Textbook of Clinical Neuropsychiatry and Behavioral Neuroscience

Textbook of Clinical Neuropsychiatry and Behavioral Neuroscience

Third edition

David P Moore MD
Associate Clinical Professor, Department of Psychiatry and Behavioral Sciences; and Associate Clinical Professor, Department of Neurosurgery (Division of Physical Medicine and Rehabilitation), University of Louisville School of Medicine, Kentucky, USA

Basant K Puri MA, PhD, MB, BChir, FRCPsych, MMath
Hammersmith Hospital and Imperial College London, UK

HODDER ARNOLD
PART OF HACHETTE LIVRE UK

First published in Great Britain in 2001 by Arnold
Second edition published in Great Britain in 2008 by Hodder Arnold
This third edition published in 2012 by
Hodder Arnold, an imprint of Hodder Education, a division of Hachette UK
338 Euston Road, London NW1 3BH

http://www.hodderarnold.com

British Library Cataloguing in Publication Data
A catalogue record for this book is available from the British Library

Library of Congress Cataloging-in-Publication Data
A catalog record for this book is available from the Library of Congress

ISBN-13 978-1-4441-2134-6

1 2 3 4 5 6 7 8 9 10

Commissioning Editor: Caroline Makepeace
Project Editor: Jenny Wright
Production Controller: Francesca Wardell
Cover Design: Amina Dudhia
Indexer: Merrall-Ross International Ltd.

Main cover image © SOVEREIGN, ISM/SCIENCE PHOTO LIBRARY
Small cover images reproduced from Graham DI, Nicoll JAR and Bone I, *Adams & Graham's Introduction to Neuropathology* 3rd edition, Hodder Arnold, 2006, with permission.

Typeset in 9.5/11.5 pt Minion by Datapage India Pvt Ltd.
Printed and bound in India by Replika Press

Dedication

This book is dedicated to our families and to the patients who have entrusted us with their care. We also wish to express our gratitude to Professors Jeffrey Cummings and Michael Trimble, whose kind words have made this third edition possible, and Dr Moore wishes to add, as he has in the first two editions, a special thanks to James W Jefferson, MD, for whose example and guidance he remains forever grateful.

'Scribere actum fidei est'

Contents

Preface

With the welcome addition of Basant Puri as a co-author, this third edition represents a major advance over prior editions: though still maintaining the clinical orientation provided by Dr Moore in the first two editions, the current edition has expanded, with Dr Puri's contributions, to cover the basic neurosciences which underpin the field of neuropsychiatry. The text is now divided into four parts. Part I, entitled 'Neuroscience Fundamentals', represents Dr Puri's main contribution, and contains chapters on functional neuroanatomy, functional neuroimaging, neuropsychology, neurophysiology, molecular genetics, psychoneuroendocrinology, and immunology. Parts II, III, and IV, contributed primarily by Dr Moore, contain not only thoroughly updated versions of all of the chapters present in the second edition, but also additional chapters which substantially broaden the coverage of clinical neuropsychiatry, including several authored by Dr Puri on clinical neuroimaging, pain, and fibromyalgia, and related disorders. Part II describes the diagnostic assessment of patients and details the interview, mental status examination, neurologic examination, and ancillary investigations; Part III provides a thorough description of the various signs, symptoms, and syndromes seen in neuropsychiatric practice; and Part IV presents virtually every single disorder seen in neuropsychiatric practice, in each instance detailing the clinical features, course, etiology, differential diagnosis, and treatment of the disorder.

The literature devoted to neuropsychiatry is vast, encompassing, as it does, both neurology and psychiatry, and the authors have attempted to cull from this tremendous reservoir the most pertinent references: although the preponderance of references are therefore from the more recent past, classic authors are never neglected, and readers will find references to the works of such physicians as Alzheimer, Binswanger, Bleuler, Hughlings Jackson, Kraepelin, and Kinnier Wilson. In all, well over six thousand references are included, thus providing the reader not only with ready access to further detail on any particular subject, but also with a window on the literature as a whole.

We are deeply indebted to the reviewers of the first two editions and to the many readers who have offered comments, critiques, and suggestions: they have enabled us to write a third edition which we believe is far stronger and more comprehensive than the first two. The specialty of neuropsychiatry continues to grow and it is our fervent hope that this third edition will not only further solidify the field but will enable the reader to both understand its underpinnings in neuroscience and to practice neuropsychiatry successfully. As with the first two editions, we invite both newcomers and established practitioners to try using this edition in their own practices, as we believe they will find it as indispensable as we do.

David P Moore
Basant K Puri
January 2012

PART I

NEUROSCIENCE FUNDAMENTALS

1

Neuroscience fundamentals

1.1 FUNCTIONAL NEUROANATOMY

In this section, aspects of functional neuroanatomy are covered, which will help readers to comprehend better later chapters of this book, and thereby aid their ability to diagnose and treat neuropsychiatric disorders.

The brain of *Homo sapiens*

It is instructive to begin by briefly reminding ourselves of our place within an evolutionary context. Rather than believing that the human body, including the human brain, sits at the pinnacle of an Aristotelian *scala naturae*, it is more accurate to recognize that there is nothing particularly superior about our place in the phylogenetic tree, as shown in Figure 1.1.

An anthropocentric view of the superiority of the anatomy of the human brain is further deflated by the realization that, while larger mammalian brains have more neocortex than smaller brains, for our species the average total cortical surface area is no more than would be expected given the average human brain volume, as shown in Figure 1.2. Moreover, in light of recent findings in molecular genetics, Butler has suggested that:

> We are now confronted with compelling evidence that ... the whole nervous system itself is organized in essentially the same way across all bilaterally symmetrical animals and is specified by the same set of regulatory genes during development. As humans we essentially share the 'same' brain not only with other primates, mammals, and craniates but also with fruit flies and octopuses.
>
> (Butler 2000)

Anteroposterior organization

There exists a general anteroposterior organization in the brain and spinal cord, whereby motor functions tend to be located anteriorly (ventrally) whereas sensory functions tend to be located posteriorly (dorsally). For example, as shown in Figure 1.3, in the formation of spinal nerves, sensory rootlets and their root ganglia are located dorsally/posteriorly, while motor rootlets are located ventrally/anteriorly. Similarly, in respect of language functions in the brain, Broca's area (motor) lies rostral to Wernicke's area (sensory). Again, the primary motor cortex (M1 or MI; Brodmann area 4) lies rostral to the primary somatosensory cortex (S1 or SI; Brodmann areas 3, 1, and 2).

This organization is common to all vertebrate central nervous systems. The cause can be traced to neurogenesis and neuronal migration during ontogeny (see Figure 1.4):

> As the neural tube differentiates rostrocaudally, it also differentiates into dorsal and ventral regions due to the actions of ... regulatory genes. After closure of the neural tube ... the underlying notochord produces the signaling factor Sonic hedgehog ... which induces the differentiation of a specific group of cells in the ventral midline of the neural plate, called the floor plate. These cells then also produce Sonic hedgehog, which promotes a ventral fate for the ventral part of the neural tube. This ventral half of the neural tube becomes the basal plate, which contains motor neurons. Other signaling molecules – bone morphogenic proteins (BMPs) – arise from the ectoderm dorsally and later from the roof plate, which comprises cells in the dorsal part of the neural tube itself. BMPs antagonize Sonic hedgehog and promote a dorsal fate. Under their influence, the dorsal part of the neural tube becomes the alar plate, which is sensory in nature.
>
> (Butler and Hodos 2005)

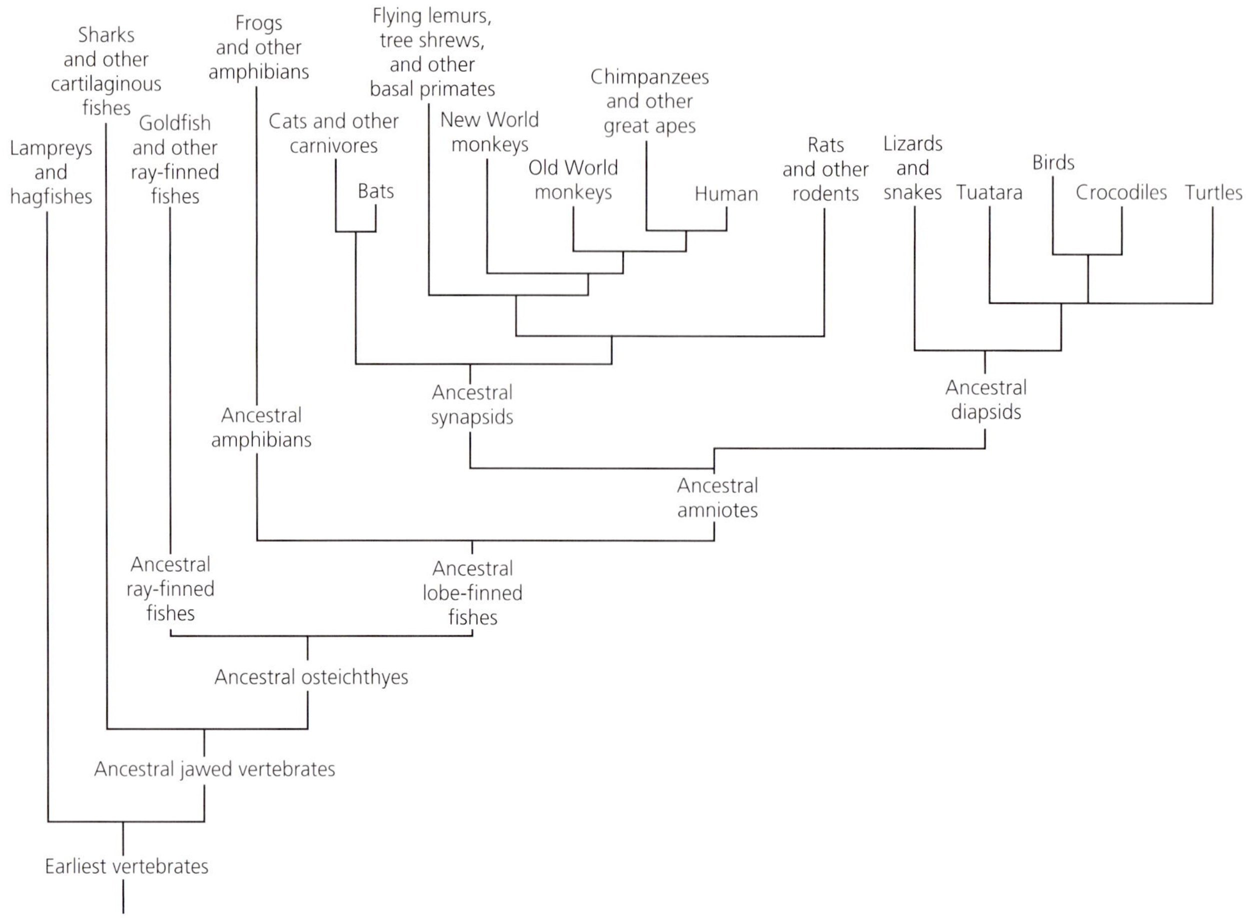

Figure 1.1 A phylogenetic tree of selected groups of vertebrates. The ancestral vertebrate groups represented at the bottom of the tree are phylogenetically older than the full set of those in the top third of the tree. Reproduced from Butler AB, Hodos W. (2005), with permission.

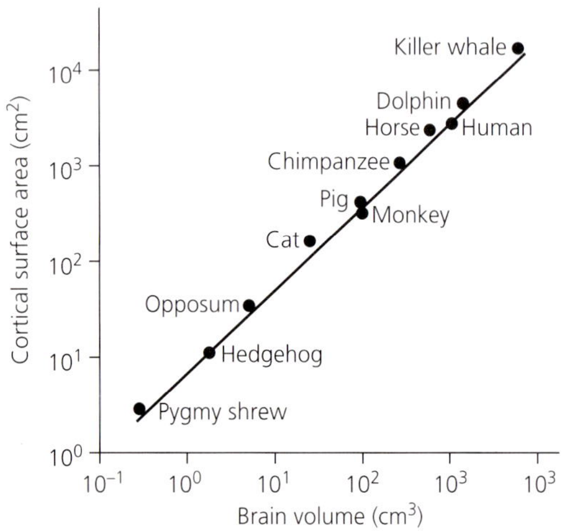

Figure 1.2 A regression line of the cortical surface area and brain volumes of 50 species of mammals, of which only 11 representative species, including humans, have been indicated. Adapted from Jerison, H.J. The study of primate brain evolution: where do we go from here? Reproduced from D. Falk and K.R. Gibson (eds) *Evolutionary Anatomy of the Primate Cerebral Cortex*. Cambridge: Cambridge University Press, 2001, with permission.

Laterality

Laterality in respect of brain functions has been found not only in human brains, but also in many other vertebrates, such as fish and birds. It is considered briefly in Chapter 1.3.

Brain lobes and major gyri and sulci

The major lobes of the brain, which are shown in Figure 1.5 ((a) lateral aspect of the left hemisphere, (b) median sagittal section of the left hemisphere, (c) view of the brain from above, and (d) view of the brain from below), are named after the approximately adjacent skull bones, namely the frontal, parietal, temporal, and occipital lobes. The frontal lobe extends rostrally from the central sulcus (the fissure of Rolando) and superiorly from the lateral fissure (the Sylvian fissure). The parietal lobe extends between the central sulcus (fissure of Rolando) and the parieto-occipital fissure (labeled as 35 in Figure 1.5b), as well as laterally as far as the lateral (Sylvian) fissure. The temporal lobe lies inferior to the lateral (Sylvian) fissure and extends posteriorly to the parieto-occipital fissure. The occipital lobe lies behind the parieto-occipital fissure. Figure 1.5 also shows the major gyri and sulci of the adult brain.

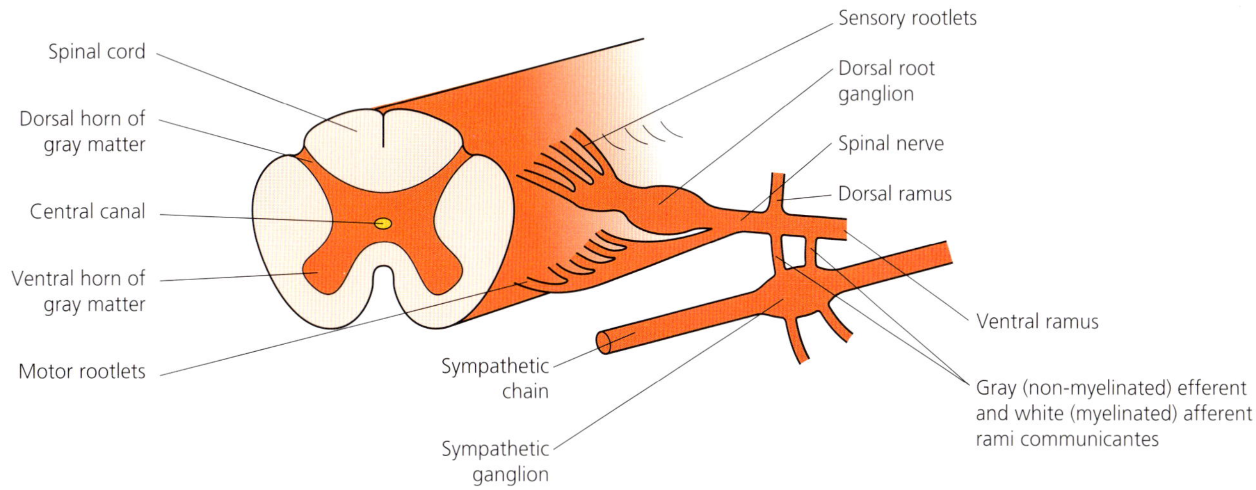

Figure 1.3 Typical spinal nerve formation and connections to the sympathetic ganglia and chains. The ventral motor rootlets unite to form a ventral motor root, joining with the dorsal sensory root which comes from the sensory rootlets and dorsal root ganglion of each spinal segment. The 'mixed' spinal nerve then divides into a dorsal and a ventral primary ramus. The dorsal rami supply the back whereas the ventral rami supply all the limbs and trunk. The sympathetic ganglia are connected to the ventral rami via white (presynaptic) rami communicantes and gray (postsynaptic) rami communicantes. Reproduced from Abrahams, Craven & Lumley, *Illustrated Clinical Anatomy* 2nd edition, Hodder Arnold, 2007 with permission.

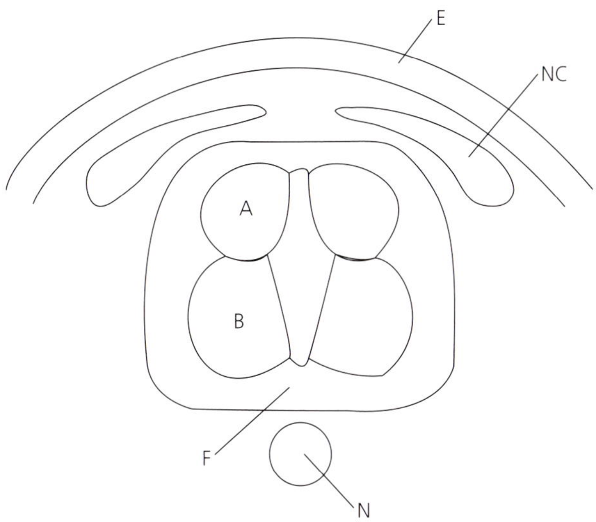

Figure 1.4 Schematic drawing of a transverse section through the dorsal part of the body of a developing craniate to show the relationships of the alar (A), basal (B), and floor (F) plates in the neural tube and the positions of the notochord (N), neural crest (NC), overlying ectoderm (E). The neural crest migrates ventrolaterally. Reproduced from Butler AB. (2000), with permission.

Cytoarchitectonic-based Brodmann areas of importance are shown in Figure 1.6.

FRONTAL LOBE

Primary motor cortex

The frontal lobe contains the primary motor cortex (M1 or MI; Brodmann area 4) just anterior to the central sulcus, occupying the precentral gyrus. Electrical stimulation of M1 is associated with contralateral muscular contraction. The somatotopic representation in the form of the homunculus was first illustrated in a paper by Penfield and Boldrey in 1937 as a combined symmetrical figure depicting motor and somatosensory features. It was then illustrated differently by Penfield and Rasmussen in their 1950 monograph as the more familiar somatotopic representation in M1, as the motor homunculus, and in S1, as the sensory homunculus, shown in Figure 1.7.

Premotor cortex

The premotor cortex (Brodmann area 6) lies immediately rostral to M1 and is divided functionally into the dorsal (PMd) and ventral (PMv) areas. PMd receives input from the dorsolateral prefrontal cortex (DLPFC) while the PMv receives input from the ventrolateral prefrontal cortex (VLPFC). The premotor cortex has important functions in both motor activity and preparing for motor activity. In addition, PMd may also be involved in decision circuits activated by free choice. Based on their finding that correlations in spike and local field potential activity between the PMd and the parietal reach region are greater when monkeys are freely making choices than when they are following instructions, Pesaran and colleagues have proposed that a decision circuit featuring a subpopulation of cells in frontal and parietal cortex may exchange information to coordinate activity between these areas; cells participating in this decision circuit may influence movement choices by providing a common bias to the selection of movement goals (Pesaran *et al.* 2008).

Frontal eye field

The frontal eye field (FEF; mainly Brodmann area 8) receives afferent connections via the thalamus from the dentate nucleus of the cerebellum, substantia nigra and superior colliculus, as well as from ipsilateral regions, including V5 (medial temporal cortex; visual), Brodmann area 7a (see Figure 1.6; parietal cortex) and the superior temporal gyrus

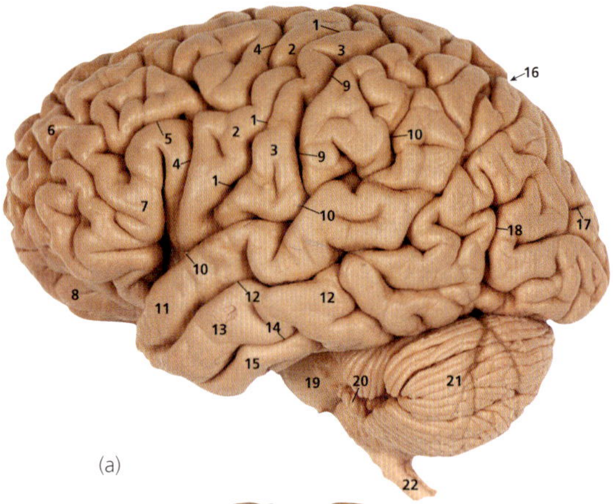

(a)

Figure 1.5a Superficial dissection of the brain viewed from the left side. From the left, with the arachnoid mater and blood vessels removed 1. Central sulcus, 2. Precentral gyrus, 3. Postcentral gyrus, 4. Precentral sulcus, 5. Inferior frontal sulcus, 6. Superior frontal gyrus, 7. Inferior frontal gyrus, 8. Orbital gyri, 9. Postcentral sulcus, 10. Lateral fissure, 11. Superior temporal gyrus, 12. Superior temporal sulcus, 13. Middle temporal gyrus, 14. Inferior temporal sulcus, 15. Inferior temporal gyrus, 16. Parieto-occipital fissure (arrowed), 17. Lunate sulcus, 18. Anterior occipital sulcus, 19. Pons, 20. Flocculus, 21. Cerebellar hemisphere, 22. Medulla oblongata;

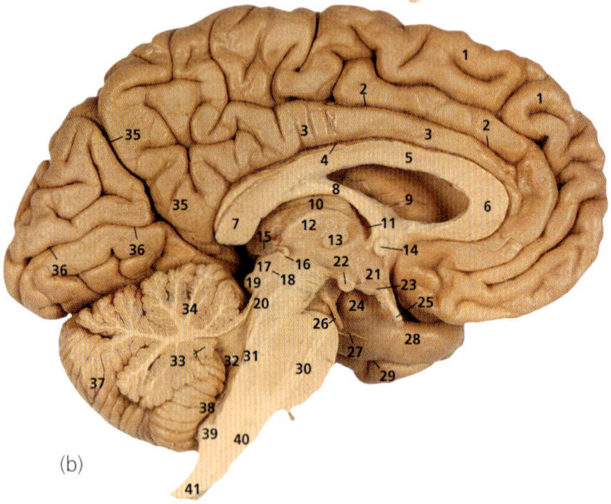

(b)

Figure 1.5b Median sagittal section of the left hemisphere. Median sagittal section. The left half, from the right, with the arachnoid mater and blood vessels removed 1. Superior frontal gyrus, 2. Cingulate sulcus, 3. Cingulate gyrus, 4. Callosal sulcus, 5. Corpus callosum - body, 6. Corpus callosum - genu, 7. Corpus callosum - splenium, 8. Fornix, 9. Caudate nucleus (head) in wall of lateral ventricle, 10. Choroid plexus, third ventricle, 11. Interventricular foramen (Monro), 12. Thalamus, 13. Massa intermedia, 14. Anterior commissure, 15. Pineal body, 16. Posterior commissure, 17. Superior colliculus, 18. Aqueduct (of Sylvius), 19. Inferior colliculus, 20. Mesencephalon, 21. Hypothalamus, 22. Mamillary body, 23. Infundibulum, 24. Uncus, 25. Optic nerve (II), 26. Oculomotor nerve (III), 27. Trochlear nerve (IV), 28. Parahippocampal gyrus, 29. Rhinal sulcus, 30. Pons, 31. Pontine tegmentum, 32. Fourth ventricle, 33. Nodulus, 34. Anterior lobe of cerebellum, 35. Parieto-occipital fissure, 36. Calcarine sulcus, 37. Cerebellar hemisphere, 38. Tonsil of cerebellum, 39. Inferior cerebellar peduncle, 40. Pyramid, 41. Medulla oblongata;

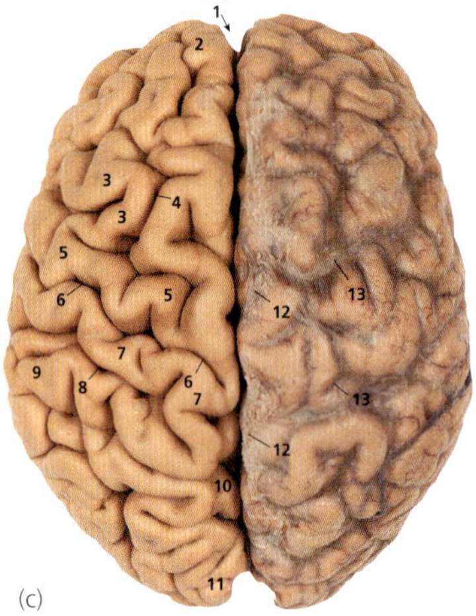

(c)

Figure 1.5c Superficial dissection of the brain viewed from above. (Left) Left cerebral hemisphere. From above, with the arachnoid mater and blood vessels removed, 1. Longitudinal cerebral fissure (arrowed), 2. Frontal pole, 3. Middle frontal gyrus, 4. Superior frontal sulcus, 5. Precentral gyrus, 6. Central sulcus, 7. Postcentral gyrus, 8. Postcentral sulcus, 9. Inferior parietal lobe, 10. Parieto-occipital fissure, 11. Occipital gyri. (Right) Right cerebral hemisphere. From above, with the arachnoid mater and blood vessels intact, 12. Arachnoid granulations, 13. Superior cerebral veins;

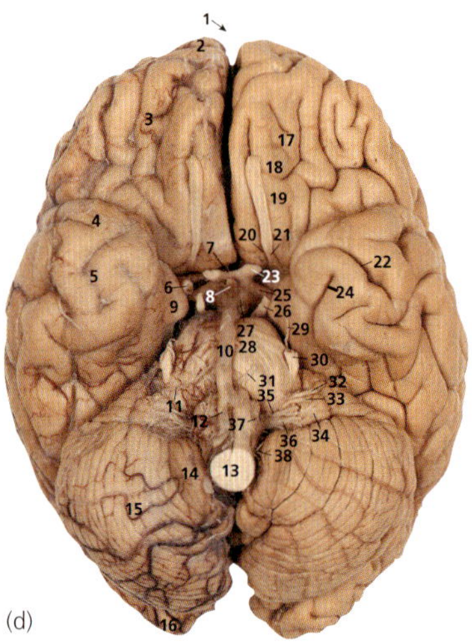

(d)

Figure 1.5d Superficial dissection of the brain viewed from below. (Left) Right cerebral hemisphere, cerebellum and brain stem. From below, with the arachnoid mater and blood vessels intact, 1. Longitudinal cerebral fissure (arrowed), 2. Frontal pole, 3. Inferior surface of frontal pole, 4. Temporal pole, 5. Inferior surface of temporal pole, 6. Internal carotid artery, 7. Optic chiasma, 8. Infundibulum, 9. Parahippocampal gyrus, 10. Basilar artery, 11. Labyrinthine artery, 12. Right vertebral artery, 13. Medulla oblongata, 14. Tonsil of cerebellum, 15. Cerebellar hemisphere, 16. Occipital pole(Right) Left cerebral hemisphere, cerebellum and brain stem. From below,with the arachnoid mater and blood vessels removed, 17. Orbital gyri, 29. Trochlear nerve (IV), 18. Olfactory bulb, 30. Trigeminal nerve (V), 19. Olfactory tract (I), 31. Abducent nerve (VI), 20. Medial olfactory stria, 32. Facial nerve (VII), 21. Lateral olfactory stria, 33. Vestibulocochlear, 22. Inferior temporal nerve (VIII)sulcus, 34. Flocculus, 23. Optic nerve (II), 35. Glossopharyngeal, 24. Collateral sulcus nerve (IX), 25. Optic tract, 36. Vagus nerve (X), 26. Oculomotor nerve (III), 37. Hypoglossal nerve, 27. Mamillary body (XII), 28. Pons, 38. Accessory nerve (XI)

Figures 1.5a-d reproduced from Ellis, Logan & Dixon, *Human Sectional Anatomy*, 2nd edition, Hodder Arnold, 2001, with permission.

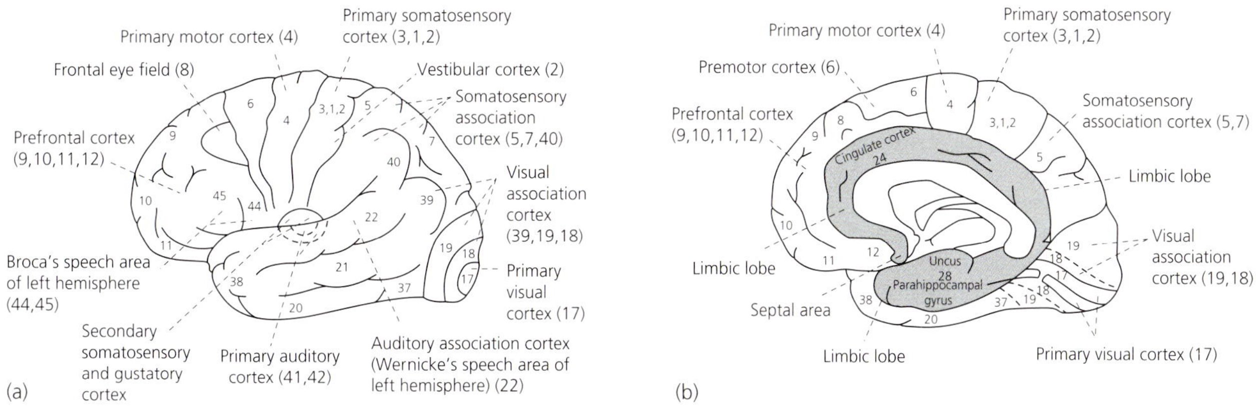

Figure 1.6 (a) Lateral and (b) medial aspects of the cerebral hemisphere showing important Brodmann areas (numbered). Reproduced from Graham DI, Nicoll JAR and Bone I, *Adams & Graham's Introduction to Neuropathology* 3rd edition, Hodder Arnold, 2006 with permission.

(auditory/language). The FEF is particularly important in controlling ocular movements; spatial selection for locating behaviurally relevant stimuli for guiding attention and eye movements appears to be computed locally in the FEF from spatially non-selective inputs (Monosov *et al.* 2008).

Prefrontal cortex

The prefrontal cortex (PFC) contains Brodmann areas 9, 10, 11, 12, 45, and 46. In the dominant (usually left) hemisphere, areas 44 and 45 comprise Broca's area (the motor speech area). The DLPFC appears to be involved in self-ordered working memory, monitoring, and the action of the central executive, while the VLPFC appears to be involved in mnemonic processing and rule representation.

Supplementary motor cortex

The supplementary motor area (SMA; M2 or MII) is on the superolateral border and medial side of the hemisphere, rostral to M1, and appears to be involved in the assembly of central motor programming of sequential voluntary movements (Roland *et al.* 1980).

It also appears to be involved in unconscious and involuntary motor control (Sumner *et al.* 2007). Indeed, the SMA has recently been shown to be involved in automatic suppression of manual motor plans (Boy *et al.* 2010).

Orbitofrontal cortex

The orbitofrontal cortex occupies the orbital gyri on the inferior surface of the frontal lobe. It is involved in processes involving the motivational or emotional value of incoming information including the representation of primary reinforcers, the representation of learnt relationships between arbitrary neutral stimuli and rewards or punishments, and the integration of this information to guide response selection, suppression and decision making (Ramnani and Owen 2004).

PARIETAL LOBE

Somatosensory cortex

The parietal lobe contains the primary somatosensory cortex (S1 or SI; Brodmann areas 3, 1, and 2) just posterior to the central sulcus, occupying the postcentral gyrus and the posterior wall of the central sulcus. The primary somatosensory cortex receives medial lemniscal, spinothalamic, and trigeminothalamic input via the thalamic ventral posterior nucleus. The sensory somatotopic representation of the contralateral half of the body on S1 as the sensory homunculus, described by Penfield and Boldrey, is shown in Figure 1.7.

There is also a sensory somatotopical representation of the body, with the head anterior and the sacral region posterior, in the secondary somatosensory cortex (S2 or SII), lying posterior to the central sulcus along the superior part of the lateral fissure. In a complex world, a sensory cue may prompt different actions in different contexts; context-dependent modulation of functional connectivity from S2 to the PFC has been demonstrated in the primate brain (Chow *et al.* 2009).

Neuronal mechanisms contributing to tactile perception have been studied using single-unit recordings from the cutaneous hand representation of primate S1 and S2 by Meftah and colleagues. S1 and S2 neurons develop a sustained change in discharge during the instruction period of a directed-attention task; selective attention to tactile stimuli have qualitatively and quantitatively greater and earlier effects in S2 than S1 (Meftah *et al.* 2002). The extent

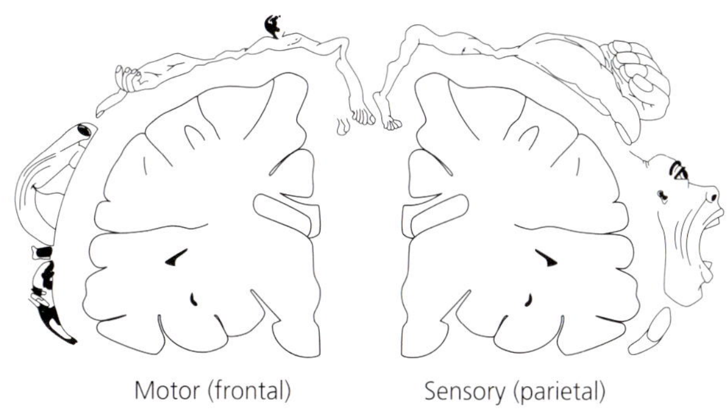

Figure 1.7 The motor and sensory homunculi. Reproduced from Graham DI, Nicoll JAR and Bone I, *Adams & Graham's Introduction to Neuropathology* 3rd edition, Hodder Arnold, 2006 with permission.

to which symbolic light cues, signaling modality (tactile, visual) to attend and discriminate, elicit changes in discharge rate during the instructed delay (ID) period of the attention task and the functional importance of this discharge have been studied. ID responses are present in about 40 percent of the neurons in S1 and S2 and are very similar in both regions in most respects (frequency, sign, latency, amplitude), suggesting a common source (Meftah *et al.* 2009). Tactile detection in humans has also been shown to be enhanced during the ID period (Meftah *et al.* 2009). Thus, ID discharge may represent a priming mechanism that prepares cortical areas to receive and process sensory inputs.

Superior and inferior parietal lobules

The somatosensory association cortex (Brodmann areas 5, 7a, 7b, and 40) is located in the superior and inferior parietal lobules. Damage to different parts of the somatosensory association cortex produces distinct somesthetic syndromes in humans (Caselli 1993).

In partially cortically blind patients, cortical activity in the parietal somatosensory association cortex has been measured, using functional magnetic resonance imaging (fMRI), before and after pairing a visual stimulus presented in the blind field with an aversive event. After pairing, self-reported negative emotional valence and blood oxygen level-dependent (BOLD) fMRI responses in somatosensory association areas were enhanced, whereby somatosensory activity predicted highly corresponding reported feelings and startle reflex amplitudes across subjects (Anders *et al.* 2004). These findings provide direct evidence that cortical activity representing physical emotional states governs emotional feelings; in other words, parietal somatosensory association cortex mediates affective blindsight.

TEMPORAL LOBE

This lies inferior to the lateral (Sylvian) fissure and extends posteriorly to the parieto-occipital fissure. The internal anatomy of the temporal lobes is described later in this chapter.

Primary auditory cortex

The primary auditory cortex (A1 or AI), corresponding to Brodmann areas 41 ± 42, receives auditory sensory input from the medial geniculate nucleus and has a tonotopic cochlear representation (low auditory frequencies anterior and high ones posterior). There is a predominant and consistent leftward functional asymmetry in A1, in right-handed individuals, during both linguistic and non-linguistic sound stimulations (Yoo *et al.* 2005).

It should be noted that, while speech perception is conventionally considered to be an auditory function, visual speech perception (e.g. of a silent video-clip of a talker saying isolated words) can occur without A1 activation (Bernstein *et al.* 2002).

Wernicke's area

The dominant temporal lobe contains Wernicke's (sensory speech) area, which is variously designated as Brodmann area 22 ± 39 ± 40, and may include part of the parietal cortex.

In comprehending complex, natural language, in addition to the 'dominant', usually left hemisphere, the right cerebral hemisphere also appears to play a role (Jung-Beeman 2005). Thus, the contralateral (non-dominant, usually right) homologue of Wernicke's area also has a role in language processing in humans, particularly in respect of lexical ambiguity resolution. The ambiguous word 'bank', for example, can be considered to have 'teller' as the dominant meaning and 'river' as a subordinate meaning. A transcranial magnetic stimulation study has shown that while (the left) Wernicke's area processes dominant meanings of ambiguous words, its contralateral homologue (the right Wernicke's area) functions as a processor of subordinate meanings of ambiguous words (Harpaz *et al.* 2009).

Middle and inferior temporal cortices

The middle temporal cortex, Brodmann area 21, has connections with different sensory modality pathways, including those related to vision, somatosensory input, and auditory input. In contrast, the inferior temporal cortex, Brodmann area 20, is particularly associated with higher processing of visual stimuli.

Induction of synaptic plasticity in surgically resected human inferior and middle temporal cortex has been studied. Synaptic strength in the human temporal cortex has been found to undergo bidirectional modifications, depending on the pattern of conditioning stimulation, and both long-term potentiation (LTP) and long-term depression can be induced; these can reverse each other, suggesting that they can act cooperatively to modify the functional state of cortical network (Chen *et al.* 1996). Thus, LTP and long-term depression are possible mechanisms for the visual memory and pattern recognition functions performed in the human temporal cortex (Chen *et al.* 1996).

OCCIPITAL LOBE

The occipital lobe is mainly made up of Brodmann areas 17, 18, and 19. Brodmann area 17, the primary visual cortex (V1, VI, or the striate cortex), receives visual sensory input from the lateral geniculate nucleus via branches (lower and upper fields) of the optic radiation. The part of the primary visual cortex above the calcarine receives a retinotopic visual input (macular to a large posterior part of V1 and peripheral to anterior V1) from the contralateral inferior visual hemifields (superior retinal quadrants) while the part below receives a similar retinotopic input from the contralateral superior visual hemifields (inferior retinal quadrants).

- V2, the second visual area, is found in much of Brodmann area 18, just concentric to area 17, and contains a retinotopic mapping of the contralateral visual hemifield.
- V3, the third visual area, is also mainly found in Brodmann area 18. It is divided into dorsal (V3d or V3), ventral (V3v or VP) and anterior (V3a) parts.
- V4, the fourth visual area, is found in Brodmann area 19. While cells in V2 and V3 tend to be orientation selective

but not selective for stimulus color or direction or movement, V4 appears to be particularly involved in the discrimination of color, as well as orientation, form, and movement (Essen and Zeki 1978).

LIMBIC SYSTEM

Papez suggested the notion of a neural network theory of emotion:

> The central emotive process of cortical origin may . . . be conceived as being built up in the hippocampal formation and as being transferred to the mammillary body and thence through the anterior thalamic nuclei to the cortex of the gyrus cinguli. The cortex of the [cingulate] gyrus may be looked on as the receptive region for the experiencing of emotion . . . Radiation of the emotive process from the gyrus cinguli to other regions in the cerebral cortex would add emotional colouring to psychic processes.

Parts of the brain involved in olfaction were also included by Papez in his circuit (Papez 1937). Many of these regions are shown as the shaded 'limbic lobe' in Figure 1.6b. Papez argued that lesions directly involving or impinging on this anatomical circuit (the Papez circuit) caused symptoms largely confined to the person's affective behavior. Thus, in rabies affecting the hippocampus and cerebellum, he noted that patients are subject to anxiety, apprehension, and paroxysms of rage and terror (MacLean 1949; Papez 1937). Papez pointed out that intact neural pathways from the mammillary bodies to the cingulate gyrus, via the anterior nuclei of the thalamus, appeared to be required for a state of vigilance and wakefulness, from which he deduced that this circuit was concerned with affective behavior (MacLean 1949; Papez 1937). Regarding a possible emotional function of the cingulate gyrus, he noted that corpus colossal tumors impinging on this gyrus are often associated with personality change, loss of affect, and a degree of somnolence and stupor (MacLean 1949; Papez 1937).

The Papez circuit was further elaborated upon by MacLean, who included the following structures (in approximately this sequence) in his version of the limbic system: olfactory bulb, olfactory tubercle, lateral olfactory stria, amygdala, stria terminalis, the septal nuclei, stria medullaris, interpeduncular nucleus, medial forebrain bundle, and mammillary body; and also a pathway from the septal nuclei via the supracallosal striae, dentate gyrus, hippocampus, and fornix back to the mammillary body. Other components of this version of the limbic system included the anterior nuclear group of the thalamus, the diagonal band of Broca, and the habenula (part of the epithalamus) (MacLean 1949).

While there is no consistent agreed set of components of the limbic system and its very concept remains controversial (McLachlan 2009), in addition to the structures suggested by Papez and MacLean, most modern definitions also tend to include the cingulate gyrus, the rest of the hypothalamus, and, at least partly, the basal ganglia.

More recently, it has been suggested that attributing emotional functions primarily to the limbic system may be inherently incorrect. For example, Pessoa has argued that:

> In attempting to localize affect in the brain, an appealing approach has been to separate the 'emotional brain' from the 'cognitive brain' . . . parcelling the brain into cognitive and affective regions is inherently problematic, and ultimately untenable for at least three reasons: first, brain regions viewed as 'affective' are also involved in cognition; second, brain regions viewed as 'cognitive' are also involved in emotion; and critically, third, cognition and emotion are integrated in the brain.
>
> (Pessoa 2008)

Basal ganglia

The basal ganglia are important in the control of movement and posture.

COMPONENTS

The basal ganglia are deep subcortical paired nuclear masses lying close to the internal capsule. A modern definition is given in Table 1.1. While the putamen and globus pallidus have traditionally been grouped together as the lentiform nucleus, in practice it is the caudate nucleus rather than the globus pallidus with which the putamen should be paired, based on cytoarchitectural and connectivity similarities; this pairing of the caudate nucleus with the putamen is phylogenetically relatively recent, and is known as the striatum (or neostriatum), in contrast to the older paleostriatum consisting of the globus pallidus.

ANATOMICAL RELATIONSHIPS

Figure 1.8 is a three-dimensional diagram showing the striatum of the left cerebral hemisphere and its relationship to nearby structures.

BASAL GANGLIA–THALAMOCORTICAL CIRCUITS

Alexander, DeLong and Strick put forward a model of five functionally segregated circuits linking the basal ganglia with the cortex in a parallel organization (Alexander *et al.* 1986),

Table 1.1 Components of the basal ganglia

- Basal ganglia
  - Corpus striatum
    - (Neo)striatum
    - Caudate nucleus
  - Putamen
  - Paleostriatum = Globus pallidus
- Amygdala ± claustrum

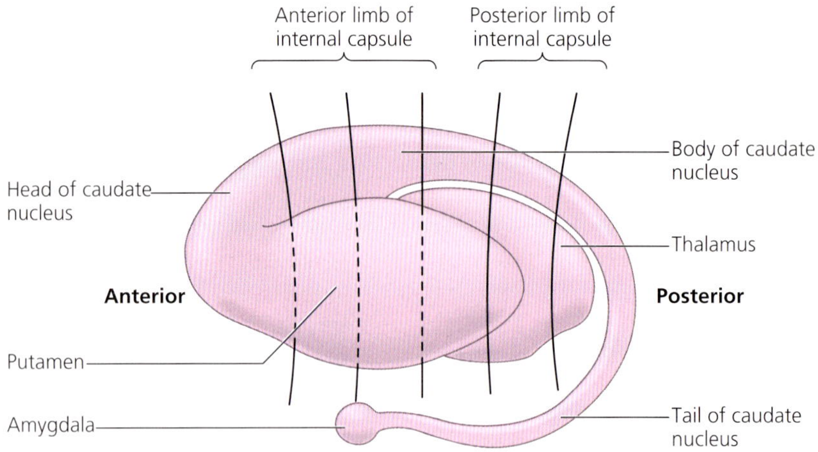

Figure 1.8 A three-dimensional view of the striatum of the left cerebral hemisphere.

which is a model supported by non-human primate studies (Middleton and Strick 2002). Figure 1.9 illustrates the five circuits, showing how each is centered on a different part of the frontal lobe. The circuits consist of partially overlapping corticostriate inputs which are progressively funneled via the pallidum (globus pallidus) and substantia nigra to the thalamus and thence back to the cortex. Within each area of connected cortico-basal ganglia structures, there are both reciprocal connections linking up regions associated with similar functions (maintaining parallel networks) and non-reciprocal connections linking up regions that are associated with different cortical-basal ganglia circuits (Haber 2003).

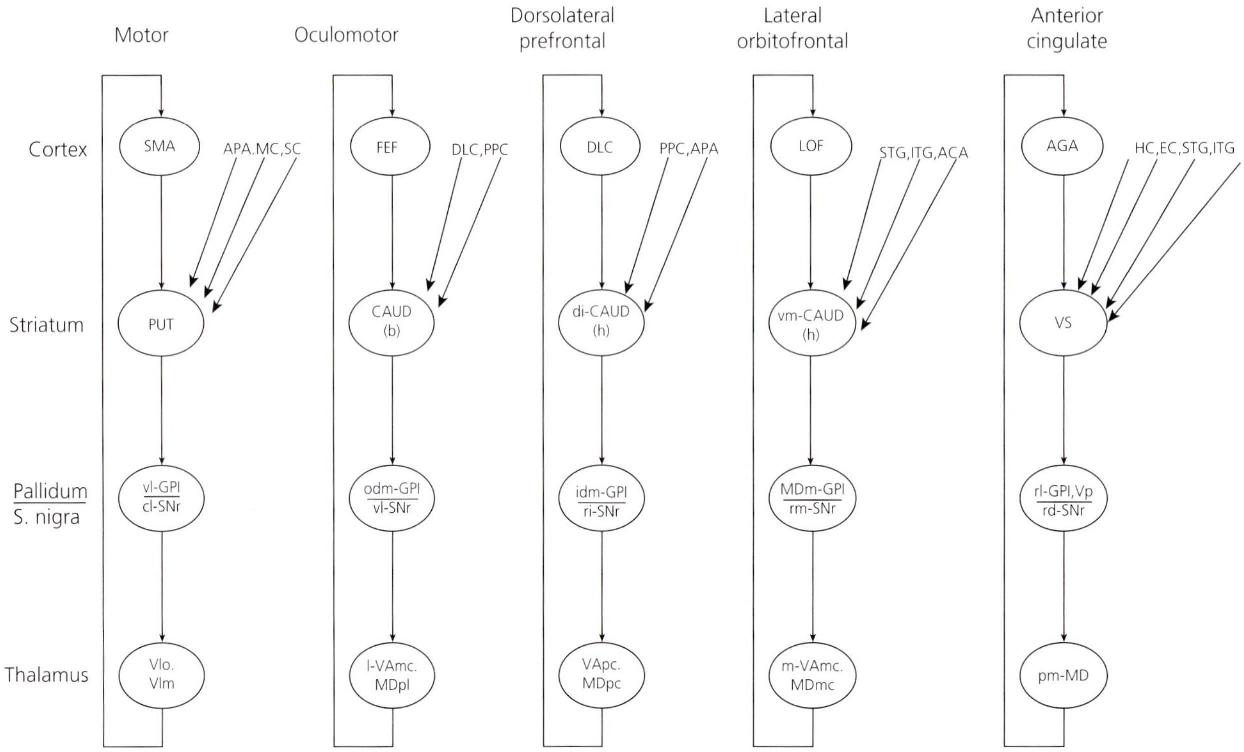

Figure 1.9 Basal ganglia-thalamocortical circuits. ACA, anterior cingulate area; APA, arcuate premotor area; CAUD, caudate; (b), body; (h), head; DLC, dorsolateral prefrontal cortex; EC, entorhinal cortex, FEF: frontal eye fields; Gpi, internal segment of globus pallidus; HC, hippocampal cortex; ITG, inferior temporal gyrus; LOF, lateral orbitofrontal cortex; MC, motor cortex; MDpl, medialis dorsalis pars paralamellaris; MDmc, medialis dorsalis pars magnocellularis; MDpc, medialis dorsalis pars parvocellularis; PPC, posterior parietal cortex; PUT, putamen; SC, somatosensory cortex; SMA, supplementary motor cortex; SNr, substantia nigra pars reticulata; STG, superior temporal gyrus; VAmc, ventralis anterior pars magnocellularis; VApc, ventralis anterior pars parvocellularis; VLm, ventralis lateralis pars medialis; VLo, ventralis lateralis pars oralis; VP, ventral pallidum; VS, ventral striatum; cl, caudolateral; cdm, caudal dorsomedial; dl, dorsolateral; l, lateral; ldm, lateral dorsomedial; m, medial; mdm, medial dorsomedial; pm, posteromedial; rd, rostrodorsal; rl, rostrolateral; rm, rostromedial; vm, ventromedial; vl, ventrolateral. Reproduced from Alexander GE, DeLong MR, Strick PL. (1986), with permission.

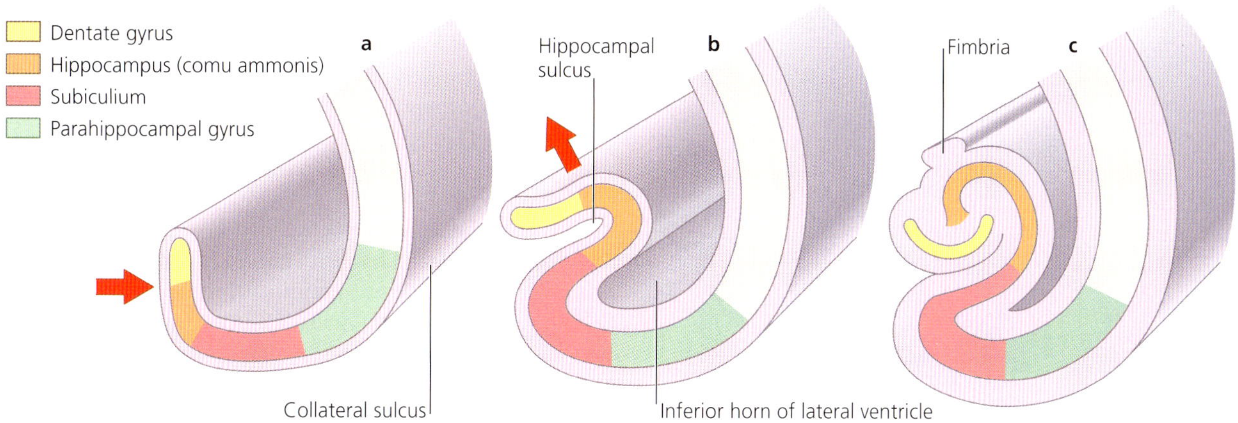

Figure 1.10 A series of coronal sections of the temporal lobe and inferior horn of the lateral ventricle illustrating the relationships between the components of the hippocampal formation. (a) to (c) show successive stages in their development. Reproduced from S. Standring (ed), *Gray's Anatomy*, 40th edition, Edinburgh, Elsevier, 2008, with permission.

Internal anatomy of the temporal lobes

COMPONENTS

There are essentially three components to the medial temporal lobe: the hippocampal formation, the amygdala and the parahippocampal cortices that cover them; the hippocampal formation refers to a constellation of structures including the subicular cortices (parasubiculum, presubiculum and subiculum proper), the hippocampus (CA1–CA3 pyramids) and the dentate gyrus (including its hilar neurons or CA4 sector) (Van Hoesen 1995). The components of the hippocampal formation are shown in Figure 1.10.

SURFACE LANDMARKS AND ANATOMICAL RELATIONSHIPS

The following sulci may be seen on the surface of the medial temporal lobe: the rhinal sulcus, which passes anteriorly and laterally and is often more prominent in great apes than in humans; the collateral sulcus, which passes posteriorly and laterally; the uncal sulcus, which separates the uncus (uncal hippocampal formation) from the parahippocampal gyrus; the hippocampal sulcus or fissure, which separates the rolled-up main body of the hippocampal formation from the posterior parahippocampal gyrus; and the sulcus semianularis, which separates the entorhinal cortex from the cortical nuclei of the amygdala (Van Hoesen 1995). They are illustrated in Figure 1.11. Figure 1.12 is a macroscopic view of the medial temporal lobe showing surface and sulcal landmarks. Interestingly, as can be seen in this figure, the surface of the entorhinal cortex often has a corrugated appearance, which is probably the result of the presence of cytochrome oxidase-reactive islands (in layer II) (Hevner and Wong-Riley 1992; Van Hoesen 1995).

CONNECTIONS

In Figure 1.13 the disposition of the cell fields CA1 to CA4 in the hippocampal formation is illustrated, as well as its different strata. Different cell fields are probably involved in different

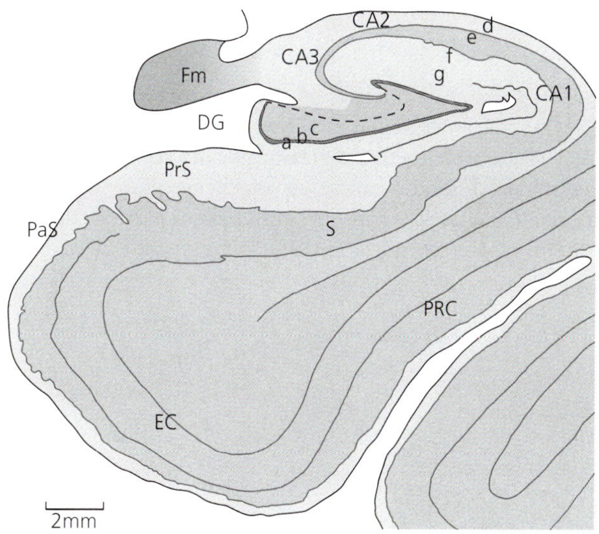

Figure 1.11 Coronal, thionin-stained section of the human hippocampal formation. Abbreviations: a, molecular layer of the dentate gyrus; b, granule cell layer of the dentate gyrus; c, plexiform layer of the dentate gyrus; CA1–3, fields of the hippocampus; d, stratum oriens layer of the hippocampus; DG, dentate gyrus; e, pyramidal cell layer of the hippocampus; EC, entorhinal cortex; f, stratum radiatum of the hippocampus; fim, fimbria; g, stratum lacunosum-molecular of the hippocampus; PaS, parasubiculum; PRC, perirhinal cortex; PrS, presubiculum; S, subiculum.

connections. For example, CA2 pyramidal cells appear to receive the hypothalamic input (from the supramammillary region of the hypothalamus). Cortical and subcortical connections of the hippocampal formation are shown in Figure 1.14, with further details being given in the legend.

Nuclei of the rat amygdala are illustrated in Figure 1.15. They may be divided into the following groups (in Figure 1.15, by different colors): the deep or basolateral group, which includes the lateral nucleus (LA), the basal nucleus (B), and accessory basal nucleus (AB); the superficial or cortical-like group, which includes the nucleus of the lateral olfactory tract

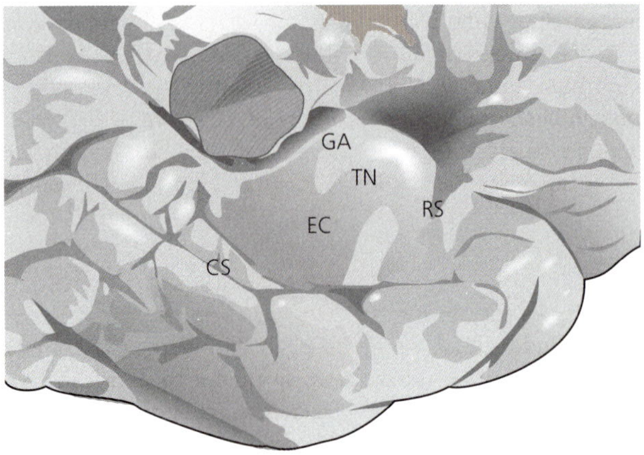

Figure 1.12 A gross view of the medial temporal lobe showing surface and sulcal landmarks. Note the corrugated appearance of the entorhinal cortex (EC) and the small elevations (known as verrucae) that define its location. CS, collateral sulcus; GA, gyrus ambiens; TN, tentorial notch; RS, rhinal sulcus. Reproduced from Van Hoesen GW. (1995), with permission.

(NLOT), the bed nucleus of the accessory olfactory tract (BAOT), the anterior and posterior cortical nucleus (CoA and CoP, respectively), and the periamygdaloid cortex (PAC); the centromedial group composed of the central (CeA) and medial (M) nuclei, and the amygdaloid part of the bed nucleus of stria terminalis (BNST); and a separate set of nuclei that do not easily fall into any of these three previous groups and include the intercalated cell masses and the amygdalo-hippocampal area (Price *et al.* 1987; Sah *et al.* 2003). The CeA has four divisions: the capsular subdivision (CeC), lateral subdivision (CeL), intermediate subdivision (CeI), and medial subdivision (CeM) (Jolkkonen and Pitkanen 1998; Sah *et al.* 2003; McDonald 1982).

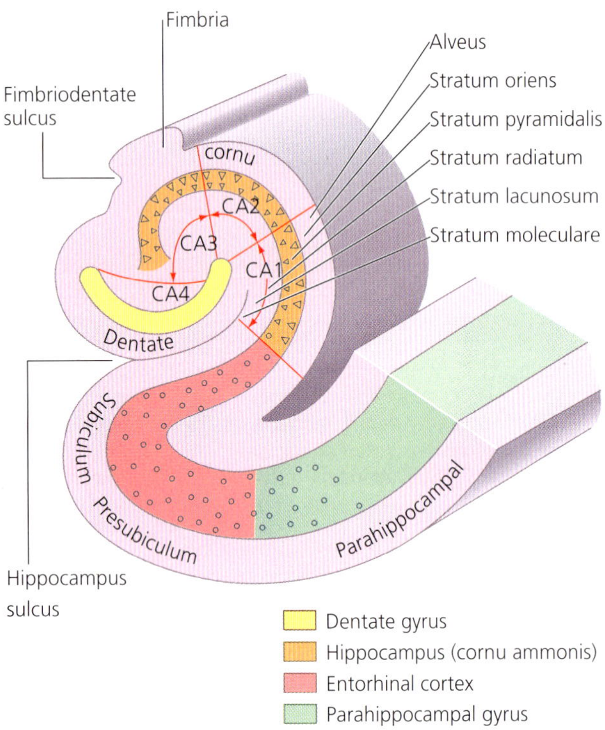

Figure 1.13 The hippocampal formation showing the disposition of the various cell fields. Redrawn from S. Standring (ed), *Gray's Anatomy*, 40th edition, Edinburgh, Elsevier, 2008, with permission.

The amygdala receives gustatory, visceral, somatosensory, auditory, visual, and olfactory inputs. It also receives afferents from the frontal cortex, perirhinal cortex, thalamus, hippocampal formation, hypothalamus, mesencephalon (dopaminergic innervation from the ventral tegmental area), and brainstem (including noradrenergic input from the locus ceruleus and serotonergic input from the raphe nuclei).

The main efferents of the amygdala are as follows. From the lateral nucleus to the prefrontal cortex, medial temporal lobe, and nucleus accumbens. From the basal nucleus to the prefrontal cortex, striatum, and medial temporal lobe. From the medial nucleus to the olfactory system, bed nucleus of the stria terminalis, hypothalamus, and thalamus. From the central nucleus to the bed nucleus of the stria terminalis, hypothalamus, mesencephalon, pons, medulla oblongata, and ascending monoaminergic and cholinergic pathways.

Major white matter pathways

The major white matter pathways of the cerebral hemispheres may be considered to fall into three major groups: association fibers, commissural fibers, and projection fibers. Papez's circuit and its relationship to the limbic system have been described above.

ASSOCIATION FIBERS

Association fibers are intra-hemispheric fibers that may be short fibers (or arcuate fibers) connecting adjacent gyri, or long fibers connecting more distant gyri, and they represent ipsilateral cortico-cortical axons. The organization of the principal association fibers is shown in Figure 1.16. Long association fibers include the cingulum, lying deep to the cingulate gyrus, between the rostral corpus callosum and the parahippocampal gyrus and adjacent temporal lobe; the superior longitudinal fasciculus, between the frontal lobe and the occipital and temporal lobes; the fronto-occipital fasciculus, lying deep to the superior longitudinal fasciculus, between the frontal pole and the occipital and temporal lobes; the inferior longitudinal fasciculus, between the occipital lobe and the temporal lobe; and the uncinate fasciculus, between the frontal lobe (Broca's motor speech area and the orbital gyri) and the temporal pole.

COMMISSURAL FIBERS

Commissural fibers are interhemispheric fibers connecting cortical areas with contralateral homologous areas. They may enter the corpus callosum, anterior commissure, or commissure of the fornix, or pass indirectly through the diencephalon or brainstem.

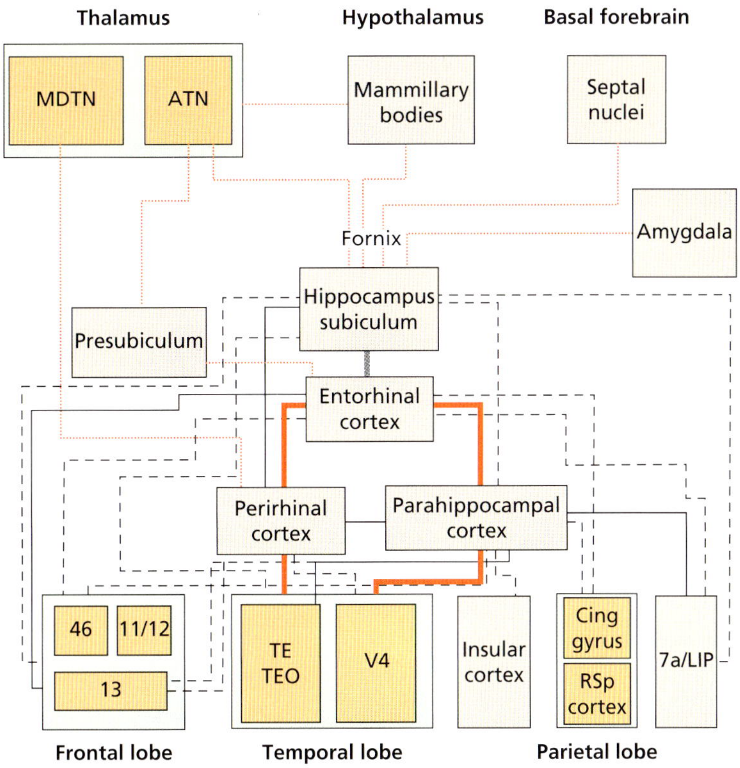

Figure 1.14 Cortical and subcortical connections of the hippocampal formation. Subcortical connections are indicated by dotted; cortical connections are indicated by solid and dashed lines. The thickness of the solid lines approximates to the strength of the connections. Most of the hippocampus's neocortical inputs come from the perirhinal and parahippocampal cortices, through the entorhinal cortex, and most of its neocortical output is through the subiculum, which also projects back to the entorhinal cortex. Both the perirhinal cortex and the parahippocampal cortex lie at the end of the ventral visual processing ('what') stream. The perirhinal cortex is crucial for the representation of complex objects, whereas the parahippocampal cortex, with its strong connections to the posterior parietal (7a/lateral intraparietal area [LIP] and retrosplenial (RSp) cortices, has a greater role in the processing of visuospatial information (from the dorsal visual processing ['where'] stream). Head-direction cells are found in the mammillary bodies, the anterior thalamic nuclei (ATN), the presubiculum and the entorhinal cortex. Some researchers have proposed that 'recollection' is dependent on the hippocampus and its links with the ATN, whereas 'familiarity' can be mediated by direct connections between the perirhinal cortex and the medial dorsal thalamic nuclei (MDTN). Cing gyrus, cingulate gyrus; TE and TEO, inferior temporal areas TE and TEO. Adapted from Bird CM, Burgess N. The hippocampus and memory: insights from spatial processing. *Nat Rev Neurosci* 2008; **9**:182–94 with permission.

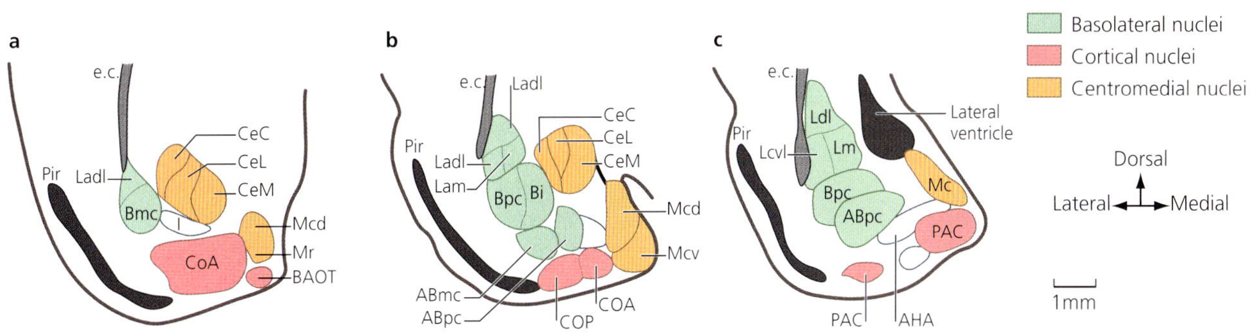

Figure 1.15 Nuclei of the rat amygdaloid complex. Coronal sections are drawn from rostral (a) to caudal (c). The different nuclei are divided into three groups as described in the text. Areas shaded green form part of the basolateral group, areas shaded pink are the cortical group, and areas shaded yellow form the centromedial group. ABmc, accessory basal magnocellular subdivision; ABpc, accessory basal parvicellular subdivision; Bpc, basal nucleus magnocellular subdivision; e.c., external capsule; Ladl, lateral amygdala medial subdivision; Lam, lateral amygdala medial subdivision; Lavl, lateral amygdala ventrolateral subdivision; Mcd, medial amygdala dorsal subdivision; Mcv, medial amygdala ventral subdivision; Mr, medial amygdala rostral subdivision; Pir, piriform cortex; s.t., stria terminalis.

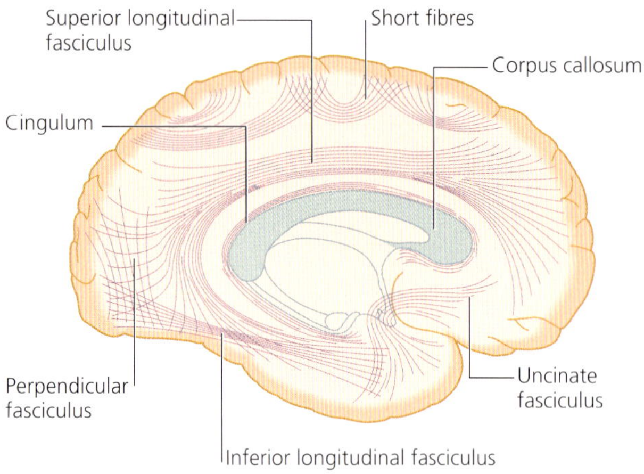

Figure 1.16 The organization of the principal association fibers projected on a sagittal section of the left cerebral hemisphere. Redrawn from S. Standring (ed.), *Gray's Anatomy*, 40th edition, Edinburgh, Elsevier, 2008, with permission.

The corpus callosum is the largest white matter pathway in the brain and is a set of commissural fibers providing both homotopic and heterotopic connections between the right and left cerebral hemispheres. The corpus callosum is shown in the median sagittal section of the left hemisphere of Figure 1.5b, from which it can be seen that it extends rostrally from the splenium, via its body, to the genu, just inferior to the callosal sulcus. Its relationship to close structures is evident from this figure. A summary of the destinations of fibers that pass through different parts of the corpus callosum is given in Table 1.2.

The anterior commissure (Figure 1.5b) crosses the midline in the lamina terminalis (helping to form the anterior wall of the third ventricle). It connects homologous neocortical and paleocortical areas. The fornix, which comprises fibers representing the hippocampal efferent system and connects the archicortex of the hemispheres with each other, is also shown in Figure 1.5b. Ventrally, close to the anterior commissure (labeled 14 in Figure 1.5b), it is divided into two parts, the precommissural part and the postcommissural part. Rostrally, the whole system arches as two posterior columns, one in each hemisphere, lying inferior to the corpus callosum and superior to the thalamus; transverse fibers connect these columns as the commissure of the fornix.

PROJECTION FIBERS

Projection fibers connect the cerebral cortex with subcortical nuclei of the brain and spinal cord, and ascend or descend in major tracts, including the fornix and the internal capsule.

Figure 1.17 illustrates the internal capsule in horizontal section, showing its main fiber components and its relationship to the lentiform complex (nucleus), to which it lies medially, and to the thalamus and caudate nucleus, to which it lies laterally. Superiorly, its fibers fan out as the corona radiata, which interdigitate with corpus callosum fibers. The division of the internal capsule into its anterior limb, genu, and posterior limb are shown in Figure 1.17, as are its ascending thalamocortical fibers and its descending cortical efferent fibers to the thalamus, brainstem, and spinal cord.

CRANIAL NERVES

The fibers of cranial nerve I, the olfactory nerve, which subserves the function of olfaction, originate from the olfactory mucosa and pass to the olfactory bulb.

The fibers of cranial nerve II, the optic nerve, which subserves the function of vision, originate from the retinal ganglion layer. The blind spot is caused by the lack of photoreceptors in the part of the retina from which this cranial nerve exits the eye. The optic nerve is ensheathed in the three meningeal layers of, from internal to external, the pia mater, arachnoid mater, and dura mater. It passes backwards to form the optic chiasma with the optic nerve from the other eye.

Cranial nerve III, the oculomotor nerve, which subserves the function of eye movement and also carries a parasympathetic innervation to the ciliary ganglion, emerges between the cerebral peduncles in the brainstem. The superior division of the nerve supplies the superior rectus and levator palpebrae superioris muscles, while the inferior division supplies the inferior oblique, inferior rectus and medial rectus, and also carries the parasympathetic innervation.

Cranial nerve IV, the trochlear nerve, which subserves the function of eye movement, emerges from the midbrain (mesencephalon) and supplies the superior oblique.

Cranial nerve V, the trigeminal nerve, which subserves the functions of general sensory input from the head and motor output to the muscles of mastication, originates from the pons. The sensory (larger) and motor roots that leave the brainstem at the middle cerebellar peduncle eventually reach the trigeminal ganglion from which the following three divisions take form: the ophthalmic division, the maxillary division, and the mandibular division. The only motor branches are those to the muscles of mastication and the mylohyoid branch of the inferior alveolar nerve to the mylohyoid muscle and the anterior belly of the digastric muscle.

Cranial nerve VI, the abducens (abducent) nerve, which subserves the function of eye movement, emerges from between the pons and the medulla oblongata and supplies the lateral rectus.

Table 1.2 Corpus callosal fiber destinations

Part of corpus callosum	Name of fibers (if applicable)	Destinations
Splenium	Forceps major	Occipital lobes
Splenium and posterior body	Tapetum	Temporal lobes
Body		Widespread neocortical areas
Genu	Forceps minor	Frontal lobes

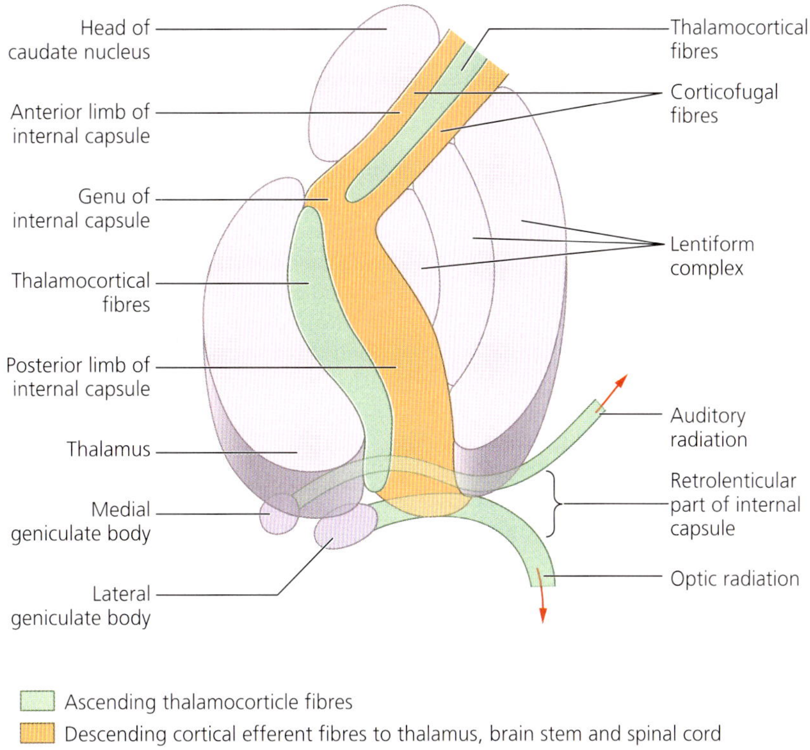

Figure 1.17 Horizontal section through the internal capsule illustrating its main fiber components.

Cranial nerve VII, the facial nerve, which subserves the functions of facial movement, taste, and lacrimal and salivary gland parasympathetic innervation, has the following major branches: greater petrosal nerve (secretomotor to the pterygopalatine ganglion); the nerve to the stapedius; chorda tympani (taste from anterior two-thirds of the tongue and parasympathetic supply to the submandibular ganglion); posterior auricular nerve (to the occipital belly of occipitofrontalis and auricular muscles); nerves to the posterior belly of the digastric and stylohyoid; five terminal motor branches from the parotid plexus (pes anserinus), namely the temporal, zygomatic, buccal, mandibular, and cervical branches, which supply the muscles of facial expression and the platysma and buccinator muscles.

Cranial nerve VIII, the vestibulocochlear nerve, has two components: the vestibular nerve, which subserves sensory functions related to balance and posture; and the cochlear nerve, which is concerned with the sensation of hearing. This cranial nerve emerges from the pontocerebellar angle and, eventually, upon entry into the petrous temporal bone (through the internal acoustic meatus), it splits into its two main components.

Cranial nerve IX, the glossopharyngeal nerve, has several functions: taste; pharyngeal sensory and motor innervation; a parasympathetic supply to the parotid gland; carotid body and sinus innervation; and a motor supply to the stylopharyngeus muscle. This cranial nerve leaves the skull through the jugular foramen. Its main branches include the carotid, carrying afferents from carotid body chemoreceptors and carotid sinus baroreceptors; the lingual branch to the pharyngeal (postsulcal) part of the tongue and to the circumvallate papillae of the oral (presulcal) part of the tongue; the muscular branch, which supplies the stylopharyngeus muscle; pharyngeal branches supplying pharyngeal sensory mucosal fibers; tonsillar branches, which, with the lesser palatine nerves, form the circulus tonsillaris from which occurs innervation of the palatine tonsil region, soft palate and oropharyngeal isthmus; and the tympanic nerve.

Cranial nerve X, the vagus nerve, has widespread functions that include sensory and motor innervation of the pharynx, larynx, palate, and esophagus; a parasympathetic supply to the viscera of the thorax and the alimentary tract as far as the splenic fissure; and sensory innervation of thoracic and abdominal viscera and the external acoustic meatus and tympanic membrane. The vagus nerve leaves the skull through the jugular foramen. Its branches in the neck include meningeal branches to the posterior cranial fossa dura mater; the auricular branch, which, after supplying a branch to the facial nerve, ultimately ends in two branches, one of which combines with the posterior auricular nerve and the other of which supplies the external acoustic meatus and part of the skin of the external ear; the pharyngeal branch, which supplies motor innervation to the pharynx; branches to the carotid body; the superior laryngeal nerve, which gives rise to the internal and external laryngeal nerves; and the recurrent laryngeal nerve.

Cranial nerve XI, the accessory nerve, which enables the head and shoulders to move as well as supplying muscles of the pharynx and larynx, comprises a smaller cranial root and a spinal root. The cranial root emerges from the skull through the jugular foramen whence it is joined briefly with the larger spinal root. After separating from the latter, the cranial root joins the tenth cranial nerve and supplies pharyngeal and laryngeal muscles (with the exception of tensor veli palatine). The spinal root supplies the sternocleidomastoid and trapezius muscles.

Cranial nerve XII, the hypoglossal nerve, enables tongue movement. It supplies all the muscles of the tongue, with the exception of palatoglossus.

Meninges

Three meninges surround the brain and spinal cord: the pia mater, the arachnoid mater, and the dura mater.

PIA MATER

This is the innermost of the three meninges. It is carried into the cerebral sulci by cerebral arteries. In the ventricular system, a double-fold of this vascular meningeal layer forms the tela choroidea, and fusion with the ependymal ventricular lining forms the choroid plexuses.

ARACHNOID MATER

The arachnoid mater lies between the pia mater and the dura mater and is named after its web-like structure. The subdural space is a potential space lying between the arachnoid mater and the dura mater, and usually contains only a film of fluid.

The subarachnoid space lies between the arachnoid mater and the pia mater and contains cerebrospinal fluid. In some regions this space is relatively large; these form subarachnoid cisterns. The subarachnoid cisterns freely communicate with each other via the subarachnoid space. The principal subarachnoid cisterns, of which the largest is the cisterna magna or cerebellomedullary cistern, are shown in the diagrammatic sagittal section of Figure 1.18. Cerebrospinal fluid passes into the cisterna magna or cerebellomedullary cistern from the fourth ventricle via the median aperture or foramen of Magendie. Cerebrospinal fluid passes into the pontine cistern from the fourth ventricle via the lateral apertures or foramina of Luschka. The subarachnoid space also continues into the orbit, surrounded by the sheath of the second cranial nerve. Enlargement of the subarachnoid space inferior to the spinal cord termination allows cerebrospinal fluid sampling via lumbar puncture.

DURA MATER

The dura mater is the outermost of the three meninges and is a firm and fibrous membrane that lines the inside of the skull. It is divided at the endosteal and meningeal layers, which separate to form venous sinuses, and also at skull foramina. The dural venous sinuses are shown in Figure 1.24.

The falx cerebri is a sickle-shaped fold of dura mater lying in the longitudinal fissure between the two cerebral hemispheres, while the crescent-shaped tentorium cerebelli is a fold of dura mater lying between the posterior cerebral hemispheres and the cerebellum.

Ventricular system

The two lateral ventricles and the third and fourth ventricles are shown in Figure 1.19. Each lateral ventricle consists of an

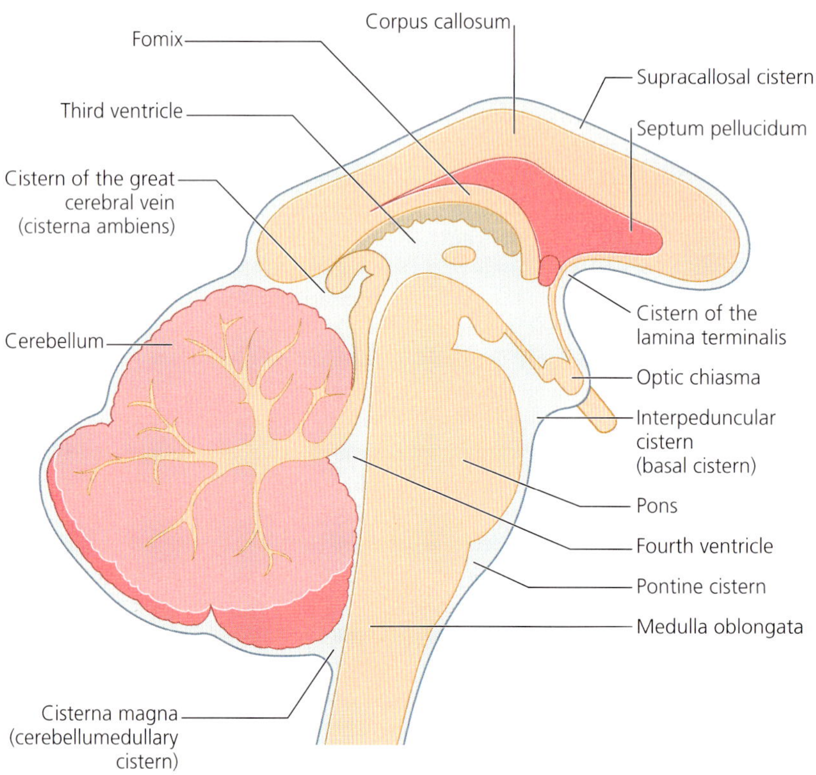

Figure 1.18 Sagittal section illustrating the principal subarachnoid cisterns.

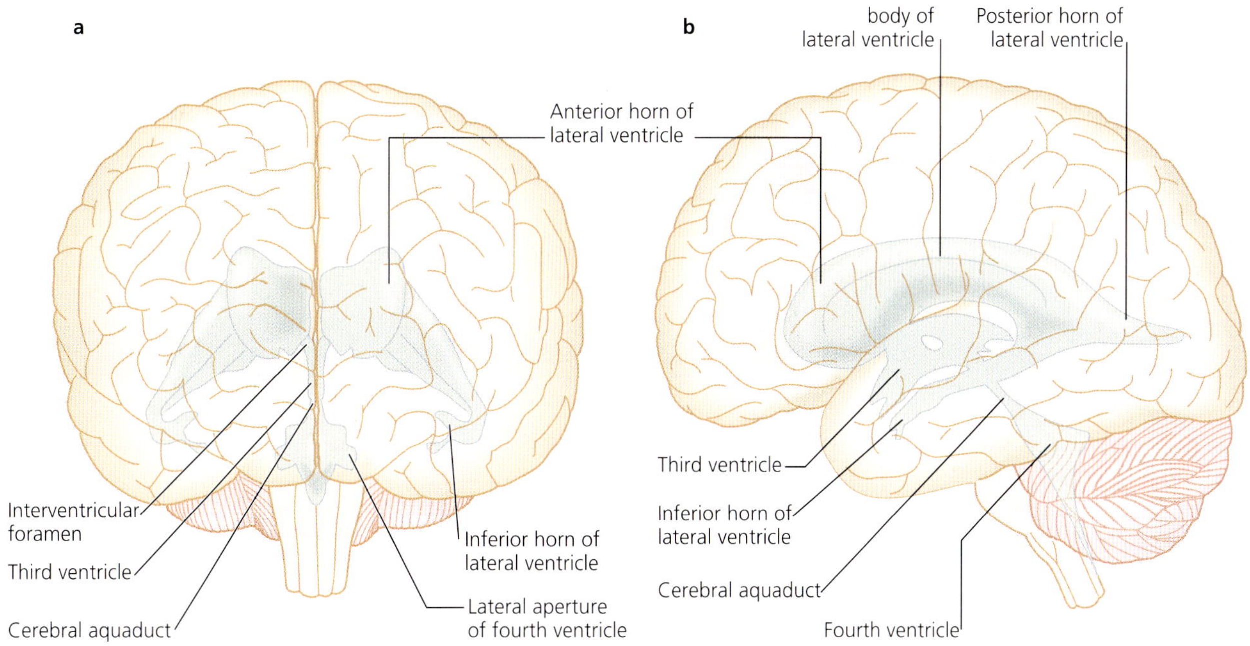

Figure 1.19 The ventricular system. (a) Anterior view; (b) left lateral view.

anterior horn (lying anterior to the interventricular foramen [of Monro]), a body, a posterior horn (in the occipital lobe), and an inferior horn (reaching the temporal lobe). The third ventricle lies between the two thalami, with the hypothalamus acting as its floor, while the fourth ventricle lies between the cerebellum and brainstem.

In the adult, the average total volume of cerebrospinal fluid is around 140 mL. It is formed by the ependymal lining of the choroid plexuses of the four ventricles. It passes from the lateral ventricles into the third ventricle via the interventricular foramen (of Monro), and from the third ventricle into the fourth via cerebral aqueduct (of Sylvius). From the fourth ventricle, the cerebrospinal fluid enters the subarachnoid space via the median aperture or foramen of Magendie (into the cisterna magna or cerebellomedullary cistern) and the lateral apertures or foramina of Luschka (into the pontine cistern), as mentioned above. In contrast to the active process of cerebrospinal fluid secretion, reabsorption is passive and takes place via the arachnoid villi of dural venous sinuses, spinal villi, and the lymphatic system.

Vascular supply and drainage

ARTERIAL SUPPLY

The brain receives its arterial supply from two pairs of arteries: the right and left internal carotid arteries, and the right and left vertebral arteries. These four arteries, which lie in the subarachnoid space, together with some of their branches, anastomose on the base of the brain to form the circulus arteriosus or circle (or polygon) of Willis, as shown in Figure 1.20. From this figure it can be seen that the circulus arteriosus surrounds the optic chiasma and the infundibulum. It usually consists, on each side, of the posterior cerebral artery (the two posterior cerebral arteries are the terminal branches of the median-placed basilar artery), the posterior communicating artery (not labeled in Figure 1.20, it passes rostrally from the posterior cerebral artery to the ipsilateral internal carotid artery), the internal carotid artery, the anterior cerebral artery, and the anterior communicating artery. Variations in the make-up and patency of the arteries of the circulus arteriosus are not uncommon.

The major arteries supplying the medial and lateral aspects of the brain are shown in Figure 1.21. The regional arterial supply of the brain is summarized in Table 1.3.

The arterial supply to the cerebral cortex from cortical branches of the anterior cerebral, middle cerebral, and posterior cerebral arteries, is shown in Figure 1.22, p. 20.

VENOUS DRAINAGE

The cerebral venous system is illustrated in Figure 1.23, p. 21, which shows the principal superficial and deep veins of the brain and also their relationship to the dural venous sinuses.

Types of cells found within the central nervous system

NEURONS

Neurons have a variety of different sizes and shapes but all have one axon, emerging from the axon hillock, and at least one dendrite; some possess many dozens of dendrites. The axon can vary in length from a few millimeters to almost a meter. The nucleus contains the prominent nucleolus, in which rRNA is transcribed (by RNA pol I), while the cytoplasm of the neuronal soma may contain large numbers

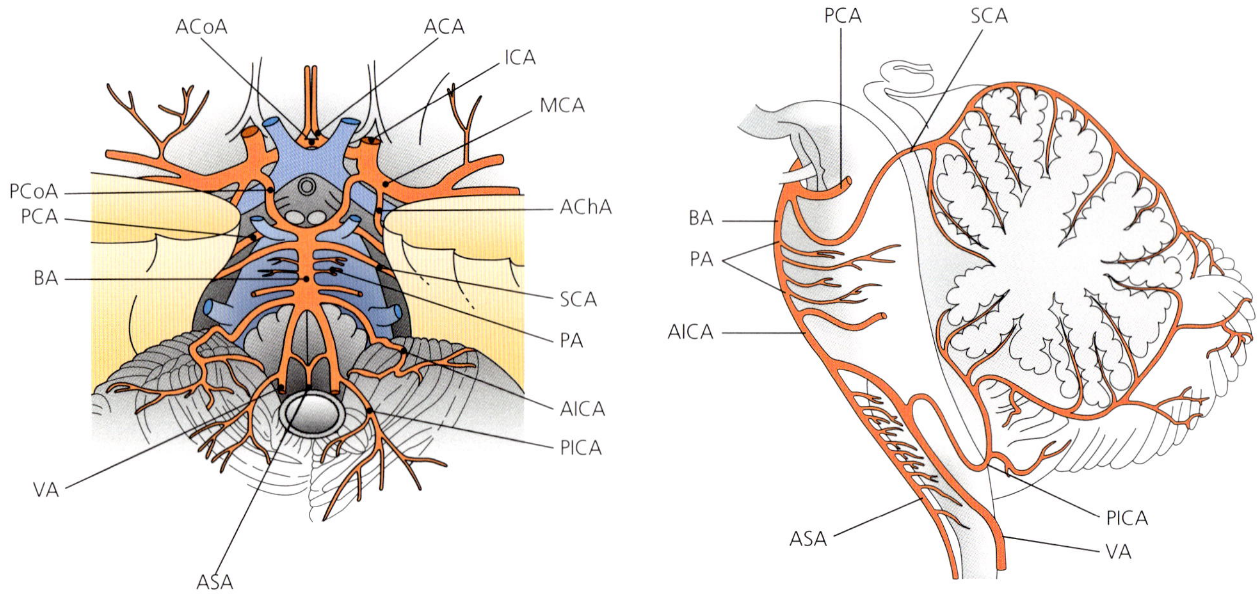

Figure 1.20 The arteries on the base of the brain. The anterior part of the left temporal lobe has been removed to display the initial course of the middle cerebral artery within the lateral fissure. Reproduced from Graham DI, Nicoll JAR and Bone I, *Adams & Graham's Introduction to Neuropathology* 3rd edition, Hodder Arnold, 2006, with permission.

of basophilic granular Nissl bodies extending into the dendrites (although typically not in the axon hillock), which are the site of protein biosynthesis, consisting of rough endoplasmic reticulum and polyribosomes. A typical neuron also contains mitochondria, which are the sites of oxidative phosphorylation; Golgi apparatus, which carry out glycosylation and phosphorylation, synthesize proteoglycans and carbohydrates, transport lipids, and create lysosomes (from endosomes by the addition of hydrolytic enzymes); and lipofuscin granules, which are membrane-bound yellow–brown organelles containing non-metabolizable remnants of lysosomal digestion.

Synaptic vesicles are the organelles of the boutons terminaux, which store neurotransmitters.

The dark tyrosine-derived pigment neuromelanin is normally found in granules in the neuronal cytoplasm of the central nervous system in the following locations: the substantia nigra, pars compacta; the locus ceruleus, from which arises the ascending noradrenergic pathway; the dorsal motor nucleus of the vagus nerve; and the pontine median raphe nucleus.

MACROGLIA

Astrocytes

Astrocytes (astroglia) are classified as being either protoplasmic or fibrous/fibrillary. Protoplasmic astrocytes are mainly found in gray matter and tend to have short processes with numerous branches. In contrast, fibrous or fibrillary astrocytes are found mainly in white matter and tend to have long processes; they form a thick felt work beneath the pia mater. Their cytoplasm tends to contain glial fibrillary acidic protein (GFAP), an intermediate filament protein.

Variants of astrocytes include Bergmann glia in the granule layer of the cerebellar cortex, and the round laminate periodic acid–Schiff-positive corpora amylacea found in subpial and periventricular white matter and in the spinal cord, particularly in subjects of advanced age.

Astrocytes have numerous functions, which may include structural functions: central nervous system repair, astrocytes having been considered to be functionally homologous to fibroblasts in the central nervous system and helping form scar tissue (astrocytic gliosis); the nourishment of neurons; aiding in neurotransmission; vasomodulation; ion concentration regulation; and promoting myelination by oligodendrocytes. Their end-feet may also play a role in the formation of the blood–brain barrier.

Oligodendrocytes

Oligodendrocytes, or oligodendroglia, possess few cell processes (hence their name) and do not have cytoplasmic filaments. Those in gray matter tend to cluster around neuronal cell bodies and are known as perineuronal satellite cells. They probably play an important role in sustaining these neurons. Those in white matter are arranged in rows between myelinated fibers and are known as interfascicular glia. These white matter oligodendrocytes are responsible for the formation of the myelin sheaths.

EPENDYMAL CELLS

Ependymal cells (ependyma) are ciliated cuboidal cells that usually form a continuous lining of the ventricular system of the brain (including its foramina and the cerebral aqueduct) and the central canal of the spinal cord (with which the ventricles are continuous). They may play a role in fluid homeostasis

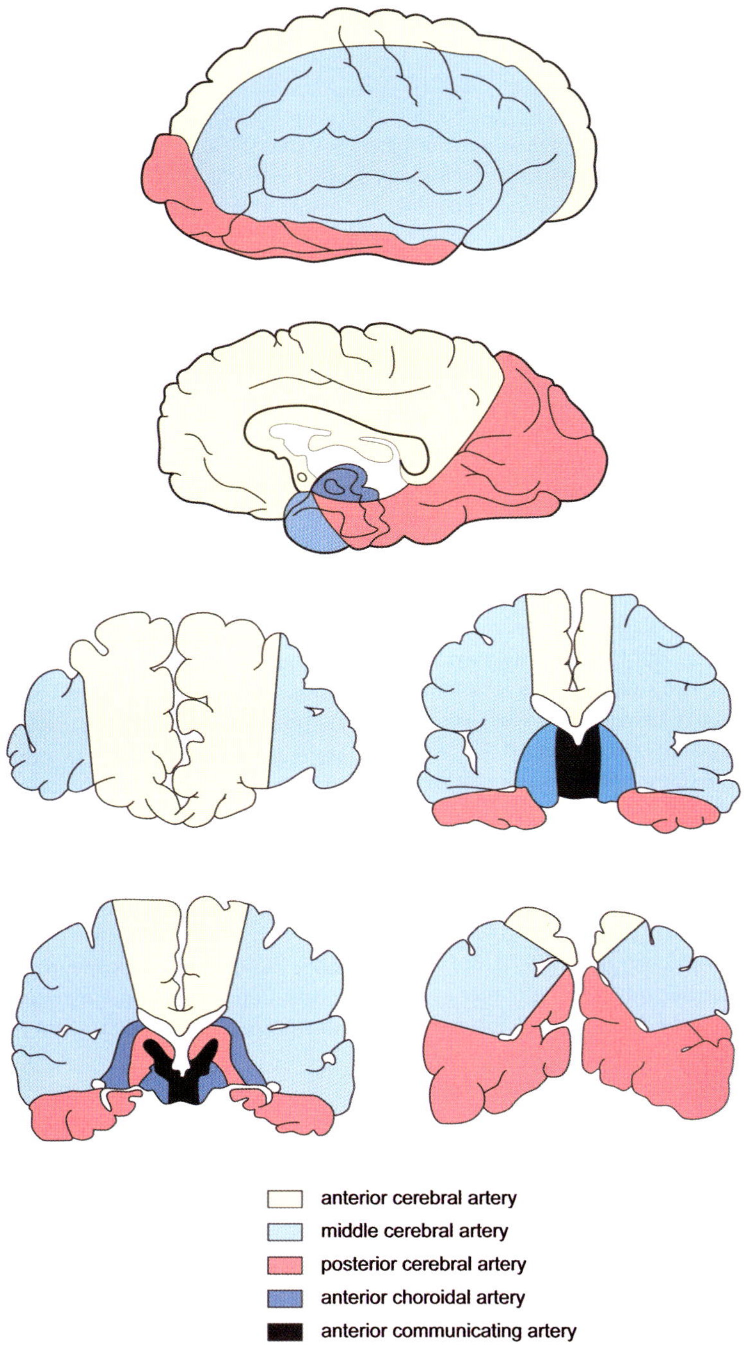

Figure 1.21 The major arteries supplying the medial (a) and lateral (b), aspects of the brain. Reproduced from Graham DI, Nicoll JAR and Bone I, *Adams & Graham's Introduction to Neuropathology* 3rd edition, Hodder Arnold, 2006, with permission.

(between cerebrospinal fluid and brain parenchyma) and in cerebrospinal fluid circulation.

CHOROID PLEXUS CELLS

The cuboidal cells that cover the fibrovascular cores of the choroid plexuses of the ventricles are non-ciliated.

MICROGLIA

Microglia are now known to be resident central nervous system immune cells which provide surveillance in the normal brain and spinal cord and become activated following tissue insult to perform a number of functions. These include proliferation, migration, phagocytosis, and secretion of cytokines, chemokines, trophic factors, nitric oxide, and other bioactive molecules and reactive species. Detailed mechanisms by which microglia detect and respond to their environment are not fully understood, but it is known that they express a number of surface receptors and ion channels, including voltage-gated sodium channels, that participate in transduction of external stimuli to intracellular responses. When activated, microglia usually become rod shaped.

Table 1.3 Regional arterial supply of the brain

Region	Arterial supply
Medulla oblongata	Branches of the vertebral, anterior spinal, posterior spinal, posterior inferior cerebellar, and basilar arteries
Pons	Basilar, anterior inferior cerebellar, and superior cerebellar arteries
Mesencephalon	Posterior cerebral, superior cerebellar, and basilar arteries
Cerebellum	Superior cerebellar, anterior inferior cerebellar, and posterior inferior cerebellar arteries
Optic chiasma	Anterior cerebral artery and branches from the internal carotid artery
Optic tract	Anterior communicating and posterior communicating arteries
Optic radiation	Branches of the middle and posterior cerebral arteries
Thalamus	Branches of the posterior communicating, posterior cerebral, and basilar arteries
Pituitary gland	Hypophysial artery (from the internal carotid)
Basal ganglia	Mainly from the striate arteries (branches from the anterior and middle cerebral arteries)
Internal capsule	Central arteries (from the circulus arteriosus and related arteries)
Cerebral cortex	Cortical branches from the anterior, middle, and posterior cerebral arteries

STEM CELLS

Stem cells are multipotent cells that can give rise to a differentiated progeny as well as self-renewed progeny; the balanced coordination of these two stem cell fates is essential for embryonic development and tissue homeostasis in the adult (Walker *et al.* 2009). Stem cells are aided by their environment, which provide support and regulatory signals; this microenvironment is the stem cell 'niche' and is the sum of all factors, cellular and molecular, that interact with and regulate the stem cell (Hellstrom *et al.* 2009; Walker *et al.* 2009). At the time of writing the following two stem cell niches have been established as existing in the adult mammalian brain: the subventricular zone (SVZ) of the lateral wall of the lateral ventricle and the subgranular zone (SGZ) of the dentate gyrus of the hippocampal formation (Hellstrom *et al.* 2009; Walker *et al.* 2009). Progenitor cells of the SVZ continually migrate rostrally to the olfactory bulb and differentiate into interneurons, which integrate into the granule cell layer and olfactory glomerular layer and serve important olfactory functions (Gheusi and Lledo 2007; Hellstrom *et al.* 2009). Postnatal neurogenesis in the dentate gyrus plays a critical role in normal cognitive development, specifically in learning and memory functions associated with the hippocampus (Hellstrom *et al.* 2009).

Corticoneurogenesis

Rakic has defined corticoneurogenesis as being the process of production, allocation, migration, and settling of neurons into proper areal and laminar positions within the cortical

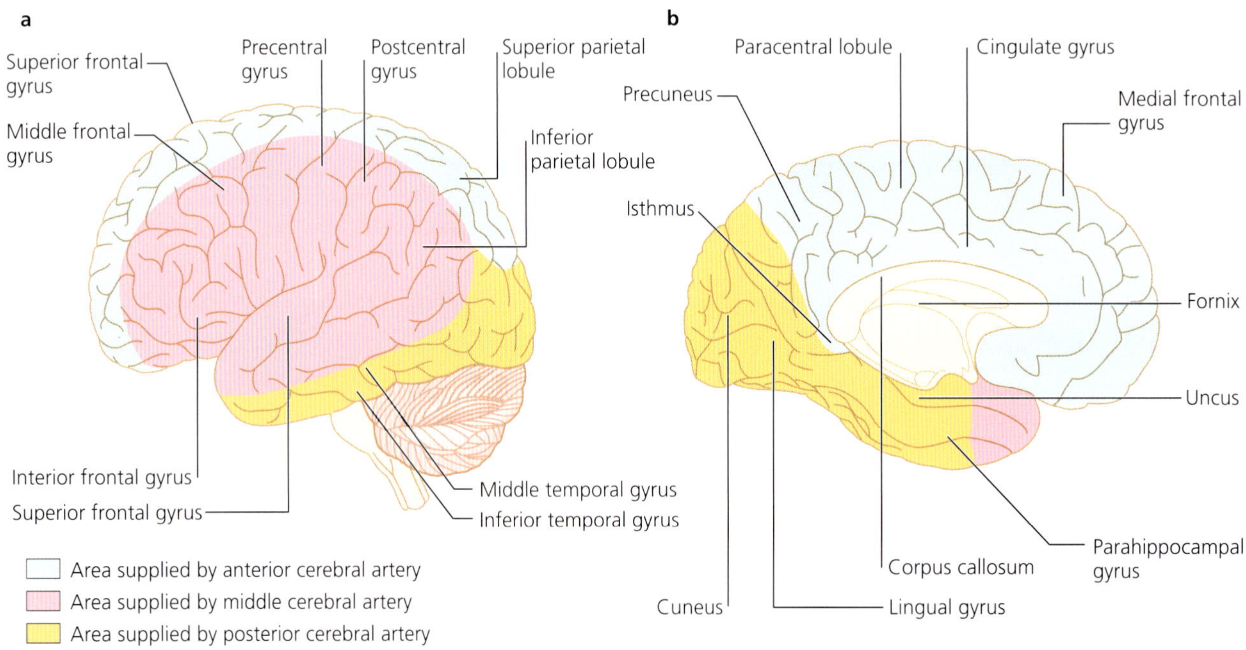

Figure 1.22 The arteries supplying the left cerebral hemisphere. (a) Lateral surface. (b) Medial surface.

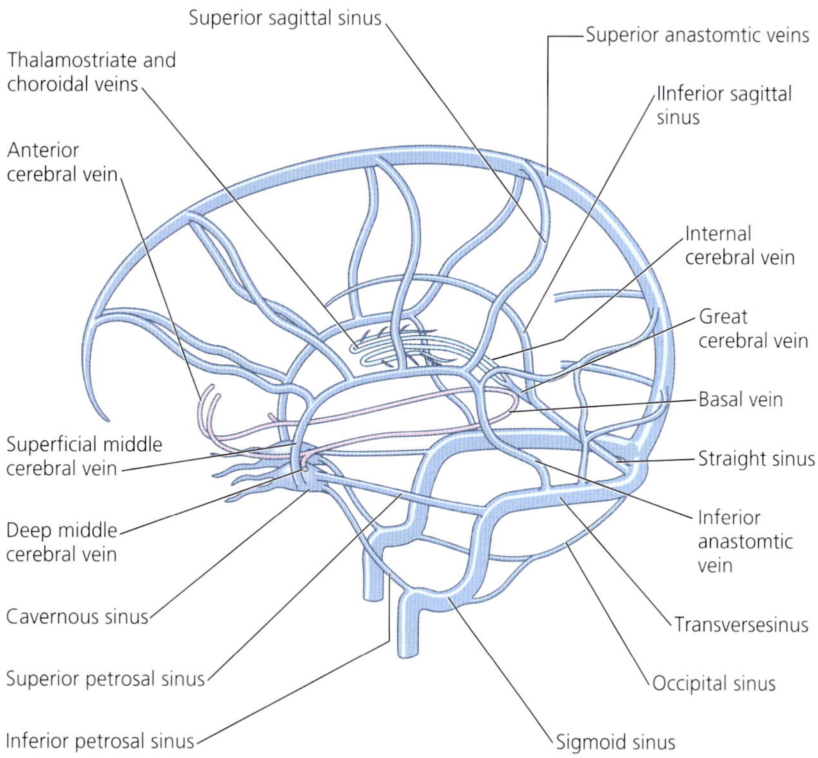

Figure 1.23 The cerebral venous system showing the principal superficial and deep veins of the brain and their relationship to the dural venous sinuses, viewed from the left side. Redrawn from S. Standring (ed.), *Gray's Anatomy*, 40th edition, Edinburgh, Elsevier, 2008, with permission.

sheet (Rakic 2006). Historically, the important finding that cortical neurons migrate from their place of origin near the cerebral ventricles (in the ventricular zone) to their final destinations in the overlying cortex was discovered towards the turn of the nineteenth century by Wilhelm His, Ramón y Cajal and others (Rakic 2006). Cajal's original drawing of Golgi-stained neonatal human visual cortex showed transient neurons, which have been named Cajal–Retzius neurons in honor of their co-discoverers (Bystron *et al.* 2008).

This pattern of corticogenesis was confirmed in the 1960s by Angevine, Sidman, Rakic, and others using the technique of tagging replicating deoxyribonucleic acid (DNA) in dividing cells of developing mammalian cerebral cortex with radiolabeled tritiated thymidine (3H-TdR); only those mature neurons which derived from cells dividing at the time of the radiolabeling will contain this radioactive marker.

MAIN STAGES AND ZONES OF THE CORTICAL WALL

The main stages in the formation of the layered structure of the human cerebral cortex are illustrated in Figure 1.24, which also shows the successive zones from the ventricular zone, adjacent to the cerebral ventricle, outwards to the marginal zone, adjacent to the pia mater. The migrating neurons stop migrating in the cortical plate, which is just deep to and adjacent to the marginal zone. Neurons migrating later end up traveling past earlier neurons (which are no longer migrating) and end up nearer the pia mater. Hence, the cortical plate has an inside-out structure.

The different zones of the cortical wall were formalized by the Boulder Committee in 1970 and based in large measure on the work of Rakic. Recently, in light of advances in our knowledge of corticoneurogenesis, the following revisions of the Boulder model have been proposed by Bystron, Blakemore, and Rakic (2008):

- A transient layer with a diverse population of neurons forms between the neuroepithelium and the pial surface of the dorsal telencephalon before the appearance of the cortical plate (CP). We suggest that the term preplate, which is already widely used, should be adopted for this layer.
- The subventricular zone appears as a distinctive proliferative layer before the emergence of the CP, earlier than previously recognized.
- There is no distinct cell-sparse layer under the pial surface before the CP forms. Thus, the term marginal zone should be used only after the appearance of the CP, to refer to the residual superficial part of the preplate, which becomes layer 1 of the mature cortex.
- The term intermediate zone (IZ) has been used in various ways. We propose that it should be reserved for the heterogeneous compartment that lies between the proliferative layers and the postmigratory cells above. The IZ contains radially and tangentially migrating cells and a thickening layer of extrinsic axons that eventually constitutes the white matter.
- The subplate (SP) is a distinct and functionally important transient layer, located directly below the cortical

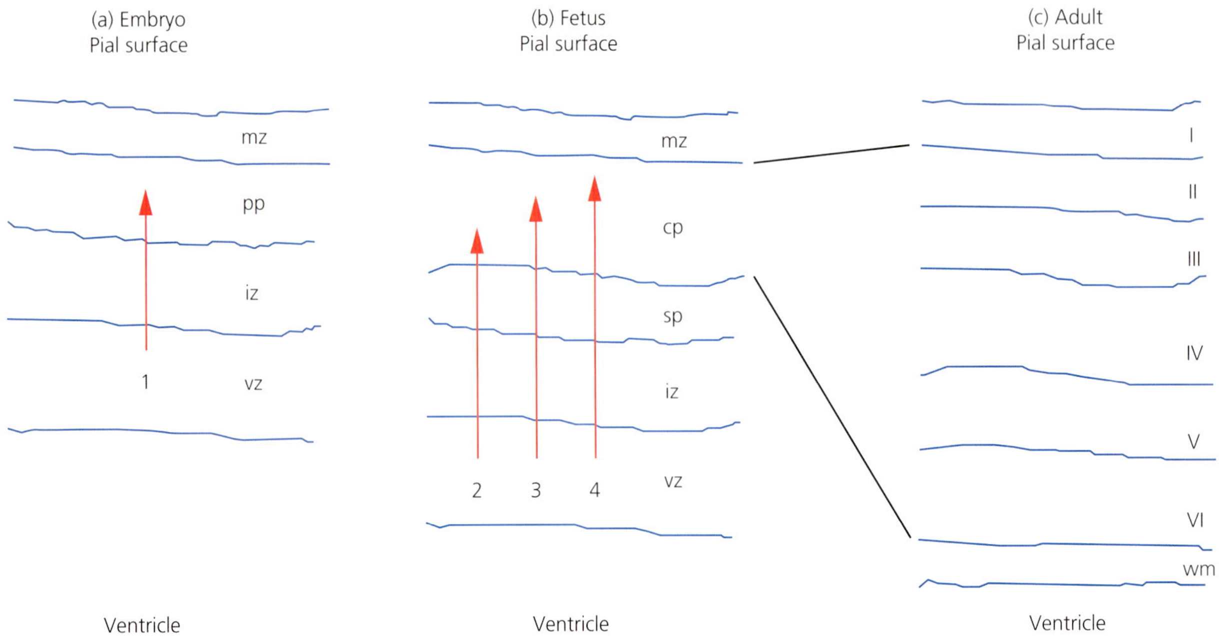

Figure 1.24 Main stages in the formation of the layered structure of the cerebral cortex. (a) The earliest migrating postmitotic neuroblasts (1) leave the ventricular zone (vz), crossing the intermediate zone (iz) to form the preplate (pp). (b) The preplate is subsequently split by the settling of later migrating neuroblasts. Successive waves of centrifugally migrating neuroblasts (2–4) establish the 'inside-out' structure of the cortical plate (cp), situated between the subplate (sp), the remnant of the inner aspect of the preplate and the marginal zone (mz). (c) Mature layered structure of the cerebral cortex, demonstrating the origin of layers II–VI from the cortical plate, whereas layer I (mz) Cajal–Retzius neurons derive largely from the outer portion of the preplate white matter (wm). Reproduced from Love S., Louis D.N. and Ellison D.W., (eds.) *Greenfield's Neuropathology*, 8th edition, Hodder Arnold, 2008, with permission.

plate, which was not recognized by the Boulder Committee. In rodents and carnivores most SP neurons are born before the first CP cells. In humans, preplate cells also contribute to the SP, but its substantial thickening at later stages probably involves the addition of later-born neurons.

The above inside-out pattern of development holds for all areas of the cerebral cortex, even though the time course varies for different cortical regions. For example, the striate cortex (Brodmann area 17, primary visual cortex, V1) develops over a longer period of time than, say, Brodmann area 24.

NEURONAL MIGRATION

Figure 1.25 is a schematic diagram of the composition of, and neuronal migration in, the developing cerebral wall. Note in particular that the radial glia appear to offer a form of scaffolding from the ventricular zone to the marginal zone of the developing cerebral cortex, along which postmitotic neuronal cells travel.

Subplate neurons are the first cortical neurons to mature. They direct migrating cortical plate neurons and targeting of their axonal projections, and are selectively vulnerable to early hypoxic–ischemic brain injury in animal models; timing of subplate neuron death determines the resulting defect in thalamocortical development: very early excitotoxic subplate neuron death results in failure of thalamocortical innervation, while later subplate neuron death interferes with the refinement of thalamocortical connections into mature circuits (Folkerth and Kinney 2008; McQuillen and Ferriero 2005).

RADIAL-UNIT HYPOTHESIS

According to Rakic (1995):

> The manner by which postmitotic cells migrate and become distributed is crucial for understanding how the neocortex forms as a sheet rather than as a lump. The waves of postmitotic neurons that are generated within the same site in the ventricular zone arrive successively at the cortical plate where they pass by each other, and become arranged radially in the form of cell stacks – named ontogenetic columns. Therefore, a radial unit consists of cells that originate from several clones but share the same birthplace, migrate along the common pathway, and settle within the same ontogenetic column.

Figure 1.26 illustrates how the developing cortical plate consists of these ontogenetic columns. The corresponding radial-unit hypothesis postulates that the size of the cerebral cortex depends on the number of contributing radial units, which, in turn, depends on the number of founder cells (Rakic 1995).

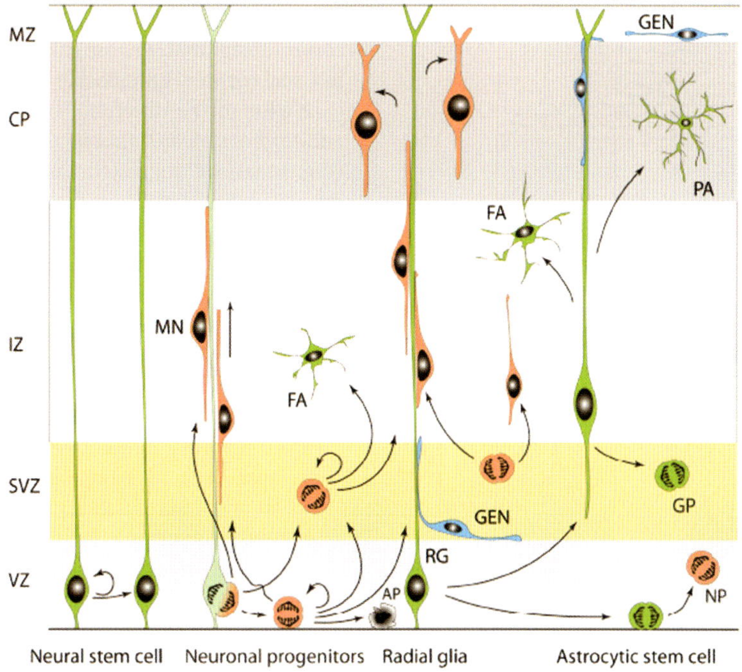

Figure 1.25 Schematic diagram of the composition, cellular events, and relationship in the developing cerebral wall, based on data obtained from species ranging from the mouse to human and non-primates. Initially, homogeneous population of neural stem cells (long fibers), which divide symmetrically, transform into radial glial cells (RG), which divide asymmetrically and over time produce migrating neurons (MN, short fibers) and dedicated neuronal progenitors with short processes (mitotic divisions) that populate the ventricular (VZ) and subventricular (SVZ) zones and produce all projection neurons, as well as the majority of interneurons in humans and a small fraction in rodents. In addition, a population of tangentially migrating neurons arrives to the dorsal telencephalon from the ganglionic eminence mostly via SVZ and marginal (MZ) zones to supply the majority of interneurons in rodents and about one-third in humans. Eventually, RG undergo apoptosis (AP) or directly or indirectly generate ependymal cells, fibrillary astrocytes (FA), protoplasmic astrocytes (PA), glial progenitors (GP), or astrocytic stem cells that retain a neurogenic potential (NP). Reproduced from Rakic P. (2006), with permission.

Neurogenesis in the adult human brain

Neurogenesis in the adult human brain was demonstrated by Eriksson and colleagues (Eriksson *et al.* 1998). They studied the dentate gyrus and the subventricular zone in humans, as these regions had been shown to generate new neurons in the postnatal and adult period in infrahuman mammalian species. Patients suffering from squamous cell carcinomas at the base of the tongue, in the larynx, or in the pharynx received bromodeoxyuridine (BrdU), dissolved in saline and given as an intravenous infusion, to assess the proliferative activity of the tumor cells, expressed as BrdU-labeling index; BrdU is a synthetic form of thymidine which has been utilized as a marker of neurogenesis in infrahuman species. The brains of the patients were studied postmortem. No signs of macroscopic or microscopic metastases were found in autopsy material from the cerebrum in any of the patients (Eriksson *et al.* 1998). Newly generated cells were found in the hippocampus; labeling with NeuN, a neuronal marker, and comparison with the adult rat brain demonstrated that these newly generated cells were indeed neurons.

Synaptogenesis and synaptic elimination in the human brain

Huttenlocher and de Courten published the results of a postmortem study in 1987 in which synaptic density in human striate cortex was determined at various ages, together with measurement of total volume of striate cortex, making it possible to estimate the total number of synapses (Huttenlocher and de Courten 1987). Their key findings included (Huttenlocher and de Courten 1987):

- Synaptogenesis in human striate cortex was found to be most rapid between ages 2 and 4 months, a time which also is critical for the development of function in visual cortex of the infant.
- Synapse elimination occurred subsequently, with loss of about 40 percent of synapses between ages 8 months and 11 years.
- Synapse numbers were held stable in adults.
- Analysis by individual cortical layers showed similar age-related changes in all strata of striate cortex, except for somewhat later synaptogenesis in cortical layers V and VI.
- Total volume of striate cortex reached adult size remarkably early, at about age 4 years.

These findings supported the hypothesis that exuberant synaptic connections are an anatomical substrate for plasticity in developing cerebral cortex (Huttenlocher and de Courten 1987).

Subsequently, Huttenlocher and Dabholkar carried out a study in which they compared the formation of synaptic contacts in human cerebral cortex in two regions: auditory cortex

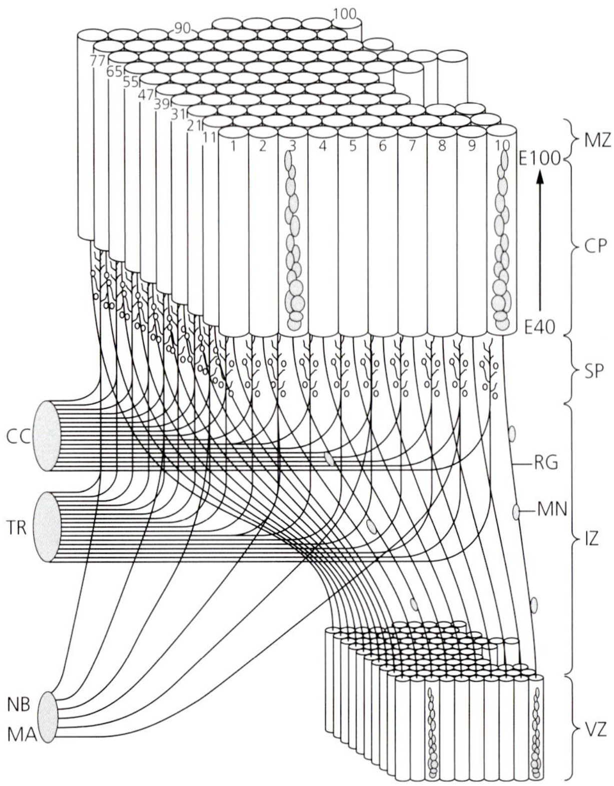

Figure 1.26 The basic developmental events and types of cell–cell interactions that occur during the early stages of corticogenesis, before formation of the final pattern of cortical connections. This diagram emphasizes radial migration, a predominant mode of neuronal movement that, in primates, underlies the elaborate columnar organization of the neocortex. After their lost division, cohorts of migrating neurons (MN) first traverse the intermediate zone (IZ) and then the subplate zone (SP), where they have an opportunity to interact with 'waiting' afferents that arrive sequentially from the nucleus basalis (NB) and monoamine (MA) subcortical centers, from the thalamic radiation (TR), and from several ipsilateral and contralateral corticocortical bundles (CC). After newly generated neurons bypass the earlier generated ones that are situated in the deep cortical layers, they settle at the interface between the developing cortical plate (CP) and the marginal zone (MZ) and, eventually, form a radial stock of cells that share a common site of origin but are generated at different times. For example, neurons that are produced between embryonic day E40 and E100 in radial unit 3 follow the same radial glial fascicle, and form ontogenetic column 3. Although some cells, presumably neurophilic in nature of their surface affinities, might detach from the cohort and move laterally, guided by an axonal bundle (for example, horizontally oriented cell leaving radial unit 3), most postmitotic cells are gliophilic, for example have affinity for the glial surface, and obey constraints strictly that are imposed by transient radial glial scaffolding (RG). This cellular arrangement preserves relationships between the proliferative mosaic of the ventricular zone (VZ), and the corresponding protomap within the SP and CP, even though the cortical surface in primates shifts considerably during the massive cerebral growth encountered in mid-gestation. Reproduced from Rakic P. (1995), with permission.

(Heschl's gyrus) and prefrontal cortex (middle frontal gyrus) (Huttenlocher and Dabholkar 1997). The combined results of this study and the previous 1987 study (Huttenlocher and de Courten 1987) of the striate cortex showed that the time course of synaptogenesis in visual cortex resembles that in auditory cortex, except that synapse elimination appears to begin earlier in visual cortex (Huttenlocher and Dabholkar 1997).

A plot of the difference between the synaptic density in auditory cortex and that in the middle frontal gyrus against conceptual age suggested earlier synaptogenesis and earlier synaptic elimination in auditory cortex compared with prefrontal cortex (Huttenlocher and Dabholkar 1997). As the authors concluded (Huttenlocher and Dabholkar 1997):

> The present data show regional differences in synaptogenesis in human neocortex both in fetal brains and postnatally up to at least age 15 months (CA [conceptual age] 700 days) in cortical layers 2 to 3, and up to at least age 3 months (CA 363 days) in the lower cortical layers. A rapid burst in synaptogenesis occurs postnatally in visual and auditory cortex. Prefrontal cortex appears to acquire synaptic junctions more slowly. By age 3.5 years, prefrontal cortex has caught up to auditory cortex. In primary visual cortex net synapse elimination appears to have started already at that age.

Cerebral plasticity

NON-HUMAN MAMMALIAN ADULT BRAIN

Reorganization of cortical representation in the adult has been demonstrated in different mammalian species. For example, in the late 1980s Robertson and Irvine examined the effect of restricted unilateral cochlear lesions on the orderly topographic mapping of sound frequency in the auditory cortex of adult guinea pigs (Robertson and Irvine 1989). These lesions, although restricted in spatial extent, resulted in a variety of patterns of histological damage to receptor cells and nerve fibers within the cochlea; nevertheless, all lesions resulted in permanent losses of sensitivity of the cochlear neural output across a limited frequency range, and 35–81 days after such damage to the organ of Corti, the area of contralateral auditory cortex in which the lesioned frequency range would normally have been represented was partly occupied by an expanded representation of sound frequencies adjacent to the frequency range damaged by the lesion (Robertson and Irvine 1989).

In the early 1990s, a demonstration of plasticity in the frequency representation of primary auditory cortex in the adult brain of owl monkeys was published by Recanzone, Schreiner, and Merzenich, who reported increased cortical area of representation of a restricted frequency range in primary auditory cortex that was correlated with performance at a frequency discrimination task (Recanzone *et al.* 1993). Monkeys trained for several weeks to discriminate small differences in the frequency of sequentially presented tonal stimuli revealed a progressive improvement in performance with training; at the end of the training period, the tonotopic organization of A1 was defined by recording multiple-unit responses at 70–258 cortical locations. These responses were compared with those derived from three normal monkeys and from two monkeys that received the same auditory stimuli but that were engaged in a tactile discrimination task. The cortical representation, sharpness of tuning, and latency of the response were greater for the behaviorally relevant frequencies of trained monkeys when compared with the same frequencies of the control monkeys; the cortical area of representation was the only studied parameter which was correlated with behavioral performance (Recanzone *et al.* 1993). This experiment showed that natural stimulation can modify the tonotopic organization of A1 in the adult primate, and that this alteration is correlated with changes in perceptual acuity.

To determine the relative contributions of intrinsic and extrinsic factors to the functional specification of different cortical regions, von Melchner, Pallas, and Sur studied ferrets, for which it was known that when retinal projections are redirected neonatally to the auditory thalamus they have visually responsive cells in auditory thalamus and cortex, form a retinotopic map in auditory cortex, and have visual receptive field properties in auditory cortex that are typical of cells in visual cortex. They found that this cross-modal projection and its representation in auditory cortex can mediate visual behavior; when light stimuli were presented in the portion of the visual field that was 'seen' only by this projection, 'rewired' ferrets responded as if they perceived the stimuli to be visual rather than auditory (von Melchner *et al.* 2000). The authors of this study deduced that the perceptual modality of a neocortical region is instructed to a significant extent by its extrinsic inputs. In addition, gratings of different spatial frequencies could be discriminated by the rewired pathway, although the grating acuity was lower than that of the normal visual pathway.

ADULT HUMAN BRAIN

Plasticity has now been demonstrated in the living human cerebral cortex, using techniques such as fMRI, positron emission tomography (PET), and MSI (magnetic source imaging), the last of which involves combining structural MRI (magnetic resonance imaging) with MEG (magnetoencephalography). Some examples are now given.

Flor, Elbert, Knecht and colleagues showed that, following upper limb amputation, phantom-limb pain is a perceptual correlate of cortical reorganization (Flor *et al.* 1995). They used MSI to study 12 men and one woman who had had an arm amputated (either post-traumatic or because of osteosarcoma). There was a large significant positive correlation between the amount of cortical reorganization and the degree of phantom-limb pain (but not non-painful phenomena); phantom-limb pain was found to explain 83 percent of the variance in cortical reorganization. These results indicated that phantom-limb pain is related to, and may be a consequence of, plastic changes in primary somatosensory cortex (Flor *et al.* 1995). Five of the subjects were relatively free of phantom-limb pain. The mean shift in the focus of cortical responsivity to facial stimulation was 0.43 cm for these five pain-free subjects but almost five times higher, at 2.05 cm, for the eight remaining subjects who suffered phantom-limb pain.

When learning to play a stringed musical instrument, it is the digits of the left hand, apart from the thumb, which are used to find notes on the strings themselves. (The right hand is used for bowing or plucking the strings.) The same group showed that the somatosensory cortical representation of the left hand of string players is expanded compared with controls; the effect was smallest for the left thumb, and no such differences were seen for the representations of the right-hand digits (Elbert *et al.* 1995). The amount of cortical reorganization in the representation of the fingering digits was correlated with the age at which the person had begun to play. These results suggest that the representation of different parts of the body in the primary somatosensory cortex of humans depends on use and changes to conform to the current needs and experiences of the individual (Elbert *et al.* 1995).

Liepert and colleagues spatially mapped the motor cortex using focal transcranial magnetic stimulation (TMS) before and after 2 weeks of treatment with constraint-induced movement therapy following stroke. Motor-output areas of the abductor pollicis brevis muscle, motor-evoked potential (MEP) amplitudes, and the location of the center of gravity of motor cortex output were studied. Changes observed after constraint-induced movement therapy included: substantially improved motor performance in all patients; an increase of motor output area size and MEP amplitudes, indicating enhanced neuronal excitability in the damaged hemisphere for the target muscles; and a considerable shift in the mean center of gravity of the motor output maps, indicating the recruitment of motor areas adjacent to the original location (Liepert *et al.* 1998). The authors concluded that even in chronic stroke patients, reduced motor cortex representations of an affected body part can be enlarged and increased in level of excitability by an effective rehabilitation procedure (Liepert *et al.* 1998). Furthermore, the cortical regions that produce abnormal slow waves, indicating dysfunctional cortex, become reduced after such rehabilitative therapy (Elbert and Rockstroh 2004).

From these and other studies demonstrating cerebral plasticity in adult humans, Elbert and Rockstroh have advanced the following general principles of cortical reorganization (Elbert and Rockstroh 2004):

- The abolition of sensory input or deafferentation, as produced by amputation or segregation of a sensory nerve, results in the 'invasion' of adjacent cortical representations of intact parts of the sensorium into the cortical representation zone of the affected sensory part.
- Increased use of a receptor pool leads to an expansion of the respective cortical representation zone.
- Training of extremity use after central nervous system injury that affected the cortical tissue representing that body part results in improved extremity function and reorganization in brain activity.
- Cortical reorganization emerges in response to massed practice of behavioural relevant tasks. This generally implies an intense training schedule of several hours a day for several successive days. The absolute amount of training seems less important than the priority of training in daily activity.
- Reorganization through synchronicity of input leads to disarrangement of representational zones.

Table 1.4 Major dopaminergic pathways of the brain

Pathway	Origin	Destination
Nigrostriatal	Substantia nigra	Corpus striatum
Mesolimbic	Ventral tegmental area	Nucleus accumbens
Mesocortical	Ventral tegmental area	Prefrontal cortex
Tuberoinfundibular	Hypothalamus	Anterior pituitary gland
A fifth pathway	Multiple sites, including the periaqueductal gray, ventral mesencephalon, hypothalamic nuclei, and lateral parabrachial nucleus	Thalamus

After Stahl SM (2008) *Stahl's Essential Psychopharmacology*, 3rd edn. Cambridge: Cambridge University Press.

Major neurochemical pathways

DOPAMINERGIC PATHWAYS

The major dopaminergic pathways of the human brain include the nigrostriatal, mesolimbic, mesocortical, and tuberoinfundibular dopaminergic pathways. These are described in Table 1.4.

The nigrostriatal dopaminergic pathway has its origin in the pars compacta of the substantia nigra of the mesencephalon (midbrain) and projects to the corpus striatum of the basal ganglia; the pars compacta is part of dopaminergic cell group A9. The nigrostriatal pathway is sometimes referred to as the mesostriatal pathway.

The origin of the mesolimbic and mesocortical dopaminergic pathways is also in the mesencephalon, in the ventral tegmental area (dopaminergic cell group A10). The mesolimbic pathway projects to the nucleus accumbens of the limbic system, while the mesocortical pathway projects to the prefrontal cortex and probably also to the cingulate and entorhinal cortices.

NOREPINEPHRINERGIC PATHWAYS

Major ascending norepinephrinergic (noradrenergic) projections are mainly from the locus ceruleus of the pons. Most ascend in the medial forebrain bundle, an exception being the tract supplying the cerebellum, which does so via the superior cerebellar peduncle. Ascending projections pass to many regions, including the thalamus, hypothalamus, amygdala, hippocampus, striatum, basal forebrain, and the cerebral cortex (Stahl 2008). The ceruleospinal pathway is the major descending noradrenergic projection to the spinal cord.

CHOLINERGIC PATHWAYS

Two major brain cholinergic (acetylcholine) pathways are of importance in neuropsychiatry. One is the brainstem cholinergic pathway, which takes its origin in the brainstem and projects to many regions, including the thalamus, hypothalamus, amygdala, hippocampus, basal forebrain, and cerebral cortex (Stahl 2008).

The second is the basal forebrain cholinergic pathway, which originates in the basal forebrain and projects widely to the prefrontal cortex, amygdala, and hippocampus (Stahl 2008).

CORTICOFUGAL GLUTAMATE SYSTEM

Corticofugal fibers project from the cerebral cortex to subcortical structures including the basal ganglia and thalamus. These fibers form the fan-shaped arrangement of white matter known as the corona radiata (*vide supra*). As the corona radiata appears as a curved linear area of low attenuation on computed tomography (CT) scans of the brain, it is also known as the centrum semiovale, particularly in the field of structural neuroimaging. As mentioned in the above section on the basal ganglia, cortical–striatal–thalamic–cortical loops are likely to subserve many important functions. These and similar loops of importance in neuropsychiatry include the DLPFC–striatal–thalamic–DLPFC loop for executive functions, the anterior cingulate cortex–striatal–thalamic–anterior cingulate cortex loop for attention, the orbitofrontal cortex–caudate–thalamic–orbitofrontal cortex loop for impulsivity, and possibly an anterior cingulate cortex–nucleus accumbens–thalamic–anterior cingulate cortex loop for some aspects of emotion. As the neurotransmitter output from the cortical pyramidal neurons that begin and end these loops is the amino acid glutamate, this whole arrangement is referred to as the corticofugal glutamate system.

SEROTONERGIC PATHWAYS

Ascending serotonergic projections originate in the brainstem and project widely, for example to the thalamus, hypothalamus, amygdala, hippocampus, nucleus accumbens, striatum, basal forebrain, and the cerebral cortex (Stahl 2008).

Descending serotonergic projections pass to the spinal cord.

The origins of the serotonergic pathways are from three groups of midline brainstem raphe nuclei: the medullary raphe group, comprising the nucleus raphe pallidus, nucleus raphe obscurus, and nucleus raphe magnus, from which project the descending bulbospinal tract into the spinal cord and the propriobulbar tract to the inferior olivary complex; the mesencephalic raphe group, comprising the nucleus raphe dorsalis, from which project ascending tracts; and the pontine raphe group, comprising the nucleus raphe pontis and nucleus centralis superior, from which project ascending tracts, the pontocerebellar tract to the cerebellum, and propriobulbar tracts to the locus ceruleus and reticular formation.

1.2 FUNCTIONAL NEUROIMAGING

An introduction is given in this section to functional neuroimaging techniques based on magnetic resonance and on radioligands. The former includes blood oxygen level-dependent (BOLD) functional magnetic resonance imaging (fMRI), magnetic resonance spectroscopy (MRS), and arterial spin labeling (ASL), while the latter includes single-photon emission computed tomography (SPECT) and positron emission tomography (PET). The emphasis is on their applications in neuropsychiatry. Detailed aspects of the physics underlying these techniques are beyond the scope of this book. Structural neuroimaging is considered in Chapter 2.4.

BOLD functional magnetic resonance imaging

In 1991, Bruce Rosen's group at Harvard Medical School published the first quantitative determination of human brain activation using MRI. This fMRI study used the measurement of cerebral hemodynamics, following the injection of the MRI contrast agent gadolinium-DTPA (Gd-DTPA). During photic stimulation, localized increases in blood volume were detected in V1 in a study involving seven subjects (Belliveau *et al.* 1991).

The following year, the same group published a non-invasive fMRI study that did not involve injections of a contrast agent (Kwong *et al.* 1992). Instead, it relied on the BOLD effect, which depends on the fact that, in contrast to oxyhemoglobin (which is diamagnetic), deoxyhemoglobin (which is paramagnetic, having unpaired electrons) may be used as an endogenous contrast agent. In theory, the greater the activation of a given brain region, the greater is its use of oxygen. This in turn leads to a greater concentration of deoxyhemoglobin. Hence greater activation is associated with greater signal intensity in BOLD fMRI. In this first human study, V1 was again shown to be activated during photic stimulation in seven volunteers, while M1 was found to be activated during repetitive contralateral hand squeezing.

BOLD fMRI rapidly established itself as a standard non-invasive functional neuroimaging technique, and it is now in widespread use.

Magnetic resonance spectroscopy

From a neuroimaging viewpoint, nuclei that can interact usefully with the strong magnetic field of an MRI scanner include protons (^{1}H), 19-fluorine, 31-phosphorus, and 13-carbon; the last of these is an isotope with an abundance of just over 1 percent in naturally occurring carbon. By exposing the brain to short pulses of radiofrequency energy at the resonance frequency of such nuclear species, some of these nuclei absorb the energy and enter a higher quantum energy state. As these nuclei return to their previous quantum energy state, they release this energy, which can be measured as an amplified signal by the MRI scanner head coil receiver (Cox and Puri 2004).

To improve the signal-to-noise ratio, multiple readings are taken. As the noise is assumed to be random, successive additions of the readings lead to relative enhancement of the true signals against a diminishing background of noise signals, which tend to cancel each other out. Furthermore, in a given molecule, all the protons do not usually resonate at the same frequency, and all the 31-phosphorus nuclei do not usually resonate at another single frequency (different from that for the protons). This is because the resonance frequencies for given nuclei are partly determined by the electron structure of the molecule. The difference between the resonance frequencies of a given nucleus and of a reference nucleus is calculated and the chemical shift of the given nucleus is then defined in terms of the ratio of this difference to the resonance frequency of the reference nucleus; the chemical shift is usually expressed in parts per million (ppm). The signals are plotted as peaks against the chemical shift, and the area under each peak is directly proportional to the concentration of the corresponding nucleus in the sample (for instance, a brain voxel) under study, while the shape of the peak(s) gives information about the electrochemical environment of the nucleus in the molecule (Cox and Puri 2004).

Figure 1.27a shows a normal brain proton magnetic resonance spectrum. As water is, by far and away, the most abundantly occurring substance containing 'free' hydrogen protons in the brain, the peak corresponding to water is much larger than that from other substances. Therefore, while acquiring proton MRS (^{1}H MRS) data, it is usually necessary to suppress this water peak, using a suitable spectroscopy sequence. Of the remaining peaks, that corresponding to *N*-acetylaspartate (NAA) is particularly prominent and may be thought of as being a neuronal marker. In Figure 1.27c it can be seen that the NAA peak is reduced in area in a glioblastoma, as a result of there being fewer normally functioning neurons in the tumor. Choline (Cho), among other functions, acts as a polar head group in phospholipid molecules. Lactate is not normally seen in the human brain (apart from the brains of some prematurely born infants) and it may become present during anaerobic respiration (Figure 1.27c). Proton MRS carried out at a (stronger magnetic field 3 tesla [T] or above) allows other peaks to be identified and is therefore more informative.

Figure 1.28a shows a spectrum from a 31-phosphorus MRS (^{31}P MRS) study of the brain in schizophrenia (Puri *et al.* 2004). The peak labeled PME (phosphomonoesters) corresponds to the phosphoethanolamine and phosphocholine peaks and indexes membrane phospholipid anabolism, while the peak labeled PDE (phosphodiesters) corresponds to the glycerol 3-phosphoethanolamine and glycerol 3-phosphocholine peaks and indexes phospholipid catabolism. Other peaks shown in Figure 1.28a are inorganic phosphate (Pi), phosphocreatine (PCr) and γ, α, and β nucleotide triphosphate (γ-, α-, and β-NTP). β-NTP indexes ATP (adenosine triphosphate). (The majority of ADP (adenosine diphosphate) is 31-phosphorus NMR-invisible, although the γ-NTP signal overlaps with signals from ADP.) The broad component from the 31-phosphorus neurospectroscopy study shown in Figure 1.28b indexes brain cell motion-restricted membrane phospholipids; this has been reported as being unchanged in schizophrenia (Puri *et al.* 2008).

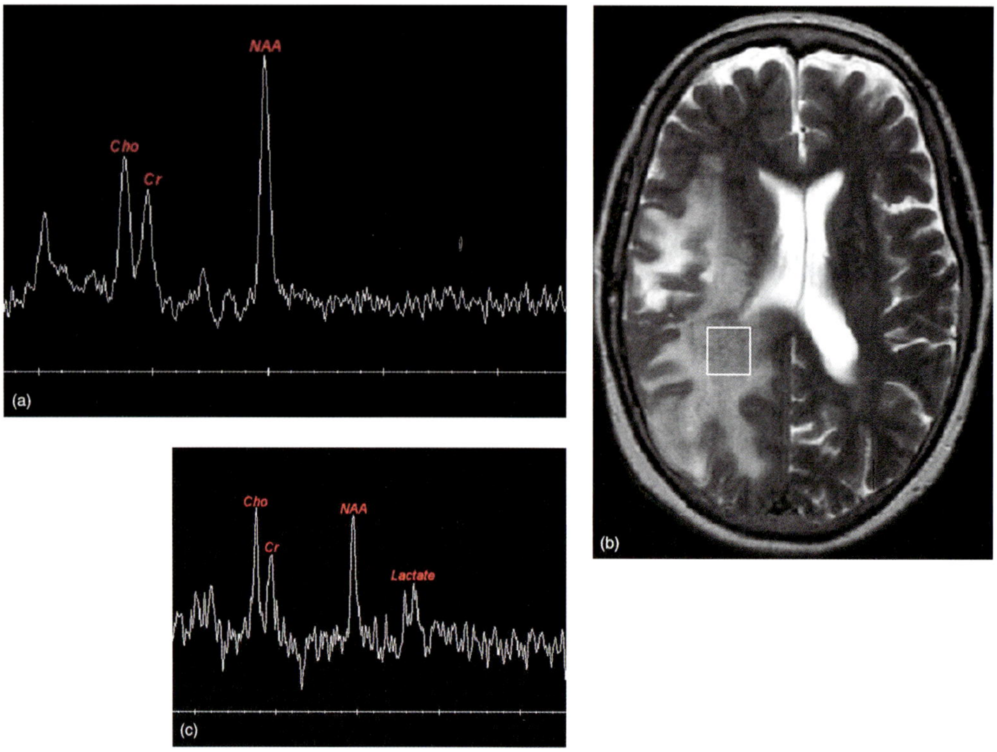

Figure 1.27 Proton magnetic resonance spectroscopy. (a) Normal proton spectrum of the brain showing the key metabolites *N*-acetylaspartate (NAA), creatine (Cr), and choline (Cho). (b) A mass lesion surrounded by vasogenic edema on a T2-weighted transverse section showing the voxel sampled in (c). (c) A typical spectrum from a glioblastoma showing reduced NAA confirming a reduction in normally functioning neurons, a raised choline indicating an increased membrane turnover, and little change in creatine. The lactate peak indicates anaerobic respiration. Reproduced from Graham DI, Nicoll JAR and Bone I, *Adams & Graham's Introduction to Neuropathology* 3rd edition, Hodder Arnold, 2006, with permission.

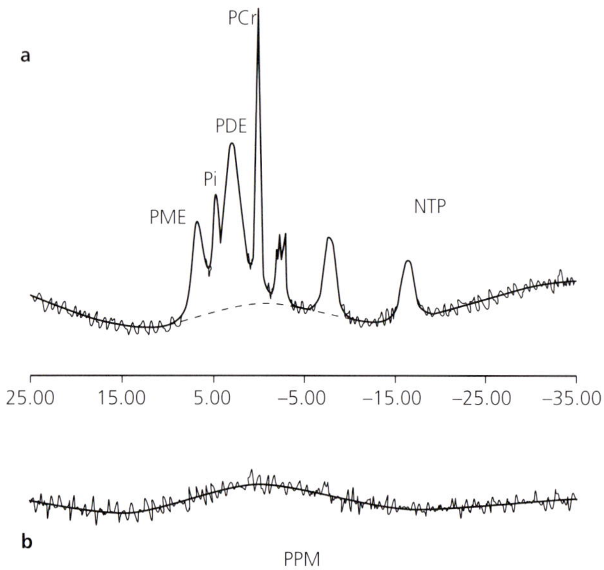

Figure 1.28 ^{31}P magnetic resonance spectroscopy. (a) Fitting seven peaks (darker line) to the ^{31}P spectrum (lighter line). The dotted line shows the baseline from the broad component, which is also shown in (b). Reproduced from Puri BK, Counsell SJ, Hamilton G *et al.* (2004), with permission.

Arterial spin labeling

Arterial spin labeling is a form of fMRI in which endogenous water molecules in arterial blood are given a special magnetization tag ('inversion') in the neck before entering the brain. The tagged images of the brain are then collected and allow cerebral blood flow to be determined. Owing to the lack of ionizing radiation, ASL is replacing PET (*vide infra*) in some studies. At the time of writing, ASL is still mainly a research tool.

Single-photon emission computed tomography

Single-photon emission computed tomography (SPECT), also known as SPET (single-photon emission tomography), is a form of functional imaging that involves the use of a radioactive ligand. For assessing cerebral blood flow, the radioisotope or radiolabeled ligand needs to be introduced into the cerebral circulation. This can be achieved by, for example, injecting a solution of the radioactive ligand into the bloodstream or by asking the subject to inhale a gaseous form of the radioactive ligand. An example of the former is the radioligand ^{99m}Tc-HMPAO (99m-technetium hexamethylpropylene amine oxime) and an example of the latter is the gas 133-xenon (^{133}Xe).

The radioactive ligand emits single γ photons, which are detected using gamma cameras. In the case of SPECT

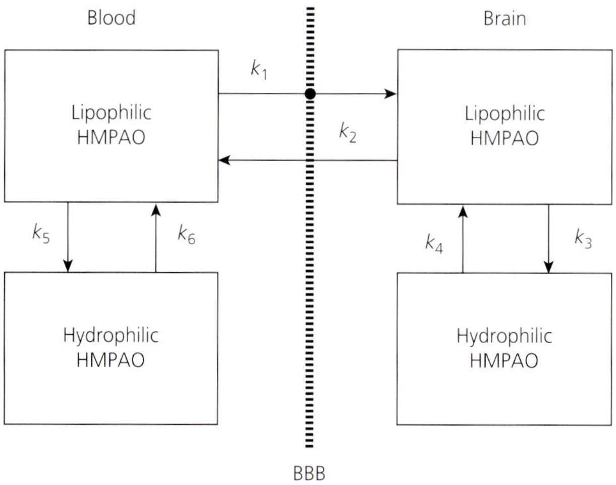

Figure 1.29 A four-compartment model describing the kinetics of ^{99m}Tc-HMPAO in the brain: k_1 to k_6 denotes constants of first-order kinetics. Lipophilic HMPAO is freely diffusible from blood to brain; hydrophilic HMPAO is non-diffusible. Since the conversion from the lipophilic to the hydrophilic form is almost irreversible $k_4 = k_6 = 0$. BBB, blood–brain barrier. Reproduced from Borch K, Greisen G. (1997), with permission.

neuroimaging, the reconstruction of the photon data may reflect cerebral blood flow or cerebral tissue binding, depending on the nature of the ligand used.

One particular advantage of SPECT is that some 99m-technetium-based ligands can be used to determine the cerebral blood flow at a time when the subject is not even in the same room as the gamma camera. For example, ^{99m}Tc-HMPAO may be injected into the bloodstream in its lipophilic form. In this form, it can readily cross the blood–brain barrier. However, having done so, it may then be converted into a hydrophilic form in a reaction that is almost irreversible (see Figure 1.29)(Borch and Greisen 1997). As the hydrophilic form does not freely cross the blood–brain barrier, one has, as it were, obtained a binding of the radioligand in the brain that corresponds to the cerebral blood flow shortly after the time of injection into the bloodstream. There is then a window of opportunity of at least half an hour, before the level of radioactivity has decayed too much, during which the subject may undergo SPECT scanning. This is useful when trying to study unpredictable symptomatology, such as auditory hallucinations or epilepsy, which occurs when the subject is at a location different from that of the gamma camera (e.g. a patient with schizophrenia may be on a hospital ward at the time when he experiences auditory hallucinations). While the spatial resolution of SPECT (of the order of 10 mm) is generally poorer than that of PET (around 5–7 mm) and fMRI, the advantage just mentioned of SPECT with radiotracers such as ^{99m}Tc-HMPAO has made it a useful neuroimaging modality in some neuropsychiatric research studies. In schizophrenia, for example, SPECT has been used to demonstrate increased blood flow in Broca's area during auditory hallucinations (McGuire *et al.* 1993), and to study abnormal saccadic distractibility (Crawford *et al.* 1996). Similarly, inter-ictal SPECT studies can be carried out in epilepsy. SPECT may have a role clinically in demonstrating early changes in cerebral blood flow in disorders such as Alzheimer's disease (Nishimiya *et al.* 2008).

Positron emission tomography

Positron emission tomography (PET) is a form of functional imaging that is currently being replaced in many centers by scanners which combine PET and multi-slice helical CT. These combined PET/CT scanners allow structural CT data to be collected rapidly and then allow the PET functional data to be registered accurately to the structural data. The subject being scanned does not need to change his position in respect of the table on which he is lying; the CT and PET components are adjacent to each other in the single-scanner gantry.

In PET neuroimaging, a positron-emitting radionuclide is introduced into the cerebral circulation; as with SPECT (*vide supra*), common methods are via intravenous administration of a radiotracer in solution or by inhalation of a gaseous radiotracer. A difference from SPECT is that in the brain the radionuclide emits a positron and a neutrino. After the positron interacts with an electron the PET scanner detects the resulting dual-photon emission; with SPECT only one photon is emitted. The PET scanner detects these two photons (which, as they travel in opposite directions, are termed antiparallel) 'in coincidence'; two suitable photons detected within a certain period of each other, the 'coincidence window', are assumed to originate from the same positron–electron interaction. Not all coincidence pairs of photons are true coincidences that relate to the image. Some are caused by photon scattering or by random single-photon emissions. These scatter and random coincidences cause background noise in the imaging.

Properties and production routes for radionuclides commonly used in PET are shown in Table 1.5. This table gives the maximum energy, E_{max}, of positrons emitted from the radionuclides. The higher the positron energy, the further the positron can travel in tissue and the poorer the spatial resolution of the corresponding PET image.

In Table 1.5, the positron-emitting radionuclides, which have low atomic numbers, that is low numbers of protons in the nucleus, can be produced in small cyclotrons. These radionuclides are ^{11}C, ^{13}N, ^{15}O, and ^{18}F.

The radiotracer ^{18}F-FDG (or just FDG), based on the radionuclide ^{18}F, is 2-[^{18}F]fluoro-2-deoxy-D-glucose and is a marker for glucose metabolism. It is a synthetic analog of the glucose molecule in which an hydroxyl group is replaced with a radioactive fluorine-18 atom. Like glucose, ^{18}F-FDG is readily transported into cells from the plasma, but, unlike glucose-6-phosphate, after intracellular phosphorylation by hexokinase, ^{18}F-FDG-6-phosphate is not acted upon by glucose-6-phosphatase and so remains trapped intracellularly, as shown in Figure 1.30. With a relatively long half-life of approximately 110 minutes and a cellular uptake similar to that of glucose, ^{18}F-FDG is a particularly good radiotracer for use in PET neuroimaging. There is a relatively high uptake, of approximately 6 percent, of administered ^{18}F-FDG by the brain. Another advantage of this radiotracer is that units are now available that allow automated synthesis from reagents

Table 1.5 Properties and production routes for positron emission tomography radionuclides

Radionuclide	Half-life	E_{max} (MeV)	Production reaction	Target	Product
^{11}C	20.4 min	0.96	$^{14}_{7}N(p,\alpha)^{11}_{6}C$	$^{nat}N_2$ gas	$^{11}CO_2$ gas
^{13}N	9.96 min	1.19	$^{16}_{8}O(p,\alpha)^{13}_{7}N$	^{nat}O water	$^{13}NH_4^+$ ion
^{15}O	2.07 min	1.72	$^{15}_{7}N(p,n)^{15}_{8}O$	$^{15}N_2$ gas	$^{15}O_2$ gas
			$^{14}_{7}N(d,n)^{15}_{8}O$	$^{nat}N_2$ gas	$^{15}O_2$ gas
^{18}F	109.8 min	0.635	$^{18}_{8}O(p,n)^{18}_{9}F$	^{18}O water	$^{18}F^-$ ion
			$^{20}_{10}Ne(d,\alpha)^{18}_{9}F$	^{nat}Ne gas	$^{18}F\text{-}F_2$ gas
^{68}Ga	68 min	1.9	$^{68}Ge \rightarrow {}^{68}Ga$	(Generator)	Ga metal
^{82}Rb	75 s	3.36	$^{82}Sr \rightarrow {}^{82}Rb$	(Generator)	RbCl

Reproduced from Barrington SF, Maisey MN and Wahl RL. *Atlas of Clinical Positron Emission Tomography*, 2nd edn. London: Hodder Arnold, 2006 with permission.

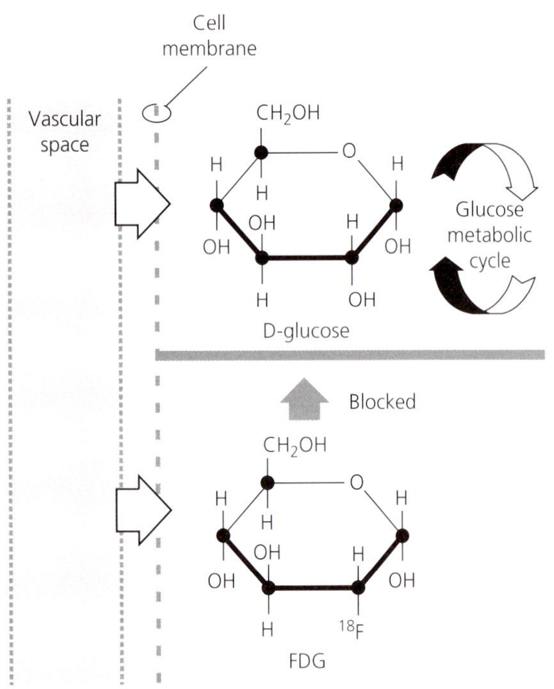

Figure 1.30 Blockage of ^{18}F-FDG from glucose metabolism after intracellular uptake. Reproduced from Dowsett, D.J, Kenny, P.A. and Johnston, R.E. *The Physics of Diagnostic Imaging*, 2nd edn, Hodder Arnold, 2006, with permission.

of quantities of ^{18}F-FDG suitable for PET imaging in only around 40 minutes.

The following techniques have been suggested to reduce physiological uptake of ^{18}F-FDG in PET neuroimaging (Barrington *et al.* 2006):

- Inject the subject in silence in a quiet darkened room.
- Avoid stimulation during the uptake period.
- Avoid any sedation for the uptake period.
- The subject should avoid any glucose-containing sedation prior to visiting the PET center.

PET neuroimaging has been widely used for psychiatric research. Specific radioligands are required when researching neurotransmission. For example, the study of dopamine (D2/D3) receptors in the brain using PET can be carried out with the dopamine receptor radioligand [^{11}C]raclopride, 5-HT_{1A} receptors can be studied with [carbonyl-^{11}C]WAY-100635 and [carbonyl-^{11}C]desmethyl-WAY-100635, cannabinoid CB_1 receptors (in non-human primates) with [^{11}C]MePPEP and [^{18}F]AM5144, and nAChR (nicotinic acetylcholine receptors) with the fluoropyridine derivatives 2-[^{18}F]FA and 6-[^{18}F]FA.

PET neuroimaging also has several clinical applications. It can be useful in determining whether or not a lesion seen on a structural brain MRI scan is a recurrence of a tumor in a patient with a history of malignancy. The use of PET to identify metabolic changes associated with a recurrent high-grade tumor is illustrated in Figure 1.31. These scans are from a patient with a history of treated glioblastoma multiforme, in whom a structural MRI scan (Figure 1.31a) showed a large mass with edema, pointing to a possible diagnosis of recurrent tumor. The PET scan (Figure 1.31b) showed increased metabolic activity in the lesion, consistent with tumor recurrence (Barrington *et al.* 2006).

The case illustrated in Figure 1.32 again shows how PET can be useful in differentiating between malignancy and, in this case, post-treatment changes. Here, while the MRI showed an enhancing region suggestive of recurrent tumor, this region did not take up ^{18}F-FDG, indicating that it was not in fact a recurrence but, probably, post-treatment fibrosis (Barrington *et al.* 2006).

PET is also useful in differentiating between intracerebral infection (commonly toxoplasmosis), which has no or low-grade ^{18}F-FDG uptake and intracerebral malignancy (commonly lymphoma) in patients with human immunodeficiency virus presenting with neurological illness (Barrington *et al.* 2006). This is illustrated in Figure 1.33, in which a focus of high ^{18}F-FDG uptake in a patient with HIV was consistent with the diagnosis of lymphoma, and Figure 1.34, in which the photopenic defect corresponding to an MRI lesion in a patient newly diagnosed with HIV was consistent with toxoplasmosis and not lymphoma (Barrington *et al.* 2006).

PET is also useful in the diagnosis of dementia. Table 1.6 shows the MRI and ^{18}F-FDG PET findings in various dementias. In addition to aiding in the differential diagnosis of dementia, ^{18}F-FDG PET is useful in differentiating between early Alzheimer's disease and benign memory loss, and also in differentiating dementia from pseudodementia or depression (Barrington *et al.* 2006).

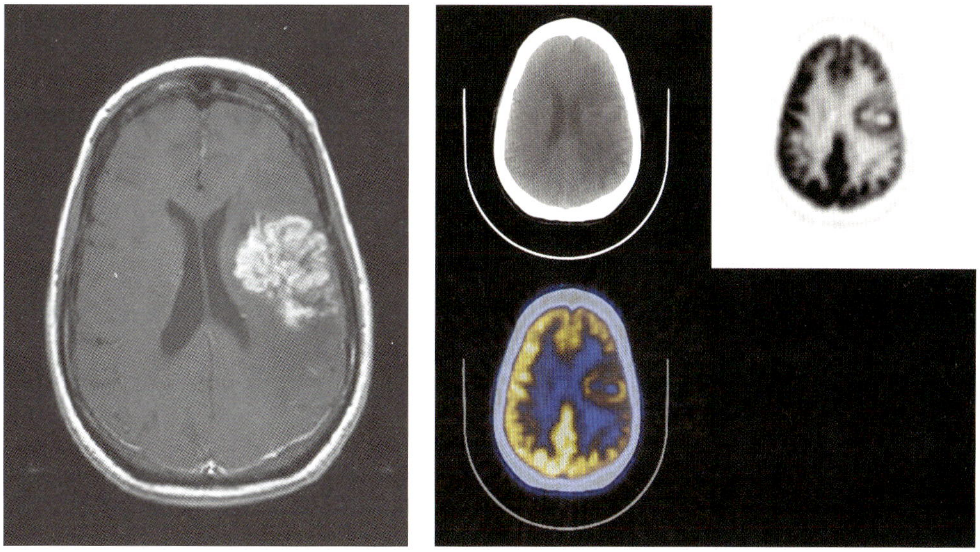

(a) (b)

Figure 1.31 (a) Magnetic resonance imaging scan showing a 5 cm mass with edema suggestive of recurrent tumor in a patient referred with a history of glioblastoma multiforme treated with surgery, chemotherapy and radiotherapy 4 years earlier. (b) Positron emission tomography scans showing increased metabolic activity associated with the mass in the inferoposterior aspect of the left frontal lobe compatible with recurrence. The diffuse decreased activity in the left frontal region is consistent with postradiation changes. Reproduced from Barrington SF, Maisey MN, Wahl RL. (2006), with permission.

1.3 NEUROPSYCHOLOGY

Assessment is considered in Part II of this book, while signs and symptoms, such as aphasia and alexia, are described in Part III. The primary focus of this section is neuropsychological testing, which may be used diagnostically, for treatment planning and review, for assessment of skills and abilities, and research. The historically important role of such testing for the detection and localization of neuropathology has now largely been superseded by advances in structural and functional neuroimaging (which are described in Chapters 1.2 and 2.4). A discussion of the theory underlying neuropsychological testing is beyond the scope of this book.

Assessment prerequisites and validity

Before neuropsychological testing is carried out, the following prerequisites should be checked (Evans 2010): the subject

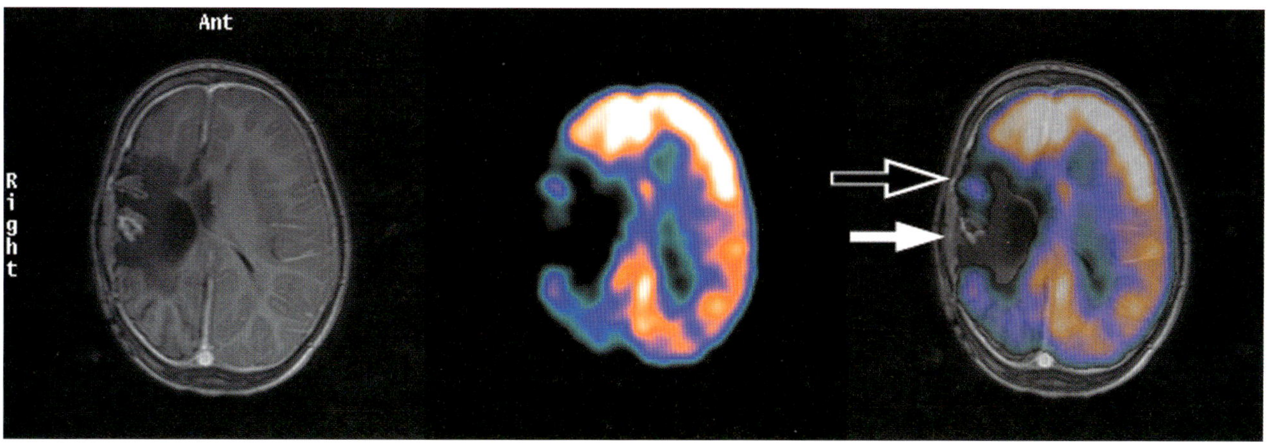

Figure 1.32 Magnetic resonance imaging (MRI) and positron emission tomography (PET) scans of the brain of a child who had undergone multiple surgery, chemotherapy, and radiotherapy for an anaplastic ependymoma. Although neurologically stable with a left hemiplegia, she complained of new headaches. The MRI scan (left) showed an enhancing region in the right parietal lobe suggestive of recurrent disease. ^{18}F-FDG PET (middle) was registered to the MRI scan (right). The region of enhancing tissue on MRI did not take up ^{18}F-FDG (solid arrow) indicating that this was likely to be scar or inflammatory tissue. The small focus of ^{18}F-FDG uptake (broken arrow) in the right parietal region corresponded to an island of normal cortex on MRI. Reproduced from Barrington SF, Maisey MN, Wahl RL. (2006), with permission.

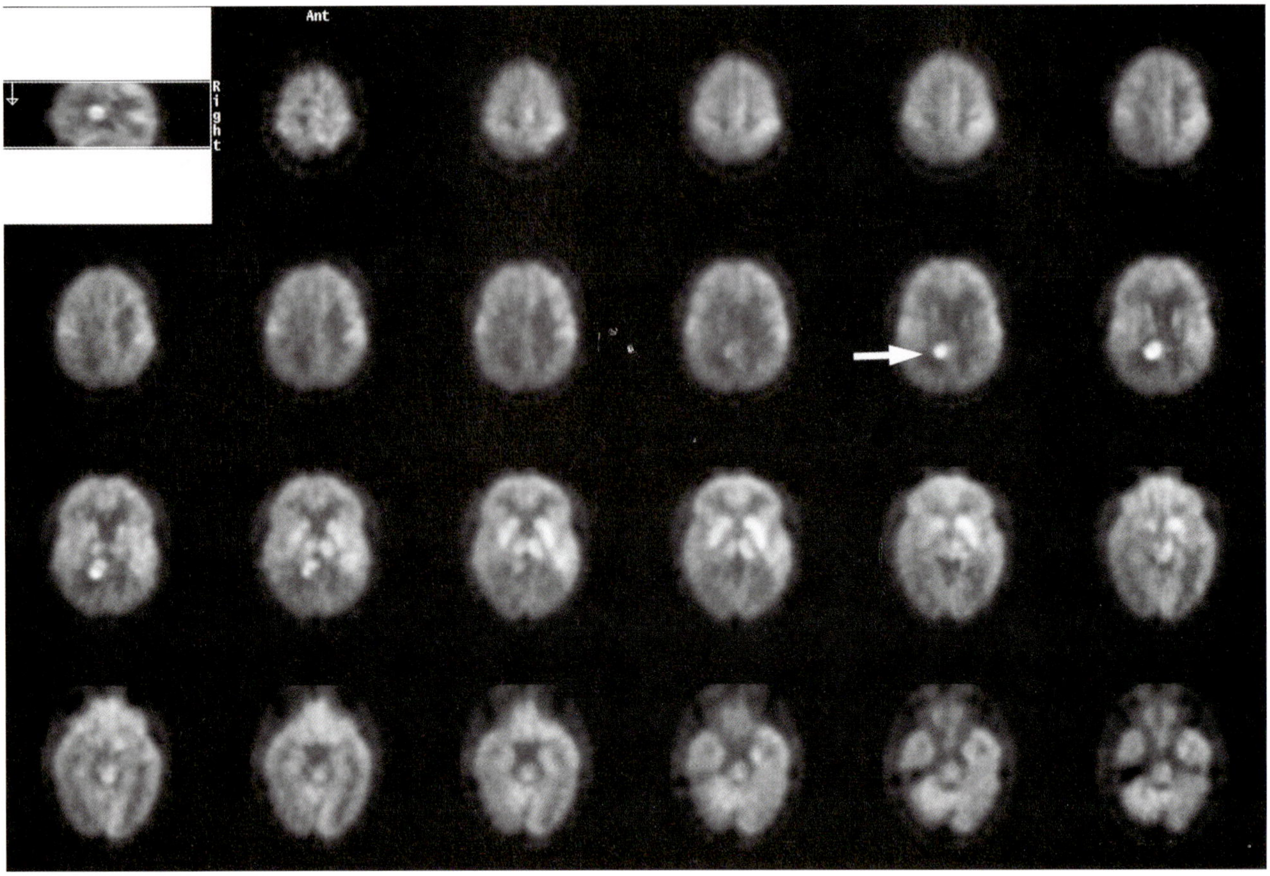

Figure 1.33 ^{18}F-fluorodeoxyglucose positron emission tomography findings from a patient with human immunodeficiency virus who was referred with a magnetic resonance imaging lesion of unknown nature in the right parietal lobe. There was a focus of intense uptake in the right parietal lobe posterior to the thalamus. Biopsy revealed lymphoma. Reproduced from Barrington SF, Maisey MN, Wahl RL. (2006), with permission.

should be able to concentrate for long enough; the subject should have adequate comprehension of the instructions; and the issue of motivation and effort, including possible motivation by hopes of secondary gain, should be considered. Notwithstanding these checks, there are several factors that may cause the results to lack validity. (Validity describes whether a test measures what it purports to measure, whereas reliability describes the level of agreement between sets of observations [Puri and Tyrer 1998]). These contra-indicators to assessment validity include (Evans 2010; Groth-Marnat 2000):

- physical problems (including pain, and sensory and motor disturbances) that affect test performance;
- psychoactive substance use;
- medication;
- psychiatric disorder;
- learning disability;
- congenital or pre-existing neurological conditions;
- the language of testing is not the first language of the subject;
- cultural test bias;
- suboptimal effort is shown by the testing;
- testing performance may vary from one session to another for many reasons. Therefore, caution should be exercised of relying on only one session.

Verbal and linguistic functions

APHASIA

Commonly available tests for aphasia are given in Table 1.7. The Boston Diagnostic Aphasia Examination, 3rd edition (BDAE-3) (Goodglass *et al.* 2000) is a well-established test battery for the assessment of aphasic syndromes in adults. It includes a short form in which fewer stimulus illustrations are shown to the subject than in the full version, and has an administration time of around 35–45 minutes. The long version contains 34 subtests, and may take up to 4 hours to administer. Domains tested include: auditory comprehension (word discrimination, body-part identification, and complex ideational material); oral expression (verbal and non-verbal agility, automatized sequences, recitation, repetition, word reading, responsive naming, animal naming, visual confrontation naming, and oral sentence reading); written language comprehension (symbolic and word discrimination, phonetic association, word–picture matching, reading sentences and paragraphs); and writing (writing mechanics, written symbol recall, written word finding, written formulation). A severity rating scale indexes the subject's communicative ability, and the BDAE-3 also allows testing of visuospatial and quantitative skills following brain injury.

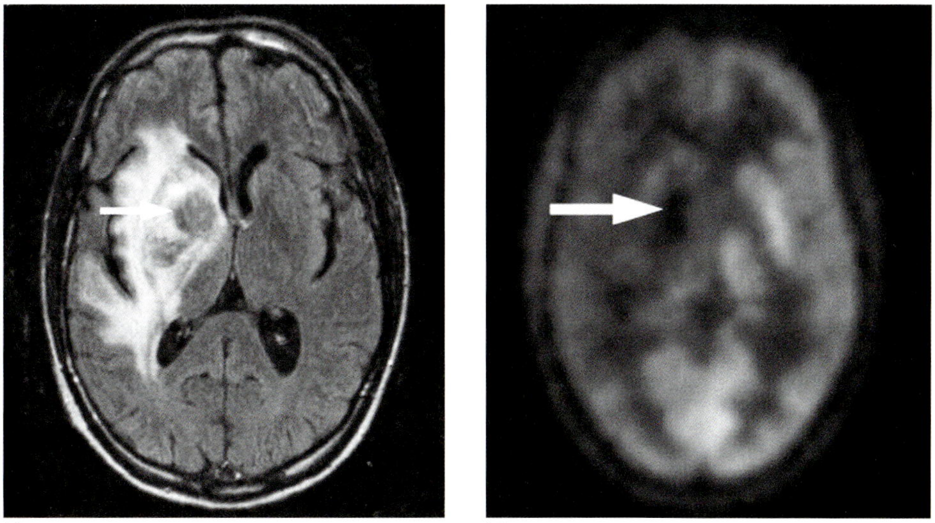

(a) (b)

Figure 1.34 This patient with newly diagnosed human immunodeficiency virus presented with a 6-week history of headache and increasing confusion associated with pyrexia. (a) Magnetic resonance imaging (MRI) showed a lesion centered on the right lentiform nucleus with surrounding edema and mass effect, consistent with toxoplasmosis or lymphoma. (b) There was a photopenic effect on positron emission tomography (PET) corresponding to the MRI lesion and consistent with toxoplasmosis and not lymphoma. The PET scan also shows diffuse reduction in the right frontoparietal region. The patient showed good clinical response to treatment with antitoxoplasmosis therapy. Reproduced from Barrington SF, Maisey MN, Wahl RL. (2006), with permission.

Table 1.6 Magnetic resonance imaging and ^{18}F-fluorodeoxyglucose positron emission tomography findings in dementias

MR imaging	FDG PET imaging
Alzheimer's disease	
Early: normal or hippocampal atrophy	Temporoparietal and cingulate hypometabolism
Advanced: frontal, parietotemporal atrophy	Temporoparietal hypometabolism with sparing of subcortical structures, primary visual and sensorimotor cortex; later also frontal hypometabolism, cortical atrophy, and thalamic separation
Multiple infarct dementia	
White matter signals and cortical and subcortical infarcts	Focal asymmetric cortical and deep hypometabolic areas
Pseudodementia	
Normal	Normal or frontal hypometabolism
Frontal dementias	
Early: normal	Frontal: frontal lobe hypometabolism
Late: frontal atrophy	Pick: frontal + temporal hypometabolism
Trauma	
Mild: normal	Focal hypometabolism
Severe: atrophy	Focal hypometabolism
HIV	
Normal	Diffuse cortical hypometabolism with sparing of deep structures
Huntington's disease	
Early: normal	Hypometabolism of caudate nucleus
Late: caudate atrophy	
Lewy body disease	
Early: nil	Alzheimer's disease-like picture but with reduced visual cortex metabolism
Late: atrophy	

HIV, human immunodeficiency virus.
Reproduced from Barrington, S.F., Maisey, M.N. and Wahl, R.L. *Atlas of Clinical Positron Emission Tomography*, 2nd edn. London: Hodder Arnold, 2006 with permission.

Table 1.7 Tests for aphasia

Test	Details
Aphasia Screening Test, 3rd edition (AST)	A screening instrument designed for severely to moderately impaired aphasic adults
Boston Diagnostic Aphasia Examination, 3rd edition (BDAE-3)	See text
Communication Abilities in Daily Living, 2nd edition (CADL-2)	Assesses functional communication skills in adults aged 18–80 years. 50 items
Mississippi Aphasia Screening Test (MAST)	Brief, repeatable screening measure for individuals with severely impaired communication/language skills, designed for serial testing to detect changes in language abilities over time
Multilingual Aphasia Examination, 3rd edition (MAE)	Brief comprehensive aphasia test battery for the age range 6–69 years. Three tests assess different aspects of oral expression (visual naming, sentence repetition, and verbal associative capacity); three tests assess oral verbal understanding; one test assesses reading comprehension; and three tests assess oral, written, and block spelling. Writing is evaluated from a written spelling test
Neurosensory Center Comprehensive Examination for Aphasia (NCCEA)	24 subtests, including four control tests
Porch Index of Communicative Ability – Revised (PICA-R)	1-hour administration time. Measures communicative ability in brain-injured adults. 18 subtests measure gestural, verbal, and graphic abilities. Multidimensional scoring system describes accuracy, responsiveness, completeness, promptness, and efficiency of response
Token Test	Designed to assess verbal comprehension of commands of increasing complexity, in children and adults. The test uses a set of 20 plastic tokens consisting of five colors, two shapes, and two sizes
Western Aphasia Battery – Revised (WAB-R)	Assessment of adults with acquired neurologic disorders. Assesses linguistic skills frequently affected by aphasia and key non-linguistic skills, and provides differential diagnosis information. Eight subtests (32 short tasks)

Table 1.8 Object naming tests

Test	Details
Boston Naming Test, 2nd edition (BNT)	See text
Graded Naming Test (GNT)	30 line drawings of objects are named in ascending order of difficulty. May take only around 10 minutes to administer, depending on the subject
Visual Naming Test (VNT)	Part of the Multilingual Aphasia Examination, containing 30 items

NAMING

Commonly available object naming tests are shown in Table 1.8. The full version of the second edition of the Boston Naming Test (BNT) (Goodglass and Kaplan 2001) is a 60-item instrument. The items are large ink drawings, starting with drawings of common items such as a bed, tree, and pencil, and progressing to less common objects such as a palette, protractor, and abacus. Adults are supposed to start at item number 30 (a harmonica) and progress towards item 60, unless a mistake is made in the first eight items (from 30 to 37) in which case the test is proceeded with in the reverse order until the subject correctly responds to eight consecutive items. If the subject makes eight consecutive mistakes, the test is stopped. Semantic cueing is permitted if the subject cannot name an item. If this does not work, then phonetic cueing is permitted. The second edition of the BNT includes a shortened version, which contains only 15 items and, if need be, recognition testing can be performed by giving subjects four choices per item. Care should be taken to factor in variation in responses related to age, education, culture, nationality, and so on. For example, some people may refer to the drawing of a mask as a 'face' or 'false face', while the drawing of a harmonica may be referred to as a 'mouth organ' or even as a 'harp' (particularly among some African American communities).

VOCABULARY

Vocabulary is examined by the vocabulary subtest of the Wechsler Adult Intelligence Scale (WAIS), described later in this chapter, and also by the Quick Test (Bowen *et al.* 1991). It is also specifically tested by the fourth edition of the Peabody Picture Vocabulary Test (PPVT-4) (Dunn and Dunn 2007), which comes in two equivalent forms containing different vocabulary items, to reduce practice effects. There are 228 stimulus words per form, presented in ascending order of difficulty, and for each stimulus the subject has to choose the correct picture from four options (which are in full color in the fourth edition). It can be administered to anyone from the age of 2.6 years onwards. For an unimpaired adult subject, testing can be carried out in just 10–15 minutes.

VERBAL FLUENCY

The Controlled Oral Word Association Test (COWAT), part of the Multilingual Aphasia Examination (Benton *et al.* 1994), requires the subject to produce as many words as he can within 1 minute that start with a given letter. This process is then repeated with two further letters. The subject is instructed to avoid word repetition, words that differ only in their endings (e.g. '-ed' is counted as '-ing' and as '-s'), numbers, and proper nouns. The three letters chosen are often FAS, CFL, or PRW, in that order; in each case, word frequency analysis indicates that the first letter should be easier than the second, and the second easier than the third. The final score is adjusted according to published age-, sex-, and education-related norms.

A well-established written word fluency test is the Thurstone Word Fluency Test (TWFT) (Thurstone 1957), in which the subject is instructed first to write down as many words as possible starting with the letter S, for which task he is given five minutes. This is followed by the task of writing down as many four-lettered words beginning with the letter C as possible during a 4-minute period.

READING

Commonly available object reading tests are shown in Table 1.9. The National Adult Reading Test (NART) (Nelson 1982; Nelson and Willison 1991) consists of the 50 words shown on the left-hand side of Table 1.9. Each of these words has a non-phonetic spelling. Performance on the NART is correlated with the premorbid intelligence quotient (IQ), making this a particularly useful test to carry out in many patients; the relationship between performance and premorbid IQ is shown on the right-hand side of Table 1.10. As the NART

Table 1.9 Reading tests

Test	Details
Gates–MacGinitie Reading Tests, 4th edition (GMRT)	Can be group administered. Suitable for the range from beginning readers upwards
Kaufman Functional Academic Skills Test (K-FAST) – reading subtest	34 items relating to everyday activities such as understanding labels on medicine containers and following directions in a recipe
National Adult Reading Test (NART)	See text
SRA Reading Index	Uses multiple-choice questions to assess picture–word association, word decoding, phrase comprehension, sentence comprehension, and paragraph comprehension

Table 1.10 The National Adult Reading Test (NART)

Words			Number of errors	Premorbid IQ estimate	Number of errors	Premorbid IQ estimate	Number of errors	Premorbid IQ estimate
Chord	Placebo	Assignate	0	131	17	110	34	89
Depot	Detente	Subtle	1	129	18	108	35	87
Bouquet	Puerperal	Gist	2	128	19	107	36	86
Capon	Gauche	Superfluous	3	127	20	106	37	85
Nausea	Leviathan	Banal	4	126	21	105	38	84
Courteous	Prelate	Cellist	5	124	22	103	39	82
Equivocal	Demesne	Zealot	6	123	23	102	40	81
Catacomb	Labile	Aeon	7	122	24	101	41	80
Thyme	Ache	Abstemious	8	121	25	100	42	79
Radix	Aisle	Idyll	9	120	26	98	43	77
Hiatus	Psalm	Aver	10	118	27	97	44	76
Procreate	Deny	Topiary	11	117	28	96	45	75
Gouge	Debt	Beatify	12	116	29	95	46	74
Simile	Rarefy	Sidereal	13	115	30	94	47	73
Quadruped	Naive	Syncope	14	113	31	92	48	71
Façade	Gaoled	Campanile	15	112	32	91	49	70
Drachm	Heir		16	111	33	90		

Reproduced from Nelson HE. The National Adult Reading Test (NART): Test Manual. Windsor: NFER-Nelson; 1982.

was originally developed in the UK and contains words particularly relevant to the British population, variations exist for other countries, such as the American NART (AMNART) and the North American Adult Reading Test.

Perception

VISUAL

A selection of commonly available tests of visual perception are shown in Table 1.11.

AUDITORY

A selection of commonly available tests of auditory perception are shown in Table 1.12.

TACTILE

A selection of commonly available tests of tactile perception are shown in Table 1.13. The Face–Hand Test is based on the finding that when light, tactile stimuli are applied simultaneously to the hand and the facial cheek, in organic mental disorders the former is often extinguished with the patient reporting being touched only on the hand (Anderson and Tranel 1989; Sivan 1992). Trials of simultaneous hand and/or cheek stimulation are conducted initially with the subject's eyes closed, for example starting with both cheeks being touched, and for each trial the subject has to specify where he was touched.

The Quality Extinction Test (QET) involves the blindfolded subject trying to identify various objects possessing different textures that are gently touched against his hands, having been allowed to see and touch these objects prior to blindfolding; taking into account the central tactile sensitivity

Table 1.11 Visual perception tests

Test	Details
Balloons Test	A quick screening test for visual inattention
Behavioural Inattention Test (BIT)	A test of unilateral visual neglect, including an objective behavioral test of everyday skills related to visual neglect
Bells Test	A cancellation test allowing for a quantitative and qualitative assessment of visual neglect in the near extrapersonal space
Closure Speed (Gestalt Completion)	A 24-item test of the 'first closure' factor – the ability to perceive an apparently disorganized/unrelated group of parts as a whole picture
Color-to-Figure Matching Test, revised (CFMT)	A nine-item test of color–object association; each drawing of common objects (e.g. cherries) needs to be matched with the most appropriate of 30 colored pencils (Caran d'Ache) placed randomly on the testing desk
Cortical Vision Screening Test (CORVIST)	Ten tests of different aspects of early visual processing by cortical centers: symbol acuity, shape discrimination, size discrimination, shape detection, hue discrimination, scattered dot counting, fragmented numbers, word reading, crowding, and face perception
Farnsworth D-15 Dichotomous Color Blindness Test	An arrangement test of color vision. An online version is now available
Hooper Visual Organization Test (VOT)	A quick (<15 min) 30-item screening instrument for measuring visual integration. Objects represented in line drawings as puzzle pieces need to be identified. Can be used from age 5 years upwards
Judgment of Line Orientation	Also known as the Benton Judgment of Line Orientation. Measures visuospatial judgment in individuals aged from 7 to 74 years
Line Bisection Test	A quick test of unilateral spatial neglect. The subject places a pencil mark in the center of a series of horizontal lines – displacement of the bisection mark to the side of a brain lesion is usually considered to be a sign of neglect
Neitz Test of Color Vision	A quick test of color vision suitable for children and adults, and individual and group administration
Perceptual Speed (Identical Forms)	Assesses the ability rapidly to compare visual configurations and identify two figures as similar or identical or to identify detail buried in distracting material
Star Cancellation Test	A quick (<5 min) screening test for unilateral spatial neglect. The subject is instructed to cancel as many as possible of 54 small stars on a page (among other distracting stimuli)
Visual Form Discrimination	Also known as the Benton Visual Form Discrimination. Measures the ability to discriminate between complex visual configurations in individuals aged from 19 to 74 years
Visual Object and Space Perception Battery (VSOP)	Eight tests each designed to assess a particular aspect of object or space perception while minimizing the involvement of other cognitive skills
Visual Search and Attention Test (VSAT)	A quick (around 6 min) test to measure attentional processes. Can be used in individuals aged 18–60 years

Table 1.12 Auditory perception tests

Test	Details
Heidelberg Phoneme Discrimination Test (HLAD)	Widely used in the diagnosis of auditory perception impairment and dyslexia in German-speaking countries
Incomplete Words	Part of the Woodcock–Johnson III Battery, in which a subject analyzes a word with missing phonemes and identifies the complete word
Putney Auditory Comprehension Screening Test (PACST)	A 60-item test of auditory comprehension in individual adults
(Seashore) Rhythm Test	A subtest of the Halsted–Reitan Battery. See later in the chapter
Sound Blending	Part of the Woodcock–Johnson III Battery requiring a subject to synthesize speech sounds to form a word
Speech Sounds Perception Test (SSPT)	A subtest of the Halsted–Reitan Battery. See later in the chapter
Wepman's Auditory Discrimination Test, 2nd edition (ADT)	A quick (5 min) test of the ability to recognize fine differences between similar phonemes. Originally developed for use in children. Involves listening to 40 pairs of words and indicating whether they are different or the same

Table 1.13 Tactile perception tests

Test	Details
Face–hand test	See text
Fingertip Number–Writing Perception	A subtest of the Halsted–Reitan Battery. See later in the chapter
Quality Extinction test	See text

pathways, this test allows discrimination between parietal and/or frontal dysfunction (Jurica *et al.* 2002; Rosen *et al.* 1984).

Cognition

ORIENTATION

The Awareness Interview is a structured interview that assesses orientation in time, place, and person, as well as awareness of acquired impairments; 'unawareness' is operationally defined as a discrepancy between the subject's description of abilities, and measurement of those abilities with neuropsychological and neurological evaluations (Anderson and Tranel 1989).

Right–left orientation can be assessed using the Benton Right–Left Orientation Test, in which the subject is asked to point to lateral body parts on verbal command.

ATTENTION AND CONCENTRATION

Continuous performance tests measure selective and sustained attention. Several are available, such as various versions of the Conners' Continuous Performance Test, which is particularly suitable for testing children and adults with attention-deficit hyperactivity disorder.

Short-term attentional capacity can be measured by the digit span, which is included in the WAIS (see below).

There are several tests of working memory. The WAIS measures the difference between digits forward and digits backward (see below). In the *n*-back task, during the serial presentation of items, such as a sequence of numbers or letters, the subject has to recall the item *n* steps back from the current item, and indicate whether the current item matches that from *n* steps before. Computerized versions of this task are part of several cognitive assessment batteries and are also available for use during fMRI studies (in which case the subject may need to respond using a finger-pad rather than orally, so as not to move his head unnecessarily during the scanning procedure). In the more difficult dual *n*-back task, the subject is presented with two different types of stimulus sequences at the same time, for example auditory and visual stimuli.

In the Stroop Color–Word Test, a common presentation is first to give the subject a page of neatly printed words, spaced into columns and rows. Each word is printed in black (on a white background) and is the name of a color. The subject is asked to read out aloud this list (from left to right, going down to the next row below when each row is completed) as quickly as possible, and the time taken is noted. This part of the test screens for any difficulty that the subject might have with reading, articulation, etc. The subject is then given another page containing the same number of words printed on a white background in the same arrangement. Again, each word is the name of a color. However, this time each word is printed in a color – a color that is not the one being described. For example, the word RED might be printed in the color blue. The subject is instructed to say out aloud as fast as possible the names of the colors in which the words are printed, and not the color described by each word. So in the example just mentioned, the correct response to articulate would be 'blue' and not 'red'. The number of mistakes is noted and difference between the longer time taken during this second task and the first task, the Stroop effect, is noted. Interactive computerized versions are available, including versions that can be used in the fMRI scanner. fMRI studies tend to show that increased activity in the anterior cingulate and dorsolateral prefrontal cortex is associated with performance of this test.

The Digit Symbol Test is a more complex test of attention and is part of the WAIS (see below).

MEMORY, CONCEPT FORMATION, AND OTHER FUNCTIONS

There are numerous tests of memory, including several memory batteries. One of the most commonly used, if not *the* most commonly used, battery is the Weschler Memory Scale, which is described later in the chapter. The Cognistat battery also assesses neurocognitive functioning, including memory, constructional ability, and reasoning/judgment, and is also described later in this chapter.

The Benton Visual Retention Test, fifth edition, assesses visual perception, memory, and visuoconstructive abilities (Sivan 1992). The subject is serially presented with ten designs, which he has to reproduce from memory. There are three test forms to diminish practice effects. It can be administered to individuals from the age of 8 years upwards. It typically takes 15–20 minutes to administer, and then another 5 minutes are required for the scoring.

The Alzheimer's Disease Assessment Scale – Cognitive Subscale (ADAS-cog), which is widely used in Alzheimer's disease trials, is part of the Alzheimer's Disease Assessment Scale (Rosen *et al.* 1984). Items include spoken language ability; spoken language comprehension; recall of test instructions; word-finding difficulty; following commands (e.g. 'make a fist'); naming objects and fingers, including objects that occur with a relatively high frequency (such as a pencil), those that occur with medium frequency (e.g. a mask), and those that occur relatively infrequently (e.g. a harmonica); constructions, in which the subject is asked to copy four geometric forms; ideational praxis, in which the subject is asked to pretend to send a letter to himself by putting a sheet of paper into an envelope (the sheet of paper needs to be folded by the subject), sealing the envelope, addressing the envelope to himself, and putting a stamp on the envelope; orientation in time, place, and person; word recall; and word recognition.

The (Mattis) Dementia Rating Scale-2 (DRS-2) (Jurica *et al.* 2002) is a revision of the original (Mattis) Dementia Rating Scale (DRS) (Mattis 1998) and is designed to measure cognitive status in adults from age 56 years upwards. There are five subscales, labeled: attention, initiation/perseveration, construction, conceptualization, and memory. The attention subscale consists of forward and backward digit span, following two-part verbal commands, following one-part verbal commands, imitating motoric gestures, visual scanning and target letter identification (two counting distraction items), reading a list of words aloud, and identifying visual designs. The initiation/perseveration subscale consists of naming items in a supermarket, naming things being worn and held by the examiner, a test of consonant perseveration (e.g. saying 'bee-key-gee' four times), a test of vowel perseveration (e.g. saying 'bee-bah-boh' four times), double alternating movements (two items), alternative tapping, and graphomotor design (reproduction of, for example, five consecutive alternating square–triangle waves; four items). The construction subscale contains six items, such as reproduction of vertical lines. The conceptualization subscale consists of identification of similar and different visual designs, identification of how objects are alike (abstract concept formation), priming inductive reasoning (generating similar objects), identification of different verbal objects, identification of how objects are alike in a multiple-choice format, and sentence initiation. The fifth and final subscale, named memory, consists of orientation in time, day, date, and situation, recall of sentence presented orally, recall of a sentence generated by the subject, verbal forced-choice recognition memory, and visual forced-choice recognition memory. In each subscale, it is the most difficult items which are presented first, so that if the individual performs well on the first one or two items, then the following items of that subscale can be omitted, thereby potentially shortening the overall testing time; it usually takes around 15–30 minutes to administer the whole test, plus another approximately 12 minutes to score.

Raven's Progressive Matrices offers a global non-verbal measure of concept formation, including abstract conceptualization (Raven *et al.* 1998). The Raven's Standard Progressive Matrices (SPM) consists of 60 items presented in five sets of 12 items. In each item, a part of a pattern is deleted. The subject has to choose the correct missing part from a choice of between six and eight possibilities per item. It can be administered to individuals or to groups, and typically takes between 20 and 45 minutes, although there is no time limit. There are various versions currently available, including the Raven's Standard Progressive Matrices Plus (SPM Plus), Colored Progressive Matrices (CPM), and Advanced Progressive Matrices. In addition to their use in clinical practice and clinical research, they find use in educational assessment and by human resources departments making staff assessments.

Verbal reasoning is tested by comprehension items in the WAIS, while a picture arrangement task in the WAIS tests a subject's ability to reason in relation to visual stimuli (see below). As mentioned later, the WAIS also tests arithmetical functioning.

Construction

In addition to construction tasks incorporated in other tests and test batteries, an important one often used is the Rey-Osterrieth Complex Figure Test (Knight and Kaplan 2003), or just Complex Figure Test (CFT) for short. One way of administering the test is as follows. First the subject is asked to copy a complex figure. Both the original figure and the subject's copy are then removed and the subject is asked to redraw the figure from memory (immediate recall). Following a further delay of 20–30 minutes, the subject is again asked to draw the figure from memory (delayed recall). Before the testing starts, the subject is not given any notice of the fact that he will be asked to draw the figure from memory. The test assesses a number of psychological functions, including construction, perceptual organization, and implicit visual memory.

The Wide Range Achievement Test 4 (WRAT4), the latest version of the WRAT that was first published in 1946, includes four subtests: word reading, sentence comprehension, spelling, and mathematical computation. Word reading evaluates letter and word decoding by assessing letter identification and word recognition. Sentence comprehension evaluates the ability to glean meaning from words and comprehend ideas and information in sentences. Spelling involves dictated spelling. Mathematical computation involves counting, number identification, simple oral problems, and written mathematical problems. The word reading and sentence comprehension subtest scores can be combined to give a reading composite score. The administration time for children aged 8 years and older, and for adults, is usually between 30 and 45 minutes, while for younger children aged 5–7 years the administration time tends to be shorter than this range.

Planning

Various tower tests can be used to assess planning. These tests tend to be named after various cities, for example, the Tower of London, Tower of Hanoi, Tower of Toronto, and the Stockings of Cambridge (SOC), the last being the equivalent test in the touch-screen computerized CANTAB software and referring to Cambridge, UK. In one example in the Tower of London task, the subject is given an apparatus with three adjacent poles; the left-hand one can hold up to three beads, the middle one two beads, and the right-hand pole one bead. Three beads of different colors are presented pre-arranged on the apparatus, and the subject is shown the final arrangement of beads that he must aim to achieve in as few moves as possible, moving just one bead at a time from one pole to another. This task can detect specific impairments of planning (as may occur following left anterior frontal lobe damage) (Shallice 1982). The task can be made more difficult, for example by including a fourth pole that can hold up to four beads, together with four beads of different colors. Computerized versions of this task are available. In the Tower of Hanoi test, there is an added complication for the subject in that now, instead of beads of different colors, he is presented with rings of different colors and different sizes; a larger-diameter ring may not be placed on top of a smaller-diameter ring. The SOC uses three adjacent computerized stockings hanging down, and colored balls (discs) are placed into the stockings, but a ball below one or more balls in a given stocking cannot be moved from that stocking until the ball(s) above it is (are) moved first.

Laterality

Based on findings such as those of the nineteenth century postmortem neuroanatomical studies of people who had suffered from language dysfunction during life by Broca, and the twentieth century split-brain studies by Sperry and his colleagues, in most people verbal functions are postulated to be associated with the left cerebral hemisphere and non-verbal, visuospatial functions with the right hemisphere. Many have gone further and ascribed to the left hemisphere such features as being associated with a more logical, analytic, and sequential way of thinking, compared with a more holistic, integrative, parallel way of thinking in the right hemisphere; such a dichotomy is probably a neuromyth. As Goswami has commented:

> . . . the left brain/right brain claims probably have their basis in the fact that there is some hemispheric specialization in terms of the localization of different skills. For example, many aspects of language processing are left-lateralized (although not . . . in blind people or in those who emigrate in later childhood to a new linguistic community). Some aspects of face recognition, in contrast, are lateralized to the right hemisphere. Nevertheless, there are massive cross-hemisphere connections in the normal brain, and both hemispheres work together in every cognitive task so far explored with neuroimaging, including language and face recognition tasks.
>
> (Goswami 2004).

Two instruments commonly used to determine the degree of laterality (how 'right-handed', ambidextrous, or 'left-handed' a person is) are the Annett Handedness Questionnaire and the Edinburgh Handedness Questionnaire.

Table 1.14 shows items from a modified version of the Annett Handedness Questionnaire (Annett 1970), which can also be supplemented by assessment of manual speed performance on a pegboard task.

The Edinburgh Handedness Questionnaire was published in 1971 (Oldfield 1971), the year after publication of Annett's questionnaire, and, as can be seen from Table 1.15, it is similar but shorter.

In the case of both questionnaires, the degree of right- or left-handedness can be expressed as a laterality quotient, calculated as $(R-L)/(R+L)$, where R refers to the number of items for which the right hand/eye/foot is preferred, and L the number for which the left is preferred.

Test batteries and rating scales

Some commonly used test batteries and rating scales are described briefly in this section.

CANTAB

The CANTAB, or Cambridge Neuropsychological Test Automated Battery, has been mentioned above in relation to the SOC. It is an automated computerized test battery offering

Table 1.14 Modified version of the Annett Handedness Questionnaire

Which hand do you use:

1. To write a letter legibly?
2. To throw a ball to hit a target?
3. To hold a racket in tennis, squash or badminton?
4. To hold a match while striking it?
5. To cut with scissors?
6. To guide a thread through the eye of a needle (or guide needle on to thread)?
7. At the top of a broom while sweeping?
8. At the top of a shovel when moving sand?
9. To deal playing cards?
10. To hammer a nail into wood?
11. To hold a toothbrush while cleaning your teeth?
12. To unscrew the lid of a jar?
    If you use the right hand for all these actions, are there any one-handed actions for which you use the left hand?
13. With which eye would you look through a telescope? (If you are not sure, roll a paper into a tube and look down it.)
14. Which foot would you use to kick a ball?

After Annett (1970).

Table 1.15 The Edinburgh Handedness Questionnaire

Surname . . . .   Given names . . . .   Date of Birth . . . .   Sex . . . . .

Please indicate your preferences in the use of hands in the following activities *by putting + in the appropriate column.* Where the preference is so strong that you would never try to use the other hand unless absolutely forced to, *put + +*. If in any case you are really indifferent *put + in both columns.*

Some of the activities require both hands. In these cases the part of the task, or object, for which hand preference is wanted is indicated in brackets.

Please try to answer all the questions, and only leave a blank if you have no experience at all of the object or task.

		LEFT	RIGHT
1	Writing		
2	Drawing		
3	Throwing		
4	Scissors		
5	Toothbrush		
6	Knife (without fork)		
7	Spoon		
8	Broom (upper hand)		
9	Striking match (match)		
10	Opening box (lid)		
i	Which foot do you prefer to kick with?		
ii	Which eye do you use when using only one?		

From Oldfield RC. The assessment and analysis of handedness: the Edinburgh inventory. *Neuropsychologia* 1971; 9:97–113 with permission.

cognitive assessment preferably with the subject having use of a touch-screen. There are 13 computerized tasks in the standard CANTAB: motor screening, big/little circle, delayed matching to sample, intradimensional/extradimensional (ID/ED) shift, matching to sample visual search, paired associated learning, pattern recognition memory, rapid visual information processing, reaction time, spatial recognition memory, spatial span, spatial working memory, and SOC.

The motor screening task is administered first. It introduces the subject to the touch-screen and, in addition, it allows the examiner to check that the subject can touch the touch-screen appropriately and that he can hear, comprehend, and follow instructions appropriately. The big/little circle task tests the ability of the subject to follow an explicit instructional rule and then his ability to reverse that rule. The ID/ED shift task tests his ability to attend to specific attributes of compound stimuli and then to shift that attention as required. Matching to sample visual search is a speed-accuracy trade-off visual search task. Paired associates learning is a delayed response procedure that tests list memory and list learning (or visuospatial conditional learning). The rapid visual information processing task tests vigilance (sustained attention) and working memory. The spatial span task tests spatial memory span while the spatial working memory task tests both spatial working memory and strategy performance. The SOC has replaced the Tower of

London task in earlier versions of CANTAB, and is described earlier in this chapter.

COGNISTAT

Cognistat, known before 1995 as the Neurobehavorial Cognitive Status Examination, provides rapid assessment in the following three general areas: consciousness level, orientation, and attention span; and in the following five major ability areas: language, constructional ability, memory, calculation skills, and reasoning and judgment. It takes just 10 minutes to screen a cognitively intact subject, and usually between 20 and 30 minutes to administer to a cognitively impaired subject. Normative data are available for those aged 12 years and older. A computerized internet-based version is available, known as the Cognistat Assessment System (CAS).

HALSTED–REITAN BATTERY (HRB)

This neuropsychological test battery is available in three versions: for adults (aged 15 years and older); for older children (aged between 9 and 14 years); and for young children (aged between 5 and 8 years).

The Aphasia Test includes the following tasks: naming common objects; spelling simple words; identifying individual numbers and letters; reading, writing, enunciating, and understanding spoken language; identifying body parts; calculating simple arithmetic problems; differentiating between right and left; and copying simple geometric shapes.

The Category Test, in which a series of stimuli are projected in front of the subject who has four answer buttons. It has seven subtests, which examine abstraction, reasoning, and organization abilities. For young children, the symbolic connotation of numbers of the answer buttons is replaced through the use of four differently colored buttons.

In the Finger Tapping Test, the subject taps as many times as possible over 10-second trial intervals. Both hands are tested and the average number of taps per 10-second interval per hand is calculated from the scores for the trials, allowing lateralized hemispheric damage to be pointed to by slowed finger tapping by the contralateral hand. Adults and older children use a manual finger tapper, while for young children a finger tapper is available, which does not traverse as large an arc as the manual tapper.

The Fingertip Number-Writing Perception Test is a tactile perception test in which the subject is asked to identify each of the numbers written using a pencil on the fingertips of the digits of each hand.

The Lateral Dominance Examination is a laterality examination involving a series of tests to determine the subject's hand preference for a unimanual task and his foot preference for a unipedal task.

In the (Seashore) Rhythm Test, the subject is asked to differentiate between 30 pairs of rhythmic beats, which are presented from a recording. It tests perception of non-verbal auditory stimuli, sustained attention, and the ability to compare different rhythms.

In the Speech–Sounds Perception Test (SSPT), the subject is asked to match a spoken sound (from a recording) to the correct choice among similar printed sounds. Among other functions, the SSPT is particularly good at testing for focused attention, with the subject being required to carry out continued discrimination over 60 items.

The Tactile Form Recognition Test requires tactile perception of objects of various shapes, such as a triangular shape. In addition to testing parietal lobe functioning, the total time required to perform this test appears to be a good screen for brain impairment.

The Tactile Performance Test (TPT) involves the blindfolded subject trying correctly to place differently shaped blocks into a board (which has different moulds into which the shapes fit correctly). It involves tactile form discrimination, kinesthesis, upper limb movement coordination, manual dexterity, an appreciation of spatial configuration, and tactile memory. Data from this test, taken together with the results of other HRB tests, including the Finger Tapping Test, can help differentiate frontal from non-frontal involvement.

In the Trail Making Test (TMT), Part A, the subject is presented with a random scattering of circles, each containing a number. One circle is marked as the one from which to start (e.g. 'Begin') and another circle as the one at which to stop ('End'). The subject must mark a trail joining circles in sequence to get from the start circle to the stop circle as quickly as possible, by joining consecutively numbered circles in ascending order. In the TMT Part B, the circles each contain either a number or a letter from the alphabet. This time the subject must mark a trail joining circles in sequence to get from the start circle to the stop circle as quickly as possible, by joining alternatively consecutively numbered and lettered circles in ascending order (e.g 1, C, 2, D, 3, E . . .). The TMT tests cognitive flexibility in addition to attention, scanning, and tracking. For young children, alternatives are available, including the Color Form Test, in which the child is asked to progress from the start to the end by alternating form and color criteria, with a reusable plastic board.

LURIA–NEBRASKA NEUROPSYCHOLOGICAL BATTERY (LNNB)

The LNNB is designed for use in individuals aged 15 years and older. It assesses a wide range of functions on clinical scales, localization scales, optional scales, and summary scales. The clinical scales assess motor functions, rhythm, tactile functions, visual functions, receptive speech, expressive speech, writing, reading, arithmetic, memory, intellectual processes, and, in the second of two forms of the LNNB, intermediate memory. The localization scales assess left frontal functioning, left sensorimotor functioning, left parieto-occipital functioning, left temporal functioning, right frontal functioning, right sensorimotor functioning, right parieto-occipital functioning, and right temporal functioning. The optional scales assess spelling, and motor writing. There are five summary scales, labeled pathognomonic, which contains items best discriminating those with brain impairment from those without; left hemisphere; right hemisphere; profile elevation; and impairment. In addition,

28 scales derived from factor analysis may reflect more specific cognitive and sensory functions.

Depending on the subject's condition, the LNNB may typically take between 1.5 and 2.5 hours to administer. The administration does not have to be in one sitting; it can be broken down into a series of short administrations. It is suitable for bedside use.

MINI-MENTAL STATE EXAMINATION (MMSE)

Folstein's MMSE is a brief test which can be used rapidly to screen for possible dementia, to estimate cognitive impairment severity, to follow cognitive changes over time, and to help differentiate dementia from delirium (Anthony *et al.* 1982; Folstein *et al.* 1975). It involves the following assessments: orientation; attention (serial subtraction or spelling a word backwards); immediate recall; short-term memory; naming common objects; following simple verbal commands; following simple written commands; writing a sentence spontaneously; and copying a figure. There is an inverse relationship between MMSE score and age. The total maximum score is 30, and a score of less than 24 tends to be considered indicative of cognitive impairment in the absence of any other cause for a low MMSE score, such as fewer than 4 years' schooling or suffering from a learning disability.

REPEATABLE BATTERY FOR THE ASSESSMENT OF NEUROPSYCHOLOGICAL STATUS (RBANS)

The RBANS is an individually administered test battery that can be used to test the neuropsychological status of adults aged from 20 to 89 years. It takes only about half an hour to administer and yields scores on the following subtests: list learning, story memory, figure copy, line orientation, picture naming, semantic fluency, digit span, coding, list recall, list recognition, story recall, and figure recall. In turn, these scores can be used to calculate various indices. A cortical–subcortical deviation score can be derived by calculating ([mean of the Attention Index and the Visuospatial Construction Index] – [mean of the Delayed Memory Index and the Language Index]); there is some evidence that this deviation score is much higher in 'cortical' dementias such as Alzheimer's disease than in 'subcortical' dementias such as Huntington's disease, with the deviation score being positive in the former and negative in the latter (Randolph *et al.* 1998).

WAIS

The WAIS, or Weschler Adult Intelligence Scale, is used to determine adult IQ scores. In 2008 the WAIS-IV was released. The following four index scores are derived: Verbal Comprehension Index (VCI), Perceptual Reasoning Index (PRI), Working Memory Index (WMI), and Processing Speed Index (PSI).

The VCI is derived from the following four tests: similarities, which assesses abstract verbal reasoning; vocabulary; information, in particular in relation to that derived from culture, such as being able to name the Prime Minister of the UK or the President of France; and comprehension, which may include asking the subject to explain the meaning of conventions and proverbs. The PRI is derived from the following five tests: block design; matrix reasoning (solving non-verbal abstract problems); visual puzzles; picture completion; and figure weights. The WMI is derived from the following three tests: digit span; arithmetic; and letter–number sequencing. The PSI is also derived from three tests: symbol search; coding, which tests visuomotor coordination; and cancellation.

The VCI and PRI can be used to derive the general ability index (GAI), while scores on all four indices (VCI, PRI, WMI, and PSI) are used to obtain the full-scale IQ (FSIQ) that follows a normal distribution with a mean of 100 and a standard deviation of 15.

Weschler Memory Scale (WMS)

In 2009 the fourth edition, the WMS-IV, was released. This now consists of the following seven subtests: spatial addition, symbol span, design memory, general cognitive screener, logical memory, verbal paired associates, and visual reproduction. The last three of these were part of the previous edition.

The spatial addition subtest is essentially an *n*-back test (see above). The subject is asked to try to remember the location of colored discs on two separate pages. He is asked to add and subtract locations and he has to be able to hold and manipulate visuospatial data.

The symbol span subtest is essentially a visual analog of the digit span test. The subject may be given a presentation with four different symbolic designs arranged from left to right. He is then shown eight designs, which contain the original four together with four new designs, and he has to choose the correct four designs (which he had previously seen) in the correct sequence (from right to left in the original).

The design memory subtest consists of four items of increasing difficulty and assesses immediate recall, delayed recall, and delayed recognition.

The cognitive screener subtest assesses temporal orientation, mental control, clock drawing, memory, inhibitory control, and verbal productivity.

The logical memory subtest involves the subject listening to stories and tests. The verbal paired associates subtest includes a combination of difficult and easy items; an example of an easy association would be Weschler's original 'baby–cries' pair from his 1945 original version of this test. For those aged between 16 and 69 years, this subtest includes 14 items, of which 10 are difficult, while for the older age group (around 65–90 years) there are just 10 items, of which six are difficult. The subtest includes a delayed free recall trial and a recognition trial. The visual reproduction subtest involves the subject attempting to draw visual designs from memory. These three subtests test immediate recall, delayed recall, recognition, and delayed contrast.

WISCONSIN CARD SORTING TEST (WCST)

The Wisconsin Card Sorting Test (WCST), also known as the Wisconsin Card Sort Test, assesses set-shifting. It was originally conducted using a set of 60 different cards containing all but four of the possible permutations of: four colors × four symbols × four numbers of printed symbols (that is, one, two, three, or four symbols). (The variable relating to the varying position of the symbol(s) on a card in the original was not counted and was eventually removed by standardizing symbol positioning.) The examiner places four further cards in front of the subject, which differ from each other in respect of color, symbols, and number. For example, these four cards might be one blue cross, two red disks (filled-in circles), three green triangles, and four yellow stars. From the pack of 60 cards, as each one is given to the subject, he must place it under one of the original four. The examiner gives immediate feedback whether the placement is correct or incorrect. The subject is not told the rule that is being followed, but once he has worked out the rule, the test continues using that rule until suddenly the rule changes, again with no notice. So, if the (unspoken) rule was to match by color, the rule might suddenly change to match by number. In addition to the number of categories correctly achieved, the perseverative errors made by the subject, in which the previous rule continues to be followed in spite of repeated negative feedback from the examiner, that is there is lack of flexibility in spite of a changed schedule of reinforcement, are also quantified. Perseverative errors have been found to occur in disorders such as schizophrenia, which is not unrelated to the finding that performance of the WCST appears to be related to activation of the left dorsolateral prefrontal cortex as well as other frontal regions and the caudate nucleus (Golberg and Weinberger 1994; Liddle *et al.* 1992). There are several versions of the WCST now available, including computerized ones which can be administered in PET and MRI (for fMRI studies) scanners and which readily score the test parameters.

1.4 NEUROPHYSIOLOGY

This section briefly reviews some basic concepts in neurophysiology.

Resting neuronal ionic concentrations

Table 1.16 shows the ionic compositions of two kinds of electrically active animal cells. As in these cases, a typical resting human neuron has a much higher extracellular sodium ion concentration (say around 145 mM) than intracellular concentration (around 15 mM), while the direction of this ratio for extracellular (around 5 mM) to intracellular (around 150 mM) potassium ions is reversed. Extracellular (around 100 mM) and intracellular (around 7 mM) chloride ion concentrations are roughly similar to those of sodium ions. It has been suggested that these extracellular animal cell ionic concentration findings reflect the composition of the primeval seawater in which animal cells first evolved; it is as if our cells are bathed in a fluid somewhat akin to ancient seawater.

Action potentials

Figure 1.35 shows how changes in the neuronal membrane permeabilities to sodium ions, P_{Na}, and potassium ions, P_K, result in the action potential.

In this figure, E_{Na} represents the equilibrium potential for sodium ions across the neuronal membrane, that is the potential at which sodium ions on either side of a membrane freely permeable to them would achieve a concentration gradient (higher extracellularly) that is equal and opposite to the electrical gradient caused by diffusion of these ions. Similarly, E_K represents the equilibrium potential for potassium ions across the neuronal membrane, that is the potential at which potassium ions on either side of a membrane freely permeable to them would achieve a concentration gradient (higher intracellularly) that is equal and opposite to the electrical gradient caused by diffusion of K^+ ions. For a given ion X, with ionic charge z, E_X can be calculated from the Nernst equation:

$$\begin{aligned} E_X &= [kT/(ze)] \ln [(\text{extracellular concentration})/ \\ &\qquad (\textit{intracellular concentration})] \\ &= [RT/(zF)] \ln [(\text{extracellular concentration})/ \\ &\qquad (\textit{intracellular concentration})] \end{aligned}$$

where $k \approx 1.381 \times 10^{-23}\,\mathrm{J\,K^{-1}}$ is Boltzmann's constant, T the temperature (in kelvin), $e \approx 1.602 \times 10^{-19}\,\mathrm{C}$ is the elementary

Table 1.16 Ionic compositions of two kinds of electrically active animal cells.

	Na^+	K^+	Cl^-
Frog muscle (resting potential c. −100 mV)			
Internal (mM)	10	124	1.5
External (mM)	109	2.3	78
Equilibrium potential (mV)	+65	−105	−100
Squid axon (resting potential c. −60 mV)			
Internal (mM)	50	400	50
External (mM)	440	20	560
Equilibrium potential (mV)	+55	−75	−60

Reproduced from Carpenter RHS. (2003), with permission.

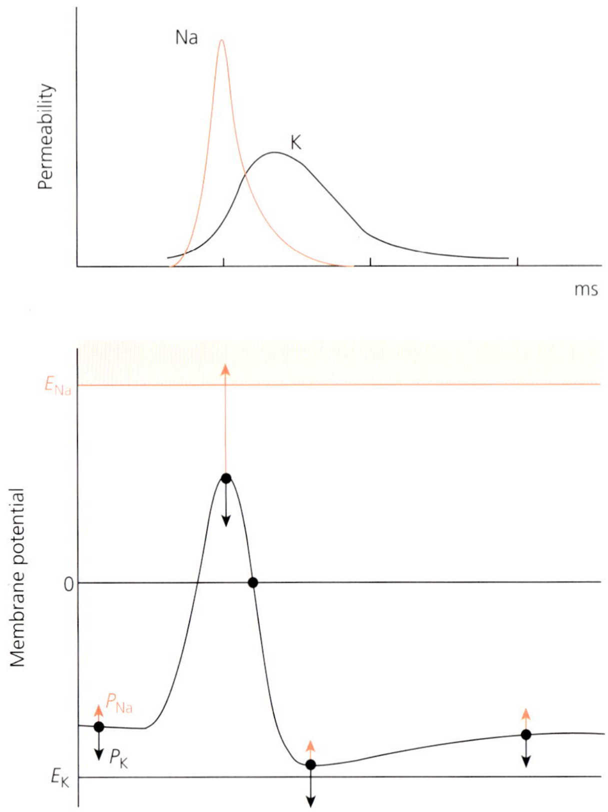

Figure 1.35 *Above* Changes in potassium and sodium permeability associated with the action potential. *Below* How these changes result in the form of the action potential itself. The arrows above and below the trace indicate roughly by their length the relative sizes of P_{Na} and P_K' pulling the potential respectively towards E_{Na} and E_K. Reproduced from Carpenter R.H.S., (2003), with permission.

charge, $R \approx 8.314\,\mathrm{J\,K^{-1}\,mol^{-1}}$ is the gas constant, and $F = 96.485 \times 10^3\,\mathrm{C\,mol^{-1}}$ is the Faraday constant (the magnitude of electric charge per mole of electrons). These changes in permeability are related to the state of voltage-gated ion channels, as shown in Figure 1.36, with initiation of an action potential following opening of neuronal membrane voltage-gated sodium ion channels in response to a depolarizing stimulus such as a region of depolarization moving into the present region from an adjacent region, or binding of a stimulatory ligand to a receptor. The absolute refractory period shown in Figure 1.36 is a period during which a further action potential cannot be initiated, while during the relative refractory period a further action potential can only be initiated by what would, at rest, be a suprathreshold stimulus.

Neuronal communication

LIGAND-GATED CHANNELS

Unlike voltage-gated channels, these membrane channels are sensitive to ligands. There are basically two ways in which membrane channels may open in response to stimuli: directly, via ionotropic mechanisms, such as the cholinergic synapse at the neuromuscular junction; and indirectly, via metabotropic mechanisms, often involving a G-protein. These types of receptor mechanisms are illustrated in Figure 1.37.

VARIABLES

Carpenter has suggested that the differences between sensory receptors and central nervous system neurons can be brought down to a consideration of just the following variables: the type of stimulus; whether the ligand-gated channel is direct or indirect, and if indirect (or metabotropic) the nature of the intermediate; what permeability changes take place; whether or not action potentials are employed; and the transmitter (etc.) released (Carpenter 2003). His neuronal ground plan is illustrated for some types of neuron in Table 1.17.

SYNAPTIC TRANSMISSION

Figure 1.38 illustrates the neuromuscular junction and some types of synapses. At the neuromuscular junction (between the motor axon and striated muscle fiber, and therefore not

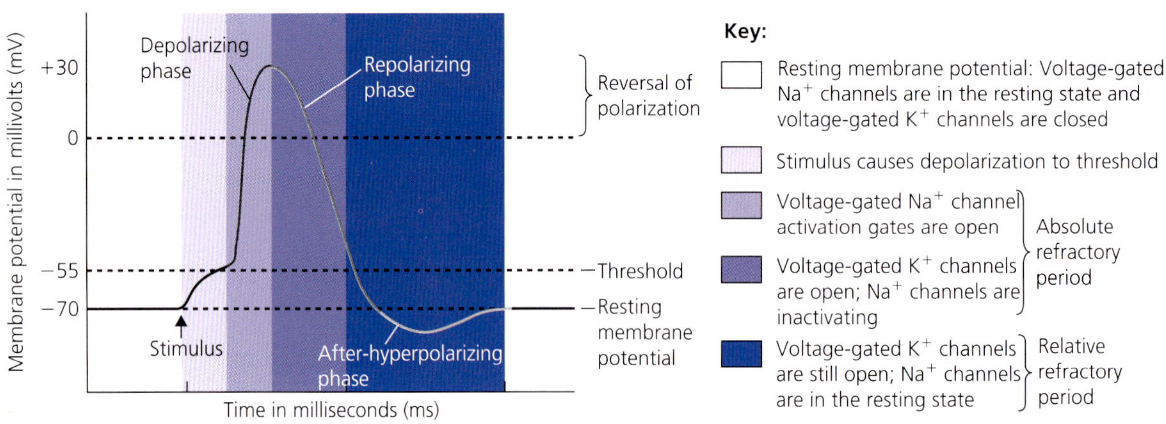

Figure 1.36 Action potential (AP) or impulse. When a stimulus depolarizes the membrane to threshold −55mV, an AP is generated. The AP arises at the trigger zone (at the junction of the axon hillock and the initial segment) and then propagates along the axon to the axon terminals. Redrawn with permission from Tortora and Derrickson (2006) *Principles of Anatomy and Physiology* 11th edn. Oxford: Wiley Blackwell.

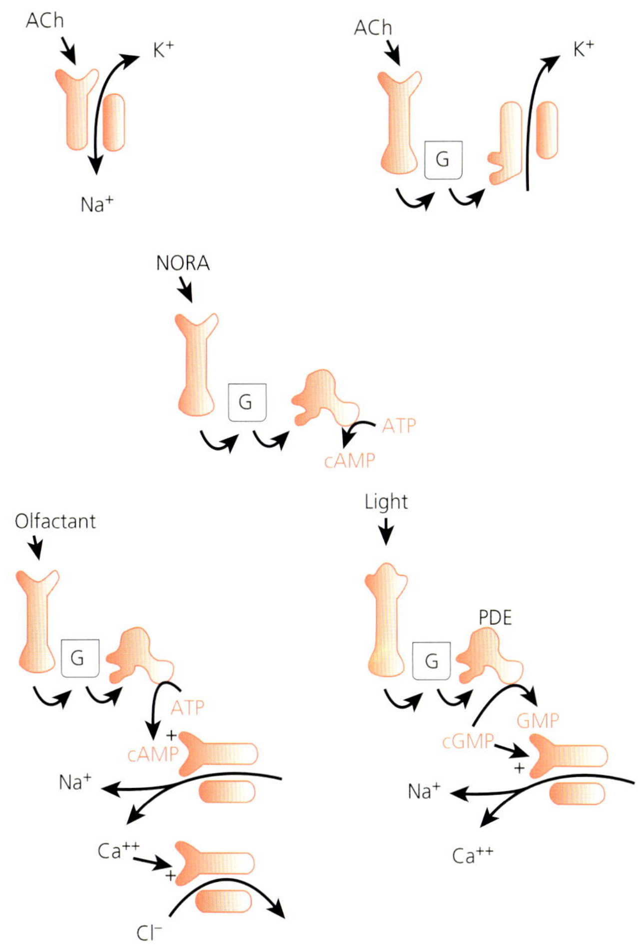

Figure 1.37 Specific examples of receptor mechanisms. *Above* Two cholinergic receptors: nicotonic (direct ligand-gated channel) *left*, and muscarinic, causing increased P_K (indirect, via G-protein) *right*. *Middle* A β-adrenergic receptor also using a G-protein, in this case resulting in cAMP. *Bottom* Two sensory receptor mechanisms, both using a G-protein, in one case (olfaction) to increase membrane permeability, and in the other (light) to decrease it. PDE, = phosphodiesterase. Reproduced from Carpenter R.H.S., (2003), with permission.

Table 1.17 Theme and variations: some types of neuron

	Nicotinic	M2 muscarinic	Photoreceptor	Olfactory receptor	Hair cell	Pacinian corpuscle
Stimulus	ACh	Ach	Light	Chemical	Mechanical	Mechanical
Indirect?	No	Yes	Yes: cGMP	Yes: cAMP, and Ca^{2+}	No	No
ΔP_{Na}?	+		−	+	+	+
ΔP_k?	+	+		+	+	+
AP_s?	Yes	(No)	No	Yes	No	Yes
Result	Mechanical	Mechanical	Glutamate?	Glutamate?	Glutamate?	Glutamate?

ACh, acetylcholine; cAMP, cyclic adenosine monophosphate; cGMP, cyclic guanosine monophosphate.
Reproduced from Carpenter RHS. (2003), with permission.

strictly speaking a synapse between two neurons), the neurotransmitter is acetylcholine. In central nervous system synapses, many other neurotransmitters and modulators act, some of which are shown in Table 1.18. Note that some of these also have hormonal properties.

Some varieties of neurotransmitter receptors are given in Table 1.19, together with typical agonists and antagonists.

Excitatory postsynaptic potentials (EPSPs) and inhibitory postsynaptic potentials (IPSPs) are integrated in the postsynaptic membrane, with temporal and spatial summation, to determine whether (postsynaptic potentials sum to a level greater than action potential threshold) or not an action potential is initiated in the postsynaptic membrane. This process is shown in Figure 1.39.

1.5 MOLECULAR GENETICS

This section briefly reviews some basic concepts in molecular genetics, a subject which presently appears destined to

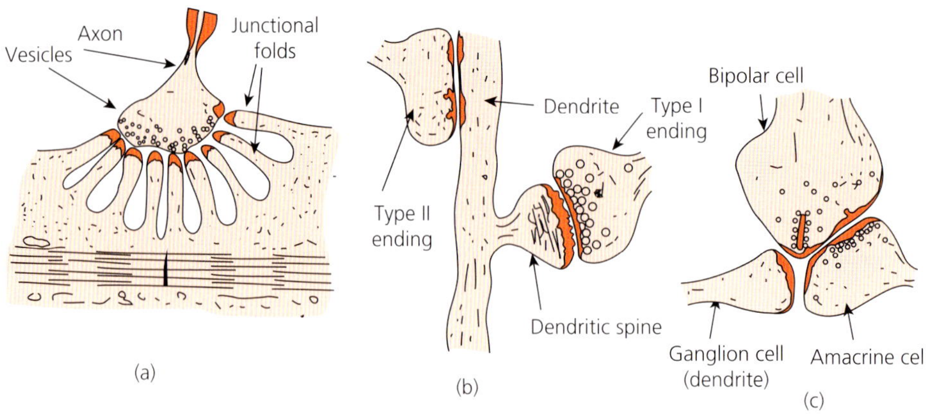

Figure 1.38 Somewhat stylized representations of some synaptic types. (a) Neuromuscular junction. (b) Two types of presynaptic axonal endings synapsing with a dendrite; the synapse on the right is with a dendritic spine. (c) A three-way synapse from the retina; the junction between bipolar and amacrine cell probably permits the transfer of information in both directions. Reproduced from Carpenter R.H.S., (2003), with permission.

Table 1.18 Examples of types of neurotransmitters and modulators

Neuropeptides (size in brackets)
β-Endorphin* (31)
Enkephalins* (5)
Oxytocin* (9)
Glucagon* (29)
Somatostatin* (14)
Substance P (11)
Vasoactive intestinal polypeptide (VIP)* (28)
(and many others)
Amino acids
Aspartate
Glutamate
Gamma-amino butyric acid (GABA)
Glycine
Monoamines
Dopamine
Noradrenaline*
5-hydroxytryptamine, serotonin (5-HT)*
Other
Acetylcholine
Nitric oxide?
Carbon monoxide?
Adenosine triphosphate (ATP)?

*Also used as a hormone.
Reproduced from Carpenter RHS. (2003), with permission.

continue helping us make increasing inroads into understanding clinical neuropsychiatric disorders.

DNA structure

Each of our 46 chromosomes is usually found in the nucleus of somatic cells, and at metaphase has a short arm, denoted p, and a long arm, denoted q. The two p and two q arms join at the centromere of the chromosome. Ultimately, the chromosome can be unwound to show the information-carrying double-stranded deoxyribonucleic acid (DNA), as shown in Figure 1.40. Each DNA strand is composed of a repeating unit of the sugar deoxyribose, a phosphate moiety, and a base. The base is usually one of four possibilities: adenine (A, a purine); cytosine (C, a pyrimidine); guanine (G, a purine); or thymine (T, a pyrimidine). A base-pairs with T, while C base-pairs with G, as shown in Figure 1.40, to give a double-stranded DNA molecule. Each haploid human genome contains approximately 3 billion base pairs (bp).

At the time of writing, only around 15 percent of human DNA appears to be incorporated into genes (exons). The true functions of the rest are presently unknown. It may play a role in helping DNA to pack into chromatin. Some, such as Dawkins, have argued that it is mostly 'junk DNA' that is a remnant of our phylogenetic past – rather as, with continued active use, a computer disk accumulates non-functional stretches of previously useful code (Dawkins 2006). Indeed, the apparently repetitious nature of this 85 per cent might appear to favor the Dawkins view, but it should be noted that there are biologists who argue strongly that the idea that this is junk DNA is incorrect (Wells 2011).

Genetic information

The transmission of protein encoding information flows as follows: transcription followed by mRNA processing followed by translation. Transcription occurs from the coding DNA strand to heteronuclear RNA (hRNA). This contains sequences between genes corresponding to DNA introns. In the next stage, mRNA processing, the introns are spliced out, a cap (a modified residue) is added to the 5′ end, and a poly-A tail to the cleaved 3′ end, to give messenger RNA (mRNA). The genetic information in mRNA may then be translated into peptide chains via ribosomes, from which proteins are made. Gene products are not confined to proteins but may also include non-translated RNA. The transmission of genetic information is illustrated in Figure 1.41.

Table 1.19 Some varieties of receptors for neurotransmitters

Neurotransmitter	Receptor type	Direct/indirect	Action	Agonists	Antagonists	
Glutamate	AMPA	Direct	↑P_{Na}, ↑P_K	AMPA	CNQX	
Glutamate	B	Indirect, ↓ cGMP and other mechanisms	↓P_{Na}	trans-ACPD	α-methyl-4-carboxy phenylglycine	Several types, some presynaptic
Glutamate	NMDA	Direct	↑P_{Ca}	NMDA	AP5	Voltage dependent
Glycine	GlyR	Direct	↑P_{Cl}		Strychnine	Mostly spinal cord
Acetylcholine	nAChR	Direct	↑P_{Na}, ↑P_K	Nicotine, carbachol	Tubocurarine	
	mAChR	Indirect	↑P_K	Muscarine	Atropine	
GABA	GABA-R_A	Direct	↑P_{Cl}	Muscimol	Picrotoxin, bicuculline	Also presynaptic
	GABA-R_B	Indirect, ↓cAMP	↑P_K	Baclofen	CGP 335348	Also presynaptic
Serotonin (5-HT)	5-HT_1	Indirect, ↓cAMP	↑P_K	(Various)	(Various)	Several subtypes, often presynaptic
	5-HT_2	Indirect	↓P_K	α-methyl-5-HT	Ketanserin	
	5-HT_3		↑P_{Na}, ↑P_K?	2-methyl-5-HT	Tropisetron	Periphery only
Dopamine	D_1	Indirect, ↑cAMP	↓P_K	Dehydrexidine	SCH 23390	
	D_2	Indirect, ↓cAMP	↑P_K	Haloperidol, spiperone	Spiroperidol	Also presynaptic
Norepinephrine	α_1-adrenoceptor	Indirect		Phenylephrine	Prazosine	
	α_2-adrenoceptor	Indirect		Clonidine	Yohimbine	Presynaptic
	β_1-adrenoceptor	Indirect		Dobutamine	Practolol	
	β_2-adrenoceptor	Indirect, cAMP		Procaterol	Butoxamine	

cAMP, cyclic adenosine monophosphate; GABA, γ-aminobutyric acid.
Reproduced from Carpenter RHS. (2003), with permission.

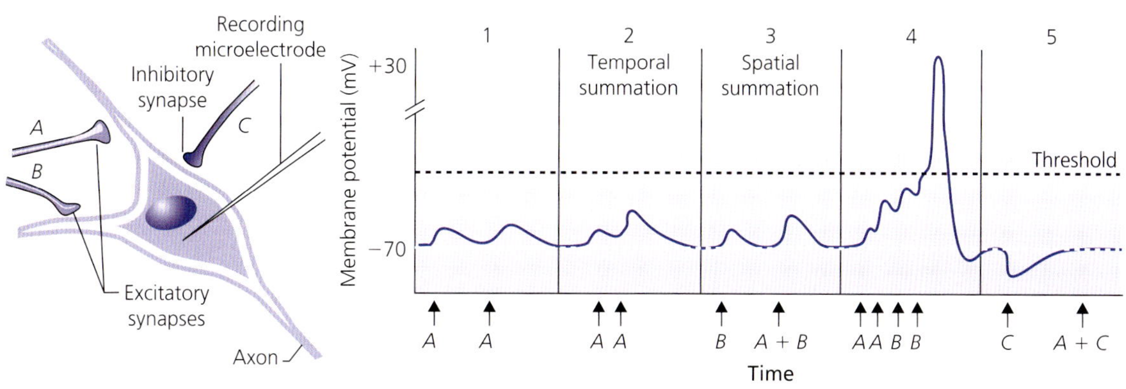

Figure 1.39 Interaction of excitatory postsynaptic potentials ($EPSP_S$) and inhibitory postsynaptic potentials ($IPSP_S$) at the postsynaptic neuron. Presynaptic neurons (A–C) were stimulated at times indicated by the arrows, and the resulting membrane potential was recorded in the postsynaptic cell by a recording microelectrode. Reproduced from Widmaier EP., Raff M. and Strang K.T. *Vander's Human Physiology: the mechanisms of body function*, 11th edition, 2008, with permission from McGraw-Hill Companies.

The DNA of genes is read in triplets. As there are four possible bases, this means that a triplet of bases can code for up to $4^3 = 64$ codons. In fact, the codons are needed to specify only 20 amino acids, plus certain regulatory or 'punctuation' signals such as 'start' and 'stop' translation of genetic information. Thus, there is redundancy in the genetic code, which is shown in Table 1.20.

Gene expression

There is clearly wide diversity in our cells. Even within one organ, the human brain, different types of neurons and glia vary in respect of morphology and function, and, indeed, protein content. And yet the DNA in the nuclei of these different brain cells is the same. Clearly there is varying gene

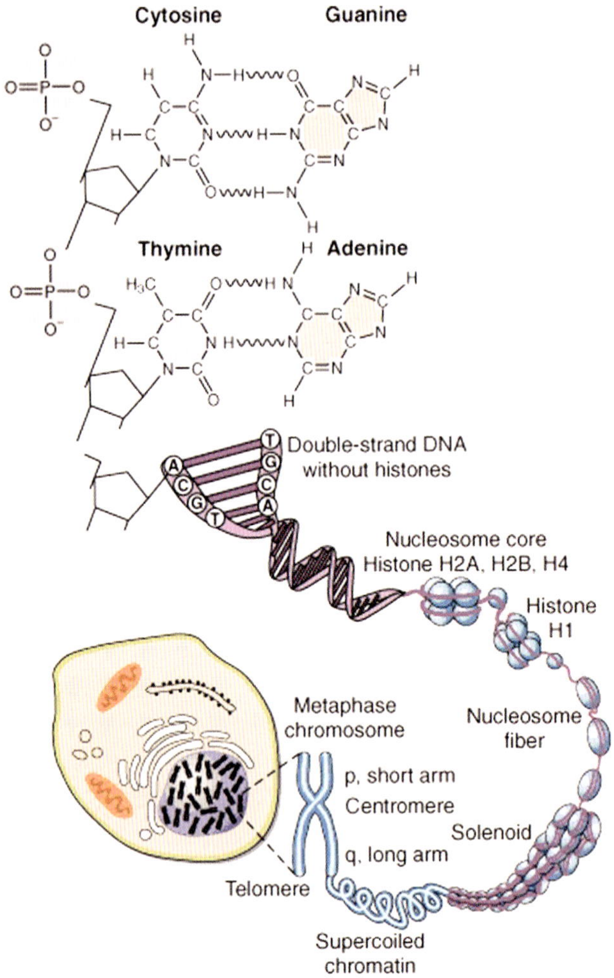

Figure 1.40 DNA and chromosomal structure. In chromatin, DNA is tightly wrapped around protein cores (such as histone). Structure of chromatin and chromosomes: chromatin is composed of double-stranded DNA that is wrapped around histone and non-histone proteins forming nucleosomes. The nucleosomes are further organized into solenoid structures. Chromosomes assume their characteristic structure, with short (p) and long (q) arms at the metaphase stage of the cell cycle. From Fauci *et al* (eds) (2008) *Harrison's Principles of Internal Medicine*, 17th edition. New York: McGraw Hill, 2008 with permission.

expression in different cells and varying expression within the same cell over time. Cells have the ability to select which genes are to be expressed, the degree of expression, and the timing of expression.

ALTERNATIVE SPLICING

The transcription stage involves the enzyme RNA polymerase, which reads the information encoded in the coding DNA strand. Note also that variation in the mRNA processing stage may occur, giving rise to alternative splicing. For example, instead of an mRNA molecule being formed from exon 1 + exon 2 + exon 3 + . . . from hRNA, one or more of these exons may be omitted, giving rise to different mRNA molecules and therefore to different gene products from the same original gene.

An example of aberrant regulation of alternative splicing of the *myomesin* 1 (*MYOM1*) gene in myotonic dystrophy type 1 (DM1) is shown in Figure 1.42; *MYOM1* is a constituent of the sarcomeric M band, suggesting its involvement in muscle impairment in patients with DM1 (Koebis *et al.* 2011).

TRANSCRIPTION INITIATION

Another method of altering gene expression involves the regulation of transcription initiation, which is comparatively well understood at the time of writing. Upstream (towards the 5′end) of the first exon of a given gene is the promoter region, to which RNA polymerase binds, shown in Figure 1.41. The DNA promoter region has a TATA box a little upstream from the start of exon 1 (see Figure 1.41), to which TATA-binding protein (TBG) binds. Further upstream are various response (or responsive) elements (REs), such as cAMP response element (CRE). RNA polymerase transcription initiation is regulated by different proteins, including specificity factors, activators, and repressors. Figure 1.41 indicates the complex way in which different extracellular signals, such as cytokines and hormones, and enhancer and silencer sequences a long way upstream of the first exon, can interact to control the initiation of transcription by RNA polymerase (type II).

One important component of the RNA polymerase II initiation complex is transcription factor TFIIB, which interacts with TBP and helps recruit RNA polymerase II into the initiation complex (Buratowski and Zhou 1993). As shown in Figure 1.43, compared with rats undergoing sham injury (anesthetized and operated on but without a cortical lesion being made), in rats experiencing traumatic brain injury (cortical incision under anesthesia) there is increased TFIIB expression in the brain (Liu *et al.* 2011).

Furthermore, as shown in Figure 1.44, the increased expression of TFIIB following traumatic brain injury is found to be associated with astrocytes and microglia; thus, TFIIB may be implicated in the proliferation of astrocytes and microglia following traumatic brain injury (Liu *et al.* 2011).

OTHER MECHANISMS

It is beyond the scope of this book to detail the numerous other complex mechanisms of regulating gene expression. They include gene silencing (post-transcription) via RNA interference and regulatory protein control of development.

Mutations

Those mutations occurring in non-gametes are somatic mutations. They are not inherited. In contrast, germ-line mutations in gametes can indeed be passed on to progeny. DNA sequence variations with a frequency of 1 percent or more are polymorphisms, and those polymorphisms affecting a single nucleotide are known as SNPs (single-nucleotide polymorphisms), pronounced 'snips'. As can be seen in Figure 1.45, point mutations (mutations involving a single

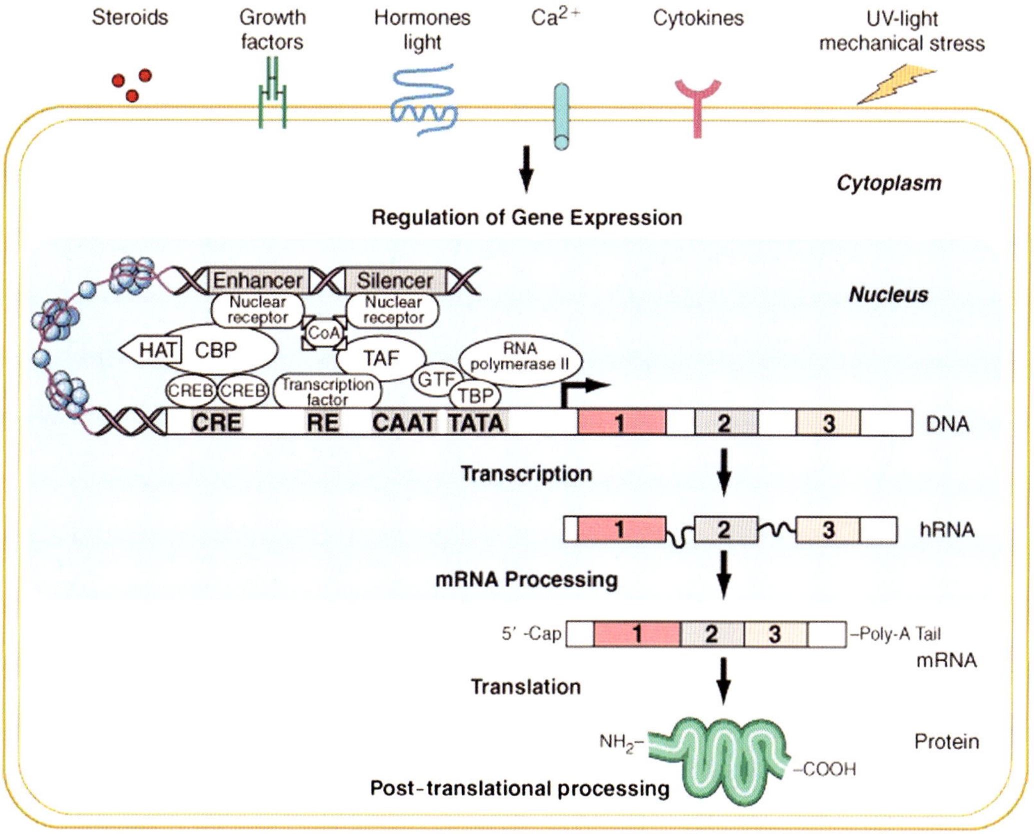

Figure 1.41 Genetic information. Flow of genetic information: multiple extracellular signals activate intracellular signal cascades that result in altered regulation of gene expression through the interaction of transcription factors with regulatory regions of genes. RNA polymerase transcribes DNA into RNA that is processed to mRNA by excision of intronic sequences. The mRNA is translated into a polypeptide chain to form the mature protein after undergoing post-translational processing. HAT, histone acetyl transferase; CBP, CREB-binding protein; CREB, cyclic AMP response element-binding protein; CRE, cyclic AMP responsive element; CoA, Co activator; TAF, TBP-associated factors; GTF, general transcription factors; TBP, TATA-binding protein; TATA, TATA box; RE, response element; NH_2, aminoterminus; COOH, carboxyterminus. From Fauci *et al* (eds) (2008) *Harrison's Principles of Internal Medicine*, 17th edition. New York: McGraw Hill, 2008, with permission

Table 1.20 The genetic code from mRNA to amino acids

Position number 1 (5′ end)	Position number 2				Position number 3 (3′ end)
	U	C	A	G	
U	Phe	Ser	Tyr	Cys	U
	Phe	Ser	Tyr	Cys	C
	Leu	Ser	STOP	STOP	A
	Leu	Ser	STOP	Trp	G
C	Leu	Pro	His	Arg	U
	Leu	Pro	His	Arg	C
	Leu	Pro	Gln	Arg	A
	Leu	Pro	Gln	Arg	G
A	Ile	Thr	Asn	Ser	U
	Ile	Thr	Asn	Ser	C
	Ile	Thr	Lys	Arg	A
	Met (STaRT)	Thr	Lys	Arg	G
G	Val	Ala	Asp	Gly	U
	Val	Ala	Asp	Gly	C
	Val	Ala	Glu	Gly	A
	Val (Met)	Ala	Glu	Gly	G

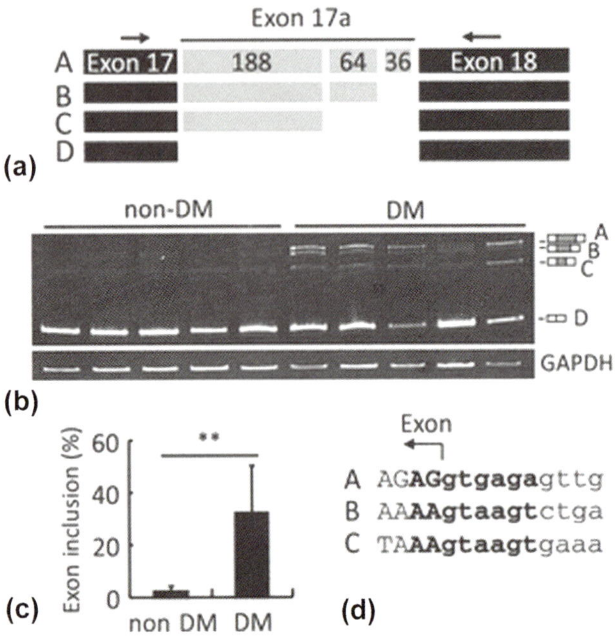

Figure 1.42 Alternative splicing of *myomesin 1* gene aberrantly regulated in myotonic dystrophy type 1 (DM1). A study of five DM1 patients and five age-matched controls. The inclusion of *MYOM1* exon 17a was increased in DM1. (a) Diagram showing the products of the alternative splicing of *MYOM1*. The number of nucleotides of exon 17a is shown: exon 17a has two internal 5′ splice sites, generating four splicing isoforms. The names of the isoforms are given on the left. Arrows represent the primer pairs used for RT-PCR. (b) Total RNA isolated from homogenized skeletal muscle tissue was subjected to reverse transcription polymerase chain reaction analysis and electrophoretically resolved on a polyacrylamide gel. The identity of the bands is indicated on the right. Glyceraldehyde-3-phosphate dehydrogenase was amplified as a loading control. (c) The histogram illustrates the inclusion of exon 17a (mean ± SD of the five individuals). The percentage of inclusions of exon 17a was calculated as the ratio of the isoforms containing full length or part of exon 17a (isoforms A, B and C) to the total spliced products. The statistical significance of the results is indicated by $^{**}P < 0.01$. All bands of interest were gel-isolated, cloned, and confirmed by sequencing. (d) Alternative 5′ splice sites of exon 17a. Reproduced from Koebis *et al.* (2011) with permission.

DNA base pair) may give rise to: a silent mutation (because of the redundancy in the genetic code); a missense mutation (a new amino acid is encoded for); a nonsense mutation; or frameshift (if the deletion or insertion is not a multiple of three – the codon size).

Examples of deletions and micro-deletion syndromes include velocardiofacial syndrome (usually associated with a hemizygous micro-deletion at 22q11.2) and Angelman's syndrome (maternal deletion of a region in 15q11 to 15q13). Unstable trinucleotide repeats occur, for example, in Huntington's disease (CAG repeat) and fragile X syndrome (CGG repeat). Inversions and translocations may be associated with schizophrenia (e.g. the inversion inv(9)(p11q13) and the translocation t(1;11)(q42;q14.3)). A copy number variant (CNV) is a structural chromosomal abnormality characterized by variation in the number of gene or DNA sequence copies, and may be inherited or caused by aberrant replication or mutations (Gill and McGrath 2010).

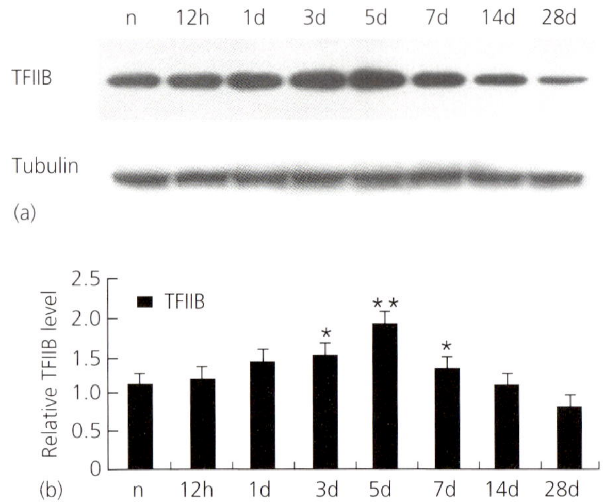

Figure 1.43 Western immunoblot analysis showing the TFIIB and (anti-rabbit) TUBULIN in the cortex surrounding the wound at various survival times after brain injury. Sample immunoblots probed for TFIIB and TUBULIN are shown above. The *bar chart* below demonstrates the ratio of TFIIB relative to TUBULIN for each time point. (a) TFIIB protein level was low in normal cortex, peaked at fifth day, and then gradually rose thereafter. *n* denotes cortical extracts from control animals which underwent the sham procedure. (b) Semiquantitative analysis (relative optical density) of the intensity of staining of TFIIB to TUBULIN for each time point. The data are means ± SEM ($n = 3$, $^{*}P < 0.05$, $^{**}P < 0.01$, significantly different from the normal groups). Reproduced from Liu *et al.* (2011) with permission.

1.6 PSYCHONEUROENDOCRINOLOGY

This section briefly reviews some basic concepts in psychoneuroendocrinology.

Hormone classification

There are several classes of hormone now recognized. Large proteins such as insulin are biosynthesized from prohormones by proteolytic processing. Small neuropeptides include gonadotrophin-releasing hormone (GnRH) and somatostatin. Catecholamines such as epinephrine are originally biosynthesized from tyrosine. The thyroid hormones thyroxine (T4) and tri-iodothryronine (T3) are biosynthesized from thyroglobulin-Tyr via iodination. The eicosanoids include prostaglandins, leukotrienes, and thromboxanes, and are derived from the n-6 long-chain polyunsaturated fatty acid dihomo-gamma-linolenic acid (e.g. series-1 prostaglandins), the n-6 long-chain polyunsaturated fatty acid arachidonic acid (e.g. series-2 prostaglandins), and the n-3 long-chain polyunsaturated fatty acid eicosapentaenoic acid (e.g. series-3 prostaglandins). Steroid hormones, such as estrogen, testosterone, and cortisol, are biosynthesized from cholesterol. Vitamin D, e.g. 1,25-dihydroxycholecalciferol, are also biosynthesized from cholesterol. Retinoids, such as retinoic acid, are derived from vitamin A. Nitric oxide may also be considered to be hormonal, and is biosynthesized from arginine.

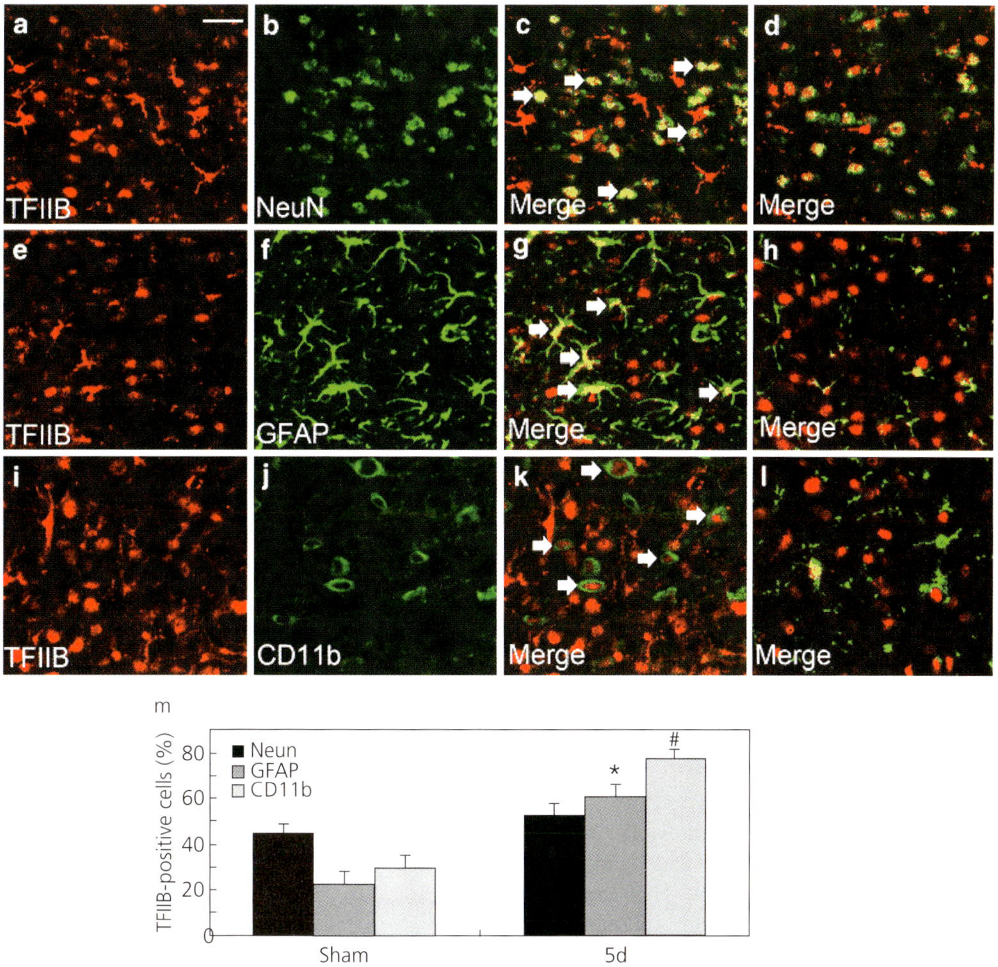

Figure 1.44 Double immunofluorescence staining for TFIIB and different phenotype-specific markers in brain cortex. In the adult rat brain cortex within 1 mm distance from the lesion site at the fifth day after traumatic brain injury (TBI), horizontal sections labeled with TFIIB (*red*, a, e, i) and different cell markers (*green*, b, f, j): neuronal marker (NeuN), astrocyte marker (GFAP), microglia marker (CD11b). The *yellow color* (*arrows*) in the merged images represented colocalization of TFIIB with different phenotype-specific markers (c, g, k). Colocalizations of TFIIB with different phenotype-specific markers in the sham group are shown in the brain cortex (d, h, l). **m** Quantitative analysis of different phenotype-specific markers positive cells expressing TFIIB (%) in sham group and 5 days after injury. For **m**: $n=3$ shams and 4 TBI 5-day post-trauma. The changes of TFIIB expression in brain cortex after TBI are significant in neurons and astrocytes ($P<0.01$). *, $^{\#}P<0.01$. Error bars represent SEM. *Scale bars* 20 μm (a–l). Reproduced from Liu *et al.* (2011) with permission.

Hormone and receptor families

Table 1.21 displays groups of families of hormones and receptors, including their effectors and signaling pathways. In this table, LH is luteinizing hormone, FSH is follicle-stimulating hormone, and TSH is thyroid-stimulating hormone; all three are glycoproteins. PTH is parathyroid hormone and PTHrP is parathyroid hormone-related peptide. Other hormones acting via GPCRs listed in Table 1.21 include ACTH (adrenocorticotropic hormone), GHRH (growth hormone-releasing hormone), CRH (corticotropin-releasing hormone), and TRH (thyrotropin-releasing hormone).

Hormones acting via receptor tyrosine kinases include insulin, EGF (epidermal growth factor), and NGF (nerve growth factor); their receptors are INS-R, ECF-R, and NGF-R respectively. IGF-I (insulin-like growth factor I) also binds to tyrosine kinase receptors.

GH (growth hormone) and PRL (prolactin) act on cytokine receptor-linked kinases.

Hormonal substances acting on serine kinase receptors include activin and MIS (müllerian-inhibiting substance or anti-müllerian hormone). The latter plays a key role in male sexual differentiation.

Membrane receptors

Figure 1.46 shows the main membrane receptors for hormones. The seven transmembrane GPCRs bind a wide variety of hormones, as mentioned above. Besides those mentioned earlier, tyrosine kinase receptors also bind several other growth factors, such as fibroblast growth factor and platelet-derived growth factor. The cytokine receptor family act via JAKs (Janus kinases) and STAT (signal transduction and activators of transcription) proteins. Serine kinase receptors act via smads.

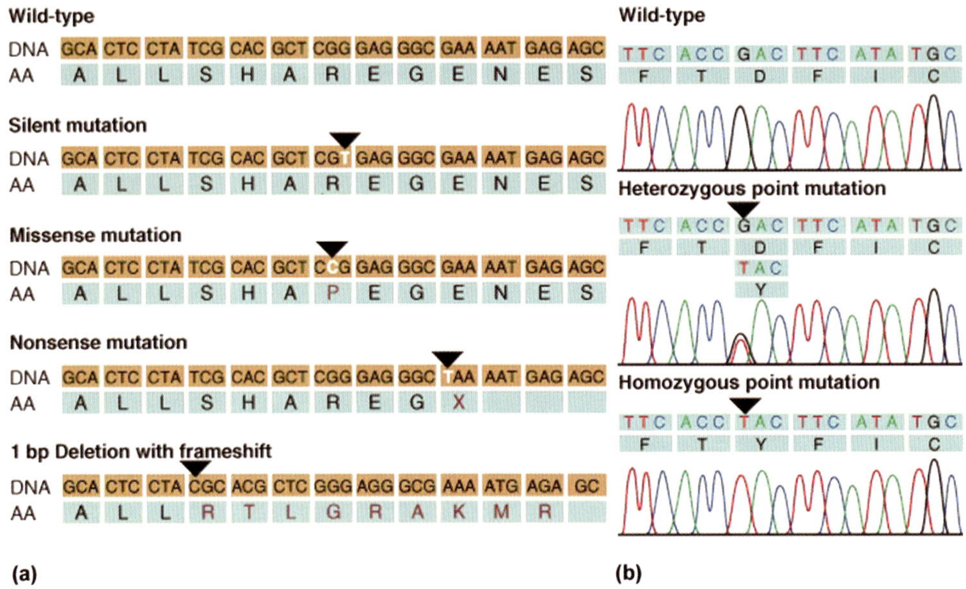

Figure 1.45 Examples of mutations. (a) Examples of mutations. The coding strand is shown with the encoded amino acid sequence. (b) Chromatograms of sequence analyses after amplification of genomic DNA by polymerase chain reaction. From Fauci *et al* (eds) (2008) *Harrison's Principles of Internal Medicine*, 17th edition. New York: McGraw Hill, 2008 with permission.

Hypothalamic–pituitary axis

The neuroendocrine axis is outlined in Figure 1.47. Neurotransmitters act on the hypothalamus, which in turn coordinates the activity of the pituitary gland. Hypothalamic stimulatory and inhibitory hormones and anterior and posterior pituitary hormones are shown in Table 1.22.

Given the large number of anterior pituitary hormones, there are many tests of anterior pituitary sufficiency, including provocative tests of pituitary reserve. These are given in Table 1.23, with interpretations based on US data.

HYPOTHALAMIC–PITUITARY–ADRENAL AXIS

The hypothalamic–pituitary–adrenal (HPA) axis is shown in Figure 1.48 and plays a key role in the reaction of the body to stress. As indicated in the figure, CRH is released from the

Table 1.21 Membrane receptor families and signaling pathways

Receptors	Effectors	Signaling pathways
G Protein-coupled seven-transmembrane (GPCR)		
β-Adrenergic LH, FSH, TSH	$G_s\alpha$, adenylate cyclase	Stimulation of cyclic AMP production, protein kinase A
Glucagon PTH, PTHrP ACTH, MSH GHRH, CRH	Ca^{2+} channels	Calmodulin, Ca^{2+}-dependent kinases
α-Adrenergic somatostatin	$G_j\alpha$	Inhibition of cyclic AMP production Activation of K^+, Ca^{2+} channels
TRH, GnRH	G_q, G_{11}	Phospholipase C, diacylglycerol, IP_3, protein kinase C, voltage-dependent Ca^{2+} channels
Receptor tyrosine kinase		
Insulin, IGF-I	Tyrosine kinases, IRS	MAP kinases, PI 3-kinase; AKT, also known as protein kinase B, PKB
EGF, NGF	Tyrosine kinases. ras	Raf, MAP kinases, RSK
Cytokine receptor-linked kinase		
GH, PRL	JAK, tyrosine kinases	STAT, MAP kinase, PI 3-kinase, IRS-1
Serine kinase		
Activin, TGF-β, MIS	Serine kinase	Smads

IP_3, inositol triphosphate; IRS, insulin receptor substrates; MAP, mitogen-activated protein; MSH, melanocyte-stimulating hormone; NGF, nerve growth factor; PI, phosphatylinositol; RSK, ribosomal S6 kinase; TGF-β, transforming growth factor β. For all other abbreviations, see text.
Reproduced from Jameson JL. Principles of endocrinology. In: Fauci AS, Braunwald E, Kasper DL, *et al.*, eds. *Harrison's Principles of Internal Medicine*, 17th edn. New York: McGraw-Hill Medical, 2008:2187–95 with permission.

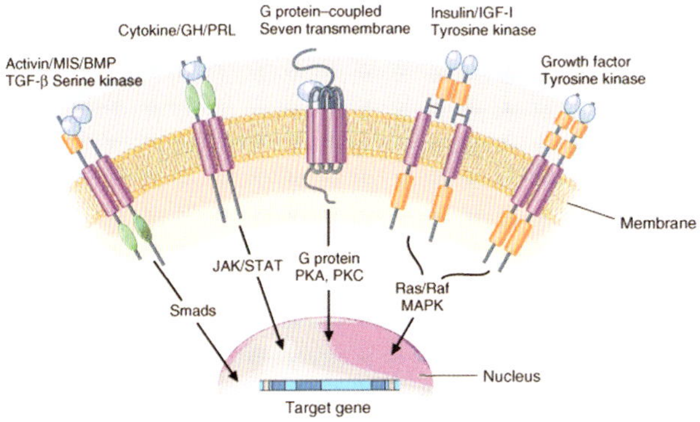

Figure 1.46 Membrane receptor signaling. MAPK, mitogen-activated protein kinase; PKA, -C, protein kinase A, C; TGF, transforming growth factor. From Fauci *et al* (eds) (2008) *Harrison's Principles of Internal Medicine*, 17th edition. New York: McGraw Hill, 2008 with permission.

paraventricular nucleus (PVN), which is a centrally located nucleus of the hypothalamus. It can be seen that there are many inputs to the PVN, including neurotransmitters such as γ-aminobutyric acid, acetylcholine, opioids, norepinehrine, and serotonin. In addition, the cytokines TNF-α (tumor necrosis factor alpha), IL-1 (interleukin 1), and IL-6 (interleukin 6) have a stimulatory effect on PVN CRH. Cushing's syndrome and adrenocortical insufficiency are considered in Chapters 17.1 and 17.2, later in the book.

HYPOTHALAMIC–PITUITARY–THYROID AXIS

The hypothalamic–pituitary–thyroid (HPT) axis is shown in Figure 1.49 and may play a role in several psychiatric disorders. Hyperthyroidism and hypothyroidism are considered in Chapters 17.3 and 17.4, later in this book.

PROLACTIN AXIS

That the hypothalamic prolactin inhibitory hormone is known to be dopamine is of clinical importance, given that most antipsychotic medication acts on the dopamine D2 receptor (D2R). Antagonism by such medication at D2R leads to inhibition of the tonic inhibition of dopamine on prolactin release and therefore to hyperprolactinemia. This can manifest as galactorrhea, gynecomastia, menstrual disturbances, reduced sperm count, and reduced libido. The role of serum prolactin level measurement in differentiating true seizures from pseudoseizures is described in Chapter 8.3.

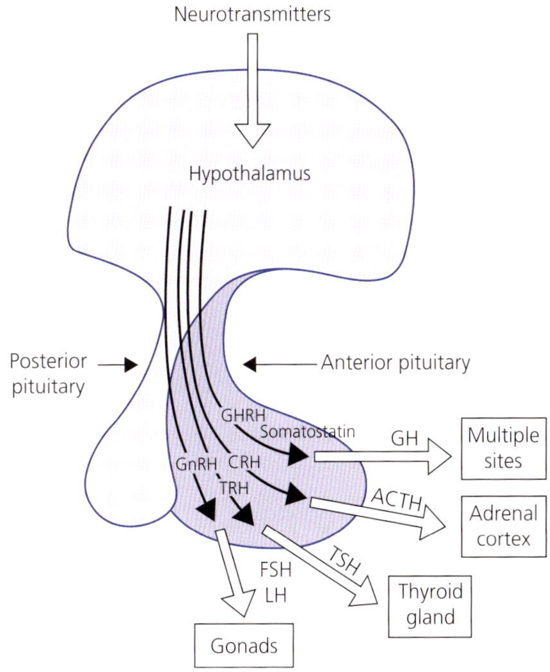

Figure 1.47 Outline of the neuroendocrine axis. ACTH, adrenocorticotropic hormone; CRH, corticotropin-releasing hormone; FSH, follicle-stimulating hormone; GH, growth hormone; GHRH, growth hormone-releasing hormone; GnRH, gonadotropin-releasing hormone; LH, luteinizing hormone; TRH, thyrotropin-releasing hormone; TSH, thyroid-stimulating hormone. From Puri BK, Treasaden IH (eds). *Psychiatry: An Evidence-based Text*. London: Hodder Arnold, 2010 with permission.

Table 1.22 Hypothalamic and pituitary hormones

Hypothalamic stimulatory hormones	Corticotropin-releasing hormone (CRH) Arginine vasopressin (AVP) Thyrotropin-releasing hormone (TRH) Growth hormone-releasing hormone (GHRH) Gonadotropin-releasing hormone (GnRH)
Inhibitory hypothalamic hormones	Dopamine Somatostatin
Anterior pituitary hormones	Adrenocorticotropic hormone (ACTH) Thyroid-stimulating hormone (TSH) Prolactin Growth hormone Follicle-stimulating hormone (FSH) Luteinizing hormone (LH)
Posterior pituitary hormones	Antidiuretic hormone (or vasopressin) Oxytocin

Reproduced from Fitzgerald and Dinan (2010) with permission.

Table 1.23 Tests of pituitary sufficiency

TESTS OF PITUITARY SUFFICIENCY	Test	Blood samples	Interpretation
GROWTH HORMONE	Insulin tolerance test: Regular insulin (0.05 – 0.15 U/kg i.v.)	–30, 0, 30, 60, 120 min for glucose and GH	Glucose <40 mg/dL; GH should be >3 μg/L
	GHRH test: 1 μg/kg i.v.	0, 15, 30, 45, 60, 120 min for GH	Normal response is GH >3 μg/L
	L-Arginine test: 30 g i.v. over 30 min	0, 30, 60, 120 min for GH	Normal response is GH >3 μg/L
	L-dopa test: 500 mg p.o.	0, 30, 60, 120 min for GH	Normal response is GH >3 μg/L
PROLACTIN	TRH test: 200–500 μg i.v.)	0, 20, and 60 min for TSH and PRL	Normal prolactin is >2 μg/L and increase >200% of baseline
ACTH	Insulin tolerance test: Regular insulin (0.05–0.15 U/kg i.v.)	–30, 0, 30, 60, 90 min for glucose and cortisol	Glucose <40 mg/dL
			Cortisol should increase and control by >7 μg/dL or to >20 μg/dL
	CRH test: 1 μg/kg ovine CRH i.v. at 0800 h	0, 15, 30, 60, 90, 120 min for ACTH and cortisol	Basal ACTH increase 2- to 4-fold and peaks at 20–100 pg/mL
			Cortisol levels >20–25 μg/dL
	Metyrapone test: metyrapone (30 mg/kg) at midnight	Plasma 11-deoxycortisol and cortisol at 0800 h; ACTH can also be measured	Plasma cortisol should be <4 μg/dL to assure an adequate response
			Normal response is 11-deoxycortisol >7.5 μg/dL or ACTH >75 pg/mL
	Standard ACTH stimulation test: ACTH 1-24 (Cosyntropin), 0.25 mg i.m. or i.v.	0, 30, 60 min for cortisol and aldosterone	Normal response is cortisol >21 μg/dL and aldosterone response of >4 ng/dL above baseline
	Low-dose ACTH test: ACTH 1-24 (Cosyntropin), 1 μg i.v.	0, 30, 60 min for cortisol	Cortisol should be >21 μg/dL
	3-day ACTH stimulation test consists of 0.25 mg ACTH 1-24 given i.v. over 8 h each day		Cortisol >21 μg/dL
TSH	Basal thyroid function tests: T_4, T_3, TSH	Basal tests	Low free thyroid hormone levels in the setting of TSH level that are not appropriately increased
	TRH test: 200–500 μg i.v.	0, 20, 60 min for TSH and PRL[a]	TSH should increase by >5 mU/L unless thyroid hormone levels are increased
LH, FSH	LH, FSH, testosterone, estrogen	Basal tests	Basal LH and FSH should be increased in postmenopausal women
			Low testosterone levels in the setting of low LH and FSH
	GnRH test: GnRH (100 μg) i.v.	0, 30, 60 min for LH and FSH	In most adults, LH should increase by 10 IU/L and FSH by 2 IU/L
			Normal responses are variable
MULTIPLE HORMONES	Combined anterior pituitary test: GHRH (1 μg/kg), CRH (1 μg/kg), GnRH (100 μg), TRH (200 μg) are given i.v.	–30, 0, 15, 30, 60, 90, 120 min for GH, ACTH, cortisol, LH, FSH, and TSH	Combined or individual releasing hormone responses must be elevated in the context of basal target gland hormone values and may not be uniformly diagnostic (see text)

response indicates lactotrope integrity.
ACTH, adrenocorticotropic hormone; CRH, corticotropin-releasing hormone; FSH, follicle-stimulating hormone; GHRH, growth hormone-releasing hormone; GnRH, gonafdotropin-releasing hormone; LH, luteinizing hormone; TRH, thyrotropin-releasing hormone; TSH, thyroid-stimulating hormone.
Reproduced from Melmed S, Jameson JL. Disorders of the anterior pituitary and hypothalamus. In: Fauci AS, Braunwald E, Kasper DL, *et al.*, eds. *Harrison's Principles of Internal Medicine*, 17th edn. New York: McGraw-Hill Medical, 2008:2195–216 with permission.
[a]Evoked PRL (prolactin).

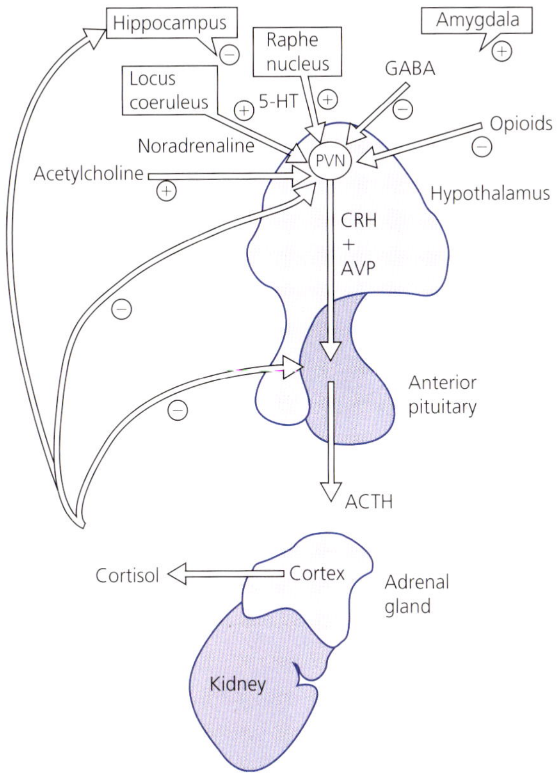

Figure 1.48 Hypothalamic–pituitary–adrenal axis. ACTH, adrenocorticotropic hormone; AVP, arginine vasopressin; CRH, corticotropin-releasing hormone; GABA, γ-aminobutyric acid; 5-HT, 5-hydroxytryptamine; PVN, paraventricular nucleus. From Puri BK, Treasaden IH (eds). *Psychiatry: An Evidence-based Text*. London: Hodder Arnold, 2010 with permission.

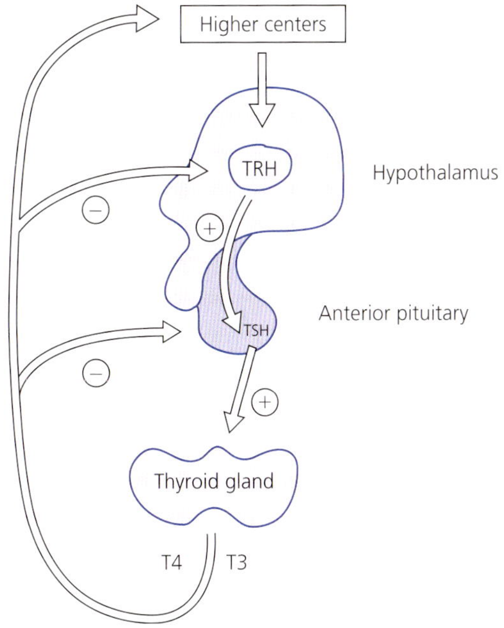

Figure 1.49 Hypothalamic–pituitary–thyroid axis. T3, tri-iodothyronine; T4, tetra-iodothyronine; TRH, thyrotropin-releasing hormone. From Puri BK, Treasaden IH (eds). *Psychiatry: An Evidence-based Text*. London: Hodder Arnold, 2010 with permission.

1.7 IMMUNOLOGY

It is becoming increasingly clear that immunology is of importance in neuropsychiatry, both in terms of research and from a clinical perspective. For example, the role of cytokines in many neuropsychiatric disorders is increasingly being recognized. Accordingly, in this section some basic concepts in immunology, cytokines, and autoimmune diseases are briefly reviewed. Clinical features of immune-related disorders of importance in neuropsychiatry are considered later in the book, particularly in Chapter 19.

Basic concepts

The innate immune system needs to be able to recognize pathogen-associated molecular patterns, for which extremely varied pattern recognition molecules are employed. Major components of the innate immune system are shown in Table 1.24.

Figure 1.50 illustrates the intercellular interactions of adaptive immune cells.

The macrophages of the central nervous system are the microglia, which act as immunological sensors, antigen-presenting cells, and cells that can release cytokines, nitric oxide, and excitatory amino acids. They engage in neuronophagia – surrounding and digesting necrotic neurons, as shown in Figure 1.51. They change their shape to become rod cells on encountering noxious stimuli such as infection. An example of activated microglia is shown in Figure 1.52.

Table 1.24 Major components of the innate immune system

Pattern recognition receptors (PRR)	C type lectins, leucine-rich proteins, scavenger receptors, pentraxins, lipid transferases, integrins
Antimicrobial peptides	α-Defensins, β-defensins, cathelin, protegrin, granulysin, histatin, secretory leukoprotease inhibitor, and probiotics
Cells	Macrophages, dendritic cells, NK cells, NK-T cells, neutrophils, eosinophils, mast cells, basophils, and epithelial cells
Complement components	Classic and alternative complement pathway, and proteins that bind complement components
Cytokines	Autocrine, paracrine, endocrine cytokines that mediate host defense and inflammation, as well as recruit, direct, and regulate adaptive immune responses

NK cells, natural killer cells.
Reproduced from Haynes BF, Soderberg KA, Fauci AS. The immune system in health and disease. In: Fauci AS, Braunwald E, Kasper DL, *et al.*, eds. *Harrison's Principles of Internal Medicine*, 17th edn. New York: McGraw-Hill Medical, 2008:2019–45 with permission.

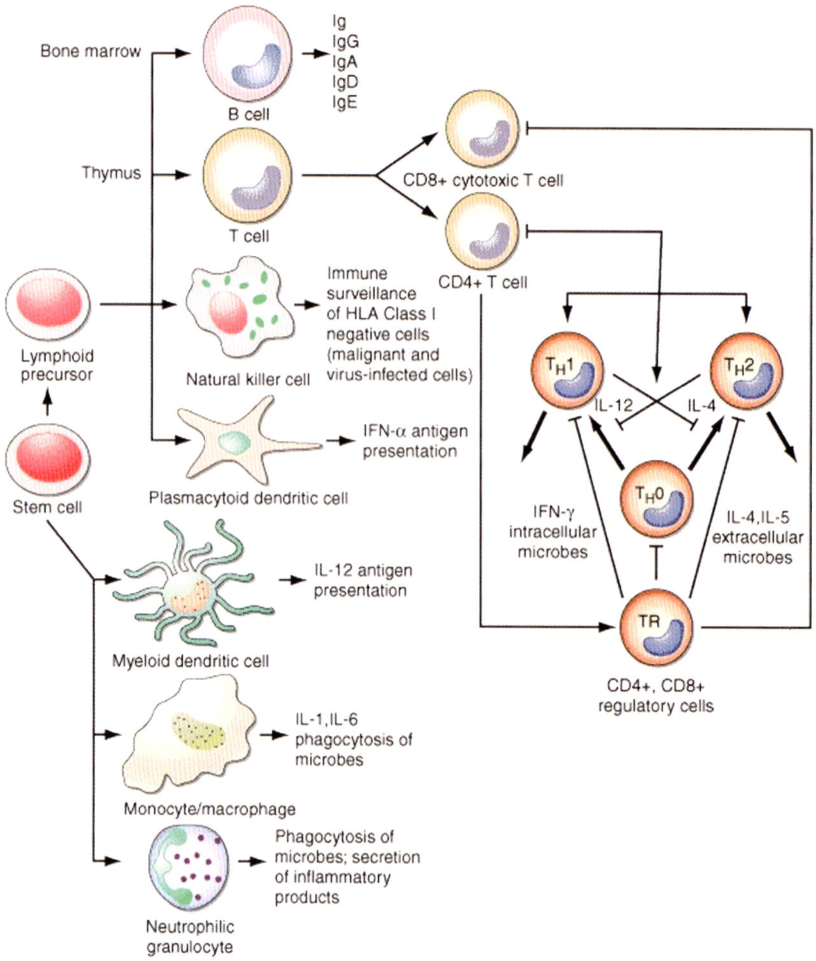

Figure 1.50 Schematic model of intercellular interactions of adaptive immune system cells. In this figure the arrows denote that cells develop from precursor cells or produce cytokines or antibodies; lines ending with bars indicate suppressive intercellular interactions. Stem cells differentiate into T cells, antigen-presenting dendritic cells, natural killer cells, macrophages, granulocytes, or B cells. Foreign antigen is processed by dendritic cells, and peptide fragments of foreign antigen are presented to CD4+ and/or CD8+ T cells. CD8+ T-cell activation leads to induction of cytotoxic T lymphocyte (CTL) or killer T-cell generation, as well as induction of cytokine-producing CD8+ cytotoxic T cells. For antibody production against the same antigen, active antigen is bound to surface immunoglobulin (sIg) within the B-cell receptor complex and drives B-cell maturation into plasma cells that secrete Ig. TH1 or TH2 CD4+ T cells producing interleukin (IL) 4, IL-5, or interferon (IFN)-γ regulate the Ig class switching and determine the type of antibody produced. CD4+, CD25+ T regulatory cells produce IL-10 and downregulate T- and B-cell responses once the microbe has been eliminated. GM-CSF, granulocyte–macrophage colony stimulating factor; TNF, tumor necrosis factor. From Fauci *et al* (eds) (2008) *Harrison's Principles of Internal Medicine*, 17th edition. New York: McGraw Hill, 2008 with permission.

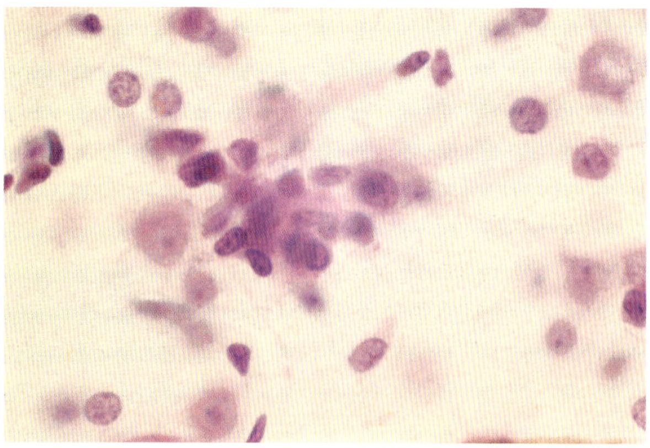

Figure 1.51 Neuronophagia. Macrophages surround necrotic neurons. Cresyl violet. Reproduced from Graham DI, Nicoll JAR and Bone I, *Adams & Graham's Introduction to Neuropathology* 3rd edition, Hodder Arnold, 2006, with permission.

Cytokines

These proteins have a relatively low relative molecular mass. They include a variety of molecular families, as depicted in Table 1.25. Their functions include roles in inflammatory responses, immune responses, and neurodevelopment.

Several studies indicate that cytokines and acute-reactive proteins are affected by the pathophysiology of mood disorders and schizophrenia, and that levels of blood inflammation-related proteins before and after treatment could act as diagnostic biomarkers of psychiatric disorders and allow treatment response to be monitored; it has also been suggested that changes in these proteins may even affect the central nervous system in such a way as to have an etiological role in neuropsychiatric disorders (Dean 2011). Dean has suggested that, further to the conclusion that

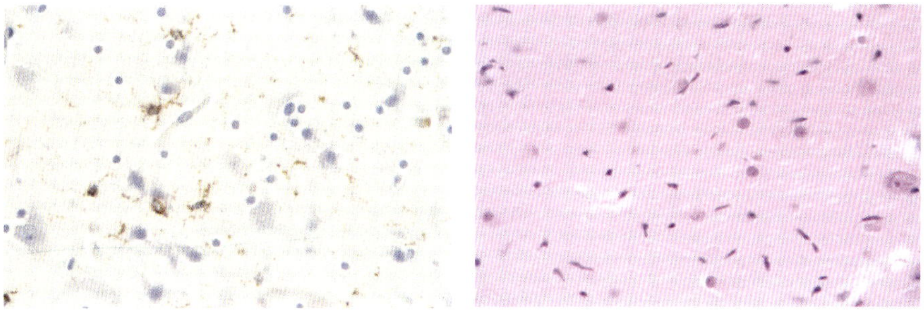

(a) (b)

Figure 1.52 (a) Microglia. Usual morphology and number in normal brain. H&E. (b) Microglia. Activated microglia in CAI sector of hippocampus in a patient who survived 4 days after cardiac arrest. H&E. Reproduced from Graham DI, Nicoll JAR and Bone I, *Adams & Graham's Introduction to Neuropathology* 3rd edition, Hodder Arnold, 2006, with permission.

Table 1.25 Major structural families of cytokines

Four α-helix-bundle family interleukins	Interleukin-2 (IL-2) subfamily: Interleukins: IL-2, IL-3, IL-4, IL-5, IL-6, IL-7, IL-9, IL-11, IL-12, IL-13, IL-15, IL-21, IL-23 Not called interleukins: colony-stimulating factor-1 (CSF1), granulocyte–macrophage colony–stimulating factor (CSF2), Flt-3 ligand, erythopoietin (EPO), thrombopoietin (THPO), leukocyte inhibitory factor (LIF) Not interleukins: growth hormone (GH1), prolactin (PRL), leptin (LEP), cardiotrophin (CTF1), ciliary neurotrophic factor (CNTF), cytokine receptor-like factor 1 (CLC or CLF) Interferon (IFN) subfamily: IFN-β, IFN-α IL-10 subfamily: IL-10, IL-19, IL-20, IL-22, IL-24 and IL-26
IL-1 family	IL-1α, (IL1A), IL-1β, (IL1B), IL-18, (IL18) and paralogs, IL-17A, IL-17B, IL-17C, IL-17D, IL-17E, IL-17F
Chemokines	IL-8, MCP-1, MCP-2, MCP-3, MCP-4, eotaxin, TARC, LARC/MIP-3α, MDC, MIP-1α, MIP-1β, RANTES, MIP-3β, I-309, SLC, PARC, TECK, GROα, GROβ, NAP-2, IP-19, MIG, SDF-1, PF4

GRO, growth-related peptide; IL, interleukin; IP, INFg-inducible protein; LARC, liver and activation-regulated chemokine; MCP, monocyte chemotatic protein; MDC, macrophage-derived chemokine; MIG, monoteine-included by IFNg; MIP, macrophage inflammatory protein; NAP, neutrophil-activating protein; PARC, pulmonary and activation-regulated chemokine; PF4, platelet factor; RANTES, regulated on activation normally T cell expressed and secreted; SDF, stromal-cell derived factor; SLC, secondary lymphoid tissue.

Reproduced from Haynes BF, Soderberg KA, Fauci AS. The immune system in health and disease. In: Fauci AS, Braunwald E, Kasper DL, *et al.*, eds. *Harrison's Principles of Internal Medicine*, 17th edn. New York: McGraw-Hill Medical, 2008:2019–45 with permission.

inflammation-related pathways are affected in chronic schizophrenia, based on changed expression in genes involved in such pathways, many of these gene expression changes could result from increased expression of transforming growth factor β1 (TGF-β1) in the central nervous system, as shown in Figure 1.53 (Dean 2011).

Autoimmune diseases

Autoantigens recognized by autoantibodies associated with several human autoimmune disorders are shown in Table 1.26. Clinical details relating to many of these disorders are given in later chapters of this book.

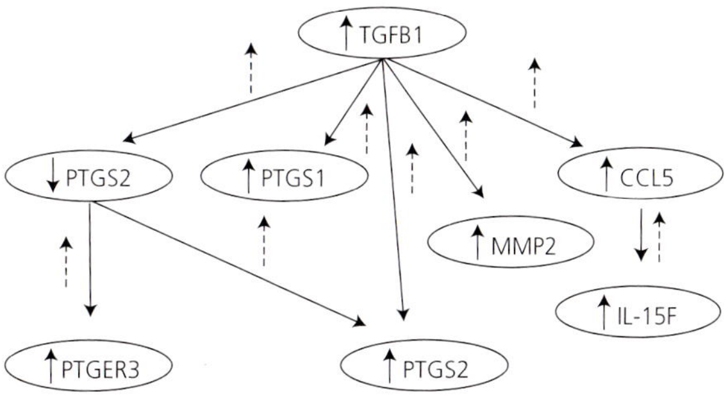

Figure 1.53 A schematic showing the genes that were identified as having changed levels of expression in the cortex of subjects with long-duration schizophrenia (direction of changes shown by solid arrows) which are also changed after the addition of exogenous transforming growth-factor β1 (TGFB1) in *in-vitro* experimental models (direction of changes shown by dotted arows). Significantly, TGFB1 alters both the expression and secretion of matrix metallopeptidase 2 (MMP2) and secreted MMP is a potent pro-inflammatory protein. CCL5, Chemokine ligand 5; C1QB, complement component 1, q subcomponent β chain; PTGS1, Prostaglandin-endoperoxide synthase 1; PTFS2, prostaglandin-endoperoxide synthase 2; PTGER3, prostaglandin E receptor 3; IL-1F5, interleukin-1F5. Adapted From Dean B. (2011), with permission.

Table 1.26 Recombinant or purified autoantigens recognized by autoantibodies associated with human autoimmune disorders.

Autoantigen	Autoimmune diseases	Autoantigen	Autoimmune diseases
Cell-or organ-specific autoimmunity			
Acetylcholine receptor	Myasthenia gravis	Insulin receptor	Type B insulin resistance, acanthosis systemic lupus erythematosus (SLE)
Actin	Chronic active hepatitis, primary biliary cirrhosis	Intrinsic factor type 1	Pernicious anemia
Adenine nucleotide translator (ANT)	Dilated cardiomyopathy, myocarditis	Leukocyte function-associated antigen (LFA-1)	Treatment-resistant Lyme arthritis
β-Adrenoreceptor	Dilated cardiomyopathy		
Aromatic L-amino acid decarboxylase	Autoimmune polyendocrine syndrome type 1 (APS-1)	Myelin-associated glycoprotein (MAG)	Polyneuropathy
Asialoglycoprotein receptor	Autoimmune hepatitis	Myelin-basic protein	Multiple sclerosis, demyelinating diseases
Bactericidal/permeability-increasing protein (Bpi)	Cystic fibrosis vasculitides	Myelin oligodendrocyte glycoprotein (MOG)	Multiple sclerosis
Calcium-sensing receptor	Acquired hypoparathyroidism		
Cholesterol side-chain cleavage enzyme (CYPIIa)	Autoimmune polyglandular syndrome-1	Myosin	Rheumatic fever
		p-80-Collin	Atopic dermatitis
Collagen type IV-α3-chain	Goodpasture syndrome	Pyruvate dehydrogenase complex-E2 (PDC-E2)	Primary biliary cirrhosis
Cytochrome P450 2D6 (CYP2D6)	Autoimmune hepatitis		
Desmin	Crohn disease, coronary artery disease	Sodium iodide symporter (NIS)	Graves' disease, autoimmune hypothyroidism
Desmoglein 1	Pemphigus foliaceus		
Desmoglein 3	Pemphigus vulgaris	SOX-10	Vitiligo
F-actin	Autoimmune hepatitis	Thyroid and eye muscle shared protein	Thyroid-associated ophthalmopathy
GM gangliosides	Guillain-Barré syndrome		
Glutamate decarboxylase (GAD65)	Type 1 diabetes, stiff man syndrome	Thyroglobulin	Autoimmune thyroiditis
Glutamate receptor (GLUR)	Rasmussen encephalitis	Thyroid peroxidase	Autoimmune Hashimoto thyroiditis
H/K ATPase	Autoimmune gastritis	Thyrotropin receptor	Graves' disease
17-α-Hydroxylase (CYP17)	Autoimmune polyglandular syndrome-1	Tissue transglutaminase	Celiac disease
21-Hydroxylase (CYP21)	Addison disease	Transcription coactivator p75	Atopic dermatitis
IA-2 (ICAS12)	Type 1 diabetes	Tryptophan hydroxylase	Autoimmune polyglandular syndrome-1
Insulin	Type 1 diabetes, insulin hypoglycemic syndrome (Hirata disease)	Tyrosinase	Vitiligo, metastatic melanoma
		Tryptophan hydroxylase	Autoimmune polyglandular syndrome-1
Systemic autoimmuniuty			
ACTH	ACTH deficiency	Histone H2A-H2B-DNA	SLE
Aminoacyl-tRNA histidyl synthetase	Myositis, dermatomyositis	IgE receptor	Chronic idiopathic urticaria
Aminoacyl-tRNA synthetase (several)	Polymyositis, dermatomyositis	Keratin	RA
Cardiolipin	SLE, anti-phospholipid syndrome	Ku-DNA-protein kinase	SLE
Carbonic anhydrase II	SLE, Sjögren syndrome, systemic sclerosis	Ku-nucleoprotein	Connective tissue syndrome
		La phosphoprotein (La 55-B)	Sjögren syndrome

Table 1.26 (*Continued*)

Autoantigen	Autoimmune diseases	Autoantigen	Autoimmune diseases
Collagen (multiple types)	Rheumatoid arthritis (RA), SLE, progressive systemic sclerosis	Myeloperoxidase	Necrotizing and crescentic glomerula nephritis (NCGN), systemic vasculitis
Centromere-associated proteins	Systemic sclerosis	Proteinase 3 (PR3)	Wegener granulomatosis, Churg–Strauss syndrome
DNA-dependent nucleosine-stimulated ATPase	Dermatomyositis	RNA polymerase I-III (RNP)	Systemic sclerosis, SLE
Fibrillarin	Scleroderma	Signal recognition protein (SRP54)	Polymyositis
Fibronectin	SLE, RA, morphea	Topoisomerase-1 (Scl-70)	Scleroderma, Raynaud syndrome
Glucose-6-phosphate isomerase	RA	Tubulin	Chronic liver disease, visceral leishmaniasis
β2-Glycoprotein I (B2-GPI)	Primary antiphospholipid syndrome		
Golgin (95, 97, 160, 180)	Sjögren syndrome, SLE, RA	Vimentin	Systemic autoimmune disease
Heat shock protein	Various immune-related disorders		
Hemidesmosomal protein 180	Bullous pemphigoid, herpes gestationis, cicatricial pemphigoid		
Plasma protein and cytokine autoimmunity			
C1 inhibitor	Autoimmune C1 deficiency	Glycoprotein IIb/IIIg and Ib/IX	Autoimmune thrombocytopenia purpura
C1q	SLE, membrane proliferative glomerulonephritis (MPGN)	IgA	Immunodeficiency associated with SLE, perniclous anemia, thyroiditis, Sjögren syndrome and chronic active hepatitis
Cyrokines (IL-1α, IL-1β, IL-6, IL-10, LIF)	RA, systemic scierosis, normal subjects		
Factor II, factor V, factor VII, factor VIII, factor IX, factor X, factor XI, thrombin vWF	Prolonged coagulation time	Oxidized LDL (OxLDL)	Atherosclerosis
Cancer and paraneoplastic autoimmunity			
Amphiphysin	Neuropathy, small-cell lung cancer	p62 (IGF-II mRNA-binding protein)	Hepatocellular carcinoma (China)
Cyclin B1	Hepatocellular carcinoma	Recoverin	Cancer-associated retinopathy
DNA topoisomerase II	Liver cancer	Ri protein	Paraneoplastic opsoclonus myoclonus ataxia
Desmoplakin	Paraneoplastic pemphigus		
Gephyrin	Paraneoplastic stiff man syndrome	βIV spectrin	Lower motor neuron syndrome
Hu proteins	Paraneoplastic encephalomyelitis	Synaptotagmin	Lambert-Eaton myasthenic syndrome
Neuronal nicotinic acetylcholine receptor	Subacute autonomic neuropathy, cancer	Voltage-gated calcium channels	Lambert-Eaton myasthenic syndrome
p53	Cancer, SLE	Yo protein	Paraneoplastic cerebellar degeneration

Reproduced from Haynes BF, Soderberg KA, Fauci AS. The immune system in health and disease. In: Fauci AS, Braunwald E, Kasper DL, *et al.*, eds. *Harrison's Principles of Internal Medicine*, 17th edn. New York: McGraw-Hill Medical, 2008:2019–45 with permission.

REFERENCES

Alexander GE, DeLong MR, Strick PL. Parallel organization of functionally segregated circuits linking basal ganglia and cortex. *Annu Rev Neurosci* 1986; **9**:357–81.

Anders S, Birbaumer N, Sadowski B *et al.* Parietal somatosensory association cortex mediates affective blindsight. *Nat Neurosci* 2004; **7**:339–40.

Anderson SW, Tranel D. Awareness of disease states following cerebral infarction, dementia, and head trauma: standardized assessment. *Clin Neuropsychol* 1989; **3**:327–39.

Annett M. A classification of hand preference by association analysis. *Br J Psychol* 1970; **61**:303–21.

Anthony JC, LeResche L, Niaz U, von Korff MR, Folstein MF. Limits of the 'Mini-Mental State' as a screening test for dementia and delirium among hospital patients. *Psychol Med* 1982; **12**:397–408.

Barrington SF, Wahl RL, Maisey M. *Atlas of Clinical Positron Emission Tomography*, 2nd edn. London: Hodder Arnold, 2006.

Belliveau JW, Kennedy DN Jr, McKinstry RC *et al.* Functional mapping of the human visual cortex by magnetic resonance imaging. *Science* 1991; **254**:716–19.

Benton AL, Hamsher K, Sivan AB. *Multilingual Aphasia Examination*, 3rd edn. Iowa: AJA Associates, **1994**.

Bernstein LE, Auer ET Jr, Moore JK *et al.* Visual speech perception without primary auditory cortex activation. *Neuroreport* 2002; **13**:311–15.

Borch K, Greisen G. ^{99m}Tc-HMPAO as a tracer of cerebral blood flow in newborn infants. *J Cereb Blood Flow Metab* 1997; **17**:448–54.

Bowen K, Mortimer A, Rooke-Ley S. Measuring IQ in schizophrenia: the 'Quick Test'. *Cogn Neuropsychiatry* 1999; **4**:81–8.

Boy F, Husain M, Singh KD, Sumner P. Supplementary motor area activations in unconscious inhibition of voluntary action. *Exp Brain Res* 2010; **206**:441–8.

Buratowski S, Zhou H. Functional domains of transcription factor TFIIB. *Proceedings of the National Academy of Sciences* 1993; **90**:5633–7.

Butler AB, Hodos W. *Comparative Vertebrate Neuroanatomy: Evolution and Adaptation*, 2nd edn. New Jersey: Wiley-Liss, 2005.

Butler AB. Chordate evolution and the origin of craniates: an old brain in a new head. *Anat Rec* 2000; **261**:111–25.

Bystron I, Blakemore C, Rakic P. Development of the human cerebral cortex: Boulder Committee revisited. *Nat Rev Neurosci* 2008; **9**:110–22.

Carpenter RHS. *Neurophysiology*, 4th edn. London: Arnold, 2003.

Caselli RJ. Ventrolateral and dorsomedial somatosensory association cortex damage produces distinct somesthetic syndromes in humans. *Neurology* 1993; **43**:762–71.

Chen WR, Lee S, Kato K, Spencer DD, Shepherd GM, Williamson A. Long-term modifications of synaptic efficacy in the human inferior and middle temporal cortex. *Proc Natl Acad Sci U S A* 1996; **93**:8011–15.

Chow SS, Romo R, Brody CD. Context-dependent modulation of functional connectivity: secondary somatosensory cortex to prefrontal cortex connections in two-stimulus-interval discrimination tasks. *J Neurosci* 2009; **29**:7238–45.

Cox IJ, Puri BK. *In vivo* MR spectroscopy in diagnosis and research of neuropsychiatric disorders. *Prostaglandins Leukot Essent Fatty Acids* 2004; **70**:357–60.

Crawford TJ, Puri BK, Nijran KS *et al.* Abnormal saccadic distractibility in patients with schizophrenia: a ^{99m}Tc-HMPAO SPET study. *Psychol Med* 1996; **26**:265–77.

Dawkins R. *The Selfish Gene*. Oxford: Oxford University Press, 2006.

Dean B. Understanding the role of inflammatory-related pathways in the pathophysiology and treatment of psychiatric disorders: evidence from human peripheral studies and CNS studies. *Int J Neuropsychopharmacology* 2011; **14**:997–1012.

Dunn LM, Dunn DM. *Peabody Picture Vocabulary Test*, fourth edition (PPVT-4). San Antonio, TX: Pearson, 2007.

Elbert T, Rockstroh B. Reorganization of human cerebral cortex: the range of changes following use and injury. *Neuroscientist* 2004; **10**:129–41.

Elbert T, Pantev C, Wienbruch C *et al.* Increased cortical representation of the fingers of the left hand in string players. *Science* 1995; **270**:305–7.

Eriksson PS, Perfilieva E, Bjork-Eriksson T *et al.* Neurogenesis in the adult human hippocampus. *Nat Med* 1998; **4**:1313–17.

Essen DC, Zeki SM. The topographic organization of rhesus monkey prestriate cortex. *J Physiol* 1978; **277**:193–226.

Evans JJ. Basic concepts and principles of neuropsychological assessment. In: Gurd JM, Kischka U, Marshall JC, eds. *The Handbook of Clinical Neuropsychology*, 2nd edn. Oxford: Oxford University Press, 2010.

Fitzgerald P, Dinan TG. The neuroendocrine system. In: Puri BK, Treasaden IH, eds. *Psychiatry: An Evidence-based Text*. London: Hodder Arnold, 2010: 385–96.

Flor H, Elbert T, Knecht S *et al.* Phantom-limb pain as a perceptual correlate of cortical reorganization following arm amputation. *Nature* 1995; **375**:482–4.

Folkerth RD, Kinney HC. Disorders of the perinatal period. In: Love S, Louis DN, Ellison DW, eds. *Greenfield's Neuropathology*, 8th edn. London: Hodder Arnold, 2008: 241–334.

Folstein MF, Folstein SE, McHugh PR. 'Mini-mental state'. A practical method for grading the cognitive state of patients for the clinician. *J Psychiatr Res* 1975; **12**:189–98.

Gheusi G, Lledo PM. Control of early events in olfactory processing by adult neurogenesis. *Chem Senses* 2007; **32**:397–409.

Gill M, McGrath J. Genetics. In: Puri BK, Treasaden IH, eds. *Psychiatry: An Evidence-based Text*. London: Hodder Arnold, 2010: 465–85.

Goldberg TE, Weinberger DR. Schizophrenia, training paradigms, and the Wisconsin Card Sorting Test redux. *Schizophr Res* 1994; **11**:291–6.

Goodglass H, Kaplan E. *Boston Naming Test*. Philadelphia: Lippincott Williams & Wilkins, 2001.

Goodglass H, Kaplan E, Barresi B. *The Boston Diagnostic Aphasia Examination*, 3rd edition (BDAE-3). Philadelphia: Lippincott, 2000.

Goswami U. Neuroscience and education. *Br J Educ Psychol* 2004; **74**:1–14.

Groth-Marnat G. *Neuropsychological Assessment in Clinical Practice: A Guide to Test Interpretation and Integration*. New York and Chichester: Wiley, 2000.

Haber SN. The primate basal ganglia: parallel and integrative networks. *J Chem Neuroanat* 2003; **26**:317–30.

Harpaz Y, Levkovitz Y, Lavidor M. Lexical ambiguity resolution in Wernicke's area and its right homologue. *Cortex* 2009; **45**:1097–103.

Haynes BF, Soderberg KA, Fauci AS. The immune system in health and disease. In: Fauci AS, Braunwald E, Kasper DL *et al.*, eds. *Harrison's Principles of Internal Medicine*, 17th edn. New York: McGraw-Hill Medical, 2008: 2019–45.

Hellstrom NA, Bjork-Eriksson T, Blomgren K, Kuhn HG. Differential recovery of neural stem cells in the subventricular zone and dentate gyrus after ionizing radiation. *Stem Cells* 2009; **27**:634–41.

Hevner RF, Wong-Riley MT. Entorhinal cortex of the human, monkey, and rat: metabolic map as revealed by cytochrome oxidase. *J Comp Neurol* 1992; **326**:451–69.

Huttenlocher PR, Dabholkar AS. Regional differences in synaptogenesis in human cerebral cortex. *J Comp Neurol* 1997; **387**:167–78.

Huttenlocher PR, de Courten C. The development of synapses in striate cortex of man. *Hum Neurobiol* 1987; **6**:1–9.

Jameson JL. Principles of endocrinology. In: Fauci AS, Braunwald E, Kasper DL *et al.*, eds. *Harrison's Principles of Internal Medicine*, 17th edn. New York: McGraw-Hill Medical, 2008: 2187–95.

Jolkkonen E, Pitkanen A. Intrinsic connections of the rat amygdaloid complex: projections originating in the central nucleus. *J Comp Neurol* 1998; **395**:53–72.

Jung-Beeman M. Bilateral brain processes for comprehending natural language. *Trends Cogn Sci* 2005; **9**:512–18.

Jurica PJ, Leitten CL, Mattis S. *Dementia Rating Scale-2 (DRS-2)*. Odessa, FL: Psychological Assessment Resources, 2002.

Knight JA, Kaplan E. *The Handbook of Rey–Osterrieth Complex Figure Usage: Clinical And Research Applications*. Lutz, FL: Psychological Assessment Resources, Inc., 2003.

Koebis M, Ohsawa N, Kino Y *et al.* Alternative splicing of myomesin 1 gene is aberrantly regulated in myotonic dystrophy type 1. *Genes Cells* 2011; **16**:961–72.

Kwong KK, Belliveau JW, Chesler DA *et al.* Dynamic magnetic resonance imaging of human brain activity during primary sensory stimulation. *Proc Natl Acad Sci U S A* 1992; **89**:5675–9.

Liddle PF, Friston KJ, Frith CD *et al.* Patterns of cerebral blood flow in schizophrenia. *Br J Psychiatry* 1992; **160**:179–86.

Liepert J, Miltner WH, Bauder H *et al.* Motor cortex plasticity during constraint-induced movement therapy in stroke patients. *Neurosci Lett* 1998; **250**:5–8.

Liu Z, Wang D, Shao B *et al.* Increased expression of transcription initiation factor IIB after rat traumatic brain injury. *Journal of Molecular Histology* 2011; **42**:265–71.

McDonald AJ. Cytoarchitecture of the central amygdaloid nucleus of the rat. *J Comp Neurol* 1982; **208**:401–18.

McGuire PK, Shah GM, Murray RM. Increased blood flow in Broca's area during auditory hallucinations in schizophrenia. *Lancet* 1993; **342**:703–6.

McLachlan RS. A brief review of the anatomy and physiology of the limbic system. *Can J Neurol Sci* 2009; **36**(Suppl 2):S84–7.

McQuillen PS, Ferriero DM. Perinatal subplate neuron injury: implications for cortical development and plasticity. *Brain Pathol* 2005; **15**:250–60.

MacLean PD. Psychosomatic disease and the visceral brain; recent developments bearing on the Papez theory of emotion. *Psychosom Med* 1949; **11**:338–53.

Mattis S. *Dementia Rating Scale (DRS)*. Odessa, FL: Psychological Assessment Resources, 1998.

Meftah el M, Shenasa J, Chapman CE. Effects of a cross-modal manipulation of attention on somatosensory cortical neuronal responses to tactile stimuli in the monkey. *J Neurophysiol* 2002; **88**:3133–49.

Meftah el M, Bourgeon S, Chapman CE. Instructed delay discharge in primary and secondary somatosensory cortex within the context of a selective attention task. *J Neurophysiol* 2009; **101**:2649–67.

Melmed S, Jameson JL. Disorders of the anterior pituitary and hypothalamus. In: Fauci AS, Braunwald E, Kasper DL *et al.*, eds. *Harrison's Principles of Internal Medicine*, 17th edn. New York: McGraw-Hill Medical, 2008: 2195–216.

Middleton FA, Strick PL. Basal-ganglia 'projections' to the prefrontal cortex of the primate. *Cereb Cortex* 2002; **12**:926–35.

Monosov IE, Trageser JC, Thompson KG. Measurements of simultaneously recorded spiking activity and local field potentials suggest that spatial selection emerges in the frontal eye field. *Neuron* 2008; **57**:614–25.

Nelson HE. *The National Adult Reading Test (NART): Test Manual.* Windsor: NFER-Nelson, 1982.

Nelson HE, Willison J, eds. *National Adult Reading Test (NART)*, 2nd edn. Windsor: NFER-Nelson, 1991.

Nishimiya M, Matsuda H, Imabayashi E, Kuji I, Sato N. Comparison of SPM and NEUROSTAT in voxelwise statistical analysis of brain SPECT and MRI at the early stage of Alzheimer's disease. *Ann Nucl Med* 2008; **22**:921–7.

Oldfield RC. The assessment and analysis of handedness: the Edinburgh inventory. *Neuropsychologia* 1971; **9**:97–113.

Papez JW. A proposed mechanism of emotion. *Arch Neurol Psychiatry* 1937; **38**:725–43.

Pesaran B, Nelson MJ, Andersen RA. Free choice activates a decision circuit between frontal and parietal cortex. *Nature* 2008; **453**:406–9.

Pessoa L. On the relationship between emotion and cognition. *Nat Rev Neurosci* 2008; **9**:148–58.

Price JL, Russchen FT, Amaral DG. *The Limbic Region. II: The Amygdaloid Complex*. New York: Elsevier Science, 1987.

Puri BK, Tyrer PJ. *Sciences Basic to Psychiatry*, 2nd edn, with contributions by Boulton M, Green J. Edinburgh: Churchill Livingstone, 1998.

Puri BK, Counsell SJ, Hamilton G *et al.* Cerebral metabolism in male patients with schizophrenia who have seriously and dangerously violently offended: a ^{31}P magnetic resonance spectroscopy study. *Prostaglandins Leukot Essent Fatty Acids* 2004; **70**:409–11.

Puri BK, Counsell SJ, Hamilton G. Brain cell membrane motion-restricted phospholipids: a cerebral 31-phosphorus magnetic resonance spectroscopy study of patients with schizophrenia. *Prostaglandins Leukot Essent Fatty Acids* 2008; **79**:233–5.

Rakic P. A small step for the cell, a giant leap for mankind: a hypothesis of neocortical expansion during evolution. *Trends Neurosci* 1995; **18**:383–8.

Rakic P. A century of progress in corticoneurogenesis: from silver impregnation to genetic engineering. *Cereb Cortex* 2006; **16**(Suppl. 1): 13–17.

Ramnani N, Owen AM. Anterior prefrontal cortex: insights into function from anatomy and neuroimaging. *Nat Rev Neurosci* 2004; **5**:184–94.

Randolph C, Tierney MC, Mohr E, Chase TN. The Repeatable Battery for the Assessment of Neuropsychological Status (RBANS): preliminary clinical validity. *J Clin Exp Neuropsychol* 1998; **20**:310–19.

Raven J, Raven JC, Court JH. *Manual for Raven's Progressive Matrices and Vocabulary Scales*, 1998 edn. Oxford: Oxford Psychologists, 1998.

Recanzone GH, Schreiner CE, Merzenich MM. Plasticity in the frequency representation of primary auditory cortex following discrimination training in adult owl monkeys. *J Neurosci* 1993; **13**:87–103.

Robertson D, Irvine DR. Plasticity of frequency organization in auditory cortex of guinea pigs with partial unilateral deafness. *J Comp Neurol* 1989; **282**:456–71.

Roland PE, Larsen B, Lassen NA, Skinhoj E. Supplementary motor area and other cortical areas in organization of voluntary movements in man. *J Neurophysiol* 1980; **43**:118–36.

Rosen WG, Mohs RC, Davis KL. A new rating scale for Alzheimer's disease. *Am J Psychiatry* 1984; **141**:1356–64.

Sah P, Faber ES, Lopez De Armentia M, Power J. The amygdaloid complex: anatomy and physiology. *Physiol Rev* 2003; **83**:803–34.

Shallice T. Specific impairments of planning. *Phil Trans R Soc B* 1982; **298**:199–209.

Sivan AB. *Benton Visual Retention Test*, 5th edition. San Antonio, TX: Psychological Corporation, 1992.

Stahl SM. *Stahl's Essential Psychopharmacology: Neuroscientific Basis and Practical Applications*, 3rd edn. Cambridge: Cambridge University Press, 2008.

Sumner P, Nachev P, Morris P *et al.* Human medial frontal cortex mediates unconscious inhibition of voluntary action. *Neuron* 2007; **54**:697–711.

Thurstone LL. *Primary Mental Abilities*. Chicago: University of Chicago Press, 1957.

Van Hoesen GW. Anatomy of the medial temporal lobe. *Magn Reson Imaging* 1995; **13**:1047–55.

von Melchner L, Pallas SL, Sur M. Visual behaviour mediated by retinal projections directed to the auditory pathway. *Nature* 2000; **404**:871–6.

Walker MR, Patel KK, Stappenbeck TS. The stem cell niche. *J Pathol* 2009; **217**:169–80.

Wells J. *The Myth of Junk DNA*. Discovery Institute, 2011.

Yoo SS, O'Leary HM, Dickey CC *et al.* Functional asymmetry in human primary auditory cortex: identified from longitudinal fMRI study. *Neurosci Lett* 2005; **383**:1–6.

PART II

DIAGNOSTIC ASSESSMENT

2

Diagnostic assessment

2.1 DIAGNOSTIC INTERVIEW

Lord Brain (1964) noted that 'in the diagnosis of nervous diseases the history of the patient's illness is often of greater importance than the discovery of his abnormal physical signs', a sentiment echoed by Russell DeJong (1979), who asserted that 'a good clinical history often holds the key to diagnosis'.

Obtaining the history, however, as noted by DeJong (1979), 'is no simple task [and] may require greater skill and experience than are necessary to carry out a detailed examination'. The acquisition of this skill is, for most, no easy matter, requiring, above all, practice and supervision. Certain points, however, may be made regarding the setting of the interview, establishing rapport, eliciting the chief complaint, the division of the interview itself into non-directive and directive portions, concluding the interview, and the subsequent acquisition of collateral history from family or acquaintances. Even these general points, however, allow exceptions depending on the clinical situation, and the physician must be flexible and prepared to exercise initiative.

Setting

The interview should ideally be conducted in a quiet and private setting, set apart from distractions and anything that might inhibit patients as they relate the history. Importantly, that means that family and friends should be excused during the interview, as patients may feel reluctant to reveal certain facts in their presence. If the interview takes place at the bedside, the physician should be seated; standing implies that time is short, and some patients, picking up on this cue, may skip over potentially valuable parts of the history in order not to waste the physician's time. In this regard, it is also important that the physician set aside a sufficient amount of time to take the history, which may range from less than half an hour in uncomplicated cases related by cooperative patients to well over an hour when the history is long and complex or the patient is unable to cooperate fully. There is debate whether the physician should take notes during the interview: some feel it is distracting, to both the patient and the physician, whereas others recommend it in order to ensure accuracy, especially when the interview is lengthy. We agree with Victor (Victor and Ropper 2001) who feels that the practice is 'particularly recommended'. The idea is not to make a transcript but simply to jot down key points and dates, and to do so in a way that allows the physician to maintain his or her attention on what the patient is saying.

Establishing rapport

DeJong (1979) noted that 'interest, understanding, and sympathy' are essential to the successful conduct of the interview: patients who experience a sense of rapport with their physicians are more likely to be truthful and forthcoming; hence, establishing rapport is of great importance.

First impressions carry great weight here: after introducing themselves, physicians should clearly relate their role in the case and then, as suggested by DeJong (1979), display 'kindness, patience, reserve, and a manner which conveys interest' throughout the interview. Provided with such a forum, most patients will, with only minor help, provide the history required to generate the appropriate differential diagnosis.

Eliciting the chief complaint

'It is well', noted Lord Brain (1964), 'to begin by asking the patient of what he complains'. The chief complaint is the epitome of the patient's illness: lacking such a focus, digressions are almost inevitable, and the history obtained may be of little diagnostic use. Thus, once introductions are out of the way the first question put by the physician should focus on what brought the patient to the hospital. Critically, as some patients may be reluctant to reveal the actual reason for their coming to the hospital, it is necessary to weigh the chief complaint offered by the patient and ask oneself whether, in fact, it sounds like a plausible reason to seek medical attention. If not, gentle probing is in order and should generally be continued until the actual chief complaint is revealed. Importantly, the physician should never accept at face value a diagnosis offered by a patient: as Bickerstaff (1980) pointed out, 'it must be made absolutely clear what the patient means by his description of his symptoms. By all means put it down in his words first, but do not be content with that . . . "Black-outs" may mean loss of consciousness, loss of vision, loss of memory, or just loss of confidence'.

Occasionally, it may not be possible to establish a chief complaint during the interview, as may occur with patients who are delirious, demented, psychotic, or simply hostile and uncooperative. In such cases, persisting overly long in the pursuit of a chief complaint may become counterproductive as patients may become resentful, and it is generally more appropriate to move on to the 'directive' portion of the interview described below, always being alert, however, to the possibility that the patient may 'slip in' the chief complaint at an unexpected moment.

The non-directive portion of the interview

Once a chief complaint has been established, the patient, as noted by Lord Brain (1964), 'should be allowed to relate the story of his illness as far as possible without interruption, questions being put to him afterwards to expand his statements and to elicit additional information'. Some patients, once asked to expand on the chief complaint, may, with little or no prompting, provide the 'perfect' history, covering each of the following essential points:

- onset, including approximate date and mode of onset (acute, gradual, or insidious)
- presence or absence of any precipitating factors
- temporal evolution of various signs and symptoms
- presence or absence of any aggravating or alleviating factors
- treatment efforts and their results
- pertinent positives and negatives
- any history of similar experiences in the past.

Most patients will require, however, either encouragement or some gentle shepherding at various times. When patients begin to falter in their history, or seem to be leaving items out, it is appropriate to encourage them to talk by asking 'open-ended' questions such as 'Tell me more about that'. Such a method is much to be preferred over the question-and-answer approach used by many. The problem with the question-and-answer approach is that many patients will lose the initiative to speak, and simply await questions from the physician, which is all well and good unless, of course, the physician fails to ask the 'right' questions, in which case potentially critical aspects of the history may remain unrevealed.

Gentle shepherding may be required in cases when patients digress or take off on a tangent. One should not, of course, rudely pull the patient back to task, but rather tactfully suggest that refocusing on the illness that prompted admission might be more appropriate.

Once the essential points have been covered, it is appropriate to summarize briefly what the patient has said in order to be sure that the history, as understood by the physician, is correct. Patients should be invited to correct any misapprehensions and once the history is complete the physician should move on to the directive portion of the interview.

The directive portion of the interview

The directive portion of the interview should be introduced to the patient as a series of perhaps 'routine' questions relating to the patient's overall health. Here, one obtains information regarding the medications that the patient is taking, allergies, the past medical history, a review of systems, the family medical history and, finally, the mental status examination (discussed in Chapter 2.2). In this regard, two points deserve special emphasis. First, when interviewing hospitalized patients it is essential to obtain an absolutely accurate list of medicines that the patient was taking at home, prior to admission: medication changes often provide the clue to otherwise puzzling syndromes, such as delirium, which may occur during the hospital stay. Second, given the increasing importance of genetics in neuropsychiatric practice, it is essential to obtain a detailed family history regarding any neuropsychiatric illness.

During the directive portion of the interview, although a question-and-answer approach is generally appropriate, the physician must always be ready to adopt a non-directive approach should the patient report a symptom or illness potentially pertinent to the chief complaint. For example, if during the review of systems the patient affirms that headaches have been present it is appropriate to stop and ask the patient to elaborate on this, with an eye towards obtaining information regarding each of the essential points described earlier.

Questions regarding alcohol/drug use and suicidal/homicidal ideation must be directly pursued if not already covered in the non-directive portion of the interview. These are, of course, delicate areas, but, if approached in a straightforward and non-judgmental way, it is remarkable how forthcoming, and indeed relieved, some patients may be at being given an opportunity to speak of them.

Concluding the interview

Once the directive portion of the interview has been completed it is appropriate to give the patient an opportunity to speak freely again. If asked whether he or she has anything else to add, many patients will offer important information that they may have either withheld or simply not recalled earlier. Asking patients whether they have anything they wish to ask the physician is also appropriate, as the patient's questions may reveal much about the concerns that brought him or her to the hospital in the first place.

Collateral history

According to Lord Brain (1964), 'the history obtained from the patient should always be supplemented, if possible, by an account of his illness given by a relative or by someone who knows him well'. This is especially the case when patients are confused or suffer from poor memory: it is remarkable how often a collateral history will change a diagnostic impression, guide further testing, or alter proposed treatments. In obtaining the collateral history, particular attention should be paid to establishing the patient's premorbid baseline ability to perform such routine activities of daily living as bathing, dressing, cooking, feeding, doing housework, shopping, driving or using public transportation, and paying bills. Inquiry should also be made regarding hobbies, such as playing cards or chess, or doing crossword puzzles. In cases characterized by cognitive deficits, the loss of these abilities may serve to establish the onset of the current illness.

Some have expressed concern that interviewing the family or acquaintances may violate patient confidentiality, but this is simply not the case, provided both that the contact knows already that the patient is in the hospital and that the physician reveals nothing about the patient while interviewing the collateral contact. No confidentiality is breached by introducing oneself as the patient's physician or by asking collateral contacts what they know about the patient.

Finally, it is also essential to review old records. This is sometimes a tedious task but, as with interviewing collateral sources, it may reveal critical information.

2.2 MENTAL STATUS EXAMINATION

The mental status examination constitutes an essential part of any neuropsychiatric evaluation and, at a minimum, should cover each of the items discussed below. Many of these may be determined during the non-directive portion of the interview; however, some, especially those concerning cognition (e.g. orientation, memory), require direct testing. As some patients may object to cognitive testing, it is important to smooth the way by indicating that these are 'routine' questions to test 'things such as memory and concentration', perhaps adding that 'patients who have had a stroke [or whatever illness the patient feels comfortable discussing] often have difficulties here'. Should patients remain uncooperative, it may at times be possible to infer their cognitive status indirectly; for example, during history taking, by asking the date of a recent event brought up by the patient.

As noted below, abnormalities on the mental status examination typically indicate the presence of one of the major syndromes, such as dementia (Chapter 6.3), delirium (Chapter 6.5), amnesia (Chapter 6.6), depression (Chapter 7.1), apathy (Chapter 7.2), mania (Chapter 7.3), anxiety (Chapter 7.5), psychosis (Chapter 8.1), and personality change (Chapter 8.2), especially the frontal lobe syndrome.

Grooming and dress

Good habits of grooming and dress may suffer in certain illnesses, sometimes with diagnostically suggestive results. In depression, hopelessness, fatigue, and anhedonia may leave patients with little interest in maintaining their appearance, with the result that grooming and dress are left in a greater or lesser degree of disarray. Manic patients, overflowing with exuberance, may truly make a spectacle of themselves with garish make-up and clothing, and all manner of decorations. Patients with psychosis, especially schizophrenia, may be quite unkempt and at times dirty, and their clothing may be bizarre, as, for example, with multiple layers and a woollen cap, even in the summer; overall dishevellment may also be seen in the frontal lobe syndrome, dementia, or delirium. Rarely, one may see evidence of the syndrome of neglect (Chapter 3.10) wherein dress and grooming suffer on only one side of the body.

General description

An overall and general description of the patient's behavior is essential, and gives room for the exercise of whatever literary talents the physician may possess.

Comments may be made on the relationship of the patient to the interviewer, noting, for example, whether the patient is cooperative or uncooperative, guarded, evasive, hostile, or belligerent. The overall quality of the relationship may also be of diagnostic importance. For example, as noted by Bleuler (1924), in schizophrenia, there is often a '*defect in … emotional rapport*' (italics in original), such that 'the joy of a schizophrenic does not transport us, and his expressions of pain leave us cold'. By contrast, in mania, as noted by Kraepelin (1921), 'the patient feels the need to get out of himself, to be on more intimate terms with his surroundings', such that the physician, willingly or not, often feels engaged, in one fashion or another, with the patient: in the case of a euphoric manic it is the rare physician who can keep from smiling, and in the case of an irritable manic most physicians will find themselves becoming, at the very least, on edge.

During the interview, one may also find evidence of certain specific personality changes. Perseveration, disinhibition, and a tendency to puerile, silly puns or jokes may suggest the frontal lobe syndrome, and in epileptics one may find evidence of the interictal personality syndrome with an overall 'viscosity' or 'adhesiveness', such that patients are unable to manage changing the subject or switching tasks. Evidence of the Kluver–Bucy syndrome may also be apparent should patients repeatedly put inedible objects in their mouths or engage in indiscriminate sexual activity.

Consideration should also be given to the overall level of the patient's verbal and motor behavior, noting whether there is either psychomotor hyperactivity or retardation.

Psychomotor hyperactivity may manifest with agitation or mere restlessness, and the activity itself may or may not be purposeful. For example, hyperactivity may be quite purposeful in mania: as pointed out by Kraepelin (1899), the manic patient 'feels the need to come out of his shell [and] to have livelier relations with those around him'. By contrast, in excited catatonia behavior is typically purposeless and bizarre: Kraepelin (1899), in commenting on this difference between mania and catatonia, noted that 'the catatonic's urge to move often takes place in the smallest space, i.e. in part of the bed, whereas the manic looks everywhere for an opportunity to be active, and runs around, occupies himself with other patients, follows the doctor and gets into all kinds of mischief'.

Psychomotor retardation may range from an almost total quietude and immobility to a mere slowing of speech and behavior. Various conditions may underlie such a change. Mere exhaustion may slow patients down, but the response to rest is generally robust. Apathetic patients, lacking in motivation, may evidence little speech or behavior; depressed patients may appear similar but here one also sees a depressed mood. In akinesia there may also be a generalized slowing of all behavior; however, here one also fails to see a depressed mood. Abulia is distinguished from akinesia by the response to supervision: in contrast with patients with apathy, depression, or akinesia, the abulic patient performs at a normal rate when supervised. Delirium may be characterized by quietude and inactivity but is distinguished by the presence of confusion and deficits in memory and orientation. Catatonia of the stuporous type may be characterized by profound immobility; however, here one typically finds distinctive associated signs, such as waxy flexibility, posturing, and negativism. Finally, one should never forget hypothyroidism, wherein, as noted by Kraepelin (1899), it may take patients 'an incredibly long time to do the simplest things, to write a letter [or] to get dressed'.

Other behavioral disturbances may occur during the interview and examination, including mannerisms, stereotypies, and echopraxia. Mannerisms represent more or less bizarre transformations of speech, gesture, or other behaviors. Stereotypies are a kind of perseveration wherein patients repeatedly engage in the same behaviors, to no apparent purpose. Echopraxia is said to be present when patients involuntarily mimic what others, such as the examining physician, do. Although each of these disturbances may be seen in schizophrenia, they may also be present in other disorders, such as dementia.

Mood and affect

Mood is constituted by an individual's prevailing emotional 'tone'. When this is within the broad limits of normal, one speaks of 'euthymia' or a euthymic mood; significant mood disturbances may tend toward depression, euphoria, anxiety, or irritability. Depressed mood, as noted by Kraepelin (1921), may be characterized by 'a profound inward dejection and gloomy hopelessness'; in contrast, euphoria, as described by Griesinger (1882), is characterized by an 'overflowing contentment', such patients being, in the words of Krapelin (1921), 'penetrated with great merriment'. Anxious patients are beset with apprehensions, may plead for help, and may complain of tremor and palpitations. Irritable patients, as noted by Kraepelin (1921), are typically 'dissatisfied, intolerant [and] fault-finding', and are often quick to react to any perceived slight or criticism.

In the case of euphoria or irritability, one should also note whether or not the mood is 'heightened', that is to say whether or not it is so abundant and at such a level that its display in strong affect is simply inevitable: for example, patients with a heightened sense of irritability may be hostile, argumentative, and uncontrollably angry, whereas other patients whose irritability is not heightened might present a picture of mere sullenness and withdrawal.

Affect has been variously defined as representing either the combination of the immediately present emotion and its accompanying expression in tone of voice, gesture, facial expression, etc. or, less commonly, as only the emotional expression itself. Although in general there is a congruence between the experienced emotion and the facial expression, disparities may arise, as in 'inappropriate affect', sensory aprosodia, emotional facial palsy, and 'emotional incontinence' (as seen in pseudobulbar palsy). In each of these conditions patients report a substantial difference between what they are feeling and what is 'showing' on their faces. This is perhaps most dramatic in emotional incontinence (or, as Wilson [1928] called it, 'pathological laughing and crying'), which is characterized by an uncontrollable affective display that occurs in the absence of any corresponding feeling. Thus 'incontinent' of affective display, patients may burst forth into laughter or tears upon the slightest of stimuli and be unable to control themselves despite the lack of any sense of mirth or sadness.

Given that, as with mood, affect may be depressed, euphoric, anxious, or irritable it may appear academic to distinguish between the two; however, disparities between mood and affect may arise. Mood is enduring, whereas affect is relatively changeable: in a sense, mood is to affect as climate is to weather. Thus, patients suffering from a depressed mood that has generally endured for weeks, months, or longer may at times during the day experience a normal, or near-normal affect, and in such cases if the physician depended solely on observation of the patient's affect and did not inquire of the overall enduring mood an important clinical finding might be missed.

Affect, in addition to being depressed, euphoric, anxious or irritable, may also be flattened or labile. Flattened (or 'blunted' as it is also known) affect is characterized by a lifeless and wooden facial expression, accompanied by an absence or diminution of feeling. As such, it may be distinguished from motor aprosodia, wherein patients do in fact still experience feelings, although speaking in a monotone as if they had none. Flattened affect is found very commonly in schizophrenia (Andreasen *et al.* 1979); it may also occur in some secondary psychoses (Cornelius *et al.* 1991) and, rarely, in dementia secondary to infarction of the mesencephalon or thalamus (Katz *et al.* 1987).

Labile affect is characterized by swift, and sometimes violent, changes in both felt and expressed emotion.

Disturbances of mood are seen in a large number of conditions, as discussed in the chapters on depression, mania, and anxiety. Furthermore, it must be stressed that changes in mood, and especially affect, are also very common in dementia and delirium. This is particularly important to keep in mind, given that effective treatment of delirium typically results in a normalization of mood without the need for treatment with antidepressants or other medications.

Incoherence and allied disturbances

Normally the thoughts we put into words are coherent, focused, and goal-directed: abnormalities here include incoherence, circumstantiality and tangentiality, and flight of ideas.

Incoherent speech is characterized by a disconnectedness and disorganization of words, phrases, and sentences such that what the patient says, to a greater or lesser degree, 'makes no sense'. Incoherence may be found in a number of different syndromes, and it is the presence of other signs and symptoms that alerts the clinician to which syndromal diagnosis should be pursued: cognitive deficits indicate the presence of dementia or delirium; heightened mood, pressure of speech, and hyperactivity suggest mania; and bizarre behavior, hallucinations, or delusions point to a psychosis, such as schizophrenia. In cases characterized primarily by incoherence but with few, if any, other abnormalities on the mental status examination, then a diagnosis of aphasia of the 'sensory' type should be considered. There has been much ink spilt on whether it is possible to reliably distinguish the incoherence seen in schizophrenia (known as 'loosening of associations') from that seen in sensory aphasia; however, of the many articles written on this subject only two studies actually compared the speech of patients with schizophrenia with that of patients with sensory aphasia secondary to stroke (Faber *et al.* 1983; Gerson *et al.* 1977), and the results, although promising, were not definitive. In general, patients with loosening of associations spoke freely and at length and, although what they said made little sense, they had no trouble in finding words. By contrast, patients with aphasia often had at least some difficulty in finding words, and their responses to questions were typically brief. Furthermore, whereas patients with loosening of associations had little or no recognition of their incoherence, the aphasic patients often seemed at least somewhat aware of their difficulty. It has been the authors' experience that these differences, although often present, are not sufficiently reliable to make the differential between loosening of associations and aphasia, and that it is much more useful to look for the presence of more or less bizarre delusions, which are typically present in any patient with loosening of associations but absent in those with aphasia.

Circumstantiality is said to be present when, perhaps in response to a question, patients take the cognitive 'long way round', traversing superfluous details and dead-ended digressions until finally getting around to the answer. In listening to such patients, the interviewer often has to suppress the urge to tell them to 'get to the point'. Tangentiality differs from circumstantiality in that the patient's thought, although coherent, takes off on a 'tangent' from the initial question, never in fact getting 'to the point'. Both of these signs are diagnostically non-specific but may be seen in the same conditions as incoherence.

Flight of ideas is, according to Kraepelin (1921), characterized by a 'sudden and abrupt jumping from one subject to another': before any given thought is fully developed, the patient's attention lights on another thought, there to stay for only a short time before moving on yet again. This differs from incoherence in that, although incomplete, the development of the subject is coherent before the patient jumps to the next. Such a flight of ideas is classic for mania.

Other disturbances of thought or speech

Poverty of thought is characterized by a dearth of thoughts: such patients, lacking anything to say, speak very little. By contrast, patients with poverty of speech may speak much. Their speech, however, is 'empty', being filled with so many stock phrases and repetitions that little is actually 'said'. Both these disturbances may be found in schizophrenia and in certain cases of aphasia.

Thought blocking is characterized by an abrupt termination of speech, sometimes in the middle of a sentence, as if the train of thought had suddenly been 'blocked'. This is not a matter of simply running out of things to say, but rather an uncanny experience wherein thoughts suddenly stop appearing. When present to a marked degree, this experience is typically accompanied by one of the Schneiderian first rank delusions, namely 'thought withdrawal'.

Pressure of speech is experienced by the patient as an 'urge to talk' that is so imperious that, as described by Kraepelin (1899), 'he cannot keep quiet for long, chatters and shouts out loud, yells, roars, bawls, whistles [and] speaks overhastily'. To be in the presence of such patients is akin to standing in front of a dam bursting with words and thoughts. Although classically seen in mania, such a disturbance may also be seen in schizophrenia, schizoaffective disorder, and, occasionally, in dementia.

Perseveration of speech is said to be present when patients either supply the same answer to successive questions or merely, and without prompting, repeat the same words or phrases over and over. This abnormality is most commonly seen in dementia or delirium. Palilalia, sometimes confused with perseveration, is characterized by an involuntary repetition of the last phrase or word of a sentence, with these repetitions occurring with increasing rapidity, but diminishing distinctness.

Echolalia is characterized by an involuntary repetition by the patient of words or sentences spoken by others, and may be seen in a large number of disorders, such as dementia, aphasia, catatonia, and Tourette's syndrome.

Obsessions are distinguished from normal thoughts by the fact that they repeatedly and involuntarily come to mind despite the fact that the patient finds them unwanted and distressing.

Hallucinations

Patients are said to be hallucinated when they experience something in the absence of any corresponding actual object; such hallucinations may occur in the visual, auditory, tactile, olfactory, or gustatory sphere. Thus, a patient who 'saw' a group of people or who 'heard' people murmuring in the next room when the room was in fact empty and silent would be considered hallucinated. Hallucinated patients may or may not retain 'insight': that is to say, they may or may not recognize that their experience is not 'real'. For example, whereas one patient might say, 'I hear some people next door, but I know that it's just my imagination and they're not really there', another might be surprised to hear that the physician did not hear them also. In cases where insight is lacking, it is generally useless to disagree with patients or try and 'talk them out of it'. As Bleuler (1924) pointed out, 'it is of no avail to try to convince the patient by his own observation that there is no one in the next room talking to him; his ready reply is that the talkers just went out or that they are in the walls or that they speak through invisible apparatus'.

Certain auditory hallucinations are included among the Schneiderian first rank symptoms, and should routinely be sought. They include audible thoughts (i.e. hearing one's own thoughts 'out loud', as if they were being spoken and as if others could also hear them), hearing voices that comment on what the patients themselves are doing, and hearing voices that argue with one another.

Although classically associated with psychosis, hallucinations are just as common in delirium and dementia, and may also occur in some cases of depression or of mania.

Delusions

A delusion, according to Lord Brain (1964), 'is an erroneous belief which cannot be corrected by an appeal to reason and is not shared by others of the patient's education and station'. Thus, whereas for a Russian in the middle part of the twentieth century to be convinced that the telephones were routinely 'bugged' would not, *prima facie*, be a delusion; for a Canadian of the twenty-first century to be so convinced, however, would be suspect. Although at times it may be difficult to decide whether or not a belief is delusional, it is in most cases quite obvious: for example, the belief that a small reptilian creature sits inside one's external auditory canal and inserts thoughts is simply not plausible in any culture.

Delusions are generally categorized according to their content or theme. Thus, there are delusions of persecution, grandeur, erotic love, jealousy, sin, poverty, and reference. Delusions of reference are said to be present when patients believe that otherwise unconnected events in some way or other refer or pertain to them. Thus, patients with a delusion of persecution who believed that they were under surveillance might, upon reading a newspaper article about undercover police, hold that the article, in fact, was a kind of 'warning' or 'message' that they could not escape.

Certain delusions are also counted among the Schneiderian first rank symptoms, and these include beliefs that one is directly controlled or influenced by outside forces, that thoughts can be withdrawn, or, alternatively, inserted, and that thoughts are being 'broadcast' such that they can be 'picked up' and known by others.

Delusions, like hallucinations, may be seen not only in psychosis, but also in delirium, dementia, mania, or depression.

Other disturbances of thought content

Phobias are fears that patients admit are irrational. Seen in the condition known as specific phobia, they may occasionally manifest during the interview, as, for example, in claustrophobia when the patient may object to the door being closed.

Depersonalization is characterized by an uncanny sense of detachment on the patient's part from what is currently going on. Patients may complain of feeling detached, as if they were not 'there', and, although doing things, were somehow removed and observing.

Obsessions consist of unwelcome and anxiety-provoking thoughts, impulses, or urges that occur repeatedly despite patients' attempts to stop them. In many cases, these obsessions may occasion compulsions, which, in some fashion or other, serve to quell the anxiety occasioned by the obsession. A familiar example would be the obsession that one had forgotten to turn off the stove, which in turn leads to a compulsion to return to the house to check and see whether or not the stove had been left on.

Level of consciousness

Note should be made of whether or not patients are alert. If not, an attempt should be made to arouse them, beginning with calling the patient's name in a normal conversational tone of voice. Should this prove unsuccessful, the next step is to call the patient's name loudly, and should that fail, to shake the patient's shoulder. Should the patient still not be fully alert, one may consider calling the patient's name loudly and also shaking the shoulder quite vigorously. Should there be no response at all at this point, or only a very sluggish one, consideration may be given to the application of painful stimuli, such as squeezing the patient's finger pad between one's thumb and forefinger, applying pressure with one's thumb to the supraorbital ridge or to the temporomandibular joint, and finally to a sternal rub.

Presence or absence of confusion

Confused patients may appear to be in a daze, and some may report feeling 'fuzzy' or 'cloudy': they have difficulty ordering their own thoughts and a similar difficulty in attending to events around them. An evocative synonym for confusion is 'clouding of the sensorium'. This is a particularly important clinical finding given that the differential diagnosis between delirium and dementia rests, in large part, on its presence or absence.

Orientation

Orientation is traditionally assessed for three 'spheres' – person, place, and time – and patients who can properly place themselves in each sphere are said to be 'oriented times three'. Orientation to person may be determined by asking patients for their full names; such orientation is only very rarely lost. Orientation to place is checked by asking patients to identify where they are, including the name of the city and of the building. In cases where patients hesitate to answer, perhaps because they are unsure, it is important to encourage them to take a guess. Should they misidentify the building, inquire further as to what kind of building it is. Some patients may betray a degree of concreteness here, for example, by replying 'a brick building' and, if they do, gently press further by offering some choices, for example, 'a hotel, hospital, or office building', and ask them to choose one. Orientation to time is determined by asking patients the date, including the day of the week, the month, day of the month, and year. If patients are oriented perfectly in all these three spheres then one may simply note 'oriented times three' in the chart. If they are not, it is critical to note their exact responses: simply noting 'oriented times two' fails to capture important information, including, as it does, the patient who believes it is 1948 and the patient who is off the date by only a few days.

In cases where patients are disoriented, it is appropriate to subsequently, and gently, state the correct orientation. This not only ensures that they have been told the correct orientation at least once, but also opens the door to the identification of the rare syndrome of reduplicative paramnesia (also known as 'delusional disorientation') wherein, for example, patients may correctly identify the name of the hospital but insist that the hospital is in a distant city.

Some authors also recommend checking orientation in a 'fourth' sphere, namely orientation to situation. This is typically determined during the non-directive portion of the interview, when it becomes clear whether or not patients recognize that they are ill, and in a hospital for treatment, etc. It is akin to 'insight' (discussed later in this chapter) and is probably appropriate.

Disorientation may be seen in delirium, dementia, amnesia, and psychosis.

Memory

Memory is discussed in detail in Chapter 6.6 and, as noted there, the most important type of memory from a clinical point of view is memory for events and facts, and it is this that is tested in the mental status examination.

Traditionally, three aspects of memory are tested: immediate, short-term, and long-term memory. Immediate recall is tested by using 'digit span'. Here, the patient is given a list of random digits, slowly, one second at a time, and then immediately asked to recall them forwards, from first to last. One starts with a list of three digits, and if the patient recalls these correctly, moves to a list four digits long, proceeding to ever longer lists until the patient either errs in recall or reaches seven digits; normal individuals can recall lists of five to seven digits in length. Once this has been accomplished, 'backward' digit recall is checked by giving a list two digits long and immediately asking the patient to recall them in reverse order. If this is done correctly one proceeds to longer lists, again until errors are made or the patient performs within the normal range of spans of three to five digits.

Short-term recall is tested by telling the patient that you will give a list of three words and that you would like him or her to memorize them because in a few minutes you will ask that they be recalled. Three unrelated words are then provided (e.g. 'cloud, tiger, daisy') and the patient is asked to repeat them once to make sure that he or she 'has' them. Once it is clear that the patient 'has' them, wait 5 minutes and then ask the patient to recall them. Importantly, during this 5-minute interval, the interviewer should stick to neutral topics (e.g. some innocuous 'review of systems' questions) and avoid any emotionally laden subjects that might upset the patient. Should patients fail to recall one or more of the three words, it is appropriate to provide a cue to see if this facilitates recall (e.g. 'something you might see up in the sky', 'an animal you might see at the zoo', and 'the name of a flower'). Normally, patients should be able to recall all three words: if they do not, in the report one should note, for example, that the patient was able to recall one out of three words after 5 minutes, and was able to get one of the other two with cueing.

Long-term memory should be checked for both personal and public events. This is often assessed informally during the non-directive portion of the interview as one ascertains whether the patient recalls what happened in the days leading up to admission, during recent holidays, or recalls where he or she worked/went to school. Recall of public events may be checked by asking about recent newsworthy events or, in a somewhat more quantitative way, by asking the patient to recall the names of the last four prime ministers or presidents.

Deficits in immediate recall are typically accompanied by confusion and generally indicate a delirium. In addition to delirium, deficits in short- and long-term memory may also be seen in dementia and amnesia. In some cases, during either testing for long-term memory or the interview, one may find evidence of confabulation, wherein the answers that patients provide are clearly false.

Abstracting ability

Abstracting ability is traditionally assessed by asking patients to interpret a proverb, such as 'Do not cry over spilled milk'. Responses to proverb testing may be 'abstract' or 'concrete', as, for example, if a patient replied, 'Well, it's already spilled'. At times, the abnormality on proverb interpretation will consist of a bizarre response instead of a concrete reply, such as 'Alien milk has no taste'. Concrete responses may be seen in delirium or dementia and typically indicate frontal lobe dysfunction. Bizarre responses suggest a psychosis, such as schizophrenia.

Calculating ability

Calculating ability is traditionally assessed with the 'serial sevens' test (Smith 1962), wherein patients are asked to subtract seven from 100, then seven from that number, and so on. In the past, physicians were encouraged to have patients persist in this task until they could go no further; however, this is a time-consuming and, for many patients, irritating task, and it has become customary to stop patients after five responses, noting in the report how many of these represented a correct subtraction (e.g. if the patient's responses were 93, 87, 80, 73, and 69, one would note that three out of five responses were correct: normally, patients of average intelligence who progressed beyond elementary school should be able to supply five correct responses). When patients are unable to provide any correct responses, it is appropriate to try a simpler subtraction task (e.g. subtracting 8 from 12, or 3 from 9), and when patients are unable to do even simple subtraction, it is appropriate to ask them to try some addition (e.g. 3 plus 5, or 7 plus 6). For some patients, rather than starting with serial sevens, it is appropriate to begin with simple subtraction, or even with simple addition: examples include patients who report doing very poorly in elementary school or patients who, as indicated by their responses to other questions, appear to be cognitively impaired.

Judgment and insight

Judgment has traditionally been assessed with test questions such as 'What would you do if you smelled smoke in a theatre?'. In many instances, however, it is appropriate to pose situations more relevant to the patient's life; thus, one might ask a police officer what should be done if a suspect refused to answer questions.

Insight, for the purposes of the mental status examination, refers not to some sophisticated appraisal of one's situation, but rather, simply, to whether or not patients recognize that they are ill or that something is wrong. This is identical to 'orientation to situation' as discussed earlier in this chapter and, if already noted, no further comment is required.

Judgment or insight may be lost in delirium, dementia, or a personality change such as the frontal lobe syndrome. Insight may also be lost when anosognosia is present, as, for example, when a patient with hemiplegia is unable to recognize the deficit.

2.3 NEUROLOGIC EXAMINATION

Bleuler (1924), in his classic *Textbook of Psychiatry*, insisted that 'a *minute physical* and *especially neurological examination* must not be omitted' (italics in original) and the reader is urged to take this admonition to heart.

Over the decades, the neurologic examination has 'thinned down' somewhat and of the dozens of abnormal reflexes that used to be *de rigueur* only a few survive today. The scheme presented here constitutes a 'middle-of-the-road' approach and, although it may be found skimpy by some, others may consider it overly detailed. We plead guilty on both accounts, but urge the reader to try this approach and then to reshape it in light of future experience and wide reading. Although, in most cases, the examination may be conducted in the order suggested here, flexibility must be maintained, especially with fatigued, agitated, or uncooperative patients. Bear in mind that even with a completely uncooperative patient, much may be gathered by a simple observation of eye and facial movements, speech, movement of the extremities, gait, etc.

For most findings, further detail on, and a consideration of, the differential diagnosis of the finding may be found in the appropriate chapter.

General appearance

In some cases, the overall appearance of the patient may immediately suggest a possible diagnosis. Examples include the moon facies of Cushing's syndrome, the puffy facial myxedema and thinning hair of hypothyroidism, and the obesity of the Bardet–Biedl and Prader–Willi syndromes or the Pickwickian syndrome.

Facial appearance, including facial dysmorphisms, may also be diagnostically suggestive, as, for example, the port wine stain of Sturge–Weber syndrome, the adenoma sebaceum of tuberous sclerosis, or the high forehead, large ears, and prognathism of fragile X syndrome.

Handedness

Inquire as to handedness and observe as patients handle implements such as a pen; if there is doubt, ask which hand the patient uses to throw a ball or which foot is used to kick with.

Pupils

The pupils are normally round in shape, regular in outline, and centered in the iris. Their diameter should be measured and their reactions to light and to accommodation should be noted. The pupillary reaction to light is tested first by shining a penlight into one eye and observing the reaction, not only of that pupil but also in terms of the consensual reaction in the opposite pupil. After a short wait, the other eye should be tested in the same fashion. The accommodation or convergence reaction is then tested by asking the patient to focus on the examiner's finger as it is slowly moved along the midline toward a spot midway between the patient's eyes: normally, as the eyes converge, both pupils undergo constriction. A preserved reaction to accommodation in the face of an absent or sluggish reaction to light is known as an Argyll Robertson pupil and is very suggestive of neurosyphilis.

While examining the pupils, the corneal limbus should also be examined for a Kayser–Fleischer ring, as seen in Wilson's disease. This is a golden-brown discoloration of the limbus, which typically begins at the 12 and 6 o'clock regions from where it gradually expands medially and laterally to eventually form a ring around the cornea.

Funduscopic examination

After examining the optic fundus for any hemorrhages or exudates, attention should be turned to the optic disk, which should be flat and sharply demarcated from the surrounding fundus. The depth of the optic cup should be noted, as should the presence or absence of venous pulsations.

Cranial nerves

CRANIAL NERVE I

The olfactory nerve is tested by first occluding one nostril and then bringing an aromatic substance, such as powdered coffee, to the patient's nostril, inquiring whether any odor is appreciated and, if so, what it is. In a pinch one may use a substance readily available at the bedside, such as toothpaste. There are also commercially available tests of the 'scratch and sniff' variety, which, although much more detailed, have not as yet found a place in routine clinical practice. Unilateral anosmia may occur secondary to compression of the olfactory bulb or tract by a tumor, such as a meningioma of the olfactory groove; bilateral anosmia may be seen in neurodegenerative diseases, such as Alzheimer's or Parkinson's disease, and may also be seen after head trauma, with rupture of the olfactory filaments as they pass through the cribriform plate.

CRANIAL NERVE II

The optic nerve is tested not only for acuity, but also for visual fields. Visual acuity may be informally tested by asking the patient to read text from a newspaper or, more formally, by use of a Snellen chart. If the patient has glasses or contact lenses, vision should be tested both with and without them. The visual fields may be assessed by confrontation testing: while facing each other, the physician and patient are separated by about a meter, each fixing vision on the other's nose; the physician then brings a small object (e.g. the tip of a reflex hammer) in from outside the patient's peripheral field, instructing the patient to say 'yes' as soon as it comes into view, and bringing the target in from all four quadrants. Importantly, in cases where the patient fails to respond to an object in one hemi-field, one must consider not only the possibility of an hemianopia, but also the possibility of visual neglect.

CRANIAL NERVES III, IV, AND VI

The oculomotor, trochlear, and abducens nerves are tested first by having the patient follow the physician's finger as it moves to either side and both upward and downward while the patient's head is kept stationary. Subsequently, patients should be asked to look to the left, to the right, then up and down. Eye movements should be full and conjugate in all directions of gaze, and without nystagmus. The oculomotor nerve also innervates the upper eyelid; thus, the presence or absence of ptosis should be noted. In cases where there is limitation of voluntary up-gaze, or, more importantly, down-gaze, one should further test the patient with the 'doll's eyes' maneuver to determine if the vertical gaze palsy is either supranuclear, nuclear, or infranuclear. To perform this, first lightly grasp the patient's head and then flex and extend it at the neck, watching how the eyes move. If eye movements are full then the lesion responsible for the voluntary vertical gaze palsy is supranuclear, as may be seen in disorders such as progressive supranuclear palsy.

CRANIAL NERVE V

The trigeminal nerve has both motor and sensory components. Masseter muscle strength is checked by lightly placing one's fingers on the patient's cheeks and then instructing the patient to bite down. Sensory testing, to both light touch and pin-prick, is checked in all three divisions, namely the ophthalmic, maxillary, and mandibular. The corneal reflex, which tests both the cranial nerves V and VII, may also be performed by lightly touching a wisp of cotton to the patient's cornea, after which there should be a bilateral blink.

CRANIAL NERVE VII

The facial nerve is first tested for voluntary facial movements by asking the patient to wrinkle the forehead and subsequently to show the teeth. In cases of unilateral voluntary facial paresis note must be made of which divisions of the facial nerve are involved: the upper (controlling forehead wrinkling), the lower (controlling elevation of the side of the mouth), or both. At times facial weakness may be quite subtle, manifesting perhaps only with a slight flattening of the nasolabial fold on one side.

After voluntary movements have been tested, the physician must then test for involuntary or 'mimetic' facial movements. This may be accomplished by telling a joke, or, if the physician is in a less humorous mood, by simply observing the patient for any spontaneous smiling. Voluntary and involuntary facial movements are quite distinct neuroanatomically, and thus both should be tested for (Hopf *et al.* 1992). Voluntary facial palsy affecting only the lower division indicates a lesion of the precentral gyrus or corticobulbar fibers, whereas emotional facial palsy (Chapter 5.8) indicates a lesion in the supplementary motor area, temporal lobe, striatum, or pons.

CRANIAL NERVE VIII

The vestibulocochlear nerve has traditionally been tested by bringing a ticking wristwatch in from a distance and noting at which point the patient first hears it. However, now that ticking watches have almost vanished, an alternative procedure (Torres-Russotto *et al.* 2009) is recommended, which involves something simpler than holding a watch, namely merely rubbing one's fingers. To perform this test, ask the patient to close his or her eyes and to indicate, by a nod of the head, whether he or she hears your fingers rubbing. Then seat yourself with your face about 15 cm from the patient's, extend your arms laterally, and, using one hand at a time, rub your thumb across your fingers. If the patient does not hear the sound, then halve the distance by bending your elbow to a 90° angle and repeat the finger rub. If the sound is still not heard, bring your hand in to about 15 cm from the patient's ear, and repeat the finger rub, and if it is still not heard, then bring your fingers as close as possible to the patient's ear without touching it, and repeat the finger rub for the last time. Each portion of this test may be accomplished with either a soft or a loud finger rub. The soft finger rub is that which can barely be heard by the physician with the arm held fully laterally, whereas the loud finger rub is the loudest sound that can be made by rubbing the fingers, *without* making them snap. It is perhaps most convenient to begin with a soft finger rub with the arm in the full lateral position, for if the patient can hear that, the test is over. Responses obviously range from an ability to hear a soft finger rub from a full lateral position to an inability to hear a loud finger rub from as close as possible to the ear. Should patients fail this latter portion, then one might amplify this test with other maneuvers, such as finger-snapping, hand-clapping, speaking in a loud voice, or even shouting.

If there are any abnormalities, both Weber and Rinne testing should be performed to determine whether the hearing loss is of the conduction or sensorineural type.

In the Weber test, a vibrating tuning fork is placed square on the midline of the patient's forehead and the patient is asked whether it sounds the same on both sides or is heard louder on one side than on the other. In the Rinne test, a vibrating tuning fork is placed against the styloid process and the patient is asked to indicate when the sound vanishes, at which point the tines of the tuning fork are immediately brought in close approximation to the ear and the patient is asked whether it can now be heard. With conductive hearing loss, the Weber lateralizes to the side with the hearing loss, and on Rinne testing, bone conduction (i.e. with the tuning fork against the styloid process) is louder than air conduction (i.e. with the tines of the fork vibrating in the air just outside the ear). With sensorineuronal loss, the Weber lateralizes to the 'good' side and, on Rinne testing, air conduction is better than bone conduction bilaterally.

CRANIAL NERVES IX AND X

The glossopharyngeal and vagus nerves are tested with the gag reflex and by observation for symmetric elevation of the palate during phonation.

CRANIAL NERVE XI

The spinal accessory nerve is tested by having patients shrug their shoulders against the resistance of the physician's hand and by turning the head to one side or the other while the physician exerts contrary pressure on the jaw.

CRANIAL NERVE XII

The hypoglossal nerve is tested first by asking the patient to open the mouth and then observing the tongue, as it rests in the oropharynx, for any atrophy or fasciculations. Once this has been accomplished, the patient is asked to protrude the tongue as far as possible, noting especially whether it protrudes past the lips and also whether it deviates to one side or the other.

Sensory testing

Elementary sensory testing involves light touch, pin-prick, and vibration. Light touch may be assessed by using a wisp of cotton, or simply by a light touch with one's finger, and pin-prick sensation is tested using a disposable safety pin. Vibratory sensation is tested by touching a vibrating tuning fork to a bony structure (such as a finger joint, the lateral malleolus, or the great toe) and asking the patient whether he or she can tell if it is vibrating; if so, the tuning fork is held in place, and the physician slowly drags his or her finger along the tines and the patient is asked to say when the vibration ceases, with the physician taking note, in a rough sort of way, of how much the tuning fork is still vibrating at that point. If there are any abnormalities in elementary sensation it is critical to determine whether or not they are bilateral. In general, it is sufficient to test sensation at both hands and both feet, reserving more detailed testing for cases in which the history suggests a more focal sensory loss.

Graphesthesia and two-point discrimination tests also constitute part of the sensory examination, but these should be used only if elementary sensation is intact. Agraphesthesia is said to be present when patients, with their eyes closed, are unable to identify letters or numerals traced on their palms by a pencil or dull pin. Two-point discrimination, as described by DeJong (1979) may be tested by 'bending a paperclip to different distances between its two points ... [starting] with the points relatively far apart ... [then] approximated until the patient begins to make errors'. As two-point discriminatory ability varies on different parts of the body (from 2–4 mm at the fingertips to 20–30 mm on the dorsum of the hand), what is most important here is to compare both sides to look for a difference.

Agraphesthesia and diminished two-point discrimination suggest a lesion in the parietal cortex; elementary sensory loss, especially to pin-prick, is also seen with parietal cortex lesions but, in addition, may occur with lesions of the thalamus, brainstem, cord, or of the peripheral nerves.

Cerebellar testing

In addition to observing the patient's gait for ataxia, as discussed below, cerebellar testing also involves finger-to-nose and heel-to-knee-to-shin testing, testing for rapid alternating movements, and observing for dysarthria.

In the finger-to-nose test, patients are instructed to keep their eyes open, extend the arm with the index finger outstretched, and then to touch the nose with the index finger. In the heel-to-knee-to-shin test, patients, while seated or recumbent, are asked to bring the heel into contact with the opposite knee and then to run that heel down the shin below the knee. In both tests one observes for evidence of dysmetria (as, for example, when the nose is missed in the finger-to-nose test) and for intention tremor, wherein, for example, there is an oscillation of the finger and hand as it approaches the target (in this case the nose, with this tremor worsening as the finger is brought progressively closer to the nose).

Rapid alternating movements also assess cerebellar function. Here, while seated, patients are asked to pronate the hand and gently slap an underlying surface (e.g. a tabletop or the patient's own thigh) and then supinate the same hand and again gently slap the underlying surface. Once they have the hang of it, patients are then asked to repeat these movements as quickly and carefully as possible. Decomposition of this movement, known as dysdiadochokinesia, if present, is generally readily apparent on this test.

Dysarthria may also represent cerebellar dysfunction and may be casually assessed by simply listening carefully to the patient's spontaneous speech, noting any evidence of slurring. In doubtful cases one may ask the patient to repeat a test phrase, such as 'Methodist Episcopal' or 'Third Riding Artillery Brigade' (DeJong 1979). Importantly, dysarthria may also be seen with lesions of the motor cortex or associated subcortical structures.

Station, gait, and the Romberg test

Station is assessed by asking patients to stand with their feet normally spaced, and observing for any sway or loss of balance. At this point, if station is adequate, one should perform the Romberg test by telling patients that you will ask them to put their feet close together, as if 'at attention', and then to close their eyes, reassuring them that you will have your hands close by and that you will not let them fall. If they are comfortable with these instructions then the test can be carried out, observing the patients for perhaps half a minute to see whether or not any swaying develops once the eyes are closed. A 'positive' Romberg test indicates a loss of position sense, as may be seen with a peripheral neuropathy or damage to the posterior columns.

Gait is tested by asking the patient to walk a straight line down a hall, then walking 'heel to toe' in a tandem walk, and, finally, if these are done adequately, by asking the patient to walk 'on the outside of your feet, like a "cowboy"'.

An ataxic gait, seen in cerebellar disorders, is wide based and staggering: steps are irregular in length, the feet are often raised high and brought down with force, and the overall course is zigzagging. In a 'magnetic' gait, as seen in hydrocephalus or bilateral frontal lesions, the feet seem stuck to the floor as if magnetized or glued to it. In a steppage gait, seen in peripheral neuropathies, the normal dorsiflexion of the feet with walking is lost and patients raise their feet high to avoid tripping on their toes. In a spastic gait, seen with hemiparetic patients, the affected lower extremity is rigid in extension and the foot is plantar flexed: with each step, the leg is circumducted around and the front of the foot is often scraped along the floor. In very mild cases of hemiparesis, the gait, to casual inspection, may not be abnormal; however, when patients walk 'on the outside' of their feet, one often sees dystonic posturing of the upper extremity on the involved side. Parkinsonian gait is described in Chapter 4.8.

Strength

Strength may, according to Lord Brain (1964), be graded as follows: 0, no contraction; 1+, a flicker or trace of movement; 2+, active movement providing that gravity is eliminated; 3+, active movement against gravity; 4+, active movement against some resistance; and 5+, full strength. In the process of assessing muscular strength one should also observe for any atrophy, fasciculations, or myotonia. Myotonia is sometimes apparent in a handshake, as patients may have trouble relaxing their grip, and may also be assessed by using a reflex hammer to lightly tap a muscle belly, such as at the thenar eminence, and watching for distinctive myotonic dimpling.

Common patterns of weakness include monoparesis if only one limb is involved, hemiparesis if both limbs on one side are weak, paraparesis if both lower extremities are weak, and quadriparesis (or, alternatively, tetraparesis) if all four extremities are weakened. In cases when strength is 0 then one speaks not of paresis but of paralysis, and uses the terms monoplegia, hemiplegia, paraplegia, or quadriplegia. When weakness is present, note should be made whether the proximal or distal portions of the limb are primarily involved; in cases of hemiparesis in which both limbs are not equally affected, the limb that is more affected should be noted.

Drift

A positive pronator drift test may be the first evidence of hemiparesis. This test, according to DeJong (1979), is accomplished by asking patients (with their eyes closed) to fully extend their upper extremities, palms up, and then maintain that position: a positive test consists of 'slow pronation of the wrist, slight flexion of the elbow and fingers, and a downward and lateral drift of the hand'.

Rigidity

Rigidity should, at a minimum, be assessed at the elbows, wrists, and knees by passive flexion and extension at the joint, with close attention to the appearance of spastic, lead pipe, or cogwheel rigidity. Spastic rigidity, seen with upper motor neuron lesions, is most noticeable on attempted extension of the upper extremity at the elbow and attempted flexion of the lower extremity at the knee. Furthermore, in spasticity, one may see the 'clasp knife' phenomenon. Here, on attempted rapid extension of the upper extremity at the elbow, an initial period of minimal resistance is quickly followed by a 'catch' of increased resistance, which, in turn, is eventually followed by a loosening, with the whole experience reminiscent of what it feels like to open the blade on a clasp knife. Lead pipe rigidity, seen in parkinsonism, is, in contrast with spastic rigidity, characterized by a more or less constant degree of rigidity throughout the entire range of motion, much as if one were manipulating a thick piece of solder. Cogwheel rigidity, also seen in parkinsonism, may accompany lead pipe rigidity or occur independently. This is best appreciated by gently holding the patient's elbow in the cup of your hand while pressing down on the patient's biceps tendon with your thumb. Once the arm is thus supported, with your other hand gradually extend the arm. When cogwheeling is present, a 'ratcheting' motion will be appreciated with your thumb, much as if there were a 'cogwheel' inside the joint.

After testing for these forms of rigidity, one should then test for gegenhalten at the elbow by repeatedly extending and flexing the arm, feeling carefully for any increasing rigidity. Evidence for this generally indicates frontal lobe damage.

Abnormal movements

Tremor (Chapter 4.1) is generally of one of three types: rest, postural, or intention. Rest tremor is most noticeable when the extremity is at rest, as, for example, when the patient is seated with the hands resting in the lap. Postural tremor becomes evident when a posture is maintained, as, for example, when the arms are held straight out in front with the fingers extended and spread. Intention tremor (as described above) appears when the patient carries out an intended action, as, for example, touching the index finger to the nose. Other forms are also possible, for example, Holmes' tremor, which has both postural and intention elements. Tremor is further characterized in terms of amplitude (from fine to coarse) and frequency (ranging from slow [3–5 cycles per second, cps] to medium (6–10 cps) to rapid (11–20 cps)).

Myoclonus (Chapter 4.2) consists, as described by Lord Brain (1964) of 'a shock-like muscular contraction' and may be focal, segmental, or generalized, occurring either spontaneously in response to some sudden stimulus (e.g. a loud noise) or as 'intention' or 'action' myoclonus that appears upon intentional movement. This is an especially valuable sign and the physician should remain alert to its occurrence throughout the interview and examination.

Motor tics (Chapter 4.3) are sudden involuntary movements that, importantly, resemble purposeful movements, such as shoulder shrugs, facial grimaces, or head jerks. Unlike myoclonus, tics, as noted by Lord Brain (1964), involve 'a number of muscles in their normal synergic relationships'.

Chorea (Chapter 4.4), according to Lord Brain (1964), is characterized by 'quasi-purposive, jerky, irregular, and non-repetitive' movements that are very brief in duration, generally erupting randomly on different parts of the body.

Athetosis (Chapter 4.5), as noted by Lord Brain (1964) 'consists of slow, writhing movements' that are generally most evident in the distal portions of a limb; they are persistent and seem to flow into one another in a serpentine fashion.

Ballismus (Chapter 4.6), which is generally unilateral, as described by DeJong (1979), consists of 'wild flaillike, writhing, twisting or rolling movements that may be intense and may lead to exhaustion'. In severe cases the flinging movements of the extremity may actually throw the patient off the chair or bed.

Dystonia (Chapter 4.7), in contrast with ballismus, consists of slow and sustained movements that variously twist or contort the involved body part. It may be focal (e.g. moving the head to one side or 'cramping' the hand), segmental (e.g. spreading to an adjacent body part, as with the head turning and the shoulder elevating), or generalized (e.g. in severe cases, creating a human 'pretzel').

Parkinsonism (Chapter 4.8), when fully developed, stamps patients with a distinctive clinical picture. Classically, a flexion posture is evident, with the patient being stooped over with the arms and legs held in flexion, and a rhythmic 'pill-rolling' rest tremor of the hands may be seen, especially with the hands resting on the lap. The face is often 'masked' and expressionless, and bradykinesia is evident in the slowness with which all movements are executed. Gait is shuffling and one may also see festination wherein the patient, as described by Lord Brain (1964), seems to hurry 'with small steps in a bent attitude, as if trying to catch up [with] his center of gravity'.

Akathisia (Chapter 4.10) is typified by an inability to keep still. If standing, patients may rock back and forth or 'march in place' and, if seated, there may be a restless fidgeting, with crossing and uncrossing of the arms or legs. In severe cases, the compulsion to move is irresistible, and patients may constantly pace back and forth. Characteristically, the restlessness is worse when lying down or seated, and most patients find some relief upon standing or moving about.

Catatonia (Chapter 4.11) of the stuporous type (Barnes *et al.* 1986) is characterized by varying degrees of immobility, mutism, and a remarkable phenomenon known as waxy flexibility (or catalepsy), wherein, as noted by Kraepelin (1899), the limbs, after being passively placed in any position, 'retain this position until they receive another impetus or until they follow the law of gravity as a result of extreme muscular fatigue'.

Asterixis (Leavitt and Tyler 1964) (Chapter 4.12) is tested for by having patients hold their upper limbs in full extension, with the hands being held in hyperextension: asterixis, if present, appears as a precipitous loss of muscle tone, such that the hands 'flap' down. When present, this may appear immediately and recur frequently, or may be delayed for up to half a minute.

Heightened startle response (Saenz-Lope *et al.* 1984) (Chapter 4.14) is often precipitated by a sudden loud noise and may go beyond being simply excessively 'jumpy'; some patients may actually be thrown to the ground during the startle.

Deep tendon reflexes

At a minimum, the following deep tendon reflexes should be tested: biceps jerk, triceps jerk, supinator jerk, knee jerk, and ankle jerk. The results may, according to DeJong (1979), be graded as 0 for absent, 1+ for present but diminished, 2+ for normal, 3+ for increased, and 4+ for markedly hyperactive. Hyperactive deep tendon reflexes may also be accompanied by ankle clonus. Testing for clonus is accomplished by placing your hand under the ball of the patient's foot and then briskly dorsiflexing the foot. When clonus is present, the foot will then briskly and spontaneously undergo plantar flexion. Keeping a light upward pressure on the ball of the foot may precipitate repetitive clonic jerking, and in some cases this may be self-perpetuating or 'sustained'.

In those cases in which patients remain so tense that their reflexes cannot be elicited, several maneuvers may render the examination possible (Bickerstaff 1980): for the upper limbs, the patient should clench his teeth tightly or while one arm is being examined he should clench the fist of the other. For the lower limbs these measures can still be used but the well-tried method of Jendressak is more reliable: the patient interlocks the flexed fingers of the two hands and pulls one against the other while the reflexes are tested.

Babinski sign

The Babinski sign, considered by DeJong (1979) as 'the most important sign in clinical neurology', may be elicited by lightly dragging a blunt object across the sole of the patient's foot: beginning at the heel, proceeding along the lateral aspect of the sole and then turning medially to cross under the ball of the foot. One then observes for the 'plantar response' of the toes, noting whether it is 'flexor' or 'extensor'. The normal response is 'flexor', wherein the toes undergo flexion. An 'extensor' response is considered abnormal and constitutes the Babinski sign, which, when fully present, consists of dorsiflexion of the great toe and fanning of the rest. The presence of the Babinski sign is a reliable indicator of damage to the corticospinal tract.

Primitive reflexes

Certain reflexes present in infancy or early childhood normally disappear. When these reappear in adult years they are known as 'primitive reflexes' (Chapter 5.6) and may indicate frontal lobe disease.

The palmomental reflex is tested for by repeatedly and rapidly dragging an object, such as the tip of a reflex hammer, across the thenar eminence: when the reflex is present, one sees, as noted by DeJong (1979) 'a wrinkling of the skin of the chin and slight retraction and sometimes elevation of the angle of the mouth'.

The snout reflex is said to be present when gentle tapping or pressure just above the patient's upper lip, in the midline, is followed by a puckering or protrusion of the lips; in advanced cases, the reflex may, as described by DeJong (1979) be elicited by merely 'sweeping a tongue blade briskly across the upper lip'.

The grasp reflex may be elicited by laying one's finger across the patient's palm such that it may be readily dragged out between the patient's thumb and index finger. If the reflex is present, the patient's fingers will grasp the physician's finger as it is slowly dragged across the palm (Walshe and Robertson 1933).

The grope reflex may be elicited by simply lightly touching the patient's hand with one's finger: when present, the patient's hand will automatically make groping movements until the physician's finger is found and grasped (Seyffarth and Denny-Brown 1948).

Aphasia and mutism

Aphasia represents a disturbance in the comprehension and/or production of spoken language. Testing involves listening to the patient's spontaneous speech, giving simple spoken commands and determining whether the patient understands them, and asking the patient to repeat a test phrase, such as 'No ifs, ands, or buts'.

As discussed in detail in Chapter 3.1, there are three basic forms of aphasia: motor (also known as expressive or Broca's), sensory (also known as receptive or Wernicke's), and global. In motor aphasia, patients are able to follow commands, but their speech, despite being coherent, is effortful, sparse, and often 'telegraphic', (i.e. lacking in prepositions and conjunctions). In sensory aphasia, in contrast, patients have a greater or lesser degree of difficulty in following oral commands, especially complex ones; furthermore, speech is quite fluent, even voluble, rather than being effortful: however, there is a greater or lesser degree of incoherence such that what the patient says makes 'no sense'. Finally, the global type of aphasia represents a combination of these two: patients have trouble following commands, speech is effortful and sparse, and what the patient says is more or less incoherent.

Each of these three types may also occur as a 'transcortical' variant, and this is said to be present when patients are able to repeat a test phrase accurately and without effort. Other variants, less common, are also possible, and these are discussed in Chapter 3.1.

Mutism (Chapter 5.1) is said to be present when there is no speech.

Alexia and agraphia

Alexia (Chapter 3.2) and agraphia (Chapter 3.3) represent, respectively, difficulties in reading and writing, and although often seen in combination with aphasia, may also appear in pure form. Testing is accomplished simply by asking the patient to read something, perhaps a headline, and then to write something, such as an address.

Aprosodia

Aprosodia (Chapter 3.7) represents a disturbance in the production or comprehension of the 'emotional' and melodic aspects of speech (Ross 1981). Thus, the patient's own speech may be monotone, lacking all prosodic elements, or the patient may have difficulty in appreciating the emotional tone of another's voice. A lack of prosody in the patient's own speech is generally apparent as the history is related; testing for the patient's ability to 'comprehend' prosody may require that patients close their eyes and then listen as the physician repeats the same neutral phrase repeatedly but with different intonations (e.g. happy, angry, or sad), asking each time what the tone was.

Apraxia

Apraxia may be ideational/ideomotor, constructional, or dressing.

Ideational and ideomotor apraxia (DeJong 1979; Heilman 1973) are tested by first asking the patient to pantomime using a common tool, such as a comb or a pair of scissors, and then, if the patient has any difficulty in performing the pantomime, by providing the tool and asking the patient to make use of it. In ideational apraxia, both miming and actual use are defective, whereas with ideomotor apraxia the patient, although unable to perform the pantomime, has no trouble correctly employing the actual tool.

Constructional apraxia is tested for by asking the patient to draw a simple figure, such as a 'stick person', or to copy a geometric design such as a cube.

Dressing apraxia is casually assessed by observing the patient put on clothing: when present, patients may put their arms in the wrong sleeve or perhaps attempt to put their shirt on backwards (Hecaen *et al.* 1956).

Agnosias

Agnosia, as discussed in Chapter 3.9, exists in various forms, including, for example, visual agnosia, tactile agnosia, and anosognosia. In each form, despite the fact that relevant elementary sensory abilities are intact, there is an inability to recognize things.

Visual agnosia, or the inability to recognize an object by sight, is tested by pointing to a common object, such as a comb, and asking patients not only to name it, but also to describe its use.

Tactile agnosia represents an inability to recognize an object by touch: with the eyes closed, the patient is given a common object, such as a key, and asked both to identify it and to describe its use.

Anosognosia is said to be present when patients fail to recognize a deficit, such as hemiparesis, or grossly minimize it, for example, by characterizing a severely hemiparetic limb as simply 'stiff'.

Other agnosias, also described in Chapter 3.9, which are generally not routinely tested for, include color agnosia, prosopagnosia (the inability to recognize faces), auditory agnosia (the inability to recognize common sounds), topographagnosia (a loss of a sense of direction), simultanagnosia (an inability to visually 'grasp' the whole of a scene to see all of its parts simultaneously), and asomatagnosia (a denial of the 'ownership' of a body part, as may be seen in some cases of hemiparesis).

Neglect

Neglect, discussed in Chapter 3.10, is characterized by an involuntary failure to attend to or notice phenomena on one side or the other, and this may be either visual or motor.

Visual neglect is tested by seating the patient squarely in front of a table, with the patient's trunk kept parallel to the edge of the table (Beschin *et al.* 1997). First, draw a line horizontally across a piece of paper, at least 15 cm long (Tegner and Levander 1991) and then place the paper directly in front of, and square to, the patient. The patient is then asked to bisect the line. Next, draw numerous short marks in a random fashion on a piece of paper, placing the paper squarely in front of the patient and asking the patient to mark or cancel out all the lines. Finally, position a blank piece of paper in front of the patient with the instruction to draw a clock face on it, with all the numbers, from 1 to 12, on the drawing. These constitute, respectively, the line bisection, line cancellation, and clock-drawing tests, and visual neglect is said to be present if the line is bisected off the midline, a significant percentage of the random lines on one side are not cancelled out, or the numerals on the clock face are bunched to one side.

Motor neglect is tested by asking the patient to perform a task that requires the use of both upper extremities, such as fastening a button: when motor neglect is present, the patient 'underutilizes' the 'neglected' side and attempts to perform the task primarily with one hand, despite the fact that with strong urging normal bilateral manual coordination is possible (Laplane and Degos 1983).

Extinction

Extinction, also discussed in Chapter 3.10, is considered a subtype of neglect, and, like neglect, may be either visual or tactile (Valler *et al.* 1994).

Visual extinction may be tested immediately after performing confrontation testing of the visual fields. While retaining the same position with respect to the patient, the physician holds both hands outstretched laterally to the edge of the peripheral fields and then simultaneously wiggles both index fingers, asking the patient to point to the finger(s) that is (are) moving. When visual extinction is present, the patient notes the motion of only one finger.

Tactile extinction may be tested during routine sensory testing. When the patient's eyes are closed, the physician instructs the patient to report which hand or hands are being touched – touching first one hand, then the other and then both simultaneously. When tactile extinction is present, only one hand will be reported as touched during simultaneous stimulation.

2.4 CLINICAL NEUROIMAGING

Computed tomography

X-ray computed tomography (CT), previously known as computed axial tomography (CAT), is a digital form of structural neuroimaging which utilizes X-ray attenuation married to a tomographic technique which separates out axial or transverse images. In general, given that magnetic resonance imaging (MRI) does not involve any ionizing radiation and permits high-resolution structural imaging, which is usually non-invasive (as contrast agents are not usually required), where available, MRI is taking over from CT in clinical neuropsychiatric practice unless there is a contraindication for the use of MRI in a particular patient (see below). There is still controversy in the area of acute stroke imaging (in which the use of an intravenous contrast

Table 2.1 Comparison of computed tomography (CT) and magnetic resonance imaging (MRI) in acute stroke imaging

	CT	MRI
Availablity	Widely available	Limited availability in most centers (especially for emergency use)
Scan duration	Short	Potentially longer
Cost	Low (standard non-contrast); high (CTA/CTP)	Very high
Radiation	Yes	No
Contrast type	Requires iodinated contrast for CT angiogram and perfusion	Gadolinium is water-soluble (risk of nephrogenic systemic fibrosis with renal insufficiency)
Sensitivity to early ischemia	Poor	High
Posterior fossa visualization	Poor (bony artefact)	Good
Brain coverage (CT perfusion)	Limited	Whole brain
Contraindications	Renal insufficiency (iodinated contrast)	Severe renal insufficiency (gadolinium-based contrast); ferromagnetic material, e.g. pacemaker, aneurysm clips; claustrophobia; medical instability

CM, computed tomography angiography; CTP, computed tomography perfusion.
After Hsia (2009) 10 questions about acute stroke imaging: CT versus MRI – the controversy. *Neurologist* 2009; **15**:105–7.

agent is usually needed for the MRI scanning) as to which of CT and MRI is better. The two are compared in Table 2.1 (Hsia 2009).

Magnetic resonance imaging

T1- AND T2-WEIGHTED IMAGING

T1 is the longitudinal time constant and indexes the rate at which the longitudinal magnetization returns to its maximum value following a radiofrequency pulse. The value of T1 varies with the magnetic field strength. At a given field strength, the T1 is longer for CSF (cerebrospinal fluid) than for gray matter, and in turn longer for gray matter than for white matter. It is particularly short for fat. (With its higher concentration of myelinated sheaths, white matter clearly has a higher density of fat than does gray matter.)

T2 is the transverse time constant and indexes the rate at which the transverse magnetization decays to its minimum value of zero following a radiofrequency pulse.

In MRI scans of the brain, if scanning sequence parameters are chosen selectively to emphasize T1 characteristics (T1-weighted scans), white matter has a higher signal intensity (i.e. it appears 'whiter') than gray matter, which in turn has a higher signal intensity than CSF. In a T2-weighted MRI scan, CSF has a higher signal intensity than gray matter or white matter, and such scans are also good for imaging cerebral edema (which also has a high signal intensity). These differences are shown in Figure 2.1, which also shows how an intravenous contrast agent (in this case, based on gadolinium) may helpfully enhance the imaging of some cerebral tumors.

RECEIVER COILS

Special receiver coils, which allow 'parallel imaging' to be carried out, are becoming available. These involve multiple imaging coils which record imaging data simultaneously using techniques such as SENSE (sensitivity encoding for MRI) and SMASH (simultaneous acquisition of spatial harmonics) so that the image acquisition time is shortened. There are also advantages from parallel imaging in terms of reducing certain types of MRI artifacts.

SAFETY

The magnetic field strengths typically used during *in vivo* human scanning are very high compared with the strength of the earth's magnetic field. For example, while the strength of the latter, at the surface of the earth, is around 30–60 μT, a field strength of 3 T inside the bore of a modern 3-T MRI scanner is 50 000–100 000 times stronger. The strength of the magnetic field falls off with increasing distance from the main MRI magnet. Special magnetic shielding is also placed around the scanner. A warning contour, usually in yellow and black, is fixed to the floor around the scanner at the distance that corresponds to a field strength of 500 μT. This corresponds to a field strength of 5 G using the previous unit of magnetic field strength (flux density) of the gauss (1 gauss, 1 G = one ten-thousandth of a tesla; that is, 1 T = 10 000 G). Therefore, this contour is often still referred to as being the '5-gauss line'. It can be very dangerous to bring freely mobile ferromagnetic materials within the 5-gauss line. Even a small piece of metal, such as a paper clip or a pen containing ferromagnetic components, can turn into an extremely dangerous, and potentially even lethal, object if it is brought within this contour – there is a strong attraction to the bore of the magnet and such an object can fly into this part of the scanner, all the time accelerating while approaching the scanner so that its momentum increases correspondingly. If a person happens to be lying in the scanner at the time, this can clearly be very dangerous.

In terms of the safety of individuals being considered for scanning, exclusions include the presence of any of the following: cardiac pacemaker; mechanical heart valve; history

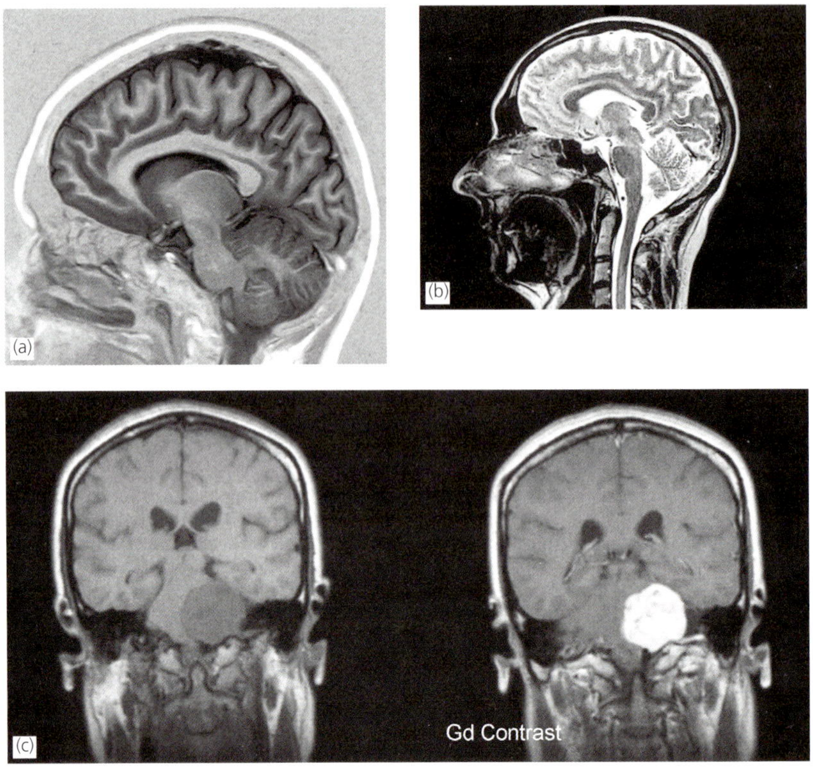

Figure 2.1 (a) Sagittal T1-weighted MRI scan of the brain. (b) Sagittal T2-weighted MRI scan of the brain. (c) A positive contrast agent (gadolinium diethylenetriamine penta-acetic acid) emphasizes the cerebral neoplasm by increasing the T1 effect. Reproduced from Dowsett, D.J, Kenny, P.A. and Johnston, R.E. *The Physics of Diagnostic Imaging*, 2nd edn, Hodder Arnold, 2006, with permission.

of a foreign body in eye (unless it is established that this is not going to be a danger, for example following careful orbital X-ray work-up); occupation as a metal worker, grinder, or welder; metallic implant, metal prosthesis, orthopedic plates, screws; shrapnel; aneurysm clip/hematostatic clip; ear implant; artificial eye; coloured contact lens; interventional radiological device; pregnancy; intrauterine contraceptive device; implantable pumps or neurostimulators; allergies – if the subject is to receive a contrast agent; a watch; any jewellery; anything in the patient's pockets (such as keys, etc.).

Claustrophobic feelings inside the scanner are a relative contraindication. Some subjects are able to go ahead after verbal reassurance. Others can do so if a second person (who should also be carefully metal-checked first) is present; this second person might sit by the subject (on a non-metallic chair) and perhaps gently hold on to a covered part of one of the subject's lower limbs during the scan.

ADVANCED STRUCTURAL IMAGING

Structural magnetic resonance neuroimaging can be used to carry out ventricular segmentation and thereby measure the volume of the ventricles of the brain. The cerebral tissue itself can be segmented into gray matter and white matter. The cortex can be analyzed for features such as its degree of folding.

Voxel-based morphometry (VBM) is based on the distribution of gray matter in the brain and, in a group, may reveal areas of gray matter loss (Duncan 2009).

It has been found that image registration is of particular use, both in terms of research and from a clinical perspective. This technique allows subtle changes in brain structure to be picked up and quantified between consecutive scans from the same subject; these scans may be several months or even years apart. This is helpful, for instance, in following the response of cerebral tumors to chemotherapy or radiotherapy. Image registration can also be used to compare groups, as shown in Figure 2.2. Here, analysis of the MRI scans from patients with Huntington's disease taking part in a randomized, double-blind placebo-controlled trial of a new intervention showed that, compared with the placebo, the active treatment was associated with significant group-level reductions in brain atrophy in the head of the caudate nucleus and the posterior thalamus (Puri *et al.* 2008) These findings showed that the active treatment was associated with significant reduction in brain atrophy, particularly in the caudate and thalamus.

FLUID-ATTENUATED INVERSION RECOVERY

FLAIR stands for fluid-attenuated inversion recovery and refers to a particular type of MRI sequence which produces a strong T2-weighting, suppresses the signal from CSF, and minimizes the contrast between gray matter and white

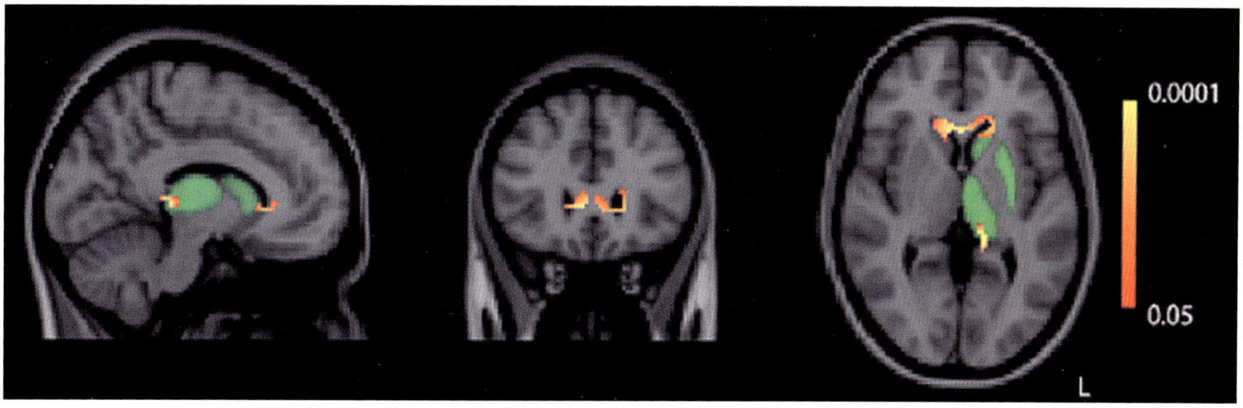

Figure 2.2 Image registration techniques were applied to structural MRI scans in a Huntington's disease clinical trial. Voxel-wise local analyses of ethyl-eicosapentaenoic acid treatment versus placebo over 12 months showed significant changes at the head of the caudate nucleus and the posterior thalamus (shown in green along with the putamen; red–yellow color bar shows the *P*-value under the null hypothesis of no change). Reproduced from Puri BK, Counsell SJ, Hamilton G (2008) Brain cell membrane motion-restricted phospholipids: a cerebral 31-phosphorus magnetic resonance spectroscopy stuffy of patients with schizophrenia, *Prostaglandins, Leukotriness and Essential Fatty Acids* **79:** 233–5 with permission.

matter, thereby producing images with significantly increased lesion-to-background CSF contrast and enhancing the visibility of lesions as well as their detectability, particularly in the peripheral subcortical and periventricular regions (De Coene *et al.* 1992; Kates *et al.* 1996) Figure 2.3 is a FLAIR MRI scan showing a hyperintense middle cerebral artery infarct.

DIFFUSION TENSOR IMAGING TRACTOGRAPHY

Diffusion tensor imaging (DTI) can be used to infer the location of white matter tracts in the brain, based on the diffusion properties of water molecules. Tractography involves calculating the fractional anisotropy (FA), which is a measure of the anisotropy of the tensor that indicates the degree of directional preference of the fiber bundles in any given voxel (Parker *et al.* 2002). Figure 2.4 shows an application of this technique in the human brain, in which the reconstructed white matter projections are shown in green.

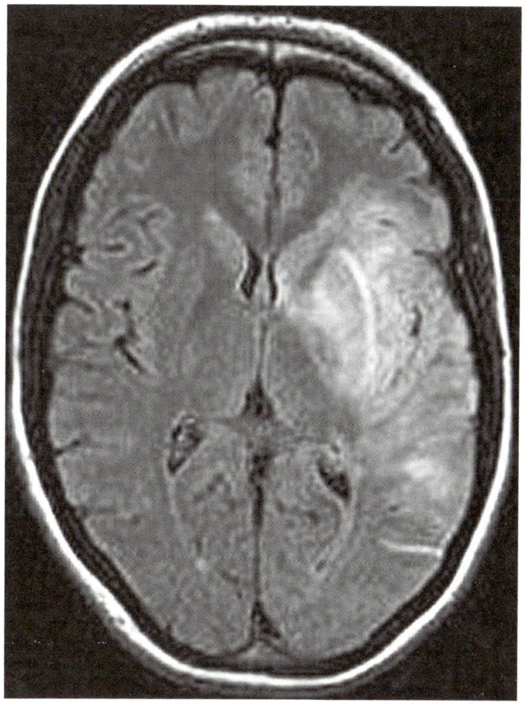

Figure 2.3 Middle cerebral artery infarct. Non-dominant hemisphere fluid attenuation inversion recovery magnetic resonance imaging (FLAIR MRI) scan showing a hyperintense infarct. Reproduced from Graham DI, Nicoll JAR and Bone I, *Adams & Graham's Introduction to Neuropathology* 3rd edition, Hodder Arnold, 2006, with permission.

2.5 ELECTROENCEPHALOGRAPHY

The existence of cerebral electrical activity was demonstrated in animals in 1875 by an English physician, Richard Caton (1875), and the first human electroencephalogram (EEG) was reported by Hans Berger in 1929 (Berger 1929). By the middle of the twentieth century, the EEG had become very important in the diagnosis of such intracranial lesions as tumors, but with the advent of CT and MRI the indications for electroencephalography have changed, and most EEGs are currently obtained in the course of the diagnosis or management of seizures or epilepsy and in the evaluation of delirium. This chapter discusses EEG instrumentation, the normal EEG, various EEG abnormalities, activation procedures (e.g. hyperventilation), normal variants, and the various artifacts that may mimic pathologic abnormalities.

As with any other diagnostic test, electroencephalography must be properly performed to yield the most useful data (Epstein *et al.* 2006a). In particular, the awake EEG should include at least 20 minutes of artifact-free recording, followed, when appropriate, by the activating procedures of hyperventilation, photic stimulation, and sleep, which should itself last an additional 20 minutes.

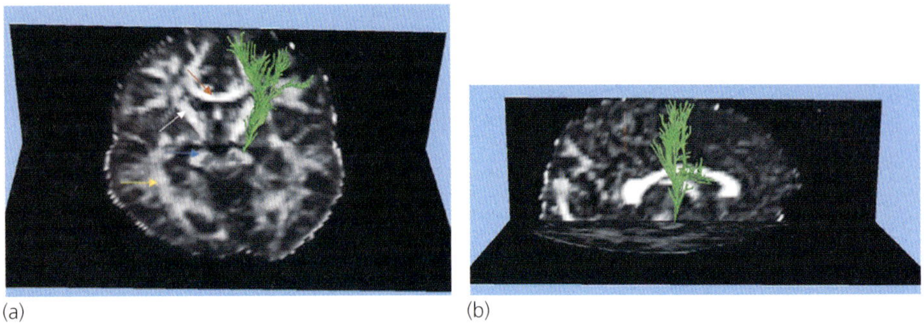

Figure 2.4 Tract reconstructions in the human corticofugal/corticospinal projections, from a start voxel in the middle of the right cerebral peduncle. Top 1 percent most likely of all possible connections shown, overlaid on fractional anisotropy (FA) maps for anatomical reference. (a) View from above and behind, showing convergence from medial and lateral motor areas. Coronal FA map bisecting the anterior portion of the corticofugal/corticospinal projections (white arrow) and corpus callosum (red arrow); near-axial FA map at the level of the peduncles (blue arrow) and inferior extent of the optic radiations (yellow arrow). (b) View from the right, showing anterior/posterior branching within the corona radiata. Axial FA map at the level of the peduncles; sagittal FA map at the plane of the midline. Reproduced from Parker *et al.* (2002) with permission.

In contrast with CT and MR scanning, there is nothing 'intuitively' obvious about an EEG tracing: whereas anyone familiar with neuroanatomy can almost immediately grasp an MRI scan, understanding an EEG, like understanding an electrocardiogram, takes some preparation.

Instrumentation

Electrodes are attached to the scalp and are connected via wires to the electroencephalography machine. Pairing of these wires, and the electrodes from which they stem, allows one to construct numerous different channels. In older, analog machines, this pairing was performed utilizing 'selector switches'; however, in the now-standard digital machines, an analog-to-digital converter allows for the creation of channels at the touch of a keyboard. Within the EEG machine itself, one finds amplifiers and filters that respectively amplify the very weak electrical signals arising from the cortex and filter out, as much as possible, electrical activity which is of little clinical interest.

The amplified and filtered electrical impulse of each channel is then used, in analog machines, to cause a deflection of the appropriate pen over a continuously moving sheet of paper, thus creating the actual EEG tracing. With digital machines, there are, of course, no pens or paper tracings; however, this terminology has stayed with us. In a standard recording, the sheet moves at a constant rate of 30 mm/s, and the sensitivity of the pen is set such that an impulse of 50 V causes a deflection of 7 mm.

The specific arrangement of electrodes on the scalp is known as an array, and the international 10–20 system described by Jasper (1958) remains a worldwide standard (Epstein *et al.* 2006b). In this system, imaginary lines are drawn on the head between specific landmarks (e.g. the nasion and inion) and the electrodes are placed along them at certain fractional intervals (i.e. either 10 percent or 20 percent of the total length of the imaginary line). These electrodes are designated with letters that refer to their location, and with numbers that indicate whether they are on the left side of the head, the right side or in the sagittal midline. Thus, F_p indicates frontopolar; F, frontal; T, temporal; O, occipital; C, central; P, parietal; and A, auricular. Odd numbers indicate the left side of the head, even numbers the right side, and zero ('z') the sagittal midline. Figure 2.5 demonstrates these placements, and Table 2.2 provides the full name for each electrode. Note, however, that clarification is needed regarding electrodes F7 and F8; although, logically, one might expect these to be called 'frontal', they are commonly referred to instead as 'anterior temporal' leads as, for the most part, they reflect activity arising from the anterior portion of the temporal lobes.

This international 10–20 system may be extended and modified by adding more electrodes (Chatrian *et al.* 1985; Epstein *et al.* 2006b), and this may be resorted to in order to improve localization or to increase spatial resolution and allow for better computed EEG analysis. Supplemental leads may also be added to better detect and localize foci in the temporal lobe. 'True' anterior temporal leads (to be distinguished from the admittedly misnamed F7 and F8 electrodes)

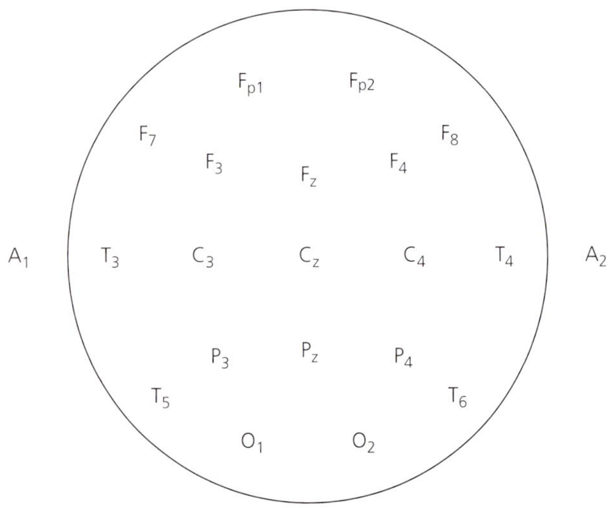

Figure 2.5 Electroencephalography (EEG) electrode placement according to the international 10–20 system (see text for details).

Table 2.2 Electrode names in the 10–20 system

Name	Position
F_{p1}, F_{p2}	Prefrontal
F_7, F_8	Anterior temporal
T_3, T_4	Mid-temporal
T_5, T_6	Posterior temporal
O_1, O_2	Occipital
F_3, F_4	Frontal
C_3, C_4	Central
P_3, P_4	Parietal
A_1, A_2	Auricular
F_z	Frontal midline
C_z	Central vertex
P_z	Parietal midline

are placed by drawing a line between the external auditory canal and the lateral canthus, and placing the electrode anterior to the external auditory canal one-third of the way forward along, and 1 cm above, this line (Homan *et al.* 1988; Silverman 1960). Nasopharyngeal leads, as the name suggests, are inserted into the nostril in order to sample the medial aspect of the temporal lobe (MacLean 1949). Sphenoidal leads are invasive, requiring a trochar to place them through the masseter muscle and up posterior to the zygomatic arch: these also attempt to sample the medial aspect of the temporal lobe (Risinger *et al.* 1989).

There is a debate over which one or combination of supplemental leads is most appropriate for detecting temporal lobe foci. The addition of anterior temporal leads provides more sensitivity than a routine 10–20 array, and it appears that anterior temporal leads are of either roughly equivalent (Sperling and Engel 1985) or even superior sensitivity (Sadler and Goodwin 1989) to nasopharyngeal leads. It is not clear how anterior temporal leads compare in sensitivity to sphenoidal leads: some studies find them equivalent (Homan *et al.* 1988; Sadler and Goodwin 1989), whereas others find sphenoidal leads superior (Sperling *et al.* 1986). Whenever temporal lobe foci are sought, at the very least 'true' anterior leads should be ordered with other supplemental leads held in reserve.

As noted earlier, the electroencephalography machine allows electrodes to be paired in various ways, and the pattern of such pairings is known as a montage. Three standard montages are recommended: a referential montage and two bipolar montages, namely a longitudinal bipolar montage and a transverse bipolar montage (Epstein *et al.* 2006c).

In a referential montage, each scalp electrode is paired with the same 'reference' electrode, usually the ipsilateral ear, producing channels such as F_7–A_1, T_3–A_1, and T_5–A_1. The scalp electrode is commonly referred to as the 'active' electrode, in contrast with the reference electrode, which is termed 'indifferent'. However, this terminology is not accurate because the ear electrode in fact picks up electrical activity arising from the temporal lobe and is thus only 'relatively' indifferent. In some instances, other electrodes, or combinations of electrodes, will be used instead of one ear; thus, the reference electrode may be found on the angle of the mandible or an 'average reference electrode' may be produced by averaging the electrical activity of a large number of scalp electrodes (Goldman 1950).

In a bipolar montage, scalp electrodes are paired in two directions – longitudinal and transverse. In a longitudinal bipolar montage, the pairings proceed ipsilaterally, from anterior to posterior, producing 'chains' of channels, such as F_{p1}–F_3, F_3–C_3, C_3–P_3, and P_3–O_1. In a transverse bipolar montage, the chain proceeds across the scalp, from left to right, for example F_7–F_3, F_3–F_z, F_z–F_4, F_4–F_8. It is appropriate to note here that in the chains of a bipolar montage one individual electrode may serve as the second electrode in one channel and the first electrode of the next: for example, in the chain noted above, containing channels F_{p1}–F_3, F_3–C_3, C_3–P_3, and P_3–O_1, note that electrode F_3 serves as the second electrode for the first channel (F_{p1}–F_3) and the first electrode for the next channel (F_3–C_3). As will be noted later in the discussion of interictal epileptiform abnormalities, the commonality of one electrode to two successive channels in a bipolar montage allows for a localization of epileptic foci.

Normal EEG

The electrical activity recorded by the EEG arises from the apical dendrites of cortical pyramidal neurons (Humphrey 1968; Purpura and Grundfest 1956). Although the electrical activity associated with an action potential is too brief to be recorded on an EEG (i.e. lasting less than 1 ms), activity derived from both inhibitory and excitatory postsynaptic potentials lasts much longer (from 15 to 200 ms) and it is this activity that is reflected in the EEG (Humphrey 1968). The electrical activity arising from one neuron is obviously too weak to affect the surface electrodes, so it is upon the summed activity of numerous neurons that the EEG depends. Furthermore, it must be borne in mind that abnormal electrical activity occurring deep below the cortex may not 'reach' the scalp electrodes (Cooper *et al.* 1965) and thus certain deep lesions, such as lacunar infarctions, may not cause any abnormality on the EEG while yet having profound clinical consequences (MacDonnell *et al.* 1988).

Electroencephalographic activity may or may not be rhythmic and it appears that rhythmicity occurs secondary to the activity of the thalamus, which acts like a pacemaker or 'conductor', exerting rhythmic control over the cortical 'orchestra', and bringing large groups of neurons into synchrony (Dempsey and Morrison 1942; Steriade *et al.* 1990). This dependence of cortical neurons upon the thalamus for rhythmic firing was demonstrated by experiments in which the destruction of the thalamus abolished rhythmic cortical activity (Jasper 1949).

The EEG consists of various waves that may differ in terms of morphology, amplitude, and duration. Thus, in terms of morphology, an individual wave may be monophasic, diphasic, triphasic, or polyphasic, depending on how many times the 'baseline' is crossed by the wave in question. Amplitude is measured in microvolts from the crest to the trough of the wave, and customarily amplitudes under 20 μV are considered low, those between 20 and 50 µV, medium, and those over 50 µV, high. Of note, however, some electroencephalographers will, rather than using this absolute scale, consider the amplitude of a given wave relative to the overall amplitude of background activity: thus, if the background activity were generally of 60 µV, a 30 µV wave, using this relative scale, might be considered low. Consequently, it is therefore critical that the electroencephalographer specifies whether an absolute or a relative scale is being used when reporting amplitude. The duration of the wave is measured in milliseconds: waves lasting less than 70 ms are referred to as 'spikes' and those lasting from 70 to 200 ms as 'sharp waves'; those lasting for over 200 ms are spoken of either as 'slow waves' or simply 'waves'.

Waves may be isolated or recurrent. If recurrent, their frequency is reported in cycles per second (Hz): by convention, frequencies less than 4 Hz are termed 'delta', those from 4 to under 8 Hz 'theta', those from 8 to 13 Hz 'alpha', and those over 13 Hz as 'beta' waves. Some electroencephalographers also use the terms 'slow' and 'fast', 'slow' referring to both delta and theta activity (i.e. anything under 8 Hz) and 'fast' referring to any activity in the beta range (i.e. over 13 Hz). Recurrent activity may also be rhythmic and regular in occurrence, or arrhythmic and irregular.

The EEG will normally have a recognizable background activity that is more or less persistent and similar throughout the recording. Upon this background, one may at times see isolated events that, for one reason or another, stand out from the background, and such events are referred to as 'transients'. Transients may, in turn, consist either of an isolated spike or wave or a 'complex' of two or more of these. Complexes themselves are further described in terms of whether they are isolated or recurrent, and, if recurrent, whether they recur irregularly or regularly. 'Spindles' comprise a specific type of transient complex, consisting of a group of rhythmic waves that gradually increase in amplitude, and then just as gradually decrease.

The normal adult awake EEG, as seen during relaxed wakefulness with the eyes closed, contains an alpha rhythm and a beta rhythm. These two terms must not be used loosely: for example, although much EEG activity may occur in the alpha *frequency*, the activity must fulfill certain other criteria to qualify as an alpha *rhythm*. In a minority of individuals, a *mu rhythm* may also be seen.

The *alpha rhythm* consists of more or less regular sinusoidal activity, ranging in amplitude from 20 to 60 µV (averaging about 50), occurring in the alpha range and most prominent posteriorly. The alpha rhythm is generally 'blocked' by eye opening, being replaced by lower amplitude and faster and much less regular activity. Although the frequency of the alpha rhythm is the same on each side, the actual waves themselves are generally out of phase. Further, there is also generally an amplitude difference between the two sides, with the left-side alpha being of lower amplitude than the right. Generally, this amplitude differential is no more than 20 percent; however, the range of normal here is wide, with some normal individuals having differentials up to 50 percent. The alpha rhythm is best seen in a state of relaxed wakefulness with the eyes closed. In a small minority of cases, variants of the alpha rhythm may occur (Goodwin 1947), wherein the frequency of the sinusoidal activity is either in the 4- to 5-Hz range ('slow' alpha variant) or 16- to 20-Hz range ('fast' alpha variant). These variants represent 'harmonics' of the more typical alpha rhythm.

The *beta rhythm* consists of bilateral beta activity of an amplitude of 30 µV or less, seen best anteriorly, which is blocked unilaterally by contralateral tactile stimulation, movement, or merely an intention to move. Although the waves are generally out of phase, the frequency is bilaterally symmetric. An amplitude variance of up to 35 percent from side to side is considered normal. Beta activity is often increased by sedatives such as benzodiazepines and barbiturates (Browne and Penry 1973; Frost *et al.* 1973; Greenblatt *et al.* 1989).

The *mu rhythm* represents another normal type of EEG activity, one that is not seen as routinely as the alpha or beta rhythms and is present in only about 10 percent of normal adults. The mu rhythm consists of theta or alpha activity (ranging from 7 to 11 Hz) that appears as long transients ('trains') lasting at least several seconds and appearing in the centroparietal region. Although these occur bilaterally, the trains are often not synchronous, with one side having a train and then losing it, and then a train appearing a little later on the opposite side. The mu rhythm is generally 50 µV or less in amplitude. The mu rhythm, like the beta rhythm, may also be unilaterally blocked by contralateral phenomena (Chatrian *et al.* 1959; Chatrian 1964) including movement (Chatrian *et al.* 1959), intention to move (Klass and Bickford 1957) and tactile stimuli (Magnus 1954).

The normal adult sleep EEG demonstrates both REM (rapid eye movement) and non-REM (NREM: non-rapid eye movement) sleep. REM sleep is, as the name suggests, characterized by rapid, saccadic, conjugate eye movements and is typically associated with dreaming. NREM sleep is generally not associated with dreaming, and during such sleep, the eyes are either still or slowly roving about. NREM sleep may further be divided into four stages: I, II, III, and IV, with each of these stages having a distinctive electroencephalographic signature (Erwin *et al.* 1984; Rechtschaffen and Kales 1968). In order to identify the various stages one must be familiar with several different transient events, including vertex sharp transients, K complexes, sleep spindles, and positive occipital sharp transients (POSTs).

Vertex sharp transients (also known as 'V waves') are intermittently occurring, bilaterally symmetric sharp waves of high amplitude (rarely more than 250 µV) seen most prominently at the vertex. K complexes are very similar to vertex sharp transients, differing only in that they generally consist of a diphasic slow wave. Sleep spindles are transients lasting from half a second to several seconds, consisting of rhythmic activity in the 11- to 14-Hz range, which, as with all spindles, demonstrates a gradual increase and decrease in amplitude, with a maximum of generally less than 50 µV. These sleep spindles occur simultaneously on both sides and, although maximal centrally, are widespread. POSTs (Vignaendra *et al.* 1974) consist of sharp waves of positive polarity seen posteriorly in the occipital regions. They are monophasic and generally of no more than 50 µV in amplitude; although they are seen bilaterally, they are not synchronous. Furthermore, they are not rhythmic and can be seen at irregular intervals of anywhere from several to one per second.

With these various transients in mind, the four sleep stages may now be defined. The onset of stage I is marked by 'alpha dropout', with slowing of the background rhythm into the theta–delta range; soon thereafter vertex sharp transients appear. Stage II is characterized by a persistence of the slowing and the vertex sharp transients, but with the appearance of K complexes, sleep spindles and POSTs. Stage III is characterized by further slowing (20–50 percent of the background activity being in the delta range), an absence of vertex sharp transients, a fading out of K complexes and sleep spindles, but a persistence of POSTs. Stage IV is identified by gross slowing (more than 50 percent delta activity), an absence of vertex sharp transients and K complexes, and only rare sleep spindles and POSTs.

The entire night's sleep typically occurs in cycles, each cycle lasting from 80 to 120 minutes. The first cycle begins as the patient drifts into stage I, progressing down through stages II and III to stage IV and thence back up through stages III and II to stage I, from which REM sleep emerges. The end of REM sleep signals the end of the first cycle and the beginning of the next. During one night's sleep, subjects normally pass through three to five of these cycles, and with each successive cycle, the amount of time spent in stage IV sleep decreases.

EEG abnormalities

The various EEG abnormalities discussed here include decreased amplitude, slowing (either focal or generalized), interictal ('epileptiform') and ictal abnormalities, periodic complexes, triphasic waves, and the burst–suppression pattern.

DECREASED AMPLITUDE

Low-amplitude EEG activity may result either from an alteration in the media between the cortex and the recording electrode or from decreased electrogenesis by the cortex (Aird and Shimuzu 1970). For example, both grease and an abnormally thick skull (e.g. in Paget's disease) act as insulators, and fluid collections, such as subgaleal, epidural, or subdural hematomas, act as 'shunts' that divert the electrical field away from the overlying electrode. Cortical electrogenesis may be reduced either because of actual destruction, as in Alzheimer's disease or tumors, or decreased neuronal activity, as in metabolic deliria or postictal states. Decreased amplitude may be either generalized or focal.

Generalized low-amplitude EEGs of from 20 to 10 μV may be seen in 5–10 percent of normal adults; an amplitude of less than 10 μV is rare in normal subjects. When the general amplitude is reduced to below 20 μV, it is helpful to be able to compare the current record with past ones, or to make serial recordings in order to determine whether the low amplitude is stable or worsening. It is also critical to ensure that the recording is made during relaxed wakefulness: tense or anxious patients, or those engaging in some more or less demanding mental activity, will have low-amplitude recordings. A generalized decrease in amplitude may be seen in conditions characterized by widespread cortical neuronal loss (e.g. Alzheimer's disease, Huntington's disease [Scott *et al.* 1972], Creutzfeldt–Jakob disease [Burger *et al.* 1972], post-anoxic encephalopathy, or AIDS dementia [Harden *et al.* 1993]) or widespread neuronal dysfunction (e.g. metabolic deliria such as hepatic or uremic encephalopathy, or other conditions such as hypothyroidism, hypothermia, uncomplicated alcohol withdrawal [Walker *et al.* 1956] or postictally after a grand mal seizure).

Focal low-amplitude EEGs may be seen in conditions that cause a unilateral increase in the media (e.g. subdural hematoma [Lusins *et al.* 1976]) and either unilateral neuronal destruction (e.g. infarction or tumor [Aird and Shimuzu 1970]) or dysfunction (e.g. transient ischemic attacks, migraine, and postictally after a partial seizure [Kaibara and Blume 1988]).

In evaluating amplitude asymmetries of the alpha rhythm, one must not forget that the left side normally has an amplitude of up to 50 percent less than the right; thus, it is only when the alpha rhythm on the left is at least 50 percent less than that on the right that one can declare with certainty that an abnormality is present. The beta rhythm is generally bilaterally symmetric, but even here an amplitude asymmetry is not unusual in normal individuals; thus, for the beta rhythm, any asymmetry must be more than 35 percent before it can be declared outside the normal range.

A unilateral reduction in amplitude of the beta rhythm indicates a frontal lesion. In general, a unilateral reduction of the alpha rhythm suggests a lesion of the underlying occipital cortex, but in the case of the alpha rhythm an amplitude reduction may also be seen with distant lesions in the frontal or parietal cortices or the ipsilateral thalamus.

Amplitude asymmetry may occasionally be spurious, as, for example, with 'breach' rhythms. Here, in conditions where the skull has been breached, for example with a burr hole or fracture (regardless of how much scar tissue has formed), an excessive amplitude is seen on the side with the breach, making the normal amplitude activity on the other side appear low by comparison (Cobb *et al.* 1979).

SLOWING

Slowing on the EEG may be either focal or generalized.

Focal slowing

Focal theta or delta slowing is generally polymorphic and non-rhythmic and may be seen in a variety of focal conditions, including infarcts and tumors (Daly and Thomas 1958; Gastaut *et al.* 1979; Gilmore and Brenner 1981), subdural hematoma (Lusins *et al.* 1976), postictally after a focal onset seizure (Gilmore and Brenner 1981), after some migraine headaches (Hockaday and Whitty 1969), and early in herpes simplex encephalitis (Upton and Gumpert 1970).

Generalized slowing

Generalized slowing appears in the theta or delta range and may be either bilaterally asynchronous or synchronous. Asynchronous generalized slowing is most commonly seen in metabolic or toxic delirium (Pro and Wells 1977; Romano and Engel 1944). Metabolic deliria accompanied by generalized asynchronous slowing include hepatic encephalopathy and uremic encephalopathy, and the deliria occurring secondary to hyperglycemia, hypoglycemia, hypernatremia, hyponatremia, hypercalcemia, or hypocalcemia. Toxic deliria associated with similar slowing include those due to phencyclidine (Stockard *et al.* 1976), phenytoin (Roseman 1961), valproate (Adams *et al.* 1978), and either carbamazepine or phenobarbital (Schmidt 1982). Generalized slowing may also be seen during alcohol intoxication (Walker *et al.* 1956) and in Wernicke's encephalopathy (Frantzen 1966). Interestingly, however, the delirium of delirium tremens, rather than slowing, is accompanied by an increase of beta activity (Kennard *et al.* 1945; Schear 1985). The delirium seen with bacterial meningitis or viral encephalitis is also marked by generalized slowing. Various dementing disorders may also be accompanied by generalized asynchronous slowing, including Alzheimer's disease (Deisenhammer and Jellinger 1974; Johannesson *et al.* 1977), Binswanger's disease (Caplan and Schoene 1978), Parkinson's disease (Neufeld *et al.* 1988), dementia with Lewy bodies (Briel *et al.* 1999), progressive supranuclear palsy (Su and Goldensohn 1973), normal pressure hydrocephalus (Wood *et al.* 1974), vitamin B12 deficiency (Walson *et al.* 1954), posthypoxic–ischemic encephalopathy (Hockaday *et al.* 1965), AIDS dementia (Harden *et al.* 1993), and Creutzfeldt–Jakob disease (Burger *et al.* 1972) (including the new-variant type [Zeidler *et al.* 1997]). Finally, generalized slowing may also be seen in the persistent vegetative state (Kulkarni *et al.* 2007).

A mild degree of generalized asynchronous slowing may also be seen as a normal variant in a small minority of subjects; furthermore, occasional scattered theta transients are not at all abnormal in normal subjects over the age of 60 years (Kooi *et al.* 1964). Generalized slowing also, of course, occurs with sleep, and thus slowing in a drowsy patient who is slipping in and out of sleep is of little significance.

Synchronous bilaterally generalized slowing typically occurs episodically, and in such cases is termed IRDA (intermittent rhythmic delta activity). In most adults, IRDA is predominantly frontal, and is termed FIRDA (frontal intermittent rhythmic delta activity) (Zurek *et al.* 1985), whereas in children IRDA is generally occipital and referred to as OIRDA (Watemberg *et al.* 2007). Although FIRDA was classically associated with deep midline lesions (Daly *et al.* 1953), it has now become clear that FIRDA is most commonly seen in metabolic and toxic deliria (Schaul *et al.* 1981), especially in patients with pre-existing ischemic lesions (Watemberg *et al.* 2002); FIRDA has also been reported in association with dementia with Lewy bodies (Calzetti *et al.* 2002), Creutzfeldt–Jakob disease (Wieser *et al.* 2004), and high-dosage antipsychotics (Koshino *et al.* 1993). There is one other, relatively rare, type of IRDA, which is restricted to the temporal areas: this is TIRDA, which, rather than being seen in delirium, most commonly occurs in patients with complex partial seizures (Normand *et al.* 1995), especially, as might be expected, in those with foci in the temporal lobes (Di Gennaro *et al.* 2003).

INTERICTAL AND ICTAL ACTIVITY

In patients with seizures or epilepsy, abnormalities seen between seizures are termed 'interictal', whereas those seen during a seizure are designated as 'ictal'.

Interictal activity

Interictal activity consists of what are known as epileptiform discharges. These interictal epileptiform discharges (IEDs) are paroxysmal transients that 'shoot' out from the background rhythm (Pedley 1980). These paroxysmal transients may consist of isolated spikes or sharp waves or may appear as complexes, such as spike-and-sharp wave, spike-and-slow wave, sharp-and-slow wave, polyspikes, or polyspike-and-wave discharges. Although epileptiform activity may be seen in a very small percentage of subjects who have never had a seizure (Ajmone-Marsan and Zivin 1970; Gibbs *et al.* 1943; Gregory *et al.* 1993; Zivin and Ajmone-Marsan 1968), and may also occur as a side-effect of certain medications (e.g. lithium [Helmchen and Kanowski 1971], clozapine [Malow *et al.* 1994] and meperidine [Kaiko *et al.* 1983]), and in patients with autism (Kim *et al.* 2006) or developmental dysphasia (Picard *et al.* 1998), they are typically present only in patients with a history of seizures. Importantly, however, although most epileptic patients will have interictal epileptiform abnormalities (Goodin and Aminoff 1984), the absence of such a finding on the first EEG does not rule out a diagnosis of epilepsy, as it may take up to four or more EEGs before an IED is 'caught' (Salinsky *et al.* 1987). Furthermore, in evaluating a patient who has had a seizure, it is important to obtain the EEG within the first 24 hours, as EEGs obtained later than this are less likely to reveal IEDs (King *et al.* 1998). It must be borne in mind, however, that in a very small minority of patients with unquestionable epilepsy routine EEGs will simply not display IEDs (Ajmone-Marsan and Zivin 1970; Martins da Silva *et al.* 1984; Morris *et al.* 1988).

Interictal epileptiform activity may be generalized, focal, or multifocal.

Generalized bisynchronous IEDs indicate the presence of one of the idiopathic generalized epilepsies, such as petit mal epilepsy (either childhood or juvenile absence epilepsy), juvenile myoclonic epilepsy, or epilepsy with tonic–clonic seizures only. In petit mal epilepsy, one sees the classic, roughly 3-Hz generalized and bisynchronous spike-and-dome discharges (Dalby 1969); in the other forms, the IEDs may be spike-and-wave, polyspike or polyspike-and-wave in character (Betting *et al.* 2006).

Focal epileptiform activity strongly suggests an underlying focal epileptogenic lesion (e.g. a tumor, scar, or area of cortical dysplasia), whereas multifocal epileptiform activity, as might be expected, suggests widespread and multiple lesions (e.g. subsequent to severe traumatic brain injury). The task of localizing focal epileptiform activity is facilitated by having in mind a spatial image of the electrical activity itself. Most paroxysmal electrical discharges are 'surface negative' (Matsuo and Knott 1977), i.e. their electrical potential is negative with regard to the normal baseline. Furthermore, most of these discharges cover a fairly wide area: although in some cases the discharge may be 'seen' under only one electrode, this is quite rare and in the vast majority of cases it covers a wider area, subtending two or more adjacent electrodes. The electrical activity itself may be visualized as a landscape, which may in turn contain various topographic features: gently rolling electrical hills and valleys represent the normal EEG background, whereas deep chasms can be likened to epileptiform discharges plunging down from the background. With this image in mind, one can understand the changes seen on either a referential or bipolar montage.

To take an example, consider an epileptiform discharge producing an electrical 'chasm' on the left hemisphere that is large enough to subtend electrodes F_3, C_3, and P_3, as illustrated in Figure 2.6. Keep in mind also that, in this example, the 'walls' of the chasm, rather than going straight up and down, instead gently slope down to the greatest electrical depth. Thus, proceeding from F_{p1} to F_3 the depth falls, from F_3 to C_3 it continues to fall to its nadir, from C_3 to P_3 it rises, and from P_3 to O_1 it continues to rise back to the surface. Assume, for the purpose of this example, that the gently rolling landscape exists at an electrical potential of $-25\,\mu V$, and that this is what electrodes F_{p1}, O_1, and A_1'see'. Furthermore, assume also that electrode F_3, being over the gently downsloping wall of the chasm, sees a potential of $-50\,\mu V$, and that electrode C_3, being over the nadir of the chasm, sees a potential of $-100\,\mu V$. Electrode P_3, being over the following wall of the chasm, sees $-50\,\mu V$, and electrode O_1 encompasses the normal landscape of $-25\,\mu V$. Depending on whether a referential or bipolar montage is used, the EEG recording of this same landscape will look quite different.

As noted earlier, in a referential recording each scalp, or active electrode, is paired with the same reference electrode, in this example the ipsilateral ear; thus, in this example, as illustrated in Figure 2.7, there are five channels to consider: F_{p1}–A_1, F_3–A_1, C_3–A_1, P_3–A_1, and O_1–A_1. Channel F_{p1}–A_1, with both electrodes 'seeing' the same potential, would register no difference; channel F_3–A_1 would see a difference of $25\,\mu V$ (i.e. looking up from a depth of -50 to the surface, which is at -25); channel C_3–A_1 would see a difference of $75\,\mu V$ (looking up from a depth of -100 to the surface at -25); channel P_3–A_1 a difference of $25\,\mu V$ (looking up from -50 to -25), and the last channel, O_1–A_1, with both electrodes seeing $-25\,\mu V$, would register 0. As seen in Figure 2.7, the greatest pen deflection is seen in the channel containing the electrode, in this example C_3, which lies over the deepest part of the electrical chasm. Thus, with referential recordings, it is the channel showing the greatest amplitude that serves to localize the focus of the electrical paroxysm.

It may be noted that the pen deflections in channels F_3–A_1, C_3–A_1, and P_3–A_1 are all positive, and this is according to the convention (Knott 1985) that whenever, in going from the first to second lead of any channel, one is 'looking' up, the pen goes up, but if one is 'looking' down the pen, likewise, goes down.

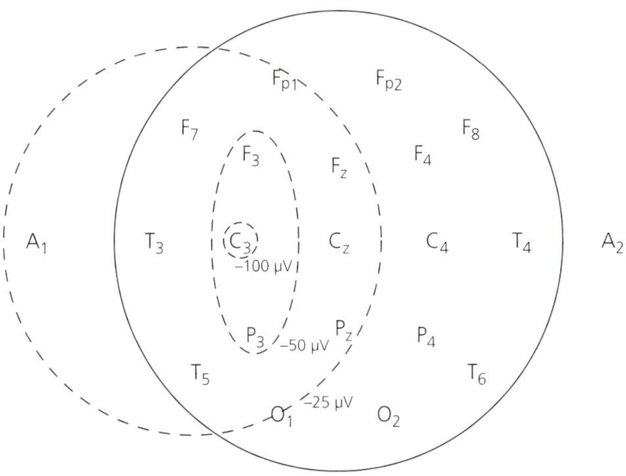

Figure 2.6 Surface-negative epileptiform discharge, of greatest extent at electrode C_3 (see text for details).

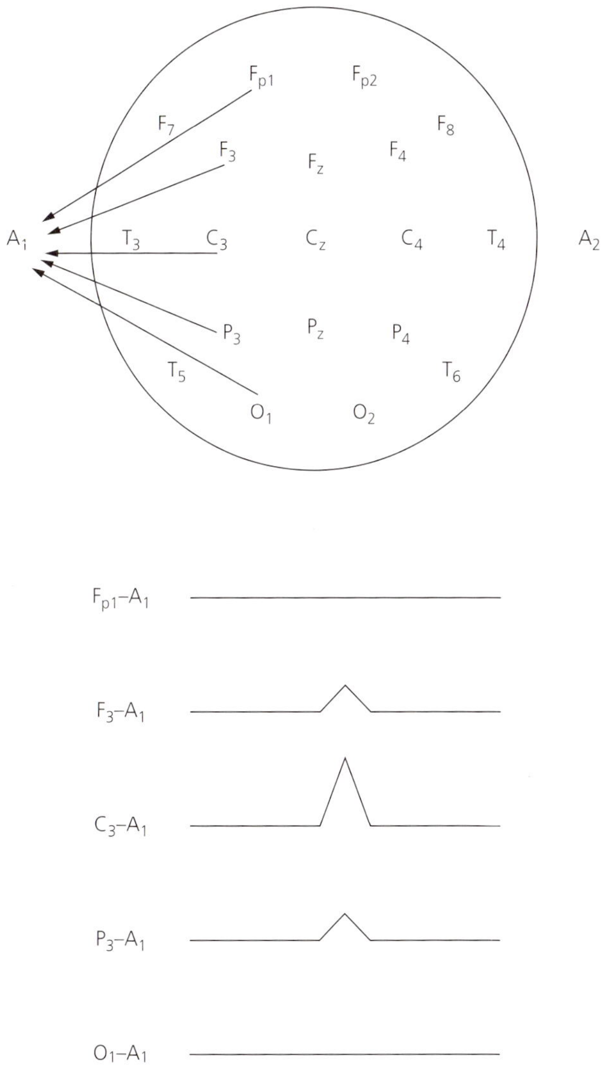

Figure 2.7 Referential recording of the epileptiform discharge shown in Figure 2.6 (see text for details).

The situation with bipolar recordings is quite different: here, it is not amplitude that is important but a phenomenon known as phase reversal (Knott 1985; Lesser 1985). Take the same example of an electrical paroxysm as used above, but this time cover it, as illustrated in Figure 2.8, with a longitudinal chain of electrodes, starting at F_{p1} and including, sequentially, F_3, C_3, P_3, and O_1. Then construct the following channels: F_{p1}–F_3, F_3–C_3, C_3–P_3, and, finally, P_3–O_1. Now consider what each channel will record. For channel F_{p1}–F_3, one looks down from F_{p1} at -25 to F_3 at -50, for a difference of $-25\,\mu V$. For the next channel, F_3–C_3, one continues to look down into the electrical chasm, now looking down from -50 to -100, for a difference of $-50\,\mu V$. At the next channel, C_3–P_3, however, something very different happens; here, standing at the nadir of the chasm at -100, one is looking 'up' to -50, for a difference of $+50\,\mu V$. Similarly, for the next channel, P_3–O_1, one continues to look up, but here from -50 to -25, for a difference of $+25\,\mu V$.

Figure 2.8 shows the various pen tracings seen for each channel. As may be noted, both channels F_{p1}–F_3 and F_3–C_3 show a downward or negative pen deflection. What happens next, however, is most critical: the next two channels, C_3–P_3 and P_3–O_1, both show an upward or positive deflection. It is apparent here that there has been a phase reversal as one goes

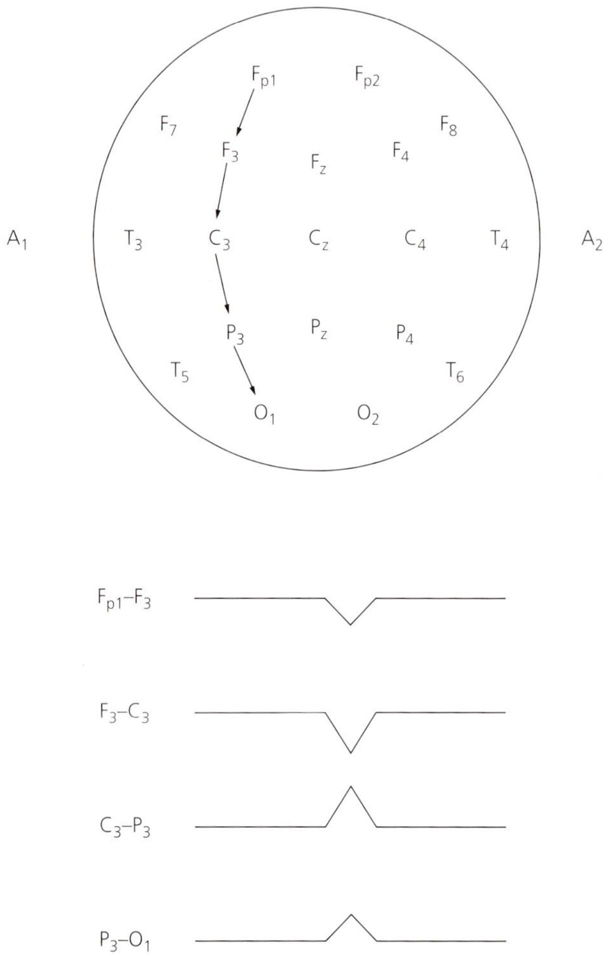

Figure 2.8 Bipolar recording of the epileptiform discharge shown in Figure 2.6 (see text for details).

from channel F_3–C_3 to channel C_3–P_3. This indicates that, in going from channel F_3–C_3 to channel C_3–P_3, one has 'crossed' over the depth of the electrical chasm; furthermore, as the electrode that both these channels have in common is C_3, it is now clear that the depth of the chasm lies under that electrode; that is to say phase reversal is seen at electrode C_3.

In some cases, focal epileptiform activity will not exhibit phase reversal with a bipolar montage. Specifically, when the focus is either proximal to the start of the chain or distal to its end, phase reversal is not possible. For example, consider a longitudinal chain linking F_{p1}, F_3, C_3, P_3, and O_1, and then imagine that the focus is located anterior to F_{p1}. In this case, all the pen deflections will be positive. Conversely, if the focus were distal to O_1, all the pen deflections would be negative.

Ictal activity

Ictal discharges consist of rhythmic activity which, unlike interictal epileptiform discharges, is sustained, at the very least, for a matter of seconds, with most ictal discharges lasting minutes. These ictal discharges may be either generalized at the onset or display a focal onset.

Ictal discharges that are generalized from the onset are typically seen in the idiopathic generalized epilepsies: myoclonic seizures are typically accompanied by polyspike or polyspike-and-wave sustained discharges, and petit mal seizures by sustained spike-and-dome discharges very similar to those seen interictally (Browne *et al.* 1974).

Ictal discharges that are focal at the onset may remain so for the duration of the seizure, and in most such cases, clinically one sees a simple partial seizure of one sort or the other. In cases when there is spread to a more or less focal area contralaterally a complex partial seizure is often seen, and in those cases when the generalization involves the entire cortex, a grand mal seizure is the typical accompaniment. Although in some cases in which there was some preceding interictal epileptiform activity the ictal discharges may resemble the interictal ones, in most cases ictal discharges are morphologically different (Blume *et al.* 1984; Geiger and Horner 1978). In general, the ictal activity is rhythmic and may occur at any frequency, from delta to beta; rarely, instead of primarily a change in frequency, one may see a change in amplitude, namely an 'electrodecremental' pattern in which the seizure is accompanied only by a paroxysmal loss of amplitude.

Remarkably, in the case of partial seizures, the surface EEG may remain normal, even in the face of indisputable seizures. In the case of simple partial seizures (Cockerell *et al.* 1996; Devinsky *et al.* 1989; Seshia and McLachlan 2005) this is seen in the majority; however, with complex partial seizures it is noted in only a very small percentage (Lieb *et al.* 1976).

PERIODIC COMPLEXES

Periodic complexes generally consist of one or more sharp waves combined with one or more slow waves that recur on a regular basis, at intervals ranging from 1 to 15 seconds, often on a background of generalized slowing. Although they may begin with a focal predominance, they fairly soon become generalized and synchronous, often with a frontal prominence. Such periodic complexes are often associated with myoclonus and are classically seen in such disorders as subacute sclerosing panencephalitis (Cobb 1966; Cobb and Hill 1950) and Creutzfeldt–Jakob disease (Aguglia *et al.* 1987; Burger *et al.* 1972; Chiofalo *et al.* 1980; Levy *et al.* 1986; Steinhoff *et al.* 1996). Importantly, although almost all patients with Creutzfeldt–Jakob disease eventually develop periodic complexes (Browne *et al.* 1986) (generally within the first 3 months [Levy *et al.* 1986]), these may be absent (Bortone *et al.* 1994; Zochodne *et al.* 1988), and this appears to be particularly the case with new-variant Creutzfeldt–Jakob disease (Will *et al.* 1996). Periodic complexes have also been noted in various other conditions, including herpes simplex viral encephalitis (Upton and Gumpert 1970), post-anoxic encephalopathy (Hockaday *et al.* 1965), Alzheimer's disease, dementia with Lewy bodies (Doran and Larner 2004; Tschampa *et al.* 2001), and baclofen intoxication (Hormes *et al.* 1988).

PERIODIC EPILEPTIFORM DISCHARGES

Periodic epileptiform discharges (PEDS), as the name clearly suggests, consist of epileptiform discharges, including either spikes or sharp waves, which in turn occur with a fairly regular periodicity, which varies from 0.5 to 5 seconds. PEDS may occur in one of three forms. In the most common form, namely periodic lateralized epileptiform discharges (PLEDS), the discharges, as the name suggests, occur in a lateralized form, on one side only. The next most common form is known as bilateral independent periodic lateralized epileptiform discharges (BIPLEDS), and here the discharges occur bilaterally; however, importantly, these bilateral discharges are not synchronous but occur independently. The least most common type consists of generalized periodic epileptiform discharges (GPEDS), and here the epileptiform discharges occur in a generalized fashion, synchronously in both hemispheres.

PLEDS (Chatrian *et al.* 1964; Chu 1980; Fitzpatrick and Lowry 2007; Garcia-Morales *et al.* 2002; Gurer *et al.* 2004; Markand and Daly 1971; Orta *et al.* 2009) are most commonly seen secondary to infarction, tumors, abscesses, intracerebral hemorrhage, herpes simplex viral encephalitis, and subdural hematoma. Other causes include Creutzfeldt–Jakob disease, posthypoxic–ischemic encephalopathy, and, interestingly, alcohol withdrawal. In cases where, as is most common, a structural lesion is responsible, PLEDS occurs ipsilateral to the lesion (Raroque *et al.* 1993); furthermore, although lesions confined to the white matter have been reported to cause PLEDS, this is very uncommon, and in almost all cases the lesion is found to affect either the gray matter alone, or the gray matter and subjacent white matter. Importantly, although the great majority of patients with PLEDS have seizures (Baykan *et al.* 2000; Orta *et al.* 2009), the periodic discharges themselves do not constitute ictal discharges: although there are exceptions (Singh *et al.* 2005), in the vast majority of cases, when it comes to seizures, PLEDS are innocent bystanders (Garcia-Morales *et al.* 2002).

BIPLEDS (de la Paz and Brenner 1981; Orta *et al.* 2009), as might be expected, occurs secondary to conditions which affect both hemispheres (Raroque *et al.* 1993), and common causes include posthypoxic–ischemic encephalopathy and herpes simplex viral encephalitis. Focal lesions, however, may also cause BIPLEDS, provided, however, that both hemispheres are involved, as has been noted with bilateral infarctions of the frontal (Nicolai *et al.* 2001) or temporal lobes (Fushimi *et al.* 2003). Seizures may occur in patients with BIPLEDS; however, this is not as common as with PLEDS, occurring in about only one-half of all cases.

GPEDS (Husain *et al.* 1999; Orta *et al.* 2009), as with BIPLEDS, occurs secondary to bihemispheric disease, and may be seen in toxic–metabolic encephalopathies, post-hypoxic–ischemic encephalopathy, and with infarctions. Seizures are less common in GPEDS than in PLEDS or BIPLEDS; however, these patients are typically critically ill, and most are in a coma.

TRIPHASIC WAVES

Triphasic waves are slow waves that, as the name indicates, possess a triphasic morphology. They are characteristically long-lasting, and typically occur in a generalized, bilaterally synchronous fashion, often with a frontal predominance. Although classically associated with hepatic encephalopathy (Summerskill *et al.* 1956), they also occur in other metabolic encephalopathies (Bahamon-Dussan *et al.* 1989; Fisch and Klass 1988; Karnaze and Bickford 1984) including uremic encephalopathy and those associated with hypernatremia, hyponatremia and hypoglycemia. Less commonly, they may occur in toxic encephalopathies, as, for example, with levetiracetam (Vulliemoz *et al.* 2009) or lithium (Kaplan and Birbeck 2006), and rarely with other disorders, including Hashimoto's encephalopathy (Henchey *et al.* 1995), post-hypoxic–ischemic encephalopathy (Bahamon-Dussan *et al.* 1989; Karnaze and Bickford 1984) the persistent vegetative state (Kulkarni *et al.* 2007), acute viral encephalitis (Rodriguez and Westmoreland 2007), and bilateral thalamic infarction (Kenangil *et al.* 2008).

BURST–SUPPRESSION

The burst–suppression pattern is characterized by bursts of generalized, bilaterally symmetric and synchronous delta activity, lasting on the order of 1–3 seconds and occurring every 3–10 seconds, in between which the background activity is suppressed to a very low amplitude or, at a normal sensitivity, to a flat line. This pattern is seen in states of severe cortical dysfunction, for example subacute sclerosing panencephalitis (Markand and Panzi 1975), viral encephalitis, or postanoxic coma (Hockaday *et al.* 1965). Although it is typically bilateral, it may occasionally be seen unilaterally, as, for example, after a very large infarction.

Activation procedures

Hyperventilation, photic stimulation, and sleep are all considered to be activation procedures in that they may activate certain abnormalities that would otherwise not be apparent on the routine EEG. In patients suspected of having reflex seizures, exposing the patient to the putative epileptogenic stimulus may also be considered in order to activate the focus.

HYPERVENTILATION

Hyperventilation is normally followed by a build-up of generalized, bilaterally symmetric and synchronous high-amplitude slow waves, maximal frontally in adults (Goldberg and Strauss 1959). Abnormalities that may occur include asymmetric or focal slowing and epileptiform abnormalities. Asymmetric or focal slowing has the same significance as spontaneous slowing, as discussed previously in this chapter. Epileptiform abnormalities are very common in the case of petit mal seizures, with most of these patients showing typical 3-Hz spike-and-dome epileptiform changes (Dalby 1969); indeed, in these patients hyperventilation is a more sensitive diagnostic procedure than a 6-hour recording (Adams and Lueders 1981). In the case of complex partial seizures, however, only a small minority of cases will display activation (Adams and Lueders 1981; Gabor and Ajmone-Marsan 1969; Miley and Forster 1977; Morgan and Scott 1970). In addition to IEDs, seizures may also be induced by hyperventilation, but this is very uncommon (Holmes *et al.* 2004), except in patients with treatment-resistant focal seizures, in whom it occurred in one-quarter of all patients (Guaranha *et al.* 2005).

Hyperventilation is generally contraindicated in patients with sickle cell disease (Protass 1978) or those with significant cerebrovascular or cardiovascular disease.

PHOTIC STIMULATION

Photic stimulation is accomplished by positioning a stroboscopic light about 30 cm from the patient's face, the eyes being either open or closed. The light is then flashed at various frequencies (e.g. 3, 5, 10, 13, 15, 17, 20, and 25 Hz), each frequency being allotted about 10 seconds. In about two-thirds of normal adults this stroboscopic illumination will produce the photic driving response, wherein bilaterally symmetric and synchronous waves appear at a frequency equal to either the stroboscopic frequency or some harmonic of it (Hughes 1960). Normally, although the photic driving is maximal occipitally, it may extend to the parietal or temporal area. Maximal photic driving is generally seen when the strobe frequency is close to the patient's normal alpha rhythm and, as with the alpha rhythm, the amplitude of the photic driving response is often lower on the left side. With very-high-frequency stroboscopic activity the resultant wave forms may resemble spikes. Abnormalities seen with photic driving include unexpected amplitude asymmetries, photomyoclonic responses, and photoparoxysmal responses. Amplitude asymmetries, wherein the left side has an amplitude of less than 50 percent that of the right, or wherein the right side has an amplitude less than that of the left, indicate a definite abnormality. The photomyoclonic response (Meier-Ewert and Broughton 1976) consists of myoclonic twitching of the eyelids and, in severe cases, myoclonus of the head and neck; it may be seen in a very small minority of normal individuals (Kooi *et al.* 1960) and more commonly in those withdrawing from alcohol or sedative hypnotics (Fisch *et al.* 1989; Gastaut *et al.* 1958). The photoparoxysmal response consists of epileptiform changes (Bickford *et al.* 1952) and is most common in those with petit mal or one of the other idiopathic generalized epilepsies (Stevens 1962; Wolf and Gooses 1986). A minority of patients with the photoparoxysmal response will experience a seizure during the photic stimulation (Seddigh *et al.* 1999).

SLEEP

Sleep may activate epileptiform activity (Sammaritano *et al.* 1991); indeed, IEDs are seen only during sleep in up to one-third of patients with complex partial seizures (Niedermeyer and Rocca 1972). The sleep portion of the recording should last at least 20 minutes and include both stages I and II of sleep. In some cases, patients will simply drift off to sleep on their own at the end of the awake recording session, whereas in others the administration of chloral hydrate will be required. Some authors recommend sleep deprivation not only to ensure that the patient falls asleep during the recording, but also in the belief that sleep deprivation *per se* is activating: this, however, is a controversial notion, supported by some (Ellingson *et al.* 1984; Leach *et al.* 2006; Mattson *et al.* 1965), but not all, studies (Pratt *et al.* 1968; Veldhuizen *et al.* 1983). In interpreting a sleep EEG it is important to distinguish epileptiform activity from normally occurring POSTs and vertex sharp waves. In addition to revealing epileptiform abnormalities, sleep recordings may, at times, reveal other abnormalities, for example the early onset of REM sleep as may be seen in narcolepsy or in alcohol or anxiolytic withdrawal (Kales *et al.* 1974).

REFLEX ACTIVATION

In cases of definite or suspected 'reflex' epilepsy, it may be appropriate, as discussed in Chapter 8.3, to expose the patient to the triggering event itself. Thus, in 'musicogenic' epilepsy, the appropriate tune may be played, and in 'reading' epilepsy, the appropriate passage read, and so on. Consideration must, of course, be given to the risk of inducing a seizure during the recording.

Normal variants

Normal variants may resemble epileptiform changes (e.g. small sharp spikes [SSS]) or pathologic slow waves (e.g. subclinical rhythmic EEG discharges of adults [SREDA]).

Small sharp spikes, also known as benign epileptiform transients of sleep (BETS), are, as the name suggests, low-amplitude, very sharp spikes (i.e. less than 50 μV in amplitude and less than 50 ms in duration) that are seen intermittently during drowsiness in both temporal areas, either intermittently or in a bilaterally synchronous fashion (Klass and Westmoreland 1985; White *et al.* 1977).

Phantom spike-and-wave (also known as 6-cps spike-and-wave) (Klass and Westmoreland 1985; Thomas and Klass 1968) is characterized by brief trains of rhythmic 6-Hz activity that are generalized, bilaterally symmetric, and synchronous, generally lasting no longer than a second or two. The name comes from the fact that the rhythmic activity is composed of a peculiar spike-and-wave complex wherein the spike is of such relatively low amplitude and brevity that, next to the much more prominent wave, it, like the 'phantom', is rarely seen.

'Occipital spikes of blind persons', an aptly named electrographic syndrome, may be seen in patients with congenital or acquired blindness and is characterized by intermittent spikes confined to the occipital area.

Subclinical rhythmic EEG discharges of adults (O'Brien *et al.* 1998; Westmoreland and Klass 1981) is characterized by lengthy trains of rhythmic theta or, less commonly, delta activity, seen best in the centroparietal areas. The trains themselves are often of abrupt onset and may be preceded by some sharp waves. They generally last about a minute and are most commonly seen with hyperventilation in the elderly.

Rhythmic mid-temporal discharges (Gibbs *et al.* 1963; Klass and Westmoreland 1985), also known as rhythmic theta bursts of drowsiness, consist of lengthy trains of rhythmic theta activity that occur in both temporal areas, either synchronously or independently. The trains themselves often last about 10 seconds but they may endure for as long as a minute. This variant was once referred to as the psychomotor variant but this terminology has been abandoned as there is no connection between rhythmic mid-temporal discharges and complex partial seizures.

Ctenoids, also known as '14- and 6-cps positive bursts' (Klass and Westmoreland 1985; Lombroso *et al.* 1966) consist of brief trains, lasting 0.5–1 second, of 6- or 14-Hz rhythmic waves that occur in the temporal regions either independently or in a bilaterally synchronous fashion. They are best seen during stages I and II of sleep, and the waves themselves have a distinctive arciform morphology.

Wicket spikes (Klass and Westmoreland 1985; Reiher and Lebel 1977) consist of brief trains of rhythmic activity at a frequency of 6–11 Hz, which are most prominent in the temporal regions where they may appear independently or in a bilaterally synchronous fashion. Like ctenoids, the waves themselves have an arciform morphology. In some cases, rather than trains, wicket spikes may occur in an isolated fashion: such isolated wicket spikes may be distinguished from IEDs by the absence of a following slow wave (Krauss *et al.* 2005).

Lambda waves (Barlow and Ciganek 1969; Evans 1953; Green 1957; Scott *et al.* 1967) are isolated, bilaterally synchronous occipital waves that occur just after saccadic eye movements made by patients as they scan a detailed scene or picture. The waves themselves are of 20–50 μV in amplitude and 200–300 ms in duration; they have a characteristic triangular or sawtooth-shaped morphology. These may be eliminated by having the patient stare at a blank sheet of paper, and this technique should be utilized and noted by the EEG technologist.

Artifacts

Artifacts may be grouped according to the EEG activity that they most closely resemble, including interictal epileptiform discharges, slow waves, or decreased amplitude. There is also a 60-cps artifact, which is usually readily identified as it does not resemble any naturally occurring EEG activity.

RESEMBLING INTERICTAL EPILEPTIFORM DISCHARGES

Drip artifact (Linnenger *et al.* 1981) reflects the electrical disturbance occurring each time a drip occurs in an intravenous line. It may appear so similar to an epileptiform discharge that its correct identification may depend on the EEG technician noting on the record when the drips occur.

Electrode pop reflects, as it were, a sudden 'spark' at an individual electrode, often occurring secondary to some impurity. Although it strongly resembles an epileptiform spike, the fact that it is restricted to one electrode betrays its artifactual nature, as true epileptiform spikes are almost always seen at more than one electrode.

Electrocardiographic (ECG) artifact represents cardiac electrical activity picked up by the EEG. In the case of a regular cardiac rhythm, its artifactual nature is immediately obvious as epileptiform spikes simply do not occur with such monotonous regularity. When irregularly occurring premature ventricular contractions are present, however, the distinction may be more difficult and indeed may depend on simultaneously recording the ECG.

Muscle artifact reflects electrical activity arising from the contraction of the scalp musculature and may appear in any one of three ways: as isolated irregularly appearing 'blips', or as 'blips' superimposed on a 'muddy' black line, the blips occurring either arrhythmically or rhythmically. The key to the identification of muscle 'blips' is their extreme brevity.

RESEMBLING SLOW WAVES

Eye movements may induce artifacts (Peters 1967). In the case of vertical movements, the artifact is seen in the frontopolar leads and, if horizontal, in the anterior temporal leads. Heuristically, the eye may be considered to be a 'battery', the cornea being positive and the retina negative: thus, whichever electrode the eye 'looks' toward becomes more positive, and whichever one the eyes 'look' away from becomes, conversely, more negative. Consider, for example, the results with a longitudinal bipolar montage when the patient engages in horizontal eye movement looking to the left: F_7 becomes more positive relative to T_3, and thus, in the channel F_7–T_3 the pen deflection is downwards; conversely, F_8 becomes more negative relative to T_4, and thus, in channel F_8–T_4, the pen deflection is up. Next, consider the situation with vertical eye movements, as occurs during a blink (Matsuo *et al.* 1975), recalling that when a blink occurs, the eyes, in Bell's phenomenon, undergo upward rotation. In this case, both frontopolar electrodes, F_{p1} and F_{p2}, become positive relative to their immediate neighbors, F_3 and F_4, and thus channels F_{p1}–F_3 and F_{p2}–F_4 both show a negative deflection. Eye movement artifacts, whether horizontal or vertical, are suggested by their bilateral synchrony. In the case of horizontal movements, a further indication is the fact that the resulting pen deflections on either side are in opposite directions. In the case of vertical eye movements occurring with blinking, the occurrence of eyelid flutter may make identification a little more difficult because the resulting artifact will appear similar to a bifrontal slow-wave focus.

Movement artifact occurs when the electrodes are actually moved, as most commonly occurs when the occipital electrodes (pressed between the patient's head and the underlying pillow) are slightly dislodged by the minimal head movements that occur with either respiration or, when present, tremor. The regularity of the resulting artifact, the identity of its frequency with the respiratory rate or the tremor frequency, and its restriction to the occipital leads all highlight its true nature.

Glossokinetic artifact occurs with tongue movement. The tip of the tongue is negative, and its up and down movement may produce slow waves in the frontal or temporal areas. Asking the patient to say 'la la la' during the recording allows for ready identification of this artifact.

Pulse artifact occurs in situations when an electrode is accidentally placed over a relatively large scalp artery, which, with every passing pulse, slightly moves the electrode resting on it. The fact that the resulting artifact occurs at a regular rate, identical to the cardiac rate, suggests its true identity: in doubtful cases, a simultaneous ECG will reveal that the pulse artifact follows, after a slight delay, every QRS complex, the delay reflecting the time required for the pulse to travel from the heart to the scalp.

Perspiration on the scalp, as may occur if the patient is febrile or anxious, both alters the resistance of the overlying electrodes and allows for some slight slippage between the electrodes and the scalp: the resulting artifact consists of very slow waves (e.g. 0.5 Hz) of very high amplitude, which occur in a generalized, bilateral, but asynchronous fashion.

RESEMBLING DECREASED AMPLITUDE

Increased electrode resistance may lead to what appears to be decreased amplitude at one electrode. Its restriction, however, to but one electrode betrays its artifactual nature as pathologic conditions capable of causing decreased amplitude are rarely so restricted in location that they will be reflected at only one electrode position.

Defective calibration of one channel may result in a decreased deflection of that channel's pen. Its isolation to one pen suggests the correct diagnosis; furthermore, the fact that the same pen continues to show decreased deflection with changing montages confirms the diagnosis.

SIXTY CYCLES PER SECOND ARTIFACT

A 60-Hz artifact occurs secondary to interference from a nearby alternating current source, typically appearing on the EEG as a thick 'muddy' line. If there is any doubt as to the source of such a 'muddy' line, lowering the paper speed to 15 mm/s will confirm the diagnosis by allowing the resolution of the line into orderly 60-Hz deflections.

2.6 LUMBAR PUNCTURE

In the not-too-distant past, lumbar puncture was a standard procedure, and it was the rare neuropsychiatric patient admitted to hospital who escaped this procedure. Although the need for this test has decreased dramatically with the advent of modern neuroimaging using CT and MRI, there are still certain clinical situations, as discussed below, in which it is vital. This chapter will discuss the indications and contraindications for lumbar puncture, its technique and complications, and the various tests, both standard and otherwise, that may be ordered.

Indications and contraindications

Lumbar puncture is indicated in cases of meningitis and acute encephalitis, and in those uncommon cases wherein subarachnoid hemorrhage is strongly suspected despite a negative CT scan. Lumbar puncture is also indicated when the following disorders are suspected: multiple sclerosis, neuromyelitis optica, lupus cerebritis, limbic encephalitis, neuroborreliosis, neurosyphilis, Creutzfeldt–Jakob disease, and benign intracranial hypertension.

Lumbar puncture is contraindicated when there is infection of any of the tissues near the site of the proposed puncture. Conditions that present a risk of bleeding during the puncture also constitute contraindications, and these include thrombocytopenia (with platelet counts below 50 000) and treatment with either heparin or warfarin, and consequently it is customary to obtain a platelet count and determine an international normalized ratio before proceeding. In cases when puncture must be performed, consideration may be given to use of platelet transfusion, protamine, or vitamin K. Evidence of increased intracranial pressure is often considered a contraindication, but this holds only where there is a risk of herniation, whether subfalcine, uncal, transtentorial, or cerebellar. Thus, the presence of mass lesions and large acute infarctions or intracranial hemorrhages generally argue against lumbar puncture. However, there are some cases of raised intracranial pressure wherein the risk of herniation is low, as, for example, in benign intracranial hypertension and some cases of subarachnoid hemorrhage, and in these instances lumbar puncture may be safely performed and indeed may be carried out therapeutically. As a rule, imaging with either CT or MRI should be performed in every patient before lumbar puncture: the only exception here would be when bacterial meningitis is strongly suspected and delay could prove fatal.

Technique

The proper technique for lumbar puncture has been reviewed by Roos (2003).The patient is placed in the left lateral decubitus position on a firm surface, with his or her back brought up to the edge of the bed or gurney. Pillows are placed under the head and neck as well as between the legs, and the back is kept strictly vertical by assuring that the right shoulder and hip are directly above the left shoulder and hip. The patient is then helped into a fetal position by flexing the head and legs, thus opening the spaces between the spinous processes. In adults, the needle should generally be inserted in the L3–L4 interspace, a spot which may be found by palpating the superior iliac crests and then drawing an imaginary line between them: the appropriate interspace generally lies just below this line. After palpating the appropriate area, the target may be marked by indenting the skin with a fingernail. After putting on a face mask, the area is then prepared with betadine and alcohol, and sterile drapes are placed. Most physicians will then infiltrate the superficial tissues with 0.5 mL of a local anesthetic; however, some, on the theory that this puncture causes as much pain as the actual lumbar puncture, will forego it. In order to reduce the risk of a postlumbar puncture headache (Evans *et al.* 2000), a 20- or 22-gauge needle should be used (higher gauge needles, up to 26 gauge, are even safer in this regard [Tourtellotte *et al.* 1972], but given their small size the flow of CSF may be quite sluggish), and the bevel of the needle should be parallel to the longitudinal axis of the patient, as this angle allows for separation, rather than laceration, of dural fibers. Atraumatic needles are even less prone to cause headache (Lavi *et al.* 2006; Strupp *et al.* 2001) but they are more difficult to use. The needle itself should be held perpendicular to the patient's back and angled slightly cephalad, generally toward the patient's umbilicus (with due regard for sagging umbilici in obese patients). The needle is then gently inserted: at a depth of between 4 and 5 cm, there is often a palpable 'give' as the dura is pierced, followed by a less obvious 'pop' when the arachnoid is punctured. Once these landmark events have transpired, the stylet should be slowly removed (rapid removal may suck a nerve rootlet into the lumen) and one should watch to see if any fluid emerges. 'Dry taps' generally indicate improper placement and, if the patient is tolerating the procedure, the stylet should be replaced, the needle gently withdrawn, and another attempt made. As fluid emerges, the patient may be allowed to relax and to slightly extend the legs. In cases when an opening pressure is desired, a manometer is attached to the needle and the pressure of water read (in centimeters). Given that compression of the abdomen may elevate the pressure, it is necessary for obese patients to relax their legs sufficiently to release any pressure on the abdomen before reading the pressure. In any case, cerebrospinal fluid (CSF) is then collected in three containers and these are then promptly sealed: for most purposes anywhere from a total of 8–15 mL is adequate; however, for some special studies (e.g. fungal cultures) up to 40 mL may be required. When sufficient fluid has been obtained, the stylet is replaced, the needle is withdrawn, and firm pressure is applied to the puncture site, followed by the application of a band aid. The patient may then be allowed to be up ad lib: bed rest at this point does not decrease the risk of a postlumbar puncture headache (Dieterich and Brandt 1985).

Recently, it has become common to perform lumbar puncture under fluoroscopic guidance, with the procedure itself performed by a radiologist.

Complications

The most common complication of a lumbar puncture is headache, which may occur in about one-quarter of all cases. The headache occurs typically a day or two after the puncture, and may last from days to a couple of weeks: it is generalized and may be either steady or throbbing in character, and is worse when sitting or standing and generally relieved by recumbancy. Treatment involves bed rest and analgesics; intravenous caffeine sodium benzoate, in a dose of 500 mg diluted in 1000 mL of normal saline and given over 1 hour, may also be used and is typically dramatically effective (Choi *et al.* 1996). If these fail, and the headache is severe, a 'blood patch' may be placed by injecting about 10 mL of the patient's blood epidurally adjacent to the puncture site: this procedure, however, should not be routine for it may cause a radiculopathy in up to one-third of all cases. The mechanism underlying the headache involves leakage of fluid through the dural hole, which allows the brain to 'sag', thus stretching pain-sensitive structures. This sagging may also stretch cranial nerves, resulting in tinnitus, deafness, or diplopia; rarely, these cranial nerve palsies may occur in the absence of headache.

Other complications include backache, radiculopathy, spinal hematoma with paraparesis, meningitis, and, as noted above, herniation.

Standard measurements

A number of determinations are considered standard and should be requested in all cases. These include the appearance of the fluid, cell count and differential, total protein, glucose, Gramstain and appropriate culture, and tests for syphilis. Measurement of the opening pressure was once standard but is now rarely required.

APPEARANCE

Normal CSF is crystal clear. A 'ground glass' appearance may occur in the presence of 400 or more red blood cells (RBCs), whereas RBC counts over 1000 may impart a pinkish cast to the fluid. A hazy fluid may occur in the presence of 200 or more white cells (WBCs), and larger WBC counts may make the fluid cloudy in appearance.

Xanthochromia, or a yellowish discoloration, may occur in any situation where the total protein is over 150 mg/dL. Such a discoloration may also occur in jaundice (serum bilrubin >6 mg/dL), cases of carotenemia and in patients treated with rifampin. Another very important cause of xanthochromia is hemolysis, wherein lysed red cells leak hemoglobin breakdown products, such as bilirubin, into the CSF. In subarachnoid hemorrhage, the xanthochromia gradually appears within 2–4 hours of the hemorrhage, and is present in over 90 percent of cases within 12 hours.

Frankly bloody CSF may occur either in cases of subarachnoid hemorrhage or after a traumatic, or 'bloody', tap. Traumatic taps occur in about 10 percent of lumbar punctures and, rather than reflecting poor technique, they generally result from the accidental nicking of a vein in the epidural plexus. These two possibilities may be differentiated by the 'three-tube test', wherein one simply watches for any change in appearance between the first, second, and third tubes collected. In subarachnoid hemorrhage, the fluid in the third tube is as bloody as that in the first, whereas in a traumatic tap there is substantial clearing of the fluid by the third tube. These two conditions may also be distinguished by centrifuging the first tube, and inspecting the supernatant. In traumatic taps the supernatant is clear, whereas in subarachnoid hemorrhage the presence of hemolyzed red cells, as noted above, creates xanthochromia. In interpreting this test, however, it must be kept in mind that in cases of subarachnoid hemorrhage xanthochromia, as noted earlier, may be absent for the first few hours.

CELL COUNT AND DIFFERENTIAL

Normal CSF is free of RBCs but may contain up to five WBCs, either lymphocytic or mononuclear. Red blood cells may be seen in subarachnoid hemorrhage and cases of either contusion or intracerebral/intraventricular hemorrhage.

A pleocytosis, that is an increased number of WBCs, is seen in a large number of conditions. Note that a polymorphonuclear predominance generally indicates a bacterial infection: although viral, fungal, or tubercular infections may initially display a polymorphonuclear predominance, a lymphocytic one is far more common, especially after the infection has passed the acute stage.

When the CSF is frankly bloody, as in subarachnoid hemorrhage or with a traumatic tap, the CSF WBC is of course elevated by white cells derived directly from the systemic circulation. As a rule of thumb, one may correct for this by subtracting one white cell from the total CSF count for every 1000 red cells.

TOTAL PROTEIN

Most of the protein in the CSF consists of albumin; other constituents include gamma-globulins, other globulins (alpha and beta) and prealbumin. The total protein in normal CSF ranges from 15 to 45 mg/dL.

When the fluid is grossly bloody, the total protein is corrected by subtracting about 1 mg for each 1000 RBCs.

GLUCOSE

Normally, CSF glucose runs anywhere from 50 to 65 percent of serum glucose, and hence normal CSF values range from 45 to 80 mg/dL. In all cases, the serum glucose should be measured, however, given that 2–4 hours are required for equilibration between serum and CSF, the timing of the blood draw varies according to the fed state of the patient: when the patient is fasting, the blood draw may be just before the puncture; when the patient has eaten, however, the blood draw should be 2–4 hours before the scheduled puncture.

Elevated CSF glucose is of no concern. Reductions in CSF glucose may occur with hypoglycemia, in bacterial or fungal meningitides, or other inflammatory disorders.

STAINS AND CULTURES

The CSF should always be Gram-stained and, whenever tuberculosis is suspected, an acid-fast stain should also be requested. An 'India ink preparation' is traditional when cryptococcal infection is suspected; however, the availability of polymerase chain reaction (PCR) assays has made this almost superfluous. Appropriate cultures should also be requested, whether bacterial, tubercular, or fungal.

TESTS FOR SYPHILIS

It is good practice to always request a CSF Venereal Disease Research Laboratory (CSF-VDRL) test with titer. Furthermore, in cases where the clinical picture is consistent with neurosyphilis but the CSF-VDRL test is negative, it is entirely appropriate to obtain a CSF fluorescent treponemal antibody (CSF-FTA) test, given that the CSF-FTA is a far more sensitive test than the CSF-VDRL (Davis and Schmitt 1989; Timmermans and Carr 2004).

OPENING PRESSURE

Measurement of the opening pressure, although once considered standard, is rarely required in modern practice: one exception consists of cases of suspected benign intracranial hypertension. The normal opening pressure ranges from 6 to 20 cm of water; values of below 5 cm may be seen with systemic dehydration, subarachnoid block, or following a prior lumbar puncture. Elevations of over 20 cm are seen in cases of benign intracranial hypertension, meningitis, cerebral venous occlusion, and with mass lesions, hemorrhages, or acute infarctions.

OTHER DETERMINATIONS

Other determinations are requested depending on one's diagnostic suspicions, and may include polymerase chain reaction (PCR) assay and testing for various antibodies and antigens: in these regards, consultation with an infectious disease specialist, if not already in the works, is always appropriate, given the rapid advances made in these fields. Other tests to consider include the immunoglobulin G (IgG) index, oligoclonal bands, myelin basic protein and the 14–3–3 protein.

Polymerase chain reaction assay represents a rapid, sensitive, and specific means for detecting various infections. Viruses detected include herpes simplex, varicella zoster, Epstein–Barr, JC, human immunodeficiency virus (HIV), and cytomegalovirus (CMV) (Steiner *et al.* 2005). Furthermore, almost all bacteria may also be identified (Poppert *et al.* 2005). Fungal and tubercular infections may also be detected.

Antibodies to arboviruses, rabies virus, *Borrelia burgdorferi* and *Coccidioides immitis* may be tested for, as may antigens specific for *Cryptococcus* and *Histoplasma*, and these should be ordered in the appropriate clinical setting.

The IgG index provides a reliable measure of increased production of IgG within the CNS. Elevations of total IgG itself may represent a false-positive, given that the immunoglobulin level in the CSF reflects not only intrathecal synthesis of globulin, but also transfer of globulins from the systemic circulation. Consequently, it is necessary to correct the CSF level by obtaining what is known as the 'IgG index'. The IgG index equals [IgG(CSF)/IgG(serum)]/[albumin(CSF)/albumin(serum)].

IgG index values above 0.65 indicate intrathecal synthesis. To calculate this, of course, one must also obtain blood for a serum protein electrophoresis, and this should be done just before the lumbar puncture.

Oligoclonal bands, that is to say anywhere from three to five bands seen on electrophoresis, are seen in a large number of autoimmune or infectious disorders, including multiple sclerosis, neuromyelitis optica, acute disseminated encephalomyelitis, systemic lupus erythematosus, limbic encephalitis, Hashimoto's encephalopathy, AIDS dementia, neuroborreliosis, and tertiary neurosyphilis.

Myelin basic protein is found in diseases characterized by myelin breakdown, such as multiple sclerosis. An increased myelin basic protein, however, is certainly not specific for multiple sclerosis, being found in various other conditions, such as encephalitis, lupus, vasculitidies, and recent infarction.

The 14–3–3 protein should be tested for in suspected cases of Creutzfeldt–Jakob disease. Although this is very sensitive for Creutzfeldt–Jakob disease, false-positives have been noted in a large number of conditions (Lemstra *et al.* 2000; Martinez-Yelamos *et al.* 2001; Van Everbroeck *et al.* 2005), such as encephalitis, recent infarction, multiple sclerosis, Hashimoto's encephalopathy, limbic encephalitis, astrocytoma, and even degenerative disorders such as Alzheimer's disease, diffuse Lewy body disease, and amyotrophic lateral scelerosis.

REFERENCES

Adams DJ, Lueders H. Hyperventilation and 6-hour EEG recording in evaluation of absence seizures. *Neurology* 1981; **31**:1175–7.

Adams DJ, Lueders H, Pippinger CH. Sodium valproate in the treatment of intractable seizure disorders: a clinical and electroencephalographic study. *Neurology* 1978; **28**:152–7.

Aguglia U, Farnarier G, Tinuper P *et al.* Subacute spongiform encephalopathy with periodic paroxysmal activities: clinical evaluation and serial EEG findings in 20 cases. *Clin Electroencephalogr* 1987; **18**:147–58.

Aird RB, Shimizu M. Neuropathological correlates of low-voltage EEG foci. *Arch Neurol* 1970; **22**:75–80.

Ajmone-Marsan C, Zivin LS. Factors related to the occurrence of typical paroxysmal abnormalities in the EEG records of epileptic patients. *Epilepsia* 1970; **11**:361–81.

Andreasen NC. Affective flattening and criteria for schizophrenia. *Am J Psychiatry* 1979; **136**:944–7.

Bahamon-Dussan JE, Celesia GG, Grigg-Damberger MM. Prognostic significance of EEG triphasic waves in patients with altered state of consciousness. *J Clin Neurophysiol* 1989; **6**:313–19.

Barlow JS, Ciganek L. Lambda responses in relation to visual evoked responses in man. *Electroencephalogr Clin Neurophysiol* 1969; **26**:183–92.

Barnes MP, Saunders M, Walls TJ *et al.* The syndrome of Karl Ludwig Kahlbaum. *J Neurol Neurosurg Psychiatry* 1986; **49**:991–6.

Baykan B, Kinay D, Gokyigit A *et al.* Periodic lateralized epileptiform discharges: association with seizures. *Seizure* 2000; **9**:402–6.

Berger H. Uber das elektrenkephalogramm des menschen. *Archiv Psychiat Nervenkr* 1929; **87**:527–70.

Beschin N, Cubelli R, Della Sala S *et al.* Left of what? The role of egocentric coordinates in neglect. *J Neurol Neurosurg Psychiatry* 1997; **63**:483–9.

Betting LE, Mory SB, Lopes-Cendes I *et al.* EEG features in idiopathic generalized epilepsy: clues to diagnosis. *Epilepsia* 2006; **47**:523–8.

Bickerstaff ER. *Neurological Examination in Clinical Practice*, 4th edn. Oxford: Blackwell Scientific, 1980.

Bickford RG, Sem-Jacobsen CW, White PT *et al.* Some observations on the mechanism of photic and photo-Metrazol activation. *Electroencephalogr Clin Neurophsysiol* 1952; **4**:275–82.

Bleuler E. *Textbook of Psychiatry* (transl. AA Brill). New York: Macmillan, 1924. Reissued by Arno Press, New York, 1976.

Blume WT, Younq GB, Lemieux JF. EEG morphology of partial epileptic seizures. *Electroencephalogr Clin Neurophysiol* 1984; **57**:295–302.

Bortone E, Bettoni L, Giorgi C *et al.* Reliability of EEG in the diagnosis of Creutzfeldt–Jakob disease. *Electroencephalogr Clin Neurophysiol* 1994; **90**:323–30.

Brain W. *Clinical Neurology*, 2nd edn. London: Oxford University Press, 1964.

Briel RC, McKeith IG, Barber WA *et al.* EEG findings in dementia with Lewy bodies and Alzheimer's disease. *J Neurol Neurosurg Psychiatry* 1999; **66**:401–3.

Browne P, Cathala F, Castaigne P *et al.* Creutzfeldt–Jakob disease: clinical analysis of a consecutive series of 230 neuropathologically verified cases. *Ann Neurol* 1986; **20**:597–602.

Browne TR, Penry JK. Benzodiazepines in the treatment of epilepsy. *Epilepsia* 1973; **14**:277–310.

Browne TR, Penry JK, Porter RJ *et al.* Responsiveness before, during and after spike–wave paroxysms. *Neurology* 1974; **24**:659–65.

Burger LJ, Rowan AJ, Goldensohn ES. Creutzfeldt–Jakob disease. An electroencephalographic study. *Arch Neurol* 1972; **26**:428–33.

Calzetti S, Bortone E, Negrotti A *et al.* Frontal intermittent rhythmic delay activity (FIRDA) in patients with dementia with Lewy bodies: a diagnostic tool? *Neurol Sci* 2002; **23**(Suppl. 2):S65–6.

Caplan LR, Schoene WL. Clinical features of subcortical arteriosclerotic encephalopathy (Binswanger disease). *Neurology* 1978; **28**:1206–15.

Caton R. The electrical currents of the brain. *British Medical Journal* 1875; **2**:278.

Chatrian GE. Characteristics of unusual EEG patterns: incidence, significance. *Electroencephalogr Clin Neurophysiol* 1964; **17**:471–2.

Chatrian GE, Petersen MC, Lazarte JE. The blocking of the rolandic wicket rhythm and some central changes related to movement. *Electroencephalogr Clin Neurophysiol* 1959; **11**:497–510.

Chatrian GE, Shaw CM, Leffman H. The significance of periodic lateralized epileptiform discharges in EEG: an electrographic, clinical and pathological study. *Electroencephalogr Clin Neurophysiol* 1964; **17**:177–93.

Chatrian GE, Lettich F, Nelson PL. Ten percent electrode system for topographic studies of spontaneous and evoked EEG activities. *Am J EEG Technol* 1985; **25**:83–92.

Chiofalo N, Fuentes A, Galvez S. Serial EEG findings in 27 cases of Creutzfeldt–Jakob disease. *Arch Neurol* 1980; **37**:143–5.

Choi A, Laurito CE, Cunningham FE. Pharmacologic management of postdural puncture headache. *Ann Pharmacother* 1996; **30**:831–9.

Chu NS. Periodic lateralized epileptiform discharges with preexisting focal brain lesions: role of alcohol withdrawal and anoxic encephalopathy. *Arch Neurol* 1980; **37**:551–4.

Cobb WA. The periodic events of subacute sclerosing panencephalitis. *Electroencephalogr Clin Neurophysiol* 1966; **21**:278–94.

Cobb WA, Hill D. Electroencephalogram in subacute progressive encephalitis. *Brain* 1950; **73**:392–404.

Cobb WA, Guiloff RJ, Cast J. Breach rhythm: the EEG related to skull defects. *Electroencephalogr Clin Neurophsyiol* 1979; **47**:251–71.

Cockerell OC, Rothwell J, Thomson PD *et al.* Clinical and physiological features of epilepsia partialis continua. Cases ascertained in the UK. *Brain* 1996; **119**:393–407.

Cooper R, Winter AL, Crown JJ *et al.* Comparison of subcortical, cortical and scalp activity using indwelling electrodes in man. *Electroencephalogr Clin Neurophysiol* 1965; **18**:217–28.

Cornelius JR, Day NL, Fabrega H *et al.* Characterizing organic delusional syndrome. *Arch Gen Psychiatry* 1991; **48**:997–9.

Dalby MA. Epilepsy and three per second spike and wave rhythms. A clinical electroencephalographic and prognostic evaluation of 346 patients. *Acta Neurol Scand* 1969; **45**(Suppl. 40):1–180.

Daly DD, Thomas JE. Sequential alterations in the electroencephalograms of patients with brain tumors. *Electroencephalogr Clin Neurophysiol* 1958; **10**:395–404.

Daly DD, Whelan JL, Bickford RG *et al.* The electroencephalogram in cases of tumor of the posterior fossa and third ventricle. *Electroencephalogr Clin Neurophysiol* 1953; **5**:203–16.

Davis LE, Schmitt JW. Clinical significance of cerebrospinal fluid tests for neurosyphilis. *Ann Neurol* 1989; **25**:50–5.

De Coene B, Hajnal JV *et al.* MR of the brain using fluid-attenuated inversion recovery (FLAIR) pulse sequences. *AJNR Am J Neuroradiol* 1992; **13**:1555–64.

Deisenhammer E, Jelinger K. EEG in senile dementia. *Electroencephalogr Clin Neurophysiol* 1974; **36**:91–3.

DeJong RN. *The Neurologic Examination*, 4th edn. New York: Harper & Row, 1979.

de la Paz D, Brenner RP. Bilateral independent periodic lateralized epileptiform discharges: clinical significance. *Arch Neurol* 1981; **38**:713–15.

Dempsey EW, Morrison RS. The production of rhythmically recurrent cortical potentials after localized thalamic stimulation. *Am J Physiol* 1942; **135**:293–300.

Devinsky O, Sato S, Kufta CV *et al.* Electroencephalographic studies of simple partial seizures with subdural electrode recordings. *Neurology* 1989; **39**:527–33.

Dieterich M, Brandt T. Is obligatory bedrest after lumbar puncture obsolete? *Eur Arch Psychiatr Neurol Sci* 1985; **235**:71–5.

Di Gennaro G, Quarato PP, Onorati P *et al.* Localizing significance of temporal intermittent rhythmic delta activity (TIRDA) in drug-resistant focal epilepsy. *Clin Neurophysiol* 2003; **114**:70–8.

Doran M, Larner AJ. EEG findings in dementia with Lewy bodies causing diagnostic confusion with sporadic Creutzfeldt–Jakob disease. *Eur J Neurol* 2004; **11**:838–41.

Duncan J. The current status of neuroimaging for epilepsy. *Curr Opin Neurol* 2009; **22**:179–84.

Ellingson RJ, Wilken K, Bennet DR. Efficacy of sleep deprivation as an activation procedure in epilepsy patients. *J Clin Neurophysiol* 1984; **2**:83–101.

Epstein CM, Bej MD, Foldvary-Schaefer N *et al.* American Clinical Neurophysiology Society. Guideline 1: Minimum technical requirements for performing clinical electroencephalography. *J Clin Neurophysiol* 2006a; **23**:86–91.

Epstein CM, Bej MD, Foldvary-Schaefer N *et al.* American Clinical Neurophysiology Society. Guideline 5: Guideline for standard electrode position nomenclature. *J Clin Neurophysiol* 2006b; **23**:107–10.

Epstein CM, Bej MD, Foldvary-Schaefer N *et al.* American Clinical Neurophysiology Society. Guideline 6: A proposal for standard montages to be used in clinical EEG. *J Clin Neurophysiol* 2006c; **23**:111–17.

Erwin CW, Somerville ER, Radtke RA. A review of electroencephalographic features of normal sleep. *Clin Neurophysiol* 1984; **1**: 253–74.

Evans CC. Spontaneous excitation of the visual cortex and association areas – lambda waves. *Electroencephalogr Clin Neurophysiol* 1953; **5**:69–74.

Evans RW, Armon C, Frohman EM *et al.* Assessment: Prevention of post-lumbar puncture headaches. *Neurology* 2000; **55**:909–14.

Faber R, Abrams R, Taylor MA *et al.* Comparison of schizophrenic patients with formal thought disorder and neurologically impaired patients with aphasia. *Am J Psychiatry* 1983; **140**:1348–51.

Fisch BJ, Klass DW. The diagnostic specificity of triphasic wave patterns. *Electroencephalogr Clin Neurophysiol* 1988; **70**:1–8.

Fisch BJ, Hauser WA, Brust JCM *et al.* The EEG response to diffuse and patterned photic stimulation during acute untreated alcohol withdrawal. *Neurology* 1989; **39**:434–6.

Fitzpatrick W, Lowry N. PLEDs: clinical correlates. *Can J Neurol Sci* 2007; **34**:443–50.

Frantzen E. Wernicke's encephalopathy. Three cases occurring in connection with severe malnutrition. *Acta Neurol Scand* 1966; **42**:426–41.

Frost JD, Carrie JRG, Borda RP *et al.* The effects of Dalmane (flurazepam hydrochloride) on human EEG characteristics. *Electroencephalogr Clin Neurophysiol* 1973; **34**:171–5.

Fushimi M, Matsubichi N, Sekine A *et al.* Benign bilateral independent periodic lateralized epileptiform discharges. *Acta Neurol Scand* 2003; **108**:55–9.

Gabor AJ, Ajmone-Marsan C. Co-existence of focal and bilateral diffuse epileptiform discharges in epileptics. *Epilepsia* 1969; **10**:453–72.

Garcia-Morales I, Garcia MT, Galan-Davila L *et al.* Periodic lateralized epileptiform discharges: etiology, clinical aspects, seizures, and evolution in 130 patients. *J Clin Neurophysiol* 2002; **19**:172–7.

Gastaut H, Trevisan C, Naquet R. Diagnostic value of electroencephalographic abnormalities provoked by intermittent photic stimulation. *Electroencephalogr Clin Neurophysiol* 1958; **10**:194–5.

Gastaut JL, Michael B, Hassan S *et al.* Electroencephalography in brain edema (127 cases of tumor investigated by cranial computerized tomography). *Clin Neurophysiol* 1979; **46**:239–55.

Geiger LR, Horner RN. EEG pattern at the time of focal seizure onset. *Arch Neurol* 1978; **35**:276–86.

Gerson SN, Benson F, Frazier SH. Diagnosis: schizophrenia versus posterior aphasia. *Am J Psychiatry* 1977; **134**:966–9.

Gibbs FA, Gibbs EL, Lennox WG. Electroencephalographic classification of epileptic patients and control subjects. *Arch Neurol Psychiatry* 1943; **50**:111–28.

Gibbs FA, Rich CL, Gibbs EL. Psychomotor variant type of seizure discharge. *Neurology* 1963; **13**:991–8.

Gilmore PC, Brenner RP. Correlation of EEG, computerized tomography, and clinical findings: a study of 100 patients with focal delta activity. *Arch Neurol* 1981; **38**:371–2.

Goldberg HH, Strauss H. Distribution of slow activity induced by hyperventilation. *Electroencephalogr Clin Neurophysiol* 1959; **11**:615.

Goldman D. The clinical use of the 'average' reference electrode in monopolar recording. *Electroencephalogr Clin Neurophysiol* 1950; **2**:209–12.

Goodin DS, Aminoff MJ. Does the interictal EEG have a role in the diagnosis of epilepsy? *Lancet* 1984; **1**:837–9.

Goodwin JE. Significance of alpha variants and their relationship to epileptiform syndrome. *Am J Psychiatry* 1947; **104**:369–79.

Green J. Some observations on lambda waves and peripheral stimulation. *Electroencephalogr Clin Neurophysiol* 1957; **9**:691–704.

Greenblatt DJ, Ehrenberg BL, Gunderman J *et al.* Pharmacokinetic and electroencephalographic study of intravenous diazepam, midazolam, and placebo. *Clin Pharmacol Ther* 1989; **45**:356–65.

Gregory RP, Oates T, Merry RTG. EEG epileptiform abnormalities in candidates for aircrew training. *Electroencephalogr Clin Neurophysiol* 1993; **86**:75–7.

Griesinger W. *Mental Pathology and Therapeutics.* New York: William Wood, 1882.

Guaranha MSB, Garzon E, Buchpiguel CA *et al.* Hyperventilation revisited: physiological effects and efficacy on focal seizure activation in the era of video-EEG monitoring. *Epilepsia* 2005; **46**:69–7.

Gurer G, Yemisci M, Saygi S *et al.* Structural lesions in periodic lateralized epileptiform discharges (PLEDS). *Clin EEG Neurosci* 2004; **35**:88–93.

Harden CL, Daras M, Tuchman AJ *et al.* Low amplitude EEGs in demented AIDS patients. *Electroencephalogr Clin Neurophysiol* 1993; **87**:54–6.

Hecaen H, Penfield W, Bertrand C *et al.* The syndrome of apractagnosia due to lesions of the minor cerebral hemisphere. *Arch Neurol Psychiatry* 1956; **75**:400–34.

Heilman KM. Ideational apraxia – a re-definition. *Brain* 1973; **96**:861–4.

Helmchen H, Kanowski S. EEG changes under lithium treatment. *Electroencephalogr Clin Neurophysiol* 1971; **30**:269.

Henchey R, Cibula J, Helveston W *et al.* Electroencephalographic findings in Hashimoto's encephalopathy. *Neurology* 1995; **45**:977–81.

Hockaday JM, Whitty CW. Factors determining the electroencephalogram in migraine: a study of 560 patients, according to clinical type of migraine. *Brain* 1969; **92**:769–88.

Hockaday JM, Potts F, Epstein E *et al.* Electroencephalographic changes in acute cerebral anoxia from cardiac or respiratory arrest. *Electroencephalogr Clin Neurophysiol* 1965; **18**:575–86.

Holmes MD, Dewaraja AS, Vanhatalo S. Does hyperventilation elicit epileptic seizures? *Epilepsia* 2004; **45**:618–20.

Homan RW, Jones MC, Rawat S. Anterior temporal electrodes in complex partial seizures. *Electroencephalogr Clin Neurophysiol* 1988; **70**:105–9.

Hopf HC, Muller-Forell W, Hopf NJ. Localization of emotional and volitional facial paresis. *Neurology* 1992; **42**:1918–23.

Hormes JT, Benarroch EE, Rodriguez M *et al.* Periodic sharp waves in baclofen-induced encephalopathy. *Arch Neurol* 1988; **45**:814–15.

Hsia AW. 10 questions about acute stroke imaging: CT versus MRI – the controversy. *Neurologist* 2009; **15**:105–7.

Hughes JR. Usefulness of photic stimulation in routine clinical electroencephalography. *Neurology* 1960; **10**:777–82.

Humphrey DR. Re-analysis of the antidromic cortical response: II. On the contribution of all discharges and PSPs to the evoked potential. *Electroencephalogr Clin Neurophysiol* 1968; **25**:421–42.

Husain AM, Mebust KA, Radtke RA. Generalized periodic epileptiform discharges: etiologies, relationship to status epilepticus, and prognosis. *J Clin Neurophysiol* 1999; **16**:51–8.

Jasper H. Diffuse projection systems: the integrative action of the thalamic reticular system. *Electroencephalogr Clin Neurophysiol* 1949; **1**:405–20.

Jasper HH. The ten-twenty electrode system of the International Federation. *Electroencephalogr Clin Neurophysiol* 1958; **10**:371–3.

Johannesson G, Brun A, Gustafson I *et al.* EEG in presenile dementia related to cerebral blood flow and autopsy findings. *Acta Neurol Scand* 1977; **56**:89–103.

Kaibara M, Blume WT. The postictal electroencephalogram. *Electroencephalogr Clin Neurophysiol* 1988; **70**:99–104.

Kaiko RF, Foley KM, Grabisnki PY *et al.* Central nervous system excitatory effects of meperidine in cancer patients. *Ann Neurol* 1983; **13**:180–5.

Kales A, Bixler EO, Tan TL *et al.* Chronic hypnotic drug use. Ineffectiveness, drug-withdrawal insomnia and dependence. *JAMA* 1974; **227**:513–17.

Kaplan PW, Birbeck G. Lithium-induced confusional states: nonconvulsive status epilepticus or triphasic encephalopathy? *Epilepsia* 2006; **47**:2071–7.

Karnaze DS, Bickford RG. Triphasic waves: a reassessment of their clinical significance. *Electroencephalogr Clin Neurophysiol* 1984; **57**:193–8.

Kates R, Atkinson D, Brant-Zawadzki *et al.* Fluid-attenuated inversion recovery (FLAIR): clinical prospectus of current and future applications. *Top Magn Reson Imaging* 1996; **8**:389–96.

Katz DI, Alexander MP, Mandell AM. Dementia following strokes in the mesencephalon and diencephalons. *Arch Neurol* 1987; **44**:1127–33.

Kenangil G, Orken DN, Yalcin D *et al.* Triphasic EEG pattern in bilateral paramedian thalamic infarction. *Clin EEG Neurosci* 2008; **39**: 185–90.

Kennard MA, Beuding E, Wortis SB. Some biochemical and electroencephalographic changes in delirium tremens. *Q J Stud Alcohol* 1945; **6**:4–14.

Kim HL, Donnelly JH, Tonay AE *et al.* Absence of seizures despite high prevalence of epileptiform EEG abnormalities in children with autism monitored in a tertiary care center. *Epilepsia* 2006; **47**:394–8.

King MA, Newton MR, Jackson GD *et al.* Epileptology of the first seizure presentation: a clinical, EEG and MRI study of 300 consecutive patients. *Lancet* 1998; **352**:1007–11.

Klass DW, Bickford RG. Observations on the rolandic arceau rhythm. *Electroencephalogr Clin Neurophysiol* 1957; **9**:570.

Klass DW, Westmoreland BF. Nonepileptogenic epileptiform electroencephalographic activity. *Ann Neurol* 1985; **18**:627–35.

Knott JR. Further thoughts on polarity, montages and localization. *J Clin Neurophysiol* 1985; **2**:63–75.

Kooi KA, Thomas MH, Mortensen FN. Photoconvulsive and photomyoclonic responses in adults. An appraisal of their clinical significance. *Neurology* 1960; **10**:113–40.

Kooi KA, Guvener AM, Tupper CJ *et al.* Electroencephalographic pattern of the temporal regions in normal adults. *Neurology* 1964; **14**:1029–35.

Koshino Y, Murata I, Murata T *et al.* Frontal intermittent rhythmic delta activity in schizophrenic patients receiving antipsychotic drugs. *Clin Electroenphalogr* 1993; **24**:13–18.

Kraepelin E. *Manic–Depressive Insanity and Paranoia*, translated by Barclay RM. Edinburgh: E&S Livingstone, 1921. Republished by Arno Press, New York, 1976.

Kraepelin E. *Psychiatry. A Textbook for Students and Physicians*, 1899, translated by Metoui H, Ayed S. Canton, MA: Science History Publications, 1990.

Krauss GL, Abdallah A, Lesser R *et al.* Clinical and EEG features of patients with EEG wicket rhythm misdiagnosed with epilepsy. *Neurology* 2005; **64**:1879–83.

Kulkarni VP, Lin K, Benbadis SR. EEG findings in the persistent vegetative state. *J Clin Neurophysiol* 2007; **24**:433–7.

Laplane D, Degos JD. Motor neglect. *J Neurol Neurosurg Psychiatry* 1983; **46**:152–8.

Lavi R, Yernitzky D, Rowe JM *et al.* Atraumatic Whitacre needle for diagnostic lumbar puncture: a randomized trial. *Neurology* 2006; **67**:1492–4.

Leach JP, Stephen LJ, Salveta C *et al.* Which electroencephalography (EEG) for epilepsy? The relative usefulness of different EEG protocols in patients with possible epilepsy. *J Neurol Neurosurg Psychiatry* 2006; **77**:1040–2.

Leavitt S, Tyler H. Studies in asterixis. *Arch Neurol* 1964; **10**:360–8.

Lemstra AW, van Meegen MT, Vreyling JP *et al.* 14–3–3 testing in diagnosing Creutzfeldt–Jakob disease. A prospective study in 112 patients. *Neurology* 2000; **55**:514–16.

Lesser RP, Luders H, Dinner DS *et al.* An introduction to the basic concepts of polarity and localization. *J Clin Neurophysiol* 1985; **2**:45–61.

Levy SR, Chiappa KH, Burke CJ *et al.* Early evolution and incidence of electroencephalographic abnormalities in Creutzfeldt–Jakob disease. *J Clin Neurophysiol* 1986; **3**:1–21.

Lieb JP, Walsh GO, Babb TL *et al.* A comparison of EEF seizure patterns recorded with surface and depth electrodes in patients with temporal lobe epilepsy. *Epilepsia* 1976; **17**:137–60.

Linnenger AW, Volow MR, Gianturco DT. Intravenous infusion artifact. *EEG Technol* 1981; **21**:167–73.

Lombroso CT, Schwartz IH, Clark DM *et al.* Ctenoids in healthy youths: Controlled study of 14- and 6-per second positive spiking. *Neurology* 1966; **16**:1152–8.

Lusins J, Jaffe R, Bender MB. Unoperated subdural hematoma: long-term follow-up study by brain scan and electroencephalography. *J Neurosurg* 1976; **44**:601–7.

MacDonnell RAL, Donnan GA, Baldin PF *et al.* The electroencephalogram in acute ischemic stroke. *Arch Neurol* 1988; **45**:520–4.

MacLean PD. A nasopharyngeal lead. *Electroencephalogr Clin Neurophysiol* 1949; **1**:110–12.

Magnus O. The cerebral alpha-rhythm ('rhythme en arceau'). *Electroencephalogr Clin Neurophysiol* 1954; **6**:349–50.

Malow BA, Reese KB, Sato S *et al.* Abnormalities during clozapine treatment. *Electroencephalogr Clin Neurophysiol* 1994; **91**: 205–11.

Markand ON, Daly DD. Pseudo-periodic lateralized paroxysmal discharges in electroencephalogram. *Neurology* 1971; **21**:975–81.

Markand ON, Panzi JG. The electroencephalogram in subacute sclerosing panencephalitis. *Arch Neurol* 1975; **32**:719–26.

Martinez-Yelamos A, Saiz A, Sanchez-Valle R *et al.* 14–3–3 protein in the CSF as prognostic marker in early multiple sclerosis. *Neurology* 2001; **57**:722–4.

Martins da Silva A, Aarts JH, Binne CD *et al.* The circadian distribution of interictal epileptiform EEG activity. *Electroencephalogr Clin Neurophysiol* 1984; **58**:1–13.

Matsuo F, Knott JR. Focal positive spikes in electroencephalography. *Electroencephalogr Clin Neurophysiol* 1977; **42**:15–25.

Matsuo F, Peters JF, Reilly EL. Electrical phenomena associated with movements of the eyelid. *Electroencephalogr Clin Neurophysiol* 1975; **38**:507–11.

Mattson RH, Pratt KL, Calverley JR. Electroencephalograms of epileptics following sleep deprivation. *Arch Neurol* 1965; **13**:310–15.

Meier-Ewert R, Broughton RJ. Photomyoclonic response of epileptic and nonepileptic subjects during wakefulness, sleep, and arousal. *Electroencephalogr Clin Neurophysiol* 1976; **23**:301–4.

Miley CM, Forster FM. Activation of complex partial seizures by hyperventilation. *Arch Neurol* 1977; **34**:371–3.

Morgan MH, Scott DF. EEG activation in epilepsies other than petit mal. *Epilepsia* 1970; **11**:255–61.

Morris HH, Dinner DS, Luders H *et al.* Supplementary motor seizures: clinical and electroencephalographic findings. *Neurology* 1988; **38**:1075–82.

Neufeld MY, Inzelberg R, Korczyn AD. EEG in demented and non-demented parkinsonian patients. *Acta Neurol Scand* 1988; **78**:1–5.

Nicolai J, van Putten MJ, Tavy DL. BIPLEDS in a kinetic mutism caused by bilateral anterior cerebral artery infarction. *Clin Neurophysiol* 2001; **112**:1726–8.

Niedermeyer E, Rocca V. The diagnostic significance of sleep electroencephalograms in temporal lobe epilepsy: a comparison of scalp and depth tracings. *Eur Neurol* 1972; **7**:119–29.

Normand MM, Wszolek ZK, Klass DW. Temporal intermittent rhythmic delta activity in electroencephalograms. *J Clin Neurophysiol* 1995; **12**:280–4.

O'Brien TJ, Sharbrough FW, Westmoreland BF *et al.* Subclinical rhythmic electrographic discharges of adults (SREDA) revisited: a study using digital EEG analysis. *J Clin Neurophysiol* 1998; **15**:493–501.

Orta DS, Chiappa KH, Quiroz AZ *et al.* Prognostic implications of periodic epileptiform discharges. *Arch Neurol* 2009; **66**:985–91.

Parker GJ, Stephan KE, Barker GJ *et al.* Initial demonstration of in vivo tracing of axonal projections in the macaque brain and comparison with the human brain using diffusion tensor imaging and fast marching tractography. *Neuroimage* 2002; **15**:797–809.

Pedley TA. Interictal epileptiform discharges: discriminating characteristics and clinical correlations. *Am J EEG Technol* 1980; **20**:101–19.

Peters JF. Surface electrical fields generated by eye movements. *Am J EEG Technol* 1967; **7**:27–40.

Picard A, Cheliout Hearaut F, Bouskraoui M *et al.* Sleep EEG and developmental dysphasia. *Dev Med Child Neurol* 1998; **40**:595–9.

Poppert S, Essig A, Stoehr B *et al.* Rapid diagnosis of bacterial meningitis by real-time PCR and fluorescence in situ hybridization. *J Clin Microbiol* 2005; **43**:3390–7.

Pratt KL, Mattson RH, Weikers NJ *et al.* EEG activation of epileptics following sleep deprivation: a prospective study of 114 cases. *Electroencephalogr Clin Neurophysiol* 1968; **24**:11–15.

Pro JD, Wells CE. The use of the electroencephalogram in the diagnosis of delirium. *Dis Nerv Syst* 1977; **38**:804–8.

Protass L. Hyperventilation in sickle cell disease. *Arch Int Med* 1978; **138**:29.

Puri BK, Bydder GM, Manku MS *et al.* Reduction in cerebral atrophy associated with ethyl-eicosapentaenoic acid treatment in patients with Huntington's disease. *J Int Med Res* 2008; **36**:896–905.

Purpura DP, Grundfest H. Nature of dendritic potentials and synaptic mechanism in cerebral cortex of cat. *J Neurophysiol* 1956; **19**:573–95.

Raroque HG, Gonzales PC, Jhaveri HS *et al.* Defining the role of structural lesions and metabolic abnormalities in periodic lateralized epileptiform discharges. *Epilepsia* 1993; **34**:279–83.

Rechtschaffen A, Kales A. *A Manual of Standardized Terminology, Technique and Scoring System for Sleep Stages of Human Subjects.* Washington DC: US Government Printing Office, 1968.

Reiher J, Lebel M. Wicket spikes: clinical correlates of a previously undescribed EEG pattern. *Can J Neurol Sci* 1977; **4**:39–47.

Risinger MW, Engel J, Van Ness PC *et al.* Ictal localization of temporal lobe seizures with scalp sphenoidal recordings. *Neurology* 1989; **39**:1288–93.

Rodriguez AJ, Westmoreland BF. Electroencephalographic characteristics of patients with West Nile virus. *J Clin Neurophysiol* 2007; **24**:386–9.

Romano J, Engel GL. Delirium, I: electroencephalographic data. *Arch Neurol Psychiatry* 1944; **51**:356–77.

Roseman E. Dilantin toxicity: a clinical and electroencephalographic study. *Neurology* 1961; **11**:912–21.

Roos KL. Lumbar puncture. *Semin Neurol* 2003; **23**:105–14.

Ross ED. The aprosdias: functional–anatomic organization of the affective components of language in the right hemisphere. *Arch Neurol* 1981; **38**:561–9.

Sadler RM, Goodwin J. Multiple electrodes for detecting spikes in partial complex seizures. *Can J Neurol Sci* 1989; **16**:326–9.

Saenz-Lope E, Herranz-Tanarro FJ, Masdeu JC *et al.* Hyperekplexia: a syndrome of pathological startle responses. *Ann Neurol* 1984; **15**:36–41.

Salinsky M, Kanter R, Dashieff RM. Effectiveness of multiple EEGs in supporting the diagnosis of epilepsy: an operational curve. *Epilepsia* 1987; **28**:331–4.

Sammaritano J, Gigli GL, Gotman J. Interictal spiking during wakefulness and sleep and the localization of foci in temporal lobe epilepsy. *Neurology* 1991; **41**:290–7.

Schaul N, Lueders H, Sachdev K. Generalized bilaterally synchronous bursts of slow waves in the EEG. *Arch Neurol* 1981; **38**:690–2.

Schear HE. The EEG pattern in delirium tremens. *Clin Electroencephalogr* 1985; **16**:30–2.

Schmidt D. The influence of antiepileptic drugs on the electroencephalogram: a review of controlled clinical studies. *Electroencephalogr Clin Neurophysiol* 1982; **36**(Suppl.):453–66.

Scott DF, Groetheysen UC, Bickford RG. Lambda responses in the human electroencephalogram. *Neurology* 1967; **17**:770–8.

Scott DF, Heathfield KWG, Toone B *et al.* The EEG in Huntington's disease: a clinical and neuropathological study. *J Neurol Neurosurg Psychiatry* 1972; **35**:97–102.

Seddigh S, Thomke F, Vogt T. Complex partial seizures provoked by photic stimulation. *J Neurol Neurosurg Psychiatry* 1999; **66**:801–2.

Seshia SS, McLachlan RS. Aura continua. *Epilepsia* 2005; **46**:454–5.

Seyffarth H, Denny-Brown D. The grasp reflex and the instinctive grasp reaction. *Brain* 1948; **71**:109–83.

Silverman D. The anterior temporal electrode and the ten-twenty system. *Electroencephalogr Clin Neurophysiol* 1960; **12**:735–7.

Smith A. The serial sevens subtraction test. *Arch Neurol* 1962; **17**: 78–80.

Singh G, Wright MA, Sander JW *et al.* Periodic lateralized epileptiform discharges (PLEDS) as the sole electrographic correlate of a complex partial seizure. *Epileptic Disord* 2005; **7**:37–41.

Sperling MR, Engel J. Electroencephalographic recordings from the temporal lobe: a comparison of ear, anterior temporal and nasopharyngeal electrodes. *Ann Neurol* 1985; **17**:510–13.

Sperling MR, Mednius JR, Engel J. Mesial temporal spikes: a simultaneous comparison of sphenoidal, nasopharyngeal and ear electrodes. *Epilepsia* 1986; **27**:81–6.

Steiner L, Budka H, Chaudhuri A *et al.* Viral encephalitis: a review of diagnostic methods and guidelines for management. *Eur J Neurol* 2005; **12**:331–43.

Steinhoff BJ, Racker S, Herrendorf G *et al.* Accuracy and reliability of periodic sharp wave complexes in Creutzfeldt–Jakob disease. *Arch Neurol* 1996; **53**:162–6.

Steriade M, Gloor P, Llinas RD *et al.* Basic mechanisms of cerebral rhythmical activities. *Electroencephalogr Clin Neurophysiol* 1990; **76**:481–508.

Stevens JR. Central and peripheral factors in epileptic discharge: clinical studies. *Arch Neurol* 1962; **7**:330–8.

Stockard JJ, Werner SS, Aalbers JA *et al.* Electroencephalographic findings in phencyclidine intoxication. *Arch Neurol* 1976; **33**: 200–3.

Strupp M, Schueler O, Straube A *et al.* 'Atraumatic' Sprotte needle reduces the incidence of post-lumbar puncture headaches. *Neurology* 2001; **57**:2310–12.

Su P, Goldensohn ES. Progressive supranuclear palsy: electroencephalographic studies. *Arch Neurol* 1973; **29**:183–6.

Summerskill WHJ, Davidson EA, Sherlock S *et al.* The neuropsychiatric syndrome associated with hepatic cirrhosis and extensive portal circulation. *Q J Med* 1956; **25**:245.

Tegner R, Levander M. The influence of stimulus properties on visual neglect. *J Neurol Neurosurg Psychiatry* 1991; **54**:882–7.

Thomas JE, Klass DW. Six-per-second spike-and-wave pattern in the electroencephalogram: a reappraisal of its clinical significance. *Neurology* 1968; **18**:587–93.

Timmermans M, Carr J. Neurosyphilis in the modern era. *J Neurol Neurosurg Psychiatry* 2004; **75**:1727–30.

Torres-Russotto D, Landau WM, Harding GW *et al.* Calibrated finger rub auditory screening test (CALFRAST). *Neurology* 2009; **72**:1595–600.

Tourtellotte WW, Henderson WG, Tucker RP *et al.* A randomized, double-blind clinical trial of post-lumbar puncture headache in normal individuals. *Headache* 1972; **12**:73–8.

Tschampa HJ, Neumnn M, Zerr I *et al.* Patients with Alzheimer's disease and dementia with Lewy bodies mistaken for Creutzfeldt–Jakob disease. *J Neurol Neurosurg Psychiatry* 2001; **71**:33–9.

Upton A, Gumpert J. Electroencephalography in diagnosis of herpes simplex encephalitis. *Lancet* 1970; **1**:650–2.

Valler G, Rusconi ML, Bignamimi L *et al.* Anatomical correlates of visual and tactile extinction in humans: a clinical CT scan study. *J Neurol Neurosurg Psychiatry* 1994; **57**:464–70.

Van Everbroeck BRJ, Boons J, Cras P. 14–3–3 Isoform detection distinguishes sporadic Creutzfeldt–Jakob disease from other dementias. *J Neurol Neurosurg Psychiatry* 2005; **76**:100–2.

Veldhuizen R, Binnie CD, Beintema DJ. The effect of sleep deprivation on the EEG in epilepsy. *Electroencephalogr Clin Neurophysiol* 1983; **55**:505–12.

Victor M, Ropper AH. *Principles of Neurology*, 7th edn. New York: McGraw Hill, 2001.

Vignaendra V, Mathews RL, Chatrian GE. Positive occipital sharp transients of sleep: relationships to nocturnal sleep cycle in man. *Electroencephalogr Clin Neurophysiol* 1974; **37**:239–46.

Vulliemoz S, Iwanowski P, Landis T. Levetiracetam accumulation in renal failure causing myoclonic encephalopathy with triphasic waves. *Seizure* 2009; **18**:376–8.

Walker A, Pescor FT, Fraser HF *et al.* Electroencephalographic changes associated with chronic alcohol intoxication and the alcohol abstinence syndrome. *Am J Psychiatry* 1956; **113**:106–14.

Walshe FMR, Robertson EG. Observations upon the form and nature of the 'grasping' movements and 'tonic innervation' seen in certain cases of lesions of the frontal lobe. *Brain* 1933; **56**:40–70.

Walson JM, Kiloh LG, Osselton JW *et al.* The electroencephalogram in pernicious anemia and subacute combined degeneration of the spinal cord. *Electroencephalogr Clin Neurophysiol* 1954; **6**: 45–64.

Watemberg N, Alehan F, Dabby R *et al.* Clinical and radiologic correlates of frontal intermittent rhythmic delta activity. *J Clin Neurophysiol* 2002; **19**:535–9.

Watemberg N, Linder I, Dabby R *et al.* Clinical correlates of occipital intermittent rhythmic delta activity (OIRDA) in children. *Epilepsia* 2007; **48**:330–4.

Westmoreland BF, Klass DW. A distinctive rhythmic EEG discharge of adults. *Electroencephalogr Clin Neurophysiol* 1981; **51**:186–91.

White JC, Langston JW, Pedley TA. Benign epileptiform transients of sleep: clarification of the small sharp spike controversy. *Neurology* 1977; **17**:1061–8.

Wieser HG, Schwarz U, Blatter T *et al.* Serial EEG findings in sporadic and iatrogenic Creutzfeldt-Jakob disease. *Clin Neurophysiol* 2004; **115**:2467–78.

Will RG, Ironside JW, Zeidler M *et al.* A new variant of Creutzfeldt–Jakob disease in the UK. *Lancet* 1996; **347**:921–5.

Wilson SA. *Modern Problems in Neurology.* London: Edward Arnold, 1928.

Wolf P, Gooses R. Relation of photosensitivity to epileptic syndromes. *J Neurol Neurosurg Psychiatry* 1986; **49**:1386–91.

Wood JW, Bartlet D, James AE *et al.* Normal pressure hydrocephalus: diagnosis and patient selection for shunt surgery. *Neurology* 1974; **24**:517–26.

Zeidler M, Stewart GE, Barraclough CR *et al.* New variant Creutzfeldt–Jakob disease: neurological features and diagnostic tests. *Lancet* 1997; **35**:903–7.

Zivin L, Ajmone-Marsan C. Incidence and prognostic significance of 'epileptiform' activity in the EEG of non-epileptic subjects. *Brain* 1968; **91**:751–79.

Zochodne DW, Young GB, McLachlan RS *et al.* Creutzfeldt–Jakob disease without periodic sharp wave complexes: a clinical, electroencephalographic and pathologic study. *Neurology* 1988; **38**:1056–60.

Zurek R, Schiemann-Delgado J, Froescher W *et al.* Frontal intermittent rhythmical delta activity and anterior bradyarrythmia. *Clin Electroencephalogr* 1985; **16**:1–10.

PART III

SIGNS, SYMPTOMS, AND SYNDROMES

3

'Cortical' signs and symptoms

3.1 APHASIA

Aphasia is a fascinating clinical phenomenon, as it represents a disruption of that most human of all characteristics, namely language. The classification of the various types of aphasia has a long and contentious history: although the approach offered here is currently dominant, further research may require a more or less substantial revision.

Clinical features

Aphasia is characterized by impairments in one or more aspects of spoken language. Assessment begins with observation of the patient's spontaneous speech, with particular attention to fluency, comprehension, and coherence and also to the presence of what is known as paraphasia – a peculiar kind of misuse of words. Following this, one tests specifically for any improvement of speech with repetition and for the patient's ability to name objects.

Fluency, or its absence, is immediately apparent. Fluent speech occurs at a normal, or perhaps even increased rate: phrases and sentences are present, and there is a normal complement of prepositions, conjunctions, adjectives, and adverbs. Non-fluent speech, by contrast, is effortful, slowed, and 'telegraphic'. Given that many readers have perhaps never seen a telegraph, some words are in order regarding this descriptive term. The cost of a telegraph was apportioned according to its length, and hence efforts were made by telegraphers to omit as many words as possible while maintaining the sense of the original message. Consequently, telegraphs often lacked prepositions, conjunctions, adverbs, and adjectives. Non-fluent aphasias share this characteristic and, as this was first described in the age of telegraphs, the term 'telegraphic' was applied.

Comprehension, or its lack, becomes apparent during interaction with the patient. For example, during the neurologic examination it may appear that the patient has trouble understanding certain commands as, for example, when one instructs the patient to take the index finger and touch the nose. If one suspects a perhaps more subtle deficit in comprehension, it is appropriate to present a more complex command, for example, 'Now I want you to touch your right knee with your left hand.'

Coherence, or, again, its lack, becomes apparent during conversation with the patient. In this regard, it is critical to engage the patient in prolonged conversation, and to avoid a 'yes–no' format that denies the patient the opportunity to speak at length: all too often, physicians come away from a brief 'yes–no' interview without any sense of how incoherent their patient actually is. Incoherent speech is characterized by the presence of such disconnectedness and disorganization of words, phrases, and sentences that what the patient says, to a greater or lesser degree, 'makes no sense'.

Paraphasias represent a specific distortion in word usage, and come in one of two forms: phonemic (also known as literal) paraphasia, and semantic (also known as verbal) paraphasia. In phonemic paraphasia, a letter or syllable is replaced or added thus producing an incorrect word: for example, rather than saying 'Put it on the table', the patient with a phonemic paraphasia may say 'Put it on the stable.' In semantic paraphasia, by contrast, an entire word is substituted: for example, rather than saying 'The book is on the table', the patient with a semantic paraphasia may say 'The book is on the mirror.'

Repetition may have a remarkable effect in some cases of aphasia. As in some cases speech improves with repetition, tell the aphasic patient that you are going to ask him or her to repeat something and then provide the patient with a test phrase, such as 'This is a hospital.' If the patient repeats this correctly and without hesitation, then give a more complex phrase, for example 'To get out of the hospital, take the elevator to the first floor and then turn right.'

Finally, the ability to name things may be tested by pointing to an object in the room, such as a lamp, or perhaps by holding up a pencil and asking the patient to name it. If the patient does so correctly, then proceed to parts of the object in question, such as the lampshade, light bulb, etc. If the patient has difficulty, then provide cues, such as 'It's something that provides light' or 'It begins with the letter L.' If cues are not helpful, tell the patient the name of the object and ask whether or not that is the correct name.

After making all these determinations, it is generally possible to classify the patient's aphasia into one of the following types (all described in detail below): motor, transcortical motor, sensory, transcortical sensory, global, transcortical mixed, conduction, pure word deafness, and anomic. Importantly, however, it must be borne in mind that this classificatory scheme is but an approximation: clinical reality often overflows the nosologic boundaries we erect, and atypical cases are not at all uncommon (Brown and Simonson 1957). One must also be prepared for surprises: for example, in bilingual patients one may see a different aphasia for each language: in one case of a native Spanish speaker who had Hebrew as a second language, there was a motor aphasia for Spanish and a sensory one for Hebrew (Silverberg and Gordon 1979).

Each of the various types of aphasia is described further below, with comments on its localizing value. Before proceeding, however, a word is in order regarding aphasia's lateralizing value. Roughly 90 percent of the general population is right-handed, and in 99 percent of these individuals the left hemisphere is dominant for language. Among left-handers, the majority also exhibit language dominance in the left hemisphere (Goodglass and Quadfasel 1954; Humphrey and Zangwill 1952), and among the remainder some form of mixed dominance is generally present. Consequently, and especially in right-handers, the presence of aphasia lateralizes the lesion to the left hemisphere. However, exceptions to this rule in right-handers have been noted, and such cases are referred to as 'crossed aphasia' (Bakar *et al.* 1996; Brown and Wilson 1973; Holmes and Sadoff 1966). Crossed aphasia has been noted for motor aphasia (Hindson *et al.* 1984; Trojanowski *et al.* 1980), transcortical motor aphasia (Ghika-Schmid and Bogousslavsky 2000), sensory aphasia (Alexander *et al.* 1989; Henderson 1983; Sweet *et al.* 1984), global aphasia (Assal *et al.* 1981), and mixed transcortical aphasia (Cappa *et al.* 1993).

MOTOR APHASIA

Motor aphasia, also known as Broca's aphasia or expressive aphasia, is characterized by non-fluent, effortful speech that is laconic, often circumlocutory, and telegraphic. Comprehension is preserved and, although one may miss hearing the expected prepositions, conjunctions, and the like, coherence is generally preserved, and paraphasias, if present, are rare. Repetition does not improve speech: naming, although effortful, is generally preserved.

Motor aphasia can be a very frustrating experience and patients tend to become irritable or depressed (Benson 1973). Asking about the patient's experience can be edifying. Try, for example, saying to the patient 'Some patients who have trouble speaking do so because they're not clear what they want to say, whereas others know exactly what they want to say but can't get the words out. Is that the kind of trouble you're having? Do you know what you want to say but can't get it out?' Patients with motor aphasia often nod their heads vigorously, relieved that the physician understands their plight. Interestingly, emotionally laden speech, such as cursing, may be relatively unaffected, and some patients may evidence a remarkably preserved ability to sing (Yamadori *et al.* 1977).

In most cases, the responsible lesion is seen to involve the posterior portion of the inferior frontal gyrus (Mohr *et al.* 1978; Naeser and Hayward 1978; Tonkonogy and Goodglass 1981). Given that most lesions extend beyond the inferior frontal gyrus, it is very common to find associated deficits, such as a right-sided hemiparesis: indeed, although noted (Henderson 1985; Masdeu and O'Hara 1983), it is very uncommon to find an isolated motor aphasia without any accompanying deficits.

Motor aphasia has also been noted with a thalamic lesion (Megens *et al.* 1992) and with infarction of the putamen and adjacent anterior limb of the internal capsule (Naeser *et al.* 1982).

TRANSCORTICAL MOTOR APHASIA

Transcortical motor aphasia is essentially identical to motor aphasia, with the exception that repetition is preserved. Transcortical motor aphasia is most often seen with lesions of the medial aspect of the left frontal lobe, as may occur with infarctions in the area of distribution of the anterior cerebral artery (Alexander and Schmitt 1980; Bogousslavsky and Regli 1990; Freedman *et al.* 1984; Racy *et al.* 1979; Rubens 1975). A syndrome similar to transcortical motor aphasia may also occur with lesions of the putamen or thalamus (Alexander and LoVerme 1980; Ghika-Schmid and Bogousslavsky 2000; McFarling *et al.* 1982).

SENSORY APHASIA

Sensory aphasia, also known as Wernicke's aphasia or receptive aphasia, is characterized by fluent speech that is more or less incoherent and contaminated by frequent paraphasias; such patients also have difficulty comprehending commands, especially complex ones. In its most severe form, one sees what as known as 'jargon aphasia', or speech that is almost totally incomprehensible. An example of this was provided by Lord Brain (Wilson 1955), who showed his patient a 'bunch of keys' and asked the patient to name it. The patient then replied: 'Indication of measurement of piece of apparatus or intimating the cost of apparatus in various forms.'

In some cases, patients may appear relatively unconcerned, despite a severe deficit. In other cases, patients may become agitated and suspicious, and some may develop delusions of persecution (Benson 1973; Singer *et al.* 1989).

Sensory aphasia may be seen with lesions involving the temporoparietal area (Naeser and Hayward 1978), especially Wernicke's area on the posterior surface of the superior temporal gyrus (Selnes *et al.* 1985); it may also be seen with lesions of the white matter subjacent to Wernicke's area (Naeser and Hayward 1978) and lesions affecting the putamen and adjacent posterior limb of the internal capsule (Naeser *et al.* 1982).

TRANSCORTICAL SENSORY APHASIA

Transcortical sensory aphasia resembles sensory aphasia, except that repetition is intact. Transcortical sensory aphasia may be seen with lesions of the parietotemporal area that spare Wernicke's area (Selnes *et al.* 1985), or with lesions of the left thalamus, principally the dorsomedial nucleus (Bogousslavsky *et al.* 1988a; Tuszynski and Petito 1988).

GLOBAL APHASIA

Global aphasia is characterized by a combination of speech that is effortful and sparse with an inability to follow complex commands. Coherence is diminished; however, this is often difficult to assess, given that many patients are reduced to single words or overlearned stock phrases. Repetition does not improve their speech.

Global aphasia is most commonly seen with very large lesions involving the frontal, parietal, and temporal cortices (Bogousslavsky 1988; Naeser and Hayward 1978) and, as such, is typically accompanied by a hemiparesis. Exceptions to this rule do occur, however, and global aphasia secondary to cortical lesions may occur without hemiparesis when there are two distinct lesions – one in the frontal, and one in the temporoparietal cortex; such a scenario has been noted with multiple embolic infarctions (Hanlon *et al.* 1996; Tranel *et al.* 1987; Van Horn and Hawes 1982). Global aphasia has also been noted with a lesion of the thalamus (Kumar *et al.* 1996) and of the putamen and adjacent anterior and posterior limbs of the internal capsule (Naeser *et al.* 1982).

TRANSCORTICAL MIXED APHASIA

This aphasia, which could also just as well be called 'transcortical global aphasia', is essentially identical to global aphasia except for the fact that speech improves with repetition.

Transcortical mixed aphasia may be seen with lesions that, in one way or other, 'isolate' the posterior portion of the superior temporal gyrus and adjacent angular gyrus from the rest of the cortex, and has been reported with watershed infarction (Bogousslavsky *et al.* 1988b) and infarction of the medial aspects of the left frontal and parietal cortices (Ross 1980).

CONDUCTION APHASIA

In conduction aphasia speech is fluent, but there is a degree of incoherence and paraphasias are present. In striking contrast with sensory aphasia, however, comprehension is preserved. Furthermore, and remarkably so in light of the preserved comprehension, patients are unable to repeat complex sentences.

Conduction aphasia is classically associated with damage to the arcuate fasciculus (Benson *et al.* 1973; Damasio and Damasio 1980; Tanabe *et al.* 1987) but has also been noted with infarction of the caudate nucleus (Godefroy *et al.* 1994).

PURE WORD DEAFNESS

Pure word deafness is a remarkable syndrome characterized by an isolated inability to comprehend spoken words and to repeat phrases. Spontaneous speech is both fluent and coherent, and there are no paraphasias. One patient commented, 'Voice comes but no words . . . There is no trouble at all with the sound. Sounds come. I can hear, but I cannot understand it' (Hemphill and Stengel 1940). Another patient, although able to recognize non-speech sounds such as telephone rings or automobile horns, could not understand spoken words: he commented, 'I can hear you talking but I can't translate it' (Kanshepolsky *et al.* 1973).

Pure word deafness has been noted with bilateral damage to the superior temporal gyrus (Coslett *et al.* 1984; Kanshepolsky *et al.* 1973) and with bilateral damage to the inferior colliculi (Meyer *et al.* 1996; Pan *et al.* 2004).

Transcortical pure word deafness, wherein patients, although unable to understand the spoken word, are yet able to repeat it, may or may not occur in pure form: a possible case was noted with a lesion of the left angular gyrus (Heilman *et al.* 1981).

ANOMIC APHASIA

Anomic aphasia is characterized by a more or less isolated inability to correctly name objects. Speech, although overall fluent and coherent, is marked by circumlocutions and a tendency to use indefinite nouns, such as 'it' or 'thing' when the correct word fails to come to mind. Comprehension is preserved.

Although cueing is not effective, if the examiner provides the correct name, the patient generally recognizes it. Furthermore, patients with anomic aphasia, though unable to name an object, are able to describe its use. For example, an anomic patient, though unable to provide the name 'pen' when the examiner held one up, would, nevertheless, be able to say that it is used for writing.

Although isolated anomic aphasia does not have strong localizing value, pathology in the temporal lobe may be suspected. Anomic aphasia may also occur in the course of a delirium and as a side-effect of certain medications, such as topiramate (Mula *et al.* 2003) and both tricyclic (Schatzberg *et al.* 1978) and monoamine oxidase inhibitor (Goldstein and Goldberg 1986) antidepressants.

ATYPICAL APHASIA

Atypical aphasias are not uncommon and simply do not fit into the categories noted above. They are particularly associated with subcortical lesions (Ciemens 1970; Damasio *et al.* 1982a).

Etiology

As noted above, the various different types of aphasia, although generally lateralizing to the left hemisphere, localize, according to the type present, to multiple different areas of the cortex and, less commonly, subcortical structures. Determining the nature of the underlying lesion is facilitated by attending to the mode of onset of the aphasia.

Aphasia of acute onset, in the vast majority of cases, occurs as part of a stroke syndrome. In this situation, it is not uncommon for the patient to initially present with mutism: over time, as perilesional edema subsides, the mutism resolves, leaving one of the classic aphasias in its wake (Pedersen *et al.* 1995). A similar syndrome may also, albeit rarely, occur secondary to appropriately situated plaques of multiple sclerosis (Achiron *et al.* 1992; Devere *et al.* 2000; Lacour *et al.* 2004; Olmos-Lau *et al.* 1977), and aphasia of paroxysmal onset may occur on an ictal basis, representing either a simple partial or a complex partial seizure. In the case of simple partial seizures, aphasia may be the sole symptom (Hamilton and Mathews 1979; Labar *et al.* 1992) or may be accompanied by some twitching of the right side of the face (Williamson *et al.* 1985); when aphasia occurs as part of a complex partial seizure (Devinsky *et al.* 1994; Knight and Cooper 1986) one also finds other typical symptoms, such as confusion or automatisms. Postictal states may also be characterized by aphasia, and this strongly suggests that the seizure focus is in the left hemisphere (Gabr *et al.* 1989; Privitera *et al.* 1991).

Aphasia of gradual onset may occur secondary to an appropriately situated mass lesion. Such a gradual onset may also occur in a condition known as primary progressive aphasia (PPA). Primary progressive aphasia represents the effects of a focal, left peri-Sylvian onset of any one of various neurodegenerative disorders (Josephs *et al.* 2006; Knibb *et al.* 2006), including Alzheimer's disease (Clark *et al.* 2003; Galton *et al.* 2000; Green *et al.* 1990; Greene *et al.* 1996), frontotemporal lobar degeneration (Mesulam *et al.* 2007; Turner *et al.* 1996), Pick's disease (Graff-Radford *et al.* 1990; Karbe *et al.* 1993; Kertesz *et al.* 1994; Wechsler *et al.* 1982), corticobasal ganglionic degeneration (Ikeda *et al.* 1996), and progressive supranuclear palsy (Knibb *et al.* 2006). The diagnosis is immediately suggested when imaging fails to reveal a mass lesion, and is confirmed clinically when, after a year or more of gradual progression of the aphasia, either cognitive deficits or a personality change gradually supervene, indicating clearly that a neurodegenenerative disorder is present (Le Rhun *et al.* 2005). Although efforts have been made to correlate the type of aphasia with a specific underlying neurodegenerative process, such correlations, at present, are not robust enough for routine clinical work, and a better approach might be to base one's diagnostic suspicions on the clinical evolution. Thus, if the aphasia were followed by a slowly progressive amnestic deficit, one might reasonably assume that Alzheimer's disease was present, while if the aphasia were followed by a personality change of the frontal lobe type, a diagnosis of frontotemporal lobar degeneration would likewise be reasonable.

In children, aphasia may also be seen as part of the Landau–Kleffner syndrome (Hirsch *et al.* 1990; Mantovani and Landau 1980; Paquier *et al.* 1992) and the syndrome of developmental dysphasia: in developmental dysphasia both motor (Sato and Dreifuss 1973) and sensory (Bartak *et al.* 1975; Cohen *et al.* 1989; Paul *et al.* 1983) types have been described.

Differential diagnosis

Motor aphasia must be distinguished from dysarthria. Dysarthric, or slurred, speech, may appear somewhat effortful, but there are no word-finding pauses or circumlocutions and, if one listens carefully one finds a normal complement of prepositions, adjectives, etc.

Sensory aphasia must be distinguished from other causes of incoherence, notably the loosening of associations seen in schizophrenia and the rambling and variably incoherent speech seen in delirium and advanced cases of dementia. Loosening of associations, in contrast with sensory aphasia (Faber *et al.* 1983; Gerson *et al.* 1977), is generally associated with more voluble speech; furthermore, patients with loosening of associations are also less concerned about their incoherence than are patients with sensory aphasia: indeed, patients with loosening often are absolutely unconcerned in this regard. Although these points may be helpful in distinguishing loosening from sensory aphasia, they represent differences in degree only, and from a clinical point of view it is far more important to look to the history and determine whether or not the speech disturbances were preceded by a gradually developing psychosis, as one typically sees in schizophrenia. The rambling and incoherent speech of delirium and advanced cases of dementia is likewise difficult to distinguish from sensory aphasia, and some might argue that the difference here is moot and go on to simply say that this finding in delirium or dementia, in fact, does represent a sensory aphasia. From a clinical point of view, however, the important point is not to resolve this debate but rather to determine whether the incoherent patient has an isolated aphasia or the syndrome of delirium or dementia, and the best way to do this is, again, to look at the clinical context of the incoherence. Thus in delirium one will find prominent confusion, while in dementia one will find a history of pre-existing cognitive deficits, such as poor short-term memory.

Pure word deafness must be distinguished from deafness, and this may be accomplished by attending to the patient's reaction to sounds other than speech. Patients with pure word deafness, although unable to comprehend spoken words, nevertheless are able to respond appropriately to other sounds, such as the ringing of a telephone or the sound of a car horn (Coslett *et al.* 1984; Kanshepolsky *et al.* 1973). Deaf patients, in contrast, fail to respond to these non-speech sounds (Bahls *et al.* 1988; Le Gros Clark and Russell 1938).

Anomia must be differentiated from visual agnosia by asking patients what the object in question is used for. While patients with anomia and those with visual agnosia all will be unable to name the object, the patient with anomia will be able to describe its use, while the patient with agnosia will not.

Treatment

In addition to treatment, if possible, of the underlying condition, speech therapy and drug treatment may be considered.

Speech therapy is generally reserved for cases of aphasia secondary to stroke or other more or less static lesions; its effectiveness in cases of aphasia secondary to disorders which, in the natural course of events tend to progress (e.g. tumors, neurodegenerative disorders), is not clear.

Although numerous drugs have been reported as effective in the treatment of aphasia, only six have been subjected to double-blinded, placebo-controlled study. Bromocriptine has been the most extensively studied, and, although the first study was positive (Bragoni *et al.* 2000), three replication efforts all ended in failure (Ashtary *et al.* 2006; Gupta *et al.* 1995; Sabe *et al.* 1995). Another dopaminergic agent, levodopa, was effective in one study (Seniow *et al.* 2009), but failed in another (Leeman *et al.* 2011). Dextroamphetamine was effective in one study (Walker-Batson *et al.* 2001), however the benefits here were not sustained over the long haul. Donepezil was effective in one study (Berthier *et al.* 2006), as was memantine in another (Berthier *et al.* 2009). Finally, piracetam has been subjected to three, relatively large, double-blinded studies (Enderby *et al.* 1994; Huber *et al.* 1997; Kessler *et al.* 2000), and in all three was superior to placebo. Based on these results, piracetam is the best choice, and, as this agent is remarkably easy to use and has a benign side-effect profile, it would be reasonable to give it a try: the dose is 2400 mg twice daily.

3.2 ALEXIA

An inability to read, despite preserved visual acuity, is known as alexia. As noted in the immediately preceding section, alexia may occur in the course of aphasia; however, in this chapter we are concerned with alexia occurring in an isolated fashion in patients who, despite their inability to read, are yet able to understand what others say to them.

Clinical features

Alexia may or may not be accompanied by agraphia, or the inability to write. Cases of alexia without agraphia are also referred to as 'pure' alexia or 'pure word blindness'.

Pure alexia is a remarkable syndrome. Although patients are unable to read words, they retain the ability to recognize letters, and because of this some are able to circumvent their deficit by utilizing a 'letter-by-letter' reading strategy: here, patients read each letter out loud, and by listening to themselves, they are able to reconstruct the word and understand it (Stommel *et al.* 1991; Warrington and Langdon 1994). This inability to read whole words, as it extends to what patients themselves have written, can be particularly frustrating. One patient (Cohen *et al.* 1976), an English teacher, in describing her situation, commented: 'I can write and I can see, but I can't read.' She went on to note that 'It was as if the writing were in a foreign language.' She attempted to cope with the deficit during class by memorizing what she was writing on the blackboard as she wrote it so she could accurately refer back to it during the ensuing discussion, and this strategy worked well unless, of course, too much time passed and the memory of what she had written had failed.

In cases of pure alexia, one often, but not always, finds a right hemianopia.

Etiology

A brief review of the visual pathways may be helpful in understanding the mechanisms involved in alexia. To begin, recall that fibers of the optic tract terminate in the lateral geniculate body of the thalamus. From the lateral geniculate body, the geniculocalcarine tract arises and proceeds to the calcarine cortex, located on the medial aspect of the ipsilateral occipital cortex. Fibers from the left calcarine cortex proceed directly anteriorly toward the left angular gyrus, whereas fibers from the right calcarine cortex must first pass forward, and then cross in the splenium of the corpus callosum, after which they proceed laterally to an eventual juncture with the fibers that originated in the left calcarine cortex. These conjoined fibers then proceed anteriorly, to terminate finally in the left angular gyrus.

It appears that the left angular gyrus subserves both reading and writing. Hence, lesions of the gyrus itself cause the syndrome of alexia with agraphia. Lesions, however, that spare the angular gyrus but which deprive it of visual afferents from both hemispheres cause 'pure' alexia.

This deprivation of the angular gyrus of visual afferents from both hemispheres may occur via a number of different mechanisms. First, and most commonly, one finds a lesion in the splenium of the corpus callosum (which severs afferents from the right occipital cortex) in combination with a lesion of the medial aspect of the left occipital cortex (which destroys afferents from the left occipital cortex) (Ajax *et al.* 1977; Damasio and Damasio 1983). This first mechanism occurs most commonly secondary to an infarction in the area of distribution of the left posterior cerebral artery, which nourishes both the splenium and the medial aspect of the left occipital cortex. Second, there may be a combination of a lesion affecting the left lateral geniculate body (thus depriving visual afferents from the left hemisphere) and one affecting the splenium (thus severing afferents from the right hemisphere) (Stommel *et al.* 1991). Third, just subjacent to the angular gyrus there may be a lesion located in the white matter, which destroys the conjoined fibers from both occipital lobes (Greenblatt 1976).

As noted earlier, cases of pure alexia may be accompanied by a right hemianopia, and this occurs with either the first or the second mechanism just described. In cases occurring via the third mechanism, however, the visual fields remain unaffected.

As in the case of spoken language, so too for written language is the left hemisphere dominant in almost all right-handers and in most left-handers also. Exceptions, however, do occur. For example, cases of 'crossed' alexia have been noted in which right-handed patients suffered alexia due to right hemisphere lesions (Fincham *et al.* 1975; Henderson *et al.* 1985). Furthermore, there is also a case report of a left-handed patient who developed alexia due to a right hemisphere lesion (Pillon *et al.* 1987).

Most cases of alexia occur on the basis of stroke, as a result of an ischemic infarction or, less commonly, an intracerebral hemorrhage. Other possible lesions include appropriately situated tumors (Turgman *et al.* 1979; Vincent *et al.* 1977) or plaques of multiple sclerosis (Day *et al.* 1987; Dogulu *et al.* 1996; Mao-Draayer and Panitch 2004).

Differential diagnosis

Aphasia is distinguished by an inability to understand the spoken word.

Neglect may include a kind of alexia (often referred to as 'hemialexia') wherein patients fail to attend to the left part of a word and thus misread it (Kinsbourne and Warrington 1962). For example, a patient with left neglect might, upon being presented with the written word 'birdhouse', read 'house'.

Dementing disorders, such as Alzheimer's disease (Arsland *et al.* 1993), may cause alexia, but this is seen in the context of other cognitive deficits, such as disorientation, short-term memory loss, etc.

The term 'alexia', by convention, refers to cases of an acquired inability to read in those who had once mastered this skill. In cases characterized by an inability, despite adequate intelligence and educational opportunity, to acquire this skill, one speaks of 'developmental dyslexia,' a condition discussed in Chapter 10.16.

Treatment

Speech therapy may also be considered in addition to treatment, where possible, of the underlying lesion.

3.3 AGRAPHIA

Agraphia, or the inability to write, may be seen, as discussed in Chapter 3.1, in conjunction with aphasia. In this section, however, we are concerned with agraphia occurring in an isolated fashion, that is to say in patients who, despite being able to speak for the most part normally, are yet unable to write.

Clinical features

Bedside testing may be accomplished by asking patients to write a brief paragraph on what prompted their admission to the hospital. When agraphia is present, two basic types of errors may occur. On the one hand, there is 'posterior' agraphia, wherein what is written may be more or less incoherent with neologisms or 'paragraphias,' similar to the spoken paraphasias discussed in Chapter 3.1. On the other hand, there is 'anterior' agraphia, wherein the written production may be 'telegraphic' in a quite literal sense, bearing a strong resemblance to telegraphs of the past century, in which, though the sense of what is written is still clear, prepositions, conjunctions, adverbs, and adjectives were lacking.

Etiology

Agraphia lateralizes to the left hemisphere, and most cases have occurred as part of a stroke syndrome. The posterior form of agraphia may occur in conjuction with alexia, and in such cases lesions are found involving the angular gyrus (Rosati and De Bastiani 1979). The posterior form of agraphia may also occur in conjunction with finger agnosia, right–left disorientation and acalculia, thus constituting Gerstmann's syndrome, discussed in Chapter 3.5. The anterior form is quite rare, but has been reported with lesions of 'Exner's area' in the posterior portion of the middle frontal gyrus (Anderson *et al.* 1990; Tohgi *et al.* 1995)

Agraphia may also be confined to the left hand in right-handed patients as part of a disconnection syndrome occurring secondary to a lesion of the corpus callosum (Yamadori *et al.* 1980).

Differential diagnosis

Poor handwriting, as may be seen with tremor, dystonia, paresis or parkinsonian micrographia, is readily apparent, provided one simply observes the patient's attempt at writing.

Apraxia may lead to disturbed handwriting; however, associated features typically suggest the correct diagnosis. In ideational apraxia, one also sees difficulty in employing other 'tools', thus patients, in addition to difficulty in utilizing a pen or pencil, will also have difficulty in handling a comb, scissors, etc. In constructional apraxia, there will be additional difficulties in copying geometric figures, drawing stick figures, etc.

Delirium and dementia may be accompanied by agraphia, but the correct syndromal diagnosis is immediately suggested by associated cognitive deficits, such as confusion, disorientation, short-term memory loss, etc.

Agraphia may also occur on a developmental basis and in these cases, it is referred to by convention as dysgraphia. Developmental dysgraphia, discussed in Chapter 10.17, is distinguished from acquired agraphia in that in developmental cases, writing skills, despite adequate intelligence and educational opportunity, were simply never acquired, whereas in acquired agraphia, these skills were acquired, only to be lost.

Treatment

Speech therapy should be considered in addition to treatment, if possible, of the underlying lesion.

3.4 ACALCULIA

Acalculia, that is to say an acquired inability to do arithmetic, may be seen as part of Gerstmann's syndrome, wherein it is accompanied by finger agnosia, right–left disorientation and agraphia, or it may occur in an isolated fashion, and it is this isolated acalculia that is dealt with here.

Clinical features

Acalculia may become obvious to patients when they find themselves unable to make change or balance their checkbooks, or may only become apparent during the mental status examination. Typically, on the mental status examination, the patient's ability to do simple, one- or two-digit addition and subtraction is tested and, if these are done well, the patient is then asked to perform 'serial 7s'. In this venerable test, the patient is asked to subtract 7 from 100, then to subtract 7 from that number and to keep on subtracting sevens until he or she can go no further: making more than two or three errors is considered abnormal (Smith 1962). If there is a clinical suspicion of acalculia and these tests are done well, it is appropriate to then test the patient's ability to do multiplication and division, as in some cases the various computational processes may be differentially affected (Lampl *et al.* 1994).

Etiology

Acalculia lateralizes to the left hemisphere and typically occurs as part of a stroke syndrome. Lesions have been noted in the parietal cortex (Lampl *et al.* 1994; Takayama *et al.* 1994) (especially the angular gyrus [Benson and Weir 1972; Bernal *et al.* 2003]), posterior inferior frontal cortex (Tohgi *et al.* 1995), striatum (Corbett *et al.* 1986), and anterior thalamus (Mendez *et al.* 2003).

Differential diagnosis

Dementia or delirium is typically accompanied by acalculia; however, here the associated cognitive deficits, such as confusion, disorientation, short-term memory loss, etc., suggest the correct syndromal diagnosis.

Perseveration, as may be seen with frontal lesions, may mimic acalculia. For example, perseveration in the 'serial 7s' test may lead to the following responses: '93, 83, 73, 63 . . .'.

In cases where patients are able to perform calculations 'in their head', but are unable to do so on paper, suspicion should be aroused for neglect or apraxia. In neglect, patients may misread written numbers (e.g. reading '69' for '1969'), and in apraxia they may be unable to properly order written numbers on the page.

Acalculia occurring on a developmental basis (as discussed in Chapter 10.18) is distinguished from acquired acalculia in that in developmental cases patients are unable to acquire mathematical skills despite adequate intelligence and educational opportunity, whereas in acquired cases this ability was developed only to be subsequently lost.

Treatment

In addition, if possible, to treatment of the underlying lesion, speech therapy may be considered. Hand-held calculators, of course, should also be utilized.

3.5 GERSTMANN'S SYNDROME

In 1924, Josef Gerstmann described a remarkable syndrome characterized by the tetrad of finger agnosia, right–left disorientation, agraphia, and acalculia (Gerstmann 1924). Although there has been controversy in the past regarding whether this occurs in 'pure' form or not, the case reports cited below indicate that, although rare, the pure syndrome does in fact occur.

Clinical features

In later papers in English, Gerstmann (1940, 1942, 1957) further described the syndrome. He noted that in finger agnosia the patient cannot name the fingers either of his or the examiner's hand, and that this deficit is usually more pronounced with regard to the index, middle, and ring fingers. Right–left disorientation may become apparent when patients are instructed, say, to touch their right knee with their left hand: patients with this sign may use the incorrect hand or may touch the ipsilateral knee. Agraphia and acalculia are tested for by having the patient write a brief paragraph and then asking him or her to perform progressively more difficult calculations.

As noted by Gerstmann (1940), it is not uncommon to see an associated right hemianopia and a degree of alexia.

Etiology

Classically, Gerstmann's syndrome, as pointed out by Gerstmann himself (1940), localizes to the angular gyrus (Gold *et al.* 1995). The syndrome has also been noted with lesions of the white matter subjacent to the angular gyrus (Mayer *et al.* 1999), and with lesions involving both the angular and the supramarginal gyri (Roeltgen *et al.* 1983; Tucha *et al.* 1997). Although most cases are due to infarction, Gerstmann's syndrome has also been noted with tumors (Tucha *et al.* 1997), trauma (Mazzoni *et al.* 1990), and subdural hematoma (Maeshima *et al.* 1998). Although left-sided lesions are by far the most common cause, Gerstmann's syndrome has been noted with a right-hemisphere lesion (Moore *et al.* 1991).

Of note, in addition to occurring on an acquired basis, Gerstmann's syndrome may also occur on a developmental basis (Benson and Geschwind 1970; PeBenito 1987; Suresh and Sebastian 2000).

Differential diagnosis

The various elements of the tetrad may occur in patients with dementia or delirium; however, here the associated cognitive deficits, such as confusion, disorientation and short-term memory loss, will suggest the correct syndromal diagnosis.

Certain neurodegenerative disorders, such as Alzheimer's disease, may, albeit rarely, present in a 'focal' fashion, with the syndrome of 'posterior cortical atrophy', which, in addition to elements of Gerstmann's syndrome, also includes Balint's syndrome, aphasia, apraxia, and neglect (Renner *et al.* 2004; Tang-Wei *et al.* 2004).

Treatment

Consideration may be given to a course of speech therapy in addition to treatment, if possible, of the underlying lesion.

3.6 HYPERGRAPHIA

Hypergraphia, as defined by Mungas (Hermann *et al.* 1988), is characterized by a tendency to excessive writing that goes beyond any social, occupational, or educational requirements.

Clinical features

In evaluating a patient's written production for evidence of hypergraphia, attention must be paid to the premorbid level of written output. Thus, for a patient who premorbidly wrote very little, the appearance of a tendency to write several pages a day would be clinically significant. The text produced is for the most part coherent, and, rather than consisting of the mere perseverative reproduction of phrases, words, or letters, displays a more or less complete working-up of various related themes.

Etiology

Hypergraphia may be seen in the interictal personality syndrome, mania, schizophrenia, and stroke.

The interictal personality syndrome, as discussed in Chapter 8.2, is seen in patients with chronic epilepsy and typically includes hypergraphia (Hermann *et al.* 1988; Okamura *et al.* 1993; Waxman and Geschwind 1974, 1975), along with deep and persistent affect, verbosity, a preoccupation with religious, ethical, or philosophical concerns, hyposexuality, and irritability.

Mania may be characterized by hypergraphia, wherein it represents the written equivalent of pressured speech. Kraepelin (1976) noted that manics may produce an 'astonishing' number of documents, all from 'the pleasure in writing'. The presence of other typical symptoms, such as increased energy, pressured speech, and a decreased need for sleep, all indicate the correct diagnosis.

Schizophrenia may also lead to hypergraphia and Kraepelin (1971) commented on the 'very numerous and monotonous' documents that may appear, all marked by neologisms, delusional thoughts, and a degree of incoherence.

Stroke, very rarely, may be characterized by hypergraphia, as has been reported with infarction of the right cingulate gyrus (Carota *et al.* 2003) and with large infarctions, involving most of the area of distribution of the right middle cerebral artery (Yamadori *et al.* 1986). In these stroke cases, however, the hypergraphia was, relative to the voluminous production seen in the other etiologies, minor, and indeed in the cases occurring with right middle cerebral artery infarctions it occurred only upon some urging from the examiner.

Differential diagnosis

Hypergraphia must be distinguished from an appropriate increase in written output and from mere perseverative writing. Increased written output is normal in professional authors and others whose personal or professional lives require writing. With perseveration, when it appears in written form, the written output itself consists of the same phrase, word, or even letter written repeatedly again and again, sometimes filling page after page. Here, there is no theme evident and no elaborative working-up of ideas. This kind of graphic perseveration has been referred to as 'automatic writing behavior' and has been noted with a right hemispheric tumor (Imamura *et al.* 1992), right hemisphere stroke (due to either infarction in the area of distribution of the right middle cerebral artery or to putaminal hemorrhage) (Evyapan and Kumral 2001), and in dementia characterized in part by a frontal lobe syndrome (Frisoni *et al.* 1993; van Vugt *et al.* 1996).

Treatment

Treatment is directed at the underlying condition.

3.7 APROSODIA

Prosody refers to the affective or emotional aspects of speech, which are conveyed by the inflection, rhythm, and tone with which patients speak (Monrad-Krohn 1947a,b); aprosodia, in turn, represents a defect either in speaking with normal prosody or in comprehending the prosody with which others speak. In a sense, as will become apparent below, aprosodia is to the right hemisphere as aphasia is to the left hemisphere. The scheme presented here follows closely that presented by Ross (Gorelick and Ross 1987; Ross 1981).

Aprosodia, although not as obvious as aphasia, may have profound effects on a patient's life. In many aspects of life, it is not so much *what* we say that counts, as it is *how* we say it, and it is this aspect that is impaired in aprosodia.

Clinical features

In assessing patients for aprosodia, attention is paid to the following aspects of speech: the presence or absence of a monotonous voice; the ability of the patient to comprehend what another is feeling by simply hearing the intonation in the other's voice; the presence or absence of a 'mismatch' between what the patient says he is feeling and the prosody with which he speaks; and, finally, for any improvement in prosodic deficits by repetition.

Monotony is said to be present when the patient's speech is devoid of inflection or changes in tone – stripped, as it were, of emotional valence. When there is doubt whether monotony is present, it may be appropriate, at some point during the interview, to ask an emotionally charged question and then to listen carefully to see whether the patient's response remains monotonous. If it does, then it is critical to ask the patient what he or she is feeling during the response. Furthermore, one may simply ask the patient if he or she has had any trouble in expressing him- or herself by tone of voice. Formal testing may be accomplished by providing a neutral phrase, such as 'I am going to the movies', and asking the patient to speak that phrase on three successive occasions with three different intonations, namely happy, angry, and sad.

Comprehension on the patient's part of the prosody with which others speak may be difficult to assess during the interview, as most physicians have been trained to keep a studied neutrality in their tone of voice. Consequently, it is necessary to ask the patient if he or she has had any difficulty in understanding what others are feeling. When there is doubt, one may also ask friends or family members of the patient whether the patient seems to have had any trouble understanding what he or she is feeling. In some cases, others may report that the only way to get across to the patient what they are feeling is to state it explicitly, rather than relying on tone of voice. Formal testing may be accomplished by telling the patient that you are going to say something (e.g. as above, 'I am going to the movies') with different tones of voice (e.g. happy, angry, or sad) and that you want him or her to report what tone you are using. In performing this test, one must stand behind the patient in order to prevent the patient from seeing facial expressions or gestures.

A 'mismatch' between what the patient is feeling and the prosody with which he or she speaks when describing feelings may become apparent during the interview. For example, after the patient has described a traumatic experience, one may ask 'How are you feeling now?' If the patient says 'I'm feeling very sad', yet there is a lilt to his or her voice, then a mismatch is present. Or, conversely, after the patient has described a gratifying experience, one may again ask how he or she is feeling. If he or she says 'Great, just great' yet the tone of voice is somber and lugubrious, then again one has demonstrated that a 'mismatch' is present. Importantly, a monotonous tone does not indicate a 'mismatch': to say that a mismatch is present, there must be a dissonance between what the patient says he or she is feeling and the tone with which he or she reports that feeling.

Repetition is tested by telling the patient that you are going to say something with different tones of voice, and that you want him or her to repeat what you said with the exact tone in which you said it. One may then proceed with the neutral phrase 'I am going to the movies', said first with a happy tone, then a sad one, and finally an angry one.

It should be noted that in aprosodia there is also often an accompanying change in gesture and facial expression. For example, if monotony is present, one may see an absence of gesturing and a lack of facial expression. This association is not invariable, however, and in many cases gesture and expression will remain intact despite a significant disturbance of prosody.

After the above assessment, it is generally possible to categorize the patient's aprosodia into one of the following categories: motor, transcortical motor, sensory, transcortical sensory, global, transcortical mixed, conduction, and pure affective deafness. Each is discussed below, with comments on its localizing value. Before proceeding, however, it is appropriate to note, as indicated earlier, that in almost all cases, aprosodia indicates a right hemisphere lesion. Cases of 'crossed' aprosodia, in which right-handed patients developed aprosodia secondary to a lesion in the left hemisphere, although reported (Darby 1993; Ross *et al.* 1989), are rare.

MOTOR APROSODIA

In motor aprosodia, speech is monotonous, but comprehension is intact and there is no 'mismatch'; repetition does not relieve the monotonous tone. The presence of motor aprosodia may be quite disconcerting to patients. One of Ross and Mesulam's (1979) patients complained that she had great difficulty in disciplining her children because when she *said* to them that she was angry they did not respond, because she did not *sound* angry. She eventually came up with a way to circumvent her motor aprosodia by adding a parenthetical statement, such as 'God damnit, I mean it' or 'I am angry and I mean it', thereby getting across what she was feeling. Of note, however, in this case, 'Even the parenthetical statement was voiced in a complete monotone.'

Motor aprosodia most commonly occurs secondary to lesions in the posterior frontal operculum (Gorelick and Ross 1987; Ross 1981; Ross and Mesulam 1979); cases have also been reported with lesions in the internal capsule (Ross *et al.* 1981) and basal ganglia (Speedie *et al.* 1993).

TRANSCORTICAL MOTOR APROSODIA

Transcortical motor aprosodia is essentially identical to motor aprosodia except that the monotonous tone of voice disappears with repetition. One case was reported secondary to medial frontal lesion (Heilman *et al.* 2004), and another to a lesion in the anterior limb of the internal capsule (Ross 1981).

SENSORY APROSODIA

In sensory aprosodia there is a normal range of intonation in the patient's spontaneous speech. Comprehension, however, is impaired, and one finds mismatches; repetition is also impaired in that when patients repeat the neutral phrase, the intonation of their speech will not be the same as that of the examiner.

Sensory aprosodia has been noted with lesions affecting the temporoparietal cortex (Darby 1993; Gorelick and Ross 1987; Ross 1981); a case has also been reported secondary to a lesion of the thalamus and adjacent posterior limb of the internal capsule (Wolfe and Ross 1987).

TRANSCORTICAL SENSORY APROSODIA

This aprosodia is similar to sensory aprosodia with the exception that repetition is performed well, with the patient accurately mimicking the examiner's tone of voice. Cases have been described secondary to a lesion of the anterior temporal cortex (Ross 1981), and the striatum and adjacent posterior limb of the internal capsule (Gorelick and Ross 1987).

GLOBAL APROSODIA

In global aprosodia, speech is monotonous and comprehension of other's prosody is poor; repetition does not improve the patient's prosodoic output. Global aprosodia has been noted with lesions affecting both the frontal operculum and the posterior temporal cortex (Darby 1993; Ross 1981).

TRANSCORTICAL MIXED APROSODIA

Transcortical mixed aprosodia, which might just as well be called transcortical global aprosodia, is essentially identical to global aprosodia with the exception that the patient's prosody improves with repetition. A case has been reported with a lesion involving the frontoparietal cortex (Ross 1981).

CONDUCTION APROSODIA

In conduction aprosodia spontaneous speech shows a normal range of intonation, and comprehension is intact. Remarkably, however, there is a 'mismatch' between the patient's statement as to how he or she is feeling and the tone in which that statement is made. Furthermore, and remarkably so in light of the preserved comprehension of prosody, patients are unable to repeat sentences with the prosody spoken by the examiner. Conduction aprosody appears to be very rare; a case was reported with a lesion involving the temporoparietal cortex (Gorelick and Ross 1987).

PURE AFFECTIVE DEAFNESS

In this condition, spontaneous speech shows a normal range of intonation and there is no mismatch present. Despite these preserved abilities however, patients are unable to comprehend the prosody with which others speak. Remarkably, however, patients are able to mimic the examiner's intonation upon repetition testing.

Cases of pure affective deafness have been reported with lesions affecting the posterior frontal and immediately subjacent temporal operculum (Gorelick and Ross 1987; Ross 1981) and also with a large lesion affecting the occipitoparietal cortex (Gorelick and Ross 1987).

Etiology

Almost all reported cases of aprosodia have occurred as part of a stroke syndrome, secondary to either ischemic infarction or, much less commonly, intracerebral hemorrhage. A case has also been reported of a gradually progressive motor aprosodia secondary to a progressive focal atrophy of the right frontal lobe (Ghacibeh and Heilman 2003). Further, there is also a report of motor aprosodia occurring paroxysmally as a simple partial seizure (Bautista and Ciampetti 2003).

Various forms of aprosodia may also be seen in dementia, as for example Alzheimer's disease (Perez-Trullen and Modrego Pardo 1996); however, here associated cognitive deficits, such as disorientation and short-term memory loss, will suggest the correct syndromal diagnosis.

Differential diagnosis

Motor aprosodia must be distinguished from flattened affect and hypophonia. Flattened affect is typically accompanied by a monotone voice; however, here there is also a 'flattening' of the patient's feelings: patients with flattened affect typically report having no feelings and this is in contrast with motor aprosodia wherein patients, although speaking in a monotone, still have feelings, sometimes quite strong ones. Hypophonia, as seen in parkinsonism, is a speech deficit characterized by whispering and low volume, which stands in contrast with the normal volume seen in motor aprosodia.

Sensory aprosodia must be distinguished from pathological laughter and crying and from inappropriate affect. Pathological laughter and crying, as discussed in Chapter 5.7, is characterized by episodes of involuntary laughing or crying (accompanied by, respectively, a mirthful or sad tone of voice) during which the patient has no feelings, except, perhaps, a sense of consternation at being unable to control the emotional display. This, superficially, is similar to the 'mismatch' seen in sensory aprosodia; however, on closer inspection there are differences. First, there is, in fact, no actual 'mismatch' in the sense of a patient feeling one thing and displaying a discordant tone of voice as he or she reports that feeling, for in emotional incontinence patients often are emotionally neutral during the episode. Second, in contrast with aprosodia, which in almost all cases is constant and more or less chronic, emotional incontinence occurs in discrete episodes, in between which there is a congruence between what the patient feels and the tone with which that feeling is reported.

Inappropriate affect is very similar to sensory aprosodia, in that in both these signs there is a mismatch between what the patient feels and the tone of voice in which that feeling is expressed. Differentiating between the two requires attention to comprehension of prosody, which is present in patients with inappropriate affect, and absent in those with sensory aprosodia.

Finally, aphasia must be clearly distinguished from aprosodia. Aphasia represents a disturbance in *what* is said, aprosodia a disturbance in *how* it is said. Consider, for example, two patients who are both grief stricken over a recent loss. The first one, having a motor aphasia, although restricted to simply repeating the word 'sad . . . sad . . . sad' over and over again, might yet say it so lugubriously that the listener has no doubt about the depth of the patient's grief. By contrast, the second patient, with a motor aprosodia, although able to say the words 'I've never felt so sad in my entire life', would say them in such a monotone that the listener might well doubt whether the patient was, in fact, really feeling any sadness.

Treatment

Speech therapy may enable patients to develop compensatory strategies, and, in the case of motor aprosodia, may actually improve patient's prosodic abilities (Rosenbek *et al.* 2006).

3.8 APRAXIA

Apraxia is said to be present when, despite preserved strength and coordination, patients are unable to carry out purposeful activities. Of the large number of different kinds of apraxia described in the literature, the following are considered in this text: ideomotor and ideational apraxia, constructional apraxia, and dressing apraxia.

Clinical features

IDEATIONAL AND IDEOMOTOR APRAXIA

Ideational and ideomotor apraxia both have to do with the use of tools, considered in a broad sense, such as scissors, combs, and knives and forks. To test for apraxia, begin by asking the patient to mimic using a pair of scissors: 'Show me how you'd use a pair of scissors to cut a piece of paper.' Next, ask the patient to mimic using a comb, and, for both the scissors and comb, ask the patient to mimic the action first with the right hand, and then with the left. Finally, ask the patient to mimic using a knife and fork, as if he or she were cutting a piece of meat. When either ideational or ideomotor apraxia is present, the patient will demonstrate significant difficulty in mimicking these actions. In severe cases, patients may not be able to execute any movements even vaguely resembling those requested, but may rather look at their hands in a perplexed sort of way. In very mild cases, however, the only abnormality may be the phenomenon of using a body part as the tool. For example, when asked to mimic using scissors, patients may move their index and middle fingers as if they were the blades of the scissors, rather than making a repetitive, squeezing sort of motion with the hand, as one would actually do when holding a pair of scissors, or, when asked to use a comb, patients may pull their fingers through their hair, as if the fingers were the teeth of the comb, rather than sweeping their hand along the head, as one would do were one holding an actual comb. In such mild cases, however, some care must be taken in interpreting the use of a body part as the tool, as many normal individuals, especially among the elderly (Peigneux and van der Linden 1999), will, in fact, do this. One way to determine if apraxia is actually present in these cases is to ask the patient to try again. For example, if the patient had used his or her fingers as if they were the blades of the scissors, one might say, 'No, you've used your fingers as if they were the blades of the scissors. Try it again this time, as if you were actually holding the scissors in your hand.' While truly apraxic patients will continue to use their body parts as the tool, normal individuals will be able to make the appropriate correction and mimic the tool use accurately.

If the patient is able to adequately mimic these actions, then it is fairly safe to assume that neither ideational nor ideomotor apraxia is present. If, however, there is significant difficulty in mimicking, further testing is in order to determine which of these two types of apraxia is present, and this is readily accomplished by giving the patient the tool in question (a pair of scissors, the patient's comb from the bedside table, and a plastic knife and fork) and asking the patient to use the tool. When ideational apraxia is present, significant difficulties persist: the scissors held in a closed fist, the comb held with the teeth facing in toward the palm, or the fork held by the tines. When ideomotor apraxia is present, however, patients *are* able to pick up the tool and use it without any significant difficulty.

Severe ideational apraxia may constitute a significant disability: patients may be reduced to eating with their hands, and grooming may suffer. Indeed, albeit rarely, such difficulties may constitute a chief complaint. In assessing the history, however, it is important to ask how much assistance patients require with grooming and feeding, as family members, even though they may have to groom and feed the patient, may not mention it. Ideomotor apraxia, however, is virtually never complained of: although it is a significant sign on the neurologic examination, patients rarely, if ever, go about attempting to mimic actions, and hence it simply does not come to light.

Before leaving this discussion of ideational and ideomotor apraxia a word of caution is in order regarding nomenclature. Though these types of apraxia have been discussed for well over a century, there is not, as yet, unanimity in their definition, and consequently it behooves the reader, when perusing articles on apraxia, to carefully examine the definition used by the authors.

CONSTRUCTIONAL APRAXIA

Constructional apraxia is said to be present when patients are unable to copy figures or to spontaneously draw them, and bedside testing is performed with pencil and paper. First, draw some simple geometric forms, such as a triangle, and ask the patient to copy them. Next, ask the patient to draw a daisy, a house, and a stick figure. In assessing the results, one is obviously not judging artistic merit, but, rather, attention is focused on any disorganization of the drawn figure.

DRESSING APRAXIA

In dressing apraxia there is, as the name implies, a significant difficulty in getting dressed, which cannot be accounted for by weakness, fatigue, or clumsiness. One patient, described by Hecaen *et al.* (1956) 'mentioned spontaneously that he had had a great deal of trouble in dressing and that he must frequently be helped by his wife. "I get puzzled in dressing. I get my clothes mixed." He said he put his arms in the wrong sleeve of his shirt or he put the back of his shirt in front.'

Etiology

IDEATIONAL AND IDEOMOTOR APRAXIA

Ideational and ideomotor apraxia both lateralize to the left hemisphere. Both may occur acutely, as part of a stroke syndrome: ideational apraxia may be seen with lesions of the parietal cortex (De Renzi and Lucchelli 1988), lentiform nucleus (De Renzi *et al.* 1986), and thalamus (De Renzi *et al.* 1986; Warren and Thompson 2000), and ideomotor apraxia may likewise occur with lesions of the parietal cortex (Heilman 1973), lentiform nucleus (Agostoni *et al.* 1983), or thalamus (Agostoni *et al.* 1983; Nadeau *et al.* 1994). Of note, both acute ideational (Watson and Heilman 1983) and ideomotor (Kazui and Sawada 1993) apraxia may also occur due to callosal infarction, and in such 'disconnection' cases, the apraxia is found only in the left hand. The existence of this phenomenon accounts for the importance of testing for apraxia with both the right and the left hand, as noted earlier.

Both ideational and ideomotor apraxia may also occur gradually with mass lesions in the same locations, or as the presenting feature of various neurodegenerative disorders such as corticobasal ganglionic degeneration (Riley *et al.* 1990; Rinne *et al.* 1994), Alzheimer's disease (Ross *et al.* 1996) and Pick's disease (Fukui *et al.* 1996).

CONSTRUCTIONAL APRAXIA

Constructional apraxia lateralizes to the right hemisphere, and may be seen as part of a stroke syndrome with lesions of the parietal cortex (Piercy *et al.* 1960) or thalamus (Motomura *et al.* 1986). With callosal infarction (Giroud and Dumas 1995), a 'disconnection' constructional apraxia may occur wherein drawing with the right hand suffers, while that with the left hand remains at the premorbid level.

DRESSING APRAXIA

Dressing apraxia, like constructional apraxia, lateralizes to the right hemisphere, and has been noted as part of a stroke syndrome due to infarction of the parietal cortex (Brain 1941; Hecaen *et al.* 1956).

Differential diagnosis

Apraxia is but one reason why patients may be unable to carry out purposeful actions, and, before the diagnosis is made, one must be certain that the difficulty is not due to weakness or poor coordination.

Ideational and ideomotor apraxia, as noted, may occur with left parietal lesions, and in many cases the infarction is large enough to also cause a sensory aphasia, which, in turn, may make it impossible for the patient to understand the instruction to mimic, or even merely to use, the tool in question, thus making formal testing very difficult. All is not lost here, however, for one may simply observe the patient during meal time, or in the morning when he or she is in front of a mirror with a comb at hand: when a moderate or severe ideational apraxia is present, the difficulty will be immediately apparent.

Constructional apraxia, to a degree, may be simulated by left neglect, as may be seen with right parietal lesions. In neglect, however, the deficient drawing is present only on the left side of the figure, whereas the right side of the figure is drawn more or less normally; by contrast, in constructional apraxia both the left and right halves of the figure are poorly drawn.

Dressing apraxia may also be simulated by left neglect, as patients may leave the left side of their dressing unattended to, with the left shirt sleeve dangling or the left shoe untied. Here, however, as with the differential with constructional apraxia, the clue to the diagnosis of neglect is the presence of adequate dressing on the right side of the body.

Treatment

Occupational therapy may assist the patient in developing compensatory strategies.

3.9 AGNOSIAS

Agnosia is a condition characterized by a failure, despite adequate sensation, to recognize phenomena. There are a number of different kinds of agnosia, and these may be grouped as described below.

Certain agnosias are distinguished by the sensory modality affected. Thus, in visual agnosia there is a failure to recognize an object by sight, in tactile agnosia by touch, and in auditory agnosia by the sound made by the object. Other agnosias are characterized by a specific kind of feature that cannot be recognized: in prosopognosia there is difficulty in recognizing faces, in topographagnosia landmarks go unrecognized and patients get lost, and in color agnosia patients cannot recognize various colors. Two other agnosias are marked by an inability to recognize certain facts: in anosognosia, patients fail to recognize certain signs and symptoms, as for example hemiparesis; in asomatognosia patients fail to recognize that a body part, for example a hemiparetic arm, belongs to them. Finally, there is a form of agnosia, namely simultanagnosia, wherein patients fail to simultaneously recognize all the objects in their view, as if one or more of them had actually disappeared. Each of these agnosias is now discussed in turn.

Visual agnosia

Visual agnosia, or the inability to recognize and name objects by sight, is a rare disorder.

CLINICAL FEATURES

In visual agnosia, patients, despite adequate visual acuity, are unable to recognize familiar objects by sight alone. Thus, if shown a pair of scissors, they will be unable to name it and they will be unable to say what it is used for. Interestingly, however, if they are given the object and allowed to handle it, they are able to recognize it by touch.

Visual agnosia may be broken down into two subtypes: apperceptive and associative. The clinical differentiation of these subtypes is based on the patient's ability, after being shown the object in question, to either draw it or pick it out from a photograph or drawing of a group of different objects (one of which is the object in question). In apperceptive visual agnosia patients can neither make a drawing of the object nor pick it out from a group of objects. In associative visual agnosia, however, patients are able to perform these tasks.

At a phenomenological level, it seems that when apperceptive subtype patients are shown an object they do not experience an image of it and, lacking such an image, have no subsequent recognition and, of course, no ability to make a drawing. In associative visual agnosia, however, an image does occur, as evidenced by the patient's ability to make a drawing of the object or to pick it out of a group of objects. Despite the presence of an image, however, there is an inability to make a connection between that image and the concept of that object, and thus a failure to recognize it, name it, or say what it is used for.

Of note, visual agnosia is typically more severe for small objects, such as a pair of scissors, than it is for large objects, such as chairs or desks, which patients are generally able to recognize and name. Furthermore, the presence of a larger object may, by providing a context, allow a patient to recognize a smaller object, which, if seen in isolation, he or she would be unable to name. For example, if shown a pair of scissors on a desk, the patient might be able to name it, whereas if shown the scissors in isolation, perhaps by placing them on the bed sheet, the patient would be unable to do so.

ETIOLOGY

Apperceptive visual agnosia has been noted with bilateral infarction of the occipital lobes, which spare the striate cortex but involve the secondary visual cortices; the adjacent temporal lobes are often also involved but only in their more posterior extent (Benson and Greenberg 1969; Ferreira *et al.* 1998; Shelton *et al.* 1994).

Associative visual agnosia may occur secondary to bilateral infarction of the medial occipitotemporal cortex and subcortical white matter, especially involving the lingual, fusiform, and parahippocampal gyri (Albert *et al.* 1979). Cases have also been reported secondary to a left unilateral occipitotemporal infarct coupled with infarction of the splenium of the corpus callosum (Feinberg *et al.* 1994). This latter mechanism represents a kind of 'disconnection' syndrome wherein, although the right occipital lobe is untouched, the lesion of the splenium disconnects it from the left hemisphere.

Theoretically, it appears reasonable to say that in the apperceptive subtype the destruction of the secondary visual cortex prevents, as it were, the 'working-up' of the raw visual experience into a coherent image. In the associative subtype, sparing of the secondary visual cortices allows for the development of an image, but destruction of the more anterior occipitotemporal cortex renders impossible an association between the image and the concepts that the patient has regarding various objects.

DIFFERENTIAL DIAGNOSIS

Both motor aphasia and anomic aphasia may, at times, occasion some differential diagnostic difficulty. In motor aphasia, some patients, though in fact having the name of the object in mind, may not be able to speak it. Here, the ability of the patient to 'circumlocute' around the aphasia, and give a description of what the object is used for, indicates that visual gnostic abilities are intact.

Anomic aphasia may be distinguished by the patient's response to being allowed to handle the object. The anomic patient, upon handling the object, will still remain unable to come up with the name, whereas the agnosic patient will recognize and name the object.

TREATMENT

Speech/language and occupational therapy may be effective, especially with regard to helping patients develop compensatory strategies.

Tactile agnosia

Tactile agnosia is characterized by an inability to recognize objects by touching and handling them, despite normal light touch, pin-prick, vibratory and two-point discriminatory sensation, and despite an ability to describe the shape of the object in question (Platz 1996).

CLINICAL FEATURES

After making sure that the patient's eyes are kept closed, place a common object, such as a key, in the patient's hand and ask the patient to identify it by touch alone. If the patient has any difficulty in doing so, ask for a description of the object. In cases of tactile agnosia, although the patient can describe the object's shape (e.g. as having a serrated edge), there is an inability to name it or to describe its function. At this point, ask the patient to take a look at the object and name it: in tactile agnosia, the patient will be able to name it immediately by sight.

Importantly, because tactile agnosia may occur unilaterally in either hand it is necessary to test both hands, using a different object each time; furthermore, it is also necessary to tell the patient to use only one hand at a time and not to palpate the object with both hands.

ETIOLOGY

Tactile agnosia generally occurs secondary to a lesion in the inferior parietal cortex (Caselli 1993), on either the right (Platz 1996) or the left (Reed *et al.* 1996).

DIFFERENTIAL DIAGNOSIS

Astereognosia is similar clinically to tactile agnosia in that patients are unable to recognize objects by touch alone; the difference lies in the fact that patients with astereognosia are unable to describe the shape of the object, whereas patients with tactile agnosia can.

Anomic aphasia may also appear similar to tactile agnosia in that anomic patients also are unable to name an object by touching it. In contrast with tactile agnosia, however, patients with anomic aphasia will remain unable to name the object when they are allowed to see it.

TREATMENT

Except in the visually impaired, treatment is rarely required.

Auditory agnosia

Auditory agnosia, or, more explicitly, environmental auditory agnosia, is a very rare condition characterized by an inability to recognize such environmental sounds as the ringing of a telephone, a knock on the door, or the honking of a horn, despite normal hearing and a normal ability to understand the spoken word (Vignolo 1982).

CLINICAL FEATURES

Auditory agnosia may come to attention when patients fail to answer the phone, or open the door, or perhaps, with potentially disastrous consequences, fail to respond to the honking of a car horn. Bedside testing may be accomplished by standing behind the patient and rustling some paper, ringing a bell, or perhaps snapping your fingers, and then asking the patient what he or she heard. In environmental auditory agnosia, patients will acknowledge hearing something but they will be unable to say what it was.

ETIOLOGY

Auditory agnosia has been reported with an infarction of the posterior portion of the right temporal lobe (Fujii *et al.* 1990); it has also been noted to evolve out of a syndrome of cortical deafness secondary to lesions affecting the auditory radiations of the retrolenticular portion of the internal capsule bilaterally (Taniwaki *et al.* 2000).

DIFFERENTIAL DIAGNOSIS

Deafness is readily distinguished in that here there is no response to any sound. Pure word deafness, as discussed in Chapter 2.1, is a form of aphasia in which patients are unable to comprehend the meaning of the word, although they are able to recognize environmental sounds and the spoken word as a word.

TREATMENT

Speech/language therapy may enable the patient to develop compensatory strategies.

Prosopagnosia

The term prosopagnosia is derived from the Greek word for face, *prosopon*, and refers to a remarkable condition in which patients are unable to recognize others by looking at their faces (Sergent and Poncet 1990).

CLINICAL FEATURES

Patients with prosopagnosia, although able to recognize faces as faces, and indeed able to describe accurately the facial features of others, are yet unable to identify the other person (Tranel *et al.* 1988). Remarkably, these patients, although unable to identify others by their facial features, may be able to identify them by other features, such as their voice, dress, or characteristic gait (Damasio *et al.* 1982b).

Some examples may help to clarify this remarkable condition. In one case, the patient, a party to a lawsuit, had come to identify his own lawyer by the setting in which the two of them generally met, namely the lawyer's office; all went well until trial, when the patient, unable to recognize people in the courtroom, went up to the opposing counsel and began to discuss confidential details of the case 'with disastrous consequences' (Pevzner *et al.* 1962). In another case, a patient had come to rely on his wife's clothing as a means of identifying her: at a party, he found himself unable to recall what she had put on and although she walked right by him he failed to recognize her (Hecaen and Angelergues 1962). In a similar vein, another patient 'failed to recognize a physician who had just examined him after the doctor had substituted his suit jacket for his white coat' (Hecaen and Angelergues 1962). Finally, and most remarkably, some patients are not able to recognize themselves when they look in the mirror (Damasio *et al.* 1980).

As might be expected, such patients, in addition to being unable to recognize old acquaintances by their facial features, are also unable to utilize facial features to recognize new acquaintances (Malone *et al.* 1982).

ETIOLOGY

Prosopagnosia, in the vast majority of cases, occurs as part of a stroke syndrome, with infarction of the fusiform gyrus: although in most cases, bilateral infarction is found (Cohn *et al.* 1986; Damasio *et al.* 1980, 1982b; Meadows 1974b; Pevzner *et al.* 1962), there are definite instances where unilateral, right infarction has caused the syndrome (Cohen *et al.* 1994; Landis *et al.* 1986, 1988; Rosler *et al.* 1997; Wada and Yamamoto 2001).

Outside of stroke, prosopagnosia has also been noted secondary to resection of the posterior portion of the right temporal lobe (Mesad *et al.* 2003) and after a right hemispherectomy (Sergent and Villemure 1989). There are also rare reports of progressive prosopagnosia occurring secondary to equally progressive atrophy of the right temporal lobe (Evans *et al.* 1995; Garrido *et al.* 2009; Joubert *et al.* 2003), forming a syndrome that might be called 'primary progressive prosopagnosia'.

Interestingly, prosopagnosia may also occur on a paroxysmal basis as a simple partial seizure: in one case, a patient with left occipitotemporal scarring had seizures characterized by a sense of 'flickering lights' followed by a brief episode of prosopagnosia (Agnetti *et al.* 1978).

Finally, there are also rare reports of prosopagnosia occurring on a developmental basis (Kress and Daum 2003).

DIFFERENTIAL DIAGNOSIS

Anomic aphasia may render patients incapable of coming up with the name of another person; in such cases, however, in contrast with prosopagnosia, there is no difficulty in recognizing the other person or in specifying the relationship that exists between the patient and the other person.

TREATMENT

Speech/language and occupational therapy may assist the patient in developing compensatory strategies.

Topographagnosia

At the outset, it should be noted that there is some debate regarding the definition of topographagnosia and, consequently, the reader, in perusing journal articles, should pay particular attention to the definition peculiar to the authors of the article. In this text, topographagnosia refers to a condition in which patients, despite adequate memory and vision, are unable to find their way in surroundings that had not previously caused any difficulty. This group of patients may be further subdivided into two types, namely one wherein the difficulty lies in an inability to recognize familiar landmarks (termed here 'landmark agnosia') and another wherein, despite a preserved ability to recognize landmarks, there is an inability to utilize these landmarks in deciding which way to proceed (herein referred to as 'topographical disorientation').

CLINICAL FEATURES

Finding one's way depends both on recognizing landmarks and on knowing which way to proceed upon encountering those landmarks. Thus, if one's accustomed route home from the office involves walking down the main street to the bank and then turning right, then in order to successfully navigate this route one must first recognize the bank as the landmark and then, second, know that at this landmark one must turn right.

In landmark agnosia, difficulty arises at the level of recognizing the bank as a landmark. Although the patient may recognize the building, and even describe it, it is not recognized as a landmark and hence the patient is as likely to walk past it as to turn.

In topographic disorientation, the situation is a little different. Here, the patient does recognize the bank as a landmark, but, having lost a 'sense of direction', the patient is unable to figure out whether to turn to the right or left, or simply proceed onward. Some examples may flesh this out. In one case, a patient 'had sudden difficulty finding her way out of a subway station she had used daily for years. She had no difficulty recognizing the surrounding buildings, but she was unable to find her way to work or to her own home despite multiple attempts' (Alsaadi *et al.* 2000). In another case, the patient: '. . . suddenly lost his way home, where he was going on foot. The buildings in front of him were familiar to him, so he could recognize them right away. However, he did not know which direction his home was from there. Relying on cues from buildings, surrounding scenery and signs, but taking several wrong turns along the way, he eventually arrived in front of his own home, and knew immediately that it was in fact his own home' (Takahashi *et al.* 1997).

ETIOLOGY

Landmark agnosia has been reported with infarction involving the medial occipitotemporal area (primarily the fusiform and lingual gyri, with some involvement of the parahippocampal gyrus), either unilaterally on the right or bilaterally (Aguirre and D'Esposito 1999). An example was also reported as a sequela to herpes simplex encephalitis wherein there was damage to both temporal lobes, with the right side being much more severely involved (McCarthy *et al.* 1996).

Topographical disorientation has also been reported with infarction of the bilateral fusiform, lingual, and parahippocampal gyri (Alsaadi *et al.* 2000), and also with infarction of the right parahippocampal gyrus (Habib and Sirigu 1987; Luzzi *et al.* 2000). Interestingly, cases have also been reported secondary to infarction of the splenium (Alsaadi *et al.* 2000) or right retrosplenial cortex (Takahashi *et al.* 1997).

Further, there are also case reports of topographagnosia occurring on a transient basis, with episodes lasting from 5 to 30 minutes (Stracciari *et al.* 1994). During these episodes, apart from the topographagnosia, patients are otherwise normal and recovery is complete. Most patients were women in their middle or later years and, although the etiology is unknown, it is speculated to be similar to that of transient global amnesia.

DIFFERENTIAL DIAGNOSIS

Patients with amnesia may lose their way in unfamiliar surroundings. The basis for this, however, lies in an inability to commit to memory features of the new environment, and this stands in contrast with topographagnosia wherein memory is intact. Furthermore, patients with amnesia typically have no trouble finding their way in environments that they have traveled in before, provided, of course, that any retrograde amnesia has not obscured their recall.

Patients with dementia or delirium often lose their way and, although this may be on the basis of memory loss (or, in the case of delirium, confusion), it can also occur in patients with relatively preserved memory, and in these cases it is perhaps appropriate to consider the topographagnosia as but one indicator of widespread cortical damage or dysfunction. The existence of other cognitive deficits, however, indicates the correct syndromal diagnosis.

TREATMENT

Speech/language and occupational therapy may assist the patient in developing compensatory strategies.

Color agnosia

CLINICAL FEATURES

In color agnosia, patients, despite normal color vision, are unable to recognize and name the color of objects (Kinsbourne and Warrington 1964; Meadows 1974a).

ETIOLOGY

Color agnosia may occur with lesions of the splenium and the medial portion of the left occipital cortex, and is often seen in association with alexia without agraphia (Geschwind and Fusillo 1966).

DIFFERENTIAL DIAGNOSIS

Color agnosia must be distinguished from acquired *color blindness* or, as it is also known, central achromatopsia (Damasio *et al.* 1980), this distinction being readily accomplished with Ishihara plates. Whereas patients with achromatopsia are unable to read the plates, patients with color agnosia are. The patient with achromatopsia exists in a world of grays; by contrast, the patient with color agnosia, although able to discern hues, cannot name them.

TREATMENT

Occupational therapy may assist in devising compensatory strategies.

Anosognosia

Anosognosia, a term coined by Babinski in 1914, is characterized by a failure to recognize neuropsychiatric deficits. Although hemiparesis is the deficit most commonly involved (and indeed anosognosia for hemiparesis is very common in the first few months after stroke [Hier 1983a,b]), other deficits, as noted below, may also go unrecognized.

CLINICAL FEATURES

The severity of the anosognosia may range from a mere tendency to minimize the extent of the deficit to a strong denial that any deficit at all exists (Baier and Karnath 2005). Upon questioning patients with anosognosia for hemiparesis, some patients may admit that there is something wrong but then go on to say that the paretic limb is only 'a bit stiff', or perhaps suffers from some 'heaviness' (Cutting 1978; Roth 1949); in other cases, 'the patient behaves as though he knew nothing about his hemiplegia, as though it had not existed, as though his paralyzed limbs were normal' (Gerstmann 1942). Even more remarkably, in some cases the patient may insist that the paralyzed limb, although motionless, is in fact moving (Feinberg *et al.* 2000).

Other deficits that may go unrecognized include hemianopia, cortical blindness, chorea, and cognitive or behavioral deficits.

Anosognosia for hemianopia after stroke was noted in over one-half of patients in one study (Celesia *et al.* 1997), a fact that emphasizes the absolute necessity of meticulous visual field testing.

Anosognosia for cortical blindness, known as Anton's syndrome, represents a most remarkable clinical phenomenon (Bergman 1957; Redlich and Dorsey 1945; Symonds and MacKenzie 1957). Patients, although blind, may insist that their vision is perhaps only slightly impaired, that the light is too dim, or they may flatly insist that there is nothing wrong with their vision at all; attempting to walk, they predictably bump into things, trip, and fall.

Abnormal movements may also go unrecognized, and this is particularly the case with chorea in Huntington's disease (Snowden *et al.* 1998), wherein patients often stoutly maintain that there is absolutely nothing wrong.

Cognitive and behavioral deficits may also be denied by patients with dementia, who may insist on keeping their own checkbooks or driving, with disastrous results.

ETIOLOGY

Anosognosia for hemiparesis is most typically seen as part of a stroke syndrome with infarction in the area of distribution of the superior division of the middle cerebral artery involving both the parietal and the frontal lobes. Although the vast majority of cases involve infarction of the right hemisphere with left hemiparesis (Cutting 1978; Roth 1949), cases have been reported of anosognosia for right hemiparesis; furthermore, it is speculated that the relative rarity of reports of anosognosia for right hemiparesis may actually be due to an associated sensory aphasia, which makes testing for anosognosia very difficult (Weinstein *et al.* 1969). Anosognosia for hemiparesis may also be seen with infarction or hemorrhage of the following areas: right posterior limb of the internal capsule and adjacent lenticular nucleus (House and Hodges 1988), right thalamus (Leibson 2000; Motomura *et al.* 1986), and the pons, on either the right or the left side (Bakchine *et al.* 1997; Evyapan and Kumral 1999).

Anosognosia for hemianopia has been noted with unilateral infarction in the area of distribution of the posterior cerebral artery (Celesia *et al.* 1997) and Anton's syndrome, with bilateral infarctions in the area of distribution of the posterior cerebral arteries.

Anognosia for chorea, as noted above, has been noted in Huntington's disease.

Cognitive and behavioral deficits are often denied in patients with Alzheimer's disease (Kashiwa *et al.* 2005; Starkstein *et al.* 1997), frontotemporal lobar degeneration (Eslinger *et al.* 2005), and traumatic brain injury (Sbordone *et al.* 1998; Flashman and McCallister 2002).

DIFFERENTIAL DIAGNOSIS

In all cases, anosognosia must be differentiated from emotionally motivated denial. This distinction must not be based on the vehemence with which patients deny any disability, as some patients with anosognosia may be quite insistent in their assertion that nothing is wrong; rather it should be based on the patient's response to a display of sympathy from others. Whereas patients with anosognosia are unmoved by sympathy, patients 'in denial' will often break down and tearfully discuss the fears aroused by their disabilities.

Anosognosia for cognitive deficits may be mimicked by a frontal lobe syndrome, wherein disinhibition and jocularity prevent patients from acting on any recognition they might have of their deficits.

TREATMENT

Patients should be seen in rehabilitation therapies and, pending resolution of the anosognosia, protective measures may be required.

Asomatognosia

In asomatognosia, patients with left hemiparesis fail to experience the hemiparetic extremity as belonging to them.

CLINICAL FEATURES

In some cases, patients, in addition to denying that the paretic arm is theirs, may go on to say that in fact the arm belongs to someone else, perhaps a family member (Brock and Merwarth 1957; Feinberg *et al.* 1990).

The presence of asomatognosia can lead to some rather bizarre interviews. One patient, asked about the hand, replied that it was '[a] hand that was left on the subway and they brought it here and they put it on me' (Feinberg *et al.* 2010). In another case (Sandifer 1946), a female patient with a left hemiplegia, when her paretic left hand was held up in front of her, indicated that it was not hers but belonged to the physician. When the physician attempted to correct her by pointing out that the hand in question had her wedding ring on it, the patient responded, 'That's my ring, you've got my ring, Doctor.'

ETIOLOGY

Most cases are due to infarction involving the right parietal cortex in the area of the supramarginal gyrus (Feinberg *et al.* 1990); cases have also been reported secondary to infarction involving the right striatum and adjacent internal capsule (Healton *et al.* 1982) and to hemorrhage in the right thalamus (Leibson 2000; Motomura *et al.* 1986). Rarely, asomatognosia may constitute the sole symptomatology of a simple partial seizure, wherein the seizure focus was in the right parietal lobe (So and Schauble 2004).

DIFFERENTIAL DIAGNOSIS

In the alien hand syndrome, although patients experience the left arm as being under the control of some 'alien' entity, they still acknowledge that the arm is theirs; furthermore, in the alien hand syndrome the left arm is quite active, whereas in asomatognosia the arm is paretic.

TREATMENT

Rehabilitation therapies may be helpful.

Simultanagnosia

Simultanagnosia, first described by Wolpert in 1924, is a remarkable syndrome in which patients, despite normal visual acuity, intermittently fail to see all parts of their environment simultaneously, with certain objects literally falling out of view, as if they had become temporarily invisible.

CLINICAL FEATURES

Simultanagnosia may present clinically either because objects 'disappear' or, conversely, 'appear'. In one case (Coslett and Saffran 1991), a patient, while walking through the apartment she had lived in for 25 years, 'fell over her dining room table' because she didn't see it as she crossed the room. The same patient reported the converse experience while watching a movie. In the movie, a character was having a heated argument, and the patient, while watching the movie, 'noted to her surprise and consternation that the character she had been watching was suddenly sent reeling across the room, apparently as a consequence of a punch thrown by a character she had never seen'.

Bedside testing may be accomplished either by asking patients to describe everything they see in their room, or by showing them a picture or photograph of a complex scene, and asking, again, for them to describe every object that they see.

ETIOLOGY

Simultanagnosia has been noted with infarction of the right temporo-occipital area (Cohen *et al.* 1994; Coslett and Saffran 1991), and with bilateral infarction of the posterior occipitoparietal area (Rizzo and Hurtig 1987; Rizzo and Robin 1990). Cases have also been reported secondary to progressive neurodegenerative conditions in the syndrome of posterior cortical atrophy (Tang-Wei *et al.* 2004).

Simultanagnosia may also occur as part of Balint's syndrome (Chapter 5.14) and, given this, all patients with simultanagnosia should be tested for optic ataxia and optic apraxia.

DIFFERENTIAL DIAGNOSIS

Hemianopia and visual neglect are experienced to only one side in contrast with simultanagnosia, wherein objects may be 'lost' from any point of the visual field.

TREATMENT

Speech/language and occupational therapy may enable patients to develop compensatory strategies.

3.10 NEGLECT

In this aptly named syndrome, patients, in one fashion or another, neglect or fail to attend to phenomena on one side, usually the left. Neglect may be either visual or motor: in visual neglect (also known as spatial neglect), as might be expected, objects are not attended to in one of the hemifields; in motor neglect, there is an 'underutilization' of the limbs on one side, as if, in a sense, the patient 'neglects' to use them.

Clinical features

Both forms of neglect are considered in turn, beginning with visual neglect. Importantly, the physician must test for both visual and motor neglect before concluding that neglect, *per se*, is, or is not, present. This is the case because visual and motor neglect may occur independently of each other (Laplane and Degos 1983).

VISUAL NEGLECT

Visual neglect may come to clinical attention in a variety of ways (Azouvi *et al.* 2002). Patients may fail to comb their hair, shave, or put on make-up on the neglected side, and food on the neglected side of a dinner plate may go uneaten. In talking with a group, patients may fail to speak with those on the neglected side, and, if patients are looking for something, they may fail to find it if it is on the neglected side, even if it is in plain view. In one case (Cherington 1974) left neglect ruined a patient's chess game: although chess pieces on the right side of the board were moved as always, those on the left, neglected, remained in position, and were easy prey for his opponent. In another case (Frantz 1950), a patient, while driving, began to run into things (such as pedestrians) on the left. Importantly, as in all cases of visual neglect, these collisions did not occur because of a hemianopia: the patient had full visual fields but simply did not attend to things to his left.

Interestingly, visual neglect may also extend to *imagined* scenes. In one study, for example, patients with neglect were instructed to imagine that they were standing on one side of a famous plaza and then describe what they saw: as might be expected, in their description of the imagined scene, they failed to speak of things on the plaza that were located on the left of the imagined scene (Bisiach and Luzzatti 1978).

Bedside testing for visual neglect may be accomplished with four 'paper and pencil' tests: line bisection, line cancellation, clock drawing, and daisy drawing. In order to obtain reliable results for these tests, the patient must be seated squarely in front of a table, the trunk parallel to the edge of the table and the piece of paper placed directly in front of the patient, such that the midline of the paper is continuous with the patient's midline. Although patients may move their head in any direction, and look in any direction, this relative position of the paper and the patient's trunk must be maintained. The reason why the position of the patient's trunk is so important in these four tests is that if the trunk is angled toward the neglected side the portion of the visual field subject to neglect will shrink, to the point at which the tests may become falsely negative (Beschin *et al.* 1997; Karnath and Hartje 1993).

In the line bisection test a single line is drawn lengthwise on the piece of paper, with the middle of the line resting at the midline of the piece of paper. The patient is then asked to make a mark directly on the middle of the line. When left neglect is present, the mark made by the patient will be to the right of the true midline. Importantly, the line should be of the order of 10 cm or longer because if the line is substantially shorter than this, there may be a 'crossover' effect, whereby the patient with left neglect will place the mark not to the right but to the left of the true midline (Anderson 1997; Tegner and Levander 1991).

In the line cancellation test (Albert 1973) the examiner places a large number of short, straight lines randomly on the surface of the piece of paper, the different lines being oriented at various and random angles, and then asks the patient to simply mark off each line. In a positive test, the proportion of lines marked off to the left of the midline will be substantially less than the proportion marked off to the right.

In the clock drawing test, the patient is asked to draw a large circle on the paper and then to put in all 12 numbers, as on a clockface. In a positive test, the numbers will be more or less 'bunched up' on the right side.

In the daisy drawing test, patients are asked to draw a daisy, and in a positive test, the petals of the daisy will end up being bunched on the right-hand side. Importantly, patients should be asked to simply 'draw a daisy' and not given instructions to 'arrange' the petals around a circle, because when patients are given such an instruction, the neglect may 'vanish' as the petals are arranged, one by one, evenly around the circle (Ishiai *et al.* 1997).

Importantly, each one of these tests must be performed before concluding that the bedside examination is negative, as some patients with neglect may 'pass' one, but not another (Azouvi *et al.* 2002; Binder *et al.* 1992; Ishiai *et al.* 1993).

Another bedside test for visual neglect involves testing for what is known as visual extinction. To test for visual extinction, establish first, with confrontation testing, that the visual fields are full. Stand directly in front of the patient, perhaps an arm's length away, and spread your arms to the sides such that your hands end up level with the patient's eyes and perhaps 30 cm in front of the patient. Next, ask the patient to stare at your nose and to point to the finger that wiggles, all the while continuing to look straight at your nose. To perform the test, wiggle one index finger at a time: if the patient points to each finger when it wiggles, then, at least on this confrontation testing, the fields are full. At this point, while maintaining the same position, wiggle both fingers simultaneously: with a positive test, the movement on the neglected side is 'extinguished' and the patient points only to the movement on the unaffected side.

MOTOR NEGLECT

Motor neglect (Laplane and Degos 1983; Triggs *et al.* 1994) may come to clinical attention in a peculiarity of gait: on the affected side, there may be reduced arm swing, and the leg, generally being underutilized and 'left behind', is seen to be 'dragged' along to keep up with the leg on the unaffected side. When patients are in a wheelchair, the arm on the affected side may dangle down from the shoulder and be dragged along passively on the floor as the wheelchair is pushed forward; for patients in bed, the affected leg may be 'left behind', still stretched out on the bed, as the patient swings the leg on the unaffected side over the edge of the bed while attempting to rise. A more subtle form of motor neglect may occur as patients neglect to chew on the neglected side, with food consequently dribbling out (Andre *et al.* 2000).

Bedside testing for motor neglect may be accomplished by asking patients to fasten a button or clap their hands: when motor neglect is present, the arm on the neglected side will either not participate in the task or do so only minimally. Importantly, this is not the result of a lack of strength or coordination: with strong urging, patients are generally able to bring the affected arm into play such that there is more or less full bimanual cooperation in the task at hand.

In observing patients during hand clapping, one may find what is known as the 'Eastchester clapping sign' (Ostrow and Llinas 2009). To test for this sign, it is important that patients first have both hands resting to the side before being asked to clap. When the sign is present the right hand will swing up toward the midline, but then stop abruptly at the midline, as if the left hand had already been correctly brought into position to make the clap. This sign, though typically present during the acute stage of stroke, generally resolves with time, and, as it does, patients may cross the midline with the right hand to try and bring it into contact with the relatively unmoving left one.

Etiology

As indicated earlier, left neglect, due to a lesion in the right hemisphere, is clinically far more prominent than right neglect. Although right neglect certainly does occur, it is less common, less severe, and, in the case of stroke, less enduring (Beis *et al.* 2004; Stone *et al.* 1991).

Although neglect may occur secondary to both cortical and subcortical lesions, in the great majority of cases, it is a lesion of the cortex that is responsible (Vallar and Perani 1986; Vallar *et al.* 1994). Cortical lesions capable of causing neglect, though most commonly found in the parietal lobe (Bender 1945; Cherington 1974; Critchley 1949; Frantz 1950) (primarily in the inferior parietal lobule [Mort *et al.* 2003; Vallar and Perani 1986]), may also be found in the frontal lobe [primarily in the posterior portion of the inferior frontal gyrus (Daffner *et al.* 1990; Heilman and Valenstein 1972; Husain and Kennard 1996; Maeshima *et al.* 1994; Stein and Volpe 1983)], and the temporal lobe (primarily the medial aspect of the parahippocampal gyrus [Mort *et al.* 2003]). Subcortical structures implicated include the caudate nucleus (Caplan *et al.* 1990; Kumral *et al.* 1999), the putamen (Hier *et al.* 1977), and the thalamus (Kumral *et al.* 1995; Motomura *et al.* 1986; Watson *et al.* 1981). Rarely, neglect has been noted with a lesion of the posterior limb of the internal capsule (Bogousslavsky *et al.* 1988c; Ferro and Kertesz 1984) and with a combined lesion of the medial aspect of the occipital lobe and the splenium (Chung Park *et al.* 2005).

Although neglect is most commonly seen as part of a stroke syndrome, it may also occur with other focal lesions (such as tumors) and may also be seen as but one sign in a dementing disorder such as Alzheimer's disease.

Differential diagnosis

Visual neglect must be distinguished from hemianopia and constructional apraxia. Hemianopic patients, of course, will fail to attend to objects in the hemianopic field, given that they simply do not see them. Confrontational testing for visual fields, done with sufficient vigor, however, reveals full fields in patients with neglect. In constructional apraxia, drawing a clock face or daisy may be very difficult; however, here one sees a random pattern of errors, rather than a 'bunching up' on the unneglected side.

Motor neglect must be distinguished from hemiparesis, and this distinction is made, as indicated earlier, on the basis of the response to vigorous encouragement. Patients with neglect eventually respond to vigorous direction, whereas patients with hemiparesis simply cannot.

Treatment

Various rehabilitation techniques are helpful for visual neglect, including prism adaptation (Frassinetti *et al.* 2002; Keane *et al.* 2006), neck muscle vibration (Schindler *et al.* 2002), and visual scanning training.

REFERENCES

Achiron A, Ziv I, Djaldetti R *et al.* Aphasia in multiple sclerosis: clinical and radiologic correlation. *Neurology* 1992; **42**:2195–7.

Agnetti V, Carreras M, Pinna L *et al.* Ictal prosopagnosia and epileptogenic damage of the dominant hemisphere. *Cortex* 1978; **14**:50–7.

Agostoni E, Coletti A, Orlando G *et al.* Apraxia in deep cerebral lesions. *J Neurol Neurosurg Psychiatry* 1983; **46**:804–8.

Aguirre GK, D'Esposito M. Topographical disorientation: a synthesis and taxonomy. *Brain* 1999; **122**:1613–28.

Ajax ET, Schenkenberg T, Kosteljanetz M. Alexia without agraphia and the inferior splenium. *Neurology* 1977; **27**:685–8.

Albert ML. A simple test of visual neglect. *Neurology* 1973; **23**: 658–64.

Albert ML, Soffer D, Silverberg R *et al.* The anatomic basis of visual agnosia. *Neurology* 1979; **29**:876–9.

Alexander MP, LoVerme SR. Aphasia after left hemispheric intracerebral hemorrhage. *Neurology* 1980; **30**:1193–202.

Alexander MP, Schmitt MA. The aphasia syndrome of stroke in the left anterior cerebral artery territory. *Arch Neurol* 1980; **37**:97–100.

Alexander MP, Fischette MR, Fischer RS. Crossed aphasias can be mirror image anomalous: case reports, review and hypothesis. *Brain* 1989; **112**:953–73.

Alsaadi T, Binder JR, Lazar RM *et al.* Pure topographic disorientation: a distinctive syndrome with varied localization. *Neurology* 2000; **54**:1864–6.

Anderson B. Pieces of the true crossover effect in neglect. *Neurology* 1997; **49**:809–12.

Anderson SW, Damasio AR, Damasio H. Troubled letters but not numbers. Domain specific cognitive impairments following focal damage in the frontal cortex. *Brain* 1990; **113**:749–66.

Andre JM, Beis JM, Morin N *et al.* Buccal hemineglect. *Arch Neurol* 2000; **57**:1734–41.

Arsland D, Larsen JP, Hoien T. Alexia in dementia of the Alzheimer's type. *Acta Neurol Scand* 1993; **88**:434–9.

Ashtary F, Janghorbani M, Chitsaz A *et al.* A randomized, double-blind trial of bromocriptine efficacy in nonfluent aphasia after stroke. *Neurology* 2006; **66**:914–16.

Assal G, Perentes E, Deruaz J-P. Crossed aphasia in a right-handed patient. *Arch Neurol* 1981; **38**:455–8.

Azouvi P, Samuel C, Louis-Dreyfus A *et al.* Sensitivity of clinical and behavioral tests of spatial neglect after right hemisphere stroke. *J Neurol Neurosurg Psychiatry* 2002; **73**:160–6.

Babinski J. Contribution a l'étude des troubles mentaux dans l'hémiplégie organique cérébrale (anosognosie). *Rev Neurol* 1914; **27**:845–7.

Bahls FH, Chatrian GE, Mesher RA *et al.* A case of persistent cortical deafness: clinical, neurophysiologic, and neuropathologic observations. *Neurology* 1988; **38**:1490–3.

Baier B, Karnath H-O. Incidence and diagnosis of anosognosia for hemiplegia. *J Neurol Neurosurg Psychiatry* 2005; **76**:358–61.

Bakar M, Kirshner HS, Wertz RT. Crossed aphasia: functional brain imaging with PET or SPECT. *Arch Neurol* 1996; **53**:1026–32.

Bakchine S, Ctassaro I, Seilhan D. Anosognosia for hemiplegia after a brainstem haematoma: a pathological case. *J Neurol Neurosurg Psychiatry* 1997; **63**:686–7.

Bartak L, Rutter M, Cox A. A comparative study of infantile autism and specific developmental receptive language disorders. I. The children. *Br J Psychiatry* 1975; **126**:127–45.

Bautista RE, Ciampetti MZ. Expressive aprosody and amusia as a manifestation of right hemisphere seizures. *Epilepsia* 2003; **44**: 466–7.

Beis JM, Keller C, Morin N *et al.* Right spatial neglect after left hemisphere stroke: qualitative and quantitative study. *Neurology* 2004; **63**:1600–5.

Bender MB. Extinction and precipitation of cutaneous sensations. *Arch Neurol Psychiatry* 1945; **54**:1–9.

Benson DF. Psychiatric effects of dysphasia. *Br J Psychiatry* 1973; **123**:555–66.

Benson DF, Geschwind N. Developmental Gerstmann syndrome. *Neurology* 1970; **20**:293–8.

Benson DF, Greenberg JP. Visual form agnosia. A specific defect in visual discrimination. *Arch Neurol* 1969; **20**:82–9.

Benson DF, Weir WF. Acalculia: acquired anarithmetria. *Cortex* 1972; **8**:465–72.

Benson DF, Sheremata WA, Bouchard R *et al.* Conduction aphasia: a clinicopathological study. *Arch Neurol* 1973; **28**:339–46.

Bergman PS. Cerebral blindness. *Arch Neurol Psychiatry* 1957; **78**:568–84.

Bernal B, Ardila A, Altman NR. Acalculia: an fMRI study with implications with respect to brain plasticity. *Int J Neurosci* 2003; **113**:1505–23.

Berthier ML, Green C, Higueras C *et al.* A randomized, placebo-controlled study of donepezil in poststroke aphasia. *Neurology* 2006; **67**:1687–9.

Berthier ML, Green C, Lara JP *et al.* Memantine and constraint-induced aphasia therapy in chronic poststroke aphasia. *Ann Neurol* 2009; **65**:577–85.

Beschin N, Cubelli R, Della Sala S *et al.* Left of what? The role of egocentric coordinates in neglect. *J Neurol Neurosurg Psychiatry* 1997; **63**:483–9.

Binder J, Marshall R, Lazar R *et al.* Distinct syndromes of hemineglect. *Arch Neurol* 1992; **49**:1187–94.

Bisiach E, Luzzatti C. Unilateral neglect of representational space. *Cortex* 1978; **14**:129–33.

Bogousslavsky J. Global aphasia without other lateralizing signs. *Arch Neurol* 1988; **45**:143.

Bogousslavsky J, Regli F. Anterior cerebral artery territory infarction in the Lausanne stroke registry: clinical and etiologic considerations. *Arch Neurol* 1990; **47**:144–50.

Bogousslavsky J, Regli F, Uske A. Thalamic infarcts: clinical syndromes, etiology, and prognosis. *Neurology* 1988a; **38**:837–48.

Bogousslavsky J, Regli F, Assal G. Acute transcortical mixed aphasia: a carotid occlusion syndrome with pia and watershed infarcts. *Brain* 1988b; **111**:631–41.

Bogousslavsky J, Miklossy J, Regli F *et al.* Subcortical neglect: neuropsychological, SPECT, and neuropathological correlations with anterior choroidal artery territory infarction. *Ann Neurol* 1988c; **23**:448–52.

Bragoni M, Altieri M, Di Piero V *et al.* Bromocriptine and speech therapy in non-fluent chronic aphasia after stroke. *Neurol Sci* 2000; **21**:19–22.

Brain WR. Visual disorientation with special reference to lesions of the right cerebral hemisphere. *Brain* 1941; **64**:244–72.

Brock S, Merwarth HR. The illusory awareness of body parts in cerebral disease. *Arch Neurol Psychiatry* 1957; **77**:366–75.

Brown JR, Simonson J. A clinical study of 100 aphasic patients: observations on lateralization and localization of lesions. *Neurology* 1957; **7**:777–83.

Brown JW, Wilson FR. Crossed aphasia in a dextral: a case report. *Neurology* 1973; **23**:907–11.

Caplan LR, Schmahmann JD, Kase CS *et al.* Caudate infarcts. *Arch Neurol* 1990; **47**:133–43.

Cappa SF, Perani D, Bressi S *et al.* Crossed aphasia: PET follow up study of two cases. *J Neurol Neurosurg Psychiatry* 1993; **56**:665–71.

Carota A, Annomi JM, Combremont P *et al.* Hypergraphia, verbal aspontaneity and post-stroke depression secondary to right cyngulate and corpus callosum infarction. *J Neurol* 2003; **250**:508–10.

Caselli RJ. Ventrolateral and dorsomedial somatosensory association cortex damage produces distinct somesthetic syndromes in humans. *Neurology* 1993; **43**:762–71.

Celesia G, Brigell MG, Vaphiades MS. Hemianopic anosognosia. *Neurology* 1997; **49**:88–97.

Cherington MS. Single case study: visual neglect in a chess player. *J Nerv Ment Dis* 1974; **159**:145–7.

Chung Park K, Jeong Y, Hwa Lee B *et al.* Left hemispatial visual neglect associated with a combined right occipital and splenial lesion: another disconnection syndrome. *Neurocase* 2005; **11**:310–18.

Ciemens VA. Localized thalamic hemorrhage: a case of aphasia. *Neurology* 1970; **20**:776–82.

Clark DG, Mendez MF, Farag E *et al.* Clinicopathologic case report: progressive aphasia in a 77-year-old man. *J Neuropsychiatr Clin Neurosci* 2003; **15**:231–8.

Cohen DN, Salanga VD, Hully W *et al.* Alexia without agraphia. *Neurology* 1976; **26**:455–9.

Cohen L, Gray F, Meyrignal C *et al.* Selective deficity of visual size perception: two cases of hemimicropsia. *J Neurol Neurosurg Psychiatry* 1994; **57**:73–8.

Cohen M, Campbell R, Yaghmai F. Neuropathological abnormalities in developmental dysphasia. *Ann Neurol* 1989; **25**:567–70.

Cohn R, Neumann MA, Wood DH. Prosopagnosia: a clinicopathological study. *Ann Neurol* 1986; **20**:177–82.

Corbett AJ, McCosker EA, Davidson OR. Acalculia following a dominant-hemisphere subcortical infarct. *Arch Neurol* 1986; **43**:964–6.

Coslett HB, Saffran E. Simultanagnosia: to see but not two see. *Brain* 1991; **114**:1523–45.

Coslett HB, Brashear HR, Heilman KM. Pure word deafness after bilateral primary auditory cortex infarcts. *Neurology* 1984; **34**:347–52.

Critchley M. Phenomenon of tactile inattention with special reference to parietal lesions. *Brain* 1949; **72**:538–61.

Cutting J. Study of anosognosia. *J Neurol Neurosurg Psychiatry* 1978; **41**:548–55.

Daffner KR, Ahern GL, Weintraub S *et al.* Dissociated neglect behavior following sequential stroke in the right hemisphere. *Ann Neurol* 1990; **28**:97–101.

Damasio AR, Damasio H. The anatomic basis of pure alexia. *Neurology* 1983; **33**:73–83.

Damasio AR, Yamada T, Damasio H *et al.* Central achromatopsia: behavioral, anatomic and physiologic aspects. *Neurology* 1980; **30**:1064–71.

Damasio AR, Damasio H, Rizzo M *et al.* Aphasia with nonhemorrhagic lesion in the basal ganglia and internal capsule. *Ann Neurol* 1982a; **39**:15–20.

Damasio AR, Damasio H, Van Hoesen GW. Prosopagnosia: anatomic basis and behavioral mechanisms. *Neurology* 1982b; **32**:331–41.

Damasio H, Damasio AR. The anatomical basis of conduction aphasia. *Brain* 1980; **30**:337–50.

Darby DG. Sensory aprosodia: a clinical clue to lesions of the inferior division of the right middle cerebral artery? *Neurology* 1993; **43**:567–72.

Day JT, Fisher AG, Mastaglia FL. Alexia with agraphia in multiple sclerosis. *J Neurol Sci* 1987; **78**:343–8.

De Renzi E, Lucchelli F. Ideational apraxia. *Brain* 1988; **111**: 1173–85.

De Renzi E, Faglioni P, Scarpa M *et al.* Limb apraxia in patients with damage confined to the left basal ganglia or thalamus. *J Neurol Neurosurg Psychiatry* 1986; **49**:1030–8.

Devere TD, Trotter JL, Cross AH. Acute aphasia in multiple sclerosis. *Arch Neurol* 2000; **57**:1207–9.

Devinsky O, Kelley K, Yacubia EMT *et al.* Postictal behavior. *Arch Neurol* 1994; **51**:254–9.

Dogulu CF, Kansu T, Karabudak R. Alexia without agraphia in multiple sclerosis. *J Neurol Neurosurg Psychiatry* 1996; **61**:528.

Enderby P, Broeckx J, Hospers W *et al.* Effect of piracetam on recovery and rehabilitation after stroke: a double-blind, placebo-controlled study. *Clin Neuropharmacol* 1994; **17**:320–31.

Eslinger PJ, Dennis K, Moore P *et al.* Metacognitive deficits in frontotemporal dementia. *J Neurol Neurosurg Psychiatry* 2005; **76**:1630–5.

Evans JJ, Heggs AJ, Antoun N *et al.* Progressive prosopagnosia associated with selective right temporal lobe atrophy: a new syndrome? *Brain* 1995; **118**:1–13.

Evyapan D, Kumral D. Pontine anosognosia for hemiplegia. *Neurology* 1999; **53**:647–9.

Evyapan D, Kumral E. Visuospatial stimulus-bound automatic writing behavior: a right hemisphere stroke syndrome. *Neurology* 2001; **56**:245–7.

Faber R, Abrams R, Taylor MA *et al.* Comparison of schizophrenic patients with formal thought disorder and neurologically impaired patients with aphasia. *Am J Psychiatry* 1983; **140**:1348–51.

Feinberg TE, Haber LD, Leeds NE. Verbal asomatognosia. *Neurology* 1990; **40**:1391–4.

Feinberg TE, Schindler RJ, Ochoa E *et al.* Associative visual agnosia and alexia without prosopagnosia. *Cortex* 1994; **30**:395–411.

Feinberg TE, Roane DM, Ali J. Illusory limb movements in anosognosia for hemiplegia. *J Neurol Neurosurg Psychiatry* 2000; **68**:511–13.

Feinberg TE, Venneri A, Simone AM *et al.* The neuroanatomy of asomatognosia and somatoparaphrenia. *J Neurol Neurosurg Psychiatry* 2010; **81**:276–81.

Ferreira CT, Ceccaldi M, Giusiano B *et al.* Separate visual pathways for perception of actions and objects: evidence from a case of apperceptive agnosia. *J Neurol Neurosurg Psychiatry* 1998; **65**:382–5.

Ferro JM, Kertesz A. Posterior internal capsule infarction associated with neglect. *Arch Neurol* 1984; **41**:422–4.

Fincham RW, Nibelink DW, Aschenbrenner CA. Alexia with left homonymous hemianopia without agraphia. *Neurology* 1975; **25**:1164–8.

Flashman LA, McCallister TW. Lack of awareness and its impact in traumatic brain injury. *Neurorehabilitation* 2002; **17**:285–96.

Frantz KE. Amnesia for left limbs and loss of interest and alteration in left fields of vision. *J Nerv Ment Dis* 1950; **112**:240–4.

Frassinetti F, Angeli V, Meneghello F *et al.* Long-lasting amelioration of visuospatial neglect by prism adaptation. *Brain* 2002; **125**: 608–23.

Freedman M, Alexander MP, Naeser MA. Anatomic basis of transcortical motor aphasia. *Neurology* 1984; **34**:409–17.

Frisoni GB, Scuatti A, Biancetti A *et al.* Hypergraphia and brain damage. *J Neurol Neurosurg Psychiatry* 1993; **56**:576–7.

Fujii T, Fukatsu R, Watabe S *et al.* Auditory sound agnosia without aphasia following a right temporal lesion. *Cortex* 1990; **26**:263–8.

Fukui T, Sugita K, Kawamura M *et al.* Primary progressive apraxia in Pick's disease: a clinicopathologic report. *Neurology* 1996; **47**:467–73.

Gabr M, Luders H, Dinner D *et al.* Speech manifestations in lateralizaton of temporal lobe seizures. *Ann Neurol* 1989; **25**:82–7.

Galton CJ, Patterson K, Xuereb JH *et al.* Atypical and typical presentations of Alzheimer's disease: a clinical, neuropsychological, neuroimaging and pathological study. *Brain* 2000; **123**: 484–98.

Garrido L, Furi N, Draganski B *et al.* Voxel-based morphometry reveals reduced grey matter volume in the temporal cortex of developmental prosopagnosics. *Brain* 2009; **132**:3443–56.

Gerson SN, Benson DF, Frazier SH. Diagnosis: schizophrenia versus posterior aphasia. *Am J Psychiatry* 1977; **134**:966–9.

Gerstmann J. Fingeragnosie: eine umschriebene storung der orientierung am eigenen korper. *Wein Klin Wschr* 1924; **37**:1010–12.

Gerstmann J. Syndrome of finger agnosia, disorientation for right and left, agraphia and acalculia. *Arch Neurol Psychiatry* 1940; **44**:398–408.

Gerstmann J. Problem of misperception of disease and of impaired body territories with organic lesions: relation to body scheme and its disorders. *Arch Neurol Psychiatry* 1942; **48**:890–913.

Gerstmann J. Some notes on the Gerstmann syndrome. *Neurology* 1957; **7**:866–9.

Geschwind N, Fusillo M. Color-naming defects in association with alexia. *Arch Neurol* 1966; **15**:137–46.

Ghacibeh GA Heilman KM. Progressive affective aprosodia and prosoplegia. *Neurology* 2003; **60**:1192–4.

Ghika-Schmid F, Bogousslavsky J. The acute behavioral syndrome of anterior thalamic infarction: a prospective study. *Ann Neurol* 2000; **48**:220–7.

Giroud M, Dumas R. Clinical and topographical range of callosal infarction: a clinical and radiological correlation study. *J Neurol Neurosurg Psychiatry* 1995; **59**:238–42.

Godefroy O, Rousseaux M, Pruvo JP *et al.* Neuropsychological changes related to unilateral lenticulostriate infarcts. *J Neurol Neurosurg Psychiatry* 1994; **57**:480–5.

Gold M, Adair JC, Jacobs DH *et al.* Right–left confusion in Gerstmann's syndrome: a model of body centered spatial orientation. *Cortex* 1995; **31**:267–83.

Goldstein DM, Goldberg RL. Monoamine oxidase inhibition-induced speech blockage. *J Clin Psychiatry* 1986; **47**:604.

Goodglass H, Quadfasel FA. Language laterality in left-handed aphasics. *Brain* 1954; **77**:521–48.

Gorelick PB, Ross ED. The aprosodias: further functional–anatomical evidence for the organization of affective language in the right hemisphere. *J Neurol Neurosurg Psychiatry* 1987; **50**:553–60.

Graff-Radford NR, Damasio AR, Hyman BT *et al.* Progressive aphasia in a patient with Pick's disease: a neuropsychological, radiologic and anatomic study. *Neurology* 1990; **40**:620–6.

Green J, Morris JC, Sandson J *et al.* Progressive aphasia: a precursor of global dementia? *Neurology* 1990; **40**:423–9.

Greenblatt SH. Subangular alexia without agraphia or hemianopia. *Brain Lang* 1976; **3**:229–45.

Greene JDW, Patterson K, Xuereb J *et al.* Alzheimer disease and nonfluent progressive aphasia. *Arch Neurol* 1996; **53**:1072–8.

Gupta MR, Micoch AG, Scolaro C *et al.* Bromocriptine treatment of nonfluent aphasia. *Neurology* 1995; **45**:2170–3.

Habib M, Sirigu A. Pure topographical disorientation: a definition and anatomical basis. *Cortex* 1987; **23**:73–85.

Hamilton NG, Mathews T. Aphasia: the sole manifestation of focal status epilepticus. *Neurology* 1979; **29**:745–8.

Hanlon RE, Lux WE, Dromerick AW. Global aphasia without hemiparesis: language profiles and lesion distribution. *J Neurol Neurosurg Psychiatry* 1996; **66**:365–9.

Hecaen H, Angelergues R. Agnosia for faces (prosopagnosia). *Arch Neurol* 1962; **7**:92–100.

Hecaen H, Penfield W, Bertrand C *et al.* The syndrome of apractagnosia due to lesions of the minor cerebral hemisphere. *Arch Neurol Psychiatry* 1956; **75**:400–34.

Healton EB, Navarro C, Bressman S *et al.* Subcortical neglect. *Neurology* 1982; **32**:776–8.

Heilman KM. Ideational apraxia – a re-definition. *Brain* 1973; **96**: 861–4.

Heilman KM, Valenstein E. Frontal lobe neglect in man. *Neurology* 1972; **22**:660–4.

Heilman KM, Rothi L, McFarling D *et al.* Transcortical sensory aphasia with relatively spared spontaneous speech and naming. *Arch Neurol* 1981; **38**:236–9.

Heilman KM, Leon SA, Rosenbek JC. Affective aprosodia from a medial frontal stroke. *Brain Lang* 2004; **89**:411–6.

Hemphill RE, Stengel E. A study on pure word deafness. *J Neurol Neurosurg Psychiatry* 1940; **3**:251–62.

Henderson VW. Speech fluency in crossed aphasia. *Brain* 1983; **106**:837–57.

Henderson VW. Lesion localization in Broca's aphasia. Implications from Broca's aphasia without hemiparesis. *Arch Neurol* 1985; **42**:1210–12.

Henderson VW, Friedman RB, Teng EL *et al.* Left hemisphere pathways in reading: inferences from pure alexia without hemianopia. *Neurology* 1985; **35**:962–8.

Hermann BP, Whitman S, Wyler AR *et al.* The neurological, psychosocial and demographic correlates of hypergraphia in patients with epilepsy. *J Neurol Neurosurg Psychiatry* 1988; **51**:203–8.

Hier DB, Davis KR, Richardson EP *et al.* Hypertensive putaminal hemorrhage. *Ann Neurol* 1977; **2**:152–9.

Hier DB, Mondlock J, Caplan LR. Behavioral abnormalities after right hemisphere stroke. *Neurology* 1983a; **33**:337–44.

Hier DB, Mondlock J, Caplan LR. Recovery of behavioral abnormalities after right hemisphere stroke. *Neurology* 1983b; **33**:345–50.

Hindson DA, Westmoreland DE, Carroll WA *et al.* Persistent Broca's aphasia after right cerebral infarction in a right-hander. *Neurology* 1984; **34**:387–9.

Hirsch E, Marescaux C, Maquet P *et al.* Landau-Kleffner syndrome: a clinical and EEG study of five cases. *Epilepsia* 1990; **31**:756–67.

Holmes JE, Sadoff RL. Aphasia due to a right hemisphere tumor in a right-handed man. *Neurology* 1966; **16**:392–7.

House A, Hodges J. Persistent denial of hardship after infarction of the right basal ganglia: a case study. *J Neurol Neurosurg Psychiatry* 1988; **51**:112–15.

Huber W, Williams K, Poeck D *et al.* Piracetam as a adjuvant to language therapy for aphasia: a randomized, double-blind placebo-controlled pilot study. *Arch Phys Med Rehabil* 1997; **78**:245–50.

Humphrey ME, Zangwill OL. Dysphasia in left-handed patients with unilateral brain lesions. *J Neurol Neurosurg Psychiatry* 1952; **15**:184–93.

Husain M, Kennard C. Visual neglect associated with frontal lobe infarction. *J Neurol* 1996; **243**:652–7.

Ikeda K, Akiyama H, Iritani S *et al.* Corticobasal degeneration with primary progressive aphasia and accentuated cortical lesion in superior temporal gyrus: case report and review. *Acta Neuropathol* 1996; **92**:534–9.

Imamura T, Yamadori A, Tsuburaya K. Hypergraphia associated with a brain tumor of the right cerebral hemisphere. *J Neurol Neurosurg Psychiatry* 1992; **54**:25–7.

Ishiai S, Sugishita M, Ichikawa T *et al.* Clock-drawing test and unilateral spatial neglect. *Neurology* 1993; **43**:106–10.

Ishiai S, Seki K, Koyama Y *et al.* Disappearance of unilateral spatial neglect following a simple instruction. *J Neurol Neurosurg Psychiatry* 1997; **63**:23–7.

Josephs KA, Duffy JR, Strand EA *et al.* Clinicopathological and imaging correlates of progressive aphasia and apraxia of speech. *Brain* 2006; **129**:1385–98.

Joubert S, Felician O, Barbeau E *et al.* Impaired configurational processing in a case of progressive prosopagnosia associated with predominant right temporal lobe atrophy. *Brain* 2003; **126**:2537–50.

Kanshepolsky J, Kelley JJ, Waggener JD. A cortical auditory disorder: clinical, audiologic and pathologic aspects. *Neurology* 1973; **23**:699–705.

Karbe H, Kertesz A, Polk M. Profiles of language impairment in primary progressive aphasia. *Arch Neurol* 1993; **50**:192–201.

Karnath HO, Hartje W. Decrease of contralateral neglect by neck muscle vibration and spatial orientation of trunk midline. *Brain* 1993; **116**:383–96.

Kashiwa Y, Kitabayashi Y, Narumoto J *et al.* Anosognosia in Alzheimer's disease: association with patient characteristics, psychiatric symptoms and cognitive deficits. *Psychiatry Clin Neurosci* 2005; **59**:697–704.

Kazui S, Sawada T. Callosal apraxia without agraphia. *Ann Neurol* 1993; **33**:401–3.

Keane S, Turner C, Sherrington C *et al.* Use of fresnel prism glasses to treat stroke patients with hemispatial neglect. *Arch Phys Med Rehabil* 2006; **87**:1668–72.

Kertesz A, Hudson L, Mackenzie IRA *et al.* The pathology and nosology of primary progressive aphasia. *Neurology* 1994; **44**:2065–72.

Kessler J, Thiel A, Karbe H *et al.* Piracetam improves activated blood flow and facilitates rehabilitation of poststroke aphasic patients. *Stroke* 2000; **31**:2112–16.

Kinsbourne M, Warrington EK. A variety of reading disability associated with right hemisphere lesions. *J Neurol Neurosurg Psychiatry* 1962; **25**:339–44.

Kinsbourne M, Warrington EK. Observations on color agnosia. *J Neurol Neurosurg Psychiatry* 1964; **27**:296–9.

Knibb J, Xuereb JH, Patterson K *et al.* Clinical and pathological characterization of progressive aphasia. *Ann Neurol* 2006; **59**:156–65.

Knight RT, Cooper J. Status epilepticus manifesting as reversible Wernicke's aphasia. *Epilepsia* 1986; **27**:301–4.

Kraepelin E. *[Dementia praecox and paraphrenia, 1916]*. Huntington, NY: Robert E. Krieger, 1971.

Kraepelin E. *[Manic-depressive insanity and paranoia, 1921]*. NY: Arno Press, 1976.

Kress T, Daum M. Developmental prosopagnosia: a review. *Behav Neurol* 2003; **14**:109–21.

Kumar R, Masih AK, Pardo J. Globl aphasia due to thalamic hemorrhage: a case report and review of the literature. *Arch Phys Med Rehabil* 1996; **77**:1312–15.

Kumral E, Kocaer T, Ertubey NO *et al.* Thalamic hemorrhage: a prospective study of 100 patients. *Stroke* 1995; **26**:964–70.

Kumral E, Evyapan D, Balkir N. Acute caudate vascular lesions. *Stroke* 1999; **30**:100–8.

Labar DR, Solomon GE, Wells CR. Aphasia as the sole manifestation of simple partial status epilepticus. *Epilepsia* 1992; **33**:84–7.

Lacour A, de Seze J, Revenco E *et al.* Acute aphasia in multiple sclerosis. *Neurology* 2004; **62**:974–7.

Lampl Y, Eshel Y, Gilad R *et al.* Selective acalculia with sparing of the subtraction process in a patient with left parietotemporal hemorrhage. *Neurology* 1994; **44**:1759–61.

Landis T, Cummings JL, Benson DF *et al.* Loss of topographic familiarity: an environmental agnosia. *Arch Neurol* 1986; **43**:132–6.

Landis T, Regard M, Bliestle A *et al.* Prosopagnosia and agnosia for noncanonical views. *Brain* 1988; **111**:1287–97.

Laplane D, Degos JD. Motor neglect. *J Neurol Neurosurg Psychiatry* 1983; **46**:152–8.

Leeman B, Laganaro M, Chetelat-Mabillard D *et al.* Crossover trial of subacute computerized aphasia therapy for anomia with the addition of either levodopa or placebo. *Neurorehabil Neural Repair* 2011; **25**:43–7.

Le Gros Clark WE, Russell WR. Cortical deafness without aphasia. *Brain* 1938; **61**:375–83.

Leibson E. Anosognosia and mania associated with right thalamic haemorrhage. *J Neurol Neurosurg Psychiatry* 2000; **68**:107–8.

Le Rhun E, Richard F, Pasquier F. Natural history of primary progressive aphasia. *Neurology* 2005; **65**:887–91.

Luzzi S, Pucci E, Di Bella P *et al.* Topographical disorientation consequent to amnesia of spatial location in a patient with right parahippocampal damage. *Cortex* 2000; **36**:427–34.

McCarthy RA, Evans JJ, Hodges JR. Topographic amnesia: spatial memory disorder, perceptual dysfunction, or category specific semantic memory impairment? *J Neurol Neurosurg Psychiatry* 1996; **60**:318–25.

McFarling D, Rothi LJ, Heilman KM. Transcortical aphasia from ischemic infarcts of the thalamus: a report of two cases. *J Neurol Neurosurg Psychiatry* 1982; **45**:107–12.

Maeshima S, Funahashi K, Ogura M *et al.* Unilateral spatial neglect due to right frontal lobe hematoma. *J Neurol Neurosurg Psychiatry* 1994; **57**:89–93.

Maeshima S, Okumura Y, Nakai K *et al.* Gerstmann's syndrome associated with chronic subdural haematoma: a case report. *Brain Inj* 1998; **12**:697–701.

Malone DR, Morris HH, Kay MC *et al.* Prosopagnosia: a double dissociation between the recognition of familiar and unfamiliar faces. *J Neurol Neurosurg Psychiatry* 1982; **45**:820–2.

Mantovani JF, Landau WM. Acquired aphasia with convulsive disorder: course and prognosis. *Neurology* 1980; **30**:524–9.

Mao-Draayer Y, Panitch H. Alexia without agraphia in multiple sclerosis: case report with magnetic resonance imaging localization. *Mult Scler* 2004; **10**:705–7.

Masdeu JC, O'Hara RJ. Motor aphasia unaccompanied by faciobrachial weakness. *Neurology* 1983; **33**:519–21.

Mayer E, Martory M-D, Pegna AJ *et al.* A pure case of Gerstmann's syndrome with a subangular lesion. *Brain* 1999; **122**:1107–20.

Mazzoni M, Pardossi L, Cantini R *et al.* Gerstmann syndrome: a case report. *Cortex* 1990; **26**:459–67.

Meadows JC. Disturbed perception of colours associated with localized cerebral lesions. *Brain* 1974a; **97**:615–32.

Meadows JC. The anatomical basis of prosopagnosia. *J Neurol Neurosurg Psychiatry* 1974b; **37**:489–501.

Megens J, van Loon J, Goffin J *et al.* Subcortical aphasia from a thalamic abscess. *J Neurol Neurosurg Psychiatry* 1992; **55**:319–22.

Mendez MF, Papasian NC, Lim GT. Thalamic acalculia. *J Neuropsychiatry Clin Neurosci* 2003; **15**:115–16.

Mesad S, Laff R, Devinsky O. Transient postoperative prosopagnosia. *Epilepsy Behav* 2003; **4**:567–70.

Mesulam M, Johnson N, Krefft TA *et al.* Progranuin mutations in primary progressive aphasia. The PPA1 and PPA3 families. *Arch Neurol* 2007; **64**:43–7.

Meyer B, Kral T, Zentner J. Pure word deafness after resection of a tectal plate glioma with preservation of wave V of brain stem auditory evoked potentials. *J Neurol Neurosurg Psychiatry* 1996; **61**:423–4.

Mohr JP, Pessin MP, Finkelstein S *et al.* Broca aphasia: pathologic and clinical. *Neurology* 1978; **28**:311–24.

Monrad-Krohn GH. Dysprosody or altered 'melody of language'. *Brain* 1947a; **70**:405–15.

Monrad-Krohn GH. The prosodic quality of speech and its disorders. *Acta Psychiatr Neurol Scand* 1947b; **22**:255–69.

Moore MR, Saver JL, Johnson KA *et al.* Right parietal stroke with Gerstmann's syndrome. Appearance on computed tomography, magnetic resonance imaging, and single-photon emission computed tomography. *Arch Neurol* 1991; **48**:432–5.

Mort DJ, Malhorta P, Mannan SK *et al.* The anatomy of visual neglect. *Brain* 2003; **126**:1986–97.

Motomura N, Yamadori A, Mori E *et al.* Unilateral spatial neglect due to hemorrhage in the thalamic region. *Acta Neurol Scand* 1986; **74**:190–4.

Mula M, Trimble MR, Thompson P *et al.* Topiramate and word-finding difficulties in patients with epilepsy. *Neurology* 2003; **60**:1104–7.

Nadeau SE, Reltgen DP, Sevush S *et al.* Apraxia due to a pathologically documented thalamic infarction. *Neurology* 1994; **44**:2133–7.

Naeser MA, Hayward RW. Lesion localization in aphasia with cranial computed tomography and the Boston diagnostic aphasia exam. *Neurology* 1978; **28**:545–51.

Naeser MA, Alexander MP, Helm-Estabrooks N *et al.* Aphasia with predominantly subcortical lesion sites: description of three capsular/putaminal aphasia syndromes. *Arch Neurol* 1982; **39**:2–14.

Okamura T, Fukai M, Yamadori A *et al.* A clinical study of hypergraphia in epilepsy. *J Neurol Neurosurg Psychiatry* 1993; **56**:556–9.

Olmos-Lau N, Ginsberg MD, Geller JB. Aphasia in multiple sclerosis. *Neurology* 1977; **27**:623–6.

Ostrow LW, Llinas RH. Eastchester clapping sign: a novel test of parietal neglect. *Ann Neurol* 2009; **66**:114–17.

Pan C-L, Kuo M-F, Hsieh S-T. Audiory agnosia caused by a tectal germinoma. *Neurology* 2004; **63**:2387–9.

Paquier PF, Van Dongen HR, Loonen CB. The Landau–Kleffner syndrome or 'acquired aphasia with convulsive disorder': long-term follow-up of six children and a review of the recent literature. *Arch Neurol* 1992; **49**:354–9.

Paul R, Cohen DJ, Caparulo BK. A longitudinal study of patients with severe developmental disorders of language learning. *J Am Acad Child Adolesc Psychiatry* 1983; **22**:525–34.

PeBenito R. Developmental Gerstmann syndrome: case report and review of the literature. *J Dev Behav Pediatr* 1987; **8**:229–32.

Pedersen PM, Jorgensen HS, Nakayama H *et al.* Aphasia in acute stroke: incidence, determinants, and recovery. *Ann Neurol* 1995; **38**:659–66.

Peigneux P, van der Linden M. Influence of ageing and educational level on the prevalence of body-part-as-objects in normal subjects. *J Clin Exp Neuropsychol* 1999; **21**:547–52.

Perez-Trullen JM, Modrego Pardo PJ. Comparative study of aprosody in Alzheimer's disease and in multi-infarct dementia. *Dementia* 1996; **7**:59–62.

Pevzner S, Bornstein B, Loewenthal M. Prosopagnosia. *J Neurol Neurosurg Psychiatry* 1962; **25**:336–8.

Piercy M, Hecaen H, de Ajuriaguerra J. Constructional apraxia with unilateral cerebral lesions – left and right sided cases compared. *Brain* 1960; **83**:225–42.

Pillon B, Bakchine S, Lhermitte F. Alexia without agraphia in a left-handed patient with a right occipital lesion. *Arch Neurol* 1987; **44**:1257–62.

Platz T. Tactile agnosia: causistic evidence and theoretical remarks on modality-specific meaning representations and sensorimotor integration. *Brain* 1996; **119**:1565–74.

Privitera MD, Morris GL, Gilliam F. Postictal language assessment and lateralization of complex partial seizures. *Ann Neurol* 1991; **30**:391–3.

Racy A, Jannotta FS, Lehner LH. Aphasia resulting from occlusion of the left anterior cerebral artery: report of a case with an old infarct in the left rolandic region. *Arch Neurol* 1979; **36**:221–4.

Redlich FC, Dorsey JF. Denial of blindness by patients with cerebral disease. *Arch Neurol Psychiatry* 1945; **53**:407–17.

Reed CL, Caselli RJ, Farah MJ. Tactile agnosia: underlying impairment and implications for normal tactile object recognition. *Brain* 1996; **119**:875–88.

Renner JA, Burns JM, Hou CE *et al.* Progressive posterior cortical dysfunction. A clinicopathologic series. *Neurology* 2004; **63**: 1175–80.

Riley DE, Lang AE, Lewis A *et al.* Cortical-basal ganglionic degneration. *Neurology* 1990; **40**:1203–12.

Rinne JO, Lee MS, Thompson PD *et al.* Corticobasal degeneration: a clinical study of 36 cases. *Brain* 1994; **117**:1183–96.

Rizzo M, Hurtig R. Looking but not seeing: attention, perception, and eye movements in simultanagnosia. *Neurology* 1987; **37**:1642–8.

Rizzo M, Robin DA. Simultanagnosia: a defect of sustained attention yields insight on visual information processing. *Neurology* 1990; **40**:447–55.

Roeltgen DP, Sevush S, Heilman KM. Pure Gerstmann's syndrome form a focal lesion. *Arch Neurol* 1983; **40**:46–67.

Rosati G, De Bastiani P. Pure agraphia: a discrete form of aphasia. *J Neurol Neurosurg Psychiatry* 1979; **42**:266–9.

Rosenbek JC, Rodriguez AD, Hieber R *et al.* Effects of two treatments for aprosodia secondary to acquired brain injury. *J Rehabil Res Dev* 2006; **43**:379–90.

Rosler A, Lanquillon S, Dippel O *et al.* Impairment of facial recognition in patients with right cerebral infarcts quantified by computer aided 'morphing'. *J Neurol Neurosurg Psychiatry* 1997; **62**:261–4.

Ross ED. Left medial parietal lobe and receptive language functions: mixed transcortical aphasia after left anterior cerebral artery infarction. *Neurology* 1980; **30**:144–51.

Ross ED. The aprosodias: functional–anatomic organization of the affective components of language in the right hemisphere. *Arch Neurol* 1981; **38**:561–9.

Ross ED, Mesulam M-M. Dominant language functions of the right hemisphere: prosody and emotional gesturing. *Arch Neurol* 1979; **36**:144–8.

Ross ED, Harney JH, de Lacoste-Utamsing C *et al.* How the brain interprets affective and propositional language into a unified behavioral function: hypothesis based on clinicoanatomic evidence. *Arch Neurol* 1981; **38**:745–8.

Ross ED, Anderson B, Morgan-Fisher A. Crossed aprosodia in strongly dextral patients. *Arch Neurol* 1989; **46**:206–9.

Ross SJM, Graham N, Stuart-Green L *et al.* Progressive biparietal atrophy: an atypical presentation of Alzheimer's disease. *J Neurol Neurosurg Psychiatry* 1996; **61**:388–95.

Roth N. Disorders of body image caused by lesions of the right parietal lobe. *Brain* 1949; **72**:89–111.

Rubens AB. Aphasia with infarction in the territory of the anterior cerebral artery. *Cortex* 1975; **11**:239–50.

Sabe L, Salvarezza F, Garcia Cuerva A *et al.* A randomized, double-blind, placebo-controlled study of bromocriptine in nonfluent aphasia. *Neurology* 1995; **45**:2272–4.

Sandifer PH. Anosognosia and disorders of body scheme. *Brain* 1946; **68**:122–37.

Sato S, Dreifuss FE. Electroencephalographic findings in a patient with developmental expressive aphasia. *Neurology* 1973; **23**:181–5.

Sbordone RJ, Seranian GD, Ruff RM. Are the subjective complaints of traumatically brain injured patients reliable? *Brain Inj* 1998; **12**:505–15.

Schatzberg AF, Cole JO, Blumer DP. Speech blockage: a tricyclic side effect. *Am J Psychiatry* 1978; **135**:600–1.

Schindler I, Kerkhoff G, Karnath H-O *et al.* Neck vibration induces lasting recovery in spatial neglect. *J Neurol Neurosurg Psychiatry* 2002; **73**:412–19.

Selnes OA, Knopman DS, Miccum N *et al.* The central role of Wernicke's area in sentence repetition. *Ann Neurol* 1985; **17**:549–57.

Seniow J, Litwin M, Litwin T *et al.* New approach to the rehabilitation of post-stroke focal cognitive syndrome: Effect of levodopa combined with speech and language therapy on functional recovery from aphasia. *J Neurol Sci* 2009; **283**:214–18.

Sergent J, Poncet M. From covert to overt recognition of faces in prosopagnosia. *Brain* 1990; **113**:989.

Sergent J, Villemure JG. Prosopagnosia in a right hemi-spherectomized patient. *Brain* 1989; **112**:975–95.

Shelton PA, Bowers D, Duara D *et al.* Apperceptive visual agnosia: a case study. *Brain Cogn* 1994; **25**:1–23.

Silverberg R, Gordon HW. Differential aphasia in two bilingual individuals. *Neurology* 1979; **29**:51–5.

Singer S, Cummings JL, Benson DF. Delusions and mood disorders in patients with chronic aphasia. *J Neuropsychiatry Clin Neurosci* 1989; **1**:40–5.

Smith A. The serial sevens subtraction test. *Arch Neurol* 1962; **17**: 78–80.

Snowden JS, Craufurd D, Griffiths HL *et al.* Awareness of involuntary movements in Huntington disease. *Arch Neurol* 1998; **55**:801–5.

So EL, Schauble BS. Ictal asomatognosia as a cause of epileptic falls. Simultaneous video EMG, and invasive EEG. *Neurology* 2004; **63**:2153–4.

Speedie LJ, Wertman E, Ta'ir J *et al.* Disruption of automatic speech following a right basal ganglia lesion. *Neurology* 1993; **43**: 1768–74.

Starkstein SE, Chemerinski E, Sabe L *et al.* Prospective longitudinal study of depression and anosognosia in Alzheimer's disease. *Br J Psychiatry* 1997; **171**:47–52.

Stein S, Volpe BT. Classical 'parietal' neglect syndrome after subcortical right frontal lobe infarction. *Neurology* 1983; **33**:797–9.

Stommel EW, Friedman RJ, Reeves AG. Alexia without agraphia associated with spleniogeniculate infarction. *Neurology* 1991; **41**:587–8.

Stone SP, Wilson B, Wroot A *et al.* The assessment of visuospatial neglect after acute stroke. *J Neurol Neurosurg Psychiatry* 1991; **54**:345–50.

Stracciari A, Lorusso S, Pazzaglia P. Transient topographical amnesia. *J Neurol Neurosurg Psychiatry* 1994; **57**:1423–5.

Suresh PA, Sebastian S. Developmental Gerstmann's syndrome: a distinct clinical entity of learning disabilities. *Pediatr Neurol* 2000; **22**:267–78.

Sweet EWS, Panis W, Levine DN. Crossed Wernicke's aphasia. *Neurology* 1984; **34**:475–9.

Symonds C, MacKenzie I. Bilateral loss of vision from cerebral infarction. *Brain* 1957; **80**:415–55.

Takahashi N, Kawamura M, Shiota J *et al.* Pure topographic disorientation due to right retrosplenial lesion. *Neurology* 1997; **49**:464–9.

Takayama Y, Sugishita M, Akiguchi I *et al.* Isolated acalculia due to left parietal lesion. *Arch Neurol* 1994; **51**:286–91.

Tanabe H, Sawada T, Inoue N *et al.* Conduction aphasia and arcuate fasciculus. *Acta Neurol Scand* 1987; **76**:422–7.

Tang-Wei DF, Graff-Radford NR, Boeve BF *et al.* Clinical, genetic, and neuropathlogic characteristics of posterior cortical atrophy. *Neurology* 2004; **63**:1168–74.

Taniwaki T, Tagawa K, Sato F *et al.* Auditory agnosia restricted to environmental sounds following cortical deafness and generalized auditory agnosia. *Clin Neurol Neurosurg* 2000; **102**:156–62.

Tegner R, Levander M. The influence of stimulus properties on visual neglect. *J Neurol Neurosurg Psychiatry* 1991; **54**:882–7.

Tohgi H, Saitoh K, Takahashi S *et al.* Agraphia and acalculalia after a left prefrontal (F1, F2) infarction. *J Neurol Neurosurg Psychiatry* 1995; **58**:629–32.

Tonkonogy J, Goodglass H. Language function, foot of the third frontal gyrus, and rolandic operculum. *Arch Neurol* 1981; **38**:486–90.

Tranel D, Biller J, Damasio H *et al.* Global aphasia without hemiparesis. *Arch Neurol* 1987; **44**:304–8.

Tranel D, Damasio AR, Damasio H. Intact recognition of facial expression, gender, and age in patients with impaired recognition of face identity. *Neurology* 1988; **38**:690–6.

Triggs WJ, Gold M, Gerstle G *et al.* Motor neglect associated with a discrete parietal lesion. *Neurology* 1994; **44**:1164–6.

Trojanowski JQ, Green RC, Levine DN. Crossed aphasia in a dextral: a clinicopathological study. *Neurology* 1980; **30**:709–13.

Tucha O, Steup A, Smely C *et al.* Toe agnosia in Gerstmann syndrome. *J Neurol Neurosurg Psychiatry* 1997; **63**:399–403.

Turgman J, Goldhammer Y, Braham J. Alexia, without agraphia, due to brain tumor: a reversible syndrome. *Ann Neurol* 1979; **6**:265–8.

Turner RS, Kenyon LC, Trojanowski JQ *et al.* Clinical, neuroimaging, and pathologic features of progressive nonfluent aphasia. *Ann Neurol* 1996; **39**:166–73.

Tuszynski MH, Petito CK. Ischemic thalamic aphasia with pathologic confirmation. *Neurology* 1988; **38**:800–2.

Vallar G, Perani D. The anatomy of unilateral neglect after right-hemisphere stroke lesions. A clinical/CT-scan correlation study in man. *Neuropsychologia* 1986; **24**:609–22.

Vallar G, Rusconi ML, Bignamini L *et al.* Anatomical correlates of visual and tactile extinction in humans: a clinical CT scan study. *J Neurol Neurosurg Psychiatry* 1994; **57**:464–70.

Van Horn G, Hawes A. Global aphasia without hemiparesis: a sign of embolic encephalopathy. *Neurology* 1982; **32**:403.

van Vugt P, Kees L, Cras P. Increased writing activity in neurological conditions: a review and clinical study. *J Neurol Neurosurg Psychiatry* 1996; **61**:510–14.

Vignolo LA. Auditory agnosia. *Phil Trans R Soc Lond Biol Sci* 1982; **298**:49–57.

Vincent FM, Sadowsky CH, Saunders RL *et al.* Alexia without agraphia, hemianopia or color-naming defect: a disconnection syndrome. *Neurology* 1977; **27**:689–91.

Wada Y, Yamamoto T. Selective impairment of facial recognition due to a haematoma restricted to the right fusiform and lateral occipital region. *J Neurol Neurosurg Psychiatry* 2001; **71**:254–7.

Walker-Batson D, Curtis S, Natarajan R *et al.* A double-blind, placebo-controlled study of the use of amphetamine in the treatment of aphasia. *Stroke* 2001; **32**:2093–8.

Warren JD, Thompson PD. Diencephalic amnesia and apraxia after left thalamic infection. *J Neurol Neurosurg Psychiatry* 2000; **68**:248.

Warrington EK, Langdon D. Spelling dyslexia: a deficit in the visual word-form. *J Neurol Neurosurg Psychiatry* 1994; **57**:211–16.

Watson RT, Heilman KM. Callosal apraxia. *Brain* 1983; **106**:391–403.

Watson RT, Valenstein E, Heilman KM. Thalamic neglect. Possible role of the medial thalamus and nucleus reticularis in behavior. *Arch Neurol* 1981; **38**:501–6.

Waxman SG, Geschwind N. Hypergraphia in temporal lobe epilepsy. *Neurology* 1974; **24**:629–36.

Waxman SG, Geschwind N. The interictal behavior syndrome of temporal lobe epilepsy. *Arch Gen Psychiatry* 1975; **32**:1580–6.

Wechsler AF, Verity A, Rosenschein S *et al.* Pick's disease: a clinical, computed tomographic, and histologic study with Golgi impregnation observations. *Arch Neurol* 1982; **39**:287–90.

Weinstein EA, Cole M, Mitchell MS *et al.* Anosognosia and aphasia. *Arch Neurol* 1969; **10**:376–86.

Williamson PD, Spencer DD, Spencer SS *et al.* Episodic aphasia and epileptic focus in the nondominant hemisphere: relieved by section of the corpus callosum. *Neurology* 1985; **35**:1069–71.

Wilson SAK. *Neurology,* 2nd edn. London, Butterworth & Co, 1955.

Wolfe GI, Ross ED. Sensory aprosodia with left hemiparesis from subcortical infarction: right hemisphere analogue of sensory-type aphasia with right hemiparesis? *Arch Neurol* 1987; **44**:668–71.

Wolpert I. Die simultanagnosie-storung der gasamtauffassung. *Z Ges Neurol Psychiatr* 1924; **93**:397–415.

Yamadori A, Osumi Y, Masuhara S *et al.* Preservation of singing in Broca's aphasia. *J Neurol Neurosurg Psychiatry* 1977; **40**:221–4.

Yamadori A, Osumi Y, Ikeda H *et al.* Left unilateral agraphia and tactile anomia: disturbances seen after occlusion of the anterior cerebral artery. *Arch Neurol* 1980; **37**:88–91.

Yamadori A, Mori E, Tabuchi M *et al.* Hypergraphia: a right hemisphere syndrome. *J Neurol Neurosurg Psychiatry* 1986; **49**:1160–4.

4

Abnormal movements

4.1 TREMOR

Tremor is a more or less rhythmic oscillatory movement, most commonly seen in the upper extremities, that occurs secondary to the rapidly alternating contraction of agonist and antagonist muscles.

Clinical features

From a clinical point of view, it is most useful to distinguish three different types of tremor, namely postural, rest, and intention; other types (e.g. Holmes' tremor) are far less common.

Postural tremor is most apparent when the patient maintains a posture, for example by extending the arms straight forward with the fingers outstretched and slightly spread apart. This is a rapid tremor, at a frequency of 5–11 Hz, and ranges in amplitude from fine (which may be almost unnoticeable to casual inspection) to coarse (which may effectively preclude even the simplest of activities, such as holding a cup of water to the lips). In essentially all cases, postural tremor is present bilaterally. Some authors also refer to this tremor as an 'action' tremor, in that it occurs when the patient makes an 'action', such as extending the arms. Not all authors are in agreement with this term, however, and it is not used further here.

Rest tremor, as the name indicates, is seen when the involved extremity is at rest, as, for example, when the hands are resting in the lap with the patient seated. This is a slow tremor, at a frequency of 3–5 Hz, and is typically 'pill-rolling' in character in that the oscillatory activity of the thumb, sweeping back and forth in proximity to the ventral surface of the fingers, resembles the movements one would make if one were rolling a pill back and forth in one's hand. As noted below, this tremor is characteristic of parkinsonian conditions, and although it may be unilateral, in most cases, with progression of the underlying condition, it becomes bilateral.

Intention tremor comes to light during finger-to-nose testing. To perform this, the patient is asked to extend the upper extremity, with the index finger extended out, and then to bring the index finger in to touch the nose. Normally the movement is smooth throughout; when intention tremor is present, however, a slow tremor, at a frequency of 3–4 Hz, develops as the target is approached, with the amplitude of the tremor increasing fairly dramatically as the target is closed in on. Some authors refer to this as a 'kinetic' tremor, indicating that it appears with movement; however, this term has not gained currency and the term 'intention' is used in this text.

Other, far less common, tremors include 'task-specific' tremors, orthostatic tremor, 'rabbit' tremor, Holmes' tremor, and 'wing-beating' tremor.

'Task-specific' tremors resemble postural tremors; however, they only appear when the patient is engaged in a specific activity, most notably writing (Bain *et al.* 1995; Klawans *et al.* 1982) but also others (Soland *et al.* 1996a) such as throwing darts or using tools.

Orthostatic tremor (Heilman 1984), as the name suggests, occurs when patients stand up. This is a very rapid tremor of the lower extremities, at a frequency of 14–16 Hz, which patients often describe as a sense of quivering, which may cause a sense of unsteadiness or even cause patients to fall. As this tremor may be invisible, it is necessary to palpate the inner aspects of the patient's thighs immediately after the patient stands up in order to assess its presence.

'Rabbit' tremor is a rest tremor that is confined to the jaw and presents an appearance similar to that of a rabbit chewing.

Holmes' tremor, named after Gordon Holmes, who first described it in 1904 (Holmes 1904), is a slow tremor (at a frequency of 2–4 Hz) of the upper and occasionally lower extremity, with rest, intentional, and postural components. The appearance of this tremor with sustained posture may be quite remarkable, with a rolling, sinuous motion of the upper arm and a 'flapping' motion of the hand. In the past, this tremor has been referred to as a 'rubal' or 'midbrain' tremor; however, these terms are discouraged because, as discussed below, this tremor may occur in the absence of involvement of the nucleus ruber or of any part of the midbrain.

'Wing-beating' tremor involves the upper extremities when they are flexed at the elbows: the tremor itself is primarily proximal and results in an oscillation of the upper extremities that appears similar to the motions a bird makes as it flaps its wings.

Etiology

The causes of tremor vary according to the type of tremor, and each is discussed in turn, beginning with the causes of postural tremor.

POSTURAL TREMOR

Of the various causes of postural tremor, perhaps the most common is drug toxicity, and common culprits include caffeine, sympathomimetics (e.g. pseudoephedrine), cocaine and stimulants, antidepressants (including, notably, the tricyclics, selective serotonin reuptake inhibitors [SSRIs], venlafaxine, and bupropion), lithium, divalproex, theophylline, and antipsychotics.

Tremor also commonly occurs on a metabolic basis, as for example with hypoglycemia (see Chapter 14.15), febrile illnesses, and end-stage chronic obstructive pulmonary disease (COPD).

Of the endocrinologic causes, although hyperthyroidism (see Chapter 17.3) is classic, Cushing's syndrome (see Chapter 17.1) may also be characterized by tremor.

When toxic, metabolic, and endocrinologic causes have been eliminated, strong consideration must be given to the possibility of alcohol withdrawal (see Chapter 28.5) or sedative-hypnotic withdrawal (see Chapter 28.6), keeping in mind that patients and family members will often collude in denying this.

Delirium may be characterized by tremor, and the presence of this sign may help in narrowing the differential of delirium. All of the causes of tremor just described may also cause delirium, and in cases associated with serotonergic antidepressants or with antipsychotics, consideration must be given, respectively, to either serotonin syndrome (see Chapter 29.5) or neuroleptic malignant syndrome (see Chapter 29.1).

Tremor may also occur in association with the syndromes of anxiety (see Chapter 7.5) or depression (see Chapter 7.1), and in such cases the diagnostic picture may become a little complicated if the syndrome in question were being treated with an antidepressant.

Finally, essential tremor (see Chapter 9.31) must be considered in cases characterized by the gradual onset of a postural tremor in middle-aged adults.

REST TREMOR

Rest tremor is a cardinal sign of parkinsonism, and of the multiple causes of parkinsonism discussed in Chapter 4.8, Parkinson's disease, dementia with Lewy bodies, and multiple system atrophy, along with treatment with first-generation antipsychotics, stand out as the most common.

INTENTION TREMOR

Intention tremor is most commonly due to disease of the cerebellar cortex, the dentate nucleus, or the outflow tracts of the dentate nucleus. Consideration should be given to degenerative conditions (e.g. alcoholic cerebellar degeneration or one of the spinocerebellar ataxias [see Chapter 9.17]), multiple sclerosis (see Chapter 19.1), and, especially in young adults, Wilson's disease (see Chapter 9.16). Severe intoxication with alcohol or sedative hypnotics (e.g. benzodiazepines) is also typified by intention tremor, and this may also constitute one of the disabling sequelae of lithium intoxication.

OTHER TYPES

Task-specific tremors may occur on an isolated, idiopathic basis, or, in the case of those occurring with writing, on an inherited basis, sometimes in association with writer's cramp (Bain *et al.* 1995).

As noted, orthostatic tremor may occur on either an idiopathic basis or secondary to other causes, such as head trauma, hydrocephalus, or, rarely, lesions of the pons (Benito-Leon *et al.* 1997).

The rabbit syndrome (discussed further in Chapter 29.4) occurs as a side-effect of antipsychotics.

Holmes' tremor, as pointed out by Holmes himself (Holmes 1904), localizes the lesion to the dentate nucleus or its outflow tracts, and has been specifically reported with lesions of the dentate nucleus (Krauss *et al.* 1995), superior cerebellar peduncle (Krack *et al.* 1994; Krauss *et al.* 1995), dorsal midbrain tegmentum (Krauss *et al.* 1995), the red nucleus and adjacent structures (Holmes 1904; Remy *et al.* 1995), the posterior thalamus (Krauss *et al.* 1992; Miwa *et al.* 1996; Tan *et al.* 2001), and the pontine tegmentum (Miyagi *et al.* 1999; Shepherd *et al.* 1997).

Wing-beating tremor, although classic for Wilson's disease, is among the presenting signs of this disease in only a small minority of cases (Starosta-Rubinstein *et al.* 1987); this tremor, rarely, has also been reported with thalamic infarction (Ghika *et al.* 1994).

Differential diagnosis

All normal individuals have a fine, postural tremor. Known as a 'physiologic tremor', although it is typically invisible to the naked eye, it may be exacerbated with stress, anxiety, or fatigue to the point where it becomes apparent.

Myoclonus is distinguished from postural tremor in that rather than being characterized by an oscillatory motion, it consists of a 'jerk' followed by a relaxation phase. In tremor, there is alternating contraction of agonist and antagonist musculature, whereas in myoclonus there is only contraction of one group. Rhythmicity may also help distinguish these two: tremor is always more or less rhythmic, whereas myoclonus, although at times rhythmic, is most often intermittent.

Asterixis appears when the arms are outstretched and the hands are dorsiflexed, and may be confused with a postural tremor. The differential rests, as with myoclonus, on the absence of oscillation. In asterixis, there is a sudden relaxation, such that the hands 'drop'; this is then followed by a slower resumption of the dorsiflexed position.

Clonic movements, as may occur during a simple partial motor seizure, may simulate postural tremor; however here, again, there is no oscillation but rather a rhythmic contraction of one muscle group followed by relaxation. Furthermore, and in contrast with postural tremor, which is almost always bilateral, simple partial motor seizures are almost always unilateral.

'Dystonic tremor' refers to a jerking motion of a dystonic limb. Although the term has currency, this is not a true tremor but more of an intermittent jerk that typically disappears when the patient relaxes and gives way to the developing dystonic spasm.

Simulated, or 'psychogenic', tremor (Deuschl *et al.* 1998), as may occur in conversion disorder, or malingering or factitious illness, is not at all uncommon, and its differential features are discussed in Chapter 8.10.

Treatment

Treatment, if possible, is directed at the underlying cause. In the case of toxic postural tremor, dose reduction is generally sufficient; however, when this is not practical, as for example with ongoing treatment with lithium, consideration may be given to symptomatic treatment with propranolol (Lapierre 1976). For primary writing tremor, consideration may be given to propranolol or primidone (Bain *et al.* 1995), botulinum injections (Singer *et al.* 2005), or the use of a writing device (Espay *et al.* 2005a). Orthostatic tremor may respond to clonazepam or gabapentin (Onofrj *et al.* 1998; Rodrigues *et al.* 2006), and Holmes' tremor may respond to carbidopa/levodopa (Remy *et al.* 1995; Samie *et al.* 1990; Tan *et al.* 2001; Velez *et al.* 2002), clonazepam (Jacob and Pratap Chand 1998), trihexyphenidyl (Krack *et al.* 1994), or levetiracetam (Striano *et al.* 2007).

4.2 MYOCLONUS

Myoclonus is an important diagnostic sign, and, in addition to close observation during the interview and examination, patients should be questioned about whether or not they have had any jerking motions. In patients with delirium, the presence of myoclonus immediately suggests the serotonin syndrome or Hashimoto's encephalopathy, while in patients with dementia its presence immediately raises the possibility of Creutzfeldt–Jakob disease.

Clinical description

Myoclonus consists of sudden, rapid muscular jerks followed by a slower relaxation phase. These myoclonic jerks may be focal, multifocal, or generalized, and they vary in amplitude from being almost imperceptible to being a gross movement that can, in severe cases, throw an arm up or knock a patient off balance. Although occasionally myoclonus may be rhythmic, in most cases myoclonic jerks occur at irregular intervals with a frequency varying from only several per day up to multiple occurrences every minute. Although in most cases myoclonus occurs spontaneously, at times one may see stimulus-sensitive myoclonus (to either a sudden touch or a loud noise); furthermore, in some cases myoclonus only occurs with an 'action', as for example when the patient extends the arms.

Etiology

The various causes of myoclonus are listed in Table 4.1. In general neuropsychiatric practice, the great importance of myoclonus lies in its usefulness in narrowing down the differential diagnosis in cases of delirium, dementia, various movement disorders, and epilepsy, and these are discussed in turn. Myoclonus may also occur as an isolated side-effect to various medications, and also in a handful of miscellaneous disorders.

ASSOCIATED WITH DELIRIUM

Toxic deliria associated with myoclonus include the serotonin syndrome, anticholinergic delirium, and bismuth intoxication, and of these the serotonin syndrome is by far the most common. Metabolic deliria include, classically, uremic encephalopathy and hepatic encephalopathy while myoclonus may also be seen in the course of hyperosmolar non-ketotic hyperglycemia. Hashimoto's encephalopathy, an autoimmune disease, is likewise classically associated with myoclonus, as is the vitamin-deficiency disorder encephalopathic pellagra, as may be seen in chronic alcoholics. Complex partial status epilepticus may present as a delirium, and in a minority of such cases, myoclonus may appear. Finally, the delirium of acute encephalitis due to arboviruses may also be characterized by myoclonus.

ASSOCIATED WITH DEMENTIA

Of the dementing disorders associated with myoclonus, the most important is Creutzfeldt–Jakob disease. Myoclonus may also be seen in Alzheimer's disease and acquired immune deficiency syndrome (AIDS) dementia, but generally only with advanced disease, long after the diagnosis has become obvious. Rare causes include subacute sclerosing panencephalitis, Whipple's disease, and dialysis encephalopathy. Finally, post-hypoxic–ischemic encephalopathy may be characterized by action myoclonus.

Table 4.1 Causes of myoclonus

Associated with delirium
Serotonin syndrome (Chapter 29.5)
Anticholinergic delirium (Chapter 29.6)
Bismuth intoxication (Chapter 14.9)
Uremic encephalopathy (Chapter 14.18)
Hepatic encephalopathy (Chapter 14.19)
Hyperosmolar non-ketotic hypergylcemia (Morres and Dire 1989)
Hashimoto's encephalopathy (Chapter 19.10)
Encephalopathic pellagra (Chapter 14.3)
Complex partial status epilepticus (Kaplan 1996)
Acute arboviral encephalitis (Kleinschmidt-DeMasters *et al.* 2004) (Chapter 15.4)

Associated with dementia
Creutzfeldt–Jakob disease (Chapter 16.1)
Alzheimer's disease (Chapter 9.1)
AIDS dementia (Chapter 15.1)
Subacute sclerosing panencephalitis (Chapter 15.12)
Whipple's disease (Chapter 15.17)
Dialysis encephalopathy (Chapter 14.13)
Posthypoxic–ischemic encephalopathy (Chapter 13.1)

Associated with movement disorders
Dementia with Lewy bodies (Chapter 9.6)
Corticobasal ganglionic degeneration (Chapter 9.8)
Progressive supranuclear palsy (Chapter 9.7)
Huntington's disease (Chapter 9.10)
Myoclonus dystonia (Asmus *et al.* 2002; Doheny *et al.* 2002; Foncke *et al.* 2006; Grimes *et al.* 2002; Roze *et al.* 2008; Schule *et al.* 2004)

Associated with epilepsy syndromes
Juvenile myoclonic epilepsy (Jain *et al.* 1998; Panayiotopoulos *et al.* 1994; Pedersen and Petersen 1998)
Degenerative epileptic syndromes (e.g. Unverricht-Lundborg disease [Koskiniemi *et al.* 1974; Magaudda *et al.* 2006])
Absence epilepsy

Medication side-effect
Gabapentin (Zhang *et al.* 2005)
Pregabalin (Huppertz *et al.* 2001)
Lamotrigine (Crespel *et al.* 2005)
Meperidine (Kaiko *et al.* 1983)
Hydromorphone (Hofmann *et al.* 2006)
SSRIs (Barucha and Sethi 1996)
Tricyclic antidepressants (Casas *et al.* 1987; DeCastro 1985; Garvey and Tollefson 1987; Lippmann *et al.* 1977)
Trazodone (Garvey and Tollefson 1987; Patel *et al.* 1988)
Non-selective monoamine oxidase inhibitors (tranylcypromine, phenelzine) (Remick *et al.* 1989)
Lithium (Caviness and Evidente 2003)
Buspirone (Ritchie *et al.* 1988)
Clozapine (Bak *et al.* 1995; Barak *et al.* 1996)
Levodopa (Klawans *et al.* 1975)
Amantadine (Matsunaga *et al.* 2001)
Trimethoprim/sulfamethoxazole (Dib *et al.* 2004)
Tardive myoclonus (Chapter 29.2)

Miscellaneous causes
Lance–Adams syndrome of postanoxic action myoclonus (Lance and Adams 1963)
Infarction of the frontoparietal cortex (Sutton and Meyere 1974) or thalamus (Kim 2001b; Lehericy *et al.* 2001)
Spinal cord lesions (De La Sayette *et al.* 1996; Hoehn and Cherington 1977; Keswani *et al.* 2002)
Paraneoplastic myoclonus (Bataller *et al.* 2001)

AIDS, acquired immune deficiency syndrome; SSRI, selective serotonin reuptake inhibitor.

ASSOCIATED WITH MOVEMENT DISORDERS

Most of the movement disorders capable of causing myoclonus may also cause a dementia; dementia with Lewy bodies is the most common. Other parkinsonian conditions in this group include corticobasal ganglionic degeneration and progressive supranuclear palsy. Huntington's disease may also cause myoclonus, but this symptom plays a negligible role in the overall clinical presentation of this disorder. Finally, there is a non-dementing inherited condition known as myoclonus dystonia. This disorder, also known as 'essential myoclonus', is inherited on an autosomal dominant basis with incomplete penetrance. Onset is in childhood or adolescence, and in addition to myoclonus there may also be cervical or task-specific dystonia (e.g. writer's cramp). Of interest, there appears to be an association between myoclonus dystonia and obsessive–compulsive disorder, depression, and alcoholism (Hess *et al.* 2007; Ritz *et al.* 2009; Saunders-Pullman *et al.* 2002).

ASSOCIATED WITH EPILEPSY SYNDROMES

Of the epileptic syndromes associated with myoclonus, the most important are juvenile myoclonic epilepsy and various degenerative conditions (e.g. Unverricht–Lundborg disease); petit mal seizures occurring in childhood- or juvenile-onset absence epilepsy may also be associated with a few myoclonic jerks of the hands.

MEDICATION SIDE-EFFECT

Of the medications capable of causing myoclonus, the anti-epileptic drugs (AEDs) gabapentin, pregabalin, and lamotrigine stand out, as do the opioids meperidine and hydromorphone. With regard to the opioids, in some cases myoclonus may appear early on in treatment; however, in others it is only after prolonged treatment that the myoclonus appears. In these cases, it is suspected that the myoclonus occurs owing to the accumulation of toxic metabolites and not secondary to the parent compound (Mercadante 1998). This same late evolution has also been noted with fluoxetine, with which years of treatment may pass before myoclonus appears. Although little needs to be said regarding other medications, some words are necessary regarding tardive myoclonus. Tardive myoclonus is a rare variant of tardive dyskinesia and, like all the tardive dyskinesia subtypes, occurs only after long exposure to antipsychotics: although tardive myoclonus may appear in isolation, it is more often seen in the setting of more common tardive movements, such as chorea.

MISCELLANEOUS CAUSES

Regarding the miscellaneous causes of myoclonus, several comments are in order. As noted earlier, myoclonus is characteristic of posthypoxic–ischemic encephalopathy; however, an episode of hypoxia may be followed by isolated myoclonus, and in such cases the term 'Lance–Adams syndrome' is used in honor of Lance and Adams, who first described it in 1963 (Lance and Adams 1963). Lesions, for example infarction of the frontoparietal cortex or the thalamus, may cause myoclonus, and months or longer may elapse between the infarct and the appearance of the abnormal movement. Lesions of the spinal cord, such as infarction, myelitis, or trauma, may cause myoclonus, which may be either focal, reflecting the segment involved, or generalized: this generalized form, often referred to as propriospinal myoclonus, typically affects the abdominal wall and extremities (Roze *et al.* 2009). Paraneoplastic myoclonus reflects autoimmune involvement of the brainstem and is often accompanied by opsoclonus (Digre 1986), thus constituting the syndrome of opsoclonus myoclonus.

Differential diagnosis

Asterixis is distinguished from myoclonus in that, rather than consisting of a jerking motion, it is characterized by a sudden drop or abrupt relaxation; furthermore, rather than occurring spontaneously, it generally appears only when the outstretched hands are held in dorsiflexion. Some authors consider asterixis to be a kind of myoclonus, namely 'negative' myoclonus, in that it consists of a sudden loss of muscle tonus rather than an active contraction. Given, however, that the etiologies of asterixis and myoclonus are not identical, it may be prudent, from a clinical point of view, to keep the two separate.

Tremor is distinguished from myoclonus by its continuous nature. In tremor, there is an ongoing and alternating contraction of agonist and antagonist muscles, producing a persistent oscillatory motion. By contrast, in myoclonus, there is a distinct relaxation phase following each jerk: after the contraction of the agonist musculature, rather than an immediately following contraction of the antagonist muscle, there is merely relaxation.

Tics are characterized by premonitory urges and by their temporary suppressibility by voluntary effort – features which are not seen in myoclonus. Furthermore, although some tics are quite simple in nature and thus may appear similar to myoclonic jerks, they may also be complex, resembling fragments of voluntary activity, a feature that, again, is not seen in myoclonus.

Choreic movements may be difficult to distinguish from myoclonic jerks. One helpful feature is the tendency of chorea to appear and disappear on different parts of the body with lightning-like fluidity; myoclonus is rarely so mercurial. Furthermore, in some cases, choreic movements may be complex and, like some tics, may resemble voluntary behaviors, thus distinguishing themselves from myoclonic jerks, which are always simple in nature.

Hypnic jerks, also known as 'sleep starts', are myoclonic jerks that occur during the transition from wakefulness to stage I sleep; they are considered a normal variant.

There is also a condition known as palatal myoclonus (Lapresle 1986); however, the name itself may be a misnomer. In this condition there is a very rapid and rhythmic movement of the palate, which may be appreciated by the patient as a kind of clicking. The movement itself is more of a tremor than anything else, and hence the name of this condition might better be 'palatal tremor'.

Treatment

Treatment is aimed, whenever possible, at the underlying cause. If symptomatic treatment is required, consideration may be given to clonazepam, levetiracetam, or valproic acid (Van Zandijcke 2003) in total daily doses of 0.5–3 mg, 1000–3000 mg, or 15–20 mg/kg, respectively. Regardless of which medication is used, the dose should be divided into two daily doses, starting at a low dose and building gradually as needed and as tolerated.

4.3 MOTOR TICS

Motor tics, although classically and most commonly seen in Tourette's syndrome, may also, as indicated below, occur as a side-effect to various medications and in a large number of other conditions.

Clinical features

Motor tics are sudden, rapid movements that, to varying degrees, resemble purposeful actions. They range from such simple tics as brow wrinkling, facial grimacing, head shaking, or shoulder shrugging to more complex activities such as gesturing or rising from a chair. In most cases, the actual tics are preceded by a 'premonitory urge' that rises in intensity until it is finally relieved by the tic's appearance. Although some patients are able to suppress the tic for a time, considerable effort is required to do so: Kinnier Wilson (Wilson 1928) noted that 'the strain in holding back is as great as the relief in letting go', and reported that one of his patients, in describing the experience, said that 'you cannot help it any more than you can sneezing'.

Although in some cases there may be 'focal' tics, with the same tic recurring uninterruptedly over time, in most cases tics are 'multifocal', with different tics occurring at different locations.

Etiology

Tics, as noted in Table 4.2, may occur in Tourette's syndrome, as a side-effect to medications, in the context of various choreiform disorders, and from a variety of miscellaneous causes.

TOURETTE'S SYNDROME

The most common cause of tics, by far, is Tourette's syndrome or a variant of Tourette's, known as chronic motor tic disorder. As noted in Chapter 9.25, the onset of Tourette's and chronic motor tic disorder is usually in childhood; however, it must also be borne in mind that Tourette's can go into remission only to recur in later life, and consequently in cases of adult-onset tics, it is critical to inquire as to a history of childhood tics (Chouinard and Ford 2000; Klawans and Barr 1985).

Table 4.2 Causes of motor tics

Tourette's syndrome
Medication side-effect
Cocaine (Cardoso and Jankovic 1993; Pascual-Leone and Dhuna 1990)
Methylphenidate (Denckla *et al.* 1976)
Pemoline (Bachman 1981; Mitchell and Mathews 1980)
Atomoxetine (Ledbetter 2005)
Lamotrigine (Lombroso 1999; Sotero de Menezes *et al.* 2000)
Carbamazepine (Robertson et *al.* 1993)
Phenobarbital (Burd *et al.* 1986)
Escitalopram (Altindag *et al.* 2005)
Sertraline (Altindag *et al.* 2005)
Clomipramine (Moshe *et al.* 1994)
Fluoxetine (Eisenhauer and Jermain 1993)
Haloperidol (Gaultieri and Patterson 1986)
Thioridazine (Gaultieri and Patterson 1986)
Clozapine (Lindenmeyer *et al.* 1995)
Tardive dyskinesia (Chapter 29.2)
Choreiform disorders
Sydenham's chorea (Chapter 19.11)
Huntington's disease (Chapter 9.10)
Neuroacanthocytosis (Chapter 9.11)
Miscellaneous causes
Autism (Ringman and Jankovic 2000)
Carbon monoxide poisoning (Ko *et al.* 2004)
Traumatic brain injury (Krauss and Jankovic 1997)
Post-encephalitis (herpes zoster [Northam and Singer 1991], varicella-zoster [Dale *et al.* 2003])
Pantothenate kinase-associated neurodegeneration (Scarano *et al.* 2002)
Infarction of the basal ganglia (Kwak and Jankovic 2002)
Behçet's syndrome (Budman and Sarcevic 2002)
With peripheral injury (Factor and Moltho 1997)

MEDICATION SIDE-EFFECT

Tics may also occur as a side-effect to medications, albeit uncommonly, and if Tourette's appears to be an unlikely cause then the patient's medication history should be scrutinized for any of the stimulants, anti-epileptics, antidepressants, or antipsychotics listed in Table 4.2, keeping in mind that in some cases a long period of time may elapse between initiation of treatment with an offending medication and the emergence of tics. As noted in Table 4.3, tics may also occur after chronic treatment with antipsychotics as a subtype of tardive dyskinesia, known as 'tardive Tourette's.'

CHOREIFORM DISORDERS

Tics may occur during the course of various choreiform disorders, such as Sydenham's chorea, Huntington's disease, and neuroacanthocytosis, and in the case of neuroacanthocytosis, eventually over one-half of patients will develop tics. Furthermore, albeit rarely, these choreiform disorders may actually present with tics, as has been noted in Sydenham's chorea (Kerbeshian *et al.* 1990), Huntington's disease (Angelini *et al.* 1998), and neuroacanthocytosis (Feinberg *et al.* 1991).

MISCELLANEOUS CAUSES

Of the other causes of tics, consideration is given to autism and to a history of carbon monoxide poisoning, traumatic brain injury, or encephalitis. Very rare causes include pantothenate kinase-associated neurodegeneration, infarction of the basal ganglia, Behçet's syndrome, and peripheral injury. In the case of tics that are secondary to peripheral injury (e.g. trauma to the neck followed by a tic of the head), the tics are persistently focal, occurring only in proximity to the site of the injury.

Differential diagnosis

Tics must be distinguished from myoclonus, chorea, hemifacial spasm, and tic douloureux. There are two differential features which distinguish tics from each of these differential possibilities, namely the premonitory urge and their temporary suppressibility, both of which are classic for tics but absent in all these other conditions.

Myoclonus may resemble a simple tic, but on close examination one will see a slow relaxation phase after a myoclonic jerk, a feature that does not occur after a tic.

Choreiform movements may be either simple in nature or somewhat complex, resembling purposeful acts. Choreiform movements, however, display an almost lightning-like fluidity, appearing and reappearing at different sites in a mercurial fashion not imitated by tics.

Hemifacial spasm is repetitive, involves only musculature innervated by the facial nerve, and is confined, as the name suggests, to one side of the face, thus clearly distinguishing it from a tic.

Tic douloureux is primarily a condition of painful spasms of the face, immediately distinguishing it from tics, which are not

Table 4.3 Causes of chorea

Secondary to medications or intoxicants
Tardive dyskinesia (Chapter 29.2)
Oral contraceptives (Gamboa *et al.* 1971; Green 1980; Nausieda *et al.* 1979)
Phenytoin (Kooiker and Sumi 1974; Nausieda *et al.* 1978)
Valproic acid (Lancman *et al.* 1994)
Gabapentin (Buetefisch *et al.* 1996; Chudnow *et al.* 1997)
Phenobarbital intoxication (Lightman 1978)
Lamotrigine (Zesiewicz *et al.* 2006)
Cocaine (Daras *et al.* 1994)
Amphetamine (Lundh and Tunving 1981)
Methylphenidate (Extein 1978; Weiner *et al.* 1978)
Pemoline (Sallee *et al.* 1989)
Levodopa (Mones *et al.* 1971)
Baclofen (both during treatment [Crystal 1990] and upon discontinuation [Kirubakaren *et al.* 1984])
Lithium intoxication (Podskalny and Factor 1996)
Neuroleptic malignant syndrome (Chapter 29.1)
Trihexyphenidyl (Nomoto *et al.* 1987)
Anticholinergic delirium (Chapter 29.6)
Fluoxetine (Bharucha and Sethi 1996)
Paroxetine (Fox *et al.* 1997)
Trazodone (McNeill 2006)
Alpha-interferon (Moulignier *et al.* 2002)
Cimetidine (Kushner 1982)
With gradual onset in late adolescence or adult years
Huntington's disease (Chapter 9.10)
Huntington disease-like syndrome type 2 (Stevanin *et al.* 2003)
Choreoacanthocytosis (Chapter 9.11)
McLeod syndrome (Danek *et al.* 2001; Jung *et al.* 2001)
Dentatorubropallidoluysian atrophy (Chapter 9.15)
Fahr's syndrome (Chapter 14.22)
Acquired hepatocerebral degeneration (Chapter 14.20)
Wilson's disease (Chapter 9.16)
Schizophrenia (Owens *et al.* 1982)
Senile chorea (Chapter 9.13)
With onset in childhood or early adolescence
Sydenham's chorea (Chapter 19.11)
Cerebral palsy (Rosenbloom 1994)
Lesch–Nyhan syndrome (Chapter 10.8)
Pantothenate kinase-associated neurodegeneration (Chapter 9.18)
Ataxia telangiectasia (Woods and Taylor 1992)
Benign hereditary chorea (Chapter 9.14)
Miscellaneous causes
Stroke (infarction of basal ganglia [Chung *et al.* 2004] or thalamus [Lera *et al.* 2000])
AIDS dementia (Pardo *et al.* 1998) (Chapter 15.1)
Toxoplasmosis (Piccolo *et al.* 1999) (Chapter 15.21)
Progressive multifocal leukoencephalopathy (Piccolo *et al.* 1999) (Chapter 15.3)
Antiphospholipid syndrome (Cervera *et al.* 1997; Paus *et al.* 2001; Sunden- Cullberg *et al.* 1998) (Chapter 19.6)
Systemic lupus erythematosus (Donaldson and Espiner 1971; Fermaglich *et al.* 1973) (Chapter 19.3)
Chorea gravidarum (Chapter 19.12)
Limbic encephalitis (Croteau *et al.* 2001; Heckmann *et al.* 1997; Kinirons *et al.* 2003; Vernino *et al.* 2002) (Chapter 19.8)
Hyperosmolar non-ketotic hyperglycemia (Chang *et al.* 1996)
Hypernatremia (Speracio *et al.* 1976)
Hyperthyroidism (Fidler *et al.* 1971; Van Uitert and Russakoff 1979)
Post-hypoglycemic (Hefter *et al.* 1993; Lai *et al.* 2004)
Delayed post-anoxic leukoencephalopathy (Davous *et al.* 1986; Hori *et al.* 1991; Schwartz *et al.* 1985) (Chapter 13.2)
'Post pump' chorea (Medlock *et al.* 1993)
B12 deficiency (Pacchetti *et al.* 2002) (Chapter 14.1)
Polycythemia vera (Gautier-Smith and Prankerd 1967)
New-variant Creutzfeldt–Jakob disease (McKee and Talbot 2003) (Chapter 16.2)
Mercury intoxication (Snyder 1972) (Chapter 14.12)
Cervical cord compression (Tan *et al.* 2002)
Paroxysmal dystonic choreoathetosis (Demirkiran and Jankovic 1995)
Paroxysmal chorea with hyperthyroidism (Fishbeck and Layzer 1979)

AIDS, acquired immune deficiency syndrome.

painful. Furthermore, the 'tic' of tic douloureux is more of a wince after the spasm of pain than a true, autonomous motor tic.

Treatment

Treatment is directed at the underlying cause. When this is ineffective or not possible, consideration may be given to the use of one of the treatments found effective in Tourette's syndrome, such as a second-generation antipsychotic.

4.4 CHOREA

Choreiform movements, classically seen in Huntington's disease, may also occur as a side-effect to various medications or as part of a host of other disorders.

Clinical features

Choreic movements consist of brief jerks, of greater or lesser complexity, which appear and disappear on different parts of the body with an amazing fluidity and rapidity, often flitting like summer lightning – they may first appear on the face, then the arm, or perhaps the leg, then again showing up in the hand. Although they are purposeless, some patients may attempt to disguise a choreiform movement by merging it into a purposeful one; thus, a choreic jerk of the arm up to the head may be melded into a purposeful movement of the hand across the top of the head, as if the movement, all along, had been meant to smooth back the hair. Gait is often disturbed in chorea, and may become lurching: when severely affected, the gait has a 'dancing and prancing' quality to it; and it is from this that the term 'chorea' (Latin for 'the dance') is derived.

Etiology

The various causes of chorea are listed in Table 4.3. These causes are divided into several groups: those secondary to medications or intoxicants (e.g. as seen in tardive dyskinesia or with oral contraceptives); those of gradual onset in late adolescence or adult years (of which Huntington's disease is the most important); those of childhood or early adolescent onset, such as Sydenham's chorea; and, finally, a miscellaneous group, of which stroke is probably the most important.

SECONDARY TO MEDICATIONS OR INTOXICANTS

Of medications capable of causing chorea, perhaps the most important group is the antipsychotics, which, after chronic use, can cause tardive dyskinesia, which in turn classically presents with chorea. Of note, and in contrast to most other causes of chorea, the chorea seen in tardive dyskinesia, although it can be generalized, tends to remain somewhat localized to the face. Oral contraceptives may cause chorea, and this is more likely to occur in females with a history of Sydenham's chorea or with either systemic lupus erythematosus or the antiphospholipid syndrome. Of the AEDs capable of causing chorea, phenytoin is most commonly involved, with the others only rarely being at fault. Of the stimulants implicated in chorea, cocaine is prominent, and chorea in these cases is often referred to as 'crack dancing'. Of note, after repeated intoxication with either cocaine (Daras *et al.* 1994) or amphetamines (Lundh and Tunving 1981), rather than subsiding with abstinence, the chorea may persist. Levodopa-induced chorea either may occur as a 'peak-dose' side-effect or, less commonly, may be seen as the blood level rises and as it falls, with an interval 'clear' of chorea during peak blood levels. Baclofen, as noted in Table 4.3, may cause chorea not only during treatment, but also with discontinuation of chronic treatment. Lithium may cause chorea, but this is only with lithium intoxication at high blood levels – although in most cases the chorea clears as the blood level falls, in some cases it may persist for months (Zorumski and Bakris 1983). The neuroleptic malignant syndrome is a rare disorder seen with an abrupt diminution of dopaminergic tone, due either to treatment with a dopamine blocker, such as an antipsychotic, or to discontinuation of a dopaminergic agent, such as levodopa: in addition to chorea, one also sees delirium, diaphoresis, fever, tachycardia, and rigidity. Anticholinergics, such as trihexyphenidyl, may cause chorea, and, as might be expected, chorea may also appear during the course of anticholinergic delirium. Regarding the other medications listed in the table, little need be said except that chorea represents a very rare side-effect to them.

WITH GRADUAL ONSET IN LATE ADOLESCENCE OR ADULT YEARS

Of the choreas with gradual onset in late adolescence or early adult years, Huntington's disease is the most common cause: most patients fall ill in their thirties with the gradual onset of a progressively worsening chorea that is eventually joined by a dementia; this is an autosomal dominant condition and sporadic cases are very rare. There are other neurodegenerative disorders that present in a more or less similar fashion to Huntington's disease, and these must be considered in the differential diagnosis. One of them is so similar, clinically, to Huntington's that it is called 'Huntington disease-like syndrome', type 2, and the differential here rests on genetic testing. Other disorders that come into the differential often have distinguishing clinical features. Thus, both choreoacanthocytosis and the X-linked McLeod syndrome are associated with elevated creatine kinase (CK) levels and acanthocytosis; choreoacanthocytosis may also cause tics, dystonia, and, classically, mutilative lip biting. Dentatorubropallidoluysian atrophy, another Huntington 'look-alike', is suggested by the presence of ataxia. Fahr's syndrome may present with chorea but this is usually accompanied by other abnormal movements, such as dystonia or parkinsonism, and computed tomography (CT) scanning will immediately reveal calcification of the basal ganglia. Acquired hepatocerebral degeneration, seen only in patients with long standing hepatic disease and a history of recurrent episodes of hepatic encephalopathy, in addition to chorea, may also cause tremor and ataxia. Wilson's disease, rarely, may present with chorea; however, other abnormal movements (e.g. dystonia, tremor) are far more common (Starosta-Rubinstein *et al.* 1987). Nevertheless, given the tragic nature of missing the diagnosis of this treatable disease, it should remain prominent on the differential. Schizophrenia may cause chorea, but this is usually quite mild and is far overshadowed, clinically, by the classic psychotic symptoms. Finally, in cases of late-onset 'pure' chorea without associated features, consideration should be given to a diagnosis of 'senile chorea'.

WITH ONSET IN CHILDHOOD OR EARLY ADOLESCENCE

Of the childhood- or early adolescent-onset choreas, Sydenham's chorea is by far the most common. Although it is often accompanied by other evidence of rheumatic fever, such as carditis or polyarthritis, the onset of the chorea may at times be delayed for so long that the other signs of rheumatic fever have already gone into remission, leaving a case of 'pure' Sydenham's chorea. The chorea is of subacute onset, over weeks, and, although typically generalized, is often most prominent in the face and upper extremities. In a small minority there may be only hemichorea. Although Sydenham's chorea may in some patients present solely with chorea, in the majority there are other neuropsychiatric features, most notably obsessions and compulsions (Swedo *et al.* 1993). In Sydenham's chorea, the chorea, in almost all cases, eventually undergoes spontaneous remission, generally within 5–9 months. The other causes listed in Table 4.3, however, are chronic. Cerebral palsy, of the 'extrapyramidal' type, may cause chorea, and its appearance may be delayed for several years after birth. Lesch–Nyhan syndrome is immediately suggested by the classic mutilative lip biting, which usually become apparent as soon as the child gets teeth. Pantothenate kinase-associated neurodegeneration may cause chorea; however, this is usually accompanied by dystonia.

Ataxia telangiectasia may also cause chorea, but this is usually preceded by ataxia. Finally, in cases of chronic childhood-onset chorea when other disorders are unlikely, one may consider the diagnosis of benign hereditary chorea.

MISCELLANEOUS CAUSES

Of the miscellaneous causes of chorea, stroke and AIDS are among the most common in a general hospital setting (Piccolo *et al.* 1999). Infarction of the basal ganglia or thalamus may cause either a hemichorea or, less commonly, a generalized chorea, and in many cases the onset of the chorea may be delayed for months (Kim 2001a). AIDS dementia may be characterized by chorea; however, chorea in the course of AIDS is more commonly due to opportunistic infections, such as, for example, toxoplasmosis or progressive multifocal leukoencephalopathy. Chorea may also occur on an autoimmune basis, either as part of the antiphospholipid syndrome or systemic lupus erythematosus. Chorea gravidarum (Latin for 'chorea of pregnant women') should obviously be suspected when chorea occurs during pregnancy, and in most of these cases the patient will have either antiphospholipid syndrome or systemic lupus erythematosus, or will have a history of Sydenham's chorea. Patients with a paraneoplastic syndrome may also present with chorea, but this is soon joined by other elements of a paraneoplastic syndrome, such as delirium, seizures, ataxia, and peripheral neuropathy. Metabolic causes of chorea include hyperosmolar non-ketotic hyperglycemia and, very rarely, hypernatremia, and both of these are readily diagnosed by a chemical survey. Hyperthyroidism may cause a persistent chorea, and the diagnosis is suggested by the associated tremor, diaphoresis, and heat intolerance. Chorea may also occur as a sequela to hypoxic coma or as part of a delayed postanoxic encephalopathy, and in children may follow cardiac surgery – the so-called 'post pump' chorea. Of the other etiologies of persistent chorea listed in Table 4.3, little more need be said except that they only very rarely cause choreiform movements. Finally, episodic paroxysmal chorea should be noted. This may occur as part of paroxysmal dystonic choreoathetosis, wherein it is typically accompanied by dystonia (discussed further in Chapter 3.7), and may also, albeit very rarely, occur as part of hyperthyroidism.

Differential diagnosis

Choreiform movements must be distinguished from myoclonus, tics, athetosis, and from the abnormal oromandibular movements seen in edentulous patients.

Myoclonic jerks are always simple in nature, and hence when choreic jerks are complex, as they sometimes may be, the differential is straightforward. Simple choreic jerks, however, may be more difficult to distinguish from myoclonic jerks. One clue is the 'mercurial' nature of chorea, with jerks appearing and disappearing, fluidly, from one part of the body to another: myoclonic jerks rarely display such lightning-like changes.

Tics, like choreic movements, may be either simple or complex, and thus complexity cannot be used to differentiate them from choreic movements. Tics, however, like myoclonus, rarely display the mercurial transience of choreic movements. Furthermore, tics are associated with a premonitory urge and are, at least partially, suppressible; these features are not characteristic of chorea.

Athetotic movements are longer lasting and more sustained in character than are choreic ones; furthermore, rather than being jerky, they are 'slow and writhing in character', as pointed out by Kinnier Wilson (Wilson 1955).

Edentulous patients may display oromandibular movements reminiscent of chorea; however, in this condition there are no abnormal movements elsewhere on the body and the tongue is not involved (Koller 1983).

Treatment

Treatment is directed at the underlying condition, as discussed in the respective chapters. In many cases this involves symptomatic treatment with an antipsychotic, and in this regard one of the second-generation antipsychotics, for example risperidone, may be considered.

4.5 ATHETOSIS

Athetosis, a term coined by Hammond in 1871 (Hammond 1871), is derived from the Greek word *athetos*, meaning 'without fixed position'.

Clinical features

Athetosis is characterized by purposeless, slow, continuously writhing movements that tend to flow, one into the other, in an unceasing, serpentine pattern. They may be unilateral or bilateral, and are typically seen in the extremities, more often upper than lower, and more often on the distal portions of the involved extremity. The movements, when seen in the hand, may mimic, in a grotesque fashion, the movements of the fingers made when playing the piano or, in some cases, may remind the observer of the hand movements made by Balinese dancers.

Etiology

The various causes of athetosis are listed in Table 4.4, beginning with cerebral palsy.

CEREBRAL PALSY

Cerebral palsy of the 'dyskinetic' type, generally due to either hypoxia or, in the past, kernicterus, may leave patients with bilateral or 'double' athetosis.

INFARCTION OF THE LENTICULAR NUCLEUS

Infarction of the lenticular nucleus may cause athetosis, and in some cases the emergence of this abnormal movement may be delayed for weeks or months after the stroke (Dooling and Adams 1975; Kim 2001a).

Table 4.4 Causes of athetosis

Cerebral palsy ('double athetosis') (Rosenbloom 1994)
Infarction of the lenticular nucleus (Carpenter 1950; Papez et al. 1938; Spiller 1920)
Associated with proprioceptive sensory (Ghika and Bogousslavsky 1997; Sharp *et al.* 1994)
Infarction of the parietal cortex or thalamus
Lesions of the cervical cord or dorsal root ganglia
Miscellaneous causes
Acquired hepatocerebral degeneration (Chapter 14.20)
Fahr's syndrome (Chapter 14.22)
Pantothenate kinase-associated neurodegeneration (Chapter 9.18)
Tiagabine
Donepezil (Tanaka *et al.* 2003)
Post-mercury intoxication (Snyder 1972)
Post-cyanide intoxication (Rosenow *et al.* 1995)

ASSOCIATED WITH PROPRIOCEPTIVE SENSORY LOSS

Severe proprioceptive sensory loss, as may be seen with infarction of the parietal cortex or thalamus, or with lesions of the cervical cord or dorsal root ganglia, may also be followed by athetosis when patients attempt to maintain a limb in a given posture. In the literature this is often referred to as 'pseudoathetosis'; however, given that clinically these cases are essentially indistinguishable from those seen with 'true' athetosis, this distinction may not be useful.

MISCELLANEOUS CAUSES

Of the miscellaneous causes of athetosis, consideration may be given to acquired hepatocerebral degeneration, Fahr's syndrome, and pantothenate kinase-associated neurodegeneration; however, in each of these conditions other features often dominate the clinical picture: in acquired hepatocerebral degeneration, chorea, tremor, and dystonia may be seen; in Fahr's syndrome, parkinsonism and calcification of the basal ganglia on CT scanning; and, in pantothenate kinase-associated neurodegeneration, dystonia. There are also single case reports of athetosis occurring as a side-effect to either tiagabine or donepezil. Finally, athetosis may occur as a sequela to intoxication with either mercury or cyanide.

Differential diagnosis

Athetosis should be distinguished from, on the one hand, chorea, and, on the other, dystonia. Chorea is marked by jerky movements which tend to appear and reappear on different parts of the body, and this is in contrast to athetotic movements which are writhing in character, sustained, and which generally remain localized to a particular limb. Dystonic movements are relatively fixed in character and lack the writhing quality of athetosis.

Although a distinction between athetosis and chorea or dystonia is often possible in well-developed cases, it must be admitted that there are numerous cases where the abnormal movement in question lies in the fuzzy borderzone between chorea and athetosis on the one hand and athetosis and dystonia on the other, and in such cases it is appropriate to consider not only the causes of athetosis listed in Table 4.4, but also the causes of chorea and dystonia outlined in Chapters 4.4 and 4.7, respectively.

Treatment

Botulinum toxin has been utilized in the treatment of double athetosis seen in cerebral palsy (Gooch and Sandell 1996); in cases of athetosis secondary to stroke, deep brain stimulation has been utilized (Katayama *et al.* 2003). It is not at all clear whether pharmacologic treatment with antipsychotics or other agents might be beneficial.

4.6 BALLISM

Ballism, or ballismus, comes originally from the Greek word *ballismos*, which means a jumping movement; it is also related to the Latin *ballista*, which refers to an ancient military machine, similar to a catapult, used for throwing large stones.

Clinical features

Ballistic movements involve the extremities – most commonly, both the upper and lower extremities are involved on one side, producing hemiballism; occasionally one may see monoballism (with involvement of only one extremity) or biballism (with bilateral, generalized involvement). The proximal musculature is preferentially involved and the movements are wild and flinging in nature, sometimes to a degree that may throw patients off balance.

Etiology

The various causes of ballism are noted in Table 4.5.

STROKE

Stroke, whether due to ischemic infarction or, less commonly, intracerebral hemorrhage, is the most common cause of ballism (Vidakovic *et al.* 1994). Although the subthalamic nucleus is most commonly involved, damage to related structures, such as the striatum, thalamus, or, very rarely, the substantia nigra may also cause ballism. In most cases a hemiballism occurs, and, as might be expected, in most of these cases, the lesion is found in the contralateral hemisphere. However, exceptions to this lateralizing rule do occur and ipsilateral hemiballism has been noted with lesions of the subthalamic nucleus (Crozier *et al.* 1996; Moersch and Kernohan 1939) or the striatum (Borgohain *et al.* 1995). Of note, transient hemiballism has also been reported secondary to a transient ischemic attack (Kim *et al.* 2003).

Table 4.5 Causes of ballism

Stroke	
Subthalamic nucleus (Carpenter 1955; Martin and Alcock 1934; Melamed *et al.* 1978; Pfeil 1952; Whittier 1947)	
Striatum (Defebvre *et al.* 1990; Kase *et al.* 1981; Schwarz and Barrows 1960; Srinivas *et al.* 1987)	
Thalamus (Antin *et al.* 1967; Dewey and Jankovic 1989; Kulisevsky *et al.* 1993)	
Substantia nigra (Caparros-Lefebvre *et al.* 1994)	
Other focal lesions	
Tumors (Glass *et al.* 1984)	
Multiple sclerosis (Riley and Lang 1988; Waubant *et al.* 1997)	
Miscellaneous causes	
Non-ketotic hyperglycemia (Chu *et al.* 2002; Lee *et al.* 1999; Lin and Chang 1994; Lin *et al.* 2001)	
Systemic lupus erythematosus (Dewey and Jankovic 1989; Vidakovic *et al.* 1994)	
Sydenham's chorea (Konagaya and Konagaya 1992)	
Limbic encephalitis (Chapter 19.8)	
Traumatic brain injury with damage to the subthalamic nucleus (Vakis *et al.* 2006)	
Hyperthyroidism (Ristic *et al.* 2004)	
Medications (phenytoin [Opida *et al.* 1978], oral contraceptives [Driesen and Wolters 1986], bupropion [de Graaf *et al.* 2003])	

OTHER FOCAL LESIONS

Other focal lesions may also cause hemiballism, and this has been noted in the subthalamic nucleus for both metastatic tumors and multiple sclerosis plaques.

MISCELLANEOUS CAUSES

Of the miscellaneous causes of ballism, the most important from a clinical point of view is non-ketotic hyperosmolar hyperglycemia, as may be seen in elderly diabetics. This syndrome may present with either generalized biballism or hemiballism; of note, in cases of hemiballism, T1-weighted magnetic resonance scanning may reveal hyperintensity of the contralateral striatum. In all cases, the ballism resolves with restoration of euglycemia; however, it may take days or weeks for full recovery to occur. Rarely, systemic lupus erythematosus, Sydenham's chorea, and limbic encephalitis may cause ballism, presumably on the basis of an autoimmune attack on the subthalamic nucleus. Traumatic brain injury is another rare cause and, in the case in question, there was damage, as might be expected, to the subthalamic nucleus. Hyperthyroidism may also cause ballism, albeit rarely, and there are isolated case reports of ballism occurring as a side-effect to phenytoin, oral contraceptives, and bupropion.

Differential diagnosis

The hallmark of ballism is its wild, flinging nature, and it is this characteristic that distinguishes it from other abnormal movements, such as dystonia, athetosis, and chorea. Dystonic movements are fixed and more or less immobile; although in athetosis there is some motility, it is slow and writhing in character. Choreic jerks, when severe, may approach the flinging character of ballistic movements, but choreic jerks appear and reappear on different parts of the body in contrast with the consistent presence of ballism on one or more limbs.

Treatment

Of the various symptomatic treatments for ballism, by far the best-established agents are the antipsychotics; among these, haloperidol (Davis 1976; Klawans *et al.* 1976) has the longest track record. Other first-generation agents to consider include perphenazine (Johnson and Fahn 1977) and chlorpromazine (Klawans *et al.* 1976). Of second-generation antipsychotics, risperidone (Evidente *et al.* 1999), olanzapine (Mukand *et al.* 2005), and clozapine (Stojanovic *et al.* 1997) have also been used. Given the better overall tolerability of the second-generation agents, a case could be made for starting with one of these, for example risperidone. Regardless of which agent is used, it is appropriate to start with a low dose (e.g. 1–2 mg of haloperidol or 0.25–0.5 mg of risperidone) and to gradually titrate the dose upward. Periodic attempts should be made to taper and, if possible, discontinue antipsychotics over the following months.

Other pharmacologic agents to consider include sertraline (Okun *et al.* 2001), topiramate (Gatto *et al.* 2004), and divalproex (Chandra *et al.* 1982; Lenton *et al.* 1981) (although not all reports are favorable for divalproex [Lang 1985]). In severe treatment-resistant cases, consideration may be given to pallidotomy (Yamada *et al.* 2004), deep brain stimulation of the thalamus (Tsubokawa *et al.* 1995), or intrathecal baclofen (Francisco 2006).

4.7 DYSTONIA

Dystonia, a term coined by Oppenheim in 1911 (Oppenheim 1911), refers to a more or less sustained contortion of one or more body parts.

Clinical features

Dystonic movements may arise either spontaneously or when a patient begins a voluntary movement. The contortion arises from the simultaneous contraction of both agonist and antagonist muscles and persists for variable periods of time. Some common varieties deserve to be mentioned. Dystonia of the cervical muscles may rotate or twist the head in one direction or the other (torticollis), pull it over to one side (lateralcollis), or forward (anterocollis), or backward (retrocollis). Dystonic contraction of the oromandibular musculature may cause a forced yawning type of movement, and when the orbicularis oculi muscles undergo dystonic contracture, blepharospasm occurs. The extraocular musculature may be involved, creating what is known as an 'oculogyric crisis', with tonic conjugate gaze deviation, generally superiorly and laterally. When the upper extremity is involved, the arm may be twisted, and in the hand the thumb may be adducted with the fingers hyperextended. Lower extremity involvement is typified by inversion and plantar flexion of the foot. The axial musculature may also be involved, resulting in variable contortions of the trunk. Finally, there are also cases where dystonia occurs only with specific movements. As noted earlier, in some cases dystonia may be precipitated by movement, and this may occur with non-specific movements such as turning the head or lifting an arm. There are, however, dystonias wherein only very specific activities, such as writing or typing, precipitate the dystonia.

Dystonias are often described with reference to the number and contiguity of body parts involved. Thus, in focal dystonia, only one body part is involved (e.g. the upper extremity alone), whereas in segmental dystonia, two adjacent parts are affected (e.g. the neck and upper extremity) – 'hemidystonia' is a term reserved for a specific segmental dystonia wherein the upper and lower extremity are both involved. In multifocal dystonia, two or more non-adjacent parts are affected; in generalized dystonia, bilateral involvement is seen.

In some cases, patients may be able to relieve a dystonia by utilizing what is known as a 'sensory trick', or *geste antagoniste*. For example, by simply placing a hand gently on the cheek, a patient with torticollis may abort the twisting motion of the neck musculature, and a patient with a generalized dystonia precipitated by walking may find relief by walking backward.

Etiology

The various causes of dystonia are listed in Table 4.6, in which they are divided into several groups. The first group, known as the primary dystonias, is composed of 'pure' dystonias that occur either idiopathically or on an inherited basis. The next group, known as the 'dystonia plus' syndromes, is also characterized by inherited disorders; however, here the dystonia is accompanied by other abnormal movements, such as parkinsonism or myoclonus. Following this are the secondary dystonias. This is a heterogeneous group, including medication-induced dystonias (e.g. antipsychotics), those occurring as part of a neurodegenerative disorder, those occurring secondary to focal lesions (e.g. infarctions of the basal ganglia), and those occurring secondary to a large number of miscellaneous causes. Finally, there is a group of paroxysmal dystonias, characterized clinically by dystonia occurring only in discrete, brief episodes. In general clinical practice, the most common causes of dystonia are one of the primary dystonias or a medication-induced dystonia, most commonly due to an antipsychotic.

PRIMARY DYSTONIAS

Of the primary dystonias, the classic example is primary torsion dystonia, or 'dystonia musculorum deformans' as it was originally named by Oppenheim (1911). This typically has an onset in childhood or adolescence and presents with a focal dystonia, often of the lower extremity, followed by gradual progression to a generalized involvement. The other primary dystonias generally have an onset in adult years and typically remain more or less focal. Idiopathic cervical dystonia, as the name clearly suggests, manifests with a cervical dystonia; Meige's syndrome is characterized by blepharospasm, Breughel's syndrome by oromandibular dystonia, and spasmodic dysphonia, again as the name suggests, by dysphonia. Rarely, only the trunk may be involved, creating the syndrome of axial dystonia. As noted above, some dystonias occur only with specific activities, and these are known as the 'task-specific dystonias': examples include writer's, typist's, and musician's cramp.

'DYSTONIA PLUS' SYNDROMES

In considering the 'dystonia plus' syndromes, by far the most important disorder to keep in mind is dopa-responsive dystonia. This disorder has an onset in childhood with dystonia of the lower extremity, and thus figures in the differential diagnosis of primary torsion dystonia. The reason why it should always be considered is because, in contrast with primary torsion dystonia, it is eminently treatable, often responding dramatically to low-dose levodopa. The other 'dystonia plus' syndromes are rare. Myoclonus dystonia, as noted in Chapter 4.2, has an onset in childhood or adolescence and is characterized by myoclonus and dystonia, often a cervical or a task-specific dystonia. Rapid-onset dystonia parkinsonism presents in adult years acutely with a combination of dystonia and parkinsonism, with symptoms escalating over days or weeks and then stabilizing. Lubag (also known as X-linked dystonia parkinsonism) is an X-linked disorder seen only in Filipino men, presenting first with dystonia but eventually joined by parkinsonism.

SECONDARY DYSTONIAS

In considering the secondary dystonias, each subgroup will be treated in turn, beginning with medication-induced dystonias.

Medication induced

Of the medications capable of causing dystonia, by far the most common are the first-generation antipsychotics, such as haloperidol and fluphenazine (Keepers *et al.* 1983; Swett 1975). These occur within 0.5–3 days of either starting the medication or substantially increasing the dose (Ayd 1961; Keepers *et al.* 1983), and are most likely to occur in young males (Boyer *et al.* 1987). Focal dystonias including oculogyric crisis, torticollis, and involvement of the upper limb are most common; in some cases, segmental spread may occur and, rarely, an acute antipsychotic-induced dystonia may become generalized. Lingual dystonia may also occur, and patients may become dysarthric and complain of a 'thick tongue'; very rarely, laryngeal dystonia with respiratory embarrassment may occur (Christodoulou and Kalaitzi 2005; Flaherty and Lahmeyer 1978). Interestingly, in patients with schizophrenia, antipsychotic-induced dystonias may be accompanied by a

Table 4.6 Causes of dystonia

Primary dystonias
Primary torsion dystonia (Chapter 9.20)
Idiopathic cervical dystonia (Chapter 9.21)
Meige's syndrome (Chapter 9.23)
Breughel's syndrome (Gilbert 1996)
Spasmodic dysphonia (Chapter 9.24)
Axial dystonia (Bhatia *et al.* 1997b)
Task-specific dystonia (Chapter 9.22)

'Dystonia plus' syndromes
Dopa-responsive dystonia (Chapter 9.19)
Myoclonus dystonia (Asmus *et al.* 2002; Doheny *et al.* 2002; Grimes *et al.* 2002; Schule *et al.* 2004)
Rapid-onset dystonia parkinsonism (Zaremba *et al.* 2004)
Lubag (Evidente *et al.* 2002)

Secondary dystonias
Medication induced
Antipsychotics
Metoclopramide (Tan *et al.* 2010)
Prochlorperazine (Olsen *et al.* 2000)
Promethazine (Tan *et al.* 2010)
Tardive dystonia (a variant of tardive dyskinesia)
Levodopa (Cubo *et al.* 2001; Kidron and Melamed 1987; Melamed 1979; Nausieda *et al.* 1980)
Bupropion (Detweiller and Harpold 2002)
Fluoxetine (Dominguez-Moran *et al.* 2001)
Sertraline (Walker 2002)
Buspirone (LeWitt *et al.* 1993)
Carbamazepine (Stryjer *et al.* 2002)
Gabapentin (Pina and Modrego 2005; Reeves *et al.* 1996)
Tiagabine (Wolanczyk and Grabowska-Grzyb 2001)
Valproic acid (Oh *et al.* 2004)
Cocaine withdrawal (Choy-Kwong and Lipton 1989)
MDMA ('Ecstasy') (Priori *et al.* 1995)
Lamivudine (Song *et al.* 2005)
α-Interferon (Atasoy *et al.* 2004)
Flunarazine (Koukoulis *et al.* 1997)
Cimetidine (Romisher *et al.* 1987)

Neurodegenerative disorders
Wilson's disease (Chapter 9.16)
Lesch–Nyhan syndrome (Chapter 10.8)
Pantothenate kinase-associated neurodegeneration (Chapter 9.18)
Corticobasal-ganglionic degeneration (Chapter 9.8)
Progressive supranuclear palsy (Chapter 9.7)
Multiple system atrophy (Chapter 9.9)
Dentatorubropallidoluysian atrophy (Chapter 9.15)
Choreoacanthocytosis (Chapter 9.11)
Huntington's disease (Chapter 9.10)
Spinocerebellar ataxia (Chapter 9.17)

Focal lesions
Infarction or hemorrhage of the basal ganglia, thalamus, or parietal cortex
Mass lesions of the basal ganglia, frontal lobe, or cerebellum

Miscellaneous causes
Traumatic brain injury (Burke *et al.* 1980; Lee *et al.* 1994; Munchau *et al.* 2000)
Fahr's syndrome (Chapter 14.22)
Delayed postanoxic leukoencephalopathy (Chapter 13.2)
Central pontine myelinolysis (Chapter 14.17)
Cyanide intoxication (Rosenow *et al.* 1995; Valenzuela *et al.* 1992)
Methanol intoxication (Quartarone *et al.* 2000) (Chapter 28.12)
Manganese intoxication (Chapter 14.6)
Postencephalitic (encephalitis lethargic [Chapter 15.6], Japanese encephalitis [Chapter15.4])
AIDS-related opportunistic infections (Factor *et al.* 2003)
Creutzfeldt–Jakob disease (Hellmann and Melamed 2002)
Tourette's syndrome (Chapter 9.25)
Cerebral palsy (Saint-Hilaire *et al.* 1991)
Cervical cord lesion (Cammarota *et al.* 1995)
Thoracic outlet syndrome (Quartarone *et al.* 1998)
Peripheral trauma (Schott 1985)
Esophageal reflux 'Sandifer's syndrome' Shahnawaz *et al.* 2001

Paroxysmal dystonias
Primary paroxysmal dystonias
Non-kinesigenic paroxysmal dystonic choreoathetosis
Kinesigenic paroxysmal dystonic choreoathetosis
Exercise-induced paroxysmal dystonic choreoathetosis

Secondary paroxysmal dystonias
Nocturnal paroxysmal dystonia
Ictal dystonia (Kuba *et al.* 2003; Newton *et al.* 1992)

Miscellaneous causes of paroxysmal dystonia
Multiple sclerosis (Waubant *et al.* 2001, Zenzola *et al.* 2001)
Fahr's syndrome (Volonte *et al.* 2001)
Traumatic brain injury (Demerkirian and Jankovic 1995)
Infarction of the basal ganglia (Demerkirian and Jankovic 1995)

AIDS, acquired immune deficiency syndrome.

transient increase of psychotic symptoms (Chiu 1989; Thornton and McKenna 1994). Although first-generation antipsychotics are most commonly at fault, dystonia has been noted, albeit rarely, with second-generation agents such as olanzapine (Alevizos *et al.* 2003) and aripiprazole (Desarkar *et al.* 2006); there are also case reports of dystonia occurring upon discontinuation of clozapine (Ahmed *et al.* 1998; Mendhekar and Duggal 2006). Furthermore, it must be borne in mind that other dopamine blockers, such as metoclopramide, may also cause dystonia. In patients who are treated chronically with antipsychotics, tardive dystonia, a variant of tardive dyskinesia, may also occur – this generally presents with a focal dystonia affecting the neck or face, which may undergo segmental spread and, rarely, become generalized (Burke *et al.* 1982; Kiriakakis *et al.* 1998; Wojick *et al.* 1991; Yassa *et al.* 1986), in which case it may be disabling (Yadalam *et al.* 1990). Interestingly, should mania occur, a pre-existing tardive dystonia may undergo substantial improvement (Kiriakakis *et al.* 1998; Yazici *et al.* 1991). As with acute dystonias, tardive dystonia may also rarely be seen with second-generation agents such as olanzapine, although it is far more common with first-generation agents (Gunal *et al.* 2001).

The most common offender of the other medications that are capable of causing dystonia is levodopa, as utilized in the treatment of Parkinson's disease – in such cases, dystonia may appear either as a peak-dose dyskinesia or, more commonly, as an end-of-dose 'wearing-off' phenomenon. The remaining medicines listed in Table 4.6 only rarely cause dystonia: most references are limited to single case reports.

Neurodegenerative disorders

Turning now to the neurodegenerative disorders capable of causing dystonia, the most important disorder to keep in mind is Wilson's disease, given that, unlike all the other disorders in this group, it is treatable. The onset of Wilson's disease occurs between childhood and early adulthood, and, in a minority of cases, it may present with dystonia (Starosta-Rubinstein *et al.* 1987) – the dystonia may consist of torticollis, dystonia of the upper or lower extremity, an oculogyric crisis (Lee *et al.* 1999), or a facial dystonia of the oromandibular type, which may leave the patient with a fixed, vacuous, wide-mouthed smile. Other symptoms and signs, including personality change, gait abnormalities, and tremor, eventually accrue. Given that treatment is available, testing for copper and ceruloplasmin levels should be performed if there is any suspicion of Wilson's disease. The occurrence of progressive dystonia in early childhood should suggest Lesch–Nyhan syndrome, especially when accompanied by mutilative lip biting. Pantothenate kinase-associated neurodegeneration presents in childhood or adolescence with progressive dystonia, usually of the upper extremity, and is eventually joined by other abnormal movements (such as chorea or tremor) and then by a dementia. The other neurodegenerative disorders listed in the table all have an onset in adult or later years, and, in addition to dystonia, typically are characterized by other abnormal movements. Thus, corticobasal ganglionic degeneration, progressive supranuclear palsy, and multiple system atrophy are characterized, in addition to dystonia, by parkinsonism, and dentatorubropallidoluysian atrophy, choreoanthocytosis, and Huntington's disease are marked by chorea. In the case of Huntington's disease, although dystonia generally appears only in far-advanced disease, there are rare cases of Huntington's disease in which the presentation was with dystonia (Louis *et al.* 2000). Finally, there are rare case reports of spinocerebellar ataxia presenting, not with the expected ataxia, but with dystonia (Nandagopal and Moorthy 2004).

Focal lesions

Of the focal lesions capable of causing dystonia, by far the most common are infarctions of either the basal ganglia (Krystkowiak *et al.* 1998; Russo 1983) or the thalamus (Lehericy *et al.* 2001) – of note is that in these cases it may take weeks, months, or, rarely, years, before the dystonia appears (Lehericy *et al.* 1996). There are also rare reports of cortical infarction causing dystonia, for example as of the parietal cortex (Burguera *et al.* 2001). Mass lesions may also cause dystonia, as has been reported for tumors of the basal ganglia (Martinez-Cage and Marsden 1984), frontal lobe (Soland *et al.* 1996b), and cerebellum (Alarcon *et al.* 2001; Krauss *et al.* 1997).

Miscellaneous causes

Of the miscellaneous secondary causes of dystonia, traumatic brain injury is perhaps the most common, and in such cases lesions of the basal ganglia are seen. Dystonia may also occur in Fahr's syndrome, but here it is generally accompanied by parkinsonism. Dystonia may also appear in a delayed postanoxic leukoencephalopathy, and there are also case reports of dystonia occurring after central pontine myelinolysis, wherein 'extrapontine' myelinolysis, involving the basal ganglia, also occurred. Other clear-cut precipitating events, rarely associated with dystonia, include intoxication with cyanide, methanol, or manganese. Encephalitis may have dystonia as a sequela, and this has been noted after encephalitis lethargica and Japanese encephalitis. AIDS may be characterized by dystonia, and this has been noted in cases of AIDS that are complicated by progressive multifocal encephalopathy and toxoplasmosis. In one case report, Creutzfeldt–Jakob disease was noted to present with dystonia. Tourette's syndrome, in addition to tics, may also, albeit rarely, cause dystonia, and cerebral palsy may also be characterized by dystonia. Remarkably, there are also reports of dystonia occurring with cervical cord lesions and with lesions outside the central nervous system, such as the thoracic outlet syndrome or with peripheral trauma: in cases of peripheral trauma, dystonia may also accompany reflex sympathetic dystrophy (van Hilten *et al.* 2001). Finally, dystonia may also occur in association with gastroesophageal reflux disease, in which case one speaks of the 'Sandifer's syndrome' – although this is generally reported only in children, adult cases have also been noted.

PAROXYSMAL DYSTONIAS

The last group in the table contains the paroxysmal dystonias, which may be divided into primary and secondary forms (Demirkiran and Jankovic 1995).

Primary paroxysmal dystonias

The primary forms are divided into three types, depending on whether the dystonia occurs spontaneously or with precipitants: spontaneously occurring dystonias are referred to as 'paroxysmal non-kinesigenic dystonic choreoathetosis'; those precipitated immediately by some movement, as 'paroxysmal kinesigenic dystonic choreoathetosis'; and those occurring only after relatively prolonged exercise as 'exercise-induced paroxysmal dystonic choreoathetosis'. Each of these primary paroxysmal dystonias has an onset in childhood or adolescence, and in all cases the attacks last generally in the order of minutes, rarely extending to hours; during the attacks, most patients will have dystonia, generally a hemidystonia or a generalized dystonia, which may be accompanied by chorea. Although sporadic cases do occur, most are familial, following a pattern consistent with autosomal dominant inheritance. Further specific details for each type follow below.

In paroxysmal non-kinesigenic dystonic choreoathetosis (Bressman *et al.* 1988; Demirkiran and Jankovic 1995; Fink *et al.* 1997; Jarman *et al.* 1997, 2000; Lance 1977; Matsuo *et al.* 1999; Mount and Reback 1940; Richards and Barnett

1968; Rosen 1964), attacks are more likely when patients are fatigued or have been indulging in alcohol, caffeine, or nicotine. A locus has been identified on chromosome 2, and there appears to be considerable genetic heterogeneity (Bruno *et al.* 2007; Chen *et al.* 2005, Spacey *et al.* 2006).

Paroxysmal kinesigenic dystonic choreoathetosis (Bruno *et al.* 2004; Demirkiran and Jankovic 1995; Kertesz 1967; Lance 1977; Stevens 1966) is suggested by the history of attacks occurring immediately upon some movement, such as walking, running, or, in some cases, simply standing up. Another distinguishing feature of this type is the fact that treatment with an AED such as carbamazepine reliably prevents attacks, often at a low dose (Wein *et al.* 1996). In this disorder, a locus has been identified on chromosome 16; however, as for non-kinesigenic cases, there is also considerable genetic heterogeneity (Valente *et al.* 2000).

In exercise-induced paroxysmal dystonic choreoathetosis (Bhatia *et al.* 1997a; Munchau *et al.* 2000), attacks occur only after fairly prolonged exercise, such as walking or cycling, and they subside fairly soon with rest.

Secondary paroxysmal dystonia

The secondary paroxysmal dystonias include the syndrome of nocturnal paroxysmal dystonia, ictal dystonia, and a miscellaneous group.

Nocturnal paroxysmal dystonia is characterized by brief attacks of dystonia that arise from non-rapid eye movement (REM) sleep and are associated with partial awakening. There are probably distinct subtypes in this syndrome, with some cases representing a parasomnia (Lee *et al.* 1985) and others representing seizures (Lugaresi and Cirignotta 1981; Lugaresi *et al.* 1986; Provini *et al.* 1999; Tinuper *et al.* 1990); in the latter subtype, the response to an AED, such as carbamazepine, is usually good.

Ictal dystonia may also be seen during otherwise typical complex partial seizures, and in such cases the dystonia fairly reliably lateralizes the epileptic focus to the contralateral hemisphere.

Miscellaneous causes of paroxysmal dystonia

Finally, the miscellaneous group includes multiple sclerosis (with plaques in the thalamus or posterior limb of the internal capsule), Fahr's syndrome, traumatic brain injury and infarction of the basal ganglia.

Differential diagnosis

Athetosis is somewhat similar to dystonia in that in both abnormal movements, contortion is present; the difference between the two lies in the degree of mobility seen. In dystonia, the contortion gradually assumes a final form and then remains fairly fixed; by contrast, in athetosis, the contortion constantly changes in a writhing, almost serpentine fashion.

Treatment

For most of the causes listed in the table, treatment is discussed in the respective chapter for the disorder in question. In the case of antipsychotic-induced dystonia, emergent treatment may be accomplished with diphenydramine, given intramuscularly or intravenously in a dose of 25–50 mg, or benztropine, given intramuscularly in a dose of 2 mg; if needed, oral benzatropine, in a similar dose, may be given on a preventive basis. For other causes where treatment is not well established, consideration may be given to empirical trials of benztropine, clonazepam, baclofen, or AEDs, such as carbamazepine; botulinum toxin also may be considered, and, as a last resort, deep brain stimulation of the globus pallidus is an option.

4.8 PARKINSONISM

Parkinsonism is a syndrome of multiple different etiologies: although Parkinson's disease is the most common cause of parkinsonism, it must always be kept in mind that it is but one of many possible causes, and the physician must dutifully consider the other causes before deciding that a patient with parkinsonism has Parkinson's disease.

Clinical features

Fully developed, classic parkinsonism is, once recognized, almost unforgettable. Patients stand in a flexion posture, bent at the waist and neck, with the arms held forward in flexion and the knees somewhat bent. A rhythmic pill-rolling rest tremor is seen, most notably in the hands. When patients attempt to do anything, bradykinesia is apparent, with almost all movements initiated and carried out slowly. The gait is shuffling and patients typically take small steps; at times the phenomenon of 'festination' is seen, wherein, with ongoing ambulation, patients begin to lean forward and take ever smaller but ever faster steps, almost as if they were trying to catch up with their forward-leaning center of gravity. Some patients may also display another gait abnormality known as 'freezing' (Giladi *et al.* 1992): here, patients become unable to initiate ambulation, as if they were 'frozen'. This curious phenomenon typically occurs at a threshold of some sort, for example a doorway or the start of a hallway.

Other motor abnormalities are evident on examination, and include rigidity, postural instability, and micrographia. Rigidity is evident in the limbs and is present with both passive extension and flexion. This rigidity may be 'lead pipe' in character, in that it is of the same severity throughout the full range of motion, or there may be what is known as 'cogwheel' rigidity. Cogwheeling is perhaps most easily appreciated at the elbow by supporting the elbow in the palm of one's hand, placing the thumb over the biceps tendon, and then, with the other hand, alternately slowly extending and flexing the forearm: when cogwheeling is present, a peculiar 'ratcheting' sensation is appreciated in the thumb over the biceps tendon, as if the extension of the forearm were proceeding one 'cog' at a time. Postural instability may be suggested by a history of falls, and may be detected using the 'pull test': to perform this test, stand several feet behind the patient and, after warning the patient what is to come, place your hands on the patient's shoulders and gently, but briskly, pull back. Typical individuals will quickly make a compensatory step back and catch their balance; those with parkinsonism, however, will fail to make compensatory movements and begin to tumble backward (in this regard, if the patient is large, it is appropriate to position yourself with your back to a wall in order to prevent a double fall). Micrographia is manifest by handwriting that becomes small and scratchy: in some cases, multiple samples of the patient's handwriting collected over months and years may provide 'graphic' evidence of a progression of the syndrome.

Other abnormalities that may be seen in classic parkinsonism include hypomimia, hypophonia, palilalia, and bradyphrenia. Hypomimia (also known as 'masked faces') is an abnormality of facial expression wherein the muscles of facial expression seem frozen and there is little blinking: Kinnier Wilson (Wilson 1955) described it as a 'starched' expression. In hypophonia, the voice becomes low, soft, and monotone; in severe cases it is reduced to an unintelligible mumbling. Palilalia is said to be present when the patient involuntarily repeats the last word of a phrase or sentence with ever-increasing rapidity and ever-increasing indistinctness. Bradyphrenia is the cognitive analog of bradykinesia and is characterized by a slowness of thought.

Of all of these signs and symptoms, the 'cardinal' ones for diagnosis are tremor, bradykinesia, rigidity, and postural instability.

Etiology

The various causes of parkinsonism are listed in Table 4.7, where they are divided into several groups. In the first group are neurodegenerative disorders, each of which causes a parkinsonism of gradual onset and progression: this group, as a whole, is by far the most common cause of parkinsonism, and of the disorders in this group, Parkinson's disease is the most frequent. The next group consists of parkinsonism

Table 4.7 Causes of parkinsonism

Neurodegenerative disorders
Parkinson's disease (Chapter 9.5)
Dementia with Lewy bodies (Chapter 9.6)
Corticobasal ganglionic degeneration (Chapter 9.8)
Multiple system atrophy (Chapter 9.9)
Progressive supranuclear palsy (Chapter 9.7)
Dentatorubropallidoluysian atrophy (Chapter 9.15)
Choreoacanthocytosis (Chapter 9.11)
Wilson's disease (Chapter 9.16)
Fahr's syndrome (Chapter 14.22)
Alzheimer's disease (Chapter 9.1)
Pantothenate kinase-associated neurodegeneration (Chapter 9.18)
Juvenile-onset Huntington's disease (Chapter 9.10)
Lubag (Evidente *et al.* 2002)
Spinocerebellar ataxia (Chapter 9.17)
Rapid onset dystonia parkinsonism (Brashear *et al.* 1996, Brashear *et al.* 1997, Brashear *et al.* 2007; Kabacki *et al.* 2005; Kramer *et al.* 1999; Pittock *et al.* 2000)
Hereditary mental depression and parkinsonism (Perry *et al.* 1975; Tsuboi *et al.* 2002)
Guamian amyotrophic lateral sclerosis–parkinsonism complex (Garruto *et al.* 1981; Hirano *et al.* 1967; Malamud *et al.* 1961)

Secondary to medications
Antipsychotics (Hardie and Lees 1988)
Metoclopramide (Indo *et al.* 1982; Sethi *et al.* 1989)
Prochlorperazine (Edelstein and Knight 1987)
Neuroleptic malignant syndrome (Chapter 29.1)
Valproic acid (Armon *et al.* 1996; Easterford *et al.* 2004; Iijima 2002)
α-Methyldopa (Strang 1966)
Lithium (Holroyd and Smith 1995)
SSRIs (e.g. fluoxetine [Ernst and Steur 1993], paroxetine [Jimenez-Jimenez *et al.* 1994])
Phenelzine (Teusink *et al.* 1984)
Calcium channel blockers (e.g. cinnarizine, flunarazine, amlodipine [Marti-Masso and Poza 1998; Sempere *et al.* 1995])
Amiodarone (Werner and Olanow 1989)
Disulfiram (de Mari *et al.* 1993; Laplane *et al.* 1992)
Budesonide (Prodan *et al.* 2006)
Cytosine arabinoside (Luque *et al.* 1987)
Kava extract (Meseguer *et al.* 2002)

Secondary to toxins and substances of abuse
Alcohol withdrawal (Carlen *et al.* 1981)
Methanol (Chapter 28.12)
MPTP (Ballard *et al.* 1985; Tetrud *et al.* 1989)
Inhalant abuse (Uitti *et al.* 1994)
Manganese (Chapter 14.16)
Cyanide poisoning (Uitti *et al.* 1985)
Diquat (Sechi *et al.* 1992)
Organophosphates (Bhatt *et al.* 1999)

Secondary to other precipitating events
Dementia pugilistica (Chapter 12.3)
Posthypoxic–ischemic encephalopathy (Chapter 13.1)
Delayed postanoxic leukoencephalopathy (Chapter 13.2)
Encephalitis lethargica (Chapter 15.6)
Arbovirus encephalitis (e.g. western equine, Japanese) (Chapter 15.4)

Miscellaneous causes
Stroke (see text for references)
Vascular parkinsonism (Chapter 11.3)
Acquired hepatocerebral degeneration (Chapter 14.20)
Hepatic encephalopathy (Federico and Zochodne 2001)
'Encephalopathic' pellagra (Chapter 14.3)
Central pontine myelinolysis (Chapter 14.17)
Multiple sclerosis (Federlein *et al.* 1997)
Systemic lupus erythematosus (Dennis *et al.* 1992)
Sjögren's syndrome (Walker *et al.* 1999)
Neurosyphilis (Sandyk 1983)

SSRI, selective serotonin reuptake inhibitor.

secondary to medications and in almost all of these cases the parkinsonism is of relatively acute onset: of all of these medications the antipsychotics, by far, are the most common offenders. Next considered are cases secondary to toxins and substances of abuse, such as methanol. Following this, there is a group of cases secondary to other, more or less obvious, precipitating events, such as repeated head trauma, or anoxia. Finally, there is a group of miscellaneous causes, prominent among which are stroke and vascular parkinsonism.

NEURODEGENERATIVE DISORDERS

Of the neurodegenerative disorders capable of causing parkinsonism, the first five, namely Parkinson's disease, dementia with Lewy bodies, corticobasal ganglionic degeneration, multiple system atrophy of the striatonigral type, and progressive supranuclear palsy (PSP) (including the two variants, PSP parkinsonism and Richardson's syndrome) are by far the most common causes, and of these Parkinson's disease and dementia with Lewy bodies account for the great majority of cases. Certain clinical features generally enable one to differentiate among them. The first feature to enquire after is whether, at onset and early on in the course, the parkinsonism was asymmetric or symmetric – Parkinson's disease, dementia with Lewy bodies, corticobasal ganglionic degeneration, and PSP parkinsonism all cause a parkinsonism of asymmetric onset. Although multiple system atrophy may be typified by asymmetric onset, the majority of cases are symmetric at onset, and in Richardson's syndrome the parkinsonism is symmetric from the start. Next, consider whether the overall clinical picture for the most part resembles the classic picture described earlier, as all of these disorders, again with the exception of Richardson's syndrome, generally do. Richardson's syndrome, in contrast to the classic picture, is characterized by an erect posture, a wide-based ataxic gait, prominent rigidity with little tremor, and extreme postural instability with, very early on, frequent, unexplained falls; finally, and very importantly, within generally 1–3 years, one sees in Richardson's syndrome a supranuclear ophthalmoplegia. Once Richardson's syndrome is ruled out, additional features enable a differentiation among the remaining disorders. Dementia with Lewy bodies is immediately suggested when the '1-year rule' is followed: in dementia with Lewy bodies, a dementia occurs within 1 year of the onset of the parkinsonism, and this is in striking contrast to the dementia of Parkinson's disease, which is delayed for at least 5 years. Furthermore, in dementia with Lewy bodies, and again early on, one typically finds both visual hallucinations and spontaneous fluctuations in mental status, which are generally brief, lasting from hours to days, and are typically characterized by confusion or drowsiness. Corticobasal ganglionic degeneration, classically, is marked by an extremely asymmetric onset, and indeed this asymmetry may persist for many, many years; other distinguishing features include apraxia and cortical sensory loss, both ipsilateral to the parkinsonism, and, in a small minority of cases, the alien hand syndrome. Multiple system atrophy of the striatonigral variant is marked by an extreme flexion posture and by evidence of other 'system' involvement, including ataxia and, early on, prominent autonomic features, such as postural hypotension (often with syncope), urinary symptoms (retention or incontinence), and erectile dysfunction.

The remaining neurodegenerative disorders in the list are far less common than those just discussed, and most also have distinctive features. Both dentatorubropallidoluysian atrophy and choreoacanthocytosis cause chorea. Wilson's disease may cause a parkinsonian rigidity, but this is generally accompanied by dysarthria or dystonia. In Fahr's syndrome, although parkinsonism may be seen in an isolated fashion, it is typically accompanied by other signs, such as ataxia (Nyland and Skre 1977) or intention tremor (Mathews 1957), and neuroimaging will reveal calcification of the basal ganglia. Alzheimer's disease may be complicated by a mild parkinsonism, but this occurs only many years after the typical dementia is well established. Adolescent-onset parkinsonism may occur with pantothenate kinase-associated neurodegeneration and in the juvenile-onset Westphal variant of Huntington's disease, but in both cases other signs are typically also present, such as dystonia in pantothenate kinase-associated neurodegeneration and either dystonia or chorea in Huntington's disease. Lubag (also known as X-linked dystonia parkinsonism) is an X-linked disorder seen only in Filipino men, which typically presents with dystonia, followed by parkinsonism. Spinocerebellar ataxia (Furtado *et al.* 2002; Shan *et al.* 2001) may present with a combination of ataxia and parkinsonism or, very rarely, with parkinsonism alone. Rapid-onset dystonia parkinsonism has a remarkable onset in adult years that is characterized by the appearance, over days or weeks, of a combination of parkinsonism and dystonia that subsequently settles into a stable chronicity. Hereditary mental depression and parkinsonism is a rare familial disorder that presents in midlife with depression and which is joined, in a matter of years, by a gradually progressive parkinsonism and, eventually, hypoventilation. Guamian amyotrophic lateral sclerosis–parkinsonism complex is a rare disorder occurring exclusively in residents of Guam, parts of Japan, and some other Pacific islands.

SECONDARY TO MEDICATIONS

Turning now to the group of medications capable of causing parkinsonism, the most frequent offenders by far are the antipsychotics. Among the antipsychotics, although first-generation agents (e.g. haloperidol or fluphenazine) are most likely to cause this side-effect, parkinsonism has also been caused by second-generation agents such as risperidone or olanzapine (Carlson *et al.* 2003). Although in the vast majority of cases the parkinsonism gradually subsides after the discontinuation of treatment, in a small minority, primarily in the elderly, it may persist indefinitely (Bocola *et al.* 1996). Other dopamine blockers, such as metoclopramide or prochlorperazine, may also cause parkinsonism, a fact often not appreciated by many physicians. When parkinsonism occurs in patients treated with dopamine blockers, the neuroleptic malignant syndrome, although rare, must also be considered: this syndrome typically presents with a delirium, followed shortly by parkinsonism and other symptoms such as coarse tremor, fever, tachycardia, labile hypertension, and diaphoresis. Although the vast majority of cases of the neuroleptic malignant syndrome are caused by treatment with antipsychotics (including clozapine [Miller *et al.* 1991]), cases have also occurred secondary to the withdrawal of dopaminergic agents such as amantadine (Cunningham *et al.* 1991; Harsch 1987) and levodopa (Sechi *et al.* 1984; Toru *et al.* 1981). Valproic acid may also cause parkinsonism, and, uniquely among medications capable of causing this side-effect, the parkinsonism here may be of delayed onset, from after 6 months to 4 years (Ristic *et al.* 2006), thus obscuring the causal role of the drug. The other medications listed in the table only rarely cause parkinsonism.

SECONDARY TO TOXINS AND SUBSTANCES OF ABUSE

The next group to consider are toxins and substances of abuse and the first to consider here is alcohol, wherein parkinsonism occurs not as a complication of alcohol intoxication but rather appears during the course of alcohol withdrawal in a very small minority of cases. In evaluating chronic alcoholics who have parkinsonism, one should also consider the possibility that the patient, unable to obtain alcohol, may have used methanol: severe cases of methanol intoxication may be complicated by putaminal necrosis with parkinsonism. The next substance to consider is methylphenyltetrahydropyridine (MPTP): this is a contaminant of illicitly prepared meperidine (Langston *et al.* 1983), and addicts who injected the contaminated meperidine developed a parkinsonism of rapid onset. A final example of substances of abuse capable of causing parkinsonism are the inhalants, wherein, very rarely, intoxication may be complicated by a parkinsonism that then persists. With regard to toxins *per se*, the classic offender is manganese. Manganese exposure may occur in manganese mines, steel mills, battery factories, or via contaminated drinking-well water and may be followed, after a variable latent interval, by a gradually progressive parkinsonism. The parkinsonism itself is characterized primarily by rigidity and bradykinesia; tremor may be present but is only rarely of the pill-rolling type. Certain additional features may also be present, including dystonia and what is known as a 'cock-walk'. Dystonic rigidity may be found in the neck or in the face, creating a vacuous, rigid grin. The cock-walk, which may be seen in up to one-third of patients, stems from a dystonic rigidity of the feet, such that patients walk on their metatarsophalangeal joints (as if in high heels), at times looking for all the world like they are imitating the walk of a rooster. Interestingly, manganese-induced parkinsonism may progress long after exposure has ceased. Another toxin to consider is cyanide, wherein exposure typically occurs after a suicidal overdose with potassium cyanide – in these cases, parkinsonism may ensue after a latent interval spanning from weeks (Rosenberg *et al.* 1989) to a year (Carella *et al.* 1988); in some cases, dystonia or athetosis may also be present (Rosenow *et al.* 1995). Finally, parkinsonism has also been reported secondary to exposure to the herbicide diquat and to organophosphate pesticides.

SECONDARY TO OTHER PRECIPITATING EVENTS

Other, more or less obvious, precipitating events include repeated head trauma, an hypoxic–ischemic insult, and encephalitis. Repeated head trauma, as may occur in boxers, may be followed, after a latent interval of 5–40 years, by the gradual evolution of a condition known as 'dementia pugilistica', comprising parkinsonism, dysarthria, ataxia, and dementia. An hypoxic–ischemic insult severe enough to cause coma may be followed by a posthypoxic–ischemic encephalopathy, which in turn may be characterized by parkinsonism, and in those who survive an anoxic insult, with or without developing a posthypoxic–ischemic encephalopathy, a delayed postanoxic leukoencephalopathy may develop, which may also be characterized by parkinsonism. Of the encephalitides associated with parkinsonism, the classic example is encephalitis lethargica, wherein parkinsonism occurred after a latent interval of 1–20 years and was often accompanied by oculogyric crises and blepharospasm; importantly, although there have been no further epidemics of encephalitis lethargica since 1928, sporadic cases still occur. Arboviral encephalitides associated with parkinsonism include western equine encephalitis, wherein parkinsonism may appear after a latent interval of a week to several months, and Japanese encephalitis, wherein parkinsonism may appear not only as a delayed sequela but also as part of the acute encephalitic syndrome.

MISCELLANEOUS CAUSES

The final group to consider is the miscellaneous one, and of the disorders in this group, stroke may be considered first. Infarctions of the basal ganglia or of the midbrain (affecting the substantia nigra), rarely, have been noted to cause a contralateral hemiparkinsonism (Boecker *et al.* 1996; Fenelon and Houeto 1997; Kulisevsky *et al.* 1995). With multiple lacunar infarctions of the basal ganglia, one may also see vascular parkinsonism: this is a 'lower-half' parkinsonism wherein symptoms are generally confined to the lower extremities, with a shuffling gait, frequent falls, and freezing.

Liver disease may also, albeit rarely, be associated with parkinsonism. Acquired hepatocerebral degeneration, seen in patients with cirrhosis, typically causes a complex movement disorder, with postural tremor, athetosis, dystonia, ataxia, and, in a substantial minority, parkinsonism with prominent rigidity and bradykinesia. Acute hepatic encephalopathy, with elevated ammonia levels, may also cause parkinsonism, but this is very rare.

Niacin deficiency causes pellagra, and this may occur in two forms. When the deficiency develops slowly, 'classic' pellagra develops, with dementia, diarrhea, and a rash on sun-exposed areas; when, however, the deficiency develops suddenly, 'encephalopathic' pellagra occurs with delirium and parkinsonism, marked primarily by rigidity. This acute onset 'encephalopathic' pellagra is not accompanied by a rash and, typically, is seen only in chronic alcoholics in developed countries.

Central pontine myelinolysis may be complicated by parkinsonism; however, as might be expected, this is not secondary to myelinolysis in the pons but to the 'extrapontine' myelinolysis that may also occur in this syndrome wherein the striatum is involved. Multiple sclerosis, rarely, may cause parkinsonism, in one case because of a midbrain plaque involving the substantia nigra. The remaining disorders in Table 4.7 are each very rare causes of parkinsonism.

Differential diagnosis

A fully developed, classic parkinsonism is distinctive and very difficult to confuse with other movement disorders. Difficulty, however, may arise when only fragments of the syndrome are present, as may be the case early on in the evolution of the syndrome. Thus, care must be taken to distinguish the rest tremor of parkinsonism from a postural tremor, as may be seen, say, in essential tremor. Care must also be taken to distinguish the shuffling gait of parkinsonism from the 'magnetic' gait seen in patients with hydrocephalus, wherein the feet seem stuck or magnetized to the floor. Further, in patients who present primarily with rigidity, a careful distinction must be made with dystonia: parkinsonian rigidity, as noted earlier, is typically accompanied by the phenomenon of cogwheeling, something not seen in dystonia; furthermore, the rigidity of parkinsonism does not contort the limb, hand, or foot, which is in contrast with dystonia, for which such contortions are the rule.

Treatment

Most of the disorders listed in the table are discussed in their respective chapters, and, as pointed out in those chapters, symptomatic treatment, if required, is generally similar to that utilized in Parkinson's disease, involving use of levodopa, direct-acting dopaminergic agents, or perhaps anticholinergics. Of note, levodopa is almost always effective in Parkinson's disease, and thus if a patient with parkinsonism fails to respond to levodopa then it is unlikely that the parkinsonism in question is occurring on the basis of Parkinson's disease. In the case of parkinsonism secondary to medications, discontinuation of the offending drug is generally in order; however, in some cases, in particular where antipsychotics are involved, this may not be possible: here, treatment with the anticholinergic benztropine is generally effective.

4.9 PURE AKINESIA

The term 'akinesia' is used in one of two senses. On the one hand, it refers merely to an extreme form of the bradykinesia seen in parkinsonism, wherein it is accompanied by other signs, such as rigidity and tremor. On the other hand, however, it refers to bradykinesia occurring in an isolated fashion in the absence of these accompanying signs, and it is this 'pure akinesia' which forms the subject of this chapter.

Clinical features

Akinesia may manifest in both motor and cognitive spheres. In motor akinesia, both the initiation and execution of actions are slowed, as if the patient were encased in molasses, and all activities may be affected, including talking, walking, or getting dressed. In cognitive (or, as it is sometimes called, 'psychic') akinesia, thoughts are likewise slow to appear and the 'stream' of thought is quite sluggish, and some patients may complain of feeling like a 'zombie'.

Etiology

In general clinical practice, the most common cause of pure akinesia is treatment with a high-potency, first-generation antipsychotic (Rifkin *et al.* 1975; Van Putten and May 1978). Acute cases of pure akinesia have also been reported with infarction of the caudate nucleus (Kumral *et al.* 1999), and the syndrome may also occur as a sequela of anoxic coma (Feve *et al.* 1993). When all of these etiologies have been ruled out, one is left with a residual group, characterized by a slowly progressive pure akinesia, and such cases have been reported secondary to progressive supranuclear palsy (the syndrome of 'pure akinesia with gait freezing', discussed in Chapter 9.7), and, extremely rarely, in atypical cases of Parkinson's disease (Quinn *et al.* 1989) and pantothenate kinase-associated neurodegeneration (Molinuevo *et al.* 2003).

Differential diagnosis

As noted earlier, the diagnosis of pure akinesia is restricted to cases where bradykinesia occurs in isolation, without any rigidity or tremor; when these accompanying signs are present, then consideration should be given to the diagnosis of parkinsonism.

Other syndromes to consider include akinetic mutism, abulia, depression, and hypothyroidism. In akinetic mutism, there is essentially a total failure to initiate any activity, whereas with akinesia, in contrast, activity does occur, albeit slowly and after a delay. In abulia, there is also a failure to initiate activity; however, here, if patients are closely supervised, activity, once initiated, proceeds at a normal rate, in contrast with akinesia wherein the execution of activity remains slow. Depression, when marked by psychomotor retardation, may present a picture quite similar to akinesia; however, here one sees other symptoms, such as depressed mood, fatigue, insomnia, etc. – symptoms that are absent in akinesia. Finally, hypothyroidism, when severe, may cause a slowing of both activity and thought, thus mimicking akinesia, although here one sees other typical symptoms and signs, such as cold sensitivity, constipation, hoarseness, hair loss, puffiness, and bradycardia.

Treatment

Akinesia secondary to antipsychotics typically responds to treatment with the anticholinergic benztropine in doses of 1–3 mg daily. Pure akinesia secondary to progressive supranuclear palsy has been reported to respond to l-threo-3, 4-dihydroxyphenylserine (Yamamoto *et al.* 1997), while a case secondary to Parkinson's disease responded to levodopa (Quinn *et al.* 1989).

4.10 AKATHISIA

Akathisia, a term coined by Ladislav Haskovec in 1901 (Haskovec 1901), is derived from the Greek, and means, literally, 'an inability to sit still'. This is a very important syndrome to recognize given that it may be disabling and that it is, in general, eminently treatable.

Clinical features

As described by multiple authors (Ayd 1961; Braude *et al.* 1983; Gibb and Lees 1986; Halstead *et al.* 1994; Sachdev and Kruk 1994), akathisia is characterized by a sense of restlessness and an inability to keep still; there is a dysphoric feeling of having to move, and, in most cases, this 'compulsion' to move is translated into action in a variety of ways. If seated, patients shift restlessly from one position to another and may tap their feet or repeatedly cross and uncross their legs or arms. If standing, patients may rock back and forth, shifting their weight alternatively from one foot to another, or may actually 'march in place'. Restless pacing may occur as patients give way to the irresistible impulsion to move about. In almost all cases, this restless urge to move is worse when seated and worst when lying down, and most patients obtain at least some relief by standing up. Coupled with this motoric restlessness, many patients also experience a 'cognitive' or 'psychic' akathisia manifest by restless thoughts: thoughts come too fast, are crowded, and may shoot about the mind 'like ping-pong balls'. In some cases, this buzzing jumble of thoughts defies expression in speech and patients may become mute.

Although mild akathisia may be tolerable to some patients, more extreme forms may become unbearable, resulting in suicide attempts (Drake and Ehrlich 1985; Rothschild and Locke 1991) or assaultive behavior (Siris 1985).

The appearance of akathisia may also exacerbate symptoms of other illnesses. For example, when antipsychotics are given to patients with Tourette's syndrome to ameliorate tics, the subsequent appearance of an akathisia may lead to an exacerbation of the tics themselves (Weiden and Bruin 1987). Furthermore, patients with schizophrenia who develop an akathisia may experience a dramatic exacerbation of their psychotic symptoms (Van Putten 1975; Van Putten *et al.* 1974). Importantly, patients with schizophrenia who experience an akathisia-mediated exacerbation of their illness may neither appear restless nor complain of restlessness if questioned (Weiden and Bruin 1987): in one of the authors' (DPM) patients with schizophrenia, an akathisia manifested with muteness and withdrawal, and it was only after treatment with propranolol that the patient was able to describe the akathetic 'chaos of thoughts' that had besieged her.

Akathisia often goes undiagnosed, especially if it is mild (Fleishman *et al.* 1994). Consequently, whenever antipsychotics, other dopamine blockers, or SSRIs are prescribed, it is important to subsequently enquire after any motoric or cognitive restlessness.

Etiology

Of the various causes of akathisia discussed below, the overwhelming majority of cases occur secondary to medications, primarily antipsychotics.

In considering antipsychotic-induced akathisia, a distinction must be drawn between *acute* akathisia, occurring soon after either initiation of treatment or a substantial dose increase, and *tardive* akathisia, which occurs only after long-term treatment. Of the antipsychotics capable of causing acute akathisia, the first-generation high-potency agents (Van Putten *et al.* 1984), such as haloperidol, are the most likely culprits, causing akathisia in half or more of patients; second-generation agents may also be at fault, but this is far less common (Kane *et al.* 2009). Typically, symptoms appear some 5–6 days after initiating treatment or substantially increasing the dose (Sachdev and Kruk 1994): exceptions to this rule, however, do occur, with some cases arising in the first day and others being delayed for up to a month (Van Putten *et al.* 1984). Tardive dyskinesia, discussed in Chapter 29.2, is symptomatically very similar to acute akathisia and appears gradually, generally after a year or more of treatment.

In all likelihood, acute akathisia induced by antipsychotics is secondary to dopamine blockade and, as might be expected, other dopamine blockers may also cause this syndrome, including droperidol (Foster *et al.* 1996), prochlorperazine (Fleishman *et al.* 1994), and metoclopramide (Fleishman *et al.* 1994; Jungmann and Schöffling 1982). Of note, tardive akathisia has also occurred with metoclopramide (Shearer *et al.* 1984). Dopamine depleters may also cause acute akathisia, and this has been noted in a substantial minority of patients treated with tetrabenazine (Kenney *et al.* 2007). SSRIs (e.g. fluoxetine [Lipinski *et* al. 1989] and paroxetine [Baldassano *et al.* 1996]) have also been associated with akathisia, and in this situation the underlying mechanism may involve decreased activity of mesencephalic ventral tegmental dopaminergic neurons secondary to serotoninergically mediated inhibition. Finally, acute akathisia has also been noted, albeit very rarely, with diltiazam (Jacobs 1983) and α-interferon (Horikawa *et al.* 1999).

Parkinson's disease, as discussed in Chapter 9.5, may cause akathisia, and here the diagnosis may be elusive given that the typical motor restlessness may be obscured by bradykinesia. Consequently, patients must be questioned closely regarding both any 'inner' sense of restlessness and any evidence of cognitive akathisia.

Finally, there is a case report of akathisia occurring secondary to bilateral necrosis of the basal ganglia after carbon monoxide intoxication (Stuppaeck *et al.* 1995).

Differential diagnosis

Restless legs syndrome may be confused with akathisia as in both cases there is motor restlessness and a tendency to pace the floor. Patients with restless legs syndrome, however, generally experience parasthesia in the legs and find some relief in rubbing their feet – characteristics not seen in akathisia; conversely, patients with akathisia often 'march in place', a sign not seen in restless legs syndrome (Walters *et al.* 1991).

Agitation (as may be seen in disorders such as schizophrenia, 'agitated' depression, mania, dementia, delirium, and traumatic brain injury) is typically accompanied by restlessness, and in cases when patients with one of these disorders are treated with an antipsychotic or an SSRI, as is very often the case, the differential question can become quite acute. A common scenario involves a patient with agitation who is given an antipsychotic, initially improves, then, days later, becomes more agitated, and the question here is whether the increasing agitation is due to the underlying disorder or to an antipsychotic-induced akathisia. In some of these cases, the appearance of a tendency to 'march in place' or of complaints of a worsening of symptoms upon being seated or lying down may suggest the correct diagnosis, but when these clues are absent the differential may not be possible on clinical grounds and a 'diagnosis by treatment response' may be in order. As noted below, akathisia responds well to treatment with various agents, and a prompt clinical improvement after initiation of, say, propranolol, would strongly suggest that the correct diagnosis is akathisia.

As noted above, patients with schizophrenia may experience an exacerbation of their psychotic symptoms owing to akathisia, and when this occurs, as it may at times, without any motor restlessness, the only clue to the diagnosis may be that the exacerbation of the psychotic symptoms occurred days after the initiation of an antipsychotic or a substantial dose increase. Should the diagnosis here be missed, a 'vicious cycle' may ensue: believing that the increase in psychotic symptoms was secondary to the underlying schizophrenia, the physician may increase the dose of the antipsychotic, thus causing a worsening of the akathisia and a worsening of the psychosis. Whenever there is doubt here, a 'diagnosis by treatment response', as just described above, is probably in order.

A similar scenario may also occur in patients with depression who are treated with an SSRI. Here, and again days or a week or more after beginning the SSRI, the patient may complain of an increase in suicidal ideation, or, in some cases, may experience suicidal ideation *de novo*. In such cases, the concurrent appearance of any restlessness, whether motoric or cognitive, should prompt an attempt at 'diagnosis by treatment response'.

Treatment

With regard to medication-induced acute akathisia, consideration may be given to either discontinuing the offending medication or substantially reducing the dose (Braude *et al.* 1983). When this is impractical or when the clinical situation demands immediate relief, any one of a large number of medications, all shown to be effective in double-blinded trials, may be considered. Propranolol remains the standard, and is generally effective in total daily doses of 20–80 mg (Adler *et al.* 1985, 1986, 1993; Lipinski *et al.* 1984); provided that the drug is well tolerated, and only partial relief has been obtained, higher doses may be utilized. Benztropine, in doses of 1–3 mg, may be utilized (Adler *et al.* 1993; DiMascio *et al.* 1976) but anticholinergic side-effects make this less attractive. Mirtazapine, in a dose of 15 mg, appears similar to propranolol (Poyurovsky *et al.* 2006), and trazodone, in a dose of 100 mg, is also effective (Stryjer *et al.* 2010). Cyproheptadine, in a dose of 8 mg b.i.d. is effective (Fischel *et al.* 2001), but sedation may prove a limiting factor. Clonazepam, in doses of 0.5–2.5 mg/day, is another alternative (Kutcher *et al.* 1989). Vitamin B6 is also effective; however, the doses required are high, ranging from 600 to 1200 mg/day (Lerner *et al.* 2004), and at these doses toxicity may ensue, with the development of a sensory neuropathy (Parry and Bredesen 1985).

The treatment of tardive akathisia is discussed in Chapter 29.2.

The treatment of akathisia secondary to Parkinson's disease is not well worked out; consideration may be given to propranolol (Adler *et al.* 1991) or one of the other agents mentioned above.

4.11 CATATONIA

The term catatonia was coined by the German psychiatrist Karl Kahlbaum in 1874 (Kahlbaum 1874), who described it in patients with schizophrenia. Since then, however, it has become quite apparent that schizophrenia, although one of the more common causes of catatonia, constitutes but one of the many etiologies of this syndrome.

Clinical features

Catatonia, as described by Kraepelin (Kraepelin 1990) and Bleuler (1924), occurs in one of two forms: stuporous (or retarded) catatonia and excited catatonia (Morrison 1973). The stuporous form is by far the most common and will be described first.

STUPOROUS CATATONIA

The cardinal signs of stuporous catatonia are immobility, waxy flexibility (also known as cataplexy or *cerea flexibalitas*), and mutism.

During the immobility of catatonic stupor, patients are quite still, either lying down or sometimes sitting. The eyes may be open or closed, and if the eyes are open, patients may either stare fixedly ahead or look about the room. Patients may not move even to relieve themselves and may become foul with urine or feces. Some may not even swallow, allowing saliva or food to dribble from their mouths; indeed, if food is placed in the mouth, there is a risk of aspiration, which may be fatal (Bort 1976). Importantly, although patients appear to lack conscious activity, most remain alert, and some, upon recovery, may evidence an accurate recall of events that occurred during the stupor. Interestingly, albeit rarely, a patient's immobility may occasionally undergo a sudden lysis: for example, if a ball is gently thrown to an immobile patient, the patient may suddenly loosen up and catch the ball.

The immobility of stuporous catatonia may or may not be accompanied by waxy flexibility. This sign derives its name from the fact that, upon passive movement of the limbs, the examiner encounters a mild, fluid rigidity similar to what one would expect upon bending a waxen object, such as a well-softened candle. In most cases, in addition to this waxy flexibility, one also finds that the patient's limbs tend to stay in whatever position the examiner places them, no matter how uncomfortable and regardless of whether or not the examiner instructs the patient to maintain the position of the limb. Bleuler (1924) recommended a bedside test for waxy flexibility that involved taking 'the patient's pulse and, as if inadvertently, "holding" his arm high and extended. Then, after taking the pulse rate. . .release the arm.' The test was considered to be positive when the patient's arm remained suspended in essentially the same position. Another bedside test involves checking for the presence of a 'psychological pillow'. Here, when the patient is lying supine in bed with the head resting on a pillow, the head is lifted slightly and the pillow removed. In a positive test, the patient's head remains in essentially the same position, as if the pillow were still present.

Mutism in catatonic stupor ranges from partial to complete. Those with only partial mutism may mumble or whisper incomprehensible words or phrases.

This syndrome of immobility (with or without waxy flexibility) and mutism may persist for hours, days, or much longer. When it does resolve, patients, now mobile and speaking, may at times display other symptoms, including posturing, negativism, automatic obedience, or the echophenomena of echolalia and echopraxia. It must be kept in mind, however, that the presence of any, or all, of these associated symptoms is neither necessary nor sufficient for the diagnosis of stuporous catatonia, and thus their presence or absence should not be accorded diagnostic weight.

Posturing is said to occur when patients automatically and spontaneously assume more or less bizarre postures, which are then maintained. For example, the arms may be spread in a cruciform position, or the head thrown back in full extension, or patients may stand, stork-like, on one leg.

Negativism does not represent mere contrariness, stubbornness, or passive aggressiveness, for in each of these phenomena, patients have not lost control and may be able to cooperate in situations that, to them, appear to be to their advantage. In contrast, negativism presents as a mulish, almost instinctual tendency to resist, which may be either passive, wherein the patient, in response to a command, simply does nothing, or active, wherein the patient does the opposite of what is requested. Negativistic patients may refuse to come to an interview, take a bath, change their clothes, take medicines, or even eat food placed in front of them. In extreme cases, negativism may extend to the patient's own urges, leading, for example, to extreme constipation with fecal impaction. Patients with active negativism may get up from a table when food is placed on it or may back out of a doorway if asked to walk through it. Typically, there is no 'reasoning' with negativistic patients, as they generally remain mute and inaccessible.

Automatic obedience represents, in a sense, the converse of negativism, in that these patients automatically do what is expected of them, without question or hesitation, regardless of whether the consequences are absurd or harmful. This is far from mere agreeableness as patients often seem to behave in a robotic or automaton-like manner.

Echo phenomena, as noted, consist of either echolalia, wherein the patient automatically, and without prompting, repeats back what the examiner said, or echopraxia, wherein, again without prompting, the patient automatically mimics the examiner's movements, gestures, or posture.

EXCITED CATATONIA

Excited catatonia is characterized by bizarre, frenzied, and purposeless hyperactivity. Uninvolved with others, almost sealed off in their own world, patients may gesticulate, march in place, loudly declaim, or engage in verbigeration, with rapid, bizarre, and senseless speech.

In rare instances, excited catatonia may undergo an evolution into a condition known as Stauder's lethal catatonia. Here there is an escalation of the hyperactivity, followed by fever, tachycardia, hypotension, leukocytosis, and, in some cases, stupor, followed by death (Castillo *et al.* 1989; Mann *et al.* 1986). The recognition of this syndrome is critical, for even when the patient is *in extremis* (with temperatures as high as 41°C [106°F]), treatment with electroconvulsive therapy may result in recovery (Aronson and Thompson 1950).

Etiology

The various causes of stuporous catatonia and of excited catatonia are listed in Table 4.8, and each is discussed in turn.

STUPOROUS CATATONIA

The overwhelming majority of cases of stuporous catatonia occur in the course of schizophrenia, a manic episode of bipolar disorder, or a depressive episode of either major depressive disorder or bipolar disorder, and in the case of depression, catatonia is more likely to occur in elderly patients (Starkstein *et al.* 1996). These diagnoses are made primarily on the basis of the clinical context within which the catatonia occurs. In the case of schizophrenia, one finds a preceding psychosis, in mania, a preceding syndrome of stages I and II of mania, and in a depressive episode, a preceding typical depressive syndrome. Allied with these conditions is the rare syndrome of periodic catatonia, characterized, as the name suggests, by recurrent episodes of catatonia. Although the nosologic status of this disorder is still not settled, recent research indicates that it may occur on a genetic basis (Stober *et al.* 2002).

Of the medications that are capable of causing stuporous catatonia, by far the most common offenders are the antipsychotics, especially high-potency first-generation agents, such as haloperidol, used in high dosage (e.g. over 20 mg) (Gelenberg and Mandel 1977). Disulfiram must also be kept in mind, as the diagnosis may be obscured by the fact that disulfiram-induced catatonia may not appear for months after the disulfiram has been administered. Catatonia secondary to ciprofloxacin, azithromycin, and levetiracetam is very rare. Although benzodiazepine treatment *per se* does not cause

Table 4.8 Causes of catatonia

Stuporous catatonia
Schizophrenia (Morrison 1973)
Depressive episode of either major depressive disorder or bipolar disorder (Barnes *et al.* 1986)
Manic episode of bipolar disorder (Abrams and Taylor 1976; Taylor and Abrams 1977)
Periodic catatonia (Gjessing 1974)
Medications
Antipsychotics (Weinberger and Wyatt 1978)
Disulfiram (Reisberg 1978; Weddington *et al.* 1980)
Ciprofloxacin (Akhtar and Ahmad 1993)
Azithromycin (Plana *et al.* 2006)
Levetiracetam (Chouinard *et al.* 2006)
Benzodiazepine withdrawal (Rosebush and Mazurek 1996)
Epileptic conditions
Complex partial seizures (Engel *et al.* 1978; Gomez *et al.* 1982; Lim *et al.* 1986; Shah and Kaplan 1980)
Post-ictal psychosis (Logsdail and Toone 1988)
Interictal psychosis (Kristensen and Sindrup 1979; Slater and Beard 1963)
Psychosis of forced normalization (Pakainis *et al.* 1987)
Viral encephalitis
Herpes simplex encephalitis (Raskin and Frank 1974)
Encephalitis lethargica (Bond 1920; Kirby and Davis 1921)
Miscellaneous conditions
Stroke (Belfer and D'Autremont 1971)
Vitamin B12 deficiency (Berry *et al.* 2003)
Wilson's disease (Davis and Borde 1993)
Systemic lupus erythematosus (Lanham *et al.* 1985; Mac and Pardo 1983)
Limbic encephalitis (Tandon *et al.* 1988)
Hepatic encephalopathy (Jaffe 1967)
Lyme disease (Pfister *et al.* 1993)
Subacute sclerosing panencephalitis (Koehler and Jakumeit 1976)
Tay–Sachs disease (Rosebush *et al.* 1995)
Thrombotic thrombocytopenic purpura (Read 1983)
Excited catatonia
Schizophrenia (Morrison 1973)
Viral encephalitis (Penn et al. 1972)

catatonia, this syndrome may occur in a very small minority of patients going through benzodiazepine withdrawal.

Epileptic conditions capable of causing stuporous catatonia include complex partial seizures, postictal psychosis, interictal psychosis, and the rare psychosis of forced normalization.

Complex partial seizures, in addition to the typical defect of consciousness, may manifest with catatonia. The diagnosis is suggested immediately by the paroxysmal onset of the disturbance, and, in most cases, by its relatively brief duration. It must be borne in mind, however, that some cases of complex partial status epilepticus with catatonia may persist for days or even longer.

Postictal psychosis is generally seen only after a flurry of grand mal or complex partial seizures, and may rarely present with catatonia. Typically, there is a lucid interval, lasting for days, between the last seizure and the onset of the psychosis, and the psychosis itself generally persists for days to months before spontaneously remitting.

Interictal psychosis is a chronic condition that evolves gradually only after many years of uncontrolled epilepsy and which may, rarely, present with catatonia.

The psychosis of forced normalization is a very rare disorder occurring in patients with uncontrolled epilepsy, which has been brought under control with aggressive treatment with AEDs, with the clinical improvement being accompanied by a 'forced normalization' of the electroencephalogram (EEG). Once again, and rarely, this may present with catatonia. The diagnosis is suggested by the seemingly paradoxical onset of a psychosis after the seizures have been stopped; further support for the diagnosis comes by comparing an EEG that has been carried out during the psychosis with one performed before aggressive anti-epileptic treatment and finding the 'normalization'.

Viral encephalitis may present with stuporous catatonia (Barnes *et al.* 1986; Kim and Perlstein 1970; Misra and Hay 1971; Wilson 1976), with the diagnosis being suggested by fever and headache, and this has been noted with herpes simplex viral encephalitis and encephalitis lethargica (although there have been no further epidemic cases of encephalitis lethargica since 1928, sporadic cases still occur; the most recent case with catatonia was reported in 2000 [Shill and Stacy 2000]).

The miscellaneous causes of stuporous catatonia are quite heterogeneous. Regarding catatonia occurring in stroke, cases have been reported with infarction of the frontal lobe. B12 deficiency, given its eminent treatability, should also be kept in mind, and the diagnostician should recall that, as with other central nervous system signs of B12 deficiency, catatonia may occur in the absence of anemia or macrocytosis (Jauhar *et al.* 2010). The remaining causes only very rarely cause the syndrome.

EXCITED CATATONIA

Excited catatonia is seen virtually exclusively in schizophrenia, and in this illness it must be kept in mind that, over long periods of time, patients may have alternating episodes of either stuporous or excited catatonia. In cases of excited catatonia, diagnostic consideration may also be given to acute viral encephalitis.

Differential diagnosis

The differential diagnosis of the stuporous and excited forms of catatonia is quite different, and thus each is treated separately.

STUPOROUS CATATONIA

Stuporous catatonia must be distinguished from stupor of other causes, akinetic mutism, abulia, and the neuroleptic malignant syndrome.

Stupor of other causes is generally associated with a decreased level of consciousness, in contrast with the alertness seen in catatonia. Other signs may also facilitate a correct differential. First, if waxy flexibility, with a tendency of the limbs to remain in whatever position they are placed, is present then this immediately indicates a diagnosis of catatonia, as this sign is not found in stupor of other causes. Second, if the eyes are closed, in catatonia one finds a resistance to opening them, while in stupor of other causes there is no resistance. If the eyes are open, in catatonia one may see saccadic movements in contrast to the roving eye-movements characteristic of stupor of other causes.

Akinetic mutism, manifesting with immobility and mutism, is almost identical to catatonic stupor. The presence of waxy flexibility (again provided it is accompanied by a tendency of the limbs to remain in whatever position they are placed) points to catatonia, but if this sign is absent, a reliable clinical distinction may be impossible. Consequently, if the history does not point to a cause associated with catatonia, one may have to resort to imaging to look for the various lesions capable of causing akinetic mutism, as discussed in Chapter 5.2.

Abulia may at first glance appear similar to catatonia, in that abulic patients, lacking any motivation or initiative, may be immobile. The diagnosis is readily apparent upon merely urging the patient to act: the abulic patient will comply, and continue to comply, with instructions (provided that supervision is ongoing), whereas with the catatonic patient there will be no response.

The neuroleptic malignant syndrome enters the differential when patients have been treated with an antipsychotic for one of the conditions which are capable of causing stuporous catatonia, as for example schizophrenia, mania, or depression. Here, although the evolution of immobility and rigidity, on the face of it, could be consistent with the development of either stuporous catatonia or the neuroleptic malignant syndrome, certain clues, all related to the rigidity, may enable a correct diagnosis. First, in stuporous catatonia the rigidity is mild, while in neuroleptic malignant syndrome it is typically severe; further, whereas in stuporous catatonia the rigidity is fluid, in neuroleptic malignant syndrome it may be cogwheel in type. Second, in stuporous catatonia one may find the limbs staying in whatever position the examiner had placed them, a finding absent in neuroleptic malignant syndrome. Finally, in neuroleptic malignant syndrome, rigidity may be accompanied by a generalized, coarse tremor or by chorea, findings which are absent in stuporous catatonia. Often fever is also cited as a key differential feature, pointing toward a diagnosis of the neuroleptic malignant syndrome, but this may not always be reliable, as some patients with stuporous catatonia may also have fever, as for example those who aspirated and developed a pneumonia.

With regard to the differential possibility of the neuroleptic malignant syndrome, however, it must be kept in mind that in some cases a reliable clinical differential may not be possible given that some very mild cases of the neuroleptic malignant syndrome may be characterized by mild rigidity, without cogwheeling and without tremor or chorea. In such instances, given the importance of avoiding the development of a severe, and potentially fatal, neuroleptic malignant syndrome, it may be prudent to discontinue the antipsychotic and see if symptoms resolve.

EXCITED CATATONIA

Excited catatonia must be distinguished from mania and simple agitation, and cases of Stauder's lethal catatonia must be distinguished from the neuroleptic malignant syndrome.

Mania, during stages I and II, is marked by hyperactivity; however, here the activity is purposeful and not bizarre. Furthermore, in stages I and II of mania one also finds symptoms such as heightened mood and pressure of speech, which are absent in excited catatonia. In some cases of stage III mania, however, the hyperactivity may fragment and lose purpose, and in these cases the differential may rest on obtaining a history of the preceding, typical stages I and II.

Agitation typically is not characterized by bizarreness, which is always seen in excited catatonia.

The possibility of a neuroleptic malignant syndrome must always be entertained in any case of excited catatonia treated with an antipsychotic where one subsequently sees the development of fever, tachycardia, and leukocytosis given that all three of these signs may be characteristic of both the neuroleptic malignant syndrome and of the evolution from excited catatonia into Stauder's lethal catatonia. In attempting to make a clinical differential between neuroleptic malignant syndrome and Stauder's lethal catatonia, one should search for features which, although typical of the neuroleptic malignant syndrome, are not characteristic of Stauder's lethal catatonia, including rigidity and an elevated blood pressure. In cases where a reliable differential is not possible on clinical grounds, it may not be clear what should be done next. Some argue for electroconvulsive therapy, given that this modality is effective for both the neuroleptic malignant syndrome and Stauder's lethal catatonia; however, as pointed out in Chapter 29.1, electroconvulsive therapy (ECT) does carry an increased risk in cases of the neuroleptic malignant syndrome. Others argue for discontinuing the antipsychotic and beginning treatment with dantrolene, and although this would be appropriate if the patient did have neuroleptic malignant syndrome, if the patient unfortunately actually had Stauder's lethal catatonia it might deprive him or her of an effective treatment. Obviously, much, and careful, clinical judgment is required in such cases.

Treatment

In cases of stuporous catatonia in which treatment of the underlying cause is ineffective or for which emergent treatment is required, consideration may be given to either lorazepam or ECT. Most cases of stuporous catatonia will show prompt, albeit temporary, improvement with parenteral lorazepam (e.g. 2 mg i.v.) (Bush *et al.* 1996; Rosebush *et al.* 1990; Salam *et al.* 1987); more lasting improvement may follow a course of ECT (Bush *et al.* 1996; Malur *et al.* 2001). Pending improvement, a careful watch must be maintained for dehydration, deep venous thrombosis with pulmonary embolism, and aspiration pneumonia.

Excited catatonia, as noted above, is seen almost exclusively in schizophrenia, and treatment proceeds with an antipsychotic, as outlined in Chapter 22.1.

4.12 ASTERIXIS

Asterixis, first described by Adams and Foley in 1953 (Adams and Foley 1953), is a very important diagnostic sign as, in most cases, it indicates a metabolic encephalopathy due to hepatic, renal, or respiratory failure.

Clinical features

Asterixis represents a precipitous loss of muscle tone (Adams and Foley 1953; Leavitt and Tyler 1964) and is typically tested for by asking patients to hold their arms straight to the front with the hands dorsiflexed as far back as possible, and holding that position for at least 30 seconds. When asterixis is present, there will be arrhythmically occurring 'flaps' of the hands downwards, followed after a brief, but distinct, moment by recovery back to the dorsiflexed position; although in most cases both hands will 'flap' down simultaneously, occasionally asterixis will be strictly unilateral, a finding with, as discussed below, considerable diagnostic significance. In cases where patients are unable to hold their arms forward, an alternative approach to eliciting asterixis involves having the patient rest the arms prone on the bed and then asking him or her to dorsiflex the hands off the bed, again holding that position for at least 30 seconds.

Table 4.9 Causes of asterixis

Metabolic encephalopathy
Hepatic encephalopathy (Chapter 14.19)
Uremic encephalopathy (Chapter 14.18)
Respiratory failure (Austen *et al.* 1957; Bacchus 1958)
Medication side-effect
Gabapentin (Babiy *et al.* 2005)
Pregabalin (Heckmann *et al.* 2005)
Phenytoin (Chi *et al.* 2000; Murphy and Goldstein 1974)
Carbamazepine (Rittmannsberger and Lebihuber 1992)
Valproate (Bodensteiner *et al.* 1981)
Lithium (Stewart and Williams 2000)
Clozapine (Rittsmannberger 1996)
Levodopa (Glantz *et al.* 1982)
Trimethoprim/sulfamethoxazole (Dib *et al.* 2004)
Ifosfamide (Meyer *et al.* 2002)
Metrizamide (Bertoni *et al.* 1981)
Infarction or intracerebral hemorrhage
Thalamus (Kim 2001a; Rio *et al.* 1995)
Frontal (Kim 2001b) or parietal (Degos *et al.* 1979) cortex
Basal ganglia (Kim 2001b)
Cerebellum (Kim 2001b)
Mesencephalon (Kim 2001b)
Pons (Peterson and Peterson 1987)
Medulla (Peterson and Peterson 1987)

Etiology

Of the various causes of asterixis listed in Table 4.9, by far the most common are the metabolic encephalopathies; asterixis occurring as a medication side-effect is somewhat less common, while infarction or intracerebral hemorrhage is an uncommon cause. From a diagnostic point of view, it is critical to keep in mind whether the asterixis is bilateral or unilateral. Asterixis occurring in the course of a metabolic encephalopathy or as a side-effect is always bilateral, while asterixis occurring as part of a stroke syndrome, although occasionally bilateral (Kim 2001b), is, in the vast majority of cases, unilateral. Thus, if a patient has unilateral asterixis, the presumption must be that it is occurring secondary to infarction or hemorrhage in one of the structures indicated below.

Of the metabolic encephalopathies, hepatic encephalopathy is so commonly associated with asterixis that, for a time, the term 'liver flap' was used as a synonym for asterixis. Uremic encephalopathy is almost always associated with asterixis, while in cases of respiratory failure it may be less common.

Of the medications capable of causing asterixis as a side-effect, the AEDs and lithium are the most frequent offenders, with the other agents only uncommonly being implicated. Metrizamide myelography may be followed by a delirium accompanied by asterixis. Of note, when asterixis is seen in a patient on valproic acid, it is important to keep in mind that it may not be the valproic acid *per se* which is at fault: valproic acid may cause hyperammonemia, and there are reports (Zaccara *et al.* 1984) of asterixis in such cases occurring secondary to the hyperammonemia.

Although infarction or intracerebral hemorrhage of a large number of structures may cause asterixis, the most commonly implicated is the thalamus (Kim 2001b; Rio *et al.* 1995): others include the cortex (primarily the frontal cortex, but occasionally also the parietal cortex), basal ganglia, cerebellum, mesencephalon, pons, and even the medulla. In cases where the structual lesion leads to unilateral asterixis, although this is generally contralateral to the lesion (Degos *et al.* 1979; Kim 2001b), exceptions do occur, including, naturally, the cerebellum (Kim 2001b), but also the pons and medulla (Peterson and Peterson 1987).

Differential diagnosis

Myoclonus is distinguished by the fact that it represents not an abrupt loss of tone with a 'flap' down, but rather an abrupt gain of tone with a resultant 'jerk' up. Tremor is distinguished by the presence of a more or less rhythmic oscillatory movement secondary to alternating contraction of agonist and antagonist musculature.

Treatment

Treatment is directed at the underlying condition; symptomatic treatment is not required.

4.13 MIRROR MOVEMENTS

Mirror movements are normal in early childhood and may persist into adult years; they may also be seen in certain disorders, for example stroke with hemiparesis.

Clinical features

Mirror movements are typically seen in the hands, and may in some cases involve the arm. They may be elicited by asking patients to perform a fine motor task with one hand, for example sequential finger–thumb apposition. As the patient performs the maneuver, simply observe the other hand for the mirrored movement. With hemiparetic patients, a simpler strategy involves telling the patient that you are going to test grip strength in one hand at a time. Then, place the second and third fingers of each of your hands into the patient's hands and ask the patient to grip with only one hand. Mirror gripping may then be appreciated in the other hand. Interestingly, such mirroring in hemiparetic patients is most commonly seen in the unaffected hand (Nelles *et al.* 1998).

Etiology

Mirror movements, as noted, are normal in young children and may persist into adult years, in which case a familial pattern may be present (Rasmussen 1993; Srour *et al.* 2009). Mirror movements may also be seen in agenesis of the corpus callosum (Schott and Wyke 1981), Kallmann's syndrome (Krams *et al.* 1999), the Klippel–Feil syndrome, Parkinson's disease (Espay *et al.* 2005b), essential tremor, and, most notably, in stroke patients with hemiparesis (Berlin 1951; Nelles *et al.* 1998).

Differential diagnosis

There is no other sign or symptom that mimics mirror movements.

Treatment

Treatment is rarely required; anecdotally, motor retraining has been successful (Cincotta *et al.* 2003).

4.14 PATHOLOGIC STARTLE

Although a startle response is a normal human reaction, in some cases it may, in one fashion or another, be exaggerated, in which case it is appropriate to speak of 'pathologic startle'.

Clinical features

The startle response is a brief, but sudden and violent, reaction to an unexpected stimulus, such as a loud noise or a bright light; in severe cases, a mere touch, as a tap on the shoulder, may be sufficient. During the startle response, the eyes blink, the face contorts in a grimace, there is flexion of the neck and trunk, and flexion and abduction of the arms; in severe cases, patients may be thrown off balance and fall.

Etiology

In pursuing the etiology of pathologic startle, the first step is to ascertain whether or not in between the episodes of pathologic startle there is a persistent state of autonomic hyperactivity, with anxiety and a greater or lesser degree of tremulousness, as this persistent state is characteristic of the most common causes of pathologic startle, namely post-traumatic stress disorder, generalized anxiety disorder, alcohol or sedative-hypnotic withdrawal, and various sympathomimetics, such as caffeine (Howard and Ford 1992).

Of those conditions wherein the intervals between episodes of pathologic startle are not characterized by persistent autonomic hyperactivity, the most common cause is hyperekplexia. This disorder, discussed in Chapter 9.32, may have an onset in infancy or later and is familial. Brainstem lesions may also cause pathologic startle, and this has been noted with pontine lesions, such as infarction (Kimber and Thompson 1997) or a plaque of multiple sclerosis (Ruprecht *et al.* 2002), and with upper medullary lesions, such as demyelinization (Della Marca *et al.* 2007); compression of either the pons (Gambardella *et al.* 1999) or medulla (Salvi *et al.* 2000) by an ectatic artery may also be at fault. Posthypoxic–ischemic encephalopathy (Brown *et al.* 1991), traumatic brain injury (Brown *et al.* 1991), and Creutzfeldt–Jakob disease (Hansen *et al.* 1998) may also be associated with pathologic startle.

Finally, there are several 'culture-bound' syndromes wherein individuals who are normally shy may, in response to an unexpected and 'startling' stimulus, engage in remarkable behavior characterized not only by pathologic startle, but also by other distinctive behaviors, such as cursing, swearing, and, at times, echolalia. These are generally considered voluntary behaviors and have been noted, famously, among French-Canadian lumberjacks of the nineteenth century (the 'Jumping Frenchmen of Maine' [Saint-Hilaire *et al.* 1986]) and in residents of Malaysia (where it is referred to as 'latah' [Bartholomew 1994]) and Siberia (where it has been termed 'myriachit' [Stevens 1965]).

Differential diagnosis

A startle reaction is a normal human response, and can be elicited in most individuals by a stimulus such as an unexpected gunshot; pathologic startle should be considered present only when the startle response is severe, frequent, and provoked by relatively minor stimuli.

Startle epilepsy, one of the reflex epilepsies discussed in Chapter 8.3, is characterized by tonic seizures precipitated by a startling stimulus. The differential here is fairly straightforward, in that the major feature here is not an exaggerated startle responses but rather a tonic seizure. Furthermore, most of these cases have an onset in childhood in multiply handicapped children who also have other seizure types.

Treatment

Treatment is directed at the underlying cause; when this is ineffective, a trial of clonazepam is in order.

REFERENCES

Abrams R, Taylor MA. Catatonia. A prospective clinical study. *Arch Gen Psychiatry* 1976; **33**:579–81.

Adams RD, Foley JM. The neurological disorder associated with liver disease. *Assoc Res Nerv Ment Dis Proc* 1953; **32**:198–237.

Adler LA, Angrist B, Peselow E *et al.* Efficacy of propranolol in neuroleptic-induced akathisia. *J Clin Psychopharmacol* 1985; **5**:164–6.

Adler LA, Angrist B, Peselow E *et al.* Controlled assessment of propranolol in the treatment of neuroleptic-induced akathisia. *Br J Psychiatry* 1986; **149**:42–5.

Adler LA, Angrist B, Weinreb H *et al.* Studies on the time course and efficacy of beta-blockers in neuroleptic-induced akathisia and the akathisia of idiopathic Parkinson's disease. *Psychopharmacol Bull* 1991; **27**:107–11.

Adler LA, Peselow E, Rosenthal M *et al.* A controlled comparison of the effects of propranolol, benzatropine and placebo on akathisia: an interim analysis. *Psychopharmacol Bull* 1993; **29**:283–6.

Ahmed S, Chengappa KN, Naidu VR *et al.* Clozapine-withdrawal-emergent dystonias and dyskinesias: a case series. *J Clin Psychiatry* 1998; **59**:472–7.

Akhtar S, Ahmad H. Ciprofloxacin-induced catatonia. *J Clin Psychiatry* 1993; **54**:115–16.

Alarcon F, Tolosa E, Munoz E. Focal limb dystonia in a patient with a cerebellar mass. *Arch Neurol* 2001; **58**:1125–7.

Alevizos B, Papageorgiou C, Christodoulou GN. Acute dystonia caused by a low dosage of olanzapine. *J Neuropsychiatr Clin Neurosci* 2003; **15**:241.

Altindag A, Yanik M, Asoglu M. The emergence of tics during escitalopram and sertraline treatment. *Int Clin Psychopharmacol* 2005; **20**:177–8.

Angelini L, Sgro V, Erba A *et al.* Tourettism as clinical presentation of Huntington's disease with onset in childhood. *Ital J Neurol Sci* 1998; **19**:383–5.

Antin SP, Prockop LD, Cohen SM. Transient hemiballism. *Neurology* 1967; **17**:1068–72.

Armon C, Shin C, Miller P *et al.* Reversible parkinsonism and cognitive impairment with chronic valproate use. *Neurology* 1996; **47**:626–35.

Aronson MJ, Thompson SV. Complications of acute catatonic excitement. *Am J Psychiatry* 1950; **107**:216–20.

Asmus F, Zimprich A, Tezenas du Montcel S *et al.* Myoclonus-dystonia syndrome: epsilon-sarcoglycan mutations and phenotype. *Ann Neurol* 2002; **52**:489–92.

Atasoy N, Ustundag Y, Konuk N *et al.* Acute dystonia during pegylated interferon alpha therapy in a case with chronic hepatitis B infection. *Clin Neuropharmacol* 2004; **27**:105–7.

Austen FK, Carmichael MW, Adams RD. Neurologic manifestations of chronic pulmonary insufficiency. *N Engl J Med* 1957; **257**:579–90.

Ayd FJ. A survey of drug-induced extrapyramidal reactions. *J Am Med Assoc* 1961; **175**:1054–60.

Babiy M, Stubblefield MD, Herklotz M *et al.* Asterixis related to gabapentin as a cause of falls. *Am J Phys Med Rehabil* 2005; **84**:136–40.

Bacchus M. Encephalopathy and pulmonary disease. *Arch Int Med* 1958; **102**:194–8.

Bachman DS. Pemoline-induced Tourette's disorder: a case report. *Am J Psychiatry* 1981; **138**:1116–17.

Bain PG, Findley LJ, Britton TC *et al.* Primary writing tremor. *Brain* 1995; **118**:1461–72.

Bak TH, Bauer M, Schaub RT *et al.* Myoclonus in patients treated with clozapine: a case series. *J Clin Psychiatry* 1995; **56**:418–22.

Baldassano CF, Truman CJ, Nierenberg A *et al.* Akathisia: a review and case report following paroxetine treatment. *Compr Psychiatry* 1996; **37**:122–4.

Ballard PA, Tetrud JW, Langston JW. Permanent human parkinsonism due to 1-methyl-4-phenyl-1,2,3,6-tetrahydropyridine (MPTP): seven cases. *Neurology* 1985; **35**:949–56.

Barak Y, Levine J, Weisz R. Clozapine-induced myoclonus: two case reports. *J Clin Psychopharmacol* 1996; **16**:339–40.

Barnes MP, Saunders M, Walls TJ *et al.* The syndrome of Karl Ludwig Kahlbaum. *J Neurol Neurosurg Psychiatry* 1986; **49**:991–6.

Bartholomew RE. Disease, disorder or deception? Latah as habit in a Malay extended family. *J Nerv Ment Dis* 1994; **182**:331–8.

Barucha KJ, Sethi DD. Complex movement disorders induced by fluoxetine. *Mov Disord* 1996; **11**:324–6.

Bataller L, Graus F, Salz A *et al.* Clinical outcome in adult-onset idiopathic or paraneoplastic opsoclonus-myoclonus. *Brain* 2001; **124**:437–43.

Belfer ML, D'Autremont CC. Catatonia-like symptomatology: an interesting case. *Am J Psychiatry* 1971; **24**:119–20.

Benito-Leon J, Rodriguez J, Orti-Pareja M *et al.* Symptomatic orthostatic tremor in pontine lesions. *Neurology* 1997; **49**:1439–41.

Berlin I. Mirror movements. Report of two cases. *Arch Neurol Psychiatry* 1951; **66**:394.

Berry N, Sagar R, Tripathi BM. Catatonia and other psychiatric symptoms with vitamin B12 deficiency. *Acta Psychiatr Scand* 2003; **108**:156–9.

Bertoni JM, Schwartzman RJ, Van Horn G *et al.* Asterixis and encephalopathy following metrizamide myelography: investigations into possible mechanisms and review of the literature. *Ann Neurol* 1981; **9**:366–70.

Bharucha KJ, Sethi KD. Complex movement disorders induced by fluoxetine. *Mov Disord* 1996; **11**:324–6.

Bhatia KP, Soland VL, Bhatt MH. Paroxysmal exercise-induced dystonia: eight new sporadic cases and a review of the literature. *Mov Disord* 1997a; **12**:1007–12.

Bhatia KP, Quinn NP, Marsden CD. Clinical features and natural history of axial predominant adult onset primary dystonia. *J Neurol Neurosurg Psychiatry* 1997b; **63**:788–91.

Bhatt MH, Elias MA, Mankodi AK. Acute and reversible parkinsonism due to organophosphate pesticide intoxication. Five cases. *Neurology* 1999; **52**:1467–71.

Bleuler E. *Textbook of Psychiatry*, 1924, translated by Brill AA. New York: Arno Press, 1976.

Bocola V, Fabbrini G, Sollecito A *et al.* Neuroleptic induced parkinsonism: MRI findings in relation to clinical course after withdrawal of neuroleptic drugs. *J Neurol Neurosurg Psychiatry* 1996; **60**:213–16.

Bodensteiner JB, Morris HH, Golden GS. Asterixis associated with sodium valproate. *Neurology* 1981; **31**:194–5.

Boecker H, Weindl A, Leenders K *et al.* Secondary parkinsonism due to focal substantia nigra lesions: a PET study with [18F]FDG and [18F]fluorodopa. *Acta Neurol Scand* 1996; **93**:387–92.

Bond ED. Epidemic encephalitis and katatonic symptoms. *Am J Insanity* 1920; **76**:261–4.

Borgohain R, Singh AK, Thadani R *et al.* Hemiballismus due to an ipsilateral striatal hemorrhage: an unusual localization. *J Neurol Sci* 1995; **130**:22–4.

Bort RF. Catatonia, gastric hyperacidity, and fatal aspiration: a preventable syndrome. *Am J Psychiatry* 1976; **133**:446–7.

Boyer WF, Bakalar NH, Lake CR. Anticholinergic prophylaxis of acute haloperidol-induced dystonic reactions. *J Clin Psychopharmacol* 1987; **7**:164–6.

Brashear A, Farlow MR, Butler I *et al.* Variable phenotype of rapid-onset dystonia–parkinsonism. *Mov Disord* 1996; **11**:151–6.

Brashear A, DeLeon D, Bressman SB *et al.* Rapid-onset dystonia–parkinsonism in a second family. *Neurology* 1997; **48**:1066–9.

Brashear A, Dubyns WB, de Carvhalo Aguiar P *et al.* The phenotypic spectrum of rapid onset dystonia-parkinsonism (RDP) and mutations in the ATP1A3 gene. *Brain* 2007; **130**:828–35.

Braude WM, Barnes TRE, Gore SM. Clinical characteristics of akathisia: a systematic investigation of acute psychiatric in-patient admissions. *Br J Psychiatry* 1983; **143**:139–50.

Bressman SB, Fahn S, Burke RE. Paroxysmal non-kinesigenic dystonia. *Adv Neurol* 1988; **50**:403–13.

Brown P, Rothwell JC, Thompson PD *et al.* The hyperekplexias and their relationship to the normal startle reflex. *Brain* 1991; **114**:1903–28.

Bruno MK, Hallett M, Gwinn-Hardy K *et al.* Clinical evaluation of idiopathic paroxysmal kinesigenic dyskinesia. New diagnostic criteria. *Neurology* 2004; **63**:2280–7.

Bruno MK, Lee H-Y, Auberger GWJ *et al.* Genotype–phenotype correlation of paroxysmal non-kinesigenic dyskinesia. *Neurology* 2007; **68**:1782–9.

Budman C, Sarcevic A. An unusual case of motor and vocal tics with obsessive-compulsive symptoms in a young adult with Behcet's syndrome. *CNS Spectr* 2002; **7**:878–81.

Buetefisch CM, Gutierrez A, Gutmann L. Choreoathetotic movements: a possible side-effect of gabapentin. *Neurology* 1996; **46**:851–2.

Burd L, Kerbeshian J, Fisher W *et al.* Anticonvulsant medications: an iatrogenic cause of tic disorders. *Can J Psychiatry* 1986; **31**:419–23.

Burguera JA, Bataller L, Valero C. Action hand dystonia after cortical parietal infarction. *Mov Disord* 2001; **16**:1183–5.

Burke RE, Fahn S, Gold AP. Delayed-onset dystonia in patients with 'static' encephalopathy. *J Neurol Neurosurg Psychiatry* 1980; **43**:789.

Burke RE, Fahn S, Jankovic J *et al.* Tardive dystonia: late-onset and persistent dystonia caused by antipsychotic drugs. *Neurology* 1982; **32**:1335–46.

Bush G, Fink M, Pertrides G *et al.* Catatonia. II. Treatment with lorazepam and electroconvulsive therapy. *Arch Psychiatr Scand* 1996; **9**:137–43.

Cammarota A, Gershank OS, Garcia S *et al.* Cervical dystonia due to spinal cord ependymoma: involvement of cervical cord segments in the pathogenesis of dystonia. *Mov Disord* 1995; **10**:500–3.

Caparros-Lefebvre D, Deleume JF, Bradai N *et al.* Ballism caused by bilateral infarction in the substantia nigra. *Mov Disord* 1994; **9**:108.

Cardoso F, Jankovic J. Cocaine-related movement disorders. *Mov Disord* 1993; **8**:175–8.

Carella F, Grassi MP, Savoiardo M *et al.* Dystonic-parkinsonian syndrome after cyanide poisoning: clinical and MRI findings. *J Neurol Neurosurg Psychiatry* 1988; **51**:1345–8.

Carlen PL, Lee MA, Jacob M *et al.* Parkinsonism provoked by alcoholism. *Ann Neurol* 1981; **9**:84–6.

Carlson CD, Cavazzoni PA, Berg PH *et al.* An integrated analysis of acute treatment-emergent extrapyramidal syndrome in patients with schizophrenia during olanzapine clinical trials: comparisons with placebo, haloperidol, risperidone or clozapine. *J Clin Psychiatry* 2003; **64**:898–906.

Carpenter MB. Ballism associated with partial destruction of the subthalamic nucleus of Luys. *Neurology* 1955; **5**:479–89.

Carpenter MR. Athetosis and the basal ganglia. *Arch Neurol* 1950; **63**:895–901.

Casas M, Garcia-Ribera C, Alvarez E *et al.* Myoclonic movements as a side-effect of treatment with therapeutic doses of clomipramine. *Int J Psychopharmacol* 1987; **2**:333–6.

Castillo E, Rubin RT, Holsboer-Trachsler E. Clinical differentiation between lethal catatonia and neuroleptic malignant syndrome. *Am J Psychiatry* 1989; **146**:324–8.

Caviness JN, Evidente VG. Cortical myoclonus during lithium exposure. *Arch Neurol* 2003; **60**:401–4.

Cervera R, Asherson RA, Font J *et al.* Chorea in the antiphospholipid syndrome. Clinical, radiologic, and immunologic characteristics of 50 patients from our clinics and the recent literature. *Medicine* 1997; **76**:203–12.

Chandra V, Wharton S, Spunt AL. Amelioration of hemiballismus with sodium valproate. *Ann Neurol* 1982; **12**:407.

Chang M-H, Li J-Y, Lee S-R *et al.* Non-ketotic hyperglycaemic chorea: a SPECT study. *J Neurol Neurosurg Psychiatry* 1996; **60**:428–30.

Chen D-H, Matsushita M, Rainier S *et al.* Presence of alanine-to-valine substitutions in myofibrillogenesis regulator 1 in paroxysmal nonkinesigenic dyskinesia. *Arch Neurol* 2005; **62**: 597–600.

Chi WM, Chua KS, Kong KH. Phenytoin-induced asterixis – uncommon or underdiagnosed? *Brain Inj* 2000; **14**:847–50.

Chiu LPW. Transient recurrence of auditory hallucinations during acute dystonia. *Br J Psychiatry* 1989; **155**:110–13.

Chouinard MJ, Nguyen DK, Clement JF *et al.* Catatonia induced by levertiracetam. *Epilepsy Behav* 2006; **8**:303–7.

Chouinard S, Ford B. Adult onset tic disorders. *J Neurol Neurosurg Psychiatry* 2000; **68**:738–43.

Choy-Kwong M, Lipton RB. Dystonia related to cocaine withdrawal: a case report and pathogenic hypothesis. *Neurology* 1989; **39**: 996–7.

Christodoulou C, Kalaitzi C. Antipsychotic drug-induced acute laryngeal dystonia: two case reports and a mini review. *J Psychopharmacol* 2005; **19**:307–11.

Chu K, Kang DW, Kim DE *et al.* Diffusion-weighted and gradient echo magnetic resonance findings of hemichorea-hemiballismus associated with diabetic hyperglycemia: a hyperviscosity syndome? *Arch Neurol* 2002; **59**:448–52.

Chudnow RS, Dewey RB, Lawson CR. Choreoathetosis as a side-effect of gabapentin therapy in severely neurologically impaired patients. *Arch Neurol* 1997; **54**:910–12.

Chung SJ, Im JH, Lee MC *et al.* Hemichorea after stroke: clinical-radiological correlation. *J Neurol* 2004; **251**:725–9.

Cincotta M, Borgheresi A, Balzini L *et al.* Separate ipsilateral and contralateral corticospinal projections in congenital mirror movements: neurophysiological evidence and significance for motor rehabilitation. *Mov Disord* 2003; **18**:1294–1300.

Crespel A, Genton P, Berrandame M *et al.* Lamotrigine associated with exacerbation or de novo myoclonus in idiopathic generalized epilepsies. *Neurology* 2005; **65**:762–4.

Croteau D, Owainati A, Dalmau J *et al.* Response to cancer therapy in a patient with a paraneoplastic choreiform disorder. *Neurology* 2001; **57**:719–22.

Crozier S, Lehericy S, Verstichel P *et al.* Transient hemiballism/hemichorea due to ipsilateral subthalamic nucleus infarction. *Neurology* 1996; **46**:267–8.

Crystal HA. Baclofen therapy may be associated with chorea in Alzheimer's disease. *Ann Neurol* 1990; **28**:839.

Cubo E, Gracies JM, Benabou R *et al.* Early morning off-medication dyskinesias, dystonia, and choreic subtypes. *Arch Neurol* 2001; **58**:1379–82.

Cunningham MA, Darby DG, Donnan GA. Controlled-release delivery of L-dopa associated with nonfatal hyperthermia, rigidity and autonomic dysfunction. *Neurology* 1991; **41**:942–3.

Dale RC, Church AJ, Heyman I. Striatal encephalitis after varicella zoster infection complicated by Tourettism. *Mov Disord* 2003; **18**:1554–6.

Danek A, Tison F, Rubio J *et al.* The chorea of McLeod syndrome. *Mov Disord* 2001; **16**:882–9.

Daras M, Koppel BS, Atos-Radzion E. Cocaine-induced choreoathetoid movements ('crack dancing'). *Neurology* 1994; **44**:751–2.

Davis EJB, Borde M. Wilson's disease and catatonia. *Br J Psychiatry* 1993; **162**:256–9.

Davis JM. Response of hemiballismus to haloperidol. *JAMA* 1976; **235**:281–2.

Davous P, Rondot P, Marion MH *et al.* Severe chorea after acute carbon monoxide poisoning. *J Neurol Neurosurg Psychiatry* 1986; **49**:206–8.

DeCastro RM. Antidepressants and myoclonus: case report. *J Clin Psychiatry* 1985; **46**:284–7.

Defebvre L, Destee A, Cassim F *et al.* Transient hemiballism and striatal infarct. *Stroke* 1990; **21**:967–8.

Degos JD, Verroust J, Bouchareine A *et al.* Asterixis in focal brain lesions. *Arch Neurol* 1979; **36**:705–7.

de Graaf L, Admiraal R, van Puijenbroeck EP. Ballism associated with bupropion use. *Ann Pharmacother* 2003; **37**:302–3.

De La Sayettte V, Schaeffer S, Querel C *et al.* Lyme neuroborreliosis presenting with propriospinal myoclonus. *J Neurol Neurosurg Psychiatry* 1996; **64**:420.

Della Marca G, Retuccia D, Mariotti P *et al.* Pathologic startle following brainstem lesion. *Neurology* 2007; **68**:47.

de Mari M, De Blasi R, Lamberti P *et al.* Unilateral pallidal lesion after acute disulfiram intoxication: a clinical and magnetic resonance imaging study. *Mov Disord* 1993; **8**:247–9.

Demirkiran M, Jankovic J. Paroxysmal dyskinesias: clinical features and classification. *Ann Neurol* 1995; **38**:571–9.

Denckla MB, Bemporad JR, MacKay MC. Tics following methylphenidate administration: a report of 20 cases. *JAMA* 1976; **235**:1349–51.

Dennis MS, Byrne EJ, Hopkinson EN *et al.* Neuropsychiatric systemic lupus erythematosus in elderly people: a case series. *J Neurol Neurosurg Psychiatry* 1992; **55**:1157–61.

Desarkar P, Thakur A, Sinha V. Aripiprazole-induced acute dystonia. *Am J Psychiatry* 2006; **163**:1112–13.

Detweiller MB, Harpold GJ. Bupropion-induced acute dystonia. *Ann Pharmacother* 2002; **36**:251–4.

Deuschl G, Koester B, Luecking CH *et al.* Diagnostic and pathophysiological aspects of psychogenic tremor. *Movement Disord* 1998; **13**:294–302.

Dewey RB, Jankovic J. Hemiballism–hemichorea. Clinical and pharmacologic findings in 21 patients. *Arch Neurol* 1989; **46**:862–7.

Dib E, Bernstein S, Benesch C. Multifocal myoclonus induced by trimethoprim–sulfamethoxazole in a patient with nocardia infection. *N Engl J Med* 2004; **350**:88–9.

Digre KB. Opsoclonus in adults – report of three cases and review of the literature. *Arch Neurol* 1986; **43**:1165–75.

DiMascio A, Bernardo DL, Greenblatt DJ *et al.* A controlled trial of amantadine in drug-induced extrapyramidal disorders. *Arch Gen Psychiatry* 1976; **33**:599–602.

Doheny DO, Brin MF, Morrison CE *et al.* Phenotypic features of myoclonus-dystonia in three kindreds. *Neurology* 2002; **59**:1187–94.

Dominguez-Moran JA, Callejo JM, Fernandex-Ruiz LC *et al.* Acute paroxysmal dystonia induced by fluoxetine. *Mov Disord* 2001; **16**:767–9.

Donaldson IM, Espiner EA. Disseminated lupus erythematosus presenting as chorea gravidarum. *Arch Neurol* 1971; **25**:240–4.

Dooling EC, Adams RD. Pathologic anatomy of posthemiplegic athetosis. *Brain* 1975; **98**:29–48.

Drake RE, Ehrlich J. Suicide attempts associated with akathisia. *Am J Psychiatry* 1985; **142**:499–501.

Driesen JJ, Wolters EC. Bilateral ballism induced by oral contraceptives. *J Neurol* 1986; **233**:379.

Easterford K, Clugh P, Kellett M *et al.* Reversible parkinsonism with normal beta-CIT-SPECT in patients exposed to sodium valproate. *Neurology* 2004; **62**:1435–7.

Edelstein H, Knight RT. Severe parkinsonism in two AIDS patients taking prochlorperazine. *Lancet* 1987; **2**:341–2.

Eisenhauer G, Jermain DM. Fluoxetine and tics in an adolescent. *Ann Pharmacother* 1993; **27**:725–6.

Engel J, Ludwig BI, Fetell M. Prolonged partial complex status epilepticus: EEG and behavioral observations. *Neurology* 1978; **28**:863–9.

Ernst NH, Steur J. Increase of parkinsonian disability after fluoxetine medication. *Neurology* 1993; **43**:211–13.

Espay AJ, Hung SW, Sanger TD *et al.* A writing device improves writing in primary writing tremor. *Neurology* 2005a; **64**:1648–50.

Espay AJ, Li JY, Johnston L *et al.* Mirror movements in parkinsonism: evaluation of a new clinical sign. *J Neurol Neurosurg Psychiatry* 2005b; **76**:1355–8.

Evidente VG, Gwinn-Hardy K, Caviness JN *et al.* Risperidone is effective in severe hemichorea/hemiballismus. *Mov Disord* 1999; **14**:377–9.

Evidente VG, Advincula J, Esteban R *et al.* Phenomenology of 'Lubag' or X-linked dystonia-parkinsonism. *Mov Disord* 2002; **17**:1271–7.

Extein I. Methylphenidate-induced choreoathetosis. *Am J Psychiatry* 1978; **135**:252–3.

Factor SA, Moltho ES. Adult-onset tics associated with peripheral injury. *Mov Disord* 1997; **12**:1052–5.

Factor SA, Troche-Panetto M, Weaver SA. Dystonia in AIDS: report of four cases. *Mov Disord* 2003; **18**:1492–8.

Federico P, Zochodne DW. Reversible parkinsonism and hyperammonemia associated with portal vein thrombosis. *Acta Neurol Scand* 2001; **103**:198–200.

Federlein J, Postert TH, Allgeier A *et al.* Remitting parkinsonism as a symptom of multiple sclerosis and the associated magnetic resonance imaging findings. *Mov Disord* 1997; **12**:1090–1.

Feinberg TE, Cianci CD, Morrow JS *et al.* Diagnostic tests for choreoacanthocytosis. *Neurology* 1991; **41**:1000–6.

Fenelon G, Houeto J-L. Unilateral parkinsonism following a large infarct in the territory of the lenticulostriate arteries. *Mov Disord* 1997; **12**:1086–90.

Fermaglich J, Streib E, Auth T. Chorea associated with systemic lupus erythematosus. *Arch Neurol* 1973; **28**:276–7.

Feve AP, Fenelon G, Wallays C *et al.* Axial motor disturbances after hypoxic lesions of the globus pallidus. *Mov Disord* 1993; **8**:321–6.

Fidler SM, O'Rourke RA, Buchsbaum HW. Choreoathetosis as a manifestation of thyrotoxicosis. *Neurology* 1971; **21**:55–7.

Fink JK, Hedera P, Mathay JG *et al.* Paroxysmal dystonic choreoathetosis linked to chromosome 2q: clinical analysis and proposed pathophysiology. *Neurology* 1997; **49**:177–83.

Fischel T, Hermesh H, Aizenberg D *et al.* Cyproheptadine versus propranolol for the treatment of acute neuroleptic-induced akathisia: a comparative double-blind study. *J Clin Psychopharmacol* 2001; **21**:612–15.

Fishbeck KH, Layzer RB. Paroxysmal choreoathetosis associated with thyrotoxicosis. *Ann Neurol* 1979; **6**:453–4.

Flaherty JA, Lahmeyer HW. Larnyngeal–pharyngeal dystonia as a possible cause of asphyxia with haloperidol treatment. *Am J Psychiatry* 1978; **135**:1414–15.

Fleishman SB, Lavin MR, Sattler M *et al.* Antiemetic-induced akathisia in cancer patients receiving chemotherapy. *Am J Psychiatry* 1994; **151**:763–5.

Foncke EMJ, Gerrits MLF, van Ruissen F *et al.* Distal myoclonus and later onset in a large Dutch family with myoclonus-dystonia. *Neurology* 2006; **67**:1677–80.

Foster PN, Stickle BR, Laurence AS. Akathisia following low-dose droperidol for antiemesis in day-case patients. *Anaesthesia* 1996; **51**:491–4.

Fox GC, Ebeid S, Vincenti G. Paroxetine-induced chorea. *Br J Psychiatry* 1997; **170**:193–4.

Francisco GE. Successful treatment of posttraumatic hemiballismus with intrathecal baclofen therapy. *Am J Phys Med Rehabil* 2006; **85**:779–82.

Furtado S, Ferrer M, Tsuboi Y *et al.* SCA-2 presenting as parkinsonism in an Alberta family. Clinical, genetic and PET findings. *Neurology* 2002; **59**:1625–7.

Gambardella A, Valentino P, Annesi G *et al.* Hyperekplexia in a patient with a brainstem vascular anomaly. *Acta Neurol Scand* 1999; **99**:255–9.

Gamboa ET, Issacs G, Harter DH. Chorea associated with oral contraceptive therapy. *Arch Neurol* 1971; **25**:112–14.

Garruto RM, Gajdusek DC, Chen KM. Amyotrophic lateral sclerosis and parkinsonism-dementia among Filipino immigrants to Guam. *Ann Neurol* 1981; **10**:341–50.

Garvey MJ, Tollefson GD. Occurrence of myoclonus in patients treated with cyclic antidepressants. *Arch Gen Psychiatry* 1987; **44**:269–72.

Gatto EM, Uribe Roca C, Raina G *et al.* Vascular hemichorea/hemiballism and topiramate. *Mov Disord* 2004; **19**:836–8.

Gaultieri CT, Patterson DR. Neuroleptic-induced tics in two hyperactive children. *Am J Psychiatry* 1986; **143**:1176–7.

Gautier-Smith PC, Prankerd TAJ. Polycythemia vera and chorea. *Acta Neurol Scand* 1967; **43**:357–64.

Gelenberg AJ, Mandel MR. Catatonic reactions to high-potency neuroleptic drugs. *Arch Gen Psychiatry* 1977; **34**:947–50.

Ghika J, Bogousslavsky J. Spinal pseudoathetosis: a rare, forgotten syndrome, with a review of old and recent descriptions. *Neurology* 1997; **49**:432–7.

Ghika J, Bogousslavsky J, Henderson J *et al.* The 'jerky dystonic unsteady hand': a delayed motor syndrome in posterior thalamic infarctions. *J Neurol* 1994; **241**:537–42.

Gibb WRG, Lees AJ. The clinical phenomenon of akathisia. *J Neurol Neurosurg Psychiatry* 1986; **49**:861–6.

Giladi N, McMahon D, Przedborski S *et al.* Motor blocks in Parkinson's disease. *Neurology* 1992; **42**:333–9.

Gilbert GJ. Brueghel syndrome: its distinction from Meige syndrome. *Neurology* 1996; **46**:1767–9.

Gjessing LR. A review of periodic catatonia. *Biol Psychiatry* 1974; **8**:23–45.

Glantz R, Weiner WJ, Goetz CG *et al.* Drug-induced asterixis in Parkinson's disease. *Neurology* 1982; **32**:553–5.

Glass JP, Jankovic J, Borit A. Hemiballism and metastatic brain tumor. *Neurology* 1984; **34**:204–7.

Gomez EA, Comstock BS, Rosario A. Organic versus functional etiology in catatonia: case report. *J Clin Psychiatry* 1982; **43**:200–1.

Gooch JL, Sandell TV. Botulinum toxin for spasticity and athetosis in children with cerebral palsy. *Arch Phys Med Rehabil* 1996; **77**:508–11.

Green PM. Chorea induced by oral contraceptives. *Neurology* 1980; **30**:1131–2.

Grimes DA, Han F, Lang AE *et al.* A novel focus for inherited myoclonus-dystonia on 18p11. *Neurology* 2002; **59**:1183–6.

Gunal DI, Onultan O, Afsar N *et al.* Tardive dystonia associated with olanzapine therapy. *Neurol Sci* 2001; **22**:331–2.

Halstead SM, Barnes TRE, Speller JC. Akathisia: prevalence and associated dysphoria in an in-patient population with chronic schizophrenia. *Br J Psychiatry* 1994; **164**:177–83.

Hammond WA. *A Treatise on Diseases of the Nervous System.* New York: Appleton-Century Crofts, 1871.

Hansen HC, Zschocke S, Sturenbur HJ *et al.* Clinical changes and EEG patterns preceding the onset of periodic sharp wave complexes in Creutzfeldt–Jakob disease. *Acta Neurol Scand* 1998; **97**:99–106.

Hardie RJ, Lees AJ. Neuroleptic-induced Parkinson's syndrome: clinical features and results of treatment with levodopa. *J Neurol Neurosurg Psychiatry* 1988; **51**:850–4.

Harsch HH. Neurologyoleptic malignant syndrome: physiological and laboratory findings in a series of nine cases. *J Clin Psychiatry* 1987; **48**:328–33.

Haskovec L. L'akathisie. *Rev Neurol* 1901; **9**:1107–9.

Heckmann JG, Lang CJG, Druschky D *et al.* Chorea resulting from paraneoplastic encephalitis. *Mov Disord* 1997; **12**:464–6.

Heckmann JG, Ulrich H, Dutsch M *et al.* Pregabalin-associated asterixis. *Am J Phys Med Rehabil* 2005; **84**:724.

Hefter H, Mayer P, Benecke R. Persistent chorea after recurrent hypoglycemia. A case report. *Eur Neurol* 1993; **33**:244–7.

Heilman KH. Orthostatic tremor. *Arch Neurol* 1984; **41**:880–1.

Hellman MA, Melamed E. Focal dystonia as the presenting sign in Creutzfeldt–Jakob disease. *Mov Disord* 2002; **17**:1097–8.

Hess CW, Raymond D, de Carbalho Aguiar P *et al.* Myoclonus-dystonia, obsessive-compulsive disorder, and alcohol dependence in SGCE mutation carriers. *Neurology* 2007; **68**:522–4.

Hirano A, Arumugasamy N, Zimmerman HM. Amyotrophic lateral sclerosis: a comparison of Guam and classical cases. *Arch Neurol* 1967; **16**:357–63.

Hoehn MM, Cherington M. Spinal myoclonus. *Neurology* 1977; **27**:942.

Hofmann A, Tangri N, Lafontaine A-L *et al.* Myoclonus as an acute complication of low-dose hydromorphone in multiple system atrophy. *J Neurol Neurosurg Psychiatry* 2006; **77**:994–5.

Holmes G. On certain tremors in organic cerebral lesions. *Brain* 1904; **27**:327–75.

Holroyd S, Smith D. Disabling parkinsonism due to lithium: a case report. *J Geriatr Psychiatry Neurol* 1995; **8**:118–19.

Hori A, Hirose G, Kataoka S *et al.* Delayed postanoxic encephalopathy after strangulation: serial neuroradiological and neurochemical studies. *Arch Neurol* 1991; **48**:871–4.

Horikawa N, Yamazaki T, Sagawa M *et al.* A case of akathisia during interferon-alpha therapy for chronic hepatitis type C. *Gen Hosp Psychiatry* 1999; **21**:134–5.

Howard R, Ford R. From the jumping Frenchmen of Maine to post-traumatic stress disorder: the startle response in neuropsychiatry. *Psychol Med* 1992; **22**:695–707.

Huppertz HJ, Feuerstein TJ, Schulze-Bonhage A. Myoclonus in epilepsy patients with anticonvulsive add-on therapy with pregabalin. *Epilepsia* 2001; **42**:790–2.

Iijima M. Valproate-induced parkinsonism in a demented elderly patient. *J Clin Psychiatry* 2002; **63**:75.

Indo T, Ando K. Metoclopramide-induced parkinsonism. Clinical characteristics of ten cases. *Arch Neurol* 1982; **39**:494–6.

Jacob PC, Pratap Chand R. Post-traumatic rubral tremor responsive to clonazepam. *Mov Disord* 1998; **13**:977–8.

Jacobs MB. Diltiazem and akathisia. *Ann Intern Med* 1983; **99**:794–5.

Jaffe N. Catatonia and hepatic dysfunction. *Dis Nerv Syst* 1967; **28**:606–8.

Jain S, Padma MV, Puri A *et al.* Juvenile myoclonic epilepsy: disease expression among Indian families. *Acta Neurol Scand* 1998; **97**:1–7.

Jarman PR, Davis MB, Hodgson SV *et al.* Paroxysmal dystonic choreoathetosis: genetic linkage studies in a British family. *Brain* 1997; **120**:2125–30.

Jarman PR, Bhatia KP, Davie C *et al.* Paroxysmal dystonic choreoathetosis: clinical features and investigation of pathophysiology in a large family. *Mov Disord* 2000; **15**:648–57.

Jauhar S, Blackett A, Sireddy P *et al.* Pernicious anemia presenting as catatonia without signs of anaemia or macrocytosis. *Br J Psychiatry* 2010; **197**:244–5.

Jimenez-Jimenez FJ, Tejeiro J, Martinez-Junquera G *et al.* Parkinsonism exacerbated by paroxetine. *Neurology* 1994; **44**:2406.

Johnson WG, Fahn S. Treatment of vascular hemiballism and hemichorea. *Neurology* 1977; **27**:634–6.

Jung HH, Hergersberg M, Kneifel S *et al.* McLeod syndrome: a novel mutation, predominant psychiatric manifestations, and distinct striatal imaging findings. *Ann Neurol* 2001; **49**:384–92.

Jungmann E, Schöffling C. Akathisia and metoclopramide. *Lancet* 1982; **2**:221.

Kabacki K, Isbruch K, Schilling K *et al.* Genetic heterogeneity in rapid onset dystonia-parkinsonism: description of a new family. *J Neurol Neurosurg Psychiatry* 2005; **76**:860–2.

Kahlbaum K. *Die Katatonie oder das Spanungsirresein.* Berlin: Hirsch, 1874.

Kaiko RF, Foley KM, Grabinski PY *et al.* Central nervous system excitatory effects of meperidine in cancer patients. *Ann Neurol* 1983; **13**:180–5.

Kane JM, Fleischhacker WW, Hansen L *et al.* Akathisia: an updated review focusing on second-generation antipsychotics. *J Clin Psychiatry* 2009; **70**:627–43.

Kaplan PW. Nonconvulsive status epilepticus in the emergency room. *Epilepsia* 1996; **37**:643–50.

Kase CS, Maulsby GO, deJuan E *et al.* Hemichorea–hemiballism and lacunar infarction in the basal ganglia. *Neurology* 1981; **31**:452–5.

Katayama Y, Yamamoto T, Kobayshi K *et al.* Deep brain and motor cortex stimulation for post-stroke movement disorders and post-stroke pain. *Acta Neurochir* 2003; **87**:121–3.

Keepers GA, Clappison VJ, Casey DE. Initial anticholinergic prophylaxis for neuroleptic-induced extrapyramidal syndromes. *Arch Gen Psychiatry* 1983; **40**:1113–17.

Kenney C, Hunter C, Jankovic J. Long-term tolerability of tetrabenazine in the treatment of hyperkinetic movement disorders. *Mov Disord* 2007; **22**:193–7.

Kerbeshian J, Burd L, Pettit R. A possible post-streptococcal movement disorder with chorea and tics. *Dev Med Child Neurol* 1990; **32**:642–4.

Kertesz A. Paroxysmal kinesigenic choreoathetosis: an entity within the paroxysmal choreoathetosis syndrome. Description of 10 cases, including 1 autopsied. *Neurology* 1967; **17**:680–90.

Keswani SC, Kossoff EH, Krauss GL *et al.* Amelioration of spinal myoclonus with levetiracetam. *J Neurol Neurosurg Psychiatry* 2002; **73**:456–9.

Kidron D, Melamed E. Forms of dystonia in patients with Parkinson's disease. *Neurology* 1987; **37**:1009–11.

Kim CH, Perlstein MA. Encephalitis with catatonic schizophrenic symptoms. *Ill Med J* 1970; **138**:503–7.

Kim JS. Delayed onset mixed involuntary movements after thalamic stroke. Clinical, radiological and pathophysiological findings. *Brain* 2001a; **124**:299–309.

Kim JS. Asterixis after unilateral stroke: lesion location of 30 patients. *Neurology* 2001b; **56**:533–6.

Kim JW, Choi SH, Kim WJ *et al.* Transient ischemic attacks presenting as hemiballism. *Mov Disord* 2003; **18**:1399–401.

Kimber TE, Thompson PD. Symptomatic hyperekplexia occurring as a result of pontine infarction. *Mov Disord* 1997; **12**:815–16.

Kinirons P, Fulton A, Keoghan M *et al.* Paraneoplastic limbic encephalitis (PLE) and chorea associated with CRMP-5 neuronal antibody. *Neurology* 2003; **61**:1623–4.

Kirby GH, Davis TK. Psychiatric aspects of epidemic encephalitis. *Arch Neurol Psychiatry* 1921; **5**:491–51.

Kiriakakis V, Bhatia K, Quinn NP *et al.* The natural history of tardive dystonia: a long-term follow-up study of 107 cases. *Brain* 1998; **121**:2053–66.

Kirubakaren V, Mayfield D, Rengachary S. Dyskinesia and psychosis in a patient following baclofen withdrawal. *Am J Psychiatry* 1984; **141**:692–3.

Klawans HL, Barr A. Recurrence of childhood multiple tic in late adult life. *Arch Neurol* 1985; **42**:1079–80.

Klawans HL, Goetz CG, Bergen D. Levodopa-induced myoclonus. *Arch Neurol* 1975; **32**:331–4.

Klawans HL, Moses H, Nausieda PA *et al.* Treatment and prognosis of hemiballismus. *N Engl J Med* 1976; **295**:1348–50.

Klawans HL, Glantz R, Tanner CM *et al.* Primary writing tremor: a selective action tremor. *Neurology* 1982; **32**:203–6.

Kleinschmidt-DeMasters BK, Marder BA, Levi ME *et al.* Naturally acquired West Nile virus encephalomyelitis in transplant recipients: clinical, laboratory, diagnostic and neuropathological features. *Arch Neurol* 2004; **61**:1210–20.

Ko SB, Ahn TB, Kim JR *et al.* A case of adult-onset tic disorder following carbon monoxide intoxication. *Can J Neurol Sci* 2004; **31**:268–70.

Koehler J, Jakumeit U. Subacute sclerosing panencephalitis presenting as Leonhard's speech-prompt catatonia. *Br J Psychiatry* 1976; **129**:29–31.

Koller WC. Edentulous orodyskinesia. *Neurology* 1983; **13**:97–9.

Konagaya M, Konagaya Y. MRI in hemiballism due to Sydenham's chorea. *J Neurol Neurosurg Psychiatry* 1992; **55**:238–9.

Kooiker JC, Sumi SM. Movement disorder as a manifestation of diphenylhydantoin intoxication. *Neurology* 1974; **24**:68–71.

Koskiniemi M, Donner M, Majuri H *et al.* Progressive myoclonus epilepsy: a clinical and histopathological study. *Acta Neurol Scand* 1974; **50**:307–32.

Koukoulis A, Herrerd JS, Gomez-Alonso J. Blepharospasm induced by flunarazine. *J Neurol Neurosurg Psychiatry* 1997; **63**:412–13.

Krack P, Deuschl G, Kapa M *et al.* Delayed onset of 'rubral tremor' 23 years after brainstem trauma. *Mov Disord* 1994; **9**:240–2.

Kraepelin E. *Psychiatry. A textbook for students and physicians, 1899, translated by Metoui H, Ayed S.*, 6th edn. Canton, MA: Science History Publications, 1990.

Kramer PL, Mineta M, Klein C *et al.* Rapid-onset dystonia-parkinsonism: linkage to chromosome 19q13. *Ann Neurol* 1999; **46**:176–82.

Krams M, Quinton R, Ashburner J *et al.* Kallmann's syndrome: mirror movements associated with bilateral corticospinal tract hypertrophy. *Neurology* 1999; **52**:816–22.

Krauss JK, Jankovic J. Tics secondary to craniocerebral trauma. *Mov Disord* 1997; **12**:776–82.

Krauss JK, Nobbe F, Wakhloo AK *et al.* Movement disorders in astrocytomas of the basal ganglia and the thalamus. *J Neurol Neurosurg Psychiatry* 1992; **55**:1162–7.

Krauss JK, Wakhloo AK, Nobbe F *et al.* MR pathological correlations of severe posttraumatic tremor. *Neurol Res* 1995; **17**:409–16.

Krauss JK, Seeger W, Jankovic J. Cervical dystonia with tumors of the posterior fossa. *Mov Disord* 1997; **12**:443–7.

Kristensen O, Sindrup EH. Psychomotor epilepsy and psychosis. III. Social and psychological correlates. *Acta Neurol Scand* 1979; **59**:1–9.

Krystkowiak P, Martinat P, Defebvre L *et al.* Dystonia after striatopallidal and thalamic stroke: clinicoradiological correlations and pathophysiological mechanism. *J Neurol Neurosurg Pyschiatry* 1998; **65**:703–8.

Kuba R, Rektor I, Brazdil M. Ictal limb dystonia in temporal lobe epilepsy: an invasive video-EEG finding. *Eur J Neurol* 2003; **10**:641–9.

Kulisevsky J, Berthier ML, Pujol J. Hemiballismus and secondary mania following right thalamic infarction. *Neurology* 1993; **43**:1422–4.

Kulisevsky J, Asuncion A, Berthier ML. Bipolar disorder and unilateral parkinsonism after a brainstem infarction. *Mov Disord* 1995; **10**:799–802.

Kumral E, Evyapan D, Balkir K. Acute caudate vascular lesions. *Stroke* 1999; **30**:100–8.

Kushner MJ. Chorea and cimetidine. *Ann Intern Med* 1982; **96**:126.

Kutcher SP, Williamson P, MacKenzie S *et al.* Successful clonazepam treatment of neuroleptic-induced akathisia in older adolescents and young adults. *J Clin Psychopharmacol* 1989; **9**:403–6.

Kwak CH, Jankovic J. Tourettism and dystonia after subcortical stroke. *Mov Disord* 2002; **17**:821–5.

Lai SL, Tseng YL, Hsu MC *et al.* Magnetic resonance imaging and single-photon emission computed tomography changes in hypoglycemia-induced chorea. *Mov Disord* 2004; **19**:475–8.

Lance JW. Familial paroxysmal dystonic choreoathetosis and its differentiation from related syndromes. *Ann Neurol* 1977; **2**: 285–93.

Lance JW, Adams RD. The syndrome of intention or action myoclonus as a sequel to hypoxic encephalopathy. *Brain* 1963; **86**:111–36.

Lancman ME, Asconape MJ, Penry JK. Choreiform movements associated with the use of valproate. *Arch Neurol* 1994; **51**:702–4.

Lang AE. Persistent hemiballismus with lesions outside the subthalamic nucleus. *Can J Neurol Sci* 1985; **12**:125–8.

Langston JW, Ballard P, Tetrud JW *et al.* Chronic parkinsonism in humans due to a product of meperidine-analog synthesis. *Science* 1983; **249**:979–80.

Lanham JG, Brown MM, Hughes GRV. Cerebral systemic lupus erythematosus presenting with catatonia. *Postgrad Med J* 1985; **61**:329–30.

Lapierre YD. Control of lithium tremor with propranolol. *Can Med Assoc J* 1976; **114**:619–20.

Laplane D, Attal N, Sauron B *et al.* Lesions of basal ganglia due to disulfiram neurotoxicity. *J Neurol Neurosurg Psychiatry* 1992; **55**:925–9.

Lapresle J. Palatal myoclonus. *Adv Neurol* 1986; **43**:265–73.

Leavitt S, Tyler H. Studies in asterixis. *Arch Neurol* 1964; **10**:360–8.

Ledbetter M. Atomoxetine associated with onset of a motor tic. *J Child Adolesc Psychopharmacol* 2005; **15**:331–3.

Lee BI, Lesser RP, Pippenger CE *et al.* Familial paroxysmal hypnogenic dystonia. *Neurology* 1985; **35**:1357–60.

Lee MS, Rinne JO, Ceballos-Baumann A *et al.* Dystonia after head trauma. *Neurology* 1994; **44**:1374–8.

Lee MS, Kim YD, Lyoo CH. Oculogyric crises as an initial manifestation of Wilson's disease. *Neurology* 1999; **52**:1714–15.

Lehericy S, Vidailhet M, Dormont D *et al.* Striatopallidal and thalamic dystonia: a magnetic resonance imaging anatamoclinical study. *Arch Neurol* 1996; **53**:241–50.

Lehericy S, Grand S, Pollak P *et al.* Clinical characteristics and topography of lesions in movement disorders due to thalamic lesions. *Neurology* 2001; **57**:1055–66.

Lenton RJ, Cofti M, Smith RG. Hemiballismus treated with sodium valproate. *BMJ* 1981; **283**:17–18.

Lera G, Scipioni O, Garcia S *et al.* A combined pattern of movement disorders resulting from posterolateral thalamic lesions of a vascular nature: a syndrome with clinico-radiologic correlation. *Mov Disord* 2000; **15**:120–6.

Lerner V, Bergman J, Statsensko N *et al.* Vitamin B6 treatment in acute neuroleptic-induced akathisia: a randomized, double-blind, placebo-controlled study. *J Clin Psychiatry* 2004; **65**:1550–4.

LeWitt PA, Walters A, Hening W *et al.* Persistent movement disorders induced by buspirone. *Mov Disord* 1993; **8**:331–4.

Lightman SL. Phenobarbital dyskinesia. *Postgrad Med J* 1978; **54**: 114–15.

Lim J, Yagnik P, Schrader P *et al.* Ictal catatonia as a manifestation of nonconvulsive status epilepticus. *J Neurol Neurosurg Psychiatry* 1986; **49**:833–6.

Lin J-J, Chang M-K. Hemiballism–hemichorea and non-ketotic hyperglycemia. *J Neurol Neurosurg Psychiatry* 1994; **57**:748–50.

Lin J-J, Lin GY, Shih C *et al.* Presentation of striatal hyperintensity on T1-weighted MRI in patients with hemiballism-hemichorea caused by non-ketotic hyperglycemia: report of seven new cases and a review of the literature. *J Neurol* 2001; **248**:750–5.

Lindenmeyer J-P, Da Silva D, Buendia A *et al.* Tic-like syndrome after treatment with clozapine. *Am J Psychiatry* 1995; **152**:649.

Lipinski JF, Zubenko GS, Cohen BM *et al.* Propranolol in the treatment of neuroleptic induced akathisia. *Am J Psychiatry* 1984; **141**: 412–15.

Lipinski JF, Mallya A, Zimmerman P *et al.* Fluoxetine-induced akathisia: clinical and theoretical implications. *J Clin Psychiatry* 1989; **50**:339–42.

Lippmann S, Moskovitz R, O'Tuama L. Tricyclic-induced myoclonus. *Am J Psychiatry* 1977; **134**:90–1.

Logsdail SJ, Toone BK. Postictal psychoses: a clinical and phenomenological description. *Br J Psychiatry* 1988; **152**:246–52.

Lombroso CT. Lamotrigene-induced tourettism. *Neurology* 1999; **52**:1191–4.

Louis ED, Anderson KE, Moskowitz C *et al.* Dystonia-predominant adult-onset Huntington disease: association between motor phenotype and age of onset in adults. *Arch Neurol* 2000; **57**:1326–30.

Lugaresi E, Cirignotta F. Hypnogenic paroxysmal dystonia: epileptic seizure or a new syndrome? *Sleep* 1981; **4**:129–38.

Lugaresi E, Cirignotta F, Montagna P. Nocturnal paroxysmal dystonia. *J Neurol Neurosurg Psychiatry* 1986; **49**:375–80.

Lundh H, Tunving K. An extrapyramidal choreiform syndrome caused by amphetamine addiction. *J Neurol Neurosurg Psychiatry* 1981; **44**:728–30.

Luque FA, Selhorst JB, Petruska P. Parkinsonism induced by high-dose cytosine arabinoside. *Mov Disord* 1987; **2**:219–22.

Mac DS, Pardo MP. Systemic lupus erythematosus and catatonia: a case report. *J Clin Psychiatry* 1983; **44**:155–6.

McKee D, Talbot P. Chorea as a presenting feature of variant Creutzfeldt–Jakob disease. *Mov Disord* 2003; **18**:837–8.

McNeill A. Chorea induced by low-dose tranzodone. *Eur Neurol* 2006; **55**:101–12.

Magaudda A, Ferlazzo E, Nguyen V-H *et al.* Unverricht–Lundborg disease: a condition with self-limited progression: long-term follow-up of 20 patients. *Epilepsia* 2006; **47**:860–6.
Malamud N, Hirano A, Kurland LT. Pathoanatomic changes in amyotrophic lateral sclerosis on Guam. *Neurology* 1961; 5:401–14.
Malur C, Pasol E, Francis A. ECT for prolonged catatonia. *J ECT* 2001; **17**:55–9.
Mann SC, Caroff SN, Bleir HR *et al.* Lethal catatonia. *Am J Psychiatry* 1986; **143**:1374–81.
Marti-Masso JF, Poza JJ. Cinnarizine-induced parkinsonism: ten years later. *Mov Disord* 1998; **13**:453–6.
Martin JP, Alcock NS. Hemichorea associated with a lesion of the corpus luysii. *Brain* 1934; **57**:504–16.
Martinez-Cage JM, Marsden CD. Hemi-dystonia secondary to localized basal ganglia tumors. *J Neurol Neurosurg Psychiatry* 1984; **47**:704–9.
Mathews WB. Familial calcification of the basal ganglia with response to parathyroid hormone. *J Neurol Neurosurg Psychiatry* 1957; **20**:172–7.
Matsunaga K, Uozumi T, Qingrui L *et al.* Amantadine-induced cortical myoclonus. *Neurology* 2001; **56**:279–80.
Matsuo H, Kamakura K, Saito M *et al.* Familial paroxysmal dystonic choreoathetosis. Clinical findings in a large Japanese family and genetic linkage to chromosome 2q. *Arch Neurol* 1999; **56**:721–6.
Medlock MD, Cruse RS, Winek SJ *et al.* A 10-year experience with postpartum chorea. *Ann Neurol* 1993; **34**:820–6.
Melamed E. Early-morning dystonia, a late effect of long-term levodopa therapy in Parkinson's disease. *Arch Neurol* 1979; **36**:308–10.
Melamed E, Korn-Lubetzk I, Reches A *et al.* Hemiballismus: detection of focal hemorrhage in subthalamic nucleus by CT scan. *Ann Neurol* 1978; **4**:582.
Mendhekar DN, Duggal HS. Isolated oculogyric crisis on clozapine discontinuation. *J Neuropsychiatry Clin Neurosci* 2006; **18**:424–5.
Mercadante S. Pathophysiology and treatment of opioid-related myoclonus in cancer patients. *Pain* 1998; **74**:5–9.
Meseguer E, Taboada R, Sanchez V *et al.* Life-threatening parkinsonism induced by Kava-Kava. *Mov Disord* 2002; **17**:195–6.
Meyer T, Ludolph AC, Munch C. Ifosfamide encephalopathy presenting with asterixis. *J Neurol Sci* 2002; **15**:85–8.
Miller DD, Sharafuddin MJA, Kathol RG. A case of clozapine-induced neuroleptic malignant syndrome. *J Clin Psychiatry* 1991; **52**:99–101.
Misra PC, Hay GG. Encephalitis presenting as acute schizophrenia. *Br Med J* 1971; **1**:532–3.
Mitchell E, Mathews KL. Gilles de la Tourette's disorder associated with pemoline. *Am J Psychiatry* 1980; **137**:1618–19.
Miwa H, Harori K, Kondo T *et al.* Thalamic tremor: case reports and implications of the tremor-generating mechanism. *Neurology* 1996; **46**:75–9.
Miyagi Y, Shima F, Ishido K *et al.* Posteroventral pallidotomy for midbrain tremor after a pontine hemorrhage. Case report. *J Neurosurg* 1999; **91**:885–8.
Moersch FP, Kernohan JW. Hemiballismus: a clinicopathologic study. *Arch Neurol Psychiatry* 1939; **41**:365–72.
Molinuevo JL, Marti MJ, Blesa R *et al.* Pure akinesia: an unusual phenotype of Hallervorden–Spatz syndrome. *Mov Disord* 2003; **18**:1351–3.
Mones RJ, Elizan TS, Siegel GJ. Analysis of L-dopa induced dyskinesias in 51 patients with parkinsonism. *J Neurol Neurosurg Psychiatry* 1971; **34**:668–73.
Morres CA, Dire DJ. Movement disorders as a manifestation of nonketotic hyperglycemia. *J Emerg Med* 1989; **7**:359–64.
Morrison JR. Catatonia. Retarded and excited types. *Arch Gen Psychiatry* 1973; **28**:515–16.
Moshe K, Iulian I, Seth K *et al.* Clomipramine-induced tourettism in obsessive-compulsive disorder: clinical and theoretical implications. *Clin Neuropharmacol* 1994; **17**:338–43.
Moulignier A, Allo S, Zittoun R *et al.* Recombinant interferon-alpha-induced chorea and frontal subcortical dementia. *Neurology* 2002; **58**:328–30.
Mount LA, Reback S. Familial paroxysmal choreoathetosis: preliminary report of an hitherto undescribed clinical syndrome. *Arch Neurol Psychiatry* 1940; **44**:841–7.
Mukand JA, Fitzsimmons C, Wennemer HK *et al.* Olanzapine for the treatment of hemiballismus: a case report. *Arch Phys Med Rehabil* 2005; **86**:587–90.
Munchau A, Mathen D, Cox T *et al.* Unilateral lesions of the globus pallidus: report of four patients presenting with focal or segmental dystonia. *J Neurol Neurosurg Psychiatry* 2000; **69**:494–8.
Murphy MJ, Goldstein MN. Diphenylhydantoin-induced asterixis. A clinical study. *JAMA* 1974; **29**:538–40.
Nandagopal R, Moorthy SG. Dramatic levodopa responsiveness of dystonia in a sporadic case of spinocerebellar ataxia type 3. *Postgrad Med J* 2004; **80**:363–5.
Nausieda PA, Koller WC, Klawans HL *et al.* Phenytoin and choreic movements. *N Engl J Med* 1978; **298**:1093–4.
Nausieda PA, Koller WC, Weiner WJ *et al.* Chorea induced by oral contraceptives. *Neurology* 1979; **29**:1605–9.
Nausieda PA, Weiner WJ, Klawans HL. Dystonic foot response of parkinsonism. *Arch Neurol* 1980; **37**:132–6.
Nelles G, Cramer SC, Schaechter JD *et al.* Quantitative assessment of mirror movements after stroke. *Stoke* 1998; **29**:1182–7.
Newton MR, Berkovic SF, Austin MC *et al.* Dystonia, clinical lateralization, and regional blood flow changes in temporal lobe seizures. *Neurology* 1992; **42**:371–7.
Nomoto M, Thompson PD, Sheehy MP *et al.* Anticholinergic-induced chorea in the treatment of focal dystonia. *Mov Disord* 1987; **2**:53–6.
Northam RS, Singer HS. Postencephalitic acquired Tourette-like syndrome in a child. *Neurology* 1991; **41**:592–3.
Nyland H, Skre H. Cerebral calcinosis with late onset encephalopathy: unusual type of pseudo-pseudohypoparathyroidism. *Acta Neurol Scand* 1977; **56**:309–25.
Oh J, Park KD, Cho HJ *et al.* Spasmodic dysphonia induced by valproic acid. *Epilepsia* 2004; **45**:880–1.
Okun MS, Riestra RA, Nadeau SE. Treatment of ballism and pseudobulbar affect with sertraline. *Arch Neurol* 2001; **58**:1682–4.
Olsen JC, Keng JA, Clark JA. Frequency of adverse reactions to prochlorperazine in the ED. *Am J Emerg Med* 2000; **18**:609–11.
Onofrj M, Thomas A, Paci C *et al.* Gabapentin in orthostatic tremor: results of a double-blind crossover with placebo in four patients. *Neurology* 1998; **51**:880–2.
Opida CL, Korthals JK, Somasunderam M. Bilateral ballismus in phenytoin intoxication. *Ann Neurol* 1978; **3**:186.
Oppenheim H. Uber eine eigenartige krampfkrankheit des kindlichen und jugendlichen alters (dysbasia lordotica progressiva, dystonia musculorum deformans). *Neurol Centrabla* 1911; **30**:1090–107.
Owens DGC, Johnstone EC, Frith CD. Spontaneous involuntary disorders of movement: their prevalence, severity, and distribution in chronic schizophrenics with and without treatment with neuroleptics. *Arch Gen Psychiatry* 1982; **39**:452–61.
Pacchetti C, Cristina S, Nappi G. Reversible chorea and focal dystonia in vitamin B12 deficiency. *N Engl J Med* 2002; **347**:295.
Pakainis A, Drake MA, John K *et al.* Forced normalization: acute psychosis after seizure control in seven patients. *Arch Neurol* 1987; **44**:289–92.

Panayiotopoulos CP, Obeid T, Tahan AR. Juvenile myolonic epilepsy: a 5-year prospective study. *Epilepsia* 1994; **35**:285–96.

Papez JW, Hertzman J, Rundles RW. Athetosis and pallidal deficiency. *Arch Neurol Psychiatry* 1938; **40**:789–9.

Pardo J, Marcos A, Bhathal H *et al.* Chorea as a form of presentation of human immunodeficiency virus-associated dementia complex. *Neurology* 1998; **50**:568–9.

Parry GJ, Bredesen DE. Sensory neuropathy with low-dose pyridoxine. *Neurology* 1985; **35**:1466–8.

Pascual-Leone A, Dhuna A. Cocaine-associated multifocal tics. *Neurology* 1990; **40**:999–1000.

Patel HC, Bruza D, Yerogani V. Myoclonus with trazodone. *J Clin Psychopharmacol* 1988; **8**:152.

Paus S, Potzach B, Risse JH *et al.* Chorea and antiphospholipid antibodies: treatment with methotrexate. *Neurology* 2001; **56**:137–8.

Pedersen SB, Petersen KA. Juvenile myoclonic epilepsy: clinical and EEG features. *Acta Neurol Scand* 1998; **97**:160–3.

Penn H, Racy J, Lapham L *et al.* Catatonic behavior, viral encephalopathy, and death. *Arch Gen Psychiatry* 1972; **27**:758–61.

Perry TL, Bratty PJA, Hansen S *et al.* Hereditary mental depression and parkinsonism with taurine deficiency. *Arch Neurol* 1975; **32**:108–13.

Peterson DI, Peterson GW. Unilateral asterixis due to ipsilateral lesions in the pons and medulla. *Ann Neurol* 1987; **22**:661–3.

Pfeil ET. Hemiballismus: hemichorea. *J Nerv Ment Dis* 1952; **116**:36–47.

Pfister H-W, Preac-Mursic V, Wilske B *et al.* Catatonic syndrome in acute severe encephalitis due to Borrelia burgdorferi infection. *Neurology* 1993; **43**:433–5.

Piccolo I, Causarano R, Sterzi R *et al.* Chorea in patients with AIDS. *Acta Neurol Scand* 1999; **100**:332–6.

Pina MA, Modrego PJ. Dystonia induced by gabapentin. *Ann Pharmacother* 2005; **39**:380–2.

Pittock SJ, Joyce C, O'Keane V *et al.* Rapid-onset dystonia-parkinsonism. A clinical and genetic analysis of a new kindred. *Neurology* 2000; **55**:991–5.

Plana MT, Blanch J, Romero S *et al.* Toxic catatonia secondary to azithromycin. *J Clin Psychiatry* 2006; **67**:492–3.

Podskalny GD, Factor SA. Chorea caused by lithium intoxication. A case report and literature review. *Mov Disord* 1996; **11**:733–7.

Poyurovsky M, Pashinian A, Weizman R *et al.* Low-dose mirtazapine: a new option in the treatment of antipsychotic-induced akathisia. A randomized, double-blind, placebo- and propranolol-controlled trial. *Biol Psychiatry* 2006; **59**:1071–7.

Priori A, Bertolasi L, Berardelli A *et al.* Acute dystonic reaction to Ecstasy. *Mov Disord* 1995; **10**:353.

Prodan CI, Monnot M, Ross ED *et al.* Reversible dementia with parkinsonian features associated with budesonide use. *Neurology* 2006; **67**:723.

Provini F, Plazzi G, Tinuper P *et al.* Nocturnal frontal lobe epilepsy. A clinical and polygraphic overview of 100 consecutive cases. *Brain* 1999; **122**:1017–31.

Quartarone A, Girlanda P, Risitano G *et al.* Focal hand dystonia in a patient with thoracic outlet syndrome. *J Neurol Neurosurg Psychiatry* 1998; **65**:272–4.

Quartarone A, Girlanda P, Vita G *et al.* Oromandibular dystonia in a patient with bilateral putaminal necrosis after methanol poisoning: an electrophysiological study. *Eur Neurol* 2000; **44**:127–8.

Quinn NP, Luthert P, Honavar M *et al.* Pure akinesia due to Lewy body Parkinson's disease: a case with pathology. *Mov Disord* 1989; **4**:85–9.

Raskin DE, Frank SW. Herpes encephalitis with catatonic stupor. *Arch Gen Psychiatry* 1974; **31**:544–6.

Rasmussen P. Persistent mirror movements: a clinical study of 17 children, adolescents and young adults. *Dev Med Child Neurol* 1993; **35**:699–707.

Read SL. Catatonia in thrombotic thrombocytopenic purpura. *J Clin Psychiatry* 1983; **44**:343–4.

Reeves AL, So EL, Sharbrough FW *et al.* Movement disorders associated with the use of gabapentin. *Epilepsia* 1996; **37**:988–90.

Reisberg B. Catatonia associated with disulfiram therapy. *J Nerv Ment Dis* 1978; **166**:607–9.

Remick RA, Froese C, Keller FD. Common side effects associated with monoamine oxidase inhibitors. *Prog Neuropsychopharmacol Biol Psychiatry* 1989; **13**:497–504.

Remy P, de Recondo A, Defer G *et al.* Peduncular 'rubral' tremor and dopaminergic innervation: a PET study. *Neurology* 1995; **45**:472–7.

Richards RN, Barnett HJ. Paroxysmal dystonic choreoathetosis. A family study and review of the literature. *Neurology* 1968; **18**: 461–9.

Rifkin AR, Quitkin F, Klein DF. Akinesia: a poorly recognized drug-induced extrapyramidal behavioral disorder. *Arch Gen Psychiatry* 1975; **32**:672–4.

Riley D, Lang AE. Hemiballism in multiple sclerosis. *Mov Disord* 1988; **3**:88–94.

Ringman JM, Jankovic J. Occurrence of tics in Asperger's syndrome and autistic disorder. *J Child Neurol* 2000; **15**:394–400.

Rio J, Montalban J, Pujadas F *et al.* Asterixis associated with anatomic cerebral lesions: a study of 45 cases. *Acta Neurol Scand* 1995; **91**:377–81.

Ristic AJ, Svetel M, Dragasevic N *et al.* Bilateral chorea-ballism associated with hyperthyroidism. *Mov Disord* 2004; **19**:982–3.

Ristic AJ, Vojvodic N, Jankovic S *et al.* The frequency of reversible parkinsonism and cognitive decline associated with valproate treatment: a study of 364 patients with different types of epilepsy. *Epilepsia* 2006; **47**:2183–5.

Ritchie EC, Bridenbaugh RH, Jabbari B *et al.* Acute generalized myoclonus following buspirone administration. *J Clin Psychiatry* 1988; **49**:242–3.

Rittsmannberger H. Asterixis induced by psychotropic drug treatment. *Clin Neuropharmacol* 1996; **19**:349–55.

Rittsmannberger H, Lebihuber F. Asterixis induced by carbamazepine therapy. *Biol Psychiatry* 1992; **32**:364–8.

Ritz K, Gerrits MCF, Foncke EMJ *et al.* Myoclonus-dystonia: clinical and genetic evaluation of a large cohort. *J Neurol Neurosurg Psychiatry* 2009; **80**:63–8.

Robertson PL, Garofalo EA, Silverstein FS *et al.* Carbamazepine-induced tics. *Epilepsia* 1993; **34**:965–8.

Rodrigues JP, Edwards DJ, Walters SE *et al.* Blinded placebo crossover study of gabapentin in primary orthostatic tremor. *Mov Disord* 2006; **21**:900–5.

Romisher S, Feller R, Dougherty J *et al.* Tagamet-induced acute dystonia. *Ann Emerg Med* 1987; **16**:1162–4.

Rosebush PI, Mazurek MF. Catatonia after benzodiazepine withdrawal. *J Clin Psychopharmacol* 1996; **16**:315–19.

Rosebush PI, Hildebrand AM, Furlong BG *et al.* Catatonic syndrome in a general hospital psychiatric population: frequency, clinical presentation, and response to lorazepam. *J Clin Psychiatry* 1990; **51**:347–53.

Rosebush PI, MacQueen GM, Clarke JTR *et al.* Late-onset Tay–Sachs disease presenting as catatonic schizophrenia: diagnostic and treatment issues. *J Clin Psychiatry* 1995; **56**:347–53.

Rosen JA. Paroxysmal choreoathetosis. *Arch Neurol* 1964; **11**:385–7.

Rosenberg NL, Myers JA, Martin WRW. Cyanide-induced parkinsonism: clinical, MRI and 6–flurodopa PET studies. *Neurology* 1989; **39**:142–4.

Rosenbloom L. Dyskinetic cerebral palsy and birth asphyxia. *Dev Med Child Neurol* 1994; **36**:285–9.

Rosenow F, Herholz K, Lanfermann H *et al.* Neurological sequelae of cyanide intoxication – the patterns of clinical, magnetic resonance imaging, and positron emission tomography findings. *Ann Neurol* 1995; **38**:825–8.

Rothschild AJ, Locke CA. Reexposure to fluoxetine after serious suicide attempts by three patients: the role of akathisia. *J Clin Psychiatry* 1991; **52**:491–3.

Roze E, Apartis E, Clot F *et al.* Myoclonus-dystonia. Clinical and electrophysiologic pattern related to SGCE mutations. *Neurology* 2008; **70**:1010–16.

Roze E, Bounolleau P, Ducreux D *et al.* Propriospinal myoclonus revisited: clinical, neurophysiologic, and neuroradiologic findings. *Neurology* 2009; **72**:1301–9.

Ruprecht K, Warmuthmetz M, Waespe W *et al.* Symptomatic hyperekplexia in a patient with multiple sclerosis. *Neurology* 2002; **58**:503–4.

Russo LS. Focal dystonia and lacunar infarction of the basal ganglia: a case report. *Arch Neurol* 1983; **40**:61–2.

Sachdev P, Kruk J. Clinical characteristics and predisposing factors in acute drug-induced akathisia. *Arch Gen Psychiatry* 1994; **51**: 963–74.

Saint-Hilaire M-H, Saint-Hilaire J-M, Granger L. Jumping Frenchmen of Maine. *Neurology* 1986; **36**:1269–71.

Saint-Hilaire M-H, Burke RE, Bressman SB *et al.* Delayed onset dystonia due to perinatal or early childhood asphyxia. *Neurology* 1991; **41**:216–22.

Salam SA, Pillai AK, Beresford TP. Lorazepam for psychogenic catatonia. *Am J Psychiatry* 1987; **144**:1082–3.

Sallee FR, Stiller RL, Perel JM *et al.* Pemoline-induced abnormal involuntary movements. *J Clin Psychopharmacol* 1989; **9**:125–9.

Salvi F, Mascalchi M, Bortolotti C *et al.* Hypertension, hyperekplexia, and pyramidal paresis due to vascular compression of the medulla. *Neurology* 2000; **55**:1381–4.

Samie MR, Selhorst JB, Koller WC. Post-traumatic midbrain tremor. *Neurology* 1990; **40**:62–6.

Sandyk R. Parkinsonism secondary to neurosyphilis. A case report. *S Afr Med J* 1983; **63**:665–6.

Saunders-Pullman R, Shriberg J, Heiman G *et al.* Myoclonus dystonia. Possible association with obsessive-compulsive disorder and alcohol dependence. *Neurology* 2002; **58**:242–5.

Scarano V, Pellechia MT, Filia A *et al.* Hallervorden–Spatz syndrome resembling a typical Tourette syndrome. *Mov Disord* 2002; **17**:618–20.

Schott GD. The relationship of peripheral trauma and pain to dystonia. *J Neurol Neurosurg Psychiatry* 1985; **48**:698–701.

Schott GD, Wyke MA. Congenital mirror movements. *J Neurol Neurosurg Psychiatry* 1981; **44**:586–99.

Schule B, Kock N, Svetl M *et al.* Genetic heterogeneity in ten families with myoclonus-dystonia. *J Neurol Neurosurg Psychiatry* 2004; **75**:1181–5.

Schwartz A, Hennerici M, Wegener OH. Delayed choreoathetosis following acute carbon monoxide poisoning. *Neurology* 1985; **35**:98–9.

Schwarz GA, Barrows LJ. Hemiballism without involvement of Luy's body. *Arch Neurol* 1960; **2**:420–34.

Sechi GP, Tanda F, Mutani R. Fatal hyperpyrexia after withdrawal of levodopa. *Neurology* 1984; **34**:249–51.

Sechi GP, Agnetti V, Piredda M *et al.* Acute and persistent parkinsonism after use of diquat. *Neurology* 1992; **42**:261–2.

Sempere AP, Duarte J, Cabezas C *et al.* Parkinsonism induced by amlopidine. *Mov Disord* 1995; **10**:115–16.

Sethi KD, Patel B, Meador KJ. Metoclopramide-induced parkinsonism. *South Med J* 1989; **82**:1581–2.

Shah P, Kaplan SL. Catatonic symptoms in a child with epilepsy. *Am J Psychiatry* 1980; **137**:738–9.

Shahnawaz M, van der Westhuizen LR, Gledhill RF. Episodic cervical dystonia associated with gastro-oesophageal reflux. A case of adult-onset Sandifer syndrome. *Clin Neurol Neurosurg* 2001; **103**:212–15.

Shan D-E, Soong B-W, Sun C-M *et al.* Spinocerebellar ataxia type 2 presenting as familial levodopa-responsive parkinsonism. *Ann Neurol* 2001; **50**:812–15.

Sharp FR, Rando TA, Greenberg SA *et al.* Pseudochoreoathetosis: movements associated with loss of proprioception. *Arch Neurol* 1994; **51**:1103–9.

Shearer PM, Bownes IT, Curran P. Tardive akathisia and agitated depression during metoclopramide therapy. *Acta Psychiatr Scand* 1984; **70**:428–31.

Shepherd GMG, Tauboll E, Bakke SJ *et al.* Midbrain tremor and hypertrophic olivary degeneration after pontine hemorrhage. *Mov Disord* 1997; **12**:432–7.

Shill HA, Stacy MA. Malignant catatonia secondary to sporadic encephalitis lethargica. *J Neurol Neurosurg Psychiatry* 2000; **69**:402–3.

Singer C, Papapetropoulos S, Spielholz NI. Primary writing tremor: report of a case successfully treated with botulinum toxin A injections and discussion of the underlying mechanism. *Mov Disord* 2005; **20**:1387–8.

Siris SG. Three cases of akathisia and 'acting out'. *J Clin Psychiatry* 1985; **46**:395–7.

Slater E, Beard SW. The schizophrenia-like psychoses of epilepsy. I. Psychiatric aspects. *Br J Psychiatry* 1963; **109**:95–112.

Snyder RD. The involuntary movements of chronic mercury poisoning. *Arch Neurol* 1972; **26**:379–81.

Soland V, Bhatia KP, Volonte MA *et al.* Focal task-specific tremors. *Mov Disord* 1996a; **11**:665–70.

Soland V, Evoy F, Rivest J. Cervical dystonia due to a frontal meningioma. *Mov Disord* 1996b; **11**:336–7.

Song X, Hu Z, Zhang H. Acute dystonia induced by lamivudine. *Clin Neuropharmacol* 2005; **28**:193–4.

Sotero de Menezes MA, Rho JM, Murphy P *et al.* Lamotrigine-induced tic disorder: report of five pediatric cases. *Epilepsia* 2000; **41**: 862–7.

Spacey SD, Adams PJ, Lam PCP *et al.* Genetic heterogeneity in paroxysmal nonkinesiogenic dyskinesia. *Neurology* 2006; **66**:1585–90.

Speracio RR, Anziska B, Schutta HS. Hypernatremia and chorea. A report of two cases. *Neurology* 1976; **26**:46–50.

Spiller WG. Acquired double athetosis (dystonia lenticularis). *Arch Neurol Psychiatry* 1920; **4**:370–86.

Srinivas K, Rao VM, Subbulakshmi N *et al.* Hemiballism after striatal hemorrhage. *Neurology* 1987; **37**:1428–9.

Srour M, Philibert M, Dion MH *et al.* Familial congenital mirror movements: report of a large 4-generation family. *Neurology* 2009; **73**:729–31.

Starkstein SE, Petracca G, Teson A *et al.* Catatonia in depression: prevalence, clinical correlates, and validation of a scale. *J Neurol Neurosurg Psychiatry* 1996; **60**:326–32.

Starosta-Rubinstein S, Young AB, Kluin K *et al.* Clinical assessment of 31 patients with Wilson's disease: correlations with structural changes on magnetic resonance imaging. *Arch Neurol* 1987; **44**:365–70.

Stevanin G, Fujigasaki H, Lebre A-S *et al.* Huntington's disease-like phenotype due to trinucleotide repeat expansion in the TBP and JPH3 genes. *Brain* 2003; **126**:1599–603.

Stevens H. 'Jumping Frenchmen of Maine'. Myriachit. *Arch Neurol* 1965; **12**:311–14.

Stevens H. Paroxysmal choreo-athetosis. *Arch Neurol* 1966; **14**: 415–20.

Stewart JT, Williams LS. A case of lithium-induced asterixis. *J Am Ger Soc* 2000; **48**:457.

Stober G, Seelow D, Ruschendork *et al.* Periodic catatonia: confirmation of linkage to chromosome 15 and further evidence for genetic homogeneity. *Hum Genet* 2002; **111**:523–30.

Stojanovic M, Sternic N, Kostic VS. Cloazapine in hemiballismus: report of two cases. *Clin Neuropharmacol* 1997; **20**:171–4.

Strang RR. Parkinsonism occurring during methyldopa therapy. *Can Med Assoc J* 1966; **95**:928–9.

Striano P, Elefante A, Coppola A *et al.* Dramatic response to levetiracetam in post-ischemic Holmes' tremor. *J Neurol Neurosurg Psychiatry* 2007; **78**:438–9.

Stryjer R, Strous RD, Bar F *et al.* Segmental dystonia as the sole manifestation of carbamazepine toxicity. *Gen Hosp Psychiatry* 2002; **42**:114–15.

Stryjer R, Rosenzcwaig S, Bar F *et al.* Trazodone for the treatment of neuroleptic-induced acute akathisia: a placebo-controlled, double-blind, crossover study. *Clin Neuropharmacol* 2010; **33**:219–22.

Stuppaeck CH, Miller CH, Ehrmann H *et al.* Akathisia induced by necrosis of the basal ganglia after carbon monoxide intoxication. *Mov Disord* 1995; **10**:229–31.

Sunden-Cullberg J, Tedroff J, Aquilonius SM. Reversible chorea in primary antiphospholipid syndrome. *Mov Disord* 1998; **13**:147–9.

Sutton GG, Meyere RF. Focal reflex myoclonus. *J Neurol Neurosurg Psychiatry* 1974; **37**:207–17.

Swedo SE, Leonard HL, Schapiro NB *et al.* Sydenham's chorea: physical and psychological symptoms of St Vitus dance. *Pediatrics* 1993; **91**:706–13.

Swett C. Drug-induced dystonia. *Am J Psychiatry* 1975; **132**:532–4.

Tabaton M, Mancardi G, Loeb C. Generalized chorea due to bilateral small, deep cerebral infarcts. *Neurology* 1985; **35**:588–9.

Tan EK, Lo YL, Chan LL *et al.* Cervical disc prolapse with cord compression presenting with choreoathetosis and dystonia. *Neurology* 2002; **58**:661–2.

Tan H, Turali G, Ay H *et al.* Rubral tremor after thalamic infarction in childhood. *Pediatr Neurol* 2001; **25**:409–12.

Tan PC, Kline PP, Vallikkannu N *et al.* Promethazine compared with metoclopramide for hyperemesis gravidarum: a randomized controlled trial. *Obstet Gynecol* 2010; **115**:975–81.

Tanaka M, Yokode M, Kita T *et al.* Donepezil and athetosis in an elderly patient with Alzheimer's disease. *J Am Geriatr Soc* 2003; **51**:889–90.

Tandon R, Walden M, Falcon S. Catatonia as a manifestation of paraneoplastic encephalopathy. *J Clin Psychiatry* 1988; **49**:121–2.

Taylor MA, Abrams R. Catatonia: prevalence and importance in the manic phase of manic-depressive illness. *Arch Gen Psychiatry* 1977; **34**:1223–5.

Tetrud JW, Langston JW, Garbe PL *et al.* Mild parkinsonism in persons exposed to 1-methyl-4-phenyl-1,2,3,6–tetrahydropyridine (MPTP). *Neurology* 1989; **39**:1483–7.

Teusink JP, Alexopoulos GS, Shamoian CA. Parkinsonism side-effects induced by a monoamine oxidase inhibitor. *Am J Psychiatry* 1984; **141**:118–19.

Thornton A, McKenna PJ. Acute dystonic reactions complicated by psychotic phenomena. *Br J Psychiatry* 1994; **164**:115–18.

Tinuper P, Cerullo A, Cirignotta F *et al.* Nocturnal paroxysmal dystonia with short-lasting attacks: three cases with evidence for an epileptic frontal lobe origin of seizures. *Epilepsia* 1990; **31**:549–56.

Toru M, Matsuda O, Makiguchi K *et al.* Single case study: neuroleptic malignant syndrome-like state following a withdrawal of anti-parkinsonian drugs. *J Nerv Ment Dis* 1981; **169**:324–7.

Tsuboi Y, Wxzolek ZK, Kusuhara T *et al.* Japanese family with parkinsonism, depression, weight loss, and cortical hypoventilation. *Neurology* 2002; **58**:1025–30.

Tsubokawa T, Katayama Y, Yamamoto T. Control of persistent hemiballismus by chronic thalamic stimulation. Report of two cases. *J Neurosurg* 1995; **82**:501–5.

Uitti RJ, Rajput AH, Ashenhurst EM *et al.* Cyanide-induced parkinsonism: a clinicopathologic report. *Neurology* 1985; **35**:921–5.

Uitti RJ, Snow BJ, Shinotoh H *et al.* Parkinsonism induced by solvent abuse. *Ann Neurol* 1994; **35**:616–19.

Vakis A, Krasoudakis A, Koutendakis D. Transient post-traumatic hemiballism. Report of a case. *J Neurosurg Sci* 2006; **50**:21–3.

Valente EM, Spacey SD, Wali GM *et al.* A second paroxysmal kinesigenic choreoathetosis locus (EKD2) mapping on 16q13–q22 indicates a family of genes which give rise to paroxysmal disorders on human chromosome 16. *Brain* 2000; **123**:2040–5.

Valenzuela R, Court J, Godoy J. Delayed cyanide induced dystonia. *J Neurol Neurosurg Psychiatry* 1992; **55**:198–9.

van Hilten JJ, van de Beek WJ, Vein AA *et al.* Clinical aspects of multifocal or generalized tonic dystonia in reflex sympathetic dystrophy. *Neurology* 2001; **56**:1762–5.

Van Putten T. The many faces of akathisia. *Compr Psychiatry* 1975; **16**:43–7.

Van Putten T, May PRA. 'Akinetic depression' in schizophrenia. *Arch Gen Psychiatry* 1978; **35**:1101–7.

Van Putten T, Mutalipassi LR, Malkin MD. Phenothiazine-induced decompensation. *Arch Gen Psychiatry* 1974; **30**:102–5.

Van Putten T, May PRA, Marder SR. Akathisia with haloperidol and thiothixene. *Arch Gen Psychiatry* 1984; **41**:1036–9.

Van Uitert RL, Russakoff LM. Hyperthyroid chorea mimicking psychiatric disease. *Am J Psychiatry* 1979; **136**:1208–10.

Van Zandijcke M. Treatment of myoclonus. *Acta Neurol Belg* 2003; **103**:66–70.

Velez M, Cosentino C, Torres L. Levodopa-responsive rubral (Holmes') tremor. *Mov Disord* 2002; **17**:741–2.

Vernino S, Tuite P, Adler CH *et al.* Paraneoplastic chorea associated with CRMP-5 neuronal antibody and lung carcinoma. *Ann Neurol* 2002; **51**:625–30.

Vidakovic A, Dragasevic N, Kostic VS. Hemiballism: report of 25 cases. *J Neurol Neurosurg Psychiatry* 1994; **57**:945–9.

Volonte MA, Perani D, Lanzi R *et al.* Regression of ventral striatum hypometabolism after calcium calcitrol therapy in paroxysmal kinesigenic choreoathetosis due to idiopathic primary hypoparathyroidism. *J Neurol Neurosurg Psychiatry* 2001; **71**:691–5.

Walker L. Setrarline-induced akathisia and dystonia misinterpreted as a panic attack. *Psychiatr Serv* 2002; **53**:1477–8.

Walker RH, Spiera H, Brin MF *et al.* Parkinsonism associated with Sjogren's syndrome: three cases and a review of the literature. *Mov Disord* 1999; **14**:262–8.

Walters AS, Hening W, Rubinstein M *et al.* A clinical and polysomnographic comparison of neuroleptic-induced akathisia and the idiopathic restless legs syndrome. *Sleep* 1991; **14**:339–45.

Waubant E, Simonetta-Moreau M, Clanet M *et al.* Left arm monoballism as a relapse in multiple sclerosis. *Mov Disord* 1997; **12**:1091–2.

Waubant E, Alize P, Tourbah A *et al.* Paroxysmal dystonia (tonic spasm) in multiple sclerosis. *Neurology* 2001; **57**:2320–1.

Weddington WW, Marks RC, Verghese JP. Disulfiram encephalopathy as a cause of the catatonic syndrome. *Am J Psychiatry* 1980; **137**:1217–19.

Weiden P, Bruin R. Worsening of Tourette's disorder due to neuroleptic-induced akathisia. *Am J Psychiatry* 1987; **144**: 504–5.

Wein T, Andermann F, Silver K *et al.* Exquisite sensitivity of paroxysmal kinesigenic choreoathetosis to carbamazepine. *Neurology* 1996; **47**:1104–5.

Weinberger DR, Wyatt RJ. Catatonic stupor and neuroleptic drugs. *JAMA* 1978; **239**:1846.

Weiner WJ, Nausieda PA, Klawans HL. Methylphenidate-induced chorea: case report and pharmacologic implications. *Neurology* 1978; **28**:1041–4.

Werner EG, Olanow CW. Parkinsonism and amiodarone therapy. *Ann Neurol* 1989; **25**:630–2.

Whittier JR. Ballism and subthalmic nucleus (nucleus hypothalamicus; corpus Luysi). Review of the literature and study of thirty cases. *Arch Neurol Psychiatry* 1947; **58**:672–92.

Wilson LG. Viral encephalopathy mimicking functional psychosis. *Am J Psychiatry* 1976; **133**:165–70.

Wilson SAK. *Modern Problems in Neurology.* London: Edward Arnold, 1928.

Wilson SAK. *Neurology*, 2nd edn. London: Butterworth, 1955.

Wojick JD, Falk WE, Fink JS *et al.* A review of 32 cases of tardive dystonia. *Am J Psychiatry* 1991; **148**:1055–9.

Wolanczyk T, Grabowska-Grzyb A. Transient dystonias in three patients treated with tiagabine. *Epilepsia* 2001; **42**:944–6.

Woods C, Taylor A. Ataxia telangiectasia in the British Isles: the clinical and laboratory features of 70 affected individuals. *Q J Med* 1992; **82**:169–79.

Yadalam KG, Korn ML, Simpson GM. Tardive dystonia: four case histories. *J Clin Psychiatry* 1990; **51**:17–20.

Yamada K, Haraa M, Goto S. Response of postapoplectic hemichorea/hemiballism to GPI pallidotomy: progressive improvement resulting in complete relief. *Mov Disord* 2004; **19**:1111–14.

Yamamoto M, Fujii S, Hatanaka Y. Result of long-term administration of L-threo-3,4-dihydrophenylserine in patients with pure akinesia as an early symptom of progressive supranuclear palsy. *Clin Neuropharmacol* 1997; **20**:371–3.

Yassa R, Nair V, Dimitry A. Prevalence of tardive dystonia. *Acta Psychiatr Scand* 1986; **73**:629–33.

Yazici O, Kantemir E, Tastaban Y *et al.* Spontaneous improvement of tardive dystonia during mania. *Br J Psychiatry* 1991; **158**:847–50.

Zaccara G, Paganini M, Campostrini R *et al.* Hyperammonemia and valproate-induced alterations of the state of consciousness. A report of 8 cases. *Eur Neurol* 1984; **23**:104–12.

Zaremba J, Mierzewska H, Lysiak Z *et al.* Rapid-onset dystonia-parkinsonism: a fourth family consistent with linkage to chromosome 19q13. *Mov Disord* 2004; **19**:1506–10.

Zenzola A, De Mari M, De Blasi R *et al.* Paroxysmal dystonia with thalamic lesion in multiple sclerosis. *Neurol Sci* 2001; **22**:391–4.

Zesiewicz TA, Sullivan KL, Hauser RA. Chorea induced by lamotrigine. *J Child Neurol* 2006; **21**:357.

Zhang C, Glenn DG, Bell WL *et al.* Gabapentin-induced myoclonus in end-stage renal disease. *Epilepsia* 2005; **46**:156–8.

Zorumski CF, Bakris GL. Choreoathetosis associated with lithium: case report and literature review. *Am J Psychiatry* 1983; **140**:1621–2.

5

Other signs and symptoms

5.1 MUTISM

The term mutism is derived from the Latin *mutus*, meaning 'inability to speak'.

Clinical features

Mute patients, as noted, do not speak; although in some cases patients may make some noises, such as grunts, there is no verbalization.

Etiology

As noted in Table 5.1, mutism may appear in several different clinical settings. Perhaps the most common one in neuro psychiatric practice is the acute onset of mutism during a stroke. Mutism may also appear gradually in far advanced neurodegenerative disorders, such as frontotemporal lobar degeneration, and is a prominent feature of several other syndromes, including akinetic mutism, catatonia, and severe cases of psychomotorically retarded depression. Finally, there is a miscellaneous group of causes including intrinsic laryngeal pathology, medications, various cerebellar lesions, and corpus callosotomy.

Stroke, when characterized by aphasia or aphemia, may cause acute mutism. Motor aphasia and transcortical motor aphasia may both, initially, present with mutism; over time, the mutism resolves, leaving the patient with the typical aphasic syndrome, as discussed in Chapter 3.1. Thus, after resolution of the mutism, the patient with motor aphasia will be left with non-fluent speech but intact comprehension, and the patient with transcortical motor aphasia will be similar but will show improvement of speech with repetition.

Aphemia is an uncommon stroke syndrome characterized by mutism but, in contrast with the aphasias, there is a preserved ability to write; furthermore, as the mutism resolves, patients are left not with an aphasia, but with a hoarse dysarthria. Aphemia, in most cases, occurs secondary to a small lesion in the posterior aspect of the left inferior frontal gyrus or the immediately subjacent white matter; rarely, however, aphemia may occur with a right-sided infarction, as for example in the supplementary motor area (Mendez 2004).

Mutism may also occur in other stroke syndromes, as for example with bilateral infarction of the lenticular nucleus and bilateral infarction of the posterior limb of the internal capsule; there is also a case report of mutism occurring with infarction of the periaqueductal gray.

Neurodegenerative disorders capable of causing dementia, such as frontotemporal lobar degeneration and Alzhei-

Table 5.1 Causes of mutism

Stroke	
As presentation of certain aphasias	
Motor aphasia (David and Bone 1984; Masdeu and O'Hara 1983)	
Transcortical motor aphasia (Alexander and Schmitt 1980; Bogousslavsky and Regli 1990)	
Aphemia (Mendez 2004; Schiff *et al.* 1983)	
Bilateral infarction of the lenticular nucleus (Evyapan Akkus 2006)	
Bilateral infarction of the posterior limb of the internal capsule (Helgason *et al.* 1988)	
Infarction of the periaqueductal gray (Esposito *et al.* 1999)	
Neurodegenerative disorders	
Frontotemporal lobar degeneration (Neary *et al.* 1993; Snowden *et al.* 1992)	
Alzheimer's disease (Mayeux *et al.* 1985)	
Primary progressive aphasia (Gorno-Tempini *et al.* 2006)	
Other syndromes characterized by mutism	
Catatonia (Kraepelin 1990)	
Depression (Benegal *et al.* 1992)	
Akinetic mutism (Cairns *et al.* 1941)	
Traumatic brain injury (Levin *et al.* 1983)	
Miscellaneous causes of mutism	
Intrinsic laryngeal pathology	
Medications	
Tacrolimus (Wijdicks *et al.* 1994)	
Cyclosporine (Valldeoriola *et al.* 1996)	
Phenytoin (Berigan and Watt 1994)	
Cerebellar lesions	
Postsurgical (Catsman-Berrevoets *et al.* 1999; Crutchfield *et al.* 1994; Van Dongen *et al.* 1994)	
Intracerebellar hemorrhage (Coplin *et al.* 1997)	
Tumor (Akil *et al.* 2006)	
Corpus callosotomy (Quattrini *et al.* 1997)	

mer's disease, may also cause a mutism of gradual onset; in this situation, however, the mutism generally occurs only at the 'tail end' of a long dementing process. Some neurodegenerative disorders may also present with aphasia, in the syndrome known as 'primary progressive aphasia', which, over time, may evolve into mutism; this is discussed further in Chapter 3.1.

Other syndromes characterized by mutism include catatonia, severe depression, and akinetic mutism, and in all three of these the mutism is accompanied by immobility. Several features allow these syndromes to be differentiated one from another. Catatonia is typically accompanied by the very distinctive finding of waxy flexibility, a sign absent in the other two syndromes. Depression of the psychomotorically retarded type, when severe, may be accompanied by mutism; however, in these cases one finds a history of a preceding, more typical depressive syndrome, with depressed mood, anergia, anhedonia, and disturbances of appetite and sleep. When a patient with mutism and immobility lacks any of these findings, then the diagnosis of akinetic mutism should be seriously considered. Finally, the syndrome of traumatic brain injury may also be characterized by mutism.

Of the miscellaneous causes of mutism, intrinsic laryngeal pathology (e.g. laryngitis, laryngeal tumors) is the most common, and in these cases the mutism is preceded by hoarseness. Medications capable of inducing mutism include tacrolimus, cyclosporine, and phenytoin. Various cerebellar lesions may also cause mutism. Most commonly, this is seen after surgical resection of a tumor, and although this is most frequently reported in children, adults may also be affected (Crutchfield *et al.* 1994). Other cerebellar lesions in adults capable of causing mutism include intracerebellar hemorrhage and tumor. Finally, transient mutism may follow corpus callosotomy.

Differential diagnosis

Aphasia, dysarthria, and hypophonia are all distinguished by the obvious fact that patients with these disorders do, in fact, speak. Although the speech may be effortful, as in aphasia, slurred, as in dysarthria, or low and monotone, as in hypophonia, it is still there.

Selective, or, as it was once called, *elective*, mutism is a condition seen in shy and withdrawn children who, although capable of speech with family members, become 'selectively' mute when with strangers or in school (Dummit *et al.* 1997; Elson *et al.* 1965; Steinhausen and Juzi 1996; Wergeland 1979).

Conversion muteness (Thorpe *et al.* 1985) and malingered muteness typically occur as isolated findings, a feature that distinguishes them from mutism occurring secondary to the causes noted under Etiology, wherein the mutism is accompanied by distinctive findings on history or examination.

Treatment

Treatment is directed at the underlying disorder; speech therapy may be attempted in cases of mutism secondary to stroke.

5.2 AKINETIC MUTISM

The name of this syndrome captures its most distinctive features, namely immobility and mutism.

Clinical features

For the most part, patients are mute and immobile, and although at first glance they appear to be in a stupor, closer observation reveals lively conjugate eye movements as patients track people in the room, such as physicians and nurses. In some cases, the mutism and akinesia may persist through all adversities, but in others, patients may respond to an especially aversive stimulus with a feeble motion of a limb or perhaps by uttering a word. Food may be eaten if placed in the mouth, but patients do not seek it, and urine and feces may be passed in the bed. Importantly, there is no rigidity, especially no waxy flexibility, and no negativism.

Some examples may serve to flesh out this definition. The first is taken from Cairns' original description of akinetic mutism (Cairns *et al.* 1941):

> In the fully developed state he makes no sound and lies inert, except that his eyes regard the observer steadily or follow the movement of objects, and they may be diverted by sound. Despite his steady gaze, which seems to give promise of speech, the patient is quite mute, or he answers only in whispered monosyllables. Oft-repeated commands may be carried out in a feeble, slow, and incomplete manner, but usually there are no movements of a voluntary character: no restless movements, struggling, or evidence of negativism. Emotional movement is almost in abeyance. A painful stimulus produces reflex withdrawal of the limb and, if the stimulus is maintained, slow, feeble, voluntary movements of the limbs may occur in an attempt to remove the source of the stimulation, but usually without tears, noise, or other manifestations of pain or displeasure. The patient swallows readily but has to be fed. Food seen may be recognized as such, but there is evidently little appreciation of its taste and other characteristics: objects normally chewed or sucked may be swallowed whole. There is total incontinence of urine and feces.

Cairns' cases, occurring with third ventricular tumors, were generally of gradual or subacute onset. Another case (Nielsen 1951), secondary to infarction of the cingulate gyri, bears detailed reporting given the remarkable abruptness of its onset. The patient was a 46-year-old woman who:

> was ironing when she suddenly stopped on the spot, complained of a severe headache, but remained standing. Her son put her to bed, where she lay motionless, without even speaking. She stared at the ceiling and did not ask for anything, not even for a drink. After 9 days she was hospitalized and on the ward she continued akinetic and mute. Under strong stimulation she did say, 'It hurts,' and 'Water,' but that was all. She was obviously conscious and took note of her environment but lay day after day motionless, not deigning to call for bed pan or food.

Etiology

Akinetic mutism most commonly occurs as a stroke syndrome, and here one typically finds infarction of the anterior cingulate gyrus, which, although generally bilateral (Barris and Schuman 1953; Faris 1969; Freemon 1971; Nielsen 1951), may at times be merely unilateral (Nagaratnam *et al.* 2004); bilateral infarction of the globus pallidus (Ure *et al.* 1998) has also been reported as a cause. Obstructive hydrocephalus (Messert *et al.* 1966) may also cause the syndrome, as may third ventricular masses (Cairns *et al.* 1941), which, although not causing hydrocephalus, compress both thalami; surgical damage to the hypothalamus (Ross and Stewart 1981) may also be at fault. Other causes include heroin vapor-induced leukoencephalopathy (Chapter 28.9) and the terminal stages of Creutzfeldt–Jakob disease (Chapter 16.1), new-variant Creutzfeldt–Jakob disease (Chapter 16.2), and acquired immune deficiency syndrome (AIDS) dementia (Chapter 15.1).

Differential diagnosis

Akinetic mutism must be distinguished from other conditions capable of causing immobility and mutism, including catatonia, depression, the persistent vegetative state, the locked-in syndrome, and stupor.

Catatonia is distinguished by the presence of waxy flexibility, a sign not seen in akinetic mutism. Depression, when severe and characterized by psychomotor retardation, may resemble akinetic mutism; however, in this instance history will reveal a typical depressive syndrome preceding the evolution of immobility and mutism, with such symptoms as depressed mood, anergia, anhedonia, and changes in sleep and appetite.

The persistent vegetative state may also appear similar to akinetic mutism; however, in this condition the eye movements – although at times seeming purposeful – are less consistently 'lively'; furthermore, in patients with the persistent vegetative state, one always finds a history of coma secondary to severe traumatic brain injury, global hypoxia, etc.

The locked-in syndrome, occurring secondary to high brainstem infarction or other lesions, such as central pontine myelinolysis, is characterized by tetraparesis, bilateral facial paresis, and mutism, and thus may appear similar to akinetic mutism. Examination of eye movements, however, will enable the correct diagnosis. In contrast with akinetic mutism, wherein extraocular movements are full, in the locked-in syndrome there is a paralysis of lateral gaze, with only vertical gaze being left intact. Furthermore, and again in contrast with akinetic mutism, in the locked-in syndrome the patient is often desperate for communication and will attempt to communicate by utilizing the remaining vertical eye movements in a sort of 'Morse code'.

Stupor of various causes may leave patients immobile and mute; however, in stupor, patients are not alert and eye movements, rather than being 'lively', are generally roving or dysconjugate.

Abulia may also be considered on the differential; however, the distinction is readily apparent upon merely instructing the patient to do something. When left undisturbed, the abulic patient is mute and immobile like the akinetic mute; however, when the abulic patient is given directions, he or she follows them, which is in contrast with the akinetic mute, who remains immobile.

Treatment

Anecdotally, bromocriptine is effective (Psarros *et al.* 2003; Ross and Stewart 1981); treatment may be initiated at 2.5 mg/day (given in two divided doses: once in the early morning and once in the early afternoon) and increased in similar increments every few days until significant improvement or limiting side-effects occur or a maximum dose of 40 mg is reached. Atomoxetine, in doses of up to 70 mg/day, was also reported as effective in one case (Kim *et al.* 2010).

5.3 STUTTERING

Stuttering, a speech dysfluency familiar to most ears, although most commonly seen on a developmental basis, may also occur on an acquired basis, as noted below.

Clinical features

The phenomenology of stuttering differs according to whether it occurs on a developmental or an acquired basis (Helm *et al.* 1978).

In developmental stuttering, patients find themselves 'blocked' as they attempt to speak the first letter or syllable of a word and have particular difficulty getting past the letters B, D, K, P, and T. Unable to surmount the sound, patients repeat it again and again, often with progressively increasing force and volume; with these ever more vigorous attempts, there may be associated features, such as facial grimacing, blinking, hissing, and fist clenching. The block is eventually overcome and there is often a cascade of words, all clearly and unhesitatingly pronounced, as if the dam had broken. Interestingly, patients with developmental stuttering may find themselves unblocked if they sing, read aloud, or are quite angry.

Although acquired stuttering may be clinically similar to developmental stuttering, it tends to differ in several ways. First, the block may be as likely to occur on the second or subsequent letter as on the first. Second, although patients may be annoyed at being blocked, and do engage in repeated attempts to make the sound, the effort to do so is rarely as vigorous or as forceful as in developmental stuttering. Finally, when a patient with acquired stuttering does get beyond the block, it is unusual to see a subsequent 'cascade' of words.

Etiology

The various causes of stuttering are listed in Table 5.2, where they are divided into developmental and acquired types.

Developmental stuttering, as noted earlier, is the most common cause of stuttering. Here, as discussed in more detail in Chapter 10.19, stuttering usually appears between the ages of 2 and 10 years, and although such developmental stuttering may persist into adulthood, most cases undergo a spontaneous remission by the early teenage years.

Of the various causes of acquired stuttering, medications are the most common: among these, the tricyclic antidepressants and selective serotonin reuptake inhibitors (SSRIs) are the most frequent offenders, followed by the antipsychotics – including both first- and second-generation agents. Alprazolam and theophylline have only rarely been reported to cause stuttering.

Stroke may be accompanied by stuttering and this has been noted with infarction of the frontal cortex, the subcortical white matter of the frontal lobe, the parietal cortex, the caudate nucleus, putamen, thalamus, pons, and cerebellum; there is also a case report of stuttering with infarction of the body of the corpus callosum. Although most cases of stuttering due to cerebral infarction occur with left-sided infarcts, there are rare cases of 'crossed' stuttering wherein stuttering occurred with a right-sided infarction in a right-handed patient (Fleet and Heilman 1985).

Parkinsonian conditions, such as Parkinson's disease and progressive supranuclear palsy, may cause *de novo* stuttering; further, in the case of Parkinson's disease, the onset of parkinsonism may be associated with a relapse of developmental stuttering (Shahed and Jankovic 2001).

Miscellaneous, and very uncommon, causes of stuttering include migraine, simple partial seizures, Landau–Kleffner syndrome, and the now vanishing, rare condition of dialysis dementia.

Table 5.2 Causes of stuttering

Developmental stuttering
Acquired stuttering
Medications (Bar *et al.* 2004; Brady 1998; Christensen *et al.* 1996; Lee *et al.* 2001; Nurnberg and Greenwald 1981)
Tricyclic antidepressants
Selective serotonin reuptake inhibitors (SSRIs)
Antipsychotics
Alprazolam
Theophylline
Stroke
Frontal cortex (Freedman *et al.* 1984) or subcortical white matter (Ludlow *et al.* 1987)
Parietal cortex (Turgut *et al.* 2002)
Caudate nucleus (Caplan *et al.* 1990)
Putamen (Ciabarra *et al.* 2000)
Thalamus (Van Borsel *et al.* 2003)
Pons (Ciabarra *et al.* 2000)
Cerebellum (Tani and Sakai 2010)
Corpus callosum (Hamano *et al.* 2005)
Parkinsonian conditions
Parkinson's disease (Koller 1983)
Progressive supranuclear palsy (Kluin *et al.* 1993; Koller 1983)
Miscellaneous causes
Migraine (Perino *et al.* 2000)
Simple partial seizures (Michel *et al.* 2004)
Landau–Kleffner syndrome (Tutuncuoglu *et al.* 2002)
Dialysis dementia (O'Hare *et al.* 1983)

Differential diagnosis

Palilalia is distinguished by the fact that the last word (or phrase) of a statement is repeated, which is in contrast with stuttering wherein typically it is the first letter or syllable which is repeated. The effortful speech of patients with motor aphasia may suggest stuttering; however, in aphasia one also finds 'telegraphic' speech, with an absence of conjugations, which stands in contrast with stuttering wherein grammar is preserved. Stuttering may also occur as a conversion symptom (Roth *et al.* 1989), and indeed is not uncommon during pseudoseizures (Vossier *et al.* 2004).

Treatment

Treatment of developmental stuttering is discussed in Chapter 10.19; whether any of these modalities would be effective in cases of acquired stuttering is not known.

5.4 PALILALIA

Palilalia, first described by Souques in 1908 (Souques 1908), is characterized, according to Kinnier Wilson, by 'the repetition of words or short phrases perhaps at increasing speed with decreasing audibility' (Wilson 1954).

Clinical features

Palilalic patients involuntarily repeat a phrase or word, and do so with increasing rapidity but with diminishing distinctness: 'One patient (Serra-Mestres *et al.* 1996), when asked "How are you today?" replied "I am very well, thank you, thank you, thank you, thank you..." up to six times with increasing speed and decreasing volume.'

Etiology

Although the classic and most common cause of palilalia is Parkinson's disease (Benke *et al.* 2000), it may also be seen in progressive supranuclear palsy (Kluin *et al.* 1993), advanced Alzheimer's disease (Cummings *et al.* 1985), Tourette's syndrome (Cardoso *et al.* 1996), and rarely as a manifestation of multi-infarct dementia (Serra-Mestres *et al.* 1996) or thalamic infarction (Yasuda *et al.* 1990). There is also a case report of palilalia occurring as part of a simple partial seizure, when it was accompanied by rhythmic head-nodding and posturing of the right upper extremity (Cho *et al.* 2009).

Differential diagnosis

Verbal perseveration is distinguished by the fact that the repeated word or phrase is spoken without the distinctive increasing rapidity and decreasing distinctness.

Stuttering is distinguished by the fact that it is the first letter of a word that is repeated, and echolalia by the fact that the word repeated is not something that the patient spoke but rather what the examiner or another person said.

Treatment

Treatment is directed at the underlying condition. In a case of palilalia occurring as part of a multi-infarct dementia, trazodone was effective (Serra-Mestres *et al.* 1996); whether this would carry over into other conditions is not known.

5.5 PERSEVERATION

Perseverative speech or behavior (Sandson and Albert 1987) is commonly seen in a variety of conditions, most notably dementia.

Clinical features

Perseverative speech generally occurs in one of two forms. In one, termed *recurrent*, patients supply the same response to repeated, but different, questions. For example, one patient, when asked where she was, replied 'Louisville;' when asked next what year it was, she perseverated in her response, and again said 'Louisville.' The other form is termed *continuous* and here, without being prompted, patients repeat the same word or phrase again and again. One of Kinnier Wilson's (Wilson 1954) patients with Pick's disease 'repeated endlessly, "I want my husband and my children and I do not know where they are".'

Perseverative behavior may involve drawing or writing: after being asked to draw a stick figure, one patient did so and then drew multiple others until the page was filled; another patient, asked to write his name, wrote out 'Paul' and then perseverated with the letter 'l', until he reached the edge of the page. Perseverative behavior may also involve other motor activities, which in turn may be simple, as for example shoulder shrugging, or complex, as demonstrated by a patient who, after being asked to get dressed, went to the closet, took out a pair of pants, then paused and hung them back up, then took them out again, and repeated the sequence until being asked to stop.

Etiology

Perseveration is common in dementia (Bayles *et al.* 1985), especially frontotemporal dementia (Thompson *et al.* 2005), Alzheimer's disease (Bayles *et al.* 2004), and dementia with Lewy bodies (Doubleday *et al.* 2002). It is commonly seen in delirium, and is also an integral part of the frontal lobe syndrome. Other conditions characterized by perseveration include sensory aphasia (Albert and Sandson 1986), autism (Rumsey *et al.* 1986), schizophrenia (Siegel *et al.* 1976), and traumatic brain injury (Hotz and Helm-Estabrooks 1995).

Differential diagnosis

Palilalia is distinguished by the fact that the verbal repetition has a distinctive character to it, in that the repeated word or phrase is said with increasing speed and decreasing distinctness.

Stereotypies, discussed in Chapter 5.28, constitute that subset of perseverative motor behavior that involves neither drawing nor writing.

Treatment

Perseverative behavior may lessen or remit with treatment of the underlying condition. In one unblinded study, bromocriptine reduced perseverative behavior in dementia (Imamura *et al.* 1998).

5.6 PRIMITIVE REFLEXES

Normal infancy is characterized by numerous reflexes that, in the usual course of events, tend to fade or disappear with development; these may, however, be found in adults, and in such cases they are known as 'primitive' reflexes or 'frontal release signs'. Although these reflexes are found in certain pathologic conditions, such as dementia, they may also, as noted below, be found in normal individuals, and hence care must be taken in their interpretation.

Clinical features

Four primitive reflexes are considered here, namely the palmomental reflex, the snout reflex, the grasp reflex, and the grope reflex.

The palmomental reflex (Blake and Kunkle 1951; Jacobs and Gossman 1980) is elicited by repeatedly and rapidly dragging an object, such as the tip of the reflex hammer, across the thenar eminence from the lateral aspect medially toward the center of the palm: the contact should be definite, and slightly disagreeable, but not painful. When the reflex is present, one sees a wrinkling of the ipsilateral *mentum*, or chin.

The snout reflex (Jacobs and Gossman 1980) is elicited by placing one's index finger just above the patient's upper lip in the midline, centered upon the philtrum, and then either gently pressing the finger in or gently tapping it with the reflex hammer: a positive reflex is indicated by a puckering or protruding of the lips.

The grasp reflex is tested for by taking hold of the patient's wrist and holding the patient's hand with the palm facing up, then laying your index finger flat across the palm of the patient's hand (with the tip of the finger pointing toward the hypothenar eminence and the base of the finger resting between the patient's thumb and index finger) and then slowly withdrawing it between the patient's thumb and index finger (Adie and Critchley 1927; Seyffarth and Denny-Brown 1948). When present, the reflex consists of the patient's fingers involuntarily closing upon the examiner's finger, as if grasping it. Once the finger has been grasped, attempts by the examiner to pull it free often result in a stronger grasp, but if the examiner ceases all effort to remove the finger, then the patient's hand typically relaxes (Walshe and Robertson 1933). If there is any doubt as to the involuntary nature of the grasp, then the test may be repeated, but only after first instructing the patient to not grab the examiner's finger. Interestingly, such a grasping reflex may be present in a hemiplegic limb, such that although the patient is unable to flex the fingers voluntarily, grasping occurs with the appropriate stimulation (Stewart-Wallace 1939).

The grope reflex (also known as 'instinctive grasping') is a truly remarkable clinical phenomenon. Testing is accomplished by simply touching the patient's hand: when the reflex is present, the patient's hand will involuntarily reach out in a groping fashion to grasp the examiner's finger (Mori and Yamadori 1985). In some cases, one may also demonstrate the 'magnet' reaction, wherein by successive touches of the patient's hand from successively different positions, one may lead the patient's hand through space, much as if it were being attracted by a magnet (Seyffarth and Denny-Brown 1948). When fully developed, actual touch may not be required to elicit the reflex, and the patient's hand may automatically begin groping at the mere sight of another object (Mori and Yamadori 1985).

The presence of a grope reflex may significantly interfere with the patient's day-to-day life. One patient, whenever passing through a doorway, found that his 'left hand clung to [the] door handle . . . with such tenacity that he had to use his right hand to free himself' (Seyffarth and Denny-Brown 1948). In another case (Stewart-Wallace 1939), the patient's right hand 'developed a habit of clutching hold of objects of its own accord, such as the lapel of his coat, the edge of his trouser pocket, or the bed-clothes. This distressed him so much that in order to prevent it, he used to wear a glove most of the time. At other times he would keep his hand in his pocket, holding a coin or a key'. Another patient (Walshe and Hunt 1936), when questioned about the experience, noted that 'I can't leave things alone. If I see anything lying near me, my hand seems to want it.'

Etiology

Primitive reflexes, as noted earlier, may be found in healthy individuals, and this is especially true of the palmomental and snout reflexes, each of which may be found in one-third to one-half of otherwise healthy elderly individuals (Jacobs and Gossman 1980; Jensen *et al.* 1983; Koller *et al.* 1982; Vreeling *et al.* 1993); consequently, finding one primitive reflex is of uncertain clinical significance. Finding more than one primitive reflex, however, is uncommon, and finding three in one patient should raise strong suspicions regarding disease of the frontal lobe (Brown *et al.* 1998; Di Legge *et al.* 2001).

Primitive reflexes, as noted, are suggestive of frontal lobe pathology, and this is especially the case with the grasp and grope reflexes (Di Legge *et al.* 2001): although they have been noted with infarction of the motor and premotor cortex, they appear to be more common with infarction of the medial aspect of the frontal lobe, particularly the cingulate gyrus in its anterior portion, and the supplemental motor area (De Renzi and Barbieri 1992; Hashimoto and Tanaka 1998; Mori and Yamadori 1985; Stewart-Wallace 1939; Walshe and Robertson 1933; Walshe and Hunt 1936).

As might be expected, primitive reflexes are found in multiple different dementing conditions (Tweedy *et al.* 1981; Waite *et al.* 1996) including frontotemporal lobar degeneration (Sjogren *et al.* 1997), Alzheimer's disease (Vreeling *et al.* 1995), multi-infarct dementia (Vreeling *et al.* 1995), AIDS dementia (Marder *et al.* 1995), and normal pressure hydrocephalus (Thomas *et al.* 2009). Parkinsonian conditions may also be characterized by primitive reflexes, including Parkinson's disease (Maertens de Noordhout and Delwaide 1988; Vreeling *et al.* 1993), dementia with Lewy bodies (Borroni *et al.* 2006), progressive supranuclear palsy (Brodsky *et al.* 2004), and corticobasal ganglionic degeneration (Rinne *et al.* 1994). Finally, these reflexes have also been noted in patients with schizophrenia (Barnes *et al.* 1995; Hyde *et al.* 2007; Lohr 1985).

Differential diagnosis

The grope reflex must be distinguished from the alien hand sign, a distinction made possible by attending to the seeming 'intent' of the moving hand. In groping, the hand reaches for whatever may be nearby, whether a doorknob or the examiner's hand, and simply grabs hold. In the alien hand sign, however, the left hand does something with the object which, importantly, is at cross-purposes with what the patient is intentionally doing with the right hand.

Treatment

Except for the grope reflex, treatment is not required; in cases when the grope reflex interferes with activities of daily living, occupational therapy may assist in developing strategies to prevent the reflex from happening.

5.7 PSEUDOBULBAR PALSY AND PATHOLOGICAL LAUGHTER AND CRYING

Pseudobulbar palsy is a syndrome occurring secondary to bilateral interruption of the corticobulbar tracts. Pathological laughter and crying, characterized by fits of laughter or crying in the absence of any corresponding feelings of mirth or sadness, although generally occurring as part of pseudobulbar palsy, may also occur independently of this syndrome, as discussed below.

Before proceeding, some words may be in order regarding three synonyms for pathological laughter and crying. 'Pseudobulbar affect', although still in use, has the disadvantage of implying a necessary connection with pseudobulbar palsy, and hence should probably be abandoned. The next synonym, 'emotional incontinence', although forcefully conveying the nature of the symptomatology, should probably also be abandoned, given that most patients find it distasteful and unacceptable. The newest synonym is 'involuntary emotional expression disorder': this, like 'pathological laughter and crying', has the advantages of neutrality regarding any association with pseudobulbar palsy and of acceptability; however, it has not gained as much currency as has 'pathological laughter and crying', which is the diagnostic term used in this text.

Clinical features

When fully developed (Langworthy and Hesser 1940; Tilney 1912), the syndrome of pseudobulbar palsy consists of dysarthria, dysphagia, a stiff and shuffling gait, an exaggerated jaw jerk and gag reflex, a variable degree of paresis of the tongue, and pathological laughter and crying; in addition, the corticospinal tracts are often also involved, as evidenced by exaggerated deep tendon reflexes and bilateral Babinski signs.

Pathological laughter and crying is characterized by either laughter or crying that occurs either spontaneously or in response to some otherwise trivial stimulus, such as someone approaching the bed (Lieberman and Benson 1977). After suffering bilateral strokes, one of Kinnier Wilson's patients:

> was seen to have a distinctly vacant, apathetic facial expression at rest. She was able to move facial muscles voluntarily on both sides, although there was the slightest weakness of the left corner of the mouth. On the slightest stimulus, even when the observer simply came to her bedside, she at once assumed a most lugubrious expression, her mouth opened widely, and a long, almost noiseless bout of weeping ensued, lasting for many seconds, even minutes, at a time.
>
> (Wilson 1924)

Another of Wilson's patients, again after bilateral strokes:

> exhibited characteristic involuntary laughing. Whatever the emotional stimulus, and however slight, he at once began to laugh and laugh loudly. Thus, on reading the war news he at once began to smile, and the more serious and anxious the news, the more he laughed. On examination, there was some voluntary facial paresis on both sides, especially the left, some dysarthria, and some dysphagia, but during the laughing the facial movements were in no way restricted. A double extensor response was present.
>
> (Wilson 1924)

Critically, these emotional displays are not accompanied by any corresponding feeling of mirth or sadness. In one case (Davison and Kelman 1939), a patient, although feeling no mirth, experienced such 'gales of laughter' that he 'felt foolish and ashamed, and had tears in his eyes, because he could not control the laughter'. The paroxysms of laughter or crying tend to last on the order of minutes and then resolve spontaneously; in some cases there may be only laughter, or only crying; however, in others one may see laughter succeeded by crying or vice versa in the same episode.

Etiology

As noted earlier, pseudobulbar palsy occurs secondary to bilateral interruption of the corticobulbar tracts. This tract originates in the cortex and descends through the genu and posterior limb of the internal capsule to the crus cerebri, the basis pontis, and, finally, the medullary pyramid (Besson *et al.* 1991). Throughout its course in the brainstem, fibers leave the tract to head toward various brainstem nuclei.

Stroke is probably the most common cause of pathological laughter and crying, and in such cases the pathological laughter and crying generally emerges after a latency period ranging from days to weeks. Such stroke-related pathological laughter and crying has specifically been noted after bilateral infarction of the posterior frontal cortex (Colman 1894; Davison and Kelman 1939; Wilson 1924), the posterior limb of the internal capsule (Helgason *et al.* 1988), and the basis pontis (Asfora *et al.* 1989; Bassetti *et al.* 1996). As might be expected, in most cases of stroke-related pathological laughter and crying, one finds a history of infarction, with the pathological laughter and crying only emerging with a subsequent infarction on the opposite side (Wilson 1924). In the majority of cases where pathological laughter and crying occur in the absence of any history of stroke, imaging will reveal evidence of an old infarction on the opposite side that is subclinical. There are, however, definite cases of pathological laughter and crying occurring with first ever infarctions and in the absence of pseudobulbar palsy, as has been reported with infarctions of the frontal lobe and of the internal capsule (Ceccaldi *et al.* 1994; Derex *et al.* 1997). Although the mechanism underlying these cases remains unknown, one may suspect congenital anomalies in the relevant corticobulbar pathways.

After stroke, the two most common causes of pathological laughter and crying are multiple sclerosis (Feinstein *et al.* 1997) and amyotrophic lateral sclerosis (Gallagher 1989; Ironside 1956; Ziegler 1930). Other disorders that are capable of causing bilateral damage to the corticobulbar tracts include Binswanger's disease (Caplan and Schoene 1978), Alzheimer's disease (Starkstein *et al.* 1995), traumatic brain injury (Tateno *et al.* 2004; Zelig *et al.* 1996), and central pontine myelinolysis (Van Hilten *et al.* 1988). Finally, pseudobulbar palsy has also been noted in Parkinson's disease (Petracca *et al.* 2009), progressive supranuclear palsy (Behrman *et al.* 1969; Menza *et al.* 1995), the cerebellar variant of multiple system atrophy (Parvizi *et al.* 2007), intrinsic brainstem tumors (Achari and Colover 1976; Cantu and Drew 1966), and with compression of the brainstem, as for example by a meningioma (Motomura *et al.* 1980).

Differential diagnosis

'Emotionalism' is not at all uncommon after strokes, but it is readily distinguished from pathological laughter and crying by the fact that 'emotional' patients experience an affect congruent with their facial expression, whereas the patient with pathological laughter and crying feels neither mirth nor sadness and is often as surprised at the emotional display, as is the observer.

Lability of affect, as may be seen in mania, is, like emotionalism, distinguished from pathological laughter and crying by the presence of an affect congruent with the facial expression.

Episodes of mirthless laughter may also occur in an isolated fashion as *le fou rire prodromique* and as a simple partial seizure in gelastic epilepsy. *Le fou rire prodromique*, as discussed in Chapter 5.9, represents a prodrome to stroke, while stroke-related pathological laughter and crying emerge after the stroke has occurred. Gelastic seizures are suggested by a history of other seizure types, such as grand mal or complex partial seizures.

Treatment

In placebo-controlled double-blind studies, several medications have been shown to be effective for pathological laughter and crying. For cases occurring secondary to infarction, one may utilize nortriptyline (in doses of 50–100 mg) (Robinson *et al.* 1993), imipramine (in doses of 10–20 mg) (Lawson and McLeod 1969), or citalopram (in doses of 10–20 mg) (Andersen *et al.* 1993): among these, citalopram is the easiest to use, and is also the most rapidly effective, with some patients responding within days. In both amyotrophic lateral sclerosis (Brooks *et al.* 2004; Pioro *et al.* 2010) and multiple sclerosis (Panitch *et al.* 2006; Pioro *et al.* 2010), a combination of quinidine and dextromethorphan is effective: in this combination, quinidine is used merely in order to inhibit the metabolism of dextromethorphan, which is the active agent. This combination is given on a twice-daily basis, and various doses have been found to be effective, including 10 mg of quinidine plus either 20 or 30 mg of dextromethorphan, or 30 mg of quinidine plus 30 mg of dextromethorphan. In multiple sclerosis, amitriptyline is also effective, in doses of 25–75 mg daily (Schiffer *et al.* 1985).

It must be stressed that although nortriptyline, imipramine, citalopram, and amitriptyline are all antidepressants, they are effective regardless of whether patients are depressed or not.

It is not clear whether the antidepressants that are effective in stroke would be effective in amyotrophic lateral sclerosis or multiple sclerosis, nor, conversely, is it clear whether the combination of quinidine and dextromethorphan would be effective in stroke-related cases. Further, it is not clear whether any of these agents would be effective in cases due to the other disorders. Clearly, some trial and error work may be necessary in some cases, and strong consideration should probably be given to citalopram, which, of all these agents, is the best tolerated.

5.8 EMOTIONAL FACIAL PALSY

Most central facial palsies are of the 'voluntary' sort, in that when patients are asked to voluntarily perform a facial maneuver, such as showing their teeth, there is a droop evident on one side; less commonly one finds an 'emotional' facial paresis wherein a 'droop' becomes evident when the patient smiles in response, say, to a joke, even though voluntary movements are intact. This emotional facial paresis has also been termed 'mimetic' or 'involuntary' facial paresis.

Clinical features

Facial palsies may be either peripheral or central. In a peripheral facial palsy, there is no movement in either the forehead or the lower face, whether on voluntary command or in response to a joke. Central facial palsies are immediately distinguished from peripheral palsies in that this palsy affects only the lower half of the face, with forehead movements being spared.

The two forms of central facial palsy, namely voluntary and emotional, may be distinguished by first noting facial movements when patients are instructed to voluntarily show their teeth and then, at some point in the examination, by closely observing facial movements when the patient spontaneously smiles, for example in response to a joke or recollection of some happy memory (Monrad-Krohn 1924).

In voluntary central facial paresis, there is a droop of one side of the face when patients are asked to show their teeth; however, when these patients are observed smiling at a joke, one sees full facial movements on both sides.

In emotional facial paresis, there is full movement bilaterally when patients are asked to show their teeth; however, when one observes these patients smiling at a joke, there is drooping of one side of the face; this produces what might be called a 'hemismile'.

Etiology

Voluntary and emotional facial paresis are dissociated because the corticobulbar fibers subserving these two functions are separate.

Corticobulbar fibers for voluntary facial movement arise in the posterior portion of the frontal cortex, descend through the corona radiata to the posterior limb of the internal capsule, and then travel through the ventral mesencephalon in the crus cerebri to the basis pontis. At this point, most of these fibers then cross to travel directly to the contralateral facial nucleus, whereas a minority of them descend further into the medulla, cross and 'loop' back up into the pons, and continue to the facial nucleus.

Corticobulbar fibers for emotional facial movements have widespread origins in the cerebrum, including the supplemental motor area on the medial aspect of the frontal lobe and temporal lobe structures (e.g. the amygdala); the thalamus also plays a role. Fibers descend in the anterior limb of the internal capsule and pass through the tegmentum of the mesencephalon into the dorsal pons. At this point, most fibers then cross to end in the contralateral facial nerve; however, a minority of fibers proceed down into the medulla, crossing in this structure and then looping back up into the pons to contact the facial nucleus.

Emotional facial paresis has been reported with contra lateral lesions, generally infarctions, of a large number of structures (Hopf *et al.* 1992), including the supplemental motor area (Gelmers 1983), temporal lobe (Remillard *et al.* 1977), thalamus (Bogousslavsky *et al.* 1988; Graff-Radford *et al.* 1984; Ross and Mathiesen 1998), anterior limb of the internal capsule (Trosch *et al.* 1990), mesencephalic tegmentum (Wilson 1924), and the dorsal pons (Hopf *et al.* 2000); a case has also been reported of emotional facial paresis with an ipsilateral medullary lesion (Cerrato *et al.* 2003).

Emotional facial paresis has also been noted in patients with chronic complex partial seizures of temporal lobe origin, and in these cases the emotional facial paresis is generally contralateral to the side with the seizure focus (Jacob *et al.* 2003). The theory here is that repetitive ictal activity leads to damage of the amygdala.

Differential diagnosis

Some individuals have a congenital 'lopsided' grin; the differential rests here on determining that the 'lopsidedness' has been present since earliest childhood.

Treatment

Specific treatment is not required.

5.9 *LE FOU RIRE PRODROMIQUE*

First described by Fere in 1903 (Fere 1903), *le fou rire prodromique* is a very rare disorder occurring as a prodrome to an ischemic infarction.

Clinical features

Le fou rire prodromique manifests as a paroxysmal, uncontrollable fit of laughter, lasting for minutes to an hour, during which there is full preservation of consciousness and, critically, no corresponding sense of mirth.

In one case (Wali 1993), a 35-year-old woman 'after a warm shower suddenly began to laugh. The laughter was inappropriate to the situation and continued for nearly 15 minutes at the end of which it became a low grade giggle. This lasted for another 15 minutes and ended abruptly when she collapsed'. On admission, she was found to have suffered a bilateral pontine infarction, and was left in a 'locked-in' state. In another case (Martin 1950), the patient's first attack occurred at his mother's funeral, and, much to his distress, recurred and recurred without any apparent cause until he collapsed and died several days later; autopsy revealed an interpeduncular aneurysm that had ruptured.

Etiology

Such prodromal laughter has been noted with infarction of the following structures: parietotemporal cortex (Lago 1998), external and extreme capsules (Kumral and Calli 2006); striatum (Carel *et al.* 1997), genu (Uzunca *et al.* 2005), or posterior limb (Ceccaldi and Milandre 1994) of the internal capsule; and the basis pontis (Assal *et al.* 2000; Gondim *et al.* 2001; Wali 1993).

Differential diagnosis

Mirthless laughter may be seen in pathological laughter and crying and during gelastic seizures. Pathological laughter and crying is distinguished by the clinical course in that in pathological laughter and crying, the episodes of mirthless laughter occur after the stroke, whereas in *le fou rire prodromique* the episodes occur beforehand. Gelastic seizures are suggested by the occurrence, at other times, of other seizure types, for example grand mal or complex partial.

Treatment

Treatment for the laughter *per se* is obviously not required; what is required, however, is that one recognizes it as the herald of stroke and proceeds accordingly.

5.10 ABULIA

The term abulia is derived from the Greek, and means 'a lack of will or drive'; synonyms include pure psychic akinesia and athymormia.

Clinical features

Abulic patients fail to experience a will to do anything because, left to themselves, no ideas, interests, or inclinations come to mind: one patient commented that there was 'a blank in my mind' (Laplane *et al.* 1984). Importantly, these patients are not depressed or bored; rather, they experience a benign sort of emptiness.

Abulic patients, thus volitionally impoverished, may be content to sit or lie quietly, doing and saying nothing, and thus present a picture of immobility and mutism. Importantly, however, if they are asked to do something or to speak, they will, after a more or less lengthy pause, respond and, provided ongoing supervision or prompting is provided, they will complete whatever it is they were asked to do. Once left to themselves, however, they rapidly lapse back into a placid quietude.

Etiology

Abulia may be seen with bilateral or unilateral infarction or hemorrhage of various subcortical structures including the caudate nucleus (Bhatia and Marsden 1994; Caplan *et al.* 1990; Kumral *et al.* 1999; Mendez *et al.* 1989), globus pallidus (Bhatia and Marsden 1994; Giroud *et al.* 1997; Laplane *et al.* 1984), thalamus (Bogousslavsky *et al.* 1991; Van Der Werf *et al.* 1999), and genu of the internal capsule (Yamanaka *et al.* 1996).

Abulia has also been noted as a sequela to carbon monoxide intoxication with bilateral necrosis of the globus pallidus (Laplane *et al.* 1984; Lugaresi *et al.* 1990).

Finally, abulia occurs commonly in schizophrenia, especially simple schizophrenia. Kraepelin (1921) noted that such patients 'experience no tediousness, have no need to pass the time... but can lie in bed unoccupied for days and weeks, stand about in corners, "stare into a hole", (or) watch the toes of their boots'.

Differential diagnosis

Akinetic mutism and catatonia enter the differential as in each of these syndromes the patients, like those with abulia, are immobile and mute. Differentiating them from abulia is quite readily done, however, by simply urging the patient to do something and providing some ongoing supervision: akinetic mutes and catatonics will not respond, whereas the abulic patient does and will continue to do so until supervision stops.

Apathy, as discussed in Chapter 7.2, may be distinguished from abulia if one attends to whether or not the patient experiences any ideas or interests. As noted earlier, abulic patients do not. Apathetic patients, however, do: what is lacking here, however, is any motivation to go on. Abulic patients, if asked whether they have any plans, will say they do not; apathetic patients, however, do have plans, although they do not 'care enough' to follow through on them.

Depression is readily distinguished from abulia by the presence of a depressed affect and mood.

Treatment

Bromocriptine, anecdotally, has been effective (Barrett 1991); treatment may be commenced at a low dose of 2.5 mg total per day, and increased in similar increments every few days until satisfactory improvement or unacceptable side-effects occur or a maximum dose of perhaps 40 mg is reached; in all cases, dosage is generally divided into two doses: one in the morning and the next in the early afternoon.

5.11 ENVIRONMENTAL DEPENDENCY SYNDROME

This rare syndrome was initially described in 1983 by Lhermitte (1983), who named it 'utilization behavior'; in a subsequent paper, however, he changed the name to the 'environmental dependency syndrome' (Lhermitte 1986).

Clinical features

In this syndrome, patients appear to lose their autonomy, becoming 'dependent' on their environment such that they feel compelled to pick up and 'utilize' whatever may come to their attention. For example, upon seeing a pair of glasses, one patient put them on, even though they were not his; another patient, upon seeing a toothbrush, picked it up and began brushing, even though there was no need to do so; and another, on seeing a bed with the sheets turned down, undressed, got in bed, and settled in as if to go to sleep. At times, the utilization behavior can be more complex. For example, in describing one of his patients, Lhermitte noted that both he and the patient:

> sat down in my office. I put some medical instruments on my desk. She immediately picked up the blood pressure gauge and very meticulously took my blood pressure... After this she took the tongue depressor and placed it in front of my mouth, which I opened, and she examined my throat... Last, she picked up the reflex tester and, to make sure she tested the ankle jerks, I knelt down on the chair. When I asked her what she thought, she said she was satisfied with my state of health.
>
> Lhermitte (1983)

Importantly, patients engage in this behavior without being invited, asked or told to, and some may even persist in the behavior despite being asked to stop.

Etiology

The syndrome has been noted most commonly with bilateral or unilateral lesions of the inferior frontal lobe (De Renzi *et al.* 1996; Lhermitte 1983, 1986; Lhermitte *et al.* 1986; Shallice *et al.* 1989); cases have also been reported with infarction of the white matter subjacent to the superior frontal gyrus (Ishihara *et al.* 2002) and with infarction of the thalamus (Eslinger *et al.* 1991; Hashimoto *et al.* 1995). As might be expected, this syndrome is also seen in frontotemporal lobar degeneration; indeed, one study found it in 80 percent of cases (Ghosh and Dutt 2010).

Differential diagnosis

One of the primitive reflexes, namely the grasp reflex, appears similar to the environmental dependency syndrome in that in both cases patients will involuntarily reach and grasp objects. The difference lies in what the patient does with the grasped object: the patient with forced grasping simply holds on, whereas the patient with the environmental dependency syndrome will utilize the object and do something with it.

The alien hand sign also appears similar to utilization behavior in that the alien hand involuntarily reaches out and does things with objects: the difference here is that what the alien hand does is at cross-purposes with what the patient is attempting to do with the 'good' hand, whereas in utilization behavior the two hands act in concert without any conflict.

Echopraxia or 'imitation behavior' is said to be present when patients involuntarily and automatically imitate what the examiner is doing. This is fundamentally different from environmental dependency, for in the environmental dependency syndrome the examiner does nothing except perhaps place objects in the patient's view.

Delirious patients may take hold of and use nearby objects, but here one also finds confusion, a sign that is absent in the environmental dependency syndrome.

Treatment

Depending on the dangerousness of objects in the patient's environment, supervision may or may not be required.

5.12 KLUVER–BUCY SYNDROME

In 1939, Kluver and Bucy (Kluver and Bucy 1939) reported striking behavioral changes in monkeys subjected to bilateral temporal lobectomy. This Kluver–Bucy syndrome has subsequently been noted in humans, wherein its most prominent clinical manifestations are hyperorality and hypersexuality.

Clinical features

In the full form of the human Kluver–Bucy syndrome, one finds a heightened, but indiscriminate, interest in nearby objects coupled with hyperorality, hypersexuality, and a certain placidity marked by an absence of fear. Patients appear to take an almost equal interest in all nearby objects and may appear restless as they reach out for one object after another on the bedside table or pace about the room, again from one thing to another. Hyperorality comes to light in that there is a strong tendency on the patients' part to put anything into their mouths, whether it is edible or not: patients may eat Styrofoam cups, drink urine out of urinals, or stuff their mouths full of tissue papers. Hypersexuality may be very problematic: patients may display no gender preference and may proposition males and females alike; there may be excessive masturbation, sometimes done publicly. The placidity of these patients is at times remarkable: they seem to have no fear, and pointing out the consequences of their behavior to them typically has no effect.

Some examples from the literature may help to fix this clinical picture in the reader's mind. In one case, a 57-year-old professor suffered a head trauma that led to a maceration of the inferior surface of both temporal lobes. After recovering from the trauma, the patient:

> ate voraciously and, in fact, had a tendency to place almost everything that came into view in his mouth. For instance, after one meal he drank a cup of tea and then ate the teabag. He wandered aimlessly and did not know the location of his room. He made inappropriate sexual advances, rather indiscriminately; both male and female attendants were cautious in his presence. In general, however, his affect was flat and unconcerned. He was readily distracted and there was a constant shifting of his attention. If restrained, he became agitated, but when his attention was diverted, he immediately calmed down.
>
> (Lilly *et al.* 1983)

In another case, a 46-year-old man had a complex partial seizure wherein:

> he was awake and alert, but almost mute and lacking facial expression. He was observed grabbing for objects on his bedside table, and he masturbated in front of the nursing staff. He also placed objects in his mouth, chewed on tissue paper, and attempted to drink from his own urine container.
>
> (Nakada *et al.* 1984)

Finally, another patient, after recovering from an episode of complex partial status epilepticus, displayed:

> a voracious appetite and indiscriminate eating habits which included paper towels, plants, styrofoam cups, and even faeces. At one point he drank urine from a catheter bag. The patient was no longer his assertive self and had become quite docile. He tended to wander about the ward touching objects or people and made inappropriate comments of a sexual nature… On the day of his death, the patient had a respiratory arrest

> after stuffing his mouth with surgical gauze. He had wandered about the ward picking up whatever he could find and putting it into his mouth.
>
> (Mendez and Foti 1997)

Kluver and Bucy (1939), as noted above, first described this syndrome in monkeys, and their descriptions are of interest. The terminology they utilized is somewhat cumbersome but is still at times seen in the literature.

The heightened interest of the monkeys (which they termed 'hypermetamorphosis') was manifest by 'an excessive tendency to take notice of and to attend and react to every visual stimulus . . . the monkey seems to be dominated by only one tendency, namely the tendency to contact every object as soon as possible [and] . . . when turned loose in a room, the monkey often behaves as if it were ceaselessly "pulled" from one object to another'.

According to Kluver and Bucy's terminology, the indiscriminate nature of this heightened interest represented a kind of 'psychic blindness' in that the monkey would approach and examine objects 'no matter whether they are very large or very small, dead or alive, edible or inedible, moving or stationary . . . The monkey seems to be just as eager to examine the tongue of a hissing snake, the mouth of a cat, feces, a wire cage or a wagon as a piece of bread'.

Hyperorality (or, as Kluver and Bucy termed it, 'oral tendencies') was manifest by a 'strong tendency to examine all objects by mouth', for example 'putting the object into the mouth, biting gently, chewing, licking, touching with the lips and smelling...the object'.

Hypersexuality was referred to as 'changes in sexual behavior' and was manifest by a general 'increase in sexual behavior' that could involve other monkeys, regardless of their gender, or masturbation.

Placidity was termed 'emotional changes' by Kluver and Bucy and was marked by 'absence of all emotional reactions...generally associated with anger and fear'. For example, 'after being attacked and bitten by another animal, (the monkey) may approach this animal again and again in an attempt to examine it'.

Etiology

Although the human Kluver–Bucy syndrome has been reported with unilateral temporal lobe damage (e.g. after a unilateral temporal lobectomy [Ghika-Schmid *et al.* 1995]), this is very rare, and, as with monkeys, the overwhelming majority of human cases occur secondary to bilateral temporal lobe damage or dysfunction, with the various causes listed in Table 5.3.

Table 5.3 Causes of the Kluver–Bucy syndrome

Bilateral ablation or injury to the temporal lobes
Temporal lobectomy (Terzian and Dalle Ore 1955)
Traumatic brain injury (Formisano *et al.* 1995; Lilly *et al.* 1983)
Late-delayed radiation encephalopathy (Lam and Chiu 1997; Thajeb 1995)
Post-herpes simplex viral encephalitis (Greenwood *et al.* 1983; Marlowe *et al.* 1975; Shoji *et al.* 1979)
Infarction (Chou *et al.* 2008; Oliveira *et al.* 1989)
Epileptic conditions
Complex partial seizure (Nakada *et al.* 1984)
Postictal, after either a grand mal seizure (Anson and Kuhlman 1993) or complex partial status epilepticus (Mendez and Foti 1997; Varon *et al.* 2003)
Neurodegenerative disorders
Pick's disease (Cummings and Duchen 1981; Mendez *et al.* 1993)
Frontotemporal dementia (Gydesen *et al.* 2002; Heutink *et al.* 1997; Mendez and Perryman 2002)
Alzheimer's disease (Kile *et al.* 2009; Mendez *et al.* 1993; Teri *et al.* 1988)
Huntington's disease (Janati 1985)
Miscellaneous causes
Adrenoleukodystrophy (Powers *et al.* 1980)
Heatstroke (Pitt *et al.* 1995)

Bilateral ablation or injury to the temporal lobes is most obvious after bilateral temporal lobectomy. Traumatic brain injury, with contusions of the inferior surfaces of both temporal lobes, may also cause the syndrome, as may a late-delayed radiation encephalopathy after irradiation, say, for a pituitary tumor or nasopharyngeal carcinoma. Herpes simplex viral encephalitis classically involves both temporal lobes, and the syndrome may appear as a sequela in those who survive. Finally, there are rare cases of infarction of both temporal lobes resulting in Kluver–Bucy syndrome.

Epileptic conditions may be characterized by Kluver–Bucy syndrome, either ictally during a complex partial seizure or postictally after either a grand mal seizure or an episode of complex partial status epilepticus.

Neurodegenerative disorders may cause Kluver–Bucy syndrome of gradual onset and progression, and this is most characteristic of those disorders such as Pick's disease and frontotemporal lobar degeneration, which preferentially involve the temporal lobes early on. Other neurodegenerative disorders, such as Alzheimer's disease and Huntington's disease, may, late in their course, also cause the syndrome.

Miscellaneous, and very rare, causes of the syndrome include adrenoleukodystrophy and heat stroke.

Differential diagnosis

The frontal lobe syndrome may enter into the differential, especially when it is marked by disinhibition, with gluttony, inappropriate sexual advances, and public masturbation. However, the gluttony of the frontal lobe syndrome is not characterized by eating inedible items, and furthermore the frontal lobe syndrome is also not characterized by a heightened and indiscriminate interest.

Treatment

There are no controlled studies of the treatment of Kluver–Bucy syndrome. Anecdotally, an overall improvement has been noted with the administration of carbamazepine (Stewart 1985), sertraline (Mendhekar and Duggal 2005), or antipsychotics (Carroll *et al.* 2001); should an antipsychotic be chosen, a second-generation agent, such as risperidone or quetiapine, may be considered. Hypersexuality has also been reduced by treatment with leuprolide (Ott 1995). As noted in one of the examples above, hyperorality may be lethal and appropriate precautions must be taken.

5.13 ALIEN HAND SIGN

The alien hand sign represents one of the most remarkable phenomena seen in neuropsychiatric practice, or for that matter, in the practice of medicine at large. Here, one of the patient's hands, almost always the left one, begins to act as if it had an independent will of its own, often engaging in a complex activity that thwarts what the patient is attempting to do with the right hand.

This sign was first described by the German physician Kurt Goldstein in 1908 (Goldstein 1908); Akelaitis, in 1941, described it in patients who had been subjected to corpus callosotomy for epilepsy, and named it 'diagnostic dyspraxia'. The current name, 'alien hand sign', is derived from a French paper by Brion and Jedynak (1972), who termed it 'le signe de la main étrangère': although the authors translated this as the 'strange hand' sign (which is indeed a more faithful translation from the French), English-speaking authors have universally favored translating it as 'alien hand sign'.

Clinical features

The alien hand sign is said to be present when one of the patient's hands (almost, as noted above, always the left one) spontaneously and autonomously acts in a way that is at definite cross-purposes to what the patient is intending to do with the right hand (Bogen 1985a). This 'intermanual conflict', as it has been termed, is not merely a matter of a clumsy or apraxic left hand 'getting in the way' or of reflexive groping or grasping by the left hand, but rather of the appearance of complex and seemingly purposeful activity of the left hand, activity that is at definite cross-purposes with the patient's consciously experienced will and intent. Patients may be astonished to see the left hand acting independently and outside their control, and may comment that it is as if the left hand had 'a mind of its own'.

Some examples of the alien hand sign will serve to flesh out this definition. The complexity of the behavior engaged in by the alien hand ranges from relatively simple activities, such as buttoning or unbuttoning, to very complex behaviors, even to the point of 'murderous' behavior. Examples of relatively simple alien hand behavior include the case of a patient, who, after an infarction of the corpus callosum, found that after he buttoned 'a shirt with the right hand the left hand would proceed to unbutton it and if pulling a cup of coffee toward him with the right hand the left hand would push it away' (Gottlieb *et al.* 1992). Similarly, subsequent to a corpus callosotomy, 'one patient was seen buttoning up his shirt with one hand while the other hand was following along behind it undoing the buttons' (Bogen 1985b). Akelaitis (1945) noted that one of his patients, also after a corpus callosotomy, 'would be putting on her clothes with her right hand and pulling them off with her left hand, opening a door or drawer with the right hand and simultaneously pushing it shut with the left hand'.

More complex behavior is illustrated in the case of a patient who, after suffering an infarction of the corpus callosum, 'told the story of herself frying a steak, turning it over in the frying pan with her right hand and, immediately after, finding herself turning it over once again with her left hand' (Barbizet *et al.* 1974; Degos *et al.* 1987). Another patient, after suffering an infarction of the anterior portion of the corpus callosum and adjacent medial aspects of both frontal lobes, found 'while playing checkers on one occasion, [that] the left hand made a move that he did not wish to make, and he corrected the move with the right hand; however, the left hand, to the patient's frustration, repeated the false move' (Banks *et al.* 1989).

Finally, there are examples of a 'murderous' alien hand. The first reported case of the alien hand sign, described by Kurt Goldstein (1908), was just such an example: the patient was a 57-year-old female with a callosal infarction whose 'left hand attempted to choke [her]... and this hand had to be pulled away. Furthermore the left hand did other unpleasant acts, such as tearing the bedclothes off the bed' (Geschwind 1981). The patient herself 'complained, "Es muss wohl ein boser Geist in der Hand sein" [There must be a devil in my hand]' (Hanakita and Nishi 1991). In another case, a patient who had suffered a callosal infarction treated her left arm 'as an alien presence with hostile motivations... she complained that the arm moved on its own and that it struck her and tried to choke her...often the hand reached for the neckline of her gown, pinched her right arm or leg, or knocked her glasses off' (Levine and Rinn 1986). Finally, there is the case of a patient who had suffered a callosal infarction (Geschwind *et al.* 1995) who 'awoke several times with her left hand choking her, and while she was awake her left hand would unbutton her gown, crush cups on her tray, and fight with the right hand while she was using the phone'.

Patients' reactions to the presence of the alien hand vary. One patient commented that 'my hands do not agree with each other' (Degos *et al.* 1987), and another that her hand 'disobeyed her' (Goldenberg *et al.* 1985), whereas a third felt 'as if someone "from the moon" were controlling her hand' (Geschwind *et al.* 1995). Patients also adopt different strategies to control the alien hand: one, 'to keep her left hand from doing mischief... would subdue it with her right hand' (Geschwind *et al.* 1995); another found the behavior 'so astonishing and uncontrollable that he occasionally hit his left [hand]... with the right one' (Leiguarda *et al.* 1989). Another patient, 'encouraged to "make friends" with the arm by talking to it... soon came to treat it as a misbehaving child, fondling and talking to it: "There, there, behave yourself now ... Do not be naughty"' (Levine and Rinn 1986).

As noted earlier, the alien hand sign is almost always found on the left: a right-handed alien hand in a right-handed person, although reported (Della Sala *et al.* 1994), is very rare.

Etiology

Once one is certain that the clinical phenomenon in question is in fact the alien hand sign, one can be reasonably assured that in all likelihood the patient has a lesion in the corpus callosum. In addition to occurring after section of the corpus callosum (Akelaitis 1941, 1945; Akelaitis *et al.* 1942; Ay *et al.* 1998; Bogen 1985a,b; Ferguson *et al.* 1985; Gazzaniga *et al.* 1962; Rayport *et al.* 1983; Smith and Akelaitis 1942; Sperry 1966; Van Waggenen and Herren 1940; Wilson *et al.* 1977), the alien hand sign has also been noted in various pathologic lesions of the corpus callosum, including infarction (Banks *et al.* 1989; Barbizet *et al.* 1974; Beukelman *et al.* 1980; Chan and Ross 1988, 1997; Chan *et al.* 1996; Degos *et al.* 1987; Geschwind *et al.* 1995; Goldberg and Bloom 1990; Goldenberg *et al.* 1985; Goldstein 1908, 1909; Hanakita and Nishi 1991; Jason and Pajurkova 1992; Nishikawa *et al.* 2001, Suwanwela and Leelacheavasit 2002; Tanaka *et al.* 1990, 1996; Watson and Heilman 1983; Watson *et al.* 1985), hemorrhage (Leiguarda *et al.* 1989; Starkstein *et al.* 1988), and tumor or angioma (Brion and Jedynak 1972). There are, however, exceptions to this rule. Thus, alien hand sign has been noted in neurodegenerative disorders, most notably corticobasal ganglionic degeneration (Ling *et al.* 2010), but also, albeit rarely, in Alzheimer's disease (Green *et al.* 1995). Focal lesions outside the corpus callosum may also cause alien hand sign, but this is probably rare. Levine and Rinn (1986) described a case secondary to infarction of the basal aspect of the temporo-occipital area, but the only imaging performed was with computed tomography (CT) scanning and the authors admitted that they could not exclude damage to either the splenium or the adjacent white matter. Gottlieb *et al.* (1992) described a case occurring with infarction of the medial aspect of the frontoparietal area. Again, only CT scanning was performed, but it appeared that the infarction was in the watershed area between the distributions of the anterior and middle cerebral arteries, and thus may truly have spared the corpus callosum. Perhaps the most convincing case comes from Dolado *et al.* (1995), who described a patient with an old infarction in the left parietal lobe who went on to develop the alien hand sign after a fresh infarction in the right parietal lobe. Finally, Creutzfeldt–Jakob disease may, very rarely, present with a combination of myoclonus and alien hand sign (MacGowan *et al.* 1997): in one case, the alien left hand 'would grab [the patient's] throat or hit her in the face and she usually limited its activity by holding it in her right hand's grasp'. In another case presented by the same authors, the alien hand, again on the left, unbuttoned the patient's blouse and removed a hair pin. In both cases, the eventual appearance of periodic complexes on electroencephalograph (EEG) and dementia suggested the correct diagnosis.

Differential diagnosis

The alien hand sign must be differentiated from the grasp and grope reflexes, asomatagnosia, the 'levitation' phenomenon, 'mirror' movements, and utilization behavior (as seen in the environmental dependency syndrome).

The grasp reflex, discussed in Chapter 5.6, is easily distinguished from the alien hand sign if one attends to what the hand does when it comes into contact with an object: the grasping hand merely holds tight and goes no further, whereas the alien hand takes hold and does something with the object. The grope reflex, also discussed in Chapter 5.6, may be a little harder to distinguish, as here the hand appears to be purposefully 'groping' for something. The distinction is again made possible by attending to what the groping hand does once it reaches the object: the groping hand, like the grasping hand, merely holds tight, again in contrast with the alien hand, which does something complex with the object. The failure to distinguish these primitive reflexes from the 'intermanual conflict' seen in alien hand sign has led to some unfortunate confusion in terminology, namely a proposal, made by Feinberg *et al.* (1992) (and further elaborated by Chan and Ross [1997]) that, rather than only one alien hand syndrome, there are in fact two: namely a 'frontal' alien hand sign and a 'callosal' alien hand sign. Close reading of these papers, however, reveals that the 'frontal' alien hand was merely grasping and groping, whereas the 'callosal' type corresponded to the clinical features noted above. Given this terminologic controversy, it is incumbent on the reader to examine closely any literature on the subject to see whether the patients described had mere grasping or groping or a true alien hand sign.

Asomatagnosia is said to be present when patients deny that a limb belongs to them. Although patients with asomatagnosia and the alien hand sign both experience the limb as foreign, the phenomena are easily distinguished in that whereas the 'foreign' limb of the patient with asomatagnosia does nothing independently, the alien hand acts 'on its own'.

The levitation phenomenon, seen in patients with parietal infarctions (Denny-Brown *et al.* 1952; Ghika *et al.* 1998) as well as in corticobasal ganglionic degeneration (Gibb *et al.* 1989; Riley *et al.* 1990) and progressive supranuclear palsy (Barclay *et al.* 1999), consists of the upper extremity spontaneously rising up or 'levitating'. This differs from the alien hand sign in that the levitated arm or hand does not do anything further, engaging in no complex behavior but simply staying 'levitated'.

Mirror movements, discussed in Chapter 4.13, are said to be present when patients who are doing something with one limb then find the contralateral limb involuntarily engaging in a more or less faithful imitation. Mirror movements may 'get in the way' and thus seem to be at cross-purposes; however, close inspection reveals that the abnormal movement does in fact 'mirror' the intended movement in the other limb and that its 'interference' is merely accidental.

Utilization behavior appears similar to the alien hand sign given that patients with utilization behavior involuntarily utilize objects. The difference between utilization behavior and the alien hand sign lies in the fact that in utilization behavior there is no intermanual conflict between the hands, which act in concert and with cooperation, whereas in the alien hand sign, the left hand acts at definite cross-purposes with the right.

Treatment

A mitt placed on the alien hand may mitigate its effects; in some cases, the entire extremity must be restrained.

5.14 BALINT'S SYNDROME

This syndrome, first described in 1909 by a Hungarian neurologist, Rezso Balint (Husain and Stein 1988), is characterized by the triad of simultanagnosia, optic ataxia, and optic apraxia.

Clinical features

Simultanagnosia, discussed further in Chapter 3.9, may be tested for by asking the patient to look at a drawing of a complex scene and asking for a description of each of the objects in the drawing: typically, patients will fail to mention one or more of the objects.

Optic ataxia is characterized by an inability to voluntarily reach and touch an object with the hand. Testing may be accomplished by holding up your index finger and asking the patient to touch it. When optic ataxia is present, the patient behaves almost as if he were blind, missing the examiner's finger by a wide margin. Interestingly, despite this inability, patients are still able to accurately touch parts of their own body.

Optic apraxia, also known as 'sticky fixation' or 'psychic paralysis of visual fixation', represents a peculiar difficulty in eye movement wherein patients are unable to voluntarily look from one fixation point to another. Testing for this may be performed by first having the patient fix his gaze on an object straight ahead and then commanding the patient to look at another object that is off to the side. When optic apraxia is present, there is a peculiar 'stickiness', as if the patient is unable to withdraw his gaze from the initial object to fix it on the next one. Interestingly, this 'stickiness' is present only with voluntary eye movements: should someone walk unexpectedly into the room, the patient may spontaneously shift his gaze without difficulty to fix it on this new object of interest.

Etiology

Most cases of Balint's syndrome occur secondary to bilateral infarctions in the parieto-occipital areas (Montero *et al.* 1982; Perez *et al.* 1996), typically watershed infarctions at the border between the areas of distribution of the middle and posterior cerebral arteries, as may occur secondary to hypotension during cardiac arrest. Another common cause of Balint's syndrome is the syndrome known as posterior cortical atrophy. In this syndrome, one may see a gradually progressive, more or less isolated Balint's syndrome; the underlying neuropathology may include corticobasal ganglionic degeneration or Alzheimer's disease (Hof *et al.* 1993; McMonagle *et al.* 2006; Mendez 2000; Renner *et al.* 2004; Tang-Wai *et al.* 2004). Balint's syndrome has also been noted in Creutzfeldt–Jakob disease (Victoroff *et al.* 1994) and progressive multifocal leukoencephalopathy (Ayuso-Peralta *et al.* 1994).

Differential diagnosis

Before deciding that Balint's syndrome is present, one must determine that the visual difficulties are not due to hemineglect, hemianopia, visual agnosia, or, with regard to optic ataxia, cerebellar dysmetria.

Treatment

Various rehabilitation techniques, anecdotally, have been helpful (Perez *et al.* 1996; Rosselli *et al.* 2001).

5.15 PHANTOM AND SUPERNUMERARY LIMBS

The phantom limb phenomenon and the supernumerary limb phenomenon are similar in that in both cases patients experience the presence of limbs which, in fact, are not present. These two phenomena differ, however, in that phantom limb follows amputation and is characterized by a re-experiencing of the limb that has been lost, whereas supernumerary limb occurs with cerebral lesions, such as infarctions, and is characterized by the experience of an extra limb, over and above the normal complement already present.

Phantom limb occurs in the vast majority of patients after amputation, and typically appears within the first few days postoperatively; in about three-quarters of cases, it is also accompanied by phantom limb pain. The supernumerary limb phenomenon, by contrast, is quite rare and is not painful.

Clinical features

Phantom limb (Carlen *et al.* 1978; Ephraim *et al.* 2005; Haber 1956; Henderson and Smythe 1948; Jensen *et al.* 1984; Shukla *et al.* 1982) is a tactile experience rather than a visual one: although patients do not see the limb, they 'feel' it and have a sense of its presence, and in some cases report being able to move it. Importantly, patients retain insight and do not, in fact, believe that a limb is actually present. Phantom limb pain is variously described as burning, electric, throbbing, cramping, tearing, or crushing, and can be quite severe. Over time, most patients experience what is known as 'telescoping': here, it is as if the phantom were being gradually absorbed back into the remaining limb, the most proximal part of the phantom disappearing first and the remaining portions telescoping after it, with the most distal portion, usually the fingers or toes, disappearing last. In some cases, the phantom may disappear entirely after months or longer, whereas in others it may be permanent.

Childhood amputations, before the age of 4 years, are only rarely followed by phantoms (Simmel 1962). The congenital absence of a limb, interestingly, may be followed in a small minority by the development of a phantom 'replacement' (Weinstein and Sersen 1961), generally around the age of 9 years (Melzak *et al.* 1997).

The phantom phenomenon, although most common after amputation of a limb, also occurs after mastectomy in one-fifth (Ackerly *et al.* 1955; Jarvis 1967) to two-thirds of women (Bressler *et al.* 1956) and may, in a small minority, be accompanied by pain. Phantom eye has also been reported after enucleation (Soros *et al.* 2003).

Supernumerary limb is characterized by the experience of having an extra arm or leg. One patient felt the extra arm lying across his abdomen (Mayeux and Benson 1979), another experienced it growing out of his forearm (Brock and Merworth 1957), and one patient, after suffering bilateral parietal lobe infarctions, experienced four legs, two on each side (Vuilleumier *et al.* 1997). Although in some cases patients may retain insight, at times this may be lost and patients may insist that, in fact, an extra limb is present (Halligan *et al.* 1993; Weinstein *et al.* 1954), and may even report seeing it (Khateb *et al.* 2009).

Etiology

The phantom limb phenomenon probably represents persistent activity of those structures, such as the parietal cortex, which maintain the body schema; phantom limb pain, in turn, appears to occur secondary to the cortical reorganization that follows amputations (Flor *et al.* 1995).

Supernumerary limbs have been noted with lesions of the parietal lobe (Vuilleumier *et al.* 1997; Weinstein *et al.* 1954), the medial aspect of the frontal cortex (McGonigle *et al.* 2002), basal ganglia (Halligan *et al.* 1993), thalamus (Miyazawa *et al.* 2004), and pons (Brock and Merwarth 1957). Although most supernumerary limbs appear on the left side, right-sided occurrence has also been noted (Mayeux and Benson 1979). Supernumerary limbs may also occur as a manifestation of a sensory simple partial seizure, a diagnosis immediately suggested by the paroxysmal onset and relatively brief duration of the experience. One patient, for a few seconds, would experience a supernumerary arm next to his actual left arm (Riddoch 1941); another had such a strong sense that an extra right arm was over his head that he asked his wife to help him pull it down (Russell and Whitty 1953).

Differential diagnosis

The phantom limb appearing after amputation is not mimicked by other conditions. It must be kept in mind, however, that, albeit rarely, not all pain experienced in the phantom limb is 'phantom limb pain' resulting from cortical reorganization. For example, one patient who experienced a non-painful phantom after amputation of a lower extremity subsequently developed disk disease with typical sciatic pain radiating down into the phantom (King 1956). In a similar vein, one patient, 5 years after undergoing bilateral below-the-knee amputations, developed restless legs syndrome in the phantoms, which was relieved by ropinirole (Vetrugno *et al.* 2010).

Treatment

The optimum treatment of phantom limb pain is not clear. Opioids are used, and provide some relief (Huse *et al.* 2001), but are certainly not universally effective. Gabapentin in high dosage was effective in one double-blind study (Bone *et al.* 2002) but not in another (Smith *et al.* 2005). Similarly, memantine, in a dose of 30 mg/day, was effective in one double-blind study (Schley *et al.* 2007) but ineffective in two others (Maier *et al.* 2003; Wiech *et al.* 2004). Infusions of salmon calcitonin were effective in one double-blinded study (Jaeger and Maier 1992). Anecdotally, patients have responded to treatment with a transcutaneous electrical nerve stimulation (TENS) unit (on the contralateral extremity) (Carabelli and Kellerman 1985; Giuffrida *et al.* 2010), deep brain stimulation (Bittar *et al.* 2005), clonazepam (Bartusch *et al.* 1996), carbamazepine (Patterson 1988), chlorpromazine (Logan 1983), and, despite an absence of depression, to fluoxetine (Power-Smith and Turkington 1993), mirtazapine (Kuiken *et al.* 2005), and to electroconvulsive therapy (ECT) (Rasmussen and Rummans 2000). In addition to pharmacologic treatment, there has also been enthusiasm for 'mirror therapy' (Chan *et al.* 2007). Here, with a suitably placed mirror within a specially constructed box, the patient sees an image of his 'good' leg moving in the same spot where the 'phantom' is experienced. Despite case reports of success with mirror therapy, however, a randomized controlled study (Brodie *et al.* 2007) found no benefit.

Regarding supernumerary limbs, there are, as of this writing, no known treatments.

5.16 DEPERSONALIZATION

Depersonalization is a common experience that is found not only in normal individuals, but also in a large number of other conditions.

Clinical features

In depersonalization, patients feel uncannily detached from themselves. Although they may continue to engage in some activity, whether it be talking, walking, or driving a car, it seems to them as if they are not really doing it but observing it being done, as if it were being carried out by a robot or an automaton. In some cases, visual distortions may occur: arms or legs may appear misshapen or shrunken. Patients may occasionally have the experience of 'floating' above the scene, watching themselves do things from a distance.

Etiology

The various causes of depersonalization are listed in Table 5.4.

Table 5.4 Causes of depersonalization

Normal response to danger (Morgan *et al.* 2001; Noyes and Kletti 1977; Noyes *et al.* 1977; Sedman 1966)
Toxic and metabolic causes
Marijuana (Mathew *et al.* 1993, Melges *et al.* 1970)
MDMA (ecstasy) (Vollenweider *et al.* 1998)
Quetiapine (Sarkar *et al.* 2001)
Fluoxetine (Black and Wojcieszek 1991)
Minocycline (Cohen 2004)
Indomethacin (Schwartz and Moura 1983)
Hypoglycemia (Chapter 14.15)
In association with migraine and epilepsy
Migraine aura (Lippman 1953)
Aura to complex partial or grand mal seizure (Devinsky *et al.* 1989; Ionasescu 1960)
Simple partial seizure (Williams 1956)
Traumatic brain injury (Grigsby and Kaye 1993; Hillbom 1960)
Panic attack (Cassano *et al.* 1989)
Depression (Noyes *et al.* 1977)
Depersonalization disorder (Chapter 25.4)

Depersonalization is most commonly seen as a normal, transient reaction to extreme danger, as may occur in combat, assaults, traumatic accidents, etc.

Of the various toxic and metabolic causes, intoxication with marijuana or MDMA (3,4-methylenedioxymethamphetamine, 'ecstasy') is most common, and in some cases patients may subsequently re-experience depersonalization as a 'flashback' when not intoxicated. Certain medications may also, albeit very rarely, cause depersonalization, including quetiapine, fluoxetine, minocycline, and indomethacin. Finally, depersonalization may constitute a prominent part of the symptomatology of hypoglycemia during its neuroglycopenic stage.

Depersonalization may also occur in association with migraine and epilepsy. Migraine headaches may be preceded by an aura of depersonalization, and the diagnosis is immediately suggested by the succeeding headache. Depersonalization may also occur as an aura to either a complex partial or grand mal seizure, and here again the diagnosis is immediately indicated by the subsequent seizure. Depersonalization may also constitute the sole symptomatology of a simple partial seizure, and here a high index of suspicion may be required to make the diagnosis. One important clue, of course, would be a history of other seizure types; however, this may not be universally present as in some cases epilepsy may be characterized only by simple partial seizures, without other seizure types: in such cases, an exquisitely paroxysmal onset of the depersonalization may be the only clue.

Traumatic brain injury may also cause depersonalization.

Panic attacks may have depersonalization as part of their symptomatology, and here the diagnosis is suggested by the associated anxiety and other autonomic symptoms such as tremor, tachycardia, and diaphoresis.

Depression may be complicated by depersonalization, and its occurrence within the setting of chronic depressive symptoms indicates the correct diagnosis.

Finally, when all other causes have been ruled out, one is left with a diagnosis, by exclusion, of depersonalization disorder. This is an idiopathic disorder, with an onset in late adolescence or early adult years, characterized by depersonalization that may either persist in a chronic waxing and waning fashion or be episodic.

Differential diagnosis

Autoscopy is a kind of visual hallucination wherein patients hallucinate a 'double' and see themselves as vividly as if looking in a mirror. Depersonalization differs from this in that there is always an 'as if' quality to the experience of depersonalization, accompanied by the sense of 'floating' over the scene that one is actually in.

Treatment

Treatment is directed toward the underlying cause.

5.17 OBSESSIONS AND COMPULSIONS

True obsessions and compulsions, once considered rare, are now known to be quite common, occurring not only in idiopathic obsessive–compulsive disorder, but also in a large number of other conditions.

Clinical features

Obsessions are involuntary, unwanted, and distressing thoughts that persistently recur despite attempts on the part of patients to stop them. Although the subject matter may at times be neutral, it more often than not concerns sexual or violent acts, acts in which the patient would never voluntarily engage. Compulsions represent an overwhelming and anxious need to do something, something that patients realize is more or less nonsensical. The compulsion is typically intimately connected with an apprehension on the part of patients that they have done something that they ought not to have done, or that they have left undone something that they ought to have done, and the compulsive act serves in some fashion either to undo what was done or to make good something that ought to have been done.

An example of an obsession is found in the case of the mother of a newborn infant, who was horrified to find herself recurrently having thoughts of 'stabbing the child, putting the infant in the microwave, or sexually abusing the newborn' (Sichel *et al.* 1993). In another case, the obsession was, *per se*, of a neutral character, consisting of 'silly poems that kept running through his consciousness': what 'appalled' the patient was the fact that he could not stop them (Mulder *et al.* 1951).

Common compulsions include compulsions to wash, to pray, or to check. Tuke (1894) described a 'young lady' who feared 'infecting herself' if she touched anything that might have germs on it, with the result that 'the act of washing her hands in the morning is repeated, immediately, again and again' such that the hands 'are quite rough and discoloured afterward', and Gordon (1950) described a patient who felt compelled to repeat his prayers again and again for fear each time that he had left a word out. Finally, Bleuler (1924), in commenting on a patient with a checking compulsion, noted that the anxious apprehension that the door had been left unlocked 'compels the patient to try it over and over again'.

Etiology

Table 5.5 lists the various causes of obsessions and compulsions.

By far the most common cause is the idiopathic disorder known as obsessive–compulsive disorder. This disorder, characterized primarily by obsessions and compulsions, has an onset in adolescence or early adult years, and generally pursues a chronic course. An episode of a major depressive disorder may also cause obsessions and compulsions; however, here the occurrence of these symptoms in the setting of a typical depressive episode suggests the correct diagnosis. Postpartum depression may also be marked by obsessions, typically, as indicated above, by horrific temptations. Schizophrenia, likewise, may be accompanied by obsessions and compulsions; however, here also it is the setting, in this case a chronic psychosis, which allows the correct diagnosis. Importantly, second-generation antipsychotic medications, as used in the treatment of schizophrenia, may also cause obsessions and compulsions as a side-effect: clozapine is most likely to do this; however, cases have also been reported secondary to olanzapine, risperidone, and quetiapine.

Obsessions and compulsions may also occur secondary to disorders characterized by abnormal movements, such as the tics of Tourette's syndrome, the chorea of Sydenham's chorea, Huntington's disease, or neuroacanthocytosis, the parkinsonism of Parkinson's disease, and dystonia, as in dopa-responsive dystonia, Meige's syndrome, and idiopathic cervical dystonia. Sydenham's chorea is of special interest here in that obsessions or compulsions are seen in the majority of patients and, interestingly, tend to appear prior to the onset of the chorea.

Of the precipitating events capable of causing obsessions and compulsions, the classic one is encephalitis lethargica, wherein these symptoms occur as chronic sequelae, often in association with oculogyric crises. Other encephalitides, such as western equine encephalitis, have also been implicated. Both posthypoxic–ischemic encephalopathy and traumatic brain injury may also leave obsessions and compulsions in their wake.

Of the miscellaneous causes, Fahr's syndrome, with calcification of the basal ganglia, stands out; infarction of the globus pallidus or the right parietal cortex has also rarely been at fault. Obsessions and compulsions may also constitute part of the symptomatology of a complex partial seizure or a simple partial seizure. In the case of a complex partial seizure, the associated defect of consciousness suggests the correct diagnosis; however, with simple partial seizures there may not be any other symptomatology besides the obsession or compulsion, and in this case the diagnosis may rest on a history of other seizure types. Finally, there is also an association of obsessions and compulsions with systemic lupus erythematosus, the velocardiofacial syndrome, and normal pressure hydrocephalus.

Table 5.5 Causes of obsessions and compulsions

Idiopathic disorders
Obsessive–compulsive disorder (Chapter 24.5)
Major depressive disorder (Chapter 23.3)
Postpartum depression (Chapter 23.6)
Schizophrenia (Byerly *et al.* 2005) (Chapter 22.1)
Medications
Clozapine (Baker *et al.* 1992)
Olanzapine (Mottard and de la Sablonniere 1999)
Risperidone (Alevizos *et al.* 2002)
Quetapine (Stamouli and Lykouras 2006)
Movement disorders
Tourette's syndrome (Chapter 9.25)
Sydenham's chorea (Chapter 19.11)
Huntington's disease (Chapter 9.10)
Neuroacanthocytosis (Chapter 9.11)
Parkinson's disease (Chapter 9.5)
Dopa-responsive dystonia (Chapter 9.19)
Meige's syndrome (Cavallaro *et al.* 2002) (Chapter 9.23)
Idiopathic cervical dystonia (Cavallaro *et al.* 2002) (Chapter 9.21)
With precipitating events
Post encephalitis lethargica (Chapter 15.6)
Post-western equine encephalitis (Mulder *et al.* 1951) (Chapter 15.4)
Posthypoxic–ischemic encephalopathy (Escalona *et al.* 1997; Laplane *et al.* 1984) (Chapter 13.1)
Traumatic brain injury (Hillbom 1960; McKeon *et al.* 1984) (Chapter 8.5)
Miscellaneous causes
Fahr's syndrome (Chapter 14.22)
Infarction of the globus pallidus (Giroud *et al.* 1997)
Infarction of the parietal cortex (Simpson and Baldwin 1995)
Complex partial (Mendez *et al.* 1996) or simple partial (Kroll and Drummond 1993) seizures
Systemic lupus erythematosus (Slatterly *et al.* 2004)
Velocardiofacial syndrome (Chapter 10.7)
Normal pressure hydrocephalus (Mendhekar *et al.* 2007)

Differential diagnosis

The adjectives 'obsessive' and 'compulsive' have been used not only to refer to true obsessions and compulsions, as described above, but also to other symptoms, and it is important to clearly distinguish 'true' obsessions and compulsions from these other symptoms. Thus, one hears of 'compulsive' spending, shopping, eating, or sex; however, one clinical feature clearly distinguishes all these from 'true' compulsions, namely the fact that patients with compulsive spending, shopping, eating, or sex enjoy the behavior they feel 'compelled' to partake in, in clear contrast with patients with 'true' compulsions who take no pleasure in their compulsive behavior, whether it be washing, praying, checking, etc.

Obsessive–compulsive personality traits must also be distinguished from obsessions and compulsions. Individuals with these personality traits are 'obsessed' with detail and feel 'compelled' to do things perfectly. These traits, however, are quite different from 'true' obsessions and compulsions in that individuals with these traits typically see nothing wrong with them (and in fact may even value them), in contrast with patients with 'true' obsessions and compulsions who see these experiences as more or less senseless and wish to be rid of them.

Perseverations, as may be seen in some patients with frontal lesions, are distinguished from compulsive behaviors by the patients' attitude toward the behavior in question. Perseverative patients can often offer no reason for their repetitive behavior and do not resist engaging in it: by contrast, patients with compulsions engage in their compulsive behavior for a definite purpose and often attempt to resist the urge. An allied symptom is 'compulsive' hoarding or collecting, as may be seen in various dementias (Hwang *et al.* 1998) and, rarely, with focal lesions of the frontal lobe (Anderson *et al.* 2005).

Treatment

Treatment is directed, if possible, at the underlying cause; if this is ineffective, consideration may be given to an empirical trial of one of the medications useful in obsessive–compulsive disorder, such as an SSRI.

5.18 REDUPLICATIVE PARAMNESIA

Reduplicative paramnesia, described in 1903 by Pick (Pick 1903) is a syndrome in which patients believe that their location has somehow been duplicated, such that, for example, the house they are in is not actually theirs or, conversely, that their house has been relocated at some distance, perhaps to another city. This syndrome has also been referred to as 'double orientation' or 'delusional disorientation'.

Clinical features

In some cases, patients may correctly identify the city they are in, but may maintain that the building they are in is not, in fact, what it appears to be. Some may admit that it perhaps looks like their home, but maintain that, in fact, it is not but rather is a replica or duplicate of their actual home (Hudson and Grace 2000).

In other cases, patients may accurately identify the building they are in, for example university hospital, but may maintain that the building, in fact, is not in the city, but has been transported or duplicated in another locale (Benson *et al.* 1976), perhaps their home town.

Etiology

Reduplicative paramnesia has been noted in schizophrenia (Joseph *et al.* 1999), mania as seen in bipolar disorder (Joseph *et al.* 1999), traumatic brain injury (Benson *et al.* 1976), Alzheimer's disease (von Gunten *et al.* 2005), and with focal lesions, as for example infarction of the right frontal lobe (Kapur *et al.* 1988), the right fusiform gyrus (Hudson and Grace 2000) or the right thalamus (Leiguarda 1983), and with a right parietal tumor (Ruff and Volpe 1981).

Differential diagnosis

Reduplicative paramnesia must be distinguished from disorientation to place. The differential here rests on the patient's reaction when corrected: the disoriented patient may either accept the correction or simply make no comment; the patient with 'delusional disorientation', however, will stoutly maintain his or her belief, and may even argue with the physician about it.

Treatment

Apart from treatment of the underlying cause, no confident recommendations can be made. Should the syndrome persist and be troubling, a case could be made for treatment with an antipsychotic.

5.19 CONFABULATION

Confabulations are false memories appearing in the context of a chronic anterograde amnesia, and these may occur in one of two forms, including 'provoked' or 'momentary' on the one hand, and 'spontaneous' or 'fantastic' on the other.

Clinical features

Provoked confabulations, which are by far the most common, appear only when the patient is 'provoked' by a question, and are generally prosaic, simple, and straightforward. For example, one patient, hospitalized for 3 days, when asked what he had done the evening before, replied that he had been out with friends at a restaurant, where everyone had enjoyed a fine meal, etc. When told the truth about what had happened, patients with provoked confabulations typically put up little resistance: the patient in question, when told that he'd been in the hospital the night before, looked thoughtful for a moment, then said that although he thought he had been out at the restaurant, he had guessed that in fact he had not, and then he let the matter rest.

Spontaneous or fantastic, confabulations, although at times appearing in response to a question, often are voiced spontaneously. Furthermore, these confabulations, unlike the prosaic provoked confabulations, have a fantastic, even at times dream-like quality to them. For example, one patient (Stuss *et al.* 1978), after the clipping of an anterior communicating artery aneurysm, during an interview:

> stood up without warning, walked to a window, gazed out at the surrounding buildings, returned to his seat, and

informed the examiner that his (the examiner's) boat had been stolen. Some weeks later, he approached another patient and told him that he had seen on television the previous night footage of the other patient's home being ripped out of the ground by a giant crane and of the patient himself running from the house. He gave equally improbable responses to routine questions about the place, his illness, and previous history.

(Stuss *et al.* 1978)

These patients, when told the truth, unlike those with provoked confabulations, will typically protest, and indeed may argue with the examiner about what had, in fact, happened.

In both types of confabulation, as noted earlier, there is an accompanying anterograde amnesia, which, as discussed in Chapter 6.6, is characterized not only by an inability to recall all out of three words after 5 minutes, but also by a variable degree of retrograde amnesia. Importantly, however, there is no confusion, nor are there any significant decrements in calculating or abstracting abilities.

Etiology

Provoked confabulations may occur in cases of chronic anterograde amnesia owing to lesions anywhere in the circuit of Papez (as discussed in Chapter 6.6), most commonly in Korsakoff's syndrome secondary to thiamine deficiency (as discussed in Chapter 14.5). Spontaneous confabulations, however, generally occur only in those cases of chronic anterograde amnesia due to bilateral basal forebrain lesions, as may occur after clipping of anterior communicating artery aneurysms (Schnider *et al.* 1996; Stuss *et al.* 1978) or traumatic brain injury (Schnider *et al.* 1996; Stuss *et al.* 1978).

Differential diagnosis

False memories may occur in a large number of other conditions, including delirium, dementia, and schizophrenia, and although a case could be made for considering these false memories to be confabulations, by convention, as indicated earlier, it is only false memories that occur in the context of a chronic anterograde amnesia that are considered to be confabulations.

Treatment

Treatment is directed at the underlying condition.

5.20 AMUSIA

Amusia, also known as amelodia, is characterized by an inability to either recognize or produce a melody, and may occur on either an acquired or a developmental basis.

Clinical features

Amusia occurs in both receptive and expressive forms.

In receptive amusia, patients are, to a greater or lesser degree, 'tone deaf'. In such a case, if the patient were to listen to a singer simply humming the tune to a song, he might be unable to recognize it, whereas if the singer actually sang the song, the lyrics would immediately allow the patient to recognize the song. As might be expected, if patients with receptive amusia listen to instrumental music they would have difficulty 'naming the tune' and might complain that the music sounded flat.

Patients with expressive amusia have great difficulty in 'carrying a tune', and if such patients hum a tune, listeners have great difficulty recognizing it; should these patients sing the lyrics, however, listeners are able to recognize the song but may comment that the singing is quite poor. Patients with expressive amusia also have great difficulty playing instruments.

In some cases, both receptive and expressive components are present, producing a global amusia.

Etiology

Most cases of acquired amusia occur secondary to infarction or hemorrhage. Receptive amusia has been noted with lesions of the temporal lobe (Mazzucchi *et al.* 1982; Sparr 2002). Expressive amusia has been noted with lesions of the superior temporal cortex (McFarland and Fortin 1982) and of the frontal cortex (Botez and Werheim 1959). Global amusia has been seen with lesions of the superior temporal cortex (Piccirilli *et al.* 2000) and also of Heschl's gyrus (Russell and Golfinos 2003). Typically, lesions are found in the non-dominant hemisphere.

Expressive amusia has also been reported as a manifestation of a simple partial seizure, with the seizure focus being found in the right temporo-occipital region (Bautista and Ciampetti 2003).

Global acquired amusia of gradual onset and slow progression has also been reported secondary to a neurodegenerative disorder characterized by bilateral atrophy of the frontotemporal cortices (Confavreux *et al.* 1992).

Developmental, or 'congenital', amusia is not uncommon and such 'tone deaf' individuals have a variable degree of difficulty in recognizing tunes and singing or playing instruments (Ayotte *et al.* 2001).

Differential diagnosis

Aprosodia is distinguished from amusia by the fact that aprosodia is related to the emotional tone with which one speaks, whether happy, angry, or sad, whereas amusia is related to the tune with which one sings.

Treatment

There are no established treatments.

5.21 FOREIGN ACCENT SYNDROME

The foreign accent syndrome, wherein patients speak with an accent foreign to their native tongue, is a rare syndrome that typically, but not always, evolves out of either a motor aphasia or an aphemia in stroke patients.

Clinical features

As noted, most patients present initially with either a motor aphasia (Christoph *et al.* 2004) (or its transcortical variant [Graff-Radford *et al.* 1986]) or an aphemic mutism (Schiff *et al.* 1983; Takayama *et al.* 1993): as the motor aphasia or the mutism resolve, patients begin to speak intelligibly but with a foreign accent. Importantly, however, more or less 'pure' cases do occur, wherein there is no preceding aphasia or mutism. Some examples of the foreign accent include a native English speaker who spoke with an Irish brogue (Seliger *et al.* 1992), another with a French accent (Hall *et al.* 2003), and another with a Chinese accent (Schiff *et al.* 1983). In one most unfortunate case, a Norwegian, who developed the syndrome during World War II, spoke in such a convincing German accent that she was suspected of being a traitor (Monrad-Krohn 1947).

Etiology

Although multiple different lesions in multiple different locations may cause this syndrome, in all cases the lesion has been in the left hemisphere.

Infarctions are typically found in the posterior–inferior portion of the frontal cortex (Schiff *et al.* 1983; Takayama *et al.* 1993) or subjacent white matter (Blumstein *et al.* 1987); cases have also been reported secondary to infarction of the left basal ganglia (Fridriksson *et al.* 2005), and there is one report of the syndrome occurring with infarction of the midportion of the body of the corpus callosum (Hall *et al.* 2003). Pure cases have been reported with infarction of the basal ganglia (Gurd *et al.* 1988) and the internal capsule (Ryallis and Whiteside 2006).

Cases have also been reported with a frontal intracerebral hemorrhage (Christoph *et al.* 2004), traumatic brain injury with a temporal intracerebral hemorrhage (Lippert-Gruener *et al.* 2005), and pure cases have further been reported with a multiple sclerosis plaque in the prerolandic white matter (Chanson *et al.* 2009) and as the initial symptom of primary progressive aphasia (Luzzi *et al.* 2008).

There are also reports of the syndrome occurring in the course of schizophrenia (Reeves and Norton 2001): in one such case, the patient, who had the delusion that he was connected with British royalty, spoke with a British accent (Reeves *et al.* 2007).

Differential diagnosis

Motor aphasia is distinguished by the characteristic effortful speech that stands in contrast with the fluent and effortless speech of patients with the foreign accent syndrome, and dysarthria does not enter into serious consideration for the simple fact that the speech of the foreign accent syndrome is not slurred. Finally, although extremely rare, consideration must also be given to the possibility that the foreign accent has occurred as a conversion symptom (Verhoeven *et al.* 2005).

Treatment

There is no known treatment.

5.22 CATAPLEXY

Cataplexy is a condition characterized by the occurrence of cataplectic attacks, that is to say episodes of a greater or lesser degree of muscle atonia.

Clinical features

Cataplectic attacks (Adie 1926; Dyken *et al.* 1996; Guilleminault *et al.* 1974; Kales *et al.* 1982; Parkes *et al.* 1975; Wilson 1928) are generally precipitated by some strong emotion, such as laughter or anger, and are characterized by the paroxysmal onset of more or less generalized weakness lasting on the order of seconds to a minute or so: subsequent recovery is rapid and complete. During a full attack, all voluntary muscle power, with the exception of the diaphragm and, at times, the extraocular muscles, is lost, and patients may fall to the ground or slump in a chair. In some cases, the muscle weakness, although generalized, may be of minor degree, and such patients may merely experience their heads lolling forward, their jaws slackening, and their knees beginning to buckle. Limited attacks, confined perhaps to an arm or leg, have also been reported. Attacks that last much longer than a minute may be joined by visual or auditory hallucinations (Van Den Hoed *et al.* 1979). Importantly, during the attack, patients remain completely alert and conscious, and afterwards are able to give a clear description of everything that happened.

Etiology

The overwhelming majority of cases of cataplexy occur secondary to narcolepsy. As discussed in Chapter 20.7, this disorder generally presents in late teenage or early adult years with narcoleptic sleep attacks; cataplectic attacks are seen in about three-quarters of patients, and although they typically occur within the first few years after the appearance of sleep attacks, in rare instances they may precede them (Parkes *et al.* 1975).

Isolated cataplectic attacks have also, very rarely, been reported to occur on an autosomal dominant (Gelardi and Brown 1967) or idiopathic basis (Van Dijk *et al.* 1991).

Cataplectic attacks, again very rarely, have also been noted with lesions, generally gliomas, of the hypothalamus (Anderson and Salmon 1977; Schwartz *et al.* 1984;

Stahl *et al.* 1980), midbrain tegmentum (Fernandez *et al.* 1995), or dorsal pontomedullary junction (D'Cruz *et al.* 1994). There are also case reports of cataplexy occurring as sequelae to encephalitis lethargica (Adie 1926; Fournier and Helguera 1934) in association with paraneoplastic limbic encephalitis (Rosenfeld *et al.* 2001) in Niemann–Pick disease (Smit *et al.* 2006) and as a side-effect of clozapine (Desarkar *et al.* 2007).

Differential diagnosis

Cataplectic attacks must be distinguished from vertebrobasilar transient ischemic attacks, syncope, and atonic seizures, and one feature helps in distinguishing all of these from cataplexy, namely that, in contrast with cataplectic attacks, none of these is typically provoked by strong emotions.

Attacks of vertebrobasilar insufficiency may be characterized by 'drop attacks' with preservation of consciousness. These are usually seen in elderly individuals, and may be accompanied by other evidence of brainstem ischemia, such as dysarthria or vertigo.

Syncopal attacks are immediately distinguished from cataplexy by loss of consciousness.

Atonic seizures may or may not be accompanied by loss of consciousness or by postictal confusion. In cases of atonic seizures with preserved consciousness and no postictal confusion, a history of other seizure types will suggest the correct diagnosis (Lipinski 1977; Pazzaglia *et al.* 1985).

Treatment

Treatment of cataplexy occurring in association with narcolepsy is discussed in Chapter 20.7; when symptomatic treatment is required for cataplexy occurring secondary to other causes, an empirical trial of these same treatments would be reasonable.

5.23 SYMPATHETIC STORM

Sympathetic storms constitute episodes of severe sympathetic hyperactivity, and are usually seen after traumatic brain injury. An often seen synonym is dysautonomia; however, this term is more often used to refer to an inherited disorder of the autonomic nervous system, and thus the more specific name, sympathetic storm, is preferable.

Clinical features

The episodes may occur either spontaneously or after some trivial precipitant, such as a change in position. During the episode one sees the acute onset of profuse diaphoresis, tachycardia, tachypnea, pupillary dilation, and, in some, rigid extensor posturing. The diaphoresis is indeed impressive, with beads of sweat dripping from the head. During the episode, patients may grimace as if in pain, and family members and other observers may become quite alarmed. The episodes themselves last from minutes to hours and terminate slowly.

Etiology

These episodes probably represent disinhibition of the hypothalamus and related structures. Although most commonly seen after severe traumatic brain injury with prominent diffuse axonal injury (Baguley *et al.* 1999; Hendricks *et al.* 2010), they may also occur in the setting of acute hydrocephalus (Rossitch and Bullard 1988) and after cardiac arrest (Diamond *et al.* 2005).

Although these episodes were initially considered by Penfield and Jasper to represent diencephalic seizures (Penfield and Jasper 1954), there is no evidence for epileptic activity during them, and anti-epileptic drugs are not effective (Boeve *et al.* 1998).

Differential diagnosis

Autonomic dysreflexia, seen in patients with spinal cord lesions above T6, manifests with diaphoresis, headache, and hypertension, and, in some cases, piloerection and flushing. The paroxysmal nature of this syndrome, and the associated diaphoresis, may suggest a sympathetic storm, but its occurrence in the setting of a spinal cord lesion will suggest the correct diagnosis.

Malignant hyperthermia, neuroleptic malignant syndrome, and Stauder's lethal catatonia may all be considered (Thorley *et al.* 2001); however, in each of these conditions, symptoms, rather than being episodic, are persistent. The setting, of course, also aids in differentiation, and one should look, respectively, for administration of an anesthetic or an antipsychotic, or pre-existing catatonia.

Treatment

The goal of treatment is prevention of future episodes, and in this regard chronic treatment with a beta-blocker, such as propranolol or labetalol (Do *et al.* 2000), constitutes a reasonable first approach. Both morphine sulfate and bromocriptine have also been used with success (Bullard 1987), as has gabapentin, in doses of 600–1800 mg/day (Baguley *et al.* 2007). Doses should be titrated to clinical effect or tolerance; in some cases a combination of agents, such as a beta-blocker plus bromocriptine, may be required. Should pharmacologic treatment fail, consideration may be given to intrathecal baclofen (Becker *et al.* 2000).

5.24 CATASTROPHIC REACTION

The catastrophic reaction, originally described by Kurt Goldstein (Goldstein 1939, 1942), is characterized by an extreme emotional reaction when patients are confronted with tasks which, since falling ill, they are no longer able to accomplish. This may not be an uncommon phenomenon, being described in 19 percent of patients with stroke (Starkstein *et al.* 1993) and 16 percent of those with Alzheimer's disease (Tilberti *et al.* 1998).

Clinical features

The reaction occurs when patients find themselves unable to accomplish a task which, prior to falling ill, would have occasioned them little or no difficulty. The task itself might be something as simple as making change, or something more formal, as for example answering questions regarding memory or calculations during a mental status examination. Finding themselves hopelessly unable to proceed, patients become frustrated, tearful, or angry; some may begin shouting or swearing, and aggressive behavior may occur. Carota *et al.* (2001) noted that when patients with expressive aphasia were given paper and pencil and asked to write, they would 'start crying, showing unwillingness, tearing the paper or even throwing the pencil and other objects toward the examiner'. Typically, this catastrophic display of emotion passes soon after patients are relieved of the burdensome task.

Etiology

There has been debate over whether the catastrophic reaction is merely an expected emotional reaction when patients find themselves failing at tasks that would have caused no difficulty in the past or whether it is perhaps related to specific pathology. Although Goldstein held to the former theory, there is some evidence to the contrary. One study (Starkstein *et al.* 1993) of stroke patients found an association between the occurrence of a catastrophic reaction and both infarction of the basal ganglia and post-stroke depression, especially when this depression occurred in the presence of subcortical frontal infarction. Furthermore, there may also be an association between the catastrophic reaction in stroke patients and the presence of a motor aphasia (Carota *et al.* 2001; Gainotti 1972), though not all studies support this (Starkstein *et al.* 1993).

Differential diagnosis

Catastrophic reactions, as defined here, are clinical events that occur in reaction to stressful demands, and thus are differentiated from more 'free-floating' agitation, as may be seen in various deliria or dementias.

Treatment

Pending improvement in the patient's abilities, environmental demands should be appropriately scaled back. Importantly, in addition to treating the underlying disorder, and any specific deficits which hinder the patient's performance, depression, if present, should be aggressively treated.

5.25 FLATTENED AFFECT

Clinical features

Flattened, or blunted, affect is characterized by a lifeless and wooden facial expression accompanied by an absence or diminution of feeling.

Etiology

Flattened affect is found very commonly in schizophrenia (Andreasen 1979); it may also occur in other psychoses (Cornelius *et al.* 1991) and, rarely, in dementia secondary to infarction of the mesencephalon or thalamus (Katz *et al.* 1987).

Differential diagnosis

Motor aprosodia, as discussed in Chapter 3.8, is characterized by a monotone voice, but in contrast with patients with flattened affect, who feel little, if anything, patients with aprosodia retain lively emotional feelings. The 'hypomimia' seen in parkinsonism is distinguished in the same way: although these patients' facial movements are more or less frozen and devoid of expression, they still may have strong feelings.

In depression, the play of emotions on the face is slowed and thus a depressed affect may appear similar to a flattened affect; however, these patients do experience a sense of sadness in contrast with patients with flattened affect, who, again, simply feel nothing.

Treatment

Treatment is directed at the underlying cause.

5.26 INAPPROPRIATE AFFECT

The term 'affect' refers both to what is felt and to what shows in facial expression. Normally, there is a congruence between these two, as, for example, when sad feelings are accompanied by a sad facial expression. In cases when there is an involuntary incongruence between these two components of affect – a mismatch, as it were, between what is felt and what shows – one speaks of inappropriate affect.

Clinical features

Examples may help fix this sign in the reader's mind. When attending his brother's funeral, one patient, while feeling nothing but grief and sadness, was noted to have a strange grin on his face; another, although reporting pleasure at receiving a gift, was noted to grimace as if in pain.

Etiology

By far the most common condition in which inappropriate affect is seen is schizophrenia (John *et al.* 2003) and the closely allied schizotypal personality disorder (Fossati *et al.* 2001). Such affect has also been noted during intoxication with hallucinogens, such as ketamine and dimethyltryptamine (Gouzoulis-Mayfrank *et al.* 2005), and in the very rare velocardiofacial syndrome (Sachdev 2002).

Differential diagnosis

Inappropriate affect, as defined here, must be distinguished from the far more common 'socially' inappropriate affect. For example, smiling at a funeral, although certainly socially inappropriate, would not be considered pathologic if the person smiling felt a sense of triumph at the death of a hated rival, for here there is no mismatch between what is felt and what is displayed.

'Nervous laughter', for example when someone laughs to cover up a feeling of sadness, although indeed representing a mismatch is distinguished by the fact that the covering laughter can be easily dispelled with a few sympathetic questions, after which the laughter is replaced by an appropriate look of sadness.

Aprosodia of the sensory type, as discussed in Chapter 3.7, is quite similar to inappropriate affect in that in both these signs there is an incongruence, or mismatch, between what the patient feels and the tone of voice with which that feeling is expressed. The difference is that in sensory aprosodia, patients also have difficulty in comprehending what others feel by listening to the tone of voice with which others speak, whereas in inappropriate affect patients retain this ability.

Treatment

Treatment is directed at the underlying condition.

5.27 MANNERISMS

Manneristic transformation of gestures or speech, or of activities such as walking or eating, although most common in schizophrenia, may occur in other disorders. The resulting mannerisms often strike others as peculiar or bizarre.

Description

Kraepelin (1907) noted that patients with mannerisms may '. . . walk with a peculiar gait, drag one foot, go in straight lines or in circles, hold their spoons at the very end, eat in a definite rhythm, and shake hands with extended fingers'. He felt that mannerisms were '. . . especially common in speech (with) . . . grunts, lisping, peculiar words, phrases and inflection'.

Bleuler (1924) commented further regarding the manneristic transformation that gestures may undergo, noting that '. . . every conceivable stilted gesture occurs. Shaking hands is done very stiffly with the hand turned or only presenting the little finger; the hand may be shot forward quickly and withdrawn just as rapidly'.

Etiology

Although mannerisms are, as noted, most commonly seen in schizophrenia (Luchins *et al.* 1992) (especially catatonic schizophrenia [Ungvari *et al.* 2005]), they may also occur in mental retardation (Leudar *et al.* 1984), autism (Sears *et al.* 1999), ketamine intoxication (Krystal *et al* 2005), and in dementia (Rabinowitz *et al.* 2004).

Differential diagnosis

Stereotypies are distinguished by their monotonous repetitiveness.

In patients treated chronically with antipsychotics, mannerisms must be distinguished from tardive dyskinesia (Granacher 1981). One clue to the differential lies in the presence or absence of a motivation for the behavior in question: mannerisms represent intentional behaviors that have undergone a bizarre transformation; the abnormal movements of tardive dyskinesias, in contrast, are involuntary and occur in the absence of any motivation.

Treatment

Treatment is directed at the underlying etiology.

5.28 STEREOTYPIES

As noted in Chapter 5.5, stereotypies constitute a subset of perseveration and are characterized by repetitive, stereotyped motor behaviors: these may range from such simple behaviors as hand flapping to complex activities such as repeatedly taking apart small machines. Importantly, although some of these stereotypies appear, on the surface, to be purposeful, patients are unable to adequately explain why they repeatedly engage in the behavior.

Clinical features

As noted, stereotypical behaviors can range from simple to complex. Simple stereotypies might consist of hand flapping, finger flicking, hand wringing or washing movements, shoulder shrugging, or body rocking. Complex stereotypies might include folding and unfolding a towel, repeatedly putting on and then taking off an article of clothing, or repeatedly sitting down and standing up. The most complex stereotypies are subsumed under the name 'punding', and these include unloading drawers or handbags, or taking apart machines (e.g. radios, clocks, or even car engines), and then sorting and arranging everything, after which there may or may not be an attempt to reassemble the item.

Stereotypies may persist for anywhere from minutes to hours, days, or even longer. Importantly, patients, provided they are not severely or profoundly retarded, are generally able, on command, to at least temporarily stop what they are doing.

Etiology

Simple stereotypies may be seen in frontotemporal lobar degeneration (Mateen and Josephs 2009; Nyatsanza *et al.* 2003), moderate to severe Alzheimer's disease, mental retardation (Bodfish *et al.* 2000) of moderate or higher grade (especially that due to Rett's syndrome ([Vignoli *et al.* 2009]), wherein hand wringing or washing movements are classic), autism (Goldman *et al.* 2009), and, albeit rarely, in children without mental retardation or autism (Mahone *et al.* 2004): in roughly half these cases the children did have either attention-deficit–hyperactivity disorder or developmental dyslexia; however, in the remainder no other significant clinical abnormalities were noted. Simple stereotypies may also be seen in schizophrenia, and some of the stereotyped choreiform movements of tardive dyskinesia (Stacy *et al.* 1993), such as shoulder shrugging, foot tapping, or body rocking, may also be subsumed under the rubric of stereotypies.

Complex stereotypies may also be seen in schizophrenia, while punding is classically associated with amphetamine (Rylander 1972) or cocaine (Brady *et al.* 1991) intoxication, or with treatment with direct-acting dopaminergic agents, most commonly in patients with Parkinson's disease (Fasano and Petrovic 2010) but also in those with restless legs syndrome (Cornelius *et al.* 2010); punding has also been noted secondary to levodopa (Evans *et al.* 2004), but this is less likely a culprit than the direct-acting agents.

Differential diagnosis

Tics are generally not repetitious, and although they can be suppressed this is done only with great effort and eventual failure. Compulsions are distinguished by the fact that there is a motivation for the behavior: if you ask a patient who repeatedly goes to and from the stove, each time touching the gas knob, why he or she is acting so, the patient with a compulsion might explain that it is because of an overwhelming anxiety that the stove might not have been turned off, whereas the patient with a stereotypy will not be able to offer a reason.

Treatment

For the stereotypies of frontotemporal lobar degeneration and Alzheimer's disease, there is no established treatment. In patients with mental retardation, low-dose antipsychotics, such as risperidone, are useful, and clomipramine is also effective (Lewis *et al.* 1996). In autism, risperidone (McDougle *et al.* 2005) and haloperidol are effective, as is clomipramine, but this is not as well tolerated as haloperidol (Remington *et al.* 2001); a recent study also suggested effectiveness for divalproex (Hollander *et al.* 2006). Behavior therapy may also be effective in the treatment of stereotypies in mental retardation and autism. The stereotypies of schizophrenia may respond to antipsychotics, and those of tardive dyskinesia may respond to the treatments outlined in Chapter 29.2. In patients with punding, consideration may be given to switching from a direct-acting dopaminergic to levodopa, which, as noted earlier, is much less likely to cause this.

5.29 ECHOLALIA AND ECHOPRAXIA

Echolalia and echopraxia represent, respectively, the repetition or mirroring of what others say or do (Ford 1989).

Clinical features

In echolalia, patients automatically repeat, parrot-like, what others have said: in some cases, only single words are repeated, whereas in others whole phrases or sentences may be echoed. In echopraxia, patients, again automatically and without being instructed to do so, will mirror what others do. For example, if the physician crosses his or her arms, the echopractic patient will do the same thing; likewise, if the physician stands, the patient will also stand. The automatic, involuntary aspect of these echophenomena becomes apparent when one instructs patients to refrain from imitating what the physician says or does: although some patients may be able to comply, this compliance is temporary at best, and very soon the echophenomena re-emerge.

Etiology

The classic cause of echophenomena is catatonia, as discussed in Chapter 4.11. Echophenomena may occur in a number of other disorders, including autism (Roberts 1989), Tourette's syndrome (Comings and Comings 1987), Pick's disease (Gustafson 1987), frontotemporal lobar degeneration (Gustafson 1987), Alzheimer's disease (Cummings *et al.* 1985), and focal lesions (De Renzi *et al.* 1996) such as infarctions or tumors, of the medial or lateral aspects of the frontal lobes, or, less commonly, of the striatum. Echolalia may also be seen in both transcortical motor (Hadano *et al.* 1998) and mixed (Mendez 2002) aphasia.

Differential diagnosis

Palilalia is distinguished from echolalia by the fact that the repetition in palilalia involves not words spoken by others but words spoken by the patient. The environmental dependency syndrome is distinguished from echopraxia by the fact that the automatic behavior of the patient with the environmental dependency syndrome represents not a mirroring of what others do but rather an involuntary utilization of objects at hand.

Treatment

Treatment is directed at the underlying cause.

5.30 HALLUCINATIONS AND DELUSIONS

Hallucinations are said to be present when patients have the experience of something that is, in fact, not actually present. Thus, a patient who hallucinates a dog in the hospital room might report seeing a dog, and even reach down to pat it, whereas others in the room see nothing.

Delusions are false beliefs that cannot be explained on the basis of either the patients' culture or religion and that persist despite evidence and reasonable argument to the contrary. Thus, in evaluating patients with false beliefs, it is necessary to take into account their culture and religion. Although a belief in zombies would not generally be considered to be a delusion in poor and uneducated Haitians, for example, any well-acculturated American who maintained such a belief would generally be considered delusional. Although some physicians make much of the 'contextual' nature of beliefs, most delusions are in practice so bizarre and unbelievable that there is no question of their status. Thus, there is but little debate about the status of such beliefs that bodies of family members are inhabited by malevolent aliens or that telepathic listening devices have been implanted inside the spleen.

A group of specific hallucinations and delusions, known as Schneiderian first-rank symptoms, is discussed separately in Chapter 5.31, and the reader is strongly encouraged to read that chapter, given the extraordinary importance of these particular hallucinations and delusions.

Clinical features

Hallucinations are categorized according to their sensory modality, while delusions may be classified according to their content.

HALLUCINATIONS

All sensory modalities are represented here, and thus there may be visual, auditory, tactile, olfactory, and gustatory hallucinations, and each is discussed in turn. From a diagnostic point of view, it is critical to determine whether the patient with an hallucination has 'insight' or not. Happily, this is not terribly difficult. 'Insight', a much belabored term, refers here not to some sophisticated level of psychological understanding, but rather simply to whether or not the patient recognizes that the hallucination is 'not real'. Sometimes simple observation will enable one to determine whether the patient experiences the hallucination as 'real': for example, should a patient report seeing a dog in the room and then ask the physician to bring it over to the bed so it could be patted, one may reasonably assume that the patient has no insight into the hallucinatory nature of the experience. In doubtful cases, further inquiry may be required. Although one might simply ask whether or not the patient thinks that the 'dog' is 'really' there in the room, such questions may offend some patients, and a more diplomatic approach is often better accepted. Thus, the patient might respond, 'Oh, there's not a dog here. I must be seeing things'. Conversely, in cases where insight has been lost, the patient might point emphatically to a corner of the room and say 'It's right over there. Can't you see it?'

Visual hallucinations range from simple to complex in form. Simple hallucinations include such phenomena as 'zigzag linear patterns' (Panayiotopoulos 1994) and 'dancing lights, rings, and circles' (Parkinson *et al.* 1952). Complex phenomena may be quite detailed: one patient, a war veteran, saw 'stretcher bearers walking past him and then the figures of nurses whom he would recognize' (Russell and Whitty 1955); another saw 'the Queen, with a bag over her arm [who] walked from the left to the centre of the ward, then vanished' (Lance 1976).

In some cases, autoscopy may occur, wherein patients hallucinate an image of themselves (Lukianowicz 1958): one 74-year-old woman 'suddenly noticed a figure seated on [her] left. "It wasn't hard to realize that it was I myself who was sitting there. I looked younger and fresher than I do now. My double smiled at me in a friendly way, as though she wanted to tell me something"' (Kolmel 1985).

'Lilliputian' hallucinations are said to be present when the hallucinated figures are quite small, much as in the novel *Gulliver's Travels* (Alexander 1926; Goldin 1955; Leroy 1922).

Palinopsia represents a peculiar kind of visual hallucination that Critchley (1951) considered to be a type of 'visual perseveration'. Here, objects originally actually seen by the patient are subsequently recurrently hallucinated out of place. For example (Michel and Troost 1980), one patient reported that 'each person she saw had the face of someone she had just seen on television. She later peeled a banana and in a few minutes saw multiple vivid images of bananas projected over the wall. She realized they were only images, and not real. Next, after putting a 20-dollar bill in her purse, she saw heaps of 20-dollar bills everywhere'. Another patient (Meadows and Munro 1977), after seeing someone costumed as Santa Claus at a Christmas party, 'noticed that a replica of the white beard of ... Santa Claus was superimposed upon the face of everyone she spoke to'.

In some cases, hallucinations occur only upon either falling asleep or awakening, and these are known respectively, as hypnagogic and hypnopompic hallucinations (Zarcone 1973).

Auditory hallucinations may range from such simple phenomena as 'grinding noises' or a noise 'like a freight train' (Cascino and Adams 1986) to complex experiences such as hearing music (Hammeke *et al.* 1983) or voices. Voices, the most compelling type of auditory hallucination, may be soft, mere whisperings, or quite clear and distinct. They may come from an internal organ or have their source outside the patient in 'the air', or perhaps electronic devices or simply the walls. Some patients seem able to ignore them, whereas others will talk back to them. There may occasionally be 'command' hallucinations, which, as noted by Kraepelin (1921), 'in certain circumstances are very precisely obeyed. They forbid the patient to eat and to speak, to work, to go to church . . .'.

Palinacousis represents the auditory counterpart to palinopsia, and here patients repeatedly hallucinate something that they had just heard (Di Dio *et al.* 2007).

Tactile hallucinations (Berrios 1982) are extraordinarily varied. Kraepelin (1921), in describing a tactually hallucinated patient, noted that he 'feels himself laid hold of, touched over his whole body, he feels tickling in his thigh and right up to his neck, pricking in his back and in his calves, a curious feeling in his neck, heat in his face'. When patients hallucinate the feeling of ants or insects crawling on the skin, one speaks of 'formication'.

Olfactory hallucinations, although at times consisting of such pleasant aromas as perfume, are generally unpleasant, even repulsive: Kraepelin (1921) noted 'a smell of sulphur; of corpses and chloride of lime, of blood, of fire, of the fumes of hell'.

Gustatory hallucinations are perhaps the least common of all types: Kraepelin (1921) reported patients who tasted 'petroleum or arsenic' in their food.

DELUSIONS

Although delusional beliefs may be extraordinarily varied in their content, certain themes are commonly found, and thus there are delusions of persecution, grandeur, and jealousy, erotomanic delusions, nihilistic delusions, delusions of sin, and delusions of reference. Each is considered in turn, beginning with the most common, namely delusions of persecution.

Delusions of persecution take the most varied of forms: patients are spied on and followed; a conspiracy has been set up, and the police, FBI, or even the mafia are involved. Neighbors and co-workers, even family members, have turned on them: poison gas surrounds them, and their food, indeed even their medicine, has been adulterated.

Capgras's syndrome (Alexander *et al.* 1979; Christodoulou 1977; Enoch 1963) represents a particular kind of delusion of persecution wherein patients believe that malevolent 'imposters' have somehow entered and taken over the bodies of familiar persons, such as family members or neighbors. In one case (Merrin and Silberfarb 1976) the patient believed that 'the "substitution" first occurred on [her] wedding day, when her husband went to the men's room and an impostor took his place'. Although others could not see anything amiss with her husband, the patient felt able to differentiate the husband from the impostor given that the impostor 'had a rotten green toenail. Once she even went so far as calling the police and demanding that her husband remove his shoe to expose the green toenail to the police'. In some cases of Capgras's syndrome, patients, in order to protect themselves, may violently turn on the 'impostor' (Thompson and Swan 1993).

In a related kind of delusion of persecution, known as the Fregoli syndrome, patients believe that the body of a stranger has somehow been invaded and taken over by a familiar person (O'Sullivan and Dean 1991).

Delusions of grandeur may involve various themes: the patient 'is "something better", born to a higher place, the "Glory of Israel", an inventor, a great singer, can do what he will' (Kraepelin 1921).

Delusions of jealousy (Soyka *et al.* 1991) fuel patients' jealousies: the infidelities of spouses or lovers are revealed and their 'playing around' is no longer hidden. Protestations of faithfulness have no effect as patients find evidence for their suspicions: the spouse is late coming home, the phone rings at odd times, the sheets are rumpled in a suggestive way.

Erotomanic delusions are characterized by a belief in an 'amorous communication with a person of much higher rank who has been the first to fall in love and the first to make advances' (Gillett *et al.* 1990). Also known as De Clerambault's syndrome (Ellis and Mellsop 1985), these delusions often impel patients to attempt contact with these supposed lovers, even to the point of stalking and kidnapping them.

Nihilistic delusions entail the belief that animate objects, such as other people, animals, or such living things as trees, are in fact dead. In a variant of this, known as Cotard's syndrome (Joseph and O'Leary 1986), patients come to believe that they themselves are dead: in one case (Cohen *et al.* 1997), a 13-year-old girl had the 'absolute conviction that she was already dead, waiting to be buried, and that she had no teeth and no hair and that her uterus was malformed'.

Delusions of sin give form to the patients' sense that they have committed unpardonable acts: they have cursed God, engaged in unspeakable practices, betrayed those close to them, and violated all their sacred principles. One of Kraepelin's patients (Kraepelin 1921) believed himself 'a damned soul, the refuse of humanity'.

Delusions of reference embody a sense that seemingly unrelated and chance events in some fashion pertain to or refer to the patient. Kraepelin (1921) noted that '[i]ndifferent remarks and chance looks, the whispering of other people, appear suspicious to the patient'. In many cases, such delusions of reference serve, as it were, to bolster or reinforce other delusions. Thus, patients with delusions of persecution, on seeing police officers talking, might immediately believe that the conversation was about them and that it was evidence that the conspiracy had begun in earnest. To take another example, patients with delusions of grandeur might, upon seeing the sun break radiantly through the storm clouds, assume that it was a sign of God's special grace for them.

Etiology

In the overwhelming majority of cases, hallucinations and delusions occur in the context of a major syndrome, such as dementia, delirium, depression, mania, and psychosis, and the first task of the diagnostician is to determine whether one of these syndromes is present. Thus, one first looks carefully for any clinically significant cognitive defects, such as memory loss, disorientation, and deficits in calculation or abstracting abilities suggestive of dementia, and then for the all-important symptom of confusion, which would indicate a delirium. Depression is suggested not only by a depressed affect and mood, but also by the presence of such historical features as anergia, anhedonia, and changes in appetite and sleep. Mania, with heightened mood, hyperactivity, pressured speech, and increased energy is hard to miss. Finally, one must ask whether a psychosis is present. In this syndrome, there are no significant cognitive defects, nor are there any significant depressive or manic symptoms. Rather, what one sees are hallucinations that are experienced without insight or delusions, or both.

Although the vast majority of cases will be found to have one of the foregoing syndromes, there yet remains a remnant of cases characterized by hallucinations experienced with insight into their hallucinatory nature, in the absence of cognitive defects, depressive or manic symptoms, or accompanying delusions, and it is to this group that we now turn.

In determining the cause of isolated hallucinations occurring with preserved insight, the differential is different for each of the different kinds of hallucinations, and thus each type is discussed in turn, beginning with visual hallucinations and proceeding to auditory, tactile, olfactory, and, finally, gustatory hallucinations.

Visual hallucinations are by far the most common type, and of the various causes of these, medications and intoxicants stand out. Of the medications capable of causing hallucinations, by far the most common are dopaminergic agents when used in the treatment of Parkinson's disease. After five or more years of treatment with levodopa, for example, visual hallucinations occur in a little over one-fifth of all patients (Friedman and Sienkiewicz 1991; Graham *et al.* 1997). The hallucinations themselves may, at times, be quite elaborate: in one case (Graham *et al.* 1997) a patient had 'hallucinations of miniature people and domestic animals . . . the figures were non-threatening, laughed and talked among themselves, and had a male leader who organized them into purposeful activities'. It is important to note that although parkinsonian patients with levodopa-induced visual hallucinations retain insight initially, over many years insight tends to be lost and the syndrome of psychosis emerges (Goetz *et al.* 2006). Other medications may also cause hallucinations with preserved insight, such as digoxin (Closson 1983) or quinidine (Fisher 1981), but given that such hallucinations are relatively uncommon with medications other than dopaminergics, caution must be utilized before drawing a causal connection.

Of the intoxicants capable of causing hallucinations, the aptly named hallucinogens are the most common. Of note, in this group, one of the earliest descriptions was provided by the eminent neurologist S. Weir Mitchell, who, in 1896 (Mitchell 1896), reported his own experiences with mescaline. Although the hallucinogen-induced hallucinations are, in most cases, fairly simple, for example in geometric forms, they may sometimes be complex (e.g. with lysergic diethylamide [LSD] [Bercel *et al.* 1956]). Hallucinogen-induced visual hallucinations may also occur as 'flashbacks', wherein long after hallucinogen intoxication the patient spontaneously re-experiences some of the visual phenomena that occurred during the intoxication (Abraham 1983; Horowitz 1969).

Partial seizures may present with visual hallucinations (Russell and Whitty 1955): in simple partial seizure the visual hallucination may constitute the entire symptomatology of the seizure, whereas in complex partial seizures it will be accompanied by some defect of consciousness.

Migraine headaches are typically preceded by an aura consisting of a visual hallucination, which tends to be simple, consisting of flashing lights or zigzagging lines (Panayiotopoulos 1994; Russell and Olesen 1996). An aura may, very rarely, persist long after the headache has cleared, sometimes for years (Liu *et al.* 1995).

Blindness, whether partial or complete, may be associated with either simple or complex visual hallucinations in what is known as the Charles Bonnet syndrome (Santhouse *et al.* 2000; White 1980). This syndrome has been noted with visual loss caused by cataracts (Bartlett 1951), glaucoma (Nesher *et al.* 2001), macular degeneration (Holroyd *et al.* 1992), and lesions of the optic nerve, chiasm, tract, and optic radiations (Lepore 1990). Although classically the hallucinations of the Charles Bonnet syndrome are said to be 'Lilliputian' in character, it in fact appears that such miniaturization occurs in only a minority of patients with this syndrome (Teunisse *et al.* 1994). The onset of the Charles Bonnet syndrome can at times be quite dramatic: in one case (White 1980), a 69-year-old man, while listening to music, 'suddenly saw a brightly coloured circus troupe burst through the window'.

Focal intracerebral lesions, usually infarctions, may cause hallucinations, and this has been noted with lesions of the occipital lobe and adjacent temporal and parietal lobes, and with lesions of the thalamus (Noda *et al.* 1993), mesencephalon, or pons.

Lesions of the occipital cortex may, of course, also cause a hemianopia, and in such cases the hallucinations tend to occur in the hemianopic field (Kolmel 1985; Vaphiades *et al.* 1996). The clinical presentation in such cases may be quite remarkable: one patient (Lance 1976) with a left hemianopia saw animals appearing 'from the left one at a time. At various times he saw dogs, goats, a lion and a horse as well as birds and butterflies. The animals would emerge from a door on the left side of the room and walk to the mid-line. If he looked to the left the animals retreated towards the door but would advance again as he looked to the front'.

Another patient, a neurologist, after suffering an infarction of the medial left occipital lobe, developed a right hemianopia in which he experienced vivid visual hallucinations: he noted that 'often there was a pony with his head cradled in my right arm' (Cole 1999).

Hallucinations occurring secondary to mesencephalic lesions are often referred to as 'peduncular' hallucinations (Lhermitte 1922, 1932) given that the lesion typically involves one of the cerebral peduncles. These peduncular hallucinations are typically complex and vivid: one patient (Geller and Bellor 1987) saw 'cats running about the floor, flowery outdoor scenes in bright purple colors, and the faces of neighbors and friends', whereas another (De La Fuente Fernandez *et al.* 1994) saw 'motorbikes . . . dogs, horses . . . and people . . . entering and driving silently around the room'. Although insight is preserved, and patients recognize the unreality of these peduncular hallucinations, their vivid character can nonetheless have a profound effect: in one case (Dunn *et al.* 1983) a patient saw 'snakes . . . that appeared from any direction' and although 'he knew the snakes were not real (nevertheless) during the examination he frequently jumped because he was pulling away from the perceived snakes'. The mesencephalic location of the responsible lesions usually entails 'neighborhood' symptoms of localizing value: thus, a right mesencephalic infarction affecting both the right cerebral peduncle and the midbrain tegmentum also caused left hemiplegia and bilateral ptosis (Geller and Bellor 1987); a left mesencephalic infarction that primarily involved the substantia nigra also caused a parkinsonian syndrome with prominent tremor on the right (De La Fuente Fernandez *et al.* 1994).

Hallucinations occurring secondary to pontine infarction have been referred to as 'Pick's visions' and have a distinctive character in that they often consist of hallucinating people walking through walls (Bing 1940).

Of the miscellaneous causes of visual hallucinations, narcolepsy (Zarcone 1973) is the most common, and in this disorder they occur on either a hypnogogic or hypnopompic basis. Guillain–Barré syndrome, at its height, may cause visual hallucinations (Cochen *et al.* 2005), possibly due to direct central nervous system involvement. Normal individuals may experience hallucinations with sleep deprivation (Kollar *et al.* 1968) or during bereavement (Grimby 1993): hallucinations seen in bereavement are often of the deceased and the grieving person often finds them comforting; they tend to clear spontaneously within a few months.

Before leaving this section on visual hallucinations, a few extra words are in order regarding palinopsia as its differential is quite limited. Palinopsia has been noted with treatment with trazodone (Hughes and Lessell 1990), mirtazapine (Ihde-Scholl and Jefferson 2001), and risperidone (Lauterbach *et al.* 2000), as a manifestation of a simple partial seizure (Muller *et al.* 1995), and with lesions of the occipital lobe (Michel and Troost 1980), occipitotemporal region (Meadows and Munro 1977), occipitoparietal region (Bender *et al.* 1968), and the parietal lobe (Critchley 1951).

Auditory hallucinations occurring with preserved insight are relatively uncommon, and may be seen as manifestations of partial seizures, with deafness or with focal intracerebral lesions.

Partial seizures may manifest with auditory hallucinations. In one case of a simple partial seizure (Williams 1956), the patient heard 'a buzz' that was quickly followed by 'music like a loudspeaker in which he heard clearly a recognized section of Veni Creator'. One of Penfield's (Penfield and Perot 1963) patients, during a complex partial seizure, 'heard voices which seemed to come from her right side' as if she were in the middle of a 'crowd of people'.

Deafness may be followed by auditory hallucinations, thus constituting, as it were, the 'auditory' equivalent of the Charles Bonnet syndrome (Griffiths 2000; Hammeke *et al.* 1983; Miller and Crosby 1979; Ross *et al.* 1975). Interestingly, the hallucinations seen here tend to be musical.

Focal intracerebral lesions, very rarely, may cause isolated auditory hallucinations, and these have been noted with lesions of the temporal cortex (Lempl *et al.* 2005), putamen (Cerrato *et al.* 2001), mesencephalon (Cascino and Adams 1986), and pontine tegmentum (Cascino and Adams 1986; Murata *et al.* 1994; Schielke *et al.* 2000).

Tactile hallucinations may be seen with intoxication with either cocaine (Siegel 1978) or stimulants (Bell 1973), and typically consist of formication.

Olfactory hallucinations may represent either the sole symptomatology of a simple partial seizure (Mauguire and Courjon 1978) or be part of the symptomatology of a complex partial seizure (Jackson and Stewart 1899), and indicate a focus in the uncus. They may also appear as a migraine aura (Fuller and Guiloff 1987; Wolberg and Ziegler 1982) or, very rarely, secondary to lesions of the olfactory bulb or tract (Kaufman *et al.* 1988; Paskind 1935).

Gustatory hallucinations with preserved insight have been reported during simple (Mulder and Daly 1952) or complex partial (Daly 1958; Hausser-Hauw and Bancaud 1987) seizures.

Differential diagnosis

As noted earlier, delusions must be distinguished from culturally or religiously sanctioned beliefs. Hallucinations, in turn, must be distinguished from illusions. Illusions are distorted experiences of actual objects, as for example the experience of a shimmering pool of water floating above a distant hot pavement, a pool that 'disappears' as one approaches closer.

Treatment

Treatment is directed, wherever possible, at the underlying cause. In the case of the Charles Bonnet syndrome, although there are no controlled studies, anecdotal reports attest to the efficacy of generally low-dose treatment with various medications including valproate (Hori *et al.* 2000), gabapentin (Paulig and Mentrup 2001), carbamazepine (Batra *et al.* 1997), donepezil (Ukai *et al.* 2004), venlafaxine (Lang *et al.* 2007), risperidone (Maeda *et al.* 2003), and olanzapine (Coletti Moja *et al.* 2005). In the case of hallucinations occurring with focal intracerebral lesions, an empirical trial of a low-dose antipsychotic may be justified.

5.31 SCHNEIDERIAN FIRST-RANK SYMPTOMS

Kurt Schneider (1887–1967) was a very influential German psychiatrist whose classic text *Clinical Psychopathology* went through multiple revisions from its first edition in 1939 up to the fifth, and last, edition, published in 1959. In this text, Schneider described a number of hallucinations and delusions that he believed were of 'first rank' importance in the diagnosis of schizophrenia. Although these Schneiderian first-rank symptoms are, indeed, most commonly seen in schizophrenia, they may also, as pointed out by Schneider himself, occur in 'diverse' other conditions, as described below.

Clinical features

The various Schneiderian first-rank symptoms, as noted in Table 5.6, may be divided into those which are auditory hallucinations and those which represent delusions.

Table 5.6 Schneiderian first-rank symptoms

Auditory hallucinations	
Audible thoughts	
Voices commenting on what the patient does	
Voices arguing with each other	
Delusions	
Delusions of passivity or influence	
Thought withdrawal	
Thought insertion	
Thought broadcasting	

Audible thoughts are said to occur when patients hear their own thoughts as if they were spoken out loud and indeed as if others might be able to hear them also. One of Schneider's (1959) patients said, 'I hear my own thoughts. I can hear them when everything is quiet.'

Voices commenting on what the patient does, keep up, as it were, a running commentary on the patient's behavior. One of Schneider's (Schneider 1959) patients 'heard a voice say, whenever she wanted to eat, "Now she is eating, here she is munching again,"' and one of Kraepelin's (1921) patients heard a voice telling her, 'Mary, you are talking nonsense, the policeman has seen you already.'

Voices arguing with each other, or, as Schneider elaborated, merely 'conversing with one another', may engage patients' attention, as if the voices were carrying on a debate about them.

Delusions of passivity or influence are, according to Schneider, said to be present when 'feelings, impulses (drives) and volitional acts . . . are experienced by the patient as the work or influence of others'. Such patients believe that their thoughts, feelings, or behavior are under the direct and unmediated control of some outside force or agency. Thus, passively played upon by these forces, patients feel as if they were robots or automatons. Patients will typically elaborate on this delusion of influence and express a belief as to the source of the influence, for example an 'electrical device', 'distant computers', or 'powerful magnets'.

Thought withdrawal represents a delusion wherein patients experience their thoughts being directly removed and withdrawn from their minds. One of Kraepelin's (1921) patients spoke of his thoughts being 'drawn off'. This is quite different from simply losing track of what one was thinking: those who lose track have a sense of having forgotten or lost something, whereas those with thought withdrawal have, as emphasized by Schneider (1959), a definite sense that some other agency or person has directly removed the thought.

Thought insertion represents the delusional belief of patients that the thoughts occurring in their minds are not their own but rather originate from, as Schneider (1959) put it, 'other people, who intrude their thoughts upon the patient'. Such inserted thoughts are quite different from obsessions: obsessions are recognized by patients as their own thoughts and as originating within them, whereas inserted thoughts are experienced as a kind of cognitive 'foreign body'.

Thought broadcasting represents the delusion that others can know what a patient is thinking without the patient in any way relating those thoughts. This 'thought diffusion' led one of Schneider's (1959) patients to complain that, 'if I think of anything, at once those opposite me know it and it is embarrassing.' As she believed that 'the doctor too knew exactly what she was thinking . . . she suggested that she would stop talking and (the physician) could just listen'. In some cases, patients will elaborate on this delusion, developing further beliefs about how such thought transfer is possible: Kraepelin (1921) noted that some patients believe that 'their thoughts are conveyed by a machine, there is a "mechanical arrangement", "a sort of little conveyance", telepathy'.

Etiology

It has, at times, been felt that Schneiderian first-rank symptoms were virtually pathognomonic of schizophrenia, occurring in virtually no other condition. The source of this belief is not clear: it certainly does not come from Schneider, who clearly stated that these first-rank symptoms could occur secondary to a 'number of diverse morbid cerebral processes' (Schneider 1959).

The various causes of the first-rank symptoms are listed in Table 5.7. Of all of these, schizophrenia is, by far, the most common cause, with first-rank symptoms being found in anywhere from one-third (Radhakrishnan *et al.* 1983) to over one-half (Tandon and Greden 1987) of such patients. Among patients with schizophrenia, it appears that thought broadcasting and thought insertion are probably most common out of all the first-rank symptoms (Mellor 1970). Schizo affective disorder is probably the next most common cause, with symptoms noted in about one-quarter of these patients (Tandon and Greden 1987). Mania, as seen in bipolar disorder, during stage II or III, may also cause first-rank symptoms, and the diagnosis here is suggested by typical manic symptoms seen in stage I, such as pressured speech, hyperactivity, increased energy, decreased need for sleep, etc. Depression, as seen in major depressive disorder, is only rarely associated with first-rank symptoms, and hence the appearance of such a symptom in a depressed patient should make one pause before giving a diagnosis of major depression; in such cases, a diagnosis of schizoaffective disorder may be more likely.

The remaining causes noted in Table 5.7, although rare, must be kept in mind. Of these, intoxications with amphetamines, cocaine, or phencyclidine are perhaps most common. Chronic, severe alcoholism may be associated with alcohol hallucinosis, which may be characterized by first-rank symptoms, and there are case reports of these symptoms occurring during benzodiazepine withdrawal and as a side-effect to fluvoxamine.

Epileptic conditions associated with first-rank symptoms include simple partial seizures and the chronic interictal psychosis. In one case of a simple partial seizure, the patient, during the ictus, had 'blurry vision, abdominal discomfort, sensation of imminent death, intense fear, and the conviction that his body (was) controlled by external forces' (Mesulam 1981). The chronic interictal psychosis, discussed further in Chapter 8.1, is seen only in patients with chronic, severe, uncontrolled epilepsy: one of Slater and Beard's (Slater and

Table 5.7 Causes of Schneiderian first-rank symptoms

Schizophrenia (Tandon and Greden 1987)
Schizoaffective disorder (Koehler and Seminario 1979)
Mania (Gonzales-Pinto *et al.* 2003; Jampala *et al.* 1989)
Depression (Tandon and Greden 1987)
Substance or medication related
Intoxication
Amphetamines (Angrist and Gershon 1970; Bell 1973; Janowsky and Risch 1979)
Cocaine (Harris and Batki 2000; Rosse *et al.* 1994)
Phencyclidine (Rosse *et al.* 1994)
Alcohol hallucinosis (Marneros 1988)
Benzodiazepine withdrawal (Roberts and Vass 1986)
Fluvoxamine (Ueda *et al.* 2003)
Epileptic conditions
Simple partial seizures (Mesulam 1981)
Chronic interictal psychosis (Kido and Yamaguchi 1989; Slater and Beard 1963)
Miscellaneous causes
Wilson's disease (Wilson 1912)
Cushing's syndrome (Trethowan and Cobb 1952)
New-variant Creutzfeldt–Jakob disease (Zeidler *et al.* 1997)
Subacute sclerosing panencephalitis (Duncalf *et al.* 1989)
Fahr's syndrome (Cummings *et al.* 1983)
Metrizamide myelography (Davis *et al.* 1986)

Beard 1963) patients believed that there was a 'pick-up' device in his body 'which transmitted thoughts from the brain, and also made the brain receive'.

The remaining miscellaneous causes listed in Table 5.7 only very rarely cause first-rank symptoms.

Treatment

Treatment is directed at the underlying condition.

REFERENCES

Abraham HD. Visual phenomenology of the LSD flashback. *Arch Gen Psychiatry* 1983; **40**:884–9.

Achari AN, Colover J. Posterior fossa tumors with pathological laughter. *JAMA* 1976; **235**:1469–71.

Ackerly W, Lhamom W, Fitts WT. Phantom breast. *J Nerv Ment Dis* 1955; **121**:177–8.

Adie WJ. Idiopathic narcolepsy: a disease sui generis, with remarks on the mechanism of sleep. *Brain* 1926; **49**:257–306.

Adie WJ Critchley M. Forced grasping and groping. *Brain* 1927; **50**:142–70.

Akelaitis AJ. Psychobiological studies following section of the corpus callosum. *Am J Psychiatry* 1941; **97**:1147–57.

Akelaitis AJ. Studies on the corpus callosum. IV. Diagnostic dyspraxia in epileptics following partial and complete section of the corpus callosum. *Am J Psychiatry* 1945; **101**:594–9.

Akelaitis AJ, Risteen WA, Herren RY *et al.* Studies on the corpus callosum. III. A contribution to the study of dyspraxia and apraxia following partial and complete section of the corpus callosum. *Arch Neurol Psychiatry* 1942; **47**:971–1008.

Akil H, Statham PF, Gotz M *et al.* Adult cerebellar mutism and cognitive-affective syndrome caused by cystic hemangioblastoma. *Acta Neurochir* 2006; **148**:597–8.

Albert ML, Sandson J. Perseveration in aphasia. *Cortex* 1986; **22**:103–15.

Alevizos B, Lykouras L, Zervas IM *et al.* Risperidone-induced obsessive-compulsive symptoms. A review of six cases. *J Clin Psychopharmacol* 2002; **22**:461–7.

Alexander MC. Lilliputian hallucinations. *J Ment Sci* 1926; **72**:187–91.

Alexander MP, Schmitt MA. The aphasia syndrome of stroke in the left anterior cerebral artery territory. *Arch Neurol* 1980; **37**:97–100.

Alexander MP, Stuss DT, Benson DF. Capgras syndrome: a reduplicative phenomenon. *Neurology* 1979; **29**:334–9.

Andersen G, Vestergaard K, Riis JO. Citalopram for post-stroke pathological crying. *Lancet* 1993; **342**:837–9.

Anderson M, Salmon MV. Symptomatic cataplexy. *J Neurol Neurosurg Psychiatry* 1977; **40**:186–91.

Anderson SW, Damasio H, Damasio AR. A neural basis for collecting behavior in humans. *Brain* 2005; **128**:201–18.

Andreasen NC. Affective flattening and criteria for schizophrenia. *Am J Psychiatry* 1979; **136**:944–7.

Angrist B, Gershon S. The phenomenology of experimentally induced amphetamine psychosis. *Biol Psychiatry* 1970; **2**:95–107.

Anson JA, Kuhlman DT. Post-ictal Kluver–Bucy syndrome after temporal lobectomy. *J Neurol Neurosurg Psychiatry* 1993; **56**:311–13.

Asfora WT, DeSalles AAF, Masamitsu ABE *et al.* Is the syndrome of pathological laughing and crying a manifestation of pseudobulbar palsy? *J Neurol Neurosurg Psychiatry* 1989; **52**:523–5.

Assal F, Valenza N, Landis T. Clinicoanatomic correlates of a Fou rire prodromique in a pontine infarction. *J Neurol Neurosurg Psychiatry* 2000; **69**:697–8.

Ay H, Buoanno S, Price BH *et al.* Sensory alien hand syndrome: case report and review of the literature. *J Neurol Neurosurg Psychiatry* 1998; **65**:366–9.

Ayotte J, Peretz I, Hyde K. Congenital amusia. A group study of adults afflicted with a music-specific disorder. *Brain* 2001; **125**:238–51.

Ayuso-Peralta L, Jimenez-Jimenez FJ, Tejeiro J *et al.* Progressive multifocal leukoencephalopathy in HIV infection presenting as Balint's syndrome. *Neurology* 1994; **44**:1339–40.

Baguley IJ, Nicholls JL, Felmingham KL *et al.* Dysautonomia after traumatic brain injury: a forgotten syndrome? *J Neurol Neurosurg Psychiatry* 1999; **67**:39–43.

Baguley IJ, Heriseanu RE, Gurka JA *et al.* Gabapentin in the management of dysautonomia following severe traumatic brain injury: a case series. *J Neurol Neurosurg Psychiatry* 2007; **78**:539–41.

Baker RW, Chengappa R, Baird JW *et al.* Emergence of obsessive compulsive symptoms during treatment with clozapine. *J Clin Psychiatry* 1992; **53**:439–42.

Banks G, Short P, Martinez AJ *et al.* The alien hand syndrome: clinical and postmortem findings. *Arch Neurol* 1989; **46**:456–9.

Bar KJ, Hager F, Sauer H. Olanzapine- and clozapine-induced stuttering. A case series. *Pharmacopsychiatry* 2004; **37**:131–4.

Barbizet J, Degos JD, Duizabo PH *et al.* Syndrome de deconnexion interhemispherique d'origine ischemique. *Rev Neurol* 1974; **130**:127–42.

Barclay CL, Bergeron C, Lang AE. Arm levitation in progressive supranuclear palsy. *Neurology* 1999; **52**:879–82.

Barnes TR, Crichton P, Nelson HE *et al.* Primitive (developmental) reflexes, tardive dyskinesia and intellectual impairment in schizophrenia. *Schizophr Res* 1995; **16**:47–52.

Barrett K. Treating organic abulia with bromocriptine and lisuride: four case studies. *J Neurol Neurosurg Psychiatry* 1991; **54**:718–21.

Barris RW, Schuman HR. Bilateral anterior cingulate gyrus lesions: syndrome of the anterior cingulate gyri. *Neurology* 1953; **3**:44–52.

Bartlett JEA. A case of organized visual hallucinations in an old man with cataract, and their relationship to the phenomenology of the phantom limb. *Brain* 1951; **79**:363–73.

Bartusch SL, Sander BJ, D'Alessio JG *et al.* Clonazepam for the treatment of lancinating phantom limb pain. *Clin J Pain* 1996; **12**:59–62.

Bassetti C, Bogousslavsky J, Barth A *et al.* Isolated infarcts of the pons. *Neurology* 1996; **46**:165–75.

Batra A, Bartels M, Wormstall H. Therapeutic options in Charles Bonnet Syndrome. *Acta Psychiatr Scand* 1997; **96**:129–33.

Bautista RE, Ciampetti MZ. Expressive aprosody and amusia as a manifestation of right hemisphere seizures. *Epilepsia* 2003; **44**:466–7.

Bayles KA, Tomoeda CK, Kaszniak AW *et al.* Verbal perseveration of dementia patients. *Brain Lang* 1985; **25**:102–16.

Bayles KA, Tomoeda CK, McKnight PE *et al.* Verbal perseveration in individuals with Alzheimer's disease. *Semin Speech Lang* 2004; **25**:335–47.

Becker R, Benes L, Sure U *et al.* Intrathecal baclofen alleviates autonomic dysfunction in severe brain injury. *J Clin Neurosci* 2000; **7**:316–19.

Behrman S, Carroll JD, Janota I *et al.* Progressive supranuclear palsy. *Brain* 1969; **92**:663–78.

Bell DS. The experimental reproduction of amphetamine psychosis. *Arch Gen Psychiatry* 1973; **29**:35–40.

Bender MB, Feldman M, Sobin AJ. Palinopsia. *Brain* 1968; **91**:321–38.

Benegal V, Hingorani S, Khanna S *et al.* Is stupor by itself a catatonic symptom? *Psychopathology* 1992; **25**:229–31.

Benke T, Hohenstein C, Poewe W *et al.* Repetitive speech phenomena in Parkinson's disease. *J Neurol Neurosurg Psychiatry* 2000; **69**:319–24.

Benson DF, Gardner H, Meadows JC. Reduplicative paramnesia. *Neurology* 1976; **26**:147–51.

Bercel NA, Travis LE, Olinger LB *et al.* Model psychoses induced by LSD-25 in normals. I. Psychophysiological investigations, with special reference to the mechanism of the paranoid reaction. *Arch Neurol Psychiatry* 1956; **75**:588–611.

Berigan T, Watt TJ. Dilantin toxicity presenting as mutism following severe head injury: case report. *Mil Med* 1994; **159**:533–4.

Berrios GE. Tactile hallucinations: conceptual and historical aspects. *J Neurol Neurosurg Psychiatry* 1982; **45**:285–93.

Besson G, Bogousslavsky J, Regli F *et al.* Acute pseudobulbar or suprabulbar palsy. *Arch Neurol* 1991; **48**:501–7.

Beukelman DR, Flowers CR, Swanson PD. Cerebral disconnection associated with anterior communicating aneurysm: implications for evaluation of symptoms. *Arch Phys Med Rehabil* 1980; **61**:18–23.

Bhatia K, Marsden CD. The behavioral and motor consequences of focal lesions of the basal ganglia in man. *Brain* 1994; **117**:859–76.

Bing R. *Compendium of Regional Diagnosis in Lesions of the Brain and Spinal Cord*, 11th edn. translated by Haymaker W. St Louis: CV Mosby, 1940.

Bittar RG, Otero S, Carter H *et al.* Deep brain stimulation for phantom limb pain. *J Clin Neurosci* 2005; **12**:399–404.

Black DW, Wojcieszek J. Depersonalizaton syndrome induced by fluoxetine. *Psychosomatics* 1991; **32**:468–9.

Blake JR, Kunkle EC. The palmomental reflex. A physiological and clinical analysis. *Arch Neurol Psychiatry* 1951; **65**:337–45.

Bleuler E. *Textbook of Psychiatry, 1924*, translated by Brill AA. New York: Arno Press, 1976.

Blumstein SE, Alexander MP, Ryalls JH *et al.* On the nature of the foreign accent syndrome; a case study. *Brain Lang* 1987; **31**: 215–44.

Bodfish JW, Symons FJ, Parker DE *et al.* Varieties of repetitive behavior in autism: comparisons to mental retardation. *J Autism Dev Disorder* 2000; **30**:237–43.

Boeve BF, Wijdicks EF, Benarroch EE *et al.* Paroxysmal sympathetic storms ('diencephalic seizures') after severe diffuse axonal head injury. *Mayo Clin Proc* 1998; **73**:148–52.

Bogen JE. The callosal syndrome. In: Heilmann KM, Valenstein E, eds. *Clinical Neuropsychology*, 2nd edn. New York: Oxford University Press, 1985a.

Bogen JE. Split-brain syndromes. In: Vinken PJ, Bruyn GW, Klawans HL, eds. *Handbook of Clinical Neurology*, 2nd edn. Amsterdam: Elsevier, 1985b.

Bogousslavsky J, Regli F. Anterior cerebral artery territory infarction in the Lausanne stroke registry: clinical and etiologic patterns. *Arch Neurol* 1990; **47**:144–50.

Bogousslavsky J, Regli F, Uske A. Thalamic infarcts: clinical syndromes, etiology, and prognosis. *Neurology* 1988; **38**:837–48.

Bogousslavsky J, Delayoe-Bischof A, Assal G. Loss of psychic self-activation with bithalamic infarction: neurobehavioral, CT, MRI and SPECT correlates. *Acta Neurol Scand* 1991; **83**:309–26.

Bone M, Critchley P, Buggy DJ. Gabapentin in postamputation phantom limb pain: a randomized, double-blind, placebo-controlled, cross-over study. *Reg Anesth Pain Med* 2002; **27**:481–6.

Borroni B, Broli M, Costanzi C *et al.* Primitive reflex evaluation in the clinical assessment of extrapyramidal syndromes. *Eur J Neurol* 2006; **13**:1026–8.

Botez MI, Werheim N. Expressive aphasia and amusia following right frontal lesion in a right-handed man. *Brain* 1959; **82**:187–202.

Brady JP. Drug-induced stuttering: a review of the literature. *J Clin Psychopharmacol* 1998; **18**:50–4.

Brady TK, Lydiard RB, Malcolm R, Ballenger JC. Cocaine-induced psychosis. *J Clin Psychiatry* 1991; **52**:509–12.

Bressler B, Cohen SI, Magnussen F. The problem of phantom breast and phantom pain. *J Nerv Ment Dis* 1956; **123**:181–7.

Brion S, Jedynak CP. Troubles de transport interhemispherique: a propos de trois observations de tumeurs du corps calleux. *Rev Neurol* 1972; **126**:257–66.

Brock S, Merwarth HR. The illusory awareness of body parts in cerebral disease. *Arch Neurol Psychiatry* 1957; **77**:366–75.

Brodie EE, Whyte A, Niven CA. Analgesia through the looking glass? A randomized controlled trial investigating the effect of viewing a 'virtual' limb upon phantom limb pain, sensation and movement. *Eur J Pain* 2007; **11**:428–36.

Brodsky H, Vuong KD, Thomas M *et al.* Glabellar and palmomental reflexes in parkinsonian disorders. *Neurology* 2004; **63**:1096–8.

Brooks BR, Thisted RA, Appel SH *et al.* Treatment of pseudobulbar affect in ALS with destromethorphan/quidine. A randomized trial. *Neurology* 2004; **63**:1364–70.

Brown DL, Smith TL, Knepper LE. Evaluation of five primitive reflexes in 240 young patients. *Neurology* 1998; **51**:322.

Bullard DE. Diencephalic seizures: responsiveness to bromocriptine and morphine. *Ann Neurol* 1987; **21**:609–11.

Byerly M, Goodman W, Acholonu W *et al.* Obsessive compulsive symptoms in schizophrenia: frequency and clinical features. *Schizophr Res* 2005; **76**:309–16.

Cairns H, Oldfield RC, Pennybacker JB *et al.* Akinetic mutism with an epidermoid cyst of the 3rd ventricle. *Brain* 1941; **64**:273–90.

Cantu RC, Drew JH. Pathological laughing and crying associated with a tumor ventral to the pons. *J Neurosurg* 1966; **24**:1024–6.

Caplan LR, Schoene WC. Clinical features of subcortical arteriosclerotic encephalopathy (Binswanger's disease). *Neurology* 1978; **28**:1206–15.

Caplan LR, Schmahmann JD, Kase CS *et al.* Caudate infarcts. *Arch Neurol* 1990; **47**:133–43.

Carabelli RA, Kellerman WC. Phantom limb pain: relief by application of TENS to contralateral extremity. *Arch Phys Med Rehabil* 1985; **66**:466–7.

Cardoso F, Veado CC, de Oliveira JT. A Brazilian cohort of patients with Tourette's syndrome. *J Neurol Neurosurg Psychiatry* 1996; **60**: 209–12.

Carel C, Albucher JF, Manelfe C *et al.* Fou rire prodromique heralding a left internal carotid artery occlusion. *Stroke* 1997; **28**:2081–3.

Carlen PL, Wall PD, Nadvorna H *et al.* Phantom limbs and related phenomena in recent traumatic amputations. *Neurology* 1978; **28**:211–17.

Carota A, Rossetti AO, Karapanayiotides T *et al.* Catastrophic reaction in acute stroke: a reflex behavior in aphasic patients. *Neurology* 2001; **57**:1902–5.

Carroll BT, Goforth HW, Raimonde LA. Partial Kluver–Bucy syndrome: two cases. *CNS Spectr* 2001; **6**:329–32.

Cascino GD, Adams RD. Brainstem auditory hallucinosis. *Neurology* 1986; **36**:1042–7.

Cassano GB, Petracca A, Perugi G *et al.* Derealization and panic attacks: a clinical evaluation on 150 patients with panic disorder/agoraphobia. *Compr Psychiatry* 1989; **30**:5–12.

Catsman-Berrevoets CE, Van Dongen HR, Mulder PG *et al.* Tumor type and size are high risk factors for the syndrome of 'cerebellar' mutism and subsequent dysarthria. *J Neurol Neurosurg Psychiatry* 1999; **67**:755–7.

Cavallaro R, Galardi G, Cavallini MC *et al.* Obsessive compulsive disorder among idiopathic focal dystonia patients: an epidemiological and family study. *Biol Psychiatry* 2002; **52**:356–61.

Ceccaldi M, Milandre L. A transient fit of laughter as the inaugural symptom of capsular-thalamic infarction. *Neurology* 1994; **44**:1762.

Ceccaldi M, Poncet M, Milandre L *et al.* Temporary forced laughter after unilateral strokes. *Eur Neurol* 1994; **34**:36–9.

Cerrato P, Imperiale D, Giraudo M *et al.* Complex musical hallucinosis in a professional musician with a left subcortical hemorrhage. *J Neurol Neurosurg Psychiatry* 2001; **71**:280–1.

Cerrato P, Imperilae D, Bergoni M *et al.* Emotional facial paresis in a patient with a lateral medullary infarction. *Neurology* 2003; **60**:723–4.

Chan BL, Witt R, Charrow AP *et al.* Mirror therapy for phantom limb pain. *N Engl J Med* 2007; **357**:2206–7.

Chan JL, Ross ED. Left-handed mirror writing following right anterior cerebral artery infarction: evidence for non-mirror transformations of motor programs by right supplementary motor area. *Neurology* 1988; **38**:59–63.

Chan JL, Ross ED. Alien hand syndrome: influence of neglect on the clinical presentation of frontal and callosal variants. *Cortex* 1997; **33**:287–99.

Chan JL, Chen RS, Ng KK. Leg manifestations in alien hand syndrome. *J Formosan Med Assoc* 1996; **45**:342–6.

Chanson JB, Kremer S, Blanc F *et al.* Foreign accent syndrome as a first sign of multiple sclerosis. *Mult Scler* 2009; **15**:1123–5.

Cho YJ, Hand SD, Song SK *et al.* Palilalia, echolalia, and echopraxia-palipraxia as ictal manifestations in a patient with left frontal lobe epilepsy. *Epilepsia* 2009; **50**:1616–19.

Chou CL, Lin YJ, Sheu YL *et al.* Persistent Kluver–Bucy syndrome after bilateral temporal lobe infarction. *Acta Neurol Taiwan* 2008; **17**:199–202.

Christensen RC, Byerly MJ, McElroy RA. A case of sertraline-induced stuttering. *J Clin Psychopharmacol* 1996; **16**:92–3.

Christodoulou GN. The syndrome of Capgras. *Br J Psychiatry* 1977; **130**:556–64.

Christoph DH, de la Freias GR, Dos Santos DP *et al.* Different perceived foreign accents in one patient after prerolandic hematoma. *Eur Neurol* 2004; **52**:198–201.

Ciabarra A, Elkino MS, Roberts JK *et al.* Subcortical infarction resulting in acquired stuttering. *J Neurol Neurosurg Psychiatry* 2000; **69**:546–9.

Closson RG. Visual hallucinations as the earliest symptom of digoxin intoxication. *Arch Neurol* 1983; **40**:386.

Cochen V, Arnulf I, Demeret S *et al.* Vivid dreams, hallucinations, psychosis and REM sleep in Guillian Barre syndrome. *Brain* 2005; **128**:2535–45.

Cohen D, Cottias C, Basquin M. Cotard's syndrome in a 15-year-old girl. *Acta Psychiatr Scand* 1997; **95**:164–5.

Cohen PR. Medication-associated depersonalization symptoms: report of transient depersonalization symptoms induced by monocycline. *South Med J* 2004; **97**:70–3.

Cole M. When the left brain is not right the right brain may be left: report of personal experience of occipital hemianopia. *J Neurol Neurosurg Psychiatry* 1999; **67**:169–73.

Coletti Moja M, Milano E, Gasverde S *et al.* Olanzapine therapy in hallucinatory visions related to Bonnet syndrome. *Neurol Sci* 2005; **26**:168–70.

Colman WS. A case of pseudo-bulbar paralysis, due to lesions in each internal capsule; degeneration of direct and crossed pyramidal tracts. *Brain* 1894; **17**:88–9.

Comings DE, Comings BG. A controlled study of Tourette syndrome, IV: obsession, compulsion, and schizoid behavior. *Am J Med Genet* 1987; **41**:782–803.

Confavreux C, Croisile B, Garassus P *et al.* Progressive amusia and aprosody. *Arch Neurol* 1992; **49**:971–6.

Coplin WM, Kim DK, Kilot M *et al.* Mutism in an adult following hypertensive cerebellar hemorrhage: nosological discussion and illustrative case. *Brain Lang* 1997; **59**:473–93.

Cornelius JR, Day NL, Fabrega H *et al.* Characterizing organic delusional syndrome. *Arch Gen Psychiatry* 1991; **48**:997–9.

Cornelius JR, Tippmann-Peikert M, Siocumb NL *et al.* Impulse control disorders with the use of dopaminergic agents in restless legs syndrome: a case–control study. *Sleep* 2010; **33**:81–7.

Critchley M. Types of visual perseveration: 'paliopsia' and 'illusory visual spread'. *Brain* 1951; **74**:267–99.

Crutchfield JS, Sawaya R, Meyers CA *et al.* Postoperative mutism in neurosurgery. Report of two cases. *J Neurosurg* 1994; **81**:115–21.

Cummings JL, Duchen LW. Kluver–Bucy syndrome in Pick disease: clinical and pathologic correlations. *Neurology* 1981; **31**:1415–22.

Cummings JL, Gosenfeld LF, Houlihan JP *et al.* Neuropsychiatric disturbances associated with idiopathic calcification of the basal ganglia. *Biol Psychiatry* 1983; **18**:591–601.

Cummings JL, Benson F, Hill MA *et al.* Aphasia in dementia of Alzheimer's type. *Neurology* 1985; **35**:394–7.

Daly D. Ictal affect. *Am J Psychiatry* 1958; **115**:97–108.

David AS, Bone I. Mutism following left hemisphere infarction. *J Neurol Neurosurg Psychiatry* 1984; **47**:1342–4.

Davis RE, Cummings JL, Malin BD *et al.* Prolonged psychosis with first-rank symptoms following metrizamide myelography. *Psychosomatics* 1986; **27**:373–5.

Davison C, Kelman H. Pathologic laughing and crying. *Arch Neurol Psychiatry* 1939; **42**:595–643.

D'Cruz OF, Vaughn BV, Gold SH *et al.* Symptomatic cataplexy in pontomedullary lesions. *Neurology* 1994; **44**:2189–91.

Degos JD, Gray F, Louarn F *et al.* Posterior callosal infarction: clinicopathological correlations. *Brain* 1987; **110**:1155–71.

De La Fuente Fernandez R, Lopez JM, Del Corral PR *et al.* Peduncular hallucinosis and right hemiparkinsonism caused by left mesencephalic parkinsonism. *J Neurol Neurosurg Psychiatry* 1994; **57**:870.

Della Sala S, Mrachetti C, Spinnler H. The anarchic hand: a fronto-mesial sign. In: Boller F, Grafman J, eds. *Handbook of Neuropsychology*, vol. 9. Amsterdam: Elsevier Science, 1994.

Denny-Brown D, Meyer JS, Horenstein S. The significance of perceptual rivalry resulting from parietal lesions. *Brain* 1952; **75**:433–71.

De Renzi E, Barbieri C. The incidence of the grasp reflex following hemispheric lesion, and its relation to frontal damage. *Brain* 1992; **115**:293–313.

De Renzi E, Cavalleri F, Facchini S. Imitation and utilization behavior. *J Neurol Neurosurg Psychiatry* 1996; **61**:396–400.

Derex L, Ostrowsky K, Nighoghossian N *et al.* Severe pathological crying after left anterior choroidal artery infarct. Reversibility with paroxetine treatment. *Stroke* 1997; **28**:1464–6.

Desarkar P, Goyan N, Khess CRJ. Clozapine-induced cataplexy. *J Neuropsychiatry Clin Neurosci* 2007; **19**:87–8.

Devinsky O, Felmann E, Burrowes K *et al.* Autoscopic phenomenon with seizures. *Arch Neurol* 1989; **46**:1080–8.

Diamond AL, Callison RC, Shokri J *et al.* Paroxysmal sympathetic storm. *Neurocrit Care* 2005; **2**:288–91.

Di Dio AS, Fields MC, Rowan AJ. Palinacousis – auditory perseveration: two cases and a review of the literature. *Epilepsia* 2007; **48**:1801–6.

Di Legge S, Di Piero V, Altieri M *et al.* Usefulness of primitive reflexes in demented and non-demented cerebrovascular patients in daily clinical practice. *Eur Neurol* 2001; **41**:104–10.

Do D, Sheen VL, Bromfield E. Treatment of paroxysmal sympathetic storm with labetalol. *J Neurol Neurosurg Psychiatry* 2000; **69**:832–3.

Dolado AM, Castrillo C, Urra DG *et al.* Alien hand sign or alien hand syndrome? *J Neurol Neurosurg Psychiatry* 1995; **59**:100–1.

Doubleday EK, Snowden JS, Varma AR *et al.* Qualitative performance characteristics differentiate dementia with Lewy bodies and Alzheimer's disease. *J Neurol Neurosurg Psychiatry* 2002; **72**:602–7.

Dummit ES, Klein RG, Tancer NK *et al.* Systematic assessment of 50 children with selective mutism. *J Am Acad Child Adolesc Psychiatry* 1997; **36**:653–60.

Duncalf CM, Kent JNG, Harbord M *et al.* Subacute sclerosing panencephalitis presenting as schizophreniform psychosis. *Br J Psychiatry* 1989; **155**:557–9.

Dunn DW, Weisberg LA, Nadell J. Peduncular hallucinations caused by brainstem compression. *Neurology* 1983; **33**:1360–1.

Dyken ME, Yamada T, Lin-Dyken DC *et al.* Diagnosing narcolepsy through the simultaneous clinical and electrophysiologic analysis of cataplexy. *Arch Neurol* 1996; **53**:456–60.

Ellis P, Mellsop G. De Clerambault's syndrome – a nosological entity? *Br J Psychiatry* 1985; **146**:90–5.

Elson A, Pearson C, Jones CD *et al.* Follow-up study of childhood elective mutism. *Arch Gen Psychiatry* 1965; **13**:182–7.

Enoch MD. The Capgras syndrome. *Acta Psychiatr Scand* 1963; **39**:437–62.

Ephraim PL, Wegener ST, MacKenzie EJ *et al.* Phantom pain, residual limb pain, and back pain in amputees: results of a national survey. *Arch Phys Med Rehabil* 2005; **86**:1910–19.

Escalona PR, Adair JC, Roberts BB *et al.* Obsessive-compulsive disorder following bilateral globus pallidus infarction. *Biol Psychiatry* 1997; **42**:410–12.

Eslinger PJ, Warner GC, Grattan LM *et al.* 'Frontal lobe' utilization behavior associated with paramedian thalamic infarction. *Neurology* 1991; **41**:450–2.

Esposito A, Demeurisse G, Alberti B *et al.* Complete mutism after midbrain periaqueductal gray lesion. *Neuroreport* 1999; **10**:681–5.

Evans AH, Katzenschlager R, Paviour D *et al.* Punding in Parkinson's disease: its relation to the dopamine dysregulation syndrome. *Mov Disord* 2004; **19**:397–405.

Evyapan Akkus D. Pure mutism due to simultaneous bilateral lenticulostriate artery territory infarction. *CNS Spectr* 2006; **11**:257–9.

Faris AA. Limbic system infarction: a report of two cases. *Neurology* 1969; **19**:91–6.

Fasano A, Petrovic I. Insights into the pathophysiology of punding reveal possible treatment strategies. *Mol Psychiatry* 2010; **15**:560–73.

Feinberg TE, Schindler RJ, Flanagan NG *et al.* Two alien hand syndromes. *Neurology* 1992; **42**:19–24.

Feinstein A, Feinstein K, Gray T *et al.* Prevalence and neurobehavioral correlates of pathological laughing and crying in multiple sclerosis. *Arch Neurol* 1997; **54**:1116–21.

Fere C. Le fou rire prodromique. *Rev Neurol* 1903; **11**:353–8.

Ferguson SM, Rayport M, Corrie WS. Neuropsychiatric observations of corpus callosum section for seizure control. In: Reeves AJ, ed. *Epilepsy and the Corpus Callosum*. New York: Plenum Press, 1985.

Fernandez JM, Sadaba F, Villaverde FJ *et al.* Cataplexy associated with midbrain lesion. *Neurology* 1995; **45**:393–4.

Fisher CM. Visual disturbances associated with quinidine and quinine. *Neurology* 1981; **31**:1569–71.

Fleet WS, Heilman KM. Acquired stuttering from a right hemisphere lesion in a right-hander. *Neurology* 1985; **35**:1343–6.

Flor H, Elbert T, Knecht S *et al.* Phantom-limb pain as a perceptual correlate of cortical reorganization following an arm amputation. *Nature* 1995; **375**:482–4.

Ford RA. The psychopathology of echophenomena. *Psychol Med* 1989; **19**:627–35.

Formisano R, Saltuari L, Gerstendbrand F. Prevalence of Kluver–Bucy syndrome as a positive prognostic feature for the remission of traumatic prolonged disturbances of consciousness. *Acta Neurol Scand* 1995; **91**:54–7.

Fossati A, Maffei C, Battaglia M *et al.* Latent class analysis of DSM-IV schizotypal personality disorder criteria in psychiatric patients. *Schizophr Bull* 2001; **27**:59–71.

Fournier JCM, Helguera RAL. Postencephalitic narcolepsy and cataplexy: muscle and motor nerves electrical inexcitability during the attack of cataplexy. *J Nerv Ment Dis* 1934; **80**:159–62.

Freedman M, Alexander MP, Naeser MA. Anatomic basis of transcortical motor aphasia. *Neurology* 1984; **34**:409–17.

Freemon FR. Akinetic mutism and bilateral anterior cerebral artery occlusion. *J Neurol Neurosurg Psychiatry* 1971; **34**:693–8.

Fridriksson J, Ryalls J, Rorden C *et al.* Brain damage and cortical compensation in foreign accent syndrome. *Neurocase* 2005; **11**:319–24.

Friedman A, Sienkiewicz J. Psychotic complications of long-term levodopa treatment of Parkinson's disease. *Acta Neurol Scand* 1991; **84**:111–13.

Fuller GN, Guiloff RJ. Migrainous olfactory hallucinations. *J Neurol Neurosurg Psychiatry* 1987; **50**:1688–90.

Gainotti G. Emotional behavior and hemispheric side of lesion. *Cortex* 1972; **8**:41–55.

Gallagher JP. Pathologic laughter and crying in ALS: a search for their origin. *Acta Neurol Scand* 1989; **80**:114–17.

Gazzaniga MS, Bogen JE, Sperry RW. Some functional effects of sectioning the cerebral commissures in man. *Proc Nat Acad Sci USA* 1962; **48**:1765–9.

Gelardi J-A M, Brown JW. Hereditary cataplexy. *J Neurol Neurosurg Psychiatry* 1967; **30**:455–7.

Geller TJ, Bellor SN. Peduncular hallucinosis: magnetic resonance imaging confirmation of mesencephalic infarction during life. *Ann Neurol* 1987; **21**:602–4.

Gelmers HJ. Non-paralytic motor disturbance and speech disorders: the role of the supplementary motor area. *J Neurol Neurosurg Psychiatry* 1983; **46**:1052.

Geschwind DH, Iacoboni M, Mega MS *et al.* Alien hand syndrome: interhemispheric motor disconnection due to a lesion in the midbody of the corpus callosum. *Neurology* 1995; **45**:802–8.

Geschwind N. The perverseness of the right hemisphere. *Behav Brain Sci* 1981; **4**:106–7.

Ghika J, Ghika-Schmid F, Bogousslasvky J. Parietal motor syndrome: a clinical description in 32 patients in the acute phase of pure parietal strokes studied prospectively. *Clin Neurol Neurosurg* 1998; **100**:271–82.

Ghika-Schmid F, Assal G, De Tribolet N *et al.* Kluver–Bucy syndrome after left anterior temporal resection. *Neuropsychologia* 1995; **33**:101–13.

Ghosh A, Dutt A. Utilization behavior in frontotemporal dementia. *J Neurol Neurosurg Psychiatry* 2010; **81**:154–6.

Gibb WRC, Luthert PJ, Marsden CD. Corticobasal degeneration. *Brain* 1989; **112**:1171–92.

Gillett T, Eminson SR, Hassanyeh F. Primary and secondary erotomania. Clinical characteristics and follow-up. *Acta Psychiatr Scand* 1990; **82**:65–9.

Giroud M, Lemesle M, Madinier G *et al.* Unilateral lenticular infarcts: radiological and clinical syndromes, etiology, and prognosis. *J Neurol Neurosurg Psychiatry* 1997; **63**:611–15.

Giuffrida O, Simpson L, Halligan PW. Contralateral stimulation, using TENS, or phantom pain: two confirmatory cases. *Pain Med* 2010; **11**:133–41.

Goetz CG, Fan W, Leurgans S *et al.* The malignant course of 'benign hallucinations' in Parkinson disease. *Arch Neurol* 2006; **63**:713–16.

Goldberg G, Bloom KK. The alien hand syndrome: localization, lateralization, and recovery. *J Phys Med Rehab* 1990; **69**:228–38.

Goldenberg G, Wimmer A, Holzner F *et al.* Apraxia of the left limb in a case of callosal disconnection: the contribution of medial frontal lobe damage. *Cortex* 1985; **21**:135–48.

Goldin S. Lilliputian hallucinations. *J Ment Sci* 1955; **101**:569–76.

Goldman S, Wang C, Salgado MW *et al.* Motor stereotypies in children with autism and other developmental disorders. *Dev Med Child Neurol* 2009; **51**:30–8.

Goldstein K. Sur Lehre der motorischen Apraxia. *J fur Psychologie und Neurologie* 1908; **11**:169–87.

Goldstein K. Der makroskopische Hirnbefund in meinem Falle von linksseitiger motorischer Apraxie. *Neurol Centralbl* 1909; **28**:898–906.

Goldstein K. *The Organism: A Holistic Approach to Biology Derived from Pathological Data in Man.* New York: American Books, 1939.

Goldstein K. *After Effects of Brain Injuries in War.* New York: Grune and Stratton, 1942.

Gondim FA, Parks BJ, Cruz-Flores S. 'Fou rire prodromique' as the presentation of pontine ischemia secondary to vertebrobasilar stenosis. *J Neurol Neurosurg Psychiatry* 2001; **71**:802–4.

Gonzales-Pinto A, van Os J, Perez de Heredia JL *et al.* Age-dependence of Schneiderian psychotic symptoms in bipolar patients. *Schizophren Res* 2003; **61**:157–62.

Gordon A. Transition of obsessions into delusions. *Am J Psychiatry* 1950; **107**:455–8.

Gorno-Tempini ML, Ogar JM, Brambati SM *et al.* Anatomical correlates of early mutism in progressive nonfluent speech. *Neurology* 2006; **67**:1849–51.

Gottlieb D, Robb K, Day B. Mirror movements in the alien hand syndrome. *Am J Phys Med Rehabil* 1992; **71**:297–300.

Gouzoulis-Mayfrank E, Heekeren K, Neukirch A *et al.* Psychological effects of (S)-ketamine and N,N-dimethyltryptamine (DMT): a double-blind, cross-over study in healthy volunteers. *Pharmacopsychiatry* 2005; **38**:301–11.

Graff-Radford NR, Eslinger PJ, Damasio AR *et al.* Nonhemorrhagic infarction of the thalamus: behavioral, anatomic, and physiologic correlates. *Neurology* 1984; **34**:14–23.

Graff-Radford NR, Cooper WE, Colsher PL *et al.* An unlearned foreign 'accent' in a patient with aphasia. *Brain Lang* 1986; **28**:86–94.

Graham JM, Gruenwald RA, Sagar HJ. Hallucinosis in idiopathic Parkinson's disease. *J Neurol Neurosurg Psychiatry* 1997; **63**:434–40.

Granacher RP. Differential diagnosis of tardive dyskinesias: an overview. *Am J Psychiatry* 1981; **38**:1288–97.

Green RC, Goldstein FC, Mirra SS *et al.* Slowly progressive apraxia in Alzheimer's disease. *J Neurol Neurosurg Psychiatry* 1995; **59**:312–15.

Greenwood R, Bhalla A, Gordon A *et al.* Behavior disturbances during recovery from herpes simplex encephalitis. *J Neurol Neurosurg Psychiatry* 1983; **46**:809–17.

Griffiths TD. Musical hallucinosis in acquired deafness. Phenomenology and brain substrate. *Brain* 2000; **123**:2065–76.

Grigsby J, Kaye K. Incidence and correlates of depersonalization following head trauma. *Brain Inj* 1993; **7**:507–13.

Grimby A. Bereavement among elderly people: grief reactions, post-bereavement hallucinations and quality of life. *Acta Psychiatr Scand* 1993; **87**:72–80.

Guilleminault C, Wilson RA, Dement WC. A study on cataplexy. *Arch Neurol* 1974; **31**:255–61.

Gurd JM, Bessell NJ, Bladon RA *et al.* A case of foreign accent syndrome, with follow-up clinical, neuropsychological and phonetic descriptions. *Neuropsychologia* 1988; **26**:237–51.

Gustafson L. Frontal lobe degeneration of non-Alzheimer type. II. Clinical picture an differential diagnosis. *Arch Gerontol Geriatr* 1987; **6**:209–23.

Gydesen S, Brown JN, Brun A *et al.* Chromosome 3 linked frontotemporal dementia (FTD-3). *Neurology* 2002; **59**:1585–94.

Haber WB. Observations on phantom-limb phenomena. *Arch Neurol Psychiatry* 1956; **75**:624–36.

Hadano K, Nakamura H, Hamanaka T. Effortful echolalia. *Cortex* 1998; **34**:67–82.

Hall DA, Anderson CA, Filley CM *et al.* A French accent after corpus callosum infarct. *Neurology* 2003; **60**:1551–2.

Halligan PW, Marshall JC, Wade DT. Three-arms: a case study of supernumerary phantom limbs after right hemisphere stroke. *J Neurol Neurosurg Psychiatry* 1993; **56**:159–66.

Hamano T, Hiraki S, Kawamur Y *et al.* Acquired stuttering secondary to callosal infarction. *Neurology* 2005; **64**:1092–3.

Hammeke TA, McQuillen MP, Cohen BA. Musical hallucinations associated with acquired deafness. *J Neurol Neurosurg Psychiatry* 1983; **46**:570–2.

Hanakita J, Nishi S. Left alien hand sign and mirror writing after left anterior cerebral artery infarction. *Surg Neurol* 1991; **35**:290–3.

Harris DL, Batki SL. Stimulant psychosis: symptom profile and acute clinical course. *Am J Addict* 2000; **9**:28–37.

Hashimoto R, Tanaka Y. Contribution of the supplementary motor area and anterior cingulate gyrus to pathological grasping phenomena. *Eur Neurol* 1998; **40**:151–8.

Hashimoto R, Yoshida M, Tanaka Y. Utilization behavior after right thalamic infarction. *Eur Neurol* 1995; **35**:58–62.

Hausser-Hauw C, Bancaud J. Gustatory hallucinations in epileptic seizures. Electrophysiological, clinical and anatomical correlates. *Brain* 1987; **110**:39–59.

Helgason C, Wilbur A, Weiss A *et al.* Acute pseudobulbar mutism due to discrete bilateral capsular infarction in the territory of the anterior choroidal artery. *Brain* 1988; **111**:507–24.

Helm NA, Butler RB, Benson DF. Acquired stuttering. *Neurology* 1978; **28**:1159–65.

Henderson WR, Smythe GE. Phantom limbs. *J Neurol Neurosurg Psychiatry* 1948; **11**:89–112.

Hendricks HT, Heeren AH, Vos PE. Dysautonomia after severe traumatic brain injury. *Eur J Neurol* 2010; **17**:1172–7.

Heutink P, Stevens M, Rizzu P *et al.* Hereditary frontotemporal dementia is linked to chromosome 17q21–q22: a genetic and clinicopathological study of three Dutch families. *Ann Neurol* 1997; **41**:150–9.

Hillbom E. After-effects of brain-injuries. *Acta Psychiatr Suppl Scand* 1960; **142**:1–195.

Hof PR, Archin N, Osmand AP *et al.* Posterior cortical atrophy in Alzheimer's disease: analysis of a new case and re-evaluation of a historical report. *Acta Neuropathol* 1993; **86**:215–23.

Hollander E, Soorya L, Wasserman S *et al.* Divalproex sodium vs, placebo in the treatment of repetitive behaviors in autism spectrum disorder. *Int J Neuropsychopharmacol* 2006; **9**:209–13.

Holroyd S, Rabins PV, Finkelstein D *et al.* Visual hallucinations in patients with macular degeneration. *Am J Psychiatry* 1992; **149**:1701–6.

Hopf HC, Muller-Forell W, Hopf NJ. Localization of emotional and volitional facial paresis. *Neurology* 1992; **42**:1918–23.

Hopf HC, Fitzek C, Mark J *et al.* Emotional facial paresis of pontine origin. *Neurology* 2000; **54**:1217.

Hori H, Terao T, Shiraishi Y *et al.* Treatment of Charles Bonnet syndrome with valproate. *Int Clin Psychopharmacol* 2000; **15**:117–19.

Horowitz MJ. Flashbacks: recurrent intrusive images after the use of LSD. *Am J Psychiatry* 1969; **126**:565–9.

Hotz G, Helm-Estabrooks N. Perseveration. Part II: A study of perseveration in closed-head injury. *Brain Inj* 1995; **9**:161–72.

Hudson AJ, Grace GM. Misidentification syndromes related to face specific area in the fusiform gyrus. *J Neurol Neurosurg Psychiatry* 2000; **69**:645–8.

Hughes MS, Lessell S. Trazodone-induced palinopsia. *Arch Ophthalmol* 1990; **108**:359.

Husain M, Stein J. Rezso Balint and his most celebrated case. *Arch Neurol* 1988; **45**:89–93.

Huse E, Larbig W, Flor H *et al.* The effect of opioids on phantom limb pain and cortical reorganization. *Pain* 2001; **90**:47–55.

Hwang JP, Tsai SJ, Yang CH *et al.* Hoarding behavior in dementia. A preliminary report. *Am J Geriatr Psychiatry* 1998; **6**:285–9.

Hyde TM, Goldberg TE, Egan MF *et al.* Frontal release signs and cognition in people with schizophrenia, their siblings and healthy controls. *Br J Psychiatry* 2007; **191**:120–5.

Ihde-Scholl T, Jefferson JW. Mirtazapine-associated palinopsia. *J Clin Psychiatry* 2001; **62**:373.

Imamura T, Takanashi M, Hattori N *et al.* Bromocriptine treatment for perseveration in demented patients. *Alzheimer Dis Assoc Disord* 1998; **12**:109–13.

Ionasescu V. Paroxysmal disorders of the body image in temporal lobe epilepsy. *Acta Psychiatr Neurol Scand* 1960; **35**:171.

Ironside R. Disorders of laughter due to brain lesions. *Brain* 1956; **79**:589–609.

Ishihara K, Nishino H, Maki T *et al.* Utilization behavior as a white matter disconnection syndrome. *Cortex* 2002; **38**:379–87.

Jackson JH, Stewart P. Epileptic attacks with a warning of a crude sensation of smell and with the intellectual aura (dreamy state) in a patient who had symptoms pointing to gross organic disease of the right temporo-sphenoidal lobe. *Brain* 1899; **32**:209.

Jacob A, Cherian PJ, Radhakrishnan K *et al.* Emotional facial paresis in temporal lobe epilepsy: its prevalence and lateralizing value. *Seizure* 2003; **12**:60–4.

Jacobs L, Gossman MD. Three primitive reflexes in normal adults. *Neurology* 1980; **30**:184–8.

Jaeger H, Maier C. Calcitonin in phantom limb pain: a double-blind study. *Pain* 1992; **48**:21–7.

Jampala VC, Taylor MA, Abrams R. The diagnostic implication of formal thought disorder in mania and schizophrenia: a reassessment. *Am J Psychiatry* 1989; **146**:459–63.

Janati A. Kluver–Bucy syndrome in Huntington's chorea. *J Nerv Ment Dis* 1985; **173**:632–5.

Janowsky DS, Risch C. Amphetamine psychosis and psychotic symptoms. *Psychopharmacology* 1979; **65**:73–7.

Jarvis JH. Post-mastectomy breast phantoms. *J Nerv Ment Dis* 1967; **144**:266–72.

Jason GW, Pajurkova EM. Failure of metacontrol: breakdown of behavioral unity after lesions of the corpus callosum and inferomedial frontal lobes. *Cortex* 1992; **28**:241–60.

Jensen JPA, Gron U, Pakkenberg H. Comparison of three primitive reflexes in neurological patients and in normal volunteers. *J Neurol Neurosurg Psychiatry* 1983; **46**:162–7.

Jensen TS, Krebs B, Nielsen J *et al.* Non-painful phantom limb phenomena in amputees: incidence, clinical characteristics and temporal course. *Acta Neurol Scand* 1984; **70**:407–14.

John JP, Khanna S, Thennarasau K *et al.* Exploration of dimensions of psychopathology in neuroleptic-naive patients with recent-onset schizophrenia/schizophreniform disorder. *Psychiatry Res* 2003; **121**:11–20.

Joseph AB, O'Leary DH. Brain atrophy and interhemispheric fissure enlargement in Cotard's syndrome. *J Clin Psychiatry* 1986; **47**:518–20.

Joseph AB, O'Leary DH, Kurland R *et al.* Bilateral anterior cortical atrophy and subcortical atrophy in reduplicative paramnesia: a case–control study of computed tomography in 10 patients. *Can J Psychiatry* 1999; **44**:685–9.

Kales A, Cadieux RJ, Soldatos CR *et al.* Narcolepsy–cataplexy. I. Clinical and electrophysiologic characteristics. *Arch Neurol* 1982; **39**: 164–8.

Kapur N, Turner A, King C. Reduplicative paramnesia: possible anatomical and neuropsychological mechanisms. *J Neurol Neurosurg Psychiatry* 1988; **51**:579–81.

Katz DI, Alexander MP, Mandell AM. Dementia following strokes in the mesencephalon and diencephalons. *Arch Neurol* 1987; **44**: 1127–33.

Kaufman MD, Lassiter KR, Shenoy BV. Paroxysmal unilateral dysosmia: a cured patient. *Ann Neurol* 1988; **24**:450–1.

Khateb A, Simon SR, Dieguez S *et al.* Seeing the phantom: A functional magnetic resonance imaging study of a supernumerary phantom limb. *Ann Neurol* 2009; **65**:698–705.

Kido H, Yamaguchi N. Clinical studies of schizophrenia-like state in epileptic patients. *Jpn J Psychiatry Neurol* 1989; **43**:433–8.

Kile SJ, Ellis WG, Olicheney JM *et al.* Alzheimer abnormalities of the amygdala with Kluver–Bucy syndrome symptoms: an amygdaloid variant of Alzheimer disease. *Arch Neurol* 2009; **66**:125–9.

Kim YW, Shin JC, An YS. Treatment of chronic akinetic mutism with atomoxetine: subtraction analysis of brain f-18 fluorodeoxyglucose positron emission tomographic images before and after medication: a case report. *Clin Neuropharmacol* 2010; **33**:209–11.

King AB. 'Phantom' sciatica. *Arch Neurol Psychiatry* 1956; **76**:72–4.

Kluin KJ, Norman LF, Berent S *et al.* Perceptual analysis of speech disorders in progressive supranuclear palsy. *Neurology* 1993; **43**:563–6.

Kluver H, Bucy PC. Preliminary analysis of functions of the temporal lobes in monkeys. *Arch Neurol Psychiatry* 1939; **42**:979–1000.

Koehler K, Seminario I. Research diagnosable 'schizo-affective' disorder in Schneiderian 'first rank' schizophrenia. *Acta Psychiatr Scand* 1979; **60**:347–54.

Kollar EJ, Namerow N, Pasnau RO *et al.* Neurological findings during prolonged sleep deprivation. *Neurology* 1968; **18**:836–40.

Koller WC. Dysfluency (stuttering) in extrapyramidal disease. *Arch Neurol* 1983; **40**:175–7.

Koller WC, Glatt S, Wilson RS *et al.* Primitive reflexes and cognitive function in the elderly. *Ann Neurol* 1982; **12**:302–4.

Kolmel HW. Complex visual hallucinations in the hemianopic field. *J Neurol Neurosurg Psychiatry* 1985; **48**:29–38.

Kraepelin E. [*Clinical Psychiatry, 1896*]. New York: MacMillan Company, 1907: 86.

Kraepelin E. [*Manic-depressive Insanity and Paranoia, 1921*]. New York: Arno Press, 1976.

Kraepelin E. [*Psychiatry. A Textbook for Students and Physicians, 1899*], 6th edn. Canton, MA: Science History Publications, 1990.

Kroll L, Drummond LM. Temporal lobe epilepsy and obsessive-compulsive symptoms. *J Nerv Ment Dis* 1993; **181**:457–8.

Krystal JH, Perry EB, Gueorguieva R *et al.* Comparative and interactive human psychopharmacologic effects of ketamine and amphetamine: implications for gluatamatergic and dopaminergic model psychoses and cognitive function. *Arch Gen Psychiatry* 2005; **62**:985–94.

Kuiken TA, Schechtman L, Harden RN. Phantom limb pain treatment with mirtazapine: a case series. *Pain Pract* 2005; **5**:356–60.

Kumral E, Calli C. External and extreme capsular stroke: clinical, topographical and etiologic patterns. *Cerebrovasc Dis* 2006; **21**:217–22.

Kumral E, Evyapan D, Balkir K. Acute caudate vascular lesions. *Stroke* 1999; **30**:100–8.

Lago A. Fou Rire Prodromique and ischemic stroke. *Stroke* 1998; **29**:1067–8.

Lam LC, Chiu HF. Kluver–Bucy syndrome in a patient with nasopharyngeal carcinoma: a late complication of radiation brain injury. *J Geriatr Psychiatry Neurol* 1997; **10**:111–13.

Lance JW. Simple formed hallucinations confined to the area of a specific visual field defect. *Brain* 1976; **99**:719–34.

Lang UE, Stogowski D, Schulze D *et al.* Charles Bonnet syndrome: successful treatment of visual hallucinations due to vision loss with selective serotonin reuptake inhibitors. *J Psycho-pharmacol* 2007; **21**:553–5.

Langworthy OR, Hesser FH. Syndrome of pseudobulbar palsy. *Arch Int Med* 1940; **65**:106–21.

Laplane D, Baulac M, Widocher D *et al.* Pure psychic akinesia with bilateral lesion of basal ganglia. *J Neurol Neurosurg Psychiatry* 1984; **47**:377–85.

Lauterbach EC, Abdlhamid A, Annandale JB. Posthallucinogen-like visual illusions (palinopsia) with risperidone in a patient without previous hallucinogen exposure: possible relation to serotonin 5-HT2A blockade. *Pharmacopsychiatry* 2000; **33**:38–41.

Lawson IR, McLeod RD. The use of imipramine ('Tofranil') and other psychotropic drugs in organic emotionalism. *Br J Psychiatry* 1969; **115**:281–5.

Lee HJ, Lee HS, Kim L *et al.* A case of risperidone-induced stuttering. *J Clin Psychopharmacol* 2001; **21**:115–16.

Leiguarda R, Starkstein S, Berthier M. Anterior callosal hemorrhage. A partial interhemispheric disconnection syndrome. *Brain* 1989; **112**:1019–37.

Leiguarda RC. Environmental reduplication associated with a right thalamic hemorrhage. *J Neurol Neurosurg Psychiatry* 1983; **46**:1154.

Lempl Y, Lorberboym M, Gilad R *et al.* Auditory hallucinations in acute stroke. *Behav Neurol* 2005; **16**:211–6.

Lepore FE. Spontaneous visual phenomena with visual loss: 104 patients with lesions of retinal and visual afferent pathways. *Neurology* 1990; **40**:444–7.

Leroy R. The syndrome of Lilliputian hallucinations. *J Nerv Ment Dis* 1922; **56**:325–33.

Leudar I, Fraser WI, Jeeves MA. Behavior disturbances and mental handicap: typology and longitudinal trends. *Psychol Med* 1984; **14**:923–35.

Levin HS, Madison CF, Bailey CB *et al.* Mutism after closed head injury. *Arch Neurol* 1983; **40**:601–6.

Levine DN, Rinn WE. Opticosensory ataxia and alien hand syndrome after posterior cerebral artery territory infarction. *Neurology* 1986; **36**:1094–7.

Lewis MH, Bodfish JW, Powell SB *et al.* Clomipramine treatment for self-injurious behavior of individuals with mental retardation: a double-blind comparison with placebo. *Am J Ment Retard* 1996; **100**:654–65.

Lhermitte F. 'Utilization behavior' and its relation to lesions of the frontal lobes. *Brain* 1983; **106**:237–55.

Lhermitte F. Human anatomy and the frontal lobes. II. Patient behavior in complex and social situations: the 'environmental dependency syndrome'. *Ann Neurol* 1986; **19**:335–43.

Lhermitte F, Pillon B, Serdaru M. Human anatomy and the frontal lobes. I. Imitation and utilization behavior: a neuropsychological study of 75 patients. *Ann Neurol* 1986; **19**:326–44.

Lhermitte J. Syndrome de la calotte du pedoncule cerebral. Les troubles psycho-sensoriels dans les lesions du mesocephale. *Rev Neurol* 1922; **38**:1359–65.

Lhermitte J. L'hallucinose pedonculaire. *Encephale* 1932; **27**:422–35.

Lieberman A, Benson DF. Control of emotional expression in pseudobulbar palsy. *Arch Neurol* 1977; **34**:717–19.

Lilly R, Cummings JL, Benson DF *et al.* The human Kluver–Bucy syndrome. *Neurology* 1983; **33**:1141–5.

Ling H, O'Sullivan SS, Holten JL *et al.* Does corticobasal degeneration exist? A clinicopathological re-evaluation. *Brain* 2010; **133**:2045–57.

Lipinski CG. Epilepsies with astatic seizures of late onset. *Epilepsia* 1977; **18**:13–20.

Lippert-Gruener M, Weinert U, Gresibach T *et al.* Foreign accent syndrome following traumatic brain injury. *Brain Inj* 2005; **19**:955–8.

Lippman CO. Hallucinations of physical disability in migraine. *J Nerv Ment Dis* 1953; **117**:345–50.

Liu GT, Scatz NJ, Galetta SL *et al.* Persistent positive visual phenomena in migraine. *Neurology* 1995; **45**:664–8.

Logan TP. Persistent phantom limb pain: dramatic response to chlorpromazine. *South Med J* 1983; **76**:1585.

Lohr JB. Transient grasp reflexes in schizophrenia. *Biol Psychiatry* 1985; **20**:172–5.

Luchins DJ, Goldman MB, Lieb M *et al.* Repetitive behaviors in chronically institutionalized schizophrenic patients. *Schizophr Res* 1992; **8**:119–23.

Ludlow CL, Rosenberg J, Salazar A *et al.* Site of penetrating brain lesion causing chronic acquired stuttering. *Ann Neurol* 1987; **22**:60–6.

Lugaresi A, Montagna P, Morreale A *et al.* 'Psychic akinesia' following carbon monoxide poisoning. *Eur Neurol* 1990; **30**:167–9.

Lukianowicz N. Autoscopic phenomena. *Arch Neurol Psychiatry* 1958; **80**:199–220.

Luzzi S, Viticchi G, Piccirilli M, Fabi K *et al.* Foreign accent syndrome as the initial sign of primary progressive aphasia. *J Neurol Neurosurg Psychiatry* 2008; **79**:79–81.

McDougle CJ, Scahill L, Aman MG *et al.* Risperidone for the core symptom domains of autism: results from the study by the autism network of the research units on pediatric psychopharmacology. *Am J Psychiatry* 2005; **162**:1142–8.

McFarland HR, Fortin D. Amusia due to right temporoparietal infarct. *Arch Neurol* 1982; **39**:725–7.

McGonigle DJ, Hanninen R, Salenius S *et al.* Whose arm is it anyway? An fMRI study of supernumerary phantom limb. *Brain* 2002; **125**:1265–74.

MacGowan DJL, Delanty N, Petito F *et al.* Isolated myoclonic alien hand as the sole presentation of pathologically established Creutzfeldt–Jakob disease: a report of two patients. *J Neurol Neurosurg Psychiatry* 1997; **63**:404–7.

McKeon J, McGuffin P, Robinson P. Obsessive-compulsive neurosis following head injury. A report of four cases. *Br J Psychiatry* 1984; **144**:190–2.

McMonagle P, Deering F, Berliner Y *et al.* The cognitive profile of posterior cortical atrophy. *Neurology* 2006; **66**:331–8.

Maeda K, Shirayama Y, Nukima S *et al.* Charles Bonnet syndrome with visual hallucinations of childhood experience: successful treatment of 1 patient with risperidone. *J Clin Psychiatry* 2003; **64**:1131–2.

Maertens de Noordhout AM, Delwaide PJ. The palmomental reflex in Parkinson's disease: comparisons with normal subjects and clinical relevance. *Arch Neurol* 1988; **45**:425–7.

Mahone EM, Bridges D, Prahme C *et al.* Repetitive arm and hand movements (complex motor stereotypies) in children. *J Pediatr* 2004; **145**:391–5.

Maier C, Derwinkel R, Mansourian N *et al.* Efficacy of NMDA-receptor antagonist memantine in patients with chronic phantom limb pain: results of a randomized double-blinded, placebo-controlled trial. *Pain* 2003; **103**:277–83.

Marder K, Liu X, Stern Y *et al.* Neurologic signs and symptoms in a cohort of homosexual men followed for 4.5 years. *Neurology* 1995; **45**:261–7.

Marlowe WB, Mancall EL, Thomas TJ. Complete Kluver–Bucy syndrome in man. *Cortex* 1975; **11**:53–9.

Marneros A. Schizophrenia first-rank symptoms in organic mental disorders. *Br J Psychiatry* 1988; **152**:625–8.

Martin JP. Fits of laughter (sham-mirth) in organic cerebral disease. *Brain* 1950; **73**:453–64.

Masdeu JC, O'Hara RJ. Motor aphasia unaccompanied by faciobrachial weakness. *Neurology* 1983; **33**:519–21.

Mateen FJ, Josephs KA. The clinical spectrum of stereotypies in frontotemporal lobar degeneration. *Mov Disord* 2009; **24**:1237–40.

Mathew RJ, Wilson WH, Humphreys D *et al.* Depersonalization after marijuana smoking. *Biol Psychiatry* 1993; **33**:431–41.

Mauguire F, Courjon J. Somatosensory epilepsy: a review of 127 cases. *Brain* 1978; **101**:307–32.

Mayeux R, Benson DF. Phantom limb and multiple sclerosis. *Neurology* 1979; **29**:724–6.

Mayeux R, Stern Y, Spanton S. Heterogeneity in dementia of the Alzheimer type: evidence of subgroups. *Neurology* 1985; **35**:453–61.

Mazzucchi A, Marchini C, Budai R *et al.* A case of receptive amusia with prominent timbre perception defect. *J Neurol Neurosurg Psychiatry* 1982; **45**:644–7.

Meadows JL, Munro SSF. Palinopsia. *J Neurol Neurosurg Psychiatry* 1977; **40**:5–8.

Melges FT, Tinklenberg JR, Hollister LE *et al.* Temporal disintegration and depersonalization during marijuana intoxication. *Arch Gen Psychiatry* 1970; **23**:204–10.

Mellor CS. First-rank symptoms of schizophrenia. I. The frequency in schizophrenics on admission to hospital. II. Differences between individual first-rank symptoms. *Br J Psychiatry* 1970; **117**:15–23.

Melzak R, Israel R, Lacroix R *et al.* Phantom limbs in people with congenital limb deficiency or amputation in early childhood. *Brain* 1997; **120**:1603–20.

Mendez MF. Corticobasal ganglionic degeneration with Balint's syndrome. *J Neuropsychiatry Clin Neurosci* 2000; **12**:273–5.

Mendez MF. Prominent echolalia from isolation of the speech area. *J Neuropsychiatry Clin Neurosci* 2002; **14**:56–7.

Mendez MF. Aphemia-like syndrome from a right supplementary motor area lesion. *Clin Neurol Neurosurg* 2004; **106**:337–9.

Mendez MF, Foti DJ. Lethal hyperoral behavior from the Kluver–Bucy syndrome. *J Neurol Neurosurg Psychiatry* 1997; **62**:293–4.

Mendez MF, Perryman KM. Neuropsychiatric features of frontotemporal dementia: evaluation of consensus criteria and review. *J Neuropsychiatry Clin Neurosci* 2002; **14**:424–9.

Mendez MF, Adams NL, Lewandowski KS. Neurobehavioral changes associated with caudate lesions. *Neurology* 1989; **39**:349–54.

Mendez MF, Selwood A, Mastri AR *et al.* Pick's disease versus Alzheimer's disease: a comparison of clinical characteristics. *Neurology* 1993; **43**:289–92.

Mendez MF, Cherrier MM, Perryman KM. Epileptic forced thinking from left frontal lesions. *Neurology* 1996; **47**:79–83.

Mendhekar DN, Duggal HS. Sertraline for Kluver–Bucy syndrome in an adolescent. *Eur Psychiatry* 2005; **20**:355–6.

Mendhekar DN, Sharma A, Inamdar A. Obsessive-compulsive disorder with hydrocephalus: effect of ventricular shunting. *J Neuropsychiatry Clin Neurosci* 2007; **19**:84.

Menza MA, Cocchiola J, Golbe LI. Psychiatric symptoms in progressive supranuclear palsy. *Psychosomatics* 1995; **36**:550–4.

Merrin EL, Silberfarb PM. The Capgras phenomenon. *Arch Gen Psychiatry* 1976; **33**:965–8.

Messert B, Henke TK, Langheim W. Syndrome of akinetic mutism associated with obstructive hydrocephalus. *Neurology* 1966; **16**:635–49.

Mesulam M-M. Dissociative states with abnormal temporal lobe EEG. *Arch Neurol* 1981; **38**:176–81.

Michel EM, Troost BT. Palinopsia: cerebral localization with computed tomography. *Neurology* 1980; **30**:887–9.

Michel V, Burbaud P, Taillard J *et al.* Stuttering or reflex seizure? A case report. *Epilept Disord* 2004; **6**:181–5.

Miller TC, Crosby TW. Musical hallucinations in a deaf elderly patient. *Ann Neurol* 1979; **5**:301.

Mitchell SW. The effects of Anhelonium lewinii (the mescal button). *BMJ* 1896; **2**:1625–9.

Miyazawa N, Hayashi M, Komiya K *et al.* Supernumerary phantom limbs associated with left hemispheric stroke: case report and review of the literature. *Neurosurgery* 2004; **54**:228–31.

Monrad-Krohn GH. On the dissociation of voluntary and emotional innervation in facial paresis of central origin. *Brain* 1924; **47**:22–35.

Monrad-Krohn GH. Dysprosody or altered 'melody of language'. *Brain* 1947; **70**:405–15.

Montero J, Pena J, Genis D *et al.* Balint's syndrome. Report of four cases with watershed parieto-occipital lesions from vertebrobasilar ischemia or systemic hypotension. *Acta Neurol Belg* 1982; **82**:270–80.

Morgan CA, Hazlett G, Wang S *et al.* Symptoms of dissociation in humans experiencing acute uncontrollable stress: a prospective investigation. *Am J Psychiatry* 2001; **158**:1239–47.

Mori E, Yamadori A. Unilateral hemispheric injury and ipsilateral instinctive grasp reaction. *Arch Neurol* 1985; **42**:485–8.

Motomura S, Tabira T, Kuroiwa Y. A clinical comparative study of multiple sclerosis and neuro-Behçet's syndrome. *J Neurol Neurosurg Psychiatry* 1980; **43**:210–13.

Mottard J-P, de la Sablonniere J-F. Olanzapine-induced obsessive-compulsive disorder. *Am J Psychiatry* 1999; **156**:799–800.

Mulder DW, Daly D. Psychiatric symptoms associated with lesions of temporal lobe. *J Am Med Assoc* 1952; **150**:173–6.

Mulder DW, Parrott M, Thaler M. Sequelae of western equine encephalitis. *Neurology* 1951; **1**:318–27.

Muller T, Buttner TH, Kuhn W *et al.* Palinopsia as sensory epileptic phenomenon. *Acta Neurol Scand* 1995; **91**:433–6.

Murata S, Naritomi H, Sawada T. Musical auditory hallucinations caused by a brainstem lesion. *Neurology* 1994; **44**:156–8.

Nagaratnam N, Nagaratnam K, Ng K *et al.* Akinetic mutism following stroke. *J Clin Neurosci* 2004; **11**:25–30.

Nakada T, Lee H, Kwee IL *et al.* Epileptic Kluver–Bucy syndrome: case report. *J Clin Psychiatry* 1984; **45**:87–8.

Neary D, Snowden JS, Mann DMA. Familial progressive atrophy: its relationship to other forms of lobar atrophy. *J Neurol Neurosurg Psychiatry* 1993; **56**:1122–5.

Nesher R, Nesher G, Epstein E *et al.* Charles Bonnet syndrome in glaucoma patients with low vision. *J Glaucoma* 2001; **10**:396–400.

Nielsen JM. The cortical components of akinetic mutism. *J Nerv Ment Dis* 1951; **114**:459–61.

Noda S, Mizoguchi M, Yamamoto A. Thalamic experiential hallucinosis. *J Neurol Neurosurg Psychiatry* 1993; **56**:1224–6.

Noyes R, Kletti R. Depersonalization in accident victims and psychiatric patients. *Compr Psychiatry* 1977; **18**:375–84.

Noyes R, Hoenk P, Kuperman S *et al.* Depersonalization in accident victims and psychiatric patients. *J Nerv Ment Dis* 1977; **164**:401–7.

Nurnberg HG, Greenwald B. Stuttering: an unusual side-effect of phenothiazines. *Am J Psychiatry* 1981; **138**:386–7.

Nyatsanza S, Shetty T, Gregory C *et al.* A study of stereotypic behaviors in Alzheimer's disease and frontal and temporal variant frontotemporal dementia. *J Neurol Neurosurg Psychiatry* 2003; **74**:1398–402.

O'Hare JA, Callaghan NM, Murnaghan DJ. Dialysis encephalopathy. Clinical, electroencephalographic, and interventional aspects. *Medicine* 1983; **62**:129–41.

Oliveira V, Ferro JM, Foreid JP *et al.* Kluver–Bucy syndrome in systemic lupus erythematosus. *J Neurol* 1989; **236**:55–6.

O'Sullivan D, Dean C. The Fregoli syndrome and puerperal psychosis. *Br J Psychiatry* 1991; **159**:274–7.

Ott BR. Leuprolide treatment of sexual aggression in a patient with dementia and the Kluver–Bucy syndrome. *Clin Neuropharmacol* 1995; **18**:443–7.

Panayiotopoulos CP. Elementary visual hallucinations in migraine and epilepsy. *J Neurol Neurosurg Psychiatry* 1994; **57**:1371–4.

Panitch HS, Thisted RA, Smith RA *et al.* Randomized controlled trial of dextromethorphan/quinidine for pseudobulbar affect in multiple sclerosis. *Ann Neurol* 2006; **59**:780–7.

Parkes JD, Baraitser M, Marsden CD *et al.* Natural history, symptoms and treatment of the narcoleptic syndrome. *Acta Neurol Scand* 1975; **52**:337–53.

Parkinson D, Rucker CW, McK Craig W. Visual hallucinations associated with tumors of the occipital lobe. *Arch Neurol Psychiatry* 1952; **68**:60–8.

Parvizi J, Joseph J, Press DZ *et al.* Pathological laughter and crying in patients with multiple system atrophy-cerebellar type. *Mov Disord* 2007; **22**:798–803.

Paskind HA. Parosmia in tumorous involvement of olfactory bulbs and nerves. *Arch Neurol Psychiatry* 1935; **33**:835–8.

Patterson JF. Carbamazepine in the treatment of phantom limb pain. *South Med J* 1988; **81**:1100–2.

Paulig M, Mentrup H. Charles Bonnet's syndrome: complete remission of complex visual hallucinations treated by gabapentin. *J Neurol Neurosurg Psychiatry* 2001; **70**:813–14.

Pazzaglia P, D'Alessandro R, Ambrosetto G *et al.* Drop attacks: an ominous change in the evolution of partial epilepsy. *Neurology* 1985; **35**:1725–30.

Penfield W, Jasper H. *Epilepsy and the Functional Anatomy of the Human Brain.* Boston: Little, Brown and Company, 1954: 383–4.

Penfield W, Perot P. The brain's record of auditory and visual experience. *Brain* 1963; **86**:595.

Perez FM, Tunkel RS, Lachmann EA *et al.* Balint's syndrome arising from bilateral posterior cortical atrophy or infarction: rehabilitation strategies and their limitation. *Disabil Rehabil* 1996; **18**:300–4.

Perino M, Famularo G, Tarroni P. Acquired transient stuttering during a migraine attack. *Headache* 2000; **40**:170–2.

Petracca GM, Jorge RE, Aclon L *et al.* Frequency and correlates of involuntary emotional expression disorder in Parkinson's disease. *J Neuropsychiatry Clin Neurosci* 2009; **21**:406–12.

Piccirilli M, Sciarma T, Luzzi S. Modularity of music: evidence from a case of pure amusia. *J Neurol Neurosurg Psychiatry* 2000; **69**:541–5.

Pick A. On reduplicative paramnesia. *Brain* 1903; **26**:260–7.

Pioro EP, Brooks BR, Cummings J *et al.* Dextromethorphan plus ultra low-dose quinidine reduces pseudobulbar affect. *Ann Neurol* 2010; **68**:693–702.

Pitt DC, Kriel RL, Wagner NC *et al.* Kluver–Bucy syndrome following heat stroke in a 12-year-old girl. *Pediatr Neurology* 1995; **13**:73–6.

Powers JM, Schaumburg HH, Gaffney CL. Kluver–Bucy syndrome caused by adrenoleukodystrophy. *Neurology* 1980; **30**:1131–2.

Power-Smith P, Turkington D. Fluoxetine in phantom limb pain. *Br J Psychiatry* 1993; **163**:105–6.

Psarros T, Zours A, Coimbra C. Bromocriptine-responsive akinetic mutism following endoscopy for ventricular neurocysticercosis. Case report and review of the literature. *J Neurosurg* 2003; **99**:397–401.

Quattrini A, Del Pesce M, Provinciali L *et al.* Mutism in 36 patients who underwent callosotomy for drug-resistant epilepsy. *J Neurosurg Sci* 1997; **41**:93–6.

Rabinowitz J, Katz JR, De Deyn PP *et al.* Behavioral and psychological symptoms in patients with dementia as a target for pharmacotherapy with risperidone. *J Clin Psychiatry* 2004; **65**:1329–34.

Radhakrishnan J, Mathew K, Richard J *et al.* Schneider's first-rank symptoms – prevalence, diagnostic use and prognostic implications. *Br J Psychiatry* 1983; **142**:557–9.

Rasmussen KG, Rummans TA. Electroconvulsive therapy for phantom limb pain. *Pain* 2000; **85**:297–9.

Rayport M, Ferguson SM, Corrie WS. Outcomes and indications of corpus callosum section for intractable seizure control. *Appl Neurophysiol* 1983; **46**:47–51.

Reeves RR, Norton JW. Foreign accent-like syndrome during psychotic exacerbations. *Neurpsychiatry Neuropsychol Behav Neurol* 2001; **14**:135–8.

Reeves RR, Burke RS, Parker JD. Characterization of psychotic patients with foreign accent syndrome. *J Neuropsychiatry Clin Neurosci* 2007; **19**:70–6.

Remillard GM, Andermann F, Rhi-Sausa A *et al.* Facial asymmetry in patients with temporal lobe epilepsy. *Neurology* 1977; **27**:109–14.

Remington G, Sloman L, Konstantareas M *et al.* Clomipramine versus haloperidol in the treatment of autistic disorder: a double-blind, placebo-controlled, crossover study. *J Clin Psychopharmacol* 2001; **21**:440–4.

Renner JA, Burns JM, Hou CE *et al.* Progressive posterior cortical dysfunction. A clinicopathologic series. *Neurology* 2004; **63**:1175–80.

Riddoch G. Phantom limbs and body shape. *Brain* 1941; **64**:197–222.

Riley DE, Lang AE, Resch L *et al.* Cortico-basal ganglionic degeneration. *Neurology* 1990; **40**:1203–12.

Rinne JO, Lee MS, Thompson PD *et al.* Cortico-basal degeneration. *Brain* 1994; **117**:1183–96.

Roberts JM. Echolalia and comprehension in autistic children. *J Autism Dev Disord* 1989; **19**:271–81.

Roberts K, Vass N. Schneiderian first-rank symptoms caused by benzodiazepine withdrawal. *Br J Psychiatry* 1986; **148**:593–4.

Robinson RG, Parikh RM, Lipsey JR *et al.* Pathological laughing and crying following stroke; validation of a measurement scale and a double-blind treatment study. *Am J Psychiatry* 1993; **150**:286–93.

Rosenfeld MR, Eichen JG, Wade DF *et al.* Molecular and clinical diversity in paraneoplastic immunity to Ma proteins. *Ann Neurol* 2001; **50**:339–48.

Ross ED, Stewart RM. Akinetic mutism from hypothalamic damage: successful treatment with dopamine agonists. *Neurology* 1981; **31**:1435–9.

Ross ED, Jossman PB, Bell B *et al.* Musical hallucinations in deafness. *JAMA* 1975; **231**:620–1.

Ross RT, Mathiesen R. Images in clinical medicine: volitional and emotional supranuclear facial weakness. *N Engl J Med* 1998; **338**:1515.

Rosse RB, Collins JP, Fay-McCarthy M *et al.* Phenomenologic comparison of the idiopathic psychosis of schizophrenia and drug-induced cocaine and phencyclidine psychoses: a retrospective study. *Clin Neuropharmacol* 1994; **17**:359–69.

Rosselli M, Ardila A, Beltran C. Rehabilitation of Balint's syndrome: a single case report. *Appl Neuropsychol* 2001; **8**:242–7.

Rossitch E, Bullard DE. The autonomic dysfunction syndrome: aetiology and treatment. *Br J Neurosurg* 1988; **2**:471–8.

Roth CR, Aronson AE, Davis LJ. Clinical studies in psychogenic stuttering of adult onset. *J Speech Hear Disord* 1989; **54**:634–46.

Ruff RL, Volpe BT. Environmental reduplication associated with right frontal and parietal lobe injury. *J Neurol Neurosurg Psychiatry* 1981; **44**:382–6.

Rumsey JM, Andreasen NC, Rapoport JL. Thought, language, commu nication and affective flattening in autistic adults. *Arch Gen Psychiatry* 1986; **43**:771–7.

Russell MB, Olesen J. A nosographic analysis of the migraine aura in a general population. *Brain* 1996; **119**:355–61.

Russell SM, Golfinos JG. Amusia following resection of Heschl gyrus glioma. Case report. *J Neurosurg* 2003; **98**:1109–12.

Russell WR, Whitty CWM. Studies in traumatic epilepsy. 2. Focal motor and somatic sensory fits: a study of 85 cases. *J Neurol Neurosurg Psychiatry* 1953; **16**:73–97.

Russell WR, Whitty CWM. Studies in traumatic epilepsy. 3. Visual fits. *J Neurol Neurosurg Psychiatry* 1955; **18**:79–96.

Ryallis J, Whiteside J. An atypical case of foreign accent syndrome. *Clin Linguist Phon* 2006; **20**:157–62.

Rylander G. Psychoses and the punding and choreiform syndromes in addiction to central stimulant drugs. *Psychiatr Neurol Neurochir* 1972; **75**:203–12.

Sachdev P. Schizophrenia-like illness in velo-cardio-facial syndrome: a genetic subsyndrome of schizophrenia? *J Psychosom Res* 2002; **53**:721–7.

Sandson J, Albert ML. Perseveration in behavioral neurology. *Neurology* 1987; **37**:1736–41.

Santhouse AM, Howard RJ, ffytche DH. Visual hallucinatory syndrome and the anatomy of the visual brain. *Brain* 2000; **123**:2055–64.

Sarkar J, Jones N, Sulivan G. A case of depersonalization-derealization syndrome during treatment with quetiapine. *J Psychopharmacol* 2001; **15**:209–11.

Schielke E, Reuter U, Hoffmnn O *et al.* Musical hallucinations with dorsal pontine lesions. *Neurology* 2000; **55**:454–5.

Schiff HB, Alexander MP, Naeser MA *et al.* Aphemia: clinical-anatomic correlations. *Arch Neurol* 1983; **40**:720–7.

Schiffer RB, Herndon RM, Rudick RA. Treatment of pathologic laughing and weeping with amitriptyline. *N Engl J Med* 1985; **312**:1480–2.

Schley M, Topfner S, Wiech K *et al.* Continuous brachial plexus blockade in combination with the NMDA receptor antagonist memantine prevents phantom limb pain in acute traumatic upper limb amputees. *Eur J Pain* 2007; **11**:299–308.

Schneider K. *Clinical Psychopathology.* London: Grune & Stratton, 1959.

Schnider A, von Daniken C, Gutbrod K. The mechanisms of spontaneous and provoked confabulations. *Brain* 1996; **119**:1365–75.

Schwartz JI, Moura RJ. Severe depersonalization and anxiety associated with indomethacin. *South Med J* 1983; **76**:679–80.

Schwartz W, Stakes JW, Jobson JA. Transient cataplexy after removal of a craniopharyngioma. *Neurology* 1984; **34**:1372–5.

Sears LL, Vest C, Mohamed S *et al.* An MRI study of the basal ganglia in autism. *Prog Neuropsychopharmacol Biol Psychiatry* 1999; **23**:613–24.

Sedman G. Depersonalization in a group of normal subjects. *Br J Psychiatry* 1966; **112**:907–12.

Seliger GM, Abrams GM, Horton A. Irish brogue after stroke. *Stroke* 1992; **23**:1655–6.

Serra-Mestres J, Shapleske J, Tym E. Treatment of palilalia with trazodone. *Am J Psychiatry* 1996; **153**:580–1.

Seyffarth H, Denny-Brown D. The grasp reflex and the instinctive grasp reflex. *Brain* 1948; **71**:109–83.

Shahed J, Jankovic J. Re-emergence of childhood stuttering in Parkinson's disease: an hypothesis. *Mov Disord* 2001; **16**:114–18.

Shallice T, Burgess PW, Schon F *et al.* The origins of utilization behavior. *Brain* 1989; **112**:1587–98.

Shoji H, Teramoto H, Satowa S *et al.* Partial Kluver–Bucy syndrome following probable herpes simplex encephalitis. *J Neurol* 1979; **221**:163–7.

Shukla GD, Sahu SC, Tripathi RP *et al.* Phantom limb: a phenomenological study. *Br J Psychiatry* 1982; **141**:54–8.

Sichel DA, Cohen LS, Dimmock JA *et al.* Postpartum obsessive compulsive disorder: a case series. *J Clin Psychiatry* 1993; **54**:156–9.

Siegel A, Harrow M, Reilly FE *et al.* Loose associations and disordered speech patterns in chronic schizophrenia. *J Nerv Ment Dis* 1976; **162**:105–12.

Siegel RK. Cocaine hallucinations. *Am J Psychiatry* 1978; **135**:309–14.

Simmel ML. Phantom experiences following amputation in childhood. *J Neurol Neurosurg Psychiatry* 1962; **25**:69–78.

Simpson S, Baldwin B. Neuropsychiatry and SPECT of an acute obsessive-compulsive syndrome patient. *Br J Psychiatry* 1995; **166**:390–7.

Sjogren M, Wallin A, Edman A. Symptomatological differences distinguish between frontotemporal dementia and vascular dementia with a dominant frontal lobe syndrome. *Int J Geriatr Psychiatry* 1997; **12**:656–61.

Slater E, Beard AW. The schizophrenia-like psychoses of epilepsy. i. Psychiatric aspects. *Br J Psychiatry* 1963; **109**:95–112.

Slatterly MJ, Dubbert BK, Allen AJ *et al.* Prevalence of obsessive-compulsive disorder in patients with systemic lupus erythematosus. *J Clin Psychiatry* 2004; **65**:301–6.

Smit LS, Lammers GJ, Catsman-Berrevoets CE. Cataplexy leading to the diagnosis of Niemann-Pick disease type C. *Pediatr Neurol* 2006; **35**:82–4.

Smith DG, Ehds DM, Hanley MA *et al.* Efficacy of gabapentin in treating chronic phantom limb and residual limb pain. *J Rehabil Res Dev* 2005; **42**:645–54.

Smith KU, Akelaitis AJ. Studies on the corpus callosum. I. Laterality in behavior and bilateral motor organization in man before and after section of the corpus callosum. *Arch Neurol Psychiatry* 1942; **47**:519–43.

Snowden JS, Neary D, Mann DMA *et al.* Progressive language disorder due to lobar atrophy. *Ann Neurol* 1992; **31**:174–83.

Soros P, Vo G, Husstedt I-W *et al.* Phantom eye syndrome. Its prevalence, phenomenology and putative mechanisms. *Neurology* 2003; **60**:1542–3.

Souques A. Palilalia. *Rev Neurol* 1908; **16**:340–3.

Soyka M, Naber G, Volcker A. Prevalence of delusional jealousy in different psychiatric disorders. *Br J Psychiatry* 1991; **158**:549–53.

Sparr SA. Receptive amelodia in a trained musician. *Neurology* 2002; **59**:1659–60.

Sperry RW. Brain bisection and mechanisms of consciousness. In: Eccles JC, ed. *Brain and Conscious Experience*. New York: Springer-Verlag, 1966.

Stacy M, Cardoso F, Jankovic J. Tardive stereotypy and other movement disorders in tardive dyskinesias. *Neurology* 1993; **43**:937–41.

Stahl SM, Layzer RB, Aminoff MJ *et al.* Continuous cataplexy in a patient with a midbrain tumor: the limp man syndrome. *Neurology* 1980; **30**:1115–18.

Stamouli S, Lykouras L. Quetiapine-induced obsessive-compulsive symptoms: a series of five cases. *J Clin Psychopharmacol* 2006; **26**:396–400.

Starkstein SE, Berthier M, Leiguarda R. Disconnection syndrome in a right-handed patient with right hemispheric speech dominance. *Eur Neurol* 1988; **28**:187–90.

Starkstein SE, Federoff JP, Price TR *et al.* Catastrophic reaction after cerebrovascular lesions: frequency, correlates, and validation of a scale. *J Neuropsychiatry Clin Neurosci* 1993; **5**:189–94.

Starkstein SE, Migliorelli R, Teson A *et al.* Prevalence and clinical correlates of pathological affective display in Alzheimer's disease. *J Neurol Neurosurg Psychiatry* 1995; **59**:55–60.

Steinhausen H-C, Juzi C. Elective mutism: an analysis of 100 cases. *J Am Acad Child Adolesc Psychiatry* 1996; **35**:606–14.

Stewart JT. Carbamazepine treatment of a patient with Kluver–Bucy syndrome. *J Clin Psychiatry* 1985; **46**:496–7.

Stewart-Wallace AM. An unusual case of the grasp reflex, with some observations on the volitional and reflex components. *J Neurol Neurosurg Psychiatry* 1939; **2**:149–53.

Stuss DT, Alexander MP, Lieberman A *et al.* An extraordinary form of confabulation. *Neurology* 1978; **28**:1166–72.

Suwanwela NC, Leelacheavasit N. Isolated corpus callosal infarction secondary to pericallosal artery disease presenting as alien hand syndrome. *J Neurol Neurosurg Psychiatry* 2002; **72**:533–6.

Takayama Y, Sugishita M, Kido T *et al.* A case of foreign accent syndrome without aphasia caused by a lesion of the left precentral gyrus. *Neurology* 1993; **43**:1361–3.

Tanaka Y, Iwasa H, Yoshuida M. Diagnostic dyspraxia: case report and movement-related potentials. *Neurology* 1990; **40**:657–61.

Tanaka Y, Yoshida A, Kawahata N *et al.* Diagnostic dyspraxia: clinical characteristics, responsible lesions and possible underlying mechanism. *Brain* 1996; **119**:859–74.

Tandon R, Greden JF. Schneiderian first-rank symptoms: reconfirmation of high specificity for schizophrenia. *Acta Psychiatr Scand* 1987; **75**:392–6.

Tang-Wai DF, Graff-Radford NR, Boeve BF *et al.* Clinical, genetic, and neuropathologic characteristics of posterior cortical atrophy. *Neurology* 2004; **63**:1168–74.

Tani T, Sakai Y. Stuttering after right cerebellar infarction: a case study. *J Fluency Disord* 2010; **35**:141–5.

Tateno A, Jorge RE, Robinson RG. Pathological laughing and crying following traumatic brain injury. *J Neuropsychiatry Clin Neurosci* 2004; **16**:426–34.

Teri L, Larson EB, Reifler BV. Behavioral disturbance in dementia of the Alzheimer's type. *J Am Ger Soc* 1988; **36**:1–6.

Terzian H, Dalle Ore G. Syndrome of Kluver and Bucy: reproduced in man by bilateral removal of the temporal lobes. *Neurology* 1955; **5**:373–80.

Teunisse RJ, Zitman FG, Raes DCM. Clinical evaluation of 14 patients with the Charles Bonnet syndrome (isolated visual hallucinations). *Compr Psychiatry* 1994; **35**:70–5.

Thajeb P. Progressive late delayed postirradiation encephalopathy with Kluver–Bucy syndrome. Serial MRI and clinico-pathological studies. *Clin Neurol Neurosurg* 1995; **97**:264–8.

Thomas RH, Bennetto L, Silva MT. Reversible grasp reflexes in normal pressure hydrocephalus. *Clin Neurol Neurosurg* 2009; **111**:387–9.

Thompson AE, Swan M. Capgras syndrome presenting with violence following heavy drinking. *Br J Psychiatry* 1993; **162**:692–4.

Thompson JC, Stopford CL, Snowden JS *et al.* Qualitative neuropsychological performance characteristics in frontotemporal dementia and Alzheimer's disease. *J Neurol Neurosurg Psychiatry* 2005; **76**:920–7.

Thorley RR, Wertsch JJ, Kingbell GE. Acute hypothalamic instability in traumatic brain injury: a case report. *Arch Phys Med Rehabil* 2001; **82**:246–9.

Thorpe LU, Keegan DL, Veeman GA. Conversion mutism: case report and discussion. *Can J Psychiatry* 1985; **30**:71–3.

Tilberti C, Sabe L, Kuzis G *et al.* Prevalence and correlates of the catastrophic reaction in Alzheimer's disease. *Neurology* 1998; **50**:546–8.

Tilney F. Pseudobulbar palsy clinically and pathologically considered with the clinical report of five cases. *J Nerv Ment Dis* 1912; **39**:505–35.

Trethowan WH, Cobb S. Neuropsychiatric aspects of Cushing's syndrome. *Arch Neurol Psychiatry* 1952; **67**:283–309.

Trosch RM, Sze G, Brass LM *et al.* Emotional facial paresis with striatocapsular infarction. *J Neurol Sci* 1990; **98**:195–201.

Tuke DH. Imperative ideas. *Brain* 1894; **17**:179–97.

Turgut N, Utku U, Balci K. A case of acquired stuttering resulting from left parietal infarction. *Acta Neurol Scand* 2002; **105**:408–10.

Tutuncuoglu S, Serdaroglu G, Kadioglu B. Landau-Kleffner syndrome beginning with stuttering: case report. *J Child Neurol* 2002; **17**:785–8.

Tweedy J, Reding M, Garcia C *et al.* Significance of cortical disinhibition signs. *Neurology* 1981; **31**:169–73.

Ueda N, Tarao T, Ohmori O *et al.* Schneiderian first-rank symptoms associated with fluvoxamine treatment: a case report. *Hum Psychopharmacol* 2003; **18**:477–8.

Ukai S, Yamamoto M, Tanaka M *et al.* Treatment of typical Charles Bonnet syndrome with donepezil. *Int Clin Psychopharmacol* 2004; **19**:355–7.

Ungvari GS, Leung SK, Ng FS *et al.* Schizophrenia with prominent catatonic features ('catatonic schizophrenia'): I. Demographic and clinical correlates in the chronic phase. *Prog Neuropsychopharmacol Biol Psychiatry* 2005; **29**:27–38.

Ure J, Faccio E, Videla H *et al.* Akinetic mutism: a report of three cases. *Acta Neurologica Scand* 1998; **98**:439–44.

Uzunca I, Utku U, Asil T *et al.* 'Fou rire prodromique' associated with simultaneous bilateral capsular genu infarction. *J Clin Neurosci* 2005; **12**:174–5.

Valldeoriola F, Graus F, Rimola A *et al.* Cyclosporine-induced mutism in liver transplant patients. *Neurology* 1996; **46**:252–4.

Van Borsel J, van der Made S, Santens P. Thalamic stuttering: a distinct clinical entity? *Brain Lang* 2003; **85**:185–9.

Van den Hoed J, Lucas EA, Dement WC. Hallucinatory experiences during cataplexy in patients with narcolepsy. *Am J Psychiatry* 1979; **136**:1210–11.

Van Der Werf YD, Weerts JGE, Jolles J *et al.* Neuropsychological correlates of a right unilateral lacunar thalamic infarction. *J Neurol Neurosurg Psychiatry* 1999; **66**:36–42.

Van Dijk JG, Lammers GJ, Blansjaar BA. Isolated cataplexy of more than 40 years duration. *Br J Psychiatry* 1991; **154**:719–21.

Van Dongen HR, Catsman-Berrevoets CE, van Mourik M. The syndrome of 'cerebellar' mutism and subsequent dysarthria. *Neurology* 1994; **44**:2040–6.

Van Hilten JJ, Buruma OJ, Kessing P *et al.* Pathologic crying as a prominent behavioral manifestation of central pontine myelinolysis. *Arch Neurol* 1988; **45**:936.

Van Waggenen WP, Herren RY. Surgical division of commisural pathways in the corpus callosum: relation to spread of an epileptic attack. *Arch Neurol Psychiatry* 1940; **44**:740–57.

Vaphiades MS, Celesia GG, Brigell MG. Positive spontaneous visual phenomena limited to the hemianopic field in lesions of central visual pathways. *Neurology* 1996; **47**:408–17.

Varon D, Pritchard PB, Wagner MT *et al.* Transient Kluver–Bucy syndrome following complex partial status epilepticus. *Epilepsy Behav* 2003; **4**:348–51.

Verhoeven J, Marien P, Engelborghs S *et al.* A foreign speech accent in a case of conversion disorder. *Behav Neurol* 2005; **16**: 225–32.

Vetrugno R, Alessandria M, D'Angelo R *et al.* 'Phantom' restless legs syndrome. *J Neurol Neurosurg Psychiatry* 2010; **81**:122–3.

Victoroff J, Ross GW, Benson DF *et al.* Posterior cortical atrophy. Neuropathologic correlations. *Arch Neurol* 1994; **51**:269–74.

Vignoli A, La Briola F, Canevini MP. Evolution of stereotypies in adolescents and women with Rett syndrome. *Mov Disord* 2009; **24**:1379–83.

Vollenweider FX, Gamma A, Liechti M *et al.* Psychological and cardiovascular effects and short-term sequelae of MDMA ('ecstasy') in MDMA-naive healthy volunteers. *Neuropsychopharmacology* 1998; **19**:241–51.

von Gunten A, Miklossy J, Suva ML *et al.* Environmental reduplicative paramnesia in a case of atypical Alzheimer's disease. *Neurocase* 2005; **11**:216–26.

Vossier DG, Haltiner AM, Schepp SK *et al.* Ictal stuttering: a sign suggestive of psychogenic nonepileptic seizures. *Neurology* 2004; **63**:516–19.

Vreeling FW, Verhey FR, Houx PJ *et al.* Primitive reflexes in Parkinson's disease. *J Neurol Neurosurg Psychiatry* 1993; **56**:1323–6.

Vreeling FW, Houx PJ, Jolles J *et al.* Primitive reflexes in Alzheimer's disease and vascular dementia. *J Geriatr Psychiatr Neurol* 1995; **8**:111–17.

Vuilleumier P, Reverdin A, Landis T. Four legs: illusory reduplication of the lower limbs after bilateral parietal lobe damage. *Arch Neurol* 1997; **54**:1543–7.

Waite LM, Broe GA, Creasey H *et al.* Neurological signs, aging, and the neurodegenerative syndromes. *Arch Neurol* 1996; **53**:498–502.

Wali GM. 'Fou rire prodromique' heralding a brainstem stroke. *J Neurol Neurosurg Psychiatry* 1993; **56**:209–10.

Walshe FMR, Hunt JH. Further observations upon grasping movements and reflex tonic grasping. *Brain* 1936; **59**:315–23.

Walshe FMR, Robertson EG. Observations upon the form and nature of the 'grasping' movements and 'tonic innervation' seen in certain cases of lesions of the frontal lobe. *Brain* 1933; **56**:40–70.

Watson RT, Heilman KM. Callosal apraxia. *Brain* 1983; **106**:391–403.

Watson RT, Heilman KM, Bowers D. Magnetic resonance imaging (MRI, NMR) scan in a case of callosal apraxia and pseudoneglect. *Brain* 1985; **108**:535–6.

Weinstein EA, Kahn RL, Malitz S *et al.* Delusional reduplication of parts of the body. *Brain* 1954; **77**:45–60.

Weinstein S, Sersen EA. Phantoms in cases of congenital absence of limbs. *Neurology* 1961; **11**:905–11.

Wergeland H. Elective mutism. *Acta Psychiatr Scand* 1979; **59**:218–28.

White NJ. Complex visual hallucinations in partial blindness due to eye disease. *Br J Psychiatry* 1980; **136**:284–6.

Wiech K, Kiefer RT, Topfner S *et al.* A placebo-controlled randomized crossover trial of the N-methyl-D-aspartate acid receptor antagonist, memantine, in patients with chronic phantom limb pain. *Anesth Analg* 2004; **98**:408–13.

Wijdicks EF, Wiesner RH, Dahlke LJ *et al.* FK506-induced neurotoxicity in liver transplantation. *Ann Neurol* 1994; **35**:498–501.

Williams D. The structure of emotions reflected in epileptic experiences. *Brain* 1956; **79**:29–67.

Wilson DH, Reeves A, Gazzaniga M. Cerebral commisurotomy for control of intractable seizures. *Neurology* 1977; **27**:708–15.

Wilson SAK. Progressive lenticular degeneration: a familial nervous disease associated with cirrhosis of the liver. *Brain* 1912; **34**:295–509.

Wilson SAK. Some problems in neurology. II. Pathological laughing and crying. *J Neurol Psychopathology* 1924; **4**:299–333.

Wilson SAK. The narcolepsies. *Brain* 1928; **51**:63–109.

Wilson SAK. *Neurology*, 2nd edn. London: Butterworth & Co, 1954.

Wolberg FL, Ziegler DK. Olfactory hallucinations in migraine. *Arch Neurol* 1982; **39**:382.

Yamanaka K, Fukuyama H, Kimura J. Abulia from unilateral capsular genu infarction: report of two cases. *J Neurol Sci* 1996; **143**:181–4.

Yasuda Y, Akiguchi I, Ino M *et al.* Paramedian thalamic and midbrain infarcts associated with palilalia. *J Neurol Neurosurg Psychiatry* 1990; **53**:797–9.

Zarcone V. Narcolepsy. *N Engl J Med* 1973; **288**:1154–66.

Zeidler M, Johnstone EC, Bamber RWK *et al.* New variant Creutzfeldt–Jakob disease: psychiatric features. *Lancet* 1997; **350**:908–10.

Zelig G, Drubach DA, Katz-Zelig M *et al.* Pathological laughter and crying in patients with closed traumatic brain injury. *Brain Inj* 1996; **10**:591–7.

Ziegler LH. Psychotic and emotional phenomena associated with amyotrophic lateral sclerosis. *Arch Neurol Psychiatry* 1930; **24**:930–6.

6

Syndromes of disturbances of consciousness and cognition

6.1 STUPOR AND COMA

Disturbances in the level of consciousness reflect dysfunction of either the ascending reticular activating system (ARAS) (Parvizi and Damasio 2003), the thalamus, or the cortex. The ARAS is a midline brainstem structure which extends from the rostral pontine tegmentum up into the mesencephalic tegmentum, where it lies ventral to the periaqueductal gray matter. Fibers from the ARAS then extend up into the thalamus, specifically to the thalamic intralaminar nuclei and the reticular nucleus, and subsequently fibers from the thalamus project to the cortex. In addition to this efferent pathway of the ARAS, there is also a subsidiary, extrathalamic pathway, which ascends from the ARAS, and then passes on through the hypothalamus toward the frontal cortex.

Before stupor or coma can occur, either the cortex must be diffusely and bilaterally compromised or there must be damage, either to the diencephalon or to the ARAS itself. It must be stressed that regardless of the nature of the insult in question, it must be *bilateral.* Unilateral lesions, in and of themselves, whether cortical, diencephalic, or brainstem, do not cause stupor or coma. Furthermore, as might be guessed, lesions of the lower pons or of the medulla, as they do not damage the ARAS, do not, again in and of themselves, cause stupor or coma.

Some examples, taken from the host of insults capable of causing stupor or coma that are discussed in the section on etiology, may serve to flesh out these concepts. An example of a diffuse bilateral cortical lesion would be hypoxic–ischemic damage (e.g. after cardiac arrest) with widespread cortical necrosis, or diffuse axonal injury after traumatic brain injury, which isolates the cortex from the ARAS below. The cortex may also be diffusely compromised by toxic or metabolic insults, including, for example, opioid overdose or hypoglycemia. As noted earlier, unilateral lesions, in and of themselves, do not cause stupor or coma; however, unilateral cerebral lesions may indirectly be pathogenic in this regard by secondarily damaging the diencephalon or brainstem. For example, a large tumor of the frontal lobe, by itself, will leave the patient awake, yet if it grows large enough to laterally displace and compress the thalamus, then stupor or coma may occur.

Diencephalic lesions capable of causing stupor or coma may include thalamic hemorrhage or infarction; however, as noted, bilateral involvement must be present, either because, in these examples, the hemorrhage was large enough to compress the opposite thalamus or, in the case of infarction, both thalami underwent infarction, as may occur when penetrating arteries to each thalamus arise from a common pedicle. Mesencephalic or rostral pontine lesions may likewise include hemorrhage or infarction affecting the midline tegmentum, which again, however, even in the brainstem, must be bilateral (Parvizi and Damasio 2003).

Clinical features

Stupor and coma are both characterized by a decreased level of consciousness. However, as any clinician knows, there is a certain variability in the definition of these terms, and consequently it is generally most useful to describe the patient's response to various stimuli. Thus, upon encountering a patient whose eyes are closed and who is either motionless or making at most seemingly purposeless movements, the first step is to call the patient's name at a normal conversational tone. If the patient does not respond, the next step is to call the patient's name loudly. If this does not work, then repeat that maneuver while shaking the patient's shoulder. If the patient still does not wake up, then again call the name loudly, and shake the shoulder vigorously.

Should this maneuver fail, then attempts should be made to arouse the patient with one or more of the following painful stimuli. First, consider applying pressure to the pad of the patient's second finger by squeezing it between your thumb and index finger. Next, one may apply pressure to the supraorbital ridge with the thumb, and finally pressure may be applied to the temporomandibular joint, again with the thumb. It is important, with each of the foregoing maneuvers, to test each side, given the possibility of hemisensory loss. Finally, if these painful stimuli fail to awaken the patient, one should perform a sternal rub by vigorous application of the knuckles of the second and third fingers. If the patient does respond to any of these maneuvers, then it is important to describe that response in some detail.

Although, as noted, there is variability in the definition of stupor and coma, most clinicians would agree that the term coma refers to a condition wherein, despite vigorous painful stimulation, the patient neither opens his eyes, nor speaks, nor makes any purposeful or semi-purposeful movement. Application of this last portion of the definition of coma, however, must be done with some care. Purposeful or semi-purposeful movements might include pulling the finger away from the physician's hand or pushing away the physician's hand from the face or sternum. These movements are often referred to as 'localizing'. Non-purposeful movements include decorticate or decerebrate posturing, as described further below, or a 'triple flexion' response upon plantar stimulation, as may be seen during the late phase of central herniation, also described further below.

The term stupor, while perhaps enjoying a less consistent definition among different physicians, may be used to refer to a condition wherein the patient does show some response to one of these maneuvers, with any one of eye opening, vocalization, or purposeful or semi-purposeful movement. In stupor, however, soon after the stimulation ceases, the patient's response likewise ceases.

Before leaving this subject of definitions, the terms 'drowsiness' and 'lethargy', often used interchangeably, may be considered. These refer to a condition wherein, with stimulation, the patient's eyes open, there is some verbal response or purposeful or semi-purposeful movement, and this response persists for a significant period of time before the eyes close once again. Clearly, the boundary between stupor and drowsiness or lethargy is a fuzzy one, pointing again to the importance of supplementing one's use of these terms with a description of what actually occurred.

Once it has been established that the patient is either stuporous or comatose, one proceeds to the physical examination. Before doing so, however, it is critical, in any cases with possible neck injury, to obtain cervical spine films, lest one inadvertently cause a cervical cord injury. The neurologic examination, attenuated as it is by the patient's inability to cooperate, can generally be accomplished in a matter of minutes, and the main aspects of this examination are described further below. In interpreting the results of the examination, one must keep in mind any pre-existing abnormalities, especially anisocoria, strabismus, cranial nerve palsies, hemiparesis, or hemianesthesia. After discussing the neurologic examination, a description is given of various structural syndromes which may cause stupor or coma, including lateral displacement of the diencephalon and two of the herniation syndromes, namely uncal herniation and central herniation. Finally, some overall comments are offered regarding the interpretation of the examination.

Before proceeding to the neurologic examination, however, certain comments are in order regarding vital signs.

An elevated temperature should always suggest an infectious process, including meningitis, encephalitis, or sepsis; temperatures rising above 41.1°C, however, indicate heat stroke. If these entities have been ruled out, then one may consider anticholinergic toxicity or the neuroleptic malignant syndrome; lesions of the hypothalamus, as may occur with traumatic brain injury, may also cause an elevated temperature; however, this should probably remain a diagnosis of exclusion.

Respiratory depression suggests overdosage with opioids, benzodiazepines, and the like, and may also be seen in myxedema. Tachypnea may be seen in diabetic ketoacidosis, sepsis, hypoxia, hepatic encephalopathy, or intoxication with aspirin or ethylene glycol. Rarely, sustained tachypnea occurs secondary to structural damage in either the lower mesencephalon or upper pons, in which case one speaks of 'neurogenic hyperventilation'. In addition to these abnormalities of the rate of respiration, one may also encounter various abnormal patterns of respiration, including Cheyne–Stokes respiration, apneustic breathing, and either cluster or ataxic respirations. Cheyne–Stokes respiration is characterized by episodes, lasting 15–20 seconds, of regular respirations which, however, gradually crescendo in depth up to a maximum, after which there is a gradual decrescendo in depth up to the end of the episode, then a period of apnea ensues, to be followed by once again a crescendo–decrescendo episode. Cheyne–Stokes respiration may be seen in congestive heart failure, hepatic or uremic encephalopathy, and also with structural damage, either to both frontal lobes (as for example in postanoxic coma) or with diencephalic lesions. Apneustic breathing is characterized by a pause at the height of inspiration, lasting several seconds, after which there is a rapid exhalation; after a few seconds of apnea, another apneustic inspiration then occurs: this pattern is seen with structural damage to the pons. Cluster and ataxic breathing are seen with structural damage at the pontomedullary junction. Cluster breathing is characterized by episodes during which the respiratory rate is irregular, with the episodes being separated by periods of apnea. Ataxic breathing, by contrast, is characterized by a persistently irregular respiratory rate.

Blood pressure may be reduced in sepsis or during intoxication with various sedative-hypnotics; however, hypotension itself generally does not cause stupor or coma. Once patients are supine, a mean arterial pressure of 60 mmHg is generally adequate to maintain cerebral perfusion. An exception to this rule might occur in cases of advanced cerebrovascular disease, with perhaps occlusion of some vessels and severe stenosis of the remaining ones. Hypertension, *per se*, although not capable of directly causing stupor or coma, may cause hypertensive encephalopathy.

At this point, the neurologic examination proper may be considered, beginning with an assessment for nuchal rigidity.

NUCHAL RIGIDITY

Presuming that the patient does not have severe cervical arthritis, resistance to neck flexion should presumptively be taken to indicate meningeal irritation, as with meningitis or subarachnoid hemorrhage. When nuchal rigidity is present, one may also, when the neck is passively flexed, see Brudzinski's sign, wherein there is flexion at the hips. One may also test for Kernig's sign by first passively flexing the hip and then, with this posture maintained, extending the leg: when the sign is present, one encounters resistance to extending the leg.

EYE OPENING

With gentle pressure from the thumbs on the upper portions of the eyelids, attempt to open the eyes, and once they are opened, release the pressure and allow them to close. In coma, there is no resistance to opening, and once they are released, the eyelids gently drift closed. In stupor, there may be slight resistance to opening, but once released, the eyelids generally drift closed. Significant resistance to eye opening, along with rapid closure once the eyelids are released, is not seen in true stupor or coma and generally indicates a case of psychogenic stupor or coma.

Rarely one may encounter cases of coma where the eyes are tonically open: this is known as Collier's sign, and may occur in cases of damage to the dorsal midbrain or to the pons.

PUPILS

In assessing the pupils, one looks to see whether they are equal bilaterally, whether they are reactive to light, both directly and consensually, and to their size. Normally, the pupils are equal; however, in a small minority of healthy individuals, a degree of anisocoria, amounting to 0.4 mm or less, may be found. Normally also, of course, the pupils are reactive to light: non-reactive (or 'fixed') pupils are always abnormal. The size of the pupils should be noted; however it is also convenient to classify pupillary size as either small (<3 mm), 'midposition' (3–4 mm), or large (>4mm).

Pupillary abnormalities may be seen in both toxic and metabolic disorders, in postictal states, and with structural damage.

In toxic or metabolic stupor or coma the pupils are equal bilaterally. As a rule, they are small (or, in opioid intoxication, pinpoint); an exception to this is anticholinergic toxicity, wherein the pupils are widely dilated. Regardless of their size, however, pupils in a toxic or metabolic stupor or coma almost always remain reactive. At times, discerning reactivity may be difficult (as with opioid or anticholinergic toxicity), but in such cases magnification and utilization of a very bright light will resolve the question. Rarely, pupillary reactivity will be lost in toxic or metabolic coma, but this is seen only in cases of profound coma, wherein, despite the most vigorous stimulation, there is no eye opening, no vocalization, and no purposeful or semi-purposeful movement.

Postictally, after a grand mal seizure, the pupils may be equally dilated, and this change may persist for up to 30 minutes.

Structural damage, as might be expected, causes different pupillary changes depending on the location of the damage; before discussing these, however, a brief review of the relevant neuroanatomy may be in order.

Dilatation of the pupils depends on the integrity of the sympathetic pupillodilator pathway. Fibers originate in the hypothalamus, then descend ipsilaterally through the lateral portions of the mesencephalic tegmentum, pons, and medulla into the cord, where they synapse in the intermediolateral gray of the first three thoracic segments. Fibers then exit via the ventral roots to enter the cervical sympathetic chain, where they synapse in the superior cervical ganglion. Fibers exit and then arise in proximity to the internal carotid artery up to the cavernous sinus; here they leave the artery and join the ophthalmic division of the fifth cranial nerve to exit via the superior orbital fissure, after which they travel via the long ciliary nerve to the pupillodilator muscle of the iris.

Constriction of the pupil in turn depends on the integrity of the parasympathetic pupilloconstrictor pathway. Fibers originating in the retina travel via the optic nerve, chiasm, and tract to the lateral geniculate body, which they pass through on their way to the tectum where they synapse in the olivary pretectal nucleus. Fibers then pass through the posterior commissure, circle around the aqueduct, and then pass, both unilaterally and bilaterally, into the mesencephalic tegmentum, where they reach the nucleus of Edinger–Westphal on both sides. Fibers from this nucleus then exit via the oculomotor nerve, and, once in the orbit, leave this nerve to synapse in the ciliary ganglion; fibers from this ganglion then innervate the pupilloconstrictor muscle of the iris.

With this anatomy in mind, the pupillary changes seen with diencephalic, pretectal, mesencephalic tegmental, or pontine damage, and with uncal herniation, are predictable.

Diencephalic lesions, by interrupting the pupillodilator pathway bilaterally at its origin, but leaving the pupilloconstrictor pathway in the mesencephalon intact, cause bilaterally equal, small, reactive pupils.

Pretectal lesions, by interrupting the pupilloconstrictor pathway bilaterally, but sparing the descending pupillodilator pathways, cause bilaterally equal, large, non-reactive pupils.

Mesencephalic tegmental lesions, by bilaterally interrupting both the pupillodilator and pupilloconstrictor pathways, cause bilaterally equal, midposition, non-reactive pupils. Here also, and in contrast with other locations, the pupils may be irregular in shape.

Pontine lesions, by bilaterally interrupting the pupillodilator pathway, cause bilaterally equal, small, reactive pupils.

Uncal herniation, by compressing or stretching the oculomotor nerve and thereby damaging the pupillodilator pathway within it, causes pupillary dilatation on that side. Although eventually pupillary reactivity on this side is also lost, early on some reactivity may be retained; however, it is sluggish compared with the other side.

EXTRAOCULAR MOVEMENTS

Examination of eye movements is of great importance in the evaluation of stupor and coma. Given that the structures involved, including the oculomotor and abducens nuclei, and the medial longitudinal fasciculi which connect them, lie in close proximity to the ascending reticular activating system, brainstem lesions capable of causing stupor or coma almost inevitably cause abnormalities in eye movements.

Before turning to some of the abnormalities that may be seen in stupor or coma, a brief, and admittedly very simplified, review of the relevant neuroanatomy may be in order.

Voluntary control of eye movements depends first on the integrity of the frontal eye fields, found within the posterior aspect of the middle frontal gyrus. From the frontal eye fields, a pathway begins that leads through the ipsilateral thalamus down into the mesencephalic tegmentum, where it crosses to the opposite side to descend in the contralateral medial longitudinal fasciculus to reach the abducens nucleus. Two groups of fibers then leave the abducens nucleus. One proceeds ventrally as the intramedullary portion of the abducens nerve. The other crosses the midline to ascend in the medial longitudinal fasciculus to the opposite oculomotor nucleus.

The oculomotor nucleus is situated in the midbrain tegmentum, just ventral to the periaqueductal gray. The oculomotor nerve proceeds anteriorly, and after exiting the midbrain along the medial aspect of the cerebral peduncle, it courses through the tentorial opening, then passes into the cavernous sinus and thence through the superior orbital fissure, after which it divides into a superior branch, which innervates the superior rectus muscle and the levator palpebrae superioris, and an inferior branch, which innervates the medial rectus and the inferior rectus muscles. In innervating these muscles, the oculomotor nerve controls not only vertical eye movement, but also adduction of the eye.

The abducens nucleus is located in the dorsal pontine tegmentum, and gives rise to the abducens nerve. This nerve then courses inferiorly and anteriorly through the pons to exit at the pontomedullary junction. It then courses superiorly over the pontine surface, then up through the tentorial opening and into the cavernous sinus, after which it joins the oculomotor nerve coursing through the superior orbital fissure. Once inside the orbit, the abducens nerve innervates the lateral rectus muscle, which in turn causes abduction of the eye.

With this anatomy in mind, the clinical abnormalities seen in stupor or coma may be considered.

To examine the extraocular movements, raise the patient's eyelids by applying gentle pressure of the thumbs to the upper portions of the lids, and observe first for any spontaneous eye movements. When spontaneous eye movements are present, they are typically of the 'roving' type; other, much less common abnormalities include 'ping pong gaze' and ocular 'bobbing' or 'dipping'. Roving eye movements are considered first.

Roving eye movements are characterized by conjugate lateral eye movements, proceeding smoothly first to one side, then promptly reversing course to smoothly reach the opposite side, whence they again reverse course, going back and forth without pause. These roving eye movements may, in some cases of light stupor, be interrupted by voluntary saccades, but this is unusual: in most cases of stupor, and in all cases of coma, saccadic eye movements are lacking. The presence of roving eye movements indicate that the brainstem from the mesencephalon to the mid pons is intact, and consequently this abnormality of eye movements generally indicates a toxic or metabolic cause. An important exception to this rule, however, occurs in the early stage of central herniation, wherein compression of the diencephalon causes stupor with roving eye movements.

Ping pong gaze, like roving eye movements, is characterized by conjugate lateral eye movements from side to side; however, in ping pong gaze, in contrast to roving eye movements, the eyes, once arrived at full excursion to a side, will pause there for a second or two before then moving to the opposite side, whence they again pause before setting off again toward the other side. Ping pong gaze, although most commonly seen in toxic or metabolic stupor or coma, may also be seen with structural lesions of the brainstem and, albeit rarely, with bilateral hemispheric lesions.

Ocular bobbing and ocular dipping are both characterized by spontaneous rhythmic vertical eye movements. In bobbing, the eyes undergo a brisk conjugate downward movement, followed by a slow 'bob' back to midposition. 'Inverse bobbing' may also be seen, wherein the eyes briskly deviate upward, and then bob slowly down to midposition. Dipping represents the converse of bobbing. In ocular dipping, there is a slow, conjugate 'dip' down, followed by a brisk return to midposition. 'Inverse dipping' may also occur, wherein the initial slow 'dip' is upward, followed by a rapid return to midposition. Both bobbing and dipping may occur with pontine lesions, in obstructive hydrocephalus, and even in some metabolic comas.

In cases where there are no spontaneous eye movements, the eyes may simply remain at midposition or there may be tonic conjugate deviation, which may be either lateral or downward.

Tonic lateral conjugate eye deviation may occur in three conditions, namely with destructive lesions of the frontal eye fields, with thalamic hemorrhage, and finally as an ictal phenomenon. The most common cause is with a frontal lesion, and here, owing to unopposed activity of the undamaged frontal lobe on the opposite side, the eyes look 'toward' the side with the lesion. With a thalamic hemorrhage, one might likewise expect the eyes to look toward the side with the hemorrhage, but here the opposite occurs, leading to the phenomenon of 'wrong way eyes' where the eyes look *away* from the side with the lesion (Fisher 1967). Ictal tonic deviation due to a seizure focus in one of the frontal lobes will cause tonic deviation away from the side with the lesion. One clue to ictal deviation is the presence of subtle 'nystagmoid' jerks which represent clonic activity. Postictally, to complicate matters further, one may see a 'Todd's paralysis' where the eyes will transiently be deviated toward the side with the inactive seizure focus.

Tonic downward conjugate deviation, although classically associated with thalamic hemorrhage, may also be seen in cases of pontine hemorrhage and also, although very rarely, in metabolic coma, as for example in hepatic encephalopathy. Of note, this downward deviation may also be accompanied by an inward deviation, creating an appearance which justifies the name 'peering at the tip of the nose sign'.

In interpreting any abnormalities, attention must be paid to the possibility of false localizing signs due to concurrent or pre-existing damage to either the oculomotor or the abducens nerve. For example, in uncal herniation, compression of the oculomotor nerve as it crosses the tentorial edge leads not only to pupillary dilatation but also, subsequently, to an oculomotor palsy, with loss of adduction on that side. To take another example, in cases characterized by increased intracranial pressure, downward movement of the brainstem may stretch the long and slender abducens nerve, leading to either a unilateral or bilateral abducens palsy. Finally, in some cases of coma due to head trauma, there may be a blowout fracture of the orbit, which in turn can result in entrapment of both the abducens and oculomotor nerves as they pass through the superior orbital fissure, leading to a paralysis of all eye movements on that side.

In cases of coma wherein there are no spontaneous eye movements, one must test for oculocephalic responses or employ cold caloric testing in order to determine whether or not the brainstem, from the mid pons to the mesencephalon, is intact.

OCULOCEPHALIC RESPONSES AND COLD CALORIC TESTING

Cases of coma accompanied by a lack of spontaneous eye movements may stem from either a toxic or metabolic cause or from structural damage to the brainstem, and oculocephalic responses or the results of cold caloric testing generally enable one to decide between these two alternatives. In toxic or metabolic coma, one or both of these tests will almost always yield normal results, whereas with structural damage, one or the other will be abnormal, and the abnormality will often disclose where in the brainstem the damage lies.

Before describing these tests, a review of the relevant anatomy is in order.

The anatomy presented in the foregoing section on extraocular movements dealt with the 'top down' control of voluntary eye movements, which began in the frontal eye fields. There is also a 'bottom up' system for control of eye movements, and this begins in the vestibular nuclei in the mid pons.

The medial and superior vestibular nuclei are located in the pontine tegmentum, lateral and dorsal to the abducens nucleus. One group of fibers from these nuclei pass up the ipsilateral medial longitudinal fasciculus to the oculomotor nucleus, thus enabling adduction of the ipsilateral eye. Another group of fibers cross the midline to the abducens nucleus, thus enabling abduction of the contralateral eye. Consequently, appropriate stimulation of the vestibular nuclei on one side leads, in the normal course of events, to conjugate lateral deviation of the eyes to the opposite side, whereas appropriate stimulation of these nuclei on both sides will lead to conjugate vertical deviation of both eyes.

With this anatomy in mind, the results of these tests may be understood. Oculocephalic responses are considered first. If these are present, then cold caloric testing is not necessary. In some cases of more severe coma, however, oculocephalic responses may be absent, and when they are, cold caloric testing, which provides a more vigorous stimulation of the vestibular nuclei, should be considered.

To test for the oculocephalic responses, grasp the patient's head with your two hands, using the thumbs to gently hold the eyelids up. The patient's head is then moved in four different directions: first it is turned to the patient's right, and then to the left; subsequently it is tilted backward, and then forward. Each movement should be brisk, but once the movement is made the patient's head should be held in that position for a few seconds in order to observe the response: after a few seconds, the eyes, if they did move, will then drift back to midposition.

When the brainstem is intact, the following responses, all involving conjugate eye movements, are typically observed: on turning the head to the right, both eyes deviate to the left; on turning the head to the left, both eyes then deviate to the right; on tilting the head backward, both eyes deviate down; and, on tilting the head forward, both eyes deviate up (this upward deviation may also be accompanied by the 'doll's eye' phenomenon, wherein the eyes also open).

With lesions of the brainstem various different abnormalities of the oculocephalic responses may be seen, depending on the location of the lesion.

With a bilateral lesion of the midbrain, both oculomotor nuclei are damaged, and hence both adduction and vertical eye movements are impossible; given that the abducens nuclei are spared, however, lateral eye movements are likewise spared. In such a case, one finds a clinical picture identical with bilateral internuclear ophthalmoplegia: on turning the head to the right, the left eye abducts but the right eye fails to adduct. Similarly, on turning the head to the left, the right eye abducts, but the left eye now fails to adduct. Finally, upon tilting the head either back or forward, there are no vertical eye movements. In addition to these abnormalities, one typically also sees a slight outward deviation of both eyes at rest, due to a failure of tonic adduction tone to both eyes.

With a bilateral dorsal mid-pontine lesion, both medial longitudinal fasciculi are destroyed, thus depriving the oculomotor nuclei of any stimulation from the vestibular nuclei while leaving the pathways from the vestibular nuclei to the abducens nuclei intact. In such a case, one sees a bilateral internuclear ophthalmoplegia: again, on turning the head to the patient's right, the left eye abducts, the right eye, however, given that the right oculomotor nucleus remains unstimulated, fails to adduct, and, on turning the head to the patient's left, the right eye will abduct but the left eye, given now that the left oculomotor is not stimulated, will not adduct. Finally, in such a case when the head is tilted, either back or forward, there will be, given again that the oculomotor nuclei remain unstimulated, no vertical eye movements. Here, however, given that the oculomotor nuclei are still intact, tonic adduction persists and consequently, at rest, the eyes will remain at midposition. As may be seen, the results of oculocephalic testing for mesencephalic and dorsal mid-pontine lesions are essentially the same, with one difference: in mesencephalic lesions, at rest the eyes show some outward deviation on both sides; by contrast, in midpontine lesions, at rest the eyes are midposition.

With a bilateral lower pontine lesion, destruction of the vestibular nuclei or their pathways deprive both the abducens nuclei and the oculomotor nuclei of any stimulation, and hence, on testing for oculocephalic responses, there are no eye movements at all.

In cases where there are no eye movements, one must proceed to cold caloric testing in order to determine, with greater certainty, whether the brainstem is intact or not.

Cold caloric testing is accomplished by drawing up 50 cc of ice water into a syringe. The head of the bed is elevated to 30 inches, and the external canal is examined to ensure that the tympanic membrane is both accessible and intact; if so, then a catheter is attached to the syringe and gently inserted in the canal to bring its end to rest at the membrane. The cold water is then infused at a rate of about 10 cc/min until either some eye movement is seen, or all 50 cc have been used up. At this point, 5 minutes are allowed to pass in order to allow any response to dissipate, after which the opposite membrane is irrigated.

The effect of cold-water irrigation of a tympanic membrane is the same as turning the head *away* from the side being irrigated. Thus, for example, when the brainstem is intact and the right tympanic membrane is irrigated with cold water, the eyes will deviate conjugately toward the side being irrigated. Of note, the conjugate deviation seen with cold caloric testing in cases of coma is tonic and lacks any nystagmus.

Although, as noted, cold caloric testing is almost always capable of eliciting appropriate eye movements in cases of toxic or metabolic coma, exceptions, although rare, do occur, as with deep coma due to intoxication with phenytoin, tricyclic antidepressants, and sedative-hypnotics. Cold caloric testing takes time, and consequently is generally deferred until appropriate imaging and any other urgent testing have been accomplished.

In interpreting either oculocephalic responses or the results of cold caloric testing, one must keep in mind some of the false localizing signs discussed in the preceding section.

CORNEAL REFLEX

The corneal reflex should be checked on both sides. Although gentle application of a wisp of cotton may be adequate, in some cases of coma a more vigorous application may be required to elicit the reflex.

The normal corneal reflex exhibits two components: first, there is closure of both eyes; second, one finds a bilateral Bell's phenomenon with upward rotation of both eyes beneath the closing lids.

The relevant neuroanatomy underlying the corneal reflex begins with the ophthalmic division of the trigeminal nerve, then the trigeminal nucleus, from whence two groups of fibers emerge: one group passes to both facial nuclei, thus enabling eye closure, whereas another passes up both medial longitudinal fasciculi to the oculomotor nuclei, thus causing Bell's phenomenon.

Given this anatomy, a normal corneal reflex bilaterally (with both components present) indicates that the brainstem, from the mid pons to the mesencephalon, is intact.

An absence of both components of the corneal reflex bilaterally as a rule indicates structural damage to the brainstem involving, at a minimum, bilateral damage to the mid pons. An exception to this rule may be found in some cases of profound toxic or metabolic coma wherein, despite an intact brainstem, the corneal reflexes are absent.

In some cases of coma due to structural damage to the brainstem, one may see a dissociation of the two components. For example, with bilateral mesencephalic damage, the oculomotor nuclei are damaged and hence Bell's phenomenon is absent: preservation of the mid pons, however, allows for eye closure.

LIMB MOVEMENTS

In evaluating limb movements, first observe the patient for any spontaneous movements, and, if these are absent, for any movements with any of the painful stimuli described earlier, up to, if necessary, a sternal rub. Should movements occur, observe carefully for any evidence of asymmetry. As a rule, hemiparesis indicates structural damage, either hemispheric or brainstem: an exception to this rule, however, may be found in a small minority of cases of metabolic stupor or coma, as for example those due to hepatic encephalopathy (Cadranel *et al.* 2001) or hypoglycemia (Hart and Frier 1998).

Spontaneous movements which are purposeful or semi-purposeful and which are symmetric as a rule indicate a toxic or metabolic cause of stupor; an exception to this rule occurs during the early diencephalic phase of central herniation.

In coma, although there are no spontaneous movements, limb movements may occur with painful stimuli, and these include both decorticate and decerebrate posturing. Decorticate posturing classically occurs with lesions at the diencephalic–mesencephalic junction, and involves flexion of the upper extremities and extension of the lower extremities. Decerebrate posturing, by contrast, involves extensor posturing of both the upper and the lower extremities, and classically occurs with mesencephalic lesions. Rarely, decorticate or decerebrate posturing may occur in coma owing to toxic (Greenberg and Simon 1982) or metabolic (e.g. hypoglycemia [Kirk *et al.* 1991] and hepatic encephalopathy [Conomy and Swash 1968]) causes. In cases of decerebrate posturing, the clue to a metabolic cause is the preservation of pupillary reactivity, something that is typically lost with structural damage to the mesencephalon.

Of note, although decorticate or decerebrate posturing typically occur only with painful stimulation, at times such posturing may appear spontaneously: in such cases, however, one typically finds a painful condition, such as bedsores, bladder distention, airway obstruction, and the like.

MUSCLE TONE

Muscle tone may be increased or decreased.

Spastic rigidity indicates damage to the corticospinal tract. Gegenhalten, manifest with increasing rigidity with rapidly repeated flexion and extension, by contrast is seen with diffuse frontal dysfunction, as may be seen in toxic or metabolic stupor or coma and in some cases of hypoxic–ischemic injury.

Flaccidity, although generally indicating a toxic or metabolic coma, may also occur in profound coma of any cause.

MYOCLONUS AND ASTERIXIS

Although myoclonus and asterixis generally indicate a toxic or metabolic cause, they may also occasionally be seen with structural damage to the frontoparietal cortex, the thalamus, or the brainstem. Testing for asterixis in a comatose patient may be accomplished by grasping the patient's ankles and then sliding them up on the bed, thereby placing the hips and knees in flexion; once this is accomplished, allow the legs to fall away from one another, and then observe for any involuntary jerky adductions (Noda *et al.* 1985).

DEEP TENDON REFLEXES

The deep tendon reflexes may be bilaterally brisk in metabolic stupor, but with the advent and deepening of coma they become progressively less active and are eventually lost. In toxic stupor or coma secondary to overdosage with opioids or hypnotic-sedatives, the reflexes are generally hypoactive.

PLANTAR RESPONSES

Bilaterally extensor plantar responses are seen in central herniation or with lesions affecting the corticospinal tracts as they pass through the mesencephalon and pons, and may also be seen in some cases of metabolic stupor or coma. Unilateral extensor plantar responses may be seen with hemispheric lesions or during the later stage of uncal herniation.

LATERAL DISPLACEMENT OF THE DIENCEPHALON, UNCAL HERNIATION, AND CENTRAL HERNIATION

At this point, the syndromes of lateral displacement of the diencephalon, uncal herniation, and central herniation are discussed. It is critical to keep these syndromes in mind when evaluating patients in stupor or coma, given their lethal nature and, in some cases, their amenability to neurosurgical treatment.

Lateral pressure upon the diencephalon, as may occur with temporal mass lesions or with basal ganglia hemorrhage, may, by direct compression of the diencephalon, cause stupor or coma, and the degree of decrement in the level of consciousness has been to shown to correlate with the extent of this lateral displacement (Ropper 1986, 1989). Of note, and with displacement due to a temporal mass, this lateral displacement may soon be followed by uncal herniation.

Uncal herniation, according to classic teaching, occurs with hemispheric mass lesions which force the uncus medially and downward into the tentorial notch, first compressing the oculomotor nerve and then the mesencephalon, eventually forcing the mesencephalon far enough laterally to compress the opposite cerebral peduncle against the opposite free edge of the tentorium. Clinically, one sees first pupillary dilatation ipsilateral to the hemisphere with the mass, followed soon thereafter by both a loss of light reactivity and stupor. With progression, an oculomotor palsy occurs, and the eye deviates laterally and the level of consciousness further decreases. With progression, compression of the contralateral cerebral peduncle leads to a hemiparesis which is ipsilateral to the hemisphere with the mass. Importantly, this entire process may transpire over hours or, more ominously, over merely minutes. In some cases, the clinical picture may depart from this classic pattern: there may be contralateral pupillary dilatation, and in some cases, owing to direct compression of the ipsilateral cerebral peduncle, there may be hemiparesis contralateral to the hemispheric mass. In the process of uncal herniation, the posterior cerebral artery may also be compressed, leading to infarction of the occipital and temporal lobes.

Central herniation (Wijdicks and Miller 1997) occurs when downward pressure upon the diencephalon causes compression and buckling of this structure. With further pressure, the brainstem is forced downward, causing a progressive, rostrocaudal compression of the mesencephalon, pons, and finally the medulla. In the process, the penetrating arteries of the mesencephalon and pons become damaged: in central herniation, although the brainstem moves down, the basilar artery, tethered to the circle of Willis above, cannot, thus leading to stretching of the penetrating arteries between their terminations in the brainstem and their origin in the basilar artery. With sufficient damage, infarction may occur or there may be slit-like hemorrhages, known as Duret hemorrhages, in the central portions of the mesencephalon or pons.

Central herniation may occur in cases of bihemispheric lesions (e.g. bilateral subdural hematomas or with viral encephalitis) or with mass lesions so situated in the frontal or parietal lobes as to exert primarily a downward vector of pressure, rather than a lateral one.

Central herniation may be divided into three rostrocaudal stages, diencephalic (which is subdivided into an early diencephalic and a late diencephalic stage), mesencephalic, pontine, and then a medullary stage. The abnormalities seen in each stage are predictable if one keeps in mind the comments made earlier regarding abnormalities of respiration, the pupils, eye movements, limb movements, and the plantar responses.

It must be stressed that recognition of the early diencephalic stage is critical, given that, if the process is reversed at this point, full recovery is possible: once the late diencephalic stage is reached, while the patient may yet live, recovery will not be full; with later stages, recovery is at best poor, and once the medullary stage is reached, death is certain. Recognition of the early diencephalic stage, however, requires a high degree of suspicion, given that, as the immediately following description makes clear, the clinical picture characteristic of this stage may be precisely mimicked by stupor due to toxic or metabolic causes.

The early diencephalic stage is characterized by stupor, and indeed this is the only stage of central herniation wherein there is any preservation of consciousness: in the late diencephalic stage, and in all the other stages of central herniation, patients are comatose. During the early diencephalic stage, respirations may be normal, often interrupted by sighing and yawning, or, in a minority, there may be Cheyne–Stokes respirations. The pupils are small and reactive, and there are spontaneous eye movements, which are roving in character; the oculocephalic responses are normal. Limb movements are purposeful or semi-purposeful in response to pain. The plantar responses are usually extensor.

The late diencephalic stage is characterized by coma. Cheyne–Stokes respirations are present. The pupils remain small and reactive. There are no spontaneous eye movements, but oculocephalic responses are preserved. With painful stimuli, there may be decorticate posturing or no movement. The plantar responses, as in the early stage, are usually extensor.

The mesencephalic stage again is characterized by coma. Respirations are typically rapid and regular, but in some cases Cheyne–Stokes respirations may still be present. The pupils now are midposition and fixed, and they may be irregular in shape. There are no spontaneous eye movements; however, the eyes may be slightly divergent. In testing for the oculocephalic responses, a pattern consistent with a bilateral internuclear ophthalmoplegia is seen; in addition, on tilting the head backward or forward, there are no vertical eye movements. In some cases, oculocephalic responses will be totally absent, and here cold caloric testing will again reveal a pattern consistent with a bilateral internuclear ophthalmoplegia. With painful stimuli, there may be decerebrate posturing or no movement. The plantar responses are extensor.

The pontine stage is characterized by persistent coma. Respirations become shallow, and may display a regular pattern or there may be apneustic or ataxic breathing. The pupils remain midposition and fixed. There are no spontaneous eye movements; oculocephalic responses are absent and cold caloric testing does not provoke any eye movement. With painful stimuli, there may be decerebrate posturing, no movement, or, in some cases, upon stimulation of the feet, there may be flexion withdrawal. The plantar responses remain extensor.

The medullary stage represents an agonal condition. Respirations remain shallow, and are often ataxic; eventually apnea and death occur.

OVERALL COMMENTS

At this point, some overall comments are in order regarding the neurologic examination in stupor or coma. The combination of the following findings indicate that the brainstem, from the mesencephalon to the mid pons, is structurally intact: preservation of pupillary reactivity; roving eye movements, or, if there are none, normal oculocephalic responses or cold caloric responses; and normal corneal reflexes. Limb movements may or may not be present with an intact brainstem: if they are present, either spontaneously or in response to painful stimuli, then they are purposeful or semi-purposeful and overall grossly symmetric. In turn, the presence of an intact brainstem indicates either a toxic or metabolic cause, or a supratentorial structural abnormality which has either diffusely compromised the cortex or which has led either to lateral displacement of the diencephalon or to the early stage of central herniation. Abnormal findings of course have the localizing values noted in the descriptions already given, and point to where in the brainstem the lesion might be. It must be kept in mind, however, that toxic or metabolic causes of coma, if severe, may, in addition to causing a deep coma, also depress the brainstem, even to the point where eye movements are lost, cold caloric testing produces no reaction, and the corneal reflexes are absent: the last abnormality to occur with deepening toxic or metabolic coma is a loss of pupillary reactivity.

Etiology

The various etiologic factors in stupor or coma are listed in Table 6.1, and are discussed further below. Before proceeding to that discussion, however, certain overall comments are in order.

Although stupor or coma may certainly occur on the basis of only one cause, often one finds mixed cases: some common examples will serve to illustrate this point. With regard to toxic cases, it is not uncommon to find stupor or coma developing in a hospitalized patient secondary to the combination of opioids and sedative-hypnotics, and the tendency of suicidal patients to overdose on a combination of agents is well known. With regard to metabolic causes, patients with pneumonia may develop sepsis, respiratory failure, and, secondary to syndrome of inappropriate anti-diuretic hormone hypersecretion (SIADH), hyponatremia. Combinations may occur involving toxic, metabolic, and structural causes: an example may be seen in some alcoholics with ethanol intoxication, hypoglycemia, a subdural hematoma, and Wernicke's encephalopathy. The list could, and does, go on, but the point is clear: the diagnostic process should not stop after one cause has been found, but should continue until at least all the common causes have been considered.

With this caveat in mind, the various etiologic factors may be considered (Forsberg *et al.* 2009). It is convenient and clinically useful to divide these into various categories. Toxic-metabolic conditions are the most common cause of stupor or coma. Common causes of structural damage come next, including stroke, tumors, hypoxic–ischemic injury, traumatic lesions, and acute hydrocephalus. Infectious causes of stupor or coma, considered next, although not common, must always be kept in mind given the potential lethality and, in some cases, treatability of meningitis and encephalitis. Finally, there is a group of miscellaneous causes.

TOXIC

The development of a toxic stupor or coma in patients hospitalized on a medical surgical floor is generally due to opioids, sedative-hypnotics, or antihistaminergic drugs, either alone or, as just stressed above, in combination. Among patients seen in the emergency room, in addition to these first

Table 6.1 Causes of stupor or coma

Toxic
Opioids
Sedative-hypnotics (especially benzodiazepines)
Antihistamines
Ethanol
Anticholinergics
Anti-epileptic drugs
Antidepressants
Antipsychotics
Serotonin syndrome
Neuroleptic malignant syndrome
Others

Metabolic
Sepsis
Respiratory failure
Carbon monoxide
Hypoglycemia
Hyperglycemia
Hyponatremia
Hypernatremia
Hypercalcemia
Hypermagnesemia
Uremic encephalopathy
Hepatic encephalopathy

Structural
Stroke
Infarction
  Large hemispheric
  Thalamic
  Hypothalamic
  Mesencephalic
  Pontine
  Cerebellar
Intracerebral hemorrhage
  Hemispheric
  Basal ganglia
  Thalamic
  Mesencephalic
  Pontine
  Cerebellar
Pituitary apoplexy
Subarachnoid hemorrhage
Intraventricular hemorrhage
Tumors
Hemispheric
Thalamic
Mesencephalic
Pontine
Cerebellar
Pineal
Pituitary
Hypoxic–ischemic injury
Traumatic lesions
Diffuse axonal injury
Epidural hematoma
Subdural hematoma
Acute hydrocephalus

Infectious
Meningitis
Acute encephalitis (including herpes simplex viral encephalitis)
Basilar meningitis
Abscess

Miscellaneous causes
Ictal coma
Wernicke's encephalopathy
Fat embolism syndrome
Multiple cholesterol emboli syndrome
Hypertensive encephalopathy
Reversible posterior leukoencephalopathy syndrome
Endocrinologic disorders
Hypothyroidism (myxedema coma)
Hyperthyroidism (thyroid storm)
Hypoadrenocorticalism (Addisonian crisis)
Autoimmune disorders
Limbic encephalitis
Lupus cerebritis
Acute disseminated encephalomyelitis
Multiple sclerosis
Heat stroke
Hypothermia
Central pontine myelinolysis
Hemiplegic migraine
Pancreatic encephalopathy

three, prime consideration is also given to ethanol and to suicide attempts with a large variety of drugs.

Whenever it is unclear, from the history or medical record, which drug is responsible, a toxicology screen should be ordered. In cases of suspected opioid overdosage, naloxone may be given, and, similarly, in suspected benzodiazepine overdose, flumazenil may be given. Overdosage of drugs with strong anticholinergic effects, such as the antihistamines diphenhydramine and hydroxyzine, and such anticholinergics as benztropine and trihexyphenidyl, may complicate interpretation of the physical examination as they cause pupillary dilatation with, in some cases, a lack of obvious reactivity. In assessing patients with high blood alcohol levels (BALs), it must be kept in mind that alcoholics may be tolerant of very high levels, even up to 400 or 500 mg, and hence finding a high BAL should not be taken as confirmation of the diagnosis of ethanol-induced stupor or coma until it is reasonably clear that the patient is not an alcoholic. Overdosage with anti-epileptic drugs has always been a concern in epileptics given the high prevalence of depression in epilepsy, but the current use of certain anti-epileptic drugs (valproic acid, carbamazepine, and lamotrigine) as mood stabilizers has made these agents much more commonly used in suicide attempts. Antidepressants used to be common culprits in suicide attempts, and tricyclic overdoses were standard fare; however, the newer antidepressants are generally remarkably safe in overdose. Some exceptions to this rule, however, must be kept in mind, including trazodone and the selective serotonin reuptake inhibitors fluvoxamine. Antipsychotic overdose-induced stupor or coma may be seen with certain antipsychotics, such as low-potency phenothiazines, clozapine, and certain second-generation agents, such as olanzapine and quetiapine; the remainder of the antipsychotics, however, are generally fairly well tolerated in overdose. With regard to the antidepressants and antipsychotics, consideration should also be given to two specific syndromes that may occur with these agents, even when taken in therapeutic doses, namely the serotonin syndrome and the neuroleptic malignant syndrome. Other medications and intoxicants to keep in mind include lithium, methanol, and ethylene glycol (both suggested by an increased anion gap), baclofen, ketamine, and phencyclidine, and finally γ-hydroxybutyrate, used not only in narcolepsy but also as the notorious 'date-rape' drug.

METABOLIC

Both sepsis and respiratory failure are common causes of metabolic stupor or coma. Carbon monoxide poisoning may also be grouped here, and is suggested by the history or exposure. Hypoglycemia and hyperglycemia (including both diabetic ketoacidosis and hyperosmolar non-ketotic hyperglycemia) should always be suspected in diabetics, and an immediate finger-stick glucose is standard. The other disorders are so common that it is appropriate to order a chemistry survey that measures not only glucose, but also sodium, calcium, magnesium, and blood urea nitrogen (BUN). In addition, one should also check the ammonia level: it must be kept in mind, however, that in some cases of hepatic encephalopathy, the ammonia level will be normal.

Certain clues suggest a diagnosis of metabolic stupor or coma. To begin with, most of these cases are of subacute onset, and, in most, one sees a delirium preceding the onset of the stupor or coma. Furthermore, seizures, myoclonus, and asterixis, although certainly possible with other causes, are, however, highly suggestive of a metabolic stupor or coma.

STRUCTURAL

Stroke, whether due to ischemic infarction, intracerebral hemorrhage, or subarachnoid or intraventricular hemorrhage, should be suspected in any case of stupor or coma of acute onset.

Ischemic hemispheric infarction may cause stupor or coma, but this occurs only with large infarctions (e.g. 'malignant' middle cerebral artery infarction), and this cause constitutes an exception to the rule that coma or stupor secondary to infarction is of acute onset. With large hemispheric infarctions, stupor or coma occurs secondary to lateral displacement of the diencephalon or to herniation, and the mass effect required to produce these syndromes occurs only after edema has developed around the infarction, a process that occurs gradually over 0.5–3 days. Acute stupor or coma, however, may occur in this situation if herniation is of the uncal type. Infarctions of the thalamus (Bogousslavsky *et al.* 1988; Kumral *et al.* 2001; Steinke *et al.* 1992), hypothalamus, mesencephalon, or pons cause stupor or coma by direct and immediate destruction of portions of the ARAS, hence in these cases the stupor or coma occurs acutely: in all these cases, as noted earlier, the area of infarction must involve the ARAS bilaterally; unilateral involvement, sparing the ARAS on one side or the other, simply is not sufficient to cause stupor or coma: in the case of thalamic or hypothalamic infarction, such bilateral involvement may occur when a pair of right and left penetrating arteries arise from a common pedicle. Cerebellar infarctions, if they do cause stupor or coma, do so by compression of the underlying brainstem, and, as in the case of large hemispheric infarctions, this process occurs only with the gradual development of sufficient edema to create a mass effect. Before leaving this discussion of infarction, a further word is in order regarding the symptoms seen with hypothalamic infarction: here, rather than coma, one typically sees only stupor, often marked by sighing and yawning.

Intracerebral hemorrhage of one of the hemispheres may cause stupor or coma, but only if it is of the large, lobar type (Massaro *et al.* 1991): here, the immediate mass effect may cause symptoms either by lateral displacement of the diencephalon or by central or uncal herniation. Intracerebral hemorrhage in the basal ganglia may also cause stupor or coma via the mechanism of lateral displacement (Massaro *et al.* 1991). Hemorrhage into the thalamus (Kumral *et al.* 1995; Steinke *et al.* 1992), mesencephalon, or pons causes stupor or coma by destruction of portions of the ARAS, and stupor or coma due to cerebellar hemorrhage may occur via several mechanisms, including direct compression of the underlying brainstem, occlusion of the fourth ventricle, acute hydrocephalus, and by what is known as 'upward' herniation, wherein the mesencephalon herniates up through the tentorial notch. Compression of the brainstem by a cerebellar intracerebral hemorrhage may be rapidly fatal, hence emergent neurosurgical consultation is required.

Pituitary apoplexy may also be mentioned here. This syndrome occurs with hemorrhage into a pituitary tumor following which there is either extravasation of blood or necrotic tissue into the subarachnoid space or direct compression of the overlying diencephalon. Clinically, one sees headache, visual field defects, and either stupor or coma.

Subarachnoid or intraventricular hemorrhage are suggested by the acute onset of stupor or coma in association with severe headache and nausea or vomiting.

Tumors of the hemispheres, diencephalon, brainstem, cerebellum, pineal gland, or pituitary may all cause stupor or coma. In the case of hemispheric tumors, stupor or coma occurs secondary to lateral displacement or to herniation, and thus only when the tumor has attained a large enough size to cause such a mass effect. Tumors of the thalamus, mesencephalon, or pons cause symptoms by direct destruction of the ARAS, while tumors of the cerebellum produce their stupor or coma, as in the case of infarctions of the cerebellum, by direct compression of the brainstem, obstruction of the fourth ventricle and resulting hydrocephalus, or by upward herniation. Pineal tumors may also cause stupor or coma: here, the mechanism involves compression of the dorsal aspect of the mesencephalon, and hence stupor or coma are often preceded by a Parinaud syndrome, with limitation of upward gaze. Pituitary tumors, such as craniopharyngiomas, may extend superiorly and compress the overlying diencephalon. In all these cases of stupor or coma caused by tumor, the onset is generally, consistent with the slow growth of the tumor, subacute or even gradual. Exceptions to this rule of gradual onset may occur in cases of intratumoral hemorrhage or, in the case of hemispheric tumors, with uncal herniation.

Hypoxic–ischemic coma may occur after cardiac arrest, during septic shock, or with massive hemorrhage; in many cases, the stupor or coma is accompanied by myoclonus.

Trauma may cause stupor or coma via a number of mechanisms, most notably diffuse axonal injury, epidural hematoma, or subdural hematoma. In some cases of head trauma, patients may be comatose at the onset, and in these cases one finds severe diffuse axonal injury. Epidural hematoma, occurring with bleeding from a lacerated middle meningeal artery, causes stupor or coma via a mass effect leading to uncal herniation. In most cases, the preceding trauma was sufficient to cause either coma or concussion, and in cases characterized by concussion one finds the classic picture of patients who initially recover consciousness, then 'talk and die', talking as the concussion clears, then dying as uncal herniation supervenes. Occasionally, the head trauma, however, although sufficient to cause laceration of the artery, has not been severe enough to cause either coma or concussion, and in these cases uncal herniation with stupor or coma occurs in patients who never experienced loss of consciousness immediately after the trauma. Subdural hematomas cause stupor or coma via a mass effect, with either lateral displacement of the diencephalon or by central or uncal herniation. Acute subdural hematomas, occurring secondary to arterial bleeding, enlarge rapidly and hence typically present acutely. Subacute or chronic subdural hematomas, being due to venous bleeding, however, present, as their names suggest, either subacutely or gradually. Although trauma sufficient to cause arterial bleeding is typically obvious by history, it must be kept in mind that, especially in the elderly or those on warfarin, trauma sufficient to cause venous bleeding may be trivial, and indeed unrecollected.

Acute hydrocephalus presents acutely or subacutely with headache, nausea or vomiting, and stupor or coma, and may, as indicated earlier, occur with cerebellar lesions which compress the fourth ventricle and basilar meningitides, which may lead to occlusion of the exit foramina of the fourth ventricle. Aqueductal compression may occur with compression of the midbrain by pineal tumors or by tumors of the mesencephalon itself, and tumors occluding the third ventricle or the interventricular foramen of Monro may also be at fault.

INFECTIOUS

Of the infectious causes of stupor or coma, meningitis and acute encephalitis must always be promptly considered in the evaluation of any patient given their lethality and also the possibility, in some cases, of effective treatment.

Meningitis classically presents with the triad of fever, stupor or coma, and nuchal rigidity. This classic presentation, however, is seen in only about two-fifths of cases, and hence a high index of suspicion is appropriate. Given the rapid course of bacterial meningitis, some clinicians advocate starting antibiotics even before lumbar puncture.

Acute encephalitis, presenting with fever, headache, and stupor or coma, may likewise have an explosive onset, and here, given the possibility of herpes simplex viral encephalitis, it has become standard to initiate presumptive treatment with acyclovir.

Other infectious causes to consider include the basilar meningitis seen with tuberculosis or fungal infections, and parenchymal abscesses. In the case of basilar meningitis, stupor or coma may occur either as a result of development of obstructive hydrocephalus or from infarction of the diencephalon or brainstem. With parenchymal abscesses, stupor or coma generally occur from a mass effect, just as with tumors.

MISCELLANEOUS CAUSES

Ictal stupor or coma is an unusual kind of seizure. Everyone is familiar with the postictal coma that follows upon a grand mal seizure, but this results, not from electrographic seizure activity, but rather from, as it is metaphorically put, 'neuronal exhaustion', and is accompanied by generalized slowing on an electroencephalogram (EEG). This postictal coma generally lasts only a matter of minutes and is followed by stupor or delirium, with these manifestations clearing up within 15–30 minutes. Ictal stupor or coma, by contrast, *is* due to electrographic seizure activity, with, rather than generalized slowing, ictal activity on EEG. Such ictal stupor or coma has been noted in three different situations: first, it may follow upon grand mal status epilepticus (DeLorenzo *et al.* 1998, Lowenstein and Aminoff 1992); second, it may evolve out of complex partial status epilepticus (Tomson *et al.* 1986); and, third, it may appear as a seizure *sui generis*, occurring independently and not preceded by any other seizure type (Tomson *et al.* 1986; Towne *et al.* 2000).

The first kind of ictal stupor or coma should be suspected whenever, after an episode of grand mal status has apparently ceased, the patient fails to regain consciousness within 15–20 minutes. Persistent coma in such cases, of course, may be due to the underlying cause of the status (e.g. hyponatremia, encephalitis, hypoxic–ischemic injury, head trauma, etc.); however, an EEG is in order before one can assume this. Furthermore, although in some cases of ictal coma there may be some subtle clonic movements, especially of the side of the mouth, there are very definitely cases of ictal coma in which patients are fully immobile, with nothing to suggest ongoing seizure activity, and consequently a lack of subtle clonic activity does not rule out ictal coma.

The second kind of ictal stupor or coma, in turn, should be suspected whenever, after a complex partial seizure, the patient, rather than regaining full alertness after a few minutes, becomes deeply stuporous and then remains that way. Here the EEG, rather than showing focal slowing, will show ongoing ictal discharges.

The third, and final, kind of ictal stupor or coma is the most difficult to recognize. In the first two types, the presence of easily recognizable seizures, whether grand mal or complex partial, prior to the onset of the stupor or coma alerts one to the true etiology. In this last type of ictal stupor or coma, however, there is no preceding grand mal or complex partial seizure: rather, the patient simply becomes either stuporous or comatose. In these cases, the sudden onset of the stupor or coma, coupled with the lack of any obvious precipitant, should alert one to the correct diagnosis, which can then be confirmed by finding generalized ictal activity on the EEG.

Wernicke's encephalopathy should be considered in any case of stupor or coma of uncertain cause, especially if there is any reason to suspect alcoholism or chronic malnutrition (De Keyser *et al.* 1985; Gibb *et al.* 1985). Wernicke's encephalopathy may not only cause coma, but in some cases also ophthalmoplegia and an absent response to oculocephalic maneuvers, or even cold caloric testing.

Fat embolism syndrome and the multiple cholesterol embolism syndrome as causes of stupor or coma should be suspected in the relevant clinical setting, namely an onset within 1–3 days after, respectively, either surgery or trauma to the long bones, or instrumentation capable of dislodging plaque material from the ascending aorta.

Hypertensive encephalopathy and the reversible posterior leukoencephalopathy syndrome are both characterized by the subacute onset of, classically, headache, seizures, visual loss, and delirium, which may evolve into stupor and then coma. Hypertensive encephalopathy is suggested by the evolution of this clinical syndrome in the setting of uncontrolled hypertension, whereas the reversible posterior leukoencephalopathy syndrome is suggested by the medication history, which reveals treatment with appropriate chemotherapeutic or immunomodulatory agents.

Endocrinologic cases of stupor or coma are generally obvious, provided one has a history adequate enough to reveal the preceding typical symptomatology of hypothyroidism, hyperthyroidism, or adrenocortical insufficiency. When an adequate history is lacking, a successful diagnosis may rest on a high index of suspicion. Myxedema coma is suggested by a myxedematous appearance, hypothermia, bradycardia, and respiratory depression. Thyroid storm is typically accompanied by fever and tachycardia, but it must be kept in mind that in cases of apathetic hyperthyroidism, these autonomic signs may be absent. Addisonian crisis presents with hypovolemic shock; if the gland itself has been destroyed, then hyponatremia and hyperkalemia will also be present.

Autoimmune disorders capable of causing stupor or coma include limbic encephalitis, lupus cerebritis, and acute disseminated encephalomyelitis (ADEM). All of these may present with delirium progressing to stupor or coma, often accompanied by seizures. When the typical setting is present (e.g. a history of cancer, lupus, or a preceding vaccination or infection, respectively) the correct diagnosis may be immediately suggested; however, the setting may be unknown, or simply elusive, and consequently these disorders should be strongly considered in any case of stupor or coma of unknown cause.

Multiple sclerosis, rarely, may cause stupor or coma with the development of a plaque in the ARAS.

The other miscellaneous causes of stupor or coma are generally easily recognized. Heat stroke and hypothermia are suggested by the patient's temperature and the history of exposure, central pontine myelinolysis by the history of a recent overly rapid correction of hyponatremia, hemiplegic migraine by the migraine, and pancreatic encephalopathy by the evidence of pancreatitis occurring anywhere from 2 to 5 days earlier.

Differential diagnosis

Stupor and coma may be mimicked by a variety of other conditions, including the locked-in state, syncope, delirium, hypersomnolence, the minimally conscious state and the vegetative state, akinetic mutism, stuporous catatonia, and, finally, psychogenic coma. Each is considered in turn.

The locked-in state may be initially confused with coma, given that patients with the locked-in state are motionless. However, attention to the fact that the eyes are open and that the patient does respond to the command to blink immediately establishes that these patients, rather than being comatose, are fully alert.

Syncope is distinguished from coma by its brevity, lasting at the most only minutes.

Delirium is distinguished from stupor by the simple fact that delirious patients, while perhaps at times appearing drowsy, are for the most part awake, and do not require constant stimulation to remain so.

Hypersomnolence is typically included in the differential diagnosis of stupor and coma; however, this is somewhat problematic. To begin with, the clinical demarcation between hypersomnolence and stupor is quite fuzzy. Hypersomnolence is often described as a condition wherein the patient appears asleep and can be aroused only with strong stimulation; once aroused, if stimulation is continued, such patients remain alert. However, once the stimulation is withdrawn, they again enter a sleep-like state. This definition, obviously, is quite similar to that for stupor. In addition to this definitional fuzziness, hypersomnolent conditions, in general, are found to occur either with excessive use of sedative-hypnotics or with lesions of either the diencephalon (particularly the hypothalamus) or the mesencephalon, and these causes, obviously, are also those implicated in many cases of stupor or coma. Given these definitional and etiologic similarities between the two conditions, it is probably appropriate, upon encountering patients whom one might consider to be hypersomnolent, to go ahead and pursue essentially the same workup that one would use for the comatose or stuporous patient. If the workup is negative, and the patient still appears somnolent, then one may consider some of the sleep disorders discussed in Chapter 20, such as the Pickwickian syndrome or the Kleine–Levin syndrome.

The minimally conscious state and the vegetative state are both distinguished from stupor and coma by the appearance of a definite sleep–wake cycle, with periods of wakefulness during which the eyes are open and the patient, at first glance, appears awake and alert. Such periods of spontaneous wakefulness are simply not seen in cases of stupor or coma.

Akinetic mutism is quickly distinguished from stupor and coma by the fact that in akinetic mutism the eyes are open; further, in akinetic mutes one sees saccadic, rather than roving, eye movements, and akinetic mutes also visually fix on objects in the room, such as a nurse or physician, and then track them as they move.

Stuporous catatonia presents with immobility and mutism, and in this condition the eyes may be open or closed. If the eyes are open, then, as in akinetic mutism, one may see spontaneous, saccadic eye movements, which would serve to make the differential. If the eyes are closed, there may be resistance to opening the eyelids, and again one may then also find saccadic movements. Other symptoms, not seen in stupor or coma, may be present in catatonia, and further serve to make the differential; these include waxy flexibility, posturing, and echo phenomena. In limiting cases, one may obtain an EEG, which will show an awake pattern.

Psychogenic coma, that is to say the appearance of coma in a patient who is, in fact, awake, may be produced intentionally, as in malingering, or without the patient's conscious intent, as in conversion disorder. The diagnosis of psychogenic coma is often entertained in cases of coma in young patients where the history, neurologic examination, toxicology and metabolic studies, and appropriate neuroimaging appear to have ruled out other causes.

Certain aspects of the neurologic examination deserve some comment in regard to this condition.

First, in assessing the level of consciousness via painful stimulation, one must keep in mind that some awake patients may be able to tolerate very noxious stimuli (Maddock *et al.* 1999), and indeed may show little, or even no, evidence of wakefulness, even after a vigorous sternal rub. In such cases, other maneuvers may enable one to determine whether the patient is, in fact, awake. The 'hand drop' test is one such maneuver (Padkin 1999). Here, while grasping the patient's wrist, one holds the patient's hand over the head, and then suddenly drops the hand. In most cases, an awake patient will pull the arm to the side to avoid a collision with the face. However, it must be kept in mind that some well-read patients, alert to this test, may allow the collision to occur; furthermore, in some cases the weight of the arm itself, during the fall, may pull it to the side. Another test involves lightly stroking the patient's eyelids: in comatose patients, this maneuver provokes no response; however, in the alert patient, this maneuver typically evokes fluttering of the lids. In limiting cases, one may resort to various maneuvers, unexpectedly performed, which may provoke a response, such as loudly clapping your hands close to the patient's ears or dousing the patient's head with ice water.

On attempted eye opening, one often encounters some resistance in psychogenic coma; further, when the eyelids are released in psychogenic coma, they tend to close rapidly, rather than drift down as in true coma. Furthermore, in psychogenic coma, when the eyelids are opened, the patient's eyes may undergo a conjugate vertical eye movement, either up or down, after which they remain fixed in that position until the eyelids are allowed to close again. Although conjugate vertical eye movements may also occur in true coma, the eyes in true coma do not remain in such a fixed position, but rather display more or less continuous movement along the vertical axis, thus demonstrating the signs of ocular bobbing or dipping, as described earlier: these signs of bobbing and dipping, to our knowledge, have never been reported in psychogenic coma.

Eye movements in psychogenic coma, if present, are generally saccadic in type, rather than roving. Exceptions to this rule, however, may occur, and one of the authors (DPM) has seen a case wherein a patient, while malingering a coma, engaged in roving eye movements. In cases of psychogenic coma where spontaneous eye movements are absent, oculocephalic testing generally likewise reveals no movements. Cold caloric testing, however, will tell the tale. In true coma, cold caloric stimulation will either provoke tonic deviation toward the irrigated ear, or, in profound coma, no eye movement at all. In psychogenic coma, however, what one finds is not tonic deviation, but rather nystagmus, with the quick phase directed away from the irrigated ear: this response to cold caloric testing indicates integrity of the overall pathway from the frontal eye fields to the mid pons, and is not seen in true coma.

Finally, on testing the corneal reflex, one typically finds at least some degree of reflex eye closure.

In limiting cases, as with catatonia, an EEG may be obtained which will show an awake recording.

Treatment

Given the potentially life-threatening nature of some causes of stupor or coma, the evaluation, including the history and physical examination, laboratory testing, and imaging, must be rapidly completed, and treatment measures must often be similarly urgently undertaken. Most physicians will then use the Glasgow coma scale to monitor the patient's progress.

The history is often sparse, and sometimes available only from ambulance personnel or nurses. The physical examination itself can often be completed within minutes, unless, of course, a cervical fracture is suspected, in which case any maneuvers which could endanger the cervical spine must be delayed pending the results of a cervical spine film. A finger-stick glucose should be immediately obtained, a pulse oximeter should be placed, and certain laboratory tests should be ordered on a stat basis, including a complete blood count (CBC), chemistry survey (including glucose, sodium, BUN, calcium, and magnesium), ammonia level, toxicology screen, and urinalysis. Other tests may be ordered depending on one's index of suspicion, including arterial blood gases, blood levels of ethanol or medications potentially involved in the case (e.g. anti-epileptic drugs, lithium, etc.), and blood cultures. A stat computed tomography (CT) scan is in order in almost all cases. Lumbar puncture is reserved for cases of suspected meningitis or encephalitis: given the necessity for emergent treatment of bacterial meningitis, some advocate performing a lumbar puncture before waiting for the results of a CT scan whenever meningitis is high on the differential list; however, the potential dangers involved here, and the fact that sometimes only minutes are required to obtain the CT scan, convince many physicians to wait for the CT results. Urgent electroencephalograms are rarely required: the one exception here is when one suspects an ictal coma.

Treatment measures in all cases include measures to ensure the airway and maintain adequate oxygenation, maintain an adequate blood pressure, and control any significant arrhythmias. If the physical examination or CT scanning suggest that lateral displacement, uncal or central herniation, or brainstem compression (e.g. by a cerebellar mass lesion) is occurring, emergent neurosurgical consultation is required: the urgency of neurosurgical consultation cannot be overstressed in cases of possible uncal herniation, which may prove to be fatal within minutes. Pending neurosurgical intervention, both hyperventilation (with an Ambu bag pending intubation) and treatment with mannitol should be instituted to decrease the overall mass effect; dexamethasone may also lower intracranial pressure, but the effects here are delayed for up to 6–12 hours.

While any metabolic abnormalities are being treated, one may also consider administering either naloxone or flumazenil in cases where opioids or benzodiazepines are implicated. In some centers, these are administered almost routinely; however, some caution is required in cases of possible addiction, given the risk of a hyperacute withdrawal syndrome with, in the case of benzodiazepine addiction, seizures. Intravenous thiamine should also be given in any case where Wernicke's encephalopathy is even a remote possibility; if at all possible, this should be done before glucose is given.

Further treatment measures depend, naturally, on the underlying cause.

6.2 PERSISTENT VEGETATIVE STATE AND THE MINIMALLY CONSCIOUS STATE

In cases of prolonged coma due to structural damage, as may be seen in such conditions as postanoxic coma or traumatic brain injury, either death ensues or, after 2–4 weeks, eye opening occurs and patients are found in a vegetative state. This vegetative state may become persistent (Jennett and Plum 1972), or some patients may emerge from it into the minimally conscious state; this minimally conscious state, in turn, may be permanent or some patients may emerge from it into a dementia of variable severity.

Clinical features and course

PERSISTENT VEGETATIVE STATE

In the vegetative state, although there is an appearance of wakefulness, there is no behavioral evidence that patients are aware of their surroundings or of themselves.

The emergence from coma is marked first by opening of the eyes, which at first may be inconsistent, but eventually becomes long-lasting, with the establishment of definite sleep–wake cycles. When the eyes are open, they may remain motionless at midposition or they may move from side to side. In some cases, there may be fixation on objects in the environment, but this is typically brief; in some cases also there may be visual tracking, but this, too, is generally brief and inconsistent. At times, there may be grunting or grimacing; smiles may occur and even tears, but these occur at random, and are not associated with any environmental events. There is no intelligible speech or mouthing of words, either spontaneously or in response to questions. Limb withdrawal to noxious stimuli may occur, as may posturing, but there is no purposeful or semi-purposeful movement, again either spontaneously or in response to command.

Respiration and blood pressure are maintained, and brainstem reflexes are normal, including the pupillary reaction, blinking to threat or bright light, the corneal reflex, and oculocephalic responses. Sucking and rooting reflexes may be elicited, as may the grasp reflex. There is incontinence of bowel and bladder.

In essence, bedside testing fails to disclose any evidence that patients have any thoughts or emotions, or any experience of the world. Recent, disturbing work, however, using functional magnetic resonance imaging (MRI) suggests that bedside testing, at least in some cases, may not always be sensitive enough to detect such phenomena. Two studies demonstrated activation of higher-level temporal cortex when patients were spoken to (Coleman *et al.* 2007; Di *et al.* 2007), and one, more remarkable, study demonstrated appropriate cerebral activation when the vegetative patient was told to *imagine* playing tennis (Owen *et al.* 2006). None of these patients demonstrated any behavioral evidence of hearing what was said, yet cortical activity indicated that they may have understood it. Although at present it is unclear what percentage of vegetative patients retain such isolated abilities, these studies raise the disturbing possibility that some patients, at least, may find themselves in the 'brain in a vat' situation that used to be the stuff of horror movies.

During the first month after emergence from coma into a vegetative state, it is difficult to make predictions about whether or not the vegetative state will become persistent. Past a month, however, the longer the vegetative state lasts, the less likely it is that the patient will emerge into a minimally conscious state, and, conversely, the more likely it becomes that the vegetative state will become persistent (Multi-Society Task Force 1994a,b). Thus, in traumatic brain injury, after a vegetative state of 3 months, about one-third will emerge into a minimally conscious state, after 6 months only about one-sixth will emerge, and after a vegetative state lasting a year, only a few percent eventually emerge. The figures for non-traumatic cases (e.g. postanoxic) are even worse: after a vegetative state lasting only a month, a mere one-sixth will emerge; after 3 months, only about one-twelfth will emerge, and in cases where the vegetative state lasts 6 months, virtually none will emerge. Given these figures, it is appropriate to withhold the diagnosis of a persistent vegetative state for 12 months after traumatic brain injury and for 3 months after non-traumatic cases. A diagnosis based on chronicity, however, must always be somewhat tentative, as, although rare, there are always exceptions to the rule; cases of emergence from a traumatic brain injury- induced vegetative state have been reported after 5 years (Avesani *et al.* 2006).

MINIMALLY CONSCIOUS STATE

In some cases of coma, patients may emerge directly into the minimally conscious state. In others, patients emerge from coma into a vegetative state, and then, after a variable period, may then emerge into the minimally conscious state.

In the minimally conscious state, patients are not only awake, with eyes open, but they also, in contrast to those in the vegetative state, display clear evidence of an awareness of their surroundings (Giacino *et al.* 2002).

This awareness of the environment may be indicated by any one of a number of different findings. Thus, there may be either sustained visual fixation or tracking. Smiling or crying may appear in response to appropriate environmental events, such as the presence of a family member. Patients may respond to questions with a 'yes' or 'no' or with an equivalent gesture, such as a 'thumbs up'. There may also be more sustained intelligible vocalizations. Patients may also follow commands, and there may be purposeful or semi-purposeful behaviors: these behaviors may include reaching for objects, or, if an object is placed in the patient's hand, movement of the hand or fingers to appropriately accommodate the object.

Over long-term follow-up, patients may either remain in the minimally conscious state or emerge into a dementia of variable severity.

Etiology

The persistent vegetative state and the minimally conscious state both stem from the same kinds of lesions, the only difference being that more severe lesions are associated with the persistent vegetative state.

Three patterns of injury may be found (Adams *et al.* 2000, Kinney and Samuels 1994). First, there may be diffuse damage to the cortex, as may occur after hypoxic–ischemic states or profound hypoglycemia. Second, there may be diffuse damage to the subcortical white matter, as may occur in diffuse axonal injury after traumatic brain injury (Graham *et al.* 2005, Jennett *et al.* 2001). Finally, there may be thalamic damage (specifically to the dorsomedial and ventral posterior nuclei [Maxwell *et al.* 2004]), as may occur with traumatic brain injury, hemorrhage, infarction, or with transtentorial herniation. By far the most common causes are traumatic brain injury and hypoxic–ischemic states, as for example after cardiac arrest.

In sum, consciousness appears to depend on the structural integrity not only of the cortex but of the thalamus and the white matter which connects these two structures, and any lesion or combination of lesions which significantly disrupts any of these structures will lead to either the persistent vegetative or minimally conscious states. Importantly, however, in these cases the brainstem and hypothalamus are spared.

Differential diagnosis

Given the profound implications of the diagnosis of either the persistent vegetative or minimally conscious state, it is imperative that patients be closely monitored and repeatedly and meticulously re-examined over long periods of time. Care must be taken to eliminate any toxic or metabolic conditions which could be influencing the patient's consciousness, and to either correct or somehow circumvent any auditory or visual impairments: in one study, blindness led to misdiagnosis in a significant minority of cases (Andrews *et al.* 1996).

Akinetic mutism is characterized by visual fixation and eye tracking, along with purposeful movements, and hence is readily distinguished from the persistent vegetative state. Distinguishing akinetic mutism from the minimally conscious state is a little more difficult: the distinction, however, may be made if one attends closely to the quality of the eye movements. In akinetic mutism, eye movements are 'lively', fixing and tracking now on one object or person, then another, all in a fashion that indicates purpose: indeed, in many cases it is almost as if patients were engaged in a 'conversation' via their eye movements. In the minimally conscious state, by contrast, although there is fixation and tracking, one fails to see the rapid, purposeful shifting of attention that is characteristic of akinetic mutism.

Some authors hold that the diagnostic terms 'minimally conscious state' and 'persistent vegetative state' also apply to the clinical end stages of such disorders as Alzheimer's disease or Creutzfeldt–Jakob disease; however, in this text these diagnoses are reserved for states that follow upon a period of prolonged coma. The end stages of such conditions as Alzheimer's disease may best, in the authors' opinion, be referred to as 'profound dementia'.

Treatment

Meticulous nursing and general medical care, with careful attention to prevention and treatment of decubiti and contractures, is essential, and in all cases nutritional support is provided via gastrostomy feedings. Beyond this, there is much debate. Claims for the effectiveness of various cognitive stimulation techniques are hard to verify. Regardless of this, it is imperative to keep in mind that patients in the minimally conscious state (and perhaps, given the functional MRI studies noted earlier, also some in the persistent vegetative state) may understand at least part of what is said by those around them: consequently, one must temper one's comments and be sure to speak as if in the presence of a fully conscious patient.

One double-blinded study (Aquilani *et al.* 2008) found that branched-chain amino acids improved cognitive functioning in both the persistent vegetative and minimally conscious states, while open work suggests zolpidem may be helpful in the persistent vegetative state (Clauss and Nel 2006), and that both amantadine (Zafonte *et al.* 1998) and apomorphine (Fridman *et al.* 2009) may improve cognition in the minimally conscious state. Given the relatively benign side-effect profile of these agents, a case may be made for trying them. Deep brain stimulation of the intralaminar thalamic nuclei has also been reported as effective in both the minimally conscious state (Schiff *et al.* 2009) and the persistent vegetative state (Yamamoto and Katayama 2005), but here the risks naturally give one pause, and extra care is absolutely necessary in obtaining informed consent from an adequate surrogate decision maker.

At some point in the care of all these patients, the question arises as to whether treatment should be provided for potentially fatal conditions, such as pneumonia or myocardial infarction; further, and especially in cases of the persistent vegetative state, the question will also arise whether nutritional support should be withdrawn. In attempting to answer these questions, one naturally first looks to whether the patient left an advance directive, expressing his or her wishes. Lacking this, one must turn to an adequate surrogate decision maker.

6.3 DEMENTIA

Dementia constitutes one of the most common syndromes seen in neuropsychiatric practice. Accurate diagnosis is critical not only for prognostic purposes, but also because a not insignificant number of the diseases capable of causing dementia are treatable.

Clinical features

Dementia is a syndrome of multiple different etiologies characterized by a global decrement in cognitive functioning occurring in a clear sensorium, without significant confusion (Cummings 1987).

An amnestic deficit is present in all cases, and, in addition, one finds one or more of the following: executive deficits, aphasia, agnosia, or apraxia

The amnestic deficit manifests most prominently with short term memory loss, as evidenced on the mental status examination by an inability to recall all out of three words after 5 minutes. In addition, patients may display confabulation, typically of the 'provoked' type. In daily life, this short-term memory loss may become apparent when patients lose their keys or wallets, or forget what people have told them; this latter situation may give rise to the common complaint by family members that the patient repeatedly asks the same question, despite already having received the answer. Long-term memory also suffers, although not as severely as does the short-term type: patients may be unable to recall public facts, such as the names of the last four presidents or biographical facts, such as where they last worked or where they went to school. Whenever memory impairment is present, one also typically finds partial or complete disorientation to time and, albeit less prominently, to place.

Executive deficits include difficulties in assessing situations, making plans, setting the plans in motion, and then monitoring the situation to see if the plans should be changed. Although there are neuropsychological tests which address these deficits, the only 'bedside' test commonly in use involves testing patient's abstracting abilities with proverb interpretation. Here, patients are asked to interpret familiar proverbs, such as 'Don't count your chickens before they hatch,' and their responses are categorized as concrete or abstract: a 'concrete' response might be 'Well, they might not all hatch,' while an abstract one might be 'Don't count on things'. In day-to-day life, the consequences of these deficits may range from trivial to disastrous. Patients may have difficulty planning a dinner party, especially if the number of guests changes at the last minute, while others may be unable to organize a budget. Family members may complain that the patient lacks 'good judgment', and in some cases patients may get involved in complex financial transactions which, at the end, may leave them penniless.

In addition to these amnestic and executive deficits, one also typically finds a degree of acalculia, and this is tested for by asking patients to do simple arithmetic, such as subtracting 8 from 12, and, if they perform well at this task, asking them to perform 'serial 7s' by first subtracting 7 from 100, then 7 from that answer, etc. Acalculia may become evident in day-to-day life when patients have trouble making change or balancing their check books.

Aphasia is typically of the sensory or anomic type. When the sensory type is present, patients, on examination, may be unable to follow complex commands, and their speech may be marked by paraphasias and a degree of incoherence. Anomic deficits may be apparent when patients are unable to name objects held up by the examiner, as for example a pen, or when, during conversation, they ask the examiner what is meant by a common word, as for example in one case, when, upon hearing the physician say 'I'm going to shine a flashlight in your eyes', the patient interrupted and asked 'What's a flashlight?'

Of the various agnosias discussed in Chapter 3.7, those most likely to occasion difficulties in daily living include anosognosia, topographagnosia, and prosopagnosia: although visual and tactile agnosia may also be present, they generally come to light only during the neurologic examination. Anosognosia is common in dementia, and patients, unaware of their deficits, may insist on continuing to drive or manage their finances, at times with disastrous results. When topographagnosia is present, patients may get lost when they drive or take a walk, and although this may initially only be apparent when they venture into new territory, eventually navigating even a familiar neighborhood may become impossible. Prosopagnosia is less common, and is typically only evident in advanced cases, when patients may fail to recognize friends or family members.

Of the apraxias discussed in Chapter 3.8, both ideational and dressing apraxia may cause significant difficulties in the activities of daily living as patients become unable to use utensils, combs, razors, and the like, or to dress themselves.

In addition to these core symptoms of dementia, most patients, at some point, will develop other symptomatologies (Savva *et al.* 2009), including the following: personality change; depression and anxiety; agitation; delusions and hallucinations; and, finally, a variety of miscellaneous disturbances, such as insomnia.

Personality change, as discussed in Chapter 8.2, may be of various types, and perhaps the most common, and most problematic, is the frontal lobe syndrome, especially when it is of the orbitofrontal type and marked by disinhibition. Here, in addition to an overall coarsening of behavior and impulsivity, sexual improprieties often loom large in the clinical picture and patients may masturbate openly, fondle others, or proposition much younger people, even children. With involvement of the temporal lobe, the personality change may manifest with repetitive behaviors, a rigid insistence on routine, a change in food preferences, or, at times, a Kluver–Bucy syndrome. Non-specific personality change may also occur, with either an accentuation of pre-existing personality traits, as for example when an overly thrifty person becomes miserly, or the emergence of new ones, as when a previously shy person becomes overly familiar.

Depression is common in dementia, and at times the diagnosis may be missed when patients become unable to accurately report their mood. Consequently it is very important to be alert to the emergence of vegetative symptoms, such as insomnia, anergia, anhedonia, and changes in appetite. Anxiety is also common, but, rather than existing in an isolated fashion, often occurs secondary to depression, thus creating the picture of an agitated depression.

Agitation becomes more frequent as the dementia progresses, and may constitute one of the most problematic of symptoms. Although at times it may occur without any obvious cause, more often than not some precipitant can be identified. There may be a catastrophic reaction, as discussed in Chapter 5.24, when patients are faced with a cognitive task beyond their abilities. Similarly, when patients are thwarted in any way, rather than simply becoming frustrated, they may erupt into an agitated rage. Pain may also cause agitation, as may 'sensory overload', as for example when there are too many visitors in the hospital, or the surroundings are too noisy.

Delusions tend to be persecutory and patients may accuse others of stealing from them. Delusions of infidelity may also occur and patients may accuse spouses of having affairs. One may also encounter the 'phantom boarder' delusion wherein patients believe that someone is hiding in the house, perhaps in the basement or attic. Hallucinations tend to be visual and may be quite complex: patients may see animals or people. Auditory hallucinations, although less common, may also occur, and patients may hear music, voices, or bells.

Of the miscellaneous disturbances seen in dementia, insomnia is perhaps the most common. Although this may occur in isolation and have its cause only in the underlying cause of the dementia, one must always look for other remediable causes, such as depression, pain, reflux, or simply a noisy environment. Wandering is also common, and while some patients may be able to find their way back, many simply wander off, at times with tragic consequences. Finally, there is the phenomenon of 'sundowning,' wherein, after dark, symptoms become worse, especially agitation and delusions and hallucinations.

Etiology

Dementia may occur secondary to any one of a large number of disorders, most of which are listed in Table 6.2, where they are grouped by etiologic category. Although the list may appear dauntingly long, in practice arriving at a 'short list' of possible causes is not that difficult. First, one looks to the age of the patient: dementia is most common in the elderly, and in this age range neurodegenerative disorders, vascular dementias, and chronic hydrocephalus rank as the most common causes. Brain tumors, of course, may occur at any age, but most of those likely to cause dementia arise in middle-aged or older adults. Next, one looks to see if the patient has a movement disorder, has suffered a clear precipitating event (e.g. head trauma), or has a history of dependence on alcohol or inhalants. Subsequently, one should determine whether or not the patient is suffering from depression: although depression is often not listed as a cause of dementia, it can, as discussed below, clearly do so, especially in the elderly, and missing the diagnosis may have tragic results. Following this, consideration should be given to either a vitamin deficiency (B12 or folic acid) or an endocrinologic disorder (i.e. hypothyroidism or hyperthyroidism), not because these are common, but because they, like depression, are eminently treatable. Infectious disorders are subsequently considered, and these are generally either obvious by history or readily diagnosed with appropriate laboratory testing. Finally one is left with a large miscellaneous group which, for the most part, contains disorders which either, in and of themselves, are rare, or which, if not rare, generally do not cause dementia.

Once a short list of diagnostic possibilities has been generated, the reader is strongly encouraged to consult the relevant chapter for each possibility, where a more 'fine-grained' differential diagnosis is presented. Further, it must be kept in mind that dementia may be multifactorial. This is especially the case in the elderly, where a combination of a neurodegenerative disorder and a vascular dementia is very common (Molsa *et al.* 1985).

Imaging and laboratory testing should of course be guided by one's diagnostic suspicions, and details are provided in the respective chapters. In most cases, however, patients will receive either a CT or MRI scan. Most clinicians will also conduct a 'screen' of laboratory tests, and not because the disorders sought for in such a screen are common causes of dementia, but rather because they are treatable and hence are not to be missed: these tests include B12 level, RBC folate level, thyroid profile, human immunodeficiency virus (HIV) (in younger adults) and fluorescent treponemal antibody.

NEURODEGENERATIVE DISORDERS

Of the neurodegenerative disorders capable of causing dementia, the most common in the elderly are Alzheimer's

Table 6.2 The causes of dementia

Neurodegenerative disorders
Alzheimer's disease (Chapter 9.1)
Dementia with Lewy bodies (Chapter 9.6)
Frontotemporal lobar degeneration (Chapter 9.3)

Vascular dementias
Multi-infarct dementia (Chapter 11-1)
Lacunar dementia (Chapter 11.2)
Binswanger's disease (Chapter 11.4)

Hydrocephalus (Chapter 21.2, 21.3)

Brain tumors (Chapter 21.2)

Movement disorders

Parkinsonism
Parkinson's disease (Chapter 9.5)
Dementia with Lewy bodies (Chapter 9.6)
Corticobasal ganglionic degeneration (Chapter 9.8)
Multiple system atrophy (Chapter 9.9)
Progressive supranuclear palsy (Chapter 9.7)
Vascular parkinsonism (Chapter 11.3)

Chorea
Huntington's disease (Chapter 9.10)
Choreoacanthocytosis (Chapter 9.11)
Dentatorubropallidoluysian atrophy (Chapter 9.15)

Ataxia
FXTAS (Chapter 9.12)
Multiple system atrophy (Chapter 9.9)
Spinocerebellar atrophy (Chapter 9.17)
Dentatorubropallidoluysian atrophy (Chapter 9.15)

Dementia occurring secondary to clear precipitants
Traumatic brain injury (Chapter 8.5)
Chronic subdural hematoma (Chapter 12.1)
Dementia pugilistica (Chapter 12.3)
Radiation encephalopathy (Chapter 12.5)
Posthypoxic–ischemic encephalopathy (Chapter 13.1)
Delayed postanoxic encephalopathy (Chapter 13.2)
Postencephalitic (Chapter 15.4, 15.5, 15.6)
Status epilepticus (Chapter 8.3)
Hypoglycemic coma (Chapter 14.15)
Dialysis encephalopathy (Chapter 14.13)

Substance induced
Alcoholic dementia (Chapter 29.8)
Inhalant-induced dementia (Chapter 29.12)

Depression (Chapter 7.1)

Vitamin deficiencies
B12 deficiency (Chapter 14.1)
Folic acid deficiency (Chapter 14.2)

Endocrinologic disorders
Hypothyroidism (Chapter 17.3)
Hyperthyroidism (Chapter 17.4)

Infectious disorders
AIDS (Chapter 15.1)
Neurosyphilis (Chapter 15.14)
Lyme disease (Chapter 15.15)

Miscellaneous causes
Multiple sclerosis (Chapter 19.1)
Collagen-vascular diseases
- Systemic lupus erythematosus (Chapter 19.3)
- Polyarteritis nodosa (Chapter 11.9)
- Wegener's granulomatosis (Chapter 11.10)
- Sjogren's syndrome (Chapter 19.4)
- Sneddon's syndrome (Chapter 19.5)

Wilson's disease (Chapter 9.16)
Fahr's syndrome (Chapter 14.22)
Creutzfeldt–Jakob disease (Chapter 16.1)
Obstructive sleep apnea (Chapter 20.8)
Valproic acid
Behçet's syndrome (Chapter 11.11)
Cerebrotendinous xananthomatosis (Chapter 9.27)
Whipple's disease (Chapter 15.17)
Myotonic muscular atrophy (Chapter 9.26)
Pellagra (Chapter 14.3)
Metachromatic leukodystrophy (Chapter 9.28)
Adrenoleukodystrophy (Chapter 9.29)
Heavy metal intoxication
- Manganese (Chapter 14.6)
- Mercury (Chapter 14.12)
- Lead (Chapter 14.11)
- Thallium (Chapter 14.7)
- Arsenic (Chapter 14.8)

Hyperviscosity syndrome
Hypereosinophilia
Hypertriglyceridemia

disease and dementia with Lewy bodies. Although some might contend that a consideration of dementia with Lewy bodies should be confined to the movement disorders group, the fact that this disorder is just as likely to present with a dementia as with parkinsonism justifies its inclusion here. Alzheimer's disease classically presents in the seventh or eighth decades with a slowly progressive amnestic syndrome. Importantly, although other cognitive deficits eventually accrue, this amnestic syndrome initially appears in isolation and remains that way for at least a year, and often much longer. Dementia with Lewy bodies presents in the same age range, but here the presentation is typically with poor attention and executive deficits, such as poor judgment and decision-making skills. Two other features also characterize dementia with Lewy bodies and further serve to differentiate it from Alzheimer's disease, namely spontaneous fluctuations in mental status, and hallucinations and delusions. The spontaneous fluctuations in mental status may last from hours to days and may be characterized by drowsiness, confusion, incoherence, or bizarre behavior. Hallucinations tend to be visual, and delusions, which are less common, tend to be of persecution. When dementia with Lewy bodies does present with dementia, parkinsonism will eventually follow, and although this may occur within the first year, it may also be delayed, often for many years.

Frontotemporal lobar degeneration, the most common neurodegenerative cause of dementia in those aged under 65 years, classically presents with a personality change, which may be of two general types: in one, there is a fairly typical frontal lobe syndrome, while in the other, known as the 'temporal variant,' a variety of changes may occur, including repetitive behaviors (e.g. hoarding), a rigid insistence on certain routines, changes in food preferences, and a Kluver–Bucy syndrome.

VASCULAR DEMENTIAS

In the evaluation of a patient with dementia, a history of repeated stroke is of great diagnostic import, and in this regard it is important to distinguish two overall groups. On the one hand, there are the large 'territorial' infarctions and intracerebral hemorrhages, both of which are associated with the entity known as multi-infarct dementia, while on the other there are lacunar infarctions, which are generally associated with the entity known as lacunar dementia. These two entities, plus Binswanger's disease, comprise what are known as the vascular dementias, and before proceeding further, some words are in order regarding the differentiation of these three vascular disorders. Although there is no doubt that each of these can occur in 'pure culture', more often than not, on imaging, one finds elements of all three, and although in some cases it is clear that one type clearly predominates, thus justifying the diagnosis of but one of these three vascular dementias, often one finds a more or less equal mixture of two or of all three types, and in such cases, the diagnostician must be content with the overarching diagnosis of 'vascular dementia'.

Multi-infarct dementia is typically characterized by a dementia that progresses in a 'stepwise' fashion, with each step corresponding to a fresh, large, cortical, or subcortical infarction or hemorrhage. Critically, in between the 'steps', the patient experiences a clinical plateau without any between-step deterioration. Examination of such patients typically reveals various focal findings, such as hemiplegia, aphasia, and apraxia, corresponding to the location of the prior infarctions. Rarely, a dementia will result from one large or strategically placed infarction in the frontal or temporal lobes: clearly, subsuming these cases under the rubric of 'multi-infarct' constitutes something of a semantic trespass, but a term equivalent to 'single-infarct dementia' has simply not found its way into current usage.

Lacunar dementia may also pursue a classic 'stepwise' course, and one typically finds a history of classic lacunar syndromes, such as pure motor stroke, pure sensory stroke, and the like. However, lacunar dementia may also be characterized by a more or less gradual cognitive decline, corresponding to lacunar infarctions in the thalamus, caudate, anterior limb of the internal capsule, and the orbitofrontal region, which, cumulatively, account for the decline.

Binswanger's disease, characterized pathologically by widespread 'white matter disease' (due to an equally widespread microangiopathy), in pure form, rather than being characterized by strokes, is typified by a gradually progressive dementia, which in turn presents with slowed thinking and various executive deficits: although an amnestic deficit may also figure early on, it is generally relatively less prominent.

HYDROCEPHALUS

Chronic hydrocephalus, including normal pressure hydrocephalus, may cause a dementia characterized by forgetfulness, slowness of thought and behavior, and apathy. The gait is somewhat wide based, shuffling, and, classically, 'magnetic', a feature characterized by difficulty in initiating steps, as if the feet were held to the floor by an underground magnet. History also typically reveals urinary urgency or incontinence.

BRAIN TUMORS

Tumors, if appropriately situated, may cause a dementia, as for example with those in the frontal lobe, anterior corpus callosum, temporal lobe, thalamus, or hypothalamus. Importantly, in these brain tumor cases, such classic focal signs as hemiparesis, hemianesthesia, aphasia, and apraxia may be lacking, and the finding of a mass on imaging may at times come as a surprise.

MOVEMENT DISORDERS

A large number of movement disorders may also cause a dementia, and, consequently, in evaluating a patient with dementia, it is particularly important to pay heed to the presence of pre-existing parkinsonism, chorea, or ataxia.

PARKINSONISM

Of the parkinsonian disorders capable of causing dementia, Parkinson's disease and dementia with Lewy bodies are by far the most common, and the time course of the evolution of dementia in these two disorders constitutes a useful differential point. In Parkinson's disease, although the vast majority of patients do become demented, this development is delayed for at least 5 years, and generally for a decade or more. In stark contrast, in cases of dementia with Lewy bodies which present with parkinsonism, a dementia will supervene in less than 1 year. Other, much less common, parkinsonian disorders capable of causing dementia include corticobasal ganglionic degeneration, multiple system atrophy (of the striatonigral variant), progressive supranuclear palsy, and vascular parkinsonism, and the differential points which distinguish these are discussed in Chapter 4.8.

CHOREA

Of the choreiform disorders which cause dementia, by far the most common is Huntington's disease; however, consideration must also be given to choreoacanthocytosis and dentatorubropallidoluysian atrophy. As discussed in

Chapter 4.4, choreoacanthocytosis is suggested by tics, dystonia, and mutilative lip biting, while the chorea of dentatorubropallidoluysian atrophy is typically accompanied by ataxia.

ATAXIA

Ataxia may constitute the presentation of various disorders capable of causing dementia, and perhaps the most common of these is FXTAS, or the fragile X-associated tremor/ataxia syndrome, an inherited disorder which presents in middle-aged or older males with a combination of ataxia and tremor. Multiple system atrophy of the olivopontocerebellar type, a sporadic disorder, may present in a similar fashion, and the differential between the two may rest on genetic testing. Spinocerebellar ataxia and dentatorubropallidoluysian atrophy are both autosomal dominant disorders presenting in early adult or middle years, and although there are clinical differences between the two, a reliable diagnosis may well depend on genetic testing.

DEMENTIA OCCURRING SECONDARY TO CLEAR PRECIPITANTS

Dementia may follow on certain precipitating events, and provided an accurate history is available, the diagnosis is relatively straightforward. The syndrome of traumatic brain injury is unmistakable; however, a chronic subdural hematoma may go unsuspected, given that, in the elderly, the trauma may have been forgotten or, on the face of it, be trivial. Dementia pugilistica is suggested by a history of repeated head trauma (e.g. as in boxers) anywhere from 5 to 40 years earlier, and is characterized by dementia in the setting of ataxia, dysarthria, and parkinsonism. Radiation encephalopathy of the late-delayed type occurring after whole-brain irradiation is generally obvious; however, here one must also keep in mind that the onset may be delayed for many years, even decades, and thus this history may have to be sought for. Posthypoxic–ischemic encephalopathy and delayed postanoxic encephalopathy are straightforward, as are cases of postencephalitic dementia. Grand mal, and, much less commonly, complex partial status epilepticus may leave patients demented; however, petit mal status does not carry this liability. Hypoglycemic coma may be followed by dementia, or, if the coma is brief, by only mild cognitive deficits: with repeated episodes, however, a dementia may eventually supervene. Dialysis encephalopathy, although now rare, should still be suspected in any chronic hemodialysis patient who develops a dementia which presents with stuttering and is characterized by myoclonus and asterixis.

SUBSTANCE INDUCED

Alcoholism may leave patients with an alcoholic dementia, and indeed this is one of the most common causes of dementia in middle-aged or older adults. Of note here is that the history, given the ubiquitous denial in patients and family members, may be difficult to obtain. Inhalant-induced dementia is generally more obvious, given not only that denial is less frequently encountered in inhalant dependence, but also because of the prominent associated cerebellar signs.

DEPRESSION

Depression, as may be seen in major depression, is a not uncommon cause of dementia (Reding *et al.* 1985). The dementia of depression used to be called a 'pseudodementia' (Caine 1981; Wells 1979), but this is inappropriate given that the dementia of depression is no more 'pseudo' than any other dementia and may indeed be very severe. Clues to the correct diagnosis include the presence of significant depressive symptoms, such as disturbances in sleep and appetite, and, above all, a history of past depressive episodes (Rabins *et al.* 1984). Furthermore, the appearance of depression in a patient with a pre-existing dementia of another cause may worsen the patient's cognitive deficits (Greenwald *et al.* 1989).

VITAMIN DEFICIENCIES

B12 deficiency is often suggested by the presence of a macrocytic anemia; however, it must always be kept in mind that B12-deficiency dementia may occur in the absence of both macrocytosis and anemia. Folic acid deficiency is a very rare cause of dementia, which likewise can occur in the absence of anemia. B12 and red blood cell folate levels constitute screening tests, which are followed up by homocysteine and methylmalonic acid levels: homocysteine levels are elevated in both B12 and folic acid deficiency, while the methylmalonic acid level is elevated in B12 deficiency alone.

ENDOCRINOLOGIC DISORDERS

Hypothyroidism typically presents with psychomotor retardation and apathy, and, in roughly one-half of all cases, a certain 'fogginess' of thought and memory, which may progress to a definite dementia. Hyperthyroidism of the 'apathetic' type, seen generally only in the elderly, is characterized, as the name indicates, by apathy, and a significant minority of these patients will also develop cognitive deficits, which may progress to a dementia: the only clues to the diagnosis may be tachycardia, congestive heart failure, or atrial fibrillation.

INFECTIOUS DISORDERS

Although acquired immune deficiency syndrome (AIDS) may present with a dementia, this is very uncommon, and in the vast majority of cases the dementia will have been preceded by a decade or more of illness characterized by lymphadenopathy, constitutional symptoms, opportunistic infections, etc. Neurosyphilis, presenting as general paresis of the insane, although once rare, has made something of a comeback in patients with AIDS, and a fluorescent treponemal antibody (FTA) test is certainly in order whenever the Argyll Robertson pupil is present, the dementia is characterized by generalized weakness and tremulousness, or in any cases where, despite a thorough work-up, the cause remains obscure. Lyme disease, in stage III, may, uncommonly, be characterized by a dementia and should be suspected in cases accompanied by a large-joint oligoarthritis or a peripheral polyneuropathy: given that stage III may not develop for more than a decade after stages I and II, one cannot count on getting a history of the earlier stages.

MISCELLANEOUS CAUSES

Multiple sclerosis eventually causes dementia in the majority of cases, and given that this development typically occurs in the setting of well-established disease, there is generally little diagnostic difficulty. It must be kept in mind, however, that the presentation of multiple sclerosis may occasionally be strongly weighted toward dementia, with little in the way of the focal findings typical of the disease, and that in some, very rare, cases, dementia may constitute the sole presentation of multiple sclerosis (Fontaine *et al.* 1994; Hotopf *et al.* 1994).

Of the collagen-vascular diseases, systemic lupus erythematosus is the most common cause. Although most cases of dementia secondary to lupus occur due to infarctions, and thus fall under the rubric of multi-infarct dementia or lacunar dementia, lupus dementia, albeit rarely, may occur secondary to a cerebritis. Polyarteritis nodosa may present in a fashion similar to lupus, and here too a dementia may occur secondary to infarctions. Wegener's granulomatosis is suggested by the combination of upper respiratory symptoms and renal failure, Sjogren's syndrome by the sicca syndrome (dry eyes and mouth) and a small joint polyarthritis, and Sneddon's syndrome by prominent livedo reticularis, which affects not only the extremities but also the trunk.

Wilson's disease, while typically presenting in late childhood or adolescence, may also present in adult years, and when dementia occurs in this disorder, it typically appears in the context of a complex movement disorder, with elements of dystonia, chorea, tremor, or parkinsonism.

Fahr's syndrome occurs secondary to prominent cerebral calcification, and may present with dementia or with a movement disorder (parkinsonism or chorea) depending on whether the calcification is primarily located at the gray–white junction or in the basal ganglia.

Creutzfeldt–Jakob disease, although rare, should be suspected in any case of rapidly progressive dementia. Although myoclonus is found in over 90 percent of cases, it may not be present early on, and hence its absence should not deter one from aggressively pursuing the diagnosis with magnetic resonance imaging, an EEG, and a lumbar puncture.

Obstructive sleep apnea very often causes cognitive deficits, and although delirium is far more likely to occur than dementia, this diagnostic possibility should still be kept in mind in any demented patient with a history of prominent snoring.

Valproic acid, albeit very rarely, may cause a combination of parkinsonism and dementia (Shill and Fife 2000), and as this evolution is delayed for 6 months to 4 years (Ristic *et al.* 2006), the connection between it and the use of valproic acid may be missed.

Many of the remaining disorders in this group have distinctive, and often highly specific, associated features. Thus, Behçet's syndrome is characterized by aphthous and genital ulcers, cerebrotendinous xanthomatosis by tendon xanthomas and early-onset cataracts, Whipple's disease by polyarthralgia and abdominal complaints (and, in a minority, by the very distinctive oculomasticatory myorhythmia), myotonic muscular dystrophy by distal extremity weakness and myotonia, pellagra by diarrhea and a rash on sun-exposed areas, adult-onset metachromatic leukodystrophy by a concurrent peripheral polyneuropathy and ataxia, and adult-onset adrenoleukodystrophy by paraparesis and adrenocortical insufficiency. Heavy metal intoxication is suggested by a history of exposure in most cases, although there are still rare cases of homicidal use of thallium or arsenic. Manganese intoxication is characterized by a personality change and parkinsonism; personality change is also seen in mercury intoxication; however, here one classically sees a tremor. Lead intoxication in adults only rarely produces a dementia, and this is always accompanied by a motor peripheral polyneuropathy with wrist or foot drop. Both thallium and arsenic intoxication are characterized by a sensorimotor peripheral polyneuropathy: in thallium intoxication, alopecia occurs, and while this may also appear with arsenic intoxication, hyperkeratosis of the palms and soles is more characteristic. All of these metals may be assayed by means of a 24-hour urine sample.

Finally, some very rare, miscellaneous causes of dementia include the hyperviscosity syndrome (as occurred in one case secondary to multiple myeloma [Mueller *et al.* 1983]), hypereosinophilia (with 92 percent eosinophils) (Kaplan *et al.* 1990), and extreme hypertriglyceridemia (with values of at least 10 times normal) (Heilman and Fisher 1974; Mas *et al.* 1985).

Differential diagnosis

Dementia must be distinguished from mild cognitive impairment, delirium, mental retardation, and amnesia.

Mild cognitive impairment is a syndrome characterized, as the name clearly suggests, by cognitive impairments that, although similar to those seen in dementia, are so mild that they cause little in the way of impairment. As noted in Chapter 6.4, this may or may not be a syndrome *sui generis* – it might just as well be thought of as a prodrome to a dementia.

Delirium is distinguished from dementia by the presence of prominent confusion; here, however, one must keep in mind that some diseases may be characterized by both dementia and intermittent delirium. In multi-infarct dementia, for example, each fresh stroke may be heralded by an episode of delirium that, once having cleared spontaneously, leaves the patient not confused but more demented. Furthermore, some diseases, albeit characterized primarily by dementia, may also cause intermittent, brief episodes of confusion, as may occur in diffuse Lewy body disease. Finally, it must be kept in mind that in dementia, owing to the underlying pathology, there is a lack of 'cognitive reserve', such that patients are likely to develop delirium in the face of insults which, in those with normal brains, would be trivial: even uncomplicated urinary tract infections may produce a profound delirium.

Mental retardation is distinguished primarily by its course: in mental retardation, the intellectual development of patients proceeds only to a certain point, at which it 'stalls', leaving patients on an intellectual 'plateau' beyond which they do not progress. Importantly, there is no decrement in intellectual ability but merely a plateauing. In contrast, in dementia there is a definite decrement from a previously acquired level of intellectual ability. Importantly, however, some disorders may cause both mental retardation and dementia: for example, in the setting of mental retardation secondary to Down's syndrome, a dementia eventually develops in most patients who survive past the age of 40 years.

Amnesia is distinguished from dementia by the restricted nature of the cognitive deficit: in amnestic disorders, one finds only a deficit in memory, whereas in dementia, in addition to a defective memory, one also finds one or more of the other cognitive deficits noted above. It must be borne in mind, however, that some dementing conditions, such as Alzheimer's disorder, may present with a pure amnesia that persists for a prolonged period before being joined by other cognitive deficits.

Treatment

Patients' liberty should be circumscribed proportionate to their reduced abilities; thus, financial affairs should usually be managed by others and guardianship may be required. Driving privileges are often retained by patients with great tenacity, but these too must eventually be withdrawn. Visiting nurses, 'meals on wheels', and adult day-care centers should each be considered, as they help patients maintain functional abilities and enable them to stay at home longer. Cognitively stimulating activities should also be encouraged (Olazaran *et al.* 2004) and patients should be encouraged to do crossword puzzles, play card games, etc., as these all help to preserve cognitive abilities.

If patients have to move, for example to a retirement or nursing home, efforts should be made to make them feel 'at home', for example by bringing in familiar photographs and, where possible, furniture, and by subscribing to the home-town newspaper. If patients are admitted to hospital, the same measures should be undertaken; furthermore, the room should have a large calendar and clock and, whenever possible, a window with a view.

The need for prosthetic devices (e.g. glasses, hearing aids, dentures, and 'quad canes') should be assessed, and medical regimens should be kept as simple as possible. Although many patients eventually require a wheelchair, ambulation should be encouraged and maintained for as long as possible. Rigorous general medical care is necessary in almost all cases, especially in light of the fact, alluded to above, that, in dementia, otherwise trivial conditions such as uncomplicated urinary tract infections, bronchitis, or anemia may have profound effects on cognition.

In addition to implementing treatment, where possible, of the underlying cause of the dementia, consideration may also be given to the symptomatic treatment of various clinical features such as agitation, delusions or hallucinations, depression, and insomnia. Although the pharmacologic treatments discussed below are applicable in most cases, it is essential to consult the relevant chapter for the cause of the dementia being treated, and for two reasons. First, in some cases, there are specific treatments. Second, and conversely, in some cases, treatments effective for some dementing disorders may have severe side-effects in others: a good example is the potentially fatal phenomenon of 'neuroleptic sensitivity' seen in patients suffering from dementia with Lewy bodies who are treated with antipsychotics.

Depression may be treated with an antidepressant, and consideration may be given to either citalopram or escitalopram, as these tend to be the best tolerated of the available agents.

Although agitation often requires pharmacologic treatment, simple psychosocial or environmental measures should be tried first. The environment should be orderly and calm, and excessive stimulation of any sort should be avoided. If patients enjoy music, selections of their choice should be allowed, and, given the obvious fact that musical tastes differ, the use of 'piped-in' music should be discouraged. Patients' 'space' should be respected, and, provided there is no danger, they should be allowed to wander where they wish. Personal privacy should also be respected, especially as regards dressing and bathing. Should redirection be necessary, a non-confrontational approach is obviously preferable, and in this regard simple distraction, as one might use with a very young child, is often effective. When these conservative measures fail, however, pharmacologic treatment should be considered. As detailed in Chapter 7.4, either risperidone or quetiapine, each in low dose, are reasonable alternatives, and, given its benign side-effect profile, consideration should also be given to citalopram.

Delusions and hallucinations may or may not require treatment, depending on whether they are causing significant distress or agitation, or are leading to behaviors that threaten the patient or others. As in the case of agitation, antipsychotics have constituted the mainstay of treatment, and consideration may be given to aripiprazole, at 10 mg (Mintzer *et al.* 2007), risperidone (0.25–2 mg), or, interestingly, to citalopram, which in one double-blinded study was equivalent in efficacy to risperidone (Pollock *et al.* 2007).

With regard to insomnia, non-pharmacologic measures should naturally be tried first. Appropriate exercise is critical, and daytime naps, if possible, should be avoided. Caffeine and other stimulants should be avoided past midday, and the bedroom should be quiet and dark. Melatonin, surprisingly, offers either little (Riemersma-van der Lek *et al.* 2008) or no (Gehrman *et al.* 2009) benefit compared with placebo, and may, indeed, increase social withdrawal during the day (Riemersma-van der Lek *et al.* 2008). In practice, many clinicians will use zolpidem; however, patients must be closely monitored for any day-time sedation or gait disturbance.

6.4 MILD COGNITIVE IMPAIRMENT

Mild cognitive impairment (MCI) (Gautier *et al.* 2006; Portet *et al.* 2006; Winblad *et al.* 2004) is a recently described syndrome seen in elderly patients that is characterized, as the name suggests, by an impairment in cognitive ability that, although clearly representing a departure from the patient's premorbid baseline, is yet not severe enough to cause significant impairment in day-to-day activities; among those aged 70 years or older, the prevalence of such impairment ranges anywhere from 14 to 18 percent.

As noted below, neuropathologic studies indicate that most of these patients suffer from an early stage of one or more of the common causes of dementia, most notably Alzheimer's disease. Given this, the question may be raised whether, from a nosologic point of view, it is useful or accurate to say that patients who suffer from mild deficits in cognitive ability have a *distinct* syndrome, namely MCI, rather than to say merely that such patients may be in the prodrome to a dementia. Indeed, from a practical point of view, it may be advantageous to say the latter, as this will immediately impel the clinician to begin a basic work-up, just as would be done were the cognitive deficits severe and numerous enough to meet the criteria for a diagnosis of dementia.

Clinical features

Cognitive impairments seen in MCI include short-term memory loss, deficits in executive or visuospatial abilities, and aphasic disturbances. Of all these impairments, short-term memory loss is the most common and, when this is seen in isolation, one speaks of the 'amnestic' type of MCI. By convention, when one of the other impairments is seen in isolation, one speaks of non-amnestic 'single-domain' MCI, and, again by convention, when two or more impairments are present, the term 'mixed-domain' MCI is used.

Cognitive impairment typically occurs gradually, and should be present for roughly a year before the diagnosis is made. In all three types, although patients or others report the cognitive impairment, which in turn is demonstrable on clinical evaluation, there is nevertheless no significant disruption in the patients' routine daily functioning.

Etiology

Autopsy work (Schneider *et al.* 2009) indicates that the most common cause of both amnestic and non-amnestic MCI is Alzheimer's disease; vascular dementia is the next most common cause, followed by dementia with Lewy bodies. This, of course, is not surprising, for these disorders constitute some of the most common causes of dementia. Of the other potentially dementing disorders discussed in Chapter 6.3 which are capable of presenting with mild cognitive deficits, several others should clearly be kept in mind, including frontotemporal lobar degeneration, hydrocephalus, alcoholic dementia, B12 deficiency, hypothyroidism, and apathetic hyperthyroidism. Depression, as noted in Chapter 6.3, can also cause dementia, and this should also be kept in mind, as it too may cause MCI (Barnes *et al.* 2006; Geda *et al.* 2006).

The question arises as to how extensive a work-up should be conducted, and, in this regard, it is reasonable, in addition to a thorough history and a detailed neurologic examination, to obtain either an MRI or CT scan along with a B12 level and a thyroid profile. In research settings, it is also customary to determine cerebrospinal fluid (CSF) levels of β-amyloid(42) and total and phosphorylated tau, and to perform either single-emission photon computed tomography (SPECT) or positron emission tomography (PET) studies, looking for the patterns suggestive of Alzheimer's disease, as discussed in Chapter 9.1. It is uncertain, however, whether these studies are justified for routine clinical work. Some argue that positive results, by indicating which patients are likely to 'convert' from the syndrome of MCI to the syndrome of dementia, have clinical value in that they allow patients to make adequate plans, and thus are justified. Others, however, note that routine clinical follow-up may be just as informative, and note further that, in the absence of any disease-modifying treatments, the earlier identification enabled by these studies lacks practical value.

Differential diagnosis

The distinction between MCI and dementia is made on the basis of both the number and severity of cognitive impairments. In dementia, in addition to an amnestic deficit, one also finds one or more other deficits, including executive and aphasic ones; furthermore, the deficits seen in dementia are of such severity as to significantly interfere with daily functioning, which is in contrast to the absence of such an effect in MCI.

Amnesia of the chronic anterograde type, discussed in Chapter 6.6, in some cases, may clinically be identical with amnestic MCI: Alzheimer's disease typically presents in this fashion, and whether the clinician categorizes such patients as having the syndrome of amnesia or amnestic MCI, from a practical point of view, makes little difference, as the work-up will be the same.

Delirium is immediately distinguished from MCI by the presence of confusion, which, by definition, is absent in MCI. The clinician should also keep in mind that certain of the toxic and metabolic factors capable of causing delirium, while not causing confusion, may also, when mild, cause similarly mild cognitive deficits. This has been shown to be the case with the chronic use of anticholinergic medications (Ancelin *et al.* 2006), and may also occur with chronic use of opioids, with mild obstructive sleep apnea or with slowly progressive, and again mild, renal or hepatic failure.

Treatment

If the work-up reveals a treatable cause (e.g. B12 deficiency), then, obviously, efforts are directed at that disorder. When Alzheimer's disease is suspected (e.g. in cases of amnestic MCI), consideration may be given to a trial of donepezil;

however, the benefits of such treatment are marginal at best (Doody *et al.* 2009).

6.5 DELIRIUM

Delirium is one of the most common neuropsychiatric disorders seen in general hospital practice: overall, roughly one-third of hospitalized patients will become delirious during the hospital stay, while among those in the intensive care unit (ICU), the figure increases to over two-thirds. Prompt diagnosis is critical as, in many cases, the underlying cause of the syndrome, untreated, carries a grave prognosis (Cameron *et al.* 1987; Pompeii *et al.* 1994; Rabins and Folstein 1982).

Clinical features

Delirium, as reviewed by Lipowski (1983, 1987, 1989), is characterized by confusion, disorientation, memory loss, and often other symptoms such as hallucinations and delusions. The syndrome is typically of acute or subacute onset, and most patients develop the full syndrome within a day or two (Levkoff *et al.* 1992).

Confusion, or, as it is also known, 'clouding of the sensorium', is the cardinal symptom of delirium and is present in all cases. Patients may appear dazed, their attention wanders, and they often seem to drift off. In most cases, a variable degree of incoherence will also be seen (Johnson *et al.* 1992; Sirois 1988). Disorientation, found in almost all cases (Morse and Litin 1971) may be for both place and time together, or merely in one of these spheres. Memory loss is typically of the short-term or anterograde type: patients are unable to remember all of three words after 5 minutes and consequently have grave difficulty in keeping track of what happens around them. Retrograde amnesia may also be present but is generally less prominent: patients may have trouble recalling clearly what happened in the days or weeks prior to the onset of the delirium.

Hallucinations are seen in about one-half of all cases (Morse and Litin 1971) and may be either visual or, less commonly, auditory (Cutting 1987; Sirois 1988). Patients may see family members in the hospital room, and bugs or small animals may hide in the sheets. Auditory hallucinations range from such simple phenomena as bells or sirens to hearing voices.

Delusions are generally fragmentary and unsystematized. One typically finds evidence of delusions of persecution (Morse and Litin 1971; Sirois 1988), and patients may report that family members are trying to get rid of them or that the medicine given by the nurse is actually poison.

The overall behavior of patients with delirium may be characterized by either agitation or quietude. Although in a minority of cases patients are either consistently hyperactive or consistently hypoactive, one sees in most a mix – patients who are now agitated, then quiet (Johnson *et al.* 1992; Liptzin and Levkoff 1992). In some cases, one may see an 'occupational delirium' wherein patients act as if they were at work (Wolff and Curran 1935): one patient, a former truck driver, was found with his hands held out in front of him as if gripping a steering wheel, busily 'driving' his bed around the ward. Sleep reversal is common (Johnson *et al.* 1992), and many patients display the phenomenon of 'sundowning', wherein their confusion markedly worsens with the coming of darkness.

Importantly, symptoms typically fluctuate in severity throughout the day, and in some, generally milder, cases, there may actually be 'lucid' intervals: these lucid intervals tend to occur in the morning, and consequently, during morning rounds, the diagnosis may be missed in such cases unless the physician makes a point of asking nursing staff how the patient does the rest of the day, especially at night, when symptoms are typically at their worst.

Etiology

For practical diagnostic purposes, it is useful, as laid out in Table 6.3, to subdivide delirium into various types. First, there is *toxic delirium*, occurring secondary to *medications* (e.g. opioids) or actual *intoxicants* (e.g. cocaine). After this, there is *metabolic delirium*, due to *specific metabolic derangements* (e.g. uremia), *systemic effects of infection* (e.g. pneumonia), or *vitamin deficiencies* (e.g. Wernicke's encephalopathy). Following this is *substance withdrawal delirium* (e.g. delirium tremens during alcohol withdrawal). Next one considers *delirium secondary to intracranial disorders* (e.g. *stroke*). At this point, attention is shifted to a syndrome known as *postoperative delirium*: this is a very common disorder, and, in most cases, represents, etiologically, a combination of toxic, metabolic, or intracranial pathologies. Finally, there is a group of *other causes*, each of them relatively uncommon, including, for example, *epileptic disorders*, *autoimmune disorders*, and *endocrinologic disorders*.

In general hospital practice, the vast majority of deliria are toxic or metabolic in character; withdrawal delirium and delirium due to intracranial disorders are somewhat less common. In working up a case, it must be kept in mind, however, that very often the etiology is multifactorial, and that the various factors involved may be simultaneous or sequential.

During the neurologic examination, it is very important to take note of the presence of certain abnormal movements, namely, postural tremor (Chapter 4.1), myoclonus (Chapter 4.2), and asterixis (Chapter 4.12), as they are very suggestive of certain etiologies. Postural tremor may be seen in toxic deliria (e.g. the serotonin and neuroleptic malignant syndromes), metabolic deliria (e.g. hypoglycemia, systemic infection), and all of the substance withdrawal deliria. Myoclonus, likewise, may be seen in toxic deliria (e.g. the serotonin syndrome), metabolic deliria (e.g. uremic encephalopathy, hyperglycemia), and certain others (e.g. Hashimoto's encephalopathy). Asterixis is highly suggestive of one of three metabolic deliria, namely uremic encephalopathy, hepatic encephalopathy, and respiratory failure.

Laboratory testing, in almost all cases, should include a complete blood count, urinalysis, and a chemistry survey (including sodium, BUN, creatinine, glucose, calcium, magnesium, and bilirubin) and an ammonia level; an antinuclear antibody (ANA) and erythrocyte sedimentation rate (ESR) test may be ordered if an autoimmune disorder is suspected,

Table 6.3 The causes of delirium

Toxic delirium

Medications

- Opioid analgesics (Eisendrath *et al.* 1987; Morisy and Platt 1986; Steinberg *et al.* 1992)
- Tramadol (Kunig *et al.* 2006)
- Anticholinergics
  - Diphenhydramine (Agostini *et al.* 2001)
  - Scopolamine (Minagar *et al.* 1999; Ziskind 1988)
  - Benztropine (De Smet *et al.* 1982)
- Commonly prescribed
  - Baclofen (Lee *et al.* 1992)
  - Baclofen withdrawal (Leo and Baer 2005)
  - Benzodiazepines (Marcantonio *et al.* 1994)
  - Amantadine (Postma and Van Tilburg 1975)
  - Dextromethorphan (Lotrich *et al.* 2005)
  - Metoclopramide (Fishbain and Rogers 1987)
- Specific syndromes
  - Serotonin syndrome (Chapter 29.5)
  - Neuroleptic malignant syndrome (Chapter 29.1)
- Dopaminergics
  - Levodopa (Chapter 9.5)
  - Direct-acting dopaminergics (Chapter 9.5)
- Cardiac medications
  - Amiodarone (Athwal *et al.* 2003)
  - Digoxin (Eisendrath and Sweeney 1987)
- Immunosuppressants
  - Tacrolimus (Buis *et al.* 2002)
  - Cyclosporine (Buis *et al.* 2002)
- Antimicrobials
  - Acyclovir (Rashiq *et al.* 1993)
  - Ganciclovir (Davis *et al.* 1990)
  - Vidarabine (Cullis and Cushing 1984)
  - Levofloxacin (Kiangkitiwan *et al.* 2008)
  - Imipenem (Fernandez-Torre *et al.* 2004)
- Anti-epileptics
  - Valproic acid (Beversdorf *et al.* 1996)
  - Carbamazepine (Horvath *et al.* 2005)
  - Gabapentin (Zhang *et al.* 2005)
- Antidepressants and mood stabilizers
  - Amitriptyline (Preskorn and Simpson 1982)
  - Bupropion (Ames *et al.* 1992)
  - Lithium (Kaplan and Birbeck 2006)
  - Valproic acid (Beversdorf *et al.* 1996)
- Miscellaneous
  - Propranolol (Topliss and Bond 1997)
  - Verapamil (Jacobsen *et al.* 1987)
  - Interleukin 2 (Illowsky *et al.* 1996)
  - Disulfiram (Laplane *et al.* 1992)
  - Bismuth (Chapter 14.9)

Intoxicants

- Cocaine (Chapter 28.2)
- Stimulants (Chapter 28.1)
- Phencyclidine (Chapter 28.4)
- Cannabis (Chapter 28.8)
- Inhalants (Chapter 28.7)
- Methanol (Chapter 28.12)

Metabolic delirium

Specific metabolic derangements

- Uremic encephalopathy (Chapter 14.18)
- Hepatic encephalopathy (Chapter 14.19)
- Respiratory failure (Austen *et al.* 1957; Bacchus 1958; Dulfano and Ishikawa 1965; Westlake *et al.* 1955)
- Obstructive sleep apnea 20.8)
- Hyponatremia (Swanson and Iseri 1958; Welti 1956)
- Hypernatremia (Jana and Romano-Jana 1973)
- Hypoglycemia (Chapter 14.15)
- Hyperglycemia (Gomez Diaz *et al.* 1996; Khardori and Soler 1984)
- Hypocalcemia (Denko and Kaelbling 1962, Hossain 1970)
- Hypercalcemia (Karpati and Frame 1964; Weizman *et al.* 1979)
- Hypomagnesemia (Fishman 1965; Hall and Joffe 1973)
- Hypermagnesemia (Clark and Bowen 1992)

Systemic effects of infection

- Sepsis (Eidelman *et al.* 1996)
- Pneumonia (Johnson *et al.* 2000)
- Urinary tract infection (Manepalli *et al.* 1990)

Vitamin deficiencies

- Wernicke's encephalopathy (Chapter 14.4)
- Encephalopathic pellagra (Chapter 14.3)

Substance withdrawal delirium

- Delirium tremens (Chapter 28.5)
- Benzodiazepine withdrawal delirium (Chapter 28.6)
- Barbiturate withdrawal delirium (Chapter 28.6)
- Gamma-hydroxybutyric acid (Snead and Gibson 2005)
- 1–4 butanediol and gamma butyrolactone (Catalano *et al.* 2001)
- Gabapentin (Pittenger and Desan 2007)

Delirium secondary to intracranial disorders

Stroke (see text for references)

- Temporal lobe
- Thalamus
- Genu of the internal capsule

Microembolism syndromes

- 'Shower' of microemboli
  - Cardiac catheterization (Karalis *et al.* 1996)
  - Coronary artery bypass grafting (McKhann *et al.* 2002; Wityk *et al.* 2001)
- Multiple cholesterol emboli syndrome (Chapter 11.17)
- Fat embolism syndrome (Chapter 11.16)

Infectious and related disorders

- Acute encephalitis (Chapter 8.8)
- Meningitis (Chapter 8.6)
- Cerebral abscess (Chapter 8.7)
- Acute disseminated encephalomyelitis (Chapter 15.11)
- Progressive multifocal leukoencephalolpathy (Chapter 15.3)

Global hypoxic–ischemic disorders

- Carbon monoxide intoxication (Chapter 13.3)
- Posthypoxic–ischemic encephalopathy (Chapter 13.1)
- Delayed postanoxic leukoencephalopathy (Chapter 13.2)

Traumatic brain injury (Chapter 8.5)

Subdural hematoma (Chapter 12.1)

Radiation encephalopathy

- Acute radiation encephalopathy (Chapter 12.5)
- Early-delayed radiation encephalopathy (Chapter 12.5)

Miscellaneous disorders

Tumors (Chapter 12.1)
  Temporal lobe
  Thalamus
  Hypothalamus
Hypertensive encephalopathy (Chapter 11.12)
Reversible posterior leukoencephalopathy (Chapter 11.13)
Thrombotic thrombocytopenic purpura (Chapter 11.15)
Central pontine myelinolysis (Chapter 14.17)

Postoperative delirium (see text for references)

Other causes
Epileptic disorders
Complex partial status epilepticus (Chapter 8.3)
Petit mal status epilepticus (Chapter 8.3)
Autoimmune disorders
Systemic lupus erythematosus (Chapter 19.3)
Polyarteritis nodosa (Chapter 11.9)
Hashimoto's encephalopathy (Chapter 19.10)
Limbic encephalitis (Chapter 19.8)
Endocrinologic disorders
Adrenocortical insufficiency (Chapter 17.2)
Cushing's syndrome (Chapter 17.1)
Hyperthyroidism (thyroid storm) (Chapter 17.3)
Miscellaneous causes
Migraine (Gardner *et al.* 1997)
Heat stroke (Yaqub and Al Deeb 1998)
Malaria (Chapter 15.19)
Hepatic porphyria (Chapter 14.21)
Pancreatitis ('pancreatic encephalopathy') (Estrada *et al.* 1979; Menza and Murray 1989)
Heavy metal intoxication
  Lead (Chapter 14.11)
  Thallium (Chapter 14.7)
  Arsenic (Chapter 14.8)
  Tin (Chapter 14.10)
Inherited disorders of urea cycle metabolism with hyperammonemia
  Ornithine transcarbamylase deficiency (Hu *et al.* 2007)
  Citrullinemia (Wong *et al.* 2007)
Celiac disease (Hadjivassiliou *et al.* 2001; Hu *et al.* 2006)
Marchiafava–Bignami disease (Chapter 29.11)
Behçet's disease (Chapter 11.11)
Granulomatous angiitis of the central nervous system (Chapter 11.8)

and a drug screen may be considered if intoxication seems likely. If there is any suspicion of pneumonia, a chest radiograph is ordered, and, if respiratory failure is suspected, arterial blood gases should be considered. Neuroimaging, preferably with MRI, is in order if an intracranial disorder is suspected. An EEG is generally not required, unless complex partial or petit mal status epilepticus is suspected. In toxic and metabolic deliria, generalized slowing is typically seen, and this may be accompanied by FIRDA (frontal intermittent rhythmic delta); triphasic waves may also be seen, especially in cases of hepatic encephalopathy. By contrast, in substance-withdrawal deliria, excess beta activity may be seen. Acute encephalitis may be accompanied by generalized slowing, while in herpes simplex encephalitis there may be focal, temporal slowing with periodic complexes. In those cases where status epilepticus is suspected, ictal activity, of course, makes the diagnosis; however, one should keep in mind that some patients may cycle between ictal and postictal states, and consequently continuous EEG monitoring may be required to capture the ictal activity. Finally, lumbar puncture should be performed whenever encephalitis or meningitis is suspected, provided, of course, that imaging does not reveal evidence of impending herniation.

These various etiologic types of delirium are now considered in turn, beginning with toxic delirium.

TOXIC DELIRIUM

Toxic delirium, as noted above, may occur owing to the toxicity of certain medications, most notably opioids, or to the effects of intoxicants *per se*, such as cocaine. Medications are considered first.

Medications

In evaluating the role that medications may play in a delirium, it is essential to have an accurate listing of all of the patient's medications, including both those taken on a regular basis and those administered as needed; furthermore, one must note when each medication was started, and, importantly, whether there have been any recent changes in dosage. In this regard, when evaluating hospitalized patients, it is also essential to obtain a list of home medications from a reliable historian, such as a close family member. Although, as noted further on, certain medications are more likely to cause delirium than others, virtually any medicine can be at fault: this is particularly the case in elderly patients or those with pre-existing brain disease, who may be exquisitely sensitive to medications which, when taken by a younger person with an intact brain, are generally innocuous. Confirmation of a medicine's etiologic role rests on demonstrating a temporal relationship between the onset of the delirium and the initiation (or substantial dosage increase) of a medicine, and the resolution of the delirium subsequent to a discontinuation (or substantial dose decrease) of the self-same medicine. In assessing this temporal relationship, one must keep in mind that although the delirium may appear fairly promptly, often within a day or so, the resolution of the delirium may be gradual, corresponding not only to the gradual 'washout' of the offending medication, but also to the diminution of the medication's pharmacodynamic effects, which may persist for some time longer. Further, one must keep in mind that should hepatic or renal failure occur, blood levels may rise to the point of toxicity, even without any change in dose.

Of the medications listed in Table 6.3, by far the most common offenders are the opioid analgesics (Lawlor *et al.* 2000; Tuma and DeAngelis 2000); although tramadol can also cause delirium, it is very much less likely to do so. Medications with anticholinergic effects, most notably diphenhydramine and scopolamine, run a close second to the opioid analgesics: although in some cases they may produce a full anticholinergic syndrome, with, as described in Chapter

29.6, mydriasis and dry skin, these additional features may not be present. Following this is a group of medicines, such as baclofen, which, although not carrying a high risk, are so commonly prescribed in some settings that they must be kept clearly in mind. With regard to baclofen, it must be remembered that delirium may occur not only with oral but also with intrathecal administration, and, in addition, that delirium may occur as a withdrawal phenomenon after long-term oral or intrathecal use: in cases of withdrawal delirium occurring after intrathecal use, it must also be kept in mind that orally administered baclofen may be ineffective in suppressing the symptomatology (Greenberg and Hendrickson 2003).

Next are two specific syndromes, namely the serotonin syndrome and the neuroleptic malignant syndrome, which, although not common, must, given their lethality, be considered. The serotonin syndrome occurs secondary to pharmacologically enhanced serotoninergic tone, which, in turn, generally requires the administration of a combination of serotoninergic agents. Clinically, the serotonin syndrome presents with a combination of delirium, myoclonus, dysarthria or ataxia, and hyper-reflexia. The neuroleptic malignant syndrome occurs secondary to any pharmacologic manipulation that leads to an abrupt diminution in dopaminergic tone. Most commonly, such a diminution occurs secondary to the use of a dopamine blocker, such as one of the first-generation antipsychotics. Second-generation antipsychotics, such as risperidone and clozapine, may also be at fault. The neuroleptic malignant syndrome will occasionally be seen secondary not to dopamine blockade but to an abrupt discontinuation of a dopaminergic agent. Thus, the neuroleptic malignant syndrome has been seen after an abrupt discontinuation of long-term treatment with levodopa or with amantadine. Clinically, within a day or two of the abrupt diminution of dopaminergic tone, patients develop varying combinations of delirium, fever, rigidity, and autonomic instability.

The remaining medications in the table are grouped under customary headings (dopaminergic medications, cardiac medications, antimetabolites, antimicrobials, anti-epileptics, antidepressants and mood stabilizers, and a miscellaneous group): for the most part these medications have only rarely been implicated in delirium and hence a fairly strong case would have to be made before ascribing a delirium to any one of them, not only by demonstrating a clear temporal relationship, but also by ruling out other, more likely causes. With regard to delirium secondary to valproate, it must be kept in mind that in some cases this is associated with elevated ammonia levels (Eyer *et al.* 2005), and that in these cases carnitine may be an effective treatment.

It must be kept in mind that this list is not presented as being comprehensive: new case reports continue to appear and both new and old medicines (even those long considered to be innocuous) continue to be implicated.

Intoxicants

Delirium secondary to intoxication with illicit substances is not uncommonly seen in the emergency room. Delirium secondary to cocaine or stimulants is suggested by agitation, hypertension, and mydriasis. Delirium secondary to phencyclidine, cannabis, inhalants, and methanol is typically accompanied by cerebellar signs, such as dysarthria and ataxia.

METABOLIC DELIRIUM

Metabolic deliria, as indicated earlier, may be due not only to specific metabolic derangements, such as uremia, but also to the systemic effects of infection and to vitamin deficiencies. Each of these types is considered in turn.

Specific metabolic derangements

The specific metabolic deliria are so common that in the evaluation of delirious patients it is appropriate to order a screening chemistry survey. In general, acute derangements are more likely to cause symptoms; gradual changes are much better tolerated.

Uremic encephalopathy is generally accompanied by both asterixis and myoclonus. In renal failure of acute onset, a BUN of over 100 mg/dL is generally sufficient to cause delirium; however, in those cases where the renal failure is gradually progressive, much higher levels may be tolerated before a delirium supervenes.

Hepatic encephalopathy may likewise be accompanied by asterixis and myoclonus, and one may, clinically, appreciate characteristic bad breath or *fetor hepaticus*. Importantly, although the ammonia level is generally elevated, cases have been reported with normal levels.

Respiratory failure, as may be seen in pneumonia or decompensated chronic obstructive pulmonary disease (COPD), generally produces a delirium when the P_aO_2 level falls to below 50 mmHg; CO_2 levels are not as reliable a guide in this regard: although acute rises to 70 mmHg may be associated with delirium, levels of 90 mmHg or more may be tolerated if reached gradually. In cases where the delirium is secondary to elevated CO_2 levels, patients may appear to be intoxicated, a phenomenon that accounts for the old term for this condition, namely 'CO_2 narcosis'.

Obstructive sleep apnea may be accompanied by delirium, not only during nocturnal awakenings but also while the patient is awake during the day, and this appears to be particularly common in stroke patients early in their recovery (Sandberg *et al.* 2001). Although the mechanism underlying the daytime delirium is not clear, it is probably related to nocturnal hypoxemia and hypercarbia.

Hyponatremia, most commonly seen, in hospital settings at least, as part of the syndrome of inappropriate antidiuretic hormone (ADH) secretion, generally is not associated with delirium until the sodium level falls to 120 mEq/L; although gradually developing falls to this level may be tolerated, falls below 110 mEq/L are generally symptomatic.

Hypernatremia, typically seen in the setting of dehydration, may, if of acute onset, cause a delirium when the sodium level reaches 160 mEq/L; gradual rises are better tolerated, and some patients may not experience delirium even with levels of 170 mEq/L.

Hypoglycemia may cause both autonomic and neuroglycopenic symptoms. Autonomic symptoms, such as anxiety, tremor, and diaphoresis, generally occur promptly as the glucose level falls below 45 mg/dL. Neuroglycopenic symptoms, such as delirium, typically do not appear until the level falls below 30 mg/dL, and, importantly, there is typically a lag of about 30 minutes until they appear, corresponding to the time required for depletion of intraneuronal glycogen. Although most cases of hypoglycemic delirium are preceded by autonomic symptoms, exceptions to this rule do occur, as for example in patients with autonomic neuropathy (as may occur in diabetes mellitus) or those treated with beta-blockers.

Hyperglycemia of sufficient degree to cause delirium is generally only found in either diabetic ketoacidosis or the hyperglycemic, hyperosmolar, non-ketotic syndrome. In diabetic ketoacidosis, delirium may appear with blood glucose levels in the range of 300–700 mg/dL, whereas in the hyperglycemic, hyperosmolar, non-ketotic syndrome, the range associated with cognitive dysfunction is higher, from 600 mg/dL up to an astounding 2000 mg/dL.

Hypocalcemia of sufficient degree to cause delirium is generally associated with muscle cramping and tetany. Acute falls to 8 mg/dL may cause symptoms, whereas gradual falls to 6 mg/dL may be well tolerated.

Hypercalcemia is typically associated with nausea, vomiting, constipation, and abdominal pain; delirium may supervene when the level rises above 16 mg/dL.

Hypomagnesemia typically does not cause symptoms until the level falls below 1.2 mg/dL; given that magnesium is required for the release of parathyroid hormone, there is often an associated hypocalcemia.

Hypermagnesemia, generally occurring only with increased magnesium intake in the presence of renal failure, may cause delirium at a level of 6 mg/dL if reached acutely, whereas levels as high as 18 mg/dL may be tolerated if reached very gradually.

Systemic effects of infection

With infection, cytokine levels rise, and it is their effect on the brain which causes delirium (Reichenberg *et al.* 2001). Although an association between sepsis and delirium is well known, it is perhaps not as well appreciated that less severe infections can also cause delirium. Indeed, in elderly or medically fragile patients, or those with pre-existing central nervous system disease (e.g. neurodegenerative disorders or stroke), pneumonia or even trivial urinary tract infections may cause delirium. This is, in fact, so common that a urinalysis should be standard whenever such patients develop a delirium.

Vitamin deficiencies

Of the vitamin deficiencies seen in malnourished patients, two are typified by delirium: namely Wernicke's encephalopathy and encephalopathic pellagra.

Wernicke's encephalopathy, occurring secondary to thiamine deficiency, although most commonly seen in alcoholics, may occur secondary to thiamine deficiency of any cause, including fasting, prolonged vomiting (as in postgastric restriction surgery), or prolonged intravenous feeding without adequate thiamine supplementation. Although most of us were taught to look for the classic 'triad' of delirium, ataxia, and nystagmus, this combination is, in fact, the exception, and most patients with Wernicke's encephalopathy present with delirium alone. The recognition of this syndrome is critical: treated promptly, patients may survive without sequelae; untreated, or with delayed treatment, patients may recover but be left with a permanent Korsakoff's syndrome.

Encephalopathic pellagra, generally seen only in malnourished alcoholics, is characterized by delirium, dysarthria, cogwheel rigidity, gegenhalten, and myoclonus.

SUBSTANCE WITHDRAWAL DELIRIUM

Withdrawal from alcohol, benzodiazepines, or barbiturates may produce a syndrome of delirium accompanied by autonomic signs such as postural tremor, diaphoresis, and tachycardia. This diagnosis, sadly, is often missed. Alcoholics in delirium tremens may deny using alcohol to excess, and their relatives are often complicit in this denial; the same holds true for those abusing benzodiazepines or barbiturates. Both benzodiazepines and barbiturates may also be taken in high doses for therapeutic reasons, and when writing admission orders it is not uncommon for physicians to not order these home medications, even though they are aware of them, on the belief that they are 'not needed'.

Gamma hydroxybutyrate is now available on prescription for the treatment of cataplexy, but it is not appreciated by many physicians that a withdrawal delirium, similar to delirium tremens, may occur upon its abrupt discontinuation. Two related compounds, namely 1–4 butanediol and γ-butyrolactone, taken illicitly, carry the same liability.

Gabapentin, used chronically and in high dosage, may also, if abruptly discontinued, cause delirium.

INTRACRANIAL DISORDERS

Various intracranial disorders may present with delirium, including *stroke*, various *microembolic syndromes* (e.g. after coronary artery bypass grafting), *infectious and related disorders*, *global hypoxic–ischemic disorders*, *traumatic brain injury*, and a group of *miscellaneous disorders*, such as tumors. Stroke will be considered first.

Stroke

In cases of delirium of acute onset wherein there is no obvious cause, consideration should be given to stroke. Delirium has been noted with infarction of the following structures: right temporal lobe (Caplan *et al.* 1986), left temporal lobe (Medina *et al.* 1974), right frontal lobe (Mori and Yamdori 1987), right (Friedman 1985) or left (Graff-Radford *et al.* 1984) thalamus, and the inferior right or left genu of the internal capsule (Tatemichi *et al.* 1992).

The diagnosis of delirium due to infarction is typically suspected when there are associated deficits, such as hemiparesis. Unfortunately for the diagnostician, in many cases, especially when it is the temporal lobe that is infarcted, there may be no obvious signs such as hemiparesis, and, although patients may indeed have other signs (such as hemianopia or various agnosias), their confusion may be so great as to preclude examination. It cannot be stressed enough that delirium in itself may at times be a 'focal' sign, indicating acute damage to one of these structures.

Microembolism syndromes

A 'shower' of microemboli to the cerebrum may cause multiple microinfarctions and present as a delirium, with or without obvious focal signs, such as hemiparesis. The emboli themselves may be composed of various materials: atherosclerotic debris may be released from plaques of the ascending aorta during cardiac catheterization or coronary bypass graft surgery; cholesterol emboli may be released during coronary or carotid catheterization to produce the multiple cholesterol emboli syndrome; and fat emboli may be released into the venous circulation with fractures of the long bones to produce the fat embolism syndrome.

Cardiac catheterization or coronary artery bypass graft surgery may be followed by an encephalopathy. In these cases, the ascending aorta is generally severely arteriosclerotic, and multiple minute emboli are released either as a catheter scrapes by or during cross-clamping of the aorta. Typically, patients are found to be delirious upon recovery from anesthesia, and diffusion-weighted MRI will reveal multiple small infarcts.

The multiple cholesterol emboli syndrome may occur after cardiac catheterization, coronary artery bypass grafting, or carotid catheterization. In this syndrome, multiple cholesterol crystals embolize from atherosclerotic plaques and eventually lodge in the brain. This syndrome, clinically, differs from that seen with embolization of atherosclerotic debris in that the onset of the encephalopathy is typically delayed for a day or so, corresponding to the time required for the development of an inflammatory response to these crystals; furthermore, one typically also finds evidence of embolization to the kidneys, with renal failure, and to the lower extremities, with livedo reticularis.

The fat embolism syndrome represents an uncommon complication of fractures of the long bones or trauma to fatty tissues; in both instances, globules of fat pass via the venous circulation first to the lungs, causing dyspnea, and then to the brain, causing multiple microinfarctions and an encephalopathy.

Infectious and related disorders

The acute appearance of a delirium in the context of headache and fever, especially if accompanied by seizures or focal signs, should immediately suggest the diagnosis of acute encephalitis, meningitis, cerebral abscess, or acute disseminated encephalomyelitis.

Acute encephalitis is generally viral and typically presents with a few days of malaise and fever, after which headache and delirium erupt over a matter of hours.

Meningitis is typically accompanied by mental status changes, and although in viral meningitis these are typically restricted to drowsiness or irritability, delirium is generally seen in bacterial cases.

Cerebral abscess may cause delirium, and this may occur not only with miliary abscesses, but also with large, solitary abscesses of the frontal or temporal lobe.

Acute disseminated encephalomyelitis is an autoimmune disorder triggered by a preceding, usually viral, infection. Within days to weeks after the infection, patients develop an illness very similar to that seen with acute encephalitis.

Another disorder to consider here is progressive multifocal leukoencephalopathy. This is an uncommon disorder, generally seen only in immunosuppressed patients, occurring secondary to infection by the JC virus; the onset, in contrast with the disorders just discussed, is generally subacute or gradual, and is characterized, initially, by progressively worsening focal signs; delirium supervenes only in a small minority of cases.

Global hypoxic–ischemic disorders

Global hypoxia or ischemia, if sustained for more than 5 minutes, will cause a variable degree of cortical necrosis. Several outcomes are possible among those who emerge from coma, including the persistent vegetative state, the minimally conscious state, or a delirium, which in turn may resolve leaving behind a dementia of variable severity. In general hospital practice, the most common causes of such global hypoxia or ischemia are cardiac arrest or severe and sustained hypotension, as may be seen intra-operatively or during sepsis.

Carbon monoxide, by binding with hemoglobin and displacing oxygen, also leads to tissue hypoxia, causing delirium and headache.

Those who recover from a global hypoxic insult are at risk for the development of a delayed postanoxic leukoencephalopathy. This is an autoimmune disorder triggered by cellular damage during the preceding hypoxic insult; clinically, within 2 days to 2 months after the initial hypoxic delirium, patients redevelop a delirium, often accompanied by incontinence, extensor plantar responses, and a movement disorder – either parkinsonism or, less commonly, chorea or dystonia.

Traumatic brain injury

Traumatic brain injury is characterized pathologically by various intracranial injuries, including diffuse axonal injury, cerebral contusions, subdural hematomas, and the like. Clinically, patients who do recover consciousness will be delirious for a variable period of time and may, eventually, be left with a dementia.

Subdural hematoma

Both subacute and chronic subdural hematoma may present with a delirium, which is often, but not always, accompanied by headache. Importantly, in the elderly, or those on warfarin or dabigatran, the initiating trauma need not be severe, and in some cases is so apparently trivial that patients may not even recall it.

Radiation encephalopathy

Radiation encephalopathy may occur in one of three forms: acute, early delayed, and late delayed. With focal irradiation of critical structures (e.g. the temporal lobe) or with whole-brain irradiation, the acute (occurring within days) and early-delayed (occurring within months) forms may present with delirium.

Miscellaneous disorders

Tumors, if appropriately situated, for example in the temporal lobe, hypothalamus, or brainstem, may cause a delirium, which, in contrast with most other deliria, will be of subacute or gradual onset, in keeping with the relatively slow growth of the responsible tumor. Focal signs may or may not be present.

Hypertensive encephalopathy may occur in the setting of an acute elevation of diastolic pressure, generally to 130 mmHg or more; patients present over a day or so with headache, nausea and vomiting, delirium, blindness (or mere visual blurring), and, in many cases, seizures.

The reversible posterior leukoencephalopathy syndrome presents in a fashion similar to hypertensive encephalopathy, and is strongly associated with use of various antimetabolites, such as tacrolimus, cyclosporine, vincristine, and others.

Thrombotic thrombocytopenic purpura (TTP), generally seen in young or middle-aged adults, is characterized by delirium, transient focal signs, and thrombocytopenia, with a platelet count below 30 000; other features include fever, renal failure, and purpura. The delirium itself typically shows a marked fluctuation in severity throughout the day.

Central pontine myelinolyis, occurring days after the overly rapid correction of chronic hyponatremia, typically presents with delirium and a flaccid quadriparesis; uncommonly, the presentation may be with delirium alone.

Postoperative delirium

Delirium is seen in a significant minority of patients after major operations. The term 'postoperative delirium', although in common usage, may be a little misleading as it appears to designate a specific entity, whereas in fact delirium seen postoperatively, rather than being an entity *sui generis*, merely results, as noted earlier, from one or more toxic, metabolic, or intracranial disorders occurring intra- or postoperatively.

Although postoperative delirium can occur in any patient, regardless of age or premorbid status, it is most commonly seen in the elderly and in those with pre-existing cognitive impairment (Litaker *et al.* 2001).

It is imperative to determine the etiology of the delirium, and in this regard it must be kept in mind that most cases are multifactorial. Toxicity secondary to medications is very common, being seen especially with meperidine and benzodiazepines (Marcantonio *et al.* 1994) and with any medications with anticholinergic properties (Mach *et al.* 1995; Tune *et al.* 1993). Metabolic factors (Yildizeli *et al.* 2005) commonly seen include uremia, hepatic failure, respiratory failure, and significant elevations or depressions of sodium or glucose. Pneumonia, line infections, or urinary tract infections may contribute to the delirium via their systemic effects. Unrecognized alcoholism or benzodiazepine dependence is not uncommon, and patients may develop withdrawal delirium; furthermore, malnourished alcoholics are at risk for Wernicke's encephalopathy.

Intracranial disorders, seen particularly after cardiac surgery, such as valve replacement or coronary artery bypass grafting, include stroke and microembolism of atherosclerotic debris or cholesterol crystals. Hypoxic–ischemic injury is not uncommon and the operative report should be inspected for any episodes of significant hypotension or blood loss.

Other causes

The other causes of delirium may be heuristically grouped into epileptic disorders, autoimmune disorders, endocrinologic disorders, and, finally, a large heterogeneous group of miscellaneous causes. Of these, the epileptic disorders are by far the most important.

Epileptic disorders

Both complex partial and petit mal status epilepticus may present with delirium. Patients with petit mal status will have a history of petit mal seizures dating back to childhood or adolescence, and consequently this diagnosis is generally not missed. In cases of complex partial status, however, the seizures may be of new onset in the setting of stroke, tumors, etc., and here a high index of suspicion may be required. The paroxysmal onset of the delirium, of course, is a strong clue, but this may have been missed by nursing personnel, and consequently an EEG is essential in such cases. As noted earlier, however, this may be falsely negative in cases where patients are cycling between seizures and postictal states, which themselves may be characterized by delirium, and hence continuous EEG monitoring may be required. It must be emphasized that complex partial status epilepticus is a not uncommon cause of delirium of unknown etiology, accounting for roughly 15 percent of such cases (Veran *et al.* 2010).

Autoimmune disorders

Of the autoimmune disorders, the most important to keep in mind is systemic lupus erythematosus. In this disease, delirium may occur secondary to infarction or to a lupus cerebritis, and this diagnosis should be considered whenever a delirium occurs in the setting of constitutional symptoms, arthralgia, rashes, pleurisy, pericarditis, and various cytopenias. Polyarteritis nodosa may cause delirium secondary to multiple cerebral infarctions and, as in lupus, the delirium generally occurs in the context of constitutional symptoms. Hashimoto's encephalopathy is a very important cause of delirium, given its treatability, and should be suspected when delirium occurs concurrent with myoclonus, ataxia, or stroke-like episodes. Limbic encephalitis, a paraneoplastic syndrome, may present with delirium alone, and, importantly, may constitute the presentation of the underlying malignancy; the diagnosis is often entertained in cases of delirium of unknown cause in middle-aged or elderly patients, especially those at risk for cancer.

Endocrinologic disorders

Endocrinologic disorders only rarely directly cause delirium. Adrenocortical insufficiency is suggested by postural dizziness and nausea and vomiting, Cushing's syndrome by moon facies, truncal obesity, and violaceous abdominal striae, and thyroid storm by proptosis, postural tremor, tachycardia, and restlessness.

Miscellaneous causes

The miscellaneous causes listed in Table 6.3 constitute rare causes of delirium but should be considered when the diagnostic work-up has been unrevealing.

Differential diagnosis

Delirium must be distinguished from dementia, sensory aphasia, amnesia, and psychosis.

Dementia is distinguished by an absence of significant confusion. As noted earlier, confusion is the cardinal symptom of delirium: both delirium and dementia are characterized by memory deficits, disorientation, etc., but only in delirium does one find prominent and persistent confusion. At times, however, one may find dementia and delirium in the same patient. Pre-existing brain damage of any sort makes patients more sensitive to various toxic and metabolic insults (Koponen and Riekkinen 1993; Koponen *et al.* 1989; Layne and Yudofsky 1971; O'Keefe and Lavan 1997), and it is not uncommon to find a patient with dementia, for example due to Alzheimer's disease or multi-infarct dementia, promptly becoming delirious upon being given an opioid analgesic or coming down with pneumonia. Furthermore, some disorders capable of causing dementia may also, at times, of themselves, cause confusion. This is perhaps most problematic, from a diagnostic point of view – in cases of Lewy body dementia, patients, once demented, are prone to develop transient episodes of confusion, lasting from hours to days, which then resolve spontaneously. Another common example is multi-infarct dementia, wherein the dementia is, as it were, punctuated by an episode of confusion occurring with each fresh infarction: as the peri-infarct edema resolves, the delirium clears, leaving the patient one cognitive 'step' further down. Finally, in the end stage of most neurodegenerative disorders, most patients gradually develop a persistent confusion: whether, nosologically, this should be considered merely an extension of the dementing process or the transition of a dementia into a delirium may be a moot point.

Sensory aphasia is characterized by incoherence, and hence patients with this condition appear similar to those delirious patients who also demonstrate incoherence. A differential point is confusion: aphasic patients, despite their incoherence, are often focused and appear attuned to events in their environment; in contrast, delirious patients, with their confusion, appear dazed and have great difficulty attending to events around them.

Amnesia, like dementia and sensory aphasia, is distinguished by an absence of confusion. In some cases, however, a single underlying pathology may initially present with a delirium, which, upon resolution, leaves the patient with an amnesia. A familiar example is thiamine deficiency, which, acutely, causes the delirium of Wernicke's encephalopathy, and which, upon resolution, leaves the patient with amnestic Korsakoff's syndrome.

Psychosis, although primarily characterized by delusions and hallucinations, may, when severe, also be characterized by confusion. A common example is the psychosis of schizophrenia: here, during periods of exacerbation, confusion may be quite severe. It is not clear whether or not one should consider this development of confusion, nosologically, as merely part of the psychosis or whether one should make a diagnosis of psychosis with superimposed delirium. Certainly, if the confusion persists after treatment of the psychosis when the delusions and hallucinations have cleared, then it would be appropriate to consider a 'dual' diagnosis and proceed with a work-up as described above.

Treatment

Concurrent with symptomatic treatment, as described below, it is imperative to pursue a vigorous diagnostic investigation and, if possible, treat the underlying cause.

Symptomatic treatment, in all cases, involves environmental measures; symptomatic pharmacologic treatment may also be required in 'noisy' deliria when agitation is prominent, and in cases where delusions or hallucinations impel patients to dangerous behaviors.

Environmental measures are aimed at enabling patients, as much as possible, to stay 'in touch' with the environment and the current situation. Large calendars and digital clocks should be kept in full view, and the importance of having a window in the room cannot be overstated (Wilson 1972). At night, the room should be as quiet as possible, and nursing procedures that can be delayed to the daylight hours should be. A night-light should be left on and a nurse's call button should be within easy reach. In some cases, for example when confused patients are apt to get out of bed and either wander off or perhaps slip and fall, sitters may be required. Restraints, although currently frowned on, may be required, and indeed may be lifesaving.

Pharmacologic treatment of delirium is discussed in detail in Chapter 7.4. As noted there, options include haloperidol, olanzapine, risperidone, and quetiapine, each of which may be used on either an as-needed basis or, should the delirium be prolonged, as a regularly scheduled dose.

6.6 AMNESIA

Memory is technically divided into two broad types: memory for how things are done (known as procedural memory) and memory for things that have happened in the past (known as declarative memory). An example of procedural memory would be one's ability to remember how to ride a bicycle or play a piano, and examples of declarative memory include one's ability to remember what happened 5 minutes earlier, a day earlier, or years ago. These two types of memory are quite distinct, and procedural memory may remain intact in cases of profound deficits in declarative memory (Cavaco *et al.* 2004). From a clinical point of view, disorders of declarative memory are most important, and this chapter focuses on these.

Defective declarative memory may or may not be accompanied by other cognitive deficits, such as confusion or deficits in abstracting or calculating abilities: when other deficits are present, consideration, as discussed below under 'differential diagnosis', should be given to a diagnosis of syndromes such as delirium or dementia. This chapter focuses on amnesia as a specific syndrome, characterized by a more or less isolated and 'pure' memory deficit.

Clinical features

New memories are formed on an ongoing basis and this process may occur either automatically or as a matter of effort, for example when one 'pays attention' to something in an attempt to 'commit' it to memory. Once memories are formed, and, as it were, 'stored', they may, with a greater or lesser degree of difficulty, be recalled. Amnesia may affect either one of these aspects of memory: when there is difficulty in forming new memories, one speaks of 'anterograde' amnesia, and when there is difficulty in summoning up memories, one speaks of 'retrograde' amnesia.

Anterograde amnesia is often also referred to as 'short-term' memory loss, and is tested for, formally, by giving patients three words to remember, having them repeat them once, and then, after 5 minutes spent on other items, coming back and asking patients to recall the three words, noting how many are accurately recalled: normally, all three are remembered. Informally, one may find during the interview that patients with short-term memory loss repeat the same question they asked minutes earlier as they have no recall of the answer that had been provided.

Retrograde amnesia is also referred to as 'long-term' memory loss and is tested for during the interview by a series of questions focusing on memories of autobiographic events successively more distant in the past. Thus, one may ask patients what they had for breakfast, what brought them to the hospital, how they had spent the past few weeks, where they lived, where they went to school, etc.; long-term memory for public events may similarly be tested by asking about current newsworthy events or by asking patients to name the last four presidents.

There are three different types of amnestic syndromes. The first is constituted by transient episodes characterized by anterograde amnesia with a variable retrograde component. The second type is chronic and characterized not only by an anterograde component but also by a retrograde one of variable severity. The third is quite rare, and is characterized primarily by a deficit in retrograde memory. Each of these is now considered in turn.

EPISODIC ANTEROGRADE AMNESIA

Episodes of anterograde amnesia typically begin abruptly and generally resolve in less than a day. During the episode, patients are unable to keep track of what is happening and are unable to recall three out of three words after 5 minutes. There may also be a retrograde component, in that patients may have difficulty recalling what happened in the minutes or hours (and sometimes months or years) just prior to the onset of the episode. Once the episode terminates, patients are once again able to keep track of ongoing events, and are able to recall three out of three words after 5 minutes; furthermore, they are able to recall events that happened up until, or just before, the episode began. Notably, however, they are unable to recall what happened while they were in the episode: looking back, it is as if there is an 'island' of amnesia, a period of time for which they have no recall.

CHRONIC ANTEROGRADE AMNESIA WITH A RETROGRADE COMPONENT

Depending on the underlying cause, as discussed further on, this type of amnesia may have either a relatively acute onset (e.g. after infarction of the temporal lobes) or a gradual one (e.g. when due to a neurodegenerative disorder such as Alzheimer's disease). Upon examination, one finds that patients are unable to keep track of ongoing events and are unable to recall three out of three words after 5 minutes. Furthermore, they will have difficulty recalling events of the more distant past events which, before falling ill, they were able to recall without difficulty. This retrograde component often exhibits a 'temporal gradient' in that the patients' recall, although perhaps quite poor for relatively recent events, becomes progressively better for events progressively more distant in the past (Albert *et al.* 1979; Seltzer and Benson 1974).

RETROGRADE AMNESIA

This is a very rare, and quite unusual, type of amnesia. Premorbidly, patients have no difficulty in remembering events of their past; once having fallen ill, however, they are unable to summon up memories of past events: one patient commented that it was as if his memories had been 'destroyed'. Remarkably, in the face of this severe retrograde memory deficit, patients generally retain anterograde capabilities and are able to keep track of ongoing events and recall all three out of three words after 5 minutes.

Etiology

The various causes of amnesia are listed in Table 6.4 and are discussed in detail below, beginning with the episodic anterograde amnesias.

EPISODIC ANTEROGRADE AMNESIA

Transient global amnesia is the prototype of episodic anterograde amnesia. This condition usually has an onset in the seventh decade of life and is characterized by the appearance of one or more episodes, lasting anywhere from 4 to 18 hours and sometimes longer, during which there is a dense anterograde amnesia coupled with a retrograde amnesia of variable duration, lasting from hours to decades. Characteristically, during the episode, patients, although not confused, may repeatedly and anxiously ask what is happen-

Table 6.4 Causes of amnesia

Episodic anterograde amnesia
Transient global amnesia (Chapter 11.18)
Pure epileptic amnesia (Chapter 8.3)
Blackouts (Chapter 28.5)
Concussion (Fisher 1966; Miller *et al.* 1982)
Transient ischemic attacks (see text for references)
'Transient tumor attacks' (see text for references)
Chronic anterograde amnesia with a retrograde component
Korsakoff's syndrome (Chapter 14.5)
Stroke (see text for references)
Temporal lobe
Thalamus
Fornix
Tumors (see text)
Limbic encephalitis (Chapter 19.8)
Neurodegenerative disorders
Alzheimer's disease (Chapter 9.1)
Pick's disease (Chapter 9.2)
Frontotemporal lobar degeneration (Chapter 9.3)
Traumatic brain injury (Chapter 8.5)
Posthypoxic–ischemic encephalopathy (Chapter 13.1)
Encephalitis (see text for references)
Status epilepticus (see text for references)
Certain neurosurgical procedures (see text for references)
Retrograde amnesia
Epileptic (see text for references)
Traumatic brain injury (see text for references)
Herpes simplex encephalitis (Chapter 15.5)
Dissociative amnesia (Chapter 25.1)

ing. Recovery is typically complete except for an 'island' of amnesia extending backwards in time from when the episode resolved to perhaps an hour or two before the episode began. In some cases, it appears that the episode is precipitated by some emotionally laden event, such as sexual intercourse or an argument.

Pure epileptic amnesia represents a partial seizure characterized solely by a combination of anterograde and variable retrograde amnesia. These amnestic episodes differ from those of transient global amnesia in that they are of hyperacute or paroxysmal onset, relatively brief, and may not be accompanied by anxious questioning on the patient's part. Furthermore, one typically finds evidence of either complex partial or grand mal seizures in the history.

Blackouts may complicate moderate or severe alcohol intoxication, and although characteristic of chronic alcoholism, they may at times be seen in healthy subjects. The patients themselves are generally not aware anything is amiss, and apart from other evidence of intoxication (e.g. dysarthria), others may not be able to discern any problem either. The next day, however, patients may find that they have no memory of the night before, and may anxiously (and often circumspectly) ask others what happened. Importantly, although alcohol is the usual culprit, blackouts may also occur with benzodiazepines, especially those of high potency, such as triazolam.

Concussion, as may occur after minor head injury or whiplash, may be accompanied by a dense anterograde and variable retrograde amnesia, and upon recovery the patient is left with the typical 'island' of amnesia, extending back from the time of recovery to, generally, either the injury itself or a short time before that. In one famous example (as reported in the New York *Daily News* of 3 August 1928), Gene Tunney, a heavyweight contender, recalled nothing of a boxing match even though he won the fight. He later decided to quit boxing before a blow could, as he put it, 'permanently hurt my brain'.

Transient ischemic attacks may be characterized in whole or in part by an episode of amnesia. This has been noted with ischemia of the left thalamus (Gorelick *et al.* 1988) or of the medial aspects of the temporal lobes, and in the latter case ischemia in other areas supplied by the posterior cerebral arteries may lead to associated signs, such as hemianopia or cortical blindness, thus providing valuable diagnostic clues (Benson *et al.* 1974).

'Transient tumor attacks' occur in association with cerebral tumors and may happen either because of transient compression of a nearby artery or because of a tumor-related seizure. In one case of a left-sided temporoparietal mass, the attack was characterized by an episode of amnesia (Lisak and Zimmerman 1977).

CHRONIC ANTEROGRADE AMNESIA WITH A RETROGRADE COMPONENT

As this type of amnesia localizes reasonably well to the circuit of Papez, a review of the relevant neuroanatomy may be helpful. The circuit of Papez involves the mamillary body, thalamus, hippocampus, and fornix. Fibers from the mamillary body project to the anterior thalamic nuclei via the mamillothalamic tract. The anterior thalamic nuclei, in turn, project to the cingulate cortex and its cingulum, which in turn projects to the entorhinal and subicular cortices on the medial aspect of the temporal lobe. Fibers from the entorhinal cortex extend, via the perforant pathway, to terminate in the dentate gyrus, from where fibers arise that project to the pyramidal cells of the hippocampus proper. In turn, hippocampal fibers become incorporated into the fornix, which also receives fibers directly from the subicular cortex. The fornix then proceeds posteriorly and superiorly, arching up under the splenium to course anteriorly under the corpus callosum, eventually turning inferiorly, as the columns of the fornix, to dive down into and through the hypothalamus, finally coming to rest in the mamillary body and thus completing the circuit.

This type of amnesia has been noted with lesions in most portions of the circuit of Papez, beginning with the mamillary bodies (Kahn and Crosby 1971; Tanaka *et al.* 1997). Amnesia has also occurred with thalamic lesions but it has been difficult to pinpoint exactly which tracts or nuclei in the thalamus are involved. Lesion studies strongly implicate the mamillothalamic tract (von Cramon *et al.* 1985; Graff-Radford *et al.* 1990; Malamut *et al.* 1992; Yoneoka *et al.* 2004), the anterior thalamic nuclei (Pepin and Auray-Pepin 1993), and perhaps the mediodorsal nucleus (Clarke *et al.* 1994; Gorelick *et al.* 1988). Turning next to the temporal lobes, temporal lobectomy may cause amnesia (Penfield and Mathieson 1974); lesions confined to the hippocampus (Duyckaerts *et al.* 1985), specifically fields CA1 and CA2 of the hippocampus (Kartsounis *et al.* 1995) or CA1 alone (Zola-Morgan *et al.* 1986), have also been responsible. Lesions of the fornix have been implicated (D'Esposito *et al.* 1995; Gaffan *et al.* 1991; Sweet *et al.* 1958), including those located beneath the splenium (Heilman and Sypert 1977) and those involving the columns (Calabrese *et al.* 1995; Hodges and Carpenter 1991).

Although these lesions have in most cases been bilateral, it appears that a properly situated unilateral lesion may also be to blame. Thus, amnesia has occurred with unilateral lesions of the thalamus, either on the left (Choi *et al.* 1983; Clarke *et al.* 1994; Gorelick *et al.* 1988; Landi *et al.* 1982) or on the right (Pepin and Auray-Pepin 1993), and with unilateral lesions of the temporal lobes, with either temporal lobectomy (Scoville and Milner 1957; Walker 1957) or infarction (Gaffan *et al.* 1991).

Specific disorders capable of producing such lesions of the circuit of Papez are now discussed, beginning with the prototype, namely, Korsakoff's syndrome.

Korsakoff's syndrome occurs as a sequela to Wernicke's encephalopathy. Wernicke's encephalopathy occurs secondary to thiamine deficiency and is characterized, pathologically, by hemorrhage in the mammillary bodies and dorsomedial nuclei of the thalamus. Clinically, patients with Wernicke's encephalopathy present with delirium (with or without nystagmus and ataxia), and after the delirium clears, patients are left with an amnesia. Of note, from a nosologic point of view, the term 'Korsakoff's syndrome' is often used to refer to all cases of chronic anterograde amnesia with a retrograde component; in this text, however, this term refers only to the form occurring secondary to thiamine deficiency.

Stroke may be characterized by amnesia, and this has been noted with infarction of the medial aspect of the temporal lobe, the thalamus, and the fornix. Infarction of the medial aspect of the temporal lobe, in the area of distribution of the posterior cerebral artery, involves, among other structures, the hippocampus, and it is from this lesion that the amnesia arises (DeJong *et al.* 1969; Victor *et al.* 1961; Zola-Morgan *et al.* 1986). The involvement of other 'downstream' structures may lead to symptoms that suggest the correct diagnosis, including hemianopia, cortical blindness, or alexia without agraphia (Benson *et al.* 1974). In the case of thalamic infarction, the stroke may be initially characterized by coma (Hodges and McCarthy 1993; Malamut *et al.* 1992) or delirium (Clarke *et al.* 1994); as alertness returns and confusion clears, the patient is left with an amnesia. In two remarkable cases, however, amnesia was the dominant and initial manifestation of the infarction (Pepin and Auray-Pepin 1993; Warren *et al.* 2000). Thalamic hemorrhage may also be at fault with, in one such case, an amnesia emerging after the clearing of a delirium (Choi *et al.* 1983). Amnesia has also been noted as a sequela of infarction of the fornices (Fukatsu *et al.* 1998; Park *et al.* 2000).

Tumors, if properly situated, may cause a progressive amnesia, which has been noted with the following: craniopharyngiomas (Kahn and Crosby 1971; Palmai *et al.* 1967; Tanaka *et al.* 1997; Williams and Pennybacker 1954), thalamic tumors (Ziegler *et al.* 1977) (which compress the overlying mamillary bodies), large tumors affecting structures such as the corpus callosum, fornices, and thalami (Sprofkin and Sciarra 1952), and a subsplenial tumor, which damage both fornices (Heilman and Sypert 1977).

Limbic encephalitis results from an autoimmune assault on the limbic system, including the hippocampus. Although most patients present with delirium, the delirium will, in a small minority, be preceded by an amnestic syndrome.

Neurodegenerative disorders may present with an amnesia that evolves very slowly into a dementia; in a sense, the amnesia represents a 'prodrome' to the dementia. This is most commonly seen with Alzheimer's disease but has also been noted with Pick's disease and frontotemporal lobar degeneration.

Traumatic brain injury characteristically causes coma, and upon recovery from the coma the patient may be left with various cognitive deficits, prominent among which may be an amnesia. The amnesia has both anterograde and retrograde components, and the duration of the anterograde amnesia generally (but not always) correlates well with the severity of the injury itself (Russell and Smith 1961; Wilson *et al.* 1994). The retrograde amnesia exhibits a temporal gradient (Levin *et al.* 1988), and as the patient experiences some recovery, the period of time covered by the retrograde amnesia gradually shrinks (Russell and Nathan 1946). Importantly, neither missile nor crush injuries are associated with much in the way of amnesia, and indeed, in some cases, there may be none at all (Russell 1951; Russell and Schiller 1949).

Posthypoxic–ischemic encephalopathy, as may occur with attempted hanging, carbon monoxide intoxication, cardiorespiratory arrest, or after inhalation anesthesia (Muramoto *et al.* 1979), may, upon the recovery of consciousness, be followed by an amnesia.

Encephalitis may have an amnesia as one of its sequelae (Rose and Symonds 1960); this is particularly true of herpes simplex encephalitis, which characteristically affects medial temporal structures.

Status epilepticus may leave an amnesia in its wake, as occurred in one case after grand mal status (Meierkord *et al.* 1997).

Certain neurosurgical procedures may have amnesia as a complication. Unilateral temporal lobectomy may be at fault (Walker 1957), but in some cases at least this appears to be because the remaining temporal lobe had already been damaged in some fashion (Penfield and Mathieson 1974). Surgery for an anterior communicating artery aneurysm may be followed by an amnesia (Abe *et al.* 1998; Lindquist and Norlen 1966; Phillips *et al.* 1987; Talland *et al.* 1967), and in such cases one finds infarction in the inferomedial frontal lobe in the areas of distribution of the small penetrating arteries arising from the anterior communicating artery (Gade 1982). Neurosurgical intervention directed toward the third ventricle, as in an attempt to remove a colloid cyst, may cause injury to the fornices with a resulting amnesia (Hodges and Carpenter 1991).

RETROGRADE AMNESIA

Pure retrograde amnesia, without any significant anterograde component, may occur on either an episodic or a chronic basis.

Episodic retrograde amnesia occurred, in one case, as a manifestation of a seizure: the patient experienced the sudden onset of an inability to recall any events of his life, but was able, even during the seizure, to keep track of ongoing events; once the seizure terminated, he was again able to recall his past (Venneri and Caffarra 1998).

Chronic retrograde amnesia may occur with lesions of the anterior portions of the temporal lobes, as may be seen in traumatic brain injury (Kroll *et al.* 1997; Markowitsch *et al.* 1993) or after herpes simplex encephalitis (Calabrese *et al.* 1996). In one traumatic brain injury case, a young woman, despite being eventually able to recall most of the events that had occurred after the head trauma, was unable to remember anything that happened before the trauma, all the way back to early childhood (Kapur *et al.* 1992). Further, after resolution of a herpes simplex encephalitis, a patient, although able to recall events that occurred subsequent to her recovery from the encephalitis, could not recall autobiographical events that occurred before the encephalitis, such as her own wedding (Tanaka *et al.* 1999).

Chronic retrograde amnesia may also occur secondary to the condition known as dissociative amnesia, and here the retrograde amnesia may be of two types. In one, known as the 'localized' type of dissociative amnesia, patients are unable to recall certain specific events of their past. In the other, patients are unable to recall any events from the time of the onset of the amnesia stretching back into the distant past: in one variation, known as the 'generalized' form, this 'blanket' of amnesia stretches back to earliest childhood and patients complain that they can recall nothing about their past, while in the other variation, known as the 'continous' form, the 'blanket' ends at some definite point, with patients being able to recall events earlier than that. Of note, history obtained from other sources will reveal that the amnesia was triggered by some emotionally traumatic event, such as rape or assault.

Differential diagnosis

Amnesia must be distinguished from delirium, dementia, and depression. Although delirium and dementia are both characterized by amnesia, here one sees additional symptoms, such as confusion in delirium and deficits in abstracting and calculating ability in dementia. Depression may mimic an amnestic condition, especially when the task of recall is effortful (Cohen *et al.* 1982; Roy-Byrne *et al.* 1986): such depressed patients essentially 'give up' on the task. The presence of other depressive symptoms, such as depressed mood, fatigue, and insomnia, suggests the correct diagnosis.

It should be noted that chronic anterograde amnesia with a retrograde component, when due to a neurodegenerative disorder such as Alzheimer's disease, is identical to the syndrome of amnestic MCI, as discussed in Chapter 6.4.

Treatment

Where available, treatment is directed at the underlying condition. Supervision of a variable degree is often required during the amnesia, and this may suffice for the transient forms. In the case of persistent amnesias, various techniques, such as the use of mnemonics and lists, may enable the patient to better navigate the temporal landscape.

6.7 MENTAL RETARDATION

Mental retardation, also known as 'intellectual disability', is characterized by a failure to progress, despite adequate educational opportunity, beyond a certain developmental point, in both an intellectual and a social sense. This is a very common disorder, seen in about 1 percent of the general population.

Clinical features

Mental retardation is divided, according to its severity, into four grades, namely mild, moderate, severe, and profound. Thus, those with mild mental retardation generally progress no further than an elementary school level, those with moderate mental retardation no further than a second-grade level, those with severe mental retardation no further than a preschool level, and those with profound mental retardation do not progress beyond the level seen in infants or very small children. By convention, these various grades of mental retardation are also demarcated by IQ level, which is presented in Table 6.5, along with the 'developmental' age at which the patient's intellectual and social progress generally stalls. Mild mental retardation is most common, seen in about 80–85 percent of cases; 10–12 percent are moderate, 3–7 percent are severe, and less than 1 percent are profoundly retarded. Among the mildly retarded, males outnumber females; however, as the grade of retardation increases, the sex ratio approaches unity. Each of these grades is described below in more detail.

Mild mental retardation may not become apparent until the child is in elementary school. These children have difficulty in learning to read, write, and do arithmetic, and at best they may progress academically to a fourth-, fifth-, or perhaps sixth-grade level. Thinking is more or less concrete, and seeing things from another's point of view or appreciating the importance of anything that lies outside of their immediate concerns is difficult. These patients fail to grasp social nuances and may appear 'immature'. Affects tend to be

Table 6.5 Grades of mental retardation

Developmental grade	Age (years)	IQ
Mild	7–11	50–55 to 65–75
Moderate	3–7	35–40 to 50–55
Severe	1–3	20–25 to 35–40
Profound	<1	20–25

exaggerated, with little or no shading: a patient may slip from joyful exuberance to profound, seemingly inconsolable, despair within moments. Judgment tends to be poor and a certain degree of gullibility is often displayed. Sudden changes or new situations often render these patients helpless and requiring direct, one-on-one supervision if they are to make it successfully through the transition. In tranquil circumstances, however, many of these patients are able to work at simple jobs and to live independently or with only minimal supervision.

Moderate mental retardation is generally apparent during preschool years. These children are often able to talk, but they have great difficulty in learning to read, write, or do arithmetic and at best may progress academically to a second-grade level. Thinking is sparse, very concrete, and limited to immediate needs. These patients fail to understand ordinary social conventions and have great difficulty in getting along with others. With close supervision, they may be able to perform simple work and live outside of an institution, perhaps in a group home.

Severe mental retardation is generally apparent during the first several years of life. Malformations may be apparent at birth and seizures are not uncommon. These infants are slow to laugh, have difficulty in imitating others, and at best acquire only limited speech. They do not learn to read or write, and at best may be able to count on their fingers. Some may be able to live with family or in closely supervised group homes; however, many are unable to survive outside of an institution.

Profound mental retardation is usually apparent during the first year of life. Malformations are common, as are seizures. These infants are generally unable to walk, and some may be unable to stand or even sit. Speech is generally not acquired, and vocalizations are generally limited to grunts, cries, or some expression of pleasure. Constant close supervision is required to maintain adequate nutrition and hygiene, and institutional care is generally required.

In addition to these specific aspects, patients with mental retardation often display a number of associated features. Aggression, impulsivity, a low frustration tolerance, and an insistence on routine may be seen in all grades. Among those with moderate and higher grades, feeding difficulties, stereotypies, and self-injurious behaviors may be seen, and among the severely and profoundly retarded, rectal digging and coprophagia may occur.

Feeding difficulties include pica, rumination, and food refusal. Food refusal at times may be related to conditions such as pharyngeal or esophageal dysmotility or to reflux, but at times appears to have no other cause than the mental retardation.

Common stereotypies include rocking, head rolling, waving or hand flapping, finger sucking, or repeated, meaningless utterances.

Self-injurious behaviors may include slapping, scratching, head banging, hair pulling, biting, and eye gouging.

Rectal digging, as noted, is generally seen only in the severely and profoundly retarded, and generally only in those who are institutionalized. Although it may be related to conditions such as hemorrhoids or pinworms, at times it seems simply to stem from the mental retardation. In addition to coprophagia, fecal smearing may also occur.

Other neuropsychiatric disorders are seen in one-half or more of patients with mental retardation (Gillberg *et al.* 1986; Gostason 1985) and may include autism, attention-deficit–hyperactivity disorder, major depressive disorder, bipolar disorder, and schizophrenia. In instances where another disorder does occur, the mental retardation itself may modify the presentation of this additional disorder, occasionally making it almost unrecognizable. Thus, depression may present merely with insomnia, weight loss, and psychomotor change, and mania may present as irritability, agitation, and sleeplessness, while schizophrenia may manifest only with bizarre behavior. The diagnosis of attention-deficit–hyperactivity disorder must be made with great care in patients with mental retardation, keeping in mind that any inattentiveness and hyperactivity must be in excess of that expected for the patient's developmental age rather than his or her chronological age.

Etiology

Although there are over a thousand different causes of mental retardation, only several dozen account for the vast majority of cases. Common causes are noted in Table 6.6, where they are divided into three groups: genetic and chromosomal abnormalities, intrauterine insults, and perinatal or postnatal factors. In the USA, by far the most common causes of mental retardation are Down's syndrome, fragile X syndrome, and fetal alcohol syndrome.

These common causes are discussed below; in working up a case, if the clinical and laboratory results fail to

Table 6.6 Common causes of mental retardation

Genetic and chromosomal abnormalities
Polygenic inheritance (Thapar *et al.* 1994)
Down's syndrome (Chapter 10.4)
Fragile X syndrome (Chapter 10.6)
Klinefelter's syndrome (Chapter 10.5)
Tuberous sclerosis (Chapter 10.2)
Sturge–Weber syndrome (Chapter 10.1)
Von Recklinghausen's syndrome (Chapter 10.3)
Prader–Willi syndrome (Chapter 10.10)
Bardet–Biedl syndrome (Chapter 10.9)
Lesch–Nyhan syndrome (Chapter 10.8)
Rett's syndrome (Chapter 10.12)
Phenylketonuria
Intrauterine insults
Fetal alcohol syndrome (Chapter 10.11)
Anti-epileptic drugs
Phentyoin (Hanson and Smith 1975)
Valproic acid (Ardinger *et al.* 1988)
Rubella (Miller *et al.* 1982)
Toxoplasmosis (Desmonts and Couvrer 1974)
Perinatal or postnatal factors
Cerebral palsy (Nicholson and Alberman 1992)
Prematurity (Collin *et al.* 1991; Volpe 1991)
Malnutrition (Chase and Martin 1970)

support one of these, then referral to a specialist is in order.

GENETIC AND CHROMOSOMAL ABNORMALITIES

Polygenic inheritance is of particular importance, especially in cases of mild mental retardation.

Down's syndrome is suggested by a characteristic appearance with narrowed palpebral fissures, epicanthal folds, a broad nasal root, and an often-protruding tongue. Importantly, as noted in Chapter 10.4, a dementia may supervene in adults with Down's syndrome, which, although most commonly due to Alzheimer's disease, may also occur secondary to hypothyroidism.

Fragile X syndrome may cause mental retardation in affected males, and is suggested by facial dysmorphism with a long, narrow face, prognathism, and, in those who have passed puberty, macro-orchidism.

Klinefelter's syndrome may also cause mental retardation and is suggested by a tall stature, small testicular size, and, in some, gynecomastia.

Tuberous sclerosis commonly causes mental retardation and is suggested by the presence of adenoma sebaceum and seizures.

Sturge–Weber syndrome is suggested by a facial port-wine stain, a contralateral hemiplegia, and seizures.

Von Recklinghausen's disease, suggested by *café au lait* spots and neurofibromas, may, in a very small minority, also cause mental retardation.

Both the Prader–Willi syndrome and the Bardet–Biedl syndrome may cause mental retardation, both being characterized by massive obesity and hypogonadism. Distinguishing characteristics for the Prader–Willi syndrome include a characteristic dysmorphism, with almond-shaped eyes, and for the Bardet–Biedl syndrome, retinitis pigmentosa and either polydactyly or syndactyly.

The Lesch–Nyhan syndrome often causes mental retardation and is suggested by choreoathetosis and self-mutilating lip or finger biting.

Rett's syndrome is seen virtually only in females and is suggested by a characteristic history: at around the age of 1.5 years, mental retardation becomes apparent, along with a characteristic stereotyped hand movement, much like hand washing. Microcephaly also gradually becomes apparent. At around the age of 3 years, there may be some partial recovery of communicative ability, and seizures often appear. Patients may then remain stable until the adult years, when scoliosis and dystonia gradually ensue.

Phenylketonuria as a cause of mental retardation is uncommon today, thanks to newborn screening. Importantly, however, adherence to the low phenylalanine diet should be lifelong in order to prevent deterioration. It is especially critical that mothers with phenylketonuria adhere strictly to the diet during pregnancy in order to prevent neurologic damage to the fetus.

INTRAUTERINE INSULTS

The fetal alcohol syndrome, a very common cause of mental retardation, is characterized by short stature, microcephaly, and a characteristic facies, with shortened palpebral fissures, epicanthal folds, a thin upper lip with a smooth philtrum, maxillary hypoplasia, and a degree of micrognathia.

Anti-epileptic drugs, notably phenytoin and valproate, taken during pregnancy may cause retardation.

Maternal infection with either rubella or toxoplasmosis may cause severe mental retardation.

PERINATAL OR POSTNATAL FACTORS

Of all the disorders of perinatal origin that may cause mental retardation, cerebral palsy is the most important. Both prematurity and malnutrition may also cause the syndrome.

Differential diagnosis

In evaluating an individual who never progressed, in an academic sense, beyond levels expected in elementary school years, consideration must first be given to whether the environment was such that, given normal intelligence, these skills could be acquired. Thus, in cases of severe deprivation or inadequate educational opportunity, the failure to acquire these skills may not reflect mental retardation at all. Observing an individual's social behavior may offer a clue here: those of normal intelligence who failed to progress academically owing to a lack of educational opportunity may yet display a keen and subtle grasp of social nuances: here, this demonstration of 'native' intelligence is a more accurate reflection of intellectual capacity than is performance on tests of academic ability.

Various other conditions may, even in the presence of adequate educational resources, prevent the acquisition of academic skills, including significant deafness or reduced visual acuity, developmental disabilities (such as developmental dyslexia or developmental dysphasia), attention-deficit–hyperactivity, schizophrenia, and significant depression.

Dementia is clearly distinguished from mental retardation by its course. In mental retardation, there is no falling off or decrement in intellectual ability from a previously acquired level: rather, there is, as noted above, a 'stalling' of development wherein patients reach a plateau significantly below that expected for their age and beyond which there is no further progress. By contrast, there is in dementia a definite falling off from a previously acquired level, with this decrement being more or less profound.

Treatment

In addition to any treatments specific for the responsible condition, several measures are generally applicable for mental retardation, regardless of the cause.

Educational and training measures are especially important for those with mild and moderate mental retardation, and family members generally will benefit from counseling; in genetic or chromosomal cases, genetic counseling should be offered.

Aggressiveness and impulsivity may respond to low-dose risperidone (Aman *et al.* 2002; Vanden Borre *et al.* 1993), lithium may also be effective (Craft *et al.* 1987; Spreat *et al.* 1989); troublesome stereotypies may respond to behavioral programs, and in some cases risperidone may also be effective. Self-injurious behaviors may also respond to behavioral programs; antipsychotics appear less responsive here, but clomipramine may be effective (Lewis *et al.* 1996).

Attention-deficit–hyperactivity may be treated with stimulants in the usual fashion (Gadow 1985; Handen *et al.* 1990; Payton *et al.* 1989); however, careful attention must be paid to any worsening of stereotypies with stimulant treatment; should this occur, alternatives, as discussed in Chapter 10.14, should be considered.

Autism, mood disorders and schizophrenia may also be treated in the usual way.

REFERENCES

Abe K, Inokawa A, Yanagihara T. Amnesia after a discrete basal forebrain lesion. *J Neurol Neurosurg Psychiatry* 1998; **65**:126–30.

Adams JH, Graham DI, Jennet B. The neuropathology of the vegetative state after an acute brain insult. *Brain* 2000; **123**:1327–38.

Agostini JV, Leo-Summers LS, Inouye SK. Cognitive and other adverse effects of diphenhydramine in hospitalized older adults. *Arch Int Med* 2001; **161**:2091–7.

Albert MS, Butters N, Levin J. Temporal gradients in the retrograde amnesia of patients with alcoholic Korsakoff's disease. *Arch Neurol* 1979; **36**:211–16.

Aman MG, De Smedt G, Derivan A *et al.* Double-blind, placebo-controlled study of risperidone for the treatment of disruptive behaviors in children with subaverage intelligence. *Am J Psychiatry* 2002; **159**:1337–46.

Ames D, Wirshing WC, Szuba MP. Organic mental disorders associated with bupropion in three patients. *J Clin Psychiatry* 1992; **53**:53–5.

Ancelin ML, Artero S, Portet F *et al.* Non-degenerative mild cognitive impairment in elderly people and use of anticholinergic drugs: longitudinal cohort study. *BMJ* 2006; **332**:455–9.

Andrews K, Murphy L, Munday R *et al.* Misdiagnosis of the vegetative state: retrospective study in a rehabilitation unit. *BMJ* 1996; **313**:13–16.

Aquilani R, Boselli M, Boschi F *et al.* Branched-chain amino acids may improve recovery from a vegetative or minimally conscious state in patients with traumatic brain injury: a pilot study. *Arch Phys Med Rehabil* 2008; **8**:1642–7.

Ardinger HH, Atkin JF, Blackston RD *et al.* Verification of the fetal valproate syndrome phenotype. *Am J Med Genet* 1988; **29**:171–85.

Athwal H, Murphy G, Chun S. Amiodarone-induced delirium. *Am J Geriatr Psychiatry* 2003; **11**:696–7.

Austen FK, Carmichael MW, Adams RD. Neurologic manifestations of chronic pulmonary insufficiency. *N Engl J Med* 1957; **257**: 579–90.

Avesani R, Gambini MG, Albertini G. The vegetative state: a report of two cases with a long-term follow-up. *Brain Inj* 2006; **20**:333–8.

Bacchus M. Encephalopathy and pulmonary disease. *Arch Int Med* 1958; **102**:194–8.

Barnes DE, Alexopoulos GS, Lopez EL *et al.* Depressive symptoms, vascular disease, and mild cognitive impairment: findings from the Cardiovascular Health Study. *Arch Gen Psychiatry* 2006; **63**:273–9.

Benson DF, Marsden CD, Meadows JC. The amnesia syndrome of posterior cerebral artery occlusion. *Acta Neurol Scand* 1974; **50**:133–45.

Beversdorf D, Allen C, Nordgren R. Valproate induced encephalopathy treated with carnitine in an adult. *J Neurol Neurosurg Psychiatry* 1996; **61**:211.

Bogousslavsky J, Regli F, Uske A. Thalamic infarcts: clinical syndromes, etiology, and prognosis. *Neurology* 1988; **38**:847–48.

Buis CI, Wiesner RH, Ruud AF *et al.* Acute confusional state following liver transplantation for alcoholic liver disease. *Neurology* 2002; **59**:601–5.

Cadranel JF, Lebiez E, Di Martino V *et al.* Focal neurological signs in hepatic encephalopathy in cirrhotic patients: an underestimated entity? *Am J Gastroenterol* 2001; **96**:515–8.

Caine ED. Pseudodementia: current concepts and future directions. *Arch Gen Psychiatry* 1981; **38**:1359–64.

Calabrese P, Markowitsch HJ, Harders AG *et al.* Fornix damage and memory: a case report. *Cortex* 1995; **31**:555–64.

Calabrese P, Markowitsch HJ, Widlitzek H *et al.* Right temporofrontal cortex as a critical locus for the ecphory of old episodeic memories. *J Neurol Neurosurg Psychiatry* 1996; **61**:304–12.

Cameron DJ, Thomas RI, Mulvihill M *et al.* Delirium: a test of DSM-III criteria on medical inpatients. *J Am Geriatr Soc* 1987; **35**: 1007–10.

Caplan LR, Kelly M, Kase CS *et al.* Infarcts of the inferior division of the right middle cerebral artery: mirror image of Wernicke's aphasia. *Neurology* 1986; **36**:1015–20.

Catalano MC, Glass JM, Catalano G *et al.* Gamma butyrolactone withdrawal syndromes. *Psychosomatics* 2001; **42**:83–8.

Cavaco S, Anderson SW, Allen JS *et al.* The scope of preserved procedural memory in amnesia. *Brain* 2004; **127**:1853–67.

Chase HP, Martin HP. Undernutrition and child development. *N Engl J Med* 1970; **282**:933–9.

Choi D, Sudarsky L, Schacter S *et al.* Medial thalamic hemorrhage with amnesia. *Arch Neurol* 1983; **40**:611–13.

Clark BA, Bowen RS. Unsuspected morbid hypermagnesemia in elderly patients. *Am J Nephrol* 1992; **12**:336–43.

Clarke S, Assal G, Bogousslavsky J *et al.* Pure amnesia after unilateral left polar thalamic infarct: topographic and sequential neurophysiological and metabolic (PET) correlations. *J Neurol Neurosurg Psychiatry* 1994; **57**:27–34.

Clauss R, Nel W. Drug induced arousal from the permanent vegetative state. *NeuroRehabilitation* 2006; **21**:23–8.

Cohen RM, Weingartner H, Smallberg SA *et al.* Effort and cognition in depression. *Arch Gen Psychiatry* 1982; **39**:593–7.

Coleman MR, Rodd JM, Davis MH *et al.* Do vegetative patients retain aspects of language comprehension? Evidence from fMRI. *Brain* 2007; **130**:2494–507.

Collin MF, Halsey CL, Anderson CL. Emerging developmental sequelae in the 'normal' extremely low birthweight infant. *Pediatrics* 1991; **88**:115–20.

Conomy JP, Swash M. Reversible decerebrate and decorticate posturing in hepatic coma. *N Engl J Med* 1968; **278**:876–9.

Craft M, Ismail IA, Krishnamurti D *et al.* Lithium in the treatment of aggression in mentally handicapped patients: a double-blind trial. *Br J Psychiatry* 1987; **150**:685–9.

Cullis PA, Cushing R. Vidarabine encephalopathy. *J Neurol Neurosurg Psychiatry* 1984; **47**:1351–4.

Cummings JL. Dementia syndrome: neurobehavioral and neuropsychiatric features. *J Clin Psychiatry* 1987; **48**(Suppl. 5):3–8.

Cutting J. The phenomenology of acute organic psychosis: comparison with acute schizophrenia. *Br J Psychiatry* 1987; **151**: 324–32.

Davis CL, Springmeyer S, Gmerek BJ. Central nervous system side-effects of ganciclovir. *N Engl J Med* 1990; **322**:933–4.

DeJong RN, Itabashi HH, Olson JR. Memory loss due to hippocampal lesions. *Arch Neurol* 1969; **20**:339–48.

De Keyser J, Deleu D, Solheid C *et al.* Coma as presenting manifestation of Wernicke's encephalopathy. *J Emerg Med* 1985; **3**:361–3.

DeLorenzo RJ, Waterhouse EJ, Towne AR *et al.* Persistent nonconvulsive status epilepticus after the control of convulsive status epilepticus. *Epilepsia* 1998; **39**:833–40.

Denko J, Kaelbling R. The psychiatric aspects of hypoparathyroidism. *Acta Psychiatr Scand* 1962; **38**(Suppl. 164):1–70.

De Smet Y, Ruberg M, Serdaru M *et al.* Confusion, dementia and anticholinergics in Parkinson's disease. *J Neurol Neurosurg Psychiatry* 1982; **45**:1161–4.

Desmonts G, Couvrer J. Congenital toxoplasmosis. *N Engl J Med* 1974; **290**:1110–16.

D'Esposito M, Verfaellie M, Alexander MP *et al.* Amnesia following traumatic bilateral fornix transection. *Neurology* 1995; **45**: 1546–50.

Di HB, Yu SM, Weng XC *et al.* Cerebral response to patient's own name in the vegetative and minimally conscious state. *Neurology* 2007; **68**:895–9.

Doody RS, Ferris SH, Salloway S *et al.* Donepezil treatment of patients with MCI: a 48-week randomized, placebo-controlled trial. *Neurology* 2009; **72**:1555–61.

Dulfano MJ, Ishikawa S. Hypercapnia: mental changes and extrapulmonary complications. *Ann Int Med* 1965; **63**:829–41.

Duyckaerts C, Derouesne C, Signoret JL *et al.* Bilateral and limited amygdalohippocampal lesions causing a pure amnestic syndrome. *Ann Neurol* 1985; **18**:314–19.

Eidelman LA, Putterman D, Putterman C *et al.* The spectrum of septic encephalopathy: definitions, etiologies and mortalities. *JAMA* 1996; **275**:470–3.

Eisendrath SJ, Sweeney MA. Toxic neuropsychiatric effects of digoxin at therapeutic serum concentrations. *Am J Psychiatry* 1987; **144**:506–7.

Eisendrath SJ, Goldman B, Douglas J *et al.* Meperidine-induced delirium. *Am J Psychiatry* 1987; **144**:1062–5.

Estrada RV, Moreno J, Martinez E *et al.* Pancreatic encephalopathy. *Acta Neurol Scand* 1979; **59**:135–9.

Eyer F, Felgenhauer N, Gempel K *et al.* Acute valproate poisoning. *J Clin Psychopharmacol* 2005; **25**:376–80.

Fernandez-Torre JL, Velasco M, Gutierrez R *et al.* Encephalopathy secondary to imipenem therapy. *Clin EEG Neurosci* 2004; **35**: 100–3.

Fishbain DA, Rogers A. Delirium secondary to metoclopramide hydrochloride. *J Clin Psychopharmacol* 1987; **7**:281–2.

Fisher CM. Concussion amnesia. *Neurology* 1966; **16**:826–30.

Fisher CM. Some neuro-ophthalmological observations. *J Neurol Neurosurg Psychiatry* 1967; **30**:383–92.

Fishman RA. Neurological aspects of magnesium metabolism. *Arch Neurol* 1965; **12**:562–9.

Fontaine B, Seilhean D, Tourbah A *et al.* Dementia in two histologically confirmed cases of multiple sclerosis: one case with isolated dementia and one case associated with psychiatric symptoms. *J Neurol Neurosurg Psychiatry* 1994; **57**:353–9.

Forsberg S, Hojer J, Enander C *et al.* Coma and impaired consciousness in the emergency room: characteristics of poisoning versus other causes. *Emerg Med J* 2009; **26**:100–2.

Fridman EA, Calvar J, Bonetto M *et al.* Fast awakening from minimally conscious state with apomorphine. *Brain Inj* 2009; **23**: 172–7.

Friedman JH. Syndrome of diffuse encephalopathy due to nondominant thalamic infarction. *Neurology* 1985; **35**:1524–6.

Fukatsu R, Yamadori A, Fujii T. Impaired recall and preserved encoding in prominent amnesic syndrome: a case of basal forebrain amnesia. *Neurology* 1998; **50**:539–41.

Gade A. Amnesia after operations on aneurysms of the anterior communicating artery. *Surg Neurol* 1982; **18**:46–9.

Gadow KD. Prevalence and efficacy of stimulant drug use with mentally retarded children and youth. *Psychopharmacol Bull* 1985; **21**:291–303.

Gaffan EA, Gaffan D, Hodges JR. Amnesia following damage to the left fornix and to other sites. *Brain* 1991; **114**:1297–13.

Gardner K, Barmada M, Ptacek LJ *et al.* A new locus for hemiplegic migraine maps to chromosome 1q31. *Neurology* 1997; **49**: 1231–1238.

Gautier S, Reisberg B, Zaudig M *et al.* Mild cognitive impairment. *Lancet* 2006; **367**:1262–70.

Geda YA, Knopman DS, Mrazek DA *et al.* Depression, apolipoprotein E genotype, and the incidence of mild cognitive impairment: a prospective cohort study. *Arch Neurol* 2006; **63**: 435–40.

Gehrman PR, Connor DJ, Martin JL *et al.* Melatonin fails to improve sleep or agitation in double-blind, randomized placebo-controlled trial of institutionalized patients with Alzheimer disease. *Am J Geriatr Psychiatry* 2009; **17**:166–9.

Giacino JT, Ashwal S, Childs N *et al.* The minimally conscious state: definition and diagnostic criteria. *Neurology* 2002; **58**: 349–53.

Gibb WR, Gorsuch AN, Lees AJ *et al.* Reversible coma in Wernicke's encephalopathy. *Postgrad Med J* 1985; **61**:607–10.

Gillberg C, Persson E, Grufman M *et al.* Psychiatric disorders in mildly and severely mentally retarded urban children and adolescents. Epidemiological aspects. *Br J Psychiatry* 1986; **149**:68–74.

Gomez Diaz RA, Rivera Moscosco R, Ramos Rodriguez R *et al.* Diabetic ketoacidosis in adults: clinical and laboratory features. *Arch Med Res* 1996; **27**:177–81.

Gorelick PB, Amico LL, Ganellen R *et al.* Transient global amnesia and thalamic infarction. *Neurology* 1988; **38**:496–9.

Gostason R. Psychiatric illness among mentally retarded: a Swedish population study. *Acta Psychiatr Scand Suppl.* 1985; 1–117.

Graff-Radford NR, Eslinger PJ, Damasio AR *et al.* Nonhemorrhagic infarction of the thalamus: behavioral, anatomic, and physiologic correlates. *Neurology* 1984; **34**:14–23.

Graff-Radford NR, Tranel D, Van GW *et al.* Diencephalic amnesia. *Brain* 1990; **113**:1–25.

Graham DI, Maxwell WL, Adams JH *et al.* Novel aspects of the neuropathology of the vegetative state after blunt head injury. *Prog Brain Res* 2005; **150**:445–55.

Greenberg DA, Simon RP. Flexor and extensor postures in sedative drug-induced coma. *Neurology* 1982; **32**:448–51.

Greenberg MI, Hendrickson RG. Baclofen withdrawal following removal of an intrathecal baclofen pump despite oral baclofen replacement. *J Toxicol Clin Toxicol* 2003; **41**:83–5.

Greenwald BS, Kramer-Ginsberg E, Marin DB *et al.* Dementia with coexistent major depression. *Am J Psychiatry* 1989; **146**: 1472–8.

Hadjivassiliou M, Grunewald RA, Lawden M *et al.* Headache and CNS white matter abnormalities associated with gluten sensitivity. *Neurology* 2001; **56**:385–8.

Hall RCW, Joffe JR. Hypomagnesemia: physical and psychiatric symptoms. *J Am Med Assoc* 1973; **224**:1749–51.

Handen BL, Beaux AM, Gosling A *et al.* Efficacy of methylphenidate among mentally retarded children with attention deficit hyperactivity disorder. *Pediatrics* 1990; **86**:922–30.

Hanson JW, Smith DW. The fetal hydantoin syndrome. *J Pediatrics* 1975; **87**:285–90.
Hart SP, Frier BM. Causes, management and morbidity of acute hypoglycaemia in adults requiring hospital admission. *QJM* 1998; **91**:505–10.
Heilman KM, Fisher WR. Hyperlipidemic dementia. *Arch Neurol* 1974; **31**:67–8.
Heilman KM, Sypert GW. Korakoff's syndrome resulting from bilateral fornix lesion. *Neurology* 1977; **27**:490–3.
Hodges JR, Carpenter K. Anterograde amnesia with fornix damage following removal of IIIrd ventricle colloid cyst. *J Neurol Neurosurg Psychiatry* 1991; **54**:633–8.
Hodges JR, McCarthy RA. Autobiographical amnesia resulting from bilateral paramedian thalamic infarction: A case study in cognitive neurobiology. *Brain* 1993; **116**:921–40.
Horvath J, Coeytax A, Jallow P *et al.* Carbamazepine encephalopathy masquerading as Creutzfeldt–Jakob disease. *Neurology* 2005; **65**:650–1.
Hossain M. Neurological and psychiatric manifestations in idiopathic hypoparathyroidism: response to treatment. *J Neurol Neurosurg Psychiatry* 1970; **33**:153–6.
Hotopf MH, Pollock S, Lishman WA. An unusual presentation of multiple sclerosis. *Psychol Med* 1994; **24**:525–8.
Hu WT, Murray JA, Greenaway MC *et al.* Cognitive impairment and celiac disease. *Arch Neurol* 2006; **63**:1440–6.
Hu WT, Kantarc OH, Meritt JL *et al.* Ornithine transcarbamylase deficiency presenting as encephalopathy during adulthood following bariatric surgery. *Arch Neurol* 2007; **64**:126–8.
Illowsky B, Yang JC, Khorsand M *et al.* Multiple cerebral lesions complicating therapy with interleukin-2. *Neurology* 1996; **47**: 417–24.
Jacobsen FM, Sack DA, James SP. Delirium induced by verapamil. *Am J Psychiatry* 1987; **144**:248.
Jana DK, Romano-Jana L. Hypernatremic psychosis in the elderly: case reports. *J Am Ger Soc* 1973; **21**:473–7.
Jennett B, Plum F. Persistent vegetative state after brain damage. A syndrome in search of a name. *Lancet* 1972; **1**:734–7.
Jennett B, Adams JH, Murray LS *et al.* Neuropathology in vegetative and severely disabled patients after head injury. *Neurology* 2001; **56**:486–90.
Johnson JC, Kerse NM, Gottleib G *et al.* Prospective versus retrospective methods of identifying patients with delirium. *J Am Geriatr Soc* 1992; **40**:316–19.
Johnson JC, Jayadevappa R, Baccash PD *et al.* Nonspecific presentation of pneumonia in hospitalized people: age effect or dementia? *J Am Geriatr Soc* 2000; **48**:1316–20.
Kahn EA, Crosby EC. Korsakoff's syndrome associated with surgical lesions involving the mammillary bodies. *Neurology* 1971; **21**: 117–25.
Kaplan PW, Birbeck G. Lithium-induced confusional states: nonconvulsive status epilepticus or triphasic encephalopathy? *Epilepsia* 2006; **47**:2071–7.
Kaplan PW, Waterbury L, Kawas C *et al.* Reversible dementia with idiopathic hypereosinophilia. *Neurology* 1990; **40**:1388–91.
Kapur N, Ellison D, Smith MP *et al.* Focal retrograde amnesia following isolated bilateral temporal lobe pathology. *Brain* 1992; **115**:73–85.
Karalis DG, Quinn V, Victor MF *et al.* Risk of catheter-related emboli in patients with atherosclerotic debris in the thoracic aorta. *Am Heart J* 1996; **131**:1149–55.
Karpati G, Frame B. Neuropsychiatric disorders in primary hyperparathyroidism: clinical analysis with review of the literature. *Arch Neurol* 1964; **10**:387–97.
Kartsounis LD, Rudge P, Stevens JM. Bilateral lesions of CA1 and CA2 fields of the hippocampus are sufficient to cause a severe amnesia syndrome in humans. *J Neurol Neurosurg Psychiatry* 1995; **59**: 95–8.
Khardori R, Soler NG. Hyperosmolar hyperglycemic nonketotic syndrome. Report of 22 cases and brief review. *Am J Med* 1984; **77**:899–904.
Kiangkitiwan B, Doppalapudi A, Fonder M *et al.* Levofloxacin-induced delirium with psychotic features. *Gen Hosp Psychiatry* 2008; **30**:381–3.
Kinney HC, Samuels MA. Neuropathology of the persistent vegetative state. A review. *J Neuropathol Exp Neurol* 1994; **53**:548–58.
Kirk MM, Hoogwerf BJ, Stoller JK. Reversible decerebrate posturing after profound and prolonged hypoglycemia. *Cleve Clin J Med* 1991; **58**:361–3.
Koponen HJ, Riekkinen PJ. A prospective study of delirium in elderly patients admitted to a psychiatric hospital. *Psychol Med* 1993; **23**:103–9.
Koponen H, Stenback U, Mattila E *et al.* Delirium among elderly persons admitted to a psychiatric hospital: clinical course during the acute stage and one-year follow-up. *Acta Psychiatr Scand* 1989; **79**:579–85.
Kroll NEA, Markowitsch HJ, Knight RT *et al.* Retrieval of old memories: the temporofrontal hypothesis. *Brain* 1997; **120**:1377–99.
Kumral E, Kocaer T, Ertubev NO *et al.* Thalamic hemorrhage. A prospective study of 100 patients. *Stroke* 1995; **26**:964–70.
Kumral E, Evyapan D, Balkir K *et al.* Bilateral thalamic infarction. Clinical, etiological, and MRI correlates. *Acta Neurol Scand* 2001; **103**:35–42.
Kunig G, Datwyler S, Eschen A *et al.* Unrecognized long-lasting tramadol-induced delirium in two elderly patients. A case report. *Pharmacopsychiatry* 2006; **39**:194–9.
Landi G, Giusti MC, Guidotti M. Transient global amnesia due to left temporal hemorrhage. *J Neurol Neurosurg Psychiatry* 1982; **45**:1062–3.
Laplane D, Attal N, Sauron B *et al.* Lesions of basal ganglia due to disulfiram. *J Neurol Neurosurg Psychiatry* 1992; **55**:925–9.
Lawlor PG, Gagnon B, Mancini IL *et al.* Occurrence, causes, and outcome of delirium in patients with advanced cancer: a prospective study. *Arch Int Med* 2000; **160**:786–94.
Layne OL, Yudofsky SC. Postoperative psychosis in cardiotomy patients: the role of organic and psychiatric features. *N Engl J Med* 1971; **284**:518–20.
Lee TH, Chen SS, Su SL *et al.* Baclofen intoxication: report of four cases and review of the literature. *Clin Neuropharmacol* 1992; **15**: 56–62.
Leo RJ, Baer D. Delirium associated with baclofen withdrawal: a review of common presentations and management strategies. *Psychosomatics* 2005; **46**:503–7.
Levkoff SE, Evans DA, Liptzin B *et al.* Delirium: the occurrence and persistence of symptoms among elderly hospitalized patients. *Arch Intern Med* 1992; **152**:334–40.
Levin HS, High WM, Eisenberg HM. Learning and forgetting during post-traumatic amnesia in head injured patients. *J Neurol Neurosurg Psychiatry* 1988; **51**:14–20.
Lewis MH, Bodfish JW, Powell SB *et al.* Clomipramine treatment for self-injurious behavior of individuals with mental retardation: a double-blind comparison with placebo. *Am J Ment Retard* 1996; **100**:654–65.
Lindquist G, Norlen G. Korsakoff's syndrome after operation on ruptured aneurysm of the anterior communicating artery. *Acta Psychiatr Scand* 1966; **42**:24–34.

Lipowski ZJ. Transient cognitive disorders (delirium, acute confusional states) in the elderly. *Am J Psychiatry* 1983; **140**:1426–36.

Lipowski ZJ. Delirium (acute confusional state). *JAMA* 1987; **258**:1789–92.

Lipowski ZJ. Delirium in the elderly patient. *N Engl J Med* 1989; **320**:578–82.

Liptzin B, Levkoff SE. An empirical study of delirium subtypes. *Br J Psychiatry* 1992; **161**:843–5.

Lisak RP, Zimmerman RA. Transient global amnesia due to a dominant hemisphere tumor. *Arch Neurol* 1977; **34**:317–18.

Litaker D, Locala J, Franco K *et al.* Preoperative risk factors for postoperative delirium. *Gen Hosp Psychiatry* 2001; **23**: 84–9.

Lotrich FE, Rosen J, Pollock BG. Dextromethorphan-induced delirium and possible methadone interaction. *Am J Geriatr Pharmacother* 2005; **3**:17–20.

Lowenstein DH, Aminoff MJ. Clinical and EEG features of status epilepticus in comatose patients. *Neurology* 1992; **42**:100–4.

Mach J, Dysken M, Richards H *et al.* Serum anticholinergic activity in hospitalized older persons with delirium: a preliminary study. *J Am Geriatr Soc* 1995; **43**:491–5.

McKhann GM, Grega MA, Borowicz LM *et al.* Encephalopathy and stroke after coronary artery bypass grafting: incidence, consequences, and prediction. *Arch Neurol* 2002; **59**:1422–8.

Maddock H, Carley S, McCluskey A. An unusual case of hysterical postoperative coma. *Anaesthesia* 1999; **54**:717–18.

Malamut BL, Graff-Radford N, Chawluk J *et al.* Memory in a case of bilateral thalamic infarction. *Neurology* 1992; **42**:163–9.

Manepalli J, Grossberg GT, Mueller C. Prevalence of delirium and urinary tract infection in a psychogeriatric unit. *J Geriatr Psychiatry Neurol* 1990; **3**:198–202.

Marcantonio ER, Juarez G, Goldman L *et al.* The relationship of postoperative delirium with psychoactive medications. *JAMA* 1994; **272**:1518 22.

Markowitsch HJ, Calabrese P, Liess J *et al.* Retrograde amnesia after traumatic injury of the fronto-temporal cortex. *J Neurol Neurosurg Psychiatry* 1993; **56**:988–92.

Mas J-L, Bousser M-G, Lacombe C *et al.* Hyperlipidemic dementia. *Neurology* 1985; **35**:1385–7.

Massaro AR, Sacco RL, Mohr JP *et al.* Clinical discriminators of lobar and deep hemorrhages: the Stroke Data Bank. *Neurology* 1991; **41**:1881–5.

Maxwell WL, Penningon K, MacKinnon MA *et al.* Differential responses in three thalamic nuclei in moderately disabled, severely disabled and vegetative patients after blunt head injury. *Brain* 2004; **127**:2470–8.

Medina JL, Rubino FA, Ross E. Agitated delirium caused by infarctions of the hippocampal formation and fusiform and lingual gyri. *Neurology* 1974; **24**:1181–4.

Meierkord H, Wieshman U, Niehaus L *et al.* Structural consequences of status epilepticus with serial magnetic resonance imaging. *Acta Neurol Scand* 1997; **96**:127–32.

Menza M, Murray GB. Pancreatic encephalopathy. *Biol Psychiatry* 1989; **25**:781–4.

Miller E, Craddock-Watson JE, Pollock TM. Consequences of confirmed maternal rubella at successive stages of pregnancy. *Lancet* 1982; **2**:781–4.

Minagar A, Shulman LM, Weiner WJ. Transderm-induced psychosis in Parkinson's disease. *Neurology* 1999; **53**:433–4.

Mintzer JE, Tune LE, Breder CD *et al.* Aripiprazole for the treatment of psychoses in institutionalized patients with Alzheimer dementia: a multicenter, randomized, double-blind, placebo-controlled assessment of three fixed doses. *Am J Geriatr Psychiatry* 2007; **15**: 918–31.

Molsa PK, Paljarvi L, Rinne JO *et al.* Validity of clinical diagnosis in dementia: a prospective clinico-pathological study. *J Neurol Neurosurg Psychiatry* 1985; **48**:1085–90.

Mori E, Yamadori A. Acute confusional state and acute agitated delirium: occurrence after infarction in the right middle cerebral artery territory. *Arch Neurol* 1987; **44**:1139–43.

Morisy L, Platt D. Hazards of high-dose meperidine. *JAMA* 1986; **255**:467–8.

Morse RM, Litin EM. The anatomy of delirium. *Am J Psychiatry* 1971; **128**:111–16.

Mueller J, Hotson JR, Langston JW. Hyperviscosity-induced dementia. *Neurology* 1983; **33**:101–3.

Multi-Society Task Force on PVS. Medical aspects of the persistent vegetative state (1). *N Engl J Med* 1994a; **330**:1499–508.

Multi-Society Task Force on PVS. Medical aspects of the persistent vegetative state (2). *N Engl J Med* 1994b; **330**:1572–9.

Muramoto O, Kuru Y, Sugishita M *et al.* Pure memory loss with hippocampal lesions: a pneumoencephalographic study. *Arch Neurol* 1979; **36**:54–6.

Nicholson A, Alberman E. Cerebral palsy – an increasing contributor to severe mental retardation. *Arch Dis Child* 1992; **67**:1050–5.

Noda S, Ito H, Umezaki H *et al.* Hip flexion-abduction to elicit asterixis in unresponsive patients. *Ann Neurol* 1985; **18**:96–7.

O'Keefe S, Lavan J. The prognostic significance of delirium in older hospital patients. *J Am Geriatr Soc* 1997; **45**:174–8.

Olazaran J, Muniz R, Reisberg B *et al.* Benefits of cognitive-motor intervention in mild to moderate Alzheimer's disease. *Neurology* 2004; **63**:2348–53.

Owen AM, Coleman MR, Boly M *et al.* Detecting awareness in the vegetative state. *Science* 2006; **313**:1402.

Padkin A. Avoiding unnecessary trauma in the differential diagnosis of coma. *Anaesthesia* 1999; **54**:1126–7.

Palmai G, Taylor DC, Falconer MA. A case of craniopharyngioma presenting as Korsakoff's psychosis. *Br J Psychiatry* 1967; **113**:619–23.

Park SA, Hahn JH, Kim JI *et al.* Memory deficits after bilateral anterior fornix infarction. *Neurology* 2000; **54**:1379–82.

Parvizi J, Damasio AR. Neuroanatomical correlates of brainstem coma. *Brain* 2003; **126**:1524–36.

Payton JB, Burkhart JE, Hersen M *et al.* Treatment of ADDH in mentally retarded children: a preliminary study. *J Am Acad Child Adolesc Psychiatry* 1989; **28**:761–7.

Penfield W, Mathieson G. Memory: autopsy findings and comments on the role of hippocampus in experiential recall. *Arch Neurol* 1974; **31**:145–54.

Pepin EP, Auray-Pepin L. Selective dorsolateral frontal lobe dysfunction associated with diencephalic amnesia. *Neurology* 1993; **43**: 733–41.

Phillips S, Sangalang V, Sterns G. Basal forebrain infarction: a clinicopathologic corelation. *Arch Neurol* 1987; **44**:1134–8.

Pittenger C, Desan PH. Gabapentin abuse, and delirium tremens upon gabapentin withdrawal. *J Clin Psychiatry* 2007; **68**: 483–4.

Pollock BG, Mulsant BH, Rosen J *et al.* A double-blind comparison of citalopram and risperidone for the treatment of behavioural and psychotic symptoms associated with dementia. *Am J Geriatr Psychiatry* 2007; **15**:942–52.

Pompeii P, Foreman M, Rudberg MA *et al.* Delirium in hospitalized older persons: outcomes and predictors. *J Am Geriatr Soc* 1994; **42**:809–15.

Portet F, Ousset PJ, Visser PJ *et al.* Mild cognitive impairment in medical practice: critical review of the concept and new diagnostic procedure. Report of the MCI working group of the European Consortium on Alzheimer's disease (EADC). *J Neurol Neurosurg Psychiatry* 2006; **77**:714–18.

Postma JU, Van Tilburg W. Visual hallucinations and delirium during treatment with amantadine (Symmetrel). *J Am Geriatr Soc* 1975; **23**:212–15.

Preskorn SH, Simpson S. Tricyclic-antidepressant-induced delirium and plasma drug concentration. *Am J Psychiatry* 1982; **139**:822–3.

Rabins PV, Folstein MF. Delirium and dementia: diagnostic criteria and fatality rates. *Br J Psychiatry* 1982; **140**:149–53.

Rabins PV, Merchant A, Nestadt G. Criteria for diagnosing reversible dementia caused by depression: validation by 2-year follow-up. *Br J Psychiatry* 1984; **144**:488–92.

Rashiq S, Briewa L, Mooney M *et al.* Distinguishing acyclovir neurotoxicity from encephalomyelitis. *J Intern Med* 1993; **234**:507–11.

Reding M, Haycox J, Blass J. Depression in patients referred to a dementia clinic: a three-year prospective study. *Arch Neurol* 1985; **42**:894–6.

Reichenberg A, Yirmiya R, Schuld A *et al.* Cytokine-associated emotional and cognitive disturbances in humans. *Arch Gen Psychiatry* 2001; **58**:445–52.

Riemersma-van der Lek RF, Swaab DF, Twisk J *et al.* Effect of bright light and melatonin on cognitive and noncognitive function in elderly residents of group care facilities: a randomized controlled trial. *JAMA* 2008; **299**:2642–55.

Ristic AJ, Vojvodic N, Jankovic S *et al.* The frequency of reversible parkinsonism and cognitive decline associated with valproate treatment: a study of 364 patients with different types of epilepsy. *Epilepsia* 2006; **47**:2183–5.

Ropper AH. Lateral displacement of the brain and level of consciousness in patients with an acute hemispheral mass. *N Engl J Med* 1986; **314**:953–8.

Ropper AH. A preliminary MRI study of the geometry of brain displacement and level of consciousness with acute intracranial masses. *Neurology* 1989; **39**:622–7.

Rose FC, Symonds CP. Persistent memory defect following encephalitis. *Brain* 1960; **83**:195–212.

Roy-Byrne PR, Weingartner H, Bierer LM *et al.* Effortful and automatic cognitive processes in depression. *Arch Gen Psychiatry* 1986; **43**:265–7.

Russell WR. Disability caused by brain wounds. *J Neurol Neurosurg Psychiatry* 1951; **14**:35–9.

Russell WR, Nathan PW. Traumatic amnesia. *Brain* 1946; **68**: 280–300.

Russell WR, Schiller F. Crushing injuries to the skull. *J Neurol Neurosurg Psychiatry* 1949; **12**:52–60.

Russell WR, Smith A. Post-traumatic amnesia in closed head injury. *Arch Neurol* 1961; **5**:4–17.

Sandberg O, Franklin KA, Bucht G *et al.* Sleep apnea, delirium, depressed mood, cognition and ADL ability after stroke. *J Am Geriatr Soc* 2001; **49**:391–7.

Savva GM, Zaccai J, Matthews FE *et al.* Prevalence, correlates and course of behavioural and psychological symptoms of dementia in the population. *Br J Psychiatry* 2009; **194**:212–19.

Schiff ND, Giacino JT, Fins JJ. Deep brain stimulation, neuroethics, and the minimally conscious state. *Arch Neurol* 2009; **66**:697–702.

Schneider JA, Arvanitakis Z, Leurgans SE *et al.* The neuropathology of probable Alzheimer disease and mild cognitive impairment. *Ann Neurol* 2009; **66**:200–8.

Scoville WB, Milner B. Loss of recent memory after bilateral hippocampal lesion. *J Neurol Neurosurg Psychiatry* 1957; **20**: 11–21.

Seltzer B, Benson DF. The temporal pattern of retrograde amnesia in Korsakoff's disease. *Neurology* 1974; **24**:527–30.

Shill HA, Fife TD. Valproic acid toxicity mimicking multiple system atrophy. *Neurology* 2000; **55**:1936–7.

Sirois F. Delirium: 100 cases. *Can J Psychiatry* 1988; **33**:375–8.

Snead OC, Gibson KM. Gamma-hydroxybutyric acid. *N Engl J Med* 2005; **352**:2721–32.

Spreat S, Behar D, Reneski B *et al.* Lithium carbonate for aggression in mentally retarded persons. *Compr Psychiatry* 1989; **30**:505–11.

Sprofkin BE, Sciarra D. Korsakoff's psychosis associated with cerebral tumor. *Neurology* 1952; **2**:427–34.

Steinberg RB, Gilman DE, Johnson F. Acute toxic delirium in a patient using transdermal fentanyl. *Anesth Analg* 1992; **75**:1014–16.

Steinke W, Sacco RL, Mohr JP *et al.* Thalamic stroke. Presentation and prognosis of infarcts and hemorrhages. *Arch Neurol* 1992; **49**: 703–10.

Swanson AG, Iseri OA. Acute encephalopathy due to water intoxication. *N Engl J Med* 1958; **258**:831–4.

Sweet WH, Talland GA, Erwin FR. Loss of recent memory following section of the fornix. *Trans Am Neurol Assn* 1958; **84**:76–82.

Talland GH, Sweet WH, Ballantine HT. Amnesic syndrome with anterior communicating artery aneurysm. *J Nerv Ment Dis* 1967; **145**: 179–92.

Tanaka Y, Miyazawa Y, Akaoka F *et al.* Amnesia following damage to the mammillary bodies. *Neurology* 1997; **48**:160–5.

Tanaka Y, Miyazawa Y, Hashimoto R *et al.* Postencephalitic focal retrograde amnesia after bilateral anterior temporal lobe damage. *Neurology* 1999; **53**:344–50.

Tatemichi TK, Desmond DW, Prohovnik I *et al.* Confusion and memory loss from capsular genu infarction: a thalamocortical disconnection syndrome? *Neurology* 1992; **42**:1966–79.

Thapar A, Gottesman II, Owen MJ *et al.* The genetic mental retardations. *Br J Psychiatry* 1994; **164**:747–58.

Tomson T, Svanborg E, Wedlund JE. Nonconvulsive status epilepticus: high incidence of complex partial status. *Epilepsia* 1986; **27**: 276–85.

Topliss D, Bond R. Acute brain syndrome after propranolol treatment. *Lancet* 1977; **2**:1133–4.

Towne AR, Waterhouse EJ, Boggs JG *et al.* Prevalence of nonconvulsive status epilepticus in comatose patients. *Neurology* 2000; **54**: 340–5.

Tuma R, DeAngelis LM. Altered mental status in patients with cancer. *Arch Neurol* 2000; **57**:1727–31.

Tune L, Carr S, Cooper T *et al.* Association of anticholinergic activity of prescribed medications with postoperative delirium. *J Neuropsychiatry Clin Neurosci* 1993; **5**:208–10.

Vanden Borre R, Vermote R, Buttiens M *et al.* Risperidone as add-on therapy in behavioral disturbances in mental retardation: a double-blind placebo–controlled cross-over study. *Acta Psychiatr Scand* 1993; **87**:167–71.

Venneri A, Caffarra P. Transient autobiographic amnesia: EEG and single-photon emission CT evidence of an organic etiology. *Neurology* 1998; **50**:186–91.

Veran O, Katnane P, Thomas P *et al.* De novo epileptic confusion in the elderly: a 1-year prospective study. *Epilepsia* 2010; **51**:1030–5.

Victor M, Angevine JB, Mancall EL *et al.* Memory loss with lesions of the hippocampal formation. *Arch Neurol* 1961; **5**:244–63.

Volpe JJ. Cognitive deficits in premature infants. *N Engl J Med* 1991; **325**:276–8.

Von Cramon DY, Hebel N, Schuri U. A contribution to the anatomical basis of thalamic amnesia. *Brain* 1985; **108**:993–1008.

Walker AE. Recent memory impairment in unilateral temporal lesions. *Arch Neurol Psychiatry* 1957; **78**:543–57.

Warren JD, Thompson PD, Thompson PD. Diencephalic amnesia and apraxia after left thalamic infarction. *J Neurol Neurosurg Psychiatry* 2000; **68**:248.

Weizman A, Eldar M, Shoenfeld Y *et al.* Hypercalcemia-induced psychopathology in malignant diseases. *Br J Psychiatry* 1979; **135**:363–6.

Wells CE. Pseudodementia. *Am J Psychiatry* 1979; **136**:895–900.

Welti W. Delirium with low serum sodium. *Arch Neurol Psychiatry* 1956; **76**:559–64.

Westlake EK, Simpson T, Kaye M. Carbon dioxide narcosis in emphysema. *Q J Med* 1955; **24**:155–73.

Wijdicks EF, Miller GM. MR imaging of progressive downward herniation of the diencephalon. *Neurology* 1997; **48**:1456–9.

Williams M, Pennybacker J. Memory disturbances in third ventricle tumours. *J Neurol Neurosurg Psychiatry* 1954; **17**:115–23.

Wilson JTL, Teasdale GM, Hadley DM *et al.* Post-traumatic amnesia: still a valuable yardstick. *J Neurol Neurosurg Psychiatry* 1994; **57**:198–201.

Wilson LM. Intensive care delirium: the effect of outside deprivation in a windowless unit. *Arch Intern Med* 1972; **130**:225–6.

Winblad B, Palmer K, Kivipelto M *et al.* Mild cognitive impairment: beyond controversies, towards a consensus: report of the International Working Group on Mild Cognitive Impairment. *J Intern Med* 2004; **256**:240–6.

Wityk RJ, Goldsborough MA, Hillis A *et al.* Diffusion- and perfusion-weighted brain magnetic resonance imaging in patients with neurologic complications after cardiac surgery. *Arch Neurol* 2001; **58**:571–6.

Wolff HG, Curran D. Nature of delirium and allied states. *Arch Neurol Psychiatry* 1935; **33**:1175–215.

Wong Y-C, Au WL, Xu M *et al.* Magnetic resonance spectroscopy in adult-onset citrullinemia. Elevated glutamine levels in comatose patients. *Arch Neurol* 2007; **64**:1034–7.

Yamamoto T, Katayama Y. Deep brain stimulation for the vegetative state. *Neuropsychol Rehabil* 2005; **15**:406–13.

Yaqub B, Al Deeb S. Heat stroke: aetiopathogenesis, neurological characteristics, treatment and outcome. *J Neurol Sci* 1998; **156**:144–151.

Yildizcli B, Ozyurtkan MO, Batirel HF *et al.* Factors associated with postoperative delirium after thoracic surgery. *Ann Thorac Surg* 2005; **79**:1004–9.

Yoneoka Y, Takeda N, Inoue A *et al.* Acute Korsakoff syndrome following mammillothalmic tract infarction. *AJNR* 2004; **25**: 964–8.

Zafonte RD, Watanabe T, Mann NR. Amantadine: a potential treatment for the minimally conscious state. *Brain Inj* 1998; **12**:617–21.

Ziegler DK, Kaufman A, Marshall HE. Abrupt memory loss associated with thalamic tumor. *Arch Neurol* 1977; **34**:545–8.

Zhang C, Glenn DG, Bell WL *et al.* Gabapentin-induced myoclonus in end stage renal disease. *Epilepsia* 2005; **46**:156–8.

Ziskind AA. Transdermal scopolamine-induced psychosis. *Postgrad Med* 1988; **84**:73–6.

Zola-Morgan S, Squire LR, Amaral DG. Human amnesia and the medial temporal region: enduring memory impairment following a bilateral lesion limited to field CA1 of the hippocampus. *J Neurosci* 1986; **6**:2950–67.

7

Syndromes of disturbances of mood and affect

7.1 DEPRESSION

Depression, as conceived of here, is a syndrome that may result from any one of a large number of underlying causes. Although the most common cause of depression is major depressive disorder, it is critical to subject each depressed patient to a thorough diagnostic evaluation before concluding that major depressive disorder, or perhaps one of the other primary mood disorders discussed below, is the cause.

Clinical features

The syndrome of depression, in its fully developed form, includes not only a depressed or irritable mood, but also other symptoms, as listed in Table 7.1, and each of these is considered in turn.

Mood is depressed or sometimes irritable; some, in addition to these symptoms, may also complain of anxiety. At times, patients may deny a depressed mood but rather complain of a sense of discouragement or perhaps lassitude and a sense of being weighted down. The patient's affect generally mirrors the mood, and there may be copious tears. At times, and especially in those with anxiety, there may be a 'pained' facial expression. Occasionally, patients may attempt to hide their depressed mood by feigning a 'happy' affect, thus creating a 'smiling depression'.

Self-esteem typically sinks during depression, and the workings of conscience may become so prominent as to create an almost insupportable burden of guilt: patients may see their sins multiply in front of them, and in reviewing their past may be blind to their accomplishments and see only their misdeeds. Pessimism settles in and patients see no hope for themselves, either now or in the future.

Suicidal ideation is typically present, and may be either passive or active. Patients may say that they would not mind if they were in a fatal accident or died of some disease. Others may actively seek death, perhaps by hanging, shooting, jumping from bridges, or taking overdoses.

Difficulty with concentration may be profound. Patients may complain that it is as if a 'fog' had settled in, and that they feel dull and heavy headed. Memory seems to fail, and patients may be unable to recall where they put things or what was said. Attempts at cognitive activities often end in failure: patients, in trying to read a book, may read the same paragraph again and again and find themselves unable to absorb or understand what they have read. Things seem too complex, and making decisions may be impossible.

The remaining symptoms, namely anhedonia, anergia, sleep disturbance, appetite change, and psychomotor change, are often referred to, collectively, as the 'vegetative' symptoms of depression, in that they represent disturbances in the basic vegetative functions critical for survival.

Anhedonia represents an inability to take pleasure in things, and patients may complain that nothing arouses or attracts them. Thus, unable to experience pleasure, patients lose interest in formerly pleasurable activities and must force themselves to get through their days, which, to them, seem desolate and lifeless. Libido is especially lost, and patients may withdraw entirely from any sexual activities.

Anergia, or a dearth of energy, may leave patients complaining of fatigue, tiredness, and exhaustion, or of being 'drained'. The anergia may be so extreme that patients are unable to complete routine tasks; some may even be unable to find the energy to get dressed.

Sleep disturbance may manifest with either insomnia or hypersomnia. Insomnia is more common, and may be a particular torment to the patient. Although some complain of what is technically known as 'initial' insomnia, that is to say trouble initiating or falling asleep, a more characteristic complaint is 'middle' insomnia, wherein patients awaken in the middle of the night, and have great difficulty in falling back asleep; some complain that their 'mind' simply will not 'shut off'. Some may also experience 'terminal' insomnia, or, as it is also known, 'early morning awakening', wherein they awaken well before the desired time of arousal and then are unable to fall back asleep. When morning finally does come, patients rise unrefreshed and exhausted, sometimes feeling as if they had

Table 7.1 Symptoms of depression

Depressed or irritable mood
Low self-esteem, guilt, pessimism
Suicidal ideation
Difficulty with concentration or forgetfulness
Anhedonia (lack of interest in formerly pleasurable activities)
Anergia (lack of energy)
Sleep disturbance (insomnia or hypersomnia)
Appetite disturbance (anorexia or increased appetite)
Psychomotor change (agitation or retardation)

not slept at all. Hypersomnia is relatively uncommon: here, patients may sleep for 12, 16, or even 18 hours a day. Remarkably, despite such extremes of sleep, patients do not feel refreshed during their waking hours.

Appetite is typically lost and this may be accompanied by an altered taste. Patients may complain that their food has no taste or has perhaps become unpalatable; some may say that their food tastes like cardboard. Although some patients may force themselves to eat, most cannot and weight loss is typical, and at times extreme. Increased appetite is relatively uncommon, but when it does occur the accompanying weight gain may be impressive.

Psychomotor change tends more to agitation than to retardation. Agitation, when slight, may be confined to a certain sense of inner restlessness. When more severe, there may be hand wringing and restless pacing: patients may complain of being unable to keep still; they may loudly lament their fate; and some may give way to wailing and miserable pleas for help. The tension experienced by these patients may be almost palpable to the observer and yet, despite their pleas, these patients cannot be comforted no matter what is done for them. Psychomotor retardation is less common. Patients may speak slowly and haltingly, and some may become mute, as if the effort to speak were simply too great; if asked, they may report that their thoughts are sluggish and come very slowly. These patients may move very little, and some may become almost completely immobile: efforts to get them up may be met with reluctance, even irritation, and some patients, if left to themselves, may neither bathe nor change their clothes.

There is debate as to how many of these symptoms must be present before a syndromal diagnosis of depression is warranted. Some authorities insist on a large number, and although such an approach yields a relatively 'pure' group, it runs the risk of ignoring patients whose symptoms, although few in number, are severe, even disabling. A 'middle of the road' approach is recommended here, reserving the diagnosis for those who, in addition to a depressed mood, also have at least three of the remaining symptoms noted in Table 7.1.

There is also debate as to the duration of symptoms before the diagnosis is given. Although it is customary to withhold the diagnosis until symptoms have persisted for about 2 weeks, some judgment is required here as there are clearly cases wherein patients have severe and numerous depressive symptoms which, however, are relatively brief in duration. Examples include the premenstrual dysphoric disorder, some medication-induced depressions, and ictal depressions. In practice, most clinicians will relax the time duration in proportion to the increasing severity and number of symptoms.

Etiology

The various causes of depression are listed in Table 7.2, in which they are divided into several groups. The first group includes the *primary mood disorders*, such as major depressive disorder: the disorders in this group account, by far, for the most cases of depression. The next group includes two of the *primary psychotic disorders*, namely schizoaffective disorder and the postpsychotic depression of schizophrenia. Next are *toxic depressions*, which may be either medication induced, for example the depression seen with high-dose prednisone, or due to abuse of substances or toxins, as may be seen in chronic alcoholism. *Metabolic depressions* are considered next, including such disorders as obstructive sleep apnea. *Medication or substance withdrawal depressions* follow, and include depressions occurring upon discontinuation of long-term treatment with anticholinergic medications or as may be seen during withdrawal from stimulants. *Endocrinologic disorders* constitute a very important cause of depression, and include such familiar conditions as hypothyroidism and Cushing's syndrome. Various *neurodegenerative and movement disorders* may cause depression, and indeed approximately one-third of patients with Parkinson's disease will be so affected. Depression may also be seen in a large number of *other intracranial disorders*, for example in the syndrome of poststroke depression. Each of these groups is considered in more detail below, beginning with the primary or idiopathic disorders.

PRIMARY MOOD DISORDERS

As noted earlier, the disorders in this group constitute, by far, the most common causes of depression. As a group, however, they constitute 'rule-out' diagnoses, and a diagnosis of one of these disorders should generally not be made until one is reasonably certain that the depressive syndrome in question is not caused by one of the other disorders in Table 7.2. It must also be kept in mind that it is not at all uncommon that in any given patient more than one disorder may be present. For example, a patient with well-established major depressive disorder, with a long history of recurrent depressions, may come down with a depression during a course of prednisone treatment, which remits shortly after treatment is discontinued. Furthermore, some depressions may be multifactorial: consider a patient, again with well-established major depressive disorder, who, shortly after treatment with metoclopramide, develops a depression that persists for weeks or longer after the metoclopramide is discontinued: in this instance, it is reasonable to assume that the medication triggered a new depressive episode of the major depressive disorder, which then persisted.

Major depressive disorder and bipolar disorder are both characterized by recurrent episodes of depression, the two disorders being distinguished by the fact that in bipolar disorder one also sees, at some point in the patient's history, a manic episode, whereas in major depressive disorder, manic

Table 7.2 Causes of depression

Primary mood disorders
Major depressive disorder (Chapter 23.3)
Bipolar disorder (Chapter 23.1)
Dysthymia (Chapter 23.4)
Premenstrual dysphoric disorder (Chapter 23.5)
Postpartum blues (Chapter 23.7)
Postpartum depression (Chapter 23.6)

Primary psychotic disorders
Schizoaffective disorder (Chapter 22.2)
Postpsychotic depression in schizophrenia (Chapter 22.1)

Toxic depressions
Medication induced
- Prednisone (Wolkowitz *et al.* 1990)
- Alpha-interferon (Fried *et al.* 2002; Krauss *et al.* 2003; Raison *et al.* 2005; Torriani *et al.* 2004)
- Beta 1b-interferon (Neilley *et al.* 1996)
- Metoclopramide (Friend and Young 1997)
- Tetrabenazine (Chapter 9.10)
- Pimozide (Bloch *et al.* 1997)
- Propranolol (Petrie *et al.* 1982; Pollack *et al.* 1985)
- Nifedipine (Hullett *et al.* 1988)
- Cimetidine (Billings *et al.* 1981)
- Ranitidine (Billings and Stein 1986)
- Subdermal estrogen–progestin (Wagner 1996; Wagner and Berenson 1994)
- Alpha-methyl dopa (DeMuth and Ackerman 1983)
- Reserpine (Jensen 1959; Quetsch *et al.* 1959)
- Levetiracetam (Mula *et al.* 2005; Wier *et al.* 2006)
- Isotretinoin (Wysowski *et al.* 2001)
- Bismuth (Chapter 14.9)

Substances of abuse or toxins
- Chronic alcoholism (Chapter 28.5)
- Lead intoxication (Chapter 14.11)

Metabolic depressions
Obstructive sleep apnea (Chapter 20.8)
Chronic hypercalcemia (Linder *et al.* 1988; Petersen 1968; Reinfrank 1961)
Vitamin B12 deficiency (Chapter 14.1)
Pellagra (Chapter 14.3)
Pancreatic cancer (Holland *et al.* 1986; McDaniel *et al.* 1995)

Medication or substance withdrawal depressions
Cholinergic rebound (Chapter 29.7)
Stimulants (Chapter 28.1, 2)
Anabolic steroids (Pope and Katz 1988)

Endocrinologic disorders
Hypothyroidism (Chapter 17.4)
Hyperthyroidism (Chapter 17.3)
Cushing's syndrome (Chapter 17.1)
Adrenocortical insufficiency (Chapter 17.2)
Hyperaldosteronism (Malinow and Lion 1979)
Hyperprolactinemia (Mattox *et al.* 1986)

Neurodegenerative and movement disorders
Parkinson's disease (Chapter 9.5)
Dementia with Lewy bodies (Chapter 9.6)
Hereditary mental depression with parkinsonism (Perry *et al.* 1975)
Huntington's disease (Chapter 9.10)
Alzheimer's disease (Chapter 9.1)
Multi-infarct dementia (Chapter 11.1)

Other intracranial disorders
Stroke (see text for references)
Traumatic brain injury (Chapter 8.5)
Multiple sclerosis (Chapter 19.1)
Epilepsy-associated depression (see text for references)
- Ictal depression
- Interictal depression

Tumors (Chapter 21.1)
Normal pressure hydrocephalus (Chapter 21.3)
Fahr's syndrome (Chapter 14.22)
Systemic lupus erythematosus (Chapter 19.3)
Limbic encephalitis (Chapter 19.8)
Tertiary neurosyphilis (Chapter 15.14)
New-variant Creutzfeldt–Jakob disease (Chapter 16.2)
Down's syndrome (Chapter 10.4)

episodes never occur. Given that bipolar disorder may commence with one, or several, episodes of depression before the first episode of mania occurs, one must, in evaluating a patient who has had only depressive episodes, allow a lengthy period of observation to pass before making a firm diagnosis of major depression. Statistically speaking, in patients with bipolar disorder, the first episode of mania will, in over 90 percent of cases, occur within 10 years of the first depressive episode or by the time five or more episodes of depression have occurred, whichever comes first (Dunner *et al.* 1976).

Although not as reliable, certain clinical characteristics of the depressive episode may also suggest whether that depressive episode is occurring on the basis of bipolar disorder or major depression. Specifically, depressive episodes of bipolar disorder are, in contrast with those of major depression, more likely to have an acute onset (over weeks rather than months) (Winokur *et al.* 1993), are more likely to be characterized by hypersomnia, increased appetite, and psychomotor retardation, and are also more likely to be accompanied by delusions and hallucinations (Guze *et al.* 1975).

Dysthymia is characterized by chronic, low-level, and generally fluctuating depressive symptoms that persist for 2 years or more.

Premenstrual dysphoric disorder is characterized by relatively brief depressions that occur with each menstrual cycle, beginning anywhere from hours to 1½ weeks before the onset of menses and remitting spontaneously 2–3 days after menstrual flow begins.

Postpartum blues is immediately suggested by its onset within the first few days postpartum and by its characteristic lability of affect.

Postpartum depression is distinguished from the postpartum blues by its later onset, with a latency of at least several weeks between delivery and the onset of the depression.

PRIMARY PSYCHOTIC DISORDERS

Schizoaffective disorder is characterized by chronic, persistent psychotic symptoms with the 'superimposition' of episodes of either depression or of mania, during either of which the pre-existing psychotic symptoms undergo a significant exacerbation. Differentiating schizoaffective disorder from bipolar disorder may be difficult (Pope *et al.* 1980), and the reader is directed to Chapter 22.2 for a discussion of this.

Postpsychotic depression in schizophrenia is seen in about one-third of patients with schizophrenia. In these patients, after a more or less complete remission of psychotic symptoms (whether occurring spontaneously or by virtue of antipsychotic treatment), a depressive episode occurs, typically within a matter of months. In contrast with schizoaffective disorder, however, there is during this postpsychotic depression *no* exacerbation of psychotic symptoms.

TOXIC DEPRESSIONS

Toxic depressions may be either medication induced or secondary to substances of abuse or actual toxins. Medications are considered first.

Of the various medications listed in Table 7.2, the most common offender is prednisone, given in relatively high doses of 40–60 mg/day or more; similar effects may be expected with other corticosteroids. α-Interferon, as used in the treatment of hepatitis C and melanoma, is another major offender, inducing depression within a few weeks in approximately one-third of all patients: the effect here is so profound that prophylactic treatment is recommended, with one study demonstrating effectiveness for paroxetine (Musselman *et al.* 2001). Depression has also been associated with β-1b-interferon, but this is comparatively rare. Metoclopramide, although producing depression in only a very small percentage of patients who take it, is, however, so commonly prescribed that it too emerges as a frequent cause. Tetrabenazine is prescribed primarily for Huntington's disease and for tardive dyskinesia, and may cause depression in up to roughly one-third of cases. The other medications listed in Table 7.2 only rarely cause depression, and before one ascribes any given depression to them it is necessary to demonstrate a close temporal relationship between the initiation of treatment and the onset of the depression and, ideally, a similarly close relationship between discontinuation and remission of the depression. This list of medicines should not be considered comprehensive, and the clinician should maintain a high index of suspicion for a medication-induced depression whenever such temporal relationships can be established, and other, more common, causes of depression appear absent.

Chronic alcoholism is commonly associated with depression, and indeed roughly two-thirds of newly admitted alcoholics will be so affected (Davidson 1995). Importantly, the symptoms typically resolve spontaneously during the first 4 weeks of abstinence (Brown *et al.* 1995).

Lead intoxication may be characterized by depression, and the diagnosis should be considered in cases wherein the depression is accompanied by a metallic taste, colicky abdominal pain, and a motor peripheral neuropathy with wrist or, less commonly, foot drop.

METABOLIC DEPRESSIONS

Obstructive sleep apnea may cause a depression marked by indecisiveness. The diagnosis is suggested by a history of prominent snoring, and by a relief of symptoms upon successful treatment of the sleep apnea.

Chronic hypercalcemia, as seen in hyperparathyroidism, may be accompanied by depression and may in some cases present with depression (Gatewood *et al.* 1975; Karpati and Frame 1964).

Both vitamin B12 deficiency and niacin deficiency may, rarely, cause depression. In one case of B12 deficiency-induced depression, the only clue was a concurrent macrocytic anemia (Fraser 1960). Chronic niacin deficiency causes pellagra, which, in one case, presented with a combination of depression and the familiar rash and diarrhea (Hardwick 1943).

Pancreatic cancer is associated with depression and may indeed present with depression (Fras *et al.* 1967; Rickles 1945). The eventual appearance of abdominal pain suggests the correct diagnosis.

MEDICATION OR SUBSTANCE WITHDRAWAL

Cholinergic rebound may occur after abrupt discontinuation of long-term treatment with anticholinergic agents, such as tricyclic antidepressants: shortly thereafter, patients fairly abruptly develop a depression accompanied by gastrointestinal symptoms, such as nausea and vomiting.

Stimulant withdrawal, for example after extended use of either amphetamines or cocaine, is typically characterized by a depression that tends to resolve within days to a few weeks.

Anabolic steroids may be taken chronically by athletes, and discontinuation may be followed by a depression that tends to remit spontaneously within weeks or months.

ENDOCRINOLOGIC DISORDERS

The endocrinologic disorders constitute a particularly important group not only because they are a relatively common cause of depression, but also because they are eminently treatable. Of them, hypothyroidism is perhaps the most important.

Hypothyroidism may cause severe depression. In addition to such features as weight gain, hair loss, dry skin, and voice change, the condition is also suggested by prominent fatigue, sluggishness, and drowsiness (Nickel and Frame 1958).

Hyperthyroidism, although generally associated with anxiety and agitation, may, conversely, cause depression in up to one-third of cases, and the correct diagnosis may be suggested by the presence of weight loss in the face of increased appetite or autonomic signs such as tremor and tachycardia. Apathetic hyperthyroidism is a condition seen in elderly patients with hyperthyroidism, and is distinguished from the more common presentation of hyperthyroidism seen in younger patients by marked apathy and the relative absence of tremor. In some cases, rather than apathy, however, one may see depression, and important clues to the correct diagnosis include atrial fibrillation or congestive heart failure.

Cushing's syndrome, as seen with pituitary or adrenal tumors, may be characterized by depression, as noted by Cushing himself (Cushing 1932), and this has been noted in from one-half to two-thirds of all cases. Suggestive features include weight gain, moon facies, hirsutism, acne, a buffalo hump, violaceous abdominal striae, hypertension, and diabetes mellitus.

Adrenocortical insufficiency, when chronic, may cause depression, and is suggested by associated features such as nausea, vomiting, abdominal pain, and postural dizziness.

Hyperaldosteronism, as may occur with adrenal tumors, very rarely, may cause depression: in one case the diagnosis was suggested by hypokalemia and weakness and cramping of the legs.

Hyperprolactinemia, as may be seen with prolactinomas, may cause depression, which, in turn, remits with treatment with bromocriptine. The diagnosis may be suggested by galactorrhea or erectile dysfunction.

NEURODEGENERATIVE AND MOVEMENT DISORDERS

Parkinsonian disorders capable of causing depression include Parkinson's disease, dementia with Lewy bodies, and a very rare disorder known as hereditary mental depression with parkinsonism. Parkinson's disease is associated with depression of varying degrees of severity in up to 40 percent of cases, and dementia with Lewy bodies in close to half. Hereditary mental depression with parkinsonism is a very rare familial disorder that presents with depression, followed, years later, by parkinsonism.

Huntington's disease causes depression in roughly one-half of all cases by the end of the second year of illness, and the suicide rate in this disorder is particularly elevated.

Alzheimer's disease, suggested by the gradual onset of a dementia with prominent amnestic features, is accompanied by prominent depressive symptoms in perhaps one-third of all cases.

Multi-infarct dementia, suggested by its stepwise course and prominent focal findings, produces depression in over one-half of all sufferers (Cummings *et al.* 1987).

OTHER INTRACRANIAL DISORDERS

Of the other intracranial disorders capable of causing depression, the most common are stroke and traumatic brain injury. Other somewhat less common causes are multiple sclerosis and epilepsy. Finally, there is a miscellaneous group that includes tumors, hydrocephalus, etc.

Poststroke depression, which can be quite severe (Lipsey *et al.* 1986), may develop in the weeks or months after stroke in roughly 40 percent of all patients. Most (Hermann *et al.* 1993; Robinson *et al.* 1983, 1984, 1985), but not all (House *et al.* 1990; MacHale *et al.* 1998; Stern and Bachman 1991), studies note that such depression is more likely with infarctions in the left frontal lobe than elsewhere. Such patients typically recover within a year (Astrom *et al.* 1993; Robinson *et al.* 1987). Poststroke depression has also been associated with infarction of the left basal ganglia (Morris *et al.* 1996; Starkstein *et al.* 1987a, 1988a).

Traumatic brain injury may be associated with depression in up to one-half of all patients during the first 2 years of illness, and, in the natural course of events, may either undergo a spontaneous remission or become chronic.

Multiple sclerosis causes depression in roughly one-quarter of all patients, and the occurrence of depression, rather than correlating with the overall degree of disability, is associated with plaque formation in specific locations, including the frontal lobes, right temporal lobe, and the left arcuate fasciculus. As with other signs of multiple sclerosis, the depression may also have a relapsing and remitting course (Dalos *et al.* 1983).

Epilepsy may be associated with depression, which may occur ictally or interictally. Ictal depressions, as discussed in Chapter 8.3, are simple partial seizures, which in turn may be brief, lasting only minutes, or, in the case of simple partial status epilepticus, quite prolonged, lasting from weeks to months. Although some patients with ictal depression may have associated symptoms which suggest the correct diagnosis, such as concurrent olfactory hallucinations, in some cases the depression may constitute the sole symptomatology of the seizure, and in such cases the best clue to the diagnosis is the paroxysmal onset of the depression, with patients reporting a sudden plunge into depression, with the full syndrome developing in seconds. Interictal depression occurs in roughly one-fifth of all patients with uncontrolled epilepsy (Guarnieri *et al.* 2009; Jones *et al.* 2005; Mendez *et al.* 1986), especially those with recurrent complex partial seizures (Indaco *et al.* 1992; Perini *et al.* 1996).

Brain tumors may cause depression, as has been noted with tumors of the frontal, and, less commonly, temporal lobes.

Normal pressure hydrocephalus, in a small minority of cases, may be characterized by prominent depression, and the diagnosis here is suggested by the concurrent urinary symptoms and 'magnetic' gait; very rarely, however, depression may constitute the presenting symptomatology of this disorder (Jones 1993).

Fahr's syndrome may, rarely, present with depression, which is eventually joined by more typical signs such as dementia (Slyter 1979) and parkinsonism (Trautner *et al.* 1988); calcification of the basal ganglia, apparent on computed tomography, strongly suggests this diagnosis.

Systemic lupus erythematosus can definitely cause depression (Dennis *et al.* 1992). What is disputed, however, is how frequently it does so: some have found depression to be common in lupus (Ainiala *et al.* 2001; Brey *et al.* 2002; Ganz *et al.* 1972; Lim *et al.* 1988; Miguel *et al.* 1994), whereas older studies have not (Guze 1967; Hugo *et al.* 1996). The diagnosis should be suspected in patients with arthralgia, rashes, and constitutional symptoms.

Limbic encephalitis may present with a depression, which is later joined by other, more typical evidence of the encephalitis, such as delirium or dementia.

Tertiary neurosyphilis may cause a dementia (general paresis of the insane – GPI) that is characterized not only by cognitive deficits, but also by depression (Dewhurst 1969;

Gomez and Aviles 1984) in almost one-fifth of all cases (Storm-Mathisen 1969).

New-variant Creutzfeldt–Jakob disease may present with depression, but eventually, over weeks or months, this will be joined by other symptoms, such as ataxia, delirium, or dementia.

Down's syndrome, immediately suggested by mental retardation and a characteristic facies, may, in adults, cause depression.

Differential diagnosis

The first item on the differential is a 'normal' or 'reactive' depression: depression, as everyone knows, is a normal reaction to the adverse events of life, especially losses, and the prototypical 'normal' depression is bereavement after the death of a loved one. The differential diagnostic question comes up, however, in some cases, about whether the depression following upon an adverse event is entirely 'normal' or perhaps is multifactorial, involving not only a 'normal' reaction but also one of the disorders capable of causing depression described earlier. Some clues should immediately suggest a multifactorial origin. First, one looks to see whether the severity of the depression is disproportionate to the severity of the adverse event: although a severe depression is not unexpected after the death of a beloved spouse, it would not be expected after a minor financial reversal. Next, one looks to duration: 'normal' depressions typically clear up within 6 months, or a year at the most. Finally, one examines the patient to see whether or not delusions, specifically delusions of guilt or poverty or nihilistic delusions, are present, as these are not found in normal depressions. Clearly, whenever one of these diagnostic clues is present, one should closely evaluate the patient to see if one of the disorders is present. A case, however, may also be made for pursuing a search for one of the disorders even when none of these 'clues' are present, given that the appearance of the depression after the adverse event might well be merely a non-causal coincidence. For example, hypothyroidism may just as well occur after the death of a loved one as at any other time.

Before leaving this discussion of 'normal' depression, some comments are in order regarding the question whether normal depressions should be treated with antidepressants. Some authors rebel at this idea, decrying the 'medicalization' of a normal human experience and asserting that the lives of people would be diminished if they were 'robbed' of the opportunity to go through the grieving 'process'. Although there is much to say for this argument, it must also be admitted that sometimes normal experiences can be intolerable, and that when this occurs, medical treatment is appropriate. To take an absolutely commonplace example, consider pain: it is a normal and a very effective teacher at times, but very few would insist that someone with a broken bone should forgo treatment with an analgesic. Normal bereavement can also become intolerable, and in such cases treatment with an antidepressant may be quite effective, as was shown in one double-blinded study of nortriptyline (Reynolds *et al.* 1999).

The next item on the differential is akinesia. Akinesia, as discussed in Chapter 4.9, is most commonly seen in patients who are treated with antipsychotics, especially first-generation agents, and is characterized by a sluggishness of thought and an overall dulling of emotion, with some patients complaining of feeling like a 'zombie'. Two points serve to differentiate this from depression: first, there is little in the way of changes in sleep or appetite, and actually little depressed mood, either; second, akinesia responds promptly to treatment with anticholinergics, such as benztropine, which have no effect on depression.

Treatment

Treatment, whenever possible, is directed at the underlying cause (e.g. BIPAP [bilevel continuous airway pressure] for obstructive sleep apnea, levothyroxine for hypothyroidism). If this is either not possible or ineffective, then consideration should be given to an antidepressant, following the guidelines discussed in the specific chapters.

7.2 APATHY

Apathy, or indifference, is very common in geriatric patients, and, as noted below, may occur secondary to a variety of causes.

Clinical features

Normally, thoughts of doing something, whether thinking a problem through, getting dressed, going to a movie, or whatever, when they appear, are invested with a greater or lesser degree of motivation to carry out the plan. In apathetic patients, however, such thoughts appear, as it were, stillborn, and without sufficient associated motivation to impel the thinker to carry them into action. In apathetic patients, there is such a pervasive sense of not 'caring' enough that many lapse into inactivity, and appear detached and indifferent.

Etiology

Apathy may appear in a more or less isolated fashion or may occur as part of another syndrome, such as dementia, the frontal lobe syndrome, or depression.

Apathy in a more or less isolated form may be seen in stroke, acquired autoimmune deficiency syndrome (AIDS) (Paul *et al.* 2005), myotonic muscular dystrophy (Rubinsztein *et al.* 1998), and hyperthyroidism in the elderly ('apathetic' hyperthyroidism [Brenner 1978]). In the case of stroke, apathy has been associated with ischemic infarction involving the posterior limb of the internal capsule (Starkstein *et al.* 1993), both thalami (Catsman-Berrevoets and von Harskamp 1988), and the right frontosubcortical circuitry (Brodaty *et al.* 2005). Apathy has also been noted as a side-effect of treatment with antipsychotics, and, with chronic use, selective serotonin reuptake inhibitors (SSRIs) (Hoehn-Saric *et al.*

1990). Finally, isolated apathy may occur in geriatric nursing home patients in whom it appears to result from chronic understimulation.

Dementia may be accompanied by apathy, and this has been noted in various neurodegenerative disorders (Levy *et al.* 1998), most notably Alzheimer's disease (Starkstein *et al.* 2001, 2005, 2006), frontotemporal dementia (Levy *et al.* 1996), Parkinson's disease (Isella *et al.* 2002; Pluck and Brown 2002), progressive supranuclear palsy (Litvan *et al.* 1996), and Huntington's disease (Burns *et al.* 1990b). Although in many of these demented patients apathy appears in association with a depressive syndrome, there is a definite, although minority, group, wherein apathy appears without any associated depressed mood, lack of energy, or other symptoms typical of a depressive syndrome. Apathy is seen commonly after traumatic brain injury, and, in these cases, is more common when the right hemisphere is involved, especially its subcortical portions (Andersson *et al.* 1999).

The frontal lobe syndrome, as discussed in Chapter 8.2, may, along with disinhibition and perseveration, also include apathy.

Depressive syndromes often include apathy as a symptom along with depressed mood, fatigue, insomnia, etc. Such syndromes may, as noted above, be seen in various dementing disorders, or more commonly may appear as a manifestation of major depressive disorder (Feil *et al.* 2003).

Psychosis, as for example schizophrenia (Roth *et al.* 2004), may also include apathy as a symptom, along with delusions, hallucinations, etc.

Differential diagnosis

Abulia, superficially, appears almost identical to apathy, in that in both cases, afflicted patients are inactive. There are, however, significant differences. In abulia, there simply are no thoughts, plans, or inclinations: the 'mental horizon' is empty and undisturbed. By contrast, thoughts etc. do occur with apathy but without sufficient associated motivation to carry the patient into action. Furthermore, when abulic patients are subject to supervision, they do carry out tasks, and typically do so at a normal rate, provided that the supervision is ongoing; by contrast, apathetic patients may shirk or withdraw when told to do something, and comply only half-heartedly, if that, with ongoing supervision.

Bradyphrenia and bradykinesia, as seen in parkinsonian conditions, may also appear similar to apathy in that, to a brief inspection, there is little activity. The differential, however, is relatively easy if one only observes the patient for a while: given enough time, the bradykinetic patient will get the job done, as here it is not a lack of motivation but rather a slowing down of all activities, whereas, by contrast, the apathetic patient will remain inactive.

Depressed mood, when accompanied by anhedonia, may appear very similar to apathy. Depressed mood, however, is a definite mood: patients feel sad, 'blue', or internally oppressed. By contrast, apathetic patients typically experience an absence of any particular mood, except, of course, for a sense of indifference. Furthermore, depressed patients who have lost their hedonic capacity may nevertheless at times feel some motivation to act but then fail to carry through as they realize that there will be no pleasure in finishing the task. By contrast, apathetic patients, lacking any motivation to act at all, simple fail to get started.

Treatment

In cases where apathy appears as part of one of the other syndromes (dementia, frontal lobe syndrome, depression, or psychosis), the first step should be to treat this underlying cause, as this may lead to relief of the apathy also, and indeed double-blinded work has demonstrated that in the case of Alzheimer's disease, use of a cholinesterase inhibitor will relieve apathy (Drijgers *et al.* 2009). Should SSRIs or antipsychotics (especially first-generation agents) be used, one must be alert to their potential, as noted earlier, to cause apathy. Should such treatment fail to relieve the apathy, or in cases where apathy is occurring in an isolated fashion, consideration may be given to symptomatic treatment aimed at apathy itself.

The most commonly used symptomatic treatment for apathy is methylphenidate. Open work has shown this to be effective in stroke (Galynker *et al.* 1997; Spiegel *et al.* 2009; Watanabe *et al.* 1995), and double-blinded studies have demonstrated its effectiveness in both Alzheimer's disease (Hermann *et al.* 2008) and among institutionalized geriatric patients (Kaplitz 1975). Methylphenidate should be started at a low dose, for example 2.5 mg in the morning and 2.5 mg in the early afternoon, and increased gradually every 2 or 3 days, in similar increments, until a response is seen, limiting side-effects occur, or a maximum dose of 60 mg/day is reached: in many elderly patients, a total daily dose of 10–20 mg generally suffices.

Apathy occurring after traumatic brain injury, in open work, responded to both bromocriptine (Powell *et al.* 1996) and selegeline (Newburn and Newburn 2005), and in one, blinded, '*N* of 1' study, to amantadine (Van Reekum *et al.* 1995).

7.3 MANIA

Mania is one of the most distinctive syndromes in neuropsychiatric practice: once seen, never forgotten. Although most commonly caused by bipolar disorder, mania may occur secondary to a host of other causes, as described below.

Clinical features

The syndrome of mania, following the elegant descriptive study of Carlson and Goodwin in 1973 (Carlson and Goodwin 1973), may be divided into three stages. Stage I mania, also known as *hypomania*, is present in all cases, and, in its fully developed form includes all the symptoms listed in Table 7.3. Most patients, over time, will enter into stage II mania, known as *acute mania*, and this transition is marked by an exacerbation of the symptoms seen in hypomania and

Table 7.3 The symptoms of hypomania

Heightened mood (either euphoric or irritable)
Increased energy
Decreased need for sleep
Pressure of speech
Flight of ideas
Distractibility
Hyperactivity

by the appearance of delusions. A minority of patients will progress further, into stage III mania, known also as *delirious mania*, and here one finds confusion, hallucinations, worsening delusions, and an overall disintegration of behavior. As the manic syndrome gradually resolves, patients tend, as it were, to retrace the symptomatic steps that marked the evolution of the syndrome: thus, a patient who reached stage III would gradually settle back into stage II, then stage I, only finally to experience a remission of symptoms. The onset of a manic syndrome in bipolar disorder tends to be fairly acute, over perhaps days to a week; the range here, however, is wide, from gradual onsets spanning months to hyperacute ones lasting hours or less. The overall duration of a manic episode of bipolar disorder ranges from weeks to months. In mania occurring secondary to other causes, the mode of onset and the duration reflect those causes.

Additional studies have largely backed up this division by Carlson and Goodwin into stage I (Abrams and Taylor 1976; Carlson and Strober 1978; Clayton *et al.* 1965; Taylor and Abrams 1973, 1977; Winokur and Tsuang 1975; Winokur *et al.* 1969), stage II (Loudon *et al.* 1977; Rosenthal *et al.* 1979, 1980; Taylor and Abrams 1973, 1977), and stage III (Black and Nasrallah 1989; Bowman and Raymond 1931; Brockington *et al.* 1980; Carlson and Strober 1978; Rosenthal *et al.* 1979, 1980; Taylor and Abrams 1973, 1977; Winokur 1984).

Each one of the three stages of mania is described further below; in several instances quotations are provided from Emil Kraepelin's (Kraepelin 1921) masterful *Manic-depressive Insanity and Paranoia.*

STAGE I MANIA: HYPOMANIA

Of the symptoms of hypomania listed in Table 7.3, a heightened mood is present in all cases and although most patients will have all of the others, some may play only a relatively minor part in the clinical picture.

Heightened mood

Heightened mood may be either predominantly euphoric or irritable.

Euphoric patients are full of jollity and cheerfulness. Although at times selfish and pompous, their mood is nevertheless quite 'infectious'. They joke, make wisecracks and delightful insinuations, and those around them often get quite caught up in the spirit, always laughing with these patients and not at them. Indeed, when physicians find themselves unable to suppress their laughter when interviewing a patient, the diagnosis of mania should be seriously considered. Self-esteem and self-confidence are greatly increased. Inflated with their own grandiosity, patients may boast of fabulous achievements and lay out plans for even grander accomplishments in the future. Such patients rarely recognize that anything is wrong with them, and although their judgment is obviously impaired they have little or no insight into their condition. Kraepelin (1921) noted that the manic patient:

> is in imperturbable good temper, sure of success, 'courageous', feels happy and merry, not rarely overflowingly so, wakes up every morning 'in excellent humor'. He sees himself surrounded by pleasant and aristocratic people, finds complete satisfaction in the enjoyment of friendship, of art, of humanity; he will make everyone happy, abolish social wretchedness, convert all in his surroundings. For the most part an exuberant unrestrained mood inclined to practical jokes of all kinds is developed. Occasionally there is developed a markedly humorous trait, the tendency to look at everything and every occurrence from the jocular side, to invent nicknames, to make fun of himself and others. A patient called himself a 'thoroughbred professional fool'; another declared that the hospital was a 'nerve-ruining institution'; a third stated that he was a 'poet, cattle-driver, author, tinker, teacher, popular reformer, chief anarchist and detective'.

Irritable patients are loud, insistent, demanding, and intolerant, and the threat of violence hangs about them as a malignant fog. Kraepelin noted that such a patient is:

> dissatisfied, intolerant, fault-finding, especially in intercourse with his immediate surroundings, where he lets himself go; he becomes pretentious, positive, regardless, impertinent and even rough, when he comes up against opposition to his wishes and inclinations; trifling external occasions may bring about extremely violent outbursts of rage. In his fury he thrashes his wife and children, threatens to smash everything to smithereens, to run amuck, to set the house on fire, abuses the 'tribe' of his relatives in the most violent language, especially when under the influence of alcohol.

Increased energy

Energy is greatly, even immensely increased: patients are on the go, busy, and involved throughout the day. They wish to be a part of life and to be more involved in the lives of those around them. They are strangers to fatigue and still quite active when their companions, exhausted, plead for sleep.

Decreased need for sleep

Decreased need for sleep typically accompanies this increased energy. The patient, as described by Kraepelin, 'cannot stay long in bed; early in the morning, even at four o'clock he gets up, he clears out lumber rooms, discharges business that was in arrears, undertakes morning walks, excursions'.

Pressure of speech

Patients with pressured speech, as noted by Kraepelin, 'talk a great deal, hastily, in loud tones, with great verbosity and prolixity'. Speech becomes imperious, incredibly rapid, and almost unstoppable, and listeners may feel veritably deluged by the torrent of words. Occasionally, patients may succumb to others' protests to stop and may be able to keep silent and withhold their speech, but such respites typically do not hold long and soon the dam bursts once again.

Pressured speech is often accompanied by pressure of thought, and patients may complain of racing thoughts. Kraepelin noted that 'thoughts come of themselves, obtrude themselves, impose upon the patients. "I cannot grasp all the thoughts which obtrude themselves," said a patient. "It is so stormy in my head," declared another, "everything goes pell-mell." "My thoughts are all tattered." "I am not master over my thoughts."'

Flight of ideas

In flight of ideas, the patients' train of thought is characterized by abrupt leaps from one topic to another. When mild, the connections between these various topics, although perhaps tenuous, may nevertheless be 'understandable' to the listener. In higher grades, however, the connections may seem to lack any logic, and may come to depend more and more on puns or word plays.

Distractibility

For distractible patients, other conversations, or events, are like glittering jewels that they must attend to, take as their own, or furiously admire, although peripheral to their present purposes. In listening to patients, one may find that a fragment of another conversation has suddenly been interpolated into their flight of ideas or there may be an absolutely abrupt change of topic triggered by a seemingly irrelevant event: one patient, in the midst of an exultant tirade, suddenly stopped and declared his unbounded admiration for the physician's tie, paused for a matter of seconds in great satisfaction, then proceeded with his previous rush of speech.

Hyperactivity

Hyperactivity, which might just as well be termed pressure of activity, follows from the patients' grandiosity and increased energy. Patients may enter into business arrangements with unbounded and completely uncritical enthusiasm. Ventures are begun, stocks are bought on a hunch, money is loaned out without collateral, and when the family fortune is spent, manic patients, undaunted, may seek to borrow more money for yet another prospect. Spending sprees are also quite typical. Clothes, furniture, and cars may be bought; the credit card is pushed to the limit, and another one is obtained; checks, without any foundation in the bank account, are written with alacrity. Kraepelin (1921) noted that such a patient is swelled by a 'need to get out of himself, to be on more intimate terms with his surroundings, to play a part'. It causes him:

> to change about his furniture, to visit distant acquaintances, to take himself up with all possible things and circumstances, which formerly he never thought about. Politics, the universal language, aeronautics, the women's question, public affairs of all kinds and their need of improvement, give him employment. A physician advertised about 'original sin, Genesis, natural selection and breeding'. Another patient drove about in a cab and distributed pictures of the saints. The patient enters into numerous engagements, suddenly pays all his business debts without it being necessary, makes magnificent presents, builds all kinds of castles in the air, and with swift enthusiasm precipitates himself in daring undertakings much beyond his powers. He has 16,000 picture post-cards of his little village printed . . .

STAGE II MANIA: ACUTE MANIA

The transition from hypomania to acute mania is marked by a severe exacerbation of the symptoms seen in hypomania and by the appearance of delusions, typically delusions of grandeur. Kraepelin noted 'The patient asserts that he is descended from a noble family, that he is a gentleman; he calls himself a genius, the Emperor William, the Emperor of Russia, Christ; he can drive out the devil. A patient suddenly cried out on the street that he was the Lord God . . .'. Delusions of persecution may also occur, especially in patients with irritability. Patients may assert that their failures are not their own but are the result of treacheries and betrayals by family members or business colleagues. They are persecuted by those jealous of their grandeur; they are pilloried and crucified by their enemies. Terrorists have set a watch on their houses and seek to destroy them before they can ascend their thrones. Occasionally, in addition to these delusions, there may be transitory hallucinations. Grandiose patients may hear a chorus of angels; persecuted patients may hear the resentful muttering of the envious crowd.

With the intensification of hypomanic symptoms and the appearance of delusions, the overall behavior of these patients may become quite extravagant. Kraepelin (1921) noted that patients may:

> run out of the house in a shirt, go to church in a petticoat, spend the night in a field of corn, give away their property, disturb the service in church by screaming and singing, kneel and pray on the street, fire a pistol in the waiting-room, put soap and soda in the food, try to force their way into the palace, throw objects out at the window. A female patient jumped into the carriage of a prince for a joke . . . A male patient appropriated the property of others in taverns. Another appeared in the court of justice in order to catch a murderer.

STAGE III MANIA: DELIRIOUS MANIA

The transition to delirious mania is marked by the appearance of confusion, more hallucinations, and an intense exacerbation of all the symptoms seen in stage II. There may be a dream-like clouding of consciousness and patients may become unaware of where they are; incoherence may appear and, as described by Kraepelin, 'Their linguistic utterances alternate between inarticulate sounds, praying, abusing, entreating, stammering, disconnected talk, in which clang-associations, senseless rhyming, diversion by external impressions, persistence of individual phrases, are recognized'. Hallucinations become prominent, and as further described by Kraepelin, 'The patient sees heaven open, full of camels and elephants, the King, his guardian-angel, the Holy Ghost; the devil has assumed the form of the Virgin Mary. The ringing of bells is heard, shooting, the rushing of water, a confused noise; Lucifer is speaking; the voice of God announces to him the day of judgment, redemption from all sins'.

With this evolution into stage III, overall behavior becomes very fragmented, and patients may 'become stupefied, confused, bewildered' and may:

> dance about, perform peculiar movements, shake their head, throw the bedclothes pell-mell, are destructive, pass their movements under them, smear everything, make impulsive attempts at suicide, take off their clothes. A patient was found completely naked in a public park. Another ran half-clothed into the corridor and then into the street, in one hand a revolver in the other a crucifix.

Etiology

The various causes of mania are listed in Table 7.4, in which they are organized into several groups. The first group is composed of *primary mood disorders*, such as bipolar disorder, which by far constitute the most common causes of mania. Following this are two *primary psychotic disorders*, namely schizoaffective disorder and postpartum psychosis, both of which may be characterized by mania. Next are *toxic* causes, including medications, for example prednisone, capable of inducing mania as a side-effect. *Metabolic* causes come next, and include both hepatic encephalopathy and uremia. *Endocrinologic* causes include Cushing's syndrome and hyperthyroidism. *Intracranial disorders* capable of causing mania are considered next, including infarctions and tumors. Mania may also occur *as part of certain dementing disorders*, for example Alzheimer's disease. Finally, there is a *miscellaneous* group of causes, including lupus and others.

PRIMARY MOOD DISORDERS

Bipolar disorder is by far the most common cause of mania and is characterized, in most cases, by recurrent episodes of mania and recurrent episodes of depression throughout the lifetime of the patient. Critically, in the intervals between these episodes, patients are either asymptomatic or experience only mild residual symptoms, tending toward either euphoria or depression. The first episode of illness may be either manic or depressive in character. In cases where the first episode is depressive, a manic episode generally occurs within either 10 years or, if there are recurrent depressive episodes, five depressive episodes, whichever comes first (Dunner *et al.* 1976). Importantly, however, although in the intervals *between* episodes there may be mild disturbances of mood, as noted above, there are *never* any psychotic symptoms.

Cyclothymia is best thought of as a very mild form of bipolar disorder. Like bipolar disorder, it is characterized by episodes of mood disturbance, but these are much milder in intensity and indeed may not bring the patient to clinical attention.

PRIMARY PSYCHOTIC DISORDERS

Schizoaffective disorder, bipolar type, possesses a distinctive overall course: patients with this illness are not only chronically psychotic, but also, in the context of this ongoing psychosis, experience episodes of mania and episodes of depression. The chronic psychosis, which leaves them with psychotic symptoms (such as delusions and hallucinations) in the intervals between mood disturbances, clearly distinguishes this illness from bipolar disorder, which is *free* of psychotic symptoms in the intervening periods.

Postpartum psychosis has an abrupt onset of between 3 days and several weeks after delivery and is, in many cases, characterized by manic symptoms. Importantly, these patients are well at other times, and although they may have recurrent postpartum psychoses after subsequent deliveries, they do not have manic symptoms *outside* the puerperium. This is the critical difference between postpartum psychosis and bipolar disorder, for although female patients with bipolar disorder may indeed have manic episodes in the puerperium, they *also* have them at other times in their lives.

TOXIC

Of all the medications listed in Table 7.4, by far the most common causes are prednisone and levodopa. Prednisone, in doses of 80 mg or more per day, may produce mania in approximately three-quarters of patients. Levodopa, as used in Parkinson's disease, in turn may cause mania in up to one-tenth of patients (Celesia and Barr 1970; O'Brien *et al.* 1971). Celesia and Barr (1970) noted that their patients became euphoric with 'excessive self-confidence, over optimism, buoyancy, lack of inhibition, exaggerated motor activity and drive. They made inappropriate jokes, were garrulous, and were often impudent'. In some cases, levodopa-induced mania may be accompanied by hallucinations (Lin and Ziegler 1976). Sometimes, the mania is so appealing that patients may abuse levodopa in order to remain manic (Giovannoni *et al.* 2000). Mania may also occur secondary to direct-acting dopaminergics, such as pramipexole and ropinirole, but this is much less common. Antidepressants are notorious for inducing mania, but this generally only occurs in patients with pre-existing bipolar disorder, and in these cases it is appropriate to consider that the antidepressant 'triggered' a manic episode. Anabolic steroid abuse, as seen in athletes such as weight-lifters and American football players, may cause mania, which may be marked by extreme irritability and violence: manic symptoms occurring in a 'bulked-up' athlete should always suggest anabolic steroid use and prompt a search for corroborating clinical evidence, such

Table 7.4 Causes of mania

Primary mood disorders Bipolar disorder (Chapter 23.1) Cyclothymia (Chapter 23.2) **Primary psychotic disorders** Schizoaffective disorder (Chapter 22.2) Postpartum psychosis (Chapter 22.4) **Toxic** Prednisone (Lyons *et al.* 1988, Minden *et al.* 1988, Wolkowitz *et al.* 1990) Anabolic steroids (Pope and Katz 1988, 1994) Oral contraceptives (Sale and Kalucy 1981) Levodopa (Ryback and Schwab 1971; Van Woert *et al.* 1971) Pramipexole (Singh *et al.* 2005) Ropinirole (Singh *et al.* 2005) Antidepressants (Shulman *et al.* 2001; Stoll *et al.* 1994) Buspirone (Liegghid and Yeragani 1988; Price and Bielfeld 1989) Alpha-interferon (Constant *et al.* 2005) Zidovudine (Maxwell *et al.* 1988; O'Dowd and McKegney 1988; Wright *et al.* 1989) Abacavir (Brouilette and Routy 2007) Clarithromycin (Abouesh and Hobbs 1998) Ciprofloxacin (Bhalerao *et al.* 2006) Isoniazid (Chaturvedi and Upadhyaya 1988; Jackson 1957) Topiramate (Jochum *et al.* 2002) Phenytoin (Patten *et al.* 1989) Zonisamide (Sullivan *et al.* 2006) Procyclidine (in high dosage) (Coid and Strang 1982) Propafenone (Jack 1985) Procarbazine (Mann and Hutchinson 1967) Disulfiram (Ceylan *et al.* 2007) Aspartame (in high dosage) (Walton 1986) Bromide (Sayed 1976) Mannitol (Navarro *et al.* 2001) Metrizamide (Kwentus *et al.* 1984) Baclofen withdrawal (Kirubakaren *et al.* 1984) Tiagabine withdrawal (Pushpal and Shamshul 2006) Reserpine withdrawal (Kent and Wilber 1982) Alpha-methyldopa withdrawal (Labbatte and Holzgang 1989) **Metabolic** Hepatic encephalopathy (Chapter 14.19) Uremia (Chapter 14.18)	**Endocrinologic** Cushing's syndrome (Chapter 17.1) Hyperthyroidism (Chapter 17.3) **Intracranial disorders** Infarctions (see text for references) Midbrain Thalamus Caudate Frontal lobe Temporal lobe Tumors (see text for references) Midbrain Hypothalamus Thalamus Frontal lobe Multiple sclerosis (Chapter 19.1) **As part of certain dementing disorders** Alzheimer's disease (Chapter 9.1) Huntington's disease (Chapter 9.10) Neurosyphilis (Chapter 15.14) Creutzfeldt–Jakob disease (Chapter 16.1) Metachromatic leukodystrophy (Chapter 9.28) Adrenoleukodystrophy (Chapter 9.29) **Miscellaneous** Traumatic brain injury (Chapter 8.5) Epileptic disorders (see text for references) Ictal mania Postictal mania Systemic lupus erythematosus (Chapter 19.3) Vitamin B12 deficiency (Chapter 14.1) Sydenham's chorea (Chapter 19.11) Chorea gravidarum (Chapter 19.12) Velocardiofacial syndrome (Chapter 10.7) Encephalitis lethargica (Chapter 15.6) Acute disseminated encephalomyelitis (Chapter 15.11) Fahr's syndrome (Chapter 14.22) Dialysis dementia (Chapter 14.13)

as gynecomastia and testicular atrophy. The other medications listed in Table 7.4 only rarely cause mania, and before attributing any given case to one of them, one would want to be able to demonstrate not only a fairly close temporal relationship between starting the medication (or substantially increasing the dose) and the onset of the mania, but also an absence of other, more common causes.

Before ending consideration of these toxic causes of mania, further comments are in order regarding the last four entries, namely baclofen, tiagabine, reserpine, and α-methyldopa. For each of these medications, mania occurs not during treatment but rather within days to a week after discontinuation of long-term treatment; here the mania occurs as a withdrawal phenomenon.

METABOLIC

Hepatic encephalopathy is often characterized by manic symptomatology, with euphoria and gregariousness. Uremia may rarely present with mania: in one case (El-Mallakh *et al.* 1987) the only sign of uremia, despite a blood urea nitrogen level of 100 mg/dL and a creatinine concentration of 5.0 mg/dL, was mania.

ENDOCRINOLOGIC

Cushing's syndrome may produce manic symptoms, which have been noted in 11 percent (Starkman *et al.* 1981) to 30 percent (Haskett 1985) of cases. Other symptoms of Cushing's syndrome, such as weight gain, hypertension, diabetes mellitus, acne, hirsutism, and easy bruising, typically accompany these manic changes (Haskett 1985) and serve to suggest the correct diagnosis.

Thyrotoxicosis may be accompanied by mania (Lee *et al.* 1991; Trzepacz *et al.* 1988), and the correct diagnosis is usually suggested by such signs as proptosis, tremor, and tachycardia. In one case, however, the only clue as to the correct diagnosis was a tachycardia of 130 beats per minute (Ingham and Nielsen 1931).

INTRACRANIAL DISORDERS

Appropriately situated infarctions, tumors, or plaques of multiple sclerosis may all cause mania and, as noted below, it appears that mania, although not of much localizing value, may have some lateralizing value.

Infarctions of the midbrain, thalamus, anterior limb of the internal capsule and adjacent head of the caudate nucleus, or the frontal or temporal lobes may all cause mania. In the case of midbrain infarction, not only has mania been seen, but also, remarkably, in these cases there were recurrent episodes of mania and of depression: this has been noted after an infarction secondary to a subarachnoid hemorrhage (Blackwell 1991) and after an ischemic lesion of the right mesencephalic–pontine junction (Kulisevsky *et al.* 1995). In this latter case, the patient, a 59-year-old woman, suddenly lost consciousness and fell to the ground. After coming to, she experienced transient dizziness and right ptosis, and then did well for about 2 weeks. She subsequently developed a severe depression accompanied by left hemiparkinsonism, both of which persisted, only to spontaneously remit after 19 months. She progressed well for the next 5 months and then developed mania characterized by elation, heightened activity and sexual interest, logorrhea, intrusiveness, and decreased sleep. Eventually she was successfully treated with lithium. Thalamic infarctions capable of causing mania may be either unilateral – on the right side – or bilateral. Regarding unilateral lesions, Cummings and Mendez (1984) reported a case of mania accompanied by left-sided hemianesthesia and slight hemiparesis, and Bogousslavsky *et al.* (1988) described a sudden onset of delirium that cleared in a few days, leaving behind mania as the sole indication of a right paramedian thalamic infarction. Kulisevsky *et al.* (1993) reported a case of mania accompanied by chorea and ballismus occurring secondary to an infarction of the ventral tier nuclei on the right. Bilateral thalamic infarction caused mania in two cases. In the first case (McGilchrist *et al.* 1993) there was an acute onset of drowsiness, followed by a cycling of mania and depression. The depression lasted for weeks and was characterized by apathy and increased eating and sleeping. The manias were brief, lasting only 1–2 days, and were characterized by elation, pressured speech, and flight of ideas. The second case (Gentilini *et al.* 1987) presented, like the first, with somnolence, which was accompanied by amnesia and a vertical gaze paresis. As the patient, a 66-year-old man, became more alert, he displayed hypersexuality and grandiosity, asserting that he had a Swiss bank account and was a general in the air force. After 7 months, the amnesia had cleared but the mania persisted: he was 'cheerful... garrulous... [and] looking for a young girl who, enticed by his wealth, would be willing to marry him'. Capsular infarction involving the anterior limb of the internal capsule and the adjacent head of the caudate nucleus has been noted to cause either depression followed within a month by mania or mania alone (Starkstein *et al.* 1990a). Cortical or subcortical white matter infarction of the right frontoparietal region (Starkstein *et al.* 1988a), bilateral orbitofrontal areas, or the right basotemporal area (Starkstein *et al.* 1990a) has been associated with mania.

Tumors affecting the mesencephalon (Greenberg and Brown 1985), the hypothalamus (via compression from an underlying craniopharyngioma [Malamud 1967] or pituitary adenoma [Alpers 1940]), the right thalamus (Stern and Dancey 1942), the right cingulate gyrus (Angelini *et al.* 1980), the inferior right frontal lobe (Gross and Herridge 1988), or both frontal lobes (Starkstein *et al.* 1988b) have all been associated with mania. As an aside, it is worth noting the remarkable ability of hypothalamic disturbances to cause mania. In one case (Alpers 1940), the patient was undergoing surgery for the removal of a craniopharyngioma; when the surgeon produced traction on the hypothalamus, 'the patient burst forth in a push of speech, quoting passages in Latin, Greek and Hebrew . . . and with every word of the [surgeon] broke into a flight of ideas'.

Multiple sclerosis has long been associated with manic symptomatology; indeed, Cottrell and Wilson (1926) noted euphoria in 63 percent of their 100 patients. Subsequent case series have reported a lower rate, varying from 42 percent (Rabins *et al.* 1986) to 26 percent (Surridge 1969), and a nationwide survey in Israel found euphoria in only 5 percent of cases (Kahana *et al.* 1971). Of all reasons underlying the admission of patients with multiple sclerosis to a psychiatric hospital, mania is, however, one of the most common: in one study, of over 2000 psychiatric admissions, ten had multiple sclerosis, and seven of these patients were admitted because of mania (Pine *et al.* 1995). Importantly, in patients with multiple sclerosis, the presence of mania correlates with brain, rather than spinal cord, involvement (Rabins *et al.* 1986).

A review of the cases of mania secondary to infarctions and tumors allows for some speculation regarding both the localizing and the lateralizing value of this syndrome. The localizing value of mania does not appear to be high, as it has been noted with lesions of the mesencephalon (Blackwell 1991; Greenberg and Brown 1985; Kulisevsky *et al.* 1995), thalamus (Bogousslavsky *et al.* 1988; Cummings and Mendez 1984; Gentilini *et al.* 1987; Kulisevsky *et al.* 1993; Liebson 2000; McGilchrist *et al.* 1993; Starkstein *et al.* 1988a; Stern and Dancey 1942), hypothalamus (Alpers 1940; Malamud 1967), anterior limb of the internal capsule and adjacent head of the caudate (Starkstein *et al.* 1990a), cingulate gyrus (Angelini *et al.* 1980), and frontal (Benjamin *et al.* 2000; Gross and Herridge 1988; Starkstein *et al.* 1988b, 1990b), and temporal lobes (Starkstein *et al.* 1990a). The lateralizing value of mania, however, may be higher. In all but one of the instances cited above, the lesion responsible for mania was either on the right side or bilateral: in only one case, namely after the excision of an arteriovenous malformation from the left frontal lobe (Benjamin *et al.* 2000), did mania occur secondary to a left-sided lesion. This impression is bolstered by some controlled studies: although not all agree

(House *et al.* 1990), it appears that when the frontal, temporal, or limbic areas of the cerebrum are involved, mania is more likely with right-sided lesions (Robinson *et al.* 1988; Starkstein *et al.* 1987b). This impression is further reinforced by a study of 19 patients who underwent hemispherectomy: of the 14 who had a right hemispherectomy, 12 became euphoric and none was depressed; of the five who had a left hemispherectomy, one was euphoric, one was depressed, and the other three showed no mood change (Sackeim *et al.* 1982).

AS PART OF CERTAIN DEMENTING DISORDERS

Alzheimer's disease may, in a small minority, be characterized by mild manic symptoms such as an elevated mood (Burns *et al.* 1990a), but this is seen only long after the dementia has become well established.

Huntington's disease, in addition to chronic chorea and dementia, may also, in a small minority of patients, cause manic symptoms, which range from simple 'excitement' (Oliver 1970) or 'euphoria' (Tamir *et al.* 1969) to a fuller syndrome with delusions of grandeur (Bolt 1970; Heathfield 1967).

Neurosyphilis, when manifesting as parenchymatous general paresis of the insane, presents with a dementia that may be marked by mania (Merritt and Springlova 1932): indeed, in one large study of 203 patients, approximately one-half were euphoric and excited (Storm-Mathisen 1969). Rarely, neurosyphilis may present with mania (Binder and Dickman 1980).

Creutzfeldt–Jakob disease is rarely characterized by mania, but in one very rare case the disease presented with mania (Lendvai *et al.* 1999).

Metachromatic leukodystrophy may, very rarely, present with mania. One patient, a 22-year-old woman (Besson 1980), was grandiose, 'spent money irresponsibly... and called out the fire brigade'; she eventually became demented after several years.

Adrenoleukodystrophy, when occurring in the adult years, often presents with a dementia, with the dementia in one case (Weller *et al.* 1992) being accompanied by mania.

MISCELLANEOUS

Of the miscellaneous disorders capable of causing mania, perhaps the most important is traumatic brain injury. Mania was found in 9 percent of 66 patients recovering from a traumatic injury over a 1-year follow-up, and there was a correlation between this sequela and damage to the inferior and polar regions of the temporal lobes, either on the left or on the right (Jorge *et al.* 1993a). In some cases, the mania may appear almost immediately after the patient recovers from the postcoma delirium (Bakchine *et al.* 1989; Bracken 1987), whereas in others there may be a latent interval, lasting up to several months, before the mania appears (Clark and Davison 1987; Nizamie *et al.* 1988). Although in most cases the mania is either persistent or eventually remits without recurrence, there may rarely be an episodic course closely resembling that seen in bipolar disorder, with alternating episodes of mania and episodes of depression (Parker 1957). The manic syndrome following traumatic brain injury may be quite classic in character (Bakchine *et al.* 1989): one patient (Bracken 1987) with euphoria, pressured speech, and flight of ideas had the grandiose delusion that he was writing a 'best seller'.

Epileptic disorders characterized by mania include both ictal and postictal mania. Ictal mania may occur as an aura or as part of a complex partial seizure. Dostoyevsky described his ecstatic aura as follows: 'the air was filled with a big noise... I felt that Heaven was going down upon the Earth and that it had engulfed me. I have really touched God... healthy people... cannot imagine the happiness which we epileptics feel during the second before our fit... for all the joys that life may bring, I would not exchange this one' (Alajouanine 1963). Complex partial seizures may themselves occasionally be characterized by manic symptoms: one patient (Mulder and Daly 1952), during her attack, 'was euphoric, talkative and pleasant. When asked how she felt, she replied "wonderful"'.

Postictal mania may be seen 1–2 days after a 'flurry' of complex partial seizures, and this may be characterized by either stage I or stage II mania (Barczak *et al.* 1988; Nishida *et al.* 2006).

Systemic lupus erythematosus may be marked by manic symptoms, and in such cases other typical symptoms, such as rash, arthralgia, pleurisy, pericarditis, nephritis, or cytopenia, suggest the diagnosis (Brey *et al.* 2002; Johnson and Richardson 1968).

Vitamin B12 deficiency may, very rarely, present with mania, as was reported in one 81-year-old man (Goggans 1984).

Sydenham's chorea may, in addition to the acute onset of chorea, be characterized by mania (often with prominent lability [Adams 1975; Gatti and Rosenheim 1969]), which may in turn be characterized by prominent hallucinations and delusions, typically of persecution (Powell 1889; Reaser 1940; Shaskan 1938; Van Der Horst 1947).

Chorea gravidarum, which essentially represents a recurrence of Sydenham's chorea during pregnancy, may also, in addition to chorea, be characterized by mania (Wilson and Preece 1932).

Velocardiofacial syndrome, a rare disorder suggested by a characteristic facial dysmorphism with hypertelorism, a large nose and micrognathia, may be characterized by mania in over one-half of all cases (Papolos *et al.* 1996).

Encephalitis lethargica, currently a very rare disorder, typically presents with headache, fever, sleep reversal, delirium, and oculomotor paralyses. In some cases, patients displayed euphoria, sometimes accompanied by lability and pressured speech (Hohman 1921).

Acute disseminated encephalomyelitis results from an autoimmune assault on the brain that is triggered by a preceding, usually viral, infection and generally has an acute onset within 2–21 days of the initial infection. Although the typical presentation is with delirium, a mania may occasionally appear. In one case (Moscovich *et al.* 1995), 2 weeks after an influenza infection, a 32-year-old woman developed mania, which remained the only indication of the underlying encephalitis for almost 2 weeks, after which she became delirious and incontinent. In another case (Paskavitz *et al.* 1995), a postmononucleosis acute disseminated encephalitis presented with a combination of mania and a grand mal seizure.

Fahr's syndrome, suggested by calcification of the basal ganglia, may, very rarely present with mania (Trautner *et al.* 1988).

Dialysis dementia, also very rarely, may be characterized by mania (Jack *et al.* 1983).

Differential diagnosis

Hypomania is a distinctive syndrome, and is very difficult to confuse with anything else; thus if the physician either sees the patient during stage I, or has a compatible history from a reliable historian, the diagnosis is generally straightforward. Difficulty can arise in cases where patients have progressed to stage II or stage III; but even here if one has a clear history of a preceding stage I, again the diagnosis remains relatively easy. However, if patients come to medical attention during stages II or III and there is no history available, then one must consider other syndromes that may cause significant agitation, namely psychosis, catatonia, and delirium.

Psychosis, that is to say a syndrome marked by delusions and hallucinations, naturally enters into the differential for patients in stages II or III. Here, the differential rests on identifying the persisting hypomanic symptoms.

Catatonia of the excited subtype is characterized by bizarre and purposeless hyperactivity, and hence may, to a degree, resemble the behavior of patients in stage III mania. In stage III mania, however, one may still see distinctive stage I symptoms; moreover, although the behavior of patients in stage III is fragmented, individual fragments may still retain some purpose. Furthermore, whereas the hyperactive catatonic tends to remain withdrawn, the hyperactive stage III manic still retains an interest in others.

Delirium is marked by confusion and incoherence, symptoms also seen in stage III mania. Here again, one must rely on identifying fragments of stage I mania, symptoms typically not seen in delirium.

It cannot be stressed enough that the easiest way to make these differentials is to obtain a 'longitudinal' view of the patient's illness by repeatedly questioning family, acquaintances, nurses, and aides until one can reliably say that the typical symptoms of hypomania were, or were not, present.

Treatment

The treatment of mania occurring secondary to bipolar disorder is discussed in Chapter 23.1, and, in those cases where agitation is prominent (as for example in stage II or stage III), discussed further in Chapter 7.4. Mania occurring in the other primary mood and psychotic disorders is discussed in the respective chapters. In cases of toxic or metabolic mania, the ideal treatment, naturally, involves eliminating or correcting the underlying cause. In some cases of steroid-induced mania, however, future treatment with steroids may be required, and in such cases it may be appropriate to utilize either prophylactic lithium (Falk *et al.* 1979; Siegal 1978) or divalproex (Abbas and Styra 1994). Treatment of mania occurring secondary to most of the other disorders discussed above may be treated with either lithium or divalproex, or with an antipsychotic, much along the lines as for bipolar disorder.

7.4 AGITATION

Agitation is extraordinarily common in general hospital and nursing home work. It is, as noted below, a non-specific symptom, being seen in a variety of conditions.

Clinical features

Agitated patients are tense and restless. If confined to bed, they may thrash about; intravenous lines and tubes may be pulled out. Ambulatory patients may pace their rooms or up and down the hall, perhaps shouting or cursing. Some may become violent, slamming doors or throwing furniture, and some, especially if attempts are made to restrain them, may become assaultive.

Etiology

Agitation, of sufficient degree to merit clinical attention, generally occurs as part of a large number of conditions, including dementia, delirium, psychosis, mania, traumatic brain injury, alcohol withdrawal, and during various intoxications, for example with stimulants. Agitation may also occur as a side-effect to various medications, for example bupropion. Pain may also cause patients to become agitated, and this is particularly the case in elderly patients with dementia.

Differential diagnosis

Anxious patients may appear quite tense but generally are not given to restless pacing, and certainly not to violent or destructive behavior.

Akathisia, seen primarily as a side-effect to antipsychotics, may appear very similar to agitation. Clues to the correct diagnosis include a worsening of symptoms when sitting or lying down and a tendency to 'march in place'.

Treatment

Environmental measures can sometimes be remarkably effective in calming an agitated patient (Alessi *et al.* 1999). Overall stimulation should be kept to a minimum, and patients should be provided with constructive and quietly engaging activities. Interactions with the patient should preferably be on a one-to-one basis and, if it is necessary to have two people with the patient, it is important to ensure that only one person does all the talking. When patients tend to roam, they should generally be allowed to do so, provided that their behavior endangers neither themselves nor others. A private room should be provided, and if that is not possible then a calm patient should be selected as a roommate; in all cases, the room should have a large clock and calendar and a window. Visitors should be screened, as in some cases certain visitors will agitate patients further; in general, there should be only one visitor at a time. Sitters are often utilized and may obviate the need for restraints, but, like visitors, sitters should be carefully screened: although some have a good 'way' with agitated patients, others may simply worsen the situation. Seclusion or restraints may at times be required and one must not be shy about ordering them, as they may at times be lifesaving.

In all instances of agitation, it is also necessary to dovetail the symptomatic treatment of agitation with other aspects of treatment of the parent syndrome, and the reader is directed to the appropriate chapter on dementia, delirium, etc.

Pharmacologic treatment is typically required: agents utilized include antipsychotics (e.g. haloperidol, risperidone, aripiprazole, olanzapine, ziprasidone, and quetiapine), antiepileptic drugs (e.g. divalproex and carbamazepine), and benzodiazepines (e.g. lorazepam). As noted above, agitation usually occurs as part of a larger syndrome and the choice of pharmacologic agent is often dictated by the syndrome within which the agitation is occurring. In the following, each of the more common syndromes is considered in turn, with recommendations for both non-emergent and emergent treatment; all of the recommendations, except where otherwise noted, are based on double-blind studies. Importantly, as noted under the concluding remarks, these recommendations are offered as guidelines only: clinical reality often dictates alternative approaches and good clinical judgment is absolutely required.

Dementia

For non-emergent care, haloperidol, olanzapine, risperidone, and quetiapine may each be considered, and with regard to effectiveness, both haloperidol (Verhey *et al.* 2006) and risperidone (Fontaine *et al.* 2003) are each equivalent to olanzapine, and risperidone, in a single-blinded study, was equivalent to quetiapine (Rainer *et al.* 2007). In a single-blinded study (Porsteinsson *et al.* 2001), divalproex was effective, but a double-blinded study found it to be no more effective than placebo (Sival *et al.* 2002). In addition, one study found trazodone to be of similar efficacy to haloperidol (Sultzer *et al.* 1997), while another (Pollock *et al.* 2007) demonstrated an equivalence between citalopram and risperidone.

In looking at more specific kinds of dementia, the effectiveness of haloperidol in Alzheimer's disease is uncertain: one study (Devanand *et al.* 1998) supported its use, whereas another (Teri *et al.* 2000) did not. Olanzapine (Street *et al.* 2000) appeared effective in a comparison with placebo, and in the monumental CATIE-AD study, olanzapine and risperidone were more effective than either quetiapine or placebo (Schneider *et al.* 2006; Sultzer *et al.* 2008): notably however, in this study, although risperidone was better tolerated than olanzapine, all three of these antipsychotics, were, overall, poorly tolerated. Carbamazepine (Olin *et al.* 2001; Tariot *et al.* 1998) appears effective but valproic acid is not (Hermann *et al.* 2008; Tariot *et al.* 2005). Trazodone does not appear to be effective (Teri *et al.* 2000).

Overall, for the non-emergent treatment of agitation in dementia, it may be best to begin with a low dose of risperidone, perhaps 0.25 or 0.5 mg/day, with a gradual titration up to a maximum of perhaps 2 mg. Should this be ineffective or not tolerated, then consideration may be given to quetiapine, beginning at 25 mg and increasing the dose gradually, if necessary, to 200 mg, or to olanzapine, beginning with a low dose of perhaps 2.5 mg and titrating gradually, if necessary, to a maximum of 10 mg. Consideration should also be given to either citalopram or trazodone.

In emergent cases, consideration may be given to intramuscular olanzapine, in a dose of 5 mg (Meehan *et al.* 2002), or intramuscular aripiprazole, in a dose of 10 mg (Rappaport *et al.* 2009), with repeat doses if needed.

Before leaving this section, some words are in order regarding the risk of death or stroke in elderly demented patients treated with antipsychotics. Although these risks are indeed increased for second-generation agents (Kryzhanovskaya *et al.* 2006), and even more so for first-generation agents (Wang *et al.* 2005), this increased risk, as with any medical treatment, must be weighed against the benefits obtained with treatment.

Delirium

The authors could not find any double-blind studies regarding the non-emergent pharmacologic treatment of delirium. In practice, patients are treated with low doses of antipsychotics (e.g. risperidone at 0.5–1 mg or haloperidol at 2–5 mg), with dosage adjustments made every day or so as needed.

Emergent treatment generally involves the use of an antipsychotic, and although in all likelihood all of the antipsychotics are probably effective in this regard, only a handful have been subjected to controlled studies. Haloperidol, a first-generation agent, is the standard (Breitbart *et al.* 1996); of second-generation agents, risperidone (Han and Kim 2004) is equally effective to haloperidol, as was olanzapine; however, the olanzapine study was an open one (Skrobik *et al.* 2004); and quetiapine has been shown to be superior to placebo (Devlin *et al.* 2010). Of these antipsychotics, haloperidol and olanzapine are both available for intramuscular use, which constitutes a distinct advantage, and haloperidol, in addition, may be administered intravenously. Treatment may be commenced with haloperidol at 1–5 mg, olanzapine at 5–10 mg, risperidone at 0.25–1 mg, or quetiapine at 25–50 mg, with repeat doses every 1–2 hours until the patient is calm, limiting side-effects occur, or one reaches a maximum dose (roughly 20 mg haloperidol, 40 mg olanzapine, 5 mg risperidone, and 200 mg quetiapine). At this point, one may continue treatment with as-needed doses only, or one may begin the patient on a regular daily dose. A regular dosage is in order when, given the underlying cause, the delirium is expected to persist for a matter of days or longer at the same intensity. In cases where a briefer duration is expected, or where there is uncertainty in this regard, it may be appropriate to continue the patient on as-needed doses only. If the decision is to place the patient on a regular daily dose, this dose, initially, should be roughly equivalent to the total amount required in p.r.n. doses, with an order for repeat doses on an as-needed basis; on subsequent days, the total daily dose should be adjusted based on the clinical situation and on the amount required in p.r.n. doses over the preceding 24 hours. Once the patient has been stabilized on this regular daily dose, and the underlying cause has either been successfully treated or undergone spontaneous improvement, attempts should be made to taper the antipsychotic.

Intravenous lorazepam is also widely used in the treatment of an agitated delirium, especially when agitation must be urgently controlled, and although it is effective in this regard, its overall effect on confusion and other cognitive deficits is negative (Breitbart *et al.* 1996); hence, an antipsychotic should be started as soon as possible.

Before leaving this discussion of the pharmacologic treatment of delirium, some special comments are in order regarding delirium in mechanically ventilated patients. Ventilated patients typically, at least initially, require sedation, and in this regard either lorazepam or midazolam are often employed. Concerns have naturally been raised, however, that the use of either of these agents may cause delirium, which in turn would make it harder to wean patients. Consequently, efforts have been undertaken to find less toxic alternatives, and it appears that dexmedetomidine, an α_1 agonist, fits the bill, as comparison studies with both lorazepam (Pandharipande *et al.* 2007) and midazolam (Riker *et al.* 2009) have demonstrated equal efficacy at sedation with, however, a lower incidence of delirium. Furthermore, in cases where mechanically ventilated non-sedated patients did become delirious, an open study (Reade *et al.* 2009) demonstrated that dexmedetomidine was more effective than haloperidol both in reducing the need for sedation and in shortening the time to successful weaning.

Psychosis

Studies regarding the treatment of agitated patients with psychosis for the most part involve patients with schizophrenia.

In non-emergent cases, one may simply begin with the antipsychotic judged most appropriate for the patient's case.

In emergent cases, rapid control of agitation has been achieved with a combination of intramuscular haloperidol and lorazepam, with the combination being more effective than use of either agent alone (Bieniek *et al.* 1998); a combination of oral risperidone and intramuscular lorazepam was also shown to be as effective as combined intramuscular treatment with haloperidol and lorazepam (Veser *et al.* 2006). Aripiprazole is of similar effectiveness to haloperidol and is better tolerated (Tran-Johnson *et al.* 2007). Olanzapine is also of similar effectiveness to haloperidol (Breier *et al.* 2002; Wright *et al.* 2001); a combination of olanzapine at a dose of 10 mg plus lorazepam as needed was less effective than simply using olanzapine alone at a higher dose of 25–30 mg (Baker *et al.* 2003). Blinded work comparing ziprasidone in doses of 20 mg (Daniel *et al.* 2001) or 10 mg (Lesem *et al.* 2001) with ziprasidone in a dose of 2 mg indicates effectiveness for the higher doses.

Overall, for emergent treatment, it seems reasonable to begin with either haloperidol (5 mg, as the concentrate or i.m.), risperidone (2 mg, as the concentrate), aripiprazole (9.75 mg, i.m.), or olanzapine (10 mg, i.m.), with repeat doses every 1–2 hours until the patient is calm, limiting side-effects occur, or maximum doses of approximately 20 mg haloperidol, 6 mg of risperidone, 39 mg of aripiprazole, or 40 mg of olanzapine are reached. In choosing which of these agents to use, one looks not only to the individual patient's treatment history for guidance but also to the side-effect profiles associated with them, and in this regard it is important to keep in mind that extrapyramidal side-effects, such as akathisia, dystonia, and parkinsonism, are far more common with haloperidol than the others (Satterthwaite *et al.* 2008). Although monotherapy is generally preferred, in cases when very rapid control is essential, intramuscular lorazepam (2 mg) may be safely given along with doses of either haloperidol or risperidone, and, although not yet demonstrated, probably also with aripiprazole or olanzapine.

Mania

In cases of mania wherein significant agitation has appeared, emergent care is almost always required, and in such cases treatment aimed at the mania itself is generally sufficient, and may proceed as outlined in Chapter 23.1. In such cases, it is appropriate to begin with a second-generation antipsychotic, such as olanzapine or aripiprazole, utilizing the scheme described above for the treatment of agitation in psychosis. As soon as the patient is cooperative enough to reliably take oral medications, a mood stabilizer, such as divalproex or lithium, should be started.

Traumatic brain injury

In non-emergent situations, a number of agents, in case reports or open studies, have been reported as effective, including quetiapine (Kim and Bijlani 2006), amitriptyline (Mysiw *et al.* 1988), sertraline (Kant *et al.* 1998), valproic acid (Chatham-Showalter and Kimmel 2000; Wroblewski *et al.* 1997), carbamazepine (Azouvi *et al.* 1999; Chatham-Showalter 1996), lithium (Glenn *et al.* 1989), and amantadine (Chandler *et al.* 1988). The only agent subjected to a double-blind trial for agitation was propranolol (Brooke *et al.* 1992), which was found to reduce the severity of agitation, but often only at high doses.

Choosing among these agents is not straightforward. In emergent situations, an antipsychotic, such as quetiapine, is often used because the dose may be rapidly escalated and the effect is usually prompt. In non-emergent situations, or in cases where an antipsychotic has reduced but not cleared the agitation, consideration should be given to one of the other agents, but here, given the lack of comparison studies, guidance must rest on accumulated clinical wisdom. Amitriptyline is often quite effective; however, anticholinergic and cardiotoxic effects make this a problematic choice. One of the authors (DPM) wishes, at this point, to note his own experience with mirtazapine (an agent that shares similar noradrenergic and serotoninergic effects with amitriptyline but lacks the anticholinergic and cardiotoxic effects), which has been often dramatically effective in reducing agitation, although often only doing so at high doses of 60–90 mg. Sertraline is also an option. Among the anti-epileptic drugs, carbamazepine is probably a better option given the increased mortality associated with divalproex in the treatment of post-traumatic epilepsy (as discussed in Chapter 8.3); carbamazepine, however, carries a significant side-effect burden. Lithium is poorly tolerated in brain-injured patients. Amantadine is a reasonable option given its other benefits in traumatic brain injury. Propranolol, although constituting the only agent with double-blinded support, is often poorly tolerated at the doses required. With all this in mind, it may be reasonable to, if not already done, start amantadine, and then add in mirtazapine or sertraline, holding carbamazepine and propranolol in reserve for treatment-resistant cases.

Alcohol withdrawal

As discussed in Chapter 28.5, agitation may be found in both mild and moderate alcohol withdrawal; however, in these degrees of withdrawal the agitation, compared with other symptoms such as tremor, is relatively mild. In severe alcohol withdrawal, that is to say delirium tremens, agitation may be extreme and in some cases may dominate the clinical picture. Successful treatment of this agitation occurs concurrent with treatment of the alcohol withdrawal itself, which is discussed at length in Chapter 28.5.

Concluding remarks

Good clinical judgment often dictates a course of treatment that differs from those recommended above. Doses must often be reduced in elderly or frail patients, and in those with significant hepatic dysfunction. Antipsychotics other than those discussed above are often used successfully, and haloperidol, given its tendency to cause extrapyramidal side-effects, is falling into disfavor. In some cases, certain medications are generally contraindicated; for example, in dementia with Lewy bodies, antipsychotics should generally be avoided given the risk of severe, and potentially fatal, parkinsonism.

7.5 ANXIETY

Pathologic anxiety occurs in two forms. In one, anxiety is more or less persistent, whereas in the other it occurs in discrete attacks.

Clinical features

Persistent anxiety tends to come on gradually, and waxes and wanes over time. The anxiety itself is typically accompanied by autonomic signs such as tremor, tachycardia, and diaphoresis. Patients complain of a sense of tremulousness, and the tremor itself is fine and postural. Tachycardic patients may complain that the heart is 'racing', and there may be palpitations. Diaphoresis may be evident when one shakes the patient's hand. The duration of this persistent form of anxiety depends on the underlying cause and may, for example, range from years or decades in the case of generalized anxiety disorder to weeks or less in alcohol withdrawal.

Anxiety attacks typically arise acutely, over minutes, and symptoms crescendo rapidly. In addition to the anxiety, which may be quite extreme, patients also typically experience a variety of other symptoms, including tremor, tachycardia, palpitations, diaphoresis, dyspnea, light-headedness, nausea, and paresthesiae. The duration of the attack, although determined by the underlying cause, is generally brief, lasting from minutes to an hour or more.

Etiology

The various causes of anxiety are listed in Table 7.5, where they are divided into those causing persistent anxiety and those causing anxiety attacks.

PERSISTENT ANXIETY

The most common cause of persistent anxiety is one of the primary anxiety disorders, namely generalized anxiety disorder. This disorder generally has an onset in adolescence or early adult years, and the characteristic anxiety tends to persist, in a waxing and waning fashion, from years to decades.

Toxic causes include caffeine and sympathomimetics, such as ephedrine and phenylpropanolamine, and ephedra alkaloids, found in many 'herbal supplements'. Others include theophylline and levodopa. Although, in general, anxiety only occurs with high doses, some patients may be quite sensitive to these medications and experience considerable anxiety at 'therapeutic' doses.

Metabolic causes include hypocalcemia, as may be seen in hypoparathyroidism, and the hypoxia and hypercarbia associated with respiratory failure, as in advanced chronic obstructive pulmonary disease.

Substance or medication withdrawals are considered next. Of the substance withdrawals, alcohol withdrawal probably constitutes one of the most common causes of persistent anxiety seen in general hospital practice, and the diagnosis is often missed, given that patients in withdrawal typically either minimize their alcohol use or deny it altogether. A similar scenario may occur in patients withdrawing from sedative/hypnotics, such as benzodiazepines. Nicotine withdrawal is typified by anxiety, irritability, and a craving for a 'smoke', and likewise may go undiagnosed, given the shame associated with smoking and patients' reluctance to admit it. Anticholinergic withdrawal, occurring after an abrupt discontinuation of drugs with strong anticholinergic properties such as benztropine or tricyclic antidepressants, may be followed by a cholinergic rebound, with anxiety, jitteriness, insomnia, and nausea.

Endocrinologic causes include hyperthyroidism and Cushing's syndrome. Hyperthyroidism classically causes chronic anxiety and may be suggested by such signs and symptoms as heat intolerance, diaphoresis, and proptosis. Cushing's syndrome, although classically associated with depression, may at times be characterized by severe anxiety. The diagnosis is suggested by such features as a 'moon' facies, buffalo hump, acne, hirsutism, violaceous abdominal striae, hypertension, and diabetes mellitus.

Intracranial disorders associated with persistent anxiety include stroke and traumatic brain injury. Poststroke anxiety occurs chronically in a minority of patients and appears to be more likely with a right hemisphere infarction (Castillo *et al.*

Table 7.5 Causes of anxiety

Persistent anxiety	
Primary anxiety disorder	
	Generalized anxiety disorder (Chapter 24.7)
Toxic	
	Caffeine (Chapter 28.11)
	Ephedrine (Sawyer *et al.* 1982)
	Phenylpropanolamine (Sawyer *et al.* 1982)
	Ephedra alkaloids (Haller and Benowitz 2000)
	Theophylline (Trembath and Boobis 1979)
	Levodopa (Celesia and Barr 1970)
Metabolic	
	Hypocalcemia (Carlson 1986; Denko and Kaelbling 1962; Lawlor 1988)
	Chronic obstructive pulmonary disease (Brenes 2003)
Substance or medication withdrawal	
	Alcohol (Chapter 28.5)
	Sedative/hypnotic (Chapter 28.6)
	Nicotine (Chapter 28.10)
	Anticholinergic (Chapter 29.7)
Endocrinologic	
	Hyperthyroidism (Chapter 17.3)
	Cushing's syndrome (Chapter 17.1)
Intracranial disorders	
	Poststroke (Chapter 8.4)
	Traumatic brain injury (Fann *et al.* 1995; Jorge *et al.* 1993b)
Anxiety attacks	
Primary anxiety disorders	
	Panic disorder (Chapter 24.1)
	Phobias (Chapter 24.2–4)
	Post-traumatic stress disorder (Chapter 24.6)
	Obsessive–compulsive disorder (Chapter 24.5)
Toxic	
	Cocaine (Louie *et al.* 1996)
	Cannabis (Chapter 28.8)
	Hallucinogens (Chapter 28.3)
	Clozapine (Bressan *et al.* 2000)
Metabolic	
	Hypoglycemia (Chapter 14.15)
Endocrinologic	
	Pheochromocytoma (Doust 1958; Modlin *et al.* 1979; Starkman *et al.* 1985; Thomas *et al.* 1966)
Miscellaneous	
	Simple partial seizures (Chapter 8.3)
	Right temporal lobe tumor (Ghadarian *et al.* 1986)
	Paroxysmal atrial tachycardia
	Angina or myocardial infarction
	Pulmonary embolus
	Parkinson's disease when 'off' (Vazquez *et al.* 1993)

1993, 1995; Starkstein *et al.* 1990b). Traumatic brain injury may be associated with anxiety in a small minority of cases.

ANXIETY ATTACKS

By far the most common causes of anxiety attacks are certain primary anxiety disorders, namely panic disorder, phobias, post-traumatic stress disorder, and obsessive–compulsive disorder. Panic disorder is the prototypical cause of anxiety attacks, which, when occurring in this disorder, are often referred to as 'panic attacks'. In this disorder, anxiety attacks typically occur first in adolescence or early adult years, and then recur, with variable frequency, over years or decades. Phobias, including specific phobia, social phobia, and agoraphobia, may all be characterized by anxiety attacks when patients are brought into close proximity to the phobic object, such as a snake for a specific phobic, speaking to an audience for the social phobic with a fear of public speaking, and venturing away from home for the agoraphobic. Post-traumatic stress disorder may also be characterized by anxiety attacks when patients are exposed to events or objects that remind them of the original trauma; thus, a combat veteran might be seized with anxiety if a combat scene appeared on a television program. Obsessive–compulsive disorder may likewise be characterized by anxiety attacks when patients attempt to resist their compulsions. Thus, patients with a hand-washing compulsion may experience unbearable anxiety if they attempt to resist the urge to wash.

Toxic causes of anxiety attacks are relatively rare, and include intoxication with cocaine, cannabis, and hallucinogens, such as LSD; anxiety attacks have also been reported secondary to treatment with clozapine.

Metabolic causes include the classic examples of hypoglycemia and hyperventilation. Hypoglycemic-induced anxiety attacks are typically seen in diabetics on insulin who miss a meal, and are suggested by the associated hunger.

Endocrinologically caused anxiety attacks may occur in pheochromocytoma, and in such cases the anxiety attack is typically accompanied not only by an elevated blood pressure, but also by headache.

Of the miscellaneous causes of anxiety attacks, perhaps the most important are simple partial seizures, which, rarely, may manifest solely with anxiety attacks (Sazgar *et al.* 2003). Although such ictal anxiety often lasts only seconds or minutes (Williams 1956), the seizure may in rare cases be very prolonged: one patient for 12 hours endured ictal fear of such a degree that she 'continually looked back over her shoulder' (McLachlan and Blume 1980). There are two clues to the correct diagnosis of these simple partial seizures: first, the exquisitely paroxysmal nature of the onset, and second, the occurrence, at other times in the patient's life, of more obvious ictal phenomena such as complex partial or grand mal seizures.

Anxiety attacks have also, very rarely, been associated with mass lesions in the right temporal lobe. Cardiopulmonary conditions associated with anxiety attacks include paroxysmal atrial tachycardia, angina or myocardial infarction, and pulmonary embolus. Paroxysmal atrial tachycardia is suggested by the almost instantaneous onset of severe tachycardia, followed immediately by severe anxiety as the patient becomes aware of the fluttering in his or her chest. The onset with tachycardia is suggestive, and the ability of the patient to terminate the attack with a Valsalva maneuver is also suggestive of the diagnosis. Angina or myocardial infarction may, in a minority, present primarily with anxiety, with chest pain or dyspnea being relatively minor symptoms. A pulmonary embolus, at the moment of its lodging in a large artery, can produce severe anxiety and dyspnea. The diagnosis is suggested by the dyspnea and by such clinical settings as prolonged immobilization or deep venous thrombophlebitis.

Finally, Parkinson's disease will, in a significant minority of levodopa-treated patients, be characterized by anxiety attacks during the 'off' interval as the effectiveness of the previous dose of levodopa wanes.

Differential diagnosis

The first task in differential diagnosis is to decide whether or not the anxiety in question is normal or pathologic. Although the experience of anxiety is a normal reaction to threatening events, the clinician should suspect pathology in cases where either the severity of the anxiety or its duration seem out of proportion to the inciting event.

Consideration should also be given to the possibility that the anxiety in question is existing as part of a larger syndrome, such as depression or delirium. Depression is commonly accompanied by anxiety, and hence in evaluating any patient with anxiety one should enquire after depressive symptoms, such as insomnia (especially middle or terminal insomnia), anergia, anhedonia, anorexia, etc., and, if they are present, then the differential for depression should be pursued, as outlined in Chapter 7.1. Delirious patients are often quite anxious, and hence the diagnostician should be alert to such symptoms as confusion, disorientation, short-term memory loss, etc., and if these are present, the differential outlined in Chapter 6.5 should be pursued.

Treatment

Treatment is directed at the underlying cause; in cases of persistent anxiety when this is not possible or ineffective, consideration may be given to symptomatic treatment with a benzodiazepine, such as lorazepam or clonazepam, or, when these are relatively contraindicated, an antidepressant or an antipsychotic, such as quetiapine. With regard to anxiety attacks, symptomatic treatment is generally not indicated: most attacks are of such brevity that they resolve spontaneously before any pharmacologic effects could be obtained.

REFERENCES

Abbas A, Styra R. Valproate prophylaxis against steroid-induced psychosis. *Can J Psychiatry* 1994; **39**:188–9.

Abouesh A, Hobbs WR. Clarithromycin-induced mania. *Am J Psychiatry* 1998; **155**:1626.

Abrams R, Taylor MA. Mania and schizoaffective disorder, manic type: a comparison. *Am J Psychiatry* 1976; **133**:1445–7.

Adams J. Psychological correlates of the course of Sydenham's chorea. Repeat evaluation of a case. *J Abnormal Child Psychol* 1975; **3**:33–40.

Ainiala H, Loukkola J, Peltola J *et al.* The prevalence of neuropsychiatric syndromes in systemic lupus erythematosus. *Neurology* 2001; **57**:496–500.

Alajouanine T. Dostoiewski's epilepsy. *Brain* 1963; **86**:209.

Alessi CA, Yoon EJ, Schnelle JF *et al.* A randomized trial of a combined physical activity and environmental intervention in nursing home residents: do sleep and agitation improve? *J Am Geriatr Soc* 1999; **37**:784–91.

Alpers BJ Personality and emotional disorders associated with hypothalamic lesions. In: Fulton JF, Ranson SW, Frantz AM, eds. *Research Publications. Association for Research in Nervous and Mental Disease*. Vol. XX. *The Hypothalamus and Central Levels of Autonomic Function*. Baltimore, MD: Williams & Wilkins, 1940.

Andersson S, Krogstad JM, Finset A. Apathy and depressed mood in acquired brain damage: relationship to lesion localization and psychophysiological reactivity. *Psychol Med* 1999; **29**: 447–56.

Angelini L, Mazzuchi A, Picciotto F *et al.* Focal lesion of the right cingulum: a case report in a child. *J Neurol Neurosurg Psychiatry* 1980; **43**:355–7.

Astrom M, Adolfsson R, Asplund K. Major depression in stroke patients: a 3-year longitudinal study. *Stroke* 1993; **24**:976–82.

Azouvi P, Jokic C, Attal N *et al.* Carbamazepine in agitation and aggressive behavior following severe closed-head injury: results of an open trial. *Brain Inj* 1999; **13**:797–804.

Bakchine S, Lacomblez L, Benoit N *et al.* Manic-like state after bilateral orbitofrontal and right temperoparietal injury: efficacy of clonidine. *Neurology* 1989; **39**:777–81.

Baker RW, Kinon BJ, Maguire GA *et al.* Effectiveness of rapid initial dose escalation of up to 40 mg per day of oral olanzapine in acute agitation. *J Clin Psychopharmacol* 2003; **23**:342–8.

Barczak P, Edmunds E, Betis T *et al.* Hypomania following complex partial seizures: a report of three cases. *Br J Psychiatry* 1988; **152**:137–9.

Benjamin S, Kirsch D, Visscher T *et al.* Hypomania from left frontal AVM resection. *Neurology* 2000; **54**:1389–90.

Besson JAO. A diagnostic pointer to adult metachromatic leukodystrophy. *Br J Psychiatry* 1980; **137**:186–7.

Bieniek SA, Ownby RL, Penalver A *et al.* A double-blind study of lorazepam versus the combination of haloperidol and lorazepam in managing agitation. *Pharmacotherapy* 1998; **18**:57–62.

Bhalerao S, Talsky A, Hansen K *et al.* Ciprofloxacin-induced manic episode. *Psychosomatics* 2006; **47**:539–40.

Billings RF, Stein MB. Depression associated with ranitidine. *Am J Psychiatry* 1986; **143**:915–16.

Billings RG, Tang SW, Rakoff VM. Depression associated with cimetidine. *Can J Psychiatry* 1981; **26**:260–1.

Binder RL, Dickman WA. Psychiatric manifestations of neurosyphilis in middle-aged patients. *Am J Psychiatry* 1980; **137**:741–2.

Black DW, Nasrallah H. Hallucinations and delusions in 1,715 patients with unipolar and bipolar affective disorders. *Psychopathology* 1989; **22**:28–34.

Blackwell MJ. Rapid-cycling manic-depressive illness following subarachnoid haemorrhage. *Br J Psychiatry* 1991; **159**:279–80.

Bloch M, Stager S, Braun A *et al.* Pimozide-induced depression in men who stutter. *J Clin Psychiatry* 1997; **58**:433–6.

Bogousslavsky J, Ferrazzini M, Regli F *et al.* Manic delirium and frontal-like syndrome with paramedian infarction of the right thalamus. *J Neurol Neurosurg Psychiatry* 1988; **51**:116–19.

Bolt JMW. Huntington's chorea in the West of Scotland. *Br J Psychiatry* 1970; **116**:259–70.

Bowman KM, Raymond AF. A statistical study of hallucinations in the manic-depressive psychoses. *Am J Psychiatry* 1931; **88**:299–309.

Bracken P. Mania following head injury. *Br J Psychiatry* 1987; **150**:690–2.

Breier A, Meehan K, Birkett M *et al.* A double-blind, placebo-controlled dose–response comparison of intramuscular olanzapine and haloperidol in the treatment of acute agitation in schizophrenia. *Arch Gen Psychiatry* 2002; **59**:441–8.

Breitbart W, Marotta R, Platt MM *et al.* A double-blind trial of haloperidol, chlorpromazine, and lorazepam in the treatment of delirium in hospitalized AIDS patients. *Am J Psychiatry* 1996; **153**:231–7.

Brenes GA. Anxiety and chronic obstructive pulmonary disease: prevalence, impact, and treatment. *Psychosom Med* 2003; **65**:963–70.

Brenner I. Apathetic hyperthyroidism. *J Clin Psychiatry* 1978; **39**:479–80.

Bressan RA, Monteiro VBM, Dias CC. Panic disorder associated with clozapine. *Am J Psychiatry* 2000; **157**:2056.

Brey RL, Holliday SL, Saklad AR *et al.* Neuropsychiatric syndromes in lupus. Prevalence using standardized definitions. *Neurology* 2002; **58**:1214–20.

Brockington IF, Wainwright S, Kendell RE. Manic patients with schizophrenic or paranoid symptoms. *Psychol Med* 1980; **10**:73–83.

Brodaty H, Sachdev PS, Withall A *et al.* Frequency and clinical, neuropsychological and neuroimaging correlates of apathy following stroke: the Sydney Stroke Study. *Psychol Med* 2005; **35**:1707–16.

Brooke MM, Patterson DR, Questad KA *et al.* The treatment of agitation during initial hospitalization after traumatic brain injury. *Arch Phys Med Rehabil* 1992; **73**:917–21.

Brouilette M-J, Routy J-P. Abacavir sulfate and mania in HIV. *Am J Psychiatry* 2007; **164**:979–80.

Brown SA, Inaba RK, Gillin JC *et al.* Alcoholism and affective disorder: clinical course of depressive symptoms. *Am J Psychiatry* 1995; **152**:45–52.

Burns A, Jacoby R, Levy R. Psychiatric phenomena in Alzheimer's disease. III. Disorders of mood. *Br J Psychiatry* 1990a; **157**: 81–6.

Burns A, Folstein S, Brandt J *et al.* Clinical assessment of irritability, aggression, and apathy in Huntington and Alzheimer disease. *J Nerv Ment Dis* 1990b; **178**:30–6.

Carlson GA, Goodwin FK. The stages of mania: a longitudinal analysis of the manic episode. *Arch Gen Psychiatry* 1973; **28**:221–8.

Carlson GA, Strober M. Affective disorders in adolescence: issues in misdiagnosis. *J Clin Psychiatry* 1978; **39**:59–66.

Carlson RJ. Longitudinal observations of two cases of organic anxiety syndrome. *Psychosomatics* 1986; **27**:529–31.

Castillo CS, Starkstein SE, Federoff JP *et al.* Generalized anxiety disorder following stroke. *J Nerv Ment Dis* 1993; **181**:100–6.

Castillo CS, Schultz SK, Robinson RG. Clinical correlates of early-onset and late-onset poststroke generalized anxiety. *Am J Psychiatry* 1995; **152**:1174–9.

Catsman-Berrevoets CE, von Harskamp F. Compulsive pre-sleep behavior and apathy due to bilateral thalamic stroke: response to bromocriptine. *Neurology* 1988; **38**:647–9.

Celesia GG, Barr AN. Psychosis and other psychiatric manifestations of levodopa therapy. *Arch Neurol* 1970; **23**:193–200.

Ceylan ME, Turkcan A, Mutlu E *et al.* Manic episode with psychotic symptoms associated with high dose disulfiram: a case report. *J Clin Psychopharmacol* 2007; **27**:224–5.

Chandler MC, Barnhill JL, Gualtieri CT. Amantadine for the agitated head-injury patient. *Brain Inj* 1988; **2**:309–11.

Chatham-Showalter PE. Carbamazepine for combativeness in acute traumatic brain injury. *J Neuropsychiatry Clin Neurosci* 1996; **8**:96–9.

Chatham-Showalter PE, Kimmel DN. Agitated symptom response to divalproex following acute brain injury. *J Neuropsychiatry Clin Neurosci* 2000; **12**:395–7.

Chaturvedi SK, Upadhyaya M. Secondary mania in a patient receiving isonicotinic acid hydrazide and pyroxidine: case report. *Can J Psychiatry* 1988; **33**:675–6.

Clark AF, Davison K. Mania following head injury: a report of two cases and a review of the literature. *Br J Psychiatry* 1987; **150**:841–4.

Clayton P, Pitts FN, Winokur G. Affective disorder. IV. Mania. *Compr Psychiatry* 1965; **3**:313–22.

Coid J, Strang J. Mania secondary to procyclidine ('Kemadrin') abuse. *Br J Psychiatry* 1982; **141**:81–4.

Constant A, Castera L, Dantzer R *et al.* Mood alterations during interferon-alfa therapy in patients with chronic hepatitis C: evidence for an overlap between manic/hypomanic and depressive symptoms. *J Clin Psychiatry* 2005; **66**:1050–7.

Cottrell SS, Wilson SAK. The affective symptomatology of disseminated sclerosis: a study of 100 cases. *J Neurol Psychopathology* 1926; **7**:1–30.

Cummings JL, Mendez MF. Secondary mania with focal cerebrovascular lesions. *Am J Psychiatry* 1984; **141**:1084–7.

Cummings JL, Miller B, Hill MA *et al.* Neuropsychiatric aspects of multi-infarct dementia and dementia of the Alzheimer type. *Arch Neurol* 1987; **44**:389–93.

Cushing H. The basophil adenomas of the pituitary body and their clinical manifestations (pituitary basophilism). *Bull Johns Hopkins Hosp* 1932; **50**:137–95.

Dalos NP, Rabins PV, Brooks BR *et al.* Disease activity and emotional state in multiple sclerosis. *Ann Neurol* 1983; **13**:573–7.

Daniel DG, Potkin SG, Reeves KR *et al.* Intramuscular (IM) ziprasidone 20 mg is effective in reducing acute agitation associated with psychosis: a double-blind, randomized trial. *Psychopharmacology* 2001; **155**:128–34.

Davidson KM. Diagnosis of depression in alcohol dependence: changes in prevalence with drinking status. *Br J Psychiatry* 1995; **166**: 199–204.

DeMuth GW, Ackerman SH. Alpha-methyldopa and depression: a clinical study and review of the literature. *Am J Psychiatry* 1983; **140**:534–8.

Denko J, Kaelbling R. The psychiatric aspects of hypoparathyroidism. *Acta Psychiatr Scand* 1962; 38(Suppl. 164):1–70.

Dennis MS, Byrne EJ, Hopkinson N *et al.* Neuropsychiatric systemic lupus erythematosus in elderly people: a case series. *J Neurol Neurosurg Psychiatry* 1992; **55**:1157–61.

Devanand DP, Marder K, Michaels KS *et al.* A randomized, placebo-controlled dose-comparison trial of haloperidol for psychosis and disruptive behaviors in Alzheimer's disease. *Am J Psychiatry* 1998; **155**:1512–20.

Devlin JW, Roberts RJ, Fong JJ *et al.* Efficacy and safety of quetiapine in critically ill patients with delirium: a prospective, multicentre, randomized, double-blind, placebo-controlled study. *Crit Care Med* 2010; **38**:419–27.

Dewhurst K. The neurosyphilitic psychoses today: a survey of 91 cases. *Br J Psychiatry* 1969; **115**:31–8.

Doust BC. Anxiety as a manifestation of pheochromocytoma. *Arch Int Med* 1958; **102**:811–55.

Drijgers RL, Aalten P, Winogrodzka A *et al.* Pharmacological treatment of apathy in neurodegenerative diseases: a systematic review. *Dement Geriatr Cogn Disord* 2009; **28**:13–22.

Dunner DL, Fleiss JL, Fieve R. The course of development of mania in patients with recurrent depression. *Am J Psychiatry* 1976; **133**:905–8.

El-Mallakh RS, Shrader SA, Widger E. Single case study: mania as a manifestation of end-stage renal disease. *J Nerv Ment Dis* 1987; **175**:243–5.

Falk WE, Mahnke MD, Poskanzer MD. Lithium prophylaxis of corticotropin-induced psychosis. *JAMA* 1979; **241**:1011–12.

Fann KR, Katon WJ, Uomoto JM *et al.* Psychiatric disorders and functional disability in outpatients with traumatic brain injuries. *Am J Psychiatry* 1995; **152**:1493–9.

Feil D, Razani J, Boone K *et al.* Apathy and cognitive performance in older adults with depression. *Int J Geriatr Psychiatry* 2003; **18**:479–85.

Fontaine CS, Hynn LS, Koch K *et al.* A double-blind comparison of olanzapine versus risperidone in the acute treatment of dementia-related behavioral disturbances in extended care facilities. *J Clin Psychiatry* 2003; **64**:726–30.

Fras I, Litin EM, Pearson JS. Comparison of psychiatric symptoms in carcinoma of the pancreas with those in some other intra-abdominal neoplasms. *Am J Psychiatry* 1967; **123**:1553–62.

Fraser TN. Cerebral manifestations of addisonian pernicious anemia. *Lancet* 1960; **2**:458–9.

Fried MW, Shiffman ML, Reddy KR *et al.* Peginterferon alfa-2a plus ribaverin for chronic hepatitis C infection. *N Engl J Med* 2002; **347**:975–82.

Friend KD, Young RC. Late-onset major depression with delusions after metoclopramide treatment. *Am J Geriatr Psychiatry* 1997; **5**: 79–82.

Galynker I, Ieronimo C, Miner C *et al.* Methylphenidate treatment of negative symptoms in patients with dementia. *J Neuro-psychiatry Clin Neurosci* 1997; **9**:231–9.

Ganz VH, Gurland BJ, Deming WE *et al.* The study of psychiatric symptoms of systemic lupus erythematosus: a biometric study. *Psychosom Med* 1972; **34**:207–20.

Gatewood JW, Organ CH, Mead BT. Mental changes associated with hyperparathyroidism. *Am J Psychiatry* 1975; **132**:129–32.

Gatti FM, Rosenheim E. Sydenham's chorea associated with transient intellectual impairment. *Am J Dis Child* 1969; **118**:915–18.

Gentilini M, De Renzi E, Crisi G. Bilateral paramedian thalamic artery infarcts: report of eight cases. *J Neurol Neurosurg Psychiatry* 1987; **50**:900–9.

Ghadarian AM, Gauthier S, Bertrand S. Anxiety attacks in a patient with a right temporal lobe meningioma. *J Clin Psychiatry* 1986; **47**:270–1.

Giovannoni G, O'Sullivan JD, Turner K *et al.* Hedonistic homeostatic dysregulation in patients with Parkinson's disease on dopamine replacement therapies. *J Neurol Neurosurg Psychiatry* 2000; **68**:423–8.

Glenn MB, Wroblewski B, Parziale J *et al.* Lithium carbonate for aggressive behavior of affective instability in ten brain-injured patients. *Am J Phys Med Rehabil* 1989; **68**:221–6.

Goggans FC. A case of mania secondary to vitamin B12 deficiency. *Am J Psychiatry* 1984; **141**:300–1.

Gomez EA, Aviles M. Neurosyphilis in community mental health clinics: a case series. *J Clin Psychiatry* 1984; **45**:127–9.

Greenberg DB, Brown GL. Single case study: mania resulting from brain stem tumor. *J Nerv Ment Dis* 1985; **173**:434–6.

Gross RA, Herridge P. A maniclike illness associated with a right frontal arteriovenous malformation. *J Clin Psychiatry* 1988; **49**:119–20.

Guarnieri R, Walz R, Hallak JE *et al.* Do psychiatric comborbidities predict postoperative seizure outcome in temporal lobe epilepsy surgery? *Epilepsy Behav* 2009; **14**:529–34.

Guze SB. The occurrence of psychiatric illness in systemic lupus erythematosus. *Am J Psychiatry* 1967; **123**:1562–70.

Guze SB, Woodrugg RA, Clayton PJ. The significance of psychotic affective disorders. *Arch Gen Psychiatry* 1975; **32**:1147–50.

Haller CA, Benowitz NL. Adverse cardiovascular and central nervous system events associated with dietary supplements containing ephedra alakaloids. *N Engl J Med* 2000; **343**:1833–8.

Han CS, Kim YK. A double-blind trial of risperidone and haloperidol for the treatment of delirium. *Psychosomatics* 2004; **45**:297–301.

Hardwick SW. Pellagra in psychiatric patients: twelve recent cases. *Lancet* 1943; **2**:43–5.

Haskett RF. Diagnostic categorization of psychiatric disturbance in Cushing's syndrome. *Am J Psychiatry* 1985; **142**:911–16.

Heathfield KW. Huntington's chorea. Investigation into the prevalence of the disease in the area covered by the North East Metropolitan Regional Hospital Board. *Brain* 1967; **90**:203–32.

Hermann M, Bartels C, Wallesch C-W. Depression in acute and chronic aphasia: symptoms, pathoanatomical–clinical correlations and functional implications. *J Neurol Neurosurg Psychiatry* 1993; **56**:672–8.

Hermann N, Rothenburg LS, Black SE *et al.* Methylphenidate for the treatment of apathy in Alzheimer disease: prediction of response using dextroamphetamine challenge. *J Clin Psychopharmacol* 2008; **28**:296–301.

Hoehn-Saric R, Lipsey JR, McLeod DR. Apathy and indifference in patients on fluvoxamine and fluoxetine. *J Clin Psychopharmacol* 1990; **10**:343–5.

Hohman LB. Epidemic encephalitis (lethargic encephalitis): its psychotic manifestations with a report of twenty-three cases. *Arch Neurol Psychiatry* 1921; **6**:295–333.

Holland JC, Korzun AH, Tross S *et al.* Comparative psychological disturbance in patients with pancreatic and gastric cancer. *Am J Psychiatry* 1986; **143**:982–6.

House A, Dennis M, Warlow C *et al.* Mood disorders after stroke and their relation to lesion location. *Brain* 1990; **113**:1113–29.

Hugo FJ, Halland AM, Spangenberg JJ *et al.* DSM-III-R classification of psychiatric symptoms in systemic lupus erythematosus. *Psychosomatics* 1996; **37**:262–9.

Hullett FJ, Potkin SG, Levy AB *et al.* Depression associated with nifedipine-induced calcium channel blockade. *Am J Psychiatry* 1988; **145**:1277–9.

Indaco A, Carrieri PB, Nappi C *et al.* Interictal depression in epilepsy. *Epilepsy Res* 1992; **12**:45–50.

Ingham SD, Nielsen JM. Thyroid psychosis: difficulties in diagnosis. *J Nerv Ment Dis* 1931; **74**:271–7.

Isella V, Melzi P, Grimaldi M *et al.* Clinical, neuropsychological, and morphometric correlates of apathy in Parkinson's disease. *Mov Disord* 2002; **17**:366–71.

Jack RA. A case of mania secondary to propafenone. *J Clin Psychiatry* 1985; **46**:104–5.

Jack RA, Rivers-Bulkeley NT, Rabin PL. Single case study: secondary mania as a presentation of progressive dialysis encephalopathy. *J Nerv Ment Dis* 1983; **171**:193–5.

Jackson SL. Psychosis due to isoniazid. *Br Med J* 1957; **2**:743–6.

Jensen K. Depression in patients treated with reserpine for arterial hypertension. *Acta Psychiatr Neurol Scand* 1959; **34**:195–204.

Jochum T, Bar KJ, Sauer H. Topiramate-induced manic episode. *J Neurol Neurosurg Psychiatry* 2002; **73**:208–10.

Johnson RJ, Richardson EP. The neurological manifestations of systemic lupus erythematosus. *Medicine* 1968; **47**:337–69.

Jones AM. Psychiatric presentation of a third ventricular colloid cyst in a mentally handicapped woman. *Br J Psychiatry* 1993; **163**:677–8.

Jones JE, Hermann BP, Barry JJ *et al.* Clinical assessment of Axis I psychiatric morbidity in chronic epilepsy: a multicenter investigation. *J Neuropsychiatry Clin Neurosci* 2005; **17**:172–9.

Jorge RE, Robinson RG, Starkstein SE *et al.* Secondary mania following traumatic brain injury. *Am J Psychiatry* 1993a; **150**:916–21.

Jorge RE, Robinson RG, Starkstein SE *et al.* Depression and anxiety following traumatic brain injury. *J Neuropsychiatry Clin Neurosci* 1993b; **5**:369–74.

Kahana E, Liebowitz U, Alter M. Cerebral multiple sclerosis. *Neurology* 1971; **21**:1179–85.

Kant R, Smith-Seemiller L, Zeiler D. Treatment of aggression and irritability after head injury. *Brain Inj* 1998; **12**:661–6.

Kaplitz SE. Withdrawn, apathetic geriatric patients responsive to methylphenidate. *J Am Geriatr Soc* 1975; **23**:271–6.

Karpati G, Frame B. Neuropsychiatric disorders in primary hyperparathyroidism. *Arch Neurol* 1964; **10**:387–97.

Kent TA, Wilber RD. Single case study: reserpine withdrawal psychosis: the possible role of denervation supersensitivity of receptors. *J Nerv Ment Dis* 1982; **170**:502–4.

Kim E, Bijlani M. A pilot study of quetiapine treatment of aggression due to traumatic brain injury. *J Neuropsychiatry Clin Neurosci* 2006; **18**:547–9.

Kirubakaren V, Mayfield D, Rengachary S. Dyskinesia and psychosis in a patient following baclofen withdrawal. *Am J Psychiatry* 1984; **141**:692–3.

Kraepelin E. *[Manic-depressive Insanity and Paranoia, 1921]* New York: Arno Press, 1976.

Krauss MR, Schafer A, Faller H *et al.* Psychiatric symptoms in patients with chronic hepatitis C receiving interferon alfa-2b therapy. *J Clin Psychiatry* 2003; **64**:708–14.

Kryzhanovskaya LA, Jeste DV, Young CA *et al.* A review of treatment emergent adverse events during olanzapine clinical trials in elderly patients with dementia. *J Clin Psychiatry* 2006; **67**:933–45.

Kulisevsky J, Berthier ML, Pujol J. Hemiballismus and secondary mania following right thalamic infarction. *Neurology* 1993; **43**:1422–4.

Kulisevsky J, Asuncion A, Berthier ML. Bipolar affective disorder and unilateral parkinsonism after a brainstem infarction. *Movement Disord* 1995; **10**:799–802.

Kwentus JA, Silverman JJ, Sprague M. Manic syndrome after metrizamide myelography. *Am J Psychiatry* 1984; **141**:700–2.

Labbatte LA, Holzgang AJ. Manic syndrome after discontinuation of methyldopa. *Am J Psychiatry* 1989; **146**:1075–6.

Lawlor BA. Hypocalcemia, hypoparathyroidism, and organic anxiety syndrome. *J Clin Psychiatry* 1988; **49**:317–18.

Lee S, Chow CC, Wing YK *et al.* Mania secondary to thyrotoxicosis. *Br J Psychiatry* 1991; **159**:712–13.

Lendvai I, Saravay SM, Steinberg MD. Creutzfeldt–Jakob disease presenting as secondary mania. *Psychosomatics* 1999; **40**:524–5.

Lesem MD, Zajecka JM, Swift RH *et al.* Intramuscular ziprasidone, 2 mg versus 10 mg, in the short-term management of agitated psychotic patients. *J Clin Psychiatry* 2001; **62**:12–18.

Levy ML, Miller BL, Cummings JL *et al.* Alzheimer disease and frontotemporal dementias. Behavioral distinctions. *Arch Neurol* 1996; **53**:687–90.

Levy ML, Cummings JL, Fairbanks LA *et al.* Apathy is not depression. *J Neuropsychiatry Clin Neurosci* 1998; **10**:314–19.

Liebson E. Anosognosia and mania associated with right thalamic hemorrhage. *J Neurol Neurosurg Psychiatry* 2000; **68**:107–8.

Liegghid NE, Yeragani VK. Buspirone-induced hypomania. *J Clin Psychopharmacol* 1988; **8**:226–7.

Lim L, Ron MA, Ormerod IEC *et al.* Psychiatric and neurological manifestations in systemic lupus erythematosus. *Q J Med* 1988; **66**:27–38.

Lin J T-Y, Ziegler DK. Psychiatric symptoms with initiation of carbidopa-levodopa treatment. *Neurology* 1976; **26**:699–700.

Linder J, Brismar K, Granberg P-O *et al.* Characteristic changes in psychiatric symptoms, cortisol and melatonin but not prolactin in primary hyperparathyroidism. *Acta Psychiatr Scand* 1988; **78**:32–40.

Lipsey JR, Spencer WC, Rabins PV *et al.* Phenomenological comparison of functional and post-stroke depression. *Am J Psychiatry* 1986; **143**:527–9.

Litvan I, Mega MS, Cummings JL *et al.* Neuropsychiatric aspects of progressive supranuclear palsy. *Neurology* 1996; **47**:1184–9.

Loudon JB, Blackburn IM, Ashworth CM. A study of the symptomatology and course of manic illness using a new scale. *Psychol Med* 1977; **7**:723–9.

Louie AK, Lannon RA, Rutzick EA *et al.* Clinical features of cocaine-induced panic. *Biol Psychiatry* 1996; **40**:938–40.

Lyons PR, Newman PK, Saunders M. Methylprednisolone therapy in multiple sclerosis: a profile of adverse effects. *J Neurol Neurosurg Psychiatry* 1988; **51**:285–7.

MacHale SM, O'Rourke SJ, Wardlaw JM *et al.* Depression and its relation to lesion location after stroke. *J Neurol Neurosurg Psychiatry* 1998; **64**:371–4.

McDaniel JS, Musselman DL, Proter MR. Depression in patients with cancer. *Arch Gen Psychiatry* 1995; **52**:89–99.

McGilchrist I, Goldstein LH, Jadresic D *et al.* Thalamo-frontal psychosis: a case report. *Br J Psychiatry* 1993; **163**:113–15.

McLachlan RS, Blume WT. Isolated fear in complex partial status epilepticus. *Ann Neurol* 1980; **8**:639–41.

Malamud N. Psychiatric disorder with intracranial tumors of the limbic system. *Arch Neurol* 1967; **17**:113–23.

Malinow KC, Lion JR. Hyperaldosteronism (Conn's disease) presenting as depression. *J Clin Psychiatry* 1979; **40**:358–9.

Mann AM, Hutchinson JL. Manic reaction associated with procarabazine hydrochloride therapy of Hodgkin's disease. *Can Med Assoc J* 1967; **97**:1350–3.

Mattox JH, Buckman MT, Bernstein J *et al.* Dopamine agonists for reducing depression associated with hyperprolactinemia. *J Reproduct Med* 1986; **31**:694–8.

Maxwell S, Scheftner WA, Kessler HA *et al.* Manic syndrome associated with zidovudine treatment. *JAMA* 1988; **259**:3406–7.

Meehan KM, Wang H, David SR *et al.* Comparison of rapidly acting intramuscular olanzapine, lorazepam, and placebo: a double-blind, randomized study in acutely agitated patients with dementia. *Neuropsychopharmacology* 2002; **26**:494–504.

Mendez MF, Cummings JL, Benson DF. Depression in epilepsy. Significance and phenomenology. *Arch Neurol* 1986; **43**:766–70.

Merritt H, Springlova M. Lissauer's dementia paralytica: a clinical and pathologic study. *Arch Neurol Psychiatry* 1932; **27**:987–1030.

Miguel EC, Rodriguez Pereira RM, de Braganca Pereira CA *et al.* Psychiatric manifestations of systemic lupus erythematosus: clinical features, symptoms, and signs of central nervous system activity in 43 patients. *Medicine* 1994; **73**:224–32.

Minden SL, Orav J, Schildkraut JJ. Hypomanic reactions to ACTH and prednisone treatment for multiple sclerosis. *Neurology* 1988; **38**:1631–4.

Modlin IM, Farndon JR, Shepherd A *et al.* Pheochromocytomas in 72 patients: clinical and diagnostic features, treatment and long-term results. *Br J Surg* 1979; **66**:456–65.

Morris RLP, Robinson RG, Raphael B *et al.* Lesion location and poststroke depression. *J Neuropsychiatr Clin Neurosci* 1996; **8**:399–403.

Moscovich DG, Singh MB, Efa FJ *et al.* Acute disseminated encephalomyelitis presenting as an acute psychotic state. *J Nerv Ment Dis* 1995; **183**:116–17.

Mula M, Trimble MR, Yuen A *et al.* Psychiatric adverse events during levetriracetam therapy. *Neurology* 2005; **61**:704–6.

Mulder DW, Daly D. Psychiatric symptoms associated with lesions of temporal lobe. *JAMA* 1952; **150**:173–6.

Musselman DL, Lawson DH, Gumnick JF *et al.* Paroxetine for the prevention of depression induced by high-dose interferon-alfa. *N Engl J Med* 2001; **344**:961–6.

Mysiw WJ, Jackson RD, Corrigan JD. Amitriptyline for post-traumatic agitation. *Am J Phys Med Rehabil* 1988; **67**:29–33.

Navarro V, Vieta E, Gasto C. Mannitol-induced acute manic state. *J Clin Psychiatry* 2001; **62**:126.

Neilley LK, Goodin DS, Goodkin DE *et al.* Side effect profile of interferon beta-1b in MS: results of an open-label trial. *Neurology* 1996; **46**:552–4.

Newburn G, Newburn D. Selegiline in the management of apathy following traumatic brain injury. *Brain Inj* 2005; **19**:149–54.

Nickel SN, Frame B. Neurologic manifestations of myxedema. *Neurology* 1958; **8**:511–17.

Nishida T, Kudo T, Inoue Y *et al.* Postictal mania versus postictal psychosis: differences in clinical features, epileptogenic zone, and brain functional changes during postictal period. *Epilepsia* 2006; **47**:2104–14.

Nizamie SH, Mizamie A, Borde M *et al.* Mania following head injury: case reports and neuropsychological findings. *Acta Psychiatr Scand* 1988; **77**:637–9.

O'Brien CP, DiGiacomo JN, Fahn J *et al.* Mental effects of high-dose levodopa. *Arch Gen Psychiatry* 1971; **24**:61–4.

O'Dowd MA, McKegney FP. Manic syndrome associated with zidovudine. *JAMA* 1988; **260**:3587–8.

Olin JT, Fox LS, Pawluczyk S *et al.* A pilot randomized trial of carbamazepine for behavioral symptoms in treatment-resistant outpatients with Alzheimer disease. *Am J Geriatr Psychiatry* 2001; **9**:400–5.

Oliver JE. Huntington's chorea in Northamptonshire. *Br J Psychiatry* 1970; **116**:241–53.

Pandharipande PP, Pun BT, Herr DL *et al.* Effect of sedation with dexmedetomidine vs lorazepam on acute brain dysfunction in mechanically ventilated patients: the MENDS randomized controlled trial. *JAMA* 2007; **298**:2644–53.

Papolos DF, Faedda GL, Veit S *et al.* Bipolar spectrum disorders in patients diagnosed with velo-cardio-facial syndrome: does a hemizygous deletion of chromosome 22q11 result in bipolar affective disorder? *Am J Psychiatry* 1996; **153**:1541–7.

Parker N. Manic-depressive psychosis following head injury. *Med J Aust* 1957; **2**:20–2.

Paskavitz JF, Anderson CA, Filley CM *et al.* Acute arcuate fiber demyelinating encephalopathy following Epstein–Barr virus infection. *Ann Neurol* 1995; **38**:127–31.

Patten SB, Klein GM, Lussier C *et al.* Organic mania induced by phenytoin: a case report. *Can J Psychiatry* 1989; **34**:457.

Paul R, Flanigan TP, Tashima K *et al.* Apathy correlates with cognitive function but not CD4 status in patients with human immunodeficiency virus. *J Neuropsychiatry Clin Neurosci* 2005; **17**: 114–18.

Perini GI, Tosin C, Carraro C *et al.* Interictal mood and personality disorders in temporal lobe epilepsy and juvenile myoclonic epilepsy. *J Neurol Neurosurg Psychiatry* 1996; **61**:601–5.

Perry TL, Bratty PJA, Hansen S *et al.* Hereditary mental depression and parkinsonism with taurine deficiency. *Arch Neurol* 1975; **32**:108–13.

Petersen P. Psychiatric disorders in primary hyperparathyroidism. *J Clin Endo Metab* 1968; **28**:1491–5.

Petrie WM, Maffucci RJ, Woosley RL. Propranolol and depression. *Am J Psychiatry* 1982; **139**:93–4.

Pine DS, Douglas CJ, Charles E *et al.* Patients with multiple sclerosis presenting to psychiatric hospitals. *J Clin Psychiatry* 1995; **56**:297–306.

Pluck GC, Brown RG. Apathy in Parkinson's disease. *J Neurol Neurosurg Psychiatry* 2002; **73**:636–42.

Pollack MH, Rosenbaum JF, Cassem NH. Brief communication: propranolol and depression revisited: three cases and a review. *J Nerv Ment Dis* 1985; **173**:118–19.

Pollock BG, Mulsant BH, Rosen J *et al.* Comparison of citalopram, perphenazine, and placebo for the acute treatment of psychosis and behavioral disturbances in hospitalized, demented patients. *Am J Psychiatry* 2007; **159**:460–5.

Pope HG, Katz DL. Affective and psychotic symptoms associated with anabolic steroid use. *Am J Psychiatry* 1988; **145**:487–90.

Pope HG, Katz DL. Psychiatric and medical effects of anabolic-androgenic steroid use: a controlled study of 160 athletes. *Arch Gen Psychiatry* 1994; **51**:375–82.

Pope HG, Lipinski JF, Cohen BM *et al.* Schizoaffective disorder': an invalid diagnosis? A comparison of schizoaffective disorder, schizophrenia and affective disorder. *Am J Psychiatry* 1980; **137**:921–7.

Powell E. Two fatal cases of acute chorea with insanity. *Brain* 1889; **12**:157–60.

Powell JH, al-Adawi S, Morgan J *et al.* Motivational deficits after brain injury: effects of bromocriptine in 11 patients. *J Neurol Neurosurg Psychiatry* 1996; **60**:416–21.

Porsteinsson AP, Tariot PN, Erb R *et al.* Placebo-controlled study of divalproex for agitation in dementia. *Am J Geriatr Psychiatry* 2001; **9**:58–66.

Price WA, Bielfeld M. Buspirone-induced mania. *J Clin Psychopharmacol* 1989; **9**:150–1.

Pushpal D, Shamshul HN. Tiagabine withdrawal-emergent mania. *Aust N Z J Psychiatry* 2006; **40**:719.

Quetsch RM, Achor RWP, Litin EM *et al.* Depressive reactions in hypertensive patients. *Circulation* 1959; **19**:366–75.

Rabins PV, Brooks BR, O'Donnell P *et al.* Structural brain correlates of emotional disorder in multiple sclerosis. *Brain* 1986; **109**:585–97.

Rainer M, Haushofer M, Pfolz H *et al.* Quetiapine versus risperidone in elderly patients with behavioural and psychological symptoms of dementia: efficacy, safety and cognitive function. *Eur Psychiatry* 2007; **22**:395–403.

Raison CL, Borisov AS, Broadwell SD *et al.* Depression during pegylated interferon-alpha plus ribaverin therapy: prevalence and presentation. *J Clin Psychiatry* 2005; **66**:41–8.

Rappaport SA, Marcus RN, Manos G *et al.* A randomized, double-blind, placebo-controlled tolerability study of intramuscular aripiprazole in acutely agitated patients with Alzheimer's, vascular, or mixed dementia. *J Am Med Dir Assoc* 2009; **10**:21–7.

Reade MC, O'Sullivan K, Bates S *et al.* Dexmedetomidine vs. haloperidol in delirious, agitated, intubated patients: a randomized, open-label trial. *Crit Care* 2009; **13**:R75.

Reaser EF. Chorea of infectious origin. *South Med J* 1940; **33**: 1324–8.

Reinfrank RF. Primary hyperparathyroidism with depression. *Arch Int Med* 1961; **108**:162–6.

Reynolds CF, Miller MD, Pasternak RE *et al.* Treatment of bereavement-related major depressive episodes in later life: a controlled study of acute and continuation treatment with nortriptyline and interpersonal psychotherapy. *Am J Psychiatry* 1999; **156**:202–8.

Rickles NK. Functional symptoms as first evidence of pancreatic disease. *J Nerv Ment Dis* 1945; **101**:566–71.

Riker RR, Shehabi Y, Bokesch PM *et al.* Dexmedetomidine vs midazolam for sedation in critically ill patients: a randomized trial. *JAMA* 2009; **301**:489–99.

Robinson RG, Kubos KL, Starr LB *et al.* Mood changes in stroke patients: relationship to lesion location. *Compr Psychiatry* 1983; **24**:555–66.

Robinson RG, Kubos KL, Starr LB *et al.* Mood disorders in stroke patients. *Brain* 1984; **107**:81–93.

Robinson RG, Lipsey JR, Bolla-Wilson K *et al.* Mood disorders in left-handed stroke patients. *Am J Psychiatry* 1985; **142**:1424–9.

Robinson RG, Bolduc P, Price TR. A two year longitudinal study of post-stroke depression: diagnosis and outcome at one and two year follow-up. *Stroke* 1987; **18**:837–43.

Robinson RG, Boston JD, Starkstein SE *et al.* Comparison of mania and depression after brain injury: causal factors. *Am J Psychiatry* 1988; **145**:172–8.

Rosenthal NE, Rosenthal LN, Stallone F *et al.* Psychosis as a predictor of response to lithium maintenance treatment in bipolar affective disorder. *J Affect Disord* 1979; **1**:237–45.

Rosenthal NE, Rosenthal LN, Stallone F *et al.* Toward the validation of RDC schizoaffective disorder. *Arch Gen Psychiatry* 1980; **37**:804–10.

Roth RM, Flashman LA, Saykin AJ *et al.* Apathy in schizophrenia: reduced frontal lobe volume and neuropsychological deficits. *Am J Psychiatry* 2004; **161**:157–9.

Rubinsztein JS, Rubinsztein DC, Goodburn S *et al.* Apathy and hypersomnia are common features of myotonic dystrophy. *J Neurol Neurosurg Psychiatry* 1998; **64**:510–15.

Ryback RS, Schwab RS. Manic response to levodopa therapy: report of a case. *N Engl J Med* 1971; **285**:788–9.

Sackeim HA, Greenberg MS, Weman AI *et al.* Hemispheric asymmetry in the expression of positive and negative emotions: neurologic evidence. *Arch Neurol* 1982; **39**:210–18.

Sale I, Kalucy P. Psychosis associated with oral contraceptive-induced chorea. *Med J Aust* 1981; **1**:79–80.

Satterthwaite TD, Wolf DH, Rosenheck RA *et al.* A meta-analysis of the risk of acute extrapyramidal symptoms with intramuscular antipsychotics for the treatment of agitation. *J Clin Psychiatry* 2008; **69**:1869–79.

Sawyer DR, Connor CS, Rumack BH. Managing acute toxicity from nonprescription stimulants. *Clin Pharm* 1982; **1**:529–33.

Sayed AJ. Mania and bromism: a case report and a look to the future. *Am J Psychiatry* 1976; **133**:228–9.

Sazgar M, Carlen PL, Wennberg R. Panic attack semiology in right temporal lobe epilepsy. *Epileptic Disord* 2003; **5**:93–100.

Schneider LS, Tariot PN, Dagerman KS *et al.* Effectiveness of atypical antipsychotic drugs in patients with Alzheimer's disease. *N Engl J Med* 2006; **355**:1525–38.

Shaskan D. Mental changes in chorea minor. *Am J Psychiatry* 1938; **95**:193–202.

Shulman RB, Scheftner WA, Mayudu S. Venlafaxine-induced mania. *J Clin Psychopharmacol* 2001; **21**:239–41.

Siegal FP. Lithium for steroid-induced psychosis. *N Engl J Med* 1978; **299**:155–6.

Singh A, Althoff R, Martineau RJ *et al.* Pramipexole, ropinirole, and mania in Parkinson's disease. *Am J Psychiatry* 2005; **162**:814–15.

Sival RC, Haffmans PM, Jansen PA *et al.* Sodium valproate in the treatment of aggressive behavior in patients with dementia: a randomized placebo controlled clinical trial. *Int J Geriar Psychiatry* 2002; **17**:579–85.

Skrobik YK, Bergeron N, Dumont M *et al.* Olanzapine vs haloperidol: treating delirium in a critical care setting. *Intensive Care Med* 2004; **30**:444–9.

Slyter H. Idiopathic hypoparathyroidism presenting as dementia. *Neurology* 1979; **29**:393–4.

Spiegel DR, Kim J, Greene K *et al.* Apathy due to cerebrovascular accidents successfully treated with methylphenidate: a case series. *J Neuropsychiatry Clin Neurosci* 2009; **21**:216–19.

Starkman MN, Schteingart DE, Schork MA. Depressed mood and other psychiatric manifestations of Cushing's syndrome: relationship to hormone levels. *Psychosom Med* 1981; **43**:3–18.

Starkman MN, Zelnick TC, Nesse RM *et al.* Anxiety in patients with pheochromocytomas. *Arch Int Med* 1985; **145**:248–52.

Starkstein SE, Robinson RG, Price TR. Comparison of cortical and subcortical lesions in the production of poststroke mood disorders. *Brain* 1987a; **110**:1045–59.

Starkstein SE, Pearlson GD, Boston J *et al.* Mania after brain injury: a controlled study of causative factors. *Arch Neurol* 1987b; **44**:1069–73.

Starkstein SE, Robinson RG, Berthier ML *et al.* Differential mood changes following basal ganglia versus thalamic lesions. *Arch Neurol* 1988a; **45**:725–30.

Starkstein SE, Boston JD, Robinson RG. Mechanism of mania after brain injury: 12 case reports and review of the literature. *J Nerv Ment Dis* 1988b; **176**:87–100.

Starkstein SE, Mayberg HS, Berthier ML *et al.* Mania after brain injury: neuroradiological and metabolic findings. *Ann Neurol* 1990a; **27**:652–9.

Starkstein SE, Cohen BS, Federoff P *et al.* Relationship between anxiety disorders and depressive disorders in patients with cerebrovascular injury. *Arch Gen Psychiatry* 1990b; **47**:785–9.

Starkstein SE, Federoff JP, Price TR *et al.* Apathy following cerebrovascular lesions. *Stroke* 1993; **24**:1625–30.

Starkstein SE, Petracca G, Chemerinski E *et al.* Syndromic validity of apathy in Alzheimer's disease. *Am J Psychiatry* 2001; **158**:872–7.

Starkstein SE, Ingram L, Garau ML *et al.* On the overlap between apathy and depression in dementia. *J Neurol Neurosurg Psychiatry* 2005; **76**:1070–4.

Starkstein WE, Jorge R, Mizrahi R *et al.* A prospective longitudinal study of apathy in Alzheimer's disease. *J Neurol Neurosurg Psychiatry* 2006; **77**:8–11.

Stern K, Dancey TE. Glioma of the diencephalon in a manic patient. *Am J Psychiatry* 1942; **98**:716–19.

Stern RA, Bachman DL. Depressive symptoms following stroke. *Am J Psychiatry* 1991; **148**:351–6.

Stoll A1, Mayer PV, Kolbrener M *et al.* Antidepressant-associated mania: a controlled comparison with spontaneous mania. *Am J Psychiatry* 1994; **151**:1642–5.

Storm-Mathisen A. General paresis: a follow-up study of 203 patients. *Acta Psychiatr Scand* 1969; **45**:118–32.

Street JS, Clark WS, Gannnon KS *et al.* Olanzapine treatment of psychotic and behavioral symptoms in patients with Alzheimer disease in nursing care facilities: a double-blind, randomized, placebo-controlled trial. The HGEU Study Group. *Arch Gen Psychiatry* 2000; **57**:968–76.

Sullivan KL, Ward CL, Zesiewicz TA. Zonisamide-induced mania in essential tremor patient. *J Clin Psychopharmacol* 2006; **26**:439–40.

Sultzer DL, Gray KF, Gunay I *et al.* A double-blind comparison of trazodone and haloperidol for treatment of agitation in patients with dementia. *Am J Geriatr Psychiatry* 1997; **5**:60–9.

Sultzer DL, Davis SM, Tariot PN *et al.* Clinical symptom responses to atypical antipsychotic medications in Alzheimer's disease: phase 1 outcomes from the CATIE-AD effectiveness trial. *Am J Psychiatry* 2008; **165**:844–54.

Surridge D. An investigation into some psychiatric aspects of multiple sclerosis. *Br J Psychiatry* 1969; **115**:749–64.

Tamir A, Whittier J, Korenys C. Huntington's chorea: a sex difference in psychopathological symptoms. *Dis Nerv Syst* 1969; **30**:103.

Tariot PN, Erb R, Podgorski CA *et al.* Efficacy and tolerability of carbamazepine for agitation and aggression in dementia. *Am J Psychiatry* 1998; **155**:54–61.

Tariot PN, Raman R, Jakimovich L *et al.* Divalproex sodium in nursing home residents with possible or probably Alzheimer disease complicated by agitation: a randomized, controlled trial. *Am J Geriatr Psychiatry* 2005; **13**:942–9.

Taylor MA, Abrams R. The phenomenology of mania: a new look at some old patients. *Arch Gen Psychiatry* 1973; **29**:520–2.

Taylor MA, Abrams R. Catatonia: prevalence and importance in the manic phase of manic-depressive illness. *Arch Gen Psychiatry* 1977; **34**:1223–5.

Teri L, Logsdon RG, Peskind E *et al.* Treatment of agitation in AD: a randomized, placebo-controlled trial. *Neurology* 2000; **55**:1271–8.

Thomas JE, Rooke ED, Kuale WF. The neurologist's experience with pheochromocytoma: a review of 100 cases. *J Am Med Assoc* 1966; **197**:754–8.

Torriani FJ, Rodriguez-Turres M, Rockstroh JK *et al.* Peginterferon alfa-2a plus ribaverin for chronic hepatitis C in HIV-infected patients. *N Engl J Med* 2004; **351**:438–50.

Tran-Johnson TK, Sack DA, Marcus RN *et al.* Efficacy and safety of intramuscular aripiprazole in patients with acute agitation: a randomized, double-blind, placebo-controlled trial. *J Clin Psychiatry* 2007; **68**:111–19.

Trautner RJ, Cummings JL, Read SL *et al.* Idiopathic basal ganglia calcification and organic mood disorder. *Am J Psychiatry* 1988; **145**:350–3.

Trembath PW, Boobis SW. Plasma theophylline levels after sustained-release aminophylline. *Clin Pharmacol Ther* 1979; **26**:654–9.

Trzepacz P, McCue M, Klein I *et al.* A psychiatric and neuropsychological study of patients with untreated Graves' disease. *Gen Hosp Psychiatry* 1988; **10**:49–55.

Van Der Horst L. Rheumatism and psychosis. *Digest Neurol Psychiatry* 1947; **15**:399–416.

van Reekum R, Bayley M, Garner S *et al.* N of 1 study: amantadine for the amotivational syndrome in a patient with traumatic brain injury. *Brain Inj* 1995; **9**:49–53.

Van Woert MH, Ambani LM, Weintraub MI. Manic behavior and levodopa. *N Engl J Med* 1971; **285**:1326.

Vazquez A, Jimenez-Jimenez FJ, Garcia-Ruiz P *et al.* 'Panic attacks' in Parkinson's disease. *Acta Neurol Scand* 1993; **87**:14–18.

Verhey FR, Verkaaik M, Lousberg R. Olanzapine versus haloperidol in the treatment of agitation in elderly patients with dementia: results of a randomized controlled double-blind trial. *Dement Geriatr Cogn Disord* 2006; **21**:1–8.

Veser FH, Veser BD, McMullan JT *et al.* Risperidone versus haloperidol, in combination with lorazepam, in the treatment of acute agitation and psychosis: a pilot, randomized, double-blind, placebo-controlled trial. *J Psychiatr Pract* 2006; **12**:103–8.

Wagner KD. Major depression and anxiety disorders associated with Norplant. *J Clin Psychiatry* 1996; **57**:152–7.

Wagner KD, Berenson AB. Norplant-associated major depression and panic disorder. *J Clin Psychiatry* 1994; **55**:478–80.

Walton RG. Seizure and mania after high intake of aspartame. *Psychosomatics* 1986; **27**:218–20.

Wang PS, Schneeweiss S, Avorn J *et al.* Risk of death in elderly users of conventional vs atypical antipsychotic medicines. *N Engl J Med* 2005; **353**:2335–41.

Watanabe MD, Martin EM, DeLeon OA *et al.* Successful methylphenidate treatment of apathy after subcortical infarcts. *J Neuropsychiatry Clin Neurosci* 1995; **7**:502–4.

Weller M, Liedtke W, Petersen D *et al.* Very-late-onset adrenoleukodystrophy: possible precipitation of demyelination by cerebral contusion. *Neurology* 1992; **42**:367–70.

Wier LM, Tavares SB, Tyrka AR *et al.* Levetiracetam-induced depression in a healthy adult. *J Clin Psychiatry* 2006; **67**:1159–60.

Williams D. The structure of emotions reflected in epileptic experiences. *Brain* 1956; **79**:29–67.

Wilson P, Preece AA. Chorea gravidarum. *Arch Int Med* 1932; **49**:671–97.

Winokur G. Psychosis in bipolar and unipolar affective illness with special reference to schizo-affective disorder. *Br J Psychiatry* 1984; **145**:236–42.

Winokur G, Tsuang MT. Elation versus irritability in mania. *Compr Psychiatry* 1975; **16**:435–6.

Winokur G, Clayton PJ, Reich T. *Manic-depressive Illness*. St Louis, MO: CV Mosby, 1969.

Winokur G, Coryell W, Endicott J *et al.* Further distinctions between manic-depressive illness (bipolar disorder) and primary depressive disorder (unipolar depression). *Am J Psychiatry* 1993; **150**: 1176–81.

Wolkowitz OM, Rubinow D, Doran AR *et al.* Prednisone effects on neurochemistry and behavior. *Arch Gen Psychiatry* 1990; **47**: 963–8.

Wright JM, Sachdev PS, Perkins RJ *et al.* Zidovudine related mania. *Med J Aust* 1989; **150**:339–41.

Wright P, Birkett M, David SR *et al.* Double-blind, placebo-controlled comparison of intramuscular olanzapine and intramuscular haloperidol in the treatment of acute agitation in schizophrenia. *Am J Psychiatry* 2001; **158**:1149–51.

Wroblewski BA, Joseph AB, Kupfer J *et al.* Effectiveness of valproic acid on destructive and aggressive behaviors in patients with acquired brain injury. *Brain Inj* 1997; **11**:37–47.

Wysowski DK, Pitts M, Beiz J. Depression and suicide in patients treated with isotretinoin. *N Engl J Med* 2001; **344**:460.

8

Other major syndromes

8.1 PSYCHOSIS

The term psychosis has been used differently by different authors, and this has led to some confusion in the literature. In this text, psychosis refers to a condition characterized by hallucinations (without insight) and/or delusions, in the absence of either significant cognitive deficits or pronounced disturbances of mood.

Clinical features

Delusions and hallucinations are discussed at length in Chapter 5.30, and readers unfamiliar with these are encouraged to consult that chapter. Although in most cases of psychosis both delusions and hallucinations are present, exceptions do occur; thus in some disorders, for example delusional disorder, one may find only delusions, whereas in the psychosis caused by levodopa in patients with parkinsonian conditions one may find only hallucinations. Critically, as discussed in Chapter 5.30, in cases characterized by only hallucinations the diagnosis of psychosis should be reserved for conditions in which insight is absent and patients react to their hallucinations as if they were real.

Depending on the cause of the psychosis, other symptoms may also be present; however, the part they play in the overall clinical picture is generally relatively minor compared with the delusions and hallucinations. Thus, there may be some incoherence, minor mood changes, anxiety, or even agitation.

Etiology

The various causes of psychosis are listed in Table 8.1, in which they are divided into several groups. The first group, composed of *idiopathic disorders*, constitutes by far the most common causes of psychosis, and of these schizophrenia is the most frequent. Next are the *toxic psychoses*, for example those seen with stimulants such as amphetamine or cocaine. *Endocrinologic psychoses*, such as the 'myxedema madness' of hypothyroidism, are considered next, followed by various *intracranial disorders* capable of causing psychosis, such as stroke or tumors. Consideration is then given to the various *epileptic psychoses* and then to *encephalitic and postencephalitic psychoses*. Finally, there is a *miscellaneous* group, including the 'megaloblastic madness' that may be seen with vitamin B12 deficiency.

IDIOPATHIC DISORDERS

Schizophrenia, discussed in Chapter 22.1, is by far the most common cause of chronic psychosis. The onset typically occurs in the late teens or early twenties, with the subacute or gradual elaboration of a psychosis characterized by varying combinations of hallucinations, delusions, incoherence, and bizarre behavior. In many cases, the symptomatology will crystallize into an enduring and recognizable subtype: paranoid, hebephrenic, catatonic, or simple. Although the symptoms gradually wax and wane over time, the illness is generally chronic and life-long, probably never going into a spontaneous and full remission.

Schizoaffective disorder, discussed in Chapter 22.2, is, like schizophrenia, characterized by a chronic psychosis: the difference is that in schizoaffective disorder one also finds recurrent episodes of either depression or mania, during either of which the chronically present psychotic symptoms undergo a significant exacerbation.

Delusional disorder, discussed in Chapter 22.3, also like schizophrenia, is characterized by a chronic psychosis: here, however, hallucinations, incoherence, and bizarre behavior are negligible or absent, with the primary or sole symptom of the illness being one or more delusions. Importantly, these

Table 8.1 Causes of psychosis

Causes of psychosis
Idiopathic disorders
Schizophrenia
Schizoaffective disorder
Delusional disorder
Postpartum psychosis
Obsessive–compulsive disorder
Body dysmorphic disorder
Borderline personality disorder
Toxic psychoses
Amphetamine
Cocaine
Hallucinogens
Phencyclidine
Cannabis
Anabolic steroids
Chronic alcoholism (alcoholic paranoia, alcoholic hallucinosis)
Neuroleptic-induced supersensitivity psychosis ('tardive psychosis')
Dopaminergics (levodopa, direct-acting dopaminergics, amantadine)
Levetiracetam
Topiramate
Vigabatrin
Phenylpropanolamine
Ephedrine
Bupropion
Fluoxetine
Disulfiram
Methysergide
Manganese intoxication
Baclofen withdrawal
Endocrinologic psychoses
Hypothyroidism
Hyperthyroidism
Cushing's syndrome
Adrenocortical insufficiency
Intracranial disorders
Stroke
Temporal lobe
Frontal lobe
Thalamus
Tumors
Temporal lobe
Corpus callosum
Frontal lobe
Parietal lobe
Multiple sclerosis
Traumatic brain injury
Neurodegenerative disorders
Huntington's disease
Dentatorubropallidoluysian atrophy
Spinocerebellar ataxia
Miscellaneous
Wilson's disease
Creutzfeldt–Jakob disease
Fatal familial insomnia
Fahr's syndrome
Aqueductal stenosis
Epileptic psychoses
Ictal psychosis
Postictal psychosis
Chronic interictal psychosis
Psychosis of forced normalization
Encephalitic and postencephalitic psychoses
Encephalitic
Herpes simplex encephalitis
Infectious mononucleosis
Encephalitis lethargica
Postencephalitic
Herpes simplex encephalitis
Encephalitis lethargica
Miscellaneous
Vitamin B12 deficiency
Neurosyphilis
Acquired immune deficiency syndrome
Systemic lupus erythematosus
Sydenham's chorea
Chorea gravidarum
Hepatic porphyria
Metachromatic leukodystrophy
Velocardiofacial syndrome
Vanishing white matter leukoencephalopathy
Subacute sclerosing panencephalitis
Prader–Willi syndrome

delusions are not bizarre but indeed have a certain plausibility to them. Certain variants of this disorder deserve special mention: parasitosis is characterized by a persistent belief that one is infested by some parasitic bug or other (Andrews *et al.* 1986; Mitchell 1989) and the olfactory reference syndrome by a delusion that one is emitting a foul odor (e.g. as in halitosis or flatus) that others detect and comment on (Videbech 1966).

Postpartum psychosis, discussed in Chapter 22.4, may appear between several days and several months postpartum and is often characterized by prominent agitation. Importantly, it is not uncommon for certain disorders, such as schizophrenia, to undergo an exacerbation postpartum, and such patients should not receive an additional diagnosis of postpartum psychosis.

The following disorders, namely obsessive–compulsive disorder, body dysmorphic disorder, and borderline personality disorder, although generally not associated with delusions and hallucinations, may at times cause these symptoms and hence are included here.

Both obsessive–compulsive disorder, discussed in Chapter 24.5, and body dysmorphic disorder (McElroy *et al.* 1993) may have psychotic subtypes wherein patients 'lose insight' and come to accept their troubling ideas as true. Thus, a patient with obsessive–compulsive disorder might come to believe that his troubling need to pray recurrently is, in fact, ordained by God, or a patient with body dysmorphic disorder may come to believe that his or her face is, in fact, deformed.

Patients with a borderline personality disorder, as discussed in Chapter 26.5, may, when under great stress, develop transient auditory hallucinations or delusions of persecution: such patients are distinguished by the chronic characteristic traits of intolerance of being alone, anger, impulsivity, and disturbed relationships.

TOXIC PSYCHOSES

Amphetamines, if taken in a sufficiently high dose, may cause a psychosis (Bell 1973; Griffith *et al.* 1972) that is typically characterized by delusions of persecution and often of reference. Hallucinations may also occur, being much more commonly auditory than visual. The psychosis typically clears within a week, but in some cases longer durations of up to 3 months have been reported (Iwanami *et al.* 1994).

Cocaine may cause a psychosis characterized by hallucinations, more often auditory than visual, and delusions of persecution and reference (Brady *et al.* 1991; Satel *et al.* 1991; Sherer *et al.* 1988). Although such a psychosis may occur in 'recreational' users (Siegel 1978), it is more characteristic of addicts, who often note that a progressively lower 'dose' becomes capable of inducing the psychosis (Brady *et al.* 1991). Although, in most cases, the psychosis clears either with the intoxication itself or shortly thereafter, it may persist in some until after the withdrawal 'crash' resolves (Satel *et al.* 1991).

Hallucinogens, such as LSD (lysergic acid diethylamide), typically cause visual hallucinations, but most patients remain aware of their unreality. In a minority, however, the intoxication will be complicated by delusions of persecution (Bercel *et al.* 1956; Kuramochi and Takahashi 1964).

Phencyclidine intoxication may render patients agitated and psychotic (Allen and Young 1978), with delusions of grandeur or persecution and auditory hallucinations. The finding of nystagmus is a very important diagnostic clue.

Cannabis intoxication, if a sufficiently high dose is taken, may be characterized by fearfulness and delusions of persecution and reference (Kroll 1975; Thacore and Shukla 1976) that may outlast the intoxication itself by a matter of days.

Anabolic steroids, as may be abused by athletes, may, in a small minority, cause a psychosis variously characterized by delusions of persecution or grandeur, delusions of reference, and auditory hallucinations (Pope and Katz 1988). The appearance of a psychosis in a 'bulked-up' young person should suggest this diagnosis.

Chronic alcoholism may be complicated by two different psychoses: alcoholic paranoia and alcohol hallucinosis. Alcoholic paranoia, as discussed in Chapter 29.10, is characterized by the gradual development of delusions, often of either jealousy or persecution. By contrast, alcohol hallucinosis, discussed in Chapter 29.9, typically appears as a sequela to an alcohol withdrawal delirium, whereas the other symptoms of the delirium tremens (DTs), such as tremor, clear, the auditory hallucinations, often accompanied by delusions of persecution, persist.

Antipsychotic-induced supersensitivity psychosis, discussed in Chapter 29.3, appears in a very small minority of patients treated with antipsychotics for a year or more and is characterized by delusions and hallucinations, which may appear either while the patient is still taking the neuroleptic or shortly after discontinuation or a significant dose reduction. This psychosis probably has an etiology similar to that of tardive dyskinesia and hence is often referred to as 'tardive psychosis'; like tardive dyskinesia, it exists as a strong reminder not to use antipsychotics chronically unless they are absolutely necessary.

Dopaminergic drugs, such as levodopa, direct-acting dopaminergic agents, or amantadine may cause a psychosis. In the case of levodopa (Celesia and Barr 1970; Fenelon *et al.* 2000; Moskovitz *et al.* 1978), the psychosis may occur either upon the initiation of treatment (Lin and Ziegler 1976) or, much more commonly, after 3–4 years (Friedman and Sienkiewicz 1991). This levodopa-induced psychosis may be characterized by hallucinations, often visual but also auditory (Fenelon *et al.* 2000; Inzelberg *et al.* 1998), and, in a minority, delusions of persecution (Graham *et al.* 1997). In the overwhelming majority of cases, patients first experience hallucinations with preserved insight: however, over many months insight is gradually lost, thus producing the syndrome of psychosis (Barnes and David 2001; Goetz *et al.* 2006). Direct-acting dopaminergic drugs may also cause psychosis but this is less common than with levodopa: bromocriptine, lergotrile (Serby *et al.* 1978), and pramipexole (Almeida and Ranjith 2006) have all been implicated. Finally, amantadine has also been implicated (Smith 2008), but this is very rare.

The remaining drugs in the list only rarely cause psychosis. These include the anti-epileptic drugs (AEDs) levetiracetam (Mula *et al.* 2003), topiramate (Kober and Gabbard 2005), vigabatrin (Sander *et al.* 1991) (in the case of levetiracetam the onset may, as with the others, occur soon after starting the drug, or may be delayed for up to 3–9 months [Kossoff *et al.* 2001; Motamedi *et al.* 2003]), the sympathomimetics phenylpropanolamine and ephedrine (Lambert 1987), the antidepressants bupropion (Golden *et al.* 1985) and fluoxetine (Mandalos and Szarek 1990), and disulfiram (Bicknell and Moore 1960) and methysergide (Cittandi and Goadsby 2005).

A toxic psychosis may also occur with chronic manganese exposure. This 'manganese madness', as discussed in Chapter 14.6, may present with a combination of parkinsonism, excitation, delusions, and hallucinations.

Finally, note should also be made of a psychosis associated with baclofen. Here, the psychosis occurs not during use of the drug but rather as a withdrawal phenomenon after chronic use. Here, about a week after discontinuation, one may see a psychosis with agitation, delusions of persecution, and hallucinations (Swigar and Bowers 1986), which, in one case, was accompanied by a complex movement disorder, with chorea, tremor, and dystonia being evident (Kirubakaren *et al.* 1984).

ENDOCRINOLOGIC PSYCHOSES

Hypothyroidism, as discussed in Chapter 17.4, may present with psychosis in a condition known as 'myxedema madness', typically characterized by delusions of persecution and reference and by hallucinations, generally auditory. The delusions of persecution may at times be so compelling that patients become assaultive (Reed and Bland 1977); in other cases, patients may be reduced to a seclusive 'mumbling' (Karnosh and Stout 1935). Pertinent clues to the correct diagnosis include slowness and a certain 'fogginess' of thought, cold intolerance, deepening of the voice, constipation, hair loss, and myxedema of the face, supraclavicular fossae, and dorsa of the hands and feet.

Hyperthyroidism may be accompanied by a psychosis, with prominent delusions of persecution: in one case, the hyperthyroid patient slashed his throat rather than let his 'persecutors' capture him (Ingham and Nielsen 1931). When the psychosis occurs in the setting of 'thyroid storm' (Bursten 1961; Greer and Parsons 1968), the prominent autonomic signs (increased temperature, tachycardia, and tremor) immediately suggest the diagnosis; however, when the responsible hyperthyroidism is milder, the diagnosis may be elusive (Hodgson *et al.* 1992).

Cushing's syndrome may be characterized by a psychosis, the diagnosis being suggested by the typical cushingoid habitus of moon facies, truncal obesity, buffalo hump, violaceous abdominal striae, and so on. One patient had classic Schneiderian first-rank symptoms, including audible thoughts, thought broadcasting, and thought insertion (Trethowan and Cobb 1952), whereas another presented with auditory hallucinations and delusions of a grandiose and religious nature (Hertz *et al.* 1955).

Adrenocortical insufficiency is suggested by abdominal complaints (nausea, vomiting, diarrhea or constipation, and abdominal pain) and orthostatic hypotension with postural dizziness. A psychosis may rarely also be seen (Cleghorn 1951; McFarland 1963).

INTRACRANIAL DISORDERS

The various intracranial disorders capable of causing psychosis include stroke, tumors, multiple sclerosis, traumatic brain injury (TBI), various heredodegenerative disorders, such as Huntington's disease, and a miscellaneous group.

Stroke

Stroke may be characterized by the fairly sudden onset of psychosis: this has been noted with infarction of the temporal lobe (Peroutka *et al.* 1982); in one case (Thompson and Nielsen 1949), the patient, a 58-year-old man, suddenly began to hear 'unusual noises which he believed were caused by wires placed in his house', and soon after, 'while in a restaurant, he suddenly declared that someone had put ground glass into his food. He then ran out of the restaurant into the street, shouting that his son-in-law had been killed after having been held captive by a gang of criminals.' Frontal lobe involvement may also be found, as in a case of infarction of the medial aspect of the right frontal lobe (Spiegel *et al.* 2009) and in a case of a ruptured aneurysm, which presented acutely with auditory hallucinations, delusions of persecution, and loosening of associations (Hall and Young 1992). Finally, thalamic infarction involving the right dorsomedial area was, in one case (Feinberg and Rapcsak 1989), associated with vivid visual hallucinations; indeed, the patient 'reached down to pat the dog' that he had hallucinated at his side.

Tumors

Tumors may present with psychosis, and, as noted in Chapter 21.1, this is most common with tumors of the temporal lobe; other locations include the corpus callosum, the frontal lobe and the parietal lobe.

Multiple sclerosis

Multiple sclerosis, as discussed in Chapter 19.1, may cause psychosis; further, in some cases psychosis may dominate the clinical picture and, albeit very, very rarely, psychosis may constitute the presentation of the disease. Rarely, multiple sclerosis may present with a psychosis.

Traumatic brain injury

Traumatic brain injury, as discussed in Chapter 8.5, may be followed by a psychosis; however, as noted in that chapter, there is some controversy over how frequently this occurs, given the possibility that psychoses following long after a TBI may, in an uncertain percentage of cases, merely represent a concurrent schizophrenia.

Neurodegenerative disorders

Of the neurodegenerative disorders capable of causing psychosis, Huntington's disease is the classic example; others include dentatorubropallidoluysian atrophy, spinocerebellar ataxia, and Wilson's disease. Of note each of these is also characterized, at some point, by abnormal movements.

Huntington's disease may cause psychosis in some 10 percent of patients (Bolt 1970; Garron 1973; Heathfield 1967), however this typically appears only after chorea has been present for 3 or more years, thus leaving little doubt as to the cause of the psychosis. Exceptions to this rule, however, do occur, and, albeit very rarely, Huntington's disease may present with psychosis (Caine and Shoulson 1983; Garron 1973; James *et al.* 1969), and in such cases the diagnosis may be suggested by a positive family history and the eventual appearance of chorea, and then confirmed by genetic testing.

Dentatorubropallidoluysian atrophy, another autosomal dominant disorder typically presents in adult years with chorea, ataxia, or both, and eventually a dementia supervenes, which may be characterized by hallucinations and delusions. Rarely, however, the presentation may be with a psychosis (Adachi *et al.* 2001), and in such cases the diagnosis, as with Huntington's disease, may be suggested by a positive family history, the eventual appearance of chorea (or ataxia), and confirmed by genetic testing.

Spinocerebellar ataxia, another autosomal dominant disorder, typically presents with a gradually progressive ataxia, and in a very small minority the ataxia may be followed by a psychosis (Chandler and Bebin 1956). Very rarely, however, the disease may present with a psychosis (Benton *et al.* 1998), and here, as in the foregoing disorders, one looks to the family history, watches for the development of a typical movement disorder, and then confirms the diagnosis with genetic testing.

Miscellaneous

Of the miscellaneous disorders capable of causing psychosis, the most important by far, given its eminent treatability, is Wilson's disease, and this diagnosis must always be considered in a young person with a psychosis occurring in the context of a movement disorder (Beard 1959; Gysin and Cooke 1950; Jackson and Zimmerman 1919), which may include varying combinations of dystonia, chorea, tremor or parkinsonism. Furthermore, it should also be clearly kept in mind that Wilson's disease not uncommonly presents with a psychosis, and, indeed, one of Kinnier Wilson's first patients presented in just this way (Wilson 1912). In such cases, the diagnosis generally rests on the eventual development of a movement disorder, and is then confirmed by copper studies: given the recessive nature of the disease, a family history is typically absent, and genetic testing, given the large number of mutations, is not practical.

Of the remaining miscellaneous disorders capable of causing psychosis, consideration may first be given to Creutzfeldt–Jakob disease which, albeit rarely, may present with a psychosis (Brown *et al.* 1984; Dervaux *et al.* 2004; Zeng *et al.* 2001); fairly rapidly, however, dementia and myoclonus supervene, and one eventually finds periodic spike-and-slow-wave complexes in the electroencephalogram (EEG) and, on magnetic resonance imaging (MRI), signal changes in the striatum and cerebral and cerebellar cortices. The rare new-variant Creutzfeldt–Jakob disease may also be considered, as this disease not uncommonly presents with psychosis (Zeidler *et al.* 1997a); within months, however, patients develop any of a variety of abnormal movements, including myoclonus, and eventually a dementia supervenes. Fatal familial insomnia, another prion disease, may also, albeit very rarely, present with a psychosis, which, true to the name of the disease, is followed by severe, intractable insomnia (Dimitri *et al.* 2006).

Fahr's syndrome may rarely present with psychosis, as in one familial case wherein the illness was manifest symptomatically with basal ganglia calcification on imaging and a chronic psychosis (Francis and Freeman 1984).

Aqueductal stenosis, one of the causes of non-communicating hydrocephalus, has also been associated with a psychosis (Roberts *et al.* 1983).

EPILEPTIC PSYCHOSES

The various psychoses seen in epilepsy may be distinguished by their relationship to the seizures experienced by the patient. Ictal psychoses are in fact seizures and are immediately suggested by their paroxysmal onset. Postictal psychoses, as the name suggests, follow seizures and, critically, are separated from the last seizure by a 'lucid' interval. The psychosis of forced normalization represents a paradoxical event in that it appears when the patient's seizures are finally brought under control with anti-epileptic treatment. Finally, chronic interictal psychosis occurs in the setting of a chronically uncontrolled seizure disorder.

Ictal psychosis may occur as a simple partial seizure or a complex partial seizure. As discussed in Chapter 8.3, some simple partial seizures may present with visual hallucinations or with delusions, including the Capgras phenomenon or Schneiderian first-rank symptoms. Complex partial status epilepticus, as also discussed in Chapter 8.3, may also cause a psychosis; however, here, in contrast to simple partial seizures, there will be some defect of consciousness.

Postictal psychosis (Briellman *et al.* 2000; D'Alessio *et al.* 2009; Devinsky *et al.* 1995; Kanner *et al.* 1996; Lancman *et al.* 1994; Leutmezer *et al.* 2003; Logsdail and Toone 1988; Nishida *et al.* 2006; Savard *et al.* 1991; Umbricht *et al.* 1995) may follow a bout of seizures, either grand mal or complex partial in type, and, critically, is separated from the last seizure by a 'lucid' interval, lasting from hours to days, during which the patient's mental status is 'clear'. The psychosis itself is characterized by delusions of persecution and hallucinations, most commonly auditory, and may last for hours or months, although most patients clear spontaneously within a matter of days; importantly this psychosis may also be characterized by violent or suicidal behavior (Kanemoto *et al.* 1999). Most patients have a long history of recurrent complex partial seizures.

Chronic interictal psychosis (Flugel *et al.* 2006; Kristensen and Sindrup 1979; Perez and Trimble 1980; Slater and Beard 1963a,b) typically occurs in the setting of uncontrolled epilepsy of a decade or more in duration. Although in some cases the epilepsy is characterized primarily by focal onset grand mal seizures, in most cases complex partial seizures, with or without secondary generalization, dominate the course (Kanemoto *et al.* 2001a). The interictal psychosis itself develops either subacutely or insidiously, and is characterized by auditory hallucinations and delusions, often of persecution or reference, all occurring in a clear sensorium. Importantly, many patients respond well to treatment with low-dose antipsychotics (Tadokoro *et al.* 2007). Of note, some patients may have both postictal psychosis and interictal psychosis (D'Alessio *et al.* 2009), and in such cases, although multiple episodes of postictal psychosis generally precede the development of the interictal psychosis (Tarulli *et al.* 2001), the reverse sequence has also been noted (Adachi *et al.* 2003), with an interictal psychosis developing first, followed by the appearance of postictal psychoses.

Psychosis of forced normalization is a rare condition, first described by Landolt (1953, 1958) and characterized by the appearance of a psychosis *after* AEDs (Pakainis *et al.* 1987) or, in one case, vagus nerve stimulation (Gatzonis *et al.* 2000), have controlled the seizure disorder and 'normalized' the EEG.

ENCEPHALITIC AND POSTENCEPHALITIC PSYCHOSES

Encephalitic

Viral encephalitis is suggested by fever, headache, and lethargy, and may in some cases present with psychosis, as has been noted with herpes simplex encephalitis (Drachman and Adams 1962; Johnson *et al.* 1972; Williams and Lerner 1978; Wilson 1976) and infectious mononucleosis (Raymond and Williams

1948). Encephalitis lethargica may present similarly (Kirby and Davis 1921; Meninger 1926; Sands 1928) and is suggested by sleep reversal and oculomotor pareses.

Postencephalitic

Viral encephalitis may also leave a psychosis in its wake, and such postencephalitic psychoses have been noted as sequelae to herpes simplex encephalitis (Rennick *et al.* 1973) and encephalitis lethargica (Fairweather 1947).

MISCELLANEOUS

A large number of miscellaneous causes of psychosis also exists, first among which is 'megaloblastic madness' due to vitamin B12 deficiency, with this colorful name being derived from the associated megaloblastic anemia. It must be borne in mind, however, that this anemia may not be present: in one case, the diagnosis became clear only when symptoms of subacute combined degeneration appeared (Smith 1929), and in another, the only evidence of vitamin B12 deficiency was the psychosis itself – there was no anemia and no evidence of spinal cord involvement (Evans *et al.* 1983).

Neurosyphilis rarely presents with a psychosis (Rothschild 1940; Schube 1934), and the diagnosis may remain elusive until other, more typical, symptoms appear, such as a dementia, pupillary changes (e.g. Argyll Robertson pupil), tremor, or seizures.

Acquired immunodeficiency syndrome (AIDS) may cause a psychosis (Bulrich *et al.* 1988; Harris *et al.* 1991), and the presence of other AIDS-related illnesses, such as thrush or *Pneumocystis* pneumonia may serve to suggest the correct diagnosis. Rarely, AIDS may present with psychosis (Thomas and Szabadi 1987), and in such cases the diagnosis may prove elusive until other, more typical, features of AIDS make their appearance.

Systemic lupus erythematosus may also cause psychosis (Brey *et al.* 2002; Devinsky *et al.* 1988a; Johnson and Richardson 1968; Lim *et al.* 1988; Miguel *et al.* 1994), and, although this usually occurs in the context of other symptoms, such as arthralgia, rash, pericarditis, or pleurisy, psychosis may rarely constitute the presenting feature of lupus (Agius *et al.* 1997).

Sydenham's chorea may rarely be complicated by a psychosis with hallucinations and delusions (Hammes 1922; Putzel 1879): the diagnosis is immediately suggested by the context of subacutely developing chorea in a child or adolescent.

Chorea gravidarum, Latin for 'chorea of pregnant women' may rarely be accompanied by psychosis (Beresford and Graham 1950; Wilson and Preece 1932). The diagnosis should be suspected in pregnant women with psychosis, chorea, and a history of Sydenham's chorea.

Hepatic porphyria typically presents in attacks accompanied by abdominal pain, often with vomiting and constipation or, less commonly, diarrhea. Rarely, such patients may also have a psychosis (Mandoki and Sumner 1994), which, in one case, was accompanied by bizarre behavior (Hirsch and Dunsworth 1955).

Metachromatic leukodystrophy, although rare, is of particular interest in that it can cause a psychosis that *very* closely resembles that caused by schizophrenia (Hyde *et al.* 1992; Muller *et al.* 1969). Indeed, in some cases, it was not initially possible to distinguish between the two disorders until other symptoms suggestive of metachromatic leukodystrophy, as for example a peripheral neuropathy (Manowitz *et al.* 1978) or a dementia (Betts *et al.* 1968a; Rauschka *et al.* 2006; Waltz *et al.* 1987) developed years later.

Velocardiofacial syndrome has also attracted great interest, as it too can cause a psychosis symptomatically quite similar to that caused by schizophrenia (Gothelf *et al.* 2007; Ivanov *et al.* 2003; Murphy *et al.* 1999). The diagnosis may be suggested by the characteristic dysmorphic facies with hypertelorism, a bulbous nose, and micrognathia.

Vanishing white matter leukoencephalopathy is an autosomal recessively inherited disorder, which, in adults, may present with a psychosis very similar to that seen in schizophrenia (Denier *et al.* 2007): the diagnosis may be suggested by evidence of bilateral corticospinal tract damage, and confirmed by finding extensive white matter disease in the cerebral hemispheres.

Subacute sclerosing panencephalitis, a vanishingly rare disease in developed countries thanks to measles vaccination, may cause psychosis and myoclonus (Cape *et al.* 1973; Salib 1988) or may present with a psychosis, the diagnosis only becoming clear with the later appearance of a delirium (Duncalf *et al.* 1989; Koehler and Jakumeit 1976).

Prader–Willi syndrome, a rare disorder characterized by massive obesity and dysmorphic facies, may rarely cause a psychosis with delusions and hallucinations (Clarke 1993).

Differential diagnosis

Delusions and hallucinations may be found in a number of other syndromes, namely dementia, delirium, depression, and mania, and thus the first task is to determine if one of these syndromes is present, and, if so, to then pursue the differential for that syndrome, as discussed in the respective chapters.

Dementia and delirium are both marked by significant cognitive deficits, such as decreased short-term memory and disorientation, and, in the case of delirium, confusion. Delusions and hallucinations are quite common in both syndromes, and in some instances, for example diffuse Lewy body disease, they may constitute a diagnostic hallmark.

Both depression and mania, when they are severe, may be characterized by delusions or hallucinations; however, in both these instances the delusions or hallucinations occur within the context of the mood syndrome, and, upon getting a reliable history, one always finds, prior to the onset of the delusions or hallucinations, a prominent and progressively worsening depressive or manic syndrome.

Finally, simulated psychoses, as seen in malingering, factitious illness, and *folie à deux* must be distinguished from 'true' psychoses. Malingerers may simulate a psychosis in order to avoid unpleasant consequences, as may occur in prisoners facing trial (Tsoi 1973). Factitious psychosis is said to occur when the simulation has the purpose of simply being a patient in the hospital (Pope *et al.* 1982). *Folie à deux* is said to occur when a person with a true psychosis, usually paranoid schizophrenia, exerts such a profound influence on close relatives or acquaintances that they come to adopt the patient's own delusional beliefs as true. Importantly, in such cases, if the 'truly' psychotic person is successfully treated or if a prolonged separation is enforced, the others gradually come to see the falseness of their beliefs (Dewhurst and Todd 1956; Kashiwase and Kato 1997; Partridge 1950; Waltzer 1963).

Treatment

Treatment is directed at the underlying cause; in cases in which that is ineffective or where symptomatic treatment is required, an antipsychotic is indicated. In general, second-generation antipsychotics are more effective and better tolerated than first-generation agents and, of the second-generation agents, risperidone, olanzapine, or quetiapine are reasonable choices. In general, and especially in the elderly or medically frail, or patients with hepatic failure, it is appropriate to 'start low and go slow' with regard to initial dose and subsequent titrations. In the case of levodopa-induced psychosis wherein dose reduction is often not feasible, good success has been had with clozapine, in low doses of 6.25–25 mg/day. In patients in whom emergent treatment is required, one may proceed as described in Chapter 7.4.

8.2 PERSONALITY CHANGE

Every individual has a definite personality structure, which, once formed in childhood and adolescence, persists in a lifelong fashion, being very resistant to any modification. Thus, the appearance of a fundamental change in personality, that is to say a far-reaching transformation of the patient's characteristic personality traits, is an ominous clinical sign and demands prompt diagnostic evaluation.

Clinical features

Personality change may either be non-specific in character, or one may see a specific type, such as the frontal lobe syndrome, or, in patients with epilepsy, the interictal personality syndrome. Regardless, however, of which kind of personality change occurs, those around the patient often make comments such as 'he's not himself any more', and indeed it may be this realization that leads family members to bring the patient to medical attention.

NON-SPECIFIC PERSONALITY CHANGE

Non-specific personality change may be characterized either by a marked exaggeration of pre-existing personality traits or by the emergence of altogether new traits, previously foreign to the patient. For example, a characteristically financially prudent person may become stingy to the point of miserliness. In another example, a previously outgoing and generous person may gradually become withdrawn and miserly, or, conversely, a premorbidly shy and timid person may become freer in personal contacts and even outgoing.

FRONTAL LOBE SYNDROME

The frontal lobe syndrome is characterized by varying combinations of disinhibition, affective change (euphoria, irritability, impulsivity, or depressed mood), perseveration, and apathy.

Disinhibited patients seem to lose regard for customs or morals, and their overall behavior becomes coarsened: they may eat with gluttony, curse with no regard for company, tell coarse and crude jokes, and neglect their personal hygiene. Inappropriate sexual advances are not uncommon, and patients may, with no hint of shame, proposition much younger individuals, even at times children. Some may engage in reckless masturbation, at the dinner table or in the front garden.

Affective changes may have some lateralizing value: euphoria is seen more often with right-sided lesions and depressed mood with left-sided lesions. The euphoria may occasionally be accompanied by *witzelsucht*, or a tendency to make simple, silly puns.

Perseveration is characterized by a tendency to repeat the same behavior over and over: examples include repeatedly uttering the same phrase, opening and closing a book, or buttoning and unbuttoning a shirt.

Apathy is characterized by a lack of motivation. Although patients may experience some urges or consider some actions, their plans, if they occur at all, often come, as it were, stillborn, and, lacking in motivation; apathetic patients may either never come to the point of action or, if they do get started, soon find themselves indifferent, after which they give up.

These clinical features often, but not always, coalesce into one of two subtypes, namely the orbitofrontal and dorsolateral frontal lobe syndromes: the orbitofrontal subtype is characterized by disinhibition and affective changes (often either euphoria or irritability) and the dorsolateral type by perseveration and apathy.

The frontal lobe syndrome is also often accompanied by what is known as the 'dysexecutive syndrome', which represents, as one might expect from the name, a disturbance in 'executive' abilities. Thus, patients with these executive deficits have difficulty in the following areas: formulating and setting goals, developing plans to meet these goals, initiating planned behavior, and, lastly, monitoring and correcting activity when it gets 'off course'. Patients with these executive deficits may not come to attention until they are faced with new, and relatively complex, situations. Thus, patients whose lives are passed in fixed routines, where habit rules the day, may have little difficulty. However, if faced with an unaccustomed task, as for example planning a formal dinner or developing a financial plan, they may find themselves unable to successfully complete the work in front of them.

Some examples may help to fix the picture of the frontal lobe syndrome. The classic case is that of Phineas Gage (Neylan 1999), who manifested disinhibition and irritability. As reported by his physician Dr Harlow (Harlow 1848), on 3 September 1848, while Gage was tamping down an explosive charge with a special 'tamping iron', the charge exploded, blowing the iron rod back up and through his skull, causing extensive frontal lobe damage. Prior to the injury, Gage had been a foreman, capable, well balanced, and an exemplary worker. Twenty years later, however, Harlow (Harlow 1868) found him to be 'irreverent, indulging at times in the grossest profanity (which was not previously his custom), manifesting but little deference for his fellows, impatient of restraint or advice when it conflicts with his desires... In this regard his mind was so radically changed... that his friends and acquaintances said he was "no longer Gage".'

Mulder *et al.* (1951) reported another case marked by disinhibition and irritability. The patient's wife complained that the patient:

> no longer cared about his appearance, that he drove through red lights, threw bills in the wastebasket remarking they were 'only bills', and frequently threatened to harm his family. He spilled his food on his clothing and on the floor, and to his family's dismay, he then picked it up and ate it. He did not wait for food to be served, but would snatch it off platters with his fingers as his wife neared the table. His sexual activity became uninhibited, and he sought intercourse with neighborhood children and prostitutes with no concern for possible consequences.

Another case (Moersch 1925) was marked by perseveration and apathy. The patient, a 54-year-old man was:

> brought to the clinic... Because of loss of bladder and rectal control, and lack of interest. The patient himself made no complaint. About three months before, a gradually increasing mental change had been observed. The patient lost his ambition and interest in work, although he had continued at his trade of carpentry until two weeks before. He had become careless in his work, would forget what he was doing, and seemed little concerned about his short-comings. For two weeks before his examination he had been content to sit aimlessly at home, or to play with his children. He voided at any time and even defecated in his clothes... During general examination, the patient was indifferent and aimless, would sit and look at a newspaper, which might be upside down. He was oriented in all spheres, and his attention might be held for a few moments when aroused. He would follow his son about in a fairly good-natured manner, but always object to being examined, saying that he was not sick. He showed considerable perseveration, repeating movements at times for long periods. For example, one evening he sat before a wash bowl for over a half hour, turning the faucets on and off.

INTERICTAL PERSONALITY SYNDROME

The interictal personality syndrome, also known as Geschwind's syndrome, is said to appear insidiously in patients with epilepsy after years of uncontrolled complex partial seizures. It should be emphasized that this is a controversial entity and that it has not as yet been possible to prove conclusively that such a specific syndrome exists. Nevertheless, the clinical impression of many is that it does occur and that it may have a profound effect on patients' lives.

The outstanding characteristic of this syndrome is a trait known variously as 'viscosity', 'adhesiveness', or 'stickiness' (Waxman and Geschwind 1975). In this, patients seem unable to break away from a train of thought or a certain emotion, thoughts and feelings plodding on and adhering to one another in a sort of viscous mass. Bleuler (Bleuler 1924), writing in the early part of the twentieth century, asserted that:

> the most conspicuous anomaly concerns the *affectivity*, which reacts to a morbid degree, and at the same time shows the peculiarity, that an existing affect lasts a long time and is difficult to divert by new impressions; *it is not merely irritability that shows itself in this manner but the other affects, as attachments, or joy, all take the same course...* In *speaking* and *writing* we have the same peculiarities: the patient does not get anywhere with his talking, not only because of its slowness, but especially because of its circumstantiality, which must depict all trivialities in repetition and in manifold expression of the same idea in different forms. Besides this the manner of speaking is verbose and clumsy, and always vague. [Italics in original.]

Bear *et al.* (Bear *et al.* 1982), in addition to this viscosity of thought and emotion, also emphasized an 'interpersonal adhesiveness' manifest in 'interpersonal clinging' and a 'tendency to draw out interpersonal encounters'. In some cases, patients, on shaking hands while saying goodbye to the physician, will simply continue to 'hang on', requiring the physician to extricate him or herself.

The viscosity of thought may have a written expression in hypergraphia, wherein patients may write voluminous amounts, far and above what is required for any social or professional purposes (Hermann *et al.* 1988), often writing about philosophic or religious concerns (Waxman and Geschwind 1974).

In addition to viscosity, patients also tend to be preoccupied with religious, ethical, or philosophical concerns and to experience hyposexuality.

Religiosity is, according to Maudsley (Maudsley 1874) 'often very notable in epileptics' and Kraepelin (Kraepelin 1902), writing in the early part of the twentieth century, noted that 'the religious content of [epileptic's] thought is another striking symptom, many patients spending a large part of their time in reading the Bible or praying aloud'. Bear *et al.* (Bear *et al.* 1982) noted the 'nascent metaphysical or cosmological preoccupations' of patients.

Hyposexuality manifests primarily as a loss of libido (Blumer 1970; Blumer and Walker 1967).

As noted earlier, the interictal personality syndrome is a controversial entity. Early attempts to validate it (Bear 1979; Bear and Fedio 1977) made use of a complex rating instrument, and subsequent attempts to replicate these earlier findings met with no (Mungas 1982; Nielsen and Kristensen 1981; Rodin and Schmaltz 1984) or only partial success (Bear

Table 8.2 Criteria for the diagnosis of the interictal personality syndrome (all five elements, A–E, must be present)

A	The patient's behavior represents an enduring change in personality
B	The disorder occurs after 3 or more years of repeated complex partial seizures
C	The patient displays viscosity, as manifested by one or more of: deep and persistent affects; verbose, overly detailed, and circumstantial speech; or hypergraphia
D	The patient displays either a preoccupation with religious, ethical, or philosophical concerns, or hyposexuality
E	There are no delusions or hallucinations

et al. 1982; Hermann and Reil 1981). Provisionally, as one of the authors (DPM) has discussed (Moore 1997), the diagnosis should probably be reserved for cases meeting the criteria listed in Table 8.2.

Etiology

The etiologies of the various types of personality change, namely the non-specific type, the frontal lobe syndrome and the interictal personality syndrome, differ, and hence each is discussed separately.

NON-SPECIFIC PERSONALITY CHANGE

The various causes of non-specific personality change are listed in Table 8.3, which divides them into three groups. The first group contains those disorders capable of causing a personality change of *gradual onset*, such as neurodegenerative disorders. The next group recognizes personality change of *subacute onset*, as may occur after stroke. Finally, there is a *miscellaneous* group of disorders, including, notably, TBI.

Table 8.3 Causes of non-specific personality change

Gradual onset
Neurodegenerative disorders
Frontotemporal lobar degeneration (Seeley *et al.* 2005)
Alzheimer's disease (Mendez *et al.* 1993)
Huntington's disease (Pflanz *et al.* 1991)
Wilson's disease (Bridgman and Smyth 1944; Dening and Berrios 1989; Starosta-Rubinstein *et al.* 1987; Walshe and Yealland 1992)
Myotonic muscular dystrophy (Delaporte 1998)
Tumors
Temporal lobe (Keschner *et al.* 1936; Strobos 1953)
Thalamus (Partlow *et al.* 1992)
Hypothalamus (Alpers 1937)
Miscellaneous
Schizophrenia (Kraepelin 1902)
Normal pressure hydrocephalus (Rice and Gendelman 1973)
Chronic subdural hematoma (Cameron 1978)
Vitamin B12 deficiency (Lindenbaum *et al.* 1988)
Limbic encephalitis (Alamowitch *et al.* 1997)
New-variant Creutzfeldt–Jakob disease (Zeidler *et al.* 1997a,b)
Mercury intoxication (O'Carroll *et al.* 1995)
Manganese intoxication (Abd El Naby and Hassanein 1965)
Subacute onset
Stroke (Stone *et al.* 2004)
Miscellaneous
Traumatic brain injury (Brooks *et al.* 1986; Thomsen 1984)
Post-grand mal status epilepticus (Oxbury and Whitty 1971, Rowan and Scott 1970)
Post-viral encephalitis (McGrath *et al.* 1997)

Gradual onset

Of the neurodegenerative disorders that may present with a personality change, perhaps the most important is frontotemporal lobar degeneration. Frontotemporal lobar degeneration may present with a frontotemporal dementia, which, in turn, is marked, at least initially, primarily by a personality change, which may be of two types: the first type, namely the frontal variant, is discussed below; the second type, namely the temporal variant, is characterized by a variety of symptoms, including stereotyped, ritualized behaviors, mental and behavioral rigidity, and changes in food preferences, typically toward sweets. Alzheimer's disease often causes a personality change; however, here, in contrast to frontotemporal lobar degeneration, the personality change typically appears only well after cognitive deficits (e.g. short-term memory loss) have become established. Huntington's disease eventually causes a personality change in over half of all cases; however, the onset is typically delayed for 2 to 3 years after the appearance of chorea. Wilson's disease may present with a personality change, characterized by lability and bizarre behavior; however, the eventual development of various abnormal movements will suggest the correct diagnosis. It is of interest that Wilson's first case (Wilson 1912), a 25-year-old woman, presented with a personality change wherein she became 'restless, unable to settle to anything, easily provoked to laughter, constantly smiling and unnaturally cheerful'. Myotonic muscular dystrophy typically presents with myotonia and weakness, and this may be followed, in a minority, by a personality change, which in turn may be marked by avoidant traits.

Tumors productive of a non-specific personality change are most often found in the temporal lobe; rarely a similar presentation may occur with tumors of the thalamus or hypothalamus.

Of the miscellaneous causes of personality change, by far the most common is schizophrenia, and any personality change occurring in adolescence or early adult years should immediately raise this diagnostic possibility. Normal pressure hydrocephalus may be suggested by incontinence (or mere urgency) and a 'magnetic' gait. Chronic subdural hematoma may, in a small minority of patients, present with a personality change, and it is important to keep in mind that the duration of the interval between the head trauma and the personality change may range from a mere 3 weeks up to as long as 2 years; furthermore, and especially in the elderly or those on anticoagulants, the inciting trauma may at times be trivial, and indeed may not be recalled by the patient. Vitamin B12 deficiency may also present with a personality change, which may or may not be accompanied by a macrocytosis or a peripheral neuropathy. Limbic encephalitis typically presents with a delirium, however in some cases a personality change may, for a time, dominate the clinical picture. New-variant Creutzfeldt–Jakob disease, a very rare prion disease, may present with a personality change; however, within a matter of months other symptoms, including, for example, ataxia and myoclonus, eventually ensue. Mercury intoxication may present with a personality change known as erethism, with emotional lability, irritability and shyness, all typically accompanied by tremor. Manganism, as may be seen in manganese miners, may present with a personality change similar to that seen in mercury intoxication, with emotional lability and irritability; however, here one may also see episodes of involuntary laughter, similar to that occurring in pseudobulbar palsy.

Subacute onset

Stroke may be followed by a non-specific personality change, and this may occur with both cortical and subcortical infarcts. By and large, patients become irritable, easily frustrated and overall less easy-going.

Miscellaneous

Traumatic brain injury, grand mal status epilepticus and viral encephalitis (e.g. herpes simplex viral encephalitis) may each have personality change as a sequela, and in the case of TBI, this may be the most disabling sequela.

FRONTAL LOBE SYNDROME

As noted in Table 8.4, the various causes may be divided into those that produce a frontal lobe syndrome of *gradual onset*, as, for example, may be seen with various neurodegenerative disorders (e.g. frontotemporal lobar degeneration), those that produce a frontal lobe syndrome of *subacute onset*, as may be seen in stroke, and a *miscellaneous* group, for example TBI. Following a discussion of these various causes, some *comments on the localizing and lateralizing value of the frontal lobe syndrome* are offered.

Table 8.4 Causes of the frontal lobe syndrome

Gradual onset
Neurodegenerative disorders
Frontotemporal dementia (Brun *et al.* 1994; Heutink *et al.* 1997; Neary *et al.* 1993)
Pick's disease (Bouton 1940; Litvan *et al.* 1997; Mendez *et al.* 1993)
Alzheimer's disease (Mega *et al.* 1996; Petry *et al.* 1988)
Amyotrophic lateral sclerosis (Massman *et al.* 1996)
Progressive supranuclear palsy (Verny *et al.* 1996))
Corticobasal ganglionic degeneration (Geda *et al.* 2007)
Spinocerebellar ataxia (Zeman *et al.* 2004)
FXTAS (fragile X-associated tremor/ataxia syndrome) (Bacalman *et al.* 2006; Bourgeois *et al.* 2007)
Metachromatic leukodystrophy (Rauschka *et al.* 2006)
Tumors
Frontal lobe (Avery 1971; Frazier 1936; Hunter *et al.* 1968; Strauss and Keschner 1935; Williamson 1896)
Corpus callosum (Alpers and Grant 1931; Beling and Martland 1919; Moersch 1925)
Miscellaneous
Alcoholic dementia (Lishman 1981)
Lacunar dementia (Ishii *et al.* 1986)
Fahr's syndrome (Lam *et al.* 2007)
Neurosyphilis (Storm-Mathisen 1969)
Subacute onset
Stroke
Frontal lobe (Logue *et al.* 1968)
Caudate nucleus (Mendez *et al.* 1989; Petty *et al.* 1996)
Thalamus (Sandson *et al.* 1991)
Mesencephalon (Adair *et al.* 1996)
Miscellaneous
Traumatic brain injury (Oder *et al.* 1992; Roberts 1976)
Postviral encephalitis (Friedman and Allen 1969; Mulder *et al.* 1951)
Cerebellar cognitive affective syndrome (Schmahmann and Sherman 1998)

Gradual onset

A frontal lobe syndrome of gradual onset may be seen with neurodegenerative disorders, tumors, and a miscellaneous group.

Of the neurodegenerative disorders capable of producing a frontal lobe syndrome, the two most important are frontotemporal lobar degeneration (when presenting as the frontal variant of frontotemporal dementia) and Pick's disease, both of which may present with a frontal lobe syndrome, with the onset of significant cognitive deficits being delayed for months or years. Alzheimer's disease may cause a frontal lobe syndrome; however, here, in contrast to frontotemporal lobar degeneration and Pick's disease, the frontal lobe syndrome typically appears only after cognitive deficits (e.g. progressive short-term memory loss) are well-established. Patients with amyotrophic lateral sclerosis may develop a frontal lobe syndrome, and, although this typically occurs only in the context of classic symptoms (e.g. upper and lower motor neuron signs), amyotrophic lateral sclerosis may, albeit rarely, present with a frontal lobe syndrome (Cavalleri and De Renzi 1994; Neary *et al.* 1990; Peavy *et al.* 1992). The other neurodegenerative disorders in Table 8.4, relative to the foregoing, only rarely cause the syndrome. Progressive supranuclear palsy (of the Richardson's subtype) presents with an atypical parkinsonism, and roughly half of these patients will develop a dementia, which, in turn, may also be characterized by a frontal lobe syndrome. Corticobasal ganglionic degeneration classically presents with a highly asymmetric parkinsonism accompanied by apraxia and cortical sensory loss, and in a minority of cases this may be followed by a frontal lobe syndrome; rarely corticobasal ganglionic degeneration may present with a frontal lobe syndrome (Bergeron *et al.* 1996). Spinocerebellar ataxia presents with a slowly progressive ataxia, and in a very small minority, this may also be followed by a frontal lobe syndrome. The fragile X-associated tremor ataxia syndrome (FXTAS), seen almost exclusively in males, presents with a gradually progressive tremor and ataxia, and in a minority of cases a frontal lobe syndrome may also develop. The adult form of metachromatic leukodystrophy may present with the frontal lobe syndrome, and it may be years before other symptoms, such as ataxia, spasticity, seizures, or a peripheral neuropathy, emerge.

Tumors capable of causing the frontal lobe syndrome, as might be expected, are found typically in the frontal lobes. Anteriorly placed tumors of the corpus callosum may also cause the syndrome; however, it is suspected that in these cases the syndrome does not occur as a result of damage to the corpus callosum but rather to lateral extension of the tumor into the adjacent medial aspects of the frontal lobes.

Of the miscellaneous disorders capable of causing the frontal lobe syndrome, alcoholic dementia is the most common, and should always be suspected in any chronic alcoholic whose behavior becomes disinhibited and coarse. It must always be kept in mind, however, that patients and family members typically stoutly deny excessive drinking. Another, relatively common, cause of the frontal lobe syndrome is lacunar dementia, occurring gradually secondary to the likewise gradual accumulation of lacunes within the thalamus, basal ganglia and inferior frontal regions, and this should always be suspected when there is a history of more typical lacunar syndromes, such as pure motor stroke. Fahr's syndrome may present with a frontal lobe syndrome, and here imaging, preferably by computed tomography (CT) scanning, reveals pronounced calcification in the basal ganglia. Finally, one must never forget general paresis of the insane (GPI), which presents with a dementia, often with frontal lobe features, anywhere from one to two decades after the initial syphilitic infection.

Subcute onset

When the frontal lobe syndrome appears subacutely, stroke should immediately be suspected. Thus, the syndrome may appear after infarction of the frontal lobe (as seen not uncommonly after subarachnoid hemorrhage [Alexander and Freedman 1984; Greene *et al.* 1995]). In addition, single, relatively large, lacunar infarctions of the caudate nucleus, thalamus and, rarely, the mesencephalon, may also produce the syndrome.

Miscellaneous

Traumatic brain injury may leave a frontal lobe syndrome in its wake, often marked by disinhibition and irritability; indeed such a personality change may represent one of the most disabling sequelae of severe TBI (Thomsen 1984). A frontal lobe syndrome may also appear as a sequela to viral encephalitis. Finally, various lesions of the posterior lobe of the cerebellum, or of the cerebellar vermis, may, as part of the 'cerebellar cognitive affective syndrome', also cause a frontal lobe syndrome, highlighting again the role the cerebellum plays in modulating frontal activity.

Comments on the localizing and lateralizing value of the frontal lobe syndrome

The frontal lobe exists in a 'circuit' involving the frontal cortex, the caudate nucleus, the lenticular nucleus, and the thalamus, which eventually sends fibers back to the frontal cortex. The frontal lobe syndrome, in general, localizes to this circuit, and may be seen with lesions of the frontal lobe (Frazier 1936; Williamson 1896), caudate nucleus (Mendez *et al.* 1989; Petty *et al.* 1996; Richfield *et al.* 1987), globus pallidus (Strub 1989), and thalamus (Sandson *et al.* 1991; Smyth and Stern 1938). The ventral tegmental area of the mesencephalon also projects to the frontal cortex and, as noted above, infarction of the mesencephalon may also cause the syndrome (Adair *et al.* 1996).

The syndrome may not have much lateralizing value: although, in most cases, the lesions are bilateral, unilateral lesions may also cause the syndrome, this having been noted with lesions of either the right or left frontal lobe (Frazier 1936; Strauss and Keschner 1935; Williamson 1896), the right caudate nucleus (Degos *et al.* 1993), and the left thalamus (Sandson *et al.* 1991; Smyth and Stern 1938).

INTERICTAL PERSONALITY SYNDROME

If indeed the interictal personality syndrome does exist as a discrete entity, it may then represent a 'functional hyperconnection' (Bear 1979; Bear and Fedio 1977) in the limbic system, which may in turn occur secondary to the 'kindling' effect of frequently repeated seizures (Adamec and Stark-Adamec 1983).

Differential diagnosis

Personality *change* must be distinguished from a personality *disorder*. Each of the various personality disorders discussed in Chapter 26 do *not* represent a change in the patient's personality make-up, but rather have been present, at least in a nascent form, since childhood or adolescence. Rather than replacing a pre-existing personality structure, the personality disorder just does, in fact, constitute the patient's life-long personality. Thus, in the history of an adult patient with a personality disorder, one finds that the various personality traits may be traced back into the patient's adolescence or childhood in a seamless and continuous fashion. By contrast, in a patient with a personality *change*, one finds a more or less distinct boundary in the history that separates the patient's original personality from that which currently exists. Relatives and friends may indicate such a change by saying that the patient is 'not himself' any more.

Dementia may be accompanied by an exaggeration of pre-existing personality traits, or by the emergence of new ones; however, the accompanying cognitive deficits indicate dementia as the primary syndrome, and the differential should then be pursued, as discussed in Chapter 6.3. Importantly, certain dementing disorders, as noted above, may present not with cognitive deficits but with a personality change, and the diagnosis may be elusive until, with further progression, cognitive deficits make their appearance. This is particularly the case, as noted above, with some of the neurodegenerative disorders, most particularly frontotemporal lobar degeneration.

Mood syndromes, namely mania and depression, may suggest the frontal lobe syndrome. The euphoria seen in mania may, superficially, appear similar to the euphoria seen in some cases of the frontal lobe syndrome; however, there are some clear differences. The euphoria of mania is heightened, full, and quite infectious, and this is in marked contrast with the shallow, silly euphoria of the frontal lobe syndrome, which lacks any infectiousness. Furthermore, in mania one also sees pressured speech, increased energy, decreased need for sleep, and hyperactivity – all symptoms lacking in the frontal lobe syndrome. A depressed mood is of course typical of depression; however, in depression, one also sees vegetative symptoms, such as anergia, anhedonia, anorexia and the like, symptoms absent in the frontal lobe syndrome.

With regard to the interictal personality syndrome, one must keep in mind that slowly growing tumors in the temporal lobe may present with epilepsy, followed, years later, by a personality change caused by the tumor itself. Thus, the appearance of a personality change in a patient with epilepsy should prompt consideration of MRI.

Treatment

If possible, treatment is directed at the underlying cause. Most patients require some form of supervision, and in some cases admission to a secure facility may be necessary. Regarding pharmacologic treatment, with the exception of utilizing carbamazepine for the disinhibition of the frontal lobe syndrome (Foster *et al.* 1989), little is known. In practice, antipsychotics (e.g. quetiapine) are also used for disinhibition, and antidepressants (e.g. a selective serotonin reuptake inhibitor, SSRI) are given when depressed mood or anxiety are prominent.

8.3 SEIZURES AND EPILEPSY

A seizure, or ictus, is paroxysmal in onset, generally brief in duration, and occurs secondary to an equally paroxysmal electrical discharge within the cerebral gray matter. Although, as detailed below, the actual symptomatology seen during a seizure and the cause of the responsible electrical discharge are both extraordinarily varied, the essential formal characteristic of the seizure, namely its 'paroxysmal-ness', remains. The term 'epilepsy' is generally reserved for cases in which there has been more than one seizure and the cause of the seizures is such that one may reasonably expect that the patient, in the absence of definitive treatment, will continue to have seizures. Thus, whereas it might not be proper to consider a patient who has had only one seizure, and that during an episode of severe hypoglycemia, to have epilepsy, it would be appropriate in the case of a patient with recurrent seizures and mesial temporal sclerosis.

Although in epilepsy the control of seizures, is, naturally, the first priority, it must be kept in mind that in cases characterized by complex partial or grand mal seizures, there are, in addition to the seizures themselves, certain postictal or interictal neuropsychiatric disorders, which, in some cases, may be more disabling than the seizures. These include postictal psychosis and interictal psychosis, both discussed in Chapter 8.1, interictal depression, discussed in Chapter 7.1, and the interictal personality disorder (also known as 'Geschwind's syndrome'), discussed in Chapter 8.2.

Clinical features

The classification of the various seizure types has changed over time, as evidenced by the evolution of criteria set forth by the International League Against Epilepsy (ILAE) in 1964 (Gastaut *et al.* 1964), 1970 (Gastaut 1970), 1981 (Commission of Classification and Terminology 1981), and 1989 (Commission of Classification and Terminology 1989). The 1981 ILAE criteria have proved the most enduring for clinical practice and, with some modification, are adhered to in this text.

Table 8.5 Types of seizures

Simple partial
Complex partial (also known as psychomotor)
Petit mal (also known as absence)
Grand mal (also known as generalized tonic–clonic)
Atonic (also known as astatic seizures or 'drop attacks')
Amnestic
Reflex
Status epilepticus

The various seizure types are listed in Table 8.5. In differentiating among these types, the first step is to determine the patient's state of consciousness during the seizure. If consciousness remains clear, that is to say the patient remains alert, with intact memory, and without any clouding or confusion, a simple partial seizure is present. If, however, consciousness is in some fashion impaired, but not entirely lost, either a complex partial or a petit mal seizure is present. The distinction between complex partial and petit mal seizures, as elaborated further in the text, is based on the overall nature of the seizure: complex partial seizures are often preceded by an aura, last several minutes, and are typically followed by some postictal confusion, while petit mal seizures occur without an aura, are very brief – lasting only seconds – and terminate abruptly, without any postictal confusion. If consciousness is entirely lost, one is typically dealing with a grand mal seizure, with its accompanying tonic–clonic activity. Atonic seizures may or may not be accompanied by an impairment or loss of consciousness but they are always characterized by an abrupt loss of muscle tone with, in most cases, a resulting fall. Amnestic seizures are unusual in that they are characterized solely by a paroxysmal amnesia in a clear consciousness. Reflex seizures are said to be present when any of the foregoing types occurs on a 'reflexive' basis, being provoked by some specific stimulus, such as hearing music or reading. Finally, each of the foregoing seizure types may also occur over a prolonged time, in which case status epilepticus is said to be present.

Although any given patient with epilepsy may experience but one type of seizure during the entire course of the illness, the history more often than not reveals different seizure types at disparate times. Thus, the course of epilepsy may be marked by varying combinations of simple partial, complex partial, and grand mal seizures (Devinsky *et al.* 1988b; Golub *et al.* 1951; Mauguire and Courjon 1978; Sperling *et al.* 1989); furthermore, patients with petit mal epilepsy often also experience grand mal seizures (Livingston *et al.* 1965; Sato *et al.* 1976). Of interest, among adults, partial seizures (either simple or complex) are more common than grand mal seizures (Gastaut *et al.* 1975; Kotsopoulos *et al.* 2002).

Importantly, in many cases, a given ictal event or seizure may in fact represent an amalgamation of two different seizure types, the first merging seamlessly into the following one. Thus, a simple partial seizure may immediately precede a complex partial seizure (Bare *et al.* 1994; Sperling *et al.* 1989) or, in other cases, a simple partial seizure may precede a grand mal seizure (Mauguire and Courjon 1978; Theodore *et al.* 1994). In such cases, it is customary no longer to refer to the ictal event that occurred in clear consciousness as a simple partial seizure but rather to speak of it as an 'aura' to the following seizure type. Furthermore, complex partial seizures may also transform without interruption into grand mal seizures (Theodore *et al.* 1994; Tinuper *et al.* 1996a), and in such cases one speaks of the complex partial seizure undergoing 'secondary generalization' into a grand mal seizure.

Each of the various seizure types noted in Table 8.5 is now considered in detail.

SIMPLE PARTIAL SEIZURES

Simple partial seizures usually last on the order of a minute or two, uncommonly extending for up to 5 minutes, and exceptionally persevering for as long as 15 minutes (Devinsky *et al.* 1988b; Mauguire and Courjon 1978). Following the 1981 ILAE criteria, they may be subdivided according to their predominant symptomatology into the following subtypes: simple partial seizures with motor signs, simple partial seizures with somatosensory or special sensory symptoms, simple partial seizures with autonomic symptoms or signs, and simple partial seizures with psychic symptoms. Furthermore, there is a group, not specified in the ILAE criteria, of miscellaneous simple partial seizures.

Simple partial seizures with motor signs

These seizures are most commonly characterized by unilateral clonic or tonic activity, or a combination of the two, such motor activity being seen most frequently in the hand, arm, or face and somewhat less so in the lower extremity (Russell and Whitty 1953). Jacksonian seizures represent a variety of simple partial seizure in which there is a 'march' of the ictal symptomatology from one part of the body to another. Such marches may begin variously in the hand or the fingers, proceed proximally to the face, and then march inferiorly; less commonly, they begin in the lower extremity. In most cases, the march is completed within a matter of minutes (Penfield and Jasper 1954; Russell and Whitty 1953). 'Versive' seizures represent a variety of motor simple partial seizure wherein there is a 'forced' tonic version of the head, and sometimes the trunk, to one side or the other (Mauguire and Courjon 1978; Russell and Whitty 1953). Motor aphasia may be seen in what the ILAE calls a 'phonatory' seizure. Such a motor aphasia may constitute the sole manifestation of the seizure (Labar *et al.* 1992) or may be accompanied by motor activity on the face. In one case (Williamson *et al.* 1985a), the seizure began with right-sided facial twitching followed by muteness: although the patient could follow commands and was still able to communicate by writing, she was unable, much to her distress, to speak. In another case (Walshe 1943), the seizure began with muteness and was 'followed by the advance of the tongue to the line of the teeth and the utterance of a rapid series of "D" sounds (D-D-D-D-D-D). The throat then "constricts", the mouth is drawn to the right and finally the head and eyes turn in a series of jerks to the right'.

Rarely, simple partial seizures with motor signs may be characterized by *bilateral* motor activity (Kanner *et al.* 1990; Tachibana *et al.* 1996), such as bilateral tonic posturing (Morris *et al.* 1988). Such seizures often lead to diagnostic uncertainty as they 'break the rule' that bilateral epileptic motor activity is always accompanied by some impairment of consciousness (Morris *et al.* 1988). In one case (Tukel and Jasper 1952): 'seizures began with a stiffness in the right arm and leg. The patient then stared and the right arm raised to the level of the chest. This was followed by the extension of both arms. There was then adversive movement of the head and eyes to the right and vocalization. The patient remained conscious throughout . . . she heard herself vocalize but could not speak.'

Although the motor behavior in these seizures may be simple, as in the foregoing example, or perhaps involve non-directed thrashing or bicycling movements of the lower extremities (Salanova *et al.* 1995), it may at times be fairly complex (Chassagnon *et al.* 2003). In one case (Weinberger 1973), the patient's 'head and neck turned to the right. Both arms moved to the right and turned rhythmically on their axis at the wrist in a fashion best described as turning two door knobs, but in opposite directions': although the patient 'was aware of what was happening to her, and could understand what people were saying. . . she could only utter a phrase such as "uh-uh-uh".' Although in almost all cases such simple partial seizures with bilateral motor signs arise from seizure foci in the supplemental motor area on the medial aspect of the frontal lobe (Salanova *et al.* 1995), exceptions do occur, as in a case in which the focus was in the parietal lobe (Bell *et al.* 1997).

Inhibitory motor simple partial seizures represent a kind of 'inverse' motor seizure, wherein, rather than seeing tonic or clonic muscle contraction, there is a paroxysmal paresis or paralysis. The hand, arm, or leg may be involved (Noachtar and Luders 1999; Russell and Whitty 1953; Villani *et al.* 2006), and in some cases hemiparesis may result (Globus *et al.* 1982; Hanson and Chodos 1978). The spread of epileptic electrical activity from the precentral gyrus to nearby areas may sometimes produce a more complex picture: in one case, a patient experienced not only ictal paresis of the right upper extremity, but also an associated motor aphasia (Lee and Lerner 1990).

Simple partial seizures with somatosensory or special sensory symptoms

Seizures with somatosensory symptoms are characterized by generally unilateral paresthesiae, numbness, pain, or a sensation of warmth or coldness. The hands and fingers are most frequently affected, followed by the face, foot, or entire upper or lower extremity (Mauguire and Courjon 1978). Ictal pain may be either diffuse (Wilkinson 1973) or localized (Young and Blume 1983, 1986) (as for example with muscle cramping in one extremity [Balkan 1995]) and may be quite severe (Russell and Whitty 1953). Although the vast majority of somatosensory seizures are unilateral, involving only one side of the body, there may rarely be bilateral involvement, for example simultaneous paresthesiae of both hands (Blume *et al.* 1992).

Somatosensory simple partial seizures may also undergo a Jacksonian 'march' (Lende and Popp 1976; Sittig 1925). These somatosensory Jacksonian marches tend to follow the same distribution as motor marches, most often beginning in the hand or fingers (Mauguire and Courjon 1978), but they are, unlike motor marches, generally quite rapid, completing their trek in a matter of seconds (Russell and Whitty 1953). One may at times encounter a kind of 'mixed' march with both somatosensory and motor components; interestingly, in such cases, one almost always finds the sensory component appearing first, followed by the motor one (Penfield and Jasper 1954; Russell and Whitty 1953, Villani *et al.* 2006).

Seizures with special sensory symptoms present with simple visual, auditory, olfactory, or gustatory hallucinations and, at times, with vertigo. Ictal visual hallucinations may consist of such crude phenomena as 'weaving patterns, zigzag lights, showers of sparks or colored clouds' (Russell and Whitty 1955). One of William Gower's patients (Holmes 1927) had simple partial seizures that 'commenced with the appearance of several small spheres, white in the center with an intermediate zone of blue and outside this a ring of red... [which] moved either at a uniform rate or in jerks to the left and downwards'. Hemianopia or blindness may also occur on an ictal basis (Barry *et al.* 1985; Russell and Whitty 1955). Ictal auditory hallucinations may consist of such phenomena as buzzing or ringing noises (Mauguire and Courjon 1978). Ictal olfactory hallucinations (Mauguire and Courjon 1978) tend to be unpleasant, such as the smell of something rotten or burning: one of Jackson's patients (Jackson and Beevor 1890) had 'a very nasty smell – "burning dirty stuff"'. Ictal gustatory hallucinations tend likewise to be unpleasant, the taste being described as foul or metallic. Ictal vertigo may be characterized either by mere giddiness or by a classic sense of rotation (Kluge *et al.* 2000; Russell and Whitty 1953).

Simple partial seizures with autonomic symptoms or signs

Ictal autonomic phenomena include an ill-defined 'rising' epigastric sensation, vomiting (Mitchell *et al.* 1983; Shukla and Mishra 1985), or diarrhea with abdominal cramping (Zarling 1984).

Simple partial seizures with psychic symptoms

These seizures may consist of a variety of experiences: 'dysmnesic' phenomena (e.g. déjà vu); 'cognitive' disturbances such as depersonalization; affective experiences such as fear; illusions such as macropsia; and complex, 'structured' hallucinations in either the auditory or the visual realms.

Dysmnesic phenomena represent a disturbance in the sense of familiarity, and include déjà vu, jamais vu, déjà entendu, and jamais entendu. In déjà vu, patients have the uncanny sense that they have already seen or experienced something that they are in fact encountering for the first time; in jamais vu, the opposite phenomenon occurs in that patients, although in the presence of something they 'intellectually' know they have experienced before, yet have the sense that it is entirely new. Déjà entendu and jamais entendu represent analagous experiences concerning not sight but hearing. A related phenomenon is the experience of prescience, wherein the seizure is characterized by a sense that the patient 'knew' beforehand what was going to happen (Sadler and Rahey 2004).

Cognitive disturbances most commonly involve depersonalization. One of Wilson's patients (Wilson 1930) noted that it was 'as if my mind were looking at myself from afar'. In another case (Daly 1958), the patient, while 'dissociated from his body' had the classic sense that he 'was looking down on the scene'. In other cases, the cognitive distortion may be difficult to categorize: one of Wilson's (Wilson 1930) patients described her seizure as an experience wherein 'her "thoughts just stopped"'.

Affective experiences may include anxiety, depression or, rarely, euphoria. Anxiety and fear have been frequently noted (Kennedy 1911; Macrae 1954a,b; Weil 1959; Williams 1956) and may be quite severe, progressing to a full anxiety attack (Alemayehu *et al.* 1995). In one case, the seizures were characterized by anxiety, palpitations, dizziness, and pallor (Rush *et al.* 1977), and in another the patient was 'afraid, perspired and did not want to be left alone' (McLachlan and Blume 1980). As might be expected, some patients may become agoraphobic on the basis of such ictal anxiety attacks: one patient, whose ictal anxiety attacks lasted of the order of a minute, 'went to the emergency room', and, fearful 'she could have spells while driving, at work, or in social situations... confined herself to home' (Weilburg *et al.* 1987). Ictal depressions 'are characterized by rather sudden let-down of mood and psychomotor retardation... from simple listlessness and apathy to agitated depression with suicidal attempts' (Weil 1959). One patient's ictal depressions were ushered in by olfactory hallucinations, like 'stuffed cabbage in a dirty outhouse', and could last for hours, during which the patient 'felt paralyzed inside and couldn't follow through an act' (Weil 1955). Euphoria may occur (Williams 1956) but in one case was followed, after 10–20 seconds, by an intense depression (Mulder and Daly 1952). Ecstasy may also occur, and may be intense: one patient commented that '[t]hanks to these experiences, I do not fear death anymore' (Picard and Craig 2009).

Illusions may include macropsia, wherein objects appear larger than they are, micropsia, wherein they appear smaller, and various illusory movements of objects (Heilman and Howell 1980; Russell and Whitty 1955). Also possible are hyperacusis or hypoacusis, wherein sounds appear louder or fainter, respectively, than they in fact are. The supernumerary limb experience may also be included here: one patient had such a strong ictal sense that an extra arm was raised above his head that he asked his wife to pull it down even though he acknowledged that he could in fact see his actual arm at his side (Russell and Whitty 1953).

Structured hallucinations are characterized by complex visual or auditory experiences. One of Russell and Whitty's (Russell and Whitty 1955) patients, wounded in the right occipital area, saw 'stretcher bearers walking past and then the figures of nurses whom he could recognize', all in the left hemifield; another, wounded in the right occipitoparietal area, 'felt that as if he was in a dream he was back in

Khartoum during the war. He saw and recognized friends around him.' In another case (Sowa and Pituck 1989) the patient saw the 'right side of people's faces missing' and 'water coming out of a clock'. Of interest, such ictal visual hallucinations may also reappear in the patient's dreams (Reami *et al.* 1991). Although, in most cases, these complex visual hallucinations occur in only one hemifield, they may at times spread to appear in the entire visual field (Russell and Whitty 1955). Polyopia may occur as may palinopsia: one patient's seizures were characterized by seeing 'non-existent pedestrians in an empty street, having seen these people minutes before in different surroundings' (Muller *et al.* 1995). Autoscopy may occur, in which patients hallucinate an image of themselves (Brugger *et al.* 1994; Devinsky *et al.* 1989a). Auditory structured hallucinations may consist of voices or music: one patient heard the same song, repeated over and over again (Wieser 1980).

Miscellaneous simple partial seizures

Various other signs and symptoms may occur on an ictal basis, including prosopagnosia (Agnetti *et al.* 1978), asomatognosia (Russell and Whitty 1955; So and Schauble 2004), and a combination of expressive aprosodia and amusia (Bautista and Ciampetti 2003).

Other miscellaneous types of simple partial seizures include unexplained urges, sexual experiences, involuntary laughing or crying, forced thoughts and, rarely, delusions. Inexplicable urges have included impulses to laugh (Sturm *et al.* 2000) or to run (Strauss 1960). Sexual experiences include strong sexual arousal (Erickson 1945) or orgasm (Reading and Will 1997; Ruff 1980).

Seizures characterized by involuntary laughing ('gelastic' seizures) or crying ('dacrystic' seizures) differ from those characterized by ictal emotion (e.g. anxiety or depression, as noted earlier) in that these patients, although laughing or crying, do not experience any associated mirth or sadness. One patient happened to see herself in the mirror in the midst of a gelastic seizure and 'was puzzled by the discrepancy between her facial expression and her feelings' (Arroyo *et al.* 1993), while another, when asked why he was laughing, could only reply 'I don't know' (Cercy and Kuluva 2009). Dacrystic seizures manifest with a sad facial expression and tears (Luciano *et al.* 1993; Marchini *et al.* 1994), in one case capping off a sensory march that began in the left leg, ascended to the left shoulder and was then succeeded by weeping (Efron 1961). Forced thoughts may occur and are quite similar to obsessions (Mendez *et al.* 1996).

Delusions noted during simple partial seizures include the Capgras phenomenon (delusion of doubles) (Kanemoto 1997), a Schneiderian first-rank symptom, namely a conviction on the patient's part 'that his body is being controlled by external forces' (Mesulam 1981), and, in one remarkable case, a delusion of sexual transformation during which the patient's voice and overall demeanor changed into that of a person of the opposite sex (Kasper *et al.* 2009).

COMPLEX PARTIAL SEIZURES

Complex partial seizures (Delgado-Escueta *et al.* 1982; Escueta *et al.* 1977; Golub *et al.* 1951; Holmes 1984; Theodore *et al.* 1983) generally last of the order of one to several minutes and may or may not be preceded by an aura. The seizure proper is characterized, in all cases, by some 'impairment' of consciousness, ranging in severity from the slightest degree of confusion to a more or less profound stupor.

Although, in some cases, the seizure manifests with this impairment of consciousness alone, one will in most cases also see a 'motionless stare' and/or automatisms. The 'motionless stare' is characterized by the abrupt appearance of a vacant stare and a complete arrest of all behavior, leaving the patient quite still and unmoving. This stare occupies only a portion of the seizure itself, and is typically either followed or, in some cases, preceded, by automatisms. Automatisms range in complexity from such simple, stereotyped behavior as lipsmacking or chewing to highly complex activity, which is, to a greater or lesser degree, 'reactive' to the environment. Such reactive automatisms may consist of a more or less faithful continuation of pre-ictal behavior or may represent behavior, which, although still reactive to the environment, represents a break with the patient's pre-ictal behavior.

After the seizure ends, most patients will display a degree of postictal confusion, lasting from one to several minutes, after which they gradually recover. Patients are subsequently totally, or sometimes only partially, amnestic for the events that occurred during the seizure.

Before discussing in detail the individual clinical aspects of complex partial seizures (i.e. aura, impairment of consciousness, motionless stare, and automatisms), some descriptions of more or less typical cases are provided to give an overall sense of what a complex partial seizure comprises. The first to be considered is Hughlings Jackson's famous patient, considered by many to be the 'paradigm of temporal lobe epilepsy' (Taylor and Marsh 1980), whom Jackson referred to as 'Dr Z' (Jackson 1889; Jackson and Colman 1898). Dr Z's first complex partial seizure was preceded by an aura of déjà vu, which was then followed by an impairment of consciousness. In Dr Z's own words:

> I was waiting at the foot of a college staircase, in the open air, for a friend who was coming down to join me. I was carelessly looking round me, watching people passing, etc., when my attention was suddenly absorbed in my own mental state, of which I know no more than that it seemed to me to be a vivid and unexpected 'recollection'; – of what, I do not know. My friend found me a minute or two later, leaning my back against the wall, looking rather pale, and feeling puzzled and stupid for the moment. In another minute or two I felt quite normal again, and was as much amused as my friend at finding that I could give no distinct account of what had happened...

In another case (Golub *et al.* 1951), the seizure was characterized by a motionless stare, followed by an impairment of consciousness and automatisms. The patient was in the physician's office when the seizure occurred. He:

> suddenly stopped talking, looked off into space, staring. He slumped down in his chair for a brief moment, then

> sat up and began to rub his abdomen with both hands. A flashlight was shined into his eyes and he turned away. He began to rummage about the desk as if looking for something. When questioned as to what he wanted, he said, 'I wanna, I wanna.' At this point he took a cigarette from his pocket, lit it and started to smoke. He then got up from his chair, walked out of the office, wandered down the hall opening all the doors saying 'I want a toilet.' Next he walked down the hall, but could not be distracted by any outside contact. He then lay down on the bed and appeared to regain contact gradually.

Aurae

As noted earlier, aurae are merely simple partial seizures that happen to evolve into complex partial seizures: they thus include all of the forms of simple partial seizure noted above. Several studies (Boon *et al.* 1991; Gupta *et al.* 1983; Kanemoto and Janz 1989; Sperling *et al.* 1989) provide a rough estimate of the frequency with which various aurae are found: the most common is a rising epigastric sensation; next come dysmnesic symptoms (déjà vu or jamais vu), affective symptoms (fear, anxiety, and depression), sensory symptoms (visual, tactile, gustatory, olfactory, or auditory hallucinations), vertigo, and nausea. Least common are illusions (macropsia, micropsia, hyperacusis, and hypoacusis) and various unclassified symptoms such as thirst or simply an 'indescribable' sensation. In some cases, the aura may be a combination of two or more symptoms: in an early report, Anderson (Anderson 1886) noted an aura compounded of auditory and visual hallucinations and déjà vu, and Hughlings Jackson (Jackson and Stewart 1899) noted a combination of an olfactory hallucination with déjà vu. Although aurae are typically remembered, amnesia for the aura may be found in approximately one-fourth of all patients upon recovering from a complex partial seizure (Schulz *et al.* 1995).

Impairment of consciousness

An impairment of consciousness (or, as it is sometimes referred to, a defect of consciousness) of one sort or another, is present in all cases. As noted by Hughlings Jackson (Taylor 1931), this impairment may involve 'all degrees of obscuration of consciousness', from a profound clouding of the sensorium to the slightest trace of confusion or inattention. Indeed, determining the presence of an impairment of consciousness may in some cases be difficult: Murray Falconer (Falconer 1954), in commenting on this, noted that a patient, during a seizure 'may carry on with what he is doing, such as playing the piano or driving a car. His performance may seem without fault, or he may betray himself [only] by ignoring the traffic light.'

Motionless stare

Although appearing in most complex partial seizures, a motionless stare is certainly not seen in all (Delgado-Escueta *et al.* 1982; Holmes 1984). Some authors have categorized complex partial seizures into two types – 'type I' referring to those seizures that are characterized by the motionless stare, and 'type II' being those without one – and have gone on to assert that, in type I seizures, the stare always precedes automatisms (Delgado-Escueta *et al.* 1982; Escueta *et al.* 1977). It appears, however, that this sequencing may not be always present, as subsequent work (Theodore *et al.* 1983) has shown that, in type I seizures, the motionless stare may be preceded by automatisms.

Automatisms

Automatisms represent behavior that is performed in a more or less automatic way, such that the patient is, to a greater or lesser degree, left in a less adaptive 'fit' with the environment. Spratling (Spratling 1902) felt that a patient in the midst of an automatism acted 'like a machine', and Kraepelin (Kraepelin 1902) described patients as acting in 'a mechanical or automatic manner'. Penfield (Penfield and Jasper 1954) echoed these impressions, noting a range in the severity of the automatism such that 'when the condition is severe, the patient acts like an automaton or a robot', but when mild, the patient 'may be cooperative and only slightly confused, but still unable to deal with new problems normally'. Automatisms are currently roughly divided into those that are stereotyped and those that are still, to some or other degree, 'reactive' to the environment.

Stereotyped automatisms generally consist of simple, purposeless behavior. Table 8.6 lists various common stereotyped automatisms as noted in studies by Delgado-Escueta *et al.* (Delgado-Escueta *et al.* 1982), Golub (Golub 1951), Holmes (Holmes 1984), and Gibbs and Gibbs (Gibbs and Gibbs 1952), roughly arranged from most (top) to least (bottom) frequent. Chewing or lipsmacking, looking around and fumbling with sheets or clothing are fairly straightforward; the other stereotyped automatisms, however, deserve some comment.

Speaking or mumbling may consist of simple, but coherent, phrases, such as 'Oh my God', or may be characterized by varying degrees of incoherence (Bell *et al.* 1990; Serafetinides and Falconer 1963; van der Horst 1953), which may at times be extreme (Gillig *et al.* 1988; Knight and Cooper 1986). In one case, the patient simply repeated neologisms such as 'exeverdedeen' over and over again (Bell *et al.* 1990).

Laughing or crying may or may not be accompanied by a sense of mirth (Yamada and Yoshida 1977) and may or may not sound natural: in one case it was 'identical' to the patient's 'natural laughter', and in another, it was 'most contagious'. More commonly, however, the laughter is unnatural (Ames and Enderstein 1975; Gascon and Lombroso 1971), not 'infectious' (Sethi and Surya 1976), and has been described

Table 8.6 Common stereotyped automatisms

Chewing or lipsmacking
Looking around
Fumbling with sheets or clothing, groping, or searching
Speaking or mumbling
Laughing or crying
Sitting or standing up
Walking or running
Thrashing or kicking; 'bicycling' movements

as cackling (Gumpert *et al.* 1970; Lehtinen and Kivalo 1965). Also, crying may or may not be accompanied by a sense of sadness (Luciano *et al.* 1993).

Running, if prominent, may allow one to speak of 'cursive' epilepsy (Chen and Forster 1973; Sethi and Surya 1976; Sisler *et al.* 1953). The running is generally uncontrollable: one patient's seizure consisted of 'howling and running' (Marsh 1978).

Bicycling movements of the lower extremities exhibit varying degrees of coordination and may or may not be accompanied by rhythmic movements of the upper extremities (Sussman *et al.* 1989; Swartz 1994).

In addition to the foregoing stereotyped automatisms, one may, rarely, see other types. In some cases, automatisms may consist of sexual activity. There may be pelvic thrusting (Geyer *et al.* 2000), genital manipulation (Leutmezer *et al.* 1999), and sexual arousal and orgasm (Remillard *et al.* 1983): in one case (Spencer *et al.* 1983), the patient's seizure consisted of lipsmacking, snorting, grimacing, thrashing about, uttering obscenities, and masturbation. Coital movements have been noted (Freemon and Nevis 1969), and in one remarkable case (Currier *et al.* 1971), a 50-year-old woman:

> began having what her daughter and husband described as 'sexual' seizures. One was as follows. The patient was sitting at the kitchen table with her daughter making out a shopping list. She stopped making the list, appeared dazed, and gradually slumped to the floor helped by her daughter. She lay on the floor on her back, lifted up her skirt, spread her knees, and elevated her pelvis rhythmically. She made appropriate vocalizations for intercourse, such as 'it feels good' and 'further, further'... Following these episodes the patient would appear confused and have no memory of them.

Other, rare, stereotyped automatisms include the Kluver–Bucy syndrome (Nakada *et al.* 1984) and stuporous catatonia with waxy flexibility (Shah and Kaplan 1980),

Reactive automatisms may, as noted earlier, consist either of behavior that represents a more or less faithful continuation of the patient's pre-ictal behavior or of behavior that, although still 'reactive' to the environment, represents a definite break with what the patient was doing before the seizure began.

Examples of 'continuation' reactive automatisms include delivering newspapers (Steegmann and Winer 1961) or continuing to drive a car (Falconer 1954). One of Hughlings Jackson's patients (Jackson 1889), if she suffered a seizure while serving tea, 'would go on pouring out but would pour out wrongly'. In another example, Forster and Liske (Forster and Liske 1963) described 'a patient, who worked in his father's shirt and pyjama factory as a sorter, [who] would, in the course of a seizure, continue to place the shirts, as they came off the line, into the same stack regardless of the size of the shirt. When the seizure ended, he would correctly place the shirts coming off the line.' In a similar example, Liddell (1953) described a 38-year-old clerk who was 'adding up a column of figures when the attack would occur. To other people he would appear to carry on as if nothing had happened. When he came to he would discover that he had written in something irrelevant or the wrong figures.'

Reactive automatisms that represent a break with pre-ictal behavior may be quite startling to observers. One patient, while walking down the street, began to 'throw away' his money (McCarthy 1900); another, a church organist, suddenly stopped playing the service music and broke out into 'hot jazz' (Forster and Liske 1963). One of Hughling Jackson's patients (Taylor 1931), in the middle of an interview, suddenly stopped responding. Jackson:

> waited a little time, and then, looking at him, I saw that he was grinning as if amused at something. Next, whilst sitting quietly in his chair, he tore a piece off a packet of prescriptions, and put it in his mouth. I took it away, but he picked up another piece from the floor and began to chew it. In about a minute more he came to himself, and then spat out into the fire a pellet of chewed paper.

In a further example, one of Penfield's patients (Penfield and Jasper 1954), a 16-year-old girl, had a complex partial seizure at a formal ball. During the seizure, she:

> proceeded to take her clothing down as though she were about to sit on the toilet. Her friends gathered round and tried to hide the performance but she seemed to come to herself and finding certain articles of apparel around her ankles, she said, 'well if they are off we had better take them off completely.' She did this and hid them in her handbag. Finally, as complete awareness returned, she was overwhelmed by shame and chagrin.

As a final example, consider a patient of Gloor's (Gloor 1975), who, during a depth electrode-monitored seizure:

> got up from his chair and began to hum and sing... He suddenly looked very cheerful, which was in marked contrast to the solemn expression which he had had prior to the attack. The radio in the room was on and some dance music was being broadcast. The patient now approached the female technician and started to dance with her to the tune coming over the radio. He continued to hum at the same time.

Before concluding this discussion of reactive automatisms, consideration must be given to the possibility of ictal violence. Violence during a complex partial seizure is a rare event, and usually only occurs with provocation, as when an attempt is made to restrain patients to protect them from harm (Delgado-Escueta *et al.* 1982). In one case (Rodin 1973), for example, a patient, during a seizure, 'suddenly lunged forward, having a bewildered and angry facial expression. An unsuccessful attempt to keep him in the chair by the attending physician resulted in making the patient angrier'; the patient then 'clenched [his] right fist... assumed a boxer-type stance, and violence [seemed] imminent. As soon as the patient was released, however, he merely got out of the chair, then sat down again and began typical fussing-type behavior with the pillow.' In other cases, patients may engage in biting, but this is generally only when someone is injudicious enough to place his or her hand near the patient's face (Tassinari *et al.* 2005).

Spontaneous unprovoked violence is rare and in most of the reported cases the patient had been accused of a crime, thus raising the question of malingering. Nevertheless, there have been reports of violence occurring during monitored seizures (Delgado-Escueta *et al.* 1981): two patients 'shouted and spat at nurses' and another 'suddenly knelt on the bed and tried to scratch the psychologist's face'. Such violence may at times be extreme: in another report of monitored violence (Ashford *et al.* 1980), the patient:

> when unrestrained in his hospital room... would jump up and run around the room. On one occasion he grabbed the drapes next to his bed and kicked out from the wall on them. On another occasion he pulled the nurse's call button out of the wall, ran into the bathroom and swung it around his head so that it was crashing against the walls. In all observations, he was locked by himself in a closed room, having no opportunity to direct aggression against people.

Other clinical manifestations during complex partial seizures

Other manifestations that may occur during a complex partial seizure include abdominal pain (Peppercorn *et al.* 1978), vomiting (Kramer *et al.* 1988; Panayiotopoulus 1988), testicular pain (York *et al.* 1979), fever (Semel 1987), unilateral dystonia (Kotagal *et al.* 1989), and various arrhthymias, including bradycardia (Britton *et al.* 2006), complete A-V block (Wilder-Smith 1992) and asystole (Kiok *et al.* 1986; Rocamora *et al.* 2003; Rugg-Gunn *et al.* 2000), which may be followed by syncope (Schufle *et al.* 2007; Smaje *et al.* 1987): in such cases, if seizures cannot be controlled, placement of a pacemaker may be required.

PETIT MAL SEIZURES

Petit mal seizures (Delgado-Escueta 1979; Penry *et al.* 1975; Sadleir *et al.* 2006), also known as absence seizures, are abrupt in onset and occur without an aura; they are very brief, lasting in the order of 10 seconds, and generally consist of an arrest of all activity accompanied by a blank stare: the seizure ends as abruptly as it began, and there is no postictal confusion or drowsiness. To the observer, it may appear that the patient had a 'blank' spell or was merely momentarily 'out of it' then 'snapped to'. In some cases, there may be some myoclonic fluttering of the eyelids and occasionally some myoclonic jerks of the hands. Some patients will also experience a partial loss of muscle tone: the head may drop forwards, or the patient may slump somewhat, but falls are unusual. Many patients will also have some simple automatisms, such as lipsmacking, chewing or fumbling, during the absence (Fuster *et al.* 1954; Penry and Dreifuss 1969; So *et al.* 1984, Sadleir *et al.* 2006; Sadleir *et al.* 2009). There may very rarely be other features, such as auditory or visual hallucinations (Guinena and Taher 1955).

A variant of petit mal, known as 'atypical absence' may be seen, most commonly in patients with mental retardation. These atypical absences are of more gradual onset and termination, tend to last longer overall, and may be associated with prominent increased muscle tone (Holmes *et al.* 1987).

GRAND MAL SEIZURES

Grand mal seizures, often referred to as generalized tonic–clonic seizures, may be preceded by an aura composed of any of the symptoms or signs seen during simple partial seizures, or may evolve out of a complex partial seizure, a process known as secondary generalization. As discussed below, under Etiology, it is critical to determine whether or not grand mal seizures are preceded by an aura or a complex partial seizure: grand mal seizures preceded by such events may be assumed to have a 'focal' onset, that is to say to be due to a more or less localized lesion, whereas those that lack such preceding events are more likely to be occurring as part of one of the idiopathic generalized epilepsies. Caution must be exercised here, however, before deciding that no 'focal' features are present, given that many patients who did, under video–EEG monitoring, clearly have an aura, will be unable, after recovering from the grand mal seizure, to recall the aura (Schulz *et al.* 1995).

Typical grand mal seizures (Theodore *et al.* 1994) begin with an abrupt loss of consciousness, often accompanied by an inarticulate 'cry': immediately there is tonic activity in all four extremities. After perhaps 15–20 seconds, the tonic activity slowly fades to be gradually replaced by regular clonic activity, which, in turn, may last anywhere from 30 seconds to 1.5 minutes. In some cases, there may be variations, the tonic phase being preceded by a few clonic jerks, or the tonic phase constituting the majority of the seizure, with only a few clonic jerks trailing behind. During the tonic activity, respirations cease and cyanosis may appear. There is often incontinence of urine, and, during the clonic phase, the tongue may be bitten. Upon cessation of the seizure proper, most patients remain in a coma or stupor for a matter of minutes. A delirium then supervenes, with prominent confusion, lasting perhaps 15–30 minutes, after which most patients fall into a deep sleep.

ATONIC SEIZURES

Atonic seizures (Gambardella *et al.* 1994; Lipinski 1977; Pazzaglia *et al.* 1985), also known as astatic seizures or 'drop attacks', occur without prodrome or aura and are characterized by a sudden loss of motor tone. In most cases, this atonus is generalized, and patients fall or slump to the ground; occasionally, however, the lack of tone may be partial, with, for example, only an abrupt drooping of the head. The atonus itself generally lasts in the order of a few seconds and may or may not be associated with a loss of consciousness. After the restoration of normal tone, most patients arise immediately, without any postictal confusion; however, others may, for a minute or two, experience a more or less profound degree of postictal confusion.

Atonic seizures generally occur only in patients who have already suffered from complex partial or grand mal seizures for many years (Gambardella *et al.* 1994; Lipinski 1977; Tinuper *et al.* 1998).

AMNESTIC SEIZURES

Amnestic seizures, also known as 'pure epileptic amnesia', are characterized solely by the appearance of amnesia. As such, they differ from complex partial seizures in that there is no impairment of consciousness, no motionless stare, and no automatic behavior. Typically, the amnesia itself is primarily of the anterograde type but, in some cases, the amnesia represents a combination of anterograde and retrograde types. Rarely, the seizure will be characterized by retrograde amnesia alone. Examples of each type follow.

Amnestic seizures of the anterograde type

These seizures (Butler *et al.* 2007; Palmini *et al.* 1992), which generally last from minutes up to an hour, are characterized by the abrupt onset of a loss of short-term memory: patients are able to recall events that occurred up to the onset of the seizure, and behave normally during the seizure itself, but subsequent to the termination of the seizure, they have no or only spotty recall of the events that transpired concurrent with the seizure itself. Palmini *et al.* (Palmini *et al.* 1992) provide some interesting examples. In one case, a waitress, during her seizure, 'was able to compute a customer's bill accurately and to bring him a correct amount of change'; later, upon recovery, 'she then realized she had a blank in her memory and could not recall her interaction with that customer'. In another case, the patient: 'was in the cafeteria at his work place, where he had to present a card and sign a register to pick up his meal. After having eaten he presented himself a second time and ate lunch twice. That same afternoon, the cook telephoned and upbraided him for having eaten two meals. He had no recollection of having done so.'

Finally, there is the case of a young woman, who, during a monitored seizure, answered a telephone, spoke with her cousin, and then went to sleep. 'The next morning she did not recall having received a phone call. When confronted with the video replay of her seizure, she was incredulous'. The investigators later contacted the cousin, who told them that 'their conversation had been entirely normal. He had not noticed anything amiss with the patient.'

Of some historical interest is the possibility that Hughlings Jackson's illustrious 'Dr Z' (Jackson 1889) may also have suffered from amnestic seizures of the anterograde type. In the midst of examining a patient, Dr Z remembered:

> taking out my stethoscope and turning away a little to avoid conversation. The next thing I recollect is that I was sitting at a writing-table in the same room, speaking to another person, and as my consciousness became more complete, recollected my patient, but saw that he was not in the room. I was interested to ascertain what had happened, and had an opportunity an hour later of seeing him in bed, with the note of a diagnosis I had made of 'pneumonia of the left base'. I gathered indirectly from conversation that I had made a physical examination, written these words, and advised him to take to bed at once. I re-examined him with some curiosity, and found that my conscious diagnosis was the same as my unconscious – or perhaps I should say, unremembered diagnosis had been. I was a good deal surprised, but not so unpleasantly as I should have thought probable.

Amnestic seizures of anterograde and retrograde types

In these seizures there is not only a defect of short-term memory, but also an inability, during the seizure itself, to recall events that occurred for a variable period of time before the onset of the seizure proper (Stracciari *et al.* 1990). Such patients may be quite alarmed at their retrograde amnesia and may ask others to fill them in (Zeman *et al.* 1998): one patient (Cole *et al.* 1987) suddenly became unable to recall whether he had gone to work that morning, long before the seizure began, and repeatedly asked others whether or not he had shown up for work. Upon termination of the seizure, the retrograde amnesia resolves, and patients are once again able to recall what events transpired up to the onset of the seizure and are also able to recall events that occurred after the termination of the seizure; those events, however, which transpired concurrent with the seizure are in almost all cases lost to memory (Deisenhammer 1981).

Amnestic seizures characterized by retrograde amnesia alone

This type of pure epileptic amnesia is very rare. In these cases, although patients are able to keep track of events during the seizure itself, they are nevertheless unable to recall events that occurred before the seizure. In one case (Venneri and Caffarra 1998), the retrograde amnesia itself was limited to only autobiographical events. Thus, during the seizure, the patient, although able to recall public events, was unable to recall personal events from her own past. Once the seizure ended, her recall of personal events was restored, and she also was able to recall being in the seizure itself and having trouble recalling those personal events.

REFLEX SEIZURES

Reflex seizures, though uncommon, are of interest, from not only a therapeutic point of view, but also a general neuropsychiatric one. Therapeutically, they offer an obvious means of seizure reduction, namely avoiding the precipitating stimulus. In general, however, their interest lies in the fact that an experience, sometimes a very complex one (as in Penfield's example below), can 'spark' off an equally complex epileptic event. Various seizure types can be reflexively induced, including simple partial seizures, complex partial seizures, grand mal and petit mal, and atonic seizures.

Reflex simple partial seizures

Reflex simple partial seizures, albeit amply reported, appear to be uncommon. Reading epilepsy is perhaps the most common, and, as the name suggests, in this form of reflex epilepsy it is the act of reading that induces seizures, which in turn may be of two types (Koutroumaniois *et al.* 1998): most commonly the seizure consists of myoclonic jaw jerking, accompanied in some by limb myoclonus, while less commonly the seizure may consist of dyslexia, which in turn may be accompanied by other symptoms, such as palinopsia (Gavaret *et al.* 2010). In eating epilepsy, eating, or the mere thought of eating, induced tingling in the tongue and hemifacial jerking in one patient (El Bouzidi *et al.* 2010).

Other stimuli may also produce various different simple partial seizures. In one patient, voluntary movement was noted to induce ictal dystonia (Falconer and Driver 1963), and in another, writing produced clonic activity in the involved upper extremity (Tanaka *et al.* 2006). In one patient, rubbing an arm produced a sensory Jacksonian march up the arm, immediately followed by a motor march up the same arm (Kanemoto *et al.* 2001b), and in others toothbrushing produced numbness or tingling in the face or tongue followed by clonic movement of the face or upper extremity (D'Souza *et al.* 2007). In another case, listening to certain kinds of music induced ictal structured visual hallucinations (Daly and Barry 1957), and ictal blindness has been induced by intermittent photic stimulation (Barry *et al.* 1985). Marchini *et al.* (Marchini *et al.* 1994) report the case of dacrystic seizures induced by speaking and, in a most unusual case (Falconer 1954; Mitchell *et al.* 1954), gazing at a safety pin was capable of precipitating a simple partial seizure characterized by a sense of profound, almost sexual, satisfaction.

Startle epilepsy (Garcia-Morales *et al.* 2009; Manford *et al.* 1996) is a relatively unique form of reflex simple partial epilepsy. The onset is in childhood or early adolescence, and most, but not all, patients have mental retardation and suffer from intractable seizures. The startle seizures themselves are induced by a startling stimulus, such as a loud noise or a sudden, unexpected touch, and the patients immediately evince a startle response, which rapidly gives way to a generalized tonic seizure, which, however, is not associated with any loss of consciousness.

Reflex complex partial seizures

Some varieties of reflex complex partial seizures have, as in the case of reflex simple partial seizures, earned specific names according to the provoking stimulus; thus one may speak of 'musicogenic', 'reading', and 'eating' epilepsy. Other stimuli, for example hearing certain people speak, may also be effective.

Musicogenic complex partial seizures, although extensively reported, are rare events (Critchley 1942; Forster 1977, Tayah *et al.* 2006). Critchley (Critchley 1937) provided an illustrative case: the patient, after hearing 'a rather loud fox-trot with a well defined tempo... suddenly became pale... [and] leaned forward in her chair with a frightened look in her eyes... and... began smacking her lips'. In some cases, it appears that only specific kinds of music are capable of inducing seizures (Newman and Saunders 1980), for example the pealing of church bells (Poskanzer *et al.* 1962), organ music (Joynt *et al.* 1962), 'a large dance band playing "swing" music' (Daly and Barry 1957), emotionally laden music (Gastaut and Tassinari 1966), and 'Italian' songs (Wieser *et al.* 1997). This last case is of interest in that, upon hearing the Italian song, the patient's complex partial seizure was preceded by an aura composed of an auditory hallucination of 'pleasing female murmuring voices, which took increasing possession of her mind', following which she had a complex partial seizure with a motionless stare.

Reading-induced complex partial seizures may occur with 'silent' reading (Gastaut and Tassinari 1966) or only upon reading aloud (Forster *et al.* 1969a). In one case (Critchley *et al.* 1959), not just any reading material would do, but rather it was reading a newspaper, especially the patient's hometown newspaper, that constituted the effective stimulus.

Eating-induced complex partial seizures may occur regardless of what is eaten (Ahuja *et al.* 1980; Fiol *et al.* 1986; Forster 1971) or may occur only with eating certain foods, in one case, for example, apples (Abenson 1969).

Other stimuli capable of inducing complex partial seizures appear to be highly idiosyncratic: thus there are case reports of seizures induced by hearing the voices of certain specific radio announcers (Forster *et al.* 1969b), by answering the telephone (Michelucci *et al.* 2004), engaging in heartfelt prayer (Glass *et al.* 2007), hearing a vacuum cleaner (Carlson and St. Louis 2004), feeling sad (Fenwick and Brown 1989), arching the back (Jacome *et al.* 1980), exercising (Sturm *et al.* 2002), or experiencing orgasm (Ozkara *et al.* 2006). More remarkably, albeit very rarely, seizures may also be induced by thinking certain thoughts, as for example, in one case, thinking about home (Martinez *et al.* 2001). Finally, there is a fascinating case reported by Penfield (Penfield and Jasper 1954) wherein it appears that it was the evocation of a memory of a specific event that served as the trigger. The history is a bit lengthy and complex but worth retelling.

The patient, when 13 years old, 'was playing with a dog that belonged to a neighbor. He remembered grabbing a stick out of the mouth of the dog and throwing it to a distance. The dog chased the stick and brought it back to him.' At the age of 17, the patient had his first seizure: 'he was in a crowd of people at the military school which he was attending [and] watched a demonstration of military tactics in which one of his classmates grabbed a rifle out of the hands of another man'. At that instant, 'there immediately came to the patient's mind' a memory of playing with the dog when he was 13, and 'in his mind he associated these two incidents and tried to put himself into the past memory. Following this, he became confused and was unable to speak for several seconds. Then he evidently lost consciousness and had a convulsive seizure.' The patient remained well until a half-year later, when 'while in a night club, he heard a man saying, "Give me my hat." He saw him grab his hat from the hat-check girl. Immediately the memory of the dog chasing the stick, and the memory of grabbing the stick from the dog's mouth, came back to him. Following that he was confused and behaved in an automatic manner.'

Reflex grand mal and petit mal seizures

Both grand mal and petit mal seizures may be induced by such intermittent photic stimulation as with electronic video games (Quirk *et al.* 1995) or watching a flickering television set (Baykan *et al.* 2005).

Reflex atonic seizures

Atonic seizure were produced reflexively in the case of one girl, whose seizures could be 'triggered by laughing if she was told jokes or watched a comedy on television' (Jacome and Risko 1984).

STATUS EPILEPTICUS

Status epilepticus is said to be present when, for at least a half-hour, the patient is either undergoing a continuous seizure or has so many closely spaced seizures that there is no time for full recovery between them. Status epilepticus may be seen with simple partial, complex partial, petit mal, grand mal, and amnestic seizures, and each is discussed in turn. Following this discussion, attention is then turned to another, and unique, form of status, namely epileptic stupor or coma.

Although most patients, upon recovery from status, are left without cognitive sequelae (Adachi *et al.* 2005), cognitive impairment, along with other sequelae, may, as noted below, be found after status, most especially after grand mal status.

Simple partial status epilepticus

Simple partial status epilepticus, although most commonly presenting with motor signs, has also been reported with psychic symptoms such as fear or structured hallucinations.

Motor simple partial status epilepticus is traditionally referred to as epilepsia partialis continua. These seizures typically consist of persistent clonic jerking involving the face, arm, or leg, and can last any time from hours to years (Cockerell *et al.* 1996; Scholtes *et al.* 1996a; Seshia *et al.* 2005); indeed in one series, the mean duration was 25 months, the longest seizure lasting 18 years (Thomas *et al.* 1977). Clonic activity may or may not persist into sleep (Cockerell *et al.* 1996; Thomas *et al.* 1977) and may be accompanied by other symptoms such as aphasia (Cockerell *et al.* 1996). Although the motor activity is, in the vast majority of cases, unilateral, there may in a small minority be bilateral involvement (Ashkenazi *et al.* 2000; Thomas *et al.* 1977). Rarely, inhibitory simple partial status epilepticus may occur, with prolonged ictal paresis (Smith *et al.* 1997).

Fear or anxiety may be the sole manifestation of simple partial status epilepticus and has been noted to last for between 10 days (Scott and Masland 1953) and 3 months (Henriksen 1973). Isolated ictal depression has also been noted; in one case (Weil 1956), the patient for 2 weeks 'felt "blue" and cried, and the smell of fat on a hot stove kept coming back'.

Structured hallucinations occurring in status have consisted of the auditory hallucination of a 'song' persisting for 3 hours (Wieser 1980) and a 5-day episode during which the patient experienced 'aphasia, acoustic hallucinations (strange sounds, music, voices), left-sided headache, and visual hallucinations (red balls in the front of the right eye)'.

Complex partial status epilepticus

Complex partial status epilepticus is always characterized by a more or less profound impairment of consciousness, may or may not be accompanied by automatisms, and generally lasts from an hour or less up to days (Cockerell *et al.* 1994; Goldensohn and Gold 1960; Mayeux and Lueders 1978; Rennick *et al.* 1969; Sheth *et al.* 2006, Tomson *et al.* 1986, 1992; Van Rossum *et al.* 1985; Williamson *et al.* 1985b). Longer episodes, however, have been reported, up to 7.5 months in one case (Roberts and Humphrey 1988) and 18 months (Cockerell *et al.* 1994) in another.

Automatisms seen in complex partial status epilepticus may be either stereotyped or reactive. Reactive automatisms include such simple behavior as feeding oneself (Jaffe 1962) or wandering off. In one case (Coats 1876), immediately after the seizure began, the patient, a craftsman, 'laid aside his tools, took off and hung up his leather apron, and after muttering some unintelligible words to the foreman, left the shop. He himself remembers nothing of this, but woke to consciousness to find himself sitting on one of the benches in George Square.' When the wandering is quite prolonged, one speaks of 'poriomania': in one case the patient, after a seizure lasting 24 hours, came to find that he had travelled 'several hundred miles' (Mayeux *et al.* 1979).

In some cases of complex partial status epilepticus, hallucinations, delusions, and bizarre behavior may be prominent (Ballenger *et al.* 1983; Ellis and Lee 1978; Mikati *et al.* 1985). In one case (Drury *et al.* 1985), the patient, an elderly clergyman, 'became agitated and paranoid and had delusional thinking with religious overtones'; two days later he was still in status, and when seen 'was tremulous, had bizarre posturing and was masturbating'. In another case (Wells 1975): 'the patient became withdrawn... refused to eat... grew restless, fitful and irritable; episodes of violent anger requiring restraints followed. She appeared to be hallucinating, crying out on several occasions, "I am God." When placed on a stretcher for transfer... she became hostile and resistant, screaming "I'm dying" and "They're taking me to the morgue".' After 2 days, she was eventually treated with an anti-epileptic drug, with prompt clearing. On recovery, she described her experience as though recovering from a bad dream.'

In other cases, stuporous catatonia has been reported (Engel *et al.* 1978): one patient (Lim *et al.* 1986) 'held his arms and legs rigidly in the air and would maintain his extremities in any bizarre posture indefinitely'.

Petit mal status epilepticus

Petit mal status epilepticus, which may occur in both childhood and the adult years (Belafsky *et al.* 1978; Michelucci *et al.* 1996; Nightingale and Welch 1982; Novak *et al.* 1971; Schwab 1953; Thompson and Greenhouse 1968; Tucker and Forster 1958; Zappoli 1955), is characterized by varying degrees of confusion, often accompanied by 'rhythmic blinking or small-amplitude myoclonic jerking of the face and arms', and generally lasts anywhere from half an hour to 2 days (Andermann and Robb 1972). In some cases, the confusion may be very slight: in the case of one 30-year-old woman seen during petit mal status (Friedlander and Feinstein 1956), 'during the first minute or so of the examination nothing unusual in her manner was noted, but as she was questioned further it became apparent that there was slowness in comprehending certain questions', the presence of myoclonus of the eyelids and eyebrows suggested the correct diagnosis.

Rarely, there may be prominent automatisms. One patient went for a walk and took a bus while in petit mal status (Vizioli and Magliocco 1953), and in another case (Bornstein *et al.* 1956) a 16-year-old girl, while playing cards in her dormitory room, 'suddenly got up, walked out without a coat, went to the airport, and boarded a plane to New York without a ticket. Upon arrival in New York she was charged as a stowaway. She was found to be confused and incoherent.'

Grand mal status epilepticus

Grand mal status epilepticus (Aminoff and Simon 1980), the most dramatic and life-threatening of all forms of status, is characterized by closely spaced grand mal seizures. In some cases, the clinical presentation may evolve into an ictal coma (discussed below), with no motor activity.

Amnestic status epilepticus

Amnestic status is rare, but has been noted, in some cases persisting for days (Zeman *et al.* 1998).

Epileptic stupor or coma

Stupor or coma may occur on an epileptic basis, and such stupor or coma may either follow upon an episode of complex partial status (Tomson *et al.* 1986) or grand mal status (DeLorenzo *et al.* 1998; Lowenstein and Aminoff 1992), or may appear *de novo* without any preceding more obvious type of seizure (Towne *et al.* 2000) to alert the diagnostician. In some parts of the literature, such cases of epileptic stupor or coma are considered a form of complex status epilepticus, while in other parts they are referred to as examples of 'non-convulsive status epilepticus': given this definitional uncertainty, the term 'epileptic stupor or coma' may be preferable, as the meaning is quite clear.

In those instances in which the epileptic stupor or coma follows upon an episode of complex partial status epilepticus, one finds that the patient, rather than remaining alert, suddenly becomes stuporous or comatose. This evolution of events should immediately suggest an epileptic stupor or coma, given that the postictal state following complex partial status epilepticus is characterized by confusion and a degree of lethargy, but not by stupor or coma.

Instances of epileptic stupor or coma, following upon grand mal status epilepticus, may be more difficult to identify, given that the postictal state following grand mal status is characterized by stupor or coma. The diagnostic clue here, however, is the duration of the stupor or coma: postictal states typically begin to resolve within a half hour or so, whereas epileptic stupor or coma will last much longer.

The final type of epileptic stupor or coma, namely that which occurs *de novo*, may, in the absence of a good history, be the most difficult to identify, given that there is no preceding complex partial or grand mal activity. One clue, of course, is the paroxysmal onset of the stupor or coma; however, if the stupor or coma began while the patient was alone, this all-important diagnostic clue will obviously be lacking.

In most cases of epileptic stupor or coma subtle motor activity may be present (Husain *et al.* 2003; Lowenstein and Aminoff 1992), thus providing a valuable diagnostic clue. There may be twitching of the corner of the mouth or the fingers or toes, or ocular abnormalities, including nystagmus, hippus or forced eye deviation. Given that these signs may be *very* subtle, prolonged and close observation is required. It must be emphasized, however, that there are undisputed cases of epileptic coma wherein there are *no* motor abnormalities at all, and patients are completely immobile (DeLorenzo *et al.* 1998; Lowenstein and Aminoff 1992).

Whenever epileptic stupor or coma is suspected, an emergent EEG is required, and here one looks for classic ictal activity, either focal or generalized: periodic lateralized epileptiform discharges (PLEDS) or bilateral independent PLEDS (BIPLEDS) simply do not count.

Sequelae of status

Enduring sequelae may be seen after simple partial, complex partial and, most especially, grand mal status epilepticus.

Simple partial status epilepticus may, albeit rarely, leave sequelae such as hemiparesis (Desbiens *et al.* 1993), which may be permanent (Borchert and Labar 1995), or a motor aphasia (Scholtes *et al.* 1996b).

Complex partial status epilepticus may also have permanent sequelae, but this appears to be uncommon. Reported sequelae, after 1–3 days of status, include amnesia, with or without other cognitive deficits (Krumholt *et al.* 1995), and in one case hemianopia and receptive aphasia after an episode that lasted 8 days (Donaire *et al.* 2006); in this latter case, autopsy revealed laminar cortical necrosis in the left hemisphere involving the occipital, parietal, and temporal cortices.

Grand mal status may be followed by either a personality change or dementia of variable severity in approximately one-fourth of patients (Oxbury and Whitty 1971, Rowan and Scott 1970); these symptoms, may, in turn, be related to a loss of hippocampal pyramidal cells (DeGiorgio *et al.* 1992).

Etiology

The multiple causes of epilepsy and seizures are listed in Table 8.7, where they are divided into several groups. The first group, the *idiopathic generalized epilepsies* (e.g. childhood absence epilepsy, with petit mal seizures), accounts for approximately one-third of all cases of epilepsy. The next three groups include seizures due to *metabolic* or *toxic* factors (such as hypoglycemia or treatment with a medication such as clozapine) or to *alcohol or sedative/hypnotic withdrawal*: in these cases, given that seizures generally do not recur if the underlying cause is treated, it is not appropriate to speak of epilepsy, a term generally reserved, as noted earlier, for patients in whom seizures, in the natural course of events, will, in all likelihood, continue to occur. The next group includes *common intracranial causes* of epilepsy, including mass lesions (e.g. tumors), cerebrovascular disorders (e.g. infarction), neuronal migration disorders (e.g. focal cortical dysplasia), mesial temporal sclerosis, and TBI. Next are seizures occurring

Table 8.7 Causes of epilepsy and seizures

Idiopathic generalized epilepsies
Childhood absence epilepsy
Juvenile absence epilepsy
Juvenile myoclonic epilepsy
Idiopathic generalized epilepsy with tonic–clonic seizures only

Metabolic
Hypoglycemia
Hyperglycemia
Hyponatremia
Hypernatremia
Hypocalcemia
Hypomagnesemia
Uremia

Toxic
Medications
- Clozapine
- Phenothiazines
- Bupropion
- Tricyclic antidepressants
- Lithium
- Tiagabine
- Baclofen
- Penicillin
- Cefipime
- Isoniazid
- Busulfan
- Ciclosporin
- Tacrolimus
- Theophylline
- Meperidine
- Ondansetron
- Bismuth

Intoxicants
- Phencyclidine
- Amphetamine
- Cocaine

Miscellaneous toxins
- Iodinated contrast dye
- Lead
- Tin
- Star fruit
- Domoic acid
- Aspartame

Alcohol or sedative/hypnotic withdrawal
Alcohol
Sedative/hypnotics

Common intracranial causes
Mass lesions
- Tumors
- Abscesses
- Vascular malformations

Cerebrovascular disorders
- Infarction
- Intracerebral hemorrhage
- Subarachnoid hemorrhage

Neuronal migration disorders
Mesial temporal sclerosis
Traumatic brain injury

Meningoencephalitis
Meningitis
Encephalitis
Acute disseminated encephalomyelitis

In association with certain dementing and neurodegenerative disorders
Multi-infarct dementia
Alzheimer's disease
Neurosyphilis
Acquired immune deficiency syndrome dementia
Progressive multifocal leukoencephalopathy
Toxoplasmosis
Dentatorubropallidoluysian atrophy
Choreoacanthocytosis
McLeod's syndrome
Wilson's disease
Spinocerebellar ataxia
Cerebrotendinous xanthomatosus
Cerebral amyloid angiopathy
Creutzfeldt–Jakob disease
Granulomatous angiitis
Metachromatic leukodystrophy
Adrenoleukodystrophy
Pantothenate kinase-associated neurodegeneration
Juvenile Huntington's disease
Subacute sclerosing panencephalitis

Congenital disorders
Mental retardation
Autism
Down's syndrome
Fetal alcohol syndrome
Fragile X syndrome
Klinefelter's syndrome
Sturge–Weber syndrome
Von Recklinghausen's disease
Rett's syndrome
Tuberous sclerosis
Prader–Willi syndrome

Miscellaneous
Systemic or autoimmune disorders
- Hypertensive encephalopathy
- Reversible posterior leukoencephalopathy syndrome
- Hashimoto's encephalopathy
- Limbic encephalitis
- Systemic lupus erythematosus
- Thrombotic thrombocytopenic purpura
- Hyperthyroidism
- Central pontine myelinolysis
- Hepatic porphyria
- Wernicke's encephalopathy
- Marchiafava–Bignami disease
- Sarcoidosis

Continued

Table 8.0 *Continued*

Behçet's syndrome	Sydenham's chorea
Wegener's granulomatosus	Precipitating events
Whipple's disease	Postanoxic encephalopathy
Celiac disease	Radiation encephalopathy
Autosomal dominant partial epilepsies	Dialysis dysequilibrium syndrome
Multiple sclerosis	Dialysis dementia
Disorders typically presenting in childhood or adolescence	Eclampsia
Rasmussen's syndrome	Postelectroconvulsive therapy
Landau–Kleffner syndrome	

due to *meningoencephalitis*, which, although not a common cause, is a diagnosis that must not be missed. The next group includes epilepsy occurring *in association with certain dementing and neurodegenerative disorders*, such as multi-infarct dementia. Following this is a group of *congenital disorders*, such as mental retardation, also associated with epilepsy. Finally, there is a large *miscellaneous* group, including such disorders as hypertensive encephalopathy.

The diagnostic evaluation is simplified by attending to the kinds of seizures that the patient has. Petit mal seizures are seen only in the idiopathic generalized epilepsies. Grand mal seizures may occur either in idiopathic generalized epilepsies or secondary to some other cause: grand mal seizures of non-focal onset suggest one of the idiopathic generalized epilepsies or a toxic or metabolic factor, whereas focal-onset grand mal seizures, that is to say those preceded by an aura or a complex partial seizure, indicate another cause. All the other seizure types, that is to say simple partial, complex partial, atonic and amnestic types, in almost all cases will be due to a focal lesion, and among adults by far the most frequent causes are found in the group of *common intracranial causes* (Currie *et al.* 1971; Dam *et al.* 1985; Jensen and Klinken 1976; King and Marsan 1977; Loiseau *et al.* 1990a; Mauguire and Courjon 1978; Rasmussen 1963, 1983; Sander *et al.* 1990; Zentner *et al.* 1995). Clearly, determining whether the seizures in question are 'focal' or 'non-focal' facilitates the diagnostic workup.

In evaluating any given patient with seizures, the possibility must be kept in mind that multiple factors may be involved. Thus, patients with epilepsy, say, due to an intracranial cause such as an infarction, may also have seizures due to hypoglycemia occurring with overvigorous treatment of diabetes mellitus. Or, to take another, not uncommon, example, seizures in an alcoholic may be due not only to alcohol withdrawal but also to lesions sustained during the traumatic brain injuries that alcoholics are so prone to (Earnest *et al.* 1988).

After each of these groups is discussed, attention is then turned to a suggested *diagnostic workup*.

IDIOPATHIC GENERALIZED EPILEPSIES

The idiopathic generalized epilepsies (IGEs) are genetically determined syndromes characterized by generalized seizures (petit mal, myoclonic, or grand mal) which usually first appear in childhood or adolescence, and which are usually, although not always, readily controlled by appropriate AEDs, such as divalproex. Although there is some controversy as to how many and what types of IGE exist, most authorities describe the following: childhood absence epilepsy; juvenile absence epilepsy; juvenile myoclonic epilepsy; and, finally, idiopathic generalized epilepsy with tonic–clonic seizures only. This is a very important group, accounting for approximately one-third of all epilepsies seen among adults.

Childhood absence epilepsy and juvenile absence epilepsy are similar in that in both disorders all patients have petit mal seizures, and most will also have grand mal seizures. These disorders differ in their age of onset, with childhood absence epilepsy appearing between the ages of 4 and 8 years, and juvenile absence epilepsy between the ages of 9 and 13 years. In all cases, the EEG shows typical interictal generalized spike and dome discharges, especially evident during hyperventilation.

Juvenile myoclonic epilepsy (Camfield and Camfield 2009; Canevini *et al.* 1992; Janz 1985) is characterized by generalized myoclonic seizures, and, in most, by either grand mal or petit mal seizures; onset is typically between the ages of 4 and 18 years. The EEG typically shows interictal generalized spike, polyspike, or polyspike and wave discharges.

Idiopathic generalized epilepsy with tonic–clonic seizures only (Camfield and Camfield 2010) is characterized, as the name suggests, by only grand mal seizures. Another name for this disorder is epilepsy with generalized tonic–clonic seizures on awakening, a synonym that calls attention to the fact that in this disorder the grand mal seizures, although able to occur at any time of the day, are most often seen early in the morning, soon after awakening. This disorder usually has an onset in adolescence or early adult years. As with juvenile myoclonic epilepsy, the EEG in this disorder typically shows interictal generalized spike, polyspike, or polyspike and wave discharges.

There is some evidence that all of these IGEs share a common genetic background, differing largely in their age of onset (Marini *et al.* 2004; Yenjun *et al.* 2003). From a general clinical point of view, the diagnosis should be suspected in all cases characterized by petit mal seizures or by non-focal grand mal seizures with onset between childhood and early adult years, and in any case when the EEG shows generalized interictal discharges of the spike and dome, spike, polyspike, or polyspike and wave types (Betting *et al.* 2006).

METABOLIC

Of the metabolic causes of seizures, hypoglycemia is perhaps most common, and although such seizures may be accompanied by autonomic signs (e.g. tremor) or delirium, it must be borne in mind that a seizure can be the presenting manifestation of hypoglycemia (Hoefer *et al.* 1946; Malouf and Brust 1985).

Hyperglycemia, occurring in the syndrome of non-ketotic hyperosmolar hyperglycemia, has been strongly associated with simple partial seizures with either sensory (Maccario *et al.* 1965) or, more commonly, motor symptoms (Grant and Warlow 1985; Hennis *et al.* 1992): in some cases, these simple partial seizures with motor symptoms display a reflex character, being induced by motion (Brick *et al.* 1989). Occasionally, simple partial status epilepticus with motor symptoms may occur (Singh *et al.* 1973; Singh and Strobos 1980). Rarely, simple partial seizures may occur with hyperglycemia during diabetic ketoacidosis (Placidi *et al.* 2001).

Hyponatremia, with levels of 120 mEq/L or less, may cause seizures, often in the context of a delirium (Swanson and Iseri 1958). Seizures secondary to hypernatremia are very rare, and generally occur only with levels of 160 mEq/L or higher (Moder and Hurley 1990).

Hypocalcemia may provoke seizures that may or may not be accompanied by other signs, such as tetany (Glaser and Levy 1960). In cases of chronic hypocalcemia secondary to hypoparathyroidism, seizures may be the presenting sign (Berger and Ross 1981) or may be preceded by other signs, for example cataracts or parkinsonism (Eraut 1974).

Hypomagnesemia, in addition to causing delirium and myoclonus, may also lead to seizures (Hall and Joffe 1973).

Uremia may be accompanied by seizures (Tyler 1968), with one study finding this to be the case in approximately one-third of uremic patients (Locke *et al.* 1961). Rarely, seizures in uremia may be caused not so much by the uremia *per se*, but by aluminum intoxication, as has occurred in some patients with renal failure who had been taking antacids chronically (Russo *et al.* 1992).

TOXIC

Toxic seizures may occur secondary to *medications*, such as clozapine, *intoxicants*, such as cocaine, and *miscellaneous toxins*, such as iodinated contrast dye.

Medications

Of the medications capable of causing seizures perhaps the most notorious is clozapine. Clozapine, an atypical antipsychotic, causes seizures overall in 1.3 percent of patients (Pacia and Devinsky 1994), with a higher incidence found in patients with a history of seizures (Wilson and Claussen 1994): the risk increases with rapid dose titration (Baker and Conley 1991; Devinsky *et al.* 1991) and with high doses, rising to approximately 10 percent in those taking more than 600 mg daily (Haller and Binder 1990). Phenothiazine antipsychotics may also cause seizures (Olivier *et al.* 1982), especially with higher doses (Logothetis 1967). Of the antidepressants, bupropion is most likely to cause seizures: among patients treated with 600 mg or more daily, some 2 percent will have seizures; importantly, seizures are particularly common in patients with a history of bulimia nervosa (Davidson 1989; Johnston *et al.* 1991). Tricyclic antidepressants may cause seizures, as has been noted with amitriptyline (Betts *et al.* 1968b) and imipramine (Petti and Campbell 1975); maprotilene (Jabbari *et al.* 1985) and clomipramine (Clomipramine Collaborative Study Group 1991) are particularly likely to cause seizures at therapeutic doses, and, in the case of amoxapine, seizures are very common in overdose (Litovitz and Troutman 1983). Lithium may cause grand mal seizures (Wharton 1969) and may increase the frequency of seizures in those with petit mal epilepsy (Moore 1981). Interestingly, tiagabine, an AED, has been associated with complex partial status epilepticus (Koepp *et al.* 2005). Baclofen, in high dosage, caused complex partial status in one case (Zak *et al.* 1994). Antibiotics known to cause seizures include penicillin (Snavely and Hodges 1984), cefipine (Dixit *et al.* 2000, Fernandez-Torre *et al.* 2005), and isoniazid (Messing *et al.* 1984). Chemotherapeutic agents associated with seizures include busulfan (Murphy *et al.* 1992), ciclosporin (Appleton *et al.* 1989), and tacrolimus (Wijdicks *et al.* 1996). Other agents include theophylline (Messing *et al.* 1984), meperidine (Kaiko *et al.* 1983), ondansetron (Singh *et al.* 2009) and bismuth: bismuth, as may be used in the treatment of peptic ulcer, is known to cause delirium with myoclonus and either partial or grand mal seizures (Supino-Viterbo *et al.* 1977); uncommonly, bismuth intoxication may present with a seizure (Molina *et al.* 1989).

Intoxicants

Intoxicants noted to cause seizures include phencyclidine, amphetamine and cocaine (Alldredge *et al.* 1989): in the case of cocaine, complex partial status epilepticus has been noted (Merriam *et al.* 1988; Ogunyemi *et al.* 1989).

Miscellaneous toxins

Of the miscellaneous toxins, perhaps the most common offender is iodinated contrast dye, given intravenously (Aurahami *et al.* 1987), intra-arterially (Vickrey and Bahls 1989), or intrathecally (Greenberg and Vance 1980; Shiozawa *et al.* 1981). Lead intoxication in adults, when acute, often presents with delirium, abdominal pain, and seizures, as may be seen in adults who drink illicit 'moonshine' whiskey made in old radiators (Morris *et al.* 1964; Whitfield *et al.* 1972) or, in one case in which adults burned discarded lead storage battery boxes for fuel (Akelaitis 1941). Acute lead intoxication in children, as can be seen in those who ingest leaded paint chips, may, when severe, cause seizures and delirium or coma (Jenkins and Mellins 1957). Tin intoxication may cause a delirium with seizures, as has been noted with trimethyl (Fortemps *et al.* 1978) and triethyl tin (Alajouanine *et al.* 1958); in one case, upon recovery from trimethyl tin intoxication, a patient subsequently developed complex partial seizures (Feldman *et al.* 1993). Star fruit is generally innocuous, however in those with renal failure, it may cause seizures (Tsai *et al.* 2005) Domoic acid intoxication, as may occur with the ingestion of mussels, is a very rare event but is noted here because the resulting neuropathology is similar to that of mesial temporal sclerosis. Domoic acid is an excitotoxin, similar to kainic acid, and in the reported case ingestion was followed by the development of hippocampal atrophy and complex partial seizures (Cendes *et al.* 1995). Finally, aspartame, taken in very high dose, has been noted, in a case report, to cause a seizure (Walton 1986).

ALCOHOL OR SEDATIVE/HYPNOTIC WITHDRAWAL

In general hospital practice, alcohol withdrawal seizures are quite common, especially among adults in their middle years (Sander *et al.* 1990); the diagnosis, however, often proves elusive as many patients (and their family members) may deny excessive alcohol use.

Alcohol withdrawal seizures are typically of the grand mal type, and are generally seen only in patients with a long history of heavy drinking (Lechtenberg and Worner 1992; Leone *et al.* 1997; Schuckitt *et al.* 1993). Seizures may occur either in an isolated fashion as 'rum fits' or in association with delirium tremens (Isbell *et al.* 1955). 'Rum fits' are generally restricted to the first 48 hours of abstinence, and, although repeated seizures, even status, may occur, most patients have only one or two seizures (Earnest *et al.* 1988; Espir and Rose 1987; Victor and Brausch 1967). Delirium tremens, marked by delirium and prominent tremor, generally appears within the first few days of abstinence, and seizures, which may either 'usher in' the delirium or occur in the midst of it, are seen in between 10 percent (Rosenbaum *et al.* 1941) and 20 percent (Lundquist 1961) of patients. Importantly, although most seizures occurring in a newly abstinent alcoholic are related to the alcohol withdrawal *per se*, other epileptogenic factors, such as head trauma, hypoglycemia, hypomagnesemia, Wernicke's encephalopathy, or meningitis may also be present (Earnest *et al.* 1988; Lechtenberg and Worner 1992).

Withdrawal from other sedative/hypnotic agents such as benzodiazepines or barbiturates may also cause grand mal seizures (Kalinowsky 1942; Levy 1984), and in the case of barbiturates, status epilepticus is not uncommon.

COMMON INTRACRANIAL CAUSES

In cases in which metabolic and toxic factors seem absent, withdrawal unlikely, and the history is not compatible with idiopathic generalized epilepsy, consideration should be given to the presence of certain intracranial disorders, the most common of which are mass lesions (e.g. tumors), cerebrovascular disorders (e.g. infarction), mesial temporal sclerosis, neuronal migration disorders, and TBI. Each of these is considered in turn.

Mass lesions

Tumors, abscesses, and vascular malformations may cause seizures, which may be simple or complex partial or grand mal.

Seizures occurring secondary to tumors are more likely to occur when the tumor involves the cerebral cortex, and less so when only the white matter or subcortical gray structures are involved (Hildebrand *et al.* 2005; Scott and Gibberb 1980). Given their frequency in adults (Dam *et al.* 1985), the clinical 'rule' that a new-onset seizure disorder in an adult is, until proven otherwise, secondary to a tumor, is a sound clinical guide.

Cerebral abscesses, which may or may not be accompanied by fever and leukocytosis, are frequently associated with seizures, not only when acute, but also among those patients who survive (Legg *et al.* 1973); in this regard it may be noted that neurocysticercosis, a common cause of seizures in developing countries, may be emerging in developed countries, such as Portugal (Monteiro *et al.* 1995) and the USA (Wallin and Kurtzke 2004).

Vascular malformations, such as arteriovenous malformations and various angiomas, are a common cause of seizures, which may indeed constitute their sole clinical expression (Kramer and Awad 1994), even in the absence of any symptomatic hemorrhage.

Cerebrovascular disorders

Simple partial, complex partial, and grand mal seizures (Labovitz *et al.* 2001) may occur after infarction (Bogousslavsky *et al.* 1992; Daniele *et al.* 1989; Kilpatrick *et al.* 1990, 1992; Richardson and Dodge 1954; So *et al.* 1996), intracerebral hemorrhage (Passero *et al.* 2002), or subarachnoid hemorrhage, and it is customary to divide these seizures into three types, depending on when they occur: (i) immediate seizures during the first day; (ii) early seizures during the following two weeks; and (iii) late seizures during the following months and years.

After cerebral infarction, immediate and early seizures are seen in less than 3 percent of patients. Late-appearing seizures are seen in a little more than 3 percent of cases, and generally appear within the first 2 years or so poststroke (Lamy *et al.* 2003): those with severe stroke seem at higher risk for late-appearing epilepsy (Lamy *et al.* 2003), and the occurrence of an immediate or early seizure may (Lamy *et al.* 2003) or may not (Lossius *et al.* 2005) increase the likelihood of a late-appearing one.

Intracerebral hemorrhage is more likely to cause seizures when the hemorrhage is lobar, rather than subcortical (Passero *et al.* 2002), and immediate or early seizures are seen in up to 25 percent of lobar cases.

Subarachnoid hemorrhage (Hart *et al.* 1981) likewise may cause immediate or early seizures in some 25 percent of cases; late-appearing seizures are seen in less than 10 percent, and appear more common in cases complicated by infarction (Claasen *et al.* 2005).

Neuronal migration disorders

Neuronal migration disorders comprise a large group of conditions, all of which result from defective neuronal migration during embryogenesis. In the normal course of embryogenesis, neurons migrate along radially oriented glial fibers from the periventricular area to the overlying cortical plate, where they come to rest in an orderly fashion, eventually resulting in a properly laminated and appropriately convoluted cortex.

Disordered migration, depending on its timing and degree, may result in a variety of morphologic changes (Raymond *et al.* 1995). In cases where migration stalls out, as it were, either at the periventricular area or in the white matter, one may find misplaced collections of neurons, known as heterotopias, in various locations (d'Orsi *et al.* 2004; Dubeau *et al.* 1995; Tassi *et al.* 2005): thus there may be periventricular nodular heterotopias (Battaglia *et al.* 2006; D'Agostino *et al.* 2002; Kothare *et al.* 1998) and what are known as 'band' or 'laminar' heteropias, wherein the heterotopia extends in a band through a greater or lesser portion of the white matter (Barkovich and Kjos 1992; d'Orsi *et al.* 2004; Ono *et al.* 1997);

in many cases one also finds an excessive number of isolated neurons scattered throughout the white matter. The overlying cortex may also be malformed, and this malformation may be either macroscopic or microscopic. Macroscopic malformations include pachygyria (a grossly thickened cortex), polymicrogyria (Leventer *et al.* 2010) (with multiple small and highly convoluted gyri), or lissencephaly (a 'smooth' cortex, with little or no gyrification visible) (Brodtkorb *et al.* 1992). Microscopic abnormalities are subsumed under the term 'focal cortical dysplasia', which, as proposed by Palmini (Palmini *et al.* 2004), includes two types: mild, or type I focal cortical dysplasia is characterized by cortical disorganization and dyslamination (Armstrong 1993; Meencke and Janz 1984), while the Taylor type, or type II is characterized not only by dyslamination but also dysmorphic neurons and balloon cells (Fauser *et al.* 2006; Lawson *et al.* 2005; Palmini *et al.* 1991; Siegel *et al.* 2005; Sisodiya *et al.* 1995; Taylor *et al.* 1971).

Various seizure types may occur secondary to disordered neuronal migration, including complex partial, simple partial, and grand mal, and, with the advent of MRI, it has become apparent that disordered neuronal migration is one of the most common causes of focal epilepsy (Barkovich *et al.* 1995; Raymond *et al.* 1995; Sisodiya *et al.* 1996): in these cases the age of onset of seizures ranges from childhood to early adult years.

Before ending this discussion of disordered neuronal migration, mention should be made of a rare condition known as hypothalamic hamartoma. In this condition one finds one or more nodules, or hamartomas, composed of otherwise normal neurons, within the hypothalamus. Classically, these present in childhood with precocious puberty and simple partial seizures of the gelastic type (Ames and Enderstein 1980; Berkovic *et al.* 1988; Breningstall 1985; Cascino *et al.* 1993a; Kuzniecky *et al.* 1997). Further experience, however, has demonstrated that the onset may be delayed until adult years, that precocious puberty may not be present, and that various other seizure types, including complex partial and grand mal, may also be seen (Boudreau *et al.* 2005; Castro *et al.* 2007; Mullatti *et al.* 2003). The interictal EEG in hypothalamic hamartoma may be normal or show interictal epileptiform discharges in the temporal or frontal areas; importantly the ictal EEG during a gelastic seizure, as is the case in most simple partial seizures, is typically unremarkable (Leal *et al.* 2003). In reviewing the MRI in cases of gelastic seizures, although finding a nodular mass in the hypothalamus is highly suggestive of a hamartoma, one must keep in mind that other lesions, such as a ganglioglioma, may appear similar (List *et al.* 1958).

Mesial temporal sclerosis

Mesial temporal sclerosis is characterized by sclerosis of one or more of the structures on the medial aspect of the temporal lobe, including the hippocampus, amygdala, and parahippocampal gyrus (Cendes *et al.* 1993a; Falconer 1971; Hudson *et al.* 1993), as illustrated in Figure 8.1. Ammon's horn (cornu ammonis) is an older term for the hippocampus, and some authors, rather than using the term 'hippocampal sclerosis', will speak of 'Ammon's horn sclerosis'.

Among patients with a seizure focus in the temporal lobe, mesial temporal sclerosis is the most common cause (Bruton 1988; Engel *et al.* 1975; Falconer 1971; Falconer *et al.* 1964; Margerison and Corsellis 1966; Sano and Malamud 1953). Given that most (but certainly not all) cases of complex partial seizures result from a focus in the temporal lobe, mesial temporal sclerosis turns out to be the most common cause of this type of seizure.

The etiology of mesial temporal sclerosis is not known with certainty. In some cases, albeit rare, it is familial (Kobayashi *et al.* 2002, 2003). With regard to the remaining large group of non-familial cases, a popular theory holds that seizures occurring secondary to other causes (e.g. childhood febrile seizures) may, via excitotoxic mechanisms, induce damage in the hippocampus, with the resulting sclerotic area then becoming epileptogenic in its own right. There is some evidence to support this. First, although certainly not all patients with mesial temporal sclerosis have a history of childhood febrile

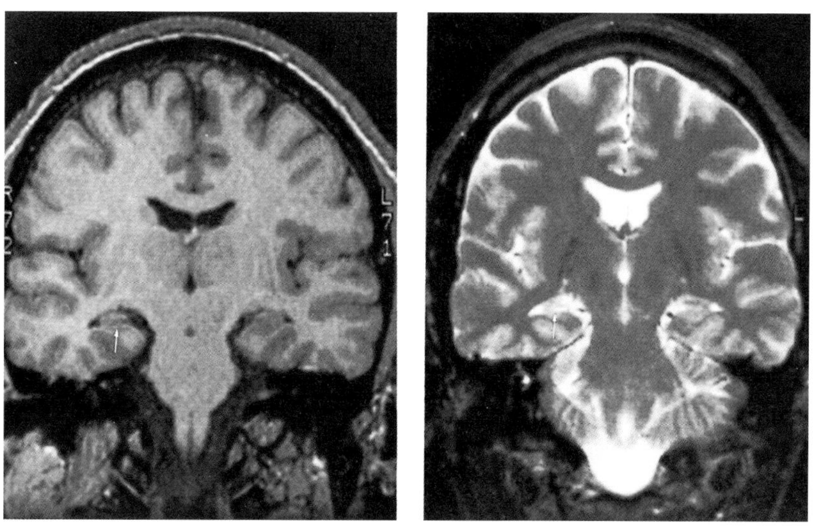

Figure 8.1 Coronal magnetic resonance imaging scan demonstrating mesial temporal sclerosis on the right. On the T1-weighted scan, atrophy of the hippocampus, indicated by the arrow, is fairly apparent, increased signal intensity being seen in the same area on the T2-weighted scan. Reproduced from Hopkins A, Shorvon S, Cascino G. *Epilepsy*, 2nd edn. London: Arnold, 1995, with permission.

seizures, such events are more common in these patients than in the general population (Adam *et al.* 1994; Cendes *et al.* 1993b; French *et al.* 1993; Saygi *et al.* 1994). Second, among patients with focal cortical dysplasia, a significant minority will also be found to have mesial temporal sclerosis (Fauser *et al.* 2006). Third, there are clearly documented cases wherein patients with normal hippocampi have developed mesial temporal sclerosis after experiencing seizures from other causes (Parmar *et al.* 2006; Worrell *et al.* 2002). Finally, although not all studies agree (Liu *et al.* 2002), there is evidence from prospective volumetric studies that progressive hippocampal atrophy occurs in patients who continue to experience complex partial or grand mal seizures (Briellmann *et al.* 2002; Fuerst *et al.* 2003; Salmenpara *et al.* 2005).

Traumatic brain injury

Seizures occurring after TBI may appear early (within the first week) or late (at any time thereafter) (Jennett 1973; Jennett *et al.* 1973). Early seizures are seen in from 2 to 15 percent of cases, and late seizures, generally appearing within the first year post injury (Mazzini *et al.* 2003), in 5 to 10 percent of cases. Several factors increase the likelihood that patients will have late seizures, including the following: the occurrence of an early seizure; the presence of contusions or intracerebral hemorrhages; any intracranial operations; and dural penetration by metal fragments or by bone (Annegers *et al.* 1980; Annegers *et al.* 1998; Ashcroft 1941; Caveness 1976; Englander *et al.* 2003; Jennett 1975; Lishman 1968; McQueen *et al.* 1983; Mazzini *et al.* 2003; Russell and Whitty 1952; Salazar *et al.* 1985).

MENINGOENCEPHALITIS

The occurrence of a grand mal or, less commonly, simple partial or complex partial seizure in the context of an illness characterized by fever, headache, and delirium should always raise the diagnostic possibility of meningitis or encephalitis (Annegers *et al.* 1988). Both arbovirus encephalitis, and herpes simplex encephalitis may present in this fashion (Haymaker 1949; Kennedy 1988; Williams and Lerner 1978), and seizures have also been noted during infectious mononucleosis (Gautier-Smith 1965; Silverstein *et al.* 1972).

In those who survive, seizures may occur as sequelae (Marks *et al.* 1992), and this appears to be especially likely in cases in which the acute phase was characterized by seizures (Annegers *et al.* 1988) and when there are other sequelae indicative of cerebral damage (Pomeroy *et al.* 1990).

In this regard, it may also be noted that acute disseminated encephalomyelitis, occurring in the days or weeks following a viral illness, may also be characterized by seizures (Paskavitz *et al.* 1995; Tenembaum *et al.* 2002).

IN ASSOCIATION WITH CERTAIN DEMENTING OR NEURODEGENERATIVE DISORDERS

Although partial and grand mal seizures may occur secondary to various dementing or neurodegenerative disorders, overall these disorders account for only a small minority of all seizure cases; furthermore, relative to other symptoms caused by these disorders, the seizures play but a minor role in the overall clinical picture.

Multi-infarct dementia may be associated with seizures at any point in its evolution (Rosenberg *et al.* 1979); in Alzheimer's disease, however, seizures typically occur only late in its course, and long after the dementia is well-established (Amatniek *et al.* 2006; Goodman 1953; Hauser *et al.* 1986; Romanelli *et al.* 1990).

Neurosyphilis, in particular general paresis of the insane (GPI), may cause either grand mal or partial seizures, most especially simple partial seizures with a Jacksonian march (Merritt and Springlova 1932; Storm-Mathisen 1969). Importantly, the Jarisch–Herxheimer reaction to penicillin treatment may also be characterized by seizures (Hahn *et al.* 1959; Zifko *et al.* 1994).

Acquired immunodeficiency syndrome dementia may, late in its course, be accompanied by seizures (Navia *et al.* 1986a). Other disorders, often seen in association with AIDS, must also be considered, including progressive multifocal leukoencephalopathy and toxoplasmosis (Navia *et al.* 1986b; Porter and Sande 1992).

Various movement disorders, typically accompanied by dementia, may also cause partial or grand mal seizures, notably the choreiform disorders dentatorubropallidoluysian atrophy (Porter *et al.* 1995; Warner *et al.* 1995), choreoacanthocytosis (Feinberg *et al.* 1991; Hardie *et al.* 1991; Lossos *et al.* 2005), which may present with complex partial seizures years before the onset of chorea (Al-Asmi *et al.* 2005), and McLeod's syndrome (Danek *et al.* 2001). In the case of seizures occurring in a young person with a movement disorder (with chorea, parkinsonism, tremor), consideration must also always be given to Wilson's disease (Dening *et al.* 1988). Spinocerebellar ataxia, characterized by a slowly progressive ataxia, in certain of its types, may also cause seizures (Grewal *et al.* 2002; Rasmussen *et al.* 2001).

Cerebrotendinous xanthomatosis, typified by a combination of either mental retardation or dementia with Achilles' tendon enlargement or cataracts, may also cause seizures (Fiorelli *et al.* 1990).

Cerebral amyloid angiopathy, classically causing lobar hematomas and, in some cases, a dementia, may also manifest with simple partial seizures, which may occur before any lobar hemorrhages and either before or concurrent with a dementia (Greenberg *et al.* 1993).

Creutzfeldt–Jakob disease, typically characterized by dementia or delirium with myoclonus, may also, albeit uncommonly and late in the course of the disease, cause partial or grand mal seizures (Brown *et al.* 1986).

Granulomatous angiitis, or isolated angiitis of the central nervous system, presents subacutely with headache, which is quite prominent, and delirium, and may, in a minority, be accompanied by seizures (Vollmer *et al.* 1993).

Certain degenerative disorders of relatively early onset, from childhood to early adult years, may also cause partial or grand mal seizures, including metachromatic leukodystrophy (Alves *et al.* 1986; Betts *et al.* 1968a; Hageman *et al.* 1995; Haltia *et al.* 1980; Lima *et al.* 2006; Rauschka *et al.* 2006), adrenoleukodystrophy (Moser *et al.* 1984), pantothenate kinase-associated neurodegeneration (Rozdilsky *et al.* 1968), juvenile Huntington's disease (Campbell *et al.* 1961; Siesling *et al.* 1997), and, especially, subacute sclerosing panencephalitis (Koehler and Jakumeit 1976; Kornberg *et al.* 1991; Prashanth *et al.* 2006).

CONGENITAL DISORDERS

Mental retardation, of severe or profound degree, is commonly associated with seizures (Steffenburg *et al.* 1996), and over one-third of patients with autism will also have seizures, which may be seen both in those with and without retardation (Danielsson *et al.* 2005; Gillberg 1991; Olsson *et al.* 1988).

Various specific congenital disorders, most associated with mental retardation, also cause seizures, with each one being marked by various distinctive features.

Down's syndrome, typified by a variable degree of mental retardation and a characteristic dysmorphism with narrowed palpebral fissures, epicanthal folds, and a small mouth, often with a large protruding tongue, may be accompanied by partial or grand mal seizures in adults (Pueschel *et al.* 1991), the proportion affected rising from about one-tenth of young adults, to about one-half of those over 50 years old (McVicker *et al.* 1994), and to over three-fourths of those who go on to develop a dementia (Lai and Williams 1989).

The fetal alcohol syndrome is characterized by short stature and a dysmorphism consisting of microcephaly, shortened palpebral fissures, epicanthal folds, maxillary hypoplasia, a thin upper lip, a flattened philtrum and micrognathia. Mental retardation may or may not be present. In one study (Bell *et al.* 2010) roughly 6 percent of patients had epilepsy, while an additional 12 percent had had at least one seizure.

Fragile X syndrome, seen generally, but not always, in males, is typified by a variable degree of mental retardation, macro-orchidism, and a characteristic dysmorphism, with a long, narrow face, prominent forehead, and large ears. A minority of these patients will also have either partial or grand mal seizures (Finelli *et al.* 1985; Wisniewski *et al.* 1985).

Klinefelter's syndrome, characterized by tall stature and eunuchoidism in postpubertal males, may, albeit rarely, be accompanied by complex partial or grand mal seizures (Tatum *et al.* 1998).

Sturge–Weber syndrome is classically characterized by a unilateral facial port-wine stain, hemiplegia contralateral to the port-wine stain, mental retardation, and seizures (Bebin and Gomez 1988; Pascual-Castroviejo *et al.* 1993). Importantly, there is an association between frequent seizures and dementia in this disorder (Lichenstein 1954; Petermann *et al.* 1958), an association that underlines the importance of seizure control.

Von Recklinghausen's disease (neurofibromatosis type I) is characterized by neurofibromas and café au lait spots: seizures may occur in 4–5 percent of sufferers (Korf *et al.* 1993; Kulkantrakorn and Geller 1998).

Rett's syndrome (Hagberg *et al.* 1983), seen virtually only in females, is suggested by a typical evolution of symptoms in childhood, resulting in microcephaly and mental retardation. In a minority of adults either partial or grand mal seizures may be seen.

Tuberous sclerosis classically presents in childhood with the triad of seizures, adenoma sebaceum, and mental retardation (Critchley and Early 1932). Seizures generally, but not always, precede the appearance of adenoma sebaceum (Alsen *et al.* 1994; Ross and Dickerson 1943), and, although generally of the 'salaam' type in infancy, manifest with typical grand mal or partial seizures by early or mid-childhood (Pampiglione and Moynahan 1976). Tuberous sclerosis may rarely present in the adult years: in one case, a 26-year-old developed adenoma sebaceum, followed, at the age of 31, by partial seizures (Kofman and Hyland 1959).

Prader–Willi syndrome is characterized by extreme obesity secondary to a ravenous hunger. Dysmorphic features are common, with a narrow head, almond-shaped eyes, a thin upper lip, thin arms, and, in males, micropenis and cryptorchidism. Perhaps one-half will also have mental retardation, generally of a mild degree. Seizures may occur in about one-fifth of these patients (Bray *et al.* 1983).

MISCELLANEOUS CAUSES

The miscellaneous causes may be divided into various groups, including: *systemic or autoimmune disorders*, such as hypertensive encephalopathy or Hashimoto's encephalopathy; the *autosomal dominant partial epilepsies; multiple sclerosis*; certain *disorders typically presenting in childhood or adolescence*, such as Rasmussen's syndrome; and, finally, a group characterized by obvious *precipitating events*, such as cerebral anoxia.

Systemic or autoimmune disorders

Various systemic or autoimmune disorders may cause partial or grand mal seizures. Hypertensive encephalopathy (Chester *et al.* 1978; Healton *et al.* 1982) and the reversible posterior leukoencephalopathy syndrome (Hinchey *et al.* 1996) both present with headache, delirium, and seizures, and should be suspected in cases of severe hypertension or treatment with immunosuppressants. Hashimoto's encephalopathy (Henchey *et al.* 1995; Shaw *et al.* 1991) and limbic encephalitis (Alamowitch *et al.* 1997) may both present with delirium and either partial or grand mal seizures; in the case of limbic encephalitis, seizures may constitute either the presentation (Brennan and Craddock 1983; Corsellis *et al.* 1968), or, rarely, the primary clinical expression of the disease (Shavit *et al.* 1999; Tsukamoto *et al.* 1993). Systemic lupus erythematosus may likewise cause partial or grand mal seizures (Devinsky *et al.* 1988a) in well over 10 percent of patients (Appenzeller *et al.* 2004). Thrombotic thrombocytopenic purpura should be suspected in cases of seizures occurring in the context of delirium and thrombocytopenia (Blum and Drislane 1996). Hyperthyroidism, rarely, may directly cause seizures, and is suggested by the accompanying autonomic symptoms (Korczyn and Bechar 1976; Jabbari and Huott 1980). Central pontine myelinolysis, suggested by the occurrence of quadriplegia or abnormal movements in the setting of rapid correction of hyponatremia, may also, rarely, cause seizures (Karp and Laureno 1993). Hepatic porphyria may be suspected when seizures occur in the context of abdominal pain and delirium (Byelsjo *et al.* 1996). Wernicke's encephalopathy, presenting with variable combinations of nystagmus, ataxia, and delirium in the setting of nutritional deficiency (as in chronic alcoholism), may, rarely, also cause seizures (Harrison *et al.* 2006). Marchiafava–Bignami disease, a very rare disorder also associated with chronic alcoholism, may present with delirium and seizures (Ironside *et al.* 1961; Koeppen and Barron 1978). Sarcoidosis is immediately suggested by typical findings on chest radiograph (Ferriby *et al.* 2001; Krumholz *et al.* 1991; Oksanen 1986). Other, rare, causes include: Behçet's syndrome, suggested by delirium accompanied by oral or genital aphthous ulcers (Aykutlu *et al.* 2002); Wegener's granulomatosus (Nishino *et al.* 1993), suggested by a combination of upper respiratory and renal disease; Whipple's disease, suggested by abdominal pain and arthralgia (Louis *et al.* 1996; Romanul *et al.* 1977); and celiac disease, also known as gluten enteropathy, suggested by abdominal pain or bloating (Chapman *et al.* 1978) or by the finding of occipital calcification (Gobbi *et al.* 1992; Tinuper *et al.* 1996b).

Autosomal dominant partial epilepsies

Autosomal dominant partial epilepsy should be suspected in cases of partial seizures, with or without grand mal seizures, which exhibit a family history consistent with autosomal dominant inheritance, and wherein the magnetic resonance (MR) scan is normal. Three, more or less distinct, types have been recognized, including autosomal dominant lateral temporal lobe epilepsy (also known as autosomal dominant partial epilepsy with auditory features), autosomal dominant nocturnal frontal lobe seizures, and familial partial epilepsy with variable foci. Autosomal dominant lateral temporal lobe epilepsy is suggested clinically by complex partial seizures occurring with an auditory aura, which may range from simple buzzing to, at times, complex auditory hallucinations, such as music or voices (Berkovic *et al.* 2004a; Brodtkorb *et al.* 2005; Hedera *et al.* 2004; Michelucci *et al.* 2003; Ottman *et al.* 1995; Poza *et al.* 1999; Winawer *et al.* 2000). Autosomal dominant nocturnal frontal lobe seizures is a syndrome distinguished by complex partial seizures that arise from sleep (Combi *et al.* 2005; Gambardella *et al.* 2000; Hayman *et al.* 1997; Hirose *et al.* 1999; Oldani *et al.* 1998; Phillips *et al.* 2000; Saenz *et al.* 1999; Scheffer *et al.* 1994, 1995; Willoughby *et al.* 2003). In both these types, the seizure type generally 'runs true' among family members; in familial partial epilepsy with variable foci, however, there is a marked heterogeneity in seizure types among family members (Berkovic *et al.* 2004b), and it is this heterogeneity that distinguishes this disorder. In addition to these three well-recognized types, there are also reports of other, less clear-cut autosomal dominant types of partial epilepsy (Brodtkorb *et al.* 2002; Depondt *et al.* 2002; Picard *et al.* 2000; Provini *et al.* 1999), and many researchers believe that what has been discovered so far represents merely the tip of the iceberg.

Multiple sclerosis

Multiple sclerosis, rarely, may cause seizures (Nicoletti *et al.* 2003; Thompson *et al.* 1993; Trouillas and Courjon 1972), and in one very rare case multiple sclerosis presented with simple partial status epilepticus with motor signs secondary to a plaque in the precentral gyrus (Spatt 1995).

Disorders typically presenting in childhood or adolescence

Certain disorders typically presenting in childhood or adolescence may also be considered, including Rasmussen's encephalitis, Landau–Kleffner syndrome, and Sydenham's chorea. Rasmussen's encephalitis typically presents with partial seizures, with or without secondary generalization; over time seizures are joined by a progressive hemiparesis, and, in some cases, a dementia (Aguilar and Rasmussen 1960; Bien *et al.* 2002; Granata *et al.* 2003; Rasmussen and Andermann 1989; Rasmussen *et al.* 1958); motor simple partial status epilepticus is not uncommon. Although pathologic studies have revealed an underlying encephalitic process consistent with an autoimmune etiology (Pardo *et al.* 2004), the nature of this autoimmune disorder remains, as yet, unknown. Although the onset is generally in childhood under the age of 10 years, adult onset, albeit rare, may also be seen (McLachlan *et al.* 1993). Landau–Kleffner syndrome is characterized by a childhood onset of aphasia with seizures, which may be either partial or grand mal in type (Hirsch *et al.* 1990; Landau and Kleffner 1957; Mantovani and Landau 1980; Paquier *et al.* 1992). Sydenham's chorea is suggested by the onset of chorea in childhood or adolescence, and may also, albeit rarely, cause seizures (Nausieda *et al.* 1980). Interestingly, and again, very rarely, epilepsy may occur as a sequela to Sydenham's chorea (Aron *et al.* 1965).

Precipitating events

Seizures may also occur after certain, obvious, precipitating events. Cerebrovascular disorders and TBI were discussed earlier; other, less common, examples include postanoxic encephalopathy (Krumholz *et al.* 1988), postradiation encephalopathy (Shewmon and Masdeu 1980), the dialysis dysequilibrium syndrome (Benna *et al.* 1981), as part of the now vanishingly rare dialysis dementia (Burks *et al.* 1976; Garrett *et al.* 1988; Lederman and Henry 1978; O'Hare *et al.* 1983), eclampsia (Manfredi *et al.* 1997) or, very rarely, as a sequela to electroconvulsive therapy (ECT) (Devinsky and Duchowny 1983; Varma and Lee 1992).

DIAGNOSTIC WORKUP

The diagnostic workup, in part, is determined by the kinds of seizures the patient has. Cases characterized by an onset in childhood or adolescence of petit mal and/or non-focal grand mal seizures probably represent one of the idiopathic generalized epilepsies, and in such cases generally an interictal EEG is indicated: if it shows the typical findings discussed below, no further workup may be required. In cases of non-focal grand mal seizures of later onset, and in all cases of focal seizures (i.e. grand mal seizures with focal features and all cases of simple partial, complex partial, atonic, or amnestic seizures) one should assume that a lesion is present and take steps to determine its location and nature (King *et al.* 1999). These steps, discussed below, include an *EEG, neuroimaging* (preferably using MRI), and a specifically focused history and examination, looking for certain distinctive *lateralizing and localizing symptoms and signs.* Lumbar puncture is reserved for cases of suspected meningitis or encephalitis. Additionally, one should always be alert to toxic or metabolic factors or substance withdrawal, which may cause grand mal seizures in otherwise normal individuals and may precipitate seizures in patients with epilepsy of any cause. If, after a thorough workup, the cause of the patient's epilepsy remains obscure, it is customary to make a diagnosis of 'cryptogenic' epilepsy. Such a diagnosis, however, should not be taken for more than it is, a mere confession of our ignorance, and the clinician should remain alert to the emergence of new clinical features that may ultimately reveal the underlying cause.

Electroencephalogram

An EEG (discussed in detail in Chapter 2.5) should be obtained in all cases. An interictal EEG is, in the vast majority of cases, sufficient, and should include hyperventilation, photic stimulation, and a sleep recording. Ictal EEGs, although helpful, are generally not required except for presurgical workups; however, in cases of reflex seizures, consideration may be given to inducing a seizure with the appropriate reflexive stimulus (Fariello *et al.* 1983).

Idiopathic generalized epilepsies are associated with generalized, bilaterally symmetric, and synchronous interictal epileptiform discharges. Patients with petit mal seizures will have classic spike and dome discharges, and patients with non-focal grand mal seizures will generally have spike and wave, polyspike, or polyspike and wave discharges.

Focal epilepsies typically display focal or multifocal interictal epileptiform discharges; however, it may require up to four routine EEGs to capture them, and even this strategy, as noted below, may fail. In cases where a focus is suspected in one of the temporal lobes, consideration should be given to the use of supplemental leads: 'true' anterior temporal leads are routinely appropriate in these cases, whereas invasive sphenoidal or nasopharyngeal leads are typically held in reserve for particularly difficult cases.

As important as the finding of interictal or ictal activity is, its absence does not rule out the diagnosis of focal epilepsy. In the case of simple partial seizures, both interictal (Devinsky *et al.* 1989b; Mauguire and Courjon 1978) and ictal (Bare *et al.* 1994; Devinsky *et al.* 1988b, 1989b) EEGs are often normal. In the case of complex partial seizures, interictal EEGs may likewise be normal (Goodin *et al.* 1990); furthermore, although ictal EEGs are more likely to be abnormal, even here the EEG may be normal during a seizure, especially when the seizure focus is on the medial surface of the frontal (Williamson and Spencer 1986; Williamson *et al.* 1985c) or temporal (Lieb *et al.* 1976) lobe. In these limiting cases when routine EEGs remain unrevealing, consideration may be given to admission to an epilepsy monitoring unit or video–EEG monitoring, or, more simply, to the utilization of a computer-assisted 24-hour ambulatory EEG.

Neuroimaging

Neuroimaging should, whenever possible, be performed using MRI rather than CT scanning, as MR scanning is far more sensitive (Franceschi *et al.* 1989; Swartz *et al.* 1992). When MR is utilized, it is important to obtain coronal images as these are superior to axial or sagittal images for the detection of mesial temporal sclerosis (Berkovic *et al.* 1991). Furthermore, it must be borne in mind that in some cases of mesial temporal sclerosis the temporal lobes and their contents may appear quite symmetric, and the only abnormality may be increased signal intensity on the T2-weighted scan (Jackson *et al.* 1994). Although MRI is more sensitive, a normal MRI examination certainly does not rule out a focal lesion, such as a small area of focal cortical dysplasia (Williamson *et al.* 1993).

In evaluating the MR scan, it must be kept in mind that, albeit rarely, seizures themselves may induce reversible changes. For example, after complex status epilepticus the mesial aspect of the temporal lobe may demonstrate increased signal intensity on diffusion-weighted imaging (Parmar *et al.* 2006; Szabo *et al.* 2005) and T2-weighted imaging (Henry *et al.* 1994; Kramer *et al.* 1987). Computed tomography scanning may also be misleading in the same situation, revealing a transient radiolucency (Kramer *et al.* 1987; Sammaritano *et al.* 1985).

Lateralizing and localizing symptoms and signs

Certain aspects of simple partial, complex partial, or focal grand mal seizures may be helpful in both lateralizing and localizing the responsible lesion, and consequently a good grasp of the clinical symptomatology of the patient's seizure is very helpful. In the case of simple partial seizures, patients are generally able to provide an accurate history, and some (but, as noted earlier, certainly not all) patients may also be able to recall an aura to a complex partial or grand mal seizure, but when it comes to complex partial seizures or grand mal seizures, one must obviously rely on a history provided by other observers. When this is not reliable, consideration may be given to asking family members to videotape seizures (Newmark 1981; Samuel and Duncan 1994) or, if this is unsuccessful, to video–EEG monitoring.

Before discussing the lateralizing and localizing features peculiar to each seizure type, mention should be made of two postictal features, either of which can be seen with any of these three seizure types and which both have strong lateralizing value. The first of these is a postictal unilateral Babinski's sign. Although it is well recognized that this may be found after a grand mal seizure, it also occurs postictally after approximately 20 percent of complex partial seizures (Walczak and Rubinsky 1994). The second is Todd's paralysis. This phenomenon, first described by Todd in 1855, is characterized by a unilateral postictal paresis, usually of the upper extremity, which may range in severity from a total paralysis to a mere drift of the upper extremity and which lasts for from seconds up to almost half an hour (Todd 1855). It may be found after simple partial seizures with motor symptomatology or complex partial or grand mal seizures (Gallmetzer *et al.* 2004; Kellinghaus and Kotagal 2004). Both of these features reliably lateralize the lesion to the contralateral hemisphere.

Simple partial seizures often display both lateralizing and localizing signs. In the case of simple partial seizures with either motor or somatosensory signs, the focus is typically, but not always (Herskowitz and Swerdlow 1972), lateralized to the hemisphere contralateral to the side on which the patient is experiencing the symptoms (Mauguire and Courjon 1978). With regard to localization to a particular lobe (Acharya *et al.* 1998; Bien *et al.* 2000; Mauguire and Courjon 1978; Morris *et al.* 1988; Nystrom 1966; Russell and Whitty 1955; Salanova *et al.* 1995), in most cases motor signs suggest the frontal lobe, somatosensory symptoms the parietal lobe, simple visual hallucinations or hemianopia the occipital lobe, structured visual hallucinations the occipitoparietal or occipitotemporal region, and most 'psychic' symptoms the temporal lobe. In cases in which more than one sort of symptom is seen, careful attention must be paid to which occurred first as this is the one with most localizing value: for example, a simple partial seizure that began with a crude visual hallucination and was immediately followed by a motor sign would indicate a focus in the occipital rather than the frontal lobe (Salanova *et al.* 1992; Williamson *et al.* 1992a).

Complex partial seizures, like simple partial seizures, may also display both lateralizing and localizing signs. Lateralizing features include: speech; unilateral ictal clonic, tonic, or dystonic movements or paresis; unilateral automatisms; possibly ictal head version; and, finally, interictal emotional facial palsy. Speech may occur either during an aura, during the seizure itself, or postictally, and may be either coherent or aphasic. Ictal speech that is coherent suggests a focus in the non-dominant hemisphere (Fakhoury *et al.* 1994; Gabr *et al.* 1989; Koerner and Laxer 1988; Marks and Laxer 1998; Yen *et al.* 1996); however, exceptions to this rule do occur (Kaiboriboon *et al.* 2006). Aphasic speech occurring either as an aura (Kanemoto and Janz 1989) or postictally (Ajmone-Marsan and Ralston 1957; Devinsky *et al.* 1994a; Fakhoury *et al.* 1994; Gabr *et al.* 1989; Koerner and Laxer 1988; Marks and Laxer 1998; Privitera *et al.* 1991; Serafetinides and Falconer 1963) suggests a focus in the dominant hemisphere. Unilateral ictal clonic, tonic or dystonic movement (Kotagal *et al.* 1989; Marks and Laxer 1998; Rusu *et al.* 2005) and unilateral paresis (Oestreich *et al.* 1995) all suggest a focus on the contralateral side. Unilateral automatisms (Janszky *et al.* 2006; Marks and Laxer 1998) or unilateral ictal eye-blinking (Benbadis *et al.* 1996) both generally indicate an ipsilateral focus. Head version at the start of a seizure is of controversial lateralizing significance: some studies (e.g. Quesney *et al.* 1990; Salanova *et al.* 1995) report that the head turns away from the side with the focus, whereas others (e.g. Geier *et al.* 1977; Ochs *et al.* 1984) have not found such a relationship. With regard to this controversy over the lateralizing significance of head version, two studies (Rheims *et al.* 2005; Wyllie *et al.* 1986) offered a potential solution by noting first that there are two types of head version: in one the head turning is casual, whereas in the other it is forced and unnatural and often accompanied by hemi-facial clonic movements on the side toward which the head is turning. Cases with forced head version lateralized to the hemisphere away from which the head turned. Interictal emotional facial palsy suggests that the focus is contralateral to the side with the palsy (Cascino *et al.* 1993b; Jacob *et al.* 2003; Remillard *et al.* 1977).

Localization of the focus of a complex partial seizure may not be as straightforward as was once thought. In the not-too-distant past, the assumption that complex partial seizures all originated from foci in the temporal lobe was so unquestioned that some authors used the terms 'complex partial epilepsy' and 'temporal lobe epilepsy' synonymously. It has become clear, however, that even though most complex partial seizures do arise from foci in the temporal lobe (Nystrom 1966), they may also occur secondary to foci in the frontal (Geier *et al.* 1977; Rasmussen 1983; Salanova *et al.* 1995; Sutherling *et al.* 1990; Williamson *et al.* 1985c), occipital (Ludwig and Marsan 1975; Salanova *et al.* 1992; Williamson *et al.* 1992a), or parietal (Cascino *et al.* 1993c; Ho *et al.* 1994; Williamson *et al.* 1992b) lobes. In cases in which the focus is at a site other than the temporal lobe, electrical activity spreads rapidly from the originating lobe to the temporal lobe, thus producing the symptomatology of the complex partial seizure itself. As suggested by Hughlings Jackson (Jackson 1894), it stands to reason that, in such cases, the aura to the complex partial seizure may indicate the lobe of origin, and this does in fact appear to be the case (Boon *et al.* 1991; Palmini and Gloor 1992), the various aurae having the same localizing value as discussed for simple partial seizures, above. Finally, before leaving the subject of localization in complex partial seizures, note should be made of the significance of ictal bilateral 'bicycling' movements. Some have held that these indicate a frontal lobe onset (Williamson *et al.* 1985c), but such movements have also been noted in complex partial seizures of temporal lobe origin (Sussman *et al.* 1989; Swartz 1994).

Grand mal seizures may also display lateralizing and localizing features. A postictal Babinski's sign or Todd's paralysis, as described above, may lateralize to the contralateral hemisphere, and any aurae may localize, as described for simple partial seizures, above.

DIFFERENTIAL DIAGNOSIS

In differentiating between seizures and other disorders capable of producing a more or less similar clinical picture, two features are generally helpful. The first to consider is the exquisitely paroxysmal onset of all seizure types, occurring over seconds: such an onset is rare in other disorders. Second, one should always, in obtaining a history, search for other seizure types, as their occurrence strongly suggests that the event in differential question is likewise a seizure. Additional differential considerations specific for each of the seizure types are discussed in turn, below, followed by a discussion of pseudoseizures.

Simple partial seizures

Transient ischemic attacks (TIAs) may mimic inhibitory motor simple partial seizures, and when these episodes of weakness occur in the elderly, the differential may be difficult: one clue exists in the rapidity with which the weakness develops: although TIAs are of sudden onset, they rarely exhibit the sort of exquisitely paroxysmal onset typical of seizures, wherein symptoms reach their peak within seconds. Transient ischemic attacks may also mimic the more typical motor simple partial seizures with clonic activity when the TIA presents in what is known as a 'limb-shaking transient ischemic attack' (Han *et al.* 2004; Persoon *et al.* 2010). These limb-shaking TIAs occur in patients with either occlusion or severe stenosis of the common or internal carotid artery, and are typically, but certainly not always, brought on by standing or any other maneuver that lowers systemic pressure to the point where perfusion of the hemisphere ipsilateral to the involved carotid artery falls below a critical level. Clinically, one sees coarse, jerky, arrhythmic movements of an arm or leg, or both, lasting generally less than 5 minutes, and often either accompanied, or followed, by a temporary weakness. The key to the differential lies in the fact that the limb-shaking fails to mimic the regular, rhythmic clonic activity seen in a simple partial seizure. The presence, of course, of a precipitating factor capable of lowering blood pressure is also, of course, an important clue, however, this is not always present, and in some cases of limb-shaking TIAs, the shaking appears to occur spontaneously.

Migraine aurae may mimic a variety of different types of simple partial seizures; however, the differential is fairly straightforward when the event is followed by a typical migraine headache. Difficulty may arise in cases of isolated migraine aurae occurring without any following headache; however, in such cases history generally reveals typical headaches.

Some further considerations come into play in cases of hallucinations or anxiety. Simple partial seizures constitute but one of many disorders capable of causing isolated hallucinations, and the differential for these events, as discussed in Chapter 5.30, should be pursued. In a similar vein, simple partial seizures, likewise, are but one cause of anxiety attacks, and the differential for these, as outlined in Chapter 7.5, should also be consulted.

Although in all these instances an EEG should be obtained, it must be kept in mind, as discussed earlier, that both the interictal and the ictal EEG is typically normal in cases characterized by simple partial seizures alone.

Complex partial seizures

Relatively brief episodes of confusion may be caused not only by complex partial seizures, but also by TIAs when the ischemia is found in one of the temporal lobes or in the thalamus. The differential here proceeds much as discussed immediately above for simple partial seizures; however, here the EEG may be more useful, as interictal epileptiform discharges are more common in cases of complex partial seizures.

Brief episodes of confusion arising from sleep may represent either a nocturnal complex partial seizure or a parasomnia such as rapid eye movement (REM) sleep behavior disorder, night terrors, or somnambulism. Rapid eye movement sleep behavior disorder may be readily identified, as patients with this disorder will, upon 'coming to', report a vivid dream that explains their behavior during the event. Night terrors and somnambulism, however, lack such distinctive clues, and may be closely mimicked by nocturnal complex seizures (Boller *et al.* 1975; Pedley and Guilleminault 1977; Plazzi *et al.* 1995); consequently, polysomnography may be required.

Brief episodes of unprovoked and uncharacteristic violence may suggest either a complex partial seizure or an entity known as intermittent explosive disorder. As discussed in Chapter 27.1, this disorder is characterized by episodes wherein patients, either without any precipitants or with only trivial ones, experience a growing sense of tension and irritability, culminating in an explosion of violence, after which they may have only a spotty memory of the event; importantly, at other times in the patients' lives, there are no significant abnormalities (e.g. no chronic irritability). Although complex partial seizures with directed aggression may naturally come into the differential, it must be kept in mind, as noted earlier, that such ictal directed aggression is extremely rare, and that in virtually all such cases there is a history of either other seizure types or of more typical complex partial seizures.

Prolonged episodes of confusion may represent complex partial status epilepticus; however, in such cases the differential becomes quite large as it encompasses the syndrome of delirium, and in such cases that differential, as discussed in Chapter 6.5, should be considered.

Finally, one must consider the possibility of a pseudoseizure, a subject considered later.

Petit mal seizures

Apart from complex partial seizures, there is little else that can even come close to mimicking typical petit mal seizures. The presence of an aura or a duration longer than 10 or 15 seconds points toward a diagnosis of a complex partial seizure; in doubtful cases, an interictal EEG will resolve the issue.

Grand mal seizures

Syncope (Aminoff *et al.* 1988; Hoefnagels *et al.* 1991; Lempert *et al.* 1994; Wieling *et al.* 2009), whether cardiac, orthostatic or vasovagal, may at times occasion some diagnostic difficulty. In syncope, although a prodrome may be lacking in the elderly, in most cases one is present, consisting of a graying of vision, dizziness, and some fuzziness of thought; in addition some two-thirds of prodromes will also be characterized by visual hallucinations, while in some one-third auditory hallucinations will occur. Following the prodrome, the gaze becomes fixed and the face exhibits a 'frozen' appearance. Although most patients will recall the prodrome, those with cardiac syncope due to asystole will not: in all cases of syncope there is some retrograde amnesia spanning several seconds, and in cases of asystole the actual faint occurs so rapidly that the prodrome is lost in the retrograde amnesia. During the faint proper, although the vast majority of patients become flaccid, in a very small minority there may be generalized tonic activity. The eyes are typically, but not always, open, and in most cases they are upturned. Myoclonus occurs in from 50 percent to almost 90 percent of cases and lasts seconds; it may be focal, multifocal, or generalized, or present as a combination of multifocal and generalized activity. In a minority of cases, there may be some vocalization, which typically consists of soft moaning. Urinary incontinence occurs in a small minority, and tongue biting may also occur, but this is rare, and when it does occur it usually involves the tip, and not the side, of the tongue. The duration of the faint proper depends on the mechanism. The duration of cardiac syncope of course depends on the duration of the arrhythmia. Orthostatic syncope generally lasts about 12 seconds, whereas vasovagal syncope, though typically lasting from 10 to 20 seconds, may persist much longer, up to, in rare instances, 5 minutes. After recovery of consciousness, there may be a brief period of confusion or fuzziness of thought, but this lasts generally no more than 30 seconds.

With this description of syncope in mind, some differential points with a grand mal seizure may be made. First, one must distinguish a presyncopal prodrome from an aura, and although there is clearly overlap here, the 'graying' out of vision is fairly specific for a prodrome. Next, one must distinguish the faint proper from a grand mal seizure, and several points aid in this regard. Although there may be some moaning during the faint proper, there is not an initial inarticulate 'cry' typical of a seizure. The occurrence of tonic or myoclonic activity during syncope naturally seems to invite confusion with a tonic–clonic seizure; however, with a reasonably good history, it is usually possible to resist this invitation. To begin with, syncope never imitates the classic sequence of events seen in a grand mal seizure, which begins with generalized tonic activity, followed by generalized clonic activity. As noted in the description of grand mal seizures, however, some seizures may be characterized by generalized tonic activity alone, and this may leave one in some doubt. Further, some seizures may also be characterized primarily, if not solely, by clonic activity, and although this too may leave one in some doubt, it is generally possible to distinguish between clonic activity and myoclonus; this, of course, is straightforward if the physician witnesses the event; however, if one must rely, as is typically the case, on a history from bystanders, it may be necessary to pantomime clonus and myoclonus in order to secure this differential point. The presence of absence of incontinence is not particularly helpful, however tongue biting is, given that in syncope it is the tip of the tongue that is snared, and not its side, as is typical of a seizure. Finally, and most importantly, one looks to how long it takes for the patient to become alert and clearheaded after the event. After syncope, this occurs in less than 30 seconds, whereas after a grand mal seizure there is a period of stupor or coma for minutes, followed by a delirium with confusion lasting anywhere from 15 to 30 minutes.

Pseudoseizures also enter into the differential, and this is discussed later.

Amnestic seizures

Amnestic seizures must be distinguished from transient global amnesia and alcoholic blackouts, a differential discussed in Chapter 5.4.

Atonic seizures

Atonic seizures may be mimicked by cataplexy, syncope and brainstem transient ischemic attacks. In this differential, a good history is essential, as in almost all cases of atonic seizures one will find a long history of other seizure types.

Cataplectic attacks are always preceded by some strong emotion and are accompanied by preservation of consciousness, and, although atonic seizures, as discussed above, may share the same characteristics, this is very rare.

Syncope may be difficult to distinguish from an atonic seizure. Keeping in mind the description of syncope provided earlier, however, will generally enable one to make an accurate differential. To begin, with the exception of cardiac syncope occurring secondary to asystole, there is typically a prodrome, something lacking in atonic seizures; in this regard, however, one must keep in mind that, in the elderly, syncope of any cause may lack a prodrome. Next, the flaccidity common to both syncope and atonic seizures develops at a different rate in these two events: in syncope there is often a 'swoon' and indeed some patients may reach for something to steady themselves, or try to lower themselves to the floor, while in a seizure the loss of tone occurs paroxysmally. Finally, although most patients with an atonic seizure become clearheaded almost immediately, just as do those with syncope, in some cases of atonic seizures there may be postictal confusion lasting a minute or two.

Brainstem TIAs which affect primarily the corticospinal tracts bilaterally may cause drop attacks with preservation of consciousness (Brust *et al.* 1979), and should be suspected in the elderly or those with multiple risk factors.

Pseudoseizures

Pseudoseizures are paroxysmal events that, rather than occurring on the basis of an electrical paroxysm in the brain, represent a simulation of a seizure, and may be seen in conversion disorder, malingering, and factitious disorder (including 'epileptic Munchausen's syndrome' [Savard *et al.* 1988]). In practice, this differential possibility comes to mind when patients present with episodes that, although resembling either grand mal or complex partial seizures, have some atypical features. After discussing some of these atypical features, consideration will be given to an appropriate workup. Before doing so, however, some comments are in order regarding the term 'pseudoseizure'. In the recent past some have argued for the abandonment of this term and the use of 'psychogenic non-epileptic seizure' in its place, primarily on the grounds that the term 'pseudoseizure', containing, as it does, the word 'pseudo' is perjorative. Although there may be some truth to this, the term 'psychogenic non-epileptic seizure' is not itself without difficulty. Although one may speak of an 'epileptic seizure', more often than not the word 'epileptic' is simply not used, given that it is assumed that a 'seizure' is an epileptic event. To speak, therefore, of a 'non-epileptic seizure' appears to involve using an oxymoron, something that invites only misunderstanding. Clearly a better term is needed, but, in the meantime, at least in this text, the term 'pseudoseizure' will be used.

The bedside clinical distinction between a pseudo-grand mal seizure and a true grand mal seizure hinges on several diagnostic points (Avbersek and Sisodiya 2010; Benbadis *et al.* 1995; Chung *et al.* 2006; Elzawahry *et al.* 2010; Oliva *et al.* 2008). Pseudo-grand mal seizures may begin with a cry, but unlike the inarticulate brief cry of a 'true' grand mal seizure, this cry may involve words and may persist long into the event. The movements seen in a pseudo-grand mal seizure may be interrupted by brief pauses, and may be thrashing and bilaterally asynchronous, and if attempts are made to restrain these movements the patient with a pseudoseizure may show active resistance; by contrast, in a grand mal seizure, there are no pauses, the movements are rhythmic and synchronous, and there is no purposeful response should one attempt to stop him or her. Eye closure, only rarely seen during grand mal seizures, is the rule in pseudoseizures. Incontinence may, albeit uncommonly, occur during pseudoesizures, however tongue biting is rare, and, if it does occur, it is the tip of the tongue that is bitten, rather than the side of the tongue, which is the site of injury in almost all cases of tongue biting during grand mal seizures. Finally, after the event most, but not all, patients with pseudosizures will rapidly become alert and clearheaded, in contrast to the postictal confusion typical of grand mal seizures; furthermore, respirations after a pseudoseizure are typically normal, in contrast to the stertorous breathing seen during the brief stupor or coma following a grand mal seizure.

Distinguishing pseudo-complex partial seizures from true complex seizures is far more difficult than is the case with grand mal seizures, in large part because there are few, if any, signs or symptoms highly specific for either type. In practice, the possibility of a pseudo-complex partial seizure comes up when the event in question is very complex, and this is the case because, as noted earlier, the automatisms seen during most true complex partial seizures are stereotyped and simple. However, as also noted earlier, some of the reactive automatisms seen during true complex seizures may be very complex indeed, and hence this differential point cannot be given too much weight.

The further workup in cases of suspected pseudoseizures should include testing for a post-event Babinski's sign, and may also include any or all of the following: appropriately timed testing for prolactin, neuron-specific enolase and, when grand mal seizures are in question, creatine kinase (CK) levels; video–EEG; neuroimaging; consideration of placebo induction, and, in selected cases, single photon emission computed tomography (SPECT) scanning.

A temporary Babinski's sign may, as noted earlier, be seen after both grand mal and complex partial seizures. As this phenomenon will generally lie outside the knowledge of most pseudoseizure patients, even medically sophisticated ones, its presence is a reliable sign of epileptic activity. Its absence, of course, is of little diagnostic import.

The serum prolactin level rises with depth-electrode stimulation of either the hippocampus or the amygdala (Gallagher *et al.* 1987; Parra *et al.* 1980; Sperling and Wilson 1986), thus serving as a marker for paroxysmal activity in these or related structures and as an indicator that the clinical event in question is a 'true' rather than a 'pseudo' seizure (Laxer *et al.* 1985; Pritchard *et al.* 1985a). Prolactin elevation is, however, not entirely specific to epileptic events, as such elevations may also be seen after hypotensive syncope (Oribe *et al.* 1996). A two-fold or greater rise in serum prolactin level is found about 15 minutes postictally in most cases of complex partial seizures with temporal lobe onset (Bauer *et al.* 1994; Sperling *et al.* 1986) and in grand mal seizures (Wyllie *et al.* 1984). However, with seizures arising from foci outside the temporal lobe, whether complex partial (Meierkord *et al.* 1992) or simple partial (Dana-Haeri *et al.* 1983; Laxer *et al.* 1985; Sperling *et al.* 1986), prolactin elevation is less likely to occur. Furthermore, after frequent seizures (Jackel *et al.* 1987; Malkowitz *et al.* 1995) or after status epilepticus (Tomson *et al.* 1989) one may find a normal prolactin level, presumably reflecting an exhaustion of the pituitary prolactin stores secondary to prolonged stimulation.

Neuron-specific enolase is a marker of neuronal injury, and, in contrast with the prolactin level, which tends to be normal after status epilepticus, the level of enolase is elevated in such situations (DeGiorgio *et al.* 1999), peaking at one (DeGiorgio *et al.* 1995) to two days (Rabinowicz *et al.* 1995). Neuron-specific enolase may at times also be elevated after single complex partial or grand mal seizures (Rabinowicz *et al.* 1996), and in such cases the peak occurs at about 6 hours (Palmio *et al.* 2008).

Creatine kinase levels rise with muscle injury, and, in cases in which the differential is between a 'true' grand mal seizure and a pseudoseizure, this should be determined. Significant rises occur in roughly 15 percent of cases within 3 hours and tend to peak at levels eight-fold or more higher than normal after a day and a half (Chesson *et al.* 1983; Libman *et al.* 1991).

A video–EEG that 'captures' a typical event occurring in the absence of any ictal EEG activity is generally considered the 'gold standard' in the diagnosis of pseudoseizures, and, although this is obviously a very important finding, it is not infallible, and for two reasons. First, the interpretation of the EEG itself is a fallible endeavor, and indeed, in one study (Benbadis *et al.* 2009) that assessed inter-rater reliability in the interpretation of video–EEG recordings, errors were made. Second, although rare, the ictal surface EEG during a complex partial seizure may be normal despite clear paroxysmal activity on depth recordings (Lieb *et al.* 1976; Morris *et al.* 1988; Williamson and Spencer 1986).

Neuroimaging, preferably by MR scanning, should be considered to determine whether or not a lesion is present that could, with some probability, be expected to cause seizures. Interpretation here must proceed with some caution, however: in some cases, epileptogenic lesions (e.g. small areas of focal cortical dysplasia) may be very difficult to detect on MRI and, conversely, not all lesions, even although they are easily detected (e.g. a small convexity meningioma), are likely to cause seizures.

Placebo induction during video–EEG monitoring, for example with saline infusion (Cohen and Suter 1982; Devinsky *et al.* 1996; Walczak *et al.* 1994), has been used as a diagnostic test for pseudoseizures on the not implausible assumption that placebos are unlikely to induce paroxysmal electrical activity. It must be borne in mind, however, that placebo may, albeit very rarely, induce 'true' seizures (Lesser *et al.* 1983). Ethical questions have been raised about utilizing such placebo induction, and one, perhaps less ethically objectionable option, is to use simple suggestion techniques, again during video–EEG monitoring, a procedure that appears to be as effective as placebo induction in precipitating a typical event (Benbadis *et al.* 2000, 2004). The suggestion techniques include photic stimulation, hyperventilation, and asking patients to count out loud as they hold their hands over their heads, informing patients ahead of time that each of these techniques may cause a seizure.

Cumulatively, the absence of Babinski's sign, normal prolactin and neuron-specific enolase levels (and CK levels when grand mal seizures are in question), and a normal ictal EEG and MR scan, coupled with a reproduction of the event in question by placebo induction would, when present in combination, argue very, very strongly for the diagnosis of pseudoseizure. It must, however, be borne in mind that some 'true' seizures will, albeit very rarely, 'slip by' all these tests. Consequently, in doubtful cases, consideration may be given to ictal SPECT scanning. Essentially, in the workup of a suspected pseudoseizure, one is attempting to 'prove a negative', that is to 'prove' that the event in question is not accompanied by a paroxysmal electrical discharge. Given that one can never definitively prove a negative, however, the diagnosis of pseudoseizure should always remain at least slightly tentative.

A further complication in the diagnostic workup of pseudoseizures is the fact that it is not uncommon for patients to have both pseudoseizures and 'true' seizures (Krumholz and Niedermeyer 1983). Thus, simply demonstrating that a patient has had a pseudoseizure does not mean that the patient does not also have epilepsy.

Breaking the diagnosis of a pseudoseizure to patients or family members requires considerable tact and diplomacy. One may wish to avoid using the term 'pseudoseizure' for fear of alienating patients, and, conversely, one may likewise wish to avoid the term 'psychogenic non-epileptic seizure', as some patients may leave the discussion with the word 'seizure' firmly in mind while the word 'non-epileptic' sinks away, such that in the future, he or she will continue to speak of having seizures, causing confusion in emergency rooms and hospitals. Sometimes it is best to say, in a quiet and non-judgmental way, that the patient is not having seizures, and that the events in question are likewise not due to any disease of the brain. When, as is virtually always the case, patients ask what is causing the event, the answer must be phrased in terms that will leave patients disposed to accept a referral for cognitive–behavioral therapy, which is effective for pseudoseizures (Goldstein *et al.* 2010; Lafrance *et al.* 2009), and this is where diplomacy and tact are most in need. The gist of the message, however, must be that it is the patient's own thoughts or emotions that are causing the event, or, if this appears too confrontational with patients, as is often the case, that 'stress' is at the root of the problem.

Treatment

This section will begin with a discussion of *non-emergent treatment*, followed by a consideration of *status epilepticus*.

NON-EMERGENT TREATMENT

In non-emergent cases, when treatment of the underlying cause is either not possible or ineffective, utilization of an AED is appropriate. The first step in choosing among the many AEDs available today is to determine the etiology, as seizures due to one of the IGEs respond differently than do focal seizures. In devising an anti-epileptic regimen for females of childbearing age, however, special care must be taken, given that monotherapy with phenytoin, carbamazepine, and especially valproic acid, or combination treatment with any agents, increases the risk of congenital malformations or cognitive impairment (Harden *et al.* 2009).

Idiopathic generalized epilepsies generally respond to treatment with valproic acid, and this is a reasonable first choice for patients with petit mal seizures (Glauser *et al.* 2010; Sato *et al.* 1982), juvenile myoclonic epilepsy (Calleja *et al.* 2001) and/or non-focal grand mal seizures (Marson *et al.* 2002) with onset from childhood to early adult years. Ethosuximide is also effective for petit mal seizures but is not for grand mal seizures and, given that most patients with childhood or juvenile-onset absence epilepsy will also, in addition to petit mal seizures, have grand mal seizures, valproic acid seems preferable. Importantly, certain other AEDs, such as phenytoin (Genton *et al.* 2000), carbamazepine (Genton *et al.* 2000; Thomas *et al.* 2006), and oxcarbazepine (Gelisse *et al.* 2004) may worsen seizure control in patients with IGEs, and therefore should be avoided. In cases in which valproic acid is only partially effective, further improvement may be obtained by adding topiramate (Biton *et al.* 1999), lamotrigine (Biton *et al.* 2005), or levetiracetam (Berkovic *et al.* 2007; Grunewald 2005; Noachtar *et al.* 2008; Rosenfeld *et al.* 2009). In patients who get little benefit from valproic acid or in whom it is poorly tolerated, consideration may be given to utilizing one of these 'add-on' drugs as monotherapy; however, it should be borne in mind that, as of this writing, there are no double-blind studies supporting this practice.

Epilepsy characterized by focal-onset seizures may respond to monotherapy with any of the following: phenytoin or carbamazepine (Heller *et al.* 1995; Mattson *et al.* 1985; Ramsey *et al.* 1983; Rowan *et al.* 2005; Simonsen *et al.* 1976; Troupin *et al.* 1977), oxcarbazepine (Beydoun *et al.* 2000; Schacter *et al.* 1999), valproic acid (Beydoun *et al.* 1997; Callaghan *et al.* 1985; Heller *et al.* 1995; Mattson *et al.* 1992), gabapentin (Rowan *et al.* 2005), topiramate (Privitera *et al.* 2003), lamotrigine (Brodie *et al.* 1995; Rowan *et al.* 2005; Saetre *et al.* 2007) and levetiracetam (Brodie *et al.* 2007; Leppik *et al.* 2003). The choice among these is guided by considerations of past response to monotherapy trials, the patient's general medical condition, potential drug–drug interactions with other medications and anticipated side-effects, and, in cases in which such considerations do not clearly tip the scales toward a particular AED, strong consideration should be given to levetiracetam, which is renally excreted, has essentially no interactions with other medications, and is generally very well tolerated.

In cases in which patients with focal-onset seizures fail to respond to monotherapy, one may choose either to try monotherapy with a different AED or to add on an AED to the one already in place. In cases in which the first AED used as monotherapy was relatively ineffective or simply not tolerated, then a switch to another monotherapy agent is reasonable; however, this 'switch' should be made gradually in order to avoid a flurry of seizures: in general, the second agent is titrated up, and, once it is in place, the original agent should be gradually tapered off. In cases, however, in which the first AED had definite benefit and was well tolerated, then it is reasonable to consider an 'add-on' strategy. Most of the 'add-on' studies have been performed with patients already on either phenytoin or carbamazepine; however, admittedly, there is no consistency over all these studies. Consideration may be given to the following as 'add-on' AEDs: oxcarbazepine (Barcs *et al.* 2000), valproic acid (Willmore *et al.* 1996), gabapentin (Fuerstein *et al.* 1994; Sivenius *et al.* 1991; US Gabapentin Study Group 1993), pregabalin (Beydoun *et al.* 2005; Elger *et al.* 2005; French *et al.* 2003), topiramate (Faught *et al.* 1996; Privitera *et al.* 1996; Yen *et al.* 2000), lamotrigine (Binnie *et al.* 1989; Boas *et al.* 1996; Jawad *et al.* 1989; Loiseau *et al.* 1990b; Matsuo *et al.* 1993; Messenheimer *et al.* 1994), lacosamide (Ben-Menachem *et al.* 2007; Chung *et al.* 2010), and levetiracetam (Boon *et al.* 2002; Cereghino *et al.* 2000; Leppik *et al.* 2003; Peltola *et al.* 2009;

Tsai *et al.* 2006), which appears to be at times very effective (Stefan *et al.* 2006). In choosing which AED to add on, the same principles discussed above regarding monotherapy come into play. In cases in which patients fail to respond to treatment with two AEDs, various options are open. First, one may try either monotherapy with a different agent or try a different combination of two agents. In some cases, patients have responded to treatment with a combination of three AEDs, but this strategy is very problematic: the risk of both pharmacokinetic and, especially, pharmacodynamic drug–drug interactions becomes so high that side-effects are often unacceptable. Furthermore, with such 'triple therapy' regimens there may be an exacerbation of the epilepsy: not uncommonly, patients on triple therapy who carry a diagnosis of 'treatment-resistant epilepsy' enjoy a substantial decrease in seizure frequency after one or more of their AEDs are discontinued.

In cases in which a patient fails to respond satisfactorily to two adequate trials of combination AED therapy, it is reasonable to assume that the patient is treatment resistant. The mechanism or mechanisms underlying such treatment resistance are not clear. Recent research has focused on the role of what are known as multidrug resistance proteins. These proteins are normally found on capillary endothelial cells and serve to maintain the blood–brain barrier by pumping out foreign substances (such as AEDs) from the interstitial fluid back into the blood: one theory holds that they are upregulated in patients with treatment resistance (Aronica *et al.* 2004). Regardless of the cause of treatment resistance, several options are open, and current thinking holds that these options should, given the consequences of uncontrolled epilepsy, be considered as soon as treatment resistance is established. The most effective option is epilepsy surgery, and this may be considered in cases of mesial temporal sclerosis (Engel 1996; Wiebe *et al.* 2001) and some cases of focal cortical dysplasia (Kral *et al.* 2003; Nobili *et al.* 2007); consideration may also be given to gamma knife radiosurgery, which, in the case of mesial temporal sclerosis, may be as effective as open surgery (Barbaro *et al.* 2009). Vagal nerve stimulation (Handforth *et al.* 1998; Salinsky *et al.* 1996) may also be considered; however, this is less effective than epilepsy surgery.

In cases in which patients have been seizure free for 2 or more years, the possibility of tapering and discontinuing AEDs may be entertained. The decision to embark on this course, however, is complex. To begin with, consideration should be given to the consequences of a recurrence of seizures and to the burden imposed by ongoing treatment. In patients with only simple partial seizures, the consequences of another seizure may not be severe, and if side-effects from current AED treatment are bothersome, then discontinuation may seem attractive; on the other hand a history of grand mal seizures and a lack of side-effects might give one pause. Another consideration, of course, is the likelihood of recurrence, and several factors may enable one to make a rough estimate in this regard (Callaghan *et al.* 1988; Cardoso *et al.* 2006; Spooner *et al.* 2006), including the presence or absence of a lesion, the frequency of seizures and the difficulty encountered in initially controlling them, the kinds of seizures experienced by the patient, and the presence of interictal epileptiform discharges (IEDs). Not unreasonably, the presence of a lesion, such as mesial temporal sclerosis, increases the risk; by contrast, patients with idiopathic generalized epilepsies appear to be at lower risk. A history of very frequent seizures coupled with great difficulty in controlling them (as evidenced by the number of trials of AEDs required to do so) suggests a more severe underlying process, and this also increases the risk of recurrence. The kinds of seizures experienced by the patient also may serve as an indicator of the severity of the underlying process. Thus, a history of complex partial seizures with secondary generalization poses a higher risk than does a history of complex partial seizures without secondary generalization or a history of non-focal grand mal seizures (as seen in the IGEs). Finally, the presence of IEDs on an EEG obtained close to the end of the 2-year seizure-free period indicates ongoing activity of the underlying disorder and suggests an increased risk of recurrence. Although some patients may fall at the extremes of these risk factors, most fall in between, and, consequently, much clinical judgment should be utilized in balancing these various risk factors. If a decision is made to taper and discontinue AEDs the process should be a slow one, generally extending over 3–6 months, and, if more than one AED is in place, only one should be tapered at a time.

In addition to the foregoing, some general measures are appropriate in all cases. First, certain non-specific stresses may aggravate epilepsy, and these should be avoided or treated. They include poor sleep, irregular dietary habits, dehydration, febrile illnesses, hyperventilation, and excessive alcohol use. Second, there are certain specific disorders that also aggravate epilepsy, including migraine, sleep apnea and, in some females, the menstrual cycle. Although, of itself, migraine probably does not cause seizures, there is evidence that migraine may be able to precipitate seizures in patients with epilepsy. In particular, it appears that a migraine aura may trigger either a partial or a grand mal seizure, the seizure being intercalated between the migraine aura and the migraine headache (Marks and Ehrenberg 1993): in such cases effective migraine prophylaxis is essential and either valproic acid or topiramate are logical choices as they are effective not only as AEDs, but also as prophylaxis against migraine. Obstructive sleep apnea may aggravate seizures, and treatment of the apnea with continuous or bilevel positive airway pressure may be very effective in reducing seizure frequency (Devinsky *et al.* 1994b). Finally, some women will experience an increase in seizure frequency either toward the end of the follicular phase or during menstrual flow (Backstrom 1976), and although oral contraceptives have been advocated in such situations (Mattson *et al.* 1986) this is controversial (Dana-Haeri and Richens 1983).

Reflex seizures, of course, constitute an invitation to the utilization of common-sense measures at avoiding the reflex stimulus itself. In cases in which this is not feasible, it may be possible, in certain cases, to blunt the provocative effect of the stimulus by repeatedly exposing the patient to that stimulus, as has been demonstrated in cases of musicogenic (Forster *et al.* 1967) and reading (Forster *et al.* 1969a) epilepsies.

Remarkably, albeit rarely, some patients have found it possible to abort partial seizures. Thus, in some cases of simple partial seizures that undergo a Jacksonian march, the seizure may be aborted by vigorously rubbing the part of the body that lies just proximal to the advancing march, and this holds true not only for motor (Russell and Whitty 1953; Symonds 1959) but also for sensory (Efron 1961; Sutherling *et al.* 1990) marches. In the case of complex partial seizures, Penfield (Penfield and Jasper 1954) noted that, when the aura consisted of some 'change in the stream of thought', the patient might be able to prevent the development of the seizure by 'forcing his thoughts with "might and main" into some other channel'. Other techniques have been reported: one patient, whose aura consisted of an olfactory hallucination, was able to prevent a complex partial seizure by the application of a 'strong odor' (Efron 1956, 1957), and another patient, whose aura consisted of an auditory hallucination of music, was able to abort the seizure by 'imagining himself fishing, his favorite leisure pursuit' (Pritchard *et al.* 1985b).

Status epilepticus

Status epilepticus is approached differently depending on the kind of seizure involved. Grand mal status epilepticus is an acutely life-threatening event that also, in those who survive, entails significant sequelae; thus it requires emergent treatment. Complex partial status epilepticus, in contrast, is not life-threatening and, although sequelae have been noted, they are uncommon and, consequently, emergency treatment may not always be required. Simple partial status, petit mal status and amnestic status are, of themselves, benign events and may be approached as a non-emergency.

In all cases of status, regardless of the seizure type, a finger stick glucose level is determined, and, if the level is low, then 50 mL of 50 percent dextrose in water (D50W) is given. Thiamine, 100 mg i.v., should also be given if there is any suspicion of alcoholism or severe malnutrition. Blood is sent for determination of glucose, sodium, potassium, calcium, magnesium, blood urea nitrogen (BUN), and, if appropriate, AED levels.

In cases requiring urgent treatment, intravenous access is obtained and lorazepam is given in a total dose of 0.07 mg/kg at a rate no faster than 2 mg/min (Alldredge *et al.* 2001; Leppik *et al.* 1983; Treiman *et al.* 1998); as a rule, fosphenytoin should then be administered in a total dose of 15–25 mg/kg at a rate of 100–150 mg/min (Browne *et al.* 1996; Lowenstein and Alldredge 1998; Ramsey and DeToledo 1996). Exceptions to this rule include the following: allergy; a history of non-response to adequate levels of phenytoin; current treatment with phenytoin with blood levels in the therapeutic range; and cases in which the grand mal seizures are occurring as part of one of the idiopathic generalized epilepsy syndromes. In cases in which one of these exceptions obtain, consideration should be given to intravenous valproate in a total dose of 15–25 mg/kg (Misra *et al.* 2006): although current recommendations suggest infusing this at a rate no faster than 20 mg/min (or over a total of 60 minutes, whichever is longer), recent work suggests that much faster infusions, at 3–10 mg/kg/min are effective and well tolerated (Boggs and Preis 2000; Limdi *et al.* 2007; Wheless *et al.* 2004). Regardless of whether fosphenytoin or valproate is given, an AED blood level should be obtained approximately 2 hours after the infusion is completed, with subsequent doses based on levels and response.

An emerging alternative to phenytoin and valproic acid is levetiracetam, which, in retrospective or open, prospective studies, has been found effective both as monotherapy (Fattouch *et al.* 2010) and in cases refractory to lorazepam (Knake *et al.* 2008) or to another AED (Berning *et al.* 2009; Moddel *et al.* 2009; Patel *et al.* 2006). If levetiracetam is utilized, a bolus of from 1500 to 2000 mg is given intravenously over 30–60 minutes, with repeat boluses as needed.

In cases refractory to the above measures, coma may be induced with midazolam, propofol or pentobarbital, and, of these three, pentobarbital appears most effective (Claassen *et al.* 2002). Finally, in cases in which drug-induced coma is ineffective, consideration, as a last resort, may be given to ECT (Cline and Roos 2007; Lisanby *et al.* 2001).

Non-emergency treatment may involve oral loading with phenytoin, or, in cases of petit mal status, valproic acid. Phenytoin may be loaded in a dose of 15–25 mg orally, and valproic acid likewise in a dose of 15–25 mg/kg (importantly, divalproex should not be used emergently, given the long time required to obtain peak blood levels). Alternatively, if the current episode of status occurred secondary to non-compliance with an AED other than phenytoin or valproic acid, and the patient had a clear history of response to that AED, one might simply load the patient with this previously effective agent.

8.4 STROKE

Stroke is defined as the more or less sudden occurrence of a neuropsychiatric deficit occurring secondary to a vascular event, such as an ischemic infarction or an intracerebral hemorrhage, and the diagnosis should be suspected in any patient with the acute onset of virtually any of the signs, symptoms or syndromes described in the preceding chapters including, most especially, weakness (such as hemiparesis), sensory changes, aphasia, agnosia, neglect, or delirium. The evaluation and care of stroke patients has undergone revolutionary advances in the recent past and, especially in acute cases, patients should generally be referred to stroke specialists.

Clinical features

Given that the clinical features of stroke are determined, in large part, by the vessel involved, this discussion will begin with a review of the *arterial supply and venous drainage of the brain.* Once this is in mind, attention is then turned to the most common cause of stroke, namely *ischemic infarction,* followed by discussions of less common causes such as *intracerebral hemorrhage* and *subarachnoid hemorrhage* and, finally, of two rare causes, namely *primary intraventricular hemorrhage* and *cerebral venous thrombosis.* Finally, attention is directed to various *sequelae of stroke,* such as dementia and poststroke depression.

ARTERIAL SUPPLY AND VENOUS DRAINAGE OF THE BRAIN

Arterial supply

In tracing the arterial supply to the brain (Tatu *et al.* 1996, 1998), one may begin at the aortic arch, from which arises first the innominate artery, which, in turn, bifurcates into the right common carotid artery and the right subclavian artery. Passing further along the aortic arch, the left common carotid artery arises and, a little after that, the aortic arch gives off the left subclavian artery. From both subclavian arteries arise the vertebral arteries. Under normal circumstances, the entire blood supply to the brain is derived from the internal carotid arteries and the vertebral arteries, as described below: in clinical work, arteries derived from the internal carotid arteries are referred to as part of the 'anterior circulation', whereas those derived from the vertebral arteries are referred to as part of the 'posterior circulation'.

Both common carotid arteries bifurcate at the level of the thyroid cartilage into an external carotid artery and an internal carotid artery (ICA). The ICA may then be divided into four segments, namely cervical, intrapetrosal, intracavernous, and supraclinoid. The cervical segment rises from the bifurcation and passes up through the neck, without giving off any branches. Upon arriving at the skull, the ICA then enters the petrous portion of the temporal bone via the foramen lacerum and passes through the temporal bone as the intrapetrosal segment. The artery emerges from the temporal bone into the cavernous sinus, and passes horizontally through the sinus close to its medial wall. This intracavernous portion bears important relations to the other occupants of the cavernous sinus, namely the third, fourth, and sixth cranial nerves, and the first and second divisions of the fifth cranial nerve. After passing through the cavernous sinus, the ICA then swings superiorly and emerges medial to the anterior clinoid process as the supraclinoid segment.

Within the cavernous sinus the ICA gives off the ophthalmic artery; upon emerging from the sinus, it then gives off the posterior communicating artery, followed by the anterior choroidal artery. The ophthalmic artery passes forward in relation to the optic nerve and enters the orbit via the optic foramen. The posterior communicating artery passes posteriorly, and forms part of the circle of Willis, described below. The anterior choroidal artery also passes posteriorly, giving off important central branches, also described below. After giving off these three branches, the internal carotid artery then bifurcates into the anterior cerebral artery (ACA) and the middle cerebral artery (MCA).

The anterior cerebral artery passes anteriorly and medially, crossing superior to the optic nerve, to reach the interhemispheric fissure. Before passing into the fissure, it is joined to its partner on the opposite side by the anterior communicating artery. After passing into the interhemispheric fissure, the ACA, traveling inferior to the rostrum and genu of the corpus callosum, gives off the medial orbitofrontal artery and the frontopolar artery. After rising over the genu, the ACA then bifurcates into a callosomarginal artery and a pericallosal artery, which extends along the body of the corpus callosum all the way to the splenium. Although this overall description of the ACA holds true in most cases, important normal variations may occur. In a few percent of cases, the ICA on one side does not give rise to an ACA, and in this situation the ACA on that side arises from the anterior communicating artery, such that both ACAs draw their blood from one ICA. In another few percent, the anterior communicating artery is absent. Another very important variation is the fact that in over one-third of individuals a given ACA will supply not only its own hemisphere, but also will send various branches to the opposite hemisphere. Finally, in perhaps 5 percent of individuals, there may be only one ACA: this azygous, or unpaired, ACA then supplies the medial aspect of both hemispheres.

The MCA passes laterally, inferior to the anterior perforated substance, to reach the lateral cerebral fissure, where the stem typically bifurcates into superior and inferior divisions. The superior division gives rise to the lateral orbitofrontal artery, the pre-Rolandic artery, Rolandic artery, anterior parietal artery, posterior parietal artery and, in about one-half of individuals, the angular artery. The inferior division in turn gives rise to the temporopolar artery, the anterior temporal artery, the middle temporal artery, the posterior temporal artery, and, in the other one-half of normal individuals, the angular artery. In a minority of cases, the MCA, rather than bifurcating, will undergo a trifurcation, or may simply divide directly into the various branches just described.

Turning now to the vertebral arteries (which have arisen from the subclavian arteries in the neck): these rise to the level of the sixth cervical vertebra where they enter the transverse foramina; subsequently, they rise through the transverse foramina of the remaining cervical vertebrae to eventually enter the cranium through the foramen magnum. Once inside the cranium, they initially ascend on either side of the medulla, and eventually merge at the junction of the medulla and pons to form the basilar artery. Before merging, however, in addition to giving off numerous small penetrating branches to the medulla, they also give rise to several large branches, namely the posterior spinal artery, the posterior inferior cerebellar artery, and the anterior spinal artery. The posterior spinal arteries move posteriorly and descend along the posterior aspect of the spinal cord. The posterior inferior cerebellar artery courses along the lateral aspect of the medulla and then reaches the inferior aspect of the cerebellum. The anterior spinal artery is formed by two branches, arising from both vertebral arteries, which meet in the midline to form this artery; the artery then descends along the anterior midline of the medulla to the cord below.

The basilar artery, arising from the junction of the two vertebral arteries, ascends along the ventral surface of the pons and onto the ventral surface of the midbrain whereupon it bifurcates into the two posterior cerebral arteries. The basilar artery gives off numerous branches. The anterior inferior cerebellar artery arises first and supplies the inferior surface of the cerebellum. Penetrating branches are given off throughout the course of the basilar artery and include paramedian, short circumferential, and long circumferential arteries. Just before the basilar artery bifurcates into the posterior cerebral arteries (PCAs) it gives off the superior cerebellar arteries, which nourish the middle and superior cerebellar peduncles and the superior surface of the cerebellum.

The PCA courses around the surface of the midbrain and passes superiorly medial to the tentorium, after which it bifurcates into a temporo-occipital artery and an internal occipital artery. The temporo-occipital artery courses laterally to supply the inferior surface of the temporal lobe and adjacent occipital lobe, including the occipitotemporal gyrus and the lingual gyrus. The internal occipital artery bifurcates into the parieto-occipital artery and the calcarine artery. The parieto-occipital artery, in addition to supplying the medial aspect of the occipital and parietal lobes, also supplies part of the splenium of the corpus callosum. The calcarine artery supplies the medial aspect of the occipital lobe, including the critically important calcarine cortex. An important normal variant of the posterior cerebral artery, seen in about 15 percent of individuals, consists of a 'fetal' pattern of circulation wherein one of the PCAs, rather than arising from the basilar artery, originates from the internal carotid artery, as it were replacing the posterior communicating artery: when such a fetal posterior communicating artery is present, the entire hemisphere is then nourished from the ICA.

The anterior, middle, and posterior cerebral arteries supply most of the cerebral cortex. Subcortical structures, including the basal ganglia and the thalamus, are supplied by various central branches, most, but not all, of which arise from the circle of Willis. The first component of the circle of Willis to consider is the posterior communicating artery, which, as noted earlier, arises from the ICA. The posterior communicating artery courses posteriorly to anastomose with the PCA. Subsequent components of the circle include the ipsilateral PCA, the contralateral PCA, the contralateral posterior communicating artery, the contralateral ACA, the anterior communicating artery, and the ipsilateral ACA, which then brings us full circle. In addition to some central branches, discussed immediately below, the circle of Willis also gives off multiple very small penetrating branches that nourish, among other structures, the basal forebrain, the hypothalamus and portions of the midbrain.

The central branches to be considered now include the following: the thalamopolar artery, the thalamoperforating artery, the thalamogeniculate artery, the posterior choroidal artery, the recurrent artery of Heubner, the lenticulostriate arteries, and the anterior choroidal artery.

The thalamopolar artery arises from the posterior communicating artery and supplies anterior and lateral portions of the thalamus, including the dorsomedial nucleus, ventral lateral nucleus, and portions of the mammillothalamic tract. In a small minority of cases, both thalamopolar arteries may arise from a common pedicle arising from one or the other posterior communicating artery.

The thalamoperforating artery arises from the PCA medial to its junction with the posterior communicating artery, and supplies the central portion of the thalamus, including the intralaminar nuclei and portions of the mammillothalamic tract. In a small minority of cases both thalamoperforating arteries may arise from a common pedicle.

The thalamogeniculate artery arises from the posterior cerebral artery lateral to the junction with the posterior communicating artery and supplies portions of the lateral geniculate body and both the ventral posterolateral and ventral posteromedial nuclei of the thalamus.

The posterior choroidal artery arises from the PCA lateral to the origin of the thalamogeniculate artery. The posterior choroidal artery bifurcates into a medial posterior choroidal artery and a lateral posterior choroidal artery. The medial posterior choroidal artery courses medially to gain the medial aspect of the thalamus, where it supplies the anterior thalamic nuclei and the choroid plexus of the third ventricle. The lateral posterior choroidal artery courses laterally and enters the choroidal fissure of the temporal lobe to supply portions of the choroid plexus of the inferior horn of the lateral ventricle and portions of the hippocampus.

The recurrent artery of Heubner arises from the ACA and gives off branches that, after arising through the anterior perforated substance, supply the ventral portion of the head of the caudate, the inferior portion of the anterior limb of the internal capsule, and the anterior portion of the putamen.

The lenticulostriate arteries arise from the stem of the MCA before it reaches the lateral cerebral fissure. These arteries rise up through the anterior perforated substance, after which they supply the superior portion of the posterior limb of the internal capsule, adjacent portions of the corona radiata, the lateral portion of the globus pallidus, and most of the putamen.

The anterior choroidal artery, as noted earlier, arises from the internal carotid artery. It courses posteriorly, lateral to the midbrain, and then turns laterally to pierce the choroidal fissure of the temporal lobe and supply portions of the choroid plexus of the inferior horn of the lateral ventricle, the uncus, and portions of the hippocampus. Before turning laterally, however, the anterior choroidal artery also gives off branches that supply the inferior portion of the posterior limb of the internal capsule, the retrolenticular portion of the posterior limb of the internal capsule and adjacent portions of the corona radiata, and portions of the lateral geniculate body. A branch also supplies a portion of the globus pallidus, and in some cases a small branch may supply part of the ventrolateral thalamus.

Venous drainage

Venous drainage of the brain is accomplished via the superficial and deep cerebral veins, which in turn drain into the dural venous sinuses.

Of the dural sinuses, the first to consider is the superior sagittal sinus, which extends posteriorly along the superior edge of the falx cerebri to reach the sinus confluens. The inferior sagittal sinus extends posteriorly along the free edge of the falx cerebri to the edge of the tentorium cerebelli, where it joins with the great vein of Galen to form the straight sinus. The straight sinus, in turn, runs posteriorly in the junction of the falx cerebri and tentorium cerebelli to eventually join the terminus of the superior sagittal sinus at the sinus confluens. The sinus confluens then gives rise to the transverse sinuses, each of which courses anteriorly along the outer edge of the tentorium cerebelli. In a majority of cases, the straight sinus enters directly into the left transverse sinus, whereas the superior sagittal sinus drains directly into the right transverse sinus. The transverse sinus, upon reaching the junction of the occipital and petrosal bones, empties into the sigmoid sinus, which curves downward to drain into the internal jugular vein.

The next dural sinus to consider is the cavernous sinus, which lies lateral to the sella turcica and its enclosed pituitary gland. The cavernous sinus receives drainage from the orbit via the superior and inferior ophthalmic veins and then, in turn, drains posteriorly by two other sinuses, the superior petrosal and inferior petrosal sinuses, which connect, respectively, with the transverse sinus and the internal jugular vein.

The superficial cerebral veins lie on the cortex and drain into the various nearby dural sinuses. Importantly, these superficial cerebral veins are interconnected by various anastomotic veins that, by providing alternative outflows, minimize the consequences of a single occlusion. Of the deep cerebral veins, the two most important are the basal vein of Rosenthal, which drains the medial aspect of the temporal lobe, and the internal cerebral vein, which drains the caudate and thalamus: both these deep cerebral veins then join to form the great vein of Galen, which, as noted earlier, drains into the straight sinus.

ISCHEMIC INFARCTION

Ischemic infarctions of the brain may be subdivided into the following groups: *large vessel syndromes*, occurring secondary to occlusion of one of the large pial vessels, such as the MCA; *low-flow (watershed) infarctions* located in the boundary zone between the areas of distribution of two large vessels, for example the MCA and ACA; and *lacunar syndromes* resulting from occlusion of central or penetrating arteries. Attention is then turned to *transient ischemic attacks (TIAs)*, which occur due to temporary occlusion of a vessel, and, finally, to so-called *'silent' infarctions.*

Large vessel syndromes

In the following paragraphs, the typical syndromes seen with occlusion of the large cerebral vessels, namely the middle, anterior, and posterior cerebral arteries, the anterior choroidal artery, and the basilar and vertebral arteries, are discussed. It must be borne in mind, however, that more often than not only fragments of these syndromes are seen. This is particularly the case with embolic infarction, wherein, rather than occluding the large vessel at or close to its origin, the embolus travels up the artery to lodge in a smaller branch. In other cases, extensive collateral circulation may blunt the ischemic challenge, thus limiting the size of the infarcted area.

Middle cerebral artery infarction may be only partial, involving only the superior division or the inferior division, or may be complete, involving both divisions. Infarction in the area of distribution of the superior division typically produces a contralateral hemiparesis and hemisensory loss, with preferential involvement of the face and upper extremity. When the right hemisphere is affected, left neglect and anosognosia are commonly seen; other signs that may appear include asomatagnosia, constructional and dressing apraxia, aprosodia, and, rarely, delirium. With involvement of the left hemisphere a motor aphasia typically occurs: in cases of left-sided infarction where the angular artery arises from the superior division, the additional involvement of Wernicke's area will produce a global aphasia; other signs that may occur include ideomotor or ideational apraxia and Gerstmann's syndrome. Infarction in the area of distribution of the inferior division generally causes a contralateral hemianopia or quadrantopia. With right-sided infarction, a delirium, often accompanied by agitation is common and prosopagnosia may also be seen. With left-sided infarction, in cases where the angular artery arises from the inferior division, one typically sees a sensory aphasia; delirium may also occur but is less common than with right-sided infarction. Complete infarction, involving both upper and lower divisions, is often a catastrophic event (Heinsius *et al.* 1998) and, in addition to a profound hemiplegia, such a 'malignant MCA infarction' may also be accompanied by stupor and vasogenic edema developing over the following 2–5 days, which may cause uncal or tentorial herniation with coma and death (Hacke *et al.* 1996). In such cases, decompressive hemicraniectomy may be life-saving.

Anterior cerebral artery infarction (Critchley 1930, Kang and Kim 2008) classically produces a contralateral hemiplegia and hemianesthesia which preferentially affect the lower extremity; when the left hemisphere is involved, a transcortical motor aphasia may also appear. Other symptoms that may appear include abulia, akinetic mutism or a frontal lobe syndrome. When the corpus callosum is infarcted, one may also see a 'disconnection syndrome' with left-sided, but not right-sided, agraphia, tactile agnosia or apraxia, and, rarely, the alien hand syndrome. In cases of bilateral infarction (e.g. as may occur with an azygous ACA), a paraplegia occurs.

Posterior cerebral artery infarction (Castaigne *et al.* 1973; Pessin *et al.* 1987) typically causes a contralateral hemianopia. If the infarction is on the left, one may also see alexia with agraphia, visual agnosia, and, if the temporal lobe is involved, a delirium. Infarctions on the right side that also involve the temporal lobe may also cause delirium, and there may also be prosopagnosia. Bilateral infarction is signaled by cortical blindness, which may or may not be accompanied by Anton's syndrome, with denial of the blindness. In cases of bilateral infarction when both temporal lobes are involved, patients may also be left with a Korsakoff's syndrome. As described earlier, central branches arise from the PCA (i.e. thalamoperforating, thalamogeniculate, and posterior choroidal) and if these are also occluded, the typical symptoms just described will be joined by the syndromes peculiar to occlusion of these central branches, as described below, under '*lacunar syndromes*'. Furthermore, when some of the very small penetrating branches to the mesencephalon are involved, there may be hemiparesis, oculomotor disturbances, and abnormal movements.

Anterior choroidal artery infarction is marked by contralateral hemiplegia and hemianopia (Helgason *et al.* 1986). Importantly, however, there is no accompanying aphasia or neglect, and generally no, or only transient, hemianesthesia, and it is the absence of these symptoms that distinguishes these infarctions from those that occur secondary to occlusion of the MCA.

Basilar artery occlusion is often a catastrophic event (Caplan 1979; von Campe *et al.* 2003), producing quadriplegia, and, if the infarction extends to the medulla, the 'locked-in syndrome', wherein patients, though still alert and cognitively intact, are able to communicate only by blinking and vertical eye movements. Occlusion of arteries nourishing the cerebellum (superior cerebellar, anterior inferior cerebellar, or posterior inferior cerebellar) may give rise to various combinations of ataxia, dysmetria, and vertigo. With occlusion of either the anterior or posterior inferior cerebellar arteries, one may also see the 'cerebellar cognitive affective syndrome', characterized by a frontal lobe syndrome (Schmahmann and Sherman 1998), described in Chapter 8.2.

Vertebral artery occlusion or occlusion of a posterior inferior cerebellar artery (or occlusion of one of the perforating branches of either of these arteries) may give rise to the classic Wallenberg or 'lateral medullary' syndrome (Kim 2003; Sacco *et al.* 1993). Infarction of the lateral medulla typically involves the following structures: inferior cerebellar peduncle, spinothalamic tract, the spinal tract and nucleus of the fifth cranial nerve, the nucleus ambiguus, vestibular nuclei, and descending sympathetic fibers. Respectively, the corresponding symptoms are: ipsilateral ataxia; contralateral hemianesthesia of the extremities; ipsilateral anesthesia of the face (which may be accompanied by pain); hoarseness and dysphagia; nausea, vomiting, and vertigo; and an ipsilateral Horner's syndrome. Occlusion of one of the originating branches of the anterior spinal artery may produce the 'medial medullary' syndrome (Kim *et al.* 1995). Here, infarction of the pyramid and of the emerging fibers of the twelfth cranial nerve give rise to an ipsilateral paresis of the tongue and a contralateral hemiparesis.

Cerebellar artery occlusion, including the posterior inferior, anterior inferior, and superior cerebellar arteries, may cause vertigo, nausea, nystagmus, or ipsilateral ataxia. Large infarctions, with attendant vasogenic edema, may, by compressing the underlying brainstem, cause stupor or coma, and in such cases emergent neurosurgical intervention may be required (Jensen and St. Louis 2005).

Low-flow (watershed) infarctions

Watershed infarctions typically present with more or less atypical fragments of some of the large vessel syndromes. Thus, ACA–MCA watershed infarctions involving the prefrontal cortex and subjacent white matter may present with brachial or crural paresis, and, when the left hemisphere is involved one may see mutism, initially, which resolves into a transcortical motor aphasia. MCA–PCA watershed infarctions involving the parieto-occipital cortex and subjacent white matter may present with hemianesthesia, hemianopia, and, if the left hemisphere is involved, a transcortical sensory aphasia. ACA–MCA infarctions involving the centrum semiovale present with a variable mixture of either of the foregoing syndromes. 'Internal' watershed infarctions in the border zone between the MCA and lenticulostriate arteries may present with hemiparesis with or without hemianesthesia (Kumral *et al.* 2004).

Lacunar syndromes

Lacunes are small cavities, ranging in size from 1 to 20 mm, which typically represent infarctions in the area of distribution of one of the central or perforating branches described earlier (Fisher 1965, 1982; Mohr 1982). Although the clinical presentation of lacunar infarctions is quite varied, depending on the location of the lacune (Arboix *et al.* 2006), certain presentations are quite distinctive and highly suggestive of lacunar infarction. These 'classic' lacunar syndromes include 'pure motor stroke', 'ataxic hemiparesis', 'dysarthria clumsy hand', 'pure sensory stroke', and 'sensorimotor stroke.'

Pure motor stroke, as the name suggests, is characterized by a hemiparesis in the absence of sensory changes or other symptoms such as aphasia or neglect. Typically, the upper and lower extremities are more or less equally involved (with or without involvement of the ipsilateral lower face) and this pattern helps distinguish hemiplegia due to lacunar infarction from hemiplegia due to middle or anterior cerebral artery infarction wherein either, respectively, the upper or lower extremity is preferentially involved. The responsible lacune in pure motor stroke may be found in the contralateral corona radiata, internal capsule, cerebral peduncle, basis pontis, or medullary pyramid. In cases where the lacune is in the brainstem, one may see contralateral cranial nerve palsies, helping to further localize it: a third nerve palsy suggests a midbrain location; a sixth nerve palsy, a pontine location; and a twelfth nerve palsy, a medullary one.

In cases when pure motor stroke is characterized by either a monoparesis (of either the upper or the lower extremity) or an isolated facial paresis, the responsible lacune may not only be due to infarction in the area of distribution of the central or perforating arteries and located in the areas just described, but may also occur secondary to infarction due to occlusion of terminal branches of the MCA (or, less commonly, the ACA), producing small infarctions in the motor cortex (Maeder-Ingvar *et al.* 2005; Paciaroni *et al.* 2005).

Ataxic hemiparesis manifests with a combination of ataxia and hemiparesis and the lacune is typically found in the corona radiata, internal capsule, or basis pontis.

Dysarthria clumsy hand presents not only with dysarthria and clumsiness of the hand but also with a lower facial paresis ipsilateral to the clumsy hand; the lacune is typically found in the corona radiata, internal capsule (near the genu), or the rostral basis pontis (Arboix *et al.* 2004).

Pure sensory stroke is characterized by hemianesthesia occurring in isolation, and in the absence of motor weakness or other symptomatology. In such cases, a lacune is typically found in the ventrolateral thalamus. In the months following a pure sensory stroke, a minority of patients may begin to experience chronic painful dysesthesiae in the involved limbs.

Sensorimotor stroke may occur secondary to infarction of either the thalamus or the posterior limb of the internal capsule. In some cases of infarction of the ventrolateral thalamus a substantial amount of edema may involve the adjacent posterior limb of the internal capsule, thus causing a contralateral hemiplegia: the hemiplegia, however, is usually relatively mild and clears as the edema resolves. In cases of infarction of the posterior limb of the internal capsule, in some instances the thalamic radiations may also undergo infarction, leading to a hemianesthesia.

Other, less 'classic' lacunar syndromes may occur with infarction of other portions of the thalamus or with lacunar infarction of the caudate nucleus.

Lacunes affecting the anterior or central portions of the thalamus (Bogousslavsky *et al.* 1988; Giannopoulos *et al.* 2006; Graff-Radford *et al.* 1985), as may occur with occlusion of the thalamopolar or thalamoperforating arteries, typically present acutely with delirium or somnolence. Over time, patients typically become alert but are left with memory deficits (reflecting involvement of the mammillothalamic tract) or a frontal lobe syndrome (with involvement of the dorsomedial nucleus); furthermore, when the lacune is on the left, aphasia may occur and, when on the right, neglect may be seen.

Caudate lacunes (Caplan *et al.* 1990; Mendez *et al.* 1989), as may occur with occlusion of the recurrent artery of Heubner, are typically characterized by frontal lobe symptoms, especially abulia, and, less commonly, by chorea. Early on there may also be a dysarthria or mild hemiparesis, both transient, and agitation or restlessness, especially when the lacune is in the right caudate nucleus.

Finally, and before leaving this discussion of lacunar syndromes, mention should be made of a condition known either as a 'giant lacune' or a 'striatocapsular infarction' (Weiller *et al.* 1990). Here, all, or almost all, of the lenticulostriate arteries on one side are occluded, resulting in a relatively large infarction involving the posterior limb of the internal capsule and the putamen. These infarctions are typically larger than 20 mm in diameter and thus, technically, do not qualify as 'lacunes'. Their discrete shape and subcortical location, however, tempt one to use the word 'lacune' in referring to them, and this linguistic tension has given rise to the compromise term 'giant lacune'. Symptomatically, as might be expected, such a giant lacune generally presents with a contralateral hemiplegia that may be accompanied, if the lesion is on the left, by a transcortical motor aphasia or, if the lesion is on the right, by left neglect. The mechanism underlying such a 'giant lacune' varies according to the anatomy of the lenticulostriate arteries in question. As noted earlier, in most cases the lenticulostriate arteries arise sequentially from along the stem of the MCA, and in these cases the responsible lesion is generally an embolus that lodges in the MCA stem proximal to the origin of the first of the lenticulostriate arteries. In such a scenario, the lack of collateral supply to the area nourished by the lenticulostriate arteries leads to infarction, whereas the cerebral cortex, being supplied with rich collaterals, may escape infarction. In other cases, all the lenticulostriate arteries arise from one common pedicle arising from the MCA stem, and in these cases the occlusion of this one pedicle, either by lipohyalinosis or encroachment of an atherosclerotic plaque, may do the job.

Transient ischemic attacks

A TIA represents the results of fully reversible neuronal ischemia due to a transient reduction in blood supply, which in turn may be due to either an embolus that fragments and passes on, or to temporary low flow past a thrombus or through a stenosed artery, as may occur during an episode of systemic hypotension. Clinically, patients may experience any of the symptoms of ischemic infarction just described, with one critical difference: after a brief period of time, the symptoms resolve completely, with a full restoration of prior functional ability. One very important kind of TIA, not mentioned earlier, is amaurosis fugax ('fleeting blindness', also referred to as transient monocular blindness), wherein the patient experiences partial or complete blindness for a matter of minutes in one eye: such episodes are highly suggestive of disease in the ipsilateral carotid artery proximal to the origin of the ophthalmic artery. Another form of TIA, also not mentioned earlier, is the 'limb-shaking TIA': here there is unilateral shaking of the upper or lower extremity, or both together, lasting from seconds to a minute. This TIA is often precipitated by sitting or standing, and is associated with occlusion or tight stenosis of either the common or internal carotid artery (Yanagihara *et al.* 1985). 'Drop attacks' may also be mentioned: here the TIA is characterized by a sudden weakness of all four extremities, with a fall to the ground. Such attacks occur with basilar artery disease and brainstem ischemia, and may herald stroke with quadriplegia (Brust *et al.* 1979).

There has been some controversy regarding the duration of symptoms seen in a TIA. In the past, it was felt that symptoms could persist for up to 24 hours; however, recent studies with diffusion-weighted imaging (DWI) MR scanning have revealed that in cases in which symptoms last much longer than 30 minutes (Inanatomia *et al.* 2004), and certainly for an hour or more (Ay *et al.* 2005), actual infarction has generally occurred. Thus, it is probably best to reserve this diagnosis for cases in which symptoms last no longer than half an hour, and in no case for more than an hour, and, if MR scanning is performed, for cases in which DWI is negative.

Transient ischemic attacks may herald either embolic or thrombotic infarctions, including lacunar infarctions. Indeed, in one study (Rothwell and Warlow 2005), approximately 20 percent of patients who had a TIA went on to have a stroke: in 17 percent of those who did develop stroke, the stroke occurred on the same day as the TIA, in 9 percent on the next day, and in 43 percent over the following week. In another study, 10.5 percent went on to have a stroke, and 51 percent of these occurred within 2 days (Johnston *et al.* 2000). These are sobering figures and serve to stress that a TIA is not a benign event but rather represents a medical emergency: patients should be immediately worked up, as described below, just as if they had already had an ischemic infarction, and then treated appropriately.

'Silent' infarctions

It is important to note that ischemic infarctions may be 'silent', that is to say they either, in and of themselves, produce no symptoms or produce symptoms of such mildness that no attention is brought to them. This may be a common occurrence; indeed in patients who have a clinical history of stroke, MR scanning with DWI reveals subsequent silent infarctions in approximately one-third of all cases within the first 5 days, with approximately one-fifth of all cases going on to have silent infarcts in the 30- to 90-day period (Kang *et al.* 2006).

Despite being individually 'silent', these infarctions should not be considered as benign, given that, with an accumulation of them, a multi-infarct or a lacunar dementia (as discussed in Chapters 11.1 and 11.2) may eventually develop.

INTRACEREBRAL HEMORRHAGE

Clinically, intracerebral hemorrhage is typified by the gradual evolution, over from 30 minutes to several hours, of a focal deficit, appropriate to the location of the hematoma, accompanied by headache and nausea.

Putaminal hemorrhages, by compressing the adjacent internal capsule, present with a dense hemiparesis, accompanied by a hemisensory loss; with left-sided hemorrhages, an aphasia may also be present. Caudate hemorrhages, by the same mechanism, may also present with hemiparesis, but this is generally less profound; a delirium may also be present. Thalamic hemorrhages typically present with a hemisensory loss; compression of the adjacent internal capsule may also cause a hemiparesis, but this is generally mild. With thalamic hemorrhage delirium may also occur, and, with left-sided hemorrhages, an aphasia. Conjugate deviation of the eyes may be seen with hemorrhage into all of these structures: with putaminal and caudate hemorrhages the eyes look toward the side of the lesion, whereas with thalamic hemorrhages the eyes look down.

Putaminal hemorrhages may rupture into the lateral ventricle, and both caudate and thalamic hemorrhages may rupture into the third ventricle, and in all these cases acute hydrocephalus may occur, with stupor or coma.

Pontine hemorrhage is characterized by quadriplegia and rapidly developing coma.

Cerebellar hemorrhage is characterized by dysmetria, ipsilateral to the hemorrhage, ataxia and nausea. As the hematoma enlarges, compression of the underlying brainstem may cause stupor, coma, and death.

Lobar intracerebral hemorrhages tend to occur in the parietal or occipital lobes, and present with appropriate focal symptoms.

Seizures occur in roughly 10 percent of cases of intracerebral hemorrhage, more commonly in cases of lobar hemorrhage.

SUBARACHNOID HEMORRHAGE

Most cases of subarachnoid hemorrhage present in a catastrophic fashion (Suarez *et al.* 2006): the eruption of blood at arterial pressure into the subarachnoid space causes a very rapid rise in intracranial pressure, and patients experience a severe generalized headache which, in almost all cases, is the worst experienced in their lives. Typically the headache reaches its maximal intensity within seconds, at the most minutes. Although some patients may remain lucid throughout, most experience the rapid onset of delirium, stupor, or coma: in those presenting in coma, there may be decerebrate rigidity, accompanied in some by clonic jerks of the extremities. Importantly, in most cases focal signs, such as hemiparesis, are absent: exceptions occur in cases where the jet of blood emerging from the bleeding point ruptures into the adjacent parenchyma, causing an enlarging intracerebral hemorrhage. Clinically obvious seizures may occur in up to one-fourth of all patients, and EEG monitoring in comatose patients further reveals electrographic seizures in almost 20 percent (Claasen *et al.* 2005).

In a minority of cases, this full picture may be preceded by transient 'sentinel' headaches occurring secondary to minor 'warning leaks' that terminate spontaneously (Ostergaard 1990).

Patients who survive the initial event are at risk for significant complications (Hijdra *et al.* 1988) over the following weeks, including rebleeding, vasospasm with cerebral infarction, and the development of hydrocephalus. Rebleeding may occur in up to 30 percent of patients and, although it is most common within the first 24 hours, the risk extends for roughly 4 weeks. Vasospasm of cerebral arteries passing through the subarachnoid blood may occur, leading to clinically evident ischemic infarction in approximately one-third of all patients (Hijdra *et al.* 1986): serial MR scans have demonstrated that approximately a further one-fifth of patients will have 'silent' ischemic infarcts (Shimoda *et al.* 2001). The infarctions themselves may be large vessel, watershed or lacunar in distribution (Weidauer *et al.* 2007). Vasospasm typically begins to occur on day 3, peaks between days 5 and 7, and then resolves over 2–4 weeks. Acute hydrocephalus, with headache and lethargy, may be seen in up to 20 percent of patients within the first hours or days, and occurs secondary to blockage, by clotted blood, of either the third or fourth ventricles or of the exit foramina of the fourth ventricle. Chronic hydrocephalus, secondary to restricted passage of cerebrospinal fluid (CSF) through the arachnoid villi, occurs in a further 20 percent of cases, with a subacute onset being seen 2–4 weeks after the initial bleed: patients present with dementia, incontinence, and gait disturbance.

Other complications include central fever, arrhythmias, neurogenic pulmonary edema, and hyponatremia. The hyponatremia, in turn, may be secondary to either the syndrome of inappropriate antidiurectic hormone secretion (SIADH), or, more commonly, cerebral salt wasting (CSW) (Rabinstein and Wijdicks 2003). Differentiating SIADH from CSW is critical, as the treatments are radically different. In SIADH there is an increased release of ADH, with consequent increased renal reabsorption of water leading to hyponatremia in the setting of volume expansion. By contrast, in CSW a release of natriuertic factors leads to renal sodium and fluid loss and a hyponatremia in a setting of volume contraction. SIADH is treated by fluid restriction, whereas CSW is treated by the administration of saline.

PRIMARY INTRAVENTRICULAR HEMORRHAGE

Primary intraventricular hemorrhage presents in a fashion similar to subarachnoid hemorrhage, with severe headache, stupor, or coma. Chronic hydrocephalus is a common sequela.

CEREBRAL VENOUS THROMBOSIS

Thrombosis of one of the dural sinuses (most commonly the superior sagittal or the transverse sinus) may or may not be followed by cerebral infarction, depending both on whether drainage of a cerebral vein is blocked by the thrombus and on whether or not the vein in question lacks the anastomotic connections that might ensure adequate venous drainage. When these unfavorable conditions are met, venous congestion of the subserved area occurs with the gradual appearance of a hemorrhagic infarction and the appearance, clinically (Bousser *et al.* 1985; Cantu and Barinagarrementeria 1993; Gosk-Bierska *et al.* 2006, Khealani *et al.* 2008), of the gradual onset of headache, focal deficits appropriate to the infarcted area, and, in a significant minority, seizures. Thrombosis of the vein of Galen, although rare, may be given special consideration here, given its clinical expression. In these cases, the thalami, which are drained by the internal cerebral veins, may undergo hemorrhagic infarction, and this may result in stupor or coma (van den Bergh *et al.* 2005) or, in milder cases, a subacute onset of dementia (Krolak-Salmon *et al.* 2006). In very rare cases, rather than undergoing infarction, the thalami may merely become edematous, and in one such case the patient, who presented with akinetic mutism, recovered completely (Gladstone *et al.* 2001).

Thrombosis of the superior sagittal sinus, by causing an elevation of intracranial pressure, may cause symptoms even in the absence of venous infarction, and patients may present with the gradual evolution of headache, vomiting, drowsiness, delirium, or stupor.

Thrombosis of the cavernous sinuses produces a distinctive syndrome with proptosis secondary to impaired venous drainage from the eye, and ophthalmoplegia, secondary to compression of the third and fourth cranial nerves found in the wall of the sinus itself.

The evolution of symptoms seen with venous infarction is very gradual, spanning days or even weeks. This leisurely onset reflects the gradual propagation of the clot and the equally gradual failure of collateral drainage.

SEQUELAE OF STROKE

The sequelae of stroke include *cognitive deficits*, *depression*, *anxiety*, and *other sequelae*, such as emotional incontinence, emotionalism, the catastrophic reaction, the frontal lobe syndrome, mania, psychosis, seizures, and central pain.

Cognitive deficits

Both dementia and amnesia may follow upon stroke.

Dementia may occur after multiple ischemic infarctions or intracerebral hemorrhages. This may occur with either cortical or white matter lesions, producing a multi-infarct dementia (discussed further in Chapter 11.1), or with lacunes, producing a lacunar dementia (discussed in Chapter 11.2). Typically, in such cases, one finds a history of repeated stroke; however, occasionally, a 'strategically' placed single infarction or hemorrhage may leave the patient with a dementia, as for example with infarcts or hemorrhages within either temporal lobe or a hemorrhage or lacune in the thalamus.

Subarachnoid hemorrhage may also be followed by a dementia, due either to chronic hydrocephalus or multiple infarctions due to vasospasm. When infarction is at fault, one may find either a relatively small number of large vessel infarcts or a myriad of 'microinfarcts' occurring secondary to spasm of small vessels. CT scanning, although quite capable of demonstrating the large infarcts, will miss the smaller ones, and hence MRI may be required.

Cerebral venous thrombosis, if accompanied by multiple venous infarctions, may also leave patients demented; in the absence of these, most patients, if they survive, do so without cognitive sequelae.

An amnestic syndrome may occur after subarachnoid hemorrhage due to microinfarctions involving the fornix as it passes through either the septal region or the hypothalamus. Such infarctions, in turn, occur secondary to damage to the delicate penetrating arteries that arise from the anterior communicating artery, damage which may occur either due to vasospasm or to their inadvertent destruction during clipping of an aneurysm arising from that artery (Alexander and Freedman 1984; Gade 1982).

Depression

In the weeks or months following stroke, roughly 40 percent of all patients will develop a depression of variable severity. As discussed in Chapter 7.1, the location of the infarct or hemorrhage plays a part here, with lesions in the anterior portions of the frontal lobes being more likely to cause depression.

Only a handful of double-blinded studies have been performed regarding the treatment of poststroke depression. Sertraline was *in*effective in one study (Murray *et al.* 2005); citalopram was effective in one study (Andersen *et al.* 1994); fluoxetine was effective in one study (Wiart *et al.* 2000), but ineffective in two (Choi-Kwon *et al.* 2006; Fruehwald *et al.* 2003); finally nortriptyline was effective in two studies (Lipsey *et al.* 1984; Robinson *et al.* 2000), and in one of these (Robinson *et al.* 2000) was also more effective than fluoxetine.

Given the frequency with which poststroke depression occurs, double-blinded studies have been performed to determine whether its occurrence may be prevented by prophylactic administration of antidepressants. In this regard, sertraline was effective in one study (Rasmussen *et al.* 2003), but not another (Almeida *et al.* 2006); escitalopram was effective in one study (Robinson *et al.* 2008); citalopram was effective in one study (Robinson *et al.* 2008); finally, fluoxetine and nortrptyline were compared with each other and placebo in one study (Narushima *et al.* 2002), and, although both were more effective than placebo, in an 18-month naturalistic follow-up, the patients who had been treated with nortriptyline were at greater risk for developing more severe depressive symptoms.

On balance, either citalopram or nortripyline are reasonable choices, both for the treatment of poststroke depression and for its prevention. Given the better tolerability of citalopram, it is probably the first choice, however should it prove ineffective, nortriptyline should be considered.

Anxiety

Chronic anxiety is seen in roughly one-fourth of stroke patients and, as discussed in Chapter 7.5, appears to be more common with right hemisphere lesions. In most cases of poststroke anxiety, however, the anxiety, rather than occurring in an isolated fashion, is part of a poststroke depression (Castillo *et al.* 1995), and in such cases an additional diagnosis should not be made. In the only double-blinded study of the treatment of poststroke anxiety, nortriptyline was found more effective than placebo (Kimura *et al.* 2003), a not-unsurprising result given that most cases of anxiety are, as just noted, found in patients with depression. It is not clear what the best course would be for cases of 'pure' poststroke anxiety, occurring in the absence of depression: although benzodiazepines are used in clinical practice, caution is needed here, as stroke patients are prone to develop cognitive deficits and lethargy secondary to these agents.

Other sequelae

Emotional incontinence, discussed further in Chapter 5.7, is characterized by displays of uncontrollable laughter or crying without any corresponding emotion and occurs secondary to bilateral interruption of the corticobulbar tracts. Nortriptyline in doses of 50–75 mg/day (Robinson *et al.* 1993) and citalopram in doses of 10–20 mg/day (Andersen *et al.* 1993) are both effective.

Emotionalism is characterized by easy tearfulness, and is distinguished from emotional incontinence by its association with a corresponding emotion of sadness. This may be seen in anywhere from one- to two-tenths of patients, and is most likely with left frontal or temporal lesions (House *et al.* 1989). Double-blinded studies indicate that both fluoxetine and sertraline (Burns *et al.* 1999) are effective.

A catastrophic reaction, as discussed further in Chapter 5.2, may be seen in close to one-fifth of patients and appears to be particularly common with infarction of the anterior left hemisphere or the left basal ganglia.

The frontal lobe syndrome, discussed in Chapter 8.2, may be seen with infarction or hemorrhage of the frontal lobes, caudate nucleus, or thalamus. Patients may present with varying combinations of disinhibition, perseveration, and affective changes.

Mania is a rare sequela to stroke and, as discussed in Chapter 7.3, may be seen after infarction of the midbrain, thalamus, anterior limb of the internal capsule, and adjacent head of the caudate nucleus, or the frontal or temporal lobes.

Psychosis, likewise, is a rare sequela, and, as discussed in Chapter 8.1, may occur with infarction of the frontal or temporal cortex, or the thalamus.

Seizures are not uncommon after stroke, and are discussed in Chapter 8.3.

Central, or 'thalamic' pain may appear in the weeks or months following infarction or hemorrhage affecting the ventrolateral thalamus. The pain is typically burning in quality, and may be particularly severe. Both amitriptyline, in a dose of 75 mg (Leijon and Boivie 1989), and lamotrigine, in a dose of 200 mg (Vestergaard *et al.* 2001), may offer relief.

Etiology

Roughly three-fourths of all cases of stroke are due to *ischemic infarction*, another one-fifth to *intracerebral hemorrhage*, and one-tenth to *subarachnoid hemorrhage*, with the remainder occurring secondary to *intraventricular hemorrhage* or *cerebral venous thrombosis*, and each of these is discussed in turn. Certain rare or unusual *other causes of stroke*, such as cerebral autosomal dominant arteriopathy with subcortical infarcts and leukoencephalopathy (CADASIL), are then discussed, followed by a *suggested workup* for the new stroke patient.

ISCHEMIC INFARCTION

Ischemic infarction occurs when arterial blood supply is reduced below that required for tissue viability, and such reductions may occur via a variety of mechanisms. First, *embolic infarctions* occur when an embolus, say from the heart, lodges in an artery, thus occluding it. Second, *thrombotic infarctions* occur when a thrombus forms inside an artery, typically on top of an ulcerated atherosclerotic plaque, causing occlusion. These two mechanisms account for most large vessel syndromes, and may also underlie certain of the lacunar syndromes. Third, *low-flow (watershed) infarctions* occur secondary, not to occlusion, but to a critical reduction in perfusion pressure. Fourth, and finally, small penetrating arteries may be subject to *lipohyalinosis*, leading to their gradual occlusion and producing a lacunar syndrome.

Embolic infarctions

Emboli may be either cardiogenic (Caplan *et al.* 1983), arising from the heart, or 'artery-to-artery' wherein they arise from a thrombus on an arterial wall and travel downstream to eventually plug a smaller caliber artery. The most common cause of cardiogenic embolic infarction is atrial fibrillation, wherein thrombi form within either the left atrium or atrial appendage. Emboli are also seen in the sick sinus syndrome; however, here the increased risk is probably due to the associated atrial fibrillation. Atrial myxomas, although rare, are very prone to fragment and undergo embolization. Valvular disease is also associated with emboli. In the case of mitral stenosis, this increased risk may be related simply to the commonly associated atrial fibrillation; however, in infective endocarditis (Anderson *et al.* 2003), Libman–Sacks endocarditis, and marantic endocarditis, actual embolization from the affected valves does occur. Prosthetic valves may be complicated by thrombus formation with embolization, and this is much more likely with mechanical than with bioprosthetic valves. Myocardial infarction is associated with thrombus formation and embolization, both acutely and in the chronic phase. Acutely, thrombi may form on the damaged endocardium, and the period of risk here extends up through the first month or so post myocardial infarction. Chronically, thrombi may form in cases characterized by ventricular aneurysm or large areas of reduced cardiac contractility. Another cardiac condition favoring thrombus formation is dilated cardiomyopathy. In all the foregoing cardiac conditions, emboli arise from the left side of the heart: there is another condition, known as 'paradoxical embolization' wherein emboli, arising, for example from a deep venous thrombosis, travel first through the right side of the heart. In these cases, one most commonly finds a patent foramen ovale; other defects allowing for such paradoxical embolization include atrial septal defect, ventricular septal defect, and pulmonary arteriovenous fistulae.

Artery-to-artery emboli typically have their origin in a thrombus atop an atherosclerotic plaque. Typical scenarios include emboli arising from the origin of the ICA and traveling downstream to occlude a branch of the MCA, or arising from the basilar artery, then traveling downstream to eventually occlude a branch of the PCA.

Emboli may also occur in association with cardiac surgery, especially valve replacement and coronary bypass grafting. Such cases are often associated with significant arteriosclerosis of the aorta, and when the aorta is clamped and unclamped, showers of embolic material may be dislodged. Mere cardiac catheterization may also cause embolic infarction, either secondary to dislodgment of a portion of an aortic plaque or secondary to thrombus formation on the tip of the catheter (Khatri and Kasner 2006).

Carotid artery dissection, although occurring across the lifespan, should be especially considered in any patient under the age of 45 years who has had an embolic infarction (Ahl *et al.* 2004). With dissection of the carotid artery and rupture of the intima, a thrombus forms, which may serve as a source for emboli (Fisher *et al.* 1978). Suggestive clinical evidence includes headache, neck pain, and, due to compression of ascending sympathetic fibers, Horner's syndrome, all occurring ipsilaterally to the infarction (Lee *et al.* 2006).

Regardless of the source of the embolus, whether it is cardiac or from an upstream artery, the sequence of events in embolic infarction is essentially the same. In all cases, the embolus is borne downstream, through arteries of progressively smaller caliber, until finally it lodges in an artery, causing its occlusion. Clinically, such plugging of an artery gives rise to a rapid onset of symptoms, over seconds to minutes. In some cases, the embolic plug remains in place; however, in many cases the embolic thrombus fragments, dislodges, and passes further downstream, either to plug yet a smaller artery or, if the fragments are sufficiently small, to pass through the capillary bed into the venous circulation. In cases when such dislodgement occurs early, before tissue infarction occurs, one may see clinical improvement, which in some cases is dramatic: the so-called 'spectacular shrinking deficit' (Minematsu *et al.* 1992), wherein the clinical symptomatology undergoes a dramatic improvement as the embolus dislodges and travels downstream. However, in cases when infarction has already occurred, the passage of the embolus downstream, rather than leading to clinical improvement, is followed by 'hemorrhagic transformation' of the infarct (Okada *et al.* 1989). Here, with the renewed presence of blood flow in the infarcted area at arterial pressure, multiple petechial hemorrhages form within the area of infarction, which, in some cases, may become confluenct. Such hemorrhagic transformation occurs in about one-third of cases, and although it generally takes places within the first 2 days, transformation may occur for up to 10 days poststroke. Importantly, there is usually little clinical change during hemorrhagic transformation: it is clinically 'silent'.

Thrombotic infarctions

Thrombotic infarctions, as noted earlier, occur when a thrombus develops on top of an atherosclerotic plaque; plaques, in turn, are most likely to occur at areas of turbulent blood flow (Moosey 1966). In the anterior circulation, this includes the origin of the common carotid artery, its bifurcation into the internal and external carotid arteries, the intracavernous portion of the internal carotid artery, the stem of the middle cerebral artery and its bifurcation, and the proximal portion of the ACA. In the posterior circulation, the likely areas include the origins of the subclavian arteries, the origin of the vertebral arteries, the proximal portion of the intracranial vertebral artery, the distal portion of the intracranial vertebral artery, the basilar artery, and the proximal portion of the PCA. The formation and enlargement of the atherosclerotic plaques is a slow process, extending over from months to years, and as this gradual encroachment of the arterial lumen occurs, collateral circulation forms to supply the downstream cerebral tissues that are being gradually deprived of their blood supply. The development of such collateral circulation explains why atherosclerotic occlusion of an artery may, even if complete, cause no symptoms at all. When, however, a thrombus forms on top of a cracked or ulcerated atherosclerotic plaque, the stage is set for infarction. One mechanism, as noted above, is embolization of a portion of the thrombus downstream to plug a smaller caliber artery. Another mechanism involves occlusion of the artery by the developing thrombus itself. Such thrombotic occlusion generally occurs over hours or a day or more, and thus most strokes due to thrombotic occlusion, in contrast to those due to embolism, have a relatively leisurely onset; this onset, in turn, may either be 'stuttering', with the overall worsening being interrupted by brief plateaus, or gradual and 'smooth' throughout. An exception to this rule is when hemorrhage occurs inside an atherosclerotic plaque: here, enlargement of the plaque into the lumen may occur rapidly, with occlusion occurring within minutes.

In cases of thrombotic infarction when traditional risk factors, such as hypertension, hyperlipidemia, and diabetes mellitus are absent, or when the patient is under 45 years old, it is appropriate to look for other causes of thrombus formation. These include deficiencies of the naturally occurring anti-thrombotic agents protein S, protein C, and anti-thrombin III, and mutations in factor V Leiden (Bertina *et al.* 1994; Nedeltchev *et al.* 2005). The anti-phospholipid syndrome should also be tested for. This syndrome, as discussed in Chapter 19.6, when fully developed, is characterized by a history of thrombophlebitis, miscarriage, stroke and heart attack, and by the presence of Libman–Sacks endocarditis, thrombocytopenia, and a false-positive Venereal Disease Research Laboratory (VDRL) test: laboratory testing reveals elevations of anti-cardiolipin antibodies (both immunoglobulin G [IgG] and IgM) and lupus anticoagulant. Consideration should also be given to carotid artery dissection. As noted above, such dissection may be followed by thrombus formation with embolization but, in some cases, rather than embolization, the thrombus may proceed to occlude the carotid artery itself.

Low-flow (watershed) infarctions

'Low-flow' infarctions differ from embolic and thrombotic infarctions in that the arteries directly serving the infarcted area are not, in fact, occluded.

Certain areas of the cerebrum lie at the juncture, as it were, of the terminal branches of the major arteries, and this location renders these areas quite vulnerable to any condition that reduces flow in two or more of these arteries. Perfusion pressure gradually declines as one proceeds further downstream along the many branchings of a major artery, and thus at the 'watershed' or 'border zone' which lies at these areas of juncture, pressure is very low. Consequently, whenever there is a substantial reduction of pressure upstream, the pressure in a border zone may fall below that required for tissue viability and infarction may occur. Such upstream reductions may occur with systemic hypotension (as may occur with cardiac arrest or as a side-effect of numerous drugs), or with severe atherosclerotic narrowing of the ICA (thus reducing flow in both the MCA and ACA), or a combination of these mechanisms.

Such 'border zone' infarctions have a characteristic location. Infarctions at the border zone of the PCA and MCA involve the occipitoparietal cortex and subjacent white matter. Those at the border zone of the MCA and ACA may involve either the anterolateral frontal cortex and subjacent white matter, or may occupy a more or less longitudinally oriented linear area within the middle and anterior portions of the centrum semiovale. Finally, there may be also so-called 'internal' border zone infarctions: these occur at the border zone between the penetrating branches of the MCA and the terminal portions of the lenticulostriate arteries, and are found in the deep white matter lateral to the lateral ventricle.

Lipohyalinosis

Lipohyalinosis is generally considered to occur secondary to hypertension and affects the central branches, described above, and the penetrating branches arising from the vertebral and basilar arteries. With occlusion of one of these small arteries, a correspondingly small infarction, known as a 'lacune' typically occurs, with production of one of the lacunar syndromes described earlier.

Although most lacunes occur secondary to lipohyalinosis, occlusion of central and penetrating arteries may also occur secondary to atherosclerosis (Caplan 1989) and, less commonly, to embolic occlusion. Atherosclerotic plaques, as they gradually enlarge, may slowly lap over the ostium of a penetrating artery, thus occluding this innocent bystander. Embolic occlusion of a small central or penetrating artery, though certainly at times occurring, is unusual given that most emboli are borne along in the mainstream of the large parent artery and simply do not make the midstream turn required to enter central or penetrating arteries, which generally arise at a more or less right angle to their parent artery.

INTRACEREBRAL HEMORRHAGE

When hemorrhage involves subcortical structures (e.g. putamen, caudate, or thalamus), the brainstem (typically the pons), or the cerebellum, bleeding has usually occurred secondary to rupture of a microaneurysm on one of the central or penetrating arteries (Cole and Yates 1967), which, in turn has usually developed on the basis of long-standing uncontrolled hypertension.

By contrast, when the hemorrhage is 'lobar', that is to say situated in one of the lobes of the cerebrum, a variety of causes may be at fault, including cerebral amyloid angiopathy, hemorrhage into a tumor, rupture of a vascular malformation (e.g. an arteriovenous malformation, a cavernous angioma, or a venous angioma), vasculitis, or in the setting of anticoagulant or thrombolytic treatment: patients treated with warfarin may experience lobar hemorrhages secondary to minor trauma and intracerebral hemorrhage is a feared complication of treatment with tissue plasminogen activator. Other possible causes include vasculitis or use of sympathomimetic agents, such as cocaine.

Intracerebral hemorrhages, in contrast to hemorrhagic transformation in an ischemic infarction, are space-occupying lesions. Expansion of the hematoma occurs not only due to bleeding from the ruptured vessel, but also due to shear forces on adjacent vessels, which may in turn rupture, with an expansion of the original hematoma. Furthermore, substantial edema also develops around the hematoma itself.

SUBARACHNOID HEMORRHAGE

Subarachnoid hemorrhage (van Gijn and Rinkel 2001) is most often due to rupture of a berry, or saccular, aneurysm: 30–40 percent of these are found on the anterior communicating artery; another 30–40 percent at the juncture of the posterior communicating artery with the internal carotid artery; 20 percent at the bifurcation of the middle cerebral artery; with the remainder either at the apex of the basilar artery or at the juncture of the posterior inferior cerebellar artery and the vertebral artery. Of note, multiple berry aneurysms are found in roughly one-fourth of patients.

Other causes include mycotic aneurysms, trauma, rupture of an arteriovenous malformation into the subarachnoid space, vasculitidies, and a condition known as perimesencephalic hemorrhage (van Gijn *et al.* 1985). This last entity is characterized by hemorrhage surrounding the midbrain and pons; symptoms are typically mild and it is suspected that the bleeding in this case, unlike all the other causes, is venous.

INTRAVENTRICULAR HEMORRHAGE

Most cases of intraventricular hemorrhage occur secondary to an obvious source, as, for example, extension of an intracerebral hemorrhage adjacent to a ventricle, or to reflux of blood into the fourth ventricle from a subarachnoid hemorrhage. Primary intraventricular hemorrhage, by contrast, is rare, and is thought to be generally due to bleeding from a small vascular malformation immediately subjacent to the ependyma.

CEREBRAL VENOUS THROMBOSIS

Cerebral venous thrombosis, as noted earlier, generally is seen as a complication of thrombosis of one of the dural sinuses. Such thrombosis, in turn, may be due to a number of causes (Gosk-Bierska *et al.* 2006; Khealani *et al.* 2008), such as Behçet's syndrome, the anti-phospholipid syndrome, deficiencies of anti-thrombin III or of proteins S or C, systemic lupus erythematosus, the puerperium, paroxysmal nocturnal hemoglobinuria, in association with certain malignancies, during treatment with oral contraceptives, and in association with certain infections: otitis or mastoiditis may lead to thrombosis of the transverse sinus, and facial or sinus infection to thrombosis of the cavernous sinus.

OTHER CAUSES OF STROKE

Other rare or relatively uncommon causes of stroke include CADASIL, moyamoya disease, vasculitidies (e.g. cranial arteritis and systemic lupus erythematosus), meningovascular syphilis, the multiple cholesterol emboli syndrome, zoster ophthalmicus, complicated migraine, sickle cell disease, and subclavian steal. CADASIL (Chapter 11.7) is suggested by a history of migraine and by the presence of leukaraiosis in the external capsule and temporal lobe. Moyamoya disease (Chiu *et al.* 1998; Mineharu *et al.* 2006, Scott and Smith 2009) is characterized by stenosis of the terminal portions of both internal carotid arteries with development of a large number of collateral vessels from the circle of Willis: the name of this disorder is Japanese for 'puff of smoke', which is the appearance generated by these numerous collaterals as seen on arteriography. Clinically, patients may present with either subcortical infarction or subcortical intracerebral hemorrhage. Moyamoya disease may be either primary or secondary to such conditions as Down's syndrome, von Recklinghausen's disease, or irradiation. Cranial arteritis (Chapter 11.5) is suggested by concurrent headache and an elevated erythrocyte sedimentation rate (ESR), and systemic lupus erythematosus (Chapter 19.3) by associated symptoms such as arthralgia and rash. Other vasculitidies to consider include Sjögren's syndrome (Chapter 19.4), suggested by dryness of the eyes and mouth, and Sneddon's syndrome (Chapter 19.5), heralded by livedo reticularis. Meningovascular syphilis (Chapter 15.15) may be characterized by both large vessel and lacunar infarctions; though often accompanied by the Argyll Robertson pupil; this is not invariable, and hence obtaining a fluorescent treponemal antibody test is suggested. The multiple cholesterol emboli syndrome (Chapter 11.17) is characterized by multiple infarctions or merely TIAs: when due to emboli arising from the ascending aorta, there may be accompanying evidence of systemic embolization, such as renal failure, livedo reticularis, or petechial hemorrhages in the skin; when the emboli arise from a plaque in the carotid system, however, the only manifestation may be stroke. Zoster ophthalmicus (Chapter 15.9) may be followed by stroke secondary to retrograde spread of the virus to the ICA or MCA with subsequent vasculitis; importantly, the zoster ophthalmicus may have resolved weeks to months earlier. Complicated migraine is suggested by the appearance of stroke in the setting of a migraine headache. Sickle cell disease is characterized by anemia and a history of abdominal crises. The subclavian steal syndrome (Hennerici *et al.* 1988) occurs secondary to stenosis of the subclavian artery proximal to the origin of the vertebral artery on either the left, or, less commonly, the right side. With flow through the subclavian artery reduced, any exercise of the ipsilateral arm, with its concomitant increased blood demand, will 'steal' blood from the vertebral artery and thus reduce flow to the brainstem. Symptoms include 'claudication' of the ipsilateral arm with exercise-induced aching, and symptoms suggestive of brainstem TIA, such as vertigo, ataxia, or diplopia: actual brainstem infarction is rare. The clue to the diagnosis is found by taking the radial pulse on both arms simultaneously: on the affected side the pulse will be delayed and reduced.

SUGGESTED WORKUP

In the initial evaluation of a patient with acute stroke, one of the primary goals is to determine the cause of the stroke and to do this rapidly. Most strokes are secondary to ischemic infarction: ischemic infarction in the area of distribution of one of the major pial vessels (either embolic, or, less commonly, thrombotic in nature) is most common, followed by lacunar infarctions and then watershed infarctions. Intracerebral hemorrhage is the next most common cause of stroke, followed by subarachnoid hemorrhage, intraventricular hemorrhage, and cerebral venous thrombosis.

Both the mode of onset and the presence or absence of severe headache help to differentiate the various causes of stroke. Strokes due to ischemic infarction may be either acute or gradual in onset but are generally not accompanied by severe headache. Embolic infarctions are generally of acute onset, with the maximal clinical deficit being reached rapidly, over minutes; by contrast thrombotic infarctions are typically of gradual or 'stuttering' onset, over hours or longer. Strokes due to intracerebral hemorrhage are relatively gradual in onset and are generally accompanied by significant headache. Strokes due to subarachnoid hemorrhage are of very acute onset, sometimes over seconds, and are typified by very severe headache; intraventricular hemorrhage shares these characteristics. Stroke due to cerebral venous thrombosis is generally of leisurely onset, over days or longer, and is generally accompanied by a more or less severe headache.

Although these clinical features are helpful, exceptions are not uncommon to these guidelines, and hence imaging is indispensable. Although MR scanning with DWI is far more sensitive than CT scanning (Chalela *et al.* 2007; Lansberg *et al.* 2000), CT scanning is generally utilized first, not only because of its speed and ready availability but also because it is at least as sensitive as is MRI for the detection of the blood characteristic of hemorrhage transformation of an ischemic infarction, intracerebral hemorrhage, subarachnoid hemorrhage, intraventricular hemorrhage, or cerebral venous thrombosis.

In ischemic infarction, the CT scan is generally normal early on. By 1 hour, some indistinctness of the gray–white boundary may occur, and in patients with occlusion of the MCA a 'hyperdense MCA sign' may be seen in about one-half of cases. Beginning 6 hours after onset, an increasing proportion of cases will demonstrate radiolucency in the area of the infarction, and up to 50 percent of cases will demonstrate this by 12 hours. If hemorrhagic transformation occurs one may see either stippled areas of radiodensity in the area of infarction, representing petechial hemorrhages, or, if the infarction is large, a confluence of these into a hematoma.

If MR scanning is performed, at a minimum T2, fluid-attenuated inversion recovery (FLAIR) and DWI sequences with apparent diffusion coefficient (ADC) mapping should be obtained. Diffusion-weighted imaging may indicate an area of ischemic infarction within minutes of the event (Hjort *et al.* 2005) and tends to remain positive for 7–14 days; T2 and FLAIR images gradually become positive and then remain so chronically. If cerebral venous thrombosis is suspected, MR venography should be ordered.

Many hospitals also offer either perfusion CT or perfusion MRI, and these procedures may be very useful in cases of ischemic infarction. In such cases, the ischemic tissue may be divided into an ischemic 'core', which is destined to undergo necrosis no matter what treatment is offered, and an ischemic 'penumbra', consisting of 'stunned' tissue, which, although not functional, is capable of recovery. Identifying such a penumbra is of great import, for it invites therapies designed to restore circulation and thus salvage the stunned tissue, and such an identification is enabled by the use of either perfusion CT or perfusion MRI (Schramm *et al.* 2004; Wintermark *et al.* 2002a,b). In this regard, perfusion CT may have an advantage as it takes only about 15 minutes and can be performed in the emergency room.

Once imaging has been accomplished, and a decision has been made whether treatment with tissue plasminogen activator (discussed below) should be offered, further laboratory evaluation is undertaken to determine the mechanism of the stroke.

In cases of ischemic infarction, the following tests should be considered: Holter monitoring, echocardiography, duplex Doppler studies of the carotid arteries, and either MR angiography or CT angiography of both the neck and the head. Holter monitoring is required to detect intermittent atrial fibrillation. Echocardiography may be either transthoracic (TTE) or transesophageal (TEE): although TEE is invasive, it is superior to TTE for imaging the left atrium, left atrial appendage, any atrial septal defects, and the aortic arch (Daniel and Mugge 1995; Sen *et al.* 2004). Duplex Doppler studies identify plaques and stenotic lesions of the carotid arteries; MR angiography and CT angiography are likewise useful in this regard and also allow imaging of the larger intracranial vessels. If an embolic source is present, it is usually demonstrated by one or more of these tests. There are, however, exceptions. In some cases, the entire emboligenic lesion, for example a small atrial thrombus, may undergo embolization, leaving nothing behind to detect. Hence, some clinical judgment is required in interpreting these tests. For example, as noted earlier, a stroke which is 'maximal' at the onset is likely embolic in nature. Furthermore, ischemic infarction in one of the distal branches of the cerebral arteries is also likely embolic: thrombotic infarctions usually occur at areas of atherosclerotic plaque formation, which, as noted earlier, are generally in the more proximal portions of the arterial tree; by contrast, smaller emboli may readily pass distally to occlude an artery further downstream. Finally, if there is more than one acute infarction, and these infarctions are in different arterial territories, then an embolic mechanism is a more likely explanation (Baird *et al.* 2000): multiple emboli, say, from a cardiac source, may course up different arteries, whereas it is unlikely that multiple different stenotic arteries would simultaneously give off emboli.

In thrombotic infarction, the underlying etiology is usually atherosclerosis, and one typically finds evidence of hyperlipidemia, diabetes mellitus, hypertension, or smoking. When these are absent, or the patient is under 45 years old, other causes must be considered, as discussed earlier, and consideration should be given to the following tests: protein S, protein C, anti-thrombin III, factor V Leiden, anti-cardiolipin antibodies (IgG and IgM), and lupus anticoagulant. The anti-phospholipid syndrome may occur on an idiopathic basis or be secondary to systemic lupus erythematosus and, consequently, in the presence of anti-cardiolipin antibodies or lupus anticoagulant, an antinuclear antibody test should also be ordered.

In patients with intracerebral hemorrhage, re-imaging with MRI (including a T2 gradient-echo sequence to detect evidence of old petechial hemorrhages in cases of suspected cerebral amyloid angiopathy) should be undertaken after a few weeks have passed, by which time enough blood will have been resorbed to allow the detection of most of the possible underlying causes.

Patients with spontaneous subarachnoid hemorrhage require arteriography to demonstrate the source of the arterial bleeding.

Differential diagnosis

Ischemic infarction may be mimicked by multiple sclerosis or a Bell's palsy. Multiple sclerosis is distinguished by imaging and CSF findings. Bell's palsy may suggest a 'pure motor' lacunar syndrome but once one recognizes that the forehead is involved, it becomes obvious that the lesion involves the facial nerve.

Intracerebral hemorrhage may be mimicked by complicated migraine; however, in complicated migraine the headache is usually delayed for from 30 to 60 minutes after the onset of focal signs, whereas in intracerebral hemorrhage, the headache evolves more or less simultaneously with focal signs. Epidural or acute subdural hematomas may likewise mimic an intracerebral hemorrhage, and, in the absence of a history of trauma the diagnosis, may depend on imaging. Hypertensive encephalopathy may closely mimic intracerebral hemorrhage, with headache, nausea and vomiting, and seizures. Finding a grossly elevated blood pressure may or may not be helpful here, as this may be common to both conditions. Delirium and visual loss favor hypertensive encephalopathy; however, here the diagnosis often depends on imaging: although there may be petechial hemorrhages in hypertensive encephalopathy, one does not see the large, well-circumscribed, dense hematoma characteristic of intracerebral hemorrhage.

Subarachnoid and primary intraventricular hemorrhage, with the associated hyperacute onset of 'worst-ever' headache, are rarely imitated by any other disorder. Both meningitis and severe migraine might be considered, but imaging will quickly resolve the issue.

Cerebral venous thrombosis may be mimicked by subacute subdural hematoma, a brain tumor (e.g. glioblastoma multiforme) or a cerebral abscess, with, again, imaging resolving the question.

Transient ischemic attacks may be mimicked, with varying degrees of faithfulness, by partial seizures, 'transient tumor attacks', transient global amnesia, and either hyper- or hypoglycemia. Inhibitory motor simple partial seizures are suggested by their exquisitely paroxysmal onset, over seconds, and by their association with other seizure types. A postictal Todd's paralysis may also enter the differential but this is immediately suggested by the history of the preceding seizure. 'Transient tumor attacks' occur in association with cerebral tumors, which are immediately apparent on imaging. Transient global amnesia (TGA) is, relative to a TIA, longlasting; furthermore, there is no heminanopia or blindness in TGA, findings that would be expected in a TIA occurring secondary to ischemia in the area of distribution of the PCAs. Finally, hyperglycemia (as may be seen in hyperosmolar non-ketotic hyperglycemia) and hypoglycemia may both present with focal signs, which resolve promptly with restoration of euglycemia.

Treatment

This section will focus first on certain aspects of the *acute treatment* of *ischemic infarction*, *intracerebral hemorrhage*, *subarachnoid hemorrhage*, *intraventricular hemorrhage*, and *cerebral venous thrombosis*, followed by certain recommended *routine measures* appropriate in most cases of stroke. It should be emphasized that the acute treatment of stroke is generally best accomplished in a specialized stroke unit.

ACUTE TREATMENT

Ischemic infarction

The acute treatment of patients with ischemic infarction may involve the use of tissue-type plasminogen activator (t-PA), also known as alteplase. Normally, plasminogen is converted by endogenous tissue plasminogen activator to plasmin, which in turn is a fibrinolytic enzyme. The administration of t-PA, by increasing fibrinolysis, may dissolve a clot (whether part of an embolus or occurring on an atherosclerotic plaque in a large vessel), thus restoring flow to cerebral tissue, which, although 'stunned' by ischemia, has not as yet undergone necrosis. Acute ischemic infarctions, as noted earlier, consist of a 'core' of tissue that has been irreversibly damaged and a surrounding 'ischemic penumbra' of tissue, which, although 'stunned' and not functioning, may still survive if blood flow is promptly restored. The window of opportunity for restoration of blood flow is narrow, measured in hours, and thus decisions must be made rapidly. Current guidelines set the outside limit for treatment with t-PA at 3 hours post onset; however, this may not be appropriate in all cases: perfusion MRI studies have demonstrated persistence of the ischemic penumbra past 3 hours (Albers *et al.* 2006; Parsons *et al.* 2002), and there is evidence that t-PA administration up to 4.5 hours after stroke onset may be beneficial (Hacke *et al.* 2008).

Even in cases of ischemic infarction when an ischemic penumbra is present not all patients should receive t-PA, given the risk of causing an intracerebral hemorrhage. Contraindications to t-PA include: a history, within the preceding 3 months, of stroke, head trauma, or myocardial infarction; major surgery within 2 weeks; a history of gastrointestinal or genitourinary bleeding within 3 weeks; a history of intracerebral hemorrhage at any time; thrombocytopenia (with a platelet count below 100 000), anticoagulation (with an international normalized ratio >1.7), or use of heparin within 48 hours (and a prolonged activated partial thromboplastin time); grossly elevated blood pressure (systolic over 185 mmHg or diastolic over 110 mmHg); and, finally, evidence of a large infarct: large infarcts, for example infarcts encompassing the entire area of distribution of the MCA, are more likely to undergo hemorrhagic transformation and thus treatment may do more harm than good.

In cases of embolic infarction when systemic t-PA fails, consideration may be given to local, intra-arterial thrombolysis or to mechanical embolectomy (Gobin *et al.* 2004).

Patients should also be given aspirin in a dose of 325 mg for the first 2 weeks, as this reduces the risk of recurrent stroke within that time frame (Chen *et al.* 2000): if t-PA is not given, aspirin should be given immediately; in cases where t-PA is utilized, aspirin may be started after 24 hours.

Some authors advocate the use of heparin in cases of thrombotic infarction, in the hope of preventing propagation of the offending thrombus. As yet, however, there is no convincing evidence for the effectiveness of heparin, and the risks attendant on its use argue against this practice.

Most patients with ischemic infarction will also have hypertension, and it cannot be stressed enough that rigid control of blood pressure is *not* indicated in the acute phase of stroke treatment (Caplan 1976). The risks of neuronal ischemia secondary to systemic hypotension are simply too great; indeed, in cases of watershed infarction, a case may be made for allowing the pressure to run a little high. Recent work has also indicated that in acute patients, lying flat in bed, as compared to sitting up, not only improves flow within the MCA territory, but may also be followed, in some cases, by prompt clinical improvement (Wojner-Alexandrov *et al.* 2005).

Obstructive sleep apnea is a common condition, and recent work indicates that obstructive apneas are associated with early clinical worsening (Iranzo *et al.* 2002). Given this, it appears appropriate to ensure, if possible, that all patients with obstructive sleep apnea receive appropriate treatment.

Once acute treatment is accomplished, preventive treatment should be instituted: in addition to control of risk factors such as diabetes mellitus, hypertension, hyperlipidemia, and to smoking cessation, consideration may be given to secondary stroke prevention with either warfarin or antiplatelet agents. Warfarin is indicated in cases of embolic infarction secondary to atrial fibrillation, atrial or ventricular thrombi, cardiomyopathy, and mechanical prosthetic valves. In other cases, or when warfarin is contraindicated, an antiplatelet agent should be utilized: these include aspirin, clopidogrel, and a time-release combination preparation of aspirin and dipyridamole. In this regard, it should be noted that clopridogrel is superior to aspirin alone (Caprie Steering Committee 1996), the time-release preparation of aspirin and dipyridamole is also superior to aspirin alone (Halkes *et al.* 2008), and that clopridogel is of equal efficacy with the time-release preparation of aspirin and dipyridamole (Sacco *et al.* 2008). Given these data, it appears reasonable, rather than using aspirin alone, to move immediately to either clopridogrel or the time-release preparation of aspirin and dipyridamole; the use of clopridogrel in combination with aspirin, however, should be avoided as it offers little advantage over using clopridogrel alone and does increase the risk of bleeding (Diener *et al.* 2004). When aspirin is employed, either alone or in the combination preparation of aspirin and dipyridamole, it is important, if at all possible, to avoid any interruption in therapy, as this may be followed by a 'rebound' *increased* risk of stroke over the following weeks (Maulaz *et al.* 2005); whether the same phenomenon occurs after discontinuation of clopidogrel is not known. Furthermore, when aspirin is used, it is important to avoid treatment with ibuprofen, which inhibits aspirin's antiplatelet effect (Catella-Lawson *et al.* 2001).

In cases when carotid artery stenosis is present at greater that 70 percent, consideration may be given to carotid endarterectomy (Chaturvedi *et al.* 2005) or carotid stenting (Yadav *et al.* 2004).

Intracerebral hemorrhage

As yet there are no specific treatments for intracerebral hemorrhage: recent work suggests a role for recombinant activated factor VII to reduce the growth of the hemorrhage (Mayer *et al.* 2005); however, this has not as yet been approved for clinical use. In cases in which the hemorrhage is causing significant herniation or compressing critical structures, treatment with dexamethasone, mannitol, or furosemide may be indicated. In emergency situations, surgery may be required to remove the clot. There is debate over whether surgery is safe in cases of cerebral amyloid angiopathy, given the widespread vascular fragility; however, on balance, even here surgery may be worth the risk. The rupture of a hemorrhage into a ventricle may be followed by hydrocephalus, and emergency ventricular drainage may be required. In cases where there is doubt as to the etiology of the bleed, as for example may occur with lobar hemorrhages, follow-up MRI is indicated to detect underlying neoplasms or vascular malformations.

Subarachnoid hemorrhage

Patients with subarachnoid hemorrhage should be treated in an intensive care unit (ICU). Various measures are undertaken (Suarez *et al.* 2006): benzodiazepines (e.g. lorazepam) are given for agitation, a 3-week course of oral nimodipine to reduce vasospasm, an AED for seizure prophylaxis, and serial transcranial Doppler studies are undertaken to monitor the development of vasospasm: should this appear, consideration is given both to 'triple-H' therapy (hypertension, hypervolemia, hemodilution) to facilitate flow through constricted arteries, and, in resistant cases, to either intra-arterial verapamil or to angioplasty. Serial CT scans are also indicated to monitor for the development of hydrocephalus, and external ventricular drainage may be required. In cases of persistent, unexplained coma, consideration should be given to an EEG, given the prevalence of electrographic seizure activity noted in such cases (Claasen *et al.* 2005).

Arteriography is generally utilized to identify the bleeding source, and when, as is usually the case, an aneurysm is identified, patients are considered for either endovascular coiling (International Subarachnoid Aneurysm Trial [ISAT] Collaborative Group 2002) or neurosurgical clipping.

Intraventricular hemorrhage

As with secondary intraventricular hemorrhage (e.g. with extension of an intracerebral hemorrhage into a ventricle), primary intraventricular hemorrhage, by producing acute hydrocephalus, may require external ventricular drainage.

Cerebral venous thrombosis

For cerebral venous thrombosis, there is debate over whether patients should be treated with heparin. Although increased hemorrhage is a risk, it appears that, in this case, the benefits obtained by preventing thrombus propagation may outweigh it. Increased intracranial pressure, as is often seen with thrombosis of the superior sagittal sinus, may require treatment with dexamethasone and mannitol.

Routine measures

Clinicians should be on the alert for urinary tract infections, deep venous thrombosis, decubiti and, especially, aspiration pneumonia: dysphagia is very common after stroke, and, although bedside testing (e.g. by simply observing a patient swallow water) is helpful in detecting aspiration, videofluorscopic examination is far more sensitive.

A close watch should be kept for any clinical deterioration. Regardless of the etiology of the stroke, worsening may occur secondary to toxic factors (e.g. injudicious use of opioids or benzodiazepines) or metabolic derangements (e.g. systemic effects of infections, electrolyte abnormalities). In the case of infarction or intracerebral hemorrhage, worsening may also occur secondary to the development of vasogenic edema, a process that gradually evolves, peaking at days 2 through 4 and then resolving over the next week or two. In cases of infarction, new strokes may occur secondary to recurrent emboli or to propagation of a thrombus. In cases of intracerebral hemorrhage where there has been intraventricular extension, and in cases of subarachnoid hemorrhage, hydrocephalus may develop. Finally, cases of subarachnoid hemorrhage may also display worsening due to rebleeding or vasospastic infarction.

Treatment with an AED may or may not be required. As discussed in Chapter 8.3, seizures may be immediate, early (within the first 2 weeks) or late (from 2 weeks to 2 years). Early and immediate seizures occur in less than 3 percent of patients with ischemic infarction, but are seen in roughly 25 percent of patients with intracerebral hemorrhage: in both infarction and intracerebral hemorrhage, seizures generally occur only in cases with cortical involvement and are rare in those with lacunar infarctions or hemorrhages confined to subcortical structures. In the case of subarachnoid hemorrhage, early and immediate seizures may be seen in roughly 25 percent of patients, and in cerebral venous thrombosis, about 15 percent. The EEG may help in predicting seizures: PLEDs and focal spikes are strongly associated with their development (Gupta *et al.* 1988; Holmes 1980). In the case of ischemic infarction, treatment with an AED is generally withheld until a seizure occurs, but in cases of intracerebral hemorrhage with cortical involvement or subarachnoid hemorrhage or cerebral venous thrombosis the risk is so high that prophylactic treatment is generally given for the first 2 weeks. Should a seizure occur after stroke, it is not clear how long treatment should continue: a prudent course would be to continue treatment until 2 years had passed without seizure. Which AED to choose is currently unclear: phenytoin has been the traditional choice, however the difficulties attendant on its use has caused it to lose some of its luster in favor of newer agents, such as levetiracetam.

Once patients are medically stable, rehabilitation measures, including, where appropriate, occupational therapy, physical therapy, and speech/language therapy, should be instituted, and consideration given to transfer to a specialized rehabilitation unit. Although the pharmacologic treatment of stroke sequelae is in its infancy, double-blinded studies regarding aphasia have demonstrated effectiveness for donepezil (Berthier *et al.* 2006), memantine (Berthier *et al.* 2009), levodopa (Seniow *et al.* 2009), and, piracetam, which, with three studies behind it (Enderby *et al.* 1994, Huber *et al.* 1997, Kessler *et al.* 2000), has the strongest support.

8.5 TRAUMATIC BRAIN INJURY

Traumatic brain injury (also known as closed head injury) occurs in the USA with a yearly incidence of approximately 150/100 000. Two age peaks are found, the first between 15 and 24 years of age, wherein motor vehicle accidents are the most common cause, and the second over the age of 64 years, wherein falls are most common. Males are more commonly affected than females at all ages, and alcohol intoxication is a very common factor, being found in from one-third to one-half of all cases (Corrigan 1995). Although multiple forms of brain injury may occur, this syndrome refers primarily to cases occurring with sudden 'acceleration–deceleration', as during a motor vehicle accident or a fall or blow to the head; trauma due to gunshot wounds and crush injuries are not considered here.

This section deals primarily with what is known as moderate or severe TBI: mild TBI, otherwise known as concussion, is dealt with in Chapter 12.4. In considering the clinical features (and their treatment) of TBI, it is convenient to divide them into two groups, namely an acute phase and a chronic phase. The acute phase, from a neuropsychiatric point of view, is often dominated by a delirium; as the confusion clears, patients gradually enter into the chronic phase, which in turn may be characterized by numerous sequelae, including cognitive deficits that may, at times, be severe enough to constitute a dementia.

Acute phase

Essentially all patients with significant TBI will initially be comatose. Among those who do survive, some will emerge from coma into a persistent vegetative or minimally conscious state (discussed in Chapter 6.2) while the remainder emerge into a delirium. This delirium, in addition to such characteristic symptoms as confusion, disorientation, and decreased short-term memory, is also often marked by hallucinations, delusions, and, especially, agitation, which is seen in the majority of cases (Rao and Lyketsos 2000; van der Naalt *et al.* 2000).

It must be borne in mind that, although the delirium in such cases is generally due to the intracranial injuries directly caused by the trauma, other factors may also be involved as the hospitalization proceeds. Toxicity from such medications as opioids, baclofen, anticholinergics, metoclopramide, and even amantadine must be considered, along with metabolic factors, such as hyponatremia, hypoglycemia, hypomagnesemia, and systemic effects of infections, such as pneumonia. Other metabolic factors to consider relate to the frequency with which alcoholism is involved, and include Wernicke's encephalopathy and delirium tremens. Consideration must also be given to the effects of global hypoxic–ischemic injury: this is very common and occurs secondary to arrhythmias, hypotension and pulmonary lesions. In those with fractures of long bones, consideration must also be given to the fat embolism syndrome.

Treatment of patients during the acute phase is, at least initially, generally undertaken in either a trauma unit or an ICU. Neurosurgical treatment may be required for evacuation of epidural or subdural hematomas or large contusions or intracerebral hemorrhages, and serial CT scans are obtained. In comatose patients, intracranial pressure monitoring is often indicated, and treatment with intravenous sedation, mannitol, and other agents may be required to reduce elevated intracranial pressure. Intubation is almost universally required, and pneumonia is a frequent occurrence. In some cases, continuous EEG monitoring may be required to detect non-convulsive status epilepticus.

Treatment of delirium, in all cases, involves simple environmental measures designed to reduce confusion. These include, whenever possible, having the patient in a quiet room, with a window. Large calendars and digital clocks should be in full view, and the nurse's call button should be readily available. Sleep is essential and consequently the room should be darkened and very quiet at night, and all non-emergency procedures (e.g. vital signs, weights, baths, laboratory testing) should be forbidden during the sleeping hours. Restraints may occasionally be required and can be life-saving. For patients prone to get out of bed unsupervised, keeping the bedrails up may be sufficient; if not, or if these are impractical, utilizing a 'low-boy' bed, surrounded by mats, may help prevent injury. In some cases, round-the-clock sitters may be required.

In cases in which these environmental measures are ineffective, pharmacologic treatment may be considered with either an antipsychotic or, in certain emergent cases, lorazepam. Antipsychotics are indicated for treatment of hallucinations or delusions, and are also effective for agitation. A second-generation agent, such as risperidone, is often used, and in practice quetiapine and olanzapine are also utilized. Initial doses should generally be low, for example 0.5–1 mg of risperidone, 12.5–25 mg of quetiapine, and 2.5–5 mg of olanzapine. The first-generation agent haloperidol is also often used, with initial doses of 2–5 mg. Repeat doses, in approximately similar milligram amounts, may then be given every hour or so until the patient is calm, limiting side-effects occur, or a maximum dose is reached: rough guidelines for dose maxima are 5 mg for risperidone, 300 mg for quetiapine, 20 mg for olanzapine, and 20 mg for haloperidol. In cases in which the patient responds satisfactorily, a regular daily dose is ordered for the next day (with the total daily dose approximately equivalent to the total required initially), divided into two or three doses. Provision is also made for further as-needed doses, with the total daily dose being adjusted according to the amount needed in as-needed doses. The eventual maintenance dose is then continued until the patient has been stable for a significant period of time, at which point it may be gradually tapered. Lorazepam is very commonly used, and, given the rapidity of its effectiveness when given intravenously, has a place in emergency situations; however, given that lorazepam may also worsen confusion, it is appropriate to substitute another agent as soon as this is practical. If lorazepam is used, one may give anywhere from 0.5 to 2 mg intravenously every hour as needed until the patient is calm, limiting side-effects (such as sedation) occur, or a maximum of approximately 12 mg is reached.

Once patients have been stabilized, general rehabilitation efforts may be started, including physical, speech, and occupational therapy. Eventually, most patients are transferred to a specialized rehabilitation facility, where these general efforts are continued.

Two rating scales have come into widespread use in the evaluation of patients with TBI: the Glasgow Coma Scale and the Rancho Los Amigos Cognitive Scale. The Glasgow Coma Scale (Teasdale and Jennett 1974) is designed for evaluating patients in the acute phase, and involves assessing three clinical features: eye opening, motor response, and verbal response, with, as noted in Table 8.8, a numerical score described for the various responses. Patients with total scores of ≤8 are said to have a severe injury, those with scores from 9 to 12, a moderate injury, and those with scores of from 13 to 15, a mild injury. The Rancho Los Amigos scale, described in Table 8.9 may also be utilized early on; however, it is often also employed as an instrument for following patients during the chronic phase. Although these scales are useful as a 'shorthand', they cannot and should not take the place of a detailed narrative description of the patient's clinical condition.

Post-traumatic seizures may occur during the acute phase, and these are discussed further below.

Table 8.8 The Glasgow Coma Scale

Eye opening	
Spontaneous	4
To speech	3
To pain	2
Nil	1
Best motor response	
Obeys commands	6
Localizes	5
Withdraws	4
Abnormal flexion	3
Extensor response	2
Nil	1
Verbal response	
Oriented	5
Confused conversation	4
Inappropriate words	3
Incomprehensible sounds	2
Nil	1

Chronic phase

As the delirium gradually clears, almost all patients will be left with one or more chronic sequelae (Rao and Lyketos 2000), and these are discussed below, beginning with cognitive deficits, which are almost universal.

COGNITIVE DEFICITS

Cognitive deficits (Levin *et al.* 1988; van Zomeren and van den Burg 1985) typically remain after the confusion of delirium clears, and these include slowed information processing, poor attention, poor concentration, poor short-term memory, impaired ability to plan and organize the day, and decreased abstracting ability. In some cases these may be quite mild and not terribly limiting; however, in others they amount to a clear, and disabling, dementia. Most patients show improvement over the first 6 months, with some further, but not as impressive, gains over the next 6 months: however, after 12 months, little further spontaneous recovery can be expected. Importantly, in assessing patients with cognitive deficits it is critical to check for the presence of depression, which, in and of itself, may cause cognitive impairment.

Table 8.9 The Ranchos Los Amigos Scale

I	*No response*: unresponsive to any stimulus
II	*Generalized response*: limited, inconsistent, and non-purposeful responses – often to pain only
III	*Localized response*: purposeful responses; may follow simple commands; may focus on presented object
IV	*Confused, agitated*: heightened state of activity; confusion and disorientation; aggressive behavior; unable to perform self-care; unaware of present events; agitation appears related to internal confusion
V	*Confused, inappropriate*: non-agitated, appears alert; responds to commands; distractible; does not concentrate on task; agitated responses to external stimuli; verbally inappropriate; does not learn new information
VI	*Confused, appropriate*: goal-directed behavior, needs cueing; can relearn old skills as activities of daily living; serious memory problems, some awareness of self and others
VII	*Automatic, appropriate*: appears appropriately oriented; frequently robot-like in daily routine; minimal or absent confusion; shallow recall; increased awareness of self and interaction with environment; lacks insight into condition; decreased judgment and problem solving; lacks realistic planning for future
VIII	*Purposeful, appropriate*: alert and oriented; recalls and integrates past events; learns new activities and can continue without supervision; independent in home and living skills; capable of driving; defects in stress tolerance, judgment and abstract reasoning persist; may function at reduced levels in society

Reprinted with permission from the Adult Brain Injury Service of the Rancho Los Amigos Medical Center, Downey, CA.

Various agents have been subjected to double-blinded studies, including methylphenidate, the cholinesterase inhibitors donepezil and rivastigmine, and amantadine. Methylphenidate is generally given in a final dose of 0.6 mg/kg (divided into a twice-a-day schedule): speed of information processing was evaluated in four studies, and three (Whyte *et al.* 1997, 2004; Wilmott and Ponsford 2009) found improvement, whereas one (Speech *et al.* 1993) did not; attention was likewise evaluated in four studies, and two (Plenger *et al.* 1996; Whyte *et al.* 2004) found improvement, one (Gualtieri and Evans 1988) a trend towards improvement, while in the last study (Speech *et al.* 1993) there was no improvement. Donepezil, in a dose of 10 mg, in one study (Zhang *et al.* 2004) improved both attention and memory. Rivastigmine has been evaluated in two studies: in the first (Silver *et al.* 2006), a dose of 3–6 mg, although failing to yield any improvement in the sample overall, did, in a secondary analysis, improve both attention and memory in those with moderate to severe memory impairment; in the second study (Tenovuo *et al.* 2009), utilizing the highest tolerated dose up to a maximum of 12 mg (average 8 mg), both speed of information processing and attention improved. Amantadine is given in a dose of 100 mg twice daily, and has been evaluated in two studies: in one (Meythaler *et al.* 2002) there was overall cognitive improvement as measured by the Folstein mini-mental status exam; in the other (Schneider *et al.* 1999), however, there was neither any overall improvement, nor was there any improvement on specific measures of attention or memory.

POST-TRAUMATIC AMNESIA

The period of post-traumatic amnesia spans, at a minimum, the period of time from the injury itself up to that point where the patient is able to again 'lay down' new memories, as evidenced by an ability to recall most events that transpired the day before being tested, or by the ability to recall three out of three words after 5 minutes: this latter criterion, of course, must be utilized with care, given that some deficit in short-term memory may persist indefinitely as part of the cognitive deficits just discussed. Thus defined, post-traumatic amnesia represents primarily an anterograde amnestic disturbance. However, in most cases there is also an associated retrograde amnesia, of variable extent: thus, patients typically are unable to recall events that occurred prior to the accident, and this 'blanket' of retrograde amnesia may encompass anywhere from the mere hours preceding the injury, up to days, weeks, or longer. Characteristically, this retrograde 'blanket', however, gradually 'shrinks' in size over time, such that most patients, although still left with the anterograde component, are eventually able to recall most events prior to the injury, even, in some cases, up to minutes before the injury itself.

ANOSOGNOSIA

Anosognosia is characterized by a failure to appreciate the severity of a deficit, or even its existence. Clinically significant anosognosia is found as a persistent symptom in almost one-half of patients (Flashman and McAllister 2002). Interestingly, the anosognosia appears selective, in that, although patients tend to acknowledge such deficits as hemiplegia, they are much less likely to appreciate their cognitive deficits (Sbordone *et al.* 1998) or the existence of mood changes or a personality change (Fahy *et al.* 1967). Distinguishing anosognosia from denial may be difficult. One helpful point is the affect accompanying the patient's response to questions about deficits: in anosognosia the affect is often bland or matter of fact, whereas in denial there may be a degree of indignation or angry protest.

Importantly, given the high prevalence of anosognosia, it is critical, in the long-term treatment of these patients, to always question others, such as family members or friends, regarding the patient's condition. Relying on the anosognosic patient's report could lead to disastrous underdiagnosing.

FOCAL SIGNS AND SYMPTOMS

Focal signs and symptoms that persist after resolution of the delirium may include not only hemiplegia and various movement disorders, but also 'cortical' signs and symptoms such as neglect and aphasia. Aphasia is particularly common, and, to a variable degree, is found in the vast majority of cases (Levin *et al.* 1976; Sarno *et al.* 1986).

AGITATION

Agitation is common, tends to fluctuate in severity, and may occur in up to two-thirds of all patients in the first few months (Nott *et al.* 2006); associated aggression may be seen in roughly one-fourth of patients (Baguley *et al.* 2006). In evaluating agitated patients, consideration must be given to the possibility that the agitation in question is not directly due to the brain injury but is rather secondary to other causes, such as pain, delusions of persecution, or akathisia.

The treatment of agitation is discussed in detail in Chapter 7.4.

SYMPATHETIC STORMS

Sympathetic storms, as described in Chapter 5.2, consist of episodes characterized by profuse diaphoresis, tachycardia, tachypnea, pupillary dilation, and, in some, rigid extensor posturing. The diaphoresis is indeed impressive, with beads of sweat dripping from the head. During the episode, patients may grimace as if in pain, and family members and other observers may become quite alarmed. The episodes themselves last from minutes to hours, and terminate slowly.

The distinction of these 'storms' from agitation is generally based on their course: the 'storms' come in discrete episodes, whereas agitation, although perhaps waxing and waning in intensity, usually does not assume an episodic course. In addition, sympathetic symptoms, such as impressive diaphoresis, although typical of sympathetic storms, are generally absent, or relatively minimal, during agitation.

Treatment with propranolol or bromocriptine usually prevents further attacks; alternatives include gabapentin and morphine.

PERSONALITY CHANGE

Personality change is very common, and indeed is found in up to two-thirds of all cases over long-term follow-up (Thomsen 1984). The change involved is reminiscent of a frontal lobe syndrome, with childish and socially inappropriate behavior and disinhibition, which may lead to sexual improprieties. There may be restlessness, emotional lability and impulsivity, and anger, and in disinhibited patients argumentativeness and assaultive behavior may occur. Some patients, however, may become withdrawn, listless, and apathetic.

Before making this diagnosis, due regard must be paid to the patient's premorbid personality, with special attention to the presence of either an antisocial or a borderline personality disorder.

The pharmacologic treatment of impulsivity, anger and assaultiveness is not well worked out. One single-blinded study (Mooney and Haas 1993) of young males found that methylphenidate in a dose of 30 mg/day was superior to placebo for anger. In practice, many such patients are treated with an antipsychotic, such as quetiapine, or an AED, such as carbamazepine.

DEPRESSION

Depression appears during the first 2 years in up to one-half of patients, and may either remit spontaneously or persist (Jorge *et al.* 1993, 2004); of the typical symptoms of depression (i.e. depressed mood, fatigue, loss of interest, loss of appetite, difficulty with concentration, and insomnia), fatigue and poor concentration are often especially problematic (Kreutzer *et al.* 2001).

Before making a diagnosis of post-TBI depression, one must first be sure that the patient does not suffer from mere emotionalism or emotional incontinence, as discussed immediately below. Furthermore it is essential to enquire whether the patient had episodes of depression before the injury, indicating that the correct diagnosis is major depressive disorder.

Treatment involves an antidepressant, and several have been studied. Desipramine, in a double-blind but methodologically problematic study (Wroblewski *et al.* 1996), was superior to placebo. Citalopram, in an open study (Rapoport *et al.* 2008) utilizing doses up to 50 mg/day appeared effective. In another double-blinded study (Lee *et al.* 2005) both sertraline, in a dose of 100 mg, and methylphenidate, in a dose of 20 mg, were superior to placebo, however in a subsequent double-blinded study of sertraline (Ashman *et al.* 2009), in doses up to 200 mg/day, no superiority was found over placebo. Given that tricyclics such as desipramine may lower the seizure threshold (Wroblewski *et al.* 1990), it appears that utilizing either sertraline or methylphenidate constitute the best options.

EMOTIONALISM

Emotionalism is relatively common after head injury (Sloan *et al.* 1992), and typically manifests with labile tearfulness. Such emotionalism must be distinguished both from depression and from emotional incontinence. The distinction from depression is made on the basis of the vegetative symptoms, such as fatigue, loss of interest, etc., which are lacking in emotionalism but, by definition, present in depression. Distinguishing emotionalism from emotional incontinence is made by attending to whether the tearfulness is associated with an experienced sense of sadness, which is present in emotionalism but absent in emotional incontinence.

Emotionalism, in one double-blinded study (Sloan *et al.* 1992), responded to fluoxetine.

EMOTIONAL INCONTINENCE

Emotional incontinence is characterized by the involuntary appearance of a sad or happy affect in the absence of any corresponding sense of sadness or mirth. Such 'empty' displays of affect may be seen as a late sequela in from 5 to 10 percent of patients (Tateno *et al.* 2004; Zeilig *et al.* 1996). This emotional incontinence must be distinguished from emotionalism, as discussed immediately above. Both citalopram and paroxetine, each in doses of from 10 to 40 mg/day, were effective in an open study (Muller *et al.* 1999), sometimes in as little as 2 or 3 days.

APATHY

Apathy may occur (Kant *et al.* 1998), and is distinct from either abulia or depression. Patients with abulia, if supervised, may complete tasks at a normal rate, whereas patients with apathy do not. Patients with depression may also complete tasks slowly; however, here, in contrast with apathy, there is a depressed mood. Apathy may respond to bromocriptine (Powell *et al.* 1996), amantadine (Van Reekum *et al.* 1995), or selegeline (Newburn and Newburn 2005).

FATIGUE

Fatigue may be present in up to one-third of patients (Hillier *et al.* 1997), and its treatment is uncertain. Modafinil, in doses up to 400 mg/day, in a double-blinded study (Jha *et al.* 2008), was not superior to placebo. Whether methylphenidate would be effective is not known.

MANIA

Mania may occur in the year following injury, but is relatively uncommon (Nizamie *et al.* 1988; Starkstein *et al.* 1990); it appears to be more likely when the polar regions of the temporal lobes have been involved, as for example by contusions. The episodes themselves may occur soon after the injury or after a latent interval lasting months. In evaluating a patient suspected of having post-traumatic mania, it is important to differentiate manic exuberance from the disinhibition that is seen in the frontal lobe syndrome: the main differential point here is that, whereas mood is heightened in mania, it is not in the frontal lobe syndrome. Furthermore, before attributing mania to head injury, a careful history is required to exclude a premorbid bipolar disorder. Treatment should generally include a mood stabilizer, such as divalproex or carbamazepine; lithium may also be considered; however, as noted earlier, patients with head injury may be especially prone to develop side-effects to this agent.

SLEEP DISTURBANCES

Sleep disturbances occur in the majority of patients (Baumann *et al.* 2007) and may include insomnia, hypersomnia, and excessive daytime sleepiness. Hypersomnia and excessive daytime sleepiness may be due to loss of hypocretin-containing neurons in the hypothalamus (Baumann *et al.* 2009) or to TBI-induced obstructive sleep apnea (Castriotta and Lai 2001; Masel *et al.* 2001), periodic limb movements (Castriotta *et al.* 2009), or narcolepsy (Lankford *et al.* 1994). Before making the diagnosis of TBI-induced sleep disturbances, however, the effect of depression (Fichtenberg *et al.* 2000) and of sedating or stimulating medications must be ruled out.

Given that obstructive sleep apnea, periodic limb movements, and narcolepsy may all respond to routine treatments (Castriotta *et al.* 2009), as discussed in Chapters 20.8, 20.12 and 20.7, respectively, it may be prudent to offer polysomnography to all post-TBI patients who complain of hypersomnia or excessive daytime sleepiness. In cases in which such specific disorders are absent, the most appropriate treatment is uncertain: modafinil, in a double-blinded study (Jha *et al.* 2008) utilizing doses up to 400 mg/day, was not superior to placebo. The treatment of insomnia is likewise uncertain: one double-blinded study (Kemp *et al.* 2004) compared melatonin, in a dose of 5 mg, with amitriptyline, in a dose of 25 mg, and found them equivalent; however, given that there was no placebo arm, it is not clear whether either agent is effective. In practice, most patients are treated with either melatonin or a hypnotic such as zolpidem.

PSYCHOSIS

Psychosis, that is to say a condition characterized by delusions and/or hallucinations in the absence of delirium or dementia, is reported, albeit rarely, as a long-term sequela to TBI (Fujii and Ahmed 2002; Sachdev *et al.* 2001); however, this assertion must be viewed with some caution. In the reported cases, latencies of 1–5 or more years are reported between the head injury and the gradual onset of the psychosis, and a family history of schizophrenia was found to be a significant risk factor (Sachdev *et al.* 2001). Given these findings, the argument may be made that the psychosis represents merely the coincidental occurrence of schizophrenia in a patient with a history of TBI. Treatment involves use of an antipsychotic, preferably one of the better tolerated second-generation agents, such as risperidone or quetiapine.

POST-TRAUMATIC SEIZURES

Post-traumatic seizures may be defined as occurring early, during the first 7 days postinjury, or late, occurring at any time thereafter. Early seizures are reported in anywhere from 2 to 15 percent of cases, and are most likely to occur within the first 24 hours. Late seizures are seen in from 5 to 10 percent of cases, and in those destined to have a late seizure, the first one usually occurs within the first year postinjury (Mazzini *et al.* 2003): the range here, however, is wide, from weeks to up to 15 years. Several features increase the risk of occurrence of a late seizure, including the following (Englander *et al.* 2003; Mazzini *et al.* 2003; McQueen *et al.* 1983; Messori *et al.* 2005): having an early seizure; the presence of contusions or intracerebral hemorrhages; intracranial operations; and dural penetration with bone or metal fragments. Importantly, although the vast majority of seizures occurring post-traumatically are due to the intracranial pathology itself, other possibilities must also be kept in mind: in the case of early seizures, metabolic causes, such as hyponatremia, hypocalcemia, or hypomagnesemia are not uncommon. Simple partial, complex partial, and grand mal seizures may all occur.

With regard to treatment, two aspects exist: prophylaxis against an early seizure, and treatment after a seizure, whether early or late, has occurred.

With regard to prophylaxis of early seizures during the first week, double-blinded studies have demonstrated that both phenytoin (Temkin *et al.* 1990) and valproate (Temkin *et al.* 1999) are more effective than placebo, and, between the two, given that there was a trend toward increased mortality with valproate, phenytoin appears preferable. A more recent single-blinded study (Szaflarski *et al.* 2010), however, compared intravenous fosphenytoin with intravenous levetiracetam and found equal effectiveness in seizure prophylaxis but an overall better tolerability for levetiracetam. Given these studies, it appears that levetiracetam is probably the best choice. Importantly, however, none of these agents have been shown to offer any prophylactic benefit after the first week, and hence, if the patient has remained seizure-free during this period, the AED should probably be discontinued.

In cases in which a seizure has occurred, whether early or late, treatment recommendations are not as clearly worked out. If a patient had an early seizure despite prophylactic treatment, some clinicians would continue the drug, on the theory that without it there might be a higher frequency of seizures, whereas others would opt for a different agent. In cases where there were no seizures during the first week, but one occurred later, perhaps much later, there are no firm guidelines regarding treatment, and consideration could be given not only to phenytoin or levetiracetam, but also to other agents, such as carbamazepine, oxcarbazepine, gabapentin, etc.

ENDOCRINOLOGIC CHANGES

Endocrinologic changes appear in a majority of patients with TBI, and may occur secondary to either hypothalamic (Crompton 1971) or pituitary damage (Edwards and Clark 1986; Salehi *et al.* 2007). Diabetes insipidus, with hypernatremia, may occur early on; other endocrinologic changes, such as hypothyroidism or gonadal failure, may not become evident for months to years after the injury. In this regard, it is prudent to check a thyroid profile in any patient who develops depression, apathy or fatigue, and a testosterone level in patients who develop decreased libido.

POST-TRAUMATIC STRESS DISORDER

At first glance, it would appear impossible for post-traumatic stress disorder (PTSD) to occur after TBI: the full syndrome of PTSD has, as a cardinal symptom, the re-experiencing of the traumatic event in dreams or intrusive memories, and, given the post-traumatic amnesia seen in TBI, such re-experiencing would be impossible. It does appear, however, that a modified form of PTSD, involving avoidance of situations similar to that involved in the TBI and heightened autonomic arousal, does occur (Warden *et al.* 1997), albeit rarely (Bombardier *et al.* 2006). In such cases it may be a matter of the patient both having being told what happened, and, after resolution of the post-traumatic amnesia, having recollection of the sometimes horrific sequelae present during the remainder of the hospital stay.

Although there are as yet no controlled studies of the treatment of PTSD following TBI, consideration may be given to those treatments discussed for PTSD in Chapter 24.6.

Etiology

In TBI, a variety of lesions may be seen (Freytag 1963; Jenkins *et al.* 1986), including diffuse axonal injury, contusions, intracerebral hemorrhage, subarachnoid hemorrhage, epidural or subdural hematomas, or infarctions. Cerebral edema accompanies most of these, and, in combination with space-occupying lesions, may cause uncal or subfalcine herniation. In some cases, hydrocephalus may appear. Fractures may or may not accompany these lesions.

Diffuse axonal injury, as discussed in Chapter 12.2, occurs secondary to the tremendous shearing and rotational forces that occur during acceleration–deceleration, and is characterized by axonal rupture or damage. Although these effects are widespread throughout the cerebrum, certain areas are most vulnerable, including the junction between the cortex and white matter, the corpus callosum, and the dorsolateral quadrants of the midbrain. In almost all cases, this diffuse axonal injury is also accompanied by diffuse vascular injury, wherein small penetrating arterioles, subjected to the same shearing and rotational forces, undergo rupture, producing widespread petechial hemorrhages.

Contusions occurring in acceleration–deceleration injuries typically occur along the inferior surfaces of the frontal and temporal lobes, as they slide along the bony protuberances at the base of the skull. In cases in which there has been a blow to the head, a contusion may form under the point of impact (the 'coup' contusion) and also, contralaterally, at the area where the brain is flung up against the inner surface of the skull (the 'contre-coup' contusion).

Intracerebral hemorrhages occur with rupture of relatively large penetrating arteries, and although these hemorrhages may be lobar in location they are most commonly seen in the basal ganglia (Katz *et al.* 1989; Macpherson *et al.* 1986). Rarely, intracerebral hemorrhages may be delayed in appearance for up to 2 days postinjury.

Subarachnoid hemorrhage may occur secondary to shearing of vessels traversing the subarachnoid space or due to leakage of blood from an area of contused or hemorrhagic cortex. In such cases, vasospasm of arteries traversing the bloody subarachnoid space may lead to ischemic infarction of subserved tissue.

Subdural hematomas occur in about one-fifth of patients, and may range in size from thin, inconsequential crescents to large, life-threatening lesions. Epidural hematomas may also occur, but are far less common.

Infarctions may occur secondary to herniations, vasospasm, arterial dissection, or, if severe hypotension occurred, via a 'watershed' phenomenon (Marino *et al.* 2006).

Cerebral edema is common, and adds considerably to the clinical expression of diffuse axonal injury, contusions, and intracerebral hemorrhages. Herniation may occur, with, as just noted, possible infarction secondary to vascular compression: with uncal herniation, such infarction may occur in the area of distribution of the posterior cerebral artery, whereas with subfalcine herniation, infarction may occur in the area of distribution of the anterior cerebral artery.

Hydrocephalus may occur via a number of mechanisms. Acute hydrocephalus may occur secondary to compression of the foramen of Monro by an expanding lesion, such as a contusion, or, when subarachnoid hemorrhage has occurred, secondary to a clot obstructing outflow from the exit foramina of the fourth ventricle. Chronic hydrocephalus, presenting weeks or months after the injury, may occur secondary to outflow obstruction at the arachnoid villi of the superior sagittal sinus.

Fractures may or may not occur, and it is important to keep in mind that there is not a good correlation between the presence of a fracture and the presence of brain damage; indeed in many cases of devastating TBI, there is no fracture at all. Linear, depressed, and compound fractures may all be seen, but of these, linear fractures, typically at the base of the skull, are most common.

In addition to these direct effects of trauma, the occurrence of hypoxia and hypotension at the scene may lead, respectively, to global anoxic brain damage or, as noted earlier, watershed infarcts.

In determining which of these injuries have occurred, it must be borne in mind that MR scanning is more sensitive than CT scanning for diffuse axonal injury, contusions, subdural and epidural hematomas, infarcts, and global anoxic injury (Mittl *et al.* 1994; Orrison *et al.* 1994).

8.6 MENINGITIS

This chapter might perhaps be more appropriately entitled 'acute meningitis', as it deals with those cases of meningitis occurring secondary to viral or bacterial infection. Meningitis, of course, may also occur secondary to tubercular, fungal, or syphilitic infections; however, these cases generally present in a gradual fashion, and are characterized by a basilar meningitis, in contrast to viral or bacterial cases, wherein inflammation involves the meninges in a generalized fashion.

Acute meningitis due to viral infection is often also referred to as 'aseptic meningitis', in contrast to the bacterial or 'septic' cases. Acute meningitis, though uncommon, must not be missed, as it constitutes a medical emergency.

Clinical features

Classically, in both viral and bacterial meningitis, onset occurs over a day or so with a triad of fever, stiff neck, and headache; the headache is often accompanied with photophobia. This triad may be enlarged to a 'quartet' of symptoms by adding mental status changes, which, initially, may consist of drowsiness or mere irritability. This full presentation is certainly not apparent in all cases: in one study of bacterial meningitis the triad was present in only 44 percent; in the same study, however, two or more symptoms of the quartet were present in 95 percent (van de Beek *et al.* 2004a). In examining patients, one should always check for Kernig's and Brudzinski's signs of meningeal irritation. Kernig's sign is elicited by flexing the leg at the thigh and knee, and then extending the leg at the knee: when the sign is present the patient will complain of pain in the back or the thigh. Brudzinski's sign is tested for by passively flexing the neck: a positive sign is indicated by involuntary flexion of the leg at the thigh and knee.

In cases of viral meningitis, the clinical picture may not evolve much further. In bacterial meningitis, however, progression is common, and one may see the rapid development of delirium, stupor, or coma, accompanied, in a minority of cases, by seizures or focal signs, such as aphasia or hemiparesis.

Emergency lumbar puncture (discussed further in Chapter 2.6) is indicated in virtually every case. A delay for imaging is only justified in cases in which stupor, coma, or focal signs have raised the possibility of increased intracranial pressure and potential herniation. In cases of bacterial meningitis, minutes count, and delays may be fatal. The cell count in viral meningitis ranges from 50 to 1000, with a lymphocytic predominance; in bacterial cases the count is higher, from 1000 to as high as 50 000, with a strong polymorphonuclear predominance. The total protein in viral cases is elevated to from 50 to 100 mg/dL, and in bacterial cases it is generally well over 100 mg/dL. The glucose in viral cases is within normal limits, whereas in bacterial cases it is reduced. The Gram stain of course is negative in viral cases, while in bacterial cases it may reveal the organism in from 60 to 90 percent. Samples are sent for viral and bacterial culture, and for appropriate polymerase chain reaction testing: in this regard, consultation with specialists in infectious diseases, if not already obtained, is essential, given the rapid development of this field. Blood cultures should also be immediately obtained.

Course

Viral meningitis is rarely fatal, and most cases run their course over a week or two. In contrast, bacterial meningitis, untreated, is generally fatal; with treatment, the overall mortality rate is roughly 20 percent, with the worst outcome being seen in cases due to *Streptococcus pneumonia*.

Upon recovery from viral meningitis, most patients, apart from some fatigue and malaise, are left with few, if any, sequelae. After bacterial meningitis, however, sequelae are common. Cognitive impairment (Hoogman *et al.* 2007; Markelbach *et al.* 2000; van de Beek *et al.* 2002) may be seen in anywhere from one-third to three-fourths of patients, and includes poor concentration, poor memory, and overall slowness of thought, in some cases leaving the patient demented. Other sequelae include seizures, focal deficits, and cranial nerve palsies.

Etiology

The majority of cases of viral meningitis are due to enteroviruses (Kupila *et al.* 2006); other organisms include type 2 herpes simplex virus, lymphocytic choriomeningitis, cytomegalovirus, mumps, and Epstein–Barr virus. Human immunovirus infection, during seroconversion, may also be associated with a meningeal picture.

In cases of bacterial meningitis, the two most common pathogens are *Streptococcus pneumonia* and *Neisseria meningitidis* (van de Beek *et al.* 2004a); other, less common pathogens include *Hemophilus influenzaa*, group B streptococcus and *Listeria monocytogenes*.

Pathologically, in all cases there is leptomeningeal inflammation, and, in the immediately subjacent cortex there is usually a degree of inflammation which is due, not to infection itself but rather to toxic effects of cytokines. In cases of viral meningitis, there may be little else; however, in bacterial cases, additional pathologic changes are common. Meningeal inflammation is more intense, and the resulting inflammation of the cortex may be extreme, leading to generalized cerebral edema, grossly increased intracranial pressure, and transtentorial herniation. Bacteria may infiltrate down perivascular spaces, and in some cases may rupture into the adjacent cortex causing an actual encephalitis. Arteritis may occur, with subsequent thrombosis and infarction. Venous inflammation may also occur, with hemorrhagic infarction. With inflammation around the exit foramina of the fourth ventricle, obstructive hydrocephalus may occur. Infiltration of cranial nerve sheathes may lead to cranial nerve palsies.

Differential diagnosis

Both acute encephalitis (Chapter 8.8) and cerebral abscess (Chapter 8.7) may mimic meningitis. Encephalitis is suggested by the prominence of mental status changes early on, and cerebral abscess by its somewhat more leisurely onset, and by the prominence of focal signs early on.

Treatment

Given the potentially catastrophic results of a failure to treat bacterial meningitis, it is essential to treat all cases of meningitis as if they were of bacterial origin, and to continue antibiotic treatment until proven otherwise. There is debate over whether one should postpone antibiotics until lumbar puncture has been performed, but, as antibiotic administration may impair the identification of organisms within the cerebrospinal fluid, most clinicians will wait until the puncture has been completed. Broad-spectrum coverage is then begun, pending identification of the organism, and, again, given the rapid progress of research, urgent consultation with specialists in infectious disease is essential. Recent research has also indicated that a 4-day course of dexamethasone, with the first dose given just before antibiotic treatment is commenced, will reduce the incidence of sequelae (van de Beek *et al.* 2004b).

8.7 CEREBRAL ABSCESS

Cerebral abscess, the subject of several recent reviews (Carpenter *et al.* 2007; Hakan *et al.* 2006; Tonon *et al.* 2006; Tseng and Tseng 2006), although common in the pre-antibiotic era, is now rare, except in the immunocompromised.

Clinical features

The classic presentation includes the triad of headache, fever, and either focal signs or seizures; this full triad, however, is found in less than one-half of all cases. In addition, there may also be delirium, stupor, or coma.

Headache is found in most cases, and may be local or generalized, persistent or intermittent, and, in some cases, quite mild.

Fever is found in up to one-half of all cases, and may actually remit as other symptoms progress.

Focal signs reflect the location of the abscess, and may include hemiparesis, apraxia, aphasia, and, with cerebellar abscesses, nystagmus or ataxia. Seizures occur in roughly one-fourth of all cases, and may be simple partial, complex partial, or grand mal in type.

Delirium may occur with either frontal or temporal abscesses. With large or multiple cerebral abscesses, increased intracranial pressure may ensue, with stupor or coma, and with cerebellar abscesses brainstem compression or upward herniation may also cause stupor or coma. Finally, should an abscess rupture into one of the ventricles, there may be an abrupt onset of stupor or coma.

Computed tomography or MRI scanning, both with contrast enhancement, has revolutionized the diagnosis of cerebral abscess. The radiologic picture evolves concurrent with the evolution of the pathologic process, described below. Initially one finds an area of edema, and as the initial cerebritis progresses, an indistinct rim of enhancement may be seen. As the capsule begins to form, this rim of enhancement becomes more pronounced, and when the capsule fully forms it becomes quite dense, clearly demarcating the abscess. Surrounding the abscess one also finds an area of edema, which may become quite large and associated with a midline shift.

Although the peripheral white count may or may not be elevated, the ESR and C-reactive protein generally are. Lumbar puncture is rarely required, and should only be undertaken with the greatest of caution.

Course

In the pre-antibiotic era, the mortality rate was high. With modern treatment, the overall rate is roughly 5 percent, with poorer outcomes seen in cases associated with stupor or coma.

Among survivors, sequelae are common, including epilepsy, dementia, and persistent focal deficits.

Etiology

Most cases are secondary to bacterial infection; in the immunocompromised, however, fungal and parasitic infections, especially with toxoplasma, enter strongly into the picture. Neurocysticercosis should also be considered in those who have resided in endemic areas.

Organisms may gain entry via one of three mechanisms: extension, direct introduction, and hematogenous spread. Extension of infection may occur from the paranasal sinuses into the frontal or temporal lobes, from otitis media or mastoiditis into the temporal lobe or the cerebellum, and from periodontal infections into the frontal lobe. Direct introduction may occur secondary to trauma or neurosurgery. Hematogenous spread is seem most often with pulmonary infection (abscess, bronchiectasis, or empyema) or bacterial endocarditis; interestingly, however, bacteremia generally does not result in cerebral abscess formation unless there is a pre-existing nidus of necrotic tissue, as may be seen in cases of acute infarction. Importantly, when hematogenous spread is the underlying mechanism, one tends to see multiple abscesses, typically at the gray–white junction in the area of distribution of the middle cerebral artery.

Abscesses begin with a localized cerebritis which expands and evolves over roughly a week's time, toward the end of which collagen deposition begins to occur at its periphery. The next stage is called the capsular one, and this generally lasts also about a week, during which necrosis occurs within the abscess, and a definite collagenous capsule forms around it. During both these stages edema surrounds the lesion. This general picture may vary, depending on the virulence of the organism and also on the immunological competence of the host.

Capsules are generally weakest on their ventricular side, and rupture into the ventricle may occur, with catastrophic results.

Differential diagnosis

When the classic triad is present, consideration must be given to acute encephalitis (especially herpes simplex viral encephalitis), meningitis, and subdural empyema. In other cases, consideration may also be given to granulomas or rapidly progressing tumors (e.g. glioblastoma multiforme).

Treatment

Infectious disease consultation should be sought urgently. Broad-spectrum coverage is instituted and maintained until the identity of the organism is established, a process that typically requires aspiration of the abscess. In cases occurring secondary to extension or hematogenous spread, the primary focus must also be adequately treated. Actual excision of the abscess is rarely required. Although controversial, some clinicians will also institute prophylactic AED treatment.

8.8 ACUTE ENCEPHALITIS

Although, technically, the term 'encephalitis' refers to any inflammation of the brain parenchyma, whether focal or diffuse, and arising from any cause, in common clinical practice to say that a patient has an encephalitis is to imply that he or she is suffering from an acute and generalized infection. The vast majority of such cases are due to viruses, with bacterial or protozoal encephalitidies being far less common (Chaudhuri and Kennedy 2002; Koskiniemi *et al.* 2001).

Acute encephalitis constitutes a medical emergency, and hence recognizing the cardinal signs of encephalitis, namely fever, headache, and delirium, is critically important.

Clinical features

In most cases, the encephalitis is preceded by a non-specific prodrome, lasting perhaps a few days, of fatigue, malaise, and fever. The encephalitis itself generally presents over anywhere from hours to days, with headache and delirium. In addition to confusion and variable disorientation, the delirium may also be accompanied by agitation, hallucinations (typically visual), and delusions. Focal findings, such as hemiparesis, aphasia, or hemianesthesia, may or may not be present, and there may also be partial or grand mal seizures. In most, but certainly not all, cases, signs of meningeal inflammation may also be present, with stiff neck or photophobia. With progression, stupor, or coma may supervene.

Imaging, preferably MRI, should be performed in all cases, and this should be done as an emergency, so as to not delay lumbar puncture. Importantly, MRI scans may be normal for the first day or two.

Lumbar puncture is indicated in virtually every case (Steiner *et al.* 2005). Glucose is normal in almost all viral infections, with the exception of a small minority of cases of herpes simplex viral encephalitis, while it is decreased in bacterial and other causes. The total protein is increased. Essentially all cases will have a pleocytosis, ranging from 5 to 10 up to 1000 white cells or more. In viral cases the pleocytosis is generally lymphocytic, whereas bacterial encephalitidies are generally marked by the presence of polymorphonuclear cells: an exception to this occurs with certain arboviral infections wherein there may be a substantial number of polymorphonuclear cells for the first day or two; a persistent polymorphonuclear predominance after this time, however, is much more characteristic of bacterial infections. Gram stain and culture are performed routinely. Polymerase chain reaction assay has revolutionized the identification of responsible organisms, and is useful in detecting herpes simplex, varicella-zoster, Epstein–Barr, mumps, and arboviruses.

Electroencephalography typically shows generalized slowing, with or without interictal epileptiform discharges. Focal slowing in the temporal areas, with or without periodic complexes, is highly suggestive of herpes simplex encephalitis.

Etiology

Although, as noted earlier, the vast majority of cases are due to viral infection, at present the responsible virus can be identified in only about one-third of cases. Herpes simplex viral encephalitis (Chapter 15.5) is the most common cause of sporadic viral encephalitis, and, given its treatability, it should always be high on the differential. Arboviral encephalitis (Chapter 15.4), occurring secondary to mosquito bites, typically occurs in an epidemic fashion. Varicella-zoster encephalitis (Chapter 15.9) occurs in the setting of a generalized zoster rash, and is typically seen only in the elderly or immunocompromised. Epstein–Barr viral encephalitis (Chapter 15.7) occurs in the setting of typical 'mono' symptoms, such as pharyngitis and cervical adenopathy, while mumps encephalitis (Chapter 15.8) occurs in the setting of a typical parotitis. During seroconversion with human immunovirus infection (Chapter 15.1), roughly one-tenth of patients develop central nervous system symptomatology, with, in some cases, an encephalitic picture. Rabies (Chapter 15.10) may appear anywhere from weeks to a year after inoculation, and, in addition to fever, headache, and delirium, patients are typically highly agitated and easily startled, and may demonstrate not only hypersalivation, but also the classic symptom of hydrophobia. Finally, although encephalitis lethargica (Chapter 15.6) no longer occurs in epidemic form, sporadic cases are seen, and here the diagnosis is suggested by finding third and sixth cranial nerve palsies.

Non-viral causes of acute encephalitis include toxoplasmosis (Chapter 15.20), generally found only in the severely immunocompromised, cerebral malaria (Chapter 15.19), and Rocky Mountain spotted fever (Chapter 15.18).

Differential diagnosis

Toxic–metabolic delirium is exceedingly common, and if it occurs in a patient with, say, pneumonia, there may be concurrent fever and headache, thus potentially mimicking the picture typical of acute encephalitis, and consequently considerable clinical judgment may be required. Certainly, if the patient is septic, one would lean toward the diagnosis of a toxic–metabolic delirium; conversely, if the systemic infectious illness were mild, say merely a urinary tract infection, and the fever, headache, and delirium were disproportionately severe, then one would obviously think strongly of performing a lumbar puncture.

Acute disseminated encephalomyelitis (ADEM), as discussed in Chapter 15.11, represents an autoimmune attack on the central nervous system, triggered by either a vaccination or an infection, generally a viral one. After a variably discrete latent interval, lasting from days to weeks, patients fall ill, and when the cerebral white matter is targeted one typically sees delirium and headache, with or without focal signs or seizures; importantly, in some cases fever may accompany these symptoms. If the illness followed closely upon a vaccination, then the diagnosis would be fairly straightforward, but if it followed upon a non-specific viral infection with fever, headache, and malaise, and if these symptoms had not cleared completely prior to the onset of the delirium, then the differential between ADEM and acute encephalitis becomes problematic. In such instances, MRI findings evident after a few days are very helpful, for in ADEM these are constituted by multiple, bilateral, large areas of increased signal intensity in the white matter, a pattern not found in encephalitis.

Meningitis, as discussed in Chapter 8.6, may mimic the picture seen in acute encephalitis, however in meningitis, a stiff neck is usually present, and quite prominent, before the appearance of a delirium.

Cerebral abscess, as discussed in Chapter 8.7, may present in a fashion quite similar to that of encephalitis, with one important exception, and that is the mode of onset: with abscesses, the full picture may take a week or more to develop, in contrast to encephalitis, which presents much more rapidly. In any case, imaging, by revealing the abscess, will immediately make the diagnosis.

The neuroleptic malignant syndrome, discussed in Chapter 29.1, may present with fever and delirium, thus suggesting encephalitis; however, here one finds that, a day or two earlier, there was a pharmacologic maneuver which abruptly decreased dopaminergic tone. Furthermore in the neuroleptic malignant syndrome one typically also sees rigidity, which is generally absent in encephalitis, with the notable exception of one of the arboviral encephalitidies, namely Japanese encephalitis. The serotonin syndrome, discussed in Chapter 29.5, may also present with fever and delirium; however, here one finds a preceding pharmacologic maneuver which increased serotoninergic tone. Furthermore, in the serotonin syndrome one may also see myoclonus, a finding typically absent in encephalitis.

Treatment

Patients should be admitted to an ICU, and infectious disease consultation should be urgently sought. In addition to general medical care, strict attention should be paid to sodium levels, as both diabetes insipidus and the syndrome of inappropriate ADH secretion may occur. Should seizures occur, patients may be treated as discussed in Chapter 8.3, and the general treatment of delirium is discussed in Chapter 6.5.

Pending the results of CSF testing, it is customary, given the devastating sequelae of herpes simplex viral encephalitis, to initiate treatment with aciclovir.

Patients who survive may or may not be left with sequelae (Arciniegas and Anderson 2004), depending on the causative organism, the severity of the infection, and the age of the patient: both the very young and the elderly are more likely to have sequelae. Focal deficits may persist, and epilepsy may ensue. There may also be personality change, dementia, and various movement disorders.

8.9 PAIN

Peripheral pain mechanisms

NOCICEPTORS

Nociceptors (pain receptors) may be divided into C-polymodal nociceptors or C-PMNs for short, and A-mechanoheat (AMH) nociceptors; the latter respond primarily to mechanical and thermal stimuli. C-PMNs respond to chemical, mechanical, and thermal stimuli.

PERIPHERAL INFLAMMATION

Tissue damage leads not just to direct stimulation of nociceptors, but also to the release of inflammatory mediators, including eicosanoids, cytokines, neurotransmitters, neuropeptides, and ATP. In turn, these inflammatory mediators act to reduce the nociceptor stimulation threshold and sensitize and stimulate the nociceptors further, as illustrated in Figure 8.2.

Further details of the events that occur during peripheral activation and sensitization, as part of the inflammatory response, and spread of this sensitization, are shown in Figure 8.3. Peripheral inflammation leads to the production of opioid receptors by the dorsal root ganglion and their transport towards the peripheral terminal, as illustrated in Figure 8.4, which also depicts the fact that these peripheral opioid receptors may be activated by endogenous opioid peptides released by monocytes, T cells, B cells, and macrophages.

PERIPHERAL TRAUMA

Figure 8.5 shows the neural mechanisms underlying the generation of pain and other symptoms associated with complex regional pain syndromes following peripheral trauma.

Dorsal horn mechanisms

Endogenous ligands that act in the dorsal horn of the spinal cord in relation to nociception include the excitatory amino acid glutamate, and substance P, as shown in Figure 8.6 (Power and Kam 2008).

A cascade of postsynaptic intracellular events is activated by *N*-methyl-D-aspartic acid receptor activation, including phospholipase C and adenylate cyclase activation, leading to cyclic AMP and diacylglycerol biosynthesis. Postsynaptic events following release of glutamate from central terminals of primary afferents in the spinal cord include *N*-methyl-D-aspartic acid receptor activation leading to calcium ion influx, which in turn activates nitric oxide synthase, which catalyzes the conversion of arginine into nitric oxide and citrulline. Increased synthesis of nitric oxide may be associated with hyperalgesia.

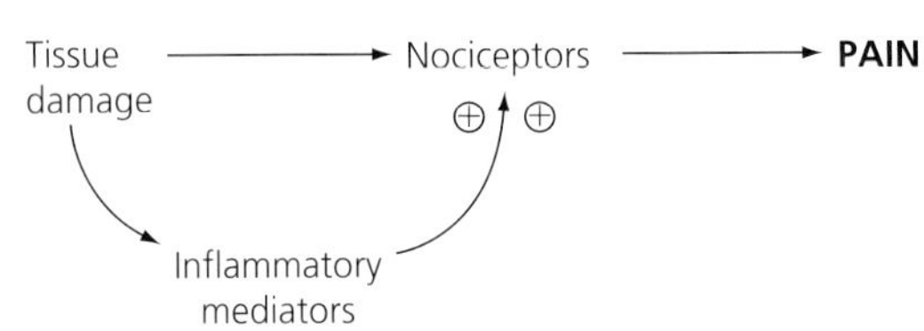

Figure 8.2 Flow diagram from tissue damage to pain. Reproduced from Power & Kam (2008), with permission.

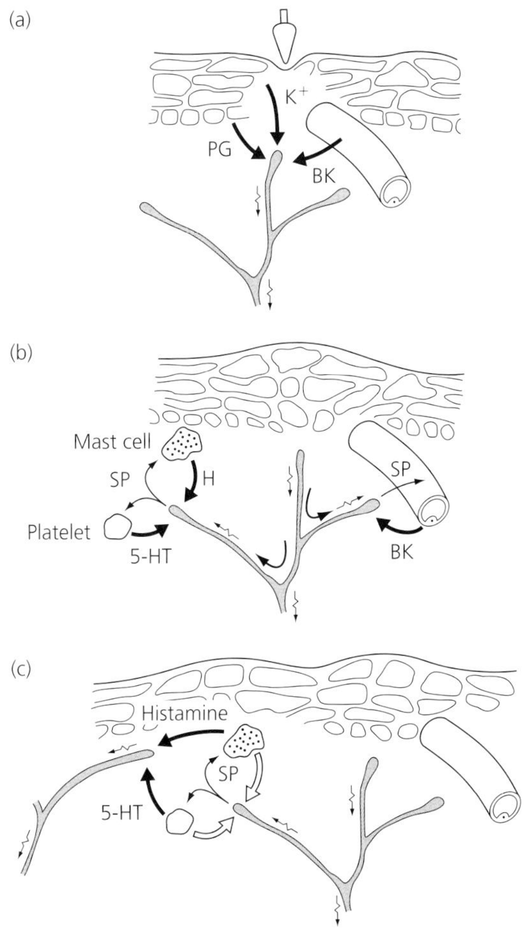

Figure 8.3 Events leading to activation, sensitization, and spread of sensitization of primary afferent nociceptor terminals. (a) Direct activation by intense pressure and consequent cell damage, leading to release of K^+ and biosynthesis of prostaglandins (PGs) and bradykinin (BK). PGs increase the sensitivity of the terminal to BK and other pain-producing substances. (b) Secondary activation. Impulses generated in the stimulated terminal propagate not only to the spinal cord but also into other terminal branches, where they induce the release of peptides such as substance P (SP), which in turn causes vasodilatation and neurogenic edema with further accumulation of BK and also the release of histamine (H) from mast cells and serotonin (5-HT) from platelets. (c) Histamine and serotonin levels rise in the extracellular space, secondarily sensitizing nearby nociceptors, leading to a gradual spread of hyperalgesia and/or tenderness. Reproduced from Fields HL. *Pain*, 1987, with permission from McGraw-Hill companies.

Spinal cord and brain pathways

Figure 8.7 shows the main ascending nociceptive pathway in the spinal cord and descending modulatory pathways arising from the mesencephalon and medulla oblongata.

8.10 SOMATOFORM DISORDERS

Somatoform disorders are characterized by complaints that either, as in the case of conversion disorder or somatization disorder, have no basis in the anatomic or physiologic facts of the matter, or, as in hypochondriasis, although perhaps having some basis in anatomy or physiology, yet are exaggerated far beyond what would be expected. Each of these is considered in turn.

Conversion disorder

Conversion disorder is characterized by symptoms that meet two criteria: first, they suggest a specific lesion of the central or peripheral nervous system; and, second, further investigation either reveals symptomatology that 'violates' the laws of anatomy or physiology, or demonstrates conclusively that the suspected lesion does not, in fact, exist. Furthermore, and again despite thorough investigation, no recognizable, conscious motive can be discovered that might account for the onset of the symptom in question. This last point is critical, as it distinguishes conversion disorder from malingering, where the motive for the symptomatology, whether it be financial gain or avoidance of some unpleasant task, is fairly obvious.

A synonym for conversion disorder is hysteria (or 'hysterical neurosis'): both terms, however, are unfortunate, as both indicate etiologic theories that have not been substantiated. 'Conversion' implies a mechanism whereby emotionally charged ideas are in some unconscious way 'converted' into 'physical' symptoms, and, although such a mechanism, as discussed below, may be operative in some cases, there is no evidence to suggest that it accounts for all or even a substantial minority of cases. 'Hysteria', an ancient term, suggests that the disorder is in some way related to having a *hystera* (Greek for 'womb'), which is clearly not the case: although females are more likely to have this disorder than males (by a factor of from 2:1 up to 10:1), it indisputably also occurs in males.

The lifetime prevalence of conversion disorder is not known with certainty: reported figures range from 0.01 to 0.5 percent of the general population.

CLINICAL FEATURES

Although conversion disorder may first appear anywhere from childhood to old age, the vast majority of patients first experience symptoms in either adolescence or early adult years.

Although in all cases the symptom is triggered off by some psychologically meaningful event, this may not always be obvious, and in some cases a thorough understanding of the patient's psychology may be necessary in order to determine what is, and is not, meaningful. Furthermore, although the onset itself is generally acute, there may subsequently be a more gradual worsening of symptoms.

Conversion symptoms may be grouped according to their symptomatology, including motor symptoms (paralysis, ataxia, tremor, or parkinsonism), sensory symptoms (anesthesia, deafness, or blindness) and pseudoseizures. Although most patients present with only one conversion symptom at a time, exceptions do occur, and in these cases one may speak of a 'mixed' presentation. Regardless of what conversion symptom is present, the distinctive feature of the symptom, as noted earlier, is that, in some way or other, it

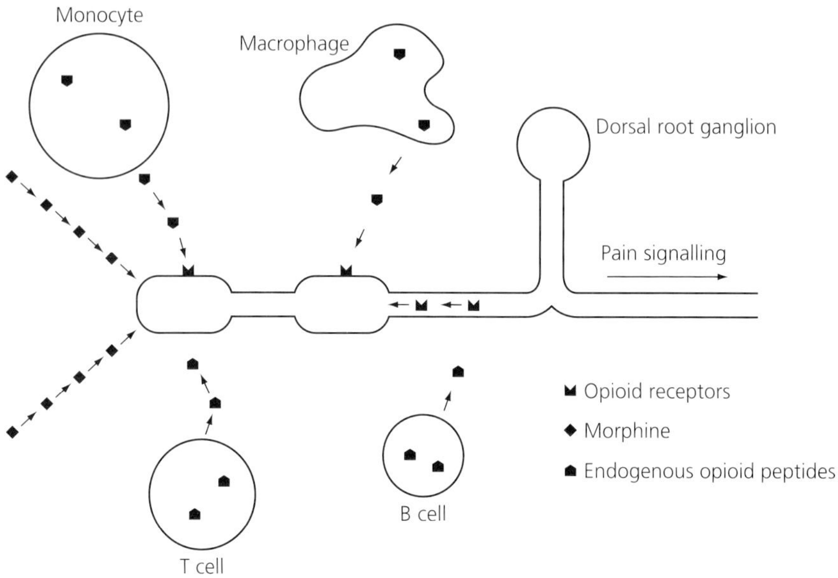

Figure 8.4 Peripheral inflammation leads to the production of opioid receptors by the dorsal root ganglion. These opioid receptors are transported towards the peripheral terminal. Peripheral opioid receptors may be activated by endogenous opioid peptides released by monocytes, T cells, B cells, and macrophages, as well as by exogenous application of morphine.

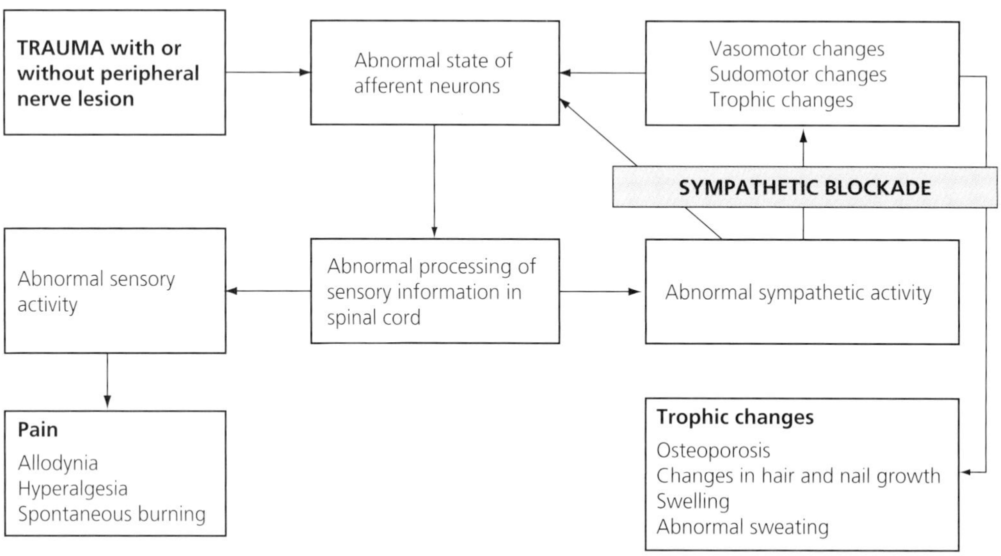

Figure 8.5 General hypothesis about the neural mechanisms underlying the generation of pain and other symptoms associated with complex regional pain syndromes following peripheral trauma with and without nerve lesions. Reproduced from Power & Kam (2008). With permission.

'violates' the laws of anatomy or physiology, and it is this 'violation' that often alerts the examiner to the conversion nature of the symptom in question (Stone *et al.* 2002). An example would be a patient who complained of an area of anesthesia that involves the entire hand, extends up to the middle of the forearm, and ends abruptly at a boundary that describes a perfect circle around the forearm. There simply is no lesion of either the central or peripheral nervous system that could conceivably produce such a pattern of anesthesia. In the following paragraphs, each of the more or less common conversion symptoms is discussed in turn, with special attention to the associated 'violations' that one may find.

Conversion paralysis may mimic *monoplegia, hemiplegia, or paraplegia.* In conversion *monoplegia* involving, for example, the upper limb, one may find a non-physiologic pattern of weakness, involving perhaps the forearm and hand with complete strength maintained in the shoulder and arm. In conversion *hemiplegia*, one may find a positive Hoover test, as described by Dr Hoover in 1908 (Hoover 1908). The Hoover test may be performed by having the patient lie supine on the bed, positioning yourself at the foot of the bed and placing one hand under the heel of the 'bad' leg and the other on top of the ankle of the 'good' leg. At this point ask the patient to lift the 'good' leg up in the air against the resistance of your hand. As the

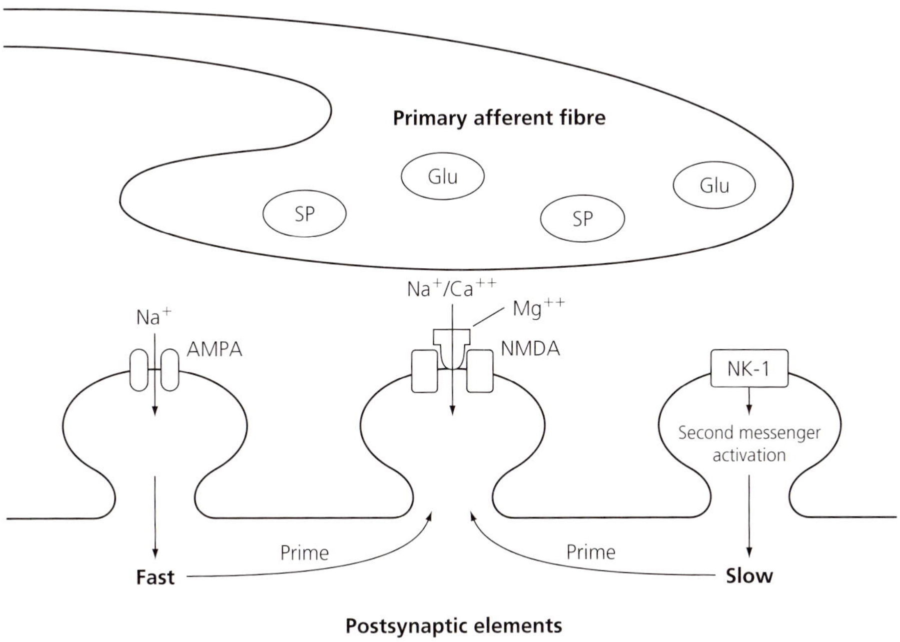

Figure 8.6 Release of glutamate (Glu) and substance P (SP) from the terminals of nociceptive primary afferents activates AMPA and neurokinin-1 (NK-1) receptors, relatively, on the postsynaptic membrane. Activation of these receptors leads to Na^+ influx at the AMPA receptor and activation of second messengers. These processes prime the NMDA receptor with removal of the Mg^{2+} 'plug' and Na^+ and Ca^{2+} influx. Reproduced from Power & Kam (2008), with permission.

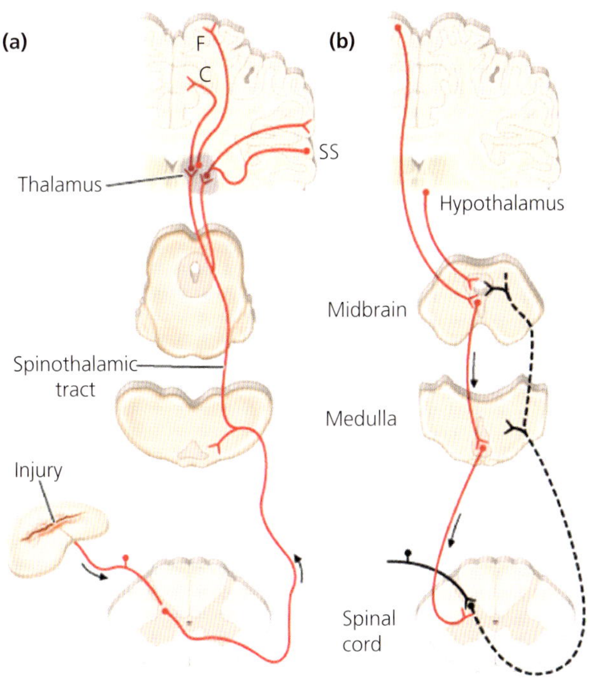

Figure 8.7 (a) This shows the main ascending nociceptive pathway to the thalamus via the spinothalamic tract. C, anterior cingulate; F, frontal insular cortex; SS, somatosensory cortex. (b) This shows descending modulatory pathways arising from the mesencephalon (with hypothalamic and frontal cortical inputs) and medulla oblongata. Reproduced from Fauci AS *et al.*, *Harrison's Principles of Internal Medicine*, 17th edition, 2008, with permission from McGraw-Hill companies.

patient attempts to lift the 'good' leg, you will feel the normal compensatory downward pressure of the contralateral leg on your hand as it lies under the heel of the 'bad' leg. Once this maneuver has been accomplished, ask the patient to rest the 'good' leg down and then switch the position of your hands, now placing one hand under the heel of the 'good' leg and placing the other hand on top of the 'bad' leg. At this point, encourage the patient to exert the greatest effort possible in raising the 'bad' leg up against the resistance of your hand. In a positive Hoover test one finds two things: first, the 'bad' leg moves little, if at all; second, one fails to feel any of the normally expected downward pressure on the hand underneath the heel of the 'good' leg, indicating that, in fact, the patient is putting no effort into the task of lifting the 'bad' leg. Recently, a similar test, designed to elicit the 'abductor sign' has been described (Sonoo 2004). In this test, with the patient with conversion hemiplegia supine in bed, stand at the foot of the bed, facing the patient, and place your hands on the lateral aspects of both of the patient's legs, maintaining a light inward pressure. At this point, ask the patient to abduct the 'bad' leg as forcefully as possible against the light pressure of your hand. In conversion paralysis, there is no movement of the 'bad' leg, and, critically, there is also no abduction of the 'good' leg either. In patients with 'true' hemiplegia one would appreciate the expected synergistic abduction of the 'good' leg, and it is the absence of this in conversion that indicates that, in fact, the patient is not putting effort into the task. Observing the patient on attempted ambulation is also helpful: when asked to walk, patients with conversion hemiparesis tend to drag the affected leg, pulling it up behind them, rather than circumducting it. Furthermore, patients with conversion hemiplegia may also display the 'wrong-way' tongue, with deviation of the protruded tongue away from the weak side, rather than toward it, as one would expect with stroke (Keane 1986). In both conversion monoplegia and hemiplegia one typically also finds symmetric deep tendon reflexes and bilaterally down-going toes. One may also find 'give-way' or 'collapsing' weakness. Here, the patient may be asked to flex the forearm after which the examiner, grasping the patient's hand, attempts to extend the forearm. In 'collapsing' weakness, one finds that an initial resistance is followed by an abrupt giving away of any resistance, as if the arm had suddenly become flaccid. Although this sign is useful, it must be borne in mind that a degree of collapsing weakness may at times be found in patients with hemiparesis due to stroke. Conversion *paraplegia* is suggested by normal reflexes, muscle tone, sensation, and sphincter function (Baker and Silver 1987). Finally, in any case of suspected conversion paralysis where the physical examination is inconclusive, one may consider transcranial magnetic stimulation motor evoked potentials, which will, in cases of conversion paralysis, demonstrate the integrity of the motor pathway from cortex to the involved muscles (Janssen *et al.* 1995).

Conversion ataxia is suggested by elaborate lurching movements of the legs, which may be accompanied by similarly exaggerated flinging of the arms. In severe cases of conversion ataxia one may find astasia–abasia, that is to say, an inability to both walk and stand. In such cases, one typically finds that, when supine on the bed, patients can perform finger-to-nose and heel-to-knee-to-shin testing adequately. Care must be taken here, however, for in some cases of lesions of the anterior vermis such a discrepancy may be found.

Conversion tremors are typically of large amplitude, and typically display characteristics (Factor *et al.* 1995) such as variability in amplitude and frequency and a diminution with distraction, as for example may occur when patients are asked to do serial sevens. Furthermore, when a weight is applied to the tremulous extremity, one often finds an exacerbation, rather than the expected diminution. In cases of unilateral tremor, the tremor may diminish when the patient is asked to perform a complex action with the contralateral extremity, for example touching the middle, then the index and finally the ring finger with the thumb. One may also attempt to elicit a phenomenon known as 'chasing the tremor'. In cases of 'true' tremor, say, of a hand, when one grasps the hand the tremor diminishes and does not appear elsewhere. In conversion tremor, however, after grasping the hand, the forearm may begin to tremble, then, if the forearm is grasped, the tremor may appear in the upper arm, as if one were 'chasing' it.

Conversion parkinsonism (Lang *et al.* 1995) may present with tremor, loss of associated arm movements, and postural instability; however, all of these may be accompanied by atypical features. Tremor may persist or even increase with action, and the arms, upon ambulation, are often held tightly against the patient's sides. Postural instability may be tested for with the standard 'pull test,' wherein one stands behind the patient, places one's hands on the patient's shoulders and then gives a quick pull back. In conversion parkinsonism patients typically regain their balance with one or two steps back, and one may also see a rapid, and quite fluid, upswing of the arms, all in contrast to cases of 'true' parkinsonism, wherein a fall back may occur after multiple steps are taken, and also wherein the associated bradykinesia prevents any rapid or fluid movements.

Conversion anesthesia is suggested by the non-physiologic boundaries of the 'anesthetic' area. Thus, patients with a 'glove and stocking' pattern may complain of a precise boundary, with the anesthesia ending abruptly at a circular boundary just above the ankles or hands, rather than displaying the gradual 'fade' seen in a peripheral polyneuropathy. Furthermore, in such cases one may find a negative Romberg test. In cases where the conversion anesthesia suggests a mononeuropathy, one finds that the anesthetic area simply does not match any known area of innervation of a peripheral nerve. Finally, in cases suggesting central sensory loss, one may see a hemianesthesia with a boundary that precisely and exactly bisects the midline (care must be taken in interpreting this finding, however, for there are rare cases of hemianesthesia secondary to thalamic lesions wherein such a precise bisection is actually seen).

Before moving on to the next conversion symptom, it is worth remarking that the oft-stated rule that conversion paralysis and anesthesia are more likely to occur on the left than the right side does not, in fact, appear to hold true.

Conversion deafness, if bilateral and complete, is suggested by observing a blink reflex to an unexpected and loud sound, thus demonstrating intactness of the brainstem.

Conversion blindness may be either monocular or bilateral. In both cases, an intact direct and consensual pupillary response demonstrates that the visual pathways to the lateral geniculate bodies are intact. In cases of bilateral conversion blindness, one finds that the patient does not sustain injury while attempting to navigate around the office or hospital room. In doubtful cases, one may have to resort to visual evoked potentials to demonstrate that the post-geniculate visual pathways are intact.

Conversion seizures, or pseudoseizures, are discussed at length in Chapter 8.3, and the reader is strongly encouraged to consult that section.

Other conversion symptoms, of course, are possible, and these include aphonia, anosmia, nystagmus, convergence spasm, and ageusia.

Demonstrating a 'violation' may at times require considerable ingenuity on the part of the examining physician, and in all cases requires a thorough and detailed neurologic examination coupled with a firm grasp of the anatomy and physiology of the nervous system. Whether or not investigation beyond the clinical examination is required, for example with MR scanning, evoked potentials or EEGs, is a question that should be decided on a case-by case basis. Certainly, if the 'violation' is questionable, such investigation should be considered.

Although conversion disorder may occur in isolation, most patients have other disorders (Binzer *et al.* 1997), most commonly either a depression or a personality disorder of the histrionic, passive–aggressive, borderline, or antisocial types. Another feature often said to occur in association with conversion symptoms is 'la belle indifference', that is to say a casual indifference to symptoms, such as blindness or hemi-anesthesia, which would normally provoke considerable alarm. Unfortunately, this is not a reliable symptom, as it may either be absent or seen in other disorders (Stone *et al.* 2006). Furthermore, it may be confused with anosognosia.

COURSE

A number of prognostic factors have been identified (Binzer and Kullgren 1998; Couprie *et al.* 1995; Mace and Trimble 1996). The following factors generally predict early recovery from a conversion symptom: good intelligence, a clearly identifiable stress, an acute onset, and early initiation of treatment, and when these factors are present well over half of all hospitalized patients recover within 2 weeks. In those who do recover early on, however, recurrences are the rule, with up to one-fourth of patients relapsing within a year. A poor prognosis is suggested by the presence of a personality disorder, an older age of onset, a failure to improve during a hospital stay, and an already chronic duration of 6 or more months. Some studies have suggested that the presence of motor symptoms and pseudoseizures also predict a chronic course, but not all are in agreement on this. These chronic cases may endure for a decade or longer, and in some cases may persist indefinitely.

ETIOLOGY

The experience of patients with a conversion symptom is quite remarkable. Although it may be clear to the examining physician that the symptom must, in some sense or other, be 'produced' by the patient, the patient is not aware of doing so: for the patient, the symptom simply appeared, and did so not on the basis of any motivation or intention that the patient was aware of.

Various psychological theories have been proposed to explain the mechanism whereby the initiating psychologically meaningful stressor results in the conversion symptom. The traditional theory holds that the stressor triggered off, or resurrected, an unconscious intrapsychic conflict, and that the symptom serves to reduce the anxiety associated with that conflict. Consider, in this regard, the case of a mild-mannered man who, shortly after an argument with his wife, developed a monoplegia of his right arm. In this case, as was eventually revealed during the course of psychodynamic psychotherapy, the man, as a young boy, harbored strong aggressive and sexual urges toward his mother, urges held only in check by an archaic and sadistic superego; whenever such urges surfaced in childhood he experienced severe anxiety. The argument with his wife resurrected this unconscious conflict, and the resulting anxiety, which itself remained largely unconscious, was in turn quelled by the paralysis, which effectively rendered him harmless. This reduction in the unconscious anxiety, in turn, is referred to as the 'primary gain' of the conversion symptom. 'Secondary gain' must be clearly distinguished from this primary gain: secondary gain, in contrast to primary gain, *is* consciously experienced, and although it is not intentionally sought, it might be quite welcome. In the case in question, the patient was disabled from work by his paralysis and became eligible for short-term disability payments. Importantly, however, secondary gain does not always occur, and it cannot be stressed enough that neither the occurrence nor the continuance of a conversion symptom is dependent on the presence of secondary gain. For example, in the case in question, the paralysis persisted long after the disability payments ceased. Another psychological theory holds that the conversion symptom results from a kind of autohypnosis, and indeed most, but not all, patients with conversion disorder are easily hypnotized. A final psychological theory invokes the concept of dissociation to explain the production of a symptom totally outside the awareness of the patient experiencing it, and the not infrequent presence of a concurrent dissociative disorder lends credence to this theory.

In addition to these psychological theories, it appears that conversion disorder is more likely to occur when various factors are present, including living in a rural setting, a lower educational status, a lower socioeconomic status, and a lack of psychological sophistication.

Although most of the work for over a century has proceeded on the assumption that the brain is normal in conversion disorder, recent work has begun to challenge this notion. For example, a functional MRI study of patients with conversion anesthesia demonstrated that sensory stimulation of the affected limb failed to activate the contralateral somatosensory cortex (Ghaffar *et al.* 2006). Similarly, in a case of conversion paralysis, when the patient attempted to move the paretic limb, positron emission tomography scanning revealed decreased activity of the contralateral primary motor cortex (Marshall *et al.* 1997). This latter finding, of course, could be explained by a simple failure to make any attempt; however a subsequent, and very elegant, study indicated otherwise. In this study (Vuilleumier *et al.* 2001), seven patients with well-documented, and strictly unilateral, conversion paralysis underwent SPECT scanning during bilateral *sensory* stimulation of both hands. In normal individuals, such passive sensory stimulation reliably activates the *motor* cortex and related subcortical structures, and the use of such sensory stimulation in this study eliminated the possibility that any differences found in SPECT scanning of the patients could be related to intentional inactivity of the involved extremity. Initial scanning, conducted when the conversion paralysis was present, revealed decreased cerebral blood flow in the striatum and thalamus contralateral to the paralyzed side; furthermore, in the four patients who recovered, repeat SPECT scanning during sensory stimulation no longer demonstrated this asymmetry. Taken together, these studies strongly suggest that the behavior of patients with conversion disorder is not only not intentional, but is also related to clearly abnormal brain activity. The nature of this abnormality, of course, remains unclear; however, it is of interest that there may also be a genetic component to conversion disorder, as indicated by a twin study which demonstrated a higher monozygotic than dizygotic concordance (Inouye 1972).

DIFFERENTIAL DIAGNOSIS

The most important differential to make, of course, is between conversion symptoms and 'true' symptoms that are, in fact, occurring on the basis of central or peripheral nervous system disease. In this regard, it must be borne in mind that, despite thorough investigation, a small minority of patients who receive the diagnosis of conversion disorder will, on follow-up, be found to have lesions missed during the initial evaluation (Binzer and Kullgren 1998; Moene *et al.* 2000). Consequently, the importance of a detailed and thorough examination, coupled with appropriate imaging and laboratory testing, cannot be overemphasized.

Conversion symptoms may occur not only in conversion disorder, but also in somatization disorder and schizophrenia. Conversion disorder is characterized solely by conversion symptoms. By contrast, in these two other disorders one sees multiple other symptoms. In somatization disorder there are multiple complaints referable to organ systems other than the central nervous system, and hence one typically hears of complaints regarding pulmonary, gastrointestinal, and musculoskeletal functioning. In schizophrenia, one sees a variable mixture of psychotic symptoms, such as delusions, hallucinations, loosened associations, and overall bizarreness.

Finally, consideration must be given to the possibility of malingering and factitious illness, and these very important possibilities are discussed in Chapter 8.11.

TREATMENT

After the diagnosis is made, one should inform the patient in a calm and quietly authoritative way that the investigation and testing indicate that the nervous system is intact and not damaged. One may then go on to add that although it is not known why these symptoms have appeared, it is known that, with time, most patients recover.

Various treatments may be considered. Cognitive–behavioral therapy appears, as discussed in Chapter 8.3, to be effective for pseudoseizures, and psychodynamic psychotherapy is effective for conversion movement disorders (Hinson *et al.* 2006) (e.g. parkinsonism); anecdotally both forms of psychotherapy also appear effective for other conversion symptoms. Hypnotherapy, though effective as monotherapy for motor conversion symptoms (Moene *et al.* 2003), did not, when added on to a comprehensive inpatient program (which utilized, among other modalities, psychodynamic psychotherapy) produce any further benefit (Moene *et al.* 2002). Physical therapy, when provided during an acute inpatient rehabilitation stay, may lead to a rapid resolution of motor symptoms (Watanabe *et al.* 1998). Finally, mention should be made of 'narcoanalysis,' which involves the intravenous administration of agents such as amobarbital or lorazepam, with or without a concurrent stimulant, in the hopes of producing an abreaction during the subsequent interview which may lead either to symptom resolution or, at the least, to data that allow a fuller understanding of the psychological processes that may have led to the appearance of the symptom. Once widely used, this technique is now fading from view, as there is little evidence for its effectiveness (Poole *et al.* 2010).

Somatization disorder

Somatization disorder, also known as Briquet's syndrome (in honor of Pierre Briquet, who first described it in 1859), is characterized by multiple complaints, referable to multiple organ systems, all occurring in the absence of any disease entity that could reasonably account for them (Perley and Guze 1962). These complaints persist chronically, and typically occasion multiple evaluations, hospitalizations, and often needless diagnostic procedures or surgeries. Conservative estimates indicate a lifetime prevalence in the general population of from 0.2 to 2 percent. Importantly, somatization disorder is far more common in females than males, by a ratio of at least 10:1, and indeed, this disorder may be rare in males (Smith *et al.* 1985).

CLINICAL FEATURES

This syndrome generally first appears in teenage years; onset beyond the age of 30 is extremely rare.

Patients tend to be excessively vague or dramatic in relating their history, often moving restlessly from one symptom to another, never lingering long enough on one symptom to give an adequately detailed account. As noted earlier, multiple complaints are heard, and these typically involve four systems, namely the nervous, genitourinary, gastrointestinal, and musculoskeletal systems.

Nervous system complaints may include headache and any of conversion symptoms discussed in the immediately preceding section of this chapter.

Genitourinary symptoms include painful symptoms, such as menstrual pain, dyspareunia or dysuria, and non-painful symptoms, such as menstrual irregularity, excessive menstrual bleeding, vomiting throughout almost the entirety of pregnancy, decreased libido, and either erectile or ejaculatory dysfunction.

Gastrointestinal symptoms include painful symptoms such as abdominal pain (which is typically poorly localized) and rectal pain, and non-painful symptoms, including bloating, nausea, vomiting, diarrhea, or multiple food intolerances.

Musculoskeletal complaints all involve pain, and include backache, arthralgia, extremity pain, and chest pain, which is typically diffuse.

Although there is debate as to how many organ systems must be involved, and now many symptoms must be present to allow for a reliable diagnosis, a landmark study (Yutzy *et al.* 1995) established the following criteria, all of which must be present: at least one conversion symptom; at least one non-painful genitourinary symptom; at least two non-painful gastrointestinal symptoms; and, finally, at least four painful symptoms, which could be from any of the four organ systems.

The large number of complaints, and the inability of the physician to pin the patient down as to details, often make the interview very frustrating for the physician, and it is typical to find chart entries indicating merely that the review of systems was 'diffusely positive'. Many physicians turn to the physical examination with some relief, hoping to determine some definite findings. Generally, however, if there are any findings, they are typically minor and not indicative of any disease or condition that could possibly account for the patient's multitudinous complaints.

When physicians attempt to reassure patients regarding the benign nature of the examination, they are often met with disbelief, if not hostility, and patients typically demand tests, and when basic tests are unremarkable, the demands persist. Some physicians may call it quits at this point, but others will proceed to invasive procedures or even to surgery. In some cases, patients have welcomed so many abdominal surgeries that they finally develop a 'battlefield abdomen'. 'Doctor shopping' is common, as exasperated physicians 'fire' their patients, or patients, dissatisfied with the diagnostic approach offered by their physicians, move on to find a more aggressive diagnostician.

In addition to these multiple complaints, depression and panic attacks are common, as is alcohol abuse or alcoholism. Personality disturbances of the borderline, histrionic, or antisocial type, are also common.

COURSE

Somatization disorder typically pursues a chronic course, with the intensity and variety of complaints fluctuating gradually over the lifespan of the patient; spontaneous remissions are very uncommon.

ETIOLOGY

It appears that somatization disorder may, in part, be inherited. A blinded family study (Guze *et al.* 1986) of first-degree relatives of patients with somatization disorder found, in female relatives, an increased prevalence of somatization disorder, and, in both female and male relatives, an increased prevalence of antisocial personality disorder. Echoing these findings, an adoption study (Bohman *et al.* 1984) demonstrated, in the fathers of females with high somatization, an increased prevalence of antisocial personality disorder. These findings are consistent with the notion that somatization disorder and antisocial personality disorder may share a common genetic background, with sex-mediated expression.

DIFFERENTIAL DIAGNOSIS

Before making the diagnosis of somatization disorder, one must, obviously, effectively rule out other disorders that could mimic the symptom profile called for in the diagnostic criteria noted earlier. The key here lies in the fact that the patient must have at least one conversion symptom, and there are only a few disorders that can produce not only a neurologic symptom but also the multiplicity of other symptoms called for in the diagnostic criteria: possibilities include systemic lupus erythematosus, multiple sclerosis, Behçet's syndrome and polyarteritis nodosa.

Conversion disorder may also be considered on the differential but is ruled out on two counts: first, rather than a multitude of symptoms, there is generally only one, or perhaps two; and, second, rather than a multitude of organ systems, only one is involved in conversion disorder, namely the nervous system.

Malingering and factitious disorder, like conversion disorder, generally are not associated with multiple complaints; furthermore, the complaints are intentionally feigned.

Hypochondriasis may also be considered, as hypochondriacal patients often have multiple complaints referable to multiple organ systems. The difference here, however, relates to the patients' attitude toward the complaint. In hypochondriasis, rather than being concerned about any suffering associated with the complaint, patients are worried about what the symptom implies, namely the presence of a serious, but undiagnosed, disease. In Briquet's syndrome, however, the focus is more on the suffering associated with the symptom.

Depression may be associated with multiple unexplained complaints, and thus can present a picture similar to somatization disorder. To complicate matters further, as noted above, patients with somatization disorder often do have concurrent depression. The key to making the differential here lies in the time course: in cases in which the complaints are secondary to depression, one finds the onset of depressed mood and associated vegetative symptoms well before the appearance of multiple complaints. In doubtful cases, it may be appropriate to attempt a 'diagnosis by treatment response' by prescribing an antidepressant and watching to see if the multiple complaints subside.

Schizophrenia may also be associated with multiple complaints, but these typically have a bizarre cast to them, and are associated with other typical psychotic symptoms, such as delusions, hallucination, etc.

TREATMENT

Overall, a conservative medical approach is clearly preferable, and ideally patients should be seen in long-term follow-up by one physician, preferably an internist or family physician.

Psychiatric consultation to the primary care physician is helpful (Smith *et al.* 1986), and it is advisable to see patients on a scheduled basis, as this may forestall exacerbations. Recent work has also demonstrated that cognitive–behavioral therapy is effective (Allen *et al.* 2006).

Hypochondriasis

In hypochondriasis (Barsky 2001), patients, on the basis of minor symptoms, come to believe, or, at the very least, strongly suspect, that they have a serious, perhaps even life-threatening, disease. Their concerns occasion multiple consultations, often with multiple physicians, and, importantly, despite negative examinations and earnest reassurances regarding their condition, these patients remain beset by their concerns. This condition probably has a lifetime prevalence of between 1 and 5 percent, and is equally common among males and females.

CLINICAL FEATURES

Although the onset of hypochondriasis may occur at any point between adolescence and old age, most patients first begin to experience their concerns in their twenties or thirties. Although in most cases there does not appear to be a precipitating event, occasionally the onset may be triggered either by observing a serious illness in an acquaintance or personally suffering one. A good example of the latter is the patient who remains an invalid, consumed by hypochondriacal concerns after recovering from a heart attack, despite reassurances from the cardiologist.

Patients come to the physician anxiously concerned that their symptoms, no matter how mild or trivial, indicate the presence of a severe disease. A mild, non-productive cough means that they have pneumonia, or perhaps lung cancer; a few palpitations indicate that the heart is about to fail; slight nausea is a sure sign that an ulcer has eaten through the stomach, and simple constipation can only mean that colon cancer has finally appeared.

Patients often present their complaints in minute and maddening detail, and many physicians will turn to the physical examination with some relief. At the end of the evaluation, it is often clear either that there is nothing wrong, or, if some condition is present, that it is mild, even trivial. Reassurance on the part of the physician however, rarely has any lasting effect, and indeed some patients may begin to have doubts before they get home. Some, indeed, and especially those who have seen other physicians, may not be reassured at all, and may demand more tests, and if the physician expresses some skepticism regarding this, they may become demanding. Predictably, 'doctor shopping' is common.

The expression of hypochondriacal concerns is not confined to hospital or doctor visits but permeates these patients' lives. They may share their worries about their health at the dinner table, the office, or at social gatherings, anxiously going from person to person until they find a sympathetic listener who will tolerate their complaints. In some cases, their complaints are so wearying that others begin to avoid these patients, who then become isolated and even more miserable. Some, paralyzed by their concerns, will opt to enter a nursing home in order to be sure that medical care is immediately available.

Depression or panic disorder is present in a majority of cases (Noyes *et al.* 1994).

COURSE

Hypochondriasis, in the majority of cases, appears to be chronic, with symptoms waxing and waning in intensity over the years (Barsky *et al.* 1998). Although it appears that spontaneous full remissions do occur, the frequency with which this occurs is not clear.

ETIOLOGY

Hypochondriasis does not appear to run in families (Noyes *et al.* 1997), and the etiology is as yet unclear. Although these patients recall having more serious illnesses in childhood and going through more emotionally traumatic events (Barsky *et al.* 1994; Noyes *et al.* 2002), there have been no prospective studies, and hence these findings may well be due to recall bias.

Based on the similarity between the persistent recurring concerns seen in hypochondriasis and obsessions seen in obsessive–compulsive disorder, there has been speculation that hypochondriasis is but one example of the 'obsessive–compulsive spectrum' disorders, but as yet proof of this is lacking.

DIFFERENTIAL DIAGNOSIS

Hypochondriasis must be distinguished from transient hypochondriacal concerns, which are very common in the general population: the notorious 'doctor's diseases' suffered by medical students are a good example. The differential here rests on the duration of hypochondriacal concerns: most transient concerns resolve spontaneously in a matter of weeks, and in any case, persist no longer than 6 months.

Depression is perhaps the most important differential to consider. Especially in the elderly, depression may manifest with hypochondriacal concerns; indeed, such patients may limit their presentation to such complaints, and not spontaneously report the accompanying vegetative symptoms, such as anergia, anhedonia, anorexia, and insomnia. These 'masked depressions' may at times be difficult to diagnose, as in some cases patients may deny feeling depressed.

The somatic subtype of delusional disorder is characterized by the presence of a delusion on the patient's part that he or she has a disease, and at first glance this disorder may appear difficult to differentiate from hypochondriasis. Two features, however, if closely attended to, will typically enable a correct differential. To begin with, assess the certainty of the patient's belief that he or she has a disease: in delusional disorder, there is no question in the patient's mind, and while some patients with hypochondriasis may also evidence such certainty, most will, at least at some point, be willing to accept some reassurance to the contrary. Next, consider the overall behavior of the patient. In delusional disorder, the presentation is often simply characterized by the belief that one has a disease, accompanied by a request for a diagnostic evaluation. The presentation of hypochondriasis, while of course also containing these elements, also is characterized by a saturation, as it were, of the entire life of the patient by hypochondriacal concerns, and this element is not typical of delusional disorder. In cases in which, however, a reliable clinical differential remains elusive, one should attempt a diagnosis by treatment response, utilizing either cognitive–behavioral therapy or an SSRI, as discussed below.

Somatization disorder is distinguished both by the nature of the patients' complaints and by the manner in which they are made. In somatization disorder, patients typically have a multitude of complaints, and, here, it is not so much a concern that the symptoms indicate a serious underlying disease as it is with the debilitating nature of the symptom itself. Thus, whereas a hypochondriacal patient complaining of constipation may admit that it is mild, and relieved with simple fiber laxatives, the patient with somatization disorder may complain that it is painful, unbearable, and impossible to overcome. Further, whereas the hypochondriacal patient typically reports complaints in precise detail, the patient with somatization disorder is for the most part vague and difficult to pin down.

Conversion disorder is likewise suggested by the nature of the complaint. In conversion disorder, the complaint always refers to the nervous system: in hypochondriasis, such complaints may also be heard, but other organ systems are more commonly implicated. Furthermore, in conversion disorder the patients' concerns are not so much related to a fear of some undiagnosed and underlying disease, as is the case in hypochondriasis, as they are to the effects of the symptom itself, such as an inability to walk because of conversion paralysis.

Malingering and factitious disorder are both distinguished by the fact that the patient's complaints are conscious and intentional lies. Neither the malingerer nor the person with factitious disorder have any sincere concern or belief that he or she is ill: in malingering and factitious disorder it is all an intentional sham in the service of some more or less obvious goal, and this is in stark contrast to hypochondriasis wherein the patient believes, to a greater or lesser degree of certainty, that he or she is, in fact, ill.

Finally, one must remain alert to the possibility that new complaints, rather than being hypochondriacal, may signal a serious underlying disease: each new complaint must be evaluated on its own merits.

TREATMENT

Cognitive–behavioral therapy is effective (Barsky and Ahern 2004), as are the SSRIs fluoxetine (Fallon *et al.* 2008) and paroxetine (Greeven *et al.* 2007), with cognitive–behavioral therapy and paroxetine being roughly equivalent in efficacy (Greeven *et al.* 2007). Should symptoms persist, it is appropriate to maintain a conservative medical approach and to see patients in regularly scheduled follow-up visits.

8.11 MALINGERING AND FACTITIOUS ILLNESS

Both malingering (LoPiccolo *et al.* 1999) and factitious illness (Krahn *et al.* 2003) are characterized by the intentional feigning of illness or the embellishment of an actual illness; they are distinguished from one another by the motive underlying this behavior. In malingering the motive is readily understandable, as for example when someone complains of a 'bad back' to get out of work or to obtain narcotics. In factitious illness, however, the motive is a little more obscure, in that the goal of these individuals is merely to be a patient in the hospital, and to assume, as it were, the 'sick role'.

Before making a diagnosis of malingering or factitious illness, one must be reasonably certain that there is no true underlying illness that could reasonably, and fully, account for the individual's complaints and symptoms. Furthermore, one must also distinguish malingering and factitious illness from conversion disorder, a distinction based primarily on the lack, in conversion disorder, of any associated intentionality. Thus, in the case of 'paralysis', the malingerer and the individual with factitious illness both consciously and intentionally feign weakness, with the goal, respectively, of either an understandable gain (e.g. winning a lawsuit) or of simply being a patient on the neurologic ward. By contrast, in the case of, say, conversion disorder, the 'paralysis' simply appears, without any planning or intention on the part of the patient.

Malingering

As noted earlier, the goal of the malingerer is to utilize a feigned or embellished illness in order to achieve a readily recognizable goal. Common scenarios include the following: an accident victim who files a lawsuit immediately after the accident; a soldier who develops an illness just before scheduled deployment to a war zone; a criminal defendant who develops amnesia or dementia, thus raising questions of competency, or psychosis, thus raising questions of culpability; or a homeless person who presents to the emergency room during inclement weather, complaining of abdominal pain, headache, or the like.

Overall, psychiatric, neurologic, or musculoskeletal disorders are often chosen by the malingerer, and the more or less sophisticated malingerer will choose from among these groups an illness that lacks definitive findings on the physical examination, laboratory tests, and imaging. Psychiatric disorders, such as PTSD, depression, psychosis and attention deficit–hyperactivity disorder, are obviously 'good' choices in this regard. Among the neurologic disorders, headache, movement disorders, and seizures may be chosen, while among musculoskeletal disorders, back pain is a common choice.

In essence, in the evaluation of possible malingering, the goal is to determine whether or not someone is engaged in conscious, deliberate, and intentional lying.

During the interview, close attention should be paid to the patient's overall demeanor. In some cases, 'conning' may be fairly obvious. Patients may come across as slick and more interested in the physician's reaction to what they are saying than in relating the history itself. For example, the experienced physician, upon hearing a malingerer's history, will often begin to feel skeptical, and no matter how hard the physician strives to present a 'neutral' facial expression, some hint of the felt skepticism may appear in his or her facial expression. Malingerers, intent on convincing the physician, rather than in relating their history, are often quick to pick up on this, and may interrupt their own story with some such exclamation as 'Look, I'm telling the truth!' or 'You don't believe me, do you?' Furthermore, malingerers, though often going to great lengths in preparing their lie, yet may find, upon questioning by the physician, that certain parts of their story are simply inconsistent, and at such points, they may become evasive, and either try to 'rehabilitate' what they had said earlier, or simply feign confusion, or perhaps offer some excuses for the inconsistencies. Finally, should it become apparent that the physician is not going to offer the desired diagnosis, or provide a prescription for the desired narcotics, benzodiazepines, or stimulants, some malingerers may become demanding or even belligerent.

The malingerer's response to the physician's request to obtain records (whether medical, military, or employment) or to interview people (whether relatives, neighbors, supervisors, or colleagues) may also be telling. Some immediately recognize that the request, if granted, will expose the lie, and thus immediately refuse. Others, who perhaps may have coached potential collateral informants into supporting the lie, may agree, but often only with reluctance, as they may realize that their 'partners in crime', no matter how well coached, may yet offer a story that varies with what the malingerer has relayed to the physician. Malingerers who do refuse, or perhaps agree but with a plea that the physician not go through with interviewing others, may offer any of a number of more or less plausible reasons: they wish to avoid embarrassment; they do not want to put others to the trouble of an interview; or, perhaps even more cloyingly, they wish to spare prospective interviewees the pain of recounting a story that also affected them.

In some cases, rather than being reluctant to allow inspection of records, malingerers may bring records with them. Here it is very important for the physician to still ask the patient's permission to request records, as some patients may be very selective in what they bring, or may even falsify original records. Should patients protest, the physician should counter with the plausible statement that what the patient does have may indeed not be complete: a continued protest from the patient strongly suggests that the requested records would reveal the lie.

The discrepancy between malingerers' behavior when they are under observation and when they believe themselves to be unobserved often also exposes the lie. Given that the goal of malingerers is to convince others that they are ill, they will generally increase the severity of symptoms when they believe themselves to be under observation, and, conversely, will relax their efforts, or even drop the act altogether when they believe themselves to be unobserved. For example, a patient malingering back pain after a 'slip and fall' accident in a grocery store, and suing the store because of it, may hobble into the office and present a picture of anguish, yet when off on vacation to another city may vigorously engage in volleyball. In a similar vein, once all prospect of gain from the feigned illness is lost, symptoms fairly promptly resolve. Thus, once a malingerer complaining of back pain lost his suit, he got 'better' in a matter of weeks.

The history provided by the patient is obviously critical, as it simply may not conform to what is known about the disorder being malingered. Sometimes the malingerers themselves may offer an unbelievable story. At other times what the malinger says may appear at least somewhat believable, and in such cases the physician may then consider asking, either during the history taking or during the review of systems, about symptoms known to be inconsistent with the disorder in question, and see if the patient endorses any of them. For example, if depression is in question, the physician may say that some cases come on gradually, over weeks or months, while others, especially more severe cases, may come on suddenly, over perhaps a day: if the patient replies that indeed his depression did come on suddenly, the evaluation is essentially over.

In cases of malingered psychiatric disease, the mental status examination may offer important clues. Thus, a patient malingering depression may simply not have a particularly depressed affect, or if he or she does, and it happens to be severe, there may be no associated psychomotor retardation or agitation. Should an attempt be made to malinger schizophrenia, the 'delusions' expressed, or the hallucinations reported, may simply fail to be consistent with what is typically seen in this disease; rather they often reflect the popular conception of insanity portrayed in the media. Mention here may also be made of a peculiar form of malingering occasionally seen in prisoners awaiting trial or sentencing, known as 'Ganser's syndrome' (Carney *et al.* 1987; Tsoi 1973). Also known as the 'nonsense syndrome', this is characterized by 'nonsense' responses to questions, which are always just off the mark or past the point. For example, if the individual is asked to add 5 plus 3, he may respond '7'; with coaching and encouragement he may give other responses, such as '6' or '9', but never the correct one. In a similar vein, if asked how many legs a horse has, the response may be '3'. Typically, although these individuals appear confused and dazed, they are generally able to find their way around the jail and to do those things that are necessary to maintain a certain degree of comfort and safety.

In cases of malingered neurologic or other general medical disorders, the physical examination is obviously essential, and, in the hands of an experienced physician, is often conclusive. When neurologic disorders are in question, the exam may turn up abnormalities which 'violate' the laws of neuroanatomy or neurophysiology, as discussed in Chapter 8.10. Imaging or laboratory testing play a similar role.

Two conditions must be distinguished from malingering. The first is factitious disorder. Factitious disorder resembles malingering in that the symptoms and signs are deliberately, consciously and intentionally feigned. Where it differs is in the goal of this conscious simulation: in malingering, the goal, whether financial gain or avoidance of punishment, is readily discernible; in factitious disorder, however, the goal, namely the desire to be a patient in a hospital, is something most people would not seek, and hence may not come to mind during the evaluation.

The second disorder to distinguish from malingering is conversion disorder, and here the differentiation is more difficult. In conversion disorder, patients present with what appears to be a neurologic disorder, and of course such a presentation may also be perpetrated by malingerers; indeed, the presentations themselves, whether of tremor, weakness, anesthesia, or seizures, may be identical. The difference lies in whether or not the symptoms are consciously and intentionally created: in malingering they are, whereas in conversion they are not. Thus, the question comes down to deciding whether the patient is lying or not, and several items may help in deciding this question. First, look to the patient's overall demeanor, as discussed earlier. Second, consider the patient's reaction to a request for records or collateral interviews: in contrast with the reluctance of most malingerers, those with conversion disorder generally readily assent. Third, look to what happens when patients believe themselves unobserved or when the putative 'goal' is no longer attainable (e.g. a lawsuit is lost): in contrast with malingerers, those with conversion disorder continue to experience symptoms when unobserved, or if, say, a lawsuit had been filed, long after the lawsuit is lost. As a corollary to this, whereas malingerers typically, if anything, intensify symptoms when under observation, in those with conversion disorder symptoms may actually lessen some during observation. In essence, these examples go to the point that in malingering, because the malingerer is 'playing' to an audience, symptoms are exquisitely sensitive to whether an audience is present or not, while in conversion disorder the symptoms are autonomous, and generally independent of environmental events. This last point deserves emphasis, especially when it comes to the issue of 'secondary gain' in conversion disorder. As discussed in the section on conversion disorder (Chapter 8.10), the traditional theory of the etiology of conversion disorder involves, in all cases, a 'primary gain', and, in most, eventually a 'secondary gain'. The primary gain is the reduction in anxiety associated with the resolution of an unconscious intrapsychic conflict, a resolution brought about by the conversion symptom, and this gain is immediate upon appearance of the symptom. The secondary gain involves potential benefits from the responses of others to the appearance of the symptom, and this may be immediate (as for example when others immediately become solicitous toward the patient) or delayed (as for example an eventual insurance settlement). Secondary gain in conversion disorder, unfortunately, is often confused with the realization of the goal of the malingerer's scam. The two, however, are quite different. Take, for example, the sudden appearance of conversion monoplegia of the right upper extremity in a man shortly after an argument with his wife. In this case, the monoplegia immediately reduced anxiety occasioned by the conflict between an aggressive urge and a fear of a brutal conscience, and this constituted the primary gain. Subsequent to the appearance of the monoplegia, however, the man was no longer able to continue his work as a laborer, and was forced to apply for disability payments. This, obviously, would have been a great gain; however, it must be stressed that the occurrence of the symptom was not related to such a potential secondary gain, and indeed, it was not even dependent on it for its continuation, as evidenced by the fact that the symptom persisted after the disability claim was denied. Contrast this with the example of a malingerer, who, shortly after a 'slip and fall' wherein he struck his arm, also developed a monoplegia, after which he filed a lawsuit alleging that he could no longer work. Here, there is no primary gain, and the potential gain, that is, winning the lawsuit, is not secondary at all, rather it is the only potential gain, and once the lawsuit is lost, there being nothing present to keep the symptom going, the monoplegia promptly resolves.

What the physician should do, once it becomes clear that malingering is present, is not clear. Some advocate a simple, but non-judgmental, discussion of the facts, and indeed some malingerers may respond favorably to this. Most, however, will not, and indeed may become even more demanding or even hostile. Regardless of what approach is taken, however, it is important to 'do no harm', and in this regard, one should not prescribe narcotics, certify non-existent illnesses, or do anything else that reinforces the patient's deception.

Factitious illness

The illnesses feigned here tend to be severe, as might be expected, given that the goal of the dissimulation is admission to the hospital. Typically, the patient arrives at the emergency room with a very convincing presentation (Reich and Gottfried 1983). Some may complain of several episodes of severe chest pain, suggesting crescendo angina. Others may report having had a 'seizure', and even bite their tongue to make the picture more convincing. Others may swallow blood and then vomit, thus simulating hematemesis, whereas others may hold the blood in their mouths and then cough, producing a picture of hemoptysis. A urine specimen may be contaminated with feldspar to mimic renal calculi, or with feces to suggest a severe urinary tract infection. More malignantly, feces may be injected to create a septic picture. Laxatives may be taken to induce diarrhea, furosemide to create hypokalemia, myelosuppressants to mimic aplastic anemia, prednisone to create Cushing's syndrome, thyroid hormone to produce hyperthyroidism, warfarin to create a coagulopathy, and either insulin or oral antidiabetic agents to produce hypoglycemia and raise the question of an insulinoma. In the case of hypoglycemia, determining simultaneous insulin and C-peptide levels is helpful to rule out surreptitious injection of insulin: in an insulinoma both are elevated, whereas when insulin is injected the C-peptide level will be normal (Grunberger *et al.* 1988).

Once admitted, these individuals may make frequent demands for narcotics, and staff are often split and played off, one against the other. Diagnostic tests are welcomed, even demanded, and as the tests become ever more invasive and dangerous, these individuals often become calmer, even content. As more and more tests come back negative or inconsistent (Wallach 1994), the complaints may change: chest pain may fail to recur, but now abdominal pain and diarrhea come to the forefront. Eventually, when told by the physician that there is 'nothing wrong', these individuals typically become angry, accuse the physician of incompetence, and leave the hospital abruptly, often against medical advice.

The majority of individuals with factitious illness are female; most are in their twenties or thirties, and most have some medical background, having worked as aides, nurses, or therapists of one sort or other. In these cases, the frequency with which hospitalization is sought varies over time, and is often related to stressful events. In most cases, the dissimulation eventually stops after a matter of years.

In a small minority of cases, however, the course is chronic and severe, and in such cases one speaks of 'Munchausen's syndrome', named after the famous German baron who traveled from city to city telling elaborate tales about himself (Asher 1951). In contrast with the typical individual with factitious illness, these individuals tend to be male and middle-aged, and to have a history of traveling from city to city with, at times, literally hundreds of hospital admissions. Also known as 'hospital hoboes' (Clarke and Melnick 1958) or 'hospital addicts' (Barker 1962), these individuals have made being a patient into a way of life.

A particularly loathsome variant of factitious illness (or Munchausen's syndrome) is that which occurs 'by proxy' (Souid *et al.* 1998). Here, a parent, typically the mother, may induce illness in a young child by any of the means noted above, and use the result as a ticket for the child's admission. These parents may, at least initially, seem appropriately concerned; however, suspicions may be aroused when staff note a certain satisfaction, even contentment, on the parent's part as the child is subjected to ever more invasive diagnostic procedures. In some cases, laboratory results may clinch the diagnosis, whereas in others it may be necessary to resort to covert video surveillance to 'catch' the parent injecting or otherwise harming the child (Hall *et al.* 2000). Although such 'proxy' factitious illness most often involves a parent and child, there have also been cases in which adults have induced illness, either in their spouses or, more commonly, their elderly parents.

Although most cases of factitious illness involve 'physical' signs and symptoms, cases may also occur where 'psychological' signs and symptoms are feigned (Gelenberg 1977; Pope *et al.* 1982; Popli *et al.* 1992). Individuals may report suicidal or homicidal ideation, or may complain of voices, visions, deep depression, or post-traumatic stress. Given that laboratory testing is generally irrelevant in such cases, unmasking the dissimulation may take a little longer; however, eventually inconsistencies become apparent. In such cases, when confronted by the physician's suspicions, these individuals may, like those with 'physical' symptoms, become indignant and demand discharge, or they may 'up the ante' by making a suicidal gesture.

As with malingering, it is not clear what is the best approach to factitious illness. Some advocate confrontation, whereas others will attempt to engage the individual in some form of psychotherapy. In cases of 'proxy' illness, legal authorities must be involved.

REFERENCES

Abd El Naby S, Hassanein M. Neuropsychiatric manifestations of chronic manganese poisoning. *J Neurol Neurosurg Psychiatry* 1965; **28**:282–8.

Abenson MH. Epileptic fits provoked by taste. *Br J Psychiatry* 1969; **115**:123.

Acharya V, Acharya J, Luders H. Olfactory epileptic auras. *Neurology* 1998; **51**:56–61.

Adachi N, Arima K, Kato M *et al.* Dentatorubral-pallidoluysian atrophy (DRPLA) presenting with psychosis. *J Neuropsychiatry Clin Neurosci* 2001; **13**:258–60.

Adachi N, Kato M, Sekimoto M *et al.* Recurrent postictal psychosis after remission of interictal psychosis: further evidence of bimodal psychosis. *Epilepsia* 2003; **44**:1218–22.

Adachi N, Kanemoto K, Muramatsu R *et al.* Intellectual prognosis of status epilepticus in adult epilepsy patients: analysis with Wechsler Adult Intelligence Scale-revised. *Epilepsia* 2005; **46**:1502–9.

Adair JC, Williamson DJG, Schwartz RL *et al.* Ventral tegmental area injury and frontal lobe disorder. *Neurology* 1996; **46**:442–3.

Adam C, Baulac M, Saint-Hilaire J-M *et al.* Value of magnetic resonance imaging-based measurements of hippocampal formation in patients with partial epilepsy. *Arch Neurol* 1994; **51**:130–8.

Adamec DE, Stark-Adamec C. Limbic kindling in animal behavior: implications for human psychopathology associated with complex partial seizures. *Biol Psychiatry* 1983; **18**:269–74.

Agius MA, Chan JW, Chung S *et al.* Role of antiribosomal P protein antibodies in the diagnosis of lupus isolated to the central nervous system. *Arch Neurol* 1997; **54**:862–4.

Agnetti V, Carreras M, Pinna L *et al.* Ictal prosopagnosia and epileptogenic damage of the dominant hemisphere. *Cortex* 1978; **14**:50–7.

Aguilar MJ, Rasmussen T. Role of encephalitis in pathogenesis of epilepsy. *Arch Neurol* 1960; **2**:663–76.

Ahl B, Bokemeyer M, Ennen JC *et al.* Dissection of the brain supplying arteries over the life span. *J Neurol Neurosurg Psychiatry* 2004; **75**:1194–6.

Ahuja GK, Mohandas S, Narayanaswamy AS. Eating epilepsy. *Epilepsia* 1980; **21**:85–9.

Ajmone-Marsan C, Ralston BL. *The Epileptic Seizure: Its Functional Morphology and Diagnostic Significance.* Springfield, IL: Charles C. Thomas, 1957.

Akelaitis AJ. Lead encephalopathy in children and adults. *J Nerv Ment Dis* 1941; **93**:313–32.

Alajouanine T, DeRobert L, Thieffry S. Etude clinique de 210 cas d'intoxication par les sels organiques d'etain. *Rev Neurol* 1958; **98**:85–96.

Alamowitch S, Graus F, Uchuya M *et al.* Limbic encephalitis and small cell lung cancer: clinical and immunological features. *Brain* 1997; **120**:923–8.

Al-Asmi A, Jansen AC, Badhwar AP *et al.* Familial temporal lobe epilepsy as a presenting feature of choreoacanthocytosis. *Epilepsia* 2005; **46**:1256–63.

Albers GW, Thijs VN, Wechsler L *et al.* Magnetic resonance imaging profile predict clinical response to early reperfusion: the diffusion and perfusion imaging evaluation for understanding stroke evolution (DEFUSE) study. *Ann Neurol* 2006; **60**:508–17.

Alemayehu S, Bergey GK, Barry E *et al.* Panic attacks as ictal manifestations of parietal lobe seizures. *Epilepsia* 1995; **36**: 824–30.

Alexander MP, Freedman M. Amnesia after anterior communicating artery aneurysm rupture. *Neurology* 1984; **34**:752–7.

Alldredge BK, Lowenstein DH, Simon RP. Seizures associated with recreational drug abuse. *Neurology* 1989; **39**:1037–9.

Alldredge BK, Gelb AM, Issacs SM *et al.* A comparison of lorazepam, diazepam, and placebo for the treatment of out-of-hospital status epilepticus. *N Engl J Med* 2001; **345**:631–7.

Allen LA, Woolfold LR, Escobar JL *et al.* Cognitive-behavioral therapy for somatization disorder: a randomized controlled trial. *Ann Intern Med* 2006; **166**:1512–18.

Allen RM, Young SJ. Phencyclidine induced psychosis. *Am J Psychiatry* 1978; **135**:1081–4.

Almeida OP, Waterreus A, Hankey GJ. Preventing depression after stroke: results from a randomized, placebo-controlled trial. *J Clin Psychiatry* 2006; **67**:1104–9.

Almeida S, Ranjith G. Using pramipexole in neuropsychiatry: a cautionary tale. *J Neuropsychiatry Clin Neurosci* 2006; **18**:556–7.

Alpers BJ. Relation of the hypothalamus to disorders of personality: report of a case. *Arch Neurol Psychiatry* 1937; **38**:291–303.

Alpers BJ, Grant FC. The clinical syndrome of the corpus callosum. *Arch Neurol Psychiatry* 1931; **25**:67–86.

Alsen G, Gillberg IC, Lindblom R *et al.* Tuberous sclerosis in Western Sweden: a population study of cases with early childhood onset. *Arch Neurol* 1994; **51**:76–81.

Alves D, Pires MM, Guimaraes A *et al.* Four cases of late onset metachromatic leukodystrophy in a family: clinical, biochemical and neuropathological studies. *J Neurol Neurosurg Psychiatry* 1986; **49**:1417–22.

Amatniek JC, Hauser WA, DelCastillo-Castenada C *et al.* Incidence and predictors of seizures in patients with Alzheimer disease. *Epilepsia* 2006; **47**:867–72.

Ames FR, Enderstein O. Ictal laughter: a case report with clinical, cinefilm, and EEG observations. *J Neurol Neurosurg Psychiatry* 1975; **38**:11–17.

Ames FR, Enderstein O. Gelastic epilepsy and hypothalamic hamartoma. *S Afr Med J* 1980; **58**:163–5.

Aminoff MJ, Simon RP. Status epilepticus. Causes, clinical features and consequences in 98 patients. *Am J Med* 1980; **69**:657–66.

Aminoff MJ, Scheinman MM, Griffin JC *et al.* Electrocerebral accompaniments of syncope associated with malignant ventricular arrhythmias. *Ann Intern Med* 1988; **108**:791–6.

Andermann F, Robb JP. Absence status: a reappraisal following review of thirty-eight patients. *Epilepsia* 1972; **13**:177–87.

Andersen G, Vestergaard K, Riis J. Citalopram for poststroke pathological crying. *Lancet* 1993; **342**:837–9.

Andersen G, Vestergaard K, Lauritzen L. Effective treatment of poststroke depression with the selective serotonin reuptake inhibitor citalopram. *Stroke* 1994; **25**:1099–104.

Anderson DJ, Goldstein LB, Wilkinson WE *et al.* Stroke location, characterization, severity, and outcome in mitral vs aortic valve endocarditis. *Neurology* 2003; **61**:1334–6.

Anderson J. On sensory epilepsy. A case of cerebral tumour, affecting the left temporo-sphenoidal lobe, and giving rise to a paroxysmal taste-sensation and dreamy state. *Brain* 1886; **8**:385–95.

Andrews E, Ballard J, Walter-Ryan WG. Monosymptomatic hypochondriacal psychosis manifesting as delusion of infestation: case studies of treatment with haloperidol. *J Clin Psychiatry* 1986; **47**:188–90.

Annegers JF, Grabow JD, Groover RD *et al.* Seizures after head trauma: a population study. *Neurology* 1980; **30**:683–9.

Annegers JF, Hauser WA, Beghi E *et al.* The risk of unprovoked seizures after encephalitis and meningitis. *Neurology* 1988; **38**:1407–10.

Annegers JF, Hauser A, Coan SP *et al.* A population-based study of seizures after traumatic brain injuries. *N Engl J Med* 1998; **383**:20–4.

Appenzeller S, Cendes F, Costallat LTL. Epileptic seizures in systemic lupus erythematosus. *Neurology* 2004; **63**:1808–12.

Appleton RE, Farrell K, Teal P *et al.* Complex partial status epilepticus associated with cyclosporin-A therapy. *J Neurol Neurosurg Psychiatry* 1989; **52**:1068–71.

Arboix A, Bell Y, Garcia-Eroles L *et al.* Clinical study of 35 patients with dysarthria-clumsy hand syndrome. *J Neurol Neurosurg Psychiatry* 2004; **75**:231–4.

Arboix A, Lopez-Grau M, Casanovas C *et al.* Clinical study of 39 patients with atypical lacunar syndromes. *J Neurol Neurosurg Psychiatry* 2006; **77**:381–4.

Arciniegas DB, Anderson CA. Viral encephalitis: neuropsychiatric and neurobehavioral aspects. *Curr Psychiatry Rep* 2004; **6**:372–9.

Armstrong DD. The neuropathology of temporal lobe epilepsy. *J Neuropathol Exp Neurol* 1993; **52**:499–506.

Aron AM, Freeman JM, Carter S. The natural history of Sydenham's chorea. *Am J Med* 1965; **38**:83–95.

Aronica E, Gorter JA, Ramkema M *et al.* Expression and cellular distribution of multidrug resistance-related proteins in the hippocampus of patients with mesial temporal lobe epilepsy. *Epilepsia* 2004; **45**:441–5.

Arroyo S, Lesser RP, Gordon B. Mirth, laughter and gelastic seizures. *Brain* 1993; **116**:757–80.

Ashcroft PB. Traumatic epilepsy after gunshot wounds in the head. *BMJ* 1941; **1**:739–44.

Asher R. Munchausen's syndrome. *Lancet* 1951; **1**:339–41.

Ashford JW, Schulz SC, Walsh GO. Violent automatism in a partial complex seizure: report of a case. *Arch Neurol* 1980; **37**:120–2.

Ashkenazi A, Kaufman Y, Ben-hur T. Bilateral focal motor status epilepticus with retained consciousness after stroke. *Neurology* 2000; **54**:976–8.

Ashman TA, Cantor JB, Gordon WA *et al.* A randomized controlled trial of sertraline for the treatment of depression in persons with traumatic brain injury. *Arch Phys Med Rehabil* 2009; **90**:733–40.

Aurahami E, Weiss-Peretz J, Cohn AF. Focal epileptic activity following intravenous contrast material injection in patients with metastatic brain disease. *J Neurol Neurosurg Psychiatry* 1987; **50**:221–3.

Avbersek A, Sisodiya S. Does the primary literature provide support for clinical signs used to distinguish psychogenic nonepileptic seizures from epileptic seizures? *J Neurol Neurosurg Psychiatry* 2010; **81**:719–25.

Avery TL. Seven cases of frontal tumor with psychiatric presentation. *Br J Psychiatry* 1971; **119**:19–23.

Ay H, Koroshetz WJ, Benner T *et al.* Transient ischemic attack with infarction: a unique syndrome. *Ann Neurol* 2005; **57**:679–83.

Aykutlu E, Baykan B, Serdaroglu P *et al.* Epileptic seizures in Behcet's disease. *Epilepsia* 2002; **43**:832–5.

Bacalman S, Farzin F, Bourgeois JA *et al.* Psychiatric phenotype of the Fragile X-associated tremor/ataxia syndrome (FXTAS) in males: newly described fronto-subcortical dementia. *J Clin Psychiatry* 2006; **67**:87–94.

Backstrom T. Epileptic seizures in women related to plasma estrogen and progesterone during the menstrual cycle. *Acta Neurol Scand* 1976; **54**:321–47.

Baguley IJ, Cooper J, Felmingham K. Aggressive behavior following traumatic brain injury: how common is common? *J Head Trauma Rehabil* 2006; **21**:45–56.

Baird AE, Lovblad KO, Schlaug G *et al.* Multiple acute stroke syndrome: marker of embolic disease? *Neurology* 2000; **54**:674–8.

Baker JH, Silver JR. Hysterical paraplegia. *J Neurol Neurosurg Psychiatry* 1987; **50**:375–82.

Baker RW, Conley RR. Seizures during clozapine therapy. *Am J Psychiatry* 1991; **148**:1265–6.

Balkan S. Painful unilateral epileptic seizure. *Acta Neurol Scand* 1995; **91**:414–16.

Ballenger CE, King DW, Gallagher BB. Partial complex status epilepticus. *Neurology* 1983; **33**:1545–52.

Barbaro NM, Quigg M, Broshek DK *et al.* A mulicenter, prospective pilot study of gamma knife radiosurgery for mesial temporal lobe epilepsy: seizure response, adverse events, and verbal memory. *Ann Neurol* 2009; **65**:167–75.

Barcs G, Walker EB, Elger CE *et al.* Oxcarbazepine placebo-controlled, dose-ranging trial in refractory partial epilepsy. *Epilepsia* 2000; **41**:1597–607.

Bare MA, Burnstine TH, Fisher RS *et al.* Electroencephalographic changes during simple partial seizures. *Epilepsia* 1994; **35**:715–20.

Barker JC. The syndrome of hospital addiction (Munchausen syndrome): a report on the investigation of seven cases. *J Ment Sci* 1962; **108**:167–82.

Barkovich AJ, Kjos BO. Gray matter heterotopias: MR characteristics and correlation with developmental and neurologic manifestations. *Radiology* 1992; **182**:493–9.

Barkovich AJ, Rowley HA, Andermann F. MR in partial epilepsy: value of high-resolution volumetric techniques. *AJNR* 1995; **16**:339–43.

Barnes J, David AS. Visual hallucinations in Parkinson's disease: a review and phenomenological survey. *J Neurology Neurosurg Psychiatry* 2001; **70**:727–33.

Barry E, Sussman NM, Bosley TM *et al.* Ictal blindness and status epilepticus amauroticus. *Epilepsia* 1985; **26**:577–84.

Barsky AJ. The patient with hypochondriasis. *N Engl J Med* 2001; **345**:1395–9.

Barsky AJ, Ahern DK. Cognitive behavior therapy for hypochondriasis: a randomized controlled trial. *JAMA* 2004; **291**:1464–70.

Barsky AJ, Wool C, Barnett MC *et al.* Histories of childhood trauma in adult hypochondriacal patients. *Am J Psychiatry* 1994; **151**:397–401.

Barsky AJ, Fama JM, Bailey ED *et al.* A prospective 4- to 5-year study of DSM-III-R hypochondriasis. *Arch Gen Psychiatry* 1998; **55**:737–44.

Battaglia G, Chiapparini L, Fransceschetti S *et al.* Periventricular nodular heterotopia: classification, epileptic history, and genesis of epileptic discharges. *Epilepsia* 2006; **47**:86–97.

Bauer J, Kaufmann P, Elger CE *et al.* Similar postictal serum prolactin response in complex partial seizures of temporal or frontal lobe onset. *Arch Neurol* 1994; **51**:645.

Baumann CR, Werth E, Stocker R *et al.* Sleep-wake disturbances 6 months after traumatic brain injury: a prospective study. *Brain* 2007; **130**:1873–83.

Baumann CR, Bassetti CL, Valko PO *et al.* Loss of hypocretin (orexin) neurons with traumatic brain injury. *Ann Neurol* 2009; **66**:555–9.

Bautista RE, Ciampetti MZ. Expressive aprosody andamuia as a manifestation of right hemisphere seizures. *Epilepsia* 2003; **44**:466–7.

Baykan B, Matur Z, Gurses C *et al.* Typical absence seizures triggered by photosensitivity. *Epilepsia* 2005; **46**:159–63.

Bear DM. Temporal lobe epilepsy: a syndrome of sensory-limbic hyperconnection. *Cortex* 1979; **15**:357–69.

Bear DM, Fedio P. Qualitative analysis of interictal behavior in temporal lobe epilepsy. *Arch Neurol* 1977; **34**:454–67.

Bear DM, Levin K, Blumer D *et al.* Interictal behavior in hospitalized temporal lobe epileptics: relationship to idiopathic psychiatric syndromes. *J Neurol Neurosurg Psychiatry* 1982; **45**:481–8.

Beard AW. The association of hepatolenticular degeneration with schizophrenia. *Acta Psychiatr Neurol Scand* 1959; **34**:411–28.

Bebin EM, Gomez ER. Prognosis in Sturge–Weber disease: comparison of unihemispheric and bihemispheric involvement. *J Child Neurology* 1988; **3**:181–4.

Belafsky MA, Carwille S, Miller P *et al.* Prolonged epileptic twilight states: continuous recordings with nasopharyngeal electrodes and videotape analysis. *Neurology* 1978; **28**:239–45.

Beling CC, Martland HS. A case of tumor of the corpus callosum and frontal lobes. *J Nerv Ment Dis* 1919; **50**:425–32.

Bell DS. The experimental reproduction of amphetamine psychosis. *Arch Gen Psychiatry* 1973; **29**:35–40.

Bell SH, Stade B, Reynolds JH *et al.* The remarkably high prevalence of epilepsy and seizure history in fetal alcohol spectrum disorders. *Alcohol Clin Exp Res* 2010; **34**:1084–9.

Bell WL, Harmes J, Logue P *et al.* Neologistic speech automatisms during complex partial seizures. *Neurology* 1990; **40**:49–52.

Bell WL, Walczak TS, Shin C *et al.* Painful generalized clonic and tonic seizures with retained consciousness. *J Neurol Neurosurg Psychiatry* 1997; **63**:792–5.

Benbadis SR, Wolgamuth BR, Goren H *et al.* Value of tongue biting in the diagnosis of seizures. *Arch Int Med* 1995; **155**:2346–9.

Benbadis SR, Kotagal P, Klem GH. Unilateral blinking: a lateralizing sign in partial seizures. *Neurology* 1996; **46**:45–8.

Benbadis SR, Johnson K, Anthony K *et al.* Induction of psychogenic nonepileptic seizures without placebo. *Neurology* 2000; **55**:1904–5.

Benbadis SR, Siegrist K, Tatum WO *et al.* Short-term outpatient EEG video with induction in the diagnosis of psychogenic seizures. *Neurology* 2004; **63**:1728–30.

Benbadis SR, LaFrance WC, Papandonatos GD *et al.* Interrater reliability of EEG-video monitoring. *Neurology* 2009; **73**:843–6.

Ben-Menachem E, Biton V, Jatuzis D *et al.* Efficacy and safety of oral lacosamide as adjunctive therapy in adults with partial-onset seizures. *Epilepsia* 2007; **48**:1308–17.

Benna P, Lacquanilti F, Triolo G *et al.* Acute neurologic complications of hemodialysis. Study of 14,000 hemodialyses in 103 patients with chronic renal failure. *Ital J Neurol Sci* 1981; **2**:53–7.

Benton CS, de Silva R, Rutledge SL *et al.* Molecular and genetic studies in SCA-7 define a broad clinical spectrum and the infantile phenotype. *Neurology* 1998; **51**:1081–6.

Bercel NA, Travis LE, Olinger LB *et al.* Model psychoses induced by LSD-25 in normals. I. Psychophysiological investigations, with special reference to the mechanism of the paranoid reaction. *Arch Neurol Psychiatry* 1956; **75**:588–611.

Beresford OD, Graham AM. Chorea gravidarum. *J Obs Gynecol Br Emp* 1950; **57**:616–25.

Berger JR, Ross DB. Reversible Parkinson syndrome complicating postoperative hypoparathyroidism. *Neurology* 1981; **31**:81–2.

Bergeron C, Pollanen MS, Weyer L *et al.* Unusual clinical presentation of cortical-basal ganglionic degeneration. *Ann Neurol* 1996; **40**:893–900.

Berkovic SF, Andermann F, Melanson D *et al.* Hypothalamic hamartomas and associated ictal laughter: evolution of the characteristic epileptic syndrome and diagnostic value of magnetic resonance imaging. *Ann Neurol* 1988; **23**:429–39.

Berkovic SF, Andermann F, Olivier A *et al.* Hippocampal sclerosis in temporal lobe epilepsy demonstrated by magnetic resonance imaging. *Ann Neurol* 1991; **29**:175–82.

Berkovic SF, Izzillo P, McMahon JM *et al.* LGI1 mutations in temporal lobe epilepsy. *Neurology* 2004a; **62**:1115–19.

Berkovic SF, Serratosa JM, Phillips HA *et al.* Familial partial epilepsy with variable foci: clinical features and linkage to chromosome 22q12. *Epilepsia* 2004b; **45**:1054–60.

Berkovic SF, Knowlton RC, Leroy RF *et al.* Placebo-controlled study of levetiracetam in idiopathic generalized epilepsy. *Neurology* 2007; **69**:1751–60.

Berning S, Boesebeck F, van Baalen A *et al.* Intravenous levetiracetam as treatment for status epilepticus. *J Neurol* 2009; **256**:1634–42.

Berthier ML, Green C, Higueras C *et al.* A randomized, placebo-controlled study of donepezil in poststroke aphasia. *Neurology* 2006; **67**:1687–9.

Berthier ML, Green C, Lara JP *et al.* Memantine and constraint-induce aphasia therapy in chronic poststroke aphasia. *Ann Neurol* 2009; **65**:577–85.

Bertina RM, Koelman BPC, Rosendall FR *et al.* Mutation in the blood coagulation factor V associated with resistance to protein C. *Nature* 1994; **369**:64–7.

Betting LE, Mory SB, Lopes-Cendes I *et al.* EEG features in idiopathic generalized epilepsy: clues to diagnosis. *Epilepsia* 2006; **47**:523–8.

Betts TA, Smith WT, Kelly RE. Adult metachromatic leukodystrophy (sulphatide lipidosis) simulating schizophrenia. *Neurology* 1968a; **18**:1140–2.

Betts TA, Kalra PL, Cooper R *et al.* Epileptic fits as a possible side-effect of amitriptyline. *Lancet* 1968b; **1**:390–2.

Beydoun A, Sackellares JC, Shu V *et al.* Safety and efficacy of divalproex sodium monotherapy in partial epilepsy: a double-blind, concentration–response design clinical trial. *Neurology* 1997; **48**:182–8.

Beydoun A, Sachdeo R, Rosenfeld WE *et al.* Oxcarbazepine monotherapy for partial-onset seizures: a multicenter, double-blind, clinical trial. *Neurology* 2000; **54**:2245–51.

Beydoun A, Uthman BM, Kugler AR *et al.* Safety and efficacy of two pregabalin regimens for add-on treatment of epilepsy. *Neurology* 2005; **64**:475–80.

Bicknell JN, Moore RA. Psychological meaning of disulfiram (Antabuse) therapy. *Arch Gen Psychiatry* 1960; **2**:661–8.

Bien CG, Benniger FO, Urbach H *et al.* Localizing value of epileptic visual auras. *Brain* 2000; **123**:244–53.

Bien CG, Widman G, Urbach H *et al.* The natural history of Rasmussen's encephalitis. *Brain* 2002; **125**:1751–9.

Binnie LD, Debets RMC, Engelsman M *et al.* Double–blind crossover trial of lamotrigene (Lamictal) as add-on therapy in intractable epilepsy. *Epilepsia Res* 1989; **4**:222–9.

Binzer M, Kullgren G. Motor conversion disorder. A prospective 2- to 5-year follow-up study. *Psychosomatics* 1998; **39**:519–27.

Binzer M, Andersen PM, Kullgren G. Clinical characteristics of patients with motor disability due to conversion disorder: a prospective control group study. *J Neurol Neurosurg Psychiatry* 1997; **63**:83–8.

Biton V, Montouris GD, Ritter F *et al.* A randomized, placebo-controlled study of topiramate in primary generalized tonic–clonic seizures. *Neurology* 1999; **52**:1330–7.

Biton V, Sackellares JC, Vuong A *et al.* Double-blind, placebo-controlled study of lamotriginein primary generalized tonic–clonic seizures. *Neurology* 2005; **65**:1737–43.

Bleuler E. *Textbook of Psychiatry, 1924.* New York: Arno Press, 1976.

Blum AS, Drislane FW. Nonconvulsive status epilepticus in thrombotic thrombocytopenic purpura. *Neurology* 1996; **47**:1079–81.

Blume WT, Jones DC, Young GB *et al.* Seizures involving secondary sensory and related areas. *Brain* 1992; **115**:1509–20.

Blumer D. Hypersexual episodes in temporal lobe epilepsy. *Am J Psychiatry* 1970; **126**:1099–106.

Blumer D, Walker AE. Sexual behavior in temporal lobe epilepsy. *Arch Neurol* 1967; **16**:37–43.

Boas J, Dam M, Friis ML *et al.* Controlled trial of lamotrigene (Lamictal) for treatment-resistant partial seizures. *Acta Neurol Scand* 1996; **54**:247–52.

Boggs J, Preis K. Successful initiation of combined therapy with valproate sodium injection and divalproex sodium extended-release tablets in the epilepsy monitoring unit. *Epilepsia* 2000; **46**:940–51.

Bogousslavsky J, Regli F, Uske A. Thalamic infarcts: clinical syndromes, etiology, and prognosis. *Neurology* 1988; **38**:837–48.

Bogousslavsky J, Martin R, Regli F *et al.* Persistent worsening of stroke sequelae after delayed seizures. *Arch Neurol Psychiatry* 1992; **49**:385–8.

Bohman M, Cloninger CR, von Knorring AL *et al.* An adoption study of somatoform disorders. III. Cross-fostering analysis and genetic relationship to alcoholism and criminality. *Arch Gen Psychiatry* 1984; **41**:872–8.

Boller F, Wright DG, Cavalieri R *et al.* Paroxysmal 'nightmares': sequel of a stroke responsive to diphenylhydantoin. *Neurology* 1975; **25**:1026–8.

Bolt JMW. Huntington's disease in the West of Scotland. *Br J Psychiatry* 1970; **116**:259–70.

Bombardier CH, Fann JR, Temkin N *et al.* Posttraumatic stress disorder symptoms during the first six months after traumatic brain injury. *J Neuropsychiatry Clin Neurosci* 2006; **18**:501–8.

Boon P, Chauvel P, Pohlmann-Eden B *et al.* Dose-response effect of levetiracetam 1000 and 2000 mg/day in partial epilepsy. *Epilepsy Res* 2002; **48**:77–89.

Boon PA, Williamson PD, Fried I *et al.* Partial seizures: an anatomo-clinical, neuropsychological, and surgical correlation. *Epilepsia* 1991; **32**:467–76.

Borchert CD, Labar DR. Permanent hemiparesis due to partial status epilepticus. *Neurology* 1995; **45**:187–8.

Bornstein M, Coddon D, Song S. Prolonged alterations in behavior associated with a continuous electroencephalographic (spike and dome) abnormality. *Neurology* 1956; **6**:444.

Boudreau EA, Liow K, Frattali CM *et al.* Hypothalamic hamartomas and seizures: distinct natural history of isolated and Pallister-Hall syndrome cases. *Epilepsia* 2005; **46**:42–7.

Bourgeois JA, Cogswell JB, Hessl D *et al.* Cognitive, anxiety and mood disorders in the fragile X-associated tremor/ataxia syndrome. *Gen Hosp Psychiatry* 2007; **29**:349–56.

Bousser M-G, Chiras J, Bories J *et al.* Cerebral venous thrombosis: a review of 38 cases. *Stroke* 1985; **16**:199–213.

Bouton SM. Pick's disease. *J Nerv Ment Dis* 1940; **91**:9–30.

Brady KT, Lydiard RB, Malcolm R *et al.* Cocaine-induced psychosis. *J Clin Psychiatry* 1991; **52**:509–12.

Bray GA, Dahms WT, Swerdhoff RS *et al.* The Prader–Willi syndrome: a study of 40 patients and a review of the literature. *Medicine* 1983; **62**:59–88.

Breningstall GN. Gelastic seizures, precocious puberty, and hypothalamic hamartoma. *Neurology* 1985; **35**:1180–3.

Brennan LV, Craddock PR. Limbic encephalopathy as a non-metastatic complication of oat cell lung cancer: its reversal after treatment of the primary lung lesion. *Am J Med* 1983; **75**:518–20.

Brey RL, Holliday SL, Saklad AR *et al.* Neuropsychiatric syndromes in lupus. Prevalence using standardized definitions. *Neurology* 2002; **58**:1214–20.

Brick JF, Gutrecht JA, Ringel RA. Reflex epilepsy and non-ketotic hyperglycemia in the elderly. *Neurology* 1989; **39**:394–9.

Bridgman O, Smyth FS. Progressive lenticular degeneration. *J Nerv Ment Dis* 1944; **99**:534–43.

Briellman RS, Kalnins RM, Hopwood MJ *et al.* TLE patients with postictal psychosis: mesial dysplasia and anterior hippocampal preservation. *Neurology* 2000; **55**:1027–30.

Briellmann RS, Berkovic SF, Syngeniotis A *et al.* Seizure-associated hippocampal volume loss: a longitudinal magnetic resonance study of temporal lobe epilepsy. *Ann Neurol* 2002; **51**:641–4.

Britton JW, Ghearing GR, Benarroch EE *et al.* The ictal bradycardia syndrome: localization and lateralization. *Epilepsia* 2006; **47**:374–84.

Brodie MJ, Richens A, Yuen AW. Double-blind comparison of lamotrigene and carbamazepine in newly diagnosed epilepsy. *Lancet* 1995; **345**:476–9.

Brodie MJ, Perucca E, Ryvlin P *et al.* Comparison of levetiracetam and controlled-release carbamazepine in newly diagnosed epilepsy. *Neurology* 2007; **68**:402–8.

Brodtkorb E, Nilsn G, Smevik O *et al.* Epilepsy and anomalies of neuronal migration: MRI and clinical aspects. *Acta Neurol Scand* 1992; **86**:24–32.

Brodtkorb E, Gu W, Nakken KO *et al.* Familial temporal lobe epilepsy with aphasic seizures and linkage to chromosome 10q22–q24. *Epilepsia* 2002; **43**:228–35.

Brodtkorb E, Michler RP, Gu W *et al.* Speech-induced aphasic seizures in epilepsy caused by LGI1 mutations. *Epilepsia* 2005; **46**:963–6.

Brooks N, Campsie L, Symington C *et al.* The five year outcome of severe blunt head injury: a relative's view. *J Neurol Neurosurg Psychiatry* 1986; **49**:764–70.

Brown P, Rodgers-Johnson P, Cathala F *et al.* Creutzfeldt–Jakob disease of long duration: clinicopathological characteristics, biotransmissability, and differential diagnosis. *Ann Neurol* 1984; **16**:295–304.

Brown P, Cathala F, Castaigne P *et al.* Creutzfeldt–Jakob disease: clinical analysis of a consecutive series of 230 neuropathologically verified cases. *Ann Neurol* 1986; **20**:597–602.

Browne TR, Kugler AR, Eldon MA. Pharmacology and pharmacokinetics of fosphenytoin. *Neurology* 1996; **46**(Suppl. 1):3–7.

Brugger P, Agosti R, Regard M *et al.* Heautoscopy, epilepsy and suicide. *J Neurol Neurosurg Psychiatry* 1994; **57**:838–9.

Brun A, Englund B, Gustafson L *et al.* Clinical and neuropathological criteria for frontotemporal dementia. *J Neurol Neurosurg Psychiatry* 1994; **57**:416–18.

Brust JC, Plank CR, Healton EB *et al.* The pathology of drop attacks: a case report. *Neurology* 1979; **29**:786–90.

Bruton CJ. *The Neuropathology of Temporal Lobe Epilepsy.* Oxford: Oxford University Press, 1988.

Bulrich N, Cooper DA, Freed A. HIV infection associated with symptoms indistinguishable from functional psychosis. *Br J Psychiatry* 1988; **152**:649–53.

Burks JS, Alfrey AC, Huddlestone J *et al.* A fatal encephalopathy in chronic hemodialysis patients. *Lancet* 1976; **1**:764–8.

Burns A, Russell E, Stratton-Powell H *et al.* Sertraline in stroke-associated lability of mood. *Int J Geriatr Psychiatry* 1999; **14**:681–5.

Bursten B. Psychoses associated with thyrotoxicosis. *Arch Gen Psychiatry* 1961; **4**:267–73.

Butler CR, Graham KS, Hodges JR *et al.* The syndrome of transient epileptic amnesia. *Ann Neurol* 2007; **61**:587–98.

Byelsjo I, Forsgren L, Lithner F *et al.* Epidemiology and clinical characteristics of seizures in patients with acute intermittent porphyria. *Epilepsia* 1996; **37**:230–5.

Caine ED, Shoulson I. Psychiatric syndromes in Huntington's disease. *Am J Psychiatry* 1983; **140**:728–33.

Callaghan N, Kenny RA, O'Neill B *et al.* A prospective study between carbamazepine, phenytoin and sodium valproate in previously untreated and recently diagnosed patients with epilepsy. *J Neurol Neurosurg Psychiatry* 1985; **48**:639–44.

Callaghan N, Garrett A, Goggin T. Withdrawal of anticonvulsant drugs in patients free of seizures for two years. A prospective study. *N Engl J Med* 1988; **318**:942–6.

Calleja S, Salas-Puij J, Robacoba R *et al.* Evolution of juvenile myoclonic epilepsy treated from the outset with sodium valproate. *Seizure* 2001; **10**:424–7.

Cameron MM. Chronic subdural hematoma: a review of 114 cases. *J Neurol Neurosurg Psychiatry* 1978; **41**:834–9.

Camfield CS, Camfield PR. Juvenile myoclonic epilepsy 25 years after seizure onset: a population-based study. *Neurology* 2009; **73**:1041–5.

Camfield P, Camfield C. Idiopathic generalized epilepsy with generalized tonic–clonic seizures (IGE-GTC): a population-based cohort with >20 year follow up for medical and social outcome. *Epilepsy Behav* 2010; **18**:61–3.

Campbell AMG, Corner B, Norman RM *et al.* The rigid form of Huntington's disease. *J Neurol Neurosurg Psychiatry* 1961; **24**:71–7.

Canevini MP, Mai R, Di Marco C *et al.* Juvenile myoclonic epilepsy of Janz: clinical observations in 60 patients. *Seizure* 1992; **1**:291–8.

Cantu C, Barinagarrementeria F. Cerebral venous thrombosis associated with pregnancy and puerperium. Review of 67 cases. *Stroke* 1993; **24**:1880–4.

Cape CA, Martinez AJ, Robertson JT *et al.* Adult onset of subacute sclerosing panencephalitis. *Arch Neurol* 1973; **28**:124–7.

Caplan LR. Positional cerebral ischemia. *J Neurol Neurosurg Psychiatry* 1976; **39**:385–91.

Caplan LR. Occlusion of the vertebral or basilar artery. *Stroke* 1979; **10**:272–82.

Caplan LR. Intracranial branch atheromatous disease. *Neurology* 1989; **39**:1246–50.

Caplan LR, Hier DB, D'Cruz I. Cerebral embolism in the Michael Reese Stroke Registry. *Stroke* 1983; **14**:530–6.

Caplan LR, Schmahmann JD, Kase CS *et al.* Caudate infarcts. *Arch Neurol* 1990; **47**:133–43.

Caprie Steering Committee. A randomized, blinded, trial of clopidogrel versus aspirin in patients at risk of ischaemic events (CAPRIE). *Lancet* 1996; **348**:1329–39.

Cardoso TAM, Coan AC, Kobayashi E *et al.* Hippocampal abnormalities and seizure recurrence after antiepileptic drug withdrawal. *Neurology* 2006; **67**:134–6.

Carlson C, St. Louis EK. Vacuum cleaner epilepsy. *Neurology* 2004; **63**:190–1.

Carney MW, Chary TK, Robotis P *et al.* Ganser syndrome and its management. *Br J Psychiatry* 1987; **151**:697–700.

Carpenter J, Stapleton S, Holliman R. Retrospective analysis of 49 cases of brain abscess and review of the literature. *Eur J Clin Microbiol Infect Dis* 2007; **26**:1–11.

Cascino GD, Andermann F, Berkovic SF *et al.* Gelastic seizures and hypothalamic hamartomas. *Neurology* 1993a; **43**:747–50.

Cascino GD, Luckstein RR, Sharbrough FW *et al.* Facial asymmetry, hippocampal pathology, and remote symptomatic seizures. *Neurology* 1993b; **43**:725–7.

Cascino GD, Hulihan JF, Sharbrough FW *et al.* Parietal lobe lesional epilepsy: electroclinical correlation and operative outcome. *Epilepsia* 1993c; **34**:522–7.

Castaigne P, Lhermitte F, Gautier J *et al.* Arterial occlusion in the vertebral-basilar system. *Brain* 1973; **96**:133–54.

Castillo CS, Schultz SK, Robinson RG. Clinical correlates of early-onset and late-onset poststroke generalized anxiety. *Am J Psychiatry* 1995; **152**:1174–9.

Castriotta RJ, Lai JM. Sleep disorders associated with traumatic brain injury. *Arch Phys Med Rehabil* 2001; **82**:1403–6.

Castriotta RJ, Atanasov S, Wilde MC *et al.* Treatment of sleep disorders after traumatic brain injury. *J Clin Sleep Med* 2009; **5**:137–44.

Castro LH, Ferreira LK, Teles LR *et al.* Epilepsy syndromes associated with hypothalamic hamartomas. *Seizure* 2007; **16**:505–8.

Catella-Lawson F, Reilly MP, Kapoor SC *et al.* Cyclooxygenase inhibitors and the antiplatelet effects of aspirin. *N Engl J Med* 2001; **345**:1809–17.

Cavalleri F, De Renzi E. Amyotrophic lateral sclerosis with dementia. *Acta Neurol Scand* 1994; **89**:391–4.

Caveness WF. Epilepsy, a product of trauma in our time. *Epilepsia* 1976; **17**:207–15.

Celesia GG, Barr AN. Psychosis and other psychiatric manifestations of levodopa therapy. *Arch Neurol* 1970; **23**:193–200.

Cendes F, Andermann F, Gloor P *et al.* Atrophy of medial structures in patients with temporal lobe epilepsy: cause or consequence of repeated seizures? *Ann Neurol* 1993a; **34**:795–801.

Cendes F, Andermann F, Dubeau F *et al.* Early childhood prolonged febrile convulsions, atrophy and sclerosis of medial structures, and temporal lobe epilepsy: an MRI volumetric study. *Neurology* 1993b; **43**:1083–7.

Cendes F, Andermann F, Carpenter S *et al.* Temporal lobe epilepsy caused by domoic acid intoxication: evidence for glutamate receptor-mediated excitotoxicity in humans. *Ann Neurol* 1995; **37**:123–6.

Cercy SP, Kuluva JE. Gelastic epilepsy and dysprosodia in a case of late-onset right frontal seizures. *Epilepsy Behav* 2009; **16**:360–5.

Cereghino JJ, Biton V, Abou-Khalil B *et al.* Levetiracetam for partial seizures: results of a double-blind, randomized clinical trial. *Neurology* 2000; **55**:236–42.

Chalela JA, Kidwell CS, Nentwich LM *et al.* Magnetic resonance imaging and computed tomography in emergency assessment of patients with suspected acute stroke: a prospective comparison. *Lancet* 2007; **369**:293–8.

Chandler JH, Bebin J. Hereditary cerebellar ataxia: olivopontocerebellar type. *Neurology* 1956; **6**:187–95.

Chapman RWG, Laidlow JM, Colin-Jones D *et al.* Increased prevalence of epilepsy in coeliac disease. *BMJ* 1978; **2**:250–1.

Chassagnon S, Minotti L, Kremer S *et al.* Restricted frontomesial epileptogenic focus generates dyskinetic behavior and laughter. *Epilepsia* 2003; **44**:859–63.

Chaturvedi S, Bruno A, Feasby T *et al.* Carotid endarterectomy – an evidence-based review. *Neurology* 2005; **65**:794–801.

Chaudhuri A, Kennedy PG. Diagnosis and treatment of viral encephalitis. *Postgrad Med J* 2002; **78**:575–83.

Chen R-C, Forster FM. Cursive epilepsy and gelastic epilepsy. *Neurology* 1973; **23**:1019–29.

Chen ZM, Sandercock P, Pan HC *et al.* Indications for early aspirin use in acute ischemic stroke: a combined analysis of 40000 randomized patients from the Chinese acute stroke trial and the international stroke trial. *Stroke* 2000; **31**:1240–9.

Chesson AL, Kasarkis EJ, Small VW. Postictal elevation of serum creatine kinase level. *Arch Neurol* 1983; **40**:315–17.

Chester EM, Agamanolis DP, Banker BQ *et al.* Hypertensive encephalopathy: a clinicopathologic study of 20 cases. *Neurology* 1978; **28**:928–39.

Choi-Kwon S, Han SW, Kwon SU *et al.* Fluoxetine treatment in poststroke depression, emotional incontinence, and anger proneness: a double-blind, placebo-controlled study. *Stroke* 2006; **37**:156–61.

Chung S, Sperling MR, Biton V *et al.* Lacosamide as adjunctive therapy for partial-onset seizures: A randomized controlled trial. *Epilepsia* 2010; **51**:958–67.

Chung SS, Gerber P, Kirlin KA. Ictal eye closure is a reliable indicator for psychogenic nonepileptic seizures. *Neurology* 2006; **66**:1730–1.

Cittandi E, Goadsby PJ. Psychiatric side effects during methysergide treatment. *J Neurol Neurosurg Psychiatry* 2005; **76**:1037–8.

Chiu D, Shedden P, Bratina P *et al.* Clinical features of Moyamoya disease in the United States. *Stroke* 1998; **29**:1347–51.

Claassen J, Hirsch LJ, Emerson RG *et al.* Treatment of refractory status epilepticus with pentobarbital, propofol, or midazolam: a systematic review. *Epilepsia* 2002; **43**:146–53.

Claasen J, Mayer SA, Hirsch LJ. Continuous EEG monitoring in patients with subarachnoid hemorrhage. *J Clin Neurophysiol* 2005; **22**:92–8.

Clarke DJ. Prader–Willi syndrome and psychoses. *Br J Psychiatry* 1993; **163**:680–4.

Clarke E, Melnick SC. The Munchausen syndrome or the problem of hospital hoboes. *Am J Med* 1958; **25**:6–12.

Cleghorn RA. Adrenal cortical insufficiency: psychological and neurological observations. *CMAJ* 1951; **65**:445–57.

Cline JS, Roos K. Treatment of status epilepticus with electroconvulsive therapy. *J ECT* 2007; **23**:30–2.

Clomipramine Collaborative Study Group. Clomipramine in the treatment of patients with obsessive-compulsive disorder. *Arch Gen Psychiatry* 1991; **48**:730–8.

Coats J. A study of two illustrative cases of epilepsy. *BMJ* 1876; **ii**:647–9.

Cockerell OC, Walker MC, Sander JWAS *et al.* Complex partial status epilepticus: a recurrent problem. *J Neurol Neurosurg Psychiatry* 1994; **57**:835–7.

Cockerell OC, Tothwell J, Marsden CD *et al.* Clinical and physiological features of epilepsia partialis continua: cases ascertained in the UK. *Brain* 1996; **116**:393–407.

Cohen RJ, Suter C. Hysterical seizures: suggestion as a provocative EEG test. *Ann Neurol* 1982; **11**:391–5.

Cole AJ, Gloor P, Kaplan R. Transient global amnesia: the electroencephalogram at onset. *Ann Neurol* 1987; **22**:721–2.

Cole F, Yates P. Intracerebral microaneurysms and small cerebrovascular lesions. *Brain* 1967; **90**:759–68.

Combi R, Dalpra L, Ferini-Strambi L *et al.* Frontal lobe epilepsy and mutations of the corticotropin-releasing hormone gene. *Ann Neurol* 2005; **58**:899–904.

Commission of Classification and Terminology of the International League Against Epilepsy. *Epilepsia* 1981; **22**:489–501.

Commission of Classification and Terminology of the International League Against Epilepsy. *Epilepsia* 1989; **30**:389–99.

Corrigan JD. Substance abuse as a mediating factor in outcome from traumatic brain injury. *Arch Phys Med Rehabil* 1995; **76**:302–9.

Corsellis JAN, Golberg GJ, Norton AR. 'Limbic encephalitis' and its association with carcinoma. *Brain* 1968; **91**:481–96.

Couprie W, Wijdicks EF, Rooijmans HG *et al.* Outcome in conversion disorder: a follow up study. *J Neurol Neurosurg Psychiatry* 1995; **58**:750–2.

Critchley M. The anterior cerebral artery, and its syndromes. *Brain* 1930; **53**:120–65.

Critchley M. Musicogenic epilepsy. *Brain* 1937; **60**:13–27.

Critchley M. Two cases of musicogenic epilepsy. *J Royal Navy Med Serv* 1942; **27**:182–4.

Critchley M, Early CJC. Tuberose sclerosis and allied conditions. *Brain* 1932; **55**:311–46.

Critchley M, Cobb W, Sears TA. On reading epilepsy. *Epilepsia* 1959; **1**:403–17.
Crompton M. Hypothalamic lesions following closed head injury. *Brain* 1971; **94**:165–72.
Currie S, Heathfield KWG, Henson RA *et al.* Clinical course and prognosis of temporal lobe epilepsy: a survey. *Brain* 1971; **97**:173–90.
Currier RD, Little SC, Suess SJ *et al.* Sexual seizures. *Arch Neurol* 1971; **25**:260–4.
D'Agostino MD, Bernasconi A, Das S *et al.* Subcortical band heterotopia (SBH) in males: clinical, imaging and genetic findings in comparison with females. *Brain* 2002; **125**:2507–22.
D'Alessio L, Giagante B, Papayannis C *et al.* Psychotic disorders in Argentine patients with refractory temporal lobe epilepsy: A case–control study. *Epilepsy Behav* 2009; **14**:604–9.
Daly D. Ictal affect. *Am J Psychiatry* 1958; **115**:97–108.
Daly DD, Barry MJ. Musicogenic epilepsy: report of three cases. *Psychosom Med* 1957; **19**:399–408.
Dam AM, Fuglsang-Fredricksen A, Svarre-Olsen U *et al.* Late-onset epilepsy: etiologies, types of seizure, and value of clinical investigation, EEG and computerized tomography scan. *Epilepsia* 1985; **26**:227–31.
Dana-Haeri J, Richens A. Effect of norethisterone on seizures associated with menstruation. *Epilepsia* 1983; **24**:377–81.
Dana-Haeri J, Trimble MR, Oxley J. Prolactin and gonadotropin changes following generalized and partial seizures. *J Neurol Neurosurg Psychiatry* 1983; **46**:331.
Danek A, Rubio JP, Rampoldi L *et al.* McLeod neuroacanthocytosis: genotype and phenotype. *Ann Neurol* 2001; **50**:755–64.
Daniel WG, Mugge A. Transesophageal echocardiography. *N Engl J Med* 1995; **332**:1268–79.
Daniele O, Mattaliano A, Tassinari CA *et al.* Epileptic seizures and cerebrovascular disease. *Acta Neurol Scand* 1989; **80**:117–22.
Danielsson S, Gillberg C, Billstedt E *et al.* Epilepsy in young adults with autism: a prospective population-based follow-up study of 120 individuals diagnosed in childhood. *Epilepsia* 2005; **46**:918–23.
Davidson J. Seizures and bupropion: a review. *J Clin Psychiatry* 1989; **50**:256–61.
DeGiorgio CM, Tomiyasu U, Gott PS *et al.* Hippocampal pyramidal cell loss in human status epilepticus. *Epilepsia* 1992; **33**:23–7.
DeGiorgio CM, Correale JD, Gott PS *et al.* Serum neuron-specific enolase in human status epilepticus. *Neurology* 1995; **45**:1134–7.
DeGiorgio CM, Heck CN, Rabinowicz AL *et al.* Serum neuron-specific enolase in the major subtypes of status epilepticus. *Neurology* 1999; **52**:746–9.
Degos J-D, da Fonseca N, Gray F *et al.* Severe frontal syndrome associated with infarcts of the left anterior cingulate gyrus and the head of the right caudate nucleus. *Brain* 1993; **116**:1541–8.
Deisenhammer E. Transient global amnesia as an epileptic manifestation. *J Neurol* 1981; **225**:289–92.
Delaporte C. Personality patterns in patients with myotonic dystrophy. *Arch Neurol* 1998; **55**:635–40.
Delgado-Escueta AV. Epileptogenic paroxysms: modern approaches and clinical correlations. *Neurology* 1979; **29**:1014–22.
Delgado-Escueta AV, Mattson RH, King L *et al.* The nature of aggression during epileptic seizures. *N Engl J Med* 1981; **305**:711–16.
Delgado-Escueta AV, Bascal FE, Treiman DM. Complex partial seizures on closed-circuit television and EEG: a study of 691 attacks in 79 patients. *Ann Neurol* 1982; **11**:292–300.
DeLorenzo RJ, Waterhouse EJ, Towne AR *et al.* Persistent nonconvulsive status epilepticus after the control of convulsive status epilepticus. *Epilepsia* 1998; **39**:833–40.
Denier C, Orgibet A, Foffi F *et al.* Adult-onset vanishing white matter leukoencephalopathy presenting as psychosis. *Neurology* 2007; **68**:1538–9.
Dening TR, Berrios GE. Wilson's disease: psychiatric symptoms in 195 cases. *Arch Gen Psychiatry* 1989; **46**:1126–34.
Dening TR, Berrios GE, Walshe JM. Wilson's disease and epilepsy. *Brain* 1988; **111**:1139–55.
Depondt C, Van Paesschen W, Matthijs G *et al.* Familial temporal lobe epilepsy with febrile seizures. *Neurology* 2002; **58**: 1429–33.
Dervaux A, Laine H, Czermak M *et al.* Creutzfeldt–Jakob disease presenting as psychosis. *Am J Psychiatry* 2004; **161**:1307–8.
Desbiens R, Berkovic S, Dubeau F *et al.* Life-threatening focal status epilepticus due to occult cortical dysplasia. *Arch Neurol* 1993; **50**:695–700.
Devinsky O, Duchowny MS. Seizures after convulsive therapy: a retrospective case series. *Neurology* 1983; **33**:921–5.
Devinsky O, Petito CK, Alonso DR. Clinical and neuropathological findings in systemic lupus erythematosus: the role of vasculitis, heart emboli and thrombotic thrombocytopenic purpura. *Ann Neurol* 1988a; **23**:380–4.
Devinsky O, Kelley K, Porter RJ *et al.* Clinical and electroencephalographic features of simple partial seizures. *Neurology* 1988b; **38**:1347–52.
Devinsky O, Felmann E, Burrowes K *et al.* Autoscopic phenomena with seizures. *Arch Neurol* 1989a; **46**:1080–8.
Devinsky O, Sato O, Kufta CV *et al.* Electroencephalographic studies of simple partial seizures with subdural electrode recordings. *Neurology* 1989b; **39**:527–33.
Devinsky O, Honigfield G, Patin J. Clozapine-related seizures. *Neurology* 1991; **41**:369–71.
Devinsky O, Kelley K, Yacubian EMT *et al.* Postictal behavior. *Arch Neurol* 1994a; **51**:254–9.
Devinsky O, Ehrenberg B, Barthlen GM *et al.* Epilepsy and sleep apnea syndrome. *Neurology* 1994b; **44**:2060–4.
Devinsky O, Abramson H, Alper K *et al.* Postictal psychosis: a case control series of 20 patients and 150 controls. *Epilepsy Res* 1995; **20**:247–53.
Devinsky O, Sanchez-Villasenor F, Vazquez B *et al.* Clinical profile of patients with epileptic and nonepileptic seizures. *Neurology* 1996; **46**:1530–3.
Dewhurst K, Todd J. The psychosis of association – folie a deux. *J Nerv Ment Dis* 1956; **124**:451–9.
Diener HC, Bogousslavsky J, Brass LM *et al.* Aspirin and clopidogrel compared with clopidogrel alone after recent ischaemic stroke of transient ischaemic attack in high-risk patients (MATCH): rardomised, double-blind, placebo-controlled trial. *Lancet* 2004; **364**: 331–7.
Dimitri D, Jehel L, Durr A *et al.* Fatal familial insomnia presenting as psychosis in an 18-year-old man. *Neurology* 2006; **67**: 363–4.
Dixit S, Kurle P, Buyan-Dent L *et al.* Status epilepticus associated with cefipime. *Neurology* 2000; **54**:2153–5.
Donaire A, Carreno M, Gomez B *et al.* Cortical laminar necrosis related to prolonged focal status epilepticus. *J Neurol Neurosurg Psychiatry* 2006; **77**:104–6.
d'Orsi G, Tinuper P, Bisulli F *et al.* Clinical features and long term outcome of epilepsy in periventricular nodular heterotopia. Simple compared with plus forms. *J Neurol Neurosurg Psychiatry* 2004; **75**:873–8.
Drachman DA, Adams RD. Herpes simplex and acute inclusion-body encephalitis. *Arch Neurol* 1962; **7**:45–63.

Drury I, Klass DW, Westmoreland BF *et al.* An acute syndrome with psychiatric symptoms and EEG abnormalities. *Neurology* 1985; **35**:911–14.

D'Souza WJ, O'Brien TJ, Murphy M *et al.* Toothbrushing-induced epilepsy with structural lesions in the primary somatosensory area. *Neurology* 2007; **68**:769–71.

Dubeau F, Tampieri E, Carpenter S *et al.* Periventricular and subcortical nodular heterotopia: a study of 33 patients. *Brain* 1995; **118**: 1273–87.

Duncalf CM, Kent JNG, Harbord M *et al.* Subacute sclerosing panencephalitis presenting as schizophreniform psychosis. *Br J Psychiatry* 1989; **155**:557–9.

Earnest MP, Feldman H, Marx JA *et al.* Intracranial lesions shown by CT scans in 259 cases of first alcohol-related seizures. *Neurology* 1988; **38**:1561–5.

Edwards OM, Clark JDA. Post-traumatic hypopituitarism: six cases and a review of the literature. *Medicine* 1986; **65**:281–90.

Efron R. The effect of olfactory stimuli in arresting uncinate fits. *Brain* 1956; **79**:267–81.

Efron R. The conditioned inhibition of uncinate fits. *Brain* 1957; **80**:251–62.

Efron R. Post-epileptic paralysis: theoretical critique and report of a case. *Brain* 1961; **84**:381–94.

El Bouzidi K, Duncan S, Whittle IR *et al.* Lesional reflex epilepsy associated with the thought of food. *Neurology* 2010; **74**:610–12.

Elger CE, Brodie MJ, Anhut H *et al.* Pregabalin add-on treatment in patients with partial seizures. A novel evaluation of flexible-dose and fixed-dose treatment in a double-blind, placebo-controlled study. *Epilepsia* 2005; **46**:1926–36.

Ellis JM, Lee SI. Acute prolonged confusion in later life as an ictal state. *Epilepsia* 1978; **19**:119–28.

Elzawahry H, Do CS, Lin K *et al.* The diagnostic utility of the ictal cry. *Epilepsy Behav* 2010; **18**:306–7.

Enderby P, Broeckx J, Hospers W *et al.* Effect of piracetam on recovery and rehabilitation after stroke: a double-blind, placebo-controlled study. *Clin Neuropharmacol* 1994; **17**:320–31.

Engel J. Surgery for seizures. *N Engl J Med* 1996; **334**:647–52.

Engel J, Driver MV, Falconer MA. Electrophysiological correlates of pathology and surgical results in temporal lobe epilepsy. *Brain* 1975; **98**:129–56.

Engel J, Ludwig BI, Fetele M. Prolonged partial complex status epilepticus: EEG and behavioral observations. *Neurology* 1978; **28**:863–9.

Englander J, Bushnik T, Duong TT *et al.* Analyzing risk factors for late posttraumatic seizures: a prospective, multicenter investigation. *Arch Phys Med Rehabil* 2003; **84**:365–73.

Eraut D. Idiopathic hypoparathyroidism presenting as dementia. *BMJ* 1974; **1**:429–30.

Erickson TC. Erotomania (nymphomania) as an expression of cortical epileptiform discharge. *Arch Neurol Psychiatry* 1945; **53**:226–31.

Escueta AV, Kunze U, Waddell G *et al.* Lapse of consciousness and automatisms in temporal lobe epilepsy: a videotape analysis. *Neurology* 1977; **27**:144–55.

Espir ML, Rose FC. Alcohol, seizures and epilepsy. *J R Soc Med* 1987; **9**:542–3.

Evans DL, Edelsohn GA, Golden RN. Organic psychosis without dementia or spinal cord symptoms in patients with vitamin B12 deficiency. *Am J Psychiatry* 1983; **140**:218–21.

Factor SA, Podskalney GD, Molho ES. Psychogenic movement disorders: frequency, clinical profile, and characteristics. *J Neurol Neurosurg Psychiatry* 1995; **59**:406–12.

Fahy TJ, Irving MJ, Millac P. Severe head injuries. A six-year follow-up. *Lancet* 1967; **2**:475–9.

Fairweather DS. Psychiatric aspects of the post-encephalitic syndrome. *J Ment Sci* 1947; **93**:201–54.

Fakhoury T, Abou-Khalil B, Peguero E. Differentiating clinical features of right and left temporal lobe seizures. *Epilepsia* 1994; **35**:1038–44.

Falconer MA. Clinical manifestations of temporal lobe epilepsy and their recognition in relation to surgical treatment. *BMJ* 1954; **2**:939–44.

Falconer MA. Genetic and related aetiological factors in temporal lobe epilepsy. *Epilepsia* 1971; **12**:13–31.

Falconer MA, Driver MV, Serafetinides EA. Seizures induced by movement: report of a case relieved by operation. *J Neurol Neurosurg Psychiatry* 1963; **26**:300–7.

Falconer MA, Serafetinides EA, Corsellis JAN. Etiology and pathogenesis of temporal lobe epilepsy. *Arch Neurol* 1964; **10**:233–48.

Fallon BA, Petkova E, Skritskaya N *et al.* A double-masked, placebo-controlled study of fluoxetine for hypochondriasis. *J Clin Psychopharmacol* 2008; **28**:638–45.

Fariello RG, Booker HE, Chun RWM *et al.* Reenactment of the triggering situation for the diagnosis of epilepsy. *Neurology* 1983; **33**:878–84.

Fattouch J, Di Bonaventura C, Casciato S *et al.* Intravenous levetiracetam as first-line treatment of status epilepticus in the elderly. *Acta Neurol Scand* 2010; **121**:418–21.

Faught E, Wilder BJ, Famsay RE *et al.* Topiramate placebo-controlled dose-ranging trial in refractory partial epilepsy using 200-, 400-, and 600-mg daily dosages. *Neurology* 1996; **46**:1684–90.

Fauser S, Huppertz H-J, Bast T *et al.* Clinical characteristics in focal cortical dysplasia: a retrospective evaluation in a series of 120 patients. *Brain* 2006; **129**:1907–16.

Feinberg TE, Cianci CD, Morrow JS *et al.* Diagnostic tests for choreoacanthocytosis. *Neurology* 1991; **41**:1000–6.

Feinberg WM, Rapcsak SZ. 'Peduncular hallucinosis' following paramedian thalamic infarction. *Neurology* 1989; **39**:1535–6.

Feldman RG, White RF, Eriator II. Triethyltin encephalopathy. *Arch Neurol* 1993; **50**:1320–4.

Fenelon G, Mahieux F, Huon R *et al.* Hallucinations in Parkinson's disease: prevalence, phenomenology and risk factors. *Brain* 2000; **123**:733–45.

Fenwick PBC, Brown SW. Evoked and psychogenic seizures. *Acta Neurol Scand* 1989; **80**:535–40.

Fernandez-Torre JL, Martinez-Martinez M, Gonzalez-Rato J *et al.* Cephalosporin-induced nonconvulsive status epilepticus: clinical and electroencephalographic features. *Epilepsia* 2005; **46**:1550–2.

Ferriby D, de Seze J, Stojkovic T *et al.* Long-term follow-up of neurosarcoidosis. *Neurology* 2001; **57**:927–9.

Fichtenberg NL, Millis SR, Mann NR *et al.* Factors associated with insomnia among post-acute traumatic brain injury survivors. *Brain Inj* 2000; **14**:659–67.

Finelli PF, Pueschel SM, Padre-Mendoza T *et al.* Neurological findings in patients with the fragile-X syndrome. *J Neurol Neurosurg Psychiatry* 1985; **45**:150–63.

Fiol ME, Leppik IE, Pretzel K. Eating epilepsy: EEG and clinical study. *Epilepsia* 1986; **27**:441–5.

Fiorelli M, Di Piero V, Bastianello S *et al.* Cerebrotendinous xanthomatosis: clinical and MRI study (a case report). *J Neurol Neurosurg Psychiatry* 1990; **53**:76–8.

Fisher CM. Lacunes: small deep cerebral infarcts. *Neurology* 1965; **15**:774–84.

Fisher CM. Lacunar strokes and infarcts: a review. *Neurology* 1982; **32**:871–6.

Fisher CM, Ojemann R, Robertson G. Spontaneous dissection of cervico-cerebral arteries. *Can J Neurol Sci* 1978; **5**:9–19.

Flashman LA, McAllister TW. Lack of awareness and its impact in traumatic brain injury. *NeuroRehabilitation* 2002; **17**:285–96.

Flugel D, Cerlignani M, Symms MR *et al.* Diffusion tensor imaging findings and their correlation with neuropsychological deficits in patients with temporal lobe epilepsy and interictal psychosis. *Epilepsia* 2006; **47**:941–4.

Forster FM. Epilepsy associated with eating. *Trans Am Neurol Assoc* 1971; **96**:106–7.

Forster FM. *Reflex Epilepsy, Behavioral Therapy and Conditioned Reflexes.* Springfield, IL: Charles C. Thomas, 1977.

Forster FM, Liske E. Role of environmental clues in temporal lobe epilepsy. *Neurology* 1963; **13**:301–5.

Forster FM, Booker HE, Gascon G. Conditioning in musicogenic epilepsy. *Trans Am Neurol Assoc* 1967; **92**:236–7.

Forster FM, Paulsen WA, Baughman FA. Clinical therapeutic conditioning in reading epilepsy. *Neurology* 1969a; **19**:717–23.

Forster FM, Hansotia P, Cleeland CS *et al.* A case of voice-induced epilepsy treated by conditioning. *Neurology* 1969b; **19**:325–31.

Fortemps E, Amand G, Bomboir A *et al.* Trimethyltin poisoning. Report of two cases. *Int Arch Occup Environ Hlth* 1978; **41**:1–6.

Foster HG, Hillbrand M, Chi CC. Efficacy of carbamazepine in assaultive patients with frontal lobe dysfunction. *Prog Neuropsychopharmacol Biol Psychiatry* 1989; **13**:865–74.

Franceschi M, Triulzi F, Ferini-Strambi L *et al.* Focal cerebral lesions found by magnetic resonance imaging in cryptogenic nonrefractory temporal lobe epilepsy patients. *Epilepsia* 1989; **30**:540–6.

Francis A, Freeman H. Psychiatric abnormality and brain calcification over four generations. *J Nerv Ment Dis* 1984; **172**:166–70.

Frazier CH. Tumor involving the frontal lobe alone: a symptomatic survey of one hundred and five verified cases. *Arch Neurol Psychiatry* 1936; **35**:525–51.

Freemon FR, Nevis AH. Temporal lobe sexual seizures. *Neurology* 1969; **19**:87–90.

French JA, Williamson PD, Thadani VM *et al.* Characteristics of medial temporal lobe epilepsy: I. Results of history and physical examination. *Ann Neurol* 1993; **34**:774–80.

French JA, Kugler AR, Robbins JL *et al.* Dose-response trial of pregabalin adjunctive therapy in patients with partial seizures. *Neurology* 2003; **60**:1631–7.

Freytag E. Autopsy findings in head injuries from blunt forces: statistical evaluation of 1,367 cases. *Arch Pathol* 1963; **75**:402–13.

Friedlander WJ, Feinstein GH. Petit mal status: Epilepsia minoris continua. *Neurology* 1956; **6**:357–62.

Friedman A, Sienkiewicz J. Psychotic complications of long-term levodopa treatment of Parkinson's disease. *Acta Neurol Scand* 1991; **84**:111–13.

Friedman HM, Allen N. Chronic effects of complete limbic lobe destruction in man. *Neurology* 1969; **19**:679–90.

Fruehwald S, Gatterbauer E, Rehak P *et al.* Early fluoxetine treatment of post-stroke depression – a three-month double-blind placebo-controlled study with an open-label long-term follow up. *J Neurol* 2003; **250**:347–51.

Fuerst D, Shah J, Shah A *et al.* Hippocampal sclerosis is a progressive disorder: a longitudinal volumetric MRI study. *Ann Neurol* 2003; **53**:413–16.

Fuerstein TJ, Sauerman W, Anhut H *et al.* Gabapentin (Neurontin) as add-on therapy in patients with partial seizures: a double-blind placebo-controlled study. *Epilepsia* 1994; **35**:795–801.

Fujii DE, Ahmed I. Characteristics of psychotic disorders due to traumatic brain injury: an analysis of case studies in the literature. *J Neuropsychiatry Clin Neurosci* 2002; **14**:130–40.

Fuster B, Castells C, Rodriguez B. Psychomotor attacks (primary automatisms) of subcortical origin. *Arch Neurol Psychiatry* 1954; **71**:455–72.

Gabapentin Study Group. US Gabapentin as add-on therapy in refractory partial epilepsy: a double-blind, placebo-controlled, parallel-group study. *Neurology* 1993; **43**:2292–8.

Gabr M, Luders H, Dinner D. Speech manifestations in lateralization of temporal lobe seizures. *Ann Neurol* 1989; **25**:82–7.

Gade A. Amnesia after operations on aneurysms of the anterior communicating artery. *Surg Neurol* 1982; **18**:46–9.

Gallagher BB, Flanigan HF, King DW *et al.* The effect of electrical stimulation of medial temporal lobe structures in epileptic patients upon ACTH, prolactin, and growth hormone. *Neurology* 1987; **37**:299–303.

Gallmetzer P, Leutezer F, Serles W *et al.* Postictal paresis in focal epilepsies: incidence, duration and causes. *Neurology* 2004; **62**:2160–4.

Gambardella A, Reutens DC, Andermann F *et al.* Late onset drop attacks in temporal lobe epilepsy. *Neurology* 1994; **44**:1074–8.

Gambardella A, Annesi G, De Fusco M *et al.* A new locus for autosomal dominant nocturnal frontal lobe epilepsy maps to chromosome 1. *Neurology* 2000; **55**:1467–71.

Garcia-Morales I, Maestu F, Perez-Jimenez MA *et al.* A clinical and magnetoencephalography study of MRI-negative startle epilepsy. *Epilepsy Behav* 2009; **16**:166–71.

Garrett PJ, Mulcahy D, Carmody M *et al.* Aluminum encephalopathy: clinical and immunologic findings. *Q J Med* 1988; **69**:775–83.

Garron D. Huntington's chorea and schizophrenia. *Adv Neurol* 1973; **1**:729–34.

Gascon GG, Lombroso CT. Epileptic (gelastic) laughter. *Epilepsia* 1971; **12**:63–76.

Gastaut H. Clinical and electroencephalographic classification of epileptic seizures. *Epilepsia* 1970; **11**:102–13.

Gastaut H, Tassinari CA. Triggering mechanisms in epilepsy: the electroclinical point of view. *Epilepsia* 1966; **7**:85–138.

Gastaut H, Caveness WF, Landolt H *et al. Epilepsia* 1964; **5**:297–306.

Gastaut H, Gastaut JL, Goncalves E *et al.* Relative frequency of different types of epilepsy: a study employing the classification of the International League Against Epilepsy. *Epilepsia* 1975; **16**:457–61.

Gatzonis SD, Stamboulis E, Siafakis A. Acute psychosis and EEG normalization after vagus nerve stimulation. *J Neurol Neurosurg Psychiatry* 2000; **69**:278–9.

Gautier-Smith PC. Neurological complications of glandular fever (infectious mononucleosis). *Brain* 1965; **88**:323–34.

Gavaret M, Guedj E, Koessler L *et al.* Reading epilepsy from the dominant temporo-occipital region. *J Neurol Neurosurg Psychiatry* 2010; **81**:710–5.

Geda YE, Boeve BF, Negash S *et al.* Neuropsychiatric features in 36 pathologically confirmed cases of corticobasal degeneration. *J Neuropsychiary Clin Neurosci* 2007; **19**:77–80.

Geier S, Bancaud J, Talairach J *et al.* The seizures of frontal lobe epilepsy: a study of clinical manifestations. *Neurology* 1977; **27**:951–8.

Gelenberg AJ. Munchausen's syndrome with a psychiatric presentation. *Dis Nerv Syst* 1977; **38**:378–80.

Gelisse P, Genton P, Kuate C *et al.* Worsening of seizures by oxcarbazepine in juvenile idiopathic generalized epilepsies. *Epilepsia* 2004; **45**:1282–6.

Genton P, Gelisse P, Thomas P *et al.* Do carbamazepine and phenytoin aggravate juvenile myoclonic epilepsy? *Neurology* 2000; **55**:1106–9.

Geyer JD, Payne TA, Drury I. The value of pelvic thrusting in the diagnosis of seizures and pseudoseizures. *Neurology* 2000; **54**:227–9.

Ghaffar O, Stains R, Feinssein A *et al.* Unexplained neurologic symptoms: an fMRI study of sensory conversion disorder. *Neurology* 2006; **67**:2036–8.

Giannopoulos S, Kostadima V, Selvi A *et al.* Bilateral paramedian thalamic infarcts. *Arch Neurol* 2006; **63**:1652.

Gibbs FA, Gibbs EC. *Atlas of Electroencephalography*, Vol. II. *Epilepsy.* London: Addison-Wesley, 1952.

Gillberg C. Outcome in autism and autistic-like conditions. *J Am Acad Child Adolesc Psychiatry* 1991; **30**:375–82.

Gillig P, Sackellares JC, Greenberg HS. Right hemisphere partial complex seizures: mania, hallucinations, and speech disturbances during ictal events. *Epilepsia* 1988; **29**:26–9.

Gladstone DJ, Silver FL, Willinsky RA *et al.* Deep cerebral venous thrombosis: an illustrative case with reversible diencephalic dysfunction. *Can J Neurol Sci* 2001; **28**:159–62.

Glaser GH, Levy LL. Seizures and idiopathic hypoparathyroidism. *Epilepsia* 1960; **4**:454–65.

Glass HC, Prieur B, Molnar C *et al.* Micturition and emotion-induced reflex epilepsy: case report and review of the literature. *Epilepsia* 2007; **47**:2180–2.

Glauser TA, Cnaan A, Shinnar S *et al.* Ethosuximide, valproic acid, and lamotrigine in childhood absence epilepsy. *N Engl J Med* 2010; **362**:790–9.

Globus M, Lavi E, Fich E *et al.* Ictal hemiparesis. *Eur Neurol* 1982; **21**:165–8.

Gloor P. Electrophysiological studies of the amygdala (stimulation and recording): their possible contribution to the understanding of neural mechanisms of aggression. In: Fields WS, Sweet WH eds. *Neural Bases of Violence and Aggression.* St Louis, MO: C.V. Mosby, 1975.

Gobbi G, Bouquet F, Greco L *et al.* Coeliac disease, epilepsy, and cerebral calcification. *Lancet* 1992; **340**:439–43.

Gobin YP, Starkman S, Duckwiler GR *et al.* MERCI 1: a phase 1 study of mechanical embolus removal in cerebral ischemia. *Stroke* 2004; **35**:2848–54.

Goetz CG, Fan W, Leurgans S *et al.* The malignant course of 'benign hallucinations' in Parkinson disease. *Arch Neurol* 2006; **63**:713–16.

Golden RN, James S, Sherer M *et al.* Psychoses associated with bupropion treatment. *Am J Psychiatry* 1985; **142**:1459–62.

Goldensohn ES, Gold AP. Prolonged behavioral disturbances as ictal phenomena. *Neurology* 1960; **10**:1–9.

Goldstein LH, Chalder T, Chigwidere C *et al.* Cognitive-behavioral therapy for psychogenic nonepileptic seizures. A pilot RCT. *Neurology* 2010; **74**:1986–94.

Golub LM, Guhlman HV, Merlis JK. Seizure patterns in psychomotor epilepsy. *Dis Nerv Syst* 1951; **12**:73–6.

Goodin DS, Aminoff MJ, Laxer KD. Detection of epileptiform activity by different noninvasive EEG methods in complex partial epilepsy. *Ann Neurol* 1990; **27**:330–4.

Goodman L. Alzheimer's disease: a clinico-pathologic analysis of twenty-three cases with a theory on pathogenesis. *J Nerv Ment Dis* 1953; **118**:97–130.

Gosk-Bierska I, Wysokinski W, Brown RD *et al.* Cerebral venous sinus thrombosis. Incidence of venous thrombosis recurrence and survival. *Neurology* 2006; **67**:814–19.

Gothelf D, Feinstein C, Thompson T *et al.* Risk factors for the emergence of psychotic disorders in adolescents with 22q11.2 deletion syndrome. *Am J Psychiatry* 2007; **164**:663–9.

Graff-Radford NR, Damasio H, Yamada T *et al.* Non-hemorrhagic thalamic infarction. *Brain* 1985; **108**:495–516.

Graham JM, Gruenwald RA, Sagar HJ. Hallucinosis in idiopathic Parkinson's disease. *J Neurol Neurosurg Psychiatry* 1997; **63**:434–40.

Granata T, Gobbi G, Spreafico R *et al.* Rasmussen's encephalitis. Early characteristics allow diagnosis. *Neurology* 2003; **60**:422–5.

Grant C, Warlow C. Focal epilepsy in diabetic non-ketotic hyperglycemia. *BMJ* 1985; **290**:1204–5.

Greenberg MK, Vance SC. Focal seizure disorder complicating iodophenylate myelography. *Lancet* 1980; **1**:312–13.

Greenberg SM, Vonsattel JPG, Stakes JW *et al.* The clinical spectrum of cerebral amyloid angiopathy: presentations without lobar hemorrhage. *Neurology* 1993; **43**:2073–9.

Greene KA, Marciano FF, Dickman CA *et al.* Anterior communicating aneurysm paraparesis syndrome: clinical manifestations and pathologic correlates. *Neurology* 1995; **45**:45–50.

Greer S, Parsons V. Schizophrenia-like psychosis in thyroid crisis. *Br J Psychiatry* 1968; **114**:1357–62.

Greeven A, van Balkom AJLM, Visser S *et al.* Cognitive behavioral therapy and paroxetin in the treatment of hypochodriasis: a randomized controlled trial. *Am J Psychiatry* 2007; **164**:91–9.

Grewal RP, Achari M, Matsuura T *et al.* Clinical features and ATTC repeat expansion in spinocerebellar ataxia type 10. *Arch Neurol* 2002; **59**:1285–90.

Griffith JD, Cavanaugh J, Held J *et al.* Dextroamphetamine: evaluation of psychotomimetic properties in man. *Arch Gen Psychiatry* 1972; **26**:97–100.

Grunberger G, Weiner JL, Silverman R *et al.* Factitious hypoglycemia due to surreptitious administration of insulin. Diagnosis, treatment, and long-term follow-up. *Ann Intern Med* 1988; **108**:252–7.

Grunewald R. Levetiracetam in the treatment of idiopathic generalized epilepsies. *Epilepsia* 2005; **46**(Suppl. 9):154–60.

Gualtieri CT, Evans RW. Stimulant treatment for the neurobehavioural sequelae of traumatic brain injury. *Brain Inj* 1988; **2**:273–90.

Guinena YH, Taher Y. Psychosensory seizures 'visual and auditory' of primary subcortical origin. *Electroencephalogr Clin Neurophysiol* 1955; **7**:425–8.

Gumpert J, Hansotia P, Upton A. Gelastic epilepsy. *J Neurol Neurosurg Psychiatry* 1970; **33**:479–83.

Gupta AK, Jeavons PM, Hughes RC *et al.* Aura in temporal lobe epilepsy: clinical and electroencephalographic correlation. *J Neurol Neurosurg Psychiatry* 1983; **46**:1079–83.

Gupta SR, Nasheedy MH, Elias D *et al.* Postinfarction seizures: a clinical study. *Stroke* 1988; **19**:1477–81.

Guze SB, Cloninger CR, Martin RL *et al.* A follow-up and family study of Briquet's syndrome. *Br J Psychiatry* 1986; **149**:17–23.

Gysin WM, Cooke ET. Unusual mental symptoms in a case of hepatolenticular degeneration. *Dis Nerv Syst* 1950; **28**:305–9.

Hacke W, Schwab S, Horn M *et al.* 'Malignant' middle cerebral artery territory infarction: clinical course and prognostic signs. *Arch Neurol* 1996; **53**:309–15.

Hacke W, Kaste M, Bluhmki E *et al.* Thrombolysis with alteplase 3 to 4.5 hours after acute ischemic stroke. *N Engl J Med* 2008; **359**:1317–29.

Hagberg B, Aicardi J, Ramos O. A progressive syndrome of autism, ataxia, and loss of purposeful hand use in girls: Rett's syndrome: report of 35 cases. *Ann Neurol* 1983; **14**:471–9.

Hageman ATM, Gabreels FJM, de Jong JGN *et al.* Clinical symptoms of adult metachromatic leukodystrophy and arylsulfatase A pseudodeficiency. *Arch Neurol* 1995; **52**:408–13.

Hahn RD, Webster B, Weickhardt G *et al.* Penicillin treatment of general paresis (dementia paralytica). *Arch Neurol Psychiatry* 1959; **81**:557–90.

Hakan T, Ceran N, Erdem I *et al.* Bacterial brain abscess: an evaluation of 96 cases. *J Infect* 2006; **52**:359–66.

Halkes PHA, Gray LJ, Bath PMW *et al.* Dipyridamole plus aspirin versus aspirin alone in secondary prevention after TIA or stroke: a meta-analysis by risk. *J Neurol Neurosurg Psychiatry* 2008; **79**: 1218–23.

Hall DE, Eubanks L, Meyyaxhagan LS *et al.* Evaluation of covert video surveillance in the diagnosis of Munchausen syndrome by proxy: lessons from 41 cases. *Pediatrics* 2000; **105**:1305–12.

Hall DP, Young SA. Frontal lobe cerebral aneurysm rupture presenting as psychosis. *J Neurol Neurosurg Psychiatry* 1992; **55**:1207–8.

Hall RCW, Joffe JR. Hypomagnesemia: physical and psychiatric symptoms. *JAMA* 1973; **224**:1749–51.

Haller E, Binder RL. Clozapine and seizures. *Am J Psychiatry* 1990; **147**:1069–71.

Haltia T, Palo J, Haltia M *et al.* Juvenile metachromatic dystrophy: clinical, biochemical, and neuropathologic studies in nine new cases. *Arch Neurol* 1980; **37**:42–5.

Hammes EM. Psychoses associated with Sydenham's chorea. *J Am Med Assoc* 1922; **79**:804–7.

Han SW, Kim SH, Kim JK *et al.* Hemodynamic changes in limb shaking TIA associated with anterior cerebral artery stenosis. *Neurology* 2004; **63**:1519–21.

Handforth A, DeGiorgio CM, Schacter SC *et al.* Vagus nerve stimulation therapy for partial-onset seizures: a randomized active-control trial. *Neurology* 1998; **51**:48–55.

Hanson PA, Chodos R. Hemiparetic seizures. *Neurology* 1978; **28**:920–3.

Harden CL, Meador KJ, Pennell PB *et al.* Practice parameter update: management issues for women with epilepsy – focus on pregnancy (an evidence-based review): teratogenesis and perinatal outcomes: report of the Quality Standards Subcommittee and Therapeutics and Technology Assessment Subcommittee of the American Academy of Neurology and American Epilepsy Society. *Neurology* 2009; **73**:133–41.

Hardie RJ, Pullon HWH, Harding AE *et al.* Neuroacanthocytosis: a clinical, haematological and pathological study of 19 cases. *Brain* 1991; **114**:13–49.

Harlow JM. Passage of an iron rod through the head. *Bost Med Surg J* 1848; **39**:389–93.

Harlow JM. Recovery from passage of an iron bar through the head. *J Mass Med Soc* 1868; **2**:327–47.

Harris MJ, Jeste DV, Gleghorn A *et al.* New-onset psychosis in HIV-infected patients. *J Clin Psychiatry* 1991; **52**:369–76.

Harrison RA, Vu T, Hunter AJ. Wernicke's encephalopathy in a patient with schizophrenia. *J Gen Int Med* 2006; **21**:1338.

Hart RG, Byer JA, Slaughter JR *et al.* Occurrence and implications of seizures in subarachnoid hemorrhage due to ruptured intracranial aneurysms. *Neurosurgery* 1981; **8**:417–21.

Hauser WA, Morris ML, Hewston LL *et al.* Seizures and myoclonus in patients with Alzheimer's disease. *Neurology* 1986; **36**:1226–30.

Haymaker W. Herpes simplex encephalitis in man. *J Neuropathol Exp Neurol* 1949; **8**:132–54.

Hayman M, Scheffer IE, Chinvarun Y *et al.* Autosomal dominant nocturnal frontal lobe epilepsy: demonstration of focal frontal onset and intrafamilial variation. *Neurology* 1997; **49**:969–75.

Healton EB, Brust JC, Feinfeld DA *et al.* Hypertensive encephalopathy and the neurologic manifestations of malignant hypertension. *Neurology* 1982; **32**:127.

Heathfield KWG. Huntington's chorea. Investigation into the prevalence of this disease in the area covered by the North East Metropolitan Regional Hospital Board. *Brain* 1967; **90**:203–32.

Hedera P, Abou-Khalil B, Cronk AE *et al.* Autosomal dominant lateral temporal epilepsy: two families with novel mutations in the LGI1 gene. *Epilepsia* 2004; **45**:218–22.

Heilman KM, Howell GJ. Seizure-induced neglect. *J Neurol Neurosurg Psychiatry* 1980; **43**:1035–40.

Heinsius T, Bogousslavsky J, van Melle G. Large infarcts in the middle cerebral artery territory. Etiology and outcome patterns. *Neurology* 1998; **50**:341–50.

Helgason C, Caplan LR, Goodwin J *et al.* Anterior choroidal artery territory infarction. *Arch Neurol* 1986; **43**:681–6.

Heller AJ, Chesterman P, Elwes RDC *et al.* Phenobarbitone, phenytoin, carbamazepine, or sodium valproate for newly diagnosed adult epilepsy: a randomized comparative monotherapy trial. *J Neurol Neurosurg Psychiatry* 1995; **58**:44–50.

Henchey R, Cibula J, Helveston W *et al.* Electroencephalographic findings in Hashimoto's encephalopathy. *Neurology* 1995; **45**: 977–81.

Hennerici M, Klemm C, Rautenberg W. The subclavian steal phenomenon: a common vascular disorder with rare neurologic deficits. *Neurology* 1988; **38**:669–73.

Hennis A, Corbin D, Fraser H. Focal seizures and non-ketotic hyperglycemia. *J Neurol Neurosurg Psychiatry* 1992; **55**:195–7.

Henriksen GF. Status epilepticus partialis with fear as clinical expression: report of a case and EEG findings. *Epilepsia* 1973; **14**:39–46.

Henry TR, Drury I, Brunberg JA *et al.* Focal cerebral magnetic resonance changes associated with partial status epilepticus. *Epilepsia* 1994; **35**:35–41.

Hermann BP, Riel P. Interictal personality and behavioral traits in temporal lobe and generalized epilepsy. *Cortex* 1981; **17**:125–8.

Hermann BP, Whitman S, Wyler AR *et al.* The neurological, psychosocial and demographic correlates of hypergraphia in patients with epilepsy. *J Neurol Neurosurg Psychiatry* 1988; **51**:203–8.

Herskowitz A, Swerdlow M. Focal seizures – a false lateralizing sign. *Dis Nerv Syst* 1972; **33**:523–5.

Hertz PE, Nadas E, Wojtkowski H. Case report: Cushing's syndrome and its management. *Am J Psychiatry* 1955; **112**:144–5.

Heutink P, Stevens M, Rizzu P *et al.* Hereditary frontotemporal dementia is linked to chromosome 17q21–q22: a genetic and clinicopathological study of three Dutch families. *Ann Neurol* 1997; **41**:150–9.

Hijdra A, Van Gijn J, Stefanko S *et al.* Delayed cerebral ischemia after aneurysmal subarachnoid hemorrhage: clinico-anatomic correlations. *Neurology* 1986; **36**:329–33.

Hijdra A, van Gijn J, Nagelkerke NJD *et al.* Prediction of delayed cerebral ischemia, rebleeding and outcome after aneurismal subarachnoid hemorrhage. *Stroke* 1988; **19**:1250–6.

Hildebrand J, Lecaille C, Perennes J *et al.* Epileptic seizures during follow-up of patients treated for primary brain tumor. *Neurology* 2005; **65**:212–15.

Hillier SL, Sharpe MH, Metzer J. Outcomes 5 years post-traumatic brain injury (with further reference to neurophysical impairment and disability). *Brain Inj* 1997; **11**:661–75.

Hinchey J, Chaves C, Appignani B *et al.* A reversible posterior leukoencephalopathy syndrome. *N Engl J Med* 1996; **334**:494–500.

Hinson VK, Weinstein S, Bernard B *et al.* Single-blind clinical trial of psychotherapy for treatment of psychogenic movement disorders. *Parkinsonism Relat Disord* 2006; **12**:177–80.

Hirose S, Iwata H, Akiyoshi H *et al.* A novel mutation of CHRNA4 responsible for autosomal dominant nocturnal frontal lobe epilepsy. *Neurology* 1999; **53**:1749–53.

Hirsch E, Marescaux C, Maquet P *et al.* Landau-Kleffner syndrome: a clinical and EEG study of five cases. *Epilepsia* 1990; **31**:756–67.

Hirsch S, Dunsworth FA. An interesting case of porphyria. *Am J Psychiatry* 1955; **111**:703.

Hjort N, Christensen S, Solling C *et al.* Ischemic injury detected by diffusion imaging in 11 minutes. *Ann Neurol* 2005; **58**:462–5.

Ho SS, Berkovic SF, Newton MR *et al.* Parietal lobe epilepsy: clinical features and seizure localization by SPECT. *Neurology* 1994; **44**:2277–84.

Hodgson RE, Murray D, Woods MR. Othello's syndrome and hyperthyroidism. *J Nerv Ment Dis* 1992; **180**:663–4.

Hoefer PFA, Guttman SA, Sands IJ. Convulsive states and coma in cases of islet cell carcinoma of the pancreas. *Am J Psychiatry* 1946; **102**:486–95.

Hoefnagels WA, Padberg GW, Overweg J *et al.* Transient loss of consciousness: the value of history for distinguishing seizure from syncope. *J Neurol* 1991; **238**:39–43.

Holmes G. Local epilepsy. *Lancet* 7 May 1927; 957–73.

Holmes GL. The electroencephalogram as a predictor of seizures following cerebral infarction. *Clin Electroencephalogr* 1980; **11**:83–6.

Holmes GL. Partial complex seizures in children: an analysis of 69 seizures in 24 patients using EEG FM radiotelemetry and videotape recordings. *Electroencephalogr Clin Neurophysiol* 1984; **57**:13–20.

Holmes GL, McKeever M, Adamson M. Absence seizures in children: clinical and electroencephalographic features. *Ann Neurol* 1987; **21**:268–73.

Hoogman M, van de Beek D, Weisfelt M *et al.* Cognitive outcome in adults after bacterial meningitis. *J Neurol Neurosurg Psychiatry* 2007; **78**:1092–6.

Hoover CF. A new sign for detection of malingering and functional paresis of the lower extremities. *J Am Med Assoc* 1908; **51**:1309–10.

Hopkins A, Shorvon S, Cascino G. *Epilepsy*, 2nd edn. London: Arnold 1995.

House A, Dennis M, Molyneux A *et al.* Emotionalism after stroke. *BMJ* 1989; **298**:991–4.

Huber W, Williams K, Poeck D *et al.* Piracetam as a adjuvant to language therapy for aphasia: a randomized, double-blind placebo-controlled pilot study. *Arch Phys Med Rehabil* 1997; **78**:245–50.

Hudson EA, Munoz DG, Miller L *et al.* Amygdaloid sclerosis in temporal lobe epilepsy. *Ann Neurol* 1993; **33**:622–31.

Hunter R, Blackwood W, Bull J. Three cases of frontal meningiomas presenting psychiatrically. *BMJ* 1968; **3**:9–16.

Husain AM, Horn GJ, Jacobson MP. Non-convulsive status epilepticus: usefulness of clinical features in selecting patients for urgent EEG. *J Neurol Neurosurg Psychiatry* 2003; **74**:189–91.

Hyde TM, Ziegler JL, Weinberger DR. Psychiatric disturbances in metachromatic leukodystrophy: insights into the neurobiology of psychosis. *Arch Neurol* 1992; **49**:401–6.

Inanatomia Y, Kimura K, Yonehara T *et al.* DWI abnormalities and clinical characteristics in TIA patients. *Neurology* 2004; **62**:376–80.

Ingham SD, Nielsen JM. Thyroid psychosis: difficulties in diagnosis. *J Nerv Ment Dis* 1931; **74**:271–7.

Inouye E. Genetic aspects of neurosis. *Int J Ment Health* 1972; **1**:176–89.

International Subarachnoid Aneurysm Trial (ISAT) Collaborative Group. International subarachnoid aneurysm trial (ISAT) on neurosurgical clipping versus endovascular coiling in 2143 patients with ruptured intracranial aneurysm: a randomized trial. *Lancet* 2002; **360**:1267–74.

Inzelberg R, Kipervasser S, Korczyn AD. Auditory hallucinations in Parkinson's disease. *J Neurol Neurosurg Psychiatry* 1998; **64**:533–5.

Iranzo A, Santamaria J, Berenguer J *et al.* Prevalence and clinical importance of sleep apnea in the first night after cerebral infarction. *Neurology* 2002; **58**:911–16.

Ironside R, Bosanquet FD, McMenemy WH. Central demyelination of the corpus callosum (Marchifava–Bignami disease): with report of a second case in Great Britain. *Brain* 1961; **84**:212–30.

Isbell H, Fraser HF, Wickler A *et al.* An experimental study of etiology of 'rum fits' and delirium tremens. *Q J Study Alcohol* 1955; **16**:1–33.

Ishii N, Nishihara Y, Imamura T. Why do frontal lobe symptoms predominate in vascular dementia with lacunes? *Neurology* 1986; **36**:340–5.

Ivanov D, Kirov G, Norton N *et al.* Chromosome 22q11 deletions, velo-cardio-facial syndrome and early onset psychosis. *Br J Psychiatry* 2003; **183**:409–13.

Iwanami A, Sugiyama A, Kuroki N *et al.* Patients with methamphetamine psychosis admitted to a psychiatric hospital in Japan. *Acta Psychiatr Scand* 1994; **89**:428–32.

Jabbari B, Huott AD. Seizures in thyrotoxicosis. *Epilepsia* 1980; **21**:91–6.

Jabbari B, Bryan GE, Marsh EE *et al.* Incidence of seizures with tricyclic and tetracyclic antidepressants. *Arch Neurol* 1985; **42**:480–1.

Jackel RA, Malkowitz D, Trivedi R *et al.* Reduction of prolactin response with repetitive seizures. *Epilepsia* 1987; **28**:588.

Jackson GD, Kuzniecky RI, Cascino GD. Hippocampal sclerosis without detectable hippocampal atrophy. *Neurology* 1994; **44**:42–6.

Jackson JA, Zimmerman SL. A case of pseudosclerosis associated with a psychosis. *J Nerv Ment Dis* 1919; **49**:5–13.

Jackson JH. On a particular variety of epilepsy ('intellectual aura'). One case with symptoms of organic brain disease. *Brain* 1889; **11**:179–207.

Jackson JH. Neurological fragments. *Lancet* 1894; **2**:182–3.

Jackson JH, Beevor CE. Case of tumour of the right temporo-sphenoidal lobe bearing on the localization of the sense of smell and on the interpretation of a particular variety of epilepsy. *Brain* 1890; **12**:346–57.

Jackson JH, Colman WJ. Case of epilepsy with tasting movements and 'dreamy state' – very small patch of softening in the left uncinate gyrus. *Brain* 1898; **21**:580–91.

Jackson JH, Stewart P. Epileptic attacks with a warning of a crude sensation of smell and with the intellectual aura (dreamy state) in a patient who had symptoms pointing to gross organic disease of the right temporo-sphenoidal lobe. *Brain* 1899; **22**:534–49.

Jacob A, Cherian PJ, Radhakrishnan K *et al.* Emotional facial paresis in temporal lobe epilepsy: its prevalence and lateralizing value. *Seizure* 2003; **12**:60–4.

Jacome DE, Risko M. Pseudocataplexy: gelastic–atonic seizures. *Neurology* 1984; **34**:1381–3.

Jacome DE, McLain W, Fitzgerald R. Postural reflex gelastic seizures. *Arch Neurol* 1980; **37**:249–51.

Jaffe R. Ictal behavior as the only manifestation of seizure disorder: case report. *J Nerv Ment Dis* 1962; **134**:470–6.

James WE, Mefferd RB, Kimbell I. Early signs of Huntington's chorea. *Dis Nerv Syst* 1969; **30**:550–9.

Janssen BA, Theiler R, Grob D *et al.* The role of motor evoked potentials in psychogenic paralysis. *Spine* 1995; **20**:608–11.

Janszky J, Fogarai A, Magalova V *et al.* Unilateral hand automatisms in temporal lobe epilepsy. *Seizure* 2006; **15**:393–6.

Janz D. Epilepsy with impulsive petit mal (juvenile myoclonic epilepsy). *Acta Neurol Scand* 1985; **72**:449–59.

Jawad S, Richens A, Goodwin G *et al.* Controlled trial of lamotrigene (Lamictal) for refractory partial seizures. *Epilepsia* 1989; **30**:356–63.

Jenkins A, Teasdale G, Hadley MD *et al.* Brain lesions detected by magnetic resonance imaging in mild and severe head injuries. *Lancet* 1986; **2**:445–6.

Jenkins CD, Mellins RB. Lead poisoning in children: a study of forty-six cases. *Arch Neurol Psychiatry* 1957; **77**:70–8.

Jennett B. Early traumatic epilepsy. *Arch Neurol* 1973; **30**:394–8.

Jennett B. Epilepsy and acute traumatic intracranial haematoma. *J Neurol Neurosurg Psychiatry* 1975; **38**:378–81.

Jennett B, Teather D, Bennie S. Epilepsy after head injury. *Lancet* 1973; **2**:652–3.

Jensen I, Klinken L. Temporal lobe epilepsy and neuropathology. *Acta Neurol Scand* 1976; **54**:391–414.

Jensen MB, St. Louis EK. Management of acute cerebellar stroke. *Arch Neurol* 2005; **62**:537–44.

Jha A, Weintraub A, Allshouse A *et al.* A randomized trial of modafinil for the treatment of fatigue and excessive daytime sleepiness in individuals with chronic traumatic brain injury. *J Head Trauma Rehabil* 2008; **23**:52–63.

Johnson KP, Rosenthal MS, Lerner PI. Herpes simplex encephalitis. *Arch Neurol* 1972; **27**:103–8.

Johnson RI, Richardson EP. The neurological manifestations of systemic lupus erythematosus. *Medicine* 1968; **47**:337–69.

Johnston JA, Lineberry CG, Archer JA *et al.* A 102-center prospective study of seizure in association with bupropion. *J Clin Psychiatry* 1991; **52**:450–6.

Johnston SC, Gress DR, Browner WS *et al.* Short-term prognosis after emergency department diagnosis of TIA. *JAMA* 2000; **284**:2901–6.

Jorge RE, Robinson RG, Arndt SV *et al.* Comparison between acute- and delayed -onset depression following traumatic brain injury. *J Neuropsychiatry Clin Neurosci* 1993; **5**:43–9.

Jorge RE, Robinson RG, Moser D *et al.* Major depression following traumatic brain injury. *Arch Gen Psychiatry* 2004; **61**:42–50.

Joynt J, Green D, Gren R. Musicogenic epilepsy. *JAMA* 1962; **179**:501–4.

Kaiboriboon K, Parent JM, Barbaro NM *et al.* Speech preservation during language-dominant, left temporal lobe seizures: report of a rare, potentially misleading finding. *Epilepsia* 2006; **47**:1343–6.

Kaiko RF, Foley KM, Grabinski PY *et al.* Central nervous system excitatory effects of meperidine in cancer patients. *Ann Neurol* 1983; **13**:180–5.

Kalinowsky LB. Convulsions in nonepileptic patients on withdrawal of barbiturates, alcohol and other drugs. *Arch Neurol Psychiatry* 1942; **48**:946–56.

Kanemoto K. Periictal Capgras syndrome after clustered ictal fear: depth-electroencephalogram study. *Epilepsia* 1997; **38**:847–50.

Kanemoto K, Janz D. The temporal sequence of aura-sensations in patients with complex partial focal seizures with particular attention to ictal aphasia. *J Neurol Neurosurg Psychiatry* 1989; **52**:52–6.

Kanemoto K, Kawasaki J, Mori E. Violence and epilepsy: a close relation between violence and postictal psychosis. *Epilepsia* 1999; **40**:107–9.

Kanemoto K, Tsuji T, Kawasaki J. Reexamination of interictal psychoses based on DSMIV psychosis classification and international epilepsy classification. *Epilepsia* 2001a; **42**:98–103.

Kanemoto K, Watanabe Y, Tsuji T *et al.* Rub epilepsy: a somatosensory evoked reflex epilepsy induced by prolonged cutaneous stimulation. *J Neurol Neurosurg Psychiatry* 2001b; **70**:541–3.

Kang D-W, Lattimore SU, Latour LL *et al.* Silent ischemic lesion recurrence on magnetic resonance imaging predicts subsequent clinical vascular events. *Arch Neurol* 2006; **63**:1730–3.

Kang SY, Kim JS. Anterior cerebral artery infarction: stroke mechanism and clinical-imaging study in 100 patients. *Neurology* 2008; **70**:2386–93.

Kanner AM, Morris HH, Dinner DS *et al.* Supplementary motor seizures mimicking pseudoseizures: some clinical differences. *Neurology* 1990; **40**:1404–7.

Kanner AM, Stagno S, Kotagal P *et al.* Postictal psychiatric events during prolonged video-electroencephalographic monitoring studies. *Arch Neurol* 1996; **53**:258–63.

Kant R, Duffy JD, Pivovarnik A. Prevalence of apathy following head injury. *Brain Inj* 1998; **12**:87–92.

Karnosh LJ, Stout RE. Psychoses of myxedema. *Am J Psychiatry* 1935; **91**:1263–74.

Karp BI, Laureno R. Pontine and extrapontine myelinolysis: a neurologic disorder following rapid correction of hyponatremia. *Medicine* 1993; **72**:359–73.

Kashiwase H, Kato M. Folie a deux in Japan – analysis of 97 cases in the Japanese literature. *Acta Psychiatr Scand* 1997; **96**:231–4.

Kasper BS, Kerling F, Graf W *et al.* Ictal delusion of sexual transformation. *Epilepsy Behav* 2009; **16**:356–9.

Katz DI, Alexander MP, Seliger GM *et al.* Traumatic basal ganglia hemorrhage: clinicopathologic features and outcome. *Neurology* 1989; **39**:862–3.

Keane JR. Wrong-way deviation of the tongue with hysterical hemiparesis. *Neurology* 1986; **36**:1406–7.

Kellinghaus C, Kotagal P. Lateralizing value of Todd's palsy in patients with epilepsy. *Neurology* 2004; **62**:289–91.

Kemp S, Biswas R, Neumann V *et al.* The value of melatonin for sleep disorders occurring post-head injury: a pilot RCT. *Brain Inj* 2004; **18**:911–19.

Kennedy F. The symptomatology of temporosphenoidal tumors. *Arch Int Med* 1911; **8**:317–50.

Kennedy PGE. A retrospective analysis of forty-six cases of herpes simplex encephalitis seen in Glasgow between 1962 and 1985. *Q J Med* 1988; **68**:533–40.

Keschner M, Bender MB, Strauss I. Mental symptoms in cases of tumor of the temporal lobe. *Arch Neurol Psychiatry* 1936; **35**:572–96.

Kessler J, Thiel A, Karbe H *et al.* Piracetam improves activated blood flow and facilitates rehabilitation of poststroke aphasic patients. *Stroke* 2000; **31**:2112–16.

Khatri P, Kasner SE. Ischemic strokes after cardiac catheterization. Opportune thrombolysis candidates? *Arch Neurol* 2006; **63**:817–21.

Khealani BA, Wasay M, Saadah M *et al.* Cerebral venous thrombosis: a descriptive multicenter study of patients in Pakistan and Middle East. *Stroke* 2008; **39**:2707–11.

Kilpatrick CJ, Davis SM, Tress BM *et al.* Epileptic seizures in acute stroke. *Arch Neurol* 1990; **47**:157–60.

Kilpatrick CJ, Davis SM, Hopper JL *et al.* Early seizures after acute stroke: risk of late seizures. *Arch Neurol* 1992; **49**:509–11.

Kim JS. Pure lateral medullary infarction: clinical-radiological correlation of 130 acute, consecutive patients. *Brain* 2003; **126**:1864–72.

Kim JS, Kim HG, Chung CS. Medial medullary syndrome: report of 18 new patients and a review of the literature. *Stroke* 1995; **26**:1548–52.

Kimura M, Tateno A, Robinson RG. Treatment of poststroke generalized anxiety disorder comorbid with poststroke depression: merged analysis of nortriptyline trials. *Am J Geriatr Psychiatry* 2003; **11**:320–7.

King DW, Marsan CA. Clinical features and ictal patterns in epileptic patients with EEG temporal lobe foci. *Ann Neurol* 1977; **2**:138–47.

King MA, Newton MR, Jackson GD *et al.* Epileptology of the first seizure presentation: a clinical, electroencephalographic, and magnetic resonance imaging study of 300 consecutive patients. *Lancet* 1999; **352**:1007–11.

Kiok MC, Terrence CF, Fromm GH *et al.* Sinus arrest in epilepsy. *Neurology* 1986; **36**:115–16.

Kirby GH, Davis TK. Psychiatric manifestations of epidemic encephalitis. *Arch Neurol Psychiatry* 1921; **5**:491–555.

Kirubakaren V, Mayfield D, Rengachary S. Dyskinesia and psychosis in a patient following Baclofen withdrawal. *Am J Psychiatry* 1984; **141**:692–3.

Kluge M, Beyenburg S, Fernandez G *et al.* Epileptic vertigo: evidence for vestibular representation in human frontal cortex. *Neurology* 2000; **55**:1906–8.

Knake S, Gruener J, Hattemer K *et al.* Intravenous levetiracetam in the treatment of benzodiazepine refractory status epilepticus. *J Neurol Neurosurg Psychiatry* 2008; **79**:588–9.

Knight RT, Cooper J. Status epilepticus manifesting as reversible Wernicke's aphasia. *Epilepsia* 1986; **27**:301–4.

Kobayashi E, Li LM, Lopes-Cendes I *et al.* Magnetic resonance imaging evidence of hippocampal sclerosis in asymptomatic, first-degree relatives of patients with familial mesial temporal lobe epilepsy. *Arch Neurol* 2002; **59**:1891–4.

Kobayashi E, D'Agostino MD, Lopes-Cendes I *et al.* Hippocampal atrophy and T2-weighted signal changes in familial mesial temporal lobe epilepsy. *Neurology* 2003; **60**:405–9.

Kober D, Gabbard GO. Topiramate-induced psychosis. *Am J Psychiatry* 2005; **162**:1542.

Koehler K, Jakumeit U. Subacute sclerosing panencephalitis presenting as Leonhard's speech-prompt catatonia. *Br J Psychiatry* 1976; **129**:29–31.

Koepp MJ, Edwards M, Collins J *et al.* Status epilepticus and tiagabine therapy revisited. *Epilepsia* 2005; **46**:1625–32.

Koeppen AH, Barron KD. Marchiafava–Bignami disease. *Neurology* 1978; **28**:290–4.

Koerner M, Laxer KD. Ictal speech, postictal language dysfunction, and seizure lateralization. *Neurology* 1988; **38**:634–6.

Kofman O, Hyland HH. Tuberous sclerosis in adults with normal intelligence. *Arch Neurol Psychiatry* 1959; **81**:557–90.

Korczyn AD, Bechar M. Convulsive fits in thyrotoxicosis. *Epilepsia* 1976; **17**:33–4.

Korf BR, Carrazana E, Holmes GL. Patterns of seizures observed in association with neurofibromatosis. I. *Epilepsia* 1993; **34**:616–20.

Kornberg AJ, Harvery AS, Shield LK. Subacute sclerosing panencephalitis presenting as simple partial seizures. *J Child Neurol* 1991; **6**:146–9.

Koskiniemi M, Rantalaiho T, Piiparinen H *et al.* Infections of the central nervous system of suspected viral origin: a collaborative study from Finland. *J Neurovirol* 2001; **7**:400–8.

Kossoff EH, Bergey GK, Freeman JM *et al.* Levetiracetam psychosis in children with epilepsy. *Epilepsia* 2001; **42**:1611–13.

Kotagal P, Luders H, Morris HH *et al.* Dystonic posturing in complex partial seizures of temporal lobe onset: a new lateralizing sign. *Neurology* 1989; **39**:196–204.

Kothare SV, VanLandingham K, Armon C *et al.* Seizure onset from periventricular nodular heterotopias: depth-electrode study. *Neurology* 1998; **51**:1723–7.

Kotsopoulos IAW, van Merode T, Kessels FGH *et al.* Systematic review and meta-analysis of incidence studies of epilepsy and unprovoked seizures. *Epilepsia* 2002; **43**:1402–9.

Koutroumaniois M, Koepp MJ, Richardson MP *et al.* The variants of reading epilepsy: a clinical and video-EEG study of 17 patients with reading-induced seizures. *Brain* 1998; **121**:1409–27.

Kraepelin E. *Clinical Psychiatry 1902.* Delmar, NY: Scholar's Facsimiles and Reprints, 1981.

Krahn LE, Li H, O'Connor MK. Patients who strive to be ill: factitious disorder with physical symptoms. *Am J Psychiatry* 2003; **160**:1163–8.

Kral T, Clusman H, Blumcke I *et al.* Outcome of epilepsy surgery in focal cortical dysplasia. *J Neurol Neurosurg Psychiatry* 2003; **74**:183–8.

Kramer DL, Awad IA. Vascular malformations and epilepsy: clinical considerations and basic mechanisms. *Epilepsia* 1994; **35**(Suppl. 6):30–43.

Kramer RE, Lesser RP, Weinstein MR *et al.* Transient focal abnormalities of neuroimaging studies during focal status epilepticus. *Epilepsia* 1987; **28**:528–32.

Kramer RE, Luders H, Goldstick LP *et al.* Ictus emeticus: an electroclinical study. *Neurology* 1988; **38**:1048–52.

Kreutzer JS, Seel RT, Gourley E. The prevalence and symptom rates of depression after traumatic brain injury: a comprehensive examination. *Brain Inj* 2001; **15**:561–2.

Kristensen O, Sindrup EH. Psychomotor epilepsy and psychosis. III: Social and psychological correlates. *Acta Neurol Scand* 1979; **59**:1–9.

Krolak-Salmon P, Montanont A, Hermier A *et al.* Thalamic venous infarction as a cause of subacute dementia. *Neurology* 2006; **58**:1689–91.

Kroll P. Psychoses associated with marijuana abuse in Thailand. *J Nerv Ment Dis* 1975; **161**:149–56.

Krumholz A, Niedermeyer E. Psychogenic seizures: a clinical study with follow-up data. *Neurology* 1983; **33**:498–502.

Krumholz A, Stern BJ, Weiss HD. Outcome from coma after cardiopulmonary resuscitation: relation to seizures and myoclonus. *Neurology* 1988; **38**:401–5.

Krumholz A, Stern BJ, Stern EG. Clinical implications of seizures in neurosarcoidosis. *Arch Neurol* 1991; **48**:842–4.

Krumholz A, Sung GY, Fisher RS *et al.* Complex partial status epilepticus accompanied by serious morbidity and mortality. *Neurology* 1995; **45**:1499–504.

Kulkantrakorn K, Geller TJ. Seizures in neurofibromatosis. 1. *Pediatr Neurology* 1998; **19**:347–50.

Kumral E, Ozdemirkiran T, Alper Y. Strokes in the subinsular territory. Clinical, topographical, and etiological patterns. *Neurology* 2004; **63**:2429–32.

Kupila L, Vuorinen T, Vainionpaa R *et al.* Etiology of aseptic meningitis and encephalitis in an adult population. *Neurology* 2006; **66**:75–80.

Kuramochi H, Takahashi R. Psychopathology of LSD intoxication: study of experimental psychosis induced by LSD-25: description of LSD symptoms in normal Oriental subjects. *Arch Gen Psychiatry* 1964; **11**:151–61.

Kuzniecky R, Guthrei B, Mountz J *et al.* Intrinsic epileptogenesis of hypothalamic hamartomas in gelastic epilepsy. *Ann Neurol* 1997; **42**:60–7.

Labar DR, Solomon GR, Wells CR. Aphasia as the sole manifestation of simple partial status epilepticus. *Epilepsia* 1992; **33**:84–7.

Labovitz DL, Hauser WA, Sacco RL. Prevalence and predictors of early seizure and status epilepticus after first stroke. *Neurology* 2001; **57**:200–6.

Lafrance WC, Miller IW, Ryan CE *et al.* Cognitive behavioral therapy for psychogenic nonepileptic seizures. *Epilepsy Behav* 2009; **14**:591–6.

Lai F, Williams RS. A prospective study of Alzheimer's disease in Down syndrome. *Arch Neurol* 1989; **46**:849–53.

Lam JS, Fong SY, Yiu GC *et al.* Fahr's disease: a differential diagnosis of frontal lobe syndrome. *Hong Kong Med J* 2007; **13**:75–7.

Lambert MT. Paranoid psychoses after abuse of proprietary cold remedies. *Br J Psychiatry* 1987; **151**:548–50.

Lamy C, Domigo V, Semah F *et al.* Early and late seizures after cryptogenic stroke in young adults. *Neurology* 2003; **60**:400–4.

Lancman ME, Craven WJ, Asconape JJ *et al.* Clinical management of recurrent postictal psychosis. *J Epilepsy* 1994; **7**:47–51.

Landau WM, Kleffner FR. Syndrome of acquired aphasia with convulsive disorder in children. *Neurology* 1957; **7**:523–30.

Landolt H. Some clinical electroencephalographic correlations in epileptic psychoses (twilight states). *Electroencephalogr Clin Neurophysiol* 1953; **5**:121.

Landolt H. Serial electroencephalographic investigations during psychotic episodes in epileptic patients and during schizophrenic attacks. In: de Haas AML, ed. *Lectures on Epilepsy.* London: Elsevier, 1958.

Lang AE, Koller WC, Fahn S. Psychogenic parkinsonism. *Arch Neurol* 1995; **52**:802–10.

Lankford DA, Wellman JJ, O'Hara C. Posttraumatic narcolepsy in mild to moderate closed head injury. *Sleep* 1994; **17**(Suppl. 8):S25–S28.

Lansberg Mg, Albers GW, Beaulieu C *et al.* Comparison of diffusion-weighted MRI and CT in acute stroke. *Neurology* 2000; **54**:1557–61.

Lawson JA, Birchansky S, Pacheo E *et al.* Distinct clinicopathologic subtypes of cortical dysplasia of Taylor. *Neurology* 2005; **64**:55–61.

Laxer KD, Mullooly JP, Howell B. Prolactin changes after seizures classified by EEG monitoring. *Neurology* 1985; **35**:31–5.

Leal AJR, Moreira A, Robalo C *et al.* Different electroclinical manifestations of the epilepsy associated with hamartomas connecting to the middle or posterior hypothalamus. *Epilepsia* 2003; **44**:1191–5.

Lechtenberg R, Worner TM. Total ethanol consumption as a seizure risk factor in alcoholics. *Acta Neurol Scand* 1992; **85**:90–4.

Lederman RJ, Henry CE. Progressive dialysis encephalopathy. *Ann Neurol* 1978; **4**:199–204.

Lee H, Lerner A. Transient inhibitory seizures mimicking crescendo transient ischemic attacks. *Neurology* 1990; **40**:165–6.

Lee H, Kim SW, Kim JM *et al.* Comparing effects of methyphenidate, sertraline and placebo on neuropsychiatric sequelae in patients with traumatic brain injury. *Hum Psychopharmacol* 2005; **20**:97–104.

Lee VH, Brown RD, Madrekar JN *et al.* Incidence and outcome of cervical artery dissection. *Neurology* 2006; **67**:1809–12.

Legg NJ, Gupta PC, Scott DF. Epilepsy following cerebral abscess. A clinical and EEG study of 70 patients. *Brain* 1973; **96**:259–68.

Lehtinen L, Kivalo A. Laughter epilepsy. *Acta Neurol Scand* 1965; **41**:255–61.

Leijon G, Boivie J. Central post-stroke pain – a controlled trial of amitriptyline and carbamazepine. *Pain* 1989; **36**:27–36.

Lempert T, Bauer M, Schmidt D. Syncope: a videometric analysis of 56 episodes of transient cerebral hypoxia. *Ann Neurol* 1994; **36**:233–7.

Lende RA, Popp AJ. Sensory Jacksonian seizures. *J Neurosurg* 1976; **44**:706–11.

Leone M, Bottacchi E, Beghi E *et al.* Alcohol use is a risk factor for a first generalized tonic–clonic seizure. *Neurology* 1997; **48**:614–20.

Leppik IE, Derivan AT, Homan RW *et al.* Double-blind study of lorazepam and diazepam in status epilepticus. *JAMA* 1983; **249**:1452–4.

Leppik IE, Biton V, Sander JWA *et al.* Levetiracetam and partial seizure subtypes: pooled data from three randomized, placebo-controlled trials. *Epilepsia* 2003; **44**:1585–7.

Lesser RP, Lueders H, Conomy JP *et al.* Sensory seizure mimicking a psychogenic seizure. *Neurology* 1983; **33**:800–2.

Leutmezer F, Serles W, Bacher J *et al.* Genital automatisms in complex partial seizures. *Neurology* 1999; **52**:1188–91.

Leutmezer F, Podreka I, Asenbaum S *et al.* Postictal psychosis in temporal lobe epilepsy. *Epilepsia* 2003; **44**:582–90.

Leventer RJ, Jansen A, Pilz DT *et al.* Clinical and imaging heterogeneity of polymicrogyria: a study of 328 patients. *Brain* 2010; **133**:1415–27.

Levin HS, Grossman RG, Kelly PJ. Aphasic disorder in patients with closed head injury. *J Neurol Neurosurg Psychiatry* 1976; **39**: 1062–70.

Levin HS, Goldstein FC, High WM *et al.* Disproportionately severe memory deficit in relation to normal intellectual functioning after closed head injury. *J Neurol Neurosurg Psychiatry* 1988; **51**:1294–301.

Levy AB. Delirium and seizures due to abrupt alprazolam withdrawal: case report. *J Clin Psychiatry* 1984; **45**:38–9.

Libman MD, Potvin L, Coupal L *et al.* Seizure v. syncope: measuring serum creatine kinase in the emergency department. *J Gen Intern Med* 1991; **6**:408–12.

Lichenstein BW. Sturge–Weber–Dimitri syndrome: cephalic form of neurocutaneous hemangiomatosus. *Arch Neurol Psychiatry* 1954; **71**:291–301.

Liddell DW. Observations on epileptic automatisms in a mental hospital population. *J Ment Sci* 1953; **99**:732–48.

Lieb JP, Walsh JO, Babb TL *et al.* A comparison of EEG seizure patterns with surface and depth electrodes in patients with temporal lobe epilepsy. *Epilepsia* 1976; **17**:137–60.

Lim J, Yagnik P, Schraeder P *et al.* Ictal catatonia as a manifestation of nonconvulsive status epilepticus. *J Neurol Neurosurg Psychiatry* 1986; **49**:833–6.

Lim L, Ron MA, Omerod IEC *et al.* Psychiatric and neurological manifestations in systemic lupus erythematosus. *Q J Med* 1988; **66**:27–38.

Lima MA, Drislane FW, Kuralnik IJ. Seizures and their outcome in progressive multifocal leukoencephalopathy. *Neurology* 2006; **66**:262–4.

Limdi Na, Knowlton RK, Cofield SS *et al.* Safety of rapid intravenous loading of valproate. *Epilepsia* 2007; **48**:478–83.

Lin J T-Y, Ziegler DK. Psychiatric symptoms with initiation of carbidopa-levodopa treatment. *Neurology* 1976; **26**:699–700.

Lindenbaum J, Healton EB, Savage DG *et al.* Neuropsychiatric disorders caused by cobalamin deficiency in the absence of anemia or macrocytosis. *N Engl J Med* 1988; **318**:1720–8.

Lipinski CG. Epilepsies with astatic seizures of late onset. *Epilepsia* 1977; **18**:13–20.

Lipsey JD, Robinson RG, Pearlson GD *et al.* Nortriptyline treatment of post-stroke depression: a double-blind study. *Lancet* 1984; **1**:297–300.

Lisanby SH, Bazl CW, Resor SR *et al.* ECT in the treatment of status epilepticus. *J ECT* 2001; **17**:210–15.

Lishman WA. Brain damage in relation to psychiatric disability after head injury. *Br J Psychiatry* 1968; **114**:373–410.

Lishman WA. Cerebral disorder in alcoholism: syndromes of impairment. *Brain* 1981; **104**:373–410.

List CF, Dowman CE, Bagchi BK *et al.* Posterior hypothalamic hamartomas and gangliogliomas causing precocious puberty. *Neurology* 1958; **8**:164–74.

Litovitz TL, Troutman WG. Amoxapine overdose: seizures and fatalities. *JAMA* 1983; **250**:1069–71.

Litvan IL, Agid Y, Sastrj N *et al.* What are the obstacles to the accurate clinical diagnosis of Pick's disease? *Neurology* 1997; **49**:62–9.

Liu RSN, Lemieux L, Bell GS *et al.* The structural consequences of newly diagnosed seizures. *Ann Neurol* 2002; **52**:573–80.

Livingston S, Torres I, Pauli L *et al.* Petit mal epilepsy. Results of a prolonged follow-up of 117 patients. *JAMA* 1965; **194**:227–32.

Locke J, Merrill JP, Tyler HR. Neurologic complications of uremia. *Arch Int Med* 1961; **108**:519–30.

Logothetis J. Spontaneous epileptic seizures and electroencephalographic changes in the course of phenothiazine therapy. *Neurology* 1967; **17**:869–77.

Logsdail SJ, Toone BK. Postictal psychoses: a clinical and phenomenological description. *Br J Psychiatry* 1988; **152**:246–52.

Logue V, Durward M, Pratt RTC *et al.* The quality of survival after rupture of an anterior cerebral aneurysm. *Br J Psychiatry* 1968; **114**:137–60.

Loiseau J, Loiseau P, Duche B *et al.* A survey of epileptic disorder in southwest France: seizures in elderly patients. *Ann Neurol* 1990a; **27**:232–7.

Loiseau P, Yuen AWC, Duche B *et al.* A randomized double-blind, placebo-controlled crossover add-on trial of lamotrigene in patients with treatment-resistant partial seizures. *Epilepsia Res* 1990b; **7**:136–45.

LoPiccolo CJ, Goodkin K, Baldewicz TT. Current issues in the diagnosis and management of malingering. *Ann Med* 1999; **31**:166–74.

Lossius MI, Ronning OM, Slapo GD *et al.* Poststroke epilepsy: occurrence and predictors: a long-term prospective controlled study. *Epilepsia* 2005; **46**:1246–51.

Lossos A, Dobson-Stone C, Monaco AP *et al.* Early clinical heterogeneity in choreoacanthocytosis. *Ann Neurol* 2005; **62**:611–14.

Louis ED, Lynch T, Daufmann P *et al.* Diagnostic guidelines in central nervous system Whipple's disease. *Ann Neurol* 1996; **40**:561–8.

Lowenstein DH, Alldredge BK. Status epilepticus. *N Engl J Med* 1998; **338**:970–6.

Lowenstein DH, Aminoff MJ. Clinical and EEG features of status epilepticus in comatose patients. *Neurology* 1992; **42**:100–4.

Luciano D, Devinsky O, Perrine K. Crying seizures. *Neurology* 1993; **43**:2113–17.

Ludwig BI, Marsan CA. Clinical ictal patterns in epileptic patients with occipital electroencephalographic foci. *Neurology* 1975; **25**:463–71.

Lundquist G. Delirium tremens: a comparative study of pathogenesis, course and prognosis with delirium tremens. *Acta Psychiatr Neurol Scand* 1961; **36**:443–66.

Maccario M, Messis CP, Vastola EF. Focal seizures as a manifestation of hyperglycemia without ketoacidosis. A report of seven cases with review of the literature. *Neurology* 1965; **15**:195–206.

Mace CJ, Trimble MR. Ten-year prognosis of conversion disorder. *Br J Psychiatry* 1996; **169**:282–8.

Macpherson P, Teasdale E, Dhakar S *et al.* The significance of traumatic hematoma in the region of the basal ganglia. *J Neurol Neurosurg Psychiatry* 1986; **49**:29–34.

Macrae D. Isolated fear: a temporal lobe aura. *Neurology* 1954a; **4**:497–505.

Macrae D. On the nature of fear, with reference to its occurrence in epilepsy. *J Nerv Ment Dis* 1954b; **120**:385–93.

McCarthy DJ. Epileptic ambulatory automatism. *J Nerv Ment Dis* 1900; **27**:143–9.

McElroy SL, Phillips KA, Keck PE *et al.* Body dysmorphic disorder: does it have a psychotic subtype? *J Clin Psychiatry* 1993; **54**:389–95.

McFarland HR. Addisons's disease and related psychoses. *Compr Psychiatry* 1963; **4**:90–5.

McGrath N, Anderson NE, Croxson MC *et al.* Herpes simplex encephalitis treated with acyclovir: diagnosis and long term outcome. *J Neurol Neurosurg Psychiatry* 1997; **63**:321–6.

McLachlan RS, Blume WT. Isolated fear in complex partial status epilepticus. *Ann Neurol* 1980; **8**:639–41.

McLachlan RS, Girvin JP, Blume WT *et al.* Rasmussen's chronic encephalitis in adults. *Arch Neurol* 1993; **50**:269–74.

McQueen JK, Blackwood DH, Harris P *et al.* Low risk of late post-traumatic seizures following severe head injury: implications for clinical trials of prophylaxis. *J Neurol Neurosurg Psychiatry* 1983; **46**:899–904.

McVicker RW, Shanks OEP, McClelland RJ. Prevalence and associated features of epilepsy in adults with Down's syndrome. *Br J Psychiatry* 1994; **164**:528–32.

Maeder-Ingvar M, van Melle G, Bogousslavsky J. Pure monoparesis. A particular stroke subgroup? *Arch Neurol* 2005; **62**:1221–4.

Malkowitz DE, Legido A, Jackel RA *et al.* Prolactin secretion following repetitive seizures. *Neurology* 1995; **45**:448–52.

Malouf R, Brust JCM. Hypoglycemia: causes, neurological manifestations and outcome. *Ann Neurol* 1985; **17**:421–30.

Mandalos GE, Szarek BL. Dose-related paranoid reaction associated with fluoxetine. *J Nerv Ment Dis* 1990; **178**:57–8.

Mandoki MW, Sumner GS. Psychiatric manifestations of hereditary coproporphyria in a child. *J Nerv Ment Dis* 1994; **182**:117–18.

Manford MRA, Fish DR, Shorvon SD. Startle provoked epileptic seizures: features in 19 patients. *J Neurol Neurosurg Psychiatry* 1996; **61**:151–6.

Manfredi M, Beltramello A, Bongiovanni LG *et al.* Eclamptic encephalopathy: imaging and pathogenetic considerations. *Acta Neurol Scand* 1997; **96**:277–82.

Manowitz P, Kling A, Kohn H. Clinical course of adult metachromatic leukodystrophy presenting as schizophrenia: a report of two living cases in siblings. *J Nerv Ment Dis* 1978; **166**:500–6.

Mantovani JF, Landau WM. Acquired aphasia with convulsive disorder: course and prognosis. *Neurology* 1980; **30**:524–9.

Marchini C, Romito D, Lucci B *et al.* Fits of weeping as an unusual manifestation of speaking: case report. *Acta Neurol Scand* 1994; **90**:218–21.

Margerison JH, Corsellis JAN. Epilepsy and the temporal lobes. *Brain* 1966; **89**:499–530.

Marini C, Scheffer IE, Crossland KM *et al.* Genetic architecture of idiopathic generalized epilepsy: clinical genetic analysis of 55 multiplex families. *Epilepsia* 2004; **45**:467–78.

Marino R, Gasparotti R, Pinelli L *et al.* Posttraumatic cerebral infarction in patients with moderate or severe head trauma. *Neurology* 2006; **67**:1165–71.

Markelbach S, Sittinger H, Schweizer I *et al.* Cognitive outcome after bacterial meningitis. *Acta Neurol Scand* 2000; **102**:118–23.

Marks DA, Ehrenberg BC. Migraine-related seizures in adults with epilepsy, with EEG correlation. *Neurology* 1993; **43**:2476–83.

Marks DA, Laxer KD. Semiology of temporal lobe seizures: value in lateralizing the seizure focus. *Epilepsia* 1998; **39**:721–6.

Marks DA, Kim J, Spencer D *et al.* Characteristics of intractable seizures following meningitis and encephalitis. *Neurology* 1992; **42**:1513–18.

Marsh GG. Neuropsychological syndrome in a patient with episodic howling and violent motor behavior. *J Neurol Neurosurg Psychiatry* 1978; **41**:366–9.

Marshall JC, Halligan PW, Fink GR *et al.* The functional anatomy of a hysterical paralysis. *Cognition* 1997; **64**:1–8.

Marson AG, Williamson PR, Clough H *et al.* Carbamazepine versus valproate monotherapy for epilepsy: a meta-analysis. *Epilepsia* 2002; **43**:505–13.

Martinez O, Reisin R, Andermann F *et al.* Evidence for reflex activation of experiential complex seizures. *Neurology* 2001; **56**:121–3.

Masel BE, Scheibel RS, Kimbark T *et al.* Excessive daytime sleepiness in adults with brain injuries. *Arch Phys Med Rehabil* 2001; **82**:1526–32.

Massman PJ, Sims J, Cooke N *et al.* Prevalence and correlates of neuropsychological deficits in amytrophic lateral sclerosis. *J Neurol Neurosurg Psychiatry* 1996; **61**:450–5.

Matsuo F, Beren D, Faught E *et al.* Placebo-controlled study of the efficacy and safety of lamotrigene in patients with partial seizures. *Neurology* 1993; **43**:2284–91.

Mattson RH, Cramer JA, Collins JF *et al.* Comparison of carbamazepine, phenobarbital, phenytoin, and primidone in partial and secondarily generalized tonic–clonic seizures. *N Engl J Med* 1985; **313**:145–51.

Mattson RH, Cramer JA, Darney PD *et al.* Use of oral contraceptives by women with epilepsy. *JAMA* 1986; **255**:238–40.

Mattson RH, Cramer JA, Collins JF *et al.* A comparison of valproate with carbamazepine for the treatment of complex partial seizures and secondarily generalized tonic–clonic seizures in adults. *N Engl J Med* 1992; **327**:765–71.

Maudsley H. *Responsibility in Mental Disease*. New York: D. Appleton, 1874.

Mauguire F, Courjon J. Somatosensory epilepsy: a review of 127 cases. *Brain* 1978; **101**:307–32.

Maulaz AB, Bezzera DC, Michel P *et al.* Effect of discontinuing aspirin therapy on the risk of brain ischemic stroke. *Arch Neurol* 2005; **62**:1217–20.

Mayer SA, Brun SC, Begtrup K *et al.* Recombinant activated factor VII for acute intracerebral hemorhage. *N Engl J Med* 2005; **352**:777–85.

Mayeux R, Lueders H. Complex partial status epilepticus: case report and proposal for diagnostic criteria. *Neurology* 1978; **28**:957–61.

Mayeux R, Alexander MP, Benson DF *et al.* Poriomania. *Neurology* 1979; **29**:1616–19.

Mazzini L, Cossa FM, Angelino E *et al.* Posttraumatic epilepsy: neuroradiologic and neuropsychological assessment of long-term outcome. *Epilepsia* 2003; **44**:569–74.

Meencke HJ, Janz D. Neuropathological findings in primary generalized epilepsy: a study of eight cases. *Epilepsia* 1984; **25**:8–21.

Mega MS, Cummings JL, Fiorello T *et al.* The spectrum of behavioral changes in Alzheimer's disease. *Neurology* 1996; **46**:130–5.

Meierkord H, Shorvon S, Lightman S *et al.* Comparison of the effects of frontal and temporal lobe partial seizures on prolactin levels. *Arch Neurol* 1992; **49**:225–30.

Mendez MF, Adams NL, Lewandowski KS. Neurobehavioral changes associated with caudate lesions. *Neurology* 1989; **39**:349–54.

Mendez MF, Selwood A, Mastri AR *et al.* Pick's disease versus Alzheimer's disease: a comparison of clinical characteristics. *Neurology* 1993; **43**:289–92.

Mendez MF, Chernier MM, Perryman KM. Epileptic forced thinking from left frontal lesions. *Neurology* 1996; **47**:79–83.

Meninger KA. Influenza and schizophrenia. *Am J Psychiatry* 1926; **82**:469–529.

Merriam AE, Medalia A, Levine B. Partial complex status epilepticus associated with cocaine abuse. *Biol Psychiatry* 1988; **23**:515–18.

Merritt HH, Springlova M. Lissauer's dementia paralytica: a clinical and pathologic study. *Arch Neurol Psychiatry* 1932; **27**:987–1030.

Messenheimer J, Ramsey RE, Willmore LJ *et al.* Lamotrigene therapy for partial seizures: a multicenter, placebo-controlled, double-blind, cross-over trial. *Epilepsia* 1994; **35**:113–21.

Messing RO, Closson RG, Simon RP. Drug-induced seizures: a 10-year experience. *Neurology* 1984; **34**:1582–6.

Messori A, Polonara G, Carle F *et al.* Predicting posttraumatic epilepsy with MRI; prospective longitudinal morphologic study in adults. *Epilepsia* 2005; **46**:1472–81.

Mesulam M-M. Dissociative states with abnormal temporal lobe EEG. *Arch Neurol* 1981; **38**:176–81.

Meythaler JM, Brunner RC, Johnson A *et al.* Amantadine to improve neurorecovery in traumatic brain injury-associated diffuse axonal injury: a pilot double-blind randomized trial. *J Head Trauma Rehabil* 2002; **17**:300–13.

Michelucci R, Rubboli G, Passarelli D *et al.* Electroclinical features of idiopathic generalized epilepsy with persisting absences in adult life. *J Neurol Neurosurg Psychiatry* 1996; **61**:471–7.

Michelucci R, Poza JJ, Sovia V *et al.* Autosomal dominant lateral temporal lobe epilepsy: clinical spectrum, new epitempin mutations, and genetic heterogeneity in seven European families. *Epilepsia* 2003; **44**:1289–97.

Michelucci R, Gardella E, de Haan G-J *et al.* Telephone-induced seizures: a new type of reflex epilepsy. *Epilepsia* 2004; **45**:280–3.

Miguel EC, Rodriguez Pereira RM, de Braganca Pereira CA *et al.* Psychiatric manifestations of systemic lupus erythematosus: clinical features, symptoms and signs of central nervous system activity in 43 patients. *Medicine* 1994; **73**:224–32.

Mikati MA, Lee WL, DeLong GR. Epileptiform encephalopathy: an unusual form of partial complex status epilepticus. *Epilepsia* 1985; **26**:563–77.

Mineharu Y, Takenada K, Yamakawa H *et al.* Inheritance pattern of familial Moyamoya disease: autosomal dominant mode and genetic imprinting. *J Neurol Neurosurg Psychiatry* 2006; **77**:1025–9.

Minematsu K, Yamaguchi T, Omae T. Spectacular shrinking deficit: rapid recovery from a major hemispheric syndrome by migration of an embolus. *Neurology* 1992; **42**:157–62.

Misra UK, Kalita J, Patel R. Sodium valproate vs phenytoin in status epilepticus: a pilot study. *Neurology* 2006; **67**:340–2.

Mitchell C. Successful treatment of chronic delusional parasittosis. *Br J Psychiatry* 1989; **155**:556–7.

Mitchell W, Falconer MA, Hill D. Epilepsy with fetishism relieved by temporal lobectomy. *Lancet* 1954; **2**:626–30.

Mitchell WG, Messenheimer JA, Greenwood RS. Abdominal epilepsy. Cyclic vomiting as the major symptom of simple partial seizures. *Arch Neurol* 1983; **40**:251–2.

Mittl RL, Grossman RI, Hiehle JF *et al.* Prevalence of MR evidence of diffuse axonal injury in patients with mild head injury and normal head CT findings. *AJNR* 1994; **15**:1583–9.

Moddel G, Bunten S, Dobis C *et al.* Intravenous levetiracetam: a new treatment alternative for refractory status epilepticus. *J Neurol Neurosurg Psychiatry* 2009; **80**:689–92.

Moder KG, Hurley DL. Fatal hypernatremia from exogenous salt intake: report of a case and review of the literature. *Mayo Clin Proc* 1990; **65**:187–94.

Moene FC, Landberg EH, Hoogduin KA *et al.* Organic syndromes diagnosed as conversion disorder: identification and frequency in a study of 85 patients. *J Psychosom Res* 2000; **49**:7–12.

Moene FC, Spinhoven P, Hoogduin KA *et al.* A randomised controlled clinical trial on the additional effect of hypnosis in a comprehensive treatment programme for in-patients with conversion disorder of the motor type. *Psychother Psychosom* 2002; **71**:66–76.

Moene FC, Spinhoven P, Hoogduin KA *et al.* A randomised controlled clinical trial of a hypnosis-based treatment for patients with conversion disorder, motor type. *Int J Clin Exp Hypn* 2003; **51**:29–50.

Moersch FP. Psychic manifestations in cases of brain tumors. *Am J Psychiatry* 1925; **81**:707–24.

Mohr JP. Lacunes. *Stroke* 1982; **13**:3–11.

Molina JA, Calandre L, Bermejo F. Myoclonic encephalopathy due to bismuth salts: treatment with dimercaprol and analysis of CSF transmitters. *Acta Neurol Scand* 1989; **79**:200–3.

Monteiro L, Nunes B, Mendonca D *et al.* Spectrum of epilepsy in neuroacanthocytosis: a long-term follow-up of 143 patients. *Acta Neurol Scand* 1995; **92**:33–40.

Mooney GF, Haas LJ. Effect of methylphenidate on brain injury-related anger. *Arch Phys Med Rehabil* 1993; **74**:153–60.

Moore DP. A case of petit mal epilepsy aggravated by lithium. *Am J Psychiatry* 1981; **138**:690–1.

Moore DP. *Partial Seizures and Interictal Disorders.* Boston, MA: Butterworth-Heinemann, 1997.

Moosey J. Morphology, sites and epidemiology of cerebral athero-sclerosis. *Res Publ Assoc Res Nerv Ment Dis* 1966; **51**:1–22.

Morris CE, Heyman A, Pozefsky T. Lead encephalopathy caused by ingestion of illicitly distilled whiskey. *Neurology* 1964; **14**:493–9.

Morris HH, Dinner DS, Luders H *et al.* Supplementary motor seizures: clinical and electroencephalographic findings. *Neurology* 1988; **38**:1075–82.

Moser HW, Moser AE, Singh I *et al.* Adrenoleukodystrophy: survey of 303 cases: biochemistry, diagnosis and therapy. *Ann Neurol* 1984; **16**:628–41.

Moskovitz C, Moses H, Klawans HL. Levodopa-induced psychosis: a kindling phenomenon. *Am J Psychiatry* 1978; **135**:669–75.

Motamedi M, Nguyen DK, Zaatreh M *et al.* Levetiracetam efficacy in refractory partial-onset seizures, especially after failed epilepsy surgery. *Epilepsia* 2003; **44**:211–14.

Mula M, Trimble MR, Yuen A *et al.* Psychiatric adverse events during levetiracetam therapy. *Neurology* 2003; **61**:704–6.

Mulder DW, Daly D. Psychiatric symptoms associated with lesions of temporal lobe. *JAMA* 1952; **150**:173–6.

Mulder DW, Parrott M, Thaler M. Sequelae of western equine encephalitis. *Neurology* 1951; **1**:318–27.

Mullatti N, Selway R, Nashef L *et al.* The clinical spectrum of epilepsy in children and adults with hypothalamic hamartoma. *Epilepsia* 2003; **44**:1310–19.

Muller D, Pilz H, Ter Meulen V. Studies on adult metachromatic leukodystrophy. I. Clinical, morphological and histochemical observations in two cases. *J Neurol Sci* 1969; **9**:567–84.

Muller T, Buttner TH, Kuhn W *et al.* Palinopsia as sensory epileptic phenomenon. *Acta Neurol Scand* 1995; **91**:433–6.

Muller U, Murai T, Bauer-Wittmund T *et al.* Paroxetine versus citalopram treatment of pathological crying after brain injury. *Brain Inj* 1999; **13**:805–11.

Mungas D. Interictal behavior abnormality in temporal lobe epilepsy. *Arch Gen Psychiatry* 1982; **39**:108–11.

Murphy CP, Harden EA, Thompson JM. Generalized seizures secondary to high dose busulfan therapy. *Ann Pharmacother* 1992; **26**:30–1.

Murphy KC, Jones LA, Owen MJ. High rates of schizophrenia in adults with velo-cardio-facial syndrome. *Arch Gen Psychiatry* 1999; **56**:940–5.

Murray V, von Arbin B, Bartfai A *et al.* Double-blind comparison of sertraline and placebo in stroke patients with minor depression and less severe major depression. *J Clin Psychiatry* 2005; **66**:708–16.

Nakada T, Lee H, Kwee IL *et al.* Epileptic Kluver–Bucy syndrome: case report. *J Clin Psychiatry* 1984; **45**:87–8.

Narushima K, Kosier JT, Robinson R. Preventing poststroke depression: a 12-week double-blind randomized treatment trial and 21-month follow-up. *J Nerv Ment Dis* 2002; **190**:296–303.

Nausieda PA, Grossman BJ, Koller WC *et al.* Sydenham chorea: an update. *Neurology* 1980; **30**:331–4.

Navia BA, Jordan BD, Price RW. The AIDS dementia complex. I. Clinical features. *Ann Neurol* 1986a; **19**:517–24.

Navia BA, Petito CK, Gold JWM *et al.* Cerebral toxoplasmosis complicating the acquired immune deficiency syndrome: clinical and neuropathological findings in 27 patients. *Ann Neurol* 1986b; **19**:224–38.

Neary D, Snowden JS, Mann DMA *et al.* Frontal lobe dementia and motor neuron disease. *J Neurol Neurosurg Psychiatry* 1990; **53**:23–32.

Neary D, Snowden JS, Mann DMA. Familial progressive atrophy: its relationship to other forms of lobar atrophy. *J Neurol Neurosurg Psychiatry* 1993; **56**:1122–5.

Nedeltchev K, der Maur TA, Georgiadis D *et al.* Ischemic stroke in young adults: predictors of outcome and recurrence. *J Neurol Neurosurg Psychiatry* 2005; **76**:191–5.

Newburn G, Newburn D. Selegeline in the management of apathy following traumatic brain injury. *Brain Inj* 2005; **19**:149–54.

Newman P, Saunders M. A unique case of musicogenic epilepsy. *Arch Neurol* 1980; **37**:244–5.

Newmark ME. Diagnosis of epilepsy with home video-cassette recorder. *N Engl J Med* 1981; **305**:769.

Neylan TC. Frontal lobe function: Mr. Phineas Gage's famous injury. *J Neuropsychiatry Clin Neurosci* 1999; **11**:281–3.

Nicoletti A, Sofia V, Biondi R *et al.* Epilepsy and multiple sclerosis in Italy: a population-based study. *Epilepsia* 2003; **44**:1445–8.

Nielsen H, Kristensen O. Personality correlates of sphenoidal EEG foci in temporal lobe epilepsy. *Acta Neurol Scand* 1981; **64**:289–300.

Nightingale S, Welch JL. Psychometric assessment in absence status. *Arch Neurol* 1982; **39**:516–19.

Nishida T, Kudo T, Inoue Y *et al.* Postictal mania versus postictal psychosis: differences in clinical features, epileptogenic zone, and brain functional changes during postictal period. *Epilepsia* 2006; **47**:2104–14.

Nishino H, Rubino FA, DeRemee RA *et al.* Neurological involvement in Wegener's granulomatosis: an analysis of 324 consecutive patients at the Mayo clinic. *Ann Neurol* 1993; **33**:4–9.

Nizamie SH, Nizamie A, Borde M *et al.* Mania following head injury: case reports and neuropsychological findings. *Acta PsychiatrScand* 1988; **77**:637–9.

Noachtar S, Luders HO. Focal akinetic seizures as documented by electroencephalography and video recordings. *Neurology* 1999; **53**:427–9.

Noachtar S, Andermann E, Meyvisch P *et al.* Levetiracetam for the treatment of idiopathic generalized epilepsy with myoclonic seizures. *Neurology* 2008; **7**:607–16.

Nobili L, Francione S, Mai R *et al.* Surgical treatment of drug-resistant nocturnal frontal lobe epilepsy. *Brain* 2007; **130**:561–73.

Nott MT, Chapparo C, Baguley IJ. Agitation following traumatic brain injury: an Australian sample. *Brain Inj* 2006; **20**:1175–82.

Novak J, Corke P, Fairley N. 'Petit mal status' in adults. *Dis Nerv Syst* 1971; **32**:245–8.

Noyes R, Kathol RG, Fisher MM *et al.* Psychiatric comorbidity among patients with hypochondriasis. *Gen Hosp Psychiatry* 1994; **16**:78–87.

Noyes R, Holt CS, Happel RL *et al.* A family study of hypochondriasis. *J Nerv Ment Dis* 1997; **185**:223–32.

Noyes R, Sturat S, Langbehn DR *et al.* Childhood antecedents of hypochondriasis. *Psychosomatics* 2002; **43**:282–9.

Nystrom HM. Relationship between location of cerebral lesion and epilepsy. A statistical study of 350 cases. *Acta Neurol Scand* 1966; **42**:191–206.

O'Carroll RE, Masterton G, Dougall N *et al.* The neuropsychiatric sequelae of mercury poisoning: the Mad Hatter's disease. *Br J Psychiatry* 1995; **167**:95–8.

Ochs R, Gloor P, Quesney F *et al.* Does head-turning during a seizure have lateralizing or localizing significance? *Neurology* 1984; **34**:884–90.

Oder W, Goldenberg G, Spatt J *et al.* Behavioral and psychosocial sequelae of severe closed head injury and regional cerebral blood flow: a SPECT study. *J Neurol Neurosurg Psychiatry* 1992; **55**:475–80.

Oestreich LJ, Berg MJ, Bachmann DL *et al.* Ictal contralateral paresis in complex partial seizures. *Epilepsia* 1995; **36**:671–5.

Ogunyemi AO, Locke GE, Kramer LD *et al.* Complex partial status epilepticus provoked by 'crack' cocaine. *Ann Neurol* 1989; **26**:785–6.

O'Hare JA, Callaghan NM, Murnaghan DJ. Dialysis encephalopathy. Clinical, electroencephalographic, and interventional aspects. *Medicine* 1983; **62**:129–41.

Okada Y, Yamaguchi T, Minematsu K *et al.* Hemorrhagic transformation in cerebral embolism. *Stroke* 1989; **20**:598–603.

Oksanen V. Neurosarcoidosis: clinical presentations and course in 50 patients. *Acta Neurol Scand* 1986; **73**:283–90.

Oldani A, Zucconi M, Asselta R *et al.* Autosomal dominant nocturnal frontal lobe epilepsy: a video-polysomnographic and genetic appraisal of 40 patients and delineation of the epileptic syndrome. *Brain* 1998; **121**:205–23.

Oliva M, Pattison C, Carino J *et al.* The diagnostic value of oral lacerations and incontinence during convulsive "seizures". *Epilepsia* 2008; **49**:962–7.

Olivier AP, Luchins DJ, Wyatt RJ. Neuroleptic-induced seizures. *Arch Gen Psychiatry* 1982; **39**:206–9.

Olsson I, Steffenberg S, Gillberg C. Epilepsy in autism and autistic-like conditions. *Arch Neurol* 1988; **45**:666–8.

Ono J, Mano T, Andermann E *et al.* Band heterotopia or double cortex in a male: bridging structures suggest abnormality of the radial glial system. *Neurology* 1997; **48**:1701–3.

Oribe E, Rohullah A, Nissenbaum E *et al.* Serum prolactin concentrations are elevated after syncope. *Neurology* 1996; **47**:60–2.

Orrison WW, Gentry LR, Stimae GK *et al.* Blinded comparison of cranial CT and MR in closed head injury evaluation. *AJNR* 1994; **15**:351–6.

Ostergaard JRL. Warning lead in subarachnoid hemorrhage. *BMJ* 1990; **301**:190–1.

Ottman R, Risch N, Hauser WA *et al.* Localization of a gene for partial epilepsy to chromosome 10q. *Nat Genet* 1995; **10**:56–60.

Oxbury JM, Whitty CWM. Causes and consequences of status epilepticus in adults: a study of 86 cases. *Brain* 1971; **94**: 733–44.

Ozkara C, Ozdemir S, Yilmaz A *et al.* Orgasm-induced seizures: a study of six patients. *Epilepsia* 2006; **47**:2193–7.

Pacia SV, Devinsky O. Clozapine-related seizures: experience with 5,629 patients. *Neurology* 1994; **44**:2247–9.

Paciaroni M, Caso V, Milia P *et al.* Isolated monoparesis following stroke. *J Neurol Neurosurg Psychiatry* 2005; **76**:805–7.

Pakainis A, Drake ME, John K *et al.* Forced normalization: acute psychosis after seizure control in seven patients. *Arch Neurol* 1987; **44**:289–92.

Palmini A, Gloor P. The localizing value of auras in partial seizures: a prospective and retrospective study. *Neurology* 1992; **42**:801–8.

Palmini A, Andermann F, Olivier A *et al.* Focal neuronal migration disorders and intractable partial epilepsy: a study of 30 patients. *Ann Neurol* 1991; **30**:741–9.

Palmini A, Najm I, Avanzini G *et al.* Terminology and classification of the cortical dysplasias. *Neurology* 2004; **62**(6 Suppl.3):S2–8.

Palmini AL, Gloor P, Jones-Gotman M. Pure amnestic seizures in temporal lobe epilepsy. *Brain* 1992; **115**:749–69.

Palmio J, Keranen T, Alapirtti T *et al.* Elevated serum neuron-specific enolase in patients with temporal lobe epilepsy: a video-EEG study. *Epilepsy Res* 2008; **81**:155–60.

Pampiglione G, Moynahan EJ. The tuberous sclerosis syndrome: clinical and EEG studies in 100 children. *J Neurol Neurosurg Psychiatry* 1976; **39**:666–73.

Panayiotopoulus CP. Vomiting as an ictal manifestation of epileptic seizures and syndromes. *J Neurol Neurosurg Psychiatry* 1988; **51**:1448–51.

Paquier PF, Van Dongen HR, Loonen CB. The Landau–Kleffner syndrome or 'acquired aphasia with convulsive disorder': long-term follow-up of six children and a review of the recent literature. *Arch Neurol* 1992; **49**:354–9.

Pardo CA, Vining SPG, Guo L *et al.* The pathology of Rasmussen syndrome: stages of cortical involvement and neuropathological studies in 45 hemispherectomies. *Epilepsia* 2004; **45**:516–26.

Parmar H, Lim S-H, Tan NCK *et al.* Acute symptomatic seizures and hippocampal damage: DWI and MRS findings. *Neurology* 2006; **66**:1732–5.

Parra A, Velasco M, Cervantes C *et al.* Plasma prolactin increase following electrical stimulation of the amygdala in humans. *Neuroendocrinology* 1980; **31**:60–5.

Parsons MW, Barber PA, Chalk J *et al.* Diffusion- and perfusion-weighted MRI response to thrombolysis in stroke. *Ann Neurol* 2002; **51**:28–37.

Partlow GD, del Carpio-ODonovan R, Melanson D *et al.* Bilateral thalamic glioma: review of eight cases with personality change and mental deterioration. *AJNR* 1992; **13**:1125–30.

Partridge M. One operation cures three people: effect of prefrontal leukotomy on a case of folie a deux et demie. *Arch Neurol Psychiatry* 1950; **54**:792–6.

Pascual-Castroviejo I, Diaz-Gonzalez G, Garcia-Melian RM *et al.* Sturge–Weber syndrome: study of 40 patients. *Pediatr Neurology* 1993; **9**:283–8.

Paskavitz JF, Anderson CA, Filley CM *et al.* Acute arcuate fiber demyelinating encephalopathy following Epstein–Barr virus infection. *Ann Neurol* 1995; **38**:127–31.

Passero S, Rocchi R, Rossi S *et al.* Seizures after spontaneous supratentorial intracerebral hemorrhage. *Epilepsia* 2002; **43**:1175–80.

Patel NC, Landan IR, Levin J *et al.* The use of levetiracetam in refractory status epilepticus. *Seizure* 2006; **15**:137–41.

Pazzaglia P, D'Alessandro R, Ambrosetto G *et al.* Drop attacks: an ominous change in the evolution of partial epilepsy. *Neurology* 1985; **35**:1725–30.

Peavy GM, Herzog AG, Rubin NP *et al.* Neuropsychological aspects of dementia of motor neuron disease: a report of two cases. *Neurology* 1992; **42**:1004–8.

Pedley TA, Guilleminault C. Episodic nocturnal wanderings responsive to anticonvulsant drug therapy. *Ann Neurol* 1977; **2**:30–5.

Peltola J, Coetzee C, Jimenez F *et al.* Once-daily extended-release levetiracetam as adjunctive treatment of partial-onset seizures in patients with epilepsy: a double-blind, randomized, placebo-controlled trial. *Epilepsia* 2009; **50**:406–14.

Penfield W, Jasper H. *Epilepsy and the Functional Anatomy of the Human Brain.* Boston, MA: Little, Brown, 1954.

Penry JK, Dreifuss FE. Automatisms associated with the absence of petit mal epilepsy. *Arch Neurol* 1969; **21**:142–9.

Penry JK, Porter RJ, Dreifuss FE. Simultaneous recording of absence seizures with video tape and electroencephalography: a study of 374 seizures in 48 patients. *Brain* 1975; **98**:427–40.

Peppercorn MA, Herzog AG, Dichter MA *et al.* Abdominal epilepsy: a case of abdominal pain in adults. *JAMA* 1978; **240**:2450–1.

Perez NM, Trimble MR. Epileptic psychosis – diagnostic comparison with process schizophrenia. *Br J Psychiatry* 1980; **137**:245–9.

Perley M, Guze B. Hysteria: the stability and usefulness of clinical criteria: a quantitative study based on a 6–8 year follow-up of 39 patients. *N Engl J Med* 1962; **266**:421–6.

Peroutka SJ, Sohmer BH, Kumer AJ *et al.* Hallucinations and delusions following a right temporo-parietal-occipital infarction. *Johns Hopkins Med J* 1982; **151**:181–5.

Persoon S, Jaap Kappelle LJ, Klijn CJ. Limb-shaking transient ischemic attacks in patients with internal carotid artery occlusion: a case-control study. *Brain* 2010; **133**:915–22.

Pessin MS, Lathi E, Cohen M *et al.* Clinical features and mechanism of occipital infarction. *Ann Neurol* 1987; **21**:290–9.

Petermann AF, Hayles AB, Dockerty MB *et al.* Encephalotrigeminal angiomatosis (Sturge–Weber disease). Clinical study of thirty-five cases. *J Am Med Assoc* 1958; **167**:2169–76.

Petry S, Cummings JL, Hill MA *et al.* Personality alterations in dementia of the Alzheimer type. *Arch Neurol* 1988; **45**:1187–90.

Petti TA, Campbell M. Imipramine and seizures. *Am J Psychiatry* 1975; **132**:538–40.

Petty RG, Bonner D, Mouratoglou V *et al.* Acute frontal lobe syndrome and dyscontrol associated with bilateral caudate infarctions. *Br J Psychiatry* 1996; **168**:237–40.

Pflanz S, Besson JAO, Ebmeier KP *et al.* The clinical manifestations of mental disorder in Huntington's disease: a retrospective case record study of disease progression. *Acta Psychiatr Scand* 1991; **83**:53–60.

Phillips HA, Marini C, Scheffer IE *et al.* A de novo mutation in sporadic nocturnal frontal lobe epilepsy. *Ann Neurol* 2000; **48**: 264–7.

Picard F, Craig AD. Epileptic ecstatic seizures: a potential window on the neural basis for human self-awareness. *Epilepsy Behav* 2009; **16**:539–46.

Picard F, Baulac S, Kahane P *et al.* Dominant partial epilepsies: a clinical, electrophysiological and genetic study of 19 European families. *Brain* 2000; **123**:1247–62.

Placidi F, Floris R, Bozzao A *et al.* Ketotic hyperglycemia and epilepsia partialis continua. *Neurology* 2001; **57**:534–7.

Plazzi G, Tinuper P, Montagna P *et al.* Epileptic nocturnal wanderings. *Sleep* 1995; **18**:749.

Plenger PM, Dixon CE, Castillo RM *et al.* subacute methylphenidate treatment for moderate to moderately severe traumatic brain injury: a preliminary double-blind placebo-controlled study. *Arch Phys Med Rehabil* 1996; **77**:536–40.

Pomeroy SL, Holmes SJ, Dodge PR *et al.* Seizures and other neurologic sequelae of bacterial meningitis in children. *N Engl J Med* 1990; **323**:1651–7.

Poole NA, Wuerz A, Agrawal N. Abreaction for conversion disorder: systematic review with meta-analysis. *Br J Psychiatry* 2010; **197**:91–5.

Pope HG, Katz DL. Affective and psychotic symptoms associated with anabolic steroid use. *Am J Psychiatry* 1988; **145**:487–90.

Pope HG, Jonas JM, Jones B. Factitious psychosis: phenomenology, family history and long-term outcome of nine patients. *Am J Psychiatry* 1982; **139**:1480–3.

Popli AP, Masand PS, Dewan MJ. Factitious disorders with psychological symptoms. *J Clin Psychiatry* 1992; **53**:315–18.

Porter NT, Meyer MA, Zimmerman AW *et al.* Molecular and clinical findings in a family with dentatorubral-pallidoluysian atrophy. *Ann Neurol* 1995; **37**:273–7.

Porter SB, Sande MA. Toxoplasmosis of the central nervous system in the acquired immunodeficiency syndrome. *N Engl J Med* 1992; **327**:1643–8.

Poskanzer DC, Brown AE, Miller H. Musicogenic epilepsy caused only by a discrete frequency band of church bells. *Brain* 1962; **85**:7–92.

Powell JH, al-Adawi S, Morgan J *et al.* Motivational deficits after brain injury: effects of bromocriptine in 11 patients. *J Neurol Neurosurg Psychiatry* 1996; **60**:416–21.

Power I, Kam P. *Principles of Physiology for the Anaesthetist.* London: Hodder Arnold, 2008.

Poza JJ, Saenz A, Martinez-Gil A *et al.* Autosomal dominant lateral temporal lobe epilepsy: clinical and genetic study of a large Basque pedigree linked to chromosome 10q. *Ann Neurol* 1999; **45**:182–8.

Prashanth LK, Taly AB, Ravi V *et al.* Adult subacute sclerosing panencephalitis: clinical profile of 39 patients from a tertiary care center. *J Neurol Neurosurg Psychiatry* 2006; **77**:630–3.

Pritchard PB, Wannamaker BB, Sagel J *et al.* Serum prolactin and cortisol levels in evaluation of pseudoepileptic seizures. *Ann Neurol* 1985a; **18**:87–9.

Pritchard PB, Holmstrom VL, Giacinto J. Self-abatement of complex partial seizures. *Ann Neurol* 1985b; **18**:265–7.

Privitera MD, Morris GL, Gilliam F. Postictal language assessment and lateralization of complex partial seizures. *Ann Neurol* 1991; **30**:391–6.

Privitera MD, Fincham R, Penry J *et al.* Topiramate placebo-controlled dose-ranging trial in refractory partial epilepsy using 600-, 800-, and 1,000-mg daily dosages. *Neurology* 1996; **46**:1678–83.

Privitera MD, Brodie MJ, Mattson RH *et al.* Topirimate, carbamazepine and valproate monotherapy: double-blind comparison in newly diagnosed epilepsy. *Acta Neurol Scand* 2003; **107**:165–75.

Provini F, Plazzi G, Tinuper P *et al.* Nocturnal frontal lobe epilepsy: a clinical and polygraphic overview of 100 consecutive cases. *Brain* 1999; **122**:1017–31.

Pueschel SM, Louis S, McKnight P. Seizure disorders in Down syndrome. *Arch Neurol* 1991; **48**:318–20.

Putzel L. Cerebral complications of chorea. *NY Med Rec* 1879; **xvi**:220–2.

Quesney LF, Constain M, Fish DR *et al.* The clinical differentiation of seizures in the parasagittal and anterolateral frontal convexities. *Arch Neurol* 1990; **47**:677–9.

Quirk JA, Fish DR, Smith SJM *et al.* First seizures associated with playing electronic screen games: a community-based study in Great Britain. *Ann Neurol* 1995; **37**:733–7.

Rabinowicz AL, Correale JD, Bracht KA *et al.* Neuron specific enolase is increased after nonconvulsive status epilepticus. *Epilepsia* 1995; **36**:475–9.

Rabinowicz AL, Correale J, Boutros RB. Neuron-specific enolase is increased after single seizures during inpatient video/EEG monitoring. *Epilepsia* 1996; **37**:122–5.

Rabinstein AA, Wijdicks EF. Hyponatremia in critically ill neurological patients. *Neurologist* 2003; **9**:290–300.

Ramsey RE, DeToledo J. Intravenous administration of fosphenytoin: options for the management of seizures. *Neurology* 1996; **46**(Suppl. 1):17–19.

Ramsey RE, Wilder BJ, Berger JR *et al.* A double-blind study comparing carbamazepine with phenytoin as initial seizure therapy in adults. *Neurology* 1983; **33**:904–10.

Rao V, Lyketsos C. Neuropsychiatric sequelae of traumatic brain injury. *Psychosomatics* 2000; **41**:95–103.

Rapoport MJ, Chan F, Lanctot K *et al.* An open-label study of citalopram for major depression following traumatic brain injury. *J Psychopharmacol* 2008; **22**:860–4.

Rasmussen A, Matsuura T, Ruano L *et al.* Clinical and genetic analysis of four Mexican families with spinocerebellar ataxia type 10. *Ann Neurol* 2001; **50**:234–9.

Rasmussen A, Lunde M, Poulsen DL *et al.* A double-blind, placebo-controlled study of sertraline in the prevention of depression in stroke patients. *Psychosomatics* 2003; **44**:216–21.

Rasmussen T. Surgical therapy of frontal lobe epilepsy. *Epilepsia* 1963; **4**:181–98.

Rasmussen T. Characteristics of a pure culture of frontal lobe epilepsy. *Epilepsia* 1983; **24**:482–93.

Rasmussen T, Andermann F. Update on the syndrome of 'chronic encephalitis' and epilepsy. *Cleve Clin J Med* 1989; **56**(Suppl. 2):181–4.

Rasmussen T, Olszewski J, Lloyd-Smith D. Focal seizures due to chronic localized encephalitis. *Neurology* 1958; **8**:435–45.

Rauschka H, Colsch B, Baumann N *et al.* Late-onset metachromatic leukodystrophy. Genotype strongly influences phenotype. *Neurology* 2006; **67**:859–63.

Raymond AA, Fish DR, Sisodiya SM *et al.* Abnormalities of gyration, heterotopias, tuberous sclerosis, focal cortical dysplasia, microdysgenesis, dysembroplastic neuroepithelial tumor and dysgenesis of the archicortex in epilepsy. Clinical, EEG and neuroimaging findings. *Brain* 1995; **118**:629–60.

Raymond RW, Williams RL. Infectious mononucleosis with psychosis. *N Engl J Med* 1948; **239**:542–4.

Reading PJ, Will RG. Unwelcome orgasms. *Lancet* 1997; **350**:1747.

Reami DO, Silva DF, Albuquerque M *et al.* Dreams and epilepsy. *Epilepsia* 1991; **32**:51–3.

Reed K, Bland RL. Masked 'myxedema madness'. *Acta Psychiatr Scand* 1977; **56**:421–6.

Reich P, Gottfried LA. Factitious disorders in a teaching hospital. *Ann Intern Med* 1983; **99**:240–7.

Remillard GM, Andermann F, Rhi-Sausa A *et al.* Facial asymmetry in patients with temporal lobe epilepsy. *Neurology* 1977; **27**:109–14.

Remillard GM, Andermann F, Testa GF *et al.* Sexual ictal manifestations predominate in women with temporal lobe epilepsy: a finding suggesting sexual dimorphism in the human brain. *Neurology* 1983; **33**:323–30.

Rennick PM, Perez-Borja C, Rodin EA. Transient mental deficits associated with recurrent prolonged epileptic clouded state. *Epilepsia* 1969; **10**:397–405.

Rennick PM, Noland DC, Bauer RB *et al.* Neuropsychologic and neurologic follow-up after herpes hominis encephalitis. *Neurology* 1973; **23**:42–7.

Rheims S, Demarquay G, Isnard J *et al.* Ipsilateral head deviation in frontal lobe seizures. *Epilepsia* 2005; **46**:1750–3.

Rice E, Gendelman S. Psychiatric aspects of normal pressure hydrocephalus. *JAMA* 1973; **223**:409–12.

Richardson EP, Dodge PR. Epilepsy in cerebral vascular disease. *Epilepsia* 1954; **3**:49–74.

Richfield EK, Twyman R, Berent S. Neurological syndrome following bilateral damage to the head of the caudate nuclei. *Ann Neurol* 1987; **22**:768–71.

Roberts AH. Sequelae of closed head injuries. *Proc R Soc Med* 1976; **69**:137–41.

Roberts JKA, Trimble MR, Robertson M. Schizophrenic psychosis associated with aqueductal stenosis in adults. *J Neurol Neurosurg Psychiatry* 1983; **46**:892–8.

Roberts MA, Humphrey PRD. Prolonged complex partial status epilepticus: a case report. *J Neurol Neurosurg Psychiatry* 1988; **51**:586–8.

Robinson RG, Parikh RM, Lipsey JR *et al.* Pathological laughing and crying after stroke: validation of a measurement scale and a double-blind treatment study. *Am J Psychiatry* 1993; **150**: 286–93.

Robinson RG, Schultz SK, Castillo C *et al.* Nortriptyline versus fluoxetine in the treatment of depression and in short-term recovery after stroke: a placebo-controlled, double-blind study. *Am J Psychiatry* 2000; **157**:658–60.

Robinson RG, Jorge RE, Moser DJ *et al.* Escitalopram and problem-solving therapy for prevention of poststroke depression: a randomized controlled trial. *JAMA* 2008; **299**:2391–400.

Rocamora R, Kurthen M, Lickfett L *et al.* Cardiac asystole in epilepsy: clinical and neurophysiologic features. *Epilepsia* 2003; **44**:179–85.

Rodin E, Schmaltz S. The Bear–Fedio personality inventory and temporal lobe epilepsy. *Neurology* 1984; **34**:591–6.

Rodin EA. Psychomotor epilepsy and aggressive behavior. *Arch Gen Psychiatry* 1973; **28**:210–13.

Romanelli MF, Morris JC, Ashkin K *et al.* Advanced Alzheimer's disease is a risk factor for late-onset seizures. *Arch Neurol* 1990; **47**:847–50.

Romanul FCA, Radvany J, Rosales RK. Whipple's disease confined to the brain: a case confirmed clinically and pathologically. *J Neurol Neurosurg Psychiatry* 1977; **40**:901–9.

Rosenbaum M, Lewis M, Piker P *et al.* Convulsive seizures in delirium tremens. *Arch Neurol Psychiatry* 1941; **45**:486–93.

Rosenberg GA, Kornfeld M, Stovring J *et al.* Subcortical arteriosclerotic encephalopathy (Binswanger's): computerized tomography. *Neurology* 1979; **29**:1102–6.

Rosenfeld WE, Benbadis S, Edrich P *et al.* Levetiracetam as add-on therapy for idiopathic generalized epilepsy syndromes with onset during adolescence: analysis of two randomized, double-blind, placebo-controlled studies. *Epilepsy Res* 2009; **85**:72–80.

Ross AT, Dickerson WW. Tuberous sclerosis. *Arch Neurol Psychiatry* 1943; **50**:233–57.

Rothschild D. Dementia paralytica accompanied by manic-depressive and schizophrenic psychoses. *Am J Psychiatry* 1940; **96**:1043–60.

Rothwell PM, Warlow CP. Timing of TIAs preceding stroke. Time window for prevention is very short. *Neurology* 2005; **64**:817–20.

Rowan AJ, Scott DF. Major status epilepticus: a series of 42 patients. *Acta Neurol Scand* 1970; **46**:573–84.

Rowan AJ, Ramsey RE, Collins JF *et al.* New onset geriatric epilepsy. A randomized study of gabapentin, lamotrigine, and carbamazepine. *Neurology* 2005; **64**:1868–73.

Rozdilsky B, Cummings JN, Huston AF. Hallervorden–Spatz disease: late infantile and adult type, report of two cases. *Acta Neuropathologica* 1968; **10**:1–16.

Ruff RL. Orgasmic epilepsy. *Neurology* 1980; **30**:1252.

Rugg-Gunn FJ, Duncan JS, Smith SJM. Epileptic cardiac asystole. *J Neurol Neurosurg Psychiatry* 2000; **68**:108–10.

Rush JL, Everett BA, Adams AH *et al.* Paroxysmal atrial tachycardia and frontal lobe tumor. *Arch Neurol* 1977; **34**:578–80.

Russell WR, Whitty CWM. Studies in traumatic epilepsy. 1. Factors influencing the incidence of epilepsy after brain wounds. *J Neurol Neurosurg Psychiatry* 1952; **15**:93–8.

Russell WR, Whitty CWM. Studies in traumatic epilepsy. 2. Focal motor and somatic sensory fits: a study of 85 cases. *J Neurol Neurosurg Psychiatry* 1953; **16**:73–97.

Russell WR, Whitty CWM. Studies in traumatic epilepsy. 3. Visual fits. *J Neurol Neurosurg Psychiatry* 1955; **18**:79–96.

Russo LS, Beace G, Sandroni S *et al.* Aluminum intoxication in undialysed adults with chronic renal failure. *J Neurol Neurosurg Psychiatry* 1992; **55**:697–700.

Rusu V, Chassoux F, Landre E *et al.* Dystonic posturing in seizures of mesial temporal origin. Electroclinical and metabolic pattern. *Neurology* 2005; **65**:1612–19.

Sacco RL, Freddo L, Bello JA *et al.* Wallenberg's lateral medullary syndrome. Clinical-magnetic resonance imaging correlations. *Arch Neurol* 1993; **50**:609–14.

Sacco RL, Diener H-C, Yusuf S *et al.* Aspirin and extended-release dipyridamole versus clopidogrel for recurrent stroke. *N Engl J Med* 2008; **359**:1238–51.

Sachdev P, Smith JS, Cathcart S. Schizophrenia-like psychosis following traumatic brain injury: a chart-based descriptive and case–control study. *Psychol Med* 2001; **31**:231–9.

Sadler RM, Rahey S. Prescience as an aura of temporal lobe epilepsy. *Epilepsia* 2004; **45**:982–4.

Sadleir LG, Farrell K, Smith S *et al.* Electroclinical features of absence seizures in childhood absence epilepsy. *Neurology* 2006; **67**:413–18.

Sadleir LG, Scheffer IE, Smith S *et al.* Automatisms in absence seizures in children with idiopathic generalized epilepsy. *Arch Neurol* 2009; **66**:729–34.

Saenz A, Galan J, Caloustian C *et al.* Autosomal dominant nocturnal frontal lobe epilepsy in a Spanish family with a SER252PHE mutation in the CHRNA4 gene. *Arch Neurol* 1999; **56**:1004–9.

Saetre E, Perucca E, Isojarvi J *et al.* An international multicenter randomized double-blind controlled trial of lamotrigine and sustained-release carbamazepine in the treatment of newly diagnosed epilepsy in the elderly. *Epilepsia* 2007; **48**:1292–302.

Salanova V, Andermann F, Olivier A *et al.* Occipital lobe epilepsy: electroclinical manifestations, electrocorticography, cortical stimulation and outcome in 42 patients treated between 1930 and 1991. *Brain* 1992; **115**:1655–80.

Salanova V, Morris HH, Van Ness P *et al.* Frontal lobe seizures: electroclinical syndromes. *Epilepsia* 1995; **36**:16–24.

Salazar AM, Jabbari B, Vance SC *et al.* Epilepsy after penetrating head injury. I. Clinical correlates: a report of the Vietnam Head Injury Study. *Neurology* 1985; **35**:1406–14.

Salehi F, Kovacs K, Scheihauer BW *et al.* Histologic study of the human pituitary gland in acute traumatic injury. *Brain Inj* 2007; **21**:651–6.

Salib EA. Subacute sclerosing panencephalitis (SSPE) presenting at the age of 21 as a schizophrenia-like state with bizarre dysmorphophobic features. *Br J Psychiatry* 1988; **152**:709–10.

Salinsky MC, Uthman BM, Ristanovic RK *et al.* Vagus nerve stimulation for the treatment of medically intractable seizures: results of a 1-year open-extension trial. *Arch Neurol* 1996; **53**:1176–80.

Salmenpara T, Kononen M, Roberts N *et al.* Hippocampal damage in newly diagnosed focal epilepsy: a prospective MRI study. *Neurology* 2005; **64**:62–8.

Sammaritano M, Andermann F, Melanson D *et al.* Prolonged focal cerebral edema associated with partial status epilepticus. *Epilepsia* 1985; **26**:334–9.

Samuel M, Duncan JS. Use of hand held video camcorder in the evaluation of seizures. *J Neurol Neurosurg Psychiatry* 1994; **57**:1417–18.

Sander JWAS, Hart YM, Johnson AL *et al.* National General Practice Study of Epilepsy: newly diagnosed epileptic seizures in a general population. *Lancet* 1990; **336**:1267–71.

Sander JWAS, Hart YM, Trimble MR *et al.* Vigabatrin and psychosis. *J Neurol Neurosurg Psychiatry* 1991; **54**:435–9.

Sands IJ. The acute psychiatric type of epidemic encephalitis. *Am J Psychiatry* 1928; **84**:975–87.

Sandson TA, Daffner KR, Carvalho PA *et al.* Frontal lobe dysfunction following infarction of the left-sided medial thalamus. *Arch Neurol* 1991; **48**:1300–3.

Sano K, Malamud N. Clinical significance of sclerosis of the cornu ammonis: ictal psychic phenomena. *Arch Neurol Psychiatry* 1953; **70**:40–53.

Sarno MT, Buonaguro A, Levita E. Characteristics of verbal impairment in closed head injury patients. *Arch Phys Med Rehabil* 1986; **67**:400–5.

Satel SL, Southwick SM, Gawin FH. Clinical features of cocaine-induced paranoia. *Am J Psychiatry* 1991; **148**:495–8.

Sato S, Dreifuss FE, Penry JK. Prognostic factors in absence seizures. *Neurology* 1976; **26**:788–96.

Sato S, White BG, Penry JK *et al.* Valproic acid versus ethosuximide in the treatment of absence seizures. *Neurology* 1982; **32**:157–63.

Savard G, Andermann F, Teitelbaum J *et al.* Epileptic Munchausen's syndrome: a form of pseudoseizures distinct from hysteria and malingering. *Neurology* 1988; **38**:1628–9.

Savard G, Andermann F, Olivier A *et al.* Postictal psychoses after partial complex seizures: a multiple case study. *Epilepsia* 1991; **32**:225–31.

Saygi S, Spencer SS, Scheyer R *et al.* Differentiation of temporal lobe ictal behavior associated with hippocampal sclerosis and tumor of the temporal lobe. *Epilepsia* 1994; **35**:737–42.

Sbordone RJ, Seyranian GD, Ruff RM. Are the subjective complaints of traumatically brain injured patients reliable? *Brain Inj* 1998; **12**:505–15.

Schacter SC, Vazquez B, Fisher RS *et al.* Oxcarbazepine: double-blind, randomized, placebo-controlled monotherapy trial for partial seizures. *Neurology* 1999; **52**:732–7.

Scheffer IE, Bhatia KP, Lopes-Cendes I *et al.* Autosomal dominant nocturnal frontal epilepsy misdiagnosed as a sleep disorder. *Lancet* 1994; **343**:515–17.

Scheffer IE, Bhatia KP, Lopes-Cendes I *et al.* Autosomal dominant nocturnal frontal lobe epilepsy: a distinctive clinical disorder. *Brain* 1995; **118**:61–73.

Schmahmann JD, Sherman JC. The cerebellar cognitive affective syndrome. *Brain* 1998; **121**:561–79.

Schneider WN, Drew-Cates J, Wong TM *et al.* Cognitive and behavioral efficacy of amantadine in acute traumatic brain injury: an initial double-blind placebo-controlled study. *Brain Inj* 1999; **13**:863–72.

Scholtes FB, Renier WO, Meinardi H. Simple partial status epilepticus: causes, treatment, and outcome in 47 patients. *J Neurol Neurosurg Psychiatry* 1996a; **61**:90–2.

Scholtes FB, Renier WO, Meinardi H. Non-convulsive status epilepticus: causes, treatment and outcome in 65 patients. *J Neurol Neurosurg Psychiatry* 1996b; **61**:93–5.

Schramm P, Schelliger PD, Klotz E *et al.* Comparison of perfusion computed tomography and computed tomography angiography source images with perfusion-weighted imaging and diffusion-weighted imaging in patients with acute stroke of less than 6 hours' duration. *Stroke* 2004; **35**:1652–8.

Schube PG. Emotional states of general paresis. *Am J Psychiatry* 1934; **91**:625–38.

Schuckitt MA, Smith TL, Anthenelli R *et al.* Clinical course of alcoholism in 636 male inpatients. *Am J Psychiatry* 1993; **150**:786–92.

Schufle SU, Bermeo AC, Alexopoulos AV *et al.* Video-electrographic and clinical features in patients with ictal asystole. *Neurology* 2007; **69**:439–41.

Schulz R, Luders HO, Noachtar S *et al.* Amnesia of the epileptic aura. *Neurology* 1995; **45**:231–5.

Schwab RS. A case of status epilepticus in petit mal. *Electroencephalogr Clin Neurophysiol* 1953; **5**:441–2.

Scott GM, Gibberb FB. Epilepsy and other factors in the prognosis of gliomas. *Acta Neurol Scand* 1980; **61**:227–39.

Scott JS, Masland RL. Occurrence of 'continuous symptoms' in epilepsy patients. *Neurology* 1953; **3**:297–301.

Scott RM, Smith ER. Moyamoya disease and moyamoya syndrome. *N Engl J Med* 2009; **360**:1226–37.

Seeley WW, Bauer AM, Miller BL *et al.* The natural history of temporal variant frontotemporal dementia. *Neurology* 2005; **64**:1384–90.

Semel JD. Complex partial status epilepticus presenting as fever of unknown origin. *Arch Int Med* 1987; **147**:1571–2.

Sen S, Laowatana S, Lima J *et al.* Risk factors for intracardiac thrombus in patients with recent ischemic cerebrovascular events. *J Neurol Neurosurg Psychiatry* 2004; **75**:1421–5.

Seniow J, Litwin M, Litwin T *et al.* New approach to the rehabilitation of post-stroke focal cognitive syndrome: Effect of levodopa combined with speech and language therapy on functional recovery from aphasia. *J Neurol Sci* 2009

Serafetinides EA, Falconer MA. Speech disturbance in temporal lobe seizures: a study in 100 epileptic patients submitted to anterior temporal lobectomy. *Brain* 1963; **86**:333–46.

Serby M, Angrist B, Lieberman A. Mental disturbances during bromocriptine and lergotrile treatment of Parkinson's disease. *Am J Psychiatry* 1978; **135**:1227–9.

Seshia SS, McLachlan RS. Aura continua. *Epilepsia* 2005; **46**:454–5.

Sethi PK, Surya R. Gelastic, quiritarian, and cursive epilepsy. *J Neurol Neurosurg Psychiatry* 1976; **39**:823–8.

Shah P, Kaplan SL. Catatonic symptoms in a child with epilepsy. *Am J Psychiatry* 1980; **137**:738–9.

Shavit YB, Graus F, Probst A *et al.* Epilepsia partialis continua: a new manifestation of anti-Hu associated paraneoplastic encephalomyelitis. *Ann Neurol* 1999; **45**:255–8.

Shaw PJ, Walls TJ, Newman PK *et al.* Hashimoto's encephalopathy: a steroid-responsive disorder associated with high anti-thyroid antibody titers – report of 5 cases. *Neurology* 1991; **41**:228–33.

Sherer MA, Kumor KM, Cone EJ *et al.* Suspiciousness induced by four-hour intravenous infusion of cocaine. Preliminary findings. *Arch Gen Psychiatry* 1988; **45**:673–7.

Sheth RD, Drazkowski JF, Sirven JI. Protracted ictal confusion in elderly patients. *Arch Neurol* 2006; **63**:529–32.

Shewmon DA, Masdeu JC. Delayed radiation necrosis of the brain contralateral to original tumor. *Arch Neurol* 1980; **37**:592–4.

Shimoda M, Takeuchi M, Tominaga J *et al.* Asymptomatic versus symptomatic infarcts from vasospasm in patients with subarachnoid hemorrhage: serial magnetic resonance imaging. *Neurosurgery* 2001; **49**:1341–8.

Shiozawa Z, Sasaki H, Ozaki Y *et al.* Epilepsia partialis continua following metrizamide cisternography. *Ann Neurol* 1981; **10**:400–1.

Shukla GD, Mishra DN. Vomiting as sole manifestation of simple partial seizure. *Arch Neurol* 1985; **42**:626.

Siegel AM, Cascino GD, Elger CE *et al.* Adult-onset epilepsy in focal cortical dysplasia of Taylor type. *Neurology* 2005; **64**:1771–4.

Siegel RK. Cocaine hallucinations. *Am J Psychiatry* 1978; **135**:309–14.

Siesling S, Vegter-van der Vlis M, Roos RAC. Juvenile Huntington's disease in the Netherlands. *Pediatr Neurol* 1997; **17**:34–43.

Silver JM, Koumaras B, Chen M *et al.* Effects of rivastigmine on cognitive function in patients with traumatic brain injury. *Neurology* 2006; **67**:748–55.

Silverstein A, Steinberg G, Nathanson M. Nervous system involvement in infectious mononucleosis. *Arch Neurol* 1972; **26**:353–8.

Simonsen N, Olsen PZ, Kuhl V *et al.* A comparative controlled study between carbamazepine and diphenylhydantoin in psychomotor epilepsy. *Epilepsia* 1976; **17**:169–76.

Singh BM, Strobos RJ. Epilepsia partialis continua associated with nonketotic hyperglycemia: clinical and biochemical profile of 21 patients. *Ann Neurol* 1980; **8**:155–60.

Singh BM, Gupta DR, Strobos RJ. Nonketotic hyperglycemia and epilepsia partialis continua. *Arch Neurol* 1973; **29**:187–90.

Singh NN, Rai A, Selhorst JB *et al.* Ondansetron and seizures. *Epilepsia* 2009; **50**:2663–6.

Sisler GC, Levy LL, Roseman E. Epilepsia cursiva: syndrome of running fits. *Arch Neurol Psychiatry* 1953; **69**:73–9.

Sisodiya SM, Free SL, Stevens JM *et al.* Widespread cerebral structural changes in patients with cortical dysgenesis and epilepsy. *Brain* 1995; **118**:1039–50.

Sisodiya SM, Stevens JM, Fish DR *et al.* The demonstration of gyral abnormalities in patients with cryptogenic partial epilepsy using three-dimensional MRI. *Arch Neurol* 1996; **53**:28–34.

Sittig O. A clinical study of sensory Jacksonian fits. *Brain* 1925; **48**:233–54.

Sivenius J, Kalviainen R, Ylinen A *et al.* Double-blind study of gabapentin in the treatment of partial seizures. *Epilepsia* 1991; **32**:539–42.

Slater E, Beard AW. The schizophrenia-like psychoses of epilepsy. I. Psychiatric aspects. *Br J Psychiatry* 1963a; **109**:95–112.

Slater E, Beard AW. The schizophrenia-like psychoses of epilepsy. V. Discussion and conclusion. *Br J Psychiatry* 1963b; **109**:143–50.

Sloan RL, Brown KW, Pentland B. Fluoxetine as a treatment for emotional lability after brain injury. *Brain Inj* 1992; **6**:315–19.

Smaje JC, Davidson C, Teasdale GM. Sino-atrial arrest due to temporal lobe epilepsy. *J Neurol Neurosurg Psychiatry* 1987; **50**:112–13.

Smith EJ. Amantadine-induced intoxication in a young healthy patient. *Am J Psychiatry* 2008; **165**:1613.

Smith GR, Monson RA, Livingston RL. Somatization disorder in men. *Gen Hosp Psychiatry* 1985; **7**:4–8.

Smith GR, Monson RA, Ray DC. Psychiatric consultation in somatization disorder. A randomized controlled study. *N Engl J Med* 1986; **314**:1407–13.

Smith LH. Mental and neurologic changes in pernicious anemia. *Arch Neurol Psychiatry* 1929; **22**:551–7.

Smith RF, Devinsky O, Luciano D. Inhibitory motor status: two new cases and a review of inhibitory motor seizures. *J Epilepsy* 1997; **10**:15–21.

Smyth GE, Stern K. Tumours of the thalamus: a clinico-pathological study. *Brain* 1938; **61**:339–74.

Snavely SR, Hodges GR. The neurotoxicity of antibacterial agents. *Ann Int Med* 1984; **101**:92–104.

So EL, Schauble BS. Ictal asomatognosia as a cause of epileptic falls. Simultaneous video, EMG, and invasive EEG. *Neurology* 2004; **63**:2153–4.

So EL, King DL, Muruin AJ. Misdiagnosis of complex absence seizures. *Arch Neurol* 1984; **41**:640–1.

So EL, Annegers JF, Hauser WA *et al.* Population-based study of seizure disorders after cerebral infarction. *Neurology* 1996; **46**:350–1.

Sonoo M. Abductor sign: a reliable new sign to detect unilateral non-organic paresis of the lower limb. *J Neurol Neurosurg Psychiatry* 2004; **75**:121–5.

Souid AK, Keith DV, Cunningham AS. Munchausen syndrome by proxy. *Clin Pediatr* 1998; **37**:497–503.

Sowa MV, Pituck S. Prolonged spontaneous visual hallucinations and illusions as ictal phenomena. *Epilepsia* 1989; **30**:524–6.

Spatt J. Epilepsia partialis continua in multiple sclerosis. *Lancet* 1995; **345**:658–9.

Speech TJ, Rao SM, Osmon DC *et al.* A double-blind controlled study of methylphenidate treatment in closed head injury. *Brain Inj* 1993; **7**:333–8.

Spencer SS, Spencer DD, Williamson PD *et al.* Sexual automatisms in complex partial epilepsy. *Neurology* 1983; **33**:527–33.

Sperling MR, Wilson CL. The effect of limbic and extralimbic electrical stimulations upon prolactin secretion in humans. *Brain Res* 1986; **371**:293–7.

Sperling MR, Pritchard PB, Engel J *et al.* Prolactin in partial epilepsy: an indicator of limbic seizures. *Ann Neurol* 1986; **20**:716–22.

Sperling MR, Lieb JP, Engel J *et al.* Prognostic significance of independent auras in temporal lobe seizures. *Epilepsia* 1989; **30**: 322–31.

Spiegel DR, Iarola R, Samuels D. A possible case of Capgras syndrome after a right anterior cerebral artery cerebrovascular accident treated successfully with mirtazapine. *J Neuropsychiatry Clin Neurosci* 2009; **20**:494.

Spooner CG, Berkovic SF, Mitchell LA *et al.* New-onset temporal lobe epilepsy in children. Lesion on MRI predicts poor seizure control. *Neurology* 2006; **67**:2147–53.

Spratling WP. Epilepsy, its etiology, pathology and treatment briefly considered. *JAMA* 1902; **38**:1126–30.

Starkstein SE, Mayberg HS, Berthier ML *et al.* Mania after brain injury: neuroradiological and metabolic findings. *Ann Neurol* 1990; **27**:652–9.

Starosta-Rubinstein S, Young AB, Kluin K *et al.* Clinical assessment of 31 patients with Wilson's disease: correlations with structural changes on magnetic resonance imaging. *Arch Neurol* 1987; **44**:365–70.

Steegmann AT, Winer B. Temporal lobe epilepsy resulting from ganglioglioma. *Neurology* 1961; **11**:406–12.

Stefan H, Wang-Tilz Y, Pauli E *et al.* Onset of action of levetiracetam; a RCT trial using therapeutic intensive seizure analysis. *Epilepsia* 2006; **47**:516–22.

Steffenburg S, Gillberg C, Steffenburg U. Psychiatric disorders in children and adolescents with mental retardation and active epilepsy. *Arch Neurol* 1996; **53**:904–12.

Steiner I, Budka H, Chaudhuri A *et al.* Viral encephalitis: a review of diagnostic methods and guidelines for management. *Eur J Neurol* 2005; **12**:331–43.

Stone J, Zeman A, Sharpe M. Functional weakness and sensory disturbance. *J Neurol Neurosurg Psychiatry* 2002; **73**:241–5.

Stone J, Townend E, Kwan J *et al.* Personality change after stroke: some preliminary observations. *J Neurol Neurosurg Psychiatry* 2004; **75**:1708–13.

Stone J, Smyth R, Carson A *et al.* La belle indifference in conversion symptoms and hysteria: systematic review. *Br J Psychiatry* 2006; **188**:204–9.

Storm-Mathisen A. General paresis: a follow-up study of 203 patients. *Acta Psychiatr Scand* 1969; **45**:118–32.

Stracciari A, Ciucci G, Bianchedi G *et al.* Epileptic transient amnesia. *Eur Neurol* 1990; **30**:176–9.

Strauss H. Paroxysmal convulsive running and the concept of epilepsia cursiva. *Neurology* 1960; **10**:341–4.

Strauss I, Keschner M. Mental symptoms in cases of tumor of the frontal lobe. *Arch Neurol Psychiatry* 1935; **33**:986–1007.

Strobos RRJ. Tumors of the temporal lobe. *Neurology* 1953; **3**:752–60.

Strub RL. Frontal lobe syndrome in a patient with bilateral globus pallidus lesions. *Arch Neurol* 1989; **46**:1024–7.

Sturm JW, Andermann F, Berkovic SF. 'Pressure to laugh': an unusual epileptic symptom associated with small hypothalamic hamartomas. *Neurology* 2000; **54**:971–3.

Sturm JW, Fedi M, Berkovic SF *et al.* Exercise induced temporal lobe epilepsy. *Neurology* 2002; **59**:1246–8.

Suarez JI, Tarr RW, Selman WR. Aneurysmal subarachnoid hemorrahge. *N Engl J Med* 2006; **354**:387–96.

Supino-Viterbo V, Sicard C, Risvegliato M *et al.* Toxic encephalopathy due to ingestion of bismuth salts: clinical and EEG studies of 45 patients. *J Neurol Neurosurg Psychiatry* 1977; **40**:748–52.

Sussman NM, Jackel RA, Kaplan LR *et al.* Bicycling movements as a manifestation of complex partial seizures of temporal lobe origin. *Epilepsia* 1989; **30**:527–31.

Sutherling WW, Risinger MW, Crandall PH *et al.* Focal functional anatomy of dorsolateral frontocentral seizures. *Neurology* 1990; **40**:87–98.

Swanson AG, Iseri OA. Acute encephalopathy due to water intoxication. *N Engl J Med* 1958; **258**:831–4.

Swartz BE. Electrophysiology of bimanual-bipedal automatisms. *Epilepsia* 1994; **35**:264–74.

Swartz BE, Tomiyasu U, Delgado-Escueta AV *et al.* Neuroimaging in temporal lobe epilepsy: test sensitivity and relationships to pathology and postoperative outcome. *Epilepsia* 1992; **33**:624–34.

Swigar ME, Bowers MB. Baclofen withdrawal and neuropsychiatric symptoms: a case report and review of other case literature. *Compr Psychiatry* 1986; **27**:394–400.

Symonds C. Excitation and inhibition in epilepsy. *Brain* 1959; **82**: 133–46.

Szabo K, Poepel A, Pohlmann-Eden B *et al.* Diffusion-weighted and perfusion MRI demonstrates parenchymal changes in complex partial status epilepticus. *Brain* 2005; **128**:1369–76.

Szaflarski JP, Sangha KS, Lindsell CJ *et al.* Prospective, randomized, single-blinded comparative trial of intravenous levetiracetam versus phenytoin for seizure prophylaxis. *Neurocrit Care* 2010; **12**:165–72.

Tachibana N, Shinde A, Ikede A *et al.* Supplementary motor area seizure resembling sleep disorder. *Sleep* 1996; **19**:811–16.

Tadokoro Y, Oshima T, Kanemoto K. Interictal psychoses in comparison with schizophrenia – a prospective study. *Epilepsia* 2007; **48**:2345–51.

Tanaka N, Sakurai K, Kamada K *et al.* Neuromagnetic source localization of epileptiform activity in patients with graphigenic epilepsy. *Epilepsia* 2006; **47**:1963–7.

Tarulli A, Devinsky O, Alper K. Progression of postictal to interictal psychosis. *Epilepsia* 2001; **42**:1468–71.

Tassi L, Colombo N, Cossu M *et al.* Electroclinical, MRI and neuropathological study of 10 patients with nodular heterotopia, with surgical outcomes. *Brain* 2005; **128**:321–37.

Tassinari CA, Tassi L, Calandra-Buonaura G *et al.* Biting behavior, aggression, and seizures. *Epilepsia* 2005; **46**:654–63.

Tateno A, Jorge RE, Robinson RG. Pathological laughing and crying after traumatic brain injury. *J Neuropsychiatry Clin Neurosci* 2004; **16**:126–34.

Tatu L, Moulin T, Bogousslavsky J *et al.* Arterial territories of human brain: brainstem and cerebellum. *Neurology* 1996; **47**:1125–35.

Tatu L, Moulin T, Bogousslavsky J *et al.* Arterial territories of human brain: cerebral hemispheres. *Neurology* 1998; **50**:1699–708.

Tatum WO, Passaro EA, Elia M *et al.* Seizure in Klinefelter's syndrome. *Pediatr Neurology* 1998; **19**:275–82.

Tayah TF, Abou-Khalil B, Gilliam FG *et al.* Musicogenic seizures can arise from multiple temporal lobe foci: intracranial EEG analyses of three patients. *Epilepsia* 2006; **47**:1402–6.

Taylor DC, Marsh SM. Hughlings Jackson's Dr. Z: the paradigm of temporal lobe epilepsy revealed. *J Neurol Neurosurg Psychiatry* 1980; **43**:758–67.

Taylor DC, Falconer MA, Bruton CJ *et al.* Focal dysplasia of the cerebral cortex in epilepsy. *J Neurol Neurosurg Psychiatry* 1971; **34**:369–87.

Taylor J, ed. *Selected Writings of John Hughlings Jackson.* London: Hodder & Stoughton, 1931.

Teasdale G, Jennett B. Assessment of coma and impaired consciousness. A practical scale. *Lancet* 1974; **2**:81–4.

Temkin NR, Dikmen SS, Wilensky AJ *et al.* A randomized, double-blind study of phenytoin for the prevention of post-traumatic seizures. *N Engl J Med* 1990; **323**:497–502.

Temkin NR, Dikmen SS, Anderson GD *et al.* Valproate therapy for prevention of posttraumatic seizures: a randomized trial. *J Neurosurg* 1999; **91**:593–600.

Tenembaum S, Chamoles N, Fejerman N. Acute disseminated encephalomyelitis. A long-term follow-up study of 84 pediatric patients. *Neurology* 2002; **59**:1224–31.

Tenovuo O, Alin J, Helenius H. A randomized controlled trial of rivastigmine for chronic sequels of traumatic brain injury – what it showed and taught? *Brain Inj* 2009; **23**:548–58.

Thacore VR, Shukla SRP. Cannabis psychosis and paranoid schizophrenia. *Arch Gen Psychiatry* 1976; **33**:383–6.

Theodore WH, Porter RJ, Penry JK. Complex partial seizures: clinical characteristics and differential diagnosis. *Neurology* 1983; **33**:1115–21.

Theodore WH, Porter RJ, Albert P *et al.* The secondarily generalized tonic–clonic seizure: a videotape analysis. *Neurology* 1994; **44**:1403–7.

Thomas CS, Szabadi E. Paranoid psychosis as the first presentation of a fulminating lethal case of AIDS. *Br J Psychiatry* 1987; **151**:693–5.

Thomas JE, Reagen TJ, Klass DW. Epilepsia partialis continua. A review of 32 cases. *Arch Neurol* 1977; **34**:266–75.

Thomas P, Valton L, Genton P. Absence and myoclonic status epilepticus precipitated by antiepileptic drugs in idiopathic generalized epilepsies. *Brain* 2006; **129**:1281–92.

Thompson AJ, Kermode AG, Moseley JF *et al.* Seizures due to multiple sclerosis: seven patients with MRI correlations. *J Neurol Neurosurg Psychiatry* 1993; **56**:1317–20.

Thompson GN, Nielsen JM. The organic paranoid syndrome. *J Nerv Ment Dis* 1949; **110**:478–96.

Thompson SW, Greenhouse HH. Petit mal status in adults. *Ann Intern Med* 1968; **68**:1271–9.

Thomsen IV. Late outcome of very severe blunt head trauma: a 10–15 year second follow-up. *J Neurol Neurosurg Psychiatry* 1984; **47**:260–8.

Tinuper P, Provini F, Marini C *et al.* Partial epilepsy of long duration: changing semiology with age. *Epilepsia* 1996a; **37**:162–4.

Tinuper P, Plazzi G, Provini F *et al.* Celiac disease, epilepsy, and occipital calcification: histopathological study and clinical outcome. *J Epilepsy* 1996b; **9**:206–9.

Tinuper P, Cerullo A, Marini C *et al.* Epileptic drop attacks in partial epilepsy: clinical features, evolution, and prognosis. *J Neurol Neurosurg Psychiatry* 1998; **64**:231–7.

Todd R. *Clinical Lectures on Paralysis, Diseases of the Brain, and Other Affections of the Nervous System.* Philadelphia, PA: Lindsay & Blakiston, 1855.

Tomson T, Svanborg E, Wedlund J-E. Nonconvulsive status epilepticus: high incidence of complex partial status. *Epilepsia* 1986; **27**:276–85.

Tomson T, Lindbom U, Nilsson BY *et al.* Serum prolactin during status epilepticus. *J Neurol Neurosurg Psychiatry* 1989; **52**:1435–7.

Tomson T, Lindbom U, Nilsson BY. Nonconvulsive status epilepticus in adults. Thirty-two consecutive patients from a general hospital population. *Epilepsia* 1992; **33**:829–35.

Tonon E, Scotton PG, Gallucci M *et al.* Brain abscess: clinical aspects of 100 cases. *Int J Infect Dis* 2006; **10**:103–9.

Towne AR, Waterhouse EJ, Boggs JG *et al.* Prevalence of nonconvulsive status epilepticus in comatose patients. *Neurology* 2000; **54**:340–5.

Treiman DM, Meyers PD, Walton NY *et al.* A comparison of four treatments for generalized convulsive status epilepticus. Veterans Affairs Status Epilepticus Cooperative Study Group. *N Engl J Med* 1998; **339**:792–8.

Trethowan WH, Cobb S. Neuropsychiatric aspects of Cushing's syndrome. *Arch Neurol Psychiatry* 1952; **67**:283–309.

Trouillas P, Courjon J. Epilepsy with multiple sclerosis. *Epilepsia* 1972; **13**:325–33.

Troupin A, Ojemann LM, Halpern L *et al.* Carbamazepine – a double-blind comparison with phenytoin. *Neurology* 1977; **27**:511–19.

Tsai JJ, Yen DJ, Hsih MS *et al.* Efficacy and safety of levetiracetam (up to 2000 mg/day) in Taiwanese patients with refractory partial seizures: a multicenter, randomized, double-blind, placebo-controlled study. *Epilepsia* 2006; **47**:72–81.

Tsai MH, Chang WN, Lui CL *et al.* Status epilepticus induced by star fruit intoxication in patients with chronic renal disease. *Seizure* 2005; **14**:521–5.

Tseng JH, Tseng MY. Brain abscess in 142 patients. Factors influencing outcome and mortality. *Surg Neurol* 2006; **65**:557–62.

Tsoi WF. The Ganser syndrome in Singapore: a report on ten cases. *Br J Psychiatry* 1973; **123**:567–72.

Tsukamoto T, Mochizuki R, Mochizuki H *et al.* Paraneoplastic cerebellar degeneration and limbic encephalitis in a patient with adenocarcinoma of the colon. *J Neurol Neurosurg Psychiatry* 1993; **56**:713–16.

Tucker WM, Forster FM. Petit mal epilepsy occurring in status. *Arch Neurol Psychiatry* 1958; **64**:823–7.

Tukel K, Jasper H. The electroencephalogram in parasaggital lesions. *Electroencephalogr Clin Neurophysiol* 1952; **4**:481–94.

Tyler RH. Neurologic disorders in renal failure. *Am J Med* 1968; **44**:734–48.

Umbricht D, Degreef G, Barr WB *et al.* Postictal and chronic psychoses in patients with temporal lobe epilepsy. *Am J Psychiatry* 1995; **152**:224–31.

Van de Beek D, Schmand B, de Gans J *et al.* Cognitive impairment in adults with good recovery after bacterial meningitis. *J Infect Dis* 2002; **186**:1047–52.

Van de Beek D, de Gans J, Spanjaard L *et al.* Clinical features and prognostic factors in adults with bacterial meningitis. *N Engl J Med* 2004a; **351**:1849–59.

Van de Beek D, de Gans J, McIntyre P *et al.* Steroids in adults with acute bacterial meningitis: a systematic review. *Lancet Infect Dis* 2004b; **4**:139–43.

Van der Horst L. Affective epilepsy. *J Neurol Neurosurg Psychiatry* 1953; **16**:25–9.

Van Der Naalt J, van Zomeren AH, Sluiter WJ *et al.* Acute behavioral disturbances related to imaging studies and outcome in mild-to-moderate head injury. *Brain Inj* 2000; **14**:781–8.

Van den Bergh MW, van der Schaaf I, van Gijn J. The spectrum of presentations of venous infarction caused by deep cerebral vein thrombosis. *Neurology* 2005; **65**:192–6.

Van Gijn J, Rinkel GJ. Subarachnoid hemorrhage: diagnosis, causes and management. *Brain* 2001; **124**:249–78.

Van Gijn J, van Dongen KJ, Vermeulen M *et al.* Perimesencephalic hemorrhage: a nonaneurysmal and benign form of subarachnoid hemorrhage. *Neurology* 1985; **35**:493–7.

Van Reekum R, Bayley M, Garner S *et al.* N of 1 study: amantadine for the amotivational syndrome in a patient with traumatic brain injury. *Brain Inj* 1995; **9**:49–53.

Van Rossum J, Groeneveld-Ockhuysen AAW, Arts RJHM. Psychomotor status. *Arch Neurol* 1985; **42**:989–93.

Van Zomeren AH, van den Berg W. Residual complaints of patients two years after severe head injury. *J Neurol Neurosurg Psychiatry* 1985; **48**:21–8.

Varma NK, Lee SI. Nonconvulsive status epilepticus following electroconvulsive therapy. *Neurology* 1992; **42**:263–4.

Venneri A, Caffarra P. Transient autobiographic amnesia: EEG and single-photon emission CT evidence of an organic etiology. *Neurology* 1998; **50**:186–91.

Verny M, Jellinger KA, Hauw J-J *et al.* Progressive supranuclear palsy: a clinicopathological study of 21 cases. *Acta Neuropathologica* 1996; **61**:427–31.

Vestergaard K, Andersen G, Gottrup H *et al.* Lamotrigine for central poststroke pain: a randomized controlled trial. *Neurology* 2001; **56**:184–90.

Vickrey BG, Bahls FH. Nonconvulsive status epilepticus following cerebral angiography. *Ann Neurol* 1989; **25**:199–201.

Victor M, Brausch J. The role of abstinence in the genesis of alcohol epilepsy. *Epilepsia* 1967; **8**:1–20.

Videbech T. Chronic olfactory paranoid syndrome. *Acta Psychiatr Scand* 1966; **42**:183–213.

Villani F, D'Amico D, Pincherle A *et al.* Prolonged focal negative motor seizures: a video-EEG study. *Epilepsia* 2006; **47**:1949–52.

Vizioli R, Magliocco EB. A case of prolonged petit mal seizures. *Electroencephalogr Clin Neurophysiol* 1953; **5**:439–40.

Vollmer TL, Guarnaccia J, Harrington W *et al.* Idiopathic granulomatous angiitis of the central nervous system: diagnostic challenges. *Arch Neurol* 1993; **50**:925–30.

von Campe G, Regli F, Bogousslavsky J. Heralding manifestations of vascular artery occlusion with lethal or severe stroke. *J Neurol Neurosurg Psychiatry* 2003; **74**:1621–6.

Vuilleumier P, Chicherio C, Assal F *et al.* Functional neuroanatomical correlates of hysterical sensorimotor loss. *Brain* 2001; **124**: 1077–90.

Walczak TS, Rubinsky M. Plantar responses after epileptic seizures. *Neurology* 1994; **44**:2191–3.

Walczak TS, Williams DT, Berten W. Utility and reliability of placebo infusion in the evaluation of patients with seizures. *Neurology* 1994; **44**:394–9.

Wallach J. Laboratory diagnosis of factitious disorders. *Arch Int Med* 1994; **154**:1690–6.

Wallin WT, Kurtzke JF. Neurocysticercosis in the United States. Review of an important emerging infection. *Neurology* 2004; **63**:1559–64.

Walshe FMR. On the mode of representation of movements in the motor cortex, with special reference to 'convulsions beginning unilaterally' (Jackson). *Brain* 1943; **66**:104–39.

Walshe JM, Yealland M. Wilson's disease: the problem of delayed diagnosis. *J Neurol Neurosurg Psychiatry* 1992; **55**:692–6.

Walton RG. Seizure and mania after high intake of aspartame. *Psychosomatics* 1986; **27**:218–20.

Waltz G, Harik SI, Kaufman B. Adult metachromatic leukodystrophy: value of computed tomographic scanning and magnetic resonance imaging of the brain. *Arch Neurol* 1987; **44**:225–7.

Waltzer H. A psychotic family – folie a douze. *J Nerv Ment Dis* 1963; **137**:67–75.

Warden DL, Labbate LA, Salazar AM *et al.* Posttraumatic stress disorder in patients with traumatic brain injury and amnesia for the event? *J Neuropsychiatry Clin Neurosci* 1997; **9**:18–22.

Warner TT, Williams LD, Walker RWH *et al.* A clinical and molecular genetic study of dentatorubropallidoluysian atrophy in four European families. *Ann Neurol* 1995; **37**:452–9.

Watanabe TK, O'Dell MW, Togliatti TJ. Diagnosis and rehabilitation strategies for patients with hysterical hemiparesis: a report of four cases. *Arch Phys Med Rehabil* 1998; **79**:1482–3.

Waxman SG, Geschwind M. Hypergraphia in temporal lobe epilepsy. *Neurology* 1974; **24**:629–36.

Waxman SG, Geschwind N. The interictal behavior syndrome of temporal lobe epilepsy. *Arch Gen Psychiatry* 1975; **32**:1580–6.

Weidauer S, Lanfemann H, Raabe A *et al.* Impairment of cerebral perfusion and infarct patterns attributable to vasospasm after aneurismal subarachnoid hemorrhage: a prospective MRI and DSA study. *Stroke* 2007; **38**:1831–6.

Weil AA. Depressive reactions associated with temporal lobe–uncinate seizures. *J Nerv Ment Dis* 1955; **121**:505–10.

Weil AA. Ictal depression and anxiety in temporal lobe disorders. *Am J Psychiatry* 1956; **113**:149–57.

Weil AA. Ictal emotions occurring in temporal lobe dysfunction. *Arch Neurol Psychiatry* 1959; **1**:101–11.

Weilburg JB, Bear DM, Sachs G. Three patients with concomitant panic attacks and seizure disorder: possible clues to the neurology of anxiety. *Am J Psychiatry* 1987; **144**:1053–6.

Weiller C, Ringelstein EB, Reiche W *et al.* The large striatocapsular infarct: a clinical and pathological entity. *Arch Neurol* 1990; **47**:1085–91.

Weinberger J. Simultaneous bilateral focal seizures without loss of consciousness. *Mt Sinai J Med* 1973; **40**:693–6.

Wells CE. Transient ictal psychosis. *Arch Gen Psychiatry* 1975; **32**:1201–3.

Wharton RM. Grand mal seizures with lithium treatment. *Am J Psychiatry* 1969; **125**:1446.

Wheless JW, Vazquez BR, Kanner AM *et al.* Rapid infusion with valproate sodium is well-tolerated in patients with epilepsy. *Neurology* 2004; **63**:1507–8.

Whitfield CL, Ch'ien LT, Whitehead JD. Lead encephalopathy in adults. *Am J Med* 1972; **52**:289–98.

Whyte J, Hart T, Schuster K *et al.* Effects of methylphenidate on attentional function after traumatic brain injury. A randomized, placebo-controlled trial. *Am J Phys Med Rehabil* 1997; **76**:440–50.

Whyte J, Hart T, Vaccaro M *et al.* Effects of methylphenidate on attention deficits after traumatic brain injury: a multidimensional, randomized, controlled trial. *Am J Phys Med Rehabil* 2004; **83**:401–20.

Wiart L, Petit H, Joseph PA *et al.* Fluoxetine in early poststroke depression: a double-blind placebo-controlled study. *Stroke* 2000; **31**:1829–32.

Wiebe S, Blume WT, Girvin JP *et al.* A randomized, controlled trial of surgery for temporal-lobe epilepsy. *N Engl J Med* 2001; **345**: 311–18.

Wieling W, Thijs RD, van Dijk N *et al.* Symptoms and signs of syncope: a review of the link between physiology and clinical clues. *Brain* 2009; **132**:2630–42.

Wieser HG. Temporal lobe or psychomotor status epilepticus. A case report. *Electroencephalogr Clin Neurophysiol* 1980; **48**: 558–72.

Wieser HG, Hungerbuhler H, Siegel AM *et al.* Musicogenic epilepsy: review of the literature and case report with ictal single photon emission computed tomography. *Epilepsia* 1997; **38**:200–7.

Wijdicks EFM, Plevak DJ, Wiesner RH *et al.* Course and outcome of seizures in liver transplant patients. *Neurology* 1996; **47**:1523–5.

Wilder-Smith E. Complete atrio-ventricular block during complex partial seizure. *J Neurol Neurosurg Psychiatry* 1992; **55**:734–6.

Wilkinson HA. Epileptic pain. An uncommon manifestation with localizing value. *Neurology* 1973; **23**:518–20.

Williams BB, Lerner AM. Some previously unrecognized features of herpes simplex encephalitis. *Neurology* 1978; **28**:1193–6.

Williams D. The structure of emotions reflected in epileptic experiences. *Brain* 1956; **79**:29–67.

Williamson PD, Spencer SS. Clinical and EEG features of complex partial seizures of extratemporal origin. *Epilepsia* 1986; **27**(Suppl. 2):46–63.

Williamson PD, Spencer DD, Spencer SS *et al.* Episodic aphemia and epileptic focus in the nondominant hemisphere: relieved by section of the corpus callosum. *Neurology* 1985a; **35**:1069–71.

Williamson PD, Spencer DD, Spencer SS *et al.* Complex partial status epilepticus: a depth-electrode study. *Ann Neurol* 1985b; **18**:647–54.

Williamson PD, Spencer DD, Spencer SS *et al.* Complex partial seizures of frontal lobe origin. *Ann Neurol* 1985c; **18**:497–504.

Williamson PD, Thadani VM, Darcey TM *et al.* Occipital lobe epilepsy: clinical characteristics, seizure spread patterns, and results of surgery. *Ann Neurol* 1992a; **31**:3–13.

Williamson PD, Boon PA, Thadani VM *et al.* Parietal lobe epilepsy: diagnostic considerations and results of surgery. *Ann Neurol* 1992b; **31**:193–201.

Williamson PD, French JA, Thadani VM *et al.* Characteristics of medial temporal lobe epilepsy. II. Interictal and ictal scalp electroencephalography, neuropsychological testing, neuroimaging, surgical results, and pathology. *Ann Neurol* 1993; **34**:781–7.

Williamson RT. On the symptomatology of gross lesions (tumours and abscesses) involving the pre-frontal regions of the brain. *Brain* 1896; **19**:346–65.

Willmore LJ, Shu V, Wallin B. Efficacy and safety of add-on divalproex sodium in the treatment of complex partial seizures. *Neurology* 1996; **46**:49–53.

Wilmott C, Ponsford J. Efficacy of methylphenidate in the rehabilitation of attention following traumatic brain injury: a randomised, crossover, double-blind, placebo controlled trial. *J Neurol Neurosurg Psychiatry* 2009; **80**:552–7.

Willoughby JO, Pope KJ, Eaton V. Nicotine as an antiepileptic agent in ADNFLE: an N-of-one study. *Epilepsia* 2003; **44**:1238–40.

Wilson LG. Viral encephalopathy mimicking functional psychosis. *Am J Psychiatry* 1976; **133**:165–70.

Wilson P, Preece AA. Chorea gravidarum. *Arch Intern Med* 1932; **49**:671–97.

Wilson SAK. Progressive lenticular degeneration: a familial nervous disease associated with cirrhosis of the liver. *Brain* 1912; **34**: 295–509.

Wilson SAK. Nervous semiology, with special reference to epilepsy. *BMJ* 1930; **2**:50–4.

Wilson WH, Claussen AM. Seizures associated with clozapine treatment in a state hospital. *J Clin Psychiatry* 1994; **55**:184–8.

Winawer MR, Ottman R, Hauser A *et al.* Autosomal dominant partial epilepsy with auditory features: defining the phenotype. *Neurology* 2000; **54**:2173–6.

Wintermark M, Reichart M, Cuisenaire O *et al.* Comparison of admission perfusion computed tomography and qualitative diffusion- and perfusion-weighted magnetic resonance imaging in acute stroke patients. *Stroke* 2002a; **33**:2025–31.

Wintermark M, Reichart M, Thiran J-P *et al.* Prognostic accuracy of cerebral blood flow measurement by perfusion computed tomography, at the time of emergency room admission, in acute stroke patients. *Ann Neurol* 2002b; **51**:417–32.

Wisniewski KE, Frency JH, Fernando S *et al.* Fragile X syndrome: associated neurological abnormalities and developmental disabilities. *Ann Neurol* 1985; **18**:665–9.

Wojner-Alexandrov AW, Garami Z, Chernyshev OY *et al.* Heads down. Flat positioning improves blood flow velocity in acute ischemic stroke. *Neurology* 2005; **64**:1354–7.

Worrell GA, Sencakova D, Jack CR *et al.* Rapidly progressive hippocampal atrophy: evidence for a seizure-induced mechanism. *Neurology* 2002; **58**:1553–6.

Wroblewski BA, McColgan K, Smith K *et al.* The incidence of seizures during tricyclic antidepressant drug treatment in a brain-injured population. *J Clin Psychopharmacol* 1990; **10**:124–8.

Wroblewski BA, Joseph AB, Cornbatt RR. Antidepressant pharmacotherapy and the treatment of depression in patients with severe traumatic brain injury: a controlled, prospective study. *J Clin Psychiatry* 1996; **57**:582–7.

Wyllie E, Luders H, MacMillan JP *et al.* Serum prolactin levels after epileptic seizures. *Neurology* 1984; **34**:1601–3.

Wyllie E, Lueders H, Morris HH *et al.* The lateralizing significance of versive head and eye movements during epileptic seizures. *Neurology* 1986; **36**:606–11.

Yadav JS, Wholex MH, Kuntz RE *et al.* Protected carotid artery stenting versus endarterectomy in high-risk patients. *N Engl J Med* 2004; **351**:1493–1501.

Yamada H, Yoshida H. Laughing attack: a review and report of nine cases. *Folia Psychiatr Neurol Japonica* 1977; **31**:129–37.

Yanagihara T, Piepgras DG, Klass DW. Repetitive involuntary movement associated with episodic cerebral ischemia. *Ann Neurol* 1985; **18**:244–50.

Yen D-J, Su M-S, Shih Y-H *et al.* Ictal speech manifestations in temporal lobe epilepsy: a video-EEG study. *Epilepsia* 1996; **37**:45–9.

Yen DJ, Yu HY, Guo YC *et al.* A double-blind, placebo-controlled study of Topiramate in adult patients with refractory partial epilepsy. *Epilepsia* 2000; **41**:1162–6.

Yenjun S, Harvey AS, Marini C *et al.* EEG in adult-onset idiopathic generalized epilepsy. *Epilepsia* 2003; **44**:252–6.

York GK, Gabor AJ, Dreyfuss PM. Paroxysmal genital pain: an unusual manifestation of epilepsy. *Neurology* 1979; **29**:516–19.

Young GB, Blume WT. Painful epileptic seizures. *Brain* 1983; **106**: 537–44.

Young GB, Blume WT. Painful epileptic seizures involving the secondary sensory area. *Ann Neurol* 1986; **19**:412.

Yutzy SH, Cloninger CR, Guze SB *et al.* DSM-IV field trial: testing a new proposal for somatization disorder. *Am J Psychiatry* 1995; **152**:97–101.

Zak R, Solomon G, Petito F *et al.* Baclofen-induced generalized nonconvulsive status epilepticus. *Ann Neurol* 1994; **36**:113–14.

Zappoli R. Two cases of prolonged twilight state with almost continuous 'wave-spikes'. *Electroencephalogr Clin Neurophysiol* 1955; **7**:421–3.

Zarling EJ. Abdominal epilepsy: an unusual cause of recurrent abdominal pain. *Am J Gastroenterol* 1984; **79**:687–8.

Zeidler M, Johnstone EC, Bamber RWK *et al.* New variant Creutzfeldt–Jakob disease: psychiatric features. *Lancet* 1997a; **350**:908–10.

Zeidler M, Stewart GE, Barraclough CR *et al.* New variant Creutzfeldt–Jakob disease: neurological features and diagnostic tests. *Lancet* 1997b; **350**:903–7.

Zeilig G, Drubach DA, Katz-Zeilig M *et al.* Pathological laughter and crying in patients with closed traumatic brain injury. *Brain Inj* 1996; **10**:591–7.

Zeman A, Stone J, Porteous M *et al.* Spinocerebellar ataxia type 8 in Scotland: genetic and clinical features in seven unrelated cases and a review of published reports. *J Neurol Neurosurg Psychiatry* 2004; **75**:459–65.

Zeman AZJ, Boniface SJ, Hodges JR. Transient epileptic amnesia: a description of the clinical and neuropsychological features in 10 cases and a review of the literature. *J Neurol Neurosurg Psychiatry* 1998; **64**:435–43.

Zeng S, Kim CH, Rahman H. Psychotic symptoms presented in familial Creutzfeldt-Jakob disease, subtype E200K. *J Clin Psychiatry* 2001; **62**:735.

Zentner J, Hufnagel A, Wolf HK *et al.* Surgical treatment of temporal lobe epilepsy: clinical, radiological, and histopathological findings in 178 patients. *J Neurol Neurosurg Psychiatry* 1995; **58**: 666–73.

Zhang L, Plotkin RC, Wang G *et al.* Cholinergic augmentation with donepezil enhances recovery in short-term memory and sustained attention after traumatic brain injury. *Arch Phys Med Rehabil* 2004; **85**:1050–5.

Zifko U, Lindner K, Wimberger D *et al.* Jarisch–Herxheimer reaction in a patient with neurosyphilis. *J Neurol Neurosurg Psychiatry* 1994; **57**:865–7.

PART IV

SPECIFIC DISORDERS

9

Neurodegenerative and movement disorders

9.1 ALZHEIMER'S DISEASE

Alzheimer's disease is the most common cause of dementia in the elderly, accounting for well over one-half of all cases in this age group. The prevalence of Alzheimer's disease increases with age: in those aged under 65 years, approximately 1 percent will be affected, whereas in those aged 65 years, some 5–10 percent will have the disease, and in those aged 85 years or older the prevalence rises to 30–40 percent. Women are slightly more likely to be affected than men.

The first known case of this disease occurred in a 51-year-old woman, Auguste D, who was initially seen by Alois Alzheimer in 1901 at the Frankfurt State Asylum. Alzheimer subsequently moved to Munich to work with Kraepelin and, after Auguste D died, he presented his findings on 3 November 1906 at the 37th annual meeting of the Southwest German Psychiatrists in Tubingen, after which he published his findings in 1907 (Alzheimer 1907). Alzheimer himself was diffident about lending his name to the disease, and it was only at the urging of Kraepelin that he agreed, thus giving us now the most famous of all medical eponyms (Amaducci *et al.* 1986; Maurer *et al.* 1997).

Clinical features

Although onsets in early or middle adult years may occur, this is rare, and in 95 percent or more of cases the onset is delayed until the age of 65 years or older, when, classically, and in the overwhelming majority of cases, patients present with a gradually progressive amnestic syndrome.

The amnestic syndrome typically presents with forgetfulness (Bowler *et al.* 1997; Didic *et al.* 1998; Linn *et al.* 1995): patients have difficulty recalling what happened earlier in the day, and the location of keys, wallets, or purses may be forgotten (Hamilton *et al.* 2009). On formal testing, patients will be unable to recall all out of three objects after 5 minutes, and, with progression, there will be disorientation to time and place. Over time, as this anterograde component of the amnesia grows more profound, it is typically joined by a progressively worsening retrograde component: patients may eventually forget where they have worked, the names of their children, the fact that they have been married, where they went to high school, etc. The fact that the amnesia is of the declarative or 'episodic' form, rather than the 'procedural' type, can make for some startling contrasts in the clinical picture. Thus, profoundly amnestic patients may still 'remember' how to play card games or musical instruments (Beatty *et al.* 1994). For example, one patient (Crystal *et al.* 1989), a professional pianist, although unable to remember the names of various musical pieces or their composers, was nevertheless still able to flawlessly play Beethoven's fifth symphony.

With progression of the amnesia, further cognitive deficits accrue to eventually complete the clinical syndrome of dementia. Attempts have been made to subdivide the course of the dementia of Alzheimer's disease into various stages, such as 'mild', 'moderate', and 'severe', and, although this has some merit, it must be kept in mind that there is substantial overlap among these stages.

The mild stage is characterized primarily by cognitive deficits: in addition to the worsening amnesia, linguistic deficits appear. There may be a degree of anomia, and patients often seem at a loss for the right word. Sentence structure becomes less complex, and speech becomes more and more characterized by stock phrases. Difficulty with calculations appear, with a concurrent difficulty in keeping tabs on finances, and executive deficits also appear with consequent problems in organizing and planning the day.

The moderate stage is marked by a worsening of these cognitive deficits, by the appearance of frankly aphasic disturbances, and by apraxic and agnostic symptoms. The aphasia is typically of the sensory type (Faber-Langendoen *et al.* 1988; Price *et al.* 1993), and speech becomes progressively more incoherent. With apraxia, patients may be unable to dress themselves or to use simple tools, such as knives and forks. Although various agnostic disturbances may be found on formal testing, three tend to stand out clinically, namely anosognosia, topographagnosia, and prosopagnosia. Anosognosia (Starkstein *et al.* 1997), very early on, may be difficult to distinguish from emotionally motivated denial, which at times can be extreme; however, over time, the emotional response fades, and patients simply and blandly do not recognize that their memory is poor or that they have any other difficulties, despite the fact that they manifestly, say, cannot dress themselves. Topographagnostic patients may get lost in their own neighborhood, or even their own house, and prosopagnostic patients, in addition to failing to recognize family members by sight, may also, and most remarkably, fail to recognize themselves, thus giving rise to the classic 'mirror sign', wherein patients, upon looking in the mirror, ask who that person is.

In the severe stage, patients are profoundly demented, often doubly incontinent, and totally dependent on others for their care; mutism may occur, or there may be palilalia or echolalia.

In addition to these cognitive and associated deficits, almost all patients will develop other syndromes, including personality change, mood disturbances, or psychotic symptoms. Although these may appear during the mild stage, they most characteristically make their appearance during the moderate stage.

Personality change may be of the non-specific type, but more commonly one finds elements of a frontal lobe syndrome, with coarseness, impulsivity, and disinhibition (Mega *et al.* 1996; Petry *et al.* 1988).

Mood changes (Mega *et al.* 1996) may include depression, apathy, anxiety, irritability, agitation, and euphoria. Depressive symptoms are fairly common, with prevalence figures noted of 14 percent (Klatka *et al.* 1996), 24 percent (Burns *et al.* 1990a), and 52 percent (Starkstein *et al.* 2005), and these symptoms tend to persist over long follow-up (Starkstein *et al.* 1997). The severity of the depression varies from mild to that encountered in the depressive episodes of a major depression (Migliorelli *et al.* 1995). Apathy may accompany depression, but it may also be seen in a pure form in roughly one-third of all patients (Starkstein *et al.* 2006). Anxiety and irritability are about as common as depression, being found in a little less than one-half of patients (Litvan *et al.* 1996a; Mega *et al.* 1996). Agitation is very common, found in over one-half of patients (Lopez *et al.* 2003). Euphoria, by contrast, is uncommon, being found in 4 percent (Burns *et al.* 1990a) to 8 percent (Mega *et al.* 1996) of cases.

Psychotic symptoms include hallucinations, either visual or auditory, and delusions. Hallucinations are noted in 10 percent (Mega *et al.* 1996) to 35 percent (Klatka *et al.* 1996) of patients and are more commonly visual than auditory (Burns *et al.* 1990b; Forstl *et al.* 1993; Ropacki and Jeste 2005). Delusions are found in 16 percent (Burns *et al.* 1990c) to 53 percent of patients (Klatka *et al.* 1996). Common delusional themes (Binetti *et al.* 1993; Burns *et al.* 1990c; Devanand *et al.* 1997; Forstl *et al.* 1993; Hirono *et al.* 1998) include theft, the Capgras phenomenon, and the phantom boarder. Patients may insist that things have been stolen or taken from them, and they may be terrified that family members are actually imposters who have come to annoy or torment them; some may insist that someone, perhaps someone malevolent, is in fact hiding in the house, perhaps in the attic or the cellar.

Other symptoms may appear as the dementia progresses, including 'frontal release' signs (such as snout and grasp reflexes), astereognosis, and agraphesthesia (Huff *et al.* 1987). Parkinsonism may be seen late in the course in a minority of cases, being typically characterized by rigidity and bradykinesia, with tremor being relatively rare (Clark *et al.* 1997; Scarmeas *et al.* 2004). Seizures may also occur in a small minority (Amatniek *et al.* 2006; Goodman 1953; Scarmeas *et al.* 2009); these tend to be grand mal and occur late in the course of the disease (Romanelli *et al.* 1990). Very late in the course, a minority may also have myoclonus (Benesch *et al.* 1993; Chen *et al.* 1991; Faden and Townsend 1976). Other possible symptoms include emotional incontinence (Starkstein *et al.* 1995), the Kluver–Bucy syndrome (Lilly *et al.* 1983), and, very rarely, hemiparesis (Jagust *et al.* 1990).

Before leaving this discussion of the symptomatology of Alzheimer's disease, some comment should be made regarding various possible atypical presentations. Although, as discussed earlier, the classic presentation is with a progressive amnestic syndrome, certain variations, although rare, do occur, and these can throw off the diagnostically unwary. Thus, Alzheimer's disease may present as one of the primary progressive aphasias, manifesting with an expressive, or, much less commonly, sensory aphasia, which gradually worsens until joined by more typical cognitive deficits (Clark *et al.* 2003; Galton *et al.* 2000; Green *et al.* 1990; Greene *et al.* 1996; Karbe *et al.* 1993; Knibb *et al.* 2006). Even more rarely, the presentation may be with a gradually worsening apraxia (Ross *et al.* 1996). Finally, and most exceedingly rarely, Alzheimer's disease may present with a psychosis, and it is of interest that this was, in fact, the presentation in Alzheimer's first patient, Auguste D, who presented, at the age of 51 years, with a delusion of jealousy regarding her husband (Maurer *et al.* 1997; Wilkins and Brody 1969a).

Magnetic resonance imaging (MRI) and computed tomography (CT) scanning reveal ventricular dilatation and cortical atrophy, especially involving the medial aspect of the temporal lobe, as illustrated in Figure 9.1; although early

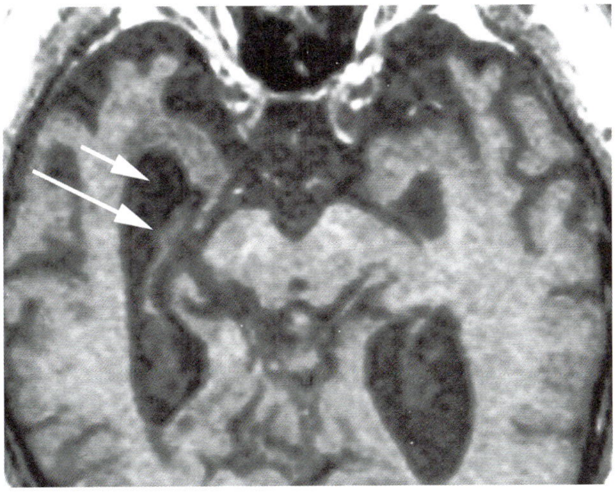

Figure 9.1 This T1-weighted magnetic resonance imaging scan demonstrates the cortical atrophy and ventricular dilation found in advanced Alzheimer's disease. Note in particular that the hippocampus, indicated by the arrows, has shrunk down to a thin remnant. Reproduced with kind permission from Gillespie and Jackson (2000)

on in the course, the degree of this change may still be within the broadly defined limits of normal for the elderly population, with progression the changes become quite pronounced.

The electroencephalogram (EEG) is typically normal in the mild stage; however, with progression, generalized theta, and eventually delta, slowing appears.

Although the cerebrospinal fluid (CSF) may display a slight elevation of total protein, routine studies are otherwise normal. Determinations of β-amyloid (42) and total and phosphorylated tau, however, are not. As noted below, Alzheimer's disease is characterized pathologically by neuritic plaques composed of an amyloid core, and by neurofibrillary tangles composed of hyperphosphorylated tau protein, and CSF studies have demonstrated both a decreased level of β-amyloid and an increased level of tau protein (Andreasen *et al.* 2001; Fagan *et al.* 2006; Galasko *et al.* 1998; Hampel *et al.* 2004; Shaw *et al.* 2009; Tapiola *et al.* 2009), and autopsy studies have demonstrated a good correlation between CSF levels of tau protein and the burden of neurofibrillary tangles (Buerger *et al.* 2006; Tapiola *et al.* 2009).

Single-photon emission computed tomography (SPECT) and fluoro-2-deoxy-D-glucose positron emission tomography (FDG-PET) scanning reveal hypoperfusion and hypometabolism, respectively, initially in the parietal cortex, then in the medial temporal lobe, and finally in most neocortical areas.

The newest imaging technique utilized in the diagnosis of Alzheimer's disease is PET scanning using the Pittsburgh binding (PiB) compound. The PiB compound binds to amyloid plaques (Ikonomovic *et al.* 2008), and, as might be expected, PET scanning with PiB reveals increased activity in neocortical areas. Unlike the foregoing diagnostic techniques, this is not as yet available for routine use.

Course

Although a small minority of patients may experience temporary plateaus, Alzheimer's disease, for the most part, is relentlessly and steadily progressive, with death occurring for most within 5–15 years. At the end, patients are vegetative, bedfast, and incontinent. Although it is generally held that cases with an early onset, before the age of 65 years, tend to run a more rapid course (Koss *et al.* 1996), not all studies agree on this (Bracco *et al.* 1994).

Etiology

Macroscopically, as illustrated in Figure 9.2, there is widespread cortical atrophy affecting primarily the temporal, parietal, and frontal lobes, with prominent sparing of the pre- and postcentral gyri; relative to the other lobes, the occipital lobe is less affected. Within the temporal lobe, the hippocampus (as illustrated in Figure 9.1) and amygdala are also very prominently involved. Subcortical and brainstem nuclei, including the nucleus basalis of Meynert (especially its cholinergic neurons) (Rogers *et al.* 1985; Whitehouse *et al.* 1981), the locus ceruleus (Mann *et al.* 1984), and the dorsal raphe nucleus (Yamamoto and Hirano 1985), also undergo significant damage.

Microscopically (Kidd 1964), there are widespread neurofibrillary tangles and neuritic plaques (also known as senile plaques) accompanied by neuronal loss (Terry *et al.* 1981). Neurofibrillary tangles are fibrillar structures found in the neuronal cytoplasm that, by electron microscopy, are seen to be composed of paired helical filaments. These paired helical filaments are composed of hyperphosphorylated tau proteins, which are one of the microtubule-associated proteins (MAPs) that ensure the integrity and stability of the cellular microtubules. Neuritic plaques are spherical extracellular structures composed of an amyloid core surrounded by 'neurites', or swollen axonal fragments. The amyloid core of the neuritic plaque is composed primarily of β-amyloid (42).

Interestingly, although the clinical severity of Alzheimer's disease correlates with the number of neurofibrillary tangles, there is little correlation with the number of neuritic plaques (Arriagada *et al.* 1992; Bennett *et al.* 2004; Bierer *et al.* 1995).

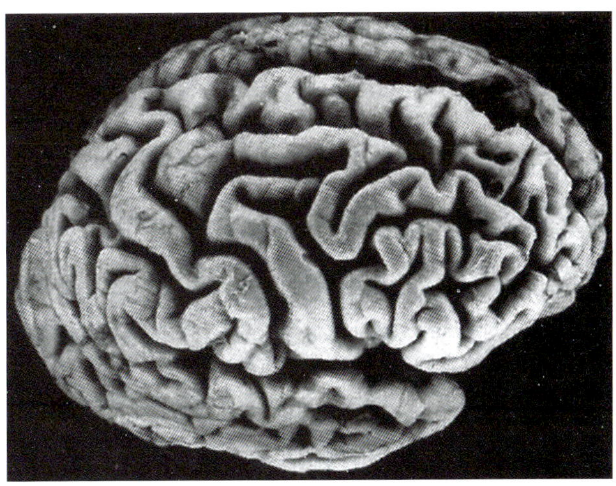

Figure 9.2 Note the relative sparing of the pre- and postcentral gyri compared with the rest of the cortex in this case of Alzheimer's disease. Reproduced with kind permission from Rees *et al.* (1996).

Furthermore, it appears that, in general, there is an orderly appearance of neurofibrillary tangles during the course of the disease, beginning first in the transentorhinal cortex and then progressing sequentially to the entorhinal cortex, hippocampus, temporal cortex, parietal and prefrontal cortex, and finally all neocortical areas (Braak and Braak 1991; Delacourte *et al.* 1999). This progression appears to supply a pathologic underpinning to the evolution of the clinical features noted above, in that damage to medial temporal structures would be expected to cause an amnesia, whereas later damage to cortical areas would account for the appearance of the additional cognitive deficits.

In addition, there appears to be a correlation between the depth of the memory loss in Alzheimer's disease and the extent of damage in the cholinergic nucleus basalis of Meynert (Neary *et al.* 1986; Rasool *et al.* 1986; Whitehouse *et al.* 1982). There is also a correlation between depression and cell loss in the superior central nucleus (Zweig *et al.* 1988) and probably the locus ceruleus (Zubenko and Moosy 1988; Zweig *et al.* 1988), although not all have replicated this finding (Hoogendijk *et al.* 1999).

Much progress has been made in the search for the etiology of Alzheimer's disease, especially with regard to genetic factors. In a small minority of cases, probably less than 1 percent, especially those of early onset before the age of 50 years, Alzheimer's disease is clearly inherited in an autosomal dominant fashion. Mutations have been identified in three genes (Janssen *et al.* 2003): the APP (amyloid precursor protein) gene on chromosome 21 (Brouwers *et al.* 2006; Chartier-Harlin *et al.* 1991), the presenilin-1 gene on chromosome 14 (Bird *et al.* 1996; Wasco *et al.* 1995), and the presenilin-2 gene on chromosome 1 (Jayadev *et al.* 2010; Levy-Lahad *et al.* 1995). The role of genetic factors in the remaining vast majority of cases of apparently sporadic Alzheimer's disease is not clear. Some (Bergem *et al.* 1997; Breitner *et al.* 1995; Gatz *et al.* 1997), but not all (Cook *et al.* 1981), studies support a higher concordance among monozygotic than dizygotic twins. Likewise, whereas some studies indicate a higher prevalence of Alzheimer's disease among the first-degree relatives of probands than among the equivalent relatives of controls, others do not. One of the reasons for these discordant results might be that, although Alzheimer's disease may be inherited, there is such great intrafamilial variability in the age of its expression that most cases among relatives are missed in cross-sectional studies. Life-table studies support this notion; indeed, studies using the life-table approach have found that the projected risk among first-degree relatives is approximately 50 percent (Mohs *et al.* 1987), just what would be expected if sporadic Alzheimer's disease was, in fact, an autosomal dominant disorder. Furthermore, and in line with this, the largest twin study to date (Gatz *et al.* 2006), which, importantly, utilized a sample of *older* twins, found an heritability of 58 percent, again what would be expected in an autosomal dominant condition.

Another gene associated with Alzheimer's disease is that for apolipoprotein E on chromosome 19. Apolipoprotein E occurs in several forms, depending on which alleles are present – ε-2, ε-3, or ε-4 – and there is a correlation between which alleles are present and the risk of Alzheimer's disease (Corder *et al.* 1993; Myers *et al.* 1996; Saunders *et al.* 1993). Thus, the risk for those homozygous for ε-4 is roughly 50 percent, while the risk for those heterozygote for ε-4/ε-3 is roughly 25 percent; interestingly, the presence of an ε-2 allele, rather than increasing the risk of Alzheimer's disease, appears to be a protective factor.

In contrast to these positive results in genetic studies, efforts to identify environmental causes have generally been unsuccessful, with possibly one exception: it does appear that a history of significant head trauma may increase the risk of Alzheimer's disease (Schofield *et al.* 1997).

The actual mechanism or mechanisms responsible for the formation of neurofibrillary tangles and neuritic plaques remain unclear. One current hypothesis (the 'amyloid cascade hypothesis') focuses on the neurotoxicity of one form of β-amyloid, the 42-amino acid form. APP is a transmembrane protein that is normally cleaved by several secretases, namely α-, β-, and γ-secretase. Depending on which secretases are involved, different fragments are produced: cleavage by α-secretase yields a non-toxic metabolite; however, when cleavage proceeds via β- and then γ-secretase, β-amyloid (42) is produced. This 42-amino acid form of β-amyloid is relatively insoluble and undergoes fibrillization to form what are known as 'diffuse' plaques. These diffuse plaques prompt an inflammatory response and are neurotoxic, and, according to the theory, this neurotoxicity leads both to the breakdown of axons, thus creating neurites that surround the 'diffuse' plaques, thereby creating classic neuritic plaques, and to the formation of neurofibrillary proteins in surviving neurons. This amyloid cascade hypothesis gains support from several quarters. First, it is well known that patients with Down's syndrome, should they survive past the age of 40 years, almost always develop Alzheimer's disease (Evenhuius 1990; Jervis 1948; Lai and Williams 1989; Olson and Shaw 1969). Down's syndrome occurs secondary to an extra chromosome 21, with a consequent extra gene for APP; this leads in turn to an overproduction of APP, which would 'start' the amyloid cascade going from the top. Second, as noted earlier, there are rare inherited cases of Alzheimer's disease that are caused by mutations in the gene for APP, which, again, could start the cascade going. Third, it appears that presenilin interacts with γ-secretase and, as noted earlier, there are also rare inherited forms of Alzheimer's disease that occur secondary to mutations in the genes for presenilin-1 or -2; conceivably, if these mutations lead to an increased activity of γ-secretase, this would lead to an overproduction of the neurotoxic 42-amino acid form of β-amyloid. Finally, it also appears that the ε-4 form of apolipoprotein E increases the rate of β-amyloid production.

Although this amyloid cascade hypothesis thus has much to recommend it, certain problems remain. For example, as noted earlier, while there is a correlation between the severity of the disease and the burden of neurofibrillary tangles, such a correlation is lacking with amyloid plaques, which is not what one would expect if plaque formation were the initiating neuropathologic event. Furthermore, it remains unclear just what the purported neurotoxic mechanism might be that leads from the presence of plaques to the formation of tangles. Although these are not insurmountable objections to the theory, they do remind us that it remains just that, a theory.

Differential diagnosis

As noted above, the typical presentation of Alzheimer's disease is characterized by the gradual onset of amnesia, with a subsequent accrual of other cognitive deficits; a personality change may occur, but usually only after cognitive deficits have become prominent, and, at least in the mild stage of the disease, there are few distinctive features. The differential diagnosis for such a dementia of gradual onset, lacking in distinctive features, as detailed in Chapter 6.3, is fairly wide, but certain considerations enable one to narrow it down.

Dementia with Lewy bodies may present in a similar fashion, but, within the first year, one sees several features that are not typical for early Alzheimer's disease, namely parkinsonism, visual hallucinations, and spontaneous confusional episodes.

Vascular dementia, including multi-infarct dementia, lacunar dementia, and Binswanger's disease figure prominently in the differential; however, in the cases of multi-infarct dementia and most instances of lacunar dementia, the course is 'step-wise', corresponding to the occurrence of new infarctions, which stands in contrast to the gradually progressive course typical of Alzheimer's disease. Some cases of lacunar dementia, and all of those of Binswanger's disease, however, do display a gradual course, and here the differential may hinge on imaging. MRI is preferable: this displays not only the large, territorial, infarctions characteristic of multi-infarct dementia, but will also display the multiple lacunar infarctions characteristic of lacunar dementia and the extensive white matter disease seen in Binswanger's disease.

Pick's disease and frontotemporal lobar degeneration both typically present with a personality change, which dominates the clinical picture for a long time before cognitive deficits appear; as noted earlier, although a personality change may occur in Alzheimer's disease, this typically has an onset only after cognitive deficits are well established.

Tumors of the frontal lobes, corpus callosum, or diencephalon may be considered, but are immediately revealed by imaging.

Cerebral amyloid angiopathy, although typically presenting with lobar intracerebral hemorrhage, may, at times, present with a gradually progressive dementia due to miliary microbleeds. The differential in such cases may depend on performing MRI with gradient recall imaging to detect these lesions.

Vitamin B12 and folate deficiency may be clinically indistinguishable from Alzheimer's disease, and, although these are uncommon, given their eminent treatability, they should always be tested for.

Major depressive disorder in the elderly may manifest with prominent cognitive deficits and thus enters the differential. Clinical features suggestive of this include the following: a history of typical depressive symptoms predating the onset of cognitive deficits; a history of prior episodes of depression; and, on mental status examination, a lack of effort on the patient's part in attempting cognitive tasks. As noted earlier, Alzheimer's disease itself may cause depression; however, in contrast to the dementia syndrome of depression, the depression seen in Alzheimer's disease occurs not before the onset of cognitive failure but only well after cognitive deficits have been thoroughly established. In cases in which a reliable history is lacking, it is appropriate to attempt a 'diagnosis by treatment response' and to prescribe an antidepressant and then observe the course of both the depressive and cognitive symptoms.

In pursuing this differential, it must also be kept in mind that many cases of dementia are of the etiologically 'mixed' type, and indeed one recent study found that among patients with autopsy-confirmed Alzheimer's disease, close to one-half also had either large territorial infarctions, dementia with Lewy bodies, or both (Schneider *et al.* 2009a). This, of course, given how common these disorders are, is not an unexpected finding; it does, however, emphasize the importance of keeping an open diagnostic mind.

Before leaving this discussion of differential diagnosis, some comments are in order regarding the place of imaging and cerebrospinal fluid studies in the workup. Although there is no debate that imaging, preferably MRI, is appropriate in virtually all cases, questions arise regarding the need for SPECT or FDG-PET scanning: although these tests do, as noted earlier, show distinctive findings in Alzheimer's disease, it is not clear how much they add to the differential workup over and above a thorough history and examination supplemented by MRI. Indeed, in the case of SPECT scanning, little, if anything, appears to be added (McNeill *et al.* 2007). Consequently, these modalities might best be held in reserve for particularly difficult cases. The same concerns hold true regarding CSF studies for tau or β-amyloid (42).

Treatment

In addition to the routine environmental measures discussed in Chapter 6.3, certain medications may offer some symptomatic relief.

Agents capable of causing modest cognitive improvement include the acetylcholinesterase inhibitors donepezil (Feldman *et al.* 2001; Greenberg *et al.* 2000; Rogers and Friedhoff 1996; Rogers *et al.* 1998), rivastigmine (Rosler *et al.* 1999), and galantamine (Rockwood *et al.* 2001). As noted above, there is a correlation between the loss of cholinergic neurons in the nucleus basalis of Meynert and memory loss, and it is probably by partly restoring cholinergic tone that the acetylcholinesterase inhibitors exert their therapeutic effect. Another agent to consider is the *N*-methyl-D-aspartic acid (NMDA) antagonist memantine (Cummings *et al.* 2006), which may be used either alone or, more commonly, as an 'add-on' drug to one of the cholinesterase inhibitors, such as donepezil (Tariot *et al.* 2004). Importantly, with treatment with a cholinesterase inhibitor or memantine, one sees not only some cognitive improvement but also some improvement in other clinical features, such as mood changes and, in some cases, delusions or hallucinations. Another agent to consider is *Ginkgo biloba*: some (Le Bars *et al.* 1997; Mazza *et al.* 2006; Oken *et al.* 1998), but not all (Schneider *et al.* 2005a), double-blinded studies have demonstrated cognitive improvement with this agent; however, the fact that standardized preparations are not available plus the multiple drug–drug interactions seen with *Ginkgo* should give one pause.

Other symptoms that may respond to pharmacologic treatment include depression, insomnia, apathy, agitation, and delusions and hallucinations. With regard to depression, despite its great importance in this disease, there have been only a handful of double-blinded studies (Thompson *et al.* 2007), and these involved the tricyclics imipramine and clomipramine and the selective serotonin reuptake inhibitors (SSRIs) fluoxetine and sertraline. Although benefits were seen with both clomipramine and sertraline, the anticholinergic effects of clomipramine make sertraline the logical choice: unfortunately, however, in a long-term extension study of sertraline, no benefits were seen over treatment with placebo (Weintraub *et al.* 2010a). Despite these discouraging results, however, it is probably reasonable to at least attempt treatment with an SSRI such as sertraline. Although the insomnia seen in Alzheimer's disease has not been subjected to double-blinded treatment studies, a reasonable approach would be to try melatonin, and, should that fail, zolpidem. Apathy does respond to methylphenidate (Hermann *et al.* 2008), but, in roughly one-quarter of patients, at the cost of significant side-effects (e.g. agitation or delusions), and hence this agent should probably be reserved for severe cases and then used only with great caution. Agitation may respond to carbamazepine (Olin *et al.* 2001; Tariot *et al.* 1998); however, the side-effect burden associated with this may give one pause, and consequently consideration may be given to citalopram (Pollock *et al.* 2002); importantly, valproic acid is not effective (Hermann *et al.* 2007). Delusions and hallucinations invite treatment with antipsychotics, and antipsychotics are also often prescribed for agitation. The most important study of these agents was the monumental 'CATIE-AD' study (Schneider *et al.* 2006). This study demonstrated an advantage of both risperidone and olanzapine over placebo for delusions and agitation, but at the cost of significant side-effects; notably, as between risperidone and olanzapine, risperidone was better tolerated. These findings, in combination with the evidence for a small, but definite, increased risk of death with antipsychotic treatment (Schneider *et al.* 2005b), make it clear that these agents should be reserved for severe cases where symptoms fail to respond to other measures; should an antipsychotic eventually be indicated, then risperidone, in low doses, is probably the best choice.

Before leaving this discussion of the treatment of Alzheimer's disease, some comments are in order regarding the putative effectiveness of some agents in preventing this disease. Epidemiologic studies suggested that chronic use of either non-steroidal anti-inflammatory drugs (Aisen and Davis 1994; in t'Veld *et al.* 2001; MacKenzie and Munoz 1998; McGeer *et al.* 1996; Stewart *et al.* 1997) or estrogens plus progestin (Scooter *et al.* 1999; Tang *et al.* 1996; Zandi *et al.* 2002) were associated with a decreased risk of developing the disease, raising great hopes. Unfortunately, however, a double-blinded study comparing either celecoxib or naproxen with placebo showed no benefit (ADAPT Research Group 2008), and, in the case of estrogens and progestin, their use by postmenopausal women, rather than protecting them, actually increased the risk of developing dementia (Shumaker *et al.* 2003).

9.2 PICK'S DISEASE

Pick's disease, first described by the neuropsychiatrist Arnold Pick in 1892 (Pick 1892), is a rare cause of dementia which typically presents with a personality change.

Before proceeding with this discussion, some words are in order regarding nosology, specifically the relationship between Pick's disease and frontotemporal lobar degeneration. Pick's disease, as noted, has been known for over a century, and its characteristic presentation with a personality change is clearly linked to its distinctive neuropathology, wherein atrophy of the frontal, and, most particularly, the temporal lobes, is associated with ballooned neurons ('Pick cells') and intraneuronal cytoplasmic inclusions ('Pick bodies') composed primarily of tau protein. Until the late twentieth century, it was believed that most early-onset dementias presenting with a personality change were cases of Pick's disease. Subsequently, however, it was found that most such cases, in fact, did not have the neuropathologic findings just described, but rather lacked Pick cells, and although they did have intracytoplasmic inclusions, these were different from Pick bodies: these cases have subsequently been grouped under the rubric of frontotemporal lobar degeneration. Many, if not most, authors then moved to subsume Pick's disease under this larger rubric of frontotemporal lobar degeneration, in effect making frontotemporal lobar degeneration a large 'umbrella' syndrome. As attractive as such 'lumping' might be, from a nosologic standpoint it may be premature. As noted in the chapter on frontotemporal lobar degeneration, there are several different neuropathologic subtypes, and it may well be that in the future, if these subtypes can be associated with distinctive clinical features, that the condition now known as frontotemporal lobar degeneration may be split into several more specific disorders. Should this prove to be the case, then 'lumping' Pick's disease into the umbrella group now may simply mean that we have to 'split' it off in the future. For the time being, then, it may be prudent to continue to treat Pick's disease in isolation, and await future developments.

Clinical features

The onset is gradual and insidious, and although most cases present in the fourth through the seventh decades, onsets as young as 21 (Lowenberg *et al.* 1939) or 25 (Coleman *et al.* 2002) years of age have been noted. In most cases, the presentation is with a personality change, typically of the frontal lobe type, with disinhibition, coarsening of behavior, and perseveration (Bouton 1940; Ferraro and Jervis 1936; Litvan *et al.* 1997a; Mendez *et al.* 1993; Munoz *et al.* 1993; Munoz-Garcia and Ludwin 1984; Nichols and Weigner 1938); elements of the Kluver–Bucy syndrome, such as hyperorality and hypersexuality, are also common (Cummings and Duchen 1981; Mendez *et al.* 1993; Munoz *et al.* 1993; Munoz-Garcia and Ludwin 1984). Some patients may also be prone to restless wandering; however, in contrast to patients with Alzheimer's disease, who tend to wander off and get lost, patients with Pick's disease tend to get back to their starting point (Mendez *et al.* 1993; Munoz *et al.* 1993). Eventually, cognitive deficits appear, such as short-term memory loss, concreteness, and difficulty with calculations; in many cases an aphasia, typically of the expressive type, will supervene, and, in a small minority, seizures may occur.

Rarely, Pick's disease may present with a slowly progressive aphasia (Graff-Radford *et al.* 1990; Kertesz *et al.* 1994; Knibb *et al.* 2006; Wechsler *et al.* 1982), amnesia (Wisniewski *et al.* 1972), or apraxia (Fukui *et al.* 1996).

Computed tomography or MRI reveals lobar atrophy (Knopman *et al.* 1989; Wechsler *et al.* 1982), typically affecting the frontal and anterior temporal lobes; in some cases the atrophy is so severe as to present a 'knife-blade' appearance.

Course

Pick's disease is relentlessly progressive, leading to a profound dementia with death within 5–10 years (Robertson *et al.* 1958), generally from an intercurrent pneumonia.

Etiology

Pick's disease is an example of 'lobar' atrophy, with the frontal and temporal lobes bearing the brunt of the disease process. Interestingly, even when the temporal lobes are very hard hit, the posterior two-thirds of the superior temporal gyrus is generally spared, often to a remarkable degree, as illustrated in Figure 9.3. Although in most cases both the frontal and the temporal lobes are clearly affected, in a minority only one lobe will be macroscopically abnormal (Sjorgen *et al.* 1952). Microscopically one sees widespread neuronal loss and gliosis. Surviving neurons may become ballooned to create the classic 'Pick cells'; within surviving neurons, 'Pick bodies' (Murayama *et al.* 1990; Rewcastle and Ball 1968; Zhukareva *et al.* 2002) may also be found: these are argentophilic intracytoplasmic inclusions containing straight or twisted filaments composed of 3R or 4R tau isoforms.

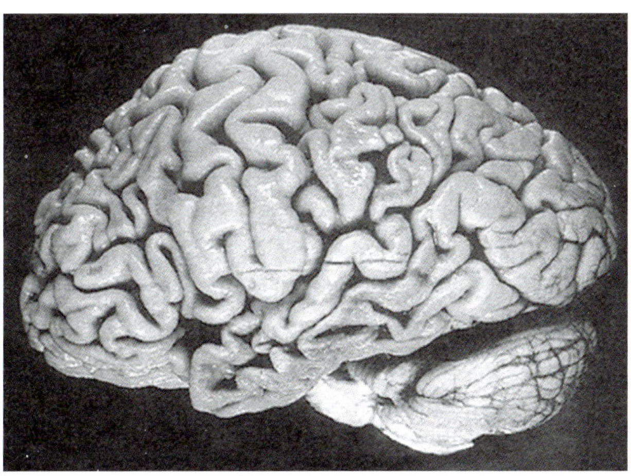

Figure 9.3 Sparing of the posterior two-thirds of the superior temporal gyrus relative to the frontal lobe and the rest of the temporal lobe in a case of Pick's disease. Reproduced with kind permission from Graham and Lantos (1996).

Although most cases of Pick's disease are sporadic (Malamud and Waggoner 1943), hereditary cases do occur (Groen and Endtz 1982; Heston *et al.* 1987) and are usually consistent with dominant inheritance (Sjorgen *et al.* 1952). Although most of these familial cases are associated with mutations in the tau gene (Bronner *et al.* 2005; Neumann *et al.* 2001), a case has also been reported secondary to a mutation in presenilin-1 (Dermaut *et al.* 2004).

Differential diagnosis

Early on in the course of the disease, the differential diagnosis includes other disorders capable of causing a personality change, as discussed in Chapter 8.2, and of these disorders the chief one on the differential is frontotemporal lobar degeneration. At present, unfortunately, there is no reliable way to distinguish Pick's disease from those cases of frontotemporal lobar degeneration which present with a frontotemporal dementia, and only large clinicopathologic studies will eventually answer the question as to whether such a differentiation is clinically possible.

Treatment

The general treatment of dementia is discussed in Chapter 6.3, of personality change in Chapter 8.2, and of the Kluver–Bucy syndrome in Chapter 5.12. There is no specific treatment for Pick's disease itself.

9.3 FRONTOTEMPORAL LOBAR DEGENERATION

Frontotemporal lobar degeneration is a relatively recently described disorder that has been shown to account for the majority of cases of dementia with onset before the age of 65 years. It is a clinically heterogeneous disorder that classically may present in one of three ways (Neary *et al.* 1998). The most common presentation is with frontotemporal dementia, a syndrome marked primarily by a personality change. The next most common is with semantic dementia, characterized primarily by a loss of word meaning. Finally, the least of the most common presentations is with a progressive non-fluent aphasia. Importantly, regardless of which presentation occurs, over time there is a clinical convergence, and the vast majority of patients will eventually display elements of all three presentations.

In addition to being clinically heterogeneous, frontotemporal lobar degeneration is also pathologically heterogeneous. As noted below, although all cases show degenerative changes in the frontal and temporal cortices, and all demonstrate inclusions in surviving neurons, the composition of these inclusions varies among patients, with some patients having tau-positive inclusions while others have tau-negative but ubiquitin-positive inclusions.

Clinical features

The onset is gradual, even insidious, and although the range in age of onset is wide, from the fourth through the eighth decade, the vast majority of patients fall ill in their late fifties or early sixties.

As just noted, the most common presentation is with a frontotemporal dementia that is marked by a personality change. This personality change (Chan *et al.* 2009; Edwards-Lee *et al.* 1997; Josephs *et al.* 2009; Piguet *et al.* 2009) in turn may be roughly divided into two variants, namely a frontal variant and a temporal variant: it must be kept in mind, however, that this division is at best a rough one and that although most patients will have predominantly one variant or the other, that admixtures are the rule (Liu *et al.* 2004). The frontal variant presents with a more or less typical frontal lobe syndrome, which may include disinhibition, coarsening of behavior, lack of empathy, a decline in personal hygiene, or mood changes, which in turn may include apathy, emotional blunting, depression, irritability, impulsivity, or elation. Of note, the environmental dependency syndrome may also occur, and indeed in one study was found in 80 percent of cases (Ghosh and Dutt 2010). The temporal variant is characterized primarily by certain peculiar behavioral changes. There may be stereotyped, almost ritualized behaviors, including counting, hand clapping, or more complex activities, such as collecting and hoarding useless items; a certain behavioral rigidity may also occur, and patients may insist on following fixed routines or taking the same route whenever traveling. A certain 'mental' rigidity may also occur, and patients may demonstrate fixed and unalterable ideas and opinions. Food preferences may also change, typically toward sweets, and some patients may gain considerable weight. Other features of the temporal variant, seen in a minority, include the Kluver–Bucy syndrome, hyper-religiosity and either hallucinations or delusions. Regardless of which variant occurs, a lack of insight is typical, and patients fail to recognize that anything is amiss or that their behavior might be offensive or troubling to others.

Semantic dementia (Hodges *et al.* 1992; Kertesz *et al.* 2010) represents a kind of aphasia characterized, as noted earlier, by what is referred to as a loss of word meaning. Here, patients no longer know what certain common words mean: it is as if the definition were stripped from the word or as if the familiar word had become part of a foreign language which the patient did not know. For example, in one case the meaning of the word 'pen' had been lost. The patient, as is often the case, during a conversation in which the physician used the word pen, interrupted and asked 'What is a pen?' Later, when the physician held up a pen and asked the patient to name the object, the patient was unable to do so. Importantly in such cases, speech is quite fluent, with no hesitations; furthermore, although patients tend to be garrulous and are prone to using stock phrases, speech remains coherent and grammatically correct, and patients, in general, have no trouble following commands, provided, of course, that the commands do not contain words the meanings of which the patient has lost.

Progressive, non-fluent aphasia represents essentially a Broca's, or motor, aphasia, and speech becomes effortful, telegraphic, and circumlocutory: in contrast with patients with semantic dementia, patients with progressive non-fluent aphasia do not ask what a word means, and although they may not be able to come up with the name of a common object held in front of them, they can say what the object is used for.

Other presentations may also occur, including prosopagnosia (Chan *et al.* 2009), topographagnosia (Chan *et al.* 2009), and an amnestic syndrome (Chan *et al.* 2009; Graham *et al.* 2005; Piguet *et al.* 2009); however, these are far less common than the classic ones just described. Importantly, these features typically also eventually appear in the context of one of the classic presentations.

As stressed earlier, over time there is a clinical convergence (Josephs *et al.* 2009; Kertesz *et al.* 2010), and, in the vast majority of cases, if patients are followed for long enough, the clinical picture contains a mixture of features, with personality change, aphasia, agnosia, and amnesia. Furthermore, in a minority of cases, either parkinsonism or a syndrome identical to that seen in amyotrophic lateral sclerosis may occur.

Magnetic resonance imaging typically reveals atrophy of either the temporal or frontal cortices, or both. Cases presenting with semantic dementia or progressive non-fluent aphasia are likely to show a predominant left temporal atrophy (Edwards-Lee *et al.* 1997), while cases presenting with frontotemporal dementia are prone to demonstrate predominantly either right temporal (Edwards-Lee *et al.* 1997) or bifrontal atrophy.

Although routine CSF studies are normal in frontotemporal lobar degeneration, total tau levels are increased and β-amyloid (42) levels are decreased, in a pattern similar to that seen in Alzheimer's disease. Notably, however, the ratio of total tau/β-amyloid-42 is lower in frontotemporal lobar degeneration than it is in Alzheimer's disease (Bian *et al.* 2008).

Course

This is a progressive disorder, and, in contrast to most other neurodegenerative disorders, is fairly rapidly so, with death occurring, on average, within 7 years; the course is even more rapid in that minority of cases marked by amyotrophic lateral sclerosis, with a fatal outcome in 3 or 4 years.

Etiology

Neuropathologically (Seelar *et al.* 2008), neuronal loss and gliosis are found in the frontal and temporal cortices, and, in a minority, the striatum, substantia nigra, or the spinal cord. In surviving neurons, intracytoplasmic inclusions are found and individual cases are characterized by one of two types of inclusions (Garcin *et al.* 2009; Hornberger *et al.* 2010; Seelar *et al.* 2008). The first type of inclusion is tau-positive; the second type is tau-negative but ubiquitin-positive. Looking further at cases with tau-negative but ubiquitin-positive inclusions, in the vast majority of cases these inclusions are found to contain the TAR DNA-binding protein-43 (TDP-43) (Garcin *et al.* 2009); in the remainder, the composition of the inclusion is either unknown or characterized by the fused in sarcoma (FUS) protein (Neumann *et al.* 2009).

Roughly one-half of all cases of frontotemporal lobar degeneration occur on a familial basis, and in roughly one-half of these this appears to be on an autosomal basis (Rohrer *et al.* 2009; Seelar *et al.* 2008). Although the genetic basis of these familial cases remains largely unknown, in a minority of cases specific mutations have been discovered. Thus, in a small minority of cases characterized by tau-positive inclusions, mutations in the gene for the microtubule-associated tau protein, *MAPT*, have been found (Seelar *et al.* 2008). Further, in cases characterized by tau-negative but ubiquitin-positive inclusions containing TDP-43, a likewise small minority are found to have mutations in the gene for progranulin, *PGN* (Rohrer *et al.* 2009; Seelar *et al.* 2008); rarely in such cases, mutations may be found in the gene for valosin-containing protein, *VCP*, or the gene for the TDR-43 protein itself, *TARDP* (Benajiba *et al.* 2009). Finally in cases characterized by tau-negative but ubiquitin-positive inclusions that do not contain TDP-43, rarely one may find mutations in the gene for chromatin-modifying protein 2B, *CHMP2B*.

Although efforts are under way to determine whether there is an association between, on the one hand, the various presentations of frontotemporal lobar degeneration and, on the other, the type of inclusion present (e.g. a possible association between semantic dementia and tau-negative, ubiquitin-positive inclusions; Kertesz *et al.* 2010), or the type of mutation present (e.g. a possible association between frontotemporal dementia and *PGN* mutations [Beck *et al.* 2008]), it currently does not seem possible to reliably predict, based on the clinical presentation, what type of microscopic pathology or mutation will be found.

Differential diagnosis

When frontotemporal lobar degeneration presents with a frontotemporal dementia, the differential includes other disorders capable of producing a gradual personality change, as discussed in Chapter 8.2, including other neurodegenerative disorders, such as Pick's disease or Alzheimer's disease, or various slowly growing tumors. At this juncture, a word is in order regarding Pick's disease. Many, if not most, authors now subsume Pick's disease into frontotemporal lobar degeneration, and consider it as but one of its causes; however, such a change may be premature from a nosologic point of view. Pick's disease, first described over a century ago, is a well-described disorder, offering a good correlation between its clinical presentation and its underlying pathology. As noted earlier, at present such a correlation has not been established between the various clinical presentations of frontotemporal lobar degeneration and the various neuropathologic underpinnings of this disorder; however, it may well be in the future that such correlations are established, and, if they are, eventually the disorder now known as frontotemporal lobar degeneration may be split into various more specific disorders, each with their own name, such as, perhaps, 'progranulin dementia'. Given the possibility of such a 'splitting' in the future, it may not, again from a nosologic point of view, at present be appropriate to 'lump' Pick's disease into frontotemporal lobar degeneration – one might have to simply 'split' it out in the future. As always in these uncertain areas, the debate between the 'lumpers' and the 'splitters' will go on until large neuropathologic studies decide the issue.

Another item on the differential for frontotemporal dementia is a recently described condition known as the 'benign' or 'phenocopy' variant of frontotemporal dementia. In this condition, patients present with a gradually progressive personality change symptomatically similar to that described above, which, however, and in contrast with the course described above, shows very little progression over long-term follow-up. Furthermore, these patients fail to display the 'convergence' of symptoms described above (Garcin *et al.* 2009; Hornberger *et al.* 2010) and develop neither any significant aphasia nor any amnesia or other cognitive deficits; they likewise do not develop parkinsonism or amyotrophic lateral sclerosis. Finally, in this condition there is little if any frontal or temporal atrophy on MRI scans (Davies *et al.* 2006; Garcin *et al.* 2009; Kipps *et al.* 2009). It is as yet unclear whether this benign variant occurs secondary to frontotemporal lobar degeneration or to some other disease process.

In cases where frontotemporal lobar degeneration presents with a progressive and relatively pure aphasic disturbance, one is, in essence, confronted with a case of the syndrome of primary progressive aphasia, and the differential here is as discussed for that syndrome in Chapter 3.1: essentially, it includes the same disorders capable of causing a gradual personality change.

In those rare instances where frontotemporal lobar degeneration presents with an amnestic syndrome, the differential is as discussed in Chapter 6.6, and the main item on the differential is Alzheimer's disease. As noted above, in most cases of frontotemporal lobar degeneration, one sees a clinical convergence, and although personality change and aphasia may be seen in Alzheimer's disease, these clinical features are much more likely to present early on in frontotemporal lobar degeneration, often within the first year or two, far earlier than one typically sees in Alzheimer's disease.

Treatment

The pharmacologic treatment of frontotemporal lobar degeneration is still in its infancy.

Regarding frontotemporal dementia, one double-blinded study compared dextroamphetamine, quetiapine, and placebo and found overall improvement with dextroamphetamine (including, surprisingly, disinhibition) but no advantage for quetiapine over placebo (Huey *et al.* 2008). Another double-blinded study (Lebert *et al.* 2004) demonstrated that trazodone ameliorated irritability, depression, and abnormal food preferences. There have been three open studies utilizing SSRIs: one (Swartz *et al.* 1997), utilizing fluoxetine, sertraline, or paroxetine, found some improvement in depression, a preference for sweets, and repetitive behaviors; another (Mendez *et al.* 2005) found that sertraline reduced repetitive behaviors; the third (Deakin *et al.* 2004), which utilized paroxetine, found not only no behavioral benefit but a cognitive worsening. Cholinesterase inhibitors have also been studied: galantamine, in a double-blinded study, was ineffective for behavioral changes (Kertesz *et al.* 2008); both rivastigmine and donepezil have been studied in an open fashion, and while rivastigmine was reported to cause behavioral improvement (Moretti *et al.* 2004), donepezil was noted to cause behavioral worsening (Mendez *et al.* 2007). Finally, memantine, in an open study (Boxer *et al.* 2009), produced transient behavioral improvement, but was associated with some cognitive decline.

Regarding treatment of aphasic disturbances, in one double-blinded study, galantamine showed a statistically non-significant trend toward improvement (Kertesz *et al.* 2008).

9.4 AMYOTROPHIC LATERAL SCLEROSIS

Amyotrophic lateral sclerosis (ALS), first clearly delineated by the French neurologist Jean-Martin Charcot in 1874, is classically characterized by a combination of upper and lower motor neuron signs reflecting involvement of, on the one hand, the corticospinal system including the pyramidal cells of the precentral gyrus, the corticospinal tract and the motor neurons of the anterior horn of the spinal cord, and, on the other, the corticobulbar system, including pyramidal cells of the precentral gyrus, the corticobulbar tracts, and brainstem nuclei, including especially the nucleus ambiguus and the hypoglossal nucleus. In addition to this classic neuropathology, it has recently become quite apparent that pathology is also found, in a substantial minority, in other areas of the cortex, including the prefrontal and temporal cortices, and that in this subgroup of patients both cognitive defects and a personality change may occur.

Amyotrophic lateral sclerosis has a lifetime prevalence of roughly 4–6 per 100 000 and is more common in men than in women by a ratio of approximately 1.5:1. In Europe, ALS is often referred to as Charcot's disease or motor neuron disease, and in the United States it is, at times, colloquially referred to as Lou Gehrig's disease, after the famous baseball player who, in 1939, was forced to retire as the disease took its toll.

Clinical features

The onset is gradual, and most patients fall ill between the ages of 40 and 70 years. Classically, patients present with signs and symptoms reflecting involvement of the corticospinal system, typically with weakness in one of the upper extremities, often in the hand, leading to difficulty buttoning clothes or using small tools; in some cases, however, the presentation may involve one of the lower extremities, and patients may begin to trip due to foot drop. Regardless, however, of whether the upper or the lower extremity is involved first, eventually all four extremities become affected. As noted earlier, amyotrophic lateral sclerosis is characterized by involvement of both upper and lower motor neurons, and this accounts for the distinctive combination of both upper and lower motor neuron signs. Thus, one finds hyperactive deep tendon reflexes and Babinski signs coexisting with muscular atrophy and fasciculations.

Subsequent to involvement of the limbs, evidence of corticobulbar involvement appears and patients develop the syndrome of pseudobulbar palsy, with dysarthria, dysphagia, and, in a significant minority, emotional incontinence, with forced laughter and crying (Gallagher 1989; Ironside 1956). With the development of these symptoms, one again finds the distinctive combination of upper and lower motor neuron signs, including a brisk jaw jerk accompanied by lingual atrophy and fasciculations. Of note, and as a variation of the typical progression from corticospinal to corticobulbar involvement, in perhaps one-fifth of all cases of ALS the presentation will be with corticobulbar symptomatology; in such cases, however, evidence of involvement of the corticospinal system eventually appears.

Although the vast majority of cases present with a combination of upper and lower motor neuron involvement, variants do occur, and in a small minority of cases the presentation will be heavily weighted toward either the upper motor neuron (producing a syndrome known as primary lateral sclerosis) or the lower motor neuron (producing the syndrome of progressive muscular atrophy). Primary lateral sclerosis (Pringle *et al.* 1992) is characterized clinically by a progressive spastic paresis, accompanied in roughly one-half of patients by a pseudobulbar palsy (Kuipers-Upmeijer *et al.* 2001). Progressive muscular atrophy, in contrast, presents with progressive weakness, atrophy, and fasciculations. In most cases, patients with primary lateral sclerosis or progressive muscular atrophy go on to develop clear clinical evidence of involvement of both upper and motor neurons (e.g. the eventual development of fasciculations in patients with primary lateral sclerosis [La Forestier *et al.* 2001]), thus leaving no doubt as to the correct diagnosis. However, in a small minority of cases, the clinical picture throughout life persistently demonstrates involvement of only the upper motor neuron or only the lower motor neuron, and these cases have raised the question of whether primary lateral sclerosis and progressive muscular atrophy might each be a disease *sui generis* and distinct from ALS. Autopsy studies, however, have indicated subclinical involvement of both the upper and lower motor neurons in these cases, thereby justifying their inclusion under the overall rubric of ALS (Brownell *et al.* 1970; Lawyer and Netsky 1953).

Although classic teaching holds that sensory changes are absent in ALS, it does appear that, late in the course of the disease, and in a minority, mild sensory loss may be demonstrated in the lower extremities.

In the context of the classic presentation of ALS, both cognitive deficits and a personality change may appear. Cognitive defects (Lomen-Hoerth *et al.* 2003; Ringholz *et al.* 2005; Rippon *et al.* 2006), of at least a mild degree, are found in roughly one-half of all cases, and these typically include deficits in executive function, such as planning or organizing tasks, and anomia; although short-term memory loss may also occur, this is a generally milder and later-occurring feature. Although, as noted, these cognitive defects are typically mild, in a very small minority, less than one-tenth of all cases, they are severe enough to constitute a dementia. Personality change may also occur (Rippon *et al.* 2006), typically in the context of these cognitive deficits, and this change is often of the frontal lobe type, with disinhibition and coarsening of behavior. It must be emphasized that these cognitive deficits and the personality change do not constitute variant presentations of ALS; rather, as noted, they occur in the context of the classic symptoms, and only after they are well established and unmistakably present.

Other symptoms seen as the disease progresses include weight loss, fatigue, and depression. The occurrence of depression deserves special note: although classic teaching holds that patients with ALS remain stoic, and even in good spirits, up to the end, it appears that one-tenth or more of patients do develop a full depressive syndrome (Rabkin *et al.* 2005; Wicks *et al.* 2007). Furthermore, the suicide rate in ALS is some six times higher than in the general population (Fang *et al.* 2008).

Electromyography typically reveals evidence of denervation, and nerve condition velocity studies are typically normal.

T2-weighted or fluid-attenuated inversion recovery (FLAIR) MRI may demonstrate a column of increased signal intensity, corresponding to the corticospinal tract: this is often subtle, and best appreciated in the motor strip, posterior limb of the internal capsule, and the crus cerebri. In cases characterized by dementia, cortical atrophy may be seen in the frontal and temporal lobes, and in the temporal lobe, increased signal intensity on T2-weighted or FLAIR imaging may be seen in the medial aspect of the white matter (Mori *et al.* 2007).

Course

Amyotrophic lateral sclerosis is almost invariably progressive, and, barring intubation, death occurs in most cases within 3–4 years, with only one-fifth of patients living longer than 5 years. Death may occur secondary to dysphagia-induced aspiration pneumonia; eventually, with involvement of the intercostal and diaphragmatic musculature, respiratory failure occurs.

Etiology

As stressed, ALS is characterized by involvement of both upper and lower motor neurons in both the corticospinal and corticobulbar systems, and at autopsy (Brownell *et al.* 1970) one finds neuronal loss and gliosis in the precentral gyrus, Wallerian degeneration of the corticospinal and corticobulbar tracts, neuronal loss and gliosis in both the brainstem nuclei and in the anterior horns of the spinal cord, and, finally, atrophy of the ventral spinal roots. Within the prefrontal and temporal cortices, neuronal loss and gliosis may also be found, especially in those cases characterized by cognitive changes (Wilson *et al.* 2001; Yoshida 2004). Within remaining neurons in these various areas, one finds, in sporadic cases, tau-negative, ubiquitin-positive inclusions, which, at least in part, are composed of the TAR DNA binding protein 43 (TDP-43) (Maekawa *et al.* 2009; Sanelli *et al.* 2007).

Ninety percent of all cases of ALS are sporadic, and although the etiology of these cases has not as yet been delineated, excitotoxicity is strongly suspected. Familial cases constitute the remaining 10 percent, and most of these are inherited in an autosomal dominant fashion: of these familial cases, roughly 20 percent are due to mutations in the gene for superoxide dismutase (*SOD1*) on chromosome 21 (Aguirre *et al.* 1999; Rosen *et al.* 1993; Siddique *et al.* 1991), 5 percent to mutations in the fused in sarcoma (*FUS*) gene on chromosome 16 (Blair *et al.* 2010), and another 5 percent to mutations in the gene for TDP-43 on chromosome 1 (Sreedharan *et al.* 2008). This last mutation bears a little more attention. As noted earlier, TDP-43 constitutes at least part of the inclusions seen in sporadic cases of ALS, and hence finding mutations in the gene for TDP-43 is most exciting: it must be stressed, however, that these mutations account for less than 1 percent of all cases of ALS, and therefore that the mechanism underlying the appearance of the TDP-43 containing inclusions in sporadic ALS must be sought elsewhere.

Differential diagnosis

Diseases capable of causing diagnostic confusion are best grouped according to whether they mimic classic ALS, primary lateral sclerosis, progressive muscular atrophy, or the dementia seen in ALS.

Classic ALS, with evidence of both upper and lower motor neuron damage, may be mimicked by cervical spondylosis and syringomyelia; rarely, a similar picture may occur on a paraneoplastic basis or secondary to lymphoma (Younger *et al.* 1991). Consideration must also be given to the very rare syndrome of benign fasciculations. In cases that present with bulbar symptomatology, consideration should be given to other causes of pseudobulbar palsy, as discussed in Chapter 5.7, and to myasthenia gravis: in myasthenia gravis, one finds ptosis and ophthalmoparesis, signs which are absent in ALS. Finally, there is a rare disorder known as amyotrophic lateral sclerosis–parkinsonism dementia complex of Guam, which is found only in Guam, parts of Japan, and other Pacific islands (Garruto *et al.* 1981; Hirano *et al.* 1967; Malamud *et al.* 1961): as the name suggests, in addition to ALS symptomatology, including dementia, one also finds a prominent parkinsonian syndrome.

Primary lateral sclerosis may be mimicked by multiple sclerosis, adrenoleukodystrophy, hereditary spastic paraplegia, and vitamin B12 deficiency.

Progressive muscular atrophy is very closely mimicked by two rare, recessively inherited disorders, namely spinal muscular atrophy type III (or the Kugelberg–Welander syndrome [Kugelberg and Welander 1956]), which presents in adolescence or early adulthood with proximal muscle weakness, and X-linked spinal muscular atrophy (or Kennedy's disease [Harding *et al.* 1982; Kennedy *et al.* 1968]), which presents in the middle years with progressive weakness and fasciculations, often accompanied by a degree of gynecomastia. Consideration may also be given to various peripheral motor neuropathies; however, these are immediately identified on nerve conduction velocity studies.

Finally, consideration must be given to the vexing question of distinguishing cases of ALS with dementia from the condition known as frontotemporal lobar degeneration. As noted in Chapter 9.3, frontotemporal lobar degeneration typically presents with a dementia symptomatically similar to that seen in ALS, and, as also noted there, in a minority of patients with frontotemporal lobar degeneration the dementia will be accompanied by upper and lower motor neuron signs identical to those seen in ALS. Cross-sectionally, therefore, it may be impossible to differentiate ALS with dementia from these cases of frontotemporal lobar degeneration. All is not lost, however, from the differential diagnostic point of view, if only one attends to the course of the illness: in ALS with dementia, the dementia, as noted earlier, only occurs after the upper and lower motor neuron signs are well established; by contrast, in frontotemporal lobar degeneration, the dementia occurs first, with the upper and lower motor neuron signs following. Although a differentiation may thus be made, things get even curiouser from an etiologic perspective, as rare kindreds (Hosler *et al.* 2000; Vance *et al.* 2006) have now been identified wherein the same mutation may, in some members, cause ALS, in others, frontotemporal lobar degeneration without upper and lower motor neuron signs, and in yet others, a combination of ALS and dementia. Although such mutations are very rare, and thus barely influence the differential task, they may offer tantalizing insights into the pathogenesis of both ALS and frontotemporal lobar degeneration.

Treatment

Meticulous general medical care goes far to ensure an optimum quality of life as the disease progresses. Physical therapy is helpful for spasticity and, when this is severe, a trial of tizanidine or baclofen may be considered. Emotional incontinence, as discussed in Chapter 5.7, may respond to a combination of 30 mg of dextromethorphan and 30 mg of quinidine, given twice daily; consideration may also be given to a trial of citalopram, nortriptyline, or amitriptyline. Fatigue may be treated with modafinil (Rabkin *et al.* 2009), and although there are no double-blinded studies of the treatment of depression in ALS, citalopram, nortriptyline, or amitriptyline are obvious choices: among these, citalopram is the best tolerated, but nortriptyline or amitripyline may appear preferable when insomnia is problematic, or when an anticholinergic effect may be desired to counteract the drooling seen with bulbar involvement; as between nortriptyline and amitriptyline, nortriptyline is generally better tolerated. As respiratory insufficiency supervenes, bilevel positive airway pressure (BIPAP) is indicated. At some point the question of a feeding tube must be considered, and in most cases this is appropriate. Eventually one must also discuss tracheostomy, which is perhaps the most difficult decision of all. Tracheostomy clearly prolongs life but, as the disease progresses, patients may become completely 'locked in'. For most patients this amounts to a 'living death' and, rather than risk this, they forgo tracheostomy. Eventually, hospice care is appropriate.

Riluzole (Bensimon *et al.* 2002) may also be considered. This agent both inhibits presynaptic release of glutamate and blocks postsynaptic glutamate receptors, and has been shown to retard the progression of the disease. At most, however, it generally adds only a few months to the patient's life.

9.5 PARKINSON'S DISEASE

Parkinson's disease, also known as idiopathic parkinsonism or paralysis agitans, although presenting with a classic parkinsonism, is also characterized by the subsequent development of significant 'non-motor' symptomatology, chief of which are dementia and depression. Furthermore, of patients treated with levodopa or direct-acting dopaminergics, a majority will develop significant neuropsychiatric side-effects, most notably visual hallucinations.

This is a common disease, with a prevalence of roughly 0.3 percent in the general population; it is slightly more common in men than in women, by a ratio of roughly 1.5:1.

Clinical features

The onset of Parkinson's disease is gradual and insidious, with symptoms generally first appearing in patients in their mid-fifties; the range, however, is wide, from 20 to 80 years.

Clinically (Hoehn and Yahr 1967; Hughes *et al.* 1993; Martin *et al.* 1973), the cardinal symptoms of the parkinsonism seen in Parkinson's disease are tremor, rigidity, and bradykinesia. In the vast majority of cases, the first symptom to appear is tremor, and almost all patients eventually develop rigidity and bradykinesia. Of note, and importantly from a differential diagnostic point of view, the tremor and rigidity usually present on one side first, and although the other side eventually becomes involved in all cases, this asymmetrical preponderance, although eventually not very marked, generally persists throughout the course of the disease.

Once fully established, Parkinson's disease leaves a stamp on patients that is, once recognized, almost unforgettable. Patients stand in a stooped 'flexion' posture, with their arms and knees in flexion. The tremor is classically a rhythmic rest tremor, at about 3–7 cycles per second (cps), and is most prominent in the hands, where it typically assumes a 'pill-rolling' character. Over time, the tremor may also be present with posture, but it typically resolves with intentional activity. In some cases, the tremor may also appear in the feet, lips, jaw, and tongue, but the head is only rarely involved: although the tremor typically resolves with sleep, in a minority it may persist (April 1966; Stern *et al.* 1968). The frequency of blinking is reduced, and the facies may become flattened or 'masked', such that although patients may be experiencing feelings normally, their expressionless face fails to convey them. Sialorrhea is also present, and there may be copious drooling. Speech is hypophonic, soft, monotonous, and lacking in emotional inflection, and there may be palilalia. Handwriting undergoes a 'micrographic' change, producing scratchy, small letters. Passive extension of the limbs reveals the rigidity, which, although often of the 'cogwheel' type, may at times be 'lead pipe' in character.

While walking there is a reduced arm swing and patients often display 'marche a petit pas', wherein they take small shuffling steps; furthermore, they often display 'festination' in which, as they walk, their steps become ever more rapid and closely spaced, to the point where a catastrophic fall forward seems almost inevitable. Patients also typically display postural instability, and although this generally occurs later in the course of the disease, it can be one of its most disabling aspects, as it predisposes to falls. One way to test for this involves the 'pull test'. In this test, one stands behind the patient and gives a sharp tug on the patient's shoulders: normally, most people regain their balance by taking at most one step backward, while in those with postural instability, it may take several steps to regain balance, and in some cases stability is not regained at all and the patient will fall backward.

Bradykinesia manifests as a slowness in virtually any activity. For example, even in cases where tremor and rigidity are relatively mild, it may take many minutes for the patient to fasten a button or type a few words. A related phenomenon is bradyphrenia, in which thoughts, although coherent and logical, move very slowly, as if stuck in molasses.

Another curious phenomenon is 'freezing': in this, patients on the brink of an intentional act suddenly become 'frozen' and unable to move at all (Giladi *et al.* 1992). For example, a patient standing in a doorway and desirous of walking down the hall may be unable to lift a foot, take a step, or move at all. Amazingly, however, such 'freezing' may be prevented by providing appropriate visual 'cues'. For example, if the hallway is marked off with pieces of tape set about one footstep apart, the patient may well be able to begin and finish the walk down the hall without much difficulty; furthermore, in some cases, patients may be able to lyse their own frozen state by simply *imagining* such cues (Morris *et al.* 1996).

Finally, akathisia may occur in Parkinson's disease (Comella and Goetz 1994) and may appear early in the course of the disease, before any pharmacologic treatment (Lang and Johnson 1987).

Other symptoms include apathy, fatigue, insomnia, various autonomic symptoms, emotional incontinence, and rapid eye movement (REM) sleep behavior disorder. Apathy is fairly common; however, although it may occur in isolation, in the vast majority of cases it is seen as part of dementia or depression (Starkstein *et al.* 2009). Fatigue is common in Parkinson's disease (Hagell and Brundin 2009) and can occur in the absence of depression or any other obvious cause. Insomnia (Gjerstad *et al.* 2007; Tandberg *et al.* 1998), often manifest with frequent awakenings throughout the night, or awakening early in the morning and being unable to get back to sleep, is also common, and although this may be related to tremor, discomfort related to not turning in bed while asleep, or to a need to use the rest room frequently, at times no obvious cause may be found, and it appears to be merely related to the disease itself. Autonomic symptoms, including constipation, orthostatic hypotension, urinary frequency or incontinence, nocturia, and erectile dysfunction are common (Verbaan *et al.* 2007), but importantly from a differential diagnostic point of view, are generally mild early on. Emotional incontinence has been reported in 7 percent (Phuong *et al.* 2009) to 17 percent (Petracca *et al.* 2009) of cases, and curiously seems limited to crying rather than laughing (Petracca *et al.* 2009): importantly, as with all cases of emotional incontinence, if these patients are asked whether they feel sad, they will report that that they do not. REM sleep behavior disorder occurs in roughly one-third of all cases (Gagnon *et al.* 2002), and may either precede or follow the onset of motor symptoms (Gagnon *et al.* 2002; Iranzo *et al.* 2005). Curiously, during the 'enacted dreams' of REM sleep behavior disorder, the motor symptomatology of Parkinson's disease lyses: one video-polysomnographic study (De Cock *et al.* 2007) disclosed 'surprisingly fast, ample, coordinated and symmetrical' movements, '*without* obvious sign of parkinsonism' (italics added).

Of all the non-motor symptoms of Parkinson's disease, dementia and depression are the most debilitating, and dementia is considered first.

Dementia is more common in patients with Parkinson's disease than in age-matched controls (Biggins *et al.* 1992), with overall prevalence figures of 11 percent (Mayeux *et al.* 1988a), 17 percent (Aarsland *et al.* 1996), 18 percent (Tison *et al.* 1995), 19 percent (Marder *et al.* 1995), 29 percent (Marttila and Rinne 1976), and 41 percent (Mayeux *et al.* 1992) being noted. The prevalence of dementia in Parkinson's disease rises with age: Mayeux *et al.* (1992) noted a prevalence of 0 percent for those aged under 50 years and 69 percent for those over 80 years. In a similar vein, over an 8-year follow-up, Aarsland *et al.* (2003) noted that 78 percent became demented. It also appears that dementia is more likely in those with more severe motor symptoms (Kuzis *et al.* 1997; Marder *et al.* 1995), in particular in those with prominent bradykinesia and rigidity (Rajput *et al.* 2009; Selikhova *et al.* 2009). Importantly, the dementia does not appear until after the motor symptoms have been well established: one study found that the onset of dementia occurred at a mean of 13 years after the appearance of motor symptoms, with a range of 6–21 years (Aarsland *et al.* 2005). It is of interest that James Parkinson, who described this disease in 1817 (Ostheimer 1922; Wilkins and Brody 1969b), maintained that dementia did not occur, an opinion that was echoed in textbooks for over one and a half centuries. This mistake of his may be forgiven, based as it was on the observation of a mere six patients, two of whom were encountered only 'casually' and one of whom was 'seen only at a distance'. The dementia itself is, at least initially, characterized primarily by mild short-term memory loss and deficits in executive functioning, with difficulties in thinking abstractly, planning, and organizing the day. Over time, however, memory deficits may become profound and other cognitive deficits may also accrue. In addition to these cognitive deficits, other symptoms, such as apathy, hallucinations, and delusions, may be seen, but in the case of psychotic symptoms these may be more related to treatment with dopaminergic agents than to the dementing process itself, as discussed further below, under Treatment.

Depressive symptoms of sufficient severity and duration to meet DSM-III criteria (American Psychiatric Association 1980) for a full depressive episode have been noted in from 3 percent (Hantz *et al.* 1994) to 5 percent (Tandberg *et al.*

1996), and, in two separate studies (Mayeux *et al.* 1984; Starkstein *et al.* 1992), to up to 21 percent of patients, and among the remainder of patients not meeting such criteria, minor depressive symptoms may be seen in up to 20 percent (Mayeux *et al.* 1986; Starkstein *et al.* 1992). Of importance from a diagnostic point of view, generalized, pervasive anxiety has also been noted in patients with Parkinson's disease; however, in the vast majority of cases (indeed in 92 percent of cases in one study [Menza *et al.* 1993]), the anxiety, rather than existing as an isolated syndrome, was merely part of a wider depressive syndrome. As might be expected, the presence of depression worsens the cognitive deficits in those who are also demented (Kuzis *et al.* 1997; Starkstein *et al.* 1992; Troster *et al.* 1995). Before leaving this discussion of depression in Parkinson's disease, some words are in order regarding the potential for misdiagnosing depression for certain non-depressive symptoms common in Parkinson's disease. Depression, among other symptoms, is characterized by depressed affect and mood, fatigue, insomnia, trouble concentrating, and in some cases psychomotor change, trending either toward psychomotor agitation or psychomotor retardation, and all of these symptoms may be mimicked or duplicated by other symptoms seen in Parkinson's disease. Thus, depressed affect may be closely mimicked by flattened facies. Both fatigue and insomnia are common in Parkinson's disease patients without depression, and trouble concentrating may be mimicked by bradyphrenia. Finally, psychomotor agitation may be mimicked by akathisia and the motoric aspect of psychomotor retardation may be mimicked by bradykinesia. The first step in making the differential is to enquire of patients whether or not they *feel* depressed: patients with depression do experience a feeling of depression, whereas those with flattened facies, although they may *appear* depressed, do not (unless, of course, one happens to have encountered a patient who, in addition to having the 'iron mask' of a flattened face, also happens to have a depression). Clearly, if a patient does not *feel* depressed, then the diagnosis of depression is effectively ruled out. Trouble concentrating may be distinguished from bradyphrenia if only one gives patients enough time to respond to a question: in those with significant trouble concentrating, time alone does not help, but bradyphrenic patients, if given enough time to wade through their 'cognitive molasses', eventually come up with the right answer (presuming, of course, that they are not developing a dementia). Psychomotor agitation may be distinguished from akathisia by attending to the patient's subjective state: in depression with psychomotor agitation, the experience is one of an anxious agitation, while in akathisia it is one of restlessness. Finally, motoric retardation may be distinguished from bradykinesia if, as with the question of bradyphrenia, one again only allows enough time. The motor retardation of depression is accompanied by a profound loss of energy and of interest, and these slowed-down patients, unless actively prodded, will, like a clock that rapidly winds down, eventually simply come to a standstill. In contrast, the patient with bradykinesia, if given enough time, will eventually complete the task.

Course

Untreated, most patients become incapacitated within 8–10 years, with death often from pneumonia; with treatment, however, survival of 15 or more years may be expected.

Etiology

Macroscopically, there is depigmentation of the substantia nigra (as illustrated in Figure 9.4) and the locus ceruleus. Microscopically (Hughes *et al.* 1993), neuronal loss is present not only in these structures but also in the dorsal raphe nucleus, the pedunculopontine nucleus, the dorsal motor nucleus of the vagus, the thalamus, the hypothalamus, the nucleus basalis of Meynert, the amygdala, and in various areas of the cortex, including the temporal, insular, and cingulate cortices. Remaining neurons typically display the hallmark of Parkinson's disease, namely the Lewy body, which, as illustrated in Figure 9.5, is an intracytoplasmic inclusion, composed of α-synuclein and ubiquitin.

Motor symptoms correlate with cell loss in the substantia nigra, and generally do not appear until 60 percent or more of these cells have been lost. Dementia in turn correlates with cell loss and Lewy bodies in the cortex (Aarsland *et al.* 2005; Apaydin *et al.* 2002; Braak *et al.* 2005; Hurtig *et al.* 2000) and the nucleus basalis of Meynert (Rogers *et al.* 1985;

Figure 9.4 Note the depigmentation, especially evident in the substantia nigra, in the case of Parkinson's disease on the left compared with a normal control on the right. Reproduced with kind permission from Stern (1990).

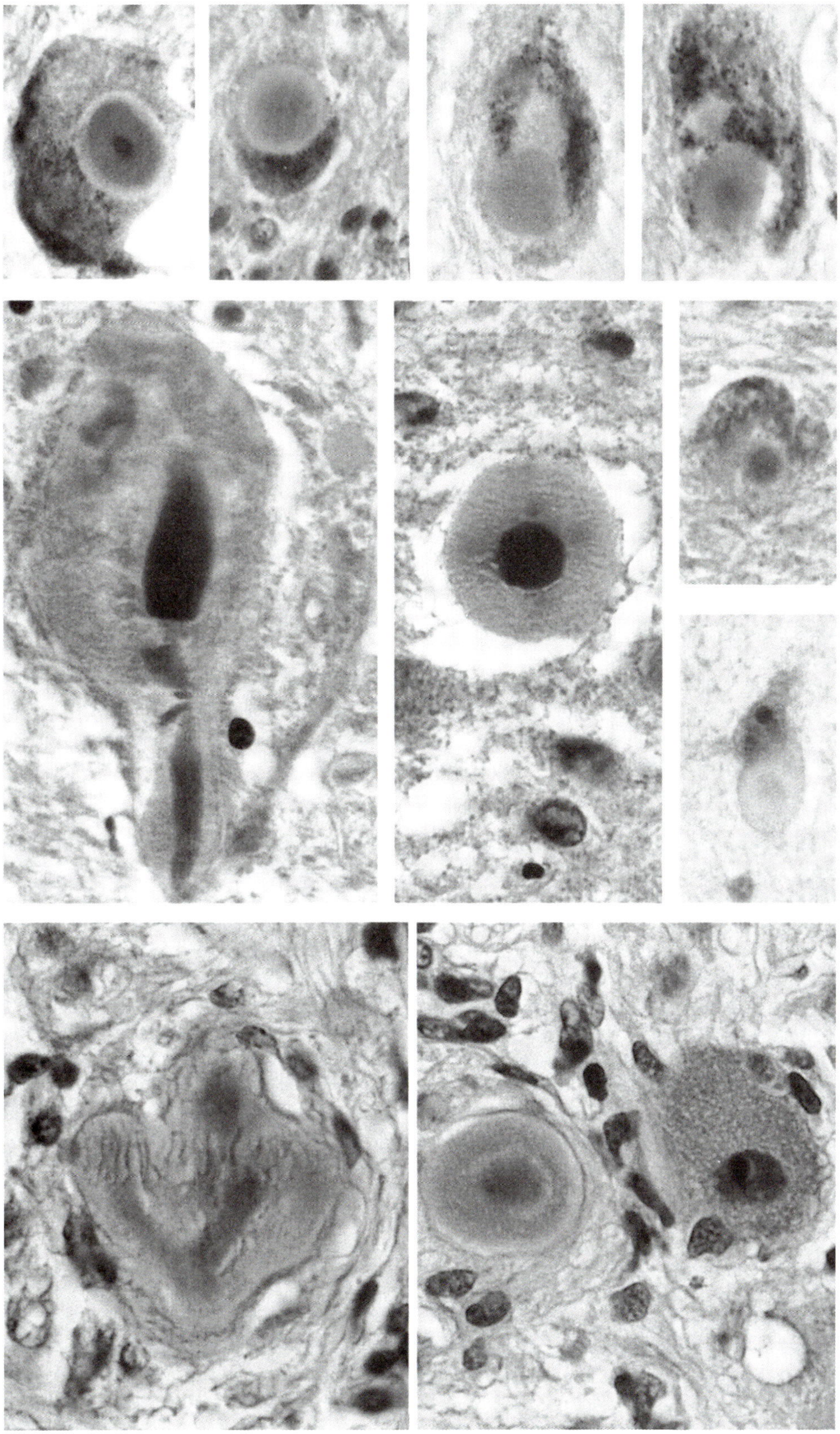

Figure 9.5 Lewy bodies in surviving neurons of the substantia nigra. Reproduced with kind permission from Stern (1990).

Whitehouse *et al.* 1983). With regard to depression, interest has naturally focused on the locus ceruleus as the main source of noradrenergic innervation, and the dorsal raphe nucleus, as the correspondingly main source of serotoninergic innervation, with some interesting results. Two autopsy studies (Chan-Palay and Asan 1989; Frisina *et al.* 2009) found a correlation between depression and cell loss in the locus ceruleus, but CSF studies looking at norepinephrine's metabolite, MHPG, gave contradictory results, with one study finding the expected reduction (Palhagen *et al.* 2009) while the other did not (Kuhn *et al.* 1996). Looking to serotonin, although an autopsy study failed to find a correlation between depression and cell loss in the dorsal raphe nucleus (Frisina *et al.* 2009), CSF studies yielded more encouraging results, with four studies (Kostic *et al.* 1987; Mayeux *et al.* 1984,1986,1988b) finding a reduction in the level of serotonin's metabolite, 5HIAA, while only one did not (Kuhn *et al.* 1996). Although clearly the case has not been proved that the pathology of the locus ceruleus and dorsal raphe nucleus cause depression in Parkinson's disease, these are intriguing results, which beg for further study.

Although the mechanism or mechanisms underlying cell loss and Lewy body formation are not known, it is theorized that, perhaps related to mitochondrial respiratory chain dysfunction, there is an increased formation of free radicals with subsequent cell damage. Genetic factors and environmental factors may both be involved. Although idiopathic Parkinson's disease has long been thought to be sporadic, recent epidemiologic work strongly suggests familial factors (Sveinbjornsdottir *et al.* 2000). Furthermore, fluorodopa PET has demonstrated not only reduced dopamine uptake in the striatum of patients but also a reduced uptake in their clinically unaffected co-twins, suggesting that the co-twins had asymptomatic disease (Piccini *et al.* 1999). Environmental toxins have long been suspected, and attention has focused on exposure to pesticides (Ascherio *et al.* 2006; Tanner *et al.* 2009), especially those containing organochlorines (Elbaz *et al.* 2009). Of interest, it also appears that cigarette smoking (Thacker *et al.* 2007), especially if it has been long term (Chen *et al.* 2010), appears to reduce the risk of developing Parkinson's disease.

Interest in genetic factors has recently been stimulated by the investigation of cases having a clear-cut, unequivocal familial basis (Feany 2004; Klein 2006; Tan and Jankovic 2006). These constitute only a small percentage of cases, no more than 10 percent, and the pattern of inheritance may be either autosomal dominant or autosomal recessive. Table 9.1 lists the various forms that have been discovered, grouping them according to whether they are dominantly or recessively inherited. As noted, these cases may be either early (under the age of 40 years) or late onset, and the parkinsonism may be typical or have atypical aspects. Two of the more common forms are PARK8, due to mutations in *LRRK*, and PARK2, due to mutations in *PRKN*. PARK8 cases may be clinically indistinguishable from idiopathic Parkinson's disease (Gaig *et al.* 2007; Healy *et al.* 2008; Papapetropoulos *et al.* 2006); PARK2 cases may also be indistinguishable or may present with a combination of parkinsonism and foot dystonia (Khan *et al.* 2003; Lucking *et al.* 2000). It is unclear at present how relevant these discoveries are to an understanding of the overwhelming majority of cases that do not exhibit a clear-cut familial basis, as these inherited cases may not have the same pathology as that described above for sporadic cases. For example, although Lewy bodies, the pathologic hallmark of Parkinson's disease, are found in some cases of PARK8 (Papapetropoulos *et al.* 2006) and PARK2 (Farrer *et al.* 2001; Pramstaller *et al.* 2005), they are not found in other cases, both of PARK8 (Funayama *et al.* 2002; Gaig *et al.* 2007) and PARK2 (Mori *et al.* 1998; Takahashi *et al.* 1994).

The place of genetic testing, at present, is uncertain. Where a strong family history exists, a case could be made for testing to allow for genetic counseling; otherwise, such testing is probably best reserved for research settings.

Differential diagnosis

Parkinson's disease constitutes only one of many different causes of parkinsonism, and the full differential for parkinsonism is discussed at length in Chapter 4.8. Of the disorders noted there, several should especially come to mind when considering a diagnosis of Parkinson's disease, including dementia with Lewy bodies, multiple system atrophy (MSA), progressive supranuclear palsy, corticobasal ganglionic degeneration, and vascular parkinsonism.

Dementia with Lewy bodies may present with a classic parkinsonism similar to that seen in Parkinson's disease, however, within a year of onset of the movement disorder; this disease also causes a dementia marked by confusional episodes and visual hallucinations. Although most patients with Parkinson's disease will also develop a dementia, this, as

Table 9.1 Genetic forms related to Parkinson's disease

Inheritance pattern	Chromosome	Locus	Gene	Protein	Onset	Parkinsonism
AD	12p11-q13	PARK8	*LRRK2*	Leucine-rich repeat kinase 2	Late	Typical
AD	4q21	PARK1	*SCNA*	Alpha-synuclein	Early to late	Typical or atypical
AD	4q21	PARK4	*SCNA*	Alpha-synuclein	Early to late	Typical or atypical
AD	4p14	PARK5	*UCH-L1*	Ubiquitin C-terminal hydrolase	Late	Typical
AD	2p13	PARK3	?	?	Late	Typical
AD	1p	PARK11	?	?	Late	Typical
AR	6q25.2–27	PARK2	*PRKN*	Parkin	Early to late	Atypical
AR	1p36	PARK6	*PINK1*	PTEN-induced putative kinase 1	Early to late	Typical
AR	1p36	PARK7	*DJ-1*	DJ-1	Early	Typical
AR	1p36	PARK9	?	?	Early	Atypical
?	1p	PARK10	?	?	Late	Typical

AD, autosomal dominant; AR, autosomal recessive.

noted above, is a late occurrence, appearing only many years after the onset of the parkinsonism, and it is this disparity in time of onset of the dementia which is most helpful in distinguishing these two disorders.

Multiple system atrophy of the striatonigral type presents with parkinsonism that is marked by a symmetric onset, an often extreme flexion posture, and tremor that tends to be jerky and dystonic, rather than rhythmic and pill rolling. Furthermore, one typically finds elements of ataxia and autonomic symptoms, or both: although autonomic symptoms, such as orthostatic hypotension, urinary frequency or incontinence, or erectile dysfunction, can also be seen in Parkinson's disease, in MSA these tend to occur earlier and to be more severe.

Progressive supranuclear palsy may cause parkinsonism, and, as noted in Chapter 9.7, this may occur in one of two types. The 'Richardson's' type is generally fairly easily distinguished from the parkinsonism seen in Parkinson's disease, as it is characterized by a symmetric onset, an extension posture, and by early postural instability with frequent, unexplained falls: postural instability and falls may of course also occur in Parkinson's disease, but this is generally a late event. Finally, in the Richardson's type, one also, within the first few years, may see a supranuclear ophthalmoplegia. The other type, referred to as 'PSP-parkinsonism' may, unfortunately, not be readily distinguished from Parkinson's disease.

Corticobasal ganglionic degeneration, when presenting with the classic 'corticobasal syndrome', like Parkinson's disease, causes a parkinsonism of asymmetric onset. However, here the onset is markedly asymmetric and, furthermore, the rigidity often has a dystonic aspect to it and is typically accompanied by cortical sensory loss and apraxia.

Vascular parkinsonism, occurring in the setting of multiple lacunar infarctions, produces a somewhat atypical parkinsonism in that tremor is often absent and the rigidity and bradykinesia are accompanied by evidence of damage to the corticospinal tracts and corticobulbar tracts, with hyperreflexia, Babinski signs, and pseudobulbar palsy.

One very important clue in the differential of parkinsonism is the presence or absence of 'levodopa responsiveness'. The parkinsonism of Parkinson's disease almost always displays a robust and sustained response to levodopa, while, by contrast, in these other disorders the response is either poor, or, if it is somewhat robust, is not sustained.

Treatment

The treatment of Parkinson's disease begins with its *motor symptoms* (and the side-effects associated with these treatments, such as hallucinations); over time, as noted earlier, *dementia* and *depression* typically appear and require their own treatments. Each of these is discussed in turn, beginning with the motor symptoms.

MOTOR SYMPTOMS

Motor symptoms may be treated with a variety of agents, including monoamine oxidase inhibitors (MAOIs; e.g. selegiline and rasagiline), anticholinergics (e.g. benztropine and trihexyphenidyl), amantadine, levodopa/carbidopa (with or without entacapone or tolcapone), and direct-acting dopamine agonists (pramipexole and ropinirole). Each of these agents will be discussed in turn, followed by a discussion of an overall treatment strategy. Treatment with levodopa or dopamine agonists eventually causes significant neuropsychiatric side-effects (e.g. visual hallucinations) in most patients, and these are discussed further below.

The MAOIs selegiline (5–10mg daily) (Myllyla *et al.* 1997; Palhagen *et al.* 1998) and rasagiline (0.5–1mg daily) (Parkinson Study Group 2002, Group 2005) are both selective against monoamine oxidase type B and, as such, inhibit the intracellular metabolism of levodopa, thus increasing the amount available in the striatum and thereby exerting a modest therapeutic effect. Although controversial, there is also some evidence that selegiline (Palhagen *et al.* 2006; Parkinson Study Group 1993; Olanow *et al.* 1995) may have a 'neuroprotective' effect and may, to a modest degree, slow the progression of the disease; although evidence for the same effect was also found for rasagaline (Parkinson Study Group 2004), a subsequent study cast doubt on this (Olanow *et al.* 2009).

Anticholinergics, although useful for tremor, have a limited effect on bradykinesia and rigidity, and in some patients may cause confusion or a memory deficit (Cooper *et al.* 1992). In light of this, they are generally reserved for cases in which tremor is prominent, with due regard for any emerging cognitive deficits.

Amantadine, in a dose of 200–300mg/day, may have a mild effect on motor symptoms, but the effect may be short-lived, with little benefit seen after a matter of months (Thomas *et al.* 2004a; Timberlake and Vance 1978). Given this limited benefit, and the side-effects seen with amantadine, routine use is probably not justified. There may, however, be a place for amantadine in the treatment of levodopa-induced dyskinesias (Verhagen Metman *et al.* 1998).

Levodopa is the most effective treatment for the motor symptomatology of Parkinson's disease, and is generally given in combination with a peripheral aromatic amino acid decarboxylase inhibitor (carbidopa or benzaride) in order to prolong its effect. The initial response to levodopa/carbidopa is generally quite gratifying, but as levodopa does not retard the progression of the disease, a dosage increase is necessary over time, and most patients eventually begin to show motor fluctuations: such fluctuations can be 'peak dose' with dyskinesias, 'end of dose' with an early wearing off of effectiveness, or unpredictable. When fluctuations appear, using lower and more closely spaced doses of levodopa may help, or one may add either tolcapone (Adler *et al.* 1998; Baas *et al.* 1997; Kurth *et al.* 1997; Rajput *et al.* 1997; Waters *et al.* 1997) or entacapone (Brooks and Sagar 2003). Both of these agents are catechol-O-methyltransferase inhibitors that extend the duration of levodopa's effect and also allow for a lowering of the levodopa dose: as between the two, although there is some suggestive evidence that tolcapone may be more effective (Entacapone to Tolcapone Switch Study Investigators 2007), there have been concerns about its hepatotoxicity, and, although serious toxicity is probably rare, and avoidable with appropriate monitoring (Olanow *et al.* 2000), the use of entacapone is easier, as laboratory testing is not required, and may leave one with a little more peace of mind; furthermore, entacapone is now available in a 'triple combination' preparation containing levodopa, carbidopa, and entacapone. Another option for control of fluctuations is to add rasagaline, which, in one study, produced as much benefit as did the addition of entacapone (Rascol *et al.* 2005).

The direct-acting dopamine agonists pramipexole (Parkinson Study Group 1997; Shannon *et al.* 1997) and ropinirole (Adler *et al.* 1997; Korczyn *et al.* 1999; Rascol *et al.* 2000) may both be used as monotherapy in early, mild Parkinson's disease. Furthermore, in patients already on levodopa who develop fluctuations, both pramipexole (Guttman *et al.* 1997; Lieberman *et al.* 1997; Pinter *et al.* 1999) and ropinirole (Lieberman *et al.* 1998) may be added on to smooth out the effect and to allow for a dose reduction. When dopamine agonists are used, patients should be cautioned that sleep attacks may occur (Avorn *et al.* 2005; Ferreira *et al.* 2000; Frucht *et al.* 1999; Hauser *et al.* 2000), and they should generally avoid driving or using hazardous machinery until, with prolonged use, it has become apparent whether or not this side-effect will occur.

The optimum overall treatment strategy is still a matter of some debate, and although the strategy recommended here is intended as a 'middle of the road' approach, not all authors will be in agreement.

In general, pharmacologic treatment should not be initiated until the motor symptomatology significantly interferes with the patient's ability to function. In mild cases one may begin with one of the MAOIs and, when tremor is prominent, some clinicians advocate the use of an anticholinergic. With disease progression, however, one must at some point add either levodopa or a dopamine agonist, and the question of which to use first has not been settled (Holloway *et al.* 2004; Parkinson Study Group 2009). Given the sometimes stunning effectiveness of levodopa, some advocate using this first, especially when a prompt response is required, for example when a patient is threatened with job loss. Others, however, with an eye toward delaying the emergence of levodopa-associated fluctuations, advocate starting with dopamine agonist monotherapy and adding levodopa only as the disease progresses.

All other things being equal, it may be prudent to begin with a dopamine agonist, increasing the dose as required until limiting side-effects occur or further dose increases are not effective, and then, with disease progression, to add levodopa. Between the regular and sustained-release preparations of levodopa, the sustained-release preparation may be preferable as it may allow for a longer duration of effect; given its slower onset of action, however, some patients may have to take a 'regular' release preparation first thing in the morning in order to 'jump start' themselves (Pahwa *et al.* 1996). When fluctuations in levodopa responsiveness begin to appear, the addition of entacapone may be considered. If one begins with levodopa rather than with dopamine agonist monotherapy, at some point one will generally wish to add a dopamine agonist to smooth out the fluctuations associated with levodopa therapy.

In treating patients with levodopa, it is important to avoid suddenly stopping it as this may, albeit rarely, precipitate a neuroleptic malignant syndrome (Sechi *et al.* 1984; Toru *et al.* 1981), and the same holds true for amantadine (Cunningham *et al.* 1991; Harsch 1987).

When motor symptoms become severe and resistant to pharmacologic treatment, consideration may be given to deep brain stimulation or to electroconvulsive therapy (ECT). Deep brain stimulation may be directed at either the subthalamic nucleus or the pars interna of the globus pallidus, and although both are clearly more effective than optimum medical treatment, they are associated with greater risk. As between the two, although subthalamic stimulation may be more effective (Deep Brain Stimulation for Parkinson's Disease Study Group 2001), this is controversial (Follett *et al.* 2010), and any increased effectiveness may be offset in cases also characterized by depression as subthalamic stimulation worsens depression, whereas pallidal stimulation alleviates it (Follett *et al.* 2010); furthermore, there is a small but definite increased risk of suicide after subthalamic stimulation (Voon *et al.* 2008). In emergent situations, or in cases where neurosurgical intervention is not possible, consideration should be given to ECT. ECT has been shown in a double-blinded study (Andersen *et al.* 1987) to improve motor symptoms regardless of whether patients are depressed or not, and the results are at times dramatic. As such, ECT may be life-saving, and although maintenance ECT may be required to prolong the benefit, this may, in selected cases, be a reasonable price to pay in light of the increased mobility.

Most patients treated with levodopa or dopamine agonists will eventually develop significant neuropsychiatric side-effects. The most common of these are visual hallucinations; others include psychosis, anxiety attacks, and certain other, less common, phenomena, including impulse control disorders, stereotypies, euphoria (with, rarely, mania), and delirium. Each of these is discussed in turn.

Visual hallucinations are very common with prolonged treatment with either levodopa or dopamine agonists: in one 6-year study, the percentage of patients with hallucinations gradually increased until, at the end of the study, fully 62 percent were experiencing them (Goetz *et al.* 2005). The hallucinations are typically complex, involving scenes, animals, or people, and may last from minutes to hours or even days; importantly, early on, patients retain insight into the hallucinatory nature of these experiences and recognize that they are not 'real' (Barnes and David 2001; Fenelon *et al.* 2000; Friedman and Sienkiewicz 1991; Graham *et al.* 1997; Inzelberg *et al.* 1998). Auditory hallucinations, or even olfactory or gustatory hallucinations, may also occur, but these are much less common (Goetz *et al.* 2005; Holroyd *et al.* 2001). Of note, although it has traditionally been taught that visual hallucinations in patients with Parkinson's disease occur, as noted here, as side-effects, recent work suggests that, in a very small minority of patients, visual hallucinations may occur before any treatment is administered (Biousse *et al.* 2004). Furthermore, it appears that they are more common in those with dementia (Fenelon *et al.* 2000), and an autopsy study revealed that cases with greater cell loss and Lewy body pathology in the amygdala were more likely to have been characterized by hallucinations (Harding *et al.* 2002a).

Psychosis is said to be present when patients either experience hallucinations without insight or develop delusions. As just noted, visual hallucinations with preserved insight are very common; however, over a matter of several years, the majority of patients who do have hallucinations will lose insight and begin to react to the visual hallucinations as if they were real (Goetz *et al.* 2006). Delusions may also occur, but are much less common (Holroyd *et al.* 2001): there may be delusions of jealousy with accusations of spousal infidelity, delusions of persecution with complaints of conspiracies, or various other delusions, as for example that things have been stolen or that people have broken into the house. In some cases, both hallucinations and delusions may occur: one patient saw 'people' in his backyard, heard them 'whispering', and was so convinced that they meant him harm that he called the police.

Treatment of hallucinations or psychosis should generally involve an attempt at dose reduction of levodopa and/or dopamine agonists. These side-effects are generally dose-responsive and in some cases it may be possible to reduce the dose sufficiently to allow for a substantial resolution of them without sacrificing control of the parkinsonism. When dose reduction is either ineffective or impractical, consideration may be given to treatment with an antipsychotic, and several have been subjected to double-blinded studies. Although quetiapine is perhaps the most commonly used agent, and was found more effective than placebo in one single-blinded study (Merims *et al.* 2006), four double-blinded studies (Kurlan *et al.* 2007; Ondo *et al.* 2005; Rabey *et al.* 2007; Shotbolt *et al.* 2009) utilizing doses of 50–200 mg/d found it to be no more effective than placebo. Olanzapine, in doses of 2.5–15 mg, was no better than placebo (Breier *et al.* 2002), less effective than clozapine, and tended to aggravate parkinsonism (Goetz *et al.* 2000). Risperidone, in a dose of 1–2 mg, was comparable to clozapine, but, in contrast to clozapine, worsened parkinsonism (Ellis *et al.* 2000). Clozapine, in doses of 6.25–50 mg, is clearly effective (Parkinson Study Group 1999; Pollak *et al.* 2004) and, rather than worsening parkinsonism, may actually improve it (Durif *et al.* 2004; Parkinson Study Group 1999). On the basis of efficacy and motor effects, clozapine is clearly preferable; however, the risk of agranulocytosis and the need for regular blood counts make most clinicians pause before recommending it. A reasonable strategy might be to try risperidone first, keeping clozapine in reserve.

Anxiety attacks may occur during motor fluctuations of the 'wearing off' type (Hillen and Sage 1996; Vazquez *et al.* 1993), and indeed such attacks, along with other 'non-motor' off symptoms (e.g. fatigue, drenching sweats, and akathisia) may be more distressing to patients than the loss of motor ability itself.

Impulse control disorders, including pathologic gambling, compulsive shopping, binge-eating and hypersexuality, may occur as side-effects of treatment in close to 15 percent of patients (Weintraub *et al.* 2010b), and although these can occur secondary to treatment with levodopa, they are more likely to appear with use of dopamine agonists (Dodd *et al.* 2005; Pontone *et al.* 2006; Voon *et al.* 2006a,b; Weintraub *et al.* 2010b). Importantly, these impulsive behaviors remit with discontinuation of treatment with the offending medication; however, it may take anywhere from days to months for this to occur (Dodd *et al.* 2005). Of interest, some patients with these impulse control disorders may also experience a withdrawal syndrome while trying to come off dopamine agonists, consisting of a craving for the drug associated with dysphoria, fatigue, diaphoresis, and orthostatic hypotension (Rabinak and Nirenber 2010). Of note, although, as described below, some patients may experience euphoria on dopaminergic agents, this is not the case here: although patients experiencing this withdrawal syndrome had experienced the impulsive behaviors, they had been neither euphoric nor 'high' at the time.

Stereotypies seen with treatment may be complex and socially disabling. Known as 'punding', these may include repeatedly taking apart and putting together machinery or engaging in other purposeless behaviors (Evans *et al.* 2004).

Euphoria may occur, and some patients may escalate the dose in order to achieve this side-effect (Giovannoni *et al.* 2000); in some cases, albeit rarely, a manic episode may occur (Celesia and Barr 1970; O'Brien *et al.* 1971).

Delirium has been noted as a side-effect (Celesia and Barr 1970; Friedman and Sienkiewicz 1991; Serby *et al.* 1978), but this is rare.

DEMENTIA

Dementia may be treated with either donepezil (Emre *et al.* 2004; Ravina *et al.* 2005) or rivastigmine (Aarsland *et al.* 2002); donepezil may be preferred as there was no worsening of parkinsonism with donepezil in contrast to rivastigmine, which did cause motor worsening, in particular an increase in tremor. Memantine has also been tested, and although one study found a benefit (Aarsland *et al.* 2009), another not only found no benefit, but also noted a distinct worsening of cognition after the drug was discontinued (Leroi *et al.* 2009): based on these results, memantine probably should not be used. The overall treatment of dementia is discussed in Chapter 6.3.

DEPRESSION

There have been only a handful of double-blinded placebo-controlled studies of the treatment of depression in Parkinson's disease. Among the SSRI antidepressants, neither paroxetine (Menza *et al.* 2009) nor sertraline (Leentjens *et al.* 2003) was effective, and, although citalopram was effective in one study (Devos *et al.* 2008), it was not in another (Wermuth *et al.* 1998). Among the tricyclics, desipramine was effective in one study (Devos *et al* 2008) and nortriptyline in two (Andersen *et al.* 1980; Menza *et al.* 2009). Of interest, omega-3 fatty acids were found effective in one study (da Silva *et al.* 2008), and of even greater interest, pramipexole has also been found effective (Barone *et al.* 2010). Transcranial magnetic stimulation (TMS) was found, in a sham-controlled study (Fregni *et al.* 2004), to be as effective as fluoxetine, but the problem here is that there was no placebo arm, and, as fluoxetine has never been shown to be more effective than placebo, it is not clear whether TMS would be. Electroconvulsive therapy, as noted earlier, is effective for motor symptoms; however, although a naturalistic study suggested it was effective for depression (Douyon *et al.* 1989), no controlled studies have been carried out. With these data in mind, it makes sense to treat patients with omega-3 fatty acids (180 mg EPA plus 120 mg DHA capsules; four capsules daily): these have virtually no side-effects and are also helpful for hyperlipidemia. If patients are not already on a direct-acting dopaminergic agent, and there are no contraindications, then pramipexole is clearly a reasonable choice as it may benefit both depression and motor symptoms, and it may be given in doses ranging from 0.125 to 1 mg three times daily. If depresssion persists, then an antidepressant is in order, and were it not for the side-effect burden associated with nortriptyline, this would be the logical choice. Given this burden, however, it may be prudent to begin with citalopram, at a dose of 20–40 mg. Some have raised concerns that SSRIs might aggravate parkinsonism, and although this has been reported, two large studies (one of paroxetine [Ceravolo *et al.* 2000] and the other of sertraline [Kulisevsky *et al.* 2008]) found at most only a mild worsening of tremor in a small minority of patients. Should citalopram not work out, then strong consideration should be given to nortriptyline: although desipramine, another tricyclic, was also effective, the side-effect burden with this agent is even worse. Nortriptyline should be started at a dose of 25 mg and increased every week or so in similar increments up to 75 mg, all given at bedtime. At this point, it is mandatory to check a blood level after steady state has been reached. Nortriptyline is the *only* antidepressant with a 'therapeutic window': levels below 50 ng/mL are ineffective and levels above 150 ng/mL are also ineffective, and this ineffectiveness at higher blood levels occurs regardless of whether there are side-effects or not – this is a 'true' therapeutic window. Before starting either an SSRI or a tricyclic, it may be prudent, should the patient be on an MAOI, to discontinue it. With an SSRI, although the risk of a serotonin syndrome is very low (Richard *et al.* 1997), it still exists, and should give one pause. With a tricyclic, there is also the risk of precipitating a hypertensive crisis.

There are certain antidepressants which should generally be avoided: mirtazapine may worsen or precipitate REM sleep behavior disorder (Onofrj *et al.* 2003), and bupropion may cause excitement, restlessness, hallucinations, and confusion (Goetz *et al.* 1984).

9.6 DEMENTIA WITH LEWY BODIES

Dementia with Lewy bodies, also known as diffuse Lewy body disease, cortical Lewy body disease, and senile dementia of the Lewy body type, first described in 1961 by Okazaki *et al.* (1961), is now recognized in both autopsy (Barker *et al.* 2002) and clinical (Rahkonen *et al.* 2003) studies to be the second most common cause of dementia in the elderly, accounting for nearly 20 percent of all cases. Clinical studies have also indicated that, in the general population, dementia with Lewy bodies is the second most common cause of parkinsonism (Pineda *et al.* 2000).

Dementia with Lewy bodies is distinctive in that it may present in one of two ways, either with a dementia or with parkinsonism: eventually, however, almost all patients will display both features.

Clinical features

The onset is gradual, generally in the seventh decade, and, as noted, the presentation may be with either dementia or parkinsonism (Byrne *et al.* 1989).

The dementia of this disease is characterized initially by difficulty in sustaining attention and by executive dysfunction, with deficits in judgment and decision making; over time, these symptoms are joined by short-term memory loss. There are two other features of the dementia with Lewy bodies that deserve note, as they are helpful, as discussed later, in distinguishing dementia with Lewy bodies from other dementing disorders: these two features are spontaneous fluctuations in mental status and early-onset hallucinations or delusions.

Fluctuations in mental status are found in over half of all patients, and tend to occur early in the course of the dementia; importantly, these appear spontaneously, without any precipitating factors (Bradshaw *et al.* 2004). Clinically (Ballard *et al.* 2001; Ferman *et al.* 2004; McKeith *et al.* 1994a), these episodes may last from hours to days, and are characterized by admixtures of drowsiness, confusion, blank staring, incoherence, and bizarre behavior.

Hallucinations occur in roughly two-thirds of all cases, and, like the spontaneous fluctuations, tend to occur early on. They are typically visual, complex, and well formed, and are experienced without insight (Ala *et al.* 1997; Ballard *et al.* 1997, 1999, 2001; Crystal *et al.* 1990; Tiraboschi *et al.* 2006; Weiner *et al.* 1996). Auditory hallucinations may also occur, but are less common. Delusions are somewhat less common than hallucinations, and delusions of persecution (Marantz and Verghese 2002) and the Capgras phenomenon (i.e. that others have been taken over by 'imposters') (Ballard *et al.* 1999) are common.

Depression may occur during the course of the dementia (Burkhardt *et al.* 1988) and indeed has been noted in up to half of all patients (Klatka *et al.* 1996).

The parkinsonism of dementia with Lewy bodies (Hely *et al.* 1996; Louis *et al.* 1997), although generally similar to the classic parkinsonism typical of Parkinson's disease, with flexion posture, bradykinesia, and rigidity, does, however, display some differences (Louis *et al.* 1995,1997), including a tendency toward a symmetric onset and less tremor; furthermore, myoclonus may be seen in close to one-fifth of cases, again in contrast with Parkinson's disease, where it is rare. Autonomic symptoms may accompany the parkinsonism (Horimoto *et al.* 2003), including orthostatic hypotension (with, in some cases, postural syncope), urinary incontinence or retention, and erectile dysfunction.

Although patients with parkinsonism of any cause are liable to get worse if treated with an antipsychotic, this liability is greatly pronounced in dementia with Lewy bodies, a phenomenon termed 'neuroleptic sensitivity' (McKeith *et al.* 1992a,b; Weiner *et al.* 1996). In some cases, the resulting exacerbation of parkinsonism may be quite profound, and fatalities have occurred; furthermore, the exacerbation, rather than resolving soon after the antipsychotic is discontinued, may persist for days or weeks.

As noted earlier, dementia with Lewy bodies may present with either a dementia or with parkinsonism. In cases that present with dementia, although the parkinsonism may appear soon thereafter, in some cases many years may pass before this occurs. In cases that present with parkinsonism, however, the accompanying dementia appears relatively early on, in almost all cases before a year has passed.

Another important clinical feature of dementia with Lewy bodies is REM sleep behavior disorder. This disorder, as discussed in Chapter 20.2, is characterized by 'enacted dreams' and is present in the vast majority of cases, often presenting years or decades before the onset of either the dementia or the parkinsonism (Boeve *et al.* 1998; Postuma *et al.* 2009).

Magnetic resonance scanning may reveal cortical atrophy and atrophy of medial temporal structures; however, the degree of atrophy tends to be relatively mild (Tam *et al.* 2005). The EEG may reveal generalized slowing with temporal theta transients (Briel *et al.* 1999); frontal intermittent delta activity (FIRDA) (Calzetti *et al.* 2002) and periodic spike and slow-wave complexes (Doran and Larner 2004) have also been reported.

Course

The course is one of gradual progression, with death on average within 7–8 years.

Etiology

Microscopically (Gibb *et al.* 1987; Gomez-Tortosa *et al.* 1999; Hansen *et al.* 1990; Lennox *et al.* 1989; Lippa *et al.* 1994), one finds neuronal loss, with Lewy bodies in surviving neurons, within the frontal, temporal, cingulate, and insular cortices, the nucleus basalis of Meynert, the substantia nigra, and the locus ceruleus and dorsal raphe nucleus. Amyloid plaques may be found in the cerebral cortex; however, neurofibrillary tangles are no more commonly found than in normal controls (Harrington *et al.* 1994).

The dementia correlates not only with the burden of Lewy bodies and plaques in the cortex, but also with the burden of Lewy bodies in the nucleus basalis of Meynert. This nucleus provides the bulk of cholinergic innervation to the cortex, and there is a good correlation between loss of cortical choline acetyltransferase activity and the severity of the dementia (Tiraboschi *et al.* 2002). Furthermore, there is also a good correlation between the occurrence of visual hallucinations and the burden of Lewy bodies in the amygdala, parahippocampus, and inferior temporal cortex (Harding *et al.* 2002b).

Although there is an increased prevalence of dementia in family members (Woodruff *et al.* 2006), as yet there is no evidence that dementia with Lewy bodies (with the distinctive combination of dementia and parkinsonism just described) runs in families, and at present it appears that the vast majority of cases occur on a sporadic basis. Nevertheless, true familial cases have been described (Galvin *et al.* 2002), and there are rare reports of cases occurring secondary to mutations at a locus on chromosome 2 (Bogaerts *et al.* 2007) and to heterozygous or homozygous duplications of the gene for α-synuclein (Ikeuchi *et al.* 2008).

Differential diagnosis

The differential considerations differ according to whether the presentation of dementia with Lewy bodies is with dementia or with parkinsonism.

In cases that present with dementia, consideration, as discussed in Chapter 6.3, must be given to certain other disorders capable of causing a dementia of gradual onset, including most notably Alzheimer's disease, Binswanger's disease, and lacunar dementia. Several features of dementia with Lewy bodies facilitate this differential. Most notable among these are the spontaneous fluctuations in mental status and the appearance of visual hallucinations: although these features of course may be seen in these other disorders, they tend to occur late in their course, whereas in dementia with Lewy bodies they are present early on, indeed sometimes at presentation. Second, REM sleep behavior disorder, common in dementia with Lewy bodies, is rare in Alzheimer's disease and is not characteristic of Binswanger's disease or lacunar dementia. Third, there is the phenomenon of neuroleptic sensitivity: in some cases of dementia with Lewy bodies, the parkinsonism is 'subclinical' and administration of an antipsychotic will provoke a totally unexpected and severe parkinsonism (McKeith *et al.* 1992c, 1994b): although parkinsonism can certainly occur with antipsychotic treatment in these other disorders, it is simply not as florid as that characteristic of neuroleptic sensitivity. Looking more specifically at the differential with Alzheimer's disease, several other factors may be helpful. First, Alzheimer's disease tends to present with an amnestic deficit, and although difficulties with short-term memory may also occur in dementia with Lewy bodies, they tend to occur later and to be relatively minor. Second, although parkinsonism may also appear in Alzheimer's disease, it tends to occur only very late in the course and to be quite mild. Finally, MRI may also be helpful (Burton *et al.* 2009; Tam *et al.* 2005), as it typically reveals prominent medial temporal atrophy in Alzheimer's disease: although medial temporal atrophy may also be seen in dementia with Lewy bodies, it tends to be mild, even subtle.

In cases that present with parkinsonism, various other neurodegenerative diseases, as discussed in Chapter 4.8, enter the differential, and of these Parkinson's disease leads the list. Although there are some differences between the parkinsonism of dementia with Lewy bodies and that of Parkinson's disease, as discussed earlier, these may not be present, or, if so, may be subtle, and consequently cannot be relied on. One feature, however, will almost always enable a differential, and that is what has come to be called the '1 year rule': as noted earlier, in cases of dementia with Lewy bodies that present with parkinsonism, a dementia will supervene, in almost all cases, within 1 year, and this is in stark contrast with Parkinson's disease, wherein, although dementia is seen in the vast majority of cases, its appearance is delayed, in almost all cases for at least 5 years.

Treatment

With regard to treatment of dementia, two double-blinded studies demonstrated the effectiveness of rivastigmine, not only for the cognitive aspects but also for hallucinations and delusions (McKeith *et al.* 2000; Wesnes *et al.* 2002); open-label studies also suggest benefit from donepezil (Thomas *et al.* 2005) and galantamine (Edwards *et al.* 2004). Memantine was also shown, in a double-blinded study (Aarsland *et al.* 2009), to cause overall improvement; however, some caution is in order here, as open work has suggested both cognitive decrement (Sabbagh *et al.* 2005) and aggravation of hallucinations and delusions (Ridha *et al.* 2005) with memantine. In cases where hallucinations and delusions fail to respond to either rivastigmine or perhaps memantine, some clinicians will utilize an antipsychotic; however, such a course should only be undertaken with great caution, given the possibility of neuroleptic sensitivity. If the decision is made to use an antipsychotic, then one of the second-generation agents, which are less likely to cause parkinsonism, should be used, and several of these have been tested in dementia with Lewy bodies. Quetiapine, in two case series (Fernandez *et al.* 2002; Takahashi *et al.* 2003) utilizing doses of 25–75 mg/day, was beneficial in the majority of cases but frequently caused sedation and orthostatic hypotension. In a subsequent double-blinded trial, however, quetiapine, in an average dose of 120 mg, was no more effective than placebo. Olanzapine, in an open study (Walker *et al.* 1999) using doses from 2.5 to 7.5 mg, caused intolerable side-effects in a minority and some benefit in the remainder. Olanzapine was also used in a large, double-blinded study (Cummings *et al.* 2002) that was designed to test its effectiveness in patients with Alzheimer's disease, but which also contained patients with dementia with Lewy bodies, and in a post-hoc analysis there was some benefit at doses of 5–10 mg, but not at 15 mg. Risperidone was reported as beneficial in a case report (Kato *et al.* 2002). Of interest, in two patients treated with low-dose clozapine (Burke *et al.* 1998), rather than any benefit, there was a behavioral deterioration.

With regard to the treatment of depression, there are no controlled studies. Consideration may be given to an antidepressant, such as an SSRI (McKeith *et al.* 2005), or, in severe and treatment-resistant cases, ECT (Rasmussen *et al.* 2003). In case reports, patients with REM sleep behavior disorder have responded well to both clonazepam and donepezil (Massironi *et al.* 2003).

The parkinsonism of dementia with Lewy bodies may respond to combination treatment with levodopa/carbidopa (Byrne *et al.* 1989); however, the response is generally not as robust as that seen in Parkinson's disease (Molloy *et al.* 2005). There is little published experience regarding the use of dopamine agonists such as pramipexole or ropinirole.

9.7 PROGRESSIVE SUPRANUCLEAR PALSY

Progressive supranuclear palsy (PSP), also known as the Steele–Richardson–Olszewski syndrome, is an uncommon disorder, with a lifetime prevalence of roughly 6 per 100 000; overall, it is somewhat more common in men than in women. Clinically, this disorder may present in one of three ways: the classic presentation, and the one first described by Steele *et al.* in 1964 (Steele *et al.* 1964), is now known as 'Richardson's syndrome', and is characterized by an atypical parkinsonism; the next most common is referred to a 'PSP-parkinsonism', and in this variant, the parkinsonism is much closer to the classic parkinsonism typical of Parkinson's disease; finally there is an uncommon variant, known as 'pure akinesia with gait freezing'.

Clinical features

The onset is gradual, typically in the seventh decade. Each of the variant presentations is described in turn.

RICHARDSON'S SYNDROME

The general clinical features of Richardson's syndrome have been described in several reports (Birdi *et al.* 2002; Collins *et al.* 1995; Litvan *et al.* 1996b; Maher and Lees 1986; Messert and Van Nuis 1966; Steele 1972; Steele *et al.* 1964). Typically, this syndrome presents with frequent, unexplained falls resulting from postural instability. Over time, an atypical parkinsonism gradually appears, characterized by a more or less symmetric onset of rigidity, generally without tremor, and an abnormal gait typified by a wide-based, ataxic stance with short, shuffling steps. Importantly, rather than the typical flexion posture seen in most cases of parkinsonism, patients with Richardson's syndrome typically display a dystonic axial rigidity, which may also affect the neck. Dystonic rigidity also affects the facial musculature, at times creating an 'astonished' appearance, as seen in Figure 9.6.

Classically, from 1 to 3 years after this atypical parkinsonism is established, one also sees a supranuclear ophthalmoplegia for vertical gaze, wherein patients most particularly have difficulty voluntarily looking down, a difficulty that may make walking down stairs particularly treacherous. This feature, however, may be delayed for many years, and in some cases may never appear (Birdi *et al.* 2002; Daniel *et al.* 1995).

Dementia occurs in about half of all patients, generally well after the parkinsonism has become established, and this is marked by frontal lobe symptomatology (Verny *et al.* 1996). Patients have difficulty in organizing and planning, and judgment and abstracting ability may be impaired. Set-shifting is likewise impaired and there may be perseveration. Forgetfulness and impaired concentration are often in evidence, and although there may be actual short-term memory loss, this may be more apparent than real, as in many cases patients are capable of recalling things if only given long enough to do so: this delayed latency to response, however, may tax the clinician's patience, as in some cases 4 or 5 minutes may be required before the correct answer is finally produced. Personality change is also present, and plays a large part in the overall picture of the dementia: there may, on the one hand, be apathy (Aarsland *et al.* 2001; Albert *et al.* 1974; Litvan *et al.* 1996a), while, on the other, a smaller percentage of patients may display disinhibition with irritability, or, rarely,

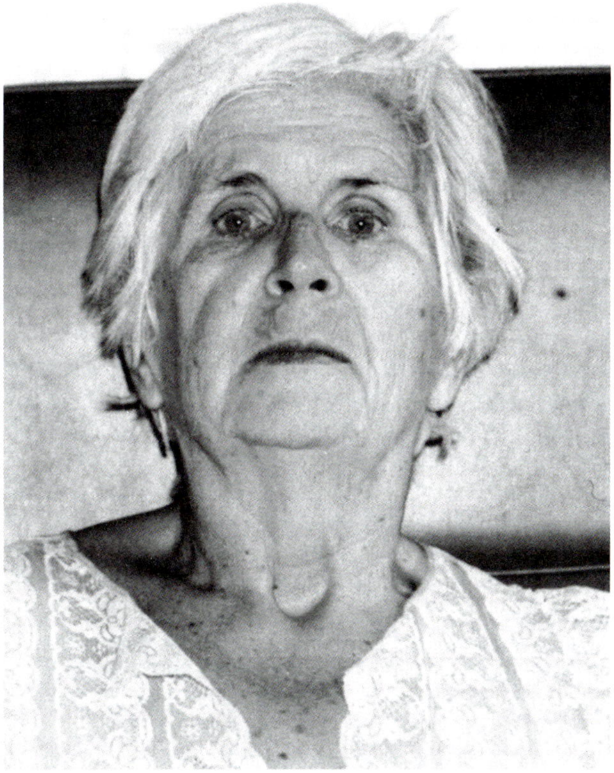

Figure 9.6 'Astonished' facial appearance in progressive supranuclear palsy. Reproduced with kind permission from Stern (1990).

euphoria (Aarsland *et al.* 2001; Albert *et al.* 1974). Over time, pseudobulbar palsy, with emotional incontinence, may also occur (Menza *et al.* 1995). Notably, such 'cortical' features as apraxia, aphasia, or agnosia are generally absent.

Although, as noted, this dementia generally appears only in the context of the atypical parkinsonism, rarely it may constitute the presenting feature of Richardson's syndrome (Davis *et al.* 1985; Kaat *et al.* 2007).

Seizures may also occur, but only in a small minority of cases (Nygaard *et al.* 1989).

Magnetic resonance imaging may reveal evidence of mesencephalic atrophy with the classic 'Hummingbird sign' (Kato *et al.* 2003): here, on midline sagittal T1-weighted scanning, the atrophy 'elongates' the area between the mesencephalon and mammillary bodies, yielding a shape similar to that of a hummingbird's head and beak. In some cases, there may also be atrophy of the frontal and temporal cortices.

The EEG may occasionally demonstrate mild, generalized slowing, but in one study roughly one-third of cases were characterized by frontal intermittent rhythmic delta activity (FIRDA) (Tashiro *et al.* 2006).

PSP-PARKINSONISM

In this variant (Williams *et al.* 2005), seen in roughly one-third of all cases of PSP, patients present with a more classic parkinsonism that is characterized by an asymmetric rigidity and tremor, a flexion posture, and short, shuffling steps without ataxia. Although postural instability with falls may also occur, this does not constitute the presentation.

PURE AKINESIA WITH GAIT FREEZING

This variant (Williams *et al.* 2007a) is the least common, and bears little resemblance to the other two. Patients present with generalized akinesia accompanied by prominent freezing. Notably absent are the following: rigidity, tremor, ophthalmoplegia, and dementia.

Course

Progressive supranuclear palsy is, as stressed by the name, progressive, and death usually occurs within 6–10 years (Golbe *et al.* 1988; Maher and Lees 1986).

Etiology

Macroscopically, there is atrophy of the globus pallidus, subthalamic nucleus, and the mesencephalon. Microscopically (Collins *et al.* 1995; Henderson *et al.* 2000; Litvan *et al.* 1996b; Steele *et al.* 1964), neuronal loss, accompanied by gliosis, is likewise seen in these structures, and, within the mesencephalon, this is prominent in the substantia nigra, locus ceruleus, red nucleus, periaqueductal gray, and superior colliculus. Similar changes are also found in the dentate nuclei, the thalamus, and the temporal and inferior frontal cortices. Within surviving neurons, one finds neurofibrillary tangles, dissimilar to those found in Alzheimer's disease (Takahashi *et al.* 1989). As might be expected, the overall burden of tau pathology is greatest in Richardson's syndrome, less in PSP-parkinsonism, and least in the syndrome of pure akinesia with gait freezing (Williams *et al.* 2007b).

Although the vast majority of cases of progressive supranuclear palsy occur on a sporadic basis, a positive family history for parkinsonism may be found in 12 percent of cases of Richardson's syndrome (Kaat *et al.* 2009); furthermore, although rare, there are also cases that are clearly inherited on an autosomal dominant basis (Brown *et al.* 1993; Rojo *et al.* 1999; Ros *et al.* 2005a,b; Stanford *et al.* 2000; de Yebenes *et al.* 1995)

Despite this lack of a clear genetic basis in most cases, an inherited susceptibility to the disease does appear to exist, correlating to which of various normally occurring tau alleles are present (Higgins *et al.* 1999; Morris *et al.* 2002).

Differential diagnosis

The differential diagnosis of Richardson's syndrome and PSP-parkinsonism includes other neurodegenerative disorders capable of causing parkinsonism, as discussed in Chapter 4.8. With regard to Richardson's syndrome, the most valuable clues are frequent, unexplained falls due to postural instability and, of course, supranuclear ophthalmoplegia, and if these are present during the first year of the illness, the diagnosis is very likely (Litvan *et al.* 1996c,1997b). Unfortunately however, as noted in

Chapter 9.8, corticobasal ganglionic degeneneration may present with a Richardson's syndrome, including a supranuclear ophthalmoplegia, and hence the diagnosis should probably always be tentative. When Richardson's syndrome is present, however, Parkinson's disease, dementia with Lewy bodies, and the striatonigral variant of MSA are very unlikely. When progressive supranuclear palsy presents with the PSP-parkinsonism type, a reliable differentiation from Parkinson's disease may have to await autopsy.

Pure akinesia with gait freezing, as discussed in Chapter 4.9, is rarely mimicked by any other condition.

Treatment

Levodopa may offer some benefit, more so in cases of PSP-parkinsonism (Williams *et al.* 2005) than in Richardson's syndrome, wherein the benefits are typically minimal and short lived (Birdi *et al.* 2002). Interestingly, in cases of Richardson's syndrome, amitriptyline, in low doses from 25 to 50 mg, improved the parkinsonism, especially regarding gait (Kvale 1982; Newman 1985). With regard to progressive akinesia with gait freezing, there is not, at present, any known treatment.

The general treatment of dementia and of the frontal lobe syndrome is discussed in Chapters 6.3 and 8.2, respectively. Donepezil offers very modest benefit for dementia, but worsens the parkinsonism, and on balance may not be beneficial (Litvan *et al.* 2001).

9.8 CORTICOBASAL GANGLIONIC DEGENERATION

Corticobasal ganglionic degeneration, also know as cortical-basal degeneration, is a rare disorder, equally common in men and women, that was first described by Rebeiz in 1968 (Rebeiz *et al.* 1968). Although the classic description of the clinical features of this disorder first provided by Rebeiz is still quite relevant, it has become apparent that corticobasal ganglionic degeneration may present in other fashions.

Clinical features

The onset is gradual, usually in the sixth or seventh decade. Classically, corticobasal ganglionic degeneration presents with an atypical parkinsonism known as the 'corticobasal syndrome.' Other presentations, however, as noted, do occur, and these include an atypical parkinsonism resembling that seen in progressive supranuclear palsy, a parkinsonism similar to that seen in Parkinson's disease, a dementia, a personality change, and a primary progressive aphasia.

The classic corticobasal syndrome (Litvan *et al.* 1997c; Riley *et al.* 1990; Rinne *et al.* 1994a; Wenning *et al.* 1998) is characterized by a strikingly asymmetric rigid-akinetic parkinsonism affecting primarily the upper extremity, which may be accompanied by dystonia (Vanek and Jankovic 2001). Ipsilateral limb apraxia and cortical sensory loss are also common, and there may be action myoclonus; tremor may also occur, but rather than a classic rest tremor, jerky, dystonic movements are seen. Another symptom consists of abnormal involuntary movements of the ipsilateral arm which, remarkably, appear voluntary: these include 'levitation' (Gibb *et al.* 1994; Riley *et al.* 1990), 'wandering' motions (Rinne *et al.* 1994a; Sawle *et al.* 1991), and, classically, an 'alien hand syndrome' (Ling *et al.* 2010), manifesting as true 'intermanual conflict' wherein the ipsilateral limb goes about, as it were, interfering with consciously voluntary acts of the normal, contralateral hand. Eventually, and gradually, the other side becomes akinetic and rigid, and roughly half of all patients will develop a dementia (Litvan *et al.* 1998) which may be marked by apathy, depression, or, rarely, delusions and hallucinations.

In addition to the atypical parkinsonism seen in this classic presentation, cases may also present with a parkinsonism similar to that seen in the Richardson's type of progressive supranuclear palsy (Ling *et al.* 2010), with early falls, a symmetric rigid–akinetic parkinsonism, and supranuclear ophthalmoplegia: in cases with supranuclear ophthalmoplegia, the main difficulty in patients with corticobasal ganglionic degeneration tends to be with voluntary horizontal eye movements, in contrast with progressive supranuclear palsy, where the main difficulty is with vertical eye movements. Cases may also present with a parkinsonism similar to that seen in Parkinson's disease (Ling *et al.* 2010), with an asymmetric (but not markedly asymmetric) parkinsonism marked by a flexion posture, rigidity and akinesia, and tremor: this tremor, however, is generally not a rest tremor, but more a jerky one.

Dementia may also constitute the presentation of corticobasal ganglionic degeneration (Bergeron *et al.* 1996; Schneider *et al.* 1997) and indeed this may be common (Grimes *et al.* 1999; Murray *et al.* 2007); rare presentations include a personality change of the frontal lobe type (Geda *et al.* 2007) or a primary progressive aphasia (Geda *et al.* 2007; Ikeda *et al.* 1996). In such cases, an atypical parkinsonism eventually appears.

Magnetic resonance imaging may reveal asymmetric cortical atrophy affecting primarily the parietal and frontal lobes.

Course

The disease is gradually progressive, with death, often from aspiration pneumonia, after an average of 8 years.

Etiology

The pathology of corticobasal ganglionic degeneration has been described in a number of papers (Gibb *et al.* 1994; Rebeiz *et al.* 1968; Riley *et al.* 1990; Rinne *et al.* 1994a). Macroscopically there is asymmetric cortical atrophy, affecting primarily the parietal lobe and the posterior portion of the frontal lobe; over time, the atrophy may spread to the contralateral side, and the temporal lobes and anterior portions of the frontal lobes may also become involved. Microscopically, changes are seen in these areas and in the

basal ganglia and substantia nigra, consisting of neuronal loss and astrocytosis; surviving neurons are large, ballooned, and achromatic and contain tau-positive filaments. Although familial cases have been described (Spillantini *et al.* 2000), the overwhelming majority of cases are sporadic.

Differential diagnosis

When the presentation is with parkinsonism, the differential, as discussed in Chapter 4.8, includes other neurodegenerative causes of parkinsonism, such as progressive supranuclear palsy, Parkinson's disease, dementia with Lewy bodies, and the striatonigral variant of MSA. Although the classic corticobasal syndrome of course strongly suggests corticobasal ganglionic degeneration, this classic corticobasal syndrome may also be caused by progressive supranuclear palsy (Ling *et al.* 2010) and by Parkinson's disease (Ling *et al.* 2010), and consequently, even when presented with cases of classic corticobasal syndrome, the diagnosis of corticobasal ganglionic degeneration, although perhaps most likely, should always be tentative. In cases where corticobasal ganglionic degeneration presents with a Richardson's parkinsonism or with a parkinsonism similar to that seen in Parkinson's disease, it is yet not clear whether an accurate diagnosis is possible by anyone other than the pathologist.

When the presentation is with dementia, a personality change, or primary progressive aphasia, only long-term observation looking for the emergence of parkinsonism will suggest the correct diagnosis, and even here some tentativeness is in order, given that dementia with Lewy bodies may also present with dementia.

Treatment

Although the parkinsonism seen with corticobasal ganglionic degeneration typically does not respond to levodopa (Rinne *et al.* 1994a), a trial is not unreasonable. Myoclonus may be treated with clonazepam. The general treatment of dementia is discussed in Chapter 6.3.

9.9 MULTIPLE SYSTEM ATROPHY

Multiple system atrophy, as the name suggests, is characterized pathologically by involvement of multiple systems. This involvement, however, does not occur evenly, and consequently the disease may present in different fashions, depending on which system is most predominantly involved (Papp and Lantos 1994; Quinn 1989). Three variants are generally recognized, namely the striatonigral variant (also known as MSA with predominant parkinsonism, or MSA-P), the olivopontocerebellar variant (also known as MSA with predominant cerebellar ataxia, or MSA-C), and the Shy–Drager variant (also known as MSA with predominant autonomic failure, or MSA-A).

This is an uncommon disease, found in 2–4/100 000 of the general population, and is somewhat more frequent in men than in women.

Clinical features

The onset is gradual, and although most patients fall ill in the sixth decade, the range is wide, from the fourth through the eighth decades. As noted, this disease may present in one of three fashions, each described below.

The striatonigral variant (Colosimo *et al.* 1995; Watanabe *et al.* 2002; Wenning *et al.* 1994,1995), seen when the striatum and substantia nigra are most involved, is characterized by the gradual onset of parkinsonism marked by bradykinesia and rigidity, often accompanied by anterocollis and extreme flexion of the trunk, sometimes amounting to classic camptocormia. Tremor is present in only about one-quarter of cases, and although a classic, rest, 'pill-rolling' tremor may occur, in most cases the tremor is postural, irregular, and jerky. Notably, in roughly three-quarters of all cases, the parkinsonism is generally symmetric. Autonomic features are seen in essentially all cases, often very early on, including postural hypotension, with or without syncope, urinary retention, frequency, or incontinence, and, in males, erectile dysfunction. Cerebellar features are found in close to half, including ataxia and dysarthria. Other features include laryngeal stridor, long-tract signs (hyperreflexia or extensor plantar responses), emotional incontinence, subtle, irregular myoclonic jerks of the fingers, and, rarely, supranuclear ophthalmoplegia (Wenning *et al.* 1995).

The olivopontocerebellar variant (Watanabe *et al.* 2002; Wenning *et al.* 1994), seen when the inferior olives, the pons, and the cerebellum are predominantly involved, presents with ataxia, intention tremor, dysarthria, scanning speech, and nystagmus. Parkinsonian features, often subtle, or autonomic features are seen in most cases; other features include reflex myoclonus (Rodriguez *et al.* 1994) and emotional incontinence, which was found in roughly one-third of all cases (Parvizi *et al.* 2007).

The Shy–Drager variant (Shy and Drager 1960; Wenning *et al.* 1994), seen when the intermediolateral gray matter of the spinal cord and the ventrolateral medulla are heavily involved, is characterized by autonomic failure, with, as noted earlier, postural hypotension, urinary symptoms, and erectile dysfunction.

Overall, of these three variants, the most common, by far, is the striatonigral variant, followed by the olivopontocerebellar one; the Shy–Drager variant, by contrast, is very uncommon. It must be emphasized, however, that, regardless of which variant predominates the clinical picture, in almost all cases, an admixture of two or all three types of symptoms is the rule: 'pure' variants are very rare.

A mild degree of cognitive impairment, primarily related to executive functioning (Robbins *et al.* 1992), is found in the majority of patients with the striatonigral and olivopontocerebellar variants, and tends to be more severe in those with the striatonigral variant (Kawai *et al.* 2008). Although many texts state that dementia does not occur, this is clearly not, in fact, the case: one clinical survey (Kitiyama *et al.* 2009) found dementia in close to one-fifth of cases, and there have been two, well-documented, autopsy-confirmed cases of dementia secondary to MSA (Critchley and Greenfield 1948; Schlossmacher *et al.* 2004).

Importantly, there is a strong association between MSA and REM sleep behavior disorder (Iranzo *et al.* 2005): one report (Plazzi *et al.* 1997) found this disorder in two-thirds of all cases and noted that it could precede the onset of the classic symptoms of MSA by years.

Magnetic resonance scanning may reveal atrophy of the striatum, pons, inferior olives, and cerebellum. In some cases, especially those with the striatonigral variant, the putamen, on T2-weighted scanning, displays both decreased signal intensity within itself and a distinctive surrounding rim of increased signal intensity. In other cases, especially those with the olivopontocerebellar variant, the basis pontis will display the 'hot cross bun sign' on axial T2-weighted images, wherein the base of the pons is marked by lines that give it the appearance of a hot cross bun seen from above.

Course

Death, often from aspiration pneumonia, occurs in cases of the striatonigral and olivopontocerebellar variants after a median of roughly 10 years, and although the overall course is marked by a steady progression, the rate of progression of the parkinsonism seen in the striatonigral variant is relatively rapid, and marked by a need for a cane or other assistive device in 3 years with eventual need for a wheelchair in about 5 years.

Etiology

Macroscopically there is variable atrophy of the cerebral cortex (particularly the frontal area), striatum (more so the putamen than the caudate), pons, inferior olives, and cerebellum (Ozawa *et al.* 2004; Wenning *et al.* 1996). Microscopically (Ozawa *et al.* 2004; Papp and Lantos 1994; Wenning *et al.* 1995), the characteristic finding in MSA consists of oligodendroglial cytoplasmic inclusions, which contain both α-synuclein and ubiquitin. Such inclusions may also be found in neurons, but overall glial pathology is more prominent. These findings, associated with cell loss and astrocytosis, are seen not only in the structures that undergo macroscopic atrophy, but also in the substantia nigra, locus ceruleus, and the intermediolateral gray of the spinal cord.

At present, although it is clear that MSA is a sporadic disorder, its cause remains unknown.

Differential diagnosis

Differential considerations vary according to the variant with which MSA presents. In cases of the striatonigral variant, consideration, as discussed in Chapter 4.8, must be given to other neurodegenerative disorders capable of causing parkinsonism, and, of these, Parkinson's disease and dementia with Lewy bodies pose the greatest differential problem. Certain features, however, may enable a correct differential. To begin with, the parkinsonism seen in the striatonigral variant is typically symmetric, and, in most cases, free of tremor, and this is in contrast to Parkinson's disease and dementia with Lewy bodies, wherein one finds an asymmetric presentation marked by tremor. Further, in the striatonigral variant, one always finds features of other system involvement. There may be ataxia, which is absent in Parkinson's disease and dementia with Lewy bodies. More commonly, one finds autonomic features, and here, given that such features may also be found in Parkinson's disease and dementia with Lewy bodies, one must pay attention both to when they occur in the course of the disease and to their severity: in multiple system atrophy they tend to both be severe and to occur early on, whereas in Parkinson's disease and dementia with Lewy bodies they are late appearing and often mild. Finally, one looks to the response to levodopa. In the striatonigral variant, although some patients do improve with levodopa, most do not, and in those who do, the response is typically transient; by contrast, the response is typically robust in Parkinson's disease and dementia with Lewy bodies.

In cases of the olivopontocerebellar variant, consideration is given to alcoholic or paraneoplastic cerebellar degeneration, and to spinocerebellar ataxia, which is distinguished by its autosomal dominant inheritance; fragile X-associated tremor/ataxia syndrome (FXTAS) may be difficult to distinguish on clinical grounds, and genetic testing may be required.

Treatment

Although, as just noted, the response to levodopa is generally neither robust nor prolonged (Wenning *et al.* 1994, 1995), it is worth a trial. In using levodopa, however, one must watch both for a potentially severe aggravation of postural hypotension and for the development of dystonic facial dyskinesias, which appear particularly common in multiple system atrophy. Interestingly, add-on treatment with paroxetine was found, as compared to placebo, to further alleviate the parkinsonism (Freiss *et al.* 2006). There is no known treatment for the cerebellar symptoms. Postural dizziness may respond to fludrocortisone or midodrine. Erectile dysfunction has been treated with sildenafil; however, this was often not tolerated because of an aggravation of hypotension (Hussain *et al.* 2001). The general treatment of dementia is discussed in Chapter 6.3. Continuous positive airway pressure (CPAP) or BIPAP may be helpful in cases of laryngeal stridor; however, tracheostomy may eventually be required.

9.10 HUNTINGTON'S DISEASE

Huntington's disease, also known as Huntington's chorea or chorea major, is an autosomal, dominantly inherited disorder that is characterized, classically, by a combination of chorea and dementia. Although the disease first arrived in America in 1630 when three women, all from Suffolk county, England, landed in Massachusetts (Vessie 1932), it was not until George Huntington's paper in 1872 (Huntington 1872) that it was clearly characterized (Brody and Wilkins 1967). Although the lifetime prevalence is not high (5–10/100 000 among white populations, and only one-third of that in black populations and one-tenth among Japanese populations),

these patients are not uncommonly seen in hospitals and clinics.

Clinical features

The onset is generally quite insidious and may occur anywhere from childhood to the eighth decade: roughly 5 percent have a juvenile or adolescent onset between the ages of 4 and 20 years, 50 percent an adult onset between 20 and 50 years, and 25 percent a late onset after the age of 50 years. As the symptomatology varies according to the age of onset, each of these age groupings is considered in turn, beginning with the most common one, namely adult-onset cases. Laboratory testing is then considered.

ADULT ONSET

Adult-onset cases (Caine and Shoulson 1983; Folstein *et al.* 1983; Kirkwood *et al.* 2001; Leroi *et al.* 2002; Marder *et al.* 2000; Paulsen *et al.* 2001; Pflanz *et al.* 1991; Shiwach 1994; Young *et al.* 1986) generally present with chorea, and this is eventually present in essentially all cases. Over time, other symptoms also occur, generally in a more or less sequential fashion, including the following: depression in the first year or two, eventually occurring in close to one-half of all cases; personality change within the first 2–3 years, eventually appearing in one-half to three-quarters of all patients; psychosis from 3 to 5 years, being seen, however, in only about one-tenth of cases; and, finally, dementia, at about 5 years, which is eventually present in almost all cases.

Chorea may initially present as fidgetiness, clumsiness, or a tendency to drop things; obvious choreiform movements are generally first visible on the face (including the forehead), from where they spread to involve the trunk and extremities. There may be facial grimacing, brow wrinkling, and blinking; upper extremity involvement may lead to shoulder shrugging, abrupt flinging of the arms, or purposeless 'piano playing' movements of the hands. In shaking the patient's hand, one may note the classic 'milkmaid' grip, wherein the patient's grasping of the examiner's hand feels as if the patient is attempting to milk a cow. Lower extremity involvement may lead to a lurching, staggering, or 'dancing and prancing' gait, which in some cases may so resemble the gait seen in alcohol intoxication that patients have been arrested for public intoxication (Lesse 1946). Importantly, these choreiform movements are quite brief, appearing and disappearing on a random basis from one location to another with lightning-like rapidity. Early on in the course of the disease, patients may attempt, with varying degrees of success, to disguise the choreic movements by merging them with purposeful movements: for example, a choreic fling of the arm up to the head may be purposefully extended to draw the fingers through the hair, as if the purpose had all along been to straighten the hair. Denial is common, and tends to worsen as the chorea progresses (Vitale *et al.* 2001). In some cases it may be quite extreme: one of the authors had a patient who, although unable to walk because of her chorea, blithely maintained that there was nothing at all amiss. Of interest, although the denial seen in Huntington's disease may certainly be emotionally mediated, there is evidence that at least in some cases it is more a matter of anosognosia (Snowden *et al.* 1998) occurring secondary to the cerebral pathology. In addition to the chorea, a majority of patients will also have dystonic movements (Louis *et al.* 1999, 2000), including internal shoulder rotation, fist clenching, knee flexion, or foot inversion: in most such instances, however, these dystonic movements are relatively mild, and so far overshadowed by the chorea, that it may take some close and dedicated observation to discern them. A small minority of patients may also have myoclonus (Carella *et al.* 1993; Thompson *et al.* 1994). The fact that both dystonia and myoclonus may appear in Huntington's disease, however, should not distract one from the overwhelmingly prominent part that chorea plays in adult-onset cases. Dysarthria often occurs, as does dysphagia, which may lead to aspiration (Leopold and Kagel 1985). The chorea eventually makes almost all purposeful activity, whether eating, dressing, or walking, almost impossible, and patients eventually become chairbound or bedridden. Curiously, at the end, the chorea may gradually disappear, to be replaced by a rigid, akinetic state (Feigin *et al.* 1995). In advanced cases, seizures may also occur.

Depression may manifest with depressed mood, loss of interest, decreased motivation and insomnia, and, interestingly, tends to lessen as the disease progresses (Paulsen *et al.* 2005). Suicidal ideation is common (Paulsen *et al.* 2005), and roughly one-quarter of all patients will attempt suicide (Farrer 1986), with completed suicides occurring in 4 percent (Schoenfeld *et al.* 1984) to 10 percent (Baliko *et al.* 2004) of all cases.

Personality change may manifest with any or all of poor judgment, impulsivity and disinhibition, apathy, and irritability and agitation, which may be quite extreme (Nance and Sanders 1996).

Psychosis typically manifests with delusions of persecution. Hallucinations may also occur, and these are typically either auditory or visual in type: olfactory and gustatory hallucinations are rare. Interestingly, Huntington's disease, albeit rarely, may present with a psychosis (Caine and Shoulson 1983).

Dementia presents with decreased concentration and poor attention, followed closely by difficulties in planning, organizational skills, and perseveration. Eventually memory begins to fail. Notably, such 'cortical' symptoms as aphasia and apraxia are generally absent. Over time the dementia may become quite profound; however, in some cases the dysarthria seen in Huntington's disease may conspire to make patients appear more demented than they actually are. Of interest, as in the case of psychosis, dementia may also, and again quite rarely, constitute the presentation of Huntington's disease (Cooper *et al.* 2006).

Other symptoms may also occur. Although, as noted earlier, depression occurs in roughly one-half of all cases, this is not the only mood change possible in Huntington's disease, and indeed roughly one-third of all patients will develop euphoria (Paulsen *et al.* 2001). Importantly, however, this euphoria is generally not accompanied by other manic

symptoms, and a full manic syndrome is in fact rare. Obsessions and compulsions have also been noted: one study (Anderson *et al.* 2001) found obsessions in roughly one-half of all cases and compulsions in about one-quarter, while another (Marder *et al.* 2000) found the combination of both in about one-fifth of patients.

LATE ONSET

As in the case of adult-onset disease, late-onset cases (Myers *et al.* 1985) also present with chorea in almost all cases. Depression, personality change, and psychosis may also occur, but tend to be much milder. Dementia also occurs, but again is typically milder, and in some late-onset cases there may be little or no cognitive decline even with a 15-year follow-up (Britton *et al.* 1995). Rarely, late-onset cases may present not with chorea, but with a levodopa-responsive parkinsonism (Reuter *et al.* 2000).

JUVENILE OR ADOLESCENT ONSET

Classically, juvenile- or adolescent-onset Huntington's disease (Bird and Paulson 1971; Campbell *et al.* 1965; Hansotia *et al.* 1968; Jervis 1979; Siesling *et al.* 1997) is characterized by what is known as the 'Westphal variant' in which, rather than chorea, there is often a rigid, akinetic state that is generally accompanied by severe dementia, often with seizures, myoclonus, and ataxia. Recent work (Ribai *et al.* 2007), however, has indicated a somewhat different presentation of juvenile- or adolescent-onset cases characterized by a movement disorder with a combination of dystonia, parkinsonism, and chorea wherein dystonia was the most prominent feature; in these cases cognitive decline, seizures, and ataxia were also noted.

LABORATORY TESTING

Computed tomography or MRI scanning may be normal early in the course of the disease; over time, however, because of atrophy of the caudate nuclei, there is dilation of the frontal horns of the lateral ventricles, yielding a characteristic 'butterfly' configuration. Cortical atrophy, especially of the frontal, parietal, and temporal lobes, may also be seen.

As noted below, Huntington's disease is a fully penetrant, autosomal dominant disease, and genetic testing is both highly specific and sensitive (International Huntington Association 1994; Kremer *et al.* 1994).

Course

Huntington's disease is relentlessly progressive. In adult-onset cases, death, often from aspiration pneumonia, occurs after anywhere from 10 to 20 years, with an average survival of 15–17 years. Late-onset cases pursue a much more leisurely course, while in juvenile- or adolescent-onset cases death typically occurs within 10 years.

Etiology

Huntington's disease is an autosomal dominant disorder: the affected gene (variously known as *IT15*, *HTT*, or *HD*) codes for a protein known as huntingtin and is located on chromosome 4. This gene normally contains anywhere from 10 to 29 CAG trinucleotide repeats, whereas in Huntington's disease 36 or more repeats are found (Kremer *et al.* 1994). Of note however, expansions to the range of 36–39 repeats may be associated with variable clinical penetrance, whereas expansions to 40 or more repeats is associated with full penetrance. Although a positive family history is found in almost all cases, spontaneous mutations, although rare (Ramos-Arroyo *et al.* 2005), do occur, and in some of these cases it appears that the patient's parent, although clinically unaffected, harbored an expansion in the 36–39 repeat range which then underwent further expansion during gametogenesis, thus passing a fully penetrant mutation to the child (Myers *et al.* 1993). Other cases with a negative family history may be due to the early death of a parent who was carrying the mutation or to uncertain parentage. Although an inverse correlation has been found between the number of repeats and both age of onset (Craufurd and Dodge 1993; Marder *et al.* 2002) and severity of chorea (Illarioshkin *et al.* 1994), there does not appear to be any correlation with the appearance of either depression or psychosis (Tsuang *et al.* 2000; Zappacosta *et al.* 1996).

Macroscopically (Mann *et al.* 1993), atrophy is noted in the caudate (as illustrated in Figure 9.7), less so in the putamen, and to a degree in the frontal, parietal, and temporal lobes. Microscopically (Vonsattel *et al.* 1985), neuronal loss and reactive astrocytosis is noted in these areas, and in the caudate nucleus, spiny neurons are lost first (Mann *et al.* 1993; Myers *et al.* 1988). Surviving neurons may display intranuclear inclusions, which contain the mutant huntingtin protein (Sapp *et al.* 1997).

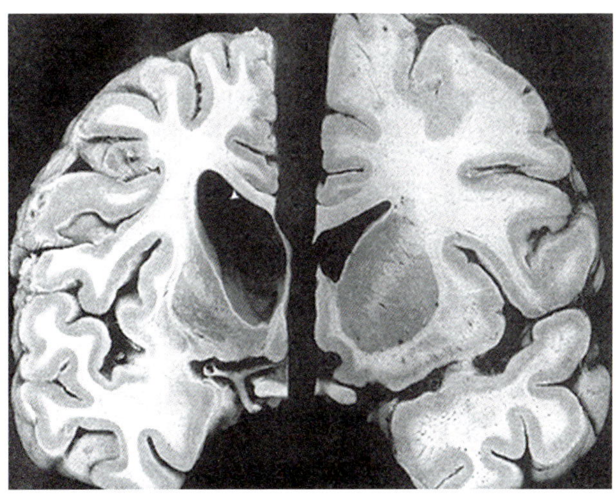

Figure 9.7 Marked atrophy of the caudate, with compensatory ventricular dilation, in a case of Huntington's disease compared with a control brain on the right. Reproduced with kind permission from Graham and Lantos (1996).

Differential diagnosis

Given that Huntington's disease is the most common cause of adult-onset progressive chorea, genetic testing is appropriate in all such cases. When testing is negative, consideration should be given to certain other disorders, discussed in Chapter 4.4, including choreoacanthocytosis, the McLeod syndrome, dentatorubropallidoluysian atrophy, spinocerebellar ataxia (specifically SCA 17), Huntington's disease-like syndrome 2, Fahr's syndrome, and senile chorea. Another disease to consider is schizophrenia, and before the advent of genetic testing, this was a much-discussed differential possibility. As noted earlier, Huntington's disease, albeit rarely, may present with a psychosis, which is then followed by the appearance of chorea. A similar scenario may occur in schizophrenia complicated by tardive dyskinesia; however, the chorea of tardive dyskinesia is different from that seen in Huntington's disease: in tardive dyskinesia the chorea is generally stereotyped and repetitive, almost never involves the forehead, and generally leaves gait relatively unaffected. This, of course, is in contrast to the chorea of Huntington's disease, which is mercurial, flitting from one body part to another in an unpredictable fashion, and which typically also affects the forehead and gait.

Treatment

Traditionally, chorea has been treated with an antipsychotic, and accumulated clinical experience attests to the effectiveness of these agents. Haloperidol was for many years the first choice, and in open work (Barr *et al.* 1988) was effective in doses of 1.5–10 mg/day; side-effects, including lethargy, akathisia and parkinsonism, are, however, common (Shoulson *et al.* 1981), and in the only double-blinded study (Leonard *et al.* 1975) utilizing haloperidol, it was no more effective than placebo. Attention, therefore, shifted to second-generation agents: risperidone, in a retrospective study (Duff *et al.* 2008), was effective in doses of 0.5–4 mg; olanzapine, in two open studies (Bonelli *et al.* 2002; Paleacu *et al.* 2002), was effective in doses of 10–30 mg; clozapine, in a double-blinded study (van Vugt *et al.* 1997), was more effective than placebo in doses up to 150 mg, but at the cost of severe side-effects; and, finally, aripiprazole (Brusa *et al.* 2009), at a mean dose of 11 mg (range 5–15 mg), in a double–blind crossover study, was as effective as tetrabenazine at a mean dose of 95 mg (range 75–125 mg), an agent discussed below. With these data in mind, it appears that aripirazole is a reasonable first choice; if another second-generation agent were to be used, then either risperidone or olanzapine may be considered, and, given its overall better tolerability, risperidone is probably preferable.

Tetrabenazine recently became available for the treatment of chorea, and has been subjected to two blinded trials. In a single-blinded study (Ondo *et al.* 2002) which used patients as their own controls, it was effective at a mean dose of 62.5 mg (range 25–150 mg). In a double-blinded study (Huntington Study Group 2006), it was more effective than placebo: although the mean dose was not given, over one-half of all the patients in the study were treated at the maximum allowable dose of 100 mg. These doses, however, appear higher than those employed in routine practice: in a long-term open follow-up study (Fasano *et al.* 2008a), the mean dose was 35 mg. As many clinicians are not familiar with tetrabenazine, some discussion may be in order regarding its pharmacology and side-effect profile. Tetrabenazine and its two active metabolites bind to the vesicular mononamine transporter type 2 (VMAT2), thus preventing uptake of the monoamines dopamine, serotonin, and norepinephrine into intraneuronal storage vesicles, leading to their depletion and to an overall reduction in dopaminergic, serotoninergic, and noradrenergic tone. Bioavailability is roughly 75 percent and is not influenced by food. Tetrabenzine is rapidly converted to its active metabolities, α-dihydrotetrabenazine (α-HTBZ) and β-dihydrotetrabenazine (β-HTBZ), which reach peak blood levels in 1–1.5 hours; both of these are in turn metabolized by CYP2D6, with a half life of 4–8 hours for α-HTBZ and 2–4 hours for β-HTBZ. Protein binding of these metabolites ranges from 60 percent to 85 percent. As might be expected, strong inhibitors of CYP2D6 (e.g. fluoxetine, sertraline, paroxetine) increase blood levels of the metabolites, and for α-HTBZ this increase is roughly threefold, while for β-HTBZ there is a ninefold increase. Another drug–drug interaction to be aware of is with MAOIs: with uptake of norepinephrine into intraneuronal storage vesicles blocked, the stage may be set for a hypertensive crisis whenever a non-selective MAOI (e.g. phenelzine or tranylcypromine), or even a selective MAOI (i.e. rasagaline at any dose and selegeline at doses over 10 mg), is employed. Side-effects are common, and may be severe: somonolence occurs in 18–31 percent and insomnia in 22 percent; 19–35 percent of patients develop depression, 15 percent anxiety, and 9 percent irritability; akathisia occurs in 20 percent and parkinsonism in 3–15 percent. When utilizing tetrabenazine, begin at 12.5 mg once daily, with increases in similar increments every week as tolerated up to a dose of 50 mg/day. At this point, if the patient is on a strong inhibitor of CYP2D6, the dose should not be increased further. If the patient is not on a CYP2D6 inhibitor, then the patient should undergo genotyping regarding CYP2D6 status: if the patient is a poor metabolizer, the dose should not be increased further; however, if the patient is an intermediate or extensive metabolizer, the dose may titrated up, again in 12.5-mg increments weekly, to a recommended maximum of 100 mg/day. Doses of 25 mg should be divided into a b.i.d. schedule, while higher doses should be divided into a t.i.d. schedule. Importantly, given the risk of depression, tetrabenazine is contraindicated in patients with depression or suicidal ideation and should only be used with great caution in those with a history of either of these.

In addition to the antipsychotics and to tetrabenazine, other agents have also been studied, with variable results. Amantadine was subjected to two double-blinded studies: at 300 mg/day it was no better than placebo (O'Suilleabhain and Dewey 2003); however, at 400 mg/day, it was (Verhagen Metman *et al.* 2002). Riluzole has also been utilized in two double-blinded studies, and at 100 mg/day it was no better than placebo (Landwehrmeyer *et al.* 2007), whereas at 200 mg/day, it was (Huntington Study Group 2003). Ethyl-eicosapentaenoic acid was studied in four double-blinded trials: two small studies (Puri *et al.* 2002; Vaddadi *et al.* 2002) found it more effective than placebo, while two larger studies (Huntington Study Group TREND-HD Investigators 2008;

Puri *et al.* 2005) found no advantage. Donepezil, in one double-blinded study (Cubo *et al.* 2006), was not superior to placebo. Open studies have found some benefit with levetiracetam (de Tommaso *et al.* 2005; Zesiewicz *et al* 2006), memantine (Ondo *et al.* 2007), and rivastigmine (de Tommaso *et al.* 2007), while additional open studies found no benefit with lamotrigine (Kramer *et al.* 1999) and valproic acid (Pearce *et al.* 1977). Finally, in one case report, deep brain stimulation of the globus pallidus was reported to yield dramatic benefit (Fasano *et al.* 2008b), while transplantation of human fetal neuroblasts into the striata did not appear to yield any lasting overall benefit (Bachoud-Levi *et al.* 2006).

With all these data in hand, either an antipsychotic or tetrabenazine appear to be reasonable options. Given, however, the considerable side-effect burden associated with tetrabenazine, and the fact that aripiprazole, a well-tolerated agent, appears as effective as tetrabenazine, it seems prudent to begin with an antipsychotic such as aripiprazole and to hold tetrabenazine in reserve. Consideration may also be given to amantadine and/or ethyl-eicosapentaenoic acid, either as monotherapy in mild cases or as 'add-on' treatment to either an antipsychotic or tetrabenazine. The place of riluzole, given the high dose utilized, is not at present clear, and consequently use of this agent might best be reserved to research centers.

With regard to the treatment of depression, although there are no blinded studies, various antidepressants have been reported as effective, including tricyclic antidepressants (Shoulson *et al.* 1981), venlafaxine (Holl *et al.* 2010), and the SSRIs fluoxetine (Como *et al.* 1997) and sertraline (Slaughter *et al.* 2001). Olanzapine, in a dose of only 5 mg, was also reported as effective (Squitieri *et al.* 2001). ECT was effective in a case series of six patients (Ranen *et al.* 1994); although one of the patients in this series showed some worsening of chorea with ECT, there is also a case report (Beale *et al.* 1997) of a non-depressed patient who was given ECT for chorea and improved. Given the overall poor tolerability of the tricyclics (Shoulson *et al.* 1981), it is probably most reasonable to begin with either venlafaxine or an SSRI.

With regard to the treatment of the personality change, the literature offers little guidance. With regard to agitation, case reports indicate effectiveness for both fluoxetine (De Marchi *et al.* 2001) and sertraline (Ranen *et al.* 1996). In practice, antipsychotics are often used not only for agitation, but also for impulsivity.

Psychosis, naturally, is treated with an antipsychotic, and of the antipsychotics used for chorea, olanzapine, in an open study (Paleacu *et al.* 2002), was effective for psychosis at a mean dose of 11 mg, with a range of 10–30 mg.

The treatment of the dementia of Huntington's disease, remarkably, has not been at all extensively researched. One open study of rivastigmine (de Tommaso *et al.* 2007) found a trend toward improvement, while a double-blinded study (Cubo *et al.* 2006) utilizing donepezil found no advantage over placebo.

Certain other treatment modalities may also be mentioned. Specialized walkers may allow patients to remain ambulatory. Dysphagia diets may reduce the risk of aspiration; however, eventually percutaneous endoscopic gastrostomy tube placement may become necessary. If seizures do occur, levetiracetam is a reasonable choice given its possible effectiveness for chorea; conversely, phenytoin, gabapentin, and divalproex should only be used with caution as they may cause chorea. Obsessions and compulsions may respond to an SSRI, as was reported in one case with sertraline (Patzold and Brune 2002). Simple euphoria does not require treatment; should mania occur, however, consideration may be given to one of the antipsychotics.

Ongoing research efforts are under way to find treatments that may retard the progression of the disease, and many patients are highly motivated to participate.

As noted earlier, genetic testing is now available, and should be offered to patients with a compatible clinical picture, and to unaffected relatives. Extensive counseling resources should be available for instances when the test is positive.

9.11 CHOREOACANTHOCYTOSIS

Choreoacanthocytosis, also known as chorea–acanthocytosis, is a rare genetic disorder, generally occurring on an autosomal recessive basis, which classically presents with a combination of chorea and acanthocytosis. As discussed further below, under Differential diagnosis, this combination places choreoacanthocytosis in a larger group of disorders, known as neuroacanthocytosis, some of whose members may present in a fashion similar to choreoacanthocytosis.

Clinical features

The onset is generally in the late twenties or early thirties but may occur anywhere from late childhood to the seventh decade. Clinically (Critchley *et al.* 1968; Feinberg *et al.* 1991; Hardie *et al.* 1991; Kartsounis and Hardie 1996; Lossos *et al.* 2005; Sakai *et al.* 1981), most cases come to attention owing to a gradually progressive chorea, which, over time, is often joined by other abnormal movements, such as dystonia, mild parkinsonism, or tics, which may be either motor or vocal. A classic symptom, seen, however, in only a minority, is self-mutilating lip or tongue biting. Importantly, this self-mutilation is outside the patient's control: one patient (Medalia *et al.* 1989), despite attempting to 'restrain herself…by placing her fingers or a folded towel in her mouth…nevertheless had bleeding and scarred lesions of her oral mucosa and lips'.

Roughly one-half of all patients will also develop a dementia, personality change, or both: the personality change has frontal lobe features, with apathy, neglect of personal hygiene, poor judgment, and, in some cases, disinhibition. Psychosis and obsessions and compulsions may also occur.

Seizures, which may be either complex partial or grand mal, occur in close to one-half of all cases, and the vast majority will also have a peripheral sensorimotor polyneuropathy.

Of note, choreoacanthocytosis, although generally presenting with chorea, may rarely present with motor and vocal tics (similar to those seen in Tourette's syndrome) (Ruiz-Sandoval *et al.* 2007), psychosis (similar to that seen in schizophrenia)

(Bruneau *et al.* 2003), obsessions and compulsions (similar to that seen in obsessive–compulsive disorder) (Walterfang *et al.* 2008), or seizures (Al-Asmi *et al.* 2005), and these atypical presentations may predate chorea by a decade or more.

Serum creatine kinase is elevated in most cases, and MRI scanning may reveal atrophy of the caudate nucleus.

Acanthocytosis, to a degree of 10 percent or more, is seen on peripheral smears; importantly, the smears must be fresh, wet preparations and, given the chance of false-negatives, at least three smears should be examined (Hardie *et al.* 1991). Genetic testing is available.

Course

The disease is gradually progressive, with death, on average, after 14 years.

Etiology

Choreoacanthocytosis occurs secondary to mutations in the *VPS13A* gene on chromosome 9q21 (formerly known as *CHAC*), which codes for a protein known as chorein. Although the vast majority of cases are inherited on an autosomal recessive basis, autosomal dominant cases have been reported (Ishida *et al.* 2009). Of note, there may be considerable phenotypic heterogeneity within the same family.

Macroscopically, there is atrophy of the caudate, putamen, and, to a lesser degree, the globus pallidus. Microscopically, neuronal loss and gliosis are seen in these structures (Hardie *et al.* 1991; Vital *et al.* 2002) and in the substantia nigra (Rinne *et al.* 1994b).

Differential diagnosis

Of the adult-onset disorders capable of causing a progressive choreiform disorder discussed in Chapter 4.4, those most likely to be confused with choreoacanthocytosis are the McLeod syndrome, Huntington's disease-like 2, and Huntington's disease itself. The McLeod syndrome (Danek *et al.* 2001) and Huntington's disease-like 2 (Walker *et al.* 2003) are both also characterized by acanthocytosis, and, along with the infantile-onset disorder abetalipoproteinemia, make up the group of disorders known as the neuroacanthocytoses. Both the McLeod syndrome and Huntington's disease-like 2 are also clinically similar to choreoacanthocytosis; however, the McLeod syndrome is X-linked while Huntington's disease-like 2 is autosomal dominant: genetic testing is available for both of these. Huntington's disease is also autosomal dominant, but is not associated with acanthocytosis, and clinically is not associated with either mutilative lip or tongue biting, or a peripheral neuropathy.

Treatment

The treatment of choreoacanthocytosis proceeds along much the same lines as that for Huntington's disease.

9.12 FXTAS

FXTAS, also known as the fragile X-associated tremor ataxia syndrome, is a recently described inherited disorder characterized by the gradual development of tremor, ataxia, and, in a minority, dementia, in middle-aged or older men. The prevalence of this disorder is as yet not known with certainty, but it may be common. As noted below, it occurs secondary to a 'premutation' in the fragile X mental retardation 1 gene (*FMR1*), and this permutation is found in 1 out of 813 males: if indeed, as again discussed below, the penetrance of this premutation rises to 75 percent in men aged 80 or more years, then FXTAS would indeed be not at all uncommon. Although this syndrome has been reported in females (Hagerman *et al.* 2004), it is far more common and severe in males.

Clinical features

The clinical symptomatology seen in males has been described in a number of papers (Greco *et al.* 2006; Hagerman *et al.* 2001; Hall *et al.* 2005). Although the age of onset ranges from the fourth to the ninth decade, most patients fall ill in the seventh decade with the gradual onset of ataxia and tremor. The tremor, although typically of the intention type, may also have a rest or postural component. Over time, a mild parkinsonism may also accrue, with rest tremor, rigidity, and bradykinesia. A peripheral neuropathy is common, and some patients may experience autonomic symptoms, such as erectile dysfunction.

Cognitive deficits eventually appear, and anywhere from one-quarter to one-half of all patients may eventually become demented: short-term memory loss, poor concentration, concreteness, and poor judgment are common. A personality change (Bacalman *et al.* 2006; Bourgeois *et al.* 2007) of the frontal lobe type may also be seen, with disinhibition, perseveration, and inappropriate jocularity; mood changes are also common, with agitation, irritability, apathy, or depression.

As noted earlier, this syndrome may also be seen in women (Chonchaiya *et al.* 2010; Hagerman *et al.* 2004), but the symptomatology is much milder; in some cases, premature ovarian failure with early menopause may also occur.

Magnetic resonance imaging may reveal both cerebral and cerebellar cortical atrophy. The most characteristic finding, however, is the 'middle cerebellar peduncle' sign: on T1-weighted scans these peduncles show decreased signal intensity, while on T2-weighted scans there is increased signal intensity (Brunberg *et al.* 2002). Although quite valuable from a diagnostic point of view, this sign is not universally present, and indeed in one study (Adams *et al.* 2007) of males with FXTAS with a mean age of 65 years, it was present in just a little over one-half.

Genetic testing is available.

Course

This disorder is relentlessly but gradually progressive: in one study of affected males, a need for a cane or some other walking aid became apparent after a mean of 15 years (Leehey *et al.* 2007).

Etiology

As might be gathered from the name of this syndrome, FXTAS bears a relationship to the fragile X syndrome: both occur as the result of a mutation in the fragile X mental retardation 1 (*FMR1*) gene, located on the long arm of the X chromosome.

Normally, the *FMR1* gene contains anywhere from 5 to 55 CGG trinucleotide repeats. The mutation in question involves an increase in the number of these repeats beyond the normal range. Increases into the range of 55 to 200 repeats have traditionally been termed a 'premutation', whereas increases above 200 repeats have been considered to constitute a 'full' mutation. Individuals with full mutations, especially males, typically develop the fragile X syndrome with mental retardation, as discussed in Chapter 10.6. Until recently, it had been thought that those with a 'premutation' were merely carriers who remained free of any symptoms. The recognition of FXTAS, however, has now made it clear that this 'premutation', rather than being benign, may cause symptoms in its own right. Indeed, among men with the premutation, the penetrance progressively rises from roughly 20 percent in those from 50 to 59 years of age up to 75 percent in those aged 80 years or above (Jacquemont *et al.* 2004).

Macroscopically, there is cerebral and cerebellar cortical atrophy. Microscopically (Greco *et al.* 2002, 2006), there is neuronal loss in these areas, and status spongiosus in the cerebellar white matter and the middle cerebellar peduncles. Intranuclear inclusions are found in both astrocytes and surviving neurons.

Differential diagnosis

Spinocerebellar ataxia is suggested by the autosomal dominant pattern of inheritance. MSA may be difficult to distinguish on clinical grounds, and the differential may rest on genetic testing.

One clue to the diagnosis rests in the family history. When the diagnosis is considered in a middle-aged or elderly man, one should inquire whether any of his grandchildren have mental retardation. If the middle-aged or elderly man in question does have FXTAS, then any daughters would also have the 'premutation'. Increases in repeat length are likely during oogenesis and, consequently, a daughter may have passed on a 'full' mutation to one of her children, who, in turn, might express the fragile X syndrome with mental retardation. As noted above, FXTAS is much less likely in females and, when it does occur, it is less severe; this is probably due to the fact that females who do have the 'premutation' also have a normal gene on the other X chromosome. In cases of middle-aged or elderly women suspected of having FXTAS, the family history may also help; however, here one looks only to the children and not the grandchildren for the telltale finding of mental retardation.

Treatment

The general treatment of dementia is described in Chapter 6.3, and of the frontal lobe-type personality change in Chapter 8.2; as yet there are no controlled studies focusing on FXTAS itself.

9.13 SENILE CHOREA

Senile chorea is a rare disorder characterized by the gradual onset of chorea late in life in the absence of a family history.

Clinical features

The onset is insidious and generally in the seventh decade. Clinically, one sees the gradual progression of a mild chorea, generally involving the upper extremities or the face (Delwaide and Desseilles 1977; Klawans and Barr 1981; Shinotoh *et al.* 1994; Varga *et al.* 1982). There are no other abnormal movements and no dementia.

Course

The disorder is very gradually progressive.

Etiology

Microscopically, neuronal loss is noted in the caudate. The mechanism is not known.

Differential diagnosis

Of the disorders capable of causing chorea, discussed in Chapter 4.4, special consideration should be given to the following (Warren *et al.* 1998): late-onset Huntington's disease, the antiphospholipid syndrome, tardive dyskinesia, and Fahr's syndrome; of these, Huntington's disease is the most probable and, hence, genetic testing is in order (Garcia Ruiz *et al.* 1997). Edentulousness should also be considered, as it may be accompanied by orofacial movements that resemble chorea (Koller 1983).

Treatment

In general, no treatment other than simple reassurance is required. In those rare cases in which the chorea becomes severe enough to interfere with functioning, consideration may be given to treatment with antipsychotics, as described for Huntington's disease.

9.14 BENIGN HEREDITARY CHOREA

This is a rare autosomal dominantly inherited disorder, characterized by a childhood-onset chorea that is generally non-progressive.

Clinical features

Onset may be anywhere from infancy to late childhood, and clinically (Behan and Bone 1977; Breedveld *et al.* 2002; Fernandez *et al.* 2001) there is a gradual development of generalized chorea that may be accompanied by a subtle degree of ataxia; importantly, there is no dementia. In some cases, hypothyroidism may also occur (Asmus *et al.* 2005).

Course

After the chorea has become established, although in most cases there is no change over time, one may see in a minority either a mild progression (Kleiner-Fisman *et al.* 2003) or, conversely, in adolescence or early adult years, a partial remission of symptoms (Fernandez *et al.* 2001).

Etiology

As noted, this is an autosomal dominant condition and mutations have been found in the thyroid transcription factor 1 gene (*TITF1*) (Asmus *et al.* 2005). Although routine gross and microscopic studies are normal, immunofluorescent studies have revealed a lack of interneurons within the striatum (Kleiner-Fisman *et al.* 2005).

Differential diagnosis

Other disorders capable of causing chorea in childhood, as discussed in Chapter 4.4, include Sydenham's chorea and cerebral palsy. The autosomal dominant pattern displayed and the course, however, are quite distinctive.

Treatment

Although there are no established treatments, case reports indicate benefit from either levodopa (Asmus *et al.* 2005) or methylphenidate (Friederich 1996); consideration may also be given to treatment with antipsychotics, as discussed for Huntington's disease.

9.15 DENTATORUBROPALLIDOLUYSIAN ATROPHY

The unwieldy name of this disorder is derived from the main structures that show pathological changes, namely the dentate nucleus, the nucleus ruber (the red nucleus), the globus pallidus, and the corpus Luysii (the subthalamic nucleus). Dentatorubropallidoluysian atrophy, also known simply as DRPLA, is a rare autosomal dominantly inherited disorder, whose clinical presentation is strongly influenced by the age of onset. Some authorities consider DRPLA to be one of the spinocerebellar ataxias; however, given that DRPLA may be characterized by chorea rather than ataxia, this seems to be inappropriate nosologic 'lumping' and the disorders are considered separately in this text.

Clinical features

The onset is gradual, regardless of whether it is early or late. Early-onset cases (Becher *et al.* 1997) present in childhood or early adolescence and are characterized by a variable combination of myoclonus, seizures, ataxia, and dementia. Late-onset cases (Becher *et al.* 1997; Iizuka *et al.* 1984; Nielsen *et al.* 1996; Villani *et al.* 1998 Warner *et al.* 1994, 1995; Wardle *et al.* 2008; Yoshii *et al.* 1998) present in early adult years and are typified by either chorea or ataxia, or a combination of the two; in a minority of cases, one may also see a mild degree of dystonia or parkinsonism. In adults, one also sees the eventual development of a dementia, which may be accompanied by prominent hallucinations and delusions: rarely, DRPLA may present with a psychosis in adult years (Adachi *et al.* 2001). A small minority of adults may also have seizures. MRI (Koide *et al.* 1997) may reveal atrophy of the cerebellum and brainstem, and, on T2-weighted scans, increased signal intensity in the globus pallidus; in adult-onset cases characterized by dementia, widespread increased signal intensity may be seen in the cerebral white matter. Genetic testing is available.

Course

DRPLA is a gradually progressive disorder.

Etiology

As noted earlier, DRPLA is an autosomal dominantly inherited disorder: mutations exist in the gene for atrophin-1, found on chromosome 12, which consist of an expansion of a normally occurring CAG trinucleotide repeat (Nagafuchi *et al.* 1994). There is considerable phenotypic heterogeneity both within and between families (Potter *et al.* 1995), and, as with most other disorders associated with such trinucleotide repeats, anticipation may be seen. At autopsy (Becher *et al.* 1997; Goto *et al.* 1982; Iizuka *et al.* 1984; Warner *et al.* 1994), neuronal loss and gliosis is found not only in the namesakes of this disorder (the dentate nuclei, red nucleus, globus pallidus, and subthalamic nucleus), but also in the striatum, substantia nigra, inferior olives, thalamus, and cerebral cortex. In surviving neurons, ubiquitinated intranuclear inclusions, composed largely of mutant atrophin-1, may be found.

Differential diagnosis

In adults, the differential is influenced by the nature of the movement disorder: when chorea is prominent, consideration is given to Huntington's disease, and when ataxia is foremost, spinocerebellar ataxia and Gerstmann–Streissler–Schencker disease enter the differential. Genetic testing is often required to make the differential.

Treatment

There is no specific treatment. The general treatment of dementia is discussed in Chapter 6.3; in cases in which chorea is problematic, treatment with antipsychotics, as described for Huntington's disease, may be considered.

9.16 WILSON'S DISEASE

Wilson's disease, also known as hepatolenticular degeneration, was first fully described by Kinnier Wilson in 1912 (Wilson 1912). It is an autosomal recessively inherited disorder characterized pathologically by copper deposition in the liver and the brain. Although relatively uncommon, occurring in perhaps 1/30 000 of the general population, it is a diagnosis which, given the availability of effective treatment, must not be missed.

Clinical features

The clinical features of Wilson's disease have been described in several large series (Akil *et al.* 1991; Dening and Berrios 1989; Machado *et al.* 2006; Starosta-Rubinstein *et al.* 1987; Taly *et al.* 2007; Walshe and Yealland 1992). Although the onset is typically between late childhood and early adulthood, averaging out at 15 years, the range is wide, from early childhood to the sixth decade. The presentation may be with a movement disorder, personality change, or psychosis, and over time most patients eventually develop a combination of these features.

The movement disorder may consist of dystonia, chorea, tremor, or parkinsonism. Dystonia may present with torticollis, dystonia of the upper or lower extremity, oculogyric crisis (Lee *et al.* 1999), or facial dystonia, with, classically, a fixed, vacuous, wide-mouthed dystonic smile. Chorea may involve either the upper or the lower extremities. The tremor may be rhythmic or irregular; at times, one may see the classic 'wing-beating' tremor in which a rhythmic elevation and lowering of the upper extremities, combined with flexion at the elbows, gives an overall appearance of a frightened bird flapping its wings. Dysarthria may also occur, and over time may become quite severe.

Personality change (Bridgman and Smyth 1944; Dening and Berrios 1989; Walshe and Yealland 1992) is generally characterized by lability, disinhibition, and, at times, bizarre behavior.

Psychosis is characterized by hallucinations and delusions (Dening and Berrios 1989; Gysin and Cooke 1950) and may be quite bizarre, with Schneiderian first-rank symptoms. Indeed, one of Wilson's (1912) patients heard 'God and the devil talking to him simultaneously' and said that he was 'influenced, willed or hypnotized to do certain things'. In one case (Davis and Borde 1993), stuporous catatonia dominated the presentation.

Other symptoms include depression, cognitive impairment, and seizures. Depression may eventually appear in roughly one-fifth of patients. Cognitive impairment may occur in a similar percentage: although this is typically mild, rarely one may see a dementia (Starosta-Rubinstein *et al.* 1987). Of note, the severe dysarthria that may develop may make cognitive testing very difficult and leave the examiner with a false impression of severe cognitive impairment. Seizures may also occur, but are uncommon, being found in only about 5 percent of cases (Dening *et al.* 1988).

Another very important sign is the Kayser–Fleischer ring. This is a golden-brown discoloration of the corneal limbus, which, although sometimes visible to the naked eye, is best seen by slit-lamp examination. Although some texts state that this Kayser–Fleischer ring is present in all cases with neuropsychiatric symptomatology, this is simply not the case, and there are well-documented instances where it is lacking (Demirkiran *et al.* 1996).

As noted earlier, copper deposition is also found in the liver, and indeed, as discussed under Etiology below, this is where pathology is first found. Thus, in all cases of neuropsychiatric Wilson's disease, hepatic pathology exists. Such pathology may range from subclinical involvement, manifested perhaps only by elevated transaminase levels, through clinical hepatitis, with nausea and abdominal pain, up to cirrhosis. In some cases, an hepatic encephalopathy may occur.

A Coombs-negative hemolytic anemia may also occur.

Magnetic resonance imaging (Sinha *et al.* 2006) may reveal cerebral cortical atrophy and ventricular dilatation. On T2-weighted or FLAIR imaging, increased signal intensity may be found in the putamen, caudate, or thalamus, and, in a small minority, one may see the virtually pathognomonic 'face of the giant panda sign' in the mesencephalon: here, there is a combination of decreased signal intensity in the red nuclei and increased signal intensity elsewhere, creating the appearance of a panda's face, with the red nuclei being its 'eyes'.

Classic laboratory serum findings include a low ceruloplasmin level, a low total copper level, and an increased free copper level. Given, however, that ceruloplasmin levels may be normal in a small minority of cases of Wilson's disease, and that, conversely, low ceruloplasmin levels may be found in various other hepatic diseases, one should always perform a 24-hour urine check, which will almost always reveal an elevated copper excretion. Should doubt still exist, a liver biopsy is in order, which will reveal not only a characteristic histopathology, but also an increased copper content.

Given the large number of possible mutations, genetic testing is not practical.

Course

Although there may be partial, temporary remissions, the overall course is one of progression, with death within 5–10 years.

Etiology

Wilson's disease is essentially a disease of copper metabolism. Normally, after intestinal absorption, copper is taken up by hepatocytes. Within hepatocytes, copper is then disposed of via two mechanisms, both of which are mediated by a protein known as ATP7B: this protein is an ATPase which, on the one hand, binds copper onto apoceruloplasmin to create ceruloplasmin, which is then secreted into the systemic circulation, and, on the other hand, incorporates copper into vesicles via which it is secreted into the bile, which constitutes the main mechanism whereby the body is rid of copper. In Wilson's disease, mutations in the gene for ATP7B impair its activity, with two results. First, less ceruloplasmin is produced, and hence serum ceruloplasmin levels are typically low. Second, copper accumulates in hepatocytes, leading to their dysfunction and 'spillage' of copper into the systemic circulation. Given that, normally, over 90 percent of serum copper is found bound in ceruloplasmin, this state of affairs leads to the curious classic serum findings noted earlier, wherein ceruloplasmin and total (bound plus free) copper levels are low, but the free copper level, reflecting the spillage from hepatocytes, is increased. This increased serum free copper then becomes deposited in the brain, primarily in a pericapillary distribution. Macroscopically (Howard and Royce 1919), there is atrophy and a brownish discoloration of the striatum, with, in advanced cases, cavitation (Wilson 1912) and a mild degree of cortical atrophy. Microscopically, there is neuronal loss and astrocytosis in the striatum (more so the putamen than the caudate) and to a lesser degree in the globus pallidus; other affected structures include the cerebral cortex (especially the frontal lobes; Barnes and Hurst 1926), thalamus, red nucleus, substantia nigra, dentate nucleus, and cerebellar cortex. Copper deposition in Descemet's membrane gives rise to the classic Kayser–Fleischer ring.

As noted earlier, Wilson's disease is an autosomal recessive disorder, and it occurs secondary to mutations in the gene for ATP7B on chromosome 13 (Bull *et al.* 1993; Chelly and Monaco 1993). Over 200 different mutations exist, and most patients are compound heterozygotes.

Differential diagnosis

Given the pleomorphic symptomatology of Wilson's disease, the differential diagnosis is large. In practice, given the availability of treatment and the consequent importance of not missing the diagnosis, it is appropriate to test for Wilson's disease in any young person with a clinical presentation consistent with it that cannot be readily and fully accounted for by another disease process. One disorder in particular deserves mention as it may mimic Wilson's disease, not only clinically, but also with respect to certain laboratory values. Hereditary ceruloplasmin deficiency, like Wilson's disease, is characterized by low ceruloplasmin and total serum copper levels, but, unlike Wilson's disease, it is characterized by a low 24-hour urinary copper level (Kawanami *et al.* 1996).

Treatment

Treatment is aimed at reducing the total copper burden. Several methods are available, including chelation with penicillamine, trientine, or tetrathiomolybdate; treatment with zinc; and limiting copper intake.

Chelation has traditionally been accomplished with penicillamine; however, this agent has multiple, severe side-effects and also, in a significant minority of patients, will either exacerbate pre-existing symptoms (Starosta-Rubinstein *et al.* 1987) or, in presymptomatic patients, cause them (Brewer *et al.* 1994; Glass *et al.* 1990). Consequently, attention has shifted to two other chelating agents, namely trientine and tetrathiomolybdate; of these, tetrathiomolybdate is as effective as trientine but less likely to cause neurologic deterioration (Brewer *et al.* 2006).

Zinc induces intestinal metallothionein, which in turn binds copper and prevents its absorption; it is less effective than chelating agents. Zinc is given in a dose of 50 mg three times daily, before meals (Hoogenraad *et al.* 1987).

Dietary restriction of copper is the least effective of the treatment measures and requires an avoidance of shellfish, legumes, nuts, grains, coffee, chocolate, and organ meats.

Initial treatment of symptomatic patients should be performed with one of the chelating agents, and of these, tetrathiomolybdate appears preferable. Unfortunately, it is not readily available, and hence some patients are started on trientine. With chelation treatment, recovery is slow, and up to 2 years may be required to see full improvement. The degree of recovery varies with the severity of symptoms before treatment; in mild cases there may be complete recovery, whereas in severe cases, some degree of residual symptomatology is to be expected. Once maximum improvement has occurred, most patients may then be switched to zinc therapy, with careful monitoring.

A final alternative is liver transplantation, which, as it restores the body's ability to handle copper, is curative. This should be considered in either treatment-resistant or fulminant cases (Bax *et al.* 1998; Stracciari *et al.* 2000).

The symptomatic treatment of personality change is discussed in Chapter 8.2, of psychosis in Chapter 8.1, and of depression in Chapter 7.1. Given that almost all patients with Wilson's disease will have some degree of hepatic failure, doses of hepatically metabolized medications must be adjusted accordingly; furthermore, these patients may be particularly liable to develop extrapyramidal side-effects to antipsychotics, and consequently agents likely to produce these, for example high-potency first-generation antipsychotics such as haloperidol, are best avoided.

All of the patient's siblings should be offered testing for copper and ceruloplasmin levels and, in doubtful cases, consideration should be given to liver biopsy.

9.17 SPINOCEREBELLAR ATAXIA

Spinocerebellar ataxia (SCA), also known as autosomal dominant cerebellar ataxia, is, as the synonym implies, an autosomal dominantly inherited disorder characterized by a slowly progressive ataxia: over time, this ataxia may be joined by various other features, including, in a small minority, personality change, depression, dementia, or psychosis. This is an uncommon disorder, found in roughly 3/100 000 of the general population, and is equally common in males and females.

Some words are required regarding the synonym for this disorder, namely autosomal dominant cerebellar ataxia or ADCA. This name was initially utilized by Harding (1982, 1984), who went on to divide ADCA into three different subtypes, namely types III, II, and I. Type III was described as being characterized solely by cerebellar signs, without any associated features. Types II and I, by contrast, did encompass associated features, the difference between type II and type I being that type II could be accompanied by pigmentary retinopathy, whereas type I cases were specifically free of this sign. The nosologic status of this proposed subdivision is, however, in doubt, on both clinical and etiologic grounds. From a clinical point of view, it appears that type III cases are exceedingly rare, perhaps (with a sufficiently long follow-up) non-existent, and that, when comparing types II and I, type I patients (that is to say, those with associated features but without pigmentary retinopathy) constitute the overwhelming majority. Furthermore, from an etiologic point of view, it appears that this subdivision does not 'cleave' nature at the genetic 'joints', in that mutations at the same gene locus may cause two different clinical types (e.g. spinocerebellar ataxia type 6 [SCA6] is associated with both type III [Ishikawa *et al.* 1999] and type I [Schols *et al.* 1998]). Given this nosologic uncertainty, it may be prudent for the clinician to refrain from worrying about subtyping cases as ADCA types III, II, or I, but rather to concentrate on making sure that one has a case of ADCA and then proceeding to identify the responsible genetic locus.

Clinical features

The clinical hallmark of SCA is the appearance of a gradually progressive cerebellar ataxia, which is generally accompanied by dysarthria and nystagmus. The onset, although generally in the early to mid-adult years, may occur anywhere from childhood to senescence. With disease progression, almost all patients will also develop one or more of the following associated features: hyperreflexia and extensor plantar responses; decreased vibratory sense, atrophy, and fasciculations; supranuclear ophthalmoplegia; tremor (including titubation), dystonia, chorea, myoclonus, or parkinsonism; or pigmentary retinopathy. Seizures may also occur; however, they are uncommon.

Personality change may occur in a small minority, and although this may be seen in the context of a dementia, it may also occur in its absence (Leroi *et al.* 2002). Dementia, often with frontal lobe features, likewise occurs in a small minority (Leroi *et al.* 2002) and has been particularly noted in SCA2 (Burk *et al.* 1999), SCA3 (Ishikawa *et al.* 2002), and SCA17 (Rolfs *et al.* 2003); rarely, as has been noted in SCA17, dementia may constitute the presenting feature of the disease (Bruni *et al.* 2004). Depression also occurs in a small minority, and notably appears to follow an episodic course (Leroi *et al.* 2002). Finally, in a very small minority, psychosis may occur (Leroi *et al.* 2002), as has been noted in SCA12 (O'Hearn *et al.* 2001).

The plenitude of these associated features should not, however, distract attention from the central feature of this syndrome, namely a progressive cerebellar ataxia; the associated features, usually few in number in any given case, generally play only a minor part in the overall clinical picture.

Genetic testing is available. Interestingly, given the wide phenotypic heterogeneity, it does not appear possible to reliably predict which SCA type is present based on the clinical picture (Schols *et al.* 1998; Tang *et al.* 2000).

Magnetic resonance scanning may reveal atrophy of the cerebellum, pons, and inferior olives (Arpa *et al.* 1999).

Etiology

As noted, SCA is an autosomal dominantly inherited syndrome, and multiple loci have been identified on multiple different chromosomes – in many instances the responsible gene has been demonstrated. The nomenclature is not particularly helpful: each locus has been assigned a number, according to when it was first identified, and at present 31 loci have been identified yielding 31 different SCAs, from SCA1 to SCA31. The most common loci, accounting for well over half of all cases, are SCA1, SCA2, SCA3 (also known as Machado–Joseph disease), SCA6, SCA7, and SCA8. Of note, many texts include DRPLA among the SCAs, but given that patients with DRPLA may have little or no ataxia, and often have prominent chorea, this inclusion may not be clinically warranted, and, in this text, DRPLA is discussed separately in Chapter 9.15.

Most of the known mutations involve CAG trinucleotide repeat expansions, as has been found for SCA1, -2, -3, -6, -7, -12, and -17; a CTG trinucleotide repeat expansion has been noted for SCA8, and a pentancleotide repeat expansion for SCA10.

Neuropathologically (Durr *et al.* 1996; Ikeuchi *et al.* 1997; Ishikawa *et al.* 1999; Jobsis *et al.* 1997; Takiyama *et al.* 1994), one finds atrophy of the cerebellum, pons, and inferior olives; in addition to these findings, associated atrophy may also be found in one or more of the following structures: the spinocerebellar tracts, Clarke's column, the globus pallidus, the subthalamic nucleus, the substantia nigra, and the cerebral cortex.

Differential diagnosis

Dentatorubropallidoluysan atrophy and Gerstmann–Straussler–Scheinker disease are both autosomal dominant disorders that may be marked by a progressive ataxia, and genetic testing is generally required to ensure an accurate differential. Friedrich's ataxia is suggested by an autosomal recessive mode of inheritance and a typically early age of onset. Multiple system atrophy of the olivopontocerebellar type is suggested by concurrent autonomic signs and by its sporadic nature. Other sporadic disorders to consider include vitamin B12 deficiency, hypothyroidism, paraneoplastic cerebellar degeneration, and alcoholic cerebellar degeneration.

Treatment

In one double-blinded study (Botez *et al.* 1996), amantadine reduced ataxia in patients with SCA1 and SCA2. The general treatment of dementia is discussed in Chapter 6.3, of personality change in Chapter 8.2, and of psychosis in Chapter 8.1.

Genetic testing should be offered to at-risk relatives.

9.18 PANTOTHENATE KINASE-ASSOCIATED NEURODEGENERATION

Pantothenate kinase-associated neurodegeneration is a rare autosomal recessively inherited disease due to mutations in the gene for panthothenate kinase. The classic presentation is with a childhood-onset progressive dystonia.

This disease was first described by Hallervorden and Spatz in 1922 (Hallervorden and Spatz 1922), and, until recently, was known as Hallervorden–Spatz syndrome. This eponym, however, should, on ethical grounds (Shevell 2003), no longer be used, as both of these physicians knowingly utilized brains in later research that had been obtained from patients killed as part of the euthanasia program of the Third Reich.

Clinical features

The onset is typically gradual and, although most patients fall ill in either childhood or in adolescence, adult-onset cases may occur. The clinical symptomatology is heavily influenced by the age of onset.

Classic childhood-onset cases (Dooling *et al.* 1974, 1980; Hayflick *et al.* 2003; Pellecchia *et al.* 2005; Swaiman 1991; Thomas *et al.* 2004b) characteristically present with a slowly progressive dystonia involving the extremities and the face: although this is typically unilateral to start, bilateral involvement eventually ensues. Dysarthria is very common, and this may progress to anarthria. Choreoathetosis may also occur, and a minority of patients may develop spasticity with hyperactive deep tendon reflexes and extensor plantar responses. Cognitive decline may also appear, and roughly a quarter of patients will become demented.

Cases with an onset in adolescence or adult years (Alberca *et al.* 1987; Hayflick *et al.* 2003; Jankovic *et al.* 1985; Pellecchia *et al.* 2005; Rozdilsky *et al.* 1968; Thomas *et al.* 2004b) may also present with dystonia, but this is generally relatively mild. Choreoathetosis and parkinsonism may appear, as may spasticity, and most will also develop dysarthria. Of interest, a small minority may present with tics, and such patients may later also develop obsessions and compulsions. Cognitive decline to dementia occurs in roughly one-third of cases, and this may be marked by a personality change with impulsivity, aggressive outbursts, and emotional lability. Rarely, adult-onset cases may present with a dementia, followed years later by a movement disorder (Cooper *et al.* 2000; Dooling *et al.* 1974; Murphy *et al.* 1989; Rozdilsky *et al.* 1968).

Magnetic resonance imaging is extremely helpful in the diagnosis, as essentially all patients will display the distinctive 'eye of the tiger sign' on T2-weighted imaging (Angelini *et al.* 1992; Hartig *et al.* 2006; Hayflick *et al.* 2003; McNeill *et al.* 2008). Here, the central portion of the globus pallidus displays increased signal intensity, while, surrounding it, there is an area of decreased signal intensity: the overall effect is of looking a tiger in the eye.

Course

Childhood-onset cases show gradual progression, with death occurring within 10–15 years; adult-onset cases tend to pursue a much more leisurely course, and some patients may remain ambulatory for decades.

Etiology

As noted earlier, this is a recessively inherited disorder, and mutations are found in the gene for pantothenate kinase, *PANK2*, located on chromosome 20 (Zhou *et al.* 2001). Macroscopically, the globus pallidus is atrophic and exhibits a rust-brown discoloration. Microscopically, iron deposition and axonal spheroids are seen not only in the globus pallidus but also in the pars reticulata of the substantia nigra and in the cerebral cortex (Dooling *et al.* 1974).

Differential diagnosis

Pantothenate kinase-associated neurodegeneration is but one member of a larger group of disorders known collectively as neurodegeneration with brain iron accumulation (NBIA). Neuropathologically, all of the members of NBIA share the same characteristic, namely brain iron accumulation within the central nervous system, primarily in the basal ganglia and the substantia nigra. Although some of these disorders have been genetically characterized (e.g. pantothenate kinase-associated neurodegeneration itself and the rare disorders aceruloplasminemia and neuroferritinopathy), it appears that the vast majority of such cases fall into a residual group that has not as yet been associated with any known mutations (Hayflick *et al.* 2003). Although the differentiation of pantothenate kinase-associated neurodegeneration from these other disorders ideally should be made with genetic testing, testing for mutations in *PANK2* is not as yet commercially available, and hence one must look to clinical features and MRI. From a clinical point of view, cases that present in childhood with a symptomatology similar to that described for the classic presentation noted above almost always, in fact, have pantothenate kinase-associated neurodegeneration. Unfortunately, however, when the clinical picture is similar to that described above for adolescent- or adult-onset cases, it may not be possible to reliably distinguish, on clinical grounds, between pantothenate kinase-associated neurodegeneration and another member of NBIA. In such cases, looking for the 'eye of the tiger' sign on T2-weighted imaging may be very helpful. As noted earlier, virtually all cases of pantothenate kinase-associated neurodegeneration, regardless of the age of onset, will have this sign, and hence it has very high sensitivity. It does not, however, have as high a specificity, as this sign has been found in a small percentage of cases of neuroferritinopathy (McNeill *et al.* 2008), and in roughly one-third of the large 'residual' group that has not as yet been genetically characterized.

In addition to these other members of NBIA, consideration should also be given to certain other disorders. In cases presenting with dystonia, consideration, as discussed in Chapter 4.7, may be given to primary torsion dystonia and to dopa-responsive dystonia, and, given the pleomorphic symptomatology possible with Wilson's disease, testing for this is also appropriate. When the presentation is primarily with parkinsonism, consideration, among the disorders discussed in Chapter 4.8, should also be given to juvenile-onset Huntington's disease. In all these disorders, however, MRI will fail to reveal the 'eye of the tiger' sign.

Treatment

Although there are no controlled trials of treatment of any of the clinical aspects of pantothenate kinase-associated neurodegeneration, consideration may be given to pharmacologic treatment as discussed for dystonia in Chapter 4.7, for parkinsonism in Chapter 4.8, and for dementia in Chapter 6.3. Deep brain stimulation of the globus pallidus may be helpful for the dystonia; however, the benefit does not appear to be sustained past a year or so (Timmerman *et al.* 2010).

9.19 DOPA-RESPONSIVE DYSTONIA

Dopa-responsive dystonia is a rare inherited syndrome that classically presents in childhood with a dystonia which, as the name indicates, is quite responsive to levodopa. Given its eminent treatability, it is a diagnosis that must not be missed.

Clinical features

Although the classic onset is in childhood, later onsets in adolescence or adult years have also been reported. Regardless of the age of onset, symptoms generally both appear and accrue gradually.

Childhood-onset cases (Deonna 1986; Harwood *et al.* 1994; Nygaard and Duvoisin 1986; Nygaard *et al.* 1990, 1991; Trender-Gerhard *et al.* 2009) generally present with an intermittent dystonia of one foot; over time the dystonia spreads, eventually involving the other lower extremity and then the upper extremities; truncal and cervical dystonia may also eventually appear. A characteristic feature of the dystonia has been termed 'sleep benefit': after a good night's sleep, the dystonia may be very mild, or even absent, only to gradually reappear and worsen as the day wears on. In some cases, a mild parkinsonism may appear, marked by bradykinesia and rigidity.

Later-onset cases (Harwood *et al.* 1994; Trender-Gerhard *et al.* 2009), although at times presenting similarly to childhood-onset cases, may also present with tremor or a relatively pure parkinsonism. Rarely, adult-onset cases have been marked by depression or obsessions and compulsions, at times without any abnormal movements (Hahn *et al.* 2001; Van Hove *et al.* 2006).

Course

In childhood-onset cases, the dystonia eventually generalizes over about 3 or 4 years; although in most cases the subsequent course is marked by a relatively stable chronicity, in a minority there may be partial remissions, which, however, are generally only temporary.

Etiology

Dopa-responsive dystonia is a genetically heterogeneous syndrome and, at present, three different diseases have been described, namely DYT5a, DYT5b, and DYT14.

DYT5a, also known as the Segawa syndrome, is an autosomal dominant disorder and constitutes the most common cause of dopa-responsive dystonia. In DYT5a there is a mutation in the gene for guanosine triphosphate cyclo-hydrolase 1 (GCH1) on chromosome 14 (Bandmann *et al.* 1998; Illarioshkin *et al.* 1998; Ohye *et al.* 1994; Steinberger *et al.* 2000). GCH1 is the rate-limiting enzyme for the synthesis of tetrahydrobiopterin, which in turn is an essential cofactor for tyrosine hydroxylase, the rate-limiting enzyme in the production of dopamine. The net result of this mutation is a reduction in the production of dopamine.

DYT5b is an autosomal recessive disorder caused by a mutation in the gene for tyrosine hydroxylase on chromosome 11 (Furukawa *et al.* 2001; Schiller *et al.* 2004). In addition to causing a dopa-responsive syndrome, these patients may display other features, including paraparesis, prosis, and seizures.

DYT14 is a very rare autosomal dominant disorder linked to a site on chromosome 14 (Grotzch *et al.* 2002); the gene has not as yet been identified.

Pathologic studies have demonstrated a reduction in melanin pigment in the substantia nigra, without cell loss, in both DYT5a (Rajput *et al.* 1994) and DYT14 (Grotzch *et al.* 2002).

Differential diagnosis

The full differential diagnosis for dystonia is discussed in Chapter 4.7: of the disorders noted there, primary torsion dystonia is highest on the differential. A key feature in making the diagnosis is the responsiveness to levodopa, and an attempt at a 'diagnosis by treatment response' is warranted in most cases of childhood-onset dystonia.

Treatment

As indicated earlier, the dystonia of dopa-responsive dystonia responds well to levodopa: treatment may be initiated with carbidopa/levodopa, 12.5/50 twice daily, and increased gradually up to 1000 mg of levodopa daily; most patients respond to doses of 200–300 mg of levodopa daily. Notably, and in contrast to other conditions treated with levodopa, such as Parkinson's disease, the response to levodopa is sustained over long periods, with little need for dose increases (Nygaard *et al.* 1991).

9.20 PRIMARY TORSION DYSTONIA

Primary torsion dystonia, also known as dystonia musculorum deformans, is a relatively rare syndrome characterized classically by the onset of a focal dystonia in childhood or adolescence that eventually generalizes.

Clinical features

Classic cases (Johnson *et al.* 1962; Marsden and Harrison 1974) are characterized by an onset of dystonia between the ages of 5 and 15 years, appearing first in one of the lower extremities. Typically, and only intermittently, the young patient may experience some dystonic inversion and plantar flexion of the foot while walking. Curiously, this dystonia is not present at rest, and it may also be absent when walking backwards or dancing. Over time, however, the dystonia becomes more frequent and begins to involve more proximal portions of the lower extremities, often with flexion at the knees and hips. With progression of the disease, the dystonia becomes more and more constant and spreads not only to the upper extremities but also to the trunk, producing lordosis and tortipelvis. In a small minority, the face may be involved, with dystonic grimacing. Although cognitive defects are not seen, it appears that these patients have a higher risk of developing major depressive disorder (Heiman *et al.* 2004).

Genetic testing is available for the most common cause of this syndrome, namely DYT1, and should be considered in any patient with an early-onset dystonia that has generalized (Brassat *et al.* 2000; Bressman *et al.* 2000).

Course

When progression does occur, it typically reaches a maximum after some 5–10 years, after which the course is more or less static. In a very small minority, partial remissions may occur, but these are generally only temporary.

Etiology

Three different autosomal dominant diseases have been identified that can cause the syndrome of primary torsion dystonia, namely DYT1, DYT6, and DYT13; the remaining cases, as yet, are considered idiopathic. Of the known causes, DYT1 was the first to be recognized and is by far the most common and the most likely to cause a classic presentation (Kramer *et al.* 1990; Ozelius *et al.* 1992).

DYT1, also known as Oppenheim's disease, occurs secondary to a mutation in the gene for torsin A on chromosome 9 (Ozelius *et al.* 1997): penetrance is low, of the order of 30–40 percent, and there can be wide intrafamilial variability in the expression of the disease. Although the underlying pathology is not known, one study found neuronal inclusion bodies in the pedunculopontine nucleus, the cuneiform nucleus, and the periaqueductal gray (McNaught *et al.* 2004).

DYT6 occurs secondary to mutations in the gene for THAP1 on chromosome 8, and tends to present with brachial dystonia, which often then generalizes (Bressman *et al.* 2009)

DYT13 is linked to chromosome 1 and also has an atypical presentation, with dystonia of an upper limb or a cervical dystonia (Bentivoglio *et al.* 1997; Valente *et al.* 2001).

Differential diagnosis

The full differential diagnosis of dystonia is discussed in Chapter 4.7. Of the disorders noted there, the most important for the differential diagnosis is dopa-responsive dystonia: given that this disorder may be clinically indistinguishable from primary torsion disorder, a 'diagnosis by treatment response', as discussed in Chapter 9.19, should be attempted by giving levodopa.

Treatment

Perhaps the best-established pharmacologic treatment is trihexyphenidyl: children and adolescents may do well with this and, in contrast to adults, may tolerate high doses (Burke *et al.* 1986). In severe cases, consideration may be given to intrathecal baclofen or to deep brain stimulation of the globus pallidus (Kupsch *et al.* 2006; Vidhailet *et al.* 2005), which appears to offer long-term benefits (Vidhailhet *et al.* 2007).

9.21 PRIMARY CERVICAL DYSTONIA

Primary cervical dystonia, also known as spasmodic torticollis or 'wry neck', is the most common of the primary focal dystonias and is somewhat more frequent in women.

Clinical features

Clinically (Chan *et al.* 1991; Jankovic *et al.* 1991; Rondot *et al.* 1991; Sorensen and Hamby 1966), the onset is generally between the ages of 20 and 60 years, with most patients falling ill in their forties. The presentation is generally with some intermittent dystonia of the neck musculature, pulling the head into a dystonic posture. The most common position is torticollis, the head being rotated to one side or the other; other positions, in order of decreasing frequency, include lateralcollis, with the head tilted to one side, retrocollis, with the head bent back, or anterocollis, with the head pulled down toward the chest. Isolated positions are the exception: most patients exhibit a combination, such as torticollis and lateralcollis. Pain is a common accompaniment.

In most cases, patients are able to use a *geste antagoniste* to temporarily relieve the dystonia (Jahanshahi 2000; Muller J *et al.* 2001). These 'tricks' consist of lightly touching a specific body part, such as the chin, face, or occiput, resulting in a prompt, but temporary, relief of the dystonic stiffness.

Over time, the dystonia becomes more constant and sustained, and will, in a minority, undergo spread to an adjacent part, such as the arm.

A tremor of the head, similar to that seen in essential tremor, is present in a substantial minority of patients.

Course

In most cases there is a gradual progression with, as noted, some segmental spread in a minority; generalization, however, does not occur. There may be remissions in some 10–20 percent of patients during the first few years (Chan *et al.* 1991; Friedman and Fahn 1986; Jayne *et al.* 1984); however, most of these patients eventually relapse within 5 years (Dauer *et al.* 1998).

Etiology

Both familial (Bressman *et al.* 1996; Chan *et al.* 1991; Jankovic *et al.* 1991) and sporadic cases occur. In some familial cases, a genetic cause has been identified, for example the DYT7 locus, mapped to chromosome 18 (Leube *et al.* 1997). The neuropathology is unknown.

Differential diagnosis

This is discussed in Chapter 4.7.

Treatment

A local injection of botulinum toxin (Greene *et al.* 1990) is effective; although some patients may benefit from treatment with trihexyphenidyl, botulinum toxin is more effective (Brans *et al.* 1996). In severe, treatment-resistant cases, consideration should be given to deep brain stimulation of the globus pallidus (Hung *et al.* 2007; Kiss *et al.* 2007).

9.22 TASK-SPECIFIC DYSTONIA

Task-specific dystonia, one of the primary dystonias, is characterized by the occurrence of a dystonia upon attempting to perform a learned task. Also known as occupational dystonia, this movement disorder is typically divided into several subtypes depending on the task that occasions the dystonia; thus, we have writer's cramp, typist's cramp, musician's cramp, etc.

Clinical features

Task-specific dystonias generally appear in adulthood between the ages of 20 and 50 years, and typically only after the patient has been engaged in the task in question for at least a number of years.

Writer's cramp (Cohen and Hallett 1988; Jedynak *et al.* 2001; Sheehy and Marsden 1982) is perhaps the most common of the task-specific dystonias and manifests as a dystonic cramping of the hand upon using a pen or pencil. Although early on in the course the dystonia may appear only after the patient has been writing for a while, with time the dystonia appears earlier and earlier until it may manifest as soon as the patient picks up the pen; furthermore, and again with time, the dystonia may spread to involve the forearm. Some patients may try and evade the cramping by writing with their non-dominant hand; however, in a minority, the dystonia will reappear on this opposite side. Musician's cramp may appear in the hands of violinists or pianists, and it may also appear in the orofacial musculature in wind-instrument players. This latter type of musician's cramp is referred to as embouchure dystonia or, colloquially, as 'loss of lip', and manifests variously with lip dystonia or dystonic jaw movements (Frucht *et al.* 2001, 2009).

Course

The dystonia gradually worsens over perhaps years and then typically remains static; remissions are unusual.

Etiology

Most cases are sporadic. Familial cases are generally consistent with an autosomal dominant mode of inheritance, and, of interest, in some familial cases different family members may manifest different subtypes of task-specific dystonia (Schmidt *et al.* 2006, 2009). Although the genetic basis of familial cases is for the most part unknown, cases have been reported as part of DYT7 with linkage to chromosome 18 (Bhidayasiri *et al.* 2005), as part of DYT1 due to mutations in the gene for torsin A on chromosome 9 (Gasser *et al.* 1998), and as an atypical presentation of dopa-responsive dystonia (Deonna *et al.* 1997).

Differential diagnosis

The differential diagnosis is discussed in Chapter 4.7.

Treatment

Writer's cramp has been treated with botulinum toxin injections (Kruisdijk *et al.* 2007; Rivest *et al.* 1991), biofeedback, retraining, splint immobilization for 4–5 weeks (Priori *et al.* 2001), and placement of a transcutaneous electrical nerve stimulation (TENS) unit (Tinazzi *et al.* 2005). Similar strategies may be considered for musician's cramp involving the hand (Jabusch *et al.* 2005).

9.23 MEIGE'S SYNDROME

Meige's syndrome (also known as essential blepharospasm or primary blepharospasm), first described by the French neurologist Henri Meige in 1910 (Tolosa and Klawans 1979), is one of the primary dystonias and is characterized, as the synonyms indicate, by blepharospasm.

Clinical features

Clinical features have been described in a number of series (Grandas *et al.* 1988; Paulson 1972; Tolosa 1981). Onset is usually in middle years, and the primary symptom is bilateral blepharospasm. Initially the blepharospasm is neither prolonged nor frequent, but over time it becomes persistent, forceful, and at times almost constant. Over a year or two, the adjacent musculature, especially the jaw and mouth, may also become involved and, in a small minority, the neck musculature may likewise be affected (Defazio *et al.* 1999). The blepharospasm may be worsened by bright light or sometimes by walking, and may be lessened by yawning, talking, or singing (Weiner and Nora 1984). *Gestes antagonistes*, such as touching the eyebrow or the temple, may relieve the blepharospasm.

Course

Symptoms tend to worsen over the first few years and then remain static. Spontaneous remissions may occur in roughly 10 percent of patients, usually within the first 5 years (Castelbuono and Miller 1998).

Etiology

A positive family history is found in only a small minority (Defazio *et al.* 2006), and the remaining cases appear sporadic. Several cases have come to autopsy; however, it is not clear how representative they are. In three patients, no clear abnormalities were found (Gibb *et al.* 1988); in one case, neuronal loss and gliosis were found in the striatum (Altrocchi and Forno 1983); and in two cases, neuronal loss and Lewy bodies were found in brainstem nuclei (Kulisevsky *et al.* 1988; Mark *et al.* 1994). It may well be that these last two cases represented atypical presentations of dementia with Lewy bodies or some variant thereof.

Differential diagnosis

The full diagnosis of dystonia is discussed in Chapter 4.7, and several of the disorders discussed there deserve special consideration. Brueghel's syndrome, another primary dystonia, is distinguished by the fact that it presents with oromandibular dystonia in the absence of blepharospasm (Gilbert 1996). Atypical presentations of primary torsion dystonia may be characterized by blepharospasm (Bressman *et al.* 1994; Ozelius *et al.* 1989). Tardive dyskinesia may present with blepharospasm, but the diagnosis is immediately suggested by the appearance of symptoms in the context of long-term treatment with antipsychotics (Weiner *et al.* 1981). Blepharospasm has also been reported secondary to lesions in the basal ganglia (Jankovic 1986) or brainstem (Jankovic and Patel 1983).

Non-dystonic disorders to consider include hemifacial spasm, distinguished by its strictly unilateral occurrence, and facial tics, distinguished by their fleeting, unsustained nature. Ocular pathology must also be considered, as inflammation of the lid, conjunctiva, or iris may all be associated with blepharospasm.

Treatment

Botulinum toxin represents standard treatment; when this fails, case reports and case series suggest a usefulness for anticholinergics (Tanner *et al.* 1982) (e.g. trihexyphenidyl or benztropine), clonazepam (Hipola *et al.* 1984), levetiracetam (Yardimci *et al.* 2006), and, interestingly, quetiapine (Reeves and Liberto 2003). In severe, treatment-resistant cases, consideration may be given to deep brain stimulation of the globus pallidus (Houser and Waltz 2005).

9.24 SPASMODIC DYSPHONIA

Spasmodic dysphonia, also at times referred to as spastic dysphonia, represents a focal dystonia of the laryngeal musculature that renders speech dysphonic.

Clinical features

Clinical features have been described in a number of reports (Aminoff *et al.* 1978; Bicknell *et al.* 1968; Pool *et al.* 1991). The onset is typically in middle or later years, and may be subacute or gradual. Dystonic spasm of the laryngeal musculature may occur either in adduction or abduction, and the quality of the dysphonia differs in these two forms. Adductor spasm, which is by far more common, renders the voice high pitched, strained, and strangulated; with abductor spasm, the voice is more breathy and whispering. Some patients may find improvement with singing, or even with shouting.

Course

After a period of progression of variable duration, the dysphonia generally remains chronic.

Etiology

Spasmodic dysphonia is considered one of the primary dystonias; the etiology is not known.

Differential diagnosis

Dysphonia may result from lesions in the central or peripheral nervous system or in the larynx itself. Dysphonia has been noted with a putaminal lesion (Lee *et al.* 1996), lesions of the nucleus ambiguus, the vagus, or recurrent laryngeal nerves, and with various intrinsic laryngeal lesions. Dysphonia has also been reported as a side-effect of valproic acid (Oh *et al.* 2004) and gabapentin (Reeves *et al.* 1996), and has also occurred as part of tardive dyskinesia (Lieberman and Reife 1989).

Treatment

Botulinum injection of the laryngeal musculature generally provides relief. In treatment-resistant cases, consideration may be given to lesioning the recurrent laryngeal nerve.

9.25 TOURETTE'S SYNDROME

Tourette's syndrome, first described by the French neurologist Georges Gilles de la Tourette in 1885 (de la Tourette 1885), is the classic cause of chronic tics. This is a not uncommon disorder, with a lifetime prevalence of about 0.05 percent, and is about three times more common in males than females.

Clinical features

The onset of Tourette's syndrome is typically with a simple tic, more often motor than verbal, and more often on the head or face than elsewhere. Although the age of onset may be anywhere from infancy to the early adult years (Marneros 1983), most patients fall ill in childhood, around the age of 7 years.

In its fully developed form, both motor and verbal tics are present, and these may be either simple or complex (Cardoso *et al.* 1996; Lang *et al.* 1993; Lees *et al.* 1984; Nee *et al.* 1980; Regeur *et al.* 1986); it has also become apparent that sensory tics, once thought to be unusual, are also present in almost all patients.

Motor tics are usually the first to appear. Simple motor tics include blinking, brow wrinkling, grimacing, and shoulder shrugging; complex motor tics may include touching, smelling, hopping, throwing, clapping, bending over, squatting, or even such very complex acts as echopraxia or copropraxia (wherein patients make obscene gestures). Motor tics usually appear first in the face or head and then spread in a caudal direction. In most cases, as noted below, before having a tic, patients first experience an urge to tic, an urge that may at times be resisted, albeit with difficulty. Furthermore, some patients are able to abort a motor tic with a *geste antagoniste*, such as placing a hand under the chin to prevent the emergence of a tic of the head (Wojcieszek and Lang 1995).

Vocal tics, like motor tics, may also be simple or complex. Simple vocal tics include snorting, hissing, coughing, throat clearing, grunting, and, classically, barking. Complex vocal tics include the utterance of words, simple phrases, or entire sentences. Echolalia or palilalia may occur and, in a minority, classic coprolalia, or involuntary swearing, may be seen.

Overall, as noted, simple motor tics are generally first to appear, and the subsequent evolution of the illness, in most cases, is generally fairly leisurely: simple vocal tics generally follow within a year or two, and subsequently complex tics, either motor or vocal, may follow.

Sensory tics occur in the majority of patients and appear to exist in two forms. In one, there is simply the experience of an itch or a tingle (Chee and Sachdev 1997), and this is seen in perhaps one-quarter of patients. In the other, the sensory tic appears more as a premonitory urge to a motor tic (Cohen and Leckman 1992; Lang 1991; Leckman *et al.* 1993), and this has been reported in over 90 percent of cases (Leckman *et al.* 1993). Remarkably, in one case a premonitory urge to itch was experienced by a patient as residing in another person, whom the patient then proceeded to scratch (Karp and Hallett 1996).

Rarely, dystonic movements, especially cervical or facial dystonias, may appear in the course of Tourette's syndrome, but not until 10–38 years have passed (Stone and Jankovic 1991).

Obsessions and compulsions are common in Tourette's syndrome, eventually appearing in nearly one-half of all patients (Frankel *et al.* 1986; Robertson *et al.* 1988); they typically begin to appear about 5 years into the course. Interestingly, compulsions experienced by patients with Tourette's syndrome often center on getting things 'just right' (Leckman *et al.* 1994).

Attention-deficit–hyperactivity disorder (ADHD) is seen in over one-half of all cases, and, when it does occur, the hyperactivity usually precedes the tics by a little over a year (Cardoso *et al.* 1996).

Course

In most cases, symptoms gradually worsen over a matter of a few years, peaking in severity around the age of 10 years; subsequently, there is a gradual and progressive remission of symptoms, such that by the age of 18 years roughly one-half of patients are left with either no, or only trivial, tics (Leckman *et al.* 1998; Pappert *et al.* 2003). In those cases where full remissions do occur, relapses may appear, sometimes decades later (Chouinard and Ford 2000; Klawans and Barr 1985). While patients are symptomatic, the frequency and severity of tics both tend to wax and wane, and, at times, there may be very brief full or partial remissions, lasting days or a week or two, after which symptoms recur.

Etiology

Tourette's syndrome is a genetic disorder: approximate concordance rates are among siblings 5 percent, among dizygotic twins 20 percent, and among monozygotic twins 50 percent, and if one accepts the presence of simple tics among twins as evidence of the disease, then the monozygotic concordance rate rises to over 75 percent (Price *et al.* 1985); furthermore, adoption work (Pauls *et al.* 1991) demonstrates that the increased risk among first-degree relatives is genetically, and not environmentally, mediated. Despite this strong evidence of a genetic etiology, as of this writing mutations in only two genes (Abelson *et al.* 2005; Ercan-Sencicek *et al.* 2010) have been identified, and in each case only one family has been identified. Although some studies have supported an autosomal dominant mode of inheritance (Eapen *et al.* 1993; Pauls and Leckman 1986; Pauls *et al.* 1990), others have indicated, in at least a minority of cases, a bilineal pattern of inheritance, with genetic contributions from both parents (Hanna *et al.* 1999; McMahon *et al.* 2003).

Although the underlying neuropathology has not as yet been clearly delineated, attention has focused on the basal ganglia, and multiple MRI studies (Bloch *et al.* 2005; Hyde *et al.* 1995; Peterson *et al.* 1993,2003; Singer *et al.* 1993) have demonstrated a reduction in size of the basal ganglia overall, while autopsy studies, although few in number, have revealed a loss of parvalbulin-positive neurons in the striatum (Kataoka *et al.* 2010) and pars externa of the globus pallidus (Kalanithi *et al.* 2005), with an increased number of such cells within the pars interna (Kalanithi *et al.* 2005).

Although the inherited mechanism underlying the neuropathology of Tourette's syndrome is not clear, one theory holds that this syndrome results from a localized disorder of neuronal migration: in the normal course of embryogenesis of the basal ganglia, cells migrate through the pars interna of the globus pallidus, to the pars externa, and then on into the striatum, and with a 'stalling out' of migration one would see the cell changes just described. Another theory, admittedly held to apply to only a small minority of cases, posits that what is inherited is a susceptibility to the development of antibasal ganglia antibodies in response to a preceding group A beta-hemolytic streptococcal pharyngitis.

This latter theory is of special interest, as it naturally suggests an etiologically based treatment approach, and consequently, it deserves further comment. Swedo and colleagues, in 1998, were the first to suggest a relationship between a preceding streptococcal pharyngitis and the development of either Tourette's syndrome or obsessive–compulsive disorder, and they coined the term PANDAS (pediatric autoimmune neuropsychiatric disorders associated with streptococcal infections) to describe this. Subsequent work has demonstrated an association between a preceding pharyngitis and both the development of Tourette's (Mell *et al.* 2005), and, over long-term follow-up, in a small minority, exacerbations of symptomatology (Kurlan *et al.* 2008). Furthermore, some cross-sectional studies of patients with Tourette's have also demonstrated elevated levels of antistreptolysin antibody levels (Church *et al.* 2003; Muller N *et al.* 2001; Rizzo *et al.* 2006). Finally, some (Church *et al.* 2003; Rizzo *et al.* 2006; Singer *et al.* 1998, 2004), but not all (Morris *et al.* 2009; Singer *et al.* 2005); cross-sectional studies have also demonstrated elevated serum levels of antibasal ganglia antibodies in patients with Tourette's. Taken together, these studies are consistent with the notion that Tourette's syndrome may, like Sydenham's chorea, be an autoimmune disorder characterized by the presence of antibasal ganglia antibodies triggered by a preceding group A beta-hemolytic streptococcal pharyngitis. Attractive as this theory may be, however, it is not as yet proven, and, in any case, at most this mechanism accounts for only a small percentage of cases.

Differential diagnosis

The full differential diagnosis for tics is provided in Chapter 4.3, and of the causes noted there, several deserve special consideration. Autism may be complicated by tics, and this diagnosis is suggested by the other features seen in autism, such as the machine-like relationships patients have with others. Sydenham's chorea, Huntington's disease, and choreoacanthocytosis may also cause tics, but these are associated with chorea, a symptom not seen in Tourette's syndrome.

Technically, a diagnosis of Tourette's syndrome should not be made until the patient has both motor tics and vocal tics. Patients with only motor or only vocal tics are said to have 'transient tic disorder' if the tics last no longer than a year, or 'chronic motor or vocal tic disorder' if they last longer than a year. This distinction may be unwarranted, however, as it appears that these 'disorders' merely represent formes frustes of the full Tourette's syndrome (Golden 1978; Kurlan *et al.* 1988).

Treatment

Until recently, the only proven treatments for Tourette's syndrome were pharmacologic; however, a recent controlled study of behavioral therapy (Piacentini *et al.* 2010) has added a new option. This study found not only that a specific form of behavior therapy, namely habit reversal therapy, led to a clinically meaningful reduction in symptoms, but, importantly, that these gains were maintained over a 6-month, post-treatment, follow-up. What remains unclear, however, is how well this behavior treatment will compare with pharmacologic treatment, and also whether this study can be replicated. Pending answers to these questions, consideration should be given to pharmacologic treatment, with any of a number of agents which have been found effective in double-blinded, placebo-controlled studies: these include antipsychotics, alpha-2 agonists, the tricyclic antidepressant desipramine, topiramate, and the direct-acting dopamine agonist pergolide.

Among the antipsychotics, consideration may be given to the second-generation agents risperidone, olanzapine, and ziprasidone, and the first-generation agents pimozide and haloperidol. The choice among these agents may be based not only on efficacy, but also side-effects. Several studies, all double-blinded (and some with a cross-over design) have compared certain of the antipsychotics on the basis of efficacy. Risperidone was equivalent to pimozide in one study (Bruggeman *et al.* 2001), but in a second study, which was crossed, risperidone was superior (Gilbert *et al.* 2004). Olanzapine was superior to pimozide in one small, but crossed, study (Onofrj *et al.* 2000). Three studies have compared pimozide with haloperidol: one found pimozide and haloperidol to be equivalent (Ross and Moldofsky 1978), one found haloperidol to be marginally superior (Shapiro *et al.* 1989), while the third, which was crossed, found a superiority for pimozide (Sallee *et al.* 1997). To date, although ziprasidone has been shown to be superior to placebo (Sallee *et al.* 2000), it has not been compared with any of the other antipsychotics. Regarding side-effects, although the second-generation agents are, overall, better tolerated than the first-generation agents, both risperidone and olanzapine are associated with weight gain, hyperlipidemia, and even diabetes mellitus, and risperidone may cause dysphoria (Margolese *et al.* 2002). Pimozide and haloperidol are less likely to cause weight gain but do carry the liabilities of akathisia (which may manifest with an increase in the frequency of tics [Weiden and Bruun 1987]), tardive dyskinesia, dysphoria (Bruun 1988), and, in children, the capacity to cause severe separation anxiety to the point where 'school phobia' sets in (Linet 1985; Mikkelsen *et al.* 1981). Furthermore, pimozide carries the further risk of prolonging the corrected QT interval, with the possibility of torsades de pointe. Based on these considerations, if an antipsychotic is chosen, a second-generation agent is probably the best choice, and among these, risperidone appears preferable. A conservative titration regimen for each of these antipsychotics involves starting at 0.125 mg of risperidone, 2.5 mg of olanzapine, 5 mg of ziprasidone, 0.5 mg of pimozide, and 0.25 mg of haloperidol, and then increasing in similar increments every week or two up to a maximum of 4 mg of risperidone, 10 mg of olanzapine, 60 mg of ziprasidone, 15 mg of pimozide, and 5 mg of haloperidol, with most patients responding to 1–3 mg of risperidone, 5–10 mg of olanzapine, 20–40 mg of ziprasidone, 1–10 mg of pimozide, and 0.25–5 mg of haloperidol.

Of the available alpha-2 agonists, both guanfacine (Scahill *et al.* 2001) and clonidine (Cohen *et al.* 1980; Leckman *et al.* 1991) are effective, and although there are no studies comparing these two agents, the lesser side-effect burden seen with guanfacine makes this agent preferable. A conservative regime involves a total daily starting dose of 0.25 mg of guanfacine or 0.025 mg of clonidine, with increases of similar increments every 1–2 weeks up to a maximum of 4 mg of guanfacine or 0.4 mg of clonidine, with most patients responding to 1–3 mg of guanfacine or 0.1–0.3 mg of clonidine. In all cases, these total daily doses should be divided into a b.i.d. schedule for guanfacine or, for clonidine, a t.i.d. or q.i.d. schedule. Of note, in cases where clonidine is used, after an optimum dose has been established, patients may be switched to a patch (Du *et al.* 2008). Importantly, in contrast to antipsychotics, where the response is fairly prompt, with the alpha-2 agonists the full response may take weeks or months to develop. Consequently, in titrating the dose, it makes sense, as soon as an initial response is seen, to pause and see how things develop before increasing the dose further. It must also be borne in mind that these agents must not be abruptly discontinued, as this may be followed by a prolonged and severe 'rebound' of tics (Leckman *et al.* 1986): a gradual tapering, over weeks, is required.

Desipramine (Spencer *et al.* 2002), a tricyclic antidepressant, may be started at a dose of 25 mg and increased every few days to 100 mg, after which the patient should be observed for a few weeks to see what sort of response emerges. Importantly, as desipramine has been associated with lethal arrhythmias in children, its use should be restricted to adults.

Topiramate (Jankovic *et al.* 2010) may be started in a dose of 25 mg and increased in similar increments every 2 weeks until a response is seen, or a maximum dose of 200 mg is reached.

Pergolide is likewise effective (Gilbert *et al.* 2003), and this finding is of great interest, as it was once held that dopamine agonists were, if anything, likely to exacerbate tics. Pergolide, however, owing to its propensity to cause a valvulopathy, has been withdrawn from the market. Open work suggests that ropinirole may also be effective (Anca *et al.* 2004), but this cannot be recommended until a double-blinded study has confirmed it.

Choosing among the antipsychotics, alpha-2 agonists, desipramine, and topiramate is not entirely straightforward. Overall, the greatest experience, by far, is with the antipsychotics and the alpha-2 agonists, and most clinicians will begin with one of these. As between these two classes, clinical experience has strongly suggested that the antipsychotics are more effective; however, a recent double-blinded study found no superiority of risperidone over clonidine (Gaffney *et al.* 2002). Other considerations may also give one pause before starting with an antipsychotic. First, the side-effect burden is generally greater with these agents. Second, the antipsychotics are ineffective for ADHD, whereas the alpha-2 agonists are. Given all these considerations, it appears reasonable to start with an alpha-2 agonist, preferably guanfacine. Should this not provide sufficient relief, then an antipsychotic, preferably risperidone, is probably in order. Desipramine may also be considered in cases marked by ADHD, as it too is effective for this, but its use should be restricted to adults. With regard to topiramate, there is such a paucity of experience with its use in Tourette's that little can be said.

As noted earlier, some cases of Tourette's may occur as part of PANDAS, and an argument may be made for treating such cases in a fashion similar to that employed for Sydenham's chorea. Before embarking on such a course of treatment, however, it is important, obviously, to be reasonably certain that the case of Tourette's in question is occurring on an autoimmune basis, and two clinical features are helpful in this regard. First, one wishes to see either an explosive onset of the Tourette's, or, in established cases, an acute, and severe, exacerbation. As noted earlier, the evolution of symptoms in most cases of Tourette's is fairly leisurely, spanning, in some cases, years: in autoimmune cases, by contrast, severe symptomatology may accrue over weeks. Second, one must establish that the onset or exacerbation is indeed related to a preceding streptococcal pharyngitis, and this may be accomplished either by a positive throat swab or by the demonstration of elevated serum antistreptolysin B or anti-hyaluronidase levels. Once these features have been clearly demonstrated, then, as discussed in Chapter 19.11, the patient should be given a course of penicillin and prednisone; in severe cases, consideration may also be given to a course of plasma exchange and immunoglobulins (Perlmutter *et al.* 1999). Subsequently, consideration should be given to prophylactic antibiotic treatment to prevent future infection, and in this regard either penicillin or azithromycin may be used (Snider *et al.* 2005).

Finally, in severe and treatment-resistant cases, consideration may be given to neurosurgical intervention, with deep brain stimulation of either the globus pallidus (Ackermans *et al.* 2006; Shahed *et al.* 2007) or the thalamus (Ackermans *et al.* 2006; Porta *et al.* 2009).

As noted earlier, ADHD is commonly seen in patients with Tourette's and, should an alpha-2 autoreceptor agonist (or desipramine in adults) not prove effective, consideration may then be given to utilizing a stimulant. Despite the fact that stimulants may cause tics (Denckla *et al.* 1976), it appears that, in practice, the risk is negligible (Gadow *et al.* 1992,1999); indeed, in the case of methylphenidate (Tourette's Syndrome Study Group 2002) and atomoxetine (Allen *et al.* 2005), tics, rather than increasing, actually decreased during treatment.

9.26 MYOTONIC MUSCULAR DYSTROPHY

This disease was first described in 1909 by Steinert (Steinert 1909) in Germany and by Batten and Gibb (1909) in England, and has had several names. Many texts refer to it as Steinert's disease, but the nomenclature of Batten and Gibb has had greater influence: their original name was myotonia atrophica, but over time this evolved to myotonia dystrophica, and finally to myotonic muscular dystrophy. The most recent name, however, for this disorder is DM1: this name was proposed after it became apparent that there was another disease (discussed below in Differential diagnosis) that was capable of producing a similar, but much milder, clinical picture (International Myotonic Dystrophy Consortium 2000). This new disorder, originally known as PROMM (for *pro*ximal *m*yotonic *m*yopathy) has subsequently been renamed DM2.

Myotonic muscular dystrophy is a not uncommon disorder, with a lifetime prevalence of about 5 per 100 000; it is equally common in males and females.

Clinical features

The onset is gradual and insidious and, although most patients fall ill in their late teens or early twenties, the range of age of onset is wide, from childhood up to the sixth decade.

The cardinal symptoms of the disease are myotonia and weakness. Myotonia may go unnoticed by the patient, or may manifest in difficulty in letting go of, for example, a doorknob, or in disengaging from a handshake. On physical examination, myotonia may be elicited by tapping the thenar eminence with a percussion hammer and observing for the characteristic muscle dimpling. Weakness appears in the distal portions of the extremities, and may eventually be accompanied by atrophy. Other signs may also accrue. Many patients have a distinctive 'myopathic facies', with frontal balding, ptosis, and wasting of the face and neck musculature; the voice often also becomes nasal and monotone. Cataracts are seen in over 90 percent of patients, and there may be deafness (Wright *et al.* 1988) and gonadal failure, with erectile dysfunction or menstrual irregularity. Cardiac conduction abnormalities and arrhythmias may occur, including atrioventricular block, fascicular block, atrial flutter or fibrillation, and ventricular premature contractions; a cardiomyopathy with congestive heart failure may also occur, but is less common. With weakness of the respiratory musculature, alveolar hypoventilation may occur.

Cognitive impairment is eventually seen in a majority of patients (Gaul *et al.* 2006; Modoni *et al.* 2004; Perini *et al.* 1989) and, although this is generally quite mild, it may progress to a dementia (Huber *et al.* 1989). Personality change may also occur and, although it tends to be mixed in type, avoidant traits may be predominant (Delaporte 1998; Winblad *et al.* 2005). Apathy appears to be more common than among comparable controls (Rubinsztein *et al.* 1998). Although depression may occur, it appears, in contrast to apathy, to be no more common among these patients than comparable controls: in considering the diagnosis of depression, one must guard against misinterpreting the expressionless 'myopathic facies' as a depressed affect (Adie and Greenfield 1924; Billings and Ravin 1941; Bungener *et al.* 1998).

Sleep apnea, predominantly obstructive in type, is seen in the vast majority of cases, and in roughly one-third may be severe (Laberge *et al.* 2009). Excessive daytime sleepiness, amounting to hypersomnia, is not uncommon (Manni *et al.* 1991) and indeed may be the presenting complaint (Hansotia and Frens 1981): in some cases it is so severe that patients may literally sleep for days (Phemister and Small 1961). Importantly, although in some cases this hypersomnolence may be accounted for by sleep apnea, in others it appears to occur independently of this (van der Meche *et al.* 1994).

As noted below, 'anticipation' may occur in this disease (Harper *et al.* 1992; Howeler *et al.* 1989), and some children may be born with 'congenital' myotonic muscular dystrophy, characterized, among other symptoms, by mental retardation (Koch *et al.* 1991).

Both CT (Avrahami *et al.* 1987) and MRI (Antonini *et al.* 2004) may reveal a mild degree of cortical atrophy. On MRI, patchy areas of increased signal intensity on T2-weighted scans may be seen in the white matter (Di Costanzo *et al.* 2002), especially in the anterior portions of the temporal lobes (Miaux *et al.* 1997).

Genetic testing is available.

Course

The disease is gradually progressive and, although those with later onsets and milder symptoms may experience a normal lifespan, cases of early onset and severe symptoms are often associated with premature death in early or middle-adult years, often from cardiac or respiratory causes.

Etiology

Myotonic muscular dystrophy is inherited in an autosomal dominant fashion, with almost 100 percent penetrance but quite variable expressivity, even within the same family (Pryse-Phillips *et al.* 1982). Mutations consist of an expansion of a normally occurring CTG triplet in the *DMPK* gene on chromosome 19, which codes for the dystrophia myotonica protein kinase (Brook *et al.* 1992). As in many inherited disorders due to triplet expansions, 'anticipation' may occur in which, in succeeding generations, and with expansion of the triplet repeat, the disease becomes more severe and has an earlier onset. Although anticipation is far more likely with maternal transmission, it has been noted in paternal cases (Nakagawa *et al.* 1994).

Within the central nervous system, neuronal heterotopias were found in the cerebral cortex (Rosman and Kakulas 1966) and neurofibrillary tangles, without senile plaques, in the limbic system (Maurage *et al.* 2005; Oyamada *et al.* 2006). In cases characterized by hypersomnolence, neuronal loss was noted in the superior central nucleus of the midbrain (Ono *et al.* 1998), and in cases marked by alveolar hyperventilation, neuronal loss was seen in the medullary reticular formation (Ono *et al.* 1996).

Differential diagnosis

The chief disorder on the differential is a disease known as PROMM, or DM2 (Day *et al.* 2003; Ricker *et al.* 1994, 1995, 1999). This was first described by Ricker *et al.* in 1994 (Ricker *et al.* 1994), and, as noted earlier, its clinical features are similar, but much milder, than those seen in myotonic muscular dystrophy; furthermore, as the acronym PROMM indicates, the muscular weakness and atrophy has a primarily proximal distribution in contrast to the distal predominance seen in myotonic muscular atrophy. Other differences include a somewhat later age of onset (in middle years), prominent myalgia, a lack of the phenomenon of anticipation, and a lack of any congenital form. PROMM occurs secondary to mutations in the gene for zinc protein finger protein 9 (*ZNF9*) on chromosome 3.

Another disorder to consider is myotonia congenita. This is an inherited, non-progressive condition characterized by myotonia and, rather than weakness and atrophy, by a degree of muscular hypertrophy: autosomal dominant cases are referred to as Thomsen's disease and recessive cases as the Becker variant. Because the phenomenon of myotonia may be the only symptom of myotonic muscular dystrophy for many years, the clinical differential between myotonic muscular dystrophy and myotonia congenita may depend on long-term follow-up, watching closely for other symptoms to appear (Maas and Paterson 1950).

Treatment

There is no specific treatment for the disease. Hypersomnolence, which, in some cases, may be the most distressing feature of the disease, has been shown in double-blinded studies to respond to modafinil (MacDonald *et al.* 2002; Wintzen *et al.* 2007); open work also suggests a usefulness for methylphenidate (van der Meche *et al.* 1994). Myotonia may respond to various medications, but given that this symptom rarely causes significant disability, treatment is generally not necessary. If, however, treatment is required, double-blinded studies have demonstrated effectiveness for mexilitine (Logigian *et al.* 2010), procainamide (Munsat 1967), imipramine (Gascon *et al.* 1989), clomipramine (Antonini *et al.* 1990), and phenytoin (Munsat 1967): given the potential pro-arrhythmic effect of all of these except phenytoin, cardiologic consultation is in order before prescribing them. Yearly electrocardiograms are required and if there is suspicion of significant arrhythmia, a 24-hour Holter monitor is appropriate; some patients may eventually require a pacemaker. As noted earlier, alveolar hypoventilation may occur, and these patients do not appear to tolerate general anesthesia well, hence surgery, if possible, should be avoided. Hearing aids and cataract surgery may be required.

Genetic counseling is appropriate and, given the variable expression of the disease, it may be appropriate to offer testing to apparently unaffected relatives.

9.27 CEREBROTENDINOUS XANTHOMATOSIS

First described by van Bogaert in 1937 (van Bogaert 1937), cerebrotendinous xanthomatosis is a very rare autosomal recessively inherited lipid storage disease that may, in roughly one-half of cases, cause a dementia. Also known as cholestanolosis, cholestanol storage disease, or van Bogaert's disease, the current name is a felicitous one, given that it draws attention to the hallmarks of the disease, namely xanthomatous deposits in the cerebrum and tendons (most particularly the Achilles tendon).

Clinical features

The overall clinical features of the disease have been described in a number of papers (Bencze *et al.* 1990; Canelas *et al.* 1983; Farpour and Mahloudji 1975; Soffer *et al.* 1995; Swanson and Cromwell 1986; Verrips *et al.* 2000a,b; Watts *et al.* 1996). Although the first symptom of the disease is a chronic diarrhea occurring in earliest childhood, patients may not come to attention until the appearance of cataracts toward the end of childhood. Subsequently, in the second or third decade of life, roughly three-quarters of all patients will develop tendon xanthomas: although these are most frequently found on the Achilles tendon, they may also occur on the patellar tendon or the extensor tendons of the arms or hands. Cognitive decline begins in the second decade, and by the third decade roughly half of all patients will develop a dementia which may be marked by agitation, depression, or hallucinations and delusions. Seizures are present in about one-half of all cases, and patients also typically develop ataxia, spasticity, and a primarily motor polyneuropathy; rarely dystonia or parkinsonism may appear.

T2-weighted or, especially, FLAIR MRI (Barkhof *et al.* 2000; De Stefano *et al.* 2001; Swanson and Cromwell 1986) reveals increased signal intensity in the dentate nuclei and adjacent cerebellar white matter, the substantia nigra, globus pallidus, and the cerebral white matter.

Serum cholestanol levels are grossly increased and cholesterol levels are either normal or decreased (Salen 1971).

Genetic testing is available (Meiner *et al.* 1994).

Course

This is a very slowly progressive disorder. In those with an early age of onset and severe symptoms, death typically occurs within 10–20 years, whereas those with later onsets and milder symptoms may experience a normal lifespan.

Etiology

Cerebrotendinous xanthomatosis occurs secondary to any of a large number of mutations in *CYP27*, a gene on chromosome 2 that codes for an enzyme known as sterol 27-hydroxylase (Verrips *et al.* 2000a). With defective activity of this enzyme, cholestanol accumulation occurs in the cornea, tendons, peripheral nerves, and in the structures with MRI signal abnormalities noted above, with neuronal loss, demyelination and reactive astrocytosis (Barkhof *et al.* 2000; Menkes *et al.* 1968); in some cases, actual xanthomas may form (Schimschock *et al.* 1968).

Differential diagnosis

Whenever dementia occurs in the setting of juvenile cataracts, chronic diarrhea, or, classically, Achilles tendon enlargement, the diagnosis of cerebrotendinous xanthomatosis is very likely. When these are not present, or are overlooked, consideration may be given to metachromatic leukodystrophy in children, and to spinocerebellar ataxia in adults.

Treatment

Chenodeoxycholic acid, at a dose of 750 mg/day, taken chronically, dramatically reduces serum cholestanol levels and leads to either a stabilization of symptoms or to a degree of remission (Berginer *et al.* 1984). Pravastatin and simvastatin, although ineffective by themselves, enhance the effect of chenodeoxycholic acid.

At-risk relatives should be offered genetic testing.

9.28 METACHROMATIC LEUKODYSTROPHY

Metachromatic leukodystrophy is a rare, recessively inherited disorder that occurs, in the vast majority of cases, secondary to mutations in the gene for arylsulfatase A, with a resulting accumulation of sulfatides in the brain, peripheral nerves, kidneys, and gallbladder. This disorder is of particular interest in that, when appearing in adolescents or adults, it may present with a psychosis, personality change, or a dementia.

Clinical features

Three forms are recognized based on the age of onset: the late infantile form presents in infants up to the age of 4 years, the juvenile form from the age of 4 to 12 years, and the adult form from the age of 12 years onwards. In the adult form, although most cases present before the age of 30 years, cases presenting as late as the seventh decade have been noted (Bosch and Hart 1978).

The late infantile form (Brain and Greenfield 1950) is characterized by gait difficulty, hypotonia or spasticity, dementia, seizures, optic atrophy, and blindness.

The juvenile form (Haltia *et al.* 1980) may present with a dementia or with a personality change that merges into a dementia. Ataxia, spasticity, and a peripheral neuropathy may also constitute the presentation, or, in cases that present with a dementia or personality change, any one or all of these features may emerge later in the course of the disease.

The adult-onset form may present with psychosis, personality change, or dementia: in those cases that present with a psychosis or a personality change, a dementia generally ensues (Hageman *et al.* 1995; Rauschka *et al.* 2006). Other symptoms and signs may also constitute presenting features, or may emerge in the context of psychosis, personality change, or dementia, including ataxia, spasticity, seizures, or a peripheral neuropathy (Rauschka *et al.* 2006). It must be emphasized that in cases which do present with psychosis, personality change, or dementia, these other symptoms and signs may not appear for years.

The psychosis seen in the adult form is very similar to that seen in schizophrenia. One patient developed bizarre delusions, auditory hallucinations, loosening of associations, and flat affect at the age of 19 years, and the diagnosis became apparent only 12 years later when a peripheral polyneuropathy was noted (Manowitz *et al.* 1978). Another, at the age of 28 years, developed 'bizarre elation, true auditory hallucinations, and poorly formulated paranoid ideas', followed by a gradually progressive dementia over the next 4 years (Betts *et al.* 1968). In a third example (Waltz *et al.* 1987), a 31-year-old man began to talk to himself and pace the floor; he was eventually fired from his job and his wife left him. By the age of 38 years he had 'poor concentration, inappropriate smiling and laughing. . .irrelevant remarks. . . with non sequiturs, restlessness and occasional auditory hallucinations'. The diagnosis was eventually suggested when nerve conduction velocity studies showed mild slowing. In some cases, the psychosis may have a distinct manic flair: one patient (Besson 1980) was 'grandiose. . .and called out the fire brigade', and another (van Bogaert and Dewulf 1939) 'had ideas of grandeur, thought he was going to become an ambassador and gave himself honorary titles'. In both these cases, the patients eventually became demented.

The personality change may be non-specific; however, a frontal lobe syndrome may be prominent (Austin *et al.* 1968; Finelli 1985; Schneider *et al.* 2009b; Wulff and Trojaborg 1985) with disinhibition, perseveration, irritability, and socially inappropriate behavior.

As might be suspected, in addition to cognitive deficits, the dementia seen in adult-onset metachromatic leukodystrophy is also often marked by delusions, hallucinations, and frontal lobe symptomatology (Alves *et al.* 1986; Bosch and Hart 1978; Hirose and Bass 1972; Hyde *et al.* 1992; Reider-Grosswasser and Bornstein 1987; Shapiro *et al.* 1994).

T2-weighted MRI reveals ventricular dilation and widespread increased signal intensity in the centrum-semiovale (Rauschka *et al.* 2006; Reider-Grosswasser and Bornstein 1987). Nerve conduction velocity studies may reveal slowing in patients who lack clinical evidence of a peripheral neuropathy; however, it must be kept in mind that such studies may be normal. Lumbar puncture may reveal normal CSF or, especially in those with late infantile or juvenile forms, an increased total protein.

An often mentioned but rarely seen finding in metachromatic leukodystrophy is non-filling of the gallbladder on cholecystography, which occurs secondary to infiltration of the gallbladder by sulfatides.

Assays of leukocytes in almost all cases will reveal decreased aryl sulfatase A activity, and assays of peripheral nerve tissue (obtained at sural nerve biopsy) or of urinary sediment will reveal increased sulfatide content. The phenomenon of metachromasia, from which this disorder derives its name, may also be observed in peripheral nervous tissue or urinary sediment. Both cresyl violet and toluidine blue undergo a chromatic metamorphosis, turning from violet or blue to brown or golden-brown, respectively, when applied to affected cells in peripheral nerves or urinary sediment.

Course

The disease is gradually progressive, with death within 2–10 years in the late infantile or juvenile forms; in the adult form, death may be delayed for up to 15 years.

Etiology

As noted, metachromatic leukodystrophy is a recessively inherited disorder, and occurs in the vast majority of cases secondary to any of a large number of mutations in the gene for arylsulfatase A, found on chromosome 22 (Barth *et al.* 1993). Arylsulfatase A catalyzes sulfatides, and with mutations in this enzyme, sulfatides accumulate within oligodendrocytes within the central nervous system, Schwann cells within the peripheral nervous system, and also within the kidneys and gallbladder. The accumulation within the peripheral nervous system and the kidneys accounts for the diagnostic metachromasia which gives the disease its name: sulfatides are positively charged, and their excessive presence reorients the negatively charged molecules of cresyl violet or toluidine blue, thereby changing their color.

Within the central nervous system, widespread demyelination is seen, and this is especially evident in the centrum semiovale. In severe cases, the white matter of the centrum semiovale may be shrunken down to a thin gliotic remnant, with only relative sparing of the U-fibers, such that there is very little intervening tissue between the depths of the sulci and the lateral ventricles. The cerebellar white matter is affected in a similar fashion, and there is also peripheral demyelination.

Rarely, rather than occurring secondary to a mutation in the gene for arylsulfatase A, metachromatic leukodystrophy may occur secondary to a mutation in a gene on chromosome 10 coding for a sulfatide activator protein (Schlote *et al.* 1991). This protein acts as an essential cofactor for arylsulfatase A and, in its absence, all of the biochemical defects seen with mutations in the gene for arylsulfatase A also accrue. Importantly, although sulfatide accumulation occurs in these cases, arylsulfatase A activity, depending on the assay used, may be normal or near normal, and this serves to underscore the necessity of determining not only arylsulfatase A activity but also of testing peripheral nerve and urinary sediment for metachromasia.

Differential diagnosis

Differential diagnostic considerations vary according to the age of the patient.

Juvenile-onset cases may be confused with adrenoleukodystrophy; however, in adrenoleukodystrophy one also sees either a hemianopia or cortical blindness, findings not seen in metachromatic leukodystrophy.

Adult-onset cases may, depending on their presentation, raise several differential considerations. Presentations with psychosis may be indistinguishable from schizophrenia until other signs and symptoms, such as ataxia or peripheral neuropathy, emerge or until a dementia supervenes. Presentations with a personality change of the frontal lobe type or with a dementia may raise the possibility of frontotemporal lobar degeneration, and the differential here may also rest on the emergence of a peripheral neuropathy or ataxia.

One final consideration is a condition known as 'pseudodeficiency' of arylsulfatase A, found in roughly 1 percent of the general population. Here, although current assays indicate arylsulfatase A deficiency, in fact this is a 'pseudodeficiency' and there is no accumulation of sulfatides, and no clinical symptomatology (Hageman *et al.* 1995). Consequently, whenever the laboratory reports a deficiency, one must always confirm a 'true' deficiency by demonstrating increased sulfatide content in peripheral tissues.

Treatment

Bone marrow transplantation may retard or halt the progression of the disease (Gorg *et al.* 2007; Kidd *et al.* 1998; Krivit *et al.* 1987; Navarro *et al.* 1996). The general treatment of psychosis is discussed in Chapter 8.1, of personality change in Chapter 8.2, and of dementia in Chapter 6.3.

9.29 ADRENOLEUKODYSTROPHY

Adrenoleukodystrophy is a rare X-linked disorder characterized pathologically by the accumulation of very long-chain fatty acids (VLCA) in the brain, spinal cord, and adrenal glands, and clinically by a variable combination of dementia, spastic paraparesis, and adrenal failure.

Clinical features

Clinically (Moser *et al.* 1984), adrenoleukodystrophy may be characterized according to either the age of onset or the predominant site of pathology. Thus, onset may be in childhood, adolescence, or adult years, and the predominant site may be cerebral, spinal, or adrenal.

Childhood-onset adrenoleukodystrophy (Moser *et al.* 1984; Schaumburg *et al.* 1975) typically presents between the ages of

3 and 10 years with the gradual onset of cerebral symptomatology, often with a personality change and visual symptoms. These patients may become withdrawn and irritable and some may develop symptoms reminiscent of ADHD. Varying degrees of hemianopia or cortical blindness may occur, followed by a dementia with, in a minority, seizures. Adrenocortical insufficiency, although occasionally present at onset, typically appears later, and is eventually present in most cases.

Adolescent-onset adrenoleukodystrophy tends to present with a combination of cerebral symptomatology (similar to that seen in the childhood form) and cord involvement, with a spastic paraparesis; adrenocortical insufficiency is also common.

Adult-onset adrenoleukodystrophy tends to present in early adult years with a variant known as adrenomyeloneuropathy, with a spastic paraparesis and adrenocortical insufficiency (Griffin *et al.* 1977). With long-term follow-up, however, cerebral involvement occurs in roughly one-fifth of cases (van Geel *et al.* 2001), marked by a dementia that may be non-specific in character, or which may be marked by manic symptoms (Weller *et al.* 1992) or a Kluver–Bucy syndrome (Powers *et al.* 1980).

Although adrenocortical insufficiency typically occurs only in concert with cerebral or cord involvement, in a very small minority of children or adolescents it may constitute the sole feature of this disease (O'Neill *et al.* 1982).

Phenotypic variability is the rule in adrenoleukodystrophy, and even members of the same family may have different presentations (Erlington *et al.* 1989). Indeed, even monozygotic twins may exhibit phenotypic heterogeneity (Sobue *et al.* 1994) to the point of one twin being affected while the other is not (Korenke *et al.* 1996).

Female heterozygotes are not immune, and indeed close to one-half will develop cord involvement with a progressive spastic paraparesis during middle adult years; notably, however, cerebral involvement and adrenocortical insufficiency are rare in females.

Computed tomography scanning in patients with cerebral involvement may reveal areas of radiolucency in the white matter: typically, these first appear in the occipital lobes and then spread anteriorly into the parietal and temporal lobes. With contrast administration, enhancement is seen at the boundary between the areas of radiolucency and normal tissue. T2-weighted MRI reveals increased signal intensity in the white matter, following the same pattern as seen on CT scans, with the same pattern of enhancement upon administration of gadolinium. CSF analysis may reveal a mild lymphocytic pleocytosis and an elevated total protein. With adrenocortical insufficiency, the laboratory values are as discussed for primary adrenocortical insufficiency in Chapter 17.2.

In all suspected cases, VLCA levels should be determined in plasma or cultured skin fibroblasts: these will be elevated in essentially all males and in some 85 percent of females. Confirmation of the diagnosis is then made by mutation analysis of the *ABCD1* gene.

Course

Childhood-onset cases generally progress fairly rapidly, with death within 5–10 years; adolescent- and adult-onset cases, especially those with the adrenomyeloneuropathy syndrome, may progress very slowly. There may rarely be periods of partial remission, only to be followed eventually by relapse (Walsh 1980).

Etiology

Adrenoleukodystrophy is an X-linked disorder (Mosser *et al.* 1994) that occurs secondary to any of a large number of different mutations in the *ABCD1* gene on the X chromosome. These mutations lead to deficient uptake of VLCFA into peroxisomes, which in turn leads to their accumulation in the brain, spinal cord, and adrenal cortex.

Within the brain (Schaumburg *et al.* 1975), one typically finds an advancing wave of demyelination that begins in the occipital lobe and then moves forward into the parietal and temporal lobes; the frontal lobes are broached in only a minority. An inflammatory response is seen at the border between the demyelination and the normal tissue, behind which much of the white matter is replaced by gliotic tissue. Importantly, the subcortical U-fibers are generally spared, as is the gray matter. Demyelination also occurs in the spinal cord, especially in the corticospinal tracts.

Differential diagnosis

In children or adolescents, a diagnosis of metachromatic leukodystrophy may be considered; however, the presence of visual symptoms and adrenocortical insufficiency are distinctive for adrenoleukodystrophy. In adults, consideration may be given to multiple sclerosis, vitamin B12 deficiency, or hereditary spastic paraplegia.

Treatment

There was much hope that reducing VLCA levels by means of 'Lorenzo's oil' would be curative; however, this hope does not seem to have been borne out. Although Lorenzo's oil does indeed reduce plasma, VLCA levels, and uncontrolled studies suggest that it may delay expression of the disease in asymptomatic males (Moser *et al.* 2005), a controlled study in symptomatic patients failed to display any benefit (van Geel *et al.* 1999). Hematopoietic stem cell transplantation, if undertaken very early in the course of the disease, may halt its progression or even lead to improvement (Shapiro *et al.* 2000). The general treatment of dementia is discussed in Chapter 6.3. Adrenocortical insufficiency is treated as described in Chapter 17.2.

9.30 KUFS' DISEASE

Kufs' disease, also known as the adult form of neuronal ceroid lipofuscinosis, is a very rare disorder that typically presents in early adult years with either seizures or dementia.

Clinical features

Kufs' disease typically presents in early adult years in one of two fashions (Berkovic *et al.* 1988; Callagy *et al.* 2000). In one, referred to as 'type A', patients present with grand mal and myoclonic seizures and a dementia or psychosis, which is often accompanied by ataxia or abnormal movements, such as athetosis or parkinsonism. In the other, 'type B', the presentation is typically with a personality change or dementia, which may also be accompanied by ataxia and other abnormal movements.

Magnetic resonance imaging may be normal or may reveal cortical atrophy.

In those with dementia, generalized slowing is seen on the EEG; when seizures are present, the EEG typically reveals generalized spike-and-wave discharges, and there may be a pronounced photosensitivity (Vadlamudi *et al.* 2003).

Diagnosis is confirmed by rectal, muscle, or skin biopsy, which reveals characteristic fingerprint or granular osmiophilic deposits on electron microscopy (Berkovic *et al.* 1988).

Course

The disease is relentlessly progressive, with death occurring on average after 12 years.

Etiology

The neuronal ceroid lipofuscinoses occur in four forms, namely infantile (CLN1), late infantile (CLN2), juvenile (CLN3), and the adult form, Kufs' disease (CLN4). Kufs' disease may occur on a sporadic or an inherited basis. Although most heritable cases occur on an autosomal recessive basis (Berkovic *et al.* 1988), families with dominant inheritance have been reported (Burneo *et al.* 2003; Nijssen *et al.* 2002). Although the genetic defect for most cases of Kufs' disease is not known, there is one case report of adult-onset neuronal ceroid lipofuscinosis occurring secondary to a mutation typically associated with the infantile form (van Diggelen *et al.* 2001).

Neuronal loss occurs, and in surviving neurons ceroid and lipofuscin deposits are seen within lysosomes, creating a typical fingerprint or granular pattern on electron microscopy.

Differential diagnosis

Kufs' disease is typically suspected in adults with a combination of myoclonus and dementia, and in such cases one must consider Creutzfeldt–Jakob disease and Hashimoto's encephalopathy. The EEG may be very helpful in making the differential here in cases with seizures, as the pattern seen in Kufs' disease is not present in these other disorders.

Treatment

The symptomatic treatment of dementia is discussed in Chapter 6.3; divalproex may be considered for seizures.

9.31 ESSENTIAL TREMOR

Essential tremor, characterized primarily by a postural tremor of the hands, is the most common cause of tremor, being found in roughly 4 percent of the general population; it is equally common in males and females.

Clinical features

Although the range in age of onset is wide, from the first to the ninth decades of life there are two peaks of incidence: one occurs in adolescence, while the second, and by far the highest, occurs in the fourth and fifth decades.

Clinically speaking (Bain *et al.* 1994; Critchley 1949; Koller *et al.* 1994; Lou and Jankovic 1991; Martinelli *et al.* 1987), in the vast majority of cases, tremor first appears in the hand, and, again in the vast majority, the onset is bilateral: in the small minority of cases with a unilateral onset, the opposite hand soon also becomes involved. The tremor, at least initially, is fine, ranging in frequency from 4 to 10 cps, and is classically a postural tremor most evident when the hands are held outstretched with the fingers spread; in a small minority, there may also be an action, or kinetic, aspect to the tremor, with an exacerbation evident with voluntary motion. Although in many cases this tremor does not cause any significant impairment, it is not necessarily benign, as in some cases it may interfere with using table utensils, drinking from a glass, or writing with a pen or pencil. Typically, the tremor is exacerbated by stress or anxiety, and also, classically, it may be suppressed by alcohol (Growdon *et al.* 1975).

In most cases, over time the tremor also becomes apparent elsewhere, including, in decreasing order of frequency, the head, the vocal cords, the chin, and, in a small minority, the feet. Head tremor is generally of the 'no–no' type (Bain *et al.* 1994), with a trembling oscillation of the head from side to side. Involvement of the vocal cords may produce a 'vocal tremor' characterized by a quavering quality (Ardran *et al.* 1966).

Although classic teaching has held that essential tremor is basically a monosymptomatic disorder, characterized by tremor alone, more recent work has demonstrated associated deficits. Depression, of at least a mild degree, may be found in roughly one-third of cases (Miller *et al.* 2007). Cognitive deficits have also been noted in those aged 65 years or older. Mild deficits, related to attentiveness, memory, and executive functioning, are very common (Benito-Leon *et al.* 2006a). More ominously, the risk of dementia is increased roughly twofold compared with matched controls (Benito-Leon *et al.* 2006b; Bermejo-Pareja *et al.* 2007): notably, however, this increased risk of dementia occurs only in those cases where essential tremor had an onset past the age of 65 years: among those patients whose tremor began earlier in life, the risk of developing dementia once they passed the age of 65 years was no higher than that found in controls.

Course

In most cases, the course is characterized by progressive worsening to a certain plateau, which may persist for years or even decades, after which there may be further progression. As the tremor worsens, it characteristically becomes of greater amplitude and slower frequency.

Etiology

Although sporadic cases do occur, most cases appear to be hereditary, and both family (Bain *et al.* 1994) and twin (Lorenz *et al.* 2004; Tanner *et al.* 2001) studies are strongly consistent with an autosomal dominant pattern of inheritance, with essentially full penetrance over long-term follow-up. Although as yet mutations have not been identified, three genetic loci have been identified, on chromosomes 2, 3, and 6 (Deng *et al.* 2007).

Although routine pathologic studies have been unremarkable (Rajput *et al.* 2004), recent work has demonstrated some very interesting findings (Axelrad *et al.* 2008; Louis *et al.* 2005, 2006a,b, 2007), suggesting that there may be two different types of pathology underlying this disorder. In the first type, found in a minority of cases, there is neuronal loss in the locus cereleus, with, in surviving neurons, Lewy bodies. In the second, more common, type, one finds cerebellar pathology, with a loss of Purkinje cells. Although speculative, these seemingly disparate findings may be reconciled as contributing to a final common pathway of reduced inhibitory output of the Purkinje cell layer. The locus ceruleus has massive stimulatory projections to the Purkinje cell layer, and with dysfunction of the locus ceruleus, and loss of this stimulation, Purkinje cell output would fall. A similar loss of output from the Purkinje cell layer would also, of course, occur with loss or damage to the Purkinje cells themselves. If these findings are replicated, and if this speculation is correct, then it may well be that essential tremor represents a syndrome composed of two, or perhaps more, inherited disorders causing pathology either in the locus ceruleus or the cerebellum.

Differential diagnosis

Essential tremor is essentially a postural tremor and, as discussed in Chapter 4.1, this must be distinguished from rest and intention tremors. Once it is clear that the patient does have a postural tremor, the differential may be pursued as outlined in that chapter, with special attention given to medication-induced tremors (e.g. sympathomimetics or caffeine), alcohol or sedative/hypnotic withdrawal, and hyperthyroidism.

Treatment

A large number of medications are effective in essential tremor. Propranolol (Winkler and Young 1974) and primidone (Koller and Royse 1986) are the mainstays, and although they are roughly equivalent in efficacy (Gorman *et al.* 1986), propranolol is generally better tolerated: propranolol may be given in doses from 80 to 240 mg/day, and primidone from 25 to 750 mg/day. Gabapentin may also be considered (Ondo *et al.* 2000), and indeed in one study (Gironell *et al.* 1999) it was as effective as propranolol: it may be given in doses of 600–3600 mg/day. Topiramate is also effective (Connor 2002; Ondo *et al.* 2006), but is generally not well tolerated. Alprazolam is likewise also effective (Gunal *et al.* 2000), but, given the high risk of physiologic dependence, should probably be held in reserve. Regardless of which agent is used, it is important to start at a low total daily dose (e.g. 40 mg propranolol, 25 mg primidone, 300 mg gabapentin) and then employ a gradual titration to determine the lowest effective dose. In severe, treatment-resistant cases, consideration should be given to thalamic deep brain stimulation (Sydow *et al.* 2003).

9.32 HYPEREKPLEXIA

Hyperekplexia, also known as hyperexplexia or 'startle disease', is a rare disorder characterized by a pathologic startle response.

Clinical features

Hyperekplexia occurs in two varieties, namely the major form and the minor form, and these differ both in age of onset and in clinical symptomatology (Andermann *et al.* 1980; Kirsten and Silfverskiold 1958; Ryan *et al.* 1992; Suhren *et al.* 1966; Tijssen *et al.* 1995).

The major form of hyperekplexia becomes apparent in earliest infancy with episodes of hypertonus. By the time that the child is able to walk, or shortly thereafter, the response typical of hyperekplexia becomes apparent: to an appropriate stimulus, often a loud, unexpected noise, patients experience an exaggerated startle response (Brown *et al.* 1991) characterized by grimacing, flexion of the neck, arms, and trunk, abduction of the arms, and generalized stiffness; many fall 'like a log' and injuries are common. Consciousness is maintained throughout and the stiffness resolves immediately, allowing the patient to stand again.

The minor form of hyperexplexia typically does not become apparent until childhood or later and consists simply of the exaggerated startle response without any associated stiffness.

Course

Symptoms typically undergo a spontaneous, albeit often not complete, remission.

Etiology

The etiology of the two forms of hyperekplexia probably differ.

Regarding the major form, although sporadic cases do occur (Saenz-Lope *et al.* 1984), most cases are inherited, and most, but certainly not all, are due to mutations in the gene for the alpha-1 subunit of the inhibitory glycine receptor

(*GLRA1*) on chromosome 5; depending on the mutation, both autosomal dominant (Andrew and Owen 1997; Shiang *et al.* 1993) and recessive (Rees *et al.* 1994; Siren *et al.* 2006; Tsai *et al.* 2004) patterns of inheritance may occur.

The etiology of the minor form is not clear. The fact that both major and minor forms may occur in the same family (Andermann *et al.* 1980; Suhren *et al.* 1966) suggests that the minor form may merely be a forme fruste of the major form; however, genetic analysis of these families, although revealing a mutation in the alpha subunit in those with the major form, failed to find mutations in those with the minor form (Tijssen *et al.* 1995).

Differential diagnosis

This is discussed in Chapter 4.14.

Treatment

Clonazepam is generally effective in reducing the severity of the startle response (Ryan *et al.* 1992); in treatment-resistant cases, consideration may be given to valproic acid (Dooley and Andermann 1989).

REFERENCES

Aarsland D, Tandberg E, Larsen JP *et al.* Frequency of dementia in Parkinson disease. *Arch Neurol* 1996; **53**:538–42.

Aarsland D, Litvan I, Larsen JP. Neuropsychiatric symptoms of patients with progressive supranuclear palsy and Parkinson's disease. *J Neuropsychiatry Clin Neurosci* 2001; **13**:42–9.

Aarsland D, Laake K, Larsen JP *et al.* Donepezil for cognitive impairment in Parkinson's disease: a randomized controlled study. *J Neurol Neurosurg Psychiatry* 2002; **72**:708–12.

Aarsland D, Andersen K, Larson JP *et al.* Prevalence and characteristics of dementia in Parkinson disease: an 8-year prospective study. *Arch Neurol* 2003; **60**:387–92.

Aarsland D, Perry R, Brown A *et al.* Neuropathology of dementia in Parkinson's disease: a prospective, community-based study. *Ann Neurol* 2005; **58**:773–6.

Aarsland D, Ballard C, Walker Z *et al.* Memantine in patients with Parkinson's disease dementia of dementia with Lewy bodies: a double-blind, placebo-controlled, multicentre trial. *Lancet Neurol* 2009; **8**:613–18.

Abelson JF, Kwan KY, O'Roak BJ *et al.* Sequence variants in SLITRK1 are associated with Tourette's syndrome. *Science* 2005; **310**:317–20.

Ackermans L, Temel Y, Cath D *et al.* Deep brain stimulation for Tourette's syndrome: two targets? *Mov Disord* 2006; **21**:709–13.

Adachi N, Arima K, Asada T *et al.* Dentatorubral-pallidoluysia atrophy (DRPLA) presenting with psychosis. *J Neuropsychiatry Clin Neurosci* 2001; **13**:258–60.

Adams JS, Adams PE, Nguyen D *et al.* Volumetric brain changes in females with fragile X-associated tremor/ataxia syndrome (FXTAS). *Neurology* 2007; **69**:851–9.

Adie WJ, Greenfield JG. Dystrophia myotonica (myotonia atrophica). *Brain* 1924; **47**:73–127.

Adler CH, Sethi KD, Hauser RA *et al.* Ropinirole for the treatment of early Parkinson's disease. *Neurology* 1997; **49**:393–9.

Adler CH, Singer C, O'Brien C *et al.* Randomized placebo-controlled study of tolcapone in patients with fluctuating Parkinson's disease treated with levodopa–carbidopa. *Arch Neurol* 1998; **55**:1089–95.

Aguirre T, Matthijs G, Robberecht W *et al.* Mutational analysis of the Cu/Zn superoxide dismutase gene in 23 familial and 69 sporadic cases of amyotrophic lateral sclerosis in Belgium. *Eur J Med Genet* 1999; **7**:599–602.

Aisen PS, Davis KL. Inflammatory mechanisms in Alzheimer's disease: implications for therapy. *Am J Psychiatry* 1994; **151**:1105–13.

Akil M, Schwartz JA, Dutchak D *et al.* The psychiatric presentations of Wilson's disease. *J Neuropsychiatry Clin Neurosci* 1991; **3**:377–82.

Ala TA, Yang KH, Sung JH *et al.* Hallucinations and signs of parkinsonism help distinguish patients with dementia and cortical Lewy bodies from patients with Alzheimer's disease at presentation: a clinicopathological study. *J Neurol Neurosurg Psychiatry* 1997; **62**:16–21.

Al-Asmi A, Jansen AC, Badhwar A *et al.* Familial temporal lobe epilepsy as a presenting feature of choreoacanthocytosis. *Epilepsia* 2005; **46**:1256–63.

Alberca RA, Chinchon I, Vadillo J *et al.* Late onset parkinsonian syndrome in Hallervorden–Spatz disease. *J Neurol Neurosurg Psychiatry* 1987; **50**:1665–8.

Albert ML, Feldman RG, Willis AL. The 'subcortical dementia' of progressive supranuclear palsy. *J Neurol Neurosurg Psychiatry* 1974; **37**:121–30.

Allen AJ, Kurlan RM, Gilbert DL *et al.* Atomoxetine treatment in children and adolescents with ADHD and comorbid tic disorders. *Neurology* 2005; **65**:1941–9.

Altrocchi PH, Forno LS. Spontaneous oral-facial dyskinesia: neuropathology of a case. *Neurology* 1983; **33**:802–5.

Alves D, Pires MM, Guimaraes A *et al.* Four cases of late onset metachromatic leukodystrophy in a family: clinical, biochemical and neuropathological studies. *J Neurol Neurosurg Psychiatry* 1986; **49**:1417–22.

Alzheimer A. Ubr eine eigenartige erkrankung der hirnrinde. *Allermeine Zeitschrift fur Psychiatrie* 1907; **64**:146–8.

Amaducci LA, Rocca WA, Schoenberg BS. Origin of the distinction between Alzheimer's disease and senile dementia: how history can clarify nosology. *Neurology* 1986; **36**:1497–9.

Amatniek JC, Hauser WA, DelCastillo-Castenada C *et al.* Incidence and predictors of seizures in patients with Alzheimer's disease. *Epilepsia* 2006; **47**:867–72.

American Psychiatric Association. *Diagnostic and Statistical Manual of Mental Disorders*, 3rd edn. Washington, DC: American Psychiatric Press, 1980.

Aminoff MJ, Dedo HH, Idebski K. Clinical aspects of spasmodic dysphonia. *J Neurol Neurosurg Psychiatry* 1978; **41**:361–5.

Anca MH, Giladi N, Korczyn AD. Ropinirole in Gilles de la Tourette syndrome. *Neurology* 2004; **62**:1626–7.

Andermann F, Keene DL, Andermann E *et al.* Startle disease or hyperekplexia: further delineation of the syndrome. *Brain* 1980; **103**:985–97.

Andersen J, Asbo E, Gulmann N *et al.* Anti-depressive treatment in Parkinson's disease. A controlled trial of the effect of nortriptyline in patients with Parkinson's disease treated with L-DOPA. *Acta Neurol Scand* 1980; **62**:210–19.

Andersen K, Baldin J, Gottfries CG *et al.* A double-blind evaluation of electroconvulsive therapy in Parkinson's disease with 'on-off' phenomenon. *Acta Neurol Scand* 1987; **76**:191–9.

Anderson KE, Louis ED, Stern Y *et al.* Cognitive correlates of obsessive and compulsive symptoms in Huntington's disease. *Am J Psychiatry* 2001; **158**:799–801.

Andreasen N, Minthon L, Davidsson P *et al.* Evaluation of CSF-tau and CSF-Abeta42 as diagnostic markers for Alzheimer disease in clinical practice. *Arch Neurol* 2001; **58**:349–50.

Andrew M, Owen MJ. Hyperekplexia: abnormal startle response due to glycine receptor mutations. *Br J Psychiatry* 1997; **170**:106–8.

Angelini L, Nardocci N, Rumi V *et al.* Hallervorden–Spatz disease: clinical and MRI study of 11 cases diagnosed in life. *J Neurol* 1992; **239**:417–25.

Antonini A, Mainero C, Romano A *et al.* Cerebral atrophy in myotonic dystrophy: a voxel-based morphometric study. *J Neurol Neurosurg Psychiatry* 2004; **75**:1611–13.

Antonini G, Vichi R, Leardi MG *et al.* Effect of clomipramine on myotonia: a placebo-controlled, double-blind, crossover trial. *Neurology* 1990; **40**:1473–4.

Apaydin H, Ahlskog JE, Parisi JE *et al.* Parkinson disease neuropathology. *Arch Neurol* 2002; **59**:102–12.

April RS. Observations on parkinsonian tremor in all-night sleep. *Neurology* 1966; **16**:720–4.

Ardran G, Kinsburne M, Rushworth G. Dysphonia due to tremor. *J Neurol Neurosurg Psychiatry* 1966; **29**:219–23.

Arpa J, Cuesta A, Cruz-Martinez A *et al.* Clinical features and genetic analysis of a Spanish family with spinocerebellar ataxia 6. *Acta Neurol Scand* 1999; **99**:44–7.

Arriagada PV, Growdon JH, Hedley-Whyte ET *et al.* Neurofibrillary tangles but not senile plaques parallel duration and severity of Alzheimer's disease. *Neurology* 1992; **42**:631–9.

Ascherio A, Chen C, Weisskopf MG *et al.* Pesticide exposure and risk for Parkinson's disease. *Ann Neurol* 2006; **60**:197–203.

Asmus F, Horber V, Pohlenz J *et al.* A novel TITF-1 mutation causes benign hereditary chorea with response to levodopa. *Neurology* 2005; **64**:1952–4.

Austin J, Armstrong D, Fouch S *et al.* Metachromatic leukodystrophy (MLD). VIII. MLD in adults: diagnosis and pathogenesis. *Arch Neurol* 1968; **18**:225–40.

Avorn J, Scneeweis S, Sudarksy LR *et al.* Sudden uncontrollable somnolence and medication use in Parkinson disease. *Arch Neurol* 2005; **62**:1242–8.

Avrahami E, Katz A, Bornstein N *et al.* Computed tomographic findings of brain and skull in myotonic dystrophy. *J Neurol Neurosurg Psychiatry* 1987; **50**:435–8.

Axelrad JE, Louis ED, Honig LS *et al.* Reduced Purkinje cell number in essential tremor: a postmortem study. *Arch Neurol* 2008; **65**:101–7.

Baas H, Beiske AG, Ghika J *et al.* Catechol-O-methyltransferase inhibition with tolcapone reduces the 'wearing off' phenomenon and levodopa requirements in fluctuating parkinsonian patients. *J Neurol Neurosurg Psychiatry* 1997; **63**:421–8.

Bacalman S, Farzin F, Bourgeois JA *et al.* Psychiatric phenotype of the Fragile X-associated tremor/ataxia syndrome (FXTAS) in males: newly described fronto-subcortical dementia. *J Clin Psychiatry* 2006; **67**:87–94.

Bachoud-Levi AC, Gaura V, Brugieres P *et al.* Effect of fetal neural transplants in patients with Huntington's disease 6 years after surgery: a long-term follow-up study. *Lancet Neurol* 2006; **5**:303–9.

Bain PG, Findley LJ, Thompson PD *et al.* A study of hereditary essential tremor. *Brain* 1994; **117**:805–24.

Baliko L, Csala B, Czopf J. Suicide in Hungarian Huntington's disease patients. *Neuroepidemiology* 2004; **23**:258–60.

Ballard C, McKeith I, Harrison R *et al.* A detailed phenomenological comparison of complex visual hallucinations in dementia with Lewy bodies and Alzheimer's disease. *Int Psychogeriatr* 1997; **9**:381–8.

Ballard C, Holmes C, McKeith I *et al.* Psychiatric morbidity in dementia with Lewy bodies: a prospective clinical and neuropathological comparative study with Alzheimer's disease. *Am J Psychiatry* 1999; **156**:1039–45.

Ballard C, O'Brien J, Gray A *et al.* Attention and fluctuating attention in patients with dementia with Lewy bodies and Alzheimer disease. *Arch Neurol* 2001; **58**:977–82.

Bandmann O, Valente EM, Holmans P *et al.* Dopa-responsive dystonia: a clinical and molecular genetic study. *Ann Neurol* 1998; **44**:649–56.

Barker WW, Luis CA, Kashuba A *et al.* Relative frequencies of Alzheimer disease, Lewy body, vascular and frontotemporal dementia, and hippocampal sclerosis in the State of Florida Brain Bank. *Alzheimer Dis Assoc Disord* 2002; **16**:203–12.

Barkhof F, Verrips A, Wesseling P *et al.* Cerebrotendinous xanthomatosis: the spectrum of imaging findings and the correlation with neuropathologic findings. *Radiology* 2000; **217**:869–76.

Barnes J, David AS. Visual hallucinations in Parkinson's disease: a review and phenomenological survey. *J Neurol Neurosurg Psychiatry* 2001; **70**:727–33.

Barnes S, Hurst W. A further note on hepatolenticular degeneration. *Brain* 1926; **49**:36–60.

Barone P, Poewe W, Albrecht S *et al.* Pramipexole for the treatment of depressive symptoms in patients with Parkinson's disease: a randomized, double-blind, placebo-controlled trial. *Lancet Neurol* 2010; **9**:573–80.

Barr AN, Fischer JH, Koller WC *et al.* Serum haloperidol concentrations and choreiform movements in Huntington's disease. *Neurology* 1988; **38**:84–8.

Barth ML, Fensom A, Harris A. Prevalence of common mutations in the arylsulfatase A gene in metachromatic leukodystrophy patients diagnosed in Britain. *Hum Genet* 1993; **91**:73–7.

Batten FE, Gibb HP. Myotonia atrophica. *Brain* 1909; **32**:187–96.

Bax RT, Hassler A, Luck W *et al.* Cerebral manifestation of Wilson's disease successfully treated with liver transplantation. *Neurology* 1998; **51**:863–5.

Beale MD, Kellner CH, Gurecki P *et al.* ECT for the treatment of Huntington's disease. *Convuls Ther* 1997; **13**:108–12.

Beatty WW, Winn P, Adams RL *et al.* Preserved cognitive skills in dementia of the Alzheimer type. *Arch Neurol* 1994; **51**:1040–6.

Becher MW, Rubinsztein DC, Leggo J *et al.* Dentatorubral and pallidoluysian atrophy (DRPLA): clinical and neuropathological findings in genetically conformed North American and European pedigrees. *Mov Disord* 1997; **12**:519–30.

Beck J, Rohrer JD, Campbell T *et al.* A distinct clinical, neuropsychological and radiological phenotype is associated with progranulin gene mutations in a large UK series. *Brain* 2008; **131**:706–70.

Behan PO, Bone I. Hereditary chorea without dementia. *J Neurol Neurosurg Psychiatry* 1977; **40**:687–91.

Benajiba L, Le Ber I, Camuzat A *et al.* TARDBP mutations in motoneuron disease with frontotemporal lobar degeneration. *Ann Neurol* 2009; **65**:470–3.

Bencze KS, Vande Polder DR, Prockop LD. Magnetic resonance imaging of the brain and spinal cord in cerebrotendinous xanthomatosis. *J Neurol Neurosurg Psychiatry* 1990; **53**:166–7.

Benesch CG, McDaniel KD, Cox C *et al.* End-stage Alzheimer's disease: Glasgow coma scale and neurologic examination. *Arch Neurol* 1993; **50**:1309–15.

Benito-Leon J, Louis ED, Bermejo-Pareja F *et al.* Population-based case–control study of cognitive function in essential tremor. *Neurology* 2006a; **66**:69–74.

Benito-Leon J, Louis ED, Bermejo-Pareja F *et al.* Elderly-onset essential tremor is azssociated with dementia. *Neurology* 2006b; **66**: 1500–5.

Bennett DA, Schneider JA, Wilson RS *et al.* Neurofibrillary tangles mediate the association of amyloid load with clinical Alzheimer disease and level of cognitive function. *Arch Neurol* 2004; **61**:378–84.

Bensimon G, Laclombez L, Delumeau JC *et al.* A study of riluzole in the treatment of advanced stage or elderly patients with amyotrophic lateral sclerosis. *J Neurol* 2002; **249**:609–15.

Bentivoglio AR, DelGrosso N, Albenese A *et al.* Non-DYT1 dystonia in a large Italian family. *J Neurol Neurosurg Psychiatry* 1997; **62**:357–60.

Bergem AL, Engedal K, Kringlen E. The role of heredity in late-onset Alzheimer disease and vascular dementia. A twin study. *Arch Gen Psychiatry* 1997; **54**:264–70.

Bergeron C, Pollanen MS, Weyer L *et al.* Unusual clinical presentation of cortical-basal ganglionic degeneration. *Ann Neurol* 1996; **40**:893–900.

Berginer VM, Salen G, Shefer S. Long-term treatment of cerebrotendinous xanthomatosis with chenodeoxycholic acid. *N Engl J Med* 1984; **311**:1649–54.

Berkovic SF, Carpenter S, Andermann F *et al.* Kufs' disease: a critical reappraisal. *Brain* 1988; **111**:27–62.

Bermejo-Pareja F, Louis ED, Benito-Leon J *et al.* Risk of incident dementia in essential tremor: a population-base study. *Mov Disord* 2007; **22**:1573–80.

Besson JAO. A diagnostic pointer to adult metachromatic leukodystrophy. *Br J Psychiatry* 1980; **137**:186–7.

Betts TA, Smith WT, Kelly RE. Adult metachromatic leukodystrophy (subacute lipidosis) simulating acute schizophrenia: report of a case. *Neurology* 1968; **18**:1140–2.

Bhidayasiri R, Jen JC, Baloh RW. Three brothers with a very-late-onset writer's cramp. *Mov Disord* 2005; **20**:1375–7.

Bian H, Van Swieten JC, Leight S *et al.* CSF biomarkers in frontotemporal lobar degeneration with known pathology. *Neurology* 2008; **70**:1827–35.

Bicknell JM, Greenhouse AH, Pesch RN. Spastic dysphonia. *J Neurol Neurosurg Psychiatry* 1968; **31**:158–61.

Bierer L, Hof P, Purohit D *et al.* Neocortical neurofibrillary tangles correlate with dementia severity in Alzheimer's disease. *Arch Neurol* 1995; **52**:81–8.

Biggins CA, Boyd JL, Harrop FM *et al.* A controlled, longitudinal study of dementia in Parkinson's disease. *J Neurol Neurosurg Psychiatry* 1992; **55**:566–71.

Billings EG, Ravin A. A psychiatric study of patients manifesting dystrophica myotonica. *Am J Psychiatry* 1941; **97**:1116–34.

Binetti G, Bianchetti A, Padovani A *et al.* Delusions in Alzheimer's disease and multi-infarct dementia. *Acta Neurol Scand* 1993; **88**:5–9.

Biousse V, Skibell BC, Watts RL *et al.* Ophthalmologic features of Parkinson's disease. *Neurology* 2004; **62**:177–80.

Bird MT, Paulson GW. The rigid form of Huntington's chorea. *Neurology* 1971; **21**:271–6.

Bird TD, Levy-Lahad E, Poorkaj P *et al.* Wide range in age of onset for chromosome 1-related familial Alzheimer's disease. *Ann Neurol* 1996; **40**:932–6.

Birdi S, Rajput AH, Fenton M *et al.* Progressive supranuclear palsy diagnosis and confounding features: report on 16 autopsied cases. *Mov Disord* 2002; **17**:1255–64.

Blair IP, Williams KL, Warraich ST *et al.* FUS mutations in amyotrophic lateral sclerosis: clinical, pathological, neurophysiological and genetic analysis. *J Neurol Neurosurg Psychiatry* 2010; **81**: 639–45.

Bloch MH, Leckman JF, Zhu HS *et al.* Caudate volumes in childhood predict symptom severity in adults with Tourette syndrome. *Neurology* 2005; **65**:1253–8.

Boeve BF, Silber MH, Ferman TJ *et al.* REM sleep behavior disorder and degenerative dementia: an association likely reflecting Lewy body disease. *Neurology* 1998; **51**:363–70.

Bogaerts V, Engelborghs S, Kumar-Singh S *et al.* A novel locus for dementia with Lewy bodies: a clinically and genetically heterogeneous disorder. *Brain* 2007; **130**:2277–91.

Bonelli RM, Niederwieser G, Tribl GG *et al.* High-dose olanzapine in Huntington's disease. *Int Clin Psychopharmacol* 2002; **17**: 91–3.

Bosch EP, Hart MN. Late adult-onset metachromatic leukodystrophy: dementia and polyneuropathy in a 63-year-old man. *Arch Neurol* 1978; **35**:475–7.

Botez MI, Botez-Marquard T, Elie R *et al.* Amantadine hydrochloride treatment in heredodegenerative ataxia: a double blind study. *J Neurol Neurosurg Psychiatry* 1996; **61**:259–64.

Bourgeois JA, Cogswell JB, Hessl D *et al.* Cognitive, anxiety and mood disorders in the fragile X-associated tremor/ataxia syndrome. *Gen Hosp Psychiatry* 2007; **29**:349–56.

Bouton SM. Pick's disease. *J Nerv Ment Dis* 1940; **91**:9–30.

Bowler JV, Eliaszaw M, Steenhuis R *et al.* Comparative evolution of Alzheimer disease, vascular dementia, and mixed dementia. *Arch Neurol* 1997; **54**:697–703.

Boxer AL, Lipton AM, Womack K *et al.* An open-label study of memantine treatment in 3 subtypes of frontotemporal lobar degeneration. *Alzheimer Dis Assoc Disord* 2009; **23**:211–17.

Braak H, Braak E. Neuropathological staging of Alzheimer-related changes. *Acta Neuropathol* 1991; **82**:239–59.

Braak H, Rub U, Jansen Steur ENH *et al.* Cognitive status correlates with neuropathologic stage in Parkinson disease. *Neurology* 2005; **64**:1404–10.

Bracco L, Gallato R, Grigoletto F *et al.* Factors affecting course and survival in Alzheimer's disease. *Arch Neurol* 1994; **51**:1213–19.

Bradshaw J, Saling M, Hopwood M *et al.* Fluctuating cognition in dementia with Lewy bodies and Alzheimer's disease is qualitatively distinct. *J Neurol Neurosurg Psychiatry* 2004; **75**:382–7.

Brain WR, Greenfield JG. Late infantile metachromatic leukodystrophy with primary degeneration of the interfascicular oligodendroglia. *Brain* 1950; **73**:291–317.

Brans JWM, Lindeboom R, Snoek JW *et al.* Botulinum toxin versus trihexyphenidyl in cervical dystonia: a prospective, randomized, double-blind controlled trial. *Neurology* 1996; **46**:1066–72.

Brassat D, Camuzat A, Vidailhet M *et al.* Frequency of the DYT1 mutation in primary torsion dystonia without family history. *Arch Neurol* 2000; **57**:333–5.

Breedveld GJ, Percy AK, MacDonald ME *et al.* Clinical and genetic heterogeneity in benign hereditary chorea. *Neurology* 2002; **59**:579–84.

Breier A, Sutton VK, Feldman PD *et al.* Olanzapine in the treatment of dopamimetic-induced psychosis in patients with Parkinson's disease. *Biol Psychiatry* 2002; **52**:438–45.

Breitner JCS, Walsh KA, Gau BA *et al.* Alzheimer's disease in the National Academy of Sciences – National Research Council of Aging Twin Veterans. III: Detection of cases, longitudinal results, and observations on twin concordance. *Arch Neurol* 1995; **52**:763–71.

Bressman SB, de Leon D, Kramer PL *et al.* Dystonia in Ashkenazi Jews: clinical characterization of a founder mutation. *Ann Neurol* 1994; **36**:771–7.

Bressman SB, Warner TT, Almasy L *et al.* Exclusion of the DYT1 locus in familial torticollis. *Ann Neurol* 1996; **40**:681–4.

Bressman SB, Sabatti C, Raymond D *et al.* The DYT1 phenotype and guidelines for diagnostic testing. *Neurology* 2000; **54**:1746–52.

Bressman SB, Raymond D, Fuchs T *et al.* Mutations in THAP1 (DYT6) in early-onset dystonia: a genetic screening study. *Lancet Neurology* 2009; **8**:441–6.

Brewer GJ, Turkay A, Yuzbasiyan-Gurkan V. Development of neurologic symptoms in a patient with asymptomatic Wilson's disease treated with penicillamine. *Arch Neurol* 1994; **51**:304–5.

Brewer GJ, Askari F, Lorincz MT *et al.* Treatment of Wilson disease with ammonium tetrathiomolybdate. IV. Comparison of tetrathiomolybdate and trientene in a double blind study of treatment of the neurologic presentation of Wilson disease. *Arch Neurol* 2006; **63**:521–7.

Bridgman O, Smyth FS. Progressive lenticular degeneration. *J Nerv Ment Dis* 1944; **99**:534–43.

Briel RC, McKeith IG, Barker WA *et al.* EEG findings in dementia with Lewy bodies and Alzheimer's disease. *J Neurol Neurosurg Psychiatry* 1999; **66**:401–3.

Britton JW, Uitti RJ, Ahlskog JE *et al.* Hereditary late-onset chorea without significant dementia: genetic evidence for substantial phenotypic variation in Huntington's disease. *Neurology* 1995; **45**:443–7.

Brody IA, Wilkins RH. Huntington's chorea. *Arch Neurol* 1967; **17**:331–3.

Bronner IF, ter Meulen BC, Azmani A *et al.* Hereditary Pick's disease with the G272V tau mutation showing predominant three-repeat tau pathology. *Brain* 2005; **128**:2645–53.

Brook JD, McCurrach ME, Harley HG *et al.* Molecular basis of myotonic dystrophy: expansion of a trinucleotide (CTG) repeat at the 39 end of a transcript encoding a protein kinase family member. *Cell* 1992; **68**:799–808.

Brooks DJ, Sagar H. Entacapone is beneficial in both fluctuating and non-fluctuating patients with Parkinson's disease: a randomized, placebo controlled, double blind, six month study. *J Neurol Neurosurg Psychiatry* 2003; **74**:1071–9.

Brouwers N, Sleegers K, Engelborghs S *et al.* Genetic risk and transcriptional variability of amyloid precursor protein in Alzheimer's disease. *Brain* 2006; **129**:2984–91.

Brown J, Lantos P, Stratton M *et al.* Familial progressive supranuclear palsy. *J Neurol Neurosurg Psychiatry* 1993; **56**:473–6.

Brown P, Rothwell JC, Thompson PD *et al.* The hyperekplexias and their relationship to the normal startle reflex. *Brain* 1991; **114**:1903–28.

Brownell B, Oppenheimer DR, Hughes JT. The central nervous system in motor neuron disease. *J Neurol Neurosurg Psychiatry* 1970; **33**:338–57.

Bruggeman R, van der Linden C, Butelaar JK *et al.* Risperidone versus pimozide in Tourette's disorder: a comparative double-blind parallel-group study. *J Clin Psychiatry* 2001; **62**:50–6.

Brunberg JA, Jacquemont S, Hagerman RJ *et al.* Fragile X premutation carriers: characteristic MR imaging findings of adult male patients with progressive cerebellar and cognitive dysfunction. *AJNR* 2002; **23**:1757–66.

Bruneau MA, Lesperance P, Chouinard S. Schizophrenia-like presentation of neuroacanthocytosis. *J Neurospcyhiatry Clin Neurosci* 2003; **15**:378–80.

Bruni AC, Takahashi-Fujigasaki J, Maltecca F *et al.* Behavioral disorder, dementia, ataxia, and rigidity in a large family with TATA box-binding protein mutation. *Arch Neurol* 2004; **61**:1314–20.

Brusa L, Oriacchio A, Moschella V *et al.* Treatment of the symptoms of Huntington's disease: preliminary results comparing aripiprazole and tetrabenazine. *Mov Disord* 2009; **24**:126–9.

Bruun RD. Subtle and unrecognized side effects of neuroleptic treatment in children with Tourette's disorder. *Am J Psychiatry* 1988; **145**:621–4.

Buerger K, Ewers M, Pirtilla T *et al.* CSF phosphorylated tau protein correlates with neocortical neurofibrillary pathology in Alzheimer's disease. *Brain* 2006; **129**:3035–41.

Bull PC, Thomas GR, Rommens JM *et al.* The Wilson disease gene is a putative copper transporting P-type ATPase similar to Menkes gene. *Nature Gen* 1993; **5**:327–37.

Bungener C, Jouvent R, Delaporte C. Psychopathological and emotional deficits in myotonic dystrophy. *J Neurol Neurosurg Psychiatry* 1998; **65**:353–6.

Burk K, Globas C, Bosch S *et al.* Cognitive deficits in spinocerebellar ataxia 2. *Brain* 1999; **122**:769–77.

Burke RE, Fahn S, Marsden CD. Torsion dystonia: a double-blind, prospective trial of high-dosage trihexyphenidyl. *Neurology* 1986; **36**:160–4.

Burke WJ, Pfeiffer RF, McComb RD. Neuroleptic sensitivity to clozapine in dementia with Lewy bodies. *J Neuropsychiatry Clin Neurosci* 1998; **10**:227–9.

Burkhardt CR, Filley CM, Kleineschmidt-DeMasters BK *et al.* Diffuse Lewy body disease and progressive dementia. *Neurology* 1988; **38**:1520–8.

Burneo JG, Arnold T, Palmer CA *et al.* Adult-onset neuronal ceroid lipofuscinosis (Kufs' disease) with autosomal dominant inheritance in Alabama. *Epilepsia* 2003; **44**:841–6.

Burns A, Jacoby R, Levy R. Psychiatric phenomena in Alzheimer's disease. III. Disorders of mood. *Br J Psychiatry* 1990a; **157**:81–6.

Burns A, Jacoby R, Levy R. Psychiatric phenomena in Alzheimer's disease. II. Disorders of perception. *Br J Psychiatry* 1990b; **157**:76–81.

Burns A, Jacoby R, Levy R. Psychiatric phenomena in Alzheimer's disease. I. Disorders of thought content. *Br J Psychiatry* 1990c; **157**:72–6.

Burton EJ, Barber R, Mukaetova-Ladinska EB *et al.* Medial temporal lobe atrophy on MRI differentiates Alzheimer's disease from dementia with Lewy bodies and vascular cognitive impairment: a prospective study with pathological verification of diagnosis. *Brain* 2009; **132**:195–203.

Byrne EJ, Lennox G, Lowe J *et al.* Diffuse Lewy body disease: clinical features in 15 cases. *J Neurol Neurosurg Psychiatry* 1989; **57**:709–17.

Caine ED, Shoulson I. Psychiatric symptoms in Huntington's disease. *Am J Psychiatry* 1983; **140**:728–33.

Callagy C, O'Neill G, Murphy SF *et al.* Adult neuronal ceroid lipofuscinosis (Kufs' disease) in two siblings in an Irish family. *Clin Neuropathol* 2000; **19**:109–18.

Calzetti S, Bortone E, Negrotti A *et al.* Frontal intermittent rhythmic delta activity (FIRDA) in patients with dementia with Lewy bodies: a diagnostic tool? *Neurol Sci* 2002; **23**(Suppl. 2):S65–6.

Campbell AMG, Corner B, Norman RM *et al.* The rigid form of Huntington's disease. *J Neurol Neurosurg Psychiatry* 1965; **24**:71–7.

Canelas HM, Quintao ECR, Scaff M *et al.* Cerebrotendinous xanthomatosis: clinical and laboratory study of 2 cases. *Acta Neurol Scand* 1983; **67**:305–11.

Cardoso F, Veado CCM, de Oliveira JT. A Brazilian cohort of patients with Tourette's syndrome. *J Neurol Neurosurg Psychiatry* 1996; **60**:209–12.

Carella F, Scaioli V, Ciano C *et al.* Adult onset myoclonic Huntington's disease. *Mov Disord* 1993; **8**:201–5.

Castelbuono A, Miller NR. Spontaneous remission in patients with essential blepharospasm and Meige syndrome. *Am J Ophthalmol* 1998; **126**:432–5.

Celesia GG, Barr AN. Psychosis and other psychiatric manifestations of levodopa therapy. *Arch Neurol* 1970; **23**:193–200.

Ceravolo R, Nuti A, Piccinni A *et al.* Paroxetine in Parkinson's disease: effects on motor and depressive symptoms. *Neurology* 2000; **55**:1216–18.

Chan D, Anderson V, Pijnenburg Y *et al.* The clinical profile of right temporal lobe atrophy. *Brain* 2009; **132**:1287–98.

Chan DB, Lang MF, Fahn S. Idiopathic cervical dystonia. *Mov Disord* 1991; **6**:119–26.

Chan-Palay V, Asan E. Alterations in catecholamine neurons of the locus coeruleus in senile dementia of the Alzheimer type and in Parkinson's disease with and without dementia and depression. *Comp Neurol* 1989; **287**:373–92.

Chartier-Harlin M, Crawford F, Houlden H *et al.* Early-onset Alzheimer's disease caused by mutations at codon 717 of the β-amyloid precursor protein gene. *Nature* 1991; **353**:844–6.

Chee KY, Sachdev P. A controlled study of sensory tics in Gilles de la Tourette syndrome and obsessive–compulsive disorder using a structured interview. *J Neurol Neurosurg Psychiatry* 1997; **62**:188–92.

Chelly J, Monaco AP. Cloning the Wilson disease gene. *Nature Gen* 1993; **5**:317–18.

Chen H, Huang X, Guo X *et al.* Smoking duration, intensity, and risk of Parkinson disease. *Neurology* 2010; **74**:878–84.

Chen J-Y, Stern Y, Sano M *et al.* Cumulative risk of developing extrapyramidal signs, psychosis or myoclonus in the course of Alzheimer's disease. *Arch Neurol* 1991; **48**:1141–3.

Chonchaiya W, Nguyen DV, Au J *et al.* Clinical involvement in daughters of men with fragile X-associated tremor ataxia syndrome. *Clin Genet* 2010; **78**:38–46.

Chouinard S, Ford B. Adult onset tic disorder. *J Neurol Neurosurg Psychiatry* 2000; **68**:738–43.

Church AJ, Dale RC, Lees AJ *et al.* Tourette's syndrome: a cross-sectional study to examine the PANDAS hypothesis. *J Neurol Neurosurg Psychiatry* 2003; **74**:602–7.

Clark CM, Ewbank D, Lerner A *et al.* The relationship between extrapyramidal signs and cognitive performance in patients with Alzheimer's disease enrolled in the CERAD study. *Neurology* 1997; **49**:70–5.

Clark DG, Mendez MF, Farag E *et al.* Clinicopathologic report: progressive aphasia in a 77-year old man. *J Neuropsychiatry Clin Neurosci* 2003; **15**:231–8.

Cohen AJ, Leckman JF. Sensory phenomena associated with Gilles de la Tourette's syndrome. *J Clin Psychiatry* 1992; **23**:319–23.

Cohen DJ, Detlor RN, Young G *et al.* Clonidine ameliorates Gilles de la Tourette's syndrome. *Arch Gen Psychiatry* 1980; **37**:1350–7.

Cohen LG, Hallett M. Hand cramps: clinical features and electromyographic patterns in a focal dystonia. *Neurology* 1988; **38**:1005–12.

Coleman LW, Digre KB, Stephenson GM *et al.* Autopsy-proven, sporadic Pick disease with onset at age 25 years. *Arch Neurol* 2002; **59**:856–9.

Collins SJ, Ahlskog JE, Parisi JE *et al.* Progressive supranuclear palsy: neuropathologically based diagnostic clinical criteria. *J Neurol Neurosurg Psychiatry* 1995; **58**:167–73.

Colosimo C, Albanese A, Hughes AJ *et al.* Some specific clinical features differentiate multiple system atrophy (striatonigral variety) from Parkinson's disease. *Arch Neurol* 1995; **52**:294–8.

Comella CL, Goetz CG. Akathisia in Parkinson's disease. *Mov Disord* 1994; **9**:545–9.

Como PG, Rubin AJ, O'Brien CF *et al.* A controlled trial of fluoxetine in nondepressed patients with Huntington's disease. *Mov Disord* 1997; **12**:397–401.

Connor GS. A double-blind placebo-controlled trial of topiramate treatment of essential tremor. *Neurology* 2002; **59**:132–4.

Cook RH, Schenck SA, Clark DB. Twins with Alzheimer's disease. *Arch Neurol* 1981; **38**:300–1.

Cooper DB, Ales G, Lange C *et al.* Atypical onset of symptoms in Huntington disease: severe cognitive decline preceding chorea or other motor manifestations. *Cogn Behav Neurol* 2006; **19**:222–4.

Cooper JA, Sagar HJ, Doherty SM *et al.* Different effects of dopaminergic and anticholinergic therapies on cognitive and motor function in Parkinson's disease. A follow-up study of untreated patients. *Brain* 1992; **115**:1701–25.

Cooper GE, Rizzo M, Jones RD. Adult-onset Hallervorden-Spatz syndrome presenting as cortical dementia. *Alzheimer Dis Assoc Disord* 2000; **14**:120–6.

Corder EH, Saunders EM, Strittmatter WJ *et al.* Gene dose of apolipoprotein II type 4 allele and the risk of Alzheimer's disease in late onset families. *Science* 1993; **261**:921–3.

Craufurd D, Dodge A. Mutation size and age of onset in Huntington's disease. *J Med Genet* 1993; **30**:1008–11.

Critchley EMR, Clark DB, Wikler A. Acanthocytosis and neurological disorder without abetalipoproteinemia. *Arch Neurol* 1968; **18**:134–40.

Critchley M. Observation on essential (heredofamilial) tremor. *Brain* 1949; **72**:113–39.

Critchley M, Greenfield JG. Olivo-ponto-cerebellar atrophy. *Brain* 1948; **71**:343–64.

Crystal HA, Grober E, Masur D. Preservation of musical memory in Alzheimer's disease. *J Neurol Neurosurg Psychiatry* 1989; **52**:1415–16.

Crystal HA, Dixon DW, Lizardi JE *et al.* Antemortem diagnosis of diffuse Lewy body disease. *Neurology* 1990; **40**:1523–8.

Cubo E, Shannon KM, Tracy D *et al.* Effect of donepezil on motor and cognitive function in Huntington disease. *Neurology* 2006; **67**:1268–71.

Cummings JL, Duchen LW. Klüver–Bucy syndrome in Pick disease: Clinical and pathologic correlations. *Neurology* 1981; **31**:1415–22.

Cummings JL, Street J, Masterman D *et al.* Efficacy of olanzapine in the treatment of psychosis in dementia with Lewy bodies. *Dement Geriatr Cogn Disord* 2002; **13**:67–73.

Cummings JL, Schneider E, Tariot PN *et al.* Behavioral effects of memantine in Alzheimer disease patients receiving donepezil treatment. *Neurology* 2006; **67**:57–63.

Cunningham MA, Darby DG, Donnan GA. Controlled-release delivery of L-dopa associated with nonfatal hyperthermia, rigidity and autonomic dysfunction. *Neurology* 1991; **41**:942–3.

Danek A, Rubio JP, Rampoldi L *et al.* McLeod neuroacanthocytosis: genotype and phenotype. *Ann Neurol* 2001; **50**:755–64.

Daniel SE, de Bruin VMS, Lees AJ. The clinical and pathological spectrum of Steele–Richardson–Olszewski syndrome (progressive supranuclear palsy): a reappraisal. *Brain* 1995; **118**:759–70.

DAPT Research Group. Cognitive function over time in the Alzheimer's Disease Anti-inflammatory Prevention Trial (ADAPT): results of a randomized, controlled trial of naproxen and celecoxib. *Arch Neurol* 2008; **65**:896–905.

Da Silva TM, Munhoz RP, Alvarez C *et al.* Depression in Parkinson's disease: a double-blind, randomized, placebo-controlled pilot study of omega-3 fatty-acid supplementation. *J Affect Disord* 2008; **111**:351–9.

Dauer WT, Burke RE, Greene P *et al.* Current concepts on the clinical features, aetiology and management of idiopathic cervical dystonia. *Brain* 1998; **121**:547–60.

Davies RR, Kipps CM, Mitchell J *et al.* Progression in frontotemporal dementia: identifying a benign behavioral variant by magnetic resonance imaging. *Arch Neurol* 2006; **63**:1627–31.

Davis EJB, Borde M. Wilson's disease and catatonia. *Br J Psychiatry* 1993; **162**:256–9.

Davis PH, Bergeron C, McLachlan DR. Atypical presentation of progressive supranuclear palsy. *Ann Neurol* 1985; **17**:337–43.

Day JW, Ricker K, Jacobsen JF *et al.* Myotonic dystrophy type 2. Molecular diagnostic and clinical spectrum. *Neurology* 2003; **60**:657–64.

Deakin JB, Rahman S, Nestor PJ *et al.* Paroxetine does not improve symptoms and impairs cognition in frontotemporal dementia: a double-blind randomized controlled trial. *Psychopharmacology* 2004; **172**:400–8.

De Cock VC, Vidailhet M, Leu S *et al.* Restoration of normal motor control in Parkinson's disease during REM sleep. *Brain* 2007; **130**:450–6.

Deep Brain Stimulation for Parkinson's Disease Study Group. Deep brain stimulation of the subthalamic nucleus or the pars interna of the globus pallidus in Parkinson's disease. *N Engl J Med* 2001; **345**:956–63.

Defazio G, Berardelli A, Abbruzzese G *et al.* Risk factors for spread of primary adult onset blepharospasm: a multicentre investigation of the Italian movement disorders study group. *J Neurol Neurosurg Psychiatry* 1999; **67**:613–19.

Defazio G, Martino D, Aniello MS *et al.* A family study on primary blepharospasm. *J Neurol Neurosurg Psychiatry* 2006; **77**:252–4.

Delacourte A, David JP, Sergeant N *et al.* The biochemical pathway of neurofibrillary degeneration in aging and Alzheimer's disease. *Neurology* 1999; **52**:1158–65.

Delaporte C. Personality patterns in patients with myotonic dystrophy. *Arch Neurol* 1998; **55**:635–40.

De la Tourette G. Etude sur une affection nerveuse, characterisee par de 'incoordination motrice, accompagne d'cholalie et de coprolalie. *Arch Neurol (Paris)* 1885; **9**:158–200.

Delwaide PJ, Desseilles M. Spontaneous buccolinguofacial dyskinesia in the elderly. *Acta Neurol Scand* 1977; **56**:256–62.

De Marchi N, Daniele F, Ragone MA. Fluoxetine in the treatment of Huntington's disease. *Psychopharmacology* 2001; **153**:264–6.

Demirkiran M, Jankovic J, Lewis RA *et al.* Neurologic presentation of Wilson disease without Kayser–Fleischer rings. *Neurology* 1996; **46**:1040–3.

Denckla MB, Bemporad JR, MacKay MC. Tics following methylphenidate administration: a report of 20 cases. *JAMA* 1976; **235**:1349–51.

Deng H, Le W, Jankovic J. Genetics of essential tremor. *Brain* 2007; **130**:1456–64.

Dening TR, Berrios GE, Walshe JM. Wilson's disease and epilepsy. *Brain* 1988; **111**:1139–55.

Dening TR, Berrios GE. Wilson's disease: psychiatric symptoms in 195 cases. *Arch Gen Psychiatry* 1989; **46**:1126–34.

Deonna T. DOPA-sensitive progressive dystonia of childhood with fluctuations of symptoms – Segawa's syndrome and possible variants. *Neuropediatrics* 1986; **17**:81–6.

Deonna T, Roulet E, Ghika J *et al.* Dopa-responsive childhood dystonia: a forme fruste with writer's cramp, triggered by exercise. *Dev Med Child Neurol* 1997; **39**:49–53.

Dermaut B, Kumar-Singh S, Engelborghs S *et al.* A novel presenilin 1 mutation associated with Pick's disease but not β-amyloid plaques. *Ann Neurol* 2004; **55**:617–26.

De Stefano N, Dotti MT, Mortilla M *et al.* Magnetic resonance imaging and spectroscopic changes in brains of patients with cerebrotendinous xanthomatosis. *Brain* 2001; **124**:121–3.

De Tommaso M, Di Fruscolo A, Sciruicchio V *et al.* Efficacy of levetiracetam in Huntington disease. *Clin Neuropharmacol* 2005; **28**:280–4.

De Tommaso M, Difruscolo O, Sciruicchio V *et al.* Two-years' follow-up of rivastigmine in Huntington disease. *Clin Neuropharmacol* 2007; **30**:43–6.

Devanand DP, Jacobs DM, Tang M-X *et al.* The course of psychopathologic features in mild to moderate Alzheimer's disease. *Arch Gen Psychiatry* 1997; **54**:257–63.

Devos D, Dujardin J, Poirot I *et al.* Comparison of desipramine and citalopram treatments for depression in Parkinson's disease: a double-blind, randomized, placebo-controlled study. *Mov Disord* 2008; **23**:850–7.

De Yebenes JG, Sarasa JL, Daniel SE *et al.* Familial progressive supranuclear palsy: description of a pedigree and review of the literature. *Brain* 1995; **118**:1095–103.

Di Costanzo A, De Salle F, Santoro L *et al.* Pattern and significance of white matter abnormalities in myotonic dystrophy type 1: an MRI study. *J Neurol* 2002; **249**:1175–82.

Didic M, Cherif AA, Gambarelli D *et al.* A permanent pure amnestic syndrome of insidious onset related to Alzheimer's disease. *Ann Neurol* 1998; **43**:526–30.

Dodd CM, Klos KS, Bower JH *et al.* Pathological gambling caused by drugs used to treat Parkinson disease. *Arch Neurol* 2005; **62**:1377–81.

Dooley JM, Andermann F. Startle disease or hyperekplexia: adolescent onset and response to valproate. *Ped Neurology* 1989; **5**:126–7.

Dooling EC, Schoene WC, Richardson EP. Hallervorden–Spatz syndrome. *Arch Neurol* 1974; **30**:70–83.

Dooling EC, Richardson EP, Davis KR. Computed tomography in Hallervorden–Spatz disease. *Neurology* 1980; **30**:1128–30.

Doran M, Larner AJ. EEG findings in dementia with Lewy bodies causing diagnostic confusion with sporadic Creutzfeldt-Jakob disease. *Eur J Neurol* 2004; **11**:838–41.

Douyon R, Serby M, Klutchbo B *et al.* ECT and Parkinson's disease revisited: a 'naturalistic' study. *Am J Psychiatry* 1989; **146**:1451–5.

Du YS, Li HF, Vance A *et al.* Randomized double-blind multicentre placebo-controlled clinical trial of clonidine adhesive patch for the treatment of tic disorders. *Aust N Z J Psychiatry* 2008; **42**:807–13.

Duff K, Beglinger LJ, O'Rourke ME *et al.* Risperidone and the treatment of psychiatric, motor, and cognitive symptoms in Huntington's disease. *Ann Clin Psychiatry* 2008; **20**:1–3.

Durif F, Debilly B, Galitzky M *et al.* Clozapine improves dyskinesias in Parkinson disease. A double-blind, placebo-controlled study. *Neurology* 2004; **62**:381–8.

Durr A, Stevanin G, Cancel G *et al.* Spinocerebellar ataxia 3 and Machado–Joseph disease: clinical, molecular, and neuropathological features. *Ann Neurol* 1996; **39**:490–9.

Eapen V, Pauls DL, Robertson MM. Evidence for autosomal dominant transmission in Tourette's syndrome – United Kingdom cohort study. *Br J Psychiatry* 1993; **162**:593–6.

Edwards KR, Hershel L, Wray L *et al.* Efficacy and safety of galantamine in patients with dementia with Lewy bodies: a 12-week interim analysis. *Dement Geriatr Cogn Disord* 2004; **17**(Suppl. 1):40–8.

Edwards-Lee T, Miller BL, Benson DF *et al.* The temporal variant of frontotemporal dementia. *Brain* 1997; **120**:1027–40.

Elbaz A, Clavel J, Rathouz PJ *et al.* Professional exposure to pesticides and Parkinson disease. *Ann Neurol* 2009; **66**:494–504.

Ellis T, Cudkowicz ME, Sexton PM *et al.* Clozapine and risperidone treatment of psychosis in Parkinson's disease. *J Neuro-psychiatry Clin Neurosci* 2000; **12**:364–9.

Emre M, Aarsland D, Albanese A *et al.* Rivastigmine for dementia associated with Parkinson's disease. *N Engl J Med* 2004; **351**:2509–18.

Entacapone to Tolcapone Switch Study Investigators. Entacapone to tolcapone switch: mulicenter double-blind, randomized, active-controlled trial in advanced Parkinson's disease. *Mov Disord* 2007; **22**:14–19.

Ercan-Sencicek AG, Stillman AA, Ghosh AK *et al.* l-histidine decarboxylas and Tourette's syndrome. *N Engl J Med* 2010; **362**:1901–8.

Erlington GM, Bateman DE, Jeffrey MJ *et al.* Adrenoleuko-dystrophy: heterogeneity in two brothers. *J Neurol Neurosurg Psychiatry* 1989; **52**:310–14.

Evans AH, Katzenshlager R, Paviour D *et al.* Punding in Parkinson's disease: its relation to the dopamine dysregulation syndrome. *Mov Disord* 2004; **19**:367–70.

Evenhuius HM. The natural history of dementia in Down's syndrome. *Arch Neurol* 1990; **47**:263–7.

Faber-Langendoen K, Morris JC, Knesvich JW *et al.* Aphasia in senile dementia of the Alzheimer type. *Ann Neurol* 1988; **23**: 365–70.

Faden AI, Townsend JJ. Myoclonus in Alzheimer's disease: a confusing sign. *Arch Neurol* 1976; **33**:278–80.

Fagan AM, Mintun MA, Mach RH *et al.* Inverse relation between in vivo amyloid imaging load and cerebrospinal fluid Abeta42 in humans. *Ann Neurol* 2006; **59**:512–19.

Fang F, Valdimarsdottir U, Furst CJ *et al.* Suicide among patients with amyotrophic lateral sclerosis. *Brain* 2008; **131**:2729–33.

Farpour H, Mahloudji M. Familial cerebrotendinous xanthomatosis: report of a new family and review of the literature. *Arch Neurol* 1975; **32**:223–5.

Farrer LA. Suicide and attempted suicide in Huntington disease: implications for preclinical testing of persons at risk. *Am J Med Genet* 1986; **24**:305–11.

Farrer M, Chan P, Chen R *et al.* Lewy bodies and parkinsonism in families with parkin mutation. *Ann Neurol* 2001; **50**:293–300.

Fasano A, Cadeddu F, Guidubaldi A *et al.* The long-term effect of tetrabenazine in the management of Huntington disease. *Clin Neuropharmacol* 2008a; **31**:313–18.

Fasano A, Mazzone P, Piano C *et al.* GPi-DBS in Huntington's disease: results on motor function and cognition in a 72-year-old case. *Mov Disord* 2008b; **23**:1289–92.

Feany MB. New genetic insights in Parkinson's disease. *N Engl J Med* 2004; **351**:1937–40.

Feigin A, Kieburtz K, Bordwell K *et al.* Functional decline in Huntington's disease. *Mov Disord* 1995; **10**:211–14.

Feinberg TE, Cianci CD, Morrow JS *et al.* Diagnostic tests for choreoacanthocytosis. *Neurology* 1991; **41**:1000–6.

Feldman H, Gauthier S, Heker J *et al.* A 24-week, randomized, double-blind study of donepezil in moderate to severe Alzheimer's disease. *Neurology* 2001; **57**:613–20.

Fenelon G, Mahieux F, Huon R *et al.* Hallucinations in Parkinson's disease: prevalence, phenomenology and risk factors. *Brain* 2000; **123**:733–45.

Ferman TJ, Smith GE, Boeve BF *et al.* DLB fluctuations: specific features that reliably differentiate DLB from AD and normal aging. *Neurology* 2004; **62**:181–7.

Fernandez HH, Trieschmann ME, Burke MA *et al.* Quetiapine for psychosis in Parkinson's disease versus dementia with Lewy bodies. *J Clin Psychiatry* 2002; **63**:513–15.

Fernandez M, Raskind W, Matsushita M *et al.* Hereditary benign chorea: clinical and genetic features of a distinct disease. *Neurology* 2001; **57**:106–10.

Ferraro A, Jervis GA. Pick's disease: clinicopathologic study with report of two cases. *Arch Neurol Psychiatry* 1936; **36**:738–67.

Ferreira JJ, Galitzky M, Montrastruc JL *et al.* Sleep attacks and Parkinson's disease treatment. *Lancet* 2000; **355**:1333–4.

Finelli PF. Metachromatic leukodystrophy manifesting as a schizophrenic disorder: computed tomographic correlation. *Ann Neurol* 1985; **18**:94–5.

Follett KA, Weaver FM, Stern M *et al.* Pallidal versus subthalamic deep brain stimulation for Parkinson's disease. *N Engl J Med* 2010; **362**:2077–91.

Folstein SE, Abbott MH, Chase GA *et al.* The association of affective disorder with Huntington's disease in a case series and in families. *Psychol Med* 1983; **13**:537–42.

Forstl H, Besthorn C, Geiger-Kabisch C *et al.* Psychotic features and the course of Alzheimer's disease: relationship to cognitive, electro-encephalographic and computerized tomography findings. *Acta Psychiatr Scand* 1993; **87**:395–9.

Frankel M, Cummings JL, Robertson MM *et al.* Obsessions and compulsions in Gilles de la Tourette's syndrome. *Neurology* 1986; **36**:378–82.

Fregni F, Santos CM, Myczkowski ML *et al.* Repetitive transcranial magnetic stimulation is as effective as fluoxetine in the treatment of depression in patients with Parkinson's disease. *J Neurol Neurosurg Psychiatry* 2004; **75**:1171–4.

Freiss E, Kuempfel T, Modell S *et al.* Paroxetine treatment improves motor symptoms in patients with multiple system atrophy. *Parkinsonism Relat Disord* 2006; **12**:432–7.

Friederich RL. Benign hereditary chorea improved on stimulant therapy. *Pediatr Neurol* 1996; **14**:326–7.

Friedman A, Fahn S. Spontaneous remissions in spasmodic torticollis. *Neurology* 1986; **36**:398–400.

Friedman A, Sienkiewicz J. Psychotic complications of long-term levodopa treatment of Parkinson's disease. *Acta Neurol Scand* 1991; **84**:111–13.

Frisina PG, Haroutunian V, Libow LS. The neuropathological basis for depression in Parkinson's disease. *Parkinsonism Relat Disord* 2009; **15**:144–8.

Frucht S, Rogers JD, Greene PE *et al.* Falling asleep at the wheel: motor vehicle mishaps in persons taking pramipexole and ropinirole. *Neurology* 1999; **52**:1908–10.

Frucht SJ. Embouchure dystonia – Portrait of a task-specific cranial dystonia. *Mov Disord* 2009; **24**:1752–62.

Frucht SJ, Fahn S, Greene PE *et al.* The natural history of embouchure dystonia. *Mov Disord* 2001; **16**:899–906.

Fukui T, Sugita K, Kawamura M *et al.* Primary progressive apraxia in Pick's disease: a clinicopathologic report. *Neurology* 1996; **47**:467–73.

Funayama M, Hasegawa K, Kowa H *et al.* A new locus for Parkinson's disease (PARK8) maps to chromosome 12p11.2-q13.1. *Ann Neurol* 2002; **51**:296–301.

Furukawa Y, Graf WD, Wong H *et al.* Dopa-responsive dystonia simulating spastic paraplegia due to tyrosine hydroxylase (TH) gene mutation. *Neurology* 2001; **56**:260–3.

Gadow KD, Nolan EE, Sverd J. Methylphenidate in hyperactive boys with comorbid tic disorder. II: Short-term behavioral effects in school settings. *J Am Acad Child Adolesc Psychiatry* 1992; **31**:462–71.

Gadow KD, Sverd J, Sprafkin J *et al.* Long-term methylphenidate therapy in children with comorbid attention-deficit hyperactivity disorder and chronic multiple tic disorder. *Arch Gen Psychiatry* 1999; **56**:330–6.

Gaffney GR, Perry PJ, Lund BC *et al.* Risperidone versus clonidine in the treatment of children and adolescents with Tourette's syndrome. *J Am Acad Child Adolesc Psychiatry* 2002; **41**:330–6.

Gagnon J-F, Bedard M-A, Fantini MC *et al.* REM sleep behavior disorder and REM sleep without atonia in Parkinson's disease. *Neurology* 2002; **59**:585–9.

Gaig C, Marti MJ, Ezquerra M *et al.* G2019S LRRK2 mutation causing Parkinson's disease without Lewy bodies. *J Neurol Neurosurg Psychiatry* 2007; **78**:626–8.

Galasko D, Chang I, Motter I *et al.* High cerebrospinal fluid tau and low amyloid beta-42 levels in the clinical diagnosis of Alzheimer disease and relation to apolipoprotein E genotype. *Arch Neurol* 1998; **55**:937–45.

Gallagher JP. Pathologic laughter and crying in ALS: a search for their origin. *Acta Neurol Scand* 1989; **80**:114–17.

Galton CJ, Patterson K, Xuereb JH *et al.* Atypical and typical presentations of Alzheimer's disease: a clinical, neuropsychological, neuroimaging and pathological study of 13 cases. *Brain* 2000; **123**:484–98.

Galvin JE, Lee SL, Perry A *et al.* Familial dementia with Lewy bodies: clinicopathologic analysis of two kindreds. *Neurology* 2002; **59**:1079–82.

Garcia Ruiz PJ, Gomez-Tortosa E, del Barrio A *et al.* Senile chorea: a multicenter prospective study. *Acta Neurol Scand* 1997; **95**:180–3.

Garcin B, Lillo P, Hornberger M *et al.* Determinants of survival in behavioral variant frontotemporal dementia. *Neurology* 2009; **73**:1656–61.

Garruto RM, Gajdusek DC, Chen KM. Amyotrophic lateral sclerosis and parkinsonism-dementia among Filipino immigrants to Guam. *Ann Neurol* 1981; **10**:341–50.

Gascon GG, Staton RD, Patterson BD *et al.* A pilot controlled study of the use of imipramine to reduce myotonia. *Am J Phys Med Rehabil* 1989; **68**:215–20.

Gasser T, Windgassen K, Bereznai B *et al.* Phenotypic expression of the DYT1 mutation: a family with writer's cramp of juvenile onset. *Ann Neurol* 1998; **44**:126–8.

Gatz M, Pedersen NL, Berg S *et al.* Heritability for Alzheimer's disease: the study of dementia in Swedish twins. *J Gerontol A Biol Sci Med Sci* 1997; **52**:17–25.

Gatz M, Reynolds CA, Fratiglioni L *et al.* Role of genes and environments for explaining Alzheimer disease. *Arch Gen Psychiatry* 2006; **63**:168–74.

Gaul C, Schmidt T, Windisch G *et al.* Subtle cognitive dysfunction in adult onset myotonic dystrophy type 1 (DM1) and type 2 (DM2). *Neurology* 2006; **67**:350–2.

Geda YE, Boeve BF, Negash S *et al.* Neuropsychiatric features in 36 pathologically confirmed cases of corticobasal degeneration. *J Neuropsychiatry Clin Neurosci* 2007; **19**:77–80.

Ghosh A, Dutt A. Utilization behavior in frontotemporal dementia. *J Neurol Neurosurg Psychiatry* 2010; **81**:154–6.

Gibb WRG, Esiri MM, Lees AJ. Clinical and pathological features of diffuse cortical Lewy body disease (Lewy body dementia). *Brain* 1987; **110**:1131–53.

Gibb WR, Lees AJ, Marsden CD. Pathological report of four patients presenting with cranial dystonia. *Mov Disord* 1988; **3**:211–21.

Gibb WR, Luthert PJ, Marsden CD. Corticobasal degeneration. A clinical study of 36 cases. *Brain* 1994; **117**:1183–96.

Giladi N, McMahon D, Przedborski S *et al.* Motor blocks in Parkinson's disease. *Neurology* 1992; **42**:333–9.

Gilbert DL, Dure L, Sethuraman G *et al.* Tic reduction with pergolide in a randomized controlled trial in children. *Neurology* 2003; **60**: 606–11.

Gilbert DL, Batterson JR, Sethuraman G *et al.* Tic reduction with risperidone versus pimozide in a randomized, double-blind, cross-over trial. *J Am Acad Child Adolesc Psychiatry* 2004; **43**:206–14.

Gilbert GJ. Brueghel syndrome: its distinction from Meige syndrome. *Neurology* 1996; **46**:1767–9.

Gillespie J, Jackson A. *MRI and CT of the Brain.* London: Arnold, 2000.

Giovannoni G, O'Sullivan JD, Turner K *et al.* Hedonistic homeostatic dysregulation in patients with Parkinson's disease on dopamine replacement therapies. *J Neurol Neurosurg Psychiatry* 2000; **68**:423–8.

Gironell A, Kulisevsky J, Barbanoj M *et al.* A randomized, placebo-controlled comparative trial of gabapentin and propranolol in essential tremor. *Arch Neurol* 1999; **56**:475–80.

Gjerstad MD, Wentzel-Larsen T, Aarsland D *et al.* Insomnia in Parkinson's disease: frequency and progression over time. *J Neurol Neurosurg Psychiatry* 2007; **78**:476–9.

Glass JD, Reich SG, DeLong MR. Wilson's disease: development of neurological disease after beginning penicillamine therapy. *Arch Neurol* 1990; **47**:595–6.

Goetz CG, Tanner CM, Klawans HL. Bupropion in Parkinson's disease. *Neurology* 1984; **34**:1092–4.

Goetz CG, Blasucci LM, Leurgans S *et al.* Olanzapine and clozapine. Comparative effects on motor function in hallucinating PD patients. *Neurology* 2000; **55**:789–94.

Goetz CG, Wuu J, Curgian LM *et al.* Hallucinations and sleep disorders in PD. Six-year prospective longitudinal study. *Neurology* 2005; **64**:81–6.

Goetz CG, Fan W, Leurgans S *et al.* The malignant course of 'benign hallucinations' in Parkinson disease. *Arch Neurol* 2006; **63**:713–16.

Golbe L, Davis P, Schoenberg B *et al.* Prevalence and natural history of progressive supranuclear palsy. *Neurology* 1988; **38**:1031–4.

Golden GS. Tics and Tourette's: a continuum of symptoms? *Ann Neurol* 1978; **4**:145–8.

Gomez-Tortosa E, Newell K, Irizarry MC *et al.* Clinical and quantitative pathologic correlates of dementia with Lewy bodies. *Neurology* 1999; **53**:1284–91.

Goodman L. Alzheimer's disease: a clinico-pathologic analysis of twenty-three cases with a theory on pathogenesis. *J Nerv Ment Dis* 1953; **118**:97–130.

Gorg M, Wilck W, Granitzny B *et al.* Stabilization of juvenile metachromatic leukodystrophy after bone marrow transplantation: a 13-year follow-up. *J Child Neurol* 2007; **22**:1139–42.

Gorman WP, Cooper R, Pocock P *et al.* A comparison of primidone, propranolol, and placebo in essential tremor, using quantitative analysis. *J Neurol Neurosurg Psychiatry* 1986; **49**:64–8.

Goto I, Tobimatsu S, Ohta M *et al.* Dentatorubropallidoluysian degeneration: clinical, neuro-ophthalmologic, biochemical and pathologic studies on autosomal dominant form. *Neurology* 1982; **32**:1395–9.

Graff-Radford NR, Damasio AR, Hyman BT *et al.* Progressive aphasia in a patient with Pick's disease: a neuropsychological, radiologic, and anatomic study. *Neurology* 1990; **40**:620–6.

Graham A, Davies R, Xuereb J *et al.* Pathologically proven frontotemporal dementia presenting with severe amnesia. *Brain* 2005; **128**:597–605.

Graham DI, Lantos PL. *Greenfield's Neuropathology*, 6th edn. London: Arnold, 1996.

Graham JM, Grunewald RA, Sagar HJ. Hallucinosis in idiopathic Parkinson's disease. *J Neurol Neurosurg Psychiatry* 1997; **63**:434–40.

Grandas F, Elston J, Quinn N *et al.* Blepharospasm: a review of 264 patients. *J Neurol Neurosurg Psychiatry* 1988; **51**:767–72.

Greco CM, Hagerman RJ, Tassone F *et al.* Neuronal intranuclear inclusions in a new cerebellar tremor/ataxia syndrome among fragile X carriers. *Brain* 2002; **125**:1760–71.

Greco CM, Berman RF, Martin RM *et al.* Neuropathology of fragile X-associated tremor/ataxia syndrome. *Brain* 2006; **129**:243–55.

Green J, Morris JC, Sandson J *et al.* Progressive aphasia: a precursor of global dementia? *Neurology* 1990; **48**:423–9.

Greenberg SM, Tennis MK, Brown LB *et al.* Donepezil therapy in clinical practice. A randomized crossover study. *Arch Neurol* 2000; **57**:94–9.

Greene JDW, Patterson K, Xuereb J *et al.* Alzheimer disease and nonfluent progressive aphasia. *Arch Neurol* 1996; **53**:1072–8.

Greene P, Kang U, Fahn S *et al.* Double-blind, placebo controlled trial of botulinum toxin injection for the treatment of spasmodic torticollis. *Neurology* 1990; **40**:1213–18.

Griffin JW, Goren E, Schaumburg H *et al.* Adrenomyeloneuropathy: a probable variant of adrenoleukodystrophy. *Neurology* 1977; **27**:1107–10.

Grimes DA, Lang AE, Bergeron CB. Dementia as the most common presentation of cortical-basal ganglionic degeneration. *Neurology* 1999; **53**:1969–74.

Groen JJ, Endtz LJ. Hereditary Pick's disease: second re-examination of a large family and discussion of other hereditary cases with particular reference to electroencephalography and computerized tomography. *Brain* 1982; **105**:443–59.

Grotzch H, Pizzolato G-P, Ghika J *et al.* Neuropathology of a case of dopa-responsive dystonia associated with a new genetic locus, DYT14. *Neurology* 2002; **58**:1839–42.

Growdon JH, Shahani BT, Young RR. The effect of alcohol on essential tremor. *Neurology* 1975; **25**:259–62.

Gunal DI, Afsar N, Bekiroglu N *et al.* New alternative agents in essential tremor therapy: double-blind placebo-controlled study of alprazolam and acetazolamide. *Neurol Sci* 2000; **21**: 315–17.

Guttman M International Pramipexole–Bromocriptine Study Group. Double-blind comparison of pramipexole and bromocriptine treatment with placebo in advanced Parkinson's disease. *Neurology* 1997; **49**:1060–5.

Gysin WM, Cooke ET. Unusual mental symptoms in a case of hepatolenticular degeneration. *Dis Nerv Syst* 1950; **11**:305–9.

Hagell P, Brundin L. Towards an understanding of fatigue in Parkinson disease. *J Neurol Neurosurg Psychiatry* 2009; **80**:489–92.

Hageman ATM, Gabreels FJM, de Jong JGN *et al.* Clinical symptoms of adult metachromatic leukodystrophy and arylsulfatase A pseudodeficiency. *Arch Neurol* 1995; **52**:408–13.

Hagerman RJ, Leehey M, Heinrichs W *et al.* Intention tremor, parkinsonism, and generalized brain atrophy in male carriers of fragile X. *Neurology* 2001; **57**:127–30.

Hagerman RJ, Leavitt BR, Farzin F *et al.* Fragile X-associated tremor/ataxia syndrome (FXTAS) in females with the FMR1 premutation. *Am J Hum Genet* 2004; **74**:1051–6.

Hahn H, Trant MR, Brownstein MJ *et al.* Neurologic and psychiatric manifestations in a family with a mutation in exon 2 of the guanosine triphosphate-cyclohydrolase gene. *Arch Neurol* 2001; **58**:749–55.

Hall DA, Berry-Kravis E, Jacquemont S *et al.* Initial diagnosis given to persons with the fragile X associated tremor/ataxia syndrome (FXTAS). *Neurology* 2005; **65**:294–301.

Hallervorden J, Spatz H. Eigenartige erkrankung im extrapyramidalen system mit besoderer beteiligung des globus pallidus und der substantia nigra: ein beitrag zu den beziehungen zwischen diesen beiden zentren. *Z Ges Neurol Psychiat* 1922; **79**:254–302.

Haltia T, Palo J, Haltia M *et al.* Juvenile metachromatic leukodystrophy: clinical, biochemical, and neuropathologic studies in nine new cases. *Arch Neurol* 1980; **37**:42–6.

Hamilton L, Fay S, Rockwood K. Misplacing objects in mild to moderate Alzheimer's disease: a descriptive analysis from the VISTA clinical trial. *J Neurol Neurosurg Psychiatry* 2009; **80**:960–5.

Hampel H, Buerger K, Zinkowski R *et al.* Measurement of phosphorylated tau epitopes in the differential diagnosis of Alzheimer disease: a comparative cerebrospinal fluid study. *Arch Gen Psychiatry* 2004; **61**:95–102.

Hanna PA, Janjua FN, Contant CF *et al.* Bilineal transmission in Tourette syndrome. *Neurology* 1999; **53**:813–18.

Hansen L, Salmon D, Galasko D *et al.* The Lewy body variant of Alzheimer's disease: a clinical and pathologic entity. *Neurology* 1990; **40**:1–8.

Hansotia P, Frens D. Hypersomnia associated with alveolar hypoventilation in myotonic dystrophy. *Neurology* 1981; **31**:1336–7.

Hansotia P, Cleeland CS, Chun RWM. Juvenile Huntington's chorea. *Neurology* 1968; **18**:217–24.

Hantz P, Caradoc-Davies G, Caradoc-Davies T *et al.* Depression in Parkinson's disease. *Am J Psychiatry* 1994; **151**:1010–14.

Hardie RJ, Pullon HWH, Harding AE *et al.* Neuroacanthocytosis: a clinical, haematological and pathological study of 19 cases. *Brain* 1991; **114**:13–49.

Harding A. The clinical features and classification of the late onset autosomal dominant cerebellar ataxias: a study of eleven families including descendants of the 'Drew family of Walworth'. *Brain* 1982; **105**:1–28.

Harding AE. *The Hereditary Ataxias and Related Disorders.* London: Churchill Livingstone, 1984.

Harding AE, Thomas PK, Baraitser M *et al.* X-linked recessive bulbospinal neuropathy: a report of ten cases. *J Neurol Neurosurg Psychiatry* 1982; **45**:1012–19.

Harding AJ, Stimson E, Hernderson JM *et al.* Clinical correlates of selective pathology in the amygdala of patients with Parkinson's disease. *Brain* 2002a; **125**:2431–45.

Harding AJ, Broe GA, Halliday GM. Visual hallucinations in Lewy body disease related to Lewy bodies in the temporal lobe. *Brain* 2002b; **125**:391–403.

Harper PS, Harley HG, Reardon W *et al.* Anticipation in myotonic dystrophy: new light on an old problem. *Am J Human Genet* 1992; **51**:10–16.

Harrington CR, Perry RH, Perry EK *et al.* Senile dementia of Lewy body type and Alzheimer type are biochemically distinct in terms of paired helical filaments and hyperphosphorylated tau protein. *Dementia* 1994; **5**:215–28.

Harsch HH. Neuroleptic malignant syndrome: physiological and laboratory findings in a series of nine cases. *J Clin Psychiatry* 1987; **48**:328–33.

Hartig MB, Hortnagel K, Garavaglia B *et al.* Genotypic and phenotypic spectrum of PANK2 mutations in patients with neurodegeneration with brain iron accumulation. *Ann Neurol* 2006; **59**:248–56.

Harwood G, Hierons R, Fletcher NA *et al.* Lessons from a remarkable family with dopa-responsive dystonia. *J Neurol Neurosurg Psychiatry* 1994; **57**:460–3.

Hauser RA, Gauger L, McDowell Anderson W *et al.* Pramipexole-induced somnolence and episodes of daytime sleep. *Mov Disord* 2000; **15**:658–63.

Hayflick SJ, Westaway SK, Levinson B *et al.* Genetic, clinical, and radiographic delineation of Hallervorden–Spatz syndrome. *N Engl J Med* 2003; **348**:33–40.

Healy DG, Falchi M, O'Sullivan SS *et al.* Phenotype, genotype and worldwide genetic penetrance of LRRK2-associated Parkinson's disease: a case–control study. *Lancet Neurol* 2008; **7**:583–90.

Heiman GA, Ottman R, Saunders-Pullman RJ *et al.* Increased risk for recurrent major depression in DYT1 dystonia mutation carriers. *Neurology* 2004; **63**:631–7.

Hely MA, Reid WGJ, Halliday GM *et al.* Diffuse Lewy body disease: clinical features in nine cases without coexistent Alzheimer's disease. *J Neurol Neurosurg Psychiatry* 1996; **60**:531–8.

Henderson JM, Carpenter K, Cartwright H *et al.* Loss of thalamic and intralaminar nuclei in progressive supranuclear palsy and Parkinson's disease: clinical and therapeutic implications. *Brain* 2000; **123**:1410–21.

Hermann N, Lanctot KL, Rothenburg LS *et al.* A placebo-controlled trial of valproate for agitation and aggression in Alzheimer's disease. *Dement Geriatr Cogn Disord* 2007; **23**:116–19.

Hermann N, Rothenburg LS, Black SE *et al.* Methylphenidate for the treatment of apathy in Alzheimer disease: prediction of response during dextroampheatmine challenge. *J Clin Psychopharmacol* 2008; **28**:296–301.

Heston LL, White JA, Mastri AR. Pick's disease. Clinical genetics and natural history. *Arch Gen Psychiatry* 1987; **44**:409–11.

Higgins JJ, Adler RI, Loveless JM. Mutational analysis of the tau gene in progressive supranuclear palsy. *Neurology* 1999; **53**:1421–4.

Hillen ME, Sage JI. Nonmotor fluctuations in patients with Parkinson's disease. *Neurology* 1996; **47**:1180–3.

Hipola D, Mateo D, Gimenez-Roldan S. Meige's syndrome: acute and chronic responses to clonazepam and anticholinergics. *Eur Neurol* 1984; **23**:474–8.

Hirano A, Arumugasamy N, Zimmerman HM. Amyotrophic lateral sclerosis: a comparison of Guam and classical cases. *Arch Neurol* 1967; **16**:357–63.

Hirono N, Mori E, Yasuda M *et al.* Factors associated with psychotic symptoms in Alzheimer's disease. *J Neurol Neurosurg Psychiatry* 1998; **64**:648–52.

Hirose G, Bass NH. Metachromatic leukodystrophy in the adult: a biochemical study. *Neurology* 1972; **22**:312–20.

Hodges JR, Patterson K, Oxbury S *et al.* Semantic dementia. *Brain* 1992; **115**:1783–1806.

Hoehn MM, Yahr MD. Parkinsonism: onset, progression and mortality. *Neurology* 1967; **17**:427–42.

Holl AK, Wilkinson L, Painold A *et al.* Combating depression in Huntington's disease: effective antidepressive treatment with venlafaxine XR. *Int Clin Psychopharmacol* 2010; **25**:46–50.

Holloway RG, Shoulson I, Fahn S *et al.* Pramipexole vs levodopa as initial treatment for Parkinson disease: a 4-year randomized controlled trial. *Arch Neurol* 2004; **61**:1044–53.

Holroyd S, Currie L, Wooten GF. Prospective study of hallucinations and delusions in Parkinson's disease. *J Neurol Neurosurg Psychiatry* 2001; **70**:734–8.

Hoogendijk WJG, Sommer IEC, Pool C *et al.* Lack of association between depression and loss of neurons in the locus coeruleus in Alzheimer disease. *Arch Gen Psychiatry* 1999; **56**:45–51.

Hoogenraad TU, Van Hattum J, Van den Hamer CJA. Management of Wilson's disease with zinc sulphate: experience in a series of 27 patients. *J Neurol Sci* 1987; **77**:137–46.

Horimoto Y, Matsumoto M, Akatsu H *et al.* Autonomic dysfunction in dementia with Lewy bodies. *J Neurol* 2003; **250**:530–3.

Hornberger M, Piguet O, Graham AJ *et al.* How preserved is episodic memory in behavioral variant frontotemporal dementia? *Neurology* 2010; **74**:472–9.

Hosler BA, Siddique T, Sapp PC *et al.* Linkage of familial amyotrophic lateral sclerosis with frontotemporal dementia to chromosome 9q21-q22. *JAMA* 2000; **284**:1664–9.

Houser M, Waltz T. Meige syndrome and pallidal deep brain stimulation. *Mov Disord* 2005; **20**:1203–5.

Howard CP, Royce CE. Progressive lenticular degeneration associated with cirrhosis of the liver (Wilson's disease). *Arch Int Med* 1919; **24**:497–508.

Howeler CJ, Busch HF, Geraedts JP *et al.* Anticipation in myotonic dystrophy: fact or fiction? *Brain* 1989; **112**:779–97.

Huber SJ, Kissel JT, Shuttleworth EC *et al.* Magnetic resonance imaging and clinical correlates of intellectual impairment in myotonic dystrophy. *Arch Neurol* 1989; **46**:536–40.

Huey ED, Garcia C, Wassermann EM *et al.* Stimulant treatment of frontotemporal dementia in 8 patients. *J Clin Psychiatry* 2008; **69**:1981–2.

Huff FJ, Boller J, Luchelli F *et al.* The neurologic examination in patients with probable Alzheimer's disease. *Arch Neurol* 1987; **44**:929–32.

Hughes AJ, Daniel SE, Blankson S *et al.* A clinicopathologic study of 100 cases of Parkinson's disease. *Arch Neurol* 1993; **50**:140–8.

Hung SW, Hamani C, Lozano AM *et al.* Long-term outcome of bilateral pallidal deep brain stimulation for primary cervical dystonia. *Neurology* 2007; **68**:4579–9.

Huntington G. On chorea. *Med Surg Reporter Philadelphia* 1872; **26**:317–21.

Huntington Study Group. Dosage effects of riluzole in Huntington's disease: a multicenter placebo-controlled study. *Neurology* 2003;**61**:1551–6.

Huntington Study Group. Tetrabenazine as antichorea therapy in Huntington disease: a randomized controlled trial. *Neurology* 2006; **66**:366–72.

Huntington Study Group TREND-HD Investigators. Randomized controlled trial of ethyl-eicosapentaenoic acid in Huntington disease: the TREND-HD study. *Arch Neurol* 2008; **6**: 1582–9.

Hurtig HI, Trojanowski JQ, Galvin J *et al.* Alpha-synuclein cortical Lewy bodies correlate with dementia in Parkinson's disease. *Neurology* 2000; **54**:1916–21.

Hussain IF, Brady CM, Swinn MJ *et al.* Treatment of erectile dysfunction with sildenafil citrate (Viagra) in parkinsonism due to Parkinson's disease or multiple system atrophy. *J Neurol Neurosurg Psychiatry* 2001; **71**:371–4.

Hyde TM, Ziegler JC, Weinberger DR. Psychiatric disturbances in metachromatic leukodystrophy: insights into the neurobiology of psychosis. *Arch Neurol* 1992; **49**:401–6.

Hyde TM, Stacey ME, Coppoa R *et al.* Cerebral morphometric abnormalities in Tourette's syndrome: a quantitative MRI study of monozygotic twins. *Neurology* 1995; **45**:1176–82.

Iizuka R, Hirayama K, Maehara KA. Dentato-rubro-pallido-luysian atrophy: a clinico-pathological study. *J Neurol Neurosurg Psychiatry* 1984; **47**:1288–98.

Ikeda K, Akiyama H, Iritani S *et al.* Corticobasal degeneration with primary progressive aphasia and accentuated cortical lesion in superior temporal gyrus: case report and review. *Acta Neuropath* 1996; **92**:534–9.

Ikeuchi T, Takano H, Koide R *et al.* Spinocerebellar ataxia type 6: CAG repeat expansion in alpha1a voltage-dependent calcium channel gene and clinical variations in Japanese population. *Ann Neurol* 1997; **42**:879–84.

Ikeuchi T, Kakita A, Shiga A *et al.* Patients homozygous and heterozygous for SNCA duplication in a family with parkinsonism and dementia. *Arch Neurol* 2008; **65**:514–19.

Ikonomovic MD, Klunk WE, Abrahamson EE *et al.* Post-mortem correlates of in vivo PiB-PET amyloid imaging in a typical case of Alzheimer's disease. *Brain* 2008; **131**:163–45.

Illarioshkin SN, Igarashi S, Onodera O *et al.* Trinucelotide repeat length and rate of progression of Huntington's disease. *Ann Neurol* 1994; **36**:630–5.

Illarioshkin SN, Markova ED, Slominsky PA *et al.* The GTP cyclohydrolase I gene in Russian families with dopa-responsive dystonia. *Arch Neurol* 1998; **55**:789–92.

International Huntington Association, World Federation of Neurology Research Group on Huntington's Chorea. Guidelines for the molecular genetics predictive test in Huntington's disease. *Neurology* 1994; **44**:1533–6.

International Myotonic Dystrophy Consortium. New nomenclature and DNA testing guidelines for myotonic dystrophy type 1 (DM1). *Neurology* 2000; **54**:1218–21.

In t'Veld BA, Ruitenberg A, Hofman A *et al.* Nonsteroidal anti inflammatory drugs and the risk of Alzheimer's disease. *N Engl J Med* 2001; **345**:1515–21.

Inzelberg R, Kipervasser S, Korczyn AD. Auditory hallucinations in Parkinson's disease. *J Neurol Neurosurg Psychiatry* 1998; **64**: 533–5.

Iranzo A, Santamaria J, Rye DB *et al.* Characteristics of idiopathic REM sleep behavior disorder and that associated with MSA and PD. *Neurology* 2005; **65**:247–52.

Ironside R. Disorders of laughter due to brain lesions. *Brain* 1956; **79**:589–609.

Ishida C, Makifuchi T, Sanik S *et al.* A neuropathological study of autosomal-dominant chorea-acanthocytosis with a mutation of VPS13A. *Acta Neuropathol* 2009; **117**:85–94.

Ishikawa A, Yamada M, Makino K *et al.* Dementia and delirium in 4 patients with Machado-Joseph disease. *Arch Neurol* 2002; **59**:1804–8.

Ishikawa K, Watanabe M, Yoshizawa K *et al.* Clinical, neuropathological, and molecular study in two families with spinocerebellar ataxia type 6 (SCA6). *J Neurol Neurosurg Psychiatry* 1999; **67**:86–9.

Jabusch HC, Zschucke D, Schmidt A *et al.* Focal dystonia in musicians: treatment strategies and long-term outcome in 144 patients. *Mov Disord* 2005; **20**:1623–6.

Jacquemont S, Hagerman RJ, Leehey MA *et al.* Penetrance of fragile X-associated tremor/ataxia syndrome in a premutation carrier population. *JAMA* 2004; **291**:460–9.

Jagust WJ, Davies P, Tiller-Borcich JK *et al.* Focal Alzheimer's disease. *Neurology* 1990; **90**:14–19.

Jahanshahi M. Factors that ameliorate or aggravate spasmodic torticollis. *J Neurol Neurosurg Psychiatry* 2000; **68**:227–9.

Jankovic J. Blepharospasm with basal ganglia lesions. *Arch Neurol* 1986; **43**:866–8.

Jankovic J, Patel S. Blepharospasm associated with brainstem lesions. *Neurology* 1983; **33**:1237–40.

Jankovic J, Kirkpatrick JB, Blomquist KA *et al.* Late-onset Hallervorden–Spatz disease presenting as familial parkinsonism. *Neurology* 1985; **35**:227–34.

Jankovic J, Leder S, Warner D *et al.* Cervical dystonia: clinical findings and associated movement disorders. *Neurology* 1991; **41**:1088–91.

Jankovic J, Jimenez-Shahed J, Brown LW. A randomized, double-blind, placebo-controlled study of topiramate in the treatment of Tourette syndrome. *J Neurol Neurosurg Psychiatry* 2010; **81**: 70–3.

Janssen JC, Beck JA, Campbell TA *et al.* Early onset familial Alzheimer's disease. Mutation frequency in 31 families. *Neurology* 2003; **60**:235–9.

Jayadev S, Leverenz JB, Steinbart E *et al.* Alzheimer's disease phenotypes and genotypes associated with mutations in presenilin 2. *Brain* 2010; **133**:1143–54.

Jayne D, Lees AJ, Stern GM. Remission in spasmodic torticollis. *J Neurol Neurosurg Psychiatry* 1984; **47**:1236–7.

Jedynak PC, Trenchant C, de Beyl DZ. Prospective clinical study of writer's cramp. *Mov Disord* 2001; **16**:494–9.

Jervis GA. Early senile dementia in mongoloid idiocy. *Am J Psychiatry* 1948; **105**:102–6.

Jervis GA. Huntington's chorea in childhood. *Arch Neurol* 1979; **23**:83–93.

Jobsis GJ, Weber JW, Barth PG *et al.* Autosomal dominant cerebellar ataxia with retinal degeneration (ADCA II): clinical and neuropathological findings in two pedigrees and genetic linkage to 3p12–p21.1. *J Neurol Neurosurg Psychiatry* 1997; **62**:367–71.

Johnson W, Schwartz G, Barbeau A. Studies on dystonia musculorum deformans. *Arch Neurol* 1962; **7**:301–13.

Josephs KA, Whitwell JL, Knopman DS *et al.* Two distinct subtypes of right temporal variant frontotemporal dementia. *Neurology* 2009; **73**:1443–50.

Kaat LD, Boon AJ, Kamphorst W *et al.* Frontal presentation in progressive supranuclear palsy. *Neurology* 2007; **69**:723–9.

Kaat LD, Boon AJ, Azmani A *et al.* Familial aggregation of parkinsonism in progressive supranuclear parkinsonism. *Neurology* 2009; **73**:95–105.

Kalanithi PS, Zheng W, Kataoka Y *et al.* Altered parvalbumin-positive neuron distribution in basal ganglia of individuals with Tourette syndrome. *Proc Natl Acad Sci USA* 2005; **102**:13307–12.

Karbe H, Kertesz A, Polk M. Profiles of language impairment in primary progressive aphasia. *Arch Neurol* 1993; **50**:192–201.

Karp BI, Hallett M. Extracorporeal 'phantom' tics in Tourette's syndrome. *Neurology* 1996; **46**:38–40.

Kartsounis LD, Hardie RJ. The pattern of cognitive impairments in neuroacanthocytosis: a frontosubcortical dementia. *Arch Neurol* 1996; **53**:77–80.

Kataoka Y, Kalanithi PS, Grantz H *et al.* Decreased number of parvalbumin and cholinergic interneurons in the striatum of individuals with Tourette syndrome. *J Comp Neurol* 2010; **518**:277–91.

Kato K, Wada T, Kawakatsu S *et al.* Improvement of both psychotic symptoms and parkinsonism in a case of dementia with Lewy bodies by the combination therapy of risperidone and L-dopa. *Prog Neuropsychopharmacol Biol Psychiatry* 2002; **26**:201–3.

Kato N, Arai K, Hatton T. Study of the rostral midbrain atrophy in progressive supranuclear palsy. *J Neurol Sci* 2003; **210**:57–60.

Kawai Y, Suenaga M, Takeda A *et al.* Cognitive impairments in multiple system atrophy: MSA-C vs MSA-P. *Neurology* 2008; **70**:1390–6.

Kawanami T, Kato T, Daimon M *et al.* Hereditary ceruloplasmin deficiency: clinicopathological study of a patient. *J Neurol Neurosurg Psychiatry* 1996; **61**:506–9.

Kennedy WR, Alter M, Sung JH. Progressive proximal spinal and bulbar muscular atrophy of late onset: a sex-linked recessive trait. *Neurology* 1968; **18**:671–80.

Kertesz A, Hudson L, Mackenzie IRA *et al.* The pathology and nosology of primary progressive aphasia. *Neurology* 1994; **44**:2065–72.

Kertesz A, Morlog D, Light M *et al.* Galantamine in frontotemporal dementia and primary progressive aphasia. *Dement Geriatr Cogn Disord* 2008; **25**:178–85.

Kertesz A, Jesso S, Harciarek M *et al.* What is semantic dementia? A cohort study of diagnostic features and clinical boundaries. *Arch Neurol* 2010; **67**:483–9.

Khan NL, Graham E, Critchley P *et al.* Parkin disease: a phenotypic study of a large case series. *Brain* 2003; **126**:1279–92.

Kidd D, Nelson J, Jones F *et al.* Long-term stabilization after bone marrow transplantation in juvenile metachromatic leukodystrophy. *Arch Neurol* 1998; **51**:98–9.

Kidd M. Alzheimer's disease: an electron microscopic study. *Brain* 1964; **87**:307–20.

Kipps CM, Hodges JR, Fryer TD *et al.* Combined magnetic resonance imaging and positron emission tomography brain imaging in behavioral variant frontotemporal dementia: refining the clinical phenotype. *Brain* 2009; **132**:2566–78.

Kirkwood SC, Su JL, Conneally P *et al.* Progression of symptoms in the early and middle stages of Huntington disease. *Arch Neurol* 2001; **58**:273–8.

Kirsten L, Silfverskiold B. A family with emotionally precipitated drop seizures. *Lancet* 1958; **2**:471–6.

Kiss ZH, Doig-Beyaert K, Eliasziw M *et al.* The Canadian multicentre study of deep brain stimulation for cervical dystonia. *Brain* 2007; **130**:2879–86.

Kitiyama M, Wada-Isoe K, Irizawa Y *et al.* Assessment of dementia in patients with multiple system atrophy. *Eur J Neurol* 2009; **16**:589–94.

Klatka LA, Louis ED, Schiffer RB. Psychiatric features in diffuse Lewy body disease: a clinicopathologic study using Alzheimer's disease and Parkinson's disease comparison groups. *Neurology* 1996; **47**:1148–52.

Klawans HL, Barr A. Prevalence of spontaneous lingual-facial-buccal dyskinesias in the elderly. *Neurology* 1981; **31**:558–9.

Klawans HL, Barr A. Recurrence of childhood multiple tics in late adult life. *Arch Neurol* 1985; **42**:1079–80.

Klein C. Implications of genetics on the diagnosis and care of patients with Parkinson disease. *Arch Neurol* 2006; **63**:328–34.

Kleiner-Fisman G, Rogaeva E, Halliday W *et al.* Benign hereditary chorea: clinical, genetic, and pathological findings. *Neurology* 2003; **54**:244–7.

Kleiner-Fisman G, Calingasan NY, Putt M *et al.* Alterations of striatal neurons in benign hereditary chorea. *Mov Disord* 2005; **20**:1353–7.

Knibb J, Xuereb JH, Patterson K *et al.* Clinical and pathological characteristics of progressive aphasia. *Ann Neurol* 2006; **59**:156–65.

Knopman DS, Christensen KJ, Schut LJ *et al.* The spectrum of imaging and neuropsychological findings in Pick's disease. *Neurology* 1989; **39**:362–8.

Koch MC, Grimm T, Harley HG *et al.* Genetic risks for children of women with myotonic dystrophy. *Am J Human Genet* 1991; **48**:1084–91.

Koide R, Onodera O, Ikeuchi T *et al.* Atrophy of the cerebellum and brainstem in dentatorubral pallidoluysian atrophy. Influence of CAG repeat size on MRI findings. *Neurology* 1997; **49**:1605–12.

Koller WC. Edentulous orodyskinesia. *Neurology* 1983; **13**:97–9.

Koller WC, Royse VL. Efficacy of primidone in essential tremor. *Neurology* 1986; **36**:121–4.

Koller WC, Busenbark K, Miner K *et al.* The relationship of essential tremor to other movement disorders: report on 678 patients. *Ann Neurol* 1994; **35**:717–23.

Korczyn AD, Brunt ER, Larsen JP *et al.* A 3-year randomized trial of ropinirole and bromocriptine in early Parkinson's disease. *Neurology* 1999; **53**:364–70.

Korenke GC, Fuchs S, Kraseman E *et al.* Cerebral adreno-leukodystrophy in only one of monozygotic twins with an identical ALD genotype. *Ann Neurol* 1996; **40**:254–7.

Koss E, Edland S, Fillenbaum G *et al.* Clinical and neuropsychological differences between patients with earlier and later onset of Alzheimer's disease: a CERAD analysis, Part XII. *Neurology* 1996; **46**:136–41.

Kostic VS, Djuricic BM, Covickovic-Sternic N *et al.* Depression and Parkinson's disease: possible role of serotonin mechanisms. *J Neurol* 1987; **234**:94–6.

Kramer B, Clark CM, Almqvist EW *et al.* Influence of lamotrigine on progression of early Huntington disease: a randomized controlled clinical trial. *Neurology* 1999; **53**:1000–11.

Kramer PL, de Leon D, Oselius L *et al.* Dystonia gene in Ashkenazi Jewish population is located on chromosome 9q32–34. *Ann Neurol* 1990; **27**:114–20.

Kremer B, Goldberg P, Andrew SE *et al.* A worldwide study of the Huntington's disease mutation: the sensitivity and specificity of measuring CAG repeats. *N Engl J Med* 1994; **330**:1401–6.

Krivit W, Lipton ME, Tsai M *et al.* Prevention of deterioration in metachromatic leukodystrophy by bone marrow transplantation. *Am J Med Sci* 1987; **249**:80–5.

Kruisdijk JJM, Koelman JHTM, Ongerboer BW *et al.* Botulinum toxin for writer's cramp: a randomized, placebo-controlled trial and 1-year follow-up. *J Neurol Neurosurg Psychiatry* 2007; **78**:264–70.

Kugelberg E, Welander L. Heredofamilial juvenile muscular atrophy simulating muscular dystrophy. *Arch Neurol Psychiatry* 1956; **75**:500–9.

Kuhn W, Muller T, Gerlach M *et al.* Depression in Parkinson's disease: biogenic amines in CSF of "de novo" patients. *J Neural Trans* 1996; **103**:1441–5.

Kuipers-Upmeijer J, de Jager AEJ, Hew JM *et al.* Primary lateral sclerosis: clinical, neurophysiological, and magnetic resonance findings. *J Neurol Neurosurg Psychiatry* 2001; **71**:615–20.

Kulisevsky J, Marti MJ, Ferrer I *et al.* Meige syndrome: neuropathology of a case. *Mov Disord* 1988; **3**:170–5.

Kulisevsky J, Pagonabarraga J, Pascual-Sedano B *et al.* Motor changes during sertraline treatment in depressed patients with Parkinson's disease. *Eur J Neurol* 2008; **15**:953–9.

Kupsch A, Beneke R, Muller J *et al.* Pallidal deep brain stimulation in primary generalized or segmental dystonia. *N Engl J Med* 2006; **355**:1978–90.

Kurlan R, Behr J, Medved L *et al.* Transient tic disorder and the spectrum of Tourette's syndrome. *Arch Neurol* 1988; **45**:1200–1.

Kurlan R, Cummings J, Raman R *et al.* Quetiapine for agitation or psychosis in patients with dementia and parkinsonism. *Neurology* 2007; **68**:1356–63.

Kurlan R, Johnson D, Kaplan EL *et al.* Streptococcal infection and exacerbations of childhood tics and obsessive–compulsive symptoms: a prospective blinded cohort study. *Pediatrics* 2008; **121**:1188–97.

Kurth MC, Adler CH, St Hilaire M *et al.* Tolcapone improves motor function and reduces levodopa requirement in patients with Parkinson's disease experiencing motor fluctuations: a multicenter, double-blind, randomized, placebo-controlled trial. *Neurology* 1997; **48**:81–7.

Kuzis G, Sabe L, Tiberti C *et al.* Cognitive functions in major depression and Parkinson disease. *Arch Neurol* 1997; **54**:982–6.

Kvale JN. Amitriptyline in the management of progressive supranuclear palsy. *Arch Neurol* 1982; **39**:387–8.

Laberge L, Begin P, Dauvilliers Y *et al.* A polysomnographic study of daytime sleepiness in myotonic dystrophy type 1. *J Neurol Neurosurgery Psychiatry* 2009; **80**:642–6.

La Forestier N, Maisonobe T, Piquard A *et al.* Does primary lateral sclerosis exist? A study of 20 patients and a review of the literature. *Brain* 2001; **124**:1989–99.

Lai F, Williams RS. A prospective study of Alzheimer's disease in Down syndrome. *Arch Neurol* 1989; **46**:849–53.

Landwehrmeyer GB, Dubois B, de Yebenes JG *et al.* Riluzole in Huntington's disease: a 3-year, randomized controlled study. *Ann Neurol* 2007; **62**:262–72.

Lang AE. Patient perception of tics and other movement disorders. *Neurology* 1991; **41**:223–8.

Lang AE, Johnson K. Akathisia in idiopathic Parkinson's disease. *Neurology* 1987; **37**:477–81.

Lang AE, Consky E, Sandor P. 'Signing tics' – insights into the pathophysiology of symptoms in Tourette's syndrome. *Ann Neurol* 1993; **33**:212–15.

Lawyer T, Netsky MG. Amyotrophic lateral sclerosis: clinico-anatomic study of 53 cases. *Arch Neurol Psychiatry* 1953; **69**:171–92.

Le Bars PL, Katz MM, Berman N *et al.* A placebo-controlled, double-blind, randomized trial of Ginkgo biloba for dementia: North American EGb Study Group. *JAMA* 1997; **278**:1327–32.

Lebert F, Stekke W, Hasenbroek C *et al.* Frontotemporal dementia: a randomized, controlled trial with trazodone. *Geriatr Cogn Disord* 2004; **17**:355–9.

Leckman JF, Ort S, Caruso KA *et al.* Rebound phenomena in Tourette's syndrome after abrupt withdrawal of clonidine. Behavioral, cardiovascular, and neurochemical effects. *Arch Gen Psychiatry* 1986; **43**:1168–76.

Leckman JF, Hardin MT, Riddle MA *et al.* Clonidine treatment of Gilles de la Tourette's syndrome. *Arch Gen Psychiatry* 1991; **48**:324–8.

Leckman JF, Walker DE, Cohen DJ. Premonitory urges in Tourette's syndrome. *Am J Psychiatry* 1993; **150**:98–102.

Leckman JF, Walker DE, Goodman WK *et al.* 'Just right' perceptions associated with compulsive behavior in Tourette's syndrome. *Am J Psychiatry* 1994; **151**:675–80.

Leckman JF, Zhang H, Vitale A *et al.* Course of tic severity in Tourette syndrome: the first two decades. *Pediatrics* 1998; **102**: 14–19.

Lee MS, Lee SB, Kim WC. Spasmodic dysphonia associated with a left ventrolateral putaminal lesion. *Neurology* 1996; **47**:827–8.

Lee MS, Kim YD, Lyoo CH. Oculogyric crisis as an initial manifestation of Wilson's disease. *Neurology* 1999; **52**:1714–15.

Leehey MA, Berry-Kravis E, Min SJ *et al.* Progression of tremor and ataxia in male carriers of the FMR1 premutation. *Mov Disord* 2007; **22**:203–6.

Leentjens AFG, Vreeling FW, Luijckx G-J *et al.* SSRIs in the treatment of depression in Parkinson's disease. *Int J Geriatr Psychiatry* 2003; **18**:552–4.

Lees AJ, Robertson M, Trimble MR *et al.* A clinical study of Gilles de la Tourette syndrome in the United Kingdom. *J Neurol Neurosurg Psychiatry* 1984; **47**:1–9.

Lennox G, Lowe J, Landon M *et al.* Diffuse Lewy body disease: correlative neuropathology using anti-ubiquitin immunocytochemistry. *J Neurol Neurosurg Psychiatry* 1989; **52**:1236–47.

Leonard DP, Kidson MA, Brown JG *et al.* A double blind trial of lithium carbonate and haloperidol in Huntington's chorea. *Aust N Z J Psychiatry* 1975; **2**:115–18.

Leopold NA, Kagel MC. Dysphagia in Huntington's disease. *Arch Neurol* 1985; **42**:57–60.

Leroi I, O'Hearn E, Marsh L *et al.* Psychopathology in patients with degenerative cerebellar diseases: a comparison to Huntington's disease. *Am J Psychiatry* 2002; **159**:1306–14.

Leroi I, Overshott R, Byrne EJ *et al.* Randomized controlled trial of memantine in dementia associated with Parkinson's disease. *Mov Disord* 2009; **24**:1217–21.

Lesse SM. Huntington's chorea: report of a case. *J Nerv Ment Dis* 1946; **104**:84–7.

Leube B, Hendgen J, Kessler KR *et al.* Sporadic focal dystonia in Northwest Germany: molecular basis on chromosome 18p. *Ann Neurol* 1997; **42**:111–14.

Levy-Lahad E, Wasco W, Poorkaj P *et al.* Candidate gene for the chromosome 1 familial Alzheimer's disease locus. *Science* 1995; **269**:973–7.

Lieberman A, Ranhosky A, Korts D. Clinical evaluation of pramipexole in advanced Parkinson's disease: results of a double-blind, placebo-controlled, parallel-group study. *Neurology* 1997; **49**:162–8.

Lieberman A, Olanow CW, Sethi K *et al.* A multicenter trial of ropinirole as adjunct treatment for Parkinson's disease. *Neurology* 1998; **51**:1057–62.

Lieberman JA, Reife R. Spastic dysphonia and denervation signs in a young man with tardive dyskinesia. *Br J Psychiatry* 1989; **154**:105–9.

Lilly R, Cummings JL, Benson DF *et al.* The human Klüver–Bucy syndrome. *Neurology* 1983; **33**:1141–5.

Linet LS. Tourette syndrome, pimozide, and school phobia: the neuroleptic separation anxiety syndrome. *Am J Psychiatry* 1985; **12**:613–15.

Ling H, O'Sullivan SS, Holten JL *et al.* Does corticobasal degeneration exist? A clinicopathological re-evaluation. *Brain* 2010; **133**:2045–57.

Linn RT, Wolf PA, Bachman DL *et al.* The 'preclinical phase' of probable Alzheimer's disease. *Arch Neurol* 1995; **52**:485–90.

Lippa CF, Smith TW, Swearer JM. Alzheimer's disease and Lewy body disease: a comparative clinicopathological study. *Ann Neurol* 1994; **35**:81–8.

Litvan I, Mega MS, Cummings JL *et al.* Neuropsychiatric aspects of progressive supranuclear palsy. *Neurology* 1996a; **47**:1184–9.

Litvan I, Mangone CA, McKee A *et al.* Natural history of progressive supranuclear palsy (Steele–Richardson–Olszewski syndrome) and clinical predictors of survival: a clinicopathologic study. *J Neurol Neurosurg Psychiatry* 1996b; **61**:615–20.

Litvan I, Agid Y, Jankovic J *et al.* Accuracy of clinical criteria for the diagnosis of progressive supranuclear palsy (Steele–Richardson Olszewski syndrome). *Neurology* 1996c; **46**:922–30.

Litvan I, Agid Y, Sastrj N *et al.* What are the obstacles for accurate clinical diagnosis of Pick's disease? *Neurology* 1997a; **49**: 62–9.

Litvan I, Campbell G, Mangone CA *et al.* Which clinical features differentiate progressive supranuclear palsy (Steele–Richardson–Olszewski syndrome) from related disorders? A clinicopathological study. *Brain* 1997b; **120**:65–74.

Litvan I, Agid Y, Goetz C *et al.* Accuracy of the clinical diagnosis of corticobasal degeneration: a clinicopathologic study. *Neurology* 1997c; **48**:119–25.

Litvan I, Cummings JL, Mega M. Neuropsychiatric features of corticobasal degeneration. *J Neurol Neurosurg Psychiatry* 1998; **65**:717–21.

Litvan I, Phipps M, Pharr VL *et al.* Randomized placebo-controlled trial of donepezil in patients with progressive supranuclear palsy. *Neurology* 2001; **57**:467–73.

Liu W, Miller BL, Kramer JH *et al.* Behavioral disorders in the frontal and temporal variants of frontotemporal dementia. *Neurology* 2004; **62**:742–8.

Logigian EL, Martens WB, Moxley RT *et al.* Mexiletine is an effective antimyotonia treatment in myotonic dystrophy type 1. *Neurology* 2010; **74**:1441–8.

Lomen-Hoerth C, Murphy J, Langmore S *et al.* Are amyotrophic lateral sclerosis patients cognitively normal? *Neurology* 2003; **60**: 1094–7.

Lopez OL, Becker JT, Sweet RA *et al.* Psychiatric symptoms vary with the severity of dementia in probable Alzheimer's disease. *J Neuropsychiatry Clin Neurosci* 2003; **15**:346–53.

Lorenz D, Frederiksen H, Moises H *et al.* High concordance for essential tremor in monozygotic twins of old age. *Neurology* 2004; **62**:208–11.

Lossos A, Dobson-Stone C, Monaco AP *et al.* Early clinical heterogeneity in choreoacanthocytosis. *Ann Neurol* 2005; **62**:611–14.

Lou J-S, Jankovic J. Essential tremor: clinical correlates in 350 patients. *Neurology* 1991; **41**:234–8.

Louis ED, Goldman JE, Powers JM *et al.* Parkinsonian features of eight pathologically diagnosed cases of diffuse Lewy body disease. *Mov Disord* 1995; **10**:188–94.

Louis ED, Klatka LA, Liu Y *et al.* Comparison of extrapyramidal features in 31 pathologically confirmed cases of diffuse Lewy body disease and 34 pathologically confirmed cases of Parkinson's disease. *Neurology* 1997; **48**:376–80.

Louis ED, Lee P, Quinn L *et al.* Dystonia in Huntington's disease: prevalence and clinical characteristics. *Mov Disord* 1999; **14**:95–101.

Louis ED, Anderson KE, Moskowitz C *et al.* Dystonia-predominant adult-onset Huntington disease: association between motor phenotype and age of onset in adults. *Arch Neurol* 2000; **57**:1325–30.

Louis ED, Honig LS, Vonsattel JPG *et al.* Essential tremor associated with focal nonnigral Lewy bodies. *Arch Neurol* 2005; **62**:1004–7.

Louis ED, Vonsattel JPG, Honig LS *et al.* Neuropathologic findings in essential tremor. *Neurology* 2006a; **66**:1756–9.

Louis ED, Vonsattel JPG, Honig LS *et al.* Essential tremor associated with pathologic change in the cerebellum. *Arch Neurol* 2006b; **63**:1189–93.

Louis ED, Faust PL, Vonsattel JP *et al.* Neuropathological changes in essential tremor: 33 cases compared with 21 controls. *Brain* 2007; **130**:3297–307.

Lowenberg K, Boyd DA, Salon DD. Occurrence of Pick's disease in early adult years. *Arch Neurol Psychiatry* 1939; **41**:1004–20.

Lucking CB, Durr A, Bonifati V *et al.* Association between early-onset Parkinson's disease and mutations in the Parkin gene. *N Engl J Med* 2000; **342**:1560–7.

Maas O, Paterson AS. Myotonia congenita, dystrophia myotonica, and paramyotonia. *Brain* 1950; **73**:318–32.

MacDonald JR, Hill JD, Tarnopolsky MA. Modafinil reduces excessive somnolence and enhances mood in patients with myotonic dystrophy. *Neurology* 2002; **59**:1876–80.

McGeer PL, Schulzer M, McGeer EG. Arthritis and anti-inflammatory agents as possible protective factors for Alzheimer's disease: a review of 17 epidemiologic studies. *Neurology* 1996; **47**:425–32.

Machado A, Chien HF, Deguti MM *et al.* Neurological manifestations of Wilson's disease: report of 119 cases. *Mov Disord* 2006; **21**:2192–6.

McKeith I, Fairbarin A, Perry R *et al.* Neuroleptic sensitivity in patients with senile dementia of Lewy body type. *BMJ* 1992a; **305**:1158–9.

McKeith IG, Perry RH, Fairbarin AF *et al.* Operational criteria for senile dementia of Lewy body type (SDLT). *Psychol Med* 1992b; **22**:911–22.

McKeith IG, Fairbairn A, Perry RH *et al.* Neuroleptic sensitivity in patients with senile dementia of Lewy body type. *BMJ* 1992c; **305**:673–8.

McKeith I, Fairbairn AF, Perry RH *et al.* The clinical diagnosis and a misdiagnosis of senile dementia of Lewy body type (SDLT). *Br J Psychiatry* 1994a; **165**:324–32.

McKeith IG, Fairbairn AF, Bothwell RA *et al.* An evaluation of the predictive validity and inter-rater reliability of clinical diagnostic criteria for senile dementia of Lewy body type. *Neurology* 1994b; **44**:872–7.

McKeith I, Del Ser T, Spano P *et al.* Efficacy of rivastigmine in dementia with Lewy bodies: a randomized, double-blind, placebo-controlled international study. *Lancet* 2000; **356**:2031–6.

McKeith IG, Dickson DW, Lowe J *et al.* Diagnosis and management of dementia with Lewy bodies: third report of the DLB consortium. *Neurology* 2005; **65**:1863–72.

MacKenzie IR, Munoz DG. Nonsteroidal anti-inflammatory drug use and Alzheimer-type pathology in aging. *Neurology* 1998; **50**:986–90.

McMahon WM, Carter AS, Fredine N *et al.* Children at risk for Tourette's disorder: child and parent diagnoses. *Am J Med Genet B Neuropsychiatr Genet* 2003; **121B**:105–11.

McNaught KSP, Kaspussin A, Jackson T *et al.* Brainstem pathology in DYT1 primary torsion dystonia. *Ann Neurol* 2004; **56**:540–7.

McNeill A, Birchall D, Hayflick SJ *et al.* T2* and FSE MRI distinguishes four subtypes of neurodegeneration with brain iron accumulation. *Neurology* 2008; **70**:1614–19.

McNeill R, Sare GM, Manoharan M *et al.* Accuracy of single-photon emission computed tomography in differentiating frontotemporal dementia from Alzheimer's disease. *J Neurol Neurosurg Psychiatry* 2007; **78**:350–5.

Maekawa S, Leigh PN, King A *et al.* TDP-43 is consistently co-localized with ubiquintinated inclusions in sporadic and Guam amyotrophic lateral sclerosis but not in familial amyotrophic lateral sclerosis with and without SOD1 mutations. *Neuropathology* 2009; **29**:672–83.

Maher ER, Lees AJ. The clinical features of the Steele–Richardson–Olszewski syndrome (progressive supranuclear palsy). *Neurology* 1986; **36**:1005–8.

Malamud N, Waggoner RW. Genealogic and clinicopathologic study of Pick's disease. *Arch Neurol Psychiatry* 1943; **50**:288–303.

Malamud N, Hirano A, Kurland LT. Pathoanatomic changes in amyotrophic lateral sclerosis on Guam. *Neurology* 1961; **5**: 401–14.

Mann D, Yates P, Marcyniuk B. A comparison of changes in the nucleus basalis and the locus coeruleus in Alzheimer's disease. *J Neurol Neurosurg Psychiatry* 1984; **47**:201–3.

Mann D, Oliver R, Snowden JS. The topographic distribution of brain atrophy in Huntington's disease and progressive supranuclear palsy. *Acta Neuropathol* 1993; **85**:553–9.

Manni R, Zucca C, Martinetti M *et al.* Hypersomnia in dystrophica myotonica: a neurophysiological and immunogenetic study. *Acta Neurol Scand* 1991; **84**:498–502.

Manowitz P, Kling A, Kohn H. Clinical course of adult metachromatic leukodystrophy presenting as schizophrenia: a report of two living cases in siblings. *J Nerv Ment Dis* 1978; **166**:500–6.

Marantz AG, Verghese J. Capgras' syndrome in dementia with Lewy bodies. *J Geriatr Psychiatry Neurol* 2002; **15**:239–41.

Marder K, Ming-Xin T, Cote L *et al.* The frequency and associated risk factors for dementia in patients with Parkinson's disease. *Arch Neurol* 1995; **52**:695–701.

Marder K, Shao H, Myers RH *et al.* Rate of functional decline in Huntington's disease. Huntington Study Group. *Neurology* 2000; **54**:452–8.

Marder K, Sandler S, Lechich A *et al.* Relationship between CAG repeat length and late-stage outcomes in Huntington's disease. *Neurology* 2002; **59**:1622–4.

Margolese HC, Annable L, Dion Y. Depression and dysphoria in adult and adolescent patients with Tourette's disorder treated with risperidone. *J Clin Psychiatry* 2002; **63**:1040–4.

Mark MH, Sage JI, Dickson DW *et al.* Meige syndrome in the spectrum of Lewy body disease. *Neurology* 1994; **44**:1432–6.

Marneros A. Adult onset of Tourette's syndrome. *Am J Psychiatry* 1983; **140**:924–5.

Marsden CD, Harrison MJG. Idiopathic torsion dystonia (dystonia musculorum deformans): a review of forty-two patients. *Brain* 1974; **97**:793–810.

Martin WE, Loewenson RB, Resch JA *et al.* Parkinson's disease: clinical analysis of 100 patients. *Neurology* 1973; **23**:783–90.

Martinelli P, Gabellini AS, Gulli MR *et al.* Different clinical features of essential tremor: a 200-patient study. *Acta Neurol Scand* 1987; **75**:106–11.

Marttila RJ, Rinne UK. Dementia in Parkinson's disease. *Acta Neurol Scand* 1976; **54**:431–44.

Massironi G, Galuzzi S, Frisoni GB. Drug treatment of REM sleep behavior disorders in dementia with Lewy bodies. *Int Psychogeriatr* 2003; **15**:377–83.

Maurage CA, Udd B, Ruchoux MM *et al.* Similar brain tau pathology in DM2/PROMM and DM1/Steinert disease. *Neurology* 2005; **65**:1636–8.

Maurer K, Volk S, Gerbaldo H. Auguste D and Alzheimer's disease. *Lancet* 1997; **349**:1546–9.

Mayeux R, Stern Y, Cote L *et al.* Altered serotonin metabolism in depressed patients with Parkinson's disease. *Neurology* 1984; **34**:642–6.

Mayeux R, Stern Y, Williams JBW *et al.* Clinical and biochemical features of depression in Parkinson's disease. *Am J Psychiatry* 1986; **143**:756–9.

Mayeux R, Stern Y, Rosenstein R *et al.* An estimate of the prevalence of dementia in idiopathic Parkinson's disease. *Arch Neurol* 1988a; **45**:260–2.

Mayeux R, Stern Y, Sano M *et al.* The relationship of serotonin to depression in Parkinson's disease. *Mov Disord* 1988b; **3**:237–44.

Mayeux R, Denaro J, Hemenegildo N *et al.* A population-based investigation of Parkinson's disease with and without dementia: relationship to age and gender. *Arch Neurol* 1992; **49**:492–7.

Mazza M, Capuano A, Bria P *et al.* Ginkgo biloba and donepezil: a comparison in the treatment of Alzheimer's disease in a randomized placebo-controlled double-blind study. *Eur J Neurol* 2006; **13**:981–5.

Medalia A, Merriam A, Sandberg M *et al.* Neuropsychological deficits in choreoacanthocytosis. *Arch Neurol* 1989; **46**:573–83.

Mega MS, Cummings JL, Fiorello T *et al.* The spectrum of behavioral changes in Alzheimer's disease. *Neurology* 1996; **46**:130–5.

Meiner V, Meiner Z, Reshef A *et al.* Cerebrotendinous xanthomatosis: molecular diagnosis enables presymptomatic detection of a treatable disease. *Neurology* 1994; **44**:288–90.

Mell LK, Davis RL, Owens D. Association between streptococcal infection and obsessive–compulsive disorder, Tourette's syndrome, and tic disorder. *Pediatrics* 2005; **116**:56–60.

Mendez MF, Selwood A, Mastr AR *et al.* Pick's disease versus Alzheimer's disease: a comparison of clinical characteristics. *Neurology* 1993; **43**:289–92.

Mendez MF, Shapira JS, Miller BL. Stereotypical movements and frontotemporal dementia. *Mov Disord* 2005; **20**:742–5.

Mendez MF, Shapira JS, McMurtay A *et al.* Preliminary findings: behavioral worsening on donepezil in patients with frontotemporal dementia. *Am J Geriatr Psychiatry* 2007; **15**:84–7.

Menkes JH, Schimschock JR, Swanson PD. Cerebrotendinous xanthomatosis: the storage of cholestanol within the nervous system. *Arch Neurol* 1968; **19**:47–53.

Menza M, Robertson-Hoffman DE, Bonapace AS. Parkinson's disease and anxiety: comorbidity with depression. *Biol Psychiatry* 1993; **34**:465–70.

Menza M, Dobkin RD, Marin H *et al.* A controlled trial of antidepressants in patients with Parkinson disease and depression. *Neurology* 2009; **72**:886–92.

Menza MA, Cocchiola J, Golbe LI. Psychiatric symptoms in progressive supranuclear palsy. *Psychosomatics* 1995; **36**:550–4.

Merims D, Balas M, Peretz C *et al.* Rater-blinded, prospective comparison: quetiapine versus clozapine for Parkinson's disease psychosis. *Clin Neuropharmacol* 2006; **29**:331–7.

Messert B, Van Nuis C. A syndrome of paralysis of downward gaze, dysarthria, pseudobulbar palsy, axial rigidity of neck and trunk and dementia. *J Nerv Ment Dis* 1966; **143**:47–54.

Miaux Y, Chiras J, Eymard B *et al.* Cranial MRI findings in myotonic dystrophy. *Neuroradiology* 1997; **39**:166–70.

Migliorelli R, Teson A, Sabe L *et al.* Prevalence and correlates of dysthymia and major depression among patients with Alzheimer's disease. *Am J Psychiatry* 1995; **152**:37–44.

Mikkelsen EJ, Detlor J, Cohen DJ. School avoidance and social phobia triggered by haloperidol in patients with Tourette's disorder. *Am J Psychiatry* 1981; **138**:1572–6.

Miller KM, Okun MS, Fernandez HF *et al.* Depression symptoms in movement disorders: comparing Parkinson's disease, dystonia and essential tremor. *Mov Disord* 2007; **22**:666–72.

Modoni A, Silvestri G, Pomponi MG *et al.* Characterization of the pattern of cognitive impairment in myotonic dystrophy type 1. *Arch Neurol* 2004; **61**:1943–7.

Mohs RC, Breitner JCS, Silverman JM *et al.* Alzheimer's disease: morbid risk among first-degree relatives approximates 50 percent by 90 years of age. *Arch Gen Psychiatry* 1987; **44**:405–8.

Molloy S, McKeith IG, O'Brien JT *et al.* The role of levodopa in the management of dementia with Lewy bodies. *J Neurol Neurosurg Psychiatry* 2005; **76**:1200–3.

Moretti R, Torre P, Antonello RM *et al.* Rivastigmine in frontotemporal dementia: an open-label study. *Drugs Aging* 2004; **21**: 931–7.

Mori H, Kondo T, Yokochi M *et al.* Pathologic and biochemical studies of juvenile parkinsonism linked to chromosome 6q. *Neurology* 1998; **51**:890–2.

Mori H, Yagishita A, Takeda T *et al.* Symmetric temporal abnormalities on MR imaging in amyotrophic lateral sclerosis with dementia. *AJNR* 2007; **28**:1511–16.

Morris CM, Pardo-Villamizar C, Gause CD *et al.* Serum autoantibodies measured by immunofluorescence confirm a failure to differentiate PANDAS and Tourette syndrome from controls. *J Neurol Sci* 2009; **276**:45–8.

Morris HR, Gibb G, Katzenschlager R *et al.* Pathological, clinical and genetic heterogeneity in progressive supranuclear palsy. *Brain* 2002; **125**:969–75.

Morris ME, Iansek R, Matyas TA *et al.* Stride length regulation in Parkinson's disease: normalization strategies and underlying mechanisms. *Brain* 1996; **119**:551–68.

Moser HW, Moser AE, Singh I *et al.* Adrenoleukodystrophy: survey of 303 cases: biochemistry, diagnosis and therapy. *Ann Neurol* 1984; **16**:628–41.

Moser HW, Raymond GV, Lu S-E *et al.* Follow-up of 89 asymptomatic patients with adrenoleukodystrophy treated with Lorenzo's oil. *Arch Neurol* 2005; **62**:1073–80.

Mosser J, Lutz Y, Stoeckel ME *et al.* The gene responsible for adrenoleukodystrophy encodes a peroxisomal membrane protein. *Hum Mol Genet* 1994; **3**:265–71.

Muller J, Wissel J, Masuhr F *et al.* Clinical characteristics of the geste antagoniste in cervical dystonia. *J Neurol* 2001; **248**: 478–82.

Muller N, Kroll B, Schwarz MJ *et al.* Increased titers of antibodies against streptococcal M12 and M19 proteins in patients with Tourette's syndrome. *Psychiatry Res* 2001; **101**:187–93.

Munoz MF, Selwood A, Mastri AR *et al.* Pick's disease versus Alzheimer's disease: a comparison of clinical abnormalities. *Neurology* 1993; **43**:289–95.

Munoz-Garcia D, Ludwin SK. Classic and generalized variants of Pick's disease: a clinico-pathological, ultrastructural, and immunocytochemical comparative study. *Ann Neurol* 1984; **16**:467–80.

Munsat TL. Therapy of myotonia. A double-blind evaluation of dipheylhydantoin, procainamide, and placebo. *Neurology* 1967; **17**:359–67.

Murayama S, Mori H, Ihara Y *et al.* Immunocytochemical and ultrastructural studies of Pick's disease. *Ann Neurol* 1990; **27**:394–405.

Murphy MA, Feldman JA, Kilburn G. Hallervorden–Spatz disease in a psychiatric setting. *J Clin Psychiatry* 1989; **50**:100–3.

Murray R, Neumann M, Forman MS *et al.* Cognitive and motor assessment in autopsy-proven corticobasal degeneration. *Neurology* 2007; **68**:1274–83.

Myers RH, Sax DS, Schoenfeld M *et al.* Late onset of Huntington's disease. *J Neurol Neurosurg Psychiatry* 1985; **48**:530–4.

Myers RH, Vonsattel JP, Stevens TJ *et al.* Clinical and neuropathologic assessment of severity in Huntington's disease. *Neurology* 1988; **38**:341–7.

Myers RH, Mac DM, Koroshetz WJ *et al.* De novo expansion of a (CAG) repeat in sporadic Huntington's disease. *Nat Genet* 1993; **5**: 174–9.

Myers RH, Schaefer EJ, Wilson PW *et al.* Apolipoprotein E epsilon-4 association with dementia in a population-based study: The Framingham Study. *Neurology* 1996; **46**:673–7.

Myllyla VV, Sotaniemi KA, Hakulinen P *et al.* Selegeline as the primary treatment of Parkinson's disease – a long-term double-blind study. *Acta Neurol Scand* 1997; **95**:211–18.

Nagafuchi S, Yanagisawa H, Sato K *et al.* Dentatorubral and pallidoluysian atrophy expansion of an unstable CAG repeat on chromosome 12p. *Nat Genet* 1994; **6**:14–18.

Nakagawa M, Yamada H, Higucki I *et al.* A case of paternally inherited congenital myotonic dystrophy. *J Med Genet* 1994; **31**:397–400.

Nance NA, Sanders G. Characteristics of individuals with Huntington disease in long-term care. *Mov Disord* 1996; **11**:542–8.

Navarro C, Fernandez JM, Dominguez C *et al.* Late juvenile metachromatic leukodystrophy treated with bone marrow transplantation: a 4-year follow-up study. *Neurology* 1996; **46**:254–6.

Neary D, Snowden J, Mann D *et al.* Alzheimer's disease – a correlative study. *J Neurol Neurosurg Psychiatry* 1986; **49**:229–34.

Neary D, Snowden JS, Gustafson L *et al.* Frontotemporal lobar degeneration: a consensus on clinical diagnostic criteria. *Neurology* 1998; **51**:1546–54.

Nee LE, Caine ED, Polinsky RJ *et al.* Gilles de la Tourette's syndrome: clinical and family study of 50 cases. *Ann Neurol* 1980; **7**: 41–9.

Neumann M, Schulz-Schaeffer W, Crowther A *et al.* Pick's disease associated with the novel tau gene mutation K369I. *Ann Neurol* 2001; **50**:503–13.

Neumann M, Rademakers R, Roeber S *et al.* A new subtype of frontotemporal lobar degeneration with FUS pathology. *Brain* 2009; **132**:2922–31.

Newman GC. Treatment of progressive supranuclear palsy with tricyclic antidepressants. *Neurology* 1985; **35**:1189–93.

Nichols IC, Weigner WC. Pick's disease – a specific type of dementia. *Brain* 1938; **61**:237–49.

Nielsen JE, Sorensen SA, Hasholt L *et al.* Dentatorubral-pallidoluysian atrophy. Clinical features of a five-generation Danish family. *Mov Disord* 1996; **11**:533–41.

Nijssen PC, Brusse E, Leyten AC *et al.* Autosomal dominant adult cerebral neuronal ceroid lipofuscinosis: parkinsonism due to both striatal and nigral dysfunction. *Mov Disord* 2002; **17**:482–7.

Nygaard TG, Duvoisin RC. Hereditary dystonia–parkinsonism syndrome of juvenile onset. *Neurology* 1986; **36**:1424–8.

Nygaard TG, Duvoisin RC, Manocha M *et al.* Seizures in progressive supranuclear palsy. *Neurology* 1989; **39**:138–40.

Nygaard TG, Trugman JM, de Yebenes JG *et al.* Dopa-responsive dystonia: the spectrum of clinical manifestations in a large North American family. *Neurology* 1990; **40**:66–9.

Nygaard TG, Marsden CD, Fahn S. Dopa-responsive dystonia: long-term treatment response and prognosis. *Neurology* 1991; **41**:174–81.

O'Brien CP, DiGiacomo JN, Fahn S *et al.* Mental effects of high-dosage levodopa. *Arch Gen Psychiatry* 1971; **24**:61–4.

Oh J, Park KD, Cho HJ. Spasmodic dysphonia induced by valproic acid. *Epilepsia* 2004; **45**:880–1.

O'Hearn E, Holmes SE, Calvert PC *et al.* SCA-12: tremor with cerebellar and cortical trophy is associated with a CAG repeat expansion. *Neurology* 2001; **56**:299–303.

Ohye T, Takahashi E, Seki N *et al.* Hereditary progressive dystonia with marked diurnal fluctuation caused by mutations in the GTP cyclohydrolase I gene. *Nat Genet* 1994; **8**:236–42.

Okazaki H, Lipkin LE, Aronson SM. Diffuse intracytoplasmic ganglionic inclusions (Lewy type) associated with progressive dementia and quadriparesis in flexion. *J Neuropathol Exp Neurol* 1961; **20**: 237–44.

Oken BS, Storzbach DM, Kaye JA. The efficacy of Ginkgo biloba on cognitive function in Alzheimer's disease. *Arch Neurol* 1998; **55**:1409–15.

Olanow CW, Hauser RA, Gauger L *et al.* The effect of deprenyl and levodopa on the progression of Parkinson's disease. *Ann Neurol* 1995; **38**:771–7.

Olanow CW Tasmar Advisory Panel. Tolcapone and hepatotoxic effects. *Arch Neurol* 2000; **57**:263–7.

Olanow CW, Rascol O, Hauser R *et al.* A double-blind, delayed-start trial of rasagaline in Parkinson's disease. *N Engl J Med* 2009; **361**:1268–78.

Olin JT, Fox LS, Pawluczyk S *et al.* A pilot randomized trial of carbamazepine for behavioral symptoms in treatment-resistant outpatients with Alzheimer disease. *Am J Geriatr Psychiatry* 2001; **9**:400–5.

Olson MI, Shaw CM. Presenile dementia and Alzheimer's disease in mongolism. *Brain* 1969; **92**:147–56.

Ondo W, Hunter C, Dat Vuong K *et al.* Gabapentin for essential tremor: a multiple-dose, double-blind, placebo-controlled trial. *Mov Disord* 2000; **15**:678–82.

Ondo WG, Tintner R, Thomas M *et al.* Tetrabenazine treatment for Huntington's disease-associated chorea. *Clin Neuropharmacol* 2002; **25**:300–2.

Ondo WG, Titner R, Voung KD *et al.* Double-blind, placebo-controlled unforced titration parallel trial of quetiapine for dopaminergic induced hallucinations in Parkinson's disease. *Mov Disord* 2005; **20**:958–63.

Ondo W, Jankovic J, Connor GS *et al.* Topiramate in essential tremor. A double-blind, placebo-controlled trial. *Neurology* 2006; **66**: 672–7.

Ondo WG, Mejia NI, Hunter CB. A pilot study of the clinical efficacy and safety of memantine in Huntington's disease. *Parkinsonism Relat Disord* 2007; **13**:453–4.

O'Neill BP, Moser HW, Saxena KM. Familial X-linked Addison disease as an expression of adrenoleukodystrophy (ALD): elevated C26 fatty acids in cultured skin fibroblasts. *Neurology* 1982; **32**:543–9.

Ono S, Kanda F, Takahashi K *et al.* Neuronal loss in the medullary reticular formation in myotonic dystrophy: a clinicopathological study. *Neurology* 1996; **46**:228–31.

Ono S, Takahashi K, Jinnai K *et al.* Loss of serotonin-containing neurons in the raphe of patients with myotonic dystrophy: a quantitative immunohistochemical study and relation to hypersomnia. *Neurology* 1998; **50**:535–8.

Onofrj M, Paci C, D'Andreamatteo G *et al.* Olanzapine in severe Gilles de la Tourette syndrome: a 5-week double-blind cross-over study vs. low-dose pimozide. *J Neurol* 2000; **247**:443–6.

Onofrj M, Luciano AL, Thomas A *et al.* Mirtazapine induces REM sleep behavior disorder in parkinsonism. *Neurology* 2003; **60**: 113–15.

Ostheimer AJ. An essay on the shaking palsy, by James Parkinson, M.D., member of the Royal College of Surgeons. *Arch Neurol Psychiatry* 1922; **7**:681–710.

O'Suilleabhain P, Dewey RB. A randomized trial of amantadine in Huntington disease. *Arch Neurol* 2003; **60**:996–8.

Oyamada R, Hayashi M, Katoh Y *et al.* Neurofibrillary tangles and deposition of oxidative products in the brain in cases of myotonic dystrophy. *Neuropathology* 2006; **26**:107–14.

Ozawa T, Pavious D, Quinn NP *et al.* The spectrum of pathological involvement of the striatonigral and olivopontocerebellar system in multiple system atrophy: clinicopathological correlations. *Brain* 2004; **127**:2657–71.

Ozelius L, Kramer PL, MosKowitz CB *et al.* Human gene for torsion dystonia located on chromosome 9q32–q34. *Neuron* 1989; **2**:1427–34.

Ozelius LJ, Kramer PL, de Leon D *et al.* Strong allelic association between the torsion dystonia gene (DYT 1) and loci on chromosome 9q34 in Ashkenazi Jews. *Am J Hum Genet* 1992; **50**: 619–28.

Ozelius LJ, Hewett JW, Page CE *et al.* The early-onset torsion dystonia gene (DYT1) encodes an ATP-binding protein. *Nat Genet* 1997; **17**:40–8.

Pahwa R, Lyons K, McGuire D *et al.* Early morning akinesia in Parkinson's disease: effect of standard levodopa/carbidopa and sustained-release levodopa/carbidopa. *Neurology* 1996; **46**:1059–62.

Paleacu D, Anca M, Giladi N. Olanzapine in Huntington's disease. *Acta Neurol Scand* 2002; **105**:441–4.

Palhagen S, Heinonen EH, Hagglund J *et al.* Selegeline delays the onset of disability in de novo parkinsonian patients. *Neurology* 1998; **51**:520–5.

Palhagen S, Heinonen E, Hagglund J *et al.* Selegeline slows the progression of the symptoms of Parkinson's disease. *Neurology* 2006; **66**:1200–6.

Palhagen S, Qi H, Martensson B *et al.* Monoamines, BDNF, IL-6 and corticosterone in CSF in patients with Parkinson's disease and major depression. *J Neurol* 2009; **257**:524–32.

Papapetropoulos S, Singer C, Ross OA *et al.* Clinical heterogeneity of the LRRK2 G2019S mutation. *Arch Neurol* 2006; **63**:1242–6.

Papp MI, Lantos PI. The distribution of oligodendroglial inclusions in multiple system atrophy and its relevance to clinical symptomatology. *Brain* 1994; **117**:235–43.

Pappert EJ, Goetz CG, Louis ED *et al.* Objective assessment of longitudinal outcome in Gilles de la Tourette's syndrome. *Neurology* 2003; **61**:936–40.

Parkinson Study Group. Effects of tocoperl and deprenyl on the progression of disability in early Parkinson's disease. *N Engl J Med* 1993; **328**:176–83.

Parkinson Study Group. Safety and efficacy of pramipexole in early Parkinson's disease: a randomized dose-ranging study. *JAMA* 1997; **278**:125–30.

Parkinson Study Group. Low-dose clozapine for the treatment of drug-induced psychosis in Parkinson's disease. *N Engl J Med* 1999; **340**:757–63.

Parkinson Study Group. A controlled trial of rasagaline in early Parkinson disease: the TEMPO study. *Arch Neurol* 2002; 59:1937–43.

Parkinson Study Group. A controlled, randomized, delayed-start study of rasagaline in early Parkinson disease. *Arch Neurol* 2004; **61**:561–6.

Parkinson Study Group. A randomized placebo-controlled trial of rasagaline in levodopa-treated patients with Parkinson disease and motor fluctuations. *Arch Neurol* 2005; **62**:241–8.

Parkinson Study Group CALM Cohort Investigators. Long-term effect of initiating pramipexole vs levodopa in early Parkinson disease. *Arch Neurol* 2009; **66**:563–70.

Parvizi J, Joseph J, Press DZ *et al.* Pathological laughter and crying in patients with multiple system atrophy – cerebellar type. *Mov Disord* 2007; **22**:798–803.

Patzold T, Brune M. Obsessive compulsive disorder in Huntington disease: a case of isolated obsessions successfully treated with sertraline. *Neuropsychiatry Neuropsychol Behav Neurol* 2002; **15**:216–19.

Pauls DL, Leckman JF. The inheritance of Gilles de la Tourette's syndrome and associated behaviors. *N Engl J Med* 1986; **315**:993–7.

Pauls DL, Pakstis AJ, Kurlan R *et al.* Segregation and linkage analysis of Tourette's syndrome and related disorders. *J Am Acad Child Adolesc Psychiatry* 1990; **29**:195–203.

Pauls DL, Leckman CL, Stevenson JM *et al.* A family study of Gilles de la Tourette syndrome. *Am J Hum Genet* 1991; **48**:154–63.

Paulsen JS, Ready RE, Hamilton JN *et al.* Neuropsychiatric aspects of Huntington's disease. *J Neurol Neurosurg Psychiatry* 2001; **71**:31–4.

Paulsen JS, Hoth KF, Nehl C *et al.* Critical periods of suicide risk in Huntington's disease. *Am J Psychiatry* 2005; **162**:725–31.

Paulson GW. Meige's syndrome. Dyskinesia of the eyelids and facial muscles. *Geriatrics* 1972; **27**:69–73.

Pearce I, Heathfield KW, Pearce MJ. Valproate sodium in Huntington chorea. *Arch Neurol* 1977; **34**:308–9.

Pellecchia MT, Valente EM, Cif L *et al.* The diverse phenotype and genotype of pantothenate kinase-associated neurodegeneration. *Neurology* 2005; **64**:1810–12.

Perini GI, Colombo G, Armani M *et al.* Intellectual impairment and cognitive evoked potentials in myotonic dystrophy. *J Nerv Ment Dis* 1989; **177**:750–4.

Perlmutter SJ, Leitman SF, Garvey MA *et al.* Therapeutic plasma exchange and intravenous immunoglobulin for obsessive–compulsive disorder and tic disorders in childhood. *Lancet* 1999; **354**:1153–8.

Peterson B, Riddle MA, Cohen DJ *et al.* Reduced basal ganglia volumes in Tourette's syndrome using 3-dimensional reconstruction techniques from magnetic resonance images. *Neurology* 1993; **43**:941–9.

Peterson BS, Thomas P, Kane MJ *et al.* Basal ganglia volumes in patients with Gilles de la Tourette syndrome. *Arch Gen Psychiatry* 2003; **60**:415–24.

Petracca GM, Jorge RE, Acion L *et al.* Frequency and correlates of involuntary emotional expression disorder in Parkinson's disease. *J Neuropsychiatry Clin Neurosci* 2009; **21**:406–12.

Petry S, Cummings JL, Hill MA *et al.* Personality alterations in dementia of the Alzheimer type. *Arch Neurol* 1988; **45**:1187–90.

Pflanz S, Besson JAO, Ebmeier KP *et al.* The clinical manifestation of mental disorder in Huntington's disease: a retrospective case record study of disease progression. *Acta Psychiatr Scand* 1991; **83**:53–60.

Phemister JC, Small JM. Hypersomnia in dystrophia myotonica. *J Neurol Neurosurg Psychiatry* 1961; **24**:173–5.

Phuong L, Garg S, Duda J *et al.* Involuntary emotional expression disorder (IEED) in Parkinson's disease. *Parkinsonism Relat Disord* 2009; **15**:511–15.

Piacentini J, Woods DW, Scahill L *et al.* Behavior therapy for children with Tourette disorder: a randomized controlled trial. *JAMA* 2010; **303**:1929–37.

Piccini P, Burn DJ, Ceravolo R *et al.* The role of inheritance in sporadic Parkinson's disease: evidence from a longitudinal study of dopaminergic function in twins. *Ann Neurol* 1999; **45**:577–82.

Pick A. Uber die beziehungen der senilin hirnatrophie zur aphasie. *Prager Med Wochenschur* 1892; **17**:165–7.

Piguet O, Hornberger M, Shelly BP *et al.* Sensitivity of current criteria for the diagnosis of behavioral variant frontotemporal dementia. *Neurology* 2009; **72**:732–7.

Pineda DA, Buritica O, Sanchez JL *et al.* Parkinsonian syndromes in Medellin (Colombia). *Rev Neurol* 2000; **31**:936–43.

Pinter MM, Pogarell O, Oertel WH. Efficacy, safety, and tolerance of the non-ergoline dopamine agonist pramipexole in the treatment of advanced Parkinson's disease: a double blind, placebo controlled, randomized, multicentre study. *J Neurol Neurosurg Psychiatry* 1999; **66**:436–41.

Plazzi G, Corsini R, Provini F *et al.* REM sleep behavior disorder in multiple system atrophy. *Neurology* 1997; **48**:1094–7.

Pollak P, Tison F, Rascol O *et al.* Clozapine in drug-induced psychosis in Parkinson's disease. A randomized, placebo controlled study with open follow up. *J Neurol Neurosurg Psychiatry* 2004; **75**: 689–95.

Pollock BG, Mulsant BH, Rosen J *et al.* Comparison of citalopram, perphenazine, and placebo for the acute treatment of psychosis and behavioral disturbances in hospitalized, demented patients. *Am J Psychiatry* 2002; **159**:460–5.

Pontone G, Williams JR, Bassett SS *et al.* Clinical features associated with impulse control disorders in Parkinson disease. *Neurology* 2006; **67**:1258–61.

Pool KD, Freeman FJ, Finitzo T *et al.* Heterogeneity in spasmodic dysphonia. Neurologic and voice findings. *Arch Neurol* 1991; **48**:305–9.

Porta M, Brambilla A, Cavanna AE *et al.* Thalamic deep brain stimulation for treatment-refractory Tourette syndrome. Two-year outcome. *Neurology* 2009; **73**:1375–80.

Postuma RB, Gagnon JF, Vendette M *et al.* Quantifying the risk of neurodegenerative disease in idiopathic REM sleep behavior disorder. *Neurology* 2009; **72**:1296–300.

Potter NT, Meyer MA, Zimmerman AW *et al.* Molecular and clinical findings in a family with dentatorubral-pallidoluysian atrophy. *Ann Neurol* 1995; **37**:273–7.

Powers JM, Schaumburg HH, Gaffney CL *et al.* Klüver–Bucy syndrome caused by adreno-leukodystrophy. *Neurology* 1980; **30**: 1131–2.

Pramstaller PP, Schlossmacher MG, Jacques TS *et al.* Lewy body Parkinson's disease in a large pedigree with 77 Parkin mutation carriers. *Ann Neurol* 2005; **58**:411–22.

Price BH, Gurvit H, Weintraud S *et al.* Neuropsychological patterns and language deficits in 20 consecutive cases of autopsy-confirmed Alzheimer's disease. *Arch Neurol* 1993; **50**:931–7.

Price RA, Kidd KK, Cohen DJ *et al.* A twin study of Tourette syndrome. *Arch Gen Psychiatry* 1985; **42**:815–20.

Pringle CE, Hudson AJ, Munoz DJ *et al.* Primary lateral sclerosis. Clinical features, neuropathology and diagnostic criteria. *Brain* 1992; **115**:495–520.

Priori A, Pesenti A, Cappellari A *et al.* Limb immobilization for the treatment of focal occupational dystonia. *Neurology* 2001; **57**: 405–9.

Pryse-Phillips W, Johnson GJ, Larsen B. Incomplete manifestations of myotonic dystrophy in a large kinship in Labrador. *Ann Neurol* 1982; **11**:582–4.

Puri BK, Bydder GM, Counsell SJ *et al.* MRI and neuropsychological improvement in Huntington disease following ethyl-EPA treatment. *Neuroreport* 2002; **13**:123–6.

Puri BK, Leavitt BR, Hayden MR *et al.* Ethyl-EPA in Huntington disease: a double-blind, randomized, placebo-controlled trial. *Neurology* 2005; **65**:286–92.

Quinn N. Multiple system atrophy – the nature of the beast. *J Neurol Neurosurg Psychiatry* 1989; **52**:78–85.

Rabey JM, Prokhorov T, Minlovitz A *et al.* Effect of quetiapine in psychotic Parkinson's disease patients: a double-blind labeled study of 3 month's duration. *Mov Disord* 2007; **22**:313–18.

Rabinak CA, Nirenberg MJ. Dopamine agonist withdrawal syndrome in Parkinson disease. *Arch Neurol* 2010; **67**:58–63.

Rabkin JG, Albert SM, Del Bene ML *et al.* Prevalence of depressive disorders and change over time in late-stage ALS. *Neurology* 2005; **65**:62–7.

Rabkin JG, Gordon PH, McElhiney M *et al.* Modafinil treatment of fatigue in patients with ALS: a placebo-controlled study. *Muscle Nerve* 2009; **39**:297–303.

Rajput AH, Gibb WRG, Zhong XH *et al.* Dopa-responsive dystonia: pathological and biochemical observations in a case. *Ann Neurol* 1994; **35**:396–402.

Rajput AH, Martin W, Saint-Hilaire M-H *et al.* Tolcapone improves motor function in Parkinsonian patients with the 'wearing-off' phenomenon: a double-blind, placebo-controlled, multicenter trial. *Neurology* 1997; **49**:1066–71.

Rajput A, Robinson CA, Rajput AH. Essential tremor course and disability. *Neurology* 2004; **62**:932–6.

Rajput AH, Voll A, Rajput ML *et al.* Course in Parkinson disease subtypes: a 39-year clinicopathologic study. *Neurology* 2009; **73**:206–12.

Rahkonen T, Eloniemi-Sulkava U, Rissanen S *et al.* Dementia with Lewy bodies according to the consensus criteria in a general population aged 75 or older. *J Neurol Neurosurg Psychiatry* 2003; **74**: 720–4.

Ramos-Arroyo MA, Moreno S, Valiente A. Incidence and mutation rates of Huntington's disease in Spain: experience of 9 years of direct genetic testing. *J Neurol Neurosurg Psychiatry* 2005; **76**:337–42.

Ranen NG, Peyser CE, Folstein SE. ECT as a treatment for depression in Huntington's disease. *J Neuropsychiatry Clin Neurosci* 1994; **6**:154–9.

Ranen NG, Lipsey JR, Treisman G *et al.* Sertraline in the treatment of severe aggressiveness in Huntington's disease. *J Neuropsychiatry Clin Neurosci* 1996; **8**:338–40.

Rascol O, Brooks DJ, Corczyn AD *et al.* A five-year study of the incidence of dyskinesia in patients with early Parkinson's disease who were treated with ropinirole or levodopa. *N Engl J Med* 2000; **342**:1484–91.

Rascol O, Brooks DJ, Melamad E *et al.* Rasagiline as an adjunct to levodopa in patients with Parkinson's disease and motor fluctuations (LARGO, Lasting effect in Adjunct therapy with Rasagiline Given Once daily, study): a randomized, double-blind, parallel-group trial. *Lancet* 2005; **365**:947–54.

Rasool C, Svendsen CN, Selkoe DJ. Neurofibrillary degeneration of cholinergic and noncholinergic neurons of the basal forebrain in Alzheimer's disease. *Ann Neurol* 1986; **20**:482–8.

Rasmussen KG, Russell JC, Kung S *et al.* Electroconvulsive therapy for patients with major depression and probably Lewy body dementia. *J ECT* 2003; **19**:103–9.

Rauschka H, Colsch B, Bamann N *et al.* Late-onset metachromatic leukodystrophy. Genotype strongly influences phenotype. *Neurology* 2006; **67**:859–63.

Ravina B, Pott M, Siderowf A *et al.* Donepezil for dementia in Parkinson's disease: a randomized double-blind, placebo controlled crossover study. *J Neurol Neurosurg Psychiatry* 2005; **76**:934–9.

Rebeiz J, Kolodney E, Richardson EJ. Corticodentatonigral degeneration with neuronal achromasia. *Arch Neurol* 1968; **18**:20–33.

Rees L, Lipsedge M, Ball C. *Textbook of Psychiatry.* London: Arnold, 1996.

Rees MI, Andrew M, Jawad S *et al.* Evidence for recessive as well as dominant forms of startle disease (hyperekplexia) caused by mutations in the alpha 1 subunit of the inhibitory glycine receptor. *Hum Mol Genet* 1994; **3**:2175–9.

Reeves AL, So EL, Sharbrough FW *et al.* Movement disorders associated with the use of gabapentin. *Epilepsia* 1996; **37**:988–90.

Reeves RR, Liberto V. Treatment of essential blepharospasm with quetiapine. *Mov Disord* 2003; **18**:1072–3.

Regeur L, Pakkenberg B, Fog R *et al.* Clinical features and long-term treatment with pimozide in 65 patients with Gilles de la Tourette's syndrome. *J Neurol Neurosurg Psychiatry* 1986; **49**:791–5.

Reider-Grosswasser I, Bornstein N. CT and MRI in late-onset metachromatic leukodystrophy. *Acta Neurol Scand* 1987; **75**:64–9.

Reuter I, Hu MT, Andrews TC *et al.* Late onset levodopa responsive Huntington's disease with minimal chorea masquerading as Parkinson plus syndrome. *J Neurol Neurosurg Psychiatry* 2000; **68**:238–41.

Rewcastle NB, Ball MJ. Electron microscopic structure of the 'inclusion bodies' in Pick's disease. *Neurology* 1968; **18**:1205–13.

Ribai P, Nguyen K, Hanhn-Barma V *et al.* Psychiatric and cognitive difficulties as indicators of juvenile Huntington's disease onset in 29 patients. *Arch Neurol* 2007; **64**:813–19.

Richard IH, Kurlan R, Tanner C *et al.* Serotonin syndrome and the combined use of deprenyl and an antidepressant in Parkinson's disease. *Neurology* 1997; **48**:1070–7.

Ricker K, Koch MC, Lehmann-Horn F *et al.* Proximal myotonic myopathy: a new dominant disorder with myotonia, muscle weakness and cataracts. *Neurology* 1994; **4**:1448–52.

Ricker K, Koch MC, Lehmann-Horn F *et al.* Proximal myotonic myopathy: clinical features of a multisystem disorder similar to myotonic dystrophy. *Arch Neurol* 1995; **52**:25–31.

Ricker K, Grimm T, Koch MC *et al.* Linkage of proximal myotonic myopathy to chromosome 3q. *Neurology* 1999; **52**:170–1.

Ridha BH, Josephs KA, Rossor MN. Delusions and hallucinations in dementia with Lewy bodies: worsening with memantine. *Neurology* 2005; **65**:481–2.

Riley DE, Lang AE, Lewis A *et al.* Cortical-basal ganglionic degeneration. *Neurology* 1990; **40**:1203–12.

Ringholz GM, Appel sh, Bradshaw M *et al.* Prevalence and patterns of cognitive impairment in sporadic ALS. *Neurology* 2005; **65**:586–90.

Rinne J, Lee MS, Thompson PD *et al.* Corticobasal degeneration: a clinical study of 36 cases. *Brain* 1994a; **117**:1183–96.

Rinne J, Daniel SE, Scaravilli F *et al.* The neuropathological features of acanthocytosis. *Mov Disord* 1994b; **9**:297–304.

Rippon GA, Scarmeas N, Gordon PH *et al.* An observational study of cognitive impairment in amyotrophic lateral sclerosis. *Arch Neurol* 2006; **63**:345–52.

Rivest L, Lees AJ, Marsden CED. Writer's cramp: treatment with botulinum toxin injections. *Mov Disord* 1991; **6**:55–9.

Rizzo R, Gulisano M, Pavone P *et al.* Increased antistreptococcal antibody titers and anti-basal ganglia antibodies in patients with Tourette's syndrome: controlled cross-sectional study. *J Child Neurol* 2006; **21**:747–53.

Robbins TW, James M, Lange KW *et al.* Cognitive performance in multiple system atrophy. *Brain* 1992; **115**:271–91.

Robertson EE, le Roux A, Brown JH. The clinical differentiation of Pick's disease. *J Ment Sci* 1958; **104**:1000–24.

Robertson MM, Trimble MR, Lees AJ. The psychopathology of Gilles de la Tourette's syndrome: a phenomenological analysis. *Br J Psychiatry* 1988; **152**:388–90.

Rockwood K, Mintzer J, Truyen L *et al.* Effects of a flexible galantamine dose in Alzheimer's disease: a randomized, controlled trial. *J Neurol Neurosurg Psychiatry* 2001; **71**:589–95.

Rodriguez ME, Artieda O, Zubieta JL *et al.* Reflex myoclonus in olivopontocerebellar atrophy. *J Neurol Neurosurg Psychiatry* 1994; **57**:316–19.

Rogers JD, Brogan D, Mirra SS. The nucleus basalis of Meynert in neurological disease: a quantitative morphological study. *Ann Neurol* 1985; **17**:163–70.

Rogers SL, Friedhoff LT. The efficacy and safety of donepezil in patients with Alzheimer's disease: results of a US multicentre randomized, double-blind, placebo-controlled study. *Dementia* 1996; **7**: 293–303.

Rogers SL, Farlow MR, Doody RS *et al.* A 24-week, double-blind, placebo-controlled trial of donepezil in patients with Alzheimer's disease. *Neurology* 1998; **50**:136–45.

Rohrer JD, Guereiro R, Vandrovcova J *et al.* The heritability and genetics of frontotemporal lobar degeneration. *Neurology* 2009; **73**:1451–6.

Rojo A, Pernaute RS, Fontan A *et al.* Clinical genetics of familial progressive supranuclear palsy. *Brain* 1999; **122**:1233–45.

Rolfs A, Koeppen AH, Bauer I *et al.* Clinical features and neuropathology of autosomal dominant spinocerebellar ataxia (SCA17). *Ann Neurol* 2003; **54**:367–75.

Romanelli MF, Morris JC, Ashkin K *et al.* Advanced Alzheimer's disease is a risk factor for late-onset seizures. *Arch Neurol* 1990; **47**:847–50.

Rondot P, Marchand MP, Dellatolas G. Spasmodic torticollis – review of 220 patients. *Can J Neurol Sci* 1991; **18**:143–51.

Ropacki SA, Jeste DV. Epidemiology of and risk factors for psychosis of Alzheimer's disease: a review of 55 studies published from 1990 to 2003. *Am J Psychiatry* 2005; **162**:2022–30.

Ros R, Garre P, Hirano M *et al.* Genetic linkage of autosomal dominant progressive supranuclear palsy to 1q31.1. *Ann Neurol* 2005a; **57**:634–1.

Ros R, Thobois S, Streichenberger N *et al.* A new mutation in the tau gene, G303V, in early-onset familial progressive supranuclear palsy. *Arch Neurol* 2005b; **62**:1444–50.

Rosen DR, Siddique T, Patterson D *et al.* Mutations in Cu/Zn superoxide dismutase gene are associated with familial amyotrophic lateral sclerosis. *Nature* 1993; **362**:59–62.

Rosler M, Anand R, Cicin-Sain A *et al.* Efficacy and safety of rivastigmine in patients with Alzheimer's disease: international randomized controlled trial. *BMJ* 1999; **318**:633–40.

Rosman NP, Kakulas BA. Mental deficiency associated with muscular dystrophy: a neuropathological study. *Brain* 1966; **89**:769–87.

Ross MS, Moldofsky H. A comparison of pimozide and haloperidol in the treatment of Gilles de la Tourette's syndrome. *Am J Psychiatry* 1978; **135**:585–7.

Ross SJM, Graham N, Stuart-Green L *et al.* Progressive biparietal atrophy: an atypical presentation of Alzheimer's disease. *J Neurol Neurosurg Psychiatry* 1996; **61**:388–95.

Rozdilsky B, Cummings JN, Huston AF. Hallervorden–Spatz disease: late infantile and adult types: report of two cases. *Acta Neuropatholog* 1968; **10**:1–16.

Rubinsztein JS, Rubinsztein DC, Goodburn S *et al.* Apathy and hypersomnia are common features of myotonic dystrophy. *J Neurol Neurosurg Psychiatry* 1998; **64**:510–15.

Ruiz-Sandoval JL, Garcia-Navarro V, Chiquete E *et al.* Choreoacanthocytosis in a Mexican family. *Arch Neurol* 2007; **64**:1661–4.

Ryan SG, Sherman SL, Terry JC *et al.* Startle disease, or hyperekplexia: response to clonazepam and assignment of the gene (STHE) to chromosome 5q by linkage analysis. *Ann Neurol* 1992; **31**:663–8.

Sabbagh MN, Hake AM, Ahmed S *et al.* The use of memantine in dementia with Lewy bodies. *J Alzheimer's Dis* 2005; **7**:285–9.

Saenz-Lope E, Herranz-Tanarro FJ, Masdeu JC *et al.* Hyperekplexia: a syndrome of pathological startle responses. *Ann Neurol* 1984; **15**:36–41.

Sakai T, Mawatari S, Hiroshi I *et al.* Choreoacanthocytosis: clues to the clinical diagnosis. *Arch Neurol* 1981; **38**:333–8.

Salen G. Cholesterol deposition in cerebrotendinous xanthomatosis. *Ann Intern Med* 1971; **75**:843–51.

Sallee FR, Nesbitt L, Jackson C *et al.* Relative efficacy of haloperidol and pimozide in children and adolescents with Tourette's disorder. *Am J Psychiatry* 1997; **154**:1057–62.

Sallee FR, Kurlan R, Goetz CG *et al.* Ziprasidone treatment of children and adolescents with Tourette's syndrome: a pilot study. *J Am Acad Child Adolesc Psychiatry* 2000; **39**:292–9.

Sanelli T, Xiao S, Horne P *et al.* Evidence that TDP-43 is not the major ubiquitinated target within the pathological inclusions of amyotrophic lateral sclerosis. *J Neuropathol Exp Neurol* 2007; **66**:1147–53.

Sapp E, Schwarz C, Chase K *et al.* Huntingtin localization in brains of normal and Huntington's disease patients. *Ann Neurol* 1997; **42**:604–12.

Saunders AM, Strittmatter WJ, Schmechel D *et al.* Association of apolipoprotein E allele epsilon 4 with late-onset and sporadic Alzheimer's disease. *Neurology* 1993; **43**:1467–72.

Sawle GV, Leenders KL, Brooks DJ *et al.* Dopa-responsive dystonia: [18F]dopa positron emission tomography. *Ann Neurol* 1991; **30**:24–30.

Scahill L, Chappell PB, Kim YS *et al.* A placebo-controlled study of guanfacine in the treatment of children with tic disorders and attention deficit hyperactivity disorder. *Am J Psychiatry* 2001; **158**:1067–74.

Scarmeas N, Hadjigeorgiou GM, Papadimitrou A *et al.* Motor signs during the course of Alzheimer disease. *Neurology* 2004; **63**:95–82.

Scarmeas N, Honig LS, Choi H *et al.* Seizures in Alzheimer disease: who, when, and how common? *Arch Neurol* 2009; **66**:992–7.

Schaumburg HH, Powers JM, Raine CS *et al.* Adrenoleukodystrophy. A clinical and pathological study of 17 cases. *Arch Neurol* 1975; **32**:577–91.

Schiller A, Wevers RA, Steenbergen GCH *et al.* Long-term course of L-dopa responsive dystonia caused by tyrosine hydroxylase deficiency. *Neurology* 2004; **63**:1524–6.

Schimschock JR, Alvord EC, Swanson PD. Cerebrotendinous xanthomatosis: clinical and pathological studies. *Arch Neurol* 1968; **18**:688–98.

Schlossmacher MG, Hamann C, Cole AG *et al.* Case records of the Massachusetts General Hospital. Case 27–2004: a 79-year-old woman with disturbances in gait, cognition, and autonomic function. *N Engl J Med* 2004; **351**:912–22.

Schlote W, Harzer K, Christomanou H *et al.* Sphingolipid activator protein 1 deficiency in metachromatic leukodystrophy with normal arylsulfatase A activity. A clinical, morphological, biochemical and immunological study. *Eur J Pediatr* 1991; **150**:584–91.

Schmidt A, Jabusch H-C, Altenmuller E *et al.* Dominantly transmitted focal dystonia in families of patients with musician's cramp. *Neurology* 2006; **67**:691–3.

Schmidt A, Jabusch HC, Altenmuller E *et al.* Etiology of musician's dystonia: familial or environmental? *Neurology* 2009; **72**:1248–54.

Schneider JA, Watts RL, Gearing M *et al.* Corticobasal degeneration: neuropathologic and clinical heterogeneity. *Neurology* 1997; **48**:959–69.

Schneider LS, DeKosky ST, Farlow MR *et al.* A randomized, double-blind, placebo-controlled trial of two doses of Ginkgo biloba extract in dementia of the Alzheimer's type. *Curr Alzheimer Res* 2005a; **2**:541–51.

Schneider LS, Dagerman KS, Insel P. Risk of death with atypical antipsychotic drug treatment for dementia: meta-analysis of randomized placebo-controlled trials. *JAMA* 2005b; **294**:1934–43.

Schneider LS, Tariot PN, Dagerman MS *et al.* Effectiveness of atypical antipsychotic drugs in patients with Alzheimer's disease. *N Engl J Med* 2006; **355**:1525–38.

Schneider JA, Arvanitakis Z, Leurgans SE *et al.* The neuropathology of probable Alzheimer disease and mild cognitive impairment. *Ann Neurol* 2009a; **66**:200–8.

Schneider A, Hasan A, Hirschel S *et al.* A novel mutation of the arylsulfatase A gene in late-onset metachromatic leukodystrophy. *J Clin Psychiatry* 2009b; **70**:1724–5.

Schoenfeld M, Myers RH, Cupples LA *et al.* Increased rate of suicide among patients with Huntington's disease. *J Neurol Neurosurg Psychiatry* 1984; **47**:1283–7.

Schofield PW, Tang M, Marder K *et al.* Alzheimer's disease after remote head injury: an incidence study. *J Neurol Neurosurg Psychiatry* 1997; **62**:119–24.

Schols L, Kruger R, Amoiridis G *et al.* Spinocerebellar ataxia type 6: genotype and phenotype in German kindreds. *J Neurol Neurosurg Psychiatry* 1998; **54**:67–73.

Scooter AJC, Bronzoua J, Witteman JCM *et al.* Estrogen use and early onset Alzheimer's disease: a population-based study. *J Neurol Neurosurg Psychiatry* 1999; **67**:779–81.

Sechi GP, Tanda F, Mutani R. Fatal hyperpyrexia after withdrawal of levodopa. *Neurology* 1984; **34**:249–51.

Seelar H, Kamphorst W, Rosso SM *et al.* Distinct genetic forms of frontotemporaldementia. *Neurology* 2008; **71**:1220–6.

Selikhova M, Williams DR, Kempster PA *et al.* A clinico-pathological study of subtypes in Parkinson's disease. *Brain* 2009; **132**: 2947–57.

Serby M, Angrist B, Lieberman A. Mental disturbances during bromocriptine and lergotrile treatment of Parkinson's disease. *Am J Psychiatry* 1978; **135**:1227–9.

Shahed J, Poysky J, Kenney C *et al.* GPi deep brain stimulation for Tourette syndrome improves tics and psychiatric comorbidities. *Neurology* 2007; **68**:159–60.

Shannon KM, Bennett JP, Friedman JH *et al.* Efficacy of pramipexole, a novel dopamine agonist, as monotherapy in mild to moderate Parkinson's disease. *Neurology* 1997; **49**:724–8.

Shapiro E, Shapiro A, Fulop G *et al.* Controlled study of haloperidol, pimozide, and placebo for the treatment of Gilles de la Tourette's syndrome. *Arch Gen Psychiatry* 1989; **46**:722–30.

Shapiro EG, Lockman LA, Knopman D *et al.* Characteristics of the dementia in late-onset metachromatic leukodystrophy. *Neurology* 1994; **44**:662–5.

Shapiro E, Krivit W, Lockman L *et al.* Long-term effect of bone-marrow transplant for childhood-onset X-linked adrenoleukodystrophy. *Lancet* 2000; **356**:713–18.

Shaw LM, Vanderstichele H, Knapik-Czajka M *et al.* Cerebrospinal fluid biomarker signature in Alzheimer's disease neuroimaging initiative subjects. *Ann Neurol* 2009; **65**:403–13.

Sheehy MP, Marsden CD. Writer's cramp – a focal dystonia. *Brain* 1982; **105**:461–80.

Shevell M. Hallervorden and history. *N Engl J Med* 2003; **348**:3–4.

Shiang R, Ryan SG, Zhu YZ *et al.* Mutations in the alpha-1 subunit of the glycine receptor cause the dominant neurologic disorder, hyperekplexia. *Nat Genet* 1993; **5**:351–8.

Shinotoh H, Calne DB, Snow B *et al.* Normal CAG repeat length in the Huntington's disease gene in senile chorea. *Neurology* 1994; **44**:2183–4.

Shiwach R. Psychopathology in Huntington's disease patients. *Acta Psychiatr Scand* 1994; **90**:241–6.

Shotbolt P, Samuel M, Fox C *et al.* A randomized controlled trial of quetiapine for psychosis in Parkinson's disease. *Neuropsychiatr Dis Treat* 2009; **5**:327–32.

Shoulson I. Huntington disease: functional capacities in patients treated with neuroleptic and antidepressant drugs. *Neurology* 1981; **31**:1333–5.

Shumaker SA, Legault C, Rapp SR *et al.* Estrogen plus progestin and the incidence of dementia and mild cognitive impairment in postmenopausal women: the Women's Health Initiative Memory Study: a randomized controlled trial. *JAMA* 2003; **289**:2651–62.

Shy GM, Drager GA. A neurological syndrome associated with orthostatic hypotension: a clinico-pathologic study. *Arch Neurol* 1960; **2**:511–27.

Siddique T, Figlewicz DA, Pericak-Vance MA *et al.* Linkage of a gene causing familial amyotrophic lateral sclerosis to chromosome 21 and evidence of genetic-locus heterogeneity. *N Engl J Med* 1991; **324**:1381–4.

Siesling S, Vegter-van der Vlis M, Roos RAC. Juvenile Huntington's disease in the Netherlands. *Pediatr Neurol* 1997; **17**:34–43.

Singer HS, Reiss AL, Brown JE *et al.* Volumetric MRI changes in basal ganglia of children with Tourette's syndrome. *Neurology* 1993; **43**:950–6.

Singer HS, Giuliano JD, Hansen BH *et al.* Antibodies against human putamen in children with Tourette syndrome. *Neurology* 1998; **50**:1618–24.

Singer HS, Loiselle CR, Lee O *et al.* Anti-basal ganglia antibodies in PANDAS. *Mov Disord* 2004; **19**:406–15.

Singer HS, Hong JJ, Yoo DY *et al.* Serum autoantibodies do not differentiate PANDAS and Tourette syndrome from controls. *Neurology* 2005; **65**:1701–7.

Sinha S, Taly AB, Ravishankar S *et al.* Wilson's disease: cranial MRI observations and clinical correlation. *Neuroradiology* 2006; **48**:613–21.

Siren A, Legros B, Crahine L *et al.* Hyperekplexia in Kurdish families: a possible GLRA1 founder mutation. *Neurology* 2006; **67**:137–9.

Sjorgen T, Sjorgen H, Lindgren AGH. Morbus Alzheimer and Morbus Pick. *Acta Psychiatr Neurol Scand* 1952; **82**(Suppl.):1–152.

Slaughter JR, Martens MP, Slaughter KA. Depression and Huntington's disease: prevalence, clinical manifestations, etiology and treatment. *CNS Spectr* 2001; **6**:306–26.

Snider LA, Lougee L, Slatterly M *et al.* Antibiotic prophylaxis with azithromycin or penicillin for childhood-onset neuropsychiatric disorders. *Biol Psychiatry* 2005; **57**:788–92.

Snowden JS, Craufurd D, Griffiths HL *et al.* Awareness of involuntary movements in Huntington disease. *Arch Neurol* 1998; **55**:801–5.

Sobue G, Ueno-Natsukari I, Okamoto H *et al.* Phenotypic heterogeneity of an adult form of adrenoleukodystrophy in monozygotic twins. *Ann Neurol* 1994; **36**:912–15.

Soffer D, Benharroch D, Berginer V. The neuropathology of cerebrotendinous xanthomatosis: a case report and a review of the literature. *Acta Neuropathol* 1995; **90**:213–20.

Sorensen BF, Hamby WB. Spasmodic torticollis: results in 71 surgically treated patients. *Neurology* 1966; **16**:867–78.

Spencer T, Biederman J, Coffey B *et al.* A double-blind comparison or desipramine and placebo in children and adolescents with chronic tic disorder and comorbid attention-deficit/hyperactivity disorder. *Arch Gen Psychiatry* 2002; **59**:649–56.

Spillantini MG, Yoshida H, Rizzini C *et al.* A novel tau mutation (N296N) in familial dementia with swollen achromatic neurons and corticobasal inclusion bodies. *Arch Neurol* 2000; **48**:939–43.

Squitieri F, Cannella M, Piorcellini A *et al.* Short-term effects of olanzapine in Huntington disease. *Neuropsychiatry Neuropsychol Behav Neurol* 2001; **14**:69–72.

Sreedharan J, Blair IP, Tripathi VB *et al.* TDP-43 mutations in familial and sporadic amyotrophic lateral sclerosis. *Science* 2008; **319**:1668–72.

Stanford PM, Halliday GM, Brooks WS *et al.* Progressive supranuclear palsy pathology caused by a novel silent mutation in exon 10 of the tau gene: expansion of the disease phenotype caused by tau gene mutations. *Brain* 2000; **123**:880–93.

Starkstein SE, Mayberg HS, Leiguarda R *et al.* A prospective longitudinal study of depression, cognitive decline and physical impairments in patients with Parkinson's disease. *J Neurol Neurosurg Psychiatry* 1992; **55**:377–82.

Starkstein SE, Migliorelli R, Teson A *et al.* Prevalence and clinical correlates of pathological affective display in Alzheimer's disease. *J Neurol Neurosurg Psychiatry* 1995; **59**:55–60.

Starkstein SE, Chemerinski E, Sabe L *et al.* Prospective longitudinal study of depression and anosognosia in Alzheimer's disease. *Br J Psychiatry* 1997; **171**:47–52.

Starkstein SE, Jorge R, Mizrahi R *et al.* The construct of minor and major depression in Alzheimer's disease. *Am J Psychiatry* 2005; **162**:2086–93.

Starkstein WE, Jorge R, Mizrahi R *et al.* A prospective longitudinal study of apathy in Alzheimer's disease. *J Neurol Neurosurg Psychiatry* 2006; **77**:8–11.

Starkstein WE, Morello M, Jorge R *et al.* The syndromal validity and nosological position of apathy in Parkinson's disease. *Mov Disord* 2009; **24**:1211–16.

Starosta-Rubinstein S, Young AB, Kluin K *et al.* Clinical assessment of 31 patients with Wilson's disease: correlations with structural changes on magnetic resonance imaging. *Arch Neurol* 1987; **44**:365–70.

Steele JC. Progressive supranuclear palsy. *Brain* 1972; **95**:693–704.

Steele JC, Richardson JC, Olszewski J. Progressive supranuclear palsy: a heterogenous degeneration involving the brain stem, basal ganglia and cerebellum, with vertical gaze and pseudobulbar palsy, vertical dystonia and dementia. *Arch Neurol* 1964; **10**:333–59.

Steinberger D, Korinthenberg R, Topka H *et al.* Dopa-responsive dystonia: mutation analysis of GCH1 and analysis of therapeutic doses of l-dopa. *Neurology* 2000; **55**:1735–7.

Steinert TH. Uber das klinische und anatomische bild des muskelschwunde der myotoniker. *Dtsch Z Nervenheilkd* 1909; **37**:58–104.

Stern G. *Parkinson's Disease.* London: Arnold, 1990.

Stern M, Roffwarg H, Duvoisin R. The parkinsonian tremor in sleep. *J Nerv Ment Dis* 1968; **147**:202–10.

Stewart WF, Kawas C, Corrada A *et al.* Risk of Alzheimer's disease and duration of NSAID use. *Neurology* 1997; **48**:626–32.

Stone LA, Jankovic J. The coexistence of tics and dystonia. *Arch Neurol* 1991; **48**:862–5.

Stracciari A, Tempestini A, Borghi A *et al.* Effect of liver transplantation on neurological manifestations in Wilson's disease. *Arch Neurol* 2000; **57**:384–6.

Suhren AL, Bruyn GW, Tuyman JA. Hyperekplexia: a hereditary startle syndrome. *J Neurol Sci* 1966; **3**:577–86.

Sveinbjornsdottir S, Hicks AA, Jonsson T *et al.* Familial aggregation of Parkinson's disease in Iceland. *N Engl J Med* 2000; **343**: 1765–70.

Swaiman KF. Hallervorden–Spatz syndrome and brain iron metabolism. *Arch Neurol* 1991; **48**:1285–93.

Swanson PD, Cromwell LD. Magnetic resonance imaging in cerebrotendinous xanthomatosis. *Neurology* 1986; **36**:124–6.

Swartz JR, Miller BL, Lesser IM *et al.* Frontotemporal dementia: treatment response to serotonin selective reuptake inhibitors. *J Clin Psychiatry* 1997; **58**:212–16.

Swedo SE, Leonhard HL, Garvey M *et al.* Pediatric autoimmune neuropsychiatric disorders associated with streptococcal infections: clinical description of the first 50 cases. *Am J Psychiatry* 1998; **154**:264–71.

Sydow O, Thobois S, Alesch F *et al.* Multicentre European study of thalamic stimulation in essential tremor: a six year follow up. *J Neurol Neurosurg Psychiatry* 2003; **74**:1387–91.

Takahashi H, Oyanagi K, Takeda S *et al.* Occurrence of 15-nm wide straight tubules in neocortical neurons in progressive supranuclear palsy. *Acta Neuropathol* 1989; **79**:233–9.

Takahashi H, Ohama E, Suzuki S *et al.* Familial juvenile parkinsonism: clinical and pathologic study in a family. *Neurology* 1994; **44**:437–41.

Takahashi H, Yoshida K, Sugita T *et al.* Quetiapine treatment of psychotic symptoms and aggressive behavior in patients with dementia with Lewy bodies: a case series. *Prog Neuropsychopharmacol Biol Psychiatry* 2003; **27**:549–53.

Takiyama T, Oyanagi S, Kawashima S *et al.* A clinical and pathologic study of a large Japanese family with Machado–Joseph disease tightly linked to the DNA markers on chromosome 14q. *Neurology* 1994; **44**:1302–8.

Taly AB, Meenakshi-Sundaram S, Sinha S *et al.* Wilson disease: description of 282 patients evaluated over 3 decades. *Medicine* 2007; **86**:112–21.

Tam CW, Burton EJ, McKeith IG *et al.* Temporal lobe atrophy on MRI in Parkinson disease with dementia: a comparison with Alzheimer disease and dementia with Lewy bodies. *Neurology* 2005; **64**:861–5.

Tan E-K, Jankovic J. Genetic testing in Parkinson's disease. *Arch Neurol* 2006; **63**:1232–7.

Tandberg E, Larsen JP, Aarsland D *et al.* The occurrence of depression in Parkinson's disease: a community-based study. *Arch Neurol* 1996; **53**:175–9.

Tandberg E, Larsen JP, Karlsen K. A community-based study of sleep disorders in patients with Parkinson's disease. *Mov Disord* 1998; **13**:895–9.

Tang B, Liu C, Shen L *et al.* Frequency of SCA1, SCA2, SCA3/MJD, SCA6, SCA7, and DRPLA CAG trinucleotide repeat expansion in patients with hereditary spinocerebellar ataxia from Chinese kindreds. *Arch Neurol* 2000; **57**:540–4.

Tang MX, Jacobs D, Stern Y *et al.* Effect of oestrogen during menopause on risk and age of onset of Alzheimer's disease. *Lancet* 1996; **348**:429–32.

Tanner CM, Glantz RH, Klawans HL. Meige disease: acute and chronic anticholinergic effects. *Neurology* 1982; **32**:783–5.

Tanner CM, Goldman SM, Lyons KE *et al.* Essential tremor in twins: an assessment of genetic vs environmental determinants of etiology. *Neurology* 2001; **57**:1389–91.

Tanner CM, Ross GW, Jewell SA *et al.* Occupation and risk of parkinsonism. *Arch Neurol* 2009; **66**:1106–13.

Tapiola T, Alafuzoff I, Herukka SK *et al.* Cerebrospinal fluid (beta)-amyloid 42 and tau proteins as biomarkers of Alzheimer-type pathologic changes in the brain. *Arch Neurol* 2009; **66**:382–9.

Tariot PN, Erb R, Podgorski CA *et al.* Efficacy and tolerability of carbamazepine for agitation and aggression in dementia. *Am J Psychiatry* 1998; **155**:54–61.

Tariot PN, Farlow MR, Grossberg GT *et al.* Memantine treatment in patients with moderate to severe Alzheimer disease already receiving donepezil: a randomized controlled trial. *JAMA* 2004; **291**:317–24.

Tashiro K, Ogata K, Goto Y *et al.* EEG findings in early-stage corticobasal degeneration and progressive supranuclear palsy: a retrospective study and literature review. *Clin Neurophysiol* 2006; **117**:2236–42.

Terry RD, Peck A, DeTeresa R *et al.* Some morphometric aspects of the brain in senile dementia of the Alzheimer type. *Ann Neurol* 1981; **10**:184–92.

Thacker EL, O'Reilly EJ, Weisskopf MG *et al.* Temporal relationship between cigarette smoking and risk of Parkinson disease. *Neurology* 2007; **68**:764–8.

Thomas A, Iacono D, Luciano AL *et al.* Duration of amantadine benefit on dyskinesia of severe Parkinson's disease. *J Neurol Neurosurg Psychiatry* 2004a; **75**:141–3.

Thomas AJ, Burn DJ, Rowan EN *et al.* A comparison of the efficacy of donepezil in Parkinson's disease with dementia and dementia with Lewy bodies. *Int J Geriatr Psychiatry* 2005; **20**:938–44.

Thomas M, Hayflick SJ, Jankovic J. Clinical heterogeneity of neurodegeneration with brain iron accumulation (Hallervorden-Spatz syndrome) and pantothenate kinase-associated neurodegeneration. *Mov Disord* 2004b; **19**:36–42.

Thompson PD, Bhatia KP, Brown P *et al.* Cortical myoclonus in Huntington's disease. *Mov Disord* 1994; **9**:633–41.

Thompson S, Hermann N, Rapoport MJ *et al.* Efficacy and safety of antidepressants for treatment of depression in Alzheimer's disease: a metaanalysis. *Can J Psychiatry* 2007; **52**:248–55.

Tijssen MA, Shiang R, van Deutekom J *et al.* Molecular genetic reevaluation of the Dutch hyperekplexia family. *Arch Neurol* 1995; **52**:578–82.

Timberlake WH, Vance MA. Four-year treatment of patients with parkinsonism using amantadine alone or with levodopa. *Ann Neurol* 1978; **3**:119–28.

Timmermann L, Pauls KA, Wieland K *et al.* Dystonia in neurodegeneration with brain iron accumulation: outcome of bilateral pallidal stimulation. *Brain* 2010; **133**:701–12.

Tinazzi M, Farina S, Bhatia K *et al.* TENS for the treatment of writer's cramp dystonia: a randomized, placebo-controlled study. *Neurology* 2005; **64**:1946–8.

Tiraboschi P, Hansen LA, Alford M *et al.* Early and widespread cholinergic losses differentiate dementia with Lewy bodies from Alzheimer disease. *Arch Gen Psychiatry* 2002; **59**:946–51.

Tiraboschi P, Salmon DP, Hansen LA *et al.* What best differentiates Lewy body from Alzheimer's disease in early-stage dementia? *Brain* 2006; **129**:729–35.

Tison F, Dartigues JF, Auriacombe S *et al.* Dementia in Parkinson's disease: population-based study in ambulatory and institutionalized individuals. *Neurology* 1995; **45**:705–8.

Tolosa ES. Clinical features of Meige's disease (idiopathic orofacial dystonia): a report of 17 cases. *Arch Neurol* 1981; **38**:147–51.

Tolosa ES, Klawans HC. Meige's disease: a clinical form of facial convulsion, bilateral and medial. *Arch Neurol* 1979; **36**: 635–7.

Toru M, Matsuda O, Makiguchi K *et al.* Single case study: neuroleptic malignant syndrome-like state following a withdrawal of anti-parkinsonian drugs. *J Nerv Ment Dis* 1981; **169**:324–7.

Tourette's Syndrome Study Group. Treatment of ADHD in children with tics: a randomized controlled trial. *Neurology* 2002; 58: 527–36.

Trender-Gerhard I, Sweeney MG, Schwingensuch P *et al.* Autosomal dominant GTPCH-1 deficient DRD: clinical characteristics and long-term outcome of 34 patients. *J Neurol Neurosurg Psychiatry* 2009; **80**:839–45.

Troster AL, Paolo AM, Lyons KE *et al.* The influence of depression on cognition in Parkinson's disease: a pattern of impairment distinguishable from Alzheimer's disease. *Neurology* 1995; **45**:672–6.

Tsai CH, Chang FC, Su YC *et al.* Two novel mutations of the glycine receptor gene in a Taiwanese hyperekplexia family. *Neurology* 2004; **63**:893–6.

Tsuang D, Almquist EW, Lipe H *et al.* Familial aggregation of psychotic symptoms in Huntington's disease. *Am J Psychiatry* 2000; **157**:1955–9.

Vaddadi KS, Soosai E, Chiu E *et al.* A randomized, placebo-controlled, double blind study of treatment of Huntington's disease with unsaturated fatty acids. *Neuroreport* 2002; **13**:29–33.

Vadlamudi L, Westmoreland BF, Klass DW *et al.* Electroencephalographic findings in Kufs disease. *Clin Neurophysiol* 2003; **114**:1738–43.

Valente EM, Bentivoglio AR, Cassetta E *et al.* DYT13, a novel primary torsion dystonia locus, maps to chromosome 1p36.13–36.32 in an Italian family with cranial-cervical or upper limb onset. *Ann Neurol* 2001; **49**:362–6.

Van Bogaert L, Schere H, Epstein E. *Une Forme Cerebrale de la Cholesterinose Generalisee.* Paris, France: Masson et Cie, 1937.

Van Bogaert L, Dewulf A. Diffuse progressive leukodystrophy in the adult with production of metachromatic degenerative products. *Arch Neurol Psychiatry* 1939; **42**:1083–97.

Van der Meche FGA, Bogaard JM, van der Sluys JCM *et al.* Daytime sleep in myotonic dystrophy is not caused by sleep apnoea. *J Neurol Neurosurg Psychiatry* 1994; **57**:626–8.

Vance C, Al-Chalabi A, Ruddy D *et al.* Familial amyotrophic lateral sclerosis with frontotemporal dementia is linked to a locus on chromosome 9p13.2–21.3. *Brain* 2006; **129**:868–76.

Van Diggelen OP, Thobois S, Tilikete C *et al.* Adult neuronal ceroid lipofuscinosis with palmitoyl-protein thioesterase deficiency: first adult-onset patients of a childhood disease. *Ann Neurol* 2001; **50**:269–72.

Vanek Z, Jankovic J. Dystonia in corticobasal degeneration. *Mov Disord* 2001; **16**:252–7.

Van Geel BM, Assies J, Haverkort EB *et al.* Progression of abnormalities in adrenomyeloneuropathy and neurologically asymptomatic X-linked adrenoleukodystrophy despite treatment with 'Lorenzo's oil'. *J Neurol Neurosurg Psychiatry* 1999; **67**:290–9.

Van Geel BM, Bezman L, Loes DJ *et al.* Evolution of phenotypes in adult male patients with X-linked adrenoleukodystrophy. *Ann Neurol* 2001; **49**:186–94.

Van Hove JLK, Steyaert J, Matthijs G *et al.* Expanded motor and psychiatric phenotype in autosomal dominant Segawa syndrome due to GTP cyclohydrolase deficiency. *J Neurol Neurosurg Psychiatry* 2006; **77**:18–23.

Van Vugt JP, Siesling S, Vergeer M *et al.* Clozapine versus placebo in Huntington's disease: a double-blind, randomized comparative study. *J Neurol Neurosurg Psychiatry* 1997; **63**:35–9.

Varga E, Sugerman AA, Varga V *et al.* Prevalence of spontaneous oral dyskinesia in the elderly. *Am J Psychiatry* 1982; **139**: 329–31.

Vazquez A, Jimenez-Jimenez FJ, Garcia-Ruiz P *et al.* 'Panic attacks' in Parkinson's disease. A long-term complication of levodopa therapy. *Acta Neurol Scand* 1993; **87**:14–18.

Verbaan D, Marinus J, Visser M *et al.* Patient-reported autonomic symptoms in Parkinson disease. *Neurology* 2007; **69**:333–41.

Verhagen Metman L, Del Dotto P, van den Munckhof P *et al.* Amantadine as treatment for dyskinesias and motor fluctuations in Parkinson's disease. *Neurology* 1998; **50**:1323–6.

Verhagen Metman L, Morris MJ, Farmer C *et al.* Huntington's disease: a randomized, controlled trial using the NMDA-antagonist amantadine. *Neurology* 2002; **59**:694–9.

Verny M, Jellinger KA, Hauw J-J *et al.* Progressive supranuclear palsy: a clinicopathological study of 21 cases. *Acta Neuropathol* 1996; **91**:427–31.

Verrips A, Hofsloot LH, Steenbergen GCH *et al.* Clinical and molecular genetic characteristics of patients with cerebrotendinous xanthomatosis. *Brain* 2000a; **123**:908–19.

Verrips A, van Engelen BGM, Wevers RA *et al.* Presence of diarrhea and absence of tendon xanthomas in patients with cerebrotendinous xanthomatosis. *Arch Neurol* 2000b; **57**:520–4.

Vessie PR. On the transmission of Huntington's chorea for 300 years – the Bures family group. *J Nerv Ment Dis* 1932; **76**:553–73.

Vidhailet M, Vercueil L, Houeto J-L *et al.* Bilateral deep brain stimulation of the globus pallidus in primary generalized dystonia. *N Engl J Med* 2005; **352**:459–67.

Vidhailhet M, Vercueil L, Houeto JL *et al.* Bilateral, pallidal, deep brain stimulation in primary generalized dystonia: a prospective 3 year follow-up study. *Lancet Neurol* 2007; **6**:223–9.

Villani F, Gellera C, Spreafilo R *et al.* Clinical and molecular findings in the first identified Italian family with dentatorubral-pallidoluysian atrophy. *Acta Neurol Scand* 1998; **98**:324–7.

Vital A, Bouillot S, Burbaud P *et al.* Chorea-acanthocytosis: neuropathology of brain and peripheral nerve. *Clin Neuropathol* 2002; **21**:77–81.

Vitale C, Pellecchia MT, Grossi D *et al.* Unawareness of dyskinesias in Parkinson's and Huntington's disease. *Neurol Sci* 2001; **22**: 105–6.

Vonsattel JP, Myers RH, Stevens TJ *et al.* Neuropathological classification of Huntington's. *disease J Neuropathol Exp Neurol* 1985; **44**:559–77.

Voon V, Hassan K, Zurowski M *et al.* Prevalence of repetitive and reward-seeking behaviors in Parkinson disease. *Neurology* 2006a; **67**:1254–7.

Voon V, Hassan K, Zurowski M *et al.* Prospective prevalence of pathologic gambling and medication association in Parkinson disease. *Neurology* 2006b; **66**:1750–2.

Voon V, Krack P, Lang AE *et al.* A multicentre study on suicide outcomes following subthalamic stimulation for Parkinson's disease. *Brain* 2008; **131**:2720–8.

Walker RH, Rasmussen A, Rudnicki D *et al.* Huntington's disease-like 2 can present as chorea-acanthocytosis. *Neurology* 2003; **61**:1002–4.

Walker Z, Grace J, Overshot N *et al.* Olanzapine in dementia with Lewy bodies: a clinical study. *Int J Geriatr Psychiatry* 1999; **14**:459–66.

Walsh PJ. Adrenoleukodystrophy: report of two cases with relapsing and remitting courses. *Arch Neurol* 1980; **37**:448–50.

Walshe JM, Yealland M. Wilson's disease: the problem of delayed diagnosis. *J Neurol Neurosurg Psychiatry* 1992; **55**:692–6.

Walterfang M, Yucel M, Walker R *et al.* Adolescent obsessive compulsive disorder heralding chorea-acanthocytosis. *Mov Disord* 2008; **23**:422–5.

Waltz G, Harik SI, Kaufman B. Adult metachromatic leukodystrophy: value of computed tomographic scanning and magnetic resonance imaging of the brain. *Arch Neurol* 1987; **44**:225–7.

Wardle M, Majounie E, Williams NM *et al.* Dentatorubral pallidoluysia atrophy in South Wales. *J Neurol Neurosurg Psychiatry* 2008; **79**:804–7.

Warner TT, Lennox GG, Janota I *et al.* Autosomal-dominant dentatorubropallidoluysian atrophy. *Mov Disord* 1994; **9**:289–96.

Warner TT, Williams LD, Walker RWH *et al.* A clinical and molecular genetic study of dentatorubropallidoluysian atrophy in four European families. *Ann Neurol* 1995; **37**:452–9.

Warren JD, Firgaira F, Thompson EM *et al.* The causes of sporadic and 'senile' chorea. *Aust N Z J Med* 1998; **28**:429–31.

Wasco W, Pettingell WP, Jondro PD *et al.* Familial Alzheimer's chromosome 14 mutations. *Nat Med* 1995; **1**:848.

Watanabe H, Saito Y, Terao S *et al.* Progression and prognosis in multiple system atrophy. An analysis of 230 Japanese patients. *Brain* 2002; **125**:1070–83.

Waters CH, Kurth M, Bailey P *et al.* Tolcapone in stable Parkinson's disease: efficacy and safety of long-term treatment. *Neurology* 1997; **49**:665–71.

Watts GF, Mitchell WD, Bending JJ *et al.* Cerebrotendinous xanthomatosis: a family study and pharmacotherapy. *Q J Med* 1996; **89**:55–63.

Wechsler AF, Verity A, Rosenschein S *et al.* Pick's disease: a clinical computed tomographic and histologic study with Golgi impregnation observation. *Arch Neurol* 1982; **39**:287–90.

Weiden P, Bruun R. Worsening of Tourette's syndrome due to neuroleptic-induced akathisia. *Am J Psychiatry* 1987; **144**:504–5.

Weiner MF, Risser RC, Cullum CM *et al.* Alzheimer's disease and its Lewy body variant: a clinical analysis of postmortem verified cases. *Am J Psychiatry* 1996; **153**:1269–73.

Weiner WJ, Nora LM. 'Trick' movements in facial dystonia. *J Clin Psychiatry* 1984; **45**:519–21.

Weiner WJ, Nausieda PA, Glantz RH. Meige syndrome (blepharospasm-oromandibulardysonia) after long-term neuroleptic therapy. *Neurology* 1981; **31**:1555–6.

Weintraub D, Rosenberg PB, Drye LT *et al.* Sertraline for the treatment of depression in Alzheimer disease: 24-week outcomes. *Am J Geriatr Psychiatry* 2010a; **18**:332–40.

Weintraub D, Koester J, Potenza MN *et al.* Impulse control disorders in Parkinson disease: a cross-sectional study of 3090 patients. *Arch Neurol* 2010b; **67**:589–95.

Weller M, Liedtke W, Petersen D *et al.* Very-late-onset adrenoleukodystrophy: possible precipitation of demyelination by cerebral contusion. *Neurology* 1992; **42**:367–70.

Wenning GK, Shlomo YB, Megalhaes M *et al.* Clinical features and natural history of multiple system atrophy: an analysis of 100 cases. *Brain* 1994; **117**:835–45.

Wenning GK, Ben-Shlomo Y, Magalhaes M *et al.* Clinicopathological study of 35 cases of multiple system atrophy. *J Neurol Neurosurg Psychiatry* 1995; **58**:160–6.

Wenning GK, Tison F, Elliott L *et al.* Olivopontocerebellar pathology in multiple system atrophy. *Mov Disord* 1996; **11**:157–62.

Wenning GK, Litvan I, Jankovic J *et al.* Natural history and survival of 14 patients with corticobasal degeneration confirmed at postmortem examination. *J Neurol Neurosurg Psychiatry* 1998; **64**:184–9.

Wermuth L, Sorensen PS, Timm S *et al.* Depression in idiopathic Parkinson's disease treated with citalopram. A placebo-controlled trial. *Nord J Psychiatry* 1998; **52**:163–9.

Wesnes KA, McKeith IG, Ferrara R *et al.* Effects of rivastigmine on cognitive function in dementia with Lewy bodies: a randomized, placebo-controlled international study using the cognitive drug research computerized assessment system. *Dement Geriatr Cogn Disord* 2002; **13**:183–92.

Whitehouse P, Price D, Clark A *et al.* Alzheimer's disease: evidence for a selective loss of cholinergic cells in the nucleus basalis. *Ann Neurol* 1981; **10**:122–6.

Whitehouse PJ, Price DL, Struble RG *et al.* Alzheimer's disease and senile dementia: loss of neurons in the basal forebrain. *Science* 1982; **215**:1237–9.

Whitehouse PJ, Hedreen JC, White CL *et al.* Basal forebrain neurons in the dementia of Parkinson disease. *Ann Neurol* 1983; **13**: 243–8.

Wicks P, Abrahams S, Masi D *et al.* Prevalence of depression in a 12-month consecutive sample of patients with ALS. *Eur J Neurol* 2007; **14**:993–1001.

Wilkins RH, Brody IA. Alzheimer's disease. *Arch Neurol* 1969a; **21**:109–10.

Wilkins RH, Brody IA. Parkinson's syndrome. *Arch Neurol* 1969b; **20**:440–5.

Williams DR, de Silva R, Paviour DC *et al.* Characteristics of two distinct clinical phenotypes in pathologically proven progressive supranuclear palsy: Richardson's syndrome and PSP-parkinsonism. *Brain* 2005; **128**:1247–58.

Williams DR, Holton JL, Strand K *et al.* Pure akinesia with gait freezing: a third clinical phenotype of progressive supranuclear palsy. *Mov Disord* 2007a; **22**:2235–41.

Williams DR, Holton JL, Strand C *et al.* Pathological tau burden and distribution distinguishes progressive supranuclear palsy-parkinsonism from Richardson's syndrome. *Brain* 2007b; **130**:1566–76.

Wilson CM, Grace GM, Munoz DG *et al.* Cognitive impairment in sporadic ALS. A pathologic continuum underlying a multisystem disorder. *Neurology* 2001; **57**:651–7.

Wilson SAK. Progressive lenticular degeneration: a familial nervous disease associated with cirrhosis of the liver. *Brain* 1912; **34**: 295–509.

Winblad S, Lindberg C, Hansen S. Temperament and character in patients with classical myotonic dystrophy type 1 (DM-1). *Neuromuscul Disord* 2005; **15**:287–92.

Winkler GF, Young RR. Efficacy of chronic propranolol therapy in action tremors of the familial, senile or essential varieties. *N Engl J Med* 1974; **290**:984–8.

Wintzen AR, Lammers GJ, van Dijk JG. Does modafinil enhance activity of patients with myotonic dystrophy? A double-blind placebo-controlled crossover study. *J Neurol* 2007; **254**:26–8.

Wisniewski HM, Coblentz JM, Terry RD. Pick's disease: a clinical and ultrastructural study. *Arch Neurol* 1972; **26**:97–108.

Wojcieszek JM, Lang AE. Gestes antagonistes in the suppression of tics: 'tricks for tics'. *Mov Disord* 1995; **10**:226–8.

Woodruff BK, Graff-Radford NR, Ferman TJ *et al.* Family history of dementia as a risk factor for Lewy body disease. *Neurology* 2006; **66**:1949–50.

Wright RB, Glantz RH, Butcher J. Hearing loss in myotonic dystrophy. *Ann Neurol* 1988; **23**:202–3.

Wulff CH, Trojaborg W. Adult metachromatic leukodystrophy: neurophysiologic findings. *Neurology* 1985; **35**:1776–8.

Yamamoto T, Hirano A. Nucleus raphe dorsalis in Alzheimer's disease: neurofibrillary tangles and loss of large neurons. *Ann Neurol* 1985; **17**:573–7.

Yardimci N, Karatas M, Kilinc M *et al.* Levetiracetam in Meige's syndrome. *Acta Neurol Scand* 2006; **114**:63–6.

Yoshida M. Amyotrophic lateral sclerosis with dementia: the clinicopathological spectrum. *Neuropathology* 2004; **24**:87–102.

Yoshii F, Tomiyasu H, Shinohara Y. Fluid attenuation inversion recovery (FLAIR) images of dentatorubropallidoluysian atrophy: case report. *J Neurol Neurosurg Psychiatry* 1998; **65**:396–9.

Young AB, Shoulson I, Penney JB *et al.* Huntington's disease in Venezuala: neurologic features and functional decline. *Neurology* 1986; **36**:244–9.

Younger DS, Rowland LP, Latov N *et al.* Lymphoma, motor neuron disease, and amyotrophic lateral sclerosis. *Ann Neurol* 1991; **29**:78–86.

Zandi PP, Carlson MC, Plassman BL *et al.* Hormone replacement therapy and incidence of Alzheimer disease in older women: the Cache County study. *JAMA* 2002; **288**:2123–9.

Zappacosta B, Monza D, Meoni C *et al.* Psychiatric symptoms do not correlate with cognitive decline, motor symptoms, or CAG repeat length in Huntington's disease. *Arch Neurol* 1996; **53**: 493–7.

Zesiewicz TA, Sullivan KL, Hauser RA *et al.* Open-label pilot study of levetiracetam (Keppra) for the treatment of chorea in Huntington's disease. *Mov Disord* 2006; **21**:1998–2001.

Zhou B, Westaway SK, Levinson B *et al.* A novel pantothenate kinase gene (PANK2) is defective in Hallervorden-Spatz syndrome. *Nat Genet* 2001; **28**:345–9.

Zhukareva V, Mann D, Pickering-Brown S *et al.* Sporadic Pick's disease: a tauopathy characterized by a spectrum of pathological tau isoforms in gray and white matter. *Ann Neurol* 2002; **51**: 730–9.

Zubenko GS, Moosy J. Major depression in primary dementia: clinical and neuropathologic correlates. *Arch Neurol* 1998; **45**:1182–6.

Zweig RM, Ross CA, Hedreen JC *et al.* The neuropathology of aminergic nuclei in Alzheimer's disease. *Ann Neurol* 1988; **24**:233–42.

10

Congenital, developmental, and other childhood-onset disorders

10.1 STURGE–WEBER SYNDROME

The Sturge–Weber syndrome, also known as the Sturge–Weber–Dimitri syndrome or encephalotrigeminal angiomatosis, was first described by William Sturge in 1879 (Sturge 1879). This is a rare disorder characterized, classically (Bebin and Gomez 1988; Chao 1959; Lichtenstein 1954; Pascual-Castroviejo *et al.* 1993, 2008; Petermann *et al.* 1958; Sujansky and Conradi 1995a,b), by a unilateral facial port-wine stain, epilepsy, hemiplegia, and mental retardation, and, on appropriate imaging, cerebral cortical calcification in the hemisphere ipsilateral to the port-wine stain.

Clinical features

The port-wine stain, also known as a nevus flammeus, is present at birth, and, at a minimum, covers the area of distribution of the first division of the trigeminal nerve, including the upper eyelid; in some cases, the stain may extend down the face to the second or even third division, and may even reach the side of the neck. Although in most cases the stain is unilateral, bilateral involvement may be seen in roughly one-third of cases; rarely there may be no port-wine stain (Aydin *et al.* 2000; Taly *et al.* 1987).

Seizures occur in 75–85 percent of patients and, although they can first appear at any time from infancy to the early adult years, the vast majority of patients will begin having seizures in the first year of life. Simple partial motor seizures are classic, and when these do occur the motor activity is seen contralateral to the hemisphere affected by calcification; complex partial seizures may also occur, and regardless of which type of partial seizure occurs, secondary generalization is common.

Mental retardation occurs in the majority of patients, and is typically seen only in those with seizures (Bebin and Gomez 1988; Sujansky and Conradi 1995a,b). Further, in those with frequent seizures, there may be a decrement in cognitive abilities (Lichtenstein 1954; Petermann *et al.* 1958).

Other symptoms include hemiparesis with hemiatrophy in roughly one-half of all cases and homonymous hemianopia in one-third to one-half of all cases, with both these symptoms occurring contralateral to the hemisphere with the calcification. An enlarged eye, present in infancy, may also occur: this is known as 'bupthalmos', which is Greek for 'ox-eye'. Glaucoma may occur in one- to two-thirds of all cases, and may present at any time from infancy to the middle years.

Patients may also suffer recurrent stroke-like episodes (Maria *et al.* 1998), which may leave lasting deficits.

Cortical calcification may not become evident on imaging until after the age of 2 years, but it becomes progressively more common with the passage of time, eventually appearing in approximately 90 percent of patients over the age of 20 years. On skull films, the calcification appears in a classic curvilinear 'trolley-track' pattern, and on computed tomography (CT) scanning, as illustrated in Figure 10.1, there is a 'serpiginous' pattern. Magnetic resonance (MR) scanning, in addition to revealing calcification, may also demonstrate leptomeningeal vascular malformations (Benedikt *et al.* 1993). Although Sturge–Weber syndrome is classically associated with unilat-

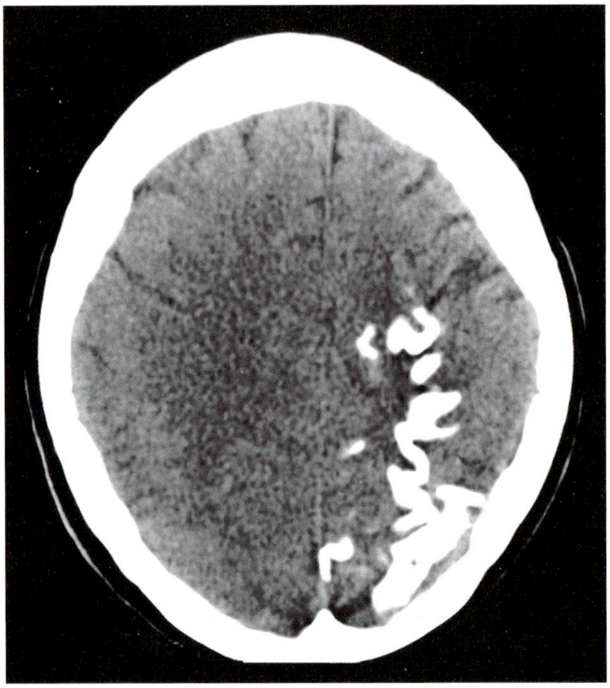

Figure 10.1 This unenhanced computed tomography scan demonstrates the serpiginous calcification seen in the Sturge–Weber syndrome. Reproduced with kind permission from Gillespie J and Jackson A. (2000), with permission.

eral cortical calcification, bilateral involvement may in fact occur in up to one-quarter of cases.

Course

In most cases, the overall clinical picture stabilizes by late childhood.

Etiology

Neuropathologically (Roizen *et al.* 1972; Weber 1929; Wohlwill and Yakovlev 1957), the hallmark of this disorder is a unilateral leptomeningeal angiomatosis, primarily venular in type, which, although often confined to the occipito-parietal area, may extend forward to the frontal area and, in some cases, may be bilateral (Bebin and Gomez 1988). In the subjacent cortex, there is calcification in the walls of the small arteries, free calcium deposits within the brain parenchyma, and neuronal loss and gliosis.

The angiomatosis is thought to be due to the failure of regression of a normally occurring vascular plexus during early embryogenesis. In turn, the vascular malformation is thought to lead to impaired venous drainage, with subsequent venous stasis, hypoxia, and, in some cases, venous thrombosis with infarction.

Sturge–Weber syndrome is sporadic in virtually all cases.

Differential diagnosis

The combination of a facial port-wine stain, seizures, and cortical calcification is pathognomonic for Sturge–Weber syndrome. In those cases in which the port-wine stain is absent, consideration should be given to celiac disease, which is also characterized by occipital calcification and seizures. It must also be borne in mind that isolated port-wine stains are not at all uncommon, and that the diagnosis of Sturge–Weber syndrome should never be made on the basis of a port-wine stain alone.

Treatment

Seizure control is essential, and vigorous treatment with anti-epileptic drugs (AEDs) should be immediately undertaken. In treatment-resistant cases, focal corticectomy or hemispherectomy should be considered (Arzimanoglou *et al.* 2000). Although not without controversy, aspirin should also be considered, as it may effectively prevent stroke-like episodes (Maria *et al.* 1998). The port-wine stain may be treated with laser surgery; glaucoma is treated in the usual way, and, in cases in which glaucoma is absent, yearly monitoring of intraocular pressure is indicated. The treatment of mental retardation is discussed in Chapter 6.7.

10.2 TUBEROUS SCLEROSIS

Tuberous sclerosis, also known as Bourneville's disease or epiloia, is a rare genetic disorder presenting, classically, with the triad of seizures, mental retardation, and a particular skin lesion known as adenoma sebaceum (Alsen *et al.* 1994; Critchley and Earl 1932; Devlin *et al.* 2006; Lagos and Gomez 1967; Pampiglione and Moynahan 1976). Although brain lesions, known as tubers, figure most prominently in this disorder, it must be borne in mind that tuberous sclerosis is a systemic disease, with additional lesions in the skin, eye, kidneys, heart, and lungs.

Clinical features

In most cases, the first sign of tuberous sclerosis is the presence of hypomelanotic macules, evident in 90 percent or more of affected infants. These macules, which are best seen with Wood's light, range in size from a few millimeters up to 3 cm, and are sometimes oval shaped, giving rise to the name 'ash leaf' spots. Additional skin lesions include adenoma sebaceum and what are known as 'shagreen patches'. Adenoma sebaceum typically appears gradually in early childhood and is present in over 90 percent of those over the age of 4 years; it consists of multiple minute facial nodules, generally arranged in a symmetrical butterfly shape over the nose, cheeks, and chin, with, typically, sparing of the upper lip. Shagreen patches are leathery-appearing areas, most frequently seen in the lumbar region.

Seizures (Chu-Shore *et al.* 2009) of various types occur in the vast majority of cases, and may have an onset anywhere from infancy to adult years Among those with an onset in infancy, infantile spasms are most common. Among the remainder, although the onset is typically in childhood or adolescence, adult-onset cases may occur (Gutowski and Murphy 1992); regardless of whether the onset in these cases is in childhood or later, the most common types of seizures are grand mal and complex partial.

Mental retardation is seen in approximately one-half of patients, and may range from mild to profound (Joinson *et al.* 2003). Furthermore, and especially in those with frequent seizures, there may be a progressive cognitive decline, thus constituting a dementia. Autism may occur in up to one-quarter of all cases (Alsen *et al.* 1994; Gutierrez *et al.* 1998; Lawlor and Maurer 1987), and is strongly associated with the presence of tubers in the temporal lobes, temporal interictal epileptiform discharges, and a history of infantile spasms (Bolton *et al.* 2002). In a very small minority of patients, hydrocephalus may occur secondary to occlusion of the foramen of Monro by a tuber or an astrocytoma, with the development of further cognitive decline accompanied by headache and gait disturbance.

Other lesions include subungual fibromas, retinal phakomas, renal angiomyolipomas and cysts, cardiac rhabdolipomas, and pulmonary cysts.

Neuroimaging reveals the distinctive tubers. Tubers typically undergo calcification and, when this occurs, they are immediately apparent on skull radiographs or CT scanning. MR imaging, in addition to revealing calcified tubers, will also demonstrate uncalcified ones, which may be missed on CT scanning. The electroencephalogram is marked by slowing and interictal epileptiform activity, which is often multifocal. Genetic testing is available.

Course

Tuberous sclerosis is a gradually progressive disease. Those with an early childhood onset, especially those with severe epilepsy, rarely survive for more than 15 years, eventually succumbing to status epilepticus or cardiac or renal complications. In adult-onset cases, the progression tends to be much slower, and the disease may be compatible with a normal lifespan.

Etiology

Roughly two-thirds of all cases of tuberous sclerosis occur secondary to mutations in one of two genes, namely *TSC1* and *TSC2*, with the remaining cases occurring secondary to unknown mutations (Au *et al.* 2007). *TSC1* is located on chromosome 9 and codes for a protein known as hamartin; *TSC2* is on chromosome 16 and codes for tuberin. Hamartin and tuberin function in concert as tumor suppressor proteins, and it is presumably a lack of this normal 'suppression' that allows for the development of the tubers and other manifestations of the disease. Approximately two-thirds of all cases represent spontaneous mutations, whereas in the remaining one-third the disease is inherited on an autosomal dominant basis.

Tubers are nodules of varying size, ranging from millimeters to 2 cm or more, and are composed of glia and enlarged 'ballooned' neuronal elements. They are typically found both in the cerebral cortex and in the subependymal white matter: in some cases, subependymal tubers may protrude into the ventricle and may be so numerous that they impart a 'candle guttering' appearance to the surface of the ventricle (Richardson 1991). Tubers, as noted, typically undergo calcification, and the calcification may be so pronounced as to produce 'brain stones' (Yakovlev and Corwin 1939). A small minority of tubers may undergo malignant transformation into astrocytomas (Goh *et al.* 2004a; Morimoto and Mogami 1986), and tubers or astrocytomas adjacent to the foramen of Monro may cause occlusion and non-communicating hydrocephalus.

Differential diagnosis

The classic triad of seizures, mental retardation, and adenoma sebaceum is pathognomonic. As noted earlier, however, about one-half of patients will have normal intelligence, and such cases may present with seizures alone in adult years (Kofman and Hyland 1959). In these cases, diagnostic suspicion may be aroused by skin lesions, such as ash leaf spots or adenoma sebaceum, or by finding tubers on imaging.

Treatment

Seizure control is imperative. Infantile spasms may respond to adrenocorticotropic hormone (ACTH) or vigabatrin. Partial or grand mal seizures may be treated with the usual AEDs, with, however, some caveats. In some cases, renal failure may occur, making the use of AEDs that undergo primary renal excretion problematic. Other AEDs, such as carbamazepine (Weig and Pollack 1993), may precipitate cardiac block in patients with cardiac lesions. In treatment-resistant cases in which there is a single, well-localized epileptogenic zone, surgical resection may be considered (Jarrar *et al.* 2004).

Obstructive hydrocephalus may respond to neurosurgical intervention. Tubers that have undergone transformation into astrocytomas may also respond to treatment with rapamycin (Franz *et al.* 2006).

Mental retardation and autism are treated in the usual fashion, as discussed in Chapters 6.7 and 10.13, respectively. When severe, adenoma sebaceum may be surgically treated, but recurrences are the rule. Genetic counseling should be offered, not only to patients but also, given that some cases, perhaps manifest only by subtle skin lesions, may go undetected, to relatives.

10.3 VON RECKLINGHAUSEN'S DISEASE (NEUROFIBROMATOSIS TYPE 1)

von Recklinghausen's disease, also known as neurofibromatosis type 1 or 'peripheral' neurofibromatosis, is a not uncommon genetic disorder and is characterized by skin lesions (such as café au lait spots), peripheral neurofibromas, and, in a minority, central nervous system tumors, such as optic nerve gliomas. Attention-deficit–hyperactivity disorder

(ADHD), developmental disabilities, and, in a small minority, mental retardation may also occur.

Clinical features

The cardinal features of von Recklinghausen's disease are café au lait spots and neurofibromas (Huson *et al.* 1988). Café au lait spots are generally present in infancy and grow in number and size throughout adolescence. Neurofibromas appear around puberty, are generally either pedunculated or sessile, and may be as large as a walnut or no bigger than a grain of sand; they typically appear on the trunk or the extremities, generally sparing the face, and range in number from a few up to literally hundreds. Large plexiform neurofibromas may occasionally occur, and these may be extremely disfiguring. Neurofibromas may be painful to strong touch and at times spontaneous neuralgic pains may occur. In a small minority, neurofibromas may appear on the spinal roots of peripheral nerves, and in such cases cord compression may occur.

Other features found in adults include Lisch nodules and axillary freckling. Lisch nodules (Lubs *et al.* 1991) are small yellow-brown spots on the iris that are at times visible only by slit-lamp examination. Axillary freckling, although found only in a minority, is virtually pathognomonic of von Recklinghausen's disease. Seizures may occur in a very small minority (Kulkantrakorn and Geller 1998).

Central nervous system tumors, such as astrocytomas, meningiomas, and, most commonly, optic nerve gliomas, may occur in a small minority (Creange *et al.* 1999; Guillamo *et al.* 2003; Listernick *et al.* 1989; Rodriguez and Barthrong 1966); further, and again in only a small minority, peripheral neurofibromas may undergo sarcomatous change, an event heralded both by an increase in size and by the occurrence or exacerbation of pain. Other malignancies seen in von Recklinghausen's disease include pheochromocytoma, Wilm's tumor, and various leukemias.

Hydrocephalus may occur in a very small minority, due either to compression of the aqueduct of Sylvius by a brainstem glioma, or to stenosis of the aqueduct by an overgrowth of glial tissue. Stroke may also occur, again in a very small minority, and this is typically associated with the development of the Moyamoya syndrome.

Other disorders found in von Recklinghausen's disease include (Hyman *et al.* 2005) ADHD in roughly one-third of patients, various developmental disabilities such as developmental dyslexia in about one-fifth, and mental retardation in a little over 5 percent of patients. Precocious puberty may occur in a small minority.

Small areas of increased signal intensity on T2-weighted scans are found in close to three-quarters of all children and adolescents with von Recklinghausen's disease (Lopes Ferraz Fiho *et al.* 2008). Of interest, some studies have found a correlation between cognitive impairment and the total number of such lesions (Feldmann *et al.* 2003; Hyman *et al.* 2003; North *et al.* 1994), but this is not a universal finding (Legius *et al.* 1995). This discrepancy, however, may be due to a failure to take into account not only the total number, but the location of such lesions, as subsequent studies have found a stronger correlation between cognitive impairment and the presence of these lesions in the thalamus rather than with their overall number (Goh *et al.* 2004; Hyman *et al.* 2007). Of interest, the number of these lesions within the gray matter structures actually decreases with age (DiMario *et al.* 1998; Hyman *et al.* 2003).

Course

Neurofibromas typically show a period of progression in early adolescence, whereas in adult years, although progression may still occur, there are typically long periods of quiescence. Pregnancy and the use of oral contraceptives may prompt renewed growth. Most patients live a normal lifespan, but exceptions may occur in those who develop any of the various tumors noted earlier.

Etiology

von Recklinghausen's disease occurs secondary to any of a large number of different mutations in a gene on chromosome 17 (Wallace *et al.* 1990) that codes for a protein known as neurofibromin (von Deimling *et al.* 1995). About half of all cases represent spontaneous mutations, whereas the other half are inherited on an autosomal dominant basis. Of note, although penetrance is near 100 percent, expressivity is quite variable, and there is considerable inter- and intra-familial phenotypic variability. Neurofibromin acts as a tumor-suppressor protein, and it is apparently a deficiency of such suppression by the abnormal protein that allows for the clinical manifestations of the disease.

Neurofibromas constitute the neuropathologic hallmark of the disease, and these may occur peripherally, where they are composed of fibroblasts and Schwann cells, or centrally, in which case one finds fibroblasts and astrocytes. Within the central nervous system, one also finds scattered glial nodules, neuronal heterotopias, and areas of cortical dysplasia (Rosman and Pearce 1967), and, in the white matter, areas of spongiosus (DiPaolo *et al.* 1995). Presumably these abnormalities account for the T2 hyperintensities seen on MR scanning and for the ADHD, developmental disabilities, and mental retardation. As noted earlier, various tumors, such as optic nerve gliomas, astrocytomas, and meningiomas, may also occur.

Differential diagnosis

The diagnosis is self-evident when numerous neurofibromas are present. When these are lacking, the diagnosis may depend on dermatologic findings, such as café au lait spots or axillary freckling. Importantly, isolated café au lait spots are not uncommon and, hence, to make the diagnosis it is necessary to find six or more of these measuring at least 0.5 cm in diameter in children and 1.5 cm in adults. It has, however, recently become apparent that the finding of six or more café au lait spots may not be specific for von Recklinghausen's disease: there is another autosomal dominant disorder, known as 'neurofibromatosis type 1-like syndrome', which presents with six or more café au lait spots (with or without axillary freckling). In this look-alike syndrome, however, there are no neurofibromas, nor are there any central nervous system lesions (Messiaen *et al.* 2009).

von Recklinghausen's disease must be differentiated from neurofibromatosis type 2, also known as 'central' neurofibromatosis. Neurofibromatosis type 2 is a very rare disorder that is clinically and genetically distinct: clinically, it presents with intracranial tumors, classically bilateral acoustic neuromas, and genetically, it results from mutations in a gene on chromosome 22 (Rouleau *et al.* 1987); cutaneous manifestations in neurofibromatosis type 2 are either absent or very scant.

Treatment

Given that surgery to neurofibromas may predispose to sarcomatous change, excision should be reserved for disfiguring peripheral neurofibromas or for central neurofibromas that are causing compression injury. Other tumors are treated in the usual fashion; bilateral optic gliomas may also be subjected to radiation treatment. ADHD, developmental disabilities, and mental retardation are treated in the usual fashion. Genetic counseling should be offered, and it should be stressed to patients with mild symptoms that, given the considerable intra-familial phenotypic heterogeneity, should they have children then those children may be severely affected by the disease.

10.4 DOWN'S SYNDROME

Down's syndrome, first described by John Langdon Down in 1866 (Down 1866), is one of the most common causes of mental retardation in the USA, and is found somewhat more frequently in males than in females. It is also known as trisomy 21; however, this synonym may not be appropriate because, although about 95 percent of cases are due to trisomy 21, the remainder, which are clinically indistinguishable, occur as a result of translocations.

Clinical features

The appearance of patients is so characteristic as to allow a diagnosis in infancy. The head is small with a flattened occiput. The palpebral fissures show a distinctive upward slant, and epicanthal folds are present. The bridge of the nose is broad, the mouth is generally small, and the tongue, which is typically enlarged, often protrudes. The patients tend to be of short stature. The hands are broad and foreshortened, and the fifth finger is often curved inward; further, the palms often display a transverse or 'Simian' crease. The first and second toes are often separated by a wide gap. The external genitalia are often small, puberty may be delayed, and fertility in males is often reduced.

Mental retardation ranges from mild to severe, with most falling within the moderate range. Social skills, however, are typically better preserved than cognitive ones, and patients with Down's syndrome tend to be affable, outgoing, and pleasant. A small minority will suffer from autism (Lund and Munk-Jorgensen 1988), and such patients may display odd or bizarre stereotypies, with anxiety and social withdrawal (Carter *et al.* 2007).

Dementia due to Alzheimer's disease occurs in adults with Down's syndrome (Collacott *et al.* 1992; Jervis 1948; Lai and Williams 1989; Lund and Munk-Jorgensen 1988; Margallo-Lana *et al.* 2007), with an ever-increasing prevalence as patients pass the age of 20 years, rising to close to 50 percent in those who live to the sixth decade and beyond. In contrast to the stable level of reduced cognitive performance characteristic of the mental retardation, there is a gradual deterioration in functioning (Wisniewski *et al.* 1985a): those with severe retardation may become more apathetic and less sociable, whereas in those with only a mild or moderate degree of retardation, the dementia may present with decreased memory, disorientation to time, and reduced verbal output (Lai and Williams 1989).

Seizures, including grand mal, complex partial, and simple partial seizures, are seen overall in roughly a tenth of all adults with Down's syndrome (McVicker *et al.* 1994; Pueschel *et al.* 1991). The age of onset, however, displays a bimodal pattern (Pueschel *et al.* 1991; Puri *et al.* 2001), with one broad peak occurring between childhood and the early adult years, and the second broad peak occurring in the mid-adult years: this second peak is associated with the onset of Alzheimer's disease, and as the years pass, the prevalence of epilepsy rises, until, by the sixth decade, over one-half of surviving patients will be affected.

Thyroid disturbances may occur in a significant minority of patients with Down's syndrome. Hypothyroidism is by far the most common problem, and this typically occurs secondary to Hashimoto's thyroiditis (Popova *et al.* 2008), with antithyroid antibodies. Hyperthyroidism may also occur, but this is seen in only a very small minority of cases: although this may also occur on the basis of Hashimoto's thyroiditis (Popova *et al.* 2008), most cases occur secondary to Grave's disease (Goday-Arno *et al.* 2009).

Depression may occur in a minority (Collacott *et al.* 1992), and in some cases this may be so severe as to cause a dementia (Warren *et al.* 1989).

Congenital heart disease, such as ventriculoseptal defect or patent ductus arteriosus, is found in a little less than one-half of all patients. Stroke secondary to ischemic infarction may occur in a small minority, and this may occur on the basis of cardiogenic emboli or, in a minority, Moyamoya syndrome (de Borchgrave *et al.* 2002; Pearson *et al.* 1985): in cases with multiple strokes, a dementia may be seen (Collacott *et al.* 1994). Obstructive sleep apnea is very common and may be associated not only with daytime sleepiness but also with a cognitive decrement (Trois *et al.* 2009). Other abnormalities seen in Down's syndrome include hearing loss, duodenal obstruction, intestinal stenosis, megacolon, leukemia, and atlanto-axial instability. This last abnormality is very important to keep in mind, as it may lead to cord compression.

Although the diagnosis can usually be reliably made on clinical grounds alone, karyotyping is indicated, not only to confirm the diagnosis but also to identify the small proportion of cases that occur secondary to a translocation.

Course

With modern care, especially cardiac surgery, the average life expectancy is roughly 50 years; once Alzheimer's disease has set in, however, most will die within 5 years (Evenhuius 1990).

Etiology

Approximately 95 percent of cases of Down's syndrome are secondary to trisomy 21 due to non-disjunction during meiosis (Petersen and Luzzatti 1965; Stoll *et al.* 1998). In almost all cases this non-disjunction occurs in the mother, and the risk for this rises dramatically with age, from about 1 in 1000 in the early twenties to almost 1 in 100 at the age of 40 years and 1 in 50 at the age of 45 years. In a very small minority, mosaicism may occur, and cases secondary to this tend to be characterized by mild symptomatology. In the remaining 5 percent of cases not due to trisomy 21, there is a translocation, generally from chromosome 21 to 14: such translocations may occur sporadically or may be inherited from either parent.

The brain is small and rounded, with a flattened occiput. The overall sulcal pattern is often simple and undeveloped, and the superior temporal gyrus is often quite hypoplastic, as illustrated in Figure 10.2. The number of neurons in the cerebral cortex is reduced (Ross *et al.* 1984), and on the remaining neurons there is typically a reduced number of dendritic spines (Suetsugu and Mehraein 1980). As noted earlier, Alzheimer's disease is common in Down's syndrome and, among patients over 40 years, senile plaques and neurofibrillary tangles (Hyman *et al.* 1995; Margallo-Lana *et al.* 2007; Schochet *et al.* 1973) are almost universal (Wisniewski *et al.* 1985b). Presumably, the appearance of Alzheimer's disease in this population is due to the fact that the gene for the amyloid precursor protein is on chromosome 21, and the extra gene leads to an overproduction of amyloid.

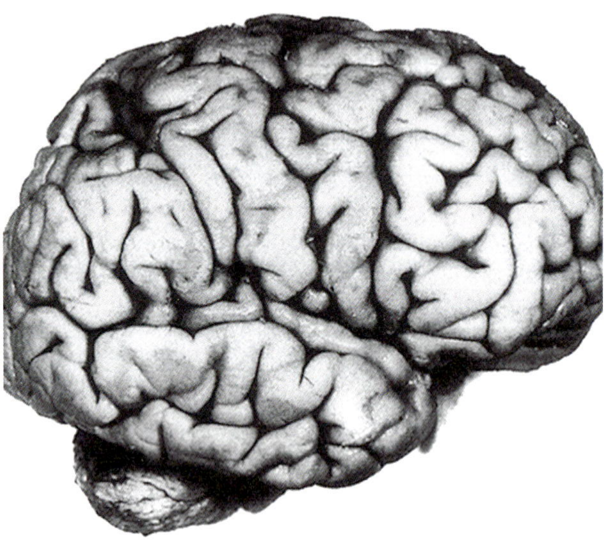

Figure 10.2 Note the marked hypoplasia of the superior temporal gyrus in this case of Down's syndrome. Reproduced with kind permission from Graham DI and Lantos PL. (1996), with permission.

Differential diagnosis

The diagnosis of Down's syndrome is evident on inspection and difficult to confuse with other causes of mental retardation. In cases in which dementia supervenes in adult years, although the most common cause is Alzheimer's disease, consideration must also be given to dementia secondary to obstructive sleep apnea, hypothyroidism, depression, and multi-infarct dementia.

Treatment

If karyotyping reveals a translocation, all first-degree relatives should be offered testing. Mothers of patients with Down's syndrome due to trisomy 21 should be informed of the high risk of Down's syndrome in future children.

The treatment of mental retardation, seizures, dementia, and depression is discussed in Chapters 6.7, 8.3, 6.3, and 7.1, respectively. When Alzheimer's disease occurs, consideration may be given to a trial of donepezil: there has been one double-blinded study of this agent in Down's syndrome, which showed a non-statistically significant trend in favor of donepezil over placebo (Prasher *et al.* 2002). Given the frequency with which hypothyroidism occurs, it is appropriate to screen patients with a thyroid profile on a periodic basis. All patients should have a CT scan of the atlanto–axial junction, and if there is evidence of instability, appropriate restrictions on contact and similar sports should be instituted.

10.5 KLINEFELTER'S SYNDROME

Klinefelter's syndrome, found only in males, occurs secondary to the presence of one or more extra X chromosomes and, when fully expressed, is characterized by tall stature, hypogonadism, and infertility. Mental retardation and other neuropsychiatric syndromes may also occur, as discussed below.

Clinical features

The classic clinical picture (Ratcliffe *et al.* 1982) of Klinefelter's syndrome becomes apparent in postpubertal males and is characterized, as noted, by tall stature, hypogonadism, and infertility. Excessive height is primarily caused by a late closure of the epiphyseal plates and results from increased leg length. Hypogonadism manifests with gynecomastia, a female escutcheon, and a small penis and testes; although most patients have a heterosexual orientation, libido is often low and erectile dysfunction may occur (Pasqualini *et al.* 1957). Moderate to severe azoospermia is present, which accounts for the infertility. Importantly, in an unclear but probably significant proportion of patients, this classic picture is not present, and some patients may come to clinical attention only during a workup for infertility or erectile dysfunction.

Although the vast majority of patients have an IQ within the normal range (Ratcliffe *et al.* 1982), mental retardation may be seen in a minority (Khalifa and Struthers 2002) and is more severe in patients with three or more X chromosomes (Forsman 1970). Further, in those who are retarded, the personality may be characterized by a sullen sort of withdrawal, with a liability to hostile outbursts (Hunter 1969). Other abnormalities include elements of developmental dysphasia (Bender *et al.* 1983; Graham *et al.* 1988), autistic traits (van Rijn *et al.* 2008), and, in a small minority of cases, alcoholism (Nielsen 1969) and bipolar disorder (Everman and Stoudemire 1994). Schizophrenia (Nielsen 1969; Pomeroy 1980; Roy 1981) or schizophrenia spectrum disorders (van Rijn *et al.* 2006) are also associated with Klinefelter's syndrome.

Tremor, similar to that seen in essential tremor, may be seen in a minority (Harlow and Gonzalez-Alegre 2009), and rarely seizures may occur (Tatum *et al.* 1998).

Other associated disorders include Hashimoto's thyroiditis, leukemia, diabetes mellitus, chronic obstructive pulmonary disease, and breast cancer.

Prepubertally, hormone levels are generally within normal limits, but after puberty abnormalities become apparent, with a low testosterone level and an elevated follicle-stimulating hormone level.

Karyotyping will reveal one or more extra X chromosomes.

Course

The evolution of the clinical picture is usually complete by adult years.

Etiology

Klinefelter's syndrome occurs secondary to non-disjunction during either spermatogenesis or oogenesis, resulting in the presence of one or more extra X chromosomes. The most common karyotype is 47, XXY; in about 10 percent of cases, mosaicism is present, with a 46, XY/47, XXY karyotype. Rarely, more severe abnormalities may occur, yielding karyotypes such as 48, XXXY or 49, XXXXY.

Although much of the symptomatology of Klinefelter's syndrome can be explained on the basis of primary hypogonadism secondary to a progressive fibrosis of the testes, it is clear that hypogonadism alone is not sufficient to explain all of the neuropsychiatric features of the disorder, such as mental retardation (Pasqualini *et al.* 1957; Wakeling 1972). MR imaging studies have demonstrated generalized cerebral cortical atrophy (Giedd *et al.* 2007; Itti *et al.* 2006), or more specifically a combination of frontal and temporal atrophy (DeLisi *et al.* 2005), or, even more specifically, left temporal atrophy (Patwardhan *et al.* 2000). The neuropathologic basis for this atrophy, however, is not known.

Differential diagnosis

When fully expressed, the clinical picture in adults is distinctive. Diagnostic difficulties may arise in partial cases, and the correct diagnosis may be revealed only incidentally during a workup for infertility or erectile dysfunction.

Treatment

Testosterone treatment improves libido and erectile function, and tends to help with energy and overall outlook (Nielsen *et al.* 1988); infertility, however, persists. Furthermore, one study found that patients treated since puberty with testosterone not only had less temporal atrophy but also better linguistic skills (Patwardhan *et al.* 2000). Gynecomastia may require surgical correction. Mental retardation and other neuropsychiatric syndromes may be treated as described in the appropriate chapters.

10.6 FRAGILE X SYNDROME

The fragile X syndrome is the most common inherited cause of mental retardation. Although the classic picture of mental retardation with a characteristic facial dysmorphism is far more common and severe in males, females may also be affected.

Clinical features

The classic syndrome (Baumgardner *et al.* 1995; Finelli *et al.* 1985; Wisniewski *et al.* 1985c), as seen in males, is characterized by facial dysmorphism, macro-orchidism, and mental retardation. Associated features include autism, ADHD, and seizures.

Dysmorphic features include a long, narrow face, prognathism, a high forehead, and large ears (De Arce and Kearns 1984). Macro-orchidism is a constant feature in postpubertal males and may also be seen in a minority during childhood (Chudley and Hagerman 1987; De Arce and Kearns 1984). Additional features found on the physical examination may include hyperextensible joints and mitral valve prolapse (Chudley and Hagerman 1987), hyperreflexia, and Babinski signs (Finelli *et al.* 1985).

Mental retardation, although ranging from mild to severe, is of moderate degree in most patients. In the small minority of cases wherein intelligence is within the normal range, one finds elements of developmental dysphasia, with both receptive and expressive deficits.

Autistic traits, such as gaze avoidance, stereotypies, and echolalia, are seen in almost all patients (Hagerman *et al.* 1986), and roughly one-quarter of patients will have the full syndrome of autism (Hagerman *et al.* 1986; Harris *et al.* 2008).

Attention-deficit–hyperactivity disorder is found in roughly one-half of all cases (Sullivan *et al.* 2006), and may be primarily inattentive, hyperactive, or of the combined type.

Seizures (Finelli *et al.* 1985; Musumeci *et al.* 1999; Sabaratnam *et al.* 2001; Wisniewski *et al.* 1985c) occur in roughly one-quarter of all cases, and are typically either complex partial or grand mal in type. These are generally readily controlled with AEDs, and the epilepsy tends to remit spontaneously by late adolescence.

Females with the fragile X syndrome tend to have much milder cases. Facial dysmorphism is absent or mild, and mental retardation is seen in only about half of the cases and likewise tends to be mild. Schizotypal traits have been noted in females (Freund *et al.* 1992; Sobosky *et al.* 1994), but the question arises as to whether these are in fact misdiagnosed autistic traits.

Genetic testing is available.

Course

Although there is some evidence that, in males, intellectual functioning may undergo a decline in late childhood or early adolescence (Dykens *et al.* 1989), it appears that, for the most part, the clinical picture becomes set in adolescence and generally remains stable thereafter.

Etiology

The fragile X syndrome occurs secondary to mutations in the *FMR1* gene on the long arm of the X chromosome (Verkerk *et al.* 1991), which codes for a protein known as the fragile X mental retardation 1 protein (FMRP). This gene normally contains a sequence of CGG trinucleotide repeats containing anywhere from 5 to 55 triplets. An expansion of this sequence to include from 55 to 200 repeats is known as a premutation, whereas expansions to over 200 triplets constitute full mutations. Patients with premutations do not develop the fragile X syndrome; however, those with full mutations do. In those with full mutations, transcription of the gene fails and levels of FMRP are low or undetectable. The relatively minor symptomatology seen in females is due to random inactivation of the X chromosome, which allows for some production of FMRP. The fragile X syndrome derives its name from the fact that when the cells of patients are cultivated in a medium deficient in thymidine and folic acid, a fragile site will be found on the long arm of the X chromosome (Sutherland 1977), which, as might be expected, corresponds to the location of the expanded CGG repeat.

Interestingly, although both female and male parents with a premutation may pass a full mutation to their children, this is far more common with female parents. The reason for this is that expansion of a premutation to a full mutation occurs readily during oogenesis but only rarely during spermatogenesis.

Magnetic resonance imaging studies have revealed a number of abnormalities. There is hypertrophy of the caudate nucleus (Gothelf *et al.* 2008), hypertrophy of the hippocampus (Reiss *et al.* 1994), and atrophy of the superior temporal gyrus (Gothelf *et al.* 2008; Reiss *et al.* 1994) and of the cerebellar vermis (Gothelf *et al.* 2008; Mostofsky *et al.* 1998).

Autopsy studies have demonstrated that, although neuronal cell counts are normal in the cortex, dendritic spines are long and tortuous in shape (Hinton *et al.* 1991; Rudelli *et al.* 1985).

Differential diagnosis

The combination of the distinctive facial dysmorphism with macro-orchidism and mental retardation immediately suggests the correct diagnosis in males.

Mention should also be made of a condition known as FXTAS, or fragile X-associated tremor/ataxia syndrome. As noted earlier, patients with premutations do not develop the fragile X syndrome. This is not to say, however, that the premutation is benign, given that those who carry this premutation are at high risk for developing FXTAS in middle or later years, with, as discussed in Chapter 9.12, tremor, ataxia, and, in a minority, dementia.

Treatment

The general treatment of mental retardation, autism, and ADHD is discussed in Chapters 6.7, 10.13, and 10.14, respectively. With regard to the treatment of ADHD when it occurs in the fragile X syndrome, there have been two double-blinded studies, one of which found effectiveness for methylphenidate (Hagerman *et al.* 1988) and the other for L-acetylcarnitine (Torrioli *et al.* 2008)

10.7 VELOCARDIOFACIAL SYNDROME

The velocardiofacial syndrome, also known as the 22q11.2 deletion syndrome, is an inherited disorder characterized by facial dysmorphism, intellectual deficits, and a number of neuropsychiatric syndromes, most notably a psychosis phenotypically similar to schizophrenia. First described in 1978 by Shprintzen *et al.* (1978), this disorder is now recognized as the most common of the microdeletion syndromes, being found in up to 1 in 4000 live births.

Clinical features

The facial dysmorphism is characterized by hypertelorism, a large, bulbous nose with a squared-off nasal root, and micrognathia. Cleft palate, which may be overt or submucous, is common, and most patients have a degree of velopharyngeal insufficiency, leading to a hypernasal voice.

About 50 percent of patients suffer from either borderline intellectual functioning or mental retardation, which is generally of mild degree (Swillen *et al.* 1997). Roughly 40 percent will also have symptomatology similar to that seen in ADHD, with varying mixtures of hyperactivity, impulsivity, and inattentiveness (Gothelf *et al.* 2004a). Autism has also been noted in roughly 20 percent (Antshel *et al.* 2007; Fine *et al.* 2005).

As these patients pass through adolescence into adult years, up to one-third will develop a psychosis phenotypically similar to that seen in schizophrenia (Bassett *et al.* 2003; Gothelf *et al.* 2007; Green *et al.* 2009; Murphy *et al.* 1999), with delusions and hallucinations. Mood disturbances may also occur and may be more frequent than psychosis: these include both manic or hypomanic episodes (Papolos *et al.* 1996) and depressive episodes (Green *et al.* 2009; Murphy *et al.* 1999). Obsessions and compulsions have also been noted in roughly one-third of teenagers (Gothelf *et al.* 2004b).

Other clinical features include cardiac defects, hypocalcemia secondary to hypoparathyroidism, and, in a small minority, seizures (Kao *et al.* 2004).

Course

The course is chronic; although some sufferers die of cardiac complications, most live a normal lifespan.

Etiology

This syndrome occurs secondary to a microdeletion at 22q11.2, which in turn may occur either sporadically or on an autosomal dominant basis with variable penetrance (McDonald-McGinn *et al.* 2001). MR imaging studies (van Amelsvoort *et al.* 2004; Bish *et al.* 2004; Campbell *et al.* 2006; Eliez *et al.* 2000; Kates *et al.* 2006; Schaer *et al.* 2006) have revealed atrophy of the cerebral cortex and white matter (with reduced gyrification of the cortex), enlargement of the right caudate, enlargement of the amygdalae, and atrophy of the thalamus. In the cerebellum, atrophy of the cortex and white matter is also seen.

Differential diagnosis

In adults, velocardiofacial syndrome must be distinguished from schizophrenia, bipolar disorder, major depressive disorder, obsessive–compulsive disorder, and ADHD. The distinctive facial dysmorphism and nasal voice, along with any associated mental retardation, hypocalcemia, or cardiac defects, may suggest the correct diagnosis; however, at times these abnormalities may be absent or subclinical, and the diagnostic possibility may only be raised when the family history suggests the syndrome in a relative, thus prompting genetic testing.

Treatment

There are no blinded treatment studies for this disorder; the neuropsychiatric syndromes seen here are treated as described in the chapters on mental retardation, ADHD, autism, schizophrenia, bipolar disorder and major depressive disorder, and obsessive–compulsive disorder.

10.8 LESCH–NYHAN DISEASE

Lesch–Nyhan disease, first described by Lesch and Nyhan in 1964 (Lesch and Nyhan 1964), is a rare X-linked recessive disorder, occurring almost exclusively in males, which may present in one of two forms, either 'classic' or 'variant'. The classic form of Lesch–Nyhan disease presents in infancy with a movement disorder, followed by mental retardation and self-mutilative lip biting. The variant form presents later, is far less severe, and lacks any self-mutilation.

Clinical features

The classic form has been described in a number of studies (Anderson and Ernst 1994; Christie *et al.* 1982; Jankovic *et al.* 1988; Jinnah *et al.* 2006; Lesch and Nyhan 1964; Nyhan 1972). The movement disorder presents towards the end of the first year of life, and although primarily dystonic in character, there may be some elements of athetosis or chorea. Mental retardation eventually becomes apparent in early childhood, and may range from severe to mild; notably, in a very small minority the IQ may be in the normal range (Mathews *et al.* 1995). The characteristic self-mutilation typically begins in early childhood, after teeth come in; later onsets, up to the age of 8 years, however, have been reported (Hatanaka *et al.* 1990). Despite being normally sensitive to pain, patients repeatedly bite at their lips, tongue, buccal mucosa, and fingers, to the point where the lips and fingers are literally bitten off in some cases. It must be clearly kept in mind that there is no anesthesia here and that the biting is involuntary: Lesch and Nyhan (1964) commented that one of their patients 'appeared terrified and screamed as if in pain during the process (of biting) and appeared happy only when restrained securely'. In a minority of cases, seizures may also occur.

The variant form (Jinnah *et al.* 2010) is characterized by a movement disorder and cognitive impairment, and is distinguished from the classic form by a lack of self-mutilation. As in the classic form, the movement disorder consists primarily of dystonia, which may range from a severe generalized form that prevents ambulation to a mild form consisting only of slight clumsiness. The dystonia is also accompanied by dysarthria, which likewise ranges in severity from severe, wherein speech may be limited to single words, to mild, with only hesitant or slow speech. Cognitive impairment may consist of mild mental retardation or merely low–normal intelligence, and in many cases there is no obvious impairment at all. Although self-mutilation, by definition, is lacking in the variant form, a minority of patients may engage in severe nail biting, a feature clearly reminiscent of the behavior seen in the classic form.

Hyperuricemia is a constant feature of this disease in both its classic and variant forms, and tophaceous gout and gouty nephropathy may appear in adolescence.

Genetic testing is available. As noted below, hypoxanthine-guanine phosphoribosyl transferase (HPRT) activity is grossly reduced or absent, and the diagnosis may also be established by measuring this enzymatic activity in erythrocytes, hair roots, or cultured skin fibroblasts.

Course

Self-mutilation may decrease, or even remit, in early adolescence (Mizuno 1986); however, other symptoms persist.

Etiology

Lesch–Nyhan disease occurs secondary to any one of a large number of different mutations in the gene for HPRT, found on the X chromosome (Davidson *et al.* 1989; Edwards *et al.* 1990; Gibbs and Caskey 1987; Mak *et al.* 2000; Yang *et al.* 1984). Although, as noted earlier, the disease occurs almost exclusively in males, female cases have been reported (van Bogaert *et al.* 1992; Ogasawara *et al.* 1989) and occur secondary to skewed inactivation of the X chromosome. Hyperuricemia is a constant feature of this disease, and this occurs secondary to disturbances in purine metabolism due to the mutation in HPRT. The hyperuricemia, however, does not explain the movement disorder, mental retardation, or self-mutilation.

Although the neuropathology of Lesch–Nyhan disease is not known, MR scanning has demonstrated mild atrophy of both the cerebral cortex and the caudate nucleus (Harris *et al.* 1998). Several studies have strongly suggested disturbances in dopaminergic functioning. Postmortem work has demonstrated reduced dopamine content in the caudate (Saito *et al.* 1999), and cerebrospinal fluid (CSF) studies have demonstrated a reduction in the level of homovanillic acid (HVA), a metabolite of dopamine (Jankovic *et al.* 1988). Positron emission tomography (PET) studies have indicated a reduction in presynaptic dopamine stores (Ernst *et al.* 1996; Wong *et al.* 1996), and postmortem work has demonstrated an increase in postsynaptic dopamine receptors (Saito *et al.* 1999). Taken together, these results are consistent with a reduction of dopamine in presynaptic striatal neurons and an expected compensatory upregulation of postsynaptic dopamine receptors.

Differential diagnosis

Although dystonia and mental retardation may be seen in a number of disorders, most notably cerebral palsy, the presence of severe self-mutilation virtually seals the diagnosis of the classic form of Lesch–Nyhan disease: although patients with other disorders may bite or hit themselves, the resulting injuries never approach the severity of those seen in Lesch–Nyhan disease.

In considering the differential of cases of dystonia and mental retardation occurring without significant self-mutilation, it is always appropriate to check for an elevated uric acid level and to look for evidence of gout or nephrolithiasis, as any one of these findings will immediately suggest the correct diagnosis.

Treatment

Allopurinol, by forestalling gouty nephropathy, may prolong life; it has, however, no effect on the central nervous system manifestations.

Lip biting may be curtailed by lip or mouth guards or by face masks; however, in severe cases, teeth extraction may be required. In cases with finger biting, restraints may be helpful. Various medications have been reported in non-blind case reports or studies to be helpful in reducing the biting, including risperidone (Allen and Rice 1996), levodopa (Jankovic *et al.* 1988, Serrano *et al.* 2008), gabapentin (McManaman and Tam 1999), S-adenosylmethionine (Glick 2006), and carbamazepine (Roach *et al.* 1996). Dramatic reductions in both dystonia and self-mutilation have been reported in two cases of deep brain stimulation of the globus pallidus (Cif *et al.* 2007; Taira *et al.* 2003).

10.9 BARDET–BIEDL SYNDROME

The Bardet–Biedl syndrome is a rare autosomal recessively inherited disorder, characterized in its fully expressed form by obesity, polydactyly or syndactyly, retinal dystrophy, and mental retardation.

Clinical features

Clinical features have been discussed in a number of papers (Beales *et al.* 1997; Green *et al.* 1989; Klein and Ammann 1969; Moore *et al.* 2005; Rathmell and Burns 1938; Roth 1947). Obesity is almost universal and tends to be of the central type. Syndactyly or polydactyly is also almost universal; when polydactyly is present, it typically manifests with an extra finger or toe, which may range from rudimentary to fully formed. Retinal dystrophy, although likewise almost universal, may not become symptomatic until later childhood or adolescence; by early adult years, however, a majority of patients will become blind. Mental retardation ranges from mild to severe, and is seen in the majority of cases. Other features include hypogenitalism in males, with a small penis and testes, menstrual irregularities in females, renal dysplasia (which may progress to renal failure in a minority), congenital heart disease, hypertension, and diabetes mellitus.

Course

Apart from retinal dystrophy and renal abnormalities, both of which are progressive, the overall clinical course remains static through adult life.

Etiology

As noted, this disorder is inherited on an autosomal recessive basis: 14 loci (*BBS1* through *BBS14*) have been identified.

Differential diagnosis

Both the Prader–Willi and Alstrom–Hallgren syndromes are characterized by obesity; however, neither of these disorders is associated with polydactyly or syndactyly. Further differentiating features include a ravenous hunger in the Prader–Willi syndrome and sensorineuronal deafness in the Alstron–Hallgren syndrome.

Treatment

The treatment of mental retardation is discussed in Chapter 6.7.

10.10 PRADER–WILLI SYNDROME

The Prader–Willi syndrome is a congenital disorder characterized by extreme hyperphagia, obesity, various dysmorphic features, and, in a majority, mild mental retardation; other neuropsychiatric features, as described below, may also be present. This is a not uncommon disorder, and is found with equal frequency in males and females.

Clinical features

The overall clinical features have been described in several papers (Bray *et al.* 1983; Burke *et al.* 1987; Butler *et al.* 1986; Dunn 1968; Greenswag 1987; Hall and Smith 1972; Robinson *et al.* 1992). This disorder presents in infancy with hypotonia and decreased oral intake. By the age of 2 years, however, a remarkable transformation occurs, in that these patients begin to display a remarkable hyperphagia.

The hyperphagia of the Prader–Willi syndrome is severe and leads to extreme obesity. It stems from a ravenous hunger: one patient literally 'took off running' and 'as soon as she could walk, she was constantly near the refrigerator, begging for food' (Zellweger and Schneider 1968). Patients often go to any lengths to satisfy this hunger and, if refrigerators and food cabinets are locked, may turn to other sources: one patient 'ate cat food, begged food from neighbors, and ate rotting chicken carcasses and other items removed from dustbins' (Clarke 1993).

Characteristic dysmorphic features include a narrow head, almond-shaped eyes, and a narrowed or tented upper lip. Micromelia is also often present, with slender arms and legs and small hands and feet. Hypogonadism is present, manifesting in males with micropenis and cryptorchidism, and in females with hypoplastic labia, a lack of breast development, and varying degrees of amenorrhea.

Mental retardation is present in the majority of patients, but is generally mild.

Patients with Prader–Willi syndrome typically display a characteristic behavioral phenotype (Akefeldt and Gillberg 1999; Hiraiwa *et al.* 2007). Skin picking, which may be severe, is classic. Behavior may become ritualized (Clarke *et al.* 2002), and patients tend to be stubborn and insistent on their routines, with temper tantrums should these be interfered with. Depressive episodes may occur (Boer *et al.* 2002; Soni *et al.* 2008), as may psychotic episodes (Soni *et al.* 2008; Vogels *et al.* 2004), which may be severe enough to require hospitalization (Clarke 1993).

Hypersomnolence is common and appears to be multifactorial. With the extreme obesity, both obstructive sleep apnea and the Pickwickian syndrome may occur (Bye *et al.* 1983; Clift *et al.* 1994; Yee *et al.* 2007); however, although these disorders may certainly contribute to the hypersomnolence, they do not account for it in many cases.

Seizures may occur in a minority of cases (Wang *et al.* 2005).

The diagnosis may be confirmed by DNA methylation analysis, as discussed below.

Course

This disorder is chronic, and many patients die prematurely of the complications of obesity.

Etiology

The Prader–Willi syndrome occurs secondary to a lack of a critical portion of the paternally derived chromosome 15, and this deficit may occur via any one of three mechanisms. The most common mechanism, occurring in roughly 75 percent of cases, involves a microdeletion on the paternally derived chromosome 15; the next most common, occurring in roughly 20 percent of cases, involves uniparental disomy, wherein both chromosome 15s are maternally derived; finally, in the remainder, various mutations may occur on the paternally derived chromosome 15. The diagnosis, as indicated earlier, may be made via methylation analysis. The methylation pattern on chromosome 15 differs for paternally and maternally derived chromosomes, and hence analysis allows one to determine whether the patient has a normal, paternally derived chromosome 15.

Importantly, Prader–Willi syndrome, in the vast majority of cases, is a sporadic disorder: the risk of recurrence in family members is less than 1 in 1000.

Although the neuropathology of the Prader–Willi syndrome has not been elucidated, MR scanning has revealed cortical atrophy in the parieto-occipital region, perisylvian polymicrogyria, and incomplete closure of the insula (Miller *et al.* 2007). The remarkable hunger seen in these patients has naturally focused attention on the hypothalamus, and one study (Swaab *et al.* 1995) of the hypothalamic paraventricular nucleus found not only an overall reduction in neuronal number, but also a specific reduction in oxytocin-containing neurons, a finding of great significance given the role of oxytocin in appetite and satiety. Further attention to the hypothalamus was also prompted by the hypersomnolence seen in the Prader–Willi syndrome; however, one study failed to find a reduction in the number of hypocretin-containing neurons (Fronczek *et al.* 2005).

Differential diagnosis

The Bardet–Biedl syndrome is distinguished by the presence of polydactyly or syndactyly.

Treatment

Growth hormone, given early in life, may improve both overall height and lean body mass (Nyunt *et al.* 2009).

Early dietary management is essential and successful (Schmidt *et al.* 2008); in some cases, however, institutionalization may be required to forestall the development of a lethal degree of obesity. In one double-blind study, fenfluramine was effective in reducing weight (Selikowitz *et al.* 1990); however, this agent is no longer available in the USA. An open study also suggested the effectiveness of risperidone in this regard (Durst *et al.* 2000). Obstructive sleep apnea and the Pickwickian syndrome are treated as discussed in Chapters 20.8 and 20.9, respectively.

The overall management of mental retardation is discussed in Chapter 6.7. Of note, open work has found that topiramate, although ineffective for weight loss in this population, does reduce skin picking (Shapira *et al.* 2004; Smathers *et al.* 2003), and in those cases with seizures, this would be a logical choice. Further case reports also suggest that selective serotonin reuptake inhibitors (SSRIs) may reduce skin picking (Hellings and Warnock 1994). Depression in these cases is typically treated with an SSRI, and should psychosis occur, risperidone is a logical choice, given its possible effectiveness for the obesity.

10.11 FETAL ALCOHOL SYNDROME

The fetal alcohol syndrome (Aragon *et al.* 2008; Clarren and Smith 1978; Jones and Smith 1973; Jones *et al.* 1974), caused by *in utero* exposure to alcohol, consists, classically, of the combination of a characteristic facial dysmorphism, mental retardation, and behavioral problems, most notably hyperactivity. This is a not uncommon disorder, being present in roughly 1 in 1000 live births, and represents one of the most common causes of mental retardation in the USA. In addition to this classic fetal alcohol syndrome, a partial syndrome may also occur, and together these are referred to as 'fetal alcohol spectrum disorders'. In some parts of the literature, the partial syndrome is also referred to as 'fetal alcohol effect'; however, this term lacks precision: some authors have used it to refer to the partial syndrome, whereas others have used it to refer to any individual sign or symptom seen in the syndrome; consequently, it has fallen out of favor.

Clinical features

The characteristic dysmorphism is seen in both the classic and the partial syndromes. It includes microcephaly, shortened palpebral fissures, epicanthal folds, maxillary hypoplasia, a thin upper lip, a flattened philtrum, and micrognathia.

In the classic syndrome, one finds, in addition to the facial dysmorphism, both a short stature and mental retardation, which, in most cases, is of mild severity. A profile similar to that found in ADHD is also common; however, here distractibility and inattentiveness are more prominent than hyperactivity.

In the partial syndrome, one again finds the characteristic dysmorphism, and short stature may also be present. The ADHD profile is also in evidence; however, mental retardation is absent. However, although intelligence is within the normal range, one typically finds symptoms reminiscent of a frontal lobe syndrome. Thus, there may be mild degrees of disinhibition, poor social judgment, perseveration, and difficulties with formulating and executing plans. Furthermore, there may be specific learning disabilities, such as developmental dyscalculia or developmental dysphasia (Larsson *et al.* 1985; Shaywitz *et al.* 1981).

Recent work has also demonstrated that seizures or epilepsy may occur in a small minority of cases of both the classic and partial syndromes (Bell *et al.* 2010).

Magnetic resonance scanning may be normal or, in severe cases, may disclose cortical atrophy, ventricular enlargement, and hypoplasia or agenesis of the corpus callosum.

Course

Although there is often some resolution of the dysmorphic features by adult years, other symptoms tend to persist (Spohr *et al.* 2007).

Etiology

Although the syndrome is clearly related to maternal alcohol ingestion, multiple uncertainties remain as to the details. For example, it is not clear whether it is the first or rather the later trimesters that constitute the greatest period of risk. Furthermore, it is also unclear whether it is the total amount of alcohol consumed that is important or whether relatively brief exposures to high levels, as may occur with binge drinking, are more toxic. Finally, there is also debate about whether it is alcohol itself that is toxic, or its metabolite, acetaldehyde.

Neuropathologic studies (Clarren *et al.* 1978; Ferrer and Galofre 1987; Konovalov *et al.* 1997; Pfeiffer *et al.* 1979; Wisniewski *et al.* 1983) have revealed numerous abnormalities (which may be present in various combinations), including microcephaly, agenesis of the corpus callosum, and hypoplasia of the cerebellar vermis; microscopically, cortical dyslamination may be present, and glial and glial–neuronal nodules may be found in the white matter and also in the leptomeninges.

Differential diagnosis

The characteristic facial dysmorphism coupled with a history of maternal alcohol use strongly suggest the diagnosis. Difficulties may arise on two counts. First, it may not be possible to establish the degree of alcohol use during pregnancy. Second, as noted earlier, the facial dysmorphism tends to fade and hence may be very subtle, or even absent, in teenagers or adults: when the diagnosis is strongly suspected but the facial dysmorphism is absent, examining childhood photographs may be very helpful.

A similar syndrome may occur in cases of *in utero* exposure to various AEDs, most notably phenytoin (Hanson *et al.* 1976) and valproic acid (Ardinger *et al.* 1988).

Treatment

Mental retardation may be treated as outlined in Chapter 6.7, and, although there are no controlled studies of the treatment of the ADHD symptoms seen in fetal alcohol spectrum disorders, consideration may be given to a trial of some of the treatments discussed in Chapter 10.14.

Prevention, of course, is critical, and females of childbearing age should be informed that binge drinking may be as dangerous as daily drinking and that there may indeed be no minimal 'threshold' of alcohol consumption below which a pregnant woman can drink without endangering the fetus.

10.12 RETT'S SYNDROME

Rett's syndrome, first described by Andreas Rett in 1966 (Rett 1966), is characterized, classically, by mental retardation, microcephaly, autistic features, and peculiar stereotypic hand movements. This is a rare disorder that occurs almost exclusively in females.

Clinical features

As described in a number of papers (Coleman *et al.* 1988; Hagberg 1989; Hagberg and Witt-Engerstrom 1986; Hagberg *et al.* 1983; Witt-Engerstrom 1990), the clinical features of Rett's syndrome may be roughly divided into four stages. It must be emphasized, however, that the delineation between various stages is not precise, and that features may overlap from one stage to another.

Stage I becomes apparent at around the age of 10 months and is characterized by a general stagnation of normal development. There may be a failure of normal weight gain, and, rather than beginning to crawl, patients may display a persistent 'bottom shuffling'. This stage persists for anywhere from a month up to a year and a half.

Stage II generally occurs around the age of 18 months and is characterized by a regression. Interestingly, although the onset of this stage is generally gradual, spanning weeks or months, it may at times occur relatively acutely, with an onset spanning days. The regression itself presents with autistic features, including a general withdrawal of contact with people and the appearance of characteristic hand stereotypies. Previously purposive hand movements are lost, to be replaced by stereotypic movements (Carter *et al.* 2010), including clapping, mouthing of the hands, and classically, hand-wringing motions, as if the patients were recurrently washing their hands. Bizarre behavior may also appear, with violent screaming, gnashing of the teeth, and nocturnal awakenings marked by peculiar laughter. Cognitive deficits also become apparent, and, characteristically, language deteriorates. Ambulation deteriorates also, and the gait becomes apraxic and marred by truncal ataxia. Most patients also display episodes of irregular respiration while awake, with periods of hyperventilation followed by apneas. During this stage, head growth decelerates, and patients gradually develop microcephaly. Stage II generally lasts about a year and a half.

Stage III typically begins around the age of 3 years and is often referred to as the 'pseudostationary' or 'plateau' stage, reflecting the fact that, for the most part, the regression seen in stage II ends and there is a period of relative stability. Patients are left with mental retardation, which is generally severe. Stereotypies persist, but often patients will display some increased interest in others. Seizures, if not already evident in stage II, now appear in roughly one-half of all patients, and these may be either complex partial or grand mal in type. This stage lasts for many years, occasionally persisting into early adult years or even middle age. Over the years, the prevalence of seizures increases, until by early adult years close to 90 percent will have epilepsy (Glaze *et al.* 2010).

Stage IV, also known as the stage of late motor deterioration, may appear at any point from late childhood to adult years, and is characterized by dystonia and lower motor neuron muscular atrophy, both most prominent in the lower extremities, and by scoliosis, which at times may be quite severe. With time, ambulation becomes progressively more difficult and many patients become wheelchair-bound.

This classic picture, although by far the most common, is not always present, and in a small minority one may find a 'variant' phenotype (Renieri *et al.* 2009). Here, patients pass through stage I into stage II, but after a few years they show substantial improvement: there is definite increased interaction with others, and stereotypies fade to be replaced by more purposive hand movements. Some cognitive ability may also be regained, and linguistic communication skills likewise improve.

As noted earlier, the vast majority of cases are seen in females. In those very rare instances in which Rett's syndrome occurs in males, it tends to be more severe (Dotti *et al.* 2002), with most patients dying in infancy (Kankirawatana *et al.* 2006).

Genetic testing is available.

Course

Stage IV may persist for decades and some patients may live into their sixties. Early death, however, is not uncommon and may occur secondary to infections or postoperative complications.

Etiology

Rett's syndrome is an X-linked dominant disorder due to any one of a large number of different mutations in the gene for methyl-CpG-binding protein 2 (*MECP2*) (Amir *et al.* 1999; Auranen *et al.* 2001). Phenotypic variation reflects not only the effects of different mutations (Dotti *et al.* 2002) but also non-random inactivation of the X chromosome (Amir *et al.* 2000).

Over 95 percent of cases of Rett's syndrome are sporadic, and in virtually all of these the *de novo* mutation occurs in the paternally derived X chromosome. The extreme rarity of cases in boys is accounted for not only by this paternal pattern of mutation but also by the fact that in boys the lethal nature of the mutation on the X chromosome is not 'balanced', as in females, by a normal X chromosome.

In the 5 percent or so of cases which are inherited, one finds a carrier state in the mother who herself was either not affected (Dayer *et al.* 2007) or affected but only minimally so (Huppke *et al.* 2006), in most cases owing to an almost completely biased pattern of X chromosome inactivation.

Neuropathologic studies have revealed microcephaly and hypopigmentation of the pars compacta of the substantia nigra (Jellinger and Seitelberger 1986; Jellinger *et al.* 1988). Microscopically, there is a loss of cortical pyramidal cells (Belichenko *et al.* 1997), and remaining neurons are generally small and closely packed (Bauman *et al.* 1995). Dendrites show decreased branching (Armstrong *et al.* 1998) and fewer dendritic spines (Chapleau *et al.* 2009). Cerebellar pathology is also present (Oldfors *et al.* 1990), with cerebellar cortical atrophy and a loss of Purkinje cells. Importantly, although gliosis may be present in those dying in adult years, it appears absent or scant in childhood.

Differential diagnosis

Rett's syndrome must be distinguished from other causes of mental retardation, and in this regard the characteristic evolution through stages I to IV is fundamental, as it is not seen in other disorders.

As an aside, some comments are in order regarding whether or not Rett's syndrome should be considered to be one of the secondary causes of autism. As noted above, during stage II, autistic features do appear, most notably hand stereotypies and a general withdrawal of social contact. Some would argue that in patients with severe mental retardation, such features merit a diagnosis of autism; however, the quality of the social withdrawal seen in Rett's syndrome may be different from that classic for autism. Specifically, patients with Rett's syndrome look at people as often as they do objects, and do not exhibit gaze avoidance (Olsson and Rett 1987); by contrast, in classic autism patients spend almost all their time looking preferentially at objects, and do exhibit gaze avoidance. This nosologic question thus comes down to whether or not one insists on a 'classic' form of social withdrawal before making the diagnosis of autism.

Treatment

The general treatment of mental retardation is discussed in Chapter 6.7. In one blinded study, in a minority, bromocriptine was followed by some improvement in cognition and overall behavior (Zappella 1990). Naltrexone should not be used, as it may cause a distinct worsening of symptoms (Percy *et al.* 1994). With regard to the treatment of seizures, although there are no blinded studies, both carbamazepine (Huppke *et al.* 2007), and, especially, levetiracetam (Specchio *et al.* 2010), appear to be reasonable choices.

10.13 AUTISM

Autism, first described by Leo Kanner in 1943 (Kanner 1943), is a chronic, lifelong disorder that is characterized, at its core, by an inability on the patient's part to form normal social relationships with others. The relationships that are formed exhibit a peculiar disturbance, in that patients act toward others as if those others were inanimate objects: there is a machine-like quality to the patient's overall behavior with others, as if the patient fails to sense the difference between the animate and inanimate. In severe cases, patients may be almost totally inaccessible, whereas in milder ones an observer might note only a peculiar awkwardness and stiltedness in the patient's social behavior. Most, but certainly not all, patients also have mental retardation.

Autism is a not uncommon disorder, with estimates as to lifetime prevalence ranging from 0.05 percent to 0.2 percent; it is far more common in males than females, by a ratio of 3–4:1. Although some recent studies have indicated an increase in the rate with which the diagnosis of autism is made, it is unclear whether this represents an actual increase in prevalence or rather better case finding.

Clinical features

Autism has an insidious onset in infancy. Parents may recall that the child cried infrequently and seemed indifferent to being held or cuddled; some parents, as noted by Kanner (Kanner 1943), were 'astonished at the children's *failure to assume at any time an anticipatory posture* to being picked up' (italics in original). Although symptoms are typically obvious by the age of 3 years, in mild cases they may escape notice, and children may not come to attention until they enter elementary school and their behavior is compared with that of their normal peers.

The full picture of autism is typically best seen in middle childhood and consists of peculiar deficits in social interaction, repetitive and stereotyped behaviors, and linguistic disturbances.

The deficits in social interaction constitute the core of autism. Autistic children prefer solitary play and, rather than playing with other children, occupy themselves with objects such as toys or machines of some sort. They seem, as noted by Kanner (1943), to be unable '*to relate themselves* in the ordinary way to people and situations' (italics in original); indeed, other people, whether adults or children, are often treated as if they were machines or perhaps mere props to be used during play. Attempts to make contact with such children generally fail. If one tries to catch their eyes, they will often exhibit 'gaze avoidance' (Richer and Coss 1976), staring fixedly at something else, perhaps something just to the side of the examiner's head. One gets the impression that a mannequin would be more satisfactory to the autistic child than an actual playmate.

The repetitive and stereotyped behaviors of autistic children may, at times, seem bizarre. There are often what are known as 'fascinations', in which the child becomes intently interested in and fascinated with certain objects, such as a piece of cloth, jewelry, or, classically, spinning things such as tops. One of Kanner's patients (Kanner 1944) spun 'toys and lids of bottles and jars by the hour…He would watch them and get severely excited and jump up and down in ecstasy'. Stereotypies such as repetitive hand flapping or finger flicking may occur, and some patients may engage in posturing or, classically, toe walking. The persistence of stereotypies can be astonishing: one child (Ornitz and Ritvo 1968) sustained 'oscillatory hand-flapping throughout the entire day'. Head banging and other self-injurious behaviors, such as self-biting, may also occur.

Other behavioral disturbances are typically seen in concert with these repetitive and stereotyped behaviors. The overall sensitivity to environmental events is often altered in autism. Some patients seem insensitive to pain, whereas others display an inordinate, often terrifying, hypersensitivity to innocuous stimuli: one of Kanner's (1943) patients 'was so afraid of the vacuum cleaner that she would not even go near the closet where it was kept, and when it was used, ran out into the garage, covering her ears with her hands'. The overall level of activity of autistic patients may also be disturbed. Some are generally hypoactive, whereas others display a generalized hyperactivity, which may be accompanied by pronounced irritability with frequent temper tantrums.

Another classic behavioral disturbance seen by Kanner (1943) in autism is an '*anxiously obsessive desire for the maintenance of sameness*' (italics in original). As noted further by Kanner (1951):

> the autistic child desires to live in a static world, a world in which no change is tolerated. The status quo must be maintained at all cost. Only the child himself may sometimes take it upon himself to modify existing combinations. But no one else may do so without arousing unhappiness and anger...The slightest change of arrangement, sometimes so minute that it is hardly perceived by others, may evoke a violent outburst of rage.

Not only arrangements but also routines and sequences must be maintained inviolate: the aisles in the grocery store must be traversed in exactly the same order and direction every time, and the exact same route must be taken to school every day.

The linguistic disturbances seen in autism include not only verbal skills, *per se*, but also defects in prosody. Some patients are mute, and in those who do speak there may be echolalia and, classically, pronomial reversals, wherein patients refer to themselves in the third person. The aprosody seen in autism is of the global type: patients not only appear insensitive to the emotional inflections and gestures used by others, but also, in their own speech, display a monotonous 'sing-song' quality.

Kanner (1944) also noted that some patients may display 'an astounding vocabulary...excellent memory for events of several years before...[and] phenomenal rote memory for poems and names'. Kanner (1944) was impressed with their 'strikingly intelligent physiognomies', believing that most, if not all, of his patients were of normal intelligence. It is now clear, however, that Kanner's original sample of 20 patients was not representative and that, in fact, about three-quarters of autistic patients have mental retardation, ranging in severity from mild to severe.

The mental retardation seen in autism is, however, peculiar in that, unlike most examples of mental retardation, the decrement in abilities is not even; indeed, in some cases one may find 'islets' of normal, or even superior, functioning in an otherwise retarded patient. For example, one autistic man with an IQ of 71, although unable to abstract on proverbs testing or do simple arithmetic, was nevertheless able to 'correctly give the day of the week of any day this century' (Hurst and Mulhall 1988). In the past, such patients have been referred to as '*idiot savants*' (Treffert 1988), and although this terminology is unfortunate, it does forcefully convey the sometimes astounding contrast between the 'islets' of superior functioning and the overall intellectual decrement.

Seizures occur in roughly one-fifth of patients and, although these are far more common in those with mental retardation, they may also occur in those of normal intelligence (Amiet *et al.* 2008). Infantile spasms may occur in younger children, whereas in older children and in adults, simple partial, complex partial, and grand mal seizures may be seen (Danielsson *et al.* 2005; Olsson *et al.* 1988).

Before leaving this discussion of the clinical features of autism, it is appropriate to comment on the putative entity known as Asperger's syndrome. As noted earlier, in some mild cases of autism, symptoms may not come to light until the child enters elementary school and, in very mild cases, recognition may be delayed until much later. Some authors believe that these cases represent a separate disorder, called Asperger's syndrome. Classically, these patients are of normal intelligence, and, unlike those with more severe autism, they do seem interested in forming relationships. This desire, however, is often foiled by an aprosodia. In conversation, they miss the non-verbal cues of others, and may continue to speak long past the point where their listeners have lost all interest; further, they may speak in a monotone, and often with a voice far too loud for the setting. They tend to be pedantic and humorless, and may speak at great length, and in great detail, about a very narrow subject. Predictably, they come off as awkward and odd. Some, compounding their symptoms, will adopt a rigid and overly formalized interpersonal style, as if incorporating an etiquette manual into one's life could somehow make up for a lack of that natural, intuitive sense of personal matters which normal people, to a greater or lesser degree, possess.

There is much debate about whether the syndrome displayed by these patients constitutes a separate entity, namely Asperger's syndrome, or rather represents merely the 'higher end' of the spectrum seen in autism. Current data appear to support the latter of these hypotheses (Witwer and Lecavalier 2008), and hence in this text such patients are considered to have autism.

Course

The overall outcome is strongly influenced by whether or not there is an associated mental retardation and, if so, its severity. With regard to autistic symptoms, there is typically some gradual improvement by adult years, and, from a prognostic point of view, the course during middle childhood is particularly important: those who attain some language and some social skills around this time have a much more favorable outcome than those who do not. No matter how great the spontaneous recovery, however, residual symptoms, such as aprosodia, social awkwardness, and a reduced awareness of social conventions, remain in adult years (Rumsey *et al.* 1985).

Etiology

Etiologic theories regarding autism have changed radically over the past few decades. This disorder was once believed to result from faulty child rearing by cold and distant parents. This theory has now been soundly discredited, and current research focuses primarily on genetics and neuropathology.

Family studies strongly support a genetic basis. Although in the general population the lifetime prevalence, as noted earlier, is no higher than 0.2 percent, among siblings of patients it rises up to close to 10 percent (Ritvo *et al.* 1989), while among monozygotic twins the concordance rate is over 60 percent (Bailey *et al.* 1995; Folstein and Rutter 1977). Genetic heterogeneity is almost assured, and linkage studies have identified loci on various chromosomes, including chromosomes 2, 3, 7, 11, 15, 19, and X.

Although MR imaging studies overall have not been conclusive, with numerous contradictory results and failures of replication, two findings appear fairly solid, namely a degree of macrocephaly and vermis hypoplasia. Neuropathologic studies have likewise not been entirely consistent; however, certain findings are of interest. Within both the cerebrum and cerebellum, there are neuronal heterotopias in the white matter, while in the cortex one finds areas of dyslamination and focal dysplasia (Wegiel *et al.* 2010). A closer look at the frontal and temporal cortices has also revealed a diminished width of the radially oriented minicolumns (Casanova *et al.* 2009). In addition, a reduced neuronal number has been noted in the cortex of the fusiform gyrus (van Kooten *et al.* 2008) and in the amygdala (Schumann and Amaral 2006), while a reduced number of Purkinje cells has been found in the cerebellar cortex (Ritvo *et al.* 1986).

Taken together, these findings are consistent with the theory that autism occurs secondary to an inherited disorder of neuronal migration, coupled with a depletion of neurons within the fusiform cortex, the amygdala, and the cerebellar cortex. The fusiform cortex is intimately related to facial recognition, and the amygdala with emotional responsiveness, and the loss of neurons in these structures is of particular interest given the gaze avoidance and deficits in social relationships seen in autism. The loss of Purkinje cells is likewise of interest given the recent recognition of the role of the cerebellum in complex behaviors (as manifest, for example, in the cerebellar cognitive affective syndrome, discussed in Chapter 8.2).

As yet, there is no firm evidence for an environmental cause, and recent concerns regarding an etiologic role of the measles/mumps/rubella vaccine appear unfounded (Madsen *et al.* 2002; Schechter and Grether 2008).

The foregoing discussion regarding etiology applies to the vast majority of cases wherein there is no obvious cause. In up to 10 percent of cases, however, autism occurs secondary to a recognizable cause. The most common of these are the fragile X syndrome, tuberous sclerosis, and the velocardiofacial syndrome, with up to one-quarter of patients with these disorders having autism; a small minority of cases of Down's syndrome may also have autism, and autistic features or traits may also be seen in Rett's syndrome and Klinefelter's syndrome. Each one of these causes of secondary autism has specific clinical features, discussed in the respective chapters, which enable its ready identification.

Differential diagnosis

As noted, mental retardation is seen in association with autism in about three-quarters of cases, and hence autism with mental retardation must be distinguished from mental retardation of other causes. At times this may be difficult, given that many patients with mental retardation of other causes will display repetitive, stereotyped behaviors, which at times may be similar to those seen in autism. A cardinal differential feature, however, is the patient's social relatedness: upon approaching a child with mental retardation of other causes, one may be greeted by a smile and an expectant posture; by contrast, the child with autism may show no more interest in the approaching physician than might be evinced for a machine.

Developmental dysphasia, especially when both expressive and receptive components are present, may be confused with autism; however, here again the quality of the patient's social relatedness enables a differentiation. The child with developmental dysphasia, although unable to communicate verbally, will still, by gesture, tone of voice, and facial expression, clearly desire social contact, in stark contrast to the child with autism.

Schizophrenia may enter the differential, especially in cases characterized by significant inaccessibility and bizarre stereotypies. Here the age of onset is helpful. As noted earlier, most cases of autism become obvious by the age of 3 years; by contrast, schizophrenia only very rarely has an onset before the age of 8 years. Furthermore, in schizophrenia one finds hallucinations and delusions – symptoms that are absent in autism.

Schizoid and schizotypal personality disorders may be confused with mild cases of autism; however, the age of onset and subsequent course enable a differential. These personality disorders have an onset in late adolescence, and then pursue a chronic, stable course. By contrast, in autism one finds the onset in childhood, with, during adolescence, some improvement.

Treatment

Treatment of autism involves behavior modification, family counseling, special education, and medication. All components are important, and most patients generally receive a combination of these.

Behavior modification programs that target specific behaviors, such as head banging or aggressive behavior, are often quite effective.

Family counseling is aimed at helping parents to adjust to their ill child, and to implement and continue behavior programs at home. In all cases, it is also important to forcefully convey to the parents that they are in no way responsible for their child's illness.

Special education classes are generally appropriate. A class that is highly structured is more effective in fostering appropriate classroom behavior and learning than a more permissive or 'open' classroom.

Medical treatment of autism, although generally playing only a modest role in the overall treatment of this disorder, should not be neglected. In the various double-blinded studies that have been performed, a number of the repetitive and stereotyped behaviors and other behavioral disturbances have been used as target symptoms, including stereotypies, self-injurious behaviors, hypersensitivity to innocuous stimuli, hyperactivity, irritability, and temper tantrums. Only one medicine has been found to be effective for all of these symptoms, and that is risperidone (McCracken *et al.* 2002; McDougle *et al.* 1998, 2005; Research Units on Pediatric Psychopharmacology Autism Network 2002; Shea *et al.* 2004) used in low doses of 1–2 mg/day; this agent was generally well tolerated, with weight gain as the main side-effect. Clonidine (Fankenhauser *et al.* 1992; Jaselskis *et al.* 1992) may also be considered, as it was found to be effective for stereotypies, hypersensitivity to innocuous stimuli, hyperactivity, and irritability; however, sedation may limit its use. Valproic acid may be considered next, and was found to be useful for both stereotypies (Hollander *et al.* 2006) and irritability (Hollander *et al.* 2010). Naltrexone may also be considered, and has been found to be useful for both hyperactivity and irritability (Felman *et al.* 1999; Willemsen-Swinkels *et al.* 1996). Antidepressants have also been used for stereotypies, and although effectiveness was found for fluvoxamine (McDougle *et al.* 1996), fluoxetine (Hollander *et al.* 2005), and clomipramine (Gordon *et al.* 1993), a trial of citalopram (King *et al.* 2009) not only found no effectiveness in this regard but also noted an *increase* in stereotypies and hyperactivity. Stimulants, including both methylphenidate (Research Units on Pediatric Psychopharmacology [RUPP] Network 2005) and atomoxetine (Arnold *et al.* 2006), are effective for hyperactivity; however, lethargy and social withdrawal may occur as side-effects. Of note, despite early concern that stimulants might increase stereotypies, one study of methylphenidate actually found a *decrease* in stereotypies (Handen *et al.* 2000).

All things considered, and all other things being equal, risperidone is probably a first choice. Importantly, however, if the patient also has epilepsy, then a trial of valproic acid is clearly in order. In planning the pharmacologic treatment of autism, it is important to adopt a methodical approach, using one agent at a time and giving each agent a 'good' trial before moving on to another: 'good' trials must involve not only adequate dosage but also adequate duration, keeping in mind that weeks or months may be required to see the effect of any given dose.

10.14 ATTENTION-DEFICIT–HYPERACTIVITY DISORDER

Attention-deficit–hyperactivity disorder (Biederman and Faraone 2005) is characterized by a variable mixture of hyperactivity, impulsivity, and inattentiveness. This is a common disorder, found in anywhere from 3 to 7 percent of school-age children, and occurs in males far more frequently than females, with reported sex ratios varying from 2:1 to 10:1 depending on the diagnostic criteria utilized. Synonyms for ADHD include minimal brain dysfunction, hyperkinetic syndrome, or, simply, hyperactivity.

Clinical features

Although symptoms are typically evident in toddlerhood, most patients do not come to clinical attention until kindergarten or early elementary school years. In infancy, one may see an unusual degree of fussiness and irritability, and during toddler years affected children may be constantly 'on the go' and forever 'into things', as if the 'motor' is never turned off. Preschoolers may be impulsive, and parents may find it almost impossible to impose discipline.

As these children enter kindergarten or elementary school, their behavior in the classroom generally brings them to attention. The clinical picture may be dominated by hyperactivity and impulsivity (in which case one speaks of ADHD 'predominantly hyperactive–impulsive type'), inattentiveness (wherein the term ADHD 'predominantly inattentive type' is used), or a combination of all three symptoms (in which case one speaks of ADHD 'combined type').

Hyperactivity may first come to attention in the classroom as these children appear incapable of remaining seated. They may squirm in their seats, fidget constantly, get up abruptly, or walk to another desk or over to a shelf. Teachers may be able to get them seated again, but the success is generally only short lived. These children may also be incessant talkers, and other students may be bothered by this. At home the hyperactivity persists, and children may restlessly go from room to room. Even favorite television shows may not be able to keep these children in one place, and asking about this point is often helpful in the attempt to make the diagnosis.

Impulsivity leads to multiple forms of 'thoughtless' behavior and may prompt teachers and parents to comment on an absence of any self-restraint. In the classroom, these children may blurt out answers even before the teacher has finished asking a question; in the lunch line they may barge ahead to get a favorite dessert; and in the playground they may rush uninvited into the middle of other children's games. At home they may be in constant conflict with siblings and neighborhood children, commandeering their toys and generally bursting into otherwise quiet activities.

Inattentiveness is typically most apparent in the classroom. These children seem incapable of paying attention to their schoolwork, and this is particularly the case whenever attention to detail is required. At times, it may appear as if it is not so much a matter of inattentiveness as easy distractibility. A noise in the hallway, a car passing by outside, or simply something on another child's desk may pull their attention away from the task at hand, as if everything had the same degree of importance as the schoolwork in front of them. Thus, incapable of giving their work the required attention, these children predictably get very poor grades. At home, parents may complain about difficulties in getting the attention of these children: sometimes it may have been necessary for parents to literally hold a child's head still in order to 'get the message through', and even with this tactic the effort may fail.

As noted below, in the natural course of events there is a gradual and spontaneous diminution of symptoms as these young patients enter adolescence and then adult years, with a consequent change in the overall clinical presentation.

In adolescence, although the grosser manifestations of hyperactivity tend to diminish and patients may be able to stay seated in the classroom, there may still be an undue amount of fidgetiness. Impulsivity may result in merely a few too many 'larks', such as 'joyrides' or unplanned drinking bouts; in more severe cases, an otherwise easily resolved confrontation with another teenager or with an authority figure may rapidly escalate into violence. Inattentiveness, abetted by normal teenage restlessness, may lead to poor grades, despite normal intelligence.

Among adults, hyperactivity tends to fade into the clinical background, being manifest in these years merely by restlessness; impulsivity likewise recedes and may fade into a mere flightiness. Inattentiveness, however, tends to persist to a significant degree, and patients may find themselves unable to advance in work situations that require sustained attention.

Course

In the natural course of events, ADHD undergoes a very gradual remission as patients pass through adolescence into adult years such that, by adult years, although roughly one-half of patients will have some symptoms, less than 5 percent will continue to have the full syndrome.

Etiology

Attention-deficit–hyperactivity disorder is clearly familial, with the concordance rate rising from first-degree relatives through dizygotic and up to monozygotic twins, and an adoption study (Sprich *et al.* 2000) lends support to an hereditary explanation for this increasing concordance. Although genetic studies have not as yet yielded definitive results, promising associations (Gizer *et al.* 2009) have been found with genes for dopamine receptors and dopamine transporters.

Structural MR imaging studies have demonstrated atrophy of the cerebral cortex (Shaw *et al.* 2009), basal ganglia (Qui *et al.* 2009), ventral striatum (Carmona *et al.* 2009), the basolateral complex of the amygdala (Plessen *et al.* 2006), the pulvinar (Ivanov *et al.* 2010), and the cerebellar vermis (Bledsoe *et al.* 2009). In contrast, hypertrophy has been noted in the head of the hippocampus (Plessen *et al.* 2006). Interestingly, and of great importance, treatment with stimulants was found to correct the atrophy found in the cerebral cortex (Shaw *et al.* 2009) and the cerebellar vermis (Bledsoe *et al.* 2009).

Taken together, these findings are consistent with the notion that ADHD occurs secondary to an inherited disturbance in the maturation of multiple areas of the brain, a disturbance which, potentially, may be corrected by treatment with stimulants.

Differential diagnosis

Attention-deficit–hyperactivity disorder may occur secondary to von Recklinghausen's disease, the fragile X syndrome, and the fetal alcohol syndrome; however, each of these disorders is readily identified on the basis of associated findings, which are discussed in the respective chapters. A similar syndrome may also occur secondary to treatment with phenobarbital. Autistic children may be quite hyperactive; however, here, in contrast with ADHD, one also sees the typical peculiar lack of relatedness with others, as if others were mere machines. This is an important differential to make, as a misdiagnosis here may lead to stimulant treatment, which may, in autistic patients, cause lethargy and social withdrawal. ADHD may also occur secondary to a rare condition of inherited resistance to thyroid hormones (Hauser *et al.* 1993), or as a sequela of traumatic brain injury (Max *et al.* 2005).

Mania is characterized by hyperactivity, impulsivity, and distractibility; however, attention to the mode and age of onset, and to the subsequent course, allow for a differential from ADHD. The onset of mania is acute or subacute, spanning from days to months, and rarely, if ever, occurs before the age of 10 years. By contrast, the onset of ADHD is very gradual, and is typically evident by the age of 7 years, if not much earlier. Further, and most importantly, mania occurs in episodes; this is in stark contrast to ADHD, which is chronic. Most episodes of mania last on the order of several weeks to several months, and although chronic mania, lasting years or decades, can occur, this is very rare. Thus, an episodic course, with spontaneous remission of symptoms, immediately favors mania and rules out ADHD.

Schizophrenia may, in some cases, be characterized by hyperactivity and impulsivity, but here attention to the age of onset and allied symptomatology immediately allows for a differential. Schizophrenia rarely has an onset before late childhood, in contrast, again, to ADHD, where the onset is typically much earlier. Further, in schizophrenia one sees delusions and hallucinations – symptoms never found in ADHD.

Depression may be characterized by agitation and inattentiveness, thus possibly engendering confusion with ADHD; however, in depression one finds that the agitation is accompanied by other 'vegetative' symptoms, such as anhedonia and changes in sleep and appetite – symptoms not seen in ADHD.

Diagnosing ADHD in patients who first present for evaluation as adults is often said to be quite difficult; however, at least in theory, it should not be. The differential diagnostic considerations include all of those just discussed, and if the diagnostician has the patient's history, from early childhood to the present, laid out, then the differential task is straightforward. The rub comes in acquiring this history, as this represents a major task, often involving interviewing not only the patient, but his or her parents or others who can provide such a history (Mannuzza *et al.* 2002).

Treatment

A large number of medications are useful in ADHD, including the following: stimulants and stimulant-like medications; certain antidepressants; lithium; and alpha-2 autoreceptor agonists. The effectiveness of these medications is reviewed below, and wherever possible, head-to-head comparator studies are included. Although the stimulants, especially methylphenidate, are the established first-line agents, other medications, including especially the antidepressant bupropion, clearly deserve attention.

The group of stimulants and stimulant-like medications include methylphenidate, mixed amphetamine salts, atomoxetine, modafinil, and dextroamphetamine. Mixed amphetamine salts were equivalent to methylphenidate (Pelham *et al.* 1999). In turn, mixed amphetamine salts were superior to atomoxetine (Wigal *et al.* 2005), and while the immediate-release preparation of methylphenidate was not superior to atomoxetine (Wang *et al.* 2007), the long-acting 'OROS' preparation *was* (Newcorn *et al.* 2008). Further, methylphenidate was superior to dextroamphetamine (Efron *et al.* 1997). All of the foregoing trials were conducted in children or adolescents. The other stimulant-like medication, modafinil (Taylor and Russo 2000), was shown, in adults, to be equivalent to dextroamphetamine. With these results in hand, it appears that either mixed amphetamine salts or the OROS preparation of methylphenidate are the top choices. Atomoxetine and modafinil represent a second tier, while the venerable dextroamphetamine represents a last choice.

Other stimulants available include dexmethylphenidate and lisdexamfetamine. Methylphenidate is a racemic preparation, and dexmethylphenidate is the dextro isomer of methylphenidate, which has been shown, in children, to be of equivalent efficacy with the racemic preparation (Muniz *et al.* 2008). Lisdexamfetamine (Adler *et al.* 2008) is a prodrug, which is converted to dextroamphetamine, and it has been shown to be superior to placebo. Given that neither of these preparations has been shown to be superior to methylphenidate, there appears, at present, to be little reason to use them.

Among the antidepressants, bupropion appears to be as effective as methylphenidate in both adults (Kuperman *et al.* 2001) and children and adolescents (Barrickman *et al.* 1995). Among the monoamine oxidase inhibitor (MAOI) drugs, studies in children and adolescents demonstrated that selegiline (Mohammadi *et al.* 2004) was equivalent to methylphenidate and that tranylcypromine (Zametkin *et al.* 1985) was equivalent to dextroamphetamine. Finally, among the tricyclics, both nortriptyline (Pringle *et al.* 2000) and desipramine (Spencer *et al.* 2002) were superior to placebo in children and adolescents. Given these results, the clear choice among the antidepressants appears to be bupropion: the MAOIs, given their dietary restrictions, are very difficult to use, and the tricyclics carry a substantial side-effect burden; furthermore, desipramine has been associated with sudden cardiac death.

In adults, lithium (Dorrego *et al.* 2002) was shown to be equivalent to methylphenidate.

Among the alpha-2 autoreceptor agonists, guanfacine (Taylor and Russo 2001), in adults, was equivalent to dextroamphetamine, while clonidine (Palumbo *et al.* 2008), in children, although more effective than placebo, was less effective than methylphenidate. Given these results, of the alpha-2 agonists, guanfacine is probably the better choice as it is generally better tolerated.

With the foregoing results in mind, it appears reasonable, barring certain complicating factors, to start with either methylphenidate or mixed amphetamine salts. The complicating factors which may dictate a change in strategy include substance abuse, depression, tic disorders, schizophrenia, and bipolar disorder.

Substance abusers, in general, should not be given stimulants, and in this group it is reasonable to begin with either bupropion or the stimulant-like medication atomoxetine, which has less abuse potential than the stimulants. When depression is present, it also makes sense to use bupropion in the hope of clearing both ADHD and depression with one medication. Tic disorders were once thought to constitute a relative contra-indication to stimulants given the possibility that tics may be exacerbated by these agents; however, recent research (discussed in Chapter 9.25) suggests that, in fact, patients with tic disorders generally do well with stimulants. In those cases, however, in which tics are exacerbated by stimulants, consideration may be given to an alpha-2 autoreceptor agonist in the hope of ameliorating both ADHD and tics. Schizophrenia constitutes a contra-indication to stimulant use as these agents exacerbate psychotic symptoms: here, either bupropion or lithium may be considered. Bipolar disorder likewise constitutes a contra-indication to stimulants given their propensity to precipitate mania, and here the obvious choice is lithium.

Regardless of which medication is chosen for children or adolescents, it is also appropriate to offer psychosocial treatments, such as parent training classes and behavior-modification programs for the classroom. It must be borne in mind, however, that medication, in particular stimulants, is more effective, typically far more effective, than psychosocial treatments.

10.15 DEVELOPMENTAL DYSPHASIA

Developmental dysphasia, also known as 'language disorder' or 'specific language impairment', is characterized by an incomplete acquisition of linguistic skills in children (Webster and Shevell 2004). In some cases, this results primarily in an expressive dysphasia, in which children, although capable of understanding what is said to them, have great difficulty in expressing their thoughts in speech. In other cases, in addition to this expressive deficit, there is also difficulty in understanding what others say, producing a clinical picture referred to as 'mixed receptive–expressive language disorder'.

Developmental dysphasia is seen in 2–4 percent of school-age children, and is two to four times more common in boys than in girls.

Clinical features

Depending on its severity, developmental dysphasia may come to clinical attention at any time from the age of 2 years up to early school years.

In the expressive form of developmental dysphasia, children, although able generally to understand what is said to them, have a greater or lesser degree of difficulty in expressing their thoughts in speech (Sato and Dreifuss 1973). In severe cases, vocabulary is restricted to simple words, and patients may be unable to speak in sentences. In milder cases, sentences may be possible, but are short, incomplete, and often telegraphic in form.

In the mixed receptive–expressive form, children, in addition to displaying the symptoms seen in the expressive form, also have a comprehension defect. Although simple, short declarative sentences may be understood, sentences containing dependent clauses or utilizing the passive voice may leave these children baffled (Bartak *et al.* 1975; Cohen *et al.* 1989; Paul *et al.* 1983). Further, and especially when they are anxious for any reason, these children may display a degree of incoherence when they speak.

Course

Although some degree of improvement may gradually occur during adolescence, for the most part this is a chronic disorder.

Etiology

Developmental dysphasia is familial and linkage has been found to loci on chromosomes 13, 16, and 19 (Bartlett *et al.* 2002; SLI Consortium 2004). In one well-studied family ('KE'), in which developmental dysphasia was inherited on a fully penetrant autosomal dominant basis, a mutation was found on chromosome 7 (Lai *et al.* 2001).

Magnetic resonance imaging has demonstrated both a reversed asymmetry of linguistic cortex (De Fosse *et al.* 2004) and, in a substantial minority of cases, polymicrogyria in the perisylvian region (Guerreiro *et al.* 2002).

One autopsy study demonstrated a dysplastic gyrus on the inferior surface of the left frontal cortex (Cohen *et al.* 1989).

Interictal epileptiform discharges on the EEG are probably common. Although they were seen on standard awake EEGs in only 6 percent of cases (Venkateswaran and Shevell 2008), standard sleep EEGs revealed a prevalence of 38 percent (Duvelleroy-Hommet *et al.* 1995), while ambulatory awake EEG demonstrated a prevalence of 50 percent (Picard *et al.* 1998).

Overall, it appears not improbable that developmental dysphasia occurs secondary to a genetically mediated disruption of normal neuronal migration to the perisylvian cortex, resulting in gyriform or dysplastic abnormalities. It is not clear whether the interictal epileptiform discharges are merely epiphenomenal, reflecting an unexpressed epileptogenic potential of dysplastic cortex, or are perhaps indicative of an epileptic process that, at least theoretically, could disturb normal development or function of the linguistic cortex.

Differential diagnosis

Deafness may simulate developmental dysphasia, and all children in whom this diagnosis is considered should have audiometry.

Severe deprivation may stunt language development; however, these children, in contrast to those with dysphasia, typically show rapid gains when placed in a linguistically stimulating environment.

Both autism and childhood-onset schizophrenia may be characterized by linguistic disturbances similar to those seen in the mixed form of developmental dysphasia; however, in these disorders other symptoms also occur (e.g. a lack of social relatedness in autism, or hallucinations and delusions in schizophrenia) that are not seen in dysphasia.

Acquired dysphasia may be seen with appropriately placed lesions, as discussed in Chapter 3.1, and may be distinguished from developmental dysphasia by its mode of onset or course. With infarction, the onset is abrupt, whereas with tumors, the course is characterized by a progressive worsening, which is in contrast to the more or less stable 'plateau' seen in developmental dysphasia. Acquired dysphasia may also be seen in the Laundau–Kleffner syndrome, but here one finds seizures. Finally, acquired dysphasia may also be seen as a side-effect of topiramate (Gross-Tsur and Shalev 2004).

Finally, there is the question of the 'late talkers', that is children whose language development, although delayed, eventually results in normal linguistic competence. At present, there is no method by which these children can be confidently distinguished from those whose development will eventually plateau at a lower than normal level, and consequently long-term clinical follow-up is required.

Treatment

Remedial education is critical and often helpful. Some practitioners recommend treatment with AEDs in cases with interictal epileptiform discharges; however, this is controversial and there are no controlled studies to support this practice.

10.16 DEVELOPMENTAL DYSLEXIA

In developmental dyslexia (Demonet *et al.* 2004), children, despite normal intelligence and adequate educational opportunity, have great difficulty in learning how to read. First described by Hinshelwood in 1896 (Hinshelwood 1896), this disorder is far more common in boys than in girls, and is seen in up to 4 percent of school-age children. Synonyms include reading disorder, specific reading disability, and congenital word blindness (Orton 1925).

Clinical features

Depending on its severity, developmental dyslexia may first come to light anywhere between the ages of 6 and 9 years as the child falls behind his or her peers in the acquisition of reading skills.

In attempting to read out loud, these young patients demonstrate a certain dysfluency as they stumble over certain words: they may skip words and go on to the next, or they may misread a word and say one that is different from that on the page. The errors in reading often involve omitting certain letters or supplying a different letter from the one printed; letter reversals often occur, such as 'd' for 'b', and in some cases, entire words are reversed, with the child, for example, saying 'pat' for the written word 'tap'. Reading comprehension is impaired, and after finally, and haltingly, reading a paragraph, the child may be unable to paraphrase it in his or her own words. In striking contrast, if the same paragraph is read out loud to the child, he or she may then be able to paraphrase it with little difficulty.

Writing is also impaired, and similar reversals may be seen. Thus, intending to write 'top', the child may write 'pot'. In some cases, entire sentences may be reversed, with the written words reading from right to left on the page. Of note, if the child is asked to take a look at what he or she has written and then asked if there are any errors, the answer is often 'no'.

Of interest, recent work has demonstrated that specific evoked-potential abnormalities in infants predict the appearance of dyslexia (Molfese 2000). This finding is of particular importance, in that it raises the possibility of identifying patients well before the actual onset of the disorder, thus enabling earlier, and perhaps more effective, therapy.

Course

Although there may be some spontaneous improvement over long periods of time, the overall natural course is marked by a chronic difficulty in reading.

Etiology

Developmental dyslexia is clearly familial, with a concordance rate of about 25 percent in dizygotic twins and of roughly 50 percent in monozygotic twins. Although genetic studies have not as yet identified specific mutations, loci have been identified on multiple chromosomes, including 1, 2, 3, 6, and 15 (Williams and O'Donovan 2006).

One autopsy study of four males revealed widespread cortical dysplasia with a marked concentration in the left perisylvian area (Galaburda *et al.* 1985). In females, although similar findings were noted, there was, in addition, widespread glial scarring (Humphreys *et al.* 1990). Furthermore, in males, dysplastic changes have also been identified in the medial geniculate body and the posterior lateral nucleus of the thalamus (Galaburda and Eidelberg 1982). Magnetic resonance imaging has also suggested a lack of normal cerebral asymmetry of the planum temporale (Hynd *et al.* 1990); however, not all studies agree on this (Rumsey *et al.* 1997).

Overall, these findings strongly suggest that developmental dyslexia occurs secondary to an inherited defect in neuronal migration to the left perisylvian area.

Differential diagnosis

Developmental dyslexia is conceived of as a more or less isolated defect; thus, the diagnosis is not made when there is a generalized cognitive deficit, as in mental retardation. Furthermore, in children who are of normal intelligence, one must rule out decreased visual acuity or a lack of adequate educational opportunity before considering the diagnosis.

Developmental dysphasia of the 'mixed receptive–expressive' form, like dyslexia, is also characterized by a comprehension deficit; however, in developmental dysphasia, this deficit is present both with regard to the spoken and written word, whereas in developmental dyslexia there is no difficulty in comprehending the spoken word, only the written one.

Acquired dyslexia in children is very rare, but may be seen with appropriately placed lesions, as discussed in Chapter 3.2. One clue to the correct diagnosis is found in the onset or course of the dyslexia: with infarctions, the onset is abrupt, and in tumors one finds a progressive worsening of the dyslexia rather than a stable 'plateau' of disability characteristic of developmental dyslexia.

Finally, there are rare cases of patients with periventricular nodular heterotopias who display a clinical picture very similar to that seen in developmental dyslexia (Chang *et al.* 2005); in such cases, however, the patients also had epilepsy, a finding not seen in developmental dyslexia.

Treatment

Remedial education is often strikingly effective.

10.17 DEVELOPMENTAL DYSGRAPHIA

Developmental dysgraphia (Deuel 1995; Gubbay and de Klerk 1995; Roeltgen and Tucker 1988), also known as developmental agraphia or 'disorder of written expression', is characterized by an impaired acquisition of the ability to write, despite normal intelligence and adequate educational opportunity. Although the precise prevalence of this disorder is not known, it is probably as common as developmental dyslexia, and, like dyslexia, it is probably more common in boys than girls.

Clinical features

This disorder usually become apparent in kindergarten or early elementary school years as the patient's ability to express him- or herself in writing falls behind that of his or her peers. In addition to misspellings, there may be deficits in the form of what is written, with poor paragraph organization, or, at the sentence level, poor grammar and punctuation and an overall 'telegraphic' style, with missing words. Penmanship may or may not be poor; at times the penmanship far outshines what is actually written. Importantly, and in stark contrast to their dysgraphia, these patients have no difficulty in reading and no difficulty in speaking.

Course

In the natural course of events, developmental dysgraphia appears to be chronic.

Etiology

Developmental dysgraphia is clearly familial (Schulte-Korne 2001; Stevenson *et al.* 1987), and a locus has been identified on chromosome 15 (Nothen *et al.* 1999).

Differential diagnosis

Acquired dysgraphia is very rare in childhood, but may be seen with appropriately placed lesions, as discussed in Chapter 3.3. Acquired cases are suggested by the mode of onset or course: with infarctions, the onset is abrupt, whereas with tumors, one would see a course marked by a progressive decline, in contrast to the stable 'plateau' of disability seen in developmental cases.

Isolated poor penmanship may be seen in developmental clumsiness or in disorders such as cerebral palsy.

Developmental dyslexia is distinguished by the presence of difficulty in reading, which is in contrast to developmental dysgraphia, where reading is unaffected. Mental retardation is distinguished by a global decrement in cognitive abilities.

Treatment

Remedial education is effective.

10.18 DEVELOPMENTAL DYSCALCULIA

Developmental dyscalculia (Shalev 2004) is characterized by a more or less complete inability, despite otherwise normal intelligence and adequate educational opportunity, to acquire arithmetic skills. This is a relatively common disorder, occurring in 5–6 percent of school-age children, and is seen with roughly equal frequency in boys and girls. Synonyms for this disorder include mathematics disorder, developmental arithmetic disorder, and specific disorder of arithmetic skills.

Clinical features

The disorder comes to light in early elementary school years, as the young patients fail to keep up with their peers in the acquisition of arithmetic skills. These skills, in ascending order of difficulty, include counting, adding, subtracting, multiplying, and dividing. In the most severe cases, the ability to count is never acquired, whereas, in the least severe cases, some ability to multiply and divide may be achieved.

Course

Roughly one-half of patients will experience considerable improvement by the early teenage years (Shalev *et al.* 2005).

Etiology

Although developmental dyscalculia is clearly familial (Shalev *et al.* 2001), the genetic basis for this is not clear. Although autopsy studies are lacking, MR imaging studies have demonstrated atrophy in both the left (Issacs *et al.* 2001) and the right (Rotzer *et al.* 2008) parietal cortex, and functional MR scanning has revealed decreased activation in the left parietal cortex (Kucian *et al.* 2006).

Differential diagnosis

Acquired dyscalculia in childhood is very rare; however, appropriately placed lesions, as discussed in Chapter 3.4, could present in such a fashion. Clues to this include both the mode of onset and the subsequent course. Dyscalculia secondary to infarction would be of acute onset, whereas with a tumor (as has been reported in one case with a left temporoparietal tumor in a child (Martins *et al.* 1999]), one would see a progressive decline in arithmetic skills, rather than the more or less stable 'plateau' characteristic of developmental disorders.

There are also cases which mimic the onset and course of developmental dyscalculia, but these can be distinguished from developmental dyscalculia by associated clinical features, as for example a characteristic facial dysmorphism in the fragile X syndrome (Shalev and Gross-Tsur 1993), petit mal seizures in petit mal epilepsy (Shalev and Gross-Tsur 1993), or by history, as in cases of extreme prematurity with very low birth weight (Issacs *et al.* 2001). Finally, mental retardation, wherein an inability to acquire arithmetic skills is common, is distinguished by the characteristic generalized cognitive deficits.

Treatment

Remedial education is effective.

10.19 DEVELOPMENTAL STUTTERING

Developmental stuttering occurs in roughly 1 percent of school-age children, and is about three times as common in boys than girls.

Clinical features

Stuttering typically first appears between the ages of 2 and 10 years, with a peak at the age of 5 years. The stuttering itself may occur with any word or, alternatively, only with certain syllables or letters. Typically, when stuttering occurs over syllables or letters, it is the first syllable or letter in a word that usually blocks speech; in this regard, 'b', 'd', 'k', 'p', and 't' are common precipitants. In attempting to speak, the patient feels 'blocked' and may appear to stumble over the word or sound, attempting to say it again and again, sometimes with increasingly explosive force. Often these attempts are accompanied by repetitive grimacing, blinking, hissing, or forceful thrusting of the head, arms, or even the trunk. After the sound or word is eventually spoken, there may be a veritable cascade of words, all correctly pronounced, until the next verbal stumbling block is encountered.

Stuttering is generally worse when patients are anxious, pressed for time, or speaking in front of a group. Interestingly, fluency may be improved or even restored if the patient reads a text, speaks in unison with others, or sings.

Course

Spontaneous remission occurs in about three-quarters of patients by early teenage years (Yairi and Ambrose 1999).

Etiology

Developmental stuttering is clearly familial, and twin studies have indicated that roughly two-thirds of cases occur on an inherited basis (Dworzynski *et al.* 2007; Felsenfeld *et al.* 2000). Linkage studies suggest loci on chromosomes 7 and 12 (Suresh *et al.* 2006), and a specific mutation has been discovered on chromosome 12 in one large extended family (Kang *et al.* 2010).

Although structural and functional imaging studies have not yielded entirely consistent results, some findings stand out. First, structural MR imaging has revealed both reduced gray matter volume in the left inferior frontal cortex (Chang *et al.* 2008) and polygyria along the left frontal operculum. Second, and as might be expected, functional imaging has also revealed decreased activation during speech in the left frontal operculum (Chang *et al.* 2009; Fox *et al.* 1996; Watkins *et al.* 2008; Wu *et al.* 1995).

Although speculative, these findings are consistent with the notion that developmental stuttering may occur secondary to an inherited disorder of neuronal migration involving the left frontal operculum.

Differential diagnosis

Acquired stuttering differs clinically from developmental stuttering in several respects: first, the stuttering is as likely to occur on the second, or later, letter or syllable, as on the first one; second, one fails to see the sort of forceful efforts characteristic of developmental stuttering; and finally, one does not see a 'cascade' of normally pronounced words after the block has been overcome. The various causes of acquired stuttering are discussed in Chapter 5.3.

Treatment

Speech therapy is generally effective. In resistant cases, consideration may be given to pharmacologic treatment. Various medications have been demonstrated to be effective in double-blind trials, including clomipramine (at a mean dose of 150 mg) (Gordon *et al.* 1995), haloperidol (at a mean dose of 3 mg) (Murray *et al.* 1977), olanzapine (in a range of from 2.5 to 5 mg) (Maguire *et al.* 2004), and risperidone (with best results seen at 0.5 mg; in some cases, higher doses were associated with a loss of effect) (Maguire *et al.* 2000). As there have been no head-to-head comparative studies of these agents, choosing among them requires clinical judgment, and in this regard the side-effects seen with the antipsychotics would argue for a trial of clomipramine first. In addition to these blinded studies, there is a very intriguing case report of a patient with developmental stuttering and epilepsy who quite unexpectedly experienced a rapid and complete remission of stuttering when treated with levetiracetam (Canevini *et al.* 2002).

REFERENCES

Adler LA, Goodman DW, Kollins SH *et al.* Double-blind, placebo-controlled study of the efficacy and safety of lisdexamfetamine dimesylate in adults with attention-deficit/hyperactivity disorder. *J Clin Psychiatry* 2008; **69**:1364–73.

Akefeldt A, Gillberg C. Behavior and personality characteristics of children and young adults with Prader–Willi syndrome: a controlled study. *J Am Acad Child Adolesc Psychiatry* 1999; **38**: 761–9.

Allen SM, Rice SN. Risperidone antagonism of self-mutilation in a Lesch–Nyhan patient. *Prog Neuropsychopharmacol Biol Psychiatry* 1996; **20**:793–800.

Alsen G, Gillberg IC, Lindblom R *et al.* Tuberous sclerosis in Western Sweden: a population study of cases with early childhood onset. *Arch Neurol* 1994; **51**:76–81.

Amiet C, Gourginkel-An I, Bouzamondo A *et al.* Epilepsy in autism is associated with intellectual disability and gender: evidence from a meta-analysis. *Biol Psychiatry* 2008; **64**:577–82.

Amir RE, Van den Veyver IB, Wan M *et al.* Rett syndrome is caused by mutations in X-linked MECP2, encoding methyl-CpG-binding protein 2. *Nat Genet* 1999; **23**:185–8.

Amir RE, Van den Veyver IB, Schultz R *et al.* Influence of mutation type and X chromosome inactivation on Rett syndrome phenotypes. *Ann Neurol* 2000; **47**:670–9.

Anderson LT, Ernst M. Self-injury in Lesch–Nyhan disease. *J Autism Dev Disord* 1994; **24**:67–81.

Antshel KM, Aneja A, Strunge L *et al.* Autistic spectrum disorders in velo-cardio facial syndrome (22q11.2 deletion). *J Autism Dev Disord* 2007; **37**:1776–86.

Aragon AS, Coriale G, Fiorentino D *et al.* Neuropsychological characteristics of Italian children with fetal alcohol spectrum disorders. *Alcohol Clin Exp Res* 2008; **32**:1909–19.

Ardinger HH, Atkin JF, Blackston RD *et al.* Verification of the fetal valproate syndrome phenotype. *Am J Med Genet* 1988; **29**:171–85.

Armstrong DD, Dunn K, Antalffy B. Decreased dendritic branching in frontal, motor and limbic cortex in Rett syndrome compared with trisomy 21. *J Neuropathol Exp Neurol* 1998; **57**: 1013–17.

Arnold LE, Aan MG, Cook AM *et al.* Atomoxetine for hyperactivity in autism spectrum disorders: placebo–controlled crossover pilot trial. *J Am Acad Child Adolesc Psychiatry* 2006; **45**:1196–205.

Arzimanoglou AA, Andermann F, Aicardi J *et al.* Sturge–Weber syndrome. Indications and results of surgery in 20 patients. *Neurology* 2000; **55**:1472–9.

Au KS, Williams AT, Roach ES *et al.* Genotype/phenotype correlation in 325 individuals referred for a diagnosis of tuberous sclerosis complex in the United States. *Genet Med* 2007; **9**: 88–100.

Auranen M, Vanhala R, Vosman M *et al.* MECP2 gene analysis in classical Rett syndrome and in patients with Rett-like features. *Neurology* 2001; **56**:611–17.

Aydin A, Cakmakci H, Kovanlikaya A *et al.* Sturge–Weber syndrome without facial nevus. *Pediatr Neurol* 2000; **22**:400–2.

Bailey A, le Couteur A, Gottesmann I *et al.* Autism as a strongly genetic disorder: evidence from a British twin study. *Psychol Med* 1995; **25**:63–77.

Barrickman LL, Perry PJ, Allen AJ *et al.* Bupropion versus methylphenidate in the treatment of attention-deficit hyperactivity disorder. *J Am Acad Child Adolesc Psychiatry* 1995; **34**:649–57.

Bartak L, Rutter M, Cox A. A comparative study of infantile autism and specific developmental receptive language disorder. I. The children. *Br J Psychiatry* 1975; **126**:127–45.

Bartlett CW, Flax JF, Logue MW *et al.* A major susceptibility locus for specific language impairment is located on 13q21. *Am J Hum Genet* 2002; **71**:45–55.

Bassett AS, Chow EWC, Abdelmalik P *et al.* The schizophrenia phenotype in 22q11 deletion syndrome. *Am J Psychiatry* 2003; **160**:1580–6.

Bauman ML, Kemper TL, Arin DM. Pervasive neuroanatomic abnormalities of the brain in three cases of Rett's syndrome. *Neurology* 1995; **45**:1581–6.

Baumgardner TL, Reiss AL, Freund LS *et al.* Specification of the neurobehavioral phenotype in males with fragile X syndrome. *Pediatrics* 1995; **95**:744–52.

Beales PL, Warner AM, Hitman GA *et al.* Bardet-Biedl syndrome: a molecular and phenotypic study of 18 families. *J Med Genet* 1997; **34**:92–8.

Bebin EM, Gomez MR. Prognosis in Sturge–Weber disease: comparison of unihemispheric and bihemispheric involvement. *J Child Neurology* 1988; **3**:181–4.

Belichenko PV, Hagberg B, Dahlstrom A. Morphological study of neocortical areas in Rett syndrome. *Acta Neuropathol* 1997; **93**:50–61.

Bell SH, Stade B, Reynolds JH *et al.* The remarkably high prevalence of epilepsy and seizure history in fetal alcohol spectrum disorders. *Alcohol Clin Exp Res* 2010; **34**:1084–9.

Bender F, Fry E, Pennington B *et al.* Speech and language development in 41 children with sex chromosome abnormalities. *Pediatrics* 1983; **71**:262–7.

Benedikt RA, Brown DC, Walker R *et al.* Sturge–Weber syndrome; cranial MR with Gd-DTPA. *AJNR* 1993; **14**:409–15.

Biederman J, Faraone SV. Attention-deficit hyperactivity disorder. *Lancet* 2005; **366**:237–48.

Bish JP, Nguyen V, Ding L *et al.* Thalamic reductions in children with chromosome 22q11.2 deletion syndrome. *Neuroreport* 2004; **15**:1413–15.

Bledsoe J, Semrud-Clikeman M, Pliszka SR. A magnetic resonance imaging study of the cerebellar vermis in chronically treated and treatment-naive children with attention-deficit/hyperactivity disorder combined type. *Biol Psychiatry* 2009; **65**:620–4.

Boer H, Holland A, Whittington J *et al.* Psychotic illness in people with Prader Willi syndrome due to chromosome 15 maternal uniparental disomy. *Lancet* 2002; **359**:135–6.

Bolton PF, Park RJ, Higgins JN *et al.* Neuro-epileptic determinants of autism spectrum disorders in tuberous sclerosis complex. *Brain* 2002; **125**:1247–55.

Bray GA, Dahms WT, Swerdloff RS *et al.* The Prader–Willi syndrome: a study of 40 patients and a review of the literature. *Medicine* 1983; **62**:59–88.

Burke CM, Kousseff BG, Gleeson M *et al.* Familial Prader–Willi syndrome. *Arch Int Med* 1987; **147**:673–5.

Butler MG, Meany FJ, Palmer CG. Clinical and cytogenetic survey of 39 individuals with Prader–Labhart–Willi syndrome. *Am J Med Genet* 1986; **23**:793–809.

Bye AM, Vines R, Fronsek K. The obesity hypoventilation syndrome and the Prader–Willi syndrome. *Aust Paediatr J* 1983; **19**:251–5.

Campbell LE, Daly E, Toal F *et al.* Brain and behavior in children with 22q11.2 deletion syndrome: a volumetric and voxel-based morphometry MRI study. *Brain* 2006; **129**:1218–28.

Canevini MP, Chifari R, Piazzini A. Improvement of a patient with stuttering on levetiracetam. *Neurology* 2002; **59**:1288.

Carmona S, Proal E, Hoekzema EA *et al.* Ventro-striatal reductions underpin symptoms of hyperactivity and impulsivity in attention-deficit/hyperactivity disorder. *Biol Psychiatry* 2009; **66**:972–7.

Carter JC, Capone GT, Gray RM *et al.* Autistic-spectrum disorders in Down syndrome: further delineation and distinction from other behavioral abnormalities. *Am J Med Genet B Neuropsychiatr Genet* 2007; **144B**:87–94.

Carter P, Downs J, Bebbington A *et al.* Stereotypical hand movements in 144 subjects with Rett syndrome from the population-based Australian database. *Mov Disord* 2010; **25**:282–8.

Casanova MF, El-BAz A, Vanbogaert E *et al.* A topographic study of minicolumnar core width by lamina comparison between autistic subjects and controls: possible minicolumnar disruption due to an anatomical element in common to multiple laminae. *Brain Pathol* 2009; **20**:451–8.

Chang BS, Ly J, Appignani B, Bodell A *et al.* Reading impairment in the neuronal migration disorder of periventricular nodular heterotopias. *Neurology* 2005; **64**:799–803.

Chang SE, Erickson KI, Ambrose NG *et al.* Brain anatomy differences in childhood stuttering. *Neuroimage* 2008; **39**:1333–44.

Chang SE, Kenney MK, Loucks TM *et al.* Brain activation abnormalities during speech and non-speech in stuttering speakers. *Neuroimage* 2009; **46**:201–12.

Chao DHC. Congenital neurocutaneous syndromes of childhood. III. Sturge–Weber disease. *J Pediatr* 1959; **55**:635–49.

Chapleau CA, Clafa GD, Lane MC *et al.* Dendritic spine pathologies in hippocampal pyramidal neurons from Rett syndrome brain and after expression of Rett-associated MECP2 mutations. *Neurobiol Dis* 2009; **35**:219–33.

Christie R, Bay C, Kaufman IA *et al.* Lesch–Nyhan disease: clinical experience with nineteen patients. *Dev Med Child Neurol* 1982; **24**:293–306.

Chu-Shore CJ, Major P, Camposano S *et al.* The natural history of epilepsy in tuberous sclerosis complex. *Epilepsia* 2009; **51**:1236–41.

Chudley AE, Hagerman RJ. Fragile X syndrome. *J Pediatrics* 1987; **110**:821–31.

Cif L, Biolsi B, Gavarini S *et al.* Antero-ventral internal pallidum stimulation improves behavioural disorders in Lesch–Nyhan disease. *Mov Disord* 2007; **22**:2126–9.

Clarke DJ. Prader–Willi syndrome and psychoses. *Br J Psychiatry* 1993; **163**:680–4.

Clarke DJ, Baer H, Whittington J *et al.* Prader–Willi syndrome, compulsive and ritualistic behaviors: the first population-based survey. *Br J Psychiatry* 2002; **180**:358–62.

Clarren SK, Smith DW. The fetal alcohol syndrome. *N Engl J Med* 1978; **298**:1063–7.

Clarren SK, Alvord EC, Sumi SM *et al.* Brain malformations related to prenatal exposure to alcohol. *J Pediatrics* 1978; **92**:64–7.

Clift S, Dahlitz M, Parkes JD. Sleep apnoea in the Prader–Willi syndrome. *J Sleep Res* 1994; **3**:121–6.

Cohen M, Campbell R, Yaghmai F. Neuropathological abnormalities in developmental dysphasia. *Ann Neurol* 1989; **25**:546–70.

Coleman M, Brubaker J, Hunter K *et al.* Rett syndrome: a survey of North American patients. *J Ment Def Res* 1988; **32**:117–24.

Collacott RA, Cooper S-A, McGrother C. Differential rates of psychiatric disorder in adults with Down's syndrome compared with other mentally handicapped adults. *Br J Psychiatry* 1992; **161**:671–4.

Collacott RA, Cooper SA, Ismail IA. Multi-infarct dementia in Down's syndrome. *J Intellect Disabil Res* 1994; **38**:203–8.

Creange A, Zeller J, Rostaing-Rigattieri S *et al.* Neurological complications of neurofibromatosis type 1 in adulthood. *Brain* 1999; **122**:473–81.

Critchley M, Earl CJ. Tuberous sclerosis and allied conditions. *Brain* 1932; **55**:311–46.

Danielsson S, Gillberg C, Billstedt E *et al.* Epilepsy in young adults with autism: a prospective population based follow-up study of 100 individuals diagnosed in childhood. *Epilepsia* 2005; **46**: 918–23.

Davidson BL, Tarle SA, Patella TD *et al.* Molecular basis of hypoxanthine-guanine phosphoribosyltransferase deficiency in ten subjects determined by direct sequencing of amplified transcripts. *J Clin Invest* 1989; **84**:342–6.

Dayer AG, Bottani A, Bouchardy I *et al.* MECP2 mutant allele in a boy with Rett syndrome and his unaffected hemizygous mother. *Brain Dev* 2007; **29**:47–50.

De Arce MA, Kearns A. The fragile X syndrome: the patients and their chromosomes. *J Med Genet* 1984; **21**:84–91.

De Borchgrave V, Saussu F, Depre A *et al.* Moyamoya disease and Down syndrome: case report and review of the literature. *Acta Neurol Belg* 2002; **102**:63–6.

De Fosse L, Hodge SM, Makris N *et al.* Language-association cortex asymmetry in autism and specific language impairment. *Ann Neurol* 2004; **56**:757–66.

DeLisi LE, Maurizo AM, Svetina C *et al.* Klinefelter's syndrome (XXY) as a genetic model for psychotic disorders. *Am J Med Genet B Neuropsychiatr Genet* 2005; **135B**:15–23.

Demonet JF, Taylor MJ, Chaix Y. Developmental dyslexia. *Lancet* 2004; **364**:247–8.

Deuel RK. Developmental dysgraphia and motor skills disorders. *J Child Neurol* 1995; **10**(Suppl. 1):6–8.

Devlin LA, Shepherd CH, Crawford H *et al.* Tuberous sclerosis complex: clinical features, diagnosis, and prevalence within Northern Ireland. *Dev Med Child Neurol* 2006; **48**:495–9.

DiMario FJ, Ramsby G. Magnetic resonance imaging lesion analysis in neurofibromatosis type 1. *Arch Neurol* 1998; **55**:500–5.

DiPaolo DP, Zimmerman RA, Rorke LB *et al.* Neurofibromatosis type 1: pathologic substrate of high-signal intensity in the brain. *Radiology* 1995; **195**:721–4.

Dorrego MF, Canevaro L, Kuzia G *et al.* A randomized, double-blind, crossover study of methylphenidate and lithium in adults with attention-deficit/hyperactivity disorder: preliminary findings. *J Neuropsychiatry Clin Neurosci* 2002; **14**:289–95.

Dotti MT, Orrico A, De Stefano N *et al.* A Rett syndrome MECP2 mutation that causes mental retardation in men. *Neurology* 2002; **58**:226–30.

Down JLH. Observation on an ethnic classification of idiots. *Clin Lect Rep Lond Hosp* 1866; **3**:259–62.

Dunn HG. The Prader–Labhart–Willi syndrome: a review of the literature and report of nine cases. *Acta Paediatr Scand Suppl* 1968; **186**:1–38.

Durst R, Rubin-Jabotinsky K, Raskin S *et al.* Risperidone in treating behavioral disturbances of Prader–Willi syndrome. *Acta Psychiatr Scand* 2000; **102**:461–5.

Duvelleroy-Hommet C, Billard C *et al.* Sleep EEG and developmental dysphasia: lack of a consistent relationship with paroxysmal EEG activity during sleep. *Neuropediatrics* 1995; **26**:14–18.

Dworzynski K, Remington A, Rijsdijk F *et al.* Genetic etiology inc cases of recovered and persistent stuttering in an unselected, longitudinal sample of young twins. *Am J Speech Lang Pathol* 2007; **16**:169–78.

Dykens EM, Hodapp RM, Ort S *et al.* The trajectory of cognitive development in males with fragile X syndrome. *J Am Acad Child Adolesc Psychiatry* 1989; **28**:422–6.

Edwards A, Voss H, Rice P *et al.* Automated DNA sequencing of the human hprt locus. *Genomics* 1990; **6**:593–608.

Efron D, Jarman F, Barker M. Methylphenidate versus dexamphetamine in children with attention deficit hyperactivity disorder: a double-blind, crossover trial. *Pediatrics* 1997; **100**:E6.

Eliez S, Schmitt JE, White CD *et al.* Children and adolescents with velocardiofacial syndrome: a volumetric MRI study. *Am J Psychiatry* 2000; **157**:409–15.

Ernst M, Zametkin AJ, Matochil JA *et al.* Presynaptic dopaminergic deficits in Lesch–Nyhan disease. *N Engl J Med* 1996; **334**: 1568–72.

Evenhuius HM. The natural history in Down's syndrome. *Arch Neurol* 1990; **47**:263–7.

Everman DB, Stoudemire A. Bipolar disorder associated with Klinefelter's syndrome and other chromosomal abnormalities. *Psychosomatics* 1994; **35**:35–40.

Fankenhauser MP, Karumanchi VC, German ML *et al.* A double-blind, placebo-controlled study of the efficacy of transdermal clondine in autism. *J Clin Psychiatry* 1992; **53**:77–82.

Feldmann R, Denecke J, Grenzebach M *et al.* Neurofibromatosis type 1. Motor and cognitive function and T2-weighted MRI hyperintensities. *Neurology* 2003; **61**:1725–8.

Felman HM, Kolmen BK, Gonzaga AM. Naltrexone and communication skills in young children with autism. *J Am Acad Child Adolesc Psychiatry* 1999; **38**:587–93.

Felsenfeld S, Kirk KM, Zhu G *et al.* A study of the genetic and environmental etiology of stuttering in a selected twin sample. *Behav Genet* 2000; **30**:359–66.

Ferrer I, Galofre E. Dendritic spine abnormalities in fetal alcohol syndrome. *Neuropediatrics* 1987; **18**:161–3.

Fine SE, Weissman A, Gerdes M *et al.* Autism spectrum disorders and symptoms in children with molecularly confirmed 22q11.2 deletion syndrome. *J Autism Dev Disord* 2005; **35**:461–70.

Finelli PF, Pueschel SM, Padre-Mendoza T *et al.* Neurological findings in patients with the fragile-X syndrome. *J Neurol Neurosurg Psychiatry* 1985; **48**:150–3.

Folstein S, Rutter M. Infantile autism: a genetic study of 21 twin pairs. *J Child Psychol Psychiatry* 1977; **18**:297–321.

Forsman H. The mental implications of sex chromosome aberrations. The Blake Marsh lecture for 1970. *Br J Psychiatry* 1970; **117**:353–63.

Fox PT, Ingham RJ, Ingham JC *et al.* A PET study of the neural systems of stuttering. *Nature* 1996; **382**:158–61.

Franz DN, Leonard J, Tudor C *et al.* Rapamycin causes regression of astrocytomas in tuberous sclerosis complex. *Ann Neurol* 2006; **59**:490–8.

Freund LS, Reiss AL, Hagerman R *et al.* Chromosome fragility and psychopathology in obligate female carriers of the fragile X chromosome. *Arch Gen Psychiatry* 1992; **49**:54–60.

Fronczek R, Lammers GJ, Balesar R *et al.* The number of hypothalamic hypocretin (orexin) neurons is not affected in Prader–Willi syndrome. *J Clin Endocrinol Metabol* 2005; **90**:5466–70.

Galaburda AM, Eidelberg D. Symmetry and asymmetry in the human posterior thalamus. II. Thalamic lesions in a case of developmental dyslexia. *Arch Neurol* 1982; **39**:333–6.

Galaburda AM, Sherman GF, Rosen GS *et al.* Developmental dyslexia: four consecutive patients with cortical abnormalities. *Ann Neurol* 1985; **18**:222–33.

Gibbs RA, Caskey CT. Identification and localization of mutations at the Lesch–Nyhan locus by ribonuclease A cleavage. *Science* 1987; **236**:303–5.

Giedd JN, Clasen LS, Wallace GL *et al.* XXY (Klinefelter syndrome): a pediatric quantitative brain magnetic resonance imaging case–control study. *Pediatrics* 2007; **119**:232–40.

Gillespie J, Jackson A. *MRI and CT of the Brain.* London: Arnold, 2000.

Gizer IR, Ficks C, Waldman ID. Candidate gene studies of ADHD: a meta-analytic review. *Hum Genet* 2009; **126**:51–90.

Glaze DG, Percy AK, Skinner S *et al.* Epilepsy and the natural history of Rett syndrome. *Neurology* 2010; **74**:909–12.

Glick N. Dramatic reduction in self-injury in Lesch–Nyhan disease following S-adenosylmethionine administration. *J Inhert Met Dis* 2006; **29**:687.

Goday-Arno A, Cerda-Esteva M, Flores-Le-Roux JA *et al.* Hyperthyroidism in a population with Down syndrome (DS). *Clin Endocrinol (Oxf)* 2009; **71**:110–14.

Goh S, Butler W, Thiele EA. Subependymal giant cell tumors in tuberous sclerosis complex. *Neurology* 2004a; **63**:1457–61.

Goh WH, Khong PL, Leung CS *et al.* T2-weighted hyperintensities (unidentified bright objects) in children with neurofibromatosis 1: their impact on cognitive function. *J Child Neurol* 2004; **19**:853–8.

Gordon CT, State RC, Nelson JE *et al.* A double-blind comparison of clomipramine, desipramine and placebo in the treatment of autistic disorder. *Arch Gen Psychiatry* 1993; **50**:441–7.

Gordon CT, Cotelingam GM, Stager S *et al.* A double-blind comparison of clomipramine and desipramine in the treatment of developmental stuttering. *J Clin Psychiatry* 1995; **56**:238–42.

Gothelf D, Presburger G, Levy D *et al.* Genetic, developmental, and physical factors associated with attention deficit hyperactivity disorder in patients with velocardiofacial syndrome. *Am J Med Genet B Neuropsychiatr Genet* 2004a; **126B**:116–21.

Gothelf D, Presburger G, Zohar AH *et al.* Obsessive–compulsive disorder in patients with velocardiofacial (22q11 deletion) syndrome. *Am J Med Genet B Neuropsychiatr Genet* 2004b; **126B**:99–105.

Gothelf D, Feinstein C, Thompson T *et al.* Risk factors for the emergence of psychotic disorders in adolescents with 22q11.2 deletion syndrome. *Am J Psychiatry* 2007; **164**:663–9.

Gothelf D, Furfaro JA, Hoeft F *et al.* Neuroanatomy of fragile X syndrome is associated with aberrant behavior and the fragile X mental retardation protein (FMRP). *Ann Neurol* 2008; **63**:40–51.

Graham DI, Lantos PL. *Greenfield's Neurology*, 6th edn. London: Arnold, 1996.

Graham JM, Bashir AS, Stark RE *et al.* Oral and written language abilities of XXY boys: implications for anticipatory guidance. *Pediatrics* 1988; **81**:795–806.

Green JS, Parfrey PS, Harnett JD *et al.* The cardinal features of Bardet-Biedl syndrome, a form of Laurence-Moon-Biedl syndrome. *N Engl J Med* 1989; **321**:1002–9.

Green T, Gothelf D, Glaser B *et al.* Psychiatric disorders and intellectual functioning throughout development in velocardiofacial (22q11.2 deletion) syndrome. *J Am Acad Child Adolesc Psychiatry* 2009; **48**:1060–8.

Greenswag LR. Adults with Prader–Willi syndrome: a survey of 232 cases. *Dev Med Child Neurol* 1987; **29**:145–52.

Gross-Tsur V, Shalev RS. Reversible language regression as an adverse effect of topiramate treatment in children. *Neurology* 2004; **62**:299–300.

Gubbay SS, de Klerk NH. A study and review of developmental dysgraphia in relation to acquired dysgraphia. *Brain Dev* 1995; **17**:1–8.

Guerreiro MM, Hage SR, Guimaraes CA *et al.* Developmental language disorder associated with polymicrogyria. *Neurology* 2002; **59**:245–50.

Guillamo J-S, Creange A, Kalifa C *et al.* Prognostic features of CNS tumors in neurofibromatosis 1 (NF1): a retrospective study of 104 patients. *Brain* 2003; **126**:152–60.

Gutierrez GC, Smalley SL, Tanguey PE. Autism in tuberous sclerosis complex. *J Autism Dev Disord* 1998; **28**:97–103.

Gutowski NJ, Murphy RP. Late onset epilepsy in undiagnosed tuberous sclerosis. *Postgrad Med J* 1992; **68**:970–1.

Hagberg BA. Rett syndrome: clinical peculiarities, diagnostic approach, and possible cause. *Pediatr Neurol* 1989; **5**:75–83.

Hagberg B, Aicardi J, Ramos O. A progressive syndrome of autism, ataxia and loss of purposeful hand use in girls: Rett's syndrome: report of 35 cases. *Ann Neurol* 1983; **14**:471–9.

Hagberg BA, Witt-Engerstrom I. Rett's syndrome: a suggested staging system for describing the impairment profile with increasing age toward adolescence. *Am J Med Genet* 1986; **24**(Suppl. 1):47–59.

Hagerman RJ, Jackson AW, Levitas A *et al.* An analysis of autism in fifty males with the fragile X syndrome. *Am J Med Genet* 1986; **23**:359–74.

Hagerman RJ, Murphy MA, Wittenberger MD. A controlled trial of stimulant medication in children with fragile X syndrome. *Am J Med Genet* 1988; **30**:377–92.

Hall BD, Smith DW. Prader–Willi syndrome. *J Pediatrics* 1972; **81**:286–93.

Handen BL, Johnson CR, Lubetsky M. Efficacy of methylphenidate among children with autism and symptoms of attention-deficit/hyperactivity disorder. *J Autism Dev Disord* 2000; **30**:245–55.

Hanson JW, Myrianthopoulos NC, Harvey MA *et al.* Risks to the offspring of women treated with hydantoin anticonvulsants, with emphasis on the fetal hydantoin syndrome. *J Pediatr* 1976; **89**:662–8.

Harlow TL, Gonzalez-Alegre P. High prevalence of reported tremor in Klinefeleter syndrome. *Parkinsonism Relat Disord* 2009; **15**:393–5.

Harris JC, Less RR, Jinnah HA *et al.* Craniocerebral magnetic resonance imaging measurement and findings in Lesch–Nyhan syndrome. *Arch Neurol* 1998; **55**:547–53.

Harris SW, Hessl D, Goodlin-Jones B *et al.* Autism profiles of males with fragile X syndrome. *Am J Ment Retard* 2008; **113**:427–38.

Hatanaka T, Higashino H, Woo M *et al.* Lesch–Nyhan syndrome with delayed onset of self-mutilation: hyperactivity of interneurons at the brainstem and blink reflex. *Acta Neurol Scand* 1990; **81**:184–7.

Hauser P, Zametkin AJ, Martinez P *et al.* Attention deficit-hyperactivity disorder in people with generalized resistance to thyroid hormone. *N Engl J Med* 1993; **328**:997–1001.

Hellings JA, Warnock JK. Self-injurious behavior and serotonin in Prader–Willi syndrome. *Psychopharmacol Bull* 1994; **30**:245–50.

Hinshelwood J. A case of dyslexia – a peculiar form of word blindness. *Lancet* 1896; **2**:1454.

Hinton VJ, Brown WT, Wisniewski K *et al.* Analysis of neocortex in three males with the fragile X syndrome. *Am J Med Genet* 1991; **41**:289–94.

Hiraiwa R, Maegaki Y, Oka A *et al.* Behavioral and psychiatric disorders in Prader–Willi syndrome: a population study in Japan. *Brain Dev* 2007; **29**:535–42.

Hollander E, Phillips A, Chaplin W *et al.* A placebo controlled cross-over trial of liquid fluoxetine on repetitive behaviors in childhood and adolescent autism. *Neuropsychopharmacology* 2005; **30**: 582–9.

Hollander E, Soorya L, Wasserman S *et al.* Divalproex sodium vs. placebo in the treatment of repetitive behaviors in autism spectrum disorders. *Int J Neuropsychopharmacol* 2006; **9**:209–13.

Hollander E, Chaplin W, Soorya L *et al.* Divalproex sodium vs placebo for the treatment of irritability in children and adolescents with autism spectrum disorders. *Neuropsychopharmacology* 2010; **35**:990–8.

Humphreys P, Kaufmann WE, Galaburda AM. Developmental dyslexia in women: neuropathological findings in three patients. *Ann Neurol* 1990; **28**:727–38.

Hunter H. A controlled study of the psychopathology and physical measurements of Klinefelter's syndrome. *Br J Psychiatry* 1969; **115**:443–8.

Huppke P, Maier EM, Warnke A *et al.* Very mild cases of Rett syndrome with skewed X inactivation. *J Med Genet* 2006; **43**:814–16.

Huppke P, Kohler K, Brockmann K *et al.* Treatment of epilepsy in Rett syndrome. *Eur J Paediatr Neurol* 2007; **11**:10–16.

Hurst LC, Mulhall DJ. Another calendar savant. *Br J Psychiatry* 1988; **152**:274–7.

Huson SM, Harper PS, Compston DAS. Von Recklinghausen neurofibromatosis: a clinical and population study in south-east Wales. *Brain* 1988; **111**:1355–81.

Hyman BT, West HL, Rebeck GW *et al.* Neuropathological changes in Down's syndrome hippocampal formation. Effect of age and apolipoprotein E genotype. *Arch Neurol* 1995; **52**:373–8.

Hyman SL, Gill DS, Shees EA *et al.* Natural history of cognitive deficits and their relationship to MRI T2-hyperintensities in NF1. *Neurology* 2003; **60**:1139–45.

Hyman SL, Shores A, North KN. The nature and frequency of cognitive deficits in children with neurofibromatosis type 1. *Neurology* 2005; **65**:1037–44.

Hyman SL, Gill DS, Shores EA *et al.* T2 hyperintensities in children with neurofibromatosis type 1 and their relationship to cognitive functioning. *J Neurol Neurosurg Psychiatry* 2007; **78**:1085–91.

Hynd GW, Semrud-Clikeman M, Lorys AR *et al.* Brain morphology in developmental dyslexia and attention deficit disorder/hyperactivity. *Arch Neurol* 1990; **47**:919–26.

Issacs EB, Edmonds CJ, Lucas A *et al.* Calculation difficulties in children of very low birthweight: a neural correlate. *Brain* 2001; **124**:1681–2.

Itti E, Gaw Gonsalo IT, Pawlikowska-Haddel A *et al.* The structural brain correlates of cognitive deficits in adults with Klinefelter's syndrome. *J Endocrinol Metabol* 2006; **91**:1423–7.

Ivanov I, Bansal R, Hao X *et al.* Morphological abnormalities of the thalamus in youths with attention deficit hyperactivity disorder. *Am J Psychiatry* 2010; **167**:397–408.

Jankovic J, Caskey TC, Stout T *et al.* Lesch–Nyhan syndrome: a study of motor behavior and cerebrospinal fluid neurotransmitters. *Ann Neurol* 1988; **33**:466–9.

Jarrar RG, Buchhalter JR, Raffel C. Long-term outcome of epilepsy surgery in patients with tuberous sclerosis. *Neurology* 2004; **62**:479–81.

Jaselskis CA, Cook EH, Fletcher KE *et al.* Clonidine treatment of hyperactive and impulsive children with autistic disorder. *J Clin Psychopharmacol* 1992; **12**:322–7.

Jellinger K, Seitelberger F. Neuropathology of Rett syndrome. *Am J Med Genet* 1986; **1**(Suppl.):259–88.

Jellinger K, Armstrong D, Zoghbi HY *et al.* Neuropathology of Rett syndrome. *Acta Neuropath* 1988; **76**:142–58.

Jervis GA. Early senile dementia in mongoloid idiocy. *Am J Psychiatry* 1948; **105**:102–6.

Jinnah HA, Visser JE, Harris JC *et al.* Delineation of the motor disorder of Lesch–Nyhan disease. *Brain* 2006; **129**:1210–17.

Jinnah HA, Ceballos-Picot I, Torres RJ *et al.* Attenuated variants of Lesch–Nyhan disease. *Brain* 2010; **133**:671–89.

Joinson O, O'Callaghan FJ, Osborne JP *et al.* Learning disability and epilepsy in an epidemiological sample of individuals with tuberous sclerosis complex. *Psychol Med* 2003; **33**:335–44.

Jones KL, Smith DW. Recognition of fetal alcohol syndrome in early infancy. *Lancet* 1973; **2**:999–1001.

Jones KL, Smith DW, Streissguth AP *et al.* Outcome in offspring of chronic alcoholic women. *Lancet* 1974; **1**:1076–8.

Kang C, Riazuddin S, Mundorff J *et al.* Mutations in the lysosomal enzyme-targeting pathway and persistent stuttering. *New Engl J Med* 2010; **362**:677–85.

Kankirawatana P, Leonard H, Ellaway C *et al.* Early progressive encephalopathy in boys and MECP2 mutation. *Neurology* 2006; **67**:164–6.

Kanner L. Autistic disturbances of affective contact. *Nervous Child* 1943; **2**:217–50.

Kanner L. Early infantile autism. *J Pediatrics* 1944; **25**:211–17.

Kanner L. The conception of wholes and parts in early infantile autism. *Am J Psychiatry* 1951; **108**:23–6.

Kao A, Mariani J, McDonald-McGinn DM *et al.* Increased prevalence of seizures in patients with a 22q11.2 deletion. *Am J Med Genet A* 2004; **129**:29–34.

Kates WR, Miller AM, Abdulsabur N *et al.* Temporal lobe anatomy and psychiatric symptoms in velocardiofacial syndrome (22q11.2 deletion syndrome). *J Am Acad Child Adolesc Psychiatry* 2006; **45**:587–95.

Khalifa MM, Struthers JL. Klinefelter syndrome is a common cause for mental retardation of unknown etiology among prepubertal males. *Clin Genet* 2002; **61**:49–53.

King BH, Hollander E, Sikich L *et al.* Lack of efficacy of citalopram in children with autism spectrum disorders and high levels of repetitive behaviour: citalopram ineffective in children with autism. *Arch Gen Psychiatry* 2009; **66**:583–90.

Klein D, Ammann F. The syndrome of Laurence–Moon–Biedl–Bardet and allied diseases in Switzerland. *J Neurol Sci* 1969; **9**:479–513.

Kofman O, Hyland HH. Tuberous sclerosis in adults with normal intelligence. *Arch Neurol Psychiatry* 1959; **81**:557–90.

Konovalov HV, Kovetsky NS, Bobryshev YV *et al.* Disorders of brain development in the progeny of mothers who used alcohol during pregnancy. *Early Hum Dev* 1997; **25**:153–66.

Kucian K, Loenneker T, Dietrich T *et al.* Impaired neural networks for approximate calculation in dyscalculic children: a functional MRI study. *Behav Brain Funct* 2006; **2**:31.

Kulkantrakorn K, Geller TJ. Seizures in neurofibromatosis 1. *Pediatr Neurology* 1998; **19**:347–50.

Kuperman S, Perry PJ, Gaffney GR *et al.* Bupropion SR vs. methylphenidate vs. placebo for attention deficit hyperactivity disorder in adults. *Ann Clin Psychiatry* 2001; **13**:129–34.

Lagos JC, Gomez MR. Tuberos sclerosis: reappraisal of a clinical entity. *Mayo Clin Proc* 1967; **42**:26–49.

Lai CS, Fisher SE, Hurst JA *et al.* A forkhead-domain gene is mutated in severe speech and language disorder. *Nature* 2001; **413**:519–23.

Lai F, Williams RS. A prospective study of Alzheimer's disease in Down syndrome. *Arch Neurol* 1989; **46**:849–53.

Larsson G, Bohlin A-B, Tunell R. Prospective study of children exposed to variable amounts of alcohol in utero. *Arch Dis Child* 1985; **60**:316–21.

Lawlor BA, Maurer RG. Tuberous sclerosis and the autistic syndrome. *Br J Psychiatry* 1987; **150**:396–7.

Legius E, Descheemaeker MJ, Steyaert J *et al.* Neurofibromatosis type 1 in childhood: correlation of MRI findings with intelligence. *J Neurol Neurosurg Psychiatry* 1995; **59**:638–40.

Lesch M, Nyhan WL. A familial disorder of uric acid metabolism and central nervous system function. *Am J Med* 1964; **36**: 561–70.

Lichtenstein BW. Sturge–Weber–Dimitri syndrome: cephalic form of neurocutaneous hemangiomatosis. *Arch Neurol Psychiatry* 1954; **71**:291–301.

Listernick R, Charrow J, Greenwald MJ *et al.* Optic gliomas in children with neurofibromatosis type 1. *J Pediatr* 1989; **114**:788–92.

Lopes Ferraz Fiho JR, Munis MP, Soares Souza A *et al.* Unidentified bright objects on brain MRI in children as a diagnostic criterion for neurofibromatosis type 1. *Pediatr Radiol* 2008; **38**:305–10.

Lubs M-LE, Bauer MS, Formas ME *et al.* Lisch nodules in neurofibromatosis type 1. *N Engl J Med* 1991; **324**:1264–6.

Lund J, Munk-Jorgensen P. Psychiatric aspects of Down's syndrome. *Acta Psychiatr Scand* 1988; **78**:369–74.

McCracken JT, McGough J, Shah H *et al.* Risperidone in children with autism and serious behavioral problems. *N Engl J Med* 2002; **347**:314–21.

McDonald-McGinn DM, Tonnesen MK, Laufer-Cahana A *et al.* Phenotype of the 22q11.2 deletion in individuals identified through an affected relative: cast a wide FISHing net!. *Genet Med* 2001; **3**:23–9.

McDougle CJ, Naylor ST, Cohen DJ *et al.* A double-blind, placebo-controlled study of fluvoxamine in adults with autistic disorder. *Arch Gen Psychiatry* 1996; **53**:1001–8.

McDougle CJ, Holmes JP, Carlson DC *et al.* A double-blind, placebo-controlled study of risperidone in adults with autistic disorder and other pervasive developmental disorders. *Arch Gen Psychiatry* 1998; **55**:633–41.

McDougle CJ, Scahill L, Aman MG *et al.* Risperidone for the core symptom domains of autism: results from the study by the Autism Network of the Research Units on Pediatric Psychopharmacology. *Am J Psychiatry* 2005; **162**:1142–8.

McManaman J, Tam DA. Gabapentin for self-injurious behavior in Lesch–Nyhan syndrome. *Pediatr Neurology* 1999; **20**:381–2.

McVicker RW, Shanks OEP, McClelland RJ. Prevalence and associated features of epilepsy in adults with Down's syndrome. *Br J Psychiatry* 1994; **164**:528–32.

Madsen KM, Huiid A, Vestergaard M *et al.* A population-based study of measles, mumps, and rubella vaccination and autism. *N Engl J Med* 2002; **347**:1477–82.

Maguire GA, Riley GD, Franklin DL *et al.* Risperidone for the treatment of stuttering. *J Clin Psychopharmacol* 2000; **20**:479–82.

Maguire GA, Riley GD, Franklin DL *et al.* Olanzapine in the treatment of developmental stuttering: a double-blind, placebo-controlled trial. *Ann Clin Psychiatry* 2004; **16**:63–7.

Mak BS, Chi CS, Tsai CR *et al.* New mutations of the HPRT gene in Lesch–Nyhan syndrome. *Pediatr Neurol* 2000; **23**:332–5.

Mannuzza S, Klein RG, Klein DF *et al.* Accuracy of adult recall of childhood attention deficit hyperactivity disorder. *Am J Psychiatry* 2002; **159**:1882–8.

Margallo-Lana ML, Moore PB, Kay DW *et al.* Fifteen-year follow-up of 92 hospitalized adults with Down's syndrome: incidence of cognitive decline, its relationship to age and neuropathology. *J Intellect Disabil Res* 2007; **51**:463–77.

Maria BL, Neufeld JA, Rosainz LC *et al.* Central nervous system structure and function in Sturge–Weber syndrome: evidence of neurologic and radiologic progression. *J Child Neurol* 1998; **13**:606–18.

Martins IP, Ferreira J, Borges L. Acquired procedural dyscalculia associated to a left parietal lesion in a child. *Child Neuropsychol* 1999; **5**:265–73.

Mathews WS, Solan A, Barabas G. Cognitive functioning in Lesch–Nyhan syndrome. *Dev Med Child Neurol* 1995; **37**:715–22.

Max JE, Schacher RJ, Levin HS *et al.* Predictors of secondary attention-deficit/hyperactivity disorder in children and adolescents 6 to 24 months after traumatic brain injury. *J Am Acad Child Adolesc Psychiatry* 2005; **44**:1041–9.

Messiaen L, Yac S, Brems H *et al.* Clinical and mutational spectrum of neurofibromatosis type 1-like syndrome. *JAMA* 2009; **302**:2111–18.

Miller JL, Couch JA, Schmalfuss I *et al.* Intracranial abnormalities detected by three-dimensional magnetic resonance imaging in Prader–Willi syndrome. *Am J Med Genet A* 2007; **143**:476–83.

Mizuno T. Long-term follow-up of ten patients with Lesch–Nyhan syndrome. *Neuropediatrics* 1986; **17**:158–61.

Mohammadi MR, Ghanizadeh A, Alaghband-Rad J *et al.* Selegeline in comparison with methylphenidate in attention deficit hyperactivity disorder children and adolescents in a double-blind, randomized trial. *J Child Adolesc Psychopharamcol* 2004; **14**:418–25.

Molfese DL. Predicting dyslexia at 8 years of age using neonatal brain responses. *Brain Lang* 2000; **72**:238–45.

Moore SJ, Green JS, Fan Y *et al.* Clinical and genetic epidemiology of Bardet-Biedl syndrome in Newfoundland: a 22-year prospective, population-based, cohort study. *Am J Med Genet A* 2005; **132**:352–60.

Morimoto K, Mogami H. Sequential CT study of subependymal giant-cell astrocytoma associated with tuberous sclerosis. *J Neurosurg* 1986; **65**:874–7.

Mostofsky SH, Mazzocco MM, Aakalu G *et al.* Decreased cerebellar posterior vermis size in fragile X syndrome: correlation with neurocognitive performance. *Neurology* 1998; **50**:121–30.

Muniz R, Brams M, Mao A *et al.* Efficacy and safety of extended-release dexmethylphenidate compared with d,l-methylphenidate and placebo in the treatment of children with attention-deficit/ hyperactivity disorder: a 12-hour laboratory classroom study. *J Child Adolesc Psychopharmacol* 2008; **18**:248–56.

Murphy KC, Jones LA, Owen MJ. High rates of schizophrenia in adults with velo-cardio-facial syndrome. *Arch Gen Psychiatry* 1999; **56**:940–5.

Murray TJ, Kelly P, Campbell L *et al.* Haloperidol in the treatment of stuttering. *Br J Psychiatry* 1977; **130**:370–3.

Musumeci SA, Hagerman RJ, Ferri R *et al.* Epilepsy and EEG findings in males with fragile X syndrome. *Epilepsia* 1999; **40**:1092–9.

Newcorn JH, Kratochvil CJ, Allen AJ *et al.* Atomoxetine and osmotically released methylphenidate for the treatment of attention deficit hyperactivity disorder: acute comparison and differential response. *Am J Psychiatry* 2008; **165**:721–30.

Nielsen J. Klinefelter's syndrome and the XYY syndrome. *Acta Psychiatr Scand Suppl* 1969; **209**:1–353.

Nielsen J, Pelsen B, Sorensen K. Follow-up of 30 Klinefelter males treated with testosterone. *Clin Genet* 1988; **33**:262–9.

North K, Joy P, Yuille D *et al.* Specific learning disability in children with neurofibromatosis type 1: significance of MRI abnormalities. *Neurology* 1994; **44**:878–83.

Nothen MM, Schulte-Korne G, Grimm T *et al.* Genetic linkage analysis with dyslexia: evidence for linkage of spelling disability to chromosome 15. *Eur Child Adolesc Psychiatry* 1999; **8**(Suppl. 3):56–9.

Nyhan WL. Clinical features of the Lesch–Nyhan syndrome. *Arch Intern Med* 1972; **130**:186–92.

Nyunt O, Harris M, Hughes I *et al.* Benefit of early commencement of growth hormone therapy in children with Prader–Willi syndrome. *J Pediatr Endocrinolol Metabol* 2009; **22**:1151–8.

Ogasawara N, Stout JT, Goto H *et al.* Molecular analysis of a female Lesch–Nyhan patient. *J Clin Invest* 1989; **84**:1024–7.

Oldfors A, Sourander P, Armstrong DL *et al.* Rett syndrome: cerebellar pathology. *Pediatr Neurol* 1990; **6**:310–14.

Olsson B, Rett A. Autism and Rett syndrome: behavioural investigations and differential diagnosis. *Dev Med Child Neurol* 1987; **29**:429–41.

Olsson I, Steffenberg S, Billberg C. Epilepsy in autism and autistic-like conditions. *Arch Neurol* 1988; **45**:666–8.

Ornitz EM, Ritvo ER. Perceptual inconstancy in early infantile autism: the syndrome of early infant autism and its variants, including certain cases of childhood schizophrenia. *Arch Gen Psychiatry* 1968; **18**:76–8.

Orton ST. Word blindness in school children. *Arch Neurol Psychiatry* 1925; **14**:581–615.

Palumbo DR, Sallee FR, Pelham WE *et al.* Clonidine for attention-deficit/hyperactivity disorder: I. Efficacy and tolerability outcomes. *J Am Acad Child Adolesc Psychiatry* 2008; **47**:180–8.

Pampiglione G, Moynahan EJ. The tuberous sclerosis syndrome: clinical and EEG studies in 100 children. *J Neurol Neurosurg Psychiatry* 1976; **39**:666–73.

Papolos DF, Faedda GL, Veit S *et al.* Bipolar spectrum disorders in patients diagnosed with velo-cardio-facial syndrome: does a hemizygous deletion of chromosome 22q11 result in bipolar affective disorder? *Am J Psychiatry* 1996; **153**:1541–7.

Pascual-Castroviejo I, Diaz-Gonzales C, Garcia-Melian RM *et al.* Sturge–Weber syndrome: study of 40 patients. *Pediatr Neurology* 1993; **9**:283–8.

Pascual-Castroviejo I, Pascual-Pascual SI, Velazquez-Fragua R *et al.* Sturge–Weber syndrome: study of 55 patients. *Can J Neurol Sci* 2008; **35**:301–7.

Pasqualini RQ, Vidal G, Bur GE. Psychopathology of Klinefelter's syndrome. Review of thirty-one cases. *Lancet* 1957; **2**:164–7.

Patwardhan AJ, Eliez S, Bender B *et al.* Brain morphology in Klinefelter syndrome: extra X chromosome and testosterone supplementation. *Neurology* 2000; **54**:2218–23.

Paul R, Cohen DJ, Caparulo BK. A longitudinal study of patients with severe developmental disorders of language learning. *J Am Acad Child Psychiatry* 1983; **22**:525–34.

Pearson E, Lenn NJ, Cail WS. Moyamoya and other causes of stroke in patients with Down syndrome. *Pediatr Neurol* 1985; **1**:174–9.

Pelham WE, Gnagy EM, Chronis AM *et al.* A comparison of morning-only and morning/late afternoon Adderall to morning-only, twice-daily, and three times-daily methylphenidate in children with attention-deficit/hyperactivity disorder. *Pediatrics* 1999; **104**:1300–11.

Percy AK, Glaze DG, Schultz RJ *et al.* Rett syndrome: controlled study of oral opiate antagonist naltrexone. *Ann Neurol* 1994; **35**:464–70.

Petermann AF, Hayles AB, Dockerty MB *et al.* Encephalotrigeminal angiomatosis (Sturge–Weber disease). Clinical study of 35 cases. *J Am Med Assoc* 1958; **167**:2169–76.

Petersen CD, Luzzatti K. The role of chromosome translocation in the recurrence risk of Down's syndrome. *Pediatrics* 1965; **35**:463–9.

Pfeiffer J, Majewski F, Fischbach H *et al.* Alcohol embryopathy and fetopathy: neuropathology of three children and three fetuses. *J Neurol Sci* 1979; **41**:125–37.

Picard A, Cheliout Heraut F, Bouskaoui M *et al.* Sleep EEG and developmental dysphasia. *Dev Med Child Neurol* 1998; **40**:595–9.

Plessen KJ, Bansal R, Zhu H *et al.* Hippocampus and amygdala morphology in attention-deficit/hyperactivity disorder. *Arch Gen Psychiatry* 2006; **63**:795–807.

Pomeroy JC. Klinefelter's syndrome and schizophrenia. *Br J Psychiatry* 1980; **136**:597–9.

Popova G, Paterson WF, Brown A *et al.* Hashimoto's thyroiditis in Down's syndrome: clinical presentation and evolution. *Horm Res* 2008; **70**:278–84.

Prasher VP, Huxley A, Haque MS *et al.* A 24-week, double-blind, placebo-controlled trial of donepezil in patients with Down syndrome and Alzheimer's disease – pilot study. *Int J Geriatr Psychiatry* 2002; **17**:270–8.

Pringle JB, Wilens TE, Biederman J *et al.* A controlled study of nortriptyline in children and adolescents with attention deficit hyperactivity disorder. *J Child Adolesc Psychopharmacol* 2000; **10**:193–204.

Pueschel SM, Louis S, McKnight P. Seizure disorders in Down syndrome. *Arch Neurol* 1991; **48**:318–20.

Puri BK, Ho KW, Singh I. Age of seizure onset in adults with Down's syndrome. *Int J Clin Pract* 2001; **55**:442–4.

Qui A, Crocetti D, Adler M *et al.* Basal ganglia volume and shape in children with attention deficit hyperactivity disorder. *Am J Psychiatry* 2009; **166**:74–82.

Ratcliffe SG, Bancroft J, Axworthy D *et al.* Klinefelter's syndrome in adolescents. *Arch Dis Child* 1982; **57**:6–12.

Rathmell TK, Burns MA. The Laurence–Moon–Biedl syndrome occurring in a brother and a sister. *Arch Neurol Psychiatry* 1938; **39**:1022–42.

Reiss AL, Lee J, Freunc L. Neuroanatomy of fragile X syndrome: the temporal lobe. *Neurology* 1994; **44**:1317–24.

Renieri A, Mari F, Mencarelli MA *et al.* Diagnostic criteria for the Zapella variant of Rett syndrome (the preserved speech variant). *Brain Dev* 2009; **31**:208–16.

Research Units on Pediatric Psychopharmacology Autism Network. Risperidone in children with autism and serious behavioral problems. *N Engl J Med* 2002; **347**:314–21.

Research Units on Pediatric Psychopharmacology (RUPP) Autism Network. Randomized, controlled, crossover trial of methylphenidate in pervasive developmental disorder with hyperactivity. *Arch Gen Psychiatry* 2005; **62**:1266–74.

Rett A. Uber ein eigenartiges hirnatrophisches syndrom bei hyperammonamie im kindersalter. *Wien Med Wochenschr* 1966; **116**:723–6.

Richardson EP. Pathology of tuberous sclerosis. Neuropathologic aspects. *Ann NY Acad Sci* 1991; **615**:128–39.

Richer JM, Coss RG. Gaze aversion in autistic and normal children. *Acta Psychiatr Scand* 1976; **53**:193–210.

Ritvo ER, Freeman BJ, Scheibel AB *et al.* Lower Purkinje cell counts in the cerebella of four autistic subjects: initial findings of the UCLA-NSAC autopsy research report. *Am J Psychiatry* 1986; **143**: 862–6.

Ritvo ER, Jorde LB, Mason-Brothers A *et al.* The UCLA-University of Utah epidemiologic survey of autism: recurrence risk estimates and genetic counseling. *Am J Psychiatry* 1989; **146**:1032–6.

Roach ES, Delgado M, Anderson L *et al.* Carbamazepine trial for Lesch–Nyhan self-mutilation. *J Child Neurol* 1996; **11**:476–8.

Robinson WP, Bottani A, Yagang X. Molecular, cytogenetic and clinical investigations of Prader–Willi syndrome patients. *Am J Hum Genet* 1992; **49**:1219–34.

Rodriguez HA, Barthrong M. Multiple primary intracranial tumors in von Recklinghausen's neurofibromatosis. *Arch Neurol* 1966; **14**:467–75.

Roeltgen DP, Tucker DM. Developmental phonological and lexical agraphia in adults. *Brain Lang* 1988; **35**:287–300.

Roizen L, Gold G, Berman HH *et al.* Congenital vascular anomalies and their histopathology in Sturge–Weber–Dimitri syndrome (nevus flammeus with angiomatosis and encephalosis calcificans). *J Neuropathol Exp Neurol* 1972; **14**:173–88.

Rosman NP, Pearce J. The brain in multiple neurofibromatosis (von Recklinghausen's disease): a suggested basis for the associated mental defect. *Brain* 1967; **90**:829–37.

Ross MH, Galaburda AM, Kemper TL. Down's syndrome: is there a decreased population of neurons? *Neurology* 1984; **34**:909–16.

Roth AA. Familial eunuchoidism: the Laurence–Moon–Biedl syndrome. *J Urology* 1947; **57**:427–42.

Rotzer S, Kucian K, Martin E *et al.* Optimized voxel-based morphometry in children with developmental dyscalculia. *Neuroimage* 2008; **39**:417–22.

Rouleau GA, Wertelecki BR, Martuza RI *et al.* Genetic linkage of bilateral acoustic neurofibromatosis to a DNA marker on chromosome 22. *Nature* 1987; **329**:246–8.

Roy A. Schizophrenia and Klinefelter's syndrome. *Can J Psychiatry* 1981; **26**:262–4.

Rudelli RD, Brown WT, Wisniewski K *et al.* Adult fragile X syndrome: clinico-neuropathologic findings. *Acta Neuropathol* 1985; **67**:289–95.

Rumsey JM, Rapoport JL, Sceery WR. Autistic children as adults: psychiatric, social, and behavioral outcomes. *J Am Acad Child Psychiatry* 1985; **24**:465–73.

Rumsey JM, Donohue BC, Brady DR *et al.* A magnetic resonance imaging study of planum temporale asymmetry in men with developmental dyslexia. *Arch Neurol* 1997; **54**:1481–9.

Sabaratnam M, Vroegop PG, Gangadhran SK. Epilepsy and EEG findings in 18 males with fragile X syndrome. *Seizure* 2001; **10**:60–3.

Saito Y, Ito M, Hanaoka S *et al.* Dopamine receptor upregulation in Lesch–Nyhan syndrome: a postmortem study. *Neuropediatrics* 1999; **30**:66–71.

Sato S, Dreifuss FE. Electroencephalographic findings in a patient with developmental expressive aphasia. *Neurology* 1973; **23**:181–5.

Schaer M, Schmitt JE, Glaser B *et al.* Abnormal patterns of cortical gyrification in velo-cardio-facial syndrome (deletion 22q11.2): an MRI study. *Psychiatry Res* 2006; **146**:1–11.

Schechter R, Grether JK. Continuing increases in autism reported to California's developmental services system: mercury in retrograde. *Arch Gen Psychiatry* 2008; **65**:19–24.

Schmidt H, Pozza SB, Bonfig W *et al.* Successful early dietary intervention avoids obesity in patients with Prader–Willi syndrome: a ten-year follow-up. *J Pediatr Endocrinol Metabol* 2008; **21**:651–5.

Schochet SS, Lampert PW, McCormick WF. Neurofibrillary tangles in patients with Down's syndrome: a light and electron-microscopic study. *Acta Neuropathol* 1973; **23**:342–6.

Schulte-Korne G. Annotation: genetics of reading and spelling disorder. *J Child Psychol Psychiatry* 2001; **42**:985–97.

Schumann CM, Amaral DG. Stereological analysis of amygdala neuron number in autism. *J Neurosci* 2006; **26**:7676–9.

Selikowitz M, Sunman J, Pendergast A *et al.* Fenfluramine in Prader–Willi syndrome: a double blind, placebo controlled trial. *Arch Dis Child* 1990; **65**:112–14.

Serrano M, Perez-Duenas B, Ormazabal A *et al.* Levodopa therapy in a Lesch–Nyhan disease patient: pathological, biochemical, neuroimaging and therapeutic remarks. *Mov Disord* 2008; **23**:1297–300.

Shalev RS. Developmental dyscalculia. *J Child Neurol* 2004; **19**:765–71.

Shalev RS, Gross-Tsur V. Developmental dyscalculia and medical assessment. *J Learn Disabil* 1993; **26**:134–7.

Shalev RS, Manor O, Kerem B *et al.* Developmental dyscalculia is a familial learning disability. *J Learn Disabil* 2001; **34**:59–65.

Shalev RS, Manor O, Gross-Tsur V. Developmental dyscalculia: a prospective six-year follow-up. *Dev Med Child Neurol* 2005; **47**:121–5.

Shapira NA, Lessig MC, Lewis MH *et al.* Effects of Topiramate in adults with Prader–Willi syndrome. *Am J Ment Retard* 2004; **109**:301–9.

Shaw P, Sharp WS, Morrison M *et al.* Psychostimulant treatment and the developing cortex in attention deficit hyperactivity disorder. *Am J Psychiatry* 2009; **166**:58–63.

Shaywitz SE, Caparulo B, Hodgson ES. Developmental language disability as a result of prenatal exposure to ethanol. *Pediatrics* 1981; **68**:850–5.

Shea S, Tungay A, Carroll A *et al.* Risperidone in the treatment of disruptive behavioral symptoms in children with autistic and other pervasive developmental disorders. *Pediatrics* 2004; **114**:634–41.

Shprintzen RJ, Goldberg RB, Lewin ML *et al.* A new syndrome involving cleft palate, cardiac anomalies, typical facies, and learning disabilities: velo-cardio-facial syndrome. *Cleft Palate J* 1978; **15**:56–62.

SLI Consortium (SLIC). Highly significant linkage to the SLI1 locus in an expanded sample of individuals affected by specific language impairment. *Am J Hum Genet* 2004; **74**:1225–38.

Smathers SA, Wilson JG, Nigro MA. Topiramate effectiveness in Prader–Willi syndrome. *Pediatr Neurol* 2003; **28**:130–3.

Sobosky WE, Hull CE, Hagerman RJ. Symptoms of schizotypal personality disorder in fragile X women. *J Am Acad Child Adolesc Psychiatry* 1994; **33**:247–55.

Soni S, Whittington J, Hollard AJ *et al.* The phenomenology and diagnosis of psychiatric illness in people with Prader–Willi syndrome. *Psychol Med* 2008; **38**:1505–14.

Specchio N, Balestri M, Striano P *et al.* Efficacy of levetiracetam in the treatment of drug-resistant Rett syndrome. *Epilepsy Res* 2010; **88**:112–17.

Spencer T, Biederaman J, Coffey B *et al.* A double-blind comparison of desipramine and placebo in children and adolescents with chronic tic disorder and comorbid attention-deficit/hyperactivity disorder. *Arch Gen Psychiatry* 2002; **59**:649–56.

Spohr HL, Willms J, Steinhausen HC. Fetal alcohol spectrum disorders in young adulthood. *J Pediatr* 2007; **150**:175–9.

Sprich S, Biederman J, Crawford MH *et al.* Adoptive and biological families of children and adolescents with ADHD. *J Am Acad Child Adolesc Psychiatry* 2000; **39**:1432–7.

Stevenson J, Graham P, Fredman G *et al.* A twin study of genetic influences on reading and spelling ability and disability. *J Child Psychol Psychiatry* 1987; **28**:229–47.

Stoll C, Alembik Y, Dott B *et al.* Study of Down syndrome in 238,942 consecutive births. *Ann Genet* 1998; **41**:44–51.

Sturge WA. A case of partial epilepsy, apparently due to a lesion of one of the vaso-motor centres of the brain. *Trans Clin Soc London* 1879; **12**:162–7.

Suetsugu M, Mehraein P. Spine distribution along the apical dendrites of the pyramidal neurons in Down's syndrome. A quantitative Golgi study. *Acta Neuropathol* 1980; **50**:207–10.

Sujansky E, Conradi S. Sturge–Weber syndrome: age of onset of seizures and glaucoma and the prognosis for affected children. *J Child Neurol* 1995a; **10**:49–58.

Sujansky E, Conradi S. Outcome of Sturge–Weber syndrome in 52 adults. *Am J Med Genet* 1995b; **57**:35–45.

Sullivan K, Hatton D, Hammer J *et al.* ADHD symptoms in children with FXS. *Am J Med Genet A* 2006; **140**:2275–88.

Suresh R, Ambrose N, Roe C *et al.* New complexities in the genetics of stuttering: significant sex-specific linkage signals. *Am J Hum Genet* 2006; **78**:554–63.

Sutherland GR. Heritable fragile sites on human chromosomes. Demonstration of their dependence on the type of tissue culture medium. *Science* 1977; **197**:265.

Swaab DJ, Purba JS, Hofman MA. Alterations in the hypothalamic paraventricular nucleus and its xytocin neurons (putative satiety cells) in Prader–Willi syndrome: a study of five cases. *J Clin Endocrinol Metab* 1995; **80**:573–9.

Swillen A, Devriendt K, Legius E *et al.* Intelligence and psychosocial adjustment in velocardiofacial syndrome: a study of 37 children and adolescents with VCFS. *J Med Genet* 1997; **34**:453–8.

Taira T, Kobayashi T, Hori T. Disappearance of self-mutilating behavior in a patient with Lesch–Nyhan syndrome after bilateral chronic stimulation of the globus pallidus internus. *J Neurosurg* 2003; **98**:414–16.

Taly AB, Nagaraja D, Das S *et al.* Sturge–Weber–Dimitri disease without facial nevus. *Neurology* 1987; **37**:1063–4.

Tatum WO, Passaro EA, Elia M *et al.* Seizure in Klinefelter's syndrome. *Pediatr Neurology* 1998; **19**:275–82.

Taylor FB, Russo J. Efficacy of modafinil compared to dextroamphetamine for the treatment of attention deficit hyperactivity disorder in adults. *J Child Adolesc Psychopharmacol* 2000; **10**:311–20.

Taylor FB, Russo J. Comparing guanfacine and destroamphetamine for the treatment of adult attention-deficit/hyperactivity disorder. *J Clin Psychopharmacol* 2001; **21**:223–8.

Torrioli MG, Vernacotola S, Peruzzi L *et al.* A double-blind, parallel, multicenter comparison of L-acetylcarnitine with placebo on the attention deficit hyperactivity disorder in fragile X syndrome boys. *Am J Med Genet A* 2008; **146**:803–12.

Treffert D. The idiot savant: a review of the syndrome. *Am J Psychiatry* 1988; **145**:563–72.

Trois MS, Capone GT, Lutz JA *et al.* Obstructive sleep apnea in adults with Down syndrome. *J Clin Sleep Med* 2009; **5**:317–23.

Van Amelsvoort T, Daly E, Henry J *et al.* Brain anatomy in adults with velocardiofacial syndrome with and without schizophrenia: preliminary results of a structural magnetic resonance imaging study. *Arch Gen Psychiatry* 2004; **61**:1085–96.

Van Bogaert P, Ceballos I, Desguerre I *et al.* Lesch–Nyhan in a girl. *J Inherit Metabol Dis* 1992; **15**:790–1.

Van Kooten IA, Palmen SJ, von Cappein P *et al.* Neurons in the fusiform gyrus are fewer and smaller in autism. *Brain* 2008; **131**:987–99.

Van Rijn S, Aleman A, Swaab H *et al.* Klinefelter's syndrome (karyotype 47,XXY) and schizophrenia-spectrum pathology. *Br J Psychiatry* 2006; **189**:459–60.

van Rijn S, Swaab H, Aleman A *et al.* Social behavior and autism traits in a sex chromosomal disorder: Klinefeleter (47XXY) syndrome. *J Autism Dev Disord* 2008; **38**:1634–41.

Venkateswaran S, Shevell M. The case against routine electroencephalography in specific language impairment. *Pediatrics* 2008; **122**:911–16.

Verkerk AJ, Pieretti M, Sutcliffe JS *et al.* Identification of a gene (FMR1) containing a CGG repeat coincident with a breakage point cluster region exhibiting length variation in fragile X syndrome. *Cell* 1991; **65**:905–14.

Vogels A, De Hert M, Descheemaeker MJ *et al.* Psychotic disorders in Prader–Willi syndrome. *Am J Med Genet A* 2004; **127A**:238–43.

von Deimling A, Krone W, Menon AG. Neurofibromatosis type 1: pathology, clinical features and molecular genetics. *Brain Pathol* 1995; **5**:153–62.

Wakeling A. Comparative study of psychiatric patients with Klinefelter's syndrome and hypogonadism. *Psychol Med* 1972; **2**:139–54.

Wallace MR, Marchuk DA, Andersen LB *et al.* Type 1 neurofibromatosis gene: identification of a large transcript disrupted in three NF1 patients. *Science* 1990; **249**:181–6.

Wang PJ, Hou JW, Sue WC *et al.* Electroclinical characteristics of seizures- comparing Prader–Willi syndrome with Angelmann syndrome. *Brain Dev* 2005; **27**:101–7.

Wang Y, Zheng Y, Du Y *et al.* Atomoxetine versus methylphenidate in paediatric outpatients with attention deficit hyperactivity disorder: a randomized, double-blind comparison trial. *Aust N Z J Psychiatry* 2007; **41**:222–30.

Warren AC, Holroyd S, Folstein MF. Major depression in Down's syndrome. *Br J Psychiatry* 1989; **155**:202–5.

Watkins KE, Smith SM, Davis S *et al.* Structural and functional abnormalities of the motor system in developmental stuttering. *Brain* 2008; **131**:50–9.

Weber FP. A note on the association of extensive haemangiomatous naevus of the skin with cerebral (meningeal) haemangioma especially cases of facial vascular naevus with contralateral hemiplegia. *Proc R Soc Med* 1929; **22**:431–42.

Webster RI, Shevell MI. Neurobiology of specific language impairment. *J Child Neurol* 2004; **19**:471–81.

Wegiel J, Kuchna I, Nowicki K *et al.* The neuropathology of autism: defects in neurogenesis and neuronal migration, and dysplastic changes. *Acta Neuropathol* 2010; **119**:755–70.

Weig SG, Pollack P. Carbamazepine-induced heart block in a child with tuberous sclerosis and cardiac rhabdomyoma: implications for evaluation and follow-up. *Ann Neurol* 1993; **34**:617–19.

Wigal SB, McGough JJ, McCracken JT *et al.* A laboratory school comparison of mixed amphetamine salts extended release (Adderall XR) and atomoxatine (Strattera) in school-aged children with attention deficit/hyperactivity disorder. *J Attent Disord* 2005; **9**:275–89.

Willemsen-Swinkels SH, Buitelaar JK, van Engeland H. The effects of chronic naltrexone treatment in young autistic children: a double-blind placebo–controlled crossover study. *Biol Psychiatry* 1996; **39**:1023–31.

Williams J, O'Donovan MC. The genetics of developmental dyslexia. *Eur J Hum Genet* 2006; **14**:681–9.

Wisniewski K, Dambska M, Sher JH *et al.* A clinical neuropathological study of the fetal alcohol syndrome. *Neuropediatrics* 1983; **14**:197–201.

Wisniewski KE, Dalton AJ, McLachlan DRC *et al.* Alzheimer's disease in Down's syndrome: clinicopathologic studies. *Neurology* 1985a; **35**:957–61.

Wisniewski KE, Wisniewski HM, Wen GY. Occurrence of neuropathological changes and dementia of Alzheimer's disease in Down's syndrome. *Ann Neurol* 1985b; **17**:278–82.

Wisniewski KE, Frency JH, Fernando S *et al.* Fragile X syndrome: associated neurological abnormalities and developmental disabilities. *Ann Neurol* 1985c; **18**:665–9.

Witt-Engerstrom I. Rett syndrome in Sweden. Neurodevelopment – disability – pathophysiology. *Acta Pediatr ScandSuppl* 1990; **369**:1–60.

Witwer AN, Lecavalier L. Examining the validity of autism spectrum disorder subtypes. *J Autism Dev Disord* 2008; 1611–24.

Wohlwill FS, Yakovlev PI. Histopathology of meningo-facial angiomatosis (Sturge–Weber's disease). *J Neuropathol Exp Neurol* 1957; **16**:341–64.

Wong DF, Harris JC, Naidu S *et al.* Dopamine transporters are markedly reduced in Lesch–Nyhan disease in vivo. *Proc Natl Acad Sci* 1996; **93**:5539–43.

Wu JC, Maguire G, Riley G *et al.* A positron emission tomography (18F) deoxyglucose study of developmental stuttering. *Neuroreport* 1995; **15**:501–5.

Yairi E, Ambrose NG. Early childhood stuttering I: persistence and recovery rates. *J Speech Lang Hear Res* 1999; **42**: 1097–112.

Yakovlev PI, Corwin W. Roentgenographic sign in cases of tuberous sclerosis of the brain (multiple 'brain stones'). *Arch Neurol Psychiatry* 1939; **40**:1030–7.

Yang TP, Patel PI, Stout JT *et al.* Molecular evidence for new mutations in the HPRT locus in Lesch–Nyhan patients. *Nature* 1984; **310**:412–14.

Yee BJ, Buchanan PR, Mahadev S *et al.* Assessment of sleep and breathing in adults with Prader–Willi syndrome: a case control series. *J Clin Sleep Med* 2007; **3**:713–18.

Zametkin A, Rapoport JL, Murphy DL *et al.* Treatment of hyperactive children with monoamine oxidase inhibitors. I. Clinical efficacy. *Arch Gen Psychiatry* 1985; **42**:962–6.

Zappella M. A double blind trial of bromocriptine in the Rett syndrome. *Brain Dev* 1990; **12**:148–50.

Zellweger H, Schneider HJ. Syndrome of hypotonia–hypomentia–hypogonadism–obesity (HHHO) or Prader–Willi syndrome. *Am J Dis Child* 1968; **115**:588–98.

11

Vascular disorders

11.1 MULTI-INFARCT DEMENTIA

Multi-infarct dementia is the traditional name given to dementia occurring secondary to multiple, generally large, cortical infarctions; classically, in addition to the dementia, such patients also have focal signs, such as aphasia, and their history is characterized by a 'step-wise' course, with successive steps further down the cognitive ladder corresponding to successive large-vessel territorial infarctions. As will be noted below, however, variations on this classic picture do occur. Thus, conceived, multi-infarct dementia is a common cause of dementia in the elderly.

Before proceeding, some comments are in order regarding the term 'vascular dementia'. Vascular dementia is an umbrella term that includes not only multi-infarct dementia but also lacunar dementia (Chapter 11.2) and Binswanger's disease (Chapter 11.4). Lacunar dementia occurs secondary to multiple subcortical lacunar infarctions, whereas Binswanger's disease appears on the basis of a gradually progressive ischemic leukoencephalopathy. Unfortunately for those diagnosticians who yearn for precision, although 'pure' cases of each of these three types of vascular dementia do occur, it is more common to find elements of two or even all three of these at the same time in any given patient.

Clinical features

The onset of multi-infarct dementia corresponds to the age of greatest risk for stroke, and hence most patients are in their sixties or older. In most cases, one finds a history of multiple strokes preceding the onset of cognitive decline. Although in some cases the dementia may be rather non-specific with mere difficulties in memory, calculations, abstractions, and the like, in the majority, other symptoms are found (Cummings *et al.* 1987; Staekenborg *et al.* 2010) including, most commonly, mood changes such as depression, irritability, agitation, apathy, or, in a small proportion, euphoria. Hallucinations and delusions occur in a small minority: hallucinations are generally visual, and delusions tend to be either of persecution or misidentification. Focal signs, as noted, are common and may include aphasia, apraxia, agnosia, neglect, and hemiparesis or hemianesthesia; late in the course a pseudobulbar palsy may also appear, with, as discussed in Chapter 5.7, emotional incontinence. Seizures may also occur.

Neuroimaging, either with computed tomography (CT) or, preferably, magnetic resonance imaging (MRI), will reveal evidence of multiple old infarctions in, as discussed below, appropriately 'strategic' locations.

Course

As noted, the classic course of multi-infarct dementia is step-wise, with successive strokes bringing the patient down yet another step into further cognitive deterioration. Importantly, many of these steps are characterized by a delirium, which gradually resolves concurrent with the resolution of peri-lesional edema. Exceptions to this rule, however, do occur. In some cases, the course may be marked by one giant step down, as in cases of multiple simultaneous infarcts or with one infarct occurring in an exquisitely strategic location.

Etiology

In most cases, the dementia occurs secondary to bilateral multiple large-vessel territorial infarctions involving the frontal, parietal, and temporal cortices (Erkinjuntti *et al.* 1988; Jayakumar *et al.* 1989; Ladurna *et al.* 1982; Liu *et al.* 1992; Tomlinson *et al.* 1970); rarely, single infarctions, for example in the temporal or frontal lobes, may be responsible (Auchus *et al.* 2002; Yoshitake *et al.* 1995). Although most cases are due to infarction, as the name 'multi-infarct' suggests, a very similar clinical picture can emerge with multiple intracerebral hemorrhages, and it is probably appropriate to lump these cases under the rubric of multi-infarct dementia. A better name might be 'multi-stroke' dementia, but the term multi-infarct has great currency and probably will not change.

The multiple causes of these infarctions (and hemorrhages) are discussed in Chapter 8.4.

Differential diagnosis

The diagnosis of multi-infarct dementia should be considered in any patient with dementia and a history of stroke. In weighing this history, however, one must take into account the location of the lesion: whereas infarcts in such cognitively strategic locations as the frontal, parietal, or temporal cortices might be expected to cause dementia, one would be hardpressed to attribute a dementia to infarctions occurring in the occipital lobes.

Lacunar dementia may also present with a history of stroke; however, here the strokes tend to be of the lacunar variety, such as pure motor stroke. Furthermore, and in contrast to multi-infarct dementia, lacunar dementia tends to be characterized by a frontal lobe syndrome.

As noted earlier, it is not uncommon to find patients with more than one vascular process underlying a dementia, and in such cases MR scanning is generally necessary to determine the various contributions of cortical infarctions, lacunar infarctions, and diffuse white matter disease. In some instances, it may not be possible to disentangle the effects of each of these separate processes, and in such cases, one may have to be content with merely making a diagnosis of vascular dementia.

Given the prevalence of Alzheimer's disease, it must also be borne in mind that it is not at all uncommon to find patients with a 'mixed' dementia, that is with a combination of multi-infarct dementia and Alzheimer's disease (Tomlinson *et al.* 1970). Such a diagnosis should be considered in cases in which the course is mixed, being composed of sequential downward steps occurring on a background of a steady, gradual decline, and also in cases where the onset of the dementia is characterized by a relatively pure and slowly progressive amnestic deficit.

Treatment

The general treatment of dementia is discussed in Chapter 6.3. Antidepressants, for example a selective serotonin reuptake inhibitor (SSRI), may be needed for depressive symptoms, and antipsychotics, such as risperidone, for hallucinations, delusions, and agitation. There is also evidence that both the cholinesterase inhibitors donepezil (Black *et al.* 2003; Wilkinson *et al.* 2003) and galantamine (Auchus *et al.* 2007) as well as memantine (Orgogozo *et al.* 2002; Wilcock *et al.* 2002) may improve overall cognitive function; however, the improvement seen is generally not impressive.

Concurrent with symptomatic treatment, steps should be taken to prevent future strokes if possible, as discussed in Chapter 8.4.

11.2 LACUNAR DEMENTIA

As discussed in Chapter 8.4, lacunes are small cavities that may be found in the thalamus, basal ganglia, and internal and external capsules, among other locations, and which occur as sequelae to infarctions in the areas of distribution of central or perforating arteries. Single lacunes may be clinically 'silent' or present with one of the classic lacunar syndromes, such as pure motor stroke. When multiple lacunes are present, one speaks of the 'lacunar state', and when significant cognitive impairment occurs on the basis of multiple lacunes (or, albeit uncommonly, on the basis of a single 'strategically' located lacune), one may speak of a 'lacunar dementia'. This lacunar dementia is one of the vascular dementias. These dementias, in turn, may be divided into two large groups, depending on the location of the vascular disease. On the one hand, cortical infarctions may cause dementia, as in multi-infarct dementia, while on the other, dementia may occur as a result of subcortical disease (Erkinjuntti *et al.* 2000), producing the two subcortical vascular dementias, namely lacunar dementia and Binswanger's disease. Although lacunar dementia may occur in a pure form, it is most commonly found in association with Binswanger's disease.

Although the prevalence of lacunar dementia is not known with any precision, the clinical impression is that it is not uncommon.

Clinical features

In addition to cognitive deficits such as slowed thinking, forgetfulness, and disorientation (Mok *et al.* 2004), etc., one also classically sees elements of the frontal lobe syndrome such as disinhibition, apathy, depression, and perseveration (Ishii *et al.* 1986; Staekenborg *et al.* 2010; Wolfe *et al.* 1990). In some cases, vascular parkinsonism may occur, as discussed in Chapter 11.3. As might be expected, there is typically also a history of lacunar syndromes, such as pure motor stroke, ataxic hemiparesis, dysarthria–clumsy hand, or pure sensory stroke. Furthermore, and in advanced cases of the lacunar state, it is common to see a psuedobulbar palsy with, as described in Chapter 5.7, emotional incontinence.

Although CT scanning may reveal some lacunes, MRI is far more sensitive and is thus strongly recommended.

Course

In most cases, lacunar dementia is a progressive condition, and the progression itself may be either 'step-wise' or more or less gradual (Ishii *et al.* 1986; Swartz and Black 2006; Yoshitake *et al.* 1995), depending on the size and location of subsequent lacunar infarctions. Larger or strategically placed lacunes may cause an obvious step down the cognitive ladder; conversely, small lacunes, although individually clinically 'silent', may, upon accumulation, cause a clinically noticeable, and more or less gradually appearing, cognitive deficit.

Etiology

Lacunar dementia is typically characterized by multiple lacunes (Esiri *et al.* 1997; van der Flier *et al.* 2005), generally a dozen or more (Ishii *et al.* 1986), scattered bilaterally in the basal ganglia (Benisty *et al.* 2009; Gold *et al.* 2005), genu (Tatemichi *et al.* 1992) and anterior limb of the internal capsule, and the thalamus (Benisty *et al.* 2009; Gold *et al.* 2005), and in all likelihood the dementia occurs secondary to interruption of the circuit running from the frontal cortex to the basal ganglia, through the internal capsule, and into the thalamus, thence back to the frontal cortex. Exceptions to this rule may occur with 'stragegically' placed lacunes, as for example in the anteromedial thalamus (Auchus *et al.* 2002; Katz *et al.* 1987; Mori *et al.* 1986; Sandson *et al.* 1991; Swartz and Black 2006; Szirmai *et al.* 2002).

Differential diagnosis

Lacunar dementia should be suspected in any patients with dementia who have a history of lacunar syndromes or multiple (or strategically placed) lacunes found on MR scanning. As noted earlier, it is not uncommon to find evidence of other vascular pathology and, in cases in which there are also large, territorial cortical infarcts or extensive white matter disease in addition to lacunes, clinical judgment must come into play in deciding how important the lacunes are in the development of the dementia. In this regard, one looks not only to the number of lacunes but also, as noted, to their location: whereas lacunes in the posterior limb of the internal capsule may not be that important, those situated in the thalamus, genu or anterior limb of the internal capsule, or the head of the caudate are more likely to cause cognitive deficits.

In cases of lacunar dementia which present in a more or less gradual fashion owing to the accumulated effects of multiple 'silent' lacunes, it may be difficult to rule out Alzheimer's disease, and again here one must look to the number and, most importantly, the location of the lacunes.

Treatment

The overall treatment of dementia, the frontal lobe syndrome, and pseudobulbar palsy is discussed in Chapters 6.3, 8.2, and 5.7, respectively. There is some evidence that donepezil (Black *et al.* 2003; Wilkinson *et al.* 2003), galantamine (Auchus *et al.* 2007), and memantine (Orgogozo *et al.* 2002; Wilcock *et al.* 2002) are helpful, but the results are not impressive. Treatment aimed at preventing future lacunar infarctions is discussed in Chapter 8.4.

11.3 VASCULAR PARKINSONISM

Vascular parkinsonism, formerly known as arteriosclerotic parkinsonism, first described by Critchley in 1929 (Critchley 1929), is characterized by an atypical parkinsonism occurring on the basis of lacunar infarctions either in the mesencephalon or basal ganglia. It is probably an uncommon disorder.

Clinical features

The onset is generally in the seventh or eighth decade. Clinically (Bruetsch and Williams 1954; Kaira *et al.* 2010; Keschner and Sloane 1931; Murrow *et al.* 1990; Tolosa and Santamaria 1984; Winikates and Jankovic 1999; Zijlmans *et al.* 2004a), the parkinsonism is primarily characterized by a gait abnormality. Although the gait is shuffling, there is no festination. Freezing is common, and patients have particular difficulty in making turns. Postural stability is significantly impaired, and falls are common. On further examination, there may be bradykinesia and rigidity; however, the rigidity is generally confined to the lower extremities, and, rather than cogwheel in type, it tends to be either lead-pipe or paratonic: further, the bradykinesia and rigidity tend to be symmetric. Notably, the face and upper extremities are relatively spared, and thus there is little in the way of facial masking or flattening, and arm swing is relatively preserved. Furthermore, tremor is either absent or minimal.

In addition to this parkinsonism, most cases are also characterized by a variable combination of corticospinal tract signs (e.g. spasticity or a Babinski sign), pseudobulbar palsy, urinary incontinence, and dementia (Bruetsch and Williams 1954; Keschner and Sloane 1931). Finally, as noted below, there is generally either no, or only a minimal, response to levodopa.

Magnetic resonance scanning reveals multiple lacunes, with at least some of them involving either the basal ganglia or the mesencephalon (Zijlmans *et al.* 1995).

Course

The overall course is generally one of progression, and may be either 'step-wise' or more or less gradual.

Etiology

Case reports have clearly demonstrated that parkinsonism can occur secondary to solitary lacunar infarctions in such strategic locations as the basal ganglia (Lazzarino *et al.* 1990) or the substantia nigra (Akyol *et al.* 2006; Hunter *et al.* 1978). In most patients, however, vascular parkinsonism occurs in the setting of multiple lacunes (Bruetsch and Williams 1954; Keschner and Sloane 1931; Murrow *et al.* 1990; Zijlmans *et al.* 2004a) that affect not only the basal ganglia or substantia nigra but also the corticospinal and corticobulbar tracts as they course through the internal capsule, thus giving rise to the corticospinal signs and pseudobulbar palsy.

Differential diagnosis

The full differential for parkinsonism is described in Chapter 4.8. Of the disorders considered there, the neurodegenerative causes of parkinsonism are most likely to be confused with vascular parkinsonism. Parkinson's disease is distinguished by the presence of tremor, cogwheel rigidity, and a good response to levodopa. Dementia with Lewy bodies is suggested by the early occurrence of a dementia marked by confusion and visual hallucinations, and multiple system atrophy by associated ataxia and autonomic failure. Progressive supranuclear palsy may appear very similar to vascular parkinsonism, and, in cases in which supranuclear ophthalmoplegia has not as yet appeared, the differential may rest on MR scanning, which reveals midbrain atrophy in progressive supranuclear palsy, in contrast to the multiple lacunes of vascular parkinsonism.

Treatment

Although the response to levodopa is generally not good (Winikates and Jankovic 1999), some patients may respond (Zijlmans *et al.* 2004b) and, hence, it is worth a try. Efforts should be made to prevent further lacunar infarctions, as discussed in Chapter 8.4; the symptomatic treatment of pseudobulbar palsy and of dementia is discussed in Chapters 5.7 and 6.3, respectively.

11.4 BINSWANGER'S DISEASE

Binswanger's disease, first described by Otto Binswanger in 1894 (Binswanger 1894), is characterized by a slowly progressive dementia occurring in the elderly on the basis of a diffuse microangiopathic ischemic leukoencephalopathy or leukoaraiosis. Thus conceived, Binswanger's disease is considered to be one of the vascular dementias (Erkinjuntti *et al.* 2000) along with multi-infarct dementia and lacunar dementia. Traditional synonyms for this disease include subcortical arteriosclerotic encephalopathy, chronic progressive subcortical encephalopathy, and encephalitis subcorticalis chronica progressiva. This is a not uncommon cause of dementia.

Clinical features

The onset is generally gradual, occurring in the sixth or later decades. Clinically (Bogucki *et al.* 1991; Caplan and Schoene 1978), as described by Binswanger himself (Blass *et al.* 1991), there is a 'slow development of the deterioration of the intellectual capacities', and patients present with a nonspecific dementia with slowness of thought, difficulty with planning or organizing the day or understanding complex events and concreteness; although eventually memory also fails, this, in the overall picture, often plays a relatively minor role. In advanced cases, a pseudobulbar palsy may occur. Minor focal signs, such as asymmetric deep tendon reflexes or a Babinski sign, may be seen, but, in uncomplicated Binswanger's disease, major, clear-cut syndromes, such as aphasia, agnosia, apraxia, or neglect, do not occur.

Computed tomography scanning reveals widespread areas of radiolucency in the periventricular white matter and centrum semiovale. Magnetic resonance scanning (De Groot *et al.* 2002; van der Flier *et al.* 2005) is far more sensitive, and on fluid-attenuated inversion recovery (FLAIR) or T2-weighted images one finds, as illustrated in Figure 11.1, multiple patchy and confluent areas of increased signal intensity in the periventricular region and the centrum semiovale, extending close to the cortex but sparing the U-fibers. In advanced cases, these patchy areas may coalesce to the point of creating a virtual 'white-out'.

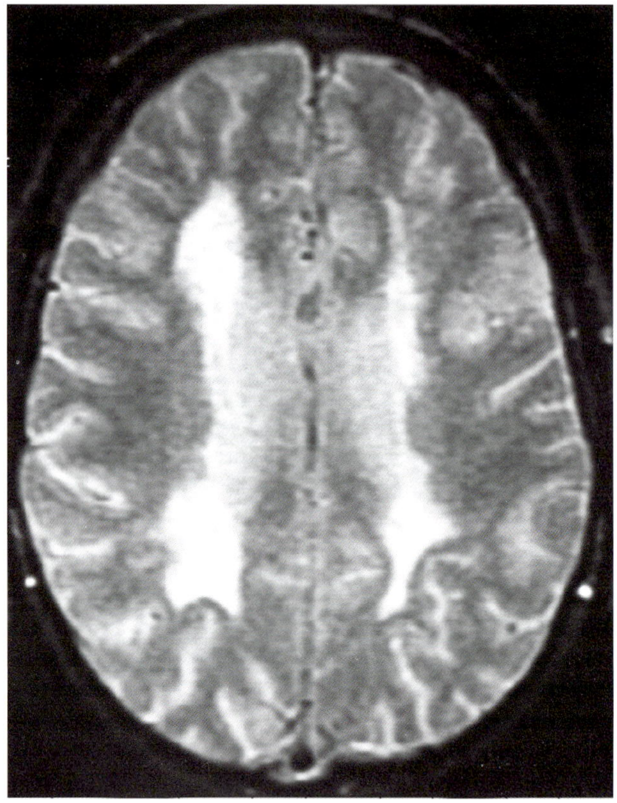

Figure 11.1 A T2-weighted magnetic resonance (MR) scan in a case of Binswanger's disease, with large confluent areas of increased signal intensity in the centrum semiovale. Reproduced from Gillespie and Jackson (2000) with permission.

Course

Binswanger's disease is characterized by a very gradual and unrelenting progression.

Etiology

The neuropathologic changes have been described in several reports (Akiguchi *et al.* 1997; Caplan and Schoene 1978; Esiri *et al.* 1997; Lin *et al.* 2000; Revesz *et al.* 1989; van Swieten *et al.* 1991; Yamanouchi 1991). At autopsy, the small penetrating medullary vessels display lipohyalinosis and, in some cases, the lumens are obliterated. There is widespread demyelination with some associated axonal loss and astrocytic gliosis, and, in severe cases, cystic changes may occur. Although it is suspected that these microangiopathic changes occur on the basis of long-standing hypertension, other factors are probably also at work, as Binswanger's disease may occur in normotensive individuals.

Differential diagnosis

Binswanger's disease is one of the dementias of gradual onset that often lack distinctive features and, as discussed in Chapter 6.3, it must therefore be distinguished from several other disorders, notably including Alzheimer's disease, dementia with Lewy bodies, and normal pressure hydrocephalus, and certain clinical features may be helpful in this differential task. Alzheimer's disease classically presents first with a slowly progressive short-term memory loss, in contrast to Binswanger's disease in which memory loss may occur late, and in any case plays a relatively minor role in the overall clinical picture. Dementia with Lewy bodies may present with a dementia, but here one finds visual hallucinations and spontaneous episodes of confusion early on – findings not seen in Binswanger's disease. Finally, in normal pressure hydrocephalus, one sees both a 'magnetic' gait and urinary urge incontinence, features that are generally not seen in Binswanger's disease.

There are two disorders that present with dementia in the context of diffuse white matter changes, namely cerebral amyloid angiopathy and CADASIL (cerebral autosomal dominant arteriopathy with subcortical infarcts and leukoencephalopathy), and these may cause some diagnostic difficulty. In cerebral amyloid angiopathy, gradient echo recall MRI may reveal evidence of 'microbleeds'; furthermore, in most cases there will eventually also be lobar intracerebral hemorrhages, which clinch the diagnosis. CADASIL is suggested by a history of migraine headaches and also by the distinctive finding of a leukoencephalopathy which is most prominent in the anterior temporal lobe and the external capsule.

As noted earlier, Binswanger's disease is one of the vascular dementias, and, given that patients may develop not only the microangiopathic changes of Binswanger's disease but also disease of the large pial vessels and the smaller central perforating vessels, it is not uncommon to find a mixture of vascular pathology; in such cases, Binswanger's disease may be complicated by territorial infarctions and lacunar infarctions. There is often a history of multiple strokes in addition to the gradually progressive dementia, and clinical judgment is required in deciding what weight to give to each of the vascular pathologies with regard to the dementia. In cases characterized by a prominent amount of all three vascular pathologies, one may have to content oneself with an 'umbrella' diagnosis of 'vascular dementia'.

Finally, in evaluating a patient with a slowly progressive, non-specific dementia, care must be taken not to place too much diagnostic weight on minor changes seen on MR scanning. Periventricular 'caps' and 'rims' are of no significance, and it is not at all uncommon to find multiple, non-confluent, small patchy areas in the centrum semiovale or periventricular areas, which are, in fact, asymptomatic. The diagnosis of Binswanger's disease should not be made unless the white matter disease is severe and extensive, approaching, ideally, a 'white-out'.

Treatment

The general treatment of dementia is discussed in Chapter 6.3; there is some evidence that donepezil (Black *et al.* 2003; Wilkinson *et al.* 2003), galantamine (Auchus *et al.* 2007), and memantine (Orgogozo *et al.* 2002; Wilcock *et al.* 2002) may be helpful, but the gains are not impressive. Control of blood pressure is important; however, one must not be overzealous in this regard, as hypotension may lead to a critical reduction in perfusion pressure in the penetrating arteries, leading to further ischemic change.

11.5 CRANIAL ARTERITIS

Cranial arteritis, also known as giant cell arteritis or temporal arteritis, is characterized pathologically by a segmental inflammation of various, primarily cranial, arteries, and clinically by symptoms related to ischemia in the areas of distribution of those arteries. Classically, this disorder presents with headache (due to inflammation of one of the branches of the external carotid artery, typically the superficial temporal artery) and either amaurosis fugax or blindness due to involvement of the ophthalmic artery or its branches.

Cranial arteritis is a relatively common disease in the elderly; although cases below the age of 50 years have been reported, this is very rare, and most patients fall ill in the latter part of the seventh decade or during the eighth decade. Of interest, this disorder is seen primarily in Caucasians, especially those of Nordic descent.

Clinical features

The symptomatology of cranial arteritis varies according to which arteries are inflamed (Wilkinson and Russell 1972). Branches of the external carotid artery are commonly involved, and, as noted above, inflammation of the superficial temporal artery is classic, causing a severe headache in the temporal region, which, although generally unilateral, can be bilateral. The headache is typically non-pulsating, and may be associated with such exquisite tenderness that patients cannot bear to let the involved side touch a pillow. To palpation, the artery is thickened and generally pulseless. In some cases the occipital artery is involved, and in these instances the headache may be localized to the neck. Involvement of the facial artery may lead to 'jaw claudication' with facial pain upon chewing, and involvement of the lingual artery may cause tongue necrosis. Of the branches of the internal carotid artery, the ophthalmic artery, as noted, is classically involved, and with occlusion of the posterior ciliary artery or the central retinal artery, unilateral blindness may occur (Caselli *et al.* 1988): this fateful development is preceded by amauosis fugax in no more than one-tenth of cases. Critically, without treatment, blindness occurs in the contralateral eye in roughly one-third of cases within 2 weeks, with some cases occurring within 24 hours.

In less than 5 percent of cases, strokes or transient ischemic attacks may occur (Caselli *et al.* 1988). In most instances, this is due to artery-to-artery embolization from foci of involvement in either the internal carotid artery or the extracranial portion of the vertebral artery. Thus, from the internal carotid artery, emboli may lodge in the middle or anterior cerebral arteries or their branches, and from foci in the extracranial vertebral artery, emboli may lodge in the posterior cerebral artery or its branches. In addition, although rare, cranial arteritis may also directly involve intracranial vessels (Solans-Laque *et al.* 2008), including the middle cerebral and anterior cerebral arteries, the basilar artery, and the posterior cerebral artery, and in such cases, in addition to artery-to-artery embolization, occlusion of these arteries may directly lead to infarction. In addition to typical stroke syndromes, some patients may develop a multi-infarct dementia (Caselli 1990) while others may develop a delirium. These latter cases are characterized either by multiple lacunar infarctions in subcortical structures (Solans-Laque *et al.* 2008) or by multiple small cortical and subcortical white matter infarctions (Caselli 1990).

Most cases of cranial arteritis are further characterized by constitutional symptoms, such as malaise, fatigue, anorexia, and a low-grade fever, and most patients will also have an associated polymyalgia rheumatica, with muscle aching and stiffness, which, although diffuse, is most prominent in the neck and shoulders.

With rare exceptions (Kansu *et al.* 1977), the erythrocyte sedimentation rate (ESR) is elevated, generally above 50 mm/min (Westergen) and often above 100 mm/min. In situations where the clinical suspicion is high yet the ESR is normal or only mildly elevated, a C-reactive protein should be ordered, as this test may be positive even when the ESR is normal. Mild anemia is common and there may be an associated mild leukocytosis. The diagnosis is confirmed by biopsy of an involved artery, almost always the superficial temporal artery: given the segmental nature of the inflammation, however, false negatives are possible, and in up to a quarter of cases the contralateral artery will have to be biopsied.

Course

The inflammatory process generally undergoes a gradual, spontaneous remission in anywhere from several months to 2 years.

Etiology

As noted earlier, cranial arteritis is characterized by a segmental inflammation, and in between these segments the artery may be normal. Within the segments, there is a granulomatous panarteritis characterized by giant cells. Intimal proliferation occurs, and there may be thrombus formation with either embolization or occlusion (Wilkinson and Russell 1972). Although the mechanism underlying this inflammation is not fully understood, an autoimmune process is strongly suspected.

As alluded to above, certain arteries are preferentially involved in cranial arteritis. The most commonly involved arteries are those derived from the external carotid artery, including the superficial temporal artery, the occipital artery, the facial artery, and the lingual artery. Both the internal carotid artery and the extracranial vertebral artery may also be involved; however, because the inflammatory process usually involves only arteries with an internal elastic lamina, intracranial branches are typically spared. As noted earlier, however, there are definite cases where these intracranial branches were involved. The ophthalmic artery, once passing extradurally and acquiring an internal elastic lamina, constitutes the one branch of the internal carotid artery which is particularly vulnerable.

Differential diagnosis

The diagnosis should be suspected in cases of amaurosis fugax, blindness, or stroke, occurring in the setting of headache, constitutional symptoms, polymyalgia rheumatica, or an elevated ESR. Systemic lupus erythematosus may be considered in the differential, but here the antinuclear antibody (ANA) test is positive. Polyarteritis nodosa may also be suspected; however, here, renal or gastrointestinal involvement will indicate the correct diagnosis.

Treatment

Urgent treatment with steroids is essential to prevent stroke or blindness, and need not wait upon biopsy, as biopsy findings generally remain positive for at least the first few days of treatment. Traditionally, prednisone is used, starting at doses of approximately 60 mg/day; however, recent work suggests that aggressive treatment with methylprednisolone at 15 mg/kg (of ideal body weight) daily for the first 3 days, followed by prednisone, induces a more rapid remission and ensures a more favorable course (Mazlumzadeh *et al.* 2006). Once symptoms are under control and the ESR has dropped, an attempt should be made to very gradually taper the dose of prednisone, keeping a close watch for any recrudescence of symptoms or for a rise in the ESR. The place of steroid-sparing agents, such as methotrexate and infliximab, is at present uncertain. Uncontrolled work suggests that adding a low dose of aspirin to prednisone reduces the risk of stroke (Nesher *et al.* 2004), and given the generally benign nature of aspirin, this appears reasonable. Importantly, a failure of the headache to respond to steroid treatment within a week argues strongly against the diagnosis.

11.6 CEREBRAL AMYLOID ANGIOPATHY

Cerebral amyloid angiopathy, also known as congophilic amyloid angiopathy, is a not uncommon disorder that is characterized classically by recurrent, spontaneous, lobar intracerebral hemorrhages.

Clinical findings

Cerebral amyloid angiopathy (Cosgrove *et al.* 1985; Gilles *et al.* 1984; Haan *et al.* 1990a; Nobuyoshi *et al.* 1984; Yoshimura *et al.* 1992) typically has an onset in the seventh or later decades, and characteristically comes to clinical attention as a result of one or more lobar intracerebral hemorrhages.

Although the basic underlying pathology of cerebral amyloid angiopathy is the same in all cases, namely the deposition of amyloid in medium and small cerebral vessels, the clinical presentation of the disease varies widely, depending on the consequences of this amyloid deposition. Thus, should a relatively large vessel burst, one will see the characteristic lobar intracerebral hemorrhage. However, should the bursting vessel be quite small, a mere 'microhemorrhage' will occur. Rather than bursting, some small vessels may become partially, or totally, occluded, leading, respectively, to either areas of white matter demyelination or to small infarctions in either the white matter or the cortex. Finally, in some cases, an autoimmune response may be mounted against the amyloid-laden vessels, leading to what has been termed the syndrome of 'cerebral amyloid angiopathy-related perivascular inflammation'. Each of these varied presentations is considered in turn.

The intracerebral hemorrhages of cerebral amyloid angiopathy are, importantly, both spontaneous and lobar in distribution, and although the temporal and occipital lobes are most favored, they may occur in any lobe. Clinically, there is a relatively slow development, over perhaps a half-hour, of headache, with or without nausea and vomiting, and a focal deficit appropriate to the location of the hemorrhage. As noted below, under Course, recurrences are the rule, and with multiple, appropriately placed hemorrhages, a multi-infarct dementia may develop.

Microhemorrhages may be asymptomatic or may present with minor focal deficits, which typically resolve. When occurring in the cortex, they may serve as seizure foci, and there may be simple or complex partial seizures, with or without secondary generalization.

With partial occlusion of small penetrating arteries, multiple, confluent, typically periventricular areas of demyelination occur, and with the gradual increase of this burden of white matter disease, a dementia may supervene, which is typically non-specific in character. As noted, multiple micro-infarctions may also occur, and these may well aggravate the developing dementia.

Cerebral amyloid angiopathy-related perivascular inflammation (Kinnecom *et al.* 2007) is a rare syndrome which, from a pathologic point of view, typically develops focally as an expanding area of vasogenic edema. Clinically, there is a subacute development, over weeks or months, of headache and delirium, with or without seizures.

Both CT and MR scanning will reveal any lobar intracerebral hemorrhages. Old microhemorrhages, however, are seen only with MR scanning utilizing gradient-echo imaging (Greenberg *et al.* 1999): these are typically scattered throughout the cerebrum, and may be seen in both the white matter and the cortex, where they are typically found at the gray–white junction. Magnetic resonance scanning is also more sensitive for the white matter disease, and is necessary for the detection of micro-infarctions. Although CT scanning may reveal the edema of the syndrome of cerebral amyloid angiopathy-related perivascular inflammation, MR scanning, using T2-weighted or FLAIR imaging, is far more sensitive.

The cerebrospinal fluid (CSF) (Verbeek *et al.* 2009) reveals reduced levels of β-amyloid-42 and β-amyloid-40 and increased levels of both total and phosphorylated tau. The cell count and glucose are normal, and the total protein is also normal except in cases of the syndrome of cerebral amyloid angiopathy-related perivascular inflammation, where it may be increased (Eng *et al.* 2004).

Course

Lobar intracerebral hemorrhages tend to recur, and indeed one study found a 2-year recurrence rate of 21 percent (O'Donnell *et al.* 2000), and hence, from a clinical point of view, there will be a history of recurrent stroke, or, and again with appropriately situated hemorrhages, a dementia that pursues a 'step-wise' course. As noted earlier, with demyelination and micro-infarctions, there may also be a dementia, and this pursues a gradually progressive course. As might be expected, in many cases the dementia seen in cerebral amyloid angiopathy reflects, to varying degree, both these pathologies, and here the clinical course of the dementia is one of steady progression punctuated by significant 'steps' down the cognitive ladder (Nobuyoshi *et al.* 1984). Finally, cases of the syndrome of cerebral amyloid angiopathy-related perivascular inflammation tend to undergo a spontaneous, but slow, resolution, and after weeks or months patients either recover essentially completely or are left with some residual symptoms.

Etiology

Amyloid deposition in the walls of medium and small cerebral arteries leads to aneurysm formation and fibrinoid degeneration (Okazaki *et al.* 1979). The aneurysms are quite fragile, and eventually burst, leading either to large lobar hemorrhages or to the smaller microhemorrhages. Fibrinoid degeneration leads to partial or complete occlusion, with, as noted earlier, either areas of demyelination (Gray *et al.* 1985) or micro-infarctions (Haglund *et al.* 2006). With regard to the syndrome of cerebral amyloid angiopathy-related perivascular inflammation, multinucleated giant cells are found surrounding vessels (Eng *et al.* 2004): as noted earlier, it is assumed that this occurs secondary to an autoimmune response mounted toward amyloid-containing vessels; however, to date, the putative antibody has not been identified.

In pathologic specimens, affected vessels are 'congophilic', in that they stain well with Congo red dye, and it is this property which prompted the synonym for this disorder, namely 'congophilic amyloid angiopathy.'

Although the vast majority of cases are sporadic, cerebral amyloid angiopathy may also occur on an autosomal dominant basis as 'hereditary cerebral hemorrhage with amyloidosis', either of the 'Dutch' (Haan *et al.* 1990a,b; Levy *et al.* 1990; Wattendorff *et al.* 1995) or 'Icelandic' (Gudmundsson *et al.* 1972; Palsdottir *et al.* 1988) type.

Differential diagnosis

The intracerebral hemorrhages of cerebral amyloid angiopathy must be distinguished from hemorrhage due to other causes. Hypertensive intracerebral hemorrhage is suggested not only by the strong association with hypertension, but by the tendency of these hemorrhages to occur in the striatum, thalamus, pons, and cerebellum, which is in striking contrast to the lobar location of the hemorrhages of cerebral amyloid angiopathy. Traumatic intracerebral hemorrhages are of course suggested by the history of trauma, keeping in mind the fact that in patients taking warfarin, the trauma need not be at all severe to cause hemorrhage. Hemorrhage into a tumor or hemorrhage secondary to rupture of a vascular malformation may be difficult to distinguish from the lobar intracerebral hemorrhages of cerebral amyloid angiopathy, and here one may have to repeat the MR scan after 2–3 months. Of note, there is one feature of cerebral amyloid angiopathy which may help to distinguish it from all these other disorders, and that is the presence of multiple microhemorrhages; although not universal, it is such a helpful sign that gradient-echo scanning should be part of the routine protocol. Furthermore, and in regard to the distinction of cerebral amyloid angiopathy from the lobar hemorrhages which may occur secondary to ruptures of vascular malformations or bleeds into tumors, one should keep in mind the fact that, in cerebral amyloid angiopathy, lobar intracerebral hemorrhages tend to recur, while the bleeds in these other situations tend to be singular events; indeed, a history of two or more spontaneous lobar intracerebral hemorrhages is strongly suggestive of cerebral amyloid angiopathy (Knudson *et al.* 2001).

The dementia of cerebral amyloid angiopathy, as noted earlier, comes in two forms, raising two differential problems. When multi-infarct dementia is the presentation, imaging will distinguish the intracerebral hemorrhages of cerebral amyloid angiopathy from the territorial infarctions which may also cause a multi-infarct dementia. When cerebral amyloid angiopathy, however, presents with a leukoencephalopathy, the distinction from Binswanger's disease may be quite difficult, and here finding evidence of old microhemorrhages may be very helpful.

The syndrome of cerebral amyloid angiopathy-related perivascular inflammation is fairly unique; however, consideration should also be given to atypical cases of hypertensive encephalopathy or the reversible posterior leukoencephalopathy syndrome, both of which occasionally may present with primarily unilateral areas of edema. Here the finding of a grossly elevated diastolic pressure or an association with any of the medicines implicated in the reversible posterior leukoencephalopathy syndrome will suggest the correct diagnosis.

Treatment

In cases of acute intracerebral hemorrhage, neurosurgical consultation should be obtained, as evacuation of the clot may be necessary. Given that aspirin (Biffi *et al.* 2010), warfarin (Rosand *et al.* 2000), and tissue plasminogen activator (Pendlebury *et al.* 1991) may all increase the risk of hemorrhage, these should be avoided.

In cases of the syndrome of cerebral amyloid angiopathy-related perivascular inflammation, success has been reported with steroids, with or without cyclophosphamide (Eng *et al.* 2004; Kinnecom *et al.* 2007).

11.7 CADASIL

CADASIL is a rare autosomal dominantly inherited arteriopathy characterized by migraine, stroke, and dementia. The name CADASIL is an acronym for *c*erebral *a*utosomal *d*ominant *a*rteriopathy with *s*ubcortical *i*nfarcts and *l*eukoencephalopathy.

Clinical features

In general, the overall clinical picture of CADASIL (Bergmann *et al.* 1996; Chabriat *et al.* 1995; Dichgans *et al.* 1998; Jung *et al.* 1995; Kim *et al.* 2006; Malandrini *et al.* 1996; Markus *et al.* 2002; Opherk *et al.* 2004; Sourander and Walinder 1977) is characterized by the cardinal manifestations of migraine, stroke, and dementia. Migraine typically appears in the third or fourth decades, and is seen in roughly one-half of all cases; it is characteristically accompanied by an aura. Stroke appears in the fifth or sixth decades and is seen in roughly three-quarters of all patients, and is typically of the lacunar type. Dementia eventually appears in the sixth decade in about a third of patients, and is characterized by executive deficits, poor attention, and forgetfulness. A pseudobulbar palsy, with emotional incontinence, accompanies the dementia in about one-half of all instances. The dementia, although generally reflecting the effects of multiple lacunar infarctions (Liem *et al.* 2007; Viswanathan *et al.* 2007), at times may also be due to a progressive leukoencephalopathy, and in rare instances CADASIL may present with a dementia secondary to the leukoencephalopathy in the absence of stroke (Filley *et al.* 1999; Mellies *et al.* 1998).

Other manifestations of CADASIL include depression, apathy, seizures, and delirium. Depression (Reyes *et al.* 2009) is seen in roughly one-half of all cases, and pure cases of apathy (Reyes *et al.* 2009), in the absence of depression, are seen in about one-sixth. Seizures occur in about one-tenth of cases, and delirium in a similar proportion. The delirium (Schon *et al.* 2003) is accompanied by seizures and fever, and may progress to coma; spontaneous recovery is seen after 1–2 weeks.

Magnetic resonance imaging reveals any prior lacunar infarctions. Furthermore, most patients will also have a diffuse leukoencephalopathy and, indeed, this may be present early on, even before the first stroke. The leukoencephalopathy is evident on either T2-weighted or FLAIR imaging: patchy, confluent areas of increased signal intensity are seen in the centrum semiovale and the periventricular white matter, and also, classically, in the external capsule and, most notably, in the white matter of the anterior temporal lobe (Markus *et al.* 2002; O'Sullivan *et al.* 2001). Recent work using gradient-echo imaging has also identified evidence for old microhemorrhages in the thalamus and basal ganglia (Choi *et al.* 2006; Lesnik Oberstein *et al.* 2001).

Definitive diagnosis rests on genetic testing. Importantly, however, commercially available testing probes for only about three-quarters of all known mutations. Consequently, if genetic testing is negative but clinical suspicion is high, one should proceed to skin biopsy: although false negatives may also occur here (Malandrini *et al.* 2007), one might pick up cases missed by genetic testing.

Course

Death, often from intercurrent infection, typically occurs in the seventh decade.

Etiology

As noted, CADASIL is an autosomal dominant disorder: mutations are found in the *Notch3* gene on chromosome 19 (Joutel *et al.* 1997; Tournier-Lasserve *et al.* 1993), over 80 different mutations are known, and *de* novo mutations, although rare, have been reported (Joutel *et al.* 2000; Liguori *et al.* 2009). Pathologically, there is concentric fibrous thickening of small penetrating arteries (Sourander and Walinder 1977), leading to both the leukoencephalopathy and the subcortical infarctions (Baudrimont *et al.* 1993). Of note, although the vast majority of strokes are ischemic, intracerebral hemorrhages within the thalamus or basal ganglia have also been reported (Choi *et al.* 2006).

Although peripheral nerves (Schroder *et al.* 1995) and skin (Ruchoux *et al.* 1994) may also be involved, symptoms referable to them are absent.

Differential diagnosis

Lacunar strokes occurring in CADASIL must be distinguished from lacunar strokes occurring on the basis of lipohyalinosis or atherosclerosis. Similarly, the dementia occurring in CADASIL must be distinguished from the dementia occurring in lacunar dementia and in Binswanger's disease. In all these instances, a positive family history is most helpful as it points to CADASIL; however, in some cases such a history may be unavailable. Other features suggesting CADASIL are migraine with aura and, most importantly, the early appearance of leukoencephalopathy in the anterior portion of the temporal lobe, which, although common in CADASIL, is most unlikely in other disorders.

Cerebral amyloid angiopathy may cause some diagnostic difficulty, as it can present with a dementia in the setting of a diffuse leukoencephalopathy with old microhemorrhages. Here, MRI evidence of lacunar infarctions or anterior temporal leukoencephalopathy point to CADASIL, and the appearance of *lobar* intracerebral hemorrhages at any time generally indicates cerebral amyloid angiopathy.

MELAS (mitochondrial encephalomyopathy, lactic acidosis, and stroke-like episodes) may present in a fashion similar to CADASIL, but is distinguished by deafness.

Treatment

At present, there is no specific treatment for CADASIL. In cases with migraine, an argument could be made to avoid triptans, given their vasospastic effect. Many clinicians will prescribe prophylactic aspirin, but there are no data to support its effectiveness in preventing stroke. Donepezil, in a double-blinded study (Dichgans *et al.* 2008), was not effective in cases of dementia. Should depression occur, an SSRI is a reasonable option, while antidepressants capable of lowering blood pressure should probably be avoided. Seizures may be treated in the usual fashion.

11.8 GRANULOMATOUS ANGIITIS OF THE CENTRAL NERVOUS SYSTEM

Granulomatous angiitis of the central nervous system is a rare disorder characterized pathologically by a granulomatous angiitis confined to the central nervous system, and clinically by headache and delirium. An often-used synonym is primary angiitis of the central nervous system; however, caution is needed here, as this term has, by some authors, been used in an 'umbrella' fashion, lumping granulomatous angiitis of the central nervous system together with certain other vasculitidies, hence the reader should examine any literature carefully to ensure that, indeed, granulomatous angiitis is the disorder in question.

Clinical features

Although the disease may present at any age, from childhood to senescence, most patients are in their thirties or forties. The onset itself is generally subacute, spanning a few weeks. Clinically (Abu-Shakara *et al.* 1994; Calabrese and Mallek 1988; Case Records 1989; Hughes and Brownell 1966; Kolodny *et al.* 1968; Koo and Massey 1988; Lie 1992; Moore 1989; Salvarani *et al.* 2007; Vollmer *et al.* 1993), patients typically have severe, generalized headache, and in this setting they develop a delirium that may be accompanied by focal deficits or, uncommonly, seizures. Although focal deficits, such as hemiparesis, may have a stroke-like onset, in most cases they appear gradually. In some cases, the spinal cord may be involved.

Magnetic resonance imaging (Alhalabi and Moore 1994; Ehsan *et al.* 1995) may initially be normal; however, over time, areas of increased signal intensity on FLAIR or T2-weighted imaging appear in the cerebrum or cerebellum, and there may be gadolinium enhancement of the meninges. The ESR is generally either normal or only very mildly elevated (Lie 1992; Moore 1989). Lumbar puncture is abnormal in 80–90 percent of cases, with a lymphocytic pleocytosis, an elevated total protein, and a normal glucose level (Vollmer *et al.* 1993). Magnetic resonance angiography and CT angiography are not helpful, given that the size of the vessels involved is generally below their level of resolution, and consequently, if the diagnosis is being seriously considered, patients come to conventional angiography. Unfortunately, however, although typically showing 'beading' (Alrawi *et al.*

1999), angiography may be falsely negative in up to a quarter of cases. The gold standard of diagnosis is a brain biopsy (Alrawi *et al.* 1999), sampling both leptomeninges and cerebrum: critically, given that biopsy may miss affected tissue (Alhalabi and Moore 1994; Lie 1992), the biopsy must be targeted at tissue which, on MRI, is clearly affected.

Course

Untreated, progressive deterioration occurs in almost all cases, and perhaps half of all patients will die within months, with the remainder surviving for a year or two.

Etiology

A granulomatous angiitis affects both small leptomeningeal vessels and small- or medium-sized parenchymal vessels, with subsequent infarctions (Cravioto and Feigin 1959). The cerebrum is most commonly affected, although the cerebellum, brainstem, and even the cord may also be involved. Although the cause is unknown, an autoimmune process, confined to the central nervous system, is suspected.

Differential diagnosis

Given the stakes involved, many authors argue that the diagnosis should not be made without a positive brain biopsy. Before proceeding to this, however, it is of course reasonable to check for other disorders which could present in a somewhat similar fashion, even to the point of a similar appearance on angiography. Hence, one must consider polyarteritis nodosa, varicella-zoster arteritis, systemic lupus erythematosus, Sjögren's syndrome, Sneddon's syndrome, Behçet's syndrome, cranial arteritis, acquired immune deficiency syndrome arteritis, and toxicity of cocaine or sympathomimetics. Another condition to consider is what has been termed 'benign angiitis of the central nervous system': this manifests in a fashion very similar to granulomatous angiitis of the central nervous system both clinically and angiographically; however, here the CSF is normal and brain biopsy reveals a lymphocytic, rather than a granulomatous, vasculitis; furthermore, this condition, as the name implies, has a favorable course, with most patients experiencing an eventual spontaneous remission.

Treatment

The general treatment of delirium is discussed in Chapter 6.5. Most patients are treated with a combination of prednisone and cyclophosphamide.

11.9 POLYARTERITIS NODOSA

Polyarteritis nodosa, also known as periarteritis nodosa, is a rare systemic vasculitis characterized pathologically by a segmental necrotizing panarteritis. Most patients present with constitutional symptoms and involvement of the kidneys, gastrointestinal tract, or muscles; in a minority there may be a peripheral neuropathy, and, in a smaller minority, central nervous system involvement with stroke, delirium, or dementia.

Clinical features

The onset is generally subacute or gradual in middle years with constitutional symptoms and evidence of involvement of multiple organs. Renal failure is common, and associated hypertension may be severe. Arthralgia and myalgia are common, as is gastrointestinal involvement with associated abdominal pain, nausea, and vomiting. Cardiac involvement may manifest with congestive heart failure or myocardial infarction. With dematologic involvement there may be rashes, skin nodules, or, rarely, livdeo reticularis. Importantly, the respiratory tract is not involved.

Nervous system involvement generally occurs only after other evidence of the disease is well established (Sigal 1987). Peripheral nervous system involvement is most common, occurring in over half of all cases, and typically manifests as a mononeuritis multiplex (Lovelace 1964). Central nervous system involvement is seen in no more than a quarter of cases, and most commonly manifests with lacunar infarctions (Reichart *et al.* 2000); larger territorial infarctions or intracerebral hemorrhages are rare. In a very small minority, delirium (Ford and Siekert 1965) or dementia (MacKay *et al.* 1950) may occur, and seizures have also been noted.

Both the peripheral white blood cell count and the ESR are elevated, and the Venereal Disease Research Laboratories (VDRL) test may be falsely positive. Magnetic resonance scanning will reveal infarctions. Definitive diagnosis is made by biopsy of an affected muscle or peripheral nerve.

Course

Although spontaneous remissions do occur, they are rare, and most cases are characterized by relentless progression, with only half of all patients surviving past 5 years.

Etiology

There is a systemic, segmental panarteritis affecting medium and small arteries. With intimal proliferation, thrombosis and occlusion of arteries may occur, and with involvement of the muscular layer, microaneurysms may form. These microaneurysms may occasionally rupture; however, they typically undergo fibrosis, thus creating nodules along the course of the artery and thereby providing the characteristic that gives the disease its name.

Involvement of the peripheral vasa nervorum leads to the mononeuritis multiplex. Within the central nervous system, involvement of small perforating arteries leads to lacunar infarctions; in those uncommon cases involving larger arteries, territorial infarctions may occur and, with rupture of an aneurysm, an intracerebral hemorrhage may be seen.

Although the mechanism underlying the arteritis is not known with certainty, deposition of immune complexes probably plays a role, and in this regard there is an association with hepatitis B antigenemia.

Differential diagnosis

In addition to suggesting polyarteritis nodosa, the occurrence of nervous system involvement in the context of a systemic illness also raises the possibility of systemic lupus erythematosus, indicated by a positive ANA, and Wegener's granulomatosis, which, unlike polyarteritis nodosa, also involves the respiratory tract.

Treatment

Treatment with corticosteroids and cyclophosphamide is recommended; consideration may also be given to antiplatelet agents (Reichart *et al.* 2000).

11.10 WEGENER'S GRANULOMATOSIS

Wegener's granulomatosis is an uncommon disease characterized pathologically by a systemic necrotizing granulomatous vasculitis, most commonly affecting the respiratory tract and kidneys. The nervous system is involved in a minority of cases: peripheral neuropathy is the most common manifestation; within the central nervous system there may be a vasculitis, intracerebral granulomas, and, rarely, a pachymeningitis.

Clinical features

Over 90 percent of patients have symptoms referable to granulomas within the respiratory tract. Upper respiratory tract involvement is most common, with sinusitis, epistaxis, or rhinorrhea; involvement of the nasal septum may lead to its collapse, resulting in a 'saddle nose' deformity, and extension of granulomatous disease to the orbit may cause proptosis. Pulmonary involvement may be asymptomatic or lead to cough or hemoptysis. Some three-quarters of patients will have renal involvement, which may manifest initially with proteinuria and microscopic hematuria.

Nervous system involvement generally occurs in the context of respiratory or renal symptomatology (Hoffman *et al.* 1992). Clinical evidence of a mononeuritis multiplex or, less commonly, a polyneuropathy, is seen in about one-fifth of patients (de Groot *et al.* 2001). Central nervous system involvement is less common, being seen in no more than a tenth of patients, and may manifest in numerous different ways (Nishino *et al.* 1993a,b; Weinberger *et al.* 1993), including delirium, dementia, stroke, seizures, and, with meningeal involvement, headache and cranial neuropathies, and rarely, endocrinologic disturbances (including diabetes insipidus; Rosete *et al.* 1991) due to hypothalamic or pitutitary granulomas.

Magnetic resonance imaging with gadolinium will reveal intracerebral granulomatous lesions, infarcts, and meningitis. The CSF may be normal or may show a lymphocytic pleocytosis and an elevated total protein.

The ESR is elevated, and over 90 percent of patients will have elevated levels of antineutrophil cytoplasmic antibodies (ANCA); although this is typically of the cANCA type, in a small minority pANCA may be present. Definitive diagnosis is made by biopsy of the lung or kidney.

Course

The disease is relentlessly progressive, with a 1-year mortality rate of roughly 80 percent.

Etiology

Within the central nervous system, several different pathologies may be found (Drachman 1963, Nishino *et al.* 1993a; Seror *et al.* 2006; Weinberger *et al.* 1993). Small, or rarely large, vessels may undergo a vasculitis and, with occlusion, infarction occurs. Granulomas may be found, and these may appear by extension from an extracranial source (e.g. a sinus or the orbit) or they may occur independently. Granulomatous involvement of the meninges, primarily the pachymeninges, may also occur, and cranial nerves may be entrapped and compressed; cranial neuropathies may also occur owing to compression of the cranial nerves in their extracranial portions by extracranial granulomas. In some cases, granulomas may appear in the hypothalamus or pituitary gland.

Although the etiology of Wegener's granulomatosis is not known, an autoimmune response, perhaps to an inhaled substance, is suspected.

Differential diagnosis

Central or peripheral nervous system involvement in the setting of respiratory tract or renal disease should raise the possibility of Wegener's granulomatosis. Sarcoidosis may be considered, but this disease has features not seen in Wegener's granulomatosis, such as erythema nodosum, lupus pernio, hypercalcemia, etc. Polyarteritis nodosa may also be considered, although in polyarteritis nodosa one does not find the respiratory tract involvement typical of Wegener's granulomatosis.

Treatment

In most cases, treatment with a combination of prednisone and cyclophosphamide is required. The symptomatic treatment of delirium and dementia is discussed in Chapters 6.5 and 6.3, respectively.

11.11 BEHÇET'S DISEASE

Behçet's disease, first described in 1937 by the Turkish dermatologist Hulusi Behçet (Behçet 1937), is characterized pathologically by a systemic, primarily venular, small-vessel vasculitis, and clinically by oral aphthous ulcers, genital ulcers, uveitis, and, in anywhere from one-tenth to one-third of patients, by evidence of central nervous system involvement. Although not uncommon in Japan and the eastern Mediterranean region, Behçet's disease is rare in Europe and the USA.

Clinical features

Behçet's disease typically presents in the twenties or thirties, and almost all patients will have oral aphthous ulcers and some form of uveitis; genital ulcers are somewhat less common. Another distinctive sign, not seen in all cases, is pathergy: here, within 1–2 days of minor skin trauma, for example phlebotomy, a large pustule forms at the site of the trauma. Other symptoms include furuncles, erythema nodosum, migratory thrombophlebitis, and a non-deforming polyarthritis. Within the context of these symptoms, evidence of central nervous system involvement may appear.

When the central nervous system is involved, a wide variety of symptoms may appear (Akman-Demir *et al.* 1993, 1999; Al-Araji *et al.* 2003; Altinors *et al.* 1987; Chajek and Fainaru 1975; Farah *et al.* 1998; Iragui and Miravi 1986; Joseph and Scolding 2007; Kidd *et al.* 1999; Motomura *et al.* 1980; Serdaroglu *et al.* 1989; Wadia and Williams 1957; Wolf *et al.* 1965). Most commonly, one finds evidence of brainstem involvement, with long-tract signs and cranial nerve palsies. When the thalamus, cerebral white matter, or frontal cortex is involved, there may be delirium, dementia, a frontal lobe syndrome (Yamada *et al.* 1978), and pseudobulbar palsy with emotional incontinence (Pallis and Fudge 1956; Motomura *et al.* 1980; Rubinstein and Urich 1963; Wechsler *et al.* 1990). Rarely, seizures may occur (Aykutlu *et al.* 2002). Should dural sinus thrombosis occur with raised intracranial pressure, there may be headache, nausea, and vomiting, and, in the very small minority of cases with associated cerebral venous thrombosis, venous infarction with stroke may ensue.

Magnetic resonance imaging (Akman-Demir *et al.* 2003) with FLAIR sequences may reveal more or less confluent areas of increased signal intensity, most commonly in the brainstem, but also in the thalamus, internal capsule, basal ganglia and frontal white matter, and cortex, and these areas may enhance with gadolinium.

The CSF (Akman-Demir *et al.* 1999; Farah *et al.* 1998) may be normal or may demonstrate a primarily lymphocytic pleocytosis and an elevated total protein level, with a normal glucose level.

Course

Most cases of Behçet's disease demonstrate an episodic course. The first attack, as noted, tends to occur in the twenties or thirties, and most attacks last in the order of weeks or a month or more, after which there is a spontaneous remission. These remissions, however, are generally not complete, and most patients are left with residual symptoms. Recurrent attacks are the rule, and after each attack the overall burden of residual symptoms increases. In some cases, this typical episodic course may evolve into one of steady, waxing, and waning progression; rarely, Behçet's disease may pursue a chronic, non-episodic course from the outset, as was seen in a case of progressive dementia with apthous and genital ulcers (Borson 1982).

Etiology

During acute attacks (Borhani Haghighi *et al.* 2007; Mirsattari *et al.* 2004), lymphocytic perivascular cuffing of small vessels, primarily venules, although found preferentially in the brainstem, is also seen in the thalamus, internal capsules, basal ganglia, and also in the cerebral white matter and cortex, wherein the frontal lobe is primarily affected. A lymphocytic meningitis may also occur, which may be associated with dural sinus thrombosis and cerebral venous thrombosis.

Although the etiology is not known, an autoimmune process is suspected. Although familial cases have been reported, the vast majority are sporadic.

Differential diagnosis

The relapsing and remitting course of Behçet's syndrome immediately suggests multiple sclerosis; however, in multiple sclerosis one does not find the characteristic apthous and genital ulcers, uveitis, or pathergy.

Treatment

During attacks, most patients are treated with a combination of steroids and either an immunosuppressant or one of the antitumor necrosis factor alpha antibodies, infliximab or rituximab. It must be emphasized that this is a rapidly evolving area, and that a familiarity with the most current literature is critical. Symptomatic treatment of delirium and dementia is discussed in Chapters 6.5 and 6.3, respectively.

11.12 HYPERTENSIVE ENCEPHALOPATHY

Hypertensive encephalopathy, a relatively common disorder, is characterized by the occurrence of delirium and headache in the setting of severe diastolic hypertension.

Before proceeding, some words are in order regarding the disorder known as reversible posterior leukoencephalopathy syndrome (RPLS). As noted in Chapter 11.13, there is a trend to include hypertensive encephalopathy as a mere part of RPLS; however, given that the mechanism of tissue injury is different in RPLS and hypertensive encephalopathy, they are treated separately in this text.

Clinical features

In the setting of acute, sustained increases in the mean arterial blood pressure, delirium and headache evolve over a matter of 1–3 days; most patients also experience nausea and vomiting, and a majority will experience bilateral visual blurring, which may progress to cortical blindness; seizures may also occur and, in a minority, focal signs such as aphasia or hemiplegia may be seen (Chester *et al.* 1978; Healton *et al.* 1982; Oppenheimer and Fishberg 1928). Untreated, patients may become comatose. Funduscopic examination may reveal papilledema or retinal hemorrhages. Acute cardiac and renal failure may accompany the cerebral symptomatology.

Computed tomography scanning may reveal bilaterally symmetric areas of hypodensity in the occipitoparietal white matter. Magnetic resonance scanning is far more sensitive and, on T2-weighted or FLAIR imaging, increased signal intensity will be seen in the same areas (Hauser *et al.* 1988), while gradient-echo imaging may reveal evidence of petechial hemorrhages (Weingarten *et al.* 1994). Diffusion-weighted imaging is generally normal unless infarction has occurred. Rarely, these MRI findings may be found in the brainstem and cerebellum (Cruz-Flores *et al.* 2004; Kanazawa *et al.* 2005).

Course

Untreated, hypertensive encephalopathy may be fatal. In cases in which blood pressure is corrected, symptoms gradually resolve over a matter of days, and white matter abnormalities seen on CT or MR scanning typically clear within a week or so. In those cases in which focal findings were noted, these may persist, and MR scanning will show persistent abnormalities consistent with infarction. With a sufficient number of strategically placed infarctions, patients may be left with a multi-infarct dementia.

Etiology

Normally, cerebral vascular autoregulation not only ensures adequate flow during periods of hypotension, but also protects delicate arterioles and capillaries from the adverse effect of elevated pressure. In hypertensive encephalopathy, this protective autoregulatory effect is lost, and with exposure to grossly elevated pressures, there is disruption of the endothelium with extravasation of fluid and plasma proteins across the now-damaged blood–brain barrier, thus creating the vasogenic edema characteristic of this disorder. Although most cases occur in the context of a grossly elevated systemic blood pressure, one must also take into account the chronic baseline blood pressure before deciding whether the current blood pressure is sufficient to cause an encephalopathy. Cerebral autoregulation is effective only along a range of mean arterial pressures, with this range extending roughly 100 mmHg. Normally, then, in non-hypertensive patients, this range covers from 50 to 150 mmHg, and hence, in a non-hypertensive patient, elevations of the mean arterial pressure above 150 mmHg may cause an encephalopathy. In patients with chronic hypertension, however, the range shifts, thereby offering more protection at these higher levels, and consequently, in the chronically hypertensive patient, greater rises in the mean arterial pressure are typically required before an encephalopathy develops. This range can only shift so far, however, and when diastolic pressure rises much above 130 mmHg autoregulation almost always fails.

In addition to vasogenic edema, further pathology may occur (Chester *et al.* 1978), including small infarctions due to the formation of fibrin thrombi within arterioles and capillaries and petechial hemorrhages due to rupture of the same structures.

Differential diagnosis

Not all deliria occurring in the setting of grossly elevated diastolic pressures occur as a result of hypertensive encephalopathy. As noted earlier, renal failure is not uncommon, and uremic encephalopathy must also be considered. Intracerebral hemorrhage may also be considered in the differential, but is generally of more acute onset and is easily recognized on neuroimaging. The reversible posterior leukoencephalopathy syndrome should be considered in cases that are clinically identical to hypertensive encephalopathy in all respects except for the fact that hypertension is either lacking or only mild.

Treatment

Hypertensive encephalopathy is a medical emergency and the pressure must be promptly lowered. Normotensive levels, however, are not the goal because, with a loss of autoregulatory capacity, a systemic pressure within normal limits may be followed by cerebral hypoperfusion and watershed infarctions (Ledingham and Rajagopalan 1979). Consequently, during acute treatment, the diastolic pressure should be lowered over a period of 2–6 hours to between 100 and 110 mmHg; once this has been accomplished, further treatment may be aimed at bringing the pressure down to normal levels in a more leisurely manner over the next few days. Acute treatment may be accomplished with intravenous sodium nicardapine or labetalol. In cases in which intravenous access is not immediately available or when the clinical situation is not as urgent, intramuscular hydralazine may be utilized. Seizures may be treated with intravenous lorazepam and an anti-epileptic drug, such as fosphenytoin, as described in Chapter 8.3. The general treatment of delirium is discussed in Chapter 6.5.

11.13 REVERSIBLE POSTERIOR LEUKOENCEPHALOPATHY SYNDROME

The reversible posterior leukoencephalopathy syndrome (RPLS), also known as posterior reversible encephalopathy syndrome (PRES), is a rare disorder occurring secondary to toxicity of chemotherapeutic or immunomodulatory agents. Clinically, hypertensive encephalopathy is identical to it, and there is a trend to expand the definition of RPLS into an 'umbrella' syndrome and include hypertensive encephalopathy under this umbrella. This may or may not be appropriate however: although both the narrowly defined RPLS and hypertensive encephalopathy involve endothelial damage, the mechanisms of this damage are different. In RPLS it is direct toxicity, whereas in hypertensive encephalopathy it is a matter of small arterioles and capillaries being exposed to grossly elevated blood pressure. In this text, therefore, RPLS is treated in the narrow sense and kept separate from hypertensive encephalopathy, which is described in Chapter 11.12.

Clinical features

The syndrome (Hinchey *et al.* 1996) presents acutely, over hours, and manifests with delirium or lethargy, accompanied typically by headache, nausea or vomiting, and seizures, which may be simple partial, complex partial, or grand mal in type. Blurring of vision is common, and may progress to cortical blindness. In some cases, focal findings, such as hemiparesis or aphasia, may also appear.

Magnetic resonance scanning with FLAIR sequences (Casey *et al.* 2000; Lamy *et al.* 2004) reveals increased signal intensity in the cerebral white matter. Importantly, although this is predominantly posterior, it often extends to the white matter of the temporal, parietal, and even frontal lobes. Furthermore, in the great majority, associated increased signal intensity may also be seen in the adjacent cortex. In a significant minority of cases, similar changes may be seen in the brainstem and cerebellum.

Diffusion-weighed imaging (Ay *et al.* 1998), although typically normal, may reveal increased signal intensity in the same areas. Importantly, however, on apparent diffusion coefficient (ADC) mapping these areas show increased signal intensity, indicating the presence of edema rather than infarction. Small areas of infarction, however, may occur within these larger areas of edema, and here one finds the typical combination of increased signal intensity on diffusion-weighted imaging in association with decreased signal intensity on ADC mapping.

Course

With prompt and adequate treatment, most symptoms resolve within days. In a minority, however, some symptoms, primarily the focal ones, may persist.

Pathology and etiology

Vasogenic edema is seen within the white matter, as indicated by both brain biopsy (Lavigne *et al.* 2004) and MRI findings. In cases with an unfavorable outcome, infarction of the white matter occurs.

This syndrome has been noted secondary to treatment with a variety of chemotherapeutic and immunomodulatory agents, including cyclosporine, tacrolimus, sirolimus, cisplatin, L-asparaginase, cytarabine, vincristine, methotrexate, cyclophosphamide, bevacizumab, interferon alpha, and immunoglobulins. Although the mechanism of toxicity is not clear, in all likelihood these agents cause damage to the endothelium. Importantly, this toxicity can occur at therapeutic doses; however, it appears that the syndrome is more likely to occur in the setting of modest elevations in diastolic pressure, for example up to 100–105 mmHg.

Differential diagnosis

The differential with hypertensive encephalopathy rests on the presence or absence of severe hypertension: although patients with the reversible posterior leukoencephalopathy syndrome may, as noted, have an elevated diastolic pressure up to 105 mmHg, pressures simply do not rise to the levels seen in hypertensive encephalopathy.

Bilateral occipital infarction, as seen in the 'top of the basilar' syndrome, is distinguished by its acute onset and the MRI findings of a large area of infarction covering both occipital lobes.

Treatment

Potentially offending medications should be discontinued, and the blood pressure should be brought within normal limits. Seizures may be treated with lorazepam and an anti-epileptic drug such as fosphenytoin, as described in Chapter 8.3. The general treatment of delirium is discussed in Chapter 6.5.

11.14 MELAS

Mitochondrial encephalomyopathy, lactic acidosis, and stroke-like episodes is a rare inherited disorder characterized by varying combinations of encephalopathy, stroke-like episodes, migrainous headaches, seizures, and deafness; as with all mitochondrial disorders, it displays a maternal pattern of inheritance. Although most cases present in childhood or adolescence, later onsets in the third and fourth decades are not uncommon, and, although rare, even later onsets have been reported.

Clinical features

As noted, the clinical presentation is quite varied. Some may present with stroke-like episodes (Iizuka *et al.* 2003), with hemiparesis, hemianopia, cortical blindness, or aphasia. These episodes are termed 'stroke-like' because, unlike most cases of stroke, they evolve gradually, over hours or days, and, on imaging, one finds abnormalities (discussed below) which do not conform to vascular territories. Delirium may accompany these episodes and may persist, only to resolve into a dementia, which, in turn, may be gradually progressive (Sharfstein *et al.* 1999). In some cases, psychosis has been noted (Apostolova *et al.* 2005). Seizures may occur, and may be simple partial (Canafoglia *et al.* 2001), complex partial (including complex partial status epilepticus; Leff *et al.* 1998), or grand mal in type. Migrainous headaches may precede, accompany, or follow any of these symptomatologies.

Hearing loss is common, and there may be an associated myopathy with proximal weakness. Diabetes mellitus and either hypo- or hyperthyroidism may also occur.

Diffusion-weighted MR scanning during stroke-like episodes reveals areas of increased signal intensity; over time, these same areas will demonstrate increased signal intensity on FLAIR imaging. Typically, abnormalities are seen in the cortex and subcortical white matter of the occipital and parietal lobes, and, as noted earlier, do not correspond to vascular territories, and indeed are not uncommonly bilateral. Furthermore, on CT scanning, one may also see calcification of the basal ganglia.

The electroencephalogram (EEG) may reveal focal slowing, generally in the occipital area, which may be accompanied by interictal epileptiform discharges or periodic lateralized epileptiform discharges (PLEDs).

The lactate level is elevated in both the serum and CSF, and the CSF total protein is elevated.

Muscle biopsy reveals ragged red fibers, and genetic testing for mutations in mitochondrial DNA may reveal characteristic mutations.

Course

Although the overall course is characterized by progression, the rate of progression, and the sequence with which various symptomatologies occur, is quite varied.

Etiology

Mitochondrial dysfunction occurs secondary to mutations in the gene for mitochondrial tRNA, with proliferation of abnormal mitochondria in vascular endothelial and smooth muscle cells. Infarcts, which, as noted earlier, do not conform to vascular territories, are found to affect both gray and white matter, primarily in the parietal and occipital lobes.

Differential diagnosis

Mitochondrial encephalomyopathy, lactic acidosis, and stroke-like episodes should be suspected in any young person with recurrent stroke, especially when accompanied by migrainous headache or hearing loss. CADASIL is high on the differential and is suggested by the MRI finding of white matter changes in the anterior portion of the temporal lobes.

Treatment

Symptomatic treatment of delirium and dementia is described in Chapters 6.3 and 6.5, respectively. Seizures may be treated with anti-epileptic drugs; however, valproic acid should be avoided as it may aggravate seizures (Lin *et al.* 2007). L-Arginine, given during stroke-like episodes, may be followed by prompt and sustained improvement (Koga *et al.* 2005).

11.15 THROMBOTIC THROMBOCYTOPENIC PURPURA

Thrombotic thrombocytopenic purpura, or TTP, is a rare, devastating disorder, typically occurring in young or middle-aged adults. It is marked by the subacute onset of delirium and thrombocytopenia (Druschky *et al.* 1998).

Clinical features

The delirium is marked by a pronounced fluctuation in the severity of symptoms throughout the day. Other symptomatology includes focal signs, such as hemiparesis or aphasia, which are typically transient, and seizures. Importantly, in addition to grand mal seizures, there may be complex partial status epilepticus, with prolonged confusion, or non-convulsive status epilepticus, with stupor or coma (Blum and Drislane 1996; Garrett *et al.* 1996).

Most patients are febrile, and widespread purpura are common. The platelet count is generally reduced below 30 000/mm^3, and there is an accompanying microangiopathic hemolytic anemia with schistocytes. Renal failure with proteinuria and azotemia is common.

Magnetic resonance scanning may be normal or reveal small areas of infarction or, in a significant minority, fully reversible cortical or subcortical areas of increased signal intensity on FLAIR imaging strongly reminiscent of the pattern seen in the reversible posterior leukoencephalopathy syndrome (Burrus *et al.* 2009).

The EEG may be normal or show generalized slowing; interictal epileptiform discharges may also be seen, as may ictal discharges during episodes of complex partial or non-convulsive status.

Course

The disorder may persist for days to months; untreated, it is typically fatal.

Etiology

Platelet thrombi are found in arterioles, capillaries, and venules both within the kidneys and throughout the cerebrum; presumably, the ongoing aggregation and disaggregation of these thrombi within the cerebrum account for the waxing and waning nature of the symptomatology which is common in this disorder. In some cases, small cortical infarctions may occur (Gruber *et al.* 2000). The consumption of platelets by these thrombi results in both the thrombocytopenia and the associated purpura.

It appears that in most cases this disorder is associated with the presence of an IgG immunoglobulin directed at an enzyme, ADAMTS13, which is responsible for cleaving von Willebrand factor.

Differential diagnosis

Disseminated intravascular coagulation is distinguished by a decreased fibrinogen level, an increase in fibrin split products, and a prolonged partial thromboplastin time.

Treatment

Plasma exchange, with or without steroids, is generally effective. With timely treatment, most patients recover, with only a minority being left with persistent deficits. In cases where confusion or stupor persists despite otherwise adequate treatment, an EEG, if not already done, is mandatory in order to rule out status.

11.16 FAT EMBOLISM SYNDROME

This syndrome represents an uncommon complication of fractures or surgery to the long bones, or trauma to fatty tissues, in which showers of fat globules pass to and through the lungs, leaving the patient in respiratory distress and with a delirium.

Clinical features

Classically, anywhere from hours to 3 days after relevant trauma or surgery, patients develop the triad of pulmonary distress, a petechial rash, and evidence of cerebral dysfunction (Dines *et al.* 1975; Jacobson *et al.* 1986), which typically takes the form of a delirium (which may progress to stupor or coma) with or without seizures and focal signs. It must be emphasized, however, that the full triad is often not present.

Although CT scanning is typically normal, MRI with diffusion-weighted or FLAIR imaging will reveal multiple, generally small, areas of increased signal intensity scattered throughout the cerebral white matter, with variable involvement of the cortex or subcortical structures (Marshall *et al.* 2004).

In a minority of cases, the chest radiograph reveals 'fluffy' infiltrates, and, likewise in a minority, fat globules may be seen in the urine.

Course

The mortality rate is as high as 10 percent; those who survive experience a variable degree of recovery over the following days or weeks.

Etiology

With fractures or orthopaedic surgery, especially when the long bones are involved, or, less commonly, during procedures (e.g. liposuction) involving trauma to fatty tissues, neutral fat is released into the venous circulation, whereupon it undergoes paradoxical embolization to the brain, either through a patent foramen ovale (Forteza *et al.* 2002), or, more commonly, pulmonary shunts (Riding *et al.* 2004). Autopsy studies (von Hochstetter and Friede 1977; Kamenar and Burger 1980) may reveal multiple petechial hemorrhages or micro-infarctions.

Differential diagnosis

In cases secondary to trauma, head trauma may also have occurred; in postoperative cases, other causes of postoperative delirium, as discussed in Chapter 6.5, must be considered. When pulmonary involvement is severe, respiratory failure may occur and global cerebral hypoxia must also be considered.

Treatment

In addition to any necessary respiratory support, seizures may be treated with anti-epileptic drugs and symptomatic treatment of delirium may be provided, as outlined in Chapter 6.5. In one case occurring in the context of a patent foramen ovale, percutaneous closure was dramatically effective (Forteza *et al.* 2002).

11.17 MULTIPLE CHOLESTEROL EMBOLI SYNDROME

In patients with severe atheromatous disease of the aorta, multiple crystals of cholesterol may break off from plaques, either spontaneously or with instrumentation, and embolize to the brain causing multiple, generally small, infarctions. Such microemboli also, of course, travel to other organs, most notably the kidneys.

Clinical features

The mode of onset appears related to whether the syndrome occurs with instrumentation or spontaneously. Cases occurring secondary to instrumentation (e.g. with coronary artery bypass grafting, cardiac catheterization, or carotid angiography [Hagiwara *et al.* 2004] tend to present acutely with a full syndrome a day or so after the instrumentation. By contrast, spontaneously occurring cases may present subacutely, with the syndrome evolving over weeks or months.

Acute cases occurring after instrumentation present with delirium (Ezzeddine *et al.* 2000), which may or may not be accompanied by focal signs, such as hemiplegia or hemianopia. Evidence of multiple emboli to other organs (Colt *et al.* 1988) includes renal failure, livedo reticularis of the lower extremities (often evident only when the patient is erect), 'blue toes', petechial hemorrhages of the skin, a peripheral mononeuritis multiplex, and abdominal pain. Conceivably, if the instrumentation in question was cerebral angiography and only the cerebral vessels were affected with atheromatous disease, with the aorta being 'clean' in this regard, the syndrome could present with central nervous system involvement in isolation, without other organ involvement. Leukocytosis, eosinophilia, and an elevated ESR may all be present. Biopsy of the skin in areas with livedo reticularis may reveal embolic particles.

Subacute cases occurring spontaneously may present with transient ischemic attacks (TIAs) or multiple cerebral infarctions; progressive renal failure is also typically present (Andreaux *et al.* 2007; Beal *et al.* 1981).

Course

In cases of acute onset, fatalities, often secondary to renal failure, are common. Patients with subacute cases typically develop severe renal failure.

Differential diagnosis

Acute cases must be differentiated from other causes of postoperative delirium, as discussed in Chapter 6.5; in this regard, the presence of livedo reticularis, 'blue toes', and renal failure constitute strong clues. Subacute cases must be distinguished from the lacunar state, and here the accompanying renal failure provides a clue.

Etiology

Showers of microemboli composed of cholesterol crystals eventually lodge in small arterioles where they provoke an inflammatory response with eventual occlusion and multiple small infarctions (Ezzeddine *et al.* 2000).

Treatment

The best treatment is prevention of atherosclerosis. Furthermore, extra caution should be exercised regarding instrumentation affecting the ascending aorta or cerebral vasculature in any patient with severe atherosclerosis. There is a case report suggesting that steroids given acutely may be beneficial (Andreaux *et al.* 2007).

11.18 TRANSIENT GLOBAL AMNESIA

Transient global amnesia, first described in the English-language literature by Fisher and Adams in 1958 (Fisher and Adams 1958), is an uncommon disorder characterized by infrequent amnestic episodes. Although the etiology of this disorder is not known, it is included in this section on vascular disorders because of the strong suspicion that, as noted below, it may result from either transient ischemia or venous congestion of medial temporal structures.

Clinical features

The overall clinical features have been described in a number of papers (Bolwig 1968; Fisher and Adams 1958, 1964; Gordon and Marin 1979; Heathfield *et al.* 1973; Hodges and Ward 1989; Kushner and Hauser 1985; Miller *et al.* 1987a; Quinette *et al.* 2006; Regard and Landis 1984; Shuttleworth and Morris 1966).

The first episode of transient global amnesia generally occurs in the sixth, seventh, or eighth decade. Episodes themselves are generally of abrupt onset, and may be associated with various precipitating events (Fisher 1982; Kushner and Hauser 1985; Quinette *et al.* 2006), such as strong emotion, sexual intercourse, pain, physical exertion, a Valsalva maneuver, and even immersion in cold water. Whether or not a precipitating event is present, patients suddenly experience an amnesia that has both retrograde and anterograde components. The retrograde component covers at least the previous few hours and in many cases may stretch much further into the past: typically, this retrograde amnesia displays a temporal gradient, such that whereas the amnesia may be quite dense for events of the very recent past, it become 'patchy' for events further back (Kritchevsky and Squire 1989). The anterograde component is fairly dense, and patients are unable to keep track of any ongoing events during the episode. Most patients, although not confused, are more or less alarmed at their state, and most will anxiously and repeatedly ask where they are and how they got to be where they are. Formal mental status testing reveals that patients are coherent, alert, retain a knowledge of who they are, and, as noted, are not confused. Although digit span is intact, patients are unable to recall any of three words after 5 minutes; furthermore, they will be unable to recall events of the recent past leading up to the onset of the episode. In essence, cognitive ability, other than memory, remains intact, and indeed some patients may engage in quite complex activity, such as playing an organ piece, during the episode itself (Byer and Crowley 1980). The neurologic examination is otherwise normal. The episode itself generally lasts anywhere from 4 to 18 hours, averaging about 6 hours, and terminates gradually.

After the episode has cleared, patients are once again able to keep track of ongoing events, and their ability to recall words after 5 minutes is fully restored. When they try and recall what happened, however, they find an 'island' of amnesia that covers not only the duration of the event itself but also any events that occurred anywhere from a few minutes to an hour or so just before the onset of the episode. Although clinically patients are thus fully restored, neuropsychological testing may reveal some subtle decrements in memory (Guillery-Girard *et al.* 2006; Steinmetz and Vroom 1972); however, not all studies agree on this (Bartsch *et al.* 2006; Uttner *et al.* 2007).

Course

Long-term follow-up of approximately 5–6 years shows that only about one-fifth to one-quarter of patients have a recurrence (Gandolfo *et al.* 1992; Hinge *et al.* 1986; Miller *et al.* 1987a), and it appears that recurrences are more likely in cases occurring secondary to precipitating events (Agosti *et al.* 2006).

Etiology

It appears that the amnesia seen in transient global amnesia represents the effects of a similarly transient dysfunction of medial temporal lobe structures. Single-photon emission computed tomography (SPECT) studies have revealed hypoperfusion of the medial aspects of both temporal lobes (Lampl *et al.* 2004), more so the left than the right, and diffusion-weighted imaging has revealed a small number of punctate areas of increased signal intensity in the hippocampus (Bartsch *et al.* 2006; Sedlaczek *et al.* 2004; Strupp *et al.* 1998); notably both of these abnormal findings clear with time. Structural MRI studies are generally normal (Bartsch *et al.* 2006); however, one high-resolution MRI study found an increased number of microcavities in the CA-1 subfield of the hippocampus, suggesting that there may be some subtle permanent damage (Nakada *et al.* 2005). Several mechanisms have been proposed to account for these changes, including epileptic activity, transient ischemia, migraine, and, recently, venous reflux. Epileptic activity appears unlikely: EEGs obtained during episodes of transient global amnesia do not display ictal activity (Miller *et al.* 1987b), and the attacks themselves, rather than ending abruptly, as is typical of a seizure, tend to clear gradually. Transient ischemia has long been an attractive hypothesis, but patients with transient global amnesia generally have few, if any, risk factors for stroke (Hodges and Warlow 1990; Lauria *et al.* 1998; Melo *et al.* 1992; Toledo *et al.* 2008) and rarely end up having stroke during follow-up (Zorzon *et al.* 1995). Migraine may play a role: there is a definite association between transient global amnesia and migraine (Crowell *et al.* 1984; Hodges and Warlow 1990; Melo *et al.* 1992; Schmidtke and Ehmsen 1998; Zorzon *et al.* 1995), and an episode of 'spreading depression of Leão' confined to medial temporal structures could, conceivably, cause amnesia. Migraine, however, would not explain the SPECT and diffusion-weighted imaging abnormalities described earlier.

The final proposed mechanism, namely venous reflux, is quite interesting. To understand the theory of venous reflux one must recall some anatomy. Normally the medial temporal structures are drained by the deep veins of Rosenthal. These veins, in turn, drain to the great vein of Galen, which drains to the straight sinus. The straight sinus, in turn, courses posteriorly to join the superior sagittal sinus at the sinus confluens. The sinus confluens gives rise to the left and right transverse sinuses, which in turn drain to the sigmoid sinuses, then into the internal jugular veins and finally into the brachiocephalic veins which merge to form the superior vena cava. Of note, and of particular importance to the venous reflux theory, in the majority of individuals, the straight sinus drains not into both transverse sinuses but only into the left transverse sinus. Given this unique drainage pattern, it is theoretically possible that venous reflux confined to the left internal jugular vein could cause venous stasis in the medial aspects of both the left *and* the right temporal lobes. Recent studies have demonstrated such reflux through incompetent valves (Maalikjy Akkawi *et al.* 2003) in the left internal jugular vein in patients with transient global amnesia, adding weight to this theory (Chung *et al.* 2006; Sander *et al.* 2000). The reflux itself could be related to a Valsalva-type maneuver, consistent with some of the precipitating events mentioned earlier, or, as indicated in one study, to intrathoracic compression of the left brachiocephalic vein as it is pressed between the aorta and the sternum (Chung *et al.* 2006). As attractive as this theory is, however, jugular venous reflux is found in no more than roughly 70 percent of cases (Agosti *et al.* 2010), indicating that other factors must be at work.

Differential diagnosis

Transient global amnesia, as discussed in Chapter 6.6, is best conceived of as one of the episodic anterograde amnesias and, as such, must be distinguished from TIAs, pure epileptic amnesia, concussion, substance-induced blackouts, and dissociative amnesia.

Transient ischemic attacks affecting the medial aspects of one or both temporal lobes may produce a syndrome quite similar to transient global amnesia, and this could occur with transient reduced blood flow in either the occipitotemporal artery or the anterior choroidal artery. In both these cases, however, one typically sees associated symptoms which are not characteristic of transient global amneisa. In the case of the occipitotemporal artery, it is usually the posterior cerebral artery that is occluded, and here, in addition to temporal lobe dysfunction, one also finds either a hemianopia or cortical blindness, and in the case of the anterior choroidal artery, proximal occlusion causes also a hemiparesis and a hemianopia. Exceptions, however, to this rule may occur when, for example, an embolus occludes only the occipitotemporal artery or only the distal portion of the anterior choroidal artery. Furthermore, transient ischemic attacks of one or both thalami may also cause a transient amnesia. The distinction between such 'pure' amnestic transient ischemic attacks and transient global amnesia may be based on the duration of the amnesia and MRI findings. Transient ischemic attacks, as noted in Chapter 8.4, last no longer than an hour, in contrast to transient global amnesia, which lasts far longer. In cases of ischemic amnesia lasting longer than an hour, actual infarction will have occurred, and here the diffusion-weighted MRI scanning will reveal a substantial area of increased signal intensity, in contrast to the few punctate areas seen in transient global amnesia.

Pure epileptic amnesia is suggested by a sudden offset of the episode, in contrast to the gradual offset of transient global amnesia, and by the fact that patients in the midst of an epileptic amnestic episode generally do not engage in the anxious questioning typical of transient global amnesia. Other distinguishing features include the occurrence of other seizure types, such as complex partial or grand mal seizures, and a pattern of frequent recurrence.

Concussion is immediately suggested by a history of head injury, which, of course, must often be gained from others. Furthermore, during concussion, patients may display a degree of confusion.

Blackouts, as seen during alcohol intoxication or intoxication with benzodiazepines, are suggested by other evidence of intoxication, such as dysarthria or ataxia, and by a lack of recognition on the part of the 'blacked-out' patient that anything is amiss and a corresponding absence of any anxious questioning.

Dissociative amnesia may be distinguished from transient global amnesia on several counts. First, the initial episode of dissociative amnesia generally occurs in the late teens or early adult years, in marked contrast with the age of onset of transient global amnesia. Second, during dissociative amnesia, some, but not all, patients may complain that they do not know who they are, in contrast to patients with transient global amnesia, who do retain personal identity. Third, although some cases of dissociative amnesia may last only hours, many last much longer, even weeks, in contrast with transient global amnesia, which typically never lasts longer than 24 hours. Finally, some, but again not all, patients with dissociative amnesia will, with psychotherapy or hypnosis, eventually be able to recall events that occurred during the episode, in marked contrast to the case of transient global amnesia, where the 'island' of amnesia is permanent.

Treatment

Apart from supervision during the episode itself, and reassurance afterwards of its overall benign nature, no treatment is generally required. However, in cases in which there are obvious precipitating factors, patients should be accordingly instructed to avoid them.

REFERENCES

Abu-Shakara M, Khraishi M, Grossman H *et al.* Primary angiitis of the CNS diagnosed by angiography. *QJ Med* 1994; **87**:351–8.

Agosti C, Akkawi NM, Borroni B *et al.* Recurrency in transient global amnesia: a retrospective study. *Eur J Neurol* 2006; **13**:986–9.

Agosti C, Akkawi NM, Borroni B *et al.* Cerebrovascular risk factors and triggers in transient global amnesia patients with and without jugular vein imcompetence: results from a sample of 243 patients. *Eur Neurol* 2010; **63**:291–4.

Akiguchi I, Tomimoto H, Suenega T *et al.* Alterations in glia and axons in the brains of Binswanger's disease patients. *Stroke* 1997; **28**:1423–9.

Akman-Demir G, Baykan-Kurt B, Serdaroglu P *et al.* Seven year follow-up of neurologic involvement in Behçet syndrome. *Arch Neurol* 1993; **53**:691–4.

Akman-Demir G, Serdaroglu P, Tasci B *et al.* Clinical patterns of neurological involvement in Behçet's disease: evaluation of 200 patients. *Brain* 1999; **122**:2171–81.

Akman-Demir G, Bahar S, Coban O *et al.* Cranial MRI in Behçet's disease: 134 examinations of 98 patients. *Neuroradiology* 2003; **45**:851–9.

Akyol A, Akyildiz UO, Tataroglu C. Vascular parkinsonism: a case of lacunar infarction localized to mesencephalic substantia nigra. *Parkinsonism Relat Disord* 2006; **12**:459–61.

Al-Araji A, Sharquie K, Al-Rawi Z. Prevalence and patterns of neurological involvement in Behçet's disease: a prospective study from Iraq. *J Neurol Neurosurg Psychiatry* 2003; **74**:608–13.

Alhalabi M, Moore PM. Serial angiography in isolated angiopathy of the central nervous system. *Neurology* 1994; **44**:1221–6.

Alrawi A, Trobe JD, Blaivas M *et al.* Brain biopsy in primary angiitis of the central nervous system. *Neurology* 1999; **53**:858–60.

Altinors N, Senveli E, Arda N *et al.* Intracerebral hemorrhage and hematoma in Behçet's disease: case report. *Neurosurgery* 1987; **21**:582.

Andreaux R, Marro B, El Khoury N *et al.* Reversible encephalopathy associated with cholesterol embolism syndrome: magnetic resonance imaging and pathological findings. *J Neurol Neurosurg Psychiatry* 2007; **78**:180–2.

Apostolova LG, White M, Moore SA *et al.* Deep white matter pathologic features in watershed regions: a novel pattern of central nervous system involvement in MELAS. *Arch Neurol* 2005; **62**:1154–6.

Auchus AP, Chen CP, Sadagar SN *et al.* Single stroke dementia: insights from 12 cases in Singapore. *J Neuro Sci* 2002; **203–4**:85–9.

Auchus AP, Brasher HR, Salloway S *et al.* Galantamine treatment of vascular dementia. *Neurology* 2007; **69**:448–58.

Ay H, Buonanno FS, Schaefer PW *et al.* Posterior leukoencephalopathy without severe hypertension: utility of diffusion-weighted MRI. *Neurology* 1998; **51**:1369–76.

Aykutlu E, Baykan B, Serdaroglu P *et al.* Epileptic seizures in Behçet disease. *Epilepsia* 2002; **43**:832–5.

Bartsch T, Alfke K, Stingele R *et al.* Selective affection of hippocampal CA-1 neurons in patients with transient global amnesia. *Brain* 2006; **129**:2874–84.

Baudrimont M, Dubas F, Joutel A *et al.* Autosomal dominant leukoencephalopathy and subcortical ischemic stroke. A clinicopathological study. *Stroke* 1993; **24**:122–5.

Beal MF, Williams RS, Richardson EP *et al.* Cholesterol embolism as a cause of transient ischemic attacks and cerebral infarction. *Neurology* 1981; **31**:860–5.

Behçet H. Uber rezidivierende, aphtose, durch ein Virus verusachte Geschwure am Mund, am Auge und an den Genitalien. *Dermatologische Wochenschrift* 1937; **105**:1152–63.

Benisty S, Gouw AA, Porcher R *et al.* Location of lacunar infarcts correlates with cognition in a sample of non-disabled subjects with age-related white-matter changes: the LADIS study. *J Neurol Neurosurg Psychiatry* 2009; **80**:478–83.

Bergmann M, Ebke M, Yuan Y. Cerebral autosomal dominant arteriopathy with subcortical infarcts and leukoencephalopathy (CADASIL): a morphological study of a German family. *Acta Neuropathol* 1996; **92**:341–50.

Biffi A, Halpin A, Towfighi A *et al.* Aspirin and recurrent intracerebral hemorrhage in cerebral amyloid angiopathy. *Neurology* 2010; **75**:693–8.

Binswanger O. Die abgrenzung des allgemeinen progressiven. *Paralysie Klin Wochenschr* 1894; **31**:1103–5, 1137–9, 1180–6.

Black S, Roman GC, Geldmacher DS *et al.* Efficacy and tolerability of donepezil in vascular dementia; positive results of a 24-week, multicenter, international, randomized, placebo-controlled trial. *Stroke* 2003; **34**:2323–30.

Blass JP, Hoyer S, Nitsch R. A translation of Otto Binswanger's article 'The delineation of the generalized progressive paralyses'. *Arch Neurol* 1991; **48**:961–72.

Blum AS, Drislane FW. Nonconvulsive status epilepticus in thrombotic thrombocytopenic purpura. *Neurology* 1996; **47**:1079–81.

Bogucki A, Janczewska E, Koszewska I *et al.* Evaluation of dementia in subcortical arteriosclerotic encephalopathy (Binswanger's disease). *Eur Arch Psychiatry Clin Neurosci* 1991; **241**:91–7.

Bolwig TG. Transient global amnesia. *Acta Neurol Scand* 1968; **44**:101–6.

Borhani Haghighi A, Sharifzad HR, Matin S *et al.* The pathological presentations of neuro-Behçet disease: a case report and review of the literature. *Neurologist* 2007; **13**:209–14.

Borson S. Behçet's disease as a psychiatric disorder: a case report. *Am J Psychiatry* 1982; **139**:1348–9.

Bruetsch WL, Williams CL. Arteriosclerotic muscular rigidity with special reference to gait disturbances. *Am J Psychiatry* 1954; **111**:332–6.

Burrus TM, Wijdicks EF, Rabinstein AA. Brain lesions are most often reversible in acute thrombotic thrombocytopenic purpura. *Neurology* 2009; **73**:66–70.

Byer JA, Crowley WJ. Musical performance during transient global amnesia. *Neurology* 1980; **30**:80–2.

Calabrese LH, Mallek JA. Primary angiitis of the central nervous system: report of 8 new cases, review of the literature, and proposal for diagnostic criteria. *Medicine* 1988; **67**:20–39.

Canafoglia L, Franceschetti S, Antozzi C *et al.* Epileptic phenotypes associated with mitochondrial disorders. *Neurology* 2001; **56**:1340–6.

Caplan LR, Schoene WC. Clinical features of subcortical arteriosclerotic encephalopathy (Binswanger disease). *Neurology* 1978; **28**:1206–15.

Caselli RJ. Giant cell (temporal) arteritis: a treatable cause of multi-infarct dementia. *Neurology* 1990; **40**:753–5.

Caselli RJ, Hunder GG, Whisnant JP. Neurologic disease in biopsy-proven giant cell (temporal) arteritis. *Neurology* 1988; **38**:352–9.

Case Records of the Massachusetts General Hospital. Case 8–1989. *N Engl J Med* 1989; **320**:514–24.

Casey SO, Sampaio RC, Michel E *et al.* Posterior reversible encephalopathy syndrome: utility of fluid-attenuated inversion recovery MR imaging in the detection of cortical and subcortical lesions. *AJNR* 2000; **21**:1199–206.

Chabriat H, Vahedi K, iba-Zizen MT *et al.* Clinical spectrum of CADASIL: a study of 7 families. *Lancet* 1995; **346**:934–9.

Chajek T, Fainaru M. Behçet's disease: report of 41 cases and a review of the literature. *Medicine* 1975; **54**:179–86.

Chester EM, Agamanolis DP, Banker BQ *et al.* Hypertensive encephalopathy: a clinicopathologic study of 20 cases. *Neurology* 1978; **28**:928–39.

Choi JC, Kang S-Y, Kang J-H *et al.* Intracerebral hemorrhages in CADASIL. *Neurology* 2006; **67**:2042–4.

Chung C-P, Hsu H-Y, Chang F-C *et al.* Detection of intracranial venous reflux in patients of transient global amnesia. *Neurology* 2006; **66**:1873–7.

Colt HG, Begg RJ, Saporito JJ *et al.* Cholesterol emboli after cardiac catheterization. Eight cases and a review of the literature. *Medicine* 1988; **67**:389–400.

Cosgrove CR, Leblanc R, Meagher-Villemure K *et al.* Cerebral amyloid angiopathy. *Neurology* 1985; **38**:625–31.

Cravioto H, Feigin I. Noninfectious granulomatous angiitis involving the central nervous system. *Neurology* 1959; **9**:599–609.

Critchley M. Arteriosclerotic parkinsonism. *Brain* 1929; **52**:23–83.

Crowell GF, Stump DA, Biller J *et al.* The transient global amnesia–migraine connection. *Arch Neurol* 1984; **41**:75–19.

Cruz-Flores S, de Assis Aquino Gondim F, Leira EC. Brainstem involvement in hypertensive encephalopathy: clinical and radiological findings. *Neurology* 2004; **62**:1417–19.

Cummings JL, Miller B, Hill MA *et al.* Neuropsychiatric aspects of multi-infarct dementia and dementia of the Alzheimer type. *Arch Neurol* 1987; **44**:389–93.

De Groot JC, De Leeuw FE, Oudkerk M *et al.* Periventricular cerebral white matter lesions predict rate of cognitive decline. *Ann Neurol* 2002; **52**:335–41.

De Groot K, Schmidt DK, Arlt AC *et al.* Standardized neurologic evaluation of 128 patients with Wegener granulomatosis. *Arch Neurol* 2001; **58**:1215–21.

Dichgans M, Mayer M, Uttner I *et al.* The phenotypic spectrum of CADASIL: clinical findings in 102 cases. *Ann Neurol* 1998; **44**:731–9.

Dichgans M, Markus HS, Salloway S *et al.* Donepezil in patients with subcortical vascular cognitive impairment: a randomized double-blind trial in CADASIL. *Lancet Neurol* 2008; **7**:310–18.

Dines DE, Burgher LW, Okazaki H. The clinical and pathologic correlation of fat embolism. *Mayo Clin Proc* 1975; **50**:407–11.

Drachman DA. Neurological complications of Wegener's granulomatosis. *Arch Neurol* 1963; **8**:145–55.

Druschky A, Erguth F, Strauss R *et al.* Central nervous system involvement in thrombotic thrombocytopenic purpura. *Eur Neurol* 1998; **40**:220–4.

Ehsan T, Hasan S, Powers JM *et al.* Serial magnetic resonance imaging in isolated angiitis of the central nervous system. *Neurology* 1995; **45**:1462–5.

Eng JA, Frosch MP, Choi K *et al.* Clinical manifestations of cerebral amyloid angiopathy-related inflammation. *Ann Neurol* 2004; **55**:250–6.

Erkinjuntti T, Haltia M, Palo J *et al.* Accuracy of the clinical diagnosis of vascular dementia: a prospective clinical and post-mortem study. *J Neurol Neurosurg Psychiatry* 1988; **51**:1037–44.

Erkinjuntti T, Inzitari D, Pantoni L *et al.* Research criteria for subcortical vascular dementia in clinical trials. *J Neural Trans* 2000; **59**(Suppl.):23–30.

Esiri MM, Wilcock GK, Morris JH. Neuropathological assessment of the lesions of significance in vascular dementia. *J Neurol Neurosurg Psychiatry* 1997; **63**:749–53.

Ezzeddine MA, Primavera JM, Rosand J *et al.* Clinical characteristics of pathologically proved cholesterol emboli to the brain. *Neurology* 2000; **54**:1681–3.

Farah S, Al-Shubaili A, Montasar A *et al.* Behçet's syndrome: a report of 41 patients with emphasis on neurological manifestations. *J Neurol Neurosurg Psychiatry* 1998; **64**:382–4.

Filley CM, Thompson LL, Sze CI *et al.* White matter dementia in CADASIL. *J Neurol Sci* 1999; **163**:163–7.

Fisher CM. Transient global amnesia: precipitating activities and other observations. *Arch Neurol* 1982; **39**:605–8.

Fisher CM, Adams RD. Transient global amnesia. *Trans Am Neurol Assoc* 1958; **83**:143–6.

Fisher CM, Adams RD. Transient global amnesia. *Acta Neurol Scand* 1964; **40**(Suppl. 9):1–83.

Ford RG, Siekert RG. Central nervous system manifestations of periarteritis nodosa. *Neurology* 1965; **15**:114–22.

Forteza AM, Rabinstein A, Koch S *et al.* Endovascular closure of a patent foramen ovale in the fat embolism syndrome: changes in the embolic patterns as detected by transcranial Doppler. *Arch Neurol* 2002; **59**:455–9.

Gandolfo C, Caponnetto C, Conti M *et al.* Prognosis of transient global amnesia: a long-term follow-up study. *Eur Neurol* 1992; **32**:52–7.

Garrett WT, Chang CW, Bleck TP. Altered mental status in thrombotic thrombocytopenic purpura is secondary to nonconvulsive status epilepticus. *Ann Neurol* 1996; **40**:245–6.

Gilles C, Brucher JM, Khoubesserian P *et al.* Cerebral amyloid angiopathy as a cause of multiple intracerebral hemorrhages. *Neurology* 1984; **34**:730–5.

Gillespie J, Jackson A. *MRI and CT of the Brain.* London: Arnold, 2000.

Gold G, Kovari E, Herrmann FR *et al.* Cognitive consequences of thalamic, basal ganglia, and deep white matter lacunes in brain aging and dementia. *Stroke* 2005; **36**:1184–8.

Gordon B, Marin OSM. Transient global amnesia: an extensive case report. *J Neurol Neurosurg Psychiatry* 1979; **42**:572–5.

Gray F, Dubas F, Roullet E *et al.* Leukoencephalopathy in diffuse hemorrhagic amyloid angiopathy. *Ann Neurol* 1985; **18**:54–9.

Greenberg SM, O'Donnell HC, Schaefer PW *et al.* MRI detection of new hemorrhages: potential marker of progression in cerebral amyloid angiopathy. *Neurology* 1999; **53**:1135–8.

Gruber O, Wittig L, Wiggins CJ *et al.* Thrombotic thrombocytopenic purpura: MRI demonstration of persistent small cerebral infarcts after clinical recovery. *Neuroradiology* 2000; **42**:616–18.

Gudmundsson G, Halgrimsson I, Jonasson TA *et al.* Hereditary cerebral haemorrhage with amyloidosis. *Brain* 1972; **95**:387–404.

Guillery-Girard B, Quinette P, Desgranges B *et al.* Long-term memory following transient global amnesia: an investigation of episodic and semantic memory. *Acta Neurol Scand* 2006; **114**:329–33.

Haan J, Lanser JBK, Zigderveld I *et al.* Dementia in hereditary cerebral hemorrhage with amyloidosis – Dutch type. *Arch Neurol* 1990a; **47**:965–7.

Haan J, Algra PR, Roos RAC. Hereditary cerebral hemorrhage with amyloidosis – Dutch type. Clinical and computed tomographic analysis of 24 cases. *Arch Neurol* 1990b; **47**:649–53.

Hagiwara N, Toyoda K, Nakayama M *et al.* Renal cholesterol emboli in patients with carotid stenosis: a severe and underdiagnosed complication following cerebrovascular procedures. *J Neurol Sci* 2004; **222**:109–12.

Haglund M, Passant U, Sjobeck M *et al.* Cerebral amyloid angiopathy and cortical microinfarcts as putative substrates of vascular dementia. *Int J Ger Psychiatry* 2006; **21**:681–7.

Hauser RA, Lacey M, Knight R. Hypertensive encephalopathy: magnetic resonance imaging demonstration of reversible cortical and white matter lesions. *Arch Neurol* 1988; **45**:1078–83.

Healton EB, Brust JC, Feinfeld DA *et al.* Hypertensive encephalopathy and the neurologic manifestations of malignant hypertension. *Neurology* 1982; **32**:127–31.

Heathfield K, Croft P, Swash M. The syndrome of transient global amnesia. *Brain* 1973; **96**:729–36.

Hinchey J, Chaves C, Appignani B *et al.* A reversible posterior leukoencephalopathy syndrome. *N Engl J Med* 1996; **334**: 494–500.

Hinge HH, Jensen TS, Kjaer M *et al.* The prognosis of transient global amnesia. Results of a multicenter study. *Arch Neurol* 1986; **43**:673–6.

Hodges JR, Ward CD. Observations during transient global amnesia: a behavioral and neuropsychological study of five cases. *Brain* 1989; **112**:595–620.

Hodges JR, Warlow CP. The aetiology of transient global amnesia. A case–control study of 114 patients with prospective follow-up. *Brain* 1990; **113**:639–57.

Hoffman GS, Kerr GS, Leavitt RY. Wegener's granulomatosis: an analysis of 158 patients. *Ann Intern Med* 1992; **116**: 488–98.

Hughes JT, Brownell B. Granulomatous giant-celled angiitis of the central nervous system. *Neurology* 1966; **16**:293–8.

Hunter R, Smith J, Thompson T *et al.* Hemiparkinsonism with infarction of the ipsilateral substantia nigra. *Neuropathol Appl Neurobiol* 1978; **4**:297–301.

Iizuka R, Sakai F, Kan S *et al.* Slowly progressive spread of the stroke-like lesions in MELAS. *Neurology* 2003; **61**:1238–44.

Iragui VJ, Miravi E. Behçet syndrome presenting as cerebrovascular disease. *J Neurol Neurosurg Psychiatry* 1986; **49**:838–9.

Ishii N, Nishihara Y, Imamura T. Why do frontal lobe symptoms predominate in vascular dementia with lacunes? *Neurology* 1986; **36**:340–5.

Jacobson DM, Terrence CF, Reinmuth OM. The neurologic manifestations of fat embolism. *Neurology* 1986; **36**:847–51.

Jayakumar PN, Taly AB, Shanmugam V *et al.* Multi-infarct dementia: a computed tomographic study. *Acta Neurol Scand* 1989; **73**:292–5.

Joseph FG, Scolding NJ. Neuro-Behçet's disease in Caucasians: a study of 22 patients. *Eur J Neurol* 2007; **14**:174–80.

Joutel A, Vahedi K, Corpechot C *et al.* Strong clustering and stereotyped nature of Notch3 mutations in CADASIL patients. *Lancet* 1997; **350**:1511–15.

Joutel A, Dodick DD, Parisi JE *et al.* De novo mutation in the Notch3 gene causing CADASIL. *Ann Neurol* 2000; **47**:388–91.

Jung HH, Bassetti C, Tournier-Lasserve E *et al.* Cerebral autosomal dominant arteriopathy with subcortical infarcts and leukoencephalopathy: a clinicopathological and genetic study of a Swiss family. *J Neurol Neurosurg Psychiatry* 1995; **59**:138–43.

Kaira S, Grosset DG, Benamer HT. Differentiating vascular parkinsonism from idiopathic Parkinson's disease: a systematic review. *Mov Disord* 2010; **25**:149–56.

Kamenar E, Burger PC. Cerebral fat embolism: a neuropathologic study of a microembolic state. *Stroke* 1980; **11**:477–84.

Kanazawa M, Sanpei K, Kasuga K. Recurrent hypertensive brainstem encephalopathy. *J Neurol Neurosurg Psychiatry* 2005; **76**: 888–90.

Kansu T, Corbett JJ, Savino P *et al.* Giant cell arteritis with normal sedimentation rate. *Arch Neurol* 1977; **34**:624–5.

Katz DI, Alexander MP, Mandell AM. Dementia following strokes in the mesencephalon and diencephalon. *Arch Neurol* 1987; **44**:1127–33.

Keschner M, Sloane P. Encephalitic, idiopathic and arteriosclerotic parkinsonism: a clinicopathologic study. *Arch Neurol Psychiatry* 1931; **25**:1011–41.

Kidd D, Stever A, Denman AM *et al.* Neurological complications in Behçet's syndrome. *Brain* 1999; **122**:2183–94.

Kim Y, Choi EJ, Choi CG *et al.* Characteristics of CADASIL in Korea. A novel cysteine-sparing Notch3 mutation. *Neurology* 2006; **66**:1511–16.

Kinnecom C, Lev MH, Wendell L *et al.* Course of cerebral amyloid angiopathy-related inflammation. *Neurology* 2007; **68**:1411–16.

Knudson KA, Rosand J, Karluk D *et al.* Clinical diagnosis of cerebral amyloid angiopathy: validation of the Boston criteria. *Neurology* 2001; **56**:537–9.

Koga Y, Akita Y, Nishioka J *et al.* L-arginine improves the symptoms of strokelike episodes in MELAS. *Neurology* 2005; **64**:710–12.

Kolodny EH, Rebeiz JJ, Caveness VS *et al.* Granulomatous angiitis of the central nervous system. *Arch Neurol* 1968; **19**:510–24.

Koo EH, Massey EW. Granulomatous angiitis of the central nervous system: protean manifestations and response to treatment. *J Neurol Neurosurg Psychiatry* 1988; **51**:1126–33.

Kritchevsky M, Squire LR. Transient global amnesia: evidence for extensive, temporally graded retrograde amnesia. *Neurology* 1989; **39**:213–18.

Kushner MJ, Hauser WA. Transient global amnesia: a case–control study. *Ann Neurol* 1985; **18**:684–95.

Ladurna G, Iliff LD, Lechner H. Clinical factors associated with ischaemic stroke. *J Neurol Neurosurg Psychiatry* 1982; **45**:97–101.

Lampl Y, Sadeh M, Loberboym M. Transient global amnesia – not always a benign process. *Acta Neurol Scand* 2004; **110**:75–9.

Lamy C, Oppenheim C, Meder JF *et al.* Neuroimaging in posterior reversible leukoencephalopathy syndrome. *J Neuroimaging* 2004; **14**:89–96.

Lauria G, Gentile M, Fassetta G *et al.* Transient global amnesia and transient ischemic attack: a community-based case–control study. *Acta Neurol Scand* 1998; **97**:381–5.

Lavigne CM, Shrier DA, Letkar M *et al.* Tacrolimus leukoencephalopathy: a neuropathologic confirmation. *Neurology* 2004; **63**:1132–3.

Lazzarino LG, Nicolai A, Toppani D. Subacute parkinsonism from a single lacunar infarct in the basal ganglia. *Acta Neurol* 1990; **12**:292–5.

Ledingham JC, Rajagopalan B. Cerebral complications in the treatment of accelerated hypertension. *Q J Med* 1979; **48**:25–41.

Leff AP, McNabb AW, Hanna MG *et al.* Complex partial status epilepticus in late-onset MELAS. *Epilepsia* 1998; **39**:438–41.

Lesnik Oberstein SAJ, van den Boom R, van Bucher MA *et al.* Cerebral microbleeds in CADASIL. *Neurology* 2001; **57**:1066–70.

Levy E, Carman MD, Fernandez Madrid IJ *et al.* Mutation of the Alzheimer's disease amyloid gene in hereditary cerebral hemorrhage, Dutch type. *Science* 1990; **248**:1124–6.

Lie JT. Primary (granulomatous) angiitis of the central nervous system: a clinicopathologic analysis of 15 new cases and a review of the literature. *Hum Pathol* 1992; **23**:164–71.

Liem MK, van den Grond J, Haan J *et al.* Lacunar infarcts are the main correlate with cognitive dysfunction in CADASIL. *Stroke* 2007; **38**:923–8.

Liguori H, Mazzei R, Ungaro C *et al.* Conventional MRI and NOTCH3 gene screening in sporadic CADASIL. *Neurology* 2009; **72**:469–71.

Lin CM, Thajeb P. Valproic acid aggravates epilepsy due to MELAS in a patient with an A3243G mutation of mitochondrial DNA. *Metabol Brain Dis* 2007; **22**:105–9.

Lin JX, Tomimoto H, Akiguchi I *et al.* Vascular cell components of the medullary arteries in Binswanger's disease brains: a morphometric and immunoelectron microscopic study. *Stroke* 2000; **31**:1838–42.

Liu CK, Miller BL, Cummings JL *et al.* A quantitative MRI study of vascular dementia. *Neurology* 1992; **42**:138–43.

Lovelace RE. Mononeuritis multiplex in polyarteritis nodosa. *Neurology* 1964; **14**:434–42.

Maalikjy Akkawi N, Agosti C *et al.* Transient global amnesia: a clinical and sonographic study. *Eur Neurol* 2003; **49**:67–71.

MacKay ME, McLardy T, Harris C. A case of periarteritis nodosa of the central nervous system. *J Ment Sci* 1950; **96**:470–5.

Malandrini A, Carrera P, Palmeri S *et al.* Clinicopathological and genetic studies of two further families with cerebral autosomal dominant arteriopathy. *Acta Neuropathol* 1996; **92**:115–22.

Malandrini A, Gaudiano C, Gambelli S *et al.* Diagnostic value of ultrastructural skin biopsy in CADASIL. *Neurology* 2007; **68**:1430–2.

Markus HS, Martin RJ, Simpson MA *et al.* Diagnostic strategies in CADASIL. *Neurology* 2002; **59**:1134–8.

Marshall GB, Heale VR, Herx L *et al.* Magnetic resonance diffusion weighted imaging in cerebral fat embolism. *Can J Neurol Sci* 2004; **31**:417–21.

Mazlumzadeh M, Hunter GG, Easley KA *et al.* Treatment of giant cell arteritis using induction therapy with high-dose glucocorticoids: a double-blind, placebo-controlled, randomized prospective clinical trial. *Arthritis Rheum* 2006; **54**:3310–18.

Mellies JK, Baumer T, Muller JA *et al.* SPECT study of a German CADASIL family: a phenotype with migraine and progressive dementia only. *Neurology* 1998; **50**:1715–21.

Melo TP, Ferro JM, Ferro H. Transient global amnesia. A case control study. *Brain* 1992; **115**:261–70.

Miller JW, Petersen RC, Metter EJ *et al.* Transient global amnesia: clinical characteristics and prognosis. *Neurology* 1987a; **37**:733–7.

Miller JW, Yanagihara T, Petersen RC *et al.* Transient global amnesia and epilepsy. Electroencephalographic distinction. *Arch Neurol* 1987b; **44**:629–33.

Mirsattari SM, McGinn GJ, Halliday WC. Neuro-Behçet disease with predominant involvement of the brainstem. *Neurology* 2004; **63**:382–4.

Mok VCT, Wong A, Lam WWM *et al.* Cognitive impairment and functional outcome after stroke associated with small vessel disease. *J Neurol Neurosurg Psychiatry* 2004; **75**:560–6.

Moore PM. Diagnosis and management of isolated angiitis of the central nervous system. *Neurology* 1989; **39**:167–73.

Mori E, Yamdori A, Mitani Y. Left thalamic infarction and disturbance of verbal memory: a clinicoanatomical study with a new method of computed tomographic stereotaxic lesion location. *Ann Neurol* 1986; **20**:671–6.

Motomura S, Tabira T, Kuroiwa Y. A clinical comparative study of multiple sclerosis and neuro-Behçet's syndrome. *J Neurol Neurosurg Psychiatry* 1980; **43**:210–13.

Murrow RW, Schweiger GD, Kepes JJ *et al.* Parkinsonism due to a basal ganglia lacunar state: clinicopathologic correlation. *Neurology* 1990; **40**:897–900.

Nakada T, Kwee IL, Fujii Y *et al.* High-field, T2 reversed MRI of hippocampus in transient global amnesia. *Neurology* 2005; **64**:1170–4.

Nesher G, Berkun Y, Mates M *et al.* Low-dose aspirin and prevention of cranial ischemic complications in giant cell arteritis. *Arthritis Rheum* 2004; **50**:1332–7.

Nishino H, Rubino FA, DeRemee RA *et al.* Neurological involvement in Wegener's granulomatosis: an analysis of 324 consecutive patients at the Mayo Clinic. *Ann Neurol* 1993a; **33**:4–9.

Nishino H, Rubino FA, Parisi JE. The spectrum of neurologic involvement in Wegener's granulomatosis. *Neurology* 1993b; **43**:1334–7.

Nobuyoshi I, Nishihara Y, Horie A. Amyloid angiopathy and lobar cerebral haemorrhage. *J Neurol Neurosurg Psychiatry* 1984; **47**:1203–10.

O'Donnell HC, Rosand J, Knudson KA *et al.* Apolipoprotein E genotype and the risk of recurrent lobar intracerebral hemorrhage. *N Engl J Med* 2000; **342**:240–5.

Okazaki H, Reagan TJ, Campbell RJ. Clinicopathologic studies of primary cerebral amyloid: angiopathy. *Mayo Clin Proc* 1979; **54**:22–31.

Opherk C, Peters N, Herzog J *et al.* Long-term prognosis and causes of death in CADASIL: a retrospective study of 411 patients. *Brain* 2004; **127**:2533–9.

Oppenheimer BS, Fishberg AM. Hypertensive encephalopathy. *Arch Intern Med* 1928; **41**:264–78.

Orgogozo JM, Rigaud AS, Stoffer A *et al.* Efficacy and safety of memantine in patients with mild to moderate vascular dementia: a randomized, placebo-controlled trial (MMM 3000). *Stroke* 2002; **33**:1834–9.

O'Sullivan M, Jarosz JM, Martin RJ *et al.* MRI hyperintensities of the temporal lobe and external capsule in patients with CADASIL. *Neurology* 2001; **56**:628–34.

Pallis CA, Fudge BJ. The neurological complications of Behçet's syndrome. *Arch Neurol Psychiatry* 1956; **75**:1–14.

Palsdottir A, Abrahamson M, Thorsteinsson L *et al.* Mutation in cystatin C gene causes hereditary brain haemorrhage. *Lancet* 1988; **2**:603–4.

Pendlebury WW, Iole ED, Tracy RP *et al.* Intracerebral hemorrhage related to cerebral amyloid angiopathy and t-PA treatment. *Ann Neurol* 1991; **29**:210–13.

Quinette P, Guillery-Girard B, Dayan J *et al.* What does transient global amnesia really mean? Review of the literature and thorough study of 142 cases. *Brain* 2006; **129**:1640–58.

Regard M, Landis T. Transient global amnesia: neuropsychological dysfunction during attack and recovery in two 'pure' cases. *J Neurol Neurosurg Psychiatry* 1984; **47**:668–72.

Reichart MD, Bogousslavsky J, Janzer RC. Early lacunar strokes complicated polyarteritis nodosa: thrombotic microangiopathy. *Neurology* 2000; **54**:883–9.

Revesz T, Hawkins CP, du Boulay EPGH *et al.* Pathological findings correlated with magnetic resonance imaging in subcortical arteriosclerotic encephalopathy (Binswanger's disease). *J Neurol Neurosurg Psychiatry* 1989; **52**:1337–44.

Reyes S, Viswanathan A, Godin O *et al.* Apathy: a major symptom in CADASIL. *Neurology* 2009; **72**:905–10.

Riding G, Daly K, Hutchinson S *et al.* Paradoxical cerebral embolization. An explanation for fat embolism syndrome. *J Bone Joint Surg Br* 2004; **86**:95–8.

Rosand J, Hylek EM, O'Donnell HC *et al.* Warfarin-associated hemorrhage and cerebral amyloid angiopathy: a genetic and pathologic study. *Neurology* 2000; **55**:947–51.

Rosete A, Cabral AR, Kraus A *et al.* Diabetes insipidus secondary to Wegener's granulomatosis: report and review of the literature. *J Rheumatol* 1991; **18**:761–5.

Rubinstein LJ, Urich H. Meningo-encephalitis of Behçet's disease: Case report with pathological findings. *Brain* 1963; **86**:151–60.

Ruchoux M-M, Chabriat H, Bousser M-G *et al.* Presence of ultrastructural arterial lesions in muscle and skin vessels of patients with CADASIL. *Stroke* 1994; **25**:2291–2.

Salvarani C, Brown RD, Calamia KT *et al.* Primary central nervous system vasculitis: analysis of 101 patients. *Ann Neurol* 2007; **62**:442–51.

Sander D, Winbeck K, Etgen T *et al.* Disturbances of venous flow patterns in patients with transient global amnesia. *Lancet* 2000; **356**:1982–4.

Sandson TA, Daffener KR, Carvalho PA *et al.* Frontal lobe dysfunction following infarction of the left-sided medial thalamus. *Arch Neurol* 1991; **48**:1300–3.

Schmidtke K, Ehmsen L. Transient global amnesia and migraine. A case control study. *Eur Neurol* 1998; **40**:9–14.

Schon F, Martin RJ, Prevett M *et al.* 'CADASIL coma': an underdiagnosed acute encephalopathy. *J Neurol Neurosurg Psychiatry* 2003; **74**:249–52.

Schroder JM, Sellhaus B, Jorg J. Identification of the characteristic vascular changes in a sural nerve biopsy of a case with cerebral autosomal dominant arteriopathy with subcortical infarcts and leukoencephalopathy (CADASIL). *Acta Neuropathol (Berl)* 1995; **89**:116–21.

Sedlaczek O, Hirsch JG, Grips E *et al.* Detection of delayed focal MR changes in the lateral hippocampus in transient global amnesia. *Neurology* 2004; **62**:2165–70.

Serdaroglu P, Yazici H, Ozdemir C *et al.* Neurologic involvement in Behçet's syndrome: a prospective study. *Arch Neurol* 1989; **46**:265–9.

Seror R, Mahr A, Ramanoelina J *et al.* Central nervous system involvement in Wegener granulomatosis. *Medicine* 2006; **85**:54–65.

Sharfstein SR, Gordon MF, Libman RB *et al.* Adult-onset MELAS presenting as herpes encephalitis. *Arch Neurol* 1999; **56**:241–3.

Shuttleworth EC, Morris CE. The transient global amnesia syndrome. *Arch Neurol* 1966; **15**:515–20.

Sigal LH. The neurologic presentation of vasculitic and rheumatologic syndromes. A review. *Medicine* 1987; **66**:157–80.

Solans-Laque R, Bosch-Gil JA, Molina-Catenario CA *et al.* Stroke and multi-infarct dementia as presenting symptoms of giant cell arteritis: report of 7 cases and review of the literature. *Medicine* 2008; **87**:335–44.

Sourander P, Walinder J. Hereditary multi-infarct dementia. *Acta Neuropathol (Berl)* 1977; **39**:247–54.

Staekenborg SS, Su T, CW van Straaten E *et al.* Behavioral and psychological symptoms in vascular dementia: differences between small- and large-vessel disease, *J Neurol Neurosurg Psychiatry* 2010;**81**:547–51.

Steinmetz EF, Vroom FQ. Transient global amnesia. *Neurology* 1972; **22**:1193–200.

Strupp M, Bruning R, Wu RH *et al.* Diffusion-weighted MRI in transient global amnesia: elevated signal intensity in the left medial temporal lobe in 7 of 10 patients. *Ann Neurol* 1998; **43**:164–70.

Swartz RH, Black SE. Anterior-medial thalamic lesions in dementia: frequent, and volume dependently associated with sudden cognitive decline. *J Neurol Neurosurg Psychiatry* 2006; **77**:1307–12.

Szirmai I, Vastagh I, Szombathelyi E *et al.* Strategic infarcts of the thalamus in vascular dementia. *J Neurol Sci* 2002; **203–204**:91–7.

Tatemichi TK, Desmond DW, Prohovnik I *et al.* Confusion and memory loss from capsular genu infarction: a thalamocortical disconnection syndrome? *Neurology* 1992; **42**:1966–79.

Toledo M, Pujadas F, Grive E *et al.* Lack of evidence for arterial ischemia in transient global amnesia. *Stroke* 2008; **39**:476–9.

Tolosa ES, Santamaria J. Parkinsonism and basal ganglia infarcts. *Neurology* 1984; **34**:1515–18.

Tomlinson BE, Blessed GE, Roth M. Observations on the brains of demented old people. *J Neurol Sci* 1970; **11**:205–42.

Tournier-Lasserve E, Joutel A, Melki J *et al.* Cerebral autosomal dominant arteriopathy with subcortical infarcts and leukoencephalopathy maps to chromosome 19q12. *Nature Genet* 1993; **3**:256–9.

Uttner I, Weber S, Freund W *et al.* Transient global amnesia – full recovery without persistent cognitive impairment. *Eur Neurol* 2007; **58**:146–51.

Van der Flier WM, van Straaten EC, Barkhof F *et al.* Small vessel disease and general cognitive function in nondisabled elderly: the LADIS study. *Stroke* 2005; **36**:2116–20.

Van Swieten JC, van den Hout JH, van Ketel BA *et al.* Periventricular lesions in the white matter on magnetic resonance imaging in the elderly. A morphometric correlation with arteriosclerosis and dilated perivascular spaces. *Brain* 1991; **141**:761–74.

Verbeek MM, Kremer BP, Rikkert MO *et al.* Cerebrospinal fluid amyloid beta(40) is decreased in cerebral amyloid angiopathy. *Ann Neurol* 2009; **66**:245–9.

Viswanathan A, Gschwendtner A, Guichard JP *et al.* Lacunar lesions are independently associated with disability and cognitive impairment in CADASIL. *Neurology* 2007; **69**:172–9.

Vollmer TL, Guarnaccia J, Harrington W *et al.* Idiopathic granulomatous angiitis of the central nervous system: diagnostic challenges. *Arch Neurol* 1993; **50**:925–30.

Von Hochstetter AR, Friede RL. Residual lesions of cerebral fat embolism. *J Neurol* 1977; **216**:227–33.

Wadia N, Williams E. Behçet's syndrome with neurological complications. *Brain* 1957; **80**:59–71.

Wattendorff AR, Frangione B, Luyendijk W *et al.* Hereditary cerebral haemorrhage with amyloidosis, Dutch type (HCHWA-D); clinicopathological studies. *J Neurol Neurosurg Psychiatry* 1995; **58**:699–705.

Wechsler B, Dell'Isola B, Vidhaillet M *et al.* MRI in 31 patients with Behçet's disease and neurological involvement: prospective study with clinical correlation. *J Neurol Neurosurg Psychiatry* 1990; **53**:793–8.

Weinberger LM, Cohen ML, Remler BF *et al.* Intracranial Wegener's granulomatosis. *Neurology* 1993; **43**:1831–4.

Weingarten K, Barbut D, Filippi C *et al.* Acute hypertensive encephalopathy: findings on spin-echo and gradient-echo MR imaging. *Am J Roentgenol* 1994; **162**:665–70.

Wilcock G, Mobius HJ, Stoffer A. A double-blind, placebo-controlled multicentre study of memantine in mild to moderate vascular dementia (MMM500). *Int Clin Psychopharmacol* 2002; **17**:297–305.

Wilkinson D, Doody R, Helme R *et al.* Donepezil in vascular dementia: a randomized, placebo-controlled study. *Neurology* 2003; **61**:479–86.

Wilkinson IMS, Russell RWR. Arteries of the head and neck in giant cell arteritis. A pathological study to show the pattern of arterial involvement. *Arch Neurol* 1972; **27**:378–91.

Winikates J, Jankovic J. Clinical correlates of vascular parkinsonism. *Arch Neurol* 1999; **56**:98–102.

Wolf SM, Schotland DL, Phillips LL. Involvement of nervous system in Behçet's syndrome. *Arch Neurol* 1965; **12**:315–25.

Wolfe N, Linn R, Babikian VL *et al.* Frontal system impairment following multiple lacunar infarcts. *Arch Neurol* 1990; **47**:129–32.

Yamada M, Kashiwamura K, Nakamura Y *et al.* On psychiatric symptoms of neuro-Behçet's syndrome. *Folia Psychiatr Neurol Jpn* 1978; **32**:191–7.

Yamanouchi H. Loss of white matter oligodendrocytes and astrocytes in progressive subcortical vascular encephalopathy of Binswanger type. *Acta Neurol Scand* 1991; **83**:301–5.

Yoshimura M, Yamanouchi H, Kazuhara S *et al.* Dementia in cerebral amyloid angiopathy: a clinicopathological study. *J Neurol* 1992; **239**:441–50.

Yoshitake T, Kiyohara Y, Kato I *et al.* Incidence and risk factors of vascular dementia and Alzheimer's disease in a defined elderly Japanese population: the Hisayama study. *Neurology* 1995; **45**:1161–8.

Zijlmans JCM, Thijssen HOM, Vogels OJM *et al.* MRI in patients with suspected vascular parkinsonism. *Neurology* 1995; **45**:2183–8.

Zijlmans JC, Daniel SE, Hughes AJ *et al.* Clinicopathological investigation of vascular parkinsonism, including clinical criteria for diagnosis. *Mov Disord* 2004a; **19**:630–40.

Zijlmans JC, Katzenschlager R, Daniel SE *et al.* The L-dopa response in vascular parkinsonism. *J Neurol Neurosurg Psychiatry* 2004b; **75**:545–7.

Zorzon M, Antonutti L, Mase G *et al.* Transient global amnesia and transient ischemic attack. Natural history, vascular risk factors, and associated conditions. *Stroke* 1995; **26**:1536–42.

12

Trauma

12.1 SUBDURAL HEMATOMA

Subdural hematoma secondary to head trauma is a not uncommon condition, especially in the elderly. It may present acutely, subacutely, or in a chronic fashion, and in the elderly it should always be included in the differential diagnosis of delirium or dementia.

Clinical features

Acute subdural hematoma occurs within several days of a moderate or severe traumatic brain injury, often in association with other injuries, including diffuse axonal injury, subarachnoid hemorrhage, contusion, and intracerebral hemorrhage. Some patients may be comatose from the time of injury, and in this case it is impossible to dissect out the part played by the subdural hematoma from the effects of the accompanying injuries. In other cases, there may be a more or less lucid interval, lasting from an hour up to several days, which then gives way to a delirium or drowsiness as the subdural hematoma expands to a critical size; stupor and coma may then supervene.

Subacute subdural hematoma presents anywhere from several days to 3 weeks after the trauma. The presentation (Black 1984) may be marked by headache, drowsiness, delirium, or a dementia, and characteristically symptoms tend to fluctuate in severity. Focal signs, such as hemiparesis or aphasia, may or may not be present, and, if they are, they are typically relatively mild. In some cases, there may be a progression to stupor, and, with appropriately situated hematomas, uncal herniation may occur, causing coma.

Chronic subdural hematomas (Arieff and Wetzel 1964; Black 1984; Gelabert-Gonzalez *et al.* 2005; Ishikawa *et al.* 2002; Ramachandran and Hegde 2007) present on average 7 weeks after the trauma, but the range is wide, from 3 weeks up to 2 years. The presentation itself may be marked by delirium, especially when it is relatively early on, or, more commonly, with a dementia; in a small minority either a personality change (Cameron 1978) or depression (Black 1984) may stand at the forefront of the clinical picture. Most patients will also complain of headache, which may or may not be ipsilateral to the hematoma. Focal signs, such as hemiparesis or aphasia, may also be present, but, as with subacute subdural hematomas, these are typically quite mild. Rarely, there may be seizures (Annegers *et al.* 1998). Importantly, as noted below, the trauma itself, especially in the elderly, may be trivial, and indeed up to one-half of all elderly patients with a chronic subdural hematoma may not recall any trauma at all (Cameron 1978; Fogelholm and Waltimo 1975).

The most common location for a subdural hematoma is over the frontoparietal convexity, where the hematoma assumes a typically convex, crescent shape. Parafalcine hematomas, adjacent to the falx cerebri, may also occur, as may tentorial hematomas and hematomas nestled in the middle cranial fossa beneath the temporal lobe. Bilateral hematomas may occur in 10–25 percent of cases. Computed tomography (CT) scanning typically provides satisfactory imaging. During the acute phase the hematoma is hyperdense relative to the underlying cortex, while during the subacute phase, it is isodense. During the chronic phase it is typically hypodense; however, at times this overall hypodensity may contain small pockets of mild hyperdensity, reflecting fresh bleeding; with contrast, the capsule surrounding the chronic subdural hematoma will typically enhance. Depending on the size of a convexity hematoma, and the degree of compression of the underlying cortex, there may or may not be a midline shift; in cases of bilateral convexity hematomas (Figure 12.1), the compressive forces even out, and, rather than a shift, one sees small, compressed lateral ventricles.

Course

Acute subdural hematomas tend to rapidly enlarge and, as noted above, may become immediately life-threatening. Subacute subdural hematomas tend to evolve over a matter of 2 or 3 weeks and either may progress to stupor or coma or may stabilize, after which there may be a greater or lesser

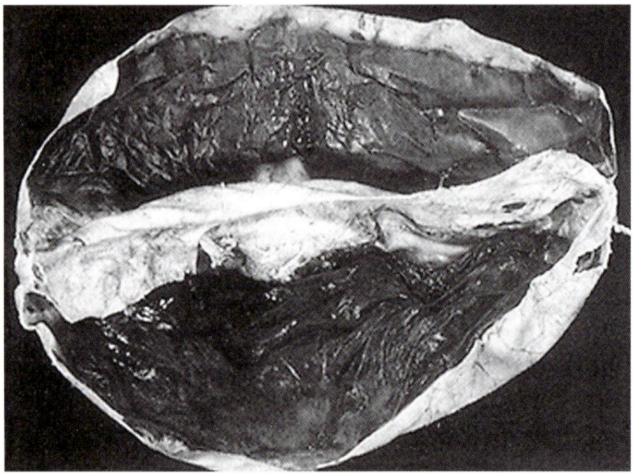

Figure 12.1 Bilateral encapsulated subdural hematomas seen over the convexities. Reproduced from Graham and Lantos (1996), with permission.

degree of gradual improvement. Chronic subdural hematomas tend to undergo a very gradual progression.

Etiology

In almost all cases, subdural hematoma occurs secondary to rupture of the fragile bridging veins which run between the cortex and the dural sinuses. The only exception to this rule occurs in a minority of cases of acute subdural hematoma in which the bleeding is arterial (Maxeiner and Wolff 2002), stemming either from rupture of a pial artery or from a 'burst' lobe, wherein an intracerebral hemorrhage bleeds into the subdural space through a cortical laceration. Regardless of the source of the bleeding, blood accumulates in the potential space between the dura and the arachnoid below. Among those who survive, the subdural hematoma undergoes a characteristic neuropathologic evolution. During the acute phase, it consists of clotted blood lying more or less freely adjacent to the cortex. In the subacute phase, if the hematoma is small, it may simply be resorbed. If it is large, however, encapsulation starts. Initially, a fibrous outer pseudomembrane begins to form, and subsequently a thin, fragile, inner pseudomembrane begins to take shape: at this point, the hematoma itself is composed of a mixture of clotted blood and thick, proteinaceous fluid. As the chronic phase is reached, capsule formation is complete, and the hematoma at this point will have resolved into a rusty, protein-rich fluid. The subsequent evolution depends on the size of the hematoma. Small hematomas may eventually be completely resorbed, leaving behind only a fibrous layer representing the collapsed capsule. Larger hematomas, however, persist, and these may either remain stable in size or progressively, and very slowly, enlarge. This enlargement, in turn, appears to be due to bleeding from the fragile capillaries found within the outer pseudomembrane (Markwalder 1981). In a small minority of cases, chronic subdural hematomas may undergo calcification.

The trauma responsible for the subdural hematoma may result from a motor vehicle accident, a blow to the head, or a fall, as may occur during grand mal seizures, or, classically, in alcoholics. Another, non-traumatic, cause of subdural hematoma is overvigorous drainage of CSF through a ventriculoperitoneal shunt, which allows the brain to sag down, thus stretching the bridging veins. Concurrent use of warfarin, aspirin, or clopidogrel naturally increases the risk, and, as noted earlier, the elderly are at particular risk for the development of a chronic subdural hematoma. This increased risk in the elderly may be explained on the basis of cerebral atrophy, which stretches the bridging veins taut, thus making them less able to accommodate to the shearing forces occurring during trauma. This phenomenon, in turn, accounts for the fact that the trauma need not be severe, and indeed may be as trivial as bumping one's head on a cabinet door.

Differential diagnosis

In the case of acute and subacute subdural hematomas, the proximity of the delirium or stupor to the head trauma immediately suggests the diagnosis; however, as noted in Chapter 8.5 on traumatic brain injury, multiple other intracranial injuries may also be present, and at times it is difficult to discern how much of the patient's clinical condition is secondary to the subdural hematoma.

With chronic subdural hematomas, given the long latency and the fact that the head trauma may have been forgotten, the diagnosis may remain obscure until, in the workup for dementia, neuroimaging reveals the encapsulated fluid collection. Importantly, however, some judgment is required here before ascribing a dementia to a chronic subdural hematoma: whereas a thick chronic hematoma that grossly compresses the underlying cortex may reasonably be held accountable for a dementia, a thin structure that barely compresses the underlying cortex might well be incidental, and in such cases it is appropriate to look for other causes of dementia.

Treatment

Symptomatic subdural hematomas generally require surgical evacuation, either by burr holes or by craniotomy. Importantly, in the case of chronic subdural hematomas, after evacuation of the clot a considerable amount of time may be required for the gradual re-expansion of the chronically compressed cortex and, hence, clinical improvement may be delayed. The symptomatic treatment of dementia and delirium is discussed in Chapters 6.3 and 6.5, respectively.

12.2 DIFFUSE AXONAL INJURY

With a sudden and severe acceleration–deceleration injury, axons and arterioles are subjected to substantial stretching, and in some cases actual shearing, leading to widespread damage. Although axons, as indicated by the name of this condition, are most susceptible to injury, the arterioles may also undergo shearing, with the production of small hemorrhages.

Clinical features

In severe or moderate cases, at the moment of injury, patients are rendered unconscious. In severe cases, patients may remain in coma and die, or emerge from coma into a persistent vegetative or minimally conscious state, as discussed in Chapter 6.2. With moderate injury, patients emerge into a delirium that eventually resolves into a dementia of variable severity, as discussed in Chapter 8.5. In mild traumatic brain injury, there may be a brief period of unconsciousness, or patients may merely be dazed; subsequently they may or may not develop a postconcussion syndrome, as discussed in Chapter 12.4.

Computed tomography scanning is often normal; in some moderate to severe cases, however, there may be a degree of generalized cerebral edema along with a few punctate or small hemorrhages scattered in the white matter. Magnetic resonance imaging (MRI) scanning is far superior to CT. On fluid-attenuated inversion recovery (FLAIR) or diffusion-weighted imaging, multiple patchy areas of increased signal intensity may be seen in the corpus callosum, centrum semiovale, internal capsule, mesencephalon and pons. Further, on gradient-echo imaging, multiple, generally small, areas of decreased signal intensity may be seen in the same areas, corresponding to hemorrhages (Pittella and Gusmao 2003). Importantly, however, the number of abnormalities seen on gradient-echo imaging is generally less than that on FLAIR or diffusion-weighted imaging, and in some cases there may be none. An emerging MRI modality, diffusion tensor imaging (DTI) fiber tractography, appears to be even more sensitive. In one study of patients with moderate to severe traumatic brain injury, reduced fractional anisotropy was found in all cases in the corpus callosum, cingulum, superior longitudinal fasciculus, corona radiata, and internal capsule (Kraus *et al.* 2007).

In cases of mild traumatic brain injury, CT scanning is almost always normal and, although MRI scanning with FLAIR, diffusion-weighted or gradient-echo imaging may show the same abnormalities seen in moderate or severe cases, they are often minimal, and may be absent. Of great interest, however, is the fact that DTI reveals reduced fractional anisotropy in essentially all cases (Kraus *et al.* 2007; Lipton *et al.* 2008; Niogi *et al.* 2008; Topal *et al.* 2008).

Course

The overall course is discussed in the chapters on the persistent vegetative and minimally conscious states, traumatic brain injury, and the postconcussion syndrome. In general, when improvement does occur, the maximum amount is seen in the first half-year. In the following half-year, some further, but less impressive, improvement may occur; however, after roughly a year, no further improvement may be expected.

Etiology

Diffuse axonal injury (Adams *et al.* 1982, 1989; Blumbergs *et al.* 1989; Ng *et al.* 1994; Strich 1956) occurs secondary to sudden acceleration–deceleration or rotation of the head. The brain, as it were, 'floats' within the skull, and is tethered to the brainstem. With such an injury, tremendous stretching, and even shearing, forces are applied to the long, delicate axons, resulting in a cascade of injury. With stretching and deformation of the axonal membrane there is an influx of calcium, which in turn activates a number of injurious enzymatic systems, leading to damage of the cytoskeleton. Fast axonal transport fails, and proteins, such as the amyloid precursor protein, accumulate at the site of injury, leading to axonal swelling, which can be demonstrated as early as 35 minutes after injury (Hortobagyi *et al.* 2007). Within hours, axonal retraction bulbs begin to form, and the portion of the axon distal to the bulb eventually undergoes Wallerian degeneration. While early studies, using silver impregnation, were able to demonstrate these bulbs in only a minority of cases (Adams *et al.* 1977), subsequent studies utilizing immunohistochemistry technique aimed at amyloid precursor protein have demonstrated axonal damage in essentially all autopsy cases (Gentleman *et al.* 1995). Such axonal damage, though widespread, is preferentially found in certain areas (Blumbergs *et al.* 1995), including the dorsolateral quadrants of the pons and midbrain, the cerebellar peduncles, the internal capsule and corona radiata, the corpus callosum, the fornix, and in the cerebral white matter. Of all these areas, it appears that damage to the brainstem is primarily responsible for death (Adams *et al.* 1977; Firsching *et al.* 2002).

Although axons are most liable to damage with this form of head injury, long, delicate penetrating arterioles may also be damaged, especially with more severe injuries (Pittella and Gusmao 2003). In such cases, widespread hemorrhages are found in the same areas demonstrating axonal injury. Although most of these hemorrhages are microscopic, in severe cases macroscopic lesions may appear. Over time, these hemorrhages are replaced by hemosiderin-laden scars.

Differential diagnosis

As noted earlier, diffuse axonal injury occurs as part of the syndrome of traumatic brain injury, and in many cases it may be very difficult to determine how great a part diffuse axonal injury plays in the overall clinical picture compared with other injuries, such as contusions, intracerebral hemorrhages, subarachnoid hemorrhage, subdural hematomas, and infarctions. In some cases, however, imaging is either normal or displays only findings consistent with diffuse axonal injury, and here one may confidently ascribe the clinical findings to this etiology.

Treatment

As yet, there is no specific treatment for diffuse axonal injury. The overall treatment of the persistent vegetative and minimally conscious states, traumatic brain injury, and the post-concussion syndrome is discussed in the relevant chapters.

12.3 DEMENTIA PUGILISTICA

Dementia pugilistica is a distinct syndrome that occurs secondary to repeated blows to the head and is characterized by dementia, ataxia, and parkinsonism. Synonyms for this disorder include 'punch drunk syndrome', 'punch drunkenness', and chronic traumatic encephalopathy.

Although dementia pugilistica is found most commonly in boxers, others may also be at risk, for example American football players (Omalu *et al.* 2005) or those who engage in chronic head-banging, as was reported in one patient with autism (Hof *et al.* 1991). In the past, it was fairly common in professional boxers, being found in close to 20 percent; however, with better medical supervision, and an overall reduced number of fights, the syndrome is becoming rare.

Clinical features

The onset of symptoms is gradual and occurs anywhere from 5 to 40 years after there has been an accumulation of a sufficient number of blows to the head, which, in the case of professional boxers, equates to perhaps a dozen or so knockouts. Thus, although some boxers may develop this disorder while they are still professionally active, in most cases symptoms are delayed until well after the boxer has left the ring.

Clinically (Corsellis 1989; Critchley 1957; Harvey and Davis 1974; Jordan 1987; McLatchie *et al.* 1987; Martland 1928; Mawdsley and Ferguson 1963), when the syndrome is fully developed, patients present with dementia, ataxia, dysarthria, and parkinsonism. The presence of ataxia and dysarthria often gives the impression of alcohol intoxication, and this accounts for the synonym 'punch drunk'. The dementia itself is non-specific, except perhaps for an undue amount of irritability.

Computed tomography or MR scanning typically displays cerebral cortical atrophy, ventricular dilatation, and an enlarged cavum septi pellucidi.

Course

Overall, the course is progressive, with a gradual worsening of symptoms.

Etiology

Neuropathologic findings are described in the classic 1973 paper by Corsellis *et al.* (1973). Macroscopically, there is generalized cerebral cortical atrophy and ventricular dilatation. The cavum septi pellucidi, as illustrated in Figure 12.2, is enlarged, and the septal walls are torn and fenestrated. Microscopically, there is widespread neuronal loss in the cerebral cortex, and in surviving neurons one finds neurofibrillary tangles, identical to those found in Alzheimer's disease (Roberts 1988; Schmidt *et al.* 2001). In contrast to Alzheimer's disease, however, in which neurofibrillary tangles are found in the deeper layers of the cortex, those in dementia pugilistica are found in the superficial layers (Hof

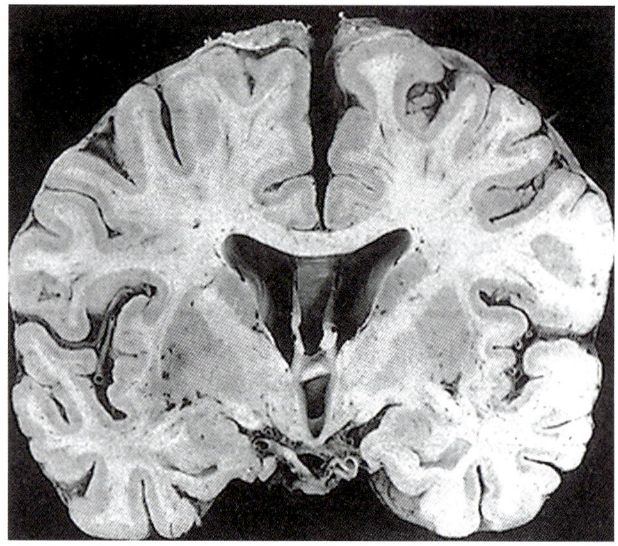

Figure 12.2 Note the wide cavum septi pellucidi in the brain of a professional boxer. Reproduced from Graham and Lantos (1996), with permission.

et al. 1992). Amyloid plaques are also found in dementia pugilistica; however, unlike the relatively discrete plaques found in Alzheimer's disease, those in dementia pugilistica are diffuse (Roberts *et al.* 1990). In addition to these findings, neuronal loss also occurs in the nucleus basalis of Meynert, the substantia nigra, and the locus ceruleus.

Although the mechanism whereby repeated blows to the head set these pathologic changes in motion is not known, it does appear that patients who carry the epsilon 4 allele of apolipoprotein E (APOE) are at greater risk (Jordan *et al.* 1997).

Differential diagnosis

Dementia pugilistica must be distinguished from dementia that occurs shortly after traumatic brain injury; such cases, discussed in Chapter 8.5, are generally due to diffuse axonal injury, acute subdural or epidural hematoma, subarachnoid hemorrhage, or intracerebral hemorrhage, and, in contrast to the course seen in dementia pugilistica, the dementia of traumatic brain injury is either static or undergoes some improvement over the first year or so. Consideration must also be given to the progressive dementia that may occur due to communicating hydrocephalus occurring secondary to traumatic subarachnoid hemorrhage, however here imaging will immediately tell the tale.

In addition to these dementias occurring secondary to trauma, consideration must also be given to the obvious fact that boxers are as likely to develop Alzheimer's disease or similar disorders as anyone else. Here, special consideration must be given to the age at onset.

Treatment

The general treatment of dementia is discussed in Chapter 6.3. Although controlled studies are lacking, it is not unreasonable to give a trial of a cholinesterase inhibitor for the dementia, and levodopa for the parkinsonism.

12.4 POSTCONCUSSION SYNDROME

Concussion is the mildest form of traumatic brain injury, and patients typically have a Glasgow Coma Scale score of 14 or 15 in the emergency room. Concussion itself is characterized by a brief loss of consciousness or merely a sense of being dazed, occurring immediately after a blow to the head or, in some cases, a whiplash injury (Miller 1982); patients also experience a variable, but short, period of amnesia, with both retrograde and anterograde components. Most recover fairly promptly; however, in a minority a postconcussion syndrome will develop. Postconcussion syndrome, also known as postconcussional disorder, is characterized by headache, fatigue, cognitive deficits, dizziness, and other symptoms.

Clinical features

When loss of consciousness does occur in concussion, it is brief, lasting at most a few minutes. Upon regaining consciousness, patients will demonstrate some confusion. As noted earlier, in some cases there will be no loss of consciousness, and here the concussion manifests with confusion alone. During this confusional state, patients appear dazed; they may ask repeatedly where they are, or what happened, and their speech may be rambling or incoherent. Some will walk away, often in a stumbling fashion, and typically in the wrong direction. This confusion is also accompanied by an amnesia, with both retrograde and anterograde components: the retrograde component typically stretches back hours before the concussion, and the anterograde component, though typically briefer, of the order of minutes, may also, in exceptional cases, likewise extend for hours (Fisher 1966; Martland 1928). Typically, the confusion clears up within minutes, and subsequently patients will be unable to recall the injury itself and the period of time immediately preceding and following it. In a very small minority of cases, less than 2 percent, a concussion will be accompanied, within a couple of seconds, by a 'concussive convulsion' (McCrory *et al.* 1997): here there is tonic extension of all four extremities, followed by symmetric myoclonic jerking of the extremities, with the episode terminating spontaneously within 2½ minutes. Although these events are styled as 'convulsions', they may or may not be epileptic in character; in any case, they do not portend the development of epilepsy.

In a minority of cases, concussion may be followed by the postconcussion syndrome (Lishman 1968; Mapothar 1937; Symonds 1962). As noted earlier, headache occurs, and this tends to be severe and may be continuous or episodic; it may be dull or throbbing, and may be exacerbated by loud noises, coughing, or sneezing. Fatigue may be constant or may become evident only when patients exert themselves. Patients may complain of difficulty concentrating and of forgetfulness, and thinking is often slowed and effortful. Dizziness may consist of mere light-headedness or there may be a true vertigo; when vertigo is present, patients may complain that it is exacerbated or precipitated by changes in position or by any sudden movements. Depression may occur and may be marked by severe insomnia. Irritability may be prominent, and patients may complain of great difficulty controlling their tempers. Anxiety may also be seen but appears less common. Other symptoms may occur, including photophobia, hyperacusis, and hyperhidrosis, which at times may be quite impressive. Many patients report that alcohol exacerbates their symptoms.

Course

Symptoms typically begin to undergo a spontaneous remission within the first week or two, and by 1 month most patients will have recovered, with only a small minority remaining symptomatic at 3 months: in those who do remain symptomatic at 3 months, the most prominent symptoms are typically fatigue, depression, insomnia, irritability, and anxiety. When symptoms persist beyond a half-year, a chronic course may be anticipated, with anywhere from 10 to 15 percent of patients remaining symptomatic at 1 year.

Etiology

Although the mechanism underlying the loss of consciousness and confusion characteristic of concussion is not known, it is speculated that, secondary to the injury, there is a brief, 'functional' deafferentation of the cortex from the underlying white matter. With regard to the postconcussion syndrome, in all likelihood diffuse axonal injury, at least in part, plays a role. Diffusion tensor imaging fiber tractography studies (Bazarian *et al.* 2007; Huang *et al.* 2009; Niogi *et al.* 2008; Smits *et al.* 2011) of the postconcussion syndrome have revealed abnormalities in fractional anisotropy in various structures, including the corpus callosum, internal capsule, inferior longitudinal fasciculus, and the uncinate fasciculus, and one of these studies (Bazarian *et al.* 2007) found a correlation between abnormal fractional anisotropy in the corpus callosum and the severity of the syndrome.

Differential diagnosis

Post-traumatic stress disorder may develop after some cases of concussion, especially those involving assaults: although the patient may not recall the actual assault itself, a memory of the terrifying build-up to the assault may remain. In such cases, however, one finds symptoms of re-experiencing the event, as in dreams or intrusive memories, and such symptoms are not found in the postconcussion syndrome.

When vertigo is prominent, consideration must be given to other causes of this symptom, including benign paroxysmal positional vertigo, labyrinthine concussion, and vertebral artery dissection with medullary infarction. Further, when cognitive deficits overshadow other symptoms, or when lethargy or confusion appears, consideration must be given to subdural hematoma or an expanding intracerebral hematoma: this is particularly important in the elderly, alcoholics, or those on warfarin, and in such cases urgent CT scanning is in order.

When symptoms persist beyond 3 months, consideration must be given to the possibility that a new disorder has supervened. Depression and generalized anxiety disorder each occur in roughly 10 percent of patients in the year following a concussion (Bryant *et al.* 2010), and the symptoms of these disorders match some of those that are most common in cases of the postconcussion syndrome that persist beyond 3 months, namely fatigue, depression, insomnia, irritability, and anxiety. It may well be, then, that the presence of these symptoms 3 or more months after concussion reflects not a persistent postconcussion syndrome, but rather the occurrence of a supervening major depressive disorder or generalized anxiety disorder. Given the similarity of the symptomatologies here, it may be that the best way to differentiate these disorders is to attempt a diagnosis by treatment response, and to treat the patient as if he or she had major depressive disorder or generalized anxiety disorder, as discussed in Chapters 23.3 and 24.7, respectively.

Malingering may occur after a concussion, and this is often suspected in cases in which litigation is in play. Sometimes in these cases the diagnostic question can be resolved only with observation after resolution of the lawsuit.

Treatment

Although concussion, *per se*, does not require any specific treatment, one must remain alert to the possibility of additional injuries, such as subdural hematomas or intracerebral hemorrhage. Thus, CT scanning should be performed in the elderly, those on warfarin or with thrombocytopenia, and alcoholics, and in any patient with repeated vomiting, worsening headache, evidence of skull fracture (looking especially for evidence of a basilar skull fracture, such as 'raccoon eyes', Battle's sign, or hemotympanum), persistent amnesia, or new-onset lethargy or confusion, or the presence of any focal signs. In other cases, patients may simply be observed for these developments over the next 24 hours, with instructions to awaken the patient every 2 hours for repeated assessments.

Treatment of the postconcussion syndrome should begin with reassurance regarding the typically benign course. Headache is treated with non-opioid analgesics. Dizziness may respond to antihistamines, but caution should be used here as these agents may exacerbate any cognitive deficits. Depressive symptoms may respond to antidepressants: one single-blind study noted that treatment with sertraline was not only effective in this regard (Fann *et al.* 2000) but was also associated with cognitive improvement (Fann *et al.* 2001). When symptoms persist beyond a month, there is some evidence that cognitive–behavioral therapy may be helpful (Al Sayegh *et al.* 2010). Alcohol should be forbidden until recovery is complete.

12.5 RADIATION ENCEPHALOPATHY

Irradiation of the central nervous system may be followed by a radiation encephalopathy, and this may occur subsequent to either whole-brain or focal irradiation. There are three different forms of radiation encephalopathy, namely acute, early delayed, and late delayed, with each form occurring on a different pathologic basis.

Clinical features

Acute radiation encephalopathy occurs within a day to a week or so of irradiation, and probably reflects a breakdown of the blood–brain barrier. Patients may experience delirium, drowsiness, ataxia, headache, nausea and vomiting, and, in a small minority, seizures (Oliff *et al.* 1978). Symptoms undergo a gradual, spontaneous resolution within days to weeks.

Early-delayed radiation encephalopathy appears subacutely anywhere from 1 to 6 months postirradiation, secondary to demyelination. In patients who received whole-brain irradiation, there may be headache, nausea, delirium, drowsiness, incontinence, and pyramidal signs. By contrast, in those subjected to focal irradiation there may be focal signs appropriate to the irradiated area. Symptoms generally resolve spontaneously over the following month or two.

Late-delayed radiation encephalopathy, which occurs secondary to a vasculopathy, presents gradually, generally within 6 months to 3 years postirradiation, with most cases occurring in a little over 1 year; in some cases, however, the latency between irradiation and the onset of symptoms may be much longer, in one case 33 years (Duffy *et al.* 1996). This late-delayed radiation encephalopathy may take one of two pathologic forms: in cases occurring secondary to whole-brain radiation, a leukoencephalopathy appears, while in those due to focal irradiation, a similarly focal area of radionecrosis appears. In cases of leukoencephalopathy, a dementia (DeAngelis *et al.* 1989; Martins *et al.* 1977; Morris *et al.* 1994) gradually appears, which is often accompanied by ataxia and urinary incontinence. Focal radionecrosis, by contrast, generally presents with focal signs appropriate to the irradiated area (Kaufman *et al.* 1990; Shewmon and Masdeu 1980). An exception to this general rule occurs in cases where the focal radionecrosis occurs in a 'strategic' area. For example, when one or both temporal lobes were damaged during irradiation of a pituitary tumor (Crompton and Layton 1961) or a nasopharyngeal carcinoma (Woo *et al.* 1988), rather than a 'focal' sign, a dementia occurred, and, in a case secondary to irradiation of a pituitary tumor, damage to the temporal lobes produced a Kluver–Bucy syndrome (Thajeb 1995). In contrast to the other two types of radiation encephalopathy, the late-delayed type does not remit spontaneously, but rather displays a progressive course.

Before leaving this section, it is also appropriate to comment on endocrinologic changes that may occur in irradiated patients (Agha *et al.* 2005; Constine *et al.* 1993; Lam *et al.* 1991). With irradiation of the hypothalamus there may be hyperprolactinemia or tertiary forms of hypothyroidism, adrenocortical insufficiency, growth hormone deficiency, or hypogonadism; with irradiation of the pituitary, one may in turn see the secondary forms of hypothyroidism, adrenocortical insufficiency, growth hormone deficiency, or hypogonadism. Such changes tend to occur very gradually, with roughly the same time course as late-delayed radiation encephalopathy; hypothyroidism is especially important to keep in mind as it may present as a dementia. Hypothalamic damage may also cause other symptomatology, as in one case of hyperphagia with severe weight gain (Christianson *et al.* 1994).

Although MRI scanning may be normal in the acute type of radiation encephalopathy, in the other two types one sees increased signal intensity on FLAIR or T2-weighted scans: in cases of whole-brain irradiation this appears diffusely in the white matter, while in cases due to focal irradiation the abnormality is likewise focal. Further, radionecrotic areas in the late-delayed type typically show intense gadolinium enhancement. Finally, in the late-delayed type due to whole-brain irradiation, cerebral atrophy may occur, along with ventricular dilation, which at times may be disproportionate to the atrophy.

Course

This is as described above.

Etiology

As indicated above, the acute form occurs secondary to a radiation-induced breakdown of the blood–brain barrier.

The early-delayed type represents demyelination (Lampert and Davies 1964), which probably occurs secondary to radiation damage of oligodendroglia. The half-life of myelin ranges from 5 to 8 weeks and, with loss or dysfunction of oligodendroglia, symptoms appear as myelin degrades in the absence of ongoing replacement. With a new 'crop' of oligodendroglia, replenishment of myelin gradually occurs and symptoms resolve.

Late-delayed radiation encephalopathy occurs on the basis of a vasculopathy (De Reuck and vander Eecken 1975; Pennybaker and Russell 1948), with hyalinization, fibrinoid necrosis, and thrombus formation within penetrating arterioles, with resulting infarction. In cases occurring secondary to whole-brain irradiation, a myriad of microinfarctions produce the leukoencephalopathy. In cases due to focal irradiation, the radionecrotic area consists of a coagulative necrosis, which may eventually resolve into an area of demyelination and scarring, with or without cystic change. The radionecrotic area is typically surrounded by edema, which in turn may be related to the release of vascular endothelial growth factor (VEGF) from damaged vessels.

Of interest, although irradiation generally affects only small vessels, rarely large arteries may be involved, and, in such cases, stroke due to large territorial infarctions occurs (Grattan-Smith *et al.* 1992).

Differential diagnosis

The history of irradiation makes the diagnosis of the acute and early-delayed forms straightforward. Difficulties, however, may arise in cases of the late-delayed form. In cases of the late-delayed form occurring secondary to whole-brain irradiation, especially if the latency between the irradiation and the onset of symptoms is prolonged, one must take into account other possibilities, including Binswanger's disease and B12 deficiency, both of which may be accompanied by similar white matter changes, and also chronic hydrocephalus. In those cases due to focal irradiation, the prime item on the differential is tumor recurrence. In such cases, routine MRI scanning is not helpful, as both the area of radionecrosis and a tumor typically have the same imaging characteristics, and consequently one must resort to other modalities. Traditionally, PET scanning (Glantz *et al.* 1991) has been utilized, as it displays decreased metabolic activity in cases of radionecrosis, and increased activity with a recurrent tumor. Perfusion-weighted MRI scanning and magnetic resonance spectroscopy, however, constitute alternative approaches. On perfusion-weighted MRI scanning, decreased blood volume will be noted with radionecrosis, whereas increased blood volume will be noted when a tumor is present. With magnetic resonance spectroscopy, increased *N*-acetylaspartate (NAA) and decreased choline will be found with radionecrosis, with the reverse pattern seen when a tumor is present. Finally, in cases where it is determined that a tumor is present, one must also take into account the fact that radiation can cause tumors (Robinson 1978), and that the tumor in question, rather than a recurrence of the original one, is of an altogether different type. Though this is a rare occurrence, both meningiomas (Brada *et al.* 1992; Soffer *et al.* 1983) and gliomas (Zampieri *et al.* 1989) have occurred secondary to radiation.

Treatment

Acute radiation encephalopathy may be treated with steroids and, indeed, and especially in cases where increased intracranial pressure is already present, it is customary to give steroids prophylactically.

Early-delayed radiation encephalopathy may show some response to either dexamethasone or prednisone.

In cases of late-delayed radiation encephalopathy with leukoencephalopathy due to whole-brain irradiation, there is no clearly established treatment. Although one large case series (Glantz *et al.* 1994) suggested stabilization with warfarin, the risks of long-term anticoagulation suggest caution until controlled studies clearly demonstrate effectiveness. Another option is shunting in cases marked by significant ventriculomegaly: one large case series (Thiessen and DeAngelis 1998) demonstrated temporary improvement lasting, on average, 6 months, in 80 percent of cases, however, it was not possible to predict who would improve on the basis of opening pressure or on the results of a 'tap test'; subsequent case reports, however, suggest that elevated pressure on intracranial monitoring may promise a good result (Perrini *et al.* 2002). As with the option of warfarin treatment, however, the risks of shunting may give one pause. In cases due to radionecrosis, however, it is clear that dexamethasone, by reducing perilesional edema, is helpful; there is also a possibility that bevacizumab, which targets VEGF, may exert a similar effect (Torcuator *et al.* 2009). Finally, consideration should be given to resection of the necrotic mass.

The symptomatic treatment of dementia and of delirium is discussed in Chapters 6.3 and 6.5, respectively.

REFERENCES

Adams H, Mitchell DE, Graham DI *et al.* Diffuse brain damage of immediate impact type. Its relationship to 'primary brain-stem damage' in head injury. *Brain* 1977; **100**:489–502.

Adams JH, Graham DI, Murray LS *et al.* Diffuse axonal injury due to non-missile injury in humans: an analysis of 45 cases. *Ann Neurol* 1982; **12**:557–63.

Adams JH, Doyle D, Ford I *et al.* Diffuse axonal injury in head injury: definition, diagnosis and grading. *Histopathology* 1989; **15**:49–59.

Agha A, Sherlock M, Brennan S *et al.* Hypothalamic–pituitary dysfunction after irradiation of nonpituitary brain tumors in adults. *J Clin Endocrinol Metabol* 2005; **90**:6355–60.

Al Sayegh A, Sandford D, Carson AJ *et al.* Psychological approaches to treatment of postconcussion syndrome: a systematic review. *J Neurol Neurosurg Psychiatry* 2010; **81**:1128–34.

Annegers JF, Hauser WA, Coan SP *et al.* A population-based study of seizures after traumatic brain injury. *N Engl J Med* 1998; **338**:20–4.

Arieff AJ, Wetzel N. Subdural hematoma following epileptic convulsion. *Neurology* 1964; **14**:731–2.

Bazarian JJ, Zhong J, Blyth B *et al.* Diffusion tensor imaging detects clinically important axonal damage after mild traumatic brain injury: a pilot study. *J Neurotrauma* 2007; **24**:1447–59.

Black DW. Mental changes resulting from subdural hematoma. *Br J Psychiatry* 1984; **145**:200–3.

Blumbergs PC, Jones NR, North JB. Diffuse axonal injury in head trauma. *J Neurol Neurosurg Psychiatry* 1989; **52**:838–41.

Blumbergs PC, Scott G, Manavis J *et al.* Topography of axonal injury as defined by amyloid precursor protein and the sector scoring method in mild and severe closed head injury. *J Neurotrauma* 1995; **12**:565–72.

Brada M, Ford D, Ashley S *et al.* Risk of second brain tumor after conservative surgery and radiotherapy for pituitary adenoma. *BMJ* 1992; **304**:1343–6.

Bryant RA, O'Donnell ML, Creamer M *et al.* The psychiatric sequelae of traumatic injury. *Am J Psychiatry* 2010; **167**:312–20.

Cameron MM. Chronic subdural hematoma: a review of 114 cases. *J Neurol Neurosurg Psychiatry* 1978; **41**:834–9.

Christianson SA, Neppe V, Hofman H. Amnesia and vegetative abnormalities after irradiation treatment: a case study. *Acta Neurol Scand* 1994; **90**:360–6.

Constine LS, Woolf PD, Cann D *et al.* Hypothalamic–pituitary dysfunction after radiation for brain tumors. *N Engl J Med* 1993; **328**:87–94.

Corsellis JAN. Boxing and the brain. *BMJ* 1989; **298**:105–9.

Corsellis JAN, Bruton CJ, Freeman-Browne D. The aftermath of boxing. *Psychol Med* 1973; **3**:270–303.

Critchley M. Medical aspects of boxing particularly from a neurological standpoint. *BMJ* 1957; **1**:357–66.

Crompton MR, Layton DD. Delayed radionecrosis of the brain following therapeutic X-irradiation of the pituitary. *Brain* 1961; **84**:85–101.

DeAngelis LM, Delattre J-Y, Posner JB. Radiation-induced dementia in patients cured of brain metastases. *Neurology* 1989; **39**:172–7.

De Reuck J, vander Eecken H. The anatomy of late radiation encephalopathy. *Eur Neurol* 1975; **13**:481–94.

Duffy P, Chari G, Cartlidge NEF *et al.* Progressive deterioration of intellect and motor function occurring several decades after cranial irradiation. *Arch Neurol* 1996; **53**:814–18.

Fann JR, Uomoto JM, Katon WJ. Sertraline in the treatment of major depression following mild traumatic brain injury. *J Neuropsychiatry Clin Neurosci* 2000; **12**:226–32.

Fann JR, Uomoto JM, Katon WJ. Cognitive improvement with treatment of depression following mild traumatic brain injury. *Psychosomatics* 2001; **42**:48–54.

Firsching R, Wolschneck D, Klein S *et al.* Brain stem lesions after head injury. *Neurol Res* 2002; **24**:145–6.

Fisher CM. Concussion amnesia. *Neurology* 1966; **16**:826–30.

Fogelholm R, Waltimo O. Epidemiology of chronic subdural haematoma. *Acta Neurochirug* 1975; **32**:247–50.

Gelabert-Gonzalez M, Iglesias-Pais M, Garcia-Allut A *et al.* Chronic subdural hematoma: surgical treatment and outcome in 1000 cases. *Clin Neurol Neurosurg* 2005; **107**:223–9.

Gentleman SM, Roberts GW, Gennarelli TA *et al.* Axonal injury: a universal consequence of fatal closed head injury? *Acta Neuropathol* 1995; **89**:537–43.

Glantz MJ, Hoffman JM, Coleman RE *et al.* Identification of early recurrence of primary central nervous system tumors by [18F] fluorodeoxyglucose positron emission tomography. *Ann Neurol* 1991; **29**:347–55.

Glantz MJ, Burger PC, Friedman AH *et al.* Treatment of radiation-induced nervous system injury with heparin and warfarin. *Neurology* 1994; **44**:2020–7.

Graham DL, Lantos PL. *Greenfield's Neuropathology*, 6th edn. London: Arnold, 1996.

Grattan-Smith PJ, Morris JG, Langlands AO. Delayed radiation necrosis of the central nervous system in patients irradiated for pituitary tumor. *J Neurol Neurosurg Psychiatry* 1992; **55**:949–55.

Harvey PKP, Davis JN. Traumatic encephalopathy in a young boxer. *Lancet* 1974; **2**:928–9.

Hof PR, Knabe R, Bovier P *et al.* Neuropathological observations in a case of autism presenting with self-injury behavior. *Acta Neuropathol* 1991; **82**:321–6.

Hof PR, Bouras C, Buee L *et al.* Differential distribution of neurofibrillary tangles in the cerebral cortex of dementia pugilistica and Alzheimer's disease. *Acta Neuropathol* 1992; **85**:23–30.

Hortobagyi T, Wise S, Cary N *et al.* Traumatic axonal damage in the brain can be detected using beta-APP immunohistochemistry within 35 min after head injury to human adults. *Neuropathol Appl Neurobiol* 2007; **33**:226–37.

Huang MX, Theilmann RJ, Robb A *et al.* Integrated imaging approach with MEG and DTI to detect mild traumatic brain injury in military and civilian patients. *J Neurotrauma* 2009; **26**:1213–26.

Ishikawa E, Yanaka K, Sugimoto K *et al.* Reversible dementia in patients with chronic subdural hematoma. *J Neurosurg* 2002; **96**:680–3.

Jordan BD. Neurologic aspects of boxing. *Arch Neurol* 1987; **44**:453–9.

Jordan BD, Relkin NR, Ravdin LD *et al.* Apolipoprotein E epsilon 4 associated with chronic traumatic brain injury in boxing. *JAMA* 1997; **298**:136–40.

Kaufman M, Swartz BE, Mandelkern M *et al.* Diagnosis of delayed cerebral radiation necrosis following proton beam therapy. *Arch Neurol* 1990; **47**:474–6.

Kraus MF, Susmaras T, Caughlin BP *et al.* White matter integrity and cognition in chronic traumatic brain injury: a diffusion tensor imaging study. *Brain* 2007; **130**:2508–19.

Lam KN, Tse VK, Wang C *et al.* Effects of cranial irradiation on hypothalamic-pituitary function – a 5-year longitudinal study in patients with nasopharyngeal cancer. *Q J Med* 1991; **78**:165–76.

Lampert PW, Davies RL. Delayed effects of radiation on the human central nervous system. *Neurology* 1964; **14**:912–17.

Lipton ML, Gellella E, Lo C *et al.* Multifocal white matter ultrastructural abnormalities in mild traumatic brain injury with cognitive disability: a voxel-wise analysis of diffusion tensor imaging. *J Neurotrauma* 2008; **25**:1335–42.

Lishman WA. Brain damage in relation to psychiatric disability after head injury. *Br J Psychiatry* 1968; **114**:373–410.

McCrory PR, Bladin PF, Berkovic SF. Retrospective study of concussive convulsions in elite Australian rules and rugby league footballers: phenomenology, aetiology, and outcome. *BMJ* 1997; **314**:171–4.

McLatchie G, Brooks N, Galbraith S *et al.* Clinical neurologic examination and computed tomographic head scanning in active amateur boxers. *J Neurol Neurosurg Psychiatry* 1987; **50**:96–9.

Mapothar E. Mental symptoms associated with head injury: the psychiatric aspects. *BMJ* 1937; **2**:1055–61.

Markwalder T-M. Chronic subdural hematomas: a review. *J Neurosurg* 1981; **54**:637–45.

Martins AN, Johnston JS, Henry JM *et al.* Delayed radiation necrosis of the brain. *J Neurosurg* 1977; **47**:336–45.

Martland HS. Punch drunk. *JAMA* 1928; **91**:1103–7.

Mawdsley C, Ferguson FR. Neurological disease in boxers. *Lancet* 1963; **2**:795–801.

Maxeiner H, Wolff M. Pure subdural hematomas: a postmortem analysis of their form and bleeding points. *Neurosrugery* 2002; **50**:503–8.

Miller CM. Whiplash amnesia. *Neurology* 1982; **32**:667–8.

Morris JG, Gratten-Smith P, Panegyres PK *et al.* Delayed cerebral radiation necrosis. *Q J Med* 1994; **87**:119–29.

Ng HK, Mahaliyana RD, Poon WS. The pathological spectrum of diffuse axonal injury in blunt head trauma: assessment with axon and myelin stains. *Clin Neurol Neurosurg* 1994; **96**:24–31.

Niogi SN, Murkherjee P, Ghajar J *et al.* Extent of microstructural white matter integrity in postconcussive syndrome correlates with impaired cognitive reaction time: a 3T diffusion tensor imaging study of mild traumatic brain injury. *AJNR* 2008; **29**:967–73.

Oliff A, Bleyer WA, Poplack DG. Acute encephalopathy after initiation of cranial irradiation for meningeal leukaemia. *Lancet* 1978; **2**:13–15.

Omalu BI, DeKosky ST, Minster RL *et al.* Chronic traumatic encephalopathy in a National Football League player. *Neurosurgery* 2005; **57**:128–34.

Pennybaker J, Russell DS. Necrosis of the brain due to radiation therapy: clinical and pathological observations. *J Neurol Neurosurg Psychiatry* 1948; **11**:183–98.

Perrini P, Scollato A, Cloffi F *et al.* Radiation leukoencephalopathy associated with moderate hydrocephalus: intracranial pressure monitoring and results of ventriculoperitoneal shunting. *Neurol Sci* 2002; **23**:237–41.

Pittella JE, Gusmao SN. Diffuse vascular injury in fatal road traffic accident victims: its relationship to diffuse axonal injury. *J Forensic Sci* 2003; **48**:626–30.

Ramachandran R, Hegde T. Chronic subdural hematomas – causes of morbidity and mortality. *Surg Neurol* 2007; **67**:367–72.

Roberts GW. Immunocytochemistry of neurofibrillary tangles in dementia pugilistica and Alzheimer's disease: evidence for common genesis. *Lancet* 1988; **2**:1456–8.

Roberts GW, Allsop D, Bruton C. The occult aftermath of boxing. *J Neurol Neurosurg Psychiatry* 1990; **53**:373–8.

Robinson RG. A second brain tumor and irradiation. *J Neurol Neurosurg Psychiatry* 1978; **41**:1005–12.

Schmidt ML, Zhukareva V, Newell KL *et al.* Tau isoform profile and phosphorylation state in dementia pugilistica recapitulate Alzheimer's disease. *Acta Neuropathol* 2001; **101**:518–24.

Shewmon DA, Masdeu JC. Delayed radiation necrosis of the brain contralateral to original tumor. *Arch Neurol* 1980; **37**:592–4.

Smits M, Houston GC, Dippel DW *et al.* Microstructural brain injury in post-concussion syndrome after minor head injury. *Neuroradiology* 2011; **53**:553–63.

Soffer D, Pittaluga S, Feiner M *et al.* Intracranial meningiomas following low-dose irradiation to the head. *J Neurosurg* 1983; **59**:1048–53.

Strich SJ. Diffuse degeneration of the cerebral white matter in severe dementia following head injury. *J Neurol Neurosurg Psychiatry* 1956; **19**:163–85.

Symonds CP. Concussion and its sequelae. *Lancet* 1962; **1**:1–5.

Thajeb P. Progressive late-delayed postirradiation encephalopathy with Kluver-Bucy syndrome. Serial MRI and clinic-pathological studies. *Clin Neurol Neurosurg* 1995; **97**:264–8.

Thiessen B, DeAngelis LM. Hydrocephalus in radiation leukoencephalopathy: results of ventriculoperitoneal shunting. *Arch Neurol* 1998; **55**:705–10.

Torcuator R, Zuniga R, Mohan YS *et al.* Initial experience with bevacizumab treatment for biopsy confirmed cerebral radiation necrosis. *J Neurooncol* 2009; **94**:63–8.

Topal NB, Hakyemez B, Erdogan C *et al.* MR imaging in the detection of diffuse axonal injury with mild traumatic brain injury. *Neurol Res* 2008; **30**:974–8.

Woo E, Lam K, Yu YL *et al.* Temporal lobe and hypothalamo-pituitary dysfunctions after radiotherapy for nasopharyngeal carcinoma: a distinct clinical syndrome. *J Neurol Neurosurg Psychiatry* 1988; **51**:1302–7.

Zampieri P, Zorat PL, Mingrino S *et al.* Radiation-associated cerebral gliomas: a report of two cases and review of the literature. *J Neurosurg Sci* 1989; **33**:271–9.

13

Hypoxic–ischemic disorders

13.1 POSTHYPOXIC–ISCHEMIC ENCEPHALOPATHY

Hypoxia or global cerebral ischemia will, if prolonged for 5 or more minutes, cause coma, and those who survive may be left with various deficits, as discussed below. Common causes include carbon monoxide poisoning, strangulation, hanging, drowning, hemorrhagic or septic shock, and, most notably, cardiac arrest.

This posthypoxic–ischemic encephalopathy is also known as postanoxic encephalopathy, posthypoxic encephalopathy or postischemic encephalopathy. Although, as noted below, etiologically 'pure' hypoxic or ischemic cases do occur, nevertheless, in the vast majority, both hypoxia and ischemia are present, thus justifying the broader term, 'hypoxic–ischemic'.

Clinical features

Among those who survive the hypoxic–ischemic event and emerge from coma, some may develop either the persistent vegetative state or the minimally conscious state, both discussed in Chapter 6.2. Others may emerge into a delirium of variable severity; after the delirium clears, some patients may recover entirely; however, most will be left with a persisting deficit, which may include dementia, amnesia, apathy, or depression. Dementia may be of variable severity (Roine *et al.* 1993), and may be accompanied by restlessness or agitation. The amnestic syndrome is characterized by an anterograde deficit, accompanied by a variable degree of retrograde amnesia (Berlyne and Strachan 1968; Broman *et al.* 1997; Medalia *et al.* 1991); although 'pure' cases have been reported (Zola-Morgan *et al.* 1986), in most instances amnesia is accompanied by other, albeit minor, cognitive deficits (Lim *et al.* 2004). Depression is common (Roine *et al.* 1993), as is apathy (Horstmann *et al.* 2010).

Action myoclonus may occur (Lance and Adams 1963; Werhahan *et al.* 1997), as may seizures. Parkinsonism, dystonia or athestosis (or a combination of these) may also develop.

Magnetic resonance (MR) imaging may reveal a variety of findings, including laminar cortical necrosis, watershed infarctions, and abnormalities in the basal ganglia or thalamus.

Course

Dementia and amnesia may show some improvement over the first half-year or so, after which these features tend to remain stably chronic. Parkinsonism and dystonia, by contrast, may show a gradual progression over many years.

Etiology

Although most cases of hypoxic–ischemic coma are, as in cardiorespiratory arrest, characterized by both hypoxia and ischemia, 'pure' instances of each type do occur. Thus, pure hypoxia may occur with strangulation or hanging, while pure global ischemia may occur with shock or severe blood loss. Hypoxic injury, if relatively mild, may cause selective hippocampal damage, and global ischemia, again if mild, may be associated only with watershed infarctions. With a combination of both mild hypoxia and mild ischemia, or in severe cases of either, more widespread damage is seen, with laminar cortical necrosis, more extensive hippocampal damage, and damage to the basal ganglia and thalamus.

Dementia, as might be expected, is associated with laminar cortical necrosis (Richardson *et al.* 1959; Weinberger *et al.* 1940), amnesia with hippocampal damage (Cummings *et al.* 1984; Muramoto *et al.* 1979, Ng *et al.* 1989), and parkinsonism and dystonia with damage to the basal ganglia.

Differential diagnosis

Both posthypoxic–ischemic encephalopathy and delayed postanoxic encephalopathy are associated with a preceding hypoxic–ischemic event. They are differentiated by attending to the time course of these two disorders. Posthypoxic–ischemic encephalopathy has an onset immediately upon recovery from coma; by contrast, in delayed postanoxic encephalopathy there is a latency period between clinical stabilization after the hypoxic–ischemic event and the onset of the encephalopathy, usually lasting for 1–2 weeks.

Treatment

The general treatment of dementia, amnesia, depression, and apathy is discussed in the appropriate chapters. A case report suggests that agitation may respond to amitriptyline (Szlabowicz and Stewart 1990), and one of the authors has seen dramatic results with high-dose mirtazapine in such cases. Myoclonus is traditionally treated with clonazepam, eventually in doses of 7–12 mg daily (Goldberb and Dorman 1976); other options include valproate (Rollinson and Gilligan 1979) and levetiracetam (Krauss *et al.* 2001). Parkinsonism may be treated with levodopa, and a case series has suggested efficacy for antipsychotics in the treatment of chorea (Park and Choi 2004).

13.2 DELAYED POSTANOXIC LEUKOENCEPHALOPATHY

After a hypoxic–ischemic event of sufficient severity to cause coma, or, in some cases, merely delirium, patients may recover completely or, in cases following coma, may develop a posthypoxic–ischemic encephalopathy, as discussed above. Regardless of the outcome after the hypoxic–ischemic event, however, an interval of more or less clinical stability will ensue. In a small minority of cases, however, this latent interval will be followed by a clinical deterioration, constituting what is known as delayed postanoxic leukoencephalopathy.

Clinical features

Clinically (Choi 1983, 2002; Courville 1957; Gottfried *et al.* 1997; Hsiao *et al.* 2004; Min 1986; Murata *et al.* 1995; Norris *et al.* 1982; Plum *et al.* 1962), the latent averages about 2–3 weeks, ranging from as little as 2 days to up to 2 months. The onset of the encephalopathy itself is fairly sudden, occurring over a matter of a day or two, and patients generally present with a combination of delirium and a movement disorder. Confusion, amnesia, apathy, irritability or agitation, and incontinence are prominent, and some patients may become mute; rarely the clinical picture may be dominated by mania (Ku *et al.* 2006) or psychosis with catatonia (Shprecher *et al.* 2009). Parkinsonism is the most common movement disorder seen, but some patients may develop dystonia or chorea. Spasticity is common, with hyper-reflexia and extensor plantar responses. Delirium may occasionally be absent, and patients may present only with a movement disorder, such as parkinsonism (Choi 2002; Goto *et al.* 1997; Grinker 1926; Klawans *et al.* 1982; Rosenberg *et al.* 1989), dystonia (Valenzuela *et al.* 1992), or chorea (Davous *et al.* 1986; Schwartz *et al.* 1985).

T2-weighted MR scanning reveals confluent areas of increased signal intensity in the centrum semiovale in those with delirium (Chang *et al.* 1992; Gottfried *et al.* 1997, Hsiao *et al.* 2004); acutely, diffusion-weighted imaging will show increased signal intensity in the same area (Kim *et al.* 2003). In patients with a movement disorder, T1-weighted scanning may reveal decreased signal intensity in the lenticular nucleus (Chang *et al.* 1992, Takahashi *et al.* 1998).

Course

In a small minority the course is fulminant, with coma and death. In most, however, the course is favorable, with patients experiencing a more or less complete recovery of cognitive abilities over 6–12 months, with only a minority being left with a residual dementia (Choi 1983; Hsiao *et al.* 2004, Min 1986; Plum *et al.* 1962; Shillito *et al.* 1936). The movement disorder, however, may persist, and in some cases may progressively worsen.

Etiology

Although in most cases of posthypoxic encephalopathy the preceding hypoxic–ischemic event was carbon monoxide intoxication, cases have also been reported after hypoxia secondary to strangulation (Dooling and Richardson 1976) or to global ischemia due to hemorrhage or severe hypotension (Mizutani *et al.* 1993). Regardless of the cause, autopsies of those who developed delirium reveal massive, symmetric, diffuse demyelination of the cerebral white matter (Plum *et al.* 1962), while, in cases characterized by a movement disorder alone, cavitary lesions are found in the basal ganglia (Dooling and Richardson 1976). Although the mechanism underlying this pathology is not known, it is strongly suspected that the preceding hypoxic–ischemic injury triggered an autoimmune response directed at both the white matter and the basal ganglia. A study (Ide and Kamijo 2008) of the cerebrospinal fluid (CSF) of patients who had recovered from an hypoxic–ischemic insult lends support to this conjecture: those patients with high CSF myelin basic protein levels went on to develop a delayed postanoxic encephalopathy, whereas those with normal levels did not: presumably the elevated myelin basic protein levels indicate significant axonal injury during the original hypoxic–ischemic event, injury, which, in turn, could plausibly prompt an immune response that subsequently went awry.

Differential diagnosis

Delayed postanoxic encephalopathy is distinguished from posthypoxic–ischemic encephalopathy by the presence of a latent interval between the hypoxic/ischemic insult and the onset of symptoms.

Treatment

Although there is no established treatment for this disorder, a case may be made, given the suspected mechanism, for acute treatment with methylprednisolone. The general treatment of delirium is discussed in Chapter 6.5; in cases characterized by persistent parkinsonism, a trial of levodopa may be considered.

13.3 CARBON MONOXIDE POISONING

Carbon monoxide poisoning may be accidental, as may occur with poorly ventilated gas, wood, or charcoal stoves, or by suicidal intent, such as when patients hook a hose to the car exhaust and then funnel it into the tightly windowed car, or simply leave the car running in an enclosed garage.

Clinical features

The onset of intoxication may be gradual or sudden, and, given that carbon monoxide is colorless and odorless, victims may be unaware of their plight. In general (Handa and Tai 2005; Sayers and Davenport 1930), although the correlation between carboxyhemoglobin level and clinical symptomatology is at best rough (Hampson and Hauff 2008), headache and delirium generally appear at a carboxyhemoglobin level between 20 and 30 percent, worsening and being joined by visual dimming, nausea, and vomiting as the level rises to 40 percent. At levels of 40–50 percent, stupor and ataxia appear, and cyanosis may be seen. When the level rises above 50 percent, coma and convulsions occur, and levels over 60 percent are often fatal. Although it is traditional to associate carbon monoxide poisoning with a cherry-red discoloration of the lips, nails and skin, this is in fact uncommon.

In those who survive, characteristic changes may be seen in the globus pallidus on MR scanning, as illustrated in Figure 13.1.

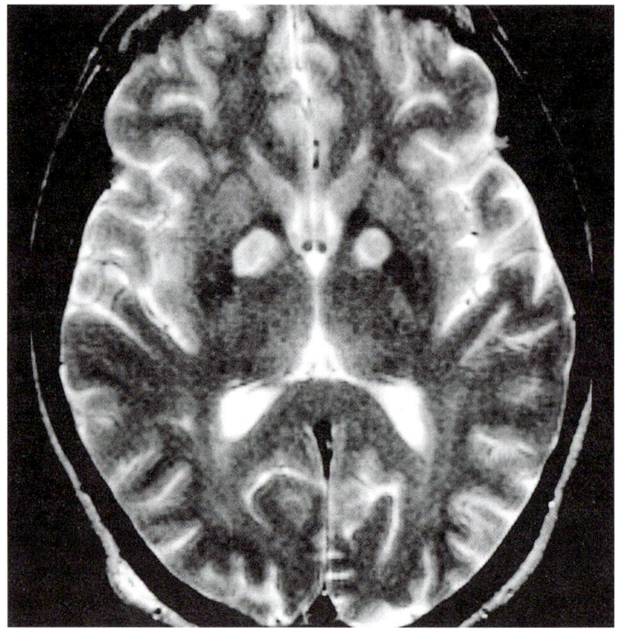

Figure 13.1 T2-weighted magnetic resonance imaging scan in a case of carbon monoxide poisoning, demonstrating increased signal intensity in the globus pallidus. Reproduced from Graham and Lantos (2002), with permission.

Course

In general, if intoxication ceases before the onset of stupor, recovery is typically complete within anywhere from hours to weeks (Smith and Brandon 1970). Should coma occur, patients may be left with a posthypoxic–ischemic encephalopathy, as discussed in Chapter 13.1. Further, in cases of coma, or even in cases characterized only by delirium or stupor, patients may, after a latent interval, develop a delayed postanoxic encephalopathy, as discussed in Chapter 13.2.

Etiology

The affinity of carbon monoxide for hemoglobin is over 200 times greater than that of oxygen and, when a high fraction of hemoglobin exists as carboxyhemoglobin, tissue anoxia supervenes. Anoxia may not, however, be the only mechanism of toxicity. Carbon monoxide also binds to mitochondrial cytochrome oxidase and thus impairs cellular respiration; furthermore, carbon monoxide also binds to areas of the central nervous system rich in iron, for example the globus pallidus and the substantia nigra. In fatal cases widespread petechial hemorrhages are found throughout the cerebrum (Finck 1966).

Differential diagnosis

The circumstances in which the patient is found usually leave little doubt as to the diagnosis; in cases of attempted suicide, however, one must keep in mind that the patient, in addition to inhaling carbon monoxide, may also, in an attempt to be certain of a fatal outcome, have taken an overdose of medications.

Treatment

The goal of treatment is to eliminate the carbon monoxide as rapidly as possible. The half-life of carboxyhemoglobin ranges from 4 to 5 hours; with inhalation of 100 percent oxygen, however, this is cut to about 1 hour, and with hyperbaric oxygen it falls to 30 minutes or less; consequently hyperbaric oxygen is preferred in virtually all cases (Weaver *et al.* 2002).

REFERENCES

Berlyne N, Strachan M. Neuropsychiatric sequelae of attempted hanging. *Br J Psychiatry* 1968; **114**:411–22.

Broman M, Rose AL, Hotson G *et al.* Severe anterograde amnesia with onset in childhood as a result of anoxic encephalopathy. *Brain* 1997; **120**:417–33.

Chang KH, Han MH, Kim HS *et al.* Delayed encephalopathy after acute carbon monoxide intoxication: MR imaging features and distribution of cerebral white matter lesions. Radiology 1992; **184**:117–22.

Choi IS. Delayed neurologic sequelae in carbon monoxide intoxication. *Arch Neurol* 1983; **40**:433–5.

Choi IS. Parkinsonism following carbon monoxide poisoning. *Eur Neurol* 2002; **48**:30–3.

Courville CB. The process of demyelination in the central nervous system. IV. Demyelination as a delayed residual of carbon monoxide asphyxiation. *J Nerv Ment Dis* 1957; **125**:534–46.

Cummings JL, Tomiyasu U, Read S *et al.* Amnesia with hippocampal lesions after cardiopulmonary arrest. *Neurology* 1984; **24**: 679–81.

Davous P, Rondot P, Marion MH *et al.* Severe chorea after acute carbon monoxide poisoning. *J Neurol Neurosurg Psychiatry* 1986; **49**:206–8.

Dooling EC, Richardson EP. Delayed encephalopathy after strangling. *Arch Neurol* 1976; **33**:196–9.

Finck PA. Exposure to carbon monoxide: review of the literature and 567 autopsies. *Milit Med* 1966; **131**:1513–39.

Goldberb MA, Dorman JD. Intention myoclonus: successful treatment with clonazepam. *Neurology* 1976; **26**:24–6.

Goto S, Kunitoku N, Suyama N *et al.* Posteroventral pallidotomy in a patient with parkinsonism caused by hypoxic encephalopathy. *Neurology* 1997; **49**:707–10.

Gottfried JA, Mayer SA, Shungu DC *et al.* Delayed posthypoxic demyelination: association with arylsulfatase A deficiency and lactic acidosis on proton MR spectroscopy. *Neurology* 1997; **49**:1400–4.

Graham DL, Lantos PL. *Greenfield's Neuropathology*, 7th edn. London: Arnold, 2002.

Grinker RR. Parkinsonism following carbon monoxide poisoning. *J Nerv Ment Dis* 1926; **64**:18–28.

Hampson NB, Hauff NM. Carboxyhemoglobin levels in carbon monoxide poisoning: do they correlate with the clinical picture? *Am J Emerg Med* 2008; **26**:665–9.

Handa PK, Tai DY. Carbon monoxide poisoning: a five year review at Tan Tock Seng Hospital, Singapore. *Ann Acad Med Singapore* 2005; **34**:611–14.

Horstmann A, Rrisch S, Jentzch RT *et al.* Rescitating the heart but losing the brain: brain atrophy in the aftermath of cardiac arrest. *Neurology* 2010; **74**:306–12.

Hsiao CL, Kuo HC, Huang CC. Delayed encephalopathy after carbon monoxide intoxication – long-term prognosis and correlation of clinical manifestations and neuroimages. *Acta Neurol Taiwan* 2004; **13**:64–70.

Ide T, Kamijo Y. Myelin basic protein in cerebrospinal fluid: a predictive marker of delayed encephalopathy from carbon monoxide poisoning. *Am J Emerg Med* 2008; **26**:908–12.

Kim JH, Chang KH, Song IC *et al.* Delayed encephalopathy of acute carbon monoxide intoxication: diffusivity of cerebral white matter lesions. *AJNR* 2003; **24**:1592–7.

Klawans HL, Stein RW, Tanner CM *et al.* A pure parkinsonian syndrome following acute carbon monoxide intoxication. *Arch Neurol* 1982; **39**:302–4.

Krauss GL, Bergin A, Kramer RE *et al.* Suppression of post-hypoxic and post-encephalitic myoclonus with levetiracetam. *Neurology* 2001; **56**:1144–5.

Ku BD, Shin HY, Kim EJ *et al.* Secondary mania in a patient with delayed anoxic encephalopathy after carbon monoxide intoxication. *J Clin Neurosci* 2006; **13**:860–2.

Lance JW, Adams RD. The syndrome of intention or action myoclonus as a sequel to hypoxic encephalopathy. *Brain* 1963; **86**:111–36.

Lim C, Alexander MP, LaFleche G *et al.* The neurological and cognitive sequelae of cardiac arrest. *Neurology* 2004; **63**:1774–8.

Medalia AA, Merriam AE, Ehrenreich JH. The neuropsychological sequelae of attempted hanging. *J Neurol Neurosurg Psychiatry* 1991; **54**:546–8.

Min SK. A brain syndrome associated with delayed neuropsychiatric sequelae following acute carbon monoxide intoxication. *Acta Psychatr Scand* 1986; **73**:80–6.

Mizutani T, Shiozawa R, Takemori S *et al.* Delayed post-anoxic encephalopathy without relation to carbon monoxide poisoning. *Intern Med* 1993; **32**:430–3.

Muramoto O, Kuru Y, Sugishita M *et al.* Pure memory loss with hippocampal lesions: a pneumoencephalographic study. *Arch Neurol* 1979; **36**:54–6.

Murata T, Itoh S, Koshino Y *et al.* Serial proton magnetic resonance spectroscopy in a patient with the interval form of carbon monoxide poisoning. *J Neurol Neurosurg Psychiatry* 1995; **58**:100–3.

Ng T, Graham DI, Adams JH *et al.* Changes in the hippocampus and the cerebellum resulting from hypoxic insults: frequency and distribution. *Acta Neuropathol* 1989; **78**:438–43.

Norris CR, Trench JM, Hook R. Delayed carbon monoxide encephalopathy: clinical and research implications. *J Clin Psychiatry* 1982; **43**:294–5.

Park S, Choi I S. Chorea following carbon monoxide poisoning. *Yonsei Med J* 2004; **45**:363–6.

Plum F, Posner JB, Hain RF. Delayed neurological deterioration after anoxia. *Arch Intern Med* 1962; **110**:56–63.

Richardson JC, Chambers RA, Heywood PM. Encephalopathies of anoxia and hypoglycemia. *Arch Neurol* 1959; **1**:178–90.

Roine RO, Kajaste S, Kaste M. Neuropsychological sequelae of cardiac arrest. *JAMA* 1993; **269**:237–42.

Rollinson RD, Gilligan BS. Post-anoxic myoclonus (Lance-Adams syndrome) responding to valproate. *Arch Neurol* 1979; **36**: 44–5.

Rosenberg NL, Myers JA, Martin WRW. Cyanide-induced parkinsonism: clinical MRI and 6-fluorodopa PET studies. *Neurology* 1989; **39**:142–4.

Sayers PR, Davenport SJ. Review of carbon monoxide poisoning. Public Health Bulletin No. 195. Washington, DC: US Government Printing Office, 1930.

Schwartz A, Hennerici M, Wegener OH. Delayed choreoathetosis following acute carbon monoxide poisoning. *Neurology* 1985; **35**:98–9.

Shillito FH, Drinker CK, Shaughnessy TJ. The problem of nervous and mental sequelae in carbon monoxide poisoning. *JAMA* 1936; **106**:669–74.

Shprecher DL, Flanigan KM, Smith AG *et al.* Clinical and diagnostic features of delayed hypoxic leukoencephalopathy. *J Neuropsychiatry Clin Neurosci* 2009; **20**:473–7.

Smith JS, Brandon S. Acute carbon monoxide poisoning – 3 years' experience in a defined population. *Postgrad Med J* 1970; **46**:65–70.

Szlabowicz JW, Stewart JT. Amitriptyline treatment of agitation associated with anoxic encephalopathy. *Arch Phys Med Rehabil* 1990; **71**:612–13.

Takahashi W, Ohnuki Y, Takizawa S *et al.* Neuroimaging on delayed postanoxic encephalopathy with lesions localized in the basal ganglia. *Clin Imaging* 1998; **22**:188–91.

Valenzuela R, Court J, Godoy J. Delayed cyanide induced dystonia. *J Neurol Neurosurg Psychiatry* 1992; **55**:198–9.

Weaver LK, Hopkins RD, Chan KJ *et al.* Hyperbaric oxygen for acute carbon monoxide poisoning. *N Engl J Med* 2002; **347**:1057–67.

Weinberger LM, Gibbon MH, Gibbon JH. Temporary arrest of the circulation to the central nervous system. II. Pathologic effects. *Arch Neurol Psychiatry* 1940; **43**:961–86.

Werhahan KJ, Brown P, Thompson PD *et al.* The clinical features and prognosis of chronic posthypoxic myoclonus. *Mov Disord* 1997; **12**:216–20.

Zola-Morgan S, Squire LR, Amaral DG. Human amnesia and the medial temporal region: enduring memory impairment following a bilateral lesion limited to the field CA1 of the hippocampus. *J Neurosci* 1986; **6**:2950–67.

14

Nutritional, toxic, and metabolic disorders

14.1 VITAMIN B12 DEFICIENCY

Vitamin B12 deficiency may lead to demyelination in the cerebrum, with, among other disturbances, dementia, or in the cord, leading to a syndrome known as subacute combined degeneration. A macrocytic anemia is also common; however, as stressed below, neuropsychiatric symptomatology may occur in patients with normal red blood cell counts and red blood cell indices. Although vitamin B12 deficiency is most commonly due to pernicious anemia, multiple other causes must also be considered.

Clinical features

Symptoms referable to the cerebrum or spinal cord tend to appear subacutely over weeks or months.

Cerebral involvement (Healton *et al.* 1991; Lindenbaum *et al.* 1988) manifests most commonly with a dementia, which may be marked by hallucinations and delusions (Lurie 1919). Personality change may also occur, and one may also see depression (Fraser 1960) or mania (Goggans 1984). Another rare manifestation is psychosis, referred to in the past as 'megaloblastic madness': in one case (Smith 1929), a 46-year-old woman developed delusions of persecution and jealousy, and was correctly diagnosed only when symptoms of subacute combined degeneration became evident; in another (Evans *et al.* 1983), a 47-year-old woman became withdrawn, guarded, and suspicious, heard the voice of God commanding her to board a spaceship, and prayed fervently, with all symptoms resolving upon B12 administration. There is also a case report of tremor and chorea occurring as a manifestation of B12 deficiency (Pacchetti *et al.* 2002).

Subacute combined degeneration (Healton *et al.* 1991; Russell *et al.* 1900) reflects demyelination of the peripheral nerves, posterior columns, and lateral corticospinal tracts. Patients present with acral parasthesiae, followed by ataxia, a positive Romberg test, and eventually spasticity: the plantar responses are generally extensor, but the deep tendon reflexes may be either increased or depressed, depending on the severity of the peripheral neuropathy. In some cases, optic neuritis and visual impairment may also occur.

Macrocytosis, with or without anemia, is common; however, as noted above, it must be stressed that both these findings may be absent. Indeed, in one large study both the red blood cell count and the mean corpuscular volume were normal in approximately one-fifth of all patients (Lindenbaum *et al.* 1988).

Magnetic resonance imaging (MRI) scanning reveals patchy, confluent areas of increased signal intensity in the centrum semiovale in patients with cerebral symptomatology (Chatterjee *et al.* 1996; Stojsavljevic *et al.* 1997). The electroencephalogram (EEG) may show generalized slowing (Watson *et al.* 1954).

It is customary to obtain a serum B12 level and, although this custom should be observed, it is also appropriate to obtain levels of both methylmalonic acid and homocysteine. In B12 deficiency, both methylmalonic acid and homocysteine levels are elevated, and this combination of findings is extremely sensitive and specific for B12 deficiency (Lindenbaum *et al.* 1988); when the serum B12 level is 'borderline', these two findings should be relied on when deciding whether intracellular B12 deficiency is present.

Before leaving this section, it is appropriate to note that clinical B12 deficiency can be precipitated by inhalation of nitrous oxide, as may occur with anaesthesia for medical or dental procedures, or in drug abusers. In cases in which there is already a 'subclinical' B12 deficiency, nitrous oxide inhalation can acutely precipitate symptoms (Beltramello *et al.* 1998; Kinsella and Green 1995; Marie *et al.* 2000); in other cases, for example in young, well-nourished drug abusers, symptoms may not occur until nitrous oxide has been abused for a long time.

Course

Most cases of B12 deficiency are gradually progressive.

Etiology

Vitamin B12, or cobalamin, is formed only by certain plant-associated bacteria, and humans generally obtain their supply indirectly by eating liver, other organ meats, beef, pork, milk, or eggs. Once ingested, cobalamin is first bound within the stomach to gastric R binder; this complex is digested by pancreatic enzymes in the duodenum and the liberated cobalamin is then bound to intrinsic factor, a glycoprotein that is secreted by gastric parietal cells. This cobalamin–intrinsic factor complex then passes to the ileum, where it is bound to a receptor on the cell wall and taken into the cell. Inside the ileal epithelial cell, cobalamin is released from intrinsic factor and bound to a carrier protein known as transcobalamin II. The cobalamin–transcobalamin II complex is then released into the systemic circulation. A substantial amount of cobalamin is stored in the liver and, because of extensive enterohepatic recirculation, years must pass before hepatic stores are depleted.

As noted earlier, the most common cause of B12 deficiency is pernicious anemia (Healton *et al.* 1991). In this disease, anti-parietal cell antibodies lead to the destruction of gastric parietal cells, with a consequent lack of intrinsic factor and a deficient uptake of ingested B12 by the ileal cells. In addition to anti-parietal cell antibodies, the majority of patients also have anti-intrinsic factor antibodies, and it is customary to test for both of these. Other antibodies may also be present in patients with pernicious anemia, including antibodies against the thyroid and the adrenal gland, and patients may develop Hashimoto's thyroiditis with hypothyroidism, or adrenocortical insufficiency.

Other causes of B12 deficiency include strict vegetarianism, severe malnutrition (as may be seen in chronic alcoholics or in anorexia nervosa [Patchell *et al.* 1994]), total or partial gastrectomy, inherited abnormalities of the R binder or intrinsic factor, achlorhydria, pancreatic insufficiency, steatorrhea or malabsorption of any cause, tapeworms, bacterial overgrowth (as may occur after a Billroth II operation), ileal diseases (e.g. Crohn's disease), ileal resection, and chronic treatment with either metformin (Andres *et al.* 2002) or omeprazole (Marcuard *et al.* 1994): this last cause is of some concern, as it is not clear whether there is something specific about omeprazole, or whether this might be a class effect, implicating all of the proton pump inhibitors as possible causes. All of the preceding causes lead to intracellular B12 deficiency because of a failure of B12 absorption, and all are associated with decreased serum B12 levels. It is important to keep in mind, however, that in addition to these causes there are also some very rare diseases characterized by intracellular B12 deficiency or malutilization, and that in these disorders the serum B12 level will be normal, but the serum methylmalonic acid and homocysteine levels will be elevated. Examples include inherited abnormalities in transcobalamin II (Turner and Talbot 2009) and abnormalities in the intracellular metabolism of cobalamin (Bodamar *et al.* 2001; Boxer *et al.* 2005; Powers *et al.* 2001). Finally, there may also be an association between acquired autoimmune deficiency syndrome (AIDS) and B12 deficiency (Herzlich and Schiano 1993; Shor-Posner *et al.* 1995).

When B12 deficiency has been confirmed by elevated methylmalonic acid and homocysteine levels, efforts should be made to determine the cause of this deficiency. In the past, the Schilling test was utilized to determine whether or not B12 absorption could be increased by the addition of hog-derived intrinsic factor: if this test was positive, then one inferred that the deficiency was due to destruction of parietal cells, as seen in pernicious anemia. With the availability of assays for anti-parietal cell and anti-intrinsic factor antibodies, however, this test is generally no longer required. Given that pernicious anemia is the most common cause of B12 deficiency, all patients should be tested for these antibodies. If these are absent, then the other causes noted above should be considered.

Neuropathologically, demyelination is seen within the centrum semiovale (Ferraro *et al.* 1945; Lurie 1919), and in the posterior columns and the lateral corticospinal tracts within the spinal cord (Russell *et al.* 1900).

Differential diagnosis

Vitamin B12 deficiency may be almost perfectly mimicked by folic acid deficiency, and the differential diagnosis rests on the results of testing for homocysteine and methylmalonic acid levels: in B12 deficiency both are elevated, whereas in folic acid deficiency the homocysteine level is elevated but the methylmalonic acid level is normal.

Subacute combined degeneration may be mimicked by cervical spondylosis or multiple sclerosis; it may also be mimicked by adult-onset adrenoleukodystrophy and by neurosyphilis, and in both of these disorders one may also find an associated dementia. Consideration should also be given to copper deficiency, as this is a readily treatable condition that may almost perfectly mimic subacute combined degeneration (Kumar 2006; Naismith *et al.* 2009).

Treatment

Traditionally, treatment involves the administration of intramuscular cyanocobalamin at a dose of 1000 μg daily for 1 week, then weekly for 4 weeks, then monthly thereafter until maximum recovery has occurred. In those rare cases in which the underlying cause has been corrected, treatment may then cease; however, given that most of the underlying causes are not treatable, chronic treatment is typically required. After patients have been treated for 2 weeks, folic acid, at a dose of 2 mg daily, should be added; importantly, folic acid should not be administered earlier than this as it may lead to an exacerbation of symptoms. Potassium levels should be checked periodically, as hypokalemia may develop early in the course of treatment with cyanocobalamin.

Although, as noted, traditional treatment involves parenteral administration of B12, recent work has indicated that oral treatment with 1 or 2 mg daily may be just as effective (Kuzminski *et al.* 1998; Rajan *et al.* 2002). Clearly, oral administration would be effective in cases of B12 deficiency due to malnutrition; however, it also appears that oral administration is also effective in pernicious anemia. The percentage of orally administered B12 that is absorbed is heavily dependent on intrinsic factor: in the presence of intrinsic factor, roughly 60 percent is absorbed, whereas in its absence, as in pernicious anemia, only about 1 percent is. With these massive oral doses, however, it appears that sufficient B12 is absorbed to allow for effective treatment. Whether oral administration would be effective in other etiologic categories, as for example achlorhydria, malabsorption, etc., however, is not clear. In practice, most physicians will begin with intramuscular treatment, and continue this until maximum recovery. Subsequently, oral administration may be started, with careful follow-up, including not only clinical evaluation, but also periodic determination of methylmalonic acid levels.

With B12 treatment improvement may not occur for months, and up to 18 months may be required for full improvement. If treatment is begun early, before axonal loss has occurred, recovery may be complete; however, if symptoms have been present for months prior to treatment, irreversible damage may already have occurred and the recovery will only be partial.

14.2 FOLIC ACID DEFICIENCY

Although there has been controversy over whether folic acid deficiency can cause disease of the central nervous system, it appears that, albeit rarely, dementia and subacute combined degeneration of the cord do in fact occur on this basis.

Clinical features

The onset of symptoms appears to be gradual over weeks or months. Reynolds *et al.* (1973) demonstrated an association of folate deficiency with dementia, and Strachan and Henderson (1967) described two very convincing cases of dementia secondary to folate deficiency, one occurring in combination with a peripheral neuropathy and the other without any other symptoms. Furthermore, Pincus *et al.* (1972) reported a convincing case of a folate-induced combination of dementia and subacute combined degeneration. Finally, Parry (1994) reported multiple cases of subacute combined degeneration secondary to folate deficiency. Folic acid deficiency, of course, is also a cause of megaloblastic anemia.

The serum homocysteine level is elevated.

Course

It appears that, in the absence of treatment, the course is progressive.

Etiology

Folic acid is a necessary factor in DNA synthesis and is found in fresh green vegetables, some fruits, yeast, kidney, and liver. Dietary deficiency, as may be seen in chronic alcoholics, is the most common cause of folic acid deficiency; intestinal malabsorption, for example in sprue, may also be a factor. Given the limited hepatic storage of folic acid, symptoms of deficiency may appear within a few months of poor oral intake or malabsorption. Certain medicines also reduce folate levels, including oral contraceptives, phenytoin, primidone, phenobarbital, carbamazepine, pyrimethamine, trimethoprim, pentamidine, sulfasalazine, and methotrexate. Marginal folic acid reserves may be rapidly depleted in conditions of increased metabolic demand, for example during pregnancy and lactation, and during the reticulocytosis seen with treatment of B12-induced anemia.

Autopsy studies are few; in two cases of subacute combined degeneration secondary to folate deficiency, findings similar to those seen in B12 deficiency were noted (Parry 1994).

Differential diagnosis

As noted in the preceding chapter, it may not be possible clinically to differentiate folic acid deficiency from B12 deficiency, and the diagnosis here rests on homocysteine and methylmalonic acid levels: in folate deficiency only homocysteine levels are increased, whereas in B12 deficiency, both homocysteine and methylmalonic acid levels are elevated.

Treatment

Parenteral treatment is almost never required, even in cases of intestinal disease, and an oral dose of 1–2 mg daily is generally sufficient. Before starting treatment, however, it is critical to rule out B12 deficiency.

14.3 PELLAGRA

Niacin, also known as vitamin B3, is a water-soluble B vitamin found in liver, poultry, fish, meat, peanuts, and yeast. Another source of this vitamin is tryptophan, which is converted endogenously into niacin.

Niacin deficiency causes pellagra, and this may occur clinically in one of two forms: acute pellagra, also known as 'encephalopathic' pellagra, is characterized by delirium, whereas the gradually developing chronic form of pellagra is characterized by dementia. In developed countries, pellagra is seen most commonly in malnourished alcoholics as the encephalopathic form; chronic pellagra, although once endemic in the American South, is now quite rare.

Clinical features

Encephalopathic pellagra presents fairly acutely, over days to a week, and, when fully developed, is characterized by delirium, dysarthria, cogwheel rigidity, gegenhalten, myoclonus, and positive grasp reflexes (Ishii and Nishihara 1981; Jolliffe *et al.* 1940; Serdaru *et al.* 1988).

Pellagra of gradual onset appears insidiously over many months, and, when fully developed, is characterized by the classic 'three Ds' of dementia, dermatitis, and diarrhea (Spivak and Jackson 1977). The dementia may present with apathy, depression, or anxiety; however, over time typical cognitive deficits, such as decreased memory and poor concentration, eventually appear; in some cases the dementia may be accompanied by delusions or hallucinations (Pierce 1924). The dermatitis is characterized initially by erythematous lesions in sun-exposed areas. Eventually, the skin becomes hyperpigmented and roughened, and it is from the Italian for rough (*pelle*) and skin (*agra*) that the disease gains its name. Diarrhea may be severe, and the fluid may be blood tinged. It must be stressed that most cases of pellagra do not, however, present with the full 'three Ds' (Spivak and Jackson 1977); some patients with pellagrous dementia may have only one of the other 'Ds', and in some cases there may be only dementia, without any rash or diarrhea. Other symptoms include stomatitis, glossitis, and a sensorimotor peripheral polyneuropathy.

MRI scanning is unremarkable; in patients with dementia, the EEG shows generalized slowing (Spivak and Jackson 1977).

Although the serum niacin level is low, a more reliable, if rarely ordered, test is a 24-hour urine for niacin metabolites.

Course

Untreated, pellagra is fatal: the encephalopathic form may be rapidly progressive, and coma and death may ensue in a matter of weeks; in the chronic form, however, one sees a more leisurely course, with death in a matter of years.

Etiology

Niacin deficiency occurs most commonly as a result of dietary deficiency. As noted earlier, in current practice in developed countries this is seen primarily in malnourished alcoholics as the encephalopathic form. The chronic form of pellagra was once endemic in the American South among those individuals who subsisted primarily on corn. As corn contains niacin in a bound, biologically less active form, and also lacks tryptophan, these individuals very gradually became niacin deficient. Since corn flour has been 'enriched' with niacin, however, the chronic form of pellagra has almost disappeared in the United States. Other causes of niacin deficiency include bowel resection, Crohn's disease (Zaki and Millard 1995), and anorexia nervosa (Jagielska *et al.* 2007).

In addition to dietary lack, pellagra has also been noted in conditions in which the normal endogenous conversion of tryptophan to niacin is, for one reason or another, impaired. Perhaps the most common example of this is in patients treated with isoniazid: in one series of patients with tuberculosis so treated, roughly 8 percent were found, at autopsy, to have pellagra (Ishii and Nishihara 1985). The normal enzymatic conversion of tryptophan to niacin is dependent on the activated form of pyridoxine (vitamin B6); isoniazid impairs the conversion of the inactive form of B6 to the active form, and by this indirect mechanism reduces the endogenous production of niacin (Ishii and Nishihara 1985). Another example is in cases of carcinoid tumor, in which the gross overutilization of tryptophan by the tumor leaves less tryptophan available for conversion to niacin.

Within the central nervous system (Hauw *et al.* 1988; Ishii and Nishihara 1981; Langworthy 1931), neurons are swollen and display chromatolysis with eccentric nuclei and a loss of Nissl substance. These chromatolytic changes may be seen in neurons of the cerebral cortex, basal ganglia, dentate nuclei, brainstem motor nuclei, dorsal root ganglia, and the anterior horn of the spinal cord.

Differential diagnosis

The acute encephalopathic form may be confused with other disorders seen in malnourished chronic alcoholics, including not only delirium tremens, but also Wernicke's encephalopathy and Marchiafava–Bignami disease. Delirium tremens is suggested by prominent tremor, and Wernicke's encephalopathy by a sixth cranial nerve palsy and ataxia; the acute form of Marchiafava–Bignami disease, a rare disorder, is in turn suggested by aphasia, ataxia, and long-tract signs. In practice, however, it must be kept in mind that these disorders often occur concurrently in the same patient, and in these cases a precise diagnosis may not be possible (Hauw *et al.* 1988; Serdaru *et al.* 1988). As such, it may be prudent to treat any malnourished delirious alcoholic not only with thiamine but also with niacin.

The chronic form of pellagra is difficult to miss when all three of the 'Ds' are present; however, as noted earlier, many patients have only two of these, and some may have only one. Consequently, a high index of suspicion is required, and chronic pellagra should always be suspected in any chronically malnourished patient who gradually develops a dementia.

Treatment

Niacin may be given orally in doses from 250 to 500 mg daily. In the encephalopathic form, the response is rapid and often robust; in the chronic form recovery is slower and may be incomplete. Once full benefit has occurred, patients may be maintained on 50–100 mg of niacin daily. In cases due to isoniazid, administration of pyridoxine, in doses of 50 mg daily, is generally sufficient, however, in some cases symptoms may persist and in these cases isoniazid must be discontinued (Burke and Hiangabeza 1977).

14.4 WERNICKE'S ENCEPHALOPATHY

Wernicke's encephalopathy occurs secondary to thiamine deficiency, and in its fully developed classic form it presents with the triad of delirium, nystagmus, and ataxia (Victor *et al.* 1971). It occurs most frequently in malnourished alcoholics and is a common cause of delirium in general hospital practice.

Before proceeding further, a word on nomenclature is in order. In the literature, one often sees the term 'Wernicke–Korsakoff syndrome'; however, this should be avoided as Wernicke's encephalopathy and Korsakoff's syndrome are, from a clinical point of view, fundamentally different. Wernicke's encephalopathy, as just noted, is characterized by delirium. Korsakoff's syndrome, although occurring as a sequela to Wernicke's encephalopathy, is an amnestic disorder in which confusion is absent. Lumping the two together under one name serves only to confuse the diagnostic picture.

Clinical features

In general the onset is subacute, spanning several days, and nystagmus is often one of the earliest signs. Occasionally, however, one may see an acute onset over hours, and this may follow a glucose load, either orally or intravenously, in a thiamine-deficient patient.

Delirium is characterized by confusion and disorientation, and is often accompanied by a degree of lethargy or drowsiness; rarely this may progress to stupor or coma (Gibb *et al.* 1985; Wallis *et al.* 1978).

Nystagmus, although generally horizontal, may at times be vertical. With progression, a bilateral and typically asymmetric sixth cranial nerve palsy may appear and patients may complain of diplopia. In extreme cases, a total ophthalmoplegia may occur.

Ataxia may be evident as an ataxia of gait or as a truncal ataxia, which, in turn, may be so severe that patients are unable to sit up in bed. Other cerebellar signs, such as intention tremor or dysarthria, are rare.

It must be emphasized that this classic triad of symptoms is the exception rather than the rule. Autopsy studies have revealed that the full triad is found in only 14–16 percent of cases (Cravioto *et al.* 1961; Harper *et al.* 1986). By far the most common presentation is with delirium alone, or with a combination of delirium and either nystagmus or ataxia (Harper *et al.* 1986).

In addition to delirium, nystagmus, and ataxia, a minority of patients will have grand mal seizures. The temperature is often decreased and there may be tachycardia and postural hypotension.

Various laboratory and imaging findings occur in Wernicke's encephalopathy, but these are rarely required for the diagnosis, and, in any case, treatment of suspected Wernicke's encephalopathy should never wait upon test results.

The EEG may show generalized slowing, but in many cases it is within normal limits. Although the cerebrospinal fluid is generally normal, an elevated total protein, rarely more than 100 mg/dL, may be found. Serum thiamine levels are low, but a more sensitive test for thiamine deficiency is a determination of red blood cell transketolase activity. As discussed below, this enzyme is dependent on thiamine, and in cases where transketolase activity is decreased, the addition of thiamine pyrophosphate will bring it closer to normal.

Magnetic resonance imaging may reveal increased signal intensity on T2-weighted, fluid-attenuated inversion recovery (FLAIR) and diffusion-weighted images of the mammillary bodies, medial aspects of the thalami, periaqueductal gray, and the midbrain tectum (Chu *et al.* 2002; Weidauer *et al.* 2003; Zuccoli *et al.* 2007, 2009); with gadolinium, enhancement may be seen in some cases in the mammillary bodies (Zuccoli *et al.* 2007, 2009).

Course

Without treatment, death occurs in a matter of weeks, and in some cases patients may die suddenly (Harper 1979). With treatment, the mortality rate is cut to roughly 10 percent, with many of these deaths occurring secondary to concurrent disorders, such as pneumonia, sepsis, or hepatic failure.

Among those who do survive, sequelae are common. With resolution of the delirium, confusion clears and roughly 80 percent of patients will be left with a Korsakoff's syndrome (Malamud and Skillicorn 1956), which, as discussed in Chapter 14.5, is characterized by an amnesia with both anterograde and retrograde components. Among patients with nystagmus, a fine degree of nystagmus will remain in roughly two-thirds, and among those with ataxia some degree of ataxia persists in two-thirds to three-quarters.

Etiology

Thiamine is found in many foods, and 1–2 mg daily is generally sufficient to meet most daily needs. Total body stores of thiamine range from 20 to 100 mg, and the half-life of thiamine ranges from 10 to 20 days; thus, with severely reduced thiamine intake, significant deficiency may appear in 1–3 months.

Once absorbed, thiamine is converted to its active form, thiamine pyrophosphate. This molecule functions as an essential co-factor for transketolase, which plays a critical role in the pentose monophosphate shunt pathway. With significant thiamine deficiency, transketolase activity is lost, and the characteristic lesions, described below, develop.

Although most cases of Wernicke's encephalopathy are seen in malnourished alcoholics, cases have also been noted in a number of other conditions (Ogershok *et al.* 2002). These include prolonged total parenteral nutrition with inadequate thiamine (Vortmeyer *et al.* 1992); hemodialysis (Jagadha *et al.* 1987); prolonged vomiting, as may occur in hyperemesis gravidarum (Gardian *et al.* 1999), with chemotherapy (Kuo *et al.* 2009) or after gastric bypass surgery (Abarbanel *et al.* 1987; Paulson *et al.* 1985; Singh and Kumar 2007); hunger strikes (Frantzen 1966); and anorexia nervosa (Handler and Perkin 1982), in one case of which the encephalopathy was precipitated by a carbohydrate binge (Peters *et al.* 2007). Importantly, two or more of these factors may at times be required to produce the encephalopathy. There are, for example, case reports of patients developing Wernicke's encephalopathy anywhere from 2 to 20 years postgastrectomy, the precipitant being a modestly decreased thiamine intake, for example secondary to a loss of appetite during an upper respiratory infection (Shimomura *et al.* 1998).

Although, generally, at least a month must pass before a significant deficiency occurs, there are exceptions to this rule. Glucose loads, either orally or parenterally, may rapidly precipitate symptoms. Furthermore, some individuals have an inherited form of transketolase that displays a decreased affinity for thiamine pyrophosphate (Blass and Gibson 1977), and these individuals are at higher risk of developing Wernicke's encephalopathy and may do so even if their thiamine reserves are only marginally depleted.

In acute cases petechial hemorrhages are seen in the gray matter adjacent to the third ventricle, aqueduct of Sylvius, and the fourth ventricle, including the dorsomedial and anterior nuclei of the thalamus, the mammillary bodies, the periaqueductal gray, the oculomotor and abducens nuclei, and the superior cerebellar vermis (Cravioto *et al.* 1961; Harper 1983; Malamud and Skillicorn 1956; Victor *et al.* 1971). In those who survive, neuronal loss and gliosis are seen in the same areas.

Differential diagnosis

The diagnosis of Wernicke's encephalopathy should be entertained in any malnourished patient who develops delirium; although from a diagnostic point of view it is comforting to also see nystagmus and ataxia, it must be forcefully kept in mind that, as noted above, these are typically not present.

Among alcoholics, differential diagnostic consideration must also be given to delirium tremens, encephalopathic pellagra, and hepatic encephalopathy. Prominent tremor suggests delirium tremens, cogwheel rigidity points to pellagra, and myoclonus suggests hepatic encephalopathy. These disorders, however, often appear concurrently with Wernicke's encephalopathy, and it is appropriate to treat all alcoholics with thiamine, as described below.

Other, rare, disorders to consider in the differential include cytomegalovirus ventriculoencephalitis, as may occur in AIDS (Torgovnik *et al.* 2000), and both Creutzfeldt–Jakob disease (Bertrand *et al.* 2009) and new-variant Creutzfeldt–Jakob disease (Stone *et al.* 2008), both of which may present subacutely with ataxia and cognitive deficits.

Treatment

Whenever Wernicke's encephalopathy is suspected, patients should emergently be given 100 mg of thiamine intravenously. Barring a severe degree of hypoglycemia, food and glucose-containing fluids should be withheld for at least several hours. Thiamine is then continued at a dose of 100 mg twice daily parenterally until substantial improvement is seen, after which patients may be continued on the same dose of oral thiamine until maximum improvement occurs.

The response to treatment is at times spectacular. Nystagmus may begin to improve within minutes, and a sixth nerve palsy may begin to resolve within hours; the improvement in delirium and ataxia is not so dramatic, and here one sees gradual resolution beginning within the first week or two, with full improvement not seen for a full 1 or 2 months. If the expected improvement does not occur, a magnesium level should be checked, as low magnesium levels have been reported to blunt the response to thiamine (Dyckner *et al.* 1985; Traviesa 1974).

On an editorial note, it is remarkable, given the devastation of Wernicke's encephalopathy and its major sequel, Korsakoff's syndrome, that governments have not mandated the addition of thiamine to alcoholic beverages. The results could be as spectacular as those that occurred when flour was 'enriched' with niacin and the plague of endemic pellagra was eradicated.

14.5 KORSAKOFF'S SYNDROME

Korsakoff's syndrome, first described in the late nineteenth century by the Russian psychiatrist S.S. Korsakoff (Victor and Yakovlev 1955), is a chronic amnestic disorder with a prominent anterograde and variable retrograde component that occurs as sequela to Wernicke's encephalopathy.

Some words are in order regarding nomenclature. As noted in Chapter 14.4, the term 'Wernicke–Korsakoff syndrome', although in common usage, should probably be abandoned as from a clinical point of view Wernicke's encephalopathy and Korsakoff's syndrome are fundamentally different: Wernicke's encephalopathy is marked by a delirium, whereas Korsakoff's syndrome is characterized by an amnesia that is not accompanied by confusion. There is an additional difficulty regarding nomenclature, and this has to do with the definition of Korsakoff's syndrome. In some texts this term is used to refer to any chronic amnestic disorder, regardless of cause. In this text, however, the term refers only to the chronic amnestic disorder that occurs as a sequela to Wernicke's encephalopathy.

Clinical features

As noted in Chapter 14.4, Wernicke's encephalopathy is characterized by delirium with or without nystagmus and ataxia. In those who survive, the delirium gradually resolves, and roughly 80 percent of patients may then be left with an amnestic disorder.

These patients, as pointed out by Korsakoff himself (Victor and Yakovlev 1955), may not, at least to casual inspection, appear ill at all. Although they may be somewhat lacking in initiative, even apathetic, they are typically able to carry on a conversation and may, when prompted, be reasonably sociable. However, formal testing reveals that, although immediate recall, as with a digit span, is intact, short-term memory, tested by asking the patient to recall three objects after 5 minutes, is severely deficient. Indeed, patients may not be able to recall a conversation they had with the physician just minutes before. A degree of disorientation to time and place inevitably accompanies this anterograde amnesia. The retrograde component of the amnesia becomes apparent on asking patients about their lives before the Wernicke's encephalopathy: answers often display a 'temporal gradient' (Albert *et al.* 1979; Seltzer and Benson 1974), such that patients, although having no or little recall of the events occurring for years before the Wernicke's encephalopathy, may have a hazy or partial recall of events occurring much earlier, and a fairly clear memory of childhood events.

In questioning patients about recent events, one often encounters the phenomenon of confabulation: here, the patient blithely makes up responses, as if to 'fill in' the amnestic gap. Thus, if a patient were asked whether he or she had ever met the physician before, the patient might respond in the affirmative and go on to talk about meeting the physician at a local tavern the week before, where they had a 'few' beers, played some pool, and the like. In addition, when one questions patients about events in the distant past that they do have some recollection of, the recollections offered often display a peculiar sequencing. In some cases there may be 'telescoping', wherein events that in fact were separated by long intervals will be recalled by the patient as if they occurred in close temporal sequence; in other cases the opposite may occur, and the patient may recall events as having been separated by long intervals, whereas in fact they occurred close to one another. Finally, in some cases the sequence of recollections offered by the patient will bear little relation to the actual sequence of events, as if memories had been tumbled together, with no respect for their actual temporal order.

Course

With long-term observation, roughly one-quarter of patients will recover, one-half will show some improvement, and one-quarter will remain unimproved (Victor *et al.* 1971). In those who do not recover, although the amnestic defect persists, the tendency to confabulate gradually subsides and eventually often remits.

Etiology

As stressed above, Korsakoff's syndrome occurs as a sequela to Wernicke's encephalopathy (Malamud and Skillicorn 1956).

Neuropathologically, there is atrophy and gliosis of the mammillary bodies and the mediodorsal and anterior nuclei of the thalamus (Halliday *et al.* 1994; Malamud and Skillicorn 1956; Victor *et al.* 1971), and it appears that the lesion in the anterior nuclei are critical for the development of the syndrome (Harding *et al.* 2000).

Differential diagnosis

Korsakoff's syndrome must be distinguished from other disorders which are capable of producing a chronic amnestic state with both anterograde and retrograde components. This differential task is fairly straightforward when one has a history of the amnestic state emerging from a Wernicke's encephalopathy, however when a good history is unobtainable, consideration must be given to these other disorders, as discussed in Chapter 6.6.

Treatment

Some form of supervision is generally required and, in severe cases, institutionalization may be necessary. Regarding pharmacologic treatment, several agents have been subjected to double-blind trials. Neither donepezil (Sahin *et al.* 2002) nor fluvoxamine (O'Carroll *et al.* 1994) was effective. Clonidine has been subjected to three double-blind trials: two, in 1980 (McEntee and Mair 1980) and 1986 (Mair and McEntee 1986), were positive, whereas the most recent, in 1993 (O'Carroll *et al.* 1993), was not. Although clonidine has not become anything like a standard treatment, it might be considered, given the otherwise hopeless nature of this disorder.

14.6 MANGANISM

Chronic exposure to manganese may be followed by the development of a personality change, an atypical parkinsonism, and, less commonly, a dementia or psychosis. Although most cases occur as a result of inhalation among manganese miners and those who work in steel or battery factories, cases have also been reported secondary to drinking contaminated well water, to prolonged intravenous total parenteral nutrition with manganese-containing solutions (Nagatomo *et al.* 1999), and to intravenous abuse of methcathinone, a stimulant drug whose illicit manufacture involves the use of potassium permanganate (Stepens *et al.* 2008).

Clinical features

The onset of symptoms is typically gradual, occurring after months or years of exposure, and patients may present with a personality change, parkinsonism, or both (Abd El Naby and Hassanein 1965).

The personality change (Abd El Naby and Hassanein 1965; Cook *et al.* 1974) typically consists of asthenia, fatigue, irritability, emotional lability, and a peculiar kind of 'incongruous' laughter, reminiscent of the emotional incontinence seen in pseudobulbar palsy: patients may smile without cause, or burst out in laughter, again for no apparent reason (Charles 1927). Insomnia or hypersomnia may accompany these changes (Abd El Naby and Hassanein 1965; Cook *et al.* 1974).

The parkinsonism (Abd El Naby and Hassanein 1965; Calne *et al.* 1994; Cook *et al.* 1974; Huang *et al.* 1989) is characterized by rigidity, bradykinesia, postural instability, and a tendency to 'freeze' and fall upon turning. Cogwheeling is often seen and, although tremor may also be present, it is generally not of the 'pill-rolling' type. The parkinsonism may also be accompanied by dystonia, often affecting the cervical musculature or the face. The most characteristic feature of manganese-induced parkinsonism is, however, a distinctive dystonic gait abnormality known as a 'cock-walk'. Here, patients walk on their metatarsophalangeal joints as if they were wearing high heels; at times, the elbows may be flexed, creating the overall appearance of the walk of a rooster. Such a 'cock-walk' has been reported in anywhere from a small minority (Cook *et al.* 1974; Huang *et al.* 1993) to up to one-third (Abd El Naby and Hassanein 1965) of patients.

Dementia may occur concurrent with the parkinsonism, and may be characterized by a marked degree of memory loss (Cook *et al.* 1974).

Psychosis, known as 'manganese madness', may occur, and is characterized by excitation, hallucinations, and delusions (Abd El Naby and Hassanein 1965).

T1-weighted MRI scanning typically reveals increased signal intensity within the globus pallidus bilaterally. Manganese levels are increased in the serum, hair, or a 24-hour urine sample.

Course

As one might expect, with ongoing exposure, symptoms gradually worsen. With cessation of exposure, however, rather than a gradual reduction of parkinsonian signs and symptoms, these actually continue to gradually worsen over the next 10 years or so (Huang *et al.* 1993,1998), after which they persist in a stable chronicity (Huang *et al.* 2007). A similar progression has been noted for the 'cock-walk' (Huang *et al.* 1997).

Etiology

Autopsy studies have revealed neuronal loss and gliosis within the basal ganglia, primarily in the pars interna of the globus pallidus, with sparing of the substantia nigra (Perl and Olanow 2007; Yamada *et al.* 1986).

Differential diagnosis

Lacking a history of manganese exposure, the differential diagnosis generally involves a consideration of other causes of parkinsonism, as discussed in Chapter 4.8, with special attention to progressive supranuclear palsy, adult-onset pantothenate kinase-associated neurodegeneration, and Wilson's disease. In all these disorders, however, one does not see the distinctive 'cock-walk'.

Treatment

The general treatment of personality change, dementia, and psychosis is outlined in Chapters 8.2, 6.3, and 8.1, respectively. If antipsychotics are required, consideration should be given to second-generation agents, such as quetiapine, in an effort to avoid exacerbating parkinsonism. Regarding the parkinsonism, although case reports suggest a usefulness for levodopa (Huang *et al.* 1993) (sometimes in high doses of up to 3 g or more daily; Mena *et al.* 1970), placebo-controlled studies cast doubt on this (Koller *et al.* 2004; Lu *et al.* 1994). In practice, however, it is reasonable to give it a trial.

Manganese is stored in bone and has a long half-life, extending to 1 month or more. Although the role of chelating agents during the first few months is not established, case reports suggest their utility (Discalzi *et al.* 2000).

14.7 THALLIUM INTOXICATON

Although thallium can be absorbed through either the lungs or the skin, industrial exposure is rare, and most cases of thallium intoxication are by ingestion. Since thallium was banned as a rodenticide, most current cases generally represent a deliberate, often homicidal, poisoning.

Clinical features

The onset of symptoms may be either acute or gradual, depending on the amount ingested.

Acute intoxication (Bank *et al.* 1972; Moore *et al.* 1993; Reed *et al.* 1963; Thompson *et al.* 1988; Wainwright *et al.* 1988) generally presents with abdominal pain, vomiting, and diarrhea. Within days or a week or so, patients develop a delirium and a polyneuropathy: initially the polyneuropathy is sensory, and typically painful; subsequently a motor component may develop, which may, in severe cases, progress to quadriplegia. Cranial neuropathies, with facial palsy or ophthalmoplegia, may also develop, and, rarely, chorea (Patterson 1975) or grand mal seizures may occur. Within 1–3 weeks patients then develop the most characteristic feature of thallium intoxication, namely a severe, generalized alopecia.

Gradual intoxication presents with dementia, a sensorimotor polyneuropathy, and alopecia; gastrointestinal symptomatology may or may not be present.

Thallium may be found in the urine, and, in long-standing cases, the hair.

Course

Acute intoxications may be fatal in up to one-tenth of cases. Those who survive generally make a gradual, more or less complete, recovery. In some cases there may be persistent cognitive deficits, which may be severe enough to produce a dementia (Reed *et al.* 1963; Thompson *et al.* 1988) or a personality change (McMillan *et al.* 1997). Gradual intoxications likewise show a variable degree of recovery.

Etiology

In acute cases there is cerebral edema, often with petechial hemorrhages; in more gradual-onset cases, and in those who have recovered, there is a variable degree of neuronal loss in the cortex, basal ganglia, and thalamus (Cavanagh *et al.* 1974). In the peripheral nerves axonal degeneration and demyelination are seen (Davis *et al.* 1981).

Differential diagnosis

Acute cases may be confused with arsenic intoxication or hepatic porphyria, or, when symptoms are confined primarily to a polyneuropathy, with the Guillain–Barré syndrome. The eventual appearance of severe alopecia, however, indicates the correct diagnosis.

Treatment

Thallium undergoes enterohepatic recirculation and, consequently, laxatives and activated charcoal should be given. Prussian blue binds to thallium in the gut, and this is part of standard treatment. Potassium displaces thallium from tissues, and thus makes it available for urinary excretion; forced diruesis also hastens excretion and hemodialysis is also effective. Chelation therapy is contra-indicated as it may be followed by a deterioration in the patient's condition (Wainwright *et al.* 1988).

14.8 ARSENIC INTOXICATION

Although elemental arsenic causes relatively little central nervous system toxicity, arsenic salts are toxic; with acute intoxication one may see a delirium, whereas with chronic, low-level exposure a dementia may occur. Arsenic salts are found in certain herbicides and rodenticides, and ingestion may occur accidentally or with suicidal or homicidal intent.

Clinical features

With acute ingestion (Jenkins 1966; Xu *et al.* 2008), patients develop nausea, vomiting, abdominal cramping and diarrhea, which may be bloody. Delirium ensues, and this may be accompanied by grand mal seizures; classically patients also have an odor of garlic on their breath. Hypotension, various arrhythmias, and renal failure may occur, and in severe cases, patients progress to coma and death. Within 1–3 weeks a peripheral polyneuropathy appears: initially this is sensory, and often painful; in many cases a motor component follows, which may progress to quadriplegia.

Chronic exposure (Jenkins 1966) to small amounts of arsenic salts may cause both a dementia and a polyneuropathy. Hyperkeratosis of the palms and soles may occur, and Mees' lines may appear, which are transverse white discolorations of the nails. Occasionally there may be a mild degree of alopecia.

Arsenic may be found in a 24-hour urine sample, and, within weeks of exposure, is also found in the hair and nails.

Course

Those who survive an episode of acute intoxication may be left with a dementia (Fincher and Koerker 1987; Freeman and Couch 1978) and a polyneuropathy (Fincher and Koerker 1987), both of which show a greater or lesser degree of improvement over long-term follow-up.

In cases of chronic intoxication, progression occurs until exposure ceases, after which a similar course is anticipated.

Etiology

In acute cases, widespread petechial hemorrhages within the white matter are seen (Hurst 1959; Russell 1937). Within the peripheral nervous system, axonal degeneration occurs.

Differential diagnosis

Acute cases may be confused with hepatic porphyria or thallium intoxication.

Treatment

Acute intoxication is treated with gastric lavage, osmotic diuresis, and intensive supportive care. Chelation therapy is appropriate for both acute and chronic cases.

14.9 BISMUTH INTOXICATION

Bismuth is found in a number of preparations and is used for the control of diarrhea and in the treatment of *Helicobacter pylori* infection. Although generally quite safe, high dosage may be followed by a delirium.

Clinical features

The onset is generally subacute, and may be marked by insomnia and mood changes, such as irritability, depression, or, uncommonly, euphoria (Supino-Viterbo *et al.* 1977). Eventually a delirium (Buge *et al.* 1981; Burns *et al.* 1974; Supino-Viterbo *et al.* 1977) supervenes, which may be accompanied by delusions and hallucinations; myoclonus is common and there may also be tremulousness and ataxia. Seizures may be seen in roughly one-third of all cases (Buge *et al.* 1981).

Computed tomography (CT) scanning may reveal hyperdensity within the cerebral cortex, basal ganglia, and cerebellum (Gardeur *et al.* 1978). The EEG may be marked by slowing and interictal epileptiform discharges (Buge *et al.* 1981).

Course

Recovery is gradual, occurring over many months, and although in most cases recovery is complete, lingering mild cognitive deficits may be seen in a minority.

Etiology

Bismuth, as indicated by the findings on CT scanning, accumulates in the gray matter.

Differential diagnosis

When the history of bismuth ingestion is lacking, the differential diagnosis expands to include other deliria associated with myoclonus, as discussed in Chapter 4.2.

Treatment

There is no established specific treatment; the general treatment of delirium is discussed in Chapter 6.4.

14.10 TIN INTOXICATION

Organotin compounds are used in the manufacture of preservatives, polymer stabilizers, and biocides, and these may gain entry by inhalation, ingestion, or transdermal absorption.

Clinical features

Within days of poisoning by trimethyl (Besser *et al.* 1987; Feldman *et al.* 1993; Fortemps *et al.* 1978; Kreyberg *et al.* 1992), triethyl (Alajouanine *et al.* 1958), or triphenyl (Wu *et al.* 1990) tin, patients develop delirium which may be associated with ataxia and seizures; in severe cases, coma eventually occurs.

Course

Although most patients survive trimethyltin intoxication (Besser *et al.* 1987; Feldman *et al.* 1993), the mortality rate after triethyltin intoxication is approximately 50 percent (Alajouanine *et al.* 1958).

Etiology

With trimethyltin intoxication, necrosis is most prominent in the hippocampus, with lesser involvement of the cerebral and cerebellar cortices (Besser *et al.* 1987; Kreyberg *et al.* 1992). In cases of triethyltin intoxication, diffuse edema and vacuolization was seen in the white matter (Alajouanine *et al.* 1958; Cossa *et al.* 1958).

Differential diagnosis

When the history of exposure is lacking the differential becomes quite wide, as discussed in Chapter 5.3.

Treatment

There is no known specific treatment.

14.11 LEAD ENCEPHALOPATHY

Toxic accumulations of lead may occur in a number of ways. Children may eat lead-based paint chips. Among adults, exposure may occur in certain occupations, for example in welders or those who work in battery factories (Menezes *et al.* 2003) or lead smelters; 'moonshine' whiskey, made with the help of old car radiators, may also be a source (Akelaitis 1941; Morris *et al.* 1964; Whitfield *et al.* 1972). Other less common sources include lead-glazed pottery (Matte *et al.* 1994), certain 'alternative' medications (Fisher and Le Couteur 2000), and retained bullets. Leaded gasoline used to be a major source, but since lead additives were banned this has essentially ceased to be a problem.

A sudden massive exposure in either children or adults may cause an acute lead encephalopathy with, among other symptoms, a delirium. Enduring, low-level exposure may produce a chronic lead encephalopathy with cognitive decline, which is often mild, and, in adults, a motor polyneuropathy.

Clinical features

Acute lead encephalopathy, in both children (Jenkins and Mellins 1957; Mellins and Jenkins 1955) and adults (Akelaitis 1941; Morris *et al.* 1964; Whitfield *et al.* 1972), is characterized by a prodrome, lasting from days to weeks, of lethargy, irritability, colicky abdominal pain and a metallic taste. Subsequently a delirium emerges, accompanied by ataxia and seizures, which may be either grand mal or focal in type. In severe cases, stupor or coma may ensue.

Chronic lead encephalopathy in children (Baghurst *et al.* 1992; Bellinger *et al.* 1987; Canfield *et al.* 2003; Needleman *et al.* 1990) presents with a very gradual cognitive decline, which may range in severity from a drop of a few IQ points to, in severe cases, a dementia. By contrast, in adults, although a cognitive decline may likewise occur, this is generally mild. The classic, and more common, manifestation of chronic lead intoxication in adults is a motor polyneuropathy that affects both the upper and lower extremities, with wrist and foot drop. Adult cases may also be characterized by colicky abdominal pain, a metallic taste, depression (Schottenfeld and Cullen 1984), and decreased libido.

Magnetic resonance imaging may reveal, on FLAIR imaging, increased signal intensity in the subcortical white matter, most especially in the occipital lobe (Atre *et al.* 2006), and CT scanning may demonstrate widespread calcification, not only in the white matter, but also in the basal ganglia (Reyes *et al.* 1986).

Whole-blood lead levels should be determined in all suspected cases. Acute lead encephalopathy may occur in children with levels of ≥ 80 μg/dL, and in adults with levels of ≥ 100 μg/dL. With regard to chronic lead encephalopathy, the levels required to produce symptoms are not as clear: it may well be that, when compatible symptoms are present, any child with a level of ≥ 20 μg/dL, and any adult with a level of ≥ 40 μg/dL should receive chelation therapy.

It is also appropriate to determine blood zinc protoporphyrin levels: although this level takes weeks to rise to its maximum, it subsequently, unlike whole-blood lead levels, remains elevated for a prolonged period. Twenty-four-hour urinary lead levels are less reliable than whole-blood levels.

In cases of acute lead encephalopathy, the cerebrospinal fluid will be under increased pressure; the total protein is often increased, but generally no higher than 100 mg/dL, and there may be a mononuclear pleocytosis of no more than 100 cells.

In both children and adults 'lead lines' may be seen at the border of the gingiva and teeth, and in childhood cases lead lines may also be seen in radiographs of the tibia or other long bones.

Course

Acute lead encephalopathy runs a rapid course, with mortality rates in children of roughly 30 percent. Those who survive may be left with dementia, seizures, or spasticity. Chronic lead encephalopathy progresses with continued exposure; with termination of exposure, there is little spontaneous improvement over time.

Etiology

In acute encephalopathy, there is widespread cerebral edema associated with petechial hemorrhages and microinfarctions (Pentschew 1965; Smith *et al.* 1960). In chronic cases, there is cortical atrophy and widespread neuronal loss.

Differential diagnosis

Colicky abdominal pain and complaints of a metallic taste in patients with delirium or dementia should always suggest lead encephalopathy, and in adults a concurrent motor polyneuropathy is likewise highly suggestive. When these symptoms are absent, or go unrecognized, the differential widens, as discussed in Chapters 6.3 and 6.5.

Treatment

Acute lead encephalopathy is a medical emergency, requiring chelation therapy and, in most cases, steroids and mannitol. Patients with chronic lead encephalopathy should likewise receive chelation.

14.12 MERCURY INTOXICATION

Three forms of mercury are potentially toxic to humans, namely elemental mercury, salts of mercury, and organic mercury.

Elemental mercury at room temperature is a liquid. When ingested, it is generally non-toxic, as it is poorly absorbed. Elemental mercury, however, readily volatizes with warming, or merely with shaking, and may be inhaled, and once inhaled, it readily passes across the alveoli and into the blood, where it is taken up into erythrocytes and oxidized to a mercuric salt. Exposure to elemental mercury generally occurs in gold miners and in certain other industries.

Salts of mercury may be either monovalent (mercurous) or divalent (mercuric), and may be absorbed through the gastrointestinal tract or transdermally. Mercury salts were once used medicinally, as in the case of calomel, also called mercurous chloride (Davis *et al.* 1974). Currently, however, most exposure occurs in those employed in the manufacture of plastics, fungicides, and electronics. In the nineteenth century, mercuric nitrate was used in the manufacture of felt, and chronic exposure among hat makers led to the 'mad hatter syndrome', made famous in Lewis Carroll's *Alice's Adventures in Wonderland.*

Organic mercury is found as methylmercury, ethylmercury (Hay *et al.* 1963), and phenylmercury (O'Carroll *et al.* 1995), and, of these, methylmercury is the most toxic. Organic mercury, especially methylmercury, is readily taken up through the gastrointestinal tract and easily crosses the blood–brain barrier. Exposure occurs generally by accidental ingestion. Perhaps the most notorious example of this is the Minamata epidemic, named after the town in Japan where it occurred (Kurland *et al.* 1960). Industrial waste containing elemental mercury was discharged into Minamata bay, where the elemental mercury was converted by bacteria into methylmercury; fish became contaminated and, when residents ate the fish, the stage was set for disaster. Other incidents have involved methylmercury used as a fungicide on seed grain. In one example, treated seed grain was sent to Iraq (Amin-Zaki *et al.* 1978; Bakir *et al.* 1980; Rustam and Hamri 1974), and although civilians had been warned not to eat the grain but only to plant it, the contaminated grain was nevertheless made into flour and bread, leading to hundreds of deaths. In another example, hogs were fed with treated seed grain, and those that then ate the hogs became ill (Snyder 1972).

Clinical features

With acute exposure to elemental mercury via inhalation, a pneumonitis may occur. Ingestion of either mercury salts or organic mercury may cause vomiting, which may be quite severe in the case of mercury salts and which may be accompanied by renal failure.

Among patients who survive the acute exposure, and in cases of chronic, low-level exposure, one or more clinical features may gradually develop, including a personality change, a dementia, various abnormal involuntary movements, visual changes, and a polyneuropathy. The personality change, known as erethism, is classically characterized by emotional lability, shyness, irritability, fatigue and insomnia. The dementia, occurring generally only with greater exposure, does not appear to have any specific distinguishing features. Of the abnormal movements seen with mercury intoxication, tremor is perhaps the most classic, and this may affect the hands, lips, and tongue. Ataxia is common and may be accompanied by an intention tremor and dysarthria. Choreoathetosis may occur, as may parkinsonian features. Visual changes include tunnel vision, various scotomata, or cortical blindness. The polyneuropathy is sensorimotor in type, and may be painful.

Although blood and urine levels of mercury will be elevated, there is not a good correlation between the degree of elevation and the clinical symptomatology. In chronic cases, neuroimaging may reveal atrophy of the cerebral cortex (most prominently the calcarine cortex) and the cerebellar cortex (Tokuomi *et al.* 1982).

Course

Over long-term follow-up, although there may be some spontaneous improvement, for the most part symptoms are chronic.

Etiology

Pathologically (Davis *et al.* 1974, 1994; Eto 2000; Hunter and Russell 1954), there is widespread neuronal loss in the cerebral and cerebellar cortices. Within the cerebral cortex, this is most pronounced within the calcarine cortex, and, to a lesser degree, the pre- and postcentral cortex, and within the cerebellar cortex, the granule cell layer is most affected, with relative sparing of the Purkinje cell layer.

Pathologically, there is widespread neuronal loss throughout the cerebral and cerebellar cortices, especially involving the calcarine cortex and, in the cerebellum, the granule cell layer.

Differential diagnosis

The diagnosis rests heavily on a history of exposure.

Treatment

Acute exposure to salts of mercury or to organic mercury may be treated with gastric lavage and activated charcoal. Chelation treatment is also indicated; however, bronchoalveolar lavage (BAL) should be avoided as it may increase mercury levels within the brain.

14.13 DIALYSIS ENCEPHALOPATHY

Chronically elevated serum levels of aluminum may cause a progressive encephalopathy. This was once a not uncommon complication of chronic hemodialysis, wherein it occurred secondary to high levels of aluminium in the dialysate fluid (Davison *et al.* 1982). With purification of the fluid, severe cases are now rare, and generally seen only in undialysed patients who happen to be taking aluminum-containing medications (Andreoli *et al.* 1984; Shirabe *et al.* 2002). It must be borne in mind, however, that even with purification of the dialysate, mild cases may still occur (Rovelli *et al.* 1988).

Clinical features

The onset of the encephalopathy is gradual, generally after 3–4 years of hemodialysis (Garrett *et al.* 1988; Lederman and Henry 1978). Clinically (Burks *et al.* 1976; Chokroverty *et al.* 1976; Garrett *et al.* 1988; Lederman and Henry 1978; O'Hare *et al.* 1983), patients present with an intermittent, stuttering kind of aphasia, joined with myoclonus and astrerixis. Over time, cognitive deficits begin to appear, with the eventual development of a dementia. Seizures may occur, and in a minority (Garrett *et al.* 1988) there may be delusions, hallucinations, and bizarre behavior (Chokroverty *et al.* 1976; Scheiber and Ziesat 1976); one case was also characterized by mania (Jack *et al.* 1983).

The EEG is almost always abnormal (Hughes and Schreeder 1980; Ladurner *et al.* 1982), showing mild generalized slowing, periodic bursts of generalized polyspike or spike and slow wave complexes and frontal intermittent rhythmic delta activity (FIRDA).

The serum aluminum level is elevated, generally above 200 μg/L.

Course

Untreated, the disease is progressive, with death on average within 6 months.

Etiology

The brain aluminum content is elevated (Alfrey *et al.* 1976) and there is widespread neuronal loss in the cerebral cortex, with, in some cases, laminar spongiform change (Shirabe *et al.* 2002; Winkelman and Ricanti 1986).

Differential diagnosis

Other disorders that are not uncommon in dialysis patients and which are capable of causing dementia include stroke and subdural hematoma. Patients on dialysis may also develop thiamine deficiency, with a resulting Wernicke's encephalopathy (Hung *et al.* 2001).

Treatment

Dialysate purity must be maintained and patients should not be given any aluminum-containing medications. Although diazepam may dramatically reduce some of the symptoms (Nadel and Wilson 1976), the effect is generally only transient (Snider *et al.* 1979), and the overall course of the disease is not altered. Chelation with deferoxamine should be considered, however, in some patients this may lead to a mobilization of aluminum from bone, with a consequent worsening of the cerebral symptomatology (Sherrard *et al.* 1988).

14.14 DIALYSIS DISEQUILIBRIUM SYNDROME

The dialysis disequilibrium syndrome (Benna *et al.* 1981; Mawdsley 1972; Peterson and Swanson 1964; Raskin and Fishman 1976; Tyler 1965) is a transient disorder occurring during, or sometimes after, a hemodialysis run. Although mild cases are common, the more severe cases, characterized by delirium or seizures, are seen in only a very small minority.

Clinical features

The syndrome usually appears several hours after the start of a dialysis run, but may sometimes be delayed for up to a day. Mild cases are characterized by headache, nausea, vomiting and muscle cramps. In more severe cases, these symptoms are joined by restlessness, agitation, myoclonus and delirium, and, in some, grand mal seizures.

The EEG typically shows generalized slowing, and MRI scanning may reveal increased signal intensity on FLAIR imaging throughout the centrum semiovale (Attur *et al.* 2008).

Course

Although the syndrome is self-limited in almost all cases, with symptoms resolving within from hours to a day or two, in rare cases there may be a progression to coma and death.

Etiology

During hemodialysis removal of urea from the plasma creates an osmotic gradient between blood and cerebrospinal fluid, with the development of cerebral edema.

Differential diagnosis

In addition to toxic and metabolic causes, consideration should be given to subdural hematoma, intracerebral hemorrhage, and non-convulsive status epilepticus.

Treatment

If symptoms are troubling, the hemodialysis run should be stopped, and in severe cases consideration may be given to administering either mannitol or hypertonic saline. Seizures may be treated with intravenous lorazepam, and the general treatment of delirium is discussed in Chapter 8.3. Future episodes may be prevented by slowing down future dialyses.

14.15 HYPOGLYCEMIA

Hypoglycemia may cause autonomic symptoms, such as anxiety and tremor, and may also cause 'neuroglycopenic' symptoms, such as delirium or seizures; when severe enough, coma may ensue and patients may be left with dementia.

Clinical features

As the blood glucose level falls below a threshold of roughly 45 mg/dL (2.5 mmol/L), autonomic symptoms appear fairly promptly, and these include anxiety, tremor, diaphoresis, and palpitations; patients may also complain of hunger, nausea, headache, and generalized weakness.

Neuroglycopenic symptoms (Hart and Frier 1998; Malouf and Brust 1985) are associated with a threshold blood glucose level of 30 mg/dL (1.67 mmol/L); however, unlike the autonomic symptoms, which tend to appear fairly promptly, neuroglycopenic symptoms generally do not appear until this threshold has been met for roughly 30 minutes. Initially, patients may experience light-headedness or depersonalization. A delirium may then ensue and this may be associated with unusual behavior: in one case (Bosboom *et al.* 1996), the patient 'was restless, opening and closing his eyes, and thrashing about with his arms and legs, occasionally hitting onlookers and spitting in their faces'; in another (Zivin 1970), the patient, a soldier, 'walked into the mess hall ... [dressed] in his underwear'. Seizures may occur, and these may be either partial or grand mal in type. A very small minority of patients will also have focal findings, such as hemiparesis (Malouf and Brust 1985) or aphasia (Shintani *et al.* 1993; Wallis *et al.* 1985), which clear with treatment.

With blood glucose levels below 10 mg/dL (0.55 mmol/L) stupor and coma may ensue (Ben-Ami *et al.* 1999).

Of note, although in most cases neuroglycopenic symptoms are preceded by autonomic symptoms, exceptions to this rule occur, and some patients may present with neuroglycopenic symptoms alone (Case Records 1988; Malouf and Brust 1985; Moersch and Kernohan 1938). This may occur in diabetic subjects who have developed a diabetic autonomic neuropathy or in patients treated with beta-blockers, which partially mask autonomic symptoms. Such a scenario may also occur in cases in which the blood glucose level drops very slowly, as for example may occur with fasting.

Electroencephalograms obtained during neuroglycopenic delirium will display generalized slowing. In cases of coma, diffusion-weighted MRI scanning may reveal areas of increased signal intensity in the cerebral cortex, white matter, and basal ganglia (Jung *et al.* 2005).

Course

Whereas autonomic symptoms respond promptly to treatment with glucose, delirium may take up to an hour to subside. Patients who have developed coma may not recover consciousness, and those who do may be left with a dementia: in such cases a greater or lesser degree of recovery may occur over the following year or two.

Etiology

Symptomatic hypoglycemia may occur in the fasting state, for example early in the morning before breakfast or in those who skip meals, or postprandially, several hours after a meal.

Fasting hypoglycemia is seen most commonly in diabetic patients on insulin or insulin secretagogues (e.g. sulfonylureas); it may also occur in patients with insulinomas (Dizon *et al.* 1999). Fasting hypoglycemia may also occur in hunger strikes or anorexia nervosa (Rich *et al.* 1990). Gluconeogenesis is one of the primary defenses in fasting states, and when this is impaired, as for example in liver disease or by the ingestion of alcohol, patients may be particularly prone to developing symptomatic hypoglycemia. Indeed, one of the most common precipitants of symptomatic hypoglycemia is an alcoholic binge (Malouf and Brust 1985): here, the combination of alcohol-inhibited gluconeogenesis with the absence of eating in favor of drinking proves to be a highly morbid combination.

Postprandial hypoglycemia may be seen in early type II diabetes, after vagotomy, gastrectomy, pyloroplasty, or gastrojejeunostomy, and in a rare (Palardy *et al.* 1989) and somewhat controversial condition known as 'functional' or 'essential' postprandial hypoglycemia.

Hypoglycemia may also be intentionally produced by malingerers who inject themselves with insulin or take high doses of insulin secretagogues (Price *et al.* 1986). Whenever this is suspected, in addition to checking the glucose level, one should also determine insulin and C-peptide levels and obtain a toxicology screen for oral agents. C-peptide is normally secreted in conjunction with insulin and, under physiologic conditions, when insulin is elevated, so too is the C-peptide level. In cases of exogenous insulin administration, however, whereas the insulin level is elevated, the C-peptide level will be normal or low (Scarlett *et al.* 1977). In cases in which the elevated insulin level is produced by an oral agent, both the insulin and C-peptide level will be increased, but the toxicology will be positive.

Autonomic symptoms reflect an immediate physiologic sympathetic and cholinergic response to hypoglycemia. Neuroglycopenic symptoms, by contrast, reflect a complex series of events triggered by intraneuronal hypoglycemia, and the half-hour or so delay in the onset of neuroglycopenic symptoms reflects the time required for the depletion of intraneuronal glucose stored as glycogen. In cases in which coma occurs, neuronal death occurs, and, if sufficient, this constitutes the cause of any persistent coma or residual dementia. In such cases pathologic studies have revealed (Auer *et al.* 1989; Kalimo and Olsson 1980; Lawrence *et al.* 1942; MacKeith and Meyer 1939) neuronal loss and gliosis in the cerebral cortex, the hippocampus (especially the dentate gyrus), and the striatum, with, in severe cases, laminar cortical necrosis.

Differential diagnosis

In cases in which the symptomatology is restricted to autonomic symptoms, the diagnosis is immediately suggested by associated hunger and the prompt relief gained by eating or drinking glucose-rich fluids. Consideration may also be given to other causes of anxiety attacks, as discussed in Chapter 7.5.

In cases characterized by autonomic symptoms followed by delirium, consideration may be given to other deliria associated with tremor, such as delirium tremens, thyroid storm, amphetamine intoxication, and either the serotonin syndrome or the neuroleptic malignant syndrome. In some cases, however, as noted above, autonomic symptoms may be lacking, and here, although the differential diagnosis becomes quite wide (as discussed in Chapter 6.5), special attention should be given to complex partial seizures.

In doubtful cases, consideration may be given to demonstrating 'Whipple's triad' by performing a glucose tolerance test. Whipple's triad consists of:

1. the presence of typical symptoms;
2. the concurrent presence of significant hypoglycemia;
3. the relief of symptoms with glucose administration.

The duration of the glucose tolerance test may range from 5 hours in cases with a postprandial pattern to up to 12 hours in those with a fasting pattern.

Treatment

If the patient is able to take fluids, a glass of orange juice or soda mixed with two or three tablespoonfuls of sugar is often adequate. Patients unable to take fluids should be given 50 mL of a 50 percent solution of glucose intravenously. In cases in which intravenous access is lacking, one may stimulate gluconeogenesis by giving glucagon in a dose of 1 mg intramuscularly, keeping in mind, of course, that glucagon is ineffective in patients with significant liver damage. Regardless of which therapy is given, patients should be closely monitored to see if repeat doses are required.

In malnourished alcoholics, an argument might be made for giving intravenous thiamine first, and then waiting for several hours before giving glucose in order to avoid the risk of precipitating a Wernicke's encephalopathy. Although this may be quite appropriate when only autonomic symptoms are present, much clinical judgment is required when neuroglycopenic symptoms appear, especially if the glucose level is trending down toward that associated with stupor or coma.

14.16 HYPERVISCOSITY SYNDROME

Significantly increased viscosity will cause microcirculatory sludging within the central nervous system, leading to, among other symptoms, a dementia. This is a very rare syndrome and is generally seen in patients with a paraproteinemia such as Waldenström's macroglobulinemia (Mehta and Singhal 2003).

Clinical features

Symptomatology may include headache, dizziness, dysarthria, and ataxia; there may also be lethargy and in some cases a dementia may occur (Mueller *et al.* 1983). Stroke has also been reported (Pavy *et al.* 1980), as have seizures. Ophthalmoscopic examination may reveal retinal engorgement and 'sausaging'.

Serum viscosity is measured in centipoises, and in most symptomatic cases the viscosity is elevated above 4 centipoises.

Course

As might be expected, the course parallels changes in serum viscosity. With progression, coma and death may occur.

Etiology

The presence of an excessive amount of variously deformed proteins, usually immunoglobulins, increases serum viscosity resulting in literal sludging of blood flow within arterioles, capillaries, and venules, in some cases leading to infarctions (Grigg *et al.* 1994).

As noted, the most common cause of hyperviscosity is Waldenström's macroglobulinemia; cases may also be seen in multiple myeloma (Camacho *et al.* 1985), cryoglobulinemia (Della Rossa *et al.* 2003), and, rarely, in rheumatoid arthritis (Bach *et al.* 1989) and disseminated breast adenocarcinoma (Grigg *et al.* 1994).

Differential diagnosis

The diagnosis should be suspected in patients with Waldenström's macroglobulinemia or any of the other conditions noted above.

Treatment

Emergent plasmapheresis should be performed until symptoms clear.

14.17 CENTRAL PONTINE MYELINOLYSIS

Central pontine myelinolysis, originally described by Adams *et al.* in 1959 (Adams *et al.* 1959), is classically characterized clinically by the development of a combination of flaccid quadriplegia and delirium, and pathologically by demyelination in the central portion of the pons, all occurring within 2–3 days of an overly rapid correction of chronic hyponatremia. Exceptions to this classic picture, as described further below, do occur, with prominent demyelination in extrapontine sites associated with delirium or with a movement disorder in the absence of quadriplegia. The existence of these exceptions has led some authors to propose alternative names for this syndrome, such as 'central pontine and extrapontine myelinolysis' or, more generally, 'osmotic demyelination syndrome'.

Clinical features

In classic cases, a flaccid symmetric quadriparesis develops within 2–3 days of rapid correction of chronic hyponatremia, accompanied by a delirium or lethargy (Karp and Laureno 1993), which in turn may be marked by both visual and auditory hallucinations (Sterns *et al.* 1986). Symptoms gradually worsen over about a week and, in severe cases, patients may be left in a 'locked-in' state, and totally paralyzed except for vertical eye movements. Over time the flaccidity resolves, to be replaced by spasticity with hyper-reflexia. Within the first day, diffusion-weighted MRI scanning may reveal increased signal intensity within the central portion of the pons, and over the following days the same area will reveal increased signal intensity on FLAIR and T2-weighted imaging and decreased signal intensity on T1-weighted imaging (Ruzek *et al.* 2004). Importantly, CT scanning is very insensitive here and may be normal despite florid symptomatology.

Exceptions to this classic picture occur when demyelination occurs in extrapontine locations. Thus, one may see delirium alone, without quadriparesis (Hadfield and Kubal 1996; Karp and Laureno 1993), or a movement disorder such as dystonia (Maraganore *et al.* 1992; Yoshida *et al.* 2000) or parkinsonism (Dickoff *et al.* 1988; Seiser *et al.* 1998; Sullivan *et al.* 2000; Tomita *et al.* 1997), again without any quadriparesis; importantly, in cases characterized by a movement disorder, the appearance of these symptoms may be delayed for weeks or months.

Course

In classic cases, death may occur within a matter of days. In those who survive, recovery may begin within a week or two of the syndrome reaching its peak intensity, with maximum recovery taking up to a year. Enduring deficits may include a variable degree of quadriparesis or dementia (Menger and Jorg 1999). In cases characterized by a movement disorder, symptoms may either resolve over many months or be chronic, or, in the case of dystonia, may actually gradually worsen.

Etiology

As noted earlier, the typical precipitant of central pontine myelinolysis is the overly rapid correction of a chronic hyponatremia (Brunner *et al.* 1990; Karp and Laureno 1993; Messert *et al.* 1979; Norenberg *et al.* 1982; Sterns *et al.* 1986). In cases of chronic hyponatremia lasting more than 2 or 3 days, intracellular osmolarity gradually falls to maintain an osmotic balance with the hypo-osmolar extracellular fluid. If the hyponatremia is corrected gradually, there is time for the intracellular osmolarity to 'catch up', so that an excessive osmotic gradient does not appear between the intracellular and extracellular fluids. However, when the hyponatremia is rapidly corrected, an osmotic gradient does occur, with a substantial fluid shift from the intracellular to the extracellular compartment. Oligodendrocytes appear to be particularly affected by this osmotic shift, and these cells, in particular those located within the pons, become damaged or die, leading to demyelination. Although such a scenario is most likely in chronic alcoholics given intravenous saline (Lampl and Yazdi 2002), other groups at risk include those in the immediate postoperative period after orthotopic liver transplantation (Estol *et al.* 1989; Lee *et al.* 2009), those suffering from severe malnutrition, and patients hyponatremic because of protracted vomiting (Dickoff *et al.* 1988) or diarrhea (Price and Mesulam 1987).

Although this mechanism appears operative for the overwhelming majority of cases, other mechanisms may also be found. For example, central pontine myelinolysis has been noted in severe burn patients who were not hyponatremic but who had rapidly developed a severe hyperosmolar state (McKee *et al.* 1988), leading, presumably, to the same kind of osmotic shift. There are also very rare cases of central pontine myelinolysis which do not appear to be related to osmotic gradients: one case apparently occurred secondary to hypokalemia (Patel *et al.* 2007).

Pathologically, in classic cases one finds symmetric demyelination within the basis pontis (Adams *et al.* 1959). This is most prominent within the central portion of the pons and spreads outward but does not involve the periphery of the pons, where a rim of intact white matter survives. This demyelination may at times extend to the pontine tegmentum and occasionally also into the mesencephalon, but the medulla is generally spared. In severe cases, cavitation may occur. Extrapontine myelinolysis, as noted, may also occur, and has been found in the centrum semiovale, corpus callosum, internal capsule, striatum, thalamus, subthalamic nucleus, and the lateral geniculate body (Wright *et al.* 1979). In pathologic material, the most common pattern is myelinolysis confined to the pons, followed by myelinolysis involving both the pons and extrapontine sites; the least common pattern consists of myelinolysis solely in extrapontine sites (Gocht and Colmant 1987).

Differential diagnosis

Classic cases must be distinguished from stroke secondary to infarction in the area of distribution of the basilar artery. Stroke is suggested by a more rapid onset and by an asymmetry of signs; furthermore, MRI scanning will reveal a lesion that, in contrast to central pontine myelinolysis, involves the periphery of the pons. In cases characterized by delirium alone, although the obvious clue would be a history of overly rapid correction of hyponatremia 2 or 3 days earlier, a high index of suspicion might be needed in alcoholics or those who have just undergone orthotopic liver transplantation, given the multitude of other etiologic factors that could account for delirium in such cases. Finally, in cases characterized by a movement disorder, an even higher index of suspicion would be required, given that the responsible correction of hyponatremia may have occurred weeks or months earlier.

Treatment

Prevention is essential, and it must be borne in mind that, when chronic, hyponatremia is often relatively well tolerated, and indeed some patients may be asymptomatic. Consequently, even in cases in which the serum sodium is as low as 115 mEq/L, it may be quite appropriate to undertake a very leisurely correction of the hyponatremia, or even to consider a brief period of mere observation. In cases, however, in which significant symptoms of hyponatremia are present, or are anticipated, and treatment with normal or hypertonic saline is required, it is essential to temper one's therapeutic enthusiasm and to limit the rise in sodium to no more than approximately 0.4 mEq/L/hour (Norenberg *et al.* 1982; Sterns *et al.* 1986) and to aim for a level of only 120 mEq/L: once that level is reached, further correction may be made at a much more leisurely pace.

Although there is no accepted treatment for central pontine myelinoysis, case reports have emerged suggesting some benefit from intravenous methylprednisolone (Sakamoto *et al.* 2007) or a combination of plasmapheresis and immunoglobulins (Saner *et al.* 2008).

Patients left with a dementia may be treated as described in Chapter 6.3, and in those with parkinsonism consideration may be given to a trial of levodopa (Sullivan *et al.* 2000).

14.18 UREMIC ENCEPHALOPATHY

In severe renal failure, patients may develop a delirium due to a condition known as uremic encephalopathy.

Clinical features

The mode of onset, as might be expected, reflects the rapidity with which the kidneys fail, and hence the onset of uremic encephalopathy may range from acute to gradual. The overall clinical picture (Mahoney and Arieff 1982; Raskin and Fishman 1976; Tyler 1968) is characterized, initially, by irritability and poor concentration, associated with lassitude and insomnia; eventually a delirium appears which is often marked by visual hallucinations. On further examination, one typically finds dysarthria, asterixis, and multifocal myoclonus; there may also be diffuse muscle twitching, and in about one-third of cases, seizures occur, which, though typically grand mal in type, may at times be focal. In one rare case, uremia presented with mania alone, without any other symptoms (El-Mallakh *et al.* 1987).

A sensorimotor polyneuropathy is common (Ropper 1993; Thomas *et al.* 1971) as is the restless legs syndrome.

The severity of the encephalopathy correlates not only with the level of the blood urea nitrogen (BUN) but also with the rapidity with which the BUN rises. For example, although in cases of acute renal failure levels of 35.7 mmol/L (100 mg/dL) are typically associated with delirium, in renal failure of very gradual onset, levels of 71.4 mmol/L (200 mg/dL) may be tolerated with few, if any, symptoms.

The EEG typically shows generalized slowing, and in some cases triphasic waves may be seen (Karnaze and Bickford 1984).

Course

Typically the symptoms of the delirium show marked fluctuations throughout the day. The overall course parallels that of the renal failure, and in severe cases coma may ensue.

Etiology

Although the level of the BUN serves as a rough guide to the presence and severity of the encephalopathy, urea itself is not the cause of the central nervous system dysfunction and in all likelihood the delirium results from an 'auto-intoxication' by as yet unidentified substances. In chronic cases that came to autopsy, an excess of protoplasmic astrocytes was found (Olsen 1961).

Differential diagnosis

Not all deliria occurring in the context of an elevated BUN represent uremic encephalopathy. Certain disorders, such as polyarteritis nodosa and systemic lupus erythematosus, may not only cause renal failure but also directly affect the central nervous system. Another example is malignant hypertension, which may cause a concurrent hypertensive encephalopathy and renal failure.

Renal failure is also associated with subdural hematoma (Fraser and Arieff 1988), hypocalcemia, and hypomagnesemia, each of which may also cause delirium. Finally, attention must be paid to toxicity from drugs excreted primarily via the kidneys.

Treatment

If the underlying cause of the renal failure is untreatable and spontaneous improvement is not expected, then uremic encephalopathy constitutes an indication for hemodialysis. With dialysis, the delirium typically resolves in a matter of days. The general treatment of delirium is discussed in Chapter 6.5, and that of seizures in Chapter 8.3. Clearly, great care must be taken in utilizing any medications, such as levetiracetam, which undergo primarily renal excretion; furthermore, in most cases of chronic renal failure, one also finds a reduced albumin level, which in turn will substantially increase the free fraction of any drug which is tightly protein-bound, as for example not only many of the antipsychotics but also phenytoin and valproic acid.

14.19 HEPATIC ENCEPHALOPATHY

With significant disease or dysfunction of hepatocytes, or with significant shunting of blood past the liver and directly into the systemic circulation, toxins normally removed by the liver reach the brain, causing an hepatic encephalopathy.

Clinical features

The mode of onset of hepatic encephalopathy mirrors the rapidity with which hepatic failure occurs, and thus may range from gradual, as in slowly progressive alcoholic cirrhosis, to acute, as in fulminant hepatic failure in some cases of viral hepatitis.

Acute-onset hepatic encephalopathy typically presents with a delirium; in gradual-onset cases, although a delirium eventually occurs, this development is often preceded by a more or less lengthy prodrome characterized by impaired judgment, difficulty with abstracting, and mood changes, which often tend toward euphoria. The delirium (Adams and Foley 1953; Fraser and Arieff 1985; Read *et al.* 1961; Summerskill *et al.* 1956) is characterized by confusion, drowsiness, sleep reversal, asterixis, and myoclonus; a small minority may have seizures, which may be partial or grand mal in type, and a minority may also have focal signs, such as hemiparesis (Cadranel *et al.* 2001). Rarely, catatonic symptoms, such as waxy flexibility, may occur (Jaffe 1967). Fetor hepaticus, a sickly-sweet musty odor to the breath, is seen in about half of the cases.

With progression of the hepatic failure, stupor and coma may ensue, and at times this may occur rapidly. In such cases there may be decorticate or decerebrate posturing (Conomy and Swash 1968) and bilateral Babinski signs.

The most important laboratory test is the ammonia level, which is elevated in the overwhelming majority of cases. Importantly, however, there is only a rough correlation between the degree of elevation and the severity of the encephalopathy (Ong *et al.* 2003); indeed, although rare, in some cases the ammonia level may be normal (Sherlock 1993).

The EEG shows generalized slowing, which is often accompanied by triphasic waves (Fisch and Klass 1988). Interictal epileptiform discharges, consisting of either focal or generalized spike and wave discharges, may be seen in a small minority (Ficker *et al.* 1997).

MRI scanning (Rovira *et al.* 2008; Thuluvath *et al.* 1995) may reveal, on T1-weighted imaging, increased signal intensity within the globus pallidus, a finding which remits with successful treatment; on FLAIR and DWI imaging, increased signal intensity may also be found within the subcortical white matter. Although MRI scanning is generally not necessary to make the diagnosis or to monitor treatment, CT scanning may be very helpful in evaluating patients who appear to be developing stupor. Computed tomography scanning in such cases may reveal evidence of cerebral edema, a finding which may predict the rapid development of coma (Wijdicks *et al.* 1995).

Course

In the absence of treatment, the course of the encephalopathy mirrors the course of the underlying hepatic failure. With treatment, provided that coma has not supervened, the recovery is generally quite good; Among cases that have progressed to coma, however, the mortality rate is high.

With repeated episodes patients may develop acquired hepatocerebral degeneration, as described in Chapter 14.20.

Etiology

As indicated earlier, hepatic encephalopathy occurs when the brain is exposed to toxins found in the portal vessels, toxins that are normally removed by the liver. Although ammonia itself is the most likely culprit here, other candidates have also been considered, including mercaptans, various aromatic amino acids, and short-chain fatty acids. In fatal cases, cerebral edema is seen, with, in some instances, transtentorial herniation (Donovan *et al.* 1998). In patients who survive encephalopathy and come to autopsy years later, one may find an increased number of Alzheimer type II astrocytes (Adams and Foley 1953).

The most common causes of hepatic failure are viral hepatitis, alcoholic cirrhosis, and drug toxicity, as for example with acetaminophen. In many cases, however, the onset of the encephalopathy may be traced to an event that increased the nitrogenous load in the gut beyond the diminished detoxifying capacity of the liver. Examples include high-protein meals, blood (as may occur with bleeding esophageal varices or peptic ulcer disease), and constipation. Other precipitants include infection, anesthesia, hypokalemia, uremia, and exposure to alcohol or sedative–hypnotics, especially benzodiazepines.

Differential diagnosis

In alcoholics, consideration must be given to delirium tremens, Wernicke's encephalopathy, hypoglycemia, encephalopathic pellagra, and subdural hematoma. The differential for delirium accompanied by asterixis is discussed in Chapter 4.12, and for delirium with myoclonus in Chapter 4.2.

Rarely, valproic acid may cause an encephalopathy associated with an elevated ammonia level (Baver 2005): treatment in such cases consists not only of discontinuing the drug but also administering carnitine (Bohan *et al.* 2001).

Treatment

In addition to treating any precipitating events, efforts are made to reduce the nitrogenous load in the gut. This may be accomplished using three methods, individually or in combination: a low-protein diet, lactulose, and treatment with rifaximin. Reducing dietary protein to 20–60 g/day is helpful, but many patients find this unpalatable. In acute situations, lactulose may be started at a dose of 20–45 mL hourly until the patient starts to have bowel movements, after which the dose may be decreased to a maintenance dose of 30–45 mL q.i.d.; in less urgent situations, one may simply start with a maintenance dose, with dosage adjustments as required to ensure three to four soft bowel movements daily. Rifaximin, which is more effective than lactulose (Mas *et al.* 2003), is a non-absorbable intestinal antibiotic that targets nitrogenous bacteria, and this may be given in a dose of 400 mg t.i.d. In urgent cases, many clinicians will use lactulose and rifaximin in combination, switching to monotherapy with the far less expensive lactulose as recovery occurs.

The development of stupor or coma dictates an additional strategy, as the accompanying cerebral edema is not responsive to treatment with lactulose and rifaximin. In such cases, mannitol is given in concert with ongoing measurement of intracranial pressure.

The general symptomatic treatment of delirium is discussed in Chapter 6.5. With regard to symptomatic pharmacologic treatment, benzodiazepines are contra-indicated, as they may worsen the delirium. If antipsychotics are utilized, low doses are appropriate, as all of these agents undergo hepatic metabolism. If seizures require treatment, consideration may be given to levetiracetam which is dependent on renal excretion.

Once patients have recovered, ongoing efforts are generally required to prevent recurrences, including the avoidance of precipitating factors and treatment with either lactulose or, if that is not effective, with the combination of lactulose and rifaximin (Bass *et al.* 2010).

14.20 ACQUIRED HEPATOCEREBRAL DEGENERATION

Acquired hepatocerebral degeneration is a rare disorder that occurs secondary to repeated or prolonged bouts of hepatic encephalopathy and is characterized by variable mixtures of cognitive deficits and abnormal movements.

Clinical features

Clinical features have been described in a number of reports (Finlayson and Superville 1981; Graham *et al.* 1970; Raskin *et al.* 1984; Victor *et al.* 1965). The onset is typically gradual and is often punctuated by repeated episodes of hepatic encephalopathy. Cognitive deficits may occur, which may be quite mild (Stracciari *et al.* 2008) or, in some cases, amount to a dementia (Spencer and Forno 2000). The movement disorder is often complex, with variable admixtures of chorea, facial grimacing, postural tremor, ataxia, dysarthria, and parkinsonism. The parkinsonism is typically of the symmetric, akinetic-rigid type, with early postural instability and no resting tremor (Burkhard *et al.* 2003). Chorea, when present, often affects the head and neck, and the resulting movements may be tic-like in character.

T1-weighted MRI scanning (Burkhard *et al.* 2003; Krieger *et al.* 1995) typically reveals increased signal intensity in the globus pallidus; similar changes may be seen in the striatum and in the substantia nigra.

Course

The overall course mirrors that of the underlying hepatic disease.

Etiology

At autopsy (Finlayson and Superville 1981; Spencer and Forno 2000; Victor *et al.* 1965), spongiform change or laminar cortical necrosis may be found in the cerebral and cerebellar cortices; spongiform change is also noted in the basal ganglia and this may be accompanied by microcavitation. Alzheimer type II astrocytes are present in the cortex and basal ganglia.

The occurrence of these changes in the context of repeated episodes of hepatic encephalopathy strongly suggests that they are toxic in nature and, of the various candidate toxins, manganese stands out. Elevated manganese levels are found in the basal ganglia and cerebral cortex (Klos *et al.* 2006; Krieger *et al.* 1995) and also in the CSF and blood (Burkhard *et al.* 2003); furthermore, one study found a correlation between the elevation in blood manganese and the increase in signal intensity on T1-weighted scanning in the globus pallidus (Krieger *et al.* 1995).

Differential diagnosis

Hepatic encephalopathy is distinguished by the presence of delirium and asterixis, alcoholic dementia by the absence of abnormal movements, and Wilson's disease by copper studies.

Treatment

Remarkably, in an as yet uncertain percentage of cases, symptoms may remit after liver transplantation (Pinarbasi *et al.* 2009; Powell *et al.* 1990; Servin-Abad *et al.* 2006; Stracciari *et al.* 2001). In cases in which transplantation is not possible, consideration may be given to treatment with branched-chain amino acids, which in case reports led to a reduction in abnormal movements and a partial resolution of MRI abnormalities (Ueki *et al.* 2002); furthermore, in cases in which parkinsonism is prominent, case series suggest an effectiveness for levodopa (Burkhard *et al.* 2003).

14.21 HEPATIC PORPHYRIA

Of the hepatic porphyrias, four can cause delirium. Three of these are inherited on an autosomal dominant basis, namely acute intermittent porphyria, variegate porphyria, and hereditary coproporphyria, whereas the fourth, delta-aminolevulinic acid (ALA) dehydratase deficiency, is an autosomal recessive disorder. Acute intermittent porphyria is found worldwide and may have a prevalence as high as 5–10 cases per 100 000 population. Variegate porphyria, although rare in the United States, may be found in Sweden and among white people in South Africa. Hereditary coproporphyria is quite rare, and ALA dehydratase deficiency is so extraordinarily rare that it is not considered further in this text.

From a clinical point of view, the three autosomal dominantly inherited hepatic porphyrias are essentially the same, with the exception that both variegate porphyria and hereditary coproporphyria may cause a photosensitive rash, a symptom not seen in acute intermittent porphyria.

Of interest, it appears that the periodic 'madness' of King George III represented attacks of hepatic porphyria (Cox *et al.* 2005; Macalpine and Hunter 1966; Macalpine *et al.* 1968).

Clinical features

The hepatic porphyrias are episodic disorders, and the first episode usually appears in late adolescence or the early adult years. The attacks themselves tend to present fairly acutely and are often precipitated by one of the factors discussed below, such as infection, menstruation, or pregnancy, or any one of a host of different medicines.

The overall symptomatology of the episodes has been reported in a number of papers (Becker and Kramer 1975; Goldberg 1959; Rowland 1961; Stein and Tschudy 1970). Classically, episodes are characterized by abdominal pain with vomiting, constipation, or, less commonly, diarrhea, accompanied, in about half of the cases, by a delirium. The delirium is often marked by affective lability, delusions of persecution, and visual hallucinations, which may be quite compelling (Cross 1956; Paredes and Jones 1959); in some cases, stupor or coma may supervene. Uncommonly, rather than a delirium there may be a psychosis with auditory hallucinations, delusions of persecution, and bizarre behavior, all occurring in a 'clear sensorium' without confusion or cognitive deficits (Hirsch and Dunsworth 1955; Mandoki and Sumner 1994). Other symptoms, seen in a minority, include a primarily motor polyneuropathy (which may progress to a quadriparesis), a cranial neuropathy (most commonly with ophthalmoplegia or facial palsy), partial or grand mal seizures, hypertension, tachycardia, and fever. A small minority of patients may also have significant hyponatremia, which in turn may be due either to a syndrome of inappropriate antidiuretic hormone (ADH) secretion or to intestinal or renal sodium loss.

As noted earlier, patients with variegate porphyria and hereditary coproporphyria may have a rash: this is a photosensitive rash that may or may not coincide with the episodes just described.

In cases characterized by central nervous system symptomatology, the EEG may reveal generalized slowing and interictal epileptiform discharges, and MRI scanning, on T2-weighted images, may reveal scattered areas of increased signal intensity within the subcortical white matter (Bylesjo *et al.* 2004; King and Bragdon 1991), without contrast enhancement or any associated changes on DWI (Maramattom *et al.* 2005).

During episodes, a 24-hour urine analysis will reveal elevated levels of both porphobilinogen and ALA in all three of these porphyrias, and the diagnosis rests on this finding. In addition to these findings, in both variegate porphyria and hereditary coproporphyria, urinary coproporphyrin will also be elevated. Distinguishing between variegate porphyria and hereditary coproporphyria requires a 24-hour stool collection for protoporphyrin and coproporphyrin. In variegate porphyria, protoporphyrin levels are higher than those of coproporphyrin, whereas in hereditary coproporphyria the converse holds true. Importantly, in between episodes all of these measurements may be normal.

Course

The duration of episodes usually ranges from days to weeks, although sometimes episodes can last much longer. Although most patients recover completely, death may occur as a result of respiratory failure due to the motor polyneuropathy, or to an arrhythmia. Repeat episodes are common and typically occur in response to a specific precipitating factor.

Etiology

All of these hepatic porphyrias occur secondary to mutations in genes that code for various enzymes involved in the heme biosynthetic pathway: acute intermittent porphyria occurs secondary to mutations in the gene for porphobilinogen deaminase (also known as hydroxymethylbilane synthase) on chromosome 11; variegate porphyria to mutations in the gene for protoporphyrinogen oxidase on chromosome 1; and hereditary coproporphyria to mutations in the gene for coproporphyrinogen oxidase on chromosome 3. Although, as noted earlier, these are all inherited on an autosomal dominant basis, penetrance is very low, and consequently the family history is typically negative.

When the heme biosynthetic pathway becomes stressed in the presence of these mutated proteins, 'upstream' proteins, such as porphobilinogen and ALA, accumulate and an episode occurs. Although the mechanism whereby this accumulation leads to cerebral pathology is unclear, one theory, consistent with the findings noted on MRI studies, posits a breakdown of the blood–brain barrier, with vasogenic edema. Autopsy studies of the brain, though generally negative, have revealed, in some cases, diffuse neuronal loss and perivascular demyelination (Suarez *et al.* 1997). Within the peripheral nervous system, axonal damage (Cavanagh and Mellick 1965; Sweeney *et al.* 1970) with associated chromatolysis of the spinal motor neurons (Hierons 1957) has also been found.

As noted earlier, most episodes are precipitated by various factors including infection, menstruation, pregnancy, fasting (or merely a low-carbohydrate diet), surgery, and a host of drugs including barbiturates, primidone, phenytoin, valproic acid, carbamazepine, clonazepam, tricyclics, sulfonamides, griseofulvin, meprobamate, glutethimide, methyprylon, ethchlorvynol, ergot derivatives, synthetic estrogens and progestins, danazol, alpha-methyldopa, chlorpropamide, and alcohol. Importantly, the following drugs appear to be safe: phenothiazines (e.g. chlorpromazine or trifluoperazine), opioid analgesics, chloral hydrate, gabapentin, levetiracetam, diazepam, aspirin, and acetaminophen.

Differential diagnosis

Both thallium and arsenic intoxication may produce a similar clinical picture; in cases characterized primarily by a motor polyneuropathy, consideration is also given to Guillain–Barré syndrome.

Treatment

Acute episodes may respond to a high carbohydrate diet or, if patients cannot take food or fluid (as is often the case), to intravenous glucose in a dose of 400 g/day. In cases resistant to these, consideration should be given to intravenous hemin (Anderson and Collins 2006) (formerly known as hematin), which suppresses the heme biosynthetic pathway and typically provides dramatic relief within a matter of days. Pending resolution of the episode, symptomatic treatment, in addition to general supportive care, may include chlorpromazine or propranolol. Chlorpromazine, in doses of 25–50 mg intramuscularly or intravenously every 4–6 hours, may not only control symptoms of the delirium but also reduce abdominal pain (Calvy *et al.* 1957). Propranolol, in doses ranging from 20 to 200 mg total daily, will control both hypertension and tachycardia (Menawat *et al.* 1979). If seizures occur and require treatment, consideration may be given to either gabapentin (Hahn *et al.* 1997; Tatum and Zachariah 1995) or levetiracetam (Paul and Meencke 2004); in emergent situations, diazepam may be given intravenously.

Prevention is critical, and this may generally be accomplished by maintaining a high-carbohydrate diet and, when possible, avoiding precipitating factors. In cases of menstrually induced attacks, consideration may be given to the administration of luteinizing hormone-releasing hormone to suppress menstruation. If episodes recur despite these measures, consideration may be given to prophylactic use of hemin (Anderson and Collins 2006).

In cases resistant to treatment with hemin, consideration should be given to liver transplantation, which may be curative (Dar *et al.* 2010).

14.22 FAHR'S SYNDROME

Fahr's syndrome is characterized by a movement disorder, generally parkinsonism, or a neuropsychiatric syndrome, most commonly dementia, or both, occurring secondary to brain calcification, which, although most common in the basal ganglia, may also be found in the dentate nuclei, thalamus, and at the gray–white junction of the cerebral and cerebellar cortices. This calcification may be either secondary to some other disorder, such as hypoparathyroidism, or occur on a primary basis as either an idiopathic or an inherited disorder, in which case one speaks of Fahr's disease. Synonyms for Fahr's disease include idiopathic striopallidodentate calcinosis and idiopathic calcification of the basal ganglia (ICBG).

Although, as noted below, asymptomatic basal ganglia calcification is a not uncommon incidental finding, especially in the elderly, Fahr's syndrome, by contrast, is rare.

Clinical features

The onset is gradual, and although cases may present with either a movement disorder or a neuropsychiatric syndrome (Manyam *et al.* 2001), most, over long-term follow-up, will be characterized by both. In the vast majority of cases, the age of onset is in the middle years: cases presenting with a psychosis, however, constitute an exception to this rule, in that such cases, when occurring on a primary basis, typically present in early adult years (Cummings *et al.* 1983).

Of the movement disorders seen in Fahr's syndrome, the most common is parkinsonism (Klawans *et al.* 1976; Manyam *et al.* 2001; Margolin *et al.* 1980; Tambyah *et al.* 1993); choreathetosis or dystonia (Larsen *et al.* 1985; Manyam *et al.* 2001; Saiki *et al.* 2007) may also occur and in a small minority there may be ataxia or dysarthria (Manyam *et al.* 2001).

Of the neuropsychiatric syndromes, the most common is dementia (Kobari *et al.* 1997; Slyter 1979; Wszolek *et al.* 2006). Personality change, which may be of the frontal lobe type (Lam *et al.* 2007; Weisman *et al.* 2007), may also occur. A psychosis, similar to that seen in schizophrenia (Chabot *et al.* 2001; Cummings *et al.* 1983; Francis and Freeman 1984), has also been reported. Other presentations include depression or mania (Trautner *et al.* 1988) and obsessions and compulsions (Lopez-Villegas *et al.* 1996). Seizures have also been noted, and in cases occurring secondary to hypoparathyroidism these may occur on the basis of hypocalcemia.

Computed tomography scanning is superior to both MRI and skull radiography and in all cases displays bilateral, symmetric, calcification in the basal ganglia; in most cases, as illustrated in Figure 14.1, one also sees bilateral calcification in other structures, such as the dentate nuclei, the thalamus, and the gray–white junction of the cerebral and cerebellar cortices.

Course

Fahr's syndrome is generally gradually progressive.

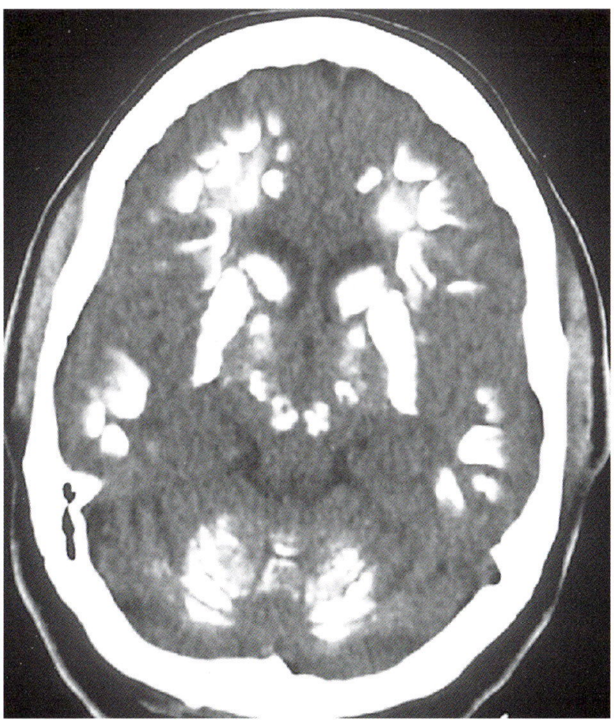

Figure 14.1 Computed tomography (CT) scan showing extensive calcium deposition in a case of Fahr's syndrome. Reproduced from Gillespie and Jackson (2000), with permission.

Etiology

The majority of cases occur on a primary basis. These may be either idiopathic (Trautner *et al.* 1988) or inherited (Boller *et al.* 1977; Kobari *et al.* 1997); of the inherited cases, although most occur on an autosomal dominant basis (Manyam *et al.* 2001), a recessive pattern has also been noted (Smits *et al.* 1983). At present, although the genetic basis of the autosomal dominant cases is not known, linkage has been demonstrated to the *IBGC1* locus on chromosome 14 in some (Geschwind *et al.* 1999), but not all (Oliveira *et al.* 2004) cases.

Secondary cases may occur on the basis of hypoparathyroidism (either surgical [Berger and Ross 1981; Klawans *et al.* 1976; Tambyah *et al.* 1993] or idiopathic [Kazis 1985; Slyter 1979]), pseudohypoparathyroidism (Mathews 1957), pseudo-pseudohypoparathyroidism (Nyland and Skre 1977), or, rarely, hyperparathyroidism (Margolin *et al.* 1980). Fahr's syndrome may also occur in concert with bone cysts of the hands and feet in the very rare autosomal recessive Nasu–Hakola disease (Klunemann *et al.* 2005; Kondo *et al.* 2002).

Differential diagnosis

In practice, the question of this diagnosis arises when, in the evaluation of patients with one of the clinical syndromes described above, neuroimaging reveals calcification. Care must be taken at this point, however, given that a degree of calcification of the basal ganglia, primarily in the globus pallidus, is not uncommon in asymptomatic individuals, being found in approximately 0.5 percent of the general population (Harrington *et al.* 1981; Koller *et al.* 1979) and up to 19 percent of 85-year-olds (Ostling *et al.* 2003). In general, then, the diagnosis should only be made in cases where the calcification is widespread, well beyond the globus pallidus, and no other clear explanation is forthcoming for the clinical syndrome in question.

Treatment

If present hypoparathyroidism should be treated as this may prompt a partial remission of symptoms (Berger and Ross 1981; Slyter 1979). Parkinsonism may respond to levodopa. The general treatment of dementia is described in Chapter 6.3. In cases characterized by depression, mania, or obsessions or compulsions, treatment may be attempted as described in Chapters 7.1, 7.3, and 5.17 respectively.

REFERENCES

Abarbanel JM, Berginer VM, Osimani A *et al.* Neurologic complications after gastric restriction surgery for morbid obesity. *Neurology* 1987; **37**:196–200.

Abd El Naby S, Hassanein M. Neuropsychiatric manifestations of chronic manganese poisoning. *J Neurol Neurosurg Psychiatry* 1965; **28**:282–8.

Adams RD, Foley JM. The neurological disorder associated with liver disease. *Assoc Res Nerv Ment Dis Proc* 1953; **32**:198–237.

Adams RD, Victor M, Mancall EL. Central pontine myelinolysis: a hitherto undescribed disease occurring in alcoholic and malnourished patients. *Arch Neurol Psychiatry* 1959; **81**:154–72.

Akelaitis AJ. Lead encephalopathy in children and adults. *J Nerv Ment Dis* 1941; **93**:313–32.

Alajouanine T, DeRobert L, Thieffry S. Etude clinique d'ensemble de 210 cas d'intoxication par les sels organiques d'etain. *Rev Neurol* 1958; **98**:85–96.

Albert MS, Butters N, Levin J. Temporal gradients in the retrograde amnesia of patients with alcoholic Korsakoff's disease. *Arch Neurol* 1979; **36**:211–16.

Alfrey AC, LeGendre GR, Kaehny WD. The dialysis encephalopathy syndrome. Possible aluminium intoxication. *N Engl J Med* 1976; **294**:184–8.

Amin-Zaki L, Majeed MA, Traxon TW *et al.* Methyl mercury poisoning in Iraqi children: clinical observations of two years. *BMJ* 1978; **1**:613–16.

Anderson KE, Collins S. Open-label study of hemin for acute porphyria: clinical practice implications. *Am J Med* 2006; **119**:801.

Andreoli SP, Bergstein JM, Sherrard DJ. Aluminum intoxication from aluminum-containing phosphate binders in children with azotemia not undergoing dialysis. *N Engl J Med* 1984; **310**:1079–84.

Andres E, Noel E, Goichot B. Metformin-associated vitamin B12 deficiency. *Arch Int Med* 2002; **162**:2251–2.

Atre AL, Shinde PR, Shinde SN *et al.* Pre- and posttreatment MR findings in lead encephalopathy. *AJNR* 2006; **27**:902–3.

Attur RP, Kandavar R, Kadavigere R *et al.* Dialysis disequilibrium syndrome presenting as a focal neurological deficit. *Hemodial Int* 2008; **12**:313–15.

Auer RN, Hugh J, Cosgrove E *et al.* Neuropathologic findings in three cases of profound hypoglycemia. *Clin Neuropathol* 1989; **8**:63–8.

Bach LA, Buchanan RR, Scarlett JD *et al.* Hyperviscosity syndrome secondary to rheumatoid arthritis. *Aust NZ J Med* 1989; **19**:710–12.

Baghurst PA, McMichael AJ, Wigg NR *et al.* Environmental exposure to lead and children's intelligence at the age of seven years – the Port Pirie Cohort Study. *N Engl J Med* 1992; **327**:1279–84.

Bakir F, Rustam H, Tikriti S *et al.* Clinical and epidemiological aspects of methylmercury poisoning. *Postgrad Med J* 1980; **56**:1–10.

Bank WJ, Pleasure DE, Suzuki K *et al.* Thallium poisoning. *Arch Neurol* 1972; **26**:456–64.

Bass NM, Mullen KD, Sanyal A *et al.* Rifaximin treatment in hepatic encephalopathy. *N Engl J Med* 2010; **362**:1071–81.

Baver MS. Fatal hepatic failure and valproate. *Am J Psychiatry* 2005; **162**:192.

Becker DM, Kramer S. The neurological manifestations of porphyria: a review. *Medicine* 1975; **56**:411–23.

Bellinger D, Leviton A, Waternaux C *et al.* Longitudinal analysis of prenatal and postnatal lead exposure and early cognitive development. *N Engl J Med* 1987; **316**:1037–43.

Beltramello A, Puppini G, Cerini R *et al.* Subacute combined degeneration of the spinal cord after nitrous oxide anaesthesia: role of magnetic resonance imaging. *J Neurol Neurosurg Psychiatry* 1998; **64**:563–4.

Ben-Ami H, Nagachandran P, Mendelson A *et al.* Drug-induced hypoglycemic coma in 102 diabetic patients. *Arch Intern Med* 1999; **159**:281–4.

Benna P, Lacquanilti F, Triolo G *et al.* Acute neurologic complications of hemodialysis. Study of 14,000 hemodialyses in 103 patients with chronic renal failure. *Ital J Neurol Sci* 1981; **2**:53–7.

Berger JR, Ross DB. Reversible Parkinson syndrome complicating postoperative hypoparathyroidism. *Neurology* 1981; **31**:881–2.

Bertrand A, Brandel JP, Grignon Y *et al.* Wernicke encephalopathy and Creutzfeldt–Jakob disease. *J Neurol* 2009; **256**:904–9.

Besser R, Kramer G, Thumler R *et al.* Acute trimethyltin limbic–cerebellar syndrome. *Neurology* 1987; **37**:945–50.

Blass JP, Gibson GE. Abnormality of a thiamine-requiring enzyme in patients with the Wernicke–Korsakoff syndrome. *N Engl J Med* 1977; **297**:1367.

Bodamar OAF, Rosenblatt DS, Appel SH *et al.* Adult-onset combined methylmalonic aciduria and homocystinuria. *Neurology* 2001; **56**:1113.

Bohan TP, Helton E, McDonald I *et al.* Effect of l-carnitine treatment for valproate-induced hepatotoxicity. *Neurology* 2001; **56**:1405–9.

Boller F, Boller M, Gilbert J. Familial idiopathic cerebral calcifications. *J Neurol Neurosurg Psychiatry* 1977; **40**:280–5.

Bosboom WMJ, Frijns CJM, Gijn J. Yelling attacks and wasted hands. *Lancet* 1996; **348**:238.

Boxer AL, Kramer JH, Johnston N *et al.* Executive dysfunction in hyperhomocystinemia responds to homocysteine-lowering treatment. *Neurology* 2005; **64**:1431–4.

Brunner JE, Redmond JM, Haggar AM *et al.* Central pontine myelinolysis and pontine lesions after rapid correction of hyponatremia: a prospective magnetic resonance imaging study. *Ann Neurol* 1990; **27**:61–6.

Buge A, Supino-Viterbo V, Rancurel G *et al.* Epileptic phenomena in bismuth toxic encephalopathy. *J Neurol Neurosurg Psychiatry* 1981; **44**:62–7.

Burke GJ, Hiangabeza T. Isoniazid-induced pellagra in a patient on vitamin B supplement. *S Afr Med J* 1977; **51**:719.

Burkhard PR, Delavelle J, Du Pasquier R *et al.* Chronic parkinsonism associated with cirrhosis: a distinct subset of acquired hepatocerebral degeneration. *Arch Neurol* 2003; **60**:521–8.

Burks JS, Alfrey AC, Huddlestone J *et al.* A fatal encephalopathy in chronic hemodialysis patients. *Lancet* 1976; **1**:764–8.

Burns R, Thomas DQ, Barron VJ. Reversible encephalopathy possibly associated with bismuth subgallate ingestion. *BMJ* 1974; **1**:220–3.

Bylesjo I, Brekke OL, Prytz J *et al.* Brain magnetic resonance imaging white-matter lesions and cerebrospinal fluid findings in patients with acute intermittent porphyria. *Eur Neurol* 2004; **51**:1–5.

Cadranel JF, Lebiz E, Di Martino V *et al.* Focal neurologic signs in hepatic encephalopathy in cirrhotic patients: an underestimated entity? *Am J Gastroenterol* 2001; **96**:515–18.

Calne DB, Chu NS, Huang CC *et al.* Manganese and idiopathic parkinsonism: similarities and differences. *Neurology* 1994; **44**:1583–6.

Calvy GL, Leeper RD, Monaco RN *et al.* Intermittent acute porphyria treated with chlorpromazine. *N Engl J Med* 1957; **256**:309–11.

Camacho J, Arnalich F, Anciones B *et al.* The spectrum of neurological abnormalities in myeloma. *J Med* 1985; **16**:597–611.

Canfield RL, Henderson CP, Cory-Slechta DA *et al.* Intellectual impairment in children with blood lead concentrations below 10 microgr per deciliter. *N Engl J Med* 2003; **348**:1517–26.

Case Records of the Massachusetts General Hospital. *N Engl J Med* 1988; **318**:1523–8.

Cavanagh JB, Mellick RS. On the nature of the peripheral nerve lesions with acute intermittent porphyria. *J Neurol Neurosurg Psychiatry* 1965; **28**:320–72.

Cavanagh JB, Fuller NH, Johnson HRM *et al.* The effects of thallium salts, with particular reference to the nervous system changes. *Q J Med* 1974; **43**:293–319.

Chabot B, Roulland C, Dollfus S. Schizophrenia and familial idiopathic basal ganglia calcification: a case report. *Psychol Med* 2001; **31**:741–7.

Charles JR. Manganese toxaemia, with special reference to the effects of liver feeding. *Brain* 1927; **50**:30–43.

Chatterjee A, Yapundich R, Palmer CA *et al.* Leukoencephalopathy associated with cobalamin deficiency. *Neurology* 1996; **46**:832–4.

Chokroverty S, Bruetman ME, Berger V *et al.* Progressive dialytic encephalopathy. *J Neurol Neurosurg Psychiatry* 1976; **39**:411–19.

Chu K, Kang DW, Kim HJ *et al.* Diffusion-weighted imaging abnormalities in Wernicke encephalopathy: reversible cytotoxic edema? *Arch Neurol* 2002; **59**:123–7.

Conomy JP, Swash M. Reversible decerebrate and decorticate posturing in hepatic coma. *N Engl J Med* 1968; **278**:876–9.

Cook DG, Fahn S, Brairt KA. Chronic manganese intoxication. *Arch Neurol* 1974; **30**:59–64.

Cossa P, Duplay J, Fischgold A *et al.* Encephalopathies toxiques au Stalinon. Aspects anatomo-cliniques et electroencephalographiques. *Rev Neurol* 1958; **98**:97–108.

Cox TN, Jack N, Lofthouse S *et al.* King George III and porphyria: an elemental hypothesis and investigation. *Lancet* 2005; **366**: 332–5.

Cravioto H, Korein J, Silberman J. Wernicke's encephalopathy: a clinical and pathological study of 28 autopsied cases. *Arch Neurol* 1961; **4**:510–19.

Cross TN. Porphyria – a deceptive syndrome. *Am J Psychiatry* 1956; **112**:1010–14.

Cummings JL, Gosenfeld LF, Houlihan JP *et al.* Neuropsychiatric disturbances associated with idiopathic calcification of the basal ganglia. *Biol Psychiatry* 1983; **18**:591–601.

Dar FS, Asai K, Haque AR *et al.* Liver transplantation for acute intermittent porphyria: a viable treatment? *Hepatobiliary Pancreat Dis Int* 2010; **9**:93–6.

Davis LE, Wands JR, Weiss SA *et al.* Central nervous system intoxication from mercurous chloride laxative. *Arch Neurol* 1974; **30**:428–31.

Davis LE, Standefer JC, Kornfeld M *et al.* Acute thallium poisoning: toxicological and morphological studies of the nervous system. *Ann Neurol* 1981; **10**:38–44.

Davis LE, Kornfield M, Mooney HS *et al.* Methylmercury poisoning: long-term clinical, radiological, toxicological, and pathological studies of an affected family. *Ann Neurol* 1994; **35**:680–8.

Davison AM, Walker GS, Oh H *et al.* Water supply aluminum concentration dialysis dementia, and effect of reverse-osmosis water treatment. *Lancet* 1982; **2**:785–7.

Della Rossa A, Tavoni A, Bombardieri S. Hyperviscosity syndrome in cryoglobuliemia: clinical aspects and therapeutic considerations. *Semin Thromb Hemost* 2003; **29**:473–7.

Dickoff DJ, Raps M, Yahr MD. Striatal syndrome following hyponatremia and its rapid correction: a manifestation of extrapontine myelinolysis confirmed by magnetic resonance imaging. *Arch Neurol* 1988; **45**:112–14.

Discalzi G, Pira E, Herrero Hernandez E *et al.* Occupational Mn parkinsonism: magnetic resonance imaging and clinical patterns following CaNa2-EDTA chelation. *Neurotoxicology* 2000; **21**:863–6.

Dizon AM, Kowaly S, Hoogwerf BJ. Neuroglycopenic and other symptoms in patients with insulinomas. *Am J Med* 1999; **106**:307–10.

Donovan JP, Schaefer DF, Shaw BW *et al.* Cerebral oedema and increased intracranial pressure in chronic liver disease. *Lancet* 1998; **351**:719–21.

Dyckner T, Ek B, Nyhlin H *et al.* Aggravation of thiamine deficiency by magnesium depletion. A case report. *Acta Med Scand* 1985; **218**:129–31.

El-Mallakh RS, Shrader SA, Widger E. Single case study: mania as a manifestation of end-stage renal disease. *J Nerv Ment Dis* 1987; **175**:243–5.

Estol CJ, Faris AA, Martinez J *et al.* Central pontine myelinolysis after liver transplantation. *Neurology* 1989; **39**:493–8.

Eto K. Minamata disease. *Neuropathology* 2000 (Suppl.); **20**:S14–9.

Evans DL, Edelsohn GA, Golden RN. Organic psychosis without anemia or spinal cord symptoms in patients with B12 deficiency. *Am J Psychiatry* 1983; **140**:218–21.

Feldman RG, White RF, Eriator II. Trimethyltin encephalopathy. *Arch Neurol* 1993; **50**:1320–4.

Ferraro A, Arieti S, English WH. Cerebral changes in the course of pernicious anemia and their relationship to psychic changes. *J Neuropathol Exp Neurol* 1945; **4**:217–39.

Ficker DM, Westmoreland BF, Sharbrough FW. Epileptiform abnormalities in hepatic encephalopathy. *J Clin Neurophysiol* 1997; **14**:230–4.

Fincher RM, Koerker RM. Long-term survival in acute arsenic encephalopathy. Follow-up using newer measures of electrophysiologic parameters. *Am J Med* 1987; **82**:549–52.

Finlayson MH, Superville B. Distribution of cerebral lesions in acquired hepatocerebral degeneration. *Brain* 1981; **104**:79–85.

Fisch BJ, Klass DW. The diagnostic specificity of triphasic wave patterns. *Electroencephalogr Clin Neurophysiol* 1988; **70**:1–8.

Fisher AA, Le Couteur DG. Lead poisoning from complementary and alternative medicine in multiple sclerosis. *J Neurol Neurosur Psychiatry* 2000; **69**:687–9.

Fortemps E, Amand G, Bomboir A *et al.* Trimethyltin poisoning. Report of two cases. *Int Arch Occup Environ Health* 1978; **41**:1–6.

Francis A, Freeman H. Psychiatric abnormality and brain calcification over four generations. *J Nerv Ment Dis* 1984; **172**:166–70.

Frantzen E. Wernicke's encephalopathy: 3 cases occurring in connection with severe malnutrition. *Acta Neurol Scand* 1966; **42**:426–41.

Fraser CL, Arieff AI. Hepatic encephalopathy. *N Engl J Med* 1985; **313**:865–73.

Fraser CL, Arieff AI. Nervous system complications in uremia. *Ann Intern Med* 1988; **109**:143–9.

Fraser TN. Cerebral manifestations of addisonian pernicious anemia. *Lancet* 1960; **2**:458–9.

Freeman JW, Couch JR. Prolonged encephalopathy with arsenic poisoning. *Neurology* 1978; **28**:853–5.

Gardeur D, Buge A, Rancurel G *et al.* Bismuth encephalopathy and cerebral computed tomography. *J Comput Assist Tomogr* 1978; **2**:436–8.

Gardian G, Voros E, Jardanhazy T *et al.* Wernicke's encephalopathy induced by hyperemesis gravidarum. *Acta Neurol Scand* 1999; **95**:196–8.

Garrett PJ, Mulcahey D, Carmody M *et al.* Aluminum encephalopathy: clinical and immunologic features. *Q J Med* 1988; **69**: 775–83.

Geschwind DH, Loginov M, Stern JM. Identification of a locus on chromosome 14q for idiopathic basal ganglia calcification (Fahr disease). *Am J Hun Genet* 1999; **65**:764–72.

Gibb WR, Gorsuch AN, Lees AJ *et al.* Reversible coma in Wernicke's encephalopathy. *Postgrad Med J* 1985; **61**:607–10.

Gillespie J, Jackson A. *MRI and CT of the Brain.* London: Arnold, 2000.

Gocht A, Colmant HJ. Central pontine and extrapontine myelinolysis: a report of 58 cases. *Clin Neuropathol* 1987; **6**:262–70.

Goggans FC. A case of mania secondary to vitamin B12 deficiency. *Am J Psychiatry* 1984; **141**:300–1.

Goldberg A. Acute intermittent porphyria. *Q J Med* 1959; **28**:183–209.

Graham DI, Adams JH, Caird FI *et al.* Acquired hepatocerebral degeneration: report of an atypical case. *J Neurol Neurosurg Psychiatry* 1970; **23**:656–62.

Grigg AP, Allardice J, Smith IL *et al.* Hyperviscosity syndrome in disseminated breast adenocarcinoma. *Pathology* 1994; **26**:65–8.

Hadfield MG, Kubal WS. Extrapontine myelinolysis of the basal ganglia without central pontine myelinolysis. *Clin Neuropathol* 1996; **15**:96–100.

Hahn M, Gildemeister OS, Krauss GL *et al.* Effects of new anticonvulsant medications on porphyrin synthesis in cultured liver cells: potential implications for patients with acute porphyria. *Neurology* 1997; **49**:97–106.

Halliday G, Cullen K, Harding A. Neuropathological correlates of memory dysfunction in the Wernicke–Korsakoff syndrome. *Alcohol Alcoholism Suppl* 1994; **2**:247–53.

Handler CE, Perkin GD. Anorexia nervosa and Wernicke's encephalopathy: an underdiagnosed association. *Lancet* 1982; **2**:771–2.

Harding A, Halliday G, Caine D *et al.* Degeneration of anterior thalamic nuclei differentiates alcoholics with amnesia. *Brain* 2000; **123**:141–54.

Harper C. Wernicke's encephalopathy: a more common disease than realized: a neuropathological study of 51 cases. *J Neurol Neurosurg Psychiatry* 1979; **42**:226–31.

Harper C. The incidence of Wernicke's encephalopathy in Australia – a neuropathological study of 131 cases. *J Neurol Neurosurg Psychiatry* 1983; **46**:593–8.

Harper CG, Giles M, Finlay-Jones R. Clinical signs in the Wernicke–Korsakoff complex: a retrospective analysis of 131 cases diagnosed at necropsy. *J Neurol Neurosurg Psychiatry* 1986; **49**:341–5.

Harrington MG, MacPherson P, McIntosh WB *et al.* The significance of the incidental finding of basal ganglia calcification on computed tomography. *J Neurol Neurosurg Psychiatry* 1981; **44**:1168–70.

Hart SP, Frier BM. Causes, management and morbidity of acute hypoglycaemia in adults requiring hospital admission. *Q J Med* 1998; **91**:505–10.

Hauw J-J, De Baecque C, Hausser-Hauw C *et al.* Chromatolysis in alcoholic encephalopathies. *Brain* 1988; **111**:843–57.

Hay WJ, Rickards AG, McMenemey WH *et al.* Organic mercurial encephalopathy. *J Neurol Neurosurg Psychiatry* 1963; **26**:199–202.

Healton EH, Savage DG, Brust JCM *et al.* Neurologic aspects of cobalamin deficiency. *Medicine* 1991; **70**:228–45.

Herzlich BC, Schiano TD. Reversal of apparent AIDS dementia complex following treatment with vitamin B12. *J Intern Med* 1993; **233**:495–7.

Hierons R. Changes in the nervous system in acute porphyria. *Brain* 1957; **80**:176–92.

Hirsch S, Dunsworth FA. An interesting case of porphyria. *Am J Psychiatry* 1955; **111**:703.

Huang C-C, Chu N-S, Lu C-S *et al.* Chronic manganese intoxication. *Arch Neurol* 1989; **46**:1104–6.

Huang C-C, Lu C-S, Chu N-S *et al.* Progression after chronic manganese exposure. *Neurology* 1993; **43**:1479–83.

Huang C-C, Chus N-S, Lu C-S *et al.* Cock gait in manganese intoxication. *Mov Disord* 1997; **12**:807–8.

Huang C-C, Chu N-S, Lu C-S *et al.* Long-term progression in chronic manganism: ten years of follow-up. *Neurology* 1998; **50**:698–700.

Huang C-C, Chu N-S, Lu C-S *et al.* The natural history of neurological manganism over 18 years. *Parkinsonism Relat Disord* 2007; **13**:143–5.

Hughes JR, Schreeder MT. EEG in dialysis encephalopathy. *Neurology* 1980; **30**:1148–54.

Hung SC, Hung SH, Tarug DC *et al.* Thiamine deficiency and unexplained neuropathology in hemodialysis and peritoneal dialysis patients. *Am J Kidney Dis* 2001; **38**:941–7.

Hunter D, Russell DS. Focal cerebral and cerebellar atrophy in a human subject due to organic mercury compounds. *J Neurol Neurosurg Psychiatry* 1954; **17**:235–41.

Hurst EW. The lesions produced in the central nervous system by certain organic arsenical compounds. *J Pathol Bacteriol* 1959; **77**:523–34.

Ishii N, Nishihara Y. Pellagra among chronic alcoholics: clinical and pathological study of 20 necropsy cases. *J Neurol Neurosurg Psychiatry* 1981; **44**:209–15.

Ishii N, Nishihara Y. Pellagra encephalopathy among tuberculous patients: its relation to isoniazid therapy. *J Neurol Neurosurg Psychiatry* 1985; **48**:628–34.

Jack RA, Rivers-Bulkeley NT, Rabin PL. Single case study: secondary mania as a presentation of progressive dialysis encephalopathy. *J Nerv Ment Dis* 1983; **171**:193–5.

Jaffe N. Catatonia and hepatic dysfunction. *Dis Nerv Syst* 1967; **28**:606–8.

Jagadha V, Deck JHN, Halliday WL *et al.* Wernicke's encephalopathy in patients on peritoneal dialysis or hemodialysis. *Ann Neurol* 1987; **21**:78–84.

Jagielska G, Tornaszewicz-Libudzic EC, Brzozowska A. Pellagra: a rare complication of anorexia nervosa. *Eur Child Adolesc Psychiatry* 2007; **16**:417–20.

Jenkins CD, Mellins RB. Lead poisoning in children. *Arch Neurol Psychiatry* 1957; **77**:70–8.

Jenkins RB. Inorganic arsenic and the nervous system. *Brain* 1966; **89**:479–96.

Jolliffe N, Bowman K, Rosenblum L *et al.* Nicotinic acid deficiency encephalopathy. *J Am Med Assoc* 1940; **114**:307–9.

Jung SL, Kim BS, Lee KS *et al.* Magnetic resonance imaging and diffusion-weighted imaging changes after hypoglycemic coma. *J Neuroimaging* 2005; **15**:193–6.

Kalimo H, Olsson Y. Effect of severe hypoglycemia on the human brain. *Acta Neurol Scand* 1980; **62**:345–56.

Karnaze DS, Bickford RG. Triphasic waves: a reassessment of their significance. *Electroencephalogr Clin Neurophysiol* 1984; **57**:193–8.

Karp BI, Laureno R. Pontine and extrapontine myelinolysis: a neurologic disorder following rapid correction of hyponatremia. *Medicine* 1993; **72**:359–73.

Kazis AD. Contribution of CT scan to the diagnosis of Fahr's syndrome. *Acta Neurol Scand* 1985; **71**:206–11.

King PH, Bragdon AC. MRI reveals multiple reversible cerebral lesions in an attack of acute intermittent porphyria. *Neurology* 1991; **41**:1300–2.

Kinsella LJ, Green R. 'Anesthesia paresthetica' nitrous oxide-induced cobalamin deficiency. *Neurology* 1995; **45**:1608–10.

Klawans HL, Lipton M, Simon L. Calcification of the basal ganglia as a cause of levodopa-resistant parkinsonism. *Neurology* 1976; **26**:221–5.

Klunemann HH, Ridha BH, Magy L *et al.* The genetic causes of basal ganglia calcification, dementia, and bone cysts. DAP12 and TREM2. *Neurology* 2005; **64**:1502–7.

Klos KJ, Ahlskog JE, Kumar N *et al.* Brain metal concentrations in chronic liver failure patients with pallidal T1 MRI hyperintensity. *Neurology* 2006; **67**:1984–9.

Kobari M, Nogawa S, Sugimoto Y *et al.* Familial idiopathic brain calcification with autosomal dominant inheritance. *Neurology* 1997; **48**:645–9.

Koller WC, Cochran JW, Klawans HL. Calcification of the basal ganglia: computerized tomography and clinical correlation. *Neurology* 1979; **29**:328–33.

Koller WC, Lyons KE, Truly W. Effect of levodopa treatment for parkinsonism in welders. *Neurology* 2004; **62**:730–3.

Kondo T, Takahashi K, Kohara N *et al.* Heterogeneity of presenile dementia with bone cysts (Nasu-Hakola disease). *Neurology* 2002; **59**:1105–7.

Kreyberg S, Torvik A, Bjorneboe A *et al.* Trimethyltin poisoning: report of a case with postmortem examination. *Clin Neuropathol* 1992; **11**:256–9.

Krieger D, Krieger S, Jansen O *et al.* Manganese and chronic hepatic encephalopathy. *Lancet* 1995; **346**:270–4.

Kumar N. Copper deficiency hyelopathy (human swayback). *Mayo Clinic Proc* 2006; **81**:1371–84.

Kuo SH, Debnam JM, Fuller GN *et al.* Wernicke's encephalopathy: an underrecognized and reversible cause of confusional state in cancer patients. *Oncology* 2009; **76**:10–18.

Kurland LT, Faro SN, Siedler H. Minamata disease. *World Neurol* 1960; **1**:370–95.

Kuzminski AM, Del Giacco EJ, Allen RH *et al.* Effective treatment of cobalamin deficiency with oral cobalamin. *Blood* 1998; **92**:1191–8.

Ladurner G, Wawschinek O, Pogglitsch H *et al.* Neurophysiological findings and serum aluminium in dialysis encephalopathy. *Eur Neurol* 1982; **21**:335–9.

Lam JS, Fong SY, Yiu GC *et al.* Fahr's disease: a differential diagnosis of frontal lobe syndrome. *Hong Kong Med J* 2007; **13**:75–7.

Lampl C, Yazdi K. Central pontine myelinolysis. *Eur Neurol* 2002; **47**:3–10.

Langworthy OR. Lesions of the central nervous system characteristic of pellagra. *Brain* 1931; **54**:291–302.

Larsen TA, Dunn HG, Jae JE *et al.* Dystonia and calcification of the basal ganglia. *Neurology* 1985; **35**:533–7.

Lawrence RD, Meyer R, Nevin S. The pathological changes in fatal hypoglycemia. *Q J Med* 1942; **11**:181–201.

Lederman RJ, Henry CE. Progressive dialysis encephalopathy. *Ann Neurol* 1978; **4**:199–204.

Lee EM, Kang JK, Yun SC *et al.* Risk factors for central pontine and extrapontine myelinolysis following orthotopic liver transplantation. *Eur Neurol* 2009; **62**:362–8.

Lindenbaum J, Healton EB, Savage DG *et al.* Neuropsychiatric disorders caused by cobalamin deficiency in the absence of anemia or macrocytosis. *N Engl J Med* 1988; **318**:1720–8.

Lopez-Villegas D, Kulisevsky J, Deus J *et al.* Neuropsychological alterations in patients with computed tomography-detected basal ganglia calcification. *Arch Neurol* 1996; **53**:251–6.

Lu C-S, Huang C-C, Chu N-S *et al.* Levodopa failure in chronic manganism. *Neurology* 1994; **44**:1600–2.

Lurie LA. Pernicious anemia with mental symptoms: observations on the variable extent and probable duration of central nervous system lesions in four autopsied cases. *Arch Neurol Psychiatry* 1919; **2**:67–109.

Macalpine I, Hunter R. The 'insanity' of King George III: a classic case of porphyria. *BMJ* 1966; **1**:65–71.

Macalpine I, Hunter R, Rimington C. Porphyria in the royal houses of Stuart, Hanover and Prussia: a follow-up study of George III's illness. *BMJ* 1968; **1**:7–8.

McEntee WJ, Mair RG. Memory enhancement in Korsakoff's psychosis by clonidine: further evidence for a noradrenergic deficit. *Ann Neurol* 1980; **7**:466–70.

McKee AC, Winkleman MD, Banker BQ. Central pontine myelinolysis in severely burned patients: relationship to serum hyperosmolality. *Neurology* 1988; **38**:1211–17.

MacKeith SA, Meyer A. A death during insulin treatment of schizophrenia; with pathological report. *J Ment Sci* 1939; **85**:96–105.

McMillan TM, Jacobson RR, Gross M. Neuropsychology of thallium poisoning. *J Neurol Neurosurg Psychiatry* 1997; **63**:247–50.

Mahoney CA, Arieff AI. Uremic encephalopathies: clinical, biochemical, and experimental features. *Am J Kidney Dis* 1982; **2**:324–36.

Mair RG, McEntee WJ. Cognitive enhancement in Korsakoff's psychosis by clonidine: a comparison with L-dopa and ephedrine. *Psychopharmacolgy* 1986; **88**:374–80.

Malamud N, Skillicorn SA. Relationship between the Wernicke and the Korsakoff syndrome. *Arch Neurol Psychiatry* 1956; **76**:585–96.

Malouf R, Brust JCM. Hypoglycemia: causes, neurological manifestations, and outcome. *Ann Neurol* 1985; **17**:421–30.

Mandoki MW, Sumner GS. Psychiatric manifestations of hereditary coproporphyria in a child. *J Nerv Ment Dis* 1994; **182**:117–18.

Manyam BV, Walters AS, Narla KR. Bilateral striopallidodentate calcinosis: clinical characteristics of patients seen in a registry. *Mov Disord* 2001; **16**:258–64.

Maraganore DM, Folger WN, Swanson JW *et al.* Movement disorders as sequelae of central pontine myelinolysis: report of three cases. *Mov Disord* 1992; **7**:142–8.

Maramattom BV, Zaldivar RA, Glynn SM *et al.* Acute intermittent porphyria presenting as diffuse encephalopathy. *Ann Neurol* 2005; **57**:581–4.

Marcuard SP, Albernaz L, Khazanie PG. Omeprazole therapy causes malabsorption of cyanocobalamin (vitamin B12). *Ann Intern Med* 1994; **120**:211–15.

Margolin D, Hammerstad J, Orwoll E *et al.* Intracranial calcification in hyperparathyroidism associated with gait apraxia and parkinsonism. *Neurology* 1980; **30**:1005–7.

Marie R-M, Le Biez E, Busson P *et al.* Nitrous oxide anesthesia-associated myelopathy. *Arch Neurol* 2000; **57**:380–2.

Mas A, Rodes J, Sunyer L *et al.* Comparison of rifaximin and lactilol in the treatment of acute hepatic encephalopathy: results of a randomized, double-blind, double-dummy, controlled clinical trial. *J Hepatol* 2003; **38**:51–8.

Mathews WB. Familial calcification of the basal ganglia with response to parathyroid hormone. *J Neurol Neurosurg Psychiatry* 1957; **20**:172–7.

Matte TD, Proops D, Palazuelos E *et al.* Acute high-dose lead exposure from beverage contaminated by traditional Mexican pottery. *Lancet* 1994; **344**:1064.

Mawdsley C. Neurological complications of haemodialysis. *Proc R Soc Med* 1972; **65**:871–3.

Mehta J, Singhal S. Hyperviscosity syndrome in plasma cell dyscrasias. *Semin Thromb Hemost* 2003; **29**:467–471.

Mellins RB, Jenkins CD. Epidemiologic and psychological study of lead poisoning in children. *J Am Med Assoc* 1955; **158**:15–20.

Mena I, court J, Fuenzalida S *et al.* Modification of chronic manganese poisoning. Treatment with L-dopa or 5-OH tryptophan. *N Engl J Med* 1970; **282**:5–10.

Menawat AS, Panwar RB, Kochar DK *et al.* Propranolol in acute intermittent porphyria. *Postgrad Med J* 1979; **55**:546–7.

Menezes G, D'Souza HS, Venkatesh T. Chronic lead poisoning in an adult battery worker. *Occup Med* 2003; **53**:476–8.

Menger H, Jorg J. Outcome of central pontine and extrapontine myelinolysis (n5 44). *J Neurol* 1999; **246**:700–5.

Messert B, Orrison WM, Quaglieri CE. Central pontine myelinolysis. *Neurology* 1979; **29**:147–60.

Moersch FP, Kernohan JW. Hypoglycemia: neurologic and neuropathologic studies. *Arch Neurol Psychiatry* 1938; **39**:242–57.

Moore D, House I, Dixon A. Thallium poisoning. Diagnosis may be elusive but alopecia is the clue. *BMJ* 1993; **306**:1527–9.

Morris CE, Heyman A, Pozersky T. Lead encephalopathy caused by ingestion of illicitly distilled whiskey. *Neurology* 1964; **14**:493–9.

Mueller J, Hotson JR, Langston JW. Hyperviscosity-induced dementia. *Neurology* 1983; **33**:101–3.

Nadel AM, Wilson WP. Dialysis encephalopathy: a possible seizure disorder. *Neurology* 1976; **26**:1130–2.

Nagatomo S, Umehara F, Hanada K *et al.* Manganese intoxication during total parenteral nutrition: report of two cases and review of the literature. *J Neurol Sci* 1999; **162**:102–5.

Naismith RT, Shepherd JB, Weihl CC *et al.* Acute and bilateral blindness due to optic neuropathy associated with copper deficiency. *Arch Neurol* 2009; **66**:1025–7.

Needleman HL, Shell A, Ballinger D *et al.* The long-term effects of exposure to low doses of lead in childhood: an 11-year follow up report. *N Engl J Med* 1990; **322**:83–8.

Norenberg MD, Leslie KO, Robertson AS. Association between rise in serum sodium and central pontine myelinolysis. *Ann Neurol* 1982; **11**:128–35.

Nyland H, Skre H. Cerebral calcinosis with late onset encephalopathy. *Acta Neurol Scand* 1977; **56**:309–25.

O'Carroll RE, Moffoot A, Ebmeir KP *et al.* Korsakoff's syndrome, cognition and clonidine. *Psychol Med* 1993; **23**:341–7.

O'Carroll RE, Moffoot APR, Ebmeier KP *et al.* Effects of fluvoxamine on cognitive functioning in the alcoholic Korsakoff syndrome. *Psychopharmacology* 1994; **116**:85–8.

O'Carroll RE, Masterton G, Dougall N *et al.* The neuropsychiatric sequelae of mercury poisoning: the Mad Hatter's disease revisited. *Br J Psychiatry* 1995; **167**:95–8.

Ogershok PR, Rahman A, Nestor S *et al.* Wernicke encephalopathy in nonalcoholic patients. *Am J Med Sci* 2002; **323**:107–11.

O'Hare JA, Callaghan NM, Murnaghan DJ. Dialysis encephalopathy. Clinical, electroencephalographic, and interventional aspects. *Medicine* 1983; **62**:129–41.

Oliveira JR, Spiteri E, Sobrio MJ *et al.* Genetic heterogeneity in familial idiopathic basal ganglia calcification (Fahr disease). *Neurology* 2004; **63**:2165–7.

Olsen S. The brain in uremia. A patho-anatomical study of brains from 104 patients dying in renal insufficiency, with reference to the influence of complicating factors, especially ischemia of the brain. *Acta Psychiatr Scand* 1961; **36**(Suppl. 156):1–128.

Ong JP, Aggarwal A, Krieger D *et al.* Correlation between ammonia levels and the severity of hepatic encephalopathy. *Am J Med* 2003; **114**:188–93.

Ostling S, Andreasson LA, Skoog I. Basal ganglia calcification and psychotic symptoms in the very old. *Int J Geriatr Psychiatry* 2003; **18**:983–7.

Pacchetti C, Cristina S, Nappi G. Reversible chorea and focal dystonia in vitamin B12 deficiency. *N Engl J Med* 2002; **347**:295.

Palardy J, Havrankova J, Lepage R *et al.* Blood glucose measurements during symptomatic episodes in patients with suspected post-prandial hypoglycemia. *N Engl J Med* 1989; **321**:1421–5.

Paredes A, Jones H. Psychopathology of acute intermittent porphyria: case report. *J Nerv Ment Dis* 1959; **129**:291–301.

Parry TE. Folate responsive neuropathy. *Presse Med* 1994; **23**:131–7.

Patchell RA, Fellows HA, Humphries LL. Neurologic complications of anorexia nervosa. *Acta Neurol Scand* 1994; **89**:111–16.

Patel SV, Parish DC, Patel RM *et al.* Resolution of MRI findings in central pontine myelinolysis associated with hypokalemia. *Am J Med Sci* 2007; **334**:490–2.

Patterson JF. Chronic thallotoxicosis: treatment of the choreiform sequelae. *South Med J* 1975; **68**:923–5.

Paul F, Meencke H-J. Levetiracetam in focal epilepsy and hepatic porphyria: a case report. *Epilepsia* 2004; **45**:559–60.

Paulson GW, Martin EW, Mojzisik C *et al.* Neurologic complications of gastric partitioning. *Arch Neurol* 1985; **42**:675–7.

Pavy MD, Murphy PL, Virella G. Paraprotein-induced hyperviscosity. A reversible cause of stroke. *Postgrad Med* 1980; **68**:109–12.

Pentschew A. Morphology and morphogenesis of lead encephalopathy. *Acta Neuropathol* 1965; **5**:133–60.

Perl DP, Olanow CW. The neuropathology of manganese-induced parkinsonism. *J Neuropathol Exp Neurology* 2007; **66**:675–82.

Peters TE, Parvin M, Petersen C *et al.* A case report of Wernicke's encephalopathy in a pediatric patient with anorexia nervosa – restricting type. *J Adolesc Health* 2007; **40**:376–83.

Peterson H, Swanson AG. Acute encephalopathy occurring during hemodialysis. *Arch Intern. Med* 1964; **113**:877–80.

Pierce LB. Pellagra: report of a case. *Am J Psychiatry* 1924; **81**:237–43.

Pinarbasi B, Kaymakoglu S, Matur Z *et al.* Are acquired hepatocerebral degeneration and hepatic myelopathy reversible? *J Clin Gastroenterol* 2009; **43**:176–81.

Pincus JH, Reynolds EH, Glaser GH. Subacute combined system degeneration with folate deficiency. *J Am Med Assoc* 1972; **221**:496–7.

Powell EE, Pender MP, Chalk JB *et al.* Improvement in chronic hepatocerebral degeneration following liver transplantation. *Gastroenterology* 1990; **98**:1079–82.

Powers JM, Rosenblatt DS, Schmidt RE *et al.* Neurological and neuropathologic heterogeneity in two brothers with cobalamin C deficiency. *Ann Neurol* 2001; **49**:396–400.

Price BH, Mesulam M-M. Behavioral manifestation of central pontine myelinolysis. *Arch Neurol* 1987; **44**:671–3.

Price WA, Zimmer B, Conway R *et al.* Insulin-induced factitious hypoglycemic coma. *Gen Hosp Psychiatry* 1986; **8**:291–3.

Rajan S, Wallache JI, Brodkin KI *et al.* Response of elevated methylmalonic acid to three dose levels of oral cobalamin in older adults. *J Am Geriatr Soc* 2002; **50**:1789–95.

Raskin NH, Fishman RA. Neurologic disorders in renal failure. *N Engl J Med* 1976; **294**:143–8, 204–10.

Raskin NH, Bredesen D, Ehrenfeld WK *et al.* Periodic confusion caused by congenital extrahepatic portocaval shunt. *Neurology* 1984; **34**:666–9.

Read AE, Laidlaw J, Sherlock S. Neuropsychiatric complications of portocaval anastamosis. *Lancet* 1961; **1**:961–4.

Reed D, Crawley J, Faro SN *et al.* Thallotoxicosis. *J Am Med Assoc* 1963; **183**:516–22.

Reyes PF, Gonzales CF, Zalewska MK *et al.* Intracranial calcification in adults with chronic lead exposure. *Am J Roentgenol* 1986; **146**:267–70.

Reynolds EH, Rothfeld P, Pincus JH. Neurological disease associated with folate deficiency. *BMJ* 1973; **2**:398–400.

Rich LM, Caine MR, Findling JW *et al.* Hypoglycemic coma in anorexia nervosa. Case report and review of the literature. *Arch Int Med* 1990; **150**:894–5.

Ropper AH. Accelerated neuropathy of renal failure. *Arch Neurol* 1993; **50**:536–9.

Rovelli E, Luciani L, Pagani C *et al.* Correlation between serum aluminum concentrations and signs of encephalopathy in a large population of patients dialyzed with aluminum-free fluids. *Clin Nephrol* 1988; **29**:294–8.

Rovira A, Alonso J, Cordoba J. MR imaging findings in hepatic encephalopathy. *AJNR* 2008; **29**:1612–21.

Rowland LP. Acute intermittent porphyria: search for an enzymatic defect with implications for neurology and psychiatry. *Dis Nerv Syst* 1961; **22(Suppl.)**:1–12.

Russell DS. Changes in the central nervous system following arsphenamine medication. *J Pathol* 1937; **45**:357–66.

Russell JSR, Batten FE, Collier J. Subacute combined degeneration of the spinal cord. *Brain* 1900; **23**:39–110.

Rustam H, Hamri T. Methyl mercury poisoning in Iraq. *Brain* 1974; **97**:499–510.

Ruzek KA, Campeau NG, Miller GM. Early diagnosis of central pontine myelinolysis with diffusion-weighted imaging. *AJNR* 2004; **25**:210–13.

Sahin HA, Gurvit IH, Bilgic B *et al.* Therapeutic effects of an acetylcholinesterase inhibitor (donepezil) on memory in Wernicke-Korsakoff's disease. *Clin Neuropharmacol* 2002; **25**:16–20.

Saiki M, Saiki S, Saiki K *et al.* Neurological deficits are associated with increased brain calcinosis, hypoperfusion and hypometabolism in idiopathic basal ganglia calcification. *Mov Disord* 2007; **22**:1027–30.

Sakamoto E, Hagiwara D, Morishita Y *et al.* Complete recovery of central pontine myelinolysis by high dose pulse therapy with methylprednisolone. *Nippon Naika Gakkai Zasshi* 2007; **96**:2291–3.

Saner FH, Koeppen S, Meyer M *et al.* Treatment of central pontine myelinolysis with plasmapheresis and immunoglobulins in liver transplant patient. *Transpl Int* 2008; **21**:390–1.

Scarlett JA, Mako ME, Rubenstein AH *et al.* Factitious hypoglycemia: diagnosis by measurement of serum C-peptide immunoreactivity and insulin-binding antibodies. *N Engl J Med* 1977; **297**: 1029–32.

Scheiber SC, Ziesat H. Brief communication: clinical and psychological test findings in cerebral dyspraxia associated with hemodialysis. *J Nerv Ment Dis* 1976; **162**:212–14.

Schottenfeld RS, Cullen MR. Organic affective illness associated with lead intoxication. *Am J Psychiatry* 1984; **141**:1425–6.

Seiser A, Schwarz S, Aichinger-Steiner MM *et al.* Parkinsonism and dystonia in central pontine and extrapontine myelinolysis. *J Neurol Neurosurg Psychiatry* 1998; **65**:119–21.

Seltzer B, Benson DF. The temporal pattern of retrograde amnesia in Korsakoff's disease. *Neurology* 1974; **24**:527–30.

Serdaru M, Hausser-Hauw C, Laplane D *et al.* The clinical spectrum of alcoholic pellagra encephalopathy. A retrospective analysis of 22 cases studied pathologically. *Brain* 1988; **111**:829–42.

Servin-Abad L, Tzakis A, Schiff ER *et al.* Acquired hepatocerebral degeneration in a patient with HCV cirrhosis: complete resolution with subsequent recurrence after liver transplantation. *Liver Transpl* 2006; **12**:1161–5.

Sherlock S. Fulminant hepatic failure. *Adv Intern Med* 1993; **38**:245–67.

Sherrard DJ, Walker JV, Boykin JL. Precipitation of dialysis dementia by deferoxamine treatment of aluminum-related bone disease. *Am J Kidney Dis* 1988; **12**:12–30.

Shimomura T, Mori E, Hirono N *et al.* Development of Wernicke–Korsakoff syndrome after long intervals following gastrectomy. *Arch Neurol* 1998; **55**:1242–5.

Shintani S, Tsuruoka S, Shiigai T. Hypoglycemic hemiplegia: a repeat SPECT study. *J Neurol Neurosurg Psychiatry* 1993; **56**:700–1.

Shirabe T, Irie K, Uchida M. Autopsy cases of aluminum encephalopathy. *Neuropathology* 2002; **22**:206–10.

Shor-Posner G, Morgan R, Wilkie F *et al.* Plasma cobalamin levels affect information processing speed in a longitudinal study of HIV-1 disease. *Arch Neurol* 1995; **52**:195–8.

Singh S, Kumar A. Wernicke encephalopathy after obesity surgery. A systematic review. *Neurology* 2007; **68**:807–11.

Slyter H. Idiopathic hypoparathyroidism presenting as dementia. *Neurology* 1979; **29**:393–4.

Smith JF, McLaurin RL, Nichols JP *et al.* Studies in cerebral oedema and cerebral swelling. 1. The changes in lead encephalopathy in children compared with those in alkyl tin poisoning in animals. *Brain* 1960; **83**:411–24.

Smith LH. Mental and neurologic changes in pernicious anemia. *Arch Neurol Psychiatry* 1929; **22**:551–7.

Smits MG, Gabreels FJ, Thijssen HO *et al.* Progressive idiopathic strio-pallido-dentate calcinosis (Fahr's disease) with autosomal recessive inheritance. Report of three siblings. *Eur Neurol* 1983; **22**:58–64.

Snider WD, DeMaria AA, Mann JD. Diazepam and dialysis encephalopathy. *Neurology* 1979; **29**:414–15.

Snyder RD. The involuntary movements of chronic mercury poisoning. *Arch Neurol* 1972; **26**:379–81.

Spencer DC, Forno LS. February 2000: Dementia with motor dysfunction in a patient with liver disease. *Brain Pathol* 2000; **10**:315–16.

Spivak JL, Jackson DL. Pellagra: an analysis of 18 patients and a review of the literature. *Johns Hopkins Med J* 1977; **140**:295–309.

Stein JA, Tschudy DP. Acute intermittent porphyria. A clinical and biochemical study of 46 patients. *Medicine* 1970; **49**:1–16.

Stepens A, Logina I, Liguts V *et al.* A parkinsonian syndrome in methcathinone users and the role of manganese. *N Engl J Med* 2008; **358**:1009–17.

Sterns RH, Riggs JE, Schochet SS. Osmotic demyelination syndrome following correction of hyponatremia. *N Engl J Med* 1986; **314**:1535–42.

Stojsavljevic N, Levic Z, Drulovic J *et al.* A 44-month clinical-brain MRI follow-up in a patient with B12 deficiency. *Neurology* 1997; **49**:878–81.

Stone R, Archer JS, Kiernan M. Wernicke's encephalopathy mimicking variant Cretuzfeldt-Jakob disease. *J Clin Neurosci* 2008; **15**: 1308–10.

Stracciari A, Guarino M, Pazzaglia P *et al.* Acquired hepatocerebral degeneration: full recovery after liver transplantation. *J Neurol Neurosurg Psychiatry* 2001; **70**:136–70.

Stracciari A, Mattarozzi K, D'Alessandro R *et al.* Cognitive functioning in chronic acquired hepatocerebral degeneration. *Metab Brain Dis* 2008; **23**:155–60.

Strachan RW, Henderson JG. Dementia and folate deficiency. *Q J Med* 1967; **36**:189–204.

Suarez JI, Cohen ML, Larkin J *et al.* Acute intermittent porphyria: clinicopathologic correlation. Report of a case and review of the literature. *Neurology* 1997; **48**:1678–83.

Sullivan AA, Chervin RD, Albin RL. Parkinsonism after correction of hyponatremia with radiological central pontine myelinolysis and changes in the basal ganglia. *J Clin Neurosci* 2000; **7**:256–9.

Summerskill WHJ, Davidson EA, Sherlock S *et al.* The neuropsychiatric syndrome associated with hepatic cirrhosis and extensive portal collateral circulation. *Q J Med* 1956; **25**:245–66.

Supino-Viterbo V, Sicard C, Risvegliato M *et al.* Toxic encephalopathy due to ingestion of bismuth salts: clinical and EEG studies of 45 patients. *J Neurol Neurosurg Psychiatry* 1977; **40**:748–52.

Sweeney VP, Pathak MA, Asbury AK *et al.* Acute intermittent porphyria. Increased ALA-synthetase activity during an acute attack. *Brain* 1970; **93**:369–80.

Tambyah PA, Ong BKC, Lee KO. Reversible parkinsonism and asymptomatic hypocalcemia with basal ganglia calcification from hypoparathyroidism 26 years after thyroid surgery. *Am J Med* 1993; **94**:444–5.

Tatum WO, Zachariah SB. Gabapentin treatment of seizures in acute intermittent porphyria. *Neurology* 1995; **45**:1216–17.

Thomas PK, Hollinrake K, Lascelles RG *et al.* The polyneuropathy of chronic renal failure. *Brain* 1971; **94**:761–80.

Thompson C, Dent J, Saxby P. Effect of thallium poisoning on intellectual function. *Br J Psychiatry* 1988; **153**:396–9.

Thuluvath PJ, Edwin D, Yue NC *et al.* Increased signals in globus pallidus in T1-weighted magnetic resonance imaging in cirrhotics are not suggestive of chronic hepatic encephalopathy. *Hepatology* 1995; **21**:440–2.

Tokuomi H, Uchino M, Imamura S *et al.* Minamata disease (organic mercury poisoning): neuroradiologic and electrophysiologic studies. *Neurology* 1982; **32**:1369–75.

Torgovnik J, Arsura EL, Lala D. Cytomegalovirus ventriculoencephalitis presenting as Wernicke's encephalopathy-like syndrome. *Neurology* 2000; **55**:1910–13.

Tomita I, Satoh H, Satoh A *et al.* Extrapontine myelinolysis presenting with parkinsonism as a sequel of rapid correction of hyponatremia. *J Neurol Neurosurg Psychiatry* 1997; **62**:422–3.

Trautner RJ, Cummings JL, Read SL *et al.* Idiopathic basal ganglia calcification and organic mood disorders. *Am J Psychiatry* 1988; **145**:350–3.

Traviesa DC. Magnesium deficiency: a possible cause of thiamine refractoriness in Wernicke-Korsakoff encephalopathy. *J Neurol Neurosurg Psychiatry* 1974; **37**:959–62.

Turner MR, Talbot K. Functional B12 deficiency. *Pract Neurol* 2009; **9**:37–41.

Tyler RH. Neurological complications of dialysis, transplantation and other forms of treatment in chronic uremia. *Neurology* 1965; **15**:1081–8.

Tyler RH. Neurologic disorders in renal failure. *Am J Med* 1968; **44**:734–48.

Ueki Y, Isozaki E, Miyazaki Y *et al.* Clinical and neuroradiological improvement in chronic acquired hepatocerebral degeneration after branched-chain amino acid therapy. *Acta Neurol Scand* 2002; **106**:113–16.

Victor M, Yakovlev PI. Korsakoff's psychic disorder in conjunction with peripheral neuritis: a translation of Korsakoff's original article with brief comments on the author and his contribution to clinical medicine. *Neurology* 1955; **5**:394–406.

Victor M, Adams RD, Cole M. The acquired 'non-Wilsonian' type of chronic hepatocerebral degeneration. *Medicine* 1965; **44**:345–96.

Victor M, Adams RD, Collins GH. *The Wernicke–Korsakoff Syndrome.* Oxford: Blackwell, 1971.

Vortmeyer AO, Hagel C, Laas R. Haemorrhagic thiamine deficient encephalopathy following prolonged parenteral nutrition. *J Neurol Neurosurg Psychiatry* 1992; **55**:826–9.

Wainwright AP, Kox WJ, House IM *et al.* Clinical features and therapy of acute thallium poisoning. *Q J Med* 1988; **69**:939–44.

Wallis WE, Willoughby E, Baker P. Coma in the Wernicke–Korsakoff syndrome. *Lancet* 1978; **2**:400–1.

Wallis WE, Donaldson I, Scott RS *et al.* Hypoglycemia masquerading as cerebrovascular disease (hypoglycemic hemiplegia). *Ann Neurol* 1985; **18**:510–12.

Watson JN, Kiloh LG, Osselton JW *et al.* The electroencephalogram in pernicious anemia and subacute combined degeneration of the spinal cord. *Electroencephalogr Clin Neurophysiol* 1954; **6**:45–64.

Weidauer S, Nichtweiss M, Lanfermann H *et al.* Wernicke encephalopathy: MR findings and clinical presentation. *Eur Radiol* 2003; **13**:1001–9.

Weisman DC, Yaari R, Hansn LA *et al.* Density of the brain, decline of the mind. An atypical case of Fahr disease. *Arch Neurol* 2007; **64**:756–7.

Whitfield CL, Ch'ien LT, Whitehead JD. Lead encephalopathy in adults. *Am J Med* 1972; **52**:289–98.

Wijdicks EF, Plevak DJ, Rakela J *et al.* Clinical and radiologic features of cerebral edema in fulminant hepatic failure. *Mayo Clin Proc* 1995; **70**:119–24.

Winkelman MD, Ricanti ES. Dialysis encephalopathy: neuropathologic aspects. *Hum Pathol* 1986; **17**:82–33.

Wright DG, Laureno R, Victor M. Pontine and extrapontine myelinolysis. *Brain* 1979; **102**:361–85.

Wszolek ZK, Baba Y, Mackenzie IR *et al.* Autosomal dominant dystonia-plus with cerebral calcifications. *Neurology* 2006; **67**:620–5.

Wu R-M, Chang Y-C, Chiu H-C. Acute triphenyltin intoxication: a case report. *J Neurol Neurosurg Psychiatry* 1990; **53**:356–7.

Xu Y, Wang Y, Zheng Q *et al.* Clinical manifestations and arsenic methylation after a rare subacute arsenic poisoning accident. *Toxicol Sci* 2008; **103**:278–84.

Yamada M, Ohno S, Okayasu I *et al.* Chronic manganese poisoning: a neuropathological study with determination of managanese distribution in the brain. *Acta Neuropathol* 1986; **70**:273–8.

Yoshida Y, Akanuma J, Tochikubo S *et al.* Slowly progressive dystonia following central pontine and extrapontine myelinolysis. *Intern Med* 2000; **39**:956–60.

Zaki I, Millard L. Pellagra complicated Crohn's disease. *Postgrad Med J* 1995; **71**:496–7.

Zivin I. The neurological and psychiatric aspects of hypoglycemia. *Dis Nerv Syst* 1970; **31**:604–7.

Zuccoli G, Gallucci M, Capellades J *et al.* Wernicke encephalopathy: MR findings at clinical presentation in twenty-six alcoholic and non-alcoholic patients. *AJNR* 2007; **28**:1328–31.

Zuccoli G, Santa Cruz D, Bertolini M *et al.* MR findings in 56 patients with Wernicke encephalopathy: nonalcoholics may differ from alcoholics. *AJNR* 2009; **30**:171–6.

15

Infectious and related disorders

15.1 ACQUIRED IMMUNODEFICIENCY SYNDROME (AIDS)

AIDS may affect the nervous system in a variety of ways. Cerebral involvement may manifest with an early mononucleosis-like syndrome, and, later, with dementia or seizures; there may also be a myelopathy with paraparesis and a peripheral sensorimotor polyneuropathy. Fatigue and depression are also common, and, rarely, mania may be seen.

Clinical features

Although nervous system manifestations, such as dementia, may constitute the presenting symptom of AIDS (Navia and Price 1987), this is distinctly uncommon, and in most cases patients already have other evidence of the illness, such as generalized lymphadenopathy, constitutional symptoms, thrush, diarrhea, shingles, various cytopenias, Kaposi's sarcoma, and *Pneumocystis* pneumonia.

Seroconversion typically occurs within 2–6 weeks of infection, and in the vast majority of cases this is accompanied by a mononucleosis-like syndrome. In some 10 percent of cases, there may also be evidence of nervous system involvement, including aseptic meningitis, cranial nerve palsies (especially affecting the fifth, seventh, and eighth cranial nerves), encephalitis (McArthur 1987; Malouf *et al.* 1990) with delirium and seizures, a transverse myelitis or a peripheral polyneuropathy. Importantly, as noted below, human immunodeficiency virus (HIV) testing may be negative at this point.

Dementia (Navia *et al.* 1986a; Price *et al.* 1988) typically appears first about 10 years after infection, generally only when the CD4 count falls below 200 cells/cm^3, and this is eventually seen in a majority of cases. The onset is typically subacute or even gradual, and initially one sees forgetfulness, poor concentration, slowed thinking, apathy, and social withdrawal. These initial symptoms are also typically accompanied by motor signs, such as clumsiness, decreased fine motor coordination, a degree of dysarthria, tremor, an unsteady gait, and lower extremity weakness, which may be accompanied by hyper-reflexia and Babinski signs. In some cases there may also be primitive reflexes, such as snout or grasp reflexes. Over time, all these symptoms and signs progress, and may be joined by hallucinations and delusions, dysphasia, dyspraxia, and seizures. Terminally (Maher *et al.* 1997), one finds akinetic mutism, myoclonus, and incontinence of bladder and bowel.

Magnetic resonance imaging (MRI) in patients with dementia generally reveals a degree of cortical atrophy and ventricular dilation, and there may also be multiple areas, some confluent, of increased signal intensity in the centrum semiovale on T2-weighted scans (Dooneief *et al.* 1992; Navia *et al.* 1986a), as illustrated in Figure 15.1. The cerebrospinal fluid (CSF) characteristically displays a mild mononuclear pleocytosis, a mildly elevated total protein level, an increased IgG index, and oligoclonal bands. The electroencephalogram (EEG) reveals generalized slowing, typically with a frontal predominance (Gabuzda *et al.* 1988).

Before leaving this discussion of dementia in AIDS, some mention must be made of the unsettled nature of the nomenclature found in the literature regarding this subject (Antinori *et al.* 2007). Thus, early on, when cognitive

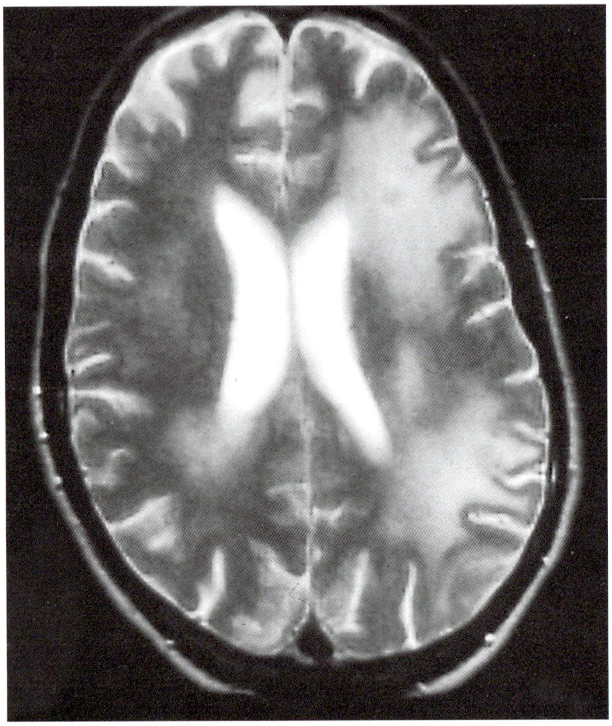

Figure 15.1 This T2-weighted magnetic resonance imaging scan shows both patchy and confluent areas of increased signal intensity in the centrum semiovale of a patient with AIDS dementia. Reproduced from Gillespie and Jackson (2000), with permission.

impairment is mild, terms such as 'HIV-associated mild cognitive disorder' and 'minor cognitive–motor disorder' have been used, depending, respectively, on whether motor signs are absent or present. When cognitive symptoms worsen to the point at which activities of daily living are clearly compromised, terms such as 'HIV-associated dementia' (or 'HIV dementia') and 'HIV-associated dementia complex' (or 'AIDS dementia complex') have been used, again depending on whether motor signs are absent or present respectively. Whether these distinctions are useful to the clinician is not clear: it may be just as appropriate to say that patients with mild cognitive symptoms are probably in a prodrome to a dementia, and then follow them up to see whether the cognitive impairment worsens.

Among other symptoms seen in AIDS, fatigue is very common, and may be debilitating. Seizures may, as noted, be seen in the context of dementia; isolated seizures, including grand mal, complex partial, and simple partial types, may also occur in a minority of cases (Pascual-Sedano *et al.* 1999; Wong *et al.* 1990), and indeed may at times constitute the presenting feature of AIDS (Holtzman *et al.* 1989). Mood disorders are common: depression occurs in roughly one-third of all cases (Bing *et al.* 2001), while mania, often marked by irritability (Kieburtz *et al.* 1991), is seen in less than one-tenth (Lyketsos *et al.* 1993). Psychosis is seen in only a few percent of cases (de Ronchi *et al.* 2000), but may, albeit rarely, constitute the presenting feature of AIDS (Harris *et al.* 1991).

The spinal cord may be affected by a vacuolar myelopathy (Navia *et al.* 1986b; Sharer *et al.* 1986), with paraparesis and sensory ataxia, and within the peripheral nervous system there may be a sensorimotor polyneuropathy (de la Monte *et al.* 1988; Morgello *et al.* 2004), which is often quite painful.

Serologic testing generally becomes positive within 2–12 weeks of the initial infection and is inevitably positive by 6 months; thus, in patients with AIDS dementia, the enzyme-linked immunosorbent assay (ELISA) test will be positive, as will the confirmatory Western blot test.

Course

In the natural course of events, AIDS dementia is relentlessly progressive, and death usually occurs within 3–6 months (Bouwman *et al.* 1998).

Etiology

HIV is an RNA virus found in blood, semen, vaginal fluid, breast milk, and colostrum, and may be spread via all these fluids. In the United States, spread occurs most commonly by homosexual contact, in particular anal intercourse; however, as the virus spreads among females, it is likely that the most common mode of transmission in the United States will eventually become heterosexual contact, as is the case in Africa. With improved screening of blood products, transfusion-related transmission is now becoming quite rare, but blood-borne transmission remains a significant problem among intravenous drug users who fail to sterilize their needles. Although the virus is also found in saliva, urine, and tears, there is as yet no convincing evidence that it can be spread by these.

Subsequent to infection there is an intense viremia followed by a vigorous cellular and humoral immune response, such that, in most cases, the viremia is substantially contained within about 3 months. The virus, however, is not eradicated but rather continues to reproduce within lymphoid tissue. Tragically, the cell most likely to be infected is the CD4+ T-lymphocyte, with the result that, over many years, there is a gradual loss of these 'helper' lymphocytes and the body's defenses are subverted to the point at which both a significant viremia again occurs and the patient becomes vulnerable to opportunistic infections.

In cases characterized by dementia, macroscopic findings include cortical atrophy, diffuse myelin pallor, and ventricular dilation. Microscopically, one finds an 'HIV encephalitis' (Brew *et al.* 1995; Budka *et al.* 1987; Petito *et al.* 1986), with multinucleated giant cells and microglial nodules scattered throughout the gray and white matter, and an 'HIV leukoencephalopathy' (Budka *et al.* 1987; Petito *et al.* 1986), with perivenular demyelination and astrocytosis, associated, in some cases, with vacuolization. There is also neuronal loss in the frontal (Everall *et al.* 1991; Wiley *et al.* 1991), occipital (Everall *et al.* 1993) and parietal (Everall *et al.* 1993) cortices, and, in remaining neurons, one finds dendritic 'simplification', with reduced dendritic branching (Masliah *et al.* 1997). Although, in all likelihood, all of these findings are related to the occurrence of dementia, there is at best only a rough correlation here, with, however, one exception, namely dendritic simplification, which does appear to be strongly associated with cognitive decline (Masliah *et al.* 1997).

The mechanisms underlying these neuropathologic changes are not clearly understood. To begin with, it is not even clear how HIV gains entry into the central nervous system: one theory holds that free virus gains entry across a disrupted blood–brain barrier at the level of the endothelium; another, known as the 'Trojan horse' theory, suggests that entry of the virus occurs as infected peripheral macrophages normally travel across the blood–brain barrier, after which they deliver their deadly cargo into the interstitial fluid. Once across the blood–brain barrier, infected macrophages coalesce into multinucleated giant cells, and infected microglia coalesce into microglial nodules. Of note, it appears that neither oligodendrocytes nor neurons are actually infected, but rather undergo damage via a variety of indirect mechanisms, including injury from cytokines released by infected macrophages (Glass *et al.* 1995), and toxicity of proteins, such as the gp120 protein, found on the surface of the virus. With oligodendrocyte loss, the leukoencephalopathy appears, and when neurons are touched, there may be both apoptosis and dendritic simplification.

Further, the mechanism or mechanisms underlying some of the other symptoms described earlier are also not fully understood. Seizures, presumably, are related to HIV encephalitis involving the cortex (Holtzman *et al.* 1989). Depression, in many cases, if not the majority, may occur secondary to major depressive disorder precipitated by the stress associated with AIDS. Mania, however, at least when occurring late in the course of AIDS (Lyketsos *et al.* 1993), may, in some as yet poorly understood fashion, occur on the basis of HIV encephalitis or leukoencephalopathy. Psychosis, especially when occurring in association with some cognitive impairment, may also occur directly as a result of HIV encephalitis or leukoencephalopathy (de Ronchi *et al.* 2000; Sewell *et al.* 1994a).

Differential diagnosis

Opportunistic infections of the central nervous system are very common in AIDS, and loom large in the differential diagnosis for seizures and dementia. Thus, seizures may occur with herpes simplex viral encephalitis, varicella-zoster encephalitis, fungal infections, progressive multifocal leukoencephalopathy, cytomegalovirus encephalitis, toxoplasmosis, tuberculosis, and syphilis. Dementia may occur with progressive multifocal leukoencephalopathy, cytomegalovirus encephalitis, toxoplasmosis, tuberculosis, and, of course, syphilis. Syphilis may also cause depression, mania, and psychosis, and it is important to keep in mind that syphilis, when occurring in the context of AIDS, may progress very rapidly. Although each of these disorders, as discussed in the relevant chapter, has certain clinical features which point to the correct diagnosis, it is appropriate, whenever the CD4 count is below 200 cells/cm^3 and the patient has a seizure or a dementia, to carry out, at a minimum, MRI scanning, as this will greatly facilitate the differential diagnosis. In doubtful cases, lumbar puncture should also be considered, keeping in mind that polymerase chain reaction (PCR) assays are available, with the exceptions of fungal infections and syphilis, for all of the opportunistic infections.

In addition to opportunistic infections, patients with AIDS are also at risk for primary central nervous system lymphoma and metastatic Kaposi's sarcoma, both of which, as mass lesions, are capable of causing seizures.

The toxicity of anti-retroviral drugs may also assume a role in the differential diagnosis. For example, abacavir and zidovudine may each cause seizures and cognitive deficits, and efavirenz may cause cognitive deficits or hallucinations.

For unclear reasons, vitamin B12 deficiency is not uncommon in patients with AIDS (Herzlich and Schiano 1993), and as it may produce a dementia or a spinal cord syndrome similar to that of vacuolar myelopathy it should always be checked for (Beach *et al.* 1992).

Patients with HIV infection are also likely to suffer from drug addiction (e.g. to stimulants or cocaine) or alcoholism, and these disorders, as discussed in Chapters 28.1, 28.2, and 28.5, respectively, may be associated with seizures, depression, mania, and psychosis.

Patients infected with HIV may also suffer from pre-existing major depressive disorder, bipolar disorder, or schizophrenia; however, these diagnoses are readily made, provided, of course, that one takes the proper amount of time to obtain an accurate history.

Before leaving this discussion of differential diagnosis, mention should also be made of the occurrence of delirium in patients with AIDS. Although delirium is very common, especially in advanced cases, it does not appear to occur on the basis of the neuropathologic changes described in the section on etiology. Thus, when delirium does occur, the differential diagnosis discussed in Chapter 6.5 must be pursued. In this regard, certain possibilities deserve special attention. First, all of the opportunistic infections, with the exception of syphilis, may cause delirium. Second, patients with AIDS are susceptible not only to *Pneumocystis* pneumonia but also to multiple other infections, and systemic effects of infection must always be taken into account.

Treatment

The advent of highly active anti-retroviral treatment (HAART) has revolutionized the treatment of AIDS, and may lead to stabilization or improvement of its central nervous system manifestations. Of note, however, in profoundly immunocompromised patients, initiation of treatment with a combination of anti-retrovirals may, in a very small percentage of cases, be followed, within weeks to months, by a neurologic immune reconstitution inflammatory syndrome ('NeuroIRIS') (McCombe *et al.* 2009), with either *de novo* signs and symptoms or a worsening of pre-existing symptomatology which had originally occurred secondary to an opportunistic infection.

Given the complex and rapidly evolving nature of antiretroviral treatment, referral to a specialist is essential. Symptomatic treatment of the various neuropsychiatric manifestations of AIDS is discussed below, and all of the agents mentioned, except as noted otherwise, were found effective in placebo-controlled, double-blinded studies.

Dementia may be approached as discussed in Chapter 6.3, and delirium as discussed in Chapter 6.5. If antipsychotics are required, low doses are in order, given that patients with AIDS seem particularly likely to develop extrapyramidal side-effects (Hriso *et al.* 1991). Of note, there has been one double-blinded study regarding the treatment of delirium, which found that low doses of either chlorpromazine or haloperidol were both effective and safe; lorazepam, however, was not effective and was associated with limiting side-effects (Breitbart *et al.* 1996).

Fatigue may be treated with a variety of agents, including modafinil (Rabkin *et al.* 2010), methylphenidate (Breitbart *et al.* 2001), dextroamphetamine (Wagner and Rabkin 2000), and testosterone cypionate (Rabkin *et al.* 2004): of all these agents, modafinil is the most benign, and should probably be tried first.

Seizures may be treated with anti-epileptic drugs, as discussed in Chapter 8.3: of the agents noted there, levetriracetam, although not specifically studied in AIDS, is, given its relatively benign side-effect profile and relative lack of drug–drug interactions, a reasonable first choice.

Depression may respond to fluoxetine (Rabkin *et al.* 1999). Other antidepressants are also used extensively, and in this regard, though not subjected to double-blind testing, mirtazapine should be considered, and for two reasons. First, it tends to boost appetite, and thus may be particularly helpful in patients with wasting. Second, as discussed in Chapter 15.3, there is suggestive evidence that mirtazapine may be beneficial in progressive multifocal leukoencephalopathy.

Mania may be treated as discussed in Chapter 7.3; although there are no blinded studies in this regard, divalproex is a reasonable choice; lithium, given the electrolyte disturbances common in AIDS patients, should probably be avoided.

Psychosis may be treated as discussed in Chapter 8.1. There has been one single-blinded study (Sewell *et al.* 1994b), done in 1994, which compared mesoridazine with haloperidol: both drugs were found to be modestly effective, but haloperidol, as might have been predicted, caused extrapyramidal side-effects in all of the participants. Neither of these agents is in common use today, and, in the authors' opinion, it is perhaps more appropriate to begin with a second-generation agent, such as risperidone or quetiapine.

Painful peripheral neuropathy may respond to gabapentin (Hahn *et al.* 2004), lamotrigine (Simpson *et al.* 2003), or smoked cannabis (Abrams *et al.* 2007), but amitriptyline does not appear to be effective (Kieburtz *et al.* 1998).

15.2 CYTOMEGALOVIRUS ENCEPHALITIS

Roughly 80 percent of adults demonstrate serologic evidence of prior infection with cytomegalovirus (CMV), and, typically during profound immunosuppression, this virus may reactivate, and undergo hematogenous spread to various organs, including the brain, where it may cause one of two syndromes: CMV encephalitis with, characteristically, a dementia, and CMV ventriculoencephalitis with a delirium. Amongst patients with CMV encephalitis (Arribas *et al.* 1996), 85 percent have AIDS, 12 percent have other causes of immunosuppresion, and 3 percent are apparently healthy. Looking further at patients who do have AIDS, although some 12 percent, at autopsy, were found to have the neuropathological features of CMV encephalitis (Arribas *et al.* 1996), only a miniority, some 2–4 percent, have symptoms attributable to this.

Clinical features

CMV encephalitis (Arribas *et al.* 1996) typically occurs in AIDS only when the CD4+ cell count falls below 100 cells/cm^3. The onset is generally subacute, and patients characteristically develop a dementia, marked by poor memory, apathy, and withdrawal; seizures may also occur. In aggressive cases, a delirium may appear. Although MRI is typically normal (Clifford *et al.* 1996), in a minority one may find, on T2-weighted or FLAIR imaging, patchy areas of increased signal intensity in the centrum semiovale.

CMV ventriculoencephalitis (Torgovnik *et al.* 2000; Kalayjian *et al.* 1993) generally occurs as a terminal event in far advanced AIDS cases, typically when the CD4+ count falls below 50 cells/cm^3. The presentation is acute, with delirium often accompanied by nystagmus, ophthalmoparesis, and ataxia. MRI scanning, on T2-weighted or FLAIR imaging, reveals increased signal intensity in a periventricular location which may be accompanied by gadolinium enhancement.

The CSF may be normal or may display a pleocytosis and an elevated total protein; in CMV ventriculoencephalitis the glucose level may be reduced (Kalayjian *et al.* 1993). Polymerase chain reaction assay for CMV, though highly specific, is positive in only roughly three-quarters of cases.

In most cases, immunocompromised patients with CMV encephalitis or ventriculoencephalitis will also have evidence of systemic infection, such as retinitis, colitis, and, importantly, adrenalitis, which may cause adrenocortical insufficiency.

In addition to infecting the brain, CMV may also infect the cord or spinal roots or peripheral nerves, causing, respectively, a myelitis, polyradiculopathy, and a mononeuritis multiplex.

Course

In the natural course of events, CMV encephalitis proceeds to death within months, while the more aggressive CMV ventriculoencephalitis leads to death in weeks.

Etiology

Cytomegalovirus encephalitis is characterized by widespread microglial nodules and inclusion-positive cytomegalic cells: although these may be found in the white matter, they are more prominent in the cerebral cortex, basal ganglia, and thalamus (Arribas *et al.* 1996; Morgello *et al.* 1987). Cytomegalovirus ventriculoencephalitis, in turn, is characterized by patchy periventricular necrosis with abundant macrophages (Arribas *et al.* 1996; Kalayjian *et al.* 1993).

Differential diagnosis

In AIDS patients the main item on the differential for CMV encephalitis is AIDS dementia itself, and on clinical grounds the differentiation between the two may not be possible. One clue, however, is the presence of evidence of systemic CMV infection, such as retinitis, colitis, or adrenalitis. CMV ventriculoencephalitis may be confused clinically with Wernicke's encephalopathy, but the MRI findings in CMV ventriculoencephalitis will make the diagnosis.

Treatment

The general treatment of dementia and delirium is discussed in Chapters 6.5 and 6.3, respectively. Ganciclovir represents a first-line treatment; foscarnet is an alternative, and in resistant cases these may be used together. In those who recover, long-term preventive treatment is required.

15.3 PROGRESSIVE MULTIFOCAL LEUKOENCEPHALOPATHY

Progressive multifocal leukoencephalopathy (PML) occurs secondary to an opportunistic infection of the central nervous system by the JC virus (Padgett *et al.* 1971). Multifocal areas of demyelination eventually appear, producing various focal signs and, in some, a personality change, dementia, or delirium. In almost all cases patients have depressed cell-mediated immunity, most commonly due to AIDS, wherein 2–5 percent of patients are afflicted.

Clinical features

The onset is typically subacute, generally over approximately 2 weeks. In most cases, patients present with a slowly progressive focal sign, such as hemianopia, aphasia, apraxia, hemisensory loss, or hemiplegia (Astrom *et al.* 1958; Krupp *et al.* 1985; Richardson 1961); rarer focal findings such as Balint's syndrome (Ayuso-Peralta *et al.* 1994) have also been reported. Over time, these initially unilateral deficits become bilateral, and many patients then go on to develop a personality change, a dementia or, rarely, a delirium. Seizures may occur in up to 20 percent of patients and may be simple partial, complex partial, or grand mal (Lima *et al.* 2006; Moulignier *et al.* 1995). Rarely, progressive multifocal leukoencephalopathy may present with a personality change (Astrom *et al.* 1958), dementia (Sellal *et al.* 1996; Zunt *et al.* 1997), or delirium (Davies *et al.* 1973). Cerebellar signs may occur but are not common (Parr *et al.* 1979). Other rare signs include quadriparesis secondary to brainstem involvement (Kastrup *et al.* 2002), and dystonia (Factor *et al.* 2003), chorea (Piccolo *et al.* 1999), or parkinsonism (Bhatia *et al.* 1996).

Magnetic resonance scanning (Guilleux *et al.* 1986; Kuker *et al.* 2006) will reveal one or more focal, well-demarcated lesions, generally in the subcortical white matter. On T1-weighted imaging these display decreased signal intensity, and on FLAIR or T2-weighted imaging, increased signal intensity is seen. Although typically there is no enhancement, in a small minority this may occur on the margins of the lesions: though this may be seen in non-HIV-infected patients (Huang *et al.* 2007), most cases are seen as part of the immune reconstitution inflammatory syndrome during treatment with HAART, as discussed further below.

The EEG may show diffuse or focal slowing.

The CSF, though characteristically normal, may occasionally reveal a mild lymphocytic pleocytosis or mild elevation in total protein. The PCR assay for JC virus is generally, but not always, positive (Hensen *et al.* 1991; Koralnik *et al.* 1999). Serum testing for antibodies to JC virus is not helpful, given, as noted below, that most adults will be positive. In doubtful cases brain biopsy may be required.

Course

In the natural course of events the disease is generally relentlessly progressive, with an overall mortality of roughly 80 percent at 9 months; rarely the course may stretch out for years, and even more rarely there may be spontaneous remissions (Price *et al.* 1983).

Etiology

Approximately 80 percent of the adult population of the United States has latent infection with the JC virus, with the virus lying dormant in bone marrow and kidney. In a very small minority of patients with profoundly depressed cell-mediated immunity, this virus reactivates and spreads to the brain. Although by far the most common cause of immunoincompetence in these patients is AIDS, PML has also been noted in patients with Hodgkin's disease, other lymphomas, leukemia, other cancers, tuberculosis, sarcoidosis, and systemic lupus erythematosus, and in patients undergoing therapeutic immunosuppression after transplantation. Progressive multifocal leukoencephalopathy may also, albeit rarely, occur in patients treated with natalizumab (Yousry *et al.* 2006; Linda *et al.* 2009) and rituximab (Carson *et al.* 2009).

Although the vast majority of cases occur in severely immunocompromised patients, there are case reports of PML occurring in patients with only minimal immunosuppression (e.g. those with alcoholic cirrhosis) (Gheuens *et al.* 2010), or, even more rarely in patients who, as best can be determined, are immunocompetent (Fermaglich *et al.* 1970; Guillaume *et al.* 2000).

Within the central nervous system, oligodendrocytes and, to a lesser degree, astrocytes are infected by the JC virus.

With destruction of oligodendrocytes, demyelination, with only relative axonal sparing, occurs, and foci of demyelination begin to appear. Within and surrounding these foci there is a variable, and typically quite slight, degree of inflammation. At least initially, these foci are generally few in number and confined to one hemisphere, most commonly in the subcortical white matter and the centrum semiovale. Over time, the foci of demyelination enlarge and become confluent, and foci begin to appear in the opposite hemisphere. In addition to cerebral involvement, foci may also occur in the brainstem, the cerebellum, and, very rarely, the cord. Although, as noted earlier, PML is primarily a disease of white matter, it has become apparent that foci of demyelination also occur in the cerebral cortex (Moll *et al.* 2008).

Differential diagnosis

In patients with AIDS, consideration must be given to AIDS dementia and primary central nervous system lymphoma. AIDS dementia is distinguished by the general lack of focal signs, and also by MRI findings: in contrast to the asymmetric, generally well-demarcated lesions seen in PML, in AIDS dementia lesions are typically symmetric and poorly demarcated. Lymphoma is distinguished primarily by MRI findings. As noted earlier, PML lesions seen on MRI generally do not enhance and, furthermore, they are not associated with perilesional edema: CNS lymphoma lesions, however, generally do demonstrate these features.

Treatment

Although there is no specific antiviral treatment directed at the JC virus, some interesting work has been done suggesting a measure whereby the ingress of the virus into oligodendrocytes may be prevented or diminished. It appears that the serotonin 5HT2A receptor mediates entry of the virus into the cell (Elphick *et al.* 2004); this receptor, in turn, is blocked by mirtazapine, and there are case reports of progressive multifocal leukoencephalopathy going into partial remission secondary to treatment with this agent (Cettomai and McArthur 2009). Although, clearly, controlled studies are needed, an argument may be made for treatment with mirtazapine, and not only on this basis: mirtazapine not only combats reduced appetite, but also relieves depression, and both these conditions are common in patients with AIDS.

In patients with AIDS, HAART, if not already in place, should be instituted, as it dramatically lengthens survival (Antinori *et al.* 2003). As noted earlier, however, in some cases HAART may be associated with the appearance of lesion enhancement. This reflects the occurrence of substantial inflammation at the margins of the lesion, and is generally considered to be part of the immune reconstitution inflammatory syndrome (IRIS) (Tan *et al.* 2009): in such cases there may be clinical worsening, and some have advocated temporary use of steroids.

In patients treated with immunosuppressants, natalizumab or rituximab, treatment, given the lethality of PML, should probably be discontinued.

Symptomatic treatment of seizures, dementia, delirium and personality change is discussed, respectively, in Chapters 8.3, 6.1, 6.5, and 8.2.

15.4 ARBOVIRUS MENINGOENCEPHALITIS

Viruses transmitted by arthropods are known as arboviruses, a term derived from the fact that they are all *ar*thropod *bo*rne. In North America, there are seven arboviruses known to cause meningoencephalitis, including six mosquito-borne viruses [eastern equine (Deresiewicz *et al.* 1997), West Nile (Davis *et al.* 2006; Debiasi and Tyler 2006), St. Louis (Wasay *et al.* 2000), western equine, Venezuelan, and La Crosse (McJunkin *et al.* 2001)] and one tick-borne virus (Powassan virus). Most cases occur in the late summer and early fall, when mosquitoes are most active. Mention should also be made of Japanese encephalitis, another mosquito-borne virus, which, although not endemic in North America, is a very common cause of meningoencephalitis in the Far East (Lewis *et al.* 1947; Solomon *et al.* 2000, 2002) and may be contracted by travelers there. Although most of these viruses tend to cause encephalitis in older adults, two of them, namely La Crosse and Japanese virus, are more likely to cause disease in children.

Clinical features

Although fulminant onsets, over hours, can occur, typically the onset is subacute, spanning days. Patients present with delirium, headache, and fever, often accompanied by meningeal signs, such as photophobia and stiff neck. Seizures and focal signs, such as hemiparesis or aphasia, may occur, and in a minority there may be hyponatremia secondary to a syndrome of inappropriate ADH (SIADH) secretion. With progression, stupor or coma may supervene, and this is particularly common with eastern equine, Japanese, and West Nile infections.

Abnormal movements may also be seen: dystonia (Kalita and Misra 2000) and parkinsonism (Pradhan *et al.* 1999) are characteristic of Japanese encephalitis, and parkinsonism may also, though less commonly, be seen in West Nile encephalitis; tremor is very characteristic of St. Louis encephalitis (Wasay *et al.* 2000). Furthermore, an acute flaccid paralysis, reminiscent of that seen in polio, may occur, and this is most characteristic of West Nile infection.

The peripheral white blood cell count may be normal or increased, and this is particularly the case in eastern equine encephalitis, where white blood cell counts up to $35\,000/cm^3$ may be seen.

Magnetic resonance imaging may be normal early on; however, eventually either T2-weighted or fluid-attenuated inversion recovery (FLAIR) imaging will reveal focal areas of increased signal intensity in the cerebral white or gray matter, and, particularly in Japanese and West Nile infections, the basal ganglia, thalamus, or upper brainstem.

The EEG typically demonstrates generalized slowing, with or without focal accentuation, and interictal epileptiform discharges are common; in some case periodic lateralized epileptiform discharges may also occur.

The most important test to perform is a lumbar puncture. The opening pressure may be increased, especially in severe cases. The cell count is typically increased, and the height of the increase parallels the overall severity of the illness: in most cases it is under 500 cells/cm^3; however, in eastern equine encephalitis the count may be as high as 3000, and in Japanese and West Nile cases up to 1000 or more. Initially, in severe cases, there may be a polymorphonuclear predominance; however, eventually lymphocytes become the most prominent type. In all cases the total protein is increased and the glucose is normal. Testing for virus-specific IgM antibodies and performing a PCR assay, again for all potentially responsible viruses, is appropriate in all instances.

Course

The mortality rate varies widely: for eastern equine encephalitis it is well over 50 percent; for Japanese encephalitis it approaches 50 percent; for West Nile, St. Louis, western equine, and Powassan encephalitis it is in the range of 10 percent; and for Venezuelan and La Crosse encephalitis it is less than 1 percent. For those who survive, the encephalitis tends to run its course within a matter of a few weeks, sometimes longer.

Once recovered from the acute illness, patients may be left with various sequelae, and both the frequency with which these occur and their severity parallel the overall mortality characteristic of each of the responsible viruses. Thus, one-third or more of those who survive eastern equine encephalitis will have sequelae, while sequelae are rare after cases of Venezuelan encephalitis. These sequelae, in turn, mirror the symptoms seen during the acute illness (Herzon *et al.* 1957; Przelomski *et al.* 1988; Smardel *et al.* 1958), and include dementia of variable severity, epilepsy, focal deficits, such as hemiplegia or aphasia, and a movement disorder: both dystonia and parkinsonism are characteristic sequelae of Japanese encephalitis; however, parkinsonism has also been noted with other viruses, including western equine virus (Mulder *et al.* 1951; Schultz *et al.* 1977). Of note, some cases of dementia may also be characterized by prominent psychotic symptoms, as has been reported after Japanese encephalitis (Richter and Shimojyo 1961).

Etiology

After the mosquito or tick bite, hematogenous spread carries the virus to the brain. Although the severity of the pathologic changes varies widely depending on the responsible virus, in general one finds widespread perivascular inflammation and areas of focal cerebritis in the cerebral cortex and white matter, the basal ganglia, thalamus, and brainstem. At times, thrombus formation may occur in the involved vessels with subsequent infarction (Leech and Harris 1977; Reyes *et al.* 1981).

Differential diagnosis

Arboviral encephalitis is but one of the many causes of acute encephalitis, and the reader is directed to Chapter 8.8, where this is discussed.

Treatment

Aggressive supportive care is required, with special attention to hyponatremia and to the development of increased intracranial pressure and herniation. Some authors advocate prophylactic use of an anti-epileptic drug, such as fosphenytoin or levetiracetam, while others hold anti-epileptic treatment in reserve until a seizure occurs. The general treatment of delirium is discussed in Chapter 6.3.

A vaccine is available for Japanese encephalitis, and travelers may wish to consider this.

15.5 HERPES SIMPLEX ENCEPHALITIS

Although herpes simplex encephalitis accounts for only about one-tenth of all cases of acute encephalitis, it is by far the most common cause of sporadic encephalitis and, given its treatability, it should be strongly considered in the differential of any adult suspected of having acute encephalitis.

Clinical features

Herpes simplex encephalitis can occur at any age but most patients are middle-aged or older (Koskiniemi *et al.* 1996).

The onset itself generally spans several days; however, the range is wide, from explosive onsets over several hours to gradual ones lasting weeks or more. In some cases the onset may be preceded be a prodrome, lasting several days, of malaise, headache, irritability, and mild fever.

Typically (Kennedy 1988; Marton *et al.* 1996; McGrath *et al.* 1997; Skoldenberg *et al.* 1984; Whitley *et al.* 1986; Williams and Lerner 1978), patients present with fever, headache, and delirium. In this setting a majority of patients will also develop focal signs, such as hemiparesis or aphasia, and approximately two-thirds will have seizures, which may be either complex partial or grand mal in type. Although meningeal signs, such as a stiff neck or photophobia, are common, they are generally not at all severe. Notably, bizarre behavior is common and, although this usually occurs in the context of the delirium, there are rare reports of the encephalitis presenting with either mania (Fisher 1996) or psychosis (Wilson 1976). Untreated, coma develops in about one-third of patients.

Computed tomography (CT) may show radiolucent areas in the medial aspects of one or both temporal lobes, but it may be normal for up to a week. Magnetic resonance imaging is much more sensitive, showing increased signal intensity on FLAIR or T2-weighted scans in the medial temporal lobe (as illustrated in Figure 15.2), although this too may be normal for the first few days; rarely, rather than temporal lobe involvement, there may only be increased signal intensity in the frontal lobes (Taylor *et al.* 2007). Diffusion-weighted imaging may show increased signal intensity in the same areas, and this abnormality may be evident earlier than that seen on T2-weighted or fluid-attenuated FLAIR imaging (Hatipoglu *et al.* 2008). Gadolinium enhancement may or may not be present.

The EEG (Upton and Gumpert 1970) is usually normal for the first few days but will eventually show theta or delta slowing in one temporal area, which may be accompanied by periodic complexes; with progression of the disease, bilateral involvement may be seen.

Although the cerebrospinal fluid may be normal very early on, abnormalities eventually occur in all cases. The cell count is increased, in the range of from 10 to 1000 cells/cm^3: although early on there may be some polymorphonuclear cells, for the most part there is a lymphocytic predominance; in some cases red blood cells may also be present. The total protein is increased, and although the glucose is typically normal, in a small minority it may be reduced. Polymerase chain reaction assay for herpes simplex virus (HSV) (Aurelius *et al.* 1991) displays a sensitivity of 97 percent, with a specificity of 100 percent, and is the diagnostic procedure of choice.

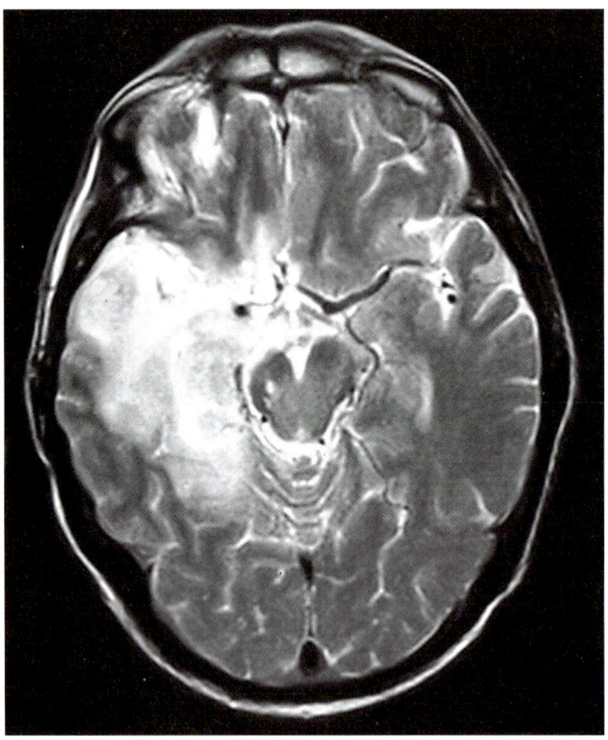

Figure 15.2 This T2-weighted magnetic resonance imaging scan shows increased signal intensity in the left temporal lobe (especially its medial aspect) of a patient with herpes simplex encephalitis. Reproduced from Gillespie and Jackson (2000), with permission.

If lumbar puncture is not possible, brain biopsy may be necessary to make a definitive diagnosis; however, as noted below, treatment should never wait upon such a procedure.

Course

Untreated, over 50 percent of patients will die in days or a few weeks. In those who survive, the clinical picture usually stabilizes in a matter of weeks; exceptionally the acute illness may be prolonged for months (Sage *et al.* 1985) or even longer (Yamada *et al.* 2003).

Although most patients have only one episode, recurrences may appear in a small minority at any time from 1 to 4 months after resolution of the initial encephalitis (Skoldenberg *et al.* 2006).

Among those who do survive, the vast majority will be left with significant sequelae (McGrath *et al.* 1997), most commonly a chronic amnestic syndrome (Hokkanen *et al.* 1996; Kapur *et al.* 1994; Young *et al.* 1992), dementia (Hokkanen *et al.* 1996; McGrath *et al.* 1997), or a personality change (Caparros-Lefebvre *et al.* 1996). Other sequelae include focal signs (e.g. aphasia or hemiplegia) (McGrath *et al.* 1997) or epilepsy (McGrath *et al.* 1997); rarely there may be a Kluver–Bucy syndrome (Greenwood *et al.* 1983; Lilly *et al.* 1983; Marlowe *et al.* 1975). There is also a report of a case in which complex motor and vocal tics occurred as a sequela (Northam and Singer 1991).

Etiology

There are two types of herpes simplex virus, namely types 1 and 2. Type 1 is responsible for the familiar 'cold sore' of herpes labialis and type 2 is responsible for genital herpes. Herpes simplex encephalitis is almost always due to type 1; although type 2 can cause encephalitis, this is seen in less than 2 percent of cases (Dennett *et al.* 1997).

Although it is not entirely clear how the virus gains entry into the central nervous system, one theory suggests that spread occurs from the trigeminal ganglion. Over 90 percent of the adult population show serologic evidence of past infection with type 1 herpes simplex, and it is known that this virus can remain latent in the trigeminal ganglion (Baringer and Swoveland 1973). Thus, virus could emerge in the ganglion either due to reactivation or, probably less commonly, due to a primary herpes labialis, with retrograde spread from the lips along the mandibular or maxillary divisions of the trigeminal nerve to the ganglion. Once present within the ganglion, the virus, regardless of how it got there, can then access the central nervous system via one of two routes. One route involves anterograde spread along the ophthalmic division to the olfactory mucosa, whereby the virus then gains access to the olfactory fila, after which it undergoes retrograde transport through the olfactory nerve to the medial aspect of the temporal lobe. The other proposed route involves anterograde transport along the mandibular and maxillary divisions to the sensory fibers supplying the dura of the anterior and middle cranial fossae, from which spread to the adjacent frontal and temporal lobes is possible.

Pathologically (Adams and Miller 1973; Esiri 1982) there is intense inflammation (which may progress to hemorrhagic necrosis) affecting initially the medial portions of the temporal lobe, with, in most cases, eventual spread to other areas, including the lateral aspects of the temporal lobe, the insula, inferior portions of the frontal lobes, and the cingulate cortex. Although involvement is typically unilateral early on, with time both temporal lobes become involved. There may be substantial edema, and both uncal and subfalcine herniation may occur. In those who survive, scarring, cavitation, and cystic change is seen in the involved areas.

As noted earlier, in a small minority of cases, recurrences may occur; however, these recurrences, rather than being due to reinfection, appear to be immunologically mediated: in such cases PCR assay for HSV is negative (Dennett *et al.* 1996), and intrathecal antibodies to HSV are absent (Dennett *et al.* 1996).

Differential diagnosis

A delirium accompanied by headache and fever immediately suggests the diagnosis of an acute encephalitis, and, as discussed in Chapter 8.8, consideration must be given not only to herpes simplex encephalitis but also to various other encephalitides, for example arboviral infections. Magnetic resonance imaging is helpful here as medial temporal lobe involvement strongly suggests herpes simplex encephalitis; however, as noted above, MRI scans may be normal early on, and consequently the diagnosis may be revealed only by PCR assay of the CSF. This takes time, however, and given the prominent mortality and devastating sequelae of herpes simplex encephalitis, and its ready treatability with acyclovir, most clinicians will presumptively diagnose Herpes simplex and institute treatment: should the diagnosis prove incorrect, the acyclovir may simply be stopped.

In those rare cases in which the presentation is with mania or psychosis, the differential is wide, as discussed in chapters 7.3 and 8.1; however, the eventual development of fever, headache, and delirium will settle the issue.

Treatment

In addition to aggressive supportive care, patients should be treated with acyclovir in a dose of 10 mg/kg intravenously every 8 hours (Skoldenberg *et al.* 1984; Whitley *et al.* 1986) for a course of 2–3 weeks. Although a retrospective study suggested that concurrent treatment with dexamethasone or methylprednisolone enhanced recovery (Kamei *et al.* 2005), this is controversial, and until the results of an ongoing double-blinded study are confirmatory, this practice cannot be recommended. Seizures may be treated with a standard anti-epileptic drug, such as fosphenytoin or levetiracetam, and it is prudent to continue treatment with an anti-epileptic until the patient has been seizure free for at least a year. The general treatment of delirium is further discussed in Chapter 6.5.

15.6 ENCEPHALITIS LETHARGICA

Encephalitis lethargica, also known as von Economo's disease or European sleeping sickness, was first described by Baron Constantin von Economo at a meeting of the Vienna Psychiatric Society in April 1917 (Dickman 2001; Wilkins and Brody 1968). This disease swept the world in an epidemic that lasted from 1917 to 1928 and although there have been no further epidemics, sporadic cases still occur (Dale *et al.* 2004; Howard and Lees 1987), making a familiarity with this disease of more than academic interest.

Clinical features

Clinically (Hohman 1921; Kirby and Davis 1921), most cases of acute encephalitis lethargica presented acutely or subacutely in childhood, adolescence, or early adult years with headache and fever, and then could assume one of four, more or less distinct, forms. The classic form, also known as the somnolent–ophthalmoplegic form, was characterized by lethargy and somnolence (which could progress to stupor or coma), third or sixth cranial nerve palsies, and delirium. The parkinsonian form was marked by a more or less profound parkinsonism, often accompanied by a degree of somnolence and also by third or sixth cranial nerve palsies; features of stuporous catatonia, such as waxy flexibility, were also at times in evidence. Next was the hyperkinetic form, marked by delirium, sleep reversal, and various hyperkinesias, including wild, flailing movements, oculogyric crises, chorea, or tics. Finally, there was the psychotic form, wherein patients presented with mania (Kirby and Davis 1921; Meninger 1926; Sands 1928), psychosis (Kirby and Davis 1921; Meninger 1926; Sands 1928), or stuporous catatonia (Bond 1920; Shill and Stacy 2000).

Magnetic resonance imaging may be normal or, on T2-weighted or FLAIR sequences, may reveal increased signal intensity in the basal ganglia, and, less frequently, the mesencephalon or thalamus (Dale *et al.* 2004); in some cases subtle, transient gadolinium enhancement may occur in the cerebral cortex (Dale *et al.* 2009).

The CSF may be normal or may reveal a lymphocytic pleocytosis and an elevated total protein; oligoclonal bands may also be present (Dale *et al.* 2004; Howard and Lees 1987).

Course

The mortality rate for acute encephalitis lethargica was about 25 percent; among those who survived, the encephalitis gradually cleared over about a month. Of great importance, neuropsychiatric sequelae, most notably postencephalitic parkinsonism, occurred in a majority of cases.

Postencephalitic parkinsonism occurred in over 50 percent of survivors after a latent interval of from 1 to over 20 years (Duvoisin and Yahr 1965). Patients gradually developed a syndrome similar to that seen in idiopathic Parkinson's disease (Rail *et al.* 1981).

Oculogyric crises (Taylor and McDonald 1928) constitute another classic sequela, and although these typically occurred in the context of postencephalitic parkinsonism, they could appear independently (McCowan *et al.* 1928). Of note, these crises were often accompanied by classic obsessions or compulsions (Jelliffee 1929), and also, at times, by palilalia (Van Bogaert 1934) or excitation (McCowan *et al.* 1928). Additional abnormal involuntary movements could appear, including dystonia or blepharospasm (Alpers and Patten 1927).

In roughly one-third of children, a syndrome reminiscent of attention-deficit–hyperactivity syndrome occurred, with hyperactivity and impulsivity (Fairweather 1947; Hohman 1922); these children also tended to be emotionally labile and disinhibited (Fairweather 1947).

In a small minority, psychosis, marked by delusions, or mania, could also occur (Fairweather 1947).

Other sequelae included dementia and a syndrome similar to narcolepsy, with both narcoleptic and cataplectic attacks (Adie 1926; Fournier and Helguera 1934).

Etiology

Autopsies of those dying in the acute stage revealed inflammation with a perivascular accumulation of lymphocytes and plasma cells in the midbrain, basal ganglia, and cortex (Buzzard and Greenfield 1919; Howard and Lees 1987). In those with sequelae, autopsy studies (Geddes *et al.* 1993; Ishii and Nakamura 1981; Rail *et al.* 1981) have revealed neuronal loss, gliosis, and, in the remaining neurons, neurofibrillary tangles similar to those seen in Alzheimer's disease, in the substantia nigra, locus ceruleus, basal ganglia, thalamus, hippocampus, and cerebral cortex.

Given that the pandemic of encephalitis lethargica coincided with the Spanish influenza epidemic, it was long suspected that encephalitis lethargica simply represented an influenza viral encephalitis. As attractive as this theory was, however, recent work, utilizing archived tissue specimens, has failed to demonstrate any evidence of direct viral infection of the brain (Lo *et al.* 2003; McCall *et al.* 2001). Consequently, attention has focused on the possibility that encephalitis lethargica may rather occur secondary to an autoimmune onslaught directed at the various structures noted earlier. Such an onslaught, in turn, could occur secondary to a mechanism similar to that seen in limbic encephalitis, or, alternatively, could be triggered by a preceding infection, similar to the case with Sydenham's chorea. In this regard, one study found anti-NMDA receptor antibodies in half of a series of recent cases (Dale *et al.* 2009), while another found evidence of a preceding streptococcal pharyngitis in 65 percent (Dale *et al.* 2004). It seems doubtful, however, that these two autoimmune mechanisms, though perhaps explaining the small collection of recent cases, could account for the epidemic that occurred in the early twentieth century that affected tens of thousands: it may well be that the Spanish influenza virus present at the time was in fact etiologic, not directly, but perhaps by acting as a trigger itself for an autoimmune response.

Differential diagnosis

Encephalitis lethargica must be distinguished from acute encephalitis, discussed in Chapter 8.8, with special attention to Japanese viral encephalitis (discussed in Chapter 15.4), which can present with parkinsonism, and herpes simplex encephalitis (discussed in Chapter 15.5), which can present with stuporous catatonia, mania, or psychosis.

With regard to sequelae, the diagnosis is fairly straightforward when they are present soon after resolution of the acute syndrome. Difficulty, however, may arise when, as in the case of postencephalitic parkinsonism, there is a prolonged latent interval between the encephalitis and the onset of the parkinsonism. In such cases, the presence of oculogyric crises is an important clue.

Treatment

Given the probable autoimmune etiology, a strong case may be made for treatment with corticosteroids in a fashion similar to that used for limbic encephalitis or Sydenham's chorea, and indeed case reports (Blunt *et al.* 1997) and case series (Dale *et al.* 2009) report good results with such measures. Postencephalitic parkinsonism may respond to levodopa, and other sequelae may be treated as outlined in the chapters for those syndromes. Whether the syndrome reminiscent of attention-deficit–hyperactivity syndrome would respond to stimulants is an open question.

15.7 INFECTIOUS MONONUCLEOSIS

Infectious mononucleosis occurs secondary to infection by Epstein–Barr virus (EBV), and, in up to a little over 5 percent of cases of infectious mononucleosis (Gautier-Smith 1965), the central or peripheral nervous system may be involved. Although most cases occur in children or adolescents, mononucleosis may also occur in adults (Fujimoto *et al.* 2003).

Clinical features

Typically there is a prodrome, lasting 1–2 weeks, of headache, malaise and fatigue, after which one sees the development of the classic syndrome of pharyngitis, fever, and cervical lymphadenopathy. Hepatosplenomegaly may also occur and, rarely, splenic rupture.

Although nervous system involvement generally develops in the context of this classic clinical picture, at times it may be the presenting feature of the disease. Encephalitis may occur, with delirium (Bergin 1960; Schlesinger and Crelinsten 1977; Schnell *et al.* 1966; Tselis *et al.* 1997), focal signs, and seizures (Doja *et al.* 2006; Freedman *et al.* 1953; Silverstein *et al.* 1972) including status epilepticus (Russell *et al.* 1985). Rare presentations of this encephalitis include an amnestic syndrome, which was also accompanied by seizures (Pommer *et al.* 1983), or a psychosis (Jarvis *et al.* 1990). Other manifestations of nervous system involvement include an aseptic meningitis, cranial nerve palsies, cerebellar ataxia, transverse myelitis, and a motor polyneuropathy.

In most cases, there is a peripheral leukocytosis with a lymphocytic predominance and atypical lymphocytes: the 'monospot' test will be positive in close to 95 percent of cases by the third week of illness. Should the diagnosis remain in doubt, one may also test for IgM antibodies to anti viral capsid antigen (VCA), which will be present in virtually 100 percent of all cases.

In delirious patients, T2-weighted or FLAIR MRI scanning may be normal or may reveal multiple areas of increased signal intensity in the cortex, white matter, basal ganglia, and thalamus, and the EEG may show generalized slowing (Doja *et al.* 2006), with, rarely, periodic complexes (Greenberg *et al.* 1982).

In encephalitic cases, the cerebrospinal fluid will be characterized by a mild lymphocytic pleocytosis, a mildly elevated total protein, a normal glucose, and, in most cases, a positive PCR assay for EBV (Weinberg *et al.* 2002).

Course

Symptoms typically resolve after about a month. Although most patients recover completely, a minority will be left with persistent fatigue (Petersen *et al.* 2006), sleepiness (Guilleminault and Mondini 1986), or depression (Cadie *et al.* 1976).

Etiology

Infectious mononucleosis is caused by the EBV and is transmitted primarily via oral secretions passed during intimate contact, such as kissing. Lymphocytes become infected and carry the virus to various organs, including the brain. Scattered autopsy reports of encephalitic cases have revealed a generalized lymphocytic and monocytic cerebritis (Roulet Perez *et al.* 1993; Sworn and Urich 1970).

Differential diagnosis

In encephalitic cases the differential diagnosis includes other acute encephalitides, as discussed in Chapter 8.8, and in this regard special attention should be paid to the possibility of the seroconversion syndrome characteristic of early HIV infection, as discussed in Chapter 15.1; consideration must also be given to acute disseminated encephalomyelitis (ADEM), as discussed in Chapter 15.11.

Treatment

As there is no specific treatment, supportive measures must suffice, and the general treatment of delirium is discussed in Chapter 6.5.

15.8 MUMPS

Although over 50 percent of patients with mumps will have a CSF pleocytosis (Russell and Donald 1958), only a minority will have clinical evidence of central nervous system involvement, with perhaps 10 percent developing a meningitis and less than 1 percent an encephalitis. Currently, thanks to widespread vaccination, mumps is quite rare in developed countries; although overall mumps is equally common in males and females, males are far more likely to develop symptomatology reflecting infection of the central nervous system.

Clinical features

Mumps typically occurs in children or adolescents, generally during the winter or spring, and presents with a prodrome of fever, myalgia, and malaise, followed within 1–7 days by the typical parotitis; males may also develop a unilateral or bilateral orchitis.

Meningitis manifests with headache, drowsiness, and a stiff neck, and typically follows the parotitis after a latency of 2–20 days; occasionally the meningitis will precede the parotitis (Levitt *et al.* 1970) and in some cases parotitis may be absent.

Encephalitis presents with an increasing fever and either delirium or stupor; seizures, ataxia, and various focal signs may also occur (Azimi *et al.* 1969; Bistrian *et al.* 1972; Finklestein 1938; Koskiniemi *et al.* 1983; Levitt *et al.* 1970).

Serum anti-mumps IgM antibody is present acutely, and acute and convalescent serum IgG antibody displays a fourfold or greater rise. The CSF typically demonstrates a lymphocytic pleocytosis, an elevated total protein and anti-mumps IgG; PCR assay may also be positive for mumps RNA.

Course

Overall, mumps usually runs its course in 3–4 weeks, and meningitis or encephalitis typically resolve after a week or two. Very rarely, the encephalitis may be chronic (Vaheri *et al.* 1982).

Although most recover from meningitis or encephalitis without sequelae, some postencephalitic patients may be left with chronic cognitive deficits, ataxia, seizures or deafness; rarely hydrocephalus, secondary to aqueductal stenosis (Thompson 1979), may occur.

Etiology

The mumps virus is spread via the respiratory route and is trophic for salivary glands, gonads, and the nervous system. Within the meninges and cerebrum there may be widespread perivascular infiltration with lymphocytes and mononuclear cells.

Differential diagnosis

Mumps encephalitis must be distinguished from other acute viral encephalitides, as discussed in Chapter 8.8; the occurrence of parotitis or orchitis constitute obvious clues. Importantly, a history of mumps or mumps vaccination effectively rules out the diagnosis, given that both of these generally confer lifelong immunity.

Treatment

There is no specific treatment for mumps; the general treatment of delirium is discussed in Chapter 6.5.

15.9 VARICELLA-ZOSTER

Infection with the varicella-zoster virus (VZV) may cause two different illnesses (Strauss *et al.* 1984). Acute infection causes chickenpox, also known as varicella. After resolution of chickenpox, the virus undergoes latency in the sensory ganglia of the cord and also in the Gasserian ganglion of the trigeminal nerve and the geniculate ganglion of the facial nerve (Mahalingam *et al.* 1990). In the aged or immunocompromised, the virus may reactivate to cause zoster, which in turn may manifest with a number of different syndromes, including the following: herpes zoster (also known as shingles) with or without a myelitis; zoster ophthalmicus, with or without an associated vasculopathy and stroke; zoster oticus (also known as the Ramsay Hunt syndrome); or a generalized zosteriform rash, with or without an encephalitis. Each of these syndromes is discussed in turn, followed by a discussion of a diagnostically challenging variant, known as zoster sine herpete.

Clinical features

Chickenpox is a disease of childhood or adolescence that is spread via respiratory droplets. The pharynx is the initial seat of the infection; subsequently a viremia occurs, resulting in a characteristic rash. Rarely, the central nervous system may be affected during the viremia, producing an encephalitis (with various symptoms, including delirium [Applebaum *et al.* 1953]), a cerebellitis (with ataxia [Johnson and Milbourn 1970]) or a meningitis.

Reactivation of the latent varicella-zoster virus may occur during the immune 'senescence' that occurs with aging (Miller 1980), during certain illnesses, such as AIDS (De La Blanchardiere *et al.* 2000; Gilden *et al.* 1988), Hodgkin's disease (McCormick *et al.* 1969) or other cancers, or during treatment with corticosteroids or various immunosuppressants, and the zoster syndrome that results is determined by the site of the reactivation.

With reactivation of virus latent within the sensory ganglia of the cord, anterograde spread of the virus down the sensory nerve eventually results in the classic dermatomal rash of herpes zoster. Typically there is a prodrome of malaise, lasting several days, followed by the occurrence of dermatomal pain; over the next 3 or 4 days, a typical vesicular rash follows. In almost all cases the rash is unilateral and typically occurs in one of the thoracic dermatomes. Resolution occurs gradually after a matter of weeks. In a very small minority of cases, the virus may spread retrogradely to the cord with the delayed and gradual appearance of myelitic symptoms (Devinsky *et al.* 1991; Hogan and Krigman 1973); involvement of the dorsal horn may lead to a segmental sensory loss whereas involvement of the ventral horn may lead to segmental weakness and amyotrophy. Occasionally the white matter of the cord may be involved and in such cases there may be a transverse myelitis or, rarely, a Brown-Séquard syndrome.

When virus latent within the Gasserian ganglion of the trigeminal nerve undergoes reactivation, anterograde spread leads to a zosteriform rash within the area of distribution of one of the divisions of the trigeminal nerve, almost always the first or ophthalmic division. The resulting herpes zoster ophthalmicus may disastrously also involve the eye. Patients who have suffered a herpes zoster ophthalmicus are also at risk for a rare complication, namely a zoster vasculopathy (Nagel *et al.* 2008). In such cases, anterograde spread of the virus occurs along the ophthalmic division to the meninges, with subsequent infection of both small, penetrating arteries and the middle and anterior cerebral arteries. Subsequently there is a latent interval, averaging 4 months (with a wide range from as little as 2 day up to 2½ years), after which infarction and stroke occur: with small-vessel involvement, typical lacunar syndromes appear, while with involvement of the middle or anterior cerebral arteries one may find either large, territorial infarctions or, alternatively, smaller infarctions in the areas of distribution of branches of these vessels. Although classically one finds a focal deficit, such as hemiparesis, contralateral to the side of the preceding zoster ophthalmicus (Hilt *et al.* 1983), this scenario, in fact, is the exception rather than the rule, and indeed one typically finds in such cases multiple, bilateral infarctions secondary to both small- and large-vessel disease (Nagel *et al.* 2008). Magnetic resonance imaging will reveal the infarctions, and the CSF will display any one or more of the following abnormalities: a mild lymphocytic pleocytosis, an elevated total protein, a positive PCR assay for viral DNA, or an elevated anti-VZV IgG level: importantly, an elevated IgG level is the most sensitive finding here, and indeed may be the *only* abnormality (Nagel *et al.* 2008); consequently it should be sought even when the fluid is acellular, the total protein is normal, and the PCR assay is negative. Segmental narrowing of the middle or cerebral arteries may be demonstrated with magnetic resonance or computed tomographic angiography, but these tests may yield normal results; conventional angiography, though far more sensitive, may also be unrevealing (Nagel *et al.* 2008).

In cases in which the geniculate ganglion is the site of reactivation, zoster oticus, (also known as the Ramsay Hunt syndrome) may occur, with a rash in the external auditory canal or the pinna, and a peripheral seventh cranial nerve palsy (Sweeney and Gilden 2001).

Rarely, and generally only in immunocompromised or severely debilitated patients, there may be a hematogenous dissemination of the reactivated virus with the development of a severe, generalized zosteriform rash, and in such cases hematogenous spread of the virus may also occur to the brain, causing an encephalitis (Applebaum *et al.* 1962; Jemsek *et al.* 1983; Krumholz and Luhan 1945). Typically in such cases, within days to weeks (or exceptionally months; Weaver *et al.* 1999) of the onset of the generalized rash, a delirium develops that may be accompanied by seizures, focal signs, ataxia, or meningismus. In such cases the EEG shows generalized slowing, and FLAIR or T2-weighted MRI will reveal multifocal areas of increased signal intensity within both the white matter and the cortex. The CSF is typically, but not always, abnormal, with a lymphocytic pleocytosis, an elevated total protein, and either a positive PCR assay for viral DNA or an elevated VZV IgG antibody titer. Although the encephalitis usually resolves in a matter of weeks, chronic courses may be seen, especially in severely immunocompromised patients, such as those with AIDS.

Before leaving this section on clinical features, note must also be made of a variant of zoster known as zoster sine herpete (Gilden *et al.* 1994). In such cases there is no rash, and hence one may have dermatomal pain, spinal cord symptoms, a vasculopathy with stroke, a facial nerve palsy, or an encephalitis with delirium without this all-important diagnostic clue (Mayo and Booss 1989).

Course

The course of zoster is dependent on the immunological competence of the patient. In immunocompetent patients all symptoms, with the exception, of course, of those due to infarction, generally resolve within a matter of weeks; in the immunocompromised, however, protracted, and at times fatal, courses may be seen.

One complication of a zoster rash, whether dermatomal or cranial, is postherpetic neuralgia, in which the pain persists beyond resolution of the rash. This neuralgic pain may be lancinating or burning in quality and may cause considerable disability; although in some cases a gradual remission may occur after many months, in others the pain proves to be chronic.

Etiology

As indicated above, zoster may be complicated by a myelitis, a cerebral vasculopathy, or an encephalitis. The myelitis results from retrograde spread from the spinal sensory ganglia. In all likelihood, as noted above, the vasculopathy occurs secondary to infection of the meninges, followed by infection of small penetrating arteries or the middle or anterior cerebral arteries: in such cases, viral particles have been found in vessel walls (Eidelberg *et al.* 1986; Linnemann and Alvira 1980; Melanson *et al.* 1996). Presumably thrombus formation occurs at the site of the infection, followed either by occlusion at that site or by artery-to-artery embolization. As noted earlier, the encephalitis occurs secondary to hematogenous seeding of the cerebrum, generally occurring only during a disseminated rash. In such cases one may find multifocal areas of cerebritis (Weaver *et al.* 1999), primarily affecting the white matter, with, at times, multiple areas of microinfarction occurring secondary to small-vessel vasculitis (Kleinschmidt-DeMasters *et al.* 1996).

Differential diagnosis

The occurrence of a myelopathy, stroke, or encephalitic symptoms (e.g. delirium, seizures, etc.) in the setting of a recent rash should immediately suggest a VZV infection. Difficulties arise, clearly, in cases of the zoster sine herpete variant, and in such cases one would have to closely question the patient and family members regarding any preceding dermatomal or facial pain. Difficulties may also arise in the case of stroke, given the long latency between the rash and the onset of the vasculopathy, and again, a careful history is most helpful.

Treatment

Immunocompetent patients with shingles (when not complicated by myelitic symptoms), zoster ophthalmicus (when not complicated by ocular involvement), and zoster oticus may be treated with oral acyclovir, valacyclovir, or famciclovir. For immunocompromised patients with these syndromes, and in all other cases, intravenous acyclovir is indicated. For patients with vasculopathy, in addition to intravenous acyclovir, some authorities also advocate concurrent use of steroids (Nagel *et al.* 2008), and a similar case could be made for cases of encephalitis occurring in the context of generalized zoster. Furthermore, in cases of vasculopathy, consideration may also be given to the administration of anti-platelet agents.

With regard to postherpetic neuralgia, a number of agents have been shown to be effective in double-blinded studies, including nortriptyline, amitriptyline, gabapentin, pregabalin, and divalproex. Importantly, with regard to nortriptyline and amitriptyline, patients get relief regardless of whether they are depressed or not. Nortriptyline and amitriptyline are equally effective, but nortriptyline is better tolerated (Watson *et al.* 1998); nortriptyline may be given in doses ranging from 10 to 75 mg, and amitriptyline in doses ranging from 10 to 100 mg. Gabapentin is roughly equivalent in effectiveness to nortriptyline but better tolerated (Chandra *et al.* 2006); it may be given in doses ranging from 1200 to 3600 mg. Pregabalin may be given in doses ranging from 150 to 600 mg (Dworkin *et al.* 2003), and divalproex in a dose of 1000 mg (Kochar *et al.* 2005); how these two agents might compare with each other or with amitriptyline, nortriptyline, or gabapentin is not clear. Based on these studies, and on the overall tolerability of each of these agents, it would seem reasonable to begin with a trial of gabapentin. Regardless, however, of which agent is used, treatment should last at least a matter of weeks before assessing the results.

15.10 RABIES

In almost all cases, rabies occurs secondary to the bite of a rabid animal, such as a dog, cat, wolf, skunk, raccoon, or bat:

in the United States, bats constitute the main vector (Noah *et al.* 1998). Rarely, cases have occurred secondary to respiratory transmission (as in spelunkers exploring bat-infested caves or in a veterinarian working with the homogenized brain of an infected animal; Conomy *et al.* 1977), or after transplantation of solid organs (Srinivasan *et al.* 2005) or corneas from infected donors (Javadi *et al.* 1996). Although very rare in developed countries, rabies continues to be a significant problem in India and parts of Asia.

Clinical features

After being bitten by an infected animal there is a latent interval of anywhere from a few weeks to a year or more, averaging 1–3 months, before symptoms appear: the duration of this latent interval appears to be related not only to the severity of the bite but also to the distance from the bitten area to the brain, with bites on the feet having the longest latency and facial bites the shortest. Importantly, however, in the majority of cases, it appears that patients do not recall the bite (Noah *et al.* 1998).

With resolution of the latent interval, there is usually a prodrome, lasting days, characterized by headache and malaise; pain or paresthesiae may also develop at the site of the original bite. Subsequently, the illness may evolve in one of two forms, namely 'furious' (or 'encephalitic') rabies or 'dumb' (or 'paralytic') rabies.

Furious rabies (Adle-Biassette *et al.* 1996; Blatt *et al.* 1938; Dupont and Earle 1965) is seen in perhaps 80 percent of cases and is characterized by restlessness, agitation, excitability, excessive startlability, hypersalivation, pharyngeal spasm, delirium, and convulsions: the delirium may be marked by bizarre behavior, including biting. The combination of hypersalivation and pharyngeal spasm may cause the patient to literally foam at the mouth, and the prospect of a worsening of the pharyngeal spasm upon drinking may prompt the patient to avoid taking fluids, thus giving rise to the classic symptom of hydrophobia.

Dumb rabies (Chopra *et al.* 1980) is characterized by a flaccid paralysis that typically begins in one limb and rapidly becomes generalized and symmetric; in such cases delirium may also occur, but is a late finding.

T2-weighted or FLAIR MRI scanning may reveal multiple foci of increased signal intensity in the gray and white matter of the cerebrum and cerebellum.

The CSF demonstrates a mild lymphocytic pleocytosis and elevated total protein; anti-rabies IgG antibodies may be found, and PCR assay may reveal rabies virus RNA.

Polymerase chain reaction assay of a skin biopsy from the back of the neck at the hairline and of saliva (testing three or more samples) is indicated in all cases, as these tests are virtually 100 percent sensitive and specific.

Course

Although survival has been reported (Porras *et al.* 1976; Willoughby *et al.* 2005), this is exceptional and the overwhelming majority of patients die within days to a couple of weeks.

Etiology

In the typical case of rabies secondary to a bite by a rabid animal, the virus slowly travels up a peripheral nerve to gain entry at the cord, after which there is rapid spread, both along axons and contiguously among adjacent neurons. Within the cord, brainstem, cerebellum, and cerebrum there are perivascular infiltrates and widespread, diffuse neuronal involvement, with particular involvement of hippocampal pyramidal cells and cerebellar Purkinje cells. Within surviving neurons cytoplasmic eosinophilic inclusions, known as Negri bodies, are found (Chopra *et al.* 1980; Dupont and Earle 1965). Once having gained entry into the central nervous system, the virus also spreads centrifugally, notably to lacrimal and salivary glands and to the skin.

Differential diagnosis

Lacking the history of exposure, the differential in the case of furious rabies includes other causes of acute encephalitis, as discussed in Chapter 8.8: certain features, however, may facilitate the diagnosis, including focal pain or paresthesia during the prodrome, hypersalivation, and, of course, hydrophobia. In cases of dumb rabies, consideration must be given to Guillain–Barré syndrome.

Treatment

Although there are case reports (Willoughby *et al.* 2005) of survival with a regimen of ribavirin, ketamine, amantadine, and midazolam-induced coma, whether such survival occurred secondary to treatment or rather represented one of the rare instances of spontaneous recovery is not certain. In any case, supportive care, at least initially, is appropriate, followed, if required, by a palliative approach. Universal precautions, with special attention to contact with saliva, are required.

Given the lack of any recognized treatment, the focus must be on prevention. In cases of animal bites, the wound should be thoroughly cleansed, followed by injection of rabies immune globulin and vaccination with human diploid cell anti-rabies vaccine.

15.11 ACUTE DISSEMINATED ENCEPHALOMYELITIS

Acute disseminated encephalomyelitis (ADEM) is a typically monophasic illness that occurs as a result of an autoimmune assault on the brain or cord, primarily involving the white matter, which is triggered by a preceding infectious illness or a vaccination. This disease was formerly called postinfectious encephalomyelitis (PIE) or, if it occurred after a vaccination, postvaccinal encephalomyelitis, and these names, although no longer current, have much to recommend them, as they refer not only to the clinical picture, namely an encephalomyelitis, but also to the etiology, namely a reaction to a preceding infection or vaccination. By contrast, the term acute disseminated encephalomyelitis encompasses only the clinical picture and, as such, it may, in the minds of some authors, perhaps inappropriately broaden the concept to include encephalomyelitides that occur on the basis of other causes.

Responsible infections (Fisher *et al.* 1983; Hart and Earle 1975; Moscovich *et al.* 1995; Paskavitz *et al.* 1995) include measles, mumps, rubella, pertussis, chickenpox, infectious mononucleosis, influenza, arboviruses, herpes simplex virus, HIV, hepatitis, scarlet fever, mycoplasma, *Borrelia*, typhoid, and, most commonly, non-specific viral upper respiratory infections. Vaccinations associated with ADEM include those for measles, mumps, rubella, chickenpox, influenza, rabies, hepatitis, typhoid, tetanus, and human papillomavirus (Wildemann *et al.* 2009). Both smallpox and smallpox vaccination (Dolgopol *et al.* 1955) were once very common causes, but with this disease now apparently eradicated they are no longer of concern.

Acute disseminated encephalomyelitis is an uncommon disease, occurring in fewer than 1:100 000 of the general population. Although most commonly seen in children, it may also occur in adolescents and adults, and, rarely, in the elderly.

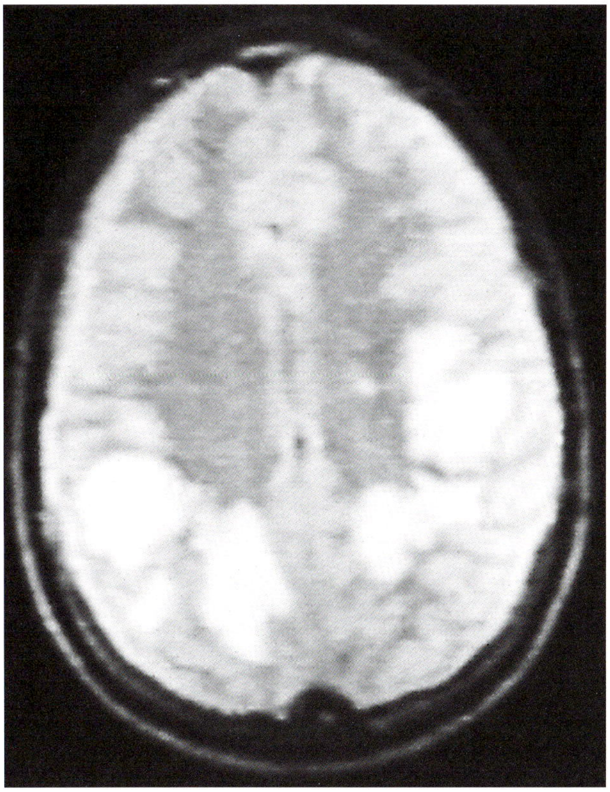

Figure 15.3 Multiple bilateral areas of increased signal intensity on a T2-weighted magnetic resonance imaging scan of a patient with acute disseminated encephalomyelitis. Reproduced from Marsden and Fowler (1998), with permission.

Clinical features

Between the responsible vaccination or infection and the onset of ADEM, there is a latent interval of variable duration, from as little as a few days to as much as a few weeks. In cases following vaccination, this latent interval is quite clear-cut, as the vaccination itself generally produces little in the way of symptoms. In cases following infection, however, this latent interval may not be clearly delineated, and indeed in many cases the symptoms of the inciting infection may not in fact have cleared before the symptoms of ADEM set in, and in such cases there is actually no 'symptom-free' interval at all.

Clinically (Hollinger *et al.* 2002; Hynson *et al.* 2001; Marchioni *et al.* 2005; Miller *et al.* 1956; Panicker *et al.* 2010; Tenembaum *et al.* 2002), the onset is generally acute, with the full picture evolving over 4–5 days. The symptomatology may be quite varied, depending on which part of the central nervous system bears the brunt of the autoimmune assault. Thus, with cerebral involvement, there may be headache, delirium, seizures, and various focal signs, such as hemiparesis; meningismus may be present, but is usually mild, and there may be fever, but again this too is generally mild; rarely, mania may occur (Moscovich *et al.* 1995; Paskavitz *et al.* 1995). In some cases marked by cerebral involvement, stupor and coma may supervene. Cerebellar involvement may cause an ataxia, and with spinal involvement there may be a transverse myelitis with paraplegia or quadriplegia. The cranial nerves (most commonly the optic nerve) may be affected, and patients may develop visual obscurations or blindness. Rarely the basal ganglia are involved with chorea or dystonia.

Magnetic resonance imaging may be normal for the first day or two; however, eventually T2-weighted or FLAIR imaging will, in cases marked by cerebral symptomatology, reveal multiple areas of increased signal intensity in the cerebral white matter (Kesserling *et al.* 1990), as illustrated in Figure 15.3: these are typically fairly large, poorly demarcated and, though bilateral, are generally asymmetric; they are most commonly found in the frontal and parietal lobes. Other structures may also be involved, including the cerebellar white matter, the brainstem, and the cord. Importantly, although these MRI findings eventually regress, they may still be evident for months following clinical improvement (O'Riordan *et al.* 1999).

The EEG, in cerebral cases, typically displays generalized slowing with or without focal accentuation.

The CSF is generally, but not always, abnormal. Findings may include a lymphocytic pleocytosis, an elevated total protein, an elevated myelin basic protein, and oligoclonal bands; the glucose level is normal. In severe cases, there may be both polymorphonuclear and red blood cells.

Course

With cerebral involvement, death may occur in 10 to 30 percent of cases; in those with cerebral involvement who do survive, and in those with cerebellar and cord involvement, symptoms gradually undergo resolution over 2–6 weeks, and in the vast majority of cases there is complete, or near complete, recovery (Hynson *et al.* 2001; Panicker *et al.* 2010).

As noted earlier, ADEM is typically a 'monophasic' illness that, in the natural course of events, does not recur unless, of course, the unfortunate patient is again exposed to one of the precipitating infections or vaccinations.

Etiology

As noted earlier, ADEM occurs secondary to an autoimmune attack on the brain or cord. Pathologically (Greenfield 1929; Hart and Earle 1975), there are multiple foci of perivenous mononuclear inflammation, surrounded by demyelination with relative axonal sparing: although these are most common in the white matter of the cerebrum, cerebellum, and cord, the gray matter may also be involved. In severe cases, known as acute hemorrhagic leukoencephalitis (Hart and Earle 1975; Russell 1955), widespread petechial hemorrhages are also seen.

Differential diagnosis

The chief items on the differential are acute viral encephalitis and multiple sclerosis.

Acute viral encephalitis, discussed in Chapter 8.8, may, during the first day or two of the illness, be indistinguishable from the cerebral form of ADEM. The eventual appearance of the typical findings on MRI scanning, however, make the diagnosis of ADEM, as these are not found in cases of viral encephalitis. The existence of a preceding illness, with a latent interval is, of course, very helpful; however, non-specific viral upper respiratory infections are very common, and viral encephalitis itself may be preceded by a prodrome with fever and malaise, and hence not too much weight can be placed on this. Naturally, a history of recent vaccination would, however, make the diagnosis of ADEM extremely likely.

The differentiation between multiple sclerosis and ADEM is discussed in Chapter 19.1.

Treatment

Methylprednisolone should be immediately started in a dose, for adults, of 250 mg intravenously every 6 hours. Although the optimum duration of treatment is not clear, it would be prudent to continue steroids for 2 weeks or until the patient has recovered, whichever comes later. The response to treatment is generally quite good; in treatment-resistant cases, consideration should be given to adding intravenous immunoglobulins or to plasmapheresis.

15.12 SUBACUTE SCLEROSING PANENCEPHALITIS

Subacute sclerosing panencephalitis (SSPE), also known as Dawson's disease, results from a reactivation of a defective measles virus and is typically characterized by dementia, myoclonus, and ataxia. This is a very rare disorder, occurring in 5–10 out of 1 000 000 with a history of measles and in approximately 1 out of 1 000 000 who receive the measles vaccine; it is three to four times more common in males than in females.

Clinical features

Over 90 percent of cases occur in childhood or adolescence; in these cases the average latency between the preceding infection or vaccination is about 8 years, and the average age of onset is about 13 years. In adults, the latency ranges from 8 to 33 years and the average age of onset is in the early twenties.

The onset itself is generally gradual, even insidious, and typically the disease evolves through three stages (Dawson 1934; Ozturk *et al.* 2002; Prashanth *et al.* 2006). In the first stage, the patient may become restless, distractable, and forgetful, and irritability and moodiness may be noted. In some cases, particularly in those with adult onset, this first stage may be characterized by a psychosis (Cape *et al.* 1973; Salib 1988) that may faithfully mimic schizophrenia, with delusions (including Schneiderian first-rank symptoms; Duncalf *et al.* 1989) and stuporous catatonia (Koehler and Jakumeit 1976). In the second stage a dementia evolves, which is accompanied by generalized myoclonus, ataxia, and seizures, which may be partial or grand mal in type; abnormal involuntary movements, such as chorea, athetosis, or dystonia, are occasionally seen at this point. In the third and final stage there is stupor and generalized rigidity, and eventually coma. It must be borne in mind that there are many exceptions to this typical picture (Risk and Haddad 1979): in some cases the onset may be relatively fulminant (Souraud *et al.* 2009), with severe symptomatology seen in a matter of months, and, furthermore, there may be considerable overlap among the various stages.

Magnetic resonance imaging may reveal cortical atrophy and, on T2-weighted scans, multiple areas of increased signal intensity in a periventricular and subcortical white matter distribution (Alkan *et al.* 2006).

The EEG in the second and third stages typically displays high-voltage periodic complexes (Praveen-Kumar *et al.* 2007), which coincide with the myoclonus; terminally a burst-suppression pattern may appear (Wulf 1982).

The CSF is generally acellular with a normal or mildly elevated total protein. The IgG level is increased, sometimes greatly so, oligoclonal bands may be found, and, most importantly, anti-measles antibodies are present.

Course

Although the course is variable, most patients pass through the three stages and die within 1–3 years; in fulminant cases, however, a fatal outcome may occur within several months. Occasionally the course is prolonged, up to a decade or more, and, rarely, there may at times be periods of relative stability (Risk *et al.* 1978), lasting many years (Cobb *et al.* 1984;

Donner *et al.* 1972; Landau and Luse 1958), which, however, inevitably give way to further progression.

Etiology

Subacute sclerosing panencephalitis occurs secondary to a reactivation of a 'dormant' measles virus that has the peculiarity of lacking a normal M protein; rather than undergoing 'budding', these defective viruses spread by means of cell fusion within the central nervous system. Pathologically there is widespread perivascular inflammation accompanied by gliosis, patchy demyelination, and neuronal loss (Ohya *et al.* 1974). Within surviving neurons inclusions are found within nuclei, which, by electron microscopy, appear similar to measles nucleocapsids.

Differential diagnosis

Subacute measles encephalitis, like SSPE, also occurs as a sequela to measles; however, in subacute measles encephalitis the latent interval between the preceding measles and the onset of the illness is much shorter than in subacute sclerosing panencephalitis, being measured in months and not years. In adult-onset cases, consideration may be given to other dementias associated with myoclonus, such as Creutzfeldt–Jakob disease.

Treatment

Various treatments have been advocated. However, in the only randomized study to date it appears that oral inosiplex was equivalent to a combination of oral inosiplex and intraventricular interferon-alpha (Gascon *et al.* 2003); the benefit, however, was not great, and the long-term outcome of treatment with inosiplex is poor (Eroglu *et al.* 2008). The general treatment of dementia is discussed in Chapter 6.3; in cases with prominent myoclonus, consideration may be given to either carbamazepine (Yigit and Sarikaya 2006), or to levetiracetam, which, in one case report, not only relieved the myoclonus but also led to a generalized clinical improvement (Becker *et al.* 2009).

15.13 SUBACUTE MEASLES ENCEPHALITIS

Subacute measles encephalitis, also known as measles inclusion body encephalitis, is a rare disorder occurring generally, but not always, in immunocompromised patients who have recently recovered from the measles. Although, given the epidemiology of measles, most cases occur in children, adult-onset cases have been reported (Croxson *et al.* 2002).

Clinical features

Clinically (Agamanolis *et al.* 1979; Aicardi *et al.* 1977), after an asymptomatic latent interval ranging from 1 to 6 months after a measles infection, there is a subacute onset of delirium over days or a week or so, accompanied, variously, by myoclonus, focal signs, and seizures, which may be either grand mal or partial in type; in some cases simple partial motor status epilepticus has been noted.

Although the CSF is generally normal, in some cases there may be a mild lymphocytic pleocytosis or a mildly elevated total protein; anti-measles antibodies may or may not be present.

Brain biopsy is typically required for definitive diagnosis.

Course

The disease is relentlessly progressive to coma and death within weeks to months.

Etiology

This disease probably occurs secondary to reactivation of a defective measles virus within neurons and oligodendrocytes. Pathologically, there is very little or no inflammation, and measles-containing inclusion bodies may be found within both the nuclei and cytoplasm of both types of cells (Aicardi *et al.* 1977; Gazzola *et al.* 1999).

Although most patients have been children with leukemia, cases have occurred in adults with Hodgkin's disease (Wolinsky *et al.* 1977), adults undergoing therapeutic immunosuppression (Gazzola *et al.* 1999), patients with AIDS (Budka *et al.* 1996), and, rarely, apparently immunocompetent adult patients (Chadwick *et al.* 1982; Croxson *et al.* 2002).

Differential diagnosis

Subacute measles encephalitis must be distinguished from other measles-related disorders. Acute measles may be complicated by an encephalitis; however, in these cases symptoms occur in the context of a measles rash. Acute disseminated encephalomyelitis, like subacute measles encephalitis, also has a latent interval but it is shorter, in the order of weeks; furthermore, there is no myoclonus in acute disseminated encephalomyelitis. Subacute sclerosing panencephalitis also has a latent interval but this is much longer, in the order of years.

Treatment

The general treatment of delirium is discussed in Chapter 6.5; anecdotally, ribavirin was beneficial.

15.14 NEUROSYPHILIS

Infection by the spirochete *Treponema pallidum* may cause primary, secondary, or tertiary syphilis. Primary syphilis typically presents with a painless chancre that remits spontaneously. Secondary syphilis may appear weeks to months after resolution of the chancre and is characterized by a widespread rash, which may or may not be accompanied by an acute syphilitic meningitis. Following resolution of the rash there is a long latency interval, lasting years, after which about 10 percent of patients go on to develop tertiary syphilis. Tertiary syphilis may manifest in a variety of organs, including the central nervous system, in which case one speaks of neurosyphilis.

Neurosyphilis may occur in any one of several forms (Merritt *et al.* 1946), including meningovascular syphilis, gummas, general paresis, and tabes dorsalis, each discussed below. Meningovascular neurosyphilis is characterized by a chronic, indolent basilar meningitis; both arteries and cranial nerves that cross the meninges may be affected and there may be infarctions and cranial nerve palsies. Gummas are granulomatous tumors that are typically found in association with meningovascular syphilis and which may vary in size, from minute, barely macroscopic lesions to large masses, in which case they may present as would any other mass lesion. General paresis is characterized pathologically by a direct invasion of the brain by the spirochete and clinically by a dementia. Tabes dorsalis is characterized by inflammation of the posterior spinal roots and softening of the posterior columns with ataxia and, in some, severe pain.

Although tertiary syphilis was common in the pre-antibiotic era, after World War II, with the widespread use of penicillin, the incidence fell dramatically to the point at which it appeared to have almost vanished in developed countries. Over the past few decades, however, it has been enjoying something of a resurgence, particularly among those with AIDS.

Clinical features

As noted, some patients with secondary syphilis may develop *acute syphilitic meningitis*. Whether or not this condition should be subsumed under the rubric of neurosyphilis is not clear, although, given that it does, of course, involve the nervous system, it appears reasonable to do so; however, many authors reserve the term 'neurosyphilis' for nervous system manifestations occurring after a long latent interval as part of tertiary syphilis. Of these traditional forms of neurosyphilis, *meningovascular neurosyphilis* is currently the most common (Conde-Sendin *et al.* 2004) and has an onset after a latency of about 6 or 7 years, with a range of 1–12 years. *Gummas* are currently quite rare and present with the same latency as meningovascular syphilis. *General paresis* has a longer latency, ranging from 10 to 20 years, and *tabes dorsalis* has potentially the longest latency, ranging from 10 to 30 years. Each of these will be discussed below, and consideration will also be given to *laboratory findings*, to the effect of concurrent *AIDS and neurosyphilis*, and to *congenital neurosyphilis*. It must be borne in mind that, although the various forms of tertiary neurosyphilis may occur in isolation, more commonly patients will have elements of two or more forms.

ACUTE SYPHILITIC MENINGITIS

In close relation to the rash of secondary syphilis, a minority of patients may develop a meningitis with typical symptoms such as malaise, headache, stiff neck, and, in some, a delirium. Although these symptoms generally undergo a spontaneous full remission, a small minority of patients may develop obstructive hydrocephalus.

MENINGOVASCULAR SYPHILIS

Although meningovascular syphilis may present with mild cognitive changes, or a personality change, the overall symptomatology is often dominated by strokes and cranial nerve palsies. The symptomatology of meningovascular syphilis depends on which arteries and which cranial nerves are affected. With involvement of large, pial arteries, typical territorial infarctions may occur (Holmes *et al.* 1984), and when small, penetrating arteries become involved, lacunar syndromes appear. The various stroke syndromes that may appear are discussed in Chapter 8.4; with multiple infarctions a multi-infarct dementia or a lacunar dementia may appear, as outlined in Chapters 11.1 and 11.2. Of the various cranial nerves that may be involved, the most common are the third and sixth, with diplopia, the seventh, with unilateral or bilateral facial palsy, and the first, with unilateral or bilateral blindness. Rarely, obstructive hydrocephalus may appear. Most cases of meningovascular syphilis will also be characterized by the Argyll Robertson pupil, in which, although pupillary constriction to direct illumination is lost, constriction upon accommodation testing is preserved.

GUMMAS

Gummas tend to occur at the base of the brain or over the convexities; occasionally they may be deep-seated. They tend to present similarly to any gradually enlarging mass lesion and, if large enough, may cause dementia (Bianchi and Frera 1957).

GENERAL PARESIS

General paresis, also known as general paresis of the insane (GPI), dementia paralytica, or paretic neurosyphilis, is not necessarily associated with meningovascular syphilis and, indeed, it may be the only manifestation of tertiary neurosyphilis. General paresis typically manifests with a dementia of gradual onset and slow progression (Gomez and Aviles 1984; Storm-Mathisen 1969). Although the dementia may be non-specific in nature, certain cases may be marked by a personality change of the frontal lobe type, by mood changes, tending either toward mania or toward depression, or by delusions and hallucinations. Rarely, general paresis may present with a psychosis (Rothschild 1940; Schube 1934). Regardless of the typology of the dementia, other signs and symptoms gradually accrue. Seizures, either partial or grand mal, are common. Dysarthria and anomia may occur, and handwriting becomes very poor. Coarse tremor is common and may be present not only in the hands but also in the lips and tongue. Very typically the facial musculature loses its tone, giving the patient a vacant, dull facial expression. The Argyll Robertson pupil, described earlier, is present in most cases. With further progression the gait becomes unsteady and a true 'general paresis' occurs, with profound widespread weakness of almost all of the voluntary musculature. The plantar responses become extensor and, unless tabes dorsalis is present, there is a generalized hyper-reflexia.

TABES DORSALIS

Tabes dorsalis, also known as locomotor ataxia, is characterized by ataxia, diminished vibratory sense, a positive Romberg test, urinary incontinence, and erectile dysfunction. There may also be 'tabetic pains' in the lower extremities that tend to be lightning-like and lancinating. 'Visceral crises' may also occur with abdominal pain, nausea, and vomiting.

LABORATORY FINDINGS

Magnetic resonance scanning may reveal a basal pachymeningitis in meningovascular syphilis, along with evidence of any concurrent infarctions. In general paresis, both cortical atrophy and ventricular dilation are seen, and if gummas are present they will immediately be apparent as space-occupying lesions.

Lumbar puncture is essential. In tertiary neurosyphilis the opening pressure is normal. There is a lymphocytic pleocytosis in the range 5–100 cells/cm^3, the total protein is elevated in the range 50–100 mg/dL, and the glucose level is normal; the IgG index is elevated and oligoclonal bands may be found. In acute syphilitic meningitis, findings are more severe: the opening pressure may be elevated, the cell count is in the range 200–400 cells/cm^3, the total protein may be as high as 200 mg/dL, and the glucose level may be reduced. Of great importance is the fact that in AIDS the CSF findings are much less dramatic, and in some cases the cell count may be normal.

The VDRL (Venereal Disease Research Laboratory) and FTA (free treponemal antibody) tests are likewise essential. The serum VDRL is positive in only about 70 percent of cases of neurosyphilis (Simon 1985), and the CSF VDRL is less sensitive: one study found a sensitivity of only 27 percent (Davis and Schmitt 1989). The serum FTA is positive in almost 100 percent of cases, whereas the CSF FTA, like the CSF VDRL, is less sensitive, being positive in about 75 percent of patients. False-positive serum VDRL tests may occur in collagen vascular diseases, such as lupus, and also in heroin addicts; generally these false-positive VDRLs are low titer (e.g. less than 1:8). False-positive serum FTAs may occur under the same circumstances but are much less common.

In practice, patients suspected of having neurosyphilis are often screened with a serum VDRL and, if this is positive, a serum FTA is then performed for confirmation. Given, however, the relative insensitivity of the serum VDRL, it is more appropriate to skip the VDRL and proceed directly to a serum FTA. If the serum FTA is negative, then neurosyphilis is effectively ruled out. If the serum FTA is positive and, again, there is a clinical suspicion of neurosyphilis, then lumbar puncture must be performed.

If the CSF VDRL is positive, then the diagnosis of neurosyphilis is almost certain. If the CSF VDRL is negative but there is other CSF evidence of infection (e.g. lymphocytic pleocytosis or an elevated total protein), then it may be prudent to obtain a CSF FTA, as this is more sensitive than the CSF VDRL. Obtaining a CSF FTA, however, is controversial, as even minute contamination of the CSF by blood may create a false-positive FTA (Davis and Sperry 1979); caution is therefore required when interpreting this result. In the rare cases in which both the CSF VDRL and the CSF FTA are negative (Ch'ien *et al.* 1970) but the clinical picture is highly suggestive of neurosyphilis and there is evidence of infection in the CSF (e.g. a pleocytosis or elevated total protein), then some clinicians will treat the patient as if neurosyphilis is present; if there is clinical improvement on follow-up, and if other CSF abnormalities resolve (e.g., the cell count falls), then it is reasonable to assume that a correct 'diagnosis by treatment response' has been made.

Given the frequency with which neurosyphilis is seen in AIDS, it is prudent to check all patients with neurosyphilis for HIV infection.

AIDS AND NEUROSYPHILIS

Some 15 percent of patients with AIDS are seropositive for syphilis and about 1 percent have neurosyphilis. Notably, the clinical presentation or neurosyphilis is altered in AIDS, primarily in that the latencies may be shorter and the evolution of the various forms may be more rapid (Johns *et al.* 1987).

CONGENITAL NEUROSYPHILIS

Infected mothers may pass the spirochete to the fetus, and congenitally infected children may or may not have the classic 'Hutchinson's triad' of cataracts, sensorineuronal deafness, and 'Hutchinson's teeth', in which the central incisors are notched, widely spaced, and often tapering. Neurosyphilis may present in these children after the normal latency periods and, in such cases, a dementia secondary to general paresis may present in the teenage years.

Course

All forms of neurosyphilis are gradually progressive; untreated, both meningovascular neurosyphilis and general paresis are fatal within 3–5 years.

Etiology

The chancre of primary syphilis represents an inflammatory response to a localized infection by *T. pallidum.* With hematogenous spread, the rash of secondary syphilis occurs, and, in about one-third of all patients, CSF findings indicate that the spirochete has gained access to the meninges (Lukehart *et al.* 1988): as noted however, only a small minority of patients with secondary syphilis develop an acute syphilitic meningitis at this point, and the remaining cases, that is to say those with a positive CSF but no neuropsychiatric symptoms, are said to have 'asymptomatic neurosyphilis'. In most cases, host defenses are eventually adequate to eradicate *T. pallidum*; however, in about 10 percent of all patients, the spirochete persists, and patients eventually develop one of the traditional forms of neurosyphilis.

In meningovascular neurosyphilis, a granulomatous basilar meningitis is present. Cranial nerves traversing these meninges may become inflamed and undergo degenerative changes, and traversing arteries develop an endarteritis that may be followed by thrombotic occlusion and infarction of subserved tissues. Obstruction of the outflow foramina of the fourth ventricle may lead to hydrocephalus.

Gummas are granulomatous tumors that probably represent 'localized' forms of meningovascular neurosyphilis.

In general paresis, cortical atrophy is prominent, more so in the frontal and temporal areas than elsewhere. The ventricles are dilated and generally display a granular ependymitis. Spirochetes are found throughout the cortex, and neurons are lost with a disruption of the normal cortical cytoarchitecture. Microglia and astrocytes are present in abundance, and there may be perivascular cuffing by lymphocytes of small penetrating vessels.

In tabes dorsalis, inflammation is present in the ventral spinal roots, and the posterior columns display atrophy and softening.

Differential diagnosis

Importantly, a history of treatment of primary or secondary syphilis with benzathine penicillin does not rule out a diagnosis of neurosyphilis, given that benzathine penicillin is not capable of consistently inducing treponemocidal levels of penicillin within the CSF (Holmes *et al.* 1984).

Meningovascular syphilis may be mimicked by other disorders capable of causing an indolent basilar meningitis, such as sarcoidosis, tuberculosis, and fungal infections. Gummas may be difficult to distinguish on clinical grounds from other granulomatous mass lesions, or from gliomas. General paresis may mimic other dementias of gradual or subacute onset, as discussed in Chapter 6.3; in this regard, particular attention must be paid to the possibility of a concurrent AIDS dementia.

There is one clinical feature that is very helpful in the differential diagnosis, and that is the Argyll Robertson pupil, which is virtually pathognomonic for neurosyphilis.

Treatment

A 2-week course of intravenous penicillin G is the treatment of choice, and this should be administered not only in patients with acute syphilitic meningitis and the various traditional forms of neurosyphilis, but also in patients with 'asymptomatic neurosyphilis'. Patients allergic to penicillin consideration should undergo desensitization. In a minority of patients, institution of antibiotic treatment will immediately be followed by what is known as the Jarisch–Herxheimer reaction, with malaise and fever, and, in some, a transient exacerbation of the clinical symptomatology (Hahn *et al.* 1959), which in one extraordinary case, characterized by a large gumma, was fatal (Zifko *et al.* 1994). This Jarisch–Herxheimer reaction probably represents an inflammatory reaction to dying spirochetes, and may be treated by a short course of steroids. Because a course of penicillin G is not universally effective (especially in patients with AIDS; Gordon *et al.* 1994), it is necessary to follow patients with serial examinations of the CSF every 3–6 months until there has been a satisfactory response. The cell count falls first, generally by 6 months, after which the total protein gradually normalizes: the VDRL titer is the last thing to fall, and overall it may take 2 years for these three abnormalities to clear. As treatment failure is most common in patients with AIDS, serial CSF examinations are particularly important in this group.

Treatment with penicillin halts the progression of the symptoms of neurosyphilis, and in the case of general paresis there may be a partial remission. Gummas may or may not respond: in treatment-resistant cases the addition of steroids may be beneficial (Fleet *et al.* 1986); however, in some cases surgical resection is required.

With regard to the dementia that may occur in meningovascular syphilis or general paresis, symptomatic treatment is discussed in Chapter 6.3. Tabetic pains may respond to carbamazepine (Gimenez-Roldan and Martin 1981).

15.15 LYME DISEASE

Lyme disease is caused by the spirochete *Borrelia burgdorferi* and is transmitted to humans by the bite of an ixodid tick (Burgdorfer *et al.* 1982). The disease itself may be loosely divided into three stages. Stage I is characterized by a rash, known as erythema chronicum migrans. Stages II and III may be characterized by involvement of the nervous system, and in such cases one speaks of neuroborreliosis: in stage II a meningitis with cranial nerve palsies and a polyradiculitis is seen, and there may also be an encephalopathy; in stage III there may be an isolated encephalopathy. In addition to these well-recognized three stages, some patients may suffer from a controversial entity known as the 'post-Lyme disease syndrome'

Although *Borrelia* is endemic in certain parts of North America, particularly the Northeast and the Midwest, Lyme disease itself is uncommon.

Clinical features

In considering the various stages of Lyme disease it must be borne in mind that there may be some overlap between them and that not all patients will experience all three stages: stage I is seen in about 75 percent and stage II nervous system involvement occurs in about 15 percent; the percentage of patients who develop stage III disease with nervous system involvement is not clear, but is probably low. Each of these stages is discussed in turn, followed by a discussion of the post-Lyme disease syndrome and finally by a consideration of laboratory testing.

STAGE I

Stage I disease, also known as 'localized disease', is, as just noted, characterized by erythema chronicum migrans, which appears anywhere from 3 days to a month after the tick bite. The rash is characterized by a gradually enlarging, roughly circular, erythematous lesion, which, as it enlarges, leaves a pale, indurated central portion, thus creating an overall 'bull's-eye' effect. The lesion may enlarge up to 5 cm in diameter, and may be joined by smaller 'satellite' lesions. In general, the rash resolves spontaneously in a matter of weeks.

STAGE II

Stage II, also known as 'early disseminated disease', typically follows stage I within weeks to months. Nervous system involvement classically manifests with the triad of meningitis, cranial neuritis, and polyradiculitis (Hansen and Lebech 1992; Pachner and Steere 1985). The meningitis presents with headache, stiff neck, and malaise. Although cranial neuritis may affect virtually any of the cranial nerves, by far the most common manifestation is a unilateral or bilateral peripheral facial nerve palsy. When radiculitis occurs it is often accompanied by significant pain. A minority of patients with central nervous system involvement may also develop an encephalopathy with somnolence, poor short-term memory, poor concentration, and irritability, and, rarely, seizures or chorea. Rarely there may be a myelitis, and some patients may also develop a peripheral neuropathy with either a mononeuritis multiplex or a primarily sensory polyneuropathy. In general, symptoms eventually, and gradually, remit over 3–18 months.

In addition to these manifestations of nervous system involvement, there may also be a polyarthralgia, and, in some 10 percent of patients, evidence of cardiac involvement, with atrioventricular conduction defects, a pericarditis or a myocarditis.

STAGE III

Stage III Lyme disease, also known as 'late disseminated disease', may appear at any time from months to up to a decade after stage I, and may be characterized by a large-joint oligoarthritis in over half of all cases and, in a minority, by an encephalopathy. The encephalopathy, though at times characterized by a dementia, is more often quite mild with symptoms similar to that seen in stage II, such as mild somnolence, poor short-term memory, poor concentration, irritability or depression, and fatigue (Halperin *et al.* 1989; Logigian *et al.* 1990; Shadick *et al.* 1994); a peripheral polyneuropathy may also be present and in a minority there may also be focal signs such as aphasia or hemiplegia; seizures have also been noted. Very rarely, stage III may also be characterized by anxiety attacks (Fallon *et al.* 1993), mania (Fallon *et al.* 1993) or psychosis with delusions and hallucinations (Hess *et al.* 1999), and, in one case, stuporous catatonia (Pfister *et al.* 1993). Although some patients may show gradual progression of the symptoms of stage III, it appears that in most cases the course is characterized by a stable chronicity.

POST-LYME DISEASE SYNDROME

This syndrome (Marques 2008) follows immediately after successful treatment of stage II Lyme disease, as indicated by a resolution of both symptoms and the CSF abnormalities, described below. The syndrome itself is characterized by fatigue, complaints of poor memory and concentration, and myalgia or arthralgia. Although this syndrome clearly resembles stage III, certain features allow a differentiation. First, the course is different: whereas there is a generally long symptom-free latent interval between the resolution of stage II and the onset of stage III, such a latent interval is absent in cases of the post-Lyme disease syndrome. Second, in stage III Lyme disease, one may see multiple symptoms not present in the post-Lyme disease syndrome, including large-joint oligoarthritis, dementia, polyneuropathy, focal signs, and seizures.

LABORATORY TESTING

Serum should be tested for both IgG and IgM anti-*Borrelia* antibodies: IgG antibodies are found in almost all patients with neurologic involvement, and IgM antibodies may be found in those with stage I and stage II disease.

The CSF in stage II disease (Pachner and Steere 1985) reveals a lymphocytic pleocytosis, ranging from 15 to 700 cells, with a median of 166. The total protein, though at times normal, is typically elevated, with a median of 79 mg/dL. The glucose level, though typically normal, may occasionally be reduced. Oligoclonal bands may be present. In all cases, an anti-*Borrelia* IgG antibody index (Blanc *et al.* 2007) should be obtained utilizing the following formula:

(CSF anti-*Borrelia* IgG/serum anti-*Borrelia* IgG)/(total CSF IgG/total serum IgG)

An antibody index ≥ 2 has a specificity for stage II disease of 97 percent and a sensitivity of 75 percent. The characteristics of the CSF in stage III have not been adequately studied; however, it appears that at least in some cases there may be a very mild pleocytosis and a mildly elevated total protein, and an elevated antibody index (Halperin *et al.* 1989). The CSF in the post-Lyme disease syndrome displays a normal cell count and total protein; although anti-*Borrelia* IgG antibodies may be present, the antibody index is normal.

Magnetic resonance imaging in stage III may reveal multifocal areas of demyelination in the subcortical and periventricular white matter.

Course

This is as noted above.

Etiology

Erythema chronicum migrans in stage I represents a localized inflammatory response to the spirochete. Subsequently, there is a spirochetemia with hematogenous dissemination to the nervous system. Stage II is characterized by a meningitis with associated inflammation of cranial nerves and spinal nerve roots, with, in some cases, a cerebral lymphocytic vasculitis (Meurers *et al.* 1990). In stage III, there may be both a small-vessel vasculitis and multifocal areas of either demyelination or cerebritis (Kobayashi *et al.* 1997; Oksi *et al.* 1996). The etiology of the post-Lyme disease syndrome is unclear.

Differential diagnosis

In cases of suspected stage II or III Lyme disease, a careful search should be made for a history of erythema chronicum migrans, as a positive finding here carries great weight. Not too much significance, however, should be placed on a negative history in this regard, and for two reasons: first, as noted earlier, this lesion is present in only about 75 percent of cases, and, secondly, of those who had the rash, no more than two-thirds will recall it. In addition to this historical detail, serum should be tested for anti-*Borrelia* IgG antibody, which will be positive in virtually all patients with stage II and stage III disease. Although a positive finding here is, of course, most important, false positives are not uncommon, and indeed in endemic areas up to 5 percent of the population will have a positive serology with no clinical evidence of disease.

The classic triad of stage II disease is rarely mimicked by any other disease process, and hence an encephalopathy in this setting will almost always be due to Lyme disease, a fact which may be confirmed by the CSF findings.

Great diagnostic difficulty, however, may arise when patients present with a mild encephalopathy long after stage I or stage II, given the non-specific character of the encephalopathic findings seen in stage III. Naturally, a concurrent oligoarthritis is supportive of the diagnosis; however, this may be absent. The MRI findings seen in stage III are likewise supportive, but these too may be absent, and may also be seen with small-vessel disease or with multiple sclerosis. The CSF findings, especially an elevated IgG antibody index, are especially supportive, but unfortunately these, too, may be absent. The upshot is that a diagnosis of stage III Lyme disease may end up being one of exclusion, and made only after ruling out the various causes of dementia discussed in Chapter 6.3.

Treatment

For stage II, intravenous ceftriaxone is typically effective; however, in a small minority, a Jarisch–Herxheimer-like reaction may develop soon after the initiation of treatment, with malaise, fever, and, in some, an exacerbation of the presenting symptomatology. Although the best approach for stage III has not been clearly worked out, ceftriaxone should also be considered: in one case of stage III characterized by dementia there was a dramatic response (Steinbach *et al.* 2005).

The treatment of the post-Lyme disease syndrome is a matter of great controversy: despite the fact that four double-blinded studies of ceftriaxone versus placebo have found no consistent benefit for ceftriaxone (Fallon *et al.* 2008; Klempner *et al.* 2001; Krupp *et al.* 2003), many patients demand a repeat course of treatment. Given the risks attendant on long-term intravenous antibiotic treatment, it would be prudent to withhold treatment, unless, of course, the CSF indicated that stage II had not completely resolved.

15.16 TUBERCULOSIS

Mycobacterium tuberculosis is an acid-fast intracellular bacillus that in most cases is spread by droplets from patients with pulmonary tuberculosis. In many patients a focus of infection may be held in check by adequate host defenses for many years or even decades until some loss of immunocompetence is followed by reactivation, as may occur in patients with AIDS, chronic alcoholism, or cancer, or after treatment with steroids or immunosuppressants. Hematogenous spread may carry the bacillus to various organs, including the brain, where it may cause a meningitis or present with one or more tuberculomas.

In North America the incidence of tuberculosis has increased dramatically over the past decades, in large part because of its occurrence in patients with AIDS; indeed, in some cases tubercular meningitis has constituted the presentation of AIDS (Barnes *et al.* 1979).

Clinical features

The meningitis seen in tuberculosis (Davis *et al.* 1993; Kennedy and Fallon 1979; Sanchez-Portocarrero *et al.* 1996; Traub *et al.* 1984; Williams and Smith 1954) presents subacutely with a prodrome, lasting 2–3 weeks, characterized by constitutional symptoms, such as headache, malaise, lethargy, and low-grade fever. Subsequently patients develop a delirium accompanied by neck stiffness, a worsening headache, and nausea and vomiting. Cranial nerves traversing the inflamed meninges may become entrapped with cranial nerve palsies, affecting especially the third, sixth, and seventh cranial nerves (McKendrick and Grose 1957). Arteries traversing the subarachnoid space may become inflamed with subsequent thrombus formation and infarction of subserved tissue (Chan *et al.* 2005): both large vessels, such as the middle cerebral artery, and small penetrating vessels may be involved and, when small vessels are involved, those supplying the basal ganglia seem particularly vulnerable, with resulting abnormal movements (Alacon *et al.* 2000) such as tremor, chorea, or dystonia. With progression of the meningitis, stupor and coma supervene, and there may be seizures. Hydrocephalus, with a dramatic worsening of the overall clinical picture, occurs in roughly one-third of cases (Chan *et al.* 2003). This is a non-communicating hydrocephalus which may occur secondary to a granular ependymitis of the aqueduct of Sylvius or, much more commonly, to obstruction of the outflow foramina of the fourth ventricle.

Tuberculomas may be single or, more commonly, multiple, may occur either in the cerebellum or cerebrum, and present as would any mass lesion, with focal signs or with seizures (Damergis *et al.* 1978); importantly tuberculomas may occur in isolation, and without evidence of meningitis, in close to one-half of all cases (Traub *et al.* 1984).

The spinal cord may also be involved. Spinal meningitis may lead to entrapment of nerve roots, and arteritis of spinal arteries may lead to cord infarction.

In cases of meningitis, MR scanning typically reveals gadolinium enhancement of the basilar meninges.

Lumbar puncture is the most important of tests, and is abnormal in virtually all cases of meningitis. The opening pressure is typically elevated. There is a pleocytosis, in the range 50–500 cells/cm^3: although there may be a polymorphonuclear predominance initially in roughly one-quarter of cases, this fairly rapidly gives way to the more characteristic lymphocytic predominance. The total protein is increased to the range 100–500 mg/dL while the glucose level is reduced in some 80 percent of cases. Staining for acid-fast bacilli may be negative on the first lumbar puncture, but with up to four serial punctures, a positive result may eventually be obtained in close to 90 percent of cases. Cultures may take weeks, but if a large amount of fluid (5–10 cm^3) is obtained, positive results may be found in roughly three-quarters of all cases. Polymerase chain reaction assay is also available, and has the advantage of yielding results within a day or two; however, although the specificity is very high, the sensitivity is not, being in the range 50–75 percent.

Chest radiography is normal in roughly half of all cases, and those abnormalities that may be present are often nonspecific. The PPD (purified protein derivative) skin test may be falsely negative, and this is particularly likely in those with AIDS (Sanchez-Portocarrero *et al.* 1996).

Course

Untreated, tubercular meningitis is generally fatal within weeks to months, and those who survive may be left with dementia, cranial nerve palsies, or focal signs (Kalita and Misra 2007). Tuberculomas generally show some progression but may eventually stabilize, in some cases undergoing calcification.

Etiology

With hematogenous spread, bacilli may infect the meninges, subpial cortex, or the ependyma, and when any of these meningeal or subpial tuberculomas rupture, bacilli spill into the subarachnoid space, prompting the development of a thick, gelatinous exudate that, as noted earlier, is typically basilar in location (Rich and McCordock 1933). Cranial nerves and arteries traversing the exudate may become inflamed with cranial nerve palsies or infarctions.

Hematogenous spread may also carry bacilli to the brain parenchyma, with the subsequent development of tuberculomas, which, like tuberculomas anywhere, may undergo caseation necrosis. As noted earlier, these tuberculomas tend to be multiple and exhibit a broad range in size, from minute to so large that they destroy an entire lobe.

Differential diagnosis

Tubercular meningitis may be confused with basilar meningitides seen with fungal infections, tertiary neurosyphilis, sarcoidosis, and meningeal carcinomatosis. Tuberculomas may be mimicked by gliomas or other mass lesions.

Treatment

Tuberculous meningitis requires urgent treatment with a multidrug regimen that typically includes isoniazid, rifampin, pyrazinamide, and ethambutol. Of grave concern is the emergence of multidrug-resistant bacilli (Frieden *et al.* 1993). When isoniazid is used, consideration should be given to supplementation with vitamin B6 to prevent the development of pellagra (Ishii and Nishihara 1985), as discussed in Chapter 14.3. Treatment is generally continued for 18–24 months, and early on in treatment, for the first 6–8 weeks, the antibiotic regimen may be supplemented with steroids. Hydrocephalus may require shunting. Tuberculomas are treated with the same drug regimen, and in most cases this is satisfactory. In some cases, biopsy may be required to secure the diagnosis, and, rarely, excision may be necessary; however this should be a last option, given the risk of spillage into the subarachnoid space, with a florid, and fatal, meningitis.

The general treatment of delirium is discussed in Chapter 6.5.

15.17 WHIPPLE'S DISEASE

Whipple's disease, first described by George Hoyt Whipple in 1907 (Whipple 1907), is a rare disorder found mostly in middle-aged white males, which occurs secondary to infection with a bacillus known as *Tropheryma whippelii.* Although Whipple's disease most commonly causes arthralgia and abdominal complaints, the nervous system may be involved with, as described below, a wide variety of signs and symptoms.

Clinical features

Whipple's disease typically presents with migratory large-joint polyarthralgia and abdominal complaints, including pain, diarrhea, and weight loss; many patients will also have a mild fever. In a minority of cases central nervous system symptomatology may occur, and although this generally occurs within the context of long-standing arthralgia or abdominal symptomatology, in a small minority, roughly 5 percent, central nervous symptomatology may constitute the presentation of Whipple's disease (Adams *et al.* 1987; Lampert *et al.* 1962; Panegyres *et al.* 2006; Pruss *et al.* 2007; Romanul *et al.* 1977).

Clinically (Gerard *et al.* 2002; Louis *et al.* 1996; Manzel *et al.* 2000; Matthews *et al.* 2005), patients typically present with a gradually progressive dementia or personality change. In most cases other symptoms are also present, and these may include upper motor neuron signs, ataxia, nystagmus, myoclonus, seizures, supranuclear ophthalmoplegia, and various manifestations of hypothalamic dysfunction including sleep disturbance (with either hypersomnolence or, less commonly, insomnia), hyperphagia, decreased libido, or diabetes insipidus. A minority of patients may also display a very distinctive abnormal movement known as oculomasticatory myorhythmia, in which pendular eye movements occur in concert with rhythmic jaw movements (Schwartz *et al.* 1986).

T2-weighted MRI may reveal multifocal areas of increased signal intensity within the cerebral cortex, the basal ganglia, thalamus, and midbrain, and in a minority these lesions may demonstrate gadolinium enhancement.

The CSF may demonstrate a variety of abnormalities. In about half, there is a modest lymphocytic pleocytosis, and half may also demonstrate a mild elevation of the total protein. Polymerase chain reaction assay is typically, but not always, positive for *T. whippelii* DNA.

The diagnosis may also be established by small bowel biopsy, which may reveal not only periodic acid–Schiff (PAS)-positive macrophages but also a positive PCR assay; in this regard, however, it must be kept in mind that, albeit rarely, central nervous system Whipple's disease can occur with a negative small bowel biopsy (Pollock *et al.* 1981).

Course

Central nervous system involvement is a grave sign in Whipple's disease; in most cases there is a steady progression, with death occurring within 6–12 months.

Etiology

Whipple's disease, as noted, occurs secondary to infection by *T. whippelii* (Raoult *et al.* 2000; Relman *et al.* 1992). Within the central nervous system (Powers and Rawe 1979; Smith *et al.* 1965), focal areas of inflammation and necrosis, along with glial scars, are found in the cerebral cortex, basal ganglia, hypothalamus, brainstem (especially the periaqueductal gray), and cerebellar cortex. PAS-positive macrophages are found in these areas, and electron microscopy may reveal the bacillus within the macrophage.

Differential diagnosis

The appearance of dementia or personality change in the context of long-standing polyarthralgia or abdominal complaints is highly suggestive; in these cases, consideration might also be given to Lyme disease or to cerebrotendinous xanthomatosis. The differential for dementia occurring with myoclonus is discussed in Chapter 6.3. Oculomasticatory myorhythmia, although present in only a minority of cases, is a particularly valuable sign as it is virtually pathognomonic for this disease.

Vitamin B12 deficiency may occur in cases characterized by severe diarrhea, making an independent contribution to central nervous system symptomatology.

Treatment

The general treatment of dementia and personality change is discussed in Chapters 6.3 and 8.2, respectively. Although there are no controlled trials of antibiotic treatment, most clinicians will begin with a 2-week course of ceftriaxone, followed by treatment with trimethoprim–sulfamethoxazole. The CSF should be monitored and treatment continued until the patient is clinically stable and the CSF has normalized; in many cases this may take up to 2 years. If antibiotics are discontinued the patient should be closely monitored as relapses may occur; in some cases indefinite treatment is required.

15.18 ROCKY MOUNTAIN SPOTTED FEVER

Rocky Mountain spotted fever (Dantas-Torres 2007) is a rare disorder occurring secondary to infection by a tick-borne obligate intracellular bacterium, *Rickettsia rickettsii*, which manifests classically with fever, rash, and delirium. In the United States, Rocky Mountain spotted fever is generally found in the Rocky Mountain area and the Appalachians, and is most common in the spring or early summer.

Clinical features

After the tick bite, which is recalled by only about three-quarters of patients, there is a latent interval of days to a couple of weeks, after which there is a fairly abrupt onset of fever, headache, malaise, and myalgia. After a few days, a characteristic maculopapular rash appears: this is initially present at the ankles and wrists but then spreads to involve the extremities and, in over 50 percent of cases, the soles and palms. In this setting, most patients become delirious or lethargic, and this may be accompanied by seizures and various focal signs; coma may supervene (Bell and Lascari 1970; Horney and Walker 1988; Kirk *et al.* 1990). Pulmonary, hepatic, and renal involvement may also occur.

The CSF typically demonstrates a mild lymphocytic pleocytosis and a mildly elevated total protein with a normal glucose; PCR assay is available.

Serologic testing may reveal anti-rickettsial IgM or, with serial testing, a fourfold or greater rise in IgG antibodies.

Course

Untreated, the mortality rate is as high as 20 percent. Among those who survive, recovery occurs gradually in a matter of weeks; various sequelae may be found, including dementia or any focal signs present during the acute illness.

Etiology

Rickettsia rickettsii undergoes hematogenous spread to the brain. Neuropathologically (Green *et al.* 1978; Miller and Price 1972) there are widespread areas of perivascular inflammation and multiple petechial hemorrhages; arterioles may become occluded with multiple microinfarctions.

Differential diagnosis

Typhus may be virtually indistinguishable from Rocky Mountain spotted fever; however, this disease is virtually extinct in North America.

Treatment

Whenever the diagnosis is suspected, emergent treatment with doxycycline or tetracycline should be instituted; in children chloramphenicol may be used.

15.19 MALARIA

Malaria occurs secondary to infection with any one of four species of the protozoon *Plasmodium*: *P. malariae*, *P. vivax*, *P. ovale*, and *P. falciparum*. Of these four species, only one, *P. falciparum*, invades the central nervous system, thereby causing cerebral malaria.

Malaria is endemic in Haiti, much of Asia and Oceania, and in tropical and subtropical areas of Africa and South America; although it has been eradicated from most of North America and Europe, cases may still be seen in those returning from travel in endemic areas who did not take adequate prophylactic medication.

Clinical features

Anywhere from 1 to 4 weeks after infection via the bite of an *Anopheles* mosquito, patients fall ill with fever, headache, malaise, and myalgia. Roughly 2 percent of those patients whose malaria has occurred secondary to *P. falciparum* may then develop cerebral malaria (Blocker *et al.* 1968; Idro *et al.* 2005), with delirium, stupor or coma, and seizures. Focal signs, such as hemiparesis, aphasia or hemianopia, may occur, but are uncommon.

Computed tomography or MRI scanning may be normal, or, in severe cases may reveal generalized cerebral edema; Magnetic resonance imaging may also, on T2-weighted or FLAIR imaging, reveal areas of increased signal intensity in the thalamus or the cerebellar white matter.

The CSF may be normal or may show a mild pleocytosis and a mildly elevated total protein.

Diagnosis is made by examination of peripheral blood smears.

Course

Cerebral malaria is fatal in perhaps one-third of all cases. Amongst those who survive a small minority may be left with epilepsy (Ngoungou and Preux 2008) or other sequelae: in children (Brewster *et al.* 1990), both dementia and various focal deficits have been noted, and in adults both dementia (Roze *et al.* 2001) and personality change or depression (Varney *et al.* 1997) have been reported.

Recently, a 'postmalaria neurological syndrome' has been described (Schnorf *et al.* 1998; Zambito Marsala *et al.* 2006). Here, after recovering from malaria there is an asymptomatic latent interval of several weeks, after which patients fall ill with various symptoms, including delirium, myoclonus, tremor, aphasia, ataxia, and seizures. Although this syndrome resolves spontaneously, treatment with corticosteroids hastens recovery.

Etiology

Erythrocytes infected with *P. falciparum* develop 'knobs' on their cell membranes, which promote the adhesion of the erythrocyte to the vascular endothelium, and within the brain, capillaries and postcapillary venules are packed with infected red cells (Oo *et al.* 1987). Petechial and ring hemorrhages are found scattered throughout the parenchyma, and in fatal cases there is generalized edema. Amongst those who survive, there may be widespread areas of focal gliosis, known as Durck's granulomas (Oo *et al.* 1987; Toro and Roman 1978).

Differential diagnosis

The differential diagnosis of acute encephalitis is discussed in Chapter 8.8. In North America and Europe, the diagnosis should be suspected in travelers who have returned from an endemic area, and in this regard it must be kept in mind that mosquitoes can gain access to airplanes during stopovers, and hence it is not necessary for travelers to actually disembark in endemic areas to become infected.

Treatment

Emergent infectious disease consultation is essential. The general symptomatic treatment of delirium is discussed in Chapter 6.5. Steroids are not recommended (Warrell *et al.* 1982), and if stupor or coma develops, mannitol should be given. Anti-epileptic drugs may be required, but they should not be given prophylactically.

15.20 TOXOPLASMOSIS

Infection by *Toxoplasma gondii* is common in birds and mammals, especially rodents, and cats serve as the definitive hosts: oocysts found in cat feces may remain viable for up to a year. Primary infection in humans occurs secondary to eating food contaminated by cat feces, or undercooked lamb, pork, or beef from infected animals, and is very common: serologic studies indicate that about one-third of adults in North America and Europe have been infected. In the vast majority of cases, host defenses are capable of rapidly controlling the infection and, during primary infection, patients are either asymptomatic or suffer a self-limited mononucleosis-like syndrome. *Toxoplasma*, however, may not be eradicated but remain latent in cysts, often in muscle.

Infection of the central nervous system is very rare in immunocompetent adults. In immunocompromised patients, however, central nervous system toxoplasmosis does occur, either during a primary infection or, more commonly, with reactivation of a latent infection. In the pre-AIDS era this was noted in patients with cancer or those undergoing therapeutic immunosuppression, but was rare (Townsend *et al.* 1975); with the advent of AIDS, however, central nervous system toxoplasmosis has become common and indeed is the most common opportunistic infection in AIDS patients, in which it typically appears only when the CD4+ count falls below $100/mm^3$.

Clinical features

Although fulminant onsets have been noted, in most cases central nervous system toxoplasmosis presents subacutely, over a matter of weeks. The presentation is variable (Navia *et al.* 1986c; Porter and Sande 1992), depending on the number and location of toxoplasma abscesses, and may include headache, fever, delirium, dementia, seizures (Pascual-Sedano *et al.* 1999), and various focal signs, including hemianopia, hemiparesis, aphasia, or, uncommonly, abnormal movements such as dystonia (Factor *et al.* 2003) or chorea (Piccolo *et al.* 1999).

Magnetic resonance imaging reveals toxoplasma abscesses as areas of increased signal intensity on T2-weighted or FLAIR imaging, which characteristically show ring enhancement with gadolinium. Typically, multiple lesions are present in the cerebrum; however, the number may vary widely, from a single large one, up to such a high number as to constitute a miliary presentation. The CSF may be normal or may display a mild lymphocytic pleocytosis and a mildly elevated total protein; occasionally the opening pressure is also elevated. Polymerase chain reaction assay for toxoplasma DNA, though demonstrating a specificity close to 100 percent, unfortunately has a low sensitivity, around 50 percent. Occasionally the organism may be detected in the CSF by staining with Wright's or Giemsa's stain.

Serologic testing for IgG anti-*Toxoplasma* antibodies may or may not be helpful. As noted earlier, up to one-third of adults in North America and Europe will be positive, and hence false-positives may be common. Furthermore, false-negative results may also occur, indeed in one large study (Porter and Sande 1992) about one-fifth of patients with AIDS and central nervous system toxoplasmosis were negative for IgG.

Course

Untreated, central nervous system toxoplasmosis is generally fatal in a matter of months.

Etiology

Although toxoplasma abscesses may occur anywhere in the brain, favored sites include the junction of the cerebral cortex and subcortical white matter, and the thalamus and basal ganglia (Navia *et al.* 1986c). These abscesses, as noted earlier, vary widely both in number and size, and in some cases they may become quite large. Rarely, an abscess may obstruct CSF outflow, causing obstructive hydrocephalus, and, also rarely, the pituitary may be involved with subsequent endocrinologic disturbances.

Differential diagnosis

Although toxoplasmosis is the most common opportunistic central nervous system infection in AIDS, it must nevertheless be distinguished from other opportunistic infections, such as tuberculosis, fungal infections, and progressive multifocal leukoencephalopathy. A solitary toxoplasma abscess must also be distinguished from primary central nervous system lymphoma. In practice, the differential here is often based on a 'diagnosis by treatment response' strategy: when lesions fail to regress within 2 weeks with adequate anti-toxoplasma therapy (Luft *et al.* 1993) then one or more of the other possibilities is strenuously investigated.

Treatment

A 2-week course of combination treatment with pyrimethamine and sulfadiazine is generally sufficient to prompt a regression of the lesions; chronic treatment, however, is typically required to prevent relapses.

15.21 FUNGAL INFECTIONS

Fungal, or mycotic, infection of the central nervous system (Mori and Ebe 1992; Scully *et al.* 2008; Walsh *et al.* 1985) may occur during the course of cryptococcosis, coccidioidomycosis, histoplasmosis, candidiasis, aspergillosis, mucormycosis, and, rarely, blastomycosis or nocardiosis. Although such fungal infection of the brain is typically found only in immunocompromised hosts, such as those with AIDS, exceptions do occur, as in cases of cryptococcosis, coccidioidomycosis or histoplasmosis, wherein immunocompetent hosts may be involved. In general, the primary site of infection is in the respiratory tract, and in most cases the fungus gains entry into the central nervous system via hematogenous spread; an exception to this rule occurs in cases of aspergillosis or mucormycosis, wherein direct invasion, often of the frontal lobe, may occur via local spread from a sinus infection.

Clinical features

Clinically, there may be a basilar meningitis, abscesses, or an arteritis, with or without aneurysm formation. The most common presentation is with a basilar meningitis, and this is discussed first.

Basilar meningitis typically presents gradually over a matter of 1–3 weeks with headache, malaise, and neck stiffness; although most patients are febrile, exceptions occur, especially in those with AIDS. Delirium and lethargy follow, and there may be seizures. With entrapment of cranial nerves there may be cranial nerve palsies, and with inflammation of arteries traversing the subarachnoid space, there may be infarctions, which are typically of the lacunar type, involving the basal ganglia. With progression, stupor and coma may supervene. Obstruction of the outflow foramina of the fourth ventricle may lead to obstructive hydrocephalus, with a dramatic worsening of the clinical picture.

Mycotic abscesses favor the deep white matter or basal ganglia: with multiple abscesses the presentation may be with a delirium, while solitary abscesses may present as would any mass lesion, with focal signs appropriate to its location; regardless of the number of abscesses, seizures may also occur.

Arteritis occurs with direct invasion of the vessel wall, and this appears particularly characteristic of aspergillosis and mucormycosis. Thrombus formation may then occur, with infarction, or, should a mycotic aneurysm occur, there may be an intracerebral hemorrhage.

Magnetic resonance imaging in cases of meningitis will reveal enhancement of the basilar meninges, and will also, of course, display any abscesses.

The CSF may be normal when pathology is confined to abscesses or granulomas. When basilar meningitis is present, the fluid may be under increased pressure. There is a moderate lymphocytic pleocytosis, typically in the range of 50–200 cells/cm^3, the total protein is elevated, generally to over 100 mg/dL, and the glucose is reduced. Of note, in patients with severe AIDS the CSF findings may not be as severe, and in some cases the pleocytosis may be quite minimal. India ink staining may reveal cryptococci, and all specimens should be tested for specific fungal antigens.

Course

In the natural course of events the basilar meningitis is fatal in roughly one-half of all cases.

Etiology

The basilar meningitis is characterized by a thick exudate at the base of the brain, which, as noted earlier, may lead to entrapment of cranial nerves, or inflammation of arteries which traverse the subarachnoid space.

Differential diagnosis

Fungal abscesses and granulomas must be distinguished from other opportunistic infections, such as toxoplasmosis or tuberculosis. A fungal basilar meningitis may be mimicked by tuberculosis, tertiary neurosyphilis, and sarcoidosis.

Treatment

Infectious disease consultation is mandatory in order to ensure selection of the proper antifungal agent. The symptomatic treatment of delirium and of seizures is discussed in Chapters 6.5 and 8.3, respectively.

REFERENCES

Abrams DI, Jay CA, Shade SB *et al.* Cannabis in painful HIV-associated sensory neuropathy: a randomized placebo-controlled trial. *Neurology* 2007; **68**:515–21.

Adams JH, Miller D. Herpes simplex encephalitis: a clinical and pathological analysis of twenty-two cases. *Postgrad Med J* 1973; **49**:393–7.

Adams M, Rhyner PA, Day J *et al.* Whipple's disease confined to the central nervous system. *Ann Neurol* 1987; **21**:104–8.

Adie WJ. Idiopathic narcolepsy: a disease *sui generis* with remarks on the mechanism of sleep. *Brain* 1926; **49**:257–306.

Adle-Biassette H, Bourhy H, Gisselbrecht M *et al.* Rabies encephalitis in a patient with AIDS: a clinicopathological study. *Acta Neuropathol* 1996; **92**:415–20.

Agamanolis DP, Tan JS, Parker DL. Immunosuppressive measles encephalitis in a patient with a renal transplant. *Arch Neurol* 1979; **36**:686–90.

Aicardi J, Goutieres F, Arsenio-Nunes HL *et al.* Acute measles encephalitis in children with immunosuppression. *Pediatrics* 1977; **55**:232–5.

Alacon F, Duenas G, Cevallos N *et al.* Movement disorders in 30 patients with tuberculous meningitis. *Mov Disord* 2000; **15**:561–9.

Alkan A, Korkmaz L, Sigirci A *et al.* Subacute sclerosing panencephalitis: relationship between clinical stage and diffusion-weighted imaging findings. *J Magn Res Imaging* 2006; **23**:267–72.

Alpers BJ, Patten CA. Paroxysmal spasm of the eyelids as a postencephalitic manifestation. *Arch Neurol Psychiatry* 1927; **18**:427–32.

Antinori A, Cingolani A, Lorenzini P *et al.* Clinical epidemiology and survival of progressive multifocal leukoencephalopathy in the era of highly active antiretroviral therapy: data from the Italian Registry Investigative Neuro AIDS (IRINA). *J Neurovirol* 2003; **9**(Suppl. 1):47–53.

Antinori A, Arendt G, Becker JT *et al.* Updated research nosology for HIV-associated neurocognitive disorders. *Neurology* 2007; **69**:1789–99.

Applebaum E, Rachelson MH, Dolgopol VB. Varicella encephalitis. *Am J Med* 1953; **15**:233–6.

Applebaum E, Krebs SI, Sunshine A. Herpes zoster encephalitis. *Am J Med* 1962; **32**:25–31.

Arribas JR, Storch GA, Clifford DB *et al.* Cytomegalovirus encephalitis. *Ann Intern Med* 1996; **125**:577–87.

Astrom KE, Mancall EL, Richardson EP. Progressive multifocal leukoencephalopathy: a hitherto unrecognized complication of chronic lymphatic leukaemia and Hodgkin's disease. *Brain* 1958; **81**:93–111.

Aurelius E, Johansson G, Skoldenberg B *et al.* Rapid diagnosis of herpes simplex encephalitis by nested polymerase chain reaction assay of cerebrospinal fluid. *Lancet* 1991; **337**:189–92.

Ayuso-Peralta L, Jimenez-Jimenez FJ, Tejeiro J *et al.* Progressive multifocal leukoencephalopathy in HIV infection presenting as Balint's syndrome. *Neurology* 1994; **44**:1339–40.

Azimi PH, Cramblett HG, Haynes RE. Mumps meningoencephalitis in children. *JAMA* 1969; **207**:509–12.

Baringer JR, Swoveland P. Recovery of herpes simplex virus from human trigeminal ganglions. *N Engl J Med* 1973; **228**:648–50.

Barnes PF, Bloch AB, Davidson PT *et al.* Tuberculosis in patients with human immunodeficiency virus. *N Engl J Med* 1979; **301**:126–30.

Beach RS, Morgan R, Wilkie F *et al.* Plasma vitamin B12 level as a potential cofactor in studies of human immunodeficiency virus type 1-related cognitive changes. *Arch Neurol* 1992; **49**:501–6.

Becker D, Patel A, Abou-Khalil BW *et al.* Successful treatment of encephalopathy and myoclonus with levetiracetam in a case of subacute sclerosing panencephalitis. *J Child Neurol* 2009; **24**: 763–7.

Bell WE, Lascari AD. Rocky Mountain spotted fever. Neurologic manifestations in the acute phase. *Neurology* 1970; **20**:841–7.

Bergin JD. Fatal encephalopathy in glandular fever. *J Neurol Neurosurg Psychiatry* 1960; **23**:69–73.

Bhatia KP, Morris JH, Frackowiak RS. Primary progressive multifocal leukoencephalopathy presenting as an extrapyramidal syndrome. *J Neurol* 1996; **243**:91–5.

Bianchi M, Frera C. A case of brain gumma. *J Neurol Neurosurg Psychiatry* 1957; **20**:133–5.

Bing EG, Burman MA, Longshore D *et al.* Psychiatric disorders and drug use among human innnunodeficiency virus-infected adults in the United States. *Arch Gen Psychiatry* 2001; **58**:721–8.

Bistrian B, Phillips CA, Kaye IS. Fatal mumps meningoencephalitis. *JAMA* 1972; **222**:478–9.

Blanc F, Jaulhal B, Fleury M *et al.* Relavance of the antibody index to diagnose Lyme neuroborreliosis among seropositive patients. *Neurology* 2007; **69**:953–8.

Blatt ML, Hoffman SJ, Schneider M. Rabies: report of twelve cases, with a discussion of prophylaxis. *JAMA* 1938; **111**:688–91.

Blocker WW, Kasil AJ, Daroff RB. The psychiatric manifestations of cerebral malaria. *Am J Psychiatry* 1968; **125**:192–6.

Blunt S, Lane RJ, Turjanski N *et al.* Clinical features and management of two cases of encephalitis lethargic. *Mov Disord* 1997; **12**:354–9.

Bond ED. Epidemic encephalitis and katatonic symptoms. *Am J Insanity* 1920; **76**:261–4.

Bouwman FH, Skolasky RL, Hes D *et al.* Variable progression of HIV-associated dementia. *Neurology* 1998; **50**:1814–20.

Breitbart W, Marotta R, Platt MM *et al.* A double-blind trial of haloperidol, chlorpromazine, and lorazepam in the treatment of delirium in hospitalized AIDS patients. *Am J Psychiatry* 1996; **153**:231–7.

Breitbart W, Rosenfeld B, Kaim M *et al.* A randomized, double-blind, placebo-controlled trial of psychostimulants for the treatment of fatigue in ambulatory patients with human immunodeficiency virus disease. *Arch Intern Med* 2001; **161**:411–20.

Brew BJ, Rosenblum M, Cronin K *et al.* AIDS dementia complex and HIV-1 brain infection: clinical-virological correlations. *Ann Neurol* 1995; **38**:563–70.

Brewster DR, Kwiatkowski D, White NJ. Neurological sequelae of cerebral malaria in children. *Lancet* 1990; **336**:1039–43.

Budka H, Costanzi G, Cristina S *et al.* Brain pathology induced by infection with the human immunodeficiency virus (HIV). A histological, immunocytochemical, and electron microscopical study of 100 autopsy cases. *Acta Neuropathol* 1987; **75**: 185–98.

Budka H, Urbanits S, Liberski PP *et al.* Subacute measles virus encephalitis: a new and fatal opportunistic infection in a patient with AIDS. *Neurology* 1996; **46**:586–7.

Burgdorfer W, Barbour AF, Hayes SF *et al.* Lyme disease – a tick-borne spirochetosis. *Science* 1982; **216**:1317–19.

Buzzard EF, Greenfield JG. Lethargic encephalitis: its sequelae and morbid anatomy. *Brain* 1919; **42**:305–38.

Cadie M, Nye FJ, Storey P. Anxiety and depression after infectious mononucleosis. *Br J Psychiatry* 1976; **128**:559–61.

Caparros-Lefebvre D, Girard-Buttaz I, Rebout S *et al.* Cognitive and psychiatric impairment in herpes simplex virus encephalitis suggest involvement of the amygdalo-frontal pathways. *J Neurol* 1996; **243**:248–56.

Cape CA, Martinez AJ, Robertson JJ *et al.* Adult onset of subacute sclerosing panencephalitis. *Arch Neurol* 1973; **28**:124–7.

Carson KR, Evens AM, Richey EA *et al.* Progressive multifocal leukoencephalopathy after rituximab therapy in HIV-negative patients: a report of 57 cases from the Research on Adverse Drug Events and Reports Project. *Blood* 2009; **113**:4834–40.

Cettomai D, McArthur JC. Mirtazapine use in human immunodeficiency virus-infected patients with progressive multifocal leucoencephalopathy. *Arch Neurol* 2009; **66**:255–8.

Chadwick DW, Martin S, Buxton PH *et al.* Measles virus and subacute neurological disease: an unusual presentation of measles inclusion body encephalitis. *J Neurol Neurosurg Psychiatry* 1982; **45**:680–4.

Chan KH, Cheung RT, Fong CV *et al.* Clinical relevance of hydrocephalus as a presenting feature of tuberculous meningitis. *QJM* 2003; **96**:643–8.

Chan KH, Cheung RT, Lee R *et al.* Cerebral infarcts complicating tuberculous meningitis. *Cerebrovasc Dis* 2005; **19**:391–5.

Chandra K, Shafiq N, Pandhi P *et al.* Gabapentin versus nortriptyline in post-herpetic neuralgia patients: a randomized, double-blind clinical trial – the GONIP Trial. *Int J Clin Pharmacol Ther* 2006; **44**:358–63.

Ch'ien L, Hathaway BM, Israel CW. Seronegative dementia paralytica: report of a case. *J Neurol Neurosurg Psychiatry* 1970; **33**:376–80.

Chopra JS, Banerjee AK, Murthy JMK *et al.* Paralytic rabies: a clinicopathological study. *Brain* 1980; **103**:789–802.

Clifford DB, Arribas JR, Storch GA *et al.* Magnetic resonance brain imaging lacks sensitivity for AIDS associated cytomegalovirus encephalitis. *J Neurovirol* 1996; **2**:397–403.

Cobb WA, Marshall J, Scaravilli F. Long survival in subacute sclerosing panencephalitis. *J Neurol Neurosurg Psychiatry* 1984; **47**:176–83.

Conde-Sendin MA, Amela-Paris R, Aladro-Benito Y *et al.* Current clinical spectrum of neurosyphilis in immunocompetent patients. *Eur Neurol* 2004; **52**:29–35.

Conomy JP, Leibovitz A, McCombs W *et al.* Airborne rabies encephalitis: demonstration of rabies virus in the human central nervous system. *Neurology* 1977; **27**:67–9.

Croxson MC, Anderson NE, Vaughn AA *et al.* Subacute measles encephalitis in an immunocompetent adult. *J Clin Neurosci* 2002; **9**:600–4.

Dale RC, Church AJ, Surtees RA *et al.* Encephalitis lethargica syndrome: 20 new cases and evidence of basal ganglia autoimmunity. *Brain* 2004; **127**:21–33.

Dale RC, Irani SR, Brilot F *et al.* N-methyl-D-asparta receptor antibodies in pediatric dyskinetic encephalitis lethargic. *Ann Neurol* 2009; **66**:704–9.

Damergis JA, Leftwich EI, Curtin JA *et al.* Tuberculoma of the brain. *JAMA* 1978; **239**:413–15.

Dantas-Torres F. Rocky Mountain spotted fever. *Lancet Infect Dis* 2007; **7**:724–32.

Davies JA, Hughes JT, Oppenheimer DR. Richardson's disease (progressive multifocal leukoencephalopathy). *Q J Med* 1973; **42**:481–93.

Davis LE, Schmitt JW. Clinical significance of cerebrospinal fluid tests for neurosyphilis. *Ann Neurol* 1989; **25**:50–5.

Davis LE, Sperry S. The CSF-FTA test and the significance of blood contamination. *Ann Neurol* 1979; **6**:68–9.

Davis LE, Rastogi KR, Lambert LC *et al.* Tuberculous meningitis in the southwest United States: a community-based study. *Neurology* 1993; **43**:1775–8.

Davis LE, DeBiasi R, Goade DE *et al.* West Nile virus neuroinvasive disease. *Ann Neurol* 2006; **60**:286–300.

Dawson JR. Cellular inclusions in cerebral lesions of epidemic encephalitis. *Arch Neurol Psychiatry* 1934; **31**:685–700.

De La Blanchardiere A, Rozenberg F, Caumes E *et al.* Neurological complications of varicella-zoster infection in adults with human immunodeficiency virus infection. *Scand J Infect Dis* 2000; **32**:263–9.

de la Monte SM, de la Gabuzda DH, Ho DD *et al.* Peripheral neuropathy in the acquired immunodeficiency syndrome. *Ann Neurol* 1988; **23**:485–92.

Debiasi RL, Tyler KL. West Nile meningoencephalitis. *Nat Clin Pract Neurol* 2006; **2**:264–75.

Dennett C, Klapper PE, Cleator GM. Polymerase chain reaction in the investigation of "relapse" following herpes simplex encephalitis. *J Med Virol* 1996; **48**:129–32.

Dennett C, Cleator GM, Klapper PE. HSV-1 and HSV-2 in herpes simplex encephalitis: a study of sixty-four cases in the United Kingdom. *J Med Virol* 1997; **53**:1–3.

Deresiewicz RL, Thaler SJ, Hsu L *et al.* Clinical and neuroradiologic manifestations of eastern equine encephalitis. *N Engl J Med* 1997; **336**:1867–74.

de Ronchi D, Faranca I, Forti P *et al.* Development of acute psychotic disorders and HIV-1 infection. *Int J Psychiatry Med* 2000; **30**: 173–83.

Devinsky O, Cho E-S, Petito CK *et al.* Herpes zoster myelitis. *Brain* 1991; **114**:1991–9.

Dickman MS. von Economo encephalitis. *Arch Neurol* 2001; **58**: 1696–8.

Doja A, Bitnun A, Ford Jones EL *et al.* Pediatric Epstein–Barr virus-associated encephalitis: 10-year review. *J Child Neurol* 2006; **21**:384–91.

Dolgopol VB, Greenberg M, Aronoff R. Encephalitis following smallpox vaccination. *Arch Neurol Psychiatry* 1955; **73**:216–23.

Donner M, Waltimo O, Poros J *et al.* Subacute sclerosing panencephalitis as a cause of chronic dementia and relapsing brain disorder. *J Neurol Neurosurg Psychiatry* 1972; **35**:180–5.

Dooneief G, Bello J, Todak G *et al.* A prospective controlled study of magnetic resonance imaging of the brain in gay men and parenteral drug users with human immunodeficiency virus infection. *Arch Neurol* 1992; **49**:38–43.

Duncalf CM, Kent JNG, Harbord M *et al.* Subacute sclerosing panencephalitis presenting as schizophreniform psychosis. *Br J Psychiatry* 1989; **155**:557–9.

Dupont JR, Earle KM. Human rabies encephalitis: a study of forty-nine fatal cases with a review of the literature. *Neurology* 1965; **15**:1023–34.

Duvoisin RC, Yahr MD. Encephalitis and parkinsonism. *Arch Neurol* 1965; **12**:227–39.

Dworkin RH, Corbin AE, Young JP *et al.* Pregabalin for the treatment of postherpetic neuralgia: a randomized, placebo-controlled trial. *Neurology* 2003; **60**:1274–83.

Eidelberg D, Sotrel A, Hooupian DS *et al.* Thrombotic cerebral vasculopathy associated with herpes zoster. *Ann Neurol* 1986; **19**:7–14.

Elphick GF, Querbes W, Jordan JA *et al.* The human polyoma virus, JCV, uses serotonin receptors to infect cells. *Science* 2004; **306**: 1380–3.

Eroglu E, Gokcil Z, Bek S *et al.* Long-term follow-up of patients with adult-onset subacute sclerosing panencephalitis. *J Neurol Sci* 2008; **275**:113–6.

Esiri MM. Herpes simplex encephalitis. An immunohistological study of the distribution of viral antigen within the brain. *J Neurol Sci* 1982; **54**:209–26.

Everall IP, Luthert PJ, Lantos PL. Neuronal loss in the frontal cortex in HIV infection. *Lancet* 1991; **337**:1119–21.

Everall IP, Luthert PJ, Lantos PL. Neuronal number and volume alterations in the neocortex of HIV infected individuals. *J Neurol Neurosurg Psychiatry* 1993; **56**:481–6.

Factor SA, Troche-Panetto M, Weaver SA. Dystonia in AIDS: report of four cases. *Mov Disord* 2003; **18**:1492–8.

Fairweather DS. Psychiatric aspects of post-encephalitic syndrome. *J Ment Sci* 1947; **93**:201–54.

Fallon BA, Nields JA, Parsons B *et al.* Psychiatric manifestations of Lyme borreliosis. *J Clin Psychiatry* 1993; **54**:263–8.

Fallon BA, Keilp JG, Corbera KM *et al.* A randomized, placebo-controlled trial of repeated IV antibiotic therapy for Lyme encephalopathy. *Neurology* 2008; **70**:992–1003.

Fermaglich J, Hardman JM, Earle KM. Spontaneous progressive multifocal leukoencephalopathy. *Neurology* 1970; **20**:479–85.

Finklestein H. Meningoencephalitis in mumps. *JAMA* 1938; **111**: 17–19.

Fisher CM. Hypomanic symptoms caused by herpes simplex encephalitis. *Neurology* 1996; **47**:1374–8.

Fisher RS, Clark AW, Wolinsky JS *et al.* Post-infectious leukoencephalitis complicating *Mycoplasma pneumoniae* infection. *Arch Neurol* 1983; **40**:109–13.

Fleet WS, Watson RT, Ballinger WE. Resolution of gumma with steroid therapy. *Neurology* 1986; **36**:1104–7.

Fournier JCM, Helguera RAL. Postencephalitic narcolepsy and cataplexy: muscles and motor nerves inexcitability during the attack of cataplexy. *J Nerv Ment Dis* 1934; **80**:159–62.

Freedman MJ, Odland LT, Cleve EA. Infectious mononucleosis with diffuse involvement of nervous system: report of a case. *Arch Neurol Psychiatry* 1953; **69**:49–54.

Frieden TR, Sterling T, Pables-Mendes A *et al.* The emergence of drug-resistant tuberculosis in New York City. *N Engl J Med* 1993; **328**:521–6.

Fujimoto H, Asaoka K, Imaizumi T *et al.* Epstein–Barr infections of the central nervous system. *Intern Med* 2003; **42**:33–40.

Gabuzda DH, Levy SR, Chiappa KH. Electroencephalography in AIDS and AIDS-related complex. *Clin Electroencephalogr* 1988; **19**:1–6.

Gascon GG International Consortium on Subacute Sclerosing Panencephalitis. Randomized treatment study of inosiplex versus combined inosiplex and intraventricular inteferon-alfa in subacute sclerosing panencephalitis (SSPE): international multicenter study. *J Child Neurol* 2003; **18**:819–27.

Gautier-Smith PC. Neurological complications of glandular fever. *Brain* 1965; **88**:323–34.

Gazzola P, Cocito L, Capello E *et al.* Subacute measles encephalitis in a young man immunocompromised for ankylosing spondylitis. *Neurology* 1999; **52**:1074–7.

Geddes JF, Hughes AJ, Lees AJ *et al.* Pathological overlap in cases of parkinsonism associated with neurofibrillary tangles. A study of recent cases of postencephalitic parkinsonism and comparison with progressive supranuclear palsy and Guamanian parkinsonism–dementia complex. *Brain* 1993; **116**:281–302.

Gerard A, Sarrot-Reynauld F, Liozon E *et al.* Neurologic presentation of Whipple disease: report of 12 cases and review of the literature. *Medicine* 2002; **81**:443–57.

Gheuens S, Pierone G, Peeters P *et al.* Progressive multifocal leukoencephalopathy in individuals with minimal or occult immunosuppression. *J Neurol Neurosurg Psychiatry* 2010; **81**:247–54.

Gilden DH, Murray RS, Wellish M *et al.* Chronic progressive varicella-zoster virus encephalitis in an AIDS patient. *Neurology* 1988; **38**:1150–3.

Gilden DH, Wright RR, Schneck SA *et al.* Zoster sine herpete, a clinical variant. *Ann Neurol* 1994; **35**:530–3.

Gillespie J, Jackson A. *MRI and CT of the Brain.* London: Arnold, 2000.

Gimenez-Roldan S, Martin M. Tabetic lightning pains: high-dosage intravenous penicillin versus carbamazepine therapy. *Eur Neurol* 1981; **20**:424–8.

Glass JD, Fedor H, Wesselingh SL *et al.* Immunocytochemical quantitation of human immunodeficiency virus in the brain: correlations with dementia. *Ann Neurol* 1995; **38**:755–62.

Gomez EA, Aviles M. Neurosyphilis in community mental health clinics: a case series. *J Clin Psychiatry* 1984; **45**:127–9.

Gordon SM, Eaton ME, George R *et al.* The response of symptomatic neurosyphilis to high-dose intravenous penicillin G in patients with human immunodeficiency virus infection. *N Engl J Med* 1994; **331**:1469–73.

Green WR, Walker DH, Cain BG. Fatal viscerotropic Rocky Mountain spotted fever. Report of a case diagnosed by immunofluorescence. *Am J Med* 1978; **64**:523–8.

Greenberg DA, Weinkle DJ, Aminoff MJ. Periodic EEG complexes in infectious mononucleosis encephalitis. *J Neurol Neurosurg Psychiatry* 1982; **45**:648–51.

Greenfield JG. The pathology of measles encephalomyelitis. *Brain* 1929; **52**:171–95.

Greenwood R, Bhalla A, Gordon A *et al.* Behavior disturbances during recovery from herpes simplex encephalitis. *J Neurol Neurosurg Psychiatry* 1983; **46**:809–17.

Guillaume B., Sindic CJ, Weber T. Progressive multifocal leukoencephalopathy: simultaneous detection of JCV DNA and anti-JCV antibodies in cerebrospinal fluid. *Eur J Neurol* 2000; **7**:101–6.

Guilleminault C, Mondini S. Mononucleosis and chronic daytime sleepiness. A long-term follow-up study. *Arch Int Med* 1986; **146**:1333–5.

Guilleux MH, Steneir RE, Young IR. MR imaging in progressive multifocal leukoencephalopathy. *AJNR* 1986; **7**:1033–5.

Hahn K, Arendt G, Braun JS *et al.* A placebo-controlled trial of gabapentin for painful HIV-associated sensory neuropathies. *J Neurol* 2004; **251**:1260–6.

Hahn RD, Webster B, Weickhardt G *et al.* Penicillin treatment of general paresis (dementia paralytica). *Arch Neurol Psychiatry* 1959; **81**:557–90.

Halperin JJ, Luft BJ, Anand AK *et al.* Lyme neuroborreliosis: central nervous system manifestations. *Neurology* 1989; **39**:753–9.

Hansen K, Lebech AM. The clinical and epidemiological profile of Lyme neuroborreliosis in Denmark 1985–1990. A prospective study of 187 patients with *Borrelia burgdorferi* specific intrathecal antibody production. *Brain* 1992; **115**:399–423.

Harris MJ, Jeste DV, Gleghorn A *et al.* New-onset psychosis in HIV-infected patients. *J Clin Psychiatry* 1991; **52**:369–76.

Hart MN, Earle KM. Haemorrhagic and perivenous encephalitis: a clinical–pathological review of 38 cases. *J Neurol Neurosurg Psychiatry* 1975; **38**:585–91.

Hatipoglu HG, Sakman B, Yukset E. Magnetic resonance and diffusion-weighted imaging findings of herpes simplex encephalitis. *Herpes* 2008; **15**:13–7.

Hensen J, Rosenblum M, Armstrong D *et al.* Amplification of JC virus DNA from brain and cerebrospinal fluid of patients with progressive multifocal leukoencephalopathy. *Neurology* 1991; **41**: 1967–71.

Herzlich BC, Schiano TD. Reversal of apparent AIDS dementia complex following treatment with vitamin B12. *J Intern Med* 1993; **233**:495–7.

Herzon H, Shelton JT, Bruyn HG. Sequelae of western equine and other arthropod-borne encephalitides. *Neurology* 1957; **7**: 535–48.

Hess A, Buchmann J, Zetti UK *et al.* *Borrelia burgdorferi* central nervous system infection presenting as an organic schizophrenia-like disorder. *Biol Psychiatry* 1999; **45**:795.

Hilt DC, Bucholz D, Krumholz A *et al.* Herpes zoster ophthalmicus and delayed contralateral hemiparesis caused by cerebral angiitis: diagnosis and management approaches. *Ann Neurol* 1983; **14**:543–53.

Hogan EL, Krigman MR. Herpes zoster myelitis: evidence for viral invasion of spinal cord. *Arch Neurol* 1973; **29**:309–13.

Hohman LB. Epidemic encephalitis (lethargic encephalitis): its psychotic manifestations with a report of twenty-three cases. *Arch Neurol Psychiatry* 1921; **6**:295–333.

Hohman LB. Post-encephalitic behavior disorders in children. *Bull Johns Hopkins Hosp* 1922; **33**:372–5.

Hokkanen L, Salonen O, Launes J. Amnesia in acute herpetic and nonherpetic encephalitis. *Arch Neurol* 1996; **53**:972–8.

Hollinger P, Sturzenegger M, Mathis J *et al.* Acute disseminated encephalomyelitis in adults: a reappraisal of clinical, CSF, EEG and MRI findings. *J Neurol* 2002; **249**:320–9.

Holmes MD, Brant-Zawadzki MM, Simon RP. Clinical features of meningovascular syphilis. *Neurology* 1984; **34**:553–6.

Holtzman DM, Kaku DA, So YT. New-onset seizures associated with human immunodeficiency virus infection: causation and clinical features in 100 cases. *Am J Med* 1989; **87**:173–7.

Horney LF, Walker DH. Meningoencephalitis as a major manifestation of Rocky Mountain spotted fever. *South Med J* 1988; **81**: 915–18.

Howard RS, Lees AJ. Encephalitis lethargica: a report of four recent cases. *Brain* 1987; **110**:19–33.

Hriso E, Kuhn T, Masdeu JC *et al.* Extrapyramidal symptoms due to dopamine-blocking agents in patients with AIDS encephalopathy. *Am J Psychiatry* 1991; **148**:1558–61.

Huang D, Cossoy M, Li M *et al.* Inflammatory progressive multifocal leukoencephalopathy in human immunodeficiency virus-negative patients. *Ann Neurol* 2007; **62**:34–9.

Hynson JL, Kornberg AJ, Coleman LT *et al.* Clinical and neuroradiologic features of acute disseminated encephalomyelitis in children. *Neurology* 2001; **56**:1308–12.

Idro R, Jenkins NE, Newton CR. Pathogenesis, clinical features, and neurological outcome of cerebral malaria. *Lancet Neurol* 2005; **4**:827–40.

Ishii N, Nishihara Y. Pellagra encephalopathy among tuberculosis patients: its relation to isoniazid therapy. *J Neurol Neurosurg Psychiatry* 1985; **48**:628–34.

Ishii T, Nakamura Y. Distribution and ultrastructure of Alzheimer's neurofibrillary tangles in postencephalitic parkinsonism of Economo type. *Acta Neuropathol* 1981; **55**:59–62.

Jarvis MR, Wasserman AL, Todd RD. Acute psychosis in a patient with Epstein–Barr virus infection. *J Am Acad Child Adolesc Psychiatry* 1990; **29**:468–9.

Javadi MA, Fayaz A, Mirdehghan SA *et al.* Transmission of rabies by corneal graft. *Cornea* 1996; **15**:431–3.

Jelliffee SE. Oculogyric crises as compulsion phenomena in postencephalitis: their occurrence, phenomenology and meaning. *J Nerv Ment Dis* 1929; **69**:59–68, 165–84, 278–97, 415–26, 531–51, 666–9.

Jemsek J, Greenberg SB, Tabor L *et al.* Herpes zoster-associated encephalitis: clinicopathologic report of twelve cases and review of the literature. *Medicine* 1983; **62**:81–97.

Johns DR, Tierney M, Felsenstein D. Alteration in the natural history of neurosyphilis by concurrent infection with the human immunodeficiency virus. *N Engl J Med* 1987; **316**:1569–72.

Johnson RT, Milbourn PE. Central nervous system manifestations of chickenpox. *Can Med Assoc J* 1970; **102**:831–6.

Kalayjian RC, Cohen ML, Bonomo RA *et al.* Cytomegalovirus ventriculoencephalitis in AIDS. A syndrome with distinct clinical and pathologic features. *Medicine* 1993; **72**:67–77.

Kalita J, Misra UK. Markedly severe dystonia in Japanese encephalitis. *Mov Disord* 2000; **15**:1168–72.

Kalita J, Misra UK, Ranjan P. Predictors of long-term neurological sequelae of tuberculous meningitis: a multivariate analysis. *Eur J Neurol* 2007; **14**:33–7.

Kamei S, Sekizawa T, Shiota H *et al.* Evaluation of combination therapy using aciclovir and corticosteroid in adult patients with herpes simplex virus encephalitis. *J Neurol Neurosurg Psychiatry* 2005; **76**:1544–9.

Kapur N, Barker S, Burrows EH *et al.* Herpes simplex encephalitis: long term magnetic resonance imaging and neuropsychological profile. *J Neurol Neurosurg Psychiatry* 1994; **57**:1334–42.

Kastrup O, Maschke M, Diener HC *et al.* Progressive multifocal leukoencephalopathy limited to the brainstem. *Neuroradiology* 2002; **44**:227–9.

Kennedy DH, Fallon RJ. Tuberculous meningitis. *J Am Med Assoc* 1979; **241**:264–8.

Kennedy PGE. A retrospective study of forty-six cases of herpes simplex encephalitis seen in Glasgow between 1962 and 1985. *Q J Med* 1988; **68**:533–40.

Kesserling J, Miller DH, Robb SA *et al.* Acute disseminated encephalomyelitis. MRI findings and the distinction from multiple sclerosis. *Brain* 1990; **113**:291–302.

Kieburtz K, Zettelmaier AE, Ketonen L *et al.* Manic syndrome in AIDS. *Am J Psychiatry* 1991; **148**:1068–70.

Kieburtz K, Simpson D, Yiannoutsos C *et al.* A randomized trial of amitriptyline and mexiletine for painful neuropathy in HIV infection. AIDS Clinical Trial Group 242 Protocol Team. *Neurology* 1998; **51**:1682–8.

Kirby GH, Davis TK. Psychiatric aspects of epidemic encephalitis. *Arch Neurol Psychiatry* 1921; **5**:491–551.

Kirk JL, Fine DP, Sexton DJ *et al.* Rocky Mountain spotted fever. A clinical review based on 48 confirmed cases, 1943–1986. *Medicine* 1990; **69**:35–45.

Kleinschmidt-DeMasters BK, Amlie-Lefond C, Gilden DH. The patterns of varicella zoster encephalitis. *Hum Pathol* 1996; **27**:927–38.

Klempner MS, Hu LT, Evans J *et al.* Two controlled trials of antibiotic treatment in patients with persistent symptoms and a history of Lyme disease. *N Engl J Med* 2001; **345**:85–92.

Kobayashi K, Mizukoshi C, Aoki T *et al.* Borrelia-burgdorferi-seropositive encephalomyelopathy: Lyme neuroborreliosis? An autopsied report. *Dement Geriatr Cogn Disord* 1997; **8**:384–90.

Kochar DK, Garg P, Bumb RA *et al.* Divalproex sodium in the management of post-herpetic neuralgia: a randomized double-blind placebo-controlled study. *QJM* 2005; **98**:29–34.

Koehler K, Jakumeit U. Subacute sclerosing panencephalitis presenting as Leonhard's speech-prompt catatonia. *Br J Psychiatry* 1976; **129**:29–31.

Koralnik IJ, Boden O, Mai VX *et al.* JC virus DNA load in patients with and without progressive multifocal leukoencephalopathy. *Neurology* 1999; **52**:253–60.

Koskiniemi M, Donner M, Pettay O. Clinical appearance and outcome in mumps encephalitis in children. *Acta Paediatr Scand* 1983; **72**:603–9.

Koskiniemi M, Piiparinen H, Mannonen L *et al.* Herpes encephalitis is a disease of middle aged and elderly people: polymerase chain reaction for detection of herpes simplex virus in the CSF of 516 patients with encephalitis. *J Neurol Neurosurg Psychiatry* 1996; **60**:174–8.

Krumholz S, Luhan JA. Encephalitis associated with herpes zoster. *Arch Neurol Psychiatry* 1945; **53**:59–67.

Krupp LB, Lipton RB, Swerdlow ML *et al.* Progressive multifocal leukoencephalopathy: clinical and demographic features. *Ann Neurol* 1985; **17**:344–9.

Krupp LB, Hyman LG, Grimson R *et al.* Study and treatment of post Lyme disease (STOP-LD). A randomized double masked clinical trial. *Neurology* 2003; **60**:1923–30.

Kuker W, Mader I, Nagele T *et al.* Progressive multifocal leukoencephalopathy; value of diffusion-weighted and contrast-enhanced magnetic resonance imaging for diagnosis and treatment control. *Eur J Neurol* 2006; **13**:819–26.

Lampert P, Tom MI, Cumings JN. Encephalopathy in Whipple's disease: a histochemical study. *Neurology* 1962; **12**:65–71.

Landau WM, Luse SA. Relapsing inclusion encephalitis (Dawson type) of eight years' duration. *Neurology* 1958; **8**:669–76.

Leech RW, Harris JC. The neuropathology of Western equine and St Louis encephalitis: a review of the 1975 North Dakota epidemic. *J Neuropathol Exp Neurol* 1977; **36**:611.

Levitt LP, Rich TA, Kinde SW *et al.* Central nervous system mumps: a review of 64 cases. *Neurology* 1970; **20**:829–34.

Lewis L, Taylor HG, Sorem MB. Japanese B encephalitis: clinical observations in an outbreak on Okinawa Shima. *Arch Neurol Psychiatry* 1947; **57**:430–63.

Lilly R, Cummings JL, Benson DF *et al.* The human Kluver–Bucy syndrome. *Neurology* 1983; **33**:1141–5.

Lima MA, Drislane FW, Koralink IJ. Seizures and their outcome in progressive multifocal leukoencephalopathy. *Neurology* 2006; **66**:262–4.

Linda H, von Heijne A, Major EO *et al.* Progressive multifocal leukoencephalopathy after natalizumab therapy. *N Engl J Med* 2009; **361**:1081–7.

Linnemann CC, Alvira MM. Pathogenesis of varicella-zoster angiitis in the CNS. *Arch Neurol* 1980; **37**:239–40.

Lo KC, Geddes JF, Daniels RS *et al.* Lack of detection of influenza genes in archived formalin-fixed, paraffin wax-embedded brain samples of encephalitis lethargic patients from 1916 to 1920. *Virchows Arch* 2003; **442**:591–6.

Logigian EL, Kaplan RF, Steere AC. Chronic neurologic manifestations of Lyme disease. *N Engl J Med* 1990; **323**:1438–44.

Louis ED, Lynch T, Kaufman P *et al.* Diagnostic guidelines in central nervous system Whipple's disease. *Ann Neurol* 1996; **40**: 561–8.

Luft BJ, Hafner R, Korzun AH *et al.* Toxoplasmic encephalitis in patients with the acquired immunodeficiency syndrome. *N Engl J Med* 1993; **329**:995–1000.

Lukehart SA, Hook EW, Baker-Zander SA *et al.* Invasion of the central nervous system by *Treponema pallidum*: implications for diagnosis and treatment. *Ann Int Med* 1988; **109**:855–62.

Lyketsos CG, Hanson AL, Fishman M *et al.* Manic syndrome early and late in the course of HIV. *Am J Psychiatry* 1993; **150**:326–7.

McArthur JC. Neurologic manifestations of AIDS. *Medicine* 1987; **66**:407–11.

McCall S, Henry JM, Reid AH *et al.* Influenza RNA not detected in archival brain tissues from acute encephalitis lethargica cases or in postencephalitic Parkinson cases. *J Neuropathol Exp Neurol* 2001; **60**:696–704.

McCombe JA, Auer RN, Maingat FG *et al.* Neurologic immune reconstitution inflammatory syndrome in HIV/AIDS: outcome and epidemiology. *Neurology* 2009; **72**:835–41.

McCormick WF, Rodnitzky RL, Schochet SS *et al.* Varicella-zoster encephalomyelitis. *Arch Neurol* 1969; **21**:559–70.

McCowan PK, Cook LC, Cantab BA. Oculogyric crises in chronic epidemic encephalitis. *Brain* 1928; **51**:285–309.

McGrath N, Anderson NE, Croxson MC *et al.* Herpes simplex encephalitis treated with acyclovir: diagnosis and long term outcome. *J Neurol Neurosurg Psychiatry* 1997; **63**:321–6.

McJunkin JE, de los Reyes EC, Irazuzta JE *et al.* La Crosse encephalitis in children. *N Engl J Med* 2001; **344**:801–7.

McKendrick GDW, Grose RJ. Tuberculous meningitis. *J Neurol Neurosurg Psychiatry* 1957; **20**:198–201.

Mahalingam R, Wellis M, Wolf W *et al.* Latent varicella-zoster viral DNA in human trigeminal and thoracic ganglia. *N Engl J Med* 1990; **323**:627–9.

Maher J, Choudri S, Halliday W *et al.* AIDS dementia complex with generalized myoclonus. *Mov Disord* 1997; **12**:593–7.

Malouf R, Jacquette G, Dobkin J *et al.* Neurologic disease in human immunodeficiency virus-infected drug abusers. *Arch Neurol* 1990; **47**:1002.

Manzel K, Tranel D, Cooper G. Cognitive and behavioral abnormalities in a case of central nervous system Whipple disease. *Arch Neurol* 2000; **57**:399–403.

Marchioni E, Ravaglia S, Piccolo G *et al.* Postinfectious inflammatory disorders. Subgroups based on prospective follow-up. *Neurology* 2005; **65**:1057–65.

Marlowe WB, Mancall EL, Thomas TJ. Complete Kluver–Bucy syndrome in man. *Cortex* 1975; **11**:53–9.

Marques A. Chronic Lyme disease: a review. *Infect Dis Clin North Am* 2008; **22**:341–60.

Marsden CD, Fowler TJ. *Clinical Neurology*, 2nd edn. London: Arnold, 1998.

Marton R, Gotlieb-Steimatsky T, Klein C *et al.* Acute herpes simplex encephalitis: clinical assessment and prognostic data. *Acta Neurol Scand* 1996; **93**:149–55.

Masliah E, Heaton RK, Marcotte TD *et al.* Dendritic injury is a pathological substrate for human immunodeficiency virus-related cognitive disorders. *Ann Neurol* 1997; **42**:963–72.

Matthews BR, Jones LK, Saad DA *et al.* Cerebellar ataxia and central nervous system Whipple disease. *Arch Neurol* 2005; **62**: 618–20.

Mayo DR, Booss J. Varicella zoster-associated neurologic disease without skin lesions. *Arch Neurol* 1989; **46**:313–15.

Melanson M, Chalk C, Georgevich L *et al.* Varicella-zoster virus DNA in CSF and arteries in delayed contralateral hemiplegia: evidence for viral invasion of cerebral arteries. *Neurology* 1996; **47**:569–70.

Meninger KA. Influenza and schizophrenia. *Am J Psychiatry* 1926; **82**:469–529.

Merritt HH, Adams RD, Solomon HC. *Neurosyphilis*. New York: Oxford University Press, 1946.

Meurers B, Kohlepp WRG, Rohrbach E *et al.* Histopathological findings in the central and peripheral nervous system in neuroborreliosis. A report of three cases. *J Neurol* 1990; **327**:113–16.

Miller AE. Selective decline in cellular immune response to varicella-zoster in the elderly. *Neurology* 1980; **30**:582–7.

Miller HG, Stanton JB, Gibbons JL. Para-infectious encephalomyelitis and related syndromes. *Q J Med* 1956; **25**:427–505.

Miller JQ, Price TR. The nervous system in Rocky Mountain spotted fever. *Neurology* 1972; **22**:561–70.

Moll NM, Rietsch AM, Ransohoff AJ *et al.* Cortical demyelination in PML and MS: similarities and differences. *Neurology* 2008; **70**:336–43.

Morgello S, Cho ES, Nielsen S *et al.* Cytomegalovirus encephalitis in patients with acquired immunodeficiency syndrome: an autopsy study of 30 cases and a review of the literature. *Hum Pathol* 1987; **18**:289–97.

Morgello S, Estanislao L, Simpson D *et al.* HIV-associated distal sensory polyneuropathy in the era of highly active antiretroviral therapy: the Manhattan HIV Brain Bank. *Arch Neurol* 2004; **61**: 546–51.

Mori T, Ebe T. Analysis of cases of central nervous system fungal infections reported in Japan between January 1979 and June 1989. *Intern Med* 1992; **31**:174–9.

Moscovich DG, Singh MB, Eva FJ *et al.* Acute disseminated encephalomyelitis presenting as an acute psychotic state. *J Nerv Ment Dis* 1995; **183**:116–17.

Moulignier A, Mikol J, Pialoux G *et al.* AIDS-associated progressive multifocal leukoencephalopathy revealed by new-onset seizure. *Am J Med* 1995; **99**:64–8.

Mulder DW, Parrott M, Thaler M. Sequelae of Western equine encephalitis. *Neurology* 1951; **1**:318–27.

Nagel MA, Cohrs RJ, Mahalingam R *et al.* The varicella zoster virus vasculopathies: clinical, CSF, imaging, and virologic features. *Neurology* 2008; **70**:853–60.

Navia BA, Price RW. The acquired immunodeficiency syndrome dementia complex as the presenting or sole manifestation of human immunodeficiency virus infection. *Ann Neurol* 1987; **44**:656–9.

Navia BA, Jordan BD, Price RW. The AIDS dementia complex. I. Clinical features. *Ann Neurol* 1986a; **19**:517–24.

Navia BA, Jordan BD, Price RW. The AIDS dementia complex. II. Neuropathology. *Ann Neurol* 1986b; **19**:525–35.

Navia BA, Petito CK, Gold JWM *et al.* Cerebral toxoplasmosis complicating the acquired immune deficiency syndrome: clinical and neuropathological findings in 27 patients. *Ann Neurol* 1986c; **19**:224–38.

Ngoungou EB, Preux PM. Cerebral malaria and epilepsy. *Epilepsia* 2008; **49**(Suppl. 6):19–24.

Noah DL, Drenzek CL, Smith JS *et al.* Epidemiology of human rabies in the United States, 1980–1996. *Ann Intern Med* 1998; **128**:922–30.

Northam RS, Singer HS. Postencephalitic acquired Tourette-like syndrome in a child. *Neurology* 1991; **41**:592–3.

Ohya T, Martinez AJ, Jabbour JT *et al.* Subacute sclerosing panencephalitis. *Neurology* 1974; **24**:211–18.

Oksi J, Kalimo H, Marttila RJ *et al.* Inflammatory brain changes in Lyme borreliosis. A report on three patients and review of the literature. *Brain* 1996; **119**:2143–54.

Oo MM, Aikawa M, Than T *et al.* Human cerebral malaria: pathological study. *J Neuropathol Exp Neurol* 1987; **46**:223–31.

O'Riordan JI, Gomez-Anson B, Moseley IF *et al.* Long term MRI follow-up of patients with post infectious encephalomyelitis: evidence for a monophasic disease. *J Neurol Sci* 1999; **167**:132–6.

Ozturk A, Gurses C, Baykan B *et al.* Subacute sclerosing panencephalitis: clinical and magnetic resonance imaging evaluation of 36 patients. *J Child Neurol* 2002; **17**:25–9.

Pachner AR, Steere AC. The triad of neurologic manifestations of Lyme disease: meningitis, cranial neuritis and radiculoneuritis. *Neurology* 1985; **35**:47–53.

Padgett BL, Walker DL, Zurheim GM *et al.* Cultivation of papova-like virus from human brain with progressive multifocal leukoencephalopathy. *Lancet* 1971; **1**:1257–60.

Panegyres PK, Edis R, Beaman M *et al.* Primary Whipple's disease of the brain: characterization of the clinical symptoms and molecular diagnosis. *QJM* 2006; **99**:609–23.

Panicker JN, Nagaraja D, Kovoor JM *et al.* Descriptive study of acute disseminated encephalomyelitis and evaluation of functional outcome predictors. *J Postgrad Med* 2010; **56**:12–6.

Parr J, Horoupian DS, Winkelman AC. Cerebellar form of progressive multifocal leukoencephalopathy (PML). *Can J Neurol Sci* 1979; **6**:123–8.

Pascual-Sedano B, Iranzo A, Marti-Fabregas J *et al.* Prospective study of new-onset seizures in patients with human immunodeficiency virus infection. *Arch Neurol* 1999; **56**:609–12.

Paskavitz JF, Anderson CA, Filley CM *et al.* Acute arcuate fiber demyelination encephalopathy following Epstein–Barr virus infection. *Ann Neurol* 1995; **38**:127–31.

Petersen I, Thomas JM, Hamilton WT *et al.* Risk and predictors of fatigue after infectious mononucleosis in a large primary-care cohort. *Q J Med* 2006; **99**:49–55.

Petito CK, Cho ES, Lemann W *et al.* Neuropathology of acquired immunodeficiency syndrome (AIDS): an autopsy review. *J Neuropathol Exp Neurol* 1986; **45**:635–46.

Pfister H-W, Preac-Mursic V, Wilske B *et al.* Catatonic syndrome in acute severe encephalitis due to *Borrelia burgdorferi* infection. *Neurology* 1993; **43**:433–5.

Piccolo I, Causarano R, Sterzi R *et al.* Chorea in patients with AIDS. *Acta Neurol Scand* 1999; **100**:332–6.

Pollock S, Lewis PD, Kendall B. Whipple's disease confined to the central nervous system. *J Neurol Neurosurg Psychiatry* 1981; **44**:1104–9.

Pommer B, Pilz P, Harrer G. Transient global amnesia as a manifestation of Epstein–Barr virus encephalitis. *J Neurol* 1983; **229**: 125–7.

Porras C, Barboza JJ, Fuenzalida E *et al.* Recovery from rabies in man. *Ann Intern Med* 1976; **85**:44–8.

Porter SB, Sande MA. Toxoplasmosis of the central nervous system in the acquired immunodeficiency syndrome. *N Engl J Med* 1992; **327**:1643–8.

Powers JM, Rawe SE. A neuropathological study of Whipple's disease. *Acta Neuropathol* 1979; **48**:223–6.

Pradhan S, Pandey N, Shashank S *et al.* Parkinsonism due to predominant involvement of substantia nigra in Japanese encephalitis. *Neurology* 1999; **53**:1781–6.

Prashanth LK, Taly AB, Ravi V *et al.* Adult subacute sclerosing panencephalitis: clinical profile of 39 patients from a tertiary care center. *J Neurol Neurosurg Psychiatry* 2006; **77**:630–3.

Praveen-Kumar S, Sinha S, Taly AB *et al.* Electroencephalographic and imaging profile in a subacute sclerosing panencephalitis (SSPE) cohort: a correlative study. *Clin Neurophysiol* 2007; **118**: 1947–54.

Price RW, Nielsen S, Horten B *et al.* Progressive multifocal leukoencephalopathy: a burnt-out case. *Ann Neurol* 1983; **13**:485–9.

Price RW, Berew B, Sidtis J *et al.* The brain in AIDS: central nervous system HIV-1 infection and AIDS dementia complex. *Science* 1988; **239**:586–91.

Pruss H, Katchanov J, Zschederlein R *et al.* A patient with cerebral Whipple disease with gastric involvement but not gastrointestinal symptoms: a consequence of local protective immunity? *J Neurol Neurosurg Psychiatry* 2007; **78**:896–8.

Przelomski MM, O'Rourke E, Grady GF *et al.* Eastern equine encephalitis in Massachusetts. *Neurology* 1988; **38**:736–9.

Rabkin JG, Wagner GJ, Rabkin R. Fluoxetine treatment for depression in patients with HIV and AIDS: a randomized, placebo-controlled trial. *Am J Psychiatry* 1999; **156**:101–7.

Rabkin JG, Wagner GJ, McElhiney MC *et al.* Testosterone versus fluoxetine for depression and fatigue in HIV/AIDS: a placebo-controlled trial. *J Clin Psychopharmacol* 2004; **24**:379–85.

Rabkin JG, McElhiney MC, Rabkin R *et al.* Modafinil treatment for fatigue in HIV/AIDS: a randomized placebo-controlled study. *J Clin Psychiatry* 2010; **71**:707–15.

Rail D, Scholtz C, Swash M. Post-encephalitic parkinsonism: current experience. *J Neurol Neurosurg Psychiatry* 1981; **44**:670–6.

Raoult D, Birg ML, La Scola B *et al.* Cultivation of the bacillus of Whipple's disease. *N Engl J Med* 2000; **342**:620–5.

Relman DA., Schmidt TM, MacDermott RP *et al.* Identification of the uncultured bacillus of Whipple's disease. *N Engl J Med* 1992; **327**:293–301.

Reyes MG, Gardner JJ, Poland JD *et al.* St Louis encephalitis: quantitative, histologic and immunofluorescent studies. *Arch Neurol* 1981; **38**:329–34.

Rich AR, McCordock HA. The pathogenesis of tuberculous meningitis. *Bull Johns Hopkins Hosp* 1933; **52**:5–37.

Richardson EP. Progressive multifocal leukoencephalopathy. *N Engl J Med* 1961; **265**:815–23.

Richter RW, Shimojyo S. Neurologic sequelae of Japanese B encephalitis. *Neurology* 1961; **11**:553–9.

Risk WS, Haddad FS. The variable natural history of subacute sclerosing panencephalitis: a study of 118 cases from the middle east. *Arch Neurol* 1979; **36**:610–14.

Risk WS, Haddad FS, Chemali R. Substantial spontaneous long-term improvement in subacute sclerosing panencephalitis: six cases from the middle east and overview of the literature. *Arch Neurol* 1978; **35**:495–502.

Romanul FCA, Radvany J, Rosales RK. Whipple's disease confined to the brain: a case confirmed clinically and pathologically. *J Neurol Neurosurg Psychiatry* 1977; **40**:901–9.

Rothschild D. Dementia paralytica accompanied by manic-depressive and schizophrenic psychoses. *Am J Psychiatry* 1940; **96**: 1043–60.

Roulet Perez E, Mader P, Cotting J *et al.* Acute fatal parainfectious cerebellar swelling in two children. A rare or an overlooked situation? *Neuropediatrics* 1993; **24**:346–51.

Roze E, Thiebault MM, Mazevet D *et al.* Neurologic sequelae after severe falciparum malaria in adult travelers. *Eur Neurol* 2001; **46**:192–7.

Russell J, Fisher M, Zivin JA *et al.* Status epilepticus and Epstein–Barr virus encephalopathy. Diagnosis by modern serologic techniques. *Arch Neurol* 1985; **42**:789–92.

Russell DS. The nosological unity of acute haemorrhagic leucoencephalitis and acute disseminated encephalomyelitis. *Brain* 1955; **78**:369–76.

Russell RR, Donald JC. The neurological complications of mumps. *BMJ* 1958; **2**:27–30.

Sage JI, Weinstein MP, Miller DC. Chronic encephalitis possibly due to herpes simplex virus: two cases. *Neurology* 1985; **35**: 1470–2.

Salib EA. Subacute sclerosing panencephalitis (SSPE) presenting at the age of 21 as a schizophrenia-like state with bizarre dysmorphophobic features. *Br J Psychiatry* 1988; **152**:709–10.

Sanchez-Portocarrero J, Perez-Cecilia E, Jimenez-Escrig A *et al.* Tuberculous meningitis: clinical characteristics and comparison with cryptococcal meningitis in patients with human immunodeficiency virus infection. *Arch Neurol* 1996; **53**:671–6.

Sands IJ. The acute psychiatric type of epidemic encephalitis. *Am J Psychiatry* 1928; **84**:975–87.

Schlesinger RD, Crelinsten GL. Infectious mononucleosis dominated by neurologic symptoms and signs. *Can Med Assoc J* 1977; **117**: 652–3.

Schnell RG, Dyck PJ, Bowie EJW *et al.* Infectious mononucleosis: neurologic and EEG findings. *Medicine* 1966; **45**:51–63.

Schnorf H, Diserens K, Schnyder H *et al.* Corticosteroid-responsive postmalaria encephalopathy characterized by motor aphasia, myoclonus, and postural tremor. *Arch Neurol* 1998; **55**:417–20.

Schube PG. Emotional states of general paresis. *Am J Psychiatry* 1934; **91**:625–38.

Schultz DR, Barthal JS, Garrett C. Western equine encephalitis with rapid onset parkinsonism. *Neurology* 1977; **27**:1095–6.

Schwartz MA, Selhorst JB, Ochs AI *et al.* Oculomasticatory myorhythmia: a unique movement disorder occurring in Whipple's disease. *Ann Neurol* 1986; **20**:677–83.

Scully EP, Baden LR, Katz JT. Fungal brain infections. *Curr Opin Neurol* 2008; **21**:347–52.

Sellal F, Mohr M, Collard M. Dementia in a 58-year-old woman. *Lancet* 1996; **347**:236.

Sewell DD, Jeste DV, Atkinson JH *et al.* HIV-associated psychosis: a study of 20 cases. *Am J Psychiatry* 1994a; **151**:237–42.

Sewell DD, jeste DV, McAdams LA *et al.* Neuroleptic treatment of HIV-associated psychosis. *Neuropsychopharmacology* 1994b; **10**:223–9.

Shadick NA, Phillis CB, Logigian EL *et al.* The long-term clinical outcomes of Lyme disease. *Ann Intern Med* 1994; **121**:560–7.

Sharer LR, Epstein LG, Cho ES *et al.* HTLV-III and vacuolar myelopathy in AIDS. *N Engl J Med* 1986; **315**:62–3.

Shill HA, Stacy MA. Malignant catatonia secondary to sporadic encephalitis lethargica. *J Neurol Neurosurg Psychiatry* 2000; **69**:402–3.

Silverstein A, Steinberg G, Nathanson M. Nervous system involvement in infectious mononucleosis. *Arch Neurol* 1972; **26**:353–8.

Simon RP. Neurosyphilis. *Arch Neurol* 1985; **42**:606–13.

Simpson DM, McArthur JC, Olney R *et al.* Lamotrigine for HIV-associated painful sensory neuropathies: a placebo-controlled trial. *Neurology* 2003; **60**:1508–14.

Skoldenberg B, Forsgren M, Alestig K *et al.* Acyclovir versus vidarabine in herpes simplex encephalitis. Randomized multicentre study in consecutive Swedish patients. *Lancet* 1984; **2**:707–11.

Skoldenberg B, Aurelius E, Hjalmarsson A *et al.* Incidence and pathogenesis of clinical relapse after herpes simplex encephalitis in adults. *J Neurol* 2006; **253**:163–70.

Smardel JE, Bailey P, Baker AB. Sequelae of the arthropod-borne encephalitides. *Neurology* 1958; **8**:873–96.

Smith WT, Frency JM, Guttsmann M *et al.* Cerebral complications of Whipple's disease. *Brain* 1965; **88**:137–50.

Solomon T, Dung NM, Kneen R *et al.* Japanese encephalitis. *J Neurol Neurosurg Psychiatry* 2000; **68**:405–15.

Solomon T, Dung NM, Kneen R *et al.* Seizures and raised intracranial pressure in Vietnamese patients with Japanese encephalitis. *Brain* 2002; **125**:1084–93.

Souraud JB, Faivre A, Waku-Kouomou D *et al.* Adult fulminant subacute sclerosing panencephalitis: pathological and molecular studies – a case report. *Clin Neuropathol* 2009; **28**:213–8.

Srinivasan A, Burton EC, Kuehnert MJ *et al.* Transmission of rabies virus from an organ donor to four transplant recipients. *N Engl J Med* 2005; **352**:1103–11.

Steinbach JP, Melms A, Skalej M *et al.* Delayed resolution of white matter changes following therapy of B burgdorferi encephalitis. *Neurology* 2005; **64**:758–9.

Storm-Mathisen A. General paresis: a follow-up study of 203 patients. *Acta Psychiatr Scand* 1969; **45**:118–32.

Strauss SE, Reinhold W, Smith HA *et al.* Endonuclease analysis of viral DNA from varicella and subsequent zoster infections in the same patients. *N Engl J Med* 1984; **311**:1362–4.

Sweeney CJ, Gilden DG. Ramsey Hunt syndrome. *J Neurol Neurosurg Psychiatry* 2001; **71**:149–54.

Sworn MJ, Urich H. Acute encephalitis in infectious mononucleosis. *J Pathol* 1970; **100**:201–5.

Tan K, Roda R, Ostrow L *et al.* PML-IRIS in patients with HIV infection: clinical manifestations and treatment with steroids. *Neurology* 2009; **72**:1458–64.

Taylor EW, McDonald CA. Forced conjugate upward movement of the eyes following epidemic encephalitis. *Arch Neurol Psychiatry* 1928; **19**:95–103.

Taylor SW, Lee DH, Jackson AC. Herpes simplex encephalitis presenting with exclusively frontal lobe involvement. *J Neurovirol* 2007; **13**:477–81.

Tenembaum S, Chamoles N, Fejerman N. Acute disseminated encephalomyelitis. A long-term follow-up of 84 pediatric patients. *Neurology* 2002; **59**:1224–31.

Thompson JA. Mumps: a cause of acquired aqueductal stenosis. *J Pediatr* 1979; **94**:923–4.

Torgovnik J, Arsura EL, Lala D. Cytomegalovirus ventriculoencephalitis presenting as a Wernicke's encephalopathy-like syndrome. *Neurology* 2000; **55**:1910–13.

Toro G, Roman G. Cerebral malaria. A disseminated vasculomyelopathy. *Arch Neurol* 1978; **35**:271–5.

Townsend JJ, Wolinsky JS, Baringer JR *et al.* Acquired toxoplasmosis: a neglected cause of treatable central nervous system disease. *Arch Neurol* 1975; **32**:335–43.

Traub M, Colchester ACF, Kingsley DPE *et al.* Tuberculosis of the central nervous system. *Q J Med* 1984; **53**:81–100.

Tselis A, Duman R, Storch GA *et al.* Epstein–Barr virus encephalomyelitis diagnosed by polymerase chain reaction: detection of the genome in the CSF. *Neurology* 1997; **48**:1351–5.

Upton A, Gumpert J. Electroencephalography in diagnosis of herpes-simplex encephalitis. *Lancet* 1970; **1**:650–2.

Vaheri A, Julkunen I, Koskiniemi M. Chronic encephalomyelitis with specific increase in intrathecal mumps antibodies. *Lancet* 1982; **2**:685–8.

Van Bogaert L. Ocular paroxysms and palilalia. *J Nerv Ment Dis* 1934; **80**:48–61.

Varney NR, Roberts SJ, Springer JA *et al.* Neuropsychiatric sequelae of cerebral malaria in Vietnam veterans. *J Nerv Ment Dis* 1997; **185**:695–703.

Wagner GJ, Rabkin R. Effects of dextroamphetamine on depression and fatigue in men with HIV: a double-blind, placebo-controlled trial. *J Clin Psychiatry* 2000; **61**:436–40.

Walsh TJ, Hier DB, Caplan LR. Fungal infections of the central nervous system: comparative analysis of risk factors and clinical signs in 57 patients. *Neurology* 1985; **35**:1654–7.

Wasay M, Diaz-Arrastia R, Suss RA *et al.* St. Louis encephalitis: a review of 11 cases in a Dallas, Tex, epidemic. *Arch Neurol* 2000; **57**:114–8.

Warrell DA, Looareesuwan S, Warrell MJ *et al.* Dexamethasone proves deleterious in cerebral malaria. A double-blind trial in 100 comatose patients. *N Engl J Med* 1982; **306**:313–19.

Wasay M, Diaz-Arrastia R, Suss RA *et al.* St Louis encephalitis. A review of 11 cases in a 1995 Dallas, Tex, epidemic. *Arch Neurol* 2000; **57**:114–18.

Watson CP, Vernich L, Chipman M *et al.* Nortriptyline versus amitriptyline in postherpetic neuralgia: a randomized trial. *Neurology* 1998; **51**:1166–71.

Weaver S, Rosenblum MK, DeAngelis LM. Herpes varicella zoster encephalitis in immunocompromised patients. *Neurology* 1999; **52**:193–5.

Weinberg A, Li S, Palmer M *et al.* Quantitative CSF PCR in Epstein–Barr virus infections of the central nervous system. *Ann Neurol* 2002; **52**:543–8.

Whipple GH. A hitherto undescribed disease characterized anatomically by deposits of fat and fatty acids in the intestinal and mesenteric lymphatic tissues. *Johns Hopkins Hosp Bull* 1907; **18**:382–91.

Whitley RJ, Alford CA, Hirsch MS *et al.* Vadarabine versus acyclovir therapy in herpes simplex encephalitis. *N Engl J Med* 1986; **314**:144–9.

Wildemann B, Jarius S, Hartmann M *et al.* Acute disseminated encephalomyelitis following vaccination against human papilloma virus. *Neurology* 2009; **72**:2132–3.

Wiley CA, Masliah E, Morey M *et al.* Neocortical damage during HIV infection. *Ann Neurol* 1991; **29**:651–7.

Wilkins RH, Brody IA. Encephalitis lethargica. *Arch Neurol* 1968; **18**:324–8.

Williams BB, Lerner AM. Some previously unrecognized features of herpes simplex encephalitis. *Neurology* 1978; **28**:1193–6.

Williams M, Smith HV. Mental disturbances in tuberculous meningitis. *J Neurol Neurosurg Psychiatry* 1954; **17**:173–82.

Willoughby RE, Tieves KS, Hoffman GM *et al.* Survival after treatment of rabies with induction of coma. *N Engl J Med* 2005; **352**: 2508–14.

Wilson LG. Viral encephalopathy mimicking functional psychosis. *Am J Psychiatry* 1976; **133**:165–70.

Wolinsky JS, Swoveland P, Johnson KP *et al.* Subacute measles encephalitis complicating Hodgkin's disease in an adult. *Ann Neurol* 1977; **1**:452–4.

Wong MC, Suite ND, Labar DR. Seizures in human immunodeficiency virus infection. *Arch Neurol* 1990; **47**:640–2.

Wulf CH. Subacute sclerosing panencephalitis: serial electroencephalographic studies. *J Neurol Neurosurg Psychiatry* 1982; **45**: 418–21.

Yamada S, Kaneyama T, Nagaya S *et al.* Relapsing herpes simplex encephalitis: pathological confirmation of viral reactivation. *J Neurol Neurosurg Psychiatry* 2003; **74**:262–4.

Yigit A, Sarikaya S. Myoclonus relieved by carbamazepine in subacute sclerosing panencephalitis. *Epileptic Disord* 2006; **8**: 77–80.

Young CA, Humphrey DM, Ghadiali EJ *et al.* Short-term memory impairment in an alert patient as a presentation of herpes simplex encephalitis. *Neurology* 1992; **42**:260–1.

Yousry TA, Major EO, Ryschkewitsch C *et al.* Evaluation of patients treated with natalizumab for progressive multifocal leukoencephalopathy. *N Engl J Med* 2006; **354**:924–33.

Zambito Marsala S, Ferracci F, Cecotti L *et al.* Post-malaria neurological syndrome: clinical and laboratory findings in one patient. *Neurol Sci* 2006; **27**:442–4.

Zifko U, Linder K, Wimberger D *et al.* Jarisch–Herxheimer reaction in a patient with neurosyphilis. *J Neurol Neurosurg Psychiatry* 1994; **57**:865–7.

Zunt JR, Tu RK, Anderson DM *et al.* Progressive multifocal leukoencephalopathy presenting as human immunodeficiency virus type 1 (HIV)-associated dementia. *Neurology* 1997; **49**:263–5.

16

Prion diseases

16.1 CREUTZFELDT–JAKOB DISEASE

Creutzfeldt–Jakob disease, also known as transmissible spongiform encephalopathy, is characterized by a fairly rapidly progressive dementia accompanied, in most cases, by myoclonus. In the past, this disease was believed to represent a 'slow virus' infection; however, it is now known that it is caused not by a virus but by a unique agent known as a prion. This is a rare disorder, occurring at a yearly rate of 1–2 cases per 1 000 000, and is equally common in males and females. About 85 percent of cases occur on a sporadic basis, 10 percent are inherited on an autosomal dominant basis, and the remainder represent iatrogenic infections, as for example with dura mater grafts.

Clinical features

On average, sporadic cases appear in the seventh decade, but the range is wide, from late teenage years to the tenth decade. Inherited cases tend to appear a bit earlier, mostly in the early sixth decade. Iatrogenic cases appear anything from 1 to 30 years after the infectious event. Although most cases appear subacutely, over weeks to months, fulminant onsets spanning only a few days have been reported.

The overall symptomatology of Creutzfeldt–Jakob disease has been described in several studies (Appleby *et al.* 2009; Brown *et al.* 1986, 1994; Collins *et al.* 2006; Cooper *et al.* 2006; Glatzel *et al.* 2005; Roos *et al.* 1973). Various presentations are possible, including dementia, mood changes, ataxia or visual changes. Dementia, in addition to memory loss, may also be characterized by aphasia and agnosia. Mood changes typically include depression and anxiety, and at times one may also see emotional lability. When ataxia is present at the onset, one speaks of the Brownell–Oppenheimer variant (Brownell and Oppenheimer 1965), and here one also typically finds dysarthria. Visual changes at the onset constitute the Heidenhain variant (Kropp *et al.* 1999), and here one may find not only hemianopia or cortical blindness, but also blurring of vision and visual hallucinations. Other, rare, presentations include psychosis (Dervaux *et al.* 2004; Zeng *et al.* 2001), aphasia (Mandell *et al.* 1989), the alien hand sign (MacGowan *et al.* 1997), or mania (Lendvai *et al.* 1999). With progression, almost all patients develop a profound dementia, often with akinetic mutism. This dementia is accompanied, in almost 90 percent of cases, by myoclonus, which in turn may be quite stimulus sensitive, as for example to touch or to sudden loud noises. Other symptoms, each seen in a small minority, include rigid–akinetic parkinsonism, upper or lower motor neuron signs, and seizures.

Magnetic resonance imaging generally discloses cortical atrophy, the progression of which may be monitored with serial scans. T2-weighted and fluid-attenuated inversion recovery (FLAIR) imaging, and, with greater sensitivity, diffusion-weighted imaging, may display increased signal intensity in the striatum and in the cerebral and cerebellar cortices in close to three-quarters of all cases (Meissner *et al.* 2009; Zerr *et al.* 2009): within the cerebral cortex the increased signal intensity follows the cortical ribbon, displaying a 'gyriform' pattern (Shiga *et al.* 2004; Tschampa *et al.* 2005).

The electroencephalogram (EEG) may show generalized slowing (Burger *et al.* 1972), and in some cases frontal intermittent rhythmic delta activity (FIRDA) has been noted (Wieser *et al.* 2004). The most characteristic EEG findings, however, are periodic spike-and-slow-wave complexes (Collins *et al.* 2006; Levy *et al.* 1986; May 1986; Steinhoff *et al.* 2004), which are eventually seen in anywhere from one-half to three-quarters of all cases. Importantly, the EEG may become abnormal only as the disease progresses, and consequently, when the initial EEG fails to reveal this finding, it may be appropriate to perform serial EEGs.

The cerebrospinal fluid (CSF) is acellular with a normal glucose; in a very small minority the total protein level may show a mild elevation. Most importantly, the 14-3-3 protein is found in anywhere from 86 to 100 percent of sporadic cases (Collins *et al.* 2006; Lemstra *et al.* 2000; Sanchez-Juan *et al.* 2006; Van Everbroeck *et al.* 2005).

Although, at present, a definitive ante mortem diagnosis may be made only by brain biopsy, this is rarely performed.

Course

This disease progresses rapidly, with death occurring, on average, within about 6 months. The overall range, however, is wide, with some dying in as little as a month and others surviving for up to 3 years (Masters and Richardson 1978; Pocchiara *et al.* 2004; Will and Mathews 1984). The rapid progression is one of the hallmarks of Creutzfeldt–Jakob disease, and often one may see a decline from week to week.

Etiology

Microscopically there is widespread spongiform change within the gray matter of the cerebral cortex, basal ganglia, thalamus, and cerebellar cortex (Masters and Richardson 1978), which in turn is accounted for by grossly swollen dendrites and axons (Beck *et al.* 1982; Chou *et al.* 1980). In a small minority, perhaps 5–10 percent of cases, plaques resembling the amyloid plaques seen in Alzheimer's disease may be found. In contrast to the plaques of Alzheimer's disease, however, these plaques are not composed of beta-amyloid but appear to be, at least in part, composed of prions (Kitamoto *et al.* 1986). There is also neuronal loss and astrogliosis, but very little, if any, inflammation. It must be emphasized that prions are found not only in the central nervous system but also in peripheral nerves (Favereaux *et al.* 2004), olfactory neuroepithelium (Tabaton *et al.* 2004; Zanusso *et al.* 2003), spleen (Glatzel *et al.* 2003), and skeletal muscle (Glatzel *et al.* 2003).

The prion protein is a normally occurring cellular protein that is coded for by the *PRNP* gene on chromosome 20. It is a constituent of the neuronal cell membrane and undergoes recycling from the exterior surface of the cell membrane into the cytoplasm, where it is digested by lysozymes. The agent responsible for Creutzfeldt–Jakob disease is a pathologic form of the prion protein. The normal prion protein exists in an alpha-helical conformation, whereas the pathologic prion protein. exists for the most part in a beta-sheet conformation (Pan *et al.* 1993). This beta-sheet conformation allows for aggregation of these pathologic prion proteins once they recycle into the cytoplasm and, as they aggregate and coalesce, a particle known as a prion is formed. Eventually, with an accumulation of prions, spongiform change occurs.

As an aside, in the literature the normal form of the prion protein is often referred to as PrPc, with the 'c' derived from the fact that these are normal '*c*ellular' components. The pathologic form, however, is often referred to as PrPsc, with the 'sc' derived from '*sc*rapie', which is the name of a disease similar to Creutzfeldt–Jakob disease that is found in sheep.

As noted earlier, Creutzfeldt–Jakob disease may occur sporadically, as an autosomal dominantly inherited disease, or iatrogenically, and each of these is now discussed in turn.

The etiology of the sporadic cases is not clear. Some believe that they represent spontaneous, age-related transformations of normal cellular prion proteins into pathogenic prion proteins, whereas others suspect that they may represent infections with a long incubation period (Galvez *et al.* 1980). The finding of prions within the olfactory neuroepithelium is intriguing in this regard, as their presence could be accounted for either by spread from the brain down the olfactory tracts and fila, in which case they might merely be seen as epiphenomenal, or by contact with airborne prions from an infected person, in which case they could represent the bridgehead of an infection. Recent work has further subdivided sporadic cases according to polymorphisms at codon 129 of *PRNP* and to the isoform of the pathologic prion protein. At codon 129, either methionine or valine may be found on either allele; thus, there are potentially three different genotypes: MM, MV, and VV. The pathologic prion protein, in turn, exists in one of two isoforms: type 1 and type 2. Parchi and colleagues (1999a), believing that in any given case only one isoform of the pathologic prion protein was present, held that all cases of sporadic disease could therefore be categorized into one of six types: MM1, MM2, MV1, MV2, VV1, and VV2. Subsequent work, however, has disclosed that in some cases one may find both type 1 and type 2 of the pathologic prion protein in the same individual patient (Cali *et al.* 2009; Puoti *et al.* 1999), thus expanding the categorization of sporadic cases into nine types: MM1, MM2, MM1–2, etc. Although there is some correlation between these types and different presentations of the disease (Cali *et al.* 2009), the correlation is rough, and it is not clear how useful this scheme is for clinical work.

Inherited cases occur on an autosomal dominant basis, secondary to any one of a large number of mutations in the *PRNP* gene (Goldfarb *et al.* 1990; Hsiao *et al.* 1991a; Owen *et al.* 1989). Importantly, in many cases mutations may occur spontaneously, and here, of course, the family history will be negative (Ladogana *et al.* 2005). Furthermore, it must also be kept in mind that these genetic forms of Creutzfeldt–Jakob disease are not fully penetrant, and up to one-half of carriers may never manifest symptoms.

Iatrogenic cases have occurred upon inadvertent exposure to tissue from patients with Creutzfeldt–Jakob disease via the following procedures: corneal transplants (Duffy *et al.* 1974; Heckmann *et al.* 1997); dura mater grafts (Brown *et al.* 2000; Hannah *et al.* 2001; Heath *et al.* 2006; Miyashita *et al.* 1991); the use of contaminated electrodes during neurosurgical procedures (Bernouilli *et al.* 1977); injections of human growth hormone (Billette de Villemeur *et al.* 1991, 1996; Brown *et al.* 2000); and the injection of human pituitary gonadotrophins (Cochius *et al.* 1992; Healy and Evans 1993); it is of especial interest that a case also occurred secondary to the use of lyophilized dura mater in the treatment of a nasopharyngeal angiofibroma, a procedure in which the dura mater remained extracranial (Antoine *et al.* 1997). The persistence of prions cannot be overemphasized: in one case (Gibbs *et al.* 1994), electrodes that had been used during neurosurgery on a patient with Creutzfeldt–Jakob disease 2 years earlier were cleaned three times and repeatedly sterilized in ethanol and formaldehyde vapor yet were still able to transmit the disease to a chimpanzee. Presumably, in these iatrogenic cases, the pathogenic prion protein acts as a template that alters the conformation of the patient's normal prion protein into the pathologic beta-sheet protein.

Differential diagnosis

A large number of other disorders may present in a fashion similar to that of Creutzfeldt–Jakob disease, that is to say with dementia accompanied by myoclonus and ataxia; however, certain clinical and laboratory features will allow for an accurate differential.

The first group of disorders to consider on the differential includes the other prion diseases. New-variant Creutzfeldt–Jakob disease is distinguished by its earlier age of onset and by the pulvinar sign, which, as noted in Chapter 16.2, is only rarely seen in Creutzfeldt–Jakob disease. In patients with Gerstmann–Sträussler–Scheinker disease both periodic complexes on the EEG and the 14-3-3 protein in the CSF are absent. Fatal familial insomnia is marked by severe insomnia, a finding, as noted in Chapter 16.4, only rarely seen in Creutzfeldt–Jakob disease, and periodic complexes are also not found.

Hashimoto's encephalopathy, limbic encephalitis, and Whipple's disease may also mimic Creutzfeldt–Jakob disease, and these diagnoses must not be missed, given their treatability. Periodic complexes are lacking in all three of these, and although the 14-3-3 protein may be found in some cases of Hashimoto's encephalopathy and limbic encephalitis, it is absent in Whipple's disease.

Lithium and bismuth intoxication must also be considered, but here the history of antecedent use will make the diagnosis.

Both Lewy body dementia and Alzheimer's disease are often included on the differential, however these disorders typically have a gradual onset, in contrast to the acute or subacute onset of Creutzfeldt–Jakob disease.

Treatment

The general treatment of dementia is discussed in Chapter 6.1. To date, the only medication shown to have any specific usefulness in a double-blinded study is flupirtine, which may slow the progression of the disease (Otto *et al.* 2004); quinacrine is not useful (Haik *et al.* 2004). In familial cases, genetic counseling should be offered.

The question inevitably arises as to what sorts of precautions should be in place to guard against transmission of the disease. Although isolation does not appear to be necessary, routine universal precautions are appropriate. It is critical to avoid contamination via CSF, blood, or biopsy specimens, and it must be borne in mind that routine sterilization procedures, including autoclaving and alcohol immersion, are *not* effective. In cases of accidental contact, consideration may be given to washing with a 1:10 solution of 5 percent common hypochlorite bleach, which is effective. Pins used for sensory testing in any patient should *never* be used twice.

16.2 NEW-VARIANT CREUTZFELDT–JAKOB DISEASE

New-variant Creutzfeldt–Jakob disease is a very rare prion disease acquired by eating meat derived from cows that had bovine spongiform encephalopathy, or 'mad cow disease'. Although most cases have occurred in the UK, at least one has been reported in the United States: this patient, however, had grown up in the UK (Belay *et al.* 2005).

Clinical features

The incubation period between the ingestion of contaminated meat and the development of the disease is probably less than 20 years. The onset of the disease is subacute and, although most patients have been in their late twenties or early thirties, the range of age of onset is wide, from early adolescence to the eighth decade.

The overall clinical features and evolution of the disease have been described in a number of large case series (Brandel *et al.* 2009; Spencer *et al.* 2002; Will *et al.* 2000; Zeidler *et al.* 1997a, b). In the majority of cases the onset is marked by 'psychiatric' features, including depression, a personality change, or psychosis, which may be marked by visual and auditory hallucinations and, in some cases, Schneiderian first-rank symptoms (Will *et al.* 2000; Zeidler *et al.* 1997a, b). In such cases, however, classic 'neurologic' features always ensue after about 6 months: these include ataxia, dysarthria, painful dysesthesiae of the limbs and face, myoclonus, dystonia or chorea, akinetic mutism, and, eventually, in all cases, dementia. In a small minority, the onset may be characterized by a combination of both psychiatric and neurologic features.

Magnetic resonance imaging, especially with FLAIR imaging, will eventually reveal the 'pulvinar sign' in almost all cases (Brandel *et al.* 2009; Collie *et al.* 2003), which, as the name indicates, is characterized by increased signal intensity in the pulvinar. In addition to this pulvinar sign, increased signal intensity may also occur in the basal ganglia and the cerebral and cerebellar cortices (Collie *et al.* 2003).

The EEG may demonstrate generalized slowing (Zeidler *et al.* 1997a). With rare exceptions (Binelli *et al.* 2006), periodic spike-and-slow-wave complexes are absent (Brandel *et al.* 2009; Zeidler *et al.* 1997a).

The CSF is acellular with a normal total protein; the 14-3-3 protein is found in up to 50 percent of cases (Brandel *et al.* 2009; Green *et al.* 2001; Sanchez-Juan *et al.* 2006; Will *et al.* 2000).

Tonsilar biopsy is both very sensitive and specific (Brandel *et al.* 2009; Hill *et al.* 1999).

Course

The disease is relentlessly progressive, with most patients dying within a little over a year, with a range of 8 to 40 months.

Etiology

As noted earlier, new-variant Creutzfeldt–Jakob disease occurs secondary to eating beef from cows that had bovine spongiform encephalopathy (Bruce *et al.* 1997); in turn, cows contract the disease by eating meal made from the offal of sheep that had a transmissible spongiform encephalopathy known as scrapie. Host factors appear to play a role in susceptibility among humans in that, with rare exceptions (Ironside *et al.* 2006), all patients have been homozygous for methionine at codon 129 of *PRNP*, the gene for normal prion proteins found on chromosome 20 (Brandel *et al.* 2009; Collinge *et al.* 1996; Will *et al.* 2000).

Within the thalamus, basal ganglia, and also the cerebral and cerebellar cortices there are widespread prion plaques surrounded by spongiform change (Will *et al.* 1996). In this disease prions are not restricted to the brain but are also found in tonsils, lymph nodes, and the spleen (Hill *et al.* 1999). It also appears that prions are present in the blood, and, alarmingly, there have been cases of new-variant Creutzfeldt–Jakob disease occurring secondary to blood transfusion (Wroe *et al.* 2006).

Differential diagnosis

In cases presenting with depression or psychosis, other disorders capable of causing these syndromes must be considered, as discussed in Chapters 7.1 and 8.1, respectively; the presence of Schneiderian first-rank symptoms narrows the diagnosis even further, as discussed in Chapter 5.31. Once neurologic features appear, the differential is similar to that described for Creutzfeldt–Jakob disease in Chapter 16.1. Importantly, the pulvinar sign, though highly suggestive of new-variant Creutzfeldt–Jakob disease, is not pathognomonic, as it has been reported, albeit very rarely, in cases of limbic encephalitis (Mihara *et al.* 2005) and in both sporadic (Meissner *et al.* 2009; Petzold *et al.* 2004) and familial (Kovacs *et al.* 2009) Creutzfeldt–Jakob disease.

Treatment

General treatment recommendations are as discussed for Creutzfeldt–Jakob disease in Chapter 16.1; although there is no established treatment for new-variant Creutzfeldt–Jakob disease, intraventricular pentosan polysulfate has shown some promise (Bone *et al.* 2008). Strenuous efforts are in place to prevent bovine spongiform encephalopathy; however, given the difficulty in monitoring the food chain, continual vigilance is necessary.

16.3 GERSTMANN–STRÄUSSLER–SCHEINKER DISEASE

Gerstmann–Sträussler–Scheinker disease is a very rare, autosomal dominantly inherited prion disease characterized classically by ataxia and, eventually, dementia.

Clinical features

The onset is subacute or gradual, and usually in mid-adult years, with a wide range from the third to the eighth decades. Classically (Barbanti *et al.* 1996; Farlow *et al.* 1989), patients present with a progressive ataxia and dysarthria, with the eventual development of a dementia; in some cases parkinsonism or long-tract signs may also occur, and, very rarely, myoclonus may appear (Takazawa *et al.* 2010).

The EEG may show generalized slowing but there are no periodic complexes; the CSF is normal and the 14-3-3 protein is absent (Sanchez-Juan *et al.* 2006).

Course

The disease is relentlessly progressive; death occurs on average within 5–7 years, with a range of 2–10 years.

Etiology

Gerstmann–Sträussler–Scheinker disease may occur secondary to any one of several mutations in the *PRNP* gene. Mutations have been noted at codons 102 (Arata *et al.* 2006; Brown *et al.* 1991; Cagnoli *et al.* 2008; Goldhammer *et al.* 1993; Hsiao *et al.* 1989; Kretzchmar *et al.* 1992; Young *et al.* 1995), 105 (Kitamoto *et al.* 1993; Yamada *et al.* 1993), 117 (Hsiao *et al.* 1991b; Nochlin *et al.* 1989), 198 (Farlow *et al.* 1989; Ghetti *et al.* 1989), and 217 (Hsiao *et al.* 1992).

Pathologically, regardless of which mutation is at fault, all cases of Gerstmann–Sträussler–Scheinker disease share a common feature, namely the presence of multicentric amyloid plaques that stain positively for prion proteins (Kitamoto *et al.* 1986; Roberts *et al.* 1986). In some cases, unicentric spiky plaques, similar to those seen in kuru, may also occur, and in others neurofibrillary tangles may be seen. Spongiform change is either minimal or absent. Although the distribution of these microscopic changes varies, the cerebellar cortex, basal ganglia, and cerebral cortex are generally involved.

Differential diagnosis

Of the disorders capable of causing ataxia and dementia discussed in Chapter 6.3, special consideration must be given to those displaying an autosomal dominant pattern of inheritance, including spinocerebellar ataxia, dentatorubropallidoluysian atrophy, and familial cases of Creutzfeldt–Jakob disease.

Treatment

There is no specific treatment; genetic counseling should be offered, and the general treatment of dementia is discussed in Chapter 6.3.

16.4 FATAL FAMILIAL INSOMNIA

Fatal familial insomnia is a rare prion disease, characterized, as the name suggests, by severe insomnia, which is typically accompanied by a dementia and prominent autonomic disturbances.

Clinical features

The overall clinical features have been described in a large number of cases series and reports (Almer *et al.* 1999; Gallassi *et al.* 1996; Krasnianski *et al.* 2008; Manetto *et al.* 1992; Medori *et al.* 1992a, b; Nagayama *et al.* 1996; Reder *et al.* 1995; Silburn *et al.* 1996; Tabernero *et al.* 2000). The onset is typically subacute or gradual and occurs in middle years, with a wide range from late adolescence to the seventh decade. Patients develop intractable insomnia, accompanied in many cases by 'oneiroid' states in which they appear confused, experience visual hallucinations, and behave as if they were acting out dreams: in one case (Almer *et al.* 1999), 'the patient performed movements of sawing with a virtual saw and stopped bewildered when told there was no saw'. Rarely, the presentation may be with a psychosis coupled with insomnia (Dimitri *et al.* 2006). Paroxysms of autonomic disturbance often occur, with hyperhidrosis, tachycardia, hypertension, and irregular respirations. Over time dementia appears, accompanied, variously, by ataxia, myoclonus, and spasticity. Importantly, although insomnia is the initial hallmark of this disease, in some cases it may appear late in the course and, rarely, it may be absent (Zerr *et al.* 1998).

The EEG typically shows generalized slowing; however, there are no periodic complexes. Magnetic resonance imaging findings may be normal; however, late in the course of the disease both cerebral and cerebellar cortical atrophy may be seen. The CSF is generally normal although in a very small minority the 14-3-3 protein may be found (Krasnianski *et al.* 2008; Sanchez-Juan *et al.* 2006).

Course

The disease is relentlessly progressive: stupor and coma eventually ensue, with death in about 1 year.

Etiology

Pathologically (Almer *et al.* 1999; Krasnianski *et al.* 2008, Lugaresi *et al.* 1986; Manetto *et al.* 1992), there is extensive neuronal loss and gliosis in the anterior and mediodorsal nuclei of the thalamus and in the inferior olives; in some cases, mild cell loss and spongiform change may also be seen in the cerebral and cerebellar cortices, and in the dorsal raphe and superior central nuclei (Almer *et al.* 1999).

Fatal familial insomnia, in almost all cases, is an autosomal dominant disease occurring secondary to mutations at codon 178 of the *PRNP* gene (Medori *et al.* 1992a, b). Rare sporadic cases have been reported (Parchi *et al.* 1999b; Scaravilli *et al.* 2000).

Differential diagnosis

In cases presenting with intractable insomnia, consideration may be given to a depressive episode of major depressive disorder; however, here one finds accompanying vegetative symptoms not seen in fatal familial insomnia; the eventual appearance of other symptoms typical of fatal familial insomnia, such as oneiroid states, paroxysmal autonomic changes, and dementia, will then seal the diagnosis.

Familial Creutzfeldt–Jakob disease may also, albeit rarely, present with intractable insomnia (Chapman *et al.* 1996; Landolt *et al.* 2006; Taratuto *et al.* 2002), and in such cases an ante mortem diagnosis may not be possible.

Treatment

There is no known specific treatment; general measures may be undertaken, as described for Creutzfeldt–Jakob disease in Chapter 16.1, and sequential trials of various hypnotics should be attempted.

16.5 KURU

Kuru is a unique human prion disease, spread by cannibalism and present only in the Fore people of New Guinea (Gajdusek 1977). With the elimination of cannibalism in the 1950s new infections stopped, but, given the long incubation period, the disease itself is still present.

Clinical features

After a long incubation period, spanning 4–50 years or more, patients gradually develop ataxia and tremor, followed, in a minority, by a dementia of relatively mild degree (Collinge *et al.* 2006, 2008).

Course

The disease is relentlessly progressive, with death occurring, on average, after about 1 year.

Etiology

Transmission of the infectious prion protein occurred with the consumption of brain during cannibalistic feasts. Pathologically (Brandner *et al.* 2008; Gajdusek and Zigas 1959; Kakulas *et al.* 1967) there is atrophy of the cerebellar vermis and flocculonodular lobe, and microscopically there is widespread spongiform change, neuronal loss, astrocytosis, and 'kuru plaques'.

Differential diagnosis

The diagnosis is only relevant in those Fore tribespeople who engaged in cannibalism.

Treatment

There is no known treatment.

REFERENCES

Almer G, Hainfellner JA, Brucke T *et al.* Fatal familial insomnia: a new Austrian family. *Brain* 1999; **122**:5–16.

Antoine JC, Michel D, Bertholon P *et al.* Creutzfeldt–Jakob disease after extracranial dura mater embolization for a nasopharyngeal angiofibroma. *Neurology* 1997; **48**:1451–3.

Appleby BS, Appleby KK, Crain BJ *et al.* Characteristics of established and proposed sporadic Creutzfeldt–Jakob disease variants. *Arch Neurol* 2009; **66**:208–15.

Arata H, Takashima H, Hirano R *et al.* Early clinical signs and imaging findings in Gerstmann–Straussler–Scheinker syndrome (Pro102-Leu). *Neurology* 2006; **66**:1672–8.

Barbanti P, Fabbrini G, Salvatore M *et al.* Polymorphism at codon 129 or codon 219 of PRNP and clinical heterogeneity in a previously unreported family with Gerstmann–Straussler–Scheinker disease (Prp–P102L mutation). *Neurology* 1996; **47**:734–41.

Beck E, Danile PM, Davey AJ *et al.* The pathogenesis of transmissible spongiform encephalopathy – an ultrastructural study. *Brain* 1982; **105**:755–86.

Belay ED, Sejvar JJ, Shieh WJ *et al.* Variant Creutzfeldt–Jakob disease death, United States. *Emerg Infect Dis* 2005; **11**:1351–4.

Bernouilli C, Siegfried J, Baumgartner G *et al.* Danger of accidental person to person transmission of Creutzfeldt–Jakob disease by surgery. *Lancet* 1977; **1**:478–9.

Billette de Villemeur T, Beuavais P, Gourmelon M *et al.* Creutzfeldt–Jakob disease in children treated with growth hormone. *Lancet* 1991; **337**:864–5.

Billette de Villemeur T, Beslys J-P, Pradel A *et al.* Creutzfeldt–Jakob disease from contaminated growth hormone extracts in France. *Neurology* 1996; **47**:690–5.

Binelli S, Agazzi P, Giaccone G *et al.* Periodic electroencephalogram complexes in a patient with variant Creutzfeldt–Jakob disease. *Ann Neurol* 2006; **59**:423–7.

Bone I, Belton L, Walker AS *et al.* Intraventricular pentosan polysulphate in human prion diseases: an observational study in the UK. *Eur J Neurol* 2008; **15**:458–64.

Brandel JP, Heath CA, Head MW *et al.* Variant Creutzfeldt-Jacob disease in France and the United Kingdom: Evidence for the same agent strain. *Ann Neurol* 2009; **65**:249–56.

Brandner S, Whitfield J, Boone K *et al.* Central and peripheral pathology of kuru: pathological analysis of a recent case and comparison with other forms of human prion disease. *Phil Trans R Soc Lond B Biol Sci* 2008; **363**:3755–63.

Brown P, Cathala F, Castaigne P *et al.* Creutzfeldt–Jakob disease: clinical analysis of a consecutive series of 230 neuropathologically verified cases. *Ann Neurol* 1986; **20**:597–602.

Brown P, Goldfarb LG, Brown WT *et al.* Clinical and molecular genetic study of a large German kindred with Gerstmann–Straussler–Scheinker syndrome. *Neurology* 1991; **41**:375–9.

Brown P, Gibbs CJ, Rodgers-Johnson P *et al.* Human spongiform encephalopathy: the National Institutes of Health series of 300 cases of experimentally transmitted disease. *Ann Neurol* 1994; **35**:513–29.

Brown P, Preece M, Brandel JP *et al.* Iatrogenic Creutzfeldt-Jacob disease at the millennium. *Neurology* 2000; **55**:1075–81.

Brownell B, Oppenheimer DR. An ataxic form of subacute presenile polioencephalopathy (Creutzfeldt–Jakob disease). *J Neurol Neurosurg Psychiatry* 1965; **28**:350–61.

Bruce ME, Will RG, Ironside JW *et al.* Transmissions to mice indicate that 'new variant' CJD is caused by the BSE agent. *Nature* 1997; **389**:498–501.

Burger LJ, Rowan J, Goldensohn E. Creutzfeldt-Jacob disease. *Arch Neurol* 1972; **26**:428–33.

Cagnoli C, Brussino A, Sbaiz L *et al.* A previously undiagnosed case of Gerstmann-Straussler-Scheinker disease revealed by PRNP gene analysis in patients with adult-onset ataxia. *Mov Disord* 2008; **23**:1468–71.

Cali I, Castellani R, Alshekhlee A *et al.* Co-existence of scrapie protein types 1 and 2 in sporadic Creutzfeldt-Jakb disease: its effect on the phenotype and prion-tye characteristics. *Brain* 2009; **132**: 2643–58.

Chapman J, Ariazoroff A, Goldfarb LG *et al.* Fatal insomnia in a case of familial Creutzfeldt–Jakob disease with the codon 200 (Lys) mutation. *Neurology* 1996; **46**:758–61.

Chou SM, Payne WN, Gibbs CG *et al.* Transmission and scanning electron microscopy of spongiform change in Creutzfeldt–Jakob disease. *Brain* 1980; **103**:885–904.

Cochius JI, Hyman N, Esiri MM. Creutzfeldt–Jakob disease in a recipient of human pituitary-derived gonadotrophin: a second case. *J Neurol Neurosurg Psychiatry* 1992; **55**:1094–5.

Collie DA, Summers DM, Sellar RJ *et al.* Diagnosing variant Creutzfeldt–Jakob disease with the pulvinar sign: MR imaging findings in 86 neuropathologically confirmed cases. *AJNR* 2003; **24**:1560–9.

Collinge J, Sidle KCL, Meads J *et al.* Molecular analysis of prion strain variation and the aetiology of 'new-variant' CJD. *Nature* 1996; **383**:685–90.

Collinge J, Whitfield J, McIntosh E *et al.* Kuru in the 21st century – an acquired human prion disease with very long incubation periods. *Lancet* 2006; **367**:2068–74.

Collinge J, Whitfield J, McKintosh E *et al.* A clinical study of kuru patients with long incubation periods at the end of the epidemic in Papua New Guinea; *Phil Trans R Soc Lond B Biol Sci* 2008; **363**:3725–39.

Collins SJ, Sanchez-Juan P, Masters CL *et al.* Determinants of diagnostic investigation sensitivities across the clinical spectrum of sporadic Creutzfeldt–Jakob disease. *Brain* 2006; **129**:2278–87.

Cooper SA, Murray KL, Will RG *et al.* Sporadic Creutzfeldt–Jakob disease with cerebellar ataxia at onset in the UK. *J Neurol Neurosurg Psychiatry* 2006; **77**:1273–5.

Dervaux A, Laine H, Czermak M *et al.* Creutzfeldt–Jakob disease presenting as psychosis. *Am J Psychiatry* 2004; **161**:1307–8.

Dimitri D, Jehel L, Durr A *et al.* Fatal familial insomnia presenting as psychosis in an 18-year-old man. *Neurology* 2006; **67**:363–4.

Duffy P, Wolf J, Collins G *et al.* Possible person to person transmission of Creutzfeldt–Jakob disease. *N Engl J Med* 1974; **290**:692–3.

Farlow MR, Yee RD, Dlouhy SR *et al.* Gerstmann–Straussler–Scheinker disease. I. Extending the clinical spectrum. *Neurology* 1989; **39**:1446–52.

Favereaux A, Quadrio I, Vital C *et al.* Pathologic prion protein spreading in the peripheral nervous system of a patient with sporadic Creutzfeldt–Jakob disease. *Arch Neurol* 2004; **61**:747–50.

Gajdusek DC. Unconventional viruses and the origin and disappearance of kuru. *Science* 1977; **197**:943–60.

Gajdusek DC, Zigas V. Clinical, pathological and epidemiological study of the central nervous system among natives of the eastern highlands of New Guinea. *Am J Med* 1959; **26**:442–69.

Gallassi R, Morreale A, Montagna P *et al.* Fatal familial insomnia: behavioral and cognitive features. *Neurology* 1996; **46**:935–9.

Galvez S, Masters C, Gajdusek C. Descriptive epidemiology of Creutzfeldt–Jakob disease in Chile. *Arch Neurol* 1980; **37**:11–14.

Ghetti B, Tagliavini F, Masters CL *et al.* Gerstmann–Straussler–Scheinker disease. II. Neurofibrillary tangles and plaques with

PrP-amyloid coexist in an affected family. *Neurology* 1989; **39**:1453–61.

Gibbs CJ, Asher DM, Kobrine A *et al.* Transmission of Creutzfeldt–Jakob disease to a chimpanzee by electrodes contaminated during neurosurgery. *J Neurol Neurosurg Psychiatry* 1994; **57**:757–8.

Glatzel M, Abela E, Maissen M *et al.* Extraneural pathologic prion protein in sporadic Creutzfeldt–Jakob disease. *N Engl J Med* 2003; **349**:1812–20.

Glatzel M, Stoeck K, Seeger H *et al.* Human prion disease. Molecular and clinical aspects. *Arch Neurol* 2005; **62**:545–52.

Goldfarb LG, Mitrova E, Brown P *et al.* Mutation in codon 200 of scrapie amyloid protein gene in two clusters of Creutzfeldt–Jakob disease in Slovakia. *Lancet* 1990; **336**:514–15.

Goldhammer Y, Gabison R, Meiner Z *et al.* An Israeli family with Gerstmann–Straussler–Scheinker disease manifesting the codon 102 mutation in the prion protein gene. *Neurology* 1993; **43**: 2718–19.

Green AJE, Thompson EJ, Stewart GE *et al.* Use of 14-3-3 protein and other brain-specific proteins in CSF in the diagnosis of variant Creutzfeldt–Jakob disease. *J Neurol Neurosurg Psychiatry* 2001; **70**:744–8.

Haik S, Brandel JP, Salomon D *et al.* Compassionate use of quinacrine in Creutzfeldt–Jakob disease fails to show significant effects. *Neurology* 2004; **63**:2413–15.

Hannah EL, Belay ED, Gambetti P *et al.* Creutzfeldt–Jakob disease after receipt of a previously unimplicated brand of dura mater graft. *Neurology* 2001; **56**:1080–3.

Healy D, Evans J. Creutzfeldt–Jakob disease after pituitary gonadotrophins. *BMJ* 1993; **307**:517–18.

Heath CA, Barver RA, Esmonde TFG *et al.* Dura mater associated Cretuzfeldt–Jakob disease: experience from surveillance in the UK. *J Neurol Neurourg Psychiatry* 2006; **77**:880–1.

Heckmann JG, Lang CJG, Petruch F *et al.* Transmission of Creutzfeldt–Jakob disease via a corneal transplant. *J Neurol Neurosurg Psychiatry* 1997; **63**:388–90.

Hill AF, Butterworth RJ, Joiner S *et al.* Investigation of variant Creutzfeldt–Jakob disease and other human prion diseases with tonsil biopsy specimens. *Lancet* 1999; **353**:183–9.

Hsiao K, Doh-ura K, Kitamoto T *et al.* A prion protein amino acid substitution in ataxic Gerstmann–Straussler–Scheinker syndrome. *Ann Neurol* 1989; **26**:137.

Hsiao K, Meiner Z, Kahana E *et al.* Mutation of the prion protein gene in Libyan Jews with Creutzfeldt–Jakob disease. *N Engl J Med* 1991a; **324**:1091–7.

Hsiao K, Cass C, Schellenberg GD *et al.* A prion protein variant in a family with the telencephalic form of Gerstmann–Straussler–Scheinker syndrome. *Neurology* 1991b; **41**:681–4.

Hsiao K, Dlouhy S, Farlow MR *et al.* Mutant prion proteins in Gerstmann–Straussler–Scheinker disease with neurofibrillary tangles. *Nat Genet* 1992; **1**:68–71.

Ironside JW, Bishop MT, Connolly K *et al.* Variant Creutzfeldt–Jakob disease: prion protein genotype analysis of positive appendix tissue samples from a retrospective prevalence study. *BMJ* 2006; **332**:1186–8.

Kakulas BA, Lecours AR, Gajdusek DC. Further observations on the pathology of Kuru (a study of two cerebra in serial section). *J Neuropathol Exp Neurol* 1967; **26**:85–97.

Kitamoto T, Tateishi J, Tashima I *et al.* Amyloid plaques in Creutzfeldt–Jakob disease stain with prion protein antibodies. *Ann Neurol* 1986; **20**:204–8.

Kitamoto T, Amano N, Terao Y *et al.* A new inherited prion disease (PrP-P105L mutation) showing spastic paraparesis. *Ann Neurol* 1993; **34**:808–13.

Kovacs GG, Horvath S, Strobel T *et al.* Genetic Creutzfeldt–Jakob disease mimicking variant Creutzfeldt–Jakob disease. *J Neurol Neurosurg Psychiatry* 2009; **80**:1410–1.

Krasnianski A, Bartl M, Sanchez Juan PJ *et al.* Fatal familial insomnia: clinical features and early identification. *Ann Neuro* 2008; **63**: 658–61.

Kretzchmar HA, Kufer P, Reithmuller G *et al.* Prion protein mutation at codon 102 in an Italian family with Gerstmann–Straussler–Scheinker syndrome. *Neurology* 1992; **42**:809–10.

Kropp S, Schultz-Schaeffer WJ, Finkenstaedt M *et al.* The Heidenhain variant of Creutzfeldt–Jakob disease. *Arch Neurol* 1999; **56**:55–61.

Ladogana A, Puopolo D, Poleggi A *et al.* High incidence of genetic human transmissible spongiform encephalopathies in Italy. *Neurology* 2005; **64**:1592–7.

Landolt H-P, Gratzel M, Blatter T *et al.* Sleep-wake disturbances in sporadic Creutzfeldt–Jakob disease. *Neurology* 2006; **66**:1418–24.

Lemstra AW, van Meegen MT, Vreyling JP *et al.* 14-3-3 testing in diagnosing Creutzfeldt–Jakob disease. A prospective study in 112 patients. *Neurology* 2000; **55**:514–16.

Lendvai I, Saravay SM, Steinberg MD. Creutzfeldt–Jakob disease presenting as secondary mania. *Psychosomatics* 1999; **40**:524–5.

Levy SR, Chiappa KH, Burke CJ *et al.* Early evolution and incidence of electroencephalographic abnormalities in Cretuzfeldt–Jakob disease. *J Clin Neurophysiol* 1986; **3**:1–21.

Lugaresi E, Medori R, Montagna P *et al.* Fatal familial insomnia and dysautonomia with selective degeneration of thalamic nuclei. *N Engl J Med* 1986; **315**:997–1003.

MacGowan DJL, Delanty N, Petito F *et al.* Isolated myoclonic alien hand as the sole presentation of pathologically established Creutzfeldt–Jakob disease: a report of two patients. *J Neurol Neurosurg Psychiatry* 1997; **63**:404–7.

Mandell AM, Alexander MP, Carpenter S. Creutzfeldt–Jakob disease presenting as isolated aphasia. *Neurology* 1989; **39**:55–8.

Manetto V, Medori R, Cortelli P *et al.* Fatal familial insomnia: clinical and pathological study of five new cases. *Neurology* 1992; **42**:312–19.

Masters CL, Richardson EP. Subacute spongiform encephalopathy (Creutzfeldt–Jakob disease) – the nature and progression of spongiform change. *Brain* 1978; **101**:333–44.

May WW. Creutzfeldt–Jakob disease. *Acta Neurol Scand* 1986; **44**: 1–32.

Medori R, Tritschler HJ, LeBlanc A *et al.* Fatal familial insomnia with a mutation at codon 178 of the prion protein gene. *N Engl J Med* 1992a; **326**:444–9.

Medori R, Montagna P, Tritschler HJ *et al.* Fatal familial insomnia: a second kindred with mutation of prion protein gene at codon 178. *Neurology* 1992b; **42**:669–70.

Meissner B, Kallenberg K, Sanchez-Juan P *et al.* MRI lesion profiles in sporadic Creutzfeldt–Jakob disease. *Neurology* 2009; **72**: 1994–2009.

Mihara M, Sgase S, Konaka K *et al.* The 'pulvinar sign' in a case of paraneoplastic limbic encephalitis associated with non-Hodgkin's lymphoma. *J Neurol Neurosurg Psychiatry* 2005; **76**:882–4.

Miyashita K, Inuzuka T, Kondo H *et al.* Creutzfeldt–Jakob disease in a patient with a cadaveric dural graft. *Neurology* 1991; **41**:940–1.

Nagayama M, Shinohara Y, Furukawa H *et al.* Fatal familial insomnia with a mutation at codon 178 of the prion protein gene: first report from Japan. *Neurology* 1996; **47**:1313–16.

Nochlin D, Sumi SM, Bird TD *et al.* Familial dementia with PrP-positive amyloid plaques: a variant of the Gerstmann–Straussler syndrome. *Neurology* 1989; **39**:910–18.

Otto M, Cepek L, Ratzka R *et al.* Efficacy of flupirtine on cognitive function in patients with CJD: A double blind study. *Neurology* 2004; **62**:714–18.

Owen F, Poulter M, Lofthouse R *et al.* Insertion in prion protein gene in familial Creutzfeldt–Jakob disease. *Lancet* 1989; **1**:51–2.

Pan K-M, Baldwin M, Nguyen J *et al.* Conversion of alpha-helices into beta-sheets features in the formation of the scrapie prion proteins. *Proc Natl Acad Sci USA* 1993; **90**:962–6.

Parchi P, Gise A, Capelleri S *et al.* Classification of sporadic Creutzfeldt–Jakob disease based on molecular and phenotypic analysis of 300 subjects. *Ann Neurol* 1999a; **46**:224–33.

Parchi P, Capellari S, Chin S *et al.* A subtype of sporadic prion disease mimicking fatal familial insomnia. *Neurology* 1999b; **52**: 1757–63.

Petzold GC, Westner I, Bohner G *et al.* False-positive pulvinar sign on MRI in sporadic Creutzfeldt–Jakob disease. *Neurology* 2004; **62**:1235–6.

Pocchiara M, Pupolo M, Croes EA *et al.* Predictors of survival in sporadic Creutzfeldt–Jakob disease and other transmissible spongiform encephalopathies. *Brain* 2004; **127**:2348–59.

Puoti G, Giaccone G, Rossi G *et al.* Sporadic Creutzfeldt–Jakob disease: co-occurrence of different types of PrP(Sc) in the same brain. *Neurology* 1999; **53**:2173–6.

Reder AT, Mednick AS, Brown P *et al.* Clinical and genetic studies of fatal familial insomnia. *Neurology* 1995; **45**:1068–75.

Roberts GW, Lofthouse R, Brown R *et al.* Prion-protein immunoreactivity in human transmissible dementias. *N Engl J Med* 1986; **315**:1231–3.

Roos R, Cajdusek DC, Gibbs CJ. The clinical characteristics of transmissible Creutzfeldt–Jakob disease. *Brain* 1973; **96**:1–20.

Sanchez-Juan P, Green A, Ladogana A *et al.* CSF tests in the differential diagnosis of Cretuzfeldt–Jakob disease. *Neurology* 2006; **67**:637–43.

Scaravilli F, Cordery RJ, Kretzchmar H *et al.* Sporadic fatal insomnia: a case study. *Ann Neurol* 2000; **48**:665–8.

Shiga Y, Miyazawa K, Sato S *et al.* Diffusion-weighted MRI abnormalities as an early diagnostic marker for Creutzfeldt–Jakob disease. *Neurology* 2004; **63**:443–9.

Silburn P, Cervenakova L, Varghese P *et al.* Fatal familial insomnia: a seventh family. *Neurology* 1996; **47**:1326–8.

Spencer MD, Knight RS, Will RG. First hundred cases of variant Creutzfeldt–Jakob disease: retrospective case note review of early psychiatric and neurological features. *BMJ* 2002; **324**:1479–82.

Steinhoff BJ, Zerr I, Glatting M *et al.* Diagnostic value of periodic complexes in Creutzfeldt–Jakob disease. *Ann Neurol* 2004; **56**:702–8.

Tabaton M, Monaco S, Cordone MP *et al.* Prion deposition in olfactory biopsy of sporadic Creutzfeldt–Jakob disease. *Ann Neurol* 2004; **55**:294–6.

Tabernero C, Polo JM, Sevillano MD *et al.* Fatal familial insomnia: clinical, neuropathological, and genetic description of a Spanish family. *J Neurol Neurosurg Psychiatry* 2000; **68**:774–7.

Takazawa T, Ikeda K, Ito H *et al.* A distinct phenotype of leg hyperreflexia in a Japanese family with Gerstmann-Straussler-Scheinker syndrome (P102L). *Intern Med* 2010; **49**:339–42.

Taratuto AL, Piccardo P, Reich EG *et al.* Insomnia associated with thalamic involvement in E220K Creutzfeldt–Jakob disease. *Neurology* 2002; **58**:362–7.

Tschampa H, Kallenberg K, Urbach H *et al.* MRI in the diagnosis of sporadic Creutzfeldt–Jakob disease: a study on inter-observer agreement. *Brain* 2005; **128**:2026–33.

Van Everbroeck BRJ, Boons J, Cras P. 14-3-3 gamma-isoform detection distinguishes sporadic Creutzfeldt–Jakob disease from other dementias. *J Neurol Neurosurg Psychiatry* 2005; **76**:100–2.

Wieser HG, Schwarz U, Blatter T *et al.* Serial EEG findings in sporadic and iatrogenic Creutzfeldt–Jakob disease. *Clin Neurophysiol* 2004; **115**:2467–78.

Will RG, Mathews WB. A retrospective study of Creutzfeldt–Jakob disease in England and Wales 1970–1979 I. Clinical features. *J Neurol Neurosurg Psychiatry* 1984; **47**:134–40.

Will RG, Ironside JW, Zeidler M *et al.* A new-variant of Creutzfeldt–Jakob disease in the UK. *Lancet* 1996; **347**:921–5.

Will RG, Zeidler M, Stewart GE *et al.* Diagnosis of new variant Creutzfeldt–Jakob disease. *Ann Neurol* 2000; **47**:575–82.

Wroe SJ, Pal S, Siddique D *et al.* Clinical presentation and pre-mortem diagnosis of variant Creutzfeldt–Jakob disease associated with blood transfusion; a case report. *Lancet* 2006; **368**:2061–7.

Yamada M, Itoh Y, Fujigasaki H *et al.* A missense mutation at codon 105 with codon 129 polymorphism of the prion protein gene in a new-variant of Gerstmann–Straussler–Scheinker disease. *Neurology* 1993; **43**:2723–4.

Young K, Jones CK, Piccardo P *et al.* Gerstmann–Straussler–Scheinker disease with mutation at codon 102 and methionine at codon 129 of PRNP in previously unreported patients. *Neurology* 1995; **45**:1127–34.

Zanusso G, Ferrari S, Cardone F *et al.* Detection of pathologic prion protein in the olfactory epithelium in sporadic Creutzfeldt–Jakob disease. *N Engl J Med* 2003; **348**:711–19.

Zeidler M, Stewart GE, Barraclough CR *et al.* New-variant Creutzfeldt–Jakob disease: neurological features and diagnostic tests. *Lancet* 1997a; **350**:903–7.

Zeidler M, Johnstone EC, Bamber RWK *et al.* New-variant Creutzfeldt–Jakob disease: psychiatric features. *Lancet* 1997b; **350**:908–10.

Zeng S, Kim CH, Rahman H. Psychotic symptoms presented in familial Creutzfeldt–Jakob disease, subtype E220K. *J Clin Psychiatry* 2001; **62**:734–5.

Zerr I, Giese A, Windl O *et al.* Phenotypic variability in fatal familial insomnia (D178N-129M) genotype. *Neurology* 1998; **51**: 1398–405.

Zerr I, Kallenberg K, Summers DM *et al.* Updated diagnostic criteria for sporadic Creutzfeldt–Jakob disease. *Brain* 2009; **132**:2659–68.

17

Endocrinologic disorders

17.1 CUSHING'S SYNDROME

Cushing's syndrome typically manifests with various neuropsychiatric features (e.g. depression or mania), classically in a patient with a 'cushingoid' habitus consisting of 'moon' facies, truncal obesity, and violaceous abdominal striae (Orth 1995). This syndrome occurs secondary to the effects of sustained hypercortisolemia, which in turn may be due either to administration of exogenous steroids (e.g. prednisone) or to endogenous overproduction of steroids.

Normally, cortisol levels are maintained within physiologic limits by the hypothalamic–pituitary–adrenal axis. The hypothalamus secretes corticotrophin-releasing hormone (CRH), which in turn stimulates the pituitary to produce adrenocorticotrophic hormone (ACTH). ACTH then stimulates the adrenal cortex to produce cortisol, which in turn exerts a negative feedback effect on both the hypothalamus and the pituitary. Endogenous overproduction of steroids may occur secondary to an ACTH-secreting pituitary adenoma, to ectopic ACTH production by a tumor (most commonly oat cell carcinoma), or to a cortisol-producing adrenal tumor.

The term 'Cushing's syndrome' refers to cases of symptomatic hypercortisolemia of any cause, whether exogenous or endogenous in origin; the term 'Cushing's disease', however, is restricted to cases caused by pituitary adenomas.

Clinical features

The mode of onset of Cushing's syndrome is dependent on the underlying cause. Cases due to exogenous steroid use are of the most rapid onset, and may appear within days. Cushing's disease secondary to a pituitary adenoma may present very gradually, over years, and cases of Cushing's syndrome secondary to ectopic ACTH production may present over months.

The neuropsychiatric features of Cushing's syndrome include depression, mania, anxiety, psychosis, and cognitive deficits, including dementia and delirium.

Depression is the most prominent neuropsychiatric feature of Cushing's syndrome (Spillane 1951), and has been noted in one-half (Haskett 1985; Jeffcoate *et al.* 1979; Kelly 1996) to three-quarters (Cohen 1980; Starkman *et al.* 1981) of all cases and indeed may constitute the presenting feature (Kelly 1996). Anxiety commonly accompanies the depression, creating the picture of an 'agitated depression'. The depression at times may be severe, with psychotic features (Cohen 1980), which may be either mood-congruent (Anderson and McHugh 1971; Maclay and Stokes 1939) or mood-incongruent (Trethowan and Cobb 1952), with Schneiderian first-rank symptoms of thought broadcasting and thought insertion. Both suicide attempts and completed suicides may occur (Jeffcoate *et al.* 1979).

Mania is less common than depression in Cushing's syndrome of endogenous origin (Haskett 1985; Jeffcoate *et al.* 1979; Kelly 1996), but the converse holds true in exogenous cases, in which mania is common (Wolkowitz *et al.* 1990).

Anxiety, in the absence of depression, has been noted in about one-tenth of patients (Kelly 1996).

Psychosis, although rare, can occur secondary to Cushing's syndrome. One patient presented with delusions of persecution, auditory and visual hallucinations, and bizarre behavior, all of which cleared with adrenalectomy (Hickman *et al.* 1961); another experienced agitation, auditory hallucinations, and religious and grandiose delusions, which again cleared with adrenalectomy (Hertz *et al.* 1955).

Cognitive deficits have been noted in over one-half of all cases (Whelan *et al.* 1980), but these are generally mild: dementia, however, may occur (Varney *et al.* 1984), and, rarely (Kelly 1996), there may be a delirium (Kawashima *et al.* 2004).

These neuropsychiatric features classically occur in the setting of a cushingoid habitus, with, as noted earlier, 'moon' facies, truncal obesity, and violaceous abdominal striae. Other features include a 'buffalo hump', acne, hirsutism, proximal myopathy, easy bruisability, hypertension, diabetes mellitus, amenorrhea, decreased libido, erectile dysfunction, and, rarely, pseudotumor cerebri. All of these features, however, take time to develop and thus may not be present in cases of Cushing's syndrome secondary to exogenous steroid use, which, as noted earlier, may develop over days.

Furthermore, in cases of Cushing's syndrome secondary to ectopic ACTH secretion by an oat cell carcinoma, one may see emaciation rather than obesity.

When Cushing's syndrome occurs secondary to exogenous steroid administration the diagnosis is fairly obvious, as symptoms appear within days of using high-dosage steroids, for example >60 mg/day prednisone. However, when endogenous steroid overproduction is suspected, laboratory testing is essential, both to confirm the diagnosis and determine the cause of the overproduction. The first step is to determine whether or not, in fact, there is an elevated cortisol level. The simplest test to perform here is a morning plasma cortisol level, which is elevated in almost all instances. An exception to this rule, however, may occur in cases due to Cushing's disease, wherein normal pulsatile secretion is maintained: if the phlebotomy happens to coincide with the nadir, then the plasma cortisol will be falsely negative. Consequently, if the plasma cortisol level is normal, but the clinical picture is strongly suggestive of the diagnosis, then a 24-hour urine for free cortisol is in order. The next step is to obtain a plasma ACTH level: this will be increased in cases of Cushing's disease and also in cases secondary to ectopic ACTH production, but will be decreased in cases due to an adrenal tumor. Finally, if the ACTH is increased, the appropriate next step is to perform a high-dose dexamethasone suppression test, which will enable one to distinguish between Cushing's disease and ectopic ACTH production due to a tumor, such as oat cell carcinoma. The test is performed by giving dexamethasone in a dose of 2 mg orally every 6 hours over 2 days: on the second day, a 24-hour urine for free cortisol is also obtained. Given that in Cushing's disease the pituitary remains sensitive to steroids, the free cortisol level will be 'suppressed' by the dexamethasone; by contrast, ectopic ACTH-secreting tumors are not sensitive to steroids, and hence in these cases the free cortisol level will not be suppressed, thus yielding a 'positive' suppression test.

Rarely, the dexamethasone test may be equivocal, and differentiating between Cushing's disease and an ectopic ACTH-secreting tumor may depend on sampling venous blood flow from the pituitary. As might be expected, in cases of pituitary tumors, the ACTH level in venous blood from either the superior petrosal sinus or the internal jugular vein will be higher than that in venous blood drawn at phlebotomy; conversely, when the source of ACTH is an ectopic tumor, this 'gradient' between superior petrosal or internal jugular blood and systemic venous blood will not be present. Sampling the superior petrosal sinus is difficult and not without risk, but is more sensitive than sampling the internal jugular vein; hence, one should begin with sampling the internal jugular vein, reserving petrosal sampling for cases in which this is negative (Doppman *et al.* 1998).

In cases in which a pituitary tumor is suspected, high-resolution magnetic resonance imaging (MRI) with gadolinium enhancement is in order, and, if positive, may obviate the need for superior petrosal or internal jugular sampling. Unfortunately, however, most ACTH-secreting pituitary tumors are microadenomas, and about 50 percent will escape detection by MRI (Kaskarelis *et al.* 2006). CT scanning of the chest and abdomen may also be in order when an ectopic ACTH-secreting tumor is suspected, and, again, if this test is positive, the need for sampling superior petrosal or internal jugular venous blood vanishes.

Course

The course is determined by the underlying cause. Once cortisol levels are returned to normal, either by stopping exogenous steroids or by virtue of treatment of endogenous causes, neuropsychiatric features gradually resolve; although such resolution may occur within weeks to days in cases secondary to relatively short-term treatment with high-dose exogenous steroids, months may be required in endogenous cases (Kelly *et al.* 1983).

Etiology

In current clinical practice, by far the most common cause of Cushing's syndrome is the administration of exogenous steroids. Before making this diagnosis, however, it is important to keep in mind that Cushing's syndrome here generally occurs only with high-dose steroids: doses below 40 mg of prednisone are toxic in less than 2 percent of cases, while doses of 80 mg or more yield symptoms in roughly 20 percent (Anonymous 1972).

Cushing's syndrome due to endogenous causes is rare: overall, about 65 percent of cases are due to Cushing's disease, 12 percent to ectopic ACTH production (most commonly oat cell carcinoma, with some cases also being seen with tumors of the thyroid, thymus, or pancreas), with the remainder being due to cortisol production by adrenal adenomas or carcinomas (Orth 1995).

Although the mechanism or mechanisms whereby sustained hypercortisolemia produces neuropsychiatric symptomatology has not been clearly elucidated, it does appear that hippocampal pathology is involved in the production of cognitive deficits. Hippocampal atrophy is associated with these deficits, and, furthermore, once the Cushing's syndrome is treated, one finds a reversal of both the atrophy and the cognitive deficits (Starkman *et al.* 2003).

Differential diagnosis

The appearance of neuropsychiatric symptomatology in a patient receiving relatively high doses of exogenous corticosteroids immediately implicates the drug. Diagnostic uncertainty may arise, however, in cases in which the condition that prompted the treatment with corticosteroids may itself affect the central nervous system, as is the case with multiple sclerosis or systemic lupus erythematosus. In such cases, a careful history, with due attention to the existence of neuropsychiatric symptoms before steroid treatment, may resolve the issue. In other cases it may be necessary to change the corticosteroid dose significantly: if the dose is increased and symptoms worsen (or, conversely, if the dose is decreased and symptoms lessen), then the steroid is implicated.

In cases of depression, mania, anxiety, psychosis, dementia, or delirium that arise in the absence of treatment with exogenous steroids, a high index of suspicion is required, with special attention being paid to the existence of a classical 'cushingoid' habitus.

Treatment

In exogenous cases, the dose of the steroid should, if possible, be reduced. In endogenous cases, the tumor should, if possible, be resected. In cases of pituitary tumors, the trans-sphenoidal approach is utilized; however, if this is not possible, radiation treatment may be considered. When the tumor cannot be treated, one may consider adrenalectomy or, alternatively, a 'chemical' adrenalectomy may be performed by administering ketoconazole, aminoglutethimide or metyrapone, each of which inhibits enzymes responsible for steroidogenesis in the adrenal glands. After adrenalectomy, whether surgical or 'chemical', daily maintenance doses of steroids will be required. Furthermore, one must be alert to the development of Nelson's syndrome, in which, owing to a lack of inhibitory feedback by cortisol, an ACTH-secreting tumor of the pituitary may develop, with the appearance of hyperpigmentation.

Symptomatic treatment of the neuropsychiatric syndromes may or may not be required. Depression may be treated as described in Chapter 7.1, and mania as in Chapter 7.3. In regard to mania due to exogenous steroids, if repeat courses of steroids are required in the future, as for example in multiple sclerosis, consideration may be given to prophylactic treatment with lithium (Falk *et al.* 1979). Anxiety, psychosis, dementia and delirium may be treated as discussed in Chapters 7.5, 8.1, 6.3, and 6.5, respectively.

17.2 ADRENOCORTICAL INSUFFICIENCY

Adrenocortical insufficiency may be primary, due to actual destruction of the adrenal gland, or secondary, due to pituitary failure or to abrupt discontinuation of long-term treatment with corticosteroids. Adrenocortical insufficiency, whether primary or secondary, may be further subdivided into acute and chronic forms. This is a rare disorder, occurring in roughly 0.00001 percent of the general population.

Before proceeding, a word is in order regarding the term 'Addison's disease'. Some authors use this term to refer to all cases of adrenocortical insufficiency, regardless of cause, whereas others restrict it to primary cases, specifically those due to autoimmune destruction of the adrenal gland. Given this unfortunate definitional ambiguity, the term is not used further in this text.

Clinical features

Acute adrenocortical insufficiency presents with nausea, vomiting, and abdominal pain, a rapidly falling blood pressure, postural dizziness, and eventually hypovolemic shock: delirium develops, followed by stupor and coma.

Chronic adrenocortical insufficiency presents gradually with fatigue, listlessness, poor concentration, anorexia, nausea, diarrhea or constipation, and abdominal pain. Depression may occur (Engel and Margolin 1941; Varadaraj and Cooper 1986) and, rarely, there may be delirium (Engel and Margolin 1941; Fang and Jaspan 1989) or psychosis (Anglin *et al.* 2006; Cleghorn 1951; McFarland 1963). Blood pressure is low and postural dizziness is common. In primary cases, with a lack of cortisol feedback on the pituitary, excessive stimulation of melanocytes by ACTH-derived melanocyte-stimulating hormone may lead to hyperpigmentation, especially prominent in sun-exposed areas and on the buccal and gingival mucosa. Chronic adrenocortical insufficiency may also be complicated by an acute episode, for example as may occur when chronic patients are subjected to a significant physiologic stress such as surgery.

Laboratory testing begins with a morning serum cortisol level, which is reduced. An ACTH level is also obtained: in primary cases ACTH levels are increased because of a lack of feedback inhibition on the pituitary, whereas in secondary cases the ACTH level is low. In cases in which ACTH levels are equivocal, further testing may be conducted with cosyntropin, a synthetic ACTH analogue, to determine whether such cases are primary or secondary. Cosyntropin is given intravenously in a dose of 0.25 mg, and cortisol levels are obtained 15 minute pre-injection and 30 and 60 minutes post-injection. In primary cases there is little or no rise in serum cortisol, whereas in secondary cases, presuming that the adrenal glands have not atrophied as a result of chronic non-stimulation, the cortisol level should undergo a significant rise.

Further laboratory abnormalities may be seen in primary cases in which destruction of the adrenal glands causes decreased mineralocorticoid levels, with resulting hyponatremia and hyperkalemia.

Course

Acute adrenocortical insufficiency is a life-threatening emergency. The course of chronic adrenocortical insufficiency is determined by the underlying cause: in the case of primary chronic adrenocortical insufficiency due to autoimmune destruction of the adrenals, there is a gradual progression of symptoms, with death occurring in perhaps 2 years.

Etiology

In developed countries, primary adrenocortical insufficiency is most commonly due to autoimmune destruction of the adrenal glands. In this disorder, other structures may also be targeted by the autoimmune process, and patients may develop Hashimoto's thyroiditis (with either hyperthyroidism or hypothyroidism), pernicious anemia with vitamin B12 deficiency, hypoparathyroidism with hypocalcemia, or diabetes mellitus. Tuberculosis is another cause of primary adrenocortical insufficiency, and indeed in underdeveloped countries it is the most common cause. Other causes include adrenoleukodystrophy, cytomegalovirus infection (as may occur in AIDS), sarcoidosis, amyloidosis, metastatic disease, and hemorrhagic infarction (as may occur during septicemia or with overly vigorous anticoagulation). With the exception of hemorrhagic infarction, all of these forms of primary adrenocortical insufficiency cause a chronic presentation.

Secondary adrenocortical insufficiency, as noted earlier, most commonly occurs due to abrupt discontinuation of long-term treatment with corticosteroids. Any patient taking supraphysiologic doses of a corticosteroid (e.g. 10 mg or more daily of prednisone) for more than 1 month will have some suppression of ACTH output from the pituitary coupled with some atrophy of the adrenal cortex. Other causes of secondary adrenocortical insufficiency include infarction, tumors or granulomas of either the pituitary or, rarely, the hypothalamus, or secondary to radiation treatment involving either of these structures. These secondary cases may present either acutely (as for example with abrupt discontinuation of steroid treatment) or chronically (as may be seen with slowly growing pituitary tumors).

Differential diagnosis

Delirium occurring in acute adrenocortical insufficiency is suggested by concurrent nausea, vomiting, and postural dizziness. In cases occurring in the course of sepsis, the delirium may be erroneously attributed to the systemic effects of the infection.

Depression, delirium, or psychosis occurring during chronic adrenocortical insufficiency are each suggested by the overall fatigue and listlessness of the patient and by the associated gastrointestinal symptomatology and postural dizziness.

Treatment

Chronic adrenocortical insufficiency is generally treated with hydrocortisone in a dose of 20–30 mg/day, with two-thirds of the dose given in the morning and one-third in the afternoon, with dose increases during periods of physiologic stress. In primary cases this should be supplemented with fludrocortisone in a dose of 0.05–0.1 mg/day.

Acute adrenocortical insufficiency is a medical emergency, requiring treatment in an intensive care unit.

17.3 HYPERTHYROIDISM

Sustained elevation of thyroid hormones typically causes a syndrome characterized by anxiety and autonomic hyperactivity; most cases are due to Graves' disease or toxic multinodular goiter. Although in routine clinical work most clinicians refer to all of these cases as 'hyperthyroidism', this practice, technically, is not correct. Strictly speaking, hyperthyroidism refers to conditions in which the thyroid gland is producing excess amounts of thyroid hormone, and the term 'thyrotoxicosis' refers to the clinical syndrome itself. Although this may seem to be splitting hairs, it does call attention to the fact that, although most cases are due to diseases that cause excess thyroid hormone production, for example Graves' disease, there are diseases, such as Hashimoto's thyroiditis, in which the elevated serum T_4 concentrations occur not because of increased production of hormone but rather because of a 'leakage' of hormone from the inflamed gland. Clinicians' habits, however, are hard to break, and in this text the term 'hyperthyroidism' is used to refer to the syndrome.

Hyperthyroidism is seen in roughly 0.5 percent of the general population and is 6–10 times more common in females as in males.

Clinical features

The age of onset varies according to the underlying etiology: Graves' disease typically appears in the twenties or thirties and toxic multinodular goiter generally has an onset in old age. Although in most cases symptoms appear gradually, over weeks or months, subacute onsets may be seen, especially in relation to various physiologic stressors.

The discussion of clinical features will begin with a description of the typical picture of hyperthyroidism, followed by a discussion of an important variant known as apathetic hyperthyroidism. Subsequently, neuropsychiatric features seen in only a minority of cases of hyperthyroidism are discussed, including depression, mania, psychosis, dementia, and delirium. Finally, 'thyroid storm' is considered.

Typically, patients are apprehensive and anxious and, although fatigued and tired, they often experience restlessness and an inability to sit still. Anxiety often stands out (Greer *et al.* 1973; MacCrimmon *et al.* 1979), and in some cases may represent the presenting complaint (Dietch 1981). Patients typically complain of diaphoresis, heat intolerance, and an increased frequency of bowel movements, and, despite an often increased appetite with increased caloric intake, there may be substantial weight loss. On examination, one finds tachycardia, widened palpebral fissures and proptosis, a fine postural tremor, and generalized hyperreflexia; there may also be a proximal myopathy. Women may complain of menstrual irregularity and men may experience erectile dysfunction. Rarely, there may be chorea (Fidler *et al.* 1971; Fishbeck and Layzer 1979; Van Uitert and Russakoff 1979), grand mal seizures (Jabbari and Huott 1980), or a motor peripheral neuropathy (Pandit *et al.* 1998).

Apathetic hyperthyroidism (Brenner 1978; Lahey 1931; Thomas *et al.* 1970) represents a peculiar variant of hyperthyroidism that is generally only seen in the elderly and which is characterized by apathy and, in some, lethargy. Remarkably, of the 'autonomic' signs and symptoms seen in typical cases, only tachycardia is common in this apathetic variant: diaphoresis, tremor, and hyperreflexia are generally absent. Many patients will also develop atrial fibrillation and congestive heart failure.

Of the neuropsychiatric features seen in hyperthyroidism, depression is the most common and is seen in roughly one-third of patients (Kathol and Delahunt 1986; Trzepacz *et al.* 1988), especially in those with apathetic hyperthyroidism (Thomas *et al.* 1970), in whom the only clue to the correct diagnosis, as just noted, may be tachycardia, atrial fibrillation, or congestive heart failure (Arnold *et al.* 1974; Brenner 1978). Depression in hyperthyroidism may be accompanied by mood-congruent delusions (Taylor 1975) or considerable agitation (Van Uitert and Russakoff 1979).

Mania is less common than depression in hyperthyroidism (Trzepacz *et al.* 1988), and in some cases may be of the 'mixed' variety (Ingham and Nielsen 1931).

Psychosis, though rare, may occur (Lazarus and Jaffe 1986; Steinberg 1994): one patient developed a delusion of jealousy and, convinced that his wife was having an affair, had her followed; when his hyperthyroidism was treated, his psychosis resolved (Hodgson *et al.* 1992).

Dementia may occur in typical hyperthyroidism (Bulens 1981; Fukui *et al.* 2001) but is rare; by contrast, a significant minority of patients with the apathetic variant will develop cognitive deficits that may be severe enough to constitute a dementia (Martin and Deam 1996). Delirium may also occur, but appears to be very rare.

Thyroid storm is a dreaded complication of hyperthyroidism and typically appears in a patient with untreated hyperthyroidism who is subjected to some significant physiologic stress, such as surgery or an infection. In this setting there is a rapid escalation of all of the typical signs and symptoms, followed by hyperthermia, delirium, stupor, and coma. In some cases seizures may occur and, rarely, thyroid storm has presented with a psychosis (Bursten 1961; Greer and Parsons 1968). Thyroid storm may also occur in the setting of apathetic hyperthyroidism, and in such cases may present with stupor or coma in the absence of any autonomic features (Ghobrial and Ruby 2002).

In almost all cases of hyperthyroidism the free T_4 level will be elevated and the thyroid-stimulating hormone (TSH) level will be reduced. Exceptions occur in cases of 'T_3 (triiodothyronine) thyrotoxicosis', in which, though the TSH level is reduced, the free T_4 level is normal: in cases in which the clinical suspicion of hyperthyroidism is high and the free T_4 is normal, the free T_3 level should be determined. Other exceptions include the very rare cases of a hypothalamic tumor secreting thyrotrophin-releasing hormone (TRH) or a pituitary adenoma secreting TSH: in both of these cases both the free T_4 and the TSH levels are elevated.

Course

The course is dictated by the underlying cause. Thyroid storm, regardless of the underlying cause, may pursue a fulminant course, with death in hours or days.

Etiology

The vast majority of cases are due to Graves' disease. Other causes include toxic multinodular goiter, toxic solitary adenoma, and the thyroiditides, including lymphocytic thyroiditis, subacute (de Quervain's) thyroiditis, and Hashimoto's thyroiditis. The thyroiditides are characterized by a 'leakage' of T_4 from the inflamed thyroid and, typically, cause a hyperthyroidism that is time limited and that may, depending on the amount of inflammatory damage and scarring, be followed by hypothyroidism. Rare causes include hypothalamic tumors, pituitary adenomas (Carlson *et al.* 1983; Wynne *et al.* 1992), inherited pituitary resistance to T_4, production of ectopic TSH by various tumors (hydatidiform mole, choriocarcinoma of the uterus, and choriocarcinoma of the testis), and production of T_4 by various tumors (struma ovarii or metastatic follicular carcinoma of the thyroid). Finally, hyperthyroidism may occur as a side-effect of amiodarone (Trip *et al.* 1991), interferon-alpha (Bohbot *et al.* 2006) and interferon-beta (Schwid *et al.* 1997) and may also be intentional, as in 'thyrotoxicosis factitia'.

Differential diagnosis

As noted earlier, anxiety is a prominent feature of hyperthyroidism, and, of the various other causes of persistent anxiety discussed in Chapter 7.5, the one that often comes to mind is generalized anxiety disorder. One clue to the differential here may be found on shaking the patient's hand: in both hyperthyroidism and generalized anxiety disorder the hand is moist with sweat, but in hyperthyroidism the skin is warm whereas in generalized anxiety disorder it is cool, yielding a 'cold and clammy' impression.

The differential diagnosis for depression is discussed in Chapter 7.1. One feature of depression secondary to typical hyperthyroidism that sets it apart from depression due to other causes is the presence of weight loss in the face of increased eating; by contrast, weight loss occurring in depression due to other causes is generally associated with anorexia. In the case of depression occurring in the context of apathetic hyperthyroidism this feature may be absent, and here a high index of suspicion, looking especially to cardiac symptoms such as tachycardia, atrial fibrillation and congestive heart failure, would be required.

Both mania and psychosis due to hyperthyroidism, as discussed in Chapters 7.3 and 8.1, respectively, are distinguished from other causes by the prominent autonomic symptoms.

Dementia occurring in the setting of typical hyperthyroidism is likewise suggested by the prominent autonomic symptoms; however, when occurring in apathetic hypothyroidism, the only clue might be the cardiac symptoms.

Treatment

Treatment is directed at the underlying cause of the hyperthyroidism, and once T_4 levels are normalized, symptoms gradually resolve. Pending this resolution, autonomic symptoms, if necessary, may be controlled with propranolol, in doses of 20–40 mg orally every 4–6 hours. Symptomatic treatment of depression and other neuropsychiatric features may or may not be required pending this resolution; if so, this may be accomplished as discussed in the relevant chapters. Importantly, thyroid storm constitutes a medical emergency and requires treatment in an intensive care unit.

17.4 HYPOTHYROIDISM

Hypothyroidism is the clinical syndrome that occurs secondary to a persistently low level of circulating thyroxine, or T_4. A synonym for this is myxedema, a name that calls attention to the mucinous edema that gives rise to the 'myxedematous' appearance of these patients, described below. Hypothyroidism is a common condition, found in close to 1 percent of the general population; it is 10–20 times more common in females than in males.

Thyroxine levels are maintained within normal limits by virtue of the hypothalamic–pituitary–thyroid axis. The hypothalamus produces TRH, which in turn stimulates the anterior pituitary to produce TSH; TSH then stimulates the thyroid gland to produce T_4, which in turn exerts a negative feedback effect on both the pituitary and the hypothalamus. The various causes of hypothyroidism may be grouped according to which element of the hypothalamic–pituitary–thyroid axis is affected. Tertiary hypothyroidism occurs as a result of hypothalamic disease (e.g. tumors) and is characterized by a low TSH and a low T_4. Secondary hypothyroidism occurs with pituitary disease (e.g. tumors or infarction) and likewise displays both a low TSH and low T_4. Primary hypothyroidism occurs with disease of the thyroid itself (e.g. Hashimoto's thyroiditis or post thyroidectomy) and, in this case, the TSH is elevated whereas the T_4 is low. Tertiary and secondary hypothyroidism are uncommon: over 90 percent of cases of hypothyroidism are of the primary type.

Clinical features

Both the age and mode of onset are determined by the underlying cause. For example, Hashimoto's thyroiditis typically appears in the late thirties or early forties, and the onset is generally gradual, over months or longer. By contrast, Graves' disease, one of the most common indications for thyroidectomy, generally appears in the late twenties or early thirties, and the hypothyroidism following thyroidectomy appears fairly acutely, within weeks.

The clinical picture includes not only typical features, such as psychomotor retardation, but also, in a majority of cases, neuropsychiatric features including cognitive deficits, depression, and psychosis.

Typically, patients develop psychomotor retardation, with slowed speech and movements; in some cases lethargy and somnolence may occur. When asked a question, up to a minute may pass before the patient responds, and the response itself, when it does come, is slow. Simple activities, such as unfastening a button, may likewise take an inordinate amount of time to complete. The overall myxedematous appearance of hypothyroid patients is often distinctive: skin becomes thickened, puffy, and even boggy, and this is particularly obvious on the face, in the supraclavicular fossae, and on the dorsal surfaces of the hands and feet. The hair becomes thin and brittle and there may be considerable hair loss; interestingly there may also be a loss of hair on the lateral thirds of the eyebrows. Other symptoms include a voice change toward hoarseness, cold intolerance, constipation, weight gain, decreased libido, erectile dysfunction, and menorrhagia. Vibratory sense may be lost and the deep tendon reflexes are often reduced; the ankle jerk is often 'hung up', with a delayed relaxation phase. Cranial nerves may be involved, with partial deafness and, rarely, a peripheral facial palsy; cerebellar ataxia ('myxedema staggers') (Jellinek and Kelly 1960) occurs in up to 20 percent of cases. Very rarely, seizures may occur (Bryce and Poyner 1992). Myxedematous infiltration may cause a carpal or tarsal tunnel syndrome; macroglossia may also occur and, if severe, may be followed by obstructive sleep apnea. Bradycardia and hypotension are common, and there may be a degree of hypothermia; pericardial and pleural effusions may also occur. Cardiomegaly may appear, and in some cases there may be congestive heart failure. In a small minority of cases, a syndrome of inappropriate antidiuretic hormone (SIADH) secretion may occur, with hyponatremia.

In severe cases of hypothyroidism a condition known as 'myxedema coma' may develop. Typically this occurs in patients with long-standing hypothyroidism who are subjected to a physiologic stress, such as surgery or a significant infection, or who are given any medication with prominent sedative effects. Stupor or coma develops, accompanied by hypothermia (which may be severe), significant bradycardia, respiratory depression, and, in a significant minority, grand mal seizures.

Cognitive deficits of variable degree occur in roughly one-half of all cases. Patients may appear apathetic and lacking in initiative, and there may be difficulty with concentration and a certain 'fogginess' of thought and memory (Nickel and Frame 1958). In an uncertain percentage of these cases, these cognitive deficits may worsen to the point where a dementia appears (Akelaitis 1936; de Fine Olivarius and Roder 1970; Uyematsu 1920), which in turn may be accompanied by delusions of persecution and auditory hallucinations.

Depression, often accompanied by anxiety, may occur in a little under than one-half of all cases; in a small minority it may be marked by psychotic features, with hallucinations and delusions (Whybrow *et al.* 1969).

Psychosis is the most uncommon neuropsychiatric manifestation of hypothyroidism, occurring in less than 5 percent of cases. Traditionally, this psychosis has been referred to as 'myxedema madness' (Asher 1949), and it is typically characterized by delusions of persecution and reference, as well as auditory hallucinations (Crowley 1940; de Fine Olivarius and Roder 1970; Karnosh and Stout 1935; Logothetis 1963; Reed and Bland 1977). One description of myxedema madness (Ziegler 1930) deserves to be quoted at length, as it illustrates the sometimes exquisite nature of the dependence of the psychosis on the level of circulating thyroid hormone. The patient had undergone radiation treatment for hyperthyroidism 3 years earlier and had subsequently done well with an appropriate dosage of supplemental thyroid hormone. At the age of 48 years, however, she became non-compliant and soon thereafter:

> she began to feel that her husband was paying attention to another woman and that he was trying to do away with her by means of gas or the electric chair. During a game at

a party in her own home during the holidays in 1928, she refused to sit in a chair designated for her, thinking it might be a plot to kill her. She also felt that her husband was trying to poison her and refused to take desiccated thyroid gland at home on account of such a belief. On several occasions subsequently, when desiccated thyroid was administered in sufficient quantity, at the repeated and urgent request of her physician, the delusions entirely disappeared and she felt so much better that she concluded that it was foolish to be taking medicine and discontinued taking it. On such occasions, the psychosis would slowly return in the same form as before.

The electroencephalogram (EEG) typically shows generalized slowing.

Thyroxine levels are available as both 'total' T_4, including both bound and free fractions, and 'free' T_4, and in all instances a free T_4 level should be obtained. This free T_4, as noted earlier, is reduced in all cases. The TSH level, however, may be elevated or decreased, depending on the cause of the hypothyroidism. In primary hypothyroidism, the TSH is increased; in both secondary and tertiary cases, however, it is reduced. Distinguishing secondary from tertiary cases may often be accomplished with an MRI scan, which often reveals appropriate pathology of the pituitary or hypothalamus. In some cases, however, the MRI scan may be inconclusive, and here a TRH stimulation test is in order. In this test, TRH is given intravenously and TSH levels are determined just before and 20 minutes after the infusion. In cases of secondary hypothyroidism, given a lack of pituitary cells capable of producing TSH, the response to exogenous TRH is blunted. By contrast, in tertiary hypothyroidism, in which a chronic lack of endogenous TRH allows for an up-regulation of TRH receptors on pituitary cells, there is an enhanced response of TSH to TRH.

Before leaving this section some words are in order regarding the condition known as 'subclinical' primary hypothyroidism. Here, although the free T_4 level is within normal limits, the TSH level is mildly elevated. Such a combination of laboratory values indicates that, although the free T_4 may be within broadly defined limits of normal, it is nevertheless below the individual patient's 'normal' level, as indicated by the rise in TSH. Although patients may not have symptoms directly related to these findings, they are significant for two reasons. First, such findings may indicate that the patient is in the very early stages of what will become clinically evident hypothyroidism and, given this, close monitoring is required. Second, such subclinically reduced free T_4 levels, although not causing symptoms *per se*, may nevertheless blunt the response to antidepressant treatment in depression occurring in either major depressive disorder (Joffe and Levitt 1992) or bipolar disorder (Cole *et al.* 2002); furthermore, there is also evidence that subclinical hypothyroidism may, in bipolar disorder, predispose to a rapid-cycling course (Cowdry *et al.* 1983).

Course

The course of hypothyroidism is determined by the underlying cause.

Etiology

As noted earlier, the various causes of hypothyroidism are divided into primary, secondary, and tertiary types.

As also noted earlier, primary hypothyroidism is by far the most common type, accounting for over 90 percent of cases. The most common cause of primary hypothyroidism is Hashimoto's thyroiditis, indicated by the presence of anti-thyroid antibodies: anti-thyroid peroxidase antibody (also known as anti-microsomal antibody) is most common, but anti-thyroglobulin antibodies may also be present. Other causes include thyroidectomy, radioactive iodine treatment, neck irradiation, iodine deficiency, and various medications, including rifampin (Takasu *et al.* 2005), ethionamide (McDonnell *et al.* 2005), and, most especially, lithium (Lindstedt *et al.* 1977), in which case the occurrence of hypothyroidism is most likely in patients who have anti-thyroid antibodies (Calabrese *et al.* 1985).

Secondary hypothyroidism may occur with infarction (e.g. post-partum Sheehan's syndrome) or tumors of the pituitary.

Tertiary hypothyroidism may occur with tumors, granulomatous disease, or infarction of the hypothalamus and after traumatic brain injury.

Hypothyroidism is also seen with treatment with either amiodarone (Trip *et al.* 1991) or carbamazepine (Isojarvi *et al.* 1995), and in such cases one may see a pattern consistent with primary hypothyroidism (i.e. with an elevated TSH) or a pattern consistent with secondary or tertiary hypothyroidism (with a normal or low TSH). Thus, in contrast to, for example, lithium-induced hypothyroidism, in which monitoring the TSH level is sufficient, in amiodarone- or carbamazepine-treated patients it is essential to monitor both the TSH and the free T_4.

Differential diagnosis

The differential diagnoses of the syndromes of dementia, depression, and psychosis are discussed in Chapters 6.3, 7.1, and 8.1, respectively. In all these cases, the chief clues to the diagnosis are the typical symptoms seen in hypothyroidism, especially the psychomotor retardation, lethargy, and the distinctive myxedematous appearance: it must be kept in mind, however, that the myxedematous appearance takes time to develop, and thus may be absent early on in cases characterized by an acute onset, as for example after thyroidectomy or pituitary infarction.

In cases of primary hypothyroidism secondary to Hashimoto's thyroiditis, it must be kept in mind that the underlying autoimmune process may cause additional disorders including primary adrenocortical insufficiency and pernicious anemia with vitamin B12 deficiency. Given that both of these disorders may cause psychosis, and that B12 deficiency, in addition, may cause depression and dementia, it is appropriate in all cases of Hashimoto's thyroiditis to check cortisol and B12 levels. Checking the cortisol level is also important because the treatment of hypothyroidism with T_4 in a patient with adrenocortical insufficiency may precipitate a life-threatening Addisonian crisis.

Cases of secondary or tertiary hypothyroidism may also be associated with other endocrinopathies, such as hyperprolactinemia or secondary adrenocortical insufficiency.

Finally, one must clearly distinguish between Hashimoto's thyroiditis and Hashimoto's encephalopathy. Both of these disorders are associated with anti-thyroid antibodies. In Hashimoto's thyroiditis, however, the thyroid alone is targeted, whereas in Hashimoto's encephalopathy, as discussed in Chapter 19.10, the thyroid is usually spared and the autoimmune onslaught is directed against the cerebrum.

Treatment

Treatment with once-daily T_4 is generally curative. In cases of hypothyroidism of relatively recent onset in young patients who are otherwise healthy and lack heart disease, one may begin with 50–75 μg daily, increasing in 25- to 50-μg increments every 2 or 3 weeks. In cases of long-standing hypothyroidism, however, or in elderly patients or those in poor health or with significant heart disease, the starting dose should be lower, in the range of 12.5–25 μg, and the increases should be in increments of 12.5–25 μg every 4–6 weeks. Serial T_4 determinations are made, and the dose should be increased until the free T_4 is within the normal range. In cases of primary hypothyroidism, TSH levels should also be obtained, with the goal of bringing TSH down to within normal limits; in this regard it must be kept in mind that the TSH level falls very slowly and that 4–8 weeks may be required at any given dose of T_4 for the TSH level to plateau. For most adult females, anywhere from 75 to 100 μg of T_4 is generally an adequate maintenance dose; in males the range is from 100 to 150 μg. In the elderly, however, the maintenance dose is generally 75 μg or less. In females given conjugated estrogens an increase in thyroid-binding globulins may decrease the free T_4, necessitating a dose increase (Arafah 2001).

It has been suggested that a combination of T_4 and T_3 produced better symptomatic relief than T_4 alone (Bunevicius *et al.* 1999); however, subsequent work has failed to confirm this (Gorzinsky-Glasberg *et al.* 2006).

Myxedema coma constitutes a medical emergency and patients should be admitted to an intensive care unit. Treatment involves giving intravenous T_4 in a dose of 300 μg, followed by daily intravenous doses of 50–100 μg. Intravenous hydrocortisone is also given, along with vigorous supportive care.

REFERENCES

Akelaitis AJE. Psychiatric aspects of myxedema. *J Nerv Ment Dis* 1936; **83**:22–36.

Anonymous. Acute adverse reactions to prednisone in relation to dosage. *Clin Pharmacol Ther* 1972; **13**:694–8.

Anderson AE, McHugh PR. Oat cell carcinoma with hypercortisolemia presenting to a psychiatric hospital as a suicide attempt. *J Nerv Ment Dis* 1971; **152**:427–31.

Anglin RE, Rosebush PI, Mazurek MF. The neuropsychiatric profile of Addison's disease: revisiting a forgotten phenomenon. *J Neuropsychiatry Clin Neurosci* 2006; **18**:450–9.

Arafah BM. Increased need for thyroxine in women with hypothyroidism during estrogen therapy. *N Engl J Med* 2001; **344**: 1743–9.

Arnold BM, Casal G, Higgins HP. Apathetic thyrotoxicosis. *Can Med Assoc J* 1974; **111**:957–8.

Asher R. Myxoedematous madness. *BMJ* 1949; **2**:555–62.

Bohbot NL, Young J, Orgiazzi J *et al.* Interferon-alfa-induced hyperthyroidism: a three-stage evolution from silent thyroiditis towards Grave's disease. *Eur J Endocrinol* 2006; **154**:367–72.

Brenner I. Apathetic hyperthyroidism. *J Clin Psychiatry* 1978; **39**: 479–80.

Bryce GM, Poyner F. Myxoedema presenting with seizures. *Postgrad Med J* 1992; **68**:35–6.

Bulens C. Neurologic complications of hyperthyroidism: remission of spastic paralysis, dementia, and optic neuropathy. *Arch Neurol* 1981; **38**:669–70.

Bunevicius R, Kazanavicius G, Zalinkevicius R *et al.* Effects of thyroxine as compared with thyroxine plus triiodothyronine in patients with hypothyroidism. *N Engl J Med* 1999; **340**:424–9.

Bursten B. Psychosis associated with thyrotoxicosis. *Arch Gen Psychiatry* 1961; **4**:267–73.

Calabrese JR, Gulledge AD, Hahn K *et al.* Autoimmune thyroiditis in manic-depressive patients treated with lithium. *Am J Psychiatry* 1985; **142**:1318–21.

Carlson HE, Linfoot JA, Burnstein GD *et al.* Hyperthyroidism and acromegaly due to a TSH- and GH-secreting tumor: lack of hormonal response to bromocriptine. *Am J Med* 1983; **74**: 915–23.

Cleghorn RA. Adrenal cortical insufficiency: psychological and neurological observations. *Can Med Assoc J* 1951; **65**:449–54.

Cohen SI. Cushing's syndrome: a psychiatric study of 29 patients. *Br J Psychiatry* 1980; **136**:120–4.

Cole P, Thase ME, Mallinger AG *et al.* Slower treatment response in bipolar depression predicted by lower pretreatment thyroid function. *Am J Psychiatry* 2002; **159**:116–21.

Cowdry RW, Wehr TA, Zis AP *et al.* Thyroid abnormalities associated with rapid-cycling bipolar illness. *Arch Gen Psychiatry* 1983; **40**:414–20.

Crowley RM. Psychoses with myxedema. *Am J Psychiatry* 1940; **96**:1105–16.

De Fine Olivarius B, Roder E. Reversible psychosis and dementia in myxedema. *Acta Psychiatr Scand* 1970; **46**:1–13.

Dietch JT. Diagnosis of organic anxiety disorders. *Psychosomatics* 1981; **22**:661–9.

Doppman JL, Oldfield EH, Nieman LK. Bilateral sampling of the internal jugular vein to distinguish between mechanisms of adrenocorticotropic hormone-dependent Cushing syndrome. *Ann Intern Med* 1998; **128**:33–6.

Engel GL, Margolin SG. Neuropsychiatric disturbances in Addison's disease and the role of impaired carbohydrate metabolism in the production of abnormal cerebral function. *Arch Neurol Psychiatry* 1941; **45**:881–4.

Falk WE, Mahnke MW, Poskanzer DC. Lithium prophylaxis of corticotrophin-induced psychosis. *JAMA* 1979; **241**:1011–12.

Fang VBS, Jaspan JB. Delirium and neuromuscular symptoms in an elderly man with isolated corticotroph deficiency syndrome completely reversed with glucocorticoid replacement. *J Clin Endocrinol Metab* 1989; **69**:1073–7.

Fidler SM, O'Rourke RA, Buschbaum HW. Choreoathetosis as a manifestation of thyrotoxicosis. *Neurology* 1971; **21**:55–7.

Fishbeck KH, Layzer RB. Paroxysmal choreoathetosis associated with thyrotoxicosis. *Ann Neurol* 1979; **6**:453–4.

Fukui T, Hasegawa Y, Takenaka H. Hyperthyroid dementia: clinico-radiological findings and response to treatment. *J Neurol Sci* 2001; **184**:81–8.

Ghobrial MW, Ruby EB. Coma and thyroid storm in apathetic thyrotoxicosis. *South Med J* 2002; **95**:552–4.

Greer S, Parsons V. Schizophrenia-like psychosis in thyroid crisis. *Br J Psychiatry* 1968; **114**:1357–62.

Greer S, Ramsay I, Bagley C. Neurotoxic and thyrotoxic anxiety: clinical, psychological and physiological measurements. *Br J Psychiatry* 1973; **122**:549–54.

Gorzinsky-Glasberg S, Fraser A, Nahsoni E *et al.* Thyroxine–triiodothyronine combination therapy versus thyroxine monotherapy for clinical hypothyroidism: meta-analysis of randomized controlled trials. *J Clin Endocrinol Metab* 2006; **91**:2592–9.

Haskett RF. Diagnostic categorization of psychiatric disturbance in Cushing's syndrome. *Am J Psychiatry* 1985; **142**:911–16.

Hertz PE, Nadas E, Wojtkowiki H. Case report: Cushing's syndrome and its management. *Am J Psychiatry* 1955; **112**:144–5.

Hickman JW, Atkinson RP, Flint LD *et al.* Transient schizophrenic reaction as a major symptom of Cushing's syndrome. *N Engl J Med* 1961; **264**:797–800.

Hodgson RE, Murray D, Woods MR. Othello's syndrome and hyperthyroidism. *J Nerv Ment Dis* 1992; **180**:663–4.

Ingham SD, Nielsen JM. Thyroid psychosis: difficulties in diagnosis. *J Nerv Ment Dis* 1931; **74**:271–7.

Isojarvi JI, Airaksinen KE, Mustonen JN *et al.* Thyroid and myocardial function after replacement of carbamazepine by oxcarbazepine. *Epilepsia* 1995; **36**:810–16.

Jabbari B, Huott AD. Seizures in thyrotoxicosis. *Epilepsia* 1980; **21**:91–6.

Jeffcoate WJ, Silverstone JT, Edwards CRW *et al.* Psychiatric manifestations of Cushing's syndrome: response to lowering of plasma cortisol. *Q J Med* 1979; **191**:465–72.

Jellinek EH, Kelly RE. Cerebellar syndrome in myxoedema. *Lancet* 1960; **2**:225–7.

Joffe RT, Levitt AJ. Major depression and subclinical (grade 2) hypothyroidism. *Psychoneuroendocrinology* 1992; **17**:215–21.

Karnosh LJ, Stout RE. Psychoses of myxedema. *Am J Psychiatry* 1935; **91**:1263–74.

Kaskarelis IS, Tsatalou EG, Benakis SV *et al.* Bilateral inferior petrosal sinuses sampling in the routine investigation of Cushing's syndrome: a comparison with MRI. *AJR* 2006; **187**:562–70.

Kathol RG, Delahunt JW. The relationship of anxiety and depression to symptoms of hyperthyroidism using operational criteria. *Gen Hosp Psychiatry* 1986; **8**:23–8.

Kawashima T, Oda M, Kund T *et al.* Metyrapone for delirium due to Cushing's syndrome induced by occult ectopic adrenocorticotropic hormone secretion. *J Clin Psychiatry* 2004; **65**:1019–20.

Kelly WF. Psychiatric aspects of Cushing's syndrome. *Q J Med* 1996; **89**:543–51.

Kelly WF, Checkley SA, Bender DA *et al.* Cushing's syndrome and depression – a prospective study of 26 patients. *Br J Psychiatry* 1983; **142**:16–19.

Lahey FH. Non-activated (apathetic) type of hyperthyroidism. *N Engl J Med* 1931; **204**:747–8.

Lazarus A, Jaffe R. Resolution of thyroid-induced schizophreniform disorder following subtotal thyroidectomy: case report. *Gen Hosp Psychiatry* 1986; **8**:29–31.

Lindstedt G, Nilsson L, Walinder J *et al.* On the prevalence, diagnosis and management of lithium-induced hypothyroidism in psychiatric patients. *Br J Psychiatry* 1977; **130**:452–8.

Logothetis J. Psychotic behavior as the initial indicator of adult myxedema. *J Nerv Ment Dis* 1963; **136**:561–8.

MacCrimmon DJ, Wallace JE, Goldberg WM *et al.* Emotional disturbances and cognitive deficits in hyperthyroidism. *Psychosom Med* 1979; **41**:331–40.

McDonnell ME, Braverman JE, Bernardo J. Hypothyroidism due to ethionamide. *N Engl J Med* 2005; **352**:2757–9.

McFarland HR. Addison's disease and related psychoses. *Compr Psychiatry* 1963; **4**:90–5.

Maclay WS, Stokes AB. Mental disorder in Cushing's syndrome. *J Neurol Neurosurg Psychiatry* 1939; **1**:110–18.

Martin FI, Deam DR. Hyperthyroidism in elderly hospitalized patients. Clinical features and treatment outcomes. *Med J Aust* 1996; **164**:200–3.

Nickel SN, Frame B. Neurologic manifestations of myxedema. *Neurology* 1958; **8**:511–17.

Orth DN. Cushing's syndrome. *N Engl J Med* 1995; **332**:791–803.

Pandit L, Shankar SK, Gayathri N *et al.* Acute thyrotoxic neuropathy – Basedow's paraplegia revisited. *J Neurol Sci* 1998; **155**: 211–14.

Reed K, Bland RC. Masked 'myxedema madness'. *Acta Psychiatr Scand* 1977; **56**:421–6.

Schwid SR, Goodman AD, Mattson DH. Autoimmune hyperthyroidism in patients with multiple sclerosis treated with interferon beta-1b. *Ann Neurol* 1997; **54**:1169–90.

Spillane JD. Nervous and mental disorders in Cushing's syndrome. *Brain* 1951; **74**:72–94.

Starkman MN, Schteingart DE, Schork MA. Depressed mood and other psychiatric manifestations of Cushing's syndrome: relationship to hormone levels. *Psychosom Med* 1981; **43**:3–18.

Starkman MN, Giordani B, Gebarski SS *et al.* Improvement in learning associated with increase in hippocampal formation volume. *Biol Psychiatry* 2003; **53**:233–8.

Steinberg PI. A case of paranoid disorder associated with hyperthyroidism. *Can J Psychiatry* 1994; **39**:153–6.

Takasu N, Takara M, Komiya I. Rifampin-induced hypothyroidism in patients with Hashimoto's thyroiditis. *N Engl J Med* 2005; **352**:518–19.

Taylor JW. Depression in thyrotoxicosis. *Am J Psychiatry* 1975; **132**:552–3.

Thomas FB, Mazzaferri EL, Skillman TG. Apathetic thyrotoxicosis: a distinctive clinical and laboratory entity. *Ann Intern Med* 1970; **72**:679–85.

Trethowan WH, Cobb S. Neuropsychiatric aspects of Cushing's syndrome. *Arch Neurol Psychiatry* 1952; **67**:283–309.

Trip MD, Wiersinga W, Plomp TA. Incidence, predictability, and pathogenesis of amiodarone-induced thyrotoxicosis and hypothyroidism. *Am J Med* 1991; **91**:507–11.

Trzepacz P, McCue M, Klein I *et al.* A psychiatric and neuropsychological study of patients with untreated Graves' disease. *Gen Hosp Psychiatry* 1988; **10**:49–55.

Uyematsu S. A case of myxedematous psychosis. *Arch Neurol Psychiatry* 1920; **3**:252–76.

Van Uitert RL, Russakoff LM. Hyperthyroid chorea mimicking psychiatric disease. *Am J Psychiatry* 1979; **136**:1208–10.

Varadaraj R, Cooper AJ. Addison's disease presenting with psychiatric features. *Am J Psychiatry* 1986; **143**:553–4.

Varney NR, Alexander B, MacIndoe JH. Reversible steroid dementia in patients without steroid psychosis. *Am J Psychiatry* 1984; **141**:369–72.

Whelan TB, Schteingart DE, Starkman MN *et al.* Neuropsychological deficits in Cushing's syndrome. *J Nerv Ment Dis* 1980; **168**: 753–7.

Whybrow PC, Prange AJ, Treadway CR. Mental changes affecting thyroid gland dysfunction: a reappraisal using objective psychological measurement. *Arch Gen Psychiatry* 1969; **20**:48–63.

Wolkowitz OM, Rubinow D, Doran AR *et al.* Prednisone effects on neurochemistry and behavior. *Arch Gen Psychiatry* 1990; **47**:963–8.

Wynne AG, Gharib H, Scheithauer BW *et al.* Hyperthyroidism due to inappropriate secretion of thyrotropin in 10 patients. *Am J Med* 1992; **92**:15–24.

Ziegler LH. Psychosis associated with myxedema. 1930; 11:20–7.

18

Myalgic encephalomyelitis, fibromyalgia, and multiple chemical sensitivity

18.1 MYALGIC ENCEPHALOMYELITIS

The nomenclature for this disorder varies. Here, we have followed the World Health Organization's neurological classification in the International Classification of Diseases (G93.3) and referred to this disorder as myalgic encephalomyelitis, but others use terms such as chronic fatigue syndrome.

Clinical features

This debilitating disease is estimated to affect around eight million people in the USA, and around a quarter of a million in the UK. It can start before adulthood, although it is more likely to start when people are in their twenties to forties. Women appear to be affected more commonly than men; the female to male ratio may be around 2:1. In severe cases, it can leave patients totally incapacitated and bedbound. Often there is a history of an influenza-like illness just before the disease manifests itself. Indeed, often patients have led very active lives before being struck down with this devastating illness.

The Revised Centers for Disease Control (CDC) criteria (Fukuda *et al.* 1994) for myalgic encephalomyelitis/chronic fatigue syndrome are in common use, particularly for clinical research into this illness. Under these criteria, any other cause for chronic fatigue must be excluded, and there should be self-reported persistent or relapsing fatigue for at least six consecutive months. Furthermore, four or more of the following symptoms need to be concurrently present for over 6 months: impaired memory or concentration; sore throat; tender cervical or axillary lymph nodes; muscle pain; multijoint pain; new headaches; unrefreshing sleep; and post-exertional malaise. Within this framework, self-reported persistent fatigue lasting 1 month or longer is defined as prolonged fatigue, and only becomes chronic fatigue after it has lasted for at least six consecutive months. While this temporal criterion helps in standardizing the disease for research purposes, in clinical practice many clinicians do not require patients to have suffered from the illness for at least 6 months before being prepared to diagnose myalgic encephalomyelitis or chronic fatigue syndrome.

An alternative set of diagnostic criteria, known as the Canadian Criteria, has been suggested by prominent researchers and clinicians involved in researching and caring for patients with myalgic encephalomyelitis (Carruthers *et al.* 2003). Among other things, the authors of the Canadian Criteria disagreed with the the fact that that the symptoms of post-exertional malaise and memory/concentration problems do not have to be present in order to make a CDC diagnosis of chronic fatigue syndrome. These Canadian Criteria were revised in 2010 (Jason *et al.* 2010). However, some have criticized both the original Canadian Criteria and the revised version (White *et al.* 2011).

More recently, a set of International Consensus Criteria for myalgic encephalomyelitis have been proposed by a group who between them have diagnosed or treated around 50 000 individuals with myalgic encephalomyelitis (Carruthers *et al.* 2011). These criteria are shown in Table 18.1.

Etiology

At the time of writing the etiology of this disease is unknown. There are many features of myalgic encephalomyelitis which point to an infectious cause. Indeed, the possibility has been raised that this disorder may be associated with XMRV (Lombardi *et al.* 2009; withdrawn), although several other groups have failed to replicate the original finding (Erlwein *et al.* 2010; Groom *et al.* 2010; Hohn *et al.* 2010; Jerome *et al.* 2011; Knox *et al.* 2011; Shin *et al.* 2011). At the time of writing, the issue of a possible etiologic role of XMRV remains controversial (see, for example, Ali *et al.* 2011; Hohn and Bannert 2011; Lombardi *et al.* 2011; Menendez-Arias *et al.* 2011; Mikovits *et al.* 2010a,b; Paprotka *et al.* 2011; Robinson *et al.* 2011; Silverman *et al.* 2011).

Table 18.1 International Consensus Criteria for myalgic encephalomyelitis

Myalgic encephalomyelitis is an acquired neurologic disease with complex global dysfunctions. Pathologic dysregulation of the nervous, immune and endocrine systems, with impaired cellular energy metabolism and ion transport are prominent features. Although signs and symptoms are dynamically interactive and casually connected, the criteria are grouped by regions of pathophysiology to provide general focus

A patient will meet the criteria for post-exertional neuroimmune exhaustion (A), at least one symptom from three neurological impairment categories (B), at least one symptom from three immune/gastrointestinal/genitourinary impairment categories (C), and at least one symptom from energy metabolism/transport impairments (D)

A post-exertional neuroimmune exhaustion (PENE pen'–e) compulsory

This cardinal feature is a pathologic inability to produce sufficient energy on demand with prominent symptoms primarily in the neuroimmune regions. Characteristics are:

1 Marked, rapid physical and/or cognitive fatigability in response to exertion, which may be minimal such as activities of daily living or simple mental tasks, can be debilitating and cause a relapse

2 Post-exertional symptom exacerbation: *e.g. acute flu-like symptoms, pain and worsening of other symptoms*

3 Post-exertional exhaustion may occur immediately after activity or be delayed by hours or days

4 Recovery period is prolonged, usually taking 24 hours or longer. A relapse can last days, weeks, or longer

5 Low threshold of physical and mental fatigability (lack of stamina) results in a substantial reduction in pre-illness activity level.

Operational notes: For a diagnosis of myalgic encephalomyelitis (ME) symptom severity must result in a significant reduction of a patient's premorbid activity level. Mild (an approximate 50 percent reduction in pre-illness activity level), moderate (mostly housebound), severe (mostly bedridden), or very severe (totally bedridden and need help with basic functions). There may be marked fluctuation of symptom severity and hierarchy from day to day or hour to hour. Consider activity, context and interactive effects. Recovery time: e.g. Regardless of a patient's recovery time from reading for ½ hour, it will take much longer to recover from grocery shopping for ½ hour and even longer if repeated the next day – if able. Those who rest before an activity or have adjusted their activity level to their limited energy may have shorter recovery periods than those who do not pace their activities adequately. Impact: e.g. An outstanding athlete could have a 50 percent reduction in his/her pre-illness activity level and is still more active than a sendentary person

B Neurological impairments

At least one symptom from three of the following four symptom categories:

1 Neurocognitive impairments

a Difficulty processing information: slowed thought, impaired concentration, *e.g. confusion, disorientation, cognitive overload, difficulty with making decisions, slowed speech, acquired or exertional dyslexia*

b Short-term memory loss: *e.g. difficulty remembering what one wanted to say, what one was saying, retrieving words, recalling information, poor working memory*

2 Pain

a Headaches: *e.g. chronic, generalized headaches often involve aching of the eyes, behind the eyes or back of the head that may be associated with cervical muscle tension; migraine; tension headaches*

b Significant pain can be experienced in muscles, muscle–tendon junctions, joints, abdomen or chest. It is non-inflammatory in nature and often migrates, *e.g. generalized hyperalgesia, widespread pain (may meet fibromyalgia criteria), myofascial or a radiating pain*

3 Sleep disturbance

a Disturbed sleep patterns: *e.g. insomnia, prolonged sleep including naps, sleeping most of the day and being awake most of the night, frequent awakenings, awaking much earlier than before illness onset, vivid dreams/nightmares*

b Unrefreshed sleep: *e.g. awaken feeling exhausted regardless of duration of sleep, day-time sleepiness*

4 Neurosensory, perceptual, and motor disturbances

a Neurosensory and perceptual: *e.g. inability to focus vision, sensitivity to light, noise, vibration, odor, taste and touch; impaired depth perception*

b Motor: *e.g. muscle weakness, twitching, poor coordination, feeling unsteady on feet, ataxia*

Notes: Neurocognitive impairments, reported or observed, become more pronounced with fatigue. Overload phenomena may be evident when two tasks are performed simultaneously. Abnormal reaction to light – fluctuation or reduced accommodation responses of the pupils with retention of reaction. Sleep disturbances are typically expressed by prolonged sleep, sometimes extreme, in the acute phase and often evolve into marked sleep reversal in the chronic stage. Motor disturbances may not be evident in mild or moderate cases but abnormal tandem gait and positive Romberg test may be observed in severe cases

C Immune, gastrointestinal, and genitourinary impairments

At least one symptom from three of the following five symptom categories:

1 Flu-like symptom may be recurrent or chronic and typically activate or worsen with exertion, *e.g. sore throat, sinusitis, cervical and/or axilllary lymph nodes may enlarge or be tender on palpitation*

2 Susceptibility to viral infections with prolonged recovery periods

3 Gastrointestinal tract: *e.g. nausea, abdominal pain bloating, irritable bowel syndrome*

4 Genitournary: *e.g. urinary urgency or frequency, nocturia*

5 Sensitivities *to food, medications, odors, or chemicals*

Notes: Sore throat, tender lymph nodes, and flu-like symptoms obviously are not specific to ME but their activation in reaction to exertion is abnormal. The throat may feel sore, dry, and scratchy. Faucial injection and crimson crescents may be seen in the tonsillar fossae, which are an indication of immune activation

D Energy production/transportation impairments: at least one symptom

1 Cardiovascular: *e.g. inability to tolerate an upright position – orthostatic intolerance, neurally mediated hypotension, postural orthostatic tachycardia syndrome, palpitations with or without cardiac arrhythmias, light-headedness/dizziness*

2 Respiratory: *e.g. air hunger, labored breathing, fatigue of chest wall muscles*

3 Loss of thermostatic stability: *e.g. subnormal body temperature, marked diurnal fluctuations; sweating episodes, recurrent feelings of feverishness with or without low-grade fever, cold extremities*

4 Intolerance of extremes of temperature

Notes: Orthostatic intolerance may be delayed by several minutes. Patients who have orthostatic intolerance may exhibit mottling of extremities, extreme pallor or Raynaud's phenomenon. In the chronic phase, moons of finger nails may recede

Pediatric considerations

Symptoms may progress more slowly in children than in teenagers or adults. In addition to post-exertional neuroimmune exhaustion, the most prominent symptoms tend to be neurological: headaches, cognitive impairments, and sleep disturbances

1 Headaches: severe or chronic headaches are often debilitating. Migraine may be accompanied by a rapid drop in temperature, shaking, vomiting, diarrhoea, and severe weakness

2 Neurocognitive impairments: difficulty focusing eyes and reading are common. Children may become dyslexic, which may only be evident when fatigued. Slow processing of information makes it difficult to follow auditory instructions or take notes. All cognitive impairments worsen with physical or mental exertion. Young people will not be able to maintain a full school program

3 Pain may seem erratic and migrate quickly. Joint hypermobility is common

Notes: Fluctuation and severity hierarchy of numerous prominent symptoms tend to vary more rapidly and dramatically than in adults

Classification

____ Myalgic encephalomyelitis

____ Atypical myalgic encephalomyelitis: meets criteria for post-exertional neuroimmune exhaustion but has two or less than required of the remaining critical symptoms. Pain or sleep disturbance may be absent in rare cases

Exclusions: *As in all diagnoses, exclusion of alternate explanatory diagnoses is achieved by the patient's history, physical examination, and laboratory/biomarker testing as indicated. It is possible to have more than one disease but it is important that each one is identified and treated.* Primary psychiatric disorders, somatoform disorder and substance abuse are excluded. *Pediatric: 'primary' school phobia*

Comorbid entities: Fibromyalgia, myofascial pain syndrome, temporomandibular joint syndrome, irritable bowel syndrome, interstitial cystitis, Rynaud's phenomenon, prolapsed mitral valve, migraines, allergies, multiple chemical sensitivities, Hashimoto's thyroiditis, Sjögren's syndrome, reactive depression. *Migraine and irritable bowel syndrome may precede ME but then become associated with it. Fibromyalgia overlaps*

Reproduced with kind permission from Carruthers BM, van de Sande MI, De Meirleir KL *et al.* Myalgic encephalomyelitis: International Consensus Criteria. *J Intern Med* 2011.

A range of biologic changes have been reported in myalgic encephalomyelitis sufferers, such as changes in autonomic nervous system functioning and neuroimaging findings.

Differential diagnosis

Hyde (2010) has reported that, of 53 consecutive patients referred to his clinic with a diagnosis of myalgic encephalomyelitis or chronic fatigue syndrome, 100 percent had missed (i.e. undiagnosed by primary or consulting doctors) measurable brain dysfunction, 98 percent had measurable significant sleep dysfunctions that included (i) lack of type 3 and 4 sleep, (ii) abnormal, absent, or significantly delayed rapid eye movement, (iii) central and peripheral apnea, (iv) restless legs syndrome, and (v) oxygen saturation that fell below 88 percent, 74 percent had measurable thyroid dysfunction, 47 percent had measurable significant arthritic, rheumatoid changes, or other indicators that were previously diagnosed as fibromyalgia; 47 percent had other missed major disease, including cardiac disease, malignant disease, vascular injuries, and autonomic nervous system dysfunction, and at least 16 percent had typical psychiatric illness but, in addition, missed physical disease.

Treatment

A full history, thorough physical examination, and full and appropriate investigations are required. In terms of treatments specifically for myalgic encephalomyelitis, these remain controversial at the time of writing. Some suggest psychologically based therapies, such as pacing, graded exercise therapy, and cognitive-behavior therapy, while those with a more organic view of myalgic encephalomyelitis tend to advocate more biologically based treatments, including nutritional supplementation (e.g. with coenzyme Q10 to help mitochondrial functioning); some of those who consider there to be good evidence of a viral etiology have advocated setting up trials to test the efficacy of antiretroviral treatments.

18.2 FIBROMYALGIA

Clinical features

Fibromyalgia is a debilitating syndrome characterized by the constant or near-constant presence of widespread pain, which is often incapacitating. It appears particularly to affect women between the ages of 25 and 50 years, with the female to male ratio overall being around 9:1. The American College of Rheumatology 1990 Criteria for the Classification of Fibromyalgia are as follows: widespread pain (for at least 3 months, on both sides of the body, both above and below the waist, and including axial skeletal pain, i.e. cervical spine or anterior chest or thoracic spine or low back pain) in combination with tenderness on digital palpation in at least 11 out of the 18 specific tender point sites shown in Figure 18.1 (Wolfe *et al.* 1990).

Muscle pain, disturbed sleep, and worsening of symptoms during emotional stress or times of temperature change are commonly occurring additional symptoms. Other clinical features and comorbid conditions that may occur include subjective swelling and numbness, chronic headaches, irritable bowel syndrome, Raynaud's phenomenon, dysmenorrhea, exercise intolerance, and weakness.

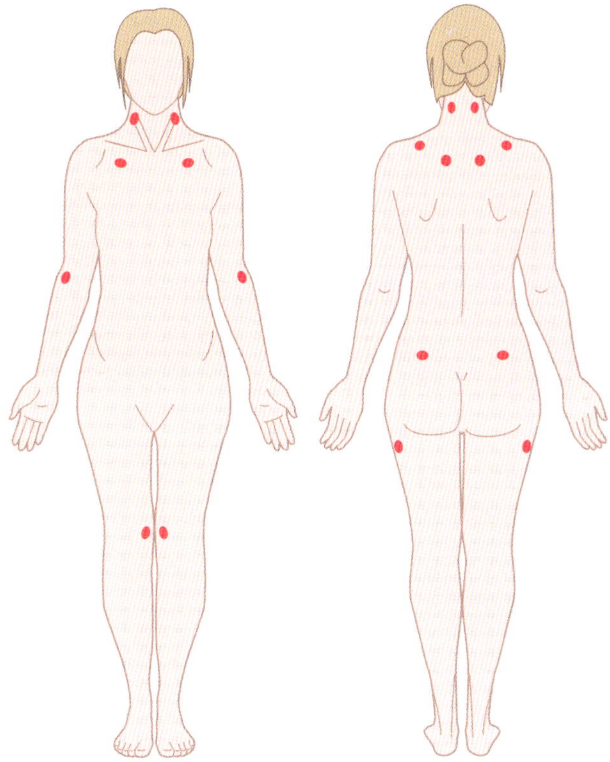

Figure 18.1 The tender points of the American College of Rheumatology 1990 Criteria for the Classification of Fibromyalgia.

Differential diagnosis

At the time of writing, there are no diagnostic tests of fibromyalgia. It may occur in patients suffering from connective tissue diseases such as rheumatoid arthritis and systemic lupus erythematosus. There appears to be an approximately 70 percent overlap of symptomatology with myalgic encephalomyelitis.

Treatment

Sleep impairment might respond to pharmacotherapy with small doses of tricyclic antidepressants (well below the psychiatric doses). Some patients respond to one or more of the following: relaxation therapy, biofeedback, injection of tender areas with steroids or local anesthetics, local heat therapy, treatment with gabapentin, and massage therapy. Several trials have suggested that sodium oxybate may aid pain, fatigue, and sleep dysfunction in fibromyalgia (Moldofsky *et al.* 2010; Russell *et al.* 2009, 2011; Scharf *et al.* 2003); however, at the time of

Table 18.2 Major classes of chemicals associated with multiple chemical sensitivity

Chemical class	Known biologic activities	Common sources/uses
Highly substituted, poly-or per-halogenated organic compounds with chlorine, bromine or fluorine atoms, e.g. DDT, DDE, lindane, hexachlorobenzene, hexachlorocyclohexanes, PCBs, aldrin, dieldrin, PBDEs, perfluorooctanoic acid polymers, and derivatives	Carcinogenic, mutagenic, kidney and liver damage, endocrine disruption	Household and agricultural pesticides as sprays and dusts, electrical insulation, flame-retardants, non-stick kitchen utensils, stain-resistant fabrics
Organophosphates, nerve agents	Nerve toxins, immune dysregulation, inhibition of key enzymes	Various pesticides in agriculture, fisheries, herbicides, engine oils
Phthalates, nonylphenol, bisphenol A and B	Endocrine disruption	Polymers, plasticizers, toys, babies' pacifiers, dialysis tubing
VOCs, aliphatic and aromatic compounds, *formaldehyde, aldehydes, esters, ketones, acids, alcohols, toluene	Disruption of brain function, nerve damage, carcinogenic	Ubiquitous in fragrances, perfumes, household goods, solvents, fuels, paints, polymers
PAHs	Carcinogenic, mutagenic	Burning fuels, exhaust fumes, power stations
Heavy metals: mercury, *lead, cadmium, arsenic, organometallics, tributyltin*	Neurotoxicity, tissue damage, endocrine disruption	Anti-fouling paints, fuels, preservatives, pesticides, electrical goods, crematoria

*Bioaccumlative and biomagnified through the food chain.
DDE, dichlorodichlorophenylethene (major metabolite of DDT); DDT, dichlorodiphenyltrichloroethane; PAHs, polyaromatic hydrocarbons; PBDEs, polybromodiphenylethers; PCBs, polychlorobiphenyls; VOCs, volatile organic compounds.
Reproduced from Puri BK and Treasaden IH (2010) *Psychiatry: An Evidence-based Text*. London: Hodder Arnold, 2010 with permission.

Table 18.3 Objective clinical signs reported in multiple chemical sensitivity

Aspects of environmental sensitivities	Commonly overlapping conditions
State of heightened reactivity to the environment	Fibromyalgia
Total allergy syndrome	Myalgic encephalomyelitis (ME), chronic fatigue syndrome
Universal allergy	Postviral fatigue syndrome
Toxicant-induced loss of tolerance (TILT)	Post-infections neuromyasthenia
Multiple chemical sensitivity(ies) (MCS)	Yuppie flu
Multiple chemical hypersensitivity (ies)	Chronic pain
Chemical intolerance(s)	Migraine
Chemical acquired immunodeficiency syndrome (AIDS)	Arthritis
Gulf War illness/syndrome (GWI/S)	Allergies
Idiopathic environmental intolerance (IEI)	Rhinitis
Environmental illness (EI)	Asthma
Environmental irritant syndrome	Food intolerance syndrome
Chemical injury/allergy	Celiac disease
Toxic injury	Irritable bowel syndrome
Tight building syndrome	Major depression
Sick building syndrome	Anxiety or panic disorder
Twentieth century disease	Hypothyrodisim
Chemical induced illness	Aerotoxic syndrome
Chemical hypersensitivity syndrome	Organophosphate poisoning
Chemophobia	Sheep dipper's flu
Electromagnetic (hyper)sensitivities/intolerance	Disorders of porphyrin metabolism
Toxic carpet syndrome	Carbon monoxide poisoning
Systemic candidiasis	
Multiple sensory sensitivity syndrome (MUSES)	

Reproduced from Puri BK and Treasaden IH (eds) *Psychiatry: An Evidence-based Text*. London: Hodder Arnold, 2010 with permission.

Table 18.4 Top 12 treatments tried in multiple chemical sensitivity

Treatment tried	n	Harmed (%)	No effect (%)	Helped (%)	Helped/harm ratio
Chemical-free living space	820	0.6	4.5	94.8	155.2
Chemical avoidance	875	0.8	4.7	94.5	118.6
Prayer	609	1.4	34.4	64.2	48.3
Meditation	423	2.8	43.3	53.8	19.2
Acupressure	308	4.5	28.3	67.2	14.9
Touch for health	75	3.8	41.8	54.4	14.3
Air filter (to prevent exposure)	786	6.0	11.8	82.1	13.7
Rotation diet	560	5.7	22.1	72.2	12.7
Acidophilus	661	4.1	44.0	52.0	12.7
Relocation	513	7.4	6.0	86.6	11.7
Reflexology	204	4.8	38.5	56.6	11.6
Personal oxygen to cope with exposures	326	7.3	14.2	78.4	10.6

Reproduced from Puri BK and Treasaden IH (eds) *Psychiatry: An Evidence-based Text*. London: Hodder Arnold, 2010 with permission.

writing, sodium oxybate has not been licensed for use in this disorder by the Food and Drug Administration.

18.3 MULTIPLE CHEMICAL SENSITIVITY

Hooper (2010) has listed the major classes of environmental chemicals which may be associated with multiple chemical sensitivity; these are listed in Table 18.2. Potential symptoms are numerous and may involve almost any body system. Objective clinical signs that have been reported in multiple chemical sensitivity are listed in Table 18.3 (Hooper 2010). The top 12 treatments tried in multiple chemical sensitivity are shown in Table 18.4; of these, the most highly rated were a chemical-free living space and avoidance of chemicals (Hooper 2010).

REFERENCES

Ali MA, Dale JK, Kozak CA, *et al.* Cohen JI. Xenotropic murine leukemia virus-related virus is not associated with chronic fatigue syndrome in patients from different areas of the US in the 1990s. *Virol J* 2011; **8**:450.

Carruthers BM, Jain AK, De Meirleir KL *et al.* Myalgic encephalomyelitis/ chronic fatigue syndrome: Clinical working case definition, diagnostic and treatment protocols. A consensus document. *J Chronic Fatigue Syndrome* 2003; **11**:7–115.

Carruthers BM, van de Sande MI, De Meirleir KL *et al.* Myalgic encephalomyelitis: International Consensus Criteria. *J Intern Med* 2011; **270**:327–38.

Erlwein O, Kaye S, McClure MO *et al.* Failure to detect the novel retrovirus XMRV in chronic fatigue syndrome. *PLoS One* 2010; **5**:e8519.

Fukuda K, Straus SE, Hickie I *et al.* The chronic fatigue syndrome: a comprehensive approach to its definition and study. International Chronic Fatigue Syndrome Study Group. *Ann Intern Med* 1994; **121**:953–9.

Groom HC, Boucherit VC, Makinson K *et al.* Absence of xenotropic murine leukaemia virus-related virus in UK patients with chronic fatigue syndrome. *Retrovirology* 2010; **7**:10.

Hohn O, Bannert N. Origin of XMRV and its demise as a human pathogen associated with chronic fatigue syndrome. *Viruses* 2011; **3**:1312–19.

Hohn O, Strohschein K, Brandt AU *et al.* No evidence for XMRV in German CFS and MS patients with fatigue despite the ability of the virus to infect human blood cells *in vitro*. *PLoS One* 2010; **5**:e15632.

Hooper M. Multiple chemical sensitivity. In: Puri BK, Treasaden IH eds. *Psychiatry: An Evidence-based Text.* London: Hodder Arnold, 2010: 793–820.

Hyde B. Mental health problems in patients with myalgic encephalomyelitis and fibromyalgia syndrome. In: Puri BK, Treasaden IH eds. *Psychiatry: An Evidence-based Text.* London: Hodder Arnold, 2010: 821–34.

Jason LA, Evans M, Porter N *et al.* The development of a revised Canadian myalgic encephalomyelitis chronic fatigue syndrome case definition. *Am J Biochem Biotechnol* 2010; **6**:120–35.

Jerome KR, Diem K, Huang ML *et al.* Xenotropic murine leukemia virus-related virus in monozygotic twins discordant for chronic fatigue syndrome. *Diagn Microbiol Infect Dis* 2011; **71**:66–71.

Knox K, Carrigan D, Simmons G *et al.* No evidence of murine-like gammaretroviruses in CFS patients previously identified as XMRV-infected. *Science* 2011; **333**:94–7.

Lombardi VC, Ruscetti FW, Das Gupta J *et al.* Detection of an infectious retrovirus, XMRV, in blood cells of patients with chronic fatigue syndrome. *Science* 2009; **326**:585–9. Withdrawn 2011.

Lombardi VC, Hagen KS, Hunter KW *et al.* Xenotropic murine leukemia virus-related virus-associated chronic fatigue syndrome reveals a distinct inflammatory signature. *In Vivo* 2011; **25**: 307–14.

Menendez-Arias L. Evidence and controversies on the role of XMRV in prostate cancer and chronic fatigue syndrome. *Rev Med Virol* 2011; **21**:3–17.

Mikovits JA, Huang Y, Pfost MA *et al.* Distribution of xenotropic murine leukemia virus-related virus (XMRV) infection in chronic fatigue syndrome and prostate cancer. *AIDS Rev* 2010a; **12**: 149–52.

Mikovits JA, Lombardi VC, Pfost MA *et al.* Detection of an infectious retrovirus, XMRV, in blood cells of patients with chronic fatigue syndrome. *Virulence* 2010b; **1**:386–90.

Moldofsky H, Inhaber NH, Guinta DR, Alvarez-Horine SB. Effects of sodium oxybate on sleep physiology and sleep/wake-related symptoms in patients with fibromyalgia syndrome: a double-blind, randomized, placebo-controlled study. *J Rheumatol* 2010; **37**:2156–66.

Paprotka T, Delviks-Frankenberry KA, Cingoz O *et al.* Recombinant origin of the retrovirus XMRV. *Science* 2011; **333**: 97–101.

Robinson MJ, Erlwein O, McClure MO. Xenotropic murine leukaemia virus-related virus (XMRV) does not cause chronic fatigue. *Trends Microbiol* 2011; **19**:525–9.

Russell IJ, Perkins AT, Michalek JE. Sodium oxybate relieves pain and improves function in fibromyalgia syndrome: a randomized, double-blind, placebo-controlled, multicenter clinical trial. *Arthritis Rheum* 2009; **60**:299–309.

Russell IJ, Holman AJ, Swick TJ *et al.* Sodium oxybate reduces pain, fatigue, and sleep disturbance and improves functionality in fibromyalgia: results from a 14-week, randomized, double-blind, placebo-controlled study. *Pain* 2011; **152**:1007–17.

Scharf MB, Baumann M, Berkowitz DV. The effects of sodium oxybate on clinical symptoms and sleep patterns in patients with fibromyalgia. *J Rheumatol* 2003; **30**:1070–4.

Shin CH, Bateman L, Schlaberg R *et al.* Absence of XMRV retrovirus and other murine leukemia virus-related viruses in patients with chronic fatigue syndrome. *J Virol* 2011; **85**:7195–202.

Silverman RH, Das Gupta J, Lombardi VC *et al.* Partial retraction. Detection of an infectious retrovirus, XMRV, in blood cells of patients with chronic fatigue syndrome. *Science* 2011; **334**: 176.

White PD, Miller A, Crawley E. Will adopting the Canadian criteria improve diagnosis of CFS? *BMJ* 2011; **343**:d4589.

Wolfe F, Smythe HA, Yunus MB *et al.* The American College of Rheumatology 1990 Criteria for the Classification of Fibromyalgia. Report of the Multicenter Criteria Committee. *Arthritis Rheum* 1990; **33**:160–72.

19

Immune-related disorders

19.1 MULTIPLE SCLEROSIS

Multiple sclerosis (MS), in 85–90 percent of cases, is characterized initially by episodes of illness and a relapsing and remitting course, with each episode occurring secondary to active plaque formation, either in the white matter or the gray matter. In the majority of cases, this initial relapsing and remitting course eventually undergoes a transformation into what is known as a secondary progressive course: here, in addition to recurring episodes, there is also a gradually progressive worsening of symptoms.

Multiple sclerosis is a fairly common disorder, occurring in about 0.1 percent of the general population, and is roughly two to three times more common in females than in males.

Clinical features

Although the range of age of onset is wide, from childhood to the seventh decade, the vast majority of patients first fall ill in their twenties or thirties. As noted (and as discussed further below, under 'Course'), in most cases, at least initially, MS pursues an episodic, relapsing, and remitting course, with each episode typically having a subacute onset, over days or a week or so. Exceptions to this rule, however, do occur, with some onsets spanning less than a day and others being quite leisurely, occurring over weeks or months. The duration of individual episodes also varies widely, from weeks to months, after which there is a gradual defervescence of symptomatology of variable degree. The severity of symptoms varies widely, and in some cases may be so mild that patients fail to recognize them as such.

The most common symptoms seen during an episode are motor or sensory in type. Spastic weakness may occur in one limb, or there may be a paraparesis, or, less commonly, a hemiparesis. Sensory symptoms, similar in distribution, may include numbness and tingling, and, in a minority, dysesthesiae or pain. Visual symptomatology is common, and may include optic neuritis with unilateral blurring of vision or blindness. Cerebellar and brainstem involvement is also common, and may produce ataxia, intention tremor, dysarthria, scanning speech, nystagmus, diplopia, and vertigo. Internuclear ophthalmoplegia may also occur, and this should be carefully sought for as it is of great diagnostic significance: although internuclear ophthalmoplegia may occur with other mesencephalic lesions, such as tumors or infarctions, these are quite rare in young adults and, hence, this finding in a young person is almost specific for MS (Keane 2005). Bladder dysfunction is quite common and has various symptoms including urgency and frequency, incontinence, or urinary retention. Bowel dysfunction is also common and most frequently presents with constipation. Sexual dysfunction is very common, with decreased libido, erectile dysfunction, or decreased vaginal lubrication. Fatigue is likewise common, and may be quite severe.

Uncommon signs and symptoms include various paroxysmal phenomena, such as hemifacial spasm, dystonic spasm of an extremity, Lhermitte's sign, trigeminal neuralgia, and lancinating pains in the extremities. Rarely one may see aphasia (Achiron *et al.* 1992; Devere *et al.* 2000; Lacour *et al.* 2004; Olmos-Lau *et al.* 1977) or seizures, which may be either partial or grand mal in type (Nicoletti *et al.* 2003).

Dementia of variable severity is eventually seen in the majority of patients (Franklin *et al.* 1989; Surridge 1969). In rare cases, dementia may constitute the sole or predominant presenting feature of MS (Bergin 1957; Hotopf *et al.* 1994; Koenig 1968; Young *et al.* 1976). In one case, for example, the only symptom in addition to the dementia was optic neuritis (Jennekens-Schinkel and Sanders 1986), and in another it was unsteady gait (Mendez and Frey 1992), and in one, very rare, case, a gradually progressive dementia constituted the *only* clinical evidence of MS (Fontaine *et al.* 1994). The anatomic basis for the dementia of MS may vary according to whether the course is still relapsing and remitting or has undergone transformation to a secondary progressive course. In cases of dementia occurring in patients with a relapsing and remitting course, there appears to be a correlation between cognitive decline and the total plaque burden, either within the white matter (Comi *et al.* 1999; Franklin *et al.* 1988; Ron *et al.* 1991) or within the gray matter (Calabrese *et al.* 2009). Conversely, dementia occurring in the context of a secondary progressive course correlates better with cortical atrophy (Staff *et al.* 2009), which, as discussed further below under 'Etiology', may, rather than reflecting plaque formation, result from a widespread inflammatory or neurodegenerative process.

Delirium has also been reported in MS (Ginsberg and Compston 1994), but it is rare.

Depression is eventually seen in perhaps one-quarter of patients (Surridge 1969). Depression may occur on a reactive basis in any debilitating disease, and MS is no exception. Certain facts, however, suggest strongly that depression in MS may also be a direct result of plaque formation. Patients with MS are more likely to experience depression than are normal control subjects (Fassbender *et al.* 1998) or those with comparably disabling neurologic diseases that generally spare the cerebrum, such as amyotrophic lateral sclerosis (Schiffer and Babigian 1984; Whitlock and Siskind 1980). Furthermore, in contrast to what one might expect if the depression were reactive, there is little or no correlation between the occurrence of depression and the extent of the patient's disability (Moller *et al.* 1994), and, regardless of the degree of overall disability, patients are more likely to experience depression when the plaques are in the cerebrum than when they are in the cerebellum or spinal cord (Rabins *et al.* 1986; Schiffer *et al.* 1983). Furthermore, a correlation has been noted between depression and the presence of plaques in right temporal (Berg *et al.* 2000; Zorzon *et al.* 2001), right frontal (Zorzon *et al.* 2001), and left frontal (Feinstein *et al.* 2004) lobes, and in the left arcuate fasciculus (Pujol *et al.* 1997).

Euphoria may also occur in MS, and this is typically of the 'bland' or non-infectious type; it has been noted in anywhere from one-quarter (Surridge 1969) to a majority of patients (Cottrell and Wilson 1926), and is correlated with cerebral rather than spinal plaque formation (Rabins *et al.* 1986). The contrast between this euphoria and the patient's actual condition can be quite dramatic: one of Kinnier Wilson's (Wilson 1955) patients, 'bedridden and unable to stand, remarked, "you will not believe me when I say I feel thundering well."' This bland euphoria is typically not accompanied by hyperactivity or pressure of speech and is often seen in concert with some degree of intellectual impairment (Surridge 1969). Although they are unusual, definite manic episodes may also occur in addition to this bland euphoria (Joffe *et al.* 1987; Schiffer *et al.* 1986); indeed, out of all of the reasons for admission to psychiatric hospital for patients with MS, mania is the most common (Pine *et al.* 1995).

Emotional incontinence, with uncontrollable laughter or crying in the absence of a corresponding affect, is seen in about one-tenth of patients (Feinstein *et al.* 1997; Surridge 1969), generally only in far-advanced cases.

Psychosis is relatively rare, and is associated with periventricular plaques within the temporal lobe (Feinstein *et al.* 1992). Even more rarely, psychosis may dominate the clinical picture (Mathews 1979; Parker 1956), and in one, exceedingly rare, case, MS *presented* with a psychosis, which was characterized by social withdrawal, 'mystic' visual hallucinations, and various delusions (Fontaine *et al.* 1994).

Magnetic resonance (MR) scanning has assumed a dominant role in the evaluation of patients with suspected MS. On axial T2, or, more especially, fluid-attenuated inversion recovery (FLAIR) imaging, most plaques appear as fairly well-circumscribed oval or circular areas of increased signal intensity, and active plaques, as demonstrated in Figure 19.1, will display gadolinium enhancement. In some cases, chronic, severe plaques may undergo cystic degeneration, and on T1-weighted images these will appear as 'black holes', with greatly reduced signal intensity. Classically, plaques are found in a periventricular distribution around the lateral ventricles: on sagittal images these periventricular plaques may appear linear, as they radiate out away from the surface of the ventricle, the so-called 'Dawson's fingers'. Plaques may also be found in the corpus callosum, the centrum semiovale, and also juxtacortically. Of interest, as noted below (under 'Etiology'), plaques are found not only in the white matter, but also in the gray matter: such plaques, however, are generally not visible on routine MR imaging (MRI) (Geurts *et al.* 2005). Plaques are also routinely found in the cerebellar white matter, cerebellar peduncles, the pons, and the cord. Most plaques measure from 3 to 15 mm in diameter; rarely one may find very large plaques, two or more centimeters in diameter. Such large plaques, when active, are often referred to as 'tumefactive' as they may resemble tumors in that they display both ring enhancement and considerable perilesional vasogenic edema (Lucchinetti *et al.* 2008).

The cerebrospinal fluid (CSF) is abnormal in almost all patients during active disease. A mild lymphocytic pleocytosis, in the range of 6–30 cells/mm^3, is seen in about one-third of cases, and the total protein is mildly elevated (rarely over 100 mg/dL) in about one-half of cases. The immuno-

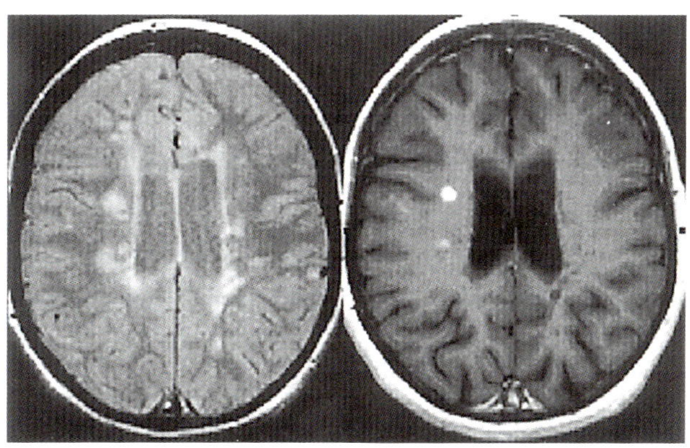

Figure 19.1 In the T2-weighted magnetic resonance imaging scan on the left, there are multiple areas of increased signal intensity corresponding to both chronic and active plaques; in the T1-weighted scan with gadolinium enhancement on the right, however, only the active lesions enhance. Reproduced with kind permission from Graham DI, Lantos PL. (1996), with permission.

globulin G (IgG) index is elevated in over two-thirds of cases, and oligoclonal bands are present in over 90 percent. The myelin basic protein is elevated in over three-quarters of cases. Of note, the 14-3-3 protein may be found in a little over one-tenth of all cases (Martinez-Yelamos *et al.* 2001).

Evoked potentials, including somatosensory, brainstem, and visual-evoked potentials, were once widely used to demonstrate lesions within the respective white matter pathways; however, with the advent of MRI, these are generally no longer required.

Course

As noted earlier, most cases of MS are characterized initially by episodes of illness, and this course, found in 85–90 percent of cases, is referred to as relapsing and remitting. After the initial episode, subsequent attacks may consist of exacerbations of symptoms seen in previous episodes, or completely new symptoms, depending on whether old plaques become reactivated or new plaques form. With the resolution of any episode, remission of symptoms is rarely perfect and most patients are left with residuals. Thus, the relapsing and remitting course demonstrates a 'step-wise' pattern. With each fresh attack, a new 'step' occurs; between successive steps, the 'baseline' of disability, reflecting the residual, although worse than the preceding baseline, remains 'flat' until the next step appears. Predicting when the next 'step' will occur is very difficult: although, on average, most patients will have an attack every 2 years or so, the range is sufficiently wide, from months up to two decades, that making accurate predictions in individual cases may not be possible until many, many years have passed and a pattern emerges. As noted earlier, the severity of symptoms seen in any particular attack varies quite widely. On the one hand, symptoms may be so mild, and residuals so minimal, that patients, after 10 years, may have little or no disability. Such cases of 'benign' MS, however, may not remain so because with a follow-up of 20 years, close to one-half will be found to have developed significant disability (Sayao *et al.* 2007). On the other hand, some cases, known as 'Marburg' variants, will be characterized by severe episodes, with multiple large plaques, and symptoms so severe as to cause, in some cases, death within weeks or months.

The course seen in relapsing and remitting cases is also influenced by pregnancy. Compared with the prepregnancy frequency of attacks, the risk of an attack falls, especially during the third trimester, then rises somewhat during the first 3 months postpartum, after which it becomes equivalent to that seen before the pregnancy (Confavreux *et al.* 1998). Furthermore, it appears that, over the long haul, women who have children after the onset of the illness tend to have less disability than those who never did (D'hooghe *et al.* 2010).

In addition to this relapsing and remitting pattern, the overall course of MS may also be characterized by a chronic, more or less steady, progression of symptoms, and this 'progressive' course may be further subdivided into a 'secondary' or 'primary' type.

Secondary progressive MS is said to occur when, in cases initially characterized by a relapsing and remitting course, the 'baseline' between successive 'steps', rather than being 'flat', displays a steady worsening. Such a secondary transformation eventually occurs in the majority of cases, typically after two decades of illness.

Primary progressive MS (Thompson *et al.* 1997) displays a fundamentally different course from relapsing and remitting MS in that in primary progressive cases there is, from the very start of the illness, a more or less steady progression of symptoms, with a gradual and relentless worsening of the baseline: in most cases this involves a progressive myelopathy. Such a course is decidedly uncommon, occurring only in roughly 10 percent of cases. Even more rarely, one may find a variant of this, known as 'progressive relapsing MS'. Here, after the relentlessly progressive course has been well established, one finds clear attacks occurring, with these 'steps' riding, as it were, on top of the ever-worsening baseline.

Symptoms of MS may undergo transient exacerbation with stress, infection, or temperature elevation (as may occur with a hot bath, fever, or exercise). These exacerbations do not reflect inflammatory plaque activity but rather impaired conduction through partially damaged axons, and thus they should not be interpreted as relapses. Interestingly, in the past this kind of transient exacerbation was used as a diagnostic test for MS: in such a 'hot bath' test, the patient was immersed in hot water and observed for the emergence of signs or symptoms.

Etiology

The underlying neuropathology of MS appears to differ relative to whether the course is relapsing and remitting or progressive.

Relapsing and remitting MS is characterized by focal areas of inflammation. In the white matter, these appear as classic plaques, which, when active, are typically characterized by an area of perivenular mononuclear inflammation with loss of oligodendrocytes and demyelination, with relative axonal sparing (Greenfield and King 1936). In a minority of cases, however, one may find primarily apoptosis of oligodendrocytes, with relatively little inflammation (Barnett and Prineas 2004). In time, these active white matter plaques resolve into chronic ones, which are composed of relatively well-circumscribed areas of demyelination and gliosis, with, in severe cases, a degree of cavitation. Focal areas of demyelination, in addition to being found in the white matter, are also found in the gray matter, including the cerebral cortex (Mainero *et al.* 2009; Peterson *et al.* 2001; Vercellino *et al.* 2005), thalamus (Vercellino *et al.* 2005), and basal ganglia (Vercellino *et al.* 2005), and indeed such areas of demyelination are found in the cortex in the vast majority of cases (Calabrese *et al.* 2010).

Secondary progressive cases of MS are characterized by diffuse inflammation which affects not only the cortex at large but also the white matter (Kutzelnigg *et al.* 2005).

Thus, although it appears that inflammation underlies the neuropathology of both relapsing and remitting cases, on the one hand, and progressive cases, on the other, it is not at present clear whether the inflammation seen in progressive cases results from the same mechanism that is active in relapsing or remitting cases, or rather represents a secondary immune response or even an inflammatory response to a supervening neurodegenerative process.

Although the cause of the inflammation seen in MS remains unknown, several lines of evidence suggest that it may represent an autoimmune disorder in genetically susceptible individuals which is triggered by a relatively common childhood viral infection.

The incidence of MS rises from that seen in the general population to that seen in first-degree relatives and finally to monozygotic twins. As noted earlier, the incidence of MS in the general population is about 0.1 percent; this rises to about 2–3 percent in first-degree relatives and up to 25–40 percent in monozygotic twins (Ebers *et al.* 1986; Mumford *et al.* 1994). Although this could conceivably be explained on the basis of a shared environment, adoption studies support a genetic cause (Ebers *et al.* 1995). Furthermore, there is significant linkage between MS and certain human leukocyte antigen (HLA) haplotypes (e.g. HLA-DRB1*1501).

The evidence for an infectious agent acting on this fertile genetic background rests on several facts. First, although found at all latitudes, MS is more common in temperate zones. Furthermore, if individuals remain in a high-risk temperate zone past the age of 15 years and then migrate to a tropical area, they 'carry' their increased risk with them (Dean 1967), indicating that the increased risk relates to some environmental exposure that is more common in temperate zones. Second, although not without controversy, it appears that there have been 'epidemics' of MS in the Faroe Islands and in Iceland, and that these epidemics may have been related to the presence of British troops stationed there during World War II. Although the nature of the presumed infection is not clear, certain evidence points toward such common viral infections as mononucleosis (DeLorenze *et al.* 2006). It must be clearly noted, however, that the deleterious effect of any such infection is not direct: there is no evidence for an actual viral infection in MS plaques. Rather, it is suspected that a childhood infection sensitizes the immune system in genetically susceptible individuals and that some event in adult life, perhaps a reactivation of a latent viral infection, triggers off an immune response that incorrectly targets oligodendrocytes and myelin.

Differential diagnosis

The differential considerations vary according to whether the illness of the patient in question is relapsing and remitting or progressive.

In relapsing and remitting cases, consideration should be given to acute disseminated encephalomyelitis, neuromyelitis optica, Behçet's syndrome, systemic lupus erythematosus, and polyarteritis nodosa.

Acute disseminated encephalomyelitis (ADEM), as discussed in Chapter 15.11, is typically seen in children or adolescents, and differs from MS both clinically and radiologically (Ketelslegers *et al.* 2010). Clinically, as ADEM is almost always a monophasic illness, history will generally fail to reveal prior episodes, whereas, by contrast, in MS one may find evidence of prior attacks. Furthermore, in ADEM, one often finds a delirium, a clinical feature which is decidedly uncommon in MS. MRI findings are also very helpful. In both ADEM and MS, one may find multiple areas of inflammation within the white matter. In ADEM, however, these lesions tend to be large and poorly demarcated, whereas in MS they tend to be smaller and well demarcated. Furthermore, in MS, one may see 'Dawson's fingers', a finding not seen in ADEM. In MS, one often is also able to find evidence of old plaques, even in cases where history fails to reveal clinical evidence of a prior attack (Rovira *et al.* 2009), and this is in marked contrast to ADEM, which, as a typically monophasic illness, will not reveal any evidence of prior inflammatory activity: in this regard, finding 'black holes' is especially indicative of MS. When the differential remains in doubt, one may have to settle for long-term follow-up: if a repeat episode occurs, then the diagnosis of ADEM is almost ruled out.

Neuromyelitis optica, as discussed in Chapter 19.2, when fully developed, is characterized by optic neuritis and transverse myelitis. Such a combination may also, of course, be seen in MS; however, in neuromyelitis optica, the transverse myelitis is longitudinally extensive, spanning three or more segments, which is in contrast with MS, where such a longitudinal extent is rarely seen. In doubtful cases, testing for neuromyelitis optica (NMO)-IgG may resolve the issue.

Behçet's syndrome, systemic lupus erythematosus, and polyarteritis nodosa may all be characterized by episodes of neuropsychiatric dysfunction; however, in each of these disorders one also finds systemic signs that are lacking in MS: in Behçet's disease there are oral and genital ulcers; and in lupus and polyarteritis nodosa, in addition to constitutional signs, one finds evidence of other organ involvement, such as the joints, kidneys, or gastrointestinal tract.

Lacunar infarctions may enter the differential, but are generally easy to rule out. First, the age of onset is generally in later years, in contrast with the younger age of onset seen in MS. Second, clinically, strokes occurring secondary to lacunar infarctions tend to present acutely, over minutes or hours, in contrast to the more subacute presentation seen with active plaques. Finally, the location of and the configuration of the lesion may help. Lacunar infarctions tend to occur in the internal capsules, an unusual location for plaques; lacunar infarctions may also occur periventricularly, as may the plaques of MS; however, infarctions rarely present the radial configuration seen in the Dawson's fingers of MS.

In cases characterized by progressive white matter disease from the onset, MS of the primary progressive type must be distinguished from vitamin B12 deficiency and adrenoleukodystrophy, a differential readily accomplished by testing for vitamin B12 and plasma long-chain fatty acid levels.

Treatment

Treatment of MS may be divided into that directed at the *acute episode*, *preventive treatments* designed to forestall future episodes or to limit secondary progression, and *symptomatic treatments*. Each of these aspects of treatment is now considered in turn, followed by some summary recommendations.

ACUTE EPISODES

Acute episodes may be treated with intravenous methylprednisolone, 250 mg every 6 hours for 5 days. Some clinicians will follow this with prednisone, 1 mg/kg/day, for 4 days, with the dose then being gradually tapered over the succeeding 2–3 weeks. Although most clinicians, when utilizing methylprednisolone, prefer the intravenous route, two studies (Barnes *et al.* 1997; Martinelli *et al.* 2009) have demonstrated that oral methylprednisolone is just as effective: this is given in a dose of 48 mg/day for week, then 24 mg/day for 1 week, and finished with an additional week at 12 mg/day. In cases resistant to steroids, consideration may be given to plasma exchange.

Before commencing treatment with steroids, however, it is important to ensure that the illness in question represents the activity of an active plaque rather than a transient exacerbation related to an old plaque precipitated by an infection, such as a urinary tract infection.

Mania may complicate a prolonged course of treatment with steroids, and this may require treatment with antipsychotics or a mood stabilizer such as lithium or valproate, as discussed in Chapter 7.3. In cases in which patients have received prior courses of steroids that were complicated by mania, it may be appropriate to consider prophylactic treatment with one of these mood stabilizers.

PREVENTIVE TREATMENTS

Preventive treatments may be considered in patients with a relapsing and remitting course, and also in those whose course is marked by secondary progression.

In relapsing and remitting MS, preventive treatment may be accomplished with several agents, including beta interferons, glatiramer, and natalizumab.

The beta interferons are immunomodulatory agents, and three are effective: interferon β-1b (Betaseron, 250 μg given subcutaneously every other day) and two preparations of interferon β-1a, one of which is given subcutaneously (Rebif, 22–44 μg subcutaneously three times weekly) and the other which is given intramuscularly (Avonex 30 μg intramuscularly once weekly). Although each of these beta interferons reduces relapse rates by roughly one-third, they may not be precisely equivalent in efficacy. Thus, although Betaseron and Rebif are equivalent (Koch-Henriksen *et al.* 2006), both of these agents appear superior to Avonex (Durelli *et al.* 2002; Panitch *et al.* 2002). Based on this, it would appear that either Betaseron or Rebif would be the logical choice; however, other considerations may give the clinician pause. To begin with, one must take into account both the fact that all three of these agents may induce the formation of neutralizing antibodies after anywhere from 6 to 18 months, which may blunt their response, and the fact that the likelihood of inducing antibodies differs among these agents, occurring in about one-third of patients on Betaseron, one-fifth of patients on Rebif, and less than one-tenth of patients on Avonex. On this count, Avonex appears preferable. One must also take into account patient preference, and on this count, many will also prefer Avonex, given that it need be given only once weekly.

Glatiramer is a mixture of polypeptides that mimics myelin basic protein, thereby presumably blunting the autoimmune assault on myelin, and is available as Copaxone, which is given in a dose of 20 mg subcutaneously once daily. Like the interferons, it reduces relapse rates by about one-third, but unlike the interferons it only rarely produces neutralizing antibodies.

The choice between one of the beta interferons, on the one hand, and glatiramer, on the other, may be based both on considerations of efficacy and patient preference. Both the beta interferons Betseron (O'Connor *et al.* 2009) and Rebif (Mikol *et al.* 2008) have been compared to Copaxone, and both were found equivalent to Copaxone, and thus on this score there appears to be no reason to choose one of these beta interferons over Copaxone. However, as noted earlier, both Betaseron and Rebif are more likely to induce neutralizing antibodies than is Copaxone, and with this in mind Copaxone might be preferable. With regard to patient preference, Copaxone may also be preferable. All of the interferons require periodic blood testing (complete blood counts and liver function tests), whereas Copaxone does not; furthermore, the overall side-effect burden is greater for the beta interferons than it is for Copaxone.

Overall, then, it may be reasonable to begin with Copaxone. This recommendation, however, must be taken as tentative only: some patients are so loathe to take daily or several times weekly injections that the once-weekly schedule of Avonex makes this agent the hands-down favorite for them. It is appropriate, therefore, to have a very detailed discussion with the patient of all the pros and cons of these agents before making a recommendation.

Regardless of which agent is chosen, it is reasonable to give it a good trial. Presuming that side-effects are tolerable, the length of the trial should be long enough to enable one to determine its efficacy. Should the initial agent prove unsatisfactory, then it is reasonable to switch: if a beta interferon had been used, then one should switch to Copaxone, and vice versa.

In cases where patients fail to respond to trials of both a beta interferon and of Copaxone, consideration should be given to natalizumab, marketed as Tysabri, which is given intravenously once monthly. This agent is a humanized monoclonal antibody directed at human α-4 integrin. This molecule exists on the surface of lymphocytes and serves to bind them to vascular cell adhesion molecules (VCAM) on vascular endothelial cells, thus allowing their transport across the vessel wall. When integrin is blocked, this 'trafficking' into the interstitial fluid is inhibited and the inflammatory response is blunted. Natalizumab appears to reduce relapse rates by roughly two-thirds (Polman *et al.* 2006), and may produce neutralizing antibodies in about one-tenth of cases. Despite this dramatic superiority over the beta interferons and glatiramer, enthusiasm for natalizumab has been tempered by the appearance of progressive multifocal leukoencephalopathy in treated patients: although this is a very rare complication, it is potentially fatal.

In cases of relapsing and remitting MS that undergo transformation to secondary progression, consideration should be given to either Betaseron (European Study Group 1998) or Avonex (Cohen *et al.* 2002) as these agents have been shown to help prevent further acute attacks; whether Rebif or Copaxone are also effective in this regard is not known. Unfortunately, however, although Betaseron and Avonex may help prevent acute attacks in this situation, they do not affect the ever-worsening baseline between attacks. Mitoxantrone (Hartung *et al.* 2002) may be effective in this regard; however, the attendant cardiotoxicity and risk of leukemia may outweigh any benefits.

Regarding primary progressive MS, there is as yet no established treatment.

SYMPTOMATIC TREATMENTS

Spasticity has traditionally been treated with baclofen, diazepam, or tizanidine; gabapentin represents a recent addition to this armamentarium (Cutter *et al.* 2000). Painful dysesthesiae, trigeminal neuralgia, or lancinating pains may respond to either carbamazepine or gabapentin. Intention tremor may be reduced by either clonazepam or, in some cases, propranolol. Bladder dysfunction is often a focus of treatment: urinary retention may respond to bethanechol, and spastic bladder with urinary urgency and frequency may be relieved by oxybutynin or tolterodine; in some cases a Foley catheter or more invasive measures may be required. Bowel dysfunction usually consists of constipation, which may be managed with a bowel program. Erectile dysfunction may be managed with a phosphodiesterase inhibitor, such as sildenafil, and decreased vaginal lubrication may be helped by lubricating agents. Seizures may be managed with standard anti-epileptic drugs. Fatigue may respond to amantadine in a dose of 100 mg b.i.d. (Krupp *et al.* 1995; Rosenberg and Appenzeller 1988); recent enthusiasm for modafinil should be tempered by a negative double-blind study (Stankoff *et al.* 2005).

Dementia is treated as discussed in Chapter 5.1. Recent work suggests that donepezil may improve cognitive functioning in MS (Krupp *et al.* 2004).

Depression should be treated with an antidepressant. Remarkably, there has been only one double-blind study in this regard, which found desipramine to be superior to placebo (Schiffer and Wineman 1990); unfortunately, as might have been predicted, this was poorly tolerated. In practice, most patients are given a selective serotonin reuptake inhibitor (SSRI) such as escitalopram.

Euphoria of the 'bland' type rarely requires any treatment. Should mania occur, however, treatment is conducted as described in Chapter 7.3.

As discussed in Chapter 5.7, emotional incontinence in MS has been shown to respond to amitriptyline or to a combination of dextromethorphan and quinidine. Whether it would respond to some of the other agents used in emotional incontinence resulting from other disorders (e.g. citalopram in vascular cases) is not clear.

Psychosis is treated as outlined in Chapter 8.1.

SUMMARY RECOMMENDATIONS

The overall treatment of patients with MS is difficult and at times quite complex, and it is often best carried out by a specialized team. Except in cases of very benign MS, most patients with relapsing and remitting MS will receive preventive treatment; whenever an episode does occur, consideration should be given to prompt treatment with steroids to lessen the chance of any permanent damage and residual symptoms. Given the number of medications involved for symptomatic treatment, the potential for drug–drug interactions and cumulative side-effects is large, and constant vigilance is required.

19.2 NEUROMYELITIS OPTICA

Neuromyelitis optica, also known as Devic's disease, although once considered a subtype of multiple sclerosis, is now known to be a disease in its own right, occurring secondary to an autoimmune attack by an IgG antibody known as NMO-IgG. Classically, it is characterized by the combination of optic neuritis and longitudinally extensive transverse myelitis (LETM), that is to say, a transverse myelitis that extends over three or more vertebral segments.

This is an uncommon disease, which is four or more times more common in females than males.

Clinical features

Although the age of onset ranges from childhood to the eighth decade, in most cases the first episode occurs in the fourth decade.

Although episodes tend to present subacutely, over days, some cases may have an explosive onset. The first episode may be characterized by either a transverse myelitis or an optic neuritis, or, occasionally, both at the same time. When optic neuritis occurs, it may be either unilateral or bilateral, with a greater or lesser degree of blindness. In some cases the optic chiasm may be primarily involved, and here one may find binasal or bitemporal field defects. In cases of transverse myelitis, either a paraplegia or a tetraplegia may occur, depending on the location of the myelitis. After the first episode subsides, recurrent episodes, as discussed below, are the rule, and eventually most patients end up with evidence for both bilateral optic neuritis and transverse myelitis.

In a minority of cases, other symptoms may occur during episodes, as for example when the hypothalamus is involved. Thus, hypersomnolence occurred in one case (Nozaki *et al.* 2009), and in another there was a delirium, with somnolence, deficits in memory and attention, delusions of persecution, and coprophagia (Viegas *et al.* 2009).

On MRI, active lesions display increased signal intensity on FLAIR or T2 imaging, and undergo enhancement with gadolinium. As might be expected, findings are generally confined to the optic nerves and chiasm, the cord, and the hypothalamus; however, a small number of lesions may also be found in the cerebral white matter. Of note, lesions in the cerebral white matter tend to appear fluffy or 'cloud like' (Ito *et al.* 2009), with indistinct borders, in contrast to the lesions of MS, which are generally well demarcated.

The cerebrospinal fluid contains an elevated protein, which may rise to over 100 mg/dL in close to half of all patients. There is a pleocytosis, often of over 50 cells: although the majority are mononuclear, five or more neutrophils are often seen. Oligoclonal bands are generally absent, being found in less than a fifth of cases.

The most important test involves a serum assay for NMO-IgG, which is positive in roughly three-quarters of all cases, and indeed may become positive before symptoms first appear (Jarius *et al.* 2008).

Classic cases present little in the way of diagnostic difficulty. Thus, when one finds a patient with optic neuritis, LETM, NMO-IgG seropositivity, and no more than a few cerebral white matter lesions, the diagnosis is secure (Wingerchuk *et al.* 2006). Difficulty may arise, however, early in the course, as for example when a patient presents with LETM alone. Here, finding NMO-IgG allows one to say with some confidence that the patient does have neuromyelitis optica, and that, over time, an episode characterized by optic neuritis will eventually occur.

Course

In most cases, neuromyelitis optica pursues a relapsing and remitting course (Wingerchuk *et al.* 1999). Overall, it is a disabling disease, and most patients are left with severe deficits after episodes, with a step-wise accrual of deficits after successive episodes. Fatalities are generally secondary to respiratory failure, which in turn occurs secondary to cervical cord lesions.

Etiology

Neuromyelitis optica is an autoimmune disease characterized by the presence of a specific IgG antibody, NMO-IgG, which binds to aquaporin-4 (AQP4) (Hinson *et al.* 2007). AQP4 is found in astrocytic foot processes, and although this is expressed throughout the central nervous system, concentrations are highest in the optic nerves and spinal cord; in a minority of individuals, high concentrations may also occur in the hypothalamus (Pittock *et al.* 2006; Poppe *et al.* 2005). During active disease phases, an intense inflammatory response, involving primarily macrophages, is seen, with destruction of oligodendrocytes and myelin in both the white and the gray matter, leading to extensive necrosis (Lucchinetti *et al.* 2002; Yanagawa *et al.* 2009).

Differential diagnosis

Early on, before the combination of optic neuritis and LETM is established, the various causes of isolated optic neuritis and transverse myelitis must be considered. Once the combination is present, however, the main item on the differential is MS. The transverse extent of the myelitis is quite helpful here, as the individual lesions of MS rarely extend the distance of three or more vertebral segments typical of neuromyelitis optica. Naturally, finding NMO-IgG in the serum settles the question.

Treatment

Acute attacks should be treated with intravenous methyl-prednisolone at 1 g/day for at least 5 days; the addition of plasma exchange appears to confer substantial further benefit (Bonnan *et al.* 2009).

Interferons and glatiramer are not effective. Although there are as yet no controlled studies regarding the prevention of future episodes, case series suggest effectiveness for cyclophosphamide, azathioprine, and mycophenolate mofetil: recent work suggests that rituximab may be superior to each of these (Jacob *et al.* 2008).

19.3 SYSTEMIC LUPUS ERYTHEMATOSUS

Systemic lupus erythematosus (SLE or, colloquially, lupus) is a systemic autoimmune disease that involves multiple organs, including the brain. Cerebral lupus (or as it is also referred to, neuropsychiatric systemic lupus erythematosus) occurs in the vast majority of cases, and may occur secondary to either a direct autoimmune assault on neurons (in which case one may speak of 'lupus cerebritis') or to one or more infarctions.

Systemic lupus erythematosus occurs in 0.015–0.05 percent of the general population. Over 90 percent of cases occur in females, and in the USA it is far more common in African-Americans, accounting for a prevalence of up to 0.4 percent in African-American females.

Clinical features

Although lupus may appear at almost any age, including in the elderly, the majority of patients fall ill between puberty and 40 years of age.

As noted, lupus is a systemic disease and in most cases cerebral lupus occurs in the setting of other symptoms, including constitutional symptoms (fatigue, fever, weight loss) and those referable to other organ systems, such as the musculoskeletal system, skin, heart, lungs, or kidneys (Johnson and Richardson 1968). Musculoskeletal symptomatology is very common and includes myalgia, arthralgia, and a non-deforming polyarthritis. Cutaneous manifestations include photosensitivity, rashes (especially a malar rash), and alopecia. Cardiac symptomatology incorporates pericarditis and Libman–Sacks endocarditis. Pulmonary involvement may manifest with pleurisy, which may or may not be accompanied by pleural effusion. Renal involvement may manifest initially with proteinuria and cellular casts; over time, renal failure may occur. Various cytopenias, including anemia, leukopenia, or thrombocytopenia, may also occur.

Cerebral lupus (Brey *et al.* 2002; Feinglass *et al.* 1976; Johnson and Richardson 1968; Sanna *et al.* 2003) most commonly manifests with a mood disorder (most commonly depression), cognitive dysfunction (including both dementia and, less commonly, delirium), and stroke. Psychosis and seizures are less common manifestations, and a movement disorder, such as chorea, is rare. Although these findings may occur independently, patients often have a mixture (Devinsky *et al.* 1988). A peripheral neuropathy may also occur in a small minority.

Depression, as noted, is the most common mood disturbance, and this at times may be accompanied by delusions and hallucinations (Ganz *et al.* 1972; Guze 1967; Hugo *et al.* 1996; Miguel *et al.* 1994). Mania, by contrast, although reported (Johnson and Richardson 1968), appears rare.

Cognitive dysfunction of mild degree may be seen in a majority of patients (Brey *et al.* 2002), but dementia, although certainly seen (Kirk *et al.* 1991; Johnson and Richardson 1968; MacNeil *et al.* 1976; Robin *et al.* 1995), is relatively uncommon (Devinsky *et al.* 1988). Delirium is even less common (Devinsky *et al.* 1988; Hugo *et al.* 1996; Miguel *et al.* 1994) and is often accompanied by hallucinations, either visual or auditory (Lief and Silverman 1960; O'Connor and Musher 1966).

Although stroke in lupus may manifest with any of the syndromes discussed in Chapter 8.4, the large-vessel syndromes are less common than the lacunar ones (Mitsias and Levine 1994).

Psychosis, as noted, is relatively uncommon (Guze 1967; Lim *et al.* 1988; Miguel *et al.* 1994). Typically, it is characterized by delusions and hallucinations; however, one may also see stuporous catatonia (Lanham *et al.* 1985; Mac and Pardo 1983). Rarely, psychosis may constitute the presenting feature of lupus (Pego-Reigosa and Isenberg 2008).

Seizures are relatively common (Devinsky *et al.* 1988) and may be partial (complex partial or simple partial) or grand mal in type (Mikdashi *et al.* 2005). They tend to appear early in the course of the disease and may either cease to occur spontaneously or, in a minority, persist indefinitely (Appenzeller *et al.* 2004).

Abnormal involuntary movements constitute a rare manifestation of lupus, and generally consist of choreiform movements, which indeed may appear very early in the course of the illness (Donaldson and Espiner 1971; Fermaglich *et al.* 1973).

Peripheral neuropathy may manifest as either a peripheral polyneuropathy (McCombe *et al.* 1987) or a mononeuritis multiplex (Hughes *et al.* 1982).

The antinuclear antibody (ANA) test is positive in approximately 95 percent of cases (Venables 1993), and the serum Venereal Disease Research Laboratories (VDRL) test may be falsely positive. As the ANA lacks specificity, however, a positive result here must be followed up by a more specific test, such as anti-native DNA (also known as anti-double-stranded DNA) or anti-Sm. During active disease, the erythrocyte sedimentation rate (ESR) is typically elevated and one or more complement levels (C3, C4, CH50) are generally decreased. Consideration should also be given to testing for the antiphospholipid syndrome, with lupus anticoagulant and anticardiolipin antibodies of both the IgG and IgM types, as this constitutes one of the major causes of stroke in lupus.

Magnetic resonance scanning may be normal or may show evidence of infarction.

The electroencephalogram (EEG) may be normal or show slowing, which may be generalized or focal (Sibley *et al.* 1992). In patients with seizures, interictal epileptiform discharges may or may not be present.

The CSF (Johnson and Richardson 1968; McLean *et al.* 1995; West *et al.* 1995) may be normal or there may be a mild lymphocytic pleocytosis or a mildly elevated total protein. An elevated IgG index and oligoclonal bands may also be seen, and indeed are found in the vast majority of cases associated with lupus cerebritis (West *et al.* 1995).

In cases characterized by infarction, one should, in addition to testing for lupus anticoagulant and anticardiolipin antibodies, also obtain an echocardiogram; when these tests are unrevealing, consideration should also be given to computed tomography (CT) or MR angiography of the neck and of the head.

Course

Overall, the intensity of autoimmune activity in lupus gradually waxes and wanes over long periods of time, sometimes becoming quite minimal. Consequently, although the sequelae of stroke, naturally, tend to persist, the severity of the other manifestations of lupus, dependent, as they are, on an underlying cerebritis, may wax and wane, and in some cases the syndrome in question (e.g. depression, delirium, psychosis) may undergo a spontaneous remission. Although SLE is in general compatible with long-term survival, the appearance of cerebral (Rubin *et al.* 1985) or renal disease is an unfavorable sign.

Etiology

Lupus is characterized by the presence of a large number of autoantibodies directed at various tissues in multiple organ systems. Although the cause is not known, it is strongly suspected that the autoimmune response occurs secondary to some environmental trigger in genetically susceptible individuals.

It must be stated at the outset that there is some uncertainty regarding the mechanism or mechanisms underlying the syndromes that make up cerebral lupus. With this caveat in mind, however, it appears that two global mechanisms may be operative (West *et al.* 1995): a direct autoimmune attack on neurons, and infarctions, which, in turn, may occur on a variety of different bases.

It is clear from autopsy studies that in some cases of cerebral lupus there is no evidence of infarction (O'Connor and Musher 1966; Tsokos *et al.* 1986), and in such cases it is suspected that the clinical symptomatology occurs secondary to antineuronal antibodies (Bluestein *et al.* 1981; Isshi and Hirohata 1998; Kelly and Denburg 1987; West *et al.* 1995), a notion supported by the finding of immune complexes in the choroid plexus (Atkins *et al.* 1972). Although a true cerebritis, with prominent inflammation of the cerebrum, is not seen, traditionally such cases have nevertheless been said to occur on the basis of a cerebritis, and this tradition is adhered to in this text.

Infarctions appear commonly in lupus and may occur on any one of several bases (Mitsias and Levine 1994). Cardio-embolic emboli may arise from Libman–Sacks endocarditis, valvulitis, or mural thrombi (Devinsky *et al.* 1988; Mitsias and Levine 1994; Tsokos *et al.* 1986). If these emboli are large, then large vessels, such as the middle cerebral or anterior cerebral artery, may be occluded, with resulting large territorial infarctions; if small, there may be widespread lacunar or microinfarctions throughout the cortex and subcortical white matter. Thrombotic infarctions of large- or medium-sized vessels may also occur, on the basis of either a fibrinoid vasculopathy (Ellis and Verity 1979; Hanley *et al.* 1992; Johnson and Richardson 1968; Malamud and Saver 1954), the antiphospholipid syndrome (Mitsias and Levine 1994), premature atherosclerosis (Mitsias and Levine 1994), or arterial dissection (Mitsias and Levine 1994). A true vasculitis may also occur (Weiner and Allen 1991), but this appears to be rare (Devinsky *et al.* 1988). Cerebral venous thrombosis may also occur but this too seems rare.

It appears that most cases of depression, mania, psychosis, and delirium occur on the basis of cerebritis; however, these syndromes may also occur with appropriately placed infarctions (e.g. depression with frontal lobe; mania with frontal or temporal lobe, thalamus, or caudate; psychosis with temporal lobe or thalamus; and delirium with temporal lobe or thalamus). Dementia may occur on the basis of cerebritis but appears more commonly owing to multiple infarctions. Seizures may likewise occur with cerebritis or with infarction. Chorea may occur with cerebritis or infarction (e.g. of the basal ganglia or thalamus) and is also associated with the presence of antiphospholipid antibodies.

Differential diagnosis

Polyarteritis nodosa, sarcoidosis, and, in older patients, cranial arteritis may all mimic lupus to a certain degree, and it is good practice to test for ANA in any case in which the neuropsychiatric syndromes described above occur in the setting of constitutional symptoms or multiorgan disease.

In evaluating neuropsychiatric syndromes in patients with established lupus, care must be taken to be sure that the syndrome is, in fact, due to lupus rather than some other cause. In this regard, special care must be taken in the evaluation of patients with depression. As discussed in Chapter 7.1, although it is not unusual for patients to become depressed in the face of adverse events, such as the occurrence of lupus, the severity of such 'normal' depressions is proportionate to the severity of the adverse event; furthermore, such 'normal' depressions are typically not accompanied by hallucinations or delusions. Consequently, the appearance of a non-psychotic depression in the context, say, of imminent renal failure might be considered normal. On the other hand, the occurrence of a severe depression in a patient whose lupus manifested only with a rash probably does not represent a normal depression, and the occurrence of a depression with hallucinations and delusions, even in patients with severe extracerebral lupus, would never represent a normal reaction. Although important in the case of depression, this concern with teasing out 'normal' reactions to stressful events does not apply to patients with mania, psychosis, or delirium, as none of these syndromes ever occurs as a 'normal' reaction.

Patients with lupus may develop renal failure with uremia, which may be accompanied by hypertension, and uremic encephalopathy or hypertensive encephalopathy may occur (Wong *et al.* 1991). Uremia may cause mania, delirium, or seizures, and hypertensive encephalopathy is characterized by delirium and seizures. Steroid treatment may also cause neuropsychiatric side-effects, including depression, mania, and psychosis. Finally, treatment with either steroids or immunosuppressants opens the way to opportunistic central nervous system infections (Futrell *et al.* 1992; Wong *et al.* 1991), which may also be associated with delirium, seizures, or focal signs.

On a final differential diagnostic note, care must be taken to distinguish naturally occurring lupus from drug-induced lupus. This syndrome is most commonly seen secondary to use of procainamide, hydralazine, or, less frequently, alpha-methyl dopa; it has also been rarely noted secondary to the use of other medications, including chlorpromazine, carbamazepine, phenytoin, and primidone. Importantly, unlike naturally occurring lupus, drug-induced lupus rarely causes cerebral symptoms. When there is doubt whether any given case of lupus is drug-induced or not, antibody levels may be helpful. Although all patients with drug-induced lupus have a positive ANA, very few will have the more specific antibodies to native DNA or Sm. Obtaining an antihistone antibody level may also be helpful: whereas patients with drug-induced lupus typically have this, it is unusual in patients with naturally occurring lupus.

Treatment

Treatment of the various features of cerebral lupus is dictated by the presumed underlying etiology, as discussed above. In this regard, a reasonable strategy is to obtain an MRI scan to determine whether or not infarction has occurred, and, if so, to then decide whether the identified infarction or infarctions could reasonably be expected to explain the patient's symptomatology. If this is the case, treatment is directed at the cause of the underlying infarction, as discussed in Chapter 8.4. In cases of infarction occurring in the context of the antiphospholipid syndrome, preventive treatment with warfarin is probably in order.

In the remaining cases, which are presumably due to a cerebritis, consideration may be given to treatment with steroids or cyclophosphamide, either individually or in combination. Initial treatment with steroids generally involves a 3- to 7-day course of methylprednisolone, 250 mg intravenously four times daily, followed by prednisone at 1 mg/kg/day, with the dose of prednisone gradually tapered over a few weeks depending on the clinical response. With regard to cyclophosphamide, some clinicians will utilize this as single-agent therapy from the start, whereas others will add it to the course of steroids, generally several days into treatment with methylprednisolone. Repeat courses of steroids or monthly infusions of cyclophosphamide are generally required to maintain remission. Remarkably, at the time of writing there has been only one double-blind study comparing steroids alone with the combination of steroids and cyclophosphamide (Barile-Fabris *et al.* 2005). This study found that the addition of cyclophosphamide to an initial course of steroids, followed by monthly cyclophosphamide, was greatly superior to initial treatment with steroids alone, followed by repeat courses of methylprednisolone. However, this study primarily included patients with seizures and excluded patients with depression, mania, psychosis, delirium, or dementia, and hence it is not clear whether the results would apply to these syndromes. In cases resistant to treatment with steroids and cyclophosphamide, consideration may be given to rituximab, which, in an open study, yielded dramatic results (Tokunaga *et al.* 2007).

Symptomatic treatment of depression, mania, psychosis, delirium, dementia, and seizures may or may not be required, and is discussed in Chapters 7.1, 7.3, 8.1, 6.5, 6.3, and 8.3, respectively.

19.4 SJÖGREN'S SYNDROME

Sjögren's syndrome is an autoimmune disease characterized by keratoconjunctivitis sicca and xerostomia (dry eyes and dry mouth: the 'sicca syndrome'), and, in a minority, by disease of the nervous system. Once thought to be rare, it is now known to occur in up to 2 percent of those aged over 60 years. It is far more common in woman than men, by a ratio of 9:1.

Clinical features

The onset is in middle or later years, and the syndrome presents, in all cases, with gradual appearance of the sicca syndrome. A majority of patients will also develop a gradually progressive, non-deforming, small-joint polyarthritis.

Nervous system involvement (Alexander *et al.* 1981, 1986, 1994; Delalande *et al.* 2004) occurs, as noted, in a minority of cases, and the peripheral nervous system is most commonly involved, with a sensory polyneuropathy, or, less commonly, a mononeuritis multiplex. Cranial neuropathy may also occur, with a trigeminal sensory neuropathy or a Bell's palsy. Central nervous system involvement may present with an optic neuritis, or cord involvement, including a transverse myelitis or a progressive myelopathy. The cerebrum may also be involved, and there may be strokes or stroke-like episodes, with hemiparesis or aphasia, or a gradually progressive dementia (Caselli *et al.* 1991, 1993; Kawashima *et al.* 1993). Aseptic meningitis may occur, and, in a very small minority, there may be seizures or parkinsonism (Walker *et al.* 1999).

T2-weighted or FLAIR MR scanning often reveals multiple small areas of increased signal intensity in the white matter (Coates *et al.* 1999) and, in cases of dementia, the increased signal intensity in the white matter may be confluent and widespread.

The CSF may be normal or may display an increased total protein, oligoclonal bands, or an increased IgG index.

Serologic abnormalities are found in almost all patients. Over 90 percent will have a positive ANA; however, the antinative DNA and anti-Sm antibodies are absent; some 75 percent will have a positive rheumatoid factor. Anti-SS-A (formerly known as anti-Ro) is found in 70–90 percent, and anti-SS-B (formerly known as anti-La) in about 60 percent. In doubtful cases, a lip biopsy will reveal lymphocytic infiltrates.

Course

Overall, the course is characterized by gradual progression.

Etiology

Sjögren's syndrome is an autoimmune disorder, and within the central nervous system there may be a lymphocytic meningitis or a lymphocytic vasculitis, primarily affecting small vessels (Caselli *et al.* 1991, 1993; de la Monte *et al.* 1983).

Differential diagnosis

The sicca syndrome may be seen in other connective tissue diseases, such as systemic lupus erythematosus, systemic sclerosis, rheumatoid arthritis, and polymyositis. In patients with involvement of the nervous system, the most important consideration is lupus, and here the differential rests on testing for antinative DNA, which is present in lupus but absent in Sjögren's syndrome.

Multiple sclerosis enters into the differential, especially in patients with optic neuritis and/or cord involvement. Here the differential rests on a lip biopsy or finding the typical serologic abnormalities mentioned above.

The sicca syndrome, of course, may also be caused by multiple different medications.

Treatment

Central nervous system involvement generally requires treatment with steroids or immunosuppressants, such as cyclophosphamide.

19.5 SNEDDON'S SYNDROME

Sneddon's syndrome, first described by Sneddon in 1965 (Sneddon 1965) is a rare disorder, more common in females than in males, that is characterized by the combination of livedo reticularis and cerebrovascular disease.

Clinical features

This disorder has been the subject of several large studies (Adair *et al.* 2001; Boesch *et al.* 2003; Frances *et al.* 1999; Stockhammer *et al.* 1993; Tourbah *et al.* 1997). The age of onset is typically in early or middle adult years, and the presentation, in almost all cases, is with livedo reticularis, which precedes other symptoms by anywhere from 6 months to 10 years. Importantly, this livedo reticularis is not confined to the lower extremities, but is also seen on the trunk.

Patients may eventually develop a slowly progressive dementia, associated with an equally progressive leukoencephalopathy involving the centrum semiovale, or they may experience recurrent cerebral infarctions or transient ischemic attacks. The infarctions are generally of the lacunar type, involving either the cortex or the subcortical white matter, and, if appropriately situated, such recurrent infarctions may make their own contribution to an evolving dementia. Seizures may occur in a small minority of cases.

Magnetic resonance imaging will reveal both the leukoencephalopathy and any areas of infarction. Echocardiography may reveal a valvulopathy, often involving the mitral valve, in close to one-half of all cases. Antiphospholipid antibodies may be found in a minority (Frances *et al.* 1999), but these are generally only transiently present. The ANA may be positive, but antinative DNA and extractable nuclear antigen (ENA) antibodies are absent (Kalashnikova *et al.* 1990).

The most important test to perform is a biopsy of the skin lesion, which will reveal the characteristic vasculopathy (Stockhammer *et al.* 1993).

Course

Sneddon's syndrome is slowly progressive, with a worsening of dementia and recurrent strokes.

Etiology

Within the central nervous system, and the skin, there is a wide-spread vasculopathy affecting small- and medium-sized arteries (Hilton and Footitt 2003; Stockhammer *et al.* 1993). This vasculopathy begins with an endothelialitis (Sepp *et al.* 1995), followed by occlusion of the lumen by a plug composed of fibrin and inflammatory cells. Endothelial proliferation follows, and eventually the vessel becomes fibrotic; in a minority of cases an arterial thrombus may be seen (Hilton and Footitt 2003). Although the mechanism underlying the initial endothelialitis is not known, an autoimmune disturbance is suspected.

In addition to this vasculopathy, there may, as noted earlier, be an accompanying valvulopathy, and it may be that this serves as a source of microemboli, which, in turn, could provide an additional basis for the occlusion of small- or medium-sized vessels (Sitzer *et al.* 1995). The extent to which such a process contributes to the overall clinical picture in Sneddon's syndrome, however, is not clear.

Although most cases are sporadic, familial cases have been noted (Pettee *et al.* 1994; Rebollo *et al.* 1983).

Differential diagnosis

The diagnosis should always be suspected in any adult with livedo reticularis (especially when it is generalized) and either stroke or dementia.

The primary antiphospholipid syndrome is distinguished from Sneddon's syndrome by the absence of livedo reticularis and by a history of deep venous thromboses, miscarriages, and, in some cases, cerebral venous infarctions.

The lacunar syndrome and Binswanger's disease are likewise distinguished by an absence of livedo reticularis. The multiple cholesterol emboli syndrome may be a little harder to distinguish, as it, like Sneddon's syndrome, may also be characterized by livedo reticularis; here, however, the livedo reticularis appears at roughly the same time as the strokes, and one finds features absent in Sneddon's syndrome, such as renal failure, petechial hemorrhages, etc.

Treatment

The optimal treatment for Sneddon's syndrome is not clear: aspirin is recommended (Frances *et al.* 1999) and there are case reports of improvement with immunosuppressive treatment (Hannon *et al.* 2010).

19.6 PRIMARY ANTIPHOSPHOLIPID SYNDROME

This is a rare syndrome, found much more commonly in females, which has an onset in early or middle adult years. It is characterized by recurrent arterial or venous thromboses, and constitutes an important cause of stroke in younger adults.

Clinical features

The overall features of this syndrome have been described in a number of reports (Asherson *et al.* 1989; Chancellor *et al.* 1991; Levine *et al.* 1990, 1995; Provenzale *et al.* 1994). Recurrent stroke secondary to ischemic infarction is the rule, and imaging may reveal either lacunar or large-vessel territorial infarctions, or both. Dural sinus thrombosis may also occur, as may venous infarction. With a multiplicity of strokes, a multi-infarct dementia may occur (Coull *et al.* 1987; Kurita *et al.* 1994). Transient ischemic attacks are common, especially amaurosis fugax.

History typically reveals evidence of deep venous thromboses, superficial thrombophlebitis, and miscarriages or prematurity, and echocardiography may reveal mitral and aortic valvular disease.

Antiphospholipid antibodies, either the lupus anticoagulant or anticardiolipin antibodies, or both, are present in every case. Anticardiolipin antibodies include both IgG and IgM types, and the IgG antibody is most strongly associated with thrombosis. Thrombocytopenia occurs in about one-third of cases, and there may be a false-positive VDRL. The ANA may be positive, but antinative DNA and anti-Sm antibodies are absent.

Course

Recurrent stroke is common.

Etiology

In this syndrome, circulating antiphospholipid antibodies attach to vascular endothelium, either arterial or venous, and induce thrombus formation (Hughson *et al.* 1993; Levine *et al.* 1990). Of interest, when arteries are targeted, it is generally only the cerebral circulation that is involved, but when veins are targeted, both the cerebral and systemic vasculature may be involved. In addition to this *in situ* thrombosis, embolic infarction may occur, with emboli arising from Libman–Sacks endocarditis of either the mitral or aortic valves. Furthermore, in patients with a patent foramen ovale, paradoxical emboli may arise from deep venous thromboses.

Differential diagnosis

The diagnosis of the primary antiphospholipid syndrome should always be suspected in any young person with stroke (Brey *et al.* 1990), especially in those with a history of deep venous thrombosis or recurrent miscarriage.

Antiphospholipid antibodies may also be found in other conditions. By far the most common cause of the secondary antiphospholipid syndrome is systemic lupus erythematosus, and here, in addition to other clinical features discussed in Chapter 19.3, the positive ANA will be accompanied by a positive antinative DNA and anti-Sm antibody. Sneddon's syndrome is also characterized by antiphospholipid antibodies; however, here one sees livedo reticularis and a leukoencephalopathy. Antiphospholipid antibodies may also occur with various malignancies and with treatment with certain drugs, including procainamide, quinidine, hydralazine, phenytoin, valproic acid, and phenothiazines. Finally, antiphospholpid antibodies may transiently occur in normal individuals – given this, it is incumbent to repeat the test after 12 weeks in order to ensure a 'true' positive.

Treatment

Long-term anticoagulation is required.

19.7 SUSAC'S SYNDROME

Susac's syndrome, first described by Susac *et al.* in 1979 (Susac *et al.* 1979), is a rare disorder which is found more commonly in females than in males, by a ratio of 3:1. The syndrome is also referred to as retinocochleocerebral vasculopathy, a term which, although cumbersome, nicely summarizes the structures involved.

Clinical features

Although the onset may occur from childhood through the eighth decade, most patients fall ill in the third or fourth decades. Classically, in fully developed cases, one sees a triad of visual changes, hearing loss, and an encephalopathy (Aubart-Cohen *et al.* 2007; Pawate *et al.* 2009).

Visual changes, occurring secondary to branch retinal arterial occlusion, may manifest initially as mere scotomata; however, with progression blindness may occur.

Hearing loss is of the sensorineuronal type, and may be accompanied by vertigo.

The encephalopathy is of subacute onset, and is marked by headache and delirium, which in turn may be accompanied by agitation and delusions of persecution.

Of note, although this classic triad eventually develops in most cases, early on, only one, or perhaps two, of the clinical elements may be present. Furthermore, in delirious patients, it may be very hard to determine whether any visual changes or hearing loss has occurred.

Magnetic resonance scanning (Susac *et al.* 2003) typically reveals multiple areas of increased signal intensity on FLAIR and T2-weighted images, scattered throughout the white and gray matter with a strong predilection for the corpus callosum. During active phases of the disease, the callosal lesions, on sagittal images, may have the appearance of 'snowballs' or assume a radial configuration; during chronic, inactive phases, the callosal lesions may appear as 'black holes', a finding almost specific for this disease.

Course

Three different courses have been observed. In some cases, there may be only one episode, which, however, may last for 1–2 years. In others, there may be a relapsing and remitting course, while in a small minority the syndrome appears to pursue a chronic waxing and waning course. The long-term outcome, in the absence of treatment, may, in the worst cases, be characterized by blindness, deafness, and dementia.

Etiology

Susac's syndrome appears to be an autoimmune disorder, with antibodies directed at endothelial cells (Jarius *et al.* 2009). A widespread microangiopathy develops, involving retinal, cochlear, and cerebral vessels, with resulting infarctions. Within the cerebrum, infarctions are seen in the white matter, especially the corpus callosum, the cortex, and deep gray structures (Heiskala *et al.* 1988; Monteiro *et al.* 1985).

Differential diagnosis

Both multiple sclerosis and systemic lupus erythematosus enter into the differential, and although both these disorders may produce the triad seen in Susac's syndrome, this triad is rare in these two disorders; furthermore, as noted earlier, the callosal abnormalities are almost specific for Susac's syndrome.

Treatment

First-line treatments include steroids and intravenous immunoglobulins. In many cases it is necessary to add an immunosuppressant, such as cyclophosphamide, or rituximab.

19.8 LIMBIC ENCEPHALITIS

Limbic encephalitis, first described by Brierly and colleagues in 1960 (Brierly *et al.* 1960), is an autoimmune disorder characterized pathologically by the presence of antineuronal antibodies and inflammatory changes in the medial aspects of the temporal lobes, and clinically, in most, but not all, cases, by delirium and seizures.

In the vast majority of cases, limbic encephalitis occurs on a paraneoplastic basis, most often in patients with small-cell lung cancer, and serum samples will be positive for typical antineuronal antibodies, such as anti-ANNA-1. Importantly, however, it must be kept in mind that limbic encephalitis is often the presenting symptom of cancer (Alamowitch *et al.* 1997; Dalmau *et al.* 1992), and in some cases the tumor itself may remain undetected for years after the onset of the encephalitis (Ahern *et al.* 1994).

Limbic encephalitis is a rare disorder, occurring in less than 0.1 percent of all patients with cancer.

Clinical features

The onset of symptoms is typically subacute, spanning days or weeks (Alamowitch *et al.* 1997).

Classically, limbic encephalitis presents with delirium marked by prominent anterograde and retrograde amnesia, often accompanied by seizures, personality change, and hallucinations and delusions (Alamowitch *et al.* 1997; Antoine *et al.* 1995; Bakheit *et al.* 1990; Gultekin *et al.* 2000; Lawn *et al.* 2003). Other less common presentations include depression (Brierly *et al.* 1960; Corsellis *et al.* 1968; Glaser and Pincus 1969), isolated amnesia (Bak *et al.* 2001; Nokura *et al.* 1997; Sutton *et al.* 2000), seizures (Corsellis *et al.* 1968), somnolence (Byrne *et al.* 1997), and catatonia (with 'confusion, stereotypy, echolalia, stiffness, verbigeration, formal thought disorder, and negativism' [Tandon *et al.* 1988]). Rare symptoms include abnormal movements, such as chorea (Croteau *et al.* 2001; Vernino *et al.* 2002), narcoleptic attacks with cataplexy (Rosenfeld *et al.* 2001), and rapid eye movement (REM) sleep behavior disorder (Iranzo *et al.* 2006).

Two recently described forms of limbic encephalitis merit some further discussion, namely that occurring secondary to anti-NMDA antibodies (Dalmau *et al.* 2007, 2008; Vitaliani *et al.* 2005), and that due to antivoltage-gated potassium channel (anti-VGKC) antibodies (Geschwind *et al.* 2008; Tan *et al.* 2008; Vincent *et al.* 2004). Although both of these forms of limbic encephalitis tend to present in a fashion similar to the classical one just described, their further evolution displays some peculiarities. In both forms, abnormal involuntary movements may be seen (in roughly two-thirds of anti-NMDA cases and one-third of anti-VGKC cases), including chorea, dystonia, or ballismus. Furthermore, in anti-VGKC cases, myoclonus and hyponatremia occur in the majority. Finally, in anti-NMDA cases, hypoventilation ensues in roughly two-thirds of cases, often necessitating intubation. The etiology of these forms also exhibits some noteworthy features. Anti-NMDA cases are seen almost exclusively in young females, and in the vast majority of these, an ovarian teratoma is found; in the few cases of males with anti-NMDA antibodies, various tumors have been noted, including testicular cancer (Eker *et al.* 2008) and Hodgkin's lymphoma (Zandi *et al.* 2009). Anti-VGKC cases may likewise occur on a paraneoplastic basis, with various different types of neoplasms being found; however, in roughly two-thirds of cases, the syndrome occurs on a non-paraneoplastic, idiopathic basis.

In cases that occur on a paraneoplastic basis, one may also see other paraneoplastic syndromes (Alamowitch *et al.* 1997), including the following: cerebellar degeneration with ataxia; brainstem encephalitis with nystagmus, ataxia, oculomotor palsies, and vertigo; opsoclonus–myoclonus; sensory neuropathy; Lambert–Eaton myasthenic syndrome; and the stiff-person syndrome.

Early in the course, MR scanning is often normal; over time, however, most cases will display increased signal intensity in the medial aspects of the temporal lobes on T2-weighted or FLAIR imaging; these abnormalities, although initially unilateral, characteristically become bilateral (Alamowitch *et al.* 1997; Bakheit *et al.* 1990; Gultekin *et al.* 2000; Lawn *et al.* 2003; Tsukamoto *et al.* 1993). This typical location of MRI abnormalities, however, may not be seen in all cases; for example, in anti-NMDA and anti-VGKC limbic encephalitis, although MRI changes may be seen in the medial aspects of the temporal lobes, they may also be found in the frontal and parietal cortices in anti-NMDA cases and in the insular, anterior cingulate, and temporal cortices of anti-VGKC cases.

The EEG is abnormal in almost all cases and typically shows temporal slowing, which, similar to the MRI findings, may initially be unilateral, only to later become bilateral. Interictal epileptiform discharges may or may not be present. As with MRI studies, however, variations may also be seen: for example, in both anti-NMDA and anti-VGKC limbic encephalitis, a generalized slowing is more common.

Although the CSF may be normal, in the vast majority of cases one will find one or more of a mild lymphocytic pleocytosis, a mildly elevated total protein, or oligoclonal bands (Alamowitch *et al.* 1997; Gultekin *et al.* 2000; Psimaras *et al.* 2010).

Although in the vast majority of cases abnormalities will be found on MRI, EEG, or CSF assay, exceptions do occur, especially early on, and in cases in which the clinical findings are strongly suggestive of the diagnosis but these tests are negative, positron emission tomography (PET) scanning should be considered: even in cases in which all other tests are negative, PET scanning may reveal focal hypermetabolism in one or both temporal lobes.

A wide variety of antineuronal antibodies have been identified (Bataller *et al.* 2007; Lai *et al.* 2009; Pittock *et al.* 2005), including ANNA-1 (also known as anti-Hu), ANNA-2 (also known as anti-Ri), ANNA-3, anti-Ma1, anti-Ma2 (also known as anti-Ta), anti-amphiphysin, anti-CRMP-5 (also known as anti-CV2), anti-AMPR, anti-NMDA, and anti-VGKC. In paraneoplastic cases, various tumors have been found, including cancer of the lung (most commonly of the small-cell type) (Alamowitch *et al.* 1997), breast, testicle (Ahern *et al.* 1994; Burton *et al.* 1988; Voltz *et al.* 1999), colon (Tsukamoto *et al.* 1993), pancreas, ovary (Nokura *et al.* 1997), thymus (Antoine *et al.* 1995; Ingenito *et al.* 1990), prostate, and bladder; cases have also been associated with lymphoma. Although there is an association between certain antineuronal antibodies and certain types of cancer (e.g. anti-ANNA-1 with lung cancer (Anderson *et al.* 1988; Alamowitch *et al.* 1997), anti-Ma2 with testicular cancer (Gultekin *et al.* 2000; Pruss *et al.* 2007; Voltz *et al.* 1999), and anti-NMDA with ovarian teratoma [Dalmau *et al.* 2008]), there is a wide overlap (Pittock *et al.* 2004, 2005); for example, anti-ANNA-1, in addition to being found in association with lung cancer, may also be found with cancer of the breast, prostate, colon, and ovary (Graus *et al.* 2001). Given this overlap, it is appropriate to test for the entire range of known antineuronal antibodies. It must also be kept in mind that new antineuronal antibodies are routinely discovered and, consequently, negative tests for known antineuronal antibodies do not rule out the diagnosis (Ances *et al.* 2005; Graus *et al.* 2008; Lai *et al.* 2009).

Once the diagnosis of limbic encephalitis has been made, it is essential to search for a tumor. In this regard, if routine investigations are unrevealing, many clinicians will undertake CT scans of the chest, abdomen, and pelvis, mammography in females, and, in young males, ultrasound of the testes. If the tumor escapes detection by these methods, consideration may be given to whole-body PET scanning.

Course

Although the overall course is one of progression, there may at times be 'plateaus', which, however, almost always give way to further decline (Alamowitch *et al.* 1997). True spontaneous remissions, although reported (Byrne *et al.* 1997), are rare, and most patients die within months to a year or more, either of complications of the limbic encephalitis or from the underlying cancer.

Etiology

Pathologically (Henson *et al.* 1965), there is a lymphocytic perivascular inflammation, with neuronal loss and gliosis within the limbic system, primarily involving medial temporal structures, such as the hippocampus and the amygdala. As noted earlier, in the vast majority of cases this inflammation occurs on a paraneoplastic basis, with antibodies raised against the cancer cross-reacting with normal neuronal tissue.

Differential diagnosis

Although limbic encephalitis immediately comes to mind in patients with known cancer who develop delirium with seizures, other disorders that are not uncommon in cancer must also be considered, including metastatic disease, Cushing's syndrome, and opportunistic infections, such as cytomegaloviral encephalitis or progressive multifocal leukoencephalopathy.

As noted earlier, in most cases, limbic encephalitis precedes other evidence of cancer and hence in most cases the differential is wider, as discussed in Chapter 6.5. Special consideration may be given to herpes simplex viral encephalitis, Hashimoto's encephalopathy, and Creutzfeldt–Jakob disease.

Treatment

Although treatment of the underlying cancer in paraneoplastic cases should be undertaken, it appears that, even with successful treatment, limbic encephalitis undergoes remission in only a minority of cases (Burton *et al.* 1988) and, consequently, other modalities must be considered. Although there are no controlled studies, various treatments, alone or in combination, are utilized, including steroids, immunosuppressants (e.g. cyclophosphamide), intravenous immunoglobulins, and plasmapheresis (Bataller *et al.* 2007). In this regard, note may be made of the sometimes dramatic response seen in anti-VGKC cases to treatment with methylprednisolone.

19.9 SARCOIDOSIS

Sarcoidosis is an uncommon disease characterized pathologically by the presence of sarcoid granulomas in multiple organ systems, including, in a minority, the nervous system, in which case one may speak of neurosarcoidosis. In addition to a granulomatous basilar meningitis, granulomas may also be found scattered throughout the cerebrum, with a special predilection for the hypothalamus and pituitary.

Sarcoidosis is somewhat more common in females than in males; in the USA it is far more common in black people than in white people, and in Europe those from northern countries are more commonly affected.

Clinical features

Although the range of the age of onset is wide, most patients fall ill in early- to mid-adult years. The onset itself is often gradual, and many cases are discovered serendipitously when a chest radiograph reveals pulmonary findings characteristic of the disease.

Although sarcoidosis may be protean in its manifestations, certain presentations deserve note. Perhaps 90 percent of patients will have pulmonary involvement, which may manifest clinically with symptoms such as cough or dyspnea, or may be asymptomatic and discovered only incidentally by chest radiograph, which may reveal bilateral hilar lymphadenopathy or a diffuse reticulonodular appearance. Other symptoms include erythema nodosum, uveitis, lupus pernio, lymphadenopathy, arthropathy, and parotid gland enlargement. Hepatic involvement occurs in almost three-quarters of patients although hepatic failure is rare. Hypercalcemia occurs in a majority of cases, and some patients may develop nephrocalcinosis and eventual renal failure.

Involvement of the nervous system occurs in anywhere from 5 to 25 percent of patients, and in a very small minority of cases it may represent the only manifestation of sarcoidosis. The overall symptomatology of neurosarcoidosis has been described in a number of reports (Allen *et al.* 2003;

Chapelon *et al.* 1990; Delaney 1977; Joseph and Scolding 2009; Manz 1983; Oksanen 1986; Sharma and Sharma 1991; Stern *et al.* 1985). With a basilar meningitis, cranial neuropathies may occur, and with obstruction of the outflow foramina of the fourth ventricle, hydrocephalus may appear; involvement of arteries as they traverse the subarachnoid space may be followed by stroke. Cerebral involvement may be characterized by multiple granulomas or by a relatively few large lesions, or even by a solitary lesion: in these cases there may be dementia, delirium, seizures, or focal signs. Hypothalamic or pituitary granulomas may present with endocrinologic syndromes. Both the cord and the peripheral nerves may also be involved.

Cranial neuropathies occur in approximately one-half of all cases and, although various of the cranial nerves may be involved (Symonds 1958), one most commonly sees a peripheral facial palsy, which may be unilateral or bilateral (Scott 1993; Sharma and Sharma 1991). The eighth cranial nerve may also be involved with deafness, as may the optic nerve or chiasm with blindness or hemianopia.

Hydrocephalus occurs in about 5 percent of cases and may present with dementia and a gait disturbance.

Stroke is rare in sarcoidosis and appears to generally present with a lacunar syndrome (Brown *et al.* 1989; Michotte *et al.* 1991), reflecting granulomatous involvement of the penetrating arteries.

Dementia does occur with cerebral involvement (Camp and Frierson 1962; Cordingly *et al.* 1981; Miller *et al.* 1988; Sanson *et al.* 1996) and may be accompanied by a frontal lobe syndrome (Hook 1954) or by delusions and hallucinations (Thompson and Checkley 1981). Although the prevalence of this syndrome is uncertain, one study found cognitive deficits of variable degree in close to 50 percent of all patients with neurosarcoidosis (Scott *et al.* 2007). Delirium has also been noted (Douglas and Maloney 1973; Silverstein and Siltzbach 1965; Wiederholt and Siekert 1965) but appears to be rare.

Seizures occur in about 15 percent of cases and may be grand mal or partial in type (Krumholz *et al.* 1991). Focal signs, such as hemiparesis, may gradually appear, and these reflect the location of any cerebral granulomas.

Endocrinologic changes have been noted in up to one-third of patients, and may consist of diabetes insipidus, hyperprolactinemia, hypothyroidism, hypogonadism, and adrenocortical insufficiency (Bihan *et al.* 2007; Scott *et al.* 1987). Hypothalamic involvement may also manifest with disturbances of appetite or with somnolence; rarely symptomatic narcolepsy may occur (Rubinstein *et al.* 1988).

Spinal cord granulomas may present with various symptomatologies, most notably paraplegia.

The peripheral nervous system is involved in up to 50 percent of patients, and may manifest with a mononeuropathy, a mononeuritis multiplex, or a primarily sensory polyneuropathy.

Magnetic resonance scanning with T2-weighted or FLAIR imaging typically reveals any basilar meningitis or macroscopic parenchymal granulomas, and with gadolinium, enhancement is seen in most of these.

The CSF is abnormal in about 50 percent of cases. Various abnormalities may occur (Joseph and Scolding 2009; Kinnman and Link 1984; McLean *et al.* 1995), including a mild lymphocytic pleocytosis, a mildly elevated total protein, and, in a minority, oligoclonal bands or an elevated IgG index; in a very small minority, the glucose level is mildly reduced. The level of CSF angiotensin-converting enzyme is elevated in approximately one-half of cases (Oksanen 1986), while the serum angiotensin-converting enzyme (SACE) level is elevated in roughly two-thirds.

Definitive diagnosis requires biopsy evidence of typical sarcoid granulomas, and in most cases this may be obtained at bronchoscopy or by lung biopsy.

Course

The course is variable (Luke *et al.* 1987; Pentland *et al.* 1985). Anywhere from one-half to two-thirds of all cases are characterized by one episode of illness, which remits spontaneously after many months; in the remainder, the course may be either relapsing and remitting or chronically progressive.

Etiology

The cardinal lesion in sarcoidosis is a non-caseating granuloma, and the neuropathology of neurosarcoidosis has been described in a number of studies (Delaney 1977; Herring and Urich 1969; Jefferson 1957). As noted above, one typically sees a granulomatous basilar meningitis, with entrapment and inflammation of cranial nerves, penetrating arteries, and, in a small minority, obstruction of the outflow foramina of the fourth ventricle. Granulomatous infiltration of the hypothalamus is very common, and further infiltration down the pituitary stalk may lead to granuloma formation in the posterior or anterior lobe of the pituitary gland. Of note, it appears that most of the endocrinologic disturbances seen in neurosarcoidosis result primarily from hypothalamic disease, with pituitary function being secondarily disturbed by the lack of releasing or inhibiting factors normally secreted by the hypothalamus (Winnacker *et al.* 1968). Parenchymal granulomas may be found not only in the white matter of the cerebrum but also in the cortex and, as noted earlier, they may range widely in size and number, from miliary lesions to large masses mimicking gliomas (Powers and Miller 1981).

Although the mechanism underlying the appearance of these granulomas is not known, it is strongly suspected that sarcoidosis represents an autoimmune disorder that is triggered in genetically susceptible individuals by an exogenous, inhaled substance.

Differential diagnosis

Neurosyphilis, tuberculosis, and fungal infections may all closely mimic sarcoidosis. Multiple sclerosis is often mentioned on the differential; however, this possibility would only arise in cases of neurosarcoidosis in which lesions were essentially restricted to the cerebral white matter.

Treatment

Active neurosarcoidosis may be treated with prednisone in a dose of approximately 1 mg/kg/day for 4–6 weeks, after which the dose may be gradually tapered over the following months, depending on the clinical evolution. In treatment-resistant cases, some clinicians will give a course of intravenous methylprednisolone (Allen *et al.* 2003), and, should that fail, an immunosuppressant (Scott *et al.* 2007) such as cyclophosphamide, azathioprine, methotrexate, or mycophenolate. Recent work has demonstrated very impressive results with the addition of infliximab to mycophenolate (Moravan and Segal 2009). Hydrocephalus may require shunting.

19.10 HASHIMOTO'S ENCEPHALOPATHY

Hashimoto's encephalopathy, first described by Lord Brain in 1966 (Brain *et al.* 1966), is an uncommon cause of delirium which occurs secondary to an autoimmune process that targets the brain.

Before proceeding further, some words are in order regarding the confusion that may exist between Hashimoto's encephalopathy and Hashimoto's thyroiditis, which are two very different diseases. The name 'Hashimoto' is generally associated not with an encephalopathy but with thyroiditis. Hashimoto's thyroiditis, first described by Hakaru Hashimoto in 1912 (Hashimoto 1912), is the most common cause of thyroiditis and is characterized by the presence of antithyroid antibodies and by a lymphocytic infiltration of the thyroid gland; affected patients may be euthyroid, transiently hyperthyroid, or, more commonly, hypothyroid. Dr Hashimoto did not note delirium in any of these patients with thyroiditis. In 1966, however, Lord Brain described a patient with thyroiditis and antithyroid antibodies who also had delirium and stroke-like episodes, and it was this association of Hashimoto's thyroiditis with delirium that prompted the new term, 'Hashimoto's encephalopathy'. Importantly, however, Hashimoto's encephalopathy does not occur secondary to hyperthyroidism or hypothyroidism. Although, as noted below, there may be elevations or reductions in free T4 in patients with Hashimoto's encephalopathy, in most cases the free T4 is within normal limits, and in those cases where it is not, the abnormality is never severe enough to directly cause any disturbance in cerebral functioning.

A recently coined synonym for Hashimoto's encephalopathy is '*s*teroid-*r*esponsive *e*ncephalopathy associated with *a*utoimmune *t*hyroiditis' or 'SREAT'. Whether this new terminology is helpful, or will gain currency, remains to be seen.

Clinical features

The clinical features have been most clearly described in two case series from the Mayo Clinic (Castillo *et al.* 2006; Sawka *et al.* 2002). Although most patients are in their forties, the age of onset varies widely, from childhood to the eighth decade; the onset itself is typically subacute, over days or perhaps weeks.

The overwhelming majority of patients have a delirium, which in most cases is accompanied by any or all of tremor, myoclonus, ataxia, or seizures: seizures may be grand mal, complex partial, or, rarely, simple partial, and grand mal status epilepticus may occur in a small minority. Stroke-like episodes are common and are typically characterized by aphasia (Bohnen *et al.* 1997; Ghika-Schmid *et al.* 1996; Henchey *et al.* 1995; Shaw *et al.* 1991; Thrush and Boddie 1974); hemiplegia or hemianesthesia may also occur. These stroke-like episodes are of brief duration, lasting in the order of hours or a day or more, and typically undergo a full remission. Very rarely Hashimoto's encephalopathy may present with a psychosis (Bostantjopoulou *et al.* 1996) or with a dementia (Galluzzi *et al.* 2002).

Magnetic resonance scanning is normal in the majority of cases; in the remainder, T2-weighted or FLAIR imaging may disclose diffusely increased signal intensity in the cerebral white matter (Bohnen *et al.* 1997) and, in a small minority, subcortical infarctions may be noted (Henchey *et al.* 1995). Computed tomography scanning is almost always normal.

The EEG shows generalized slowing in over 90 percent of cases, and this may be accompanied by other changes in a small minority, including triphasic waves and interictal epileptiform discharges (Schauble *et al.* 2003).

The CSF typically, but not always, displays various abnormalities. An elevated total protein is most common, and in a small minority there may be a mild lymphocytic pleocytosis. Rarely there may be oligoclonal bands or the 14-3-3 protein (Hernandez Echebarria *et al.* 2000).

A small minority may have a mildly elevated ANA or ESR. Thyroid indices (Henchey *et al.* 1995; Shaw *et al.* 1991) are normal in roughly one-half of all cases; in a little less than one-half of cases, the thyroid-stimulating hormone is mildly elevated, with or without a mild reduction in free T_4 levels, and, in a very small minority, the free T_4 may be elevated. It must be emphasized, however, that when abnormalities are present, they are generally quite mild and could not even begin to account for the clinical symptomatology.

Elevations of antithyroid antibodies, either antithyroid peroxidase or antithyroglobulin, are present in all cases. Although in the vast majority of cases both of these are elevated, exceptions do occur and patients may have elevation of only one; consequently, both should be routinely tested for. Typically the levels of these antibodies are elevated tenfold or greater; importantly, however, there is no correlation between the degree of elevation and the severity of the clinical syndrome.

Course

Although the course of Hashimoto's encephalopathy has not been clearly delineated, it appears to be an episodic disease. The episodes themselves tend to persist for anywhere from weeks up to 6 months, after which there is generally a remission. Repeat episodes can occur; however, it is not clear whether this is the case for all, or even most, patients, nor is it clear how long the intervals are between episodes.

Etiology

Routine neuropathologic studies may be normal (Oide *et al.* 2004), display widespread demyelination (Mahad *et al.* 2005), or, in the majority of cases, reveal a widespread lymphocytic vasculitis involving both arterioles and venules, accompanied by microglial activation and gliosis (Castillo *et al.* 2006; Chong *et al.* 2003; Doherty *et al.* 2002; Duffey *et al.* 2003; Nolte *et al.* 2000).

Although the mechanism underlying these neuropathologic changes has not been clearly elucidated, the association of Hashimoto's encephalopathy with high levels of serum antithyroid antibodies and the excellent response to treatment with steroids strongly suggests an autoimmune etiology, and several studies support this notion. First, serum levels of antibodies to α-enolase, found on endothelial cells, are elevated in Hashimoto's encephalopathy (Yoneda *et al.* 2007). Second, in one patient a serum antineuronal antibody was identified which was found to bind to cortical neurons (Oide *et al.* 2004). Third, IgG found in the CSF was found to bind to both vessels and neurons (Gini *et al.* 2008), and finally, in one study, antithyroid peroxidase antibodies were identified in CSF, which in turn were found to bind to astrocytes (Blanchin *et al.* 2007).

Although it is tempting to conclude that antithyroid antibodies constitute the main portion of the autoimmune onslaught against the brain and its vasculature, certain facts may give one pause. First, although all patients do have elevated serum levels of antithyroid antibodies, there is little correlation between these levels and the overall clinical picture, and second, there is likewise little correlation between falls in these levels and the response to steroid treatment; indeed, in some cases the level of antithyroid antibodies may rise, despite a good response to treatment. It may well be, therefore, that antithyroid antibodies, although ubiquitous in this disorder, constitute but one part of a more heterogeneous autoimmune response.

Differential diagnosis

Of the many disorders capable of causing delirium, discussed in Chapter 6.5, several deserve special attention, including delirium tremens, the serotonin syndrome, viral encephalitis, limbic encephalitis, and Creutzfeldt–Jakob disease, which may closely mimic Hashimoto's encephalopathy (Doherty *et al.* 2002).

Furthermore, it must be kept in mind that the mere finding of antithyroid antibodies in a patient with delirium does not clinch the diagnosis. First, it is possible that the patient in question has a delirium of some other cause and a coincident Hashimoto's thyroiditis. Second, antithyroid antibodies may constitute an incidental finding in otherwise normal individuals, as has been noted in about 3 percent of children and adolescents (Kabelitz *et al.* 2003) and 8 percent of those aged over 60 years (Roti *et al.* 1992): in such individuals, however, the level of antithyroid antibody is generally far below the tenfold or more rises seen in Hashimoto's encephalopathy.

Treatment

Although there are no blind treatment studies, most patients are given 1000 mg/day intravenous methylprednisolone for 5–7 days, followed by 60–100 mg/day oral prednisone in a tapering dose over the following weeks (Castillo *et al.* 2006). In most cases the response is prompt, within days, and most patients do well with tapering. In some cases, however, prolonged treatment with steroids is required, and it appears also that some patients are resistant to steroids and require treatment with immunosuppressants (azathioprine, methotrexate, or cyclophosphamide), plasma exchange, or intravenous immunoglobulins. Although intuitively it makes sense to monitor levels of antithyroid antibodies to gauge treatment response, in practice this is not useful, for, as noted earlier, in some cases levels may actually rise despite a good clinical response.

The general treatment of delirium is discussed in Chapter 6.5; seizures may be treated as described in Chapter 8.3.

19.11 SYDENHAM'S CHOREA

Sydenham's chorea, also known as Saint Vitus' dance, rheumatic chorea, or chorea minor, is one of the major manifestations of rheumatic fever (Anonymous 1992; Bland and Jones 1951,1952), occurring in about one-quarter of all cases (Cardoso *et al.* 1997). In addition to chorea, these patients also display other neuropsychiatric features, most notably obsessions and compulsions.

Because of the widespread treatment of streptococcal pharyngitis with penicillin, Sydenham's chorea is currently uncommon; it occurs more frequently in females than in males, with a ratio of approximately 2:1.

Clinical features

The diagnosis of rheumatic fever is made according to the 'Jones criteria' (Anonymous 1992), which demand the presence of (i) laboratory evidence of recent pharyngeal infection with group A beta-hemolytic streptococci and (ii) either two 'major' criteria or one 'major' criterion plus two 'minor' criteria. Major criteria include carditis, migratory polyarthritis, Sydenham's chorea, subcutaneous nodules, and erythema marginatum. Minor criteria include fever, arthralgia, a prolonged PR interval on electrocardiography, and either an elevated ESR or an elevated C-reactive protein level. Acceptable laboratory evidence of recent pharyngeal infection includes a positive throat culture or an elevated anti-DNAase B, antistreptolysin O, or antihyaluronidase titer; of these three tests, anti-DNAase B tends to remain positive for the longest period of time, in some cases up to 6 months after the pharyngitis.

Although most of the clinical manifestations of rheumatic fever appear about 10 days after the pharyngitis (Taranta 1959), Sydenham's chorea is an exception in that the latency between the pharyngitis and the onset of the chorea is, on average, in the order of 2 or 3 months (Taranta 1959; Taranta and Stollerman 1956). Consequently, it is not unusual to see a case of 'pure' chorea (Feinstein and Spagnuola 1962; Taranta and Stollerman 1956) in which the other manifestations of rheumatic fever have already undergone a full remission, leaving the chorea as the sole manifestation of the disease.

Sydenham's chorea generally presents subacutely, over several weeks (McCullogh 1938; Nausieda *et al.* 1980), in children aged 8–15 years (Cardoso *et al.* 1997; Kilic *et al.* 2007; Thayer 1906); later onsets, however, are certainly possible – as late, indeed, as the ninth decade (Goyal and Williams 1967). The onset itself is typically characterized by symptoms reminiscent of attention-deficit–hyperactivity disorder (Ridel *et al.* 2010), such as restlessness, fidgetiness, irritability, and emotional lability; choreiform movements, if present at this stage, are mild and evanescent (Diefendorf 1912; Gerstley *et al.* 1935). When the chorea does settle in, it is usually generalized but most prominent in the limbs and face; alternatively, one may find hemichorea in up to 30 percent of cases (Abt and Levinson 1916; Nausieda *et al.* 1980; Zomorrodi and Wald 2006); rarely, rather than chorea, there may be a profound weakness, known as 'chorea mollis'.

Neuropsychiatric features are very common during Sydenham's chorea and include obsessions and compulsions, tics, delirium, depression, and, rarely, mania or psychosis. In addition to these features, seizures (Ch'ien *et al.* 1978; Nausieda *et al.* 1980), either complex partial or grand mal, may be seen in a small minority, along with various signs, such as a positive Babinski reflex (Ganji *et al.* 1988).

Obsessions and compulsions occur most notably in Sydenham's chorea (Swedo *et al.* 1989): one prospective study reported them in 70 percent of patients (Asbahr *et al.* 1998), another in 82 percent (Swedo *et al.* 1993). The course of these obsessions and compulsions is of interest. Although they tend to peak in severity along with the worsening of the chorea and to remit before the chorea does, in fact they generally make their appearance before the chorea sets in (Swedo *et al.* 1993); importantly, it appears that in cases of rheumatic fever, it is only those patients who develop Sydenham's chorea who develop obsessions and compulsions; those without chorea remain free of them (Asbahr *et al.* 1998).

Tics, including vocal tics (de Teixeira *et al.* 2009), may also occur during Sydenham's chorea (Creak and Guttmann 1935), but appear to be uncommon.

Delirium is seen in fewer than 10 percent of patients (Nausieda *et al.* 1980), but may be profound (Diefendorf 1912).

Depression appears to be common during Sydenham's chorea: one recent study found it in close to one-half of all cases (Ridel *et al.* 2010).

Mania is a rare manifestation of Sydenham's chorea (Abt and Levinson 1916; MacKenzie 1887) and may present as either pure mania or mixed mania (Bradley 1904; Ebaugh 1926; Gay 1889; Lewis and Minski 1935; Powell 1889; Reaser 1940; Shaskan 1938) or, rarely, be coupled with a depression (Abt and Levinson 1916; Haskell 1914).

Psychosis, with hallucinations and delusions (Teixeira *et al.* 2007), may occur in a small minority (Hammes 1922) and may symptomatically resemble the psychosis seen in schizophrenia (Leys 1946; Putzel 1879).

On average, an episode of Sydenham's chorea gradually remits after 5–9 months, with, however, a wide range, from 1 week up to 2 years or more (Aron *et al.* 1965; Kilic *et al.* 2007; Lessof 1958; Swedo *et al.* 1993); very rarely Sydenham's' chorea may be chronic, even lifelong (Gibb *et al.* 1985). Although there may be some very mild residual chorea, especially evident when the patient is under stress (Lessof 1958; Swedo *et al.* 1993), in most cases recovery is essentially complete. The mortality rate is less than 1 percent (Abt and Levinson 1916; Bussiere and Rhea 1926; Lessof and Bywaters 1956).

Although MRI is generally unrevealing (Zomorrodi and Wald 2006), T2-weighted or FLAIR imaging may, in a very small minority, display increased signal intensity in the caudate and in the cerebral white matter (Castillo *et al.* 1999; Emery and Vieco 1997; Moreau *et al.* 2005; Robertson and Smith 2002).

The EEG is abnormal in the majority of cases, demonstrating either posterior slowing or, in a minority, sharp waves or interictal epileptiform discharges (Ch'ien *et al.* 1978).

Course

The overall course of Sydenham's chorea parallels that of the underlying rheumatic fever, and if there are recurrences of rheumatic fever, the patient is at risk for another episode of Sydenham's chorea. Indeed, it appears that with repeated bouts of rheumatic fever, the likelihood that future episodes of rheumatic fever will be characterized by Sydenham's chorea increases (Aron *et al.* 1965; Bland and Jones 1951; Jones and Bland 1935). Although most recurrences of Sydenham's chorea occur after about 2 years (Nausieda *et al.* 1980; Schwartzman *et al.* 1948), in some cases a very long interval may separate individual episodes, in one case up to 52 years (Gibb and Lees 1989).

Although most cases of recurrent episodes of Sydenham's chorea are associated with laboratory evidence of a recent streptococcal pharyngitis, there are exceptions and some cases may appear in the absence of such indicators (Berrios *et al.* 1985; Korn-Lubetzki *et al.* 2004); whether this implies that other factors may be involved in recurrences or that the preceding infection was of such low intensity as to leave no laboratory trace is not clear.

In addition to relapses of chorea consequent upon recurrent group A beta-hemolytic pharyngitis, patients who have recovered from Sydenham's chorea are also at risk for other sequelae. Females with a history of Sydenham's chorea may develop chorea gravidarum during pregnancy (Beresford and Graham 1950; Wilson and Preece 1932a) and are also at increased risk for developing chorea during treatment with oral contraceptives (Nausieda *et al.* 1983). There is also suggestive evidence that some cases of obsessive–compulsive disorder (Swedo 1994; Swedo *et al.* 1994), Tourette's syndrome (Kerbeshian *et al.* 1990), and even schizophrenia (Casanova *et al.* 1995; Wilcox and Nasrallah 1986,1988) occur as sequelae.

Etiology

Autopsy studies of patients dying during acute Sydenham's chorea have revealed two kinds of damage: vasculitic and encephalitic. Vasculitic changes affecting small vessels (Buchanen 1941; Buchanen *et al.* 1942; Poynton and Holmes 1906; Van Bogaert and Bertrand 1932; Winkelman and Eckel 1932) have been noted in the cerebral cortex and basal ganglia; encephalitic changes (Bradley 1904; Buchanen *et al.* 1942; Colony and Malamud 1956; Coombs 1912; Gordon and Norman 1934; Greenfield and Wolfsohn 1922; Ziegler 1927), with inflammation and neuronal loss, have been found throughout the cerebral cortex and the basal ganglia, without any corresponding vasculitis. Autopsies of patients who died from unrelated causes long after recovering from the chorea have revealed evidence of old endarteritis (Benda 1949) and patchy gliosis and neuronal loss (Lange *et al.* 1976).

These changes occur secondary to an autoimmune attack on the cerebrum and cerebral vasculature, which in turn is triggered by the preceding group A beta-hemolytic streptococcal pharyngitis. Antibasal ganglia antibodies are found in the serum of 95 percent of cases (Church *et al.* 2002), and these antibodies, in turn, have been shown to target both the caudate 2nd subthalamic nuclei (Husby *et al.* 1976). Recent work has refined this picture even further, demonstrating that these antibodies are directed at lysoganglioside found on the cell membrane and intracellular tubulin (Kirvan *et al.* 2007).

Differential diagnosis

Of the various causes of chorea in childhood or adolescence discussed in Chapter 4.4, only toxicity (e.g. from phenytoin) is capable of causing a self-limited episode of chorea, while all the others (e.g. infarction of the basal ganglia, Wilson's disease, benign hereditary chorea) cause chronic chorea.

Treatment

Patients should be given penicillin VK in a dose of 500 mg twice daily for 10 days, regardless of whether there is any clinical evidence of an active pharyngitis, as it is critical to eradicate all of the offending streptococci. Erythromycin may be substituted in patients allergic to penicillin. Various treatments have been proposed, including steroids (e.g. prednisone), valproic acid, carbamazepine, haloperidol, and plasma exchange or intravenous immunoglobulins. Of these, only prednisone has been subjected to a double-blind trial.

Treatment with steroids makes sense in that it targets the underlying mechanism of the disease and could conceivably, if effective, prevent the occurrence of some of the sequelae noted above, such as obsessive–compulsive disorder. In the double-blind study that demonstrated the effectiveness of prednisone, patients were given 2 mg/kg/day for 4 weeks, after which the dose was gradually tapered (Paz *et al.* 2006).

In contrast to steroids, valproic acid, carbamazepine, and haloperidol all represent purely symptomatic treatments. One open study found valproic acid to be superior to carbamazepine, which in turn was superior to haloperidol (Pena *et al.* 2002); another open study, however, found valproic acid and carbamazepine to be equally effective (Genel *et al.* 2002). Valproic acid and carbamazepine are given in customary doses; haloperidol has been given in doses ranging from 1 to 4 mg/day (Axley 1981; Shenker *et al.* 1973).

A reasonable strategy would be to initiate treatment with prednisone and observe the patient. In cases in which there is a response but it is slow and the chorea is of such severity as to threaten the patient, consideration may then be given to adding a symptomatic treatment, beginning with valproic acid. In cases in which there is little or no response to prednisone, symptomatic treatment may also, of course, be considered, but some authors would recommend further etiologic treatment, with consideration given to a course of intravenous methylprednisolone (Cardoso *et al.* 2003) or to treatment with plasma exchange or intravenous immunoglobulins. Indeed, in one controlled, but open, comparative study, both plasma exchange and immunoglobulins provided fuller and more rapid relief than did prednisone (Garvey *et al.* 2005).

Patients should also be treated with injections of penicillin G benzathine, using a dose of 1.2 million units every 3–4 weeks, to prevent recurrences of streptococcal pharyngitis. For children or adolescents, such treatment should probably continue for 5 years or until the age of 21 years, whichever comes later. For adults, the decision must be individualized, with special attention given to those at risk of contracting further streptococcal pharyngitides, such as teachers or pediatricians.

19.12 CHOREA GRAVIDARUM

As a result of the endocrinologic changes occurring during pregnancy, women with certain disorders may develop chorea. This chorea gravidarum is a rare disorder, occurring in anywhere from 0.001 (Wilson and Preece 1932a) to 0.003 percent (Beresford and Graham 1950) of pregnancies. In the past, the most common of the pre-existing disorders was Sydenham's chorea; with the greatly decreased incidence of this disorder, however, other conditions, such as the antiphospholipid syndrome and systemic lupus erythematosus, have become more important.

Clinical features

The onset of the chorea is usually during the first half of pregnancy, and symptoms resolve either toward the end of the third trimester or during the puerperium. In cases occurring in women with a history of Sydenham's chorea, anywhere from 7 (Wilson and Preece 1932b) to 26 percent (Thiele 1935) may also experience hallucinations or delusions; furthermore, mania or delirium may also occur (Wilson and Preece 1932b).

Course

With subsequent pregnancies there may be recurrences, and this appears to be especially common in those with a history of Sydenham's chorea (Wilson and Preece 1932a).

Etiology

Sydenham's chorea greatly increases the chances of chorea gravidarum, which may occur in anywhere from 4 percent (Beresford and Graham 1950) to an astounding 26 percent (Wilson and Preece 1932a) of patients with a history of this condition. Chorea gravidarum has also been noted in association with the anti-phospholipid syndrome (Omdal and Roalso 1992) and systemic lupus erythematosus (Wolf and McBeath 1985). There has been one autopsy case, which revealed neuronal loss and gliosis within the caudate nucleus (Ichikawa *et al.* 1980).

Differential diagnosis

Of the multiple causes of chorea discussed in Chapter 4.4, several may occur coincidentally with pregnancy. For example, women with schizophrenia being treated with antipsychotics may stop taking the antipsychotic upon learning of a pregnancy, only to then go on to develop tardive dyskinesia; another example might be the coincidental onset of a neurodegenerative disease, such as Huntington's disease.

Treatment

Although termination of pregnancy is followed by a resolution of symptoms, chorea gravidarum, given its essentially benign nature, is generally not considered an indication for abortion. If symptomatic treatment is required, consideration may be given to an antipsychotic, such as haloperidol (Patterson 1979) or risperidone; however, in most cases, given the possible teratogenicity of antipsychotics, it is prudent to simply let the disease run its course.

REFERENCES

Abt IA, Levinson A. A study of two-hundred twenty-six cases of chorea. *JAMA* 1916; **67**:1342–7.

Achiron A, Ziv I, Djaldetti R *et al.* Aphasia in multiple sclerosis: clinical and radiologic correlation. *Neurology* 1992; **42**:2195–7.

Adair JC, Digre KB, Swanda RM *et al.* Sneddon's syndrome: a cause of cognitive decline in young adults. *Neuropsychiatry Neuropsychol Behav Neurol* 2001; **14**:197–204.

Ahern GL, O'Connor M, Dalmau J *et al.* Paraneoplastic temporal lobe epilepsy with testicular neoplasm and atypical amnesia. *Neurology* 1994; **44**:1270–4.

Alamowitch S, Graus F, Uchuya M *et al.* Limbic encephalitis and small cell lung cancer: clinical and immunological features. *Brain* 1997; **120**:923–8.

Alexander EL, Malinow K, Lejewski JE *et al.* Primary Sjogren's syndrome with central nervous system disease mimicking multiple sclerosis. *Ann Intern Med* 1986; **104**:323–30.

Alexander EL, Ranzenbach MR, Kumar AJ *et al.* Anti-Ro(SS-A) autoantibodies in central nervous system disease associated with Sjogren's syndrome (CNS-SS): clinical, neuroimaging, and angiographic correlates. *Neurology* 1994; **44**:899–908.

Alexander GE, Provost TT, Stevens MB *et al.* Sjogren syndrome: central nervous system manifestations. *Neurology* 1981; **31**:1391–6.

Allen RK, Sellars RE, Sandstrom PA. A prospective study of 32 patients with neurosarcoidosis. *Sarcoidosis Vasc Diffuse Lung Dis* 2003; **20**:118–25.

Ances BM, Vitaliani R, Taylor RA *et al.* Treatment-responsive limbic encephalitis identified by neuronal antibodies: MRI and PET correlates. *Brain* 2005; **128**:1764–77.

Anderson NE, Rosenblum MK, Graus F *et al.* Autoantibodies in paraneoplastic syndromes associated with small-cell lung cancer. *Neurology* 1988; **38**:1391–8.

Anonymous Guidelines for the diagnosis of rheumatic fever: Jones criteria, 1992 update. *JAMA* 1992; **268**:2069–73.

Antoine JC, Honnorat J, Anterion CT *et al.* Limbic encephalitis and immunological perturbations in two patients with thymoma. *J Neurol Neurosurg Psychiatry* 1995; **58**:706–10.

Appenzeller S, Cendes F, Costallat LT. Epileptic seizures in systemic lupus erythematosus. *Neurology* 2004; **63**:1808–12.

Aron AM, Freeman JM, Carter S. The natural history of Sydenham's chorea. *Am J Med* 1965; **38**:83–95.

Asbahr FR, Negrao AB, Gentil V *et al.* Obsessive-compulsive and related symptoms in children and adolescents with and without chorea: a prospective 6-month study. *Am J Psychiatry* 1998; **155**:1122–4.

Asherson RA, Khamashta MA, Ordis-Ros J *et al.* The 'primary' antiphospholipid syndrome: major clinical and serological features. *Medicine* 1989; **68**:366–74.

Atkins CJ, Kondon JJ, Quismorio FP *et al.* The choroid plexus in systemic lupus erythematosus. *Ann Intern Med* 1972; **76**: 65–72.

Aubart-Cohen F, Klein I, Alexandra JF *et al.* Long-term outcome in Susac syndrome. *Medicine* 2007; **96**:93–102.

Axley J. Rheumatic chorea controlled with haloperidol. *J Pediatr* 1981; **72**:1216–17.

Bak TH, Antoun N, Balan KK *et al.* Memory lost, memory regained: neuropsychological findings and neuroimaging in two cases of paraneoplastic limbic encephalitis with radically different outcomes. *J Neurol Neurosurg Psychiatry* 2001; **71**:40–7.

Bakheit AMO, Kennedy PGE, Behan PO. Paraneoplastic limbic encephalitis: clinico-pathological correlation. *J Neurol Neurosurg Psychiatry* 1990; **53**:1084–8.

Barile-Fabris L, Ariza-Andraca R, Olguin-Ortega L *et al.* Controlled clinical trial of cyclophosphamide versus IV methylprednisolone in severe neurological manifestations in systemic lupus erythematosus. *Ann Rheum Dis* 2005; **64**:620–5.

Barnes D, Hughes RA, Morris RW *et al.* Randomised trial of oral and intravenous methylprednisolone in acute relapses of multiple sclerosis. *Lancet* 1997; **349**:902–6.

Barnett MH, Prineas JW. Relapsing and remitting multiple sclerosis: pathology of the newly forming lesions. *Ann Neurol* 2004; **55**: 458–60.

Bataller L, Kleopa KA, Wu GF *et al.* Autonomic limbic encephalitis in 39 patients: immunophenotypes and outcomes. *J Neurol Neurosurg Psychiatry* 2007; **78**:381–5.

Benda CA. Chronic rheumatic encephalitis, torsion dystonia, and Hallervorden–Spatz disease. *Arch Neurol Psychiatry* 1949; **61**:137–63.

Beresford OD, Graham AM. Chorea gravidarum. *J Obs Gynaecol Br Emp* 1950; **57**:616–25.

Berg D, Supprian T, Thomas J *et al.* Lesion patterns with multiple sclerosis and depression. *Mult Scler* 2000; **6**:156–62.

Bergin JD. Rapidly progressive dementia in disseminated sclerosis. *J Neurol Neurosurg Psychiatry* 1957; **20**:285–92.

Berrios X, Quesney F, Morales A *et al.* Are all recurrences of 'pure' Sydenham chorea true recurrence of acute rheumatic fever? *J Pediatr* 1985; **107**:867–72.

Bihan H, Christozova V, Dumas JL *et al.* Sarcoidosis: clinical, hormonal, and magnetic resonance imaging (MRI) manifestations of hypothalamic-pituitary disease in 9 patients and review of the literature. *Medicine* 2007; **86**:259–68.

Blanchin S, Coffin C, Viader F *et al.* Anti-thyroperoxidase antibodies from patients with Hashimoto's encephalopathy bind to cerebellar astrocytes. *J Neuroimmunol* 2007; **192**:13–20.

Bland EF, Jones TD. Rheumatic fever and rheumatic heart disease: a twenty-year report on 1000 patients followed since childhood. *Circulation* 1951; **4**:836–43.

Bland EF, Jones TD. The natural history of rheumatic fever: a 20 year perspective. *Ann Intern Med* 1952; **37**:1006–26.

Bluestein HG, Williams GW, Stebeg AD. Cerebrospinal fluid antibodies to neuronal cells: association with neuropsychiatric manifestations of systemic lupus erythematosus. *Am J Med* 1981; **70**:240–6.

Boesch SM, Plorer AL, Auer AJ *et al.* The natural course of Sneddon's syndrome: clinical and magnetic resonance imaging findings in a prospective six year observation study. *J Neurol Neurosurg Psychiatry* 2003; **74**:542–4.

Bohnen NIJ, Parnell KJ, Harper CM. Reversible MRI findings in a patient with Hashimoto's encephalopathy. *Neurology* 1997; **49**:246–7.

Bonnan M, Valentino R, Olindo S *et al.* Plasma exchange in severe spinal attacks associated with neuromyelitis optica spectrum disorder. *Mult Scler* 2009; **15**:487–92.

Bostantjopoulou S, Zafiriou D, Katsarou Z *et al.* Hashimoto's encephalopathy: clinical and laboratory findings. *Funct Neurol* 1996; **11**:247–51.

Bradley IA. A case of chorea insaniens with autopsy. *Am J Insanity* 1904; **60**:777–85.

Brain L, Jellinek EH, Ball K. Hashimoto's disease and encephalopathy. *Lancet* 1966; **2**:512–14.

Brey RL, Hart RG, Sherman DG *et al.* Antiphospholipid antibodies and cerebral ischemia in young people. *Neurology* 1990; **40**:190–6.

Brey RL, Holliday SL, Saklad AR *et al.* Neuropsychiatric syndromes in lupus. Prevalence using standardized definitions. *Neurology* 2002; **58**:1214–20.

Brierly JB, Corsellis JAN, Hierons R *et al.* Subacute encephalitis of later adult life. *Brain* 1960; **83**:357–68.

Brown MM, Thompson AJ, Wedzicha JA *et al.* Sarcoidosis presenting with stroke. *Stroke* 1989; **20**:400–5.

Buchanen DN. Pathologic change in chorea. *Am J Dis Child* 1941; **62**:443–4.

Buchanen DN, Walker AK, Case TJ. The pathogenesis of chorea. *J Pediatr* 1942; **20**:555–75.

Burton GE, Bullard DE, Walther PJ *et al.* Paraneoplastic limbic encephalopathy with testicular carcinoma. A reversible neurologic syndrome. *Cancer* 1988; **62**:2248–51.

Bussiere HC, Rhea LJ. Acute rheumatic fever and chorea in children. *Can Med Assoc J* 1926; **16**:35–40.

Byrne T, Mason WP, Posner JB *et al.* Spontaneous neurologic improvement in anti-Hu associated encephalomyelitis. *J Neurol Neurosurg Psychiatry* 1997; **62**:276–8.

Calabrese M, Agosta F, Rinaldi F *et al.* Cortical lesions and atrophy are associated with cognitive impairment in relapsing–remitting multiple sclerosis. *Arch Neurol* 2009; **66**:1144–50.

Calabrese M, Rocca MA, Atzori M *et al.* A 3-year magnetic resonance imaging study of cortical lesions in relapse-onset multiple sclerosis. *Ann Neurol* 2010; **67**:376–83.

Camp WA, Frierson JG. Sarcoidosis of the central nervous system. *Arch Neurol* 1962; **7**:432–41.

Cardoso F, Eduardo C, Silva AP *et al.* Chorea in 50 consecutive patients with rheumatic fever. *Mov Disord* 1997; **12**:701–3.

Cardoso F, Maia D, Cunningham MC *et al.* Treatment of Sydenham chorea with corticosteroids. *Mov Disord* 2003; **18**:1374–7.

Casanova MF, Crapanzano KA, Mannheim G *et al.* Sydenham's chorea and schizophrenia: a case report. *Schizophr Res* 1995; **16**:73–6.

Caselli RJ, Scheithauer BW, Bowles CA *et al.* The treatable dementia of Sjögren's syndrome. *Ann Neurol* 1991; **30**:98–101.

Caselli RJ, Scheithauer BW, O'Duffy JD *et al.* Chronic inflammatory meningoencephalitis should not be mistaken for Alzheimer's disease. *Mayo Clin Proc* 1993; **68**:846–53.

Castillo M, Kwock L, Arbelaez A. Sydenham's chorea: MRI and proton spectroscopy. *Neuroradiology* 1999; **41**:943–5.

Castillo P, Woodruff B, Caselli R *et al.* Steroid-responsive encephalopathy associated with autoimmune thyroiditis. *Arch Neurol* 2006; **63**:197–202.

Chancellor AM, Cull RE, Kilpatrick DC *et al.* Neurological disease associated with anticardiolipin antibodies in patients without systemic lupus erythematosus: clinical and immunological features. *J Neurol* 1991; **238**:401–7.

Chapelon C, Ziza JM, Piette JC *et al.* Neurosarcoidosis: signs, course and treatment in 36 confirmed cases. *Medicine* 1990; **69**:261–76.

Ch'ien LT, Economides AH, Lemmi H. Sydenham's chorea and seizures: clinical and electroencephalographic studies. *Arch Neurol* 1978; **35**:382–5.

Chong JY, Rowland LP, Utiger RD. Hashimotos's encephalopathy: syndrome or myth? *Arch Neurol* 2003; **60**:164–71.

Church AJ, Cardoso F, Dale RC *et al.* Anti-basal ganglia antibodies in acute and persistent Sydenham's chorea. *Neurology* 2002; **59**:227–31.

Coates T, Slavotinek JP, Rischmueller M *et al.* Cerebral white matter lesions in primary Sjögren's syndrome: a controlled study. *J Rheumatol* 1999; **26**:1301–5.

Cohen JA, Cutter GR, Fischer JS *et al.* Benefit of interferon beta-1a on MSFC progression in secondary progressive MS. *Neurology* 2002; **59**:679–87.

Colony HS, Malamud N. Sydenham's chorea: a clinicopathologic study. *Neurology* 1956; **6**:672–6.

Comi G, Rovaris M, Falautano M *et al.* A multiparametric study of frontal lobe dementia in multiple sclerosis. *J Neurol Sci* 1999; **171**:135–44.

Confavreux C, Hutchinson M, Hours MM *et al.* Rate of pregnancy-related relapse in multiple sclerosis. *N Engl J Med* 1998; **339**:285–91.

Coombs C. Cerebral rheumatism. *Practitioner* 1912; **88**:99–106.

Cordingly G, Navarro C, Brust JCM *et al.* Sarcoidosis presenting as senile dementia. *Neurology* 1981; **31**:1148–51.

Corsellis JAN, Goldberg GJ, Norton AR. Limbic encephalitis and its association with carcinoma. *Brain* 1968; **91**:481–96.

Cottrell SS, Wilson SAK. The affective symptomatology of disseminated sclerosis: a study of 100 cases. *J Neurol Psychopathology* 1926; **7**:1–30.

Coull BM, Bourdette DM, Goodnight SH *et al.* Multiple cerebral infarctions and dementia associated with anticardiolipin antibodies. *Stroke* 1987; **18**:1107–12.

Creak M, Guttmann E. Chorea, tics, compulsive utterances. *J Ment Sci* 1935; **81**:834–9.

Croteau d, Owainati A, Dalmau J *et al.* Response to cancer therapy in a patient with paraneoplastic choreiform disorder. *Neurology* 2001; **57**:719–22.

Cutter NC, Scott DD, Johnson JC *et al.* Gabapentin effect on spasticity in multiple sclerosis: a placebo-controlled, randomized trial. *Arch Phys Med Rehabil* 2000; **81**:164–9.

Dalmau J, Graus F, Rosenblum MK *et al.* Anti-Hu-associated paraneoplastic encephalomyelitis/sensory neuropathy. A clinical study of 71 patients. *Medicine* 1992; **71**:59–72.

Dalmau J, Tuzun E, Wu H-Y *et al.* Paraneoplastic anti-N-methyl-D-aspartate receptor encephalitis associated with ovarian teratoma. *Ann Neurol* 2007; **61**:25–36.

Dalmau J, Gleichman AJ, Hughes EG *et al.* Anti-NMDA-receptor encephalitis: case series and analysis of the effects of antibodies. *Lancet Neurol* 2008; **7**:1091–8.

Dean G. Annual incidence, prevalence and mortality in white South African-born and in white immigrants to South Africa. *BMJ* 1967; **2**:724–30.

Delalande S, de Seze J, Fauchais AL *et al.* Neurologic manifestations in primary Sjögren syndrome: a study of 82 patients. *Medicine* 2004; **83**:280–91.

de la Monte SM, Hutchins GM, Gupta PK. Polymorphous meningitis with atypical mononuclear cells in Sjögren's syndrome. *Ann Neurol* 1983; **14**:455–61.

Delaney P. Neurologic manifestations in sarcoidosis. *Ann Intern Med* 1977; **87**:336–45.

DeLorenze GN, Munger KL, Lennette ET *et al.* Epstein–Barr virus and multiple sclerosis. Evidence of association from a prospective study with long-term follow-up. *Arch Neurol* 2006; **63**: 839–44.

de Teixeira AL, Cardoso F, Maia DP *et al.* Frequency and significance of vocalizations in Sydenham's chorea. *Parkinsonism Relat Disord* 2009; **15**:62–3.

Devere TD, Trotter JL, Cross AH. Acute aphasia in multiple sclerosis. *Arch Neurol* 2000; **57**:1207–9.

Devinsky O, Petito CK, Alonso DR. Clinical and neuropathological findings in systemic lupus erythematosus: the role of vasculitis, heart emboli, and thrombotic thrombocytopenic purpura. *Ann Neurol* 1988; **23**:380–4.

D'hooghe MB, Nagels G, Uitdehaag BM. Long-term effects of childbirth in MS. *J Neurol Neurosurg Psychiatry* 2010; **81**:38–41.

Diefendorf AR. Mental symptoms of acute chorea. *J Nerv Ment Dis* 1912; **39**:161–72.

Doherty CP, Schlossmacher M, Torres N *et al.* Hashimoto's encephalopathy mimicking Creutzfeldt–Jakob disease: brain biopsy findings. *J Neurol Neurosurg Psychiatry* 2002; **73**:601–2.

Donaldson IM, Espiner EA. Disseminated lupus erythematosus presenting as chorea gravidarum. *Arch Neurol* 1971; **25**:240–4.

Douglas AC, Maloney AFJ. Sarcoidosis of the central nervous system. *J Neurol Neurosurg Psychiatry* 1973; **36**:1024–33.

Duffey P, Yee S, Rein IN *et al.* Hashimoto's encephalopathy: postmortem findings after fatal status epilepticus. *Neurology* 2003; **61**:1124–6.

Durelli L, Verdun E, Barbero P *et al.* Every-other-day interferon beta-1b versus once-weekly interferon beta-1a for multiple sclerosis: results of a 2-year prospective randomized multicentre study (INCOMIN). *Lancet* 2002; **359**:1453–60.

Ebaugh FG. Neuropsychiatric aspects of chorea in children. *JAMA* 1926; **87**:1083–8.

Ebers GC, Bulman DE, Sodovnik AD *et al.* Population-based studies of multiple sclerosis in twins. *N Engl J Med* 1986; **315**:1638–42.

Ebers GC, Sadovnik AD, Risch NJ. A genetic basis for familial aggregation in multiple sclerosis. Canadian Collaborative Study Group. *Nature* 1995; **377**:150–1.

Eker A, Saka E, Dalmau J *et al.* Testicular teratoma and anti-N-methyl-D-aspartate receptor-associated encephalitis. *J Neurol Neurosurg Psychiatry* 2008; **79**:1082–3.

Ellis SG, Verity MA. Central nervous system involvement in systemic lupus erythematosus: a review of neuropathologic findings in 57 cases. *Semin Arthritis Rheum* 1979; **8**:212–21.

Emery ES, Vieco PT. Sydenham chorea: magnetic resonance imaging reveals permanent basal ganglia injury. *Neurology* 1997; **48**:531–3.

European Study Group on Interferon Beta-1b in Secondary Progressive MS. Placebo-controlled multicentre randomized trial of interferon beta-1b in treatment of secondary progressive multiple sclerosis. *Lancet* 1998; 352:1491–7

Fassbender K, Schmidt R, Mofsner R *et al.* Mood disorders and dysfunction of the hypothalamic–pituitary-adrenal axis in multiple sclerosis. *Arch Neurol* 1998; **55**:66–72.

Feinglass EJ, Arnett FC, Dorsch CA. Neuropsychiatric manifestations of systemic lupus erythematosus: diagnosis, clinical spectrum, and relationship to other features of the disease. *Medicine* 1976; **55**:323–39.

Feinstein A, du Boulay G, Ron MA. Psychotic illness in multiple sclerosis. A clinical and magnetic resonance imaging study. *Br J Psychiatry* 1992; **161**:680–5.

Feinstein A, Feinstein K, Gray T *et al.* Prevalence and neurobehavioral correlates of pathological laughing and crying in multiple sclerosis. *Arch Neurol* 1997; **54**:1116–21.

Feinstein A, Roy P, Lobaugh N *et al.* Structural brain abnormalities in multiple sclerosis patients with major depression. *Neurology* 2004; **62**:586–90.

Feinstein AR, Spagnuola M. The clinical pattern of acute rheumatic fever. *Medicine* 1962; **41**:279–305.

Fermaglich J, Streib E, Auth T. Chorea associated with systemic lupus erythematosus. *Arch Neurol* 1973; **28**:276–7.

Fontaine B, Seilhean D, Tourbah A *et al.* Dementia in two histologically confirmed cases of multiple sclerosis: one case with isolated dementia and one case associated with psychiatric symptoms. *J Neurol Neurosurg Psychiatry* 1994; **57**:353–9.

Frances C, Papo T, Wechsler B *et al.* Sneddon syndrome with or without antiphospholipid antibodies. A comparative study in 46 patients. *Medicine* 1999; **78**:209–19.

Franklin GM, Heaton RK, Nelson LM *et al.* Correlations of neuropsychological and MRI findings in chronic-progressive multiple sclerosis. *Neurology* 1988; **38**:1826–9.

Franklin GM, Nelson LM, Filley CM *et al.* Cognitive loss in multiple sclerosis: case reports and review of the literature. *Arch Neurol* 1989; **46**:162–7.

Futrell N, Schultz LR, Millikan C. Central nervous system disease in patients with systemic lupus erythematosus. *Neurology* 1992; **42**:1649–57.

Galluzzi S, Geroldi C, Zanetti O *et al.* Hashimoto's encephalopathy in the elderly: relationship to cognitive impairment. *J Geriatr Psychiatry Neurol* 2002; **15**:175–9.

Ganji S, Duncan MC, Frazier E. Sydenham's chorea: clinical, CT scan, and evoked potential studies. *Clin Electroencephalogr* 1988; **19**:114–22.

Ganz VH, Gurland BJ, Deming WE *et al.* The study of the psychiatric symptoms of systemic lupus erythematosus: a biometric study. *Psychosom Med* 1972; **34**:207–20.

Garvey MA, Snider LA, Leitman SF *et al.* Treatment of Sydenham's chorea with intravenous immunoglobulin, plasma exchange or prednisone. *J Child Neurol* 2005; **20**:424–9.

Gay W. Chorea insaniens. *Brain* 1889; **12**:150–6.

Genel F, Arslanoglu S, Uran N *et al.* Sydenham's chorea: clinical findings and comparison of the efficacies of sodium valproate and carbamazepine regimens. *Brain Dev* 2002; **24**:73–6.

Gerstley JR, Wile SA, Falstein E *et al.* Chorea: is it a manifestation of rheumatic fever? *J Pediatr* 1935; **6**:42–50.

Geschwind MD, Tan KM, Lennon VA *et al.* Voltage-gated potassium channel autoimmunity mimicking Creutzfeldt–Jakob disease. *Arch Neurol* 2008; **65**:1341–6.

Geurts JJ, Bo L, Pouwels PJ *et al.* Cortical lesions in multiple sclerosis: combined postmortem MR imaging and histopathology. *AJNR* 2005; **26**:572–7.

Ghika-Schmid F, Ghika J, Regli F *et al.* Hashimoto's myoclonic encephalopathy: an underdiagnosed treatable condition? *Mov Disord* 1996; **11**:555–62.

Gibb WR, Lees AJ, Scadding JW. Persistent rheumatic chorea. *Neurology* 1985; **35**:101–2.

Gibb WRG, Lees AJ. Tendency to late recurrence following rheumatic chorea. *Neurology* 1989; **39**:999.

Gini B, Lovato L, Cianti R *et al.* Novel autoantigens recognized by CSF IgG from Hashimoto's encephalitis revealed by a proteonomic approach. *J Neuroimmunol* 2008; **196**:153–8.

Ginsberg L, Compston DA. Acute encephalopathy: diagnosis and outcome in patients at a regional neurological unit. *QJ Med* 1994; **87**:169–80.

Glaser G, Pincus JH. Limbic encephalitis. *J Nerv Ment Dis* 1969; **149**:59–67.

Gordon RG, Norman RM. A case of acute toxic chorea. *J Neurol Psychopathology* 1934; **15**:313–19.

Goyal BK, Williams BJ. Sydenham's chorea in an octogenarian. *Gerontologica Clinica* 1967; **9**:176–81.

Graham DI, Lantos PL. *Greenfield's Neuropathology*, 6th edn. London: Arnold, 1996.

Graus F, Keime-Guibert F, Rene R *et al.* Anti-Hu-associated paraneoplastic encephalomyelitis: analysis of 200 patients. *Brain* 2001; **124**:1138–48.

Graus F, Saiz A, Lai M *et al.* Neuronal surface antigen antibodies in limbic encephalitis: clinical–immunologic associations. *Neurology* 2008; **71**:930–6.

Greenfield JG, King LS. Observations on the histopathology of the cerebral lesions in disseminated sclerosis. *Brain* 1936; **59**:445–58.

Greenfield JG, Wolfsohn JM. The pathology of Sydenham's chorea. *Lancet* 1922; **2**:603–6.

Gultekin SH, Rosenfeld MR, Voltz R *et al.* Paraneoplastic limbic encephalitis: neurological symptoms, immunological findings and tumour association in 50 patients. *Brain* 2000; **123**:1481–94.

Guze SB. The occurrence of psychiatric illness in systemic lupus erythematosus. *Am J Psychiatry* 1967; **123**:1562–70.

Hammes EM. Psychoses associated with Sydenham's chorea. *JAMA* 1922; **79**:804–7.

Hanley JG, Walsh NMG, Sangalang V. Brain pathology in systemic lupus erythematosus. *J Rheumatol* 1992; **19**:732–41.

Hannon PM, Kuo SH, Strutt AM *et al. Clin Neurol Neurosurg* 2010; **112**:544–7.

Hartung HP, Gonsetta R, Konig N *et al.* Mitoxantrone in progressive multiple sclerosis: a placebo-controlled, double-blind, randomised, multicentre trial. *Lancet* 2002; **360**:2018–25.

Hashimoto H. Zur kenntnis der lymphomatosan veranderung der schilddruse (struma lymphomatosa). *Arch Klin Chir* 1912; **97**:219–48.

Haskell RH. Mental disturbance associated with acute articular rheumatism. *Am J Insanity* 1914; **71**:361–81.

Heiskala H, Somer H, Kovanen J *et al.* Microangiopathy with encephalopathy, hearing loss and arteriolar occlusions: two new cases. *J Neurol Sci* 1988; **8**:239–50.

Henchey R, Cibula J, Helveston W *et al.* Electroencephalographic findings in Hashimoto's encephalopathy. *Neurology* 1995; **45**:977–81.

Henson RA, Hofman HL, Ulrich H. Encephalomyelitis with carcinoma. *Brain* 1965; **88**:449–64.

Hernandez Echebarria LE, Saiz A, Graus F *et al.* Detection of 14-3-3 protein in the CSF of a patient with Hashimoto's encephalopathy. *Neurology* 2000; **54**:1539–40.

Herring AB, Urich H. Sarcoidosis of the central nervous system. *J Neurol Sci* 1969; **9**:405–22.

Hilton DA, Footitt D. Neuropathological findings in Sneddon's syndrome. *Neurology* 2003; **60**:1181–2.

Hinson SR, Pittock SJ, Lucchinetti CF *et al.* Pathogenic potential of IgG binding to water channel extracellular domain in neuromyelitis optica. *Neurology* 2007; **69**:2221–31.

Hook O. Sarcoidosis with involvement of the nervous system. *Arch Neurol Psychiatry* 1954; **71**:554–75.

Hotopf MH, Pollock S, Lishman WA. An unusual presentation of multiple sclerosis. *Psychol Med* 1994; **24**:525–8.

Hughes RAC, Cameron JS, Hall SM *et al.* Multiple mononeuropathy as the initial presentation of SLE. *J Neurol* 1982; **228**:239–47.

Hughson MD, McCarty GA, Sholer CM *et al.* Thrombotic cerebral angiopathy in patients with the antiphospholipid syndrome. *Mod Pathol* 1993; **6**:644–53.

Hugo FJ, Halland AM, Spangenberg JJ *et al.* DSM-III-R classification of psychiatric symptoms in systemic lupus erythematosus. *Psychosomatics* 1996; **37**:262–9.

Husby G, van de Rijn I, Zabriskie JB *et al.* Antibodies reacting with cytoplasm of subthalamic and caudate neurons in chorea and acute rheumatic fever. *J Exp Med* 1976; **144**:1094–110.

Ichikawa K, Kim RC, Givelbar H *et al.* Chorea gravidarum. Report of a fatal case with neuropathological observations. *Arch Neurol* 1980; **37**:429–32.

Ingenito GG, Berger JR, David NJ *et al.* Limbic encephalitis associated with benign thymoma. *Neurology* 1990; **40**:382.

Iranzo A, Graus F, Clover L *et al.* Rapid eye movement sleep disorder and potassium channel antibody-associated limbic encephalitis. *Ann Neurol* 2006; **59**:178–82.

Isshi K, Hirohata S. Differential roles of the anti-ribosomal P antibody and antineuronal antibody in the pathogenesis of central nervous system involvement in systemic lupus erythematosus. *Arthritis Rheum* 1998; **41**:1819–27.

Ito S, Mori M, Makino T *et al.* "Cloud-like enhancement" is a magnetic resonance imaging abnormality specific to neuromyelitis optica. *Ann Neurol* 2009; **66**:425–8.

Jacob A, Weinshenker BG, Violich I *et al.* Treatment of neuromyelitis optica with rituximab: retrospective analysis of 25 patients. *Arch Neurol* 2008; **65**:1443–8.

Jarius S, Aboul-Enein F, Waters P *et al.* Antibody to aquaporin-4 in the long-term course of neuromyelitis optica. *Brain* 2008; **131**:3072–80.

Jarius S, Meumayer B, Wandinger KP *et al.* Anti-endothelial serum antibodies in a patient with Susac's syndrome. *J Neurol Sci* 2009; **285**:259–61.

Jefferson M. Sarcoidosis of the central nervous system. *Brain* 1957; **80**:540–56.

Jennekens-Schinkel A, Sanders EACM. Decline of cognition in multiple sclerosis: dissociable artifacts. *J Neurol Neurosurg Psychiatry* 1986; **49**:1354–60.

Joffe RT, Lippert GP, Gray TA *et al.* Mood disorder and multiple sclerosis. *Arch Neurol* 1987; **44**:376–8.

Johnson RT, Richardson EP. The neurological manifestations of systemic lupus erythematosus: a clinical–pathological study of 24 cases and review of the literature. *Medicine* 1968; **47**:337–69.

Jones TD, Bland EF. Clinical significance of chorea as a manifestation of rheumatic fever: a study in prognosis. *JAMA* 1935; **105**:571–7.

Joseph FG, Scolding NJ. Neurosarcoidosis: a study of 30 new cases. *J Neurol Neurosurg Psychiatry* 2009; **80**:297–304.

Kabelitz M, Liesenkotter KP, Stach B *et al.* The prevalence of anti-thyroid antibodies and autoimmune thyroiditis in children and adolescents in an iodine replete area. *Eur J Endocrinol* 2003; **148**:301–7.

Kalashnikova LA, Noasonov EL, Kushekbaeva AE *et al.* Anticardiolipin antibodies in Sneddon's syndrome. *Neurology* 1990; **40**:464–7.

Kawashima N, Shindo R, Kohno M. Primary Sjögren's syndrome with subcortical dementia. *Intern Med* 1993; **32**:561–4.

Keane JR. Internuclear ophthalmoplegia. Unusual causes in 114 of 410 patients. *Arch Neurol* 2005; **62**:714–17.

Kelly MC, Denburg JA. Cerebrospinal fluid immunoglobulins and neuronal antibodies in neuropsychiatric systemic lupus erythematosus and related conditions. *J Rheumatol* 1987; **14**:740–4.

Kerbeshian J, Burd L, Pettit R. A possible post-streptococcal movement disorder with chorea and tics. *Dev Med Child Neurol* 1990; **32**:642–4.

Ketelslegers IA, Neuteboom RF, Boon M *et al.* A comparison of MRI criteria for diagnosing pediatric ADEM and MS. *Neurology* 2010; **74**:1412–15.

Kilic A, Unuvar E, Tatli B *et al.* Neurologic and cardiac findings in children with Sydenham chorea. *Pediatr Neurol* 2007; **36**: 159–64.

Kinnman J, Link H. Intrathecal production of oligoclonal IgM and IgG in CNS sarcoidosis. *Acta Neurol Scand* 1984; **69**:97–106.

Kirk A, Kertesz A, Polk MJ. Dementia with leukoencephalopathy in systemic lupus erythematosus. *Can J Neurol Sci* 1991; **18**:344–8.

Kirvan CA, Cox CJ, Swedo SE *et al.* Tubulin is a neuronal target of autoantibodies in Sydenham's chorea. *J Immunol* 2007; **178**:7412–21.

Koch-Henriksen N, Sorensen PS, Christensen T *et al.* A randomized study of two interferon-beta treatments in relapsing–remitting multiple scelerosis. *Neurology* 2006; **66**:1056–60.

Koenig H. Dementia associated with the benign form of multiple sclerosis. *Trans Am Neurol Assoc* 1968; **93**:227–31.

Korn-Lubetzki I, Brand A, Steiner I. Recurrence of Sydenham chorea: implications for pathogenesis. *Arch Neurol* 2004; **61**: 1261–4.

Krumholz A, Stern BJ, Stern EG. Clinical implications of seizures in neurosarcoidosis. *Arch Neurol* 1991; **48**:842–4.

Krupp LB, Coyle PK, Doescher C *et al.* Fatigue therapy in multiple sclerosis: results of a double-blind, randomized, parallel trial of amantadine, pemoline, and placebo. *Neurology* 1995; **45**: 1956–61.

Krupp LB, Christodoulou C, Melville P *et al.* Donepezil improved memory in multiple sclerosis in a randomized clinical trial. *Neurology* 2004; **63**:1579–85.

Kurita A, Hasunuma T, Mochio S *et al.* A young case with multi-infarct dementia associated with lupus anticoagulant. *Intern Med* 1994; **33**:373–5.

Kutzelnigg A, Lucchinetti CF, Stadelmann C *et al.* Cortical demyelination and diffuse white matter injury in multiple sclerosis. *Brain* 2005; **128**:2705–12.

Lacour A, de Seze J, Revenco E *et al.* Acute aphasia in multiple sclerosis. *Neurology* 2004; **62**:974–7.

Lai M, Hughes EG, Peng X *et al.* AMPA receptor antibodies in limbic encephalitis alter synaptic receptor location. *Ann Neurol* 2009; **65**:424–34.

Lange H, Thorner G, Hopf A *et al.* Morphometric studies of the neuropathological changes in choreatic disease. *J Neurol Sci* 1976; **28**:401–25.

Lanham JG, Brown MM, Hughes GRV. Cerebral systemic lupus erythematosus presenting with catatonia. *Postgrad Med J* 1985; **61**:329–30.

Lawn ND, Westmoreland BF, Kiely MJ *et al.* Clinical, magnetic resonance imaging, and electroencephalographic findings in paraneoplastic limbic encephalitis. *Mayo Clinic Proc* 2003; **78**:1363–8.

Lessof MH. Sydenham's chorea. *Guys Hospital Reports* 1958; **107**:185–206.

Lessof MH, Bywaters EGL. The duration of chorea. *BMJ* 1956; **1**: 1520–3.

Levine SR, Deegan MJ, Futrell N *et al.* Cerebrovascular and neurologic disease associated with antiphospholipid antibodies: 48 cases. *Neurology* 1990; **40**:1181–9.

Levine SR, Brey RL, Sawaya KL *et al.* Recurrent stroke and thrombo-occlusive events in the antiphospholipid syndrome. *Ann Neurol* 1995; **38**:119–24.

Lewis A, Minski L. Chorea and psychosis. *Lancet* 1935; **1**:536–8.

Leys D. Rheumatic encephalopathy. *Edinburgh Med J* 1946; **53**:444–9.

Lief VF, Silverman T. Psychosis associated with lupus erythematosus disseminatus. *Arch Gen Psychiatry* 1960; **3**:608–11.

Lim L, Ron MA, Ormerod IEC *et al.* Psychiatric and neurological manifestations in systemic lupus erythematosus. *Q J Med* 1988; **66**:27–38.

Lucchinetti CF, Mandler RN, McGavern D *et al.* A role for humoral mechanisms in the pathogenesis of Devic's neuromyelitis optica. *Brain* 2002; **125**:1450–61.

Lucchinetti CF, Gavrilova RH, Metz I *et al.* Clinical and radiographic spectrum of pathologically confirmed tumefactive multiple sclerosis. *Brain* 2008; **131**:1759–75.

Luke RA, Stern BJ, Krumholz A *et al.* Neurosarcoidosis: the long-term clinical course. *Neurology* 1987; **37**:461–3.

Mac DS, Pardo MP. Systemic lupus erythematosus and catatonia: a case report. *J Clin Psychiatry* 1983; **44**:155–6.

McCombe PA, McLeod JG, Pollard JG *et al.* Peripheral sensorimotor and autonomic neuropathy associated with systemic lupus erythematosus. *Brain* 1987; **110**:533–50.

McCullogh H. Encephalitis rheumatica: chorea minor of Sydenham. *J Pediatr* 1938; **13**:741–7.

McLean BN, Miller D, Thompson D. Oligoclonal banding of IgG in CSF; blood–brain barrier function, and MRI findings in patients with sarcoidosis, systemic lupus erythematosus, and Behçet's disease involving the nervous system. *J Neurol Neurosurg Psychiatry* 1995; **58**:548–54.

MacKenzie S. Report on inquiry: II. Chorea. *BMJ* 1887; **1**:425–36.

MacNeil A, Grennan DM, Ward D *et al.* Psychiatric problems in systemic lupus erythematosus. *Br J Psychiatry* 1976; **128**:442–5.

Mahad DJ, Staugaitis S, Ruggieri P *et al.* Steroid-responsive encephalopathy associated with autoimmune thyroiditis and primary CNS demyelination. *J Neurol Sci* 2005; **228**:3–5.

Mainero C, Benner T, Radding A *et al.* *In vivo* imaging of cortical pathology in multiple sclerosis using ultra-high field MRI. *Neurology* 2009; **73**:941–8.

Malamud N, Saver G. Neuropathologic findings in disseminated lupus erythematosus. *Arch Neurol Psychiatry* 1954; **71**:723–31.

Manz HJ. Pathobiology of neurosarcoidosis and clinicopathologic correlation. *Can J Neurol Sci* 1983; **10**:50–5.

Martinelli V, Rocca MA, Annovazzi P *et al.* A short-term randomized MRI study of high-dose oral vs intravenous methylprednisolone in MS. *Neurology* 2009; **73**:1842–8.

Martinez-Yelamos A, Saiz A, Sanchez-Valle R *et al.* 14-3-3 protein in the CSF as prognostic marker in early multiple sclerosis. *Neurology* 2001; **57**:722–4.

Mathews WB. Multiple sclerosis presenting with acute remitting psychiatric symptoms. *J Neurol Neurosurg Psychiatry* 1979; **42**:859–63.

Mendez MF, Frey WH. Multiple sclerosis dementia. *Neurology* 1992; **42**:696.

Michotte A, Dequenne P, Jacobovitz D *et al.* Focal neurological deficit with sudden onset as the first manifestation of sarcoidosis: a case report with MRI follow-up. *Eur Neurol* 1991; **31**:376–9.

Miguel EC, Rodriguez Pereira RM, de Braganca Pereira CA *et al.* Psychiatric manifestations of systemic lupus erythematosus: clinical features, symptoms, and signs of central nervous system activity in 43 patients. *Medicine* 1994; **73**:224–32.

Mikdashi J, Krumholz A, Handwerger B. Factors at diagnosis predict subsequent occurrence of seizures in systemic lupus erythematosus. *Neurology* 2005; **64**:2102–7.

Mikol DD, Barkhof F, Chang P *et al.* Comparison of subcutaneous interferon beta-1a with glatiramer acetate in patients with relapsing multiple sclerosis (the REbif vs Glatiramer Acetate in Relapsing MS Disease [REGARD] study): a multicentre, randomized, parallel, open-label trial. *Lancet Neurol* 2008; **7**:864–6.

Miller DH, Kendall BE, Barter S *et al.* Magnetic resonance imaging in central nervous system sarcoidosis. *Neurology* 1988; **38**:378–83.

Mitsias P, Levine SR. Large cerebral vessel occlusive disease in systemic lupus erythematosus. *Neurology* 1994; **44**:385–93.

Moller A, Wiedemann G, Rohde U *et al.* Correlates of cognitive impairment and depressive mood disorder in multiple sclerosis. *Acta Psychiatr Scand* 1994; **89**:117–21.

Monteiro ML, Swanson RA, Coppeto JR *et al.* A microangiopathic syndrome of encephalopathy, hearing loss, and retinal arteriolar occlusions. *Neurology* 1985; **35**:1113–21.

Moravan M, Segal BM. Treatment of CNS sarcoidosis with infliximab and mycophenolate mofetil. *Neurology* 2009; **72**:337–40.

Moreau C, Devos D, Delmaire C *et al.* Progressive MRI abnormalities in late recurrence of Sydenham's chorea. *J Neurol* 2005; **252**:1341–4.

Mumford CJ, Wood NW, Kellar-Wood HF *et al.* The British Isles Survey of multiple sclerosis in twins. *Neurology* 1994; **44**:11–15.

Nausieda PA, Grossman BJ, Koller WC *et al.* Sydenham chorea: an update. *Neurology* 1980; **30**:331–4.

Nausieda PA, Bielauskas LA, Bacon LD *et al.* Chronic dopaminergic sensitivity after Sydenham's chorea. *Neurology* 1983; **33**:750–4.

Nicoletti A, Sofia V, Biondi R *et al.* Epilepsy and multiple sclerosis in Sicily: a population-based study. *Epilepsia* 2003; **44**:1445–8.

Nokura K, Yamamoto H, Okawara Y *et al.* Reversible limbic encephalitis caused by ovarian teratoma. *Acta Neurol Scand* 1997; **95**:367–73.

Nolte KW, Undehaun A, Sieker H *et al.* Hashimoto's encephalopathy: a brainstem vasculitis? *Neurology* 2000; **54**:769–70.

Nozaki H, Shimohata T, Kanbayashi T *et al.* A patient with anti-aquaporin 4 antibody who presented with recurrent hypersomnia, reduced orexin (hypocretin) level anymmetrical hypothalamic lesions. *Sleep Med* 2009; **10**:253–5.

O'Connor JF, Musher DM. Central nervous system involvement in systemic lupus erythematosus. *Arch Neurol* 1966; **14**:157–64.

O'Connor P, Filippi M, Aronson B *et al.* 250 microgr or 500 microgr interferon beta-1b versus 20 mg glatiramer acetate in relapsing–remitting multiple scelrosis: a prospective, randomized, multicentre study. *Lancet Neurol* 2009; **8**:889–97.

Oide T, Tokuda T, Yazaki M *et al.* Anti-neuronal antibody in Hashimoto's encephalopathy: neuropathological, immunochemical, and biochemical analysis of two patients. *J Neurol Sci* 2004; **217**:7–12.

Oksanen V. Neurosarcoidosis: clinical presentations and course in 50 patients. *Acta Neurol Scand* 1986; **73**:283–90.

Olmos-Lau N, Ginsberg MD, Geller JB. Aphasia in multiple sclerosis. *Neurology* 1977; **27**:623–6.

Omdal R, Roalso S. Chorea gravidarum and chorea associated with oral contraceptives – diseases due to antiphospholipid antibodies? *Acta Neurol Scand* 1992; **86**:219–20.

Panitch H, Goodin DS, Francis G *et al.* Randomized, comparative study of interferon beta-1a treatment regimens in MS: the EVIDENCE trial. *Neurology* 2002; **59**:1496–506.

Parker N. Disseminated sclerosis presenting as schizophrenia. *Med J Aust* 1956; **1**:405–7.

Patterson JF. Treatment of chorea gravidarum with haloperidol. *South Med J* 1979; **72**:1220–1.

Pawate S, Agarwal A, Moses H *et al.* The spectrum of Susac's syndrome. *Neurol Sci* 2009; **30**:59–64.

Paz JA, Silva CA, Marques-Dias MJ. Randomized double-blind study with prednisone in Sydenham's chorea. *Pediatr Neurol* 2006; **34**:264–9.

Pego-Reigosa JM, Isenberg DA. Psychosis due to systemic lupus erythematosus: characteristics and long-term outcome of this rare manifestation of the disease. *Rheumatology* 2008; **47**:1498–502.

Pena J, Mora E, Cardozo J *et al.* Comparison of the efficacy of carbamazepine haloperidol and valproic acid in the treatment of children with Sydenham's chorea: clinical follow-up of 18 patients. *Arq Neuropsiquitr* 2002; **60**:374–7.

Pentland B, Mitchell JD, Cull RE *et al.* Central nervous system sarcoidosis. *Q J Med* 1985; **56**:457–65.

Peterson JW, Bo L, Mork S *et al.* Transected neuritis, apoptotic neurons, and reduced inflammation in cortical multiple sclerosis lesions. *Ann Neurol* 2001; **50**:389–400.

Pettee AD, Wasserman BA, Adams NL *et al.* Familial Sneddon's syndrome: clinical, hematologic, and radiographic findings in two brothers. *Neurology* 1994; **44**:399–405.

Pine DS, Douglas CJ, Charles E *et al.* Patients with multiple sclerosis presenting to psychiatric hospitals. *J Clin Psychiatry* 1995; **56**:297–306.

Pittock SJ, Kryzer TJ, Lennon VA. Paraneoplastic antibodies coexist and predict cancer, not neurological syndrome. *Ann Neurol* 2004; **56**:715–19.

Pittock SJ, Lucchinetti CF, Parisi JE *et al.* Amphiphysin autoimmunity: paraneoplastic accompaniments. *Ann Neurol* 2005; **58**: 96–107.

Pittock SJ, Weinshenker BG, Lucchinetti CF *et al.* Neuromyelitis optica brain lesions localized at sites of high aquaporin 4 expression. *Arch Neurol* 2006; **63**:964–8.

Polman CH, O'Connor PW, Havrdova E *et al.* A randomized, placebo-controlled trial of natalizumab for relapsing multiple sclerosis. *N Engl J Med* 2006; **354**:899–910.

Poppe AY, Lapierre Y, Melancon D *et al.* Neuromyelitis optica with hypothalamic involvement. *Mult Scler* 2005; **11**:617–21.

Powell E. Two fatal cases of acute chorea with insanity. *Brain* 1889; **12**:157–60.

Powers WJ, Miller EM. Sarcoidosis mimicking glioma: case report and review of intracranial sarcoid mass lesions. *Neurology* 1981; **31**:907–10.

Poynton FJ, Holmes GM. A contribution to the study of chorea. *Lancet* 1906; **2**:982.

Provenzale JM, Heinz ER, Ortel TL *et al.* Antiphospholipid antibodies in patients without systemic lupus erythematosus: neuroradiologic findings. *Radiology* 1994; **192**:531–7.

Pruss H, Voltz R, Flath B *et al.* Anti-ta-associated paraneopastic encephalitis with occult testicular intratrabecular germ cell neoplasia. *J Neurol Neurosurg Psychiatry* 2007; **78**:651–2.

Psimaras D, Carpentier AF, Rossi C *et al.* Cerebrospinal fluid study in paraneoplastic syndromes. *J Neurol Neurosurg Psychiatry* 2010; **81**:42–5.

Pujol J, Bello J, Deus J *et al.* Lesions in the left arcuate fasciculus region and depressive symptomatology in multiple sclerosis. *Neurology* 1997; **49**:1105–10.

Putzel L. Cerebral complications of chorea. *Med Rec* 1879; 220–2.

Rabins PV, Brooks BR, O'Donnell P *et al.* Structural brain correlates of emotional disorder in multiple sclerosis. *Brain* 1986; **109**: 585–97.

Reaser EF. Chorea of infectious origin. *South Med J* 1940; **33**:1324–8.

Rebollo M, Val JF, Garijo F *et al.* Livedo reticularis and cerebrovascular lesions (Sneddon's syndrome). Clinical, radiological and pathological features in eight cases. *Brain* 1983; **106**:965–79.

Ridel KR, Lipps TD, Gilbert DL. The prevalence of neuropsychiatric disorders in Sydenham's chorea. *Pediatr Neurol* 2010; **42**:243–8.

Robertson WC, Smith CD. Sydenham's chorea in the age of MRI: a case report and review. *Pediatr Neurol* 2002; **27**:6–7.

Robin C, Gonnaud P-M, Durieu I *et al.* Demence lupique progressive: deux cas avec anticorps antiphospholipides ou non. *Rev Neurol* 1995; **151**:699–707.

Ron MA, Callanan MM, Warrington EK. Cognitive abnormalities in multiple sclerosis: a psychometric and MRI study. *Psychol Med* 1991; **21**:59–68.

Rosenberg GA, Appenzeller O. Amantadine, fatigue and multiple sclerosis. *Arch Neurol* 1988; **45**:1104–6.

Rosenfeld MR, Eichen JG, Wade DF *et al.* Molecular and clinical diversity in paraneoplastic immunity to Ma proteins. *Ann Neurol* 2001; **50**:339–48.

Roti E, Gardini F, Minelli R *et al.* Prevalence of anti-thyroid peroxidase antibodies in serum in the elderly: comparison with other tests for anti-thyroid antibodies. *Clin Chem* 1992; **38**:88–92.

Rovira A, Swanton J, Tintore M *et al.* A single, early magnetic resonance imaging study in the diagnosis of multiple sclerosis. *Arch Neurol* 2009; **66**:587–92.

Rubin LA, Urowitz MB, Gladman DD. Mortality in systemic lupus erythematosus: the bimodal pattern revisited. *Q J Med* 1985; **55**:87–98.

Rubinstein I, Gray TA, Moldofsky H *et al.* Neurosarcoidosis associated with hypersomnolence treated with corticosteroids and brain irradiation. *Chest* 1988; **94**:205–6.

Sanna G, Bertolaccini ML, Cuadrado MJ *et al.* Neuropsychiatric manifestations in systemic lupus erythematosus: prevalence and association with antiphospholipid antibodies. *J Rheumatol* 2003; **30**:985–92.

Sanson M, Duyckaerts C, Thibault J-L *et al.* Sarcoidosis presenting as late-onset dementia. *J Neurol* 1996; **243**:484–7.

Sayao AL, Devonshire V, Tremlett H. Longitudinal follow-up of "benign" multiple sclerosis at 20 years. *Neurology* 2007; **68**:496–500.

Sawka AM, Fatourechi V, Boeve BF *et al.* Rarity of encephalopathy associated with autoimmune thyroiditis: a case series from Mayo Clinic from 1950 to 1996. *Thyroid* 2002; **12**:227–8.

Schauble B, Castillo PR, Boeve BF *et al.* EEG findings in steroid-responsive encephalopathy associated with autoimmune thyroiditis. *Clin Neurophysiol* 2003; **114**:32–7.

Schiffer RB, Babigian H. Behavioral disorders in multiple sclerosis, temporal lobe epilepsy and amyotrophic lateral sclerosis. An epidemiologic study. *Arch Neurol* 1984; **41**:1067–9.

Schiffer RB, Wineman NM. Antidepressant pharmacotherapy of depression associated with multiple sclerosis. *Am J Psychiatry* 1990; **147**:1493–7.

Schiffer RB, Caine ED, Bamford KA *et al.* Depressive episodes in patients with multiple sclerosis. *Am J Psychiatry* 1983; **140**:1498–500.

Schiffer RB, Wineman M, Weitkamp LR. Association between bipolar affective disorder and multiple sclerosis. *Am J Psychiatry* 1986; **143**:94–5.

Schwartzman J, McDonald DH, Derillo L. Sydenham's chorea: report of 140 cases and review of the recent literature. *Arch Pediatr* 1948; **65**:6–24.

Scott IA, Stocks AE, Saines N. Hypothalamic/pituitary sarcoidosis. *Aust NZ J Med* 1987; **17**:243–5.

Scott TF. Neurosarcoidosis: progress and clinical aspects. *Neurology* 1993; **43**:8–12.

Scott TF, Yandora K, Valeri A *et al.* Aggressive therapy for neurosarcoidosis. *Arch Neurol* 2007; **64**:691–6.

Sepp N, Zelger B, Schuler G *et al.* Sneddon's syndrome – an inflammatory disorder of small arteries followed by smooth muscle proliferation. Immunohistochemical and ultrastructural evidence. *Am J Surg Pathol* 1995; **19**:448–53.

Sharma OP, Sharma AM. Sarcoidosis of the nervous system: a clinical approach. *Arch Int Med* 1991; **151**:1317–21.

Shaskan D. Mental changes in chorea minor. *Am J Psychiatry* 1938; **95**:193–202.

Shaw PJ, Walls TJ, Newman PK *et al.* Hashimoto's encephalopathy: a steroid-responsive disorder associated with high anti-thyroid antibody titers – report of 5 cases. *Neurology* 1991; **41**:228–33.

Shenker DM, Grossman HJ, Klawans HL. Treatment of Sydenham's chorea with haloperidol. *Dev Med Child Neurol* 1973; **15**:19–24.

Sibley JT, Olszynski WP, DeCoteau WE *et al.* The incidence and prognosis of central nervous system disease in systemic lupus erythematosus. *J Rheumatol* 1992; **19**:47–52.

Silverstein A, Siltzbach LE. Neurologic sarcoidosis: a study of 18 cases. *Arch Neurol* 1965; **12**:1–11.

Sitzer M, Sohngen D, Siebler M *et al.* Cerebral microembolism in patients with Sneddon's syndrome. *Arch Neurol* 1995; **52**:271–5.

Sneddon JB. Cerebrovascular lesions and livedo reticularis. *Br J Dermatol* 1965; **77**:180–5.

Stankoff B, Waubant E, Confarveux C *et al.* Modafinil for fatigue in MS. A randomized placebo-controlled double-blind study. *Neurology* 2005; **64**:1139–43.

Staff NP, Lucchinetti CF, Keegan BM. Multiple sclerosis with predominant, severe cognitive impairment. *Arch Neurol* 2009; **66**:1139–43.

Stern BJ, Krumholz A, Johns C *et al.* Sarcoidosis and its neurological manifestations. *Arch Neurol* 1985; **42**:909–17.

Stockhammer G, Felber SR, Zelger B *et al.* Sneddon's syndrome: diagnosis by skin biopsy and MRI in 17 patients. *Stroke* 1993; **24**:685–90.

Surridge D. An investigation into some psychiatric aspects of multiple sclerosis. *Br J Psychiatry* 1969; **115**:749–64.

Susac JO, Hardman JM, Selhorst JB. Microangiopathy of the brain and retina. *Neurology* 1979; **29**:313–16.

Susac JO, Murtagh FR, Egan RA *et al.* MRI findings in Susac's syndrome. *Neurology* 2003; **61**:1783–7.

Sutton I, Winer J, Rowlands D *et al.* Limbic encephalitis and antibodies to Ma2: a paraneoplastic presentation of breast cancer. *J Neurol Neurosurg Psychiatry* 2000; **69**:266–8.

Swedo SE. Sydenham's chorea: a model for childhood autoimmune neuropsychiatric disorders. *JAMA* 1994; **272**:1788–91.

Swedo SE, Rapoport JL, Cheslow DL *et al.* High prevalence of obsessive–compulsive symptoms in patients with Sydenham's chorea. *Am J Psychiatry* 1989; **146**:246–9.

Swedo SE, Leonard HL, Schapiro MB *et al.* Sydenham's chorea: physical and psychological symptoms of St Vitus dance. *Pediatrics* 1993; **91**:706–13.

Symonds C. Recurrent multiple cranial nerve palsies. *J Neurol Neurosurg Psychiatry* 1958; **21**:95–100.

Tan KM, Lennon VA, Klein CJ *et al.* Clinical spectrum of voltage-gated potassium channel autoimmunity. *Neurology* 2008; **70**:1883–90.

Tandon R, Walden M, Falcon S. Catatonia as a manifestation of paraneoplastic encephalopathy. *J Clin Psychiatry* 1988; **49**:121–2.

Taranta A. Relation of isolated recurrences of Sydenham's chorea to preceding streptococcal infection. *N Engl J Med* 1959; **260**: 1204–10.

Taranta A, Stollerman GH. Relationship of Sydenham's chorea to infection with Group A streptococcus. *Am J Med* 1956; **20**:170–5.

Teixeira AL, Maia DP, Cardoso F. Psychosis following acute Sydenham's chorea. *Eur Child Adolesc Psychiatry* 2007; **16**:67–9.

Thayer WS. An analysis of 808 cases of chorea with special reference to cardiovascular manifestations. *J Am Med Assoc* 1906; **47**: 1352–4.

Thiele R. Clinical observations on chorea of pregnancy. *Arch Neurol Psychiatry* 1935; **39**:1077.

Thompson C, Checkley S. Short term memory deficit in a patient with cerebral sarcoidosis. *Br J Psychiatry* 1981; **139**:160–1.

Thompson AJ, Polmon CH, Miller DH *et al.* Primary progressive multiple sclerosis. *Brain* 1997; **120**:1085–96.

Thrush DC, Boddie HG. Episodic encephalopathy associated with thyroid disorders. *J Neurol Neurosurg Psychiatry* 1974; **37**:697–700.

Tokunaga M, Saito K, Kawabata D *et al.* Efficacy of rituximab (anti-CD20) for refractory systemic lupus erythematosus involving the central nervous system. *Ann Rheum Dis* 2007; **66**:470–5.

Tourbah A, Piette JC, Iba-Zizen MT *et al.* The natural course of cerebral lesions in Sneddon syndrome. *Arch Neurol* 1997; **54**:53–60.

Tsokos GC, Tsokos M, LeRiche NGH *et al.* A clinical and pathological study of cerebrovascular disease in patients with systemic lupus erythematosus. *Semin Arthritis Rheum* 1986; **16**:70–8.

Tsukamoto T, Mochizuki R, Mochizuki H *et al.* Paraneoplastic cerebellar degeneration and limbic encephalitis in a patient with adenocarcinoma of the colon. *J Neurol Neurosurg Psychiatry* 1993; **56**:713–16.

Van Bogaert L, Bertrand I. Hemorrhagic affection of cortico-neostriatal site revealed clinically by acute and fatal chorea. *J Neurol Psychopathology* 1932; **13**:1–13.

Venables PJW. Diagnosis and treatment of systemic lupus erythematosus. *BMJ* 1993; **307**:663–6.

Vercellino M, Plano F, Votta B *et al.* Grey matter pathology in multiple sclerosis. *J Neuropathol Exp Neurol* 2005; **64**:1101–7.

Vernino S, Tuite P, Adler CH *et al.* Paraneoplastic chorea associated with CRMP-5 neuronal antibody and lung carcinoma. *Ann Neurol* 2002; **51**:625–30.

Viegas S, Eir A, Esiri M *et al.* Symptomatic, radiological and pathological involvement of the hypothalamus in neuromyelitis optica. *J Neurol Neurosurg Psychiatry* 2009; **80**:679–82.

Vincent A, Buckley C, Schott JM *et al.* Potassium channel antibody-associated encephalopathy: a potentially immunotherapy-responsive form of limbic encephalitis. *Brain* 2004; **127**:701–12.

Vitaliani R, Mason W, Ances B *et al.* Paraneoplastic encephalitis, psychiatric symptoms, and hypoventilation in ovarian teratoma. *Ann Neurol* 2005; **58**:594–604.

Voltz R, Gultekin H, Rosenfeld MR *et al.* A serologic marker of paraneoplastic limbic and brain-stem encephalitis in patients with testicular cancer. *N Engl J Med* 1999; **340**:1788–95.

Walker RH, Spiera H, Brin MF *et al.* Parkinsonism associated with Sjögren's syndrome: three cases with a review of the literature. *Mov Disord* 1999; **14**:262–8.

Weiner DK, Allen NB. Large vessel vasculitis of the central nervous system in systemic lupus erythematosus: report and review of the literature. *J Rheumatol* 1991; **18**:748–51.

West SG, Emlen W, Wener MH *et al.* Neuropsychiatric lupus erythematosus: a 10-year prospective study on the value of diagnostic tests. *Am J Med* 1995; **99**:153–63.

Whitlock FA, Siskind MM. Depression as a major symptom of multiple sclerosis. *J Neurol Neurosurg Psychiatry* 1980; **43**:861–5.

Wiederholt WC, Siekert RG. Neurological manifestations of sarcoidosis. *Neurology* 1965; **15**:1147–54.

Wilcox JA, Nasrallah HA. Sydenham's chorea and psychosis. *Neuropsychobiology* 1986; **15**:13–14.

Wilcox JA, Nasrallah HA. Sydenham's chorea and psychopathology. *Neuropsychobiology* 1988; **19**:6–8.

Wilson P, Preece M. Chorea gravidarum. *Arch Intern Med* 1932a; **49**:471–533.

Wilson P, Preece M. Chorea gravidarum. *Arch Intern Med* 1932b; **49**:671–97.

Wilson SAK. *Neurology*, 2nd edn. London: Butterworth & Co, 1955.

Wingerchuk DM, Hogancamp WF, O'Brien PC *et al.* The clinical course of neuromyelitis optica (Devic's syndrome). *Neurology* 1999; **53**:1107–14.

Wingerchuk DM, Lennon VA, Pittock SJ *et al.* Revised diagnostic criteria for neuromyelitis optica. *Neurology* 2006; **66**:1485–9.

Winnacker JL, Becker KL, Katz S. Endocrine aspects of sarcoidosis. *N Engl J Med* 1968; **278**:483–92.

Winkelman NW, Eckel JL. The brain in acute rheumatic fever. *Arch Neurol Psychiatry* 1932; **28**:844–70.

Wolf RE, McBeath JG. Chorea gravidarum in systemic lupus erythematosus. *J Rheumatol* 1985; **12**:992–3.

Wong KL, Woo EKW, Yu YL *et al.* Neurologic manifestations of systemic lupus erythematosus. *Q J Med* 1991; **81**:857–70.

Yanagawa K, Kawachi I, Toyoshima Y *et al.* Pathologic and immunologic profiles of a limited form of neuromyelitis optica with myelitis. *Neurology* 2009; **73**:1628–37.

Yoneda M, Fujii A, Ito A *et al.* High prevalence of serum autoantibodies against the amino terminal of alpha-enolase in Hashimoto's encephalopathy. *J Neuroimmunol* 2007; **185**:195–200.

Young AC, Saunders J, Ponsford JR. Mental change as an early feature of multiple sclerosis. *J Neurol Neurosurg Psychiatry* 1976; **39**:1008–13.

Zandi MS, Irani SR, Follows G *et al.* Limbic encephalitis associated with antibodies to the NMDA receptor in Hodgkin lymphoma. *Neurology* 2009; **73**:2039–40.

Ziegler LH. The neuropathological findings in a case of acute Sydenham's chorea. *J Nerv Ment Dis* 1927; **65**:273–81.

Zomorrodi A, Wald ER. Sydenham's chorea in western Pennsylvania. *Pediatrics* 2006; **117**:e675–9.

Zorzon M, de Masi R, Nasuelli D *et al.* Depression and anxiety in multiple sclerosis. A clinical and MRI study in 95 subjects. *J Neurol* 2001; **248**:416–21.

20

Sleep disorders

20.1 SOMNAMBULISM

Isolated episodes of sleepwalking are common in children, occurring in about 15 percent (Kales *et al.* 1987). Recurrent, frequent episodes, however, are not normal, and in these instances one speaks of somnambulism or, as it is also known, 'sleepwalking disorder'. Somnambulism is not common, being found in 2–3 percent of children and far fewer adolescents or adults.

Clinical features

Somnambulism generally has an onset in childhood, at some time between the ages of 3 and 6 years; onsets in adolescence are uncommon, and in adult years are rare.

Sleepwalking arises from non-rapid eye movement (NREM) sleep: in children, episodes generally occur in the first third of the night (Kales and Kales 1974), whereas in adults episodes may occur during the second and final thirds (Kavey *et al.* 1990). Typically (Kales *et al.* 1966), the patient sits up in bed; the eyes may be closed or, more often, open and, if open, the patient may look about the room with a blank stare. Some patients merely engage in simple, stereotyped behavior, such as fumbling with pyjamas or sheets, but most will get out of bed and begin to walk. Although some patients may bump into furniture or walls, many are able to navigate in such a way as to avoid obstacles. Some patients may simply wander, whereas others may attempt to climb out of windows or go down stairs. Though uncommon, injuries, and even violent behavior, may occur, especially in adults (Kavey *et al.* 1990, Schenck *et al.* 1989). Rarely, patients may engage in complex activities such as eating, writing, or even driving a car (Schenck and Mahowald 1995). Patients may be mute or mumble incoherently; some may respond to requests with a few simple words.

Attempts to redirect patients or lead them back to bed may or may not be met with success; some patients actively resist any attempts to interfere with them, even to the point of combativeness. If patients are awakened, there is typically a few minutes of confusion before normal consciousness is restored.

Most episodes end spontaneously within 15–30 minutes; some patients may make it back to bed, whereas others will lie down on a couch, or even the floor, and resume sleep. Upon awakening the next day, patients are either amnestic for the episode or, at most, have only a patchy recall of the events that transpired. Importantly, there is little, if any, dreaming associated with the episode: if patients are awakened during the episode, or questioned the next morning, they may recall at best a fragmentary kind of dream-like event (Oudiette *et al.* 2009a).

Childhood-onset cases are generally not associated with other neuropsychiatric disturbances; by contrast, some (Kales *et al.* 1980a; Sours *et al.* 1963), but not all (Parkes 1986), authors have found an association between adult-onset cases and various abnormal personality traits.

Polysomnography reveals the onset of sleepwalking in stage IV or, at times, stage III sleep. If polysomnography is required, it is useful to precede it with between 25 (Zadra *et al.* 2008) and 38 (Joncas *et al.* 2002) hours of sleep deprivation, as this greatly increases the chances of catching an episode. Importantly, there are no ictal discharges and no interictal epileptiform discharges.

Course

If the onset is in early childhood, a remission is likely by early adolescence (Kales *et al.* 1980a); later-onset cases, however, may persist into adult years. In cases where remission does occur in adolescence, recurrences, though rare, may occur in adult years.

Overall, the frequency of sleepwalking in somnambulism increases during febrile illnesses, with sleep disruption (as may be seen with shift work [Driver and Shapiro 1993]), and, as noted above, with sleep deprivation. Recent work has also demonstrated, in both children (Guilleminault *et al.* 2003) and adults (Guilleminault *et al.* 2005), a strong association between episodes and partial or complete obstructive sleep apneas.

Etiology

Somnambulism probably occurs on a hereditary basis: it is clearly familial (Kales *et al.* 1980b) and the concordance rate is higher for monozygotic than for dizygotic twins (Bakwin 1970; Hublin *et al.* 1997).

Differential diagnosis

At first glance, rapid eye movement (REM) sleep behavior disorder may appear quite similar to somnambulism; however, the differential is easily made by awakening the patient and asking if he or she were dreaming during the episode. In REM sleep behavior disorder there is always a dream present and the patient's behavior during the episode is 'explained' by the dream content; by contrast, in somnambulism, there is either no dreaming or, at best, only a fragment, which could not 'explain' the behavior seen during the episode.

Night terrors are distinguished from sleepwalking by both the extreme terror evident in the sleeping patient and the lack of any actual walking about.

Sleep drunkenness may resemble sleepwalking; however, the episodes of sleep drunkenness are seen in the morning as the patient struggles to awaken, in contrast to sleepwalking in which episodes occur during the night and are not associated with awakening.

Nocturnal complex seizures may, as discussed further in Chapter 8.3, be difficult to distinguish from episodes of sleepwalking. A history of complex partial or other types of seizures during waking hours is helpful, but in limiting cases polysomnography may be required.

Sleepwalking may also occur on a secondary basis. Thus, sleepwalking may be caused by various medications, including zolpidem (Hogue and Chesson 2009), metoprolol (Hensel and Pillman 2008), paroxetine (Kawashima and Yamada 2003), bupropion (Khazaal *et al.* 2003), mirtazapine (Yeh *et al.* 2009), olanzapine (Kolivakis *et al.* 2001), quetiapine (Hafeez and Kalinowski 2007), and the combination of antipsychotics with lithium (Charney *et al.* 1979). Sleepwalking may also occur during hyperthyroidism (Ajlouni *et al.* 2005).

Treatment

Patients and family members should be reassured regarding the essentially benign nature of somnambulism, and common sense precautions, such as locking windows or doors and removing potentially dangerous objects, should be taken to reduce the risk of any injury. Should these measures fail, consideration may be given to 'anticipatory awakenings' (Tobin 1993). Here, one monitors the patient to determine when episodes are most likely to occur and then awakens the patient just beforehand; in many cases this may provide lasting relief after only a few nights. Consideration should also be given to polysomnography to determine whether sleep apneas are occurring and associated with episodes, for if they are, treatment of the sleep apnea may eliminate the episodes (Guilleminault *et al.* 2003, 2005). In limiting cases, medications may be utilized. The only blind study carried out found diazepam, in a dose of 10 mg nightly, to be effective in adults (Reid *et al.* 1984). Anecdotally, clonazepam (Kavey *et al.* 1990, Schenck and Mahowald 1996), alprazolam (Schenck and Mahowald 1996) and imipramine (Cooper 1987) are also effective.

20.2 REM SLEEP BEHAVIOR DISORDER

Rapid eye movement sleep behavior disorder is an uncommon, and most remarkable, condition in which patients, while asleep and dreaming, literally 'act out' their role in the dream, with the bedroom serving as the 'stage'. As noted below, under 'etiology' it occurs most commonly either on an idiopathic basis or secondary to a parkinsonian condition or to narcolepsy: in cases occurring on an idiopathic basis or associated with a parkinsonian condition, it generally has an onset in middle or later years, is far more common in males, and may be quite severe; by contrast, in cases associated with narcolepsy, it has an onset in late adolescence or early adult years, is roughly equally common in males and females, and tends to be mild.

Clinical features

Episodes occur during REM sleep, typically in the middle or last third of the night, and occur with variable frequency, from multiple episodes nightly to isolated episodes occurring every few months.

Schenck and colleagues have comprehensively described this condition (Schenck and Mahowald 1990; Schenck *et al.* 1986, 1987, 1989). Violent or potentially dangerous behavior is common, and patients may suffer bruises, lacerations, or even fractures; in one case (Dyken *et al.* 1995) a subdural hematoma was sustained. Patients are generally difficult to awaken and, if they do come to full consciousness, they relate a vivid dream that, in retrospect, clearly provides the context for their behavior.

Some examples will help to convey a sense of the remarkable phenomenology of this disorder. In one case (Schenck *et al.* 1986) a 67-year-old man described his dream and what he found when he woke up:

> I was a halfback playing football, and after the quarterback received the ball from the center he lateraled it sideways to me and I'm supposed to go around end and cut back over tackle and – this is very vivid – as I cut back over tackle there is this big 280-pound tackle waiting, so I, according to football rules, was to give him my shoulder

and bounce him out of the way, supposedly, and when I came to I was standing in front of our dresser and I had knocked lamps, mirrors, and everything off the dresser, hit my head against the wall and my knee against the dresser.

Damage to property is not the only danger here: during one incident (Schenck *et al.* 1989), the patient 'was awakened one night by his wife's yelling as he was choking her. He was dreaming of breaking the neck of a deer he had just knocked down'; during another (Culebras and Moore 1989), a 70-year-old man 'dreamed that an alligator was trying to get into his car, and in order to prevent it, he held the animal's snout with great force...' After finally being awakened by his wife calling him, he found himself 'strongly grabbing her arm'.

Although violent behavior is common, nonviolent behavior may also occur (Oudiette *et al.* 2009b), such as eating, drinking, singing or dancing: one patient during an episode introduced himself and then reached out and moved his hand as if shaking someone else's; later, although he had not smoked for years, he positioned his hand as if he were holding a cigarette, and then made movements of his cheeks and lips as if drawing in, and then breathing out, the smoke.

Importantly, although the dream content, and resulting behavior, are often aggressive, the waking behavior of these patients is not characterized by any increased aggressiveness (Fantini *et al.* 2005).

Polysomnography is diagnostic and reveals REM sleep during the episode, without, obviously, the expected atonia.

Course

In the natural course of events, REM sleep behavior disorder is chronic.

Etiology

Rapid eye movement sleep behavior disorder most commonly occurs either on an idiopathic basis or secondary to certain parkinsonian conditions or to narcolepsy; other, relatively rare, causes are noted below.

Idiopathic REM sleep behavior is diagnosed when the disorder occurs in an isolated fashion, and this is generally seen in middle or later years. This diagnosis, however, must be tentative, given that a large percentage of these patients will eventually develop a parkinsonian condition (Iranzo *et al.* 2006a); indeed, in one study (Postuma *et al.* 2009) this occurred in roughly one-half of participants over a 12-year follow-up. For example, dementia with Lewy bodies may follow 6–8 years later, and Parkinson's disease some 12 years later (Schenck *et al.* 1996a; Tan *et al.* 1996).

Parkinsonian conditions capable of causing REM sleep behavior disorder include dementia with Lewy bodies (Boeve *et al.* 2003a), Parkinson's disease (Olson *et al.* 2000), and the striatonigral variant of multiple system atrophy (Boeve *et al.* 2003a; Plazzi *et al.* 1997).

Rapid eye movement sleep behavior disorder also occurs in narcolepsy (Schenck and Mahowald 1992), being found in between one-third (Nightingale *et al.* 2005) and two-thirds (Knudsen *et al.* 2010) of cases. Notably, in cases of narcolepsy, REM sleep behavior disorder is far more common in cases with cataplexy than in those without (Knudsen *et al.* 2010; Nightingale *et al.* 2005).

Rare causes of REM sleep behavior disorder include Alzheimer's disease (Schenck *et al.* 1996b), spinocerebellar ataxia type 3 (Iranzo *et al.* 2003), limbic encephalitis with anti-voltage-gated potassium channel antibodies (Iranzo *et al.* 2006b), and certain focal lesions. Focal lesions capable of causing REM sleep behavior disorder have all been found in the pontine tegmentum, and include paramedian infarction (Xi and Luning 2009), plaques of multiple sclerosis (Plazzi and Montagna 2002; Tippmann-Peikert *et al.* 2006), and tumor (Zambelis *et al.* 2002). Various medications have also been reported to either cause REM sleep behavior disorder or aggravate pre-existing cases, including antidepressants (Frauscher *et al.* 2010) (e.g. fluoxetine [Schenck *et al.* 1992] and mirtazapine [Onofrj *et al.* 2003]) and selegiline (Louden *et al.* 1995).

Although the precise structures involved in the genesis of REM sleep behavior disorder are not known, attention has focused on brainstem nuclei and the hypothalamus. In cases occurring secondary to diffuse Lewy body disease, neuronal loss and Lewy bodies have been noted in both the locus ceruleus and the substantia nigra (Turner *et al.* 2000). This, of course, is not unexpected in diffuse Lewy body disease; however, in cases in which REM sleep behavior disorder was the *only* manifestation of Lewy body disease, this too was where the pathology was found (Uchiyama *et al.* 1995). Other investigators, however, have cast doubt on this localization, finding only minimal changes in these nuclei in a patient with diffuse Lewy body disease who also had REM sleep behavior disorder (Boeve *et al.* 2007).

Diffuse projections occur from the hypothalamus to the locus ceruleus, and there is evidence to suggest that loss of hypocretin-1-containing neurons in the hypothalamus is associated with REM sleep behavior disorder. Specifically, cerebrospinal fluid (CSF) levels of hypocretin-1 are reduced in narcoleptics with REM sleep behavior disorder as compared with narcoleptics without this disorder (Knudsen *et al.* 2010).

Differential diagnosis

Somnambulism is readily differentiated from REM sleep behavior disorder by simply asking patients upon awakening whether or not they had been dreaming: somnambulists will either have no dream to recall or, at most, a few dream-like fragments, whereas those with REM sleep behavior disorder will recall a vivid dream, which, as noted earlier, will 'explain' their behavior. If there is any doubt, polysomnography will reveal episodes arising from NREM sleep in somnambulism and from REM sleep in REM sleep behavior disorder.

Complex partial seizures arising from sleep may, likewise, be distinguished from REM sleep behavior disorder by asking about dreaming, which is absent in the patient having a seizure. Furthermore, one typically also finds a history of other types of seizures, such as grand mal seizures, or of complex partial seizures during the day. In doubtful cases, polysomnography may reveal ictal activity during the episode. Importantly, not too much weight should be placed on finding interictal epileptiform discharges, as these are not uncommon incidental findings in elderly patients with REM sleep behavior disorder (Manni *et al.* 2006).

Obstructive sleep apnea may at times enter into the differential. In some cases of obstructive sleep apnea, patients, during arousals from an apneic episode, may appear confused and engage in complex activity (Iranzo and Santamaria 2005). Polysomnography may be required here and reveals that these episodes do not arise from REM sleep.

Treatment

Although there are no blinded studies of the treatment of REM sleep behavior disorder, clonazepam (Schenck and Mahowald 1990), in doses ranging from 0.5 to 2 mg nightly, has become the accepted first-line treatment. Melatonin, in doses ranging from 3 to 12 mg, constitutes an acceptable alternative (Boeve *et al.* 2003b), and an argument may be made that it should be tried first as it does not carry with it any risk of falls. In cases in which both of these agents are either ineffective or not well tolerated, consideration may be given to donepezil (Ringman and Simmons 2000); pramipexole was also reported effective in one report (Schmidt *et al.* 2006); however, follow-up work has cast serious doubts on its utility (Fantini *et al.* 2003, Kummu *et al.* 2008). Pending effective treatment, the bedroom should be made as safe as possible.

20.3 NIGHTMARE DISORDER

Occasional nightmares, occurring at a frequency of no more than once every few weeks, are not abnormal. Nightmares occurring on a weekly, or more frequent, basis may be seen in a variety of conditions, discussed in the section on 'etiology', below. In cases, however, where frequent nightmares occur in the absence of any clear cause, the diagnosis of nightmare disorder, an idiopathic condition, should be considered. Although the actual prevalence of nightmare disorder is not known, it may be common in childhood; its appearance in adult years, however, is probably uncommon, if not rare.

Clinical features

The nightmares of nightmare disorder typically first appear after an emotionally troubling event.

During a nightmare (Kales *et al.* 1980c), patients typically lie quite still; rarely is there much movement at all, and one never sees any thrashing about or complex behavior. During the nightmare, patients may be chased, attacked, tortured, or preyed upon by any number of unspeakable apparitions. Typically, as the fear crescendos, patients awaken with a cry of fright and are shaky, mildly diaphoretic, and tachycardic. Within moments, full alertness and orientation are achieved, and patients are then able to provide a vivid and emotional account of the frightening dream. Although some patients are able to go directly back to sleep, most, fearful of another nightmare, have some difficulty in this regard and may lie awake for half an hour or more.

Nightmares arise from REM sleep, typically in the last half of the night.

Course

As noted above, the nightmares in nightmare disorder occur at least on a weekly basis and in some cases may occur multiple times a night. Fever, fatigue, emotional stress, and watching frightening shows before bed may all aggravate this condition. Although in most cases the disorder eventually remits, in a minority it may be lifelong.

Etiology

Apart from a suggestion that frequent nightmares may be inherited (Hublin *et al.* 1999), little is known about the etiology of nightmare disorder.

Differential diagnosis

Night terrors may be distinguished from nightmares in that night terrors are associated with overt signs of fright while the patient is asleep (e.g. crying out), arise from NREM sleep in the first third of the night, and are not associated with any clear dream recall; by contrast, nightmares are not associated with any movement, arise from REM sleep in the last half of the night, and are clearly and readily recalled.

Nocturnal panic attacks are easily distinguished from nightmares in that the panic attacks are not associated with frightening dreams; although patients with nocturnal panic attacks awaken into panic, they do not awaken out of a frightening dream.

Frequent nightmares are very common in post-traumatic stress disorder, and may also occur in borderline, schizotypal or schizoid personality disorder and during a depressive episode of a major depressive disorder or bipolar disorder, in the course of schizophrenia, or during delirium of any cause.

Various medications may cause nightmares, including tricyclic antidepressants, selective serotonin reuptake inhibitors (SSRIs), antipsychotics, levodopa, and beta-blockers. Nightmares may also occur during the 'REM rebound' seen during withdrawal from sedative–hypnotics or alcohol, or after discontinuation of long-term treatment with any medication capable of causing REM suppression, such as antidepressants.

Very rarely, nightmares may constitute the only symptomatology of a seizure (Boller *et al.* 1975). Clues to this would be a history of other seizure types and a lack of any other reasonable explanation. Polysomnography may not be helpful here, as ictal electroencephalograms (EEGs) during simple partial seizures are typically normal. In doubtful cases one may consider attempting a 'diagnosis by treatment response' to an anti-epileptic drug.

Treatment

Avoidance of fatigue, frightening stories or shows, and, whenever possible, emotional stress, along with prompt treatment of any febrile illness, are all in order.

Behavioral treatments, such as densensitization or dream rehearsal, appear to be effective (Germain and Nielsen 2003, Kellner *et al.* 1992; Neidhardt *et al.* 1992). When such treatment is not feasible or is ineffective, one may consider suppressing REM sleep with a benzodiazepine, such as diazepam, or using cyproheptadine in a dose of 16–24 mg at bedtime (Harsch 1986). Consideration may also be given to prazosin, which is effective in the treatment of nightmares seen in post-traumatic stress disorder.

20.4 SLEEP TERRORS

Sleep terrors, also known as sleep terror disorder, night terrors, or pavor nocturnus, occurs in 1–4 percent of children and is somewhat more common in boys than girls.

Clinical features

The onset is generally in childhood, between the ages of 4 and 12 years; rarely the onset may be delayed until early adult years.

Attacks (Fisher *et al.* 1973a; Kales and Kales 1974) arise from stage III or IV NREM sleep in the first third of the night, typically before the first episode of REM sleep. Patients typically scream or cry out in terror, sit bolt upright, and appear dazed; the heart rate is increased, sometimes greatly so, and respirations are rapid and panting. The sheets may be grasped and patients may cry out for help. Attempts to awaken the child are generally unsuccessful, and the episode generally runs its course in a matter of minutes. Afterwards, patients may simply fall back into sleep or they may awaken; those who do awaken, although recalling the sense of terror, find either no memory of a dream or merely fragments of one. Remarkably, in contrast to their parents, who are generally quite shaken at witnessing the episode, the children themselves are generally able to fall back to sleep without any difficulty.

Otherwise, typical attacks may rarely occur during daytime naps, thus constituting a 'pavor diurnus'.

Course

The frequency of attacks varies widely, from daily or weekly attacks to widely spaced attacks occurring at monthly or longer intervals. In childhood-onset cases attacks eventually cease, generally after about 4 years or so; uncommonly, they may persist into adolescence or even adult years (DiMario and Emery 1987). In adult-onset cases the course may be chronic. Anxiety, stress, fatigue, and irregular sleep habits may all increase the frequency of attacks.

Etiology

Sleep terrors may well occur on a hereditary basis: they run in families (Kales *et al.* 1980b), and the concordance rate is higher among monozygotic than dizygotic twins (Nguyen *et al.* 2008).

Differential diagnosis

Nightmares are quite different from night terrors. Nightmare sufferers typically have a vivid recall of the nightmare and are reluctant to go back to sleep for fear of having another one; in contrast, patients with night terrors have little or no recall of any dream and, if awakened, can fall back asleep with little anxiety. Furthermore, nightmares arise from REM sleep typically in the second half of the night, whereas night terrors arise from NREM sleep in the first third of the night.

Nocturnal panic attacks are often included in the differential, but these are also quite different from night terrors. In nocturnal panic attacks, patients awaken from sleep into the panic attack and are awake and alert during the attack, with a full recall of it afterward. By contrast, patients with night terrors stay asleep during the attack and have little, if any, recall of it. Most patients with nocturnal panic attacks will also have typical attacks during waking hours (Mellman and Uhde 1989a), and this history, of course, is very helpful; exceptions do occur, however, and some patients with panic disorder may have only nocturnal attacks (Mellman and Uhde 1990). Polysomnography is not very helpful in this differential, given that nocturnal panic attacks arise from NREM sleep (Mellman and Uhde 1989b), just as sleep terrors do.

Nocturnal complex partial seizures may closely resemble sleep terrors. In one case (Tinuper *et al.* 1990), a teenager had nocturnal seizures characterized by 'episodes of sudden arousal, screaming, grimacing, and violent, repetitive movements of the trunk and all four limbs' lasting for 5–10 minutes. The occurrence of seizures during waking hours, or a family history of epilepsy, is an important clue. In doubtful cases, polysomnography will be required.

Finally, there is a case report of adult-onset night terrors occurring secondary to a thalamic tumor (Di Gennaro *et al.* 2004).

Treatment

Parents should be reassured regarding the benign nature of night terrors, and, in most cases, as the episodes do not appear to bother patients much, this is all that is required. Should suppression of the episodes be necessary, open studies and case reports suggest effectiveness for clonazepam (0.5–1 mg) (Schenck and Mahowald 1996), diazepam (2.5–10 mg) (Fisher *et al.* 1973b), imipramine (25–50 mg) (Burstein *et al.* 1983; Cooper 1987), paroxetine (Wilson *et al.* 1997), trazodone (Balon 1994) and *L*-5-hydroxytryptophan (Bruni *et al.* 2004).

20.5 NOCTURNAL HEAD BANGING

Nocturnal head banging, also known as jactatio nocturna capitis, is characterized by repetitive head banging or bumping as the child falls asleep. This is a common disorder in infants, seen in about 15 percent.

Clinical features

Head banging usually begins around the age of 9 months and typically occurs during stage II sleep (Dyken *et al.* 1997); the head movements are rhythmic, occurring several times a minute, with the head repeatedly making contact with the bed, bedrails, or headboard (Kravitz *et al.* 1960). The contact itself is generally not very forceful, and injuries are rare.

Course

The vast majority of cases resolve by middle childhood (Abe *et al.* 1984); persistence into adult years, though reported (Anderson *et al.* 2006), is very rare.

Etiology

The etiology is not known.

Differential diagnosis

Nocturnal head banging must be distinguished from head banging seen during wakefulness, as may occur in mental retardation, autism, and schizophrenia. There is also a case report of acquired nocturnal head banging occurring after traumatic brain injury (Drake 1986).

Treatment

In most cases treatment is not required. Anecdotally, behavior therapy and nightly clonazepam (0.25–0.5) are both effective.

20.6 ENURESIS

In the normal course of events, 90 percent or more of children become dry during the night by the age of 5 or 6 years. Persistent, recurring bedwetting past this age is considered abnormal and is termed enuresis (Forsythe and Redmond 1974) or, more precisely, nocturnal enuresis. Enuresis is nearly twice as common in males as females.

The vast majority of cases of enuresis occur on an idiopathic basis; secondary causes of enuresis, such as diabetes mellitus, are relatively uncommon.

Clinical features

The achievement of nocturnal continence of urine is a normal developmental event, and, in most cases of idiopathic enuresis, this developmental milestone is simply never attained at the expected age. Enuresis due to other causes may have an onset at any time, from middle childhood to adult years, depending on the underlying cause.

In idiopathic enuresis, bedwetting, though perhaps more common in the first half of the night, may occur at any time. Children may or may not awaken during the bedwetting; if they do, it is always after urination has begun. Although wetting tends to be more common in NREM sleep and is generally not associated with dreaming (Pierce *et al.* 1961), it may occur during REM sleep (Mikkelsen *et al.* 1980; Neveus *et al.* 1999).

In idiopathic enuresis there are no associated symptoms. In other cases, however, one may find polyuria, dribbling, or dysuria, depending on the underlying cause.

Course

In the natural course of events, idiopathic enuresis typically undergoes a spontaneous remission: by the age of 12 years, only 3 percent of children are still bedwetting, and by adult years the figure drops to 1 percent (Forsythe and Butler 1989; Forsythe and Redmond 1974).

Etiology

In about two-thirds of cases, idiopathic enuresis is inherited on an autosomal dominant basis (von Gontard *et al.* 2001). Although it is not exactly clear what is inherited, several mechanisms have been proposed, including delays in the normal neuromuscular maturation that allows for continence, a smaller than normal bladder capacity, or either a reduced secretion of vasopressin or a decreased sensitivity of renal tubule cells to vasopressin. There is no evidence for any association with personality variables or particular methods of toilet training.

Enuresis may also occur secondary to diabetes mellitus, diabetes insipidus, urinary tract infections, cystic medullary disease, sickle cell disease, bladder or urethral obstructions, spastic bladder (as in cerebral palsy), compression of the bladder by pelvic masses or impacted stool, obstructive sleep apnea, nocturnal seizures, and treatment with various drugs, including sedatives, lithium, risperidone, clozapine and SSRIs.

Differential diagnosis

In mental retardation of moderate or greater degree, a developmental age of 4 or more years may simply never be attained and, hence, in the normal course of events nocturnal continence does not occur.

Some authors include awake wetting in children over the age of 5 or 6 years under the rubric of enuresis; however, this may not be appropriate as in these cases the wetting is usually intentional or secondary to a resistance on the child's part to make the trip to the bathroom, as may be seen in young children who are reluctant to leave their friends on the playground and simply cannot 'hold it'.

Treatment

Idiopathic enuresis may respond to a number of different treatments. Behavioral treatments should probably be tried first. Caffeinated beverages are eliminated and, except for ice chips for thirst or small sips of water for toothbrushing, fluids are withheld for the 3 hours leading up to bedtime, and the bladder is emptied just before going to bed. If the child remains dry through the night, a reward, perhaps a 'gold star' or a small present, is given the next morning. If bedwetting does occur the child should strip the bed but parents should take care of cleaning the sheets and bed and there should be no punishment. Many children respond favorably to this program in about a month. If it is unsuccessful, it should be supplemented with an 'enuresis alarm' (Forsythe and Butler 1989). Use of these alarms, which are inexpensive devices triggered by minute amounts of urine, is generally followed by gradual improvement over weeks or months. Once dryness has been maintained for a few months, the program and, if utilized, the alarm, may be discontinued; should a relapse occur, a repeat course may be given.

If behavioral measures fail, medications may be considered; of the various medications shown to be effective in double-blinded trials, imipramine (Fritz *et al.* 1994) and desmopressin (Schulman *et al.* 2001) are both considered first-line treatments. Overall, imipramine takes longer to work and tends to cause more side-effects; desmopressin, although easier to use, is not without liabilities, however, in that it carries a small risk of hyponatremia, with delirium or seizures (Dehoorne *et al.* 2006; Odeh and Oliven 2001). Once continence has been achieved with either imipramine or desmopressin, treatment should be continued until continence has been maintained for several months, after which the medication may be tapered over the following 3 months; relapses may be treated with reinstitution of the previously effective regimen.

Treatment of secondary enuresis is directed at the underlying cause.

20.7 NARCOLEPSY

Narcolepsy is characterized by narcoleptic attacks and, in most cases, cataplexy; sleep paralysis and either hypnagogic or hypnopompic hallucinations may also occur in a minority. In addition to this classic tetrad, a minority of patients will also experience episodes of 'automatic behavior'.

Narcolepsy occurs in 0.025–0.05 percent of the general population and is equally common in males and females.

Clinical features

The basic clinical features of narcolepsy have been described in a number of reports (Adie 1926; Kales *et al.* 1982; Parkes *et al.* 1975; Wilson 1928). Of the classic tetrad of symptoms, narcoleptic attacks occur in all patients, cataplexy in about three-quarters, sleep paralysis in one-third, and hypnagogic or hypnopompic hallucinations in about one-third; only about 1 in 10 patients experience the full tetrad. In over 90 percent of cases, the first manifestation of the illness is a narcoleptic attack and, although this may appear anywhere from childhood to the middle years, most patients fall ill in their late teens or early adult years.

Narcoleptic attacks are ushered in by an overwhelming and irresistible desire to sleep; although such attacks are most likely to occur in situations conducive to drowsiness, such as long-distance driving or sitting through a lecture or meeting, they can occur at any time, even during otherwise lively and engaging conversations at the dinner table. Undisturbed, patients may sleep for minutes or even up to half an hour, after which they awaken, feeling more or less refreshed. Importantly, waking patients from a narcoleptic attack is not at all difficult and may be accomplished by a light touch or simply by calling the patient's name. Distinctively, the narcoleptic attack itself consists of REM sleep (Dement *et al.* 1964, 1966; Hishikawa and Kaneko 1965; Hishikawa *et al.* 1968), and most patients will be able to recall dreams upon awakening. Narcoleptic attacks are most frequent in the afternoon or evening and may occur anywhere from once to dozens of times per day. Nocturnal sleep is often broken and some patients may complain of insomnia; indeed, for most patients, the total 24-hour sleep time is actually not increased.

Although cataplectic attacks may appear before narcoleptic attacks, this is uncommon, and in the vast majority of cases the first appearance of cataplexy is delayed, often for several years. These attacks may be either generalized or focal. Typically, the cataplectic attack is precipitated by some strong emotion, such as laughter, fear, anger, or a sudden surprise. In the generalized form, all voluntary muscle power (except for that of the diaphragm and, at times, the extraocular muscles) is diminished or lost, and the head droops forward, the jaw sags, the knees buckle, and patients may sink and collapse to the floor. In the focal type, the sudden muscle weakness is confined to one part, for example the neck musculature, with consequent head droop, the forearm and hand musculature, with items being dropped, or the extraocular muscles, with diplopia. Most attacks last for about a minute. During the attack, patients, even if fully paralyzed, remain conscious and alert, and are able, upon recovery, to give a full description of the event. In some cases, cataplectic attacks may be prolonged, lasting 5 minutes or more, and during such lengthy attacks patients may experience vivid visual hallucinations. Very prolonged attacks, lasting 20 minutes or more, are referred to as 'status cataplecticus', and these may be precipitated by the sudden discontinuation of certain medications, such as fluoxetine (Poryazova *et al.* 2005). The frequency with which cataplexy occurs varies widely, from a mere handful of attacks through the patient's lifetime to multiple attacks daily.

Sleep paralysis may occur upon either falling asleep or awakening. Although fully conscious, patients find themselves unable to move. Most attacks last only a minute or so, and some may be accompanied by visual hallucinations. Importantly, although patients appear to the observer to be sound asleep, they may nevertheless be easily awakened by simply calling their names or lightly touching them.

Hypnagogic hallucinations appear upon falling asleep, being generally visual and quite vivid and complex, as if the patient were dreaming while still awake. Hypnopompic hallucinations are quite similar and appear upon awakening.

In addition to the classic tetrad just described, perhaps one-third or so of patients will also experience episodes of 'automatic behavior'. During these episodes, patients appear to be half-asleep and, although they may continue to engage in complex behavior, such as driving a car, there is generally a decrement in the quality of behavior: if writing, patients may write a gibberish scrawl, and, if speaking, they may engage in incoherent muttering. Importantly, upon coming to full alertness, these patients generally have no recall of what they did during the event; thus, a patient who drove automatically may, upon 'coming to', have no idea of how many exits were passed.

Other sleep disorders may also be found in patients with narcolepsy. As noted in Chapter 20.2, from one- to two-thirds of patients will eventually develop REM sleep behavior disorder; a small minority will also have periodic limb movements of sleep and sleep apnea.

The clinical diagnosis of narcolepsy may be confirmed with the multiple sleep latency test (MSLT). In narcolepsy, REM sleep occurs soon after sleep onset (Rechtschaffen *et al.* 1963), and the MSLT is designed to capture this. The results, however, must be interpreted in light of the overall clinical picture, as false negatives may occur. False positives may also be found in various conditions, including psychomotorically retarded, hypersomnic depressions, sleep apnea, and alcohol or sedative/hypnotic withdrawal.

Some clinicians will also test patients to see if they are positive for the human leukocyte antigen (HLA)-DQB1*0602 haplotype. This haplotype is found in about 20 percent of the general population but in almost all patients with narcolepsy; consequently, whereas a negative test for this haplotype argues against the diagnosis, a positive test is of little diagnostic value.

Course

Although in a small minority there may be temporary or, rarely, permanent remissions, for the vast majority of patients narcolepsy is a chronic, lifelong disease.

Etiology

The various clinical features of narcolepsy each seem to represent an 'intrusion' of REM sleep into waking life. The narcoleptic attack itself appears to be nothing but an episode of REM sleep, and the other features appear to represent a fragment of REM sleep, namely atonia, with or without associated dream imagery.

Although the mechanism underlying this phenomenon has not as yet been completely worked out, current theory holds that narcolepsy occurs secondary to an autoimmune-mediated destruction of hypocretin-containing neurons within the hypothalamus; the autoimmune process, in turn, is probably triggered by an infection in genetically susceptible individuals.

One aspect of this theory has been conclusively demonstrated. Neuropathologic studies have demonstrated both gliosis and a loss of hyocretin-containing neurons in the hypothalamus (Blouin *et al.* 2005; Crocker *et al.* 2005; Thannickal *et al.* 2009). Furthermore, as might be expected, hypocretin levels within the CSF are also decreased (Dauvilliers *et al.* 2003; Ebrahim *et al.* 2003; Mignot *et al.* 2002). The search for auto-antibodies, though revealing some tantalizing results, has not, however, as yet turned up the responsible antibody.

This neuropathology does not occur on an inherited basis. As yet, only one case has been found to be secondary to a mutation in the gene for hypocretin, and the case itself was atypical (Peyron *et al.* 2000). Furthermore, the prevalence of narcolepsy among first-degree relatives of patients is low, on the order of only a few percent (Chen *et al.* 2007; Guilleminault *et al.* 1989). Genetics, however, does play a role: as noted earlier, the HLA-DQB1*0602 haplotype, though present in only 20 percent of the general population, is found in almost all patients with narcolepsy. The presence of this haplotype could indicate either a genetically determined susceptibility to certain infections, or a propensity to develop an autoimmune response to those infections. Of interest in this regard, a recent report found evidence for an increased incidence of group B streptococcal infections in patients with narcolepsy as compared with controls (Aran *et al.* 2009).

Differential diagnosis

Excessive daytime sleepiness may be seen in a number of other conditions, including sleep apnea, the Pickwickian syndrome and primary hypersomnia; however, in these conditions the excessive sleepiness, rather than occurring in the discrete, brief attacks characteristic of narcolepsy, rather waxes and wanes throughout the day; furthermore, cataplexy is not present in any of these conditions, and thus once it appears the differential question is resolved.

True sleep attacks, however, may be seen in a number of other conditions. Thus, they may occur as a side-effect of direct dopamine agonists, such as bromocriptine, pergolide, pramipexole, and ropinirole (Ferreira *et al.* 2000; Frucht *et al.* 1999; Hauser *et al.* 2000). Furthermore, they may be seen with lesions of the hypothalamus, including radiation necrosis (Dempsey *et al.* 2003), tumor (Aldrich and Naylor 1989), sarcoid granuloma (Aldrich and Naylor 1989), vascular malformation (Clavelou *et al.* 1995), infarction (Scammell *et al.* 2001) and lesions characteristic of acute disseminated encephalomyelitis (Gledhill *et al.* 2004) and both multiple sclerosis and neuromyelitis optica (Kanbayashi *et al.* 2009). Sleep attacks have also been reported with a brainstem glioma (Stahl *et al.* 1980), as a symptom of limbic encephalitis (Rosenfeld *et al.* 2001), and as a sequela to traumatic brain injury (Ebrahim *et al.* 2005) and encephalitis lethargica (Adie 1926; Fournier and Helguera 1934).

Although cases of isolated cataplexy have been noted in both multiple sclerosis, with involvement of the pons and medulla, and with a ponto-medullary tumor (D'Cruz *et al.* 1994), and also on an idiopathic basis (van Dijk *et al.* 1991), the occurrence of cataplexy outside of narcolepsy is exceedingly rare. In evaluating such cases, one must rule out drop attacks, atonic seizures and syncope. Drop attacks, as discussed in Chapter 8.4, represent vertebro-basilar transient ischemic attacks (TIAs), and should be suspected in the elderly. Atonic seizures, as discussed in Chapter 8.3, occur only in those with a long-standing history of complex partial or grand mal seizures, and of course should be suspected in those with such a history. Syncope, in turn, is distinguished from cataplexy by a loss of consciousness during the syncopal attack, a feature which contrasts with the preservation of consciousness seen in cataplectic attacks.

Treatment

Activities such as driving or operating hazardous machinery should be prohibited until narcoleptic and cataplectic attacks have been brought under control.

Narcoleptic attacks may be partially eliminated by brief scheduled naps (Roehrs *et al.* 1986; Rogers *et al.* 2001), judiciously spread out during the day. In most cases, however, pharmacologic treatment is required, and this generally involves the use of modafinil or methylphenidate. Modafinil is currently preferred and may be started at a dose of 200 mg once daily in the morning; in some cases, a total of 400 mg may be required (Broughton *et al.* 1997; US Modafinil in Narcolepsy Multicenter Study Group 1998, 2000), generally divided into a morning and early afternoon dose (Schwartz *et al.* 2005a). Methylphenidate (Mitler *et al.* 1986a) may be used as a second-line agent and is given in a total daily dose of 20–60 mg, equally divided into a morning and early afternoon schedule. Tolerance to methylphenidate may occur, and in these cases the drug should be gradually tapered over a few days followed by a 'drug holiday' for another day or two, after which it may be gradually restarted, with a resumption of effect. Importantly, neither modafinil nor methylphenidate are effective against cataplexy. Alternatives to modafinil or methylphenidate include selegiline (in a dose of 20–40 mg) (Hublin *et al.* 1994; Mayer *et al.* 1995) and sodium oxybate (Black and Houghton 2006).

Cataplectic attacks may be prevented by selegiline (Hublin *et al.* 1994) and by sodium oxybate (Black and Houghton 2006). Open studies or case reports also suggest usefulness for clomipramine (Schacter and Parkes 1980), fluvoxamine (Schacter and Parkes 1980), fluoxetine (Frey and Darbonne 1994), citalopram (Thirumalai and Shubin 2000), escitalopram (Sonka *et al.* 2006), and venlafaxine (Guilleminault *et al.* 2000).

Overall, the vast majority of patients are treated with either modafinil or methylphenidate. When cataplexy requires treatment it would appear logical to add either selegiline or sodium oxybate, given that they have support from double-blinded studies; however, in practice, these are not often used. Selegiline, in the doses required, is no longer a selective monoamine oxidase B (MAO-B) inhibitor, and hence the low tyramine diet is required, making this problematic. Sodium oxybate has high abuse potential and must be taken in the middle of the night, and is rarely used. In practice, when cataplexy does require treatment, most patients seem to do quite well on one of the antidepressants, such as venlafaxine.

All of the foregoing treatments are merely symptomatic. Although, of course, there is no way to replace the lost hypocretin-containing neurons in the hypothalamus, the theory positing an autoimmune basis for their destruction has prompted the use of intravenous immunoglobulins early in the course of the disease in the hopes of blunting the putative autoimmune assault, with one report noting prolonged improvement (Dauvilliers 2006) while another found only a temporary reprieve (Valko *et al.* 2008).

20.8 SLEEP APNEA

Sleep apnea, sometimes also referred to as 'breathing-related sleep disorder', may manifest in one of three different types, namely obstructive, central, and mixed. The common denominator in all of these types is the appearance of frequent apneic episodes during sleep, and complaints of either daytime sleepiness or, less commonly, insomnia. Cognitive deficits (including delirium and dementia) and depression are also common.

During episodes of obstructive sleep apnea the oropharyngeal airway closes and, despite ongoing vigorous inspiratory efforts of the diaphragmatic and intercostal musculature, air flow at the mouth and nares ceases. By contrast, in central sleep apnea, cessation of air flow occurs not because of any obstruction but because of a lack of inspiratory effort due to inappropriate relaxation of the diaphragmatic and intercostal musculature. In the mixed type of sleep apnea, there is, as the name suggests, a combination of the preceding two types: initially one sees a lack of inspiratory effort, as in the central type; this is soon followed by inspiratory effort of the diaphragmatic and intercostal musculature, which, owing to oropharyngeal obstruction, is not followed by inspiration.

Of these three types, obstructive sleep apnea is the most common, followed by the mixed type; pure central types are relatively rare.

Sleep apnea is a common disorder, found in anywhere from 2 to 4 percent of the adult population over 40 years; it is at least twice as common in males as in females.

Clinical features

Although sleep apnea can occur at any age, it generally appears in middle years.

Apneic episodes last any time from 10 seconds up to 2 minutes, and may occur anywhere from 30 to several hundred times a night. In addition to episodes of complete apnea, patients will often also have hypopneic episodes, in which, although some inspiration occurs, it is at least 50 percent less than that seen with a normal breath. The severity of sleep apnea is often quantified by the 'apnea index' or the 'apnea–hypopnea index', which is equal to either the total number of apneic episodes per night or the total number of apneic plus hypopneic episodes per night divided by the number of hours spent in sleep. At a minimum, one wishes to see an apnea index of greater than 5 or an apnea–hypopnea index greater than 10.

In obstructive sleep apnea episodes (Strollo and Rogers 1996; Whyte *et al.* 1989), the patient, often an obese, middle-aged man with a short, thick neck, presents with a chief complaint of daytime sleepiness, which may or may not be accompanied by complaints of broken sleep. Almost universal is a history of frequent loud snoring, which may have prompted equally loud complaints from the patient's bed partner. Upon observing these patients while they sleep, for example while making rounds early in the morning, one may see a characteristic episode: oral and nasal airflow ceases despite increasingly vigorous diaphragmatic and intercostal muscular activity, until finally the obstruction resolves with a loud, gasping snort, at which point the patient may or may not awaken; should awakening occur it lasts only seconds, after which sleep resumes.

In central sleep apnea, patients tend to complain of insomnia and restless sleep, and there may or may not be any daytime drowsiness. As snoring is typically absent, there may be no complaints from the patient's bed partner, who may be unaware of the problem. The typical episode of central sleep apnea is far less dramatic than an obstructive one, as patients with central sleep apnea simply stop breathing: the chest and diaphragm are relaxed and there is no air flow. Eventually, inspiratory effort occurs with easy inspiration, and, at this point, the patient often has a transient awakening.

Mixed apneic episodes pursue a biphasic course. After an initial period of central apnea an inspiratory effort begins but is met by oropharyngeal obstruction, causing a period of obstructive apnea. Clinically, these patients resemble more the obstructive than the central type, and one often hears complaints of snoring and daytime sleepiness.

In addition to complaints of daytime sleepiness or insomnia, patients with sleep apnea typically also experience other symptoms (Guilleminault *et al.* 1977), including headache, sleep drunkenness, during which patients experience a brief period of confusion as they struggle to awaken in the morning, enuresis, and erectile dysfunction, which may be quite common (Goncalves *et al.* 2005).

Cognitive deficits are common. There may be difficulty with concentration and patients may complain of feeling 'fuzzy' or 'wooden-headed' during the day; there may also be more or less severe inattentiveness, short-term memory loss and perseveration (Feuerstein *et al.* 1997). Mechanical ability is impaired, and traffic accidents are more frequent in patients with obstructive sleep apnea than in the general public (George *et al.* 1987). In an uncertain percentage of patients, delirium may occur (Dyken *et al.* 2004; Lee 1998; Munoz *et al.* 1998; Sandberg *et al.* 2001; Whitney and Gannon 1996); in general hospital practice this has not been at all uncommon. Rarely, rather than delirium there may be a dementia (Scheltens *et al.* 1991; Steiner *et al.* 1999).

Depressive symptoms are also common (Millman *et al.* 1989; Schwartz and Karatinos 2007, Schwartz *et al.* 2005b), with depressed mood, indecisiveness, low self-esteem, helplessness, tearfulness, and, in some cases, suicidal ideation.

During episodes of sleep apnea, hypercapnia and hypoxia occur, and cyanosis may be seen. With frequent episodes of hypercapnia, pulmonary hypertension may occur, leading to cor pulmonale. Systolic hypertension is common. Arrhythmias may occur during episodes, including sinus bradycardia, sinus tachycardia, sinus arrest, atrial flutter, premature ventricular contractions and ventricular tachycardia.

Sleep apnea may also aggravate epilepsy (Malow *et al.* 2000) and increase the risk of stroke and death (Yaggi *et al.* 2005).

Polysomnography should be considered in all cases of suspected sleep apnea, not only to document its presence and the type of sleep apneic episodes but also to determine the presence of arrhythmias. Often, overnight oximetry is performed as a screening test; however, this is not as sensitive as polysomnography.

Course

Sleep apnea is chronic.

Etiology

Obstructive episodes may occur secondary to various conditions, either singly or in combination, including the following: hypertrophy of the adenoids or tonsils; micrognathia (in which the tongue falls back to occlude the airway); lingual hypertrophy (as may be seen in hypothyroidism, acromegaly, or Down's syndrome); myotonic muscular atrophy; and, most commonly, obesity. Occasionally, no obvious cause is found for the obstruction, and in such cases it is suspected that there is a failure of the brainstem mechanisms responsible for maintaining the patency of the airway during sleep.

Central episodes are seen in association with obesity and congestive heart failure and may also occur in multiple system atrophy and with medullary lesions, for example infarctions, tumors, multiple sclerosis, or syringobulbia.

Differential diagnosis

Morbidly obese patients may be afflicted not only with sleep apnea but also with the Pickwickian syndrome, and the differential between obstructive sleep apnea alone and the combination of obstructive sleep apnea plus the Pickwickian syndrome, as discussed in Chapter 20.9, rests on waking arterial blood gases, which reveal a normal PCO_2 in sleep apnea alone, but an elevated PCO_2 when the Pickwickian syndrome is present.

Narcolepsy is distinguished by the fact that in this condition daytime sleepiness comes in discrete attacks, in contrast to the chronic, waxing and waning sleepiness seen in sleep apnea.

Treatment

In obstructive sleep apnea, correction of the underlying cause, such as obesity (Smith *et al.* 1985), may be followed by significant relief. In very mild cases, some relief may also be gained by having patients sleep on their sides, a position that favors airway patency (Cartwright *et al.* 1985, 1991). In most cases, however, a continuous positive airway pressure (CPAP) or bi-level positive airway pressure (BIPAP) device is required, and the relief gained with these devices may be pronounced: not only daytime sleepiness but also cognitive deficits (including delirium or dementia), depression, erectile dysfunction, and enuresis are all improved or cleared. Pending acquisition of one of these devices, patients are often treated with nocturnal oxygen via nasal cannulae: the results here, however, are generally not impressive.

In those who cannot tolerate CPAP or BIPAP, or in those in whom they are ineffective, the use of an oral appliance or surgery may be contemplated. Oral appliances serve to keep the jaw advanced, thus preventing the tongue from occluding the airway. Surgical options include uvulopalatopharyngoplasty or uvuloplasty, and, in severe and limiting cases, tracheostomy.

In all cases where treatment is not fully effective, consideration should be given to the use of modafinil, which, in doses of 200–400 mg daily, may partially relieve daytime sleepiness (Kingshott *et al.* 2001; Black and Hirshkowitz 2005).

Further, in cases where depression is not fully relieved, consideration may be given to either protriptyline, 10–20 mg at bedtime (Brownell *et al.* 1982), or paroxetine 20 mg at bedtime (Kraiczi *et al.* 1999), not only because these are effective antidepressants, but also because they have some effect, albeit a very modest one, in the treatment of obstructive sleep apnea itself.

In central sleep apnea, acetazolamide, in a dose of 250 mg four times daily, may provide some relief (White *et al.* 1982); however, its effectiveness often wanes. In severe cases, diaphragmatic pacing may be required.

The treatment of mixed episodes is similar to that of the obstructive type.

Regardless of the type of sleep apnea present, care must be taken to avoid alcohol (Scrima *et al.* 1982) or any medication that might depress respiratory drive, including sedative–hypnotics (Mendelson *et al.* 1981), anxiolytics, sedating antidepressants, opioids, etc. Of note, sildenafil, which may be prescribed for the erectile dysfunction seen in obstructive sleep apnea, may also worsen the apnea (Roizenblatt *et al.* 2006), and consequently care must be taken in its use.

20.9 PICKWICKIAN SYNDROME (OBESITY–HYPOVENTILATION SYNDROME)

The constellation of morbid obesity, a ruddy complexion, and somnolence has long been recognized and was exemplified by the overweight boy in Charles Dickens's *The Pickwick Papers*. Following the lead given by Dickens, Burwell *et al.* (1956) coined the term 'Pickwickian syndrome'; a more recent name is 'obesity–hypoventilation syndrome' and, although this is gaining currency, it lacks the color and flair of the eponym. As noted below, the extreme obesity in this syndrome leads to waking hypoventilation.

Clinical features

Patients are morbidly obese and often have a ruddy complexion; they are typically somnolent and lethargic and have difficulty paying attention or concentrating (Burwell *et al.* 1956; Drachman and Gumnit 1962; Nowbar *et al.* 2004; Sieker *et al.* 1955; Ward and Kelsey 1962). Arterial blood gases drawn while patients are awake reveal significant hypercapnia and hypoxemia; erythrocytosis may also occur, as may pulmonary hypertension and cor pulmonale. Although, as might be expected given the morbid obesity, most patients will also have obstructive sleep apnea, this is not inevitable, and a small minority will have the Pickwickian syndrome alone, occurring in isolation without any significant sleep apnea (Kessler *et al.* 2001).

Course

In general, this parallels the course of the underlying obesity. These morbidly obese patients are prone to venous stasis and deep venous thrombosis, and any acute worsening of their clinical status should prompt a search for pulmonary emboli.

Etiology

The burden of excess adipose tissue encircling the chest and also pushing up the diaphragm from the obese abdomen below leads to chronic alveolar hypoventilation, with resulting chronic hypercapnia.

Differential diagnosis

Morbidly obese patients may have the obstructive sleep apnea syndrome alone, a combination of obstructive sleep apnea and the Pickwickian syndrome, or the Pickwickian syndrome alone, and the differential may be made with polysomnography and waking arterial blood gases. Polysomnography reveals an apnea index of 5 or more (or an apnea–hypopnea index of 10 or more) in both obstructive sleep apnea alone and in the combination of obstructive sleep apnea and the Pickwickian syndrome, whereas only a minimal number of events is seen in patients with the Pickwickian syndrome alone. Waking arterial blood gases reveal a normal PCO_2 in obstructive sleep apnea alone, whereas in both the combination of obstructive sleep apnea and the Pickwickian syndrome, and the Pickwickian syndrome alone, one finds an elevated PCO_2. The key here is the presence of an elevated PCO_2 while the patient is awake: presuming that other causes of an elevated PCO_2 have been ruled out (e.g. chronic obstructive pulmonary disease [COPD, or 'blue bloaters'], respiratory depression due to sedating medications, myxedema), this finding, in a morbidly obese patient, strongly suggests that the Pickwickian syndrome is present, either in combination with obstructive sleep apnea or in isolation.

Treatment

Weight loss is critical and effective (Chiang *et al.* 1980), and, pending this, BIPAP should be utilized, as it leads to a reduction in daytime hypercapnia (Piper *et al.* 2008); alcohol, sedative–hypnotics, and any other medications that may reduce respiratory drive, such as antihistamines or opioids, should be avoided. In cases with significant hypoxemia, one may be tempted to use supplemental oxygen, but this must be approached only with extreme caution: these patients typically run on oxygen drive alone, and a correction of hypoxemia may be followed by a sudden rise in PCO_2. Consideration may also be given to a trial of either modafinil or methylphenidate, as used in obstructive sleep apnea.

20.10 KLEINE–LEVIN SYNDROME

The Kleine–Levin syndrome, first described by Kleine in 1925 (Kleine 1925) and Levin in 1936 (Levin 1936), is a rare disorder characterized by episodes of hypersomnia and associated symptoms, the most notable of which is hyperphagia. The first episode typically occurs in adolescence, and the disorder is far more common in males than females.

Clinical features

The clinical features of the Kleine–Levin syndrome have been described in a number of case series (Arnulf *et al.* 2008; Critchley 1962; Critchley and Hoffman 1942; Dauvilliers *et al.* 2002; Gadoth *et al.* 2001) and a recent comprehensive review (Arnulf *et al.* 2005). As noted, this is an episodic disorder, and although the age of onset of the first episode ranges from childhood through geriatric years, in roughly 80 percent of cases the first episode occurs in adolescence. Although in the majority of cases the first episode is preceded by an infection, often viral, subsequent episodes generally occur without any precipitating factors. The episodes themselves generally last in the order of 2 weeks; however, the range is wide, from days up to 3 months.

Episodes may be ushered in by a brief prodrome, lasting perhaps days, of malaise, headache, and lethargy. During the episode proper, all patients experience hypersomnia, often sleeping 18 or more hours per day. During waking hours, between two-thirds and three-quarters of patients will also experience hyperphagia. Cognitive changes, ranging from mere impaired concentration up to confusion and disorientation, are seen in almost all cases. Mood changes are seen in over half of all patients and typically consist of depression. Hypersexuality occurs in roughly one-half of patients and may manifest with exhibitionism, unwelcome sexual advances, and frequent, and at times public, masturbation. Delusions and hallucinations may appear in a small minority, as may unusual behaviors such as persistent humming and singing.

As noted, hypersomnia and hyperphagia constitute the primary symptomatology seen during an episode. Levin (1936) noted that

> the patient sleeps excessively day and night, in extreme instances waking only to eat and go to the toilet. He can always be roused. When roused he is usually irritable and wants to be left alone so that he can go back to sleep. He is abnormally hungry and eats excessively.

The hyperphagia seen during the episode, though often favoring sweets, may be indiscriminate, with patients eating whatever is at hand (Critchley 1962), begging for food from other patients (Garland *et al.* 1965), or indulging in unusual food preferences (Will *et al.* 1988).

Of the possible cognitive changes, impaired concentration and sluggish, slowed thought are most common, often accompanied by impaired decision making. The overwhelming majority will also experience one or more of confusion, decreased short-term memory, and disorientation to time or place.

Mood changes tend towards depression, often tinged with irritability; some may become agitated. Critchley (1962) noted that 'in the majority ... the patient when awake is intensely irritable and resentful ... a few patients are actually hostile in their behavior ... in his attitude toward others he may be uninhibited, insolent, and quarrelsome'.

Hypersexuality may be very problematic: one patient masturbated in public (Fernandez *et al.* 1990), others 'to the point of bleeding' (Arnulf *et al.* 2008), and one patient continued his 'vigorous attempts at masturbation' during the interview, refusing laboratory testing unless 'the house physician was prepared to "come into bed and give me a feel over"'(Garland *et al.* 1965).

Delusions, which are typically of persecution, and hallucinations, which may be either auditory or visual, are seen in a small minority (Arnulf *et al.* 2008).

Unusual behaviors may also be seen. Some authors have commented that patients would 'sing inappropriately' (Chiles and Wilkus 1976); in one case the patient 'frequently burst into song' but sang the same song over and over again (Garland *et al.* 1965). In other cases patients may pace, wring their hands, tear out their hair, or engage in body rocking.

As noted earlier, most episodes last in the order of 2 weeks. Upon recovery, most patients are more or less amnestic for the events that occurred during the episode (Critchley 1962; Levin 1936), and some may experience a residual mood disturbance (Critchley 1962), tending towards either depression (Gallinek 1954) or elation (Gilbert 1964), which passes within a week or so.

Magnetic resonance scanning is normal, as is assay of the CSF. In the majority of cases the electroencephalogram (EEG) displays diffuse, mild slowing during an episode, with at times an accentuation of this in the temporal or frontotemporal areas. There are no ictal or interictal epileptiform discharges, and in between episodes the EEG normalizes. Polysomnography and multiple sleep latency testing may reveal sleep-onset REM, frequent awakenings, and, most commonly, decreased delta sleep (Huang *et al.* 2008; Pike and Stores 1994; Reynolds *et al.* 1980, 1984).

Course

Subsequent to recovery from the first episode, it appears that in most, if not all, cases there will be subsequent episodes. Although the intervals between subsequent episodes varies widely, ranging from a few weeks up to 6 years, on average, at least early on in the course, the interval is about 6 months. Over long-term follow-up, however, in most cases the intervals between episodes gradually lengthen, and eventually spontaneous remission occurs, typically, in those with adolescent onset, in early adult years.

In the intervals between episodes, although it appears that the vast majority of patients return to normal, there is some suggestive evidence that in some cases there may be some residual quarrelsomeness and slightly reduced academic ability (Sagar *et al.* 1990).

Etiology

Only a few cases have come to autopsy: in one (Carpenter *et al.* 1982) there was widespread microgliosis in the thalamus, with little neuronal loss; in another (Fenzi *et al.* 1993) there was microgliosis not only in the thalamus but also in the midbrain; and in a final one (Koerber *et al.* 1984) there was mild depigmentation of the substantia nigra and locus ceruleus, without evidence of Lewy bodies. The role of the thalamus in this disorder is further highlighted by a single photon emission computed tomography (SPECT) study (Huang *et al.* 2005), which found hypoperfusion of the thalami in seven out of seven patients tested, with resolution of this abnormality upon recovery.

Although the mechanism underlying these changes is not clear, the frequency with which the first episode follows an infectious illness has suggested an autoimmune basis (Dauvilliers *et al.* 2002), with the infection triggering an immune attack on diencephalic or mesencephalic structures.

Differential diagnosis

The main item on the differential is a depression. Depressive episodes of either major depressive disorder or bipolar disorder may, in a minority of cases, be characterized by hypersomnia and increased appetite; however, the increased eating seen in depression does not lead to the sort of indiscriminate eating seen in Kleine–Levin syndrome; furthermore, one does not see hypersexuality in depression.

Treatment

There are no controlled studies regarding either acute or preventive treatment.

With regard to acute treatment, some success has been reported with modafinil (Arnulf *et al.* 2008; Dauvilliers *et al.* 2002), methylphenidate (Arnulf *et al.* 2008) and amphetamine (Arnulf *et al.* 2008; Gallinek 1962). Importantly, these agents were effective for hypersomnia alone, with the other symptoms being left untouched.

Regarding preventive treatment, aimed at lengthening the interval between episodes, success has been noted with lithium (Arnulf *et al.* 2008; Dauvilliers *et al.* 2002; Goldberg 1983; Muratori *et al.* 2002; Ogura *et al.* 1976; Poppe *et al.* 2003), valproic acid (Arnulf *et al.* 2008), carbamazepine (Dauvilliers *et al.* 2002; El Hajj *et al.* 2009) and gabapentin (Itokawa *et al.* 2009). Of these agents, lithium has the best track record (Arnulf *et al.* 2005), and should probably be tried first.

20.11 RESTLESS LEGS SYNDROME

The restless legs syndrome, also known as Ekbom's syndrome, is a common disorder, with a lifetime prevalence of about 5 percent. As the name implies, it is characterized by the experience of restlessness in the legs, an experience so disagreeable and uncomfortable that it keeps patients from falling asleep. This disorder occurs in two forms: a primary form, which is hereditary, and a secondary form, which occurs on the basis of iron deficiency anemia, uremia, and a host of other disorders, described below in the section on etiology.

Clinical features

In both primary and secondary forms the onset is typically gradual. Primary forms generally first appear in early adult years, whereas the age of onset of the secondary forms is determined by the underlying condition; for the most part, however, the onset of the secondary form tends to be in middle or later years.

Clinically (Ekbom 1960; Montplaisir *et al.* 1997), patients report that when they lie down, or even merely sit down, they feel a restlessness deep within the legs, especially in the calves; some may also experience formication, with a sense of crawling under the skin, or an aching discomfort. Often the experience is so uncomfortable that patients feel impelled to get up and walk about, which brings some relief. Over time, the restlessness may begin to involve the upper extremities.

At night, falling asleep may be almost impossible, and patients may either try and lay still and bear the discomfort or spend hours out of bed, pacing about. Typically, symptoms lessen by early morning hours and patients may then be able to get some sleep.

The majority of patients will also experience periodic limb movements of sleep (Montplaisir *et al.* 1997), described in Chapter 20.12.

Course

Primary restless legs syndrome is generally chronic, and symptoms may either wax and wane in intensity over time or progressively worsen. The course of the secondary form is determined by the underlying cause.

Etiology

The primary, or idiopathic, form of restless legs syndrome is hereditary, and occurs in both autosomal dominant and recessive forms (Levchenko *et al.* 2006; Trenkwalder *et al.* 1996; Winkelmann *et al.* 2000, 2002); several loci have also been identified (Kemlink *et al.* 2009). Although the neuropathology of restless legs syndrome remains uncertain, attention has been directed toward the substantia nigra and the putamen. The effectiveness of dopaminergic agents in the treatment of restless legs syndrome and the similarity of restless legs syndrome to the syndrome of akathesia produced by dopamine-blocking agents both suggest that these structures may be involved. The pathology here, however, appears to be functional: neuropathologic studies have revealed neither cell loss within the substantia nigra nor any tau or synucelin pathology within these cells (Pittock *et al.* 2004). Neuropathology does, however, reveal a deficiency of transferrin receptors on neuromelanin cells (Connor *et al.* 2003, 2004), and both magnetic resonance imaging (MRI) (Allen *et al.* 2001) and transcranial sonographic (Godau *et al.* 2008; Schmidauer *et al.* 2005) studies have demonstrated a lack of iron in the substantia nigra, while within the CSF there is a reduction in ferritin levels (Earley *et al.* 2000). Neuropathology further reveals both an increase in tyrosine hydroxylase within neuromelanin cells and a reduction in the number of D_2 receptors within the putamen (Connor *et al.* 2009). These findings may all be connected by positing an intracellular iron deficiency. A deficiency of transferrin receptors could lead to a reduced iron uptake into the neuromelanin cells. Reduced intracellular iron then prompts an upregulation of tyrosine hydroxlase activity, with a consequent increased production of dopamine; the increased dopamine release within the putamen would then lead to a downregulation of the postsynaptic D_2 receptors within that structure. It must be stressed, however, that this posited intracellular iron deficiency is due to a failure to take up iron into the cells: there is no evidence for any role for systemic iron deficiency in the primary form of restless legs syndrome.

Secondary restless legs syndrome, by contrast, is associated with iron deficiency anemia (Rangarajan and D'Souza 2007), and, indeed, this is one of the most common causes of the secondary form. Other causes include uremia (Winkelmann *et al.* 1996, 2000), chronic hemodialysis (Kim *et al.* 2008), Parkinson's disease (Krishnan *et al.* 2003, Peralta *et al.* 2009), infarction of the basal ganglia, internal capsule or thalamus (Lee *et al.* 2009), multiple sclerosis (Manconi *et al.* 2008), spinal cord lesions (Hartmann *et al.* 1999), spinal anesthesia with either bupivacaine or mepivacaine (Hogl *et al.* 2002), rheumatoid arthritis (Salih *et al.* 1994), and pregnancy (Manconi *et al.* 2004). Various medications may also either cause or exacerbate restless legs syndrome: mirtazapine is the most common offender, followed by SSRIs, venlafaxine and duloxetine (Rottach *et al.* 2008). Some studies have also found an association between restless legs syndrome and sensory polyneuropathy (Gemignani *et al.* 2006; Polydefkis *et al.* 2000); however, a recent, well-done study revealed that whereas there was an association with various hereditary polyneuropathies, the acquired polyneuropathies (e.g. diabetic polyneuropathy) were innocent in this regard (Hattan *et al.* 2009).

Differential diagnosis

The chief differential possibility is akathisia. In akathisia, as described in Chapter 4.10, patients experience a restlessness when seated or lying down and may feel impelled to get up; hence, they present a picture similar to that of the restless legs syndrome. Two features, however, enable a differentiation to be made (Walters *et al.* 1991). First, when patients with the restless legs syndrome get up they tend to pace about, whereas those with akathisia often 'march in place'. Second, patients with the restless legs syndrome often get some relief by massaging their calves, a maneuver that is useless in akathisia.

Painful sensory polyneuropathies, as, for example, certain cases of diabetic polyneuropathy, may make it hard to fall asleep; however, here, the pain is present throughout the day, whether lying down, sitting or standing, in contrast to the discomfort present in some cases of restless legs syndrome, which is generally most prominent in the evening, and is relieved by standing and walking about.

Treatment

In secondary cases the underlying cause should, if possible, be treated, as this may bring relief. In this regard, it is appropriate to keep in mind that even mild iron deficiency may exacerbate symptoms in patients who suffer from the primary form of restless legs syndrome. Thus, a ferritin level should be checked in all cases, and, if it is low, oral iron should be given (Wang *et al.* 2009).

In secondary cases in which treatment of the underlying cause is either not possible or ineffective, and in all primary cases, various medications may be considered. The most commonly used are dopaminergic agents, including both levodopa and the direct-acting dopaminergic agents pramipexole and ropinirole. Gabapentin constitutes a viable option, and is considered next. Other, much less commonly used, options include clonazepam, oxycodone, carbamazepine and clonidine.

Levodopa, in combination with carbidopa (Benes *et al.* 1999; Brodeur *et al.* 1988), may be started at a dose of 100 mg levodopa and 25 mg carbidopa, which may be doubled if needed after 3–7 days. In many cases, the effect of levodopa may wear off early in the night, allowing for a re-emergence of symptoms, and in such cases, combining immediate-release levodopa/carbidopa with a sustained-release preparation may provide longer relief. Of the direct-acting dopaminergic agents that are effective in the restless legs syndrome, pramipexole (Jama *et al.* 2009; Montplaisir *et al.* 1999) and ropinirole (Bogan *et al.* 2006) are currently used. Starting doses are 0.125 mg of pramipexole or 0.25 mg of ropinirole, and the dose may be increased in similar increments every 3–7 days up to a maximum of 0.75 mg of pramipexole or 4 mg of ropinirole. All of these dopaminergic agents may be either given as a single dose 1–3 hours before bedtime or divided into two doses, given in the early evening and then at bedtime. Importantly, treatment with dopaminergic agents may be associated with the side-effect of 'augmentation': this phenomenon, seen in roughly half of all patients treated with either levodopa (Hogl *et al.* 2010) or the direct-acting dopaminergic agents (Ondo *et al.* 2004; Winkelman and Johnston 2004), appears after several months, and is characterized by both an earlier appearance and a worsening of the symptoms of restless legs. Treatment with direct-acting dopaminergic agents may also cause other side-effects: pathologic gambling occurs in a small minority (Driver-Dunckley *et al.* 2007), and rarely psychosis may emerge, as has been reported with ropinirole (Stoner *et al.* 2009).

Gabapentin (Garcia-Borreguero *et al.* 2002) may be started at a low dose of 300–400 mg and increased every few days in similar increments up to a maximum of 3600 mg; the dose should be divided into an early evening dose, equal to one-third of the total, and a bedtime dose of the remainder. Most patients respond to a total daily dose of 1800 mg.

Clonazepam (Saletu *et al.* 2001) may be started at a dose of 0.5 mg 1 hour before bedtime, with similar increments every 3–7 days up to a maximum of 2 mg. Carbamazepine (Telstad *et al.* 1984) may be started at a dose of 100 mg at bedtime, and increased in similar increments every week up to 300 mg. Oxycodone (Walters *et al.* 1993) is effective in a dose of about 15 mg at bedtime, and clonidine in a dose of 0.025–0.05 mg (Wagner *et al.* 1996).

Although most texts recommend trying a dopaminergic agent first, consideration should be given to a trial of gabapentin, and for two reasons. First, in the only double-blinded comparison of gabapentin vs. a dopaminergic agent, gabapentin was superior to ropinirole (Saletu *et al.* 2010). Second, gabapentin, unlike the dopaminergic agents, does not cause augmentation. If, however, a dopaminergic agent is chosen, either pramipexole or ropinirole should be tried first, as clinical experience suggests that the direct-acting dopaminergic agents are more effective than levodopa. Of the other agents, clonazepam is probably the best tolerated.

20.12 PERIODIC LIMB MOVEMENTS IN SLEEP

Periodic limb movements in sleep is characterized by multiple jerkings of one or both legs during sleep. Although this disorder has traditionally been referred to as nocturnal myoclonus, this is actually a misnomer because, as pointed out below, the abnormal movements seen in this disorder are not truly myoclonic in character but, when fully developed, resemble more a 'triple flexion' response.

Patients may or may not be awakened by the abnormal movements; if they are awakened and complain of either insomnia or of daytime sleepiness, then it is customary to speak not of 'periodic limb movements in sleep' but rather 'periodic limb movement disorder' – this, in turn, may occur either on an idiopathic basis or secondary to various other disorders. Periodic limb movement disorder is common, seen in at least 4 percent of the general population.

Clinical features

The onset of the disorder may occur at any time from early adult years to old age.

The jerking movement itself may occur in one or both of the lower extremities. Upon observation, one sees dorsiflexion at the ankle accompanied in most cases by dorsiflexion of the great toe; in many cases these movements are accompanied by flexion at the knee and hip, thus mimicking a classic triple flexion response (Coleman *et al.* 1980; Smith 1985; de Weerd *et al.* 2004). The jerkings evolve over anywhere from 0.5 to 4 seconds, and repeat every 20–40 seconds during episodes that last from minutes to an hour or more. Episodes occur during NREM sleep and may recur throughout the night. The jerkings may or may not be accompanied by an awakening and, if they are, patients may complain of either insomnia or daytime sleepiness. Despite these complaints, patients themselves may be unaware of the jerkings, and an accurate history of these nocturnal events may depend on their description by a bed partner, who may complain of being repeatedly 'kicked' during the night.

A minority of patients will also have the restless legs syndrome.

Course

Idiopathic periodic limb movement disorder appears to be chronic; the course of the secondary form is determined by the underlying cause. Of note, antidepressants may exacerbate symptoms, as has been noted with tricyclics (Ware *et al.* 1984), SSRIs, and venlafaxine (Yang *et al.* 2005).

Etiology

The idiopathic form appears to be inherited, and an association has been found with a common variant on chromosome 6 (Stefansson *et al.* 2007). Although the mechanism underlying the abnormal movements is not known, their strong resemblance to a Babinski response suggests that they result from a lack of normal supraspinal inhibition (Smith 1985).

Secondary forms have been associated with congestive heart failure (Skomro *et al.* 2009), chronic hemodialysis (Rijsman *et al.* 2004), alcohol withdrawal (Gann *et al.* 2002), Parkinson's disease (Wetter *et al.* 2000), lacunar infarctions in the corona radiata (Kang *et al.* 2004), striatum (Sechi *et al.* 2008) or pons (Kim *et al.* 2003), and spinal cord lesions (Lee *et al.* 1996; Nogues *et al.* 2000).

Differential diagnosis

Isolated jerkings, occurring at a frequency of up to five per hour, may be an incidental finding on polysomnography (Mendelson 1996) and are not associated with any symptoms.

Hypnic jerks (Oswald 1959), also known as 'sleep starts', are a normal accompaniment of the transition into sleep. They differ from the jerkings seen in periodic limb movements in that they are very brief, typically involve all four extremities, and occur only as the individual is falling asleep.

Myoclonus, discussed in Chapter 4.2, is distinguished from the jerkings of periodic limb movements by its 'shock-like' rapid onset, in contrast to the relatively leisurely evolution of the jerking movement.

Treatment

Medications effective in periodic limb movement disorder are similar to those used in restless legs syndrome. Thus, immediate-release levodopa/carbidopa (Benes *et al.* 1999;

Brodeur *et al.* 1988; Kaplan *et al.* 1993) is effective, and should the effect prove too short lived, this immediate-release preparation may be given in combination with a sustained-release preparation. The direct-acting dopaminergic agents pramipexole (Jama *et al.* 2009; Montplaisir *et al.* 1999) and ropinirole (Bliwise *et al.* 2005) are also effective, as are gabapentin (Garcia-Borreguero *et al.* 2002), clonazepam (Mitler *et al.* 1986b; Ohanna *et al.* 1985; Peled and Lavie 1987), and oxycodone (Walters *et al.* 1993). The choice among these medications, and their method of use, is similar to that noted for restless legs syndrome in Chapter 20.11; of note, 'augmentation', as may be seen with dopaminergic agents in the restless legs syndrome, has not been reported in periodic limb movement disorder.

20.13 PAINFUL LEGS AND MOVING TOES

As the name implies, this syndrome is characterized by pain in the legs (which may be quite severe) and involuntary movements of the toes, all resulting in insomnia. This is a rare but potentially devastating condition.

Clinical features

The syndrome (Alvarez *et al.* 2008; Dressler *et al.* 1994; Spillane *et al.* 1971) generally has an onset in the sixth or seventh decade. In most cases pain appears first and, although symptoms may begin unilaterally, bilateral involvement eventually ensues. As noted by Spillane *et al.* (1971), the pain varies 'in intensity from discomfort to a pain of great severity... an ache, an intense pressure, a tightness, a feeling that the toes were pulling or being pulled, a throbbing, bursting, crushing... [or] a deep burning' and the movements consist of a 'sinuous clawing and re-straightening, fanning and circular movements of the toes'. The effect of the toe movements can be remarkable: one of Spillane *et al.*'s patients 'was surprised to see that the toes were actually moving "as though they were playing a piano on their own"'. The symptoms are not relieved by walking about and insomnia can be severe (Montagna *et al.* 1983). When sleep does come, the abnormal movements cease.

Course

In most cases this syndrome appears to be chronic.

Etiology

The syndrome has been noted secondary to lesions of the cord, posterior lumbar roots, and peripheral nerves, and with trauma to the back or feet (Dressler *et al.* 1994; Ikeda *et al.* 2004; Montagna *et al.* 1983; Nathan 1978; Pla *et al.* 1996; Schott 1981). Interestingly, lesions or trauma need not be bilateral; unilateral lesions may be followed initially by an ipsilateral onset, but eventually the contralateral extremity becomes involved.

Differential diagnosis

The restless legs syndrome is distinguished by an absence of pain and abnormal movements, and by the characteristic relief obtained by walking about. Reflex sympathetic dystrophy is distinguished by the lack of abnormal movements.

Treatment

There are no controlled treatment studies. Case reports indicate relief with gabapentin and clonazepam, and with epidural block, lumbar sympathectomy, and utilization of transcutaneous electrical nerve stimulation (TENS).

20.14 CIRCADIAN RHYTHM SLEEP DISORDER

The onset and duration of sleep, as is the case with most biological rhythms, is under the control of the suprachiasmatic nucleus of the hypothalamus, which functions as the body's 'internal clock'. Under its influence, most people begin to feel sleepy in the evening, go to sleep between the hours of 2000 and 2400, sleep for 7 or 8 hours, and then awaken, generally feeling refreshed. Social demands for work and other functions are built around this biologically determined schedule.

Whenever there is a mismatch between social demands for sleep and wakefulness and this biologically determined rhythm, one speaks of a circadian rhythm sleep disorder. Such mismatches may occur in one of two ways: either the timing of social demands changes or the internal 'clock' changes. With the first kind of change, one may see the 'jet lag' and 'shift work' types of circadian rhythm sleep disorder, and with the second kind of change one may find the 'delayed sleep phase', the 'advanced sleep phase', the 'free running' and the 'irregular sleep–wake' types of circadian rhythm sleep disorder.

As a whole, circadian rhythm sleep disorders are very common. 'Jet lag' occurs routinely in air travelers and some 7 million Americans are exposed to the deleterious effects of shift work. The delayed sleep phase type occurs in roughly 7 percent of adolescents; the free-running and irregular sleep–wake types, by contrast, are very uncommon.

Clinical features

The jet lag type occurs with air travel that crosses five or more meridians. With eastward flight, the traveller's internal clock is phase-delayed relative to the local schedule, and travelers tend to feel awake long past the local 'bedtime', with resulting insomnia. With westward travel, which is generally better tolerated, the traveller's internal clock is phase-advanced relative to the local time, and travelers find themselves sleepy quite early in the evening.

The shift work type occurs when individuals, normally accustomed to working during the day and sleeping at night, are shifted to night-time work. Although a minority of individuals adjust to this change fairly rapidly, most do not, and, under the influence of their internal clock, they feel sleepy during the night while at work and have trouble sleeping during their daytime off-hours.

The delayed sleep phase type is characterized by an inability to fall asleep at socially approved times. Patients often stay awake until the early morning hours and are then unable to awaken early enough in the morning to get to school or work on time. Importantly, if such 'night owls' are allowed to 'sleep in' they get a normal amount of sleep and awaken refreshed. In most cases, this type has an onset in adolescence.

The advanced sleep phase type, which is generally restricted to the elderly, is characterized by an urge to go to sleep very early in the evening. Such individuals then awaken some 7 or so hours later, in the very early morning. These 'morning larks' awaken refreshed and get a start on the day long before their companions.

In the free-running type (also known as the 'non-24-hour' or 'non-entrained' type) there is a failure of entrainment of the internal clock to the normally occurring 24-hour schedule. Given that the internal clock, when allowed to 'run free', has a cycle length some 15–30 minutes longer than 24 hours, the internally generated sleep–wake cycle tends to 'drift' forward more and more each day, and such patients experience progressively worsening insomnia, accompanied by daytime sleepiness, until their free-running internal clock finally completes enough cycles to bring it into alignment with the normal environmental 24-hour cycle.

In the irregular sleep–wake type the urge to sleep seems to come at random, with no clear-cut relationship to any cycle, whether environmental or internal.

Etiology

The jet lag and the shift work types of circadian rhythm sleep disorder occur secondary to the environmental changes, which create a mismatch between the newly enforced sleep–wake schedule and the ongoing workings of the internal clock.

The delayed sleep phase type may have a genetic basis. Associations have been found with polymorphisms of the 'circadian clock gene' *hPER3* (Pereira *et al.* 2005), polymorphisms of the gene encoding an enzyme that is responsible for phosphorylation of clock proteins (Takano *et al.* 2004), and polymorphisms of the gene coding for the rate-limiting enzyme responsible for melatonin synthesis (Hohjoh *et al.* 2003). Uncommonly, this type may occur after closed head injury (Ayalon *et al.* 2007; Quinto *et al.* 2000).

The advanced sleep phase type occurs most commonly on a sporadic basis in the elderly, and, in these cases, is presumed to be secondary to age-related changes in the suprachiasmatic nucleus or related structures. Rarely, the syndrome may occur on a familial basis, with autosomal dominant inheritance (Reid *et al.* 2001); in some such cases, mutations have been found in *hPER2* (Xu *et al.* 2005).

The free-running type generally occurs in patients who are blind secondary to lesions affecting the eyes, optic nerves or optic chiasm. Anatomically, entrainment of the suprachiasmatic nucleus to the environmental light–dark schedule is dependent on fibers of the retino-hypothalamic tract, which arise in the retina and then traverse the optic nerves to the optic chiasm from which they ascend into the hypothalamus. With bilateral lesions of any of these structures, the internal clock is allowed to 'run free' (Sack *et al.* 1992). Less commonly, this syndrome may occur in sighted patients, and such cases have been noted after head trauma (Boivin *et al.* 2003) and in otherwise normal individuals with disorders of melatonin metabolism (McArthur *et al.* 1996, Nakamura *et al.* 1997).

The irregular sleep–wake type is generally seen only in patients with diffuse brain disease, whether congenital or acquired (e.g. after traumatic brain injury [Ayalon *et al.* 2007] or in patients with Alzheimer's disease [Witting *et al.* 1990]), and in such cases it is presumed that the suprachiasmatic nucleus has been damaged or destroyed.

Differential diagnosis

The diagnosis of the jet lag and shift work types of circadian rhythm sleep disorder is usually self-evident. In the other types, routine history taking alone generally suffices; however, in some cases it may be necessary to have the patient or a caretaker keep a 'sleep log' for a couple of weeks; polysomnography is rarely required.

In evaluating patients suspected of having the delayed sleep phase, free-running or irregular sleep–wake types, consideration should be given to the effects of alcohol or stimulants, and to the presence of various disorders that routinely disturb sleep, such as depression, anxiety, mania, and schizophrenia.

Treatment

Both bedtime melatonin and appropriately timed exposure to bright light are useful in most of these disorders. If melatonin is used, the immediate-release type should be prescribed, in doses ranging from 3 to 6 mg.

The jet lag type is approached by rigidly adhering to the local bedtime; an additional strategy is to adjust one's internal clock before departure by adhering progressively closer, over 2 or 3 days, to the anticipated bedtime at one's destination. In cases of jet lag triggered by eastward travel, either melatonin (Beaumont *et al.* 2004) or zolpidem (Jamieson *et al.* 2001) may be used to induce sleep; in this regard, one study (Suhner *et al.* 2001) found zolpidem to be more effective than melatonin; furthermore, it was found that the combination of the two agents caused confusion the next morning. Over the next few days, caffeine may also be used to increase alertness (Beaumont *et al.* 2004).

The shift work type is treated by use of bright light during night-time work and use of sunglasses or goggles during daytime hours (Crowley *et al.* 2003; Czeisler *et al.* 1990). Melatonin does not appear to be robustly effective (Cavallo *et al.* 2005; Sadeghniiat-Haghighi *et al.* 2008); however, it is worth a try. Many individuals also utilize hypnotics, such as zolpidem; although there are no data to support this use, it is probably preferable to drinking oneself to sleep. Use of modafinil during the day is marginally helpful (Czeisler *et al.* 2005).

The delayed sleep phase type may be treated with a combination of modalities including administration of melatonin sometime between the hours of 1900 and 2100 (Kayumov *et al.* 2001), dim lights in the evening, rigid adherence to a normal wake-up time, and exposure to bright light for 2 or 3 hours after arising (Watanabe *et al.* 1999). Another method of resetting the internal clock involves what is known as 'chronotherapy' (Czeisler *et al.* 1981). Here, patients advance their sleep time progressively by 3 hours every day until, after 5 or 6 days (depending on when their original sleep time was), their new sleep time coincides with environmental demands, which are then rigidly adhered to. Although effective, this procedure is difficult to implement.

The advanced sleep phase type rarely requires treatment, as it rarely causes significant problems; in limiting cases, evening bright light treatment may be considered.

The free-running type in blind patients may be treated with melatonin (Sack *et al.* 2000). In sighted patients, and in those whose blindness is due to retrochiasmal lesions, morning bright-light treatment may be considered.

Anecdotally, the irregular sleep–wake type has been improved with melatonin (Pillar *et al.* 1998).

20.15 PRIMARY INSOMNIA

Primary insomnia occurs in one of two types: psychophysiologic and idiopathic. The psychophysiolgic type is common, occurring in 1 to 10 percent of the general population and up to 25 percent in the elderly. The idiopathic type, by contrast, is rare.

Clinical features

Psychophysiologic insomnia (Frankel *et al.* 1976; Hauri and Fisher 1986) typically has an onset in middle or later years, and generally, but not always, follows upon some stressful life event. Idiopathic insomnia (Hauri and Olmstead 1980), by contrast, has an onset in childhood.

In both types, patients typically have trouble falling asleep, experience multiple awakenings, and have an overall reduced total sleep time. Most patients are fretful about their insomnia and, while lying in bed, may find themselves worrying constantly over their inability to achieve sleep.

Course

The psychophysiologic type of primary insomnia remits in about 50 percent of patients within a year; in the remainder, it runs a chronic course. The idiopathic type appears to be chronic.

Etiology

The onset of the psychophysiologic type of primary insomnia is typically associated with the emergence of a self-perpetuating vicious cycle. After a stressful life event, most individuals will experience some insomnia; however, in the vast majority, this is transient. In those who do go on to develop chronic insomnia, however, a certain anxiety appears over whether or not sleep will come. This anxiety then makes sleep less likely to occur and a vicious cycle is established in which a failure to sleep engenders more anticipatory anxiety, which in turn makes the insomnia worse. Once this pattern is firmly established, the bed, rather than being seen as a place for relaxation and restoration, becomes an anxiety-provoking stimulus in itself. Interestingly, in such cases patients may sleep better on the couch or in a hotel.

Various neurophysiologic changes have also been found in the psychophysiologic type of primary insomnia. A SPECT study (Smith *et al.* 2002) revealed hypoperfusion in the basal ganglia and frontal, parietal and occipital cortices, and magnetic resonance spectroscopy revealed globally reduced cerebral gamma-aminobutyric acid (GABA) levels (Winkelman *et al.* 2008). Further, two studies (Hajak *et al.* 1995; Riemann *et al.* 2002a) revealed lower nocturnal plasma melatonin levels. Of interest in this regard, treatment with doxepin in one study (Hajak *et al.* 1996) was associated not only with relief of insomnia but also with a rise in plasma melatonin levels.

The etiology of the idiopathic type of primary insomnia is less well understood. One study found an increase in REM sleep, and an association with 'soft' neurologic signs (Hauri and Olmstead 1980) suggesting congenital brain damage.

Differential diagnosis

Chronic insomnia may occur secondary to a multitude of other conditions. Certain other sleep disorders, including restless legs syndrome, periodic leg movement disorder and the syndrome of painful legs and moving toes, routinely cause insomnia. Depression, as seen in major depressive disorder, dysthymia or bipolar disorder, is perhaps the most common cause of insomnia, and one should always inquire after other 'vegetative' symptoms, such as anergia, anhedonia, and changes in appetite: attention should also be payed to the presence of prominent 'middle insomnia', wherein patients awaken in the middle of the night and are unable to get back to sleep for a prolonged period, as this is strongly associated with depression. Generalized anxiety disorder, post-traumatic stress disorder, and schizophrenia must also be considered.

Caffeine and other stimulants taken too late in the day cause insomnia, and insomnia is universal and often very severe and long-lasting in alcohol withdrawal. Furthermore, chronic insomnia is often found in hyperthyroidism and in various painful conditions, such as heartburn or rheumatoid arthritis.

Finally, one must consider whether the patient suffers from 'paradoxical insomnia' (also known as 'sleep state misperception') or is merely an otherwise normal 'short sleeper'. Paradoxical insomnia is said to exist in cases in which, despite often bitter complaints of insomnia, polysomnography reveals normal sleep. Before making this diagnosis, however, it must be borne in mind that some patients with psychophysiologic insomnia sleep better when they are away from home. When this is suspected, it may be appropriate to perform polysomnography at the patient's home before making the diagnosis. 'Short sleepers' are individuals who, despite getting little sleep, awake refreshed and have no complaints.

Treatment

Good sleep hygiene is essential. Naps should not be taken, and patients should get some exercise every day. Caffeine and other stimulants should be reserved for morning use only. Evenings should be reserved for relaxing activities, the bedroom should be darkened and quiet, and the bed should be reserved for sleep or sexual activity. If sleep does not come, patients should do something else, perhaps reading, until drowsiness occurs. Whether insomnia occurs or not, the wake-up time should be strictly adhered to.

Should insomnia persist despite good sleep hygiene, consideration may be given to cognitive behavioral therapy, which is not only effective acutely (Edinger *et al.* 2001, 2009) but also confers enduring benefits (Backhaus *et al.* 2001). Should cognitive behavioral therapy be ineffective or impractical, pharmacologic treatment may be considered.

Zolpidem and eszopiclone are the most widely used agents for primary insomnia, and are roughly of equivalent efficacy. When prescribing zolpidem, however, one must watch for the emergence of both somnambulism (as noted in Chapter 20.1) and a variant of this, wherein patients engage in eating (Morgenthaler and Silber 2002).

Melatoninergic agents are also widely used. Melatonin itself was found ineffective in a dose of 1 mg (Almeida Montes *et al.* 2003), but effective when a dose of 3 mg was used (Zhdanova *et al.* 2001). Ramelteon, a melatonin agonist, is clearly effective (Erman *et al.* 2006, Mayer *et al.* 2009).

The tricyclic doxepin has been extensively studied. Doses of 25–50 mg were effective, but in a small minority of cases discontinuation of this agent was followed by a severe rebound insomnia (Hajak *et al.* 2001). Interestingly, lower doses of 1–6 mg are also effective, in both adult (Roth *et al.* 2007) and geriatric (Scharf *et al.* 2008) patients. Trimipramine, another tricyclic antidepressant, is also effective in doses of 100–200 mg at bedtime (Hohagen *et al.* 1994, Riemann *et al.* 2002b), and neither appears to suppress REM sleep nor cause rebound insomnia. Trazodone, although widely used in doses of 25–100 mg, may not be effective: in one study (Walsh *et al.* 1988) that compared 50 mg of trazodone with placebo and 10 mg of zolpidem, after 1 week trazodone was no more effective than placebo, whereas zolpidem continued to show superiority over placebo.

Choosing among these various agents requires considerable clinical judgment. Given the excellent tolerability and low price of melatonin, starting with this agent, at a dose of 3–6 mg in the evening, is a reasonable first choice; if this proves ineffective, ramelteon should be considered. Should melatoninergic agents fail, consideration may be given to zolpidem or eszopiclone; doxepin and trimipramine, though reasonable alternatives, are not widely used.

The foregoing discussion of pharmacologic treatment concerns the psychophysiologic type of primary insomnia; in cases of the idiopathic type, melatonin, in a dose of 5 mg, was effective (Smits *et al.* 2001).

20.16 PRIMARY HYPERSOMNIA

Primary insomnia, more commonly known as idiopathic hypersomnia, is an uncommon condition characterized by hypersomnia and hypersomnolence.

Clinical features

The clinical features have been described in several large case series (Anderson *et al.* 2007; Bassetti and Aldrich 1997; Vernet and Arnulf 2009). The onset is gradual, typically occurring in late adolescence. Patients typically complain of both excessive night-time sleep and frequent, long daytime naps, with the resulting total hypersomnia averaging some 11–12 hours. Importantly, sleep, whether during the night, or with naps, is non-restorative, and patients often have difficulty awakening. Sleep drunkenness may occur in roughly one-third of cases, and hypnogogic hallucinations and sleep paralysis may each occur in roughly one-quarter; importantly, there are no cataplectic attacks. During waking hours patients often complain of hypersomnolence and many have to fight to stay awake.

The MSLT may reveal a reduced sleep latency; however, there are no sleep-onset REM periods.

Course

Although partial remission may eventually occur in roughly one-quarter of patients (Bassetti and Aldrich 1997), the overall course is one of chronicity, with only about one-tenth of patients enjoying an eventual full remission (Anderson *et al.* 2007).

Etiology

In roughly one-third of cases a positive family history may be obtained, with the rest apparently being sporadic. Although a disturbance in hypothalamic functioning is suspected, CSF studies of hypocretin levels have yielded mixed results, with reports of both decreased (Ebrahim *et al.* 2003) and normal (Dauvilliers *et al.* 2003) levels.

Differential diagnosis

Although some authorities subsume not only chronic but also intermittent forms of hypersomnia under the rubric of 'primary hypersomnia', in this text the diagnosis of primary hypersomnia is reserved for chronic cases only. Consequently, the first task in differential diagnosis is to determine whether the hypersomnia is chronic or occurs in episodes. Episodes of hypersomnia may occur in the Kleine–Levin syndrome, in association with menstruation (Sachs *et al.* 1982), with intermittent use of sedating medications, and, most importantly, in certain depressive illnesses. Depressive episodes of major depressive disorder, and, most especially, bipolar disorder may be characterized by severe hypersomnia, and in taking the history one must be alert to other vegetative symptoms and, of course, to any history of mania.

Once it is established that the patient indeed has chronic hypersomnia, other sleep disorders must be distinguished, including sleep apnea, the Pickwickian syndrome, restless legs syndrome, periodic limb movement disorder, and painful legs and moving toes. Narcolepsy is also often considered; however, the naps seen in narcolepsy, unlike those of primary hypersomnia, are brief and refreshing, and patients may be easily aroused from them; furthermore, one sees sleep-onset REM periods on the MSLT and, in most cases, cataplectic attacks, findings not seen in primary hypersomnia.

Excessive daytime sleepiness may also be seen in other disorders, including hypothyroidism, myotonic muscular dystrophy, the Prader–Willi syndrome and as a sequela to traumatic brain injury or infectious mononucleosis. Excessive sleepiness has also been noted with lesions of the hypothalamus (Eisensehr *et al.* 2003), thalamus (Bassetti *et al.* 1996), or pons (Ganji *et al.* 1996). Chronic use of sedating medications, such as benzodiazepines, tricyclic antidepressants, certain anti-epileptic drugs, antihistamines, and opioids must also be considered on the differential.

Finally, there are individuals known as 'long sleepers' who require more than 8 hours of sleep a night. In contrast to patients with primary hypersomnia, however, these individuals awaken refreshed and are not subject to unrefreshing naps during the day.

Treatment

Although there are no controlled studies, case series (Anderson *et al.* 2007; Bassetti and Aldrich 1997) report success with the same agents used in the treatment of narcolepsy, including modafinil and methylphenidate. Interestingly, in two cases with subclinical hypothyroidism (with normal thyroxine but elevated thyroid stimulating hormone levels), treatment with 25 μg/day of thyroxine was followed by significant improvement (Shinno *et al.* 2009).

REFERENCES

Abe K, Oda N, Amatomi M. Natural history and predictive significance of head-banging, head-rolling and breath-holding spells. *Dev Med Child Neurol* 1984; **26**:644–8.

Adie WJ. Idiopathic narcolepsy: a disease sui generis; with remarks on the mechanism of sleep. *Brain* 1926; **49**:257–306.

Ajlouni KM, Ahmad AT, El-Zaheri MM *et al.* Sleepwalking associated with hyperthyroidism. *Endocr Pract* 2005; **11**:5–10.

Aldrich MS, Naylor MW. Narcolepsy associated with lesions of the diencephalon. *Neurology* 1989; **39**:1505–8.

Allen R, Barker PB, Wehrl F *et al.* MRI measurements of brain iron in patients with restless legs syndrome. *Neurology* 2001; **56**:263–5.

Almeida Montes LG, Ontiveros Uribe MP, Cortes Sotres J *et al.* Treatment of primary insomnia with melatonin: a double-blind, placebo-controlled, crossover study. *J Psychiatry Neurosci* 2003; **28**:191–6.

Alvarez MV, Driver-Dunckley EE, Caviness JN *et al.* Case series of painful legs and moving toes: clinical and electrophysiologic observations. *Mov Disord* 2008; **23**:2062–6.

Anderson KN, Smith IE, Shneerson JM. Rhythmic movement disorder (head banging) in an adult during rapid eye movement sleep. *Mov Disord* 2006; **21**:866–7.

Anderson KN, Pilsworth S, Sharples LD *et al.* Idiopathic hypersomnia: a study of 77 cases. *Sleep* 2007; **30**:1274–81.

Aran A, Lin L, Nevsimalova S, Plazzi G *et al.* Elevated anti-streptococcal antibodies in patients with recent narcolepsy onset. *Sleep* 2009; **32**:979–63.

Arnulf I, Zeitzer JM, File J *et al.* Kleine–Levin syndrome: a systematic review of 186 cases in the literature. *Brain* 2005; **128**:2763–76.

Arnulf I, Lin L, Gadoth N *et al.* Kleine–Levin syndrome: a systematic study of 108 patients. *Ann Neurol* 2008; **63**:482–93.

Ayalon L, Borodkin K, Dishon L *et al.* Circadian rhythm sleep disorders following mild traumatic brain injury. *Neurology* 2007; **68**: 1136–40.

Backhaus J, Hohagen F, Voderholzer U *et al.* Long-term effectiveness of a short-term cognitive-behavioral group treatment for primary insomnia. *Eur Arch Psychiatry Clin Neurosci* 2001; **251**:35–41.

Bakwin HI. Sleepwalking in twins. *Lancet* 1970; **2**:446–7.

Balon R. Sleep terror disorder and insomnia treated with trazodone: a case report. *Ann Clin Psychiatry* 1994; **6**:161–3.

Bassetti C, Aldrich MS. Idiopathic hypersomnia. A series of 42 patients. *Brain* 1997; **120**:1423–35.

Bassetti C, Mathis J, Gugger M *et al.* Hypersomnia following paramedian thalamic stroke: a report of 12 patients. *Ann Neurol* 1996; **39**:471–80.

Beaumont M, Batejat D, Pierard C *et al.* Caffeine or melatonin effects on sleep and sleepiness after rapid eastward transmeridian travel. *J Appl Physiol* 2004; **96**:50–8.

Benes H, Kurella B, Kummer J *et al.* Rapid onset of action of levodopa in restless legs syndrome: a double-blind, randomized, multicenter, crossover trial. *Sleep* 1999; **22**:1073–81.

Black JE, Hirshkowitz M. Modafinil for treatment of residual excessive sleepiness in nasal continuous positive airway pressure-treated obstructive sleep apnea/hypopnea syndrome. *Sleep* 2005; **28**: 464–71.

Black J, Houghton WC. Sodium oxybate improves excessive daytime sleepiness in narcolepsy. *Sleep* 2006; **29**:939–46.

Bliwise DL, Freeman A, Ingram CD *et al.* Randomized, double-blind, placebo-controlled, short-term trial of ropinirole in restless legs syndrome. *Sleep Med* 2005; **6**:141–7.

Blouin AM, Thannickal TC, Worley PF *et al.* NARP immunostaining of human hypocretin (orexin) neurons. Loss in narcolepsy. *Neurology* 2005; **65**:1189–92.

Boeve BF, Silber MH, Parisi JE *et al.* Synucleinopathy pathology and REM sleep behavior disorder plus dementia or parkinsonism. *Neurology* 2003a; **61**:40–5.

Boeve BF, Silber MH, Ferman TJ. Melatonin for treatment of REM sleep behavior disorder in neurologic disorders: results in 14 patients. *Sleep Med* 2003b; **4**:281–4.

Boeve BF, Dickson DW, Olson EJ *et al.* Insights into REM sleep behavior disorder pathophysiology in brainstem-predominant Lewy body disease. *Sleep Med* 2007; **8**:60–4.

Bogan RK, Fry JM, Schmidt MH *et al.* Ropinirole in the treatment of patients with restless legs syndrome: A US-based randomized, double-blind, placebo-controlled clinical trial. *Mayo Clin Proc* 2006; **81**:17–27.

Boivin DB, James FO, Santo JB *et al.* Non-24-hour sleep-wake syndrome following a car accident. *Neurology* 2003; **60**:1841–3.

Boller F, Wright DG, Cavalieri R *et al.* Paroxysmal 'nightmares'. Sequel of a stroke responsive to diphenylhydantoin. *Neurology* 1975; **25**:1026–8.

Brodeur C, Monplaisir J, Godbout R *et al.* Treatment of restless legs syndrome and periodic movements during sleep with l-dopa: a double-blind, controlled study. *Neurology* 1988; **38**:1845–8.

Broughton RJ, Fleming JAE, George CFP *et al.* Randomized, double-blind, placebo-controlled crossover trial of modafinil in the treatment of excessive daytime sleepiness in narcolepsy. *Neurology* 1997; **49**:444–51.

Brownell LG, West P, Sweatman P *et al.* Protriptyline in obstructive sleep apnea: a double-blind trial. *N Engl J Med* 1982; **307**: 1037–42.

Bruni O, Ferri R, Miano S *et al.* L-5-hydroxytryptophan treatment of sleep terrors in children. *Eur J Pediatr* 2004; **163**:402–7.

Burstein A, Burstein A, Freehold NJ. Treatment of night terrors with imipramine. *J Clin Psychiatry* 1983; **44**:82.

Burwell CS, Robin ED, Whaley RD *et al.* Extreme obesity associated with alveolar hypoventilation: a Pickwickian syndrome. *Am J Med* 1956; **21**:811–18.

Carpenter S, Yassa R, Ochs R. A pathologic basis for Kleine–Levin syndrome. *Arch Neurol* 1982; **39**:25–8.

Cartwright RD, Lloyd S, Lilie J *et al.* Sleep position training as treatment for sleep apnea syndrome: a preliminary study. *Sleep* 1985; **8**:87–94.

Cartwright RD, Ristanovic R, Diaz F *et al.* A comparative study of treatments for positional sleep apnea. *Sleep* 1991; **14**:546–52.

Cavallo A, Ris MD, Succop P *et al.* Melatonin treatment of pediatric residents for adaptation to night shift work. *Ambul Pediatr* 2005; **5**:172–7.

Charney DS, Kales A, Soldatos CR *et al.* Somnambulistic-like episodes secondary to combined lithium-neuroleptic treatment. *Br J Psychiatry* 1979; **135**:418–24.

Chen L, Fong SY, Lam CW *et al.* The familial risk and HLA susceptibility among narcolepsy patients in Hong Kong Chinese. *Sleep* 2007; **30**:851–6.

Chiang ST, Lee PY, Liu SY. Pulmonary function in a typical case of Pickwickian syndrome. *Respiration* 1980; **39**:105–13.

Chiles JA, Wilkus RJ. Behavioral manifestations of the Kleine–Levin syndrome. *Dis Nerv Syst* 1976; **37**:646–8.

Clavelou P, Tournilhac M, Vidal C *et al.* Narcolepsy associated with arteriovenous malformation of the diencephalon. *Sleep* 1995; **18**:202–5.

Coleman RM, Pollack CP, Weitzman ED. Periodic movements in sleep (nocturnal myoclonus): relationship to sleep disorders. *Ann Neurol* 1980; **8**:416–21.

Connor JR, Boyer PJ, Menzies SL *et al.* Neuropathological examination suggests impaired brain iron acquisition in restless legs syndrome. *Neurology* 2003; **61**:304–9.

Connor JR, Wang XS, Patton SM *et al.* Decreased transferrin receptor expression by neuromelanin cells in restless legs syndrome. *Neurology* 2004; **62**:1563–7.

Connor JR, Wang XS, Allen RP *et al.* Altered dopaminergic profile in the putamen and substantia nigra in restless leg syndrome. *Brain* 2009; **132**:2403–12.

Cooper AJ. Treatment of co-existent night-terrors and somnambulism in adults with imipramine and diazepam. *J Clin Psychiatry* 1987; **48**:209–10.

Critchley M. Periodic hypersomnia and megaphagia in adolescent males. *Brain* 1962; **85**:627–56.

Critchley M, Hoffman HL. The syndrome of periodic somnolence and morbid hunger (Kleine–Levin syndrome). *BMJ* 1942; **1**: 137–9.

Crocker A, Espana RA, Papadopoulou M *et al.* Concomitant loss of dynorphin NARP and orexin in narcolepsy. *Neurology* 2005; **65**:1184–8.

Crowley SJ, Lee C, Tseng CY *et al.* Combinations of bright light, scheduled dark, sunglasses, and melatonin to facilitate circadian entrainment to night shift work. *J Biol Rhythms* 2003; **18**:513–23.

Culebras A, Moore JT. Magnetic resonance findings in REM sleep behavior disorder. *Neurology* 1989; **39**:1519–23.

Czeisler CA, Richardson GS, Coleman RM *et al.* Chronotherapy: resetting the circadian clocks of patients with delayed sleep phase insomnia. *Sleep* 1981; **4**:1–21.

Czeisler CA, Johnson MP, Duffy JF *et al.* Exposure to bright light and darkness to treat physiologic maladaptation to night work. *N Engl J Med* 1990; **322**:1253–9.

Czeisler CA, Walsh JK, Roth T *et al.* Modafinil for excessive daytime sleepiness associated with shift-work sleep disorder. *N Engl J Med* 2005; **353**:476–86.

Dauvilliers Y. Follow-up of four narcolepsy patients treated with intravenous immunoglobulins. *Ann Neurol* 2006; **60**:153.

Dauvilliers Y, Mayer G, Lecendreux M *et al.* Kleine–Levin syndrome. An autoimmune hypothesis based on clinical and genetic analyses. *Neurology* 2002; **59**:1739–45.

Dauvilliers Y, Bauman CR, Carlander B *et al.* CSF hypocretin-1 levels in narcolepsy, Kleine–Levin syndrome, and other hypersomnias and neurological conditions. *J Neurol Neurosurg Psychiatry* 2003; **74**:1667–73.

D'Cruz OF, Vaughn BV, Gold SH *et al.* Symptomatic cataplexy in pontomedullary lesions. *Neurology* 1994; **44**:2189–91.

Dehoorne JL, Raes AM, van Laecke E *et al.* Desmopressin toxicity due to prolonged half-life in 18 patients with nocturnal enuresis. *J Urol* 2006; **176**:754–7.

Dement WC, Rechtschaffen AR, Gulevich GD. A polygraphic study of the narcoleptic sleep attacks. *Electroencephalogr Clin Neurophysiol* 1964; **17**:608–9.

Dement WC, Rechtschaffen AR, Gulevick GD. The nature of the narcoleptic sleep attack. *Neurology* 1966; **16**:18–33.

Dempsey OJ, McGeoch P, de Silva RN *et al.* Acquired narcolepsy in an acromegalic patient who underwent pituitary irradiation. *Neurology* 2003; **61**:537–40.

Di Gennaro G, Autret A, Mascia A *et al.* Night terrors associated with thalamic lesion. *Clin Neurophysiol* 2004; **115**:2489–92.

DiMario FJ, Emery ES. The natural history of night terrors. *Clin Pediatr* 1987; **26**:505–11.

Drachman DB, Gumnit R. Periodic alteration of consciousness in the 'Pickwickian syndrome'. *Arch Neurol* 1962; **6**:471–7.

Drake ME. Jactatio nocturna after head injury. *Neurology* 1986; **36**:867–8.

Dressler D, Thompson PD, Marsden CD. The syndrome of painful legs and moving toes. *Mov Disord* 1994; **9**:13–21.

Driver HS, Shapiro CM. Parasomnias. *BMJ* 1993; **306**:921–4.

Driver-Dunckley ED, Noble BN, Hentz JG *et al.* Gambling and increased sexual desire with dopaminergic medications in restless legs syndrome. *Clin Neuropharmacol* 2007; **30**:249–55.

Dyken ME, Lin-Dyken DC, Seaba P *et al.* Violent sleep-related behavior leading to subdural hematoma. *Arch Neurol* 1995; **52**: 318–21.

Dyken ME, Lin-Dyken DC, Yamada T. Diagnosing rhythmic movement disorder with video-polysomnography. *Pediatr Neurol* 1997; **16**:37–41.

Dyken ME, Yamada T, Glenn CL *et al.* Obstructive sleep apnea associated with cerebral hypoxemia and death. *Neurology* 2004; **62**:491–3.

Earley CJ, Connor JR, Beard JL *et al.* Abnormalities in CSF concentrations of ferritin and transferrin in restless legs syndrome. *Neurology* 2000; **54**:1698–700.

Ebrahim IO, Sharieff MK, de Laly S *et al.* Hypocretin (orexin) deficiency in narcolepsy and primary hypersomnia. *J Neurol Neurosurg Psychiatry* 2003; **74**:127–30.

Ebrahim IO, Peacock KW, Williams AJ. Posttraumatic epilepsy-two case reports and a mini review. *J Clin Sleep Med* 2005; **1**: 153–6.

Edinger JD, Wohlgemuth WK, Radtke RA *et al.* Cognitive behavioral therapy for treatment of chronic primary insomnia: a randomized controlled trial. *JAMA* 2001; **285**:1856–64.

Edinger JD, Olsen MK, Stechuchak KM *et al.* Cognitive behavioural therapy for patients with primary insomnia or insomnia associated predominantly with mixed psychiatric disorders: a randomized clinical trial. *Sleep* 2009; **32**:499–510.

Eisensehr I, Noachter S, von Schlippenbach C *et al.* Hypersomnia associated with bilateral posterior hypothalamic lesions. A polysomniographic case study. *Eur Neurol* 2003; **49**:169–72.

Ekbom KA. Restless legs syndrome. *Neurology* 1960; **10**:868–73.

El Hajj T, Nasreddin W, Korri H *et al.* A case of Kleine–Levin syndrome with a complete and sustained response to carbamazepine. *Epilepsy Behav* 2009; **15**:391–2.

Erman M, Seiden D, Zammit G *et al.* An efficacy, safety, and dose–response study of ramelteon in patients with chronic primary insomnia. *Sleep Med* 2006; **7**:17–24.

Fantini ML, Gagnon J-F, Filipini D *et al.* The effects of pramipexole in REM sleep behavior disorder. *Neurology* 2003; **61**:1418–20.

Fantini ML, Cona A, Clerici S *et al.* Aggressive dream content without daytime aggressiveness in REM sleep behavior disorder. *Neurology* 2005; **65**:1010–15.

Fenzi F, Simonati A, Crosato F *et al.* Clinical features of Kleine–Levin syndrome with localized encephalitis. *Neuropediatrics* 1993; **24**:292–5.

Fernandez J-M, Lara I, Gila L *et al.* Disturbed hypothalamic–pituitary axis in idiopathic recurring hypersomnia syndrome. *Acta Neurol Scand* 1990; **82**:361–3.

Ferreira JJ, Galitzky M, Montrastruc JL *et al.* Sleep attacks and Parkinson's disease treatment. *Lancet* 2000; **355**:1333–4.

Fisher C, Kahn E, Edwards A *et al.* A psychophysiological study of nightmares and night terrors. I. Physiological aspects of the stage 4 night terror. *J Nerv Ment Dis* 1973a; **157**:75–98.

Feuerstein C, Naegele B, Pepin JL *et al.* Frontal-lobe related cognitive functions in patients with sleep apnea syndrome before and after treatment. *Acta Neurol Belg* 1997; **97**:96–107.

Fisher C, Kahn E, Edwards A *et al.* A psychophysiological study of nightmares and night terrors: the suppression of stage 4 night terrors with diazepam. *Arch Gen Psychiatry* 1973b; **28**: 252–9.

Forsythe WI, Butler RJ. Fifty years of enuretic alarms. *Arch Dis Child* 1989; **64**:879–85.

Forsythe WI, Redmond A. Enuresis and spontaneous cure rate of 1129 enuretics. *Arch Dis Child* 1974; **49**:259–63.

Fournier JCM, Helguera RAL. Postencephalitic narcolepsy and cataplexy: muscle and motor nerves electrical inexcitability during the attack of cataplexy. *J Nerv Ment Dis* 1934; **80**:159–62.

Frankel BL, Coursey RD, Buchbinder R *et al.* Recorded and reported sleep in chronic primary insomnia. *Arch Gen Psychiatry* 1976; **33**:615–23.

Frauscher B, Gschliesser V, Brandauer E *et al.* REM sleep behavior disorder in 703 sleep-disorder patients: the importance of eliciting a comprehensive sleep history. *Sleep Med* 2010; **11**:167–71.

Frey J, Darbonne C. Fluoxetine suppresses human cataplexy: a pilot study. *Neurology* 1994; **44**:707–9.

Fritz GK, Rockney RM, Yeung AS. Plasma levels and efficacy of imipramine treatment for enuresis. *J Am Acad Child Adolesc Psychiatry* 1994; **33**:60–4.

Frucht S, Rogers JD, Greene PE *et al.* Falling asleep at the wheel: motor vehicle mishaps in persons taking pramipexole and ropinirole. *Neurology* 1999; **52**:1908–10.

Gadoth N, Kesler A, Vainstein G *et al.* Clinical and polysomnographic characteristics of 34 patients with Kleine–Levin syndrome. *J Sleep Res* 2001; **10**:337–41.

Gallinek A. Syndrome of episodes of hypersomnia, bulimia and abnormal mental states. *JAMA* 1954; **154**:1081–3.

Gallinek A. The Kleine–Levin syndrome: hypersomnia, bulimia, and abnormal mental states. *World Neurology* 1962; **3**:235–43.

Ganji SS, Ferriss GS, Rao J *et al.* Hypersomnia associated with a focal pontine lesion. *Clin Electroencephalogr* 1996; **27**:52–6.

Gann H, Feige B, Fesihi S *et al.* Periodic limb movements during sleep in alcohol dependent patients. *Eur Arch Psychiatry Clin Neurosci* 2002; **252**:124–9.

Garcia-Borreguero D, Larrosa O, de la Llave Y *et al.* Treatment of restless legs syndrome with gabapentin. A double-blind cross-over study. *Neurology* 2002; **59**:1573–9.

Garland H, Sumner D, Fourman P. The Kleine–Levin syndrome: some further observations. *Neurology* 1965; **15**:1161–7.

Gemignani F, Brindani F, Negrotti A *et al.* Restless legs syndrome and polyneuropathy. *Mov Disord* 2006; **21**:1254–7.

George CF, Nickerson PW, Janley PJ *et al.* Sleep apnoea patients have more automobile accidents. *Lancet* 1987; **2**:447.

Germain A, Nielsen T. Impact of imagery rehearsal treatment on distressing dreams, psychological distress, and sleep parameters in nightmare patients. *Behav Sleep Med* 2003; **1**:140–54.

Gilbert GJ. Periodic hypersomnia and bulimia: the Kleine–Levin syndrome. *Neurology* 1964; **14**:844–50.

Godau J, Wevers AK, Gaenslen A *et al.* Sonographic abnormalities of brainstem structures in restless legs syndrome. *Sleep Med* 2008; **9**:782–9.

Gledhill RF, Bartel PR, Yoshida Y *et al.* Narcolepsy caused by acute disseminated encephalomyelitis. *Arch Neurol* 2004; **61**: 758–60.

Goldberg MA. The treatment of Kleine–Levin syndrome with lithium. *Can J Psychiatry* 1983; **28**:491–3.

Goncalves MA, Guilleminault C, Ramos E *et al.* Erectile dysfunction, obstructive sleep apnea syndrome and nasal CPAP treatment. *Sleep Med* 2005; **6**:333–9.

Guilleminault C, Eldridge FL, Tilkian A *et al.* Sleep apnea syndrome due to upper airway obstruction: a review of 25 cases. *Arch Int Med* 1977; **137**:296–300.

Guilleminault C, Mignot E, Grumet FC. Familial patterns of narcolepsy. *Lancet* 1989; **2**:1376–9.

Guilleminault C, Aftab FA, Daradeniz D *et al.* Problems associated with switch to modafinil – a novel alerting agent in narcolepsy. *Eur J Neurol* 2000; **7**:381–4.

Guilleminault C, Palombini L, Pelayo R *et al.* Sleepwalking and sleep terrors in prepubertal children: what triggers them? *Pediatrics* 2003; **111**:17–25.

Guilleminault C, Kirisoglu C, Bao G *et al.* Adult chronic sleepwalking and its treatment based on polysomnography. *Brain* 2005; **128**:1062–9.

Hafeez ZH, Kalinowski CM. Somnambulism induced by quetiapine: two case reports and a review of the literature. *CNS Spectr* 2007; **12**:910–12.

Hajak G, Rodenbeck A, Staedt J *et al.* Nocturnal plasma melatonin levels in patients suffering from chronic primary insomnia. *J Pineal Res* 1995; **19**:116–22.

Hajak G, Rodenbeck A, Adler L *et al.* Nocturnal melatonin secretion and sleep after doxepin administration in chronic primary insomnia. *Pharmacopsychiatry* 1996; **29**:187–92.

Hajak G, Rodenbeck A, Voderholzer U *et al.* Doxepin in the treatment of primary insomnia: a placebo-controlled, double-blind, polysomnographic study. *J Clin Psychiatry* 2001; **62**:453–63.

Harsch HH. Cyproheptadine for recurrent nightmares. *Am J Psychiatry* 1986; **143**:1491–2.

Hartmann M, Pfister R, Pfadenhauer K. Restless legs syndrome associated with spinal cord lesions. *J Neurol Neurosurg Psychiatry* 1999; **66**:688–9.

Hattan E, Chalk C, Postuma RB. Is there a higher risk of restless legs syndrome in peripheral neuropathy? *Neurology* 2009; **72**:955–960.

Hauri P, Fisher J. Persistent psychophysiologic (learned) insomnia. *Sleep* 1986; **9**:38–53.

Hauri P, Olmstead E. Childhood-onset insomnia. *Sleep* 1980; **3**:59–65.

Hauser RA, Gauger L, McDowell Anderson W *et al.* Pramipexole-induced somnolence and episodes of daytime sleep. *Mov Disord* 2000; **15**:658–63.

Hensel J, Pillmann F. Late-life somnambulism after therapy with metoprolol. *Clin Neuropharmacol* 2008; **31**:248–50.

Hishikawa Y, Kaneko Z. Electroencephalographic study on narcolepsy. *Electroencephalogr Clin Neurophysiol* 1965; **18**:249–59.

Hishikawa Y, Nanno H, Tachibana M *et al.* The nature of sleep attack and other symptoms of narcolepsy. *Electroencephalogr Clin Neurophysiol* 1968; **24**:1–10.

Hogl B, Frauscher B, Seppi K *et al.* Transient restless legs syndrome after spinal anesthesia. A prospective study. *Neurology* 2002; **59**:1705–7.

Hogl B, Garcia-Borreguero D, Kohnen D *et al.* Progressive development of augmentation during long-term treatment with levodopa in restless legs syndrome: results of a prospective multi-center study. *J Neurol* 2010; **257**:230–7.

Hogue R, Chesson AL. Zolpidem-induced sleepwalking, sleep related eating disorder, and sleep-driving: fluorine-18-fluorodeoxyglucose positron emission tomography analysis, and a literature review of other unexplained clinical effects of zolpidem. *J Clin Sleep Med* 2009; **5**:471–6.

Hohagen R, Monteero RF, Weiss E *et al.* Treatment of primary insomnia with trimipramine: an alternative to benzodiazepine hypnotics? *Eur Arch Psychiatry Clin Neurosci* 1994; **244**:65–72.

Hohjoh H, Takasu M, Shishikura K *et al.* Significant association of the arylalkylamine N-acetyltransferase (AA-NAT) gene with delayed sleep phase syndrome. *Neurogenetics* 2003; **4**:151–3.

Huang YS, Guilleminault C, Kao PF *et al.* SPECT findings in the Kleine–Levin syndrome. *Sleep* 2005; **28**:955–60.

Huang YS, Lin YH, Guilleminault C. Polysomnography in Kleine–Levin syndrome. *Neurology* 2008; **70**:795–801.

Hublin C, Partinen M, Heinonen EH *et al.* Selegiline in the treatment of narcolepsy. *Neurology* 1994; **44**:2095–101.

Hublin C, Kaprio J, Partinen M *et al.* Prevalence and genetics of sleepwalking: a population-based twin study. *Neurology* 1997; **48**:177–81.

Hublin C, Kaprio J, Partinen M *et al.* Nightmares: familial aggregation and association with psychiatric disorders in a nationwide twin cohort. *Am J Med Genet* 1999; **88**:329–36.

Ikeda K, Deguchi K, Touge T *et al.* Painful legs and moving toes syndrome associated with herpes zoster myelitis. *J Neurol Sci* 2004; **219**:147–50.

Iranzo A, Santamaria J. Severe obstructive sleep apnea/hypopnea mimicking REM sleep behavior disorder. *Sleep* 2005; **28**:203–6.

Iranzo A, Munoz E, Santamaria J *et al.* REM sleep behavior disorder and vocal cord paralysis in Machado–Joseph disease. *Mov Disord* 2003; **18**:1179–83.

Iranzo A, Molinuevo JL, Santamaria J *et al.* Rapid-eye-movement sleep behavior disorder as an early marker for a neurodegenerative disorder: a descriptive study. *Lancet Neurol* 2006a; **5**:572–7.

Iranzo A, Graus F, Clover L *et al.* Rapid eye movement sleep behavior disorder and potassium channel antibody-associated limbic encephalitis. *Ann Neurol* 2006b; **59**:178–82.

Itokawa K, Fukui M, Ninomiya M *et al.* Gabapentin for Kleine–Levin syndrome. *Intern Med* 2009; **48**:1183–5.

Jama L, Hirvonen K, Partinen M *et al.* A dose-ranging study of pramipexole for the symptomatic treatment of restless legs syndrome: polysomnographic evaluation of periodic leg movements and sleep disturbance. *Sleep Med* 2009; **10**:630–6.

Jamieson AO, Zammit GK, Rosenberg RS *et al.* Zolpidem reduces the sleep disturbance of jet lag. *Sleep Med* 2001; **5**:423–30.

Joncas S, Zadra A, Paquet J *et al.* The value of sleep deprivation as a diagnostic tool in adult sleepwalkers. *Neurology* 2002; **58**: 936–40.

Kales A, Kales JD. Sleep disorders: recent findings in the diagnosis and treatment of disturbed sleep. *N Engl J Med* 1974; **290**:487–99.

Kales A, Jacobson A, Paulson MJ *et al.* Somnambulism: psychophysiological correlates. *Arch Gen Psychiatry* 1966; **14**:586–94.

Kales A, Soldatos CR, Caldwell AB *et al.* Somnambulism: clinical characteristics and personality patterns. *Arch Gen Psychiatry* 1980a; **37**:1406–10.

Kales A, Soldatos CR, Bixler EO *et al.* Hereditary factors in sleepwalking and nightmares. *Br J Psychiatry* 1980b; **137**:111–18.

Kales A, Soldatos CR, Caldwell AB *et al.* Nightmares: clinical characteristics and personality patterns. *Am J Psychiatry* 1980c; **137**:1197–201.

Kales A, Cadieux RJ, Soldatos CR *et al.* Narcolepsy–cataplexy. I. Clinical and electroencephalographic characteristics. *Arch Neurol* 1982; **39**:164–8.

Kales A, Soldatos CR, Kales JD. Sleep disorders: insomnia, sleepwalking, night terrors, nightmares and enuresis. *Ann Int Med* 1987; **106**:582–92.

Kanbayashi T, Shimohata T, Nakashima I *et al.* Symptomatic narcolepsy in patients with neuromyelitis optica and multiple sclerosis: new neurocehmical and immunological implications. *Arch Neurol* 2009; **66**:1563–6.

Kang SY, Sohn YH, Lee IK *et al.* Unilateral periodic limb movements in sleep after supratentorial cerebral infarction. *Parkinsonism Relat Disord* 2004; **10**:429–31.

Kaplan PW, Allen RP, Buchholz DW *et al.* A double-blind, placebo-controlled study of the treatment of periodic limb movements in sleep using carbidopa/levodopa and propoxyphene. *Sleep* 1993; **16**:7171–23.

Kavey NB, Whyte J, Resor SR *et al.* Somnambulism in adults. *Neurology* 1990; **40**:749–52.

Kawashima T, Yamada S. Paroxetine-induced somnambulism. *J Clin Psychiatry* 2003; **64**:483.

Kayumov L, Brown G, Jindal R *et al.* A randomized, double-blind, placebo-controlled crossover study of the effect of exogenous melatonin on delayed sleep phase syndrome. *Psychosom Med* 2001; **63**:40–8.

Kellner R, Neidhardt J, Krakow B *et al.* Changes in chronic nightmares after one session of desensitization or rehearsal instructions. *Am J Psychiatry* 1992; **149**:659–63.

Kemlink D, Polo O, Frauscher B *et al.* Replication of restless legs syndrome loci in three European populations. *J Med Genet* 2009; **46**:315–18.

Kessler R, Chaouat A, Shinkewitch P *et al.* The obesity–hypoventilation syndrome revisited: a prospective study of 34 consecutive cases. *Chest* 2001; **120**:369–76.

Khazaal Y, Krenz S, Zullino DF. Bupropion-induced somnambulism. *Addict Biol* 2003; **8**:359–62.

Kim JS, Lee SB, Park SK *et al.* Periodic limb movement during sleep developed after pontine lesion. *Mov Disord* 2003; **18**:1403–5.

Kim JM, Kwon HM, Lim CS *et al.* Restless legs syndrome in patients on hemodialysis: symptom severity and risk factors. *J Clin Neurol* 2008; **4**:153–7.

Kingshott RN, Vennelle M, Coleman EL *et al.* Randomized, double-blind, placebo-controlled crossover trial of modafinil in the treatment of residual excessive daytime sleepiness in the sleep apnea–hypopnea syndrome. *Am J Respir Crit Care Med* 2001; **163**:918–23.

Kleine W. Periodische schlafsucht. *Monatsschriff fur Psychiatrie und Neurologie* 1925; **57**:285–320.

Knudsen S, Gammeltoft S, Jennum PJ. Rapid eye movement sleep behavior disorder in patients with narcolepsy is associated with hypocretin-1 deficiency. *Brain* 2010; **133**:568–79.

Koerber RK, Torkelson R, Haven G *et al.* Increased cerebrospinal fluid 5-hydroxytryptamine and 5-hydroxyindoleacetic acid in Kleine–Levin syndrome. *Neurology* 1984; **34**:1597–600.

Kolivakis TT, Margolese HC, Beauclair L *et al.* Olanzapine-induced somnambulism. *Am J Psychiatry* 2001; **158**:1158.

Kraiczi H, Hedner J, Dahlof P *et al.* Effect of serotonin reuptake inhibition on breathing during sleep and daytime symptoms in obstructive sleep apnea. *Sleep* 1999; **22**:61–7.

Kravitz H, Rosenthall V, Teplitz Z *et al.* A study of head-banging in infants and children. *Dis Nerv Syst* 1960; **21**:203–8.

Krishnan PR, Bhatia M, Behari M. Restless legs syndrome in Parkinson's disease: a case–controlled study. *Mov Disord* 2003; **18**:181–5.

Kummu H, Iranzo A, Carrasco E *et al.* Lack of effects of pramipexole on REM sleep behavior disorder in Parkinson disease. *Sleep* 2008; **31**:1418–21.

Lee JW. Recurrent delirium associated with obstructive sleep apnea. *Gen Hosp Psychiatry* 1998; **20**:120–2.

Lee MS, Choi YC, Lee SH *et al.* Sleep-related periodic leg movements associated with spinal cord lesions. *Mov Disord* 1996; **11**:719–22.

Lee SJ, Kim JS, Song IU *et al.* Poststroke restless legs syndrome and lesion location: anatomical considerations. *Mov Disord* 2009; **24**:77–84.

Levchenko A, Provost S, Montplaisir J-Y *et al.* A novel autosomal dominant restless legs syndrome locus maps to chromosome 20p13. *Neurology* 2006; **67**:900–1.

Levin M. Periodic somnolence and morbid hunger: a new syndrome. *Brain* 1936; **59**:494–504.

Louden MB, Morehead MA, Schmidt HS. Activation by selegiline (Eldepryl) of REM sleep behavior disorder in parkinsonism. *W V Med J* 1995; **91**:101.

McArthur AJ, Levy AJ, Sack RL. Non-24 hour sleep-wake syndrome in a sighted man: circadian rhythm studies and efficacy of melatonin treatment. *Sleep* 1996; **19**:544–53.

Malow BA, Levy K, Maturen K *et al.* Obstructive sleep apnea is common in medically refractory epilepsy patients. *Neurology* 2000; **55**:1002–7.

Manconi M, Govoni V, De Vito A *et al.* Restless legs syndrome and pregnancy. *Neurology* 2004; **63**:1065–9.

Manconi M, Ferini-Strambi L, Filippi M *et al.* Multicenter case–control study on restless legs syndrome in multiple sclerosis: the REMS study. *Sleep* 2008; **31**:944–52.

Manni R, Terzaghi M, Zambrelli E *et al.* Interictal, potentially misleading, epileptiform EEG abnormalities in REM sleep behavior disorder. *Sleep* 2006; **29**:934–7.

Mayer G, Ewert Meier K, Hephata K. Selegiline hydrochloride treatment in narcolepsy: a double-blind, placebo-controlled study. *Clin Neuropharmacol* 1995; **18**:306–19.

Mayer G, Wang-Weigand S, Roth-Schechter B *et al.* Efficacy and safety of 6-month nightly ramelteon administration in adults with chronic primary insomnia. *Sleep* 2009; **32**:351–60.

Mellman TA, Uhde TW. Sleep panic attacks: new clinical findings and theoretical implications. *Am J Psychiatry* 1989a; **146**: 1204–7.

Mellman TA, Uhde TW. Electroencephalographic sleep in panic disorder: a focus on sleep-related panic attacks. *Arch Gen Psychiatry* 1989b; **46**:178–89.

Mellman TA, Uhde TW. Patients with frequent sleep panic: clinical findings and response to medication treatment. *J Clin Psychiatry* 1990; **51**:513–16.

Mendelson WB. Are periodic leg movements associated with clinical sleep disturbance? *Sleep* 1996; **19**:219–23.

Mendelson WB, Garnett D, Gillin JC. Single case study: flurazepam-induced sleep apnea syndrome in a patient with insomnia and mild sleep-related respiratory changes. *J Nerv Ment Dis* 1981; **169**: 261–4.

Mignot E, Lammers GJ, Ripley B *et al.* The role of cerebrospinal fluid hypocretin measurement in the diagnosis of narcolepsy and other hypersomnias. *Arch Neurol* 2002; **59**:1553–62.

Mikkelsen EJ, Rapoport JL, Nee L *et al.* Childhood enuresis. I. Sleep patterns and psychopathology. *Arch Gen Psychiatry* 1980; **37**:1139–44.

Millman RP, Fogel BS, McNamara ME *et al.* Depression as a manifestation of obstructive sleep apnea: reversal with nasal continuous positive airway pressure. *J Clin Psychiatry* 1989; **50**:348–51.

Mitler MM, Shafor R, Hajdukovich R *et al.* Treatment of narcolepsy: objective studies on methylphenidate, pemoline, and protriptyline. *Sleep* 1986a; **9**:260–4.

Mitler MM, Browman CP, Menn SJ *et al.* Nocturnal myoclonus: treatment efficacy of clonazepam and temazepam. *Sleep* 1986b; **9**:385–92.

Montagna P, Cirignotta F, Sacquegna T *et al.* Painful legs and moving toes' associated with polyneuropathy. *J Neurol Neurosurg Psychiatry* 1983; **46**:399–403.

Montplaisir J, Boucher S, Poirier G *et al.* Clinical, polysomnographic and genetic characteristics of restless legs syndrome: a study of 133 patients diagnosed with new standard criteria. *Mov Disord* 1997; **12**:61–5.

Montplaisir J, Nicolas A, Denesle R *et al.* Restless legs syndrome improved by pramipexole: a double-blind randomized trial. *Neurology* 1999; **52**:938–43.

Morgenthaler TI, Silber MH. Amnestic sleep-related eating disorder associated with zolpidem. *Sleep Med* 2002; **3**:323–7.

Munoz X, Marti S, Sumalla J *et al.* Acute delirium as a manifestation of obstructive sleep apnea syndrome. *Am J Respir Crit Care Med* 1998; **158**:1306–7.

Muratori R, Bertini N, Masi G. Efficacy of lithium treatment in Kleine–Levin syndrome. *Eur Psychiatry* 2002; **17**:232–3.

Nakamura K, Hashimoto S, Honma S *et al.* Daily melatonin intake resets circadian rhythms of a sighted man with non-24 hour sleep–wake syndrome who lacks the nocturnal melatonin rise. *Psychiatry Clin Neurosci* 1997; **51**:121–7.

Nathan PW. Painful legs and moving toes: evidence on the site of the lesion. *J Neurol Neurosurg Psychiatry* 1978; **41**:934–9.

Neveus T, Stenberg A, Lackgren G *et al.* Sleep of children with enuresis: a polysomnographic study. *Pediatrics* 1999; **103**:1193–7.

Neidhardt EJ, Krakow B, Kellner R. The beneficial effects of one treatment session and recording of nightmares on chronic nightmare sufferers. *Sleep* 1992; **15**:470–3.

Nguyen BH, Perusse D, Paquet J *et al.* Sleep terrors in children: a prospective study of twins. *Pediatrics* 2008; **122**:1164–7.

Nightingale S, Orgill JC, Ebrahim IO *et al.* The association between narcolepsy and REM behavior disorder (RBD). *Sleep Med* 2005; **6**:253–8.

Nogues M, Cammarota A, Leiguarda R *et al.* Periodic limb movements in syringomyelia and syringobulbia. *Mov Disord* 2000; **15**:113–19.

Nowbar S, Burkart KM, Gonzales R *et al.* Obesity-associated hypoventilation in hospitalized patients: prevalence, effects, and outcome. *Am J Med* 2004; **116**:58–9.

Odeh M, Oliven A. Coma and seizures due to severe hyponatremia and water intoxication in an adult with intranasal desmopressin therapy for nocturnal enuresis. *J Clin Pharmacol* 2001; **41**:582–4.

Ogura C, Okuma T, Nakazawa A. Treatment of periodic somnolence with lithium carbonate. *Arch Neurol* 1976; **33**:143.

Ohanna N, Peled R, Rubin AH *et al.* Periodic leg movements in sleep: effect of clonazepam treatment. *Neurology* 1985; **35**:408–11.

Olson EJ, Boeve BF, Silber MH. Rapid eye movement sleep behavior disorder: demographic, clinical and laboratory findings in 93 cases. *Brain* 2000; **123**:331–9.

Ondo W, Romanyshyn J, Vuong KD *et al.* Long-term treatment of restless legs syndrome with dopamine agonists. *Arch Neurol* 2004; **61**:1393–7.

Onofrj M, Luciano AL, Thomas A *et al.* Mirtazapine induces REM sleep behavior disorder (RBD) in parkinsonism. *Neurology* 2003; **60**: 113–15.

Oswald I. Sudden bodily jerks on falling asleep. *Brain* 1959; **82**: 92–103.

Oudiette D, Leu S, Pottier M *et al.* Dreamlike mentations during sleepwalking and sleep terrors in adults. *Sleep* 2009a; **32**:1621–7.

Oudiette D, De Cock VC, Lavault S *et al.* Nonviolent elaborate behaviors may also occurin REM sleep behavior disorder. *Neurology* 2009b; **7**:551–7.

Parkes JD. The parasomnias. *Lancet* 1986; **2**:1021–5.

Parkes JD, Baraitser M, Marsden CD *et al.* Natural history of symptoms and treatment of the narcoleptic syndrome. *Acta Neurol Scand* 1975; **52**:337–53.

Peled R, Lavie P. Double-blind evaluation of clonazepam on periodic leg movements in sleep. *J Neurol Neurosurg Psychiatry* 1987; **50**:1679–81.

Peralta CM, Frauscher B, Seppi K *et al.* Restless legs syndrome in Parkinson's disease. *Mov Disord* 2009; **24**:2076–80.

Pereira DS, Tufik S, Louzada FM *et al.* Association of the length polymorphism in the human Per3 gene with the delayed sleep phase syndrome: does latitude have an influence upon it? *Sleep* 2005; **28**:29–32.

Peyron C, Faraco J, Rogers W *et al.* A mutation in a case of early onset narcolepsy and a generalized absence of hypocretin peptides in human narcoleptic brains. *Nat Med* 2000; **6**:991–7.

Pierce CM, Whitman RM, Gay ML. Enuresis and dreaming: experimental studies. *Arch Gen Psychiatry* 1961; **4**:166–70.

Pike M, Stores G. Kleine–Levin syndrome: a cause of diagnostic confusion. *Arch Dis Child* 1994; **71**:355–7.

Pillar G, Etzioni A, Shahar E *et al.* Melatonin treatment in an institutionalized child with psychomotor retardation and an irregular sleep–wake pattern. *Arch Dis Child* 1998; **79**:63–4.

Piper AJ, Wang D, Yee BJ *et al.* Randomised trial of CPAP vs bilevel support in the treatment of obesity hypoventilation syndrome without severe nocturnal desaturation. *Thorax* 2008; **63**: 395–401.

Pittock SJ, Parrett T, Adler CH *et al.* Neuropathology of primary restless legs syndrome: absence of specific tau- and alpha-synucelin pathology. *Mov Disord* 2004; **19**:695–9.

Pla MER, Dillingham TR, Spellman LT *et al.* Painful legs and moving toes associated with tarsal tunnel syndrome and accessory soleus muscle. *Mov Disord* 1996; **11**:82–6.

Plazzi G, Corsini R, Provini F *et al.* REM sleep behavior disorders in multiple system atrophy. *Neurology* 1997; **48**:1094–7.

Plazzi G, Montagna P. Remitting REM sleep behavior disorder as the initial sign of multiple sclerosis. *Sleep Med* 2002; **3**:437–9.

Polydefkis M, Allen RP, Hauer P *et al.* Subclinical sensory neuropathy in late-onset restless legs syndrome. *Neurology* 2000; **55**:1115–21.

Poppe M, Freibel D, Reuner U *et al.* The Kleine–Levin syndrome – effects of treatment with lithium. *Neuropediatrics* 2003; **34**:113–19.

Poryazova R, Siccoli M, Werth E *et al.* Unusually prolonged rebound cataplexy after withdrawal of fluoxetine. *Neurology* 2005; **65**:967–8.

Postuma RB, Gagnon JF, Vendette M *et al.* Quantifying the risk of neurodegenerative disease in idiopathic REM sleep behavior disorder. *Neurology* 2009; **72**:1296–300.

Quinto C, Gellido C, Chokroverty S *et al.* Posttraumatic delayed sleep phase syndrome. *Neurology* 2000; **54**:250–2.

Rangarajan S, D'Souza GA. Restless legs syndrome in Indian patients having iron deficiency anemia in a tertiary care hospital. *Sleep Med* 2007; **8**:247–51.

Rechtschaffen A, Wolpert EA, Dement WC *et al.* Nocturnal sleep of narcoleptics. *Electroencephalogr Clin Neurophysiol* 1963; **15**:599–609.

Reid KJ, Chang AM, Dubocovich ML *et al.* Familial advanced sleep phase syndrome. *Arch Neurol* 2001; **58**:1089–94.

Reid WH, Haffke EA, Chu CC. Diazepam in intractable sleepwalking: a pilot study. *Hillside J Clin Psychiatry* 1984; **6**:49–55.

Reynolds CF, Black RS, Coble P *et al.* Similarities in EEG sleep findings for Kleine–Levin syndrome and unipolar depression. *Am J Psychiatry* 1980; **137**:116–18.

Reynolds CF, Kupfer DJ, Christiansen CL *et al.* Multiple sleep latency test findings in Kleine–Levin syndrome. *J Nerv Ment Dis* 1984; **172**:41–4.

Riemann D, Klein T, Todenbeck A *et al.* Nocturnal cortisol and melatonin secretion in primary insomnia. *Psychiatry Res* 2002a; **113**:17–27.

Riemann D, Voderholzer U, Cohrs S *et al.* Trimipramine in primary insomnia: results of a polysomnographic double-blind study. *Pharmacopsychiatry* 2002b; **35**:165–74.

Rijsman RM, de Weerd AW, Stam CJ *et al.* Periodic limb movement disorder and restless legs syndrome in dialysis patients. *Nephrology* 2004; **9**:353–61.

Ringman JM, Simmons JH. Treatment of REM sleep behavior disorder with donepezil; a report of three cases. *Neurology* 2000; **55**: 870–1.

Roehrs TA, Zorick FJ, Wittig RM *et al.* Alerting effects of naps in patients with narcolepsy. *Sleep* 1986; **9**:194–9.

Rogers AE, Aldrich MS, Liu X. A comparison of three different sleep schedules for reducing daytime sleepiness in narcolepsy. *Sleep* 2001; **24**:385–91.

Roizenblatt S, Guilleminault C, Poyares D *et al.* A double-blind, placebo-controlled, crossover study of sildenafil in obstructive sleep apnea. *Arch Int Med* 2006; **166**:1763–7.

Rosenfeld MR, Eichen JG, Wade DF *et al.* Molecular and clinical diversity in paraneoplastic immunity to Ma proteins. *Ann Neurol* 2001; **50**:339–48.

Roth T, Rogowski R, Hull S *et al.* Effiacy and safety of doxepin 1, 3, and 6 mg in adults with primary insomnia. *Sleep* 2007; **30**:1555–61.

Rottach KG, Schaner BM, Kirch MH *et al.* Restless legs syndrome as side effect of second generation antidepressants. *J Psychiatr Res* 2008; **43**:70–5.

Sachs S, Persson HE, Hagenfeldt K. Menstruation-related periodic hypersomnia: a case study with successful treatment. *Neurology* 1982; **32**:1376–9.

Sack RL, Lewy AJ, Blood ML *et al.* Circadian rhythm abnormalities in totally blind people: incidence and clinical significance. *J Clin Endocrinol Metabol* 1992; **75**:127–34.

Sack RL, Brandes RW, Kendall AR *et al.* Entrainment of free-running circadian rhythms by melatonin in blind people. *N Engl J Med* 2000; **343**:1070–7.

Sadeghniiat-Haghighi K, Aminian O, Pouryaghoub G *et al.* Efficacy and hypnotic effects of melatonin in shift-work nurses: double-blind, placebo-controlled crossover trial. *J Circadian Rhythms* 2008; **6**:10.

Sagar RS, Khandelwal SK, Gupta S. Interepisodic morbidity in Kleine–Levin syndrome. *Br J Psychiatry* 1990; **157**:139–41.

Saletu M, Anderer P, Saletu-Zyhlarz G *et al.* Restless legs syndrome (RLS) and periodic limb movement disorder (PLMD): acute placebo-controlled sleep laboratory studies with clonazepam. *Eur Neuropsychopharmacol* 2001; **11**:153–61.

Saletu M, Anderer P, Saletu-Zyhlarz GM *et al.* Comparative placebo-controlled polysomnographic and psychometric studies on the acute effects of gabapentin versus ropinirole in restless legs syndrome. *J Neural Trans* 2010; **117**:463–473.

Salih AM, Gray RE, Mills KR *et al.* A clinical, serological and neurophysiological study of restless legs syndrome in rheumatoid arthritis. *Br J Rheumatology* 1994; **33**:60–3.

Sandberg O, Franklin KA, Bucht G *et al.* Sleep apnea, delirium, depressed mood, cognition and ADL ability after stroke. *J Am Geriatr Soc* 2001; **49**:391–7.

Scammell TE, Nishino S, Mignot E *et al.* Narcolepsy and low CSF orexin (hypocretin) concentration after a diencephalic stroke. *Neurology* 2001; **56**:1751–3.

Schacter M, Parkes JD. Fluvoxamine and clomipramine in the treatment of cataplexy. *J Neurol Neurosurg Psychiatry* 1980; **43**:171–4.

Scharf M, Rogowski R, Hull S *et al.* Efficacy and safety of doxepin 1 mg, 3 mg, and 6 mg in elderly patients with primary insomnia: a randomized, double-blind, placebo-controlled crossover study. *J Clin Psychiatry* 2008; **69**:1557–64.

Scheltens P, Visscher F, Van Keimpema ARJ *et al.* Sleep apnea syndrome presenting with cognitive impairment. *Neurology* 1991; **41**:155.

Schenck CH, Bundlie SR, Ettinger MG *et al.* Chronic behavioral disorders of human REM sleep: a new category of parasomnia. *Sleep* 1986; **9**:293–308.

Schenck CH, Bundlie SR, Patterson AL *et al.* Rapid eye movement sleep behavior disorder. A treatable parasomnia affecting older adults. *JAMA* 1987; **257**:1786–9.

Schenck CH, Milner DM, Hurwitz TD *et al.* A polysomnographic and clinical report on sleep-related injury in 100 adult patients. *Am J Psychiatry* 1989; **146**:1166–73.

Schenck CH, Mahowald MW. Polysomnographic, neurologic, psychiatric, and clinical outcome report on 70 consecutive cases with REM sleep behavior disorder (RBD): sustained clonazepam efficacy in 89.5 percent of 57 treated patients. *Cleve Clin J Med* 1990; **57**(Suppl.):S9–23.

Schenck CH, Mahowald MW. Motor dyscontrol in narcolepsy: rapid-eye-movement (REM) sleep without atonia and REM sleep behavior disorder. *Ann Neurol* 1992; **32**:3–10.

Schenck CH, Mahowald MW, Kim SW *et al.* Prominent eye movements during NREM sleep and REM sleep behavior disorder associated with fluoxetine treatment of depression and obsessive–compulsive disorder. *Sleep* 1992; **15**:226–35.

Schenck CH, Mahowald MW. A polysomnographically documented case of adult somnambulism with long-distance automobile driving and frequent nocturnal violence: parasomnia with continuing danger as a noninsane automatism? *Sleep* 1995; **18**: 765–72.

Schenck CH, Mahowald MW. Long-term, nightly benzodiazepine treatment of injurious parasomnias and other disorders of disrupted nocturnal sleep in 170 adults. *Am J Med* 1996; **100**:333–7.

Schenck CH, Bundlie SR, Mahowald MW. Delayed emergence of a parkinsonian disorder in 38% of 29 older men initially diagnosed with idiopathic rapid eye movement sleep behavior disorder. *Neurology* 1996a; **46**:388–93.

Schenck CH, Garcia-Rill E, Skinner RD *et al.* A case of REM sleep behavior disorder with autopsy-confirmed Alzheimer's disease: postmortem brain stem histochemical analyses. *Biol Psychiatry* 1996b; **40**:422–5.

Schmidauer C, Sojer M, Seppi K *et al.* Transcranial ultrasound shows migral hypoechogenicity in restless legs syndrome. *Ann Neurol* 2005; **58**:630–4.

Schmidt MH, Koshal VB, Schmidt HS. Use of pramipexole in REM sleep behavior disorder: results from a case series. *Sleep Med* 2006; **7**:418–23.

Schott GD. Painful legs and moving toes': the role of trauma. *J Neurol Neurosurg Psychiatry* 1981; **44**:344–6.

Schulman SI, Stokes A, Salzman PM. The efficacy and safety of oral desmopressin in children with primary nocturnal enuresis. *J Urol* 2001; **166**:2427–31.

Schwartz DJ, Karatinos G. For individuals with obstructive sleep apnea, institution of CPAP therapy is associated with an amelioration of symptoms of depression which is sustained long term. *J Clin Sleep Med* 2007; **3**:631–5.

Schwartz JRL, Feldman NT, Bogan RK. Dose effects of modafinil in sustaining wakefulness in narcolepsy patients with residual evening sleepiness. *J Neuropsychiatry Clin Neurosci* 2005a; **17**:405–12.

Schwartz DJ, Kohler WC, Katarinos G. Symptoms of depression in individuals with obstructive sleep apnea may be amenable to treatment with continuous positive airway pressure. *Chest* 2005b; **128**:1304–9.

Scrima L, Broudy M, Nay KN *et al.* Increased severity of obstructive sleep apnea after bedtime alcohol ingestion: diagnostic potential and proposed mechanism of action. *Sleep* 1982; **5**:318–28.

Sechi G, Agnetti V, Galistu P *et al.* Restless legs syndrome and periodic limb movements after ischemic stroke in the right lenticulostriate region. *Parkinsonism Relat Disord* 2008; **14**:157–60.

Shinno H, Inami Y, Inagaki T *et al.* Successful treatment with levothyroxine for idiopathic hypersomnia patients with subclinical hypothyroidism. *Gen Hosp Psychiatry* 2009; **31**:190–3.

Sieker HO, Estes EH, Kelser GA *et al.* Cardiopulmonary syndrome with extreme obesity. *J Clin Invest* 1955; **34**:916.

Skomro R, Silva R, Alves R *et al.* The prevalence and significance of periodic leg movements during sleep in patients with congestive heart failure. *Sleep Breath* 2009; **13**:43–7.

Smith PL, Gold AR, Meyers DA *et al.* Weight loss in mildly to moderately obese patients with obstructive sleep apnea. *Ann Intern Med* 1985; **103**:850–5.

Smith RC. Relationship of periodic movements in sleep (nocturnal myoclonus) and the Babinski sign. *Sleep* 1985; **8**:239–43.

Smith MT, Perlis ML, Changazi VU *et al.* Neuroimaging of NREM sleep in primary insomnia: a Tc-99-HMPAO single photon emission computed tomography study. *Sleep* 2002; **25**:325–35.

Smits MG, Nagtegaal EE, van der Heijden J *et al.* Melatonin for chronic sleep onset insomnia in children: a randomized placebo-controlled trial. *J Child Neurol* 2001; **16**:86–92.

Sonka K, Kemlink D, Pretl M. Cataplexy treated with escitalopram – clinical experience. *Neurol Endocrinol Lett* 2006; **27**:174–6.

Sours JA, Frumkin P, Indermill RR. Somnambulism: its clinical significance and dynamic meaning in late adolescence and adulthood. *Arch Gen Psychiatry* 1963; **9**:400–13.

Spillane JD, Nathan PW, Kelly RE *et al.* Painful legs and moving toes. *Brain* 1971; **94**:541–56.

Stahl SM, Layzer RB, Aminoff MJ *et al.* Continuous cataplexy in a patient with a midbrain tumor: the limp man syndrome. *Neurology* 1980; **30**:1115–18.

Stefansson H, Rye DB, Hicks A *et al.* A genetic risk factor for periodic limb movements in sleep. *N Engl J Med* 2007; **357**: 639–47.

Steiner MC, Ward MJ, Ali NJ. Dementia and snoring. *Lancet* 1999; **353**:204.

Stoner SC, Dahmen MM, Makos M *et al.* An exploratory retrospective evaluation of ropinirole-associated psychotic symptoms in an outpatient population treated for restless legs syndrome or Parkinson's disease. *Ann Pharmacother* 2009; **43**:1426–32.

Strollo PJ, Rogers RM. Obstructive sleep apnea. *N Engl J Med* 1996; **334**:99–104.

Suhner A, Schlagenhauf P, Hofer I *et al.* Effectiveness and tolerability of melatonin and zolpidem for the alleviation of jet lag. *Aviat Space Environ Med* 2001; **72**:638–46.

Takano A, Uchiyama M, Kajimura N *et al.* A missense variation in human casein kinase I epsilon gene that induces functional alteration and shows an inverse association with circadian rhythm sleep disorders. *Neuropsychopharmacology* 2004; **29**:1901–9.

Tan A, Selgado M, Fahn S. Rapid eye movement sleep behavior disorder preceding Parkinson's disease with therapeutic response to levodopa. *Mov Disord* 1996; **11**:214–16.

Telstad W, Sorensen O, Larsen S *et al.* Treatment of restless legs syndrome with carbamazepine: a double blind study. *Br Med J (Clin Res Ed)* 1984; **288**:444–6.

Thannickal TC, Nienhuis R, Siegel JM. Localized loss of hypocretin (orexin) cells in narcolepsy without cataplexy. *Sleep* 2009; **32**:993–8.

Thirumalai SS, Shubin RA. The use of citalopram in resistant cataplexy. *Sleep Med* 2000; **1**:313–16.

Tinuper P, Cerullo A, Cirignotta F *et al.* Nocturnal paroxysmal dystonia: three cases with evidence for an epileptic frontal lobe origin of seizures. *Epilepsia* 1990; **31**:549–56.

Tippmann-Peikert M, Boeve BF, Keegam BM. REM sleep behavior disorder initiated by acute brainstem multiple sclerosis. *Neurology* 2006; **66**:1277–9.

Tobin JD. Treatment of somnambulism with anticipatory awakening. *J Pediatr* 1993; **122**:426–7.

Trenkwalder C, Seidel VC, Gasser T *et al.* Clinical symptoms and possible anticipation in a large kindred of familial restless legs syndrome. *Mov Disord* 1996; **11**:389–94.

Turner RS, D'Amato CJ, Chervin RD *et al.* The pathology of REM sleep behavior disorder with comorbid Lewy body dementia. *Neurology* 2000; **55**:1730–2.

Uchiyama M, Isse K, Tanaka K *et al.* Incidental Lewy body disease in a patient with REM sleep behavior disorder. *Neurology* 1995; **45**:709–12.

US Modafinil in Narcolepsy Multicenter Study Group. Randomized trial of modafinil for the treatment of pathological somnolence in narcolepsy. *Ann Neurol* 1998; 43:88–97.

US Modafinil in Narcolepsy Multicenter Study Group. Randomized trial of modafinil as a treatment for the excessive daytime sleepiness of narcolepsy. *Neurology* 2000; 54:1166–75.

Valko PO, Khatami R, Baumann CR *et al.* No persistent effect of intravenous immunoglobulins in patients with narcolepsy with cataplexy. *J Neurol* 2008; **255**:1900–3.

van Dijk JG, Lemmers GJ, Blansjaar BA. Isolated cataplexy of more than 40 years' duration. *Br J Psychiatry* 1991; **159**:719–21.

Vernet C, Arnulf I. Idiopathic hypersomnia with and without long sleep time: a controlled series of 75 patients. *Sleep* 2009; **32**:753–9.

von Gontard A, Scaumberg H, Hollmann E *et al.* The genetics of enuresis: a review. *J Urol* 2001; **166**:2438–43.

Wagner ML, Walters AS, Coleman RG *et al.* Randomized, double-blind, placebo-controlled study of clonidine in restless legs syndrome. *Sleep* 1996; **19**:52–8.

Walsh JK, Erman MK, Erwin JR *et al.* Subjective hypnotic efficacy of trazodone and zolpidem in DSM-III-R primary insomnia. *Hum Psychopharmacol* 1988; **13**:191–8.

Walters AS, Hening W, Rubinstein M *et al.* A clinical and polysomnographic comparison of neuroleptic-induced akathisia and the idiopathic restless legs syndrome. *Sleep* 1991; **14**:339–45.

Walters AS, Wagner ML, Hening WA *et al.* Successful treatment of idiopathic restless legs syndrome in a randomized double-blind trial of oxycodone versus placebo. *Sleep* 1993; **16**:327–32.

Wang J, O'Reilly B, Venkataraman R *et al.* Efficacy of oral iron in patients with restless legs syndrome and low-normal ferritin: a randomized, double-blind, placebo-controlled study. *Sleep Med* 2009; **10**:973–5.

Ward WA, Kelsey WM. The Pickwickian syndrome: a review of the literature and report of a case. *J Pediatr* 1962; **61**:745–50.

Ware JC, Brown FU, Moorad PJ *et al.* Nocturnal myoclonus and tricyclic antidepressants. *Sleep Res* 1984; **13**:72.

Watanabe T, Kajimura N, Kato M *et al.* Effects of phototherapy in patients with delayed sleep phase syndrome. *Psychiatry Clin Neurosci* 1999; **53**:231–3.

de Weerd AW, Rijsman RM, Brinkley A. Activity patterns of leg muscles in periodic limb movement disorder. *J Neurol Neurosurg Psychiatry* 2004; **75**:317–19.

Wetter TC, Callado-Seidel V, Pollmacher T *et al.* Sleep and periodic leg movement patterns in drug-free patients with Parkinson's disease and multiple system atrophy. *Sleep* 2000; **23**:361–7.

White DP, Zwillich CW, Pickett CK *et al.* Central sleep apnea. Improvement with acetazolamide therapy. *Arch Int Med* 1982; **142**:1816–19.

Whitney JF, Gannon DE. Obstructive sleep apnea presenting as acute delirium. *Am J Emerg Med* 1996; **14**:270–1.

Whyte KF, Allen MB, Jeffrey AA *et al.* Clinical features of the sleep apnoea/hypopnoea syndrome. *Q J Med* 1989; **72**:659–66.

Will RG, Young JPR, Thomas DJ. Kleine–Levin syndrome: report of two cases with onset of symptoms precipitated by head trauma. *Br J Psychiatry* 1988; **152**:410–12.

Wilson SAK. The narcolepsies. *Brain* 1928; **51**:63–109.

Wilson SJ, Lillywhite AR, Potokar JP *et al.* Adult night terrors and paroxetine. *Lancet* 1997; **350**:185.

Winkelman JW, Johnston L. Augmentation and tolerance with long-term pramipexole treatment of restless legs syndrome (RLS). *Sleep Med* 2004; **5**:9–14.

Winkelman J, Buxton OM, Jensen JE *et al.* Reduced brain GABA in primary insomnia: preliminary data from 4T proton magnetic resonance spectroscopy (1H-MRS). *Sleep* 2008; **31**:1499–506.

Winkelmann JW, Chertow GM, Lazarus JM. Restless legs syndrome in end-stage renal disease. *Am J Kidney Dis* 1996; **28**:372–8.

Winkelmann J, Wetter TC, Collado-Seidel V *et al.* Clinical characteristics of the hereditary restless legs syndrome in a population of 300 patients. *Sleep* 2000; **23**:597–602.

Winkelmann J, Muller-Myhsok B, Wittchen H-U *et al.* Complex segregation analysis of restless legs syndrome provides evidence of an autosomal dominant mode of inheritance in early age at onset families. *Ann Neurol* 2002; **52**:297–302.

Witting W, Kwa IH, Eikelemboom P *et al.* Alterations in the circadian rest-activity rhythm in aging and Alzheimer's disease. *Biol Psychiatry* 1990; **27**:563–72.

Xi Z, Luning W. REM sleep behavior disordere in a patient with pontine stroke. *Sleep Med* 2009; **10**:143–6.

Xu Y, Padiath QS, Shapiro RE *et al.* Functional consequences of a CKIdelta mutation causing familial advanced sleep phase syndrome. *Nature* 2005; **434**:640–4.

Yaggi HK, Concato J, Kernan WM *et al.* Obstructive sleep apnea as a risk factor for stroke and death. *N Engl J Med* 2005; **353**:2034–45.

Yang C, White DP, Winkelman JW. Antidepresssants and periodic limb movements of sleep. *Biol Psychiatry* 2005; **58**:510–14.

Yeh YW, Chen CH, Feng HM *et al.* New onset somnambulism associated with different dosage of mirtazapine: a case report. *Clin Neuropharmacol* 2009; **32**:232–3.

Zadra A, Pillon M, Montplaisir J. Polysomnographic diagnosis of sleepwalking: effects of sleep deprivation. *Ann Neurol* 2008; **63**:513–19.

Zambelis T, Paparrigopoulos T, Soldatos CR. REM sleep behavior disorder associated with a neurinoma of the left pontocerebellar angle. *J Neurol Neurosurg Psychiatry* 2002; **72**:821–2.

Zhdanova IV, Wurtman RJ, Regan MM *et al.* Melatonin treatment for age-related insomnia. *J Clin Endocrincol Metabol* 2001; **86**: 4727–30.

21

Brain tumors and hydrocephalus

21.1 BRAIN TUMORS

Neoplastic brain tumors (DeAngelis 2001) may be broadly divided into two types, namely those neoplasms that are primary to the brain, such as gliomas or meningiomas, and those that represent metastases from systemic cancers, such as lung or breast cancer.

Clinical features

Although brain tumors may occur at any age, most patients are middle-aged or older. The onset itself ranges from acute to insidious, depending in large part on the aggressiveness of the tumor involved. Certain gliomas, such as glioblastoma multiforme, may present rapidly over several weeks or months, whereas some meningiomas may attain a large size without ever causing symptoms (Olivero *et al.* 1995) and may indeed be found incidentally on imaging for other reasons.

The symptomatology of brain tumors varies according to their location, size, the extent of peri-tumoral vasogenic edema, and the appearance of increased intracranial pressure: there is generally little room for expansion within the intracranial vault and, with growth of a tumor and, especially, expansion of vasogenic edema, there is an inevitable rise in overall intracranial pressure, which may eventually produce symptoms in its own right.

The overall symptomatology seen with brain tumors may be divided into the following domains: headache; non-focal symptoms; focal signs and specific syndromes, such as dementia or personality change; and seizures. Each of these is considered in turn below.

Headache (Forsyth and Posner 1993) may be generalized or have a unilateral predominance, in which case it may have some lateralizing value. The headache itself tends to be dull and may be severe. It is typically worsened by recumbancy, a Valsalva manuever (e.g. coughing, straining) or merely by bending over. Although classically it is worst in the morning, such diurnal variation is not common. In most cases the headache reflects stretching of pain-sensitive structures; however, should increased intracranial pressure occur, the headache may worsen and be joined, classically, by projectile vomiting.

Non-focal symptoms (Davison and Demuth 1945, 1946; McKendree and Feinier 1927) often reflect increased intracranial pressure and may include dizziness, 'fuzziness' of thought, and somnolence. In cases where the increase in intracranial pressure approaches cerebral perfusion pressure, a Valsalva manuever, or, again, merely bending over, may increase pressure to the point where cerebral perfusion essentially halts, producing, variously, worsening headache, nausea, vomiting, visual dimming, ataxia, and generalized weakness. In some cases, syncope may occur, which in turn may, as with any form of syncope, be accompanied by some clonic jerks. Overall, provided that the stimulus for the increased pressure passes, symptoms abate in a few minutes.

Focal signs and specific syndromes typically reflect compression of brain tissue by the tumor mass or peri-tumoral edema. Traditional focal signs, such as hemiplegia, aphasia, apraxia, and hemianopia, may occur and may serve to both lateralize and localize the tumor; compression or stretching of cranial nerves may result in appropriate cranial nerve palsies. Specific syndromes seen with tumors include, most commonly, dementia, personality change and depression; other specific syndromes, seen in a small minority, include delirium, amnesia, mania, and psychosis. Each of these is discussed in turn.

Dementia is classically seen with tumors of the frontal lobe (Keschner *et al.* 1936; Sachs 1950) or corpus callosum (Alpers and Grant 1931; Moersch 1925), and in such cases it is often accompanied by apathy, dullness, and somnolence (Williamson 1896) or by a frontal lobe syndrome (Frazier 1936). Tumors of the temporal lobe (Keschner *et al.* 1936), thalamus and hypothalamus (Alpers 1937; Liss 1958; Strauss and Globus 1931) may also cause dementia, and with hypothalamic tumors one often sees additional symptoms (Beal *et al.* 1981), such as hypersomnolence, weight gain, or diabetes insipidus.

Personality change is seen with tumors of the frontal lobe in the vast majority of cases (Direkze *et al.* 1971; Keschner *et al.* 1936; Strauss and Keschner 1935), with tumors of the temporal lobe in roughly half (Keschner *et al.* 1936), and in an unclear percentage of tumors of the corpus callosum

(Alpers and Grant 1931; Moersch 1925); uncommonly tumors of the thalamus or hypothalamus may also cause a personality change. Although the personality change may be non-specific, in cases of frontal lobe tumors one classically sees an accompanying frontal lobe syndrome (Avery 1971).

Depression is seen in close to one-half of all cases, and may occur secondary to tumors in a variety of locations (Keschner *et al.* 1936; Wellisch *et al.* 2002), including especially the frontal lobes, somewhat less so the temporal lobe, and far less so with the parietal and occipital lobes.

Delirium may occur with tumors of the temporal lobe (Keschner *et al.* 1936), thalamus, or hypothalamus (Alpers 1940).

Amnesia, with isolated short-term memory loss, may be seen with tumors that impinge on any part of the circuit of Papez, for example the fornix (Heilman and Sypert 1977; Sprofkin and Sciarra 1952), mammilary bodies (Kahn and Crosby 1971; Palmai *et al.* 1967), or thalamus (Ziegler *et al.* 1977).

Mania is a very uncommon symptom (Keschner *et al.* 1936; Strauss and Keschner 1935), but has been noted with tumors of the frontal lobes (Starkstein *et al.* 1988), thalamus (Stern and Dancy 1942), hypothalamus (Alpers *et al.* 1940), and mesencephalon (Greenberg and Brown 1985). Simple euphoria is far more common than mania (Keschner *et al.* 1936; Strauss and Keschner 1935), and has been noted in a substantial minority of cases with frontal lobe tumors, and somewhat less so in cases with temporal lobe tumors.

Psychosis may occur with tumors, most commonly of the temporal lobe (Gal 1958; Keschner *et al.* 1936; Malamud 1967; Strobos 1953; Tucker *et al.* 1986); other locations include the frontal lobe (Strauss and Keschner 1935), the corpus callosum (Murthy *et al.* 1997), and, rarely, the parietal lobe (Keschner *et al.* 1936).

Finally, a few words are in order regarding tumors located in the hypothalamus. As noted earlier, these may present with dementia, personality change, delirium, amnesia, or mania. Other symptoms may also be seen, including diabetes insipidus, anorexia with profound weight loss (Heron and Johnston 1976; White *et al.* 1977), and hyperphagia with extreme weight gain (Fulton and Bailey 1929; Liss 1958), which may, in rare instances, be accompanied by episodic rage (Haugh and Markesbery 1983; Reeves and Plum 1969).

Seizures may constitute the presentation of a tumor, and are seen early on in roughly one-third of cases; over time, eventually two-thirds of patients will have seizures, which may be simple partial, complex partial, or grand mal in type. Of note, seizures are particularly common with slower-growing tumors, such as oligodendrogliomas or low-grade astrocytomas. They are less common with metastatic tumors and high-grade astrocytomas: in the case of astrocytomas, one study (Lote *et al.* 1998) found that 95 percent of grade I astrocytomas were associated with seizures, with the percentage gradually declining as the malignancy worsened, with 49 percent of cases of grade IV tumors (glioblastoma multiforme) having seizures. As discussed in Chapter 8.3, seizures may have considerable lateralizing and localizing value.

With growth of the tumor and enlargement of the area of peri-tumoral edema, the clinical picture evolves, with worsening of initial symptoms and addition of new ones. In some cases, hydrocephalus may occur, with symptoms as discussed in Chapter 21.2. In other cases, there may be acute clinical exacerbations due to either intratumoral hemorrhage or infarction secondary to arterial compression.

Magnetic resonance (MR) scanning should, if possible, be obtained in all cases, and, again if possible, this should be with gadolinium enhancement. Computed tomography (CT) scanning, again, if possible, with enhancement, is an alternative, but far less sensitive, technique.

Although some types of tumors, for example meningiomas, may be fairly confidently diagnosed on the basis of their clinical presentation, course, and imaging characteristics, in most cases definitive diagnosis must wait upon examination of tissue obtained either by stereotaxic biopsy or craniotomy.

Lumbar puncture, although not routine, may be appropriate when certain tumors are suspected, such as primary central nervous system lymphoma or leptomeningeal carcinomatosis.

Although in most cases of metastatic disease the systemic cancer is already known, in a minority of cases, perhaps up to one-quarter, the metastasis represents the presentation of the systemic cancer, and, consequently, in evaluating patients with a brain tumor who do not apparently have systemic cancer, this possibility must always be kept in mind. One clue to the metastatic nature of the disease is the number of tumors and their location. Whereas primary tumors, with the exception of primary central nervous system lymphoma, are generally singular, metastatic disease generally manifests with two or more lesions. Further, metastatic tumors are typically found at the gray–white junction. When metastatic disease is suspected, a diligent search must be made for a primary tumor and this may include CT of the chest, abdomen, and pelvis, mammography, testicular ultrasound, and, in limiting cases, total body positron emission tomography (PET) scanning.

Course

The natural course varies widely, depending on the malignancy of the tumor itself, ranging from as little as months in the case of glioblastoma multiforme up to a decade or more with meningiomas or low-grade gliomas.

Etiology

As noted earlier, brain tumors may be either primary to the central nervous system or metastatic; of these two broad types, metastatic tumors are far more common than primary ones.

PRIMARY TUMORS

Of the primary brain tumors, *gliomas* and *meningiomas* constitute the vast majority of cases. *Primary central nervous system lymphoma*, once rare, has become increasingly common, both in immunocompromised and immunocompetent patients. Other primary brain tumors, seen in a small minority, include *neuromas*, *medulloblastoma*, *gangliocytoma*, *pituitary adenoma*, *craniopharyngioma*, *pineal tumors*, *hemangioblastoma*, and *colloid cyst of the third ventricle*. Each of these is considered in turn.

Gliomas include astrocytomas, oligodendrogliomas, and ependymomas. Astrocytomas are by far the most common type and may be divided into four grades according to their malignancy, namely grades I, II, III, and IV. Grades I (pilocytic) and II (diffuse) astrocytoma, or 'low-grade' astrocytomas, are slow-growing, whereas grades III (anaplastic) and IV (glioblastoma multiforme), or 'malignant' or 'high-grade' astrocytomas, tend to be rather aggressive: this is especially the case with glioblastoma multiforme, which is the most common of the gliomas. Astrocytomas tend to occur in the white matter of the frontal, temporal, or parietal lobes, or in the corpus callosum, but may also be found in the cerebellum or brainstem. On MR scanning, low-grade astrocytomas have decreased signal intensity on T1-weighted imaging and display homogeneous increased signal intensity on T2-weighted imaging; there may or may not be enhancement. High-grade astrocytomas may appear heterogeneous on T1- and T2-weighted imaging, and typically undergo enhancement, which, especially in the case of glioblastoma multiforme, may be ring-shaped. Over long periods of time, low-grade astrocytomas may undergo malignant transformation. A variant form of glioma is known as gliomatosis cerebri. This is a very aggressive growth characterized not by a discrete mass but by a widespread infiltration of one or both hemispheres of the cerebrum, primarily of the white matter; patients typically present with delirium, headache, and seizures, and T2-weighted MR scanning may reveal diffusely increased signal intensity in the white matter, which may or may not display a degree of enhancement.

Oligodendrogliomas arise from oligodendrocytes and are generally indolent growths (Wilkinson *et al.* 1987) found typically in the white matter of the frontal or temporal lobes, or in the corpus callosum, which often present with seizures; on MR scanning they present as non-enhancing masses with increased signal intensity on T2-weighted images, and on CT scans one may appreciate calcification. A very small minority of oligodendrogliomas may be anaplastic, and these forms do undergo enhancement.

Ependymomas, although most commonly seen in children, may present in early adult years. These tumors arise from ependymal cells of the fourth, third, or lateral ventricles, and may produce symptoms either by causing obstructive hydrocephalus or by extending into the adjacent brain parenchyma; imaging characteristics are similar to oligodendrogliomas and, like these tumors, a very small minority of ependymomas may also be anaplastic.

Meningiomas are very slow-growing tumors that arise from arachnoidal cells and which have an attachment to the dura. These are well-demarcated, extra-axial tumors that produce symptoms by compression of the subjacent brain parenchyma, from which they are clearly separated, as illustrated in Figure 21.1. Although they may or may not be readily discernible on T2-weighted scans, they do undergo homogeneous enhancement. There are several favored locations. Meningiomas of the falx cerebri, by compression of the medial aspects of the frontal or parietal cortices, may cause dementia or paraparesis. Lateral convexity meningiomas may also cause dementia, along with various other focal signs, such as hemiplegia or aphasia. Olfactory groove meningiomas may cause anosmia, blindness, and, by upward extension against the frontal lobe,

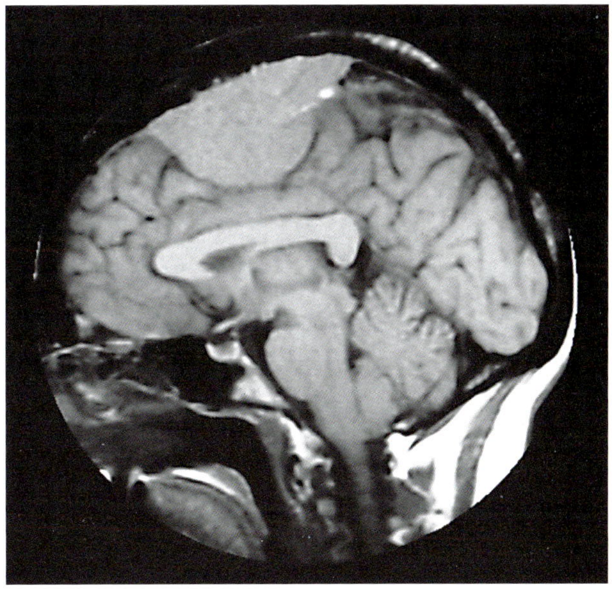

Figure 21.1 A sagittal T1-weighted magnetic resonance scan demonstrates the clear demarcation between a meningioma and the surrounding tissue. Reproduced with kind permission from Gillespie J and Jackson A. *MRI and CT of the Brain*. London: Arnold, 2000.

dementia; should they attain a size capable of causing increased intracranial pressure, a Foster Kennedy syndrome may occur, with ipsilateral anosmia, ipsilateral optic atrophy, and contralateral papilledema. Suprasellar meningiomas may cause a bitemporal hemianopia and pituitary failure, and meningiomas of the sphenoid ridge may present with extraocular nerve palsies and proptosis. All meningiomas may also be associated with seizures. Rarely, rather than presenting as a circumscribed mass, meningiomas may present 'en plaque' as a sheath-like structure, and, again, also rarely, anaplastic or atypical meningiomas may occur.

Primary central nervous system lymphoma, once rare, has recently shown an increasing incidence, and this appears to be the case not only in immunocompromised patients, such as transplant patients (Schneck and Penn 1970) and those with acquired immune deficiency syndrome (AIDS) (Feiden *et al.* 1993; Lang *et al.* 1989), but also in immunocompetent elderly patients. These tumors may be single or multiple and typically show bright, homogeneous enhancement (Lai *et al.* 2002). Although most are found in the cerebrum, often in a periventricular location, they may also occur in the cerebellum or brainstem. They often seed into the cerebrospinal fluid (CSF), and a suspicion of primary central nervous system lymphoma is one situation where a lumbar puncture should be seriously considered.

Neuromas, although capable of arising from cranial nerves V, VII, IX, or X, are by far most commonly associated with cranial nerve VIII, in which case they are referred to as acoustic neuromas. These acoustic neuromas constitute the most common cause of a cerebellopontine angle tumor and typically present with hearing loss, accompanied in most cases by dysequilibrium and tinnitus. With growth of the tumor and compression of cranial nerves V and VII, there may be facial numbness and a peripheral facial palsy; rarely, a trigeminal neuralgia may occur (Harner and Laws 1983). In cases in which the cerebellum is compressed, there may be nystagmus and ataxia.

Medulloblastomas, although generally seen only in children, may occasionally occur in adults. These are typically found in the midline cerebellum and often protrude into the fourth ventricle, causing hydrocephalus.

Gangliocytomas (Kernohan *et al.* 1932) are composed of neural elements, and gangliogliomas (Morris *et al.* 1993) of both neural and glial elements. These are rare, indolent tumors, generally found in the temporal, frontal, or parietal cortices, which typically present with seizures.

Pituitary adenomas may be subclassified according to either their size or their endocrinologic status. Macroadenomas are larger than 1 cm, whereas microadenomas, which are more common, are smaller. Endocrinologically, more than 80 percent of adenomas are secretory, with the remainder being non-productive. Pituitary adenomas may cause symptoms by either compression of adjacent tissue or secondary to the secretion of various hormones. With compression of adjacent pituitary tissue there may be pituitary failure, with, for example, hypothyroidism or adrenocortical insufficiency. Lateral extension of a macroadenoma into the adjacent cavernous sinus may cause an oculomotor palsy or facial numbness in the areas of the first or second divisions of the trigeminal nerve. In turn, upward extension of a macroadenoma may impinge on the optic chiasm, with a bitemporal hemianopia; on the hypothalamus, with diabetes insipidus; or, in extreme cases, on the frontal lobe, with cognitive changes and seizures. Endocrinologic changes seen with secretory tumors most commonly involve hyperprolactinemia, with amenorrhea in females and gynecomastia and erectile dysfunction in males. Excess growth hormone and adrenocorticotropic hormone (ACTH) secretion are the next most common changes and may cause acromegaly or Cushing's disease, respectively. Gonadotropin and thyroid-stimulating hormone (TSH) excesses are rare. Importantly, in about 10 percent of secreting adenomas, two or more hormones may be excessively produced. In addition, in some cases of macroadenomas there may be 'kinking' of the pituitary stalk, and in such cases dopamine delivery to the posterior pituitary may be impaired with a resultant secondary hyperprolactinemia. Magnetic resonance scanning is typically positive with macroadenomas; however, microadenomas with a diameter of less than 3 mm may escape detection. In rare instances, pituitary adenomas may undergo hemorrhage or infarction, producing the syndrome of 'pituitary apoplexy' with severe headache, diplopia, blindness, and, critically, an acute case of adrenocortical insufficiency, which may be fatal.

Craniopharyngiomas are lobulated, calcified tumors that, although most commonly occurring in children, may occur in patients of any age. They typically arise from the junction of the infundibulum and pituitary gland, and may produce a variety of symptoms, depending on which direction they grow in. With upward extension and impingement of the hypothalamus, there may be diabetes insipidus and obesity, and with compression of the third ventricle, hydrocephalus may occur. Compression of the optic chiasm may cause a bitemporal hemianopia, and with downward extension various forms of pituitary failure may appear.

Pineal tumors include not only pinealomas but also germ cell tumors, gliomas, and teratomas. These tumors cause symptoms primarily by compression of the adjacent quadrigeminal plate and the underlying aqueduct of Sylvius, with a Parinaud syndrome (limitation of upgaze and an Argyll Robertson pupil) and obstructive hydrocephalus, respectively.

Hemangioblastomas (Boughey *et al.* 1990) are cystic masses found in the cerebellum, which typically contain an enhancing nodule on MR scanning. They are found in children or young adults, usually present with cerebellar symptoms, and may occur as part of von Hippel–Lindau disease.

Colloid cysts of the third ventricle (Nitta and Symon 1985) are rare cystic masses that may cause dementia (Kelly 1951), either by compression of the surrounding thalamus (Faris and Terrence 1989; Lobosky *et al.* 1984) or as a result of obstructive hydrocephalus, as illustrated in Figure 21.2. Interestingly, by virtue of a 'ball-valve' effect, in which the foramen of Monro is intermittently occluded, these tumors may also cause intermittent elevations of intraventricular pressure. Thus, patients may experience intermittent headache, unsteadiness, confusion, or drop attacks, and report that these symptoms are posture dependent.

METASTATIC TUMORS

Metastases may arise from various different primary tumors and almost always reach the brain via hematogenous spread. By far the most common source is lung cancer, primarily the non-small-cell type, which is closely followed by breast cancer; renal and colorectal cancer are less common, and

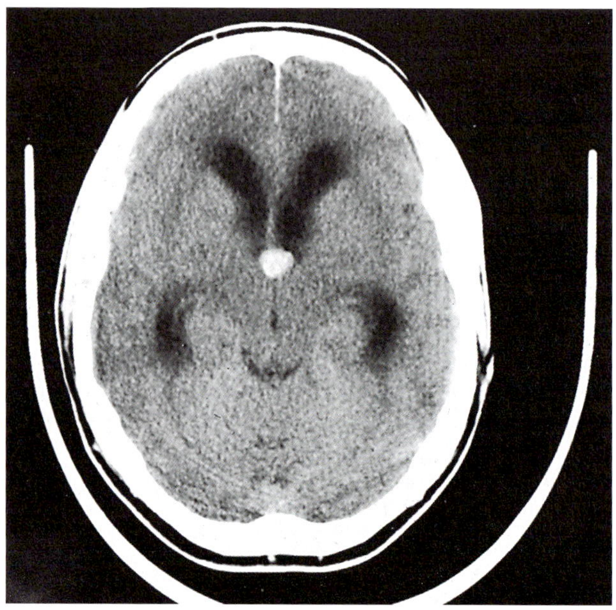

Figure 21.2 This unenhanced computed tomography scan demonstrates a colloid cyst, which, by obstructing the foramen of Monro, has caused obstructive hydrocephalus with dilation of the lateral ventricles. Reproduced with kind permission from Gillespie J and Jackson A. *MRI and CT of the Brain.* London: Arnold, 2000.

melanoma even less so. Other, far less common, sources include the thyroid, bladder, testis, and ovary; rarely, cancer of the pancreas, prostate, or uterus may also metastasize to the brain.

Two-thirds of metastases eventually settle in the brain parenchyma, whereas the remainder lodge either in the leptomeninges (where they create a condition known as leptomeningeal carcinomatosis) or in the dura. Parenchymal metastases may be found throughout the cerebrum (Delattre *et al.*1988a); they are found in the hemispheres in about three-quarters of cases, in the cerebellum in about one-eighth, and in the brainstem or thalamus in the remainder; of those lodging in the hemispheres, all lobes may be involved, including, in descending order, the frontal, parietal, occipital, and temporal lobes.

Metastases to the parenchyma generally appear as homogenously enhancing masses, often with considerable peri-tumoral edema; in high-grade malignancies, however, central necrosis may lead to a ring-enhancing pattern. Although solitary lesions may occur, multiple tumors, as noted earlier, are more common, and indeed at autopsy one may find widespread, microscopic foci.

Leptomeningeal carcinomatosis occurs when tumor cells that are widely distributed throughout the subarachnoid space spread into subjacent tissue, leading to delirium and cranial nerve palsies; obstruction of CSF outflow may also lead to hydrocephalus. Magnetic resonance scanning typically reveals meningeal enhancement, and tumor cells may be found with lumbar puncture; this may have to be repeated multiple times, however, to avoid a false-negative result. Although spread to the leptomeninges may occur with almost every type of primary tumor, it is, as noted earlier, a relatively uncommon occurrence. The exception occurs with systemic lymphoma or leukemia, which, unique among the primaries, tend to preferentially seed into the leptomeninges.

Dural metastases may occur via direct spread to the dura or by virtue of an inward extension of a calvarial metastasis. Symptoms are produced by virtue of compression of subjacent tissues and may include focal signs and seizures.

Differential diagnosis

In evaluating a mass lesion, consideration must be given not only to primary or metastatic neoplasms but also to sarcoid granulomas, syphilitic gummas, tuberculomas, and bacterial abscesses. Furthermore, in patients immunocompromised by AIDS or chemotherapy, opportunistic infections enter the differential, including toxoplasmosis and fungal infections.

Care must also be taken to not misattribute pre-existing conditions to the effects of a tumor. In the elderly, dementia from other causes is particularly common, and in any adult, mania may be due to bipolar disorder, depression may be due to major depressive disorder, and psychosis may be due to schizophrenia.

In patients with systemic cancer, toxic and metabolic deliria are common (Clouston *et al.* 1992), and consideration must also be given to the possibility of a paraneoplastic limbic encephalitis, as discussed in Chapter 19.8.

With high-dose steroid treatment, as discussed in Chapter 17.1, mood changes, tending towards mania rather than depression, are very common, and in a small minority one may also see psychosis or delirium. With radiation treatment, a radiation encephalopathy, as discussed in Chapter 12.5, may occur, and treatment with chemotherapeutic agents may cause delirium or seizures, as has been noted with ifosfamide, procarbazine, vincristine, the combination of 5-fluorouracil and levamisole, and methotrexate (when given intrathecally or, in high doses, intravenously). Bevacizumab and various chemotherapeutic agents may also cause the reversible posterior leukoencephalopathy syndrome, as discussed in Chapter 11.13.

Treatment

Treatment may involve any one or a combination of surgery, radiation treatment, chemotherapy, and use of dexamethasone and anti-epileptic drugs: neurosurgical and neuro-oncologic consultation is essential in all cases. It must be borne in mind, however, that not all tumors require treatment. Indolent, asymptomatic tumors, such as small meningiomas or low-grade gliomas, may merely be monitored clinically and with serial MR scans.

Surgery may be curative in cases of single, well-demarcated, accessible lesions, such as meningiomas or low-grade gliomas. With large, highly infiltrative tumors, such as glioblastoma multiforme, cure is not to be expected, but debulking of the tumor may improve the patient's overall functioning and prolong his or her life. Focal irradiation may be considered in patients who are not good surgical candidates or in whom the tumor is surgically inaccessible, and is also used post-surgically in cases where total excision is not possible. Finally, for malignant primary tumors, and for metastatic tumors, chemotherapy or bevacizumab may also be considered.

Dexamethasone may be dramatically effective in cases characterized by extensive peri-tumoral edema: results may be seen within hours, with maximal effect within 1–3 days.

An anti-epileptic drug should be started after a seizure occurs. Given that enzyme-inducing anti-epileptics, such as phenobarbital, phenytoin and carbamazepine, may reduce the effectiveness of chemotherapeutic agents, they are best avoided; furthermore, they tend to carry a high side-effect burden, and phenytoin, when used in conjunction with radiation treatment, is quite prone to cause severe rashes (Delattre *et al.*1988b). Divalproex may be considered, but it is a second choice in patients undergoing chemotherapy, as it tends to increase the toxicity of these agents. Gabapentin, lamotrigine, and levetiracetam are all reasonable first choices, and, of these, levetiracetam is preferable, given its ease of use and low side-effect burden. Although anti-epileptic drugs are clearly indicated in cases where a seizure has occurred, there is no evidence that they are useful prophylactically, except post-craniotomy, where there is some evidence for a prophylactic benefit from levetiracetam (Lim *et al.* 2009).

In cases characterized by pituitary involvement, hormone replacement therapy is often required after surgery or radiotherapy. Prolactinomas, unique among the pituitary adenomas, may be treated medically, with bromocriptine.

The general treatment of dementia, delirium, psychosis, depression, personality change, and mania is discussed in Chapters 6.3, 6.5, 8.1, 7.1, 8.2, and 7.3, respectively. Importantly, if antipsychotics or antidepressants are used, agents with low toxicity are obviously preferable: of the antipsychotics, the second-generation agents risperidone, ziprasidone, aripiprazole, and quetiapine constitute reasonable choices, and of the antidepressants, either citalopram or escitalopram should be considered, because of not only their excellent tolerability but also their lack of effect on the cytochrome P450 enzyme system. In cases of mania, lithium is generally a second choice, given its side-effect burden. If chemotherapy is not being utilized, then divalproex is a reasonable choice for mania; otherwise, one of the second-generation antipsychotics should be considered. Finally, in cases of glioblastoma multiforme, an open study (Meyers *et al.* 1998) demonstrated that methylphenidate, in a dose of 10 mg twice daily, improved cognitive function, gait, energy, and motivation. Regardless of what medication is used, in all non-emergent situations the general rule of 'start low and go slow' applies especially to patients with brain tumors.

21.2 HYDROCEPHALUS

The overall volume of CSF is about 140 mL: about 20 mL is contained within the lateral ventricles, 5 mL within the third and fourth ventricles, and the remainder found in the subarachnoid space surrounding the brain and cord. Cerebrospinal fluid is produced by the choroid plexus at a rate of about 20 mL/hour: although some is produced by the choroid plexus in the roof of the third and fourth ventricle, the vast majority derives from the choroid plexus in the lateral ventricles. Cerebrospinal fluid produced within the lateral ventricles normally flows through the foramina of Monro into the third ventricle and then via the aqueduct of Sylvius into the fourth ventricle, from which it exits via the foramina of Magendie and Luschka into the subarachnoid space surrounding the brainstem. From here, the CSF circulates around the cord and brain, and finally arrives at the cerebral convexity, where it exits the subarachnoid space via the villi of the arachnoid granulations into the superior saggital sinuses.

Hydrocephalus is characterized by an enlargement of one or more of the ventricles as a result of an increase in CSF pressure. It may be usefully divided into two types: non-communicating (or 'obstructive') and communicating. In non-communicating hydrocephalus, the overall 'communication' between the ventricular system and the brainstem subarachnoid space is partially or totally obstructed at some point, thus hindering or totally preventing the egress of CSF from one or more of the ventricles into the subarachnoid space. By contrast, in communicating hydrocephalus, this 'communication' is unimpeded and CSF flows freely from the ventricular system into the brainstem subarachnoid space.

Non-communicating hydrocephalus may occur with obstruction of CSF flow at various locations. With obstruction of one of the foramina of Monro, only one of the lateral ventricles enlarges, whereas with obstruction of both foramina, both lateral ventricles become enlarged. Obstruction at the aqueduct of Sylvius is followed by enlargement of the third and both lateral ventricles, and obstruction at the exit foramina of Magendie and Luschka entails expansion of all four ventricles.

Communicating hydrocephalus typically occurs as a result of an obstruction at the level of the arachnoid villi, and is associated with enlargement of all of the ventricles. Very rarely, communicating hydrocephalus may occur not because of an obstruction at any point but rather because of a grossly excessive production of CSF, which overwhelms the drainage system.

Although hydrocephalus may occur at any age, from infancy to senescence, in this text only adult or later-onset cases are considered.

Clinical features

From a clinical point of view it is very useful to divide hydrocephalus into two forms, namely acute and chronic, depending on their mode of onset.

Acute hydrocephalus is a form of non-communicating hydrocephalus in which there is a complete, or near-complete, obstruction of CSF flow. This form of hydrocephalus is characterized clinically by a rapid onset of symptoms, over days, hours, or even minutes. Patients present with headache, stupor, and vomiting, and, without treatment, coma and death may rapidly ensue.

Chronic hydrocephalus may represent either a communicating or a non-communicating condition: when it occurs as a result of non-communicating hydrocephalus, one finds only a partial obstruction. Clinically (Gustafson and Hagberg 1978; Harrison *et al.* 1974), chronic hydrocephalus typically presents gradually, or even insidiously, and is characterized by a dementia marked by forgetfulness, apathy, and a generalized slowing of thought and behavior; rarely, akinetic mutism may occur (Messert *et al.* 1966). A gait disturbance also occurs, and this may either precede or follow the onset of the dementia. The gait may be shuffling, apraxic, or 'magnetic' in type; in this last type of gait, it appears as if the patient's feet are 'stuck' to the floor, as if held there by magnets. Urinary urgency or frequency or urinary incontinence may also occur. On examination there may be generalized hyper-reflexia and bilaterally positive Babinski signs.

Although ventriculomegaly is evident on both CT and MR scanning (as illustrated in Figure 21.3), MR scanning is preferred as it is more likely to reveal the underlying cause of any obstructive hydrocephalus. In addition, on T2-weighted or fluid-attenuated inversion recovery (FLAIR) images one typically sees evidence of transependymal flow of CSF from the ventricles into the immediately subjacent white matter, producing a hazy rim of increased signal intensity in a periventricular distribution.

Lumbar puncture is indicated in most cases; exceptions include cases where imaging reveals an obvious cause (e.g. a tumor). In cases of non-communicating hydrocephalus, the opening pressure will be normal. In cases of communicating hydrocephalus, the opening pressure may or may not be elevated, depending on the etiology.

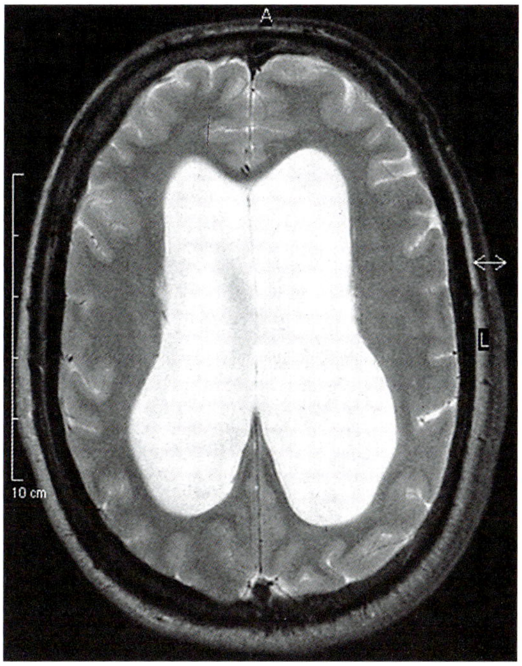

Figure 21.3 Massive enlargement of the lateral ventricles in a case of communicating hydrocephalus, as demonstrated on a T2-weighted magnetic resonance scan. Reproduced with kind permission from Gillespie J and Jackson A. *MRI and CT of the Brain*. London: Arnold, 2000.

Course

Acute hydrocephalus is a catastrophic event, with a rapid evolution of symptoms. Chronic hydrocephalus, by contrast, is marked by a very slow progression of symptoms as the pressure of the CSF very slowly increases. In most cases, however, a new 'equilibrium' is eventually reached between CSF production and outflow, and in this situation, although the ventricles remain under increased pressure, they do not undergo any further expansion. When this development occurs, the previously 'active' hydrocephalus is said to have undergone 'arrest', and such 'arrested' cases of hydrocephalus clinically manifest a plateau of symptoms, with a subsequently stable clinical course.

Etiology

In non-communicating hydrocephalus the obstruction may occur at various sites and may be caused by various different lesions. Thus, the foramina of Monro may be obstructed by a tumor, such as an astrocytoma, or by a colloid cyst of the third ventricle, and the third ventricle may be occluded by a tumor (Riddoch 1936). The aqueduct of Sylvius may be stenotic (Nag and Falconer 1966; Wilkinson *et al.* 1966) or may suffer post-infectious scarring; it may also be compressed by adjacent tumors, such as pineal tumors. The fourth ventricle may be compressed by cerebellar lesions, such as tumors, hemorrhages, or infarctions. The exit foramina of Magendie and Luschka may be occluded by scarring, as may occur after an episode of viral or bacterial meningitis, or in the course of an indolent basilar meningitis, as may be seen in meningovascular syphilis, tuberculosis, or fungal infections; scarring and obstruction may also occur after a subarachnoid hemorrhage, either spontaneous or as may be seen with traumatic brain injury.

Communicating hydrocephalus almost always occurs secondary to impaired outflow of CSF through the arachnoid granulations. In the middle-aged or elderly this impaired outflow is most commonly associated with a normal opening pressure, and this condition, customarily referred to as 'normal-pressure hydrocephalus', is discussed in Chapter 21.3. As an aside, it should be clear that the term 'normal-pressure hydrocephalus', used in this fashion, is technically a misnomer, given that a normal opening pressure is also seen with cases of non-communicating, or obstructive, hydrocephalus: the term, however, is firmly ingrained as referring only to cases of communicating hydrocephalus, and hence confusion is unlikely to occur. Communicating hydrocephalus may also occur with impaired outflow through the arachnoid granulations due to leptomeningeal carcinomatosis or extremely elevated total protein values, as may be seen in some cases of spinal tumor or polyneuritis. Finally, there are very rare cases of communicating hydrocephalus occurring secondary to an overproduction of CSF by a choroid plexus papilloma (Eisenberg *et al.* 1974).

Differential diagnosis

The first task in making the differential diagnosis is to distinguish 'true' hydrocephalus from a condition known as 'hydrocephalus *ex vacuo*'. In true hydrocephalus, ventricular enlargement occurs as the result of an increase in pressure within the ventricles. By contrast, in hydrocephalus *ex vacuo*, ventricular enlargement occurs not because of any pressure increase but merely secondary to shrinkage of the surrounding brain parenchyma, as may be seen in normal ageing or in various neurodegenerative conditions such as Alzheimer's disease. There are two imaging clues that allow one to differentiate between these two forms of ventricular enlargement. First, and most importantly, one should look at the size of the cortical sulci and determine whether or not these are enlarged. In true hydrocephalus, although there may be some enlargement of the sulci, the ventricular enlargement is disproportionately much greater, often markedly so. By contrast, in hydrocephalus *ex vacuo*, there is always sulcal enlargement, and the degree of sulcal enlargement is proportionate to the degree of ventricular enlargement. Next, one should examine the pattern of enlargement of the lateral ventricles: typically the frontal horns undergo disproportionate enlargement and appear 'ballooned', and pressure on the corpus callosum causes both thinning and upward bowing.

In cases where hydrocephalus is 'truly' present, one must keep in mind that, especially in the elderly, the individual clinical picture may be occurring secondary to a combination of hydrocephalus and one or more of the more common causes of dementia, such as Alzheimer's disease or Binswanger's disease, and this differential is dealt with in the following chapter on normal-pressure hydrocephalus.

Treatment

Treatment consists of shunting, or, in highly selected cases, endoscopic third ventriculosotomy (Farin *et al.* 2006). Although ventriculo-atrial shunting is occasionally performed, ventriculo-peritoneal shunts are standard, and in either case, a programmable valve is utilized. Importantly, although shunting is typically most effective early in the course of hydrocephalus, it may also provide benefit in 'arrested' cases (Larsson *et al.* 1999). In shunted patients, one should keep a watch for complications, which are not uncommon. These include intracerebral hemorrhage with shunt placement, infections, catheter migration, mechanical failure and what is termed 'overdrainage'. In overdrainage, an excessive amount of fluid is withdrawn, and with the resulting intracranial hypotension, the brain literally sags down, with potentially significant consequences. With excessive traction on the bridging veins between the cortical surface and overlying dura, bleeding may occur, with the formation of a subdural hematoma. Further, the base of the brain may undergo compression, resulting, variously, in nausea, vomiting, lethargy, or delirium.

21.3 NORMAL-PRESSURE HYDROCEPHALUS

The general subject of hydrocephalus is discussed in the preceding chapter, and the reader is encouraged to review this. Normal-pressure hydrocephalus, discussed in this chapter, is a kind of chronic communicating hydrocephalus wherein the opening pressure at lumbar puncture is within normal limits. As discussed further below, under 'Etiology', normal-pressure hydrocephalus may occur secondary to known causes, such as subarachnoid hemorrhage, or on an idiopathic basis. Classically (Adams *et al.* 1965; Gallassi *et al.* 1991; Hill *et al.* 1967), it presents with the triad of gait disturbance, dementia, and urinary urgency or incontinence.

Clinical features

The onset is typically quite gradual in the latter part of the sixth decade or beyond. Of the classic triad, gait disturbance typically occurs first, followed by dementia, with urinary symptoms typically appearing last.

With the gait disturbance, patients may walk with short steps on a somewhat widened base, and sometimes there is a degree of shuffling, but the distinctive feature is what is often referred to as a 'magnetic' gait. Here, patients have difficulty initiating steps, as if their feet are held in place on the floor by a large magnet; some may complain that it feels as if their feet are 'glued to the floor'.

The dementia of normal-pressure hydrocephalus is typically characterized by forgetfulness, slowness of thought and action, apathy, and indifference. Rarely the clinical picture may be dominated by a personality change (Rice and Gendelman 1973) or depression (Pujol *et al.* 1989), and even more rarely by aggressiveness (Crowell *et al.* 1973; Sandyk 1984) or mania (Kwentus and Hart 1987). Importantly, 'cortical' signs, such as aphasia, agnosia, or neglect, are absent; furthermore, although patients, as noted, are forgetful, it is not until late in the course that clear-cut, prominent anterograde amnesia becomes apparent.

Urinary urgency and incontinence may initially be embarrassing to patients; however, with progression of the hydrocephalus this embarrassment gives way to indifference and a lack of concern.

On examination, in addition to the gait abnormality, there may be additional findings, including hyperreflexia, the Babinski sign, snout and grasp reflexes, paratonia, and, in a small minority, a postural hand tremor. Importantly, when these additional findings are present, they are bilaterally symmetric. Finally, there may also be postural instability, with an associated history of falls.

Findings on MR scanning are as discussed in the preceding chapter. As noted earlier, the opening pressure at lumbar puncture is within normal limits.

Course

In most cases there is a gradual progression of symptoms, and some patients may eventually develop akinetic mutism and quadriparesis. Often, however, the hydrocephalus becomes arrested, and symptoms then 'plateau out'.

Etiology

As discussed in the preceding chapter, normal-pressure hydrocephalus occurs as a result of impaired outflow of CSF through the arachnoid granulations (Borgesen and Gjerris 1987). Such impaired outflow, in a minority of cases, occurs secondary to subarachnoid hemorrhage or meningitis. In the remaining, idiopathic cases, the cause of the impaired outflow is either unknown or, in a substantial minority of cases (Bech *et al.* 1997,1999), is related to arachnoid fibrosis of uncertain cause. Although, as stressed earlier, the opening pressure at lumbar puncture is normal, it appears that there are episodic, mild elevations of intraventricular pressure, typically occurring at night (Packard 1982). Presumably, with expansion of the frontal horns, there is compression and stretching of the adjacent white matter tracts, which in turn leads to the symptoms seen in the disease.

Differential diagnosis

The question of normal-pressure hydrocephalus typically arises when, in evaluating patients with a gait disorder or dementia, imaging reveals hydrocephalus which is of the 'true' rather than the *ex vacuo* type, as discussed in the preceding chapter. Given, however, that elderly patients may have other disorders that could also cause a gait disorder or dementia more or less similar to those seen in normal-pressure hydrocephalus, the next step is to attempt to determine whether or not a concurrent disorder is present, and, if so, what contribution it makes to the overall clinical picture. Of the many disorders that could make such a contribution, the most important are parkinsonian conditions, Alzheimer's disease, and Binswanger's disease.

Of parkinsonian conditions, Parkinson's disease and dementia with Lewy bodies are most common. The gait disorder seen in these conditions bears some similarity to that seen in normal-pressure hydrocephalus; however, in parkinsonism, the gait is with a narrow base, rather than the wide base seen in normal-pressure hydrocephalus. Furthermore, in parkinsonism one typically sees a rest tremor and cogwheel rigidity, whereas in normal-pressure hydrocephalus tremor, if present, is postural, and rigidity, again if present, is of the paratonic type. Both Parkinson's disease and dementia with Lewy bodies may also cause a dementia that may be roughly similar to that seen in normal-pressure hydrocephalus; however, in dementia with Lewy bodies, the dementia is typically accompanied by spontaneous, brief episodes of delirium and by visual hallucinations, features not seen in normal-pressure hydrocephalus.

The dementia of Alzheimer's disease is characterized, at least early on, by an amnestic syndrome which overshadows any other cognitive deficits, and this clearly differentiates it from the dementia seen in normal-pressure hydrocephalus; furthermore, there is typically, again at least early on, no gait disorder in Alzheimer's disease. However, it must be borne in mind that Alzheimer's disease and normal-pressure hydrocephalus often coexist; indeed biopsies taken during shunt insertion have revealed Alzheimer's disease pathology in roughly one-quarter of all cases of normal-pressure hydrocephalus (Bech *et al.* 1997, 1999). Consequently, one may see a mixed clinical picture (e.g. a gait disorder typical of normal-pressure hydrocephalus, followed by a dementia that presented with a disproportionate amount of amnesia), and the question then arises whether such patients should be shunted. The answer appears to be affirmative: although the response to shunting may not be as robust as in cases of pure normal hydrocephalus (Hamilton *et al.* 2010), it is nevertheless substantial (Bech *et al.* 1999; Golomb *et al.* 2000).

The dementia of Binswanger's disease is quite similar to that seen in normal-pressure hydrocephalus, and such patients, if they also happen to have multiple lacunar infarcts (as is often the case), may develop a vascular parkinsonism. In such cases, MR scanning will reveal white matter disease not only in a periventricular location but also extensively throughout the centrum semiovale. Presuming, then, that imaging confirms both 'true' hydrocephalus and the white matter disease characteristic of Binswanger's disease, the same question regarding shunting occurs as with concurrent Alzheimer's disease, and the answer is the same: shunting should still be considered in these cases (Bech *et al.* 1999).

Treatment

Ventriculoperitoneal shunting should be considered in all cases; however, selecting patients for shunting may at times be difficult. Certainly, in cases in which the classic triad is definitely present and imaging reveals unequivocal hydrocephalus, one should strongly consider surgery. However, these features may not all be present and in such cases one may consider a number of ancillary tests, including the 'tap test', prolonged external lumbar drainage, and the 'infusion' test. Of these, the tap test is most commonly utilized. In this test, a detailed mental status is performed, the patient's gait is videotaped, and 30–50 mL of CSF is withdrawn. Approximately 30–60 minutes later, the mental status exam and videotaped gait testing is repeated. Although this test has reasonably high specificity and sensitivity, false negatives may occur, and if the test is negative, but clinical suspicion is high, then consideration should be given to prolonged external ventricular drainage. In this test, after the initial mental status and videotaped gait testing, roughly 50 mL of fluid is withdrawn daily for 3–5 days, and concurrently one repeats the testing on at least a daily basis. Although this test may pick up some cases missed by the tap test, external ventricular drainage is not without risk, and probably should not be resorted to unless clinical suspicion is fairly high. The infusion test represents the last alternative to consider. Here, Ringer's solution or artificial CSF is infused into the lumbar cistern and outflow resistance measured: increased resistance indicates the presence of impaired egress of CSF through the arachnoid villi, and thus confirms the diagnostic suspicion of normal-pressure hydrocephalus. It must be borne in mind, however, that none of these tests is completely accurate in predicting the response to shunting, and they may yield false negatives in this regard. Consequently, even when these tests are all negative, one may still consider shunting in clinically appropriate cases.

Among patients who do respond, the gait disturbance generally improves first, typically over the first week or so; dementia improves next, but full improvement may take months. Urinary incontinence is the last symptom to improve, and typically shows the least improvement.

REFERENCES

Adams RD, Fisher CM, Hakim S *et al.* Symptomatic occult hydrocephalus with 'normal' cerebrospinal fluid pressure. *N Engl J Med* 1965; **273**:117–26.

Alpers BJ. Relation of the hypothalamus to disorders of personality: report of a case. *Arch Neurol Psychiatry* 1937; **38**:291–303.

Alpers BJ. Personality and emotional disorders associated with hypothalamic lesions. In: Fulton JF, Ranson SW, Frantz AM, eds., *Research Publications. Association for Research in Nervous and Mental Disease.* Vol. XX. *The Hypothalamus and Central Levels of Autonomic Function.* Baltimore: William & Wilkins, 1940.

Alpers BJ, Grant FC. The clinical syndrome of the corpus callosum. *Arch Neurol Psychiatry* 1931; **25**:67–86.

Avery TL. Seven cases of frontal tumor with psychiatric presentation. *Br J Psychiatry* 1971; **119**:19–23.

Beal MF, Kleinman GM, Ojemann RG *et al.* Gangliocytoma of third ventricle: hyperphagia, somnolence, and dementia. *Neurology* 1981; **31**:1224–8.

Bech RA, Juhler M, Waldemar G *et al.* Frontal brain and leptomenigeal biopsy specimens correlated with cerebrospinal outflow resistance and B-wave activity in patients suspected of normal-pressure hydrocephalus. *Neurosurgery* 1997; **40**:497–502.

Bech RA, Waldemar G, Gjerris F *et al.* Shunting effects in patients with idiopathic normal pressure hydrocephalus: correlation with cerebral and leptomeningeal biopsy findings. *Acta Neurochir* 1999; **141**:633–9.

Borgesen SE, Gjerris F. Relationships between intracranial pressure, ventricular size, and resistance to CSF outflow. *J Neurosurg* 1987; **67**:535–9.

Boughey AM, Fletcher NA, Harding AE. Central nervous system haemangioblastoma: a clinical and genetic study of 52 cases. *J Neurol Neurosurg Psychiatry* 1990; **53**:644–8.

Clouston PD, DeAngelis LM, Posner JB. The spectrum of neurological disease in patients with systemic cancer. *Ann Neurol* 1992; **31**:268–73.

Crowell RM, Tew JM, Mark VH. Aggressive dementia associated with normal pressure hydrocephalus. *Neurology* 1973; **23**:461–4.

Davison C, Demuth EL. Disturbances in sleep mechanism: a clinicopathologic study. I. Lesions at the cortical level. *Arch Neurol Psychiatry* 1945; **53**:399–406.

Davison C, Demuth EL. Disturbances in sleep mechanism: a clinicopathologic study. III. Lesions at the diencephalic level (hypothalamus). *Arch Neurol Psychiatry* 1946; **55**:111–25.

DeAngelis LM. Brain tumors. *N Engl J Med* 2001; **344**:114–23.

Delattre JY, Krol G, Thaler HT *et al.* Distribution of brain metastases. *Arch Neurol* 1988a; **45**:741–4.

Delattre JY, Safai B, Posner JB. Erythema multiforme and Stevens–Johnson syndrome in patients receiving cranial irradiation and phenytoin. *Neurology* 1988b; **38**:194–8.

Direkze M, Bayliss SG, Cutting JC. Primary tumors of the frontal lobe. *Br J Clin Pract* 1971; **25**:207–13.

Eisenberg HM, McComb JG, Lorenzo AV. Cerebrospinal fluid overproduction and hydrocephalus associated with choroid plexus papilloma. *J Neurosurg* 1974; **40**:381–5.

Farin A, Aryan HE, Ozgur BM *et al.* Endoscopic third ventriculostomy. *J Clin Neurosci* 2006; **13**:663–70.

Faris AA, Terrence CF. Limbic system symptomatology associated with colloid cyst of the third ventricle. *J Neurol* 1989; **236**:60–1.

Feiden W, Bise K, Steude U *et al.* The stereotactic biopsy diagnosis of focal intracerebral lesions in AIDS patients. *Acta Neurol Scand* 1993; **87**:228–33.

Forsyth PA, Posner JB. Headaches in patients with brain tumors: a study of 111 patients. *Neurology* 1993; **43**:1678–83.

Frazier CH. Tumor involving the frontal lobe alone: a symptomatic survey of one hundred and five verified cases. *Arch Neurol Psychiatry* 1936; **35**:525–71.

Fulton JF, Bailey P. Tumors in the region of the third ventricle: their diagnosis and relation to pathological sleep. *J Nerv Ment Dis* 1929; **69**:1–25.

Gal P. Mental symptoms in cases of tumor of the temporal lobe. *Am J Psychiatry* 1958; **115**:157–60.

Gallassi R, Morreale A, Montagna P *et al.* Binswanger's disease and normal-pressure hydrocephalus: clinical and neuro-radiological comparison. *Arch Neurol* 1991; **48**:1156–9.

Golomb J, Wisoff J, Miller DC *et al.* Alzheimer's disease comorbidity in normal pressure hydrocephalus: prevalence and shunt response. *J Neurol Neurosurg Psychiatry* 2000; **68**:778–81.

Greenberg DB, Brown GL. Single case study: mania resulting from brain stem tumor. *J Nerv Ment Dis* 1985; **173**:434–6.

Gustafson L, Hagberg B. Recovery in hydrocephalic dementia after shunt operation. *J Neurol Neurosurg Psychiatry* 1978; **41**:940–7.

Hamilton R, Patel S, Lee EB *et al.* Lack of shunt response in suspected idiopathic normal pressure hydrocephalus with Alzheimer's disease pathology. *Ann Neurol* 2010; **68**:535–40.

Harner SG, Laws ER. Clinical findings in patients with acoustic neurinoma. *Mayo Clin Proc* 1983; **58**:721–8.

Harrison MJG, Robert CM, Uttley D. Benign aqueductal stenosis in adults. *J Neurol Neurosurg Psychiatry* 1974; **37**:1322–8.

Haugh RM, Markesbery WR. Hypothalamic astrocytoma: syndrome of hyperphagia, obesity and disturbances of behavior and endocrine and autonomic function. *Arch Neurol* 1983; **40**:560–3.

Heilman KM, Sypert GW. Korakoff's syndrome resulting from bilateral fornix lesion. *Neurology* 1977; **27**:490–3.

Heron GB, Johnston DA. Hypothalamic tumor presenting as anorexia nervosa. *Am J Psychiatry* 1976; **133**:580–2.

Hill ME, Lougheed WM, Barnett HJM. A treatable form of dementia due to normal-pressure, communicating hydrocephalus. *Can Med Assoc J* 1967; **97**:1309–11.

Kahn EA, Crosby EC. Korsakoff's syndrome associated with surgical lesions involving the mammillary bodies. *Neurology* 1971; **21**:117–25.

Kelly R. Colloid cysts of the third ventricle: analysis of twenty-nine cases. *Brain* 1951; **74**:23–65.

Kernohan JW, Learmonth JR, Doyle JB. Neuroblastomas and gangliocytoma of the central nervous system. *Brain* 1932; **55**:287–310.

Keschner M, Bender MB, Strauss I. Mental symptoms in cases of tumor of the temporal lobe. *Arch Neurol Psychiatry* 1936; **35**:572–96.

Kwentus JA, Hart RP. Normal pressure hydrocephalus presenting as mania. *J Nerv Ment Dis* 1987; **175**:500–2.

Lai R, Rosenblum MK, DeAngelis LM. Primary CNS lymphoma: a whole-brain disease? *Neurology* 2002; **59**:1557–62.

Lang W, Miklossy J, Deruaz JP *et al.* Neuropathology of the acquired immune deficiency syndrome (AIDS): a report of 135 consecutive autopsy cases from Switzerland. *Acta Neuropathol* 1989; **77**:379–90.

Larsson A, Stephenson H, Wikkelso C. Adult patients with 'asymptomatic' and 'compensated' hydrocephalus benefit from surgery. *Acta Neurol Scand* 1999; **99**:81–90.

Lim DA, Tarapore P, Chang E *et al.* Safety and feasibility of switching from phenytoin to levetiracetam monotherapy for glioma-related seizure control following craniotomy: a randomized phase II pilot study. *J Neurooncol* 2009; **93**:349–54.

Liss L. Pituicytoma, a tumor of the hypothalamus. *Arch Neurol Psychiatry* 1958; **80**:567–76.

Lobosky JM, Vangilder JC, Damasio AR. Behavioral manifestations of third ventricular colloid cysts. *J Neurol Neurosurg Psychiatry* 1984; **47**:1075–80.

Lote K, Stenwig AE, Skullerud K *et al.* Prevalence and prognostic significance of epilepsy in patients with gliomas. *Eur J Cancer* 1998; **34**:98–102.

McKendree CA, Feinier L. Somnolence: its occurrence and significance in cerebral neoplasms. *Arch Neurol Psychiatry* 1927; **17**:44–56.

Malamud N. Psychiatric disorder with intracranial tumors of limbic system. *Arch Neurol* 1967; **17**:113–23.

Messert B, Henke TK, Langheim W. Syndrome of akinetic mutism associated with obstructive hydrocephalus. *Neurology* 1966; **16**:635–49.

Meyers CA, Weitzner MA, Valentine AD *et al.* Methylphenidate therapy improves cognition, mood, and function of brain tumor patients. *J Clin Oncol* 1998; **16**:2522–7.

Moersch FP. Psychic manifestations in cases of brain tumors. *Am J Psychiatry* 1925; **81**:707–24.

Morris HH, Estes ML, Gilmore R *et al.* Chronic intractable epilepsy as the only symptom of primary brain tumor. *Epilepsia* 1993; **34**:1038–43.

Murthy P, Jayakumar PN, Sampat S. Of insects and eggs: a case report. *J Neurol Neurosurg Psychiatry* 1997; **63**:522–3.

Nag TK, Falconer MA. Non-tumoral stenosis of the aqueduct in adults. *BMJ* 1966; **2**:1168–70.

Nitta M, Symon L. Colloid cysts of the third ventricle: a review of 36 cases. *Acta Neurochir* 1985; **76**:99–104.

Olivero WC, Lister JR, Elwood PW. The natural history and growth rate of asymptomatic meningiomas. *J Neurosurg* 1995; **83**: 222–4.

Packard JD. Adult communicating hydrocephalus. *Br J Hosp Med* 1982; **27**:35–44.

Palmai G, Taylor DC, Falconer MA. A case of craniopharyngioma presenting as Korsakoff's psychosis. *Br J Psychiatry* 1967; **113**:619–23.

Pujol J, Leal S, Fluvia X *et al.* Psychiatric aspects of normal pressure hydrocephalus. *Br J Psychiatry* 1989; **154**(Suppl. 4):77–80.

Reeves AG, Plum F. Hyperphagia, rage and dementia accompanying a ventromedial hypothalamic neoplasm. *Arch Neurol* 1969; **20**:616–24.

Rice E, Gendelman S. Psychiatric aspects of normal pressure hydrocephalus. *JAMA* 1973; **223**:409–12.

Riddoch G. Progressive dementia, without headache or changes in the optic discs, due to tumors of the third ventricle. *Brain* 1936; **59**:225–33.

Sachs E. Meningiomas with dementia as the first and presenting feature. *J Ment Sci* 1950; **96**:998–1007.

Sandyk R. Aggressive dementia in normal-pressure hydrocephalus. *S Afr Med J* 1984; **65**:114.

Schneck SA, Penn I. Cerebral neoplasms associated with renal transplantation. *Arch Neurol* 1970; **22**:226–33.

Sprofkin BE, Sciarra D. Korsakoff's psychosis associated with cerebral tumor. *Neurology* 1952; **2**:427–34.

Starkstein SE, Boston JD, Robinson RG. Mechanism of mania after brain injury: 12 case reports and review of the literature. *J Nerv Ment Dis* 1988; **176**:87–100.

Stern K, Dancey TE. Glioma of the diencephalon in a manic patient. *Am J Psychiatry* 1942; **98**:716–19.

Strauss I, Globus JH. Tumor of the brain with disturbance in temperature regulation. *Arch Neurol Psychiatry* 1931; **25**:506–22.

Strauss I, Keschner M. Mental symptoms in cases of tumor of the frontal lobe. *Arch Neurol Psychiatry* 1935; **33**:986–1007.

Strobos RRJ. Tumors of the temporal lobe. *Neurology* 1953; **3**:752–60.

Tucker GJ, Price TRP, Johnson VB *et al.* Phenomenology of temporal lobe dysfunction: a link to atypical psychosis: a series of cases. *J Nerv Ment Dis* 1986; **174**:348–56.

Wellisch DK, Kaleita TA, Freeman D *et al.* Predicting major depression in brain tumor patients. *Psychooncology* 2002; **11**:230–8.

White JH, Kelly P, Dorman K. Clinical picture of atypical anorexia nervosa associated with a hypothalamic tumor. *Am J Psychiatry* 1977; **134**:323–5.

Wilkinson HA, Lemay M, Drew JH. Adult aqueductal stenosis. *Arch Neurol* 1966; **15**:643–8.

Wilkinson IM, Anderson JR, Holmes AE. Oligodendroglioma: an analysis of 42 cases. *J Neurol Neurosurg Psychiatry* 1987; **50**:304–12.

Williamson RT. On the symptomatology of gross lesions (tumors and abscesses) involving the pre-frontal regions of the brain. *Brain* 1896; **19**:346–65.

Ziegler DK, Kaufman A, Marshall HE. Abrupt memory loss associated with thalamic tumor. *Arch Neurol* 1977; **34**:545–8.

22

Schizophrenia and other psychotic disorders

22.1 SCHIZOPHRENIA

Schizophrenia is a chronic, more or less debilitating, psychosis that occurs in about 1 percent of the general population and which is equally common in males and females. This illness was first noted by Morel in 1860 (Anonymous 1954), who referred to it as *démence précoce.* A full description of the disease, however, had to await the efforts of Emil Kraepelin. Kraepelin, who latinized the name to dementia praecox, was a German psychiatrist of the late nineteenth and early twentieth centuries, whose work remains a guiding force for modern psychiatry. The current name for the disease, schizophrenia, was coined by Eugen Bleuler, a Swiss psychiatrist who amplified Kraepelin's original descriptions. Another guiding light in the elucidation of the disease was the German psychiatrist Kurt Schneider, who isolated certain symptoms, now known as Schneiderian first-rank symptoms, which, although not pathognomonic of the disease, are very, very suggestive.

Clinical features

The age at onset of schizophrenia, although generally falling in the late teens or early twenties, may range from late childhood to the seventh decade (Brodaty *et al.* 1999; Grahame 1984). A prodrome may or may not be present; in some cases, the premorbid personality may have been completely normal, whereas in others, peculiarities may have been present for years or even decades (Walker and Lewine 1990). In cases in which the prodrome begins in childhood, history may reveal introversion and peculiar interests. In cases in which the prodrome appears in teenage years or later, well after the patient's personality has formed, family members may recall a stretch of time when the patient 'changed' and was no longer 'himself'; prior interests and hobbies may have been abandoned and replaced by a certain irritable seclusiveness or perhaps suspiciousness.

The mode of onset ranges from acute to gradual. Acute onsets occur over weeks or months and may be initially marked by perplexity or depressive symptoms; patients may recognize that something is going wrong, and some may make desperate attempts to bring order and structure into a life that is rapidly fragmenting. By contrast, gradual onsets, which may span months or a year or more, may not be particularly disturbing to the patient; there may be fleeting, whispering auditory hallucinations, vague intimations, or strange occurrences.

Although the symptomatology of schizophrenia may be quite varied, in most cases one sees *hallucinations*, *delusions*, *disorganized speech*, and *catatonic* or *bizarre behavior. Negative symptoms* (e.g. flattening of affect) are also often seen and, as with all typologies, there is also a category of miscellaneous symptoms, the most important of which is transient disturbances of mood. Generally, based on the overall constellation of symptoms, one may classify any given case of schizophrenia into one of several *subtypes*, namely paranoid, catatonic, disorganized (also known as hebephrenic), or simple schizophrenia; in a large minority, however, there appears to be a mixture of these subtypes, and in such cases one speaks of 'undifferentiated' schizophrenia. Each of these symptoms and subtypes is now considered in some detail.

HALLUCINATIONS

Hallucinations may occur in the auditory, visual, gustatory, olfactory, and tactile realms (Mueser *et al.* 1990).

Auditory hallucinations are most common in schizophrenia and, as noted by Kraepelin (1919), it is '*the hearing of voices*' (italics in original) that is by far most 'peculiarly characteristic' of schizophrenia. Voices may come from inside the body or perhaps the air; sometimes they are sent by electronic devices or emerge from the walls or furniture. They often speak in short phrases and may at times manifest as commands, which patients may or may not be able to resist. Certain auditory hallucinations, included among the Schneiderian first-rank symptoms, although not specific for schizophrenia, are highly suggestive: these include voices that repeat the patient's thoughts, voices that comment on what the patient is doing, and voices that argue with each other. For most patients the voices sound as real as the voice of another person, and they may talk back to them or argue with them. At times when the voices are loud or unpleasant, patients may try and drown them out by listening to music or watching television. In addition to voices, patients may also hear sounds, such as the ringing of bells, footsteps, or tapping on the walls or windows.

Visual hallucinations, although common, play a much less prominent role in the overall symptomatology than auditory hallucinations. Often they are poorly formed and indistinct and perhaps seen only fleetingly, 'out of the corner of the eye'. In some cases, however, they may be detailed, vivid, and compelling: strange people walk the halls; the devil appears in violent red, straight ahead; heads may float through the air; and reptilian creatures may crawl the floors.

Gustatory and olfactory hallucinations are uncommon but may at times be quite compelling. Patients may smell poison gas and inspect the air ducts to find the source. Tastes, often foul and bitter, may appear on the tongue, and patients may become convinced that their drinks have been fouled by poisons.

Tactile hallucinations may also occur and may be quite varied. Patients may feel electric currents course over their bodies; they may complain that fluids are poured on them at night or that they are pricked by needles from behind; in some cases they may experience movements deep inside, such as crampings and twistings.

DELUSIONS

Delusions are almost universal in schizophrenia and, although their content is extremely varied, certain themes stand out, including delusions of persecution, delusions of grandeur, delusions of reference, and a number of Schneiderian first-rank symptoms. These false beliefs may either grow slowly in the patient or occur suddenly, as if in an enlightenment. Although some patients may entertain some lingering doubts as to the veracity of these beliefs, for most they are as self-evidently true as any other belief. Occasionally patients may argue with others about these beliefs, and even attempt to convince others of their truth, but more often they do not press their case on the unbeliever. Most patients hold multiple delusions; these are often not well elaborated and are often poorly coordinated with each other, and they may even be contradictory. An exception to this rule is seen in the paranoid subtype of schizophrenia, in which one may find a certain degree of systematization of the patient's delusions into a more or less coherent corpus of beliefs.

Delusions of persecution are particularly common. Patients may believe that there is a conspiracy against them, for example that the police have coordinated their efforts with co-workers or neighbors or that perhaps the mafia or certain underground organizations are involved. Patients may believe that they are being followed, that their telephone conversations are being listened to, and that their mail is being cleverly opened. Some patients may endure these persecutions stoically, whereas others may engage in what to them appears to be a justifiable self-defense and fight back. Some may flee the area, seeking safety in another town or state. In some cases, if the delusion of persecution is a bizarre one, the response may be proportionately bizarre: one patient, believing that 'rays' were being sent through the ceiling to kill him, proceeded to cover the ceiling in aluminum foil in order to protect himself.

Delusions of grandeur are also common and may coexist with delusions of persecution. Patients may believe that they have developed great inventions and that others persecute them out of envy. Some believe that they have been elected by God, that millions of dollars are held privately for them, that heads of state secretly await their advice on foreign affairs. Although some patients may harbor these beliefs quietly, others may feel compelled to make an announcement. One patient took out a full-page advertisement in the newspaper in which he described his plans for 'world peace' through alliances with 'extraterrestrial beings' that had chosen him as their emissary.

Delusions of reference are intimately related to delusions of persecution and to delusions of grandeur, and serve, as it were, to reinforce them. Here, patients believe that chance events, rather than being innocuous and unrelated to them, in fact bear special meaning and pertain specifically to them. Newspaper headlines seem to be a kind of code, which only the patient can decipher, informing him or her that the time is near; street lights which blink on are a sign to those pursuing the patient that the time has come to 'move in'. The patient with grandiose delusions may hear church bells on Sunday morning and know that they serve as an announcement of his or her glory. For patients with delusions of reference, at times all things seem pregnant with meaning: there are no more chance occurrences, no accidents, and no coincidences.

Schneiderian first-rank symptoms may comprise delusions, including thought broadcasting, thought withdrawal, thought insertion, and delusions of influence, control, or passivity. In thought broadcasting, patients believe that thoughts can leave the head without being spoken or written, and that others may 'pick up' these thoughts directly; some patients say that it works 'like a radio' and they may feel no need to speak their thoughts as they assume that others, perhaps including the doctors, have already picked up the 'broadcast'. Thought withdrawal is characterized by the experience of having one's thoughts directly withdrawn from the mind. In such cases, patients suddenly become bereft of thoughts and are left with blank minds; some may elaborate on the experience and speak of electrical or magnetic devices that remove the thoughts. Patients who experience thought withdrawal while speaking may exhibit the sign known as 'thought blocking'. Here, in the middle of speaking, patients abruptly cease talking and become silent; this happens because they just as abruptly find themselves with no thoughts to speak. Thought insertion represents the converse of thought withdrawal, and in this experience patients believe that thoughts that are not their own have been 'inserted' directly into their minds; these 'alien' thoughts exist as a kind of cognitive 'foreign body', and patients may elaborate on the experience and believe that strange devices, perhaps using magnets, are responsible for their appearance. Delusions of influence, control or passivity are characterized by the belief on the patient's part that their thoughts, feelings, or actions are somehow directly influenced or controlled by some outside force or agency, and that they have somehow become like robots or passive automatons, without any independence of will. Some, again, may elaborate on these beliefs and speak of being under the influence of a spell or perhaps of an electric or magnetic machine, or perhaps a distant computer.

Other delusions may also occur. Indeed, any imaginable belief may be held, no matter how fantastic: angels live in the patient's nose; a small person is seated at a chair in the external auditory canal and whispers to the patient; parents have risen from the grave, and so on.

DISORGANIZED SPEECH

In considering the sign of disorganized speech we are concerned not with the content of speech, which may be composed of delusions, but rather with the form of speech. This 'formal thought disorder' is most often referred to as 'loosening of associations', but may sometimes be termed 'derailment'. Thought and speech become illogical and incoherent; thoughts are juxtaposed that have no conceivable connection and family members may complain that the patient 'does not make sense': there is a lack of 'goal-directedness', and thoughts, lacking such an organizing goal, become disparate and unconnected. At its most extreme, loosening of associations produces a 'word salad' in which successive spoken words have no more inherent connection with each other than do the tossed leaves of a salad. In one case, a patient, after being asked about the previous day's activities, replied, 'The sun bestrides the mouse doctor. In the morning, if you wish. Twenty-five dollars is a lot of money! Large faces and eyes. Terrible smells. Rats in the socket. Can there be darkness? Oh, if you only knew!' Here, any inner connection between the various ideas and concepts is lost; it is as if they come at random. Typically, patients are unconcerned with their incoherence; if asked to explain what they mean, they make little, if any, effort at clarification.

In some cases, rather than loosening of associations one may see mere tangentiality or circumstantiality. In tangentiality, in response to a question, thoughts proceed off on a tangent and patients never get around to giving an answer; in circumstantiality, the responses are circuitous and patients take a long 'round about' path before they finally 'get to the point'.

Neologisms may also occur, in which, in the course of speaking, the patient may use a word that seems to have no meaning. For example, one patient, when asked if he wanted anything, replied, 'Yes, please, some bufkuf.' When asked the meaning of 'bufkuf', the patient responded, 'Oh, you know', and made no further attempt to define or explain it.

CATATONIC OR BIZARRE BEHAVIOR

Catatonia may occur in one of two forms, namely stuporous catatonia and excited catatonia. Both forms may be seen in schizophrenia and, indeed, it is not uncommon to see individual patients with the catatonic subtype of schizophrenia exhibit both forms at different times (Morrison 1973). In stuporous catatonia one sees immobility, catalepsy, and mutism, which may be joined by posturing, echolalia or echopraxia, negativism, or automatic obedience, whereas in excited catatonia one sees bizarre, frenzied, purposeless behavior. All of these symptoms are discussed in detail in Chapter 4.11.

In addition to catatonic symptoms there may be other bizarre aspects to schizophrenia, including mannerisms, bizarre affect, and an overall disorganization and disintegration of behavior.

Mannerisms, as discussed in Chapter 5.27, represent bizarre or odd caricatures of gestures, speech, or behavior. In a manneristic gesture, the patient may offer a hand to shake with the fingers splayed out, or the fingers may intermittently writhe in a peculiar, contorted way. In manneristic speech, cadence, modulation or volume may be erratic and dysmodulated. One patient spoke in a 'sing-song' voice and, in another, random syllables were accented in a pompous way. Overall behavior may undergo manneristic transformation; one patient walked in a stiff-legged fashion, rigidly swinging only one arm with each step.

Bizarre affect may manifest in a variety of ways. In some cases facial expression appears theatrical, wooden, or under a peculiar constraint, for example patients may report feeling joy but the rapturous facial expression may appear brittle, tenuous, and disconnected. In other cases there may be inappropriate affect. Here, as described in Chapter 5.26, one finds a disconnection between what the patient feels and what 'shows' on his face, as, for example, in the case of a patient who, although feeling sadness at the death of a parent, was seen to involuntarily snicker.

There may also be a bizarre disintegration of the patient's overall behavior and demeanor, and it is this sign that often makes these patients 'stand out' in public. Patients become untidy and may neglect to clean their clothes or to bathe. Dress and grooming may become bizarre: several layers of clothing may be worn, even when it is hot outside, and bits of string or cloth may festoon jackets or shirts. Patients may cut off their eyebrows or make deep gashes in their arms or legs. Some seem almost analgesic: one patient bit out part of his biceps muscle; another eviscerated himself, 'just to see' what was inside. In some cases the bizarre behavior may be in response to delusions, as, for example, when a patient wrapped his legs in aluminum foil to 'keep the bugs off'.

NEGATIVE SYMPTOMS

'Negative', or 'defect', symptoms (Andreasen 1982; Andreasen and Olson 1982; Andreasen *et al.* 1990a) include flattening of affect, poverty of thought or speech, and avolition.

Flattening of affect, also known as mere 'blunting' of affect when less severe, is, as discussed further in Chapter 5.25, characterized by a lifeless and wooden facial expression accompanied by a corresponding dearth or diminution of all feelings.

Poverty of speech and poverty of thought are often referred to collectively as 'alogia'. Poverty of speech is said to be present when patients, although speaking a normal amount, seem to 'say' very little: there is a dearth of meaningful content and speech is often composed of stock phrases. In poverty of thought, by contrast, patients speak little, essentially because there is a wide-ranging and far-reaching impoverishment of their entire thinking: patients may complain that their heads are 'empty' and that, simply, nothing comes to mind.

Avolition, referred to by Kraepelin as 'annihilation of the will', is allied to poverty of thought in that avolitional patients simply do not experience any impulses, desires, stirrings, or inclinations; if left undisturbed they may spend hours or days in quietude, doing nothing.

MISCELLANEOUS SYMPTOMS

Of the miscellaneous symptoms seen in schizophrenia, perhaps the most important are transient disturbances of mood, which may tend towards depression, mania, or anxiety. Patients may complain of depressive symptoms, such as feeling depressed, being tired, or having trouble sleeping; some may demonstrate some euphoria and increased energy and talkativeness, whereas others may complain of feeling anxious and tremulous. Indeed, at first glance these symptoms may seem to dominate the clinical picture; however, on a closer and wider look one finds that they are transient, lasting only hours or days, are mild overall, and, relative to other symptoms, such as hallucinations and delusions, play only a very minor role in the overall clinical picture. Agitation may also be seen, and this may occur either as a non-specific part of an exacerbation of the disease or as a reaction to delusions of persecution or threatening voices.

SUBTYPES

Classifying patients as to subtype is useful not only because the various subtypes pursue different courses with different prognoses, as noted further below, but also because knowledge of a patient's subtype diagnosis may allow the clinician to predict how the patient will act in any given situation.

After discussing the paranoid, catatonic, disorganized, simple, and undifferentiated subtypes of schizophrenia, some comments will also be offered on an alternative mode of subdividing schizophrenia, namely into 'reactive' and 'process' types.

Paranoid schizophrenia tends to have a somewhat later onset, sometimes as late as in middle years, and is characterized primarily by hallucinations and delusions; disorganized speech, catatonic or bizarre behavior, and negative symptoms are either absent or relatively minor. Hallucinations are generally auditory and delusions are generally of persecution and reference. In paranoid schizophrenia, more so than in any other subtype, the delusions tend to be systematized and, on first glance, even plausible. Voices may warn patients that their supervisors are plotting against them. Patients may begin to suspect that people are talking about them, perhaps laughing at them behind their backs. Newspaper headlines pertain to them; the Central Intelligence Agency, or perhaps the Federal Bureau of Investigation, may be involved. At times patients may appeal to the authorities for help, but often they suffer their persecutions in rigid silence; occasionally they may try to escape, perhaps by moving to another area, or they may turn on their supposed attackers, sometimes violently. Often, allied with delusions of persecution, there may also be delusions of grandeur. Patients believe that they are being persecuted not for some trivial reason; they suspect that others know that they have developed great inventions. Rarely, grandiose delusions may be more prominent than persecutory ones, and they may even dominate the clinical picture. One patient believed himself to be the anointed of God; he heard trumpets proclaiming his advent and was prepared to announce himself to the world.

Catatonic schizophrenia, as the name obviously indicates, is dominated by catatonic symptoms. As described in Chapter 4.11, catatonia occurs in both stuporous and excited forms and, although some patients with catatonic schizophrenia may demonstrate only one form throughout the course of the illness, in most cases, as noted earlier, these two forms are seen to alternate in the same patient. The duration of these forms is quite variable, ranging from hours on one extreme to months or years on the other. The transition from one form to another may be quite unpredictable and, at times, quite sudden; in one case a chronically stuporous patient, without any warning, suddenly jumped from his bed, screamed incoherently, and paced agitatedly from one wall to another, only to lapse into immobility and muteness an hour later.

Disorganized schizophrenia, also known classically as hebephrenic schizophrenia, tends to have an earlier onset than the other subtypes and develop very slowly. Although hallucinations and delusions are present, they generally play a minor role and the clinical picture is dominated by disorganized speech and bizarre behavior. Overall, the behavior of these patients seems at times to represent a caricature of childish silliness. Senselessly, they may busy themselves, first with this and then with that, generally to no purpose and often with silly, shallow giggling. At times they may be withdrawn and inaccessible. When delusions are at all prominent, they tend to be hypochondriacal in nature and very unsystematized. In some cases there may also be disorganized speech, with marked loosening of associations to the point of a fatuous, drivelling incoherence.

Simple schizophrenia (Black and Boffeli 1989; Kendler *et al.* 1994a), also known as 'simple deteriorative disorder', has the earliest age of onset, often appearing in childhood, with a very slow and insidious onset, stretching over years. Hallucinations, delusions, and disorganized speech are sparse, and, indeed, are, for the most part, absent, and the clinical picture is dominated by negative symptoms. Over the years, these patients fall away from any acquaintances that they may have had and often become distant and emotionally dead. They may appear shiftless and some may accuse them of being 'lazy'. Few thoughts, desires, or inclinations disturb them and they may appear quite content to lie in bed or sit in a darkened room all day. For the most part they do little to attract the attention of others and may pass their lives in homeless shelters.

Undifferentiated schizophrenia is said to be present when the clinical picture of any given case does not fit well into any one of the foregoing subtypes. This is not uncommonly the case, and it also appears that, over long periods of time, the clinical picture, which initially did fit a particular subtype, may gradually change and become less distinctive. This transition from a recognized specific subtype to an undifferentiated presentation appears to be most common with the catatonic and disorganized subtypes; by contrast, the paranoid and simple subtypes tend to 'run true'.

As noted earlier, in addition to these classic subtypes there is an alternative proposal for classifying schizophrenia, which divides the illness into two types: 'reactive' schizophrenia, also known as 'good prognosis' or 'type I' schizophrenia, and 'process' schizophrenia, also known as 'poor prognosis' or 'type II' schizophrenia. In reactive schizophrenia, the premorbid personality tends to be normal and the onset, which is marked by depression and perplexity, is acute and occurs generally in adult years, often following some obvious social or personal stress; hallucinations and delusions, with some speech disorganization, dominate the clinical picture. In process schizophrenia, by contrast, the premorbid personality is often abnormal and the onset is insidious, often in childhood or adolescence, and without any recognizable precipitants; negative symptoms tend to dominate the clinical picture. As might be expected, the overall prognosis is favorable for 'reactive' cases and quite poor for 'process' ones.

Although this reactive–process dichotomy is useful, many patients do not fit neatly into one type or the other but rather demonstrate a mix of features. Furthermore, it is not clear whether this scheme represents an advance over the classic subtyping; indeed, one might say that the 'reactive' types represent paranoid schizophrenia whereas the 'process' types include disorganized and simple schizophrenia.

Course

Schizophrenia is a chronic, lifelong disease and most patients suffer considerable disability throughout their lives (McGlashan 1984; Tsuang *et al.* 1979). In most cases, the disease exhibits one of two courses: an overall waxing and waning course or a chronic, slow progression.

The waxing and waning course is characterized by exacerbations and partial remissions. The pattern of these changes is often quite irregular, as are the durations of the exacerbations and partial remissions, which may range from weeks to months or even years. In some cases, after a partial remission patients may develop a depressive episode, commonly referred to as 'post-psychotic depression' (Mandel *et al.* 1982); this is an important development as the risk of suicide is higher during this time. Importantly, this sustained depressive episode must not be confused with the frequently observed, transient, minor depressive symptoms that may occasionally accompany an exacerbation of psychotic symptoms. Overall, this waxing and waning course may persist throughout the life of the patient or, in many cases, it may give way, after anywhere from 5 to 20 years, to a stable chronicity.

The chronic progressive course may be evident from the onset of the disease, as, for example, in the simple subtype, or may become evident only after the initial onslaught has settled some. Over long periods of time, there is a very gradual progression, after which many patients eventually 'burn out' with a subsequent stable chronicity.

As noted earlier, the subtype diagnosis may enable some predictions to be made as to course (Fenton and McGlashan 1991; Kendler *et al.* 1994b). Thus, the paranoid and catatonic subtypes tend to pursue a waxing and waning course and, of these two, the overall prognosis appears to be better for the paranoid subtype. The disorganized and simple subtypes, by contrast, tend to demonstrate a slowly progressive course.

Schizophrenia may have a fatal outcome. Patients with catatonic schizophrenia of the stuporous type may die of aspiration pneumonia or extensive decubiti. Suicide occurs in about 10 percent of patients (Tsuang 1978); overall, about one-third will make a suicide attempt (Allebeck *et al.* 1987).

Before leaving this discussion of the course of schizophrenia it is appropriate to consider whether or not, in the natural course of events and in the absence of treatment, schizophrenia ever undergoes a full, complete, and spontaneous remission. Certainly, in cases that exhibit a waxing and waning course, the partial remissions may be far reaching and, to casual inspection, it may appear that the patient has recovered. Closer inspection, however, almost always reveals lingering residual symptoms in these 'recovered' patients, such as fleeting hallucinations, odd beliefs, mannerisms, or a certain poverty of thought. Consequently, in perusing the literature on the course of schizophrenia, one must pay careful attention to the definition used by the authors for 'recovery' or 'remission', for often it is a broad one that looks more to social functioning than to a true and complete absence of all symptoms. In this regard one may recall Bleuler's opinion (Bleuler 1950) that a full cure, a *restitutio ad integrum*, never occurs; although he saw 'far reaching improvements' to the point of 'social restitution', he was always able, upon careful examination, to 'see distinct signs of the disease'.

Etiology

Both computed tomographic (CT) and magnetic resonance imaging (MRI) studies have amply demonstrated the presence of ventricular dilation and cortical atrophy, most prominently in the temporal cortex. Furthermore, several studies have also demonstrated a correlation between the degree of atrophy seen in the left temporal cortex and the severity of such psychotic symptoms as loosening of associations and auditory hallucinations (Barta *et al.* 1990; Hirayasu *et al.* 2000; Menon *et al.* 1995; Shenton *et al.* 1992). Autopsy studies support the results of neuroimaging, demonstrating a reduced volume in the medial temporal lobe structures (Bogerts *et al.* 1985, 1990) and a prominence of ventricular enlargement in the temporal horn (Crow *et al.* 1989).

Microscopic studies have demonstrated the presence of an excessive number of interstitial neurons within the white matter (Akbarian *et al.* 1993a,b, 1996; Rioux *et al.* 2003) and neuronal disarray within the temporal lobe, specifically the hippocampus (Conrad *et al.* 1991). Strikingly, and importantly, gliosis is absent (Bogerts *et al.* 1985; Bruton *et al.* 1990; Roberts *et al.* 1987).

Although the mechanism underlying these anatomic changes is not known with certainty, it is strongly suspected that they represent a disorder of neuronal migration. In the normal course of development, neurons migrate along radial glial fibers from the ventricular area through the embryonic white matter to the overlying cortical subplate, where they come to rest in an orderly fashion to create the laminated cortex. Furthermore, and again normally, a small number of these neurons fail to migrate through the white matter and remain there as interstitial neurons. The findings noted above of an increased number of interstitial neurons and neuronal disarray in the cortex are consistent with the hypothesis of a failure of normal migration in schizophrenia.

The mechanism underlying such a failure of normal migration, although not clear, may involve one or more environmental insults acting on a genetically determined vulnerability. As noted earlier, schizophrenia occurs in about 1 percent of the general population; however, among first-degree relatives the prevalence rises to about 10 percent. Further, whereas the concordance rate in dizygotic twins is about 15 percent, among monozygtic twins it rises to about 50 percent. Furthermore, adoption studies have made clear that the increased prevalence in first-degree relatives reflects genetic and not environmental factors (Kendler and Gruenberg 1984; Kety 1987; Tienari *et al.* 2003). These impressive figures have naturally prompted a search for the gene or genes involved; however, despite intensive effort, no unequivocal results have been obtained, and the thinking is that the genetics of schizophrenia are complex, and probably involve the effects of multiple different susceptibility genes. Of course, genetics cannot explain the entire picture or one would expect a much higher concordance rate in monozygotic twins, and consequently one must look to environmental factors. Several have been proposed, including obstetrical complications (Kendell *et al.* 1996), maternal malnutrition (Susser and Lin 1992), and *in utero* exposure to certain viral illnesses (Murray 1994), for example influenza (Brown *et al.* 2004). All of these factors can cause disorders of neuronal migration, and it may be that what is inherited in schizophrenia is not the disease *per se* but rather a defect that renders the fetal brain particularly vulnerable to certain of these factors.

If, indeed, the neuropathology of schizophrenia represents a disorder of neuronal migration, one must also explain why the onset of the disease is delayed until the late teenage or early adult years. One hypothesis is that the phenotypic expression of the disease is dependent upon an interaction between the fixed neuronal migration defect and the normally evolving neuroanatomic changes seen during adolescence. During normal childhood and adolescence there is a progressive and selective 'pruning' of dendrites, and, according to this hypothesis, the ability of the fixed neuronal migration defect to express itself clinically may have to wait until a specific degree and type of 'pruning' has 'cleared the way'.

Although this neurodevelopmental theory of the etiology of schizophrenia has much to recommend it, the case is certainly not proven, and readers are encouraged to watch the literature.

Differential diagnosis

Of the multiple other different causes of a chronic psychosis discussed in Chapter 8.1, only a few play a large role in the differential diagnosis of schizophrenia, and these include bipolar disorder, major depressive disorder, schizoaffective disorder, delusional disorder, alcoholic psychoses, several personality disorders, and two putative disorders known as schizophreniform disorder and brief psychotic disorder.

Bipolar disorder, discussed in Chapter 23.1, is characterized by episodes of mania and depression, whereas major depressive disorder, as noted in Chapter 23.3, is marked by depressive episodes alone. Both manic episodes, as seen in bipolar disorder, and depressive episodes, as seen in either bipolar disorder or major depressive disorder, may be characterized by psychotic symptoms.

Manic episodes, as discussed in Chapter 7.3, typically show a more or less orderly progression through three stages: stage I is characterized by manic symptoms alone; in stage II, manic symptoms intensify and are joined by delusions and hallucinations; in stage III, manic symptoms may often fade and the clinical picture is often dominated by delusions, hallucinations, disorganized speech, bizarre behavior, and, at times, stuporous catatonia. Naturally, if one sees a patient in stage II or stage III, and has no history, a reliable differentiation between mania and schizophrenia might not be possible. If, however, one does have a reliable history, the differential diagnosis is fairly straightforward. In mania, classic manic symptoms (e.g. heightened mood, increased energy, and decreased need for sleep) occur, during stage I, in the absence of psychotic symptoms. By contrast, in schizophrenia one finds that the course is marked by the presence of hallucinations before the onset of any mood symptoms; furthermore, as noted earlier, any manic symptoms seen in schizophrenia are transient and relatively mild. Certain other features may also help in the differential diagnosis between mania and schizophrenia. In mania, euphoria is typically 'infectious' and deeply felt, and this contrasts with the euphoria seen, for example, in hebephrenic schizophrenia, which is shallow and silly and which, rather than being infectious, typically leaves the interviewer unmoved. Furthermore, the hyperactivity of mania is typically outgoing and extroverted, and this is in marked contrast to the behavior of the excited catatonic patient, who typically avoids contact. Finally, in cases in which the heightened mood is one of irritability rather than euphoria, the manic patient is typically 'on the attack', which contrasts with the irritable paranoid schizophrenic, who is more 'on guard': although both irritable manics and irritable paranoid schizophrenics are dangerous, the manic patient is recklessly so, in contrast to the paranoid schizophrenic, who may become violent only if approached in what appears, to the patient at least, to be a potentially hostile manner.

Depression, as discussed in Chapter 7.1, may likewise, at its depth, be characterized by hallucinations and delusions, and this may be seen in depressive episodes of either bipolar disorder or major depressive disorder. As with manic symptoms, however, the differential is fairly straightforward providing that one has a reliable history: in depression characterized by hallucinations and delusions, these symptoms only occur well after the typical depressive symptoms have become well established and severe; by contrast, in schizophrenia, delusions and hallucinations precede the advent of depressive symptoms. Furthermore, the depressive symptoms seen in schizophrenia are generally transient and are typically not severe. The nature of the delusions seen in depression may also be helpful. In depression, the delusions are 'mood congruent' in that they 'make sense' in light of the prevailing mood. Thus, patients with depression may come to believe that they have committed unpardonable sins or that their insides are drying up and dying, as is only fitting for such miserable sinners. By contrast, in schizophrenic patients who experience depressive symptoms, the delusions are often 'mood incongruent' and indeed often bear no conceivable relationship to depression, as, for example, in the case of a schizophrenic who complained of feeling depressed and then went on to talk about the 'telepathy' that made his toe tingle. Finally, consideration must be given to differentiating a severe depressive episode from a catatonic stupor. In both conditions one may see immobility and mutism; however, with prolonged observation one may chance to see a sudden 'lysis' of catatonic stupor, into either relatively normal mobility or a brief period of excitation; such a lysis, occurring over seconds or minutes, simply is not seen in depression.

Schizoaffective disorder, as described further in Chapter 22.2, is an illness characterized by the chronic presence of psychotic symptoms, such as hallucinations, delusions, or disorganized speech, and by intermittent episodes of either mania or depression, during which the pre-existing psychotic symptoms undergo an exacerbation. The differential point that allows a distinction from schizophrenia is that the mood disturbances, whether manic or depressed, are full, severe, and sustained, generally lasting at least weeks, in marked contrast to the mood changes that may be seen in schizophrenia, which are fragmentary, mild, and transient.

Delusional disorder, as noted in Chapter 23.3, is characterized by delusions, and thus may be confused with paranoid schizophrenia. The differential here rests on the degree of plausibility and systematization of the delusions and on the absence of other symptoms. Delusional disorder is marked by delusions that are often quite plausible and typically very well systematized, and by the absence of other symptoms, in particular disorganized speech and bizarre behavior. By contrast, although there may be some plausibility and systematization to the delusions seen in paranoid schizophrenia, close inspection typically reveals some fragmentation and inconsistencies and almost always the presence of other symptoms.

Alcoholic psychoses, namely alcoholic hallucinosis and alcoholic paranoia, discussed in Chapters 28.9 and 28.10, respectively, may be characterized by delusions and hallucinations, and thus mimic paranoid schizophrenia. The differential here rests on the history: if the psychosis in question occurred only after many years of alcoholism with repeated episodes of delirium tremens, then a case may be made for an alcoholic psychosis; in cases, however, in which delusions or hallucinations occurred early on, perhaps in adolescence or early adult years, and only after a relatively brief drinking career, then one would be hard pressed to explain them on the basis of alcoholism.

Several personality disorders, namely those of the paranoid, borderline, and schizoid and schizotypal types, may occasionally offer some diagnostic difficulties. Patients with paranoid personality disorder are chronically distrustful and on guard, quick to take offense and to read malevolence into what others do, and prone to harbor deep, long-standing resentments; under great stress they may develop delusions of persecution and thus may resemble patients with paranoid schizophrenia. In patients with paranoid schizophrenia, however, close inspection will reveal other symptoms, such as delusions of other types (e.g. delusions of grandeur) or hallucinations, which are not seen in the personality disorder; furthermore, in paranoid schizophrenia one often also sees some lack of full systematization of the delusions, in contrast to the personality disorder in which logic is preserved. Borderline personality disorder is characterized by a chronic instability in all aspects of the patient's life, accompanied by intense loneliness and dramatic and stormy emotions; when under great stress, these patients may at times also develop delusions, either of persecution or reference, and may also experience auditory hallucinations. The distinction from schizophrenia rests on the stormy instability that characterizes these patients' lives and on the fact that the psychotic symptoms appear only at times of great stress; this is in contrast to schizophrenia, in which they are present throughout the course, even during stress-free periods. Schizoid and schizotypal personality disorders are both characterized by an aloofness and detachment, with, in schizotypal personality disorder, the addition of eccentricities and odd beliefs, and each may thus mimic simple schizophrenia. The differential here rests on the overall course: in the personality disorder there is no deterioration, whereas in simple schizophrenia one sees a very slow progression.

Schizophreniform disorder and brief psychotic disorder (also known as brief reactive psychosis) are both characterized by symptoms that are similar to those seen in schizophrenia; where they differ is in their supposed course. Patients who experience a full, complete, and spontaneous remission within 1 month are said to have brief psychotic disorder, whereas those whose illness lasts longer than 1 month but less than 6 months are said to have schizophreniform disorder. However, there is a debate whether such disorders actually exist. Certainly, there are patients with schizophrenia who are treated with antipsychotics early in the course of the illness and who experience a complete, antipsychotic-induced remission of symptoms; however, in these cases, if treatment is discontinued, symptoms gradually recur. What is at issue here is whether there are, in fact, cases in which symptoms spontaneously and completely undergo an absolutely complete remission, which then persists over long-term follow-up. This issue has not been conclusively settled.

Treatment

In almost all cases, treatment involves the use of an antipsychotic drug. These agents may be broadly divided into two different categories, namely first generation and second generation or, as they are often also termed, typical and atypical agents.

First-generation agents are further subdivided into 'high-potency' (haloperidol, fluphenazine, perphenazine, trifluoperazine, and thiothixene), 'low-potency' (chlorpromazine and thioridazine), and 'medium-potency' (loxapine and molindone) drugs. High-potency drugs require lower milligram doses and are more likely to cause extrapyramidal side-effects (e.g. parkinsonism, dystonia, akathisia, akinesia). Low-potency drugs require higher doses and are less likely to cause extrapyramidal side-effects, but are prone to cause sedation, hypotension, and anticholinergic effects. Medium-potency drugs, as might be expected, fall in between regarding both milligram dosage and side-effects. Of these first-generation agents, both haloperidol and fluphenazine are available in intramuscular 'depot' formulations. All other things being equal, of the first-generation agents, haloperidol is probably a reasonable first choice.

Second-generation agents include clozapine, olanzapine, risperidone, quetiapine, aripiprazole, ziprasidone, paliperidone, and iloperidone. Choosing among these agents is not straightforward. With regard to effectiveness, clozapine is clearly superior; however, its side-effect profile, especially the risk of agranulocytosis, limits its use to treatment-resistant cases. Of the other agents, although there is controversy here, it appears that olanzapine may have an edge in terms of overall effectiveness over risperidone, which in turn may have an edge over the others (Leucht *et al.* 2009; Stroup *et al.* 2007). The advantage of olanzapine, however, is severely tempered by its tendency to cause metabolic derangements, including weight gain, hyperlipidemia, and diabetes. Both risperidone and quetiapine may also cause these metabolic derangements, but are much less likely to do so; aripiprazole and ziprasidone stand out in that they are not associated with metabolic changes. As with the first-generation agents, some of the second-generation agents are also available in depot preparations, including olanzapine, risperidone, and paliperidone. Overall, and all other things being equal, if a second-generation agent is used, it may be reasonable to start with risperidone; however, again, it must be acknowledged that this is an area of great controversy.

In deciding which antipsychotic, whether first or second generation, to prescribe, the first step is to obtain an accurate treatment history, and this may require not only questioning the patient but also reviewing records and interviewing family members. If there is a history of a good response combined with good tolerability, then it makes sense to use the same drug again. In treatment-naive patients, or in cases in which prior treatments were unsatisfactory, other considerations come into play. Although not without controversy (Jones *et al.* 2006), it appears that, overall, second-generation agents are more effective than first-generation ones; furthermore, and very importantly, they are unquestionably far less likely to cause tardive dyskinesia and the extrapyramidal side-effects seen with the high-potency first-generation agents. Consequently, a second-generation agent appears preferable. It must be emphasized, however, that the choice of an agent is not simple or straightforward, and often multiple trials of different agents must be performed before a regimen is found that is reasonably effective and well tolerated. Furthermore, consideration must be given to the problem of non-compliance, which is very common in schizophrenia; in such cases, utilization of one of the depot preparations should be strongly considered.

Regardless of which oral agent is chosen, it is important that one gives it an 'adequate' trial before moving on to another, in terms of not only dose but also duration. In general, assuming an adequate dose is used, one should observe the patient for at least several weeks to get a reasonable idea as to response. In cases characterized by significant agitation, one may consider adding adjunctive lorazepam to the antipsychotic, and continuing this until the agitation has passed, after which the patient may be continued on the antipsychotic alone. If the overall initial response is only partial, but otherwise promising, one may elect to observe the patient for another month, and, if the response is good, then move to maintenance treatment, as described below. If the response to the first agent is poor, or the side-effects are unacceptable, then a trial with a different agent should be considered. Should patients fail to get a good response to adequate trials of two agents, one may be dealing with a treatment-resistant case. In such cases several options are available, including a trial of clozapine, trials of other agents, or simply 'living with' a less than optimal treatment regimen. As regards effectiveness, clozapine is clearly head and shoulders above all other antipsychotics in treatment-resistant cases, and indeed may yield some of the most gratifying treatment responses in all of medical practice; however, its side-effect profile gives pause to many patients and physicians. If clozapine is not an option, some clinicians will try trial after trial of different agents, hoping to find one that 'works'; provided that patients have the fortitude, this is not an unreasonable strategy, as in some cases patients will simply have idiosyncratic and unpredictable 'good' responses to one agent but not others. In some cases, however, patients will opt to stay with a regimen that, although perhaps providing less relief than is hoped for, is at least tolerable.

Maintenance treatment is appropriate in almost all cases. Initially, patients should be maintained on a dose that is similar to, if not identical to, that utilized during initial treatment. Once patients are stable, cautious dose adjustments may be considered every 3–4 months. As noted earlier, in many cases the course of schizophrenia is characterized by gradually occurring exacerbations and partial remissions, and in such cases it is appropriate to try to 'titrate' the dose to the underlying severity of the disease, always seeking the lowest possible dose consistent with acceptable symptomatic control. This serves to reduce not only cost and side-effect burden but also the risk of tardive dyskinesia. In a minority of cases, the underlying course is so favorable, and the partial spontaneous remissions so profound, that it may be possible to taper the dose to almost nothing, at which point some patients and physicians may consider stopping treatment. This is a difficult decision. Schizophrenia is a chronic disease and, although far-reaching spontaneous partial remissions do occur, exacerbations may be expected at some point in the future. Consequently, it is necessary to continue seeing patients in regular follow-up visits and to discuss with them, and with family, the importance of calling immediately should significant symptoms recur.

In both initial and maintenance treatment phases there are two side-effects that must always be kept in mind, namely akinesia and akathisia. Akinesia, as noted in Chapter 4.9, may leave patients psychomotorically retarded and, in the unwary clinician, may prompt a misdiagnosis of depression (King *et al.* 1995; Van Putten and May 1978). Akathisia, described in Chapter 4.10, classically renders patients restless, but at times may manifest only with an exacerbation of psychotic symptoms (Van Putten 1975; Van Putten *et al.* 1974); unless this is properly diagnosed, a 'vicious cycle' may occur, in which the exacerbation is mistakenly attributed to the underlying schizophrenia, prompting an increased dose of the antipsychotic with a consequent worsening of the psychosis. Although these side-effects are classically seen with high-potency first-generation agents, they may also, albeit rarely, occur with second-generation ones.

Post-psychotic depression may occur after psychotic symptoms have partially remitted, either spontaneously or by virtue of antipsychotic treatment. These sustained depressions must be treated as they carry a significant risk of suicide. Treatment may be accomplished with an antidepressant, such as a selective serotonin reuptake inhibitor (SSRI), or, in severe cases, with electroconvulsive therapy (ECT). Of interest, ECT may also be effective in the acute treatment of catatonic schizophrenia.

In addition to treatment with antipsychotics and routine supportive care, many patients will also require extensive assistance in gaining housing and sheltered employment; social skills training and cognitive–behavioral treatment may also be beneficial. Insight-oriented or psychodynamically oriented psychotherapy is generally contraindicated, as it may make patients worse.

Hospitalization is required for most patients at some point in their illness and, in many cases, repeated admissions occur. In some cases involuntary hospitalization is required and may be life-saving. Partial hospitalization or 'day hospitals' may enable some severely ill patients to be maintained in the community.

22.2 SCHIZOAFFECTIVE DISORDER

The term 'schizoaffective' has had many definitions since it was first coined by Kasanin in 1933. As conceived of here, schizoaffective disorder is characterized by chronic, unremitting psychotic symptoms, similar to those seen in schizophrenia, upon which are superimposed full episodes of either depression or mania, during which the pre-existing psychotic symptoms undergo an exacerbation. This is an uncommon disorder, being found in roughly 0.3 percent of the general population.

Clinical features

The onset is typically in the late teens or early twenties, and, viewed over time, this disorder, as suggested above, appears to represent a superimposition of a mood disorder, such as major depressive disorder or bipolar disorder, upon schizophrenia. Thus, these patients typically present with a psychosis similar to that described in the preceding chapter for schizophrenia, and the symptoms (e.g. hallucinations, delusions, disorganized speech, etc.) persist in a chronic, generally lifelong fashion. Periodically, however, these chronic symptoms are joined by either a depressive episode or a manic episode (with symptoms as described in Chapters 7.1 and 7.3, respectively). Characteristically, whenever a depressive or manic episode does occur, the chronically persistent psychotic symptoms become more severe, only to eventually return to their previous level of severity once the mood episode has run its course.

Course

Although it is clear that the psychotic symptoms remain chronic, the frequency with which episodes of depression or mania occur has not been well studied. In some cases it appears that only depressive episodes occur, and here one speaks of schizoaffective disorder, depressed type; in others one sees both depressive and manic episodes or manic episodes alone, and in these cases one speaks of schizoaffective disorder, bipolar type. The overall outcome of this disorder, although not as favorable as that of the mood disorders, is better than that seen in schizophrenia (Jager *et al.* 2004).

Etiology

Family studies (Angst *et al.* 1979; Cohen *et al.* 1972; Kendler *et al.* 1995; Lauren *et al.* 2005; Pope *et al.* 1980) suggest that schizoaffective disorder may either 'run true' or result from a 'double loading' of genetic susceptibilities for both schizophrenia and either major depressive disorder or bipolar disorder.

Differential diagnosis

Schizoaffective disorder must be distinguished from both schizophrenia and mood disorders.

In schizophrenia, as noted in the preceding chapter, one may see mood disturbances, but these differ fundamentally from the episodes of depression or mania seen in schizoaffective disorder. The mood changes seen in schizophrenia are transient, fragmentary, and generally mild, whereas those seen in a depressive or manic episode are sustained, pervasive, and severe, typically enduring at least for weeks. Post-psychotic depression (McGlashan and Carpenter 1976), as seen in some cases of schizophrenia, is distinguished by the fact that there is no exacerbation of psychotic symptoms during the depression.

In major depressive disorder or bipolar disorder, episodes of mood disturbance may, when severe, be accompanied by delusions and hallucinations; however, the differential may be easily made if one simply attends to the overall course of the illness. In mood episodes of major depressive disorder or bipolar disorder, psychotic symptoms occur only within the context of the mood episode, generally at their height, and are not present in the intervals between mood episodes. By singular contrast, in schizoaffective disorder, psychotic symptoms are *always* present: they are present during the mood episodes and also in the intervals between these episodes.

Treatment

As might be expected, the treatment of schizoaffective disorder borrows heavily from the treatments for schizophrenia and for either bipolar disorder or major depressive disorder. Given that all patients with schizoaffective disorder, whether of the bipolar type or the depressive type, are chronically psychotic, in almost all cases chronic treatment with an antipsychotic is indicated. To date, only one of the second-generation agents has been specifically tested in schizoaffective disorder and found effective, and that is paliperidone (Canuso *et al.* 2010). Other antipsychotics have been utilized in studies which looked at combined populations of patients with schizoaffective disorder and schizophrenia, and these studies, though partially flawed by the use of such a combination, did report effectiveness for aripirazole (Potkin *et al.* 2003), ziprasidone (Simpson *et al.* 2004), risperidone (Volavka *et al.* 2002) and olanzapine (Volavka *et al.* 2002). Whether paliperidone has an advantage over these agents is unclear; however, it would appear unlikely. In some cases, monotherapy with an antipsychotic alone may effectively treat manic or depressive episodes; however, in others, combination treatment is required, and in most cases this involves adding on a mood stabilizer. Both lithium (Carman *et al.* 1981) and carbamazepine (Okuma *et al.* 1989) have double-blinded support in this regard, and a long-term open study (Greil *et al.* 1997), although finding overall equivalence, did note a preferential response of mania to lithium and depression to carbamazepine. Divalproex also appears effective in open work (Bogan *et al.* 2000); however, there are no double-blinded studies to support this. All other things being equal, lithium is probably the first choice. When depression is present, and resists treatment with a mood stabilizer, consideration may be given to an antidepressant, and in this regard, citalopram has some support from a double-blinded study (Zisook *et al.* 2009). Adding an antidepressant, however, must be done with caution: if the patient has well-established schizoaffective disorder, depressed type, then it might be safe, but if the patient has schizoaffective disorder, bipolar type, then there is a risk of precipitating a manic episode. Consequently, antidepressants should probably be used only in cases where patients are already on a mood stabilizer.

22.3 DELUSIONAL DISORDER

Delusional disorder is a chronic disorder characterized by the gradual appearance of one or more delusions that eventually elaborate into a coherent system (Kendler 1980, 1988). In contrast to the delusions seen in schizophrenia, the delusions of delusional disorder have a certain plausibility, and the eventual delusional system is within itself quite logical.

The traditional name for this disorder, as originally bestowed by Kraepelin, was paranoia, and although this term is still seen in the literature, delusional disorder is probably a better name, for two reasons. First, it emphasizes the cardinal aspect of this disorder, namely the presence and prominence of delusions. Second, it avoids the unfortunate association of paranoia with persecution, and reminds us that delusions of persecution are but one of many types of delusions seen in this disorder.

This is an uncommon disorder, with a lifetime prevalence of roughly 0.2 percent.

Clinical features

The onset is generally in early to mid-adult years (Winokur 1977). Generally, these patients experience a lengthy prodrome, with evanescent delusions or perhaps intimations that things are not 'right'. Eventually, persistent and clear-cut delusions occur, and the illness begins to assume its definitive form. This transition to active illness may at times be sudden, and patients may experience a sort of 'epiphany' in which lurking suspicions and concerns suddenly crystallize into beliefs; often, however, the transition from prodrome to active illness is gradual, almost furtive.

Although the most common type of delusion seen in delusional disorder is that of persecution, other themes may be prominent: jealousy, grandiosity, erotic longing, litigiousness, and bodily concerns may all occur. Regardless of which delusion is most prominent, however, one often also sees delusions of reference. Furthermore, as noted earlier, the delusions experienced by these patients often fit together quite logically, and the entire corpus of beliefs is well systematized. Often, one finds that all of the delusions appear to stem from one delusional 'premise' and that this premise often has a certain plausibility to it.

Traditionally, delusional disorder has been divided into several subtypes, depending on the type of delusion present (Winokur 1977); thus, there are persecutory, grandiose, erotomanic, jealous, and somatic subtypes, and these are described further below. However, this division is somewhat artificial given that, although one type of delusion will dominate the picture, two or more may be present, thus creating a 'mixed' picture.

In the persecutory subtype the dominant delusion is one of being persecuted or conspired against. Patients feel singled out; they are watched and followed. Delusions of reference typically appear, and patients may, for example, believe that people on the street talk about them. Patients may move to another city to avoid their persecutors, and may feel safe for a while, but eventually their persecutors catch up with them. These patients may at times be dangerous and may attack others in what, to them, appears to be justified self-defence (Kennedy *et al.* 1992).

The litigious subtype may be the most difficult to diagnose as the initial delusion may appear very plausible and the ensuing delusions may have an almost unassailable logic to them. During the onset of the illness, patients are typically involved in legal proceedings that go badly for them. A lawsuit may be lost or, if won, the award may appear to the patient to be too low. Patients become convinced that someone is at fault: their attorneys are incompetent; the judges were biased; the juries were prejudiced. Patients may pore over trial manuscripts until, finally, some irregularity, no matter how minor, is found. New attorneys are then hired and appeals are filed, and a series of legal proceedings is embarked upon. With each failed legal proceeding, patients may become more convinced that the legal system as a whole is conspiring in the denial of justice. Eventually, the disparity between the magnitude of the patient's sense of being unjustly harmed and the trivial, insignificant nature of the original inciting insult to their sense of justice brings to light the pathologic nature of their behavior.

In the grandiose subtype, the dominant theme of grandiosity may come to light in a variety of ways. Patients may believe themselves to be secret 'captains of industry' whose advice is sought by leaders in the financial community. Others may entertain delusions of high descent. One man believed that his mother was an heiress who, out of shame for her out-of-wedlock pregnancy, gave him up for adoption to a poor family; the patient, believing that an inheritance would soon be his, quit his job as he felt no need for his income any more. Others may believe themselves to be great, although unrecognized, inventors, and toil on in their homes, littering their walls with fantastic diagrams and sketches of their magnificent creations.

In the erotomanic subtype (also known as de Clerambault's syndrome), patients come to believe that they are loved by someone else, generally someone of much higher social station. The clinical picture is dominated by the patient's belief that the imagined lover, for some reason or other, cannot openly express the love. For example, one woman believed that the mayor was in love with her and was unable to tell her this openly as he was married. She saw him at a political rally and he turned his gaze from her, a move she interpreted as evidence that he could not bear the unrequited longing he surely must have felt had he allowed himself to look at her. The newspaper hinted at a strain in his marriage, and the patient believed that, had it not been for the mayor's high moral character, he would immediately have divorced his wife. This attitude of hopeful expectation lasted for years, throughout which the patient occasionally sent 'secret' letters to the mayor at his office. Eventually, unable to wait longer she began stalking the mayor's wife, and called him to say that she would soon 'dispose' of his wife, thus freeing him.

In the jealous subtype, the patient becomes convinced that his or her spouse or lover is being unfaithful. The patient not only 'sees' evidence for the infidelity but may seek it out. The spouse is a few minutes late getting home from work and the patient believes that only passionate lovemaking could have caused the delay. Sheets and underclothing are inspected for telltale stains; telephone conversations are listened in on; a private detective may be hired. The patient may insist that the spouse stay at home, and at times spouses may become virtual prisoners in their own homes.

In the somatic subtype patients believe, despite reassurances to the contrary from their physicians, that they have a serious disease. Although the 'symptoms' may be mild, perhaps a minor headache, patients are convinced that the underlying disease is severe, perhaps a brain tumor. Two atypical variations on this subtype deserve mention, namely the olfactory reference syndrome (Videbech 1966) and parasittosis (Andrews *et al.* 1986; Mitchell 1989). In the olfactory reference syndrome patients are convinced that they are emanating a foul odor from the mouth or some other orifice, and may anxiously ask others if they smell it also. In parasitosis patients believe that the skin is infested: they may complain that they feel bugs crawling and may dig under the skin to find them.

In addition to delusions, some patients with delusional disorder may also have hallucinations, but these generally play only a minor role in the clinical picture and are consistent with the delusions. For example, a persecuted patient may hear a voice warning him that his life is in danger or an erotomanic patient may hear voices whispering caressing words. Mood and affect may be unremarkable or may show changes, again consistent with the delusions: the persecuted patient may become quite irritable and the grandiose patient may experience a shallow, contented euphoria.

Overall, regardless of subtype, the behavior of these patients may be quite normal in areas of their lives that are not touched by their delusions. For example, a patient with the persecutory subtype who believed that co-workers at the factory were conspiring against him, but who was free of delusional concerns regarding all others, led an unremarkable life at home and sang in the choir at church, took his children to baseball games, and was considered a 'good neighbor'.

Course

Over long-term follow-up (Opjordsmoen 1988; Opjordsmoen and Rettersol 1991), it appears that spontaneous remission occurs in roughly one-third of cases, while in the remainder a chronic course is seen.

Etiology

Delusional disorder is familial, and appears to 'run true' in that there is no familial relationship with schizophrenia (Kendler *et al.* 1982, 1985; Watt 1985; Winokur 1985).

Differential diagnosis

Of the multiple causes of psychosis discussed in Chapter 8.1, several figure highly in the differential diagnosis of delusional disorder, including schizophrenia, body dysmorphic disorder and alcoholic paranoia.

Schizophrenia is distinguished on two counts, namely the lack of systematization and the presence of other symptoms. As noted, in delusional disorder the various delusions are logically connected into a well-systematized corpus of beliefs. By contrast, in schizophrenia there is always some lack of connectedness among the various delusions, which at times may be flatly contradictory. Furthermore, in schizophrenia one sees other symptoms, such as bizarre delusions, prominent hallucinations and speech disorganization. Importantly, given that early in its course schizophrenia, especially the paranoid subtype, may be characterized by only delusions, long-term follow-up is necessary before making the diagnosis of delusional disorder.

Body dysmorphic disorder is characterized by a chronic concern on the patient's part that he or she, in some fashion, is ugly or misshapen. In the vast majority of cases, these concerns are not delusional, and patients admit that their concerns, though persistently nagging, are unfounded. In a small minority, however, delusional transformation occurs and the patient becomes unshakeably convinced. At this point, a reliable clinical differentiation from the somatic subtype of delusional disorder may not be possible, and given this it is reasonable to proceed to a diagnosis by treatment response and treat the patient with fluoxetine or clomipramine.

Alcoholic paranoia may be characterized by chronic delusions of persecution or jealousy, and given that alcoholics are often adept at hiding other evidence of their drinking, a careful history obtained from collateral sources is imperative.

Finally, consideration should also be given to a non-psychotic disorder, namely hypochondriasis. Hypochondriacal patients, like those with the somatic subtype of delusional disorder, have a persistent unfounded concern that they are ill; however, hypochondriacal patients, unlike those with delusional disorder, are not delusional and do accept reassurance from the physician.

Treatment

Treatment generally involves the use of an antipsychotic (Manschreck and Khan 2006); however, in most cases, gaining the patient's agreement to take medication may be quite difficult. These patients see nothing wrong with themselves or their beliefs, and if told that what they believe may not be true, they may be quite offended. Consequently, great tact and diplomacy are required, and the subject of medication may often have to be approached obliquely; whereas the persecuted patient might never accept a prescription for 'psychosis', he might well agree to 'try' a medicine to help him maintain his composure in the face of all the tribulations he is facing. An antipsychotic may be chosen using the scheme discussed in Chapter 22.1: importantly, one should choose a medication with a low propensity for side-effects and start at a low dose, as patients with delusional disorder are typically prone to seize on the occurrence of a side-effect, no matter how trivial, as a reason to never take medication of any sort again. The somatic subtype may constitute an exception to the foregoing as there are case reports of this subtype responding to antidepressants such as clomipramine (Wada *et al.* 1999) or paroxetine (Hayashi *et al.* 2004).

Occasionally, hospitalization may be required to protect others, for example in the persecutory or erotomanic subtypes.

22.4 POSTPARTUM PSYCHOSIS

Postpartum psychosis, also known as puerperal psychosis, as conceived of in this text, is a disorder characterized by the occurrence of psychosis exclusively in the puerperium and never at other times in the patient's life. As noted below in the discussion on differential diagnosis, this strictly defined disorder must be clearly differentiated from other disorders, which, though perhaps presenting with psychosis in the puerperium, are also characterized by psychotic symptomatology occurring at other times in the woman's life. The actual prevalence of strictly defined postpartum psychosis is not known: although 1–2 out of every 1000 deliveries is complicated by a psychosis occurring within days to weeks, most of these cases can be accounted for by other illnesses, most notably bipolar disorder.

Clinical features

Symptoms generally have an abrupt onset, anywhere from days to a few weeks after delivery (Brockington *et al.* 1981; Munoz 1985). Clinically (Bagedahl-Strindlund 1986; Brockington *et al.* 1981; Hadley 1941; Kumar *et al.* 1983; Munoz 1985; Oosthuizen *et al.* 1995), patients present with a combination of manic symptoms and delusions and hallucinations. The mood is often heightened, tending towards either irritability or euphoria, and is often accompanied by considerable agitation. Delusions often center on the baby, who may be seen as either an evil or a divine presence; other delusions, including Schneiderian first-rank symptoms, may also occur. Hallucinations tend to be auditory, and at times command hallucinations may occur, compelling the patient to harm the infant. Often, in addition to these symptoms, there may be a degree of confusion or disorientation. Suicide may occur, and infanticide is seen in roughly 4 percent of cases.

Course

In the natural course of events, symptoms undergo a gradual, spontaneous, and full remission after a matter of weeks or months. Close to one-third of patients will have another episode should they have another child (Davidson and Robertson 1985).

Etiology

Strictly defined postpartum psychosis does not appear to be genetically related to bipolar disorder, schizoaffective disorder, or schizophrenia (Schopf and Rust 1994a). Although the etiology is not known, it is strongly suspected that postpartum psychosis occurs secondary to the effects of the profound endocrinologic changes characteristic of the puerperium (Ahokas *et al.* 2000a,b).

Differential diagnosis

The main disorders on the differential diagnosis include bipolar disorder, schizoaffective disorder, and schizophrenia. The first step in pursuing the differential, then, would be to obtain a reliable history. With regard to bipolar disorder, there is a clear increased risk of mania in the puerperium (Bratfos and Haug 1966); however, in most of these cases there will be a history of prior episodes of either mania or depression. With regard to schizoaffective disorder and schizophrenia the task is easier, as there is a history of chronic psychotic symptoms, which would be present not only during the pregnancy, but in most cases also well before. Importantly, however, in cases where there is no history of psychosis prior to the puerperium, one must resist the temptation to confidently diagnose postpartum psychosis: bipolar disorder, schizoaffective disorder, and schizophrenia each can have an onset in the puerperium (Schopf and Rust 1994b), and consequently long-term follow-up is required in order to establish two facts: first, that the current psychosis undergoes a full, complete and spontaneous remission within weeks to months, and, second, that there is either no recurrence, or, if there is, that such recurrence occurs *only* postpartum.

Certain other disorders, though only rarely causing a psychosis in the puerperium, must also be considered. Eclampsia may cause delirium with psychotic features; however, here one finds hypertension and proteinuria. Dural sinus thrombosis may occur postpartum, and if the transverse sinus is involved there may be venous infarction of the temporal lobe, with resulting delirium (Dhasmana *et al.* 2010). Certain tumors, such as meningiomas, may undergo considerable enlargement during pregnancy, and may present with psychosis in the puerperium (Khong *et al.* 2007). Finally, bromocriptine (Canterbury *et al.* 1987), as may be used to suppress lactation, may also cause a toxic psychosis.

Treatment

Patients should be emergently hospitalized. Traditionally, treatment has involved an antipsychotic, combined, in many cases, with a mood stabilizer. Of the antipsychotics, a second-generation agent is preferable, and this may be effective as monotherapy. When manic symptoms are prominent, however, the addition of either divalproex or lithium may be very helpful, and given the rapid onset of action of divalproex (when given in a loading dose), this may be preferable. Recent work, however, suggests an alternative approach: in an open study (Ahokas *et al.* 2000b), estradiol, 1 mg four times a day, yielded impressive results. If one does choose to try estradiol, however, it may be prudent to start initially with an antipsychotic, then add on the estradiol: if there is a rapid and complete remission, then one could discontinue the antipsychotic and see if the estradiol was effective as monotherapy. In severe and treatment-resistant cases, consideration should be given to ECT (Reed *et al.* 1999). Once a drug-induced remission has been achieved, treatment should continue for at least several months: at this point, provided that the patient is completely symptom free, the drug may be gradually tapered.

As patients begin to improve, attempts should be made to gradually guide them into appropriate interactions with their babies: these visits, however, should always be closely monitored until patients have fully recovered.

Subsequent to recovery, patients should be counselled regarding the risk of recurrence after future pregnancies. If patients do become pregnant again, close monitoring is required postpartum, and a case may also be made for prophylactic use of lithium (Austin 1992; Stewart 1988; Stewart *et al.* 1991) or whichever other agent was effective during the earlier episode, with treatment beginning either immediately post-delivery or, in some highly selected cases, shortly before anticipated delivery.

REFERENCES

Ahokas A, Aito M, Turianinen S. Association between oestradiol and puerperal psychosis. *Acta Psychiatr* 2000a; **101**:167–9.

Ahokas A, Aito M, Rimon R. Positive treatment effect of estradiol in postpartum psychosis: a pilot study. *J Clin Psychiatry* 2000b; **61**:166–9.

Akbarian S, Bunney WE, Potkin SG *et al.* Altered distribution of nicotinamide-adenine dinucleotide phosphate-diaphorase cells in frontal lobe of schizophrenics implies disturbances of cortical development. *Arch Gen Psychiatry* 1993a; **50**:169–77.

Akbarian S, Vinuela A, Kim JJ *et al.* Distorted distribution of nicotinamide-adenine dinucleotide phosphate-diaphorase neurons in temporal lobe of schizophrenics implies anomalous cortical development. *Arch Gen Psychiatry* 1993b; **50**:178–87.

Akbarian S, Kim JJ, Potkin SG *et al.* Maldistribution of interstitial neurons in prefrontal white matter of the brains of schizophrenic patients. *Arch Gen Psychiatry* 1996; **53**:425–36.

Allebeck P, Varla A, Kristjansson E *et al.* Risk factor for suicide among patients with schizophrenia. *Acta Psychiatr Scand* 1987; **76**:414–19.

Andreasen NC. Negative symptoms in schizophrenia: definition and reliability. *Arch Gen Psychiatry* 1982; **39**:784–8.

Andreasen NC, Olson S. Negative versus positive schizophrenia: definition and validation. *Arch Gen Psychiatry* 1982; **39**:789–94.

Andreasen NC, Flaum M, Swayze VW *et al.* Positive and negative symptoms in schizophrenia: a critical reappraisal. *Arch Gen Psychiatry* 1990a; **47**:615–21.

Andrews E, Ballard J, Walter-Ryan WG. Monosymptomatic hypochondriacal psychosis manifesting as delusions of infestation: case studies of treatment with haloperidol. *J Clin Psychiatry* 1986; **47**:188–90.

Angst J, Felder W, Lohmeyer B. Schizoaffective disorders. I. Results of a genetic investigation. *J Affect Disord* 1979; **1**:139–53.

Anonymous. Historical notes: earliest use of the term dementia praecox. *Am J Psychiatry* 1954; 111:470.

Austin MP. Puerperal affective psychosis: is there a case for lithium prophylaxis? *Br J Psychiatry* 1992; **161**:692–4.

Bagedahl-Strindlund M. Postpartum mental illness: timing of illness onset and its relation to symptoms and sociodemographic characteristics. *Acta Psychiatr Scand* 1986; **74**:490–6.

Barta PE, Pearlson GD, Powers RE *et al.* Auditory hallucinations and smaller superior temporal gyral volume in schizophrenia. *Am J Psychiatry* 1990; **146**:1457–62.

Black DW, Boffeli TJ. Simple schizophrenia: past, present and future. *Am J Psychiatry* 1989; **146**:1267–73.

Bleuler E. *Dementia Praecox or the Group of Schizophrenias.* New York: International Universities Press, 1950.

Bogan AM, Brown ES, Suppes T. Efficacy of divalproex therapy for schizoaffective disorder. *J Clin Psychopharmacol* 2000; **20**:520–2.

Bogerts B, Meertz E, Schonfeldt-Bausch R. Basal ganglia and limbic system pathology in schizophrenia: a morphometric study of brain volume and shrinkage. *Arch Gen Psychiatry* 1985; **42**:784–91.

Bogerts B, Falkai P, Haupts M *et al.* Post-mortem volume measurements of limbic system and basal ganglia structures in chronic schizophrenics. Initial results from a new brain collection. *Schizophren Res* 1990; **3**:295–301.

Bratfos O, Haug JO. Puerperal mental disorders in manic–depressive females. *Acta Psychiatr Scand* 1966; **42**:285–94.

Brockington IF, Cernik KF, Schofield EM *et al.* Puerperal psychosis: phenomena and diagnosis. *Arch Gen Psychiatry* 1981; **38**:829–33.

Brodaty H, Sachdev P, Rose N *et al.* Schizophrenia with onset after age 50 years. I. Phenomenology and risk factors. *Br J Psychiatry* 1999; **175**:410–15.

Brown AS, Begg MD, Gravenstein S *et al.* Serologic evidence of prenatal influenza in the etiology of schizophrenia. *Arch Gen Psychiatry* 2004; **61**:774–80.

Bruton CJ, Crow TJ, Frith CD *et al.* Schizophrenia and the brain: a prospective clinico-pathological study. *Psychol Med* 1990; **20**:285–304.

Canterbury RJ, Haskins B, Kahn N *et al.* Postpartum psychosis induced by bromocriptine. *South Med J* 1987; **80**:1463–4.

Canuso CM, Lindenmayer JP, Kosik-Gonzalez C *et al.* A randomized, double-blind, placebo-controlled study of 2 dose ranges of paliperidone extended-release in the treatment of subjects with schizoaffective disorder. *J Clin Psychiatry* 2010; **71**:587–98.

Carman JS, Bigelow LB, Wyatt RJ. Lithium combined with neuroleptics in chronic schizophrenia and schizoaffective patients. *J Clin Psychiatry* 1981; **42**:124–8.

Cohen SM, Allen MG, Pollin W *et al.* Relationship of schizoaffective psychosis to manic–depressive psychosis and schizophrenia: findings in 15,909 veteran pairs. *Arch Gen Psychiatry* 1972; **26**:539–46.

Conrad AJ, Abebe T, Austin R *et al.* Hippocampal pyramidal cell disarray in schizophrenia as a bilateral phenomenon. *Arch Gen Psychiatry* 1991; **48**:413–17.

Crow TJ, Ball J, Bloom SR *et al.* Schizophrenia as an anomaly of development of cerebral asymmetry: a post-mortem study and a proposal concerning the genetic basis of the disease. *Arch Gen Psychiatry* 1989; **46**:1145–50.

Davidson J, Robertson E. A follow-up study of postpartum illness, 1946–1978. *Acta Psychiatr Scand* 1985; **71**:451–7.

Dhasmana DJ, Brockington IF, Roberts A. Post-partum transverse sinus thrombosis presenting as acute psychosis. *Arch Womens Ment Health* 2010; **13**:365–7.

Fenton WS, McGlashan TH. Natural history of schizophrenia subtypes. I. Longitudinal study of paranoid, hebephrenic and undifferentiated schizophrenia. *Arch Gen Psychiatry* 1991; **48**:969–77.

Grahame PS. Schizophrenia in old age (late paraphrenia). *Br J Psychiatry* 1984; **145**:493–5.

Greil W, Ludwig-Mayerhofer W, Erazo N *et al.* Lithium vs carbamazepine in the maintenance treatment of schizoaffective disorder: a randomized study. *Eur Arch Psychiatry Clin Neurosci* 1997; **247**:42–50.

Hadley HG. A case of puerperal psychosis recovering from four attacks. *J Nerv Ment Dis* 1941; **94**:540–1.

Hayashi H, Oshino S, Ishikawa J *et al.* Paroxetine treatment of delusional disorder, somatic subtype. *Hum Psychopharmacol* 2004; **19**:351–2.

Hirayasu Y, McCarley RW, Salisbury DF *et al.* Planum temporale and Heschl gyrus volume reduction in schizophrenia: a magnetic resonance imaging study of first-episode patients. *Arch Gen Psychiatry* 2000; **57**:692–9.

Jager M, Bottlender R, Strauss A *et al.* Fifteen-year follow-up of ICD-10 schizoaffective disorders compared with schizophrenia and affective disorders. *Acta Psychiatr Scand* 2004; **109**:30–7.

Jones PB, Barnes TR, Davies L *et al.* Randomized controlled trial of the effect on quality of life of second- vs. first-generation antipsychotic drugs in schizophrenia: Cost Utility of the Latest Antipsychotic Drugs in Schizophrenia Study (CUTLASS 1). *Arch Gen Psychiatry* 2006; **63**:1079–87.

Kendell RE, Juszczak E, Cole SK. Obstetric complications and schizophrenia: a case–control study based on standardized obstetric records. *Br J Psychiatry* 1996; **168**:556–61.

Kendler KS. The nosological validity of paranoia (simple delusional disorder): a review. *Arch Gen Psychiatry* 1980; **37**:699–706.

Kendler KS. Kraepelin and the diagnostic concept of paranoia. *Compr Psychiatry* 1988; **29**:4–11.

Kendler KS, Gruenberg AM. An independent analysis of the Danish Adoption Study of Schizophrenia. IV. The relationship between psychiatric disorders as defined by DSM-III in the relatives and adoptees. *Arch Gen Psychiatry* 1984; **41**:555–64.

Kendler KS, Gruenberg AM, Strauss JS. An independent analysis of the Copenhagen sample of the Danish Adoption Study of Schizophrenia. III. The relationship between paranoid psychosis (delusional disorder) and the schizophrenia spectrum disorders. *Arch Gen Psychiatry* 1982; **38**:985–7.

Kendler KS, Masterson CC, Davis KL. Psychiatric illness in first-degree relatives of patients with paranoid psychosis, schizophrenia and medical illness. *Br J Psychiatry* 1985; **147**:524–31.

Kendler KS, McGuire M, Gruenberg AM *et al.* An epidemiologic, clinical and family study of simple schizophrenia in County Roscommon, Ireland. *Am J Psychiatry* 1994a; **151**:27–34.

Kendler KS, McGuire M, Gruenberg AM *et al.* Outcome and family study of the subtypes of schizophrenia in the West of Ireland. *Am J Psychiatry* 1994b; **151**:849–56.

Kendler KS, McGuire M, Gruenberg AM *et al.* Examining the validity of DSM-III-R schizoaffective disorder and its putative subtypes in the Roscommon family study. *Am J Psychiatry* 1995; **152**:755–64.

Kennedy HG, Kemp LI, Dyer DE. Fear and anger in delusional (paranoid) disorder: the association with violence. *Br J Psychiatry* 1992; **160**:488–92.

Kety SS. The significance of genetic factors in the etiology of schizophrenia: results from the national study of adoptees in Denmark. *J Psychiatr Res* 1987; **21**:423–9.

Khong SY, Leach J, Greenwood C. Meningioma mimicking puerperal psychosis. *Obstetr Gynecol* 2007; **109**:515–16.

King DJ, Burke M, Lucas RA. Antipsychotic drug-induced dysphoria. *Br J Psychiatry* 1995; **167**:480–2.

Kumar R, Issacs S, Meltzer E. Recurrent post-partum psychosis: a model for prospective clinical investigation. *Br J Psychiatry* 1983; **142**:618–20.

Kraepelin E. *Dementia Praecox and Paraphrenia*, Huntington, 1919. New York: Robert E. Krieger, 1971.

Lauren TM, Labouriau R, Licht RW *et al.* Family history of psychiatric illness as a risk factor for schizoaffective disorder: A Danish register-based cohort study. *Arch Gen Psychiatry* 2005; **62**:841–8.

Leucht S, Kornossa K, Rummel-Kluge C *et al.* A meta-analysis of head-to-head comparisons of second-generation antipsychotics in the treatment of schizophrenia. *Am J Psychiatry* 2009; **166**:152–63.

McGlashan TH. The Chestnut Lodge Follow-Up Study. II. Long-term outcome of schizophrenia and the affective disorders. *Arch Gen Psychiatry* 1984; **41**:586–601.

McGlashan TH, Carpenter WT. Postpsychotic depression in schizophrenia. *Arch Gen Psychiatry* 1976; **33**:231–9.

Mandel MR, Severe JB, Schooler NR *et al.* Development and prediction of postpsychotic depression in neuroleptic-treated schizophrenia. *Arch Gen Psychiatry* 1982; **39**:197–203.

Manschreck TC, Khan NL. Recent advances in the treatment of delusional disorder. *Can J Psychiatry* 2006; **51**:114–19.

Menon RR, Barta PE, Aylward EH *et al.* Posterior superior temporal gyrus in schizophrenia: grey matter changes and clinical correlates. *Schizophren Res* 1995; **16**:127–35.

Mitchell C. Successful treatment of chronic delusional parasittosis. *Br J Psychiatry* 1989; **155**:556–7.

Morrison JR. Catatonia: retarded and excited types. *Arch Gen Psychiatry* 1973; **28**:39–41.

Mueser KT, Bellack AS, Brady EU. Hallucinations in schizophrenia. *Acta Psychiatr Scand* 1990; **82**:26–9.

Munoz RA. Postpartum psychosis as a discrete entity. *J Clin Psychiatry* 1985; **46**:182–4.

Murray RM. Neurodevelopmental schizophrenia: the rediscovery of dementia praecox. *Br J Psychiatry* 1994; **165**(Suppl. 25):6–12.

Okuma T, Yamashita I, Takahashi R *et al.* A double-blind study of adjunctive carbamazepine versus placebo on excited states of schizophrenic and schizoaffective disorders. *Acta Psychiatr Scand* 1989; **80**:250–9.

Oosthuizen P, Russouw H, Roberts M. Is puerperal psychosis bipolar mood disorder? A phenomenological comparison. *Compr Psychiatry* 1995; **36**:77–81.

Opjordsmoen R. Long-term course and outcome in delusional disorder. *Acta Psychiatr Scand* 1988; **78**:576–86.

Opjordsmoen R, Rettersol N. Delusional disorder: the predictive validity of the concept. *Acta Psychiatr Scand* 1991; **84**:250–4.

Pope HG, Lipinski JF, Cohen BM *et al.* 'Schizoaffective disorder', an invalid diagnosis? A comparison of schizoaffective disorder, schizophrenia and affective disorder. *Am J Psychiatry* 1980; **137**:921–7.

Potkin SG, Saha AR, Kujawa WH *et al.* Aripiprazole, an antipsychotic with a novel mechanism of action, and risperidone vs placebo in patients with schizophrenia and schizoaffective disorder. *Arch Gen Psychiatry* 2003; **60**:681–90.

Reed P, Sermin N, Appleby L *et al.* A comparison of clinical response to electroconvulsive therapy in puerperal and non-puerperal psychoses. *J Affect Dis* 1999; **54**:255–60.

Rioux L, Nissanov J, Lauber K *et al.* Distribution of microtubule-associated protein MAP2-immunoreactive interstitial neurons in the parahippocampal white matter in subjects with schizophrenia. *Am J Psychiatry* 2003; **160**:149–55.

Roberts GW, Colter N, Lofthouse RM *et al.* Is there gliosis in schizophrenia? Investigation of the temporal lobe. *Biol Psychiatry* 1987; **22**:1459–68.

Schopf J, Rust B. Follow-up and family study of postpartum psychosis. Part III: Charactristics of psychoses occurring exclusively in relation to childbirth. *Eur Arch Psychiatry Clin Neurosci* 1994a; **244**:138–140.

Schopf J, Rust B. Follow-up and family study of postpartum psychosis. Part I: Overview. *Eur Arch Psychiatry Clin Neurosci* 1994b; **244**:101–11.

Shenton ME, Kikinis R, Jolesz FA *et al.* Abnormalities of the left temporal lobe and thought disorder in schizophrenia: a quantitative magnetic resonance imaging study. *N Engl J Med* 1992; **327**:604–12.

Simpson GM, Glick ID, Weiden PJ *et al.* Randomized, controlled, double-blind multicenter comparison of the efficacy and tolerability of zioprasidone and olanzapine in acutely ill inpatients with schizophrenia or schizoaffective disorder. *Am J Psychiatry* 2004; **161**:1837–47.

Stewart DE. Prophylactic lithium in postpartum affective psychosis. *J Nerv Ment Dis* 1988; **176**:485–9.

Stewart DE, Klompenhouwe JL, Kendell RE *et al.* Prophylactic lithium in puerperal psychosis. The experience of three centres. *Br J Psychiatry* 1991; **158**:393–7.

Stroup TS, Lieberman JA, McEvoy JP *et al.* Effectiveness of olanzapine, quetipaine, and risperidone in patients with chronic schizophrenia after discontinuing perphenazine: a CATIE study. *Am J Psychiatry* 2007; **164**:415–27.

Susser ES, Lin SP. Schizophrenia after prenatal exposure to the Dutch Hunger Winter of 1944–1945. *Arch Gen Psychiatry* 1992; **49**:983–8.

Tienari P, Wynne LC, Laksy K *et al.* Genetic boundaries of the schizophrenic spectrum: evidence from the Finnish adoptive family study. *Am J Psychiatry* 2003; **160**:1587–94.

Tsuang MT. Suicide in schizophrenics, manics, depressives, and surgical controls: a comparison with general population suicide mortality. *Arch Gen Psychiatry* 1978; **35**:153–5.

Tsuang MT, Woolson RF, Fleming JA. Long-term outcome of major psychoses. I. Schizophrenia and affective disorders compared with psychiatrically symptom-free surgical controls. *Arch Gen Psychiatry* 1979; **36**:1295–304.

Van Putten T. The many faces of akathisia. *Compr Psychiatry* 1975; **16**:43–7.

Van Putten T, May PRA. 'Akinetic depression' in schizophrenia. *Arch Gen Psychiatry* 1978; **35**:1101–7.

Van Putten T, Multalipassi LR, Malkin MD. Phenothiazine-induced decompensation. *Arch Gen Psychiatry* 1974; **30**:102–5.

Videbech T. Chronic olfactory paranoid syndromes. *Acta Psychiatr Scand* 1966; **42**:183–213.

Volavka J, Czobor P, Sheitman B *et al.* Clozapine, olanzapine, risperidone, and haloperidol in the treatment of patients with chronic schizophrenia and schizoaffective disorder. *Am J Psychiatry* 2002; **159**:255–162.

Wada T, Kawakatsu S, Nadaoka T *et al.* Clomipramine treatment of delusional disorder, somatic type. *Int Clin Psychopharmacol* 1999; **14**:181–3.

Walker E, Lewine RJ. Prediction of adult-onset schizophrenia from childhood home movies of the patients. *Am J Psychiatry* 1990; **147**:1052–6.

Watt JAG. The relationship of paranoid states to schizophrenia. *Am J Psychiatry* 1985; **142**:1456–8.

Winokur G. Delusional disorder (paranoia). *Compr Psychiatry* 1977; **18**:511–21.

Winokur G. Familial psychopathology in delusional disorder. *Compr Psychiatry* 1985; **26**:241–8.

Zisook S, Kasckow JW, Golshan S *et al.* Citalopram augmentation for subsyndromal symptoms of depression in middle-aged and older outpatients with schizophrenia and schizoaffective disorder: a randomized controlled trial. *J Clin Psychiatry* 2009; **70**:562–71.

23

Mood disorders

23.1 BIPOLAR DISORDER

Bipolar disorder is characterized by the occurrence of at least one manic or mixed-manic episode during the patient's lifetime; almost all patients will also have one or more depressive episodes at other times. In the intervals between these episodes, most patients return to their normal state of well-being. Thus, bipolar disorder is properly considered an 'episodic', 'periodic', or 'cyclic' illness, with patients 'cycling' up into a manic or mixed-manic episode, then returning to normal and cycling 'down' into a depressive episode, from which they likewise eventually more or less fully recover.

Bipolar disorder has a lifetime prevalence of roughly 2 percent, and is equally common in males and females.

A synonym for bipolar disorder is 'manic–depressive illness, circular type'; however, this terminology is gradually fading from use. In the past it was believed that patients with what is now termed bipolar disorder and patients with major depressive disorder actually suffered from the same illness, namely manic–depressive illness, which merely manifested in different forms. Those who had both manic episodes and depressive episodes during their lifetimes were considered to have the 'circular' form, whereas those who had only depressive episodes during their lifetimes were considered to have the 'depressive' form. Given that it is now clear that bipolar disorder and major depressive disorder are different diseases, it may be appropriate to leave the term 'manic–depressive illness' to history.

Clinical features

The onset of bipolar disorder is heralded by the appearance of a first episode of illness, which may be manic, depressive, or mixed-manic. In general, most patients have their first episode in their late teens or early twenties, and by the age of 50 years over 90 percent of patients will have had their first episode. The range of age at onset is, however, wide, from as young as 11 years (McHarg 1954) up to the eighth decade (Charron *et al.* 1991), even as late as the age of 79 years (Summers 1983). Each type of episode is described in turn.

MANIC EPISODE

Manic episodes are often preceded by a prodrome, lasting from a few days to a few months, of mild and often indistinct manic symptoms. At times, however, there may be little or no prodrome and the episode may appear quite abruptly; when this occurs, patients often unaccountably wake up in the middle of the night full of energy and vigor, thus manifesting the so-called 'manic alert'.

The overall symptomatology of mania has been well described (Abrams and Taylor 1981; Black and Nasrallah 1989; Bowman and Raymond 1931; Brockington *et al.* 1980; Carlson and Goodwin 1973; Kraepelin 1921; Lipkin *et al.* 1970; Rosenthal *et al.* 1979, 1980; Stevens 1904; Winokur 1984). As discussed in Chapter 7.3, mania may be divided into three stages: hypomania or stage I mania; acute mania or stage II mania; and delirious mania or stage III mania. All patients who enter a manic episode experience hypomania and most progress to acute mania; however, only a minority eventually reach delirious mania. By convention, patients whose manic episodes never pass beyond hypomania are said to have 'bipolar II disorder', whereas those who, during at least one episode of mania, pass beyond hypomania are said to have 'bipolar I disorder'.

The rapidity with which patients pass from hypomania through acute mania and on to delirious mania varies from a week to a few days to, rarely, hours; indeed, in hyperacute onsets, patients may already have passed through hypomania before being brought to medical attention. The duration of an entire manic episode varies from the extremes of only a few days up to many years, or even a decade (Wertham 1929). On average, however, most episodes of mania last several months. In general, once the peak of the episode is reached, symptoms gradually subside and, after remission finally occurs, many patients, looking back over what they did, often feel guilt and remorse.

Hypomania, or stage I mania, is characterized by the cardinal manic symptoms of heightened mood, increased energy and decreased need for sleep, pressure of speech and flight of ideas, and pressure of activity (Abrams and Taylor 1976a; Beigel and Murphy 1971a; Clayton *et al.* 1965; Loudon *et al.* 1977; Taylor and Abrams 1973; Winokur and Tsuang 1975; Winokur *et al.* 1969). The heightened mood may be one of either euphoria or irritability, or a mixture of the two, and is often quite labile. Euphoric patients are in great good cheer and wish to share their immense enjoyment with others; they are often full of jokes, puns, and wisecracks, and their humor is often irresistibly infectious to those around them. Indeed, it is the rare physician who can resist at least inwardly smiling when in the presence of a euphoric manic. Irritable manics, by contrast, are irascible, fault-finding, and accusatory, and when their intemperate demands are not immediately met, they may erupt into a tirade of curses and threats, and indeed may become violently assaultive. Increased energy leaves these patients strangers to fatigue and in little need of sleep. Pressured speech is rapid and voluble. Patients have much to say, their thoughts come rapidly and race pell-mell, and in extreme cases they cannot speak fast enough to express them. Although patients may, with great urging, be able momentarily to dam up their words, such respites, when an interviewer may be able to get a few words in, are but transient events before the dam bursts and the interviewer is again inundated with a torrent of words. Such pressured speech is also typically characterized by flight of ideas, in which patients' interests change abruptly from one subject to another, successive subjects having little in common with each other. Pressure of activity impels patients to be ever on the go and perpetually involved in schemes, plans, projects, and activities, activities in which they also often seek to involve others. Patients may also demonstrate distractibility, in which their attention changes mercurially from one subject to another. As might be expected, hypomanic patients often become involved in impetuous and ill-considered ventures: there may be spending sprees; intense, injudicious, and often sexual, relationships; and ruinous business ventures. Attempts to reason with such patients, and to bring them back to some good judgment, are typically in vain. Hypomanic patients rarely see anything wrong with themselves; indeed, they often opine that if only others saw as they did, and partook of their confidence, all would be well.

Acute, or stage II, mania is characterized by an intensification of all of the symptoms seen in hypomania and by the appearance of delusions. The mood becomes extraordinarily heightened and labile, and irritability may be quite pronounced, with unpredictable assaults and tirades. Energy seems boundless, and the pressure of activity and speech may begin to fragment the overall behavior; patients may shout, then cry, hop on the floor or race to the nurses' station, making one demand, then an opposite one, and be completely incapable of channelling themselves towards any one overall purpose. Delusions are typically either of grandeur or of persecution, according to the mood of the patient. Euphoric patients may announce their divinity or lavish listeners with promises to share their great wealth; irritable patients may accuse others of irrationally thwarting and persecuting them. Patients may also hear voices.

Delirious, or stage III, mania represents the height of mania and is characterized by a sometimes startling metamorphosis. The cardinal symptoms of mania may fade, and speech and behavior may become profoundly fragmented (Bond 1980). Loosening of associations may occur, and patients are often confused; some may become mute. Hallucinations and delusions abound and, in addition to delusions of grandeur or persecution, one may also see bizarre delusions, including Schneiderian first-rank symptoms (Jampala *et al.* 1989). Catatonic stupor may appear, with immobility, waxy flexibility, and bizarre posturing (Abrams and Taylor 1976b; Taylor and Abrams 1977).

DEPRESSIVE EPISODE

As discussed in Chapter 7.1, depression manifests with depressed or irritable mood, low self-esteem or guilt, pessimism, difficulty with concentration and forgetfulness, anhedonia, anergia, sleep and appetite disturbances, and psychomotor change. In depression seen in bipolar disorder there tends to be an increased need for sleep and increased appetite and psychomotor retardation, which may be profound (Beigel and Murphy 1971b; Hartmann 1968; Johnson 1984; Mitchell *et al.* 2001). Delusions and hallucinations may occur (Guze *et al.* 1975) and are 'mood congruent', in that they seem fitting when viewed in the context of the patient's overall mood. Patients may believe themselves to be the worst of sinners and that they are to be taken into imprisonment or to execution. Voices may accuse and condemn them, urging them on to suicide. Depressive episodes of bipolar disorder tend to come on subacutely, over several weeks (Casper *et al.* 1985; Winokur *et al.* 1993), and last 6 months or so; the range here, however, is wide, from as little as weeks up to years.

MIXED MANIC EPISODE

Mixed mania (Himmelhoch *et al.* 1976; Kotin and Goodwin 1972; McElroy *et al.* 1992) is a variant of mania in which there is a strong admixture of depressive symptoms. In some cases, manic and depressive symptoms may rapidly alternate, and in others they may exist simultaneously. Euphoric patients, singing and proclaiming their glory and beneficence, may suddenly be thrust into the profoundest of despair, weeping, bereft of all hope and energy, and intensely suicidal. Mixtures of manic and depressive symptoms may present a startling clinical picture: one patient strode through the ward, shouting unstoppably that he was the greatest of sinners and would die of unspeakable tortures; another, weeping uncontrollably with a look of utter despair, proclaimed to feel wonderful, at peace and transcendently happy.

Mixed manic episodes are relatively uncommon, being found in a little over 10 percent of cases, and tend to last longer than straight manic ones.

Course

As noted earlier, bipolar disorder is an episodic disorder and its course is characterized in most patients by the intermittent appearance, over the lifespan, of episodes of illness, in between which most patients experience 'euthymic' intervals during which they more or less return to their normal state of health. Both the duration of the euthymic intervals and the sequencing of episodes varies widely among patients.

The duration of the euthymic interval varies from as little as a few hours or days (Bunney *et al.* 1972; Sitaram *et al.* 1978) up to years, or even decades. In contrast to this interpatient variability, however, one may often find a remarkable intrapatient regularity, and indeed in some patients the euthymic intervals are so regular that it is possible to predict, even to the month, when the next episode will occur. Occasionally, one may also see a seasonal pattern, with manic episodes more likely in the spring or early summer, and depressive ones in the autumn or winter. Manic episodes in patients with bipolar disorder may also become 'entrained' to certain biologic events, such as the puerperium (Bratfos and Haug 1966; Viguera *et al.* 2000) or the premenstruum (D'Mello *et al.* 1993).

The total number of episodes experienced by a patient depends, of course, not only on the duration of the euthymic interval but also on the duration of the episodes themselves. On one extreme, the euthymic interval may be so long that patients have only a couple of episodes in their lifetimes; indeed, in the natural course of events, some patients may die before they were 'scheduled' for another episode, and thus they end up having only one episode in their entire life. On the other extreme, patients with very brief episodes and brief intervals may have literally hundreds of episodes per year (Bunney and Hartmann 1965; Jenner *et al.* 1967); such an extremely high frequency of episodes, however, is very rare. Patients with more than four episodes per year are termed 'rapid cyclers' (Bauer *et al.* 1994; Dunner *et al.* 1977). Interestingly, it appears that, in some instances, rapid cycling is associated with subclinical hypothyroidism (Bauer *et al.* 1990; Cowdry *et al.* 1983) (as may be induced by treatment with lithium [Terao 1993]).

The sequence of episodes is also quite variable. It is rare to find patients whose courses are characterized by regularly alternating manic and depressive episodes. Most patients experience either a preponderance of manic or a preponderance of depressive episodes throughout their lives. Thus, to look at two extremes, whereas one patient may have six episodes of depression and only one of mania throughout his life, another might have a dozen episodes of mania but only one of depression. Indeed, albeit very rarely, one may encounter 'unipolar' manic patients who have only manic episodes throughout their lives, never experiencing a depressive one (Pfohl *et al.* 1982; Shulman and Tohen 1994).

Bipolar disorder may present with either a depressive or a manic episode. Importantly, in cases in which the first episode is depressive, it appears that, in over 90 percent of cases, a manic episode will ensue within either 10 years or a total of five episodes of depression, whichever comes first (Dunner *et al.* 1976).

As noted earlier, during the intervals between episodes, most patients are euthymic and free of mood symptoms. In about one-quarter of cases, however, the intervals may be 'colored' by very mild mood symptoms, and the 'direction' or 'polarity' of this coloring correlates with the preponderance of episodes. Thus, a patient who tends to have very mild, 'sub-hypomanic' symptoms during the interval is more likely to have manic than depressive episodes, whereas a patient whose intervals are clouded by minor depressive symptoms is more likely to have depressive than manic episodes.

Occasionally, one may find cases in which certain events, pharmacologic or otherwise, may more or less reliably precipitate a manic episode. These include serotoninergic agents, such as tryptophan or 5-hydroxytryptophan; noradrenergic or dopaminergic agents, such as cocaine, amphetamine, methylphenidate, various sympathomimetics, levodopa, or direct-acting dopaminergics such as bromocriptine; alcohol or sedative–hypnotic withdrawal, or sudden discontinuation of clonidine; and treatment with steroids, such as prednisone. Most notably, as discussed further below, antidepressants may also precipitate mania. Other agents capable of inducing mania include herbal preparations, such as St John's wort (Moses and Mallinger 2000), and phototherapy.

Suicide occurs in 10–20 percent of patients with bipolar disorder, and although most suicides tend to occur during a depressive episode, patients in the uncommon mixed-manic episodes are actually at highest risk.

Etiology

Bipolar disorder has a definite genetic component. The incidence of bipolar disorder is roughly fivefold higher among the first-degree relatives of probands than among the general population, and the concordance rate among dizygotic twins is in the range of 20–30 percent, while in monozygotic twins it rises to 60–80 percent. Furthermore, adoption studies have demonstrated that the prevalence of bipolar disorder is several times higher among the biologic parents of patients with bipolar disorder than among their adoptive parents (Mendlewicz and Rainer 1977). Although genetic studies have offered tantalizing clues, replication of positive findings has been difficult. In all likelihood, multiple genes on multiple different chromosomes are involved, each conferring a susceptibility to the disease.

Endocrinologic changes strongly suggest hypothalamic disturbances: the dexamethasone suppression test (DST) is generally positive (Rush *et al.* 1997; Watson *et al.* 2004), and the thyroid-stimulating hormone (TSH) response to thyrotropin-releasing hormone (TRH) stimulation is blunted (Extein *et al.* 1980). It also appears that there may be a disturbance of cholinergic transmission in bipolar disorder: the infusion of physostigmine reliably precipitates depression in patients who are currently manic (Janowsky *et al.* 1973), and the latency to rapid eye movement (REM) sleep upon infusion of arecoline is shortened in bipolar patients compared with control subjects.

Although neuropathologic studies have indicated glial and neuronal loss in the dorsolateral prefrontal cortex and the subgenual cingulated cortex, the endocrinologic changes just noted strongly suggest hypothalamic involvement, and indeed within the hypothalamus overall neuronal loss has been noted in the paraventricular nucleus (Manaye *et al.* 2005), with, however, an increased number of corticotropin-releasing hormone (CRH)-containing cells (Bao *et al.*

2005). Furthermore, within the brainstem, disturbances have been noted in the locus ceruleus and the dorsal raphe nucleus (Baumann and Bogerts 2001). Although speculative, taken together these findings are consistent with the notion that bipolar disorder represents an inherited disturbance of the structure or function of hypothalamic and brainstem nuclei.

Differential diagnosis

In considering a diagnosis of bipolar disorder, the first step is to ensure that the patient either has had a manic episode or is in the midst of one. As noted in Chapter 7.3, hypomania is a distinctive syndrome and is difficult to confuse with anything else. Difficulties arise, however, when one either lacks this history or happens to see the patient when the stage of hypomania has already been passed and the patient is now in acute mania or delirious mania. Acute mania may be confused with the syndrome of psychosis, given the presence of delusions and hallucinations; however, in acute mania the cardinal manic symptoms (e.g. heightened mood, increased energy, etc.) are still prominent and offer a clue. In delirious mania, however, these cardinal symptoms, as noted above, may fade from the picture and, at this point, in addition to a syndromal diagnosis of psychosis, one may also entertain syndromal diagnoses of catatonia or delirium. It must be emphasized that the easiest and best way to make a correct syndromal diagnosis of mania is to obtain an accurate history. This may be laborious at times given that patients in acute mania or delirious mania are generally unable to provide a reliable history, and consequently one may have to contact friends, family members, or co-workers; the best diagnostic strategy is to establish a typical clinical evolution of symptoms, from normalcy to hypomania and then on to acute and perhaps delirious mania.

Once the syndromal diagnosis of mania is established, the next step is to determine the cause of the mania. Although bipolar disorder is by far the most common cause of mania, multiple other etiologies, as discussed in Chapter 7.3, are possible. Of the disorders discussed there, the idiopathic ones, namely cyclothymia, schizoaffective disorder, and postpartum psychosis, figure most prominently on the differential. Cyclothymia is, in all likelihood, merely a forme fruste of bipolar disorder (Akiskal *et al.* 1977), and patients experience recurrent episodes of very mild hypomania and correspondingly very mild depression. Schizoaffective disorder is immediately distinguished by the course. In schizoaffective disorder the intervals between episodes of mania or depression are marked by psychotic symptoms, such as delusions, hallucinations, and loosening of associations, and this is in stark contrast to the intervals of bipolar disorder, in which such psychotic symptoms are *never* seen. Postpartum psychosis is characterized by a psychosis, often with prominent manic symptoms, occurring in the postpartum period, and is distinguished from bipolar disorder by its course. In postpartum psychosis, symptoms occur only in the postpartum period, whereas in bipolar disorder, although episodes of mania may occur postpartum, they are also seen at other times in the patient's life. Other disorders noted in Chapter 7.3, although less common, must also be considered, and the reader is encouraged to review that chapter.

Treatment

The treatment of bipolar disorder involves acute and continuation treatment for manic and mixed-manic episodes and for depressive episodes, and prophylactic treatment to prevent them, and each is discussed in turn.

MANIC AND MIXED-MANIC EPISODES

Acute treatment of a manic or mixed-manic episode almost always involves the administration of one of the mood stabilizers (lithium, divalproex, or carbamazepine; lamotrigine has not been shown to be effective in the acute treatment of mania). Although lithium has by far the longest track record, divalproex has certain advantages: first, it appears to be more effective than lithium for mixed mania (Swann *et al.* 1997); second, and more importantly, when divalproex is rapidly loaded (Hirschfeld *et al.* 2003) results are often seen in a matter of days, in contrast to the 'lag' period of 1–2 weeks between starting lithium and seeing substantial results. Carbamazepine is less well studied than either lithium or divalproex, and though it may be roughly equivalent to lithium (Ceron-Litvac *et al.* 2009), it appears less effective than divalproex (Vasudev *et al.* 2000). Although, based on these data, divalproex appears to be the first choice, consideration should also be given to the patient's treatment history: for example, if the history indicated an overall excellent response to lithium, both for acute mania and for prophylaxis, then of course a trial of lithium in the current case would be most appropriate. Although some patients may respond to monotherapy with lithium or divalproex alone, a substantial minority will not, and in such cases combination therapy with one of these mood stabilizers plus a second-generation antipsychotic will often turn the tide, as has been demonstrated with the addition of quetiapine (Sachs *et al.* 2004), risperidone (Sachs *et al.* 2002), aripirazole (Vieta *et al.* 2008a), and olanzapine (Houston *et al.* 2009). In practice, however, given the often pressing need to bring mania under control, most clinicians will either rapidly add one of these antipsychotics or simply begin treatment with the combination. It is also possible to treat mania utilizing monotherapy with one of the antipsychotics (as has been shown for quetiapine [Bowden *et al.* 2005], risperidone [Khanna *et al.* 2005], aripiprazole [Vieta *et al.* 2005], olanzapine [Perlis *et al.* 2006], and ziprasidone [Keck *et al.* 2003]); however, this approach should be discouraged, and for two reasons: first, as just noted, the combination of a mood stabilizer and an antipsychotic is more effective; second, as noted below, almost all patients should eventually be transitioned to prophylactic treatment, and as this almost inevitably involves using one of the mood stabilizers, it makes sense to have one of these in place.

Most manic patients require admission to a locked unit. Stimulation, including visitors, phone calls, and mail, should be kept to an absolute minimum, as they routinely exacerbate symptoms. In some cases seclusion is required, and certain patients, still possessed of some insight, may demand seclusion, as they know that the reduced stimulation of the seclusion room will allow for some reduction in their symptoms.

Continuation treatment is designed to prevent the recurrence of symptoms once they have been brought under control during the acute phase of treatment. Generally, this is accomplished by continuing the regimen that was effective during the acute phase, and doing so for the anticipated duration of the manic episode. In some cases, however, if a combination of a mood stabilizer plus an antipsychotic were utilized it may be possible to discontinue the antipsychotic and maintain the patient on the mood stabilizer alone; should symptoms recur, the antipsychotic may simply be restarted. As noted earlier, most manic episodes last on average from several months; however, given the wide variability here, the best guide is the individual patient's past history. Of note, if the patient declines prophylactic treatment and wishes to stop all medications, it is critical to hold off on this until it is reasonably certain that the manic episode has undergone a spontaneous remission. This is sometimes difficult, especially if one does not have a reliable history regarding the length of earlier episodes. Certainly, if the patient is having *any* manic symptoms, no matter how mild, treatment should be continued, and in general it is best to hold off on discontinuing treatment until the patient's personal life is fairly stable. Furthermore, if the patient had been on lithium, the dose should be tapered gradually over a few weeks, as it appears that abrupt discontinuation of lithium may predispose to a recurrence of mania (Baldessarini *et al.* 1996, 1997). Although this effect has not been demonstrated for divalproex or carbamazepine, prudence dictates following a similar strategy.

DEPRESSIVE EPISODES

In some cases, depression may respond to monotherapy with lithium (Goodwin *et al.* 1972), divalproex (Ghaemi *et al.* 2007), or carbamazepine (Post *et al.* 1986). There is also some evidence that lamotrigine may likewise be effective (Geddes *et al.* 2009), but this has been disputed (Calabrese *et al.* 2008). Often, however, the response is not robust, and alternatives include quetiapine (Calabrese *et al.* 2005; McElroy *et al.* 2010; Young *et al.* 2010), the combination of olanzapine plus fluoxetine (Brown *et al.* 2006) or, traditionally, the combination of one of the mood stabilizers (lithium, divalproex, or carbamazepine) plus an antidepressant. This last option, namely the combination of a mood stabilizer with an antidepressant, deserves further comment. To begin with, although some studies (Gijsman *et al.* 2004) support the effectiveness of such a combination, a more recent study (Sachs *et al.* 2007) casts doubt on it. Second, there is a risk that an antidepressant, even when administered concurrently with a mood stabilizer, may precipitate a manic episode (Leverich *et al.* 2006), and this risk seems particularly high with venlafaxine (Leverich *et al.* 2006) and the tricyclics (Gijsman *et al.* 2004). This risk seems particularly high in cases where the depression is 'contaminated' with any manic symptoms, no matter how mild (Frye *et al.* 2009). Thus, before adding an antidepressant, it is critical to closely examine the patient for such easily missed symptoms as distractability or racing thoughts, and to always remember that even a hint of mania *is* mania. With all these data in mind, a reasonable course would be to start with lithium or divalproex. Although the response to these agents is often incomplete, the fact that most patients will be treated prophylactically with one of these makes their utilization at this point logical. If the response is incomplete, then consideration could be given to adding quetiapine or the combination of olanzapine plus fluoxetine: as between these two add-ons, quetiapine is probably preferable, given the higher side-effect burden associated with olanzapine. A final option would be to add an antidepressant, and given the higher risk of a switch into mania with venlafaxine or a tricyclic, a selective serotonin reuptake inhibitor (SSRI) or bupropion would probably be an appropriate first choice. Should patients fail to respond to one of these combination strategies, one should, before trying another combination, first check thyroid status, as even 'chemical' hypothyroidism, manifest only with an elevated TSH, will blunt the antidepressant response (Cole *et al.* 2002). Should all strategies fail, or should the patient be *in extremis*, electroconvulsive therapy (ECT) should be considered.

Once depressive symptoms have been relieved, treatment should be continued to prevent a reappearance of symptoms. If the depression has responded to a mood stabilizer alone, one may then simply continue it. If quetiapine had been used, then one could either discontinue it, or, if it were being used as part of a prophylactic regimen (discussed further below), it could be continued. If an antidepressant has been utilized, however, the ongoing risk of an antidepressant-precipitated mania will dictate an attempt at some point to discontinue the antidepressant and see if the patient can be maintained on the mood stabilizer alone. Deciding when to discontinue an antidepressant is not straightforward. Certainly, if a reliable history indicates that the patient's depressive episodes routinely last for a more or less definite and predictable length of time, one could discontinue the antidepressant if that time has elapsed. Lacking this guidance, however, one may wish to wait until the patient has been euthymic for a significant period of time, at least weeks, before attempting a discontinuation; should symptoms recur, then one could simply restart the antidepressant. If a combination of olanzapine and fluoxetine had been used, although it is not clear whether the same reasoning applies, and whether an attempt should be made to discontinue the fluoxetine, prudence suggests such a course.

PROPHYLACTIC TREATMENT

Prophylactic treatment should be seriously considered in every case. Many patients may balk at the idea of indefinite treatment, and in some cases, they may have a point. Thus, if prior episodes have been mild, and widely spaced, occurring 5 or more years apart, then the burden of ongoing treatment, especially if side-effects are prominent, may outweigh the benefits of prevention. On the other hand, if prior episodes had been severe, or closely spaced, with symptom-free intervals of less than 2 years, then the benefits of treatment in all likelihood outweigh any other considerations. Some patients, though, may continue to demur, and promise to contact the physician as soon as an episode starts. Although such a strategy may work for depressions, it may fail miserably when it comes to manic episodes: one of the first things to go out the window when mania makes its appearance is good judgment.

If prophylaxis is decided on, then once the continuation phase of treatment of either mania or depression has been successfully completed, many patients may be maintained on monotherapy with a mood stabilizer alone, and in this regard lithium, though more effective for the prevention of mania than depression, has the longest track record (Geddes *et al.* 2004) of all the mood stabilizers for the prevention of both mania and depression, and constitutes the first choice. Either divalproex (Bowden *et al.* 2000) or carbamazepine (Ceron-Litvac *et al.* 2009), of course, may also be considered, and if there is a history of a good prophylactic effect with either of these, then it would constitute a reasonable choice. Some clinicians will also consider utilizing lamotrigine (Bowden *et al.* 2003) in cases characterized by frequent depressive episodes; however, it must always be kept in mind that lamotrigine is not effective in the prevention of mania, and hence cannot be used as an overall monotherapy agent. Should 'breakthrough' symptoms of mania occur it is critical to determine thyroid status, as even 'chemical' hypothyroidism, with a normal free T4 but an elevated TSH, may blunt the effectiveness of a mood stabilizer (Cowdry *et al.* 1983). Should breakthrough mania or breakthrough depression occur despite normal good compliance, consideration may then be given to combination therapy with lithium and divalproex together (Solomon *et al.* 1997) or to combination therapy with either lithium or divalproex plus one of the antipsychotics, quetiapine (Suppes *et al.* 2009; Vieta *et al.* 2008a), risperidone (Macfadden *et al.* 2009), aripiprazole (Vieta *et al.* 2008b), olanzapine (Tohen *et al.* 2004), or ziprasidone (Bowden *et al.* 2010).

Before leaving this discussion of prophylaxis, some comments are in order regarding the potential emerging role of antipsychotics as monotherapy agents for prophylaxis. Recent work has demonstrated that both aripiprazole (Keck *et al.* 2007) and olanzapine (Tohen *et al.* 2005) are effective in this regard, and hence there may be a temptation to utilize one of these in prophylaxis rather than the traditional scheme just described. At present, however, it may be more prudent to stick with the traditional scheme, which has decades of experience behind it, and await on the results of further research before switching to monotherapy with an antipsychotic.

23.2 CYCLOTHYMIA

Cyclothymia is a chronic mood disorder characterized by the sequential appearance of periods characterized by either very mild hypomanic symptoms or equally mild depressive ones. Lifetime prevalence figures vary widely among studies: a conservative estimate would place the figure at 0.5–1 percent; it is probably equally common among males and females.

Clinical features

The onset is usually quite insidious, occurring in adolescence or early adult years.

These mild manic and depressive periods typically alternate in an irregular fashion, and last for days to weeks. Although in most cases these periods are separated one from another by a euthymic interval, lasting from hours up to 2 months, in a minority there may be no euthymic interval, and patients may undergo continuous cycling.

During manic periods, patients may be euphoric (or merely 'over-happy') and enthusiastic, or at times irritable. Energetic, talkative and overactive, these patients tend to become over-involved with projects or with people. In gifted individuals this may result in heightened productivity; however, in those with average talents or those with predominantly irritable mood, conflicts with others or authority abound.

During depressed periods, patients are moody, irritable, and unduly sensitive to criticism; they may be withdrawn and taciturn. Fatigue and apathy hamper their efforts, and modest changes in appetite and sleep may be seen, with a tendency towards hypersomnia.

Course

The course is chronic, extending for at least 2 years, and in some may be lifelong. Over long-term follow-up, a third or more of patients will develop either a full manic or a full depressive episode (Akiskal *et al.* 1977).

Etiology

Cyclothymia runs in the same families as does bipolar disorder (Vazquez *et al.* 2008), and probably represents a mild form of this.

Differential diagnosis

The full differential for manic or hypomanic symptoms is discussed in Chapter 7.3. Of the disorders mentioned there, special attention should be paid to the differential between cyclothymia and bipolar disorder. In cases where cycling has been present for less than 2 years, the diagnosis of cyclothymia should be considered provisional: should a full manic episode, or a full depressive episode occur before 2 years have passed, then the correct diagnosis is bipolar disorder (either bipolar I or bipolar II, respectively). In cases, however, where the cycling has been present for 2 or more years and either a full manic or a full depressive episode occur, then, by convention, one makes two diagnoses: namely cycylothymia and bipolar disorder.

Borderline personality disorder may also enter the differential. The affective instability experienced by patients with cyclothymia may take a toll on personal relations, and their lives may at times be quite stormy, a feature which will suggest borderline personality disorder, especially if the prevailing mood is one of irritability rather than euphoria. Some features may enable a correct differential. First, sustained euphoria, as may occur in cyclothymia, is not seen in borderline personality disorder. Second, whereas in cyclothymia the periods of hypomania and depression occur autonomously, and independently of any life events, in borderline personality disorder mood changes are often exquisitely tied to changes in personal relations, especially when rejection or abandonment occur, or are merely perceived as likely.

Treatment

Although intuitively it makes sense to treat cyclothymia with the same agents as are used in bipolar disorder, there have been no placebo-controlled studies to test this intuition. Open studies suggest the usefulness of lithium (Peselow *et al.* 1982), valproic acid (Jacobsen 1993), and quetiapine (Bisol and Lara 2010). Importantly, antidepressants should not be used unless the patients are first 'covered' with a mood stabilizer or an antipsychotic, given the risk of precipitating a full manic episode (Akiskal *et al.* 1977).

23.3 MAJOR DEPRESSIVE DISORDER

Major depressive disorder, often referred to simply as major depression, is characterized by the occurrence of one or more depressive episodes during the patient's lifetime. Synonyms for this disorder include unipolar affective disorder, melancholia, and manic–depressive illness, depressed type. 'Unipolar' highlights the critical difference between major depressive disorder and bipolar disorder, namely the fact that the patient with major depressive disorder has episodes that go towards only one 'pole', namely the depressive one, whereas the patient with bipolar disorder, in addition to depressive episodes, will also have episodes that go to the other, or manic, pole. Melancholia is the most ancient term for this disorder, and comes from the Greek word meaning 'black bile', reflecting the Greek's humoral theory of this disorder. Manic–depressive illness, depressed type, the final synonym for this disorder, is problematic for two reasons. First, as originally conceived by Kraepelin, manic–depressive illness was a disorder that subsumed what today are recognized as two disorders, namely major depressive disorder and bipolar disorder. Kraepelin felt that these two disorders were merely variants or subtypes of one overarching disorder, but recognizing the clinical differences of these variants he spoke of a 'depressive' type and a 'circular' type. Given the separateness of major depressive disorder and bipolar disorder, however, it may no longer be appropriate to use the overarching term. The second reason has to do with common usage. In the USA, when clinicians use the term 'manic–depressive illness', most are referring to bipolar disorder, and hence to use it to refer to major depressive disorder may lead to confusion.

Major depressive disorder is a very common disorder, with a lifetime prevalence of roughly 15 percent; it is twice as common in females as males.

Clinical features

Major depression is characterized by the occurrence of episodes of depression, in between which patients return to more or less normal functioning. Although the first episode of depression generally appears in the mid-twenties, the range in age at onset is wide, from childhood to the ninth decade.

Depressive episodes tend to appear gradually, even insidiously. Typically, there is a long prodrome, often lasting months, characterized by fleeting and often indefinite symptoms such as moodiness, anxiety, or fatigue. Furthermore, when the depressive episode finally does settle in, the various symptoms often appear haltingly and with differing severities, and it is consequently rare that a patient can date the onset with any sort of precision. This is not to say, however, that acute and obvious onsets are not seen. They do occur, and some patients may describe a rapid fall from emotional well-being into a full depressive episode in as little as a week or two. Such acute onsets, however, are the exception rather than the rule. The duration of the depressive episode varies widely: in perhaps one-quarter of cases spontaneous remissions occur in a matter of months, while in roughly one-half of cases, anywhere from 0.5 to 1 year must pass before remission spontaneously occurs. In the remaining one-quarter of cases, the depressive episode proves to be chronic, lasting years, or, in exceptional cases, decades.

At times depressive episodes may appear to be precipitated by a stressful life event, typically a serious loss, such as the death of a loved one, divorce, or the loss of a job. At times, however, close enquiry will reveal that the stressful life event, rather than actually precipitating the depressive episode, was itself caused by the depressive episode. For example, a married person, in the midst of a long prodrome, may be sufficiently irritable to cause the spouse to leave. In such cases, although the patient may blame the depression on the separation, in fact it was the depressive prodrome that caused the separation. This is not to say, of course, that independent precipitating events do not at times trigger depressive episodes, for they do. However, it appears that in the group of patients with precipitated depressive episodes, subsequent episodes tend to become independent and to occur autonomously without any precipitating events (Brown *et al.* 1994; Frank *et al.* 1994).

As discussed in Chapter 7.1, depressive episodes are characterized by a number of symptoms, including depressed or irritable mood; low self-esteem, guilt or pessimism; suicidal ideation; difficulty with concentration or forgetfulness; anhedonia; anergia; sleep disturbance; appetite disturbance; and psychomotor change. Patients may also describe an overall diurnal variation in the severity of their depression, with symptoms being more intense in the morning. Delusions and hallucinations may also occur, as may certain other symptoms, such as anxiety attacks.

Mood is typically depressed but may be primarily irritable; some patients may also complain of anxiety. In some cases, however, although possessed of a depressed affect, patients may deny feeling depressed but may rather speak of a sense of discouragement, lassitude, or a feeling of being weighed down. Occasionally one may see patients who 'put on' a 'happy face', and feign a normal affect despite experiencing a depressed mood. Such a 'smiling depression' may mislead the diagnostician who fails to specifically enquire after the patient's mood.

Self-esteem typically sinks and the workings of conscience become prominent. Patients may consider themselves worthless and as having never done anything of value; in looking over their past, they see their sins multiplied. Indeed, in reviewing the past, patients seem 'blind' to any accomplishments and fix only on their misdeeds and shortcomings, which, as they recall them, may become magnified to heinous proportions. Some patients may give way to rumination, in which their failings and defects repeat themselves again and again in a litany of hopelessness. Pessimism is common and patients see no hope for the future; to them there are no prospects and all appears bleak.

Suicidal ideation is almost always present. At times this may be merely passive and patients may wish aloud that they might die of some disease or accident. Conversely, it may be active, and patients may consider hanging or shooting themselves, jumping from bridges, or overdosing on their medications. Often, and seemingly paradoxically, the risk of suicide is greatest as patients begin to recover. Still seeing themselves as worthless and hopeless sinners, these patients, now with some relief from fatigue, may find themselves with enough energy to carry out their suicidal plans. The overall suicide rate in major depressive disorder is about 4 percent; among those with depressive episodes severe enough to prompt hospitalization, however, the rate rises to about 9 percent.

Difficulty with concentration (Roy-Byrne *et al.* 1986) may be particularly troubling. Patients may complain of a dull heavy-headedness, as if they are 'in a fog'; attempts to read may be especially painful as patients read and reread the same paragraph but never find themselves understanding what they read. Short-term memory may become difficult, and patients may be unable to recall where they put their keys or what was said just minutes before. Making decisions, even simple ones, may become an almost insuperable task; everything appears too complicated, with too many possibilities and choices. In some cases of severe depression in the elderly, cognitive decrements may be severe enough to constitute a dementia (Rabins *et al.* 1984), which may be particularly severe. One 76-year-old became disoriented to time, was unable to recall any items after 3 minutes, and presented cachectic and curled up in a fetal position (McAllister and Price 1982), and a 66-year-old was disoriented to time and place, confused, and incontinent (Kramer 1982); both recovered with adequate antidepressant treatment.

Anhedonia manifests with a lack of interest in formerly pleasurable activities; sports, hobbies, and the like no longer arouse patients, and, if they force themselves to partake, they take no pleasure in such activities and merely go through the motions. Libido is routinely lost and there is no pleasure in sexual activity.

Anergia manifests with a dearth of energy, and patients may complain of feeling tired, fatigued, lifeless, or drained. Occasionally, patients may complain of having too much energy; however, on closer questioning one finds rather that patients have a 'nervous energy', more akin to agitation: such 'increased energy' is useless to the patient and is always overshadowed by a sense of imminent and impending exhaustion.

Sleep disturbance tends towards insomnia, which may be a torment to the patient. Although many complain of what is technically known as 'initial' insomnia, or trouble falling asleep, the most characteristic kind of insomnia in depression comes later in the night as either 'middle' or 'terminal' insomnia. In middle insomnia the patient awakens in the middle of the night for no particular reason and then has great difficulty falling back to sleep, often lying awake for an hour or more before sleep finally comes. Terminal insomnia, also known as 'early morning awakening', comes later in the night, and, here, once awake, the patient cannot fall back to sleep at all. As they lie awake, many patients experience ruminations or restless, unproductive thoughts. When the morning finally does come, patients find themselves unrefreshed and exhausted, as if they had not slept at all. Rarely, rather than experiencing insomnia, patients with major depressive disorder may complain of hypersomnia, wherein they sleep excessively, sometimes for up to 18 hours.

Appetite is routinely lost and many patients lose weight, sometimes in substantial amounts (Stunkard *et al.* 1990). Food may lose its taste or become unpalatable, and some patients may complain that food tastes like cardboard or leaves them nauseated. Uncommonly, patients may experience an increased appetite and may gain weight.

Psychomotor change tends toward agitation. When slight, this agitation may be experienced as a mere inner restlessness. When more severe, however, there may be hand-wringing and incessant pacing: patients may complain that they cannot keep still; they may lament their fate out loud and give way to wailing and repetitive pleas for help. The tension of these patients may be extreme, almost palpable, and yet, despite their piteous pleas for help, they cannot be comforted, no matter what is done for them. Psychomotor retardation, although uncommon, may also occur. In this, thoughts come slowly and sluggishly, if at all, and speech, when it occurs, is slow and halting. Some patients may sit immobile for hours, and if asked to do something may complain that they cannot and that it is too difficult. In severe cases patients may not move for any reason; they may go without bathing or changing their clothes, and some may defecate or urinate in the chair or on the bed.

Delusions and hallucinations may occur in about 15 percent of the depressive episodes of major depressive disorder, most commonly in cases in which the depressive symptoms are quite severe (Lykouras *et al.* 1986; Maj *et al.* 1990). Delusions are mood congruent in that, in some extreme, fantastic way, they are appropriate to the patients' moods and views of themselves. Guilt may be extreme and patients may confess to unspeakable sins: they have poisoned their children; family or friends are imprisoned for crimes that they, the patients, have committed. They may believe that they are condemned to Hell, that they have only hours to live. Delusions of persecution are common and have a peculiar twist to them. In contrast to patients with schizophrenia, who have delusions of persecution and protest that they are innocent victims, depressive patients with delusions of persecution typically feel that they deserve their persecutions for their miserable sins and shortcomings. Delusions of poverty and nihilistic delusions may also occur and are entirely consistent with these patients' views of themselves as worthless and hopeless. They may believe that they are without any funds and are completely bankrupt and unable to pay any bills, and that their families will go destitute. Those with nihilistic delusions may believe that they are near death: their insides have turned to dust or concrete; their brains have shrivelled up; the heart has dried up for lack of blood. In extreme cases patients may believe that they are dead, and some may believe that all are dead, and that death has finally achieved complete dominion over the world. Auditory hallucinations may also occur and generally reflect the patients' delusions. Voices may accuse them of crimes or sins, or announce that their well-deserved punishment is at hand. Visual hallucinations occasionally occur, and patients may see corpses or accusatory spirits. In some cases patients may develop stuporous catatonia (Starkstein *et al.* 1996).

Other symptoms seen during a minority of depressive episodes include anxiety attacks (Van Valkenburg *et al.* 1984) and what have been called 'anger attacks', in which generally hostile and irritable patients occasionally experience episodes of violence and autonomic arousal (Fava *et al.* 1993). Obsessions and compulsions may also appear: a particularly common obsession is the 'horrific temptation' to use a knife or gun to kill a loved one, and a consequent overwhelming necessity to rid the house of all potential weapons and to avoid any contact with them outside the house.

Course

Major depressive disorder is a relapsing and remitting illness (Thase 1990), characterized in most patients by the recurrence of depressive episodes throughout their lives, in between which they return to a more or less normal mood. The duration of the interval between successive episodes ranges widely, from as little as 1 year to up to decades, with an overall average of about 5 years. In general, however, with repeated episodes the interval between episodes tends to shorten, and the episodes themselves tend to lengthen, such that, over a very long period of time, successive episodes may eventually 'merge' to create a chronic, non-remitting condition.

Recently, much attention has been focused on patients whose depressive episodes seem entrained to the changing seasons. In patients with this seasonal pattern of illness, depressive episodes appear to occur far more commonly in the autumn or winter than in the spring or summer. Of note, in such cases of 'seasonal affective disorder' where the depression is of mild to moderate severity, although the pharmacologic treatments discussed below are effective, one may also consider phototherapy with bright morning light (Lam *et al.* 2006).

Etiology

Hereditary factors clearly play a role: amongst first-degree relatives, the risk of major depressive disorder is increased roughly threefold, and the approximate concordance rate amongst dizygotic twins is 20 percent, while among monozygotic twins it rises to 30 percent. To date, however, genetic studies have not identified genes or loci that may be confidently associated with this illness, indicating in all likelihood that, from a genetic point of view, this is a complex disorder, involving multiple genes and multiple modes of inheritance.

Hereditary factors, although clearly important, do not appear to provide a complete account, and environmental factors also seem to play a role. Among the various environmental events proposed, it appears that early childhood loss may be the most important. Events may also serve as precipitants for episodes in adult life; however, as noted earlier, the importance of precipitants fades with successive episodes to the point where, over long periods of time, episodes become, as it were, autonomous.

There is abundant evidence for endocrinologic changes in depression, all of which point to disturbances in the hypothalamus. Within the hypothalamus, for example, the level of messenger RNA for CRH in the paraventricular nucleus is elevated (Raadsheer *et al.* 1995), as is the cerebrospinal fluid (CSF) level of CRH (Nemeroff *et al.* 1984); consistent with this, the low-dose DST test is generally positive (Carroll *et al.* 1968, 1976). The thyroid axis also shows disturbances: the CSF level of TRH is elevated (Banki *et al.* 1988), and, consistent with this, the response of TSH to exogenous TRH is blunted (Prange *et al.* 1972).

Abnormalities in brainstem structures responsible for REM sleep are suggested by the fact that REM sleep latency is reduced in depression (Hauri *et al.* 1974; Rush *et al.* 1986), and by the response to cholinergic agents. Normally, REM sleep may be induced by cholinergic stimulation, and in patients with major depressive disorder the latency to REM sleep with such cholinergic agents as arecoline (Gillin *et al.* 1991) and donepezil (Perlis *et al.* 2002) is shorter than in control subjects.

The undoubted success of antidepressant medications has focused attention on biogenic amines. Given that all antidepressants have effects on either noradrenergic or serotoninergic functioning, it appears reasonable to assume that there is a complementary disturbance in these amines in patients with major depressive disorder. Despite enormous research efforts, however, it has been difficult to isolate definite abnormalities here, with two notable exceptions. First, if patients with major depressive disorder are given alpha-methylparatyrosine (an agent which, by blocking tyrosine hydroxylase, acutely depletes norepinephrine stores), they rapidly develop depressive symptoms, and this holds true not only for those patients who are in drug treatment-induced remission, but also for those whose depressive episode had undergone spontaneous remission and who were no longer taking an antidepressant (Berman *et al.* 1999; Hasler *et al.* 2008). Second, if patients are given a mixed amino-acid drink lacking in tryptophan (a manuever which leads to rapid depletion of intraneuronal tryptophan, and hence a reduction in serotonin) they also rapidly become depressed, and this, as with alpha-methyparatyrosine, occurs regardless of whether patients are in a treatment-induced (Aberg-Wistedt *et al.* 1998; Delgado *et al.* 1990) or spontaneous (Neumeister *et al.* 2004; Smith *et al.* 1997) remission.

Relatively speaking, neuropathologic studies are in their infancy in this disorder. Although some studies have demonstrated decreases in the density of glia or neurons in the dorsolateral frontal cortex and subgenual cingulate gyrus, the endocrinologic changes just noted suggest rather that the primary pathology may rest in the hypothalamus and indeed autopsy studies have demonstrated, in the paraventricular nucleus of the hypothalamus, both an overall reduction in neuronal number (Manaye *et al.* 2005) and a relative increase in the number of CRH-containing neurons (Bao *et al.* 2005).

Integrating all of the foregoing findings into a coherent theory is problematic and involves some speculation. With this caveat in mind, however, it may be reasonable to propose that major depressive disorder represents an interaction between certain environmental events and an inherited abnormality of noradrenergic, serotoninergic, or cholinergic functioning, of variable degree, associated with subtle neuropathologic changes in the hypothalamus, dorsolateral frontal cortex, subgenual cingulate gyrus, and mesencephalon.

Differential diagnosis

Before even considering a diagnosis of major depressive disorder, one must first be sure that the patient does indeed suffer from the syndrome of depression, and is not merely experiencing depressive symptoms that are either too few or of too brief a duration to qualify for such a syndromal diagnosis, as discussed in Chapter 7.1. Once it is reasonably certain that the patient does suffer from the syndrome of depression, the next step is to pursue the differential diagnosis of depression, as discussed, again, in Chapter 7.1. One of the greatest errors in medicine is to assume that any patient with depression has, *ipso facto*, major depressive disorder: because of this error, not only other mood disorders, but also multiple toxic, metabolic, and endocrinologic disorders all go undiagnosed.

Of the other mood disorders, the two that loom largest on the differential are dysthymia and bipolar disorder. Dysthymia, which indeed may represent merely a forme fruste of major depressive disorder, is distinguished by the mildness of its symptoms and their chronicity. Bipolar disorder is distinguished by the history of a manic episode, and in this regard one must *never* rely on the patient as a reliable historian, but must *always* question friends or family: it is remarkable how, during a depression, patients are absolutely unable to recall a manic episode which, in the minds of other informants, and indeed in the mind of patients *once they have recovered from the depression*, burns with vivid intensity.

Treatment

The overall pharmacologic treatment of major depressive disorder is conveniently divided into three phases: acute treatment designed to initially relieve symptoms during a depressive episode; continuation treatment to prevent the re-emergence of symptoms; and maintenance or prophylactic treatment aimed at preventing the occurrence of subsequent episodes. Although this text focuses on pharmacologic treatment, consideration may also be given, in mild to moderate cases, to psychotherapy (e.g. cognitive–behavioral therapy).

ACUTE TREATMENT

Before initiating treatment a decision must be made whether or not to admit the patient. Indications for hospitalization include the following: significant suicide risk; depressive symptoms of such severity as to preclude independent functioning at home and work; significant concurrent illness that requires inpatient monitoring; and a need for acute treatment with ECT.

Acute treatment generally begins with the selection of an antidepressant medication from one of the following: an SSRI (e.g. citalopram or its S-enantiomer, escitalopram), mirtazapine, venlafaxine (or its metabolite, desvenlafaxine), duloxetine, bupropion, a tricyclic (e.g. nortriptyline), or trazodone. Monoamine oxidase inhibitors (MAOIs) are often added to this list; however, as noted below, there are so many serious potential drug–drug interactions with this group that they are rarely used. Several considerations come into play when making this selection, including clinical effectiveness, a personal or family history of response to a particular agent, anticipated side-effects, potential drug–drug interactions, and, finally, lethality in overdose.

With regard to clinical effectiveness, in mild to moderate depressions, although there may be some superiority of venlafaxine or tricyclics over an SSRI, the difference among these agents is modest at best, and hence monotherapy with any one of the agents noted above would constitute, from the point of view of effectiveness, a reasonable first choice. When it comes to severe depressions, however, it may be more prudent to begin with either monotherapy with venlafaxine (Poirier and Boyer 1999) or a tricyclic (Perry 1996), or to use one of the following combinations: an SSRI plus mirtazapine (Blier *et al.* 2009, 2010), mirtazapine plus venlafaxine (Blier *et al.* 2010), mirtazapine plus bupropion (Blier *et al.* 2010), or the combination of the antipsychotic risperidone plus any of the antidepressants (Reeves *et al.* 2008).

A personal history of a good response to a particular agent is a good predictor of future response, and a family history of a good response may also predict a good response; however, this relationship is not as robust.

Side-effects sometimes loom large in the patient's overall reaction to an antidepressant. Where weight gain is a concern, SSRIs and bupropion are good choices, and certainly preferable, in this regard, to mirtazapine and tricyclics. Sexual side-effects (e.g. erectile dysfunction, decreased vaginal lubrication, decreased libido) are common with most antidepressants, with the notable exceptions of mirtazapine and bupropion. Orthostatic hypotension, of particular concern in the elderly, is unlikely with SSRIs, mirtazapine, venlafaxine, and bupropion, but common with tricyclics and trazodone. Cardiac arrhythmias may be induced by tricyclics, but SSRIs are particularly benign in this regard. The seizure threshold may be reduced by venlafaxine, tricyclics, and bupropion, and in this regard, particular attention must be given to any history of bulimia nervosa, as such patients appear to be particularly at risk for seizures if treated with bupropion. Sedation may be problematic with mirtazapine, tricyclics (with the exception of nortriptyline), and trazodone, but is generally negligible with the other agents.

Drug–drug interactions are a constant concern, and in this regard the SSRIs (especially citalopram and escitalopram) are the 'cleanest' of agents. Conversely, there are so many drug–drug interactions with MAOIs, some of them lethal, that these agents generally constitute a last choice.

Lethality in overdose is always a concern in prescribing for depressed patients, and in this regard the SSRIs are by far the safest agents, followed by mirtazapine, venlafaxine, duloxetine, and bupropion; the tricyclics and MAOIs are by far the most dangerous of the antidepressants.

All other things being equal, and lacking clear guidance from a personal or strong family history, in cases of mild to moderate depression, one may begin with monotherapy with an SSRI, mirtazapine, venlafaxine, duloxetine, or bupropion. In cases of severe depression, monotherapy, with either venlafaxine or nortriptyline, may also be considered, or one may move immediately to combination therapy, utilizing one of the combinations discussed earlier. Regardless of which agent or combination of agents is chosen, it must be given an 'adequate' trial, which involves not only an 'adequate' dose but also an 'adequate' duration. Even with an average or above-average dose, one should not expect to see any improvement for the first week or two; by the third week, however, if there has been no improvement at all it is unlikely that the patient will get a response, at least at the dose used (Quitkin *et al.* 1996; Szegedi *et al.* 2009). If there is improvement by the third week, however, the patient should then be advised that it may take up to 3 months to see a full response.

In cases in which there has been little or no response by 3 weeks, one should, if not already done, check the free T_4 and TSH levels. Hypothyroidism, even if only 'chemical' (with a normal free T_4 but an elevated TSH), blunts the response to antidepressants (Joffe and Levitt 1992) and must be corrected. If the patient is euthyroid, and the side-effect burden is readily tolerable, and the depression is of only mild or moderate severity, one may at this point consider simply increasing the dose: with the sole exception of nortriptyline (which has a true 'therapeutic window') it appears that for most other antidepressants the dose–response curve is linear, and thus it is a reasonable strategy to gradually push the dose to the maximum recommended, or to tolerability. Another option, again in cases of mild to moderate severity which have not responded to monotherapy, is to switch to a different monotherapy regimen. In this regard, however, if the patient failed treatment with an SSRI, rather than merely switching to another SSRI, one should try an agent from another class (Papakostas *et al.* 2010), such as mirtazapine, venlafaxine, or bupropion.

Patients who fail to respond to two or more adequate monotherapy trials should be considered to be treatment resistant, and the same probably holds true for patients who failed to respond to initial combination treatment. At this point, if the patient has not already been tried on a combination, then consideration should be given to the following combination strategies: mirtazapine plus an SSRI (Carpenter *et al.* 2002); olanzapine plus fluoxetine (Trivedi *et al.* 2009); risperidone (Keitner *et al.* 2009; Mahmoud *et al.* 2007) or aripiprazole (Berman *et al.* 2007; Marcus *et al.* 2008) or quetiapine (Bauer *et al.* 2009, McIntyre *et al.* 2007) plus any of a variety of different antidepressants; and, finally, lithium (Crossley and Bauer 2007) added to an SSRI or a tricyclic. One other traditional strategy has involved adding triiodothyronine to an antidepressant; however, the evidence for this is weak (Aronson *et al.* 1996). As there are no head-to-head studies comparing these various combination strategies in treatment-resistant depression, the choice of a combination is often made on the basis of tolerability, and in this regard combinations utilizing olanzapine or lithium should probably be last choices. If the depression remains treatment resistant to one combination, another might be tried, or one could consider transcranial magnetic stimulation (Bares *et al.* 2009; George *et al.* 2010; O'Reardon *et al.* 2007). Some physicians may attempt serial trials with different combinations; however, if the depression proves completely resistant, strong consideration should be given to ECT, which is the most effective treatment available (Geddes 2003).

Although the foregoing schema for the acute treatment of depression is in general applicable to most cases, exceptions do occur. First, in severe cases requiring hospitalization, most physicians will immediately move to treatment with one of the combinations utilized for treatment-resistant cases. Second, for the highly suicidal patient, or in other life-threatening cases (e.g. in severely debilitated elderly patients), immediate use of ECT should be strongly considered.

CONTINUATION TREATMENT

Once acute treatment has effected a remission of depressive symptoms, continuation treatment is required. It must be kept in mind that antidepressants, rather than 'curing' depression, merely suppress symptoms, and consequently, if treatment is discontinued before the depressive episode has run its course, symptoms will gradually recur. In general, continuation treatment involves simply a continuation of the regimen that was effective for acute treatment. In a minority of cases it may be possible to scale back treatment without any loss of effectiveness (e.g. in monotherapy cases, allowing a dose reduction, or in combination therapy, discontinuing one of the agents); however, this must be carried out carefully given the risk of relapse. If ECT were required, one should consider either 'continuation ECT' or antidepressant treatment, for example with either an SSRI (e.g. paroxetine) or a combination of nortriptyline and lithium.

Continuation treatment should persist for the duration of the depressive episode. In cases in which the patient has a history of depressive episodes of fairly uniform and discrete duration, this may be used as a guide for determining the length of continuation treatment. In most cases, however, such guidance is not available, either because the current episode is the first or because prior episodes were of such indistinct onset and offset that a reliable estimation of their duration is not possible. In these cases, continuation treatment should probably be performed until the patient has been symptom free for 6 consecutive months (Altamura and Percudani 1993; Kupfer *et al.* 1992; Prien and Kupfer 1986; Reimherr *et al.* 1998). This guideline is based on the assumption that no treatment is perfect and that, given the waxing and waning nature of symptoms, one will find that, when symptoms rise to a 'peak', they 'break through' the antidepressant 'barrier' to the point where they cause distress. Furthermore, the longer a patient goes without such a breakthrough, the more likely it becomes that the underlying episode has finally gone into a spontaneous remission, and it appears that 6 months represents a reasonable duration for this to occur. When the time does come to discontinue treatment, this should be done gradually, preferably tapering off the current regimen over a month or two.

MAINTENANCE TREATMENT

Given that the majority of patients will eventually have another episode, it is reasonable to discuss maintenance or preventive treatment. Prevention should be strongly considered in cases in which past history suggests that, in the natural course of events, the euthymic interval between episodes is less than 2 years, or in cases in which, regardless of the frequency of episodes, the episodes themselves have been so severe that one would not wish to risk a relapse, no matter how far in the future that might occur. If patients do elect for maintenance therapy one may simply continue the antidepressant regimen used in continuation therapy. In cases in which ECT was required for continuation therapy it is not clear whether ongoing ECT is required, and one may opt for an antidepressant regimen.

23.4 DYSTHYMIA

Dysthymia is characterized by chronic, low-level depressive symptoms, and is a common condition, with a lifetime prevalence of 2.5–6 percent; it is more common in females than males.

Clinical features

Dysthymia typically has an insidious onset in childhood or teenage years, less commonly in adulthood. Customarily, those with an onset at or before the age of 21 are said to have an 'early onset' whereas those with an onset past the age of 21 are said to have a 'late onset'.

Patients with dysthymia complain of a depressed mood, and in some cases they may assert that they have been depressed all their lives; some may also complain of anxiety, and in these cases a certain amount of irritability or querulousness may be found. Self-esteem and self-confidence are low, and patients typically have a pessimistic outlook. Thinking is difficult, and patients may complain of feeling heavy-headed or slow and of not being able to concentrate. Irresolution is common, and decisions may be postponed again and again. Energy is often low, and patients may fatigue easily. Insomnia, or occasionally hypersomnia, may also appear, and, likewise, appetite may be either decreased, or, occasionally, increased. No matter how numerous the depressive symptoms are, however, they always remain low level and relatively mild, and are never disabling: patients typically 'soldier on' and shoulder the burden of their lives in never ending toil.

Course

Dysthymia pursues a chronic course, with symptoms persisting for at least 2 years. Over time, the severity of symptoms typically undergoes gradual fluctuations, and in some cases there may actually be periods of euthymia, but these are brief, generally lasting only days (Dunner 1999), and never more than a month. In the majority of cases, after 2 or more years have passed, patients will develop a full depressive episode, identical to those found in major depressive disorder: eventually symptoms subside again, however, back into the low-level depressive symptoms typical of the disorder.

Etiology

There is considerable debate whether dysthymia represents a disorder *sui generis*, or merely represents a form of major depressive disorder: family studies offer support for both hypotheses (Klein *et al.* 1995).

Differential diagnosis

The full differential for depression is discussed in Chapter 7.1, and the reader is referred to that chapter. Special consideration should be given to the differential between dysthymia and major depressive disorder, bipolar disorder, and cyclothymia.

In cases where low-level depressive symptoms have been present for less than 2 years, the diagnosis of dysthymia must be provisional: should a full depressive episode appear before 2 years have passed, then the correct diagnosis, rather than dysthymia, is major depressive disorder, and in such cases the preceding low-level symptoms are seen as merely a prodrome to the full depressive episode. In cases where the low-level depressive symptoms have, however, persisted for 2 or more years, and then a full depressive episode supervenes, one speaks of a 'double depression', and makes two diagnoses: dysthymia and major depressive disorder.

The appearance of manic symptoms, no matter how mild and evanescent, at any time in the patient's life rules out the diagnosis of dysthymia, in favor of bipolar disorder, cyclothymia, or one of the other conditions capable of causing manic symptoms, as discussed in Chapter 7.3.

Treatment

Given the close relationship of dysthymia to major depressive disorder, it makes intuitive sense, in choosing a treatment for dysthymia, to use any of those found effective in major depressive disorder. Whether this is justified, however, is not clear, as there have been only a few placebo-controlled studies regarding the treatment of dysthymia. The SSRIs fluoxetine (Vanelle *et al.* 1997) and sertraline (Ravindran *et al.* 1999, 2000; Thase *et al.* 1996) are effective, as are the tricyclics imipramine (Thase *et al.* 1996) and desipramine (Miller *et al.* 2001). Psychotherapy has also been utilized, but group cognitive behavioral therapy was less effective than sertraline, and no more effective than placebo in one study (Ravindran *et al.* 1999), while in two other studies interpersonal psychotherapy was less effective than sertraline (Browne *et al.* 2002; Markowitz *et al.* 2005).

Given that SSRIs are better tolerated than tricyclics, it makes sense, all other things being equal, to start, perhaps, with sertraline. If this is ineffective, then consideration might be given to imipramine or desipramine, but given their poor tolerability, many clinicians might opt for one of the other antidepressants found useful in major depressive disorder.

23.5 PREMENSTRUAL DYSPHORIC DISORDER

Premenstrual dysphoric disorder, formerly known as late luteal phase dysphoric disorder, is characterized by recurrent, relatively brief, depressive episodes that are tightly entrained to the events of the menstrual cycle. This is a common disorder, occurring in roughly 5 percent of all menstruating females. An older synonym for premenstrual dysphoric disorder is 'premenstural syndrome'; however, this is a problematic term, as it is often used to refer to minimal, even trivial, symptoms occurring during the premenstruum, and, as such, is said to be found in up to 80 percent of all menstruating females.

Clinical features

The onset of this disorder is generally between menarche and the late twenties, and its most remarkable aspect is the timing of the depressive episodes, which have a fairly abrupt onset, anywhere from 3 to 10 days before menstruation begins, and then just as rapidly remit, 2–3 days after menstrual flow commences.

During the depressive episode, the mood (Bloch *et al.* 1997) is variously depressed, sad, anxious, or, classically, irritable; often there is considerable lability with inexplicable crying spells or unwonted anger, and many patients complain of feeling 'out of control'. Patients may have difficulty with concentration or paying attention to things, and may lose interest in work, hobbies, or sexual activity. There may be prominent fatigue, and some may become lethargic. Sleep disturbance is common and may be towards either insomnia or hypersomnia. Appetite may also change and is often increased with associated cravings for sweets or chocolates. Other symptoms may also occur, including headache, mastalgia, bloating (particularly with swelling of the hands and feet), clumsiness, nausea, and constipation.

Some patients, concerned over their irritability, may voluntarily isolate themselves, and family members may soon learn to 'leave well enough alone' during this part of the menstrual cycle.

Course

Over long periods of time the depressive episodes may become more severe and may appear progressively earlier in the menstrual cycle. Eventually, with menopause, the episodes cease to occur.

Etiology

Premenstrual dysphoric disorder appears to be inherited, with the concordance rate rising from 28 percent for dizygotic twins to 55 percent for monozygotic twins (Condon 1993), and recent research (Huo *et al.* 2007) has implicated an association between this disorder and normal variations in the estrogen receptor alpha gene. Additional research has also strongly implicated changes in serotonergic function. Fenfluramine, a serotoninergic agent, normally stimulates prolactin secretion, and the prolactin response to this agent in premenstrual dysphoric disorder is blunted (Fitzgerald *et al.* 1997). Furthermore, symptoms may be exacerbated either by administration of a serotonin-blocking agent (Roca *et al.* 2002) or by tryptophan depletion (Menkes *et al.* 1994). A positron emission tomography (PET) study (Eriksson *et al.* 2006) also revealed deficient trapping of tryptophan within raphe nuclei during the luteal phase in patients. Finally, whereas serotoninergic agents (e.g. SSRIs) are effective in this disorder, non-serotoninergic antidepressants (e.g. desipramine [Freeman *et al.* 1999] and bupropion [Pearlstein *et al.* 1997]) are not.

Overall, although speculative, it may be reasonable to hypothesize that in premenstrual dysphoric disorder there is an inherited disturbance of serotoninergic functioning, probably in mesencephalic or hypothalamic structures, that is episodically triggered by the dramatic hormonal changes seen during the transition from the follicular to the luteal phase of the normal menstrual cycle.

Differential diagnosis

The key to making this diagnosis is to monitor the patient over at least several menstrual cycles to establish the classic onset and duration of the episodes and, very importantly, to demonstrate that during the follicular phase patients are *symptom free*. Essentially, there is no other disorder that mimics this remarkable pattern.

Of the multiple other causes of depression discussed in Chapter 7.1, major depressive disorder and dysthymia must be clearly ruled out. In these disorders, depressive symptoms may undergo an exacerbation during the luteal phase, and if this exacerbation is sufficiently severe, patients may mistakenly report that they feel 'better' during the follicular phase, thus lulling the diagnostician into a mistaken conclusion. Close examination in these cases, however, reveals that, although patients do indeed feel better during the follicular phase, this is only relatively so, and in fact the depressive symptoms are present throughout both phases of the menstrual cycle, in stark contrast to the pattern seen in premenstrual dysphoric disorder.

Dysmenorrhea, characterized by cramps, headaches, and bloating, is easily distinguished by its course, beginning not before, but after the commencement of menstrual flow.

Treatment

Selective serotonin reuptake inhibitors, including fluoxetine (20 mg) (Cohen *et al.* 2002; Pearlstein and Stone 1994; Pearlstein *et al.* 1997; Steiner *et al.* 1995), paroxetine (20 mg) (Landen *et al.* 2007), and sertraline (50–150 mg) (Freeman *et al.* 1999, 2004; Yonkers *et al.* 1997; Young *et al.* 1998), constitute the mainstay of treatment, and these may be taken either chronically on a daily basis or on a timed basis relative to the timing of the episode itself, such that they are started just before the anticipated onset of symptoms and then stopped several days into menstrual flow. Venlafaxine (50–200 mg) (Freeman *et al.* 2001), taken chronically, is another option. Hormonal treatments may also be considered, and these include either an oral contraceptive containing 20 μg of estradiol and 3 mg of drosperenone (Pearlstein *et al.* 2005), or the use of gonadotropin-releasing hormones (Wyatt *et al.* 2004). Other options include alprazolam (1–4 mg) (Freeman *et al.* 1995), taken during the luteal phase, and a herbal preparation, *Vitex agnus-castus* (Atmaca *et al.* 2003; Schellenberg 2001).

In cases where women desire an oral contraceptive for contraceptive purposes, then the estradiol–drosperenone preparation is a logical first choice. When oral contraception is not desired, or when the estradiol–drosperenone preparation fails, then an SSRI should be tried. Alprazolam, given the risk of neuroadaptation, should probably be held in reserve. Gonadotropin-releasing hormone treatment should probably also be held in reserve, given the associated long-term side-effects. The place of *Vitex agnus-castus* is simply not as yet established.

23.6 POSTPARTUM DEPRESSION

In the strict sense used here, postpartum depression is characterized by the occurrence of depressive episodes in only the postpartum period (Cooper and Murray 1995); as such it must be distinguished from cases of major depressive disorder, for example, in which depressive episodes, although at times occurring in the postpartum period, also occur at other times in the patient's life. Thus, strictly defined, the prevalence of this disorder is not certain: most texts quote lifetime prevalence rates of anywhere from 10 to 25 percent; however, these studies include not only patients with postpartum depression in this strict sense, but also cases where the patient, in addition to depression occurring in the postpartum period, also had had depressions at other times.

Clinical features

The onset of the depressive episode is anywhere from several weeks to several months post partum. Mood is depressed and often accompanied by a considerable amount of anxiety (Hendrick *et al.* 2000). Self-esteem falls, particularly in regard to the patient's estimate of her abilities as a mother. There may be poor concentration, anhedonia, fatigue, initial insomnia, and anorexia. Obsessions may occur and typically involve 'horrific temptations' to do harm to the baby (Wisner *et al.* 1999). Rarely, infanticide may occur, and, as with any depression, suicide remains a risk.

Course

Although most patients experience a spontaneous remission within months or years, the depression may be chronic in a minority. Those who do recover are at increased risk for another episode after a subsequent pregnancy.

Etiology

Postpartum depression may occur secondary to an unusual sensitivity of mood-regulating central nervous system structures to the profound endocrinologic changes that occur postpartum. One particularly interesting study (Bloch *et al.* 2000) looked at two groups of women: both groups had a history of depressive episodes, but in one they were confined to the postpartum period, whereas in the other, depressive episodes occurred at times outside this period. All patients were euthymic at the time of the study, and all were given supraphysiologic doses of estradiol and progesterone for 8 weeks, after which these hormones were abruptly discontinued. In the group of women with a history of postpartum depression in the strict sense, the hormone withdrawal was followed by a depressive episode in the majority, whereas in the group with a history of depression outside the postpartum period, there were no depressions.

Differential diagnosis

Of the many causes of depression discussed in Chapter 7.1, the two most important are major depressive disorder and bipolar disorder. Patients with major depressive disorder or bipolar disorder may have depressive episodes in the postpartum period; however, as stressed earlier, these patients also have episodes at other times. In cases in which a depressive episode occurring in the postpartum period represents the first ever episode in the patient's life, long-term follow-up will be required to determine whether any future episodes occur and, if so, whether they are unrelated to pregnancy or are confined to the postpartum period. It must be stressed that a careful search for a history of mania is imperative here, given that some patients may be treated with antidepressants which, of course, may precipitate mania in bipolar patients who happen to be in a depressive episode.

Other disorders to consider include the postpartum blues, hypothyroidism, and Sheehan's syndrome. The postpartum blues is distinguished by its early onset, within days, and rapid resolution, within 2 weeks; there are also symptomatic differences here, with the postpartum blues being marked by prominent lability. Hypothyroidism, especially that due to

Hashimoto's thyroiditis, is not uncommon post partum, and determining a free T_4 level and a TSH level should be part of the work-up. Sheehan's syndrome is suggested by a failure of lactation, persistent amenorrhea, and, often, by loss of pubic and axillary hair.

Treatment

The acute treatment of postpartum depression may involve cognitive–behavioral therapy or interpersonal therapy, and one double-blinded, placebo-controlled study found transdermal estrogen to be effective (Gregoire *et al.* 1996). Regarding the use of antidepressants, there is a remarkable paucity of double-blinded studies: one (Appleby *et al.* 1997) found fluoxetine more effective than placebo; another (Wisner *et al.* 2006) found sertraline equivalent to nortriptyline – however, as there was no placebo arm in this study, it may well be that they are both equally ineffective in this regard.

Some patients, once recovered and either pregnant or considering becoming pregnant, may be interested in preventive treatment, and two double-blinded, placebo-controlled studies have been conducted: one (Wisner *et al.* 2004) found sertraline to be effective, while the other (Wisner *et al.* 2001) found nortriptyline to be ineffective.

23.7 POSTPARTUM BLUES

The postpartum blues, also known as maternity blues or baby blues, is a common disturbance, seen in well over half of all women. Given this very high prevalence, although most authors consider this to be a non-pathologic disturbance, many sufferers find it totally foreign and out of character.

Clinical features

Clinically (Pitt 1973; Rohde *et al.* 1997; Yalom *et al.* 1968), the onset is acute, usually within the first few days postpartum. Mood may be depressed, irritable or fearfully anxious. Crying spells are frequent, and there may be a striking lability of affect; crying spells may come and go with remarkable rapidity, and at times the patient may actually be laughing and claim to feel happy with her delivery, yet be absolutely unable to stop the tears cascading down past her smile. There may also be minor degrees of difficulty with concentration, fatigue, and insomnia. Symptoms tend to peak within a couple of days and then gradually undergo a full remission by the end of the second postpartum week.

Course

Although the postpartum blues may recur after subsequent pregnancies, they tend to be less severe.

Etiology

Although the postpartum blues is almost certainly related to the profound hormonal and neurophysiologic changes present in the immediate postpartum period, as yet it has not been possible to develop a unified theory. One study noted a relationship with lower allopregnanolone levels (Nappi *et al.* 2001); another study found a relationship with lower postpartum progesterone levels (Harris *et al.* 1994); however, another study did not confirm this (Kuevi *et al.* 1983). Disturbances in tryptophan metabolism have also been suggested: two studies found lower levels in relation to postpartum blues (Bailara *et al.* 2006; Kohl *et al.* 2005), and one (Bailara *et al.* 2006) of these also noted a reduced ratio of tryptophan to other large neutral amino acids; notably, however, treatment with tryptophan is not effective (Harris 1980). Finally, two studies have noted an association between an increased number of platelet alpha-2 autoreceptors and the occurrence of the blues (Best *et al.* 1988; Metz *et al.* 1983).

Differential diagnosis

Postpartum depression is distinguished by its later onset (usually at least several weeks post-partum), longer duration (at least months), and the absence of lability. It must be borne in mind, however, that, like any other postpartum women, patients with the postpartum blues may go on to develop a postpartum depression, and hence any persistence of symptoms beyond a couple of weeks should prompt a diagnostic re-evaluation.

Treatment

Given the brevity of the syndrome, treatment with antidepressants is not indicated, as a spontaneous remission may be anticipated before an antidepressant could be expected to take effect. Support, reassurance, and assistance are generally sufficient; in some cases a brief course of treatment with a benzodiazepine, such as lorazepam, may be considered, but it must be kept in mind that these drugs do appear in the breast milk.

REFERENCES

Aberg-Wistedt A, Hasselmark L, Stain-Malmgren R *et al.* Serotonergic 'vulnerability' in affective disorder: a study of the tryptophan depletion test and relationships between peripheral and central serotonin indexes in citalopram responders. *Acta Psychiatr Scand* 1998; **97**:374–8.

Abrams R, Taylor MA. Manic and schizoaffective disorder, manic type: a comparison. *Am J Psychiatry* 1976a; **133**:1445–7.

Abrams R, Taylor MA. Catatonia: a prospective clinical study. *Arch Gen Psychiatry* 1976b; **33**:579–81.

Abrams R, Taylor MA. Importance of schizophrenic symptoms in the diagnosis of mania. *Am J Psychiatry* 1981; **138**:658–61.

Akiskal HS, Djenderedjian AM, Rosenthal RH *et al.* Cyclothymic disorder: validating criteria for inclusion in the bipolar affective group. *Am J Psychiatry* 1977; **134**:1227–33.

Altamura AC, Percudani M. The use of antidepressants for long-term treatment of recurrent depression: rationale, current methodologies, and future directions. *J Clin Psychiatry* 1993; **54**(Suppl. 8):29–37.

Appleby L, Warner R, Whitton A *et al.* A controlled study of fluoxetine and cognitive behavioral counselling in the treatment of postnatal depression. *BMJ* 1997; **314**:932–6.

Aronson R, Offman HJ, Joffe RT *et al.* Triiodothyronine augmentation in the treatment of refractory depression. A meta-analysis. *Arch Gen Psychiatry* 1996; **53**:842–8.

Atmaca M, Kumru S, Tezcan E. Fluoxetine versus agnus castus extract in the treatment of premenstrual dysphoric disorder. *Hum Psychopharmacol* 2003; **1**:191–5.

Bailara KM, Henry C, Lestage J *et al.* Decreased brain tryptophan availability as a partial determinant of post-partum blues. *Psychoneuroendocrinology* 2006; **31**:407–13.

Baldessarini RJ, Tondo L, Faeda GL *et al.* Effects of the rate of discontinuing lithium maintenance treatment in bipolar disorder. *J Clin Psychiatry* 1996; **57**:441–8.

Baldessarini RJ, Tondo L, Floris G *et al.* Reduced morbidity after gradual discontinuation of lithium treatment for bipolar I and II disorders: a replication study. *Am J Psychiatry* 1997; **154**:551–3.

Banki CM, Bissette G, Arato M *et al.* Elevation of immunoreactive CSF TRH in depressed patients. *Am J Psychiatry* 1988; **145**: 1525–31.

Bao A-M, Hestiantoro A, Van Someren EJW *et al.* Colocalization of corticotropin-releasing hormone and oestrogen receptor-alpha in the paraventricular nucleus of the hypothalamus in mood disorders. *Brain* 2005; **128**:1301–13.

Bares M, Kopecek M, Novak T *et al.* Low frequency (1-Hz), right prefrontal repetitive transcranial stimulation (rTMS) compared with venlafaxine ER in the treatment of resistant depression: a double-blind, single-centre, randomized study. *J Affect Disord* 2009; **118**:94–100.

Bauer MS, Whybrow PC, Winokur A. Rapid cycling bipolar affective disorder. I. Association with grade I hypothyroidism. *Arch Gen Psychiatry* 1990; **47**:427–32.

Bauer MS, Calabrese J, Dunner DL *et al.* Multisite data reanalysis of the validity of rapid cycling as a course modifier for bipolar disorder in DSM-IV. *Am J Psychiatry* 1994; **151**:506–15.

Bauer M, Pretorius HW, Constant EL *et al.* Extended-release quetiapine as adjunct to an antidepressant in patients with major depressive disorder: results of a randomized, placebo-controlled study. *J Clin Psychiatry* 2009; **70**:540–9.

Baumann B, Bogerts B. Neuroanatomical studies on bipolar disorder. *Br J Psychiatry* 2001; **41**(suppl.):142–7.

Beigel A, Murphy DL. Assessing clinical characteristics of the manic state. *Am J Psychiatry* 1971a; **128**:688–94.

Beigel A, Murphy DL. Unipolar and bipolar affective illness: differences in clinical characteristics accompanying depression. *Arch Gen Psychiatry* 1971b; **24**:215–20.

Berman RM, Narasimhan M, Miller HL *et al.* Transient depressive relapse induced by catecholamine depletion: potential phenotypic vulnerability marker? *Arch Gen Psychiatry* 1999; **56**: 395–403.

Berman RM, Marcus RN, Swanink R *et al.* The efficacy and safety of aripiprazole as adjunctive therapy in major depressive disorder: amulticenter, randomized, double-blind, placebo-controlled study. *J Clin Psychiatry* 2007; **68**:843–53.

Best NR, Wiley M, Stump K *et al.* Binding of tritiated yohimbine to platelets in women with maternity blues. *Psychol Med* 1988; **18**:837–42.

Bisol LW, Lara DR. Low-dose quetiapine for patients with dysregulation of hyperthymic and cyclothymic temperaments. *J Psychopharmacol* 2010; **24**:421–4.

Black DW, Nasrallah H. Hallucinations and delusions in 1,715 patients with unipolar and bipolar affective disorders. *Psychopathology* 1989; **22**:28–34.

Blier P, Gobbi G, Turcotte JE *et al.* Mirtazapine and paroxetine in major depression: a comparison of monotherapy versus their combination from treatment initiation. *Eur Neuropsychopharmacol* 2009; **19**:457–65.

Blier P, Ward HE, Tremblay P *et al.* Combination of antidepressant medications from treatment initiation for major depressive disorder: a double-blind randomized study. *Am J Psychiatry* 2010; **167**:281–8.

Bloch M, Schmidt PJ, Rubinow DR. Premenstrual syndrome: evidence of symptom stability across cycles. *Am J Psychiatry* 1997; **154**:1741–6.

Bloch M, Schmidt PJ, Danaceau M *et al.* Effects of gonadal steroids in women with a history of postpartum depression. *Am J Psychiatry* 2000; **157**:924–30.

Bond TC. Recognition of acute delirious mania. *Arch Gen Psychiatry* 1980; **37**:553–4.

Bowden CL, Calabrese JR, McElroy SL *et al.* A randomized, placebo-controlled 12-month trial of divalproex and lithium in treatment of outpatients with bipolar I disorder. *Arch Gen Psychiatry* 2000; **57**:481–9.

Bowden CL, Calabrese JF, Sachs G *et al.* A placebo-controlled 18-month trial of lamotrigine and lithium maintenance in recently manic of hypomanic patients with bipolar I disorder. *Arch Gen Psychiatry* 2003; **60**:392–400.

Bowden CL, Grunze H, Mullen J *et al.* A randomized, double-blind, placebo-controlled efficacy and safety study of quetiapine or lithium as monotherapy for mania in bipolar disorder. *J Clin Psychiatry* 2005; **66**:111–21.

Bowden CL, Vieta E, Ice KS *et al.* Ziprasidone plus a mood stabilizer in subjects with bipolar I disorder: a 6-month, randomized, placebo-controlled, double-blind trial. *J Clin Psychiatry* 2010; **71**:130–7.

Bowman KM, Raymond AF. A statistical study of hallucinations in the manic–depressive psychoses. *Am J Psychiatry* 1931; **88**:299–309.

Bratfos O, Haug JO. Puerperal mental disorders in manic–depressive females. *Acta Psychiatr Scand* 1966; **42**:285–94.

Brockington IF, Wainwright S, Kendell RE. Manic patients with schizophrenic or paranoid symptoms. *Psychol Med* 1980; **10**:73–83.

Brown GW, Harris TO, Hepworth C *et al.* Life events and endogenous depression: a puzzle re-examined. *Arch Gen Psychiatry* 1994; **51**:525–34.

Brown EB, McElroy SL, Keck PE *et al.* A 7-week, randomized, double-blind trial of olanzapine/fluoxetine combnation versus lamotrigine in the treatment of bipolar I depression. *J Clin Psychiatry* 2006; **67**:1025–33.

Browne G, Steiner M, Roberts J *et al.* Sertraline and/or interpersonal psychotherapy for patients with dysthymic disorder in primary care: a 6-month comparison with longitudinal 2-year follow-up of effectiveness and costs. *J Affect Disord* 2002; **68**:317–30.

Bunney WE, Hartmann EL. Study of a patient with a 48-hour manic-depressive cycle. *Arch Gen Psychiatry* 1965; **12**:611–18.

Bunney WE, Murphy DL, Goodwin FK *et al.* The 'switch process' in manic-depressive illness. *Arch Gen Psychiatry* 1972; **27**:295–302.

Calabrese JR, Keck PE, Macfadden W *et al.* A randomized, double-blind, placebo-controlled trial of quetiapine in the treatment of bipolar I or II depression. *Am J Psychiatry* 2005; **162**:1351–60.

Calabrese JR, Huffman RF, White RL *et al.* Lamotrigine in the acute treatment of bipolar depression: results of five double-blind, placebo-controlled clinical trials. *Bipolar Disord* 2008; **10**:323–33.

Carlson GA, Goodwin FK. The stages of mania: a longitudinal analysis of the manic episode. *Arch Gen Psychiatry* 1973; **28**:221–8.

Carpenter LL, Yasmin S, Pierce LH. A double-blind, placebo-controlled study of antidepressant augmentation with mirtazapine. *Biol Psychiatry* 2002; **51**:183–8.

Carroll BJ, Martin FI, Davis B. Pituitary-adrenal function in depression. *Lancet* 1968; **2**:1373–4.

Carroll BJ, Curtis GC, Mendels J. Neuroendocrine regulation in depression. *Arch Gen Psychiatry* 1976; **33**:1039–44.

Casper RC, Redmond E, Katz MM *et al.* Somatic symptoms in primary affective disorder: presence and relationship to the classification of depression. *Arch Gen Psychiatry* 1985; **42**:1098–104.

Ceron-Litvac D, Soares BG, Geddes J *et al.* Comparison of carbamazepine and lithium in treatment of bipolar disorder: a systematic review of randomized controlled trials. *Hum Psychopharmacol* 2009; **24**:19–28.

Charron M, Fortin L, Paquette I. *De novo* mania among elderly people. *Acta Psychiatr Scand* 1991; **84**:503–7.

Clayton P, Pitts FN, Winokur G. Affective disorders. IV. Mania. *Compr Psychiatry* 1965; **3**:313–22.

Cohen LS, Mier C, Brown EW *et al.* Premenstrual daily fluoxetine for premenstrual dysphoric disorder: a placebo-controlled, clinical trial using computerized diaries. *Obstet Gynecol* 2002; **100**:435–44.

Cole P, Thase ME, Mallinger AG *et al.* Slower treatment response in bipolar depression predicted by lower pretreatment thyroid function. *Am J Psychiatry* 2002; **159**:116–21.

Condon JT. The premenstrual syndrome: a twin study. *Br J Psychiatry* 1993; **162**:481–6.

Cooper PJ, Murray L. Course and recurrence of postnatal depression. Evidence for the specificity of the diagnostic concept. *Br J Psychiatry* 1995; **166**:191–5.

Cowdry RW, Wehr TA, Zis AP *et al.* Thyroid abnormalities associated with rapid-cycling bipolar illness. *Arch Gen Psychiatry* 1983; **40**:414–20.

Crossley NA, Bauer M. Acceleration and augmentation of antidepressants with lithium for depressive disorders: two meta-analyses of randomized, placebo-controlled trials. *J Clin Psychiatry* 2007; **68**:935–40.

Delgado PL, Charney DS, Price LH *et al.* Serotonin function and the mechanism of antidepressant action: reversal of antidepressant-induced remission by rapid depletion of plasma tryptophan. *Arch Gen Psychiatry* 1990; **47**:411–18.

D'Mello DA, Pinheiro AL, Lalinet-Michaud M. Premenstrual mania: two case reports. *J Nerv Ment Dis* 1993; **181**:330–1.

Dunner DL. Duration of periods of euthymia in patients with dysthymic disorder. *Am J Psychiatry* 1999; **156**:1992–3.

Dunner DL, Fleiss JL, Fieve RR. The course of development of mania in patients with recurrent depression. *Am J Psychiatry* 1976; **133**:905–8.

Dunner DL, Patrick V, Fieve RR. Rapid cycling manic depressive patients. *Compr Psychiatry* 1977; **18**:561–6.

Eriksson O, Wall A, Marteinsdottir I *et al.* Mood changes correlate to changes in brain serotonin precursor trapping in women with premenstrual dysphoria. *Psychiatry Res* 2006; **146**:107–16.

Extein I, Pottash AL, Gold MS *et al.* Differentiating mania from schizophrenia by the TRH test. *Am J Psychiatry* 1980; **137**:981–2.

Fava M, Rosenbaum JF, Pava JA *et al.* Anger attacks in unipolar depression. I. Clinical correlates and response to fluoxetine treatment. *Am J Psychiatry* 1993; **150**:1158–63.

Fitzgerald M, Malone KM, Li S *et al.* Blunted serotonin response to fenfluramine challenge in premenstrual dysphoric disorder. *Am J Psychiatry* 1997; **154**:556–8.

Frank E, Anderson B, Reynolds CF *et al.* Life events and the research diagnostic criteria endogenous subtype: a confirmation of the distinction using the Bedford College methods. *Arch Gen Psychiatry* 1994; **51**:519–24.

Freeman EW, Rickels K, Sondheimer SJ *et al.* A double-blind trial of oral progesterone, alprazolam and placebo in the treatment of severe premenstrual syndrome. *JAMA* 1995; **27**:51–4.

Freeman EW, Rickels K, Sondheimer SJ *et al.* Differential response to antidepressants in women with premenstrual syndrome/premenstrual dysphoric disorder: a randomized controlled trial. *Arch Gen Psychiatry* 1999; **56**:932–9.

Freeman EW, Rickels K, Yonkers KA *et al.* Venlafaxine in the treatment of premenstrual dysphoric disorder. *Obstet Gynecol* 2001; **98**: 737–44.

Freeman EW, Rickels K, Sondheimer SJ *et al.* Continuous or intermittent dosing with sertraline for patients with severe premenstrual syndrome or premenstrual dysphoric disorder. *Am J Psychiatry* 2004; **161**:343–51.

Frye MA, Helleman G, McElroy SL *et al.* Correlates of treatment-emergent mania associated with antidepressant treatment in bipolar depression. *Am J Psychiatry* 2009; **166**:154–7.

Geddes J. Efficacy and safety of electroconvulsive therapy in depressive disorders: a systematic review and meta-analysis. *Lancet* 2003; **361**:799–808.

Geddes JR, Burgess S, Hawton K *et al.* Long-term lithium therapy for bipolar disorder: systematic review and meta-analysis of randomized controlled trials. *Am J Psychiatry* 2004; **161**: 217–22.

Geddes JF, Calabrese JR, Goodwin GM. Lamotrigine for treatment of bipolar depression: independent meta-analysis and meta-regression of individual patient data from five randomized trials. *Br J Psychiatry* 2009; **194**:4–9.

George MS, Lisanby SH, Avery D *et al.* Daily left prefrontal transcranial magnetic stimulation therapy for major depressive disorder: a sham-controlled randomized trial. *Arch Gen Psychiatry* 2010; **67**:507–16.

Ghaemi SN, Gilmer WS, Goldberg JF *et al.* Divalproex in the treatment of acute bipolar depression: a preliminary double-blind, randomized, placebo-controlled study. *J Clin Psychiatry* 2007; **68**: 1840–4.

Gijsman HJ, Geddes JR, Rendell JM *et al.* Antidepressants for bipolar depression: a systematic review of randomized, controlled trials. *Am J Psychiatry* 2004; **161**:1537–47.

Gillin JC, Sutton L, Ruiz C *et al.* The cholinergic rapid eye movement induction test with arecoline in depression. *Arch Gen Psychiatry* 1991; **48**:264–70.

Goodwin FC, Murphy DL, Dunner DL *et al.* Lithium response in unipolar versus bipolar depression. *Am J Psychiatry* 1972; **129**: 44–7.

Gregoire AJ, Kumar L, Everitt B *et al.* Transdermal oestrogen for treatment of severe postnatal depression. *Lancet* 1996; **347**: 930–3.

Guze SB, Woodruff RA, Clayton PJ. The significance of psychotic affective disorders. *Arch Gen Psychiatry* 1975; **32**:1147–50.

Harris B. Prospective trial of l-tryptophan in maternity blues. *Br J Psychiatry* 1980; **137**:233–5.

Harris B, Lovett L, Newcombe RG *et al.* Maternity blues and major endocrine changes: Cardiff puerperal mood and hormone study II. *BMJ* 1994; **308**:949–53.

Hartmann E. Longitudinal studies of sleep and dream patterns in manic–depressive patients. *Arch Gen Psychiatry* 1968; **19**:312–29.

Hasler G, Fromm S, Carlson PJ *et al.* Neural response to catecholamine depletion in unmedicated subjects with major depressive disorder in remission and healthy subjects. *Arch Gen Psychiatry* 2008; **65**:521–31.

Hauri PJ, Chernik D, Hawkins D *et al.* Sleep of depressed patients in remission. *Arch Gen Psychiatry* 1974; **31**:386–91.

Hendrick V, Altshuler L, Strouse T *et al.* Postpartum and non postpartum depression: differences in presentation and response to pharmacologic treatment. *Depress Anxiety* 2000; **11**:66–72.

Himmelhoch JM, Mulla D, Neil JF *et al.* Incidence and significance of mixed affective states in a bipolar population. *Arch Gen Psychiatry* 1976; **33**:1062–6.

Hirschfeld RM, Baker JD, Wozniak P *et al.* The safety and early efficacy of oral-loaded divalproex versus standard titration divalproex, lithium, olanzapine, and placebo in the treatment of acute mania associated with bipolar disorder. *J Clin Psychiatry* 2003; **64**:841–6.

Houston JP, Tohen M, Degenhardt EK *et al.* Olanzapine–divalproex combination versus divalproex monotherapy in the treatment of bipolar mixed episodes: a double-blind, placebo-controlled study. *J Clin Psychiatry* 2009; **70**:1540–7.

Huo L, Straub RE, Roca C *et al.* Risk for premenstrual dysphoric disorder is associated with genetic variation in ESR1, the estrogen receptor alpha gene. *Biol Psychiatry* 2007; **62**:925–33.

Jacobsen FM. Low-dose valproate: a new treatment for cyclothymia, mild rapid cycling disorders, and premenstrual syndromes. *J Clin Psychiatry* 1993; **54**:229–34.

Jampala VC, Taylor MA, Abrams R. The diagnostic implication of formal thought disorder in mania and schizophrenia: a reassessment. *Am J Psychiatry* 1989; **146**:459–63.

Janowsky DS, El-Yousof MK, David JM *et al.* Parasympathetic suppression of manic symptoms by physostigmine. *Arch Gen Psychiatry* 1973; **28**:542–7.

Jenner FA, Gjessing LR, Cox JR *et al.* A manic depressive psychotic with a persistent forty-eight hour cycle. *Br J Psychiatry* 1967; **113**: 895–910.

Joffe RT, Levitt AJ. Major depression and subclinical (grade 2) hypothyroidism. *Psychoneuroendocrinology* 1992; **17**:215–21.

Johnson J. Stupor: review of 25 cases. *Acta Psychiatr Scand* 1984; **70**:370–7.

Keck PE, Versiani M, Potkin S *et al.* Ziprasidone in the treatment of acute bipolar mania: a three-week, placebo-controlled, double-blind, randomized trial. *Am J Psychiatry* 2003; **160**:741–8.

Keck PE, Calabrese JR, McIntyre RS *et al.* Aripiprazole monotherapy for maintenance therapy in bipolar I disorder: a 100-week, double-blind study versus placebo. *J Clin Psychiatry* 2007; **68**: 1480–91.

Keitner GI, Garlow SJ, Ryan CE *et al.* A randomized, placebo-controlled trial of risperidone augmentation for patients with difficult-to-treat unipolar, non-psychotic major depression. *J Psychiatr Res* 2009; **43**:205–14.

Khanna S, Vieta E, Lyons B *et al.* Risperidone in the treatment of acute mania: double-blind, placdebo-controlled study. *Br J Psychiatry* 2005; **187**:229–34.

Klein DN, Riso LP, Donaldson SK *et al.* Family study of early-onset dysthymia. Mood and personality disorders in relatives of out-patients with dysthymia and episodic major depression and normal controls. *Arch Gen Psychiatry* 1995; **52**:487–96.

Kohl C, Walch T, Huber R *et al.* Measurement of tryptophan, kynurenine and neopterin in women with and without postpartum blues. *J Affect Disord* 2005; **86**:135–42.

Kotin J, Goodwin FK. Depression during mania: clinical observations and theoretical implications. *Am J Psychiatry* 1972; **129**:679–86.

Kraepelin E. *Manic–Depressive Insanity and Paranoia*, 1921. New York: Arno Press, 1976.

Kramer BA. Depressive pseudodementia. *Compr Psychiatry* 1982; **23**:538–44.

Kuevi V, Causon R, Dixson AF *et al.* Plasma amine and hormone changes in 'post-partum blues'. *Clin Endocrinol* 1983; **19**:39–46.

Kupfer DJ, Frank E, Perel JM *et al.* Five-year outcome for maintenance therapies in recurrent depression. *Arch Gen Psychiatry* 1992; **49**:769–73.

Lam RW, Levitt AJ, Levitan RD *et al.* The Can-SAD study: a randomized controlled trial of the effectiveness of light therapy and fluoxetine in patients with winer seasonal affective disorder. *Am J Psychiatry* 2006; **163**:805–12.

Landen M, Nissbrandt H, Allgulander C *et al.* Placebo-controlled trial comparing intermittent and continuous paroxetine in premenstrual dysphoric disorder. *Neuropsychopharmacology* 2007; **32**:153–61.

Leverich GS, Altshuler LL, Frye MA *et al.* Risk of switch in mood polarity to hypomania or mania in patients with bipolar depression during acute and continuation trials of venlafaxine, sertraline and bupropion as adjuncts to mood stabilizers. *Am J Psychiatry* 2006; **163**:232–9.

Lipkin KM, Dyrud J, Meyer GD. The many faces of mania; therapeutic trial of lithium carbonate. *Arch Gen Psychiatry* 1970; **22**:262–7.

Loudon JB, Blackburn IM, Ashworth CM. A study of the symptomatology and course of manic illness using a new scale. *Psychol Med* 1977; **7**:723–9.

Lykouras E, Malliargas D, Christdoulou GN *et al.* Delusional depression: phenomenology and response to treatment. *Acta Psychiatr Scand* 1986; **73**:324–9.

McAllister TW, Price TRP. Severe depressive pseudodementia with and without dementia. *Am J Psychiatry* 1982; **139**:626–9.

McElroy SL, Keck PE, Pope HG *et al.* Clinical and research implications of the diagnosis of dysphoric or mixed mania. *Am J Psychiatry* 1992; **149**:1633–44.

McElroy SL, Weisler RH, Chang W *et al.* A double-blind, placebo-controlled study of quetiapine and paroxetine as monotherapy in adults with bipolar depression (EMBOLDEN II). *J Clin Psychiatry* 2010; **71**:163–74.

McHarg JF. Mania in childhood: report of a case. *Arch Neurol Psychiatry* 1954; **72**:531–9.

McIntyre A, Gendron A, McIntyre A. Quetiapine adjunct to selective serotonin reuptake inhibitors or venlafaxine in patients with major depression, comorbid anxiety, and residual depressive symptoms: a randomized, placebo-controlled pilot study. *Depress Anxiety* 2007; **24**:487–94.

Macfadden W, Alphs L, Haskins JT *et al.* A randomized, double-blind, placebo-controlled study of maintenance treatment with adjunctive risperidone long-acting therapy in patients with bipolar I disorder who relapse frequently. *Bipolar Disord* 2009; **11**:827–39.

Mahmoud RA, Pandina GJ, Turkoz I *et al.* Risperidone for treatment-refractory major depressive disorder: a randomized trial. *Ann Intern Med* 2007; **147**:593–602.

Maj M, Pirozzi R, DiCaprio EL. Major depression with mood-congruent psychotic features: a distinct diagnostic entity or a more severe subtype of depression? *Acta Psychiatr Scand* 1990; **82**:439–44.

Manaye KF, Lei DL, Tizabi Y *et al.* Selective neuron loss in the paraventricular nucleus of hypothalamus in patients suffering from major depression and bipolar disorder. *J Neuropathol Exp Neurol* 2005; **64**:224–9.

Marcus RN, McQuade RD, Carson WH *et al.* The efficacy and safety of aripiprazole as adjunctive teharpy in major depressive disorder: a second multicenter, randomized, double-blind, placebo-controlled trial. *J Clin Psychopharmacol* 2008; **28**:156–65.

Markowitz JC, Kocsis JH, Bleiberg KL *et al.* A comparative trial of psychotherapy and pharmacotherapy for 'pure' dysthymic patients. *J Affect Disord* 2005; **89**:167–75.

Mendlewicz J, Rainer JD. Adoption study supporting genetic transmission in manic–depressive illness. *Nature* 1977; **268**:327–9.

Menkes DB, Coates DC, Fawcett JP. Acute tryptophan depletion aggravates premenstrual syndrome. *J Affect Disord* 1994; **32**:37–44.

Metz A, Stump K, Cowen PJ *et al.* Changes in platelet alpha 2-autoreceptor binding post partum: possible relation to maternity blues. *Lancet* 1983; **1**:495–8.

Miller NL, Kocsis JH, Leon AC *et al.* Maintenance desipramine for dysthymia: a placebo-controlled study. *J Affect Disord* 2001; **64**:231–7.

Mitchell PB, Wilhelm K, Parker G *et al.* The clinical features of bipolar depression: a comparison with matched major depressive patients. *J Clin Psychiatry* 2001; **62**:212–16.

Moses EL, Mallinger AG. St. John's wort: three cases of possible mania induction. *J Clin Psychopharmacol* 2000; **20**:115–17.

Nappi RE, Petraglia F, Luisi S *et al.* Serum allopregnanolone in women with postpartum 'blues'. *Obstet Gynecol* 2001; **97**:77–80.

Nemeroff CB, Widerlow E, Bissete G *et al.* Elevated concentrations of CSF corticotropin-releasing factor-like immunoreactivity in depressed patients. *Science* 1984; **226**:1342–4.

Neumeister A, Nugent AC, Waldeck T *et al.* Neural and behavioral responses to tryptophan depletion in unmedicated patients with remitted major depressive disorder and controls. *Arch Gen Psychiatry* 2004; **61**:765–73.

O'Reardon JP, Solvason HB, Janicak PG *et al.* Efficacy and safety of transcranial magnetic stimulation treatment of major depression: a multisite randomized controlled trial. *Biol Psychiatry* 2007; **62**:1208–16.

Papakostas GI, Mischoulon D, Shyu I *et al.* S-adenosyl methionine (SAMe) augmentation of serotonin reuptake inhibitors for antidepressant nonresponders with major depressive disorder: a double-blind, randomized clinical trial. *Am J Psychiatry* 2010; **167**:942–8.

Pearlstein TB, Stone AB. Long-term fluoxetine treatment of late luteal phase dysphoric disorder. *J Clin Psychiatry* 1994; **55**:332–5.

Pearlstein TB, Stone AB, Lund SA *et al.* Comparison of fluoxetine, bupropion, and placebo in the treatment of premenstrual dysphoric disorder. *J Clin Psychopharmacol* 1997; **17**:261–6.

Pearlstein TB, Bachmann GA, Zacur HA *et al.* Treatment of premenstrual dysphoric disorder with a new drosperenone-containing oral contraceptive formulation. *Contraception* 2005; **72**:414–21.

Perlis ML, Smith MT, Orff HJ *et al.* The effects of orally administered cholinergic agonists on REM sleep in major depression. *Biol Psychiatry* 2002; **51**:457–62.

Perlis RH, Baker RW, Zarate CA *et al.* Olanzapine versus risperidone in the treatment of manic or mixed states in bipolar I disorder: a randomized, double-blind trial. *J Clin Psychiatry* 2006; **67**:1747–53.

Perry PJ. Pharmacotherapy for major depression with melancholic features: relative efficacy of tricyclic versus selective serotonin reuptake inhibitor antidepressants. *J Affect Disord* 1996; **39**:1–6.

Peselow ED, Dunner DL, Fieve RR *et al.* Lithium prophylaxis of depression in unipolar, bipolar II, and cyclothymic patients. *Am J Psychiatry* 1982; **139**:747–52.

Pfohl B, Vaquez N, Nasrallah H. Unipolar vs. bipolar mania: a review of 247 patients. *Br J Psychiatry* 1982; **141**:453–8.

Pitt B. Maternity blues. *Br J Psychiatry* 1973; **122**:431–3.

Poirier MF, Boyer P. Venlafaxine and paroxetine in treatment-resistant depression. Double-blind, randomized comparison. *Br J Psychiatry* 1999; **175**:12–6.

Post RM, Unde TW, Roy-Byrne PP *et al.* Antidepressant effects of carbamazepine. *Am J Psychiatry* 1986; **143**:29–34.

Prange AJ, Lara PP, Wilson IC *et al.* Effect of thyrotropin-releasing hormone in depression. *Lancet* 1972; **2**:999–1002.

Prien RF, Kupfer DJ. Continuation drug therapy for major depression episodes: how long should it be maintained? *Am J Psychiatry* 1986; **143**:18–23.

Quitkin FM, McGrath PJ, Stewart JW *et al.* Chronological milestones to guide drug change. When should clinicians switch antidepressants? *Arch Gen Psychiatry* 1996; **53**:785–92.

Raadsheer FC, van Heerikhuize JJ, Lucassen PJ *et al.* Corticotropin-releasing hormone mRNA in the paraventricular nucleus of patients with Alzheimer's disease and depression. *Am J Psychiatry* 1995; **152**:1372–6.

Rabins PV, Merchant A, Nestadt G. Criteria for diagnosing reversible dementia caused by depression: validation by 2-year follow-up. *Br J Psychiatry* 1984; **144**:488–92.

Ravindran AV, Anisman H, Merali Z *et al.* Treatment of primary dysthymia with group cognitive therapy and pharmacotherapy: clinical symptoms and functional impairments. *Am J Psychiatry* 1999; **156**:1608–17.

Ravindran AV, Guelfi JD, Lane RM *et al.* Treatment of dysthymia with sertraline: a double-blind, placebo-controlled trial in dysthymic patients without major depression. *J Clin Psychiatry* 2000; **61**:821–7.

Reeves H, Batr S, May RS *et al.* Efficacy of risperidone augmentation to antidepressants in the management of suicidality in major depressive disorder: a randomized, double-blind, placebo-controlled pilot study. *J Clin Psychiatry* 2008; **69**:1228–336.

Reimherr FW, Amsterdam JD, Quitkin FM *et al.* Optimal length of continuation therapy in depression: a prospective assessment during long-term fluoxetine treatment. *Am J Psychiatry* 1998; **155**:1247–53.

Roca CA, Schmidt PJ, Smith MJ *et al.* Effects of metergoline on symptoms in women with premenstrual dysphoric disorder. *Am J Psychiatry* 2002; **159**:1876–81.

Rohde LA, Busnello E, Wolf A *et al.* Maternity blues in Brazilian women. *Acta Psychiatr Scand* 1997; **95**:231–5.

Rosenthal NE, Rosenthal LN, Stallone F *et al.* Psychosis as a predictor of response to lithium maintenance treatment in bipolar affective disorder. *J Affect Disord* 1979; **1**:237–45.

Rosenthal NE, Rosenthal LN, Stallone F *et al.* Toward the validation of RDC schizoaffective disorder. *Arch Gen Psychiatry* 1980; **37**: 804–10.

Roy-Byrne PR, Weingartner H, Bierer LM *et al.* Effortful and automatic cognitive processes in depression. *Arch Gen Psychiatry* 1986; **43**:265–7.

Rush AJ, Erman MK, Giles DE *et al.* Polysomnographic findings in recently drug free and clinically remitted depressed patients. *Arch Gen Psychiatry* 1986; **43**:878–84.

Rush AJ, Giles DE, Schlesser MA *et al.* Dexamethasone response, thyrotropin-releasing hormone stimulation, rapid eye movement latency and subtypes of depression. *Biol Psychiatry* 1997; **41**: 915–28.

Sachs GS, Grossman F, Ghaemi SN *et al.* Combination of mood stabilizer with risperidone or haloperidol for treatment of acute mania: a double-blind, placebo-controlled comparison of safety and efficacy. *Am J Psychiatry* 2002; **159**:1146–54.

Sachs GS, Chengappa KN, Suppes T *et al.* Quetiapine with lithium or divalproex for the treatment of bipolar mania: a randomized, double-blind, placebo-controlled study. *Bipolar Disord* 2004; **6**:213–23.

Sachs GS, Nieenberg AA, Calabrese JR *et al.* Effectiveness of adjunctive antidepressant treatment for bipolar depression. *N Engl J Med* 2007; **356**:1711–22.

Schellenberg R. Treatment for the premenstrual syndrome with agnus castus fruit extract: prospective, randomized, placebo controlled study. *BMJ* 2001; **322**:134–7.

Shulman KI, Tohen M. Unipolar mania revisited: evidence from an elderly cohort. *Br J Psychiatry* 1994; **164**:547–9.

Sitaram N, Gillen JC, Bunney WE. The switch process in manic–depressive illness: circadian variation in time of switch and sleep and manic ratings before and after switch. *Acta Psychiatr Scand* 1978; **58**:267–78.

Smith KA, Fairburn CG, Cowen PJ. Relapse of depression after rapid depletion of tryptophan. *Lancet* 1997; **349**:915–19.

Solomon DA, Ryan CE, Keitner GI *et al.* A pilot study of lithium carbonate plus divalproex sodium for the continuation and maintenance treatment of patients with bipolar I disorder. *J Clin Psychiatry* 1997; **58**:95–9.

Starkstein SE, Petraca G, Teson A *et al.* Catatonia in depression: prevalence, clinical correlates, and validation of a scale. *J Neurol Neurosurg Psychiatry* 1996; **60**:326–32.

Steiner M, Steinberg S, Stewart D *et al.* Fluoxetine in the treatment of premenstrual dysphoria. *N Engl J Med* 1995; **332**: 1529–34.

Stevens JW. Manic–depressive insanity, with the report of a typical case. *J Nerv Ment Dis* 1904; **31**:513–25.

Stunkard AJ, Fernstrom MH, Price A *et al.* Direction of weight gain in recurrent depression: consistency across episodes. *Arch Gen Psychiatry* 1990; **47**:857–60.

Summers WK. Mania with onset in the eighth decade: two cases and a review. *J Clin Psychiatry* 1983; **44**:141–3.

Suppes T, Vieta E, Brecher M *et al.* Maintenance treatment for patients with bipolar I disorder: results from a North American study of quetiapine in combination with lithium or divalproex (trial 127). *Am J Psychiatry* 2009; **166**:476–88.

Swann AC, Bosden CL, Morris D *et al.* Depression during mania. Treatment response to lithium or divalproex. *Arch Gen Psychiatry* 1997; **54**:37–42.

Szegedi A, Jansen WT, van Willigenburg AP *et al.* Early improvement in the first 2 weeks as a predictor of treatment outcome in patients with major depressive disorder: a meta-analysis including 6562 patients. *J Clin Psychiatry* 2009; **70**:344–53.

Taylor MA, Abrams R. The phenomenology of mania: a new look at some old patients. *Arch Gen Psychiatry* 1973; **29**:520–2.

Taylor MA, Abrams R. Catatonia: prevalence and importance in the manic phase of manic-depressive illness. *Arch Gen Psychiatry* 1977; **34**:1223–5.

Terao T. Subclinical hypothyroidism in recurrent mania. *Biol Psychiatry* 1993; **38**:853–4.

Thase ME. Relapse and recurrence in unipolar major depression: short-term and long-term approaches. *Compr Psychiatry* 1990; **51**(Suppl. 26):51–7.

Thase ME, Fava M, Halbreich U *et al.* A placebo-controlled, randomized clinical trial comparing sertraline and imipramine for the treatment of dysthymia. *Arch Gen Psychiatry* 1996; **53**:777–84.

Tohen M, Chengappa KN, Suppes T *et al.* Relapse prevention in bipolar I disorder: 18-month comparison of olanzapine plus mood stabiliser v. mood stabiliser alone. *Br J Psychiatry* 2004; **184**: 37–45.

Tohen M, Greil W, Calabrese JR *et al.* Olanzapine versus lithium in the maintenance treatment of bipolar disorder: a 12-month, randomized, double-blind, controlled clinical trial. *Am J Psychiatry* 2005; **162**:1281–90.

Trivedi MH, Thase ME, Osuntokum O *et al.* An integrated analysis of olanzapine/fluoxetine combination in clinical trials of treatment-resistant depression. *J Clin Psychiatry* 2009; **70**:387–96.

Van Valkenburg C, Winokur G, Behar D *et al.* Depressed women with panic attacks. *J Clin Psychiatry* 1984; **45**:367–9.

Vanelle JM, Attar-Levy P, Poirier MF *et al.* Controlled efficacy study of fluoxetine in dysthymia. *Br J Psychiatry* 1997; **170**:345–50.

Vazquez GH, Kahn C, Schiavo CE *et al.* Bipolar disorders and affective temperaments: a national family study testing the 'endophenotype' and 'subaffective' theses using the TEMPS-A Buenos Aires. *J Affect Disord* 2008; **108**:25–32.

Vasudev K, Goswami U, Kohli K. Carbamazepine and valproate monotherapy: feasibility, relative safety and efficacy, and therapeutic drug monitoring in manic disorder. *Psychophramocolgy* 2000; **150**:15–23.

Vieta E, Bourin M, Sanchez R *et al.* Effectiveness of aripiprazole v. haloperidol in acute bipolar mania: a double-blind, randomized, comparative 12-week trial. *Br J Psychiatry* 2005; **187**:235–42.

Vieta E, Suppes T, Eggens I *et al.* Efficacy and safety of quetiapine in combination with lithium or divalproex for maintenance of patients with bipolar I disorder (international trial 126). *J Affect Disord* 2008a; **109**:251–63.

Vieta E, T'joen C, McQuade RD *et al.* Efficacy of adjunctive aripiprazole to either valproate or lithium in bipolar mania patients partially nonresponsive to valproate/lithium monotherapy: a placebo-controlled study. *Am J Psychiatry* 2008b; **165**:1316–25.

Viguera AC, Nonacs R, Cohen LS *et al.* Risk of recurrence of bipolar disorder in pregnant and nonpregnant women after discontinuing lithium maintenance. *Am J Psychiatry* 2000; **157**:179–84.

Watson S, Gallagher P, Ritchie JC *et al.* Hypothalamic–pituitary–adrenal axis function in patients with bipolar disorder. *Br J Psychiatry* 2004; **184**:496–502.

Wertham FI. A group of benign psychoses: prolonged manic excitements. *Am J Psychiatry* 1929; **86**:17–78.

Winokur G. Psychosis in bipolar and unipolar affective illness with special reference to schizo-affective disorder. *Br J Psychiatry* 1984; **145**:236–42.

Winokur G, Tsuang MT. Elation versus irritability in mania. *Compr Psychiatry* 1975; **16**:435–6.

Winokur G, Clayton PJ, Reich T. *Manic Depressive Illness*. St. Louis, MO: Mosby, 1969.

Winokur G, Coryell W, Endicott J *et al.* Further distinction between manic–depressive illness (bipolar disorder) and primary depressive disorder (unipolar depression). *Am J Psychiatry* 1993; **150**:1176–81.

Wisner KL, Peindel KS, Gigliotti T *et al.* Obsessions and compulsions in women with postpartum depression. *J Clin Psychiatry* 1999; **60**:17–80.

Wisner KL, Perel JM, Peindel KS *et al.* Prevention of postpartum depression: a randomized clinical trial. *J Clin Psychiatry* 2001; **62**:82–6.

Wisner KL, Perel JM, Peindel KS *et al.* Prevention of postpartum depression: a pilot randomized clinical trial. *Am J Psychiatry* 2004; **161**:1290–2.

Wisner KL, Hanusa BH, Perel JM *et al.* Postpartum depression: a randomized trial of sertraline versus nortriptyline. *J Clin Psychopharmacol* 2006; **26**:353–60.

Wyatt KM, Dimmock PW, Ismail KM *et al.* The effectiveness of GnRHa with and without 'add-back' therapy in treating premenstrual syndrome: a meta analysis. *BJOG* 2004; **111**:585–93.

Yalom ID, Lunde DT, Moos RH *et al.* Postpartum blues' syndrome: a description and related variables. *Arch Gen Psychiatry* 1968; **18**:16–27.

Yonkers KA, Halbreich U, Freeman E *et al.* Symptomatic improvement of premenstrual dysphoric disorder with sertraline treatment: a randomized controlled trial. *JAMA* 1997; **278**: 983–8.

Young SA, Hurt PH, Benedek DM *et al.* Treatment of premenstrual dysphoric disorder with sertraline during the luteal phase: a randomized, double-blind, placebo-controlled crossover trial. *J Clin Psychiatry* 1998; **59**:76–80.

Young AH, McElroy SL, Bauer M *et al.* A double-blind, placebo-controlled study of quetiapine and lithium monotherapy in adults in the acute phase of bipolar depression (EMBOLDEN I). *J Clin Psychiatry* 2010; **71**:150–62.

24

Anxiety disorders

24.1 PANIC DISORDER

Panic disorder is characterized by the repeated occurrence of discrete anxiety attacks, which, in the context of this disorder, are referred to as 'panic attacks'. This is a common disorder, with a lifetime prevalence of roughly 3 percent, and is two to three times more common in females than males.

Clinical features

The first panic attack generally occurs in late adolescence or early adult years; however, the range is wide, from childhood to the fifth decade.

The panic attack itself usually comes on acutely, often within a minute, and symptoms crescendo rapidly. Anxiety is typically severe, and patients may report the classic 'sense of impending doom', as if some catastrophic event were looming; some fear that they are having a heart attack, others that they are about to 'go crazy'. Uncommonly, however, one may see attacks in which the feeling of anxiety is mild relative to other symptoms, and rarely patients may deny any significant anxiety at all: although the existence of these 'panic attacks without panic' was initially controversial, there is no doubt that they do in fact occur (Russell *et al.* 1991). Other symptoms include tremor, tachycardia, palpitations, chest pain, dyspnea, dizziness, nausea, diaphoresis, and acral parasthesiae; rarely, one may see hemianesthesia, macropsia, or microspia (Coyle and Sterman 1986). Chest pain may be most alarming to patients, and, as this pain may at times radiate to the left shoulder or left side of the neck, it may likewise cause some alarm in emergency room physicians. The attack itself generally lasts about 5–15 minutes, but may persist for up to an hour; symptoms then remit over a few minutes, after which patients may feel 'shaken', drained, and apprehensive for up to a few hours. Although most attacks occur during waking hours, some patients may have nocturnal attacks (Mellman and Uhde 1989a), and, in a small minority, attacks may occur only nocturnally (Mellman and Uhde 1990). Nocturnal attacks arise from non-rapid eye movement (NREM) sleep (Mellman and Uhde 1989b) and often awaken the patient, who typically has trouble falling back asleep and is able to recall the panic the next morning (Hauri *et al.* 1989).

In most cases, panic attacks occur spontaneously, without any precipitating factors, and patients complain that they 'come out of the blue' and strike without warning.

Course

The frequency with which panic attacks occur varies widely among patients, from the extremes of multiple attacks daily to only one attack every few months. The overall course of the disorder appears to follow one of two patterns. In one, the frequency of attacks may vary over years or decades, with the patient never experiencing any prolonged attack-free interval. In the other, one does see prolonged attack-free intervals, and in patients with this pattern one may speak of panic disorder with an episodic course; the 'episodes', characterized by recurrent attacks, are separated from each other by prolonged intervals in which no spontaneous attacks occur.

Over time, and with repeated attacks, most patients begin to develop a chronic, anxious apprehension that another attack may be just around the corner. This 'anticipatory anxiety' may induce patients to avoid situations in which, should they have an attack, they might not be able to find immediate help; thus, patients may avoid travelling by plane or boat, or even on limited-access highways. Should this anticipatory anxiety become very severe, patients may develop the syndrome of agoraphobia, described in the next chapter.

Abuse of alcohol or benzodiazepines is a serious risk. Some patients may use these to quell their anticipatory anxiety, whereas others will consume them during the attack itself, in the mistaken belief that oral administration of these agents will result in pharmacologic blood levels before the attack runs its course. In cases in which chronic use of these agents leads to neuroadaptation, any attempt to stop them may lead to a withdrawal state that, in turn, may precipitate further attacks, thus setting up a vicious cycle.

Etiology

Panic disorder is probably hereditary: as the degree of consanguinity rises from unrelated persons to first-degree relatives, then to dizygotic and finally monozygotic twins, so too does the risk of having panic disorder (Crowe *et al.* 1983; Noyes *et al.* 1986; Torgerson 1990). Genetic studies, however, have not as yet yielded robust findings, and neither the mode of inheritance nor candidate loci have been clearly demonstrated.

The discovery of 'panicogens' has been one of the major fruits of the research efforts into panic disorder. Panicogens are substances which, though innocuous in normal control subjects, reliably produce panic attacks in patients with panic disorder. Importantly, these panicogen-induced attacks are identical to spontaneously occurring attacks and, furthermore, may be prevented by the same medicines that are effective in the treatment of panic disorder. Panicogens include inhalation of 5 or 35 percent carbon dioxide (Gorman *et al.* 1988), intravenous sodium lactate (Cowley and Arana 1990; Liebowitz *et al.* 1984), caffeine (Charney *et al.* 1985), yohimbine (Charney *et al.* 1992), isoproterenol (Pohl *et al.* 1988), intravenous meta-chlorophenylpiperazine (Kahn and Wetzler 1991), flumazenil (Nutt *et al.* 1990) (although not all studies concur with this [Strohle *et al.* 1998]), and cholecystokinin tetrapeptide (Bradwejn *et al.* 1991).

Various neurotransmitters have also been implicated, including norepinephrine, serotonin, and gamma-aminobutyric acid (GABA). Norepinephrine is implicated by the panicogenic effects of caffeine, yohimbine, and isoproterenol. Serotonin involvement is suggested by the response to the panicogen meta-chlorophenylpiperazine and by studies involving manipulation of serotonin levels. For example, depletion of the serotonin precursor tryptophan increases the effectiveness of panicogens such as flumazenil (Bell *et al.* 2002), whereas the administration of l-5-hydroxytryptophan, a serotonin precursor, will blunt the panicogenic effect of carbon dioxide inhalation (Schruers *et al.* 2002). GABAergic involvement is suggested by the panicogenic effect of flumazenil and also by studies that have demonstrated reduced levels of GABA within the occipital cortex (Goddard *et al.* 2001a).

Integrating all of these findings into a coherent theory regarding the pathophysiology of panic disorder is problematic and does require some speculation. With this caveat in mind it appears plausible to say that panic disorder results from an inherited disturbance in the metabolism of norepinephrine, serotonin, and/or GABA in one or more of the central nervous system structures that subserve the experience of anxiety. Candidate structures include the locus ceruleus, the dorsal raphe nucleus, the parahippocampal gyrus, hippocampus, and amygdala. The locus ceruleus is noradrenergic and the dorsal raphe nucleus is serotoninergic, and both send fibres to a large number of structures, including the parahippocampus, hippocampus, and amygdala, structures that are rich in GABA receptors. Stimulation of the parahippocampus, hippocampus, and, especially, the amygdala is well known to produce anxiety. If one considers the amygdala as being the 'final common pathway' in the development of anxiety, then dysfunction of any one of these upstream structures, or of the amygdala itself, could produce an attack. It has been proposed that one possible 'trigger' for the activation of this circuitry is an abnormal sensitivity of brainstem structures to disturbances in the acid–base balance (Klein 1993). Such a sensitivity, in turn, could explain the panicogenic effects of lactate infusion and carbon dioxide inhalation.

Before leaving this section on etiology, some words are in order regarding mitral valve prolapse. Although there is a clear association between this disorder and panic disorder (Katerndahl 1993), it is probably not etiologic in a direct sense. In all likelihood the association is probably secondary to some, as yet unidentified, common factor that underlies both disorders.

Differential diagnosis

As noted earlier, the term 'panic attack' refers to an anxiety attack that happens to occur secondary to panic disorder, and, consequently, in pursuing the differential diagnosis of panic disorder one must consider the various other causes of anxiety attacks, as discussed in Chapter 7.5.

Occasionally, an otherwise normal individual will have an isolated anxiety attack, and, after thorough investigation, no clear cause may be found. In such cases one is tempted to make a diagnosis of panic disorder; however, by convention, this diagnosis should probably be withheld until subsequent attacks have occurred and one is able to demonstrate that, at least at some point in the course, a frequency of once monthly or more has been observed.

Anxiety attacks may also be seen as part of the depressive episodes of either major depressive disorder or bipolar disorder; if their occurrence is confined solely within the limits of the depressive episode, an additional diagnosis of panic disorder is not warranted. However, if one can demonstrate that the anxiety attacks preceded the onset of the depressive episode, or persisted beyond the resolution of the episode, then the additional diagnosis is appropriate.

Of the other causes of anxiety attacks noted in Chapter 7.5, some of the most common are simple phobia, social phobia, post-traumatic stress disorder, and obsessive–compulsive disorder. In all of these disorders, however, the anxiety attacks are precipitated. For example, if the simple phobic has to approach a snake, the social phobic public speaking, the post-traumatic patient a situation reminiscent of the original trauma, or the obsessive–compulsive a contaminated object, a severe anxiety attack may indeed occur. If, however, these precipitating situations may be avoided, these patients remain free of attacks.

Of the less common causes of anxiety attacks noted in Chapter 7.5, special consideration should be given to myocardial infarction or angina, paroxysmal atrial tachycardia, hypoglycemia, hyperventilation, and simple partial seizures. As noted earlier, panic attacks may be accompanied by chest pain, which may radiate to the neck or shoulder, hence suggesting a diagnosis of cardiac disease, and in these cases the general medical setting is very helpful. Clearly, if the patient is elderly and has known risk factors and no history of anxiety attacks, then one would lean towards a diagnosis of coronary artery disease; by contrast, if the patient is young and lacks risk factors, one might be inclined to lean towards a diagnosis of panic disorder. An episode of paroxysmal atrial tachycardia may occasionally prompt considerable anxiety in the sufferer. However, here one finds a hyperacute onset of the tachycardia, over a second or so, in contrast to the build-up of a panic attack, which occurs over a minute or so; furthermore, in paroxysmal atrial tachycardia a valsalva manuever may terminate the attack, whereas such a manuever has no effect on a panic attack. Hypoglycemia should be suspected in the case of a diabetic who has missed a meal, and in whom the attack is associated with hunger; prompt relief with glucose confirms the suspicion. Hyperventilation is suggested by the prominent dyspnea and by relief with re-breathing through a paper bag. Simple partial seizures are suggested by an exquisitely paroxysmal onset, over seconds, and by a history in most cases of other, more obvious seizure types, such as complex partial seizures or grand mal seizures.

Treatment

Pharmacologic treatment of panic disorder has as its goal the prevention of future panic attacks, and the first step is to choose one of the various medications that are effective in this regard, including antidepressants (various selective serotonin reuptake inhibitors [SSRIs], venlafaxine, and the tricyclic antidepressants imipramine and clomipramine) and various benzodiazepines. Cognitive–behavioral therapy (Barlow *et al.* 2000; Beck *et al.* 1992) is also effective and constitutes a reasonable alternative to pharmacologic treatment; should patients fail to respond, then one of the agents discussed below may be tried.

Among the antidepressants, the following SSRIs, with their average effective doses, may be used: fluoxetine (20 mg) (Michelson *et al.* 2001), paroxetine (40 mg) (Ballenger *et al.* 1998; Oehrberg *et al.* 1995), sertraline (50–150 mg) (Dohl *et al.* 1998; Londborg *et al.* 1998), citalopram (20 mg) (Wade *et al.* 1997), escitalopram (10 mg) (Stahl *et al.* 2003), and fluvoxamine (200 mg) (Asnis *et al.* 2001). Venlafaxine (Pollack *et al.* 2007) is used in a dose of 225 mg, imipramine (Mavissakalian and Perel 1989) in a dose of 150–200 mg, and clomipramine (Modigh *et al.* 1992) in a dose of 50–150 mg. As all of these are of approximately equal effectiveness, the choice among them is generally based on side-effect profile and ease of use, and with these criteria as guides, an SSRI is probably the first choice (with the exception of fluvoxamine, whose side-effects and many problematic drug–drug interactions make this a second choice). Regardless of which antidepressant is chosen, it is important to start with a low dose, as initiation of treatment at a full dose may cause a 'flurry' of panic attacks. Thus, one should start with anywhere from one-tenth to one-third of the full dose (breaking up tablets, if need be), followed by upward titration in similar increments every week or so. Importantly, however, although most patients require, and eventually tolerate, a full dose, there is a small minority of patients who do not tolerate anything more than a minimal dose, which they do benefit from; consequently, if patients do not tolerate a dose escalation, one option is to continue with a low dose and see if they respond. Once an optimum dose has been reached, it may take up to 3 months to see a good prophylactic effect.

Among the benzodiazepines, choices include clonazepam (Rosenbaum *et al.* 1997; Tesar *et al.* 1991), alprazolam (Ballenger *et al.* 1988), lorazepam (Schweizer *et al.* 1990), and diazepam (Noyes *et al.* 1996). Each should be started at a low dose, with the understanding that dose increases may be required to prevent attacks. Total daily starting doses and potential final doses are as follows: clonazepam, starting at 0.5–1.5 mg, increasing to 1–4 mg; aprazolam, starting at 0.75–1.5 mg, increasing to 1.5–6 mg; lorazepam, starting at 1–2 mg, increasing to 2–6 mg; and diazepam, starting at 4–10 mg and increasing to 10–60 mg. In general, the total daily dose should be divided into a twice- or thrice-daily schedule. Clearly, given the wide ranges of starting and potential final doses here, considerable clinical judgment is required. Importantly, however, and unlike the antidepressants, these benzodiazepines are immediately effective once the 'right' dose has been reached; consequently, if patients continue to have attacks, this is an indication that their current dose will not be effective and that the dose should promptly be increased.

Many clinicians, concerned about the possibility of neuroadaptation with benzodiazepines, and the risk of precipitating a flurry of attacks if the patient runs out of medication or tapers the dose too rapidly (Pecknold *et al.* 1988; Rickels *et al.* 1993a), lean towards trying an antidepressant first. However, in some cases, especially those characterized by very frequent attacks and considerable anticipatory anxiety, some clinicians will 'split the difference' and start the patient on a combination of a benzodiazepine and an antidepressant (e.g. clonazepam and sertraline [Goddard *et al.* 2001b]), with plans to gradually taper the benzodiazepine once enough time has passed for the antidepressant to take effect.

Before leaving this discussion of the pharmacologic treatement of panic disorder, mention should be made of inositol. This is a naturally occurring isomer of glucose that is normally converted to inositol 1,4,5-triphosphate, which in turn functions as an intracellular second messenger, and doses of 6000–9000 mg twice daily have been found to be superior to placebo and comparable to fluvoxamine in preventing panic attacks (Palatnik *et al.* 2001). Although experience with inositol is relatively limited, it may be a reasonable option in cases of intolerance to other agents or for patients who prefer 'natural' remedies.

24.2 AGORAPHOBIA

Agoraphobia is derived from the Greek meaning, literally, a fear of the marketplace or, more generally, a fear of open spaces. However, on close questioning, what becomes apparent is that patients suffering from agoraphobia do not in fact fear a particular place or situation, but rather the possibility of becoming ill or incapacitated in some fashion or other and not being able to either escape or get immediate help.

As discussed further below, there are probably two kinds of agoraphobia (Horwarth *et al.* 1993; Lelliott *et al.* 1989). In one type, seen in the vast majority of cases, agoraphobia represents a complication of panic disorder, and what patients fear is being incapacitated by a panic attack while away from home and safety; in the other type, there is no history of panic disorder, and patients seem unable to clearly delineate what they fear might happen.

The lifetime prevalence of agoraphobia ranges from 1 to 5 percent; it is probably two to four times more common in females than males.

Clinical features

The onset is generally in the twenties or thirties; in those cases in which agoraphobia represents a complication of panic disorder, anywhere from as little as a few days to up to a year or more may elapse between the occurrence of the first panic attack and the onset of the agoraphobia.

A large number of places and situations may be feared and avoided (Page 1994), including travelling on airplanes or trains, travelling across bridges or through tunnels, and, in extreme cases, simply waiting in long lines, and the prospect of being caught in such a situation may fill patients with a catastrophic, yet difficult to describe, sense of dread (Goisman *et al.* 1995). Some patients become completely housebound and literally may never set foot outside the house for years. Interestingly, in many cases patients are able to temporarily overcome their agoraphobia if they take their 'safety' with them by travelling with a good and trusted friend.

Course

Agoraphobia is a chronic disorder and tends to gradually worsen over long periods of time.

Etiology

In cases occurring secondary to panic disorder, the normal anticipatory anxiety about having another panic attack becomes so severe and pervasive that patients cannot bear venturing into 'unsafe' situations. In those cases that occur in the absence of panic disorder, patients, although describing well the dread they feel over being in agoraphobic situations, cannot clearly explain, either to themselves or others, what it is they fear might happen, and the etiology in these cases is simply not known.

Differential diagnosis

In social phobia of the generalized subtype (also known as social anxiety disorder), patients may fear and avoid many different situations. The difference here is that social phobics fear that they will do something that embarrasses or humiliates them, whereas agoraphobics fear that something will happen to them, such as a panic attack.

In illnesses characterized by delusions of persecution, patients may avoid going out for fear that people will talk about them, spy on them, assault them, etc. This may be seen in schizophrenia, schizoaffective disorder, the persecutory subtype of delusional disorder, and depressive episodes of either major depressive disorder or bipolar disorder.

Treatment

Agoraphobia may be treated either with cognitive–behavioral therapy or a behavioral program of graded exposure, in which the patient gradually and sequentially takes progressively greater 'steps' towards and into the feared situation. Critically, however, in those cases occurring secondary to panic disorder, one must first treat the panic disorder and render the patient free of panic attacks. The occurrence of even one panic attack can, and often does, destroy any gains made by cognitive–behavioral or exposure therapy.

24.3 SPECIFIC (SIMPLE) PHOBIA

The patient with a specific phobia, or, as it was formerly known, a simple phobia, experiences extreme anxiety upon approaching something that for others arouses little or no apprehension. Although adult patients acknowledge the irrationality of their fear, they go out of their way to avoid the feared object. First described by Hippocrates, this condition is quite common, occurring in at least 10 percent of the general population; it is about twice as frequent in females as in males.

In the past it was fashionable to subdivide specific phobia according to the feared object or situation, and thus one reads of arachnophobia, acrophobia, claustrophobia, etc. For the most part, however, this subdividing added little to our understanding of the disorder, with one probable exception, namely blood-injury phobia, which, as noted below, may be a unique specific phobia.

Clinical features

The age of onset of specific phobia ranges from childhood to early adult years: animal phobias and blood-injury phobia tend to first appear in childhood, whereas the other phobias may first appear at any point from childhood to adult years (Marks and Gelder 1966).

A wide range of objects or situations may come to be feared, including snakes, spiders, heights, being in closed spaces, darkness, storms, and the sight of blood. Although most patients with specific phobia have only one phobia, in a minority two or more may be present.

On approaching the feared object or situation, or even upon simply imagining doing so, patients experience the acute onset of an anxiety attack, characterized by anxiety, tremor, tachycardia, diaphoresis, and piloerection; in some cases, these symptoms may be accompanied by depersonalization. Some patients may be able to steel themselves and stay nearby, but for the most part the fear and anxiety is so great that they must escape, no matter how humiliating or embarrassing such behavior might be for them. Importantly, as soon as patients can get away, the anxiety rapidly subsides and patients, although perhaps a little 'shaken' by the experience, promptly return to normal.

Blood-injury phobia, as indicated earlier, may be unique. A common example is found in patients who are phobic about having a venipuncture (Chapman *et al.* 1993). If these patients can force themselves to hold still for the phlebotomist, they typically experience a biphasic symptomatology. Initially, there is anxiety and tachycardia, similar to that seen in the anxiety attack just described; soon after, however, these sympathetic symptoms give way to a parasympathetic response, with hypotension and either pre-syncope or actual fainting (Curtis and Thyer 1983). It is this second phase that gives blood-injury phobia its unique status.

Course

Phobias with an onset in childhood tend to remit within months or a year or so. In cases of childhood-onset phobias that persist beyond a year or so, and in cases characterized by later onsets in adolescence or early adult years, the course tends to be chronic.

Etiology

Specific phobias run in families (Fyer *et al.* 1990), although it is not clear whether this reflects environmental or genetic factors, or both. Some authors believe that children become phobic by virtue of modeling their parents' behavior; however, this theory begs the question as to the source of the parents' phobias. Other authors suggest that phobias represent a vestigial remnant from our ancient evolutionary history, when an instinctual avoidance of certain objects, such as snakes, could have conferred considerable survival value. In support of this evolutionary view is the fact that certain monkeys who have never had any contact with a snake nevertheless react with extreme fear upon seeing one.

In the case of blood-injury phobia, it appears that there is an underlying chronic dysfunction of the autonomic nervous system, which predisposes these patients to the development of vaso-vagal syncope (Accurso *et al.* 2001; Donadio *et al.* 2007).

Differential diagnosis

Specific phobia must be distinguished from social phobia of the circumscribed type, certain cases of obsessive–compulsive disorder, and post-traumatic stress disorder. Social phobia of the circumscribed type may be distinguished from specific phobia by the fact that, in social phobia, what is feared is humiliation or embarrassment in certain situations, rather than the situation *per se*. For example, the patient with a circumscribed social phobia of public speaking, although terrified at the prospect of speaking when others are present, may, when rehearsing the speech in private, have little or no anxiety: here, it is the apprehension of embarrassing oneself when speaking, rather than speaking itself, which the patient fears. This is in contrast to the patient with a specific phobia, for example of snakes, who reacts with anxiety upon seeing the snake regardless of whether this occurs in public or in private. Certain cases of obsessive–compulsive disorder may also cause diagnostic concern, for example when patients have a fear of contamination and avoid various objects or situations, such as shaking hands or using a public restroom. In these cases, however, the reaction to the feared object or situation persists long after contact ceases, for example when the patient with obsessive–compulsive disorder washes his or her hands repeatedly; this is in contrast to specific phobia, in which patients return to normal promptly upon disengaging from the phobic object or situation. Post-traumatic stress disorder may be confused with specific phobia when post-traumatic patients report avoiding certain situations. Here, however, one finds that the avoided situations have a specific meaning for patients in that they are reminiscent of an earlier traumatic event; this is in contrast to social phobia, in which patients admit that the fear is 'irrational' and not based on any earlier event.

Treatment

Both desensitization and cognitive–behavioral therapy are effective. The place of pharmacologic therapy of specific phobias is not entirely clear. Many patients are prescribed benzodiazepines on a chronic basis; however, there is no research support for this and it runs the risk of neuroadaptation. Brief, 'one time' use of benzodiazepines may, however, have a place, as, for example, in the pre-treatment of a claustrophobic patient before a magnetic resonance imaging (MRI) study in a 'closed' system or for the traveler with a fear of flying who simply cannot miss the flight. There is also one study that found paroxetine to be superior to placebo (Benjamin *et al.* 2000).

24.4 SOCIAL PHOBIA

In social phobia, patients fear that if they attempt to do certain things in public they will appear inept, foolish, or inadequate, and thus suffer shame, humiliation, or embarrassment. Feeling this way, patients, although admitting that their fears are perhaps groundless, nevertheless become intensely anxious on approaching these situations and may go to great lengths to avoid them.

Importantly, there are two subtypes of social phobia, namely the circumscribed subtype and the generalized subtype (Heimberg *et al.* 1990; Mannuzza *et al.* 1995). The circumscribed subtype is characterized by fears of only one or two specific social situations, whereas the generalized subtype is characterized by fears that span multiple situations. The generalized subtype is also known as 'social anxiety disorder'; however, the reader should use caution in examining the literature, as some authors use the phrase 'social anxiety disorder' to refer to both subtypes.

This is a common disorder, occurring in anywhere from up to 13 percent of the adult population; it is more frequent in females than males.

Clinical features

The onset of social phobia ranges from late childhood to the early adult years, with most patients falling ill in their mid-teens (Marks and Gelder 1966). Pre-morbidly, in some cases there is a history of shyness and easy blushing. The actual onset itself is heralded by the first wave of irrational anxiety over doing something in public.

In the generalized subtype, patients fear multiple situations, including, for example, answering questions in class, asking others out for dates, or attending meetings; in some cases patients may have a global fear of interacting socially in any situation.

In the circumscribed subtype there may be a fear of public speaking, or fears that one will tremble when writing in public, misplay notes when giving a musical performance, choke when eating in public, or, in males, be unable to urinate in a public restroom. Regardless of the specific circumscribed fear involved, it is important to note that it is not the activity itself that is feared but rather the performance of that activity in public. Thus, the patient with a fear of public speaking who is paralyzed into muteness upon approaching the podium with an audience present may nevertheless be able to make the speech flawlessly on rehearsal in an empty hall.

If patients do approach the phobic situation (or, at times, merely imagine doing so), they experience an anxiety attack of variable severity, with anxiety, tremor, tachycardia, diaphoresis, and dyspnea. In some cases patients may be able to endure the anxiety and proceed with the activity at hand; however, in many cases the anxiety is simply 'too much' and patients may refuse to go on. One variation on this reaction is known as erythrophobia. Here, what patients are concerned about is excessive blushing in public. In some cases there may actually be an obvious, scarlet blush; however, in others the blush is actually not noted by others but is simply experienced by the patient as a sense of flushing or uncomfortable warmth.

Course

Social phobia appears to be a chronic disorder.

Etiology

The etiology of the two types of social phobia may not be the same, and hence they are considered separately.

Generalized social phobia, which has been subjected to the most study, appears to be familial (Fyer *et al.* 1993) and to be associated with disturbances in various neurotransmitter systems, including serotonin, endogenous benzodiazepines, and dopamine. With regard to serotonin, several abnormalities have been noted, including an aggravation of symptoms with depletion of the serotonin precursor tryptophan (Argyropoulos *et al.* 2004), a decrease in the serotonin-1A binding potential on both platelets and neurons in the amygdala and cingulate cortex (Lanzenberger *et al.* 2007), and an augmented cortisol response to the serotoninergic agent fenfluramine (Tancer *et al.* 1994). In looking at endogenous benzodiazepine function, there appears to be a reduced number of platelet benzodiazepine receptors (Johnson *et al.* 1998). Finally, both a reduced density of dopamine receptors in the striatum (Schneier *et al.* 2000) and a decrease in striatal dopamine reuptake sites (Tiihonen *et al.* 1997a) have also been noted.

Regarding the circumscribed type of social phobia, little can be said except that it is suspected to reflect a disturbance in noradrenergic functioning.

Differential diagnosis

Agoraphobia may be distinguished from the generalized type of social phobia by the source of the patient's fear about going out in public. In the agoraphobic, what is feared is having a panic attack, regardless of whether anyone is around or not; this is in contrast to social phobics who fear that they will do something humiliating or embarrassing.

Specific phobia may be readily distinguished from social phobia of the circumscribed type by asking patients whether they can engage in the fearful behavior when in private; the patient with a specific phobia, for example of snakes, is no more able to approach the snake in private than in public, whereas the patient with a circumscribed social phobia, for example of speaking in public, may, as noted earlier, go through a rehearsal of the speech without anxiety, provided that there is no audience.

Similar to the generalized type of social phobia, body dysmorphic disorder is characterized by a fear of humiliation or embarrassment in public; however, in body dysmorphic disorder the fear arises from the patient's belief that he is in some way misshapen or deformed, rather than from a concern that he might suffer embarrassment because of something he does.

Certain conditions, by their very symptoms, may occasion some embarrassment, and in such cases a separate diagnosis of social phobia is not made. Examples include the tremor of essential tremor or of Parkinson's disease, and developmental stuttering.

One must also distinguish between normal fears and social phobia. Most individuals experience some apprehension or 'stage fright' when first approaching certain situations; however, in contrast to social phobia, the fear is not overwhelming and, importantly, diminishes after a few 'dress rehearsals'.

Treatment

The treatment of the generalized and circumscribed subtypes is different, and thus each is discussed separately, beginning with the generalized subtype.

A large number of medications are effective in the generalized type of social phobia, including various antidepressants, clonazepam, and two anti-epileptic drugs, gabapentin and pregabalin. Effective antidepressants include sertraline 50–200 mg/day (Katzelnick *et al.* 1995), paroxetine 20–40 mg/day (Allgulander 1999; Baldwin *et al.* 1999; Stein *et al.* 1998), escitalopram 10–20 mg/day (Kasper *et al.* 2005), fluvoxamine 150 mg/day (Stein *et al.* 1999; van Vliet *et al.* 1994), venlafaxine 75–225 mg/day (Liebowitz *et al.* 2005), and phenelzine 60–90 mg/day (Heimberg *et al.* 1998; Liebowitz *et al.* 1992; Versiani *et al.* 1992). The benzodiazepine clonazepam 0.5–3.0 mg/day (Davidson *et al.* 1993) is also effective, as are the anti-epileptics gabapentin 900–3600 mg/day (Pande *et al.* 1999) and pregabalin 600 mg/day (Pande *et al.* 2004). All other things being equal, one of the antidepressants (with the exception of phenelzine, which, given the difficulty in its use, should be held in reserve) is a reasonable first choice. Although experience with gabapentin and pregabalin is limited in social phobia, they might be a reasonable second choice. Although clonazepam is often used, the potential for neuroadaptation may give one pause. Cognitive–behavioral therapy is also effective, but may be slightly less so than treatment with phenelzine (Heimberg *et al.* 1998). Intuitively it makes sense to use medications and cognitive–behavioral therapy in combination, however, of the three studies (Blanco *et al.* 2010; Blomhoff *et al.* 2001; Davidson *et al.* 2004) that looked at this question, two (Blomhoff *et al.* 2001; Davidson *et al.* 2004) found no advantage for the combination.

Circumscribed social phobia, especially fears of speaking, writing, or giving a musical performance, may be treated with a beta-blocker such as propranolol, given on an as-needed basis in anticipation of the public performance (Brantigan *et al.* 1982; Hartley *et al.* 1983; James *et al.* 1977; Liden and Gottfries 1974). The dose of propranolol ranges from 20 to 60 mg, and the drug appears to work not so much by reducing anxiety *per se* as by reducing autonomic symptoms, such that patients no longer 'shake like a leaf'. Desensitization is also effective, and, interestingly, the effectiveness of exposure therapy for a public speaking phobia is enhanced by pre-treatment with 50 mg of d-cylcoserine (Guastella *et al.* 2008; Hofmann *et al.* 2006).

24.5 OBSESSIVE–COMPULSIVE DISORDER

Obsessive–compulsive disorder is characterized by obsessions and compulsions (Rasmussen and Tsuang 1986). It is a common disorder, with a lifetime prevalence of 2–3 percent.

Clinical features

Most patients fall ill in adolescence or early adult years; although onset in childhood is not rare, onset past the age of 40 years is quite rare.

The vast majority of patients experience both obsessions and compulsions (Foa *et al.* 1995), and the majority of patients will experience multiple obsessions and compulsions. In a very small minority of patients, however, only obsessions may occur, and, likewise, in a very small minority there will be only compulsions; very rarely one finds patients with a subtype known as primary obsessive slowness.

Obsessions are unwanted, intrusive, and troubling thoughts, impulses or images that repeatedly appear despite patients' attempts to stop them. In most cases, these give rise to a compulsion to do something: a typical example would be the appearance of a thought that one had touched something dirty followed by an anxious need to clean oneself. A minority of obsessions, however, though anxiety provoking, are not associated with a compelling need to do anything, and these obsessions often center on sexual or violent themes. Thus, a minister found himself repeatedly thinking of bestiality, while a neurologist, whenever interviewing patients, repeatedly experienced an urge to question his patients about the shape of their bowel movements.

Compulsions may, as just noted, arise from obsessions, and in some way or other serve to quell the anxiety associated with these obsessions. Compulsions are often categorized according to what patients feel compelled to do: thus, for example, there are checkers, washers, counters, arrangers, and hoarders. One checker, described by Tuke in 1894 (Tuke 1894), after going upstairs to bed, was beset with an obsessive concern that he had failed to turn off the electric lights. Despite being almost certain that he had, in fact, turned it off, his anxiety over the possibility that he had not rose to the point where had to go downstairs and check on it. Once he did, and found out that he had turned it off, he went back upstairs, only to experience again the doubt that he had left if on, which compelled him again to go downstairs and check on it. 'Washers' are obsessed with the thought that they have been contaminated in some way, perhaps by dirt, 'germs', or fecal material, and feel compelled to wash their hands, no matter how clean they may be or how many times they have already been washed; some may wash to the point of rawness, yet still feel unclean. 'Counters' feel compelled to count to a certain number and do it repeatedly, whereas 'arrangers', under the compulsion to arrange things 'just so', may spend much time arranging and then rearranging things on the desk or in the closet, always finding afterwards on inspection, however, that some critical point of symmetry has been missed. 'Hoarders' feel compelled to not discard things, with the result that almost everything may be saved, including not only clothing, books, furniture and the like, but also unusual, even bizarre, items, such as nail clippings, urine, and feces. These counters and arrangers and hoarders are sometimes at a loss to explain what they fear might happen should they fail to act; it is as if a there is a nameless dread motivating them. Although most compulsive behaviors are visible and evident to others, some patients may at times engage in 'silent' or private compulsions: one patient, obsessed with the thought that he had failed to say his prayers correctly, repeatedly and silently said the Lord's prayer until he was certain the recitation had been done perfectly.

Almost all patients with compulsions, at least initially, recognize the 'senselessness' or irrationality of their compulsions and will attempt to resist them (Stern and Cobb 1978); over the years, however, many will give in, carrying out the compulsions without even a token show of resistance. In a small minority of cases, patients may lose insight into the irrationality of their compulsive behavior and become convinced, to a delusional degree, that their behavior is in fact reasonable and appropriate (Eisen and Rasmussen 1993; Insel and Akiskal 1986). Such patients with psychotic obsessive–compulsive disorder, now finding their compulsions reasonable, pursue them with purpose rather than resisting them. One patient (Gordon 1950), after 25 years of resisting the urge to repeat his prayers for fear that he had left out a word, eventually came to believe that he was in fact a sinner and that God had given him the compulsion to specifically ensure that he said the prayers with the perfection that salvation and eternal life demanded; subsequently, he freely gave himself to the repeating of his prayers whenever he had the slightest doubt that he had left anything out.

Primary obsessive slowness is characterized by the transformation of routine daily activities into lengthy, meticulous rituals. Dressing or preparing a meal may take literally hours; the sequence and form of each step in the process is carefully and scrupulously observed, and any deviation leads to an accretion of severe anxiety. Only when completely satisfied that the job has been perfectly done can the patient move on to the next task of the day. Patients with primary obsessive slowness typically have had, or continue to have, other compulsions, such as checking or washing.

Course

Although in a small minority of cases a spontaneous remission may occur, in the natural course of events obsessive–compulsive disorder is chronic. In most chronic cases, symptoms very gradually wax and wane in intensity; in a small minority, however, there may be a progressive downhill course until patients' lives are literally consumed by their obsessions and compulsions.

Etiology

Obsessive–compulsive disorder is familial (Grabe *et al.* 2006; Pauls *et al.* 1995), and genetic research, although inconclusive, is promising (Shugart *et al.* 2006); in some cases it appears that obsessive–compulsive disorder may represent a phenotypic variant of the same genes responsible for Tourette's syndrome (Apter *et al.* 1993; Frankel *et al.* 1986). Although neuropathologic studies are lacking, positron emission tomography (PET) studies (Baxter *et al.* 1987, 1992; Benkelfat *et al.* 1990; Perani *et al.* 1995; Saxena *et al.* 2002) have demonstrated increased metabolic activity in the orbitofrontal cortex and caudate nucleus. Furthermore, several lines of evidence suggest a disturbance in serotoninergic functioning: all of the medicines effective in this disorder are serotoninergic, and meta-chlorophenylpiperazine, a mixed agonist/antagonist at postsynaptic serotonin receptors, generally increases the severity of symptoms (Broocks *et al.* 1998; Hollander *et al.* 1992).

In a minority of childhood-onset cases it appears that obsessive–compulsive disorder occurs on an autoimmune basis, via a mechanism similar to that seen in Sydenham's chorea, and such patients are said to have one of the 'PANDAS' or '*p*ediatric *a*utoimmune *n*europsychiatric *d*isorders *a*ssociated with *s*treptococcal infections' (Swedo *et al.* 1998). In these cases, as in Sydenham's chorea, a preceding beta-hemolytic streptococcal pharyngitis initiates an autoimmune attack on the basal ganglia, resulting in obsessions and compulsions. Whether this group of patients should be considered to have obsessive–compulsive disorder or a phenocopy is a nosologic question that has not as yet been answered. In any case, PANDAS should be suspected in cases in which either the onset of the disorder or exacerbations are closely related to streptococcal pharyngitides, and the work-up should proceed as outlined for Sydenham's chorea in Chapter 19.11.

Importantly, there is no relationship between obsessive–compulsive disorder and obsessive–compulsive personality disorder. Although in the past, the personality disorder was felt to provide the 'fertile' soil for the development of obsessive–compulsive disorder, there is no evidence for this, and patients with obsessive–compulsive disorder are no more likely to have a pre-existing obsessive–compulsive personality disorder than any other personality disorder (Baer *et al.* 1990; Black *et al.* 1993; Joffe *et al.* 1988).

Differential diagnosis

Of the multiple different causes of obsessions and compulsions discussed in Chapter 5.17, the most common are major depressive disorder and schizophrenia. Distinguishing depressive episodes of major depressive disorder from obsessive–compulsive disorder may at times be difficult: patients with obsessive–compulsive disorder may become quite demoralized and appear depressed; however, in obsessive–compulsive disorder the obsessions and compulsions appear first, well before the demoralization, whereas in depressive episodes the depressive symptoms are well in evidence before any obsessions or compulsions appear. Schizophrenia may be distinguished from psychotic obsessive–compulsive disorder likewise on the basis of the course of the illness: in schizophrenia psychotic symptoms are well in evidence prior to the onset of any obsessions or compulsions.

One other disorder to consider is simple hoarding. Simple hoarders may, like those with compulsive hoarding, also hang on to things; however, certain features enable a correct differentiation (Pertusa *et al.* 2008). First, compulsive hoarders save things out of a compulsion to do so, whereas simple hoarders save things for sentimental or practical reasons: a compulsive hoarder might not be able to explain his anxious need to save every scrap of paper; in contrast, the simple hoarder may save objects associated with a past love, or perhaps check stubs that might be useful should an audit ever occur. Second, compulsive hoarders often save bizarre things, such as feces or urine, objects which simple hoarders do not save. Third, and finally, compulsive hoarders typically experience other compulsions, whereas simple hoarders are free of such symptoms.

Treatment

Effective therapy involves the use of a serotoninergic medication or either the behavioral techniques of exposure and response prevention (Foa *et al.* 1984; Steketee *et al.* 1982) or cognitive–behavioral therapy; most patients do best with a combination of medication and one of these two techniques (Foa *et al.* 2005).

Serotoninergic drugs effective in obsessive–compulsive disorder include clomipramine, venlafaxine, and the SSRIs. Clomipramine is effective in the dose range 200–250 mg/day (Clomipramine Collaborative Study Group 1991), venlafaxine in a dose of 300 mg (Denys *et al.* 2003), fluoxetine (Bergeron *et al.* 2002; Pigott *et al.* 1990) in doses of 20–60 mg/day (Tollefson *et al.* 1994), fluvoxamine (Freeman *et al.* 1994; Hollander *et al.* 2003a; Koran *et al.* 1996) in doses of 100–300 mg/day, sertraline in doses of 50–200 mg/day (Bergeron *et al.* 2002; Greist *et al.* 1995; Krong *et al.* 1999), paroxetine (Hollander *et al.* 2003b; Zohar and Judge 1996) in doses of 40–60 mg/day, and escitalopram in a dose of 20 mg/day (Stein *et al.* 2007). Most clinicians will initiate treatment with an SSRI. Importantly, regardless of which medication is used, 6 weeks must be allowed to see an initial response, and a full response may not be seen for 3 months.

When an initial trial of medication fails, several strategies are available. First, one may consider an increased dose, for example increasing the dose of sertraline up to 400 mg (Ninan *et al.* 2006). If dosage increase is either not tolerated or ineffective, then one may consider switching from one agent to monotherapy with another, for example from an SSRI to clomipramine or venlafaxine. Another option is to add an antipsychotic, such as risperidone (1–3 mg) (Li *et al.* 2005; McDougle *et al.* 2000) or haloperidol (2–5 mg) (Li *et al.* 2005; McDougle *et al.* 1994, 2000), and as between these, risperidone is probably the treatment of first choice, as it is better tolerated than haloperidol (Li *et al.* 2005). Importantly, these antipsychotics are effective only as augmenting agents and are not useful in monotherapy. In severe and treatment-resistant cases, consideration may be given to cingulotomy (Baer *et al.* 1995; Dougherty *et al.* 2002; Hay *et al.* 1993), anterior capsulotomy (Ruck *et al.* 2008), or deep brain stimulation of the anterior capsule (Goodman *et al.* 2010; Nuttin *et al.* 2008).

In acute cases of obsessive–compulsive disorder occurring as part of PANDAS, it is tempting to consider utilizing the same sort of acute treatment as is recommended for Sydenham's chorea; although prednisone has not been studied in this regard, one study did report benefit from plasma exchange (Perlmutter *et al.* 1999). In chronic cases one might also consider preventive treatment with benzathine penicillin.

24.6 POST-TRAUMATIC STRESS DISORDER

Post-traumatic stress disorder (PTSD) is a common disorder, occurring in 7–9 percent of the general population, and, though occasionally seen after relatively minor emotionally traumatic events, is typically seen only after severe trauma, such as combat (Lee *et al.* 1995), rape (Bownes *et al.* 1991), torture (Basoglu *et al.* 1994; Ramsey *et al.* 1993), incarceration in concentration camps (Kinzie *et al.* 1984), mass shootings (North *et al.* 1994), earthquakes (Goenjian *et al.* 1994), or fire, as in the famous Coconut Grove disaster in Boston in 1943 (Adler 1943). Subsequent to the trauma, patients, in various fashions, re-experience the event over and over again; there is also a general withdrawal from present life, and patients tend to be anxious and easily startled.

Clinical features

As trauma may occur at any age, so too may PTSD; however, given that the most common precipitants for PTSD, such as combat and rape, occur in early adult years, most cases have an onset in the twenties. Symptoms may appear either acutely, within days or weeks after the trauma, or, in roughly one-quarter of cases, in a delayed fashion, after a latency of months to many years (Smid *et al.* 2009; Watson *et al.* 1988); indeed, in one case of PTSD secondary to combat in the Second World War, the syndrome was delayed for three decades (Van Dyke *et al.* 1985). In cases of delayed onset, the latency period is generally, but not always, marked by dysphoria and a tendency to avoid situations reminiscent of the original trauma (Smid *et al.* 2009). Although delayed onsets typically occur subacutely, occasionally the latent interval will end abruptly if the patient experiences a new trauma similar to the original one. Although it is not possible to predict with any degree of certainty who, after a trauma, will go on to develop PTSD, it appears that a history of dissociation during the trauma portends a greater risk (Birmes *et al.* 2003; Koopman *et al.* 1994; Shalev *et al.* 1996): such patients, in the hours or days after the trauma, may appear 'in a daze' with, in some cases, derealization, depersonalization, or dissociative amnesia.

Clinically (Gersons and Carlier 1992), in one fashion or another, patients become numb to the world around them. Events that used to arouse interest now leave them unaffected and unmoved; they may complain of feeling 'dead inside' or of having no feelings at all, and some may appear listless and detached.

The experience of the trauma lives on in these patients, and they typically have intense, intrusive, and vivid memories of it. Nightmares are common and, unlike most nightmares, these have little of the fantastic in them; rather they tend to stick to the persistently disturbing facts. At times the waking recollections of the trauma may be more compelling and vivid than the patients' actual surroundings, and patients may experience 'flashbacks' in which they act as if the original trauma were actually occurring; in extreme cases, these flashbacks may be characterized by visual and auditory hallucinations that recreate the original trauma, thus creating the picture of 'psychotic PTSD' (Hamner *et al.* 1999).

Situations that remind patients of the trauma tend to be avoided. Veterans may refuse to view war movies, and Second World War concentration camp survivors may avoid anything German. If unavoidably trapped in such a situation, patients become intensely anxious and some may have an anxiety attack.

Patients tend to be anxious, tense, and easily startled. Although often fatigued, they try to remain alert, as if on guard against some fresh onslaught. There may be difficulty with concentration, insomnia, and lability of mood; irritability is common, and patients may become enraged with little or no provocation.

Over time, a full depressive episode may occur, and alcohol abuse or alcoholism may also occur, in which case one typically sees a florid exacerbation of the symptoms of PTSD.

Course

In roughly one-half of all cases, symptoms undergo a gradual and spontaneous remission within a year or so, and this appears to be more common in cases with an acute onset shortly after the original trauma. In the remainder, a chronic course occurs, and this appears likely when either symptoms have persisted beyond a year or the onset was delayed. When the course is chronic, symptoms tend to wax and wane over years or decades.

Etiology

At present, the neuropathology of PTSD is unclear. There has been only one autopsy study (Bracha *et al.* 2005), and this demonstrated a reduction in neurons in the locus ceruleus in patients as compared with controls. Magnetic resonance scanning has revealed hippocampal atrophy in some (Bremner *et al.* 1995, 2003; Lindauer *et al.* 2006; Villarreal *et al.* 2002; Wang *et al.* 2010), but not all (Bonne *et al.* 2001; Jatzko *et al.* 2006; Woodward *et al.* 2006), studies, and there has been controversy whether this atrophy occurs secondary to the stress-related hormonal changes seen in PTSD or rather precedes the development of PTSD.

The role of genetics in PTSD is also unclear. One widely quoted twin study (True *et al.* 1993) reported a genetic contribution; however, subsequent twin work (Stein *et al.* 2002a) has suggested that what is inherited is not so much a diathesis towards developing PTSD *per se*, but rather a disposition to get involved in the sorts of high-stakes endeavors that increase the risk of traumatic exposure. Although these studies leave the question unsettled, a unique MRI study of the hippocampus may shed some further light on this issue. This study (Gilbertson *et al.* 2002) looked at two sets of monozygotic twin pairs: in the first set, one of the twins had been exposed to combat and had developed PTSD whereas the co-twin had not been exposed to combat and did not have PTSD; in the second set, one of the twins had been exposed to combat and had *not* developed PTSD and the co-twin did not have PTSD. As might be expected from the MRI studies noted earlier, in the first set, the twins who had been exposed to combat and had developed PTSD also had hippocampal atrophy: what was surprising was that the co-twins of those who had developed PTSD also had hippocampal atrophy. In the second set of twins, that is to say the set wherein none of the twins developed PTSD, there was no hippocampal atrophy. What this intriguing study shows is that hippocampal atrophy is not only associated with PTSD but that it is present *before* combat exposure, and probably so on the basis of inheritance.

In contrast with the neuropathologic, imaging, and genetic studies, there are clear endocrinologic and biochemical abnormalities in PTSD. The autonomic symptomatology seen in this disorder, such as a heightened startle response (Butler *et al.* 1990; Shalev *et al.* 1992) and sensitivity to stimuli reminiscent of the original trauma (Pitman *et al.* 1987), immediately suggests disturbances of catecholamines, and indeed both cerebrospinal fluid (CSF) (Geracioti *et al.* 2001) and 24-hour urine levels (Young and Breslau 2004) of norepinephrine are increased; furthermore, intravenous administration of yohimbine, a noradrenergic agonist, will increase the severity of symptoms (Southwick *et al.* 1993). Of the endocrinologic studies carried out on PTSD, the hypothalamic–pituitary–adrenal axis has been most extensively studied and shows unique abnormalities. Cerebrospinal fluid levels of corticotropin-releasing hormone (CRH) are increased (Baker *et al.* 1999, 2005), and this is similar to what is seen in major depressive disorder. However, unlike in major depressive disorder where the dexamethasone test is positive (i.e. there is a non-suppression of cortisol), in PTSD the dexamethasone test is strongly positive (Goenjian *et al.* 1996; Yehuda *et al.* 1993, 2004) (i.e. there is an intense suppression of cortisol).

Overall, though speculative, it seems reasonable, based on these findings, to theorize that PTSD occurs secondary to an inherited abnormality of brainstem mononaminergic structures and/or the hippocampus and related limbic structures, that predisposes to the development, in the face of an emotionally traumatic event, of a chronic state of autonomic hyperarousal. Such a 'stress–diathesis' model can account for two facts. First, some individuals, even in the face of a severe stress (e.g. horrendous combat) do not develop PTSD, and it may well be that such individuals are fortunate to have inherited a normal or near-normal hippocampus and related structures. Second, although initially doubted, it is clear that some individuals may, after relatively minor trauma, develop full-blown PTSD (Feinstein and Dolan 1991; McFarlane 1989), and in such individuals the stress, though minor, is sufficient to precipitate the disorder because these individuals may have inherited gross abnormalities in the same structures.

Differential diagnosis

After a significant trauma, rather than developing PTSD, individuals may develop a major depressive disorder, and the resulting depressive episode may at times be difficult to distinguish from the symptoms seen in PTSD. Certain features, however, may enable a differential diagnosis to be made. Patients with depression have a depressed mood rather than a sense of numbness or detachment; furthermore, patients with depression may experience ruminations, which are heavy, leaden recollections of misfortune, thus standing in contrast to the starkly vivid and mercurial recollections and flashbacks seen in PTSD. As noted earlier, patients with PTSD may develop a depressive episode, but in these cases one sees clear-cut PTSD preceding the onset of the depression.

Malingerers and those with factitious illness, as discussed in Chapter 8.11, may feign PTSD. In those feigning combat-related PTSD, a review of military records may reveal the lie; in cases, however, in which an actual trauma did occur, for example a motor vehicle accident, the diagnosis may have to wait upon resolution of pending litigation, after which the 'PTSD' magically resolves.

Treatment

Cognitive–behavioral therapy appears to be effective. Certain antidepressants are also effective, including phenelzine (Frank *et al.* 1988), imipramine (Frank *et al.* 1988), amitriptyline (Davidson *et al.* 1990), venlafaxine (Davidson *et al.* 2006), mirtazapine (Davidson *et al.* 2003), and the SSRIs fluoxetine (Connor *et al.* 1999; Martenyi *et al.* 2002; van der Kolk *et al.* 1994), paroxetine (Marshall *et al.* 2001; Tucker *et al.* 2001), and sertraline (Brady *et al.* 2000; Davidson *et al.* 2001), (although not all studies are in agreement regarding the effectiveness of fluoxetine [Martenyi *et al.* 2007] or sertraline [Freidman *et al.* 2007]). Of the antidepressants, the SSRIs have the longest track record, and one of these is generally used first; consideration, however, may also be given to mirtazapine or venlafaxine; the tricyclics amitriptyline and imipramine are generally not as well tolerated, and phenelzine is difficult to use. Regardless of which antidepressant is chosen, a high dose may be required, and at least 6 weeks should be allowed to pass before assessing any effects. In cases in which the antidepressant is either poorly tolerated or ineffective, consideration may be given to switching to a different antidepressant or, in patients on an SSRI, to augmentation with either risperidone (1–2 mg) (Bartzokis *et al.* 2005; Reich *et al.* 2004) or olanzapine (10 mg) (Stein *et al.* 2002b); of these two antipsychotics, risperidone is generally better tolerated over the long haul and should probably be tried first. Another medication to consider is prazosin. This alpha-1 antagonist has been shown in double-blind work to reduce the nightmares seen in PTSD (Raskind *et al.* 2003, 2007), and may be started at 1 mg in the evening, with the dose increased in similar increments every few days until patients obtain relief or limiting side-effects occur. There is also suggestive work indicating that daytime prazosin may provide further relief. (Taylor *et al.* 2006). A non-blind study also suggests that cyproheptadine, in a night-time dose of 4–12 mg, may reduce nightmares (Gupta *et al.* 1998).

As noted earlier, alcohol abuse or alcoholism not uncommonly accompany PTSD, and when this is the case it is critical to treat these, either first or concomitantly with the PTSD.

24.7 GENERALIZED ANXIETY DISORDER

Generalized anxiety disorder is characterized by a persistent, 'free-floating' anxiety, which is accompanied by autonomic symptoms such as tremor. Although the lifetime prevalence of this disorder has been estimated to be as high as 5 percent, there has been some doubt expressed regarding this; as noted below, it may be difficult to distinguish generalized anxiety disorder from depression, and such a misdiagnosis may have inflated the prevalence figures.

Clinical features

The onset is gradual and may occur in either teenage or early adult years.

Clinically (Anderson *et al.* 1984; Hoehn-Saric *et al.* 1989; Marten *et al.* 1993; Nisita *et al.* 1990; Starcevic *et al.* 1994), patients experience a chronic, pervasive sense of anxious apprehension and tension. They frequently worry about the future and are easily startled. Some complain of a sense of shakiness, and there may be a mild degree of a fine postural tremor of the hands. There is occasionally tachycardia, and the skin is often cold and clammy. Insomnia may occur, and many complain of nausea and headache. Some patients can offer no explanation for their symptoms, whereas others will offer 'reasons' and blame their state on some life event or other. Observation over time, however, reveals that the anxiety is in fact autonomous and 'free-floating'.

Course

This appears to be a chronic disorder, with symptoms waxing and waning in intensity over the years or decades. In some cases there appear to be spontaneous remissions; however, it is not clear in what percentage of these cases relapses eventually occur.

Etiology

Both family (Newman and Bland 2006; Noyes *et al.* 1987) and twin studies (Hettema *et al.* 2001; Kendler *et al.* 1992) suggest a genetic role in generalized anxiety disorder. Although neuropathologic studies are as yet lacking, MRI (Elkin *et al.* 2009) has revealed not only alterations in the size of amygdalar subregions but also changes in the connectivity of the amygdala with limbic structures and the fronto-parietal cortex. Further, and as might be expected from both the symptomatology of this disorder and its responsiveness to to antidepressants and benzodiazepines, several findings suggest abnormalities both in GABAergic and noradrenergic functioning. GABAergic dysfunction is indicated by the finding of reduced benzodiazepine receptor sites, not only on peripheral blood lymphocytes (Rocca *et al.* 1998) but also in the left temporal pole (Tiihonen *et al.* 1997b). Noradrenergic dysfunction is suggested by several facts. To begin with, plasma norepinephrine levels are increased in generalized anxiety disorder (Sevy *et al.* 1989) and there is a reduced number of alpha-2 adrenoreceptor sites on platelets (Cameron *et al.* 1990; Sevy *et al.* 1989). Further, and as might be expected, the growth hormone response to the alpha-2 agonist clonidine is blunted (Abelson *et al.* 1991) in generalized anxiety disorder.

Differential diagnosis

By far, the most important item on the differential is depression, particularly as seen in dysthymia or in depressive episodes of major depressive disorder. In both cases, patients may complain of anxiety and insomnia; however, in depression one sees additional symptoms, not typically seen in generalized anxiety disorder, such as crying spells, anergia, anhedonia, and changes in appetite.

Of the other causes of persistent anxiety discussed in Chapter 7.5, special attention should be paid to caffeine intoxication, chronic obstructive pulmonary disease, hyperthyroidism, and, especially, alcoholism.

Treatment

Antidepressants, benzodiazepines, pregabalin, and buspirone have all been shown effective in double-blind studies. Of the antidepressants, duloxetine and venlafaxine are roughly equivalent in efficacy (Nicolini *et al.* 2009); venlafaxine may have an edge over escitalopram (Bose *et al.* 2008), but is equivalent to paroxetine (Kim *et al.* 2006), which, in turn, is equivalent to sertraline (Ball *et al.* 2005): with these data in mind, if an antipdepressant is to be utilized, though the SSRIs are overall better tolerated, consideration might be given to trying either duloxetine or venlafaxine first; although imipramine (Rickels *et al.* 1993b) and trazodone (Rickels *et al.* 1993b) are also effective, the side-effect burden associated with each of these makes them clear second choices. Among the benzodiazepines, consideration may be given to alprazolam (Elie and Lamontagne 1984), lorazepam (Laakmann *et al.* 1998), and diazepam (Rickels *et al.* 1993b). Pregabalin is also a very viable alternative: it may have an edge over venlafaxine (Montgomery *et al.* 2006) and is as effective as the benzodiazepines alprazolam and lorazepam (Lydiard *et al.* 2010). Buspirone is less effective than venlafaxine (Davidson *et al.* 1999), but is equivalent in efficacy to sertraline (Mokhber *et al.* 2010) and to the benzodiazepines alprazolam (Enkelmann 1991), lorazepam (Laakmann *et al.* 1998), and diazepam (Ross and Matas 1987).

Choosing among these various agents is not exactly straightforward. Given the almost immediate onset of action of the benzodiazepines, in contrast to the weeks required to see an effect with the other agents, there is a temptation to start with a benzodiazepine. However, the high risk of neuroadaptation with the benzodiazepines argues strongly for starting with one of the other agents: as among these, although a recommendation to start with venlafaxine or pregabalin is not unreasonable, probably any of the other agents could be used. Regardless of which agent is tried first, a trial at adequate dose and duration is necessary. Regarding dosage, the antidepressants should be used in doses equivalent to those for the treatment of depression; of the benzodiazepines, alprazolam is used in a dose of 1–6 mg, lorazepam 2–4 mg, and diazepam 15–25 mg; pregabalin is given in doses of 300–600 mg; and buspirone is utilized in a dose of 15–30 mg. In all cases, treatment is started at a low dose and titrated up; subsequently, the agent should be given for an adequate duration: for the benzodiazepines the response, once a therapeutic dose has been reached, is prompt; for the other agents weeks or more may be required to see a therapeutic effect.

In cases where patients fail to respond, a reasonable course would be to attempt monotherapy with another agent; for example, if an antidepressant failed, one might consider pregabalin or buspirone. In practice, many clinicians will opt for combination treatment, say with an antidepressant plus a benzodiazepine; however, there are no blinded studies to support this approach.

REFERENCES

Abelson JL, Glitz D, Cameron OG *et al.* Blunted growth hormone response to clonidine in patients with generalized anxiety disorder. *Arch Gen Psychiatry* 1991; **48**:157–62.

Accurso V, Winnicki M, Shamsuzzaman AS *et al.* Predisposition to vasovagal syncope in subjects with blood/injury phobia. *Circulation* 2001; **104**:903–7.

Adler A. Neuropsychiatric complications in victims of Boston's Coconut Grove disaster. *JAMA* 1943; **123**:1098–101.

Allgulander C. Paroxetine in social anxiety disorder: a randomized placebo-controlled study. *Acta Psychiatr Scand* 1999; **100**:193–8.

Anderson JD, Noyes R, Crowe RR. A comparison of panic disorder and generalized anxiety disorder. *Am J Psychiatry* 1984; **141**:572–5.

Apter A, Pauls DL, Bleich A *et al.* An epidemiologic study of Gilles de la Tourette's syndrome in Israel. *Arch Gen Psychiatry* 1993; **50**:734–8.

Argyropoulos SV, Hood SD, Adrover M *et al.* Tryptophan depletion reverses the therapeutic effect of selective serotonin reuptake inhibitors in social anxiety disorder. *Biol Psychiatry* 2004; **56**:503–9.

Asnis GM, Hameedi FA, Goddard AW *et al.* Fluvoxamine in the treatment of panic disorder: a multi-center, double-blind, placebo-controlled study in outpatients. *Psychiatry Res* 2001; **103**:1–14.

Baer L, Jenike MA, Ricciardi JN *et al.* Standardized assessment of personality disorders in obsessive–compulsive disorder. *Arch Gen Psychiatry* 1990; **47**:826–30.

Baer L, Rauch SL, Ballantine HT *et al.* Cingulotomy for intractable obsessive–compulsive disorder: prospective long-term follow-up of 18 patients. *Arch Gen Psychiatry* 1995; **52**:384–92.

Baker DG, West SA, Nicholson WE *et al.* Serial CSF corticotropin-releasing hormone levels and adrenocortical activity in combat veterans with posttraumatic stress disorder. *Am J Psychiatry* 1999; **156**:585–8.

Baker DG, Ekhator NN, Kasckow JW *et al.* Higher levels of basal serial CSF cortisol in combat veterans with posttraumatic stress disorder. *Am J Psychiatry* 2005; **162**:992–4.

Baldwin D, Bobes J, Stein DJ *et al.* Paroxetine in social phobia/social anxiety disorder: randomized, double-blind, placebo-controlled study. *Br J Psychiatry* 1999; **175**:120–6.

Ball SG, Kuhn A, Wall D *et al.* Selective serotonin reuptake inhibitor treatment for generalized anxiety disorder: a double-blind, prospective comparison between paroxetine and sertraline. *J Clin Psychiatry* 2005; **66**:94–9.

Ballenger JC, Burows CD, DuPont RL *et al.* Alprazolam in panic disorder and agoraphobia: results from a multicenter trial. I. Efficacy in short-term treatment. *Arch Gen Psychiatry* 1988; **45**:413–22.

Ballenger JC, Wheadon DE, Steiner M *et al.* Double-blind, fixed-dose, placebo-controlled study of paroxetine in the treatment of panic disorder. *Am J Psychiatry* 1998; **155**:36–42.

Barlow DH, Gorman JM, Shear MK *et al.* Cognitive–behavioral therapy, imipramine, or their combination for panic disorder: a randomized controlled trial. *JAMA* 2000; **283**:2529–36.

Bartzokis G, Lu PH, Turner J *et al.* Adjunctive risperidone in the treatment of chronic combat-related posttraumatic stress disorder. *Biol Psychiatry* 2005; **57**:474–9.

Basoglu M, Paker M, Paker O *et al.* Psychological effects of torture: a comparison of tortured with nontortured political activists in Turkey. *Am J Psychiatry* 1994; **151**:76–81.

Baxter LR, Phelps ME, Mazziotta JC *et al.* Local cerebral glucose metabolic rates in obsessive–compulsive disorder: a comparison with rates in unipolar depression and in normal controls. *Arch Gen Psychiatry* 1987; **44**:211–18.

Baxter LR, Schwartz JM, Bergman KS *et al.* Caudate glucose metabolic rate changes with both drug and behavior therapy for obsessive–compulsive disorder. *Arch Gen Psychiatry* 1992; **49**:681–9.

Beck AT, Sokol L, Clark DA *et al.* A crossover study of focused cognitive therapy for panic disorder. *Am J Psychiatry* 1992; **149**:778–83.

Bell C, Forshall S, Andover M *et al.* Does 5-HT restrain panic? A tryptophan depletion study in panic disorder patients recovered on paroxetine. *Psychopharmacol* 2002; **16**:5–14.

Benjamin J, Ben-Zion IZ, Karbofsky E *et al.* Double-blind placebo-controlled pilot study of paroxetine for specific phobia. *Psychopharmacology* 2000; **149**:194–6.

Benkelfat C, Nordahl TE, Semple WE *et al.* Local cerebral glucose metabolic rates in obsessive–compulsive disorder: patients treated with clomipramine. *Arch Gen Psychiatry* 1990; **47**:840–8.

Bergeron R, Ravindran AV, Chaput Y *et al.* Sertraline and fluoxetine treatment of obsessive–compulsive disorder: results of a double-blind, 6-month treatment study. *J Clin Psychopharmacol* 2002; **22**:148–54.

Birmes P, Brunet A, Carreras D *et al.* The predictive power of peritraumatic dissociation and acute stress symptoms for post-traumatic stress symptoms: a three-month prospective study. *Am J Psychiatry* 2003; **160**:1337–9.

Black DW, Noyes R, Pfohl B *et al.* Personality disorder in obsessive–compulsive volunteers, well comparison subjects, and their first-degree relatives. *Am J Psychiatry* 1993; **150**:1226–32.

Blanco C, Heimberg RG, Schneir FR *et al.* A placebo-controlled trial of phenelzine, cognitive behavioral group therapy, and their combination for social anxiety disorder. *Arch Gen Psychiatry* 2010; **67**:286–95.

Blomhoff S, Haug TT, Hellstrom K *et al.* Randomised controlled general practice trial of sertraline, exposure therapy and combined treatment in generalized social phobia. *Br J Psychiatry* 2001; **179**:23–30.

Bonne O, Brandes D, Gilboa A *et al.* Longitudinal MRI study of hippocampal volume in trauma survivors with PTSD. *Am J Psychiatry* 2001; **158**:1248–51.

Bose A, Korotzer A, Gommoll C *et al.* Randomized placebo-controlled trial of escitalopram and venlafaxine XR in the treatment of generalized anxiety disorder. *Depress Anxiety* 2008; **25**:854–61.

Bownes IT, O'Gorman EC, Sayers A. Assault characteristics and posttraumatic stress disorder in rape victims. *Acta Psychiatr Scand* 1991; **83**:27–30.

Bracha HS, Garcia-Rill E, Mrak RE *et al.* Postmortem locus coeruleus neuron count in three American veterans with probable or possible war-related PTSD. *J Neuropsychiatry Clin Neurosci* 2005; **17**:503–9.

Bradwejn J, Koszycki D, Shriqui C. Enhanced sensitivity to cholecystokinin tetrapeptide in panic disorder. Clinical and behavioral findings. *Arch Gen Psychiatry* 1991; **48**:603–10.

Brady K, Pearlstein T, Asnis GM *et al.* Efficacy and safety of sertraline treatment of posttraumatic stress disorder: a randomized controlled trial. *JAMA* 2000; **283**:1837–44.

Brantigan CO, Brantigan TA, Joseph N. Effect of beta blockade and beta stimulation on stage fright. *Am J Med* 1982; **72**:88–94.

Bremner JD, Randall P, Scott TM *et al.* MRI-based measurement of hippocampal volume in patients with combat-related posttraumatic stress disorder. *Am J Psychiatry* 1995; **152**:973–81.

Bremner JD, Vythilingam M, Vermetten E *et al.* MRI and PET study of deficits in hippocampal structure and function in women with childhood sexual abuse and posttraumatic stress disorder. *Am J Psychiatry* 2003; **160**:924–32.

Broocks A, Pigott TA, Hill JL *et al.* Acute intravenous administration of ondansetron and m-CPP, alone and in combination, in patients with obsessive–compulsive disorder (OCD): behavioral and biological results. *Psychiatry Res* 1998; **79**:11–20.

Butler RW, Braff DL, Rausch J *et al.* Physiological evidence of exaggerated startle response in a subgroup of Vietnam veterans with combat-related posttraumatic stress disorder. *Am J Psychiatry* 1990; **147**:1308–12.

Cameron OG, Smith CB, Lee MA *et al.* Adrenergic status in anxiety disorders: platelet alpha-2-adrenergic receptor binding, blood pressure, pulse and plasma catecholamines in panic and generalized anxiety disorder patients and in normal subjects. *Biol Psychiatry* 1990; **28**:2–20.

Chapman TF, Fyer AJ, Mannuzza S *et al.* A comparison of treated and untreated simple phobia. *Am J Psychiatry* 1993; **150**:816–18.

Charney DS, Henninger GR, Jatlow PI. Increased anxiogenic effects of caffeine in panic disorders. *Arch Gen Psychiatry* 1985; **42**:233–43.

Charney DS, Woods SW, Krystal JH *et al.* Noradrenergic neuronal dysregulation in panic disorder: the effects of intravenous yohimbine and clonidine in panic disorder patients. *Acta Psychiatr Scand* 1992; **86**:273–82.

Clomipramine Collaborative Study Group. Clomipramine in the treatment of patients with obsessive–compulsive disorder. *Arch Gen Psychiatry* 1991; 48:730–8.

Connor KM, Sutherland SM, Tupler LA *et al.* Fluoxetine in post-traumatic stress disorder. *Br J Psychiatry* 1999; **175**:17–22.

Cowley DS, Arana GW. The diagnostic utility of lactate sensitivity in panic disorder. *Arch Gen Psychiatry* 1990; **47**:277–84.

Coyle PK, Sterman AB. Focal neurologic symptoms in panic attacks. *Am J Psychiatry* 1986; **143**:648–9.

Crowe RR, Noyes R, Pauls DL *et al.* A family study of panic disorder. *Arch Gen Psychiatry* 1983; **40**:1065–9.

Curtis GC, Thyer B. Fainting on exposure to phobic stimuli. *Am J Psychiatry* 1983; **140**:771–4.

Davidson J, Potts N, Richichi E *et al.* Treatment of social phobia with clonazepam and placebo. *J Clin Psychopharmacol* 1993; **13**:423–8.

Davidson J, Baldwin D, Stein DJ *et al.* Treatment of posttraumatic stress disorder with venlafaxine extended release: a 6-month randomized controlled trial. *Arch Gen Psychiatry* 2006; **63**:1158–65.

Davidson JR, DuPont RL, Hedges D *et al.* Efficacy, safety and tolerability of venlafaxine extended release and buspirone in outpatients with generalized anxiety disorder. *J Clin Psychiatry* 1999; **60**:528–35.

Davidson JR, Weisler RH, Butterfield MI *et al.* Mirtazapine vs placebo in posttraumatic stress disorder: a pilot trial. *Biol Psychiatry* 2003; **53**:188–91.

Davidson JR, Foa EB, Huppert JD *et al.* Fluoxetine, comprehensive cognitive behavioral therapy, and placebo in generalized social phobia. *Arch Gen Psychiatry* 2004; **61**:1005–13.

Davidson JRT, Kudler HS, Smith R *et al.* Treatment of posttraumatic stress disorder with amitriptyline and placebo. *Arch Gen Psychiatry* 1990; **47**:259–66.

Davidson JRT, Rothbaum BO, van der Kolk BA *et al.* Multicenter, double-blind comparison of sertraline and placebo in the treatment of posttraumatic stress disorder. *Arch Gen Psychiatry* 2001; **58**:485–92.

Denys D, van der Wee N, van Megen HJGM *et al.* A double blind comparison of venlafaxine and paroxetine in obsessive–compulsive disorder. *J Clin Psychopharmacol* 2003; **23**:568–75.

Dohl RB, Wolkow RM, Clary CM. Sertraline in the treatment of panic disorder: a double-blind multicenter trial. *Am J Psychiatry* 1998; **155**:1189–95.

Donadio V, Liguori R, Elam M *et al.* Arousal elicits exaggerated inhibition of sympathetic nerve activity in phobic syncope patients. *Brain* 2007; **130**:1653–7.

Dougherty DD, Baer L, Cosgrove GR *et al.* Prospective long-term follow-up of 44 patients who received cingulotomy for treatment-refractory obsessive–compulsive disorder. *Am J Psychiatry* 2002; **159**:269–75.

Eisen JL, Rasmussen SA. Obsessive–compulsive disorder with psychotic features. *J Clin Psychiatry* 1993; **54**:373–9.

Elie R, Lamontagne Y. Alprazolam and diazepam in the treatment of generalized anxiety. *J Clin Psychopharmacol* 1984; **4**:125–9.

Elkin A, Prater KE, Schatzberg AF *et al.* Disrupted amygdalar subregion functional connectivity and evidence of a compensatory network in generalized anxiety disorder. *Arch Gen Psychiatry* 2009; **66**:1361–72.

Enkelmann R. Alprazolam versus buspirone in the treatment of outpatients with generalized anxiety disorder. *Psychopharmacology* 1991; **105**:428–32.

Feinstein A, Dolan R. Predictors of post-traumatic stress disorder following physical trauma: an examination of the stressor criterion. *Psychol Med* 1991; **21**:85–91.

Foa EB, Steketee G, Grayson JB *et al.* Deliberate exposure and blocking of obsessive compulsive rituals: immediate and long-term effect. *Behav Ther* 1984; **15**:450–72.

Foa EB, Kozak MJ, Goodman WK *et al.* DSM-IV field trial: obsessive–compulsive disorder. *Am J Psychiatry* 1995; **152**:90–6.

Foa EB, Liebowitz MR, Kozak MJ *et al.* Randomized, placebo-controlled trial of exposure and ritual prevention, clomipramine, and their combination in the treatment of obsessive–compulsive disorder. *Am J Psychiatry* 2005; **162**:151–61.

Frank JB, Kosten TR, Giller EL *et al.* A randomized clinical trial of phenelzine and imipramine for posttraumatic stress disorder. *Am J Psychiatry* 1988; **145**:1289–91.

Frankel M, Cummings JL, Robertson MM *et al.* Obsessions and compulsions in Gilles de la Tourette's syndrome. *Neurology* 1986; **36**:378–82.

Freeman CP, Trimble MR, Deakin JF *et al.* Fluvoxamine versus clomipramine in the treatment of obsessive–compulsive disorder: a multi-center, randomized, double-blinded, parallel group comparison. *J Clin Psychiatry* 1994; **55**:301–5.

Freidman MJ, Marmar CR, Baker DG *et al.* Randomized, double-blind comparison of sertraline and placebo for posttraumatic stress disorder in a Department of Veterans Affairs setting. *J Clin Psychiatry* 2007; **68**:711–20.

Fyer AJ, Mannuzza S, Gallops MS *et al.* Familial transmission of simple phobias and fears: a preliminary report. *Arch Gen Psychiatry* 1990; **47**:252–6.

Fyer AJ, Mannuzza S, Chapman TF *et al.* A direct interview family study of social phobia. *Arch Gen Psychiatry* 1993; **50**:286–93.

Geracioti TD, Baker DG, Ekhatur NN *et al.* CSF norepinephrine concentrations in posttraumatic stress disorder. *Am J Psychiatry* 2001; **158**:1227–30.

Gersons BPR, Carlier IVE. Post-traumatic stress disorder: the history of a recent concept. *Br J Psychiatry* 1992; **161**:742–8.

Gilbertson MW, Shenton ME, Ciszewski A *et al.* Smaller hippocampal volume predicts pathologic vulnerability to psychological trauma. *Nat Neurosci* 2002; **5**:1242–7.

Goddard AW, Mason GF, Almai A *et al.* Reductions in occipital cortex GABA levels in panic disorder detected with 1h-magnetic resonance spectography. *Arch Gen Psychiatry* 2001a; **58**:556–61.

Goddard AW, Brouette T, Almai A *et al.* Early coadministration of clonazepam with sertraline for panic disorder. *Arch Gen Psychiatry* 2001b; **58**:681–6.

Goenjian AK, Najarian LM, Pynoos RS *et al.* Posttraumatic stress disorder in elderly and younger adults after the 1988 earthquake in Armenia. *Am J Psychiatry* 1994; **151**:895–901.

Goenjian AK, Yehuda R, Pynoos RS *et al.* Basal cortisol dexamethasone suppression of cortisol and MHPG in adolescents after the 1988 earthquake in Armenia. *Am J Psychiatry* 1996; **153**:929–34.

Goisman RM, Warshaw MG, Steketee GS *et al.* DSM-IV and the disappearance of agoraphobia without a history of panic disorder: new data on a controversial diagnosis. *Am J Psychiatry* 1995; **152**:1438–43.

Goodman WK, Foote KD, Greenberg BD *et al.* Deep brain stimulation for intractable obsessive–compulsive disorder: pilot study using a blinded, staggered-onset design. *Biol Psychiatry* 2010; **67**:535–42.

Gordon A. Transition of obsessions into delusions. *Am J Psychiatry* 1950; **107**:455–8.

Gorman JM, Fyer MR, Goetz R *et al.* Ventilatory physiology of patients with panic disorder. *Arch Gen Psychiatry* 1988; **45**:31–9.

Grabe HJ, Ruhrmann S, Ettelt S *et al.* Familiarity of obsessive–compulsive disorder in nonclinical and clinical subjects. *Am J Psychiatry* 2006; **163**:1986–92.

Greist J, Chouinard G, DuBoff E *et al.* Double-blind parallel comparison of three dosages of sertraline and placebo in outpatients with obsessive–compulsive disorder. *Arch Gen Psychiatry* 1995; **52**:289–95.

Guastella AJ, Richardson R, Lovibond PF *et al.* A randomized controlled trial of d-cylcoserine enhancement of exposure therapy for social anxiety disorder. *Biological Psychiatry* 2008; **63**:544–9.

Gupta S, Popli A, Bathurst E *et al.* Efficacy of cyproheptadine for nightmares associated with posttraumatic stress disorder. *Compr Psychiatry* 1998; **39**:160–4.

Hamner MB, Frueh BC, Ulmer HG *et al.* Psychotic features and illness severity in combat veterans with chronic posttraumatic stress disorder. *Biol Psychiatry* 1999; **45**:846–52.

Hartley LR, Ungapen S, Davie I *et al.* The effect of beta adrenergic blocking drugs on speakers' performance and memory. *Br J Psychiatry* 1983; **142**:512–17.

Hauri PJ, Friedman M, Ravaruis CL. Sleep in patients with spontaneous panic attacks. *Sleep* 1989; **12**:323–37.

Hay P, Sachdev P, Cummings S *et al.* Treatment of obsessive–compulsive disorder by psychosurgery. *Acta Psychiatr Scand* 1993; **87**:197–207.

Heimberg RG, Hope DA, Dodge CS *et al.* DSM-III-R subtypes of social phobia: comparison of generalized social phobics and public speaking phobics. *J Nerv Ment Dis* 1990; **178**:172–9.

Heimberg RG, Liebowitz MR, Hope DA *et al.* Cognitive behavioral group therapy vs. phenelzine therapy for social phobia. *Arch Gen Psychiatry* 1998; **55**:1133–41.

Hettema JM, Prescott CA, Kendler KS. A population-based twin study of generalized anxiety disorder in men and women. *J Nerv Ment Dis* 2001; **189**:413–20.

Hoehn-Saric R, McLeod DR, Zimmerli WD. Somatic manifestations in women with generalized anxiety disorder. *Arch Gen Psychiatry* 1989; **46**:1113–19.

Hofmann SG, Meuret AE, Smits JA *et al.* Augmentation of exposure therapy with d-cycloserine for social anxiety disorder. *Arch Gen Psychiatry* 2006; **63**:298–304.

Hollander E, DeCaria CM, Nitescu A *et al.* Serotoninergic function in obsessive–compulsive disorder: behavioral and neuroendocrine responses to m-chlorophenylpiperazine and fenfluramine in patients and healthy volunteers. *Arch Gen Psychiatry* 1992; **49**:21–8.

Hollander E, Koran LM, Goodman WK *et al.* A double-blind, placebo-controlled study of the efficacy and safety of controlled-release fluvoxamine in patients with obsessive–compulsive disorder. *J Clin Psychiatry* 2003a; **64**:640–7.

Hollander E, Allen A, Steiner M *et al.* Acute and long-term treatment and prevention of relapse of obsessive–compulsive disorder with paroxetine. *J Clin Psychiatry* 2003b; **64**:1113–21.

Horwarth E, Lish JD, Johnson J *et al.* Agoraphobia without panic: clinical reappraisal of an epidemiologic finding. *Am J Psychiatry* 1993; **150**:1496–501.

Insel TR, Akiskal HS. Obsessive–compulsive disorder with psychotic features: a phenomenologic analysis. *Am J Psychiatry* 1986; **143**:1527–33.

James IM, Griffith DNW, Pearson RM *et al.* Effect of oxyprenolol on stage-fright in musicians. *Lancet* 1977; **2**:952–4.

Jatzko A, Rothenhofer S, Schmitt A *et al.* Hippocampal volume in chronic posttraumatic stress disorder (PTSD): MRI study using two different evaluation methods. *J Affect Disord* 2006; **94**:121–6.

Joffe RT, Swinson RP, Regan JJ. Personality features of obsessive–compulsive disorder. *Am J Psychiatry* 1988; **145**:1127–9.

Johnson MR, Marazziti D, Brawman-Mintzer O *et al.* Abnormal peripheral benzodiazepine receptor density associated with generalized social phobia. *Biol Psychiatry* 1998; **43**:306–9.

Kahn RS, Wetzler S. *m*-Chlorophenylpiperazine as a probe of serotonin function. *Biol Psychiatry* 1991; **30**:1139–66.

Kasper S, Stein DJ, Loft H *et al.* Escitalopram in the treatment of social anxiety disorder: randomized, placebo-controlled, flexible-dosage study. *Br J Psychiatry* 2005; **186**:222–6.

Katerndahl DA. Panic and prolapse: meta-analysis. *J Nerv Ment Dis* 1993; **181**:539–44.

Katzelnick DJ, Kobak KA, Greist JH *et al.* Sertraline for social phobia: a double-blind, placebo-controlled crossover study. *Am J Psychiatry* 1995; **152**:1368–71.

Kendler KS, Neale MC, Kessler RC *et al.* Generalized anxiety disorder in women: a population-based twin study. *Arch Gen Psychiatry* 1992; **49**:267–72.

Kim TS, Pae CU, Yoon SJ *et al.* Comparison of venlafaxine extended release versus paroxetine for treatment of patients with generalized anxiety disorder. *Psychiatry Clin Neurosci* 2006; **60**:347–51.

Kinzie JD, Fredrickson RH, Ben R *et al.* Posttraumatic stress disorder among survivors of Cambodian concentration camps. *Am J Psychiatry* 1984; **141**:645–50.

Klein DF. False suffocation alarms, spontaneous panics, and related conditions: an integrative hypothesis. *Arch Gen Psychiatry* 1993; **50**:306–17.

Koopman C, Classen C, Spiegel D. Predictors of posttraumatic stress symptoms among survivors of the Oakland/Berkeley, Calif., firestorm. *Am J Psychiatry* 1994; **151**:888–94.

Koran LM, McElroy SL, Davidson JR *et al.* Fluvoxamine versus clomipramine for obsessive–compulsive disorder: a double-blind comparison. *J Clin Psychopharmacol* 1996; **16**:121–9.

Krong MH, Apter J, Asnis G *et al.* Placebo-controlled multicenter study of sertraline treatment for obsessive–compulsive disorder. *J Clin Psychopharmacol* 1999; **19**:172–6.

Laakmann G, Schule C, Lorkowski G *et al.* Buspirone and lorazepam in the treatment of generalized anxiety disorder in outpatients. *Psychopharmacology* 1998; **136**:357–66.

Lanzenberger RR, Mitterhauser M, Spindelegger C *et al.* Reduced serotonin-1A receptor binding in social anxiety disorder. *Biol Psychiatry* 2007; **61**:1081–9.

Lee KA, Vaillant GE, Torrey WC *et al.* A 50-year prospective study of the psychological sequelae of World War II combat. *Am J Psychiatry* 1995; **152**:516–22.

Lelliott P, Marks I, McNamee G *et al.* Onset of panic disorder with agoraphobia. *Arch Gen Psychiatry* 1989; **46**:1000–8.

Li X, May RS, Tolbert LC *et al.* Risperidone and haloperidol augmentation of serotonin reuptake inhibitors in refractory obsessive–compulsive disorder: a crossover study. *J Clin Psychiatry* 2005; **66**:736–43.

Liden S, Gottfries CG. Beta-blocking agents in the treatment of catecholamine-induced symptoms in musicians. *Lancet* 1974; **2**:529.

Liebowitz MR, Fyer AJ, Gorman JM *et al.* Lactate provocation of panic attacks. I. Clinical and behavioral findings. *Arch Gen Psychiatry* 1984; **41**:764–70.

Liebowitz MR, Schneier F, Campeas R *et al.* Phenelzine versus atenolol in social phobia: a placebo-controlled comparison. *Arch Gen Psychiatry* 1992; **39**:290–300.

Liebowitz MR, Gelenberg AJ, Munjack D. Venlafaxine extended release vs placebo and paroxetine in social anxiety disorder. *Arch Gen Psychiatry* 2005; **62**:190–8.

Lindauer RJ, Olff M, van Meijel EP *et al.* Cortisol, learning, memory and attention in relation to smaller hippocampal volume in police officers with posttraumatic stress disorder. *Biol Psychiatry* 2006; **59**:171–7.

Londborg PD, Wolkow R, Smith WT *et al.* Sertraline in the treatment of panic disorder. *Br J Psychiatry* 1998; **173**:54–60.

Lydiard RB, Rickels K, Herman B *et al.* Comparing efficacy of pregabalin and benzodiazepines in treating the psychic and somatic symptoms of generalized anxiety disorder. *Int J Neuropsychopharmacol* 2010; **13**:229–41.

McDougle CJ, Goodman WK, Leckman JF *et al.* Haloperidol addition to fluvoxamine-refractory obsessive–compulsive disorder: a double-blind, placebo-controlled study in patients with and without tics. *Arch Gen Psychiatry* 1994; **51**:302–8.

McDougle CJ, Epperson CN, Pelton GH *et al.* A double-blind, placebo-controlled study of risperidone addition in serotonin reuptake inhibitor-refractory obsessive–compulsive disorder. *Arch Gen Psychiatry* 2000; **57**:794–801.

McFarlane AC. The aetiology of post-traumatic morbidity: predisposing, precipitating and perpetuating factors. *Br J Psychiatry* 1989; **154**:221–8.

Mannuzza S, Schneier FR, Chapman TF *et al.* Generalized social phobia: reliability and validity. *Arch Gen Psychiatry* 1995; **52**:230–7.

Marks IM, Gelder MG. Different ages of onset in varieties of phobia. *Am J Psychiatry* 1966; **123**:218–21.

Marshall RD, Beebe KL, Oldham M *et al.* Efficacy and safety of paroxetine treatment for chronic PTSD: a fixed-dose, placebo-controlled study. *Am J Psychiatry* 2001; **158**:1982–8.

Marten PA, Brown TA, Barlow DH *et al.* Evaluation of the ratings comprising the associated symptom criteria of DSM-III-R generalized anxiety disorder. *J Nerv Ment Dis* 1993; **181**:676–82.

Martenyi F, Brown EB, Zhang H *et al.* Fluoxetine versus placebo in posttraumatic stress disorder. *J Clin Psychiatry* 2002; **63**:199–206.

Martenyi F, Brown EB, Caldwell CD. Failed efficacy of fluoxetine in the treatment of posttraumatic stress disorder: results of a fixed-dose placebo-controlled study. *J Clin Psychopharmacol* 2007; **27**:166–70.

Mavissakalian MR, Perel JM. Imipramine dose–response relationship in panic disorder with agoraphobia: preliminary findings. *Arch Gen Psychiatry* 1989; **46**:127–31.

Mellman TA, Uhde TW. Sleep panic attacks: new clinical findings and theoretical implications. *Am J Psychiatry* 1989a; **146**:1204–7.

Mellman TA, Uhde TW. Electroencephalographic sleep in panic disorder: focus on sleep-related panic attacks. *Arch Gen Psychiatry* 1989b; **46**:178–89.

Mellman TA, Uhde TW. Patients with frequent sleep panic: clinical findings and response to medication treatment. *J Clin Psychiatry* 1990; **51**:513–16.

Michelson D, Allgulander C, Dantendorfer K *et al.* Efficacy of usual antidepressant dosing regimens of fluoxetine in panic disorder: randomized, placebo-controlled trial. *Br J Psychiatry* 2001; **179**:514–18.

Modigh K, Westberg P, Eriksson E. Superiority of clomipramine over imipramine in the treatment of panic disorder: a placebo-controlled trial. *J Clin Psychopharmacol* 1992; **12**:251–61.

Mokhber N, Azapazhooh MR, Khaejdaluee M *et al.* Randomized, single-blind, trial of sertraline and buspirone for treatment of elderly patients with generalized anxiety disorder. *Psychiatry Clin Neurosci* 2010; **64**:128–33.

Montgomery SA, Tobias K, Zornberg GL *et al.* Efficacy and safety of pregabalin in the treatment of generalized anxiety disorder: a 6-week multicenter, randomized, double-blind, placebo-controlled comparison of pregabalin and venlafaxine. *J Clin Psychiatry* 2006; **67**:771–82.

Newman SC, Bland RC. A population-based family study of DSM-III generalized anxiety disorder. *Psychol Med* 2006; **36**:1275–81.

Nicolini H, Bakish D, Duenas H *et al.* Improvement of psychic and somatic symptoms in adult patients with generalized anxiety disorder: examination from a duloxetine, venlafaxine extended-release and placebo-controlled trial. *Psychol Med* 2009; **39**:267–76.

Ninan PT, Koran LM, Kiev A *et al.* High-dose sertraline strategy for nonresponders to acute treatment of obsessive–compulsive disorder; a multicenter double-blind trial. *J Clin Psychiatry* 2006; **67**:15–22.

Nisita C, Petracca A, Akiskal HS *et al.* Delimitation of generalized anxiety disorder: clinical comparisons with panic and major depressive disorder. *Compr Psychiatry* 1990; **31**:409–15.

North CS, Smith EM, Spitznagel EL. Posttraumatic stress disorder in survivors of a mass shooting. *Am J Psychiatry* 1994; **151**:82–8.

Noyes R, Crowe RR, Harris EL *et al.* Relationship between panic disorder and agoraphobia. A family study. *Arch Gen Psychiatry* 1986; **43**:227–32.

Noyes R, Clarkson C, Crow RR *et al.* A family study of generalized anxiety disorder. *Am J Psychiatry* 1987; **144**:1019–24.

Noyes R, Burows GD, Reich JH *et al.* Diazepam versus alprazolam for the treatment of panic disorder. *J Clin Psychiatry* 1996; **57**:349–55.

Nutt DJ, Glue P, Lawson C *et al.* Flumazenil provocation of panic attacks: evidence for altered benzodiazepine receptor sensitivity in panic disorder. *Arch Gen Psychiatry* 1990; **47**:917–25.

Nuttin BJ, Gabriels LA, Cosyns PR *et al.* Long-term electrical capsular stimulation in patients with obsessive–compulsive disorder. *Neurosurgery* 2008; **62**(Suppl 3):966–7.

Oehrberg S, Christiansen PE, Behnke K *et al.* Paroxetine in the treatment of panic disorder: a randomized, double-blind, placebo-controlled study. *Br J Psychiatry* 1995; **167**:374–9.

Page AC. Distinguishing panic disorder and agoraphobia from social phobia. *J Nerv Ment Dis* 1994; **182**:611–17.

Palatnik A, Frolov K, Fux M *et al.* Double-blind, controlled, crossover trial of inositol versus fluvoxamine for the treatment of panic disorder. *J Clin Psychopharmacol* 2001; **21**:335–9.

Pande AC, Davidson JRT, Jefferson JW *et al.* Treatment of social phobia with gabapentin: a placebo-controlled study. *J Clin Psychopharmacol* 1999; **19**:341–8.

Pande AC, Feltner DE, Jefferson JW *et al.* Efficacy of the novel anxiolytic pregabalin in social anxiety disorder: a placebo-controlled, multicenter study. *J Clin Psychopharmacol* 2004; **24**:141–9.

Pauls DL, Alsobrook JP, Goodman W *et al.* A family study of obsessive–compulsive disorder. *Am J Psychiatry* 1995; **152**:76–84.

Pecknold JC, Swinson RP, Krich K *et al.* Alprazolam in panic disorder and agoraphobia: results from a multicenter trial. III. Discontinuation effects. *Arch Gen Psychiatry* 1988; **45**:429–36.

Perani D, Colombo C, Bressi S *et al.* [18F]FDG PET study in obsessive–compulsive disorder: a clinical/metabolic correlation study after treatment. *Br J Psychiatry* 1995; **166**:244–50.

Perlmutter SJ, Leitman SF, Garvey MA *et al.* Therapeutic plasma exchange and intravenous immunoglobulin for obsessive–compulsive disorder and tic disorders in childhood. *Lancet* 1999; **354**:1153–8.

Pertusa A, Fullana MA, Singh S *et al.* Compulsive hoarding: OCD symptom, distinct clinical syndrome, or both? *Am J Psychiatry* 2008; **165**:1289–98.

Pigott TA, Pato MT, Bernstein SE *et al.* Controlled comparisons of clomipramine and fluoxetine in the treatment of obsessive–compulsive disorder: behavioral and biological results. *Arch Gen Psychiatry* 1990; **47**:926–32.

Pitman RK, Orr SP, Forgue DF *et al.* Psychophysiologic assessment of post-traumatic stress disorder imagery in Vietnam combat veterans. *Arch Gen Psychiatry* 1987; **44**:970–5.

Pohl R, Yeragani VK, Balon R *et al.* Isoproterenol-induced panic attacks. *Biol Psychiatry* 1988; **24**:891–902.

Pollack M, Mangano R, Entsuah R *et al.* A randomized controlled trial of venlafaxine ER and paroxetine in the treatment of outpatients with panic disorder. *Psychopharmacology* 2007; **194**:233–42.

Ramsey R, Gorst-Unsworth C, Turner S. Psychiatric morbidity in survivors of state violence including torture. *Br J Psychiatry* 1993; **162**:55–9.

Raskind MA, Peskind ER, Kanter ED *et al.* Reduction of nightmares and other PTSD symptoms in adult veterans by prazosin: a placebo-controlled study. *Am J Psychiatry* 2003; **160**:371–3.

Raskind MA, Peskind EF, Hoff DJ *et al.* A parallel group placebo controlled study of prazosin for trauma nightmares and sleep disturbances in combat veterans with post-traumatic stress disorder. *Biol Psychiatry* 2007; **61**:928–34.

Rasmussen SA, Tsuang MT. Clinical characteristics and family history in DSM-III obsessive–compulsive disorder. *Am J Psychiatry* 1986; **143**:317–22.

Reich DB, Winternitz S, Hennen J *et al.* A preliminary study of risperidone in the treatment of posttraumatic stress disorder related to childhood abuse in women. *J Clin Psychiatry* 2004; **65**:1601–6.

Rickels K, Schweizer E, Weiss S *et al.* Maintenance drug treatment for panic disorder. II. Short- and long-term outcome after drug taper. *Arch Gen Psychiatry* 1993a; **50**:61–8.

Rickels K, Downing R, Schweizer E *et al.* Antidepressants for the treatment of generalized anxiety disorder. A placebo-controlled comparison of imipramine, trazodone, and diazepam. *Arch Gen Psychiatry* 1993b; **50**:884–95.

Rocca P, Beoni AM, Eva C *et al.* Peripheral benzodiazepine receptor messenger RNA is decreased in lymphocytes of generalized anxiety disorder patients. *Biol Psychiatry* 1998; **43**:767–73.

Rosenbaum JF, Moroz G, Bowden CL *et al.* Clonazepam in the treatment of panic disorder with or without agoraphobia: a dose–response study of efficacy, safety and discontinuance. *J Clin Psychopharmacol* 1997; **17**:390–6.

Ross CA, Matas M. A clinical trial of buspiroine and diazepam in the treatment of generalized anxiety disorder. *Can J Psychiatry* 1987; **32**:351–5.

Ruck C, Karlsson A, Steele JD *et al.* Capsulotomy for obsessive–compulsive disorder: long-term follow-up of 25 patients. *Arch Gen Psychiatry* 2008; **65**:914–21.

Russell JL, Kushner MG, Beitman BD *et al.* Nonfearful panic disorder in neurology patients validated by lactate challenge. *Am J Psychiatry* 1991; **148**:361–4.

Saxena S, Brody AL, Ho ML *et al.* Differential cerebral metabolic changes with paroxetine treatment of obsessive–compulsive disorder vs major depression. *Arch Gen Psychiatry* 2002; **59**:250–61.

Schneier FR, Liebowitz MR, Abi-Dargham A *et al.* Low dopamine D(2) receptor binding potential in social phobia. *Am J Psychiatry* 2000; **157**:457–9.

Schruers K, van Diest R, Overbeek T *et al.* Acute l-5-hydroxytryptophan administration inhibits carbon dioxide-induced panic in panic disorder patients. *Psychiatry Res* 2002; **113**:237–43.

Schweizer E, Pohl R, Balon R *et al.* Lorazepam vs. alprazolam in the treatment of panic disorder. *Pharmacopsychiatry* 1990; **23**:90–3.

Sevy S, Papadimitriou GN, Surmont DW *et al.* Noradrenergic function in generalized anxiety disorder, major depressive disorder, and healthy subjects. *Biol Psychiatry* 1989; **25**:141–52.

Shalev AY, Orr SP, Peri T *et al.* Physiologic responses to loud tones in Israeli patients with posttraumatic stress disorder. *Arch Gen Psychiatry* 1992; **49**:870–5.

Shalev AY, Peri T, Cangetti L *et al.* Predictors of PTSD in injured trauma survivors: a prospective study. *Am J Psychiatry* 1996; **153**: 219–25.

Shugart YY, Samuels J, Willour VL *et al.* Genomewide linkage scan for obsessive–compulsive disorder: evidence for susceptibility loci on chromosomes 3q, 7p, 15q, and 6q. *Mol Psychiatry* 2006; **11**:736–70.

Smid GE, Mooren TT, van der Mast RC *et al.* Delayed posttraumatic stress disorder: systematic review, meta-analysis, and meta-regression analysis of prospective studies. *J Clin Psychiatry* 2009; **70**:1572–82.

Southwick SM, Krystal JH, Morgan CA *et al.* Abnormal noradrenergic function in post-traumatic stress disorder. *Arch Gen Psychiatry* 1993; **50**:31–7.

Stahl SM, Gergel I, Li D. Escitalopram in the treatment of panic disorder; a randomized, double-blind, placebo-controlled trial. *J Clin Psychiatry* 2003; **64**:1322–7.

Starcevic V, Fallon S, Uhlenhuth EH. The frequency and severity of generalized anxiety disorder symptoms: toward a less cumbersome conceptualization. *J Nerv Ment Dis* 1994; **182**:80–4.

Stein DJ, Anderson EW, Tonnoir B *et al.* Escitalopram in obsessive–compulsive disorder: a randomized, placebo-controlled, paroxetine-referenced, fixed-dose, 24-week study. *Curr Med Res Opin* 2007; **23**:701–11.

Stein MB, Liebowitz MD, Lydiard RB *et al.* Paroxetine treatment of generalized social phobia (social anxiety disorder). *JAMA* 1998; **280**:708–13.

Stein MB, Fyer AJ, Davidson JRT *et al.* Fluvoxamine treatment of social phobia (social anxiety disorder): a double-blind, placebo-controlled study. *Am J Psychiatry* 1999; **156**:756–60.

Stein MB, Jang KL, Taylor S *et al.* Genetic and environmental influences on trauma exposure and posttraumatic stress disorder symptoms: a twin study. *Am J Psychiatry* 2002a; **159**:1675–21.

Stein MB, Kline NA, Matloff JL. Adjunctive olanzapine for SSRI-resistant combat-related PTSD: a double-blind, placebo-controlled study. *Am J Psychiatry* 2002b; **159**:1777–9.

Steketee F, Foa EB, Grayson JB. Recent advances in the behavioral treatment of obsessive–compulsives. *Arch Gen Psychiatry* 1982; **39**:1365–71.

Stern RS, Cobb JP. Phenomenology of obsessive–compulsive neurosis. *Br J Psychiatry* 1978; **132**:233–9.

Strohle A, Kellner M, Yassouridis A *et al.* Effect of flumazenil in lactate-sensitive patients with panic disorder. *Am J Psychiatry* 1998; **155**:610–12.

Swedo SE, Leonard HL, Garvey M *et al.* Pediatric autoimmune neuropsychiatric disorders associated with streptococcal infections: clinical description of the first 50 cases. *Am J Psychiatry* 1998; **155**:264–71.

Tancer ME, Mailman RB, Stein MB *et al.* Neuroendocrine responsivity to monoaminergic system probe in generalized social phobia. *Anxiety* 1994; **1**:216–23.

Taylor FB, Lowe K, Thompson C *et al.* Daytime prazosin reduces psychological distress to trauma specific cues in civilian trauma posttraumatic stress disorder. *Biol Psychiatry* 2006; **59**:577–81.

Tesar GE, Rosenbaum JF, Pollack MH *et al.* Double-blind, placebo-controlled comparison of clonazepam and alprazolam for panic disorder. *J Clin Psychiatry* 1991; **52**:69–76.

Tiihonen J, Kuikka J, Bergstrom K *et al.* Dopamine reuptake site densities in patients with social phobia. *Am J Psychiatry* 1997a; **154**:239–42.

Tiihonen J, Kuikka J, Rasanen P *et al.* Cerebral benzodiazepine receptor binding and distribution in generalized anxiety disorder: a fractal analysis. *Mol Psychiatry* 1997b; **2**:463–71.

Tollefson GD, Rampey AH, Potvin JH *et al.* A multicenter investigation of fixed-dose fluoxetine in the treatment of obsessive–compulsive disorder. *Arch Gen Psychiatry* 1994; **51**:559–67.

Torgerson S. Comorbidity of major depression and anxiety disorders in twin pairs. *Am J Psychiatry* 1990; **147**:1199–202.

True WR, Rice J, Eisen SA *et al.* A twin study of genetic and environmental contributions to liability for posttraumatic stress symptoms. *Arch Gen Psychiatry* 1993; **50**:257–64.

Tucker P, Zaninelli R, Yehuda R *et al.* Paroxetine in the treatment of chronic posttraumatic stress disorder: results of a placebo-controlled, flexibile dosage trial. *J Clin Psychiatry* 2001; **62**:860–8.

Tuke DH. Imperative ideas. *Brain* 1894; **17**:179–97.

van der Kolk BA, Dreyfuss D, Michaels M *et al.* Fluoxetine in posttraumatic stress disorder. *J Clin Psychiatry* 1994; **55**:517–22.

Van Dyke C, Zilberg NJ, McKinnon JA. Posttraumatic stress disorder: a thirty-year delay in a World War II veteran. *Am J Psychiatry* 1985; **142**:1070–3.

van Vliet IM, den Boer JA, Westenberg HG. Psychopharmacological treatment of social phobia: a double blind placebo controlled study with fluvoxamine. *Psychopharmacology* 1994; **115**:128–34.

Versiani M, Nardi AE, Mundim FD *et al.* Pharmacotherapy of social phobia: a controlled study with moclobemide and phenelzine. *Br J Psychiatry* 1992; **161**:353–60.

Villarreal G, Hamilton DA, Petropoulos H *et al.* Reduced hippocampal volume and total white matter volume in posttraumatic stress disorder. *Biol Psychiatry* 2002; **52**:119–25.

Wade AG, Lepola U, Koponen HJ *et al.* The effect of citalopram in panic disorders. *Br J Psychiatry* 1997; **170**:549–53.

Wang Z, Neylan TC, Mueller SG *et al.* Magnetic resonance imaging of hippocampal subfields in posttraumatic stress disorder. *Arch Gen Psychiatry* 2010; **67**:296–303.

Watson CG, Kucala T, Manifold V *et al.* Differences between post traumatic stress disorder patients with delayed and undelayed onset. *J Nerv Ment Dis* 1988; **176**:568–72.

Woodward SH, Kaloupek DG, Streeter CC *et al.* Hippocampal volume, PTSD, and alcoholism in combat veterans. *Am J Psychiatry* 2006; **163**:674–81.

Yehuda R, Southwick SM, Krystal JH *et al.* Enhanced suppression of cortisol following dexamethasone administration in posttraumatic stress disorder. *Am J Psychiatry* 1993; **150**:83–6.

Yehuda R, Golier JA, Hallian SL *et al.* The ACTH response to dexamethasone in PTSD. *Am J Psychiatry* 2004; **161**:1397–403.

Young EA, Breslau N. Cortisol and catecholamines in posttraumatic stress disorder: an epidemiologic community study. *Arch Gen Psychiatry* 2004; **61**:394–401.

Zohar J, Judge R. Paroxetine versus clomipramine in the treatment of obsessive–compulsive disorder. *Br J Psychiatry* 1996; **169**:468–74.

25

Dissociative disorders

25.1 DISSOCIATIVE AMNESIA

Dissociative amnesia, also known as psychogenic amnesia, is an amnestic disorder characterized by episodes of amnesia for emotionally traumatic events. This disorder is more common in females than males, and may occur, to varying degrees of severity, in up to 2 percent of the general population.

Clinical features

The onset of dissociative amnesia is usually in the late teens or early adult years, and patients may present in one of two ways.

The first presentation to consider, often referred to as the 'localized' type of dissociative amnesia, involves patients who, though at the time of evaluation have no difficulty with recalling recent events, complain that they cannot recall certain events in the more distant past, which, through other sources, are known to have occurred. These events are always of an emotionally traumatic type, such as rape, torture, kidnapping, or assault, or perhaps less severe sorts, such as tempestuous arguments with a spouse. Patients have little difficulty in recalling events that occurred before the emotionally traumatic one, and likewise can recall events that occurred after the emotional trauma had passed: the period of time that contained the traumatic event, however, is empty and exists as a kind of 'island' of amnesia. This first type is by far the most common of the two presentations.

The second presentation involves patients who, during the evaluation, complain that they cannot recall anything that happened from the time of the evaluation stretching back to some point in the past. The duration of this period of retrograde amnesia varies widely, from decades down to weeks or merely days. This type of dissociative amnesia is divided into two subtypes, namely 'generalized' and 'continuous'. In the generalized subtype the retrograde amnesia stretches back to earliest childhood, and patients complain that they cannot remember anything about their lives, and indeed in some of these cases patients may complain that they do not even know their own names. In the continuous type the retrograde amnesia does not stretch back indefinitely, but ends at a certain point, with patients being able to recall events prior to that. Importantly, regardless of the subtype of this presentation, one also finds, again from independent sources, that an emotionally traumatic event preceded the onset of the amnesia. For example, a relative may report that the patient seemed normal, and had no difficulty with memory, up until a savage rape occurred, soon after which the patient was no longer able to remember the rape, nor events that occurred for a variable period of time before the rape itself occurred. In addition to the retrograde amnesia, these patients will also, from the time of evaluation forward, experience a degree of anterograde amnesia, such that, upon successive evaluations, they complain that they are unable to recall events that occurred after their initial presentation. This anterograde component may be of variable duration, lasting for from hours to weeks (Kanzer 1939), but eventually the amnesia ends and patients are once again able to keep track of and recall events of their recent past.

Course

The overall natural course of dissociative amnesia has not been well studied. Thus, in the absence of treatment, in the localized type it is not clear how often patients will eventually be able to recall the emotionally traumatic events, nor is it clear whether they will become amnestic for emotionally traumatic events that, at some point earlier in their lives, they were able to recall. Further, and again in the absence of treatment, in the generalized and continuous types, it is not clear how great the risk is for a recurrence in the future should additional emotionally traumatic events occur.

Etiology

From a psychological point of view, the amnesia of dissociative amnesia probably occurs on the basis of repression or dissociation precipitated by the emotionally traumatic event. From a physiological point of view, positron emission tomography (PET) studies have demonstrated various abnormalities, such as decreased metabolism in the right inferolateral prefrontal cortex in patients suffering from the continuous subtype (Brand *et al.* 2009).

Differential diagnosis

The full differential for amnestic disorders is discussed in Chapter 6.6, and, in pursuing this differential, it is useful to consider each type of dissociative amnesia separately.

Patients who present with the localized type of amnesia, though capable of recalling the recent past, have difficulty recalling more distant events. As such, they are similar to patients who suffered from any of the conditions listed in Table 6.3 of Chapter 6.6 that can cause an episode of transient anterograde amnesia in the past, such as a concussion, transient global amnesia or an episode of pure epileptic amnesia; in this regard one may also add to this differential patients who suffered from complex partial seizures in the past, as these patients, likewise, will not be able to recall anything that happened while the seizure was occurring. Each of these conditions is associated with differential points, discussed in that chapter, which enable their recognition.

Patients who present with either the generalized or the continuous subtypes of dissociative amnesia present with a difficulty in recalling events that stretch back from the time of presentation towards the distant, sometimes far distant, past, and also a difficulty in keeping track of present events. As such they must be distinguished from patients who are *currently* in any one of the conditions capable of causing episodic anterograde amnesia, as many of these are also associated with a retrograde component. Thus, a patient with transient global amnesia may have both difficulty recalling the past and difficulty keeping track of present events. In addition, they must be distinguished from patients who are suffering from conditions capable of causing permanent retrograde and anterograde amnesia, such as Korsakoff's syndrome, and these are also listed in Table 6.3 and discussed in the accompanying text. One final point to consider here is the oft-quoted clinical rule that patients with an amnestic disorder who also are unable to recall their names are always suffering from dissociative amnesia. As with most clinical rules, although this is true in the vast majority of cases, there are exceptions, as for example a patient, described in Chapter 6.6, who, during a partial seizure, forgot everything, including her name.

Before leaving this discussion of differential diagnosis, two other differential points should be highlighted. First, as indicated earlier, dissociative amnesia is always precipitated by an emotionally traumatic event. By contrast, although other amnestic disorders may occasionally be preceded by emotional events (e.g. transient global amnesia), they more often than not occur in an emotionally neutral setting. Second, patients with dissociative amnesia may, either in the natural course of events or with treatment, eventually be able to recall the emotionally traumatic event and recover their memory. Again by contrast, with only rare exceptions (e.g. in some cases of epilepsy), patients suffering from other forms of amnesia are never able to recall the events in question.

Treatment

Although there are no controlled studies, psychotherapy, with or without hypnosis, appears effective. In the past the 'Amytal Interview', involving intravenous infusion of amobarbital, was popular, and although it does facilitate the recovery of lost memories, it appears to offer no advantage over psychotherapy, and is rarely used today.

25.2 DISSOCIATIVE FUGUE

In dissociative fugue, also known as psychogenic fugue, patients suddenly become amnestic for much of their lives and engage in purposeful travel, at times establishing a new identity (Coons 1999). This is probably a rare disorder, and may be equally common in males and females.

Clinical features

An episode of fugue generally appears quite suddenly, and is often immediately preceded by some emotionally traumatic event. Patients leave off what they are doing and begin to travel: some may go only cross-town, while others may travel great distances, even to distant countries. When travel does cease, patients carry on as if nothing were amiss: some may find a place to live, get a job and basically set up a new life; generally this new life is simple and unassuming, with little about it to call attention to the patient. When questioned, however, patients will be unable to recall most, or even all, of their past; furthermore, some, but not all, either will be uncertain about their names or will adopt a new name (Riether and Stoudemire 1988).

Episodes of fugue may last for from days to months or even longer, and they tend to end abruptly, and when they do, patients subsequently, though able to recall events which led up to the onset of the fugue, will be unable to recall what they did during the fugue itself.

Course

Although the natural course of fugue has not been well-studied, it appears that in some cases, there is only one episode of fugue in the patient's entire life, whereas in others, recurrences may be found.

Etiology

As the name suggests, dissociative fugue occurs on the basis of dissociation triggered by a preceding emotionally traumatic event.

Differential diagnosis

Dissociative amnesia is distinguished by the lack of purposeful travel: though these amnestic patients may wander, there is no apparent purpose to the wandering, and they do not end up establishing new lives for themselves. In dissociative identity disorder, one does see the assumption of a new identity; however, in dissociative identity disorder there is typically more than one identity, with fairly rapid shifting back and forth between different identities.

In complex partial seizures characterized by reactive automatisms, patients may travel, all the while engaging in complex activity. Here, however, there is no assumption of a new life, and certainly not of a new identity. Furthermore, in most cases one will also find a history of other, more obvious, seizure types, such as grand mal seizures.

Treatment

The treatment approach here is the same as that described for dissociative amnesia.

25.3 DISSOCIATIVE IDENTITY DISORDER

Dissociative identity disorder, also known as multiple personality disorder, is characterized by the appearance, generally in late adolescence or early adult years, of one or more additional personalities (Boon and Draijer 1993). Each of these is different from that present in the patient from childhood up to the onset of the disorder, and each, at separate times, may take control of the patient's behavior.

The prevalence of this disorder is a matter of much debate: although figures ranging from 0.1 to 1.5 percent of the general population are cited, some authors believe that it is rare, and that almost all cases are iatrogenic (Piper and Mersky 2004a,b). It is anywhere from five to nine times as common in females as males.

Clinical features

The onset is heralded by the appearance of a new personality structure in, as just noted, late adolescence or early adult years; over long periods of time additional personalities, or 'alters', may also emerge. The transition, or 'switch', from one personality to another, although occasionally taking from hours up to an entire day, is generally quite abrupt, occurring in seconds to minutes. The new personality presents as a coherent and cohesive structure, with different attitudes, habits, facial expressions, posture, gestures, tone of voice and even handwriting, from the original one. Indeed, were an observer prevented from recognizing the patient by sight, he or she would have little reason to doubt that, at separate times, different persons were indeed under observation. Typically, these alters will engage in behaviors that are quite foreign to the 'original' or 'primary' personality. For example, a patient who, in the original personality, was reserved and inhibited, may, in one of the alters, be outgoing and promiscuous.

Generally, the original, or primary, personality will have no direct knowledge of any of the alters, and the existence of these alters may come to the attention of the primary personality in one of two ways. First, the primary personality may, in looking back, realize that there are gaps in memory, or 'lost time': these gaps, in turn, are found to have been occupied, as it were, by an alter. Second, the primary personality may find indirect evidence of the activity of an alter: for example, one primary, prim and restrained, in going through her dresser, found a host of sexual paraphernalia.

The lives of patients with dissociative identity disorder are often chaotic, and suicidal behavior is common. These patients are also likely to have other disorders, such as mood disorders and substance abuse or dependence.

Course

Although this is a chronic, lifelong disorder, over the decades the frequency of 'switching' tends to decrease.

Etiology

As the name of this disorder indicates, dissociation is felt to underlie the development of alternate personalities, and this dissociation, in turn, is related to the occurrence, in virtually all patients, of severe and protracted abuse, often sexual in nature, during childhood. Of interest, quantitative electroencephalograms (EEGs) (Lapointe *et al.* 2006) and PET scans (Reinders *et al.* 2006) reveal different findings with different alters in the same patient.

Patients with dissociative identity disorder are readily hypnotizable, and some authors believe that many, if not most, cases are iatrogenic, with new personalities being suggested by therapists.

Differential diagnosis

In dissociative amnesia, although patients may complain of blank spots or 'lost time' in their memory, there is no appearance of a different personality. In dissociative fugue, some patients, during the fugue, may adopt a new name; however, in fugue only one new personality is present, and there is no rapid shifting back and forth. Furthermore, in fugue, patients often travel great distances before developing a new identity, in contrast to dissociative identity disorder patients, who typically remain in the same locale, indeed in the same house.

Borderline personality, like dissociative identity disorder, is characterized by chaotic behavior, and many borderlines have very unstable personality structures. Indeed, it is in these cases that iatrogenic formation of new personalities may occur.

Complex partial seizures are commonly listed on this differential; however, this is not an issue, as these patients do not assume a new identity.

Finally, one must consider malingering. Given the enormous media interest in multiple personality disorder, persons wishing to evade responsibility for illegal activity may plead this disorder as an excuse for their behavior.

Treatment

Although there are no controlled studies, most patients are seen in individual psychotherapy. Hospitalization may be necessary to protect patients from unacceptable behaviors by alters, or by suicide attempts; however, long stays are to be avoided as patients may worsen during them; partial hospitalization, though, may be quite beneficial. Although pharmacologic treatment does not influence the alters *per se*, it may be very useful for comorbid conditions. Overall, the treatment of these patients is very difficult, and referral to a specialist is generally in order.

25.4 DEPERSONALIZATION DISORDER

Depersonalization disorder (Baker *et al.* 2003; Simeon *et al.* 1997, 2003a) is a rare disorder, occurring with roughly equal frequency in males and females, in which the symptom of depersonalization occurs on an idiopathic basis.

Clinical features

The onset is generally in mid to late adolescence; although early adult-onset cases may occur, onsets in middle or later years probably do not occur.

In depersonalization, patients find themselves, as it were, detached from themselves. Though they continue to go about their lives, they experience themselves as observing their activities, as if an automaton were actually performing them. In some cases, patients will report that it is as if they were observing themselves from above, as if they were looking down from several feet above; further, some patients may also report that their extremities appear distorted, enlarged, or shrunken.

Frequently, depersonalization is accompanied by derealization. Here, though patients clearly perceive their surroundings, those surroundings appear as if they were unreal, with objects appearing strange, dead, lifeless, and without substance.

Course

In the majority of cases, symptoms, though waxing and waning in intensity, are chronic. In the remaining minority, an episodic course appears.

Etiology

Although this disorder is currently included within the dissociative disorders group, there is debate whether the phenomenon of dissociation accounts for the clinical picture. From a physiologic point of view, there is both an inverse correlation between the severity of dissociation and plasma norepinephrine levels (Simeon *et al.* 2003b) and a decreased galvanic skin response to unpleasant visual stimuli (Sierra *et al.* 2006). Further, PET scanning has revealed abnormalities in both the temporal and parietal lobes (Simeon *et al.* 2000).

Differential diagnosis

The full differential for the symptom of depersonalization is discussed in Chapter 5.16, and the diagnosis of depersonalization disorder should not be made until these other causes have been ruled out.

Treatment

Depersonalization disorder often appears treatment resistant. Although psychotherapy is often engaged in, there are no controlled trials demonstrating its effectiveness. Anxiolytics, antipsychotics, and antidepressants, particularly selective serotonin reuptake inhibitors (SSRIs), are often prescribed, but again there are no controlled trials demonstrating their effectiveness; indeed, the only controlled trials of these agents, one comparing clomipramine and desipramine (Simeon *et al.* 1998) and one fluoxetine and placebo (Simeon *et al.* 2004), failed to find any convincing benefit.

REFERENCES

Baker D, Hunter E, Lawrence E *et al.* Depersonalization disorder: clinical features of 204 cases. *Br J Psychiatry* 2003; **182**:428–33.

Boon S, Draijer N. Multiple personality disorder in the Netherlands: a clinical investigation of 71 patients. *Am J Psychiatry* 1993; **150**:489–94.

Brand M, Eggers C, Reinhold N *et al.* Functional brain imaging in 14 patients with dissociative amnesia reveals right inferolateral prefrontal hypometabolism. *Psychiatry Res* 2009; **174**:32–9.

Coons PM. Psychogenic or dissociative fugue: a clinical investigation of five cases. *Psychol Rep* 1999; **84**:881–6.

Kanzer M. Amnesia: a statistical study. *Am J Psychiatry* 1939; **96**: 711–16.

Lapointe AR, Crayton JW, DeVito R *et al.* Similar or disparate brain patterns? The intra-personal EEG variability of three women with multiple personality disorder. *Clin EEG Neurosci* 2006; **37**: 235–42.

Piper A, Mersky H. The persistence of folly: critical examination of dissociative identity disorder. Part I. The excesses of an improbable concept. *Can J Psychiatry* 2004a; **49**:592–600.

Piper A, Mersky H. The persistence of folly: critical examination of dissociative identity disorder. Part II. The defence and decline of multiple personality or dissociative identity disorder. *Can J Psychiatry* 2004b; **49**:678–83.

Reinders AA, Nijenhuis ER, Quak J *et al.* Psychobiological characteristics of dissociative identity disorder: a symptom provocation study. *Biol Psychiatry* 2006; **60**:730–40.

Riether AM, Stoudemire A. Psychogenic fugue states: a review. *South Med J* 1988; **81**:568–71.

Sierra M, Senior C, Phillips ML *et al.* Autonomic response in the perception of disgust and happiness in depersonalization disorder. *Psychiatry Res* 2006; **145**:225–31.

Simeon D, Gross S, Guralnik O *et al.* Feeling unreal: 30 cases of DSM-III-R depersonalization disorder. *Am J Psychiatry* 1997; **154**: 1107–13.

Simeon D, Stein DJ, Hollander E. Treatment of depersonalization disorder with clomipramine. *Biol Psychiatry* 1998; **44**:302–3.

Simeon D, Guralnik O, Hazlett DA *et al.* Feeling unreal: a PET study of depersonalization disorder. *Am J Psychiatry* 2000; **157**:1782–8.

Simeon D, Knutelska M, Nelson D *et al.* Feeling unreal: a depersonalization disorder update of 117 cases. *J Clin Psychiatry* 2003a; **64**:990–7.

Simeon D, Guralnik O, Knutelska M *et al.* Basal norepinephrine in depersonalization disorder. *Psychiatry Res* 2003b; **121**:93–7.

Simeon D, Guralnik O, Schmeidler J *et al.* Fluoxetine therapy in depersonalization disorder: randomized controlled trial. *Br J Psychiatry* 2004; **185**:31–6.

26

Personality disorders

26.1 PARANOID PERSONALITY DISORDER

The patient with paranoid personality disorder is characteristically distrustful and on guard, quick to take offense and read malevolence into what others do, and prone to harbor deep, long-standing resentments. This disorder occurs in 0.5–2 percent of the general population, and appears to be more common among males than females.

Clinical features

The coalescence of paranoid traits occurs during adolescence.

As far as these patients are concerned, most others are ill intentioned or hostile; consequently, being suspicious and watchful seems only natural, even prudent. Keenly observing those around them, these patients are prone to see insults where none exist, and are quick to take offense where none was ever intended. A less than respectful greeting becomes an attack; a passerby who accidentally steps off the sidewalk onto the patient's lawn is clearly testing the limits of the patient's tolerance and forbearance.

Relations with neighbors or co-workers are generally tense and brittle. To others, these patients appear to have a 'chip on the shoulder'. They question motives and intentions and are quick to feel betrayed. Grudges form, and these patients may wait intently for their chance to exact some form of revenge. They may file lawsuits that are subsequently pursued with righteous tenacity.

Intimacy or cooperation are almost impossible for these patients to achieve, because for them every relationship seems to be one of either dominance or submission. They are keenly aware of rank and authority, and the closer they get to others the more important it becomes for them to be in the dominant position. Only around the clearly submissive can these patients begin to relax, but even these relationships may not last long because they come to despise such abject weakness and refuse to associate with such weaklings any further.

At times these patients may settle in to a sort of truce with those around them. To the outside world they seem cold and humorless. Left to themselves they may become preoccupied with machines, computers, or dossiers of various sorts. They may take great pride in their independence and self-sufficiency and look with disdain on those with weaker hearts who allow themselves to depend on others.

Such uneasy truces, at times, however, may be impossible for these people to maintain. Under great stress they may pass from having mere suspicions to actually developing transient delusions of persecution. What they perceive as insults and assaults can no longer be borne, and violence may ensue.

Course

Paranoid personality disorder appears to be a chronic, lifelong condition.

Etiology

Paranoid personality disorder is over-represented among the first-degree relatives of patients with schizophrenia (Kendler and Gruenbeerg 1984; Kendler *et al.* 1993), and it is suspected that the personality disorder exists as a 'forme fruste' of paranoid schizophrenia.

Differential diagnosis

The persecutory subtype of delusional disorder may resemble paranoid personality disorder; however, in delusional disorder, delusions are chronic whereas in paranoid personality disorder they are transient or may not occur at all. Furthermore, the patient with delusional disorder does not tend to display the constant vigilance and pervasive mistrust seen in paranoid personality disorder.

Paranoid schizophrenia may superficially mimic paranoid personality disorder; however, in schizophrenia one sees not only chronic delusions or hallucinations but also a greater or lesser degree of incoherence or illogical speech, characteristics that are not seen in paranoid personality disorder. It must be borne in mind, however, that in some cases paranoid personality traits may constitute a lengthy prodrome to schizophrenia (Rodriguez Solano and Gonzalez De Chavez 2000), and consequently long-term diagnostic follow-up is in order.

In borderline personality disorder one may see deep mistrust and anger; however, on close inspection, patients with borderline personality disorder, in contrast to those with paranoid personality disorder, desperately want a relationship.

Treatment

These patients rarely seek treatment themselves. They see nothing wrong with themselves and would interpret any suggestion that they seek treatment as an insult or as part of a movement to discredit them. A studied courtesy on the part of the physician may ease these patients somewhat. Low-dose antipsychotics may be helpful, but it is critical to avoid side-effects, as patients with paranoid personality disorder rarely tolerate them. Should a patient tolerate oral forms of risperidone, paliperidone, or haloperidol, consideration might be given to using one of the long-acting injectable forms of these medicines.

26.2 SCHIZOID PERSONALITY DISORDER

Schizoid personality disorder is marked by detachment, social isolation, and emotional coldness. This is probably an uncommon disorder, seen in less than 1 percent of the general population, and is more common in males.

Clinical features

Although some schizoid traits may be evident in late childhood, the full clinical picture generally becomes established only in adolescence.

Patients with schizoid personality disorder are characteristically aloof, detached, and socially isolated. They take little or no pleasure in social relationships and experience little interest in being with others or having sexual relations with others. Lacking any desire or interest in relationships, and indeed being more comfortable by themselves, these patients are generally found engaged in solitary pursuits, often involving machines, computers or mechanical routines.

If forced into conversation or social activities, these patients are typically awkward, socially inept, and seemingly incapable of grasping the nuances or subtleties of normal social interchange.

Occasionally, if under great stress, these patients may experience delusions or hallucinations. These phenomena, however, are generally brief, lasting less than a day, and resolve spontaneously.

Course

This is a chronic, lifelong disorder.

Etiology

Schizoid personality disorder is more common among relatives of patients with schizophrenia (Kendler *et al.* 1993; Thaker *et al.* 1993), and is often considered to be one of the 'schizophrenia spectrum' disorders. Like schizophrenia, it is also associated with *in utero* malnutrition (Hoek *et al.* 1996).

Differential diagnosis

Distinguishing adults with autism from adults with schizoid personality disorder may be very difficult on clinical grounds. The age at onset, however, is distinctive. Although patients with mild cases of autism may not come to clinical attention until they reach elementary school, a careful history will reveal a clear onset in very early childhood, well before the onset typical for schizoid personality disorder.

Schizotypal personality disorder is distinguished by the chronic presence of magical thinking and peculiarities or oddities of speech and behavior.

Paranoid personality disorder is distinguished by the presence of a pervasive suspiciousness and mistrust.

Schizophrenia, with the exception of the simple subtype, is distinguished by the chronic presence of hallucinations, delusions or loosening of associations. The simple subtype of schizophrenia is distinguished by its gradual downhill course, which contrasts with the stable course of schizoid personality disorder.

Treatment

These patients generally do not do well in any psychotherapy that involves interpretation or emotional interchange. Within the context of a calmly supportive doctor–patient relationship, however, they may respond to gentle suggestions regarding work or limited social exchanges. Antipsychotics are of little or no benefit.

26.3 SCHIZOTYPAL PERSONALITY DISORDER

Patients with schizotypal personality disorder are distant and aloof from others; their thoughts and behavior tend towards the peculiar, and others often view them as odd or eccentric (Dickey *et al.* 2005; Jacobsberg *et al.* 1986). Schizotypal personality disorder is not rare and may indeed be present in up to 3 percent of the general population. It may be more common in males than females.

Clinical features

This disorder typically has an onset in adolescence.

These patients have few friends, and tend to remain isolated and aloof, even when with people, such as family members, whom they have known for a long time. Their affect tends to be constricted, and they rarely display any humorousness. At times, albeit with hesitation, they may express their opinions, and when they do, certain peculiarities become evident. To begin with, their communications tend to be vague and overelaborated. Further, they may express odd beliefs, such as in telepathy or clairvoyance, and they may suspect that others have been talking about them. Their overall behavior, likewise, tends to be odd, and they may gain a reputation for eccentricity.

Occasionally, if under great stress, these patients may transiently become delusional. The delusions tend to be either persecutory or hypochondriacal. They may have to move because forces have gathered in the attic. They may consult their internist for bizarrely characterized aches and pains. After the stress has passed and the delusions have faded, these patients may be embarrassed to admit that they have ever thought that way.

Course

Schizotypal personality disorder is chronic and lifelong.

Etiology

Schizotypal personality disorder appears to run true in families (Baron *et al.* 1985; Kendler *et al.* 1993), and may represent a 'forme fruste' of schizophrenia: the frequency of schizotypal personality disorder is increased in the first-degree relatives of patients of schizophrenia (Kendler and Gruenberg 1984; Kendler *et al.* 1993), and the converse is also true, namely that the rate of schizophrenia is increased in the first-degree relatives with schizotypal personality disorder (Siever *et al.* 1990; Thaker *et al.* 1993). Although there are no neuropathologic studies of this disorder, magnetic resonance imaging (MRI) has revealed atrophy of the superior temporal gyrus (Goldstein *et al.* 2009) and both reduced size and abnormal shape of the caudate nucleus (Levitt *et al.* 2009).

Differential diagnosis

Schizoid personality disorder shares the social isolation seen in schizotypal personality disorder, but in schizoid personality disorder one fails to see the odd beliefs and eccentricities of behavior typical of schizotypal personality disorder.

Autism is characterized by social isolation and odd behavior, yet here the course is fundamentally different. The onset of autism is in early childhood, after which one typically finds, during adolescence, some improvement. By contrast, schizotypal personality disorder typically has an onset in adolescence, after which the course is chronic, with, essentially, no improvement.

Most subtypes of schizophrenia are readily distinguished from schizotypal personality disorder by the chronic presence of delusions and hallucinations. Although delusions may occasionally appear in schizotypal personality disorder, this is only during periods of great stress; by contrast, in schizophrenia, delusions are present throughout, both in stressful and stress-free times. Difficulty may arise, however, in cases of the simple subtype of schizophrenia, for in this subtype psychotic symptoms may be transient and minimal. Here, observation over a long course may be required to make the differential: in simple schizophrenia, one sees a gradual deterioration, whereas in schizotypal personality disorder the overall course is stable. Finally, it must be borne in mind that symptomatology similar to that seen in schizotypal personality disorder may also constitute a prodrome to schizophrenia (Rodriguez Solano and Gonzalez De Chavez 2000), and consequently long-term follow-up is required before a confident diagnosis can be made.

Females with the fragile X syndrome may display some schizotypal traits (Sobesky *et al.* 1994). When mental retardation or a facial dysmorphism is present, the diagnosis may be relatively easy, but many females with fragile X may lack these features, and here only a high index of suspicion will lead to appropriate genetic testing.

Treatment

Within the context of a long-term psychotherapeutic relationship, patients should be diplomatically directed towards work that demands little in the way of personal contact, individual initiative, or abstract thinking. Should the patient attempt an intimate relationship, contact with the patient's partner is in order to help correct the inevitable distortions that the patient will develop about the relationship.

Low doses of antipsychotics are helpful, as has been demonstrated in a case series of patients treated with olanzapine (mean dose ~10 mg) (Keshaven *et al.* 2004) and in double-blind studies of risperidone (2 mg) and thiothixene (mean dose ~10 mg) (Goldberg *et al.* 1986).

26.4 ANTISOCIAL PERSONALITY DISORDER

In the distant past, this disorder was known as 'moral insanity' (Cleckley 1941), a name that hints at the fundamental aspect of this disorder. At times, 'sociopathy' and 'psychopathy' are used as synonyms, from which are derived the sobriquets 'sociopath' and 'psychopath': in recent literature, the term psychopathy is also used to denote a more severe form of antisocial personality disorder.

Antisocial personality disorder is much more common in males than females and is much more prevalent among the transient populations of impoverished inner cities. Roughly 3 percent of all American males have this disorder, in contrast to about 1 percent of females. As might be expected, upwards of 50 percent of all prison inmates are found to have this disorder.

Clinical features

The onset of this disorder is always in childhood (Robins 1978). In boys, a pattern of antisocial behavior is often apparent before the age of eight. In girls, the pattern may not become fully established until puberty. The antisocial tendencies may appear insidiously, as when the child gradually commits more and more minor infractions, or at times dramatically, as when a child 'senselessly' murders a playmate or family member.

A peaceful, yet not oppressive, society rests upon the restraint that each person exercises over those urges and impulses that might lead to unnecessary harm to others. That restraint, in turn, rests primarily on three things: an effective conscience, a respect for the law and social norms (or, failing that, an effective fear of consequences), and a capacity to sympathize with the suffering of others. To varying degrees, sociopaths lack these and lack them to such a degree that they repeatedly, in most areas of their lives, fail to maintain their conduct in conformity with the rules.

Lacking in conscience, respect for the law, and a fear of consequences, sociopaths, when seized by impulse or temptation, are easily swept past the bounds set by law and custom. They are irritable with authority and seem comfortable when aggressive. If caught, they rarely admit to being at fault, and blame is cast on others: the police, judges, parents, or society at large. They have little recognition that most of those who are successful in life are so primarily because they do conform to social norms; rather, the sociopath sees the successful citizen as simply a better 'con', as someone who is 'smart' enough to 'get away with it'.

Sociopaths lack any enduring sympathy or fellow-feeling. Although at times they may evidence some affection for others, such feelings pale rapidly when the other person is found to be in the way of the sociopath's wants. The capacity for empathy, however, may be quite good: sociopaths may be quite capable of understanding the suffering of others – what is lacking is the capacity to feel that suffering. Sociopaths evidence a remarkable degree of selfishness and egocentricity; others are manipulated as if they had no more inherent value than a pawn on a chessboard, exploited when it serves the sociopath's purpose, discarded or sacrificed when no longer useful. Genuine shame, guilt, or remorse are foreign emotions to sociopaths. They may feign guilt when such an act serves to get others off their backs, yet once surveillance is lifted, this imposture is dropped.

Although deficient in morality, sociopaths may otherwise evidence a high development of other faculties. Indeed, they may show a fine and discriminating taste in food, music, literature, and the arts. Their feelings, in so far as they concern themselves, may be complex and highly developed; they may be remarkably intelligent. Such gifted sociopaths become experts at 'conning' others; they may come across as caring, delightful, and personally attractive. So long as they get what they want, they seem immune to depression or anxiety. Some successful politicians have been excellent sociopaths.

In practice, however, most sociopaths, like the rest of the population, have only a modicum of these gifts or endowments. Thus, limited in their ability to deceive others, they come across, at best, as 'slick' rather than cunning, and as manipulative rather than deftly diplomatic. Unable to successfully pursue a Machiavellian course through life, they repeatedly come into open conflict with society, breaking laws and failing to live up to everyday responsibilities.

Telling the truth is a matter of convenience for sociopaths; if a lie would prove more useful, that is what is heard. This may progress to the point of using aliases or of lying in almost all situations. Being faithful in a relationship, whether married or not, is, likewise, a matter of convenience. If the partner no longer pleases him or commits the mistake of making too many demands, the sociopath may simply discard the relationship and move on to another. Marriages rarely last for more than a year. If children become a burden, they be neglected or even abandoned. Money that should properly go for food, clothing, or shelter is used for more immediate pleasures, such as gambling, alcohol, or drugs.

Financial obligations are disregarded as soon as they crimp the sociopath's lifestyle. Bills are left unpaid, debts are left outstanding, and alimony and child support payments become overdue.

The normal practice of staying at one job and living in one place is often too confining for sociopaths. Should it not suit them, they simply may not go to work; quitting one job after another or being fired for insubordination is common. Sociopaths may go for months or longer with no work and seem unconcerned: 'moving on' is a common theme. Some sociopaths travel from town to town and state to state; roots are not put down, and being transient becomes a way of life.

Reckless, violent, and often cruel behavior is common. Those who get in the way may be assaulted; fights are very common. Children and wives are beaten, and murder may be committed. As noted below, in the discussion on treatment, it is important to distinguish between impulsive violence and predatory violence. Sociopaths who are prone to anger, have a 'short fuse', and harbor resentments and the desire for revenge often commit violence on the spur of the moment, with little premeditation. In contrast, predatory types typically feel little and set about their violence with premeditation and often careful planning.

When others have what sociopaths want, theft, fraud, and embezzlement are seen. Most sociopaths have a long history of arrests. Abuse of alcohol, cocaine, stimulants, opioids, and other substances is common, and when this occurs whatever restraints or craftiness the sociopath had may disappear. Promiscuity is often found; some sociopaths boast of an incredible number of sexual partners. Female sociopaths often lack the foresight to use contraceptives and multiple pregnancies are common. Suicide threats are often heard, and though at times suicide may occur, more often than not these patients do not seriously act on these threats, and they pass as soon as the sociopath's demands are met.

The expression of sociopathy is influenced by age, and the childhood history of adult sociopaths shows some typical features. As children, sociopaths often start fights and at times would use weapons. Often they acquired a 'reputation' in the neighborhood, and other children were advised to keep away from them. Bullying or cruelty to younger or smaller playmates, and cruelty, often savage, to animals is common. Early sexual activity is also common, often voluntary, sometimes not.

Overall, sociopathic boys and girls seem wild and undisciplined; punishment has little lasting effect on them, and they seemed unmoved by kindness. They often simply run away when efforts are made to enforce conventional rules. Most are recurrently truant from school; indeed, a history of satisfactory academic adjustment militates strongly against this diagnosis. Stealing and extortion are common; property may be destroyed just for the pleasure of it. Fire setting is also common. Habitual lying for no purpose other than to escape deserved punishment or to get one's undeserved way is very common.

Course

Although sociopathic tendencies and attitudes are chronic and lifelong, the frequency of actual antisocial acts changes with time. Highest in late adolescence, the frequency gradually decreases, so the majority of patients seem to 'burn out' by middle years (Black *et al.* 1995). After that, the frequency of antisocial acts is quite low. Some patients eventually become amusing raconteurs; many, however, become hypochondriacally preoccupied with minor aches and pains.

Etiology

Although the underlying neuropathology of this disorder is not known, MRI studies have demonstrated atrophy in the prefrontal cortex (Raine *et al.* 2000), the frontopolar, anterior temporal and insular cortices (de Oliveira-Souza *et al.* 2008), and the amygdala (Yang *et al.* 2009). Further, cerebrospinal (CSF) studies (Soderstrom *et al.* 2001, 2003) have demonstrated a positive correlation between antisocial traits and an elevated ratio of the metabolites of dopamine (homovanillic acid [HVA]) and serotonin (5-hydroxyindoleacetic acid [5-HIAA]). This elevated HVA–5-HIAA ratio in turn could reflect either increased dopaminergic tone, resulting in increased inhibition of serotoninergic neurons, or a primary reduction in serotoninergic tone, or perhaps a combination of these. Of interest, and in keeping with the presence of genetic factors, decreased CSF 5-HIAA has also been found in newborns when one of the biologic parents had antisocial personality disorder (Constantino *et al.* 1997).

Both family (Cadoret *et al.* 1985) and twin (Lyons *et al.* 1995) studies indicate the importance of genetic factors, and although environmental factors, such as being raised by parents with antisocial traits or alcoholism or growing up in socioeconomically deprived areas (Cadoret *et al.* 1990), may also be important, the fact that adoptees, raised apart from their sociopathic biologic parents, also display an increased risk of antisocial personality disorder (Cadoret *et al.* 1990) strongly supports the notion that antisocial personality disorder is an inherited disease. Attention-deficit disorder with hyperactivity may also play a role, as children with this disorder are more likely to develop antisocial personality disorder (Simonoff *et al.* 2004).

It may well be that antisocial personality disorder arises from an interplay of, on the one hand, inherited neuroanatomical and neurophysiologic defects or a history of hyperactivity, and, on the other, adverse environmental events. Atrophy of limbic structures may impair both the development of conscience and the development of a capacity for sympathy, while deficient serotoninergic tone may predispose patients towards impulsive behavior. Attention-deficit disorder and hyperactivity may, in addition, so disrupt the normal relations between child and parent that the developmental formation of conscience simply does not fully occur. Adverse environmental features during childhood may likewise impair this development. If parents have antisocial traits and condone this behavior in children, or if home life is sufficiently chaotic, say, because of parental alcoholism or drug addiction, that consistent discipline simply does not occur, then children, as they grow up, simply will not acquire the basics of conscience.

Differential diagnosis

A criminal record, no matter how long, does not of itself qualify a person for a diagnosis of antisocial personality disorder. In simple criminality, one finds either that the onset of criminal behavior was well beyond puberty or that the antisocial behavior, unlike the case in sociopathy, fails to pervade all aspects of the person's life. In sociopathy, selfishness and exploitation are evident in the patient's relationships with family, friends, and society at large. By contrast, the person with simple criminality may have normal, indeed exceptionally loving, relationships with family or children.

By convention, those under the age of 18 are not given the diagnosis of antisocial personality disorder even though they have the signs and symptoms described earlier. The merit in this lies in the fact that in most of these adolescents the signs and symptoms remit before adult years (Robins 1978).

Those with a narcissistic personality disorder are exploitative and manipulative, yet here, in contrast with antisocial personality disorder, the purpose is to feed their vanity and sense of glory, not simply to achieve financial gain. Indeed, the narcissist may have a deep respect for law and custom and take pride in following them.

Patients with borderline personality disorder are often quite impulsive, and, thus driven by impulse, may commit antisocial acts. Here, however, in contrast to the sociopath, the borderline patient suffers from a guilty conscience, experiences true remorse, and feels sympathy for others.

Patients with mental retardation may have a history of antisocial behavior dating back to childhood, yet here, in contrast with sociopathy, the patient simply lacks the cognitive ability to appreciate rules and consequences. The sociopath recognizes them, but simply chooses not to abide by them.

Manic patients, especially irritable manics, are notorious for their antisocial behavior. However, the antisocial behavior here is clearly episodic, being confined within the episodes of mania themselves.

Personality change may be characterized by antisocial behavior; however, here the premorbid history shows a non-sociopathic personality structure. Such personality change may be seen with frontal infarction (Meyers *et al.* 1992) or in the context of frontotemporal dementia (Mendez *et al.* 2005). The key to a correct diagnosis in such cases is found in the history: in antisocial personality disorder one finds a string of antisocial behaviors stretching back at least into adolescence, whereas in personality change the antisocial behaviors represent a break from the premorbid, non-sociopathic, personality structure.

Differentiating alcoholism, cocaine addiction, or stimulant or opioid addiction from sociopathy may be very difficult. Lying, manipulation, and crime in the service of ensuring an uninterrupted supply of alcohol or drugs is quite common. In cases where the alcohol or substance use disorder began relatively late in life, the consequent late age at onset of antisocial behavior stands in contrast to the childhood onset seen in sociopathy. In cases where the substance use began very early in life, or where a reliable early history is not available, diagnostic certainty may not be possible until a long period of abstinence ensues. If the antisocial phenomena clear up with abstinence, then sociopathy in all likelihood is not present. If, however, the antisocial behavior and attitudes persist into the sobriety, then both diagnoses should be made.

Treatment

In general, sociopaths present for treatment only to escape consequences or because they are under court order. Once freed from these concerns, almost all drop out of treatment.

No evidence supports any long-term effectiveness of individual or family psychotherapy or for behavior therapy in antisocial personality disorder. Some evidence suggests that long-term stays in well-organized, strict, and tightly structured residential treatment centers may help. For those sociopaths who have alcoholism, intense involvement in Alcoholics Anonymous may not only help their alcoholism but also be followed by a partial remission of the sociopathy.

In cases characterized by impulsive violence, medical treatment should be considered. Lithium (Sheard *et al.* 1976), phenytoin (Barratt *et al.* 1997; Stanford *et al.* 2001, 2005), divalproex (Hollander *et al.* 2003) (or valproate [Stanford *et al.* 2005]), and carbamazepine (Stanford *et al.* 2005) are all effective in reducing impulsive violence. These medications, however, are of no use in cases where the violence is predatory, planned, and 'cold blooded'.

Debate exists as to how to treat the sociopath who also evidences residual symptoms of childhood attention-deficit–hyperactivity disorder. Some clinicians recommend methylphenidate; however, the risk of abuse is so high among sociopaths that a more prudent choice would be bupropion or, and especially in cases characterized by impulsive violence, lithium.

By and large, however, sociopathy remains resistant to most currently available treatments. For the dangerous sociopath, incarceration until the disorder 'burns out' may be the only safe alternative.

26.5 BORDERLINE PERSONALITY DISORDER

Patients with a borderline personality disorder have been aptly described as 'stably unstable'. Their attitudes and feelings, about both themselves and others, are subject to dramatic, at times violent, change. Their relationships with others are intense and generally stormy, and a clinging dependency may alternate with enraged attacks. Their histories often reveal a life of tumult and chaos.

The lifetime prevalence of borderline personality disorder is estimated at 1–2 percent; although females far outnumber males in clinical samples, the sex ratio in the general population is probably much closer to 1:1.

Clinical features

Symptoms are often apparent by early teenage years; by late teenage years, the pattern of instability is generally set.

These patients experience a pervasive sense of loneliness and emptiness. They cannot count on feeling well; their mood might plummet unpredictably into depression and despair, or they may be seized by an unreasoning irritability or anger. Desperate to borrow some order and comfort from others, they fear abandonment, perhaps more than anything else. They do not tolerate being alone; weekends without assurance of companionship may offer a terrifying prospect. Yet, at the same time, they are fearful of engulfment: lacking any stable sense of themselves they fear, in a very literal way, losing themselves in others, as if in joining with another they lose their own boundaries and dissolve into, or become subsumed by, the person they wish to be with.

These patients tend to have very strong feelings and opinions about others. They tend to perceive others as either all good or all bad. Those who are seen as all good may be idealized, even worshipped. They are seen as a balm, a source of comfort and salvation, and the relationship is seen by the patient as essential and vital to his or her existence. In such cases, the patient often clings to the other person and fiercely resents any perceived threats to the relationship. Should, however, there be any threat of rejection or abandonment, this loving adoration may shatter, and the patient may then come to see the other person as all bad. Now the patient sees the other person as a pretense, a sham, and as hollow and without substance. Thus, disappointed, the patient may become intensely angry and impulsive: the other person may find him- or herself the target of aggressive and enraged outbursts, and the patient may engage in suicidal behavior.

Predictably, these patients rarely enjoy any stable relationships: others can rarely tolerate the intense and unpredictable changes in them. Idealization and adoration may be a seductive lure for some, but often the intense demandingness of the patient gives the other person pause. Such a pause is often taken as an abandonment or a threat of one, prompting the rage just described. Eventually, the relationship can no longer bear these extremes and the patient, once again, is left alone.

Thus, alone and unable to 'fill the emptiness', these patients are often bitter, cynical, and sarcastic. Some may give up entirely in their quest for a relationship and set out to establish a structure of self-sufficiency. Inevitably, however, such an edifice is fragile, intensely brittle, and maintained only with enormous effort. Interestingly, many of these patients form strong attachments to 'transitional objects', such as teddy bears, and, indeed, may bring them to the hospital (Cardasis *et al.* 1997).

When desperate, these patients may become very impulsive and may go to any extreme to fill the void. Intoxication may be sought, and alcohol is often the preferred intoxicant. Promiscuous behavior is common. At times, the thrill of daredevil recklessness may be sought.

The thought of suicide is often a constant companion. In despair, patients yearn for death as relief, and even when calm, it may be viewed as a very rational alternative. Suicidal behavior is common, and at times it may appear manipulative. Some patients have slashed their wrists and forearms so many times as to create a veritable 'ladder' of scars. Relatively innocuous overdoses are common. At times, however, the intent is serious, and up to 10 percent of patients may die at their own hands.

At times patients may mutilate themselves with no interest in killing themselves. When asked why they slashed or gouged themselves they may have difficulty offering a reason. Some do it to relieve tension, whereas others use it as a way to inflict a feeling, to convince themselves that they are still, in fact, alive.

At times, certain patients may display transient psychotic symptoms. Voices may be heard and are generally critical; delusions of reference or persecution may occur and may be accompanied by guardedness. Other, allied symptoms may appear, including depersonalization, derealization, or ideas of reference. Often these symptoms may be precipitated by extreme stress, most typically by the fear, or the actual fact of, abandonment.

Patients with a borderline personality disorder often have other disorders. Alcohol abuse or alcoholism are not uncommon; and the same holds true for dysthymia and major depressive disorder. Panic attacks may also occur. Likewise, bulimia nervosa may develop and, when it does, kleptomania is also often in evidence.

Course

Although this is a chronic disorder, most patients experience a diminution in the intensity of symptoms, especially impulsivity, as they reach middle years.

Etiology

There is little doubt that borderline patients, as children, were subjected to neglect and abuse, especially sexual abuse. Theoretically, such experiences, by interfering with the development of stable object relations in childhood, could have prevented the subsequent development both of the capacity to form stable relations in adulthood and of the capacity to form those stable intrapsychic structures necessary for the control of affect.

There may also be a genetic contribution to borderline personality disorder, as suggested by studies involving twins (Torgersen *et al.* 2000) and first-degree relatives (Baron *et al.* 1985); however, these studies did not control for the effect of shared environment, and until adoption studies are performed, this question will remain undecided.

Although there are no neuropathologic studies of borderline personality disorder, MRI studies have shown smaller volumes of the hippocampus and amygdala (as reviewed in Nunes *et al.* 2009), and functional MRI (fMRI) studies in turn have demonstrated hyperreactivity of the amygdala to emotionally laden stimuli (Minzenberg *et al.* 2007).

Differential diagnosis

Patients with histrionic personality disorder, like those with borderline personality disorder, often lead emotionally stormy lives, yet there is a qualitative difference here. In the histrionic, the storm is, at base, an act, whereas in the borderline the emotional tumult is authentic and deeply felt. Consequently, the histrionic will fairly rapidly get over it and carry on, whereas the borderline is often consumed and disabled by it.

Patients with narcissistic personality disorder may become enraged if others turn their backs; however, here, in contrast to the borderline, the rage is often expressed by an icy disdain, rather than an overt display. Furthermore, in the narcissist, the issue is loss of admiration, rather than the threat of desolate loneliness faced by the borderline patient.

Cyclothymia and bipolar II disorder, given their intense affective changes and the destructive effect of these on relationships, may be difficult to distinguish from borderline personality disorder. Certain qualitative differences in the affective changes of these disorders, however, may allow a differential. To begin with, most patients with cyclothymia and bipolar II disorder will, at least at times, experience sustained euphoria, a symptom that is not, at least in a sustained fashion, seen in borderline personality disorder. In some cases of cyclothymia or bipolar II disorder, however, euphoria is minimal or absent, and rather one sees sustained irritability. In such cases, however, one finds that the irritability occurs autonomously, in contrast to the situation in borderline personality disorder, where the irritability is reactive, occurring in response to environmental events, typically a perceived or actual abandonment.

Schizophrenia may enter the differential diagnosis in cases of borderline personality disorder marked by prominent psychotic symptoms. In schizophrenia, however, the psychotic symptoms are sustained and autonomous, in contrast to the case of borderline personality disorder where they are generally transient, and occur in reaction to events in the patient's life. Furthermore, with the exception of the paranoid subtype, one finds in schizophrenia other symptoms, such as incoherence and bizarre behavior, which are not seen in borderline personality disorder.

Treatment

These patients are perhaps the most difficult of all to treat. Even in the most skilful hands, progress, at best, evolves slowly and haltingly. Setbacks are the rule, and patients often become extraordinarily demanding, enraged, or suicidal. When a substance use disorder, such as alcoholism, is present, it is critical to treat this, as little progress can be expected otherwise.

Most patients are seen in individual psychotherapy. Perhaps the best-researched form of psychotherapy for borderline patients is dialectical behavior therapy; however, in a recent trial, this approach was no better than psychoanalytically-oriented psychotherapy combined with appropriately symptom-targeted medications (McMain *et al.* 2009). Regardless of what form of therapy is utilized, definite structure and firm limit setting are required, for without these a deterioration in the patient's condition is almost inevitable. The length of appointments must be strictly adhered to, and during the session a more than average amount of direction and support is required. Often firm limits must be placed on telephone calls; in addition, fees must be scrupulously collected.

Hospitalizations, which often follow suicidal ideation or behavior, should be kept as short as possible and highly structured. The focus should be on practical planning and not on the deeper aspects of the patient's emotional life. Partial hospitalization is also utilized, and with a psychoanalytically oriented program, improvement is maintained long after discharge (Bateman and Fonagy 2001).

The tendency for borderline patients to see people either as all good or as all bad, by virtue of its effect on any developing relationship, often leaves those involved in the patient's care split into one of two camps. Not uncommonly, clinical personnel in a hospital are split by the patient into two groups: those who are drawn to the patient and are inclined to 'go the extra mile', and those who see the patient as manipulative and demanding and recommend immediate 'therapeutic' discharge. Often these two clinical camps find themselves at war. Indeed, when one finds the staff in civil war, often one finds the war began after a particular borderline patient was admitted.

Medications play a definite role in the treatment of borderline personality disorder; however, in planning the psychopharmacologic treatment of patients with borderline personality disorder it is essential, given the chronicity of this disorder, to take the long view: sequential treatment trials of one agent at a time, if possible, are ideal, with the goal of finding the best regimen for any particular patient. Antipsychotics, mood stabilizers, and antidepressants may all be considered, and are discussed below. Importantly, the use of anxiolytics should be approached with great caution, given the possibility of disinhibition with these agents, as has been demonstrated in the case of alprazolam (Cowdry and Gardner 1988).

Psychotic and allied symptoms, as might be expected, respond to antipsychotics, and this has been demonstrated for aripiprazole (Bellino *et al.* 2008; Nickel *et al.* 2006), thiothixene (Goldberg *et al.* 1986; Serban and Siegel 1984), and trifluoperazine (Cowdry and Gardner 1988); importantly with each of these agents, one sees not only a reduction in psychotic and allied symptoms, but also an overall, general improvement. Importantly, it appears that not all antipsychotics, however, are beneficial. For example, haloperidol was found beneficial for psychotic symptoms in one study (Soloff *et al.* 1986), but an attempt to replicate this by the same group failed (Soloff *et al.* 1993). Additionally, olanzapine, in terms of overall improvement, was likewise found beneficial in one study (Bogenschultz and Nurnberg 2004), but not another (Schulz *et al.* 2008).

Affective dyscontrol has been successfully approached with mood stabilizers, antipsychotics, and antidepressants. Anger may respond to topiramate (Nickel *et al.* 2004) or fluoxetine (Salzman *et al.* 1995). Impulsivity may respond to lamotrigine (Reich *et al.* 2009) or aripiprazole (Bellino *et al.* 2008). Impulsive aggressive behavior may respond to divalproex (Hollander *et al.* 2001, 2003, 2005). Impulsive aggressive behavior may also respond to olanzapine (Zanarini *et al.* 2004), but some further work casts doubt on this. In one study, all patients received dialectical behavior therapy; half received olanzapine while the other half received placebo, and, in this study, olanzapine was beneficial (Soler *et al.* 2005). However, in a subsequent study that used a similar design, olanzapine was no more beneficial than placebo (Linehan *et al.* 2008). Impulsive aggressive behavior responded to fluoxetine in one study (Coccaro and Kavoussi 1997); however, when fluoxetine was compared with placebo in patients receiving dialectical behavior therapy, there was no benefit (Simpson *et al.* 2004). Overall affective dyscontrol leading to significant behavioral dyscontrol improved with carbamazepine in two studies (Cowdry and Gardner 1988; Gardner and Cowdry 1986), but not a third (de la Fuente and Lotstra 1994).

26.6 HISTRIONIC PERSONALITY DISORDER

Patients with a histrionic personality disorder are dramatic, overly emotional, and often quite colorful. They seek incessantly to keep themselves the center of attention, and also to involve others in their drama. Others are often put off and find these patients shallow, insincere, and unconvincing.

Histrionic personality disorder used to be called 'hysterical personality disorder'; however, this is an unfortunate term because the possession of a uterus is by no means a prerequisite for this disorder. Indeed, though females far outnumber males in clinical samples, the sex ratio is probably much closer to 1:1 in the general population. Overall in the general adult population the prevalence is roughly 2 percent.

Clinical features

Patients with a histrionic personality disorder strive to create a lasting impression by their dress, speech, and behavior. Females tend to dress seductively, provocatively, even exhibitionistically; males, on the other hand, tend towards the 'macho' image with striking hair styles, 'tough' or revealing clothing, and often an excess of jewelry. The speech of these patients is emotional and expressive and is often accompanied by broad or exaggerated gestures. Their exciting and melodramatic stories may, at first, be engaging; however, pinning these people down as to details is often impossible. Indeed, in relating a pivotal event, as for example the death of a loved one, such a patient may not be able to get beyond saying 'most horrible' or 'I was crushed, just crushed'. The overall behavior of these patients is marked by crisis and the melodramatic. Indeed, if their current situation lacks these elements, their actions may rapidly create them. They are highly emotional and often effusive. Mere acquaintances are greeted as if they were long lost friends; there may be tears and exclamations; the hand shake is too strong, too 'sincere' and heart felt; the embrace is too close and too long. Temper is always a problem, and should these patients be thwarted, should they not get their way, they may become explosively angry or threatening or, conversely, they may withdraw into a devastated pout that seems to beg for sympathy and understanding.

Friends and acquaintances often find themselves part of the melodrama. These patients are attracted to causes, fads, and often seek to have others play a supporting role in their quest. Romantic entanglements are common. Females seem unable to stop flirting, and males may seem restlessly 'on the make'. Promiscuity, though common, is not universal.

Those with a histrionic personality disorder are rarely able to maintain a stable, enduring, and meaningful relation with another, and relationships tend to be short-lived and often stormy. Eventually, others often find these patients shallow, unsubstantial, and lacking in any honest emotion. Thus, no longer engaged, others often prefer to end the relationship, much as they might wish to turn off a bad soap opera on the television.

Course

Histrionic personality disorder is a chronic lifelong disorder, and although symptoms tend to lessen somewhat with age, true remission does not appear to occur.

Etiology

Although this disorder appears to be familial (Torgersen *et al.* 2000), the basis of the aggregation is not known.

Differential diagnosis

Although narcissistic and borderline personality often enter into the differential, certain features allow for an accurate differentiation. Narcissistic patients, like histrionic patients, seek to be the center of attention. Narcissists, however, are never as emotional or dramatic as are histrionic patients, and would never allow themselves the kind of effusive displays or angry outbursts so common in histrionic personality disorder. Borderline patients often lead lives marked by crisis and tumult; however, in borderline personality disorder, emotions are experienced deeply and genuinely, in contrast to the shallow experience in histrionics: the crisis of the borderline patient is real, and such patients cannot simply walk 'off stage' like the histrionic can.

Briquet's syndrome (somatization disorder) is typified by histrionic behavior; however, here the drama centers on only one theme, namely the patient's concern about having multiple other symptoms and illnesses. By contrast, the drama of the patient with the histrionic personality disorder covers a wide range of subjects.

In cyclothymia, patients may go from one flamboyant crisis to another. Here, however, one sees pressured speech, racing thoughts, and truly heightened mood, all of which are not part of histrionic personality disorder.

Treatment

There are no controlled studies of the treatment of histrionic personality disorder. Anecdotally, patients improve with individual or group psychotherapy.

26.7 NARCISSISTIC PERSONALITY DISORDER

Patients with a narcissistic personality disorder (Ronningstam and Gunderson 1990) see themselves as superior to others, regardless of their actual achievements in life. They feel entitled to admiration and expect that others will act accordingly. Relationships are valued only insofar as they enhance these patients' self-esteem: they have little capacity for sympathy nor do they experience any true interest in the well-being of others.

This personality disorder is rare, occurring in less than 1 percent of the general population; it is somewhat more common among males than females.

Clinical features

These patients have a lofty opinion of themselves. They are often preoccupied with daydreams or fantasies of glory, power, and idealized love, and whatever accomplishments they do have are typically showcased and exaggerated. They view themselves as above the need for help, and thus rarely, if ever, permit themselves to seek assistance from others. They have difficulty tolerating any defects in their persons, and grooming and dress may capture a large amount of time.

Despite this lofty opinion, these patients seem to require the admiration of others. Provided that others remain fawning, they are tolerated, even welcomed, but should others offer any criticism, or should the accomplishments of others shine more brightly than the patients', then these patients may become intensely envious, even quietly enraged. Patients may then attempt to undercut others, or belittle and devalue the others' accomplishments. Should such a strategy fail they may counter-attack with sarcasm or simply withdraw from the field with a cold, regal disdain and contempt for their opponents.

These patients do not have truly reciprocal relationships with others. They may be socially correct, even polite, yet they are perceived as frosty and distant. They seem to lack a capacity for sympathy or intimacy. Should an acquaintance be in need, their only motive for offering help would be to demonstrate their own power. These patients do not make 'anonymous' gifts of time or money and are rarely willing to sacrifice anything of their own. Predictably, others often feel exploited and manipulated by narcissistic patients. They feel as if they are regarded as mere pawns in a narcissist's life and, often in short order, come to regard the narcissist as abrasive and arrogant.

Course

This disorder is chronic and lifelong; with middle years, however, the clinical picture may change. These patients are often intolerant of the inevitable decline that comes with age and often develop lingering, mild depressive symptoms.

Etiology

Although narcissistic personality disorder appears to be familial (Torgersen *et al.* 2000), little else is known with any certainty regarding its etiology.

Differential diagnosis

Four other personality disorders may, at times, enter into the differential diagnosis: antisocial, histrionic, obsessive–compulsive, and borderline.

In antisocial personality disorder one often sees vanity and always sees a persistent exploitation of others. Here, however, the purpose of the exploitation is for financial or material gain. This contrasts with narcissistic personality disorder, wherein financial gain may not even be an issue, so long as the narcissists' sense of personal glory is enhanced.

Those with histrionic personality disorder are notoriously vain and anxious to be at the center of attention. Here, however, one finds a willingness to depend on others and to become strongly emotionally involved with them. This contrasts with narcissistic personality disorder, wherein a cool distance is kept between the narcissist and others, the narcissist being no more involved with others than with a trinket, which is discarded without sentiment as soon as it loses its luster.

Those with obsessive–compulsive personality disorder often evidence an air of superiority and conduct themselves with condescension toward others. Here, however, the smug sense of superiority derives from a perceived adherence to a rigid code of conduct, which others, most regrettably, fall short of. In contrast, in narcissistic personality disorder the grandiose superiority seems to rest on nothing. One sees no smugness about narcissists; they counter-attack, sometimes with viciousness, should a stone of criticism approach the fragile vessel of their self-esteem.

Those with borderline personality disorder may affect an air of regal indifference and often tend to either devalue or idolize others. Here, however, longer observation reveals intense and dramatic swings in their self-conception and their mood. Furthermore, one finds an almost frighteningly intense yearning for relationships with others, as if the borderline wishes to fuse with them. This contrasts with narcissistic personality disorder, wherein the narcissist's behavior remains remarkably stable over time and rarely indicates an interest in friendship or love.

Treatment

Patients with narcissistic personality disorder have been treated with individual, group, and family therapy, and each method has anecdotally been reported as successful. However, no controlled outcome studies have been carried out.

26.8 OBSESSIVE–COMPULSIVE PERSONALITY DISORDER

Patients with obsessive–compulsive personality disorder are typically rigid and perfectionistic. Their lives are orderly and carefully controlled and they give undue attention to details, rules, and schedules. These patients rarely display any emotion and seem to march through life with stultifying, regimental thoroughness.

In Europe, this disorder is often referred to as anancastic personality disorder, and this name gives the distinct advantage of not allowing for any confusion between this personality disorder and obsessive–compulsive disorder, a disorder which, as noted below, may have no relation at all with obsessive–compulsive personality disorder.

Obsessive–compulsive personality disorder occurs in 1–2 percent of the general population.

Clinical features

These patients appear to be saturated with scruples; moral rectitude seems second nature. Their lives are scheduled and orderly; perfectionism and cleanliness characterize their behavior. Details demand strict attention, and lists and timetables are extensively used. Emotions are handled with intense care; affect is often constricted and tight, and displays of emotion, especially anger, are avoided. Decisions are extremely difficult to make, and patients are generally unwilling to make one, unless they are absolutely certain that their decision is the one and only correct one.

Work typically occupies a large portion of these patients' lives, and their dedication to work characteristically leads them to neglect both their families and any leisure pursuits. Despite this dedication to work, however, some patients may remain remarkably unproductive, as their commitment to perfection, down to the last detail, may keep them from ever being satisfied that the project in question is ready for display. Furthermore, and complicating efforts otherwise to complete a task, these patients are typically unwilling to delegate work to anyone else, unless they can be absolutely certain that the other person will follow all their rules, down to the last exacting detail, and thus not make any mistakes.

These patients are often stubborn. They resent and resist any attempts of others to control them. Demands are met with a rigid defiance, as if any sign of weakness or relaxation on the patient's part would mean the destruction of autonomy and integrity. They are often frugal and parsimonious, at times even miserly.

The relationships that these patients form are often strained, stilted, and tense. They apply their own inflexible standards to others and insist that things are done the 'correct' way. Their work and family lives are often formal and rigidly conformist. They seek to exercise a tyranny of rectitude over others and may become quite angry when others dare to stray from the selected path.

Course

Although this is a chronic disorder, some patients appear to relax as they enter middle and later years.

Etiology

Obsessive–compulsive personality disorder tends to run in families (Torgesen *et al.* 2000) and these patients tend to be the oldest in their sibships. It is not clear, however, whether this personality disorder is secondary to rigid upbringing or to a congenitally determined temperamental style.

Differential diagnosis

As noted earlier, some may confuse obsessive–compulsive disorder with obsessive–compulsive personality disorder. The differential is fairly easy in most instances, given that almost all patients with obsessive–compulsive disorder will have true obsessions or compulsions, whereas these symptoms are absent in obsessive–compulsive personality disorder. The one exception to this differential rule occurs in a subtype of obsessive–compulsive disorder, namely primary obsessional slowness. Patients with primary obsessional slowness lack true obsessions or compulsions, and rather present with such a ritualization of daily routines that their days are consumed in completing them. This symptom, of course, is similar to the attention to detail seen in obsessive–compulsive personality disorder; however, patients with primary obsessional slowness lack the other symptoms seen in the personality disorder.

Another source of confusion between obsessive–compulsive personality disorder and obsessive–compulsive disorder was the once-current belief that the personality disorder existed, as it were, as the soil out of which obsessive–compulsive disorder sprang. Research has demonstrated, however, that obsessive–compulsive personality disorder is not present in the history of the vast majority of patients with obsessive–compulsive disorder (Mancebo *et al.* 2005).

Narcissistic personality disorder may be confused with obsessive–compulsive personality disorder, given that in both disorders patients are concerned with work, accomplishment, and perfection. The difference here lies in the motive for this concern. In narcissistic personality disorder, work is done to achieve recognition and fame, whereas in obsessive–compulsive personality disorder work is typically done out of a sense of duty.

Treatment

In contrast with patients with other personality disorders, patients with obsessive–compulsive personality disorder may be aware of the suffering their behavior engenders and may seek treatment. Psychotherapy, such as cognitive–behavioral therapy, or more psychoanalytically-oriented therapy, may be helpful.

26.9 PASSIVE–AGGRESSIVE PERSONALITY DISORDER

Patients with passive–aggressive personality disorder automatically, almost instinctively, see any demand, request, or expectation, no matter how fair or reasonable, as unwarranted, unjust, and excessive. They see themselves as unfairly burdened, singled out, and put upon. They live in resentment towards those who make demands on them, yet do not become openly hostile or angry. Rather, by all manner of sabotage and covert action, they seek to bring those who make demands to frustration, failure, and dismay.

This personality disorder has also been called 'negativistic personality disorder', and although there has been debate whether it exists as an identifiable syndrome, recent research suggests that it is still with us, and clearly recognizable (Hopwood *et al.* 2009). It probably occurs in 1–2 percent of the general population, and may be more common in males than females.

Clinical features

History often reveals that adults with passive–aggressive personality disorder displayed these traits in childhood. By late adolescence, the traits gradually become entrenched.

These patients believe that they are being unfairly treated by the world, and that any demands or expectations placed on them are unreasonable or unduly burdensome. Seeing the world thus, they tend to be sullen and chronically resentful of those who are, in one fashion or another, in positions of authority over them. Although at times they may voice their complaints or show obvious disrespect, in general they avoid open display of aggression, and rather resort to passive disobedience or resistance.

Such passive–aggressive behavior is more likely to occur in settings where demands are routinely made, such as in the military or in the workplace. When asked to do something, these patients may begin the task, but soon engage in all manner of procrastination, dawdling and foot-dragging, along the way, engaging in inefficiencies and making minor mistakes. If taken to task they may complain that the instructions they had been given were unclear, or may assert that they 'forgot' certain steps. Later, they may privately complain to fellow workers about their supervisors, but should others offer useful suggestions about how to get the work done with a minimum of effort, these suggestions are irritably brushed aside. If ever caught in their efforts to 'throw a wrench into the works', they may plead all manner of excuses and promise to do better later. Eventually, should they be threatened with termination, or time in the brig, rather than immediately mending their ways, they may become stubbornly argumentative.

Such passive–aggressive behavior, though more common in the workplace or military, may also occur at home. For example, the expectations of spouses regarding household tasks may be met with the same passive–aggressiveness, and patients may refuse to pick up after themselves, help with chores, or even keep up their personal hygiene.

Course

Passive–aggressive personality disorder is chronic and may not 'burn out' as do some other personality disorders.

Etiology

Although this personality disorder is familial, it is not clear whether this represents the influence of genetics or shared environment (Czajkowski *et al.* 2008).

Differential diagnosis

As with all personality disorders, the traits seen in passive–aggressive personality are pervasive throughout all aspects of the patient's life, and hence before making this diagnosis one must ascertain that passive–aggressiveness does indeed characterize the patient's behavior in all settings. Thus, an individual who resented his boss and displayed passive–aggressive behavior at work, yet who had an exemplary military record and was helpful and considerate at home, would obviously not qualify for the diagnosis.

Dysthymia, if of early onset, chronic, and characterized by irritability, may leave patients feeling burdened by life's demands and resentful; however, although these patients may grumble about their work, and often fail to complete it, they do not engage in passive–aggressiveness.

Negativism, as seen in catatonia, may present as a mulish resistance to any demands. Other signs, such as waxy flexibility, stupor, and the like, suggest, however, the correct diagnosis.

Treatment

Passive–aggressive patients rarely seek treatment on their own, but only end up there at the insistence of others. Medications are not helpful, and psychotherapy, though anecdotally at times effective, is usually sabotaged in a characteristically passive–aggressive way.

REFERENCES

Baron M, Gruen R, Asnis L *et al.* Familial transmission of schizotypal and borderline personality disorders. *Am J Psychiatry* 1985; **142**:927–34.

Barratt ES, Stanford MS, Felthous AR *et al.* The effects of phenytoin on impulsive and premeditated aggression: a controlled study. *J Clin Psychopharmacol* 1997; **17**:341–9.

Bateman A, Fonagy P. Treatment of borderline personality disorder with psychoanalytically oriented partial hospitalization: an 18-month follow-up. *Am J Psychiatry* 2001; **158**:36–42.

Bellino S, Paradiso E, Bogetti F. Efficacy and tolerability of aripiprazole augmentation in sertraline-resistant patients with borderline personality disorder. *Psychiatry Res* 2008; **161**:206–12.

Black DW, Baumgard CH, Bell SE. A 16 to 45 year follow up of 71 men with antisocial personality disorder. *Compr Psychiatry* 1995; **36**:130–40.

Bogenschultz MP, Nurnberg H. Olanzapine versus placebo in the treatment of borderline personality disorder. *J Clin Psychiatry* 2004; **65**:104–9.

Cadoret RJ, O'Gorman TW, Troughton E *et al.* Alcoholism and antisocial personality disorder. Interrelationships, genetic and environmental factors. *Arch Gen Psychiatry* 1985; **42**:161–7.

Cadoret RJ, Troughton E, Bagford J *et al.* Genetic and environmental factors in adoptee antisocial personality disorder. *European Arch Psychiatry Neurol Sci* 1990; **239**:231–240.

Cardasis W, Hochman JA, Silk KR. Transitional objects and borderline personality disorder. *Am J Psychiatry* 1997; **154**:250–5.

Cleckley HM. *The Mask of Sanity.* Mosby: St. Louis, 1941.

Coccaro EF, Kavoussi RJ. Fluoxetine and aggressive behavior in personality-disordered subjects. *Arch Gen Psychiatry* 1997; **54**:1081–8.

Constantino JN, Morris JA, Murphy DL. CSF 5-HIAA and family history of antisocial personality disorder in newborns. *Am J Psychiatry* 1997; **154**:1771–1773.

Cowdry RW, Gardner DL. Pharmacotherapy of borderline personality disorder. Alprazolam, carbamazepine, trifluoperazine, and tranylcypromine. *Arch Gen Psychiatry* 1988; **45**:111–19.

Czajkowski N, Kendler KS, Jacobson JC *et al.* Passive–aggressive (negativistic) personality disorder: a population-based twin study. *J Person Disord* 2008; **22**:109–22.

de la Fuente JM, Lotstra F. A trial of carbamazepine in borderline personality disorder. *Eur Neuropsychopharmacol* 1994; **4**:479–86.

de Oliveira-Souza R, Hare RD, Bramati IE *et al.* Psychopathy as a disorder of the moral brain: fronto-temporo-limbic grey matter reductions demonstrated by voxel-based morphometry. *Neuroimage* 2008; **40**:1202–13.

Dickey CC, McCarley RW, Niznikiewicz MA *et al.* Clinical, cognitive, and social characteristics of a sample of neuroleptic-naive persons with schizotypal personality disorder. *Schizophr Res* 2005; **78**:297–308.

Gardner DL, Cowdry RW. Positive effects of carbamazepine on behavioral dyscontrol in borderline personality disorder. *Am J Psychiatry* 1986; **143**:519–22.

Goldberg SC, Schulz SC, Schulz PM *et al.* Borderline and schizotypal personality disorders treated with low-dose thiothixene vs placebo. *Arch Gen Psychiatry* 1986; **43**:680–6.

Goldstein KE, Hazlitt EA, New AS *et al.* Smaller superior temporal gyrus volume specificity in schizotypal personality disorder. *Schizophr Res* 2009; **112**:14–23.

Hoek HW, Susser E, Buck KA *et al.* Schizoid personality disorder after parental exposure to famine. *Am J Psychiatry* 1996; **153**:1637–9.

Hollander E, Allen A, Lopez RC *et al.* A preliminary double-blind, placebo-controlled trial of divalproex sodium in borderline personality disorder. *J Clin Psychiatry* 2001; **62**:199–203.

Hollander E, Tracy KA, Swann AC *et al.* Divalproex in the treatment of impulsive aggression: efficacy in cluster B personality disorders. *Neuropsychopharmacology* 2003; **28**:1186–97.

Hollander E, Swann AC, Coccaro EF *et al.* Impact of trait impulsivity and state aggression on divalproex versus placebo response in borderline personality disorder. *Am J Psychiatry* 2005; **162**:621–4.

Hopwood CJ, Morey LC, Markowitz JC *et al.* The construct validity of passive–aggressive personality disorder. *Psychiatry* 2009; **72**:256–67.

Jacobsberg LB, Humowitz P, Barasch A *et al.* Symptoms of schizotypal personality disorder. *Am J Psychiatry* 1986; **143**:1222–7.

Kendler KS, Gruenberg AM. An independent analysis of the Danish Adoption Study of Schizophrenia. VI. The relationship between psychiatric disorders as defined by DSM-III in the relatives and adoptees. *Arch Gen Psychiatry* 1984; **41**:555–64.

Kendler KS, McGuire M, Gruenberg AM *et al.* The Roscommon Family Study. III. Schizophrenia-related personality disorders in relatives. *Arch Gen Psychiatry* 1993; **50**:781–8.

Keshaven M, Shad M, Soloff P *et al.* Efficacy and tolerability of olanzapine in the treatment of schizotypal personality disorder. *Schizophr Res* 2004; **71**:97–101.

Levitt JJ, Styner M, Niethammer M *et al.* Shape abnormalities of caudate nucleus in schizotypal personality disorder. *Schizophr Res* 2009; **110**:127–39.

Linehan MM, McDavid JD, Brown MZ *et al.* Olanzapine plus dialectical behavior therapy for women with high irritability who meet criteria for borderline personality disorder: a double-blind, placebo-controlled study. *J Clin Psychiatry* 2008; **69**:999–1005.

Lyons MJ, True WR, Eisen SA *et al.* Differential heritability of adult and juvenile antisocial traits. *Arch Gen Psychiatry* 1995; **52**:906–915.

McMain SF, Links PS, Gnam WH *et al.* A randomized trial of dialectical behavior therapy versus general psychiatric management for borderline personality disorder. *Am J Psychiatry* 2009; **166**:1365–74.

Mancebo MC, Eisen JL, Grant JE *et al.* Obsessive–compulsive personality disorder and obsessive–compulsive disorder: clinical characteristics, diagnostic difficulties and treatment. *Ann Clin Psychiatry* 2005; **17**:197–204.

Mendez MF, Chen AK, Shapira JS *et al.* Acquired sociopathy and frontotemproral dementia. *Dement Geriatr Cogn Disord* 2005; **20**:99–104.

Meyers CA, Berman SA, Scheibel RS *et al.* Case report: acquired antisocial personality disorder associated with unilateral left orbital frontal damage. *J Psychiatry Neurosci* 1992; **17**:121–5.

Minzenberg MJ, Fan J, New AS *et al.* Fronto-limbic dysfunction in response to facial emotion in borderline personality disorder: an event-related fMRI study. *Psychiatry Res* 2007; **155**:231–43.

Nickel MK, Nickel C, Mitterlehner FO *et al.* Topiramate treatment of aggression in female borderline personality disorder patients: a double-blind, placebo-controlled study. *J Clin Psychiatry* 2004; **65**:1515–19.

Nickel MK, Muehlbacher M, Nickel C *et al.* Aripiprazole in the treatment of patients with borderline personality disorder: a double-blind, placebo-controlled study. *Am J Psychiatry* 2006; **163**:833–8.

Nunes PM, Wenzel A, Borges KT *et al.* Volumes of hippocampus and amygdale in patients with borderline personality disorder: a meta-analysis. *J Pel Disord* 2009; **23**:333–45.

Raine A, Lencz T, Bihrle S *et al.* Reduced prefrontal gray matter volume and reduced autonomic activity in antisocial personality disorder. *Arch Gen Psychiatry* 2000; **57**:119–29.

Reich DB, Zanarini MC, Bieri KA. A preliminary study of lamotrigine in the treatment of affective irritability in borderline personality disorder. *Int Clin Psychopharmacol* 2009; **24**:270–5.

Robins LN. Sturdy childhood predictors of adult antisocial behavior: replications from longitudinal studies. *Psychol Med* 1978; **8**: 611–622.

Rodriguez Solano JJ, Gonzalez De Chavez M. Premorbid personality disorders in schizophrenia. *Schizophr Res* 2000; **4**:37–44.

Ronningstam E, Gunderson J. Identifying criteria for narcissistic personality disorder. *Am J Psychiatry* 1990; **147**:918–22.

Salzman C, Wolfson AN, Schatzberg A *et al.* Effect of fluoxetine on anger in symptomatic volunteers with borderline personality disorder. *J Clin Psychopharmacol* 1995; **15**:23–9.

Schulz SC, Zanarini MC, Bateman A *et al.* Olanzapine for the treatment of borderline personality disorder: variable dose 12-week randomized double-blind placebo-controlled study. *Br J Psychiatry* 2008; **193**:485–92.

Serban G, Siegel S. Response of borderline and schizotypal patients to small doses of thiothixene and haloperidol. *Am J Psychiatry* 1984; **141**:1455–8.

Sheard MH, Marini JL, Bridges CI *et al.* The effect of lithium on impulsive aggressive behavior in man. *Am J Psychiatry* 1976; **133**:1409–13.

Siever LJ, Silverman JM, Horvath TB *et al.* Increased morbid risk for schizophrenia-related disorders in relatives of schizotypal personality disordered patients. *Arch Gen Psychiatry* 1990; **47**:634–40.

Simonoff E, Elader J, Holmshaw J *et al.* Predictors of antisocial personality. Continuities from childhood to adult life. *Br J Psychiatry* 2004; **184**:118–27.

Simpson EB, Yen S, Costello E *et al.* Combined dialectical behavior therapy and fluoxetine in the treatment of borderline personality disorder. *J Clin Psychiatry* 2004; **65**:379–85.

Sobesky WE, Hull CE, Hagerman RJ. Symptoms of schizotypal personality disorder in fragile X women. *J Am Acad Child Adolesc Psychiatry* 1994; **33**:247–55.

Soderstrom H, Blennow K, Manhem A *et al.* CSF studies in violent offenders. I. 5-HIAA as a negative and HVA as a positive predictor of psychopathy. *J Neural Trans* 2001; **108**:869–78.

Soderstrom H, Blannow K, Sjodin AK *et al.* New evidence for an association between the CSF HVA:5-HIAA ratio and psychopathic traits. *J Neurol Neurosurg Psychiatry* 2003; **74**:918–21.

Soler J, Pascual JC, Campins J *et al.* Double-blind, placebo-controlled study of dialectical behavior therapy plus olanzapine for borderline personality disorder. *Am J Psychiatry* 2005; **162**:1221–4.

Soloff PH, George A, Nathan RS *et al.* Progress in pharmacotherapy of borderline disorders. A double-blind study of amitriptyline, haloperidol and placebo. *Arch Gen Psychiatry* 1986; **43**:691–7.

Soloff PH, Cornelius J, George A *et al.* Efficacy of phenelzine and haloperidol in borderline personality disorder. *Arch Gen Psychiatry* 1993; **50**:377–85.

Stanford MS, Houston RJ, Mathias CW *et al.* A double-blind placebo-controlled crossover study of phenytoin in patients with impulsive aggression. *Psychiatry Res* 2001; **103**:193–203.

Standford MS, Helfritz LE, Conklin SM *et al.* A comparison of anticonvulsants in the treatment of impulsive aggression. *Exp Clin Psychopharmacol* 2005; **13**:72–7.

Thaker G, Adami H, Moran M *et al.* Psychiatric illnesses in families of subjects with schizophrenia-spectrum personality disorders: high morbidity risks for unspecified functional psychoses and schizophrenia. *Am J Psychiatry* 1993; **150**:66–71.

Torgersen S, Lygren S, Oien PA *et al.* A twin study of personality disorders. *Compr Psychiatry* 2000; **41**:416–25.

Yang Y, Raine R, Narr KL *et al.* Localization of deformations within the amygdala in individuals with psychopathy. *Arch Gen Psychiatry* 2009; **66**:986–94.

Zanarini MC, Frankenbur FR, Parachini EA. A preliminary trial of fluoxetine, olanzapine, and the olanzapine–fluoxetine combination in women with borderline personality disorder. *J Clin Psychiatry* 2004; **65**:903–7.

27

Impulse control disorders

27.1 INTERMITTENT EXPLOSIVE DISORDER

In intermittent explosive disorder (Coccaro *et al.* 1998, 2004; McElroy *et al.* 1998), formerly known as episodic dyscontrol syndrome, patients experience recurrent, brief episodes of aggression, which are out of character for them, and for which they feel remorse. Once thought to be rare, it now appears that this disorder, which may be equally common in males and females, occurs in roughly 6 percent of the general population.

Clinical features

The onset of the disorder is heralded by the first episode of aggressiveness, and this usually occurs between mid-teenage years and the middle twenties. Episodes themselves are ushered in by a rising sense of tension, after which patients, despite efforts to restrain themselves, erupt into violence. Those around the patient may be assaulted, or the patient's wrath may be turned on furniture or other property. Attempts at restraining the patient are met with violent resistance. The episode itself may last anywhere from a half-hour to an hour or more, and remits spontaneously. Afterwards, the patient may appear spent, and some may have only a spotty memory for what happened. Remorse and guilt follow upon surveying the destruction. These episodes may have no precipitating event at all, or, if one is present, it may appear absurdly trivial in light of the ensuing violence. Furthermore, the violence itself is neither premeditated, nor does it have any discernible goal, such as intimidating others.

Importantly, these episodes are out of character for the patient: in the intervals between episodes, one fails to see any persistent impulsivity or aggressive tendencies.

Course

The frequency with which episodes occur varies widely among patients from only several per year to weekly. Over the decades following the first episode, the frequency lessens, and spontaneous, permanent remission may occur.

Etiology

Although the similarity between these episodes and complex partial seizures suggests that the episodes may, in fact, be seizures, electroencephalogram (EEG) studies, though revealing focal or generalized slowing in a minority, have not demonstrated any epileptiform discharges (Drake *et al.* 1992). Recent research, prompted by findings of serotoninergic dysfunction in patients with impulsivity, has also found such abnormalities in intermittent explosive disorder: for example, not only is there is a blunted prolactin response to fenfluramine (Coccaro *et al.* 2010), but this abnormality also predicts a good response to treatment with fluoxetine (Coccaro *et al.* 1997).

Differential diagnosis

Episodic violence may occur in a host of other disorders, the most common of which are schizophrenia, the manic phase of bipolar disorder, antisocial personality disorder, and borderline personality disorder. In these two personality disorders one typically finds that the intervals between the episodes of violence are characterized by irritability or impulsivity, or, in those cases of antisocial personality disorder not characterized by generalized irritability of impulsivity, one finds that the violence was premeditated and purposeful.

Other disorders in which episodic violence may occur are intoxication (with alcohol, cocaine, or amphetamine, and phencyclidine or ketamine), anabolic steroid abuse, traumatic brain injury, Tourette's syndrome, and during depressive episodes of major depressive disorder ('anger attacks'). Delirium, dementia, mental retardation, and personality change (especially of the frontal lobe type) may all be associated with disinhibition or lack of self-restraint, which in turn may be followed by violence on the most trivial of provocations. Finally, there are rare reports of episodic violence associated with hypothalamic lesions (Tonkonogy and Geller 1992), and, as discussed in Chapter 8.3, there are likewise rare cases of complex partial seizures characterized by directed violence.

Malingerers may claim this diagnosis in order to escape legal responsibility for their violent behavior. Here a careful history will reveal either premeditation or a clear goal underlying the violence.

Treatment

A variety of treatment options are available. In one double-blind study, fluoxetine, in a dose of 20 mg daily, was superior to placebo, showing benefit in as little as 2 weeks (Coccaro *et al.* 2009). Phenytoin, in a double-blind discontinuation study of patients who had benefited in a prior open study, was likewise superior to placebo (Maletzky and Klotter 1974). Open work has also suggested benefit from carbamazepine (Mattes 1990), high-dose propranolol (Jenkins and Maruta 1987; Mattes 1990), and metoprolol (Mattes 1985).

Given these results, it may be prudent to begin with fluoxetine, and, in nonresponders, to add either phenytoin or carbamazepine, holding the beta-blockers in reserve. Some clinicians, reasoning by analogy from their effectiveness in the treatment of impulsive violence in antisocial personality disorder, will also consider using divalproex or lithium.

27.2 KLEPTOMANIA

Patients with kleptomania (Aboujaoude *et al.* 2004; Grant and Kim 2002; McElroy *et al.* 1991) repeatedly steal items, not for gain, but rather to placate an irrational but irresistible urge.

Although thought to be uncommon, the prevalence of kleptomania is not known, as there have been no epidemiologic studies; amongst those arrested for shoplifting, however, less than 5 percent have kleptomania. Traditionally, kleptomania has been considered far more common in females; however, this may reflect a judicial sentencing bias, with females being referred for treatment, and males being incarcerated.

Clinical features

Although the age at onset ranges from childhood to middle years, most patients fall ill in their late teens or early twenties.

Characteristically, patients give little forethought to the stealing, and there is very little in the way of planning. Typically, patients experience a restless urge to steal, which may increase over hours or days; in some cases, however, the urge may strike suddenly as the patient walks down the aisle of a store. As the object is stolen, and then hurriedly stuffed into a pocket or purse, patients experience a wave of relief or pleasure. Importantly, patients typically either do not need the object, or, if they do, they actually have the money to purchase it. Furthermore, after leaving the store, patients typically either throw the object away or hide it at home. Most experience shame or guilt over their behavior, and make attempts to resist the urge in the future; attempts which, however, are generally fruitless. Eventually, arrest, and, in some, incarceration, occurs.

Patients with kleptomania typically have one or more other disorders, including major depressive disorder, bipolar disorder, substance use disorders, pathologic gambling, and, notably, bulimia nervosa.

Course

The overall course is chronic: in most cases, the frequency of theft gradually waxes and wanes over decades, without any true remissions; in a minority, however, there is an episodic course, with long intervals free of thievery. When the disorder is active, patients engage in theft generally twice weekly (McElroy *et al.* 1991).

Etiology

Although the family histories (Grant 2003; McElroy *et al.* 1991) of patients with kleptomania are characterized by mood disorders, alcoholism, and other impulse control disorders, there is as yet no evidence that kleptomania, *per se*, is familial. There has been one imaging study, which revealed decreased fractional anisotropy in the frontal lobes (Grant *et al.* 2006).

Differential diagnosis

Ordinary theft, as may occur in antisocial personality disorder, or in anyone in financial straits, is distinguished by the presence of gain as a motive, the absence of an irresistible impulse, and the presence of planning and forethought. Ordinary thieves may also malinger a plea of kleptomania to avoid punishment.

Patients with schizophrenia may steal in response to delusions or command hallucinations, and manic patients may take items that do not belong to them because, as far as they are concerned, they own the store anyway. Patients with dementia or mental retardation may take items and fail to pay for them, often because they do not realize they have to. There are also rare case reports of kleptomaniac behavior occurring after head trauma (Aizer *et al.* 2004), a tumor involving the right orbitofrontal region (Nyffeler and Regard 2001), and in depressed patients treated with a selective serotonin reuptake inhibitor (SSRI) (Kindler *et al.* 1997).

Treatment

Naltrexone, 50–150 mg per day, was superior to placebo in a double-blind study (Grant *et al.* 2009a). Behavior therapy has also been reported as helpful in some cases.

27.3 PYROMANIA

Pyromania (Grant and Kim 2007) is characterized by repeated fire-setting, which is motivated solely by a fascination with fire. This disorder is more common in males, and is rare: amongst those charged with repeated fire-setting, only 1–3 percent were so diagnosed (Lindberg *et al.* 2005; Ritchie and Huff 1999).

Clinical features

The onset is usually in the late teens. The fire-setting itself is not impulsive, and indeed there is often considerable forethought and planning. The sense of anticipation prior to setting the fire may be quite acute, and afterwards the patient may experience a considerable sense of relief. Often the patient may be found in the crowd of onlookers, observing with delight his handiwork.

These patients are fascinated with fires: they take delight in watching them, and may regularly follow fire engines; some may even become volunteer firefighters.

Course

Pyromania appears to be chronic, and fires tend to be set, on average, every month or two.

Etiology

Little is known regarding the etiology of pyromania. One study of 20 arsonists, all of whom had borderline personality disorder, found low cerebrospinal fluid (CSF) levels of 3-methoxy-4-hydroxyphenylglycol and 5-hydroxyindoleacetic acid levels (Virkkunen *et al.* 1987): whether similar findings would be noted in 'pure' pyromaniacs without this personality disorder is not known. In another study, three juvenile fire-setters were found to have a photoparoxysmal response on the EEG, and all three responded to anti-epileptic drugs (Milrod and Urion 1992).

Differential diagnosis

Simple arson is suggested by the presence of an understandable motive, such as revenge or financial gain, and may be seen in otherwise normal individuals or in patients with antisocial personality disorder. In other cases (Lindberg *et al.* 2005; Ritchie and Huff 1999), one finds additional disorders, including other personality disorders, mental retardation, schizophrenia, bipolar disorder, or, most frequently, alcoholism, with alcohol intoxication at the time of the fire-setting.

Treatment

There are no controlled treatment studies; anecdotally, both behavior and cognitive–behavioral therapy are useful. Consideration might also be given to obtaining an EEG, and, if a photoparoxysmal response were found, to providing a trial of an anti-epileptic.

27.4 PATHOLOGIC GAMBLING

Pathologic gambling (Stinchfield 2003) is characterized by ever-increasing gambling, to the point of financial or social ruin. Though currently classified as an impulse control disorder, the phenomenology of pathologic gambling is much more akin to an addiction. This is a relatively common disorder, being found in up to 3 percent of the population in the USA, and is far more common in males than females.

Clinical features

In males, the onset is usually in adolescence, whereas in females it tends to be in early adult years.

The onset is generally protracted, over several years, with the frequency of gambling increasing insidiously. Occasionally, however, one may see an acute onset, coinciding with a chance 'big win' that seems to light a fire in the patient.

For these patients, gambling becomes the single most important motivating factor in their lives, and former pursuits are neglected, unless, of course, they can be brought into the service of more gambling. Over time, larger and larger bets are required to satisfy the urge to make a big win. These patients seem unable to walk away from a loss and call it quits: they have an incessant, heart-breaking hope that, given just 'one more chance', they will be able to recoup all their losses. Indeed, in the excitement of such a desperate hope, they may literally 'bet the farm', and stake their house or all their savings on one more card game, horse race, or roll of the die. Those who do attempt to stop typically find themselves more or less powerless over the incessant craving for the excitement of the gamble. For most, the tension is unbearable, and in order to 'feed the habit', loans and taxes may go unpaid, and theft, embezzlement or fraud may be resorted to.

When gambling, these patients are often overconfident and extremely energetic. When not engaged in some 'high stakes' endeavor, however, they tend to be bored and irritable. Heavy losses may leave them feeling depressed, but this generally gives way to a hope that the next bet will pay off.

Other disorders often accompany pathologic gambling, including alcoholism, stimulant dependence, and bipolar disorder.

Course

In about two-thirds, the course is chronic, while in the remainder one sees an episodic course, with long intervals free of problematic gambling; over long periods of time, eventually one-third will eventually experience a spontaneous remission (Slutske 2006).

Etiology

There are clearly disturbances in biogenic amines in this disorder. Central nervous system noradrenergic tone is increased, as indicated by higher CSF levels of norepinephrine and its metabolite, 3-methoxy-4-hydroxyphenylglycol (Roy *et al.* 1988), and by a blunted growth hormone response to clonidine (Pallanti *et al.* 2006), as would be expected due to down-regulation of post-synaptic norepinephrine receptors. Serotoninergic dysfunction is indicated by a blunted growth hormone response to sumatriptan (Pallanti *et al.* 2010a), and an enhanced prolactin response to meta-chlorophenylpiperazine (Pallanti *et al.* 2010b). Dopaminergic dysfunction was indicated by a decrease in CSF levels of dopamine, associated with an increase in the levels of its metabolites, homovanillic acid and 3,4-dihydroxyphenylacetic acid (Bergh *et al.* 1997).

Functional magnetic resonance imaging (MRI) has demonstrated that in pathologic gamblers, as compared with controls, viewing a videotape of others gambling is associated with decreased metabolic activity in the frontal and orbitofrontal cortex, thalamus, and basal ganglia (Potenza *et al.* 2003), while actually engaging in gambling results in an opposite pattern, with increased metabolic activity now seen in the inferior frontal cortex and the thalamus (Miedl *et al.* 2010).

Pathologic gambling is clearly familial (Black *et al.* 2006), and genetic studies have revealed associations with different alleles in genes for dopamine receptors (da Silva Lobo *et al.* 2007).

Differential diagnosis

Both the 'social' gambler and the professional gambler are distinguished from the pathologic gambler by their ability to 'walk away' from a big loss, thereby keeping their losses from becoming at all unmanageable.

Manic patients may gamble away a fortune with breath-taking speed. The associated symptoms of mania, however, clearly indicate the correct diagnosis.

Direct-acting dopaminergic agents, such as ropinirole and pramipexole, as discussed in the chapters on Parkinson's disease and restless legs syndrome, may induce a clinical picture very similar to that of pathologic gambling.

Treatment

Both cognitive–behavioral therapy (Ladouceur *et al.* 2001) and imaginal desensitization (Grant *et al.* 2009b) are effective; Gamblers Anonymous, a self-help program modeled after Alcoholics Anonymous, may also be effective, but this is a demanding program, and the overall response rate is low (Stewart and Brown 1988).

Selective serotonin reuptake inhibitors, bupropion, lithium, valproic acid, and topiramate have all been utilized, but have not stood the test of large, double-blinded studies. Naltrexone, in a dose of 50 mg per day, however, appears effective (Grant *et al.* 2008), and a recent, very interesting study, also demonstrated effectiveness for acetylcysteine (Grant *et al.* 2007).

27.5 TRICHOTILLOMANIA

Trichotillomania (Christenson *et al.* 1991) is characterized by chronic hair pulling leading to substantial hair loss, generally from the scalp. There is currently debate regarding its lifetime prevalence: a conservative estimate would be around 1 percent of the general population. It is up to 10 times more common in females than males.

Clinical features

Trichotillomania generally has an onset in early teenage years. Classically, patients experience an urge to pull out their hair. Though they may attempt to resist the urge, most eventually give in, and as they pull the hair they experience a sense of relief or pleasure. Recently, however, it has become apparent that a considerable portion of patients pull their hair in a 'semi-automatic' fashion, almost outside their awareness. Although the resulting hair loss, as noted, is generally on the scalp, loss may also occur from the eyebrows, eyelashes, pubic, or axillary hair. After pulling out hair, many patients will mouth it, and some will swallow it, resulting, in a small minority of cases, in bezoar formation.

On the scalp, the resulting hair loss is often patchy: the patches themselves are poorly demarcated, and the overall appearance is often of a 'moth eaten' scalp. Within the patch itself, there is typically a mixture of normal hair and short, broken ones; importantly, there is no evidence of inflammation. Most patients are embarrassed by the hair loss, and will go to great lengths to conceal it.

Course

The course may be chronic or episodic. In chronic courses, the intensity and frequency of hair-pulling slowly waxes and wanes over the decades. In episodic cases, there are, between lengthy episodes of hair-pulling, long intervals during which hair-pulling is absent.

Etiology

Trichotillomania runs in families, and the concordance rate is higher in monozygotic than in dizygotic twins (Novak *et al.* 2009). MRI studies have not yielded consistent results.

Differential diagnosis

Tinea capitis, systemic lupus erythematosus, secondary syphilis, and alopecia areata may all cause patchy hair loss. In all these conditions, however, one does not find normal hairs in the patches, and inflammation, either to casual inspection or on biopsy, is present.

Patients with schizophrenia or mental retardation may pull out their hair, but here the associated symptoms indicate the correct diagnosis.

Obsessive–compulsive disorder may be considered on the differential, and indeed many patients with trichotillomania, in describing their urges, will speak of a 'compulsion' to pull out their hair. This 'compulsion' to pull out hair in trichotillomania is, however, quite different than the true compulsions seen in obsessive–compulsive disorder. In trichotillomania, the act itself may be associated with pleasure, whereas in obsessive–compulsive behavior, the compulsive act, though perhaps accompanied by a reduction in anxiety, is not, in itself, pleasurable. Furthermore, whereas hair-pulling is, as it were, an end unto itself, most compulsions are not; rather, they are instrumental to some other purpose: for example, in compulsive hand-washing, the hand-washing, rather than being a goal in itself, is merely a way to reach the true goal, which is to get rid of germs.

Treatment

Although clomipramine is more effective than placebo, it is less effective than cognitive–behavioral therapy (Ninan *et al.* 2000); fluoxetine is no better than placebo (Streichenwein and Thornby 1995). Of interest, a recent study found that *N*-acetylcysteine was more effective than placebo (Grant *et al.* 2009c).

REFERENCES

Aboujaoude E, Gamel N, Koran LM. Overview of kleptomania and phenomenological description of 40 patients. *Prim Care Companion J Clin Psychiatry* 2004; **6**:244–7.

Aizer A, Lowengrup K, Dannon PN. Kleptomania after head injury: two case reports and combination treatment strategies. *Clin Neuropharmacol* 2004; **27**:211–15.

Bergh C, Eklund T, Sodersten P *et al.* Altered dopamine function in pathological gambling. *Psychol Med* 1997; **27**:473–5.

Black DW, Monahan PO, Temkit M *et al.* A family study of pathological gambling. *Psychiatry Res* 2006; **141**:295–303.

Christenson GA, Mackenzie TB, Mitchell JE. Characteristics of 60 chronic hair pullers. *Am J Psychiatry* 1991; **148**:365–70.

Coccaro EF, Kavoussi RJ, Hauger RL. Serotonin function and antiaggressive response to fluoxetine: a pilot study. *Biol Psychiatry* 1997; **42**:546–52.

Coccaro EF, Kavoussi RJ, Berman ME *et al.* Intermittent explosive disorder-revised: development, reliability, and validity of research criteria. *Compr Psychiatry* 1998; **39**:368–76.

Coccaro EF, Schmidt CA, Samuels JF *et al.* Lifetime and 1-month prevalence rates of intermittent explosive disorder in a community sample. *J Clin Psychiatry* 2004; **65**:820–4.

Coccaro EF, Lee RJ, Kavoussi RJ. A double-blind, randomized, placebo-controlled trial of fluoxetine in patients with intermittent explosive disorder. *J Clin Psychiatry* 2009; **70**:653–62.

Coccaro EF, Lee R, Kavoussi RJ. Aggression, suicidality, and intermittent explosive disorder: serotoninergic correlates in personality disorder and healthy control subjects. *Neuropsychopharmacology* 2010; **35**:435–44.

Da Silva Lobo DS, Vallada HP, Knight J *et al.* Dopamine genes and pathological gambling in discordant sibships. *J Gambl Stud* 2007; **23**:421–33.

Drake ME, Hietter SA, Pakalnis A. EEG and evoked potentials in episodic-dyscontrol syndrome. *Neuropsychobiology* 1992; **26**:125–8.

Grant JE. Family history and psychiatric comorbidity in persons with kleptomania. *Compr Psychiatry* 2003; **44**:437–41.

Grant JE, Kim SW. Clinical characteristics and associated psychopathology of 22 patients with kleptomania. *Compr Psychiatry* 2002; **43**:378–84.

Grant JE, Kim SW. Clinical characteristics and psychiatric comorbidity of pyromania. *J Clin Psychiatry* 2007; **68**:1717–22.

Grant JE, Correia S, Brennan-Kohn T. White matter integrity in kleptomania: a pilot study. *Psychiatry Res* 2006; **147**:233–7.

Grant JE, Kim SW, Odlaug BL. N-acetyl cysteine, a glutamate-modulating agent, in the treatment of pathological gambling: a pilot study. *Biol Psychiatry* 2007; **62**:652–7.

Grant JE, Kim SW, Hartman BK. A double-blind, placebo-controlled study of the opiate antagonist naltrexone in the treatment of pathological gambling urges. *J Clin Psychiatry* 2008; **69**:783–9.

Grant JE, Kim SW, Odlaug BL. A double-blind, placebo-controlled study of the opiate antagonist, naltrexone, in the treatment of kleptomania. *Biol Psychiatry* 2009a; **65**:600–6.

Grant JE, Donahue CB, Odlaug BL *et al.* Imaginal desensitization plus motivational interviewing for pathological gambling: randomized controlled trial. *Br J Psychiatry* 2009b; **195**:266–7.

Grant JE, Odlaug BL, Kim SW. N-acetylcysteine, a glutamate modulator, in the treatment of trichotillomania. A double-blind, placebo-controlled study. *Arch Gen Psychiatry* 2009c; **66**: 756–63.

Jenkins SC, Maruta R. Therapeutic use of propranolol for intermittent explosive disorder. *Mayo Clin Proc* 1987; **62**:204–14.

Kindler S, Dannon PN, Iancu I *et al.* Emergence of kleptomania during treatment for depression with serotonin selective reuptake inhibitors. *Clin Neuropharmacol* 1997; **20**:126–9.

Ladouceur R, Sylvain C, Boutin C *et al.* Cognitive treatment of pathological gambling. *J Nerv Ment Dis* 2001; **189**:774–80.

Lindberg N, Holi MM, Tani P *et al.* Looking for pyromania: characteristics of a consecutive sample of Finnish male criminals with histories of recidivist fire-setting between 1973 and 1992. *BMC Psychiatry* 2005; **5**:47.

McElroy SL, Pope HG, Hudson JL *et al.* Kleptomania: a report of 20 cases. *Am J Psychiatry* 1991; **148**:652–7.

McElroy SL, Soutullo CA, Beckman DA *et al.* DSM-IV intermittent explosive disorder: a report of 27 cases. *J Clin Psychiatry* 1998; **59**:203–10.

Maletzky BM, Klotter J. Episodic dyscontrol: a controlled replication. *Dis Nerv Syst* 1974; **35**:175–9.

Mattes JA. Metoprolol for intermittent explosive disorder. *Am J Psychiatry* 1985; **142**:1108–9.

Mattes JA. Comparative effectiveness of carbamazepine and propranolol for rage outbursts. *J Neuropsychiatry Clin Neurosci* 1990; **2**: 159–64.

Miedl SF, Fehr T, Meyer G *et al.* Neurobiological correlates of problem gambling in a quasi-realistic blackjack scenario as revealed by fMRI. *Psychiatry Res* 2010; **181**:165–73.

Milrod LM, Urion DK. Juvenile fire setting and the photoparoxysmal response. *Ann Neurol* 1992; **32**:222–3.

Ninan PT, Rothbaum BO, Marsteller FA *et al.* A placebo-controlled trial of cognitive–behavioral therapy and clomipramine in trichotillomania. *J Clin Psychiatry* 2000; **61**:47–50.

Novak CE, Keuthen NJ, Stewart SE *et al.* A twin concordance study of trichotillomania. *Am J Med Genet B Neuropsychiatr Genet* 2009; **150B**:944–9.

Nyffeler T, Regard M. Kleptomania in a patient with a right frontolimbic lesion. *Neuropsychiatry Neuropsychol Behav Neurol* 2001; **14**:73–6.

Pallanti S, Bernardi S, Quercioli L *et al.* Serotonin dysfunction in pathological gamblers: increased prolactin response to oral m-CPP versus placebo. *CNS Spectr* 2006; **11**:956–64.

Pallanti S, Bernardi S, Allen A *et al.* Noradrenergic function in pathological gambling: blunted growth hormone response to clonidine. *J Psychopharmacol* 2010a; **24**:847–53.

Pallanti S, Bernardi S, Allen A *et al.* Serotonin function in pathological gambling: blunted growth hormone response to sumatriptan. *J Psychopharmacol* 2010b; **24**:1802–9.

Potenza MN, Steinberg MA, Skudlarski P *et al.* Gambling urges in pathological gambling: a functional magnetic resonance imaging study. *Arch Gen Psychiatry* 2003; **60**:828–36.

Ritchie EC, Huff TG. Psychiatric aspects of arsonists. *J Forensic Sci* 1999; **44**:733–40.

Roy A, Adinoff B, Lamparski D *et al.* Pathological gambling. A psychobiological study. *Arch Gen Psychiatry* 1988; **45**:369–73.

Slutske WS. Natural recovery and treatment-seeking in pathological gambling: results of two U.S. national surveys. *Am J Psychiatry* 2006; **163**:297–302.

Stewart RM, Brown RI. An outcome study of Gamblers Anonymous. *Br J Psychiatry* 1988; **152**:284–8.

Stinchfield R. Reliability, validity, and classification accuracy of a measure of DSM-IV diagnostic criteria for pathological gambling. *Am J Psychiatry* 2003; **160**:180–2.

Streichenwein SM, Thornby JI. A long-term, double-blind, placebo-controlled trial of the efficacy of fluoxetine for trichotillomania. *Am J Psychiatry* 1995; **152**:1192–6.

Tonkonogy JM, Geller JL. Hypothalamic lesions and intermittent explosive disorder. *J Neuropsychiatry Clin Neurosci* 1992; **4**:45–50.

Virkkunen M, Nuutila A, Goodwin FK *et al.* Cerebrospinal fluid monoamine metabolite levels in male arsonists. *Arch Gen Psychiatry* 1987; **44**:241–7.

28

Substance use disorders

28.1 STIMULANTS

Of the many stimulants that are used for intoxication, including methamphetamine, amphetamine, dextroamphetamine, and even methylphenidate, by far the most commonly utilized is methamphetamine. All of these stimulants may be taken orally or, after crushing, intravenously, and freebase methamphetamine, known as 'ice', may be either smoked or intranasally 'snorted'. Methamphetamine has a half-life of 6–12 hours, and urine toxicology may be positive for up to 24 hours. Importantly, urine toxicology may yield not only false negatives, but also, in the presence of selegiline or bupropion, false positives.

Clinical features

The speed of onset of intoxication varies with the preparation used. With intravenous use of any of the stimulants, intoxication begins within seconds, and a similarly rapid onset occurs when methamphetamine is smoked; with 'snorted' methamphetamine, intoxication is delayed for 5 or more minutes, while with oral use of any of the stimulants, the delay amounts to 20–30 minutes. Mild intoxication (Hollister and Gillespie 1970) is characterized by increased energy, varying degrees of elation, increased self-confidence, and talkativeness. With more severe intoxication there may be agitation and some bizarre behavior: often patients take a particular interest in things mechanical, and hours may be spent first taking apart, and then trying to put back together, various items, such as clocks, radios, and televisions. Visual hallucinations, typically of insects ('crank bugs'), may occur, and these may be associated with the tactile hallucination of formication, leading patients to dig at their skin to 'get' the bugs. Auditory hallucinations may likewise occur, and some patients may develop delusions of persecution. In severe intoxication a delirium may ensue, with confusion, incoherence, disorientation, and severe agitation. Abnormal movements, such as bruxism, chorea (Lundh and Tunving 1981), and, in some cases, generalized dystonia, may occur. In almost all cases of intoxication, from mild to severe, there is evidence of adrenergic hyperactivity, with mydriasis, tachycardia, and elevated blood pressure; in more severe cases the pupils may become poorly reactive, tachycardia and hypertension may become severe, the temperature may rise to hyperthermic levels, and there may be profuse diaphoresis. In the natural course of events, intoxication subsides gradually over 12–24 hours.

Additional symptoms may also be seen during intoxication. Cardiac symptomatology includes various arrhythmias, cardiac arrest, anginal pain, and myocardial infarction. Pulmonary symptomatology includes pulmonary edema and hemoptysis. As noted earlier, hyperthermia may occur, and in these cases there may be rhabdomyolysis with acute renal failure. Cerebrovascular pathology may also occur. Stroke may occur secondary to ischemic infarction due to vasospasm or a vasculitis (Bostwick 1981); ischemic infarction may also occur on an embolic basis in patients with a myocardial infarction or in those who have resorted to intravenous use and who developed an infective endocarditis. With severe hypertension, stroke may also occur secondary to intracerebral hemorrhage (Harrington *et al.* 1983), subarachnoid hemorrhage, or, rarely, intraventricular hemorrhage. With grossly elevated blood pressure, a hypertensive encephalopathy may also occur. Finally, grand mal seizures may occur (Alldredge *et al.* 1989).

With chronic use of stimulants, both tolerance and withdrawal may occur. With the development of tolerance, patients require ever larger doses to achieve euphoria, in some cases up to several grams daily. A withdrawal syndrome (Kramer *et al.* 1967; Watson *et al.* 1972) may appear with abstinence after long-term use, and is characterized by irritability, fatigue, suicidal ideation, and sleep disturbance, either hypersomnia or, less commonly, insomnia. This 'crash', as the withdrawal syndrome is often referred to, typically undergoes considerable clearing within days or a week or so; however, mild symptomatology may linger for weeks or months. A craving for the drug may be particularly severe during the withdrawal, and may persist, to a varying degree of severity, long into abstinence.

As noted earlier, some cases of intoxication are characterized by delusions of persecution and auditory hallucinations, and although these typically clear along with other symptoms of the intoxication, in some cases, after repeated episodes of intoxication, they may persist. This persistent psychosis has been most studied in methamphetamine addicts, and is characterized not only by delusions of persecution and auditory hallucinations (Angrist and Gershon 1970; Bell 1973; Derlet *et al.* 1989; Ellinwood 1967; Griffith *et al.* 1972; Iwanami *et al.* 1994), but also by Schneiderian first-rank symptoms (Janowsky and Risch 1979), such as thought broadcasting, and also visual hallucinations. Although this persistent psychosis typically clears within weeks, with chronic use the psychosis may be protracted, enduring for months to years into abstinence (Akiyama 2006). Furthermore, patients with a persistent psychosis also display the phenomenon of sensitization (Akiyama 2006): here, after the psychosis clears, even small doses of a stimulant may trigger the psychosis again, which then persists for a long period. Such patients may at times be quite dangerous and some may resort to murdering others whom they thought were after them (Ellinwood 1971).

Additional complications of intravenous use include acquired immune deficiency syndrome (AIDS) and hepatitis B and C.

Course

Recreational use of stimulants is not uncommon, and such occasional use generally causes few, if any, consequences. Some patients may develop an abusive pattern of use, however, and continue to seek intoxication despite suffering social or legal consequences. Addiction is said to occur when a craving develops for the stimulant, accompanied by the phenomena of tolerance and withdrawal; such patients may use stimulants on a daily basis, in relatively modest doses, or display a 'binge' pattern (Kramer *et al.* 1967), in which they use ever-escalating doses until, after a matter of days or more, they either run out of money or become so debilitated that they have to stop, after which they suffer through a withdrawal syndrome before going on yet another binge.

Etiology

All stimulants promote release of both dopamine and norepinephrine, and it appears that the euphoria seen with them occurs secondary to dopamine release within the ventral striatum (Drevets *et al.* 2001), while the autonomic symptomatology is mediated by enhanced noradrenergic tone (Nurnberger *et al.* 1984). Methamphetamine also partially blockades reuptake of dopamine and norepinephrine by presynaptic neurons; however, this effect is relatively minor.

Differential diagnosis

Intoxication with cocaine may be clinically indistinguishable from stimulant intoxication, and the differential may rest on history or drug screening.

Intoxication may also suggest mania: here toxicology is of course helpful, and observation in a controlled environment will reveal a resolution of symptoms within a day. Withdrawal may suggest depression; however, here the onset of the symptoms shortly after intoxication, and their resolution within a week of abstinence, will confirm the diagnosis of withdrawal.

The persistent psychosis is suggested by its onset during a binge. Difficulties may arise, however, in cases of a protracted psychosis, especially when the history of methamphetamine use is lacking. Here the differential broadens to include many of the causes of psychosis discussed in Chapter 8.1, including, most especially, paranoid schizophrenia.

Treatment

Intoxication, if mild, may be managed with simple observation. When agitation becomes severe, patients may be treated with intravenous or intramuscular lorazepam, in doses of 2–4 mg, with repeat doses every 5 minutes if given intravenously, and every hour if given intramuscularly. Should lorazepam fail to be fully effective, an antipsychotic, such as ziprasidone, may be given intramuscularly in a dose of 10 mg, with repeat doses every hour. Restraints, if possible, should be avoided in order to reduce the risk of rhabdomyolysis. When delusions or hallucinations require treatment, or in cases of delirium, though lorazepam may be helpful, more often than not an antipsychotic will be required. Cardiac and pulmonary complications, hyperthermia and rhabdomyolysis are all treated in the usual fashion; when hypertension requires treatment, however, beta-blockers should be avoided in favor of an agent such as phentolamine. Stroke may be treated as outlined in Chapter 8.4.

Withdrawal typically does not require specific treatment; however, if suicidal ideation is prominent, then hospitalization will be required until the risk has passed.

The overall treatment of stimulant addiction has as its goal abstinence. Although hospitalization may be required to break the pattern of use, the emphasis is on outpatient treatment. Cognitive–behavioral treatment, contingency management and referral to 12-step programs such as Cocaine Anonymous or Narcotics Anonymous all represent viable alternatives, and are typically used in combination.

28.2 COCAINE

Several different preparations of cocaine are available illegally. Cocaine hydrochloride is a white powder that may be insufflated ('snorted') into the nasal passages where it is absorbed through the nasal mucosa; it is also water soluble and thus may be dissolved and injected intravenously. Cocaine hydrochloride is destroyed by heat and is thus not suitable for smoking; it may, however, be treated with sodium bicarbonate and then either extracted with ether to yield a 'free base' preparation or warmed to create a 'rock' of 'crack' cocaine. Both the free base and crack preparations evaporate with heating and thus may be smoked.

Cocaine has a half-life of 0.5–1 hour. There are two main metabolites, both pharmacologically inactive: ecgonine methyl ester and benzoylecgonine, which has a half-life of 5–8 hours. In cases of acute intoxication, serum toxicology for cocaine is appropriate; however, when conducting urine toxicology screens to detect recent use in non-intoxicated patients the short half-life of cocaine makes urine testing impractical. In such a situation, however, toxicologic screening for benzoylecgonine is practical, as this metabolite may be found in the urine for 2 or 3 days, or, in heavy users, up to 2 weeks. Importantly, in some cases the level of this metabolite may fluctuate, such that a 'negative' urine test may occasionally be followed by a 'positive' one (Burke *et al.* 1990).

Finally, although cocaine, as noted, has a short half-life, the presence of alcohol may prolong any intoxication. This is because a portion of cocaine, in the presence of alcohol, undergoes a transesterification reaction to produce cocaethylene, which, though somewhat less potent, has a much longer half-life.

Clinical features

The onset of intoxication varies according to the preparation used, occurring within 20–30 minutes after snorting, and after only several minutes with intravenous use or smoking; in turn, the intensity of the intoxication parallels the rapidity of onset, and is much more pronounced with intravenous use or smoking. The duration of intoxication ranges from 30 to 60 minutes after snorting, while after intravenous use or smoking, it shrinks to 5–20 minutes. During intoxication (Kleber and Gawin 1984), patients become euphoric, hyperalert, talkative, and grandiose. Hyperactivity is common, and with higher doses agitation, at times severe, may occur (Fischman *et al.* 1976). Some patients may experience visual hallucinations, and these are typically of insects, which are often referred to as 'cocaine bugs'; these hallucinations may be accompanied by tactile hallucinations of bugs crawling beneath the skin, and some patients may excoriate themselves in an attempt to 'get' them. Uncommonly, there may be chorea ('crack dancing') (Daras *et al.* 1994) or tics (Pascual-Leone and Dhuna 1990). With mild intoxication libido increases, and in males there may be delayed ejaculation; with more severe intoxication, however, there may be erectile dysfunction. In severe intoxication, especially after intravenous use or smoking, a delirium may occur, with confusion, incoherence, lability, and delusions and hallucinations. Other symptoms and signs include mydriasis, hypertension, headache, nausea and vomiting, and tachycardia.

Additional symptoms may also occur during intoxication. Cardiac symptoms include angina, myocardial infarction (Virmani *et al.* 1988), various arrhythmias, and, rarely, cardiac arrest (Hsue *et al.* 2007). Pulmonary symptomatology includes dyspnea, hemoptysis, and pulmonary edema; pneumothorax and pneumomediastinum may also occur in cases where patients engage in a Valsalva maneuver while smoking in order to prolong the effect. Hyperthermia may occur, in all likelihood secondary to the combined effect of agitation and vasoconstriction of dermal vessels, which prevents heat dissipation. Rhabdomyolysis (Roth *et al.* 1988) may occur in a very small minority: in such cases, the intoxication is generally characterized not only by severe agitation but also by hyperthermia, and some patients progress to acute renal failure. Cerebrovascular pathology may also occur. Stroke secondary to ischemic infarction most commonly occurs as a result of vasospasm or increased platelet aggregation; such infarctions, in turn, may occur in the areas of distribution (Daras *et al.* 1991; Levine *et al.* 1990) of the middle, posterior, or anterior cerebral arteries or the basilar artery. Ischemic infarction may also, albeit rarely, occur secondary to a true vasculitis (Fredericks *et al.* 1991; Krendel *et al.* 1990): in these cases infarction may occur in the area of distribution of both large and small vessels, and, importantly, multiple infarctions may occur over the following weeks. Of note, ischemic infarction may also occur on a cardioembolic basis because of infectious endocarditis in intravenous users or myocardial infarction with thrombus formation, and these embolic events may or may not coincide with intoxication itself. Stroke may also occur as a result of intracerebral hemorrhage (Levine *et al.* 1990; Nolte *et al.* 1996) occurring secondary to severe hypertension during intoxication. In such cases, the location of the hemorrhage provides a further clue as to the mechanism: when the hemorrhage occurs in the caudate, putamen, thalamus, cerebellum, or pons, it probably reflects rupture of a penetrating artery; by contrast, when the hemorrhage is lobar, it more often reflects rupture of a pre-existing vascular malformation, and in such cases follow-up angiography is in order. Subarachnoid hemorrhage may also occur (Lichtenfeld *et al.* 1986; Nolte *et al.* 1996), and in these cases a pre-existing aneurysm is often present, again suggesting the need for angiography. Finally, stroke may also occur secondary to intraventricular hemorrhage (Levine *et al.* 1990). Severe hypertension may also cause a hypertensive encephalopathy, with cortical blindness, delirium, and seizures. Seizures may also occur in the absence of hypertensive encephalopathy: in one study (Koppel *et al.* 1996), some 3 percent of 1900 cocaine-related admissions were related to seizures, and although most of these were seen in patients with pre-existing epilepsy who had been either sleep deprived or non-compliant with their anti-epileptic drug regimen, in close to 1 percent of the 1900 cases seizures occurred secondary to cocaine toxicity alone.

Shortly after resolution of the intoxication, most patients will experience a 'crash', lasting from hours to a day, characterized by fatigue, depressed mood, irritability, and anxiety. Given that this 'crash' may occur with first-time use of cocaine as well as after widely spaced repeat intoxications, it may not be appropriate to consider it a withdrawal syndrome. Clear-cut withdrawal, however, does occur with chronic use and indeed may appear after only a few days of heavy use. The withdrawal symptoms (Weddington *et al.* 1990) include not only those seen in the 'crash' but also anhedonia, hyperphagia, insomnia, and, often, suicidal ideation; furthermore, during withdrawal there is typically a tense craving for more cocaine. This withdrawal reaches a maximum of severity within a few days and then gradually remits over days or weeks. Rarely, dystonia may appear during withdrawal (Choy-Kwong and Lipton 1989).

Tolerance may develop rapidly with repeated use; indeed, during a binge of intravenous use or smoking, tolerance may appear within a day.

After approximately 2 or more years of frequent cocaine use, intoxications may become characterized by delusions of persecution and of reference, and by auditory hallucinations (Brady *et al.* 1991; Rosse *et al.* 1994; Satel *et al.* 1991; Sherer *et al.* 1988). Although initially these symptoms tend to resolve shortly after the intoxication resolves (Brady *et al.* 1991), over time, and with repeated episodes, they appear with lower doses and also tend to last much longer (Bartlett *et al.* 1997), in some cases creating a persistent psychosis that may endure for weeks (Manschreck *et al.* 1987), despite abstinence.

Additional complications of cocaine use include a perforated nasal septum in patients who snort, and, in those who use cocaine intravenously, AIDS, hepatitis B and C, and, as noted earlier, infectious endocarditis.

Course

Recreational use of cocaine, that is occasional intoxications without consequences, is generally seen only with 'snorting'. Abusive use, with legal, social, and medical consequences, may also occur with snorting but is more common when cocaine is taken intravenously or smoked. Cocaine is one of the most, perhaps the most, addictive substances in the world, and craving may develop rapidly, leading to chronic, frequent use and the development of tolerance and withdrawal.

When abuse or addiction does set in, the pattern of cocaine use may be either continuous or episodic. Continuous use is characterized by daily intoxication, via snorting, injection, or smoking. Episodic use is far more common, and is characterized by 'binges' during which cocaine may be injected or smoked very frequently, sometimes every 10 or 15 minutes: with such frequent use, however, the euphoria of the intoxication becomes progressively briefer and the crashes progressively more severe, until finally either exhaustion or a lack of funds ends the binge. Although binges may last for days, on average they endure only about half a day. The interval between binges varies widely, from as little as a day or two in addicts to up to weeks or months in the abuser.

Etiology

Within the central nervous system cocaine both inhibits the reuptake and facilitates the release of monoamines by presynaptic neurons. Although both serotonin and norepinephrine are involved, it appears that the euphoriant effects of cocaine are related to the increased concentration of dopamine at the terminals of the mesolimbic and mesocortical dopaminergic pathways.

Differential diagnosis

A clinical differentiation of cocaine intoxication from stimulant intoxication may not be possible, and the differential often rests on history or a drug screen.

Intoxication may suggest mania; however, the extreme brevity of intoxication, relative to the duration of mania, immediately indicates the correct diagnosis, which may then be confirmed by toxicology. Withdrawal may suggest depression; however, the onset of symptoms at the end of a binge, and their subsequent resolution with abstinence, will suggest the correct diagnosis.

A persistent cocaine-induced psychosis is suggested by the onset during a binge, and the diagnosis is again confirmed by a resolution of symptoms during abstinence.

Treatment

Given the brevity of intoxication, in most cases simple observation is sufficient. In cases characterized by severe agitation, consideration may be given to intravenous lorazepam or to intramuscular use of a second-generation antipsychotic, such as ziprasidone or olanzapine; a second-generation agent may also be utilized in those cases of delirium which demand symptomatic treatment. Cardiac and pulmonary complications, hyperthermia, and rhabdomyolysis are all treated in the usual fashion. Hypertension should be treated with an agent such as phentolamine, and not with beta-blockers. The optimal treatment of stroke due to ischemic infarction, in general, may proceed along the lines laid out in Chapter 8.4.

Psychosis, as noted earlier, may eventually complicate intoxication, and indeed, over time, may persist long after the intoxication subsides. Although there are no controlled studies here, most clinicians will opt for treatment with a second-generation antipsychotic.

With regard to cocaine withdrawal, there are no established pharmacologic treatments. Although in most cases simple observation on an outpatient basis suffices, the presence of significant suicidal ideation may mandate inpatient care.

The overall goal of treatment of cocaine abuse or addiction is abstinence from cocaine, and established treatment modalities include cognitive–behavioral therapy, 12-step programs such as Cocaine Anonymous or Narcotics Anonymous, and contingency management programs, which may provide for payment for 'clean' urines or various consequences for 'dirty' ones. At present, there is no clearly established pharmacologic treatment for cocaine dependence. Multiple agents have been tested, but as yet none of these has emerged as clearly reliable. Agents that have been tested include amantadine, bromocriptine, cabergoline, carbidopa/levodopa, bupropion, methylphenidate, modafinil, tiagabine, gabapentin, carbamazepine, valproic acid, topiramate, baclofen, and disulfiram. One bright spot on the horizon is a vaccine against cocaine (Martell *et al.* 2009), but this is in the early stages of development.

28.3 HALLUCINOGENS

The hallucinogens (also known as psychotomimetics) can be roughly divided into two groups, namely the indolealkylamines and the phenylethylamines. Of the indolealkylamines, the prototype is lysergic acid diethylamide (LSD), which, though typically synthesized, may be found, in small amounts, in morning glory seeds. Other indolealkylamines include psilocybin (found in 'magic mushrooms'), the synthetic agents dimethyltryptamine (DMT), and 5-methoxy-*N,N*-diisopropyltryptamine ('foxy'), and ergine (a molecule in the same family as LSD), which is found in Hawaiian baby woodrose. Among the phenylethylamines, mescaline is the prototype, and this is joined by numerous synthetic agents, including dimethoxymethylamphetamine (DOM), dimethoxyamphetamine (DMA), methylenedioxyamphetamine (MDA), 3,4-methylenedioxy-*N*-ethylamphetamine (MDEA or 'eve'), and the very popular 3,4-methyle nedioxymethamphetamine (MDMA or 'ecstasy').

In addition to the foregoing hallucinogens, all of which exert their effects on serotoninergic transmission, consideration may also be given to salvinorin A, an opioid antagonist, which is found in a member of the mint family known as *Salvia divinorum.*

With the exception of DMT, which must be insufflated, smoked, or injected (Strassman *et al.* 1994a,b), all of the hallucinogens are active orally and are usually taken in this fashion.

Although in the past LSD was the most commonly used hallucinogen, its use has now been surpassed by MDMA, psilocybin, and *Salvia divinorum*; furthermore, although the vast majority of hallucinogens are illegal, certain agents are both legal and readily obtained through Internet sales, including LSD found in morning glory seeds, ergine found in Hawaiian baby woodrose, and *Salvia divinorum.*

Clinical features

Intoxication begins within 20–60 minutes after oral ingestion and is characterized by a profound, 'cosmic' sense, without any drowsiness or sedation (Bercel *et al.* 1956; Freedman 1968; Hollister *et al.* 1960; Isbell *et al.* 1956; Ungerleider *et al.* 1966). Patients may experience vivid memories, and commonplace events may appear exceedingly meaningful. Visual illusions and hallucinations are common, and a minority may experience synesthesia, wherein, for example, they may 'see' sounds. Importantly, patients maintain insight during intoxication and recognize the hallucinations as being unreal. Mild elevations of temperature, pulse and blood pressure may occur, along with mydriasis, fine tremor, hyperreflexia, and poor coordination. In a minority of cases the intoxication may turn into a 'bad trip', with the development of severe anxiety, which may be accompanied by delusions of persecution and of reference (Kuramochi and Takahashi 1964). Most intoxications last in the order of 4–12 hours; exceptions include psilocybin and DMT, which produce shorter intoxications lasting 2–6 hours.

Intoxication with MDMA ('ecstasy') deserves special mention for several reasons. First, unlike the other hallucinogens, intoxication is characterized by an initial, often quite intense, 'rush', which is then followed by a heightened sense of empathy and connectedness with others. Next, it has become immensely popular, especially among young adults who may use it during all-night parties known as 'raves'. Furthermore, in addition to the possibility of a 'bad trip', MDMA toxicity may present in other fashions. There may be agitation and hyperthermia, with, in some cases, rhabdomyolysis and acute renal failure; this agitation and hyperthermia may also be accompanied by delirium, rigidity, and pronounced hyperreflexia, thus creating a picture similar to that seen in the serotonin syndrome (Demirkiran *et al.* 1996). Severe hyponatremia due to syndrome of inappropriate antidiuretic hormone secretion (SIADH) and/or fluid overload, may also occur, with seizures, delirium, stupor, or coma (Hartung *et al.* 2002). Cardiac symptoms may accompany the foregoing or may occur independently, and include myocardial infarction, various arrhythmias, and cardiac arrest. There are also rare case reports of stroke occurring secondary to ischemic infarction (Manchanda and Connolly 1993), subarachnoid hemorrhage (Lee *et al.* 2003), or intracerebral hemorrhage (Hughes *et al.* 1993). Finally, there is accumulating evidence that chronic MDMA use may cause persistent deficits in attention, decision-making ability, and short-term memory (Bolla *et al.* 1998; Gouzoulis-Mayfrank *et al.* 2000; Kish *et al.* 2010; Zakzanis and Young 2001).

Tolerance develops rapidly to the intoxicating effects of all of the hallucinogens (with the sole exception of DMT), and most patients must wait a few days before they can again achieve intoxication (Isbell *et al.* 1956). Interestingly, despite the occurrence of tolerance, withdrawal phenomena do not occur.

A small minority of patients may develop certain sequelae after intoxication, including flashbacks, mood changes, or a psychosis.

Flashbacks (Abraham 1983; Frosch *et al.* 1965; Horowitz 1969) may occur in up to one-quarter of patients, during which, although not intoxicated, patients re-experience one or more of the symptoms that occurred during a prior intoxication. Classically, patients experience visual hallucinations, which may be complex or quite simple, or may consist of 'trailing', a kind of palinopsia in which after-images follow along after patients divert their gaze away from an object. As with the hallucinations occurring during intoxication, patients retain insight and recognize the flashbacks as unreal. Flashbacks may occur spontaneously or may be precipitated by moving into a darkened area or by the use of alcohol, cannabis, or an antipsychotic drug. Flashbacks may be infrequent or occur multiple times daily, and they may remit spontaneously after a few weeks or months or persist indefinitely. When flashbacks occur frequently, chronically, and cause distress, by convention one speaks of 'hallucinogen persisting perception disorder'.

Mood changes may appear shortly after intoxication resolves, generally within a few days, and may consist of depression or, less commonly, mania. These mood changes are typically quite brief, lasting in the order of days, and only rarely weeks, after which patients return to their baseline.

Psychosis (McGuire *et al.* 1994) may emerge with resolution of the intoxication; it is generally characterized by one or more of the hallucinations seen during intoxication (with the critical difference that patients now believe the hallucinations are real) or by delusions. Although this 'hallucinogen-induced psychotic disorder' generally remits spontaneously after a few days, in a small minority it may persist for weeks, months, or longer.

Course

Recreational use of hallucinogens is very common among adolescents and young adults; it is typically confined to occasional use without consequences at parties or social gatherings. Hallucinogen abuse, that is repeated use despite significant consequences, is relatively rare. Neither craving nor, as pointed out earlier, withdrawal occur, and addiction is not seen.

Etiology

With the exception of *Salvia divinorum*, which produces blockade of postsynaptic kappa-opioid receptors, the intoxication of all of the other hallucinogens is mediated by stimulation of postsynaptic 5-HT_{2A} serotonin receptors. MDMA, as noted earlier, may also be associated with persistent cognitive deficits, and this in turn is associated with a reduction in presynaptic serotonin transporters (Kish *et al.* 2010).

Differential diagnosis

Mild intoxication with phencyclidine may be quite similar to that seen with hallucinogens, especially MDMA, but is suggested by nystagmus, dysarthria, and ataxia.

Treatment

Simple 'bad trips' may generally be managed by observation in a calm, quiet, and darkened room. In some cases lorazepam may be indicated for severe anxiety; antipsychotics are generally not required, and in the case of MDMA intoxication, should, given the potential for rigidity, be avoided. Additional measures are indicated for MDMA intoxication. A complete blood count (CBC), basic metabolic profile, creatinine kinase (CK) level, urinalysis, and electrocardiogram (ECG) should be obtained, along with intravenous access. Intravenous lorazepam, 1–2 mg, or midazolam, again in doses of 1–2 mg, is given for agitation, with repeat doses as required. If severe hypertension persists despite resolution of agitation, nitroprusside or phentolamine may be administered. Hyperthermia, again if persisting despite resolution agitation, may be treated with ice-water baths or cooling blankets. In those instances where a serotonin syndrome-like picture emerges, consideration may also be given to treatment with cyproheptadine, as discussed in Chapter 29.5. In cases characterized by significant hyponatremia, a serum and urine osmolarity should be obtained, and treatment with hypertonic saline may be required.

Flashbacks generally do not require treatment; however, when these are frequent, severe and troubling, consideration may be given to clonazepam in a dose of 2 mg/day (Lerner *et al.* 2003). Postintoxication mood changes may generally be managed with simple observation, on either an inpatient or an outpatient basis. Postintoxication psychosis may be treated with an antipsychotic.

28.4 PHENCYCLIDINE AND KETAMINE

Phencyclidine and ketamine are both arylcyclohexylamines that were developed as 'dissociative' anesthetics. Although the frequent occurrence of postoperative psychosis with phencyclidine led to its abandonment in medical practice, ketamine is still used, both as an anesthetic and as an analgesic. Both drugs are used as intoxicants, and although the use of phencyclidine has dropped since its peak in the late 1970s, the use of ketamine in this regard appears to be on the rise. Phencyclidine is also known as PCP, 'angel dust', 'hog', and 'peace pill', and ketamine may be referred to as 'K', 'special K', 'vitamin K', 'cat valium', 'kat', or 'kit-kat'.

Clinical features

Intoxication with phencyclidine may be roughly characterized as mild, moderate, or severe; although ketamine intoxication is similar to that seen with phencyclidine, it is typically relatively mild (Krystal *et al.* 1994; Weiner *et al.* 2000). Both drugs are typically smoked in laced cigarettes; ingestion or intranasal snorting are alternate routes used in a small minority, while intravenous or intramuscular injection is rare.

Mild intoxication with phencyclidine (Javitt and Zukin 1991; Luby *et al.* 1959; McCarron *et al.* 1981a,b; Meyer *et al.* 1959; Pearlson 1981) is characterized by euphoria and a peculiar sense of detachment or dissociation. Patients may feel as if they are floating, and their bodies may, to them, appear misshapen. With somewhat higher doses, agitation appears, with violent behavior, which at times may be extreme. Bizarre behavior may also occur, as may hallucinations and delusions: of note, the delusions seen in phencyclidine intoxication are often of grandeur, and patients may believe they are religious figures or that they have incredible powers. In a minority, stuporous catatonia may also be seen. Patients may complain of nausea, vomiting, or vertigo, and examination may disclose dysarthria, ataxia, and, classically, nystagmus, which may be rotatory, horizontal, or vertical. Additional signs include increased deep tendon reflexes, decreased pin-prick sensation in the extremities, and, in a small minority, myoclonus. The temperature, pulse, and blood pressure are all increased.

Moderate intoxication (McCarron *et al.* 1981a,b) is signalled by the appearance of a delirium and abnormal movements, including facial grimacing, posturing, stereotypies, dystonia, and opisthotonus; grand mal seizures may also occur (Alldredge *et al.* 1989). Agitation may be extreme, and with further elevation in temperature, a small percentage of patients may develop rhabdomyolysis with renal failure.

Severe intoxication is characterized by stupor or coma (McCarron *et al.* 1981a,b). In stupor, patients may be quite still or may evidence random, purposeless movements; myoclonus is frequent and the deep tendon reflexes are greatly increased. With even higher doses coma supervenes; this is a kind of 'coma vigil' in which the eyes are open and the patient appears vigilant, but shows no response to pain or to anything else. The temperature rises further, as does the blood pressure, and hypertensive encephalopathy, though rare, may occur (Eastman and Cohen 1975). In a very small minority respiratory failure or cardiac arrest may occur. The electroencephalogram (EEG) in phencyclidine-induced coma (Fariello and Black 1978; Stockard *et al.* 1976) demonstrates generalized, non-reactive, theta activity, which may be interrupted by brief periodic, or pseudoperiodic, bursts of delta activity.

Ketamine intoxication generally lasts from 4 to 6 hours, consistent with its half-life of 1–3 hours. Phencyclidine intoxication, by contrast, is prolonged, lasting from a mere half-day up to a half-week or even longer. This longer duration seen with phencyclidine is accounted for by its longer half-life, which averages about 24 hours, and by its highly lipophilic nature. This lipophilicity leads to uptake into adipocytes, from which it gradually leaks back into the bloodstream. Both these agents undergo extensive hepatic metabolism: in the case of ketamine, norketamine is produced, which is far less potent, and in the case of phencyclidine, the metabolites are inactive. Only about 5–10 percent of phencyclidine is excreted unchanged in the urine.

Qualitative testing for phencyclidine is generally performed with urine samples, which may remain positive for up to a week after a single dose, and up to a month in those who have used the drug chronically and heavily. False negatives may occur, as may false positives, especially with venlafaxine, dextromethorphan, and ketamine. Urine toxicology for ketamine or its metabolites is less widely available; positive results may be found for up to 3 or 4 days.

Although tolerance to both phencyclidine and ketamine may occur, this is uncommon. Whether or not a withdrawal syndrome occurs is not clear: although some patients may complain of a 'let down', characterized by dysphoria and irritability, this generally does not last much longer than a few days, and as it may occur after one-time use, it may not be proper to style it as withdrawal.

A minority of patients may experience a persistent psychosis (Allen and Young 1978) after resolution of other symptoms of intoxication. This psychosis is characterized by delusions of grandeur, persecution or influence, and may be accompanied by a degree of agitation and hostility. Although symptoms typically resolve within days to weeks, in chronic, heavy users they may persist for months.

Course

Occasional recreational use of either phencyclidine or ketamine, without consequences, is not uncommon in late adolescence or early adult years. Prolonged and repeated use, despite medical or social consequences, appears uncommon, and such abuse of these drugs tends to resolve while patients are still in their twenties. Whether or not dependence occurs is not clear. As noted, tolerance may occur, but the existence of true withdrawal is debatable; furthermore, it is not at all clear whether actual craving occurs.

Etiology

Both phencyclidine and ketamine act as non-competitive antagonists at cation channels within the *N*-methyl-d-aspartic acid (NMDA) receptor complex, and, although numerous other receptors are also affected, it is this action at the NMDA receptor that appears to be responsible for the intoxicating effect of these drugs. Furthermore, phencyclidine, but not ketamine, at high blood levels, also blocks reuptake of norepinephrine and dopamine.

Differential diagnosis

Hallucinogen intoxication, especially with MDMA ('ecstasy') may resemble mild intoxication with phencyclidine or ketamine; however, hallucinogen intoxication lacks some of the distinctive characteristics of arylcyclohexylamine intoxication, including ataxia, dysarthria, and, most notably, nystagmus.

The persistent psychosis seen in some patients, if chronic, may be confused with paranoid schizophrenia; however, the onset shortly after intoxication will generally enable a correct differential. The diagnostic problem, however, may become more difficult in cases where patients with schizophrenia take phencyclidine, but the onset of the exacerbation of symptoms in the context of intoxication will again enable a correct diagnosis.

Treatment

Intoxicated patients should be observed in a supervised setting. Patients should be separated from unnecessary stimulation, and isolation in a quiet, dimly lit, well-protected room with non-intrusive monitoring is often best; unlike hallucinogen-intoxicated patients, arylcyclohexylamine-intoxicated patients do not respond well to verbal support, and cannot be 'talked down'. Agitated patients may be treated with 2–4 mg of parenteral lorazepam, with repeat doses as needed; although restraints may be required to even get close to highly agitated patients, they should be used sparingly, for when patients struggle against them rhabdomyolysis may appear, or, if already present, may worsen (Lahmeyer and Stock 1983). In some cases, especially those with prominent delusions and hallucinations, supplemental treatment with a parenteral antipsychotic, such as haloperidol, may be indicated (Giannini *et al.* 1984). Dystonia, if severe, and opisthotonus may be treated with intravenous lorazepam, and seizures may be treated with intravenous lorazepam and, if repetitive, fosphenytoin. Vigorous general medical care may be required for hyperthermia, hypertension, and rhabdomyolysis.

In the case of phencyclidine, consideration may be given to continuous gastric lavage with activated charcoal. Phencyclidine is a weak base which undergoes ionization in the stomach and cannot be absorbed until it passes into the alkaline bowel. Furthermore, phencyclidine is actively secreted into the stomach, and the concentration of phencyclidine within the stomach may be 50 times higher than that in the plasma. However, despite the inviting target that gastric phencyclidine presents, the risk of aspiration probably outweighs any benefits. Consideration may also be given to acidification of the urine, a maneuver which would trap phencyclidine in the urine, thus hastening its clearance. Although a double-blind study (Giannini *et al.* 1987) demonstrated that acidification does indeed shorten intoxication, the possibility that the metabolic acidosis could worsen rhabdomyolysis has dampened enthusiasm for this technique.

Treatment of the persistent psychosis typically involves the use of an antipsychotic. It must be kept in mind, however, that antipsychotics are often not very effective in this psychosis, and hence long-term hospitalization may be required until the psychosis runs its course.

28.5 ALCOHOL

This chapter deals with the various disorders and phenomena associated with alcohol use, including alcohol intoxication, blackouts, pathologic intoxication, tolerance to alcohol, alcohol withdrawal, delirium tremens, alcohol withdrawal seizures, and various other disorders often seen in association with alcohol use, such as head trauma. The full panoply of these is often seen in alcoholism, and this disorder is also discussed.

Clinical features

ALCOHOL INTOXICATION

Alcohol intoxication of mild degree is characterized by euphoria, talkativeness, and a degree of disinhibition; in some patients there may be some irritability or dysphoria rather than euphoria. In moderate intoxication, behavior becomes coarse and thinking is slow and unclear. There is facial flushing, conjunctival injection, dysarthria, nystagmus, and ataxia. With severe intoxication there is drowsiness, stupor, and disabling ataxia; coma may ensue, with respiratory depression and death.

The blood alcohol level (BAL) is customarily reported in milligrams/deciliter or, as it is often worded, milligrams percent (mg %). In an alcohol-naive subject, mild intoxication is seen at 100 mg %, moderate intoxication at 200 mg %, and severe intoxication at 300 mg %; in the alcohol-naive patient, levels of 400 mg % generally cause coma, and levels of approximately 500 mg % cause respiratory depression.

In general, in a 70-kg subject, the rapid consumption of 15 mL of pure, 100 percent ethanol will elevate the BAL by about 15–20 mg %, and this amount of pure ethanol is generally found in one mixed drink, one can of beer, or one glass of wine. In those with normal hepatic function, ethanol is metabolized at a rate of 10–15 mL/hour. Of interest, females tend to become intoxicated with a smaller amount of ingested alcohol than males, and this may be because of a reduction in gastric alcohol dehydrogenase activity, thus allowing a greater percentage of the ingested alcohol to escape this initial metabolic step (Frezza *et al.* 1990).

After the intoxication has passed, patients may experience a 'hangover', which may last from hours up to a day. This is characterized by headache, malaise, dysphoria, nausea, mild tremulousness, and diaphoresis.

BLACKOUTS

Blackouts may occur during moderate or severe alcohol intoxication and consist of transient episodes of anterograde amnesia, lasting anywhere from minutes to days, depending on how long the BAL remains high (Goodwin 1971; Goodwin *et al.* 1969a). During a blackout, patients' behavior may, to casual inspection, not appear to be changed: patients may recall what they were doing just before the blackout started and may also be able to keep track of ongoing events during the blackout sufficiently well that they are able to keep up a conversation, play cards, etc. If, however, short-term memory is tested during the blackout, one finds that patients are unable to recall anything that happened much more than 5 minutes earlier. Furthermore, once the blackout ends, the time period covered by the blackout remains an 'island of amnesia' to patients, who can recall little or nothing of the events that transpired during the blackout.

Should blacked-out patients fall asleep during the blackout, they may, upon awakening the next day, anxiously ask acquaintances what they did the night before. Those who are still awake when the blackout abruptly ends may be quite startled at their situation; one patient (Goodwin *et al.* 1969b) 'found himself dancing with no recollection of what he had been doing during the previous six hours'. A variant of blackouts, known as 'brownouts', may occur in which the amnesia is not complete and patients can recall some of what happened (Tamerin *et al.* 1971).

PATHOLOGIC INTOXICATION

Pathologic intoxication (Perr 1986), also known as 'alcohol idiosyncratic intoxication', although long written off (Banay 1944; May and Ebaugh 1953), is a controversial diagnosis (Coid 1979). Putatively, patients, after only a small amount of alcohol, undergo a dramatic change, becoming uncharacteristically irritable and often violent. One study was able to reproduce these symptoms in patients thought to have suffered pathologic intoxication (Maletzky 1976), whereas another was not (Bach-y-Rita *et al.* 1971): if, indeed, pathologic intoxication does exist, it is probably rare.

TOLERANCE TO ALCOHOL

Tolerance to alcohol typically develops gradually with chronic, repeated intoxications, and greater and greater amounts must be drunk to achieve the desired intoxication. Thus, whereas in the alcohol-naive patient levels of 100 mg % typically cause intoxication, those with tolerance may need to reach a level of 300 mg % before they begin to 'feel' the alcohol; indeed, in some cases alcohol-tolerant patients may sustain levels of 500 mg % without loss of consciousness (Adachi *et al.* 1991; Minion *et al.* 1989). Interestingly, late in the course of alcoholism, some patients may fairly rapidly 'lose' their tolerance. In such cases, patients who had previously been able to consume a liter of alcohol and still be standing may now find themselves hopelessly and severely intoxicated after only a few drinks.

ALCOHOL WITHDRAWAL

Alcohol withdrawal (Isbell *et al.* 1955) may be roughly divided into mild, moderate, and severe forms: mild and moderate degrees of withdrawal are considered here while severe withdrawal, also known as delirium tremens, is considered next. Although mild withdrawal may be seen in anyone who engages in heavy, prolonged drinking, moderate withdrawal is seen only in alcohol abusers or alcoholics, while delirium tremens occurs only in alcoholics. Regardless of the form of withdrawal, symptoms generally appear anywhere from 4 to 12 hours after the BAL has fallen below the patient's 'threshold' for intoxication. Importantly, this implies that for some patients, such as alcoholics who have developed tolerance, withdrawal symptoms may appear while the patient is still drinking, provided that the alcohol consumption has 'slowed down' sufficiently.

Mild withdrawal, colloquially known as the 'shakes', has, as its characteristic symptom, a rapid, fine, postural tremor seen most prominently in the outstretched hands. Additional symptoms include anxiety, easy startability, agitation, insomnia, nausea and vomiting, and mild, non-sustained elevations in pulse and systolic blood pressure. In moderate degrees of withdrawal, all of the symptoms seen in mild withdrawal are intensified and the elevations in blood pressure and pulse become constant, with the pulse climbing to over 120 beats per minute (bpm) in some cases. Additional symptoms include transient confusion and fleeting visual or auditory hallucinations.

Most patients recognize that a drink will 'solve' their problem, and many will take some 'hair of the dog that bit you'. This does offer temporary relief of the shakes but of course threatens to set off a vicious cycle.

Mild and moderate alcohol withdrawal generally peaks within a couple of days and then gradually settles over the next 2 or 3 days: among recreational or abusive drinkers, the symptoms may eventually fully remit within a week; however, among alcoholics, residual symptoms may persist for months.

DELIRIUM TREMENS

Delirium tremens (Lundquist 1961; Nielsen 1965; Rosenbaum *et al.* 1941), also known as alcohol withdrawal delirium or simply the 'DTs', is the most severe form of alcohol withdrawal and is generally seen only in alcoholics who have been drinking heavily for many years and who have had repeated episodes of alcohol withdrawal: amongst such patients the risk of developing the DTs rises with each of the following features (Palmstierna 2001): a concurrent infectious illness (such as pneumonia); tolerance to alcohol (as evidenced by the appearance of withdrawal symptoms while the BAL is over 100 mg %); a history of alcohol withdrawal seizures; and, finally, a history of the DTs. Clinically, patients first develop a moderately severe alcohol withdrawal syndrome. Subsequently, anywhere from hours to a week later, but generally after 2–3 days, there is a dramatic exacerbation of the typical alcohol withdrawal symptoms, and in the context of this exacerbation, a delirium emerges. Agitation may become extreme, and the tremor becomes coarse and involves not only the upper extremities, but now also the legs, and in some cases the eyelids and tongue. This 'trembling delirium', in addition to confusion, disorientation, and short-term memory loss, is also marked by constant and prominent hallucinations and delusions. Visual hallucinations are most common, and they tend to be extremely vivid and complex, typically involving animals or insects: dogs may circle the bed; rats eat at the toes; bugs crawl on the arms and face and the patient may try and swat them away. Tactile hallucinations may accompany the visual ones, and the patient may feel animals tearing at the flesh or bugs biting into the skin. Auditory hallucinations may occur. Patients may hear bells, whistles, or alarms, and at times voices, and when voices do occur they tend to be critical and persecutory. Delusions are common and tend to be persecutory and referential. Patients may believe that staff are conspiring against them to have them killed and may believe that chance conversations refer especially to them.

Delirium tremens is fatal in over 5 percent of cases, with patients succumbing to concurrent illnesses such as pneumonia, hepatic failure, cardiac arrhythmias, or hypovolemic shock. Those who survive generally experience a gradual remission of symptoms within a few days; however, in some cases the delirium may persist much longer, at times up to 2 weeks.

Rarely, one may see a variant of delirium tremens known as delirium tremens sine tremore (trembling delirium without tremor). Here, the tremor and other autonomic symptoms classically associated with delirium tremens are either minimal or absent, and patients present with a 'quiet' delirium, seemingly undisturbed while all the time experiencing confusion, hallucinations, and delusions.

ALCOHOL WITHDRAWAL SEIZURES

Alcohol withdrawal seizures, also known as 'rum fits', are an uncommon accompaniment of the alcohol withdrawal syndrome, seen only in a few percent of alcoholics (Isbell *et al.* 1955; Schuckit *et al.* 1993); typically they occur only after many years of heavy drinking and repeated episodes of alcohol withdrawal. They may occur within hours to 2 days after either a cessation of alcohol use or a substantial reduction in the amount consumed, and typically consist of grand mal seizures, which may, in about one-quarter of cases, have focal features. Most patients have only one seizure; occasionally patients may experience two or three, or even half a dozen, and very rarely status epilepticus may occur. Once patients do have alcohol withdrawal seizures, they are more likely to in the future, should they go through withdrawal again (Booth and Blow 1993).

OTHER DISORDERS

Other disorders often accompany alcohol intoxication, especially in patients with alcoholism. Falls and motor vehicle accidents are common, and traumatic brain injury (Chapter 8.5), especially with subdural hematoma (Chapter 12.1), is not uncommon. Vitamin deficiencies may occur: thiamine deficiency may cause a Wernicke's encephalopathy (Chapter 14.4), and the much less commonly seen niacin deficiency may cause the acute encephalopathic form of pellagra (Chapter 14.3). Among alcoholics, recurrent attacks of alcoholic hepatitis may lead to hepatic cirrhosis and, in some, to a hepatic encephalopathy (Chapter 14.19). Alcoholics are also prone to infection, and aspiration pneumonia is not uncommon; bacterial meningitis may also occur but this is relatively rare. Hypoglycemia may occur, not only due to malnourishment but also to the direct inhibitory effect alcohol has on hepatic gluconeogenesis; hypomagnesemia may also be seen. Gastritis, bleeding esophageal varices, and pancreatitis may likewise occur. Chronic hyponatremia may be seen in some cases, and its overly rapid correction may cause central pontine myelinolysis (Chapter 14.17).

Patients who have survived a Wernicke's encephalopathy may be left with the chronic amnesia of Korsakoff's syndrome (Chapter 14.5). Long-term, chronic alcoholism may also be associated with the development of various other disorders, including Marchiafava–Bignami disease (Chapter 29.11), alcoholic cerebellar degeneration, alcoholic polyneuropathy, alcoholic myopathy, alcoholic dementia (Chapter 29.8), alcoholic paranoia (Chapter 29.10), and alcohol hallucinosis (Chapter 29.9). As an aside, it is important to note at this point that some authors use the term 'alcohol hallucinosis' to refer to the transient, fleeting hallucinations that may occur in some cases of moderate alcohol withdrawal, whereas in this text, as discussed in Chapter 29.9, the term refers to a condition seen only in alcoholics, which is characterized by hallucinations emerging during delirium tremens, which, however, persist after resolution of the delirium, and indeed remain chronic. Finally, consideration must be given to alcoholic depression. Roughly two-thirds of alcoholics admitted to hospital will have a depressive syndrome (Davidson 1995), and although symptoms may be severe, the prognosis, with abstinence, is good, with symptoms spontaneously remitting generally within a month (Brown *et al.* 1995).

Course

Recreational or 'social' use of alcohol is almost ubiquitous in the Western world, but the intoxications here are generally mild, and, although a significant minority of recreational drinkers may experience blackouts and some may suffer head trauma, the other consequences of alcohol use described above do not occur. Alcohol abuse is characterized by recurrent intoxication despite significant medical, social, and legal consequences, and these patients, in addition to episodes of intoxication of moderate degree, commonly experience blackouts and may develop a degree of tolerance and experience mild or even moderate forms of alcohol withdrawal; these patients, however, do not have a craving for alcohol and, albeit with some effort, are able to either moderate their drinking or stop altogether. Alcoholism emerges from alcohol abuse and is signalled by the development of craving, with an inability to 'leave it alone', and by clear-cut tolerance and frequent episodes of moderate alcohol withdrawal or delirium tremens; alcoholics are also prone to alcohol withdrawal seizures, and all of the other disorders described earlier, such as head trauma and Wernicke's encephalopathy.

Alcoholism is a common disorder, seen in about 10 percent of the adult population in the USA, and is much more frequent in males than females. In most cases of alcoholism, excessive alcohol use begins in the late teens or early twenties, and progressively and gradually worsens until the full clinical picture is reached after a decade or two; exceptions to this pattern do occur, however, and in some cases the onset may be delayed until middle years or the progression may be unusually rapid. Among alcoholics, drinking becomes the primary need in life, and patients continue to drink despite disastrous consequences. Denial is ubiquitous, and almost all alcoholics will strenuously deny that they have a problem with drinking or they will attempt to rationalize it in one way or another. Most alcoholics, at least initially, will try at times to control their drinking, perhaps after a hospital stay, incarceration for driving while intoxicated, or the angry departure of a spouse, and some may experience some success at this, only to soon find themselves again hopelessly intoxicated. The overall course of alcoholism may be episodic or chronic. The episodic course is characterized by more or less lengthy episodes, or 'binges', during which there are frequent, closely spaced intoxications; these binges may last anywhere from weeks to months, and in between them patients generally remain abstinent. The chronic course may be apparent from the onset of the alcoholism or may supervene upon a prior episodic course. In this chronic course there are repeated, more or less closely spaced, intoxications without any intervening periods of sobriety of any significant length. Regardless of whether the course is episodic or chronic, the great majority of alcoholics continue to drink despite losing jobs, friends, families, and their health, and indeed they continue to do so until they are either institutionalized or dead. Exceptions to this dismal prognosis, however, do occur. Roughly 20 percent of alcoholics are able to stop on their own; however, these patients generally have mild cases, and enjoy good family support and have a stable work and social life. There are also rare cases of 'burnt out' alcoholics who simply stop after many decades of severe alcoholism: in these cases, one typically finds that patients lost their tolerance for alcohol and were unable to take even one drink without becoming intoxicated.

Etiology

Alcohol intoxication may reflect not only increases in membrane fluidity, with consequent changes in ion channel function, but also alcohol-mediated sensitization of gamma-aminobutyric acid (GABA) receptors and inhibition of NMDA glutamate receptors; furthermore, in the presence of alcohol, dopaminergic activity in the meso-limbic pathway is increased. With repeated, frequent intoxications, it is suspected that there is a chronic downregulation of GABA receptors and an upregulation of NMDA receptors, and that it is these changes that predispose to the development of the alcohol withdrawal syndrome and delirium tremens. The mechanisms underlying blackouts, pathologic intoxication, and alcohol withdrawal seizures are not clear. Although the etiology of alcoholism has not as yet been fully elucidated, it is clear from both twin (Kendler *et al.* 1997) and adoption (Cloninger *et al.* 1985) studies that genetics plays a significant role.

Differential diagnosis

Alcohol intoxication may be mimicked by intoxication with various other substances, but certain signs allow for a differential. In the case of intoxication with benzodiazepines (or other sedative–hypnotics) and ethylene glycol, there is no odor of alcohol on the breath. Both isopropyl alcohol and methanol intoxication may be accompanied by an odor of alcohol; however, in these intoxications there is prominent nausea and vomiting and, furthermore, in methanol intoxication there may be bilateral dimming or loss of vision. Consideration must also be given to some of the other disorders that may accompany intoxication, especially in cases in which intoxicated patients fail to sober up within the expected time frame, including especially head trauma, infection, hypoglycemia, and hypomagnesemia.

Blackouts must be distinguished from certain other causes of episodic anterograde amnesia, as discussed in Chapter 6.6. Transient global amnesia and pure epileptic amnesia are distinguished by the absence of other signs of intoxication, such as dysarthria. Concussion may occur during intoxication, but the evidence of head injury is generally obvious.

Alcohol withdrawal may be clinically indistinguishable from withdrawal from benzodiazepines or other sedative–hypnotics, and the diagnosis therefore rests on the history of substance use, keeping in mind that many patients will also use these other agents in addition to alcohol. Hypoglycemia, not uncommonly seen in intoxicated alcoholics, may produce a similar picture (Fredericks and Lazor 1963; de Moura *et al.* 1967) and, as noted below, a blood glucose determination should be standard.

Delirium tremens may be clinically indistinguishable from benzodiazepine or sedative–hypnotic withdrawal delirium, and the differential here rests on the history of substance use. Other disorders capable of causing a 'trembling delirium' with marked autonomic signs include the serotonin syndrome, the neuroleptic malignant syndrome, thyroid storm, and hypoglycemia. In some cases, alcohol withdrawal may be accompanied by a delirium secondary to one of the other disorders seen in alcoholics, thus producing a clinical picture that may be easily passed off as delirium tremens rather than an etiologically multifactorial delirium; consideration should therefore be given to the presence of Wernicke's encephalopathy, hepatic encephalopathy, traumatic brain injury, hypomagnesemia, encephalopathic pellagra, Marchiafava–Bignami disease, meningitis, and pneumonia.

Alcohol withdrawal seizures may be mimicked by seizures occurring secondary to many of the conditions commonly seen in association with either alcohol abuse or alcoholism, including subdural hematoma, Wernicke's encephalopathy, hepatic encephalopathy, meningitis, hypoglycemia, and hypomagnesemia. Such mimics should be strongly suspected whenever seizures occur outside the 'window' for alcohol withdrawal seizures, which, as noted earlier, encompasses the period which begins several hours after withdrawal symptomatology emerges and ends no later than 2 days after the withdrawal symptomatology emerged. Patients with epilepsy are especially likely to have a seizure within a day of intoxication; however, here the history of seizures outside this time frame will suggest the correct diagnosis.

Other disorders found in association with chronic alcoholism may occasionally present some differential diagnostic problems, and this is especially the case with alcoholic paranoia, alcohol hallucinosis, and alcoholic depression. The first two of these are discussed in Chapters 29.10 and 29.9, respectively. Alcoholic depression must be distinguished from other causes of depression, most especially from major depressive disorder. This distinction from major depressive disorder may be made by reference to the course of the current depression, and to any history of previous depression during periods of abstinence. As noted earlier, alcoholic depression generally remits spontaneously within a month, and consequently any depression lasting much longer than that is probably not an alcoholic depression. In looking at history, if one finds that the patient suffered from depression during earlier periods of sobriety and both that the earlier depression occurred outside this 1-month period and that the current depression is symptomatically similar to this earlier one, then one may generally assume that the current depression, again, is not an alcoholic one. Overall, in pursuing the differential between major depressive disorder and alcoholic depression, it is probably better to err on the side of major depressive disorder: a depressive episode of major depressive disorder is one of the most common causes for non-compliance with the treatments for alcoholism, and a failure to treat the depression could have disastrous consequences.

Treatment

ALCOHOL INTOXICATION

Alcohol intoxication is generally managed by observation and routine general medical supportive care. An estimate should be made as to the time of the last drink and a BAL should be obtained, after which one may calculate the time required for the alcohol to wash out. With severe intoxication, one must be prepared for the development of respiratory failure and the need for intubation.

In older, debilitated, or alcoholic patients, or those with a history of seizures, moderate withdrawal, or delirium tremens, routine laboratory tests should be obtained, including a complete blood count, electrolytes, glucose, magnesium, ammonia, bilirubin, liver enzymes, stool for occult blood, and an electrocardiogram; if there is any suspicion of pneumonia or head trauma, a chest radiograph or computed tomography scan, respectively, should also be obtained. Patients should also receive 100 mg of thiamine parenterally, followed by 100 mg daily of parenteral or oral thiamine; glucose and food should, if possible, be withheld until 2 hours have passed from the administration of the initial dose of parenteral thiamine.

BLACKOUTS

Blackouts are managed by simple observation until serial mental status examinations have demonstrated a recovery of short-term memory function.

PATHOLOGIC INTOXICATION

Pathologic intoxication is likewise managed by observation; however, here, seclusion and, at times, restraints may be required until the intoxication has passed.

ALCOHOL WITHDRAWAL

Alcohol withdrawal of mild degree may or may not require pharmacologic treatment. Provided that the tremor and other autonomic symptoms are not too uncomfortable and do not pose a threat to the patient (e.g. in cases of cardiac disease or epilepsy) and the risk of both alcohol withdrawal seizures and of progression to a moderate degree of withdrawal or to delirium tremens is low, then patients may be left to simply 'shake it out', as the discomfort associated with this may provide a valuable instructive lesson. In other cases, and during moderate withdrawal, pharmacologic treatment is generally indicated, and benzodiazepines remain the agents with the longest track record, and are preferred by most clinicians.

For mild withdrawal, a 'symptom-triggered approach' (Daeppen *et al.* 2002) may be utilized, wherein only as-needed doses are given, every 4–6 hours, of diazepam (5–10 mg), chlordiazepoxzide (25–50 mg), lorazepam (1–2 mg), or oxazepam (15–20 mg). This approach requires frequent and reliable monitoring, but if this is available, the overall dose may be titrated fairly precisely to the severity of symptoms, and the course of treatment lasts only a few days. Some clinicians prefer diazepam or chlordiazepoxzide, given their long half-lives and active metabolites, on the theory that this allows for a 'self-tapering' once the last as-needed dose has been given. These two benzodiazepines, however, undergo extensive hepatic metabolism, and in cases characterized by hepatic failure, most clinicians will opt for either lorazepam or oxazepam, as neither of these have any active metabolites, but rather, after glucuronidation, undergo direct renal excretion. This symptom-triggered approach may be utilized not only on an inpatient basis, but also, provided patients are reliable and not likely to abuse the benzodiazepine, on an outpatient basis. Moderate alcohol withdrawal may also be treated with a symptom-triggered approach; however, given that moderate cases may be complicated by transient confusion or hallucinations, it is essential that medications be administered by someone else, and if such a reliable person is not available, then inpatient care is generally indicated.

This symptom-triggered approach, though having much to recommend it, does carry the potential liability of allowing autonomic symptoms to occur, thus posing a risk to patients with cardiac disease, epilepsy, a history of alcohol withdrawal seizures or a current alcohol withdrawal seizure, or a history of delirium tremens or of any of the risk factors for the development of delirium tremens discussed earlier. In such cases it may be prudent to modify the symptom-triggered approach to ensure that autonomic symptoms do not occur. On day one, with this modified approach, the patient is given as-needed doses of one of the benzodiazepines every 2–4 hours. On the morning of day two, the patient is placed on a regular dose, roughly equivalent to the total amount given on day one, with this regular dose divided into four equal doses; the patient is also allowed as-needed doses. This procedure is then repeated on each successive day until no as-needed doses are required. Subsequently, the total daily dose may then be reduced in decrements of one-third to one-fifth, such that the tapering occurs over 3–4 days. In general, either lorazepam or oxazepam should be used in this modified approach in order to avoid 'over-shooting' with long-acting agents, and, given that lorazepam is available for injection, it may be preferable. Although this modified approach takes longer (anywhere from 4 to 7 days), and typically entails a higher overall dose of benzodiazepine, reducing the risk of seizures or of delirium tremens may make it advisable. Given the complexity of this modified approach, it generally should not be utilized on an outpatient basis.

Although benzodiazepines, as noted earlier, have the longest track record, consideration may also be given to either carbamazepine (Malcolm *et al.* 1989; Stuppacek *et al.* 1992) or divalproex (Reoux *et al.* 2001), both of which appear superior to oxazepam. In otherwise healthy patients with normal hepatic function, one may begin with carbamazepine in a dose of 200 mg three or four times daily or with divalproex in a total daily loading dose of 20 mg/kg, divided into two or three doses, with subsequent dosage adjustments made on the basis of clinical response, side-effects, and blood levels. As both of these agents may take a day or two to quell the symptoms of alcohol withdrawal, most clinicians will combine their use with symptom-triggered lorazepam until symptoms are completely controlled. Generally, after 3 days at the most, no further lorazepam will be required, and the patient may then be continued on the anti-epileptic drug alone for about a week in total, that is to say, long enough for the alcohol withdrawal to run its natural course, after which it may be rapidly tapered and discontinued.

At present, it is not clear when this combination approach, utilizing an anti-epileptic drug and lorazepam, should take the place of treatment with a full course of benzodiazepine monotherapy. Certainly, in cases of mild withdrawal, where symptom resolution is expected in a matter of days, it makes little sense to use an anti-epileptic drug, as the withdrawal will almost have run its course before the anti-epileptic drug could become effective, and thus the use of anti-epileptic drugs should generally be restricted to cases of moderate withdrawal. Further, given that lorazepam, as noted below, is effective both in preventing recurrent withdrawal seizures and in treating delirium tremens, whereas the anti-epileptic drugs have not been shown to be effective in these regards, many clinicians feel safer using lorazepam. Others, noting that the combination strategy provides coverage with lorazepam until symptoms are completely controlled, and the advantage of getting away from potentially intoxicating substances as soon as possible, find the combination strategy acceptable. One of the authors (DPM) has also successfully utilized both carbamazepine and divalproex as an 'add-on' drug to a benzodiazepine when the benzodiazepine alone has not succeeded in controlling symptoms. Clearly, although more work is needed here, the option of an anti-epileptic drug should not be forgotten.

In addition to these two anti-epileptic drugs, other agents have also been utilized for alcohol withdrawal. Baclofen, 10 mg three times daily, in a double-blinded study (Addolorato *et al.* 2002) was superior to placebo, and in a rater-blinded study (Addolorato *et al.* 2006) was superior to diazepam 0.5–0.75 mg/kg/day. Gabapentin, 1200 mg/day, was effective in one study (Myrick *et al.* 2009) but not another (Bonnet *et al.* 2003). Finally, one small study (Krupitsky *et al.* 2007) found topiramate, lamotrigine, and memantine were each superior to placebo. None of these agents, however, has entered mainstream treatment.

Finally, if not already done, patients should receive thiamine as described earlier.

DELIRIUM TREMENS

Delirium tremens represents a medical emergency. In addition to routine measures used for delirium, as discussed in Chapter 6.5, patients should be treated with intravenous lorazepam in doses of 2–4 mg every 5–10 minutes until they are lightly sedated, and massive doses, up to 200 mg/day, may be required to accomplish this. Provision is then made for additional as-needed doses should symptoms reappear. The next day, patients should be treated with a regular dose of lorazepam, roughly equivalent to that required during the first day, with this regular dose divided into six equal doses, again with provision for as-needed doses. This procedure is repeated daily until no further as-needed doses are required. At this point, the total daily dose may be reduced in decrements equivalent to 20–10 percent of the total amount, thus allowing for discontinuation of the lorazepam anywhere from 5 to 10 days later. In many cases, lorazepam, although capable of calming patients and quelling tremor and other autonomic symptoms, does not control hallucinations and delusions, and in these cases consideration may be given to adding an antipsychotic. Some authors urge caution here, given that antipsychotics do lower the seizure threshold; however, when the patients' reaction to these symptoms creates a dangerous situation (e.g. attempting to escape the hospital to avoid their 'persecutors'), the use of an antipsychotic may be justified. Should this be the case, one may proceed as described in Chapter 6.5. Occasionally patients do not respond to lorazepam, even in massive doses, and in such cases it is not clear what the next step should be. Some authors recommend switching to intravenous phenobarbital, beginning with 120–240 mg, with repeat as-needed doses every 1–2 hours; however, if this approach is utilized, one must be prepared for respiratory depression and the need for intubation. One of the authors (DPM) has successfully utilized carbamazepine as an add-on drug to lorazepam in such situations.

In addition to the routine laboratory tests described earlier, a careful search should be made for other illnesses, such as pneumonia, pancreatitis, gastrointestinal bleeding, hepatic failure, as one or more of these is usually present. Diaphoresis, vomiting, and diarrhea may cause dehydration, and massive fluid replacement may be required.

In those rare cases of delirium tremens sine tremore, treatment with benzodiazepines may or may not be required, depending on whether any autonomic symptoms are present. Antipsychotics, likewise, may or may not be required, depending again on whether patients' reactions to the hallucinations and delusions threaten them.

ALCOHOL WITHDRAWAL SEIZURES

Alcohol withdrawal seizures, as noted earlier, occur in the context of an alcohol withdrawal syndrome anywhere from hours to 2 days after the BAL has fallen below the patient's threshold for intoxication. If patients are not already being aggressively treated with a benzodiazepine, they should be, as a double-blinded study has demonstrated that the administration of as little as 2 mg of lorazepam intravenously will reduce the risk of a second seizure over the next 6 hours (D'Onofrio *et al.* 1999). In cases of status epilepticus, fosphenytoin may be given. Importantly, none of the anti-epileptic drugs, including carbamazepine and divalproex, have any prophylactic effect against alcohol withdrawal seizures.

OVERALL GOALS

Overall, the goal of treatment of alcoholism is abstinence. Although attempts have been made to enable alcoholics to continue drinking in a 'controlled' fashion, these are not reliable and cannot be recommended: the goal of abstinence must be stated to alcoholics clearly, starkly, and unmistakably. Whether the same applies to alcohol abusers is not clear; however, given the risk that alcohol abuse may evolve into alcoholism or that, if it does not, further drinking may get out of control with disastrous consequences, it is prudent to advance the same goal to alcohol abusers.

Although some alcoholics are able to stop drinking by an extraordinary act of will, this, as noted earlier, is an uncommon event, and generally confined to mild cases. The vast majority will continue to drink unless they receive help. In such cases, various psychosocial methods are helpful and may be offered. Pharmacologic treatment of alcoholism, utilizing naltrexone, acamprosate, topiramate, gabapentin, baclofen, or disulfiram, although at times helpful, is for the most part adjunctive only and cannot replace psychosocial treatments.

Various psychotherapies, such as cognitive–behavioral therapy, have been utilized, and in some cases are successful. A referral to Alcoholics Anonymous (AA), however, should be considered, regardless of whether patients are referred to psychotherapy. Many patients are also hospitalized in residential treatment facilities, and this may be necessary to effect a period of sobriety of sufficient length to allow patients to begin to think clearly again; during such stays patients are often engaged in various psychotherapies and, importantly, are often taken to AA meetings.

Alcoholics Anonymous is the oldest treatment approach to alcoholism and, if participated in fully, has an excellent success rate. The key word here is 'fully': no treatment works unless patients are compliant, and AA is no exception. Patients must be told that much will be expected of them but that, if they persist, they will become sober. AA meetings are available around the world, and in the USA contact may be made by simply calling directory assistance in virtually any city. Patients should be instructed to attend meetings as frequently as possible, ideally attending 'ninety meetings in ninety days', and to get an AA 'sponsor'. Should patients protest that they do not have the time to go to a meeting every day, they may be gently, but clearly, reminded that they have a debilitating and potentially fatal disease, and that the time spent at meetings will pay off not only in keeping them alive but also in salvaging what is left of their careers and relationships.

Of the medications to be considered, naltrexone, acamprosate, topiramate, gabapentin, and baclofen reduce either the urge to drink or the intoxication that occurs with drinking, whereas disulfiram, by contrast, induces a toxic 'Antabuse reaction' if patients do drink, thus adding another motivation to remain sober. Naltrexone, utilized only in patients with normal hepatic function, may be given in a dose of 50 mg orally daily or in a long-acting intramuscular preparation (Garbutt *et al.* 2005) in a dose of 380 mg once monthly. Acamprosate, used only in patients with normal renal function, is given in a dose of 666 mg three times daily; topiramate is titrated to a dose of 200–300 mg daily, gabapentin to a dose of 300 mg twice daily or 1500 mg nightly, and baclofen to a dose of 10 mg three times daily. Disulfiram is started at 500 mg daily, with the dose reduced to 250 mg daily after 1 or 2 weeks; patients should be given a graphic description of the toxic reaction that they may expect should they ingest even a minuscule amount of alcohol, and, given the potential toxicity of disulfiram, treatment is generally maintained for only a matter of months.

Choosing among these medications is not straightforward. Naltrexone (Anton *et al.* 2006) has been standard. Acamprosate, though superior to placebo in several studies (Kranzler and Gage 2008), is less effective than naltrexone (Anton *et al.* 2006; Morley *et al.* 2006). Topiramate is relatively new on the scene, and in addition to its superiority over placebo (Johnson *et al.* 2003, 2007), it also appears superior to naltrexone (Baltieri *et al.* 2008). Gabapentin (Brower *et al.* 2008; Furieri and Nakamura-Palacios 2007) and baclofen (Addolorato *et al.* 2007) are also relatively new on the scene, and, although superior to placebo, have not been compared with the other agents. With these data in hand, although naltrexone is certainly worthy of consideration, a case could clearly be made for starting with topiramate. Disulfiram should be reserved for patients who are highly motivated to remain abstinent.

Although these medications may be helpful, it is imprudent to allow patients to believe that taking one of them will 'cure' their alcoholism. Although there are rare exceptions, pharmacologic treatment alone cannot be used as a substitute for psychosocial treatments, but must, as indicated earlier, be considered adjunctive only. Alcoholism is a chronic disease and hence relapses are to be expected, and these occur most frequently in the first 6 months of treatment. Relapses, or 'slips', therefore, although certainly undesirable, should not be taken as an indication of failure but rather as an indication for patients to redouble their efforts.

In addition to psychosocial and pharmacologic treatment, it is essential to treat certain common comorbid conditions, such as panic disorder, schizophrenia, bipolar disorder, and major depressive disorder, as each of these, if untreated, may predispose to a resumption of drinking. As a general rule, in prescribing medications for these or any other conditions, potentially intoxicating drugs, such as benzodiazepines or opioids, should be avoided as they may trigger a desire to drink: exceptions to this rule are few.

28.6 SEDATIVES, HYPNOTICS, AND ANXIOLYTICS

The sedatives, hypnotics, and anxiolytics comprise a large group of agents, often collectively referred to as the 'sedative–hypnotics', each of which has an effect similar to that of alcohol. Representative agents include benzodiazepines, non-benzodiazepine hypnotics (eszopiclone, zolpidem, and zaleplon), and the barbiturates. Although most commonly used in conjunction with alcohol or other substances, such as opioids, stimulants, or cocaine, they may at times be used in isolation. Although these drugs, in the vast majority of cases, are taken orally, some addicts will crush, dissolve, and filter them for intravenous use.

Table 28.1 lists the members of this large group which are most likely to be abused, classifying them, according to their half-lives, as short acting (generally less than 6 hours), intermediate acting (6–18 hours), or long acting (generally over 24 hours). This classification is useful as it allows one to make a rough prediction as to when withdrawal, withdrawal seizures, or withdrawal delirium are likely to occur.

The popularity of these various agents has changed over time. The barbiturates, once commonly abused, have been supplanted by the benzodiazepines, among which alprazolam, lorazepam, and diazepam are most popular.

Clinical features

As indicated, the clinical features of sedative–hypnotic use are similar to those of alcohol, and thus one may see

Table 28.1 Sedative–hypnotics grouped by duration of effect

Name	Equivalent dose (mg)	Phenobarbital conversion constant
Short acting		
Triazolam	0.25	120
Ezsopiclone	2	15
Zolpidem	10	3
Zaleplon	10	3
Intermediate acting		
Alprazolam	1	30
Lorazepam	2	15
Temazepam	15	2
Chlordiazepoxide	25	1.2
Oxazepam	30	1
Long acting		
Clonazepam	2	15
Diazepam	10	3
Flurazepam	15	2
Phenobarbital	30	1
Butalbital	100	0.33
Secobarbital	100	0.33
Amobarbital	100	0.33
Pentobarbital	100	0.33

sedative–hypnotic intoxication, blackouts, tolerance, withdrawal, withdrawal seizures, and withdrawal delirium. Each of these is discussed below in turn.

Mild, sedative–hypnotic intoxication with barbiturates (Curran 1938, 1944; Isbell *et al.* 1950) and benzodiazepines is characterized by euphoria, a degree of affective lability, and disinhibition. With moderate intoxication, reaction time is slowed, lethargy and drowsiness appear, and patients often develop nystagmus, dysarthria, and ataxia; falls may occur, with possible head trauma. With severe intoxication stupor or coma may occur, with respiratory depression and death. Although severe intoxication is not uncommonly seen with overdosage of barbiturates, it is rare with benzodiazepines unless the benzodiazepine overdose has been accompanied by use of alcohol, opioids, or other sedative–hypnotics. Although urine toxicology screens are definitely appropriate, it is important to keep in mind that some tests will fail to screen for certain drugs (e.g. alprazolam), and consequently one must contact the lab to determine precisely what is screened for.

Sedative–hypnotic blackouts are quite similar to those seen with alcohol; although possible with long-acting benzodiazepines such as diazepam (Wolkowitz *et al.* 1987), they are more common with short-acting, high-potency benzodiazepines (Scharf *et al.* 1987) such as triazolam (Ewing *et al.* 1988; Greenblatt *et al.* 1991; Morris and Estes 1987).

Tolerance may appear with long-term use and at times may be quite remarkable: for example some patients may end up taking hundreds of milligrams of diazepam daily, with little or no evidence of sedation.

Sedative–hypnotic withdrawal, as seen with benzodiazepines (Busto *et al.* 1986; Covi *et al.* 1973; Hollister *et al.* 1961; Juergens and Morse 1988; Murphy *et al.* 1984, 1989; Rickels *et al.* 1990; Shader and Greenblatt 1993; Tyrer *et al.* 1983) and barbiturates (Fraser *et al.* 1958; Isbell *et al.* 1950), is characterized by anxiety, irritability, insomnia, tremor, tachycardia, diaphoresis, nausea and vomiting, and postural hypotension; there are also rare reports of stuporous catatonia occurring as a manifestation of benzodiazepine withdrawal (Rosebush and Mazurek 1996). The onset of the withdrawal syndrome varies according to the half-life of the agent: roughly speaking, for short-acting agents withdrawal starts in less than a day; for intermediate-acting agents, 2–3 days; and for long-acting agents, 2–6 days. For certain very-long-acting agents, such as diazepam or phenobarbital, the blood level may fall so slowly that there is, in effect, a 'self-tapering', with, in many cases, a substantially less severe withdrawal syndrome. The duration of withdrawal likewise varies with the half-life of the agent. Roughly speaking, for short- and intermediate-acting agents, symptoms peak within 1–3 days and persist for 1–2 weeks, whereas for long-acting agents the peak arrives in 5–7 days and the syndrome may persist for up to 2 or 3 weeks. As with alcohol withdrawal, some patients may experience a protracted withdrawal syndrome (Ashton 1984; Shader and Greenblatt 1993), lasting for from weeks to months, which is characterized by lingering, low-level symptoms such as anxiety, irritability, insomnia, increased sensitivity to light and sound, and parasthesiae.

Sedative–hypnotic withdrawal seizures typically occur within the context of withdrawal symptomatology and, although these may occur with benzodiazepines (e.g. alprazolam [Breier *et al.* 1984; Noyes *et al.* 1986]), they are much more common with barbiturates, in which case there is a significant risk of multiple seizures and status epilepticus.

Sedative–hypnotic withdrawal delirium, noted with benzodiazepines (such as alprazolam [Levy 1984; Zipursky *et al.* 1985], lorazepam, and triazolam [Heritch *et al.* 1987; Martinez-Cano *et al.* 1995]) and barbiturates (Fraser *et al.* 1958; Isbell *et al.* 1950), emerges out of a withdrawal syndrome and is characterized by an intensification of those symptoms, accompanied by confusion, disorientation, agitation, hallucinations, and persecutory delusions. In the natural course of events, the delirium tends to clear in anywhere from days to a couple of weeks.

Before leaving this discussion, some words are in order regarding the non-benzodiazepine hypnotics. Although these agents are considerably safer than the benzodiazepines, there is no doubt that, for example, zolpidem can cause withdrawal seizures (Cubala and Landowski 2007) and both tolerance and withdrawal (Cavallaro *et al.* 1993).

Course

Recreational use of these agents, particularly the benzodiazepines, is common among adolescents and young adults. Sedative–hypnotic abuse is said to occur when patients continue to seek intoxication despite experiencing blackouts and social or legal consequences, and the onset of addiction is heralded by the development of craving, tolerance, and withdrawal phenomena, such as a withdrawal syndrome or seizures. As indicated earlier, concurrent use of other substances is more the rule than the exception: benzodiazepines may be used by alcoholics to treat the shakes, by opioid addicts to lessen the dysphoria and insomnia of withdrawal, and by cocaine or stimulant addicts to 'smooth out' an intoxication.

Etiology

All of these agents, by somewhat different mechanisms, enhance chloride flux across neuronal membranes when GABA receptors are stimulated. Presumably, with prolonged use, downregulation of these receptors occurs, with the consequent development of tolerance and withdrawal phenomena.

Differential diagnosis

Sedative–hypnotic intoxication is clinically indistinguishable from alcohol, isopropyl alcohol, and methanol intoxication except for the fact that sedative–hypnotic-intoxicated patients do not have an odor of alcohol on their breath. Ethylene glycol intoxication, which also lacks an odor of alcohol, is distinguished by an increased anion gap. In cases in which patients fail to recover from an intoxication within the expected time frame, other disorders, for example traumatic brain injury, should be considered.

Sedative–hypnotic blackouts must be distinguished from other causes of episodic anterograde amnesia, as discussed in Chapter 6.6.

Sedative–hypnotic withdrawal is basically indistinguishable from alcohol withdrawal, and the diagnosis must rest on an accurate history of substance use, a BAL and urine toxicology, keeping in mind that many patients will have a 'mixed' picture of both sedative–hypnotic and alcohol withdrawal.

Although seizures occurring during sedative–hypnotic withdrawal may, obviously, be due to this withdrawal, *per se*, consideration must also be given to head trauma, alcohol withdrawal, and some of the other causes of seizures seen in alcoholics, as discussed under 'differential diagnosis' in the immediately preceding chapter on alcoholism.

Sedative–hypnotic withdrawal delirium may be indistinguishable from delirium tremens and the diagnosis may rest on an accurate history, keeping in mind that, as for withdrawal, a 'mixed' picture may be present.

Treatment

Sedative–hypnotic intoxication typically requires only observation and general medical support. In cases of benzodiazepine intoxication in which respiratory depression threatens, flumazenil may be given (Brogden and Goa 1991); however, caution is needed here in cases in which tolerance has developed, as flumazenil may precipitate both a hyperacute withdrawal syndrome and withdrawal seizures.

Sedative–hypnotic blackouts require only observation until serial mental status examinations have revealed a restoration of short-term memory, and the intoxication itself has resolved.

Sedative–hypnotic withdrawal may be pharmacologically managed with one of three approaches. The simplest is to use the same drug the patient was on. A second approach is to switch to diazepam. Finally, consideration may also be given to treatment with phenobarbital. Each of these approaches will be discussed in turn, followed by a discussion of how to choose among them.

If the choice is to withdraw the patient using the same drug, one may simply reinstate the same average daily dose, with provision for as-needed doses, and once the patient is stabilized, the drug may be tapered off. One difficulty with this approach is the fact that many patients will either underestimate or overestimate their average daily dose. Underestimates, of course, are taken care of by the as-needed doses; however, when an overestimate is made, one runs the risk of overdosing the patient. Given this concern, many clinicians will start with a total daily dose (divided, as appropriate, given the half-life of the drug in question, into more or less multiple doses) that deliberately is below the patient's estimate, knowing full well that as-needed doses will be required.

The second approach, namely switching to diazepam, is, as discussed below, appropriate only when the drug in question is a benzodiazepine. Furthermore, if the benzodiazepine in question is alprazolam, one should probably not switch to diazepam, as there are anecdotal reports of alprazolam withdrawal not responding to diazepam (Zipursky *et al.* 1985). The first step in utilizing this approach is to determine how much diazepam to give, and the column headed 'Equivalent dose' in Table 28.1 may be used to establish a rough equivalency; thus, if the patient had been taking 40 mg total per day of lorazepam, approximately 200 mg of diazepam would be required. The estimated diazepam dose requirement, however, must be taken as only a rough guide, and for two reasons. First, there is the problem of underestimation and overestimation, discussed above. Second, the drug's metabolism must be taken into account. For example, in a patient with liver disease, the metabolism of lorazepam, which is dependent on glucuronidation alone, would not be greatly disturbed, whereas that of diazepam, which is heavily dependent on hepatic metabolism, would be. In such a case, if the 'estimated' dose of diazepam were given, an overdose situation might ensue. Given these concerns, as with the first option, most clinicians will start with a dose of diazepam that is deliberately low, and then make up for any shortcoming with as-needed doses. Once the patient is stabilized, then the dose of diazepam is tapered off.

The last option, namely switching to phenobarbital, may be accomplished by arriving at a dose of phenobarbital equivalent to the total daily dose of the drug in question, and this may be obtained utilizing the figures supplied in the column entitled 'Phenobarbital conversion constant' in Table 28.1. The total daily dose of phenobarbital = (total daily dose of drug in question) × (phenobarbital conversion constant for the drug in question). For example, if the patient had been using a total of 20 mg of lorazepam daily, the total daily dose of phenobarbital would equal 20 × 15, or 300 mg, and this total daily dose is divided into three equal ones. This approach, of course, yields only a rough figure, and the same concerns occur here regarding the problem of underestimation and overestimation. This problem, however, is particularly acute in the case of phenobarbital, given that, unlike the case with the benzodiazepines, overdoses with phenobarbital may be fatal. Given this, it is prudent to underestimate the total daily dose and utilize as-needed doses of 30 mg every 2 or 3 hours, and, before each dose, to examine the patient closely for evidence of intoxication, such as nystagmus, dysarthria, or ataxia, and if there is such evidence, then withhold treatment until the intoxication passes, after which treatment is recommenced at a suitably lower dose.

Once the patient has been stabilized for 2 or 3 days, regardless of which approach is utilized, tapering may be commenced. The rate of tapering depends on the setting: if the patient is hospitalized, the dose may be tapered at roughly 20–25 percent per day; with outpatients, however, a slower taper is appropriate, at a rate of no more than 10 percent per day.

The overall choice among these three approaches depends, in large part, on what the patient had been on, and of the many possible different clinical scenarios, most patients will fall into one of the following five: one benzodiazepine; two or more benzodiazepines; a benzodiazepine plus alcohol; a barbiturate alone; and, finally, a barbiturate plus alcohol.

In cases of withdrawal from only one benzodiazepine, the first approach, namely managing the withdrawal with that particular benzodiazepine, though attractive, may have drawbacks when the benzodiazepine in question is short acting. When short-acting agents, such as triazolam, are utilized, there is a risk of 'mini-withdrawals' during the day, and although this risk may be mitigated with frequent observation during a hospital stay, their occurrence in outpatients typically makes compliance very, very difficult. In such a situation, consideration should be given to the second approach, namely switching to diazepam, as diazepam's long half-life typically allows a 'self-tapering' without the risk of any mini-withdrawals.

In cases of withdrawal from two or more benzodiazepines, patients should, in general, be switched to diazepam, as stabilizing and then tapering two benzodiazepines is simply unnecessarily difficult and complex. Naturally, if diazepam is used, the dose should be based on the sum of the doses required for each of the benzodiazepines.

In cases of withdrawal from a benzodiazepine plus alcohol, which is indeed the most common of these five scenarios, proper management depends on whether or not the patient is dependent on alcohol, as is usually the case. In such a case, one should utilize the modified symptom-targeted approach for alcohol withdrawal, as discussed in Chapter 28.5. If, however, the patient had been using alcohol only recreationally, then one may proceed as if the patient were on only one benzodiazepine, or two or more benzodiazepines, as just discussed.

In cases of withdrawal from a barbiturate alone, one may conduct the withdrawal utilizing the first approach, namely with the barbiturate in question, or one may utilize the third approach and switch the patient to phenobarbital. In cases of barbiturate dependence, no attempt should be made to manage the withdrawal with benzodiazepines, as these are simply not effective.

Finally, in cases where the patient is taking a barbiturate plus alcohol one should withdraw the patient utilizing the third approach, with phenobarbital alone. Here, one first calculates the phenobarbital dose required for the barbiturate in question and then adds that to the phenobarbital dose required to cover the alcohol; this figure in turn is calculated using the formula discussed above, utilizing a phenobarbital conversion constant of 1 for each milliliter of pure ethanol (e.g. if the patient were drinking a liter of '100 proof' whiskey on a regular basis, he would be drinking 500 mL of pure ethanol, and the required total daily dose of phenobarbital would equal 500×1, or 500 mg).

Sedative–hypnotic withdrawal seizures should be treated by rapidly controlling the withdrawal with intravenous medications, such as lorazepam, diazepam, or phenobarbital.

Sedative–hypnotic withdrawal delirium demands vigorous treatment of the withdrawal syndrome, utilizing intravenous medications, with the goal of producing a light degree of sedation. Should hallucinations and delusions persist in a troubling fashion, an antipsychotic, as described in Chapter 6.5, may be required. Once symptoms have been brought under control, the dose may be gingerly tapered at a rate of from 10 to 25 percent daily.

Overall, the goal of treatment in the case of abuse or addiction is abstinence. Those addicted to sedative–hypnotics alone, or to a combination of a sedative–hypnotic and alcohol, may do well in Alcoholics Anonymous.

28.7 INHALANTS (SOLVENTS)

The volatile ingredients of many readily available products are often inhaled for intoxication. These include airplane or model glue, paint thinner, kerosene, gasoline, fingernail polish remover, the propellants in aerosol sprays and spray paints, and typewriter correction fluid. Each of these products contains various mixtures of aliphatic and aromatic hydrocarbons, some of which may be halogenated; of all of these intoxicating hydrocarbons, toluene appears to be the most significant.

Clinical features

Intoxication is typically obtained either by soaking a rag in the volatile substance and holding it to the face or by placing the substance in a plastic or paper bag and then inhaling; when a bag is used it may leave a telltale circular rash on the face. The intoxication (Evans and Raistrick 1987) occurs rapidly and is characterized by a dreamy euphoria, drowsiness, dizziness, dysarthria, diplopia, nystagmus, and ataxia. Some may also experience confusion, delusions, or hallucinations (which may be either visual or, less commonly, auditory), or irritability and impulsivity. The intoxication is generally short lived, however, many users will take repeated 'hits', and extend the intoxication for several hours. Although most intoxications are benign, fatalities may occur secondary to cardiac arrhythmias (King *et al.* 1985; Steffee *et al.* 1996) or, if a plastic bag is placed over the head, suffocation; in some cases vomiting may occur during the intoxication, and aspiration may follow. In severe intoxications, convulsions, coma and respiratory depression may ensue (King *et al.* 1985). Other, uncommon, complications of intoxication include rhabdomyolysis, hepatitis and renal failure (Gupta *et al.* 1991; Taverner *et al.* 1988).

Toluene may be detected in the blood for 4–10 hours; toluene is also metabolized to hippuric acid, and although this may be detected in the urine a little longer, this test may yield both false positives and false negatives.

Both tolerance and withdrawal may occur: withdrawal (Evans and Raistrick 1987; Watson 1979) occurs within 1–2 days of abstinence and is characterized by irritability, sweating, tremulousness, and insomnia, all of which generally remit within a matter of days.

With long-term use a dementia may occur (discussed in Chapter 29.12), as may a motor peripheral polyneuropathy (Altenkirch *et al.* 1977; King *et al.* 1985b), sensorineuronal hearing loss, and cerebellar ataxia, which, in a case report, responded to treatment with amantadine (Deleau and Hanssens 2000).

Course

Occasional, recreational use of inhalants is not at all uncommon among adolescents. Abuse and dependence are far less common.

Etiology

Although the mechanism whereby volatile hydrocarbons produce intoxication in humans is not known with certainty, it does appear that these agents both enhance GABAergic transmission and inhibit NMDA-mediated transmission.

Differential diagnosis

Intoxication with alcohol or sedative–hypnotics may yield a somewhat similar clinical picture. The odor of solvents on clothing or skin may be a clue, as may a rash on the face.

Treatment

Simple observation is generally sufficient. Patients who also abuse or are dependent on alcohol or opioids may be referred to Alcoholics Anonymous or Narcotics Anonymous; the optimal treatment of those who are solely involved with inhalants is not clear.

28.8 CANNABIS

The name 'cannabis' comes from the Greek word for hemp and refers specifically to the flowering tops of the hemp plant, *Cannabis sativa*. In the USA, the four most common preparations of cannabis are marijuana, sinsemilla, hashish, and hashish oil. Marijuana (also known as 'grass', 'pot', 'reefer', 'weed' or 'Mary Jane') is simply a dried collection of the flowers and nearby leaves and sprouts of the hemp plant and, although at times ingested, it is usually rolled into a cigarette and smoked. Sinsemilla, in turn, is a dried collection of only the flowers. Hashish is composed of the resin scraped from the leaves and flowers of the plant, while hashish oil is an ethanol extract of hashish. The active ingredient in all these preparations is the delta-9 isomer of tetrahydrocannabinol (THC), and the potency of these preparations rises from marijuana on through to hashish oil, which is by far the most potent.

Clinical features

The onset of intoxication occurs within minutes of smoking; however, after oral intake, effects may be delayed for hours. Clinically (Allentuck and Bowman 1942; Bromberg 1934; Clark and Nakashima 1968; Clark *et al.* 1970; Klonoff *et al.* 1973; Melges 1976) there is a dreamy sense of well-being, an unusual heightening of the senses, and a feeling that time is slowing down or disintegrating (Melges *et al.* 1970). Thinking becomes slowed and patients often develop a heightened sense of the ridiculous, laughing and giggling at otherwise prosaic things; in some cases depersonalization may occur (Mathew *et al.* 1993). Typically, intoxication is accompanied by conjunctival injection, dry mouth, increased appetite, mild ataxia, mild tachycardia, and a combination of increased supine blood pressure and orthostatic hypotension. In general, symptoms undergo substantial resolution after 3 or 4 hours. THC or its metabolites, however, may be detected in the urine for 2–6 days in infrequent users, and in chronic heavy users the urine may remain positive for several weeks.

In a minority of cases of intoxication, complications may occur, including anxiety, psychosis, and delirium.

Anxiety may occur during otherwise unremarkable intoxications and may at times crescendo to constitute an anxiety attack (Bromberg 1934), with tremor, tachycardia, and palpitations: typically the anxiety resolves, as does the intoxication.

Psychosis (Kroll 1975; Mathers and Ghodse 1992; Talbott and Teague 1969; Weil 1970) may also occur during intoxication and patients may develop delusions of persecution, which may be accompanied by auditory or visual hallucinations. Patients may become quite agitated in the midst of this, and some will flee the scene or seek safety in some other way. This 'cannabis-induced psychotic disorder' generally outlasts the intoxication itself and indeed may persist for days. Before leaving this discussion of psychosis, mention should also be made of the possible occurrence of a chronic psychosis secondary to cannabis use. This is a controversial notion: although it does appear that chronic cannabis use is associated with the occurrence of non-affective psychosis later in life (McGrath *et al.* 2010), it is not clear whether cannabis is a direct causative factor, or merely increases the risk for the development of schizophrenia.

Delirium (Chopra and Smith 1974; Palsson *et al.* 1982) may also occur during an intoxication, but generally only when very high doses have been taken. Patients become confused, agitated, and at times incoherent; delusions and hallucinations may also occur. This delirium may either clear as the intoxication does, or may persist for up to a few days.

Tolerance to cannabis can develop (D'Souza *et al.* 2008) and is manifest by a decreased euphoric response and a diminution of the tachycardia and elevated supine blood pressure normally seen during intoxication.

Withdrawal (Duffy and Millin 1996; Haney *et al.* 1999; Mendelson *et al.* 1984) may occur in tolerant patients, and typically appears anywhere from 1 to 2 days after the resolution of intoxication. Symptoms are typically mild and consist of a craving for the drug, anxiety, irritability, restlessness, a fine tremor, diaphoresis, and insomnia. Withdrawal symptoms reach a peak intensity within a few days and then resolve over roughly a week's time.

Before leaving this discussion of the clinical features of cannabis use, some mention may be in order regarding a putative entity referred to as the 'amotivational syndrome'. This syndrome is said to occur after long-term heavy use of cannabis and to be characterized by a lack of motivation and sense of purpose, and by a tendency towards passivity and aimless 'drifting' through life. Whether or not such a syndrome actually exists, however, is controversial, and to date there is no convincing evidence for it.

Course

Recreational use of cannabis is extremely common among adolescents and young adults, and most of these individuals either stop entirely or greatly reduce their use as they begin to assume adult responsibilities. Cannabis abuse is marked by frequent intoxication despite social and legal consequences, and dependence is heralded by the onset of tolerance and withdrawal. Overall, abuse and dependence are far less common than recreational use, occurring in no more than a few percent of adolescents and young adults in the USA.

Etiology

The principal intoxicant in cannabis, as noted earlier, is the delta-9-tetrahydrocannibinol (Isbell *et al.* 1967), which is highly lipid soluble and readily crosses the blood–brain barrier. There are two endogenous cannabinoid receptors, namely CB2 and CB1, and the CB1 receptor is found within the central nervous system, primarily within the cerebral cortex, hippocampus, basal ganglia, and cerebellar cortex.

Although environmental factors play a role in the development of dependence, genetic factors also clearly play a role (Lynskey *et al.* 2002), and recent data suggest that different polymorphisms in the gene for CB1 may confer different risks for the development of dependence (Hopfer *et al.* 2006).

Differential diagnosis

Uncomplicated intoxication with cannabis may be mimicked by intoxication with alcohol, sedative–hypnotics, inhalants, and opioids; more to the point, however, is the fact that many cannabis users also partake of these substances, particularly alcohol, and hence a mixed intoxication is not uncommon. Here, as with intoxications marked by anxiety attacks, psychosis or delirium, a drug screen is typically necessary.

Treatment

Uncomplicated intoxication generally requires only observation until the intoxication has passed. Anxiety, if troubling, may be relieved by oral diazepam, in a dose of 10 mg. Psychosis, if problematic, may be treated with an antipsychotic, such as 1–2 mg of risperidone or 50 mg of quetiapine, with repeat doses as needed. Delirious patients should be closely monitored until the delirium has passed, and may be treated as outlined in Chapter 6.5. Cannabis withdrawal generally does not require specific treatment: although oral delta-9-tetrahydrocannabinol does reduce withdrawal symptoms (Budney *et al.* 2007), it is questionable whether this is justified.

The overall goal of treatment in abusers and addicts is abstinence, and although various forms of psychotherapy, including cognitive–behavioral therapy and motivational enhancement therapy (with or without contingency management with vouchers) have shown success, many adolescents and young adults simply see nothing wrong with using cannabis, and often drop out of treatment.

28.9 OPIOIDS

Opium is derived from the juice of the poppy plant, and within opium two opiates are found, namely morphine and codeine. The term opioid is broader, and includes not only the two naturally occurring opiates, but also their semisynthetic and synthetic derivates, including fentanyl, heroin, oxycodone, hydrocodone, hydromorphone, meperidine, pentazocine, methadone, and buprenorphine: these last two derivatives, although often used in the treatment of opioid withdrawal and addiction, may also be used for intoxication (Torrens *et al.* 1993). Of all the opioids, oxycodone and heroin are the most commonly used for intoxication.

With the exceptions of fentanyl and heroin, all of these drugs may be taken orally (or, in the case of buprenorphine, sublingually). Although some patients may utilize the oral (or sublingual) route, most strongly favor the parenteral, most especially the intravenous, route, either by utilizing parenteral preparations where available or, when only pills are available, crushing and dissolving them and then filtering the solution, often with cigarette filters, to yield a more or less adulterated preparation which is then injected. Occasionally heroin may be snorted or smoked, or, in the process known as 'chasing the dragon', heated, with subsequent inhalation of the heroin vapor.

Of note, buprenorphine comes in two preparations. One contains only buprenorphine, in both a sublingual and parenteral form. The other contains a combination of buprenorphine and naloxone, and is available only in a sublingual form. The rationale for this combination preparation lies in the fact that whereas buprenorphine is absorbed sublingually, naloxone is not. Consequently, when the combination is taken sublingually, the effect of buprenorphine is not blocked by naloxone, but if the combination is crushed and taken parenterally, buprenorphine's effect is blocked by the antagonist.

Clinical features

Intoxication with opioids is intensely seductive. Within moments of intravenous injection, the user may be rewarded by an intense 'rush'. The body is suffused with warmth, and orgasmic sensations may be experienced. In less than a minute the rush tends to pass, to be replaced by a drowsy, vaguely euphoric feeling that is accompanied by difficulty with concentration and dysarthria. The pupils are generally constricted, peristalsis is slowed, and constipation ensues; urinary hesitancy or retention may also occur. When taken orally, the onset of action is slower, occurring over 0.5–1 hour, and the intoxication itself lacks the 'rush' seen with intravenous use; otherwise, however, orally-induced intoxication is the same.

Meperidine and pentazocine intoxication have distinctive features. Meperidine is metabolized to normeperidine, and this metabolite may cause agitation, tremor, mydriasis, increased deep tendon reflexes, and, occasionally, myoclonus or seizures (Kaiko *et al.* 1983). Pentazocine intoxication, when high doses are utilized, may be accompanied by dysphoria, anxiety, hallucinations, and bizarre thoughts, along with dizziness and diaphoresis (Challoner *et al.* 1990).

The overall duration of intoxication may be roughly divided into short, medium, and long. Short-duration intoxication, lasting 1–2 hours, occurs with fentanyl. Medium-duration intoxications, lasting 3–6 hours, is seen with morphine, codeine, heroin, oxycodone, hydrocodone, hydromorphone, meperidine, and pentazocine. Long-duration intoxications, lasting anywhere from 12 to 48 hours, are seen with methadone and buprenorphine.

Overdoses of opioids are not uncommon. The purity of 'street' heroin varies widely, and the unwary user may unintentionally take a lethal amount; furthermore, some addicts, after a period of abstinence during which tolerance is lost, may resume use with a dose that, although prior to the loss of tolerance produced only intoxication, is now sufficient to cause overdose. Overdose itself is characterized by stupor or coma, accompanied by hypotension and respiratory depression. Pupils are initially pinpoint; however, with the advent of cerebral anoxia, mydriasis appears. Pulmonary edema and seizures may occur, and death is usually due to respiratory arrest. Those who survive may be left with a posthypoxic encephalopathy or sequelae of watershed infarctions.

Urine toxicology may be positive for 12–36 hours for drugs causing short- or medium-duration intoxications, and up to several days for those causing long-duration intoxications. It is critical to determine exactly which drugs the toxicology screen covers, as many of these drugs, in particular fentanyl, are not routinely covered. False positives may occasionally be seen when patients have eaten foods, such as bagels, which have poppy seeds in them.

Tolerance may develop to almost all of the effects of opioids (with the exception of miosis and constipation) and addicts may progressively increase their doses to obtain intoxication, sometimes to stunning levels of a gram or more of morphine.

Withdrawal symptomatology may occur after only a matter of weeks of daily use, and the severity of withdrawal increases with both longer duration of use and higher doses. Mild withdrawal is characterized by a sense of uneasiness and a craving for the drug; with progression to moderately severe withdrawal, restlessness appears along with an intensified craving, accompanied by yawning, lacrimation, rhinorrhea, and diaphoresis. Severe withdrawal follows and is characterized by extreme dysphoria, agitation, and waves of goose flesh that may be so severe as to resemble the skin of a plucked turkey, an appearance that prompted the phrase 'going cold turkey'. Intense bone and muscle pain, especially in the back, arms, and legs, also occurs, and patients may engage in seemingly involuntary kicking movements, a phenomenon that gave rise to another synonym for opioid withdrawal, namely 'kicking the habit'. Insomnia may be severe. The pupils are dilated and the temperature, pulse, and blood pressure are all increased. Nausea, vomiting, abdominal cramping, and diarrhea occur, and the resulting fluid loss may be so severe as to cause circulatory collapse.

Withdrawal from drugs causing short- and medium-duration intoxications usually begins within 6–12 hours, peaks in 1–3 days, and then generally subsides over a week or so. Withdrawal from either methadone or buprenorphine typically begins in 1–3 days, peaks in 3–8 days, and then subsides in 2–3 weeks. In heavy users, although the most prominent signs of withdrawal subside, a protracted withdrawal syndrome (Martin and Jasinski 1969) may persist for up to half a year, which is characterized by dysphoria, anhedonia, insomnia and craving.

Other disorders may accompany opioid abuse or addiction. Intravenous use brings the risk of bacteremia with pulmonary abscess, endocarditis, cerebral abscess, cerebral mycotic aneurysm, meningitis, osteomyelitis, and tetanus. Furthermore, particulates, if large enough, may lodge in small arteries and cause stroke, and there are also rare cases of a myelopathy occurring after intravenous heroin use. Parenteral users often share needles and are thus at risk for AIDS, syphilis, and hepatitis. Furthermore, the presence of particulates in the injected fluid (as may occur when cigarette filters are used) may lead to pulmonary fibrosis, pulmonary hypertension, and cor pulmonale. Particulates may also collect in regional lymph nodes causing a chronic lymphadenopathy with edema, especially of the hands.

Although most patients who use opioids parenterally stick to the intravenous route, some will inject the drug intramuscularly or, in the practice known as 'skin popping', subcutaneously, and these practices carry distinct liabilities. With intramuscular use a myositis may occur, which may undergo ossification. Those who engage in skin popping may start out that way, only to progress to intravenous use. More commonly, however, skin popping represents a terminal stage wherein all the patient's veins have become sclerotic. Skin popping may be followed by cellulitis or ulceration, which may eventually be disfiguring.

Patients who 'chase the dragon' and inhale heroin vapor may develop a leukoencephalopathy (Gacouin *et al.* 2003; Kriegstein *et al.* 1999; Ropper and Blair 2003) with variable mixtures of quadriparesis, akinetic mutism, and ataxia. Of note, there is anecdotal evidence that the leukoencephalopathy may respond to treatment with coenzyme Q (Gacouin *et al.* 2003; Kriegstein *et al.* 1999).

Illegally manufactured 'street' meperidine may be contaminated with a by-product, methyl-phenyl-tetrahydropyridine (MPTP), which in turn may cause a chronic parkinsonism (Langston *et al.* 1983).

Course

Long-term recreational use of opioids is relatively uncommon and most long-term recreational users pass fairly rapidly to abuse, and generally on to addiction. Although some addicts, notably physicians who may have ready access to opioids, are able to maintain their social positions, most quickly lose whatever gainful employ they may have had and turn to crime to support their addiction. Those who become deeply involved in the drug 'subculture' are liable to meet violent death at the hands of others; suicide attempts are also not uncommon, and those who do survive often end up losing all in their pursuit of the drug.

Etiology

The intoxicant effects of opioids are mediated by their binding to mu and, to a much lesser extent, kappa receptors within the central nervous system. Although genetic factors account for roughly one-half of the liability for opioid addiction, environmental factors are also clearly important, including peer group pressure and, naturally, availability. The presence of a conduct disorder or antisocial personality disorder also increases the risk.

Differential diagnosis

Opioid intoxication may be partially mimicked by intoxication with alcohol, sedative–hypnotics or inhalants; however, these intoxications generally lack the miosis characteristic of most opioid intoxications; in doubtful cases an appropriate urine toxicology will resolve the issue. It must be borne in mind, however, that many patients will use other substances in addition to opioids: cocaine may be used to reduce sedation, and alcohol or sedative–hypnotics may be employed to ease the pain of withdrawal.

Treatment

Intoxication, if mild, may require only simple observation. Whenever there is any evidence of overdose (e.g. stupor or respiratory depression), naloxone should be given in a dose of 0.4–2 mg intravenously, with repeat doses every 3–5 minutes until the patient is out of danger. Although in most cases a failure to respond to 10 mg argues strongly against opioid intoxication, exceptions do occur, especially with intoxications due to buprenorphine, a drug which bonds with exceptional avidity to the mu receptor. Whenever naloxone is administered to patients, such as abusers or addicts, who are tolerant to opioids, care must be taken to avoid 'overshooting' and precipitating a hyperacute withdrawal syndrome. Indeed, given that such hyperacute withdrawals may be marked by agitation and combativeness, consideration may be given to applying restraints in such cases either before administering naloxone or as soon thereafter as possible.

Although withdrawal may be attempted on either an inpatient or outpatient basis, the odds of success with inpatient withdrawal are much higher, in the order of 80 percent, as compared with no more than 20 percent for outpatient withdrawal. When inpatient withdrawal is attempted, this, ideally, should be performed on a secure unit, and, given the intense drug craving seen during withdrawal, patients should be confined to the ward until the withdrawal has run its course; visitation, if allowed at all, must be closely and continuously supervised. If outpatient withdrawal is undertaken, it is critical to provide only sublethal doses of any medications utilized, and to see patients frequently, either in the office or via home visits.

Currently, there are three traditional approaches that remain standard for withdrawal: withdrawal with opioids, symptomatic treatment with medications such as the alpha-2 agonists clonidine and lofexidine, and, finally, 'cold turkey' withdrawal.

Withdrawal utilizing an opioid may be accomplished with methadone, buprenorphine, or, if the patient had been using another legally available drug (e.g. oxycodone), with that agent. Treatment is generally commenced as withdrawal symptoms appear. Methadone may be started in a dose of 5–10 mg intramuscularly or 10–20 mg orally, with repeat doses every 4 hours as needed to suppress symptoms. Most patients are stabilized on a dose ranging from 20 to 40 mg orally daily, after which the total daily dose may be reduced in decrements of 5–10 mg daily. Buprenorphine may be given sublingually in an initial dose of 4–6 mg, with repeat doses as needed every 2 hours until symptoms are suppressed, a process that generally requires anywhere from 8 to 32 mg; once the patient is stabilized, the dose may be gradually tapered in daily decrements of 2–4 mg. Importantly, in the case of buprenorphine, treatment should not be initiated until withdrawal has unequivocally started. This is because buprenorphine, as a partial agonist, may precipitate withdrawal in patients addicted to other opioids.

Symptomatic treatment generally involves using either clonidine (Charney *et al.* 1981; Gold *et al.* 1978; Jasinski *et al.* 1985) or lofexidine (Beam *et al.* 1996; Carnwath and Hardman 1998; Lin *et al.* 1997; Yu *et al.* 2008). Although neither of these alpha-2 agonists are as effective as an opioid in suppressing withdrawal, they do help, especially with regard to insomnia, nausea and vomiting, and diarrhea. Although these two are roughly equivalent with regard to efficacy, clonidine is much more likely to cause hypotension. Once withdrawal begins, clonidine is given in a dose ranging from 0.1 to 0.3 mg, with repeat doses every 2–3 hours until maximum effect is obtained. The initial total dose is then given on a daily basis in four divided doses, with provision for further as-needed doses, after which the total daily dose is titrated up based on how much is given in as-needed doses until no further as-needed doses are required; most patients are stabilized by a total daily dose ranging from 0.6 to 2.4 mg. The patient is then generally 'covered' by this final dose for about a week to a week and a half, that is to say for the expected duration of the withdrawal, after which the dose may be tapered over 3 or 4 days and then discontinued. Lofexidine is generally started at a total daily dose of 0.4–0.6 mg, in two or three divided doses, with the dose gradually increased on a daily basis until symptoms are controlled, with most patients being stabilized at doses of 1.6–3.2 mg. As with clonidine, the patient is then covered for the expected duration of the withdrawal, after which the dose is tapered over 3 or 4 days. This tapering of clonidine and lofexidine is important in order to avoid rebound hypertension. Other symptomatic treatments to consider include the following: ibuprofen for bone and muscle pain; trazodone for insomnia; prochlorperazine or ondansetron for nausea and vomiting; dicyclomine for abdominal cramping; and either diphenoxylate or bismuth subsalicylate for diarrhea. Finally, consideration should also be given to two newly recognized symptomatic treatments, namely tramadol and buspirone. Tramadol, given its agonist effect at mu receptors, is a logical candidate, and a double-blinded study indicated that it was more effective than placebo (Lofwall *et al.* 2007). Tramadol may be given in doses of 200–400 mg daily, and once symptoms are controlled, the dose may be tapered over the following 4 or 5 days. Tramadol, however, is not without risk, as some patients, albeit rarely, may experience seizures, while those taking selective serotonin reuptake inhibitors (SSRIs) or other serotoninergic medications will be at risk for a serotonin syndrome. Buspirone, in a remarkable double-blind study (Buydens-Branchey *et al.* 2005), was as effective in opioid withdrawal as methadone. Patients were treated with a total dose of 30 or 45 mg of buspirone, and this was maintained throughout a 12-day withdrawal period.

'Cold turkey' withdrawal, as the name implies, involves going through withdrawal without any pharmacologic support at all.

Choosing among these treatment options for withdrawal is not entirely straightforward, with two exceptions. First, pregnant females should be withdrawn with opioids, as any of the other options risks fetal death, and indeed the tapering schedule should be prolonged. Second, in cases likely to progress to severe withdrawal, debilitated patients or those with significant cardiac or pulmonary disease should also be withdrawn with opioids, given that these patients are at risk for death. In other cases, however, any one of these options is potentially viable. Most patients facing severe withdrawal will opt for treatment with opioids. Symptomatic treatment, however, is a viable option, and indeed some patients, wishing never to take opioids again, may prefer it. If symptomatic treatment is decided on, in general one of the two alpha-2 agonists should be used, and a case could also be made for combining this with buspirone. The other symptomatic treatments may then be used on an as-needed basis. Finally, one may consider the oldest form of withdrawal, namely 'cold turkey'. Although this may be appropriate for some cases of mild, or even moderate, withdrawal, very, very few patients will choose this once severe withdrawal supervenes.

Before leaving this discussion of the treatment of withdrawal, some words are in order regarding what is known as 'ultra-rapid' detoxification. Although details differ among various programs, in most cases patients are first 'covered' with clonidine and an anti-emetic and then either sedated or anesthetized. In this state they are then given naloxone and naltrexone and a hyperacute withdrawal is produced. Consciousness is then allowed to return and patients are continued on the naltrexone and, temporarily, on the clonidine and anti-emetic. Although these programs are heavily advertised and have a certain appeal, several considerations should give one pause. First, upon regaining consciousness, patients are in withdrawal, and although the withdrawal may be brief, lasting only 3 or 4 days, it can be severe. Second, the procedure is not without risk, and deaths have occurred. Finally, there is no evidence that patients going through ultra-rapid withdrawal are any more likely to achieve long-term abstinence than those withdrawing with opioids or symptomatic care. With all these caveats in mind, the general conclusion has been that ultra-rapid withdrawal should not be recommended.

Overall, in some cases the goal is abstinence from all opioids, whereas in others the goal is to reduce the frequency of intoxication by providing maintenance treatment with either methadone or buprenorphine. Whichever goal is chosen, it is critical to provide ongoing psychosocial treatment, which may include drug counseling, motivational enhancement therapy, cognitive–behavioral therapy, or contingency management.

Abstinence, although very difficult to achieve, is clearly preferable, and patients desiring this option may, in addition to the psychosocial treatment just mentioned, also be referred to Narcotics Anonymous; many also require a lengthy inpatient stay, not only to get over withdrawal but also to begin their rehabilitation efforts. Consideration may also be given to treatment with naltrexone in those with normal hepatic function: this blocks opioid-induced euphoria, and thus effectively prevents the overwhelmingly reinforcing 'rush' that patients get if they 'slip'. Naltrexone may be given orally in a dose of 50 mg/day, or via a long-acting intramuscular form in a dose of 380 mg once monthly.

In the USA, methadone maintenance is available only in specially licensed facilities. In some cases indefinite maintenance is anticipated, and such patients are generally treated with doses of 80–120 mg of methadone daily. In other cases consideration is given to eventual abstinence, and here patients are initially maintained on doses ranging from 20 to 60 mg; once patients have consistently tested drug-free for a year or so, attempts are then made to reduce the dose of methadone in decrements of 5–10 percent of the total daily dose every week until the drug can be discontinued.

Buprenorphine maintenance, given the less stringent controls imposed on its use, is becoming popular, and is conducted by individual physicians on an outpatient basis utilizing the combination product containing buprenorphine and naloxone. Most patients are managed with doses ranging from 16 to 32 mg three times weekly. As with methadone, some patients are continued indefinitely on the drug, whereas in other cases attempts may be made to eventually taper it in weekly increments of 2 mg.

Maintenance with either methadone or buprenorphine clearly 'works' in that such maintained patients are less likely to become intoxicated with illicitly obtained opioids or to engage in the criminal behavior that often accompanies addiction. Both methadone and buprenorphine, however, as noted earlier, may produce intoxication, and supplies obtained through maintenance programs have been used for this purpose.

28.10 NICOTINE

Although nicotine is the substance in question, this chapter might just as well be entitled 'tobacco', for tobacco is the only 'vehicle' by which abuse or addiction to nicotine occurs. Of the various ways that tobacco is used, cigarette smoking is by far the most clinically important; although both chewing tobacco and smoking cigars or pipes can cause addiction, their numbers pale in comparison to cigarettes. With each cigarette smoked, anywhere from 1 to 3 mg of nicotine is delivered (Benowitz and Henningfield 1994).

Clinical features

Intoxication with nicotine is mild: especially with the first cigarette of the day, there is a sense of satisfaction and a reduction in any irritability. With this intoxication there may be a mild tachycardia, elevation of blood pressure, and an increase in peristaltic activity, and in some cases there may be palpitations; tobacco-naive patients may also experience nausea and vomiting. Overall, appetite decreases and frequent smokers may lose some weight.

Tolerance occurs rapidly and, by the end of a day spent smoking, there is little effect from a cigarette. Such tolerance, however, rapidly decreases, such that by the next day intoxication may again be achieved. Despite these daily fluctuations, however, a chronic tolerance does develop. Whereas a tobacco-naive patient may, as noted above, experience toxic nausea and vomiting with one cigarette, the nicotine addict may be able to smoke dozens of cigarettes a day without any immediate adverse effects.

Withdrawal symptoms (Hughes and Hatsukami 1986; Hughes *et al.* 1991a) may appear within anywhere from hours to a day of abstinence. There is a restless craving for a cigarette, and patients become tense and irritable. Headache, difficulty concentrating, and insomnia may occur, as may increased appetite with, in some, substantial weight gain. Withdrawal generally peaks within 1–3 days and then gradually subsides over a matter of 2–3 weeks; in some, however, mild withdrawal symptoms may persist for months, and in many cases the craving for a cigarette may recur intermittently for years, often at times of stress.

Other disorders may appear in association with tobacco use, and these occur not as a result of the effects of nicotine itself but rather of the by-products of tobacco. Smoked tobacco produces over 4000 different compounds, in both gaseous and particulate form. Gaseous components include carbon monoxide and hydrogen cyanide, whereas particulates contain a substance known as 'tar', which, for the most part, contains polycyclic aromatic hydrocarbons. With chronic smoking, patients are at risk for cancer of the mouth, larynx, and lung, chronic obstructive pulmonary disease (COPD), coronary artery disease, cerebrovascular disease, peripheral vascular disease, Raynaud's phenomenon, gingivitis, gastroesophageal reflux, cancer of the esophagus, and peptic ulcer disease. Smoking during pregnancy may cause spontaneous abortion, abruptio placentae, and low birth weight.

Smoking may also influence drug metabolism. Polycyclic aromatic hydrocarbons induce the metabolism of various drugs, lowering blood levels by 30–40 percent: affected drugs include haloperidol, clozapine, nortriptyline, valproic acid, and propranolol. Thus, if a smoker taking one of these drugs becomes abstinent, toxicity may occur, and vice versa.

Course

Recreational use of nicotine is rare, occurring in less than 5 percent of all smokers: those who start smoking usually either stop in short order or go on to fairly rapidly develop nicotine addiction (with craving, tolerance, and withdrawal) and persistent use, despite the development of one or more of the other disorders associated with tobacco use.

Etiology

Although the mechanisms that determine which patients will stop and which will go on to develop addiction are not clearly understood, heritability clearly plays a role, accounting for over 50 percent (Kendler *et al.* 2000).

Differential diagnosis

There is generally no diagnostic difficulty: in those who deny smoking, but who appear to be doing so, one may screen for cotinine, a metabolite of nicotine, in urine, saliva, or blood, with such screens remaining positive for up to 2 weeks.

Treatment

Patients should choose a 'quit date', which ideally should fall during a relatively stress-free time in the not-too-distant future. Patients should be instructed to stop smoking on that day and to avoid, if possible, situations or gatherings where smoking is likely to occur. One or more psychosocial measures should also be instituted. Brief, scheduled, telephone follow-ups are helpful, and the first one should be 2 or 3 days after the quit date, when withdrawal is peaking. Some patients may do well with this approach, supplemented with brief, and again scheduled, follow-up visits. Should patients resume smoking, however, cognitive–behavioral therapy should be considered. In addition to psychosocial measures, all patients should be offered pharmacologic treatment, which may include nicotine, varenicline, bupropion, and nortriptyline. Importantly, both bupropion and nortriptyline are effective regardless of whether or not patients are depressed.

Nicotine is available in patches and as a lozenge, gum, inhaler, and nasal spray. Patches typically contain 21, 14 and 7 mg of nicotine: the 21 mg size is intended for those who smoke roughly a pack a day of cigarettes. The patches are generally well tolerated; however, some patients, if they leave them on overnight, may experience insomnia and distressing dreams. Both the lozenge and gum are available in 2 and 4 mg sizes: the 2 mg size is intended for those smoking less than a pack a day, while those smoking more than this should use the 4 mg size, and patients, initially, should take from one to two pieces of lozenge or gum on an as-needed basis every hour. Between these two preparations, the lozenge is much easier to use. The inhaler is difficult to use, as one must puff on it for close to 20 minutes to absorb the same amount as is absorbed from a 2 mg-size lozenge or gum. The nasal spray delivers 0.5 mg to each nostril, and although it produces the highest levels of nicotine, it is irritating and most patients find it simply unacceptable. In practice, the best results are often obtained by using a patch supplemented by as-needed pieces of lozenge or gum. The overall duration of treatment with nicotine replacement therapy is not settled. Although initial recommendations were to continue treatment until the patient had been abstinent from tobacco for 2 weeks, and then gradually taper off the total nicotine dose over the following 4 weeks, an argument may clearly be made for much longer, even indefinite, treatment should patients not be able to tolerate the taper without 'slipping': it is the by-products of tobacco that are dangerous, not the nicotine itself.

Varenicline is a partial agonist at $\alpha_4\beta_2$ nicotinic acetylcholine receptors, and reduces both nicotine intoxication and craving. Treatment is begun 1 week before the quit date with 0.5 mg/day for 3 days, then 0.5 mg twice daily for 3 days, and finally 1 mg twice daily thereafter, with treatment continued for 3–12 months. Varenicline, although generally safe, is not without side-effects. Insomnia and distressing dreams are not uncommon. More importantly, varenicline may also, albeit rarely, cause assaultive behavior or homicidal ideation (Moore *et al.* 2010), suicide or suicidal ideation (Tonstad *et al.* 2010), mania in those with bipolar disorder (Kohen and Kremen 2007), and visual hallucinations (Raidoo and Kutscher 2009). Visual hallucinations have also occurred upon discontinuation of varenicline (Laine *et al.* 2009), as has delirium with prominent visual and auditory hallucinations (May and Rose 2010).

Bupropion is started 1 week before the quit date at 150 mg daily and increased 3 days later to 150 mg twice daily, and then continued for 3–12 months. Although, like varenicline, bupropion is also generally safe, it does lower the seizure threshold, and may induce seizures in patients with a history of bulimia nervosa.

Nortriptyline is started 1 week before the quit date at 25 mg daily, increased to 50 mg daily after 3 days, and then, if tolerated, to 75 mg 3 days later, after which it is continued for 3–12 months. Although studies measuring blood levels have not been performed, it would not be unreasonable to check a blood level after a week or so of the full dose and to make appropriate adjustments based on the results. Nortriptyline is likewise generally safe; however, anticholinergic side-effects not uncommonly make it unacceptable to patients.

All things considered, it is appropriate to offer either varenicline or nicotine replacement therapy, utilizing both a patch plus as-needed lozenge or gum to all patients. Between these two options, up until recently varenicline seemed preferable; however, the side-effects discussed earlier are making many clinicians lean towards nicotine replacement therapy. If nicotine replacement therapy is chosen, then consideration may also be given to enhancing the effect of nicotine by combining it with either bupropion (Jorenby *et al.* 1999) or nortriptyline (Prochazka *et al.* 2004), and between these two, bupropion is preferable, as it is superior to nortriptyline for smoking cessation (Wagena *et al.* 2005), has fewer side-effects, and may also prevent the weight gain often seen during withdrawal.

Without treatment, only 5–10 percent of patients will achieve 1 year of abstinence, whereas with treatment the figure rises to roughly 25 percent. Dismal as this figure may appear, it must be kept in mind that 'slips' are the rule, and indeed it typically takes five or six attempts before abstinence is achieved, with some 40–50 percent of smokers eventually reaching this goal.

28.11 CAFFEINE

Caffeine is the most commonly used substance in the Western world. A cup of coffee contains about 100 mg of caffeine, a cup of tea about 50 mg, and over-the-counter 'stimulant', analgesic, and 'cold' remedies may contain anywhere from 25 to 200 mg.

Clinical features

In caffeine-naive patients, about 100 mg of caffeine produces an increased sense of alertness and decreased fatigue. At doses ranging from 200 to 500 mg, however, intoxication (Hughes *et al.* 1991b; Rapoport *et al.* 1981) occurs, with anxiety, restlessness, jitteriness, tremulousness, muscle twitching, tachycardia, palpitations, and insomnia. In severe cases, with doses of roughly 1000 mg, there may be anxiety attacks, agitation, and, albeit very rarely, a delirium (Stillner *et al.* 1978) or a psychosis (Mackay and Rollins 1989). Finally, with clear overdoses of up to 5000 mg, seizures, ventricular arrhythmias, shock and death may ensue (Curatolo and Robertson 1983). Most intoxications clear within 6–12 hours.

In addition to these clear-cut episodes of intoxication, patients may also develop a chronic intoxication secondary to ongoing use of relatively low doses. Here, one finds chronic anxiety associated with a mild, fine, postural tremor and insomnia.

Tolerance to caffeine may appear after only 2 or 3 weeks' use of 300 mg or more on a daily basis (Evans and Griffiths 1992). In cases in which tolerance has occurred, withdrawal symptoms (Griffiths *et al.* 1990; Hughes *et al.* 1991b; Juliano and Griffiths 2004; Silverman *et al.* 1992; Strain *et al.* 1994) may appear anywhere from 12 to 24 hours after the last dose, with headache, poor concentration, drowsiness, fatigue, anxiety, irritability, and depressed mood. Withdrawal symptomatology tends to peak within a day or two, with symptoms then gradually resolving over days to a week or so.

Other disorders associated with chronic caffeine use include gastroesophageal reflux disease, peptic ulcer, fibrocystic disease, and hypertension; caffeine may also precipitate anxiety attacks seen in panic disorder (Boulenger *et al.* 1984; Charney *et al.* 1985).

Course

Recreational use of caffeine is extraordinarily common. Abusive use, that is continued use despite recurrent intoxication or the aggravation of some of the other disorders seen in association with caffeine use, is not uncommon. Whether or not true dependence actually occurs is a matter of some contention. The diagnosis certainly comes to mind in cases where there is tolerance and withdrawal associated with continued use despite adverse consequences and an inability to stop despite a desire to do so, and there are certainly a large number of cases which meet these criteria (Hughes *et al.* 1998; Strain *et al.* 1994). The contentious issue resolves around whether true craving actually develops, as this, for many, is a *sine qua non* for the diagnosis of dependence. It may well be that the continued use occurs not because of a craving, but rather to relieve withdrawal symptoms.

Etiology

The intoxicating effects of caffeine appear to be mediated by its competitive blockade of central nervous system adenosine receptors. There is also evidence that a dependence-like pattern of use is genetically determined (Kendler and Prescott 1999).

Differential diagnosis

Recurrent episodes of caffeine intoxication, when characterized by anxiety attacks, may suggest panic disorder; however, the history of excessive use of caffeine immediately indicates the correct diagnosis. Chronic, mild intoxication may suggest generalized anxiety disorder, akathisia, or hyperthyroidism. Here, abstinence from caffeine will make the diagnosis (Greden 1974).

Treatment

Severe cases of intoxication may require intensive supportive care. In dependence-like cases where patients are unable to stop 'cold turkey' the dose may be gradually decreased in daily increments equivalent to 10 percent of the average daily dose.

28.12 METHANOL

Methanol, also known as methyl alcohol or 'wood alcohol', is found in 'canned heat' preparations such as 'Sterno', in certain solvents and paint thinners, and as a 'denaturant' added to ethanol, which is then sold tax-free as 'denatured alcohol' for use as a cleaner. Methanol is at times used by desperate alcoholics when no other source of intoxication is at hand; methanol intoxication may also occur accidentally with inhalation of fumes, absorption through the skin, or inadvertent ingestion. This is an extremely toxic substance, and fatalities may occur with ingestion of only 30 cc^3.

Methanol is metabolized first via alcohol dehydrogenase to formaldehyde and then via aldehyde dehydrogenase to formic acid, which is the ultimate cause of the devastating sequelae of methanol intoxication: formic acid is not only directly toxic to neuronal mitochondria but also produces a severe systemic acidosis. Importantly, this is the same metabolic pathway used by ethanol, a fact of considerable importance regarding not only the evolution of methanol intoxication but also one of the traditional treatments of this intoxication.

Clinical features

Clinically (Bennet *et al.* 1953; Erlanson *et al.* 1965; Hovda *et al.* 2005; Kaplan 1962; Paasma *et al.* 2007; Wood and Buller 1904), the intoxication seen with methanol evolves in a biphasic fashion. The initial response, occurring in response to the methanol itself, is characterized by euphoria, headache, and nausea, all accompanied by an odor of alcohol on the breath. Subsequently, as formic acid begins to accumulate, the second phase emerges anywhere from hours to 2 days after ingestion. This second phase may be characterized by delirium, restlessness, dizziness, vomiting, and, classically, bilateral blurring or dimming of vision. With severe intoxication, seizures, respiratory depression, and coma may supervene. Methanol levels are generally above 20 mg/dL, and a metabolic acidosis is present with an increased anion gap reflecting the presence of formic acid. Untreated, methanol intoxication may be fatal in up to one-third of cases; those who survive generally recover from the intoxication in 1–3 days, but may be left with one or more sequelae, such as visual loss (Wood and Buller 1904), parkinsonism (Guggenheim *et al.* 1971; McLean *et al.* 1980), or dementia (McLean *et al.* 1980; Scrimgeour 1980). Magnetic resonance imaging may, on fluid-attenuated inversion recovery (FLAIR) sequences, reveal increased signal intensity in the putamen, thalamus and cerebral white matter, and with gradient echo recall sequences, hemorrhage may be found in the same areas.

Course

Repeated ingestion of methanol is uncommon, as most alcoholics learn their lesson.

Etiology

Secondary to the direct toxicity of formic acid, there is widespread necrosis and petechial hemorrhage involving the cortex and putamen (Erlanson *et al.* 1965; Karayel *et al.* 2010; McLean *et al.* 1980) and the retina and optic nerve (Benton and Calhoun 1953; Sharpe *et al.* 1982).

Differential diagnosis

Ethanol and isopropanol intoxication are distinguished by the absence of features such as visual loss and delirium, and ethylene glycol intoxication by the absence of an odor of alcohol.

A confusing diagnostic picture may emerge when both methanol and ethanol are consumed, as may occur when denatured alcohol is ingested. As noted earlier, ethanol inhibits the metabolism of methanol to formic acid, and hence the evolution of the second phase of methanol intoxication may be delayed until the ethanol is cleared, after which the remaining methanol is converted to formic acid.

Treatment

Methanol intoxication constitutes a medical emergency. If patients are seen within 2 hours of ingestion, gastric lavage may be performed. Throughout treatment one must monitor pH, bicarbonate levels, the anion gap, and methanol levels: initial methanol levels above 20 mg/dL are considered toxic, and levels above 50 mg/dL are potentially life threatening.

The cornerstone of treatment rests on delaying the transformation of methanol to formic acid. In the past this was accomplished by giving ethanol, which binds preferentially to the enzymes responsible for the metabolism of methanol to formic acid, thus preventing an overly rapid accumulation of formic acid. A more recent, and far better, option involves the administration of fomepizole, a drug that inhibits alcohol dehydrogenase (Brent *et al.* 2001). Folic acid hastens the excretion of formic acid and may be given in doses of 50 mg intravenously every 6 hours. In cases in which the foregoing measures are ineffective, hemodialysis may be utilized to remove formic acid.

28.13 ISOPROPANOL

Isopropanol, also known as isopropyl alcohol, is found in aftershave lotions, hand lotions, hair tonics, and in 'rubbing alcohol'. Among desperate alcoholics it may be known as 'rubby dubby', with reference to the source in rubbing alcohol, or as 'blue heaven', the latter name derived from the blue coloring often added to rubbing alcohol. In part, isopropyl alcohol is metabolized via alcohol dehydrogenase to acetone.

Clinical features

Intoxication (Lacouture *et al.* 1983; Rich *et al.* 1990) is characterized by mild euphoria, an odor of alcohol on the breath, dizziness, and ataxia, and may also be accompanied by headache, nausea, vomiting, and, in some cases, hematemesis. With high doses, severe intoxication may occur, with coma and respiratory depression. Isopropanol, as noted, is converted to acetone, leading to both acetonemia and acetonuria. The intoxication generally passes within 12 hours.

Tolerance may occur: whereas 250 mL may constitute a lethal dose in a non-tolerant patient, alcoholics may be able to consume this much with little untoward effect. Although withdrawal probably occurs, this has not been documented.

Course

Isopropanol is generally taken by alcoholics for want of something better and, once alcohol is available, its use is abandoned.

Etiology

The mechanism whereby isopropanol produces intoxication is probably similar to that for ethanol.

Differential diagnosis

The prominent nausea, with, in some cases, hematemesis, helps to distinguish isopropanol from ethanol or methanol intoxication; furthermore, the absence of dimming of vision argues against methanol intoxication. Ethylene glycol intoxication is suggested by an absence of the odor of alcohol on the breath.

Treatment

If patients are seen within 2 hours of ingestion, gastric lavage may be performed. Hemodialysis (Freireich *et al.* 1967) should be considered in cases of severe intoxication or when the isopropanol level is over 400 mg/dL.

REFERENCES

Abraham HD. Visual phenomenology of the LSD flashback. *Arch Gen Psychiatry* 1983; **40**:884–9.

Adachi J, Mizoi Y, Fukunaga T *et al.* Degrees of alcohol intoxication in 117 hospitalized cases. *J Stud Alcohol* 1991; **52**:448–53.

Addolorato G, Caputo F, Capristo E *et al.* Baclofen efficacy in reducing alcohol craving and intake: a preliminary double-blind randomized controlled study. *Alcohol Alcohol* 2002; **37**:504–8.

Addolorato G, Leggio L, Abenavoli L *et al.* Baclofen in the treatment of alcohol withdrawal symptoms: a comparative study vs diazepam. *Am J Med* 2006; **119**(276):e13–18.

Addolorato G, Leggio L, Ferruli A *et al.* Effectiveness and safety of baclofen for maintenance of alcohol abstinence in alcohol-dependent patients with liver cirrhosis: randomized, double-blind controlled study. *Lancet* 2007; **370**:1915–22.

Akiyama K. Longitudinal clinical course following pharmacological treatment of methamphetamine psychosis which persists after long-term abstinence. *Ann N Y Acad Sci* 2006; **1074**:125–34.

Alldredge BK, Lowenstein DH, Simon RP. Seizures associated with recreational drug abuse. *Neurology* 1989; **39**:1037–9.

Allen RM, Young SJ. Phencyclidine induced psychosis. *Am J Psychiatry* 1978; **135**:1081–4.

Allentuck S, Bowman SM. The psychiatric aspects of marijuana intoxication. *Am J Psychiatry* 1942; **99**:248–50.

Altenkirch H, Mager J, Stoltenburg G *et al.* Toxic polyneuropathies after sniffing a glue thinner. *J Neurol* 1977; **214**:137–52.

Angrist B, Gershon S. The phenomenology of experimentally induced amphetamine psychosis. *Biol Psychiatry* 1970; **2**:95–107.

Anton RF, O'Mally SS, Ciraulo DA *et al.* Combined pharmacotherapies and behavioural interventions for alcohol dependence: the COMBINE study: a randomized controlled trial. *JAMA* 2006; **295**:2003–17.

Ashton H. Benzodiazepine withdrawal: an unfinished story. *BMJ* 1984; **288**:1135–40.

Bach-y-Rita G, Lion JR, Climent CE *et al.* Episodic dyscontrol: a study of 130 violent patients. *Am J Psychiatry* 1971; **127**:1473–8.

Baltieri DA, Daro FR, Ribeiro PL *et al.* Comparing topiramate with naltrexone in the treatment of alcohol dependence. *Addiction* 2008; **103**:2035–44.

Banay RS. Pathologic reaction to alcohol. I. Review of the literature and original case reports. *Q J Stud Alcohol* 1944; **4**:580–605.

Bartlett E, Hallin A, Chapman B *et al.* Selective sensitization to the psychosis-inducing effects of cocaine: a possible marker for addiction relapse vulnerability? *Neuropsychopharmacology* 1997; **16**:77–82.

Beam J, Gossop M, Strang J. Randomised double-blind comparison of lofexidine and methadone in the in-patient treatment of opiate withdrawal. *Drug Alcohol Depend* 1996; **43**:87–91.

Bell DS. The experimental reproduction of amphetamine psychosis. *Arch Gen Psychiatry* 1973; **29**:35–40.

Bennet IL, Cary FM, Mitchell GL *et al.* Acute methyl alcohol poisoning: a review based on experience in an outbreak of 323 cases. *Medicine* 1953; **32**:431–63.

Benowitz NL, Henningfield JE. Establishing a nicotine threshold for addiction. The implications for tobacco regulation. *N Engl J Med* 1994; **331**:123–5.

Benton ED, Calhoun FP. The ocular effects of methyl alcohol poisoning. Report of a catastrophe involving 320 persons. *Arch Ophthalmol* 1953; **36**:1677–85.

Bercel NA, Travis LE, Olinger LB *et al.* Model psychoses induced by LSD-25 in normals. I. Psychophysiological investigations, with special reference to the mechanism of the paranoid reaction. *Arch Neurol Psychiatry* 1956; **75**:588–611.

Bolla KI, McCann UD, Ricaurte GA. Memory impairment in abstinent MDMA ('Ecstasy') users. *Neurology* 1998; **51**:1532–7.

Bonnet U, Banger M, Leweke FM *et al.* Treatment of acute alcohol withdrawal with gabapentin: results from a controlled two-center trial. *J Clin Psychopharmacol* 2003; **23**:514–19.

Booth BM, Blow FC. The kindling hypothesis: further evidence from a U.S. national study of alcoholic men. *Alcohol Alcohol* 1993; **28**:593–8.

Bostwick DG. Amphetamine induced cerebral vasculitis. *Hum Pathol* 1981; **12**:1031–3.

Boulenger JP, Uhde TW, Wolff EA *et al.* Increased sensitivity to caffeine in patients with panic disorder: preliminary evidence. *Arch Gen Psychiatry* 1984; **41**:1067–71.

Brady KT, Lydiard RB, Malcolm R *et al.* Cocaine-induced psychosis. *J Clin Psychiatry* 1991; **52**:509–12.

Breier A, Charney DS, Nelson JC. Seizures induced by abrupt discontinuation of alprazolam. *Am J Psychiatry* 1984; **141**:1606–7.

Brent J, McMartin K, Phillips S *et al.* Fomepizole for the treatment of methanol poisoning. *N Engl J Med* 2001; **344**:424–9.

Brogden RN, Goa KL. Flumazenil. A reappraisal of its pharmacological properties and therapeutic efficacy as a benzodiazepine antagonist. *Drugs* 1991; **42**:1061–89.

Bromberg W. Marijuana intoxication. *Am J Psychiatry* 1934; **91**:303–30.

Brower KJ, Myra Kim H, Strobbe S *et al.* A randomized double-blind pilot trial of gabapentin versus placebo to treat alcohol dependence and comorbid insomnia. *Alcohol Clin Exp Res* 2008; **32**:1429–38.

Brown SA, Inaba RK, Gillin JC *et al.* Alcoholism and affective disorder: clinical course of depressive symptoms. *Am J Psychiatry* 1995; **152**:45–52.

Budney AJ, Vandrey RG, Hughes JR *et al.* Oral delta-9-tetrahydrocannibinol suppresses cannabis withdrawal symptoms. *Drug Alcohol Depend* 2007; **86**:22–9.

Burke WM, Ravi NV, Dhopesh V *et al.* Prolonged presence of metabolite in urine after compulsive cocaine use. *J Clin Psychiatry* 1990; **51**:145–8.

Busto U, Sellers EM, Naranjo CA *et al.* Withdrawal reactions after long-term therapeutic use of benzodiazepines. *N Engl J Med* 1986; **315**:854–7.

Buydens-Branchey L, Brancey M, Reel-Brander C. Efficacy of buspirone in the treatment of opioid withdrawal. *J Clin Psychopharmacol* 2005; **25**:230–6.

Carnwath T, Hardman J. Randomised double-blind comparison of lofexidine and clonidine in the out-patient treatment of opiate withdrawal. *Drug Alcohol Depend* 1998; **50**:251–4.

Cavallaro R, Regazzetti MG, Covelli G *et al.* Tolerance and withdrawal with zolpidem. *Lancet* 1993; **342**:374–5.

Challoner KR, McCarron MM, Newton EJ. Pentazocine (Talwin) intoxication: report of 57 cases. *J Emerg Med* 1990; **8**:67–74.

Charney DS, Sternberg DE, Kleber HD *et al.* The clinical use of clonidine in abrupt withdrawal from methadone: effects on blood pressure and specific signs and symptoms. *Arch Gen Psychiatry* 1981; **38**:1273–7.

Charney DS, Heninger GR, Jatlow PI. Increased anxiogenic effects of caffeine in panic disorders. *Arch Gen Psychiatry* 1985; **42**:233–43.

Chopra GS, Smith JW. Psychotic reactions following marijuana use in East Indians. *Arch Gen Psychiatry* 1974; **30**:24–7.

Choy-Kwong M, Lipton RB. Dystonia related to cocaine withdrawal: a case report and pathogenic hypothesis. *Neurology* 1989; **39**:996–7.

Clark LD, Nakashima EN. Experimental studies of marijuana. *Am J Psychiatry* 1968; **125**:379–84.

Clark LD, Hughes R, Nakashima EN. Behavioral effects of marijuana. *Arch Gen Psychiatry* 1970; **23**:193–8.

Cloninger CR, Bohman M, Sigvardsson S *et al.* Psychopathology in adopted-out children of alcoholics. The Stockholm Adoption Study. *Recent Dev Alcohol* 1985; **3**:37–51.

Coid J. Mania a potu: a critical review of pathological intoxication. *Psychol Med* 1979; **9**:709–19.

Covi L, Lipman RS, Pattison JH *et al.* Length of treatment with anxiolytic sedatives and response to their sudden withdrawal. *Acta Psychiatr Scand* 1973; **49**:51.

Cubala WJ, Landowski J. Seizure following sudden zolpidem withdrawal. *Prog Neuropsychopharmacol Biol Psychiatry* 2007; **31**:539–40.

Curatolo PW, Robertson D. The health consequences of caffeine. *Ann Intern Med* 1983; **98**:641–53.

Curran FJ. The symptoms and treatment of barbiturate intoxication and psychosis. *Am J Psychiatry* 1938; **95**:73–85.

Curran FJ. Current views on the neuropsychiatric effects of barbiturates and bromides. *J Nerv Ment Dis* 1944; **100**:142–69.

Daeppen JB, Gache P, Landry U *et al.* Symptom-triggered vs fixed-schedule doses of benzodiazepines for alcohol withdrawal: a randomized treatment trial. *Arch Int Med* 2002; **162**:1117–21.

Daras M, Tuchman AJ, Marks S. Central nervous system infarction related to cocaine abuse. *Stroke* 1991; **22**:1320–5.

Daras M, Koppel BS, Atos-Radzion E. Cocaine-induced choreoathetoid movements ('crack dancing'). *Neurology* 1994; **44**:751–2.

Davidson KM. Diagnosis of depression in alcohol dependence: changes in prevalence with drinking status. *Br J Psychiatry* 1995; **166**:199–204.

Deleau D, Hanssens Y. Cerebellar dysfunction in chronic toluene abuse: beneficial response to amantadine hydrochloride. *J Toxicol Clin Toxicol* 2000; **38**:37–41.

Demirkiran M, Jankovic J, Dean JM. Ecstasy intoxication: an overlap between serotonin syndrome and neuroleptic malignant syndrome. *Clin Neuropharmacol* 1996; **19**:157–64.

De Moura MC, Correla JP, Madeira F. Clinical alcohol hypoglycemia. *Ann Int Med* 1967; **66**:893–905.

Derlet RW, Rice P, Horowitz BZ *et al.* Amphetamine toxicity: experience with 127 cases. *J Emerg Med* 1989; **7**:157–61.

D'Onofrio G, Rathlev NK, Ulrich AS *et al.* Lorazepam for the prevention of recurrent seizures related to alcohol. *N Engl J Med* 1999; **340**:915–19.

Drevets WC, Gautier C, Price JC *et al.* Amphetamine-induced dopamine release in human ventral striatum correlates with euphoria. *Biol Psychiatry* 2001; **49**:81–96.

D'Souza DC, Ranganathan M, Braley G *et al.* Blunted psychotomimetic and amnestic effects of delta-9-tetrahydrocannabinol in frequent uses of cannabis. *Neuropsychopharmacology* 2008; **33**:2505–16.

Duffy A, Millin R. Case study: withdrawal syndrome in adolescent chronic cannabis users. *J Am Acad Child Adol Psychiatry* 1996; **35**:1618–21.

Eastman JW, Cohen SN. Hypertensive crisis and death associated with phencyclidine poisoning. *JAMA* 1975; **231**:1270–1.

Ellinwood EH. Amphetamine psychosis. I. Description of the individuals and the process. *J Nerv Ment Dis* 1967; **144**:273–83.

Ellinwood EH. Assault and homicide associated with amphetamine abuse. *Am J Psychiatry* 1971; **127**:1170–5.

Erlanson P, Fritz H, Hagstam KE *et al.* Severe methanol intoxication. *Arch Med Scand* 1965; **177**:393–408.

Evans SM, Griffiths RR. Caffeine tolerance and choice in humans. *Psychopharmacology* 1992; **108**:51–9.

Evans AC, Raistrick D. Phenomenology of intoxication with toluene-based adhesives and butane gas. *Br J Psychiatry* 1987; **150**:769–73.

Ewing JA, Elliot WJ, Maio LD *et al.* You don't have to be a neuroscientist to forget everything with triazolam but it helps. *JAMA* 1988; **259**:350–2.

Fariello RG, Black JA. Pseudoperiodic bilateral EEG paroxysms in a case of phencyclidine intoxication. *J Clin Psychiatry* 1978; **39**:579–81.

Fischman MW, Schuster CR, Resnekov L *et al.* Cardiovascular and subjective effects of intravenous cocaine administration in humans. *Arch Gen Psychiatry* 1976; **33**:983–9.

Fraser HF, Wiler A, Essig CF *et al.* Degree of physical dependence induced by secobarbital or pentobarbital. *JAMA* 1958; **166**:126–9.

Fredericks EJ, Lazor MZ. Recurrent hypoglycemia associated with acute alcoholism. *Ann Int Med* 1963; **59**:90–4.

Fredericks RK, Lefkowitz DS, Challa VR *et al.* Cerebral vasculitis associated with cocaine abuse. *Stroke* 1991; **22**:1437–9.

Freedman DX. On the use and abuse of LSD. *Arch Gen Psychiatry* 1968; **18**:330–47.

Freireich AW, Cinqu TJ, Xanthaky G *et al.* Hemodialysis for isopropanol poisoning. *N Engl J Med* 1967; **277**:699.

Frezza M, di Padova C, Pozzato G *et al.* High blood alcohol levels in women: the role of decreased gastric alcohol dehydrogenase activity and first-pass metabolism. *N Engl J Med* 1990; **322**:95–9.

Frosch WA, Robbins ES, Stern M. Untoward reactions to lysergic acid diethylamide (LSD) resulting in hospitalization. *N Engl J Med* 1965; **273**:1235–9.

Furieri FA, Nakamura-Palacios EM. Gabapentin reduces alcohol consumption and craving: a randomized, double-blind, placebo-controlled trial. *J Clin Psychiatry* 2007; **68**:1691–700.

Gacouin AA, Lavoue S, Signouret T *et al.* Intensive spongiform leukoencephalopathy after inhalation of heated heroin. *Intensive Care Med* 2003; **29**:1012–15.

Garbutt JC, Kranzler HR, O'Malley SS *et al.* Efficacy and tolerability of long-acting injectable naltrexone for alcohol dependence: a randomized controlled trial. *JAMA* 2005; **293**:1617–25.

Giannini AJ, Eighan MS, Loiselle RH *et al.* Comparison of haloperidol and chlorpromazine in the treatment of phencyclidine psychosis. *J Clin Pharmacol* 1984; **24**:202–4.

Giannini AJ, Loiselle RH, DiMarzio LR *et al.* Augmentation of haloperidol by ascorbic acid in phencyclidine intoxication. *Am J Psychiatry* 1987; **144**:1207–9.

Gold MS, Redmond DE, Kleber HD. Clonidine blocks acute opiate withdrawal symptoms. *Lancet* 1978; **2**:599.

Goodwin DW. Two species of alcoholic 'blackout'. *Am J Psychiatry* 1971; **127**:1665–70.

Goodwin DW, Crane JB, Guze SB. Alcoholic 'blackouts': a review and clinical study of 100 alcoholics. *Am J Psychiatry* 1969a; **126**:191–8.

Goodwin DW, Crane JB, Guze SB. Phenomenological aspects of the alcoholic 'blackout'. *Br J Psychiatry* 1969b; **115**:1033–8.

Gouzoulis-Mayfrank E, Daumann J, Tuchtenhagen F *et al.* Impaired cognitive performance in drug free users of recreational ecstasy (MDMA). *J Neurol Neurosurg Psychiatry* 2000; **68**:719–25.

Greden JF. Anxiety or caffeinism: a diagnostic dilemma. *Am J Psychiatry* 1974; **131**:1089–92.

Greenblatt DJ, Harmatz JS, Shapiro L *et al.* Sensitivity to triazolam in the elderly. *N Engl J Med* 1991; **324**:1691–8.

Griffith JD, Cavanaugh J, Held J *et al.* Dextroamphetamine: evaluation of psychotomimetic properties in man. *Arch Gen Psychiatry* 1972; **26**:97–100.

Griffiths RR, Evans SM, Heishman SJ *et al.* Low-dose caffeine physical dependence in humans. *J Pharmacol Exp Ther* 1990; **255**:1123–32.

Guggenheim MA, Courch JR, Weinberg W. Motor dysfunction as a permanent complication of methanol intoxication: presentation of a case with a beneficial response to levodopa treatment. *Arch Neurol* 1971; **24**:550–4.

Gupta RK, van der Meulen J, Johny KV. Oliguric acute renal failure due to glue-sniffing. Case report. *Scand J Urol Nephrol* 1991; **25**:247–50.

Haney M, Ward AS, Comer SD *et al.* Abstinence symptoms following oral THC administration to humans. *Psychopharmacology* 1999; **141**:385–94.

Harrington H, Heller HA, Dawson D *et al.* Intracerebral hemorrhage and oral amphetamine. *Arch Neurol* 1983; **40**:503–7.

Hartung TK, Schofield E, Short AI *et al.* Hyponatremic states following 3,4-methylenedioxymethamphetamine (MDMA, 'ecstasy') ingestion. *Q J Med* 2002; **95**:431–7.

Heritch AJ, Capwell R, Roy-Byrne PR. A case of psychosis and delirium following withdrawal from triazolam. *J Clin Psychiatry* 1987; **48**:168–9.

Hollister LE, Gillespie HK. Marijuana, ethanol, and dextroamphetamine: mood and mental function alterations. *Arch Gen Psychiatry* 1970; **23**:199–203.

Hollister LE, Prusmack JJ, Paulsen JAS *et al.* Comparison of three psychotropic drugs (psilocybin, JB-329, IT-290) in volunteer subjects. *J Nerv Ment Dis* 1960; **131**:428–34.

Hollister LE, Motzenbecker FP, Degan RO. Withdrawal reactions from chlordiazepoxide (Librium). *Psychopharmacology* 1961; **2**:63–8.

Hopfer CJ, Young SE, Purcell S *et al.* Cannabis receptor haplotype associated with fewer cannabis dependence symptoms in adolescents. *Am J Genet B Neuropsychiatr Genet* 2006; **141**:895–901.

Horowitz MJ. Flashbacks: recurrent intrusive images after the use of LSD. *Am J Psychiatry* 1969; **126**:565–9.

Hovda KE, Hunderi OH, Tajjord AB *et al.* Methanol outbreak in Norway: epidemiology, clinical features and prognostic signs. *J Intern Med* 2005; **258**:181–90.

Hsue PY, McManus D, Selby V *et al.* Cardiac arrest in patients who smoke crack cocaine. *Am J Cardiol* 2007; **99**:822–4.

Hughes JR, Hatsukami D. Signs and symptoms of tobacco withdrawal. *Arch Gen Psychiatry* 1986; **43**:289–94.

Hughes JR, Gust SW, Skoog K *et al.* Symptoms of tobacco withdrawal: a replication and extension. *Arch Gen Psychiatry* 1991a; **48**:52–9.

Hughes JR, Higgins ST, Bickel WK *et al.* Caffeine self-administration, withdrawal, and adverse effects among coffee drinkers. *Arch Gen Psychiatry* 1991b; **48**:611–17.

Hughes JC, McCabe M, Evans RJ. Intracranial hemorrhage associated with ingestion of 'ecstasy'. *Arch Emerg Med* 1993; **10**:372–4.

Hughes JR, Oliveto AH, Liguori A *et al.* Endorsement of DSM-IV dependence criteria among caffeine users. *Drug Alcohol Depend* 1998; **52**:99–107.

Isbell H, Altschul S, Kornetsky CH *et al.* Chronic barbiturate intoxication. *Arch Neurol Psychiatry* 1950; **64**:1–28.

Isbell H, Fraser HF, Wickler A *et al.* An experimental study of etiology of 'rum fits' and delirium tremens. *Q J Stud Alcohol* 1955; **16**:1–33.

Isbell H, Belleville RE, Fraser HF *et al.* Studies on lysergic acid diethylamide (LSD-25). I. Effects in former morphine addicts and development of tolerance during chronic intoxication. *Arch Neurol Psychiatry* 1956; **76**:468–78.

Isbell H, Gorodetzky CW, Jasinski D *et al.* Effects of (–) 9-trans-tetrahydrocannabinol in man. *Psychopharmacologia* 1967; **11**:184–8.

Iwanami A, Sugiyama A, Kuroki N *et al.* Patients with methamphetamine psychosis admitted to a psychiatric hospital in Japan. *Acta Psychiatr Scand* 1994; **89**:428–32.

Janowsky DS, Risch C. Amphetamine psychosis and psychotic symptoms. *Psychopharmacology* 1979; **65**:73–7.

Jasinski DR, Johnson RE, Kocher TR. Clonidine in morphine withdrawal: differential effects on signs and symptoms. *Arch Gen Psychiatry* 1985; **42**:1063–6.

Javitt DC, Zukin SR. Recent advances in the phencyclidine model of schizophrenia. *Am J Psychiatry* 1991; **148**:1301–8.

Johnson BA, Ait-Daoud N, Bowden CL *et al.* Oral topiramate for treatment of alcohol dependence: a randomized controlled trial. *Lancet* 2003; **361**:1677–85.

Johnson BA, Rosenthal N, Capece JA *et al.* Topiramate for treating alcohol dependence; a randomized controlled trial. *JAMA* 2007; **298**:1641–51.

Jorenby DE, Leischow SJ, Nides MA *et al.* A controlled trial of sustained-release bupropion, a nicotine patch, or both for smoking cessation. *N Engl J Med* 1999; **340**:685–91.

Juergens SM, Morse RM. Alprazolam dependence in seven patients. *Am J Psychiatry* 1988; **145**:625–7.

Juliano LM, Griffiths RR. A critical review of caffeine withdrawal: empirical validation of symptoms and signs, incidence, severity and associated features. *Psychopharmacology* 2004; **176**:1–29.

Kaiko RF, Foley KM, Grabinski PY *et al.* Central nervous system excitatory effects of meperidine in cancer patients. *Ann Neurol* 1983; **13**:180–5.

Kaplan K. Methyl alcohol poisoning. *Am J Med Sci* 1962; **244**:170–4.

Karayel F, Turan AA, Sav A *et al.* Methanol intoxication: pathological changes of central nervous system (17 cases). *Am J Forensic Med Pathol* 2010; **31**:34–6.

Kendler KS, Prescott CA, Neale MC *et al.* Temperance board registration for alcohol abuse in a national sample of Swedish male twins, born 1902 to 1949. *Arch Gen Psychiatry* 1997; **54**:176–84.

Kendler KS, Prescott CA. Caffeine intake, tolerance, and withdrawal in women: a population-based twin study. *Am J Psychiatry* 1999; **156**:223–8.

Kendler KS, Thornton LM, Pedersen NL. Tobacco consumption in Swedish twins reared apart and reared together. *Arch Gen Psychiatry* 2000; **57**:886–92.

King GS, Smialek JE, Trouthman WG. Sudden deaths in adolescents resulting from inhalation of typewriter correction fluid. *JAMA* 1985; **253**:1604–9.

King PJ, Morris JG, Pollard JD. Glue sniffing neuropathy. *Aust N Z J Med* 1985b; **15**:293–9.

Kish SJ, Lerch J, Furukawa Y *et al.* Decreased cerebral cortical serotonin transporter binding in ecstasy users: a positron emission tomography/[(11)C]DASB and structural brain imaging study. *Brain* 2010; **133**:1779–97.

Kleber HD, Gawin FH. The spectrum of cocaine abuse and its treatment. *J Clin Psychiatry* 1984; **45**(Suppl. 2):18–23.

Klonoff H, Low M, Marcus A. Neuropsychological effects of marijuana. *Can Med Assoc J* 1973; **108**:150–7.

Kohen I, Kremen N. Varenicline-induced manic episode in a patient with bipolar disorder. *Am J Psychiatry* 2007; **164**:1269–70.

Koppel BS, Samkoff L, Daras M. Relation of cocaine use to seizures and epilepsy. *Epilepsia* 1996; **37**:875–8.

Kramer JC, Fischman VS, Littlefield DC. Amphetamine abuse patterns and effects of high doses taken intravenously. *JAMA* 1967; **201**:305–9.

Kranzler HR, Gage A. Acamprosate efficacy in alcohol-dependent patients: summary of results from three pivotal trials. *Am J Addict* 2008; **17**:70–6.

Krendel DA, Ditter SM, Frankel MR. Biopsy-proven cerebral vasculitis associated with cocaine abuse. *Neurology* 1990; **40**:1092–4.

Kriegstein AR, Shungu DC, Millar WS *et al.* Leukoencephalopathy and raised brain lactate from heroin vapor inhalation ('chasing the dragon'). *Neurology* 1999; **53**:1765–73.

Kroll P. Psychoses associated with marijuana abuse in Thailand. *J Nerv Ment Dis* 1975; **161**:149–56.

Krystal JH, Karper LP, Selbyl JP *et al.* Subanesthetic effects of the noncompetitive NMDA antagonist, Ketamine, in humans. Psychotomimetic, perceptual, cognitive, and neuroendocrine responses. *Arch Gen Psychiatry* 1994; **51**:199–214.

Kuramochi H, Takahashi R. Psychopathology of LSD intoxication: study of experimental psychosis induced by LSD-25: description of LSD symptoms in normal oriental subjects. *Arch Gen Psychiatry* 1964; **11**:151–61.

Krupitsky EM, Rudenko AA, Burakov AM *et al.* Antiglutaminergic strategies for ethanol detoxification: comparison with placebo and diazepam. *Alcohol Clin Exp Res* 2007; **31**:604–11.

Lacouture PG, Watson S, Abrams A *et al.* Acute isopropyl alcohol intoxication: diagnosis and management. *Am J Med* 1983; **75**:680–4.

Lahmeyer HW, Stock PG. Phencyclidine intoxication, physical restraint and acute renal failure. *J Clin Psychiatry* 1983; **44**:184–5.

Laine P, Marttila J, Lindeman S. Hallucinations in the context of varenicline withdrawal. *Am J Psychiatry* 2009; **166**:619–20.

Langston JW, Ballard P, Tetrud JW *et al.* Chronic parkinsonism in humans due to a product of meperidine-analog synthesis. *Science* 1983; **219**:979–80.

Lee GY, Gong GW, Vrodos N *et al.* 'Ecstasy'-induced subarachnoid haemorrhage: an under-reported neurological complication? *J Clin Neurosci* 2003; **10**:705–7.

Lerner AG, Gelkopf M, Skladman I *et al.* Clonazepam treatment of lysergic acid diethylamide-induced hallucinogen persisting perception disorder with anxiety features. *Int Clin Psychopharmacol* 2003; **18**:101–5.

Levine SR, Brust JC, Futrell N *et al.* Cerebrovascular complications of the use of the 'crack' form of alkaloidal cocaine. *N Engl J Med* 1990; **323**:699–704.

Levy AB. Delirium and seizures due to abrupt alprazolam withdrawal: case report. *J Clin Psychiatry* 1984; **45**:38–9.

Lichtenfeld J, Rubin DB, Feldman RS. Subarachnoid hemorrhage precipitated by cocaine snorting. *Arch Neurol* 1986; **41**:223–4.

Lin SK, Strang J, Su LW *et al.* Double-blind randomised controlled trial of lofexidine versus clonidine in the treatment of heroin withdrawal. *Drug Alcohol Depend* 1997; **48**:127–33.

Lofwall MR, Walsh SL, Bigelow GE *et al.* Modest opioid withdrawal suppression efficacy of oral tramadol in humans. *Psychopharmacology* 2007; **194**:381–93.

Luby ED, Cohen BD, Rosenbaum G *et al.* Study of a new schizophrenomimetic drug – Sernyl. *Arch Neurol Psychiatry* 1959; **81**:363–9.

Lundh H, Tunving K. An extrapyramidal choreiform syndrome caused by amphetamine addiction. *J Neurol Neurosurg Psychiatry* 1981; **44**:728–30.

Lundquist G. Delirium tremens: a comparative study of pathogenesis, course and prognosis. *Acta Psychiatr Neurol Scand* 1961; **36**:443–66.

Lynskey MT, Heath AC, Nelson EC *et al.* Genetic and environmental contributions to cannabis dependence in a national young adult twin sample. *Psychol Med* 2002; **32**:195–207.

McCarron MM, Schulze BW, Thompson GA *et al.* Acute phencyclidine intoxication: clinical patterns, complications, and treatment. *Ann Emerg Med* 1981a; **10**:290–7.

McCarron MM, Schulze BW, Thompson GA *et al.* Acute phencyclidine intoxication: incidence of clinical findings in 1,000 cases. *Ann Emerg Med* 1981b; **10**:237–42.

McGrath J, Welham J, Scott J *et al.* Association between cannabis use and psychosis-related outcomes using sibling pair analysis in a cohort to young adults. *Arch Gen Psychiatry* 2010; **67**:440–7.

McGuire PK, Cope H, Fahy TA. Diversity of psychopathology associated with use of 3,4-methylenedioxyamphetamine ('Ecstasy'). *Br J Psychiatry* 1994; **165**:391–5.

McLean DR, Jacobs H, Mielke BW. Methanol poisoning: a clinical and pathological study. *Ann Neurol* 1980; **8**:161–7.

Mackay DC, Rollins JW. Caffeine and caffeinism. *J R Nav Med Serv* 1989; **75**:65–7.

Maletzky BM. The diagnosis of pathological intoxication. *Q J Stud Alcohol* 1976; **37**:1215–28.

Malcolm R, Ballenger JC, Sturgis ET *et al.* Double-blind controlled trial comparing carbamazepine to oxazepam treatment of alcohol withdrawal. *Am J Psychiatry* 1989; **146**:617–21.

Manchanda S, Connolly MJ. Cerebral infarction in association with Ecstasy abuse. *Postgrad Med J* 1993; **69**:874–5.

Manschreck TC, Allen DF, Neville M. Freebase psychosis: cases from a Bahamian epidemic of cocaine abuse. *Compr Psychiatry* 1987; **28**:555–64.

Martell BA, Oson FM, Poling J *et al.* Cocaine vaccination for the treatment of cocaine dependence in methadone-maintained patients: a randomized, double-blind, placebo-controlled efficacy trial. *Arch Gen Psychiatry* 2009; **66**:1116–23.

Martin WA, Jasinski DR. Physiological parameters of morphine dependence in man – tolerance, early abstinence and protracted abstinence. *J Psychiatr Res* 1969; **7**:7–9.

Martinez-Cano H, Vela-Bueon A, de Iceta M *et al.* Benzodiazepine withdrawal syndrome seizures. *Pharmacopsychiatry* 1995; **28**:257–62.

Mathers DC, Ghodse AH. Cannabis and psychotic illness. *Br J Psychiatry* 1992; **161**:648–53.

Mathew RJ, Wilson WH, Humphreys D *et al.* Depersonalization after marijuana smoking. *Biol Psychiatry* 1993; **33**:431–41.

May AC, Rose D. Varenicline withdrawal-induced delirium with psychosis. *Am J Psychiatry* 2010; **167**:720–1.

May PRA, Ebaugh FG. Pathological intoxication, alcoholic hallucinosis, and other reactions to alcohol: a clinical study. *Q J Stud Alcohol* 1953; **14**:200–27.

Melges FT. Tracking difficulties and paranoid ideation during hashish and alcohol intoxication. *Am J Psychiatry* 1976; **133**:1024–8.

Melges FT, Tinklenberg JR, Hollister LE *et al.* Temporal disintegration and depersonalization during marijuana intoxication. *Arch Gen Psychiatry* 1970; **23**:204–10.

Mendelson JH, Mello NK, Lex BW *et al.* Marijuana withdrawal syndrome in a woman. *Am J Psychiatry* 1984; **141**:1289–90.

Meyer JS, Greifenstein F, Devault M. A new drug causing symptoms of sensory deprivation. *J Nerv Ment Dis* 1959; **129**:54–61.

Minion GE, Slovid CM, Boutiette L. Severe alcohol intoxication: a study of 204 consecutive patients. *Clin Toxicol* 1989; **27**:375–84.

Moore TJ, Glenmullen J, Furberg CD. Thoughts and acts of aggression/violence toward others reported in association with varenicline. *Ann Pharmacother* 2010; **44**:1389–94.

Morley KC, Teesson M, Reid SC *et al.* Naltrexone versus acamprosate in the treatment of alcohol dependence: a multi-centre, randomized, double-blind, placebo-controlled trial. *Addiction* 2006; **101**:1451–62.

Morris HH, Estes ML. Traveler's amnesia: transient global amnesia due to triazolam. *JAMA* 1987; **258**:945–6.

Murphy SM, Owen RT, Tyrer PJ. Withdrawal symptoms after six weeks' treatment with diazepam. *Lancet* 1984; **2**:1389.

Murphy SM, Owen RT, Tyrer PJ. Comparative assessment of efficacy and withdrawal symptoms after 6 and 12 weeks' treatment with diazepam or buspirone. *Br J Psychiatry* 1989; **154**:529–34.

Myrick H, Malcolm R, Randall PK *et al.* A double-blind trial of gabapentin versus lorazepam in the treatment of alcohol withdrawal. *Alcohol Clin Exp Res* 2009; **33**:1582–8.

Nielsen J. Delirium tremens in Copenhagen. *Acta Psychiatr Scand* 1965; **41**(Suppl):1–92.

Nolte KB, Brass LM, Fletterick CF. Intracranial hemorrhage associated with cocaine abuse: a prospective autopsy study. *Neurology* 1996; **46**:1291–6.

Noyes R, Perry PJ, Crowe RR *et al.* Seizures following withdrawal of alprazolam. *J Nerv Ment Dis* 1986; **174**:50–2.

Nurnberger JL, Simmons-Aling S, Kessler L *et al.* Separate mechanisms for behavioral, cardiovascular, and hormonal responses to dextroamphetamine in man. *Psychopharmacology* 1984; **84**:200–4.

Paasma R, Hovda KE, Tikkerberi A *et al.* Methanol mass poisoning in Estonia: outbreak in 154 patients. *Clin Toxicol* 2007; **45**:152–7.

Palmstierna T. A model for predicting alchol withdrawal delirium. *Psychiatr Serv* 2001; **52**:820–3.

Palsson A, Thulin SO, Tunving K. Cannabis psychosis in south Sweden. *Acta Psychiatr Scand* 1982; **66**:311–21.

Pascual-Leone A, Dhuna A. Cocaine-associated multifocal tics. *Neurology* 1990; **40**:999–1000.

Pearlson GD. Psychiatric and medical syndromes associated with phencyclidine (PCP) abuse. *Johns Hopkins Med J* 1981; **148**:25–33.

Perr IN. Pathological intoxication and alcohol idiosyncratic intoxication. I. Diagnostic and clinical aspects. *J Forensic Sci* 1986; **31**:806–11.

Prochazka AV, Kick S, Steinbrunn C *et al.* A randomized trial of nortriptyline combined with transdermal nicotine for smoking cessation. *Arch Int Med* 2004; **164**:2229–33.

Raidoo BM, Kutscher EC. Visual hallucinations with varenicline: a case report. *J Med Case Reports* 2009; **3**:7560.

Rapoport JL, Jensvold M, Elkins R *et al.* Behavioral and cognitive effects of caffeine in boys and adult males. *J Nerv Ment Dis* 1981; **169**:726–32.

Reoux JP, Saxon AJ, Malte CA *et al.* Divalproex sodium in alcohol withdrawal: a randomized double-blind placebo-controlled trial. *Alcohol Clin Exp Res* 2001; **25**:1324–9.

Rich J, Scheife RT, Katz N *et al.* Isopropyl alcohol intoxication. *Arch Neurol* 1990; **47**:322–4.

Rickels K, Schweizer E, Case G *et al.* Long-term therapeutic use of benzodiazepines. I. Effects of abrupt discontinuation. *Arch Gen Psychiatry* 1990; **47**:899–907.

Ropper Ah, Blair R. Symmetric deep cerebellar lesions after smoking heroin. *Arch Neurol* 2003; **60**:1605–6.

Rosebush PI, Mazurek MF. Catatonia after benzodiazepine withdrawal. *J Clin Psychopharmacol* 1996; **16**:315–19.

Rosenbaum M, Lewis M, Piker P *et al.* Convulsive seizures in delirium tremens. *Arch Neurol Psychiatry* 1941; **45**:486–93.

Rosse RB, Collins JP, Fay-McCarthy M *et al.* Phenomenologic comparison of the idiopathic psychosis of schizophrenia and drug-induced cocaine and phencyclidine psychoses: a retrospective study. *Clin Neuropharmacol* 1994; **17**:349–69.

Roth D, Alarcon FJ, Fernandez JA. Acute rhabdomyolysis associated with cocaine intoxication. *N Engl J Med* 1988; **319**:673–7.

Satel SL, Southwick SM, Gawin FH. Clinical features of cocaine-induced paranoia. *Am J Psychiatry* 1991; **148**:495–8.

Scharf MB, Saskin P, Fletcher K. Benzodiazepine-induced amnesia: clinical laboratory findings. *J Clin Psychiatry* 1987; Monograph **5**: 14–17.

Scrimgeour EM. Outbreak of methanol and isopropanol poisoning in New Britian, Papua New Guinea. *Med J Aust* 1980; **2**:36–8.

Schuckit MA, Smith TL, Anthenelli R *et al.* Clinical course of alcoholism in 636 male inpatients. *Am J Psychiatry* 1993; **150**:786–92.

Shader RI, Greenblatt DJ. Use of benzodiazepines in anxiety disorders. *N Engl J Med* 1993; **328**:1398–405.

Sharpe JA, Hostovsky M, Bilbao JM *et al.* Methanol optic neuropathy: a histopathological study. *Neurology* 1982; **32**:1093–100.

Sherer MA, Kumor KM, Cone EJ *et al.* Suspiciousness induced by four-hour intravenous infusion of cocaine. Preliminary findings. *Arch Gen Psychiatry* 1988; **45**:673–7.

Silverman K, Evans SM, Strain EC *et al.* Withdrawal syndrome after the double-blind cessation of caffeine consumption. *N Engl J Med* 1992; **327**:1109–14.

Steffee CH, Davis GJ, Nicol KK. A whiff of death: fatal volatile solvent inhalation abuse. *South Med J* 1996; **89**:879–84.

Stillner V, Popkin MK, Pierce CM. Caffeine-induced delirium during prolonged competitive stress. *Am J Psychiatry* 1978; **135**:855–6.

Stockard JJ, Werner SS, Aalbers JA *et al.* Electroencephalographic findings in phencyclidine intoxication. *Arch Neurol* 1976; **33**:200–3.

Strain EG, Mumford GK, Silverman K *et al.* Caffeine dependence syndrome: evidence from case histories and experimental evaluations. *JAMA* 1994; **272**:1043–8.

Strassman RJ, Qualls CR. Dose–response study of N,N-dimethyltryptamine in humans. I. Neuroendocrine, autonomic, and cardiovascular effects. *Arch Gen Psychiatry* 1994a; **51**:85–97.

Strassman RJ, Qualls CR, Uhlenhuth EH *et al.* Dose-response of *N,N*-dimethyltryptamine in humans. II. Subjective effects and preliminary results of a new rating scale. *Arch Gen Psychiatry* 1994b; **51**:98–108.

Stuppacek CH, Pycha R, Miller C *et al.* Carbamazepine versus oxazepam in the treatment of alcohol withdrawal: a double-blind study. *Alcohol Alcoholism* 1992; **27**:153–8.

Talbott JA, Teague JW. Marijuana psychosis. *JAMA* 1969; **210**:299–302.

Tamerin JS, Weiner S, Poppen R *et al.* Alcohol and memory: amnesia and short-term memory function during experimentally induced intoxication. *Am J Psychiatry* 1971; **128**:1659–64.

Taverner D, Harrison DJ, Bell GM. Acute renal failure due to interstitial nephritis induced by 'glue sniffing' with subsequent recovery. *Scott Med J* 1988; **33**:246–7.

Tonstad S, Davies S, Flammer M *et al.* Psychiatric adverse events in randomized, double-blind, placebo-controlled trials of varenicline: a poked analysis. *Drug Saf* 2010; **33**:289–301.

Torrens M, San L, Cami J. Buprenorphine versus heroin dependence: comparison of toxicologic and psychopathologic characteristics. *Am J Psychiatry* 1993; **150**:822–4.

Tyrer HR, Owen R, Dawling S. Gradual withdrawal of diazepam after long-term therapy. *Lancet* 1983; **1**:1402–6.

Ungerleider JT, Fisher DD, Fuller M. The dangers of LSD. *JAMA* 1966; **197**:389–92.

Virmani R, Robinowitz M, Smialek JE *et al.* Cardiovascular effects of cocaine: an autopsy study of 40 patients. *Am Heart J* 1988; **115**:1068–76.

Wagena EJ, Knipschild PG, Huibers MJ *et al.* Efficacy of bupropion and nortriptyline for smoking cessation among people at risk of or with chronic obstructive pulmonary disease. *Arch Int Med* 2005; **165**:2286–92.

Watson JM. Glue sniffing: two case reports. *Practitioner* 1979; **222**:845–7.

Watson R, Hartmann E, Schildkraut JJ. Amphetamine withdrawal: affective state, sleep patterns, and MHPG excretion. *Am J Psychiatry* 1972; **129**:263–9.

Weddington WW, Brown BS, Haertzen CA *et al.* Changes in mood, craving, and sleep during short-term abstinence reported by male cocaine addicts. *Arch Gen Psychiatry* 1990; **47**:861–8.

Weil AJ. Adverse reactions to marijuana. *N Engl J Med* 1970; **282**:997–1000.

Weiner AL, Vieira L, McKay CA *et al.* Ketamine abusers presenting to the emergency department: a case series. *J Emerg Med* 2000; **18**:447–51.

Wolkowitz OM, Weingartner H, Thompson K *et al.* Diazepam-induced amnesia: a neuropharmacological model of an 'organic amnesic syndrome'. *Am J Psychiatry* 1987; **144**:25–9.

Wood CA, Buller F. Poisoning by wood alcohol. Cases of death and blindness from Columbian spirits and other methylated preparations. *JAMA* 1904; 43:972–7, 1058–62, 1117–23, 1213–21.

Yu E, Miotto K, Akerele E *et al.* A phase 3 placebo-controlled, double-blind, multi-site trial of the alpha-2-adrenergic agonist, lofexidine, for opioid withdrawal. *Drug Alcohol Depend* 2008; **97**:158–68.

Zakzanis KK, Young DA. Memory impairment in abstinent MDMA ('Ecstasy') users: a longitudinal investigation. *Neurology* 2001; **56**:966–9.

Zipursky RB, Baker RW, Zimmer B. Alprazolam withdrawal delirium unresponsive to diazepam: case report. *J Clin Psychiatry* 1985; **46**:344–5.

29

Medication- and substance-induced disorders

29.1 NEUROLEPTIC MALIGNANT SYNDROME

The neuroleptic malignant syndrome is a rare and potentially fatal disorder characterized, in its full expression, by the tetrad of delirium, rigidity, fever, and autonomic instability, which occurs secondary to an abrupt diminution of dopaminergic tone, due to either the use of a dopamine-blocking agent, such as an antipsychotic, or, much less commonly, the discontinuation of a dopaminergic agent, such as levodopa.

Before proceeding further, some words are in order regarding the name of this syndrome. This syndrome was first described in association with the use of antipsychotics, and at the time these agents were known as neuroleptics, and hence this malignant syndrome was called the neuroleptic malignant syndrome. With the shift in coinage from 'neuroleptic' to 'antipsychotic' it might have been appropriate to rename the syndrome as 'antipsychotic malignant syndrome'; however, this did not occur. As noted above, this syndrome can also occur secondary to discontinuation of a dopaminergic agent, most commonly in patients with Parkinson's disease. Some authors have suggested carving out this group of patients and, for them, renaming the syndrome in a variety of ways, including 'neuroleptic malignant-like syndrome', 'malignant syndrome in Parkinson disease', or 'parkinsonism hyperpyrexia syndrome'. However, these terms have never gained currency, and hence we are left with the name 'neuroleptic malignant syndrome', which, despite its inadequacies, appears destined to persist.

Clinical features

The onset is usually within a day or two of the diminution in dopaminergic tone; exceptionally, the syndrome may appear within an hour or, at the other extreme, be delayed for weeks (Keck *et al.* 1987).

Classically (Addonizio *et al.* 1987; Caroff 1980; Kellam 1987; Pope *et al.* 1986; Rosebush and Stewart 1989; Velamoor *et al.* 1994), the syndrome, as noted, is characterized by a tetrad of delirium, rigidity, fever, and autonomic instability. Although in most cases the evolution of the syndrome follows this order, exceptions do occur, and some cases may present first with rigidity, fever, or autonomic instability. The delirium may become quite profound, and patients may also develop stuporous catatonia (Koch *et al.* 2000); eventually stupor or coma may supervene. Rigidity may be of either the lead pipe or cogwheel type, is typically generalized, and may likewise become profound, to the point of compromising chest wall movement. In some cases the rigidity may be accompanied by a generalized, coarse, tremor or, less commonly, dystonia or chorea. Fever is generally over 38 °C (100.4 °F) and autonomic instability manifests with pallor, diaphoresis, tachycardia, tachypnea, and elevated blood pressure, which may be quite labile.

Rhabdomyolysis may occur (Jones and Dawson 1989), with myoglobinuria and, in some cases, acute renal failure. Other complications include dehydration, pulmonary emboli, aspiration pneumonia, respiratory failure, and disseminated intravascular coagulation.

The white blood cell count is almost always elevated, typically to 15 000 cells/mm^3, and sometimes to as high as 40 000 cells/mm^3. The creatine phosphokinase level is likewise increased, typically to over 1000 units/L, and sometimes to as high as 60 000 units/L. Lactate dehydrogenase, serum glutamic oxaloacetic transaminase, and alkaline phosphatase levels are also often elevated.

Magnetic resonance (MR) scanning is typically normal, while the electroencephalogram (EEG) typically reveals generalized slowing. The cerebrospinal fluid, though characteristically normal, may, in perhaps one-third of cases, demonstrate a mildly elevated total protein level.

Although this classic picture is the rule, exceptions do occur, and atypical cases have been reported wherein the full tetrad is not present.

Course

The mortality rate is between 10 and 20 percent. Those who survive generally recover within a week or two, although an exception to this rule occurs in the case of the neuroleptic malignant syndrome occurring secondary to treatment with long-acting depot preparations of the antipsychotics fluphenazine, haloperidol, risperidone, or paliperidone, in which the syndrome is prolonged for at least a month.

Although the vast majority of patients who survive do so without sequelae, exceptions do occur (Adityanjee Sajatovic and Munshi 2005) and some patients may be left with cognitive deficits, rigidity, or dystonia.

Etiology

As noted earlier, the syndrome occurs secondary to an abrupt diminution of dopaminergic tone, and this most commonly occurs secondary to either initiation of treatment with an antipsychotic or a substantial dose increase (Kellam 1990): importantly, this appears to be an idiosyncratic reaction, and in some instances the syndrome has occurred within hours of the *first* dose of an antipsychotic. Although most cases have occurred secondary to treatment with first-generation agents, such as haloperidol, the neuroleptic malignant syndrome has also been seen with second-generation agents, such as risperidone (Levin *et al.* 1996; Meterissian 1996), olanzapine (Levenson 1999; Suh *et al.* 2003), quetiapine (Sing *et al.* 2002), aripiprazole (Brunelle *et al.* 2007), ziprasidone (Ozen *et al.* 2007), and clozapine (Anderson and Powers 1991; Das Gupta and Young 1991; Miller *et al.* 1991; Sachdev *et al.* 1995). Other dopamine blockers may also cause the syndrome, such as metoclopramide (Friedman *et al.* 1987) and promethazine (Mendhekar and Andrade 2005). The syndrome has also occurred secondary to treatment with the antidepressant amoxapine (Taylor and Schwartz 1988); however, in all likelihood the cause here is not amoxapine but one of its metabolites, loxapine, which is also a first-generation antipsychotic. There are also reports of the syndrome occurring after an antidepressant was added to a stable dose of an antipsychotic, for example with the addition of venlafaxine to trifluoperazine (Nimmagadda *et al.* 2000) or paroxetine to olanzapine (Kontaxakis *et al.* 2003). Finally, and very rarely, the neuroleptic malignant syndrome has also been reported as occurring secondary to treatment with venlafaxine (Lu *et al.* 2006) or paroxetine (Tanii *et al.* 2006) alone, in the absence of a neuroleptic: there is some doubt about these reports, however, given the sometimes difficult differential between the neuroleptic malignant syndrome and the serotonin syndrome, discussed further below.

The neuroleptic malignant syndrome has also been seen upon the cessation of treatment with not only levodopa (Friedman *et al.* 1984; Gibb and Griffith 1986; Keyser and Rodnitzky 1991; Sechi *et al.* 1984; Toru *et al.* 1981) but also amantadine (Cunningham *et al.* 1991; Harsch 1987) and a combination of levodopa and bromocriptine (Figa-Talamanca *et al.* 1985). In addition, the neuroleptic malignant syndrome has been reported secondary to the use of the dopamine-depleting drug tetrabenazine (Ossemann *et al.* 1996).

Although the mechanism by which a diminution of dopaminergic tone produces the syndrome is not known with certainty, suspicion naturally falls on the meso-limbic and nigro-striatal pathways, with reductions in dopaminergic tone in the limbic system leading to delirium and reductions in the striatum to rigidity; similarly, a reduction in dopaminergic tone in the hypothalamus could account for the fever and autonomic instability. There have been only a handful of autopsy studies: although one (Horn *et al.* 1988) revealed necrotic changes in the anterior and lateral hypothalamic nuclei, another (Jones and Dawson 1989) failed to reveal anything but non-specific agonal changes.

Differential diagnosis

A particularly difficult differential task may occur in patients with Parkinson's disease who develop a significant infection, for example a pneumonia, and who, during their hospital stay, are made nil per os (NPO). In such cases a full tetrad of symptoms may occur: delirium may develop secondary to systemic effects of the infection, rigidity due to a lack of levodopa, and fever along with tachycardia due to the infection. Ideally, from a differential diagnostic point of view, one might merely observe the patient while antibiotic treatment is under way, and if the delirium, fever, and tachycardia all resolved, then the diagnosis of neuroleptic malignant syndrome would be ruled out. Obviously, however, given the stakes involved, such a course of action is untenable, and in such cases one treats the patient as if a neuroleptic malignant syndrome had occurred, with prompt administration of levodopa via a nasogastric tube.

A similarly difficult differential problem may occur in patients being treated with antipsychotics who develop a significant infection. Here, along with fever and tachycardia, a delirium may likewise develop secondary to systemic effects of the infection, and if the delirium were characterized by agitation, the dose of the antipsychotic might be increased to treat that: should an acute 'extrapyramidal' side-effect of parkinsonism accompany that dose increase, the full tetrad may appear. In such cases, one might, as in the example noted above, simply wait to see what happened with antibiotic treatment; however, again, given the stakes involved, a prompt reduction in dose of the antipsychotic, or, to be on the safe side, its discontinuation, would be in order.

Heat stroke may present with delirium, fever, tachycardia, and tachypnea; however, the occurrence of these symptoms in the proper setting of an elevated ambient temperature, and the presence, rather than diaphoresis, of a hot, dry, and flushed skin immediately suggest the correct diagnosis. The absence of any maneuvers that reduce dopaminergic tone is also, obviously, an important diagnostic point. Differential diagnostic difficulties, however, may arise in cases where the patient had been on a low-potency first-generation antipsychotic agent. These agents may reduce sweating, and although the patient may not have rigidity, the picture may resemble a partial neuroleptic malignant syndrome suitably altered by the anticholinergic effects of the low-potency agent. In such cases, it would be prudent to institute treatment measures appropriate to both heat stroke and the neuroleptic malignant syndrome.

The serotonin syndrome, as discussed in Chapter 29.5, may present with the full tetrad of delirium, rigidity, fever, and autonomic instability. Clearly, a medication history is of importance here, as the diagnosis of the serotonin syndrome can be made only in patients exposed to serotoninergic agents. However, there are cases where a serotoninergic agent may be added to a regimen already including an antipsychotic, making the differential problematic. In such cases, the presence of signs and symptoms typical for the serotonin syndrome, and absent in the neuroleptic malignant syndrome are particularly helpful, and these include the following: nausea, vomiting and diarrhea; myoclonus; and clonus.

Other disorders capable of producing a full tetrad of symptoms include arboviral encephalitis (secondary to Japanese encephalitis or West Nile encephalitis), abrupt withdrawal of either oral (Turner and Gainsborough 2001) or intrathecal (Coffey *et al.* 2002) baclofen, and severe cocaine intoxication (Daras *et al.* 1995). A partial syndrome, with delirium, fever and autonomic instability (and, in some cases rhabdomyolysis) may also occur with methylenedioxymethamphetamine (MDMA) intoxication.

Finally, consideration must be given to Stauder's lethal catatonia, a syndrome described well before the advent of antipsychotics. As discussed in Chapter 4.11, this syndrome develops in the setting of excited catatonia in patients with catatonic schizophrenia. These patients become progressively more agitated, and then develop stupor, fever, tachycardia, and leukocytosis. As most patients with catatonic schizophrenia are already on an antipsychotic, the differential possibility of a partial neuroleptic malignant syndrome naturally arises. In such cases, if the dose of the antipsychotic had been stable, then one might, should stupor, fever, and tachycardia occur, lean towards a diagnosis of Stauder's lethal catatonia. If, however, the dose of the antipsychotic had been recently increased, then a diagnosis of neuroleptic malignant syndrome becomes a distinct possibility. Further, in cases where rigidity supervenes, the diagnosis of neuroleptic malignant syndrome becomes far more likely, because rigidity is not part of the syndrome of Stauder's lethal catatonia.

Treatment

The neuroleptic malignant syndrome constitutes a medical emergency. Intensive supportive care is required, with particular attention to fluid and electrolyte balance; adequate hydration must be assured to reduce the risk of renal failure. Cooling blankets may be required, as may ventilatory support. In addition to these measures it is critical to restore dopaminergic tone as quickly as possible. Thus, if a neuroleptic agent is at fault it should be discontinued, and if a dopaminergic agent has been stopped it should be restarted. Another strategy includes the use of bromocriptine and/or dantrolene (Granato *et al.* 1983; Guze and Baxter 1985; May *et al.* 1983; Mueller *et al.* 1983). Although these agents have not been assessed in controlled trials, anecdotal reports support their use. Bromocriptine is given orally, by nasogastric tube if necessary, in doses ranging from 2.5 to 20 mg three times a day, and dantrolene is given intravenously in a dose of 1–2 mg/kg, which is repeated as necessary to a maximum of 10 mg/kg per 24 hours. Recent double-blind work in cases secondary to withdrawal of dopaminergic agents in patients with Parkinson's disease demonstrated that the addition of intravenous methylprednisolone (1000 mg daily) to combined treatment with bromocriptine (7.5 mg daily) and dantrolene (75 mg daily) for a 3-day course not only further reduced the severity of the syndrome but also shortened its overall course (Sato *et al.* 2003). Although bromocriptine has the longest track record in restoring dopaminergic tone, consideration may also be given to subcutaneous apomorphine (Lattanzi *et al.* 2006; Wang and Hsieh 2001) which may be given in a dose of 2 mg every 3 hours for 2 days, then every 6 hours for an additional 2 days (Lattanzi *et al.* 2006), or to amantadine (McCarron *et al.* 1982), given by nasogastric tube in a total daily dose of 300–400 mg, given in two divided doses. Finally, there are case reports of the neuroleptic malignant syndrome responding to parenteral lorazepam (Yacoub and Francis 2006), in doses of 1–2 mg every 4–6 hours, to a maximum of 8 mg per day.

Regardless of which regimen is used, treatment should be continued for at least a week after symptoms have resolved, after which the dose may be tapered: an exception to this rule, of course, occurs in cases secondary to depot antipsychotics, and in such cases weeks or months of treatment may be required before the antipsychotic eventually washes out.

In treatment-resistant cases, consideration should be given to electroconvulsive therapy (ECT) (Davis *et al.* 1991; Hermesh *et al.* 1987; Troller and Sachdev 1999). However, despite its often dramatic effectiveness, this option should be approached with caution, as some patients may develop ventricular fibrillation, cardiac arrest, or status epilepticus.

In many cases of the neuroleptic malignant syndrome occurring secondary to an antipsychotic, patients require ongoing treatment. In such cases it has been found possible to reinstitute treatment with an antipsychotic (Rosebush *et al.* 1989) provided that one waits for at least 2 weeks after the neuroleptic malignant syndrome has entirely cleared and then reintroduces the antipsychotic cautiously, starting with a low dose and increasing very gradually. Although there are case reports of the successful reinstitution of treatment with the same agent that caused the syndrome, prudence dictates using a different antipsychotic. Thus, if a high-potency first-generation agent was used, one should probably choose either a low-potency first-generation agent or a second-generation one. If the syndrome has occurred secondary to a second-generation agent, then one might consider an alternative second-generation agent with a statistically lower chance of producing extrapyramidal side-effects.

29.2 TARDIVE DYSKINESIA

Tardive dyskinesia is a movement disorder that occurs secondary to chronic treatment with a dopamine-blocking drug, which, in the vast majority of cases, is an antipsychotic. In contrast to the more common typical 'extrapyramidal' side-effects seen with these agents, however, tardive dyskinesia does not remit promptly when the offending medication is discontinued, but rather persists for prolonged periods, and may indeed be permanent.

The most typical, and by far the most common, form of tardive dyskinesia is a choreiform one, often with bucco–lingual–masticatory movements. Less commonly, one encounters dystonic and akathetic forms, and rarely one may see tics, tremor, myoclonus, or pain. Although some authors treat these less common forms as separate entities, this distinction appears unwarranted. Another nosologic dispute centers on the question of whether there is any fundamental difference between those cases that are permanent and those in which the abnormal movements remit spontaneously within a few months of the discontinuation of the antipsychotic. Some have referred to the latter form as 'withdrawal dyskinesia', but this distinction, too, may be unwarranted: in all likelihood, tardive dyskinesia, like many disorders, varies in severity, with some cases demonstrating permanent symptomatology and others displaying a spontaneous remission.

The overall prevalence of tardive dyskinesia in those treated chronically with first-generation antipsychotics is in the order of 15–20 percent (Kane and Smith 1982; Woerner *et al.* 1991), while among those treated with second-generation agents, it is about half that (Woods *et al.* 2010).

Clinical features

Although some cases of tardive dyskinesia have been reported after only a month of treatment with an antipsychotic, this is quite rare; in general, at least 6 months are required, and often one or more years pass before the abnormal movements appear. An exception to this rule is seen in the elderly, in whom the latency between initiation of treatment and the appearance of abnormal movements may be much shorter (Kane and Smith 1982; Saltz *et al.* 1991).

The mode of onset may be either acute or gradual. An acute onset may occur when the antipsychotic is discontinued or the dose rapidly decreased, in which case an abrupt 'unmasking' occurs with an equally abrupt appearance of abnormal movements. Conversely, if the dose is maintained, or decreased only slightly, the abnormal movements may make their appearance gradually or insidiously.

As noted earlier, tardive dyskinesia may manifest with chorea, dystonia, akathisia, tics, tremor, myoclonus, or pain, and each of these presentations is considered in turn.

Choreiform movements in tardive dyskinesia are most commonly found in the lower part of the face (Burke *et al.* 1982; Kang *et al.* 1986), less so in the extremities, and only rarely in the trunk, and range in severity from mild to disabling (Kennedy 1969). Classically, one sees bucco–linguo–masticatory movements in the face, with pursing or puckering of the lips, chewing motions, and repetitive tongue protrusions. Facial grimacing may also be seen, but this, importantly, spares the forehead. Extremity involvement may present with shoulder-shrugging or a restless piano-playing movement of the fingers and hands; in the lower extremities there may be foot-tapping. Truncal involvement may manifest with axial to-and-fro rocking or with pelvic thrusting. Uncommonly, respiratory dyskinesias may occur, with irregular, grunting respirations (Chiang *et al.* 1985; Ivanovitch *et al.* 1993; Rich and Radwany 1994; Yassa and Lal 1986a). Critically, the choreiform movements of tardive dyskinesia tend to be repetitive and stereotypied (Stacy *et al.* 1993); furthermore, they tend to wax and wane in severity throughout the day, and they disappear in sleep. Some patients are able to suppress them voluntarily, but such suppression is at best temporary and they inevitably reappear.

Dystonic movements (Burke *et al.* 1982; Kang *et al.* 1986; Kiriakakis *et al.* 1998; Sachdev 1993a; Wojcik *et al.* 1991; Yassa *et al.* 1986) typically appear focally and most commonly involve the face or neck, followed by the upper extremity, the trunk, and the lower extremity. Over a long period of time, segmental spread to adjacent body parts may occur, and, rarely, the dystonia may become generalized (Burke *et al.* 1982; Wojcik *et al.* 1991). Facial involvement may manifest with blepharospasm (Ananth *et al.* 1988; Glazer *et al.* 1983; Weiner *et al.* 1981), oromandibular dystonia (Tan and Jankovic 2000), or oculogyric crises (FitzGerald and Jankovic 1989; Sachdev 1993b). As with chorea, the severity of tardive dystonia ranges from mild to disabling (Yadalam *et al.* 1990). Importantly, many patients with tardive dystonia will also have choreiform movements (Sachdev 1993a; Wojcik *et al.* 1991).

Akathisia occurring on a tardive basis is symptomatically very similar to that occurring as an acute extrapyramidal side-effect of an antipsychotic, with restlessness, marching in place, and 'restless' thoughts (Burke *et al.* 1989; Dufresne and Wagner 1988; Lang 1994; Sachdev and Loneragan 1991).

Tics may be either focal or multifocal, and at times are so extensive as to mimic the picture seen in Tourette's syndrome (Bharucha and Sethi 1995; Stahl 1980), thus prompting the term 'tardive tourettism'. Other, rare manifestations of tardive dyskinesia include tremor (Stacy and Jankovic 1992), myoclonus (Little and Jankovic 1987), and pain (Ford *et al.* 1994), and all of these typically present in the context of choreiform movements. The tremor, though primarily of the postural type, may also be seen at rest and with movement; myoclonus was noted in one case in the face. When pain occurs, it is most common in the mouth or genital area, and has a burning quality.

Course

The course of tardive dyskinesia has been most thoroughly studied with reference to the choreiform type. In situations in which the antipsychotic is continued at a constant dose, there is a gradual worsening of symptoms; although in most cases the severity eventually reaches and stays at a plateau, in a minority one sees a relentless progression (Gardos *et al.* 1988, 1994; Glazer *et al.* 1984; Yagi and Itoh 1987). Should the dose of the antipsychotic be substantially increased, there is a 'masking' of symptoms; however, this is temporary and they eventually reappear. Conversely, if the dose is decreased, symptoms worsen. If the antipsychotic is discontinued there is typically a pronounced worsening of symptoms, which is followed, however, over the succeeding weeks or months, by at least some diminution of symptoms, after which one of two eventualities may ensue (Glazer *et al.* 1990). In some cases, perhaps one-third, abnormal movements continue to lessen and, after many months or a year or more, they go into a spontaneous remission. In the remainder, however, only a partial remission occurs, after which the abnormal movements persist indefinitely. This chronic course is more likely in the elderly and in those who have been treated with very high doses.

Interestingly, choreiform movements in tardive dyskinesia are worsened by anticholinergics (Greil *et al.* 1984), such as benztropine (Reunanen *et al.* 1982), and also may vary with mood. Patients who develop a depressive syndrome typically experience a worsening of symptoms (Sachdev 1989), whereas during mania there may be a partial remission (de Potter *et al.* 1983; Yazici *et al.* 1991); symptom improvement has also been noted during catatonic stupor (Assmann and van Woerkom 1987).

Etiology

Although the vast majority of cases of tardive dyskinesia occur secondary to treatment with antipsychotics, cases have also been reported with the dopamine-blocker metoclopramide (Sewell and Jeste 1992; Sewell *et al.* 1994), and secondary to chronic treatment with the antidepressant amoxapine (Huang 1986; Thornton and Stahl 1984); in this latter case, however, it is probably not amoxapine that is at fault but rather one of its metabolites, loxapine, which is an antipsychotic.

Given that all of the drugs capable of causing tardive dyskinesia have one thing in common, namely a blockade of postsynaptic dopamine receptors, and given that the risk of tardive dyskinesia increases with higher doses (Morganstern and Glazer 1993) and a longer duration of treatment (Glazer *et al.* 1993; Kane *et al.* 1988), it seems reasonable to suppose that it is some effect of the chronic dopamine blockade that is etiologic in this disorder, and, indeed, this is the basis for the dopamine upregulation hypothesis. According to this hypothesis, with chronic dopamine blockade there is a gradually worsening compensatory upregulation of the postsynaptic receptors to the point at which, if the dose of the dopamine blocker is reduced, the 'unmasking' of these upregulated receptors leads to a hyperdopaminergic state and the production of the abnormal movements. Further support for a disturbance in dopamine transmission is provided by the response to anticholinergics in tardive dyskinesia. Dopaminergic and cholinergic systems exist in a balance in the basal ganglia, such that an increase in dopaminergic tone may be mimicked by a reduction in cholinergic tone and vice versa. Given this, one would predict that, in patients with tardive dyskinesia, a reduction in cholinergic tone, as might occur with the administration of an anticholinergic medication, would increase the abnormal movements, and this is generally what happens (Klawans and Rubovits 1974).

As attractive as this dopamine theory is, it does not account for several important findings. First, the fact that acute antipsychotic-induced parkinsonism can co-exist with tardive dyskinesia (Richardson and Craig 1982) argues against any generalized hyperdopaminergic state, as theoretically this should bring a relief of the parkinsonism. Second, some cases of the dystonic subtype of tardive dyskinesia may, as noted below, be relieved, rather than worsened, by anticholinergic agents. Finally, if it were simply a matter of the compensatory upregulation of the postsynaptic dopamine receptors, then providing that the dose of a neuroleptic were held constant, with no dose decrease, one should not see any abnormal movements as there would be no 'unmasking', and thus no hyperdopaminergic state. Unfortunately for the theory, however, tardive dyskinesia does in fact emerge while patients continue at the same dose, a fact strongly suggesting that some other process, in addition perhaps to a progressive upregulation, is at work.

The nature of this other process is not clear. Theories include disturbances in gamma-aminobutyric acid (GABA)-ergic transmission, glutamate toxicity, and progressive neuronal destruction by free radicals produced in the course of neuroleptic treatment. This last theory is of some interest given the evidence, noted below, for the treatment efficacy of vitamin E, an antioxidant.

Differential diagnosis

In evaluating a patient with abnormal movements the diagnosis of tardive dyskinesia cannot be considered unless the patient has undergone chronic treatment with an antipsychotic or other dopamine-blocker such as metoclopraminde. The overall differential diagnoses of chorea, dystonia, akathisia, and tics are discussed in Chapters 4.4, 4.7, 4.10, and 4.3, respectively; of the disorders discussed in these chapters, several deserve special consideration.

Choreiform movements may occur in both Huntington's disease and schizophrenia, and both of these disorders are typically treated with antipsychotics. With regard to Huntington's disease, genetic testing is of course diagnostic; however, certain clinical features may also allow a differentiation to be made, including the presence of forehead chorea and a 'dancing and prancing' gait, both of which may be seen in Huntington's disease but not in tardive dyskinesia. The nature of the chorea itself provides another differential point: in Huntington's disease it is extremely transient, flickering from one part of the body to another in an almost random pattern, whereas in tardive dyskinesia the chorea tends to be stereotypied, repetitive, and persistently recurring in the same area. Schizophrenia may at times be characterized by choreiform movements, as pointed out by Kraepelin in the early part of the twentieth century (Kraepelin 1971 [1919]) and confirmed by subsequent investigators (Farran-Ridge 1926; Mettler and Crandall 1959; Owens *et al.* 1982; Yarden and Discipio 1971); these choreiform movements, however, are mild, generally involve the face, and are present before any treatment with antipsychotics.

Both dystonia and akathisia may occur as *acute* extrapyramidal side-effects of antipsychotics but are readily distinguished from tardive dystonia or tardive akathisia by their course: acute side-effects occur shortly after the initiation or dose increase of an antipsychotic and subside with a dose decrease or discontinuation; by contrast, tardive dystonia and akathisia come on only after a long period of treatment and worsen when the dose is decreased or the antipsychotic is discontinued.

Tourette's syndrome typically has an onset in childhood, and the presence of tics before treatment with antipsychotics makes the diagnosis straightforward. As noted in Chapter 9.25, however, after a full or partial remission in late adolescence, there may be a relapse decades later, and if this relapse should occur in a patient being treated with an antipsychotic for another reason, diagnostic confusion may ensue; in such cases it is critical to obtain a childhood history, with special reference to the symptomatology of the prior tics to allow a comparison with the current ones.

Treatment

The best 'treatment' is prevention, and chronic antipsychotic treatment should not be undertaken for those disorders for which other medications, with better side-effect profiles, would be as effective.

When tardive dyskinesia does first appear, a decision must be made whether ongoing treatment with an antipsychotic is required, carefully weighing the risk of worsening dyskinesia against the risk of relapse. In the case of schizophrenia, the balance often tips towards continuing antipsychotic treatment. If treatment is continued, efforts, if not already in place, should be made to keep the dose as low as possible, consistent with adequate symptomatic control. In the past it was felt that intermittent 'drug holidays' for patients with schizophrenia might reduce the risk of tardive dyskinesia, but subsequent studies have suggested that drug holidays may actually *increase* the risk (Goldman and Luchins 1984; van Harten *et al.* 1998; Jeste *et al.* 1979). In cases due to treatment with a first-generation antipsychotic, consideration should be given to switching to a second-generation agent; with such a switch, adequate symptom control is generally either maintained or improved and the risk of worsening tardive dyskinesia with further treatment is lessened.

In cases in which treatment cannot be discontinued, or in cases in which discontinuation is possible but symptoms fail to go into remission, various medical treatments may be considered, all of which have been tested under double-blind conditions. Of all these medical treatments, vitamin E, in total daily doses of 1200–1600 international units (IU)/day (divided into three doses), has been most extensively studied: out of 12 trials, it was effective in eight (Adler *et al.* 1993, 1998; Akhtar *et al.* 1993; Dabiri *et al.* 1994; Elkashef *et al.* 1990; Lohr and Caligiuri 1996; Lohr *et al.* 1988; Zhang *et al.* 2004), but no more effective than placebo in four (Adler *et al.* 1999; Dorevitch *et al.* 1997; Egan *et al.* 1992; Shriqui *et al.* 1992). Melatonin, in a dose of 2 mg (Shamir *et al.* 2000) was ineffective, but at a dose of 10 mg (Shamir *et al.* 2001) it was superior to placebo. Vitamin B6 (Lerner *et al.* 2001) was effective at a high dose of 300 mg; levetiracetam (Woods *et al.* 2008) in doses ranging from 500 to 1000 mg; piracetam (Libov *et al.* 2007) in a dose of 3600 mg; clonazepam (Thaker *et al.* 1990) in doses from 2 to 4.5 mg; the dopamine depleter tetrabenazine (Ondo *et al.* 1999) in doses ranging from 25 to 150 mg; and, finally, branched-chain amino acids (Richardson *et al.* 2003). This last option, from a pharmacologic point of view, works in a fashion similar to tetrabenazine: it involves administration of a combination of valine, isoleucine, and leucine (given in a ratio of 3:3:4) at a dose of 222 mg/kg three times daily: this is followed by a fall in the plasma levels of the aromatic amino acids tyrosine and tryptophan, with, presumably, a fall in central nervous system levels of dopamine and serotonin.

Other medical treatments have also been reported effective. In the case of tardive dystonia, anecdotal reports support treatment with high-dose anticholinergic agents (Burke *et al.* 1982), such as 20 mg trihexyphenidyl (Fahn 1983); however, in cases in which the dystonia is accompanied by choreiform movements, these may worsen. Anecdotal reports also suggest that tardive akathisia may respond to propranolol (Yassa *et al.* 1988); however, high doses may be required.

When medical treatment fails, other options may be considered. Botulinum toxin, in a single-blinded study (Slotema *et al.* 2008), was effective for orofacial dyskinesias, and open work (Tarsy *et al.* 1997) suggests its benefit for dystonia. Deep brain stimulation of the globus pallidus was effective in a double-blinded study (Damier *et al.* 2007), but for obvious reasons constitutes a last resort. Anecdotally, ECT is effective for tardive dyskinesia (Adityanjee 1990), especially in cases in which the patient is depressed (Hay *et al.* 1990).

29.3 SUPERSENSITIVITY PSYCHOSIS

Supersensitivity psychosis represents a very rare side-effect of chronic treatment with antipsychotic drugs. As such, it bears a strong etiologic relationship to tardive dyskinesia, and indeed this syndrome is also referred to as 'tardive psychosis'. This is, for some, a controversial diagnosis. As noted below (see 'Differential diagnosis'), it may be difficult to distinguish supersensitivity psychosis from an exacerbation of schizophrenia, and indeed some observers have held that all reported cases of supersensitivity psychosis represent merely such exacerbations. However, there are cases of supersensitivity psychosis in patients with bipolar disorder treated with antipsychotics (Steiner *et al.* 1990), and there is one well-described case (Moore 1986) in a patient chronically treated with an antipsychotic for major depressive disorder, who had *never* been psychotic, yet who developed a psychosis immediately after the antipsychotic was discontinued.

Clinical features

After a year or more of treatment with an antipsychotic, the syndrome emerges. This emergence may be fairly abrupt if the antipsychotic is discontinued or there is a rapid dose reduction; in other cases a more gradual onset may occur, for example when the dose of the antipsychotic is held constant or only gradually and modestly decreased.

Clinically (Chouinard and Jones 1980; Kirkpatrick *et al.* 1992; Moncrieff 2006; Steiner *et al.* 1990), patients develop a psychosis characterized by delusions and hallucinations. Importantly, tardive dyskinesia may or may not accompany this development.

Course

Although the course has not been well studied, it appears that, in cases in which the antipsychotic is discontinued, symptoms very gradually lessen; the percentage of cases that go on to full remission or to a stable chronicity is not known.

Etiology

Although the etiology is not known, it is speculated that chronic treatment with antipsychotics causes an upregulation of postsynaptic dopamine receptors within the limbic system leading to a chronic hyperdopaminergic state. Of note, although almost all described cases have occurred secondary to chronic treatment with antipsychotics, cases have also been described secondary to chronic treatment with the dopamine-blocker metoclopramide (Lu *et al.* 2002) and also with reserpine (Kent and Wilber 1982). This last case is of interest for two reasons. First, this patient, like the one described earlier who had depression, had never been psychotic: reserpine had been prescribed for hypertension, and thus the prompt occurrence of psychosis after discontinuation of the reserpine could not be explained on the basis of a worsening of a pre-existing psychotic illness. Second, the fact that chronic treatment with reserpine was followed by upregulation of dopamine receptors provides support to the notion that supersensitivity psychosis does occur secondary to such a mechanism.

Differential diagnosis

The diagnosis should be entertained only in patients who have been treated chronically with an antipsychotic, another form of dopamine blocker, or a dopamine depleter.

As most patients treated chronically with antipsychotics have schizophrenia, the main differential for supersensitivity psychosis is an exacerbation of the schizophrenia. For example, consider the case of a patient with schizophrenia who had been doing well and whose antipsychotic dose had been substantially reduced, after which there was a reappearance of hallucinations or delusions. The question here is whether this appearance of hallucinations and delusions represents merely the schizophrenia or whether the patient now has two disorders, namely schizophrenia and supersensitivity psychosis. One helpful differential point here lies in the course of the development of the hallucinations and delusions: exacerbations of schizophrenia, although they can be hyper-acute, are generally more leisurely affairs, spanning weeks or months; by contrast, supersensitivity psychosis appears fairly promptly, often within days.

Patients with major depressive disorder or with bipolar disorder may also be treated chronically with antipsychotics, and the emergence of delusions and hallucinations in such patients may or may not represent a supersensitivity psychosis. The guiding differential point here is whether the delusions and hallucinations emerge in the context of a severe depression or mania. If they do, then they may well be simply part of the expression of the underlying mood disorder; if, however, the patient was euthymic during the emergence of the psychosis, then the diagnosis of supersensitivity psychosis is obviously more likely.

Treatment

The best treatment is prevention, and antipsychotics should not be used chronically in conditions that may respond to other agents.

In cases in which it is possible to discontinue the antipsychotic, this should be carried out, in the hope of achieving a gradual spontaneous remission. In cases in which ongoing treatment is required, or when discontinuation is not followed by a spontaneous remission, consideration may be given to increasing the dose of the antipsychotic to suppress symptoms; in such instances, if a first-generation agent has been at fault it is probably appropriate to switch to a second-generation one. Whether or not some of the treatments effective for tardive dyskinesia, such as vitamin E, are effective here is simply not known; however, given the benign nature of these treatments a trial would be appropriate.

29.4 RABBIT SYNDROME

The rabbit syndrome is an uncommon movement disorder that generally occurs secondary to chronic treatment with an antipsychotic (Chiu *et al.* 1993; Yassa and Lal 1986b); although first-generation agents are the most common causes, cases have been reported with second-generation agents such as risperidone (Hoy and Alexander 2002) and aripiprazole (Caykoylu *et al.* 2010).

Clinical features

The onset is gradual, usually after a year or more of antipsychotic treatment. Clinically (Decina *et al.* 1990; Deshmukh *et al.* 1990; Todd *et al.* 1983), the syndrome is characterized by a rhythmic rest tremor of the jaw, with a frequency and amplitude such that the appearance is that of a rabbit chewing. Importantly, there is no involvement of the tongue nor is there any tremor elsewhere, or any associated rigidity or bradykinesia.

Course

With continued antipsychotic treatment the tremor becomes more pronounced; with discontinuation of the antipsychotic the tremor gradually decreases and typically remits.

Etiology

Apart from the fact that this disorder generally occurs secondary to antipsychotic treatment, the etiology is not known. Although some have speculated that it represents a form of tardive dyskinesia, the fact that it responds to treatment with anticholinergics argues for its inclusion among the 'extrapyramidal' side-effects commonly seen with antipsychotics. However, to complicate matters further, there are rare reports of the rabbit syndrome occurring secondary to treatment with antidepressants, such as imipramine (Fornazzari *et al.* 1991) and citalopram (Parvin and Swartz 2005).

Differential diagnosis

Essential tremor can produce a tremor of the jaw and chin; however, the tremor here is generally more rapid than the 'rabbit chewing' movement seen in the rabbit syndrome; furthermore, in essential tremor one also finds a tremor in the hands, something that is absent in the rabbit syndrome.

Parkinsonism may also cause a jaw tremor; however, in parkinsonism this is accompanied by tremor of the hands and, at some point, rigidity and bradykinesia – findings absent in the rabbit syndrome.

Tardive dyskinesia may be considered, but the bucco–lingual–masticatory movement with repetitive tongue protrusion is fundamentally different from the chewing motion of the rabbit syndrome, in which the tongue is spared. Furthermore, anticholinergics worsen tardive dyskinesia, whereas, as noted below, they are of benefit in the rabbit syndrome.

Treatment

If possible, the antipsychotic should be discontinued to allow for a spontaneous resolution of the syndrome. If, as is often the case, this is not possible, treatment with an anticholinergic agent, such as benztropine (Todd *et al.* 1983) or trihexyphenidyl (Wada and Yamaguchi 1992) is effective.

29.5 SEROTONIN SYNDROME

The serotonin syndrome is a potentially fatal complication arising from any pharmacologic maneuver that abruptly increases serotoninergic tone within the central nervous system.

Clinical features

The onset occurs within hours or days of one of the pharmacologic maneuvers noted below (see 'Etiology').

In its full expression (Bodner *et al.* 1995; Boyer and Shannon 2005; Feighner *et al.* 1990; Mason *et al.* 2000; Sternbach 1991), the serotonin syndrome presents with a delirium, often accompanied by restlessness or agitation. Nausea, vomiting, and diarrhea are present. The temperature, pulse, and blood pressure are all increased, and there may be diaphoresis and mydriasis. Tremor, shivering, and myoclonus are present, along with rigidity; the deep tendon reflexes are increased, especially in the lower extremities, the plantar responses are extensor and there is bilateral clonus. The white blood cell count is increased, as is the creatine phosphokinase level.

It must be emphasized that this full expression, however, is not the rule, and most cases present with only partial syndromes.

In severe cases, delirium may give way to stupor or coma, and the temperature may rise well over 38 °C (100.4 °F). Rhabdomyolysis may occur, with renal failure, as may shock and disseminated intravascular coagulation.

Course

Although potentially fatal, most patients recover, and if nothing else is done besides discontinuing the offending medications, the syndrome gradually remits within anywhere from a day to a week. Longer courses, however, may be seen when the offending agent has a long half-life, as, for example, with fluoxetine.

Etiology

This syndrome occurs secondary to any of a large number of pharmacologic maneuvers, all of which have in common the effect of abruptly increasing serotoninergic tone within the central nervous system. Although monotherapy with serotoninergic agents can be at fault, this is most unusual, and in general what one has to be concerned about are combinations of serotoninergic agents, and of these combinations, the one with by far the highest risk is a non-selective strong monoamine oxidase inhibitor (MAOI) (e.g. phenelzine and tranylcypromine) and a selective serotonin reuptake inhibitor (SSRI), and this combination should not be used. Importantly, the combination of a selective strong MAOI (i.e. selegiline and rasagaline) and an SSRI, though also capable of causing the serotonin syndrome, carries a far lower risk. This lower risk, however, is obtained only at doses below 10 mg of selegiline and 1 mg of rasagaline: at higher doses these MAOIs become non-selective, and hence carry a higher risk. Mention here should also be made of the antibiotic linezolid: although this is a non-selective MAOI, it has only weak activity in this regard, and hence carries no greater risk than the selective MAOIs.

Other combinations of some concern include the following: a strong non-selective MAOI with a tricyclic or with venlafaxine, and an SSRI with a tricyclic. With regard to the tricyclics, special attention should be paid to the use of clomipramine, as it has strong serotoninergic effects. Although these combinations can cause the syndrome, this is quite uncommon, and indeed the combination of a strong, non-selective MAOI and a tricylic is one of the recommended treatments for treatment-resistant depression.

Various other combinations have also been reported to cause the syndrome, but in general these combinations are quite safe. They include an MAOI with duloxetine, cyclobenzaprine, or meperidine; an SSRI in combination with any one of the following: a triptan (e.g. sumatriptan), a second-generation antipsychotic (e.g. risperidone, olanzapine, or quetiapine), an opioid (e.g. tramadol, fentanyl, hydromorphone, meperidine, or pentazocine), carbamazepine, trazodone, mirtazapine, bupropion, buspirone, and, finally, levodopa; and lithium with a tricyclic, venlafaxine, or a triptan. Unusual combinations include trazodone with venlafaxine, buspirone, or bupropion; buspirone and bupropion; and cyclobenzaprine and duloxetine.

Monotherapy with serotoninergic agents can cause the syndrome, but at therapeutic doses this is quite rare; it has been reported for SSRIs, tricyclics, venlafaxine, mirtazapine, and trazodone. With overdoses of SSRIs, however, the syndrome is common.

Before leaving this discussion, mention should be made of tryptophan. Tryptophan, a serotonin precursor, has been banned in the USA but may be found in various 'supplements', some of which may not be labeled as such, and the syndrome has occurred with the combination of tryptophan and an MAOI, an SSRI, or a tricyclic.

Differential diagnosis

The occurrence of a full or partial syndrome in the setting of any of the pharmacologic maneuvers noted above, under 'Etiology', should suggest the serotonin syndrome. A particularly difficult differential, however, may arise in cases where patients had been on an antipsychotic and a serotoninergic agent had been added on, because the neuroleptic malignant syndrome and the serotonin syndrome may both cause delirium, rigidity, fever, tachycardia, and hypertension. In such cases, certain features, such as nausea, vomiting, diarrhea, myoclonus, and clonus, all point towards the serotonin syndrome, as these features are lacking in the neuroleptic malignant syndrome.

When the medication history is lacking, consideration must be given to encephalitis and to various other disorders capable of causing delirium with myoclonus, including uremic encephalopathy, hepatic encephalopathy, hyperosmolar non-ketotic hyperglycemia, Hashimoto's encephalopathy, baclofen withdrawal, complex partial status epilepticus, and encephalopathic pellagra.

Treatment

In addition to discontinuing all potentially offending medications and providing vigorous supportive care, parenteral lorazepam should be used for agitation. In mild cases, these measures may be sufficient. In moderate or severe cases, however, specific treatment with cyproheptadine (Graudins *et al.* 1998; Lappin and Auchincloss 1994; McDaniel 2001), a serotonin antagonist, should be undertaken. Cyproheptadine may be given in a dose of 4–8 mg every 2 hours, or in a loading dose of 12 mg followed by 2 mg every 2 hours, until symptoms are controlled or a maximum dose of 32 mg is reached. Most patients respond within a day, and some within hours; additional doses may or may not be required, depending, in part, on the half-lives of the medications at fault. In severe cases sedation, paralysis, and intubation may be required.

29.6 ANTICHOLINERGIC DELIRIUM

In sufficient dosage, any drug with anticholinergic properties that crosses the blood–brain barrier may cause a delirium. Old age and medical frailty increase the risk, and anticholinergic toxicity is a prominent cause of delirium in hospitalized patients (Han *et al.* 2001), especially in postoperative cases (Tune *et al.* 1981).

Of some literary interest is the possibility that the Reverend Dimmesdale, of Nathaniel Hawthorne's *The Scarlet Letter*, may himself have succumbed to an herbal preparation with anticholinergic properties, administered by the good Doctor Chillingworth (Khan 1984).

Clinical features

Clinically (Itil and Fink 1966), there is delirium and restlessness or agitation, often accompanied by visual hallucinations, and patients may report dry mouth, blurry vision, and constipation. On examination, the temperature and pulse are elevated, the skin is typically dry and flushed (at times to a scarlet hue), the pupils are dilated, the bowel sounds decreased, and the deep tendon reflexes are brisk; rarely, one may see myoclonus and choreoathetosis (Holinger and Klawans 1976). Urinary retention may occur and, in severe cases, there may be seizures, coma, respiratory depression, and death.

Course

In the natural course of events, provided that the offending medication is discontinued, there is a gradual remission of symptoms, consistent with the half-life of the anticholinergic in question.

Etiology

Any of a large number of drugs with anticholinergic properties may, if given in sufficient dose, cause a delirium (Tune *et al.* 1981); although in most cases relatively high doses are required, individual sensitivity varies widely, and in some patients, especially the elderly, seemingly innocuous doses can have devastating results. Anticholinergically active drugs to consider include the following: atropine, scopolamine (Vonderahe 1929; Ziskind 1988), and homatropine ophthalmic drops (Tune *et al.* 1992); anticholinergic antiparkinsonian agents (De Smet *et al.* 1982) such as benztropine, biperidin, and trihexyphenidyl (Porteous and Ross 1956); tricyclic antidepressants (Goodwin 1983; Preskorn and Simpson 1982); antihistamines such as diphenhydramine (Tejera *et al.* 1994) and hydroxyzine; low-potency first-generation antipsychotics; cyclobenzaprine (Engel and Chapron 1993); and oxybutynin and tolteridone (Tsao and Heilman 2003). Delirium has also been noted with 'recreational' intoxication with Angel's trumpet (*Datura stramonium*) (Hall *et al.* 1977; Oberndorfer *et al.* 2002). Of all of these agents, diphenhydramine stands out (Agostino *et al.* 2001), not so much because of its inherent toxicity but because of the frequency with which it is used in hospitalized patients.

Differential diagnosis

Heat stroke is clinically very similar to anticholinergic delirium, with delirium; skin that is hot, dry, and flushed; tachycardia; and fever. However, in heat stroke the temperature is often at 41.1 °C (106 °F), whereas in simple anticholinergic delirium it rarely rises above 40 °C (104 °F). Certainly, the diagnosis of heat stroke should not be considered unless the ambient temperature is quite high; however, in some cases of high ambient temperature when patients are taking anticholinergics, one may be confronted with an etiologically mixed picture in which the anticholinergic, by reducing sweating, sets the stage for the dramatic temperature increases seen in heat stroke.

Treatment

The anticholinergic should be stopped and, if ingestion is recent, consideration may be given to gastric lavage or activated charcoal. General supportive measures include intravenous fluids and, if the temperature is significantly elevated, cooling blankets. If symptomatic treatment of hallucinations or agitation is required, an antipsychotic may be utilized; however, those with anticholinergic effects (all first-generation agents and both olanzapine and quetiapine) should obviously be avoided in favor of agents innocent in this regard, such as risperidone, ziprasidone, and aripiprazole. Seizures may be treated with lorazepam.

In emergent situations, treatment with physostigmine (Beaver and Gavin 1998; Burns *et al.* 2000; Duvoisin and Katz 1968; Stern 1983) may be considered; however, as noted below, this carries some risk. Physostigmine may be given intravenously in a dose of 0.5–2 mg, at a rate no faster than 1 mg/min, with repeat doses every 5–10 minutes until the patient is out of danger. Given the half-life of most anticholinergics, repeat doses of physostigmine are often required, initially every 30–60 minutes. A failure to respond to physostigmine essentially rules out a diagnosis of anticholinergic delirium. Physostigmine is not a benign treatment and patients may develop bronchospasm, seizures, bradycardia, or asystole, and hence treatment should be undertaken only under cardiac monitoring with intravenous atropine close at hand. Furthermore, in cases of tricyclic overdose, physostigmine has no effect on the development of arrhythmia, which is the main concern in this situation. With all these considerations in mind, many clinicians will opt for conservative treatment and simply let the offending agent 'wash out', allowing the delirium to resolve on its own. In severe cases, however, with significant temperature elevations, seizures, coma, or respiratory depression, treatment is justified.

29.7 CHOLINERGIC REBOUND

The abrupt discontinuation of medications with significant anticholinergic properties may, in about one-third of cases, be followed by a syndrome known as cholinergic rebound, or cholinergic overdrive.

Clinical features

The onset of cholinergic rebound is in large part determined by the half-life of the drug that has just undergone abrupt discontinuation; thus, in the case of most tricyclic antidepressants, the onset may be anticipated within 36–48 hours.

Clinically (Dilsaver 1989; Dilsaver *et al.* 1983a; Petti and Law 1981), patients present with depressed mood, anxiety, malaise, and insomnia; typically, there is also nausea and abdominal cramping, with, at times, vomiting and diarrhea.

Course

In the natural course of events, symptoms subside over 1–3 days.

Etiology

With chronic treatment with any drug possessing anticholinergic properties there is a compensatory upregulation of postsynaptic acetylcholine receptors; if the drug is abruptly discontinued and there is insufficient time for a gradual 'downregulation', the increased number of receptors is 'unmasked', resulting in an increase in cholinergic transmission and the symptoms seen in this disorder (Dilsaver *et al.* 1983b; Luchins *et al.* 1980). This has been noted with tricyclic antidepressants (Dilsaver *et al.* 1983a; Wolfe 1997), low-potency first-generation antipsychotics, the second-generation antipsychotic clozapine (Delassus-Guenault *et al.* 1999), atropine, scopolamine, and antiparkinsonian anticholinergics, such as benztropine, biperidin, and trihexyphenidyl.

Differential diagnosis

The appearance of depressed mood and insomnia shortly after stopping a tricyclic antidepressant may suggest a relapse of depression; however, the abruptness of the onset of symptoms is inconsistent with a relapse of depression, which would not be expected for at least a matter of weeks after stopping an antidepressant.

Treatment

The best treatment is prevention, and medications with strong anticholinergic effects should be tapered over 3 or 4 days. In cases in which rebound does occur, some patients may elect to simply wait it out. When symptoms are severe, however, one may restart the original medication, or, if this is not feasible, use another anticholinergic medication (e.g. benztropine in a dose of 1–2 mg), and adjust the dose to suppress symptoms, after which a gradual tapering may be undertaken (Wolfe 1997).

29.8 ALCOHOLIC DEMENTIA

Alcoholic dementia, also known as alcohol-related dementia or alcohol-induced persisting dementia, is one of the most common causes of dementia and indeed may account for approximately one-fifth of all cases among the institutionalized elderly (Carlen *et al.* 1994). Among chronic alcoholics, approximately 10 percent will develop this dreaded complication.

Clinical features

Alcoholic dementia presents insidiously, generally after decades of alcoholism. Clinically (Lee *et al.* 1979; Lishman 1981), patients typically present with a personality change, with frontal lobe features: there is a general coarsening of personality and a heedless disregard for social conventions; judgment becomes poor and many patients become apathetic. Over time cognitive deficits gradually accrue, including short-term memory loss, concreteness, and difficulties in understanding complex situations.

Computed tomography (Carlen *et al.* 1981; Gurling *et al.* 1984; Harper *et al.* 1985) and MRI (Jernigan *et al.* 1991; Schroth *et al.* 1988) reveal both cortical atrophy and ventricular dilation, and there is a correlation between the extent of atrophy and the severity of the dementia (Carlen *et al.* 1981; Carlsson *et al.* 1979).

Course

With continued drinking the dementia progresses and may become profound; with abstinence a variable degree of recovery may be expected over about a 6-month period (Grant *et al.* 1984; Hambidge 1990), after which any remaining deficits tend to persist indefinitely.

Etiology

Alcohol, in all likelihood, is directly toxic to the white matter and perhaps also to cortical neurons. Autopsy studies have demonstrated a reduction in brain weight (Harper and Blumbergs 1982; Torvik *et al.* 1982) with overall cerebral atrophy (Harper and Krill 1985; Lynch 1960; Neuberger 1957), primarily due to loss of white matter (Harper *et al.* 1985; Jensen and Pakkenberg 1993); there is also some evidence for neuronal loss, particularly in the superior frontal cortex (Harper *et al.* 1987; Krill *et al.* 1997). Of note, it appears that some of these changes may be reversible: with sobriety, MR scanning has revealed re-expansion of the brain (Pfefferbaum *et al.* 1995; Schroth *et al.* 1988; Zipursky *et al.* 1989), primarily as a result of enlargement of the white matter (Agartz *et al.* 2003).

Differential diagnosis

The diagnosis of alcoholic dementia, given the rampant denial present in alcoholics and their family members, is often missed. In cases in which the history of alcoholism is clear, one should also bear in mind that cognitive deficits associated with withdrawal may persist for some time, and consequently the diagnosis of alcoholic dementia should probably be tentative only until a month or more of sobriety has been maintained.

Given that most cases of alcoholic dementia present in middle years or later, consideration must be given to other common causes of dementia in this age group. Of the various causes discussed in Chapter 6.3, certain deserve special attention, including Alzheimer's disease, Binswanger's disease, lacunar dementia, and, especially, frontotemporal lobar degeneration. Although each of these has more or less distinctive features, they all share one feature in common, which distinguishes them from alcoholic dementia, and that is their progressive nature, which contrasts with the initial improvement and subsequent stable chronicity characteristic of alcoholic dementia, provided, of course, that alcohol use has ceased. Another disorder to consider, given the falls so common in active alcoholism, is chronic subdural hematoma.

Alcoholic dementia must also be distinguished from Korsakoff's syndrome; however, this should not be difficult. As discussed in Chapter 14.5, Korsakoff's syndrome is characterized by anterograde and variably severe retrograde amnesia, and lacks the personality change and concreteness seen in alcoholic dementia.

Treatment

Adequate nutrition, including thiamine and niacin, and, above all, abstinence are essential. The overall treatment of alcoholism is discussed in Chapter 28.5. In cases in which patients are unable to participate successfully in rehabilitative efforts, institutionalization may be required.

29.9 ALCOHOL HALLUCINOSIS

Alcohol hallucinosis, also known as alcohol-induced psychotic disorder with hallucinations, is seen only in chronic alcoholics, typically after one or more decades of severe, heavy drinking. This psychosis is an uncommon disorder and may be more common in men than women.

Clinical features

The onset is typically abrupt, over a matter of days, and generally occurs in the context of delirium tremens. Clinically (Soyka 1990; Victor and Hope 1958), the principal symptom of alcohol hallucinosis is auditory hallucinations. These are often extremely vivid and clear, and the patient has no doubt as to their reality. For the most part they are critical, deprecatory, and often persecutory. Generally, more than one voice is heard, and, curiously, the voices often talk among themselves. Visual hallucinations may also occur, but these are far less prominent than auditory hallucinations. Delusions of persecution and reference often accompany the auditory hallucinations, and are generally congruent with them. Patients may believe that others are plotting against them, or that the police are following them. Occasionally there may be Schneiderian first-rank delusions, such as thought-broadcasting or delusions of influence (Soyka 1990). Patients are often constrained and very watchful, and tend to be irritable and querulous.

Course

With continued drinking, symptoms persist and may worsen. With abstinence, a gradual remission of symptoms of variable extent may occur over the following weeks or months. Should patients commence drinking again, symptoms typically recur, and, with another period of abstinence, the remission is generally not as substantial. Eventually, with repeated relapses, there may be a chronic persistence of symptoms, even with long-sustained sobriety.

Etiology

Although the etiology of alcohol hallucinosis is not clear, it does appear that the risk for developing this disorder rises in direct proportion to the severity of the alcoholism and, more importantly, to the frequency with which alcohol withdrawal and delirium tremens occurs. With this in mind, some investigators have proposed that alcohol hallucinosis is the end result of repeated 'kindling' within the temporal lobes. Importantly, alcohol hallucinosis is not etiologically related to paranoid schizophrenia (Schuckit and Winokur 1971).

Differential diagnosis

Delirium tremens may cause auditory hallucinations, and the fact that alcohol hallucinosis often has an onset in the course of delirium tremens sets the stage for some diagnostic difficulty. The question, however, may be resolved by observing the patient during enforced abstinence: in cases in which delirium tremens alone is present, all symptoms, including auditory hallucinations, gradually resolve; in cases, however, in which alcohol hallucinosis has appeared, the auditory hallucinations will persist despite resolution of other symptoms of delirium tremens, such as confusion, disorientation, and tremor.

Alcoholic paranoia, discussed in the following chapter, is distinguished by the prominence of delusions of persecution in the relative absence of any hallucinations.

Of the other causes of psychosis discussed in Chapter 8.1, the one that occasions the most diagnostic difficulty is paranoid schizophrenia. Patients with schizophrenia may also develop alcoholism; however, in these cases the psychosis generally occurs either before the onset of the alcoholism or relatively early on, in contrast to alcohol hallucinosis, which occurs only after a decade or more of heavy drinking. Certain symptoms may also enable a differential diagnosis to be made: loosening of associations, bizarre delusions, and bizarre behavior, although common in paranoid schizophrenia, are not seen in alcohol hallucinosis.

Treatment

Abstinence is essential, and the overall treatment of alcoholism is discussed in Chapter 28.5. Unfortunately, the very symptoms that are characteristic of alcohol hallucinosis often make it impossible for patients to participate in rehabilitative efforts, and hence additional treatment is generally required. Traditionally, antipsychotics are utilized (Soyka *et al.* 2007), and the choice among these may be made based on the same principles outlined in Chapter 22.1 for schizophrenia. Of interest, however, there is a double-blinded study (Aliyev and Aliyev 2008) demonstrating the effectiveness of valproate over placebo when given early on in the course of the disorder. Whether valproate is more effective than an antipsychotic, however, is an open question. Regardless of what agent is used it may be possible, given the natural course of the disorder, to taper the dose after the patient has been symptom free and sober for a prolonged interval.

29.10 ALCOHOLIC PARANOIA

Alcoholic paranoia, also known as alcohol-induced psychotic disorder with delusions, like alcohol hallucinosis, is seen only in chronic alcoholics, and only after many years of severe, heavy drinking. The prevalence of this disorder is a matter of some debate: although traditionally it is thought to be rare, clinical experience suggests that it is not uncommon.

Clinical features

The onset is gradual, and symptoms appear without any direct connection with either alcohol withdrawal or delirium tremens. Clinically (Albert *et al.* 1995), patients gradually develop either delusions of persecution or of jealousy. Those with delusions of persecution may believe that the police are following them or that intruders are hiding in the house; patients may peek out of windows to try and catch a glimpse of the police they are sure are 'staking out' the house, or they may feel compelled to go from room to room, perhaps with a weapon, looking for intruders. Those with delusions of jealousy typically suspect their spouse or lover of infidelity, and they may follow them or look for clues of the suspected romantic encounters. Occasionally there may be some hallucinations but these never dominate the clinical picture; indeed, if they do occur they typically play a very minor role, for example the persecuted patient may hear footsteps outside or the jealous patient may smell an unaccustomed cologne or perfume.

Course

In general, with continued drinking there is a gradual worsening of symptoms until eventually a 'plateau' of severity is reached, after which symptoms remain stable, despite ongoing drinking. With abstinence, symptoms gradually lessen over months to up to 2 years, and then either go into remission or settle into a stable, low-level chronicity.

Etiology

The mechanism by which this disorder occurs is not clear.

Differential diagnosis

Alcohol hallucinosis, discussed in the preceding chapter, is distinguished by its abrupt onset during alcohol withdrawal or delirium tremens, and by the prominence of hallucinations.Of the other causes of psychosis discussed in Chapter 8.1, delusional disorder most closely resembles alcoholic paranoia. In delusional disorder, however, the history of chronic alcoholism is lacking.

Treatment

Abstinence is essential, and the overall treatment of alcoholism is discussed in Chapter 28.5. Treatment with antipsychotics should be considered, especially in cases in which delusions of persecution hinder the patient's efforts to engage in rehabilitation. The choice of antipsychotic is based on the same principles as outlined for schizophrenia in Chapter 22.1; given the natural history of alcoholic paranoia, attempts should eventually be made to taper the dose after the patient has been symptom free and sober for at least a few months.

29.11 MARCHIAFAVA–BIGNAMI DISEASE

Marchiafava–Bignami disease is a very rare disorder that was first described by Doctors Marchiafava and Bignami (Marchiafava and Bignami 1903) in 1903 amongst Italian alcoholics. Traditionally, this disorder has been considered one of the alcohol-related diseases, and consequently it is included in this book with other substance-induced illnesses. As noted below, however, Marchiafava–Bignami disease, rather than occurring secondary to the direct toxic effects of alcohol, may in fact be a vitamin-deficiency disease, and if this does prove to be the case, it will have to be regrouped with the nutritional disorders.

Clinical features

Depending on the mode of onset, clinical symptomatology and MRI findings, Marchiafava–Bignami disease may roughly be divided (Heinrich *et al.* 2004) into an acute and subacute form.

The acute form (Bohrod 1942; Chang *et al.* 1992; Heinrich *et al.* 2004; Ironside *et al.* 1961; Kamaki *et al.* 1993; Kawamura *et al.* 1985; Menegon *et al.* 2005; Rosa *et al.* 1991) evolves fairly rapidly, over days or weeks, and is characterized by delirium progressing to stupor and, in most cases, coma. Before the onset of stupor and coma, additional findings may also be noted, including dysarthria, ataxia, seizures, and callosal disconnection symptoms (Lechevalier *et al.* 1977; Lhermitte *et al.* 1977), such as left-sided apraxia and agnosia. Fluid-attenuated inversion recovery (FLAIR), T2-weighted and diffusion-weighted MRI typically reveal increased signal intensity within the middle tier of the corpus callosum along its entire extent, including the genu, body, and splenium; in some cases these signal abnormalities may extend symmetrically into the white matter on either side of the corpus callosum.

The subacute form (Heinrich *et al.* 2004; Mayer *et al.* 1987) presents in a more leisurely fashion, and though also characterized by delirium with dysarthria, ataxia, seizures, and disconnection symptoms, it does not progress to stupor or coma. Furthermore, on MRI, abnormalities are generally confined to the corpus callosum, which, instead of being involved in its full extent, rather demonstrates more or less focal involvement of the genu, body, or splenium, or perhaps combinations of only two of these areas.

Course

Prior to modern imaging, when diagnosis rested primarily on autopsy findings, it was assumed that most cases were fatal. With MRI-based diagnosis, however, it is apparent that the outlook is much better. In the acute form a fatal outcome is seen in only about one-fifth of cases, while the remainder experience a variable degree of recovery over the following weeks and months (Heinrich *et al.* 2004). The subacute form has an even more favorable course, with no fatalities and a far better cognitive outcome (Heinrich *et al.* 2004), with, in some cases, apparent full recovery (Heinrich *et al.* 2004; Helenius *et al.* 2001).

T1-weighted MRI obtained long after recovery typically reveals cystic lesions corresponding to the abnormalities seen acutely; in most cases one also sees a variable degree of callosal atrophy.

Etiology

Pathologically (Bohrod 1942; Ironside *et al.* 1961; Poser 1973), demyelination with relative axonal sparing is found in the central portion of the corpus callosum, sparing the dorsal and ventral areas, and this may extend laterally in a symmetric fashion to involve adjacent areas of the centrum semiovale. Demyelination may also be seen in the anterior and posterior commissures and the middle cerebellar peduncles. Cystic changes are seen in the same areas.

The cause of this demyelination is not clear. Originally it was believed that there was an association with the consumption of cheap red wine by Italian men, as most of the original cases fit this description; however, it has become quite clear that this disease may occur in alcoholics who consume whiskey, beer, or white wine, and also in non-Italians (Ironside *et al.* 1961). Indeed, there are also rare reports of the disease occurring in association with severe malnutrition in non-alcoholics (Leong 1979), including one case occurring during the course of a severe case of anorexia nervosa (Tao *et al.* 2007).

Given the high prevalence of severe alcoholism and the rarity of Marchiafava–Bignami disease, and the occurrence of this disease in malnourished non-alcoholics, it appears that direct toxicity of alcohol may well not be the major factor, and that a deficiency of one or of a combination of vitamins may underlie the development of the demyelination, perhaps in genetically predisposed individuals.

Differential diagnosis

Marchiafava–Bignami disease must be distinguished from several other disorders seen in chronic alcoholism or other malnourished states, including Wernicke's encephalopathy, encephalopathic pellagra, and hepatic encephalopathy. Though myoclonus might point toward encephalopathic pellagra and hepatic encephalopathy, and disconnection signs to Marchaifava–Bignami disease, in the vast majority of cases the diagnosis is not even suspected until MRI reveals the characteristic findings in the corpus callosum. Consideration might also be given to stroke occurring secondary to infarction in the area of distribution of the anterior cerebral artery; however, in these cases the entire width of the corpus callosum is involved, rather than merely the middle tier.

Treatment

Abstinence and adequate nutrition are critical, and it is appropriate to give thiamine, pyridoxine, niacin, and folic acid. There is also one case report, in a patient with anorexia nervosa, of a response to intravenous steroids (Tao *et al.* 2007).

29.12 INHALANT-INDUCED DEMENTIA

Inhalant intoxication (discussed in Chapter 28.7), if chronic and repeated, may cause a dementia.

Clinical features

The onset of dementia is gradual in the setting of ongoing, chronic inhalant use. Clinically (Fornazzari *et al.* 1983; Grabski 1961; Hormes *et al.* 1986; Knox and Nelson 1966; Lazar *et al.* 1983; Lewis *et al.* 1981; Rosenberg *et al.* 1988), patients become apathetic and concrete, and have difficulty with memory; in general, 'cortical' signs, such as aphasia, are lacking. In concert with the dementia, cerebellar signs are very common, including ataxia, dysarthria, titubation, intention tremor, and nystagmus. A minority of patients may also have spasticity. In some cases there may also be a peripheral neuropathy (Korobkin *et al.* 1975).

Computed tomography scanning reveals cerebral and cerebellar cortical atrophy (Fornazzari *et al.* 1983; Hormes *et al.* 1986; Lazar *et al.* 1983); in addition to the atrophy, T2-weighted MRI also reveals diffuse, increased signal intensity in the cerebral and cerebellar white matter, with, in some cases, decreased signal intensity in the thalami or basal ganglia (Aydin *et al.* 2002; Filley *et al.* 1990; Rosenberg *et al.* 1988).

Course

With continued use a progression occurs; with abstinence there may be a gradual, but generally only partial, remission.

Etiology

Autopsy studies have revealed both cerebral and cerebellar atrophy, with widespread demyelination (Al-Hajri and Del Bigio 2010; Escobar and Aruffo 1980; Kornfeld *et al.* 1994; Rosenberg *et al.* 1988).

Differential diagnosis

Consideration may be given to other disorders capable of causing dementia in combination with ataxia, as discussed in Chapter 6.3; however, the history of chronic inhalant use is hard to miss.

Treatment

Abstinence is essential and may require institutionalization. There is no specific treatment.

REFERENCES

Addonizio G, Susman VL, Roth SD. Neuroleptic malignant syndrome: review and analysis of 115 cases. *Biol Psychiatry* 1987; **22**:1004–20.

Adityanjee , Jayaswal SK, Chan TM *et al.* Temporary remission of tardive dystonia following electroconvulsive therapy. *Br J Psychiatry* 1990; **156**:433–5.

Adityanjee Sajatovic M, Munshi KR. Neuropsychiatric sequelae of neuroleptic malignant syndrome. *Clin Neuropharmacol* 2005; **28**:197–204.

Adler LA, Peselow E, Rotrosen J *et al.* Vitamin E in the treatment of tardive dyskinesia. *Am J Psychiatry* 1993; **150**:1405–7.

Adler LA, Edson R, Lavori P *et al.* Long-term treatment effects of vitamin E for tardive dyskinesia. *Biol Psychiatry* 1998; **43**:868–72.

Adler LA, Rotrosen J, Edson R *et al.* Vitamin E treatment for tardive dyskinesia. *Arch Gen Psychiatry* 1999; **56**:863–41.

Agartz I, Brag S, Franck J *et al.* MR volumetry during acute alcohol withdrawal and abstinence: a descriptive study. *Alcohol Alcohol* 2003; **38**:71–8.

Agostino JV, Leo-Summers LS, Inouye SK. Cognitive and other adverse effects of diphenhydramine use in hospitalized older patients. *Arch Int Med* 2001; **161**:2091–7.

Akhtar S, Jajor TR, Kumar S. Vitamin E in the treatment of tardive dyskinesia. *J Postgrad Med* 1993; **39**:124–6.

Al-Hajri Z, Del Bigio MR. Brain damage in a large cohort of solvent abusers. *Acta Neuropathol* 2010; **119**:435–45.

Albert A, Mirza S, Mirza KAH *et al.* Morbid jealousy in alcoholics. *Br J Psychiatry* 1995; **167**:668–72.

Aliyev ZN, Aliyev NA. Valproate treatment of acute alcohol hallucinosis: a double-blind, placebo-controlled study. *Alcohol Alcohol* 2008; **43**:456–9.

Ananth J, Edelmuth E, Dargan B. Meige's syndrome associated with neuroleptic treatment. *Am J Psychiatry* 1988; **145**:513–15.

Anderson ES, Powers PS. Neuroleptic malignant syndrome associated with clozapine use. *J Clin Psychiatry* 1991; **52**:102–4.

Assmann VCCA, van Woerkom TCAM. Disappearance of tardive dyskinesia during catatonic stupor. *Acta Psychiatr Scand* 1987; **76**:217–18.

Aydin K, Sencer S, Demir T *et al.* Cranial MR findings in chronic toluene abuse by inhalation. *AJNR* 2002; **23**:1173–9.

Beaver KM, Gavin TJ. Treatment of acute anticholinergic poisoning with physostigmine. *Am J Emerg Med* 1998; **16**:505–7.

Bharucha KJ, Sethi KD. Tardive Tourettism after exposure to neuroleptic therapy. *Mov Disord* 1995; **10**:791–3.

Bodner RA, Lynch T, Lewis L *et al.* Serotonin syndrome. *Neurology* 1995; **45**:219–23.

Bohrod MG. Primary degeneration of the corpus callosum (Marchiafava's disease): report of the second American case. *Arch Neurol Psychiatry* 1942; **47**:465–73.

Boyer EW, Shannon M. The serotonin syndrome. *N Engl J Med* 2005; **352**:1112–20.

Brunelle J, Guigueno S, Gouin P *et al.* Aripiprazole and neuroleptic malignant syndrome. *J Clin Psychopharmacol* 2007; **27**:212–14.

Burke RE, Fahn S, Jankovic J *et al.* Tardive dyskinesia: late-onset and persistent dystonia caused by antipsychotic drugs. *Neurology* 1982; **32**:1335–46.

Burke RE, Kang UK, Jankovic J *et al.* Tardive akathisia: an analysis of clinical features and response to open therapeutic trials. *Mov Disord* 1989; **4**:157–75.

Burns MJ, Linden CH, Graudins A *et al.* A comparison of physostigmine and benzodiazepines for the treatment of anticholinergic poisoning. *Ann Emerg Med* 2000; **35**:374–81.

Carlen PL, Wilkinson DA, Wotzman G *et al.* Cerebral atrophy and functional deficits in alcoholics without clinically apparent liver disease. *Neurology* 1981; **31**:377–85.

Carlen PL, McAndrews MP, Weiss RT *et al.* Alcohol-related dementia in the institutionalized elderly. *Alcohol Clin Exp Res* 1994; **18**:1330–4.

Carlsson C, Claeson L-E, Karlsson K-I *et al.* Clinical, psychometric and radiological signs of brain damage in chronic alcoholism. *Acta Neurol Scand* 1979; **60**:85–92.

Caroff SN. The neuroleptic malignant syndrome. *J Clin Psychiatry* 1980; **41**:79–83.

Caykoylu A, Ekinci O, Kuloglu M *et al.* Aripiprazole-induced rabbit syndrome: a case report. *J Psychopharmacol* 2010; **24**:429–31.

Chang KH, Cha SH, Han MH *et al.* Marchiafava–Bignami disease: serial changes in corpus callosum on MRI. *Neuroradiology* 1992; **34**:480–2.

Chiang E, Pitts WM, Rodriguez-Garcia M. Respiratory dyskinesia: review and case reports. *J Clin Psychiatry* 1985; **46**:232–4.

Chiu HFK, Lan LCW, Chung DWS. Prevalence of the rabbit syndrome in Hong Kong. *J Nerv Ment Dis* 1993; **181**:264–5.

Chouinard G, Jones BD. Neuroleptic-induced supersensitivity psychosis: clinical and pharmacologic characteristics. *Am J Psychiatry* 1980; **137**:16–21.

Coffey RJ, Edgar TS, Francsco GE *et al.* Abrupt withdrawal from intrathecal baclofen: recognition and management of a potentially life-threatening syndrome. *Arch Phys Med Rehabil* 2002; **83**:1479.

Cunningham MA, Darby DG, Donnan GA. Controlled-release delivery of L-dopa associated with nonfatal hyperthermia, rigidity and autonomic dysfunction. *Neurology* 1991; **41**:942–3.

Dabiri LM, Pasta D, Darby JK *et al.* Effectiveness of vitamin E for the treatment of long-term tardive dyskinesia. *Am J Psychiatry* 1994; **151**:925–6.

Damier P, Thobois S, Witjas T *et al.* Bilateral deep brain stimulation of the globus pallidus to treat tardive dyskinesia. *Arch Gen Psychiatry* 2007; **64**:170–6.

Daras M, Kakkouras L, Tuchman AJ *et al.* Rhabdomyolysis and hyperthermia after cocaine abuse: a variant of the neuroleptic malignant syndrome? *Acta Neurol Scand* 1995; **92**:161–5.

Das Gupta K, Young A. Clozapine-induced neuroleptic malignant syndrome. *J Clin Psychiatry* 1991; **52**:105–7.

Davis JM, Janicak PG, Sakkas P *et al.* Electroconvulsive therapy in the treatment of the neuroleptic malignant syndrome. *Convulsive Ther* 1991; **7**:111–20.

Decina P, Caracci G, Spichio PL. The rabbit syndrome. *Mov Disord* 1990; **5**:263–6.

Delassus-Guenault N, Jegouzo A, Odou P *et al.* Clozapine–olanzapine: a potentially dangerous switch. A report of two cases. *J Clin Pharmacol Ther* 1999; **24**:191–5.

de Potter RW, Linkowski P, Mendelwicz J. State-dependent tardive dyskinesia in manic–depressive illness. *J Neurol Neurosurg Psychiatry* 1983; **46**:666–8.

Deshmukh DK, Joshi VS, Agarwal MR. Rabbit syndrome – a rare complication of long-term neuroleptic medication. *Br J Psychiatry* 1990; **157**:293.

De Smet Y, Ruberg M, Serdaru M *et al.* Confusion, dementia and anticholinergics in Parkinson's disease. *J Neurol Neurosurg Psychiatry* 1982; **45**:1161–4.

Dilsaver SC. Antidepressant withdrawal syndromes: phenomenonology and pathophysiology. *Acta Psychiatr Scand* 1989; **79**:113.

Dilsaver SC, Feinberg M, Greden JF. Antidepressant withdrawal symptoms treated with anticholinergic agents. *Am J Psychiatry* 1983a; **140**:249–51.

Dilsaver SC, Kronfol Z, Sackellares JC *et al.* Antidepressant withdrawal syndrome: evidence supporting the cholinergic overdrive hypothesis. *J Clin Psychopharmacol* 1983b; **3**:157–64.

Dorevitch A, Lerner V, Shalfman M *et al.* Lack of effect of vitamin E on serum creatine phosphokinase in patients with long-term tardive dyskinesia. *Int Clin Psychopharmacol* 1997; **12**:171–3.

Dufresne RL, Wagner RL. Antipsychotic-withdrawal akathisia versus antipsychotic-induced akathisia: further evidence for the existence of tardive akathisia. *J Clin Psychiatry* 1988; **49**:435–8.

Duvoisin RC, Katz R. Reversal of central anticholinergic syndrome in man by physostigmine. *JAMA* 1968; **206**:1963–5.

Egan MF, Hyde TM, Albers GW *et al.* Treatment of tardive dyskinesia with vitamin E. *Am J Psychiatry* 1992; **149**:773–7.

Elkashef AM, Ruskin PE, Bacher N *et al.* Vitamin E in the treatment of tardive dyskinesia. *Am J Psychiatry* 1990; **147**:505–6.

Engel PA, Chapron D. Cyclobenzaprine-induced delirium in two octogenarians. *J Clin Psychiatry* 1993; **54**:39.

Escobar A, Aruffo C. Chronic thinner intoxication: clinico-pathologic report of a human case. *J Neurol Neurosurg Psychiatry* 1980; **43**:986–94.

Fahn S. High dosage anticholinergic therapy in dystonia. *Neurology* 1983; **33**:1255–61.

Farran-Ridge C. Some syndromes referable to the basal ganglia occurring in dementia praecox and epidemic encephalitis. *J Ment Sci* 1926; **72**:513–23.

Feighner JP, Boyer WF, Tyler DL *et al.* Adverse consequences of fluoxetine–MAOI combination therapy. *J Clin Psychiatry* 1990; **51**:222–5.

Figa-Talamanca L, Gualandi C, Di Meo L *et al.* Hyperthermia after discontinuance of levodopa and bromocriptine therapy: impaired dopamine receptors a possible cause. *Neurology* 1985; **35**:258–61.

Filley CM, Heaton RK, Rosenberg NL. White matter dementia in chronic toluene abuse. *Neurology* 1990; **40**:532–4.

FitzGerald PM, Jankovic J. Tardive oculogyric crises. *Neurology* 1989; **39**:1434–7.

Ford B, Greene P, Fahn S. Oral and genital tardive pain syndromes. *Neurology* 1994; **44**:2115–19.

Fornazzari L, Wilkinson DA, Kapur BM *et al.* Cerebellar, cortical and functional impairment in toluene abusers. *Acta Neurol Scand* 1983; **67**:319–29.

Fornazzari L, Ichise M, Remington G *et al.* Rabbit syndrome, antidepressant use, and cerebral perfusion SPECT scan findings. *J Psychiatry Neurosci* 1991; **16**:227–9.

Friedman J, Feinberg SS, Feldman RG. A neuroleptic malignant-like syndrome due to L-dopa withdrawal. *Ann Neurol* 1984; **16**:126–7.

Friedman LS, Weinrauch LA, D'Elia JA. Metoclopramide-induced neuroleptic malignant syndrome. *Arch Int Med* 1987; **147**:1495–7.

Gardos G, Cole JO, Haskell D *et al.* The natural history of tardive dyskinesia. *J Clin Psychopharmacol* 1988; **8**(Suppl. 4):31–3.

Gardos G, Casey D, Cole JO *et al.* Ten year outcome of tardive dyskinesia. *Am J Psychiatry* 1994; **151**:836–41.

Gibb W, Griffith D. Levodopa withdrawal syndrome identical to neuroleptic malignant syndrome. *Postgrad Med J* 1986; **62**: 59–60.

Glazer WM, Moore DC, Hansen TC *et al.* Meige syndrome and tardive dyskinesia. *Am J Psychiatry* 1983; **140**:798–9.

Glazer WM, Moore DC, Schooler NR *et al.* Tardive dyskinesia: a discontinuation study. *Arch Gen Psychiatry* 1984; **41**:623–7.

Glazer WM, Morgenstern H, Schooler N *et al.* Predictors of improvement in tardive dyskinesia following discontinuation of neuroleptic medication. *Br J Psychiatry* 1990; **157**:585–92.

Glazer WM, Morgenstern H, Doucette JT. Predicting the long-term risk of tardive dyskinesia in outpatients maintained on neuroleptic medication. *J Clin Psychiatry* 1993; **54**:133–9.

Goldman MB, Luchins DJ. Intermittent neuroleptic therapy and tardive dyskinesia: a literature review. *Hosp Comm Psychiatry* 1984; **35**:1215–19.

Goodwin CD. Case report of tricyclic-induced delirium at a therapeutic drug concentration. *Am J Psychiatry* 1983; **140**:1517–18.

Grabski DA. Toluene sniffing producing cerebellar degeneration. *Am J Psychiatry* 1961; **118**:461–2.

Granato JE, Stern BJ, Ringel J *et al.* Neuroleptic malignant syndrome: successful treatment with dantrolene and bromocriptine. *Ann Neurol* 1983; **14**:89–90.

Grant I, Adams KM, Reed R. Aging, abstinence, and medical risk factors in the prediction of neuropsychologic deficit among long-term alcoholics. *Arch Gen Psychiatry* 1984; **41**:710–18.

Graudins A, Stearman A, Chan B. Treatment of serotonin syndrome with cyporoheptadine. *J Emerg Med* 1998; **16**:615–19.

Greil W, Haag H, Rossnagl G *et al.* Effect of anticholinergics on tardive dyskinesia: a controlled discontinuation study. *Br J Psychiatry* 1984; **145**:304–10.

Gurling HMD, Reveley MA, Murray RM. Increased cerebral ventricular volume in monozygotic twins discordant for alcoholism. *Lancet* 1984; **1**:986–8.

Guze BH, Baxter LR. Current concepts: neuroleptic malignant syndrome. *N Engl J Med* 1985; **313**:163–6.

Hall RCW, Popkin MK, McHenry LE. Angel's trumpet psychosis: a central nervous system anticholinergic syndrome. *Am J Psychiatry* 1977; **134**:312–14.

Hambidge DM. Intellectual impairment in male alcoholics. *Alcohol Alcoholism* 1990; **25**:555–9.

Han L, McCusker J, Cole M *et al.* Use of medications with anticholinergic effect predicts clinical severity of delirium symptoms in older medical inpatients. *Arch Int Med* 2001; **161**:1099–105.

Harper CG, Blumbergs PC. Brain weights in alcoholics. *J Neurol Neurosurg Psychiatry* 1982; **45**:838–40.

Harper CG, Krill JJ. Brain atrophy in chronic alcoholic patients – a quantitative pathological study. *J Neurol Neurosurg Psychiatry* 1985; **48**:211–17.

Harper CG, Krill JJ, Holloway RL. Brain shrinkage in chronic alcoholics: a pathological study. *BMJ* 1985; **290**:501–4.

Harper CG, Kril J, Daly J. Are we drinking our neurones away? *Br Med J (Clin Res Ed)* 1987; **294**:534–6.

Harsch HH. Neuroleptic malignant syndrome: physiological and laboratory findings in a series of nine cases. *J Clin Psychiatry* 1987; **48**:328–33.

Hay DP, Hay L, Blackwell B *et al.* ECT and tardive dyskinesia. *J Geriatr Psychiatr Neurol* 1990; **3**:106–9.

Heinrich A, Runge U, Khaw AV. Clinicoradiologic subtypes of Marchiafava–Bignami disease. *J Neurol* 2004; **251**:1050–9.

Helenius J, Tatlisumak T, Soinne L *et al.* Marchiafava–Bignami disease: two cases with favorable outcome. *Eur J Neurol* 2001; **8**:269–72.

Hermesh H, Aizenburg D, Weizman A. A successful electroconvulsive treatment of neuroleptic malignant syndrome. *Acta Psychiatr Scand* 1987; **75**:237–9.

Holinger PC, Klawans HL. Reversal of tricyclic-overdosage-induced central anticholinergic syndrome by physostigmine. *Am J Psychiatry* 1976; **133**:1018–23.

Hormes JT, Filley CM, Rosenberg NL. Neurologic sequelae of chronic solvent vapor abuse. *Neurology* 1986; **36**:698–702.

Horn E, Lach B, Lapierre Y *et al.* Hypothalamic pathology in the neuroleptic malignant syndrome. *Am J Psychiatry* 1988; **145**: 617–20.

Hoy JS, Alexander B. Rabbit syndrome secondary to risperidone. *Pharmacotherapy* 2002; **22**:513–15.

Huang CC. Persistent tardive dyskinesia associated with amoxapine therapy. *Am J Psychiatry* 1986; **143**:1069–70.

Ironside R, Bosanquet FD, McMenemy WH. Central demyelination of the corpus callosum (Marchiafava–Bignami disease): with report of a second case in Great Britain. *Brain* 1961; **84**:212–30.

Itil T, Fink M. Anticholinergic drug-induced delirium: experimental modification, quantitative EEG, and behavioral correlations. *J Nerv Ment Dis* 1966; **143**:492–507.

Ivanovitch M, Glantz R, Bone RC *et al.* Respiratory dyskinesia presenting as acute respiratory distress. *Chest* 1993; **103**:314–16.

Jensen GB, Pakkenberg B. Do alcoholics drink their neurons away? *Lancet* 1993; **342**:1201–4.

Jernigan TL, Butters N, Ditraglia G *et al.* Reduced cerebral grey matter in alcoholics using MRI. *Alcoholism Clin Exp Res* 1991; **15**:418–27.

Jeste DV, Potkin SG, Sinha S *et al.* Tardive dyskinesia – reversible and persistent. *Arch Gen Psychiatry* 1979; **36**:585–90.

Jones EM, Dawson A. Neuroleptic malignant syndrome: a case report with post-mortem brain and muscle pathology. *J Neurol Neurosurg Psychiatry* 1989; **52**:1006–9.

Kamaki M, Kawamura M, Moriya H *et al.* 'Crossed homonymous hemianopia' and a 'crossed left hemispatial neglect' in a case of Marchiafava–Bignami disease. *J Neurol Neurosurg Psychiatry* 1993; **56**:1027–32.

Kane JM, Smith JM. Tardive dyskinesia: prevalence and risk factors, 1959–1979. *Arch Gen Psychiatry* 1982; **39**:473–81.

Kane JM, Woerner M, Lieberman J. Tardive dyskinesia: prevalence, incidence and risk factors. *J Clin Psychopharmacol* 1988; **8**:S52–S56.

Kang UJ, Burke RE, Fahn S. Natural history and treatment of tardive dystonia. *Mov Disord* 1986; **1**:193–208.

Kawamura M, Shiota J, Yagishita T *et al.* Marchiafava–Bignami disease: computed tomographic scan and magnetic resonance imaging. *Ann Neurol* 1985; **18**:103–4.

Keck PE, Pope HG, McElroy SL. Frequency and presentation of neuroleptic malignant syndrome: a prospective study. *Am J Psychiatry* 1987; **144**:1344–6.

Kellam AMP. Neuroleptic malignant syndrome, so-called. A survey of the world literature. *Br J Psychiatry* 1987; **150**:752–9.

Kellam AMP. The (frequently) neuroleptic malignant syndrome. *Br J Psychiatry* 1990; **157**:169–73.

Kennedy PF. Chorea and phenothiazines. *Br J Psychiatry* 1969; **115**:103–4.

Kent TA, Wilber RD. Reserpine withdrawal psychosis: the possible role of denervation supersensitivity of receptors. *J Nerv Ment Dis* 1982; **170**:502–4.

Keyser DL, Rodnitzky RL. Neuroleptic malignant syndrome in Parkinson's disease after withdrawal or alteration of dopaminergic therapy. *Arch Int Med* 1991; **151**:794–6.

Khan JA. Occasional notes: atropine poisoning in Hawthorne's The Scarlet Letter. *N Engl J Med* 1984; **311**:414–17.

Kiriakakis V, Bhatia K, Quinn NP *et al.* The natural history of tardive dystonia: a long-term follow-up study of 107 cases. *Brain* 1998; **121**:2053–66.

Kirkpatrick B, Alphs L, Buchanen RW. The concept of supersensitivity psychosis. *J Nerv Ment Dis* 1992; **180**:265.

Klawans HL, Rubovits R. Effect of cholinergic and anticholinergic agents on tardive dyskinesia. *J Neurol Neurosurg Psychiatry* 1974; **37**:941–7.

Knox JW, Nelson JR. Permanent encephalopathy from toluene inhalation. *N Engl J Med* 1966; **275**:1494–6.

Koch M, Chandragiri S, Rizvi S *et al.* Catatonic signs in neuroleptic malignant syndrome. *Compr Psychiatry* 2000; **41**:73–5.

Kontaxakis VP, Havaki-Kontaxaki BJ, Pappa DA *et al.* Neuroleptic malignant syndrome after addition of paroxetine to olanzapine. *J Clin Psychopharmacol* 2003; **23**:671–2.

Kornfeld M, Moser AB, Moser HW *et al.* Solvent vapor abuse leukoencephalopathy. Comparison to adrenoleukodystrophy. *J Neuropathol Exp Neurol* 1994; **53**:389–98.

Korobkin R, Asbury AK, Sumner AJ *et al.* Glue-sniffing neuropathy. *Arch Neurol* 1975; **32**:158–62.

Kraepelin E. *Dementia Praecox and Paraphrenia.* Huntington, NY: Robert E. Krieger, 1971.

Krill JJ, Halliday GM, Svoboda MD *et al.* The cerebral cortex is damaged in chronic alcoholics. *Neuroscience* 1997; **79**:983–98.

Lang AE. Withdrawal akathisia: case reports and a proposed classification of chronic akathisia. *Mov Disord* 1994; **9**:188–92.

Lappin RI, Auchiscloss EL. Treatment of the serotonin syndrome with cyproheptadine. *N Engl J Med* 1994; **331**:1021–2.

Lattanzi L, Mungai F, Romano A *et al.* Subcutaneous apomorphine for neuroleptic malignant syndrome. *Am J Psychiatry* 2006; **163**:1450–1.

Lazar RB, Ho SU, Melen O *et al.* Multifocal central nervous system damage caused by toluene abuse. *Neurology* 1983; **33**:1337–40.

Lechevalier B, Anderson JC, Morin P. Hemispheric disconnection syndrome with a 'crossed avoiding' reaction in a case of Marchiafava–Bignami disease. *J Neurol Neurosurg Psychiatry* 1977; **40**:483–97.

Lee K, Moller L, Hardt F *et al.* Alcohol induced brain damage and liver damage in young males. *Lancet* 1979; **2**:759–61.

Leong ASV. Marchiafava–Bignami disease in a non-alcoholic Indian male. *Pathology* 1979; **11**:241–9.

Lerner V, Miodownik C, Kapstan A *et al.* Vitamin B6 in the treatment of tardive dyskinesia: a double-blind, placebo-controlled, crossover study. *Am J Psychiatry* 2001; **158**:1511–14.

Levenson JL. Neuroleptic malignant syndrome after the initiation of olanzapine. *J Clin Psychopharmacol* 1999; **19**:477–8.

Levin GM, Lazowick AL, Powell HS. Neuroleptic malignant syndrome with risperidone. *J Clin Psychopharmacol* 1996; **16**:192–3.

Lewis JD, Moritz D, Mellis LP. Long-term toluene abuse. *Am J Psychiatry* 1981; **138**:368–70.

Lhermitte F, Marteau R, Serdaru M *et al.* Signs of interhemispheric disconnection in Marchiafava–Bignami disease. *Arch Neurol* 1977; **34**:254–7.

Libov I, Miodowni C, Bersudsky Y *et al.* Efficacy of piracetam in the treatment of tardive dyskinesia in schizophrenic patients: a randomized, double-blind, placebo-controlled crossover study. *J Clin Psychiatry* 2007; **68**:1031–7.

Lishman WA. Cerebral disorder in alcoholism: syndromes of impairment. *Brain* 1981; **104**:373–410.

Little JT, Jankovic J. Tardive myoclonus. *Mov Disord* 1987; **2**:307–11.

Lohr JB, Caligiuri MP. A double-blind placebo-controlled study of vitamin E in the treatment of tardive dyskinesia. *J Clin Psychiatry* 1996; **57**:167–73.

Lohr JB, Cadet JL, Lohr MA *et al.* Vitamin E in the treatment of tardive dyskinesia: the possible involvement of free radical mechanisms. *Schizophr Bull* 1988; **14**:291–6.

Lu ML, Pan JJ, Teng HW *et al.* Metoclopramide-induced supersensitivity psychosis. *Ann Pharmacother* 2002; **36**:1387–90.

Lu TC, Chu PL, Wu CS *et al.* Neuroleptic malignant syndrome after the use of venlafaxine in a patient with generalized anxiety disorder. *J Formos Med Assoc* 2006; **105**:90–3.

Luchins DJ, Freed WJ, Wyatt RJ. The role of cholinergic supersensitivity in the medical symptoms associated with withdrawal of antipsychotic drugs. *Am J Psychiatry* 1980; **137**:1395–8.

Lynch MJG. Brain lesions in chronic alcoholism. *Arch Pathol* 1960; **69**:342–53.

McCarron MM, Boettger ML, Peck JJ. A case of neuroleptic malignant syndrome successfully treated with amantadine. *J Clin Psychiatry* 1982; **43**:381–2.

McDaniel WW. Serotonin syndrome: early management with cyproheptadine. *Ann Pharmacother* 2001; **35**:870–3.

Marchiafava E, Bignami A. Sopra un alterazione del corpo callsos osservata in sogetti alcoolisti. *Riv Patol Nerv Mentale* 1903; **8**:544–9.

Mason PJ, Morris VA, Balcezak TJ. Serotonin syndrome. Presentation of 2 cases and review of the literature. *Medicine* 2000; **79**:201–9.

May EDC, Morris SW, Stewart RM *et al.* Neuroleptic malignant syndrome: response to dantrolene sodium. *Ann Intern Med* 1983; **98**:183–4.

Mayer JM, De Liege P, Netter JM *et al.* Computerized tomography and nuclear magnetic resonance imaging in Marchiafava–Bignami disease. *J Neuroradiol* 1987; **14**:152–8.

Mendhekar DN, Andrade CR. Neuroleptic malignant syndrome with promethazine. *Aust N Z J Psychiatry* 2005; **39**:310.

Menegon P, Sibon I, Pachia C *et al.* Marchiafava–Bignami disease: diffusion-weighted MRI in corpus callosum and cortical lesions. *Neurology* 2005; **65**:475–7.

Meterissian GB. Risperidone-induced neuroleptic malignant syndrome: a case report and review. *Can J Psychiatry* 1996; **41**:52–4.

Mettler FA, Crandall A. Neurologic disorders in psychiatric institutions. *J Nerv Ment Dis* 1959; **128**:148–59.

Miller DD, Sharafuddin MJA, Kathol RG. A case of clozapine-induced neuroleptic malignant syndrome. *J Clin Psychiatry* 1991; **52**:99–101.

Moncrieff J. Does antipsychotic withdrawal provoke psychosis? Review of the literature on rapid onset psychosis (supersensitivity psychosis) and withdrawal-related relapse. *Acta Psychiatr Scand* 2006; **114**:3–13.

Moore DP. Tardive psychosis. *J Ky Med Assoc* 1986; **84**:351–3.

Morganstern H, Glazer WM. Identifying risk factors for tardive dyskinesia among the long-term outpatients maintained with neuroleptic medications. *Arch Gen Psychiatry* 1993; **50**:723–33.

Mueller PS, Vester JW, Fermaglich J. Neuroleptic malignant syndrome – successful treatment with bromocriptine. *JAMA* 1983; **249**:386–8.

Neuberger KT. The changing neuropathological picture of chronic alcoholism. *Arch Pathol* 1957; **63**:1–6.

Nimmagadda SR, Ryan DH, Atkin SL. Neuroleptic malignant syndrome after venlafaxine. *Lancet* 2000; **354**:289–90.

Oberndorfer S, Grisold W, Hinterholzer G *et al.* Coma with focal neurological signs caused by Datura stramonium intoxication in a young man. *J Neurol Neurosurg Psychiatry* 2002; **73**:458–9.

Ondo WG, Hanna PA, Jankovi J. Tetrabenazine treatment for tardive dyskinesia: assessment by randomized videotape protocol. *Am J Psychiatry* 1999; **156**:1279–81.

Ossemann M, Sindic CJM, Laterre C. Tetrabenazine as a cause of neuroleptic malignant syndrome. *Mov Disord* 1996; **11**:95.

Owens DGC, Johnstone EC, Frith CD. Spontaneous involuntary disorders of movement: their prevalence, severity and distribution in chronic schizophrenics with and without treatment with neuroleptics. *Arch Gen Psychiatry* 1982; **39**:452–61.

Ozen ME, Yumru M, Savas HA *et al.* Neuroleptic malignant syndrome induced by ziprasidone on the second day of treatment: a case report. *World J Biol Psychiatry* 2007; **8**:133–4.

Parvin MM, Swartz CM. Dystonic rabbit syndrome from citalopram. *Clin Neuropharmacol* 2005; **28**:289–91.

Petti TA, Law W. Abrupt cessation of high-dose imipramine treatment in children. *JAMA* 1981; **246**:768–9.

Pfefferbaum A, Sullivan EV, Mathalon DH *et al.* Longitudinal changes in magnetic resonance imaging brain volumes in abstinent and relapsed alcoholics. *Alcohol Clin Exp Res* 1995; **19**:1177–91.

Pope HG, Keck PE, McElroy SL. Frequency and presentation of neuroleptic malignant syndrome in a large psychiatric hospital. *Am J Psychiatry* 1986; **143**:1227–33.

Porteous HB, Ross DN. Mental symptoms in parkinsonism following benzhexol hydrochloride therapy. *BMJ* 1956; **2**:138–40.

Poser CM. Demyelinization in the central nervous system in chronic alcoholism, central pontine myelinolysis, and Marchiafava–Bignami's disease. *Ann NY Acad Sci* 1973; **215**:373–81.

Preskorn SH, Simpson S. Tricyclic-antidepressant-induced delirium and plasma drug concentrations. *Am J Psychiatry* 1982; **139**:822–3.

Reunanen M, Kaarnen P, Vaisanen E. The influence of anticholinergic treatment on tardive dyskinesia caused by neuroleptic drugs. *Acta Psychiatr Scand* 1982; 65(Suppl. 90):278–9.

Rich MW, Radwany SM. Respiratory dyskinesia: an underrecognized phenomenon. *Chest* 1994; **105**:1826–32.

Richardson MA, Craig TJ. The coexistence of Parkinsonism-like symptoms and tardive dyskinesia. *Am J Psychiatry* 1982; **139**:341–3.

Richardson MA, Vevans ML, Read LL *et al.* Efficacy of branched-chain amino acids in the treatment of tardive dyskinesia. *Am J Psychiatry* 2003; **160**:1117–24.

Rosa A, Demiati M, Cartz L *et al.* Marchiafava–Bignami disease, syndrome of interhemispheric disconnection, and right-handed agraphia in a left-hander. *Arch Neurol* 1991; **48**:986–8.

Rosebush P, Stewart T. A prospective analysis of 24 episodes of neuroleptic malignant syndrome. *Am J Psychiatry* 1989; **146**:717–25.

Rosebush PI, Stewart TED, Gelelberg AJ. Twenty neuroleptic rechallenges after neuroleptic malignant syndrome in 15 patients. *J Clin Psychiatry* 1989; **50**:295–8.

Rosenberg NL, Kleineschmidt-DeMasters BK, Davis KA *et al.* Toluene abuse causes diffuse central nervous system white matter changes. *Ann Neurol* 1988; **23**:611–14.

Sachdev PS. Depression-dependent exacerbation of tardive dyskinesia. *Br J Psychiatry* 1989; **155**:253–5.

Sachdev P. Clinical characteristics of 15 patients with tardive dystonia. *Am J Psychiatry* 1993a; **150**:498–500.

Sachdev P. Tardive and chronically recurrent oculogyric crises. *Mov Disord* 1993b; **8**:93–7.

Sachdev P, Loneragan C. The present status of akathisia. *J Nerv Ment Dis* 1991; **179**:381–91.

Sachdev P, Kruk J, Kneebone M *et al.* Clozapine-induced neuroleptic malignant syndrome: review and report of new case. *J Clin Psychopharmacol* 1995; **15**:365–71.

Saltz BL, Woerner MG, Kane JM *et al.* Prospective study of tardive dyskinesia incidence in the elderly. *JAMA* 1991; **266**:2402–6.

Sato Y, Asoh T, Metoki N *et al.* Efficacy of methylprednisolone pulse therapy on neuroleptic malignant syndrome in Parkinson's disease. *J Neurol Neurosurg Psychiatry* 2003; **74**:574–6.

Schroth G, Naegele T, Klose U *et al.* Reversible brain shrinkage in abstinent alcoholics, measured by MRI. *Neuroradiology* 1988; **30**:385–9.

Schuckit MA, Winokur G. Alcohol hallucinosis and schizophrenia: a negative study. *Br J Psychiatry* 1971; **119**:549–53.

Sechi GP, Tanda F, Mutani R. Fatal hyperpyrexia after withdrawal of levodopa. *Neurology* 1984; **34**:249–51.

Sewell DD, Jeste DV. Metoclopramide-associated tardive dyskinesia: an analysis of 67 cases. *Arch Fam Med* 1992; **1**:271–8.

Sewell DD, Kodsi AB, Caligiuri MP *et al.* Metoclopramide and tardive dyskinesia. *Biol Psychiatry* 1994; **36**:630–2.

Shamir E, Barak Y, Plopsky I *et al.* Is melatonin treatment effective for tardive dyskinesia? *J Clin Psychiatry* 2000; **61**:556–8.

Shamir E, Varak Y, Shalman I *et al.* Melatonin treatment for tardive dyskinesia. A double-blind, placebo-controlled, crossover study. *Arch Gen Psychiatry* 2001; **58**:1049–52.

Shriqui CL, Bradwejn J, Annable L *et al.* Vitamin E in the treatment of tardive dyskinesia: a double-blind placebo-controlled study. *Am J Psychiatry* 1992; **149**:391–3.

Sing KJ, Ramaekers GMGI, van Harten PN. Neuroleptic malignant syndrome and quetiapine. *Am J Psychiatry* 2002; **159**:149–50.

Slotema CW, van Harten PN, Bruggeman R *et al.* Botulinum toxin in the treatment of orofacial tardive dyskinesia: a single blind study. *Prog Neuropsychopharmacol Biol Psychiatry* 2008; **32**:507–9.

Soyka M. Psychopathological characteristics in alcohol hallucinosis and paranoid schizophrenia. *Acta Psychiatr Scand* 1990; **81**:255–9.

Soyka M, Taschner B, Clausius N. Neuroleptic treatment of alcohol hallucinosis: case series. *Pharmacopsychiatry* 2007; **40**:291–2.

Stacy M, Jankovic J. Tardive tremor. *Mov Disord* 1992; **7**:53–7.

Stacy M, Cardoso F, Jankovic J. Tardive stereotypy and other movement disorders in tardive dyskinesias. *Neurology* 1993; **43**:937–41.

Stahl SM. Tardive Tourette syndrome in an autistic patient after long-term neuroleptic administration. *Am J Psychiatry* 1980; **137**:1267–9.

Steiner W, Laporta M, Chouinard G. Neuroleptic-induced supersensitivity psychosis in patients with bipolar affective disorder. *Acta Psychiatr Scand* 1990; **81**:437–40.

Stern TA. Continuous infusion of physostigmine in anticholinergic delirium: case report. *J Clin Psychiatry* 1983; **44**:463–4.

Sternbach H. The serotonin syndrome. *Am J Psychiatry* 1991; **148**:705–13.

Suh H, Bronson B, Martin R. Neuroleptic malignant syndrome and low-dose olanzapine. *Am J Psychiatry* 2003; **160**:796.

Tan E-K, Jankovic J. Tardive and idiopathic oromandibular dystonia: a clinical comparison. *J Neurol Neurosurg Psychiatry* 2000; **68**:186–90.

Tanii H, Ichihashi K, Inoue K *et al.* Possible neuroleptic malignant syndrome related to concomitant treatment with paroxetine and alprazolam. *Prog Neuropsychopharmacol Biol Psychiatry* 2006; **30**:1176–8.

Tao H, Kitigawa N, Kato Y *et al.* A case of anorexia nervosa with Marchiafava–Bignami disease that responded to high-dose intravenous corticosteroid administration. *Psychiatry Res* 2007; **156**:181–4.

Tarsy D, Kaufman D, Sethi KD *et al.* An open-label study of botulinum toxin A for treatment of tardive dystonia. *Clin Neuropharmacol* 1997; **20**:90–3.

Taylor NE, Schwartz HI. Neuroleptic malignant syndrome following amoxapine overdose. *J Nerv Ment Dis* 1988; **176**:249–51.

Tejera CA, Saravay SM, Goldman E *et al.* Diphenhydramine-induced delirium in elderly hospitalized patients with mild dementia. *Psychosomatics* 1994; **35**:399–402.

Thaker GK, Nguyen JA, Strauss ME *et al.* Clonazepam treatment of tardive dyskinesia: a practical GABAmimetic strategy. *Am J Pscyhiatry* 1990; **147**:445–51.

Thornton NE, Stahl SM. Case report of tardive dyskinesia and parkinsonism associated with amoxapine therapy. *Am J Psychiatry* 1984; **141**:704–5.

Todd R, Lippmann S, Manshadi M *et al.* Recognition and treatment of rabbit syndrome, an uncommon complication of neuroleptic therapies. *Am J Psychiatry* 1983; **140**:1519–20.

Toru M, Matsuda O, Makiguchi K *et al.* Single case study: neuroleptic malignant syndrome-like state following withdrawal of antiparkinsonian drugs. *J Nerv Ment Dis* 1981; **169**:324–7.

Torvik A, Lindboe CF, Rogde S. Brain lesions in alcoholics. A neuropathological study with clinical correlations. *J Neurol Sci* 1982; **56**:233–48.

Troller JN, Sachdev PS. Electroconvulsive treatment of neuroleptic malignant syndrome: a review and report of cases. *Aust N Z J Psychiatry* 1999; **33**:650–9.

Tsao JW, Heilman KM. Transient memory impairment and hallucination associated with tolteridone use. *N Engl J Med* 2003; **349**: 2274–5.

Tune LE, Dainloth NF, Holland A *et al.* Association of postoperative delirium with raised serum levels of anticholinergic drugs. *Lancet* 1981; **2**:651–3.

Tune LE, Bylsma FW, Hilt DC. Anticholinergic delirium caused by topical homatropine ophthalmic solution: confirmation by anticholinergic radioreceptor assay in two cases. *J Neuropsychiatry Clin Neurosci* 1992; **4**:195–7.

Turner MR, Gainsborough N. Neuroleptic malignant-like syndrome after abrupt withdrawal of baclofen. *J Psychopharmacol* 2001; **15**:61–3.

van Harten PN, Hoek HW, Matroos GE *et al.* Intermittent neuroleptic treatment and risk of tardive dyskinesia: Curacao extrapyramidal syndromes study III. *Am J Psychiatry* 1998; **155**:565–7.

Velamoor VR, Norman RMG, Caroff SN *et al.* Progression of symptoms in neuroleptic malignant syndrome. *J Nerv Ment Dis* 1994; **182**:168–73.

Victor M, Hope JM. The phenomenon of auditory hallucinations in chronic alcoholism. *J Nerv Ment Dis* 1958; **126**:451–8.

Vonderahe AR. Lilliputian hallucinations: report of a case of hyoscine poisoning. *Arch Neurol Psychiatry* 1929; **22**:585–8.

Wada Y, Yamaguchi N. The rabbit syndrome and antiparkinsonian medication in schizophrenic patients. *Neuropsychobiology* 1992; **25**:149–52.

Wang H-C, Hsieh Y. Treatment of neuroleptic malignant syndrome with subcutaneous apomorphine therapy. *Mov Disord* 2001; **16**:765–7.

Weiner WJ, Nausieda PA, Glantz RH. Meige syndrome (blepharospasm–oromandibular dystonia) after long-term neuroleptic therapy. *Neurology* 1981; **31**:1555–6.

Woerner M, Kane JM, Lieberman JA *et al.* The prevalence of tardive dyskinesia. *J Clin Psychopharmacol* 1991; **11**:34–42.

Wojcik JD, Falk WE, Fink JS *et al.* A review of 32 cases of tardive dystonia. *Am J Psychiatry* 1991; **148**:1055–9.

Wolfe RM. Antidepressant withdrawal reactions. *Am Fam Physician* 1997; **56**:455–62.

Woods SW, Saksa JR, Baker CB *et al.* Effects of levetiracetam on tardive dyskinesia: a randomized, double-blind, placebo-controlled study. *J Clin Psychiatry* 2008; **69**:546–54.

Woods SW, Mergenstern H, Saksa JR *et al.* Incidence of tardive dyskinesia with atypical versus conventional antipsychotic medications: a prospective cohert study. *J Clin Psychiatry* 2010; **71**:463–74.

Yacoub A, Francis A. Neuroleptic malignant syndrome induced by atypical neuroleptics and responsive to lorazepam. *Neuropsychiatr Dis Treat* 2006; **2**:235–40.

Yadalam KG, Korn ML, Simpson GM. Tardive dystonia: four case histories. *J Clin Psychiatry* 1990; **51**:17–20.

Yagi G, Itoh H. Follow-up study of 11 patients with potentially reversible tardive dyskinesia. *Am J Psychiatry* 1987; **144**:1596–8.

Yarden PE, Discipio WJ. Abnormal movements and prognosis in schizophrenia. *Am J Psychiatry* 1971; **128**:317–23.

Yassa R, Lal S. Respiratory irregularity and tardive dyskinesia: a prevalence study. *Acta Psychiatr Scand* 1986a; **73**:506–10.

Yassa R, Lal S. Prevalence of the rabbit syndrome. *Am J Psychiatry* 1986b; **143**:656–7.

Yassa R, Nair V, Dimitry A. Prevalence of tardive dystonia. *Acta Psychiatr Scand* 1986; **73**:629–33.

Yassa R, Iskander H, Nastase C. Propranolol in the treatment of tardive akathisia: a report of two cases. *J Clin Psychopharmacol* 1988; **8**:283–5.

Yazici O, Kantemir E, Tastaban Y *et al.* Spontaneous improvement of tardive dystonia during mania. *Br J Psychiatry* 1991; **158**:847–50.

Zhang XY, Zhou DF, Cao LY *et al.* The effect of vitamin E treatment on tardive dyskinesia and blood superoxide dismutase: a double-blind, placebo-controlled trial. *J Clin Psychopharmacol* 2004; **24**:83–6.

Zipursky RB. Lim KC, Pfefferbaum A. MRI study of brain changes with short-term abstinence from alcohol. *Alcohol Clin Exp Res* 1989; **13**:664–6.

Ziskind AA. Transdermal scopolamine-induced psychosis. *Postgrad Med* 1988; **84**:73–6.

Index

Note:
Abbreviations:
ADHD – attention-deficit hyperactivity disorder
EEG – electroencephalography
PTSD – post-traumatic stress disorder
SLE – systemic lupus erythematosus
Page number in *italics*, and in **bold**, refer to pages on which figures and tables appear respectively.
vs. denotes differential diagnosis